2016 IEEE Applied Power Electronics Conference and Exposition (APEC 2016)

Long Beach, California, USA
20-24 March 2016

Pages 1468-2222

IEEE Catalog Number: CFP16APE-POD
ISBN: 978-1-4673-9551-9

Copyright © 2016 by the Institute of Electrical and Electronic Engineers, Inc
All Rights Reserved

Copyright and Reprint Permissions: Abstracting is permitted with credit to the source. Libraries are permitted to photocopy beyond the limit of U.S. copyright law for private use of patrons those articles in this volume that carry a code at the bottom of the first page, provided the per-copy fee indicated in the code is paid through Copyright Clearance Center, 222 Rosewood Drive, Danvers, MA 01923.

For other copying, reprint or republication permission, write to IEEE Copyrights Manager, IEEE Service Center, 445 Hoes Lane, Piscataway, NJ 08854. All rights reserved.

***This publication is a representation of what appears in the IEEE Digital Libraries. Some format issues inherent in the e-media version may also appear in this print version.

IEEE Catalog Number: CFP16APE-POD
ISBN (Print-On-Demand):   978-1-4673-9551-9
ISBN (Online):   978-1-4673-9550-2
ISSN: 1048-2334

Additional Copies of This Publication Are Available From:

Curran Associates, Inc
57 Morehouse Lane
Red Hook, NY 12571 USA
Phone: (845) 758-0400
Fax: (845) 758-2633
E-mail: curran@proceedings.com
Web: www.proceedings.com

TECHNICAL PAPERS

Session T01: Three-Phase AC-DC Converters
Location: 101A
March 22, 2016 8:30 - 12:00
Session Chairs: Gerry Moschopoulos, *Western University, Canada*
Patrick Wheeler, *University of Nottingham*

Hardware Implementation and Characterization of SiC-Based Hybrid Three-Phase Rectifier Employing Third Harmonic Injection 1
M. Makoschitz, *Technische Universität Wien, Austria*
M. Hartmann, *Schneider Electric SE, Austria*
H. Ertl, *Technische Universität Wien, Austria*

Voltage Oriented Control of the Three-Level Vienna Rectifier using Vector Control Method 9
Jeevan Adhikari, *National University of Singapore, Singapore*
Prasanna IV, *National University of Singapore, Singapore*
S.K. Panda, *National University of Singapore, Singapore*

Compensation of Neutral Point Deviation in 3-Level NPC Converter Under Unbalanced Grid Conditions 17
Kyungsub Jung, *Chungbuk National University, Korea, South*
Yongsug Suh, *Chungbuk National University, Korea, South*

High Power Factor Modular Polyphase AC/DC Converters with Galvanic Isolation based on Resistor Emulators 25
Javier Sebastián, *Universidad de Oviedo, Spain*
Ignacio Castro, *Universidad de Oviedo, Spain*
Diego G. Lamar, *Universidad de Oviedo, Spain*
Aitor Vázquez, *Universidad de Oviedo, Spain*
Kevin Martín, *Universidad de Oviedo, Spain*

Reduced Duty-Cycle Loss and Output Inductor Current Ripple in a ZVS Switched Three-Phase Isolated PWM Rectifier 33
Jahangir Afsharian, *Ryerson University, Canada*
Dewei David Xu, *Ryerson University, Canada*
Tao Zhao, *Ryerson University, Canada*
Bing Gong, *Murata Power Solution, Canada*
Zhihua Yang, *Murata Power Solution, Canada*

Analysis, Design, and Evaluation of Three-Phase Three-Wire Isolated AC-DC Converter Implemented with Three Single-Phase Converter Modules 38
Laszlo Huber, *Delta Products Corporation, United States*
Misha Kumar, *Delta Products Corporation, United States*
Milan M. Jovanović, *Delta Products Corporation, United States*
Dinggang Ping, *Delta Electronics Shanghai Co., Ltd., China*
Gang Liu, *Delta Electronics Shanghai Co., Ltd., China*

Startup Procedure for Three-Phase Three-Wire Isolated AC-DC Converter Implemented with Three Single-Phase Converter Modules 46

Misha Kumar, *Delta Products Corporation, United States*
Laszlo Huber, *Delta Products Corporation, United States*
Milan M. Jovanović, *Delta Products Corporation, United States*
Dinggang Ping, *Delta Electronics Shanghai Co., Ltd., China*
Gang Liu, *Delta Electronics Shanghai Co., Ltd., China*

Control of a Single-Stage Three-Phase Boost Power Factor Correction Rectifier 54

Ayan Mallik, *University of Maryland, United States*
Bryan Faulkner, *Virginia Polytechnic Institute and State University, United States*
Alireza Khaligh, *University of Maryland, United States*

A Bidirectional Single-Stage Three-Phase Rectifier with High-Frequency Isolation and Power Factor Correction 60

Bruno Ricardo de Almeida, *Universidade Federal do Ceará, Brazil*
Demercil de Souza Oliveira Jr., *Universidade Federal do Ceará, Brazil*
Paulo P. Praça, *Universidade Federal do Ceará, Brazil*

Session T02: High Frequency and Fast-Response DC-DC Converters
Location: 104A
March 22, 2016 8:30 - 12:00
Session Chairs: Olivier Trescases, *University of Toronto*
Jeff Nilles, *Texas Instruments*

A 5 MHz, 12 V, 10 A, Monolithically Integrated Two-Phase Series Capacitor Buck Converter 66

Pradeep S. Shenoy, *Texas Instruments Inc., United States*
Orlando Lazaro, *Texas Instruments Inc., United States*
Ramanathan Ramani, *Texas Instruments Inc., United States*
Mike Amaro, *Texas Instruments Inc., United States*
Wlodek Wiktor, *Texas Instruments Inc., United States*
Joseph Khayat, *Texas Instruments Inc., United States*
Brian Lynch, *Texas Instruments Inc., United States*

A 10-MHz Isolated Class-Φ_2 Synchronous Resonant DC-DC Converter 73

Yuan Zhou, *Nanjing University of Aeronautics and Astronautics, China*
Zhiliang Zhang, *Nanjing University of Aeronautics and Astronautics, China*
Xue-Wen Zou, *Nanjing University of Aeronautics and Astronautics, China*
Zhou Dong, *Nanjing University of Aeronautics and Astronautics, China*
Xiaoyong Ren, *Nanjing University of Aeronautics and Astronautics, China*

865 MHz Switching-Speed Step-Down DC-DC Power Converter for Envelope Tracking 79

Vivek Mehrotra, *Teledyne Scientific Company, United States*
Andrea Arias, *Teledyne Scientific Company, United States*
Joshua Bergman, *Teledyne Scientific Company, United States*
Charles Neft, *Teledyne Scientific Company, United States*
Miguel Urteaga, *Teledyne Scientific Company, United States*
Berinder Brar, *Teledyne Scientific Company, United States*

Current Parking Regulator for Zero Droop/Overshoot Load Transient Response 86

Sudhir S. Kudva, *Nvidia Corporation, United States*
William J. Dally, *Nvidia Corporation, United States*
Thomas H. Greer III, *Nvidia Corporation, United States*
C. Thomas Gray, *Nvidia Corporation, United States*

A 5MHz, 24V-to-1.2V, AO^{2}T Current Mode Buck Converter with One-Cycle Transient Response and Sensorless Current Detection for Medical Meters 94

Xugang Ke, *University of Texas at Dallas, United States*
Joseph Sankman, *Texas Instruments Inc., United States*
Dongsheng Ma, *University of Texas at Dallas, United States*

Capacitively-Aided Switching Technique for High-Frequency Isolated Bus Converters 98

Seungbum Lim, *Massachusetts Institute of Technology, United States*
Alex J. Hanson, *Massachusetts Institute of Technology, United States*
Juan A. Santiago-González, *Massachusetts Institute of Technology, United States*
David J. Perreault, *Massachusetts Institute of Technology, United States*

A 10 MHz, 48-to-5V Synchronous Converter with Dead Time Enabled 125 ps Resolution Zero-Voltage Switching 106

Alexander Barner, *Robert Bosch GmbH, Germany*
Jürgen Wittmann, *Hochschule Reutlingen, Germany*
Thoralf Rosahl, *Robert Bosch GmbH, Germany*
Bernhard Wicht, *Hochschule Reutlingen, Germany*

Plug-and-Play Electronic Capacitor for VRM Applications 111

Or Kirshenboim, *Ben-Gurion University of the Negev, Israel*
Alon Cervera, *Ben-Gurion University of the Negev, Israel*
Bar Halivni, *Ben-Gurion University of the Negev, Israel*
Eli Abramov, *Ben-Gurion University of the Negev, Israel*
Mor Mordechai Peretz, *Ben-Gurion University of the Negev, Israel*

Adaptive Voltage Positioning (AVP) Design of Multi-Phase Constant On-Time I^2 Control for Voltage Regulators with Ramp Compensations 118

Kuang-Yao Cheng, *Texas Instruments Inc., United States*
Yipeng Su, *Texas Instruments Inc., United States*

Session T03: Microgrids and Hybrid Systems
Location: 104B
March 22, 2016 8:30 - 12:00
Session Chairs: Yunwei Li, *University of Alberta*
Joesep Guerrero, *Aalborg University*

Reactive Power Support Capabilities of Nonsynchronous Interconnection Systems in Microgrid Applications 125

Yong-Duk Lee, *University of Connecticut, United States*
Sung-Yeul Park, *University of Connecticut, United States*

Zero Standby Power High Efficiency Hot Plugging Outlet for 380VDC Power Delivery System 132

Kai Tan, *North Carolina State University, United States*
Chang Peng, *North Carolina State University, United States*
Pengkun Liu, *North Carolina State University, United States*
Xiaoqing Song, *North Carolina State University, United States*
Alex Q. Huang, *North Carolina State University, United States*

Design of Control System for Smooth Mode Transfer in Smart Microgrid Application 138
Mingzhi Gao, *Zhejiang University, China*
Canhui Zhang, *Zhejiang University, China*
Maohang Qiu, *Zhejiang University, China*
Min Chen, *Zhejiang University, China*
Aron Levy, *Technology Dynamics Inc., United States*

Resonance Propagation Modeling and Analysis of AC Filters in a Large-Scale Microgrid ... 143
Yusi Liu, *University of Arkansas, United States*
Chris Farnell, *University of Arkansas, United States*
H. Alan Mantooth, *University of Arkansas, United States*
Juan Carlos Balda, *University of Arkansas, United States*
Roy A. McCann, *University of Arkansas, United States*
Cheng Deng, *University of Arkansas, United States*

A New Bidirectional DC-DC Converter for Fuel Cell, Solar Cell and Battery Systems 150
Ankur Patel, *Vicor Corporation, United States*

A Multiport Isolated DC-DC Converter 156
Yan-Kim Tran, *École Polytechnique Fédérale de Lausanne, Switzerland*
Drazen Dujic, *École Polytechnique Fédérale de Lausanne, Switzerland*

A Seamless Transfer Control Method with High Load Sharing Performance for Modular ESS ... 163
Jung-Hoon Ahn, *Sungkyunkwan University, Korea, South*
Won-Yong Sung, *Sungkyunkwan University, Korea, South*
Chang-Yeol Oh, *Sungkyunkwan University, Korea, South*
Byoung-Kuk Lee, *Sungkyunkwan University, Korea, South*
Yun-Sung Kim, *Dongahelecomm Corporation, Korea, South*

A Plug-and-Play Ripple Mitigation Approach for DC-Links in Hybrid Systems 169
Sinan Li, *University of Hong Kong, Hong Kong*
Albert T.L. Lee, *University of Hong Kong, Hong Kong*
Siew-Chong Tan, *University of Hong Kong, Hong Kong*
S.Y. Ron Hui, *University of Hong Kong, Hong Kong*

Active Control of Low Frequency Common Mode Voltage to Connect AC Utility and 380 V DC Grid 177
Fang Chen, *Virginia Polytechnic Institute and State University, United States*
Rolando Burgos, *Virginia Polytechnic Institute and State University, United States*
Dushan Boroyevich, *Virginia Polytechnic Institute and State University, United States*
Xuning Zhang, *Virginia Polytechnic Institute and State University, United States*

Session T04: Control Strategies for Inverters and Motor Drives
Location: 103C
March 22, 2016 8:30 - 12:00
Session Chairs: Bilal Akin, *Univeristy of Texas, Dallas*
Babak Nahid-Mobarakeh, *University of Lorraine*

A Three-Level Space Vector Modulation Scheme for Paralleled Two Converters to Reduce Zero-Sequence Circulating Current and Common Mode Voltage 185
Zhongyi Quan, *University of Alberta, Canada*
Yunwei Li, *University of Alberta, Canada*

Nonlinearity Analysis and Linear Modulation Method for Two Level Voltage Source Inverter with Low Switching to Operating Frequency Ratio 193

Yongjae Lee, *Seoul National University, Korea, South*
Jung-Ik Ha, *Seoul National University, Korea, South*

Synchronization Strategies in Cascaded H-Bridge Multi Level Inverters for Carrier based Sinusoidal PWM Techniques 199

Saroj Kumar Sahoo, *Indian Institute of Technology Kharagpur, India*
Tanmoy Bhattacharya, *Indian Institute of Technology Kharagpur, India*

Design and Implementation of a Sinusoidal Flux Controller for Core Loss Measurements 207

Burak Tekgun, *University of Akron, United States*
Ali R. Boynuegri, *University of Akron, United States*
Md Asif Mahmood Chowdhury, *University of Akron, United States*
Yilmaz Sozer, *University of Akron, United States*

Implementation of Deadbeat-Direct Torque and Flux Control for Synchronous Reluctance Machines to Minimize Loss Each Switching Period 215

Michael Saur, *Universität der Bundeswehr München, Germany*
Francisco Ramos, *Universität der Bundeswehr München, Germany*
Aday Perez, *Universität der Bundeswehr München, Germany*
Dieter Gerling, *Universität der Bundeswehr München, Germany*
Robert D. Lorenz, *University of Wisconsin at Madison, United States*

Addressing the Unbalance Loading Issue in Multi-Drive Systems with a DC-Link Modulation Scheme for Harmonic Reduction 221

Yongheng Yang, *Aalborg University, Denmark*
Pooya Davari, *Aalborg University, Denmark*
Firuz Zare, *Danfoss Power Electronics A/S, Denmark*
Frede Blaabjerg, *Aalborg University, Denmark*

Input Current Interharmonics in Adjustable Speed Drives Caused by Fixed-Frequency Modulation Techniques 229

Hamid Soltani, *Aalborg University, Denmark*
Pooya Davari, *Aalborg University, Denmark*
Poh Chiang Loh, *Aalborg University, Denmark*
Frede Blaabjerg, *Aalborg University, Denmark*
Firuz Zare, *Danfoss Power Electronics A/S, Denmark*

Low-Frequency Voltage Ripples in the Flying Capacitors of the Nested Neutral-Point-Clamped Converter 236

Amer M.Y.M. Ghias, *University of Sharjah, U.A.E.*
Josep Pou, *University of New South Wales, Australia*
Salvador Ceballos, *TECNALIA, Spain*
Vassilios G. Agelidis, *University of New South Wales, Australia*

DC Bus Capacitor Discharge of Permanent Magnet Synchronous Machine Drive Systems for Hybrid Electric Vehicles 241

Ziwei Ke, *Oregon State University, United States*
Julia Zhang, *Oregon State University, United States*
Michael W. Degner, *Ford Motor Company, United States*

Session T05: Si Devices and Power Module Packaging
Location: 101B
March 22, 2016 8:30 - 12:00
Session Chairs: Iulian Nistor, *Corporate Research, ABB Inc.*
Brian Rowden,

C_{OSS} Hysteresis in Advanced Superjunction MOSFETs 247
J.B. Fedison, *Enphase Energy, Inc., United States*
M.J. Harrison, *Enphase Energy, Inc., United States*

Compact Electrothermal Models for Unbalanced Parallel Conducting Si-IGBTs 253
Roozbeh Bonyadi, *University of Warwick, United Kingdom*
Olayiwola Alatise, *University of Warwick, United Kingdom*
Ji Hu, *University of Warwick, United Kingdom*
Zarina Davletzhanova, *University of Warwick, United Kingdom*
Yeganeh Bonyadi, *University of Warwick, United Kingdom*
Jose Ortiz-Gonzalez, *University of Warwick, United Kingdom*
Li Ran, *University of Warwick, United Kingdom*
Philip Mawby, *University of Warwick, United Kingdom*

General 3D Lumped Thermal Model with Various Boundary Conditions for High Power IGBT Modules 261
Amir Sajjad Bahman, *Aalborg University, Denmark*
Ke Ma, *Aalborg University, Denmark*
Frede Blaabjerg, *Aalborg University, Denmark*

Improved 6.5kV FREEMD-Pair based on SiC JFET and Si IGBT 269
Xiaoqing Song, *North Carolina State University, United States*
Alex Q. Huang, *North Carolina State University, United States*
Chang Peng, *North Carolina State University, United States*
Liqi Zhang, *North Carolina State University, United States*

On the Comparative Assessment of 1.7 kV, 300 a Full SiC-MOSFET and Si-IGBT Power Modules 276
Muhammad Nawaz, *ABB Corporate Research, Sweden*
Kalle Ilves, *ABB Corporate Research, Sweden*

Suppression of Reverse Recovery Ringing 3.3kV/450A Si/SiC Hybrid in Low Internal Inductance Package Next High Power Density Dual; nHPD^2 283
Katsuaki Saito, *Hitachi Europe Ltd., United Kingdom*
Daisuke Kawase, *Hitachi Power Semiconductor, Ltd., Japan*
Masamitsu Inaba, *Hitachi Power Semiconductor, Ltd., Japan*
Keiichi Yamamoto, *Hitachi Power Semiconductor, Ltd., Japan*
Katsunori Azuma, *Hitachi Power Semiconductor, Ltd., Japan*
Seiichi Hayakawa, *Hitachi Power Semiconductor, Ltd., Japan*

New Layout Concepts in MW-Scale IGBT Modules for Higher Robustness during Normal and Abnormal Operations 288
Paula Diaz Reigosa, *Aalborg University, Denmark*
Francesco Iannuzzo, *Aalborg University, Denmark*
Stig Munk-Nielsen, *Aalborg University, Denmark*
Frede Blaabjerg, *Aalborg University, Denmark*

Design, Package, and Hardware Verification of a High Voltage Current Switch 295
Ankan De, *North Carolina State University, United States*
Adam Morgan, *North Carolina State University, United States*
Vishnu Mahadeva Iyer, *North Carolina State University, United States*
Haotao Ke, *North Carolina State University, United States*
Xin Zhao, *North Carolina State University, United States*
Kasunaidu Vechalapu, *North Carolina State University, United States*
Subhashish Bhattacharya, *North Carolina State University, United States*
Douglas C. Hopkins, *North Carolina State University, United States*

Investigation of Short Circuit in a IGBT Power Module with Three-Level Neutral Point Clamped Type 2 (NPC2, T-NPC, Mixed Voltage) Topology 303
Kevin Lenz, *Danfoss Silicon Power, Germany*
Vladan Jerinic, *Danfoss Silicon Power, Germany*
Reiner Hinken, *Danfoss Silicon Power, Germany*

Session T06: DC-DC Converter Control
Location: 102AB
March 22, 2016 8:30 - 12:00
Session Chairs: Sombuddha Chakraborty, *Texas Instruments*
Rafael Pena Alzola, *University of British Columbia*

Closed-Loop Design and Time-Optimal Control for a Series-Capacitor Buck Converter 308
Timur Vekslender, *Ben-Gurion University of the Negev, Israel*
Ofer Ezra, *Ben-Gurion University of the Negev, Israel*
Yevgeny Bezdenezhnykh, *Ben-Gurion University of the Negev, Israel*
Mor Mordechai Peretz, *Ben-Gurion University of the Negev, Israel*

Unified Constant On/Off-Time Hybrid Compensation for Fast Recovery in Digitally Current-Mode Controlled Point-of-Load Converters 315
K. Hariharan, *Indian Institute of Technology Kharagpur, India*
Santanu Kapat, *Indian Institute of Technology Kharagpur, India*
Siddhartha Mukhopadhyay, *Indian Institute of Technology Kharagpur, India*

Digital Implementation of Adaptive Synchronous Rectifier (SR) Driving Scheme for LLC Resonant Converters 322
Chao Fei, *Virginia Polytechnic Institute and State University, United States*
Fred C. Lee, *Virginia Polytechnic Institute and State University, United States*
Qiang Li, *Virginia Polytechnic Institute and State University, United States*

Digital Synchronous Rectification Controller for LLC Resonant Converters 329
Maryam S. Amouzandeh, *University of Toronto, Canada*
Behzad Mahdavikhah, *University of Toronto, Canada*
Aleksandar Prodić, *University of Toronto, Canada*
Brent McDonald, *Texas Instruments Inc., United States*

A Novel Adaptive Synchronous Rectification Method for Digitally Controlled LLC Converters ... 334
Fan Wang, *Texas Instruments Inc., United States*
Brent A. McDonald, *Texas Instruments Inc., United States*
Jeff Langham, *Texas Instruments Inc., United States*
Bo Fan, *Texas Instruments Inc., China*

Influence of the ADC Zero Bin on the Performance of an Integrated DC-DC Converter 339
S. Vesti, *Infineon Technologies Austria AG, Austria*
M. Agostinelli, *Infineon Technologies Austria AG, Austria*
H. Koltsov, *Infineon Technologies Austria AG, Austria*
S. Marsili, *Infineon Technologies Austria AG, Austria*

Improved Current-Mode Control with Single-Cycle Load Transient 343
Virginia Li, *Virginia Polytechnic Institute and State University, United States*
Pei-Hsin Liu, *Virginia Polytechnic Institute and State University, United States*
Qiang Li, *Virginia Polytechnic Institute and State University, United States*
Fred C. Lee, *Virginia Polytechnic Institute and State University, United States*

A Mixed-Signal Ripple-Based Controller for a 16 V, 10 MHz Integrated Buck Converter 350
Sergii Tkachov, *Infineon Technologies Austria AG, Austria*
Matteo Agostinelli, *Infineon Technologies Austria AG, Austria*

**New Control Concept for Soft-Switching Flyback Converters with Very High
Switching Frequency** 355
A.M. Connaughton, *Technische Universität Graz, Austria*
K. Krischan, *Technische Universität Graz, Austria*
K.K. Leong, *Infineon Technologies AG, Austria*
A. Muetze, *Technische Universität Graz, Austria*

Session T07: Solar Energy Systems
Location: 104C
March 22, 2016 8:30 - 12:00
Session Chairs: Babak Fahimi, *UT- Dallas*
Morgan Kiani, *Texas Christian University*

**Analysis, Modeling and Control of an Interleaved Isolated Boost Series Resonant
Converter for Microinverter Applications** 362
Luciano A. Garcia-Rodriguez, *University of Arkansas, United States*
Cheng Deng, *University of Arkansas, United States*
Juan Carlos Balda, *University of Arkansas, United States*
Andrés Escobar-Mejía, *Universidad Tecnologica de Pereira, Colombia*

**Benchmarking of Constant Power Generation Strategies for Single-Phase Grid-
Connected Photovoltaic Systems** 370
Ariya Sangwongwanich, *Aalborg University, Denmark*
Yongheng Yang, *Aalborg University, Denmark*
Frede Blaabjerg, *Aalborg University, Denmark*
Huai Wang, *Aalborg University, Denmark*

**Advanced Slip Mode Frequency Shift Islanding Detection Method for Single Phase Grid
Connected PV Inverters** 378
Bahador Mohammadpour, *Queen's University, Canada*
Majid Pahlevani, *Queen's University, Canada*
Sajjad Makhdoomi Kaviri, *Queen's University, Canada*
Praveen Jain, *Queen's University, Canada*

Direct MPPT Control of PWM Converters for Extreme Transient PV Applications 386
Ignacio Galiano Zurbriggen, *University of British Columbia, Canada*
Francisco Paz, *University of British Columbia, Canada*
Martin Ordonez, *University of British Columbia, Canada*

Feeding Partial Power into Line Capacitors for Low Cost and Efficient MPPT of Photovoltaic Strings 392
Ali Elrayyah, *Qatar Environment and Energy Research Institute, Qatar*
Mohammed Badawey, *University of Akron, United States*
Yilmaz Sozer, *University of Akron, United States*

Single Phase Cascaded H5 Inverter with Leakage Current Elimination for Transformerless Photovoltaic System 398
Xiaoqiang Guo, *Yanshan University, China*
Xiaoyu Jia, *Yanshan University, China*
Zhigang Lu, *Yanshan University, China*
Josep M. Guerrero, *Aalborg University, Denmark*

Optimal Low Switching Frequency Pulse Width Modulation of Current-Fed Three-Level Inverter for Solar Integration 402
Gnana Sambandam Kulothungan, *National University of Singapore, Singapore*
Akshay K. Rathore, *National University of Singapore, Singapore*
Amarendra Edpuganti, *National University of Singapore, Singapore*
Dipti Srinivasan, *National University of Singapore, Singapore*

Low Leakage Current Single-Phase PV Inverters with Universal Neutral-Point-Clamping Method 410
Liwei Zhou, *Shandong University, China*
Feng Gao, *Shandong University, China*

Modular Subpanel Photovoltaic Converter System: Analysis and Control 417
Yuan Li, *Sichuan University / Northeastern University, China*
Yue Zheng, *Northeastern University, United States*
Su Sheng, *Northeastern University, United States*
Brad Scandrett, *PowerFilm, Inc., United States*
Brad Lehman, *Northeastern University, United States*

Session T08: Advanced Converter for Power Systems used in Transportation
Location: 103AB
March 22, 2016 8:30 - 12:00
Session Chairs: Omer Onar, *Oak Ridge National Laboratory*
  Khurram Afridi, *University of Colorado, Boulder*

Integrated DC-DC Converter Design for Electric Vehicle Powertrains 424
Saeed Anwar, *University of Tennessee, United States*
Weimin Zhang, *University of Tennessee, United States*
Fred Wang, *University of Tennessee, United States*
Daniel J. Costinett, *University of Tennessee, United States*

A 1 MHz Bi-Directional Soft-Switching DC-DC Converter with Planar Coupled Inductor for Dual Voltage Automotive Systems 432
Chenhao Nan, *Arizona State University, United States*
Raja Ayyanar, *Arizona State University, United States*

A Bridgeless Totem-Pole Interleaved PFC Converter for Plug-In Electric Vehicles 440
Yichao Tang, *University of Maryland, United States*
Weisheng Ding, *University of Maryland, United States*
Alireza Khaligh, *University of Maryland, United States*

Stability Analysis of Hybrid AC/DC Power Systems for More Electric Aircraft 446
Mehdi Karbalaye Zadeh, *Norwegian University of Science and Technology, Norway*
Roghayeh Gavagsaz-Ghoachani, *Université de Lorraine, France*
Babak Nahid-Mobarakeh, *Université de Lorraine, France*
Serge Pierfederici, *Université de Lorraine, France*
Marta Molinas, *Norwegian University of Science and Technology, Norway*

On the Concept of the Multi-Source Inverter 453
Lea Dorn-Gomba, *McMaster University, Canada*
Pierre Magne, *McMaster University, Canada*
Clement Barthelmebs, *McMaster University, Canada*
Ali Emadi, *McMaster University, Canada*

**Time-Domain Analysis of a Wide-DC-Range Series Resonant Dual-Active-Bridge
Bidirectional Converter with a New Passive Auxilliary Circuit** 460
Alireza Safaee, *Queen's University, Canada*
Praveen Jain, *Queen's University, Canada*
Alireza Bakhshai, *Queen's University, Canada*

A New High Capacity Compact Power Modules for High Power EV/HEV Inverters 468
Seiichiro Inokuchi, *Mitsubishi Electric Corporation, Japan*
Shoji Saito, *Mitsubishi Electric Corporation, Japan*
Arata Izuka, *Mitsubishi Electric Corporation, Japan*
Yuki Hata, *Mitsubishi Electric Corporation, Japan*
Shinji Hatae, *Mitsubishi Electric Corporation, Japan*
Toshiya Nakano, *Powerex, Inc., United States*
Eric R. Motto, *Powerex, Inc., United States*

**Modular Pet, Two-Phase Air-Cooled Converter Cell Design and Performance Evaluation
with 1.7kV IGBTs for MV Applications** 472
Frederick Kieferndorf, *ABB Switzerland Ltd, Switzerland*
Uwe Drofenik, *ABB Switzerland Ltd, Switzerland*
Francesco Agostini, *ABB Switzerland Ltd, Switzerland*
Francisco Canales, *ABB Switzerland Ltd, Switzerland*

**A Phase Shift Full Bridge based Reconfigurable PEV Onboard Charger with Extended
ZVS Range and Zero Duty Cycle Loss** 480
Haoyu Wang, *ShanghaiTech University, China*

Session T09: Gate Drives, Failure Analysis, and Protection
Location: 102C
March 22, 2016 8:30 - 12:00
Session Chairs: Zhiliang Zhang, *Nanjing University of Aeronautics and Astronautics*
Indumini Ranmuthu, *Texas Instruments*

Series Arc Fault Detection Method based on Statistical Analysis for DC Microgrids 487
Gab-Su Seo, *Seoul National University, Korea, South*
Jung-Ik Ha, *Seoul National University, Korea, South*
Bo-Hyung Cho, *Seoul National University, Korea, South*
Kyu-Chan Lee, *Smart Power Supply Co., Ltd., Korea, South*

Arc Welding Inverter with Embedded Digital Active EMI Controller 493
Junpeng Ji, *Xi'an Jiaotong University, China*
Wenjie Chen, *Xi'an Jiaotong University, China*
Xu Yang, *Xi'an Jiaotong University, China*

A Thermo-Sensitive Electrical Parameter with Maximum dI_C/dt during Turn-Off for High Power Trench/Field-Stop IGBT Modules 499
Yuxiang Chen, *Zhejiang University, China*
Haoze Luo, *Zhejiang University, China*
Wuhua Li, *Zhejiang University, China*
Xiangning He, *Zhejiang University, China*
Jun Ma, *Shanghai Electric, China*
Guodong Chen, *Shanghai Electric, China*
Ye Tian, *Shanghai Electric, China*
Enxing Yang, *Shanghai Electric, China*

A Software Frequency Response Analysis Method to Monitor Degradation of Power MOSFETs in Basic Single-Switch Converters 505
Serkan Dusmez, *University of Texas at Dallas, United States*
Manish Bhardwaj, *Texas Instruments Inc., United States*
Lei Sun, *University of Texas at Dallas, United States*
Bilal Akin, *University of Texas at Dallas, United States*

A New Capacitance Estimation Method of Supercapacitor Bank using a Bank Impedance and Current Injection 511
Junwon Lee, *Chungnam National University, Korea, South*
Hyunsik Jo, *Chungnam National University, Korea, South*
Hanju Cha, *Chungnam National University, Korea, South*

Gate Driver Design for 1.7kV SiC MOSFET Module with Rogowski Current Sensor for Shortcircuit Protection 516
Jun Wang, *Virginia Polytechnic Institute and State University, United States*
Zhiyu Shen, *Virginia Polytechnic Institute and State University, United States*
Christina Dimarino, *Virginia Polytechnic Institute and State University, United States*
Rolando Burgos, *Virginia Polytechnic Institute and State University, United States*
Dushan Boroyevich, *Virginia Polytechnic Institute and State University, United States*

2 MHz High-Density Integrated Power Supply for Gate Driver in High-Temperature Applications 524

Remi Perrin, *Université Claude Bernard Lyon 1, France*
Bruno Allard, *Université Claude Bernard Lyon 1, France*
Cyril Buttay, *Université Claude Bernard Lyon 1, France*
Nicolas Quentin, *Université Claude Bernard Lyon 1, France*
Wenli Zhang, *Virginia Polytechnic Institute and State University, United States*
Rolando Burgos, *Virginia Polytechnic Institute and State University, United States*
Dushan Boroyevich, *Virginia Polytechnic Institute and State University, United States*
Philippe Preciat, *Labinal Power Systems, France*
Donatien Martineau, *Labinal Power Systems, France*

Design Consideration of Gate Driver Circuits and PCB Parasitic Parameters of Paralleled E-Mode GaN HEMTs in Zero-Voltage-Switching Applications 529

Juncheng Lu, *Kettering University, United States*
Hua Bai, *Kettering University, United States*
Alan Brown, *Hella Corporate Center USA Inc., United States*
Matt McAmmond, *Hella Corporate Center USA Inc., United States*
Di Chen, *GaN Systems Inc., Canada*
Julian Styles, *GaN Systems Inc., Canada*

A Gate Driver of SiC MOSFET for Suppressing the Negative Voltage Spikes in a Bridge Circuit 536

Qi Zhou, *Shandong University, China*
Feng Gao, *Shandong University, China*

Session T10: Control of AC-DC Converters
Location: 102AB
March 23, 2016 8:30 - 10:10
Session Chairs: Tsorng-Juu Liang, *National Cheng-Kung University (Taiwan)*
Laszlo Balogh, *Fairchild Semiconductor*

Interleaved Boost based AC/DC Bidirectional Converter with Four Quadrant Power Control based on One-Cycle Controller (OCC) 544

Snehal Bagawade, *Queen's University, Canada*
Praveen Jain, *Queen's University, Canada*

A New Control Scheme to Improve Load Transient Response of Single Phase PWM Rectifier with Auxiliary Current Injection Circuit 552

Naga Brahmendra Yadav Gorla, *National University of Singapore, Singapore*
Sandeep Kolluri, *National University of Singapore, Singapore*
Pritam Das, *National University of Singapore, Singapore*
Sanjib Kumar Panda, *National University of Singapore, Singapore*

Active Capacitor with Ripple-Based Duty Cycle Modulation for AC-DC Applications 558

Ching-Chieh Yang, *National Taiwan University, Taiwan*
Yang-Lin Chen, *National Taiwan University, Taiwan*
Yaow-Ming Chen, *National Taiwan University, Taiwan*

Novel Approach to Current-Mode Control in DCM/CCM Boundary Boost PFC 564

Giovanni Gritti, *STMicroelectronics, Italy*
Claudio Adragna, *STMicroelectronics, Italy*

Reducing the Switching Frequency Variation Range for CRM Buck PFC Converter by Variable On-Time Control 572

Xiaoping Wang, *Nanjing University of Science and Technology, China*
Kai Yao, *Nanjing University of Science and Technology, China*
Junfang Zhang, *Nanjing University of Science and Technology, China*

Session T11: GaN-Based DC-DC Converters
Location: 104A
March 23, 2016 8:30 - 10:10
Session Chairs: Alexis Kwasinski, *University of Pittsburgh*
Regan Zane, *Utah State*

High Efficiency 20-400 MHz PWM Converters using Air-Core Inductors and Monolithic Power Stages in a Normally-Off GaN Process 580

Alihossein Sepahvand, *University of Colorado at Boulder, United States*
Yuanzhe Zhang, *University of Colorado at Boulder, United States*
Dragan Maksimović, *University of Colorado at Boulder, United States*

Thermal Evaluation of Chip-Scale Packaged Gallium Nitride Transistors 587

David Reusch, *Efficient Power Conversion Corporation, United States*
Johan Strydom, *Efficient Power Conversion Corporation, United States*
Alex Lidow, *Efficient Power Conversion Corporation, United States*

Over 300kHz GaN Device based Resonant Bidirectional DCDC Converter with Integrated Magnetics 595

Gang Liu, *Fudan University, China*
Dan Li, *Fudan University, China*
Yungtaek Jang, *Delta Products Corporation, United States*
Jianqiu Zhang, *Fudan University, China*

Effective Control & Software Techniques for High Efficiency GaN FET based Flexible Electrical Power System for Cube-Satellites 601

Ashish Shrivastav, *North Carolina State University, United States*
Shikhar Singh, *IBM, United States*
Anirudh Mahajan, *North Carolina State University, United States*
Subhashish Bhattacharya, *North Carolina State University, United States*

A 98.8% Efficient Bidirectional Full-Bridge Isolated DC-DC GaN Converter 609

Rakesh Ramachandran, *University of Southern Denmark, Denmark*
Morten Nymand, *University of Southern Denmark, Denmark*

Session T12: Electric Machines
Location: 101A
March 23, 2016 8:30 - 10:10
Session Chairs: Bilal Akin, *Univeristy of Texas, Dallas*
Bulent Sarlioglu, *University of Wisconsin - Madison*

Comparison of Lateral- and Cylindrical-Stator Electrical Machines for High-Speed Direct-Drive Applications in Confined Spaces 615

Arda Tüysüz, *ETH Zürich, Switzerland*
Johann W. Kolar, *ETH Zürich, Switzerland*

Novel Contactless Axial-Flux Permanent-Magnet Electromechanical Energy Harvester 623
Michael Flankl, *ETH Zürich, Switzerland*
Arda Tüysüz, *ETH Zürich, Switzerland*
Ivan Subotic, *Liverpool John Moores University, United Kingdom*
Johann W. Kolar, *ETH Zürich, Switzerland*

Design of Rare-Earth Free Five-Phase Outer-Rotor IPM Motor Drive for Electric Bicycle 631
Md. Zakirul Islam, *University of Akron, United States*
Seungdeog Choi, *University of Akron, United States*

Transverse Flux Machines with Rotary Transformer Concept for Wide Speed Operations without using Permanent Magnet Material 638
Iftekhar Hasan, *University of Akron, United States*
Md Wasi Uddin, *University of Akron, United States*
Yilmaz Sozer, *University of Akron, United States*

Field Oriented Modeling and Control of Six Phase, Open-Delta Winding, Interior Permanent Magnet Synchronous Machines Considering Current Unbalance and Zero Sequence Currents 643
Murat Senol, *RWTH Aachen University, Germany*
Michael Schubert, *RWTH Aachen University, Germany*
Georges Engelmann, *RWTH Aachen University, Germany*
Rik W. De Doncker, *RWTH Aachen University, Germany*
Thorben Grosse, *RWTH Aachen University, Germany*
Kay Hameyer, *RWTH Aachen University, Germany*

Session T13: Advances in Magnetics
Location: 101B
March 23, 2016 8:30 - 10:10
Session Chairs: Matthew Wilkowski, *Enpirion*
Charles Sullivan, *Dartmouth*

Passive Integration using FMLF Technique for Integrated Boost Resonant Converters 651
Cheng Deng, *University of Arkansas, United States*
Luciano Andres Garcia Rodriguez, *University of Arkansas, United States*
Juan Zou, *Xiangtan University, China*
Juan Carlos Balda, *University of Arkansas, United States*

Magnetic Characterization Technique and Materials Comparison for Very High Frequency IVR 657
Dongbin Hou, *Virginia Polytechnic Institute and State University, United States*
Fred C. Lee, *Virginia Polytechnic Institute and State University, United States*
Qiang Li, *Virginia Polytechnic Institute and State University, United States*

Large-Signal Power Circuit Characterization of On-Silicon Coupled Inductors for High Frequency Integrated Voltage Regulation 663
S. Kulkarni, *Tyndall National Institute, Ireland*
Z. Pavlovic, *Tyndall National Institute, Ireland*
S. Kubendran, *Tyndall National Institute, Ireland*
C. Carretero, *Universidad de Zaragoza, Spain*
N. Wang, *Tyndall National Institute, Ireland*
C. O'Mathuna, *Tyndall National Institute / University College Cork, Ireland*

Point-of-Load Inductor with High Swinging and Low Loss at Light Load 668

Ting Ge, *Virginia Polytechnic Institute and State University, United States*
Khai Ngo, *Virginia Polytechnic Institute and State University, United States*
Jim Moss, *Texas Instruments Inc., United States*

Iron Loss Evaluation of Three-Phase Inductor for Three-Phase PWM Inverter 676

Hiroaki Matsumori, *Tokyo Metropolitan University, Japan*
Toshihisa Shimizu, *Tokyo Metropolitan University, Japan*
Koushi Takano, *Iwatsu Test Instrument Corporation, Japan*
Ishii Hitoshi, *Iwatsu Test Instrument Corporation, Japan*

Session T14: System Design and Layout for Improved Performance
Location: 102C
March 23, 2016 8:30 - 10:10
Session Chairs: Jeff Nilles, *Texas Instruments*
                    Ernie Parker, *Crane Aerospace & Electronics*

CMOS Gate Drive IC with Embedded Cross Talk Suppression Circuitry for SiC Devices 684

Jeffery Dix, *University of Tennessee, United States*
Zheyu Zhang, *University of Tennessee, United States*
Benjamin J. Blalock, *University of Tennessee, United States*

Optimal Design of a Voltage Regulator based Resonant Switched-Capacitor Converter IC 692

Eli Abramov, *Ben-Gurion University of the Negev, Israel*
Alon Cervera, *Ben-Gurion University of the Negev, Israel*
Mor Mordechai Peretz, *Ben-Gurion University of the Negev, Israel*

Novel Highly Integrated Current Measurement Method for Drive Inverters 700

N. Langmaack, *Technische Universität Braunschweig, Germany*
G. Tareilus, *Technische Universität Braunschweig, Germany*
M. Henke, *Technische Universität Braunschweig, Germany*

**A Novel DBC Layout for Current Imbalance Mitigation in SiC MOSFET Multichip
Power Modules** 704

Helong Li, *Aalborg University, Denmark*
Stig Munk-Nielsen, *Aalborg University, Denmark*
Szymon Bęczkowski, *Aalborg University, Denmark*
Xiongfei Wang, *Aalborg University, Denmark*

**A Double-End Sourced Multi-Chip Improved Wire-Bonded SiC MOSFET Power
Module Design** 709

Miao Wang, *Ohio State University, United States*
Fang Luo, *Ohio State University, United States*
Longya Xu, *Ohio State University, United States*

Session T15: Modeling of AC Energy Converters and Systems
Location: 104B
March 23, 2016 8:30 - 10:10
Session Chairs: Jaber Abu Qahouq, *The University of Alabama*
Xiongfei Wang, *Aalborg University*

Comparing Extended Kalman Filter and Particle Filter for Estimating Field and Damper Bar Currents in Brushless Wound Field Synchronous Generator for Stator Winding Fault Detection and Diagnosis 715
Sivakumar Nadarajan, *National University of Singapore, Singapore*
S.K. Panda, *National University of Singapore, Singapore*
Bicky Bhangu, *Rolls-Royce Singapore Pte. Ltd., Singapore*
Amit Kumar Gupta, *Rolls-Royce Singapore Pte. Ltd., Singapore*

Analytical Determination of Conduction Power Losses for Active Neutral-Point-Clamped Multilevel Converter 720
Vahid Dargahi, *Clemson University, United States*
Arash Khoshkbar Sadigh, *Extron Electronics, United States*
Keith Corzine, *Clemson University, United States*

Multifrequency Small-Signal Model of Voltage Source Converters Connected to a Weak Grid for Stability Analysis 728
Xing Li, *Huazhong University of Science and Technology, China*
Hua Lin, *Huazhong University of Science and Technology, China*

A New Approach to Control the Modified LinVerter for High Frequency Applications 733
Peyman Farhang, *University of Southern Denmark, Denmark*
Stefan Mátéfi-Tempfli, *University of Southern Denmark, Denmark*

Small-Signal Terminal Characteristics Modeling of Three-Phase Boost Rectifier with Variable Fundamental Frequency 739
Zeng Liu, *Xi'an Jiaotong University, China*
Jinjun Liu, *Xi'an Jiaotong University, China*
Dushan Boroyevich, *Virginia Polytechnic Institute and State University, United States*

Session T16: Manufacturing, Test, and Reliability
Location: 103C
March 23, 2016 8:30 - 10:10
Session Chairs: Jim Marinos, *Payton Group*
Brian Narveson, *Narveson Innovative Consulting*

Reliability Analysis of a High-Efficiency SiC Three-Phase Inverter for Motor Drive Applications 746
Juan Colmenares, *KTH Royal Institute of Technology, Sweden*
Diane-Perle Sadik, *KTH Royal Institute of Technology, Sweden*
Patrik Hilber, *KTH Royal Institute of Technology, Sweden*
Hans-Peter Nee, *KTH Royal Institute of Technology, Sweden*

RCP Evaluation of Electrolytic Capacitor Degradation for SMPS Failure Prediction 754
Hiroshi Nakao, *Fujitsu Laboratories Ltd., Japan*
Yu Yonezawa, *Fujitsu Laboratories Ltd., Japan*
Yoshiyasu Nakashima, *Fujitsu Laboratories Ltd., Japan*
Fujio Kurokawa, *Nagasaki University, Japan*

Modular Test System Architecture for Device, Circuit and System Level Reliability Testing 759
Roland Sleik, *Kompetenzzentrum Automobil- und Industrieelektronik GmbH, Austria*
Michael Glavanovics, *Kompetenzzentrum Automobil- und Industrieelektronik GmbH, Austria*
Sascha Einspieler, *Kompetenzzentrum Automobil- und Industrieelektronik GmbH, Austria*
Annette Muetze, *Technische Universität Graz, Austria*
Klaus Krischan, *Technische Universität Graz, Austria*

EMI Noise Cancelation by Optimizing Transformer Design without Need for the Traditional Y-Capacitor 766
Yongjiang Bai, *Xi'an Jiaotong University, China*
Wenjie Chen, *Xi'an Jiaotong University, China*
Ruirui He, *Xi'an Jiaotong University, China*
Dan Zhang, *Silergy Corp., China*
Xu Yang, *Xi'an Jiaotong University, China*

Manufacturing, Assembly and Production Qualifications of High Density, High Reliability POL DC-DC Converters 772
Fariborz Musavi, *CUI Inc., United States*

Session T17: Soft-Switching Converters in Renewable Energy Systems
Location: 104C
March 23, 2016 8:30 - 10:10
Session Chairs: Khurram Afridi, *University of Colorado at Boulder*
Katherine Kim, *Ulsan NIST*

Power Flow Control and ZVS Analysis of Three Limb High Frequency Transformer based Three-Port DAB 778
Ritwik Chattopadhyay, *North Carolina State University, United States*
Subhashish Bhattacharya, *North Carolina State University, United States*

A Novel Multi-Input Converter using Soft-Switched Single-Switch Input Modules with Integrated Power Factor Correction Capability for Hybrid Renewable Energy Systems 786
Sanjida Moury, *York University, Canada*
John Lam, *York University, Canada*
Vineet Srivastava, *Cistel Technology Inc., Canada*
Ron Church, *Cistel Technology Inc., Canada*

Analysis and Design of Impulse Commutated ZCS Three-Phase Current-Fed Push-Pull DC/DC Converter 794
Radha Sree Krishna Moorthy, *National University of Singapore, Singapore*
Akshay Kumar Rathore, *National University of Singapore, Singapore*

ZCS Resonant Converter based Parallel Balancing of Serially Connected Batteries String 802
Ilya Zeltser, *Rafael Advanced Defense Systems Ltd., Israel*
Or Kirshenboim, *Ben-Gurion University of the Negev, Israel*
Nadav Dahan, *Ben-Gurion University of the Negev, Israel*
Mor Mordechai Peretz, *Ben-Gurion University of the Negev, Israel*

A Novel Topology of High Voltage and High Power Bidirectional ZCS DC-DC Converter based on Serial Capacitors 810

Lejia Sun, *Xi'an Jiaotong University, China*
Fang Zhuo, *Xi'an Jiaotong University, China*
Feng Wang, *Xi'an Jiaotong University, China*
Tianhua Zhu, *Xi'an Jiaotong University, China*

Session T18: Solid State Lighting
Location: 103AB
March 23, 2016 8:30 - 10:10
Session Chairs: Jim Spangler, *Spangler Prototype Inc*
Nan Chen, *ABB*

Control Scheme for TRIAC Dimming High PF Single-Stage LED Driver with Adaptive Bleeder Circuit and Non-Linear Current Reference 816

Weizhong Ma, *Hangzhou Dianzi University, China*
Xiaogao Xie, *Hangzhou Dianzi University, China*
Yang Han, *Hangzhou Dianzi University, China*
Hao Deng, *Hangzhou Dianzi University, China*

Three Phase Converter with Galvanic Isolation based on Loss-Free Resistors for HB-LED Lighting Applications 822

Ignacio Castro, *Universidad de Oviedo, Spain*
Diego G. Lamar, *Universidad de Oviedo, Spain*
Manuel Arias, *Universidad de Oviedo, Spain*
Javier Sebastián, *Universidad de Oviedo, Spain*
Marta M. Hernando, *Universidad de Oviedo, Spain*

A ZV-ZCS Electrolytic Capacitor-Less AC/DC Isolated LED Driver with Continous Energy Regulation 830

John Lam, *York University, Canada*
Nader A. El-Taweel, *York University, Canada*

High Efficiency and Power Density GaN-Based LED Driver 838

Eric Faraci, *Texas Instruments Inc., United States*
Michael Seeman, *Texas Instruments Inc., United States*
Bin Gu, *Texas Instruments Inc., United States*
Yogesh Ramadass, *Texas Instruments Inc., United States*
Paul Brohlin, *Texas Instruments Inc., United States*

A Novel LED Drive System based on Matrix Rectifier 843

Baoping Shi, *Nanjing University of Aeronautics and Astronautics, China*
Bo Zhou, *Nanjing University of Aeronautics and Astronautics, China*
Jiadan Wei, *Nanjing University of Aeronautics and Astronautics, China*
Xianhui Qin, *Nanjing University of Aeronautics and Astronautics, China*
Yuanyu Yang, *Nanjing University of Aeronautics and Astronautics, China*
Bing Liu, *Nanjing University of Aeronautics and Astronautics, China*

Session T19: Resonant and Soft Switching DC-DC Converters
Location: 101A
March 23, 2016 14:00 - 17:30
Session Chairs: Mahshid Amirabadi, *Northeastern University*
Ray Orr, *Solantro*

LLC Synchronous Rectification using Coordinate Modulation 848
Mehdi Mohammadi, *University of British Columbia, Canada*
Navid Shafiei, *University of British Columbia, Canada*
Martin Ordonez, *University of British Columbia, Canada*

Low Parasitics Planar Transformer for LLC Resonant Battery Chargers 854
Mohammad Ali Saket, *University of British Columbia, Canada*
Navid Shafiei, *University of British Columbia, Canada*
Martin Ordonez, *University of British Columbia, Canada*
Marian Craciun, *Delta-Q Technologies Corporation, Canada*
Chris Botting, *Delta-Q Technologies Corporation, Canada*

New Symmetrical Bidirectional L3C Resonant DC-DC Converter with Wide Voltage Range 859
Minjae Kim, *Seoul National University of Science and Technology, Korea, South*
Shinyoung Noh, *Seoul National University of Science and Technology, Korea, South*
Sewan Choi, *Seoul National University of Science and Technology, Korea, South*

Influence of the Junction Capacitance of the Secondary Rectifier Diodes on Output Characteristics in Multi-Resonant Converters 864
Stefan Ditze, *Fraunhofer Institute for Integrated Systems and Device Technology, Germany*
Thomas Heckel, *Fraunhofer Institute for Integrated Systems and Device Technology, Germany*
Martin März, *Fraunhofer Institute for Integrated Systems and Device Technology, Germany*

A Triple Active Bridge DC-DC Converter Capable of Achieving Full-Range ZVS 872
Ling Jiang, *University of Tennessee, United States*
Daniel Costinett, *University of Tennessee, United States*

A Novel High Gain Step-Up Resonant DC-DC Converter for Automotive Application 880
Fei Shang, *Illinois Institute of Technology, United States*
Mahesh Krishnamurthy, *Illinois Institute of Technology, United States*
Alexander Isurin, *Vanner Inc., United States*

Series Injection Enabled Full ZVS Light Load Operation of a 15kV SiC IGBT based Dual Active Half Bridge Converter 886
Awneesh Tripathi, *North Carolina State University, United States*
Sachin Madhusoodhanan, *North Carolina State University, United States*
Krishna Mainali, *North Carolina State University, United States*
Kasunaidu Vechalapu, *North Carolina State University, United States*
Subhashish Bhattacharya, *North Carolina State University, United States*

Soft Switching for Half Bridge Current Doubler for High Voltage Point of Load Converter in Data Center Power Supplies 893
Yutian Cui, *University of Tennessee, United States*
Weimin Zhang, *University of Tennessee, United States*
Leon M. Tolbert, *University of Tennessee, United States*
Daniel J. Costinett, *University of Tennessee, United States*
Fred Wang, *University of Tennessee, United States*
Benjamin J. Blalock, *University of Tennessee, United States*

An Algorithm to Analyze Circulating Current for Multi-Phase Resonant Converter 899
Hongliang Wang, *Queen's University, Canada*
Yang Chen, *Queen's University, Canada*
Zhiyuan Hu, *Queen's University, Canada*
Laili Wang, *Queen's University, Canada*
Tianshu Liu, *Queen's University, Canada*
Wenbo Liu, *Queen's University, Canada*
Yan-Fei Liu, *Queen's University, Canada*
Jahangir Afsharian, *Murata Power Solutions, Canada*
Zhihua Yang, *Murata Power Solutions, Canada*

Session T20: Control Applications and Modulation Schemes
Location: 102C
March 23, 2016 14:00 - 17:30
Session Chairs: Masoud Karimi Ghartemani, *Mississippi state University*
Paul Bauer, *University of Lorraine*

A Simple Active Damping Method for Active Power Filters 907
Huawei Yuan, *Tsinghua University, China*
Xinjian Jiang, *Tsinghua University, China*

Simultaneous Voltage and Current Compensation of the 3-Phase Electric Spring with Decomposed Voltage Control 913
Shuo Yan, *University of Hong Kong, Hong Kong*
Tianbo Yang, *University of Hong Kong, Hong Kong*
C.K. Lee, *University of Hong Kong, Hong Kong*
Siew-Chong Tan, *University of Hong Kong, Hong Kong*
S.Y. Ron Hui, *University of Hong Kong / Imperial College London, Hong Kong*

Self-Synchronization Operation of Global Synchronous Pulsewidth Modulation with Communication Fault Tolerant and Simplified Calculation Capabilities 921
Tao Xu, *Shandong University, China*
Feng Gao, *Shandong University, China*

Design Considerations and Predictive Direct Current Control of Active Regenerative Rectifiers for Harmonic and Current Ripple Reduction 928
Alberto Berzoy, *Florida International University, United States*
A.A.S. Mohamed, *Florida International University, United States*
Osama Mohammed, *Florida International University, United States*

A Robust Controller for Medium Voltage AC Collection Grid for Large Scale Photovoltaic Plants based on Medium Frequency Transformers 936
Bahaa Hafez, *Texas A&M University, United States*
Prasad Enjeti, *Texas A&M University, United States*
Shehab Ahmed, *Texas A&M University at Qatar, Qatar*

Optimal Low Switching Frequency Pulse Width Modulation of Current-Fed Five-Level Inverter for Solar Integration 943
Gnana Sambandam Kulothungan, *National University of Singapore, Singapore*
Akshay K. Rathore, *National University of Singapore, Singapore*
Amarendra Edpuganti, *National University of Singapore, Singapore*
Dipti Srinivasan, *National University of Singapore, Singapore*

Design and Implementation of D-Σ Digital Controlled Multi-function Inverter to Achieve APF, Active Power Injection and Rectification 951
T.-F. Wu, *National Tsing Hua University, Taiwan*
H.-C. Hsieh, *National Chung Cheng University, Taiwan*
L.-C. Lin, *National Tsing Hua University, Taiwan*
C.-H. Chang, *National Tsing Hua University, Taiwan*

Operation and Analysis of an Improved Transformerless Unified Power Flow Controller 959
Yang Liu, *Michigan State University, United States*
Shuitao Yang, *Michigan State University / Ford Motor Company, United States*
Fang Zheng Peng, *Michigan State University, United States*

Design Consideration of Converter based Transmission Line Emulation 966
Bo Liu, *University of Tennessee, United States*
Shuoting Zhang, *University of Tennessee, United States*
Sheng Zheng, *University of Tennessee, United States*
Yiwei Ma, *University of Tennessee, United States*
Fred Wang, *University of Tennessee, United States*
Leon M. Tolbert, *University of Tennessee, United States*

Session T21: Advances in Wide BandGap Devices
Location: 104A
March 23, 2016 14:00 - 17:30
Session Chairs: Doug Hopkins, *North Carolina State University*
Alex Huang, *North Carolina State University*

Short-Circuit Characterization of 10 kV 10A 4H-SiC MOSFET 974
Emanuel-Petre Eni, *Aalborg University, Denmark*
Szymon Bęczkowski, *Aalborg University, Denmark*
Stig Munk-Nielsen, *Aalborg University, Denmark*
Tamas Kerekes, *Aalborg University, Denmark*
Remus Teodorescu, *Aalborg University, Denmark*

Record-Low 10mΩ SiC MOSFETs in TO-247, Rated at 900V 979
Vipindas Pala, *Wolfspeed, A Cree Company, United States*
Gangyao Wang, *Wolfspeed, A Cree Company, United States*
Brett Hull, *Wolfspeed, A Cree Company, United States*
Scott Allen, *Wolfspeed, A Cree Company, United States*
Jeffrey Casady, *Wolfspeed, A Cree Company, United States*
John Palmour, *Wolfspeed, A Cree Company, United States*

Performance Evaluation of Multiple Si and SiC Solid State Devices for Circuit Breaker Application in 380VDC Delivery System 983
Kai Tan, *North Carolina State University, United States*
Pengkun Liu, *North Carolina State University, United States*
Xijun Ni, *North Carolina State University, United States*
Chang Peng, *North Carolina State University, United States*
Xiaoqing Song, *North Carolina State University, United States*
Alex Q. Huang, *North Carolina State University, United States*

Evaluation of High Voltage Cascode GaN HEMTs in Parallel Operation 990

He Li, *Ohio State University, United States*
Xuan Zhang, *Ohio State University, United States*
Lucheng Wen, *Ohio State University, United States*
John Alex Brothers, *Ohio State University, United States*
Chengcheng Yao, *Ohio State University, United States*
Ke Zhu, *Ohio State University, United States*
Jin Wang, *Ohio State University, United States*
Liming Liu, *ABB Inc., United States*
Jing Xu, *ABB Inc., United States*
Joonas Puukko, *ABB Inc., United States*

A New Driving Concept for Normally-On GaN Switches in Cascode Configuration 996

Bernhard Zojer, *Infineon Technologies Austria AG, Austria*

Avoiding Divergent Oscillation of Cascode GaN Device Under High Current Turn-Off Condition 1002

Weijing Du, *Virginia Polytechnic Institute and State University, United States*
Xiucheng Huang, *Virginia Polytechnic Institute and State University, United States*
Fred C. Lee, *Virginia Polytechnic Institute and State University, United States*
Qiang Li, *Virginia Polytechnic Institute and State University, United States*
Wenli Zhang, *Virginia Polytechnic Institute and State University, United States*

Temperature-Dependent Turn-On Loss Analysis for GaN HFETs 1010

Edward A. Jones, *University of Tennessee, United States*
Fred Wang, *University of Tennessee, United States*
Daniel Costinett, *University of Tennessee, United States*
Zheyu Zhang, *University of Tennessee, United States*
Ben Guo, *United Technologies Research Center, United States*

Analysis of Parasitic Elements of SiC Power Modules with Special Emphasis on Reliability Issues 1018

Diane-Perle Sadik, *KTH Royal Institute of Technology, Sweden*
Juan Colmenares, *KTH Royal Institute of Technology, Sweden*
Hans-Peter Nee, *KTH Royal Institute of Technology, Sweden*
Konstantin Kostov, *Acreo Swedish ICT AB, Sweden*
Florian Giezendanner, *Alstom Power Sweden AB, Sweden*
Per Ranstad, *Alstom Power Sweden AB, Sweden*

Static and Dynamic Characterization of GaN HEMT with Low Inductance Vertical Phase Leg Design for High Frequency High Power Applications 1024

Nidhi Haryani, *Virginia Polytechnic Institute and State University, United States*
Xuning Zhang, *Virginia Polytechnic Institute and State University, United States*
Rolando Burgos, *Virginia Polytechnic Institute and State University, United States*
Dushan Boroyevich, *Virginia Polytechnic Institute and State University, United States*

Session T22: Motor Drive Design and Inverter Topologies
Location: 101B
March 23, 2016 14:00 - 17:30
Session Chairs: Yingying Kuai, *Caterpillar Inc.*
Jin Wang, *The Ohio State University*

A Family of Single-Phase Current Source Converters with Double Outputs 1032
Louelson A. Costa, *Universidade Federal de Campina Grande, Brazil*
Maurício B.R. Corrêa, *Universidade Federal de Campina Grande, Brazil*
Montiê A. Vitorino, *Universidade Federal de Campina Grande, Brazil*
Gutemberg G. Dos Santos, *Universidade Federal de Campina Grande, Brazil*
Darlan A. Fernandes, *Universidade Federal da Paraíba, Brazil*

**Multiple-Output Boost Resonant Inverter for High Efficiency and Cost-Effective
Induction Heating Applications** 1040
Hector Sarnago, *Universidad de Zaragoza, Spain*
Oscar Lucia, *Universidad de Zaragoza, Spain*
José M. Burdío, *Universidad de Zaragoza, Spain*

Development of 2-kW Interleaved DC-Capacitor-Less Single-Phase Inverter System 1045
Runruo Chen, *Michigan State University, United States*
Hulong Zeng, *Michigan State University, United States*
Deepak Gunasekaran, *Michigan State University, United States*
Yunting Liu, *Michigan State University, United States*
Fang Z. Peng, *Michigan State University, United States*

Single Stage Transformer Isolated High Frequency AC Link based Open End Drive 1051
Srikant Gandikota, *University of Minnesota, United States*
Ned Mohan, *University of Minnesota, United States*

**A Quasi-Z-Source Integrated Multi-Port Power Converter with Reduced Capacitance for
Switched Reluctance Motor Drives** 1057
Fan Yi, *University of Texas at Dallas, United States*
Wen Cai, *University of Texas at Dallas, United States*

**A Fault-Tolerant Topology of T-Type NPC Inverter with Increased Thermal
Overload Capability** 1065
Jiangbiao He, *Marquette University, United States*
Nathan Weise, *Marquette University, United States*
Lixiang Wei, *Rockwell Automation, United States*
Nabeel A.O. Demerdash, *Marquette University, United States*

**A Novel Analysis and Design Method of Phase Lead Filters in Repetitive Controllers for
Pulse-Width Modulated Inverters** 1071
Shunfeng Yang, *Nanyang Technological University, Singapore*
Peng Wang, *Nanyang Technological University, Singapore*
Yi Tang, *Nanyang Technological University, Singapore*
Michael Zagrodnik, *Rolls-Royce Singapore Pte. Ltd., Singapore*
Xiaolei Hu, *Nanyang Technological University, Singapore*
King Jet Tseng, *Nanyang Technological University, Singapore*

Research on the Filter of Load Side Converter in BDFG based Ship Shaft Power Generation System 1078

Meilin Wang, *Huazhong University of Science and Technology, China*
Hua Lin, *Huazhong University of Science and Technology, China*
Hongbin Yang, *Huazhong University of Science and Technology, China*
Xingwei Wang, *Huazhong University of Science and Technology, China*

Investigation of Common Mode Current Related DC-Bus Overvoltage in Multiple Converter Systems 1084

Jiangbiao He, *Rockwell Automation, United States*
Zoran Vrankovic, *Rockwell Automation, United States*
Patrick E. Ozimek, *Rockwell Automation, United States*
Craig Winterhalter, *Rockwell Automation, United States*

Session T23: Modeling of Magnetic Circuits and Systems
Location: 102AB
March 23, 2016 14:00 - 17:30
Session Chairs: Ed Herbert,
Jin Ye, *San Francisco State University*

High Frequency AC Inductor Analysis and Design for Dual Active Bridge (DAB) Converters .. 1090

Zhe Zhang, *Technical University of Denmark, Denmark*
Michael A.E. Andersen, *Technical University of Denmark, Denmark*

A Comprehensive Assessment of PM Motor Topology Impact on Magnet Defect Fault Signatures 1096

Mohsen Zafarani, *University of Texas at Dallas, United States*
Taner Goktas, *University of Texas at Dallas, United States*
Bilal Akin, *University of Texas at Dallas, United States*

High Frequency Modeling for Transformer Common Mode Noise Coupling Path based on Multiconductor Transmission Line Theory 1102

Peipei Meng, *Wuhan University of Technology, China*
Xiangming Zhang, *Naval University of Engineering, China*

Leakage Flux Modelling of Multi-Winding Transformer using Permeance Magnetic Circuit ... 1108

Min Luo, *École Polytechnique Fédérale de Lausanne, Switzerland*
Drazen Dujic, *École Polytechnique Fédérale de Lausanne, Switzerland*
Jost Allmeling, *Plexim GmbH, Switzerland*

Modeling Magnetic Devices using SPICE: Application to Variable Inductors 1115

J. Marcos Alonso, *Universidad de Oviedo, Spain*
Gilberto Martínez, *Continental Automotive R&D, Mexico*
Marina Perdigão, *Universidade de Coimbra, Portugal*
Marcelo Cosetin, *Universidade Federal de Santa Maria, Brazil*
Ricardo N. do Prado, *Universidade Federal de Santa Maria, Brazil*

Investigation of a Thermal Model for a Permanent Magnet Assisted Synchronous Reluctance Motor 1123

Joseph Herbert, *University of Akron, United States*
A.K.M. Arafat, *University of Akron, United States*
Guo-Xiang Wang, *University of Akron, United States*
Seungdeog Choi, *University of Akron, United States*

Design Procedure for Multi-Phase External Rotor Permanent Magnet Assisted Synchronous Reluctance Machines 1131

Sai Sudheer Reddy Bonthu, *University of Akron, United States*
Seungdeog Choi, *University of Akron, United States*

Applicability and Limitations of an M2Spice-Assisted "Planar-Magnetics-in-the-Circuit" Simulation Approach 1138

Samantha J. Gunter, *Massachusetts Institute of Technology, United States*
Minjie Chen, *Massachusetts Institute of Technology, United States*
Stephanie A. Pavlick, *Massachusetts Institute of Technology, United States*
Rose A. Abramson, *Massachusetts Institute of Technology, United States*
Khurram K. Afridi, *University of Colorado at Boulder, United States*
David J. Perreault, *Massachusetts Institute of Technology, United States*

Session T24: Inverter/Converter Control
Location: 103C
March 23, 2016 14:00 - 17:30
Session Chairs: Siavash Pakdelian, *UMass Lowell*
Behrooz Mirafzal, *Kansas State University*

Solution of Input Double-Line Frequency Ripple Rejection for High-Efficiency High-Power Density String Inverter in Photovoltaic Application 1148

Xiaonan Zhao, *Virginia Polytechnic Institute and State University, United States*
Lanhua Zhang, *Virginia Polytechnic Institute and State University, United States*
Rachael Born, *Virginia Polytechnic Institute and State University, United States*
Jih-Sheng Lai, *Virginia Polytechnic Institute and State University, United States*

Fractional-Order Phase Lead Compensation for Multi-Rate Repetitive Control on Three-Phase PWM DC/AC Inverter 1155

Zhichao Liu, *University of South Carolina, United States*
Bin Zhang, *University of South Carolina, United States*
Keliang Zhou, *University of Glasgow, United Kingdom*

A Robust Modified Model Predictive Control (MMPC) based on Lyapunov Function for Three-Phase Active-Front-End (AFE) Rectifier 1163

M. Parvez, *University of Malaya, Malaysia*
S. Mekhilef, *University of Malaya, Malaysia*
Nadia M.L. Tan, *Universiti Tenega Nasional, Malaysia*
Hirofumi Akagi, *Tokyo Institute of Technology, Japan*

Adaptive Reference Model Predictive Control for Power Electronics 1169

Yun Yang, *University of Hong Kong, Hong Kong*
Siew-Chong Tan, *University of Hong Kong, Hong Kong*
Shu-Yuen Ron Hui, *Imperial College London, United Kingdom*

Power Switch Lifetime Extension Strategies for Three-Phase Converters 1176

Serkan Dusmez, *University of Texas at Dallas, United States*
Enes Ugur, *University of Texas at Dallas, United States*
Bilal Akin, *University of Texas at Dallas, United States*

Current Controller Modeling for an Interleaved Boost with Voltage Multiplier Cells for PV Applications 1183
Alessandro Pevere, *Katholieke Universiteit Leuven, Belgium*
Urmimala Chatterjee, *Katholieke Universiteit Leuven, Belgium*
Johan Driesen, *Katholieke Universiteit Leuven, Belgium*

New Active Capacitor Voltage Balancing Method for Five-Level Stacked Multicell Converter 1191
Arash Khoshkbar Sadigh, *Extron Electronics, United States*
Vahid Dargahi, *Clemson University, United States*
Keith Corzine, *Clemson University, United States*

Gate Signal Jitter Elimination and Noise Shaping Modulation for High-SNR Class-D Power Amplifiers 1198
M. Mauerer, *ETH Zürich, Switzerland*
A. Tüysüz, *ETH Zürich, Switzerland*
J.W. Kolar, *ETH Zürich, Switzerland*

Analysis and Compensation of Inverter Nonlinearity for Three-Level T-Type Inverters 1206
Hyeon-Sik Kim, *Seoul National University, Korea, South*
Yong-Cheol Kwon, *Seoul National University, Korea, South*
Seung-Jun Chee, *Seoul National University, Korea, South*
Seung-Ki Sul, *Seoul National University, Korea, South*

Session T25: Topics in Renewable Energy Systems I
Location: 104B
March 23, 2016 14:00 - 17:30
Session Chairs: Fei Gao, *University of Technology of Belfort-Montbéliard*
Kent Wanner, *John Deere*

Front-End Isolated Quasi-Z-Source DC-DC Converter Modules in Series for Photovoltaic High-Voltage DC Applications 1214
Yushan Liu, *Texas A&M University at Qatar, Qatar*
Haitham Abu-Rub, *Texas A&M University at Qatar, Qatar*
Baoming Ge, *Texas A&M University, United States*

Analysis of Non Detection Zone for Multiple Distributed PCS based on Equivalent Single PCS using Reactive Power Approach 1220
Byeong-Heon Kim, *Seoul National University, Korea, South*
Seung-Ki Sul, *Seoul National University, Korea, South*

Optimal Power Scheduling for a Grid-Connected Hybrid PV-Wind-Battery Microgrid System .. 1227
Adriana Luna, *Aalborg University, Denmark*
Nelson Diaz, *Aalborg University, Denmark*
Mehdi Savaghebi, *Aalborg University, Denmark*
Juan C. Vásquez, *Aalborg University, Denmark*
Josep M. Guerrero, *Aalborg University, Denmark*
Kai Sun, *Tsinghua University, China*
Guoliang Chen, *Shanghai Solar Energy & Technology Co., Ltd., China*
Libing Sun, *Shanghai Solar Energy & Technology Co., Ltd., China*

High Efficiency Power Converter for a Doubly-Fed SOEC/SOFC System 1235
Kevin Tomas-Manez, *Technical University of Denmark, Denmark*
Alexander Anthon, *Technical University of Denmark, Denmark*
Zhe Zhang, *Technical University of Denmark, Denmark*

A Hierarchical Active Balancing Architecture for Li-Ion Batteries 1243
Han-Dong Gui, *Nanjing University of Aeronautics and Astronautics, China*
Zhiliang Zhang, *Nanjing University of Aeronautics and Astronautics, China*
Dong-Jie Gu, *Nanjing University of Aeronautics and Astronautics, China*
Yang Yang, *Nanjing University of Aeronautics and Astronautics, China*
Zhouyu Lu, *Nanjing University of Aeronautics and Astronautics, China*
Yan-Fei Liu, *Queen's University, Canada*

A Series-DG based Autonomous Islanding Microgrid 1249
Beihua Liang, *Tianjin University, China*
Yun Wei Li, *University of Alberta, Canada*
Jinwei He, *Tianjin University, China*
Chengshan Wang, *Tianjin University, China*

An Enhanced Droop Control Scheme for Resilient Active Power Sharing in Paralleled Two-Stage PV Inverter Systems 1253
Hongpeng Liu, *Harbin Institute of Technology, China*
Yongheng Yang, *Aalborg University, Denmark*
Xiongfei Wang, *Aalborg University, Denmark*
Poh Chiang Loh, *Aalborg University, Denmark*
Frede Blaabjerg, *Aalborg University, Denmark*
Wei Wang, *Harbin Institute of Technology, China*
Dianguo Xu, *Harbin Institute of Technology, China*

Voltage Closed-Loop Virtual Synchronous Generator Control of Full Converter Wind Turbine for Grid-Connected and Stand-Alone Operation 1261
Yiwei Ma, *University of Tennessee, United States*
Liu Yang, *University of Tennessee, United States*
Fred Wang, *University of Tennessee, United States*
Leon M. Tolbert, *University of Tennessee, United States*

DC Voltage Ripple Quantification for a Flywheel-Battery based Hybrid Energy Storage System 1267
Christopher R. Lashway, *Florida International University, United States*
Ahmed T. Elsayed, *Florida International University, United States*
Osama A. Mohammed, *Florida International University, United States*

Session T26: Electric Vehicle Charging Systems
Location: 104C
March 23, 2016 14:00 - 17:30
Session Chairs: Jim Spangler, *Spangler Prototype Inc*
Hadi Malek, *Ford*

Adaptive Loss Reduction Charging Strategy Considering Variation of Internal Impedance of Lithium-Ion Polymer Batteries in Electric Vehicle Charging Systems 1273
Nari Kim, *Sungkyunkwan University, Korea, South*
Jung-Hoon Ahn, *Sungkyunkwan University, Korea, South*
Dong-Hee Kim, *Sungkyunkwan University, Korea, South*
Byoung-Kuk Lee, *Sungkyunkwan University, Korea, South*

A Pulse Width Modulated LLC Type Resonant Topology Adpated to Wide Output Voltage Range 1280

Haoyu Wang, *ShanghaiTech University, China*

A Series Resonant Circuit for Voltage Equalization of Series Connected Energy Storage Devices 1286

Yanqi Yu, *University of British Columbia, Canada*
Raed Saasaa, *University of British Columbia, Canada*
Wilson Eberle, *University of British Columbia, Canada*

Implementation of 3.3-kW GaN-Based DC-DC Converter for EV On-Board Charger with Series-Resonant Converter that Employs Combination of Variable-Frequency and Delay-Time Control 1292

Yungtaek Jang, *Delta Products Corporation, United States*
Milan M. Jovanović, *Delta Products Corporation, United States*
Juan M. Ruiz, *Delta Products Corporation, United States*
Misha Kumar, *Delta Products Corporation, United States*
Gang Liu, *Delta Electronics Shanghai Co., Ltd., China*

Dual Active Bridge-Based Full-Integrated Active Filter Auxiliary Power Module for Electrified Vehicle Applications with Single-Phase Onboard Chargers 1300

Ruoyu Hou, *McMaster University, Canada*
Ali Emadi, *McMaster University, Canada*

All-SiC Inductively Coupled Charger with Integrated Plug-In and Boost Functionalities for PEV Applications 1307

M. Chinthavali, *Oak Ridge National Laboratory, United States*
O.C. Onar, *Oak Ridge National Laboratory, United States*
S.L. Campbell, *Oak Ridge National Laboratory, United States*
L.M. Tolbert, *Oak Ridge National Laboratory, United States*

Switching Condition and Loss Modeling of GaN-Based Dual Active Bridge Converter for PHEV Charger 1315

Lingxiao Xue, *Virginia Polytechnic Institute and State University, United States*
Dushan Boroyevich, *Virginia Polytechnic Institute and State University, United States*
Paolo Mattavelli, *Università degli Studi di Padova, Italy*

Analysis of Cascaded Multi-Output-Port Converter for Wireless Plug-In Hybrid/On-Board EV Chargers 1323

Erdem Asa, *Hevo Power Inc. / New York University, United States*
Kerim Colak, *Istanbul Ulasim A.S., Turkey*
Dariusz Czarkowski, *New York University, United States*

Comparative Analysis of High Step-Down Ratio Isolated DC/DC Topologies in PEV Applications 1329

Zhiqing Li, *ShanghaiTech University, China*
Haoyu Wang, *ShanghaiTech University, China*

Session T27: Utility Interface and Inverter Applications
Location: 103AB
March 23, 2016 14:00 - 17:30
Session Chairs: Akshay Kumar Rathore, *Concordia University*
Yichao Tang, *Texas Instruments*

DC to Single-Phase AC Voltage Source Inverter with Power Decoupling Circuit based on Flying Capacitor Topology for PV System 1336
Hiroki Watanabe, *Nagaoka University of Technology, Japan*
Keisuke Kusaka, *Nagaoka University of Technology, Japan*
Keita Furukawa, *Nagaoka University of Technology, Japan*
Koji Orikawa, *Nagaoka University of Technology, Japan*
Jun-Ichi Itoh, *Nagaoka University of Technology, Japan*

GaN FET and Hybrid Modulation based Differential-Mode Inverter 1344
Sudip K. Mazumder, *NextWatt LLC, United States*
Ankit Gupta, *University of Illinois at Chicago, United States*
Shirish Raizada, *University of Illinois at Chicago, United States*
Harshit Soni, *University of Illinois at Chicago, United States*
Nikhil Kumar, *University of Illinois at Chicago, United States*
Paromita Mazumder, *NextWatt LLC, United States*
Parijat Bhattachaarjee, *NextWatt LLC, United States*

Thermal and Electrical Co-Design of a Modular High-Density Single-Phase Inverter using Wide-Bandgap Devices 1350
Steven Chung, *University of Toronto, Canada*
Miad Nasr, *University of Toronto, Canada*
David Guirguis, *University of Toronto, Canada*
Masafumi Otsuka, *University of Toronto, Canada*
Shahab Poshtkouhi, *University of Toronto, Canada*
David K.W. Li, *University of Toronto, Canada*
Vishal Palaniappan, *University of Toronto, Canada*
David Romero, *University of Toronto, Canada*
Cristina Amon, *University of Toronto, Canada*
Ray Orr, *Solantro Semiconductor, Canada*
Olivier Trescases, *University of Toronto, Canada*

Reactive Power Compensation with Improvement of Current Waveform Quality for Single-Phase Buck-Type Dynamic Capacitor 1358
Xinwen Chen, *Huazhong University of Science and Technology, China*
Ke Dai, *Huazhong University of Science and Technology, China*
Chen Xu, *Huazhong University of Science and Technology, China*
Ziwei Dai, *Huazhong University of Science and Technology, China*
Li Peng, *Huazhong University of Science and Technology, China*

Circulating Current Reduction for a D-Σ Digital Controlled Transformerless UPS 1364
T.-F. Wu, *National Tsing Hua University, Taiwan*
T.-H. Shiu, *National Tsing Hua University, Taiwan*
P.-H. Lin, *National Tsing Hua University, Taiwan*
L.-C. Lin, *National Tsing Hua University, Taiwan*
J.-W. Huang, *Industrial Technology Research Institute, Taiwan*

A Multi-Function Three-Level Dynamic Voltage Corrector with Wide Correction Range and Short Circuit Fault Isolation 1371

Jiankun Cao, *Nanjing University of Aeronautics and Astronautics, China*
Pengling Ding, *Nanjing University of Aeronautics and Astronautics, China*
Haichun Liu, *Nanjing University of Aeronautics and Astronautics, China*
Shaojun Xie, *Nanjing University of Aeronautics and Astronautics, China*

Effects and Analysis of Minimum Pulse Width Limitation on Adaptive DC Voltage Control of Grid Converters 1376

Bo Sun, *Aalborg University, Denmark*
Ionut Trintis, *Aalborg University, Denmark*
Stig Munk-Nielsen, *Aalborg University, Denmark*
Josep M. Guerrero, *Aalborg University, Denmark*

Improved Three-Phase Micro-Inverter using Dynamic Dead Time Optimization and Phase-Skipping Control Techniques 1381

S. Milad Tayebi, *University of Central Florida, United States*
Xianmin Mu, *University of Central Florida, United States*
Issa Batarseh, *University of Central Florida, United States*

Correcting Current Imbalances in Three-Phase Four-Wire Distribution Systems 1387

Vinson Jones, *University of Arkansas, United States*
Juan Carlos Balda, *University of Arkansas, United States*

Session T28: Isolated DC-DC Converters
Location: 104A
March 24, 2016 8:30 - 11:20
Session Chairs: Dragan Maksimovic, *UC Boulder*
Zhong Ye, *Texas Instruments*

New Design Methdology for Megahertz-Frequency Resonant DC-DC Converters using Impedance Control Network Architecture 1392

Yushi Liu, *University of Colorado at Boulder, United States*
Ashish Kumar, *University of Colorado at Boulder, United States*
Jie Lu, *University of Colorado at Boulder, United States*
Dragan Maksimovic, *University of Colorado at Boulder, United States*
Khurram K. Afridi, *University of Colorado at Boulder, United States*

Dual Voltage Regulations of Single Switch Flyback Converter using Variable Switching Frequency 1398

Jin-Woong Kim, *Seoul National University, Korea, South*
Jung-Ik Ha, *Seoul National University, Korea, South*

On-Chip PLL-Based Methods for Synchronizing Active Switches Across the Isolation Boundary in DC-DC Converters 1403

Shahab Poshtkouhi, *University of Toronto, Canada*
Miad Fard, *University of Toronto, Canada*
Olivier Trescases, *University of Toronto, Canada*

An Isolated Soft-Switching Buck-Boost Converter Utilizing Two Transformers and Embedded Bidirectional Switches on Secondary-Side for Wide Voltage Applications 1410
Tingting Liu, *Nanjing University of Aeronautics and Astronautics, China*
Hongfei Wu, *Nanjing University of Aeronautics and Astronautics, China*
Yan Xing, *Nanjing University of Aeronautics and Astronautics, China*
Kai Sun, *Tsinghua University, China*

Effect of Transformer Design on Operation of Fundamental Duty Modulation for Dual-Active-Bridge Converter 1416
Wooin Choi, *Seoul National University, Korea, South*
Moonhyun Lee, *Seoul National University, Korea, South*
Bo-Hyung Cho, *Seoul National University, Korea, South*

A High Step-Up Bidirectional Isolated Dual-Active-Bridge Converter with Three-Level Voltage-Doubler Rectifier for Energy Storage Applications 1424
Xiaohai Zhan, *Nanjing University of Aeronautics and Astronautics, China*
Hongfei Wu, *Nanjing University of Aeronautics and Astronautics, China*
Yan Xing, *Nanjing University of Aeronautics and Astronautics, China*
Hongjuan Ge, *Nanjing University of Aeronautics and Astronautics, China*
Xi Xiao, *Tsinghua University, China*

Digitized Self-Oscillating Loop for Piezoelectric Transformer-Based Power Converters 1430
Marzieh Ekhtiari, *Technical University of Denmark, Denmark*
Thomas Andersen, *Technical University of Denmark, Denmark*
Zhe Zhang, *Technical University of Denmark, Denmark*
Michael A.E. Andersen, *Technical University of Denmark, Denmark*

Session T29: Multilevel Converters
Location: 101A
March 24, 2016 8:30 - 11:20
Session Chairs: Maryam Saeedifard, *Georgia Tech*
Julia Zhang, *Oregon State University*

An Isolated Topology for Reactive Power Compensation with a Modularized Dynamic-Current Building-Block 1437
Hao Chen, *Georgia Institute of Technology, United States*
Anish Prasai, *Varentec, Inc., United States*
Deepak Divan, *Georgia Institute of Technology, United States*

Design and Control of a Compact MMC Submodule Structure with Reduced Capacitor Size using the Stacked Switched Capacitor Architecture 1443
Yuan Tang, *University of Warwick, United Kingdom*
Minjie Chen, *Massachusetts Institute of Technology, United States*
Li Ran, *University of Warwick, United Kingdom*

Fundamental Frequency Sorting Strategy for Capacitor Voltage Balance of Modular Multilevel Converters with Phase Disposition PWM 1450
Kun Wang, *Zhejiang University, China*
Yan Deng, *Zhejiang University, China*
Wenyu Li, *Zhejiang University, China*
Hao Peng, *Zhejiang University, China*
Guipeng Chen, *Zhejiang University, China*
Xiangning He, *Zhejiang University, China*

Active Voltage Balancing Control for 10kV Three-Level Converter using Series-Connected HV-IGBTs 1456
Shiqi Ji, *Tsinghua University, China*
Ting Lu, *Tsinghua University, China*
Zhengming Zhao, *Tsinghua University, China*
Hualong Yu, *Tsinghua University, China*
Fred Wang, *University of Tennessee, United States*

Average-Value Model of Modular Multilevel Converters Considering Capacitor Voltage 1462
Heya Yang, *Zhejiang University, China*
Yuxiang Chen, *Zhejiang University, China*
Wuhua Li, *Zhejiang University, China*
Xiangning He, *Zhejiang University, China*
Wei Sun, *China Electric Power Research Institute, China*
Yongning Chi, *China Electric Power Research Institute, China*
Yan Li, *China Electric Power Research Institute, China*

New Submodule Circuits for Modular Multilevel Current Source Converters with DC Fault Ride through Capability 1468
Xinyu Yu, *Tsinghua University, China*
Yingdong Wei, *Tsinghua University, China*
Qirong Jiang, *Tsinghua University, China*

Voltage and Power Balance Control Strategy for Three-Phase Modular Cascaded Solid Stated Transformer 1475
Zhiyu Zhang, *Zhejiang University, China*
Hengyang Zhao, *Zhejiang University, China*
Shihang Fu, *Zhejiang University, China*
Jianjiang Shi, *Zhejiang University, China*
Xiangning He, *Zhejiang University, China*

Session T30: Multilevel and Matrix Converters for Motor Drives
Location: 102C
March 24, 2016 8:30 - 11:20
Session Chairs: SeonHwan Hwang, *Kyungnam University, Korea*
Xiaohu Liu, *GE*

New Flying-Capacitor-Based Multilevel Converter with Optimized Number of Switches and Capacitors Controlled with a New Logic-Form-Equation based Active Voltage Balancing Technique 1481
Vahid Dargahi, *Clemson University, United States*
Arash Khoshkbar Sadigh, *Extron Electronics, United States*
Keith Corzine, *Clemson University, United States*

New Low-Cost Five-Level Active Neutral-Point Clamped Converter 1489
Hongliang Wang, *Queen's University, Canada*
Lei Kou, *Queen's University, Canada*
Yan-Fei Liu, *Queen's University, Canada*
Paresh C. Sen, *Queen's University, Canada*
Sucheng Liu, *Anhui University of Technology, China*

Medium Voltage (≥ 2.3 kV) High Frequency Three-Phase Two-Level Converter Design and Demonstration using 10 kV SiC MOSFETs for High Speed Motor Drive Applications .. 1497

Sachin Madhusoodhanan, *North Carolina State University, United States*
Krishna Mainali, *North Carolina State University, United States*
Awneesh Tripathi, *North Carolina State University, United States*
Kasunaidu Vechalapu, *North Carolina State University, United States*
Subhashish Bhattacharya, *North Carolina State University, United States*

Novel Three Phase Multi-Level Inverter Topology with Symmetrical DC-Voltage Sources . 1505

Ahmed Salem, *Aswan University, Egypt*
Emad M. Ahmed, *Aswan University, Egypt*
Mahrous Ahmed, *Aswan University, Egypt*
Mohamed Orabi, *Aswan University, Egypt*

A 2 kW, Single-Phase, 7-Level, GaN Inverter with an Active Energy Buffer Achieving 216 W/in^3 Power Density and 97.6% Peak Efficiency 1512

Yutian Lei, *University of Illinois at Urbana-Champaign, United States*
Christopher Barth, *University of Illinois at Urbana-Champaign, United States*
Shibin Qin, *University of Illinois at Urbana-Champaign, United States*
Wen-Chuen Liu, *University of Illinois at Urbana-Champaign, United States*
Intae Moon, *University of Illinois at Urbana-Champaign, United States*
Andrew Stillwell, *University of Illinois at Urbana-Champaign, United States*
Derek Chou, *University of Illinois at Urbana-Champaign, United States*
Thomas Foulkes, *University of Illinois at Urbana-Champaign, United States*
Zichao Ye, *University of Illinois at Urbana-Champaign, United States*
Zitao Liao, *University of Illinois at Urbana-Champaign, United States*
Robert C.N. Pilawa-Podgurski, *University of Illinois at Urbana-Champaign, United States*

Indirect Matrix Converter based Open-End Winding AC Drives with Zero Common-Mode Voltage 1520

Saurabh Tewari, *MTS Systems Corporation, United States*
Ranjan K. Gupta, *First Solar, Inc., United States*
Apurva Somani, *Dynapower Company LLC, United States*
Ned Mohan, *University of Minnesota, United States*

Precharging Strategy for Soft Startup Process of Modular Multilevel Converters based on Various SM Circuits 1528

Jiangchao Qin, *Arizona State University, United States*
Suman Debnath, *Oak Ridge National Laboratory, United States*
Maryam Saeedifard, *Georgia Institute of Technology, United States*

Session T31: System Design Techniques for Reduced EMI
Location: 101B
March 24, 2016 8:30 - 11:20
Session Chairs: John Vigars, *Allegro Microsystems*
Doug Hopkins, *North Carolina State University*

Conducted EMI Analysis and Filter Design for MHz Active Clamp Flyback Front-End Converter 1534

Xiucheng Huang, *Virginia Polytechnic Institute and State University, United States*
Junjie Feng, *Virginia Polytechnic Institute and State University, United States*
Fred C. Lee, *Virginia Polytechnic Institute and State University, United States*
Qiang Li, *Virginia Polytechnic Institute and State University, United States*
Yuchen Yang, *Virginia Polytechnic Institute and State University, United States*

EMC Investigation of a Very High Frequency Self-Oscillating Resonant Power Converter 1541
Jeppe A. Pedersen, *Technical University of Denmark, Denmark*
Arnold Knott, *Technical University of Denmark, Denmark*
Michael A.E. Andersen, *Technical University of Denmark, Denmark*

Numerical Optimization of Passive Line Filter Components for Suppression of Electromagnetic Interference (EMI) 1547
Carsten Henkenius, *Universität Paderborn, Germany*
Norbert Fröhleke, *Universität Paderborn, Germany*
Joachim Böcker, *Universität Paderborn, Germany*
Heiko Figge, *Delta Energy Systems GmbH, Germany*

Electromagnetic Noise Coupling and Mitigation for Fast Response On-Die Temperature Sensing in High Power Modules 1554
Chengcheng Yao, *Ohio State University, United States*
Pengzhi Yang, *Ohio State University, United States*
Mingzhi Leng, *Ohio State University, United States*
He Li, *Ohio State University, United States*
Lixing Fu, *Ohio State University, United States*
Jin Wang, *Ohio State University, United States*
Ke Zou, *Ford Motor Company, United States*
Chingchi Chen, *Ford Motor Company, United States*

Ultra-Low Inductance Vertical Phase Leg Design with EMI Noise Propagation Control for Enhancement Mode GaN Transistors 1561
Xuning Zhang, *Virginia Polytechnic Institute and State University, United States*
Zhiyu Shen, *Virginia Polytechnic Institute and State University, United States*
Nidhi Haryani, *Virginia Polytechnic Institute and State University, United States*
Dushan Boroyevich, *Virginia Polytechnic Institute and State University, United States*
Rolando Burgos, *Virginia Polytechnic Institute and State University, United States*

Decoupling of Interaction between WBG Converter and Motor Load for Switching Performance Improvement 1569
Zheyu Zhang, *University of Tennessee, United States*
Fred Wang, *University of Tennessee, United States*
Leon M. Tolbert, *University of Tennessee, United States*
Benjamin J. Blalock, *University of Tennessee, United States*
Daniel J. Costinett, *University of Tennessee, United States*

Control and Characterization of Electromagnetic Emissions in Wide Band Gap based Converter Modules for Ungrounded Grid-Forming Applications 1577
Robert Cuzner, *University of Wisconsin at Milwaukee, United States*
Rasoul Hosseini, *University of Wisconsin at Milwaukee, United States*
Andrew Lemmon, *University of Alabama, United States*
James Gafford, *Mississippi State University, United States*
Michael Mazzola, *Mississippi State University, United States*

Session T32: Modeling of DC Energy Converters and Systems
Location: 102AB
March 24, 2016 8:30 - 11:20
Session Chairs: Santanu Kapat, *IIT Kharagpur*
Sombuddha Chakraborty, *Texas Instruments*

A Practical Switching Time Model for Synchronous Buck Converters 1585
Yuan Rao, *Texas Instruments Inc., United States*
Surinder P. Singh, *Texas Instruments Inc., United States*
Taisuke Kazama, *Texas Instruments Inc., United States*

**Off-Line Identification of Digitally Controlled Power Converters using an Analog
Frequency Response Analyzer** 1591
Marco Meola, *Zentrum Mikroelektronik Dresden AG, Germany*
Anthony Kelly, *Altera Corporation, Ireland*

**Extended Wide-Load Range Model for Multi-Level DC-DC Converters and a Practical
Dual-Mode Digital Controller** 1597
Nenad Vukadinović, *University of Toronto, Canada*
Aleksandar Prodić, *University of Toronto, Canada*
Brett A. Miwa, *Maxim Integrated, United States*
Cory B. Arnold, *Maxim Integrated, United States*
Michael W. Baker, *Maxim Integrated, United States*

Burst Mode Control and Switched-Capacitor Converters Losses 1603
Michael Evzelman, *Utah State University, United States*
Regan Zane, *Utah State University, United States*

Equivalent Circuit Modeling of LLC Resonant Converter 1608
Shuilin Tian, *Virginia Polytechnic Institute and State University, United States*
Fred C. Lee, *Virginia Polytechnic Institute and State University, United States*
Qiang Li, *Virginia Polytechnic Institute and State University, United States*

Small Signal Modeling of the Hysteretic Modulator with a Current Ripple Synthesizer 1616
Yi Huang, *Intersil Corporation, United States*
Chun Cheung, *Intersil Corporation, United States*

A Black-Box Modeling Approach for DC Nanogrids 1624
A. Francés, *Universidad Politécnica de Madrid, Spain*
R. Asensi, *Universidad Politécnica de Madrid, Spain*
O. García, *Universidad Politécnica de Madrid, Spain*
R. Prieto, *Universidad Politécnica de Madrid, Spain*
J. Uceda, *Universidad Politécnica de Madrid, Spain*

Session T33: Gate Drive Techniques

Location: 103C
March 24, 2016 8:30 - 11:20
Session Chairs: Christopher Bridge, *SIMPLIS Technologies*
Martin Ordonez, *University of British Columbia*

Design and Evaluation of Isolated Gate Driver Power Supply for Medium Voltage Converter Applications 1632
Krishna Mainali, *North Carolina State University, United States*
Sachin Madhusoodhanan, *North Carolina State University, United States*
Awneesh Tripathi, *North Carolina State University, United States*
Kasunaidu Vechalapu, *North Carolina State University, United States*
Ankan De, *North Carolina State University, United States*
Subhashish Bhattacharya, *North Carolina State University, United States*

General-Purpose Clocked Gate Driver (CGD) IC with Programmable 63-Level Drivability to Reduce IC Overshoot and Switching Loss of Various Power Transistors 1640
Koutarou Miyazaki, *University of Tokyo, Japan*
Seiya Abe, *Kyushu Institute of Technology, Japan*
Masanori Tsukuda, *Kyushu Institute of Technology, Japan*
Ichiro Omura, *Kyushu Institute of Technology, Japan*
Keiji Wada, *Tokyo Metropolitan University, Japan*
Makoto Takamiya, *University of Tokyo, Japan*
Takayasu Sakurai, *University of Tokyo, Japan*

An Integrated SiC CMOS Gate Driver 1646
Matthew Barlow, *University of Arkansas, United States*
Shamim Ahmed, *University of Arkansas, United States*
H. Alan Mantooth, *University of Arkansas, United States*
A. Matt Francis, *Ozark Integrated Circuits, Inc., United States*

Digital Active Gate Drives using Sequential Optimization 1650
Daniel J. Rogers, *University of Oxford, United Kingdom*
Boris Murmann, *Stanford University, United States*

One Adaptive Turn-Off Method for PFC Converter with Voltage Spike Limitation 1657
Qunfang Wu, *Nanjing University of Aeronautics and Astronautics, China*
Qin Wang, *Nanjing University of Aeronautics and Astronautics, China*
Lan Xiao, *Nanjing University of Aeronautics and Astronautics, China*
Jialin Xu, *Nanjing University of Aeronautics and Astronautics, China*
Hongxu Li, *Nanjing University of Aeronautics and Astronautics, China*

A Digital Implementation for PWM Phase-Frequency Synchronization in SMPS Systems 1663
Luca Bizjak, *Infineon Technologies Austria AG, Austria*
Emanuele Bodano, *Infineon Technologies Austria AG, Austria*
Ante Gotovac, *Infineon Technologies Austria AG, Austria*
Sergii Tkachov, *Infineon Technologies Austria AG, Austria*

A High Accuracy and High Bandwidth Current Sense Circuit for Digitally Controlled DC-DC Buck Converters 1670
David Stack, *Altera Corporation, Ireland*
Anthony Kelly, *Altera Corporation, Ireland*
Thomas Conway, *University of Limerick, Ireland*

Session T34: Energy Storage Systems

Location: 104B
March 24, 2016 8:30 - 11:20
Session Chairs: Wei Qiao, *University of Nebraska Lincoln*
Yilmaz Sozer, *University of Akron*

Modular Multilevel Dual Active Bridge DC-DC Converter with ZVS and Fast DC Fault Recovery for Battery Energy Storage Systems 1675
Yuxiang Shi, *Florida State University, United States*
Rui Li, *Florida State University, United States*
Hui Li, *Florida State University, United States*

An Analytical Framework to Design a Dynamic Frequency Control Scheme for Microgrids using Energy Storage 1682
Ajit A. Renjit, *Ohio State University, United States*
Feng Guo, *NEC Laboratories America, Inc., United States*
Ratnesh Sharma, *NEC Laboratories America, Inc., United States*

Comparative Evaluation of $LiFePO_4$ Cell SOC Estimation Performance with ECM Structure and Noise Model/Data Rejection in the EKF for Transportation Application 1690
Hyun-jun Lee, *Soongsil University, Korea, South*
Joung-hu Park, *Soongsil University, Korea, South*
Jonghoon Kim, *Chosun University, Korea, South*

A Power Sharing Scheme for Series Connected Offshore Wind Turbines in a Medium Voltage DC Collection Grid 1695
Michael T. Daniel, *Texas A&M University, United States*
Prasad N. Enjeti, *Texas A&M University, United States*

Fault Ride-Through Performance Evaluation of an Interleaved Grid-Connected Converter Employing Low Switching Frequency 1702
Lorand Bede, *Aalborg University, Denmark*
Ghanshyamsinh Gohil, *Aalborg University, Denmark*
Mihai Ciobotaru, *University of New South Wales, Australia*
Tamas Kerekes, *Aalborg University, Denmark*
Remus Teodorescu, *Aalborg University, Denmark*
Vassilios G. Agelidis, *University of New South Wales, Australia*

Analysis of Two Charging Modes of Battery Energy Storage System for a Stand-Alone Microgrid 1708
Jongmin Jo, *Chungnam National University, Korea, South*
Hanju Cha, *Chungnam National University, Korea, South*

Proposition and Experimental Verification of a Bi-Directional Isolated DC/DC Converter for Battery Charger-Discharger of Electric Vehicle 1713
Ryota Kondo, *Mitsubishi Electric Corporation, Japan*
Yusuke Higaki, *Mitsubishi Electric Corporation, Japan*
Masaki Yamada, *Mitsubishi Electric Corporation, Japan*

Session T35: Topics on Inductive and Capacitive Wireless Power Transfer
Location: 104C
March 24, 2016 8:30 - 11:20
Session Chairs: Chris Mi, *San Diego State University*
Omer Onar, *Oak Ridge National Laboratory*

A CLLC-Compensated High Power and Large Air-Gap Capacitive Power Transfer System for Electric Vehicle Charging Applications 1721
Fei Lu, *University of Michigan at Ann Arbor, United States*
Hua Zhang, *Northeastern Polytechnical University, China*
Heath Hofmann, *University of Michigan at Ann Arbor, United States*
Chris Mi, *San Diego State University, United States*

A Large Air-Gap Capacitive Power Transfer System with a 4-Plate Capacitive Coupler Structure for Electric Vehicle Charging Applications 1726
Hua Zhang, *Northwestern Polytechnical University, China*
Fei Lu, *University of Michigan at Ann Arbor, United States*
Heath Hofmann, *University of Michigan at Ann Arbor, United States*
Weiguo Liu, *Northwestern Polytechnical University, China*
Chris Mi, *San Diego State University, United States*

Dynamic Wireless Power Transfer System for Electric Vehicles to Simplify Ground Facilities – Power Control and Efficiency Maximization on the Secondary Side – 1731
Katsuhiro Hata, *University of Tokyo, Japan*
Takehiro Imura, *University of Tokyo, Japan*
Yoichi Hori, *University of Tokyo, Japan*

Uniform-Gain Frequency Tracking of Wireless EV Charging for Improving Alignment Flexibility 1737
Yabiao Gao, *University of Georgia, United States*
Antonio Ginart, *University of Georgia / Sonnenbatterie GmbH, United States*
Kathleen Blair Farley, *Southern Company Services, Inc., United States*
Zion Tsz Ho Tse, *University of Georgia, United States*

Design and Optimization of a Multi-Coil System for Inductive Charging with Small Air Gap .. 1741
Christopher Joffe, *Fraunhofer Institute for Integrated Systems and Device Technology, Germany*
Andreas Roßkopf, *Fraunhofer Institute for Integrated Systems and Device Technology, Germany*
Stefan Ehrlich, *Fraunhofer Institute for Integrated Systems and Device Technology, Germany*
Christian Dobmeier, *Fraunhofer Institute for Integrated Systems and Device Technology, Germany*
Martin März, *Fraunhofer Institute for Integrated Systems and Device Technology, Germany*

Core Design for Better Misalignment Tolerance and Higher Range of Wireless Charging for HEV 1748
Mostak Mohammad, *University of Akron, United States*
Sangshin Kwak, *Chung-ang University, Korea, South*
Seungdeog Choi, *University of Akron, United States*

A 25 kW Industrial Prototype Wireless Electric Vehicle Charger 1756
Mariusz Bojarski, *Hevo Power Inc., United States*
Erdem Asa, *Hevo Power Inc. / New York University, United States*
Kerim Colak, *Istanbul Ulasim A.S., Turkey*
Dariusz Czarkowski, *New York University, United States*

Session T36: Wireless Power Transfer

Location: 103AB
March 24, 2016 8:30 - 11:20
Session Chairs: Sriram Jala Reddy, *Ford Motors*
Michael Masquelier, *WAVE*

Full-Bridge Series Resonant Multi-Inverter Featuring New 900-V SiC Devices for Improved Induction Heating Appliances 1762
Mario Pérez-Tarragona, *Universidad de Zaragoza, Spain*
Héctor Sarnago, *Universidad de Zaragoza, Spain*
Óscar Lucía, *Universidad de Zaragoza, Spain*
José M. Burdío, *Universidad de Zaragoza, Spain*

A Novel Phase Control of Single Switch Active Rectifier for Inductive Power Transfer Applications 1767
Kerim Colak, *Istanbul Ulasim A.S., Turkey*
Erdem Asa, *Hevo Power Inc. / New York University, United States*
Dariusz Czarkowski, *New York University, United States*

Optimal Shaped Dipole-Coil Design and Experimental Verification of Inductive Power Transfer System for Home Applications 1773
Duy T. Nguyen, *Korea Advanced Institute of Science and Technology, Korea, South*
Eun S. Lee, *Korea Advanced Institute of Science and Technology, Korea, South*
Byeung G. Choi, *Korea Advanced Institute of Science and Technology, Korea, South*
Chun T. Rim, *Korea Advanced Institute of Science and Technology, Korea, South*

A Novel Time-Sharing Current-Fed ZCS High Frequency Inverter-Applied Resonant DC-DC Converter for Inductive Power Transfer 1780
Kyohei Konishi, *Kobe University, Japan*
Tomokazu Mishima, *Kobe University, Japan*
Mutsuo Nakaoka, *University of Malaya, Malaysia*

Optimization of Coils for Magnetically Coupled Resonant Wireless Power Transfer System based on Maximum Output Power 1788
Dan Jiang, *Nanjing University of Aeronautics and Astronautics, China*
Yong Yang, *Nanjing University of Aeronautics and Astronautics, China*
Fuxin Liu, *Nanjing University of Aeronautics and Astronautics, China*
Xinbo Ruan, *Nanjing University of Aeronautics and Astronautics, China*
Xuling Chen, *Nanjing University of Aeronautics and Astronautics, China*

Online Regulation of Receiver-Side Power and Estimation of Mutual Inductance in Wireless Inductive Link based on Transmitter-Side Electrical Information 1795
Jeff Po Wa Chow, *City University of Hong Kong, Hong Kong*
Henry Shu-Hung Chung, *City University of Hong Kong, Hong Kong*
Chun Sing Cheng, *City University of Hong Kong, Hong Kong*

Dynamic Period Switching of PRS-PWM with Run-Length Limiting Technique for Spurious and Ripple Reduction in Fast Response Wireless Power Transmission 1802
Takahiro Moroto, *Keio University, Japan*
Toru Kawajiri, *Keio University, Japan*
Hiroki Ishikuro, *Keio University, Japan*

Session T37: Single-Phase AC-DC Converters

Location: 102AB
March 24, 2016 14:00 - 17:30
Session Chairs: Dusty Becker, *Emerson Network Power*
Pritam Das, *National University of Singapore*

A Flyback AC/DC Converter using Power Semiconductor Filter for Input Power Factor Correction 1807
Chung-Pui Tung, *City University of Hong Kong, Hong Kong*
Henry Shu-Hung Chung, *City University of Hong Kong, Hong Kong*

Reducing the Variation Range of the Switching Frequency for CRM Boost PFC Converter by Injecting 3rd Harmonic into the Input Current 1815
Yi Wang, *Nanjing University of Science and Technology, China*
Kai Yao, *Nanjing University of Science and Technology, China*

A Sustained Increase of Input Current Distortion in Active Input Current Shapers to Eliminate Electrolytic Capacitor for Designing AC to DC HB-LED Drivers for Retrofit Lamps Applications 1823
D.G. Lamar, *Universidad de Oviedo, Spain*
M. Arias, *Universidad de Oviedo, Spain*
A. Rodriguez, *Universidad de Oviedo, Spain*
J. Sebastian, *Universidad de Oviedo, Spain*
A. Fernandez, *European Space Agency, Netherlands*
J.A. Villarejo, *Universidad de Cartagena, Spain*

Reduced Current Stress Bridgeless Cuk PFC Converter with New Voltage Multiplier Circuit ... 1831
Yi-Hung Liao, *National Penghu University of Science and Technology, Taiwan*

Implementation of Multi-Level Bridgeless PFC Rectifiers for Mid-Power Single Phase Applications 1835
Trong Tue Vu, *Eisergy Ltd., Ireland*
George Young, *Eisergy Ltd., Ireland*

US Mains Stacked Very High Frequency Self-Oscillating Resonant Power Converter with Unified Rectifier 1842
Jeppe A. Pedersen, *Technical University of Denmark, Denmark*
Mickey P. Madsen, *Technical University of Denmark, Denmark*
Jakob D. Mønster, *Technical University of Denmark, Denmark*
Thomas Andersen, *Technical University of Denmark, Denmark*
Arnold Knott, *Technical University of Denmark, Denmark*
Michael A.E. Andersen, *Technical University of Denmark, Denmark*

Digital-Based Interleaving Control for GaN-Based MHz CRM Totem-Pole PFC 1847
Zhengyang Liu, *Virginia Polytechnic Institute and State University, United States*
Zhengrong Huang, *Virginia Polytechnic Institute and State University, United States*
Fred C. Lee, *Virginia Polytechnic Institute and State University, United States*
Qiang Li, *Virginia Polytechnic Institute and State University, United States*

A Novel AC-to-DC Adaptor with Ultra-High Power Density and Efficiency 1853
Yan-Cun Li, *Virginia Polytechnic Institute and State University, United States*
Fred C. Lee, *Virginia Polytechnic Institute and State University, United States*
Qiang Li, *Virginia Polytechnic Institute and State University, United States*
Xiucheng Huang, *Virginia Polytechnic Institute and State University, United States*
Zhengyang Liu, *Virginia Polytechnic Institute and State University, United States*

A Single-Stage Single-Phase Isolated AC-DC Converter based on LLC Resonant Unit and T-Type Three-Level Unit for Battery Charging Applications 1861
Yikai Gao, *University of Texas at Dallas, United States*
Wen Cai, *University of Texas at Dallas, United States*
Fan Yi, *University of Texas at Dallas, United States*

Session T38: Non-Isolated DC-DC Converters
Location: 101A
March 24, 2016 14:00 - 17:30
Session Chairs: Pradeep Shenoy, *Texas Instruments*
Juan Rivas-Davila, *Stanford*

DC-DC Power Converter Controller for SOC Balancing of Paralleled Battery System 1868
Jaber A. Abu Qahouq, *University of Alabama, United States*
Lin Zhang, *University of Alabama, United States*
Yuan Cao, *University of Alabama, United States*
Bharat Balasubramanian, *University of Alabama, United States*

Ultra-Step-Up DC-DC Converter with Integrated Autotransformer and Coupled Inductor ... 1872
Yam P. Siwakoti, *Aalborg University, Denmark*
Frede Blaabjerg, *Aalborg University, Denmark*
Poh Chiang Loh, *Aalborg University, Denmark*

Optimal Dynamic Phase Add/Drop Mechanism in Multiphase DC-DC Buck Converters 1878
Anandha Ruban T T, *Texas Instruments India Pvt. Ltd., India*
Preetam Tadeparthy, *Texas Instruments India Pvt. Ltd., India*
Sankaran Aniruddhan, *Indian Institute of Technology Madras, India*
Vikram Gakhar, *Texas Instruments India Pvt. Ltd., India*
Muthusubramanian Venkateswaran, *Texas Instruments India Pvt. Ltd., India*

A Universal Self-Calibrating Dynamic Voltage and Frequency Scaling (DVFS) Scheme with Thermal Compensation for Energy Savings in FPGAs 1882
Shuze Zhao, *University of Toronto, Canada*
Ibrahim Ahmed, *University of Toronto, Canada*
Carl Lamoureux, *University of Toronto, Canada*
Ashraf Lotfi, *Altera Corporation, United States*
Vaughn Betz, *University of Toronto, Canada*
Olivier Trescases, *University of Toronto, Canada*

Morphing Switched-Capacitor Step-Down DC-DC Converters with Variable Conversion Ratio 1888
Song Xiong, *University of Hong Kong, Hong Kong*
Ying Huang, *University of Hong Kong, Hong Kong*
Siew-Chong Tan, *University of Hong Kong, Hong Kong*
Shu-Yuen Ron Hui, *University of Hong Kong, Hong Kong*

Compact Modular Switched-Capacitor DC/DC Converters with Exponential Voltage Gain 1894
Ying Huang, *University of Hong Kong, Hong Kong*
Song Xiong, *University of Hong Kong, Hong Kong*
Siew-Chong Tan, *University of Hong Kong, Hong Kong*
Shu-Yuen Ron Hui, *University of Hong Kong, Hong Kong*

Study and Implementation of a High Step-Up Voltage DC-DC Converter using Coupled-Inductor and Cascode Techniques 1900
Tsorng-Juu Liang, *National Cheng Kung University, Taiwan*
Yung-Ting Huang, *National Cheng Kung University, Taiwan*
Jian-Hsing Lee, *National Cheng Kung University, Taiwan*
Lo Pang-Yen Ting, *National Cheng Kung University, Taiwan*

20 mV Input, 4.2 V Output Boost Converter with Methodology of Maximum Output Power for Thermoelectric Energy Harvesting 1907
Taichi Ogawa, *Toshiba Corporation, Japan*
Takeshi Ueno, *Toshiba Corporation, Japan*
Takayuki Miyazaki, *Toshiba Corporation, Japan*
Tetsuro Itakura, *Toshiba Corporation, Japan*

Clarification of Relationship between Current Ripple and Power Density in Bidirectional DC-DC Converter 1911
Hoai Nam Le, *Nagaoka University of Technology, Japan*
Koji Orikawa, *Nagaoka University of Technology, Japan*
Jun-Ichi Itoh, *Nagaoka University of Technology, Japan*

Session T39: Inverter Applications and Technologies
Location: 101B
March 24, 2016 14:00 - 17:30
Session Chairs: Ali Khajehoddin, *University of Alberta*
Wen Cai, *University of Texas, Dallas*

Grid-Voltage Feedforward based Control for Grid-Connected LCL-Filtered Inverter with High Robustness and Low Grid Current Distortion in Weak Grid 1919
Jinming Xu, *Nanjing University of Aeronautics and Astronautics, China*
Qiang Qian, *Nanjing University of Aeronautics and Astronautics, China*
Shaojun Xie, *Nanjing University of Aeronautics and Astronautics, China*
Binfeng Zhang, *Nanjing University of Aeronautics and Astronautics, China*

Evaluation of PV Frequency-Watt Function for Fast Frequency Reserves 1926
J. Neely, *Sandia National Laboratories, United States*
J. Johnson, *Sandia National Laboratories, United States*
J. Delhotal, *Sandia National Laboratories, United States*
S. Gonzalez, *Sandia National Laboratories, United States*
M. Lave, *Sandia National Laboratories, United States*

A Systematic Design Method and Verification for a Zero-Ripple Interface for PV/Battery-to-Grid Applications 1934
Suvankar Biswas, *University of Minnesota, United States*
Ned Mohan, *University of Minnesota, United States*
William Robbins, *University of Minnesota, United States*

Grid-Voltage-Feedforward Active Damping for Grid-Connected Inverter with LCL Filter 1941
Minghui Lu, *Aalborg University, Denmark*
Xiongfei Wang, *Aalborg University, Denmark*
Frede Blaabjerg, *Aalborg University, Denmark*
S.M. Muyeen, *Petroleum Institute, U.A.E.*
Ahmed Al-Durra, *Petroleum Institute, U.A.E.*
Siyu Leng, *Petroleum Institute, U.A.E.*

A High Power Density Single-Phase Inverter using Stacked Switched Capacitor Energy Buffer 1947
Colin McHugh, *University of Colorado at Boulder, United States*
Sreyam Sinha, *University of Colorado at Boulder, United States*
Jeffrey Meyer, *University of Colorado at Boulder, United States*
Saad Pervaiz, *University of Colorado at Boulder, United States*
Jie Lu, *University of Colorado at Boulder, United States*
Fan Zhang, *University of Colorado at Boulder, United States*
Hua Chen, *University of Colorado at Boulder, United States*
Hyeokjin Kim, *University of Colorado at Boulder, United States*
Usama Anwar, *University of Colorado at Boulder, United States*
Ashish Kumar, *University of Colorado at Boulder, United States*
Alihossein Sepahvand, *University of Colorado at Boulder, United States*
Scott Jensen, *University of Colorado at Boulder, United States*
Beomseok Choi, *University of Colorado at Boulder, United States*
Daniel Seltzer, *University of Colorado at Boulder, United States*
Robert Erickson, *University of Colorado at Boulder, United States*
Dragan Maksimovic, *University of Colorado at Boulder, United States*
Khurram K. Afridi, *University of Colorado at Boulder, United States*

A Novel Single-Stage Dual-Active Bridge based Isolated DC-AC Converter 1954
Shiladri Chakraborty, *Indian Institute of Technology Kharagpur, India*
Souvik Chattopadhyay, *Indian Institute of Technology Kharagpur, India*

Ultra-Low Ripple Inverters for Distributed Generation Applications 1962
Ang Shen, *Missouri University of Science and Technology, United States*
Pourya Shamsi, *Missouri University of Science and Technology, United States*
Mehdi Ferdowsi, *Missouri University of Science and Technology, United States*

A 15 kV SiC MOSFET Gate Drive with Power Over Fiber based Isolated Power Supply and Comprehensive Protection Functions 1967
Xuan Zhang, *Ohio State University, United States*
He Li, *Ohio State University, United States*
John A. Brothers, *Ohio State University, United States*
Jin Wang, *Ohio State University, United States*
Lixing Fu, *Texas Instruments Inc., United States*
Mico Perales, *MH GoPower Co., Ltd., Taiwan*
John Wu, *MH GoPower Co., Ltd., Taiwan*

A 15-kV Class Intelligent Universal Transformer for Utility Applications 1974
Jih-Sheng Lai, *Virginia Polytechnic Institute and State University, United States*
Wei-Han Lai, *Enertronics, Inc., United States*
Seung-Ryul Moon, *Virginia Polytechnic Institute and State University, United States*
Lanhua Zhang, *Virginia Polytechnic Institute and State University, United States*
Arindam Maitra, *Electric Power Research Institute, United States*

Session T40: Modeling, Modulation and Control of Motor Drive
Location: 102C
March 24, 2016 14:00 - 17:30
Session Chairs: Jin Wang, *The Ohio State University*
River-TinHo Li, *ABB*

Modulation Technique for Common Mode Voltage Reduction in a Matrix Converter Drive Operating with High Voltage Transfer Ratio 1982
Varsha Padhee, *Rockwell Automation, United States*
Ashish Kumar Sahoo, *University of Minnesota, United States*
Ned Mohan, *University of Minnesota, United States*

Soft-Switched Discontinuous Pulse-Width Pulse-Density Modulation Scheme 1989
Arash Rahnamaee, *University of Illinois at Chicago, United States*
Alireza Mojab, *University of Illinois at Chicago, United States*
Hossein Riazmontazer, *University of Illinois at Chicago, United States*
Sudip K. Mazumder, *University of Illinois at Chicago, United States*
Milos Zefran, *University of Illinois at Chicago, United States*

A Novel Flux Estimator based on SOGI with FLL for Induction Machine Drives 1995
Rende Zhao, *China University of Petroleum, China*
Zhen Xin, *Aalborg University, Denmark*
Poh Chiang Loh, *Aalborg University, Denmark*
Frede Blaabjerg, *Aalborg University, Denmark*

Performance Characterization of Random Pulse Width Modulation Algorithms in Industrial and Commercial Adjustable Speed Drives 2003
Kevin Lee, *Eaton Corporation, United States*
Guangtong Shen, *Purdue University, United States*
Wenxi Yao, *Zhejiang University, China*
Zhengyu Lu, *Zhejiang University, China*

Stability Analysis and Controller Synthesis for Digital Single-Loop Voltage-Controlled Inverters 2011
Xiongfei Wang, *Aalborg University, Denmark*
Poh Chiang Loh, *Aalborg University, Denmark*
Frede Blaabjerg, *Aalborg University, Denmark*

High Efficiency, Hybrid Selective Harmonic Elimination Phase-Shift PWM Technique for Cascaded H-Bridge Inverters to Improve Dynamic Response and Operate in Complete Normal Modulation Indices 2019
Amirhossein Moeini, *University of Florida, United States*
Zhao Hui, *University of Florida, United States*
Shuo Wang, *University of Florida, United States*

Implementation and Experimental Validation of Efficiency Improvement in PMSM Drives through Switching Frequency Reduction 2027
Parag Kshirsagar, *United Technologies Research Center, United States*
Krishnan Ramu, *Virginia Polytechnic Institute and State University, United States*

Sensorless Speed Control of Symmetrical Triple-Star Nine-Phase Interior Permanent Magnet Machines 2035
Olorunfemi Ojo, *Tennessee Technological University, United States*
Medhi Ramezani, *Tennessee Technological University, United States*

Mitigation of Common-Mode Noise in Wide Band Gap Device based Motor Drives 2043
Sneha Narasimhan, *Rockwell Automation, United States*
Saurabh Tewari, *MTS Systems Corporation, United States*
Eric Severson, *University of Minnesota, United States*
Rohit Baranwal, *University of Minnesota, United States*
Ned Mohan, *University of Minnesota, United States*

Session T41: Gate Drivers and Integrated Packaging
Location: 103C
March 24, 2016 14:00 - 17:30
Session Chairs: Qiang Li, *Virginia Tech*
Jean-Luc Schanen, *Ecole Nationale Supérieure de l'Energie*

A High-Efficient Driving Isolated Drive-by-Microwave Half-Bridge Gate Driver for a GaN Inverter 2051
Shuichi Nagai, *Panasonic Corporation, Japan*
Yasufumi Kawai, *Panasonic Corporation, Japan*
Osamu Tabata, *Panasonic Corporation, Japan*
Songbaek Choe, *Panasonic Corporation, Japan*
Noboru Negoro, *Panasonic Corporation, Japan*
Tesuzo Ueda, *Panasonic Corporation, Japan*

Sensing Gallium Nitride HEMT Junction Temperature using Gate Drive Output Transient Properties 2055
He Niu, *University of Wisconsin at Madison, United States*
Robert D. Lorenz, *University of Wisconsin at Madison, United States*

Design and Application of a 1200V Ultra-Fast Integrated Silicon Carbide MOSFET Module ... 2063
Suxuan Guo, *North Carolina State University, United States*
Liqi Zhang, *North Carolina State University, United States*
Yang Lei, *North Carolina State University, United States*
Xuan Li, *North Carolina State University, United States*
Wensong Yu, *North Carolina State University, United States*
Alex Q. Huang, *North Carolina State University, United States*

Active Gate Charge Control Strategy for Series-Connected IGBTs 2071
Fan Zhang, *Xi'an Jiaotong University, China*
Xu Yang, *Xi'an Jiaotong University, China*
Yu Ren, *Xi'an Jiaotong University, China*
Ying Chen, *Xi'an Jiaotong University, China*
Ruifeng Gou, *Xi'an XD Power Systems Co., LTD, China*

A MV Intelligent Gate Driver for 15kV SiC IGBT and 10kV SiC MOSFET 2076
Awneesh Tripathi, *North Carolina State University, United States*
Krishna Mainali, *North Carolina State University, United States*
Sachin Madhusoodhanan, *North Carolina State University, United States*
Akshat Yadav, *North Carolina State University, United States*
Kasunaidu Vechalapu, *North Carolina State University, United States*
Subhashish Bhattacharya, *North Carolina State University, United States*

Linear Temperature Sensors in High-Voltage GaN-HEMT Power Devices 2083
Richard Reiner, *Fraunhofer Institute for Applied Solid State Physics, Germany*
Patrick Waltereit, *Fraunhofer Institute for Applied Solid State Physics, Germany*
Beatrix Weiss, *Fraunhofer Institute for Applied Solid State Physics, Germany*
Matthias Wespel, *Fraunhofer Institute for Applied Solid State Physics, Germany*
Dirk Meder, *Fraunhofer Institute for Applied Solid State Physics, Germany*
Michael Mikulla, *Fraunhofer Institute for Applied Solid State Physics, Germany*
Rüdiger Quay, *Fraunhofer Institute for Applied Solid State Physics, Germany*
Oliver Ambacher, *Fraunhofer Institute for Applied Solid State Physics, Germany*

An Innovative Power Module with Power-System-in-Inductor Structure 2087
Laili Wang, *Sumida Corporation, Canada*
Doug Malcolm, *Sumida Corporation, Canada*
Yan-Fei Liu, *Queen's University, Canada*

Thermal Analysis of a Magnetic Packaged Power Module 2095
Laili Wang, *Sumida Corporation, Canada*
Doug Malcolm, *Sumida Corporation, Canada*
Wenbo Liu, *Queen's University, Canada*
Yan-Fei Liu, *Queen's University, Canada*

Analysis of a Low-Inductance Packaging Layout for Full-SiC Power Module Embedding Split Damping 2102
Yu Ren, *Xi'an Jiaotong University, China*
Xu Yang, *Xi'an Jiaotong University, China*
Fan Zhang, *Xi'an Jiaotong University, China*
Linlin Tan, *Xi'an Jiaotong University, China*
Xiangjun Zeng, *Xi'an Jiaotong University, China*

Session T42: Component Modeling
Location: 103AB
March 24, 2016 14:00 - 17:30
Session Chairs: Sheldon Williamson, *University of Ontario Institute of Technology*
Abhijit Pathak, *Infineon/IR*

Comprehensive Parametric Analyses of Thermally Aged Power MOSFETs for Failure Precursor Identification and Lifetime Estimation based on Gate Threshold Voltage 2108
Serkan Dusmez, *University of Texas at Dallas, United States*
Bilal Akin, *University of Texas at Dallas, United States*

Modeling and Design Guidelines of High Density Power Inductor for Battery Power Unit 2114
Zhigang Dang, *University of Alabama, United States*
Jaber A. Abu Qahouq, *University of Alabama, United States*

Degradation of Low Voltage Metal Oxide Varistors in Power Supplies 2122
Dawood Talebi Khanmiri, *Northeastern University, United States*
Roy Ball, *Mersen USA, United States*
Jerry Mosesian, *Mersen USA, United States*
Brad Lehman, *Northeastern University, United States*

Characterization and Modeling of SiC MOSFET Body Diode 2127
Kang Peng, *University of South Carolina, United States*
Soheila Eskandari, *University of South Carolina, United States*
Enrico Santi, *University of South Carolina, United States*

A Simple Behavioral Electro-Thermal Model of GaN FETs for SPICE Circuit Simulation 2136
Liyao Wu, *Georgia Institute of Technology, United States*
Maryam Saeedifard, *Georgia Institute of Technology, United States*

Decomposition and Electro-Physical Model Creation of the CREE 1200V, 50A 3-Ph SiC Module 2141
Adam J. Morgan, *North Carolina State University, United States*
Yang Xu, *North Carolina State University, United States*
Douglas C. Hopkins, *North Carolina State University, United States*
Iqbal Husain, *North Carolina State University, United States*
Wensong Yu, *North Carolina State University, United States*

A Three-Legged MATLAB/Simulink Transformer Model using a Fictitious Delta Winding 2147
Thomas A. Nondahl, *Rockwell Automation, United States*
Jingbo Liu, *Rockwell Automation, United States*
Peter B. Schmidt, *Rockwell Automation, United States*

A Lifetime Prediction Method for LEDs Considering Mission Profiles 2154
Xiaohui Qu, *Southeast University, China*
Huai Wang, *Aalborg University, Denmark*
Xiaoqing Zhan, *City University of Hong Kong, Hong Kong*
Frede Blaabjerg, *Aalborg University, Denmark*
Henry Shu-Hung Chung, *City University of Hong Kong, Hong Kong*

Enhanced Li-Ion Battery Modeling using Recursive Parameters Correction 2161
Jae-Gu Kim, *Sungkyunkwan University, Korea, South*
Jung-Hoon Ahn, *Sungkyunkwan University, Korea, South*
Byoung-Kuk Lee, *Sungkyunkwan University, Korea, South*

Session T43: Grid and Utility Interface
Location: 104A
March 24, 2016 14:00 - 17:30
Session Chairs: Manish Bhardwaj, *Texas Instruments*
Nan Chen, *ABB*

Robust Sensorless Control of Grid Connected Converters with LCL Line Filters using Frequency Adaptive Observers as AC Voltage Estimators 2167
Vlatko Miskovic, *Danfoss Drives, United States*
Vladimir Blasko, *United Technologies Research Center, United States*
Thomas Jahns, *University of Wisconsin at Madison, United States*
Robert Lorenz, *University of Wisconsin at Madison, United States*
Haojiong Zhang, *Danfoss Drives, United States*

Active Stabilization of Direct Matrix Converter Input Side Filter through Grid Current Control 2175
Martin Leubner, *Technische Universität Dresden, Germany*
Nico Remus, *Technische Universität Dresden, Germany*
Marc Stübig, *Technische Universität Dresden, Germany*
Wilfried Hofmann, *Technische Universität Dresden, Germany*

Impedance-Based Stability Analysis of Single-Phase Inverter Connected to Weak Grid with Voltage Feed-Forward Control 2182
Jiangfeng Wang, *Nanjing University of Aeronautics and Astronautics, China*
Jianhui Yao, *Nanjing University of Aeronautics and Astronautics, China*
Haibing Hu, *Nanjing University of Aeronautics and Astronautics, China*
Yan Xing, *Nanjing University of Aeronautics and Astronautics, China*
Xiaobin He, *Shanghai Institute of Space Power-Sources, China*
Kai Sun, *Tsinghua University, China*

New Configuration of Dynamic Voltage Restorer for Medium Voltage Application 2187
Arash Khoshkbar Sadigh, *Extron Electronics, United States*
Vahid Dargahi, *Clemson University, United States*
Keith Corzine, *Clemson University, United States*

Studies on the Clustered Voltage Balancing Mechanism for Cascaded H-Bridge STATCOM ... 2194
Daorong Lu, *Nanjing University of Aeronautics and Astronautics, China*
Haibing Hu, *Nanjing University of Aeronautics and Astronautics, China*
Yan Xing, *Nanjing University of Aeronautics and Astronautics, China*
Xiaobin He, *Shanghai Institute of Space Power-Sources, China*
Kai Sun, *Tsinghua University, China*
Jianhui Yao, *Nanjing University of Aeronautics and Astronautics, China*

Design of a Fast Response Time Single-Phase PLL with DC Offset Rejection Capability ... 2200
Abhijit Kulkarni, *Indian Institute of Science, India*
Vinod John, *Indian Institute of Science, India*

Four New Applications of Second-Order Generalized Integrator Quadrature Signal Generator 2207
Zhen Xin, *Aalborg University, Denmark*
Rende Zhao, *China University of Petroleum, China*
Xiongfei Wang, *Aalborg University, Denmark*
Poh Chiang Loh, *Aalborg University, Denmark*
Frede Blaabjerg, *Aalborg University, Denmark*

Three-Phase Multiple Harmonic Sequence Detection based on Generalized Delayed Signal Superposition 2215
Yong Lu, *Xi'an Jiaotong University, China*
Guochun Xiao, *Xi'an Jiaotong University, China*
Xiongfei Wang, *Aalborg University, Denmark*
Frede Blaabjerg, *Aalborg University, Denmark*

Hybrid Modelling and Control of Single-Phase Grid-Connected NPC Inverters 2223
Xingda Yan, *University of Southampton, United Kingdom*
Zhan Shu, *University of Southampton, United Kingdom*
Suleiman M. Sharkh, *University of Southampton, United Kingdom*

Session T44: Topics in Renewable Energy Systems II
Location: 104B
March 24, 2016 14:00 - 17:30
Session Chairs: Akshay Kumar Rathore, *Concordia University*
Yichao Tang, *Texas Instruments*

Stability Criterion and Controller Parameter Design of Radial-Line Renewable Systems with Multiple Inverters 2229
Wenchao Cao, *University of Tennessee, United States*
Xuan Zhang, *University of Tennessee, United States*
Yiwei Ma, *University of Tennessee, United States*
Fred Wang, *University of Tennessee, United States*

Stability Analysis and Improvement of Solid State Transformer (SST)-Paralleled Inverters System using Negative Impedance Feedback Control 2237
Qing Ye, *Florida State University, United States*
Hui Li, *Florida State University, United States*

Compensator-Less Structures for Droop Control of Single Phase Inverters in a Flexible Microgrid 2245
Onkar Vitthal Kulkarni, *Indian Institute of Technology Bombay, India*
Suryanarayana Doolla, *Indian Institute of Technology Bombay, India*
B.G. Fernandes, *Indian Institute of Technology Bombay, India*

Comparative Evaluation of the Loss and Thermal Performance of Advanced Three Level Inverter Topologies 2252
Alexander Anthon, *Technical University of Denmark, Denmark*
Zhe Zhang, *Technical University of Denmark, Denmark*
Michael A.E. Andersen, *Technical University of Denmark, Denmark*
Grahame Holmes, *RMIT University, Australia*
Brendan McGrath, *RMIT University, Australia*
Carlos Teixeira, *RMIT University, Australia*

Dual Buck Inverter with Series Connected Diodes and Single Inductor 2259
Liwei Zhou, *Shandong University, China*
Feng Gao, *Shandong University, China*

Magnetic Integration of the Harmonic Filter Inductor for Dual-Converter Fed Open-End Transformer Topology 2264
Ghanshyamsinh Gohil, *Aalborg University, Denmark*
Lorand Bede, *Aalborg University, Denmark*
Remus Teodorescu, *Aalborg University, Denmark*
Tamas Kerekes, *Aalborg University, Denmark*
Frede Blaabjerg, *Aalborg University, Denmark*

Mechanism Analysis and Mitigation of Instability in Grid-Connected Voltage Source Inverter with LCL Filters based on Terminal Impedance 2272
Teng Liu, *Xi'an Jiaotong University, China*
Zeng Liu, *Xi'an Jiaotong University, China*
Jinjun Liu, *Xi'an Jiaotong University, China*
Qingyun Dou, *Xi'an Jiaotong University, China*

Seven-Switch Five-Level Active Neutral-Point Clamped Converter and Optimal Modulation Strategy 2278

Hongliang Wang, *Queen's University, Canada*
Lei Kou, *Queen's University, Canada*
Yan-Fei Liu, *Queen's University, Canada*
Paresh C. Sen, *Queen's University, Canada*
Sucheng Liu, *Anhui University of Technology, China*

A Simple Variable Step Size Method for Maximum Power Point Tracking using Commercial Current Mode Control DC-DC Regulators 2286

Su Sheng, *Northeastern University, United States*
Brad Lehman, *Northeastern University, United States*

Session T45: Envelope Tracking and Resonant Conversion
Location: 104C
March 24, 2016 14:00 - 17:30
Session Chairs: Brian Zahnstecher, *PowerRox*
Davide Giacomini, *Infineon*

Envelope Tracking GaN Power Supply for 4G Cell Phone Base Stations 2292

Yuanzhe Zhang, *University of Colorado at Boulder, United States*
Johan Strydom, *Efficient Power Conversion Corporation, United States*
Michael de Rooij, *Efficient Power Conversion Corporation, United States*
Dragan Maksimović, *University of Colorado at Boulder, United States*

Envelope Tracking Power Supply for Volume-Sensitive Low-Power Applications based on a Resonant Switched-Capacitor Converter 2298

Alon Cervera, *Ben-Gurion University of the Negev, Israel*
Mor Mordechai Peretz, *Ben-Gurion University of the Negev, Israel*

A Passive-Impedance-Matching Concept for Multi-Phase Resonant Converter 2304

Hongliang Wang, *Queen's University, Canada*
Yang Chen, *Queen's University, Canada*
Yan-Fei Liu, *Queen's University, Canada*

LLC Converter with Auxiliary Switch for Hold Up Mode Operation 2312

Yang Chen, *Queen's University, Canada*
Hongliang Wang, *Queen's University, Canada*
Yan-Fei Liu, *Queen's University, Canada*
Jahangir Afsharian, *Murata Power Solutions, Canada*
Zhihua Yang, *Queen's University, Canada*

A Common Capacitor Multi-Phase LLC Resonant Converter 2320

Hongliang Wang, *Queen's University, Canada*
Yang Chen, *Queen's University, Canada*
Zhiyuan Hu, *Queen's University, Canada*
Laili Wang, *Queen's University, Canada*
Yajie Qiu, *Queen's University, Canada*
Wenbo Liu, *Queen's University, Canada*
Yan-Fei Liu, *Queen's University, Canada*
Jahangir Afsharian, *Murata Power Solutions, Canada*
Zhihua Yang, *Murata Power Solutions, Canada*

LLC Resonant Converter Design for Bendable Power Converter 2328
Kwun Yuan Godwin Ho, *University of Hong Kong, Hong Kong*
M.H. Bryan Pong, *University of Hong Kong, Hong Kong*
Shu-Yuen Ron Hui, *University of Hong Kong, Hong Kong*

Design Consideration of MHz Active Clamp Flyback Converter with GaN Devices for Low Power Adapter Application 2334
Xiucheng Huang, *Virginia Polytechnic Institute and State University, United States*
Junjie Feng, *Virginia Polytechnic Institute and State University, United States*
Weijing Du, *Virginia Polytechnic Institute and State University, United States*
Fred C. Lee, *Virginia Polytechnic Institute and State University, United States*
Qiang Li, *Virginia Polytechnic Institute and State University, United States*

A New Capacitor Voltage Balancing Control for Hybrid Modular Multilevel Converter with Cascaded Full Bridge 2342
Mahendra B. Ghat, *Indian Institute of Technology Bombay, India*
Anshuman Shukla, *Indian Institute of Technology Bombay, India*
Richa Mishra, *Indian Institute of Technology Bombay, India*

Sensorless Scheduling of the Modular Multilevel Series-Parallel Converter: Enabling a Flexible, Efficient, Modular Battery 2349
Stefan M. Goetz, *Duke University, United States*
Zhongxi Li, *Duke University, United States*
Angel V. Peterchev, *Duke University, United States*
Xinyu Liang, *North Carolina State University, United States*
Chengduo Zhang, *North Carolina State University, United States*
Srdjan M. Lukic, *North Carolina State University, United States*

Session D01: AC-DC Converters
Location: Poster Area
March 24, 2016 11:30 - 14:00
Session Chairs: Nathan Weise, *Marquette*
Daniel Costinett, *University of Tennessee-Knoxville*

An Input Current Calculation Switching Driver for High Power-Factor and Phase-Cut Dimmer Compatibility 2355
Hyunchul Eum, *Fairchild Semiconductor International, Inc., Korea, South*
Youngjong Kim, *Fairchild Semiconductor International, Inc., Korea, South*
Kuohsien Huang, *Fairchild Semiconductor International, Inc., Taiwan*

High Frequency Range Conducted Common-Mode Noise Suppression in SMPS 2360
Jinping Zhou, *Delta Electronics Shanghai Co., Ltd., China*
Yicong Xie, *Delta Electronics Shanghai Co., Ltd., China*
Min Zhou, *Delta Electronics Shanghai Co., Ltd., China*

Improved Medium Voltage AC-DC Rectifier based on 10kV SiC MOSFET for Solid State Transformer (SST) Application 2365
Qianlai Zhu, *North Carolina State University, United States*
Li Wang, *North Carolina State University, United States*
Liqi Zhang, *North Carolina State University, United States*
Wensong Yu, *North Carolina State University, United States*
Alex Q. Huang, *North Carolina State University, United States*

Suppression of Circulating Current in Parallel Operation of Three-Level Converters 2370
Young-Kwang Son, *Seoul National University, Korea, South*
Seung-Jun Chee, *Seoul National University, Korea, South*
Younggi Lee, *Seoul National University, Korea, South*
Seung-Ki Sul, *Seoul National University, Korea, South*
Changjin Lim, *LG Electronics, Korea, South*
Sungjae Huh, *LG Electronics, Korea, South*
Jaeyoon Oh, *LG Electronics, Korea, South*

Hybrid Bridgeless DCM SEPIC Rectifier Integrated with a Modified Switched Capacitor Cell ... 2376
Paulo Junior Silva Costa, *Universidade Federal de Santa Catarina, Brazil*
Telles Brunelli Lazzarin, *Universidade Federal de Santa Catarina, Brazil*
Carlos Henrique Illa Font, *Universidade Tecnológica Federal do Paraná, Brazil*

LCL Filter Design for Three-Phase Two-Level Power Factor Correction using Line Impedance Stabilization Network 2382
Alireza Kouchaki, *University of Southern Denmark, Denmark*
Morten Nymand, *University of Southern Denmark, Denmark*

Sensorless Current Rebuilding Strategy in a Single Phase Bridgeless PFC 2389
Felipe López, *Universidad de Cantabria, Spain*
Paula Lamo, *Universidad de Cantabria, Spain*
Alberto Pigazo, *Universidad de Cantabria, Spain*
F.J. Azcondo, *Universidad de Cantabria, Spain*

A Compact Electrolytic-Free Two-Stage Universal Input Offline LED Driver 2395
Saad Pervaiz, *University of Colorado at Boulder, United States*
Ashish Kumar, *University of Colorado at Boulder, United States*
Khurram K. Afridi, *University of Colorado at Boulder, United States*

Session D02: DC-DC Converters I
Location: Poster Area
March 24, 2016 11:30 - 14:00
Session Chairs: Charles Sullivan, *Dartmouth*
Mahshid Amirabadi, *Northeastern University*

Design Methodology for a High Insulation Voltage Power Transmission Function for IGBT Gate Driver 2401
Sokchea Am, *Grenoble Institute of Technology, France*
Pierre Lefranc, *Grenoble Institute of Technology, France*
David Frey, *Grenoble Institute of Technology, France*
Mahmoud Ibrahim, *Grenoble Institute of Technology, France*

Optimized Design of GaN Switching Capacitor based Envelope Tracking Power Supply for Satellite Applications 2409
Qian Jin, *Nanjing University of Aeronautics and Astronautics, China*
M. Vasić, *Universidad Politécnica de Madrid, Spain*
O. Garcia, *Universidad Politécnica de Madrid, Spain*
P. Alou, *Universidad Politécnica de Madrid, Spain*
J.A. Oliver, *Universidad Politécnica de Madrid, Spain*
J.A. Cobos, *Universidad Politécnica de Madrid, Spain*

An Isolated High Step-Up Converter with Continuous Input Current and LC Snubber 2415

K.I. Hwu, *National Taipei University of Technology, Taiwan*
W.Z. Jiang, *National Taipei University of Technology, Taiwan*
Y.T. Yau, *National Taipei University of Technology, Taiwan*

Output-Inductor-Less Full-Bridge Converter with SiC-MOSFETs for Low Noise and ZVS Operation 2422

Kazuhide Domoto, *Nagasaki University, Japan*
Yoichi Ishizuka, *Nagasaki University, Japan*
Seiya Abe, *Kyushu Institute of Technology, Japan*
Tamotsu Ninomiya, *Green Electronics Research Institute, Kitakyushu, Japan*

Reduction Technique of Leakage Flux Effects on GaN-HEMTs in 5 MHz / 100 W Isolated DC-DC Converters 2430

Akinori Hariya, *Nagasaki University, Japan*
Tomoya Koga, *Nagasaki University, Japan*
Ken Matsuura, *TDK Corporation, Japan*
Hiroshige Yanagi, *TDK-Lambda Corporation, Japan*
Satoshi Tomioka, *TDK-Lambda Corporation, Japan*
Yoichi Ishizuka, *Nagasaki University, Japan*
Tamotsu Ninomiya, *City of Kitakyushu, Japan*

A High-Voltage Level Shifter with Sub-Nano-Second Propagation Delay for Switching Power Converters 2437

Ahmed Abdelmoaty, *Ohio State University, United States*
Mohammad Al-Shyoukh, *TSMC Inc., United States*
Ayman Fayed, *Ohio State University, United States*

Dual-Output, Three-Level GaN-Based DC-DC Converter for Battery Charger Applications 2441

Ren Ren, *Nanjing University of Aeronautics and Astronautics, China*
Bo Liu, *University of Tennessee, United States*
Edward A. Jones, *University of Tennessee, United States*
Fred Wang, *University of Tennessee, United States*
Zheyu Zhang, *University of Tennessee, United States*
Daniel Costinett, *University of Tennessee, United States*

Quadruple Active Bridge DC-DC Converter as the Basic Cell of a Modular Smart Transformer 2449

Levy F. Costa, *Christian-Albrechts-Universität zu Kiel, Germany*
Giampaolo Buticchi, *Christian-Albrechts-Universität zu Kiel, Germany*
Marco Liserre, *Christian-Albrechts-Universität zu Kiel, Germany*

Analytical Model of a Phase-Shift Controlled Three-Level Zero-Voltage Switching Converter 2457

Cas Bakker, *Prodrive Technologies, Netherlands*
Bas Vermulst, *Technische Universiteit Eindhoven, Netherlands*
Anton Driessen, *Prodrive Technologies, Netherlands*

High Efficiency Design for ISOP Converter System with Dual Active Bridge DC-DC Converter 2465

Masaki Sato, *Nagasaki University, Japan*
Kazuhide Domoto, *Nagasaki University, Japan*
Yoichi Ishizuka, *Nagasaki University, Japan*
Masahiro Yamaguchi, *Tohoku University, Japan*
Shinya Manabe, *RICOH Electronic Devices Co., Ltd., Japan*
Hiizu Okubo, *RICOH Electronic Devices Co., Ltd., Japan*
Atsushi Itagaki, *Ryowa Electronics Co., Ltd., Japan*

Wide Input Range Power Converters using a Variable Turns Ratio Transformer 2473

Ziwei Ouyang, *Technical University of Denmark, Denmark*
Michael A.E. Andersen, *Technical University of Denmark, Denmark*

Design Approaches for Fast Supercapacitor Chargers for Applications like SCATMA, SRUPS 2479

Nicoloy Gurusinghe, *University of Waikato, New Zealand*
Nihal Kularatna, *University of Waikato, New Zealand*
W. Howell Round, *University of Waikato, New Zealand*
D. Alistair Steyn-Ross, *University of Waikato, New Zealand*

Stack Multiphase Asymmetrical Half-Bridge Topology Offering Advance Performance and Efficiency 2485

Trong Tue Vu, *Eisergy Ltd., Ireland*
George Young, *Eisergy Ltd., Ireland*

Session D03: DC-DC Converters II
Location: Poster Area
March 24, 2016 11:30 - 14:00
Session Chairs: Jason Stauth, *Dartmouth*
Yan-Fei Liu, *Queens*

Design of a Novel APWM Half-Bridge DC-DC Resonant Converter with Load-Independent Soft-Switching and Reduced Circulating Current 2491

Kawsar Ali, *National University of Singapore, Singapore*
Sandeep Kolluri, *National University of Singapore, Singapore*
Naga Brahmendra Yadav Gorla, *National University of Singapore, Singapore*
Pritam Das, *National University of Singapore, Singapore*
Sanjib Kumar Panda, *National University of Singapore, Singapore*

A Low-Volume Hybrid Step-Down DC-DC Converter based on the Dual use of Flying Capacitor 2497

S.M. Ahsanuzzaman, *University of Toronto, Canada*
Yingxian Ma, *University of Toronto, Canada*
Abrar Ahmed Pathan, *University of Toronto, Canada*
Aleksandar Prodić, *University of Toronto, Canada*

Fractional Pulse Skipping in Digitally Controlled DC-DC Converters for Improved Light-Load Efficiency and Power Spectrum 2504

Bipin Chandra Mandi, *Indian Institute of Technology Kharagpur, India*
Santanu Kapat, *Indian Institute of Technology Kharagpur, India*
Amit Patra, *Indian Institute of Technology Kharagpur, India*

A New Compact and High Efficiency Resonant Converter 2511
Sheng-Yang Yu, *Texas Instruments Inc., United States*

A 10-MHz eGaN FETs based Isolated Class-Φ_2 DCX 2518
Xuewen Zou, *Nanjing University of Aeronautics and Astronautics, China*
Zhiliang Zhang, *Nanjing University of Aeronautics and Astronautics, China*
Zhou Dong, *Nanjing University of Aeronautics and Astronautics, China*
Yuan Zhou, *Nanjing University of Aeronautics and Astronautics, China*
Xiaoyong Ren, *Nanjing University of Aeronautics and Astronautics, China*
Qianhong Chen, *Nanjing University of Aeronautics and Astronautics, China*

**Multi-Level Capacitor Clamped DC-DC Multiplier/Divider with Variable and Fractional
Voltage Gain – An (n/m)X DC-DC Converter** 2525
Deepak Gunasekaran, *Michigan State University, United States*
Liang Qin, *Wuhan University, China*
Ujjwal Karki, *Michigan State University, United States*
Yuan Li, *Sichuan University, China*
Fang Z. Peng, *Michigan State University, United States*

**Multi-Mode Quasi-Z-Source Series Resonant DC/DC Converter for Wide Input Voltage
Range Applications** 2533
Dmitri Vinnikov, *Ubik Solutions LLC, Estonia*
Andrii Chub, *Tallinn University of Technology, Estonia*
Indrek Roasto, *Ubik Solutions LLC, Estonia*
Liisa Liivik, *Tallinn University of Technology, Estonia*

Hybrid Serial-Output Converter for Integrated LED Lighting Applications 2540
T. McRae, *University of Toronto, Canada*
A. Prodić, *University of Toronto, Canada*
G. Lisi, *Texas Instruments Inc., United States*
W. McIntrye, *Texas Instruments Inc., United States*
A. Aguilar, *Texas Instruments Inc., United States*

Analysis and Modeling of a Modular ISOP Full Bridge based Converter with Input Filter ... 2545
P. Zumel, *Universidad Carlos III de Madrid, Spain*
E. Oña, *Universidad Carlos III de Madrid, Spain*
C. Fernandez, *Universidad Carlos III de Madrid, Spain*
M. Sanz, *Universidad Carlos III de Madrid, Spain*
A. Lazaro, *Universidad Carlos III de Madrid, Spain*
A. Barrado, *Universidad Carlos III de Madrid, Spain*
A. Vazquez, *Universidad de Oviedo, Spain*
D.G. Lamar, *Universidad de Oviedo, Spain*

Wide-Input High Power Density Flexible Converter Topology for DC-DC Applications 2553
Parth Jain, *University of Toronto, Canada*
Aleksandar Prodić, *University of Toronto, Canada*
Alexander Gerfer, *Würth Elektronik eiSos GmbH & Co. KG, Germany*

High Efficiency LLC Converter Design for Universal Battery Chargers 2561
Navid Shafiei, *University of British Columbia, Canada*
Ali Arefifar, *University of British Columbia, Canada*
Mohammad Ali Saket, *University of British Columbia, Canada*
Martin Ordonez, *University of British Columbia, Canada*

A New High Power Density Modular Multilevel DC-DC Converter with Localized Voltage Balancing Control for Arbitrary Number of Levels 2567
Ahmed Morsy, *Texas A&M University, United States*
Yong Zhou, *Texas A&M University, United States*
Prasad Enjeti, *Texas A&M University, United States*

Design and Control of a Fault Tolerant Soft Switching DC-DC Converter for High Power High Voltage Applications 2573
Tao Li, *Rensselaer Polytechnic Institute, United States*
Leila Parsa, *Rensselaer Polytechnic Institute, United States*

Accurate Parametric Steady State Analysis and Design Tool for DC-DC Power Converters 2579
Mohammad Daryaei, *University of Alberta, Canada*
Mohammad Ebrahimi, *University of Alberta, Canada*
S. Ali Khajehoddin, *University of Alberta, Canada*

Analysis of Multi-Output Half-Wave Semi-Synchronous Rectifier with a Uniform Magnetic Field Transmitter 2587
Erdem Asa, *Hevo Power Inc. / New York University, United States*
Kerim Colak, *Istanbul Ulasim A.S., Turkey*
Dariusz Czarkowski, *New York University, United States*

High Gain QZS DC/DC Converter with Coupled Inductor 2592
Rafael V. Silva, *Universidade Federal do Ceará, Brazil*
Antônio A.A. Freitas, *Universidade Federal do Ceará, Brazil*
Marcus R. Castro, *Universidade Federal do Ceará, Brazil*
Fernando L.M. Antunes, *Universidade Federal Rural do Semi-Árido, Brazil*
Edilson M. Sá Jr., *Universidade Federal do Ceará, Brazil*

Session D04: Utility Interface
Location: Poster Area
March 24, 2016 11:30 - 14:00
Session Chairs: Ali Khajehoddin, *University of Alberta*
Julia Zhang, *Oregon State University*

A Power Decoupling Method with Small Capacitance Requirement based on Single-Phase Quasi-Z-Source Inverter for DC Microgrid Applications 2599
Dingyi He, *University of Texas at Dallas, United States*
Wen Cai, *University of Texas at Dallas, United States*
Fan Yi, *University of Texas at Dallas, United States*

Operation Analysis of High Efficiency Grid Connected Bi-Directional Power Conversion System for Various Storage Battery Systems with Bi-Directional Switch Circuit Topology 2607
Go Yamada, *Panasonic Corporation, Japan*
Takaaki Norisada, *Panasonic Corporation, Japan*
Fumito Kusama, *Panasonic Corporation, Japan*
Keiji Akamatsu, *Panasonic Corporation, Japan*
Masakazu Michihira, *Kobe City College of Technology, Japan*

Fault Tolerant Control of MMC with Redundant Sub-Modules based on Carrier Phase Shift Modulation 2613

Kai Li, *Tsinghua University, China*
Zhengming Zhao, *Tsinghua University, China*
Liqiang Yuan, *Tsinghua University, China*
Sizhao Lu, *Tsinghua University, China*
Bing Pan, *State Grid Smart Grid Research Institute, China*
Zhengang Lu, *State Grid Smart Grid Research Institute, China*

A New Topology of Multilevel VSC Converter for Hybrid HVDC Transmission System 2620

Jae-Jung Jung, *Seoul National University, Korea, South*
Shenghui Cui, *RWTH Aachen University, Germany*
Seung-Ki Sul, *Seoul National University, Korea, South*

Performance of Solid State Transformers Under Imbalanced Loads in Distribution Systems 2629

Tao Yang, *University College Dublin, Ireland*
Ronan Meere, *University College Dublin, Ireland*
Cathal O'Loughlin, *University College Dublin, Ireland*
Terence O'Donnell, *University College Dublin, Ireland*

Steady-State Analysis of Modular Multilevel Converter (MMC) Under Unbalanced Grid Conditions 2637

Xiaojie Shi, *University of Tennessee, United States*
Yalong Li, *University of Tennessee, United States*
Zhiqiang Wang, *University of Tennessee, United States*
Bo Liu, *University of Tennessee, United States*
Leon M. Tolbert, *University of Tennessee, United States*
Fred Wang, *University of Tennessee, United States*

Design and Control of a Compensated Submodule Testing Scheme for Modular Multilevel Converter 2645

Yuan Tang, *University of Warwick, United Kingdom*
Li Ran, *University of Warwick, United Kingdom*
Olayiwola Alatise, *University of Warwick, United Kingdom*
Philip Mawby, *University of Warwick, United Kingdom*

A Voltage Independent Islanding Detection Method and Low Voltage Ride through of a Two-Stage PV Inverter 2652

Partha Pratim Das, *Indian Institute of Technology Kharagpur, India*
Souvik Chattopadhyay, *Indian Institute of Technology Kharagpur, India*
Shiladri Chakraborty, *Indian Institute of Technology Kharagpur, India*

Low Cost and High Efficiency Topology for Flexible Integration of Multi-PV and Batteries in Resonant-Based Converters 2660

Ali Elrayyah, *Qatar Environment and Energy Research Institute, Qatar*

Real-Time Integrated Model of a Micro-Grid with Distributed Clean Energy Generators and their Power Electronics 2666

Weiqiang Chen, *University of Connecticut, United States*
Ali M. Bazzi, *University of Connecticut, United States*
James Hare, *University of Connecticut, United States*
Shalabh Gupta, *University of Connecticut, United States*

Minimization of Inter-Module Leakage Current in Cascaded H-Bridge Multilevel Inverters for Grid Connected Solar PV Applications 2673
V.V.S. Pradeep Kumar, *Indian Institute of Technology Bombay, India*
B.G. Fernandes, *Indian Institute of Technology Bombay, India*

Effect of Grid Inductance on Grid Current Quality of Parallel Grid-Connected Inverter System with Output LCL Filter and Closed-Loop Control 2679
Wooyoung Choi, *University of Wisconsin at Madison, United States*
Woongkul Lee, *University of Wisconsin at Madison, United States*
Bulent Sarlioglu, *University of Wisconsin at Madison, United States*

Small Signal Modeling and Control of a Grid Tied Converter without a Syncronization Unit 2687
Subhajyoti Mukherjee, *Missouri University of Science and Technology, United States*
Pourya Shamsi, *Missouri University of Science and Technology, United States*
Mehdi Ferdowsi, *Missouri University of Science and Technology, United States*

Bridgeless SEPIC PFC Converter for Low Total Harmonic Distortion and High Power Factor ... 2693
Yasemin Onal, *Bilecik Seyh Edebali University, Turkey*
Yilmaz Sozer, *University of Akron, United States*

Effectiveness of Pareto-Front Analysis Applied to the Design of a Single-Phase PFC Rectifier 2700
Mahmoud Ibrahim, *Eaton Corporation, France*
Luc Gonnet, *Eaton Corporation, France*
Pierre Lefranc, *Grenoble Institute of Technology, France*
David Frey, *Grenoble Institute of Technology, France*
Jean-Paul Ferrieux, *Grenoble Institute of Technology, France*
Sokchea Am, *Grenoble Institute of Technology, France*

State Space Analysis and Duty Cycle Control of a Switched Reactance based Center-Point-Clamped Reactive Power Compensator 2706
Pankaj Kumar Bhowmik, *University of North Carolina at Charlotte, United States*
Somasundaram Essakiappan, *University of North Carolina at Charlotte, United States*
Madhav Manjrekar, *University of North Carolina at Charlotte, United States*

A SiC-Based Power Converter Module for Medium-Voltage Fast Charger for Plug-In Electric Vehicles 2714
Srdjan Srdic, *North Carolina State University, United States*
Chi Zhang, *North Carolina State University, United States*
Xinyu Liang, *North Carolina State University, United States*
Wensong Yu, *North Carolina State University, United States*
Srdjan Lukic, *North Carolina State University, United States*

Shunt Active Power Filter based on Cascaded Transformers Coupled with Three-Phase Bridge Converters 2720
Gregory A. de Almeida Carlos, *Universidade Federal de Campina Grande, Brazil*
Cursino B. Jacobina, *Universidade Federal de Campina Grande, Brazil*
João Paulo R. Méllo, *Universidade Federal de Campina Grande, Brazil*
Euzeli C. dos Santos Jr., *Indiana University - Purdue University, United States*

Independent DC Link Voltage Control of Cascaded Multilevel PV Inverter 2727
Qingyun Huang, *North Carolina State University, United States*
Wensong Yu, *North Carolina State University, United States*
Alex Q. Huang, *North Carolina State University, United States*

New Active Damping Method for LCL Filter Resonance based on Two Feedback System 2735

Mahmoud A. Gaafar, *Kyushu University, Japan*
Gamal M. Dousoky, *Minia University, Egypt*
Masahito Shoyama, *Kyushu University, Japan*

Static Synchronous Generator Model for Investigating Dynamic Behaviors and Stability Issues of Grid-Tied Inverters 2742

Liansong Xiong, *Xi'an Jiaotong University, China*
Xiaokang Liu, *Xi'an Jiaotong University, China*
Feng Wang, *Xi'an Jiaotong University, China*
Fang Zhuo, *Xi'an Jiaotong University, China*

Session D05: Motor Drives and Inverters: Modeling and Control I
Location: Poster Area
March 24, 2016 11:30 - 14:00
Session Chairs: Liming Liu, *ABB Inc.*
Thomas Gietzold, *United Technologies Aerospace Systems*

Initial Orientation and Sensorless Starting Strategy of Wound-Rotor Synchronous Starter/Generator 2748

Jichang Peng, *Northwestern Polytechnical University, China*
Weiguo Liu, *Northwestern Polytechnical University, China*
Jinhao Meng, *Northwestern Polytechnical University, China*
Tao Meng, *Northwestern Polytechnical University, China*
Guangzhao Luo, *Northwestern Polytechnical University, China*

A Novel Method for Polarity Detection of Non-Salient PMSMs in Initial Position Estimation 2754

Bing Liu, *Nanjing University of Aeronautics and Astronautics, China*
Bo Zhou, *Nanjing University of Aeronautics and Astronautics, China*
Jiadan Wei, *Nanjing University of Aeronautics and Astronautics, China*
Long Wang, *Nanjing University of Aeronautics and Astronautics, China*
Tianheng Ni, *Nanjing University of Aeronautics and Astronautics, China*

A Speed Adaptive Sensorless Flux Observer for the Induction Motor Drive using Sylvester Criterion Design 2759

Mihai Comanescu, *Penn State Altoona, United States*

Discontinuous PWM for Low Switching Losses in Indirect Matrix Converter Drives 2764

Yeongsu Bak, *Ajou University, Korea, South*
Kyo-Beum Lee, *Ajou University, Korea, South*

Model Predictive Control for Extended Kalman Filter based Speed Sensorless Induction Motor Drives 2770

Jie Li, *Xi'an University of Technology, China*
Li-Heng Zhang, *Xi'an University of Technology, China*
Ying Niu, *Xi'an University of Technology, China*
Hai-Peng Ren, *Xi'an University of Technology, China*

Research on Excitation Control Methods for the Two-Phase Brushless Exciter of Wound-Rotor Synchronous Starter/Generators in the Starting Mode 2776
Ningfei Jiao, *Northwestern Polytechnical University, China*
Weiguo Liu, *Northwestern Polytechnical University, China*
Tao Meng, *Northwestern Polytechnical University, China*
Jichang Peng, *Northwestern Polytechnical University, China*
Shuai Mao, *Northwestern Polytechnical University, China*

A High Performance Speed Regulator Design for AC Machines 2782
Adil Khurram, *American University of Sharjah, U.A.E.*
Habibur Rehman, *American University of Sharjah, U.A.E.*
Shayok Mukhopadhyay, *American University of Sharjah, U.A.E.*

Zero-Sequence Current Suppression for Open-End Winding Induction Motor Drive with Resonant Controller 2788
Hajime Kubo, *Meidensha Corporation, Japan*
Yasuhiro Yamamoto, *Meidensha Corporation, Japan*
Takeshi Kondo, *Meidensha Corporation, Japan*
Kaushik Rajashekara, *University of Texas at Dallas, United States*
Bohang Zhu, *University of Texas at Dallas, United States*

Optimized Control of High-Performance Servo-Motor Drives in the Field-Weakening Region 2794
Jack Bermingham, *Moog Ireland Ltd, Ireland*
Gerard O'Donovan, *Moog Ireland Ltd, Ireland*
Ray Walsh, *Moog Ireland Ltd, Ireland*
Michael Egan, *University College Cork, Ireland*
Gordon Lightbody, *University College Cork, Ireland*
John G. Hayes, *University College Cork, Ireland*

Motor Current Reference Generation for Reducing Motor Currents in Drive Systems with Single-Phase Diode Rectifier and Small DC-Link Capacitor 2801
Young-Ho Chae, *Seoul National University, Korea, South*
Jung-Ik Ha, *Seoul National University, Korea, South*

A Simple Double Mapping based SVPWM Method for Balancing DC-Link Capacitor Voltages of Five-Level Diode-Clamped Converters 2806
Aparna Saha, *University of Akron, United States*
Ali Elrayyah, *Qatar Environment and Energy Research Institute, Qatar*
Yilmaz Sozer, *University of Akron, United States*

Session D06: Motor Drives and Inverters: Modeling and Control II
Location: Poster Area
March 24, 2016 11:30 - 14:00
Session Chairs: Bulent Sarlioglu, *University of Wisconsin - Madison*
Yichao Tang, *Texas Instruments*

Capacitor-Clamped Inverter based Transient Suppression Method for Azimuth Thruster Drives 2813
Shantha Gamini Jayasinghe, *Australian Maritime College, University of Tasmania, Australia*
Viknash Shagar, *Australian Maritime College, University of Tasmania, Australia*
Hossein Enshaei, *Australian Maritime College, University of Tasmania, Australia*
Danyal Mohammadi, *Boise State University, United States*
Mahinda Vilathgamuwa, *Queensland University of Technology, Australia*

Active Common-Mode Voltage Reduction in a Fault-Tolerant Three-Phase Inverter 2821
Danyal Mohammadi, *Boise State University, United States*
Said Ahmed-Zaid, *Boise State University, United States*

Power Cycling Lifetime Improvement of Three-Level NPC Inverters with an Improved DPWM Method 2826
Jiangbiao He, *Marquette University, United States*
Lixiang Wei, *Rockwell Automation, United States*
Nabeel A.O. Demerdash, *Marquette University, United States*

Synchronous Optimal Pulsewidth Modulation Digital Implementation Concept for Multilevel Converters 2833
Jackson Lago, *Universidade Federal de Santa Catarina, Brazil*
Marcelo Lobo Heldwein, *Universidade Federal de Santa Catarina, Brazil*

Analytical Determination of Conduction Losses for Modified Flying Capacitor Multicell Converters 2840
Vahid Dargahi, *Clemson University, United States*
Arash Khoshkbar Sadigh, *Extron Electronics, United States*
Keith Corzine, *Clemson University, United States*

Comparison of Electrical Losses in an Inverter-Fed Five-Phase and Three-Phase Permanent Magnet Assisted Synchronous Reluctance Motor 2847
Akm Arafat, *University of Akron, United States*
Seungdeog Choi, *University of Akron, United States*

A Hybrid Adaptive Observer for the Speed and Flux Estimation of Induction Motors 2855
Mihai Comanescu, *Penn State Altoona, United States*

Determination of CM Choke Parameters for SiC MOSFET Motor Drive based on Simple Measurements and Frequency Domain Modeling 2861
Di Han, *University of Wisconsin at Madison, United States*
Casey Morris, *University of Wisconsin at Madison, United States*
Woongkul Lee, *University of Wisconsin at Madison, United States*
Bulent Sarlioglu, *University of Wisconsin at Madison, United States*

An Improved Model Predictive Current Control of Permanent Magnet Synchronous Motor Drives 2868
Yongchang Zhang, *North China University of Technology, China*
Sugu Gao, *North China University of Technology, China*
Wei Xu, *Huazhong University of Science and Technology, China*

Analysis of Magnet Defect Faults in Permanent Magnet Synchronous Motors through Fluxgate Sensors 2875
Taner Goktas, *University of Texas at Dallas, United States*
Kun Wang Lee, *University of Texas at Dallas, United States*
Mohsen Zafarani, *University of Texas at Dallas, United States*
Bilal Akin, *University of Texas at Dallas, United States*

Session D07: Motor Drives and Inverters: Topologies
Location: Poster Area
March 24, 2016 11:30 - 14:00
Session Chairs: Amirnaser Yazdani, *Ryerson University*
Babak Nahid-Mobarakeh, *University of Lorraine*

Performance Comparison of Transfer Switch Topologies in Switched-Doubly-Fed Machine Drives 2881
Arijit Banerjee, *Massachusetts Institute of Technology, United States*
Steven B. Leeb, *Massachusetts Institute of Technology, United States*
James L. Kirtley, *Massachusetts Institute of Technology, United States*

Multilevel Converter Topologies for High-Power High-Speed Switched Reluctance Motor: Performance Comparison 2889
Devendra Patil, *University of Texas at Dallas, United States*
Shiliang Wang, *University of Texas at Dallas, United States*
Lei Gu, *University of Texas at Dallas, United States*

Bidirectional Magnetically Coupled T-Source Inverter for Extra Low Voltage Application 2897
Thomas Baier, *Friedrich-Alexander-Universität Erlangen-Nürnberg, Germany*
Bernhard Piepenbreier, *Friedrich-Alexander-Universität Erlangen-Nürnberg, Germany*

Active Virtual Ground: Single Phase Grid-Connected Voltage Source Inverter Topology ... 2905
River Tin-Ho Li, *ABB China Ltd., China*
Carl Ngai-Man Ho, *University of Manitoba, Canada*

Design and Evaluation of 30kVA Inverter using SiC MOSFET for 180°C Ambient Temperature Operation 2912
Feng Qi, *Ohio State University, United States*
Miao Wang, *Ohio State University, United States*
Longya Xu, *Ohio State University, United States*
Bo Zhao, *State Grid Corporation of China, China*
Zhe Zhou, *State Grid Corporation of China, China*
Xizhou Ren, *State Grid Corporation of China, China*

A DC to Three-Phase Boost-Buck Inverter with Stored Energy Modulation and a Tiny DC Link Capacitor 2919
Mahima Gupta, *University of Wisconsin at Madison, United States*
Giri Venkataramanan, *University of Wisconsin at Madison, United States*

Drive Circuits for Ultra-Fast and Reliable Actuation of Thomson Coil Actuators used in Hybrid AC and DC Circuit Breakers 2927
Chang Peng, *North Carolina State University, United States*
Alex Huang, *North Carolina State University, United States*
Iqbal Husain, *North Carolina State University, United States*
Bruno Lequesne, *E-Motors Consulting, LLC, United States*
Roger Briggs, *Energy Efficiency Research, LLC, United States*

Improved Transformerless Dual Buck Inverters with Buffer Inductors 2935
Liwei Zhou, *Shandong University, China*
Feng Gao, *Shandong University, China*

A 99% Efficiency SiC Three-Phase Inverter using Synchronous Rectification 2942
Shan Yin, *Nanyang Technological University, Singapore*
K.J. Tseng, *Nanyang Technological University, Singapore*
C.F. Tong, *Nanyang Technological University, Singapore*
Rejeki Simanjorang, *Rolls-Royce Singapore Pte. Ltd., Singapore*
C.J. Gajanayake, *Rolls-Royce Singapore Pte. Ltd., Singapore*
Amit K. Gupta, *Rolls-Royce Singapore Pte. Ltd., Singapore*

Comparison and Evaluation of Common Mode EMI Filter Topologies for GaN-Based Motor Drive Systems 2950
Casey T. Morris, *University of Wisconsin at Madison, United States*
Di Han, *University of Wisconsin at Madison, United States*
Bulent Sarlioglu, *University of Wisconsin at Madison, United States*

Analysis of Thermal Cycling Stress on Semiconductor Devices of the Modular Multilevel Converter for Drive Applications 2957
Xiangyu Han, *Georgia Institute of Technology, United States*
Qichen Yang, *Georgia Institute of Technology, United States*
Liyao Wu, *Georgia Institute of Technology, United States*
Maryam Saeedifard, *Georgia Institute of Technology, United States*

Fault Tolerant Topologies of Five-Level Active Neutral-Point-Clamped Converters 2963
Jun Li, *ABB Inc., United States*

Session D08: Advanced Components and Devices
Location: Poster Area
March 24, 2016 11:30 - 14:00
Session Chairs: Abhijit Pathak, *Infineon/IR*
Doug Hopkins, *North Carolina State University*

Dynamic Characterization of the Input and Reverse Transfer Capacitances in Power MOSFETs under High Current Conduction 2969
Cristino Salcines, *Universität Stuttgart, Germany*
Ingmar Kallfass, *Universität Stuttgart, Germany*
Hisao Kakitani, *Keysight Technologies International, Japan*
Atsushi Mikata, *Keysight Technologies International, Japan*

Medium Voltage Power Switch based on SiC JFETs 2973
Xueqing Li, *United Silicon Carbide, Inc., United States*
Hao Zhang, *United Silicon Carbide, Inc., United States*
Peter Alexandrov, *United Silicon Carbide, Inc., United States*
Anup Bhalla, *United Silicon Carbide, Inc., United States*

Numerical Model and Experimental Study on Comparison of Semiconductor Pulsed Power Devices 2981
Lin Liang, *Huazhong University of Science and Technology, China*
Changdong Chen, *Huazhong University of Science and Technology, China*
Fang Luo, *Ohio State University, United States*

A Normalization Procedure of DC-Side Stray Inductance for High-Speed Switching Circuit 2986
Masato Ando, *Tokyo Metropolitan University, Japan*
Keiji Wada, *Tokyo Metropolitan University, Japan*

Thermal Network Parameter Identification of IGBT Module based on the Cooling Curve of Junction Temperature 2992

Xiong Du, *Chongqing University, China*
Tengfei Li, *Chongqing University, China*
Jun Zhang, *Chongqing University, China*
Heng-Ming Tai, *University of Tulsa, United States*
Pengju Sun, *Chongqing University, China*
Luowei Zhou, *Chongqing University, China*

Design and Evaluation of High Current PCB Embedded Inductor for High Frequency Inverters 2998

Mehrdad Biglarbegian, *University of North Carolina at Charlotte, United States*
Neel Shah, *University of North Carolina at Charlotte, United States*
Iman Mazhari, *University of North Carolina at Charlotte, United States*
Johan Enslin, *University of North Carolina at Charlotte, United States*
Babak Parkhideh, *University of North Carolina at Charlotte, United States*

Prognosis of Wire Bond Lift-Off Fault of an IGBT based on Multisensory Approach 3004

Moinul Shahidul Haque, *University of Akron, United States*
Jeihoon Baek, *Korean Rail Research Institute, Korea, South*
Joseph Herbert, *University of Akron, United States*
Seungdeog Choi, *University of Akron, United States*

Electrical Parasitics and Thermal Modeling for Optimized Layout Design of High Power SiC Modules 3012

Amir Sajjad Bahman, *Aalborg University, Denmark*
Frede Blaabjerg, *Aalborg University, Denmark*
Atanu Dutta, *University of Arkansas, United States*
Alan Mantooth, *University of Arkansas, United States*

Calculation of Losses in PCB Windings for Multi-Coil Contactless Charging Systems 3020

J. Serrano, *Universidad de Zaragoza, Spain*
J. Acero, *Universidad de Zaragoza, Spain*
I. Lope, *BSH Home Appliances Group, Spain*
C. Carretero, *Universidad de Zaragoza, Spain*
J.M. Burdío, *Universidad de Zaragoza, Spain*
R. Alonso, *Universidad de Zaragoza, Spain*

Design of Efficient Loads for Domestic Induction Heating Applications by Means of Non-Magnetic Thin Metallic Layers 3026

Jesús Acero, *Universidad de Zaragoza, Spain*
Claudio Carretero, *Universidad de Zaragoza, Spain*
Rafael Alonso, *Universidad de Zaragoza, Spain*
José Miguel Burdío, *Universidad de Zaragoza, Spain*

A New Evaluation Circuit with a Low-Voltage Inverter Intended for Capacitors used in a High-Power Three-Phase Inverter 3032

Kazunori Hasegawa, *Kyushu Institute of Technology, Japan*
Ichiro Omura, *Kyushu Institute of Technology, Japan*
Shin-Ichi Nishizawa, *Kyushu Institute of Technology / National Institute of Advanced Industrial Science and Technology, Japan*

Energy Absorption Capability of Low Voltage Metal Oxide Varistors in AC and Impulse Currents 3038

Dawood Talebi Khanmiri, *Northeastern University, United States*
Roy Ball, *Mersen USA, United States*
Craig McKenzie, *Mersen USA, United States*
Brad Lehman, *Northeastern University, United States*

Optimization and Experimental Validation of Medium-Frequency High Power Transformers in Solid-State Transformer Applications 3043

M.A. Bahmani, *Chalmers University of Technology, Sweden*
T. Thiringer, *Chalmers University of Technology, Sweden*
M. Kharezy, *SP Technical Research Institute of Sweden, Sweden*

Evaluation of Core Loss in Magnetic Materials Employed in Utility Grid AC Filters 3051

Remus Beres, *Aalborg University, Denmark*
Xiongfei Wang, *Aalborg University, Denmark*
Frede Blaabjerg, *Aalborg University, Denmark*
Claus Leth Bak, *Aalborg University, Denmark*
Hiroaki Matsumori, *Tokyo Metropolitan University, Japan*
Toshihisa Shimizu, *Tokyo Metropolitan University, Japan*

A Novel Gate Assisted Circuit to Reduce Switching Loss and Eliminate Shoot-Through in SiC Half Bridge Configuration 3058

Shan Yin, *Nanyang Technological University, Singapore*
K.J. Tseng, *Nanyang Technological University, Singapore*
C.F. Tong, *Nanyang Technological University, Singapore*
Rejeki Simanjorang, *Rolls-Royce Singapore Pte. Ltd., Singapore*
C.J. Gajanayake, *Rolls-Royce Singapore Pte. Ltd., Singapore*
Amit K. Gupta, *Rolls-Royce Singapore Pte. Ltd., Singapore*

Session D09: System Design Considerations for Power Electronics
Location: Poster Area
March 24, 2016 11:30 - 14:00
Session Chairs: John Vigars, *Allegro Microsystems*
Ernie Parker, *Crane Aerospace & Electronics*

Methods to Enhance the Thermal Performance of a 3D Power Package 3065

Jonathan Noquil, *Texas Instruments Inc., United States*
Ozzie Lopez, *Texas Instruments Inc., United States*
Tianyi Luo, *Lehigh University, United States*

Highly Reliable and Cost Effective Thick Film Substrates for Power LEDs 3069

Paul Gundel, *Heraeus Deutschland GmbH & Co. KG, Germany*
Ryan Persons, *Heraeus Deutschland GmbH & Co. KG, Germany*
Melanie Bawohl, *Heraeus Deutschland GmbH & Co. KG, Germany*
Mark Challingsworth, *Heraeus Deutschland GmbH & Co. KG, Germany*
Christoph Czwickla, *Heraeus Deutschland GmbH & Co. KG, Germany*
Virginia Garcia, *Heraeus Deutschland GmbH & Co. KG, Germany*
Christina Modes, *Heraeus Deutschland GmbH & Co. KG, Germany*
Ilias Nikolaidis, *Heraeus Deutschland GmbH & Co. KG, Germany*
Jessica Reitz, *Heraeus Deutschland GmbH & Co. KG, Germany*
Caitlin Shahbazi, *Heraeus Deutschland GmbH & Co. KG, Germany*
Torsten Nowak, *Fraunhofer-Institut für Zuverlässigkeit und Mikrointegration, Germany*

Design and Evaluation of SiC-Based High Power Density Inverter, 70kW/Liter, 50kW/kg ... 3075
Koji Yamaguchi, *IHI Corporation, Japan*

An Improved Automatic Layout Method for Planar Power Module ... 3080
Puqi Ning, *Chinese Academy of Sciences, China*
Xuhui Wen, *Chinese Academy of Sciences, China*
Yaohua Li, *Chinese Academy of Sciences, China*
Xiongxuan Ge, *Chinese Academy of Sciences, China*

Practical Implementation Schemes of Motor Speed Measurement by Magnetic Encoder on Electric Power Steering Applications ... 3086
Jae-Hyun Lee, *Hyundai Mobis, Korea, South*

Low-Cost Input Impedance Estimator of DC-to-DC Converters for Designing the Control Loop in Cascaded Converters ... 3090
M. Sanz, *Universidad Carlos III de Madrid, Spain*
A. Lázaro, *Universidad Carlos III de Madrid, Spain*
M. Bermejo, *Universidad Carlos III de Madrid, Spain*
D. López del Moral, *Universidad Carlos III de Madrid, Spain*
P. Zumel, *Universidad Carlos III de Madrid, Spain*
C. Fernández, *Universidad Carlos III de Madrid, Spain*
A. Barrado, *Universidad Carlos III de Madrid, Spain*

On-Chip High Performance Magnetics for Point-of-Load High-Frequency DC-DC Converters ... 3097
Dragan Dinulovic, *Würth Elektronik eiSos GmbH & Co. KG, Germany*
Mahmoud Shousha, *Würth Elektronik eiSos GmbH & Co. KG, Germany*
Martin Haug, *Würth Elektronik eiSos GmbH & Co. KG, Germany*
Alexander Gerfer, *Würth Elektronik eiSos GmbH & Co. KG, Germany*
Mike Wens, *MinDCet NV, Belgium*
Jef Thone, *MinDCet NV, Belgium*

Effects of Auxiliary Source Connections in Multichip Power Module ... 3101
Helong Li, *Aalborg University, Denmark*
Stig Munk-Nielsen, *Aalborg University, Denmark*
Szymon Bęczkowski, *Aalborg University, Denmark*
Xiongfei Wang, *Aalborg University, Denmark*
Emanuel-Petre Eni, *Aalborg University, Denmark*

Session D10: Modeling and Simulation
Location: Poster Area
March 24, 2016 11:30 - 14:00
Session Chairs: Marco Meola, *ZMD AG*
Mehdi Ferdowsi, *Missouri University of Science & Technology*

Modelling Technique Utilizing Modified Sigmoid Functions for Describing Power Transistor Device Capacitances Applied on GaN HEMT and Silicon MOSFET ... 3107
H.L. Yeo, *Nanyang Technological University, Singapore*
K.J. Tseng, *Nanyang Technological University, Singapore*

Design and Precise Modeling of a Novel Digital Active EMI Filter ... 3115
Junpeng Ji, *Xi'an Jiaotong University, China*
Wenjie Chen, *Xi'an Jiaotong University, China*
Xu Yang, *Xi'an Jiaotong University, China*

Development of a Hybrid Emulation Platform based on RTDS and Reconfigurable Power Converter-Based Testbed 3121
Shuoting Zhang, *University of Tennessee, United States*
Yiwei Ma, *University of Tennessee, United States*
Liu Yang, *University of Tennessee, United States*
Fred Wang, *University of Tennessee, United States*
Leon M. Tolbert, *University of Tennessee, United States*

Online Temperature Estimation for Phase Change Composite – 18650 Lithium Ion Cells based Battery Pack 3128
Mohamad Salameh, *Illinois Institute of Technology, United States*
Ben Schweitzer, *AllCell Technologies, United States*
Peter Sveum, *AllCell Technologies, United States*
Said Al-Hallaj, *AllCell Technologies, United States*
Mahesh Krishnamurthy, *Illinois Institute of Technology, United States*

Modeling and Fault Diagnosis of Inter-Turn Short Circuit for Five-Phase PMSM based on Particle Swarm Optimization 3134
Jianwei Yang, *Northwestern Polytechnical University, China*
Manfeng Dou, *Northwestern Polytechnical University, China*
Zhiyong Dai, *Northwestern Polytechnical University, China*
Dongdong Zhao, *Northwestern Polytechnical University, China*
Zhen Zhang, *Northwestern Polytechnical University, China*

Comprehensive Modeling, Testing, and Experimental Validation of Ultracapacitor Open Circuit Voltage Characteristics 3140
Amandeep Singh, *University of Ontario Institute of Technology, Canada*
Najath Abdul Azeez, *University of Ontario Institute of Technology, Canada*
Sheldon S. Williamson, *University of Ontario Institute of Technology, Canada*

Novel SPICE Model for Common Mode Choke Including Complex Permeability 3146
Katsuya Nomura, *Toyota Central R&D Labs., Inc., Japan*
Naoto Kikuchi, *Toyota Central R&D Labs., Inc., Japan*
Yoshitoshi Watanabe, *Toyota Central R&D Labs., Inc., Japan*
Shuntaro Inoue, *Toyota Central R&D Labs., Inc., Japan*
Yoshiyuki Hattori, *Toyota Central R&D Labs., Inc., Japan*

Session D11: Control I
Location: Poster Area
March 24, 2016 11:30 - 14:00
Session Chairs: Bilal Akin, *Univeristy of Texas, Dallas*
Brian Zahnstecher, *PowerRox LLC*

Analysis and Design of Capacitive Power Transmission System Employing Out-of-Band Wireless Feedback Link 3153
Sung-Jin Choi, *University of Ulsan, Korea, South*
Hee-Su Choi, *University of Ulsan, Korea, South*

Introducing Fourier-Based Modeling and Control of Active-Bridge Converters 3158
B.J.D. Vermulst, *Technische Universiteit Eindhoven, Netherlands*
J.L. Duarte, *Technische Universiteit Eindhoven, Netherlands*
C.G.E. Wijnands, *Technische Universiteit Eindhoven, Netherlands*
E.A. Lomonova, *Technische Universiteit Eindhoven, Netherlands*

A Stability Analysis and Efficiency Improvement of Synchronverter 3165
Prasanna Piya, *Mississippi State University, United States*
Masoud Karimi-Ghartemani, *Mississippi State University, United States*

Compensation of Switching Dead-Time Effects in Voltage-Fed PWM Inverters using FPGA-Based Current Oversampling 3172
Bastian Weber, *Leibniz Universität Hannover, Germany*
Tobias Brandt, *Leibniz Universität Hannover, Germany*
Axel Mertens, *Leibniz Universität Hannover, Germany*

Control Strategy of High Power Converters with Synchronous Generator Characteristics for PMSG-Based Wind Power Application 3180
Yuzhi Zhang, *University of Arkansas, United States*
Haoyan Liu, *University of Arkansas, United States*
H. Alan Mantooth, *University of Arkansas, United States*

Phase Compensation, ZVS Operation of Wireless Power Transfer System based on SOGI-PLL 3185
Pingan Tan, *Xiangtan University, China*
Haibing He, *Xiangtan University, China*
Xieping Gao, *Xiangtan University, China*

A Novel Low-Cost Online State of Charge Estimation Method for Reconfigurable Battery Pack 3189
Ni Lin, *University of Nebraska at Lincoln, United States*
Song Ci, *University of Nebraska at Lincoln, United States*
Dalei Wu, *University of Tennessee at Chattanooga, United States*

Effect of Decoupling Terms on the Performance of PR Current Controllers Implemented in Stationary Reference Frame 3193
Sizhan Zhou, *Xi'an Jiaotong University, China*
Jinjun Liu, *Xi'an Jiaotong University, China*

Fuzzy Predictive DTC of Induction Machines with Reduced Torque Ripple and High Performance Operation 3200
Alberto Berzoy, *Florida International University, United States*
Osama Mohammed, *Florida International University, United States*
Johnny Rengifo, *Universidad Simon Bolivar, Venezuela*

Session D12: Control II
Location: Poster Area
March 24, 2016 11:30 - 14:00
Session Chairs: Martin Ordonez, *University of British Columbia*
Jiangbiao He, *GE Global Research*

Fixed-Frequency Generalized Peak Current Control (GPCC) for Inverters 3207
Mohammad Ebrahimi, *University of Alberta, Canada*
S. Ali Khajehoddin, *University of Alberta, Canada*

Improved Control Strategy of 1 MHz LLC Converter for High Frequency Resolution 3213
Hwa-Pyeong Park, *Ulsan National Institute of Science and Technology, Korea, South*
Jee-Hoon Jung, *Ulsan National Institute of Science and Technology, Korea, South*

Bumpless Control for Reduced THD in Power Factor Correction Circuits 3219
Joel Steenis, *Microchip Technology, United States*
Alex Dumais, *Microchip Technology, United States*

Mixed-Signal Hysteretic Internal Model Control of Buck Converters for Ultra-Fast Envelope Tracking 3224
V. Inder Kumar, *Indian Institute of Technology Kharagpur, India*
Santanu Kapat, *Indian Institute of Technology Kharagpur, India*

A Continuous Actor-Critic Maximum Power Point Tracker Applied to Low Power Wind Turbine Systems 3231
J.L. Wattes, *Universidade Federal do Ceará, Brazil*
A.J.S. Dias Jr., *Universidade Federal do Ceará, Brazil*
A.P.S. Braga, *Universidade Federal do Ceará, Brazil*
P.P. Praça, *Universidade Federal do Ceará, Brazil*
A.U. Barbosa, *Universidade Federal do Ceará, Brazil*
D.S. de Souza Oliveira Jr., *Universidade Federal do Ceará, Brazil*

Multi-Band Mixed-Signal Hysteresis Current Control for EMI Reduction in Switch-Mode Power Supplies 3237
Arindam Mandal, *Indian Institute of Technology Kharagpur, India*
V. Inder Kumar, *Indian Institute of Technology Kharagpur, India*
Santanu Kapat, *Indian Institute of Technology Kharagpur, India*

A Parabolic Current Control based Digital Current Control Strategy for High Switching Frequency Voltage Source Inverters 3243
Lanhua Zhang, *Virginia Polytechnic Institute and State University, United States*
Rachael Born, *Virginia Polytechnic Institute and State University, United States*
Xiaonan Zhao, *Virginia Polytechnic Institute and State University, United States*
Jih-Sheng Jason Lai, *Virginia Polytechnic Institute and State University, United States*
Hongbo Ma, *Southwest Jiaotong University, China*

Finite Control Set Model Predictive Control of Dual-Output Four-Leg Indirect Matrix Converter Under Unbalanced Load and Supply Conditions 3248
Ozan Gulbudak, *University of South Carolina, United States*
Enrico Santi, *University of South Carolina, United States*

A Silicon Carbide Integrated Circuit Implementing Nonlinear-Carrier Control for Boost Converter Applications 3255
Richard Kyle Harris, *University of Tennessee, United States*
Benjamin M. McCue, *University of Tennessee, United States*
Benjamin D. Roehrs, *University of Tennessee, United States*
Charles Roberts II, *University of Tennessee, United States*
Benjamin J. Blalock, *University of Tennessee, United States*
Daniel J. Costinett, *University of Tennessee, United States*
Kouros Sariri, *Frequency Management International, United States*
George Megyei, *Frequency Management International, United States*
Cheng-Po Chen, *GE Global Research, United States*
Avinash Kashyap, *GE Global Research, United States*
Reza Ghandi, *GE Global Research, United States*

A New Current Mode Constant on Time Control with Ultrafast Load Transient Response 3259
Syed Bari, *Virginia Polytechnic Institute and State University, United States*
Qiang Li, *Virginia Polytechnic Institute and State University, United States*
Fred C. Lee, *Virginia Polytechnic Institute and State University, United States*

A Web-Based Tool for Compensation Design of Power Converters using Hybrid Optimization 3266

Srikanth Pam, *Texas Instruments Inc., India*
Yudhister Satija, *Texas Instruments Inc., India*
Pradeep Chawda, *Texas Instruments Inc., United States*
Makram Mansour, *Texas Instruments Inc., United States*
Robert Hanrahan, *Texas Instruments Inc., United States*
Jeff Perry, *Texas Instruments Inc., United States*

Second Order Sliding Mode Controlled Point of Load Power Supply 3273

Prasanta K. Achanta, *University of Colorado at Boulder, United States*
David C. Jones, *University of Colorado at Boulder, United States*
Dragan Maksimovic, *University of Colorado at Boulder, United States*
Serhii M. Zhak, *Linear Technology Corporation, United States*
Brett Miwa, *Maxim Integrated, United States*
Cory Arnold, *Maxim Integrated, United States*

Vibration and Torque Ripple Reduction of Switched Reluctance Motors through Current Profile Optimization 3279

Cong Ma, *University of Nebraska at Lincoln, United States*
Liyan Qu, *University of Nebraska at Lincoln, United States*
Rakesh Mitra, *Nexteer Automotive, United States*
Prerit Pramod, *Nexteer Automotive, United States*
Rakib Islam, *Nexteer Automotive, United States*

Modified Predictive Current Control of Neutral-Point Clamped Converter with Reduced Switching Frequency 3286

Dinto Mathew, *Indian Institute of Technology Bombay, India*
Anshuman Shukla, *Indian Institute of Technology Bombay, India*
Santanu Bandyopadhyay, *Indian Institute of Technology Bombay, India*

Implicit Finite Control Set Model Predictive Current Control for Modular Multilevel Converter based on IPA-SQP Algorithm 3291

Hamed Nademi, *ABB AS, Norway*
Lars Einar Norum, *Norwegian University of Science and Technology, Norway*

Resolution Requirements to Avoid Limit Cycling in LLC Resonant Converter 3297

Shadi Dashmiz, *University of Toronto, Canada*
Behzad Mahdavikhah, *University of Toronto, Canada*
Aleksandar Prodić, *University of Toronto, Canada*
Brent McDonald, *Texas Instruments Inc., United States*

Session D13: Renewable Energy Systems I
Location: Poster Area
March 24, 2016 11:30 - 14:00
Session Chairs: Akshay Kumar Rathore, *Concordia University*
Xiaoqiang Guo, *Yanshan University, China*

Reduction of Storage Capacity in DC Microgrids using PV-Embedded Series DC Electric Springs 3302

Ming-Hao Wang, *University of Hong Kong, Hong Kong*
Siew-Chong Tan, *University of Hong Kong, Hong Kong*
Shu-Yuen Ron Hui, *University of Hong Kong, Hong Kong*

A Vector Control Strategy of Grid-Connected Brushless Doubly Fed Induction Generator based on the Vector Control of Doubly Fed Induction Generator 3310

Sheng Hu, *Wuhan University of Technology, China*
Guorong Zhu, *Wuhan University of Technology, China*

An Energy Router based on Multi-Winding High-Frequency Transformer 3317

Xianzhuo Liu, *Tsinghua University, China*
Zedong Zheng, *Tsinghua University, China*
Kui Wang, *Tsinghua University, China*
Yongdong Li, *Tsinghua University, China*

Noise Suppression of the DWT-Based MRA on Mother Wavelet and Decomposition Level Optimization for a Robust Adaptive SOC Estimator in Multi-Cell Battery String 3322

Jonghoon Kim, *Chosun University, Korea, South*
Chang Yoon Chun, *Seoul National University, Korea, South*
Woonki Na, *California State University, Fresno, United States*

A Feedforward Control based Power Decoupling Scheme for Voltage-Controlled Grid-Tied Inverters 3328

Baojin Liu, *Xi'an Jiaotong University, China*
Zeng Liu, *Xi'an Jiaotong University, China*
Jinjun Liu, *Xi'an Jiaotong University, China*
Teng Wu, *Xi'an Jiaotong University, China*
Shike Wang, *Xi'an Jiaotong University, China*

Light Load Efficiency Improvement of Solar Farms Three-Phase Two-Stage Module Integrated Converter 3333

Ahmadreza Amirahmadi, *University of Central Florida, United States*
Utsav Somani, *University of Central Florida, United States*
Mahmood Alharbi, *University of Central Florida, United States*
Charlie Jourdan, *University of Central Florida, United States*
Issa Batarseh, *University of Central Florida, United States*

Switching System Stability Analysis of DC Microgrids with DBS Control 3338

Na Zhi, *Xi'an University of Technology, China*
Hui Zhang, *Xi'an University of Technology, China*
Xi Xiao, *Tsinghua University, China*

A Grid-Connected WECS with Power Limiting Control 3346

Jéssica Santos Guimarães, *Universidade Federal do Ceará, Brazil*
Demercil de Souza Oliveira Jr., *Universidade Federal do Ceará, Brazil*
Juliano de Oliveira Pacheco, *Universidade Federal do Ceará, Brazil*
Paulo P. Peixoto, *Universidade Federal do Ceará, Brazil*

Overshoot Control of the Electromagnetic Torque during Fault Recovery for an SCIG with a STATCOM 3353

Zahra Mahmoodzadeh, *Washington State University, United States*
Mehrdad Yazdanian, *Washington State University, United States*
Hooman Ghaffarzadeh, *Washington State University, United States*
Ali Mehrizi-Sani, *Washington State University, United States*

A Self-Adaptive Power Balance Control Strategy for PV Inverters in Islanded Microgrids 3358
Zhenxiong Wang, *Xi'an Jiaotong University, China*
Hao Yi, *Xi'an Jiaotong University, China*
Fang Zhuo, *Xi'an Jiaotong University, China*
Zhigang Zhang, *Xi'an Jiaotong University, China*

High Performance ZVT with Bus Clamping Modulation Technique for Single Phase Full Bridge Inverters 3364
Yinglai Xia, *Arizona State University, United States*
Raja Ayyanar, *Arizona State University, United States*

Small AC Signal Droop based Secondary Control for Microgrids 3370
Teng Wu, *Xi'an Jiaotong University, China*
Zeng Liu, *Xi'an Jiaotong University, China*
Jinjun Liu, *Xi'an Jiaotong University, China*
Baojin Liu, *Xi'an Jiaotong University, China*
Shike Wang, *Xi'an Jiaotong University, China*

Mode Transition Control Strategy for Multiple Inverter based Distributed Generators Operating in Grid-Connected and Stand-Alone Mode 3376
Onkar Vitthal Kulkarni, *Indian Institute of Technology Bombay, India*
Suryanarayana Doolla, *Indian Institute of Technology Bombay, India*
B.G. Fernandes, *Indian Institute of Technology Bombay, India*

An Autonomous Power Management Strategy based on DC Bus Signaling for Solid-State Transformer Interfaced PMSG Wind Energy Conversion System 3383
Rui Gao, *North Carolina State University, United States*
Iqbal Husain, *North Carolina State University, United States*
Alex Q. Huang, *North Carolina State University, United States*

An Isolated Buck-Boost Type High-Frequency Link Photovoltaic Microinverter 3389
Shiladri Chakraborty, *Indian Institute of Technology Kharagpur, India*
Souvik Chattopadhyay, *Indian Institute of Technology Kharagpur, India*

Energy Management and Stabilization of a Hybrid DC Microgrid for Transportation Applications 3397
Mehdi Karbalaye Zadeh, *Norwegian University of Science and Technology, Norway*
Louis-Marie Saublet, *Université de Lorraine, France*
Roghayeh Gavagsaz-Ghoachani, *Université de Lorraine, France*
Babak Nahid-Mobarakeh, *Université de Lorraine, France*
Serge Pierfederici, *Université de Lorraine, France*
Marta Molinas, *Norwegian University of Science and Technology, Norway*

A Low-Cost Solar Micro-Inverter with Soft-Switching Capability Utilizing Circulating Current 3403
Xiaohu Liu, *GE Global Research, United States*
Mohammed Agamy, *GE Global Research, United States*
Dong Dong, *GE Global Research, United States*
Maja Harfman-Todorovic, *GE Global Research, United States*
Luis Garces, *GE Global Research, United States*

Session D14: Renewable Energy Systems II
Location: Poster Area
March 24, 2016 11:30 - 14:00
Session Chairs: Haoyu Wang, *Shanghai Tech University*
Robert Pilawa-Podgurski, *University of Illinois at Urbana-Champaign*

Design and Stability Analysis for an Autonomous DC Microgrid with Constant Power Load ... 3409
Qianwen Xu, *Nanyang Technological University, Singapore*
Xiaolei Hu, *Nanyang Technological University, Singapore*
Peng Wang, *Nanyang Technological University, Singapore*
Jianfang Xiao, *Nanyang Technological University, Singapore*
Leonardy Setyawan, *Nanyang Technological University, Singapore*
Changyun Wen, *Nanyang Technological University, Singapore*
Lee Meng Yeong, *Rolls-Royce Singapore Pte. Ltd., Singapore*

MPC-SVM Method for Vienna Rectifier with PMSG used in Wind Turbine Systems 3416
June-Seok Lee, *Korea Railroad Research Institute, Korea, South*
Yeongsu Bak, *Ajou University, Korea, South*
Kyo-Beum Lee, *Ajou University, Korea, South*
Frede Blaabjerg, *Aalborg University, Denmark*

An Equivalent Circuit Model for State of Energy Estimation of Lithium-Ion Battery 3422
Kaiyuan Li, *Nanyang Technological University, Singapore*
King Jet Tseng, *Nanyang Technological University, Singapore*

Distributed Optimal Control of Reactive Power and Voltage in Islanded Microgrids 3431
Yanbo Wang, *Aalborg University, Denmark*
Xiongfei Wang, *Aalborg University, Denmark*
Zhe Chen, *Aalborg University, Denmark*
Frede Blaabjerg, *Aalborg University, Denmark*

New Start-Up Scheme for HF Transformer Link Photovoltaic Inverter 3439
Abhijit Kulkarni, *Indian Institute of Science, India*
Vinod John, *Indian Institute of Science, India*

Analysis and Improvement of Harmonic Quasi Resonant Control for LCL-Filtered Grid-Connected Inverters in Weak Grid 3446
Qiang Qian, *Nanjing University of Aeronautics and Astronautics, China*
Jinming Xu, *Nanjing University of Aeronautics and Astronautics, China*
Shaojun Xie, *Nanjing University of Aeronautics and Astronautics, China*
Lin Ji, *Nanjing University of Aeronautics and Astronautics, China*

Model Predictive Control Method to Reduce Common-Mode Voltage and Balance the Neutral-Point Voltage in Three-Level T-Type Inverter 3453
Xiangyang Xing, *Shandong University, China*
Alian Chen, *Shandong University, China*
Zicheng Zhang, *Shandong University, China*
Jie Chen, *Shandong University, China*
Chenghui Zhang, *Shandong University, China*

Convergence Analysis of Distributed Control for Operation Cost Minimization of Droop Controlled DC Microgrid based on Multiagent 3459
Chendan Li, *Aalborg University, Denmark*
Juan C. Vásquez, *Aalborg University, Denmark*
Josep M. Guerrero, *Aalborg University, Denmark*

A Novel Model Predictive Control Algorithm to Suppress the Zero-Sequence Circulating Currents for Parallel Three-Phase Voltage Source Inverters 3465
Zicheng Zhang, *Shandong University, China*
Alian Chen, *Shandong University, China*
Xiangyang Xing, *Shandong University, China*
Chenghui Zhang, *Shandong University, China*

Design of Dynamic Voltage Restorer and Active Power Filter for Wind Power Systems Subject to Unbalanced and Harmonic Distorted Grid 3471
Woei-Luen Chen, *Chang Gung University, Taiwan*
Meng-Jie Wang, *Chang Gung University, Taiwan*

Dynamic Variable Coupling Analysis and Modeling of Proton Exchange Membrane Fuel Cells for Water and Thermal Management 3476
Daming Zhou, *Université de Technologie de Belfort-Montbéliard, France*
Elena Breaz, *Université de Technologie de Belfort-Montbéliard, France*
Alexandre Ravey, *Université de Technologie de Belfort-Montbéliard, France*
Fei Gao, *Université de Technologie de Belfort-Montbéliard, France*
Abdellatif Miraoui, *Université de Technologie de Belfort-Montbéliard, France*
Ke Zhang, *Northwestern Polytechnical University, China*

Voltage and Frequency Control of Electric Spring based Smart Loads 3481
Yun Yang, *University of Hong Kong, Hong Kong*
Siew-Chong Tan, *University of Hong Kong, Hong Kong*
Shu-Yuen Ron Hui, *University of Hong Kong, Hong Kong*

Second Harmonic Current Compensator with Improved One-Cycle-Control 3488
Li Zhang, *Nanjing University of Aeronautics and Astronautics, China*
Xinbo Ruan, *Nanjing University of Aeronautics and Astronautics, China*
Xiaoyong Ren, *Nanjing University of Aeronautics and Astronautics, China*

Frequency Adaptive Control of a Smart Transformer-Fed Distribution Grid 3493
Zhi-Xiang Zou, *Christian-Albrechts-Universität zu Kiel, Germany*
Giovanni De Carne, *Christian-Albrechts-Universität zu Kiel, Germany*
Giampaolo Buticchi, *Christian-Albrechts-Universität zu Kiel, Germany*
Marco Liserre, *Christian-Albrechts-Universität zu Kiel, Germany*

A Synchronization Scheme for Single-Phase Grid-Tied Inverters under Harmonic Distortion and Grid Disturbances 3500
Lenos Hadjidemetriou, *University of Cyprus, Cyprus*
Elias Kyriakides, *University of Cyprus, Cyprus*
Yongheng Yang, *Aalborg University, Denmark*
Frede Blaabjerg, *Aalborg University, Denmark*

Series-Parallel Connection of Low-Voltage Sources for Integration of Galvanically Isolated Energy Storage Systems 3508

Ramy Georgious, *Universidad de Oviedo, Spain*
Jorge Garcia, *Universidad de Oviedo, Spain*
Angel Navarro, *Universidad de Oviedo, Spain*
Sarah Saeed, *Universidad de Oviedo, Spain*
Pablo Garcia, *Universidad de Oviedo, Spain*

Saturation Controller-Based Direct Power Control for Doubly-Fed Induction Generator 3514

Chun Wei, *University of Nebraska at Lincoln, United States*
Zhe Zhang, *Nexteer Automotive, United States*
Wei Qiao, *University of Nebraska at Lincoln, United States*
Liyan Qu, *University of Nebraska at Lincoln, United States*

Inductance-Simulating Control for DFIG-Based Wind Turbine to Ride-Through Grid Faults 3521

Donghai Zhu, *Huazhong University of Science and Technology, China*
Xudong Zou, *Huazhong University of Science and Technology, China*
Yong Kang, *Huazhong University of Science and Technology, China*
Lu Deng, *Wuhan NARI Limited Company of State Grid Electric Power Research Institute, China*
Qingjun Huang, *State Key Laboratory of Disaster Prevention & Reduction for Power Grid Transmission and Distribution Equipment, China*

Session D15: Transportation Power Electronics
Location: Poster Area
March 24, 2016 11:30 - 14:00
Session Chairs: Ted Bohn, *Argonne National Labs*
Khurram Afridi, *University of Colorado, Boulder*

Misalignment Effect on Efficiency of Wireless Power Transfer for Electric Vehicles 3526

Yabiao Gao, *University of Georgia, United States*
Antonio Ginart, *University of Georgia / Sonnenbatterie GmbH, United States*
Kathleen Blair Farley, *Southern Company Services, Inc., United States*
Zion Tsz Ho Tse, *University of Georgia, United States*

Genetic Algorithm Design of a 3D Printed Heat Sink 3529

Tong Wu, *University of Tennessee, United States*
Burak Ozpineci, *Oak Ridge National Laboratory, United States*
Curtis Ayers, *Oak Ridge National Laboratory, United States*

Evaluation of Power Flow Control for an All-Electric Warship Power System with Pulsed Load Applications 3537

J. Neely, *Sandia National Laboratories, United States*
L. Rashkin, *Sandia National Laboratories, United States*
M. Cook, *Sandia National Laboratories, United States*
D. Wilson, *Sandia National Laboratories, United States*
S. Glover, *Sandia National Laboratories, United States*

Reduced Active Switch AC to DC Rectifier with High Frequency Isolation for Electric Vehicle Chargers 3545

José Juan Sandoval, *Texas A&M University, United States*
Taeyong Kang, *Texas A&M University, United States*
Prasad Enjeti, *Texas A&M University, United States*

A Wide Bandgap Device based Multilevel Switched-Capacitor Converter 3553
Diogo Cesar Santos de Moura, *North Dakota State University, United States*
Boris Curuvija, *North Dakota State University, United States*
Dong Cao, *North Dakota State University, United States*

Session D16: Power Topologies, Distribution, and Control
Location: Poster Area
March 24, 2016 11:30 - 14:00
Session Chairs: Tiefu Zhao, *Eaton*
Xiaonan Lu, *Argonne National Laboratory*

Novel Circulating Current Suppression Strategy for MMC based on Quasi-PR Controller 3560
Shengbao Geng, *Shanghai Jiao Tong University, China*
Yiliang Gan, *Shanghai Jiao Tong University, China*
Yungui Li, *Shanghai Jiao Tong University, China*
Lijun Hang, *Shanghai Jiao Tong University, China*
Guojie Li, *Shanghai Jiao Tong University, China*

Assymmetric Duty-Cycle Phase-Shift Modulation for Power Management in Double Half-Bridge Inverter with Partly Coupled Inductive Loads 3566
C. Carretero, *Universidad de Zaragoza, Spain*
H. Sarnago, *Universidad de Zaragoza, Spain*
O. Lucia, *Universidad de Zaragoza, Spain*
J. Acero, *Universidad de Zaragoza, Spain*
J.M. Burdío, *Universidad de Zaragoza, Spain*

Control Implementation for a Wide Voltage Range High Efficiency Power Supply Utilizing Low Voltage MOSFETs 3570
Werner Konrad, *Technische Universität Graz, Austria*
Gerald Deboy, *Infineon Technologies AG, Austria*
Annette Muetze, *Technische Universität Graz, Austria*

A Single-Phase Dual Frequency Inverter based on Multi-Frequency Selective Harmonic Elimination 3577
Chongwen Zhao, *University of Tennessee, United States*
Daniel Costinett, *University of Tennessee, United States*
Brad Trento, *University of Tennessee, United States*
Daniel Friedrichs, *Medtronic, United States*

Grid Connected DC Distribution Network Deploying High Power Density Rectifier for DC Voltage Stabilization 3585
Danillo B. Rodrigues, *Universidade Federal do Triângulo Mineiro, Brazil*
Paulo R. Silva, *Universidade Federal de Uberlândia, Brazil*
Gustavo B. Lima, *Universidade Federal do Triângulo Mineiro, Brazil*
Ernane A.A. Coelho, *Universidade Federal de Uberlândia, Brazil*
Luiz C.G. Freitas, *Universidade Federal de Uberlândia, Brazil*

Even-Harmonic Repetitive Control for Circulating Current Suppression in Modular Multilevel Converters 3591

Shunfeng Yang, *Nanyang Technological University, Singapore*
Peng Wang, *Nanyang Technological University, Singapore*
Yi Tang, *Nanyang Technological University, Singapore*
Michael Zagrodnik, *Rolls-Royce Singapore Pte. Ltd., Singapore*
Xiaolei Hu, *Nanyang Technological University, Singapore*
King Jet Tseng, *Nanyang Technological University, Singapore*

A New DSC-PLL using Recursive Discrete Fourier Transform for Robustness to Frequency Variation 3598

Jaedo Lee, *Korea Institute of Nuclear Safety, Korea, South*
Hanju Cha, *Chungnam National University, Korea, South*

A Four-Quadrant Modulation Technique for Cascaded Multilevel Inverters to Extend Solution Range for Selective Harmonic Elimination/Compensation 3603

Hui Zhao, *University of Florida, United States*
Shuo Wang, *University of Florida, United States*

Online Battery Impedance Spectrum Measurement Method 3611

Jaber A. Abu Qahouq, *University of Alabama, United States*

Analysis and Control of a Reduced Switch Converter for Active Magnetic Bearings 3616

Dong Jiang, *Huazhong University of Science and Technology, China*
Parag Kshirsagar, *United Technologies Research Center, United States*

A Novel Balanced Winding Topology to Mitigate EMI without the Need for a Y-Capacitor 3623

Yongjiang Bai, *Xi'an Jiaotong University, China*
Xu Yang, *Xi'an Jiaotong University, China*
Xinlei Li, *Silergy Corp., China*
Dan Zhang, *Silergy Corp., China*
Wenjie Chen, *Xi'an Jiaotong University, China*

Topology and Control Strategy for Accelerated Lifetime Test Setup of DC-Link Capacitor of Wind Turbine Converter 3629

Youngjong Ko, *Christian-Albrechts-Universität zu Kiel, Germany*
Holger Jedtberg, *Christian-Albrechts-Universität zu Kiel, Germany*
Giampaolo Buticchi, *Christian-Albrechts-Universität zu Kiel, Germany*
Marco Liserre, *Christian-Albrechts-Universität zu Kiel, Germany*

Voltage Droop Compensation based on Resonant Circuit for Generalized High Voltage Solid-State Marx Modulator 3637

Hiren Canacsinh, *Instituto Superior de Engenharia de Lisboa, Portugal*
Luís M. Redondo, *Instituto Superior de Engenharia de Lisboa, Portugal*
J. Fernando Silva, *Instituto Superior Técnico, Portugal*
Beatriz Borges, *Instituto Superior Técnico, Portugal*

Four H-Bridge based Shunt Active Power Filter for Three-Phase Four Wire System 3641

Edgard L.L. Fabricio, *Universidade Federal da Paraíba, Brazil*
Cursino B. Jacobina, *Universidade Federal de Campina Grande, Brazil*
Gregory A.A. Carlos, *Universidade Federal de Campina Grande, Brazil*
Maurício B.R. Correa, *Universidade Federal de Campina Grande, Brazil*

High-Frequency AC Distributed Power Delivery System 3648
Mengqi Wang, *University of Michigan at Dearborn, United States*
Qingyun Huang, *North Carolina State University, United States*
Wensong Yu, *North Carolina State University, United States*
Alex Q. Huang, *North Carolina State University, United States*

Effect of the Capacitance Distribution on the Output Impedance of the Half-Wave Cockcroft-Walton Voltage Multiplier 3655
Liran Katzir, *Tel Aviv University, Israel*
Doron Shmilovitz, *Tel Aviv University, Israel*

Session D17: Emerging and Renewable Power
Location: Poster Area
March 24, 2016 11:30 - 14:00
Session Chairs: Katherine Kim, *Ulsan NIST*
Dimitri Torregrossa, *EPFL*

A Cost Effective High Performance LED Driver Powered by Electronic Ballasts 3659
Jianwen Shao, *STMicroelectronics, United States*
Thomas Stamm, *STMicroelectronics, United States*

Model Predictive Control of Z-Source Four-Leg Inverter for Standalone Photovoltaic System with Unbalanced Load 3663
Sertac Bayhan, *Gazi University, Turkey*
Mohamed Trabelsi, *Texas A&M University at Qatar, Qatar*
Haitham Abu-Rub, *Texas A&M University at Qatar, Qatar*

Efficiency Optimization of an Integrated Wireless Power Transfer System by a Genetic Algorithm 3669
Rosario Pagano, *Integrated Device Technology Inc., United States*
Siamak Abedinpour, *Integrated Device Technology Inc., United States*
Angelo Raciti, *Università degli Studi di Catania, Italy*
Salvatore Musumeci, *Università degli Studi di Catania, Italy*

Loss Analysis of a High Efficiency GaN and Si Device Mixed Isolated Bidirectional DC-DC Converter 3677
Fei Xue, *North Carolina State University, United States*
Ruiyang Yu, *North Carolina State University, United States*
Alex Q. Huang, *North Carolina State University, United States*

Dynamic Efficiency Tracking Controller for Reconfigurable Four-Coil Wireless Power Transfer System 3684
Yuan Cao, *University of Alabama, United States*
Zhigang Dang, *University of Alabama, United States*
Jaber A. Abu Qahouq, *University of Alabama, United States*
Evan Phillips, *University of Alabama, United States*

Wireless Power and Data Transfer System for Smart Bridge Sensors 3690
Yujin Jang, *Korea Advanced Institute of Science and Technology, Korea, South*
Jung Kyu Han, *Korea Advanced Institute of Science and Technology, Korea, South*
Shin Young Cho, *Korea Advanced Institute of Science and Technology, Korea, South*
Gun-Woo Moon, *Korea Advanced Institute of Science and Technology, Korea, South*
Ji-Min Kim, *Korea Advanced Institute of Science and Technology, Korea, South*
Hoon Sohn, *Korea Advanced Institute of Science and Technology, Korea, South*

Inrush Transient Current Analysis and Suppression of Photovoltaic Grid-Connected Inverters during Voltage Sag 3697

Zhongyu Li, *China University of Petroleum, China*
Rende Zhao, *China University of Petroleum, China*
Zhen Xin, *Aalborg University, Denmark*
Josep M. Guerrero, *Aalborg University, Denmark*
Mehdi Savaghebi, *Aalborg University, Denmark*
Peide Li, *Shandong Jinan Power Equipment Factory Co., LTD, China*

A Highly Reliable Single-Stage Converter for Electric Vehicle Applications 3704

S.A.Kh. Mozaffari Niapour, *Northeastern University, United States*
Mahshid Amirabadi, *Northeastern University, United States*

Simple and Efficient Low Power Photovoltaic Emulator for Evaluation of Power Conditioning Systems 3712

Jesus Gonzalez-Llorente, *Universidad Sergio Arboleda, Colombia*
Andres Rambal-Vecino, *Universidad Sergio Arboleda, Colombia*
Luciano A. Garcia-Rodriguez, *University of Arkansas, United States*
Juan C. Balda, *University of Arkansas, United States*
Eduardo I. Ortiz-Rivera, *University of Puerto Rico at Mayaguez, Puerto Rico*

Data Transmission Method without Additional Circuits in Bidirectional Wireless Power Transfer System 3717

Yeongrack Son, *Seoul National University, Korea, South*
Jung-Ik Ha, *Seoul National University, Korea, South*

Improved Impedance Source Inverter for Hybrid/Electric Vehicle Application with Continuous Conduction Operation 3722

Thilak Senanayake, *University of Tsukuba, Japan*
Ryuji Iijima, *University of Tsukuba, Japan*
Takanori Isobe, *University of Tsukuba, Japan*
Hiroshi Tadano, *University of Tsukuba, Japan*

New Submodule Circuits for Modular Multilevel Current Source Converters with DC Fault Ride Through Capability

Xinyu Yu, Yingdong Wei, and Qirong Jiang

Department of Electrical Engineering
Tsinghua University
Beijing, 100084, China
Email: yuxy08@126.com, {qrjiang, wyd}@tsinghua.edu.cn

Abstract— **In this paper, two types of new submodule (SM) circuits are proposed for modular multilevel current source converter (MMCSC) with dc fault ride through capability, namely, type I diagonal bridge SM (DBSM) and type II DBSM. The topologies and possible operation states of the two types of DBSMs are presented. Each type of DBSM is capable of providing bipolar voltages in unidirectional current. Replacing the half bridge SMs (HBSMs) in the conventional modular multilevel converter (MMC) with either type of DBSMs, the MMC can operate as a current source converter during normal operation, and is capable of not only blocking the dc fault current, but also functioning as a static synchronous compensator (STATCOM) to support the ac grid to ride through dc faults. Compared with the other MMC topologies with dc fault ride through capability, the proposed DBSM-based MMCSC is cost-effective, as it saves on the use of a significant number of semiconductor devices. The operation principle, control strategy and dc fault ride-through scheme of the proposed DBSM-based MMCSCs are presented. Simulation results in the PSCAD/EMTDC environment demonstrate the validity and characteristics of the DBSM-based MMCSC.**

Keywords—modular multilevel current source converters (MMCSC); diagonal bridge submodule (DBSM); dc fault ride through capability

I. INTRODUCTION

Compared with traditional voltage-source converters, such as the 2-level converter, the neutral-point-clamped multilevel converter, and the cascaded H-bridge multilevel converter, the modular multilevel converter (MMC) offers significant advantages such as highly modular structure, reduced harmonic distortion of ac-side currents and dc current, low switching frequency, and low power loss [1-2]. As a result, the MMC has become one of the most promising converter topology for HVDC transmissions.

However, the conventional MMC which adopts half-bridge SMs (HBSM-MMC) suffers from poor dc fault performance. When a pole-to-pole dc fault occurs, high currents may damage the freewheeling diodes [3]. In addition, a dc short-circuit fault of the MMC could lead to voltage instability in the connected ac system [4-6]. Therefore, the dc fault ride through scheme

has become one of the major concerns for MMC-based HVDC systems [3], [6-10]. During a dc fault condition, the MMC is expected to be able to protect the converter itself by clearing the fault current quickly, and support the ac grid by working as a static synchronous compensator (STATCOM) [3], [6].

So far the study of the dc fault ride through scheme of the MMC mainly relies on modifying the SM topology or converter topology [8-10], because the technology of high-voltage dc breakers is not mature yet [7]. With the replacement of the HBSMs with full bridge SMs (FBSMs) or clamp double SMs (CDSMs) [8], the MMC can clear the dc fault current. However, a large number of IGBTs are added, which results in increased cost and power loss. Modifying the converter topology, the hybrid multilevel converters such as the alternative arm converter (AAC) [9] and hybrid cascaded modular multilevel converters (HCMC) [10] are capable of working as a STACOM to support the ac grid during dc faults. However, the size of dc filters are much larger, and additional control methods and peripheral circuits are needed to handle the IGBT series operation [11], [12]. The current source converter handles the dc fault very well, and several papers studied the possible topologies of modular multilevel current source converters (MMCSCs) that adopt current-source SMs. Two inductor based current-source SMs are proposed in [13], and a current-source H bridge SM is proposed in [14]. However, additional inductors are employed in these current source SM circuits, which results in a significant increase in power loss, volume, and cost.

In this paper, two types of diagonal bridge SMs (DBSMs), namely, type I DBSM and type II DBSM are proposed. The operation states of the DBSMs are studied, and each type of DBSM is found to be able to provide bipolar voltages in unidirectional current. Therefore, the DBSMs can operate as current-source SMs. Compared with FBSMs and CDSMs, DBSMs use fewer IGBTs. Unlike the aforementioned current source SMs, the DBSMs are free of inductors, which reduces loss and cost. Adopting the DBSMs ensures that the DBSM-based MMCSC can operate as a current source converter during normal operation, and is capable of not only blocking the dc fault current, but also working as a STATCOM to

---

This work was supported by the National High Technology Research and Development Program of China (863 program, No 2012AA050401) and Power Electronics Science and Education Development Program of Delta Environmental & Educational Foundation.

support the ac grid to ride through dc faults. The rest of the paper are organized as follows:

Section II presents the topologies and operating states of the two proposed type I and type II DBSMs. By adopting the DBSMs in place of the HBSMs, the topology of the MMCSC adopting DBSMs (DBSM-MMCSC) is also presented. In Section III, the operation principles and control strategies of the DBSM-MMCSC under both normal condition and dc-fault condition are analyzed. In section IV, the number of SMs, capacitors, and power devices of the DBSM-based MMCSC are calculated, and are compared with other MMCs with dc fault ride through capability. Section V demonstrates the validity and characteristics of the DBSM- MMCSC by simulation results in the PSCAD/EMTDC environment. Section VI summarizes the whole paper and makes conclusions.

II. TOPOLOGIES

The topologies of the type I DBSM and type II DBSM are shown in Figure 1. Each type of DBSM consists of 2 IGBTs, 2 diodes and a capacitor. Unlike the HBSMs, the IGBTs in the DBSMs are placed in diagonal position. The possible operation states of the DBSMs are shown in Table 1. The current paths of the DBSMs in all these states are shown in Figure 2. Attention should be paid that the anti-parallel diodes of the IGBTs in Figure 1 are not essential and can be eliminated.

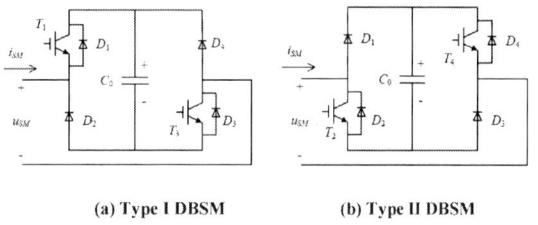

(a) Type I DBSM　　　　**(b) Type II DBSM**

Figure 1: Topologies of the DBSMs.

Table 1 and Figure 2 show that the type I DBSM can

provide bipolar voltages in negative arm current, while the type II DBSM can provide bipolar voltages in positive current. This is different from the operation states of a HBSM, since the HBSM can provide non-negative voltages in bidirectional currents. As a result, the HBSM can operate as a voltage source SM while the DBSMs can operate as current-source SMs. Compared with existed current-source SMs [13], [14], the proposed DBSMs doesn't contain inductors, thus can save cost and loss.

Replacing the HBSMs in the conventional HBSM-MMC with either type of the DBSMs can obtain a DBSM-based MMCSC (DBSM-MMCSC), as shown in Figure 3. Similar with the conventional MMC, the DBSM-MMCSC is also composed of three phases, each consists of a positive arm (upper arm) and a negative arm (lower arm). Each arm is formed with N series-connected DBSMs and an inductor. In a 2-terminal HVDC system, the two terminals should employ different types of DBSMs.

TABLE 1
POSSIBLE OPERATION STATES OF THE DBSMS

	SM state	T_1	T_2	T_3	T_4	i_{SM}	u_{SM}
Type I DBSM	Positive	1	--	1	--	<0	U_c
	Bypassing	1	--	0	--	<0	0
	Negative	0	--	0	--	<0	$-U_c$
Type II DBSM	Positive	--	0	--	0	>0	U_c
	Bypassing	--	0	--	1	>0	0
	Negative	--	1	--	1	>0	$-U_c$

III. OPERATION PRINCIPLE AND CONTROL STRATEGY

A. Operation Principle

The circuit of phase $x(x=a,\ b,\ c)$ of the MMCSC adopting the type I DBSM (DBSM-I-MMCSC) is taken as an example as shown in Fig. 4 to explain the operation principle of the proposed MMCSC. As shown in Fig. 4, all arms of the DBSM-I-MMCSC are conducing and working as wave shaping circuits (WSCs) during normal operation, similar to the conventional MMC. However, because the type-I DBSM

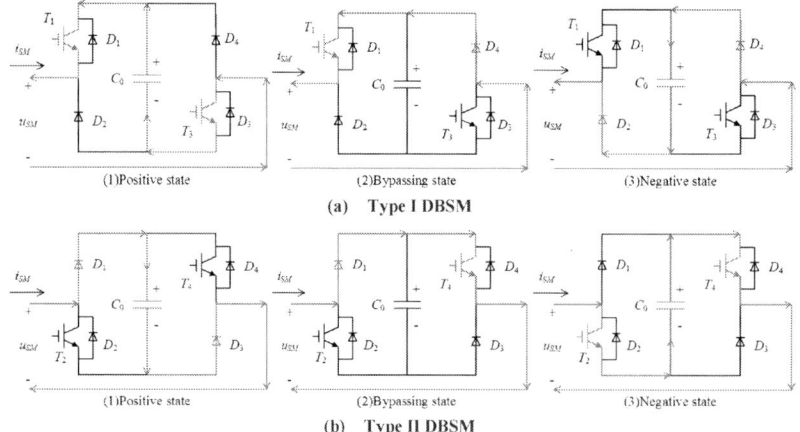

Figure 2: Current paths of the DBSMs.

978-1-4673-9551-9/16 $31.00 © 2016 IEEE　　　　1469

provides bipolar voltages only in negative current, the arm current should be always negative. The upper and lower arm currents of phase x can be expressed as follows:

$$\begin{cases} i_{xp}(t) = \dfrac{1}{3}i_{cirx}(t) + \dfrac{1}{2}i_x(t) = \dfrac{1}{3}i_{dc} + \dfrac{1}{2}I_m \sin(\omega t) \\ i_{xn}(t) = \dfrac{1}{3}i_{cirx}(t) - \dfrac{1}{2}i_x(t) = \dfrac{1}{3}i_{dc} - \dfrac{1}{2}I_m \sin(\omega t) \end{cases} \quad (1)$$

where $i_{xp}(t)$ and $i_{xn}(t)$ denote the upper and lower arm current of phase x, respectively, $i_{cirx}(t)$ denotes the circulating current of phase x, i_{dc} denotes the dc-link current, and I_m denotes the amplitude of the ac-side currents. Equation (1) can be satisfied by controlling the circulating currents and ac-side currents of the DBSM-I-MMCSC. To make sure the arm currents are always negative, it can be derived from (1) that:

$$\frac{1}{3}i_{dc} \le \frac{1}{2}I_m \quad (2)$$

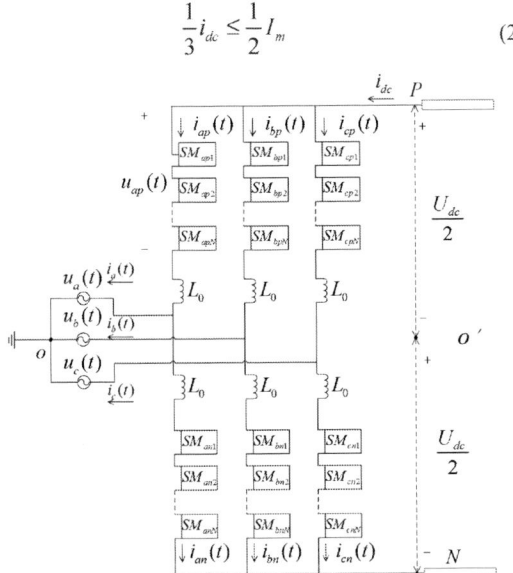

Figure 3: Topology of the DBSM-based MMCSC.

Because the arm currents are always negative, i_{dc} is also always negative. Consequently, the three-phase DBSM-I-MMCSC operates as a current-source converter during normal operation. To ensure the energy of each arm is balanced in one cycle, the following equations should be satisfied:

$$\begin{cases} P_{xp} = \dfrac{1}{T}\int_0^T u_{xp}(t)i_{xp}(t)dt = 0 \\ P_{xn} = \dfrac{1}{T}\int_0^T u_{xn}(t)i_{xn}(t)dt = 0 \end{cases} \quad (3)$$

where $u_{xp}(t)$ and $u_{xn}(t)$ denote the upper and lower arm voltage, respectively, and $u_{xp}(t)$ and $u_{xn}(t)$ satisfy:

$$\begin{cases} u_{xp}(t) = \dfrac{1}{2}U_{dc} - U_m \sin(\omega t + \varphi) - L_0 \dfrac{di_{xp}(t)}{dt} \\ u_{xn}(t) = \dfrac{1}{2}U_{dc} + U_m \sin(\omega t + \varphi) - L_0 \dfrac{di_{xn}(t)}{dt} \end{cases} \quad (4)$$

From (1)-(4) it can be derived that the dc-link voltage should be smaller than the amplitude of ac phase voltage:

$$U_{dc} \le U_m \quad (5)$$

From (5) it can be deduced that each arm of the DBSM-I-MMC employs fewer SMs than the HBSM-MMC with the same ac voltage. Besides, when the MMCSC operates under normal operation, U_{dc} can vary from $-U_m$ to $+U_m$. By reversing the dc-link voltage, the power flow can be reversed, therefore the proposed MMCSC is able to transmit bidirectional power.

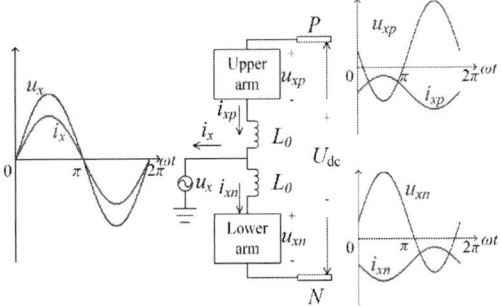

Figure 4: One phase of the DBSM-I-MMCSC showing the arm voltages and arm currents.

B. Mathematic Model

In order to control the ac-side currents as well as the dc-link current, an accurate mathematic model of the proposed MMCSC should be set up. From (1)(4) it can be deduced that the ac-side currents and circulating currents satisfy the following equations:

$$\begin{cases} u_{comx}(t) = \dfrac{1}{2}U_{dc} - L_0 \dfrac{di_{cirx}(t)}{dt} \\ u_{diffx}(t) = u_x(t) + \dfrac{L_0}{2} \dfrac{di_x(t)}{dt} \end{cases} \quad (6)$$

where u_{comx} and u_{diffx} represents the common component and differential component of the upper and lower arm voltages of phase x, respectively, and satisfy the following equations:

$$\begin{cases} u_{comx} = \dfrac{1}{2}[u_{xp}(t) + u_{xn}(t)] \\ u_{diffx} = \dfrac{1}{2}[u_{xn}(t) - u_{xp}(t)] \end{cases} \quad (7)$$

By utilizing the abc-dq transformation, the second equation of (6) can be transformed into dq axis:

$$\begin{cases} L_0 \dfrac{di_d(t)}{dt} = \omega L_0 i_q(t) + u_{diffd}(t) - u_d(t) \\[2mm] L_0 \dfrac{di_q(t)}{dt} = -\omega L_0 i_d(t) + u_{diffq}(t) - u_q(t) \end{cases} \quad (8)$$

where $i_d(t)$ and $i_q(t)$ are transformed from $i_a(t)$, $i_b(t)$ and $i_c(t)$, $u_{diffd}(t)$ and $u_{diffq}(t)$ are transformed from $u_{a*}(t)$, $u_{b*}(t)$, and $u_{c*}(t)$, and $u_{d*}(t)$ and $u_{q*}(t)$ are transformed from $u_a(t)$, $u_b(t)$, and $u_c(t)$. Therefore, (8) is the mathematic model of ac-side currents, while the mathematic model of circulating currents and dc-link current can be expressed as:

$$\begin{cases} u_{coma}(t) = \dfrac{1}{2} U_{dc} - L_0 \dfrac{di_{cira}(t)}{dt} \\[2mm] u_{coma}(t) = \dfrac{1}{2} U_{dc} - L_0 \dfrac{di_{cira}(t)}{dt} \\[2mm] u_{coma}(t) = \dfrac{1}{2} U_{dc} - L_0 \dfrac{di_{cira}(t)}{dt} \\[2mm] i_{dc}(t) = i_{cira}(t) + i_{cirb}(t) + i_{circ}(t) \end{cases} \quad (9)$$

C. Control Strategy

It can be seen that (8) is similar to the mathematical model of HBSM-MMC [15], therefore the inner current control method in dq axis proposed in [15] can be simply transplanted to control ac-side currents in of the DBSM-MMC. Similar to the HBSM-MMC, the reference values of $i_d(t)$ and $i_q(t)$, i.e. $i_{d*}(t)$ and $i_{q*}(t)$, can be obtained by outer loop controllers. For example, $i_{d*}(t)$ can be obtained by the active power controller or the dc-link voltage controller, while $i_{q*}(t)$ can be obtained by the reactive power controller [16].

On the other hand, the circulating current of phase x can be controlled by a PI controller according to (9):

$$u_{comx*}(t) = \frac{1}{2} U_{dc} - [k_{p1} \Delta i_{cirx}(t) + \frac{k_{p1}}{T_{i1}} \int_0^t \Delta i_{cirx}(\tau) d\tau] \quad (10)$$

where $u_{comx*}(t)$ denotes the reference voltage of $u_{comx}(t)$, while $\Delta i_{cirx}(t)$ satisfies the following equation:

$$\Delta i_{cirx}(t) = i_{cirx}(t) - i_{cirx*}(t) = i_{cirx} - \frac{1}{3} i_{dc} \quad (11)$$

where $i_{cirx*}(t)$ denotes the reference value of $i_{cirx}(t)$, and is set to be $i_{dc}/3$ so that there will be no harmonic components in the circulating currents.

From the ac-side current controller and circulating current controller above, the reference value of u_{comx} and u_{diffx}, i.e. u_{comx*} and u_{diffx*} can be obtained. The reference value of each arm voltage can be calculated as follows:

$$\begin{cases} u_{xp*} = u_{comx*} - u_{diffx*} \\[1mm] u_{xn*} = u_{comx*} + u_{diffx*} \end{cases} \quad (12)$$

By utilizing proper modulation schemes, such as the nearest level modulation scheme [16], the numbers of DBSMs in different operation states can be calculated. Since the DBSMs provide bipolar voltages like FBSMs do, the sorting method for FBSMs proposed in [17] can be simply transplanted to balance the capacitor voltages in each arm of the DBSM-MMCSC, and the operation states of each DBSM can be obtained. The whole control diagram of the proposed DBSM-MMCSC during normal operation is shown in Figure 5.

D. DC Fault Management

Equation (5) indicates that the DBSM-MMCSC is able to work not only when the dc-link voltage is reversed, but also when the dc-link voltage drops to zero. When a pole-to-pole short circuit happens on the dc side, the arm currents are still under control, therefore the MMCSC can operate as a STATCOM, and provide reactive power to support the ac grid. The control strategy of the DBSM-MMCSC doesn't have to change during dc fault. As a matter of fact, the control strategy of the DBSM-MMCSC during normal operation is also suitable for dc-fault STATCOM operation, as long as the reference value of i_d is set to be zero.

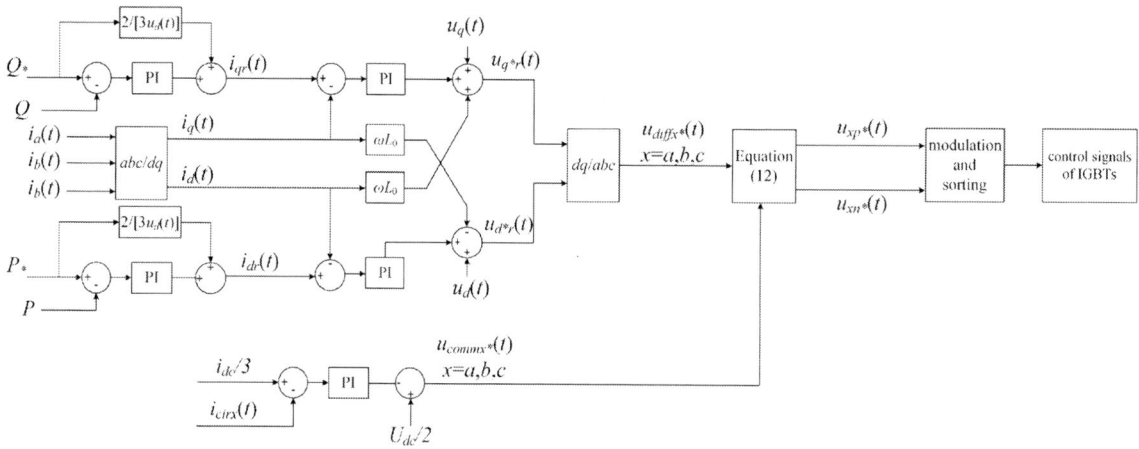

Figure 5: Control diagram of the DBSM-MMCSC.

TABLE II
COMPARISON OF DIFFERENT MULTILEVEL CONVERTERS

Parameters	HBSM-MMC	CDSM-MMC	FBSM-MMC	Hybrid FBSM -MMC	DBSM-MMC
DC-link voltage/kV	40	40	40	40	20
AC line-line RMS voltage/kV	24.5	24.5	24.5	24.5	24.5
U_c/kV	1.7	1.7	1.7	1.7	1.7
SM type	HBSM	CDSM	FBSM	FBSM and HBSM	DBSM
No. of SMs	144	72	144	72 FBSMs and 72 HBSMs	108
No. of IGBTs in total	288	360	576	432	216
No. of diodes in total*	288	504	576	432	216
No. of capacitors in SMs	144	144	144	144	108
DC fault blocking capability	No	Yes	Yes	Yes	Yes
DC-fault STATCOM capability	No	Yes[19]	Yes	Yes	Yes

*: The number of anti-parallel diodes are included.

Besides, by blocking all the IGBTs in each DBSM, the DBSMs can work in blocking state. When all the DBSMs are blocked, the arm currents will charge the SM capacitors and drop to zero rapidly. When the arm currents drops to zero, the SM capacitors provide enough reverse voltages, thus all diodes are reverse biased, and all arm currents will not reignite. As a result, the dc-fault current can be cleared. That is to say, the DBSM-MMCSC is able to work as a STATCOM during a pole-to-pole dc fault, and is able to clear the dc fault by blocking all IGBTs.

IV. COMPARSONS

In this part, the number of power electronic devices employed by the DBSM-MMC, HBSM-MMC, CDSM-MMC, FBSM-MMC and hybrid FBSM-MMC [18] are compared. Suppose the amplitudes of ac-side voltages of the converters above are all U_m, and the rated SM capacitor voltage is U_c. Therefore the dc-link voltage of MMCs should be larger than $2U_m$, and the minimum number of SMs in each arm satisfies:

$$N_{MMC} = round(2U_m / U_c) \qquad (13)$$

Where $round(X)$ denotes the nearest integer of X.

On the other hand, from (5) it can be deduced that the dc-link voltage of the DBSM-MMC is smaller than U_m, therefore the maximum number of SMs in each arm satisfies:

$$N_{MMCSC} = round(1.5U_m / U_c) \approx 0.75 N_{MMC} \qquad (14)$$

That is to say, the MMCSC employs fewer SMs in each arm. Considering that different types of SMs consist of different numbers of IGBTs and diodes, the total number of IGBTs and diodes employed in the converters above are compared in Table II.

As compared in Table II, the MMCSC provides the best dc fault ride through performance with the smallest number of IGBTs. The number of IGBTs employed in the DBSM-MMCSC is even smaller than HBSM-MMC, which is unable to deal with dc-link pole-to-pole fault. Therefore, the proposed DBSM-MMCSC can be an economic option in long-distance HVDC overhead-line transmission.

V. SIMULATION

A detailed model of a unilateral DBSM-I-MMCSC system is simulated in PSCAD/EMTDC. The specifications of the simulated system are listed in Table III. Two cases are simulated as follows to verify the operation principle and control strategy of the proposed DBSM-MMCSC:

Case 1: The DBSM-I-MMCSC transmits an active power of -6MW and a reactive power of 2Mvar under normal condition before t=0.4s. At t=0.4s, the reference value of reactive power steps to -2MVar. At t=0.5s, the reference value of active power steps to -3MW.

Case 2: The DBSM-I-MMCSC transmits an active power of -6MW and a reactive power of 2Mvar under normal condition before t=0.4s. At t=0.4s, a pole-to-pole dc fault occurs on the dc port, then the DBSM-I-MMCSC is switched to dc fault blocking mode, and all IGBTs of the DBSM-MMCSC are switched off. At t=0.5s, the DBSM-I-MMCSC is switched to dc-fault STATCOM mode, and provides a reactive power of -4MVar to support the ac grid.

TABLE III
PARAMETERS OF THE DBSM-MMC SIMULATION MODEL

Parameter	Value
Inductance of each arm	25mH
Capacitance of each DBSM	2000μF
Number of CDSMs in each arm	8
Rated capacitor voltage	5kV
Rated dc-link voltage	20kV
Rated ac-side line to line RMS voltage	30kV
Rated capacity of the DBSM-MMCSC	10MVA

Figure 6 shows the simulation results of case 1. As shown in Figure 6, the active power and reactive power are controlled precisely, and the ac-side currents are with little distortion. Besides, the dc-link current is controlled to be constant, while the dc-link voltage varies with the active power that is being transmitted. Moreover, the capacitor voltages value. The simulation results show that the DBSM-MMCSC is able to work as a current-source converter during normal operation, and the reactive power can be controlled decoupled with the active power.

978-1-4673-9551-9/16 $31.00 © 2016 IEEE

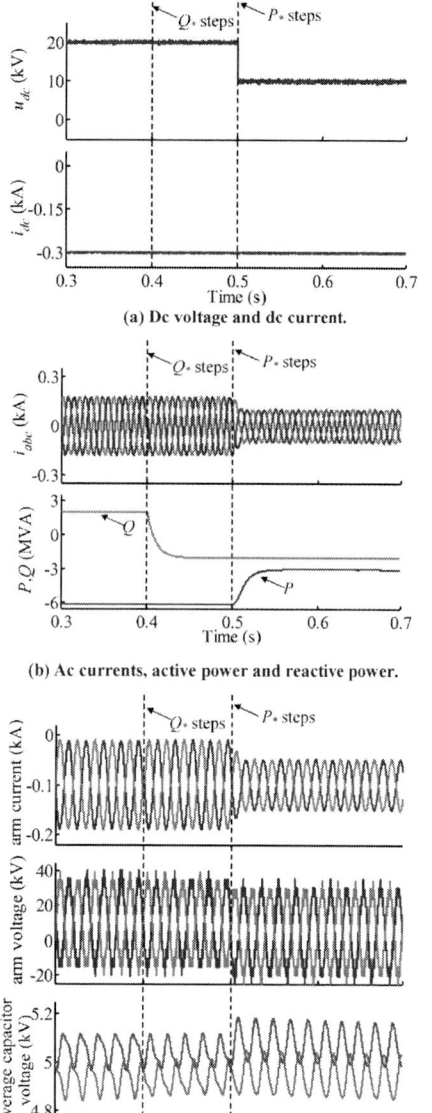

(a) Dc voltage and dc current.

(b) Ac currents, active power and reactive power.

(c) Arm currents, arm voltages and capacitor voltages of phase a.

Figure 6: Simulation results of case 1.

Figure 7 shows the simulation results of case 2. As shown in Figure 7, the DBSM-MMCSC can clear the dc fault current by switching off all the IGBTs. Besides, after being switched to the dc-fault STATCOM mode, the DBSM-MMCSC can provide reactive power to support the ac grid, and the reactive power can be controlled precisely. What's more, the capacitor voltages are all maintained around the nominated value. The simulation results verify the dc fault ride through capability of the DBSM-MMCSC.

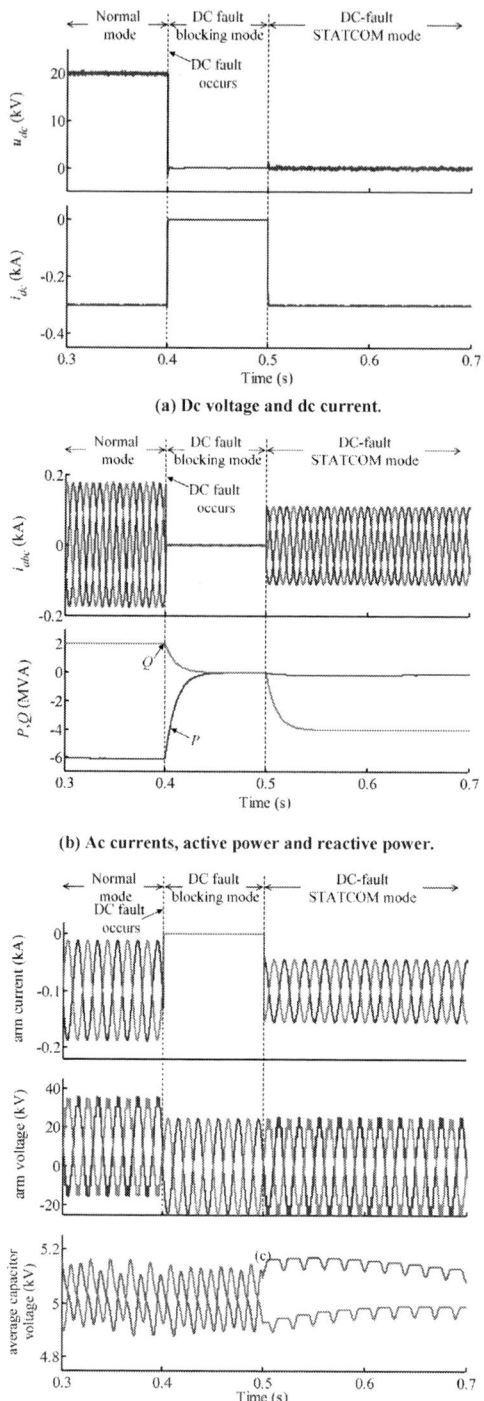

(a) Dc voltage and dc current.

(b) Ac currents, active power and reactive power.

(c) Arm currents, arm voltages and capacitor voltages of phase a.

Figure 7: Simulation results of case 2.

VI. CONCLUSION

In this paper, two types of DBSMs are proposed for MMCSCs with dc fault ride through capability. The topologies and operation states of the proposed DBSMs are presented. The operation principle of the DBSM-MMCSC is analyzed, and the control strategies during normal operation and dc fault are designed. Simulation results show that the DBSM-MMCSC can operate as a current source converter during normal operation, and can block dc fault current and provide reactive power to support the ac grid to ride through dc faults. Compared with the existing modular multilevel voltage source converters, the proposed DBSM-MMCSC employs smaller number of SMs and IGBTs. Therefore, the DBSM-based MMCSC can be an economic option for long-distance HVDC transmission using overhead lines.

REFERENCES

[1] A. Lesnicar and R.Marquardt, "An innovative modularmultilevel converter topology suitable for a wide power range," in Proc. IEEE Power Tech Conf., Bologna, Italy, 2003, vol. 3, pp. 23–26.

[2] M. Saeedifard and R. Iravani, "Dynamic performance of a modular multilevel back-to-back HVDC system," IEEE Trans. Power Del., vol.25, no. 4, pp. 2903–2912, Oct. 2010.

[3] S. Norrga, X. Li, and L. Angquist, "Converter topologies for HVDC grids," in Proc. IEEE EnergyCon 2014, Dubrovnik, Croatia, May 2014, pp. 1634–1641.

[4] D. T. Oyedokun, K. A. Folly and S. P. Chowdhury, "Effect of converter DC fault on the transient stability of a multi-machine power system with HVDC transmission lines," in Proc. AFRICON 2009, pp. 1-6, 23-25 Sept. 2009.

[5] Y. Zhuang, R.W. Menzies, O.B. Nayak, and H.M. Turanli, "Dynamic performance of a STATCON at an HVDC inverter feeding a very weak AC system", IEEE Trans. Power Del., vol.11, no.2, pp.958-964, Apr. 1996.

[6] G. Tang, Z. Xu, and Y. Zhou, "Impacts of three MMC-HVDC configurations on ac system stability under dc line faults," IEEE Trans. Power Systems, vol.29, no.6, pp.3030-3040, Nov. 2014.

[7] C. M. Franck, "HVDC circuit breakers: A review identifying future research needs," IEEE Trans. Power Del., vol. 26, no. 2, pp. 998–1007,Apr. 2011.

[8] R. Marquardt, "Modular multilevel converter: An universal concept for HVDC-networks and extended dc-bus-applications," in Proc. Int. Power Electron. Conf., 2010, pp. 502–507.

[9] M. M. C. Merlin, T. C. Green, P. D. Mitcheson, D. Trainer, W. Critchley, R. Crookes, and F. Hassan, "The Alternate Arm Converter: A New Hybrid Multilevel Converter With DC-Fault Blocking Capability," IEEE Trans. Power Del., vol. 29, no. 1, pp. 310–317, 2014.

[10] G. P. Adam, K. H. Ahmed, S. J. Finney, K. Bell, and B. W. Williams, "New breed of network fault-tolerant voltage-source-converter HVDC transmission system," IEEE Trans. Power Syst., vol. 28, no. 1, pp. 335–346, Feb. 2012.

[11] T.-V. Nguyen, P. Jeannin, E. Vagnon, D. Frey, and J.-C. Crebier, "Series connection of IGBT," in Proc. IEEE 25th Applied Power Electronics Conference and Exposition, Feb. 2010, pp. 2238-2244.

[12] S. Ji, T. Lu, Z. Zhao, H. Yu, L. Yuan, S. Yang, and C. Secrest, "Physical model analysis during transient for series-connected HVIGBTs," IEEE Trans. Power Electron., vol. 29, no. 11, pp. 5727–5737, Nov. 2014.

[13] J. Liang, A. Nami, F. Dijkhuizen, P. Tenca, and J. Sastry, "Current source modular multilevel converter for HVDC and FACTS," presented at the EPE ECCE Eur., Lille, France, Sep. 3–5, 2013.

[14] M. A. Perez, R. Lizana, C. Azocar, J. Rodriguez, and B. Wu, "Modular multilevel cascaded converter based on current source H-Bridges cells, " presented at the IEEE Ind. Electron. Soc. Annu. Conf., Montreal, QC, Canada, Oct. 2012.

[15] Y. Zhao, X. Hu, G. Tang, Z. He, "A study on MMC model and its current control strategies," Power Electronics for Distributed Generation Systems (PEDG), 2010 2nd IEEE International Symposium on , vol., no., pp.259,264, 16-18 June 2010.

[16] M. Guan, Z. Xu, and H. Chen, "Control and modulation strategies for modular multilevel converter based HVDC System," in Proc. Conf. IEEE Ind. Electron. Soc., 2011, pp. 849–854.

[17] Yinglin Xue; Zheng Xu; Qingrui Tu, "Modulation and Control for a New Hybrid Cascaded Multilevel Converter With DC Blocking Capability," IEEE Transactions on Power Delivery, , vol.27, no.4, pp.2227,2237, Oct. 2012.

[18] S. Cui, S. Kim, J. Jung, and S. Sul, "Principle, control and comparison of modular multilevel converters (MMCs) with DC short circuit fault ride-through capability," in Proc. IEEE 29th Applied Power Electronics Conference and Exposition, Mar. 2014, pp. 610-616.

[19] X. Yu, Y. Wei, and Q. Jiang, "STATCOM Operation Scheme of the CDSM-MMC during a Pole-to-pole DC Fault," IEEE Trans. Power Del., in press.

Voltage and Power Balance Control Strategy for Three-phase Modular Cascaded Solid Stated Transformer

Zhiyu Zhang, Hengyang Zhao, Shihang Fu, Jianjiang Shi and Xiangning He

College of Electrical Engineering, Zhejiang University, Hangzhou, Zhejiang 310027, China

Email: zhiyuzhang@zju.edu.cn, sunspirit@zju.edu.cn, bellnorth@zju.edu.cn, jianjiang@zju.edu.cn, hxn@zju.edu.cn

Abstract—The three-phase modular cascaded solid stated transformer (SST) is an important element in the micro-grid systems as its outstanding adventures compared with conventional power transformer. It consists of three stages, the three-phase cascaded modular rectifier stage, the dual active bridge (DAB) converter stage and the three-phase inverter stage. Three output-paralleled DAB converters offer three dc-buses, which forming the interfaces of renewable energy. However, unbalanced modular voltage or transferred power causes overvoltage or overcurrent issues, which increase the stress of the semiconductor switches and even the instability of SST system. Focusing on the imbalance issues, this paper proposes a systematic control strategy for the three-phase modular cascaded SST. A d-q vector-based common-duty-ratio controller is applied to the rectifier stage aiming at balancing the modular current in each phase. Voltage controller for DAB stage achieves the balanced dc-link and dc-bus voltage, and proportional-resonant controller is applied to inverter stage. The effect of this proposed control method has been verified theoretically. In addition, a scaled-down experimental prototype is built to prove the performance.

Keywords—*cascaded H-bridge rectifier; dual active bridge converter; three-phase solid stated transformer*

I. INTRODUCTION

With increased concerns about the energy consumption and environment protection, more and more distributed renewable energy resources (DRER), such as photovoltaic cells, small wind turbines and fuel cells, have been integrated into the power grid in the form of distributed generation (DG) [1]-[12]. These DRER based DG systems are normally interfaced to the grid through dedicated power electronics converters, advanced communication technologies and the distributed energy storage devices (DESD), forming the micro-grids [2]. They work as the integrations of DRER and DESD for small area power systems in the smart grid, which regulate and distribute energy in more flexible and credible ways [6] [7]. Thus, micro-grids are being extensively investigated as a promising solution to electrical energy challenges in 21st century [8]-[10]. Hence, an intelligent transformer with advanced controller is becoming indispensable in micro-grids [12].

The power electronic converter known as the solid state transformer (SST) [13] [14], has outstanding adventures compared with conventional power transformer [15] [16]. With the development of the power electronics technologies, especially the power semiconductor devices, SST has got widespread concern and received a great number of results in the past decades [15]-[24]. The types of SST topologies can be divided into three categories: ac-ac single-stage, ac-dc-ac two-stage and ac-dc-dc-ac three-stage structure [23]. Although the three-stage SST has more complex circuits and lower efficiency, its good control characteristics make it more suitable, potential and flexible for energy distribution and management. In addition, modular multilevel cascaded converters are commonly employed at the SST rectifier stage in order to reduce the rectifier switching voltage stress and match the grid medium-voltage level. However, unbalanced modular voltage or transferred power cause overvoltage or overcurrent issues, which increase the stress of the semiconductor switches and even the instability of SST system.

Different control strategies were proposed for the same single-phase three-stage modular multilevel cascaded SST topology [17]-[20]. A voltage and power balance controller was proposed in [17] consists of a voltage balance controller for cascaded H-bridge rectifier and an average power calculator for dc-dc converters. In [18], a 3-D space modulation with voltage balancing capability was proposed for the cascaded seven-level rectifier in SST. However, the control schemes for dc-dc stage in [17] [18] cost a lot of calculation, high hardware requirements and was difficult to implement in Engineering. Shi et al. [19] proposed a system strategy, however, the mentioned function of bi-directional power transmission was not achieved by experiment. A current sensorless power balance strategy was proposed for DAB converters in [20]. However, the bi-directional power flow, which is indispensable in the renewable energy field, was not considered and verified in that paper. An enormous amount of research effort goes into single-phase SST and several researches discuss on the three-phase SST [21]-[24]. In [21], it proposed a hysteresis-based control method for three-phase non-cascaded topology and was not suitable for high voltage input. In [22], 15 kV SiC N-IGBTs was used in SST converter, which greatly contributed to the engineering application in medium voltage. Several SST strategies had been discussed in [23], while that topology without 400V dc bus was not suitable for DESD and DRER. Wu et al. [24] proposed an abc/dq0 control scheme for the three-phase modular multilevel cascaded, the three-phase input voltages were coupled together and may affect system stability.

This work is sponsored by the National Nature Science Foundation of China(51277162).

As shown in Figure 1, the three-phase SST consists of three stages, the three-phase cascaded modular rectifier stage, the dual active bridge (DAB) converter stage and the three-phase inverter stage. Rectifier stage with multilevel cascaded structure, allows the use of low voltage module to fit for the grid medium-level voltage input. In each phase, the output-paralleled DAB converters offer the dc-bus, which forming the interfaces of DESD and DRER. Three single-phase 220V ac voltages, differed 120 degree from each other, form the three-phase output voltage.

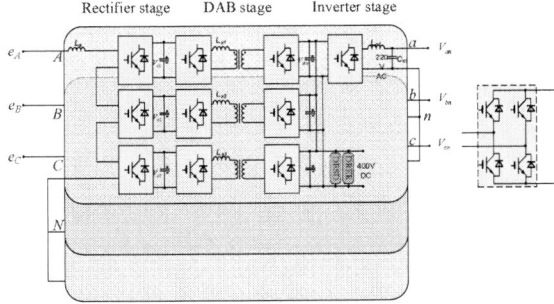

Figure 1: The topology of three-phase SST.

This paper presents a systematic control strategy for the three-phase cascaded H-Bridge multilevel converter-based SST topology. The remainder of this paper is organized as follows. Section II focuses on the modeling and controller of three-phase SST. Section III focuses on modular voltage and power balance analysis. Section IV focuses on the experiment results of a scaled-down prototype.

II. SST CIRCUIT ANALYSIS

A. Three-phase Modular Cascaded H-Bridge Rectifier Modeling and Controller Analysis

In order to enhance reliability and stability of SST system, each phase dc-bus is separated as shown in Figure 1. Hence, each phase decoupled control strategy is suitable for this topology. In case of phase A, cascaded rectifier equivalent circuit is shown in Figure 2. Referred to it, each rectifier module ac side is connected in series, the ac voltage supported between nodes "a" and "n" is the sum of the voltages supported by each module, which can be written as follows:

$$e_{an} = e_{ab} + e_{bc} + e_{cn} \tag{1}$$

Assuming $V_{a1} = V_{a2} = V_{a3} = V_d$, which is the rectifier dc-link reference voltage, each module can produce three different voltage levels: $\pm V_d$ and 0. With reference to upper rectifier module #1, $e_{ab} = V_d$ is achieved with switches T_{11} and T_{14} on, with T_{12} and T_{13} on, and if either T_{11} and T_{13} or T_{12} and T_{14} are on. The legs are complementarily off in each of the three modules. The remaining rectifier modules can operate similarly. Therefore, seven distinct voltage levels can appear on the ac input terminal between the points "a" and "n".

Based on the above analysis of converter switching state, average differential equations of the cascaded H-bridge rectifier can be described as follows:

$$\begin{cases} L_a \dfrac{di_a}{dt} = e_A - R_s i_a - e_{ab} - e_{bc} - e_{cn} \\ C_{ai} \dfrac{d_a V_{ai}}{dt} = i_{ai} - \dfrac{V_{ai}}{Z_{ai}} \end{cases} \tag{2}$$

where e_A is grid voltage; i_a is grid current; L_a is input inductance; $V_{ab} = d_{a1}V_{a1}$, $V_{bc} = d_{a2}V_{a2}$, $V_{cn} = d_{a3}V_{a3}$, and V_{ai} is the dc-link capacitor voltage of rectifier module #i; $i_{ai} = d_{ai}i_a$ is the output current of rectifier module #i, and d_{ai} is the SPWM duty cycle of module #i; Z_{ai} corresponds to equivalent input-impedance of the down-stream DAB module #i.

According to (2), it can be observed that the power transferred by each rectifier module is related to its output current i_{ai} and equivalent impedance Z_{ai}, with i_{ai} depending on the duty cycle of rectifier module #i and Z_{ai} not decided by the module .

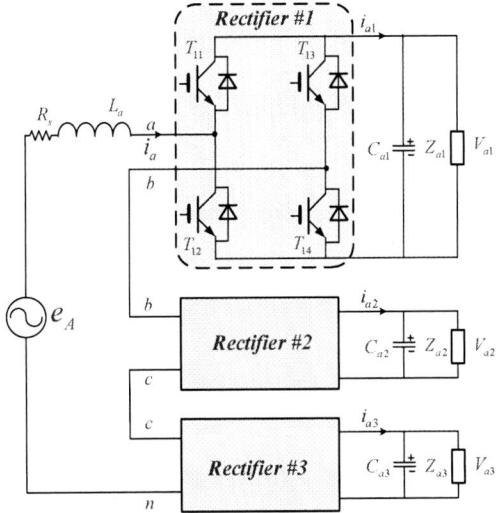

Figure 2: Modular cascaded rectifier equivalent circuit.

Illustrated with example phase A, a single-phase d-q vector-based controller in Figure 3 is applied to each phase rectifier respectively, which generates the same SPWM duty cycles for each H-bridge. The following d-q transformation is used in the controller:

$$\begin{bmatrix} x_d \\ x_q \end{bmatrix} = T \cdot \begin{bmatrix} x_a \\ x_m \end{bmatrix} = \begin{bmatrix} \sin\theta & -\cos\theta \\ \cos\theta & \sin\theta \end{bmatrix} \cdot \begin{bmatrix} x_a \\ x_m \end{bmatrix} \tag{3}$$

where T is the d-q transformation matrix; $\theta = wt = 2\pi f_{La}t$; f_{La} is the grid voltage frequency; x_a are the variables of phase A; x_m are the variables of an imaginary phase M, which is 90 degree lagging phase A.

From (3), small signal model can be obtained and decoupled signal model can be derived. Detailed modeling analysis can be illustrated in [17]-[19]. As shown in Figure 3, only the rectifier module #1 is closed-loop controlled, and the SPWM signals for module #2 and module #3 are directly derived from that for module #1 by shifting the high-frequency carriers' phase of 120° and 240°, respectively, which aims at producing the seven-level staircase voltage on the ac side of the cascaded rectifier. Hence, the common-duty-ratio control scheme realizes the simplified and decoupled control for the modular multilevel cascaded rectifier.

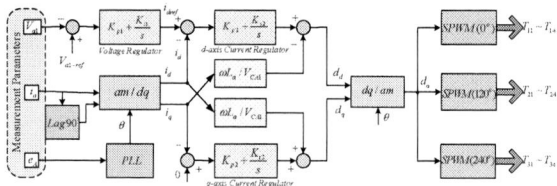

Figure 3: Controller for modular multilevel cascaded rectifier.

B. DAB Converter Modeling and Controller Analysis

DAB offers low switching losses, high power density, high efficiency, and configuration symmetry that allows seamless control of bidirectional power flow. It consists of a high voltage H-bridge, a low voltage H-bridge and a high frequency transformer as shown in Figure 4(a). According to DAB modal analysis, the power transferred by DAB module #a1 is given as follows:

$$p_{a1} = \frac{nV_{a1}V_{dca}D_{a1}(1-D_{a1})}{2f_sL_{a1}} \quad (4)$$

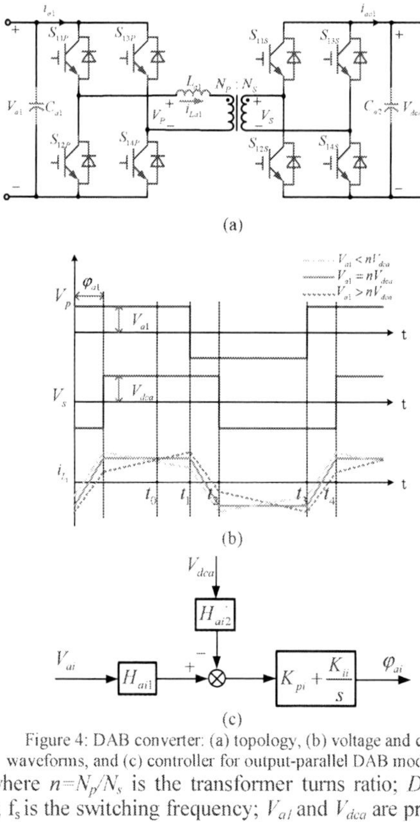

where $n = N_p/N_s$ is the transformer turns ratio; D_{a1} is duty ratio; f_s is the switching frequency; V_{a1} and V_{dca} are primary and secondary dc voltages; and L_{a1} is inductance of DAB module #a1.

Active power flows from the bridge with leading phase angle to that with a lagging phase angle. Detailed analysis of the DAB working process, with the main theoretical voltage and current waveforms shown in Figure 4(b), has been illustrated in [25]-[29]. In theory, full range ZVS can be achieved on the time $V_{a1}=nV_{dca}$. In other words, if the ratio of input to output voltage in DAB module varies from the DAB transformer turns ratio, which results in the narrowing of soft-switching range. The full range ZVS conditions are based on the ideal case. In fact, due to the switch junction capacitor, the switching process time will make it impossible for switching device to achieve ZVS under the condition of small DAB load current.

The PI regulators generate independently their phase-shift signals to eliminate the voltage imbalance caused by mismatched DAB parameters. As shown in Figure 4(c), DAB #a dc bus voltage V_{dca} multiplied by the feedback coefficient H_{ai2} is compared with input-voltage V_{ai} multiplied by feed-forward coefficient H_{ai1}. Hence, V_{dca} equals to $V_{a1}\cdot H_{a11}/H_{a12}$, V_{a2} equals to $V_{dca}\cdot H_{a22}/H_{a21}$ and V_{a3} equals to $V_{dca}\cdot H_{a32}/H_{a31}$.

C. Three-phase Inverter Controller Analysis

The decoupled three-phase SST makes it possible to combine three single-phase inverter to form the three-phase inverter. Three-phase inverter controller is shown in Figure 5.

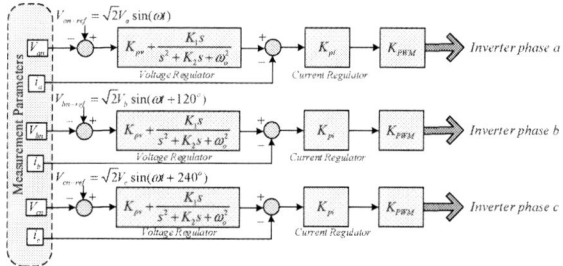

Figure 5: Three-phase inverter controller

where V_{ai}, V_{bn} and V_{cn} are the output voltages; $V_{an\text{-}ref}$, $V_{bn\text{-}ref}$ and $V_{cn\text{-}ref}$ are the output voltage references; i_a, i_b and i_c are the output currents of inverter phase a, b and c, respectively.

The three inverter controllers are implemented by one DSP28335. In digital controller, it is easy to generate three output voltage references spaced 120 degree, respectively. Proportional-resonant (PR) controller is applied to the voltage outer loop to track voltage ac reference in zero static error and proportional controller is used in inner current loop to improve the dynamic response. K_{pv} is the proportional parameter of voltage controller; w_o is the resonant frequency, which means the controlled frequency; the gain of resonant frequency refers to the ratio of K_1 and K_2; K_{pi} is the proportional parameter of current controller.

III. VOLTAGE AND POWER BALANCE ANALYSIS

A systematic control scheme is applied for the modular SST, which include three d-q vector-based common-duty-ratio

Figure 4: DAB converter: (a) topology, (b) voltage and current waveforms, and (c) controller for output-parallel DAB modules.

978-1-4673-9551-9/16 $31.00 © 2016 IEEE

controllers for the three-phase modular cascaded rectifier stage and the voltage feedback based controller for the output-parallel DAB converter stage. The two controllers cooperate with each other to realize voltage and power balance, even in the case of module mismatched parameters.

A. Voltage and Power Balance Analysis

This common-duty-ratio rectifier controller in Figure 3 ensures that the three rectifier modules in each phase share the same SPWM duty cycle. According to the analysis in SECTION II-A modular currents are the same and V_{a1} refers to the dc-link voltage reference. The sine grid current in phase with grid voltage can be implemented by the zero q-component reference.

$$i_{a1} = i_{a2} = i_{a3} = d_a i_a \qquad (5)$$

Rectifier controller generates the same modular current greatly in theory as shown in (5). It makes it possible to design DAB controller without current sampling as DAB average current can not be measured easily due to the current high ripple.

For the three output-paralleled DAB converters, the PI regulators generate their phase-shift signals independently. As shown in Figure 4(c), DAB #a dc bus voltage V_{dca} multiplied by the feedback coefficient H_{a12} is compared with input-voltage V_{a1} multiplied by feed-forward coefficient H_{a11}. Hence, V_{dca} equals to $V_{a1} \cdot H_{a11}/H_{a12}$, V_{a2} equals to $V_{dca} \cdot H_{a22}/H_{a21}$ and V_{a3} equals to $V_{dca} \cdot H_{a32}/H_{a31}$. On the condition that $H_{a11}/H_{a12}=H_{a21}/H_{a22}=H_{a31}/H_{a32}=H_{va}$, the following equation can be achieved:

$$V_{a1} = V_{a2} = V_{a3} = V_{dca} / H_{va} \qquad (6)$$

It is illustrated that the dc-link voltages of the rectifier modules are only decided by their corresponding voltage feedback coefficient ratios. If the three-phase feedback coefficient ratios are the same, the balanced modular dc-link voltages can be achieved. Consequently, voltage balance can be achieved similarly with the same feedback coefficient ratios in phase B and C, respectively. As shown in Figure 2, the power transferred by each rectifier module is determined by its output currents and voltage.

$$P_{ai} : P_{bi} : P_{ci} = d_{ai} / H_{va} : d_{bi} / H_{vb} : d_{ci} / H_{vc} \qquad (7)$$

In equation (7), the unbalanced power is mainly caused by rectifier PWM duty cycle and feedback coefficient ratio. For each phase the three modular PWM duty cycles are the same and the feedback coefficient ratios is selected the same. Hence, it is concluded that voltage and power balance in rectifier and DAB modules can be well achieved by the proposed control strategy.

B. Balance Analysis on Missmatched Conditions

PWM duty cycle is affected by delay time caused by switching devices and gate-driving circuitries in the real circuit. In this paper, it is neglected since the influence can not be measured easily in quantity.

Feedback coefficient ratio is achieved by sampling and digital calculation in DSP controller.

Considering a general situation, where there is 10% difference in the parameters H_v and PWM duty cycles are the same, the power being transferred by each rectifier modules in steady-state can be derived as follows:

$$P_{ai} : P_{bi} : P_{ci} = H_{va} : H_{vb} : H_{vc} = 1 : 0.9 : 1.1 \qquad (8)$$

From (8), even with little difference in voltage feedback coefficient ratios, the power transferred by each converter module will not be diverged seriously. Hence, voltage and power balance can be realized on mismatched parameters.

IV. EXPERIMENTAL RESULTS

A 15kVA scaled-down prototype is designed and built to verify the systematic control strategy. The control schemes for the prototype are implemented in DSP TMS320F28335. The SST prototype experimental parameters are shown in TABLE I.

As shown in Figure 1, the three-phase SST system consists of three single-phase SST. SST ac-dc stage consists of three H-bridge rectifiers with their ac sides connected in series. The dc voltage of each rectifier module then feeds its DAB module. Each DAB module incorporates galvanic isolation via a high-frequency transformer. The three DAB modules are output connected in parallel to form a 400V dc-bus, with the inverter and an emulated 5kW DRER connected. The grid voltage of the SST prototype is three-phase 220V/50Hz. The dc voltage of each rectifier module is 130V. The dc output of each DAB module is 400V. The ac voltage of the inverter is three-phase 220V/50Hz.

TABLE I. SST PROTOTYPE EXPERIMENTAL PARAMETERS

Circuit parameters	
Rated RMS grid voltage: E_a, E_b, E_c	220V
Dc-link voltage reference: $V_{ai\ ref}$, $V_{bi\ ref}$, $V_{ci\ ref}$	130V
Dc-link voltage capacitor: C_{ai}, C_{bi}, C_{ci}	2mF
Rectifier switching frequency: f_s	15kHz
Line input inductance: L_a, L_b, L_c	1mH
Dc-bus voltage reference: V_{dca}, V_{dcb}, V_{dcc}	400V
Dc-bus voltage capacitor: C_{ao2}, C_{bo2}, C_{co2}	1mF
DAB transformer turns: n	13:40
Voltage coefficient ratio: H_{va}, H_{vb}, H_{vc}	40/13
DAB transformer inductance: L_{ai}, L_{bi}, L_{ci}	30uH
Switching frequency for DAB: f_s	30kHz
Inverter filter inductor: L_{oi}	1mH
Inverter filter capacitor: C_{oi}	10uH
Switching frequency for Inverter: f_s	15kHz
Controller parameters	
Rectifier voltage control parameter: K_{p1}	10
Rectifier voltage control parameter: K_{i1}	20
Rectifier current control parameter: K_{p2}	1
Rectifier current control parameter: K_{i2}	20
DAB voltage control parameter: K_p	0.7
DAB voltage control parameter: K_i	50
Inverter Voltage control parameter: K_{pv}	3
Inverter Voltage control parameter: K_1	10
Inverter Voltage control parameter: K_2	18000
Inverter Voltage control parameter: w_o	100π
Inverter Voltage control parameter: K_{pi}	0.12

Figure 6 show the experimental results in three-phase SST. Figure 6(a) and Figure 6(b) shows three-phase grid voltage and current. Sinusoidal grid current is well achieved and in phase with grid voltage, hence, unity power factor is achieved.

Figure 6(c) shows the grid voltage, current, seven-level voltage and inverter output voltage of phase A. Waveforms of phase B and C are similar with Figure 6(c). In Figure 6(d), V_{a1} equals to 130V, while V_{a2} and V_{a3} equal to 131V and 130V. Average DAB input current I_{a1}, I_{a2} and I_{a3} equals to 12.4A. The experimental waveforms illustrate that the three dc-link voltages and DAB input currents in phase A are balanced. Figure 6(e) shows that three-phase output voltages get the same peak voltage 311V and the spaced 120 degree is greatly achieved. From Figure 6(f), the SST has a good performance in realizing optimal power flow between the renewable energy source and power grid through the 400V dc-bus, while the photovoltaic cells of 2.5 kW is being connected to the SST at local load of 1 kW.

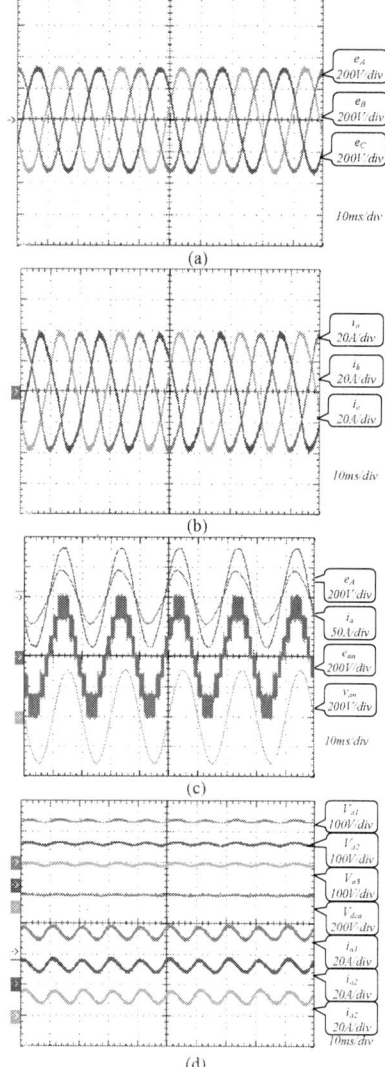

(a)

(b)

(c)

(d)

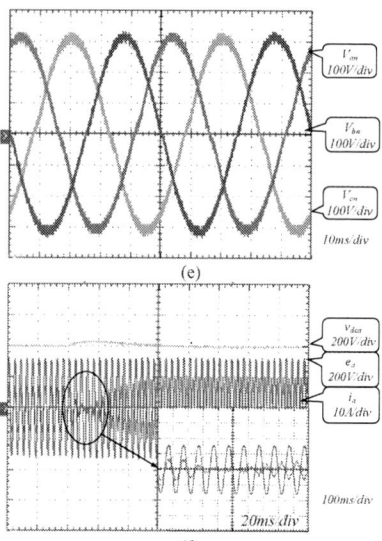

(e)

(f)

Figure 6: (a)Three-phase grid voltage, (b)three-phase grid current, (c)grid voltage, current, seven-level voltage and inverter output voltage of phase A, (d)three rectifier dc-link voltages, dc-bus voltage and three DAB input current, (e)three-phase inverter output voltage and (f)bidirectional experimental results.

The experimental results, in accordance with the theoretic analysis in Section III, demonstrate that the voltage and power balance in each module are well achieved by the systematic control strategy for the three-phase SST system. In addition, other performances such as low-distortion grid current and unity power factor, as well as seamless control of bidirectional power flow between the power grid and the dc-bus, are realized.

V. CONCLUSIONS

In order to achieve voltage and power balance for each converter module reliably, a three-phase SST topology is studied and a systematic control strategy is applied for the three-phase modular cascaded SST, which include a single-phase d-q vector-based common-duty-ratio controller for the cascaded active rectifier stage, a voltage feedback based controller for the output-parallel DAB converter stage and a single-phase based three-phase PR controller for three-phase inverter. This control scheme is proved to work well in realizing voltage and power balance, even in the case of module mismatched parameters. Finally, the stability and other performances of the SST control scheme are substantiated by a 15kW scaled-down experimental prototype.

ACKNOWLEDGMENT

Thanks are due to J.Shi and X.He for assistance with both the theoretical guidance and experiments. The authors thank Zhejiang University Power Electronics Laboratory for the experimental devices supply. And also, this work is sponsored by the National Nature Science Foundation of China (51277162). Thanks for the research funding support.

REFERENCES

[1] J. M. Guerrero, J. C. Vasquez, J. Matas, L. G. de Vicuna, and M. Castilla, "Hierarchical control of droop-controlled ac and dc microgrids - a general approach toward standardization," IEEE Trans. Ind. Electron., vol. 58, no. 1, pp.158-172, Jan. 2011.

[2] Y. Li, and C.-N. Kao, "An accurate power control strategy for power-electronics-interfaced distributed generation units operation in a low-voltage multibus microgrid," IEEE Trans Power Electron, vol. 24, no. 12, pp. 2977-2988, Dec. 2009.

[3] S. Chakraborty, M. D. Weiss, and M. G. Simoes, "Distributed intelligent energy management system for a single-phase high-frequency AC microgrid," IEEE Trans Ind. Electron, vol. 54, no. 1, pp. 97-109, Feb. 2007.

[4] H. Kakigano, Y. Miura, and T. Ise, "Low-voltage bipolar-type dc microgrid for super high quality distribution," IEEE Trans Power Electron, vol. 25, no. 12, pp. 3066-3075, Dec. 2010.

[5] J. M. Guerrero, J. C. Vasquez, J. Matas, M. Castilla, and L. G. de Vicuna, "Control strategy for flexible microgrid based on parallel line-interactive UPS systems," IEEE Trans. Ind. Electron., vol. 56, no. 3, pp. 726-736, Mar. 2009.

[6] N. C. Ekneligoda and W. W. Weaver, "Game-theoretic communication structures in microgrids," IEEE Trans. Power Del., vol. 27, no. 4, pp. 2334-2341, Oct. 2012.

[7] J. M. Guerrero, M. Chandorkar, T.-L. Lee, and P. C. Loh, "Advanced control architectures for intelligent microgrids—Part I: Decentralized and hierarchical control," IEEE Trans. Ind. Electron., vol. 60, no. 4, pp. 1254–1262, Apr. 2013.

[8] Ravichandran, A., et al. The critical role of microgrids in transition to a smarter grid: A technical review. in Transportation Electrification Conference and Expo (ITEC), 2013 IEEE. 2013.

[9] Hossain, M.J., et al. Distributed control scheme to regulate power flow and minimize interactions in multiple microgrids. in PES General Meeting | Conference & Exposition, 2014 IEEE. 2014.

[10] J. M. Guerrero, J. C. Vasquez, J. Matas, M. Castilla, L. G. de Vicuna, and M. Castilla "Hierarchical Control of Droop-Controlled AC and DC Microgrids - A General Approach Toward Standardization," IEEE Trans. Ind. Electron., vol. 58, no. 1, pp. 158-171, Jan. 2011.

[11] Lee, J., et al., Distributed Energy Trading in Microgrids: A Game Theoretic Model and Its Equilibrium Analysis. IEEE Transactions on Industrial Electronics, 2015. 62(6): p. 1-1

[12] Wang, J. L. Duarte, and M. A. M. Hendrix, "Grid-Interfacing Converter Systems With Enhanced Voltage Quality for Micro-grid Application - Concept and Implementation," IEEE Trans. Power Electron., vol. 26, no. 12, pp. 3501-3513, Dec. 2011.

[13] W. McMurray, "Power converter circuits having a high-frequency link," U.S. Patent 3517300, June 23, 1970.

[14] J. L. Brooks, "Solid state transformer concept development," in Naval Material Command. Port Hueneme, CA: Civil Eng. Lab., Naval Construction Battalion Center, 1980.

[15] M. Kang, P. N. Enjeti, and Ira J. Pitel, "Analysis and design of electronic transformers for electric power distribution system," IEEE Trans. Power Electron., vol. 14, no. 6, pp. 1133-1141, Nov. 1999.

[16] E. R.Ponan, S. D. Sudhoff, and D. L. Glover, "A power electronic-based distribution transformer," IEEE Trans. Power Delivery, vol.17, no.2, pp. 537-543, Apr. 2002.

[17] T. Zhao, G. Wang, S. Bhattacharya, and A. Q. Huang, "Voltage and Power Balance Control for a Cascaded H-Bridge Converter-Based Solid-State Transformer," IEEE Transactions on Power Electronics, vol. 28, pp. 1523-1532, 2013.

[18] X. She, A.Q. Huang, and G. Y. Wang, "3-D Space Modulation With Voltage Balancing Capability for a Cascaded Seven-Level Converter in a Solid-State Transformer," IEEE Trans. Power Electron., Vol. 26 , No. 12 , pp. 3778-3789, Dec. 2011.

[19] J. Shi, W. Gou, H. Yuan, T. Zhao, and A. Q. Huang, "Research on voltage and power balance control for cascaded modular solid-state transformer," IEEE Trans. Power Electron., vol. 26, no. 4, pp. 1154-1166 , Apr. 2011.

[20] X. She, A. Q. Huang and X. Ni, "Current Sensorless Power Balance Strategy for DC/DC Converters in a Cascaded Multilevel Converter Based Solid State Transformer," IEEE Transactions on Power Electronics, vol. 29, pp. 17-22, 2014.

[21] Wang Dan, Mao Chengxiong PROCEEDINGS. Since the balance of Electronic Power Transformer China CSEE, 2007,27 (6): 77-83.

[22] Kadavelugu, A., et al., Medium voltage power converter design and demonstration using 15 kV SiC N-IGBTs. 2015, IEEE. p. 1396 - 1403.

[23] Wang, X., et al., Control and Experiment of an H-Bridge Based Three-Phase Three-Stage Modular Power Electronic Transformer. Power Electronics, IEEE Transactions on, 2015. PP(99): p. 1-1.

[24] J.Wu, J.Shi and Z.Zhang, "Research on Voltage and Power Balance Control for Threephase Cascaded Modular Solid-State Transformer," Journal of Power Supply, vol.13, no.2, pp. 17-26, Mar. 2015.

[25] H. Bai, and C. Mi, "Eliminate reactive power and increase system efficiency of isolated bidirectional dual-active-bridge dc-dc converters using novel dual-phase-shift control," IEEE Trans. Power Electron., vol. 23, no.6, pp. 2905-2914, Nov. 2008.

[26] H. J. Chiu and L. W. Lin, "A bidirectional dc-dc converter for fuel cell electric vehicle driving system," IEEE Trans. Power Electron., vol. 21, no. 4, pp. 950-958, Jul. 2006.

[27] D. H. Xu, C. H. Zhao, and H. F. Fan, "A PWM plus phase-shift control bidirectional dc-dc converter," IEEE Trans. Power Electron., vol. 19, no. 3, pp. 666-675, May 2004.

[28] L. Zhu, "A novel soft-commutating isolated boost full-bridge ZVS-PWM dc-dc converter for bidirectional high power application," IEEE Trans. Power Electron., vol. 21, no. 4, pp. 422-429, Mar. 2006.

[29] G. G. Oggier, G. O. Garcia, and A. R. Oliva, "Switching control strategy to minimize dual active bridge converter Losses," IEEE Trans Power Electron, vol. 24, no. 7, pp. 1826-1838, July 2009.

978-1-4673-9551-9/16 $31.00 © 2016 IEEE

New Flying-Capacitor-Based Multilevel Converter with Optimized Number of Switches and Capacitors Controlled with a New Logic-Form-Equation Based Active Voltage Balancing Technique

Vahid Dargahi[1], *Student Member, IEEE*, Arash Khoshkbar Sadigh[2], *Member, IEEE*, and Keith Corzine[1], *Senior Member, IEEE*

[1] *Microgrid and Power Electronics Laboratory, Holcombe Department of Electrical and Computer Engineering, Clemson University, Clemson, SC 29634, USA*

[2] *Extron Electronics, Anaheim, CA 92805, USA*

Abstract - The flying-capacitor-based multicell converter is one of the well-known breeds of multilevel power converters. This paper proposes a new flying-capacitor-based multilevel converter to minimize the number of flying capacitors (FCs) and power switches. The advantage of the proposed FC-based multilevel converter in comparison with the conventional flying capacitor multicell converter is the fewer number of FCs. In comparison with the stacked multicell converter, the proposed multilevel converter requires less number of the semiconductor switches. In order to balance the voltage of the FCs in proposed multilevel converter, a new active voltage balancing method which is fully implemented utilizing logic-form equations is presented. The proposed voltage balancing method, measures output current and FC voltages to generate switching states to produce the required output voltage level as well as balance FC voltages at their reference values. The output voltage of the proposed multilevel converter controlled with suggested active voltage balancing method can be modulated with any pulse-width-modulation (PWM) method such as phase-shifted-carrier PWM (PSC-PWM) or level-shifted-carrier PWM (LSC-PWM). Simulation results and experimental measurements of proposed FC-based multilevel are presented to verify performance of the proposed converter, and its novel switching and modulation strategy which is based on the active voltage balancing method.

Index Terms – Active voltage balancing, Flying capacitor multicell converter; multilevel converter; level-shifted-carrier; phase-shifted-carrier; PWM; renewable energy integration.

I. INTRODUCTION

The neutral point clamped (NPC) [1], cascaded H-bridge (CHB) [2], modular multilevel converter (MMC) [3], flying capacitor multicell (FCM) [4], and active NPC (ANPC) [5] inverters are the most widely used topologies of multilevel converters. The substantial advantages concerning multilevel converters are improved dynamics, high modularity, reduced line harmonics, fault tolerant, and extended operating and power ranges [6]–[9].

Among the multilevel topologies, the FCM converter and its sub-topology breed referred as the stacked multicell (SM) converter are finding increased attention in the industry and academia due to advantages such as the natural balancing of flying capacitor (FC) voltages as a result of the redundancies present in phase switching states, transformer-less operation, and even distribution of the switching stress among power transistors [10]–[16]. As number of voltage levels increases, amount of energy stored in FCs, voltage rating of FCs, and diversity in their voltage ratings profoundly impact converter size and price. To negate this issue and to decrease voltage diversity as well as price in FC-based multilevel converter,

DFCM converter [17] or its improved structure (I-DFCM) [18] are considered as the cost-effective alternate topologies. Nevertheless, DFCM converter breed might not be able to fulfill the restricted requirements of the high-power market deservedly because of requiring isolated dc voltage sources in each phase. Hence, this paper proposes a novel FC-based multilevel converter to connect the medium-voltage dc link fed from a wind farm or a large-scale solar plant to power grid. The advantage of the proposed multilevel converter in comparison with conventional topology of FCM converter is a lower number of bulky FCs. Besides, in comparison with classical SM converter, proposed multilevel converter needs fewer power switches (IGBTs/diodes) and FCs. Moreover, a new active voltage balancing method is presented to control the proposed converter.

II. PROPOSED FC-BASED MULTILEVEL CONVERTER

A four-cell five-level FCM and SM converters are shown in Figure 1 and Figure 2, respectively. As shown in Figure 1, a five-level FCM converter requires eight IGBTs/diodes with voltage rating of $E/4$ and three FCs with voltage rating of $E/4$, $2E/4$, and $3E/4$. In order to reduce the number of FCs and their voltage rating diversity, an SM converter has been proposed in the literature. As shown in Figure 2, a five-level SM converter needs two FCs with voltage rating of $E/4$ which is advantageous in comparison with five-level FCM converter. However, five-level SM converter requires twelve IGBTs/diodes with voltage rating of $E/4$ which is 50% more than those in five-level FCM converter which is considered as disadvantage. In other words, FCM converter is optimized regarding the number of power switches while SM converter is optimized regarding the number of FCs and diversity of their voltage ratings. This paper proposes a novel FC-based multilevel converter which is superior to the FCM converter regarding the number of FCs and superior to SM converter regarding the number of IGBTs/diodes and FCs. Figure 3 illustrates five-level proposed FC-based multilevel converter which has 2 FCs with voltage ratings of $E/4$ and $E/2$. The state of power switches for the five-level proposed FC-based multilevel converter are also illustrated in Table. 1. Figure 3 shows a reduction in the number of the FCs and their voltage rating in proposed multilevel converter as compared to FCM converter. Also, the five-level proposed FC-based multilevel converter has eight IGBTs/diodes with voltage rating of $E/4$. This fact proves a decrease in number of IGBTs/diodes in proposed multilevel in comparison with SM converter. Table 2 illustrates comparison between FCM, SM, and proposed

978-1-4673-9551-9/16 $31.00 © 2016 IEEE

FC-based multilevel converter to generate the same output voltage with an identical number of levels and peak-to-peak voltage value. General configuration of $n + 1$-cell $2n + 1$-level proposed converter is illustrated in Figure 5.

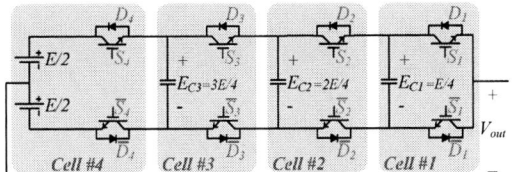

Figure 1. Four-cell five-level FCM converter.

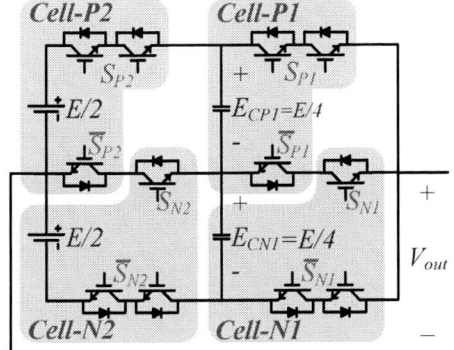

Figure 2. Four-cell five-level SM converter.

In Table 2, two cases of A and B are considered; in case A, voltage ripple percentage is the same in all FCs while in case B, voltage ripple amplitude (in volt) is the same in all FCs. Case A and B are illustrated in Figure 4. Furthermore, the amount of stored energy in all FCs of each converter is illustrated in Figure 6 considering both cases of A and B. In Figure 2 and Figure 3, some of the IGBTs are drawn in pairs to clearly illustrate that their voltage rating are two times of other IGBTs voltage rating. This does not mean that they should be in pair in reality. One IGBT with higher voltage rating can be utilized instead of a one pair of IGBTs. For instance in Figure 3, if the S1 and S2 are 3.3kV for a specific application, a 6.5kV IGBT can be used for S3. However, it is possible in practice to balance the voltage stress between two IGBTs connected in series which is discussed in [19]–[23]. It is worth mentioning that in FC-based multilevel converters it is desirable to have the same percentage of the voltage ripple across the FCs resulting in the same factor of capacitance-voltage for all FCs [24]–[26]. This means that C3 and C2 are the series connection of the three and two capacitors of C1, respectively, in order to have the same percentage of voltage ripple in the five-level FCM converter illustrated in Figure 1. Likewise in the five-level proposed converter as depicted in Figure 3, C2 is the series connection of two capacitors of C1 in order to have the same percentage of the voltage ripple. This fact is considered in Table 2 to calculate the number of required capacitors in each of the FCM, SM and proposed multilevel converters. As it can be pointed out, the proposed multilevel converter is superior to both of the FCM and SM converters regarding the number of FCs and the number of IGBTs/diodes.

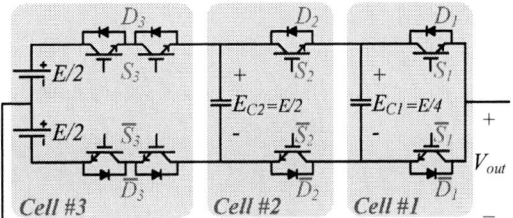

Figure 3. Three-cell five-level proposed FC-based multilevel converter.

Table 1. States of power switches and charging/discharging process of flying capacitors in five-level proposed FC-based multilevel converter.

Output Voltage	State of switches (S_1, S_2, S_1)	ΔE_{C2}	ΔE_{C1}	i_{out}	Number of States
+0.5E	(1 , 1 , 1)	0	0	whatever	1
+0.25E	(1 , 1 , 0)	0	+	$i_{out} > 0$	2
		0	-	$i_{out} < 0$	
	(1 , 0 , 1)	+	-	$i_{out} > 0$	
		-	+	$i_{out} < 0$	
0	(1 , 0 , 0)	+	0	$i_{out} > 0$	2
		-	0	$i_{out} < 0$	
	(0 , 1 , 1)	-	0	$i_{out} > 0$	
		+	0	$i_{out} < 0$	
-0.25E	(0 , 0 , 1)	0	-	$i_{out} > 0$	2
		0	+	$i_{out} < 0$	
	(0 , 1 , 0)	-	+	$i_{out} > 0$	
		+	-	$i_{out} < 0$	
-0.5E	(0 , 0 , 0)	0	0	whatever	1

III. ACTIVE VOLTAGE BALANCING OF FLYING CAPACITORS IN THE PROPOSED FC-BASED MULTILEVEL CONVERTER

To elaborate the proposed active voltage balancing which is fully implemented using logic-form equation, the required parameters are defined in Table 3 and in following equations accordingly to facilitate the determination of switching states for power switches.

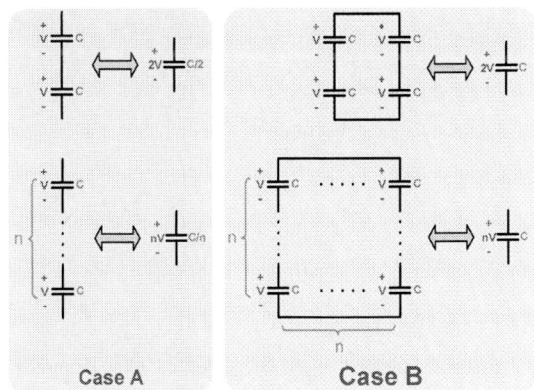

Figure 4. Capacitor connection for case A and B considered in Table 2

978-1-4673-9551-9/16 $31.00 © 2016 IEEE   1482

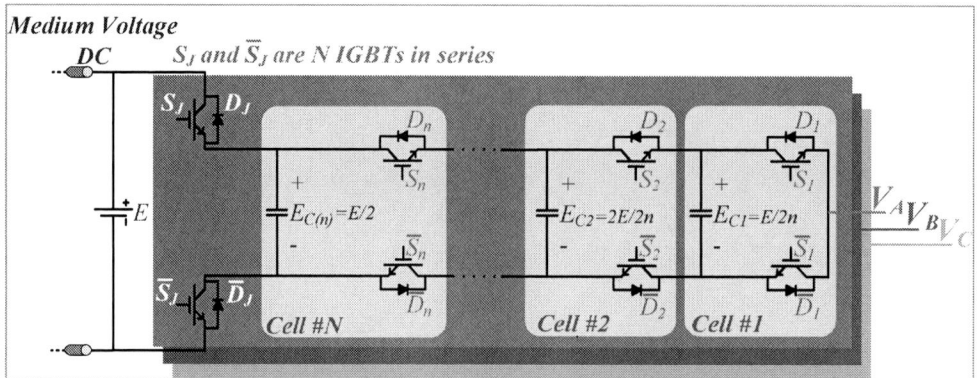

Figure 5. $N+1$-cell $2N+1$-level proposed FC-based multilevel converter.

Table 2. Comparison between single-phase FCM, SM, and proposed multilevel converters to generate the same staircase output voltage, *i.e.*, $2n+1$-level with peak-to-peak voltage of E (* **Case – A:** Voltage ripple percentage is the same in all FCs.   * **Case – B:** Voltage ripple amplitude (in volt) is the same in all FCs).

Multilevel Converter	No. of Switches With the Same Voltage Rating	No. of Cells	No. of Flying Capacitors	Voltage Rating of Capacitors (per unit)	Overall No. of Identical Capacitors Case – A*	Overall No. of Identical Capacitors Case – B*
FCM	4n	2n	2n-1	One set: from 1 to $2n$-1 (increasing 1 pu)	$\sum_{i=1}^{i=2n-1} C_i = \sum_{i=1}^{i=2n-1}\left(\frac{i}{n}C_{pu}\right)$ $=\frac{(4n-2)}{2}C_{pu}$	$\sum_{i=1}^{i=2n-1} C_i = \sum_{i=1}^{i=2n-1}\left[\left(\frac{i}{n}\right)^2 C_{pu}\right]$ $=\frac{(4n-2)(4n-1)}{6n}C_{pu}$
SM	6n	2n	2n-2	Two sets: from 1 to n-1 (increasing 1 pu)	$2\times\sum_{i=1}^{i=n-1} C_i = 2\times\sum_{i=1}^{i=n-1}\left(\frac{i}{n}C_{pu}\right)$ $=\frac{(2n-2)}{2}C_{pu}$	$2\times\sum_{i=1}^{i=n-1} C_i = 2\times\sum_{i=1}^{i=n-1}\left[\left(\frac{i}{n}\right)^2 C_{pu}\right]$ $=\frac{(2n-2)(2n-1)}{6n}C_{pu}$
Proposed Converter	4n	n+1	n	One set: from 1 to n (increasing 1 pu)	$\sum_{i=1}^{i=n-1} C_i = \sum_{i=1}^{i=n}\left(\frac{i}{n}C_{pu}\right)$ $=\frac{(n+1)}{2}C_{pu}$	$\sum_{i=1}^{i=n} C_i = \sum_{i=1}^{i=n}\left[\left(\frac{i}{n}\right)^2 C_{pu}\right]$ $=\frac{(n+1)(2n+1)}{6n}C_{pu}$

$$I = \begin{cases} 1 & i_{out}(t) \geq 0 \\ 0 & i_{out}(t) < 0 \end{cases} \qquad (1)$$

$$\Delta V_{C_j} = \frac{E_{C_j}}{E_{C_j}\big|_{Reference}} - 1 \qquad (2)$$

$$\Delta_{C_j} = \begin{cases} 1 & \Delta V_{C_j} \geq 0 \\ 0 & \Delta V_{C_j} < 0 \end{cases} \qquad (3)$$

$$Y_{ij} = \begin{cases} 1 & |\Delta V_{C_i}| \geq |\Delta V_{C_j}| \\ 0 & |\Delta V_{C_i}| < |\Delta V_{C_j}| \end{cases} \qquad (4)$$

where variable I determines direction of load current, and E_{C_j} and $E_{C_j}\big|_{Reference}$ are measured and reference voltages of capacitor C_j at each sampling moment, respectively.

Balancing the FC voltages at their reference values is a concern of utmost importance in FC-based converters since it results in a safe operation of switches (IGBTs/diodes) and consequently, the proper function of converter. In FC-based converters, voltage stress of IGBTs/diodes is limited to its safe operation range only if FC voltages are balanced at their reference values. Therefore, it is essential to ensure that FC voltages are balanced at their reference values which require redundancies to generate desired voltage level at the output. Switching states of power switches in five-level FC-based proposed multilevel converter are listed in Table 1 wherein voltage variation of each FC, *i.e.*, $\Delta E_{Cx} = E_{Cx}(t+\Delta t) - E_{Cx}(t)$, is determined for each switching state by taking into account the direction of the output current. As it is shown, all levels except $+0.5E$ and $-0.5E$ have redundancies which can be used to charge or discharge the FCs in order to balance their voltage at their reference value. So, states of switches should be selected properly according to direction of output current and the voltage of FCs in order to generate the desired output voltage level as well as to balance FC voltages. In this paper, an active voltage balancing method is presented for proposed FC-based multilevel converter.

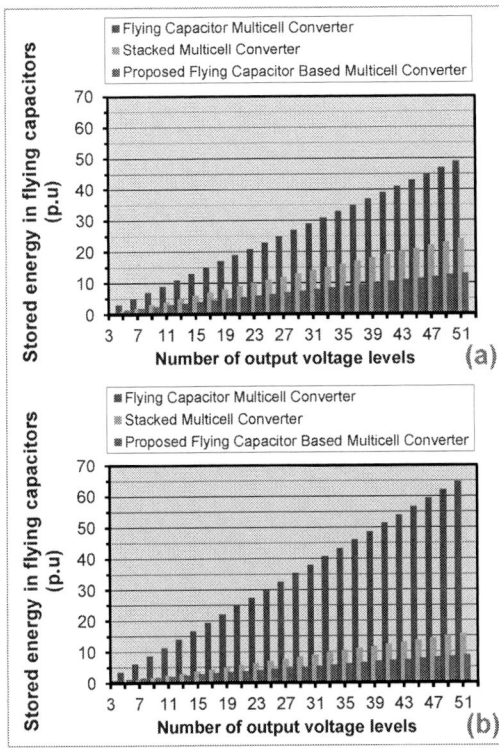

Figure 6. Amount of stored energy in FCs in FCM, SM, and the proposed converters considering capacitor connection of: (a) **Case A**; (b) **Case B**. (**Case – A**: Voltage ripple percentage is the same in all FCs. **Case – B**: Voltage ripple amplitude (in volt) is the same in all FCs.)

As depicted in Figure 7, a sinusoidal reference waveform (V_Ref) is intersected with the triangle carriers to generate staircase reference waveform. It is worth mentioning that the triangle carriers can be arranged based on any type of PWM methods like as PSC or LSC. LSC-PWM switching strategy includes the phase-disposition (PD), as represented in Figure 7, phase-opposition-disposition (POD), and alternate-phase-opposition-disposition (APOD) [27]. Due to a superior line-to-line harmonic performance of the LSC-PWM method, this switching strategy is used in this paper for proposed active voltage balancing technique [28]–[31].

Table 3. Definition of the required variables for proposed active voltage balancing method.

Level (per unit)	L_{-2}	L_{-1}	L_0	L_{+1}	L_{+2}
+2	0	0	0	0	1
+1	0	0	0	1	0
0	0	0	1	0	0
-1	0	1	0	0	0
-2	1	0	0	0	0

$$P_{C2} = Y_{21} + L_0 \qquad (5)$$

$$P_{C1} = \overline{P_{C2}} \qquad (6)$$

According to the explained procedure to balance the FC voltages, following logic-form equations must be applied to control the proposed multilevel converter. The switching state of power switch S_1 is as follows:

$$S_1 = L_{+2} + X_1.L_{+1} + X_0.L_0 + X_1.L_{-1} \qquad (7)$$

where:

$$X_0 = \overline{I}.\overline{\Delta}_{C2} + I.\Delta_{C2} \qquad (8)$$

$$X_1 = I.P_{C2}.\overline{\Delta}_{C2} + \overline{I}.P_{C2}.\Delta_{C2} + I.P_{C1}.\Delta_{C1} + \overline{I}.P_{C1}.\overline{\Delta}_{C1} \qquad (9)$$

The switching state of power switch S_2 is as follows:

$$S_2 = L_{+2} + X_2.L_{+1} + X_0.L_0 + X_2.L_{-1} \qquad (10)$$

where:

$$X_2 = I.P_{C2}.\Delta_{C2} + \overline{I}.P_{C2}.\overline{\Delta}_{C2} + I.P_{C1}.\overline{\Delta}_{C1} + \overline{I}.P_{C1}.\Delta_{C1} \qquad (11)$$

The switching state of power switch S_3 is as follows:

$$S_3 = \overline{S}_1.L_0 + L_{+1} + L_{+2} \qquad (12)$$

IV. SIMULATION RESULTS

In this section, simulation results are illustrated to verify the performance and dynamic behavior of the proposed FC-based converter and effectiveness of its new active voltage balancing technique. Three-phase five-level (line-to-neutral) proposed converter is utilized to inject/absorb reactive power into/from power grid and inject active power harvested from renewable energy, such as solar PV or wind energy, into electric grid. The main circuit parameters used in simulations is abridged in Table 4. Line-to-neutral and line-to-line output voltages of the converter, power grid voltages, and converter injected currents (increased 10 times) in all three phase and in steady states are shown in Figure 8 where the converter injects 700 kW and 900kVAR into electric power grid. The converter is controlled with LSC-PWM switching method wherein triangle-carriers' frequency is *1.95 kHz*. Figure 9 illustrates FCs' voltages of converter in both transient and steady states. As it is shown, the FCs' voltages are balanced properly at their reference values, *i.e.*, 2.25kV and 4.5kV, validating the proposed active voltage balancing method. To investigate the dynamic behavior of the proposed FC-based multilevel converter and its ability to balance voltages across FCs with undesired initial voltages, the proposed converter is tested with undesired initial voltages across the FCs. As it is depicted in Figure 10 which reflects the transient and steady states of the FC voltages, proposed active voltage balancing method regulates the FC voltages correctly.

Table 4. Main parameters of the simulated grid-tied proposed converter.

System Parameters	Values
Nominal grid rms voltage (line-line), V_{gnd}	4.16 kV
System frequency	50 Hz
dc link voltage (E)	9 kV
Capacitance of FCs ($C1$, $C2$) in five-level converter	2000 *u*F - 1000 *u*F
PD LSC-PWM carrier frequency	1.95 kHz
Converter output *LC* filter	3 mH ; 10 μF

978-1-4673-9551-9/16 $31.00 © 2016 IEEE

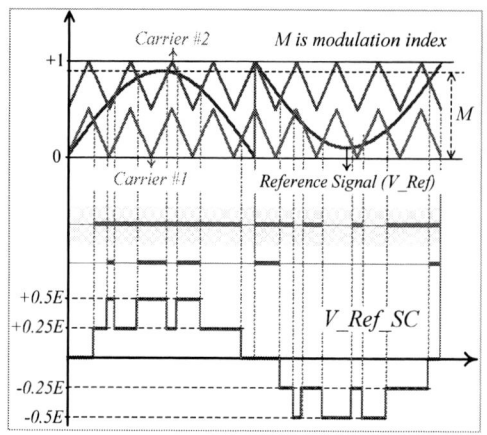

Figure 7. Phase-disposition LSC-PWM strategy to produce the staircase output voltage reference (V_Ref_SC).

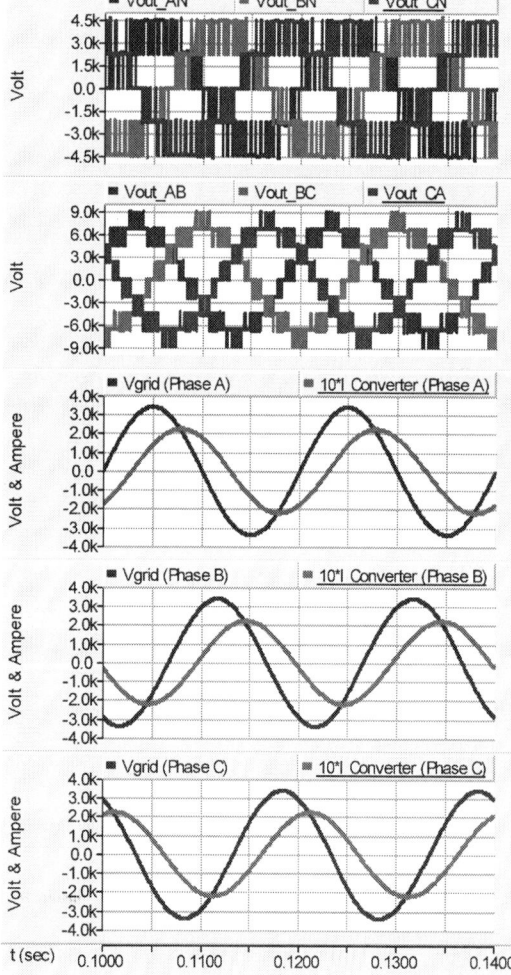

Figure 8. Converter output voltage, grid voltage and converter injected current (increased 10 times) in five-level (line-to-neutral) grid-tied proposed FC-based multilevel converter.

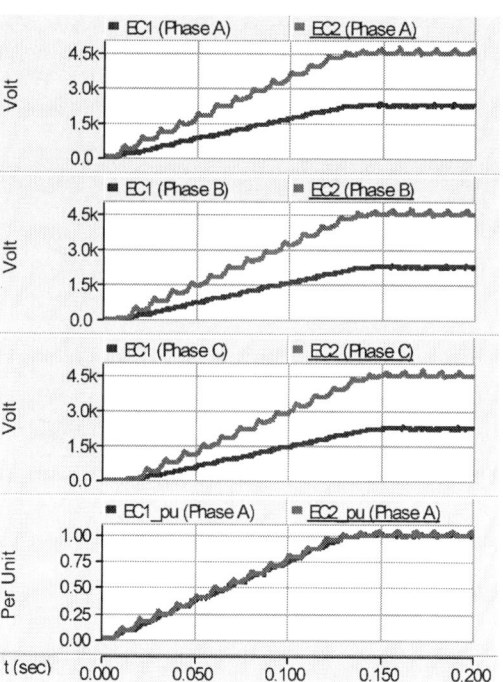

Figure 9. Voltage of FCs in five-level (line-to- neutral) grid-tied proposed FC-based multilevel converter.

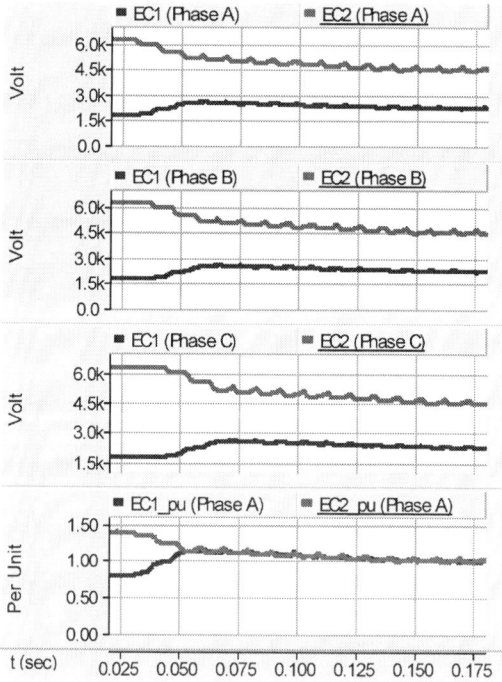

Figure 10. Voltage balancing of FCs with undesired initial voltage in five-level (line-to- neutral) grid-tied proposed FC-based multilevel converter.

The dynamic performance and stability of the proposed FC-based multilevel converter under difference modulation indices ranging from 1 to 0.1 with intervals of 0.1 is depicted in Figure 11 where the resistive-inductive load is $10\,\Omega$ and

42 mH. As it is demonstrated, the proposed converter and its active voltage balancing method do not have any restriction regarding value of modulation index. Besides, performance of the proposed FC-based converter under different output voltage frequencies from 35Hz to 105Hz with intervals of 10Hz is shown in Figure 12 where resistive-inductive load is $10\,\Omega$ and 42 mH. As it is shown, FC voltages get stabilized and balanced at their reference values even if the switching frequency is not tuned to the output voltage frequency. This substantiates performance of proposed multilevel converter modulated under suggested active voltage balancing method which is fully implemented using only logic-form equations for variable-frequency applications such as medium-voltage high-power adjustable-speed ac drives.

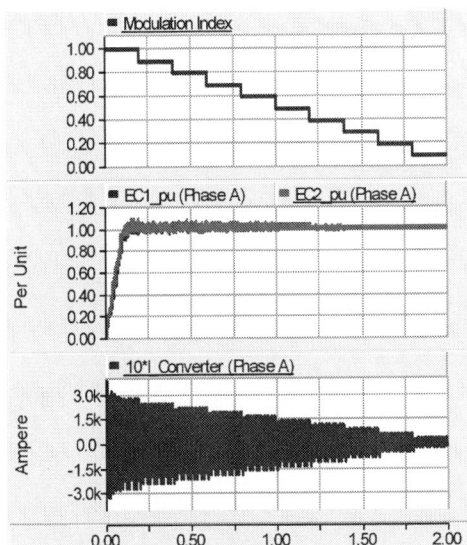

Figure 11. Dynamic performance of the proposed FC-based multilevel converter under different modulation indices.

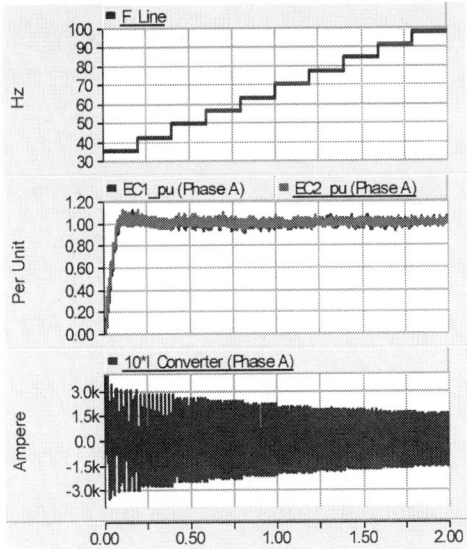

Figure 12. Dynamic performance and stability of the proposed FC-based multilevel converter under different output voltage frequencies.

V. EXPERIMENTAL RESULTS

To verify proposed FC-based multilevel converter and its active voltage balancing switching pattern, a laboratory scale prototype of proposed FC-based converter is built utilizing IXYS IXGH48N60B3D1 600V 48A discrete IGBT modules. The TMS320F28335 DSP has been used to modulate the proposed FC-based multilevel converter controlled with the new active voltage balancing method. The gate driver ICs for IGBTs are IR2184 with the gate resistance of $R_G = 10\Omega$. The DCP021515 IC is used to provide required isolated dc voltages for gate drive circuitry of IGBT modules. The dc link voltage is 600 V and the resistive-inductive load is 13 Ω and 42 mH. It is worth mentioning that both of the voltage and current are scaled down 15 times in comparison with the simulation results to match each other. All the voltages and currents are measured with probe of 10:1.

The measured output voltage, load current, and the FC voltages attained from prototype system of a five-level (line-to-neutral) single-phase proposed FC-based converter are shown in Figure 13. As shown in Figure 13, the FCs voltages are stabilized at their reference values, *i.e.*, 150 and 300V, respectively, which confirm the performance of the proposed active voltage balancing technique which is implemented using only logic-form equations. Data of both output voltage and current waveforms are saved as csv file from scope and then, analyzed in Matlab as a digital signal to obtain their frequency spectrums which are portrayed in Figure 14. As shown in Figure 14(a), output voltage of five-level proposed FC-based multilevel converter has harmonic clusters around integer multiples of 1.95 kHz wherein switching frequency is 1.95 kHz. As shown in Figure 14(b), the converter output current is almost a pure sinusoidal waveform; consequently, it contains only the fundamental component in its frequency spectrum. In order to investigate dynamic performance of the proposed active voltage balancing method, the main dc link voltage is increased from 300V to 600V and also, decreased from 600V to 300V as shown in Figure 15 which illustrates transient of FCs voltages balanced at their reference values.

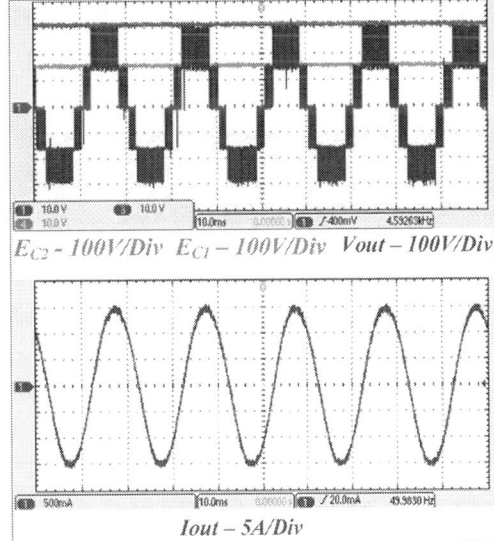

Figure 13. Experimental measurements of the output voltage, output current and FCs voltages in the five-level proposed FC-based multilevel converter.

978-1-4673-9551-9/16 $31.00 © 2016 IEEE

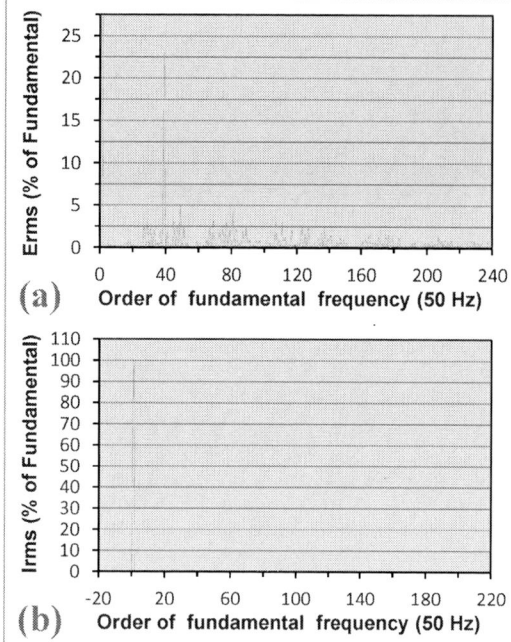

(a)

(b)

Figure 14. Frequency spectrum in the five-level (line-to-neutral) proposed FC-based multilevel converter. (a) output voltage; (b) output current.

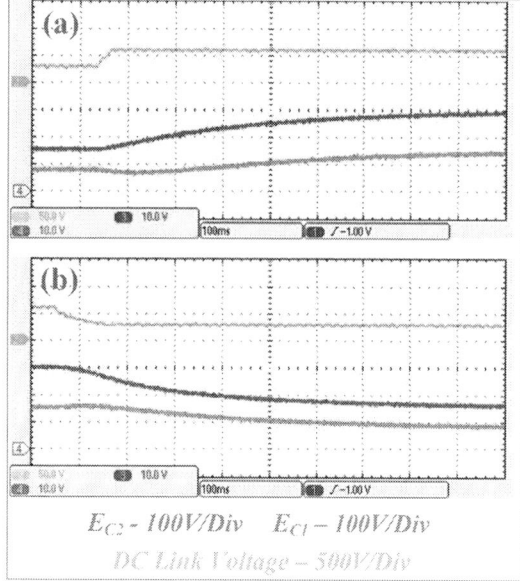

(a)

(b)

$E_{C2} - 100V/Div$ $E_{C1} - 100V/Div$

DC Link Voltage – 500V/Div

Figure 15. Experimental measurements of transient of FCs voltages in the five-level (line-to-neutral) proposed FC-based multilevel converter during change of main dc link voltage: (a) from 300V to 600V; (b) from 600V to 300V

VI. CONCLUSION

A new flying-capacitor-based multilevel power converter is proposed in this paper. The proposed converter requires the same number of IGBTs/diodes in comparison with the conventional flying capacitor multicell converter to generate an identical output voltage. However, the proposed converter

requires a fewer number of flying capacitors. Moreover, in order to generate an identical output voltage, the proposed multilevel converter requires two thirds of IGBTs/diodes and almost half of the flying capacitors required in a stacked multicell converter. This paper also presented a new active voltage balancing method for modulating the proposed FC-based multilevel converter to stabilize the capacitor voltages at their target values. The proposed active voltage balancing technique is implemented based on the logic-form equations and modulates the output voltage based on PD-LSC PWM in order to optimize the THD of the line-to-line output voltage as well as to balance the capacitor voltages. Additionally, the dynamic performance and stability of the proposed flying-capacitor-based converter under different modulation indices and different output voltage frequencies were perused. It was concluded that proposed converter and its voltage balancing method do not have any restriction regarding the value of the modulation index and output voltage frequency. Finally, in order to verify the feasibility of the proposed converter and performance of presented active voltage balancing method, both simulation results and experimental measurements were provided.

REFERENCES

[1] J. Rodriguez, S. Bernet, P. K. Steimer, and I. E. Lizama, "A Survey on Neutral-Point-Clamped Inverters," *IEEE Trans. Ind. Electron.*, vol. 57, no. 7, pp. 2219–2230, Jul. 2010.

[2] M. Malinowski, K. Gopakumar, J. Rodriguez, and M. A. Pérez, "A Survey on Cascaded Multilevel Inverters," *IEEE Trans. Ind. Electron.*, vol. 57, no. 7, pp. 2197–2206, Jul. 2010.

[3] H. Akagi, "Classification, terminology, and application of the modular multilevel cascade converter (MMCC)," *IEEE Trans. Power Electron.*, vol. 26, no. 11, pp. 3119–3130, 2011.

[4] A. K. Sadigh, V. Dargahi, and K. A. Corzine, "New Multilevel Converter Based on Cascade Connection of Double Flying Capacitor Multicell Converters and Its Improved Modulation Technique," *IEEE Trans. Power Electron.*, vol. 30, no. 12, pp. 6568–6580, Dec. 2015.

[5] D. Andler, R. Alvarez, S. Bernet, and J. Rodriguez, "Experimental Investigation of the Commutations of a 3L-ANPC Phase Leg Using 4.5-kV–5.5-kA IGCTs," *IEEE Trans. Ind. Electron.*, vol. 60, no. 11, pp. 4820–4830, 2013.

[6] A. Khoshkbar-Sadigh, V. Dargahi, and K. Corzine, "Analytical Determination of Conduction Power Losses in Flying-Capacitor-Based Active Neutral-Point-Clamped Multilevel Converter," *IEEE Trans. Power Electron.*, pp. 1–1, 2015.

[7] M. Aleenejad, H. Iman-Eini, and S. Farhangi, "Modified space vector modulation for fault-tolerant operation of multilevel cascaded H-bridge inverters," *IET Power Electron.*, vol. 6, no. 4, pp. 742–751, Apr. 2013.

[8] V. Dargahi and S. Dargahi, "Analytical modelling of single-phase stacked multicell multilevel converters exploiting Kapteyn (Fourier–Bessel) series," *IET Power Electron.*, vol. 6, no. 6, pp. 1220–1238, Jul. 2013.

[9] V. Dargahi, A. Khoshkbar Sadigh, M. Abarzadeh, S. Eskandari, and K. Corzine, "A New Family of Modular Multilevel Converter Based on Modified Flying-Capacitor Multicell Converters," *IEEE Trans. Power Electron.*, vol. 30, no. 1, pp. 138–147, 2015.

[10] A. M. Y. M. Ghias, J. Pou, V. G. Agelidis, and M. Ciobotaru, "Initial Capacitor Charging in Grid-Connected Flying Capacitor Multilevel Converters," *IEEE Trans. Power Electron.*, vol. 29, no. 7, pp. 3245–3249, Jul. 2014.

[11] A. M. Y. M. Ghias, J. Pou, M. Ciobotaru, and V. G. Agelidis, "Voltage-balancing method using phase-shifted PWM for the flying capacitor multilevel converter," *IEEE Trans. Power Electron.*, vol. 29, no. 9, pp. 4521–4531, Sep. 2014.

[12] R. H. Wilkinson, T. A. Meynard, and H. du Toit Mouton, "Natural Balance of Multicell Converters: The General Case," *IEEE Trans. Power Electron.*, vol. 21, no. 6, pp. 1658–1666, Nov. 2006.

[13] R. H. Wilkinson, T. A. Meynard, and H. du Toit Mouton, "Natural Balance of Multicell Converters: The Two-Cell Case," *IEEE Trans. Power Electron.*, vol. 21, no. 6, pp. 1649–1657, Nov. 2006.

[14] W. Qian, H. Cha, F. Z. Peng, and L. M. Tolbert, "55-kW Variable 3X DC-DC Converter for Plug-in Hybrid Electric Vehicles," *IEEE Trans. Power Electron.*, vol. 27, no. 4, pp. 1668–1678, Apr. 2012.

[15] L. Shi, B. P. Baddipadiga, M. Ferdowsi, and M. L. Crow, "Improving the Dynamic Response of a Flying-Capacitor Three-Level Buck Converter," *IEEE Trans. Power Electron.*, vol. 28, no. 5, pp. 2356–2365, May 2013.

[16] S. Thielemans, A. Ruderman, B. Reznikov, and J. Melkebeek, "Improved Natural Balancing With Modified Phase-Shifted PWM for Single-Leg Five-Level Flying-Capacitor Converters," *IEEE Trans. Power Electron.*, vol. 27, no. 4, pp. 1658–1667, Apr. 2012.

[17] A. K. Sadigh, S. H. Hosseini, M. Sabahi, and G. B. Gharehpetian, "Double Flying Capacitor Multicell Converter Based on Modified Phase-Shifted Pulsewidth Modulation," *IEEE Trans. Power Electron.*, vol. 25, no. 6, pp. 1517–1526, Jun. 2010.

[18] V. Dargahi, A. Khoshkbar Sadigh, M. Abarzadeh, M. R. A. Pahlavani, and A. Shoulaie, "Flying Capacitors Reduction in an Improved Double Flying Capacitor Multicell Converter Controlled by a Modified Modulation Method," *IEEE Trans. Power Electron.*, vol. 27, no. 9, pp. 3875–3887, Sep. 2012.

[19] R. Withanage and N. Shammas, "Series connection of Insulated Gate Bipolar Transistors (IGBTs)," *IEEE Trans. Power Electron.*, vol. 27, no. 4, pp. 2204–2212, 2012.

[20] Y. Abe and K. Maruyama, "Multi-series Connection of High-Voltage IGBTs," *Fuji Electr. J.*, vol. 8, pp. 1–4, 2002.

[21] S. Hong, V. Chitta, and D. A. Torrey, "Series connection of IGBT's with active voltage balancing," *IEEE Trans. Ind. Appl.*, vol. 35, no. 4, pp. 917–923, 1999.

[22] D. Ning, X. Tong, M. Shen, and W. Xia, "The experiments of voltage balancing methods in IGBTs series connection," in *Asia-Pacific Power and Energy Engineering Conference, APPEEC*, 2010.

[23] S. Zheng, Y. Wang, X. Wu, Z. Qian, and F. Z. Peng, "The voltage sharing of commercial IGBTS in series with passive components," in *IEEE Energy Conversion Congress and Exposition: Energy Conversion Innovation for a Clean Energy Future, ECCE 2011, Proceedings*, 2011, pp. 3008–3012.

[24] J.-S. Lai and F. Z. Peng, "Multilevel converters-a new breed of power converters," *IEEE Trans. Ind. Appl. Appl.*, vol. 32, no. 3, pp. 509–517, 1996.

[25] J. Rodríguez, J. S. Lai, and F. Z. Peng, "Multilevel inverters: A survey of topologies, controls, and applications," *IEEE Trans. Ind. Electron.*, vol. 49, no. 4, pp. 724–738, 2002.

[26] B. Wu, "High-Power Converters and AC Drives, Chapter 9," 2006, pp. 179–186.

[27] A. KhoshkbarSadigh and M. Barakati, *Modeling and Control of Sustainable Power Systems, , Chap. 11: Topologies and Control Strategies of Multilevel Converters*. Springer, 2012.

[28] M. Aleenejad, P. Moamaei, H. Mahmoudi, and R. Ahmadi, "Unbalanced Selective Harmonic Elimination for fault-tolerant operation of three phase multilevel Cascaded H-bridge inverters," in *2015 IEEE Applied Power Electronics Conference and Exposition (APEC)*, 2015, pp. 1589–1594.

[29] M. Aleenejad, R. Ahmadi, and P. Moamaei, "Selective harmonic elimination for cascaded multicell multilevel power converters with higher number of H-Bridge modules," in *2014 Power and Energy Conference at Illinois (PECI)*, 2014, pp. 1–5.

[30] V. Dargahi and A. Shoulaie, "Capacitors voltage balancing modeling in three phase flying capacitor converters with booster," in *2012 3rd Power Electronics and Drive Systems Technology (PEDSTC)*, 2012, pp. 103–108.

[31] M. Aleenejad, R. Ahmadi, and P. Moamaei, "A modified selective harmonic elimination method for fault-tolerant operation of multilevel cascaded H-bridge inverters," in *2014 Power and Energy Conference at Illinois (PECI)*, 2014, pp. 1–5.

New Low-Cost Five-Level Active Neutral-Point Clamped Converter

Hongliang Wang, *Senior Member, IEEE*, Lei Kou,
Yan-Fei Liu, *Fellow, IEEE* and Paresh C. Sen, *Life Fellow,IEEE*
Department of Electrical and Computer Engineering
Queen's University
Kingston, Canada
hongliang.wang@queensu.ca, kou.lei@queensu.ca,
yanfei.liu@queensu.ca, senp@queensu.ca

Sucheng Liu
School of Electrical and Information Engineering
Anhui University of Technology
Ma' anshan, China
liusucheng@gmail.com

Abstract—Multilevel converters are a popular solution for medium-voltage and high-power applications, including renewable energy conversion. Five-level active neutral-point-clamped converter (5L-ANPC) is one of the most advantageous topologies among five-level multilevel converters. A six-switch 5L-ANPC (6S-5L-ANPC) topology is proposed, which can achieve low-cost because several fast recovery diodes are employed to replace the body-diodes and active switches. However, the power branches including the diode can only provide active current path. Special modulation strategy is proposed to achieve reactive power operation. The simulation analysis shows the proposed 6S-5L-ANPC is suitable for unity power factor application such as photovoltaic (PV) application. A 500W single-phase inverter prototype is built to verify the validity and flexibility of proposed topology and modulation.

I. INTRODUCTION

Multilevel converters (or inverters) have been used for power conversion in high-power applications such as medium voltage grid (2.3KV, 3.3KV, or 6.9KV) to reduce the switch voltage stress, and photovoltaic (PV) application to reduce the filter size [1, 2]. Compared to two-level voltage source inverters, the advantages of multilevel inverters are lower voltage stress, higher efficiency, smaller filter size and lower common-mode voltage [3].

There are three traditional multilevel topologies [4-8]: the neutral-point-clamped (NPC) type [4, 5], flying-capacitor (FC) type [6], and cascaded H-bridge (CHB) type [7-8]. Many five-level NPC topology has been derived in [9]. NPC

type generates the voltage levels from the neutral point voltage by adopting diodes. The drawback is the increased number of switching devices when voltage level increases. FC type outputs the voltage level by summing the flying-capacitor voltage. However, higher voltage level leads to more flying-capacitors and the complexity of control strategy to balance the voltages of each flying-capacitor is then increased. The CHB multilevel inverters use series-connected H-bridge cells with an isolated dc voltage sources connected to each cell [10]. Similarly, to have more output levels, more cells are needed. This will lead to impracticality of this type of topology since more DC sources are required.

Active-neutral-point-clamped (ANPC) which is one of the multilevel topology has been proposed in Refs [11-15]. Three 5L-ANPC topologies are shown in Fig.1. The ANPC type converter combines the features of NPC and FC topology. The ANPC topologies is receiving more and more attentions nowadays because of high efficiency and multi-level output. In this paper, a novel six-switch five-level ANPC (6S-5L-ANPC) inverter topology is proposed. Compared to traditional ANPC topologies, the proposed topology adopts only six active semiconductor switches, greatly reducing the volume of system. This paper is organized as follows: Section II describes working principles of proposed topology; Section III discusses the modulation strategy of 6S-5L-ANPC topologies; Section IV and V show the simulation and experimental results and Section VI gives the conclusion.

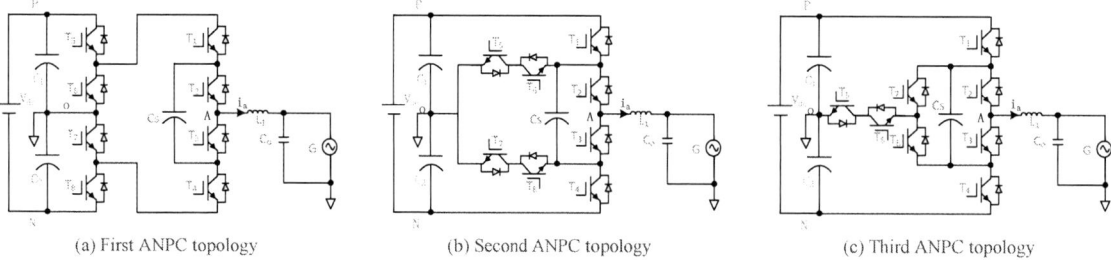

(a) First ANPC topology   (b) Second ANPC topology   (c) Third ANPC topology

Fig. 1. Five-level ANPC topology

II. Operational Principles of Proposed 6S-5L-ANPC Inverter

A. Introduction of 6S-5L-ANPC Inverter

To increase the overall efficiency and improved output waveform, the five-level ANPC inverters are good choice for renewable energy harvesting. Additionally, based on half-bridge, the 5L-ANPC inverters guarantee no leakage current generation. Therefore, they are suitable for transformer-less type photovoltaic (PV) system. For grid-connection application, the inverter output current is required to be in phase with grid voltage. In this situation, reactive current paths can be ignored, which means some active switches can be replaced by fast recovery diodes in order to increase the efficiency. Based on this, a novel 5L-ANPC inverter topology is proposed, which is composed of six switches (T_1 to T_6), two discrete-diodes (D_{F1}, D_{F2}) and one flying-capacitor (C_S). The configuration of six-switch 5L-ANPC (6S-5L-ANPC) inverter is shown in Fig.2.

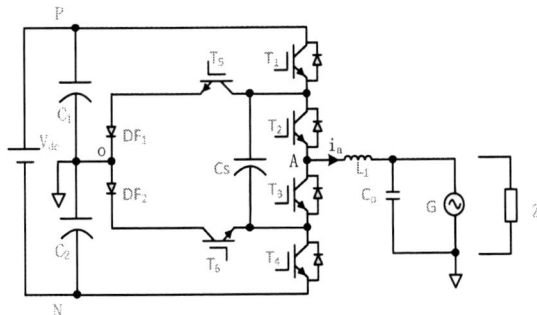

Fig. 2. Configuration of 6S-5L-ANPC inverter

As can be seen from Fig.2, two DC capacitors, C_1 and C_2, are connected in series to DC-link. The complexity of DC-link capacitor voltages balancing is reduced compared to NPC and FC type converters which need four DC capacitors in series. Additionally, in contrast to traditional 5L-ANPC topologies (as shown in Fig.1) which need eight active switches and other types of five-level inverters which require

more, the proposed 6S-5L-ANPC inverter only has six active switches and two discrete diodes. The reduced number of active switches sacrifice some reactive current paths. However, the proposed topology is capable of operating under reactive power condition with special modulation method, which will be discussed in the following section.

B. Operation of 6S-5L-ANPC Topology

DC voltage is defined as V_{dc}. The 6S-5L-ANPC converter consists of eight switching states that generate five-level voltage levels at the output based on capacitor voltages (DC voltage and FC voltage), as shown in Table I. Five output voltage levels $+V_{dc}/2$, $+V_{dc}/4$, 0, $-V_{dc}/4$ and $-V_{dc}/2$ are achieved by summing the flying-capacitor voltage and DC-link capacitor voltage. Fig.3 shows the specific eight switching states (mode A to H) and current paths (active power branch in red line and reactive power branch in green line).

In Table I, it is observed that some of the switching states are redundant in generating certain output voltage level: mode B and mode C are redundant switching states to generate $+V_{dc}/2$; similarly, (D, E) and (F, G) are redundant states to generate 0 and $-V_{dc}/2$, respectively. Although the redundant states (B, C) and (F, G) generate the same output voltage level, their effect on the FC voltage is opposite to each other due to the change in the direction of FC current. This leads to the possibility of regulating the FC voltage to a constant value ($V_{dc}/4$).

In Fig.3, it is observed that among eight switching states, four modes (C, D, E, F) allow unidirectional current-flow path due to the presence of discrete diode. Therefore, appropriate selection of switching states under reactive power operation is very important, which also increases the complexity of modulation. Therefore, the proposed 6S-5L-ANPC converter is suitable for active power application (e.g. PV grid-connection application). The following section will cover the specific modulation strategy for proposed topology.

TABLE I. Switching States, Output Voltage and Voltage of Flying-Capacitor of 6S-5L-ANPC Inverter

Switching state	Switch number						Output voltage V_{Ao}	Flying capacitor C_S	
	T_1	T_2	T_3	T_4	T_5	T_6		$i_a>0$	$i_a<0$
A	1	1	0	0	0	1	$+V_{dc}/2$	--	--
B	1	0	1	0	0	1	$+V_{dc}/4$	Charge	Discharge
C	0	1	0	0	0	1	$+V_{dc}/4$	Discharge	Charge
D	0	0	1	0	0	1	$+0$	--	--
E	0	1	0	0	1	0	-0	--	--
F	0	0	1	0	1	0	$-V_{dc}/4$	Discharge	Charge
G	0	1	0	1	1	0	$-V_{dc}/4$	Charge	Discharge
H	0	0	1	1	1	0	$-V_{dc}/2$	--	--

978-1-4673-9551-9/16 $31.00 © 2016 IEEE

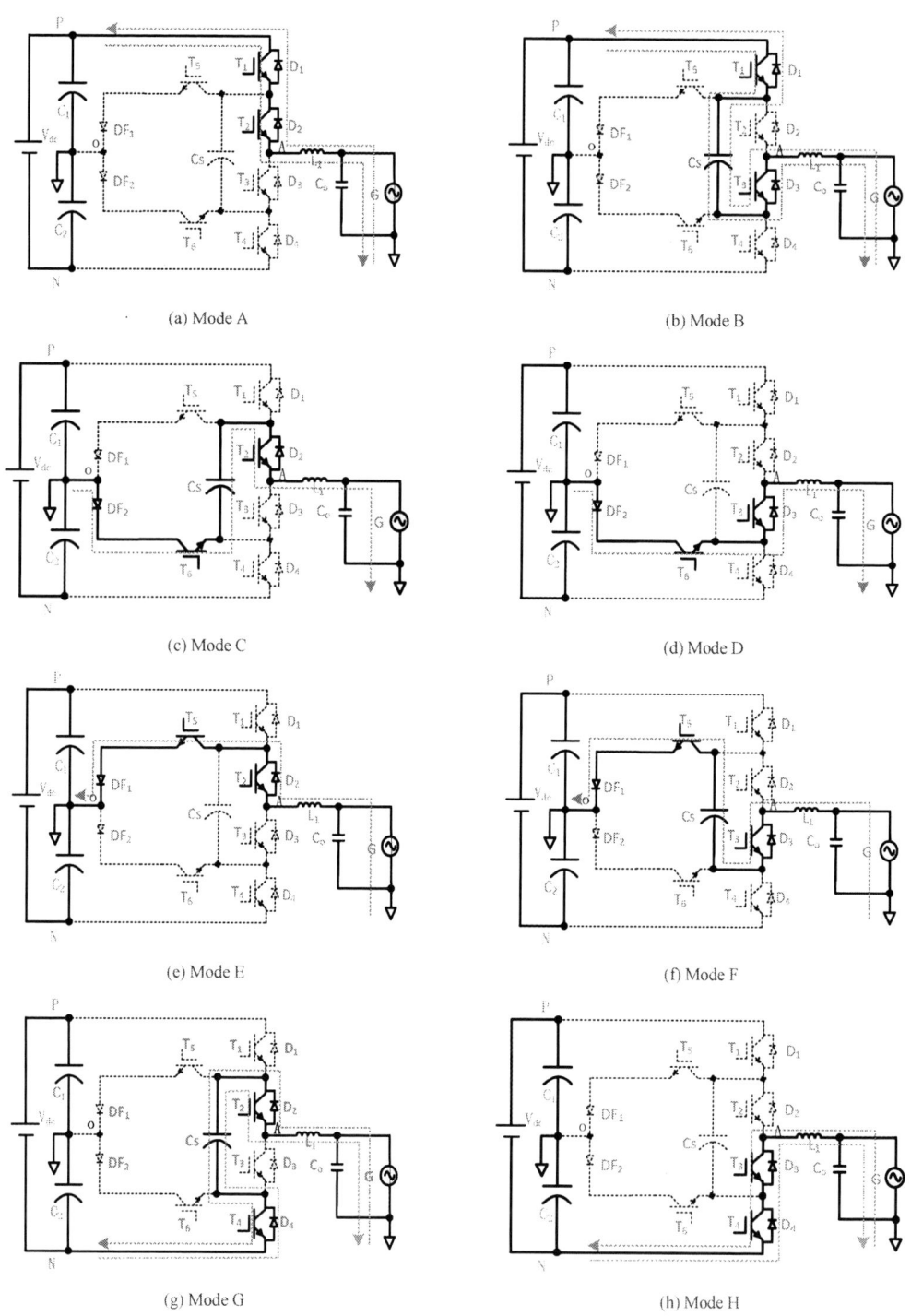

Fig. 3. Eight switching modes and current paths for 6S-5L-ANPC

III. Modulation Strategy

As can be seen from Fig.3, four switching modes allow unidirectional current path due to the adoption of fast recovery diode. Therefore, the modulation method for 6S-5L-ANPC inverter is different from traditional ANPC topologies.

Selection of zero output states: both zero output modes (D and E) belong to unidirectional current path states: the current in mode D can only flow from left to right (positive) while mode E allows only negative current. So selection of zero output states is based on the direction of output current.

Among two pairs of redundant states which output $V_{dc}/2$ level, mode C ($+V_{dc}/2$) and F ($-V_{dc}/2$) allow unidirectional current-flow path. These two states can only be used when output voltage and current are in same direction. Therefore,

under reactive power factor condition, when directions of current and voltage are different, we can only use mode B ($+V_{dc}/2$) and E ($-V_{dc}/2$) to achieve $V_{dc}/2$ output level. This will result in the continuous drop of flying-capacitor voltage because both two modes are discharging the flying-capacitor. Consequently, if power factor is low which leads to wide region of output current and voltage in opposite direction, a large flying-capacitor voltage ripple will be generated. In this case, a large value of flying-capacitor is selected to reduce the voltage ripple.

From the analysis above, it is concluded that the complexity of modulation is increased when inverter is working under reactive power condition. Therefore, only reactive power modulation strategy is discussed. The diagram of modulation for 6S-5L-ANPC inverter is shown in Fig.4.

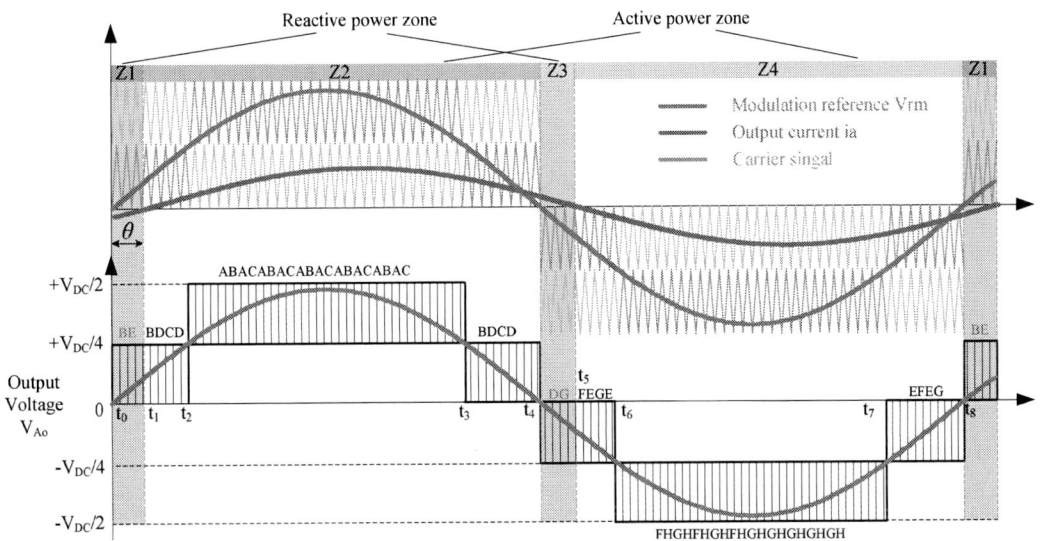

Fig. 4. PWM modulation for 6S-5L-ANPC inverter ($cos\theta > \sqrt{3}/2$)

From Fig.4, it can be observed that during a whole grid cycle, there are four zones according to whether the output current and voltage are in same direction: Z1 and Z3 are reactive power zones while Z2 and Z4 are active power zones.

As mentioned before, the selection of two redundant switching states generation zero output is based on the direction of output current. So from t_1 to t_5, the current is positive and mode D is chosen. Similarly, from t_0 to t_1 and t_5 to t_8, mode E is selected due the negative output current.

Reactive power zone Z1 [from t_0 to t_1]: inverter is outputting zero and $+V_{dc}/4$. In this region, only mode B can be used to generate $+V_{dc}/4$ output because voltage and current are in opposite direction. For zero output, mode E is

selected due the negative current. So, (B, E) states combination is obtained.

Active power zone Z2 [From t_1 to t_2 and t_3 to t_4]: the output voltage is switched between zero and $+V_{dc}/4$. Because it is in the active power zone, two redundant $+V_{dc}/4$ states can be used. This gives an opportunity to regulate the voltage across flying-capacitor. When actual FC voltage is lower than reference FC voltage, charging state B is chosen; when greater than reference FC voltage, mode C is selected to discharge flying-capacitor. When outputting zero voltage level, mode D is chosen. Consequently, the switching state sequence of (B, D, C, D) is achieved.

[From t_2 to t_3]: the output reference voltage is between $+V_{dc}/4$ and $+V_{dc}/2$. Mode A is required to generate $+V_{dc}/2$ output level. Similarly, in active power zone, appropriate

978-1-4673-9551-9/16 $31.00 © 2016 IEEE

selection of redundant switching states (B, C) is necessary. In this way, sequence of (B, A, C, A) guarantees flying-capacitor voltage balancing and inverter output.

Reactive power zone Z3 [from t_4 to t_5]: output current is negative, so mode D is needed to output zero level. Mode G is selected to generate -V_{dc}/4 output. In this region, switching states sequence (D, G) is acquired.

Active power zone Z4 [From t_5 to t_8]: similar to zone Z2, mode D and mode H are selected for zero and -V_{dc}/2 output levels. When outputting -V_{dc}/4 voltage levels, redundant switching states (F, G) are employed alternately to keep flying-capacitor balanced. Consequently, during t_5 to t_6 and t_7 to t_8, switching state sequence (F, E, G, E) is achieved; from t_6 to t_7, switching state sequence (F, H, G, H) is adopted.

According the description of modulation strategy for 6S-5L-ANPC inverter topology, it is obtained that in reactive power zone, the flying-capacitor cannot be regulated. A higher value of flying-capacitor can be adopted in low power factor case. In conclusion, the proposed 6S-5L-ANPC inverter is suitable for active power application (e.g. PV grid-connection) and high power factor application.

IV. SIMULATION VERIFICATION

In order to testify the effectiveness of modulation strategy especially under reactive power condition, simulation verification has been carried out using MATLAB/Simulink. The simulation has been done in two cases: unity power factor operation and reactive power case. The power level of inverter system in the simulation is 500W. The system parameters are shown in Table II.

TABLE II.   SYSTEM PARAMETERS

Input voltage	400V	Grid voltage	110V RMS
DC capacitor	2000μF	Grid frequency	60Hz
Flying capacitor	310μF	Output current	4.6A RMS
Output filter inductor	1.6mH	Switching frequency	15kHz

A. Unity power factor operation

To investigate the effectiveness of proposed topology and modulator, waveforms of inverter output voltage V_{Ao}, flying-capacitor voltage, grid voltage and output current are obtained from simulation, as shown in Fig.5.

Fig.5 (a) shows the five-level inverter output. Fig.5 (b) shows the flying-capacitor voltage. From calculation, 310 uF is chosen for flying-capacitor value. As can be seen, the FC voltage is kept at 100 volts, and 2 volts voltage variation is applied on the flying-capacitor. So voltage ripple in this case is 2%. Fig.5 (c) shows the gird-voltage and inverter output current. Under unity power factor, the current and voltage are in phase. The output current is sinusoidal wave without distortion.

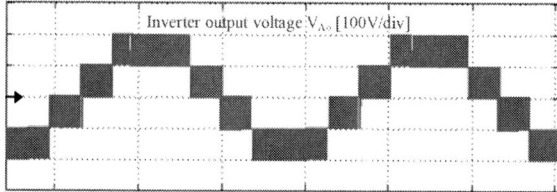

(a)Inverter output voltage waveform

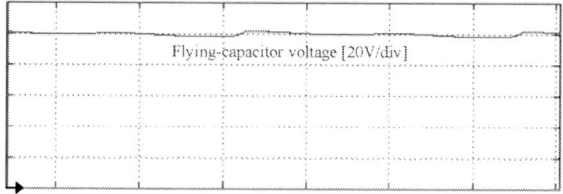

(b)Flying capacitor voltage waveform

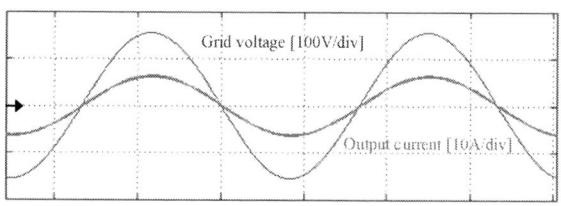

(c)Grid voltage and output current waveform

Fig. 5.   Active power operation: simulation waveforms of inverter output voltage, flying-capacitor voltage, grid voltage, output current

B. Reactive Power operation ($cos\theta = 0.9$)

To verify the proposed modulation method in section III, simulation under high power factor case ($cos\theta = 0.9$) has been carried out. Waveforms of inverter output voltage V_{Ao}, flying-capacitor voltage, grid voltage and output current are obtained from simulation are shown in Fig.5.

Compared to the output voltage in Fig.5 (a), the output voltage under reactive power condition is almost the same. As can been seen from Fig. 6 (a), a small voltage spike happens near the output current zero-crossing point. This is due to the switching states transition: the output current is changing from mode D to E (or from mode E to D).

In Fig.6 (b), it is observed that flying-capacitor voltage ripple is increased compared to unity power factor case. As can been seen, the large voltage ripples occur at reactive power zone (in Fig. 6 (c), the region where grid voltage and output current are in opposite direction). The waveform of FC voltage verify the analysis in section III: the FC voltage cannot be regulated in reactive power zone. Appropriate selection of flying-capacitor value can keep the voltage ripple within acceptable range. In my case, power level is 500W (input voltage is 400V and output current is 4.6A RMS) and 310 uF flying-capacitor is selected to achieve maximum 6 volts (6%) ripple voltage.

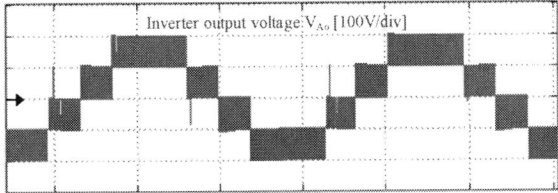

(a)Inverter output voltage waveform

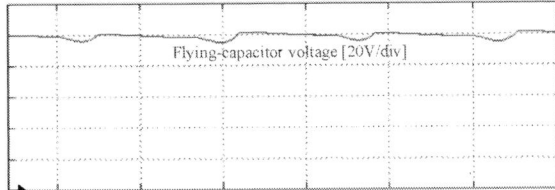

(b)Flying capacitor voltage waveform

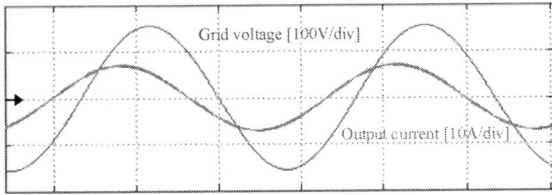

(c)Grid voltage and output current waveform

Fig. 6. Reactive power operation $cos\theta = 0.9$: simulation waveforms of inverter output voltage, flying-capacitor voltage, grid voltage

From the simulation results in two cases, it can be concluded that with proposed modulation method applied on proposed 6S-5L-ANPC inverter, the system is capable of operating under both active power condition and reactive power condition.

V. EXPERIMENTAL RESULTS

To verify the effectiveness of proposed topology and its modulation strategy. A 500W single-phase 6S-5L-ANPC inverter grid-connection experimental prototype is implemented, as shown in Fig.7. The system includes main circuit, DSP and FPGA control board, DC source, output filter and measurement instruments.

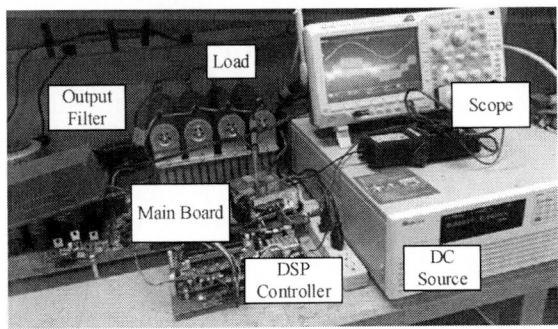

Fig. 7. Experimental prototype

The control board employs a combination of the Texas Instruments TMS320F28335 control card and the Altera Cyclone IV EP4CGX22 FPGA card to provide powerful real-time mathematical calculations and control functions. The experimental parameters are the same as parameters used in simulation, shown in Table II.

Fig.8 and Fig.9 show the experimental results under unity power factor condition. Fig.8 shows inverter output voltage, flying-capacitor voltage, grid voltage and output current: channel 1 is the output bridge voltage; channel 2 is $110V_{rms}$ grid voltage; channel 3 is the flying-capacitor voltage, which is balanced at 100 volts; channel 4 is the output current, which is sinusoidal without distortion and in phase with grid voltage in this case. Fig.9 shows two DC-link capacitors voltages, flying-capacitor voltage and output current: channel 1 is lower DC-link capacitor voltage and channel 2 is upper DC-link capacitor voltage, which both have a line-frequency fluctuation. The measured flying-capacitor ripple voltage is 3V (3%) and DC-link capacitor ripple voltage is 8V (4%).

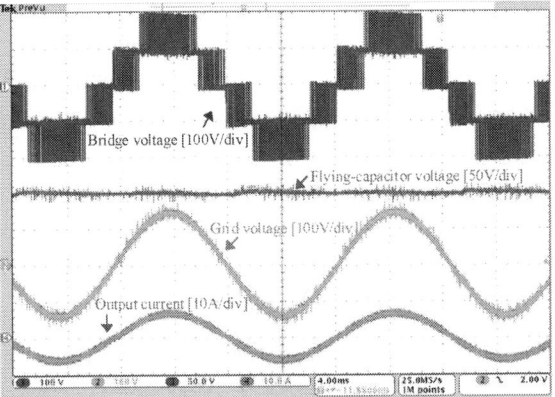

Fig. 8. Experimental results under unity power factor condition: waveforms of inverter bridge voltage, flying-capacitor voltage, grid voltage and output current

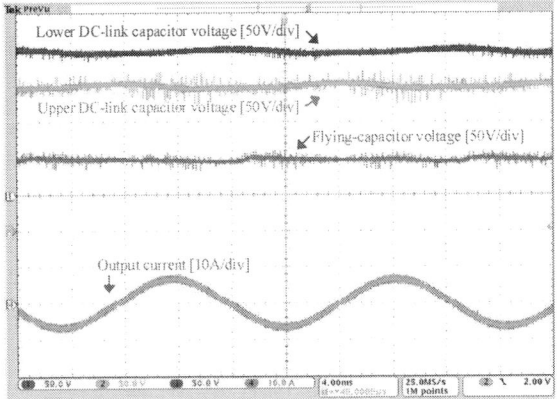

Fig. 9. Experimental results under unity power factor condition: waveforms of lower DC-link capacitor votlage, upper DC-link capacitor voltage, grid voltage and output current

978-1-4673-9551-9/16 $31.00 © 2016 IEEE

To testify the proper system operation under reactive power condition, experimental works are carried out. The power factor is selected as $cos\theta = 0.9$. Fig.10, Fig.11 and Fig.12 show the experimental results.

Fig.10 shows inverter output voltage, flying-capacitor voltage, grid voltage and output current. It is observed that the waveform of inverter output voltage V_{Ao} is same as one achieved in the simulation: small voltage spikes occur around output current zero-crossing points. The flying-capacitor voltage is also balanced at 100 volts. Under 0.9 power factor, the output current and grid voltage has a 25 degree phase shift. In this situation, the inverter still produces good quality current waveform without distortion.

Fig.11 shows two DC-link capacitors voltages, flying-capacitor voltage and output current. The DC-link capacitors voltages waveforms in this situation are almost the same as one under active power condition. The measured DC-link capacitor ripple voltage is also 8V (4%).

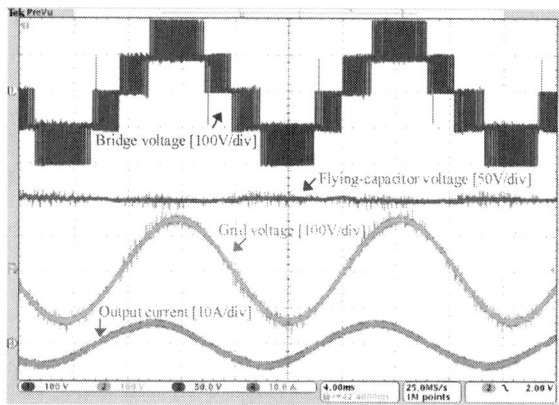

Fig. 10. Experimental results under reacitve power operation $cos\theta = 0.9$: waveforms of inverter bridge voltage, flying-capacitor voltage, grid voltage and output current

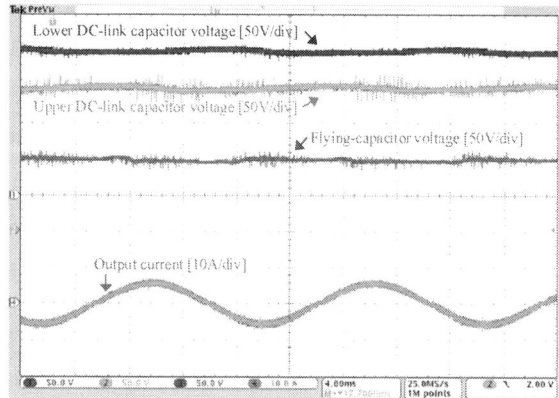

Fig. 11. Experimental results under unity power factor $cos\theta = 0.9$: waveforms of lower DC-link capacitor votlage, upper DC-link capacitor voltage, grid voltage and output current

Fig.12 shows lower DC-link capacitor voltage, flying-capacitor voltage, grid voltage and output current. It can be seen that large FC voltage ripple happen during the time when grid voltage and output current are in opposite direction. The voltage drop in reactive power zone is 7 volts (7%). The experimental results are consistent with analysis in section III and simulation results in section IV.

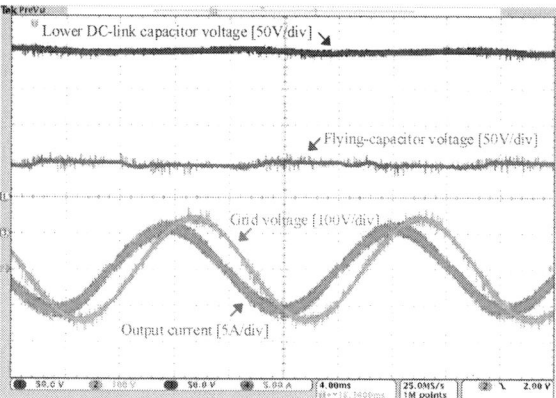

Fig. 12. Experimental results under unity power factor $cos\theta = 0.9$: waveforms of lower DC-link capacitor votlage, flying-capacitor voltage, grid voltage and output current

VI. CONCLUSION

A novel six-switch five-level flying-capacitor based ANPC converter has been proposed. Compared to traditional 5L-ANPC converters, it reduces two active semiconductor switches. The working principles and switching states are presented. The specific modulation strategy of 6S-5L-ANPC inverter under reactive power operation has been investigated. Issues related to the balancing of flying-capacitor voltage and reactive power operation are discussed. The simulation and experimental verifications have been carried out in unity power factor condition and reactive power case to demonstrate the effectiveness of proposed topology and modulation method.

REFERENCE

[1] P Hammond. A New Approach to Enhance Power Quality for Medium Voltage AC Drives. IEEE Trans. on Industry Applications, 1997, 33(1): 202-208.

[2] G Beinhold, R Jakob and M Nahrstaedt. A New Range of Medium Voltage Multilevel Inverter Drives with Floating Capacitor Technology[C]. In: Conference Proceedings of the 9th European Conference on Power Electronics, EPE 2001, Austria, CD-ROM.

[3] J. Rodriguez, J. S. Lai, and F. Z. Peng, "Multilevel inverters: A survey of topologies, controls, and applications," IEEE Trans. Ind. Electron. Vol. 49, no. 4, pp. 724–738, Aug. 2002.

[4] A. Nabae, I. Takahashi, and H. Akagi, "A new neutral-point clamped PWM inverter," IEEE Trans. Ind. Applicant., vol. IA-17, pp. 518–523,Sept./Oct. 1981.

[5] J. Rodriguez, S. Bernet, B. Wu, J. O. Pontt, and S. Kouro, "Multilevel voltage-source-converter topologies for industrial medium-voltage drives," IEEE Trans. Ind. Electron., vol. 54, no. 6, pp. 2930–2945,Dec. 2007.

[6]   F. Richardeau, P. Baudesson, and T. A. Meynard, "Failure-tolerance and remedial strategies of a PWM multi-cell inverter," *IEEE Trans. Power Electron.*, vol. 17, no. 6, pp. 905–912, Nov. 2002.

[7]   J. I. Leon, S. Kouro, S. Vazquez, R. Portillo, L. G. Franquelo, J. M.Carrasco, and J. Rodriguez, "Multidimensional modulation technique for cascaded multilevel converters," IEEE Trans. Ind. Electron., vol. 58, no. 2,pp. 412–420, Feb. 2011.

[8]   Kagarlu,M.F and Babaei.E" A Generalized Cascaded Multilevel Inverter Using Series Connection of Sub-multilevel Inverters," *IEEE Trans. Power Electron.* Vol. 28, no. 2, pp. 625–636, Aug. 2013.

[9]   H. Wang, Y.-F. Liu, and P. C. Sen, "A neutral point clamped multilevel topology flow graph and space NPC multilevel topology," in Energy Conversion Congress and Exposition (ECCE), 2015 IEEE, 2015, pp. 3615-3621.

[10] Kouro. S, Malinowski. M, Gopakumar. K, Pou. J, Franquelo. L, "Recent Advances and Industrial Applications of Multilevel Converters," IEEE Trans. Ind. Electron., vol. 57, no. 8,pp. 2553-2580, Feb. 2010.

[11] P.Barbosa, P. Steimer, J. Steinke, J. meysenc, *"Active Neutral-point-clamped Multilevel Converter" Power Electronics Specialists Conference, 2005. PESC' 05. IEEE 36ᵗʰ 16-16 June 2005. (2001), 2001, pp. 2296-2301.*

[12] *T. Brückner, S. Bernet, and H. Güldner, "The active NPC converter and its loss-balancing control," IEEE Trans. Ind. Electron., vol. 52, no. 3, pp. 855–868, 2005.*

[13] *L.A. Spepa, P.M. Brbosa, P.K. Steimer, and J. W. Kolar, "Five-Level virtual –flux direct power control for the active neutral-point-clamped multilevel inverter," in Proc. IEEE PESC, Jun. 2008, PP. 1668-1674.*

[14] K.Wang, Y.Li and Z. Zheng, "A neutral-point potential balancing algorithm for five-level ANPC converters," in Proc. ICEMS, Aug. 2011, pp. 1-5.

[15] Soeiro.T.B, Carballo. R, Moia. J, Garcia. G. O, Heldwein. M. L, "Three-phase five-level active-neutral-point-clamped converters for medium voltage applications," in Power Electronics Conference (COBEP), Brazilian, 2013, pp. 85

Medium Voltage ($\geq$ 2.3 kV) High Frequency Three-Phase Two-Level Converter Design and Demonstration using 10 kV SiC MOSFETs for High Speed Motor Drive Applications

Sachin Madhusoodhanan, Krishna Mainali, Awneesh Tripathi, Kasunaidu Vechalapu, Subhashish Bhattacharya
Department of Electrical and Computer Engineering, North Carolina State University
Email: sachin, kmainal, aktripat, kvechal, sbhatta4@ncsu.edu

Abstract—High speed variable frequency motor drives are required for marine applications, compressors for oil and gas industries, wind energy generation systems etc. Traditionally, low voltage high speed motor drives are used in such applications. This results in large currents at high power levels leading to large copper loss in the motor winding. Therefore, medium voltage (MV) drives are being considered. The silicon (Si) based MV drives need gears to increase the speed due to low switching frequency operation of Si devices in the converter. Gears reduce both efficiency and power density. With the development of 10 kV SiC MOSFET, high switching frequency at MV is possible, which has enabled the scope of high power density MV direct drive variable speed controlled motors. In this paper, the design of a three-phase, 2-level, $\geq$ 2.3 kV MV, high frequency converter based on 10 kV SiC MOSEFT is explained. Performance analysis is presented along with experimental demonstration.

I. INTRODUCTION

Medium voltage (MV), high speed motor drives are suitable for applications which demand small size and weight of the motor drive system with high efficiency [1]–[4]. MV drives are popular compared to traditional low voltage drives due to better efficiency for the same power level and higher power density. For example, with MV drive, the copper loss in the motor coils is small due to small rms current for same power level [1]–[4]. Till recently, such MV drives are composed of multilevel converters made of silicon (Si) IGBTs [5], [6]. The Si IGBTs are limited in voltage to 6.5 kV and are highly inefficient if switched at more than 1 kHz [7], [8]. Thus, a Si based MV drive needs higher order multilevel converter for meeting the blocking voltage requirement [3]–[8]. Neutral point clamped (NPC) multilevel converters need large number of devices and diodes which reduces the power density. NPC converters also suffer from leakage balancing issues which are critical at MV [8]. Alternatively, cascaded multilevel converters can be used for attaining higher effective converter switching frequency (f_{sw}) to run the motor at very high speed [3], [4]. This is because, the speed of the motor is proportional to the fundamental frequency (f_m) of the converter, and high f_m is attainable only with high f_{sw}. With practical limitations in the number of cascaded converter modules, due to dc bus voltage balancing issues and lower power density, the motor speed attainable in such MV drives are not high enough to achieve very small size of the motor [9], [10]. Therefore, mechanical gears are used for increasing the speed. This reduces efficiency, in addition to making the system bulky, and the need for regular maintenance of the gear due to wear and tear. This calls for an alternative approach.

Recently developed 10 kV silicon carbide (SiC) MOSFETs are getting attention for MV converters [7], [8], [11]–[15]. These SiC MOSFETs can be switched at higher switching frequencies ($\geq$ 10 kHz) from MV levels with smaller switching loss compared to Si IGBTs [7], [8]. Also, due to very small specific on-resistance, the conduction loss is very small [11]–[15]. Hence, these can be used in a simple three-phase, 2-level topology switching at 10–20 kHz, and used for high speed MV motor drives with $\geq$ 2.3 kV ac voltage and 6 kV dc bus voltage. Single phase converter topologies based on 10 kV SiC MOSFETs are reported in [12], [13]. In this paper, a three-phase converter using 10 kV/10 A SiC MOSFET co-pack modules is developed and experimentally demonstrated for high speed motor drive. The test setup schematic of the demonstrated converter is shown in Fig. 1. The 10 kV/10 A SiC MOSFET co-pack modules from Cree, packaged by Powerex are used. A boost converter based on 15 kV/20 A SiC IGBT from Cree is used as the source for the 6 kV dc bus voltage. The 3-phase converter is switched using sine-triangle pulse width modulation (SPWM) and the load currents are filtered using inductors of 180 mH per phase.

The paper is organized as follows: Section II describes

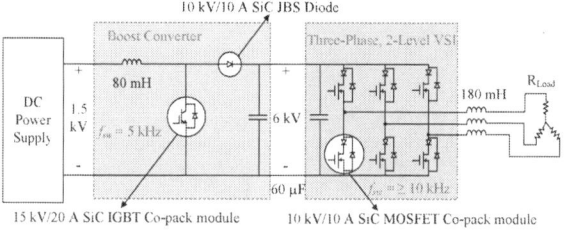

Fig. 1. Test setup schematic for the three-phase, 2-level converter based on 10 kV/10 A SiC MOSFETs

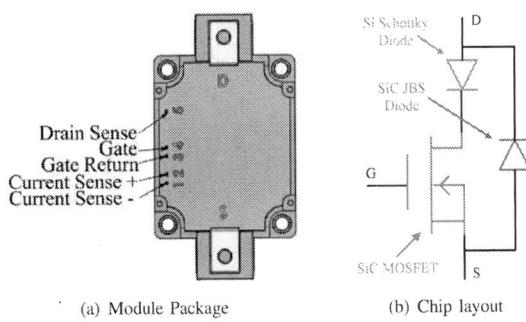

| (a) Module Package | (b) Chip layout |

Fig. 2. 10 kV/10 A SiC MOSFET co-pack module

the package and gives the detailed loss analysis of the 10 kV/10 A SiC MOSFET co-pack module along with the testing of the gate driver. Section III explains the module heat-run tests and power density estimation of the converter. Section IV covers the three-phase converter hardware development and the experimental validation at high f_m along with efficiency estimation. Section V concludes the paper.

II. Device Characterization and Gate Driver

A. 10 kV/10 A SiC MOSFET Module Details

The package of the 10 kV/10 A SiC MOSFET co-pack module used in the converter is shown in Fig. 2 (a). The package size is 4.41" X 2.36" X 0.79" including the height of the control pins. Drain (D) and source (S) are the power terminals. The Kelvin connection for the gate signal is marked as 'Gate'. There is a current sensing resistance in series with the MOSFET chip between 'Gate Return' and source. The chip layout inside the package is shown in the schematic in Fig. 2 (b). The 10 kV/10 A SiC MOSFET die size is 8.11 X 8.11 mm^2. The anti-parallel diode with size 10.6 X 8.3 mm^2 in the co-pack module is a 10 kV/10 A SiC junction barrier schottky (JBS) diode. There is a 4.8 X 4.8 mm^2 Si Schottky diode in series with the SiC MOSFET to cancel out the body diode of the SiC MOSFET.

B. Forward V-I Characteristics of 10 kV/10 A SiC MOSFET Co-pack Module

The forward characteristics of the co-pack module is captured using the curve tracer Tektronix 370A for nine values of gate voltages (v_{gs}) from 0 to 16 V. From the V-I plot given in Fig. 3, it can be seen that the drain-source forward voltage drop (v_{ds}) is 6.5 V at 9 A drain-source current (i_{ds}) when v_{gs} equals 16 V. The operating v_{gs} for the three-phase converter is 20 V. As seen from the plot, the V-I curves are matching for higher v_{gs} and hence, the curve is similar at $v_{gs} = 20$ V. The Si diode in the module also contributes to the voltage drop in the on-state. The forward V-I curves are important in the thermal design of the converter, as they are used for calculating the device conduction loss.

C. Switching Characteristics of 10 kV/10 A SiC MOSFET Co-pack Module

The turn-on (E_{on}) and turn-off (E_{off}) energy losses of the co-pack module are measured experimentally using a clamped

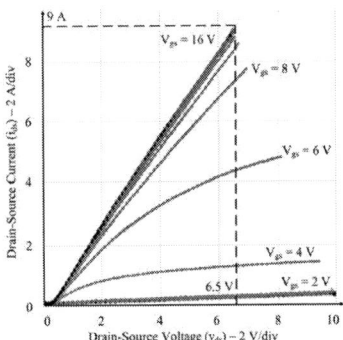

Fig. 3. 10 kV/10 A SiC MOSFET co-pack module V-I characteristics

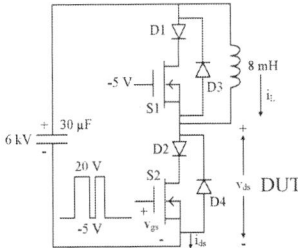

Fig. 4. DPT circuit schematic for switching characterization

inductive double pulse test (DPT) circuit [8]. Fig. 4 shows the circuit schematic with parameters. The gate signal to the top module MOSFET $S1$ is kept low (-5 V) while the double pulse signal is applied on the gate of the bottom module MOSFET $S2$. The bottom co-pack module is the device under test (DUT). During the first pulse, the DUT current i_{ds} rises to the required level. The energy losses E_{off} and E_{on} are measured during the first turn-off pulse and second turn-on pulse, respectively. The current level is decided by the dc bus voltage, inductor value and the width of the pulses. The anti-parallel diode $D3$ in the top module is used as the free-wheeling diode for the inductor current. This represents the operating scenario in the three-phase converter with motor load. The top module total package capacitance affects the switching loss of the bottom module, and vice-versa, in a two-level pole. When the diode $D3$ is conducting, the voltage across the top module is very small and hence, its capacitance is high. Thus, when the bottom module turns on, the high dv/dt causes a spiky current to flow through it due to this high capacitance of the top module. This increases the switching loss in the bottom module compared to a discrete free-wheeling diode due to the relatively higher package capacitance of the top module. A discrete free-wheeling diode in place of the top module (no $D1$ and $S1$) as explained in [8] is only helpful to estimate the loss of the standalone packages. Therefore, an arrangement shown in Fig. 4 using the entire half-bridge is used for switching loss characterization in this paper. A hot plate below the module base plate maintains the junction temperature of the DUT, since there is no significant power loss in the module in a DPT.

978-1-4673-9551-9/16 $31.00 © 2016 IEEE   1498

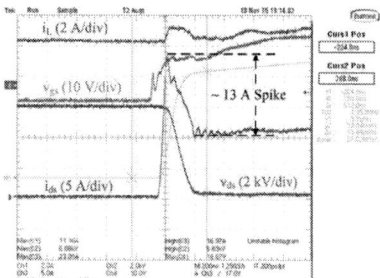

Fig. 5. 10 kV/10 A SiC MOSFET turn-on waveforms at 6 kV, 10 A, 125°C

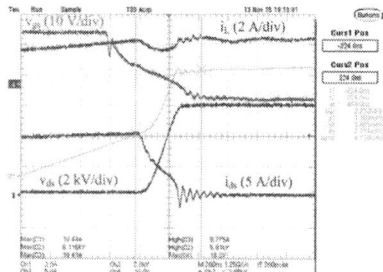

Fig. 6. 10 kV/10 A SiC MOSFET turn-off waveforms at 6 kV, 10 A, 125°C

Figs. 5 and 6 show the turn-on and turn-off characteristics of the 10 kV/10 A module, respectively, at 6 kV blocking voltage, 10 A current and 125°C junction temperature (θ_j). Both the turn-on and turn-off gate resistances used are 14.7 Ω based on a trade-off between switching loss and dv/dt. The turn-on and turn-off energy losses are found to be 13.84 mJ and 2.114 mJ, respectively. This results in a total switching power loss of about 160 W per module at 10 kHz. This is reasonably small and hence, the MOSFET based converter can be pushed to higher switching frequencies. The spiky current (~13 A) in bottom module due to the capacitance of the top module is indicated in Fig. 5 and is considered in E_{on} calculation. Figs. 7 and 8 give the E_{on} and E_{off} values, respectively, at different operating points measured using DPT. It can be seen from Fig. 7 that, E_{on} decreases with temperature for the same current and voltage. This is due to the decrease in threshold voltage with temperature and hence, faster turn-on reducing E_{on}. From Fig. 8, E_{off} is found to be comparatively very small and varying only with blocking voltage. E_{off} variation with temperature is negligible while it slightly decreases with current. The turn-on and turn-off dv/dts at 125°C are 36 $kV/\mu s$ and 27 $kV/\mu s$, respectively, while they reduce to 29 $kV/\mu s$ and 26 $kV/\mu s$, respectively, at 25°C. Hence, the turn-on dv/dt increase while the turn-off dv/dt almost remain constant with increase in temperature. The variation in turn-on dv/dt with temperature at 6 kV and 10 A is shown in Fig. 9. The maximum dv/dt value is considered for designing the shield needed to suppress the generated EMI.

D. Gate driver Isolated Power Supply and High Side Test

The gate drivers used in the DPT and also for the six 10 kV SiC MOSFETs in the three-phase converter are shown

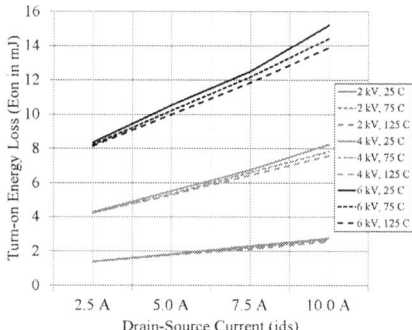

Fig. 7. E_{on} variation with blocking voltage, current, θ_j

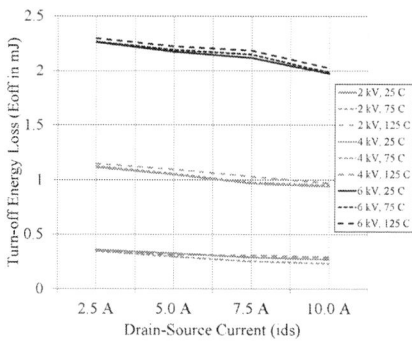

Fig. 8. E_{off} variation with blocking voltage, current, θ_j

in Fig. 10 (a). The design of the gate driver is similar to that for the 15 kV SiC IGBT discussed in [16]. The top board is the isolated power supply board generating +20 V for turn-on and -5 V for turn-off. The bottom board is the gate driver logic circuit board. The main requirements are low coupling capacitance of the isolation transformer in the power supply and high voltage isolation. The coupling capacitance is measured to be 3.5 pF at 9.5 MHz and 13.5 pF at 110 MHz as shown in the frequency response curve in Fig. 10 (b). Multiple layers of Kapton tape are used for more than 20 kV dc insulation. The emitter of the top module in each pole of the three-phase converter is exposed to high voltage swing. Thus, it is required to test the gate driver on the high side for high voltage isolation capability. For this, a boost-buck topology

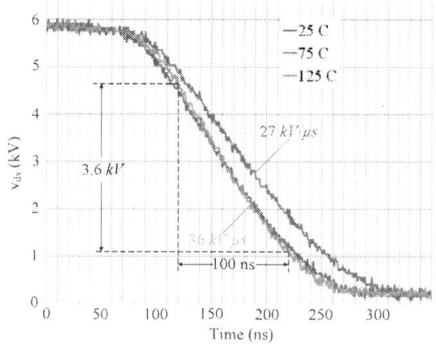

Fig. 9. Turn-on dv/dt variation with θ_j at 6 kV, 10 A

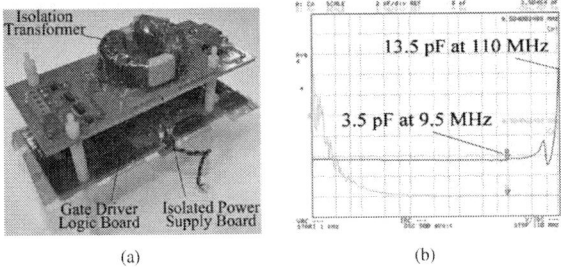

(a)   (b)

Fig. 10. (a) Gate driver for the 10 kV SiC MOSFET (b) Coupling capacitance vs frequency of the gate driver power supply isolation transformer

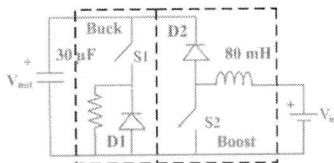

Fig. 11. Boost-buck converter setup for gate driver test

shown in the schematic in Fig. 11 is developed using the two poles of the three-phase converter. Two MOSFET modules in the poles are replaced by 20 kV SiC JBS diodes D1 and D2 with similar packaging as the MOSFET modules. The 'Boost' section is switched at 10 kHz with 0.75 duty ratio while the 'Buck' section is switched at 10 kHz with 0.25 duty ratio and without filter. The input and output voltages V_{in} and V_{out} are 1.5 kV and 6 kV, respectively, and the power is 6 kW. The gate driver under test drives the buck-MOSFET module in position S1. Fig. 12 shows the waveforms indicating that the gate driver is operating reliably under high dv/dt and 6 kV swing of the emitter of the high side device S1. The buck diode (D1) voltage (Ch3) shows this voltage swing.

III. THERMAL ANALYSIS OF THE 10 KV SIC MOSFET MODULES

A. Single Chip 10 kV/10 A SiC MOSFET Module

For understanding the thermal performance of the device, continuous heat run tests are carried out for 15 mins, when the device thermals are assumed to reach steady state. The heat sink temperature adjacent to the device base plate (θ_h) is

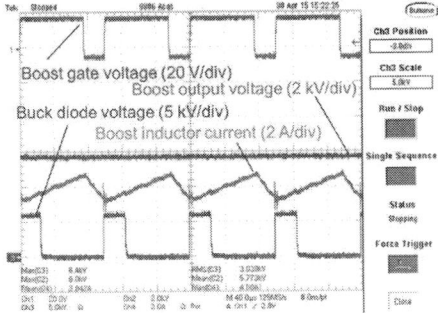

Fig. 12. Boost-buck converter waveforms

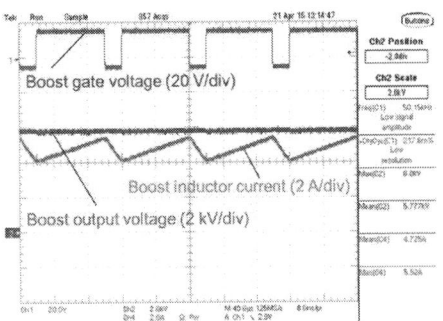

Fig. 13. Boost converter waveforms

TABLE I
10 KV/10 A SIC MOSFET THERMAL CONDUCTIVITY

Layer	Thickness (mm)	Thermal Conductivity $(W/m/^\circ C)$
SiC Drift	0.12	380
SiC Bulk	0.31	380
Solder	0.076	33
DBC Copper	0.305	393
AlN	1.016	170
DBC	0.305	393
Solder	0.127	33

monitored using a Flir thermal camera with emissivity setting adjusted to 0.9. Based on small assumptions, design thumb rules and experimental data, the thermal performance of the module is estimated. The heat sink and the cooling system are finalized based on a trade-off between the device separation for proper isolation and the maximum utilization of thermal path for faster heat dissipation. The heat sink used per pole has dimensions 7.087" X 4.921" X 5.346" from Wakefield Solutions (part no. 345-1046-ND). Forced air cooling is used with one fan-set each, at the inlet and the outlet of the heat sink. Each fan-set is composed of two fans running at 115 CFM coupled on the same axis to increase the CFM. Heat-run tests are carried out both using the boost-buck circuit in Fig. 11 as well a simple boost converter using only the boost section in Fig. 11. This helps to understand the thermal performance of the high side device as well as the low side device in the three-phase converter. Fig. 13 shows the boost converter waveforms at 6 kV output dc voltage, 7 kW and switching at 10 kHz. The condition under high side test is the same as in Fig. 12.

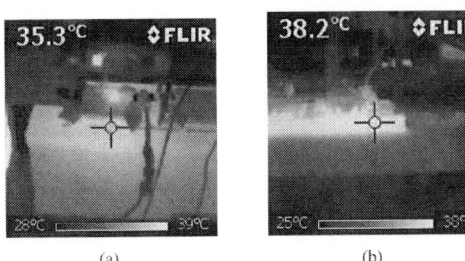

(a)   (b)

Fig. 14. (a) Boost converter and (b) Boost-buck converter thermal images

978-1-4673-9551-9/16 $31.00 © 2016 IEEE   1500

Table. I shows the thickness of different layers and their thermal conductivity values in the 10 kV/10 A SiC MOSFET co-pack module. The thermal resistance from junction to base plate (Rth_{jc}) is $0.23°C/W$ (chip area of 8.11 X 8.11 mm^2) for the 10 kV/10 A SiC MOSFET. The thermal resistances Rth_{tg} and Rth_{hs} for the thermal grease and heat sink are $0.05°C/W$ and $0.11°C/W$ (value for forced air-cooling from data-sheet), respectively. Fig. 14 (a) shows the boost converter thermal image after 15 mins. The temperature on the heat sink near to the device base plate is θ_h = 35.3°C. The ambient temperature θ_a is 25°C. Therefore, the loss in the co-pack module is calculated to be roughly 94 W using (1). Based on this, and assuming that the loss is contributed only by the SiC MOSFET chip, the junction temperature θ_j of the MOSFET is estimated to be 61.52°C using (2). This is very small compared to the maximum operating temperature of 175°C allowed for the 10 kV SiC MOSFET chip. In reality, the Si diode also contributes to this loss and hence, θ_j is smaller than 61.52°C. These calculations are possible because the anti-parallel diode does not carry current in the module S2 in the boost converter.

$$P_{loss} = \frac{(\theta_h - \theta_a)}{Rth_{hs}} \quad (1)$$

$$\theta_j = \theta_h + P_{loss}(Rth_{jc} + Rth_{tg}) \quad (2)$$

Fig. 14 (b) shows the thermal image of the boost-buck converter indicating that θ_h has settled at 38.2°C near to the boost device. So the θ_j of the MOSFET is estimated to be 71.8°C using the same assumptions. In this case, the heat sink temperature has increased, due to both boost and buck devices switching on heat sinks which are tied together. This condition closely resembles that in the three-phase converter. These experimental data show that f_{sw} can be increased further. If both boost and buck in Fig. 11 are switched at 20 kHz, the power loss in the buck-module S1 is 240 W (based on (1) and switching loss $\gg$ conduction loss). This increases θ_j of the SiC MOSFET to 118.6°C, which is reasonable. Hence, f_{sw} is increased to 20 kHz in the three-phase converter to increase f_m up to 1 kHz as explained later. To have some safety margin, this is considered as the best operating point for the converter. Hence, the DPT results are captured up to 125°C.

For cross checking the calculation, the DPT data from Figs. 7 and 8 are used for the case with θ_j equal to 125°C (approx.), along with conduction loss data from Fig. 3. The switching (P_{sw}) and conduction (P_{cond}) losses are 225 W and 17 W, respectively, based on the operating conditions at 20 kHz and using (3). The total loss of 242 W is similar to that estimated by heat-run tests.

$$P_{loss} = f_{sw}(E_{on} + E_{off}) + P_{cond} \quad (3)$$

The total weight of the three-phase converter power hardware excluding the control power supply and filter inductor is 50 kg (110 lbs). Considering 60% de-rating for the device voltage and current, the maximum operating power for the designed three-phase converter using 10 kV/10 A SiC MOSFET is

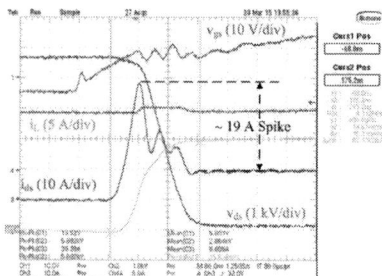

Fig. 15. 10 kV/20 A SiC MOSFET turn-on waveforms at 6 kV, 10 A, 25°C

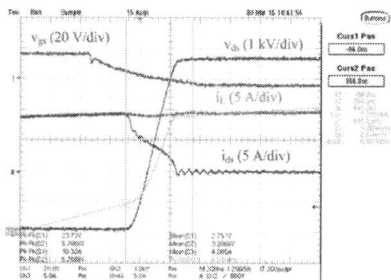

Fig. 16. 10 kV/20 A SiC MOSFET turn-off waveforms at 6 kV, 10 A, 25°C

21 kW. Therefore, the power density is 420 W/kg (191 W/lbs). Considering volume (29" X 16" X 11"), the power density is 4.11 $W/inch^3$ (or 0.25 MW/m^3). The volume is decided by the spacing constraints due to high voltage isolation requirement between the devices on the heat sink.

B. Two Chip 10 kV/20 A SiC MOSFET Module

10 kV/20 A modules using the same size package is used to double the power density of the three-phase converter. Two 10 A chips are connected in parallel to increase the current rating. Since the MOSFET chips have positive temperature coefficients, parallel operation with equal current sharing is achievable. The series Si diode is omitted from this package. Figs. 15 and 16 show the turn-on and turn-off characteristics of the 10 kV/20 A module, respectively, at 6 kV blocking voltage, 10 A current and room temperature (25°C). As discussed before, for the 10 kV SiC MOSFETs, the loss does not vary much with temperature and therefore, the loss data at 25°C is considered for analysis. The turn-on and turn-off energy

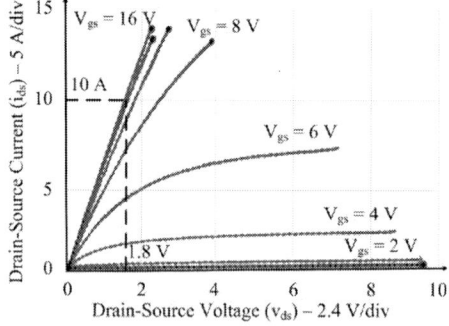

Fig. 17. 10 kV/20 A SiC MOSFET co-pack module V-I characteristics

losses are 15.11 and 2.87 mJ, respectively, with 14.7 Ω turn-on and turn-off gate resistances. From the V-I characteristics in Fig. 17, the forward drop is 1.8 V at 10 A. Compared to the 10 A module, the forward drop reduces due to reduction in the on-state resistance after paralleling the chips. The total loss in the module when switched in a boost converter at 6 kV dc bus, 20 kHz and 10 A average current through the boost inductor is roughly 370 W. Considering $\theta_a = 25°C$ and half the total power loss dissipated through one chip, θ_j per chip reaches 97.15°C which is smaller than 108.3°C estimated for the one 10 A chip module. Therefore, paralleling the MOSFET chips reduces the maximum θ_j per chip for double the total current through the module (4.7 A in Fig. 13 compared to 10 A considered here). This is because, the total switching loss does not increase much while the total conduction loss reduces by more than half. Hence, the power density can be increased by a factor of more than two by paralleling two chips with same package dimensions and eliminating the series Si diode.

A multi-chip 120 A half-bridge module is reported in [12] with dimensions 7.48" X 5.51" X 1.89". That is only 5 times increase in device volume compared to the 20 A module for 6 times more power. Based on the analysis in this paper, it is expected that the power density of the three-phase converter using this module will increase enormously. In the next section, the converter demonstration is done using 10 A modules for understanding the impact of high dv/dt.

IV. THREE-PHASE CONVERTER

A. Converter Hardware

A sandwich bus bar is developed for the 3-phase converter based on copper sheets as shown in Fig. 18. The bus bar stray inductance L_{stray} is calculated to be less than 15 nH. Small value of L_{stray} is required to protect the devices from the $L_{stray}.di/dt$ voltage spikes. The positive and negative copper plates are separated from each other using FR4 insulation. The corners are rounded and smoothed to avoid corona discharge at higher dc bus voltages. All sides are covered with Kapton tape insulation (not shown in Fig. 18). One heat sink is used per pole. The three heat sinks are electrically connected together. Four low ESL (30 nH) film capacitors from ICAR of 120 μF, 3.5 kV rating, are connected in series in the dc bus. The MV, high speed motors have very low winding inductance. Therefore, external filters are required between the converter and the motor to filter out the high magnitude switching frequency ripple voltage present in the converter line voltage. This is required to prevent the motor insulation breakdown. In the experimental setup of the presented three-phase converter, 180 mH inductors are used in each phase. Since the devices are switching at very high dv/dt, the parasitic capacitance C_p across the filter needs to be small in order to reduce the $C_p.dv/dt$ generated spikes in the motor currents. Fig. 19 shows the frequency response plots of the filter. At 28.572 kHz resonant frequency, the value of C_p is calculated to be 222 pF in each phase. Series connection of nine 20 mH inductors are used in each phase to minimize C_p. The converter hardware

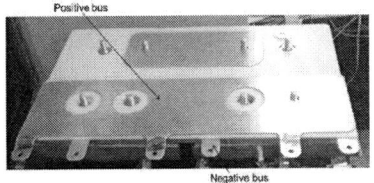

Fig. 18. Three-phase converter bus-bar arrangement

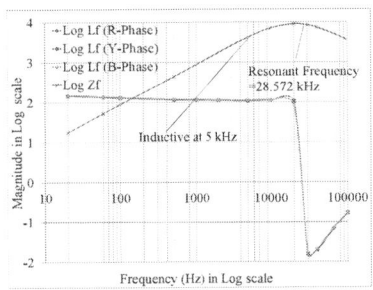

Fig. 19. Inductance and impedance vs frequency for the L-filter (in log scale)

setup along with the probes used for measurement are shown in Fig. 20. Six gate drivers are used in total and they are mounted on top of the corresponding modules. Bottom of each gate driver board is covered with Nomex paper for insulating the logic ICs from high dv/dt generated noise interference.

B. Experimental Validation

The converter is experimentally demonstrated under various operating conditions. Figs. 21 to Fig. 23 show the converter waveforms with f_{sw} = 10 kHz. The converter is tested up to f_m = 400 Hz at 5 kV dc bus voltage using f_{sw} = 10 kHz. For higher values of f_m, f_{sw} is increased to 20 kHz. Higher f_m waveforms are shown in Figs. 24 to Fig. 27 with 3 kV dc bus voltage and up to 1 kHz (Fig. 25). The dv/dt generated noise can be seen superimposed with the converter currents in all the waveforms. The current spikes are more visible with increase in f_m as the time-scale becomes smaller in the waveforms. The zoomed in current waveforms in

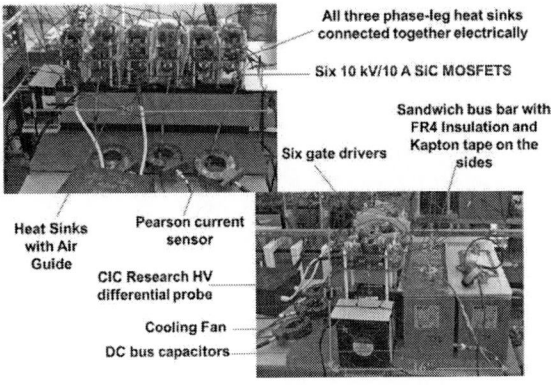

Fig. 20. Three-phase converter hardware setup

978-1-4673-9551-9/16 $31.00 © 2016 IEEE

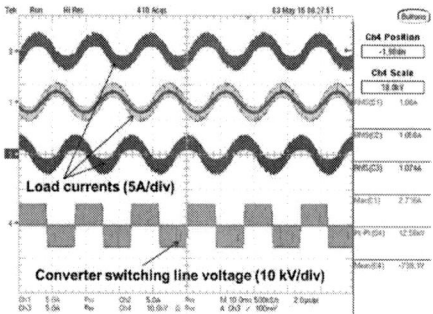

Fig. 21. Three-phase converter waveforms at 6 kV dc bus voltage, 3 kV ac line voltage rms, $f_{sw} = 10$ kHz, 5 kW load and $f_m = 60$ Hz

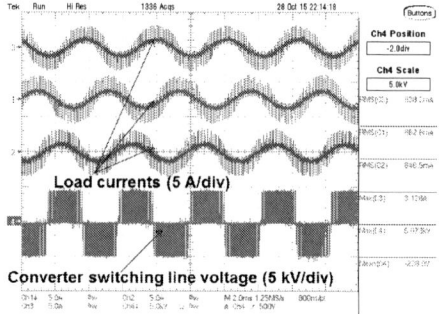

Fig. 22. Three-phase converter waveforms at 5 kV dc bus voltage, 2.6 kV ac line voltage rms, $f_{sw} = 10$ kHz, 3.8 kW load and $f_m = 240$ Hz

Fig. 26 show the spikes clearly at 20 kHz switching frequency in every 50 μs interval during device switching. The filter inductor voltage in Fig. 27 shows the 20 kHz, 6 kV peak-peak switching voltage ripple superimposed on the 1 kHz fundamental frequency voltage drop. For any 6-pole motor, $f_m = 1$ kHz results in a speed of 20,000 RPM at MV level which is not possible with any Si based converter.

C. Three-Phase Converter Loss Analysis

The experimentally measured switching and conduction loss data of the 10 kV/10 A SiC MOSFET are used for the loss analysis. The loss data captured experimentally in Section II are used in the thermal model of PLECS simulation software to estimate the loss of the three-phase converter under various

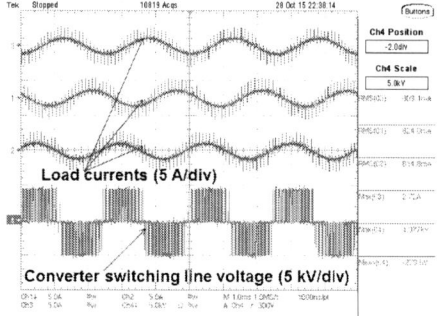

Fig. 23. Three-phase converter waveforms at 5 kV dc bus voltage, 2.6 kV ac line voltage rms, $f_{sw} = 10$ kHz, 3.7 kW load and $f_m = 400$ Hz

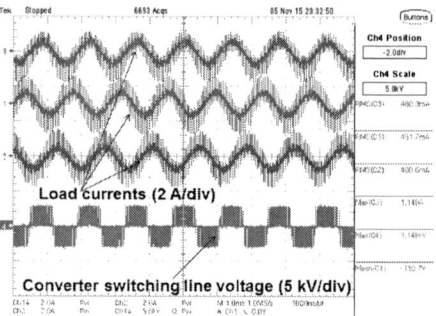

Fig. 24. Three-phase converter waveforms at 3 kV dc bus voltage, 900 V ac line voltage rms, $f_{sw} = 20$ kHz, 1.45 kW load and $f_m = 720$ Hz

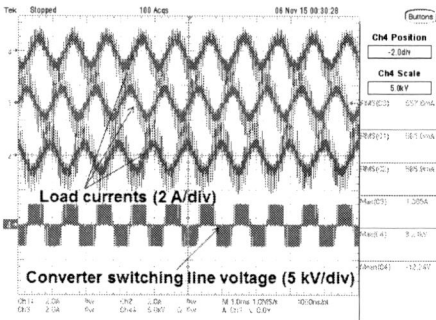

Fig. 25. Three-phase converter waveforms at 3 kV dc bus voltage, 900 V ac line voltage rms, $f_{sw} = 20$ kHz, 1.45 kW load and $f_m = 1000$ Hz

conditions. Fig. 28 shows the variation of the converter loss with load at $f_m = 1$ kHz, 6 kV dc bus voltage and 3 kV ac line voltage. Loss variations at both 10 kHz and 20 kHz f_{sw} are plotted. The loss does not vary with f_m and therefore, only loss values at $f_m = 1$ kHz are given. The conduction loss is small compared to the switching loss at both 10 kHz and 20 kHz f_{sw}. This shows good forward characteristics of the 10 A module. The switching loss and therefore, the total loss almost doubles when f_{sw} is increased from 10 kHz to 20 kHz. At $f_{sw} = 20$ kHz and 20 kVA load, the total loss is estimated to be 695 W, which gives an efficiency of 96.64 %. This efficiency at a power density of 4.11 $W/inch^3$ is reasonable.

V. CONCLUSIONS

In this paper, the design and experimental validation of a MV, high frequency, three-phase, two-level converter based on 10 kV/10 A SiC MOSFET co-pack module are presented. The application of the converter is in high speed MV motor drives for marine applications, compressors for oil and gas industries, wind energy generation systems etc. The design of the converter is discussed including gate driver, bus bar, filter and dc bus capacitor. The 10 kV/10 A SiC MOSFET co-pack module characterization and heat-run tests are discussed for loss analysis, and the thermal design of the converter. The power density improvement with a two-chip 20 A module is explained. The experimental results up to 1 kHz fundamental and 20 kHz switching frequencies are also given.

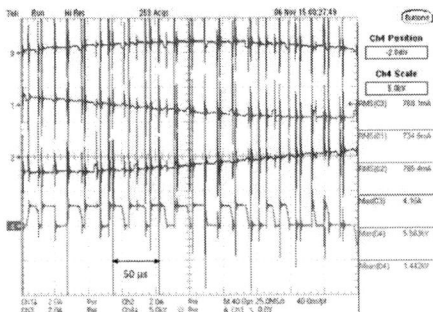

Fig. 26. Zoomed waveforms in Fig. 25 at $f_{sw} = 20$ kHz

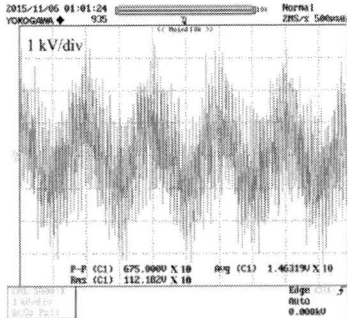

Fig. 27. Filter inductor voltage drop at 3 kV dc bus voltage, 900 V ac line voltage rms, 20 kHz f_{sw}, 1.45 kW load and 1000 Hz f_m

ACKNOWLEDGMENT

The information, data, or work presented herein was funded in part by the Office of Energy Efficiency and Renewable Energy (EERE), U.S. Department of Energy, under Award Number DE-EE0006521 with North Carolina State University, PowerAmerica Institute. This work made use of FREEDM ERC shared facilities supported by NSF under award no. EEC-0812121.

DISCLAIMER

The information, data, or work presented herein was funded in part by an agency of the United States Government. Neither the United States Government nor any agency thereof, nor any of their employees, makes any warranty, express or implied, or assumes any legal liability or responsibility for the accuracy, completeness, or usefulness of any information,

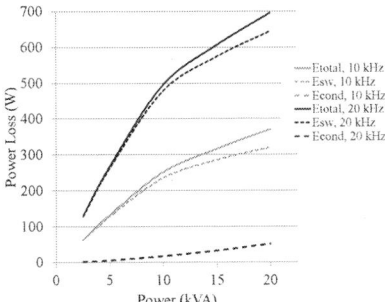

Fig. 28. Loss variation with load at $f_m = 1$ kHz, 6 kV dc bus voltage and 3 kV ac line voltage

apparatus, product, or process disclosed, or represents that its use would not infringe privately owned rights. Reference herein to any specific commercial product, process, or service by trade name, trademark, manufacturer, or otherwise does not necessarily constitute or imply its endorsement, recommendation, or favoring by the United States Government or any agency thereof. The views and opinions of authors expressed herein do not necessarily state or reflect those of the United States Government or any agency thereof.

REFERENCES

[1] E. Cengelci, P. N. Enjeti, and J. W. Gray, "A new modular motor-modular inverter concept for medium-voltage adjustable-speed-drive systems," *IEEE Trans. Ind. Applicat.*, Vol. 36, no. 3, pp. 786-796, Aug. 2002.

[2] J. Dai, S. W. Nam, M. Pande, and G. Esmaeili, "Medium-voltage current-source converter drives for marine propulsion system using a dual-winding synchronous machine," *IEEE Trans. Ind. Applicat.*, Vol. 50, no. 6, pp. 3971-3976, Nov./Dec. 2014.

[3] R. Teodorescu, F. Blaabjerg, J. K. Pedersen, E. Cengelci, and P. N. Enjeti, "Multilevel inverter by cascading industrial VSI," *IEEE Trans. Ind. Electron.*, Vol. 49, no. 4, pp. 832-838, Aug. 2002.

[4] G. Baoming, F. Z. Peng, A. T. de Almeida, and H. Abu-Rub, "An effective control technique for medium-voltage high-power induction motor fed by cascaded neutral-point-clamped inverter," *IEEE Trans. Ind. Electron.*, Vol. 57, no. 8, pp. 2659-2668, Aug. 2010.

[5] D. Krug, S. Bernet, S. S. Fazel, K. Jalili, and M. Malinowski, "Comparison of 2.3-kV medium-voltage multilevel converters for industrial medium-voltage drives," *IEEE Trans. Ind. Electron.*, Vol. 54, no. 6, pp. 2979-2992, Dec. 2007.

[6] J. A. Sayago, S. Bernet, and T. Bruckner, "Comparison of medium voltage IGBT-based 3L-ANPC-VSCs," in *proc. 2008 IEEE Power Electronics Specialist Conference*, Rhodes, pp. 851-858, June 2008.

[7] S. Madhusoodhanan, K. Hatua, S. Bhattacharya, S. Leslie, S. H. Ryu, M. Das, A. Agarwal, and D. Grider, "Comparison study of 12 kV n-type SiC IGBT with 10 kV SiC MOSFET and 6.5 kV Si IGBT based on 3L-NPC VSC Applications," in *proc. 2012 IEEE Energy Conversion Congress and Exposition*, Raleigh, NC, pp. 310-317, Sept. 2012.

[8] S. Madhusoodhanan et al., "Solid state transformer and MV grid tie applications enabled by 15 kV SiC IGBTs and 10 kV SiC MOSFETs based multilevel converters," *IEEE Trans. Ind. Applicat.*, Vol. 51, no. 4, pp. 3343-3360, Jul./Aug. 2015.

[9] D. Jiang, J. Xue, F. Wang, and M. H. Kao, "High density modular multilevel cascade converter for medium-voltage motor drive," in *proc. 2011 IEEE Electric Ship Technologies Symposium*, Alexandria, VA, pp. 482-485, Apr. 2011.

[10] J. Gong, L. Xiong, F. Liu, and X. Zha, "A regenerative cascaded multilevel converter adopting active front ends only in part of cells," *IEEE Trans. Ind. Applicat.*, Vol. 51, no. 2, pp. 1754-1762, Mar./Apr. 2015.

[11] J. Wanget et al., "Characterization, modeling and application of 10 kV SiC MOSFET," *IEEE Trans. Electron. Devices*, Vol. 55, no. 8, pp. 1798-1806, Aug. 2008.

[12] M. K. Das et al., "10 kV, 120 A SiC half H-bridge power MOSFET modules suitable for high frequency, medium voltage applications," in *proc. 2011 IEEE Energy Conversion Congress and Exposition*, Phoenix, AZ, pp. 2689-2692, Sept. 2011.

[13] G. Wang et al., "Comparisons of 6.5kV 25A Si IGBT and 10-kV SiC MOSFET in solid-state transformer application," in *proc. 2010 IEEE Energy Conversion Congress and Exposition*, Atlanta, GA, pp. 100-104, Sept. 2010.

[14] C. DeMarino et al., "10 kV, 120 A SiC MOSFET modules for a power electronics building block (PEBB)," in *proc. 2014 IEEE Workshop on Wide Bandgap Power Devices and Applications (WiPDA)*, Knoxville, TN, pp. 55-58, Oct. 2014.

[15] S. Ryu et al., "10-kV, 123-m/spl Omega//spl middot/cm^2 4H-SiC power DMOSFETs," *Electron Device Letters*, Vol. 25, no. 8, pp. 556-558, Aug. 2004.

[16] A. Kadavelugu and S. Bhattacharya, "Design considerations and development of gate driver for 15 kV SiC IGBT," in *proc. 2014 IEEE Applied Power Electronics Conference and Exposition*, Fort Worth, TX, pp. 1494-1501, Mar. 2014.

978-1-4673-9551-9/16 $31.00 © 2016 IEEE

Novel Three Phase Multi-Level Inverter Topology with Symmetrical DC-Voltage Sources

Ahmed Salem[1], Emad M. Ahmed[1a], Mahrous Ahmed[1] and Mohamed Orabi[1b], IEEE Senior Member

[1]APEARC, Faculty of Engineering, Aswan University, Aswan 81542, Egypt

[a]eelbakoury@apearc.aswu.edu.eg, [b]orabi@ieee.org

Abstract— In this paper, a novel three phase modular multi-level inverter (*MMLI*) is proposed. The proposed inverter consists of primary cell and repetitive modular cells which are connected in series arrangement with the primary cell. Therefore, the proposed topology is able to get more output voltages levels number by adding extra modular cells. Both the sinusoidal pulse width modulation (*SPWM*) and staircases modulation are effectively executed. The proposed inverter is distinguished by several advantages such as: reduction in the number of semiconductor power switches, reduced Dc-voltage sources count, high utilization factor of the used Dc-voltage sources, and the control execution simplicity. Accordingly, the installation cost and size are reduced. It is simulated using MATLAB software package-tool. In addition, a prototype is developed and examined, to verify both control techniques and performance of the topology. Moreover, experimental results are provided to authenticate the simulation results and it show high similarity with it.

Keywords—multilevel invereter; reduced compnants count; sinsudal pulse width modulation; low frequency operation.

I. Introduction

In latest years, the worldwide energy consumption has increased in high rates, driven mainly by growing demand in developing countries for both industrial and residential fields, and also by emergence of new promising markets. In order to overcome the world excess energy demands, the world trend is replacing the traditional energy sources (*TES*) which based on non-sustainability sources such as fossil fuels and nuclear fission by utilizing new efficient and clean energy resources, that are based on natural sources, such as sunlight power, wind power, and geothermal power, in addition to ocean energy, and bioenergy. All these energy sources decrease carbon emission, and it classified as renewable energy sources (*RES*) [1].

One of the leading advances in the *RES* management systems is the enhancement in power converters, which have the main role to convert the raw generated energy to both loads and utility grid. Accordingly, the conventional interfacing converters as DC-AC inverters, DC-DC converters topologies …etc. are replaced by new sophisticated topologies such as multi-level inverters (*MLIs*) or single-stage buck-boost inverters like quasi-Z-source inverters (*QZSIs*). Their target is to increase system reliability, efficiency, power-transfer, life time, and in addition to reduce overall system cost and size. Besides the others DC-AC converters, *MLIs* are characterized

by various features such as low *THD* in the inverter outputs, Low *EMI*, and low *dv/dt* across power switches. These features pushing MLIs to acquire a great attention as a single-stage inverter [2] – [5].

Many topologies were addressed in the last decade focusing on minimizing the required number of components per voltage levels and decreasing the *THD* in MLIs. The author in [6] presented a multilevel DC link (*MLDCL*) topology. It consists of a group of basic cells connected in series configuration. Each cell produces *E* or *0* voltage across the connected cells. The required number of active switches for (*m*) output voltage levels is (*m+3*) for the *MLDCL* inverters. However, this topology requires an increase in the number of components compared to the conventional topologies in addition to high voltage stresses. The authors in [7] presented a topology named transistor-clamped H-bridge (*TCHB*). The primary cell can produce a five-levels per pole in the output voltage (±*E*, ± (1/2)*E*, 0). However, it suffers also from the increased components counts, requirement of electrolytic capacitors, and complex control methodology. While, the authors in [8] presented three-phase asymmetrical multi-level cascade inverter. The output voltage levels synthesized by series connected cells like in [6]. Using two cells configuration, it produces four levels per pole. However, instead of using H-bridge to obtain the opposite voltage polarities as in [6], it uses simply the phase shift relationship between the three legs, by subtracts each leg's voltage with the neighboring one to produce the line voltage. The same subtraction idea was presented in [9].

Obviously, almost all of addressed topologies suffering from increased number of components counts and usage of electrolytic capacitors as floating power sources which add more complicated problems in the control system. On the other hand, introducing new topology that can solve the stated challenges is highly recommended.

This paper introduces a new circuit with the features of reduced components count compared with the conventional and the addressed *MLIs* topologies in the literature with keeping the same pole voltage levels number. The paper is organized as follow: the proposed three phase *MLI* topology is introduced in section II. The modulation scheme for the proposed *MLI* is presented in section III. Simulation and experimental results are provided in section IV. A comparison between the proposed topology and the others *MLIs* in the literature and a conclusion are presented in sections V, VI, respectively.

978-1-4673-9551-9/16 $31.00 © 2016 IEEE

II. PROPSED MLI TOPOLGY CONFIGRATION

A. n-level configration

The proposed topology configuration for generating n-levels across pole voltage is demonstrated in Fig. 1. Where the poles voltage is defined as the voltage difference between points *A* or *B* or *C* and point *0*. The proposed *MLI* topology shared the DC voltages sources for different inverter legs power switches. And it consists from repeated cells in cascaded configuration; the first cell is considered as the primary/main cell, which the other repeated modular cells are connected to it.

As shown in Fig. 1 the primary cell constructed by using twelve power switching devices and two DC voltage sources. While the modular cells have same construction of the primary cell, except that its three legs are connected together from one side (named as *0*) while the other sides of the three legs are isolated from each other, to add the connecting-capability for the modular cell to by connected with the primary cell. In order to compute the required components for the proposed inverter as function of output voltage levels number, equation (1) is provided. In (1) both of modular cells number (N_{MC}), isolated DC voltage sources count (N_{DC}), and the utilized switching devices count (N_{SW}) are related to output voltage levels number (n).

$$\left. \begin{array}{l} N_{MC} = 0.5(n-3) \\ N_{DC} = n-1 \\ N_{SW} = 6(n-1) \end{array} \right\} \qquad (1)$$

For example, to generate five-levels across the output voltage ($n=5$), one modular cells ($N_{MC} = 0.5(5-3)$) is required to be connected with the primary cell, in addition to four DC voltage sources ($N_{DC} = 5-1$). Also, a twenty-four power switches ($N_{SW} = 6(5-1)$) are necessary.

B. Three-level configration

In order to investigate the proposed topology, a scaled-down circuit is deduced from the general *n-level'* configuration, to produce only three-levels across the pole voltage, it depicted in Fig. 2. Also, according to (1), it is constructed by using only primary cell as no need for adding extra modular cells. In other words, it requires two DC voltage sources beside twelve switching devices to produce three voltage levels *0*, *E*, and *2E*, where *E* is the DC voltage source value.

The operation of the proposed inverter can be clarified as following: to get zero voltage between points *A* and *0*, the power switch which is titled S_4 should be turned on while the others switches S_1, S_2, and S_3 are required to be in off-states. On the other hand, to generate voltage difference between A and 0 equal to *E* voltage, both of S_2 and S_3 are needed to be turned on while both of S_1 and S_4 should be turned off. Finally, in order to generate *2E* voltage across pole voltage (V_{A0}), only S_1 is turned on, however S_2, S_3, and S_4 all are turned off. These different operation modes are summarized in Table I.

In contrast to, the three allowed switching states (i.e. state 1, state 2, and state 3) that are listed in Table I, there are three forbidden switching states which are not allowed in switching algorithms for the proposed *MLI*, in order to prevent short-circuit occurrence across the utilized DC voltage sources.

These on-state switching combinations are: A) S_1, S_2, and S_3 in on-state. B) S_1, and S_4 in on-state. C) S_2, S_3, and S_4 in on-state.

III. PROPOSED TOPOLOGY MODULATION SCHEMES

In order to produce a semi-like sinusoidal voltages waveforms across output terminals, two modulation schemes are proposed and implemented for the scaled down circuit which is shown in Fig. 2, and they are described in following subsections:

A) Low frequency modulation scheme:

The low frequency modulation is preferred in *MLI* switching algorithms due to reducing the switching losses as the frequency of the switching pulses is very low. The low frequency scheme is depended on Table II which lists the different operating modes for the proposed *MLI* to produce five-level across the line voltages V_{AB}, V_{BC}, and V_{CA}. Actually, Table II lists twelve operation modes (*M1* to *M12*) and it is based on Table I. By driving the proposed *MLI* switches according to the different switching states which are listed in Table II, a balancing three-phase output voltages are achieved.

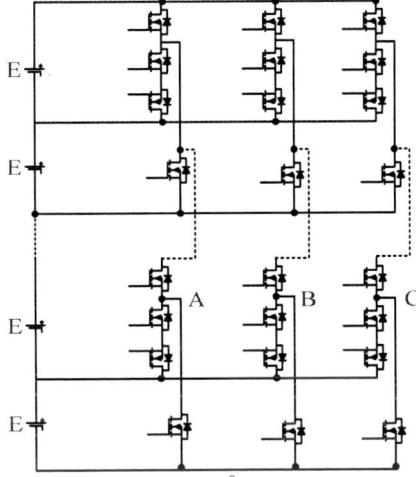

Fig. 1 The proposed three-phase *MLI* configuration for *n*-level.

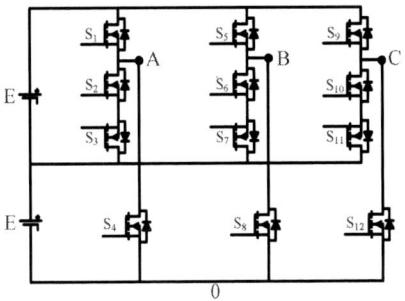

Fig. 2. Proposed three-level *MLI* topology.

978-1-4673-9551-9/16 $31.00 © 2016 IEEE

TABLE I. THE DIFFERENT SWITCHING STATES AND THE CORRESPONDING POLE VOLTAGE (V_{AO}).

Switching states	Switch				Pole voltage (V_{A0})
	S_1	S_2	S_3	S_4	
State 1	ON	OFF	OFF	OFF	2E
State 2	OFF	ON	ON	OFF	E
State 3	OFF	OFF	OFF	ON	0

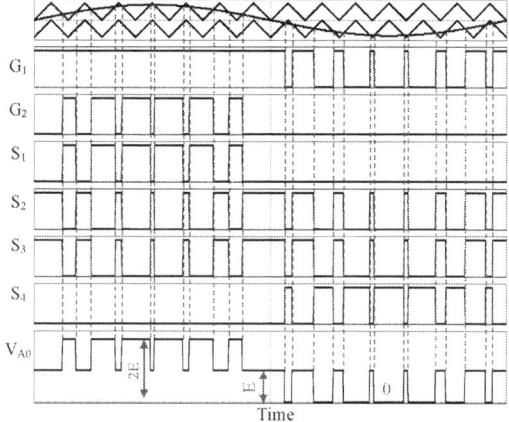

Fig. 3. SPWM switching pattern and the pole voltage (V_{A0}) of the proposed *MLI* topology.

B) *Sinusoidal pulse width modulation (SPWM):*

In addition to low frequency modulation, the suggested inverter is modulated by using pulse-width modulation (*PWM*) based scheme. The proposed switching scheme uses three signals in sinusoidal shape, they are shifted from each other by *120°*, and named modulation signals. Besides multi-carrier signals with higher frequency compared to sinusoidal signals which has a frequency equals to *50* Hz. The modulation signals are sampled by using the carrier signals in order to obtain on/off switching pulses with variable width, in distinguish manner to achieved finally sinusoidal voltages waveforms in the output of the inverter.

The proposed *SPWM* scheme is categorized under phase disposition (*PD*) modulation scheme. Where all carriers are in phase and having same amplitudes, but they are shifted in their levels. So as to generate *n*-levels across the output terminals, a (*n-1*) carrier signals are necessary. Fig. 3 shows the switching pulses generation for leg *A*. As the case-study circuit (shown in Fig. 2.) generates three-level, so only two carrier signals are required as depicted in Fig. 3. To generate switching pulses that are belong to legs *B* and *C*, the same procedure is applied, but the modulation signals should be shifted by *-120°*, *120°* respectively. After comparing the modulation waveforms with the two carrier signals, two new signals G_1 and G_2 will be generated. These signals are considered as a train of on/off pulses, but to obtain the gate drives switching signals, a several logical operations should be carried on these two signals as described in (2). Where S_1, S_2, S_3, and S_4 are the switching signals that are applied to the switching devices in leg *A*. also, it is notable from both of Fig. 3 and (2) that, the switching pulses for S_2 and S_3 are matched. Because S_2 and S_3 are turned on/off together.

$$\left.\begin{array}{l} S_1 = G_2 \\ S_2 = G_1 \times \overline{G_2} \\ S_3 = G_1 \times \overline{G_2} \\ S_4 = \overline{G_1} \end{array}\right\} \qquad (2)$$

TABLE II. SWITCHING STATES OF THE PROPOSED TOPOLOGY

Mode	V_{AB}	V_{BC}	V_{CA}	S_1	S_2	S_3	S_4	S_5	S_6	S_7	S_8	S_9	S_{10}	S_{11}	S_{12}
M1	E	-2E	E	OFF	ON	ON	OFF	OFF	OFF	OFF	ON	ON	OFF	OFF	OFF
M2	2E	-2E	0	ON	OFF	OFF	OFF	OFF	OFF	OFF	ON	ON	OFF	OFF	OFF
M3	2E	-E	-E	ON	OFF	OFF	OFF	OFF	OFF	OFF	ON	OFF	ON	ON	OFF
M4	2E	0	-2E	ON	OFF	OFF	OFF	OFF	OFF	OFF	ON	OFF	OFF	OFF	ON
M5	E	E	-2E	ON	OFF	OFF	OFF	OFF	ON	ON	OFF	OFF	OFF	OFF	ON
M6	0	2E	-2E	ON	OFF	OFF	OFF	ON	OFF	OFF	OFF	OFF	OFF	OFF	ON
M7	-E	2E	-E	OFF	ON	ON	OFF	ON	OFF	OFF	OFF	OFF	OFF	OFF	ON
M8	-2E	2E	0	OFF	OFF	OFF	ON	ON	OFF	OFF	OFF	OFF	OFF	OFF	ON
M9	-2E	E	E	OFF	OFF	OFF	ON	ON	OFF	OFF	OFF	OFF	ON	ON	OFF
M10	-2E	0	2E	OFF	OFF	OFF	ON	ON	OFF	OFF	OFF	ON	OFF	OFF	OFF
M11	-E	-E	2E	OFF	OFF	OFF	ON	OFF	ON	ON	OFF	ON	OFF	OFF	OFF
M12	0	-2E	2E	OFF	OFF	OFF	ON	OFF	OFF	OFF	ON	ON	OFF	OFF	OFF

IV. SIMULATION AND EXPERIMENTAL RESULTS

Based on the three-level case-study circuit, simulation model is developed using MATLAB/Simulink environment, in order to investigate the proposed MLI topology and to implement its two modulation schemes. Beside the simulation study of the proposed topology, the experimental verification is conducted to validate the operation of the proposed inverter for the two aforesaid switching schemes low frequency modulation and *SPWM*. The metal–oxide–semiconductor field-effect transistor (*MOSFET*) of type *AOT22N50L* is used to construct the laboratory inverter power stage. While the control algorithms are executed by using the digital signal processor (*DSP*) of type *TMS320F28335*. Fig. 4 shows the proposed *MLI* experimental setup, which includes two DC-voltage sources (*E*) everyone having *80* volt, twelve switching devices, measurement tools (i.e. current probe, voltage probes, and oscilloscope), *DSP* controller, and three-phase resistive-inductive load. The

978-1-4673-9551-9/16 $31.00 © 2016 IEEE

switching frequency (F_S) is chosen to be *3* kHz, while the modulation index (M_I) is equal to *0.95*.

Figure. 5 shows three voltages waveforms V_{A0}, V_{B0}, and V_{C0}, these three voltages are called pole voltages, they are considered as the key elements in operation of the proposed inverter. Because, both of the line voltages (V_{AB}, V_{BC}, and V_{CA}) and load-phase voltages (V_{AN}, V_{BN}, and V_{CN}) are synthesized from it as it specified in (3) and (4). According to (3) and (4), three-levels (*0, E, 2E*) across the pole voltages will generates five-levels (*2E, E, 0, -E, -2E*) in line voltages. In addition to seven-levels (*0, 2E/3, E, 4E/3, -2E/3, -E,* and *-4E/3*) in load-phase voltage for low frequency modulation and nine-levels (*0, E/3, 2E/3, E, 4E/3, -E/3, -2E/3, -E,* and *-4E/3*) for *SPWM* scheme

$$V_{AB} = V_{AO} - V_{BO} \qquad (3)$$

$$\begin{bmatrix} \mathbf{V_{AN}} \\ \mathbf{V_{BN}} \\ \mathbf{V_{CN}} \end{bmatrix} = \frac{1}{3} \times \begin{bmatrix} 2 & -1 & -1 \\ -1 & 2 & -1 \\ -1 & -1 & 2 \end{bmatrix} \times \begin{bmatrix} V_{AO} \\ V_{BO} \\ V_{CO} \end{bmatrix} \qquad (4)$$

The two additional voltage levels across the load-phase voltages waveforms are '*E/3*' and '*-E/3*'. These extra voltage levels are created across the three-phase terminals as a result of applying the *SPWM* switching scheme. In other words, a new voltage combinations between V_{A0}, V_{B0}, and V_{C0} are appeared. While, it are not appeared in the pole voltages for low frequency switching scheme. Furthermore, both of Fig. 6 and Fig. 7 show the simulation and experimental waveforms for three-phase line voltages (V_{AB}, V_{BC}, and V_{CA}) and load-phase voltages (V_{AN}, V_{BN}, and V_{CN}) respectively. It is obviously from the above Figs, that the experimental results for the proposed *MLI* when it is modulated with low frequency modulation and

SPWM scheme, are well matched with the obtained simulation results under the same conditions.

Furthermore, Fig. 8 shows the frequency spectrum for line voltage V_{AB} before filtering. It is noticeably that, the harmonic contents are centered on switching frequency multiples (f_S= 3 kHz). And the *THD* value is about *39.3%*. It is decreasing due to the increasing of modulation signal (M_I) amplitude as illustrated in Fig. 9.

V. COMPARISON OF THE PROPOSED TOPOLOGY WITH OTHER MLIs

In order to shows the impact of the proposed *MLI* topology on reducing the utilized components count, it is compared to the presented topologies in [6]-[16].

Fig. 4. Experimental setup of the proposed *MLI*.

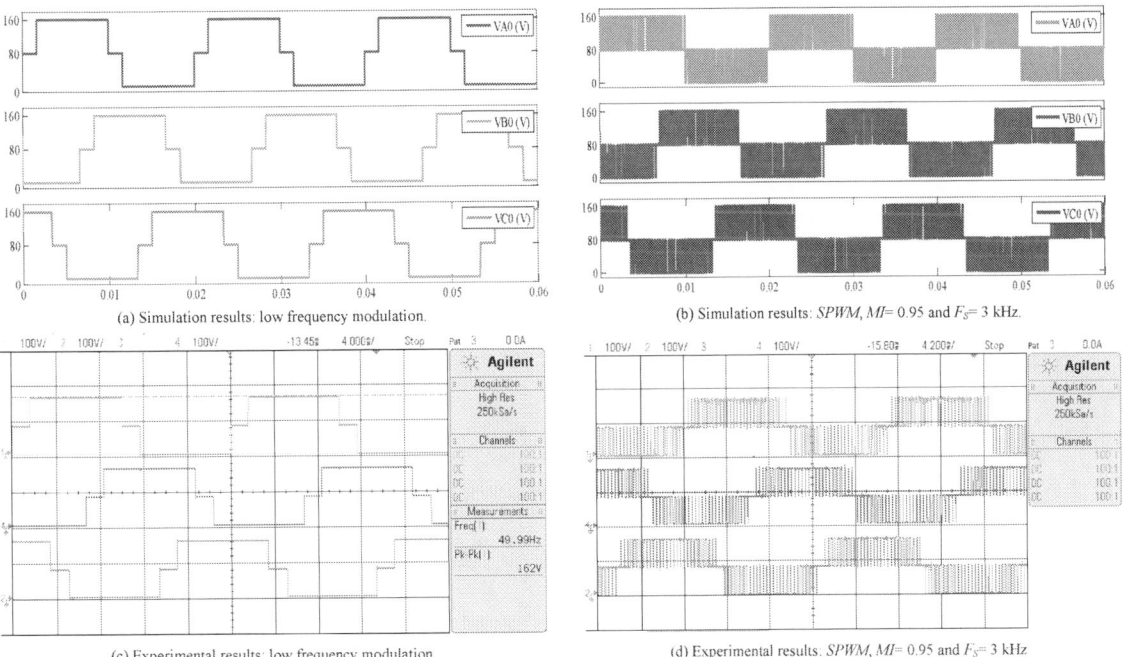

(a) Simulation results: low frequency modulation.

(b) Simulation results: *SPWM, MI*= 0.95 and F_S= 3 kHz.

(c) Experimental results: low frequency modulation.

(d) Experimental results. *SPWM, MI*= 0.95 and F_S= 3 kHz

Fig. 5. Three-phase Pole voltages V_{A0}, V_{B0}, and V_{C0}

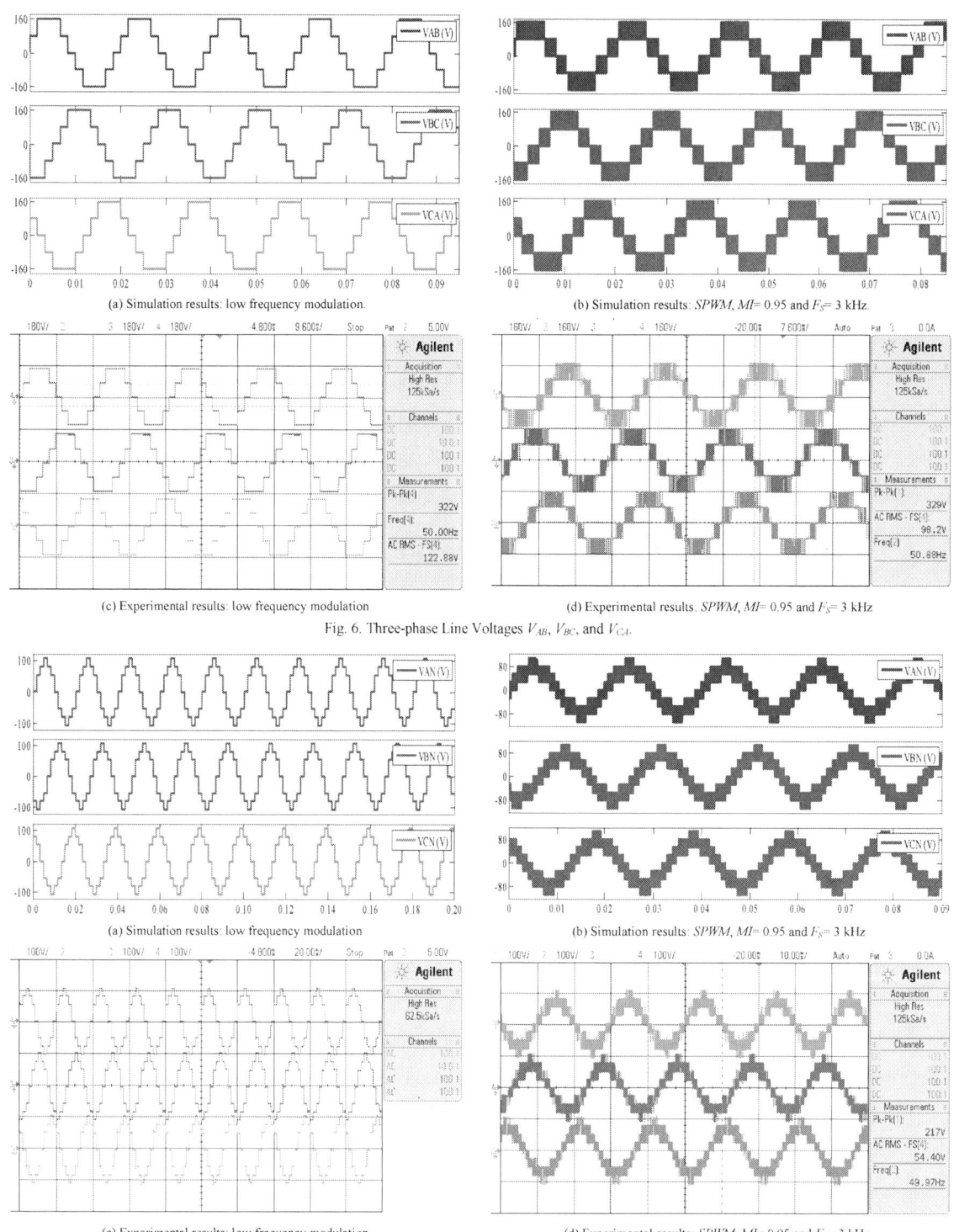

(a) Simulation results: low frequency modulation.

(b) Simulation results: *SPWM*, *MI*= 0.95 and F_S= 3 kHz.

(c) Experimental results: low frequency modulation

(d) Experimental results: *SPWM*, *MI*= 0.95 and F_S= 3 kHz

Fig. 6. Three-phase Line Voltages V_{AB}, V_{BC}, and V_{CA}.

(a) Simulation results: low frequency modulation

(b) Simulation results: *SPWM*, *MI*= 0.95 and F_S= 3 kHz

(c) Experimental results: low frequency modulation.

(d) Experimental results: *SPWM*, *MI*= 0.95 and F_S= 3 kHz.

Fig. 7. The Load-phase Voltages V_{AN}, V_{BN}, and V_{CN}.

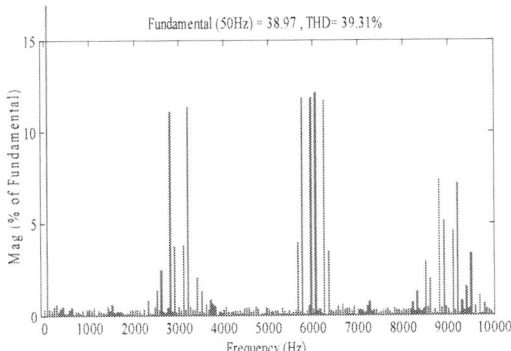

Fig. 8. Fast Fourier transform (*FFT*) analysis of inverter line voltage V_{AB}

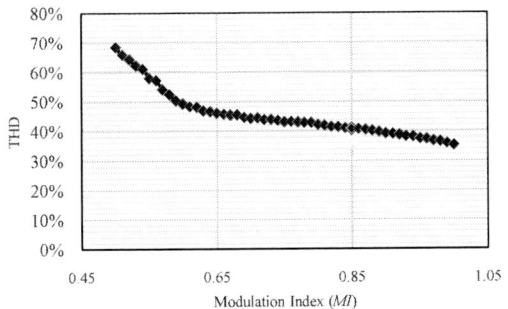

Fig. 9. *THD* of the line voltage V_{AB} against modulation index (*MI*).

TABLE III. COMPARISON BETWEEN PROPOSED TOPOLOGY AND OTHER MLIs TOPOLOGIES

Presented In	Voltage -levels	Dc Sources	Switches	Diodes	Capacitors	Transformers	$F_{C/L}$
[6]	3	3	18	0	0	0	7.0
[8]	3	6	12	0	0	0	6.0
[9]	3	1	9	12	2	0	8.0
[10]	3	1	24	0	6	0	10.3
[11]	3	6	12	0	0	0	6.0
The half-bridge based cell [12]	3	1	12	0	6	0	6.3
The clamp-double[12]	3	1	15	6	6	0	9.3
The three-level FC [12]	3	1	12	0	6	0	6.3
The three-level NPC [12]	3	1	12	6	6	0	8.3
Alternative Active 3-L NPC [13]	3	1	18	0	6	0	8.3
T-connected NPC [13]	3	1	12	0	6	0	6.3
[14]	3	6	9	3	0	0	6.0
[15]	3	1	28	0	0	3	10.6
[16]	3	4	12	0	0	0	5.3
[17]	3	4	12	0	0	0	5.3
Proposed topology	3	2	12	0	0	0	4.6

The comparison strategy is based on a comparative tool named 'components per level factor ($F_{C/L}$)', this factor was presented in [16] and used to distinguish between the different *MLIs* topologies, by compute the total components count that are necessary for each topology to produce same output voltage levels. The comparison is summarized in tabulated

format and is shown in Table III. Also, Table III shows the dominant features of the proposed topology against the others topologies, as it utilize only twelve power switches and two Dc-voltage sources. Also, both of power diodes and electrolytic capacitors are not existed. Which make the control algorithms simpler, due to removing the additional procedures for handling the capacitors voltage balancing issues.

VI. CONCLUSIONS

A novel three-phase *MLI* topology is proposed. It is modulated by using two switching schemes: low frequency modulation and *SPWM*. The proposed topology power and control stages are simulated and studied via MATLAB/Simulink software. Also, a prototype is constructed and tested to realize the proposed topology and validate its performance. The suggested inverter compared with the other *MLIs* topologies have Low-components count (equal to *14*). Both of electrolytic capacitors and power diodes are not used so their limitation (i.e. increasing inverter size, reducing life time of the system, adding more complexity in the control procedures) are avoided.

ACKNOWLEDGMENT

This work is partially funded by the Academy of Scientific Research & technology under Egypt-Tunis Collaboration project no. 39-13-A2 entitled by "Smart PV Micro-Grid System with Advanced Energy Management Control".

REFERENCES

[1] Renewables 2015 Global Status Report, avaliable online: http://www.ren21.net/wp-content/uploads/2015/07/REN12-GSR2015_Onlinebook_low1.pdf , 2015.

[2] S. Kouro, J. I. Leon, D. Vinnikov, and L. G Franquelo, "Grid-Connected Photovoltaic Systems: An Overview of Recent Research and Emerging PV Converter Technology, " *IEEE Mag. Ind. Electron.*,vol. 9, no. 1, pp.47-61, Mar. 2015.

[3] S. Park, F. Kang, M. Lee, and C. Kim, "A new single-phase five level PWM inverter employing a deadbeat control scheme," *IEEE Trans.Power Electron.*, vol. 18, no. 18, pp. 831–843, May. 2003.

[4] G. Su, "Multilevel DC-Link Inverter", *IEEE Trans.Ind. Applications*, vol. 41, no. 3, may/june 2005.

[5] M. Calais, L. Borle, and V. Agelidis, "Analysis of multicarrier PWM methods for a single-phase five-level inverter," in *Proc. Power Electronics Specialists Conference*, vol. 3, pp. 1173–1178, Jun. 2001.

[6] S. N. Rao, D. V. A. Kumar, and C. S. Babu, "New multilevel inverter topology with reduced number of switches using advanced modulation strategies," *In Proc. International Conference on Power, Energy and Control* , pp. 693,699, Feb. 2013.

[7] N. A. Rahim, M. F. M. Elias, and W. P. Hew, "Transistor-Clamped H-Bridge Based Cascaded Multilevel Inverter With New Method of Capacitor Voltage Balancing," *IEEE Trans. Ind. Electron.*, vol. 60, no. 8, pp. 2943-2956, Aug. 2013

[8] H. Belkamel, S. Mekhilef, A. Masaoud, and M. A. Naeim, "Novel three-phase asymmetrical cascaded multilevel voltage source inverter," *IET Power Electron.*, vol. 6, no. 8, pp. 1696-1706, Sept. 2013.

[9] E. A. Mahrous, N. A. Rahim, and W. P. Hew, "Three-phase three-level voltage source inverter with low switching frequency based on the two-level inverter topology," *IET Electric Power Appl.*, vol. 1, no. 4, pp. 637-641, Jul. 2007.

[10] K. Ilves, F. Taffner, S. Norrga, A. Antonopoulos, L. Harnefors, and H. P. Nee, "A Submodule Implementation for Parallel Connection of Capacitors in Modular Multilevel Converters," *IEEE Trans. Power Electron.*, vol. 30, no. 7, pp. 3518-3527, Jul. 2015.

978-1-4673-9551-9/16 $31.00 © 2016 IEEE

[11] A. Salem, E. M. Ahmed, M. Orabi, and A. B. Abdelghani, "Novel three-phase multilevel voltage source inverter with reduced no. of switches," *In Proc. International Renewable Energy Congress*, pp. 1-5, Mar. 2014.

[12] S. Debnath, Q. Jiangchao, B. Bahrani, M. Saeedifard, and P. Barbosa, "Operation, Control, and Applications of the Modular Multilevel Converter: A Review," *IEEE Trans. Power Electron.*, vol. 30, no. 1, pp. 37-53, Jan. 2015.

[13] A. Nami, J. Liang, F. Dijkhuizen, and G. D. Demetriades, "Modular Multilevel Converters for HVDC Applications: Review on Converter Cells and Functionalities," *IEEE Trans. Power Electron.* vol. 30, no. 1, pp. 18-36, Jan. 2015.

[14] A. Salem, E. M. Ahmed, M. Ahmed, M. Orabi, and A. B. Abdelghani, "Reduced Switches Based Three-Phase Multi-Level Inverter for Grid Integration," *In Proc. International Renewable Energy Congress*, pp. 1-6, Mar. 2015.

[15] S. Essakiappan, H. S. Krishnamoorthy, P. Enjeti, R. S. Balog, and S. Ahmed, "Multilevel Medium-Frequency Link Inverter for Utility Scale Photovoltaic Integration," *IEEE Trans. Power Electron.*, vol. 30, no. 7, pp. 3674-3684, Jul. 2015.

[16] M. Hasan, S. Mekhilef, and M. Ahmed, "Three-phase hybrid multilevel inverter with less power electronic components using space vector modulation," *IET Power Electron.*, vol. 7, no. 5, pp. 1256-1265, May. 2014.

[17] A. Salem, E. M. Ahmed, M. Orabi, and M. Ahmed, "New Three-Phase Symmetrical Multilevel Voltage Source Inverter,"*IEEE Journal on Emerging and Selected Topics in Circuits and Systems*, vol. 5, no. 3, pp. 430-442, Sept. 2015.

A 2 kW, Single-Phase, 7-Level, GaN Inverter with an Active Energy Buffer Achieving 216 W/in³ Power Density and 97.6% Peak Efficiency

Yutian Lei, Christopher Barth, Shibin Qin, Wen-chuen Liu, Intae Moon, Andrew Stillwell,
Derek Chou, Thomas Foulkes, Zichao Ye, Zitao Liao and Robert C.N. Pilawa-Podgurski
University of Illinois at Urbana-Champaign, Urbana, Illinois, USA
E-mail: pilawa@illinois.edu

Abstract—High efficiency and compact single phase inverters are desirable in many applications such as solar energy harvesting and household appliances. This paper presents a 2 kW, 60 Hz, 450 V_{DC} to 240 V_{RMS} power inverter, designed and tested subject to the specifications of the Google/IEEE Little Box Challenge. The inverter features a 7-level flying capacitor multilevel converter, with low-voltage GaN switches operating at 120 kHz, the highest switching frequency to date at this power level. The inverter also includes an active buffer for twice-line-frequency power pulsation decoupling, which reduces the required capacitance by a factor of eight compared to conventional passive decoupling capacitor, while maintaining an efficiency above 99%. The inverter prototype is a self-contained box that achieves a high power density of 216 W/in³ and a peak overall efficiency of 97.6% while meeting the constraints including input current ripple, load transient, thermal and EMC specifications.

I. INTRODUCTION

Recent efforts in industry and academia have emphasized the reduction in the physical dimensions of inverters and rectifiers for use in residential and commercial grid-interfaced applications, such as solar energy harvesting. A smaller and lighter inverter can reduce the size of the overall system, and the associated cost of installation and maintenance. In addition to high power density, it is also preferable to maintain high efficiency.

A common inverter topology is an H-bridge operating with pulse width modulation (PWM). Compared to non-PWM inverters, which usually have a large harmonic content in the output waveform, the PWM H-bridge pushes the undesired frequency content up near the switching frequency. It is common to leverage a high switching frequency such that the output filter size can be significantly reduced. Yet, the high frequency switching actions result in high power loss, which limits the maximum switching frequency in practice. Flying-capacitor multilevel (FCML) converters (also known as multilevel buck converters in dc-dc applications) allow the use of low voltage switches and can be used to further increase the switching frequency and reduce the inductor size [1], [2]. Although FCML converters have received attention in medium (1 kV - 70 kV) and high (>70 kV) voltage applications, they have found limited success for the low voltage range around a few hundred volts. Existing efforts either has a small number of levels or switch at a low frequency (less than

10 kHz). This paper is the first demonstration of a 7-level FCML converter implemented using GaN switches operating at a high frequency of 120 kHz. It will be shown that in the targeted application space (several hundreds of volts and several kilowatts), FCML converters can exhibit significantly higher performance in terms of power density and efficiency, compared to H-bridge converters.

Another challenge in the single-phase inverter involves the twice-line-frequency power pulsation. On the ac side of the inverter, the product of the sinusoidal output voltage and current results in a twice-line-frequency (120 Hz) power ripple. On the dc side, a constant power draw is desirable. This instantaneous power mismatch requires energy to be stored by the converter. This energy storage requirement cannot be reduced by increasing the switching frequency, as it depends only on the power output and line frequency. This work incorporates an active energy buffer architecture that utilizes a partial processing technique to drastically reduce the required buffering capacitor values compared to conventional passive decoupling capacitors, while maintaining a very high efficiency [3], [4].

In order to establish a common baseline to compare this inverter with other and future work, the proposed inverter is designed and built according to the specifications outlined in the Google/IEEE Little Box Challenge [5]. The inverter prototype is a stand-alone box, including auxiliary converters (to power the control and gate drivers), EMC filters (to meet FCC Class B requirements) and cooling fans. With a rectangular dimensions of 4.02 in × 2.42 in × 0.95 in and a total volume of 9.26 in³, the experimentally verified power density is 216 W/in³. A peak efficiency of 97.6% is achieved, including the power losses from control and cooling fan.

The paper is organized as follows. Section II provides the theory of operation and hardware implementation of the 7-level flying capacitor multilevel converter. Section III presents the operation and implementation of the energy buffer. Section IV provides the experimental results. Finally, concluding thoughts are given in Section V.

II. FLYING-CAPACITOR MULTILEVEL CONVERTER

The overall architecture of the single-phase inverter is shown in Fig. 1. Apart from the main multilevel converter

978-1-4673-9551-9/16 $31.00 © 2016 IEEE

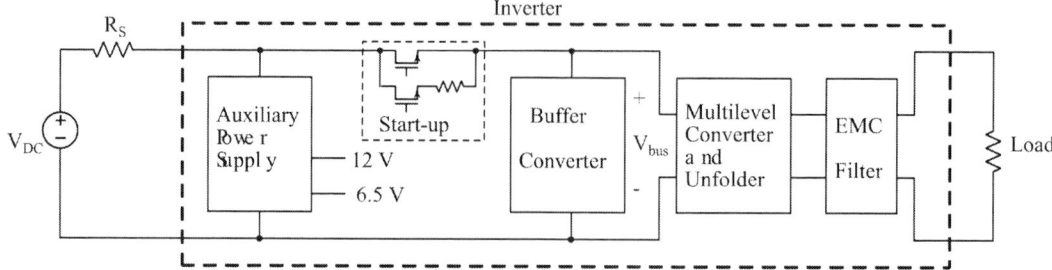

Fig. 1. Full system overview.

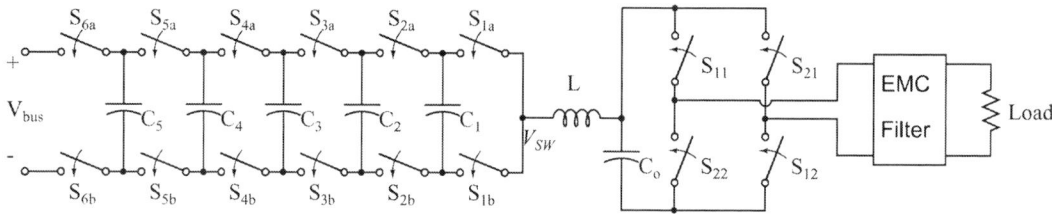

Fig. 2. Schematic drawing of the 7-level flying-capacitor multilevel inverter.

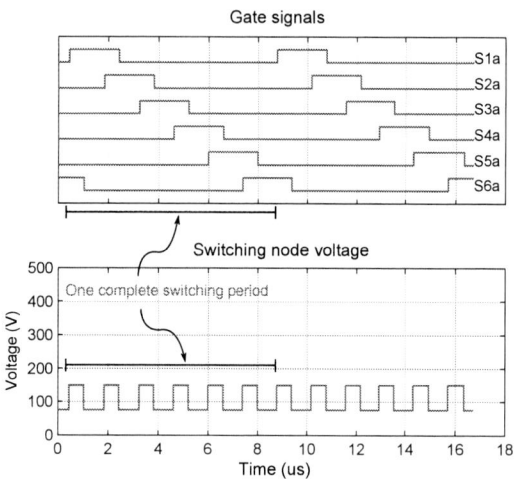

Fig. 3. Gate signals (with high represents on) and simulated switching node voltage (V_{sw}) at the switching frequency.

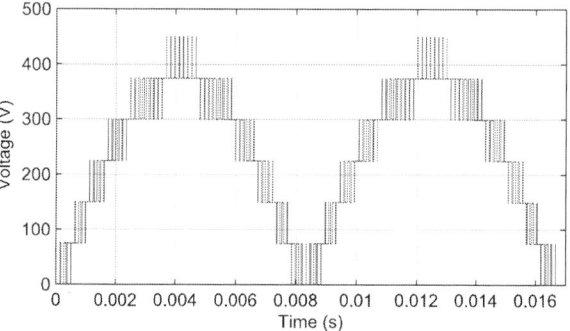

Fig. 4. Simulated switching node voltage (V_{sw}) in $\frac{1}{60}$ of a second. Note that the switching frequency is reduced for better illustration.

and the buffer converter, there are also an auxiliary converter, which generates 12 V and 6.5 V from 450 V for control and fan power, a start-up circuitry to limit the inrush current, and filters for electromagnetic emission compliance. This section presents the details of the multilevel converter block while the buffer converter block is presented in Section III.

A. Theory of operation

The schematic drawing of the dc-ac conversion stage is shown in Fig. 2. It consists of a 7-level FCML converter and an H-bridge unfolder. The FCML converter produces a rectified

sinusoidal output between 0 V and V_{bus}, while the unfolder flips the polarity of the output every 1/120 second to produce a true AC output.

For an FCML converter with N levels, there are $(N-1)$ pairs of switches, each with an ideal voltage rating of $\frac{V_{\text{bus}}}{N-1}$. There are also $(N-2)$ flying capacitors with ratings of $\frac{V_{\text{bus}}}{N-1}$, $\frac{2V_{\text{bus}}}{N-1}$, ..., $\frac{(N-2)V_{\text{bus}}}{N-1}$, respectively. In this work, the FCML converter is operated with phase-shifted pulse width modulation (PSPWM) [1]. In this control scheme, all of the "a" switches have a duty ratio of D, and all of the "b" switches have a duty ratio of $1-D$. The average output voltage is given by DV_{bus}. Each adjacent switch turns on and off with a phase shift of $\frac{360}{N-1}$ degrees. Figure 3 shows the gate signals for the "a" switches with a duty ratio of 25%. It can be seen that each switch has a phase lag of 60 degrees from the previous one. In one complete switching period of each switch, the

978-1-4673-9551-9/16 $31.00 © 2016 IEEE   1513

switching node voltage has six pulses. In addition, the pulses are in-between two intermediate voltage levels set by the flying capacitors. By modulating the duty ratio, the switching node voltage can be made to follow a rectified sinusoidal waveform, as shown in Fig. 4.

A major advantage of the FCML converter is the reduction in the required output filter inductor size. Comparing to a conventional two-level or three-level converter, the PWM operation is done between two voltage levels that are only $\frac{V_{bus}}{N-1}$ apart, where N is the number of levels. In addition, the effective pulse frequency seen by the filter inductor is $(N-1)f_{sw}$, where f_{sw} is the switching frequency of each switch. At the same time, the current ripple in the filter inductor is directly proportional to the pulse voltage levels across the inductor and inversely proportional to the pulse frequency. Thus, on an analytical level, given a certain inductor current ripple, an FCML converter can have an output inductor and capacitor that are $(N-1)^2$ times smaller than a two-level converter [6].

A common practical implementation issue associated multi-level converters is the capacitor voltage imbalance, where the actual capacitor voltages are deviated from the ideal levels. Imbalanced capacitor voltages increase the drain-source voltages across the switches, which can lead to switch failure if the rated blocking voltage is exceeded. Efforts are made towards a precise generation of the PWM gate signals using the micro-controller, and a symmetrical board layout. As such, this work can rely on the natural balance of the PSPWM scheme [9]. Active capacitor voltage balancing is possible by monitoring the capacitor voltages and change the switching patterns to selectively charge or discharge one or more flying capacitors [7], [8]. Yet, these methods can be difficult to apply to inverters with a high number of levels, high switching frequency and small capacitor values, due to the high bandwidth sensing and control required.

B. Hardware implementation

The top, side and bottom view of the FCML inverter are shown in Fig. 5. To the left is the start-up circuitry in order to charge up the flying capacitors and bus capacitors to their steady-state values. The flying capacitors and inductors are on the bottom side of the PCB, and are aligned in the center. To the right on top of the PCB are the common-mode and differential-mode EMI filters. A small common-mode choke that fits on the right edge on the back is not shown. The GaN switches are placed in the center, distributed on three red daughter boards. The use of modular interchangeable daughter boards facilitate the assembling and debugging of the converter. The total thickness of the inverter is 10.3 mm and the tallest component is the inductor at 7.5 mm. This low profile is enabled by the drastically reduced inductor size with the multilevel structure as discussed previously. A complete component listing of the inverter is given in Table I.

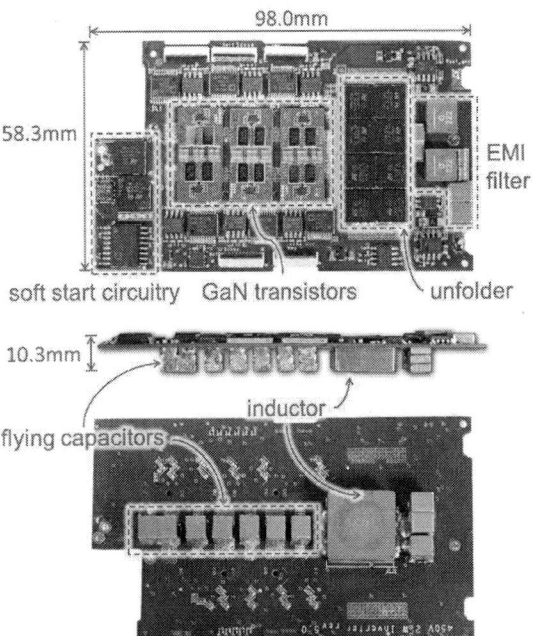

Fig. 5. Annotated photographs of the FCML inverter board.

III. 120 Hz INPUT CURRENT RIPPLE COMPENSATION

The most common method to achieve twice-line-frequency power decoupling is to connect a passive capacitor bank across the dc bus. To meet the stringent ripple requirement on bus voltage and dc input current, a large volume of capacitors are required. Such a capacitor bank is typically formed by electrolytic capacitors due to their high energy density, but their high power loss, poor ripple current capability and short lifetime becomes a significant constraint [10]. Ceramic or film capacitors are preferable from an efficiency and reliability per-spective, but requires even large volume, as their capacitance density is generally significantly lower than that of electrolytic capacitors.

Many active decoupling approaches have been proposed in the literature to reduce the required capacitance and even allow for the use of ceramic or film capacitors [11]–[15]. One common approach is a full ripple port converter [11], [12], where energy storage capacitors are interfaced to the dc bus through a power converter. The capacitors operate with a large voltage swing to buffer more energy, while the dc bus voltage is regulated by the ripple port converter. To achieve this functionality, the converter needs to process a large portion of the full power on average (i.e., $\frac{2}{\pi}P_{out}$, where P_{out} is the average output power of the inverter) and to withstand high voltage stress (i.e., the full dc bus voltage at least). As a result, the ripple port converter can have a significant negative impact on the overall inverter efficiency. Moreover, the added volume (especially the magnetic components) can be very large and

978-1-4673-9551-9/16 $31.00 © 2016 IEEE

TABLE I
COMPONENT LISTING OF THE INVERTER BOARD

Component	Part number	Parameters
GaN switches (S_{1a}, S_{1b} to S_{6a}, S_{6b})	EPC 2033	150 V, 7 mΩ
GaN gate driver	Texas Instruments LM5113	100 V half-bridge
Unfolder MOSFETs ($S_{11}, S_{12}, S_{21}, S_{22}$)	STMicroelectronics STL57N65M5 $\times 2$	650 V, 69 mΩ
Unfolder gate driver	Fairchild FAN73932MX	600 V half-bridge
Flying capacitors ($C_1 - C_5$)	TDK C5750X6S2W225K250KA $\times 3$	2.2 μF
Inductor	Vishay IHLP6767GZER220M11	22 μH
Digital isolators	Silicon Labs Si8423BB-D-IS	
Power isolators	Analog Devices ADUM5210	

even completely offsets the volume reduction from smaller energy storage capacitors.

In this work we use a series-stacked buffer architecture that overcomes the aforementioned problems. This architecture achieves very high efficiency and power density with a low complexity circuit while tightly regulating the dc bus voltage. Details on this buffer architecture are presented in [3], [4], and this section only provides an overview.

A. Theory of operation

The schematic of the active energy buffer is shown in Fig. 6. In this architecture, a full-bridge converter is connected in series with a main energy storage capacitor, C_1. Unlike conventional dc bus capacitors, C_1 is operated with a relatively large voltage ripple, for example at 25% of rated voltage. Since V_{C1} has relatively large ripple, the energy utilization is significantly increased compared to the dc bus capacitors, and much smaller capacitance is needed (eight times reduction in this design). To maintain a constant dc bus voltage, the voltage of C_3 is controlled such that its ac component is of the same magnitude but opposite sign to that of V_{C1}, as shown in Fig. 7. A support capacitor C_2 is used to maintain a certain voltage (higher than V_{C3}) to ensure the correct operation of the full-bridge converter, while C_3 is only a small filter capacitor to absorb switching frequency ripples.

A key advantage of this architecture is that the full-bridge converter only processes a fraction of the total output power, as C_1 provides the bulk of the power buffering. While the entire buffer architecture buffers 2 kW in this application, the full-bridge converter only processes 100 W on average, as shown in Fig. 7, so that the efficiency penalty on the overall system is small. V_{C3} has only the ripple magnitude of V_{C1}, and thus the full-bridge converter experiences low voltage stress. A small-size inductor and low-voltage, high-speed transistors can be used for a small buffer converter size. A current hysteresis control scheme is implemented to match the buffer current with the inverter current, such that the voltage waveform in Fig. 7 follows naturally.

A considerable challenge in this architecture is that the losses in the full-bridge converter will drain capacitor C_2 over time. A loss compensation scheme which exploits the phase difference between the buffer current and bus voltage is devised to supply additional energy to capacitor C_2 to

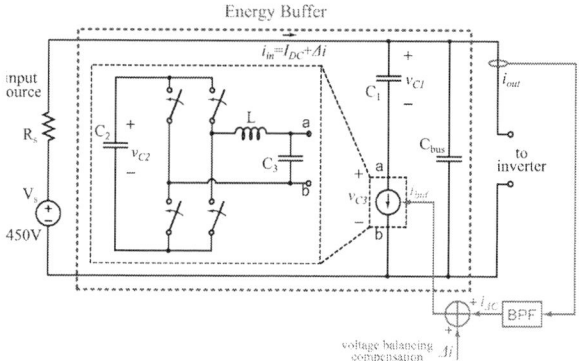

Fig. 6. Schematic drawing of the active energy buffer with high level control flow diagram.

compensate for this loss. In our proposed method, the small bus voltage ripple is utilized to deliver a net positive energy into the converter, so that the voltage of C_2 can be maintained. A detailed explanation of the control scheme is presented in [4].

B. Hardware implementations

One practical issue with sizing ceramic capacitors for energy buffering is that their capacitance decreases in a non-linear fashion as the dc bias increases. It is common for the capacitance to reduce to one quarter of the nominal value when biased at the rated dc voltage. Thus, care must be taken to not over or underestimate the required capacitor volume. The experiment based methodology presented in [16] is followed to help determine precisely the energy storage capability of the ceramic capacitors.

Annotated photographs of the buffer converter board is shown are Fig. 8. The full bridge converter and the sensing circuitry are placed on top, together with the micro-controller, which controls both the FCML inverter and the energy buffer. The capacitors C_1 and C_2 on the bottom side are placed such that the inverter flying capacitors and inductors can fit inside. This allows the inverter board to stack on top of the buffer board with minimal unused space in-between, as shown in Fig. 9. The auxiliary converter to power both the buffer converter and the inverter from the 450 V are placed

978-1-4673-9551-9/16 $31.00 © 2016 IEEE

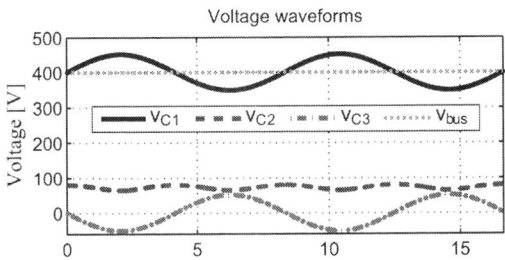

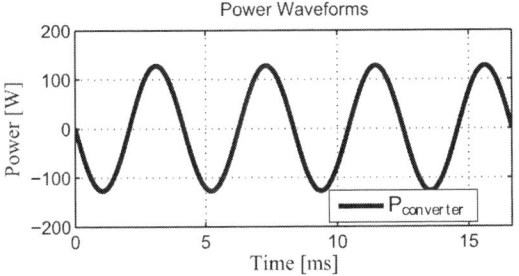

Fig. 7. Waveforms illustrating the operation of the active energy buffer architecture: voltages (top) and converter power (bottom).

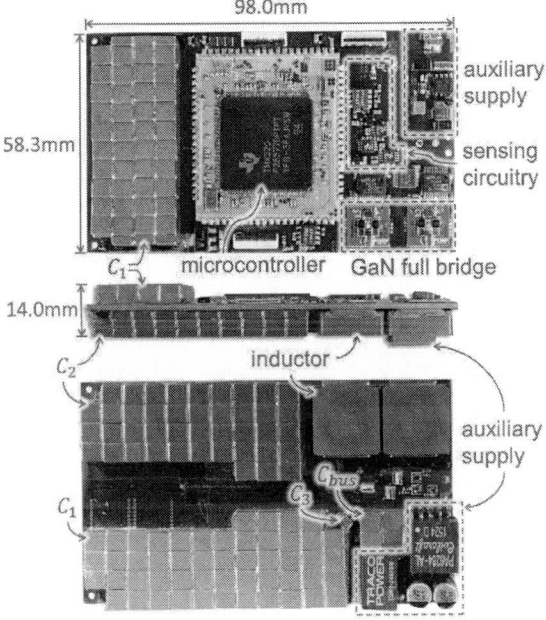

Fig. 8. Annotated photographs of the active energy buffer board.

in the bottom right corner. A component listing of the buffer converter is shown in Table II.

IV. EXPERIMENTAL RESULTS

The box enclosure is shown in Fig. 10. With a dimension of 102.1 mm× 61.4 mm×24.2 mm (4.02 in ×2.42 in ×

Fig. 9. Photograph of the FCML inverter board and energy buffer fit together.

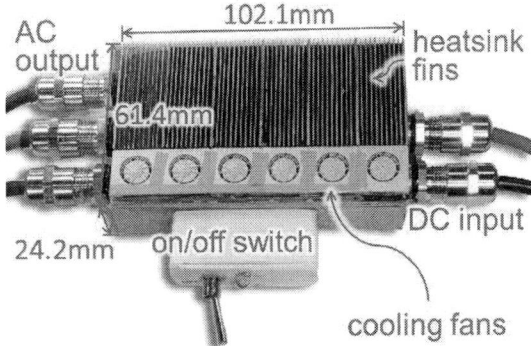

Fig. 10. Photograph of the inverter prototype inside the heat-sink and enclosure.

0.95 in) and a total volume of 15.2 cm^3 (9.24 in^3), the inverter prototype achieves a power density of 132 W/cm^3 (216 W/in^3). Table III displays the high-level performance metrics of prototype. For all metrics listed, the converter meets or exceeds the specifications required by the Little Box Challenge [5].

The individual efficiencies of the FCML inverter and the active energy buffer are measured with Yokogawa WT310 digital power meters, and are plotted in Fig. 11. It can be seen that the energy buffer alone is able to achieve a high efficiency of 99%, thanks to the partial power processing architecture. For the inverter, the peak conversion efficiency is 98.6% at about half load and the efficiency at full load is above 98%. The efficiency of the inverter at light load drops to about 92%, without any special light load control. Since the output inductance is designed with a small current ripple, the light load efficiency of the converter can be improved by reducing the switching frequency, if so desired. Also shown in Fig. 11 is the efficiency after including all power losses from control and cooling fans. A 450 V to 12 V fly-back converter and a 12 V to 6.5 V converter provide power for both the cooling fans, which consume about 5 W, and the control and gate driving circuits, which dissipate about 4 W. They represent a constant power loss independent of the load conditions. The overall light load efficiency can be further improved by turning-off the fan when the output power is below a preset threshold.

TABLE II
COMPONENT LISTING FOR THE ACTIVE ENERGY BUFFER

Component	Part number	Parameters
GaN FETs	EPC 2016C	100 V, 16 mΩ
Capacitors (C_1)	TDK C5750X6S2W225K250KA × 239	450 V, 2.2 μF
Capacitors (C_2)	TDK CKG57NX7R2A106M500JH × 126	100V, 15 μF
Inductor	Vishay IHLP6767GZER470M11 × 2	47 μH
Power isolators	Analog Devices ADUM5210	
Microcontroller	Texas Instruments TMX320F28377D	

TABLE III
KEY PERFORMANCE SPECIFICATIONS

Parameters	Values
Rated power	2 kVA
Volume	9.52 in^3
Power density	216 W/in^3
Rated input voltage	450 V
Rated output voltage	240 V_{RMS}
Efficiency (CEC Method)	97.0%
Efficiency at rated power	97.4%
Load power factor	0.7 – 1.0
Voltage THD at rated power	0.3%
Input current ripple at rated power	15%
Case temperature at rated power	57 °C

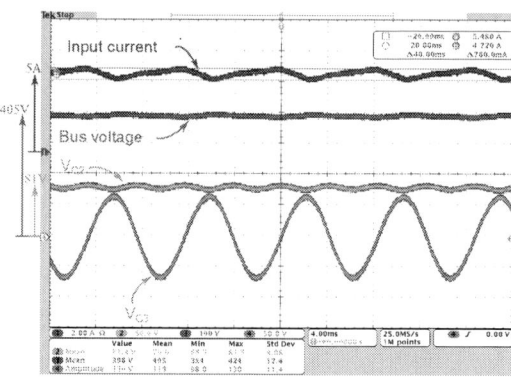

Fig. 12. Waveforms showing active energy buffer operation.

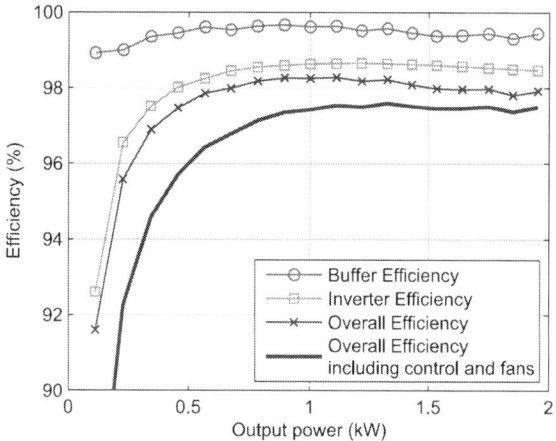

Fig. 11. Converter efficiency at different power levels.

The full load operation of the energy buffer is shown in Fig 12. At full load, the input current ripple is approximately 760 mA, which is 15% of the average input current. The voltage of C_2 and C_3 as well as the bus voltage are also shown. To generate a resultant bus voltage with a very small ripple, C_3 has a large voltage ripple in order to counter the voltage ripple of C_1 (not shown). The large voltage swing on C_1 is the key to the volume reduction of the buffer capacitors.

The operation of the 7-level inverter can be seen in Fig. 13, which shows the switching node voltage as well as the output voltage and current. The output voltage is 240 V RMS and the output current is 8.3 A RMS at full load. The switching node has a unipolar 120 Hz envelope as well as a high frequency PWM between two smaller voltage levels. The average flying capacitor voltages are monitored with a National Instruments data acquisition system (PXIe-1073), and are plotted in Fig. 14. With a PSPWM control scheme and no active capacitor voltage balancing, the capacitor voltages are evenly distributed across the input voltage range. This is the first demonstration that the FCML converter is able to self-balance capacitor voltages at this high number of levels and switching frequencies, with aggressive flying capacitor sizing. The maximum voltages across the GaN switches are measured and tabulated in Table IV. The deviations of capacitor voltages from ideal values are small and the largest increase is approximately 12%.

TABLE IV
DRAIN-SOURCE VOLTAGES ACROSS THE SWITCHES

Switch	S_6	S_5	S_4	S_3	S_2	S_1
Measured at 20% load (V)	78	84	63	81	66	79
Measured at 100% load (V)	70	76	51	71	51	76

The response of the inverter during a load transient are presented in Fig. 15 and Fig. 16. Figure 15 shows the input current, bus voltage, V_{C2} and V_{C3} during a load step-down

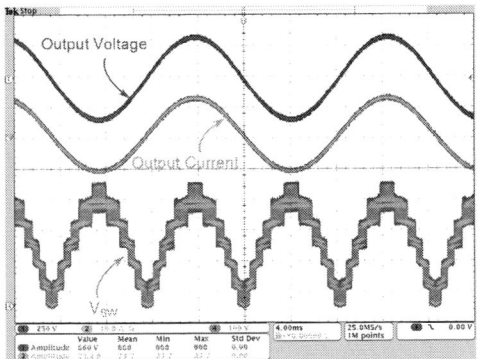

Fig. 13. Waveforms showing the output voltage, output current and the switching node voltage (V_{SW}) of the 7-level inverter at full load.

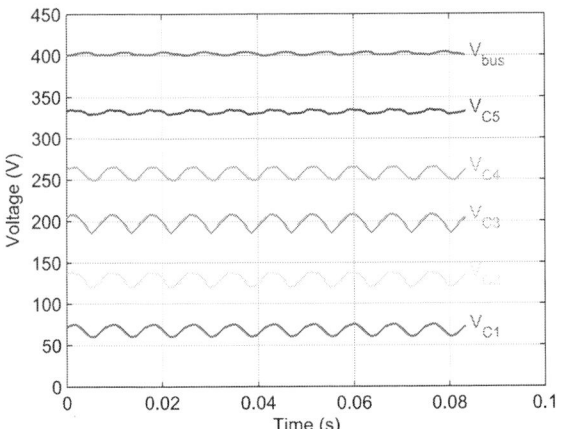

Fig. 14. Capacitor voltages of the 7-level inverter during full load operation.

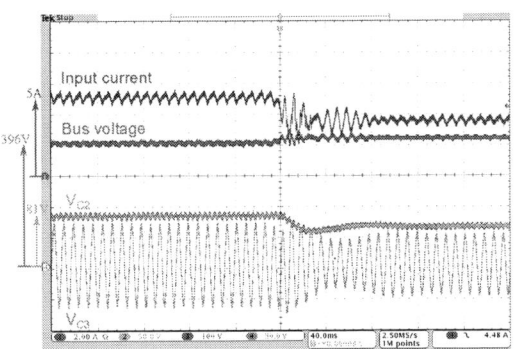

Fig. 15. Energy buffer operation during a load step-down from 100% to 75%.

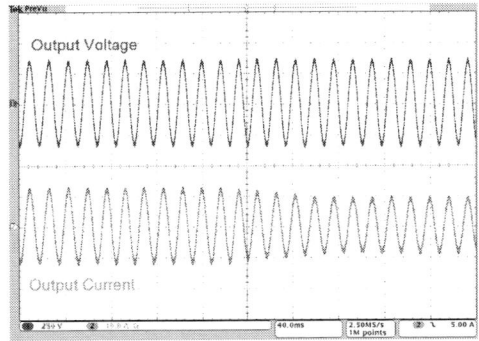

Fig. 16. Inverter operation during a load step-down from 100% to 75%.

from 100% to 75%. The input current ripple comes back within the 20% specification in 80 ms after the load change. Figure 16 shows the output voltage and output current of the inverter during the same load step. The output voltage is not impacted by the change in load current.

At full power, the inverter incurs approximately 40 W of power loss. To dissipate this amount of heat, the top of the enclosure is machined with fins that are 2 mm tall. A total of six radial fans are placed on the edge of the top, blowing air across the fins. A thermal image is taken with a thermal camera after the converter reaches thermal steady-state at full power operation. It can be seen that the maximum temperature is at 57 °C.

For grid-interfaced converters, the electro-magnetic interference (EMI) emissions must be filtered to meet the required limits set by regulatory bodies. To date, EMI filter design considerations for FCML converters have not been discussed thoroughly in previous literature. In this work, an EMI filter has been designed so that the conducted emissions of the converter passes the FCC Part 15 Class B requirements [5].

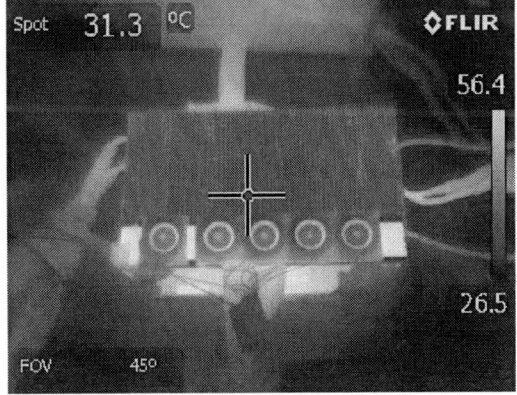

Fig. 17. Thermal image of the inverter operating at full power (2kW).

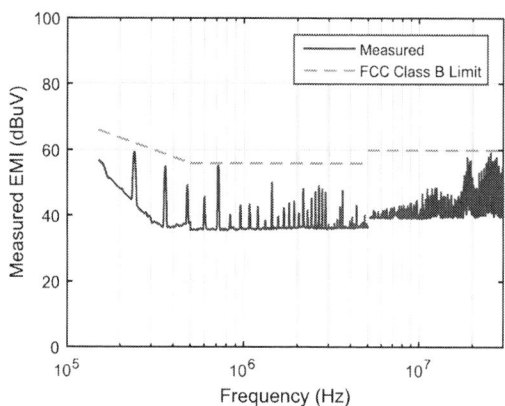

Fig. 18. Conducted EMI measurement at full power (2kW).

Compared to conventional PWM H-bridge converters, the filter requirements for multilevel converters are much smaller. This reduction in EMI filter size is a direct result of the switching nodes of the multilevel inverter experiencing voltage steps with only an amplitude of $\frac{V_{bus}}{N-1}$ and at a frequency of $(N-1)f_{sw}$. Thus, this beneficial property of multilevel inverters reduces both the volume of the main output filter inductor, and the EMI filter. The conducted EMI measurements made with a Tektronix RSA5126A real-time signal analyzer are depicted in Fig. 18. For the entire 150 kHz to 30 MHz measurement range, the inverters conducted EMI emission is lower than the Class B limit. The peak emission closest to the limit is at 720 kHz as expected, which is six times the switching frequency of 120 kHz. Parasitics in the filtering elements, such as the equivalent series inductance (ESL) of filter capacitors and inter-winding capacitance of filter inductors, cause the emissions to rise significantly after 10 MHz.

V. CONCLUSIONS

This paper has presented a 2 kW, 450 V_{DC} to 240 V_{RMS} single-phase inverter. The inverter consists of a 7-level flying-capacitor multilevel converter, with GaN transistors switching at 120 kHz, which is the highest switching frequency achieved to date. The multilevel inverter is complemented by a series-stacked buffer converter for twice-line-frequency ripple compensation. The active energy buffer achieves ultra high efficiency of 99% while reducing the required capacitor volume by a factor of eight. The combined inverter prototype successfully demonstrated a 216 W/in^3 power density and a peak overall efficiency of 97.6%, while meeting all the specifications of the Little Box Challenge.

ACKNOWLEDGEMENT

We would like to thank Texas Instruments, NASA, Army Research Laboratory, and TDK for their support. We would also like to acknowledge Google for the Google Little Box Academic Award that supported this work.

REFERENCES

[1] T. Meynard and H. Foch, "Multi-level conversion: high voltage choppers and voltage-source inverters," in *Power Electronics Specialists Conference, 1992. PESC '92 Record., 23rd Annual IEEE*, pp. 397–403 vol.1, Jun 1992.

[2] V. Yousefzadeh, E. Alarcon, and D. Maksimovic, "Three-level buck converter for envelope tracking applications," *IEEE Transactions on Power Electronics*, vol. 21, pp. 549–552, Mar. 2006.

[3] S. Qin, Y. Lei, C. Barth, W.-C. Liu, and R. Pilawa-Podgurski, "Architecture and control of a high energy density buffer for power pulsation decoupling in grid-interfaced applications," in *Control and Modeling for Power Electronics (COMPEL), 2015 IEEE 16th Workshop on*, pp. 1–8, July 2015.

[4] S. Qin, Y. Lei, C. Barth, W.-C. Liu, and R. C. Pilawa-Podgurski, "A high-efficiency high energy density buffer architecture for power pulsation decoupling in grid-interfaced converters," in *Energy Conversion Congress and Exposition (ECCE), 2015 IEEE*, pp. 149–157, Sept 2015.

[5] "Little box challenge detailed inverter specifications, testing procedure, and technical approach and testing application requirements for the little box challenge." [Online] https://www.littleboxchallenge.com/pdf/LBC-InverterRequirements-20150717.pdf.

[6] Y. Lei and R. Pilawa-Podgurski, "An analytical method to evaluate flying capacitor multilevel converters and hybrid switched-capacitor converters for large conversion ratios," in *Control and Modeling for Power Electronics (COMPEL), 2015 IEEE 16th Workshop on*, pp. 1–7, July 2015.

[7] A. Shukla, A. Ghosh, and A. Joshi, "Capacitor voltage balancing schemes in flying capacitor multilevel inverters," in *Power Electronics Specialists Conference, 2007. PESC 2007. IEEE*, pp. 2367–2372, June 2007.

[8] M. Khazraei, H. Sepahvand, K. Corzine, and M. Ferdowsi, "A generalized capacitor voltage balancing scheme for flying capacitor multilevel converters," in *Applied Power Electronics Conference and Exposition (APEC), 2010 Twenty-Fifth Annual IEEE*, pp. 58–62, Feb 2010.

[9] X. Yuan, H. Stemmler, and I. Barbi, "Self-balancing of the clamping-capacitor-voltages in the multilevel capacitor-clamping-inverter under sub-harmonic pwm modulation," *Power Electronics, IEEE Transactions on*, vol. 16, pp. 256–263, Mar 2001.

[10] H. Wang and F. Blaabjerg, "Reliability of capacitors for dc-link applications in power electronic converters - an overview," *Industry Applications, IEEE Transactions on*, vol. 50, pp. 3569–3578, Sept 2014.

[11] R. Wang, F. Wang, D. Boroyevich, R. Burgos, R. Lai, P. Ning, and K. Rajashekara, "A high power density single-phase pwm rectifier with active ripple energy storage," *Power Electronics, IEEE Transactions on*, vol. 26, pp. 1430–1443, May 2011.

[12] P. Krein, R. Balog, and M. Mirjafari, "Minimum energy and capacitance requirements for single-phase inverters and rectifiers using a ripple port," *Power Electronics, IEEE Transactions on*, vol. 27, pp. 4690–4698, Nov 2012.

[13] H. Hu, S. Harb, N. Kutkut, Z. Shen, and I. Batarseh, "A single-stage microinverter without using eletrolytic capacitors," *Power Electronics, IEEE Transactions on*, vol. 28, pp. 2677–2687, June 2013.

[14] M. Chen, K. Afridi, and D. Perreault, "Stacked switched capacitor energy buffer architecture," *Power Electronics, IEEE Transactions on*, vol. 28, pp. 5183–5195, Nov 2013.

[15] H. Wang, H.-H. Chung, and W. Liu, "Use of a series voltage compensator for reduction of the dc-link capacitance in a capacitor-supported system," *Power Electronics, IEEE Transactions on*, vol. 29, pp. 1163–1175, March 2014.

[16] C. B. Barth, I. Moon, Y. Lei, S. Qin, C. Robert, and Pilawa-Podgurski, "Experimental evaluation of capacitors for power buffering in single-phase power converters," in *Energy Conversion Congress and Exposition (ECCE), 2015 IEEE*, pp. 6269–6276, Sept 2015.

Indirect Matrix Converter Based Open-End Winding AC Drives With Zero Common-Mode Voltage

Saurabh Tewari[*], Ranjan K. Gupta[†], Apurva Somani[‡], and Ned Mohan[§], *Life Fellow, IEEE*

Email: tewari@umn.edu

[*]MTS Systems Corporation, Eden Prairie, MN 55344

[†]First Solar, Bridgewater, NJ 08807

[‡]Dynapower Company LLC, South Burlington, VT 05403

[§]Department of Electrical and Computer Engineering, University of Minnesota, Minneapolis, MN 55455

Abstract—Common-mode voltage (CMV) generated by semi-conductor switching causes stray currents and mechanical failure in modern drive systems. Solutions employed to attenuate or isolate the common-mode voltage (CMV) require additional components, and may still fail to eliminate the detrimental effects. Matrix converter based open-end winding drives, when modulated using synchronous vectors, do not generate CMV to begin with. Additionally, these drives do not rely upon a large DC capacitor that is used in the state-of-the-art systems; and are therefore expected to be more compact and reliable. This paper will present prototypes of two distinct indirect matrix converter based open-end winding drives that eliminate output common-mode voltage, provide high voltage transfer ratio (up to 1.5), and allow input power factor control. These indirect drives have the additional advantages of clamp circuit elimination, lower voltage stress on the devices, naturally intelligent commutation, and natural low-voltage ride-through integration over their direct matrix converter counterpart. Experimental evidence of the voltage transfer ratio and input power factor control will be provided. Compared to 2-level and 3-level inverters, significant reduction in the CMV induced shaft voltage and ground currents will be shown. An optimal third-order grid filter applicable to all matrix converter based drives will also be discussed. This filter will be used with the presented drives to validate its superior performance.

I. INTRODUCTION

PWM inverter drives allow superior and energy-efficient control of electric motors. Currently, the state-of-the-art drive systems are based on voltage source inverters (VSIs) with two-level and three-level space vector pulse width modulated (SVPWM) VSIs [1] being, by far, the dominant. Operation of a VSI-based drive is conditional upon stable DC input voltages. Torque transients and ride-through specifications in motor drives necessitate the use of large DC capacitors at the DC-link, few tens of microfarads per unit output current being a common practice in the industry. Electrolytic capacitors, while inexpensive and compact, compromise the reliability. Equivalent film capacitors are larger and costlier, but more robust [2].

Matrix Converters (MCs) [3]–[7] are AC-AC PWM converters that do not require capacitors to absorb the load torque ripple, but only to filter the switching frequency harmonics in the input current. Therefore the filter can be constructed with high-quality film capacitors at low cost and volume penalty [8]–[10]. On the other hand, back-to-back VSI based drives

outperform MCs in the voltage transfer ratio, input power factor control (possible in conventional MCs with further degradation of the voltage transfer ratio), and ride-through capability.

However, both approaches, VSI-based and MC-based, result in low- and high-frequency common-mode components in the output voltage. The bearing currents attributed to drive-generated high-frequency common-mode voltage (CMV) have been conclusively linked to failure due to pitting and fluting in the bearing races [11]–[15], and to shaft voltage buildup that may pose more immediate risks in hazardous environments [16], [17].

To counter the effects of high-frequency CMV, passive measures like shaft grounding brushes, common-mode chokes, insulated bearings and electrostatically shielded motor (ESIM) [12]; and active measures like cancelers and compensators [18], [19] can be implemented. Completely passive solutions add components and may still fall short of satisfactory performance and/or shift the detrimental effects elsewhere [18]–[20], or may prove to be expensive and difficult to implement [12], [20]. Active measures [18], [19] require some passive components and operation of the active devices in the linear region leading to additional power dissipation and thermal management requirements.

Therefore, considerable research has been devoted to mitigating common-mode voltage at the PWM converter itself. Reduction in the common-mode voltage generated by the two-level inverter has been reported [21]. Compared to the standard two-level SVPWM VSI, a lower common-mode voltage is inherent to multilevel inverters [22]. Further mitigation in multilevel inverters [23], [24] has been demonstrated. Theoretically complete elimination of the common-mode voltages for multilevel inverters has also been demonstrated [25]–[28]. However, the aforementioned strategies that provide complete elimination do so by using only a limited set of available vectors, compromising either, or both, the voltage transfer ratio and the output voltage harmonic distortion.

Research in conventional matrix converter modulation with the aim of mitigating CMV has also been reported [29]–[38]. It should be noted that these strategies only mitigate, not eliminate, the common-mode voltage. Additionally, the shortcomings of matrix converters against back-to-back VSIs remain.

978-1-4673-9551-9/16 $31.00 © 2016 IEEE

Yet another approach to common-mode voltage elimination has been open-end winding (OE Wdg.) drives wherein the stator windings are supplied from both ends. Several VSI-derived implementations have been reported in the literature [39]–[43]. However, the VSI-derived OE Wdg. drives still require a large capacitor to maintain the DC-link voltage same as the wye- or delta-connected two- or multilevel inverter drives.

Gupta *et al.* proposed a direct matrix converter (DMC) based OE Wdg. drive [44] that eliminated output CMV, and did not require large reactive components for operation. It also overcame the voltage transfer ratio and input power factor limitation of conventional MCs. Low-voltage ride-through can be implemented by addition of a grid isolator similar to [45]. More recently, DMC OE Wdg. drives have been studied for n-phase machines [46]–[48].

In this paper, the authors demonstrate two distinct *indirect* matrix converter based open-end winding drives that use synchronously rotating vectors to eliminate common-mode output voltage. Voltage transfer ratio up to 1.5 and continuous power factor control at full voltage transfer ratio is possible same as [44]. Compared to [44], following additional features are unique to the presented drives:

1) Intelligent commutation is possible without the need to sense current direction;
2) No clamp circuit is necessary for protection or to aid commutation;
3) Power semiconductor devices do not switch a voltage higher than 86.6% of the peak line-to-line voltage;
4) To implement a low-voltage ride-through (LVRT) strategy similar to [45], no additional switches are required.

In addition to explaining and demonstrating the drives' operation, this paper will also optimize a third-order grid current filter that outperforms the second order LC filter common in matrix converter applications.

II. PRINCIPLE OF OPERATION

Fig. 1 shows the block diagrams of the presented drives with the filter components shown. The proposed drives utilize three converters — a front-end converter, and two load-end converters that supply the machine terminals. All three converters are identical three-level inverters and could use either of the two popular three-level inverter structures: diode-clamped I-type as shown in Fig. 1(a), or T-type as shown in Fig. 1(b). In contrast, the DMC based OE Wdg. drive of [44] utilizes two identical direct matrix converters, but the switch count in all three drives, the two presented here and the one reported in [44], is same (36 IGBTs).

A. Front-end converter

Let v_a, v_b, v_c be the three input voltages. Let,

$$v_{\max} = \max\{v_a, v_b, v_c\}$$
$$v_{\min} = \min\{v_a, v_b, v_c\} \qquad (1)$$
$$v_{\mathrm{mid}} = v_a + v_b + v_c - v_{\max} - v_{\min}$$

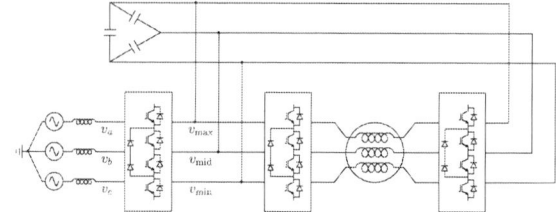

(a) I-type indirect MC OE Wdg. drive

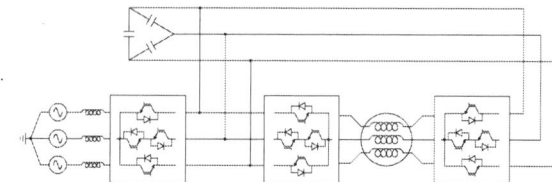

(b) T-type indirect MC OE Wdg. drive

Fig. 1. Presented indirect matrix converter (IMC) open-end winding drives based on (a) I-type and (b) T-type three-level inverter topologies.

The front-end converter arranges the input grid voltages by their instantaneous value:

$$\{v_a, v_b, v_c\} \xrightarrow{\text{front-end}} \{v_{\max}, v_{\mathrm{mid}}, v_{\min}\} \qquad (2)$$

The voltages $\{v_{\max}, v_{\mathrm{mid}}, v_{\min}\}$ generated by the front-end are shown in Fig. 2(a). These are also three-phase voltages alternating between positive sequence and negative sequence. Let the input voltage sector be determined by Table I where $\vec{v}_{abc}$ is the input voltage space vector:

$$\vec{v}_{abc} = v_a + v_b\, e^{j2\pi/3} + v_c\, e^{-j2\pi/3} \qquad (3)$$

The synchronous vector $\vec{v}_{\mathrm{xdn}}$ can be defined by a space vector transform on the voltages $(v_{\max}, v_{\mathrm{mid}}, v_{\min})$:

$$\vec{v}_{\mathrm{xdn}} = v_{\max} + v_{\mathrm{mid}}\, e^{j2\pi/3} + v_{\min}\, e^{-j2\pi/3} \qquad (4)$$

Note that the letters 'x', 'd', 'n' in the subscript of $\vec{v}_{\mathrm{xdn}}$ represent 'max', 'mid' and 'min' respectively. Other synchronous vectors can be defined similarly. The synchronous vectors $\{\vec{v}_{\mathrm{xdn}}, \vec{v}_{\mathrm{dnx}}, \vec{v}_{\mathrm{nxd}}\}$ rotate counterclockwise (CCW) in the odd sectors of the input voltage and the complementary set $\{\vec{v}_{\mathrm{xnd}}, \vec{v}_{\mathrm{ndx}}, \vec{v}_{\mathrm{dxn}}\}$ rotates clockwise (CW) in these odd sectors. In the even sectors, the sets reverse their direction of rotation. This is illustrated in Fig. 2(b) for the first set $\{\vec{v}_{\mathrm{xdn}}, \dots\}$. In summary, the output voltages of the front-end converter form a set of CCW rotating vectors and a set of CW

$\angle \vec{v}_{abc}$	Sector
$[0, \pi/3)$	I
$[\pi/3, 2\pi/3)$	II
$[2\pi/3, \pi)$	III
$[\pi, 4\pi/3)$	IV
$[4\pi/3, 5\pi/3)$	V
$[5\pi/3, 2\pi)$	VI

TABLE I
DEFINITION OF THE INPUT VOLTAGE SECTORS.

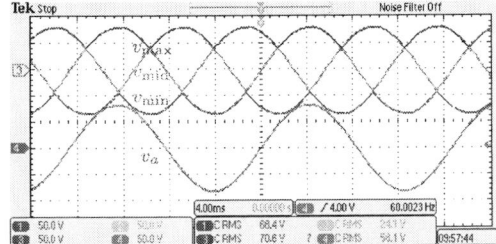

(a) 'max', 'mid', 'min' voltages generated by the front-end converter and the phase 'a' voltage at no-load. The input voltage is $\sim 100\,\mathrm{V}$.

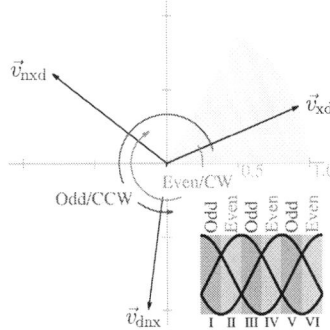

(b) Vector diagram of the front-end outputs

Fig. 2. Front-end operation. (a) shows the time-domain waveforms of the front-end converter and (b) shows the vector diagram of the front-end outputs.

rotating vectors that are available to the load-end converters for output voltage synthesis.

B. Output voltage synthesis

Let $\{\vec{v}_1, \vec{v}_2, \vec{v}_3\}$ be a set of synchronous vectors available for synthesis. In an open-end winding machine, the same set is available at both ends. $\{\vec{v}_1, \vec{v}_2, \vec{v}_3\}$ could be the set of CCW rotating vectors, or the set of CW rotating vectors. However, the direction of rotation of all three vectors in the set $\{\vec{v}_1, \vec{v}_2, \vec{v}_3\}$ must be the same.

Let $\vec{v}_{1,2} = \vec{v}_1 - \vec{v}_2$ etc. Applying $\vec{v}_1$ at one end and applying $\vec{v}_2$ at the other end of an open-end winding machine is equivalent to applying a longer vector $\vec{v}_{1,2}$ at the terminals of an equivalent wye-connected machine. Therefore, the vectors $\{\vec{v}_{1,2}, \vec{v}_{2,3}, \vec{v}_{3,1}, \vec{v}_{2,1}, \vec{v}_{3,2}, \vec{v}_{1,3}\}$ are *effectively* available for the load voltage synthesis in an open-end winding machine. This is illustrated in Fig. 3.

Any target vector lying in the convex hull of the vectors $\{\vec{v}_{1,2}, \ldots\}$ can be synthesized. Since the angle of a given target vector $(:= \vec{v}^*)$ is general, the magnitude of the target vector must be less than or equal to the radius of the incircle of the convex hull (the hexagon in Fig. 3). The radius of the incircle is 1.5 times the magnitude of vectors $\vec{v}_1, \vec{v}_2, \vec{v}_3$, and the magnitude of the vectors $\vec{v}_1, \vec{v}_2, \vec{v}_3$ is equal to the magnitude of the input voltage space vector $\vec{v}_{abc}$ — therefore, an output voltage vector with a magnitude up to 1.5 times the magnitude of the input voltage vector can be synthesized.

Let the duty ratios for which the vectors $\vec{v}_1, \vec{v}_2, \vec{v}_3$ are applied at the first set of the machine terminals A_1, B_1, C_1

be $\lambda_{11}, \lambda_{12}, \lambda_{13}$; and similar duty ratios at the other set of machine terminals A_2, B_2, C_2 be $\lambda_{21}, \lambda_{22}, \lambda_{23}$. The target vector $\vec{v}^*$ is synthesized as:

$$\vec{v}^* = \underbrace{\lambda_{11}\vec{v}_1 + \lambda_{12}\vec{v}_2 + \lambda_{13}\vec{v}_3}_{=\vec{v}_{A_1 B_1 C_1}} - \underbrace{(\lambda_{21}\vec{v}_1 + \lambda_{22}\vec{v}_2 + \lambda_{23}\vec{v}_3)}_{=\vec{v}_{A_2 B_2 C_2}}$$

(5)

The decomposition of the above equation into real and imaginary parts yields two equations. The conditions that $\sum_{1,2,3} \lambda_{1i} = 1, \sum_{1,2,3} \lambda_{2j} = 1$, yield two more. Four equations, however, are insufficient to determine six duty ratios.

Let the target vector $\vec{v}^*$ lie in the $\pi/3$ sector formed by the vectors $\vec{v}_{12}$ and $\vec{v}_{13}$ ($\angle\vec{v}^* - \angle\vec{v}_1 \in [-\pi/6, \pi/6)$). Let us impose an additional constraint that within this sector, $\vec{v}^*$ will be synthesized by a convex combination $(\sigma_1, \sigma_2, \sigma_3)$ of a zero vector and the two nearest available vectors $\vec{v}_{12}, \vec{v}_{13}$ [49], [50]:

$$\begin{aligned}
\vec{v}^* &= \sigma_1\vec{0} + \sigma_2\vec{v}_{1,2} + \sigma_3\vec{v}_{1,3} \\
&= \sigma_1\vec{v}_1 + \sigma_2\vec{v}_1 + \sigma_3\vec{v}_1 - (\sigma_1\vec{v}_1 + \sigma_2\vec{v}_2 + \sigma_3\vec{v}_3) \\
&= \vec{v}_1 - (\sigma_1\vec{v}_1 + \sigma_2\vec{v}_2 + \sigma_3\vec{v}_3)
\end{aligned}$$

(6)

The zero vector in (6) was chosen to be $\vec{v}_{11}$ which is equivalent to choosing $\lambda_{11} = 1, \lambda_{12} = \lambda_{13} = 0$ in (5). The equation (6) has a unique solution due to imposing the additional constraint that the output voltage vector will be synthesized using adjacent vectors — this choice of the synthesizing vector eliminates the switching losses in one of the two load-end converters ($\lambda_{11} = 1, \lambda_{12} = \lambda_{13} = 0$). The synthesis of the target vector is illustrated in the vector diagram of Fig. 4.

C. Input power factor control

The load current lags the applied voltage in a typical motor load. Assuming sinusoidal steady state, the load current space

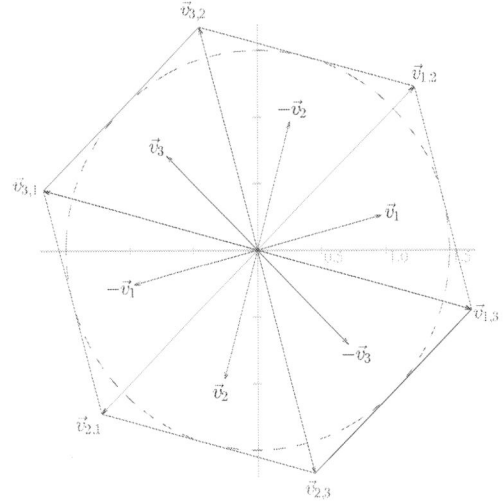

Fig. 3. Vectors available for output voltage synthesis.

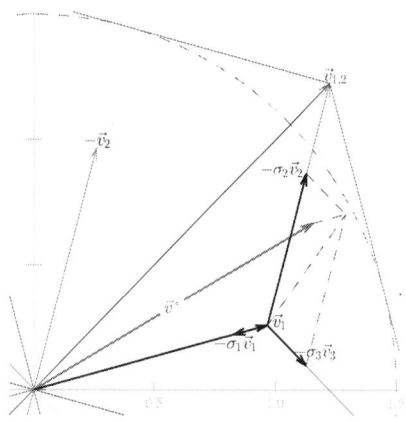

Fig. 4. Output voltage synthesis: $\vec{v}^* = \vec{v}_1 - (\sigma_1\vec{v}_1 + \sigma_2\vec{v}_2 + \sigma_3\vec{v}_3)$.

vector can be written as:

$$\vec{i}_{ABC} = \frac{\vec{v}^*}{|Z|} \angle - \phi \qquad \phi > 0 \tag{7}$$

where $|Z|\angle\phi$ is the equivalent load impedance. In the previous subsection, the set $\{\vec{v}_1, \vec{v}_2, \vec{v}_3\}$ could have been either of the two oppositely rotating sets of vectors formed by the front-end output voltages. Let $(\vec{v}_1, \vec{v}_2, \vec{v}_3) = (\vec{v}_{\text{xdn}}, \vec{v}_{\text{dnx}}, \vec{v}_{\text{nxd}})$. It can be shown that for balanced input voltages and load currents,

$$\vec{i}_{\text{xdn}} = m^2 \frac{\vec{v}_{\text{xdn}} \angle - \phi}{|Z|} \tag{8}$$

where m is the voltage transfer ratio $|\vec{v}^*|/|\vec{v}_{abc}|$. The input current vector $\vec{i}_{abc}$ can be calculated based on (8) and Table I. The normalized input voltages and currents (switching cycle averaged values) in different sectors of the input voltage are shown in Fig. 5 for modulation using the set $\{\vec{v}_{\text{xdn}}, \vec{v}_{\text{dnx}}, \vec{v}_{\text{nxd}}\}$.

Equation (8)/Table I and Fig. 5 reveal that when the load voltage is synthesized from CCW rotating vectors ($\{\vec{v}_{\text{xdn}}, \ldots\}$ in sectors I, III, V), the input current power factor is equal to the load power factor, i.e. the input current lags the input

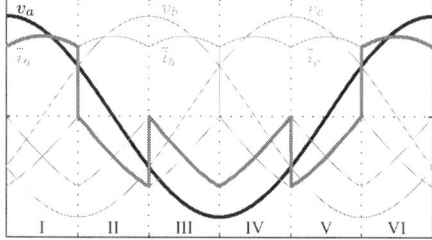

Fig. 5. Normalized input voltages and currents (simulated) for modulation using only one set of the available vectors. The bar above the currents denotes the switching cycle average.

voltage for an inductive load; whereas when the load voltage is synthesized from CW rotating vectors ($\{\vec{v}_{\text{xdn}}, \ldots\}$ in sectors II, IV, VI), the input current power factor is opposite of the load power factor. Let the CCW and CW rotating sets of vectors, and the corresponding duty ratios be defined as:

$$\mathbf{V}_{\text{pos}} = \begin{cases} [\vec{v}_{\text{xdn}} \ \vec{v}_{\text{dnx}} \ \vec{v}_{\text{nxd}}]^T & \text{odd sectors} \\ [\vec{v}_{\text{xnd}} \ \vec{v}_{\text{ndx}} \ \vec{v}_{\text{dxn}}]^T & \text{even sectors} \end{cases} \tag{9}$$

$$\mathbf{V}_{\text{neg}} = \begin{cases} [\vec{v}_{\text{xnd}} \ \vec{v}_{\text{ndx}} \ \vec{v}_{\text{dxn}}]^T & \text{odd sectors} \\ [\vec{v}_{\text{xdn}} \ \vec{v}_{\text{dnx}} \ \vec{v}_{\text{nxd}}]^T & \text{even sectors} \end{cases} \tag{10}$$

$$\mathbf{\Lambda}_{1,\text{pos}} = \begin{cases} [\lambda_{1,\text{xdn}} \ \lambda_{1,\text{dnx}} \ \lambda_{1,\text{nxd}}] & \text{odd sectors} \\ [\lambda_{1,\text{xnd}} \ \lambda_{1,\text{ndx}} \ \lambda_{1,\text{dxn}}] & \text{even sectors} \end{cases} \tag{11}$$

$$\mathbf{\Lambda}_{1,\text{neg}} = \begin{cases} [\lambda_{1,\text{xnd}} \ \lambda_{1,\text{ndx}} \ \lambda_{1,\text{dxn}}] & \text{odd sectors} \\ [\lambda_{1,\text{xdn}} \ \lambda_{1,\text{dnx}} \ \lambda_{1,\text{nxd}}] & \text{even sectors} \end{cases} \tag{12}$$

$$\mathbf{\Lambda}_{2,\text{pos}} = \begin{cases} [\lambda_{2,\text{xdn}} \ \lambda_{2,\text{dnx}} \ \lambda_{2,\text{nxd}}] & \text{odd sectors} \\ [\lambda_{2,\text{xnd}} \ \lambda_{2,\text{ndx}} \ \lambda_{2,\text{dxn}}] & \text{even sectors} \end{cases} \tag{13}$$

$$\mathbf{\Lambda}_{2,\text{neg}} = \begin{cases} [\lambda_{2,\text{xnd}} \ \lambda_{2,\text{ndx}} \ \lambda_{2,\text{dxn}}] & \text{odd sectors} \\ [\lambda_{2,\text{xdn}} \ \lambda_{2,\text{dnx}} \ \lambda_{2,\text{nxd}}] & \text{even sectors} \end{cases} \tag{14}$$

Let $\alpha : \alpha \in [0, 1]$ be a power factor control parameter. The target vector $\vec{v}^*$ can be synthesized using both sets of vectors:

$$\vec{v}^* = \alpha(\mathbf{\Lambda}_{1,\text{pos}} - \mathbf{\Lambda}_{2,\text{pos}})\mathbf{V}_{\text{pos}} + \\ (1-\alpha)(\mathbf{\Lambda}_{1,\text{neg}} - \mathbf{\Lambda}_{2,\text{neg}})\mathbf{V}_{\text{neg}} \tag{15}$$

By varying α, the input power factor angle can be controlled continuously in $[-\phi, \phi]$ where ϕ is the load power factor angle.

D. Common-mode elimination principle

Equation (15) expresses the target output vector as a convex combination of all vectors available for synthesis. These vectors are $\{\vec{v}_{\text{xdn}}, \vec{v}_{\text{dnx}}, \vec{v}_{\text{nxd}}\}$ and $\{\vec{v}_{\text{xnd}}, \vec{v}_{\text{ndx}}, \vec{v}_{\text{dxn}}\}$. All of these vectors are obtained by a space vector transform on triplets elements of which are $v_{\text{max}}, v_{\text{mid}}, v_{\text{min}}$. If the input voltages are balanced, that is, $\sum_{a,b,c} v_i = 0$, it follows from (1) that $\sum_{\text{max,mid,min}} v_i = 0$, which further means that $\sum_{A_1, B_1, C_1} v_i = 0$ and $\sum_{A_2, B_2, C_2} v_i = 0$.

Thus, the common-mode voltages generated at both ends of the machine are identically equal to zero for modulation with rotating voltage vectors.

E. Switching, commutation, and clamp circuit elimination

The phase legs of the two indirect MC OE Wdg. drives presented in this paper are shown in Fig. 6. Every converter (one front-end and two load-end) consists of three such legs with the voltages $v_{\text{max}}, v_{\text{mid}}, v_{\text{min}}$ connected at the voltage terminals and either an input or an output phase connected at the current terminal of the three-level inverter phase leg. The switches are numbered according to their position in the phase leg and the input/output phase they control — e.g. the phase legs in Fig. 6 control the load terminal A_1.

1) Front-end converter: The input voltages are sensed and the sign of the input line-to-line voltages is used to define the

978-1-4673-9551-9/16 $31.00 © 2016 IEEE

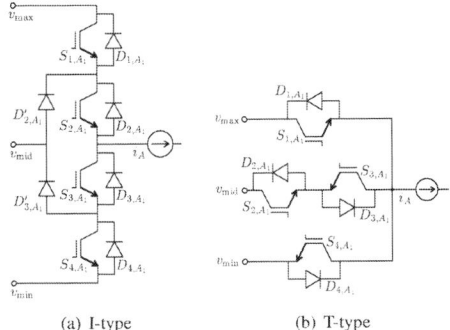

(a) I-type   (b) T-type

Fig. 6. Phase legs of the presented drives: (a) I-type and (b) T-type.

status signals:

$$s_{ab} = \begin{cases} 1 & v_{ab} \geq 0 \\ 0 & v_{ab} < 0 \end{cases}$$

$$s_{bc} = \begin{cases} 1 & v_{bc} \geq 0 \\ 0 & v_{bc} < 0 \end{cases} \quad (16)$$

$$s_{ca} = \begin{cases} 1 & v_{ca} \geq 0 \\ 0 & v_{ca} < 0 \end{cases}$$

The status signals are used to generate the switching signals q for the topmost and the bottommost switches of the front-end phase legs. The switching signals for the remaining switches are computed by logical inversion:

$$\begin{aligned} q_{1,a} &= s_{ab}\overline{s}_{ca} & q_{1,b} &= s_{bc}\overline{s}_{ab} & q_{1,c} &= s_{ca}\overline{s}_{bc} \\ q_{4,a} &= \overline{s}_{ab}s_{ca} & q_{4,b} &= \overline{s}_{bc}s_{ab} & q_{4,c} &= \overline{s}_{ca}s_{bc} \\ q_{2,a} &= \overline{q}_{4,a} & q_{2,b} &= \overline{q}_{4,b} & q_{2,c} &= \overline{q}_{4,c} \\ q_{3,a} &= \overline{q}_{1,a} & q_{3,b} &= \overline{q}_{1,b} & q_{3,c} &= \overline{q}_{1,c} \end{aligned} \quad (17)$$

Since the switches of the front-end switch only at the zero crossings of the line-to-line voltages, ZVS is inherent to the front-end converter. Furthermore, the switching transitions occur only at the line frequency and therefore the front-end converter does not incur any switching losses.

2) Load-end converters: Having computed the elements of the Λ vectors in (15), the series of the λ coefficients is compared with a carrier signal to generate the switching signals for the individual switches. Since a total of six vectors are applied at either set of the load terminals, multiple vector sequences are possible. Among all the possible sequences, certain sequences eliminate the $v_{max} \leftrightarrow v_{min}$ transitions at the current terminal of the phase legs of Fig. 6. One such sequence is illustrated in Fig. 7. Using the vector sequence of Fig. 7 reduces the switching stress since the semiconductor devices never switch the peak line-to-line voltage. It is noteworthy that in the direct MC based OE Wdg. drive of [44], the switches must switch the peak line-to-line voltage. Utilizing this intelligent vector sequence may also reduce the voltage rating of the IGBTs used in the I-type IMC OE Wdg. topology (Figs. 1(a) and 6(a)), thus reducing the conduction losses.

The switching signals of the switches at positions 3 and 2 are the inverted switching signals (with some blanking time) of the switches at positions 1 and 4 respectively, similar to conventional three-level inverters. Due to this strategy, the

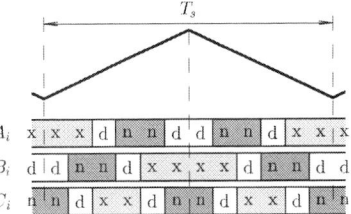

Fig. 7. An intelligent vector sequence that minimizes the voltage switched: $\vec{v}_{xdn}, \vec{v}_{xnd}, \vec{v}_{dnx}, \vec{v}_{ndx}, \vec{v}_{nxd}, \vec{v}_{dxn}, \vec{v}_{dxn}, \vec{v}_{nxd}, \vec{v}_{ndx}, \vec{v}_{dnx}, \vec{v}_{xnd}, \cdots$

current always commutates directly to the incoming voltage terminal. Therefore 4-step commutation is achieved without having to sense the current direction which is susceptible to noise [6].

In addition to the above, the freewheeling diodes of the IGBTs always ensure a path into the voltage terminals of a phase leg. Since the filter capacitors are placed at the $v_{max}, v_{mid}, v_{min}$ terminals (Fig. 1), these capacitors can act as the clamp capacitance in the event of commutation failure. No additional clamp capacitance is therefore necessary in the presented indirect topologies unlike their direct matrix converter counterpart.

III. FILTER OPTIMIZATION

LC filter with a damping resistor is commonly used in matrix converters for filtering the switching current. The damping resistor degrades the high-frequency (HF) attenuation. We employ a series RL branch connected in parallel to the main inductor to provide the necessary damping while offering sufficient impedance to the HF currents [51]. The filter circuit at the switching frequency is shown in Fig. 8.

Let L_f, C_f be the main filter inductor and the filter capacitor. Let the damping resistor be r_d and the auxiliary inductor placed in the damping branch be L_d. We optimize r_d for a given L_d/L_f such that the amplification at the damped natural frequency is minimized. Let,

$$\begin{aligned} n &= L_d/L_f \\ \omega_0 &= 1/\sqrt{L_f C_f} \end{aligned} \quad (18)$$

The transfer function of the converter input current to the filtered grid current in the switching frequency circuit of Fig. 8 is:

$$\begin{aligned} G(s) &:= \frac{i_{grid}(s)}{\tilde{i}_{sw}(s)} \\ &= \frac{sL_f(n+1) + r_d}{s^3 n L_f^2 C_f + s^2 L_f C_f r_d + s L_f(n+1) + r_d} \end{aligned} \quad (19)$$

The transfer function $G(s)$ has three poles and one zero as opposed to two poles and one zero in the corresponding transfer function for the second-order filter. Therefore the high-frequency attenuation provided by $G(s)$ is $-40\,$dB/decade.

It is found that for a given n $(= L_d/L_f)$, $\exists \omega^* : |G(j\omega^*)|$

is independent of $r_d \forall r_d \geq 0$:

$$\omega^* = \sqrt{\frac{2(n+1)}{2n+1}}\,\omega_0 \tag{20}$$
$$|G(j\omega^*)| = 2n+1$$

The next step in optimizing r_d is to find r_d such that the maximum amplification at the damped resonant frequency is equal to the value above, i.e. $\|G(s, r_d)\|_\infty = |G(j\omega^*)|$. The steps are omitted for brevity; the final results is:

$$r_d = \omega_0 L_f \left(\frac{n + \frac{1}{2}}{n(n+1)^2} \right)^{-\frac{1}{2}} \tag{21}$$

A. Practical designs

The practical design of the proposed optimal third-order filter is based on the choice of the switching frequency and the desired attenuation. The initial steps are identical to a conventional second-order filter design. In a second-order filter, the damping resistor is chosen based on the damping coefficient and the power loss. In the proposed filter, the maximum amplification is determined from (20), and then the damping resistor is calculated from (21). The current rating of the auxiliary inductor and the power lost in the damping resistor are implicit in this procedure. The design procedure is ultimately iterative based on the simulation performance and component availability.

Table II and Fig. 9 show a comparison of the proposed third-order filter against a conventional second-order filter. Both filters achieve similar attenuation at the switching frequency (10 kHz for a line frequency of 60 Hz). It is seen that the proposed filter is more compact, less lossy, and provides better attenuation of the higher frequencies ($f > 10$ kHz).

IV. Results

Prototypes of the two indirect matrix converter based open-end winding drives, I-type and T-type, were constructed to experimentally validate these drives. Figs. 10(a) and 10(b) show the input and the output voltage and current waveforms of the I-type and the T-type drive respectively. Sinusoidal input and output currents are observed. These results also prove the voltage transfer ratio and the input power factor control capabilities of the drives presented.

The prototype of the I-type IMC OE Wdg. drive uses the second-order filter of Table II whereas the prototype of the T-type IMC OE Wdg. drive uses the more compact third-order filter (Table II) designed to provide the same attenuation of the switching frequency as the former. The spectra of the input currents in Figs. 10(a) and 10(b) confirm that both filters provide adequate attenuation of the switching frequency; thus validating the proposed optimal third-order input filter design.

In Fig. 11, one of the presented drives (I-type) is compared to conventional two- and three-level VSIs at the same operating point. Significant improvement in the output common-mode voltage related phenomena, namely shaft voltage and ground currents, is demonstrated by the presented drive.

V. Conclusion

The indirect matrix converter based open-end winding drives presented in this paper combine the common-mode elimination and high voltage transfer ratio of the direct matrix converter drive with the robust three-level inverter structure. The indirect topologies retain all features of the direct topology and offer further advantages of naturally intelligent commutation, lower switching voltage stress, clamp circuit elimination and possibly lower device voltage ratings.

A comparison of the indirect MC OE Wdg. drives against the state-of-the-art on criteria other than the HF common-mode performance is the topic of a future publication. Research on integrating the LVRT strategy proposed in [45] with the presented drives as well as the implementation of direct torque control using these drives is underway.

References

[1] D. G. Holmes and T. A. Lipo, *Pulse width modulation for power converters: principles and practice*. John Wiley & Sons, 2003, vol. 18.

[2] S. Parler and L. R. Macomber, "Power film capacitors versus aluminum electrolytic capacitors for dc link applications," Cornell Dubilier.

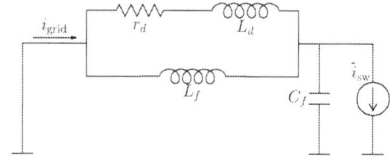

Fig. 8. Filter topology and the equivalent circuit at the switching frequency.

	3rd order		2nd order	
	Design	Prototype	Design	Prototype
L_f	0.0261	0.95 mH	0.0300	1.5 mH
C_f	0.1631	10.75 μF(Δ)	0.3333	18 μF(Δ)
L_d	0.0083	330 μH	--	--
r_d	0.5547	8 Ω	0.6000	10 Ω
P_{loss}	0.0013		0.0017	

TABLE II

THE DESIGN VALUES ARE EXPRESSED IN PU. THE PROTOTYPES ARE FOR A 208 V, 3 kVA SYSTEM.

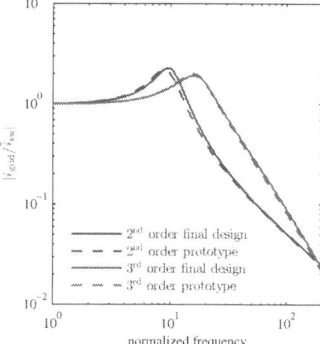

Fig. 9. Filter response against conventional: the proposed filter exhibits superior attenuation of higher frequencies.

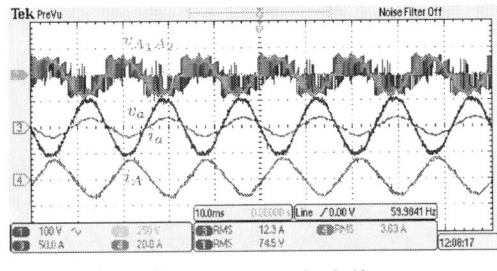

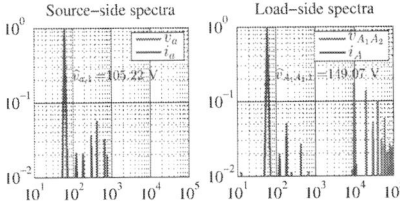

(a) I-type topology of Fig. 1(a) with ACIM load and ∼upf operation at $f_{\text{out}} = 60$ Hz

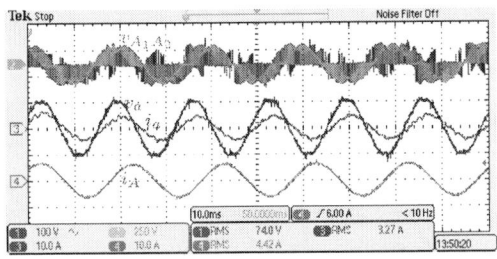

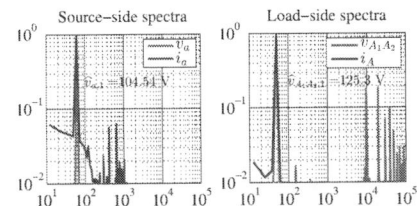

(b) T-type topology of Fig. 1(b) with RL load and lagging input current at $f_{\text{out}} = 50$ Hz

(c) Prototypes of I-type (left) & T-type (right) implementations

Fig. 10.   For all cases $f_{\text{sw}} \approx 10$ kHz, load $V/f = 200/60$ V/Hz, and: v_a is the input voltage, i_a is the input current, $v_{A_1 A_2}$ is the load voltage, i_A is the load current. Voltage transfer ratio in (a) = 1.42, and (b) = 1.20. The x-axis in the spectra is the frequency (Hz), the y-axis is the normalized amplitude.

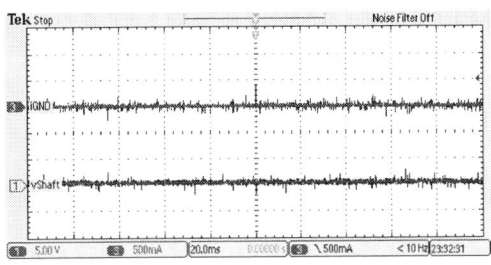

(a) I-type topology of Fig. 1(a)

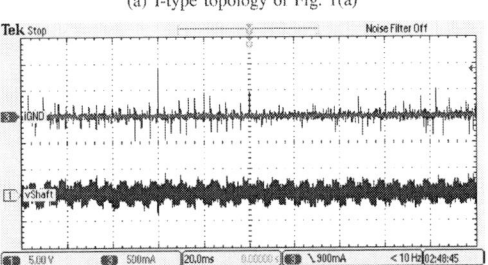

(b) Conventional three-level NPC VSI

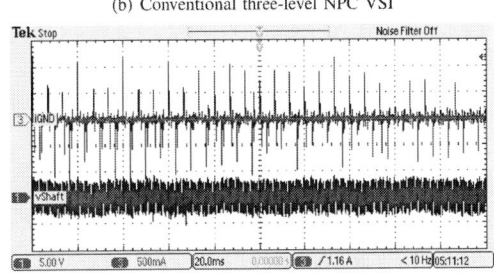

(c) Conventional two-level VSI

Fig. 11.   Ground current (top) and shaft voltage (bottom) for different converters driving an ACIM at ∼166.67 V, 50 Hz ($V/f = 200/60$ V/Hz). $f_{\text{sw}} \approx 10$ kHz for all cases. The presented drive's results in (a) show remarkable improvement over state-of-the-art shown in (b),(c).

[6] P. Wheeler, J. Rodriguez, J. Clare, L. Empringham, and A. Weinstein, "Matrix converters: a technology review," *Industrial Electronics, IEEE Transactions on*, vol. 49, no. 2, pp. 276–288, Apr 2002.

[7] J. Rodriguez, M. Rivera, J. Kolar, and P. Wheeler, "A review of control and modulation methods for matrix converters," *Industrial Electronics, IEEE Transactions on*, vol. 59, no. 1, pp. 58–70, Jan 2012.

[8] P. Nielsen, F. Blaabjerg, and J. Pedersen, "New protection issues of a matrix converter: design considerations for adjustable-speed drives," *Industry Applications, IEEE Transactions on*, vol. 35, no. 5, pp. 1150–1161, Sep 1999.

[9] S. Bernet, S. Ponnaluri, and R. Teichmann, "Design and loss comparison of matrix converters, and voltage-source converters for modern ac drives," *Industrial Electronics, IEEE Transactions on*, vol. 49, no. 2, pp. 304–314, Apr 2002.

[10] T. Friedli, J. Kolar, J. Rodriguez, and P. Wheeler, "Comparative evaluation of three-phase ac-ac matrix converter and voltage dc-link back-to-back converter systems," *Industrial Electronics, IEEE Transactions on*, vol. 59, no. 12, pp. 4487–4510, Dec 2012.

[11] S. Chen, T. Lipo, and D. Fitzgerald, "Source of induction motor bearing currents caused by pwm inverters," *Energy Conversion, IEEE Transactions on*, vol. 11, no. 1, pp. 25–32, Mar 1996.

[12] J. Erdman, R. Kerkman, D. Schlegel, and G. Skibinski, "Effect of pwm inverters on ac motor bearing currents and shaft voltages," *Industry Applications, IEEE Transactions on*, vol. 32, no. 2, pp. 250–259, Mar 1996.

[13] S. Chen, T. Lipo, and D. Fitzgerald, "Modeling of motor bearing currents in pwm inverter drives," *Industry Applications, IEEE Transactions on*, vol. 32, no. 6, pp. 1365–1370, Nov 1996.

[3] A. Alesina and M. Venturini, "Solid-state power conversion: A fourier analysis approach to generalized transformer synthesis," *Circuits and Systems, IEEE Transactions on*, vol. 28, no. 4, pp. 319–330, Apr 1981.

[4] ——, "Analysis and design of optimum-amplitude nine-switch direct ac-ac converters," *Power Electronics, IEEE Transactions on*, vol. 4, no. 1, pp. 101–112, Jan 1989.

[5] L. Huber and D. Borojevic, "Space vector modulated three-phase to three-phase matrix converter with input power factor correction," *Industry Applications, IEEE Transactions on*, vol. 31, no. 6, pp. 1234–1246, Nov 1995.

978-1-4673-9551-9/16 $31.00 © 2016 IEEE

[14] D. Busse, J. Erdman, R. Kerkman, D. Schlegel, and G. Skibinski, "System electrical parameters and their effects on bearing currents," *Industry Applications, IEEE Transactions on*, vol. 33, no. 2, pp. 577–584, Mar 1997.

[15] S. Chen and T. Lipo, "Bearing currents and shaft voltages of an induction motor under hard- and soft-switching inverter excitation," *Industry Applications, IEEE Transactions on*, vol. 34, no. 5, pp. 1042–1048, Sep 1998.

[16] F. Wang, "Motor shaft voltages and bearing currents and their reduction in multilevel medium-voltage pwm voltage-source-inverter drive applications," *Industry Applications, IEEE Transactions on*, vol. 36, no. 5, pp. 1336–1341, Sep 2000.

[17] M. Melfi, F. Ladonne, and D. Ankele, "Can a shaft brush be safely applied on a motor in a class i hazardous location?" in *Petroleum and Chemical Industry Technical Conference (PCIC), 2014 IEEE*, Sept 2014, pp. 85–90.

[18] S. Ogasawara, H. Ayano, and H. Akagi, "An active circuit for cancellation of common-mode voltage generated by a pwm inverter," *Power Electronics, IEEE Transactions on*, vol. 13, no. 5, pp. 835–841, Sep 1998.

[19] M. Di Piazza, G. Tine, and G. Vitale, "An improved active common-mode voltage compensation device for induction motor drives," *Industrial Electronics, IEEE Transactions on*, vol. 55, no. 4, pp. 1823–1834, April 2008.

[20] S. Bell, T. Cookson, S. Cope, R. Epperly, A. Fischer, D. Schlegel, and G. Skibinski, "Experience with variable-frequency drives and motor bearing reliability," *Industry Applications, IEEE Transactions on*, vol. 37, no. 5, pp. 1438–1446, Sep 2001.

[21] A. Hava and E. Un, "A high-performance pwm algorithm for common-mode voltage reduction in three-phase voltage source inverters," *Power Electronics, IEEE Transactions on*, vol. 26, no. 7, pp. 1998–2008, July 2011.

[22] J. Rodriguez, J.-S. Lai, and F. Z. Peng, "Multilevel inverters: a survey of topologies, controls, and applications," *Industrial Electronics, IEEE Transactions on*, vol. 49, no. 4, pp. 724–738, Aug 2002.

[23] H.-J. Kim, H.-D. Lee, and S.-K. Sul, "A new pwm strategy for common-mode voltage reduction in neutral-point-clamped inverter-fed ac motor drives," *Industry Applications, IEEE Transactions on*, vol. 37, no. 6, pp. 1840–1845, Nov 2001.

[24] A. Gupta and A. Khambadkone, "A space vector modulation scheme to reduce common mode voltage for cascaded multilevel inverters," *Power Electronics, IEEE Transactions on*, vol. 22, no. 5, pp. 1672–1681, Sept 2007.

[25] K. Ratnayake and Y. Murai, "A novel pwm scheme to eliminate common-mode voltage in three-level voltage source inverter," in *Power Electronics Specialists Conference, 1998. PESC 98 Record. 29th Annual IEEE*, vol. 1, May 1998, pp. 269–274 vol.1.

[26] H. Zhang, A. von Jouanne, S. Dai, A. Wallace, and F. Wang, "Multilevel inverter modulation schemes to eliminate common-mode voltages," *Industry Applications, IEEE Transactions on*, vol. 36, no. 6, pp. 1645–1653, Nov 2000.

[27] M. Renge and H. Suryawanshi, "Five-level diode clamped inverter to eliminatecommon mode voltage and reduce dv/dt inmedium voltage rating induction motor drives," *Power Electronics, IEEE Transactions on*, vol. 23, no. 4, pp. 1598–1607, July 2008.

[28] P. Rajeevan and K. Gopakumar, "A hybrid five-level inverter with common-mode voltage elimination having single voltage source for im drive applications," *Industry Applications, IEEE Transactions on*, vol. 48, no. 6, pp. 2037–2047, Nov 2012.

[29] H. J. Cha and P. Enjeti, "An approach to reduce common-mode voltage in matrix converter," *Industry Applications, IEEE Transactions on*, vol. 39, no. 4, pp. 1151–1159, July 2003.

[30] R. Vargas, U. Ammann, J. Rodriguez, and J. Pontt, "Predictive strategy to reduce common-mode voltages on power converters," in *Power Electronics Specialists Conference, 2008. PESC 2008. IEEE*, June 2008, pp. 3401–3406.

[31] F. Bradaschia, M. Cavalcanti, E. Ibarra, F. Neves, and E. Bueno, "Generalized pulse-width-modulation to reduce common-mode voltage in matrix converters," in *Energy Conversion Congress and Exposition, 2009. ECCE 2009. IEEE*, Sept 2009, pp. 3274–3281.

[32] T. Nguyen and H.-H. Lee, "Modulation strategies to reduce common-mode voltage for indirect matrix converters," *Industrial Electronics, IEEE Transactions on*, vol. 59, no. 1, pp. 129–140, Jan 2012.

[33] M. Rivera, J. Rodriguez, J. Espinoza, and B. Wu, "Reduction of common-mode voltage in an indirect matrix converter with imposed sinusoidal input/output waveforms," in *IECON 2012 - 38th Annual*

Conference on IEEE Industrial Electronics Society, Oct 2012, pp. 6105–6110.

[34] T. Nguyen and H.-H. Lee, "A new svm method for an indirect matrix converter with common-mode voltage reduction," *Industrial Informatics, IEEE Transactions on*, vol. 10, no. 1, pp. 61–72, Feb 2014.

[35] X. You, S. Liu, H. Abu-Rub, B. Ge, X. Jiang, and F. Peng, "A new space vector modulation strategy to reduce common-mode voltage for quasi-z-source indirect matrix converter," in *Energy Conversion Congress and Exposition (ECCE), 2014 IEEE*, Sept 2014, pp. 1064–1069.

[36] J. Espina, C. Ortega, L. de Lillo, L. Empringham, J. Balcells, and A. Arias, "Reduction of output common mode voltage using a novel svm implementation in matrix converters for improved motor lifetime," *Industrial Electronics, IEEE Transactions on*, vol. 61, no. 11, pp. 5903–5911, Nov 2014.

[37] H.-N. Nguyen and H.-H. Lee, "An enhanced svm method to drive matrix converters for zero common-mode voltage," *Power Electronics, IEEE Transactions on*, vol. 30, no. 4, pp. 1788–1792, April 2015.

[38] V. Padhee, A. Sahoo, and N. Mohan, "Svpwm technique with varying dc-link voltage for common mode voltage reduction in an indirect matrix converter," in *Energy Conversion Congress and Exposition (ECCE)*, Sep 2015, p. to appear.

[39] M. Baiju, K. Mohapatra, R. Kanchan, and K. Gopakumar, "A dual two-level inverter scheme with common mode voltage elimination for an induction motor drive," *Power Electronics, IEEE Transactions on*, vol. 19, no. 3, pp. 794–805, May 2004.

[40] S. Lakshminarayanan, G. Mondal, P. Tekwani, K. Mohapatra, and K. Gopakumar, "Twelve-sided polygonal voltage space vector based multilevel inverter for an induction motor drive with common-mode voltage elimination," *Industrial Electronics, IEEE Transactions on*, vol. 54, no. 5, pp. 2761–2768, Oct 2007.

[41] B. Reddy and V. Somasekhar, "A dual inverter fed four-level open-end winding induction motor drive with a nested rectifier-inverter," *Industrial Informatics, IEEE Transactions on*, vol. 9, no. 2, pp. 938–946, May 2013.

[42] V. Somasekhar, B. Venugopal Reddy, and K. Sivakumar, "A four-level inversion scheme for a 6 n -pole open-end winding induction motor drive for an improved dc-link utilization," *Industrial Electronics, IEEE Transactions on*, vol. 61, no. 9, pp. 4565–4572, Sept 2014.

[43] J. Kalaiselvi and S. Srinivas, "Bearing currents and shaft voltage reduction in dual-inverter-fed open-end winding induction motor with reduced cmv pwm methods," *Industrial Electronics, IEEE Transactions on*, vol. 62, no. 1, pp. 144–152, Jan 2015.

[44] R. Gupta, K. Mohapatra, A. Somani, and N. Mohan, "Direct-matrix-converter-based drive for a three-phase open-end-winding ac machine with advanced features," *Industrial Electronics, IEEE Transactions on*, vol. 57, no. 12, pp. 4032–4042, Dec 2010.

[45] D. Orser and N. Mohan, "A matrix converter ride-through configuration using input filter capacitors as an energy exchange mechanism," *Power Electronics, IEEE Transactions on*, vol. 30, no. 8, pp. 4377–4385, Aug 2015.

[46] A. Iqbal, R. Alammari, H. Abu-Rub, and S. Ahmed, "Pwm scheme for dual matrix converters based five-phase open-end winding drive," in *Industrial Technology (ICIT), 2013 IEEE International Conference on*, Feb 2013, pp. 1686–1690.

[47] S. Ahmed, H. Abu-Rub, and Z. Salam, "Investigation of space vector modulated dual matrix converters feeding a seven phase open-end winding drive," in *Industrial Electronics Society, IECON 2014 - 40th Annual Conference of the IEEE*, Oct 2014, pp. 3305–3310.

[48] S. Ahmed, H. Abu-Rub, Z. Salam, M. Rivera, and O. Ellabban, "Generalized carrier based pulse width modulation technique for a three to n-phase dual matrix converter," in *Industrial Electronics Society, IECON 2014 - 40th Annual Conference of the IEEE*, Oct 2014, pp. 3298–3304.

[49] S. Tewari, R. Gupta, and N. Mohan, "Three-level indirect matrix converter based open-end winding ac machine drive," in *IECON 2011 - 37th Annual Conference on IEEE Industrial Electronics Society*, Nov 2011, pp. 1636–1641.

[50] S. Tewari, R. Gupta, A. Somani, and N. Mohan, "A new sinusoidal input-output three-phase full-bridge direct power converter," in *Industrial Electronics Society, IECON 2013 - 39th Annual Conference of the IEEE*, Nov 2013, pp. 4824–4830.

[51] R. Erickson, "Optimal single resistors damping of input filters," in *Applied Power Electronics Conference and Exposition, 1999. APEC '99. Fourteenth Annual*, vol. 2, Mar 1999, pp. 1073–1079 vol.2.

978-1-4673-9551-9/16 $31.00 © 2016 IEEE

Precharging Strategy for Soft Startup Process of Modular Multilevel Converters Based on Various SM circuits

Jiangchao Qin
School of Electrical, Computer
and Energy Engineering
Arizona State University
Tempe, AZ 85287-5706
Email: jqin@asu.edu

Suman Debnath
Oak Ridge National Laboratory
Oak Ridge, TN 37831
Email: suman2k42000@gmail.com

Maryam Saeedifard
School of Electrical and
Computer Engineering
Georgia Institute of Technology
Atlanta, GA 30332–0250
Email: maryam@ece.gatech.edu

Abstract— The modular multilevel converter (MMC) has become one of the most promising converter technologies for medium/high-power applications, specifically for high-voltage direct current (HVDC) transmission systems. One of the technical challenges associated with the operation and control of the MMC-based system is to precharge the submodule (SM) capacitors to their nominal voltages during the startup process. In this paper, considering various SM circuits, a general precharging strategy is proposed for the MMC-based systems under ac- and dc-side startup conditions. The proposed startup method does not require any additional feedback control loop, extra measurements, and/or auxiliary power supplies. Based on the developed startup method, the charging current is controllable by adjusting the changing rate of the number of blocked and bypassed SM capacitors. Performance of the proposed strategy for various MMCs is evaluated based on time-domain simulation studies in the PSCAD/EMTDC software environment.

I. INTRODUCTION

The modular multilevel converter (MMC) has become the most attractive converter topology for medium/high voltage applications, especially for voltage-sourced converter high-voltage direct current (HVDC) transmission systems because of its modularity and scalability [1], [2].

One of the main technical challenges associated with the control and operation of the MMC systems is to smoothly precharge the submodules (SM) capacitors to their nominal voltages during the converter startup process. Considering various MMC systems based on various SM circuits, which have been investigated to block the dc-side short-circuit faults [1], [3]–[7], precharging process becomes even more complicated. For the MMC-based drive systems, precharging the capacitors during startup process is carried out by absorbing power from the dc-side voltage source, while the MMC-HVDC system is charged from the ac grid.

The existing startup methods are mainly based on using additional power supplies and/or complex control methods while only considering one type of SM circuit topology. In [8] and [9], the proposed startup methods only consider half-bridge MMC systems and charge the SM capacitors from the dc-side voltage source. In [10], [11], an auxiliary voltage source is employed to charge the SM capacitors individually, which adds to the system complexity and cost. In [12], [13], closed-loop startup control methods are applied to charge the SM capacitors from ac and dc sides. However, those closed-loop methods only consider the MMC based on half-bridge SMs and employ additional PI controllers. Reference [14] investigates the startup issue of a clamp-double SM-based MMC-HVDC system and proposes a grouping sequentially controlled charge method. However, the method is only applicable to the MMC with clamp-double SMs. Furthermore, additional control effort is needed to group and charge the SM capacitors.

In this paper, a general precharging strategy is proposed to charge the SM capacitors of the MMC system during a startup process. Based on the proposed precharging strategy, several startup processes are analyzed for various MMCs based on different SM circuits under different startup conditions, i.e., ac- and dc-side startups. The proposed method is based on adjustment of the number of blocked and bypassed SM capacitors in conjunction with the conventional capacitor voltage sorting algorithm, without using any auxiliary power supplies and/or additional feedback control loop. By using the proposed procedure, the SM capacitors can be charged smoothly to their nominal voltages and the inrush charging current can be limited. The proposed strategy is also applicable to black start applications [10]. The effectiveness of the proposed startup method for a 21-level MMC system, based on various SM circuits, is presented. The studies are carried out based on time-domain simulations in the PSCAD/EMTDC environment under ac- and dc-side startup processes.

II. THE PROPOSED PRECHARGING STRATEGY

A schematic diagram of an MMC is shown in Fig. 1. The MMC consists of two arms per each phase where each arm comprises N series-connected, nominally identical, half-bridge (HB) submodules (SMs), and a series-connected inductor. The details of operation of the MMC has been described in [2] and is not repeated here.

978-1-4673-9551-9/16 $31.00 © 2016 IEEE

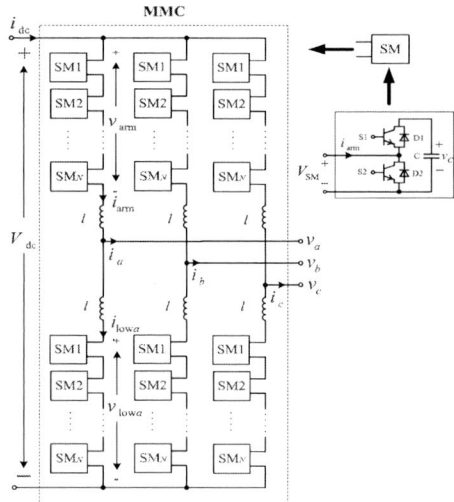

Fig. 1. Schematic representation of an MMC.

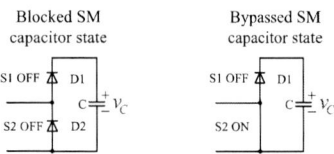

Fig. 2. The equivalent circuit of an HB SM under blocked and bypassed SM capacitor states.

To start up an MMC, two phases need to be considered:

- Uncontrollable precharging (Phase I): During this phase, the SMs are not controllable and all the switches are blocked. The charging current from either the ac or the dc side flows through the anti-parallel diodes to charge the SM capacitors to an initial voltage. Since the system is uncontrollable, to limit the inrush charging current, a current-limiting resistor arranged in the dc side, ac side, or the arm loops is employed [9], [12], [14].
- Controllable precharging (Phase II): Subsequent to an initial voltage built-up across each SM capacitor in Phase I, the SMs can be controlled to be in the inserted, bypassed, or blocked state.

For the proposed precharging strategy, two operating states are defined:

- Blocked SM capacitor (charging state): In an HB SM, both S1 and S2 are off, and its equivalent circuit is shown in Fig. 2. During this state, the SM capacitor is only charged when the arm current is positive.
- Bypassed SM capacitor (bypassing state): In an HB SM, S1 is off and S2 is on, and its equivalent circuit is shown in Fig. 2. Ideally, the SM capacitor voltage is kept unchanged.

The idea of the proposed precharging strategy is based on control of the number of blocked and bypassed SM capacitors

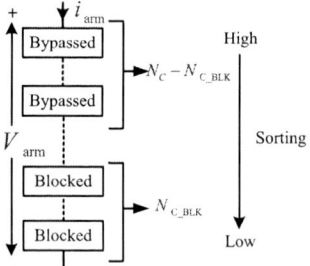

Fig. 3. The proposed precharging strategy for an arm.

in conjunction with the conventional SM capacitor voltage sorting algorithm so that all SM capacitors can be charged to their nominal voltages V_{cref}. When the MMC is connected to a dc voltage source on its dc side or to an ac voltage source on its ac side, a steady-state voltage V_{arm} is established across each arm, as shown in Fig. 3. Each SM capacitor is charged to a certain voltage level, which is usually less than its nominal value during Phase I. For the HB SM, the blocked SM capacitors are charged when the arm current is positive. When the arm current is negative, the charging current flows through the anti-parallel diode of S2. Thus, the positive arm voltage can charge the capacitors during the blocked SM capacitor state. During Phase II, assuming all SMs are controllable, n_{C_BLK} SM capacitors out of N_C SM capacitors of each arm are blocked and $(N_C - n_{C\_\mathrm{BLK}})$ SM capacitors are bypassed during startup process to charge the SM capacitors. N_C is the total number of SM capacitors in each arm, where $N_C = N$ for the HB-MMC system. As shown in Fig. 3, to charge the capacitors to their nominal voltage V_{cref}, the steady-state value of n_{C_BLK} is given by

$$n_{C\_\mathrm{BLK}} = N_{C\_\mathrm{BLK}} = \frac{V_{\mathrm{arm}}}{V_{\mathrm{cref}}}, \qquad (1)$$

where V_{arm} depends on various operating conditions and will be discussed in the following sections. To balance the capacitor voltages/energy during the startup precharging process, the conventional sorting algorithm is employed [15], [16]. In each arm, $(N_C - n_{C\_\mathrm{BLK}})$ SM capacitors with the highest voltages are bypassed while n_{C_BLK} SM capacitors with the lowest voltages are blocked during the startup precharging process. After startup process, N_{C_BLK} SM capacitors support the charging voltage V_{arm} while the sorting algorithm maintains the capacitor voltages/energy of all SMs at the same level.

Besides the HB SM, other SM circuits including the full-bridge (FB), the unipolar-voltage full-bridge (UFB), the clamp-double (CD), the three-level/five-level cross-connected (3LCC/5LCC) SMs can be also used for the purpose of dc-fault blocking [1], [7], [17]. Considering all these SM circuits, a general startup precharging procedure is developed and described in Fig. 4. Subsequent to Phase I, the SM capacitors are precharged to an initial steady-state voltage and the SMs become controllable. In case of the fault-blocking SMs, i.e., the FB, UFB, CD, 3LCC and 5LCC SMs, their conducting

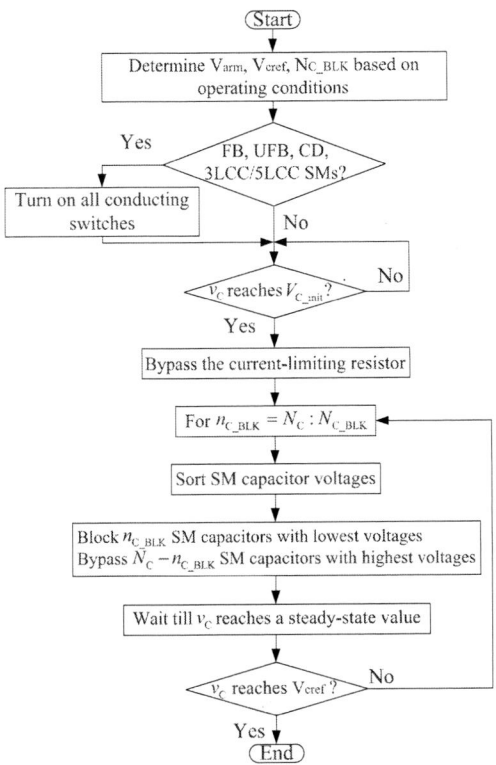

Fig. 4. Flowchart of the proposed startup strategy of the MMC.

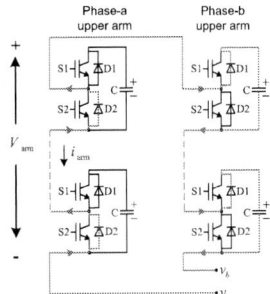

Fig. 5. HB-MMC charged from ac side.

switches are turned on. In this way, all SMs become equivalent to the HB SMs and controlled as either in the blocked or the bypassed state. After all capacitor voltages reach their steady-state value, the current-limiting resistor is bypassed and the number of the blocked SM capacitors n_{C_BLK} is controlled from N_C to N_{C_BLK} smoothly while, in conjunction with the capacitor voltage sorting algorithm, the number of bypassed SM capacitors is changed from zero to $(N_C - N_{C\_BLK})$. The changing rate of n_{C_BLK} can limit the inrush charging current. The steady-state value of n_{C_BLK} is determined by (1) to charge the SM capacitors to their nominal voltages.

III. STARTUP PROCEDURES FOR VARIOUS MMC CONFIGURATIONS

A. DC-Side Startup

Due to a positive dc-bus voltage on the dc side of the MMC in Fig. 1, a positive arm voltage is generated, which is equal to $V_{\mathrm{arm}} = \frac{V_{dc}}{2}$. Regardless of the SM circuit topology, the charging current is positive, charging the SM capacitor. To charge the SM capacitor to its nominal voltage, the steady-state number of the blocked SM capacitors is

$$N_{C\_BLK} = \frac{V_{\mathrm{arm}}}{V_{\mathrm{cref}}} = \frac{V_{dc}}{2V_{\mathrm{cref}}}. \quad (2)$$

Before starting up, the initial steady-state voltage for each SM capacitor is

$$V_{C\_\mathrm{init}} = \frac{V_{\mathrm{dc}}}{2N_C}, \quad (3)$$

which is only half of the nominal capacitor voltage.

If the dc-bus voltage is not at its nominal value, the MMC can be still started up by choosing appropriate N_{C_BLK} based on (2). Thus, by the proposed startup strategy, a black start [10] can be realized with a low dc-bus voltage (i.e., $V_{\mathrm{dc}} = 2V_{\mathrm{cref}}$ and $N_{C\_BLK} = 1$).

B. AC-Side Startup

When the ac side of the MMC is connected to an ac grid/voltage source, the arm voltage is determined by the ac-side line-to-line voltage and the SM circuit topology.

1) Startup Procedure for the HB-MMC: For the HB-MMC, when v_{ab} is positive, the current flows from phase-a upper arm to phase-b upper arm. Due to the HB SM circuit, the current flows through the anti-parallel diodes of the SMs in phase-a upper arm, charging the SM capacitors of phase-b upper arm, as shown in Fig. 5. Thus, the maximum available charging voltage in each arm is $V_{\mathrm{arm}} = V_{\mathrm{LL}}$, where V_{LL} is the amplitude of the ac-side line-to-line voltage.

To charge the SM capacitors to their nominal voltages, the steady-state number of blocked SM capacitors is

$$N_{C\_BLK} = \frac{V_{\mathrm{LL}}}{V_{\mathrm{cref}}}. \quad (4)$$

Before starting up, the initial steady-state voltage is $V_{C\_\mathrm{init}} = \frac{V_{\mathrm{LL}}}{N_C}$. By choosing appropriate N_{C_BLK}, a smaller ac-side voltage source can still charge the SM capacitors to their nominal voltage.

2) Startup Procedure for the FB-MMC: For the FB-MMC, during Phase I, the ac-side current charges the SM capacitors in both phase-a and phase-b arms, as shown in Fig. 6. Thus, during Phase I, when blocking all SMs, the maximum available charging voltage across each arm is $\frac{V_{\mathrm{LL}}}{2}$. The steady-state SM capacitor voltage during Phase I is

$$V_{C\_\mathrm{init\_I}} = \frac{V_{\mathrm{LL}}}{2N_C}. \quad (5)$$

978-1-4673-9551-9/16 $31.00 © 2016 IEEE

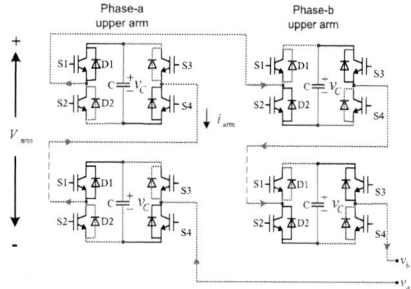

Fig. 6. FB-MMC charged from ac side.

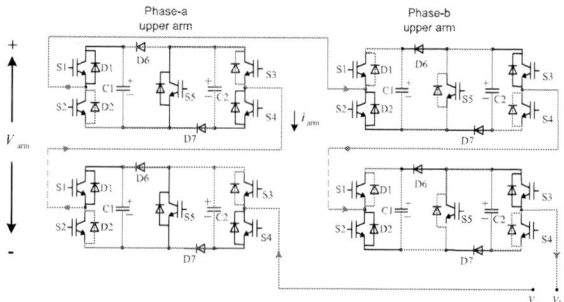

Fig. 7. CD-MMC charged from ac side.

During Phase II, when all SMs become controllable, the conducting switch (i.e., $S4$ of Fig. 6) is turned on. The ac-side current flows through the anti-parallel diodes of one arm and charge the SM capacitors in the other arm, which has similar operating characteristics of the HB-MMC. Consequently, the initial steady-state SM capacitor voltage during Phase II becomes $V_{C\_init} = \frac{V_{LL}}{N_C}$, which is the same as that of the HB-MMC. Thus, the steady-state number of the blocked SM capacitors can be chosen the same as that of the HB-MMC, i.e., $N_{C\_BLK} = \frac{V_{LL}}{V_{cref}}$. For the UFB, 3LCC, and 5LCC SMs in [7], similar procedures are applied.

3) Startup Procedure for the CD-MMC: For the CD-MMC, during Phase I, the ac-side current charges the SM capacitors in both phase-a and phase-b arms, as shown in Fig. 7. However, when the arm current is negative, every two capacitors connected in parallel in each SM are charged. The equivalent number of SM capacitors in charging loop is $\frac{N_C}{2} + N_C = 1.5N_C$. Thus, the initial steady-state SM capacitor voltages during Phase I is

$$V_{C\_init\_I} = \frac{V_{LL}}{1.5N_C}. \tag{6}$$

During Phase II, when all SMs become controllable, the conducting switch (i.e., $S5$ of Fig. 7) is turned on and the initial steady-state SM capacitor voltages during Phase II become $V_{C\_init} = \frac{V_{LL}}{N_C}$, which is the same as that of the HB-MMC. Therefore, the startup procedure for the CD-MMC is

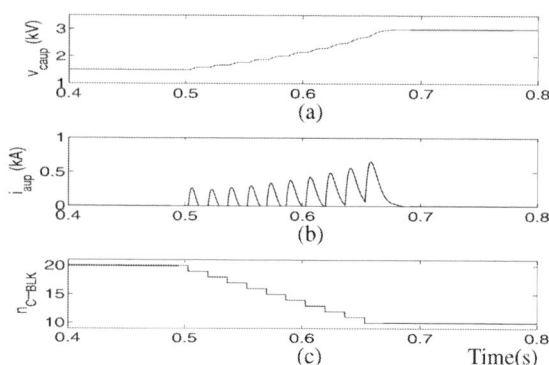

Fig. 8. Dc-side startup process of the HB-MMC system: (a) SM capacitor voltages of the upper arm of phase-a, (b) phase-a upper arm current, and (c) the commanded number of blocked SM capacitors within each arm.

similar to that of the FB-MMC.

IV. STUDY RESULTS

In this section, performance of the proposed strategy for various MMC systems with different SM circuits is evaluated based on simulation studies in the PSCAD/EMTDC environment. The study system parameters are the same as those used in [7].

A. Startup Process of an MMC from the Dc Side

Figure 8 shows the startup process of the HB-MMC system from its dc side. During Phase I, the SMs are uncontrollable and the current-limiting resistor is inserted to limit the charging current. The SM capacitor voltages are charged to their steady-state voltage $V_{C\_init} = \frac{V_{dc}}{2N_C} = 1.5$ kV at $t = 0.5$ s. During Phase II, the SMs are controllable and the current-limiting resistor is bypassed. From $t = 0.5$ s, the number of blocked SM capacitors is being dynamically changed from $N_C = 20$ to $N_{C\_BLK} = \frac{V_{dc}}{2V_{cref}} = \frac{60kV}{2 \times 3kV} = 10$ to charge the SM capacitors to their nominal voltages, as shown in Figs. 8(a) and (c). The arm currents are limited by the changing rate of n_{C_BLK}, as shown in Fig. 8(b). As expected, the startup process of the MMCs based on the SMs with fault-blocking capability is similar to that of the HB-MMC.

The proposed strategy is also applicable for the black start of the MMC systems. Figure 9 shows the black start of the HB-MMC system from dc side. The dc side is supplied by a low dc voltage source, i.e., $V_{dc} = 2V_{cref} = 6$ kV. Based on (2), $N_{BLK} = 1$. As shown in Fig. 9, due to the low dc-bus voltage, a longer time is needed to charge the SM capacitors to their nominal voltages.

B. Startup Process of an MMC from Ac Side

In Fig. 10, the ac-side startup process of the HB-MMC system is shown. During Phase I (prior to $t = 1$ s), assuming the uncontrollable SMs, the current-limiting resistor is inserted to limit the charging current. The SM capacitor voltages are charged to their steady-state voltage $V_{C\_init} = \frac{V_{LL}}{N_C} = 2.1$ kV. After Phase I, the SMs are controllable and the current-limiting

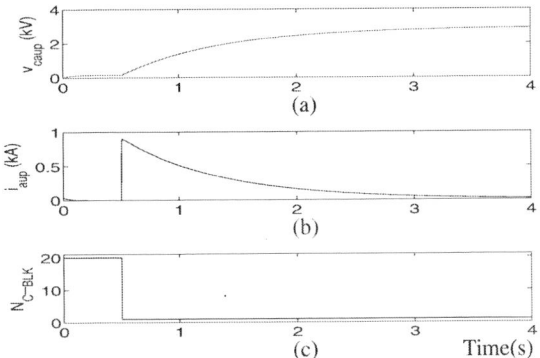

Fig. 9. Black start of the HB-MMC system: (a) SM capacitor voltages of the upper arm of phase-a, (b) phase-a upper arm current, and (c) the commanded number of blocked SM capacitors within each arm.

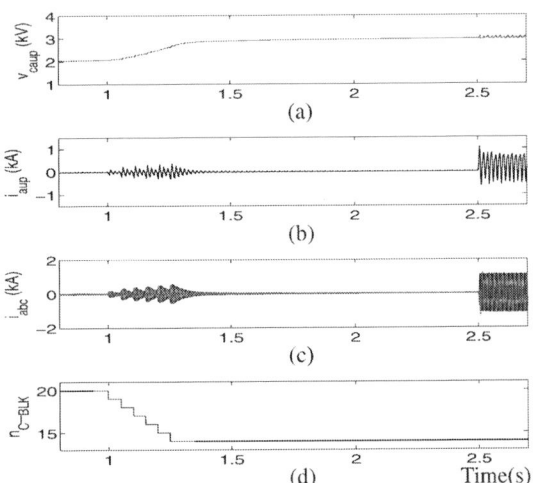

Fig. 10. Ac-side startup process of the HB-MMC system: (a) SM capacitor voltages of the upper arm of phase-a, (b) phase-a upper arm current, (c) ac-side three-phase currents, and (d) the commanded number of blocked SM capacitors within each arm.

resistor is bypassed at $t = 1$ s. The number of blocked SM capacitors is being changed from $N_C = 20$ to $N_{C\_BLK} = 14$ to charge the SM capacitors to their nominal voltages, as shown in Figs. 10(a) and (d). The arm and ac-side currents are limited by the changing rate of n_{C_BLK}, as shown in Figs. 10(b) and (c). After the startup process, the MMC starts to transfer power between ac and dc side at $t = 2.5$ s.

The ac-side startup process of the FB-MMC system is illustrated in Fig. 11. During Phase I (prior to $t = 0.5$ s), the SMs are uncontrollable and the SM capacitor voltages are charged to their steady-state voltage $V_{C\_init\_I} = \frac{V_{LL}}{2N_C} = 1.1$ kV. After Phase I, the SMs are controllable and all conducting switches are turned on at $t = 0.5$ s. Consequently, the SM capacitors are charged to their new steady-state voltage $V_{C\_init} = \frac{V_{LL}}{N_C} = 2.1$ kV at $t = 1$ s, which is the same as that of the HB-MMC. At $t = 1$ s, the current-limiting resistor is

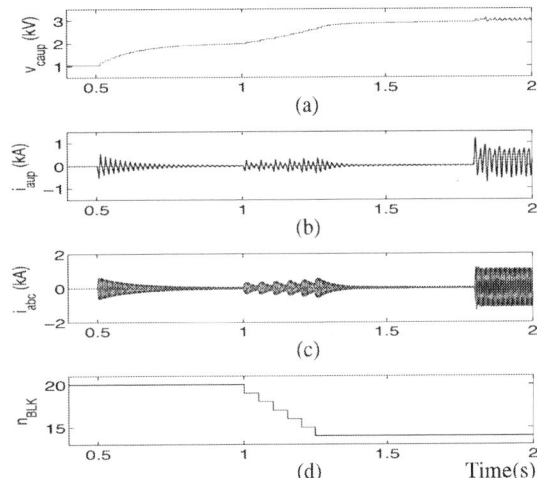

Fig. 11. Ac-side startup process of the FB-MMC system: (a) SM capacitor voltages of the upper arm of phase-a, (b) phase-a upper arm current, (c) ac-side three-phase currents, and (d) the commanded number of blocked SM capacitors within each arm.

Fig. 12. SM capacitor voltages of the upper arm of phase-a during ac-side startup process of the CD-MMC system.

bypassed, and the number of blocked SM capacitors is being dynamically changed from $N_C = 20$ to $N_{C\_BLK} = 14$ to charge the SM capacitors to their nominal voltages, as shown in Fig. 11. The arm and ac-side currents are limited by the changing rate of n_{C_BLK}. After the startup process, the MMC starts to transfer power between its ac and dc side at $t = 1.8$ s. For the UFB, 3LCC, and 5LCC SM-based MMC systems, the startup process is similar to that of the FB-MMC.

The startup process of the CD-MMC is also similar to that of the FB-MMC. However, the initial steady-state capacitor voltage is different. For the CD-MMC, $V_{C\_init\_I} = \frac{V_{LL}}{1.5N_C} = 1.4$ kV, as shown in Fig. 12

V. CONCLUSION

In this paper, a generalized precharging strategy is proposed for the MMC-based systems built upon various SM circuits under ac- and dc-side startup conditions. The steady-state SM capacitor voltages of various MMC systems during startup process are analyzed. The proposed startup strategy can smoothly charge the SM capacitors without using any additional feedback control loop, extra measurements, and/or auxiliary power supplies. Moreover, the proposed procedure is also applicable for the black start application. Effectiveness of the proposed strategy for various MMCs under ac- and dc-side startup conditions, is evaluated based on time-domain simulation studies in the PSCAD/EMTDC software environment. The

978-1-4673-9551-9/16 $31.00 © 2016 IEEE

study results demonstrate the proposed strategy can charge the SM capacitors smoothly and limit the inrush charging current during the startup process for various MMC configurations under different startup conditions.

REFERENCES

[1] R. Marquardt, "Modular multilevel converter: An universal concept for HVDC-networks and extended DC-bus-applications," in *International Power Electronics Conf.*, Jun. 2010, pp. 502–507.

[2] S. Debnath, J. Qin, B. Bahrani, M. Saeedifard, and P. Barbosa, "Operation, control, and applications of the modular multilevel converter: A review," *IEEE Trans. Power Electron.*, vol. 30, no. 1, pp. 37–53, 2015.

[3] R. Marquardt, "Modular multilevel converter topologies with dc-short circuit current limitation," in *Power Electronics and ECCE Asia (ICPE ECCE), 2011 IEEE 8th International Conference on*, May 2011, pp. 1425–1431.

[4] T. Modeer, H.-P. Nee, and S. Norrga, "Loss comparison of different sub-module implementations for modular multilevel converters in HVDC applications," in *Power Electronics and Applications (EPE 2011), Proceedings of the 2011-14th European Conference on*, 2011, pp. 1–7.

[5] G. Adam, K. Ahmed, S. Finney, K. Bell, and B. Williams, "New breed of network fault-tolerant voltage-source-converter HVDC transmission system," *IEEE Trans. Power Syst.*, vol. 28, no. 1, pp. 335–346, Feb 2013.

[6] A. Nami, J. Liang, F. Dijkhuizen, and G. Demetriades, "Modular multilevel converters for HVDC applications: Review on converter cells and functionalities," *IEEE Trans. Power Electron.*, vol. 30, no. 1, pp. 18–36, 2015.

[7] J. Qin, M. Saeedifard, A. Rockhill, and R. Zhou, "Hybrid design of modular multilevel converters for HVDC systems based on various submodule circuits," *IEEE Trans. Power Del.*, vol. 30, no. 1, pp. 385–394, 2015.

[8] K. Shi, F. Shen, D. Lv, P. Lin, M. Chen, and D. Xu, "A novel start-up scheme for modular multilevel converter," in *Energy Conversion Congress and Exposition (ECCE), 2012 IEEE*, Sept 2012, pp. 4180–4187.

[9] A. Das, H. Nademi, and L. Norum, "A method for charging and discharging capacitors in modular multilevel converter," in *IECON 2011 - 37th Annual Conference on IEEE Industrial Electronics Society*, Nov 2011, pp. 1058–1062.

[10] R. Marquardt and A. Lesnicar, "New concept for high voltage-modular multilevel converter," in *IEEE Power Electronics Specialists Conf.*, Aachen, Germany, 2004.

[11] J. Xu, C. Zhao, B. Zhang, and L. Lu, "New precharge and submodule capacitor voltage balancing topologies of modular multilevel converter for VSC-HVDC application," in *Power and Energy Engineering Conference (APPEEC), 2011 Asia-Pacific*, March 2011, pp. 1–4.

[12] B. Li, D. Xu, Y. Zhang, R. Yang, G. Wang, W. Wang, and D. Xu, "Closed-loop precharge control of modular multilevel converters during start-up processes," *IEEE Trans. Power Electron.*, vol. 30, no. 2, pp. 524–531, Feb 2015.

[13] Y. Yang, Q. Ge, M. Lei, X. Wang, X. Yang, and R. Gou, "Precharging control strategies of modular multilevel converter," in *Electrical Machines and Systems (ICEMS), 2013 International Conference on*, Oct 2013, pp. 1842–1845.

[14] Y. Xue, Z. Xu, and G. Tang, "Self-start control with grouping sequentially precharge for the C-MMC-Based HVDC system," *IEEE Trans. Power Del.*, vol. 29, no. 1, pp. 187–198, Feb 2014.

[15] S. Rohner, S. Bernet, M.Hiller, and R. Sommer, "Modelling, simulation and analysis of a modular multilevel converter for medium voltage applications," in *IEEE International Conference on Industrial Technology,ICIT*, 2010, pp. 775–782.

[16] M. Saeedifard and R. Iravani, "Dynamic performance of a modular multilevel back-to-back HVDC system," *IEEE Trans. Power Del.*, vol. 25, no. 4, pp. 2903–2912, Oct. 2010.

[17] A. Nami, L. Wang, F. Dijkhuizen, and A. Shukla, "Five level cross connected cell for cascaded converters," in *Proc. European Conf. Power Electronics and applications (EPE)*, 2013.

978-1-4673-9551-9/16 $31.00 © 2016 IEEE

Conducted EMI Analysis and Filter Design for MHz Active Clamp Flyback Front-end Converter

Xiucheng Huang, Junjie Feng, Fred C. Lee, Qiang Li, and Yuchen Yang

Center for Power Electronics Systems, the Bradley Department of Electrical and Computer Engineering
Virginia Polytechnic Institute and State University, Blacksburg, VA, 24061, USA

Abstract— **High frequency is the major catalyst for size reduction in the advancement of power conversion technology. It is essential to understand the EMI characteristic of high frequency converter since the EMI filter typically occupies one third of total system volume. This paper presents an insight view of CM/DM noise transformation in Flyback converter for low power adapter applications. The so-called mixed-mode noise is DM noise transformed from CM noise due to unbalanced impedance when the diode-bridge is off. The transformer shielding technique can effectively reduce the CM noise, and therefore it can significantly reduce the magnitude of CM transformed noise. A 65W (19V/3.3A) prototype of active clamp flyback front-end converter operating around 1MHz with GaN devices is developed to verify the analysis. A single stage EMI filter is designed accordingly with much smaller size compared with industry practice.**

Keywords—Gallium nitride device, active clamp flyback, soft-switching, common-mode noise, differential-mode noise, transformation, EMI filter

I. INTRODUCTION

High efficiency and high density are the key drivers and metrics for the advancement of power conversion technology. The pushing for higher switching frequency is the major catalyst for performance improvement and size reduction. The emerging GaN device, with much improved figures of merit, opens the door for operating frequency well into the MHz range [1-8]. However, Electromagnetic Interference (EMI) problems is more challenging due to high dv/dt induced by GaN devices. Fully understanding the EMI mechanism and designing the filter accordingly are also important from efficiency and density perspective.

Flyback front-end converters are widely used in low power offline application due to its simplicity and low cost. Several literatures [6-8] have demonstrated flyback converter with GaN devices operating over 1MHz in order to reduce passive components volume. Usually, an RCD clamp circuit is necessary to dissipate the leakage energy and suppress the voltage spike when the switch is turned off. However, the leakage energy is proportional to the switching frequency and it is quite considerable at MHz frequency range. Moreover, the voltage ringing causes high voltage slew rate which has significant impact on EMI noise, especially at 10~30MHz range. The active clamp flyback converter can clamp the voltage without any ringing and recycle the transformer leakage energy [9-11]. Zero-voltage-switching (ZVS) for both main switch and clamping switch can be realized by proper design as well, which is beneficial for efficiency and EMI. The modeling and analysis of conducted EMI noise of flyback

front-end converter are illustrated in [12-15]. The common-mode (CM) noise is majorly coupled through the parasitic capacitance of the transformer and the differential-mode (DM) noise is determined by switching current ripple and input impedance. Except for the CM and DM noise, there is additional mixed-mode noise which is transformed from unbalanced CM current. This mix-mode noise also contributes to the total DM noise when it is measured using spectrum separator. While the mix-mode noise phenomenon has been described and some analysis are reported in their studies, the fundamental mechanism by which the unbalanced current is induced has not been revealed. The detailed analysis of CM/DM noise transformation is shown in section II.

On the other hand, shielding technique is used to block the CM noise path from either side of transformer [16, 17]. The traditional flyback transformer is in hand-made fashion and it is complicated to insert a shielding. With much higher switching frequency, PCB winding based transformer is feasible and it is easier to integrate the shielding layer as well as precisely control the parasitics. Conventional theory only expects shielding has impact on CM noise reduction, however, it also helps reduce the total DM noise by reducing the part of mix-mode noise. The impact of shielding on EMI noise is illustrated in section III.

With better understanding of EMI noise mechanism and impact of shielding on total noise, EMI filter can be designed effectively. High frequency operation and shielding have significant impact on EMI filter design. The corner frequency of CM and DM filter shifts to higher frequency which requires smaller CM and DM choke. One-stage filter can be used to achieve the required attenuation over whole conducted EMI noise testing frequency range (150kHz ~ 30MHz). The detailed filter design is discussed in section IV.

A MHz active clamp flyback converter is built to verify the theoretical analysis and filter design. The experimental results are shown in section V. The prototype can achieve much higher density compared with industry practice due to much higher switching frequency and smaller EMI filter. By utilizing GaN devices and PCB winding transformer, the prototype can achieve very high efficiency which is very important for thermal consideration in adapter application.

II. ANALYSIS OF CM/DM NOISE TRANSFORMATION

To better understand the CM/DM transformation mechanism, the DM and CM noise propagation path of flyback converter is shown in Fig. 1. The DM noise source is the switching current ripple of the transformer primary winding, which is

978-1-4673-9551-9/16 $31.00 © 2016 IEEE

represented as I_S. Majority of the noise current is bypassed by DC bus capacitor and the remaining noise current flows through the diode bridge (either diode or junction capacitance of the diode) and Line Impedance Stabilization Network (LISN). The current in these two paths is determined by impedance.

The primary switches induce high dv/dt which dominates the overall CM noise magnitude, and it is represented as a noise source V_S. Compared to the primary to the power earth parasitic capacitance C_{PE}, the interwinding capacitance of the flyback transformer C_{PS} and the secondary ground to power earth parasitic capacitance C_{SE} are much larger and it is the major coupling path of the CM noise. The CM noise current flows through the LISN, diode bridge (either diode or junction capacitance of the diode), and back to the noise source.

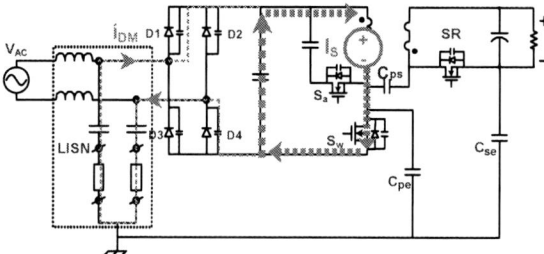

(a) DM noise path

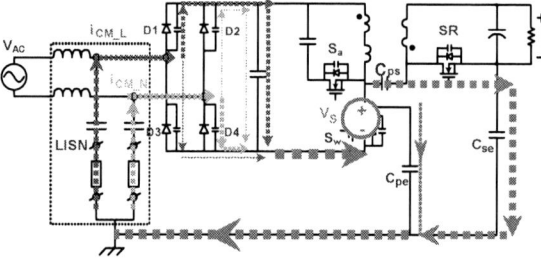

(b) CM noise path

Fig. 1 EMI noise path of active clamp flyback converter

It has been identified that CM/DM noise transformation occurs when the diode-bridge is off [12-15]. Using zero-span mode of a spectrum analyzer, the magnitude of a selected EMI frequency can be displayed with respect to time [12]. Fig. 2 shows the time-domain waveform of the switching frequency noise in spectrum analyzer based on a 65W MHz flyback front-end converter. The diode bridge on and off interval are clearly marked. The magnitude of DM noise in diode-on interval was expected to be higher than diode-off interval due to lower impedance when diode is on. However, the measured DM noise magnitude in diode-off interval is much higher than diode-on interval, especially in the short period right after or before diode is on. The explanation of CM/DM noise transformation in [12, 13, 15] is that one of the four diode is forward bias by CM current and causes CM noise path impedance unbalance. This is actually not an accurate description, the fundamental reason of this phenomenon is illustrated as follows.

As mentioned above, the majority of CM noise current flows through Cps and Cse to earth and LISN. Then the CM noise current go through L and N lines separately. The L line current goes through the junction capacitors of D1 and D3 during the diode-off period. Similarly, the N line current goes through D2 and D4 junction capacitors, and then the two current merges together back to the noise source.

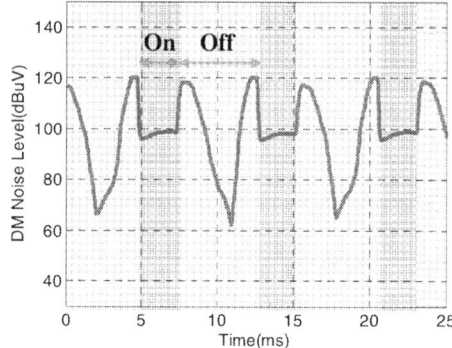

Fig. 2 DM noise of switching frequency using zero-span mode in spectrum analyzer

Since the CM noise current is induced by switch in every switching instant, there is switching frequency current charging and discharging the junction capacitors of the diode bridge, and the high frequency voltage ripple of diode voltage is observed during diode off period, as shown in Fig. 3. The zoom in waveform at t_1 instant at where the diode bridge is just off during positive line cycle is shown in Fig.4. The voltage ripple induced by CM noise current is about 10V.

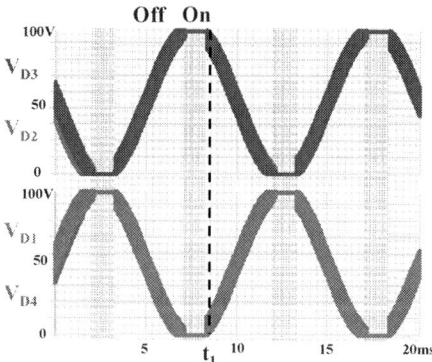

Fig. 3 Diode bridge voltage waveforms in line cycle

As shown in Fig. 4, V_{D1} and V_{D4} are much smaller than V_{D2} and V_{D3} at t_1 instant, therefore, the impedance of junction capacitor of D1 and D4 is much lower compared with D2 and D3 according to the nonlinearity of diode junction capacitance relationship with reverse voltage which is shown in Fig. 5 [18]. As a result, the CM noise current mainly go through D1 and D4. More importantly, the 10V voltage ripple impacts the junction capacitance significantly when the reverse voltage is low according to Fig.5. In other words, there is a great discrepancy between the impedance of junction capacitance of

978-1-4673-9551-9/16 $31.00 © 2016 IEEE   1535

D1 and D4 due to nonlinear junction capacitance. Consequently, the CM noise current in L line and N line is different due to unbalanced impedance, as illustrated by Fig. 6. The difference between CM noise current in L line and N line can be measured as DM noise, which means CM noise transformed into DM noise in such a manner. The impedance of diode junction capacitance is more balanced in the middle of diode-off period, therefore the magnitude of DM noise gradually reduces due to less CM/DM transformation.

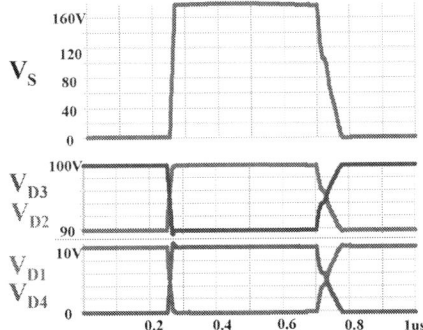

Fig. 4 Diode bridge voltage waveform at t1 instant

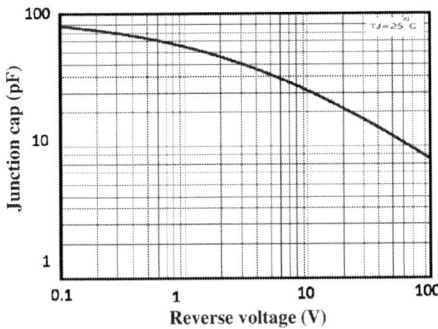

Fig. 5 Diode junction capacitance characteristics

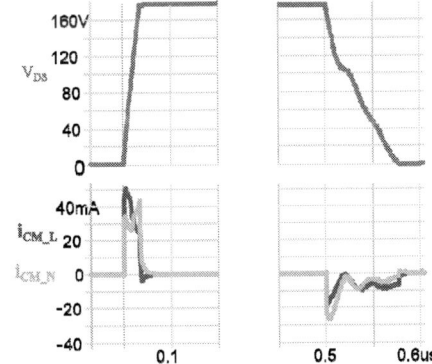

Fig. 6 CM noise current in L line and N line

It is obvious that the DM filter will be overdesigned according to Fig. 2 without understanding the mechanism of CM/DM noise transformation.

III. REDUCTION OF CM/DM TRANSFORMATION WITH SHIELDING TECHNIQUE

There are several way to minimize the CM/DM noise transformation. The most straight-forward method is to parallel sufficient large capacitors (several nF) with each diode of the bridge to make impedance balanced and therefore eliminate the transformation. Adding X-cap (C_X in Fig. 14 and Fig. 15) also can balance the impedance and suppress the transformed noise [12-15]. However, both of these method try to solve the problem passively with extra components. A more effective way is try to reduce the CM noise source current.

A. Analysis of shielding impact on CM/DM noise

It is well known that shielding technique can block the CM noise from either side of transformer [16]. The CM noise propagation path with shielding is shown in Fig. 7. The CM noise current path through interwinding capacitance is cut off and the amplitude of the remaining noise current is small due to small parasitic capacitance C_{pe}. As a result, the magnitude of the DM noise which is transformed from CM noise is reduced. Moreover, the switching frequency voltage ripple of diode in diode-off period caused by CM noise current also decreases. The difference of junction capacitance of diode is much smaller and this makes the impedance more balanced.

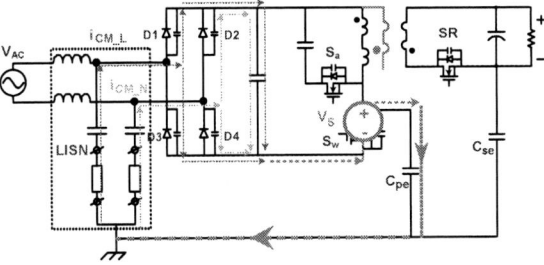

Fig. 7 CM noise path with shielding

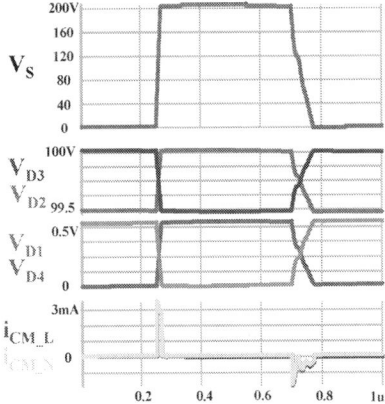

Fig. 8 Detailed diode-bridge voltage waveform and CM noise current in L line and N line right after diode bridge is off

978-1-4673-9551-9/16 $31.00 © 2016 IEEE   1536

The detailed switching frequency ripple on diode-bridge is shown in Fig. 8. The magnitude of the voltage ripple is reduced to 0.5V. The L line and N line CM noise current reduces to 3mA and the difference becomes very small due to more balanced impedance.

The measured CM/DM noise with shielding is shown in Fig. 9. The CM noise reduces by 27dB at switching frequency and the shielding is effective in the whole testing frequency range (up to 30MHz). This is an important feature since the high frequency noise is difficult to be attenuated by filters due to the poor characteristic of filter at that frequency range. The DM noise reduces by 20dB at all frequency range due to less CM noise and CM/DM transformation. As illustrated by the time-domain waveform shown in Fig. 10, DM noise in diode-off interval is smaller than diode on interval.

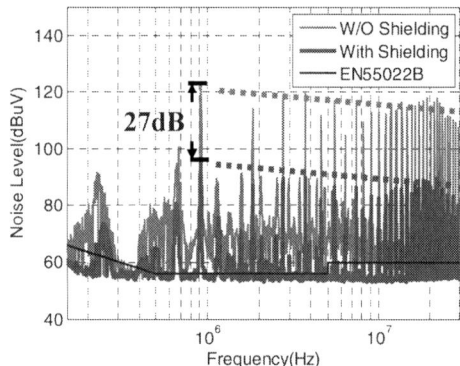

(a) CM noise measurement w/ and w/o shielding

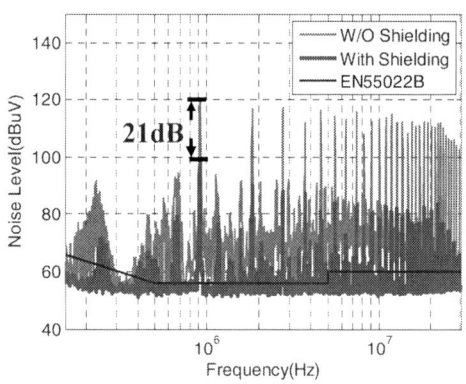

(b) DM noise measure w/ and w/o shielding

Fig. 9 EMI noise measurement

B. Design consideration of shielding layer

The complete shielding can be easily implemented in the PCB winding based transformer [16]. Fig. 11 shows the circuit diagram of flyback converter with shielding and Fig. 12 shows the transformer structure with shielding. The shielding is connected to the primary ground. Therefore the CM noise current coming from the primary noise source is circulating within the primary side. Making the shielding layer exactly the

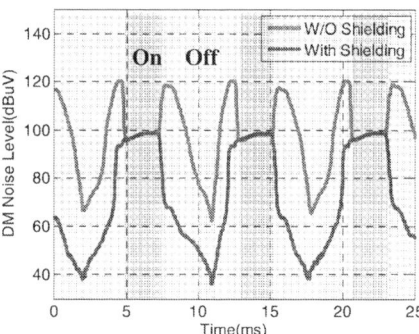

Fig. 10 DM noise of switching frequency using zero-span mode in spectrum analyzer

same as the secondary side winding generates no voltage difference between secondary side winding and shielding, which means no CM current flowing from either side. Rotating the shielding layer with any angle creates displacement current circulating between secondary side winding and shielding, but it does not create CM current from either side [16].

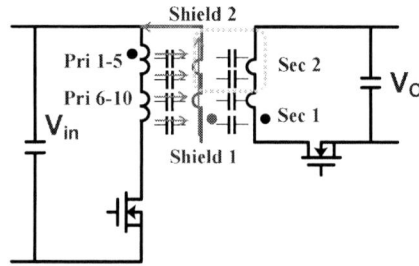

Fig. 11 Flyback transformer with shielding

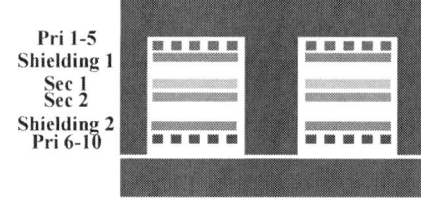

(a) Transformer winding cross section view

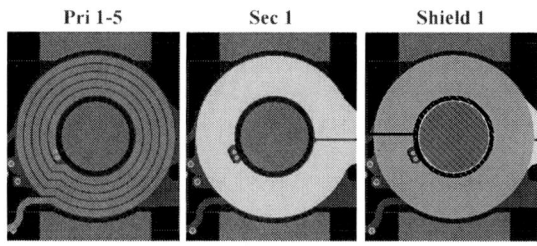

(b) detailed PCB layout view

Fig. 12 Flyback transformer winding structure with shielding

One of the general concern of adding shielding layer to the flyback transformer is the possible winding loss increase. The

3D FEA simulation is carried out to evaluate the impact of the shielding layer on the total transformer winding loss as shown in Fig. 13. It shows that the current distribution in the primary and secondary winding almost remains the same and the current density in shielding layer is also very small. The total winding loss with the shielding layer is 1.35W and it is only 0.05W larger than the case without shielding layer, which is negligible loss increase.

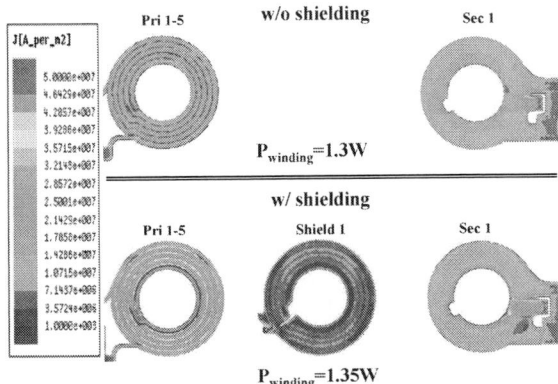

Fig. 13 3D FEA simulation of flyback transformer with and without shielding layer

IV. EMI FILTER DESIGN OF MHz ACTIVE CLAMP FLYBACK CONVERTER

It is essential to understand the mechanism of CM/DM noise transformation and the impact of shielding on the total EMI noise in order to design the EMI filter with required attenuation [19, 20].

The typical EMI filter for adapter application is shown in Fig. 14. L_{CM2} is the major CM choke with large number of turns to attenuate the switching frequency and harmonic frequency noise. The parasitic capacitance of the L_{CM2} is relatively large and the impedance at high frequency turns to be capacitive. L_{CM1} is usually made in a small toroid core with few turns to deal with very high frequency range (10MHz~30MHz) CM noise. The leakage inductance of L_{CM1} and L_{CM2} is not sufficient to attenuate the DM noise due to low switching frequency. An additional L_{DM} is required to suppress the DM noise. Therefore, three chokes in total are needed to attenuate the total EMI noise which makes the EMI filter quite large. Generally speaking, the EMI filter occupies at least one fourth of the total system volume.

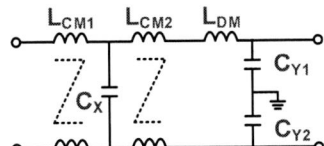

Fig. 14 EMI filter structure for conventional adapter application

It is always true that the desired corner frequency of the CM and DM filter increases with switching frequency when switching frequency is higher than 150kHz. The corner frequency can be calculated by the desired attenuation at switching frequency according to the conducted EMI standard EN55022B. Due to MHz operation and effectiveness of the designed shielding layer, the simple one-stage filter can be used to attenuate the total EMI noise, as shown in Fig. 15. It is assumed that the attenuation of the one-stage filter is 40dB/dec after the corner frequency. Based on the measurement results shown in Fig. 8, the corner frequency of CM and DM increases to around 100kHz which is much higher than current industry practice. The maximum Y-cap that can be used is limited by the leakage current defined in IEC60950. Two 1nF Y-cap is used in the prototype design. Then L_{CM} can be calculated according to the CM filter corner frequency. L_{DM} is the leakage inductance of CM choke and C_X is calculated based on DM filter corner frequency. The parameters of the EMI filter is summarized in Table I.

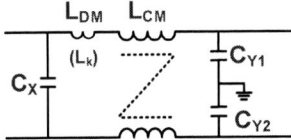

Fig. 15 One stage EMI filter

TABLE I. PARAMETERS OF THE EMI FILTER

Parameters	L_{CM}	L_{DM}	$C_{Y1}=C_{Y2}$	C_X
Value	1.4mH	0.03mH	1nF	130nF

The core materials for CM/DM chokes in current products have high permeability at few hundreds of kHz and drops sharply at frequency above 500kHz. They are good enough for the converter operates below few hundreds of kHz. However, these kinds of materials are not suitable for MHz application. The ferrite material 3E6 from Ferroxcube is shown as an example. The high initial permeability provides high inductance at frequency range from tens of kHz to few hundreds of kHz. The permeability drops sharply above 500kHz which may result in insufficient attenuation at high frequency.

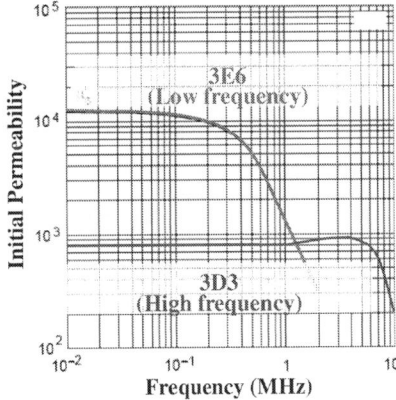

Fig. 16 Core material comparison

978-1-4673-9551-9/16 $31.00 © 2016 IEEE

The desired features for high frequency chokes includes relative high initial permeability and relative stable permeability over the frequency range from 1MHz to 5MHz. Based on the material survey, the ferrite materials 3D3 from Ferroxcube is chosen for the prototype design. The permeability curve is also shown in Fig.16. It clearly shows that 3D3 has flat permeability over wide frequency range which is quite critical to suppress the harmonic frequency noise.

To reduce the volume and footprint of the EMI filter, a smaller toroid core is preferred. Two stacked TC9.5/4.8/3.2 toroid cores are selected. The turns number for each winding is 40 and AWG 28 solid wire is used for the two windings. The basic parameters are listed in Table I. The parasitic capacitance is around 15pF which is acceptable in this design. The prototype of the CM/DM choke is shown in Fig. 17. The volume is $10 \times 10 \times 8$ mm^3.

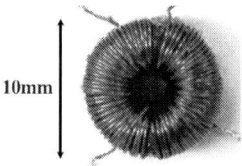

Fig. 17 Picture of CM/DM choke for MHz flyback converter

V. EXPERIMENTAL RESULTS AND DISCUSSIONS

In order to verify the EMI analysis and filter design of MHz flyback converter, a 65W (19.5V/3.3A) prototype is developed as shown in Fig. 18.

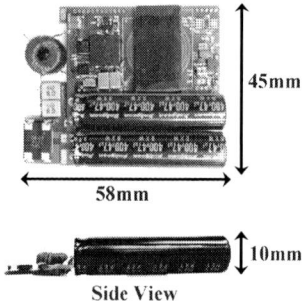

Fig. 18 1MHz 65W active clamp flyback converter prototype

GaN devices are used for both primary and secondary switches to reduce conduction and switching losses. The 600V GaN HEMT from Transphorm is used as the primary main switch and active clamp switch. The 150V eGaN from EPC is used as the synchronous rectifier. The PCB board is 6-layer which includes two layers shielding as shown in Fig. 11. The power density excludes the case is over 40W/in^3, which is at least two times higher than the state-of-the-art products.

Fig. 19 shows that ZVS is achieved for primary switches and there is no voltage spike across the main switch which is beneficial for EMI noise reduction.

Fig. 20 shows the CM/DM noise spectrum with peak mode measurement under 110V$_{AC}$ input full load output condition which is the worst case for the prototype design. The red

curves are the results with shielding but without filters. The blue curves are the final results with EMI filter. It clearly shows that the blue curves are already lower than the EN55022B standard.

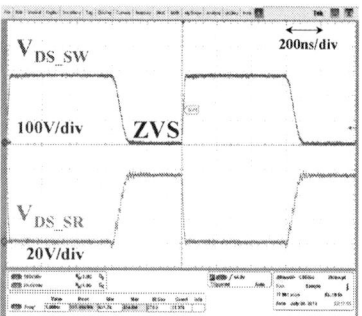

Fig. 19 Prototype experiment waveforms

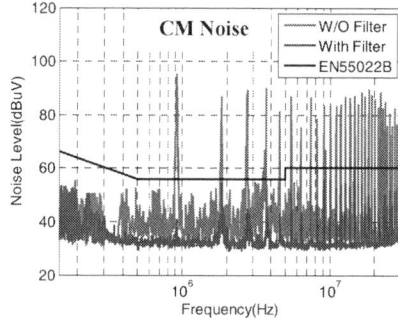

(a) CM noise measurement

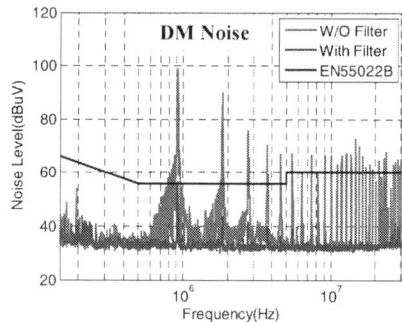

(b) DM noise measurement

Fig. 20 Prototype EMI noise measurement with and without filter

The measured full load efficiency at 90V$_{AC}$ input is 92.7% which is 1.5% higher than the state-of-the-art product. The efficiency improvement is accomplished by several design aspects as listed below.

- GaN devices minimize the conduction and switching loss.

- The active clamp circuit recycles the transformer leakage energy and realizes soft switching for primary switches.

- High frequency operation reduces the core volume and winding turns number.

- Simplified EMI filter reduces conduction loss.

VI. Conclusion

This paper presents the insight view of the CM/DM noise transformation in flyback front-end converter. The impedance unbalance of diode-bridge caused by switching frequency CM noise current is the fundamental mechanism. The transformer shielding technique can effectively reduce the CM noise, and therefore it can significantly reduce the magnitude of CM transformed noise. Based on fully understanding of the conducted EMI mechanism, an effective one-stage EMI filter is designed and validated by experiment. The prototype can achieve very high efficiency and high density with MHz switching frequency operation. The authors are continuing working on the optimization of the EMI filter design in order to minimize the volume and the parasitic capacitance.

It is worthwhile to point out that the analysis and design concept are also applicable to other topologies for offline applications.

Acknowledgment

The information, data, or work presented herein was funded in part by the Office of Energy Efficiency and Renewable Energy (EERE), U.S. Department of Energy, under Award Number DE-EE0006521 with North Carolina State University, PowerAmerica Institute.

This work was supported in part by the Power Management Consor-tium in CPES, Virginia Tech.

This work was conducted with the use of GaN device samples by Transphorm and EPC of the CPES Industry Consortium Program.

Disclaimer

The information, data, or work presented herein was funded in part by an agency of the United States Government. Neither the United States Government nor any agency thereof, nor any of their employees, makes any warranty, express or implied, or assumes any legal liability or responsibility for the accuracy, completeness, or usefulness of any information, apparatus, product, or process disclosed, or represents that its use would not infringe privately owned rights. Reference herein to any specific commercial product, process, or service by trade name, trademark, manufacturer, or otherwise does not necessarily constitute or imply its endorsement, recommendation, or favoring by the United States Government or any agency thereof. The views and opinions of authors expressed herein do not necessarily state or reflect those of the United States Government or any agency thereof.

References

[1] B. Hughes, J. Lazar, S. Hulsey, D. Zehnder, D. Matic, and K. Boutros, "GaN HFET switching characteristics at 350V-20A and synchronous boost converter performance at 1MHz," in *proc. IEEE APEC*, 2012, pp 2506-2508.

[2] J. Delaine, P. Olivier, D. Frey, and K. Guepratte, "High frequency DC-DC converter using GaN device," in *proc. IEEE APEC*, 2012, pp 1754-1761.

[3] X. Huang, Z. Liu, Q. Li, and F C. Lee, "Evaluation and application of 600 V GaN HEMT in cascode structure," *IEEE Trans. on Power Electron.*, vol. 29, no. 5, pp.2453-2461, May. 2014.

[4] X. Huang, Z, Liu, F. C. Lee, and Q. Li, "Characterization and enhancement of high-votlage cascode GaN devices," *IEEE Trans. on Electron Devices*, vol. 62, no. 2, pp.270-277, Feb. 2015.

[5] X. Zhang, C.Yao, X. Lu, E. Davidson, M. Sievers, M. J. Scott, P. Xu, and J. Wang, "A GaN transistor based 90W AC/DC adapter with a buck-PFC stage and an isolated Quasi-switched-capacitor DC/DC stage," in *proc. IEEE APEC*, 2014, pp 109-116.

[6] X. Huang, W. Du, F. C. Lee, and Q. Li, "A novel driving scheme for synchronous rectifier in MHz CRM flyback converter with GaN devices, " in *proc. IEEE ECCE*, 2015, pp 5089-5095.

[7] Z. Zhang, K. D. T. Ngo, and J. L. Nilles, "A 30-W flyback converter operating at 5 MHz." in *proc.IEEE APEC*, 2014, pp 1415-1421.

[8] T. Labella, B. York, C. Hutchens, and J. S. Lai, "Dead time optimization through loss analysis of an active-clamp flyback converter utilizing GaN devices," in *proc.IEEE ECCE*, 2012, pp 3882-3889.

[9] C. T. Choi, C. K. Li, and S. K. Kok, "Control of an active clamp discontinuous conduction mode flyback converter," in Proc. *IEEE Power Electronics and Drive Systems Conf.*, 1999, vol. 2, pp. 1120-1123.

[10] R. Watson, F. C. Lee, and G. Hua, "Utilization of an Active-Clamp Circuit to achieve Soft Switching in Flyback Converters," *IEEE Trans. on Power Electron.*, vol.11, pp162-169, January 1996.

[11] J. Zhang, X. Huang, X. Wu, and Z. Qian, "A high efficiency flyback covnerter with new active clamp technique," *IEEE Trans. on Power Electron.*, vol.25, No. 7, pp1775-1785, Jul. 2010.

[12] D. Zhang; D. Chen; D. Sable; "Non-intrinsic differential mode noise caused by ground current in an off-line power supply," in *proc. IEEE PESC*, 1998, pp.1131-1133

[13] S. Qu; D. Chen; "Mixed-mode EMI noise and its implications to filter design in offline switching power supplies," *IEEE Trans. on , Power Electron.*, vol.17, no.4, pp.502-507, Jul. 2002

[14] J. Meng, W. Ma, "A new technique for modeling and analysis of mixed-mode conducted EMI noise," *IEEE Trans. on Power Electron.*, vol. 19, no. 6, pp.1679-1687, Nov. 2004.

[15] H. Hsieh; J. Li; D. Chen; "Effects of X Capacitors on EMI Filter Effectiveness," *IEEE Trans on Ind. Electron.*, vol.55, no.2, pp.949-955, Feb. 2008.

[16] Y. Yang; D. Huang; F C. Lee; Q. Li; "Transformer shielding technique for common mode noise reduction in isolated converters," in *proc. IEEE ECCE*, 2013, pp.4149-4153.

[17] L. Xie; X. Ruan; Q. Ji; Z. Ye, "Shielding-Cancelation Technique for Suppressing Common-Mode EMI in Isolated Power Converters," *IEEE Trans. on Ind. Electron.*, vol.62, no.5, pp.2814-2822, May 2015

[18] Comchip SMD Diode Specialist, "Low VF SMD Bridge Rectifiers" Z4DGP4046L-HF datasheet.

[19] M. Nave, Power line filter design for switched-mode power supply, Van Nostrand Reinhold, NewYork, 1991.

[20] F. Shih; Y. Chen; Y. Wu; Y. Chen, "A procedure for designing EMI filters for AC line applications," *IEEE Trans on Power Electron.*, vol.11, no.1, pp.170,181, Jan 1996

EMC Investigation of a Very High Frequency Self-oscillating Resonant Power Converter

Jeppe A. Pedersen*, Arnold Knott*, Michael A. E. Andersen*
*Technical University of Denmark
Richard Petersens Plads, building 325
2800 Kongens Lyngby
Denmark
Email: {jarpe, akn, ma}@elektro.dtu.dk

Abstract—This paper focuses on the electromagnetic compatibility (EMC) performance of a Very High Frequency (VHF) converter and how to lower the emissions. To test the EMC performance a VHF converter is implemented with a Class-E inverter and a Class-DE rectifier. The converter is designed to deliver 3 W to a 60 V LED, it has a switching frequency of 37 MHz and achieves an efficiency of 80%. For an LED driver to be used on the consumer market it has to fulfil the standard regarding EMC emissions. The conducted emission is often used as a reason to increase the switching frequency to the VHF range to avoid the regulations. This converter shows to be well below the levels for conducted emission even without filtering. For the radiated emissions the converter is above the limits without input and output filters. Several designs with different ways to lower the emissions are implemented and the different layouts and filtering are compared and discussed.

I. INTRODUCTION

In most Switch Mode Power Supplies (SMPS) the passive components set the limit for the minimum physical size. The passive components scale inversely with frequency and hence increasing the switching frequency reduces the physical size of the power supply. The switching frequency is typically limited to a few megahertz due to switching losses that increase linearly with frequency. In resonant converters that utilize soft switching the switching losses can be neglected. When increasing the switching frequency to the Very High Frequency (VHF) range (30 MHz - 300 MHz) resonant converters are used [1]–[5]. Converter operating in the VHF range often consists of an inverter and a rectifier. This paper describes the electromagnetic compatibility (EMC) performance of a DC-DC converter with a Class-E Inverter described in [6], [7] and with a Class-DE rectifier described in [8]–[12]. The converter is galvanic isolated by two resonant capacitors (C_{res1} and C_{res2}) placed in series with the resonant inductor and in the ground return path separating the inverter from the rectifier. The schematic of the full converter is seen in Fig. 1. The converter uses a self-oscillating gate drive [13] shown in Fig. 2. This gate drive is also used for VHF converters in [14], [15].

VHF converters have been a hot topic in recent years and with the development of wide bandgap devices this field will grow. There has not been any documentation of soft switching VHF resonant converters EMC performance. Only the EMC performance of a hard switched VHF converter is described in [16]. This subject is of great importance for industrial

applications since products must comply with the standard for EMC emissions both in radiated and conducted mode. The standard for conducted emissions stops at 30 MHz which makes VHF converter attractive since the switching frequency is no longer within the measured frequency range. The radiated emissions is measured from 30 MHz and up and the presented converter shows that the radiated emissions needs attention when designing VHF converters.

II. THEORY

The converter described in this paper is made with a Class-E inverter and a Class-DE rectifier. When designing resonant converters the rectifier [11] is designed first. Then the inverter [14] is designed to match the input impedance of the rectifier. The load is a 60 V LED and due to the high output voltage a Class-DE rectifier is chosen since the peak voltage of the diode is equal to the output voltage. The design of a Class-DE rectifier is described in [11] and the input impedance of the Class-DE rectifier for $D = 0.25$ is:

$$Z_{IN} = \frac{1}{2 \cdot \pi} \cdot R_L \qquad (1)$$

The design for the Class-E inverter is described in [14] where the circuit components needed to obtain zero voltage switch (ZVS) is found from the load impedance. Ideally the Class-E inverter and Class-DE rectifier have no rapid transients since they are soft switching. Soft switching is known to have less harmonic content at the switch node and hence better EMC performance [17].

In power electronics the current loops are kept small to avoid large magnetic fields on the PCB, it also ensures that the ground return inductance is kept as small as possible. Ground bounce can create common mode emissions because it has a capacitive coupling to other parts of the circuit. The impedance of the traces in the circuit is usually considered inductive [18] and therefore the impedance increase with frequency which again increases ground bounce. For the PCB itself to become a good antenna at a given frequency it has to be above a quarter of a wavelength and since the PCB is usually smaller than this within the VHF range it is the cables that contribute most to the radiation. One of the means to reduce the radiated emissions is to use an image plane as described in [19]. The image plane can be used to lower the impedance for traces in the circuit

978-1-4673-9551-9/16 $31.00 © 2016 IEEE

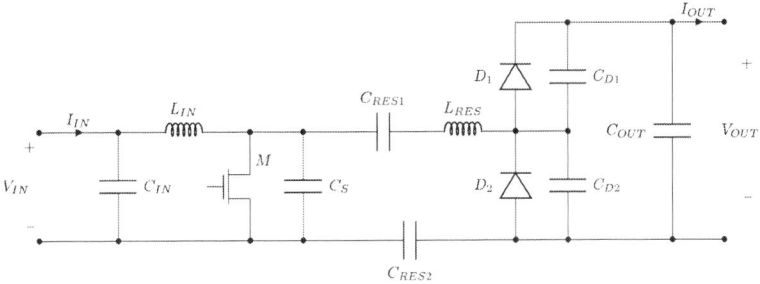

Fig. 1. Schematic of the converter used in this paper, with a Class-E inverter and a Class-DE rectifier.

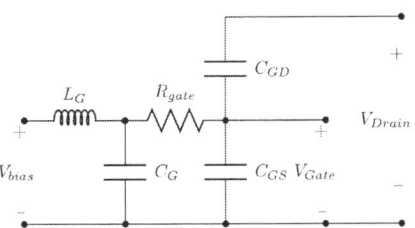

Fig. 2. Schematic of the gate drive with MOSFET parasitics.

Fig. 3. Picture of the converters labeled from A-I.

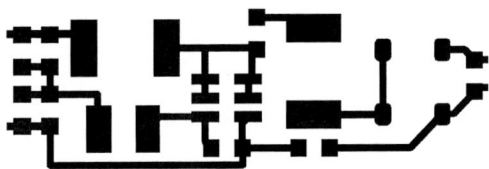

Fig. 4. PCB layout with small traces.

and together with proper filtering it can create a low impedance return path. If the PCB layout is not enough to suppress the emissions input and output filtering can be implemented on the PCB. Filtering can be used to create a low impedance path for differential and common mode signals and a high impedance path to the cables ensuring that the high frequency signals stays within the PCB.

III. IMPLEMENTED CONVERTERS

The converter used in this paper is designed to deliver 3 W for a 60 V LED with a switching frequency of 37 MHz and the input voltage is 30 V DC. The component list is found in Table II. The MOSFET used in this design is a IRF5802. This is a 150 V device and was chosen due to its low parasitic capacitances. The diodes used in the rectifier is a MBR0560 which is a 60 V schottky diode. The converter is achieving an efficiency of 80% at 30 V input.

Different versions of the converter are implemented as shown in Fig. 3 to test the EMC performance of different layouts. The PCBs are made with different input and output orientation on the PCB, converter A-C is made with input and

TABLE I. COMPONENT VALUES FOR THE VHF CONVERTER.

Component	Value
C_{IN}	1 µF
L_{IN}	1 µH
C_{S1}	18 pF
C_{RES1}	100 pF
C_{RES2}	4.7 nF
L_{RES}	1.2 µH
C_{D1}	7 pF
C_{D2}	7 pF
C_{OUT}	1 µF
L_G	82 nH
C_G	33 pF

output is in each end of the PCB and converter D-I are made with input and output close together on one side of the PCB. Converter B-C and F-I is made with image plane. Input and output filters are implemented on H-I and in addition to this I is implemented with an EMC shield connected to the image plane.

There is also made a comparison between having narrow traces between the components as shown in Fig. 4 and having polygons fill up the PCB with copper as shown in Fig. 5. The PCBs with small traces will add inductance to the wires and this increases ground bounce and can create resonances in the circuit. The PCB with large polygon traces will have less inductance in the traces and better thermal cooling of the components as the copper helps spread the heat. However due to the larger copper planes the capacitive coupling within the circuits and to the surroundings increases.

The converters H and I have an input and output filter which is shown in Fig. 6, the filters is connected on both the input and output to minimize the emissions from the board. The filter is made with 1008AF inductors from Coilcraft and capacitors with the values shown in table II. The two inductors is placed to filter both common mode and differential mode signals. The capacitors C_{f1} and C_{f6} are only filtering differential mode signals, C_{f6} is also shown in Fig. 1 as the input (C_{IN}) and output (C_{OUT}) capacitors. The capacitors

978-1-4673-9551-9/16 $31.00 © 2016 IEEE   1542

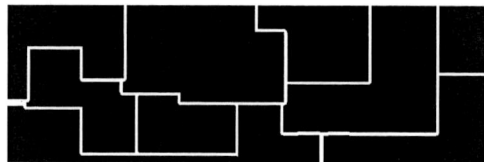

Fig. 5.   PCB layout with large copper polygons.

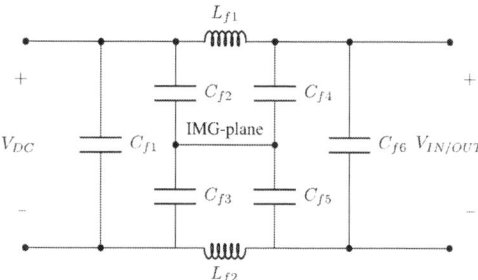

Fig. 6.   Schematic of the filter used, the V_{IN-OUT} is placed at the input and output of the VHF converter.

C_{f2}-C_{f5} are connected to the image plane to filter the common mode signals. By connecting them to the image plane they create a low impedance path for the signals at high frequencies. In this way the signals is kept within the PCB and hereby minimise the emissions. For the last converter I the EMC shield is also connected to the image plane with vias all around the edge of the shield.

IV.   EMC PERFORMANCE

The converter implemented is designed for an LED therefore the Cispr 15 standard should be followed. This standard however refers to the limits from Cispr 22 (EN 55022) so these limits are used in this paper. One of the reasons for increasing the frequency into the VHF range is to avoid the conducted EMC requirements that only goes to 30 MHz. Conducted measurements using a LISN network have been made on the converters and the results without filter is shown in Fig. 7. It is clear from the measurements that the noise floor is increased at the lower frequencies. This is caused by the self-oscillating gate drive drifting a little up and down in frequency. The measurement indicates that the conducted emission is not an issue for VHF converters.

The radiated emissions is more interesting and radiated measurements have been made inside an EMC chamber with an antenna placed 3 m from the converter. The chamber and

TABLE II.   COMPONENT VALUES FOR THE INPUT AND OUTPUT FILTER.

Component	Value	Type
L_{f1}	3.3 µH	Ferrite
L_{f2}	3.3 µH	Ferrite
C_{f1}	1 µF	X5R
C_{f2}	680 pF	C0G
C_{f3}	680 pF	C0G
C_{f4}	680 pF	C0G
C_{f5}	680 pF	C0G
C_{f6}	1 µF	X5R

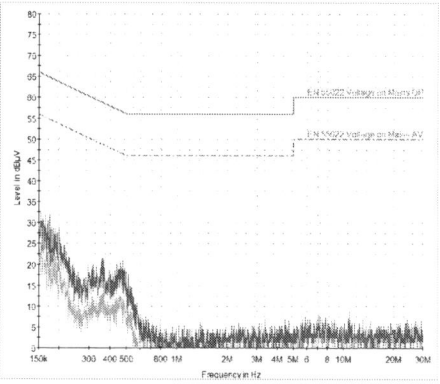

(a) Conducted emission reference.

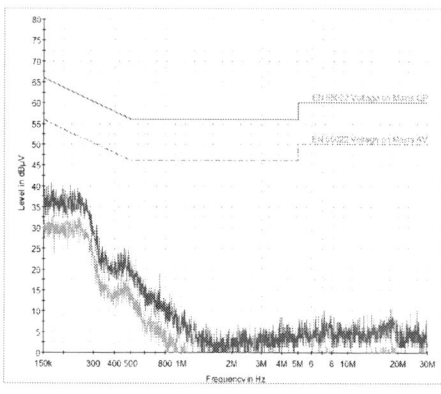

(b) Conducted emission of a converter without filter (converter E).

Fig. 7.   Conducted measurements with limits from EN 55022 (blue is peak and green is average).

antenna is not fully calibrated however it can be used to compare the performance of different layouts. All measurements are done with the same setup and at same voltages. A reference measurement is shown in Fig. 8(a) and is compared to the basic converter A that is expected to have the highest emission. The measurement of converter A is shown in Fig. 8(b). The converter is clearly radiating at the fundamental and all the harmonic frequencies. The drain voltage in this converter is going from 0 V to a peak voltage of $V_{DSpeak} = 3.6 \cdot V_{in} = 108$ V in a quarter of a period which at 37 MHz is 6.75 ns. This results in a high $\frac{\delta V}{\delta t}$ which can generates high radiated emissions if not filtered properly.

As mentioned before an image plane is one of the methods used to reduce radiated emissions. Two identical boards one with and one without an image plane is tested. The image plane is connected to ground on the primary side with a via placed next to the input connector. As seen in Fig. 9 the image plane is damping the radiated emissions as expected, the only filter components used in this setup is the input and output capacitors C_{IN} and C_{OUT} so some radiation is expected. It seems that the image plane is working well and reduces the

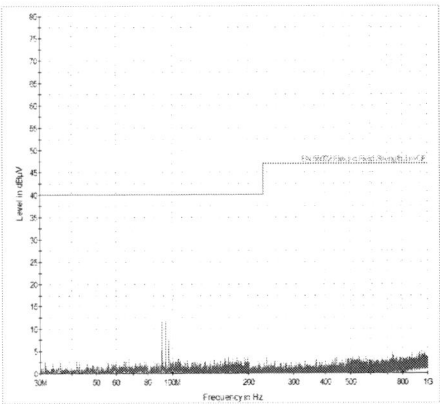

(a) Radiated emmisions reference.

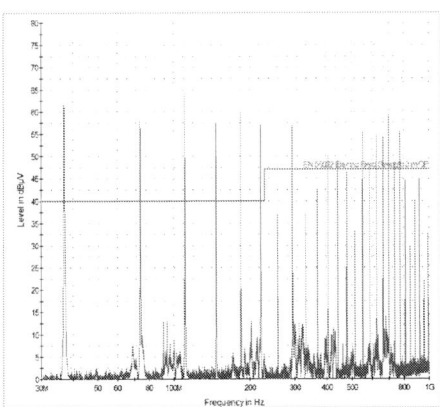

(b) Radiated emmisions, the converter has input and output in each end of the PCB and no image plane (converter A).

Fig. 8.   Radiated emmisions reference and basic converter.

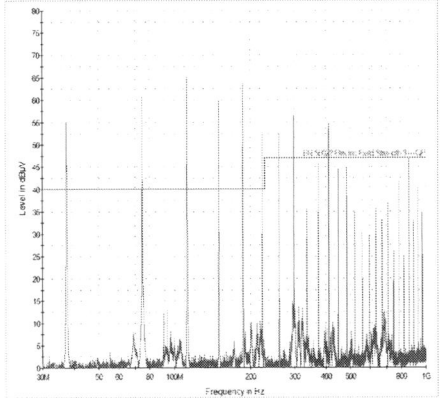

(a) Radiated emissions, the converter has input and output in the same end of the PCB and no image plane (converter E).

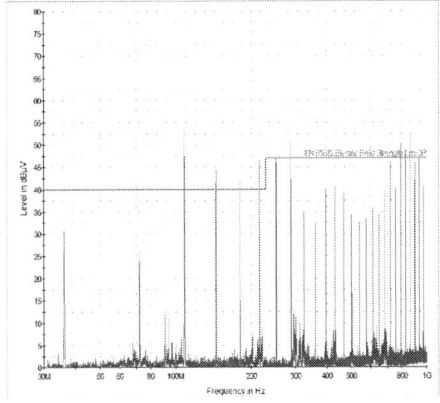

(b) Radiated emissions, the converter has input and output in the same end of the PCB and an image plane (converter G).

Fig. 9.   Radiated emissions measured on converters with and without image plane.

emissions.

The standard recommendation is to place the input and output in the same side of the PCB. The measurements on two converters with different placement of input and output are shown in Fig. 10. It is clear from the measurement that having the input and output next to each other helps lowering the emissions. In this case it keeps the current loop smaller in the power stage which seems to have a positive effect on the fundamental and second harmonic.

The two types of trace layouts shown in Fig. 4 and 5 is measured. First on converter D and E that is both implemented without an image plane, the results are shown in Fig. 11. Here there is only a small difference between the two PCB layouts. Only at frequencies above 400 MHz the polygons are performing better which indicates that it lowers the trace inductance. The same test was performed on two PCB's implemented with an image plane. The emissions from the PCB with thin traces (converter F) is shown in Fig. 10(b) and the emissions with large polygons (converter G) is shown in Fig.

9(b). The polygon implementation seems to lower inductance in the traces thereby reducing the radiated emissions when there is an image plane. Because of the better performance all the converters with additional filtering have been implemented with polygon traces. Placing two large copper areas closely above each other can add capacitance to the circuit. This is not considered a problem in this implementation as the areas are relatively small.

Measurement was performed on a converter with filters on the input and output this is shown in Fig. 12(a). The filters placed on the input and output of the converter is reducing the radiated emission at the high frequencies. However there is still much harmonic content at the frequencies below 400 MHz. In figure 12(b) the measurements on a converter with the same filters and an EMC shield is shown. The EMC shield is surrounding the converter to minimize the capacitive coupling to the cables and surroundings. Adding the EMC

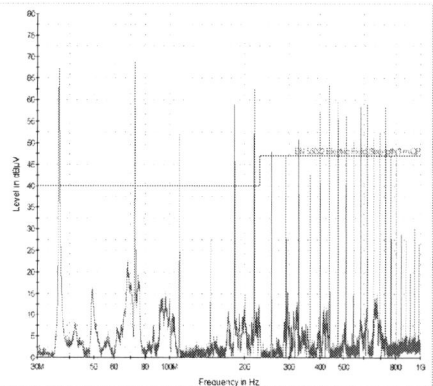

(a) Radiated emmisions, the converter has input and output in each end of the PCB and an image plane (converter B).

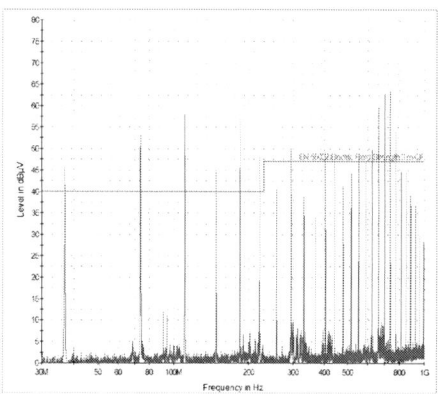

(b) Radiated emmisions, the converter has input and output in the same end of the PCB and an image plane(converter F).

Fig. 10. Radiated emmisions for converters with different input and output layout.

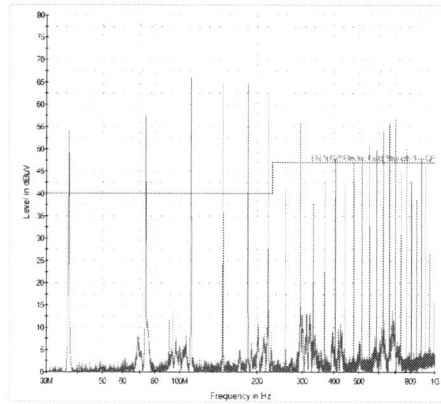

(a) Radiated emmisions, the converter has input and output in the same end of the PCB, no image plane and the thin traces (converter D).

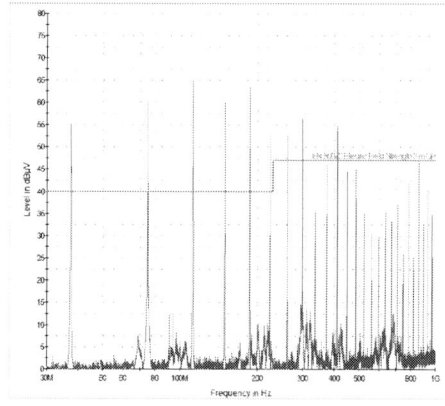

(b) Radiated emmisions, the converter has input and output in the same end of the PCB, no image plane and polygon traces (converter E).

Fig. 11. Radiated emmisions for thin vs polygons traces and no image plane.

shield reduces the radiated emissions significant and only the fundamental and first harmonics are present. By adding the EMC shield the capacitive coupling across the filter is limited and most of the radiated emissions are eliminated.

To summarize the EMC performance of this VHF converter seems to have no problems with the conducted EMC, and therefore the main focus is on the radiated emissions. As for the radiated part the converter has a high level of radiation if left without any filtering besides the input and output capacitor. By implementing the image plane the emissions are lowered. Placing the input and output connecters on the same side also helps reducing the emissions. The polygon layout on the PCB seems to have a positive effect together with an image plane. By adding additional filtering on the input and output the emissions are reduced further. The EMC shield prevents the electric field from coupling across the filters.

V. CONCLUSION

This paper describes a VHF converter implemented with different layouts to investigate the EMC challenges when increasing the switching frequency to the VHF range. The converter has a switching frequency of 37 MHz and an efficiency of 80%. The conducted EMC test shows that the converter has no problem in the frequency range from 150 kHz to 30 MHz. The challenge with VHF converters is to lower the radiated emissions. This paper shows that the image plane can be used to reduce the radiated emission to some extent and filtering and proper shielding can be used to lower the emissions.

REFERENCES

[1] T. Andersen, S. Christensen, A. Knott, and M. A. E. Andersen, "A VHF class E dc-dc converter with self-oscillating gate driver," in *Applied*

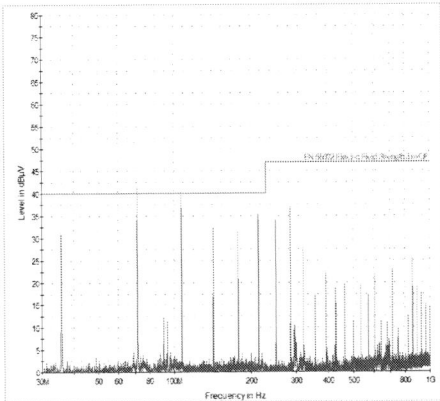

(a) Radiated emmisions, the converter has input and output in the same end of the PCB, no image plane, polygon traces and filters on the input and output (converter H).

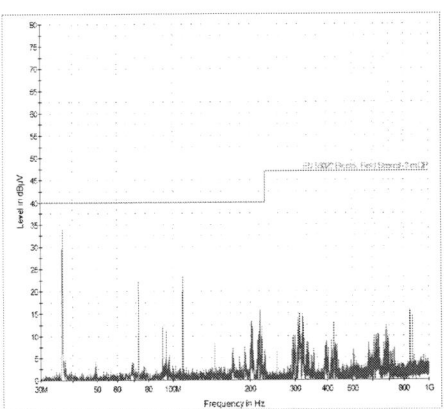

(b) Radiated emmisions, the converter has input and output in the same end of the PCB, no image plane, polygon traces, filters on the input and output and an EMC sield (converter I).

Fig. 12. Radiated emmisions for converters with filters, and with and without EMC shield.

Power Electronics Conference and Exposition (APEC), 2011 Twenty-Sixth Annual IEEE, 2011, pp. 885–891.

[2] M. Kovacevic, A. Knott, and M. Andersen, "A VHF interleaved self-oscillating resonant sepic converter with phase-shift burst-mode control," in Applied Power Electronics Conference and Exposition (APEC), 2014 Twenty-Ninth Annual IEEE, March 2014, pp. 1402–1408.

[3] M. Madsen, A. Knott, and M. Andersen, "Very high frequency half bridge dc/dc converter," in Applied Power Electronics Conference and Exposition (APEC), 2014 Twenty-Ninth Annual IEEE, March 2014, pp. 1409–1414.

[4] M. K. Song, M. Dehghanpour, J. Sankman, and D. Ma, "A VHF-level fully integrated multi-phase switching converter using bond-wire inductors, on-chip decoupling capacitors and dll phase synchronization," in Applied Power Electronics Conference and Exposition (APEC), 2014 Twenty-Ninth Annual IEEE, March 2014, pp. 1422–1425.

[5] W. Cai, Z. Zhang, X. Ren, and Y.-F. Liu, "A 30-MHz isolated push-pull VHF resonant converter," in Applied Power Electronics Conference and Exposition (APEC), 2014 Twenty-Ninth Annual IEEE, March 2014, pp. 1456–1460.

[6] R. Redl, B. Molnar, and N. Sokal, "Class E resonant regulated dc/dc power converters: Analysis of operations, and experimental results at 1.5 MHz," Power Electronics, IEEE Transactions on, vol. PE-1, no. 2, pp. 111–120, April 1986.

[7] M. Kazimierczuk and J. Jozwik, "Resonant dc/dc converter with class-E inverter and class-E rectifier," Industrial Electronics, IEEE Transactions on, vol. 36, no. 4, pp. 468–478, 1989.

[8] D. Kazimierczuk, Marian K. Czarkowski, Resonant Power Converters, 2nd ed. Wiley, 2011.

[9] K. Fukui and H. Koizumi, "Analysis of half-wave class DE low dv/dt rectifier at any duty ratio," Power Electronics, IEEE Transactions on, vol. 29, no. 1, pp. 234–245, Jan 2014.

[10] D. Hamill, "Class DE inverters and rectifiers for dc-dc conversion," in Power Electronics Specialists Conference, 1996. PESC '96 Record., 27th Annual IEEE, vol. 1, 1996, pp. 854–860 vol.1.

[11] K. Fukui and H. Koizumi, "Half-wave class DE low dv/dt rectifier," in Circuits and Systems (APCCAS), 2012 IEEE Asia Pacific Conference on, Dec 2012, pp. 69–72.

[12] L. Raymond, W. Liang, and J. Rivas, "Performance evaluation of diodes in 27.12 MHz class-D resonant rectifiers under high voltage and high slew rate conditions," in Control and Modeling for Power Electronics (COMPEL), 2014 IEEE 15th Workshop on, June 2014, pp. 1–9.

[13] M. Madsen, J. Pedersen, A. Knott, and M. Andersen, "Self-oscillating resonant gate drive for resonant inverters and rectifiers composed solely of passive components," in Applied Power Electronics Conference and Exposition (APEC), 2014 Twenty-Ninth Annual IEEE, March 2014, pp. 2029–2035.

[14] J. Pedersen, M. Madsen, A. Knott, and M. Andersen, "Self-oscillating galvanic isolated bidirectional very high frequency dc-dc converter," in Applied Power Electronics Conference and Exposition (APEC), 2015 IEEE, March 2015, pp. 1974–1978.

[15] J. Mønster, M. Madsen, J. Pedersen, and A. Knott, "Investigation, development and verification of printed circuit board embedded air-core solenoid transformers," in Applied Power Electronics Conference and Exposition (APEC), 2015 IEEE, March 2015, pp. 133–139.

[16] C. Delepaut, J. Wolf, N. Le Gallou, F. Leroy, O. Deblecker, and F. Dualibe, "VHF switching dc/dc converter: Electromagnetic emissions assessment," in Electromagnetic Compatibility, Tokyo (EMC'14/Tokyo), 2014 International Symposium on, May 2014, pp. 856–859.

[17] H. Chung, S. Hui, and K. Tse, "Reduction of power converter EMI emission using soft-switching technique," Electromagnetic Compatibility, IEEE Transactions on, vol. 40, no. 3, pp. 282–287, Aug 1998.

[18] T. Williams, EMC for Product Designers. Elsevier Science, 2011.

[19] R. German, H. Ott, and C. Paul, "Effect of an image plane on printed circuit board radiation," in Electromagnetic Compatibility, 1990. Symposium Record., 1990 IEEE International Symposium on, Aug 1990, pp. 284–291.

978-1-4673-9551-9/16 $31.00 © 2016 IEEE

Numerical Optimization of Passive Line Filter Components for Suppression of Electromagnetic Interference (EMI)

Carsten Henkenius[*], Norbert Fröhleke[*], Joachim Böcker[*] and Heiko Figge[§]

[*] University of Paderborn
Power Electronics and Electrical Drives
33098 Paderborn, Germany
Email: henkenius@lea.upb.de
[§] Delta Energy Systems (Germany) GmbH
59494 Soest, Germany
Email: heiko.figge@delta-es.com

Abstract—Objective functions and constraints for electric losses and component volume are derived to gain an optimal design of a line filter for grid rectifiers with power factor correction (PFC). The gained characteristics described by these models are valid for coupled inductors with toroidal cores and conventional capacitors. A modified noise predicting technique for a Totem-Pole rectifier is then used to calculate transfer functions of a given filter structure to obtain voltage and current noise levels. Beside geometrical and physical constraints the noise levels must not exceed specific standards defining the admissible sets of parameters for the following optimization. Application of a multi-objective optimization approach yields to an optimal set of construction parameters.

I. INTRODUCTION

Most PFC rectifiers contain passive line filters for high frequency disturbances shaping regarding electromagnetic compatibility (EMC) and residual current (RC) device limitations. Beside parasitic interferences these filters mainly consist of resistive, capacitive and inductive components whereas the latter occupy a remarkable amount of space in SMPS. Furthermore the resistive characteristic of the line filter caters for power losses. In practice, the most common design procedures for passive line filters typically apply try and error methods in order to fulfill EMC standards but they may not result in optimized design regarding minimum component volume, costs or losses. With the increasing demand on SMPS with high power density and high efficiency at low costs the optimization of line filter circuits regarding power losses and component volume becomes focus of industrial and scientific research.

Several publications have been released about line filter design optimization of PFC rectifiers. [1] shows a line filter optimization of a PFC rectifier where the boost inductor is considered as part of the line filter. It is shown that the costs for line filter components mainly depend on the switching frequency but proceeds discontinuously because as the switching frequency increases more harmonics of the boost inductor current reach the lower frequency limit of the EMC-

standards (150 kHz). It is mentioned, that both differential and common mode noise modeling is necessary since their propagation paths affect each other.

Both types of noise are considered in [2] where an optimization approach is used to create nomographs. Depending on the required attenuation at the stop frequency (frequency of the first harmonic above the lower frequency limit of the EMC standard), the common mode and differential mode impedance of the converter, the nominal filter current, the optimal filter structure and filter component values are selected.

[3], [4] present a multi-objective optimization of the line filter components regarding filter losses and volume where the boost inductor is utilized as part of the input filter. It is shown that for a fixed switching frequency the filter volume increases almost linearly with an increasing boost current ripple. [3] depicts that the line filter often allocates the highest amount of space within the framework of the PFC rectifier for all optimized design results.

Above mentioned publications indicate that the largest amount of space within a line filter framework is occupied by the filter chokes and investigations solely concentrate either on common mode or differential mode aspects.

For this reason this paper presents a method to design a line filter choke for a PFC rectifier with a fixed line filter circuit that applies multi-objective optimization techniques to obtain trade-off solutions between component losses and volume visualized by the (P-V) Pareto-Front. This method is bridging the gap outlined above and is exemplarily explained by analyzing a Totem-Pole rectifier like shown in Fig.1 with line filter containing idealized capacitors and coupled windings on a toroidal core.

The PFC rectifier consists of a boost inductor L, an output capacitor C and two half bridges. One bridge contains the switches S_1 and S_2 which are actively switched with high frequency. The second bridge involves line commutated diodes D_1 and D_2. The capacitors C_{ap}, C_{1p} and C_{2p} are used for modeling of parasitic effects. The rectifier is connected in

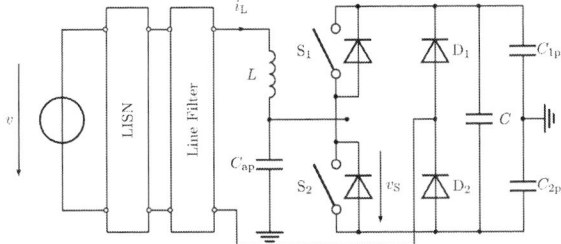

Fig. 1. Totem-Pole-Rectifier with line filter and LISN

series to a line filter circuit, a line impedance stabilization network (LISN) and the sinusoidal grid voltage v.

The paper is organized in four sections. Section II presents the derivation of the computation models for component losses, volume and temperature. Furthermore an excursion investigates common mode and differential mode choke integration within a single electrical component. Section III shows the formulation of the loss and volume optimization problem concerning technical, geometrical, electrical and thermal constraints. The design method is evaluated by an optimization of an existing common mode. Section IV depicts the (P-V) Pareto-Front for a specific line filter design and some parameter studies. The paper ends with a conclusion and an outlook in section V.

II. LINE FILTER MODELING

Coupled windings for passive line filters are typically distinguished between common mode chokes and differential mode chokes. The differential mode noise (DM) is influenced by the effective DM inductance. The same is valid for the common mode noise (CM). So they are designed separately depending on the noise levels. In order to reduce the required amount of space for both chokes it is a legitimate question whether it is possible to combine both inductances within one electrical component [5]. One important parameter is the coupling factor between the two windings. This factor is generally affected by the geometrical arrangement of the two windings and the core material. Since line filter inductors mostly evolve toroidal core shapes, a short excursion will show an analysis of how the coupling coefficient is influenced by the core material in case of two coupled windings placed on a toroidal shaped core with ferromagnetic material.

1) Investigation of coupling coefficient: The inductive component of a passive line filter considered in this paper includes two windings which are placed on a toroidal core like shown in Fig.2. Assuming homogeneous, isotropic core material and two symmetrically wrapped windings ($L_1 = L_2 = L$) the coupling coefficient k between the two windings is defined as the ratio of

$$k = \frac{M}{L} \tag{1}$$

where M describes the mutual inductance and L is the self inductance. Equation 1 can be solved by calculating the stored

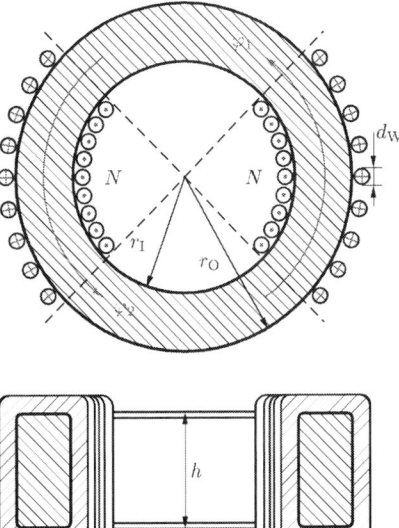

Fig. 2. Two windings attached on a toroidal core

energy when currents flow through the windings and excite the magnetic field. The energy can be determined by using

$$W = \frac{1}{2}LI_1^2 + MI_1I_2 + \frac{1}{2}LI_2^2 \tag{2}$$

where I_1 is the current through the first winding and I_2 is the current through the second winding. Now two mental experiments are conducted. In a first step sets $I_2 = 0$. Then the self inductance is simply calculated by

$$L = \frac{2W_1}{I_1^2} \tag{3}$$

where W_1 is the stored energy caused by the excitation of current I_1. In a second next step both currents are assumed to be equal $I_1 = I_2 = I$. Therefore the stored energy is

$$W_2 = LI^2 + MI^2 \tag{4}$$

and the mutual inductance is therefore

$$M = \frac{W_2}{I^2} - L = \frac{W_2 - 2W_1}{I^2}. \tag{5}$$

To determine the coupling factor k it is necessary to calculate the partial energies. These are either gained from numerical simulation or from analytical modeling. In this case an analytical approach is preferred by solving the Laplace equation

$$\nabla^2 A = 0 \tag{6}$$

on a two dimensional surface. The vector potential $\vec{A} = A(\varrho, \varphi)\vec{e}_z$ is represented by ϱ, φ and z in cylindrical coordinates. Fig.3 depicts a cross sectional sketch of the toroidal core. To simplify this approach it is assumed that the diameter

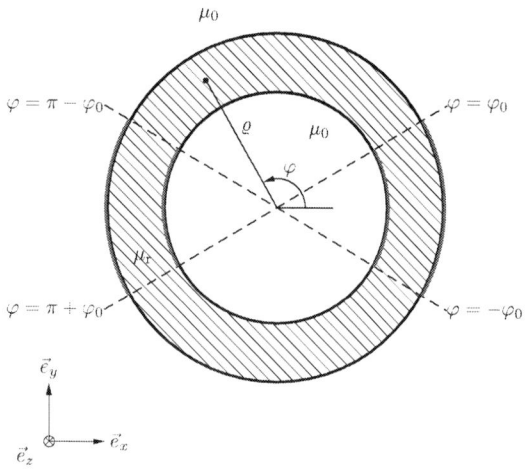

Fig. 3. Sectional drawing of the toroidal core

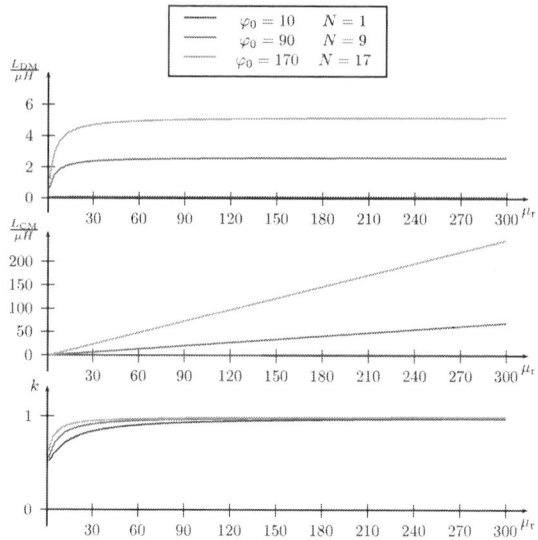

Fig. 4. Progression of the effective inductances and the coupling coefficient

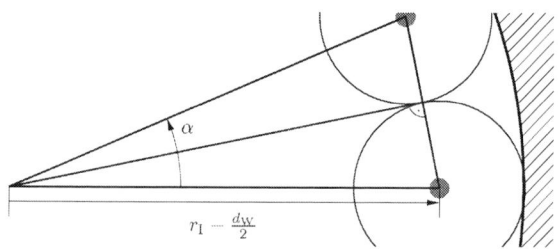

Fig. 5. Zooming of the sectional drawing

of the wire d_{W} is small compared to the inner radius r_{I} and the resulting current filaments are thus densely wound around the core. Therefore the excitation of the magnetic field is modeled by surface currents $\vec{k} = k(\varphi)\vec{e}_z$ in the inner and outer winding (red lines in Fig.3).

Using the calculated vector potential and the given surface currents the stored energy is

$$W_{1,2} = \frac{1}{2} \sum_{\mathrm{i}} \int_{A_{\mathrm{i}}} \vec{A}_{1,2} \vec{k}_{1,2} da_{\mathrm{i}} \tag{7}$$

where the indexes mark the two experiments. Note, that the shown winding configuration is referred to as common mode choke because the magnetic fluxes φ_1 and φ_2 caused by common mode currents have the same positive directions. For this configuration the effective inductance is

$$L_{DM} = 2(L - M) \tag{8}$$

for DM and

$$L_{CM} = \frac{L + M}{2} \tag{9}$$

for CM. Fig.4 shows a numerical evaluation of the modeling approach for an outer radius $r_{\mathrm{O}} = 34$ mm, an inner radius $r_{\mathrm{I}} = 19$ mm, a wire diameter of $d_{\mathrm{W}} = 3$ mm and a core depth of $h = 25$ mm. The bottom graph shows k for different numbers of turns N. It does never exceed unity but gets saturated for high values of the relative permeability μ_{r}. That affects L_{DM} like shown in the top graph because although μ_{r} increases L_{DM} saturates. This becomes clear when looking at equation 8. The inductance L increases with μ_{r} as $L \propto \mu_{\mathrm{r}}$ but the difference $1 - k$ gets closer and closer to zero at the same time. So further increasing of μ_{r} does not increase L_{DM}. This is an interesting fact for the inductor design because for ferromagnetic core material with $\mu_{\mathrm{r}} > \approx 200$ the effective differential mode inductance is only influenced by geometrical parameters.

A. Volume and loss modeling

The volume of the choke is divided into three parts

$$V_{\mathrm{C}} = \pi h (r_{\mathrm{O}}^2 - r_{\mathrm{I}}^2) \tag{10a}$$

$$V_{\mathrm{W}_1} = 4 N \alpha d_{\mathrm{W}} (r_{\mathrm{O}}^2 - r_{\mathrm{I}}^2 + 2 d_{\mathrm{W}}(r_{\mathrm{O}} - r_{\mathrm{I}})) \tag{10b}$$

and

$$V_{\mathrm{W}_2} = 4 N \alpha d_{\mathrm{W}} h (r_{\mathrm{O}} + r_{\mathrm{I}}) \tag{10c}$$

where V_{C} is the core volume, V_{W_1} is the winding volume on top and bottom of the core and V_{W_2} is the winding volume on the left and right side of the core. The angle between two wires at r_{I} is calculated by

$$\alpha = \arcsin\left(\frac{d_{\mathrm{W}}}{2r_{\mathrm{I}} - d_{\mathrm{W}}}\right) \tag{11}$$

like depicted in Fig.5. Summarizing all partial component volumes the total volume V is calculated by

$$V = V_{\mathrm{C}} + V_{\mathrm{W}_1} + V_{\mathrm{W}_2}. \tag{12}$$

978-1-4673-9551-9/16 $31.00 © 2016 IEEE

The winding losses of both windings are calculated as

$$P = R I_{\text{rms}}^2 \tag{13}$$

where I_{rms} is the root mean square (RMS) of the current through the windings and R is the resistance calculated by

$$R = \frac{N \, l_{\text{W}}}{\kappa \, \pi \, d_{\text{W}}^2} (1 + \alpha_{\text{m}} (\vartheta_{\text{L}} - \vartheta_{\text{E}})). \tag{14}$$

ϑ_{L} is the surface temperature, ϑ_{E} is the air temperature of the environment, κ is the specific conductivity and l_{W} is the length of the wire. The change of resistance caused by change of the wire temperature is considered by a linear term with the temperature coefficient α_{m}.

B. Thermal modeling

For non forced cooling the heat dissipation is described by convectional and radiated heat transfer [6]. The convectional heat dissipation is generally described by

$$P_{\text{c}} = \alpha_{\text{c}} A_{\text{c}} (\vartheta_{\text{L}} - \vartheta_{\text{E}}) \tag{15}$$

where P_{c} is the heat conduction, A_{c} is the effective surface area. α_{c} is a dimensionless number which is calculated by Nusselt, Prandtl and Rayleigh numbers for free convectional heat transfer through air. α_{c} varies for vertical and horizontal surfaces. The radiated heat dissipation is described by

$$P_{\text{r}} = \sigma_{\text{r}} \epsilon_{\text{r}} A_{\text{r}} (\vartheta_{\text{L}}^4 - \vartheta_{\text{E}}^4) \tag{16}$$

where P_{r} is the heat radiation, σ_{r} is the Boltzmann constant, ϵ_{r} is a dimensionless number for wire isolation and A_{r} is the effective surface area. In this paper to the sum of both effective areas

$$A = A_{\text{c}} + A_{\text{r}}. \tag{17}$$

is referred to as effective cooling area A. Note, that temperatures ϑ_{L} and ϑ_{E} are measured in Kelvin.

C. Electrical modeling

Following the analysis results in [7] and [8] CM emission is mainly caused by the capacitors C_{ap}, C_{1p}, and C_{2p} and high temporal changes of the voltage waveform v_{S} across the switch S_2. [9] and [4] show that the current waveform through the boost inductor i_{L} is mainly responsible for DM emission. Similar to the noise source modeling approaches of [7] the rectifier is approximated by an ideal voltage source v_{D} and the components L_3, C_{tp} and C_{bp} where $v_{\text{D}} = v_{\text{S}}$, $L_3 = L$, $C_{\text{tp}} = C_{\text{ap}}$ and $C_{\text{bp}} = C_{1p} + C_{2p}$ like shown in Fig.6.

The LISN circuit arrangement of the components R_{SN}, C_{SN} and L_{SN} defines a high-pass characteristic for the impedance

$$\underline{Z}_{\text{SN}} = \frac{V_{\text{N}}}{I_{\text{SN}}}. $$

Noise voltage levels are obtained by measuring the voltage drop v_N across the resistor R_{SN} of the LISN.

The line filter circuit consists of two coupled windings represented by the components L_1, L_2, R_1 and R_2, the X-Capacitors C_1 and C_2 and the Y-Capacitors C_3 and C_4.

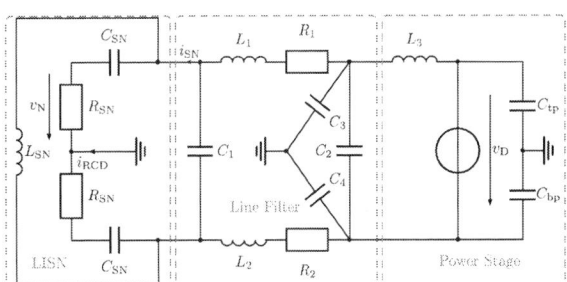

Fig. 6. Proposed circuit for modeling

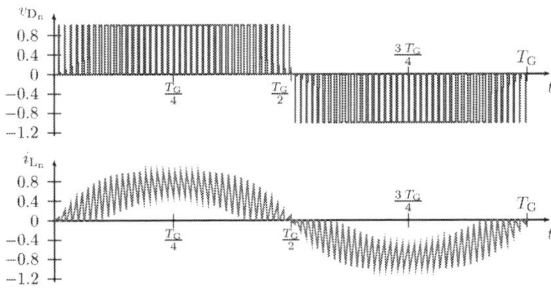

Fig. 7. Exemplary voltage waveform across switch S_2

The signal waveform of the excitation voltage v_{D} can be obtained in two ways. If the power stage is already designed these parameters can be directly measured and transferred for further analysis or it is obtained from simulation in time domain. Fig.7 depicts exemplary simulation results of the normalized waveforms

$$v_{\text{D}_{\text{n}}} = \frac{v_{\text{S}}}{\|v_{\text{S}}\|_{\max}} \text{ and } i_{\text{L}_{\text{n}}} = \frac{i_{\text{L}}}{\|i_{\text{L}}\|_{\max}}$$

for one period T_{G} of the grid voltage V. Note, that especially the typical waveform of the current is valid only if the voltage across the capacitor C_2 is nearly constant during a single switching interval. This means, that the transfer function for the rectifier input voltage (IV)

$$\underline{Gv}_{\text{ID}} = \frac{V_{\text{I}}}{V_{\text{D}}}. $$

needs to have sufficient damping at the switching frequency.

Transfer functions are gained from Modified Nodal Analysis (MNA) in frequency domain for the equivalent circuit shown in Fig.6 used together with the spectrum of the excitation voltage waveform to determine CM, DM, RC and IV.

III. MULTI-OBJECTIVE OPTIMIZATION

The design parameters used for the design optimization are

$$x = \{r_{\text{O}}, r_{\text{I}}, h, N, d_{\text{W}}\} \tag{18}$$

like depicted in Fig.2. They are used to calculate L_1, L_2, R_1 and R_2 in Fig.6. The capacitance values of the line filter $C_{1,\ldots,4}$ are fixed. Target of the optimization is to find a minimum value for

$$F(x) = \beta\,\widehat{P}(x) + (1-\beta)\,\widehat{V}(x) \tag{19}$$

where $\widehat{P}$ describes the normalized losses and $\widehat{V}$ is the normalized volume of the two windings with the toroidal core. The factor β is the design priority parameter. Normalization is used to obtain two objectives of the same size. Like shown in [10] the easiest way is to scale and shift the objectives from one extrema $[P_{\min}, V_{\max}]$ to the other $[P_{\max}, V_{\min}]$. These values are the base for the normalization

$$\widehat{P} = \frac{P - P_{\min}}{P_{\max} - P_{\min}} \tag{20a}$$

$$\widehat{V} = \frac{V - V_{\min}}{V_{\max} - V_{\min}}. \tag{20b}$$

For $\beta = 0 \rightarrow V = V_{\min}$ and for $\beta = 1 \rightarrow P = P_{\min}$. The values in between represent design trade-offs.

The constraints are defined as

$$0 > -x + x_{\min}, \tag{21a}$$

$$0 > x - x_{\max}, \tag{21b}$$

$$0 > r_{\mathrm{I}} - r_{\mathrm{O}} + r_{\Delta}, \tag{21c}$$

$$0 \geq \vartheta_{\mathrm{L}} - \vartheta_{\mathrm{L,max}}, \tag{21d}$$

$$0 \geq \alpha - \pi + \alpha_{\Delta}, \tag{21e}$$

$$0 \geq |V_{\mathrm{N}}(\omega)| - V_{\mathrm{NORM}}(\omega) + V_{\Delta}, \tag{21f}$$

$$0 \geq |I_{\mathrm{RCD}}(\omega)| - I_{\mathrm{NORM}}(\omega), \tag{21g}$$

$$0 \geq \left|Gv_{\mathrm{ID}}(\omega_{\mathrm{s}})\right| - Gv_{\max}(\omega_{\mathrm{s}}). \tag{21h}$$

$x_{\min}$ is a set of minimum parameters with $x_{\min} > 0$. Parameters and quantities indexed by Δ represent safety margins.

The optimization routine to solve this non-linear constrained problem contains a combination of metaheuristic computation and line search strategy. It starts with the differential evolution (DE) algorithm [11]. If the DE routine converges to a reasonable solution the trust-region-reflective algorithm is used for further optimization.

IV. Evaluation

To demonstrate the performance of the design approach presented in the previous sections an already designed common mode choke of a passive line filter was optimized. In the following the already designed common mode choke is referred to as device under test (DUT). The filter structure is shown in Fig.6. Boundary conditions are a nominal power of $P_{\mathrm{N}} = 3.6\,\mathrm{kW}$ with a mains supply voltage of $V = 230\,\mathrm{V}$ and a grid frequency of $f_{\mathrm{G}} = 50\,\mathrm{Hz}$. The switching frequency of the PFC rectifier is $f_{\mathrm{S}} = 60\,\mathrm{kHz}$ with a DC-link voltage of $V_{\mathrm{DC}} = 400\,\mathrm{V}$. The boost inductor was given with $L_3 \approx 67\,\mu\mathrm{H}$ whereby the ratio of continuous current mode (CCM) to discontinuous current mode (DCM) operation time is approximately 60:40 at full load condition. The DUT contains a ferromagnetic core with toroidal shape and two windings. The measured parameters of the DUT are given in table I.

TABLE I
PARAMETERS OF THE DUT

Parameter	Value
r_{O}	32.5 mm
r_{I}	19 mm
h	25 mm
N	11
d_{W}	2x2 mm
L	1.6 mH
R	5.5 mΩ
A	130.21 cm^2
V	75.3 cm^3
κ_{Cu}	56 10^6 S/m

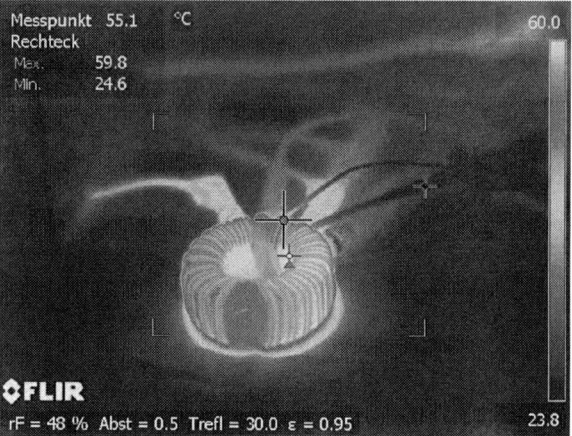

Fig. 8. Thermal image of the choke

The optimization starts with the definition of parameter values for the based calculation models described in section II. Fig.8 shows a thermographic view of the DUT when it is thermally excited by a direct current of 25 A. The radiation factor $\epsilon_{\mathrm{r}} = 0.95$ is shown at the bottom line. The temperature is measured in two different ways for better accuracy of the validation (thermographic camera and thermocouple for multimeter). The results of the thermal model after parameter definition are shown in the graphs of Fig.9. The graph on top shows the measured and calculated temperature of the DUT and the measured temperature of the environment. The maximum deviation between the computation model and the measured temperature is $\vartheta_{\mathrm{diff,max}} = 2.5^\circ\mathrm{C}$. The bottom graph shows temperature dependency of the resistance of the copper wire. The measurements are used to fit the temperature coefficient of the linear term in equation 14 with $\alpha_{\mathrm{m}} = 4.3\,10^{-3}$.

The next step is the definition of the constraints (equations 21a-21h). The geometrical constraints for the optimization parameters are given in table II. The temperature limit of the common mode choke is set to $\vartheta_{\mathrm{L,max}} = 125^\circ\mathrm{C}$. Noise and RC limits are specified in [12] and [13] respectively. The margin for the noise measurement is set to $V_{\Delta} = 6\,\mathrm{dB\mu V}$.

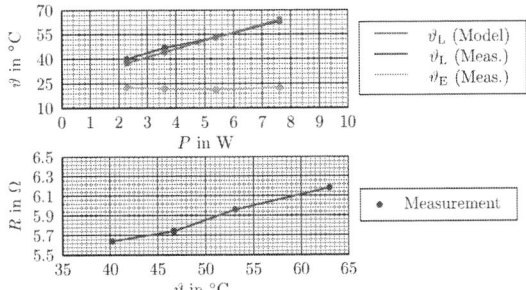

Fig. 9. Validation of the thermal model

TABLE II
PARAMETER CONSTRAINTS

Parameter	Value
$r_{O,min}, r_{I,min}$	10 mm
h_{min}	15 mm
$d_{W,min}$	1 mm
$d_{W,max}$	3 mm
N_{min}	10
r_Δ	10 mm
α_Δ	10°

A. Pareto-Front

The optimization procedure leads to different optimal designs results depending on the chosen priority parameter β like shown in Fig.10. It can be seen that the DUT of the previous section is not part of the Pareto-Front so it is not an optimal design solution. The Pareto-Front seems to have a hyperbolical behavior but there exist a wide range where the difference quotient of the Pareto-Points

$$\frac{\Delta P}{\Delta V}$$

is close to zero. The point where the losses do not exceed the losses of the DUT significantly is chosen as the optimal design result. Limiting constraints are the minimum height h_{min}, the minimum difference between the outer and inner radii r_Δ

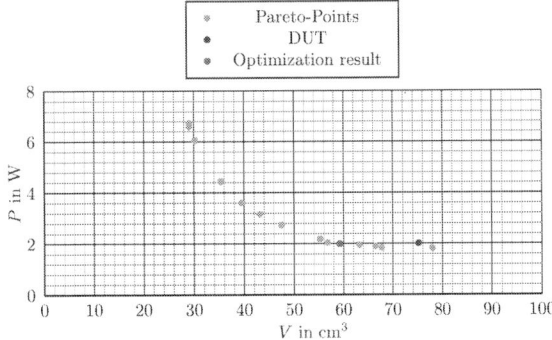

Fig. 10. Pareto-Front and optimization result

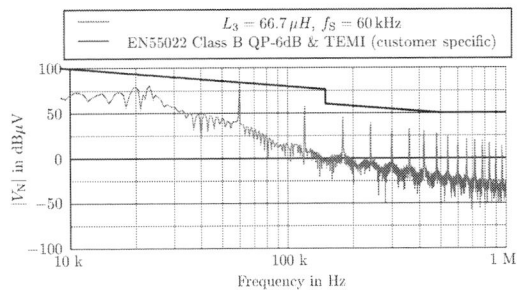

Fig. 11. EMC Standard and simulated noise spectrum of the optimization result

and the allowed noise measurement like shown in Fig.11. The parameter set of the optimal solution and the deviations compared to the parameters of the DUT are given in table III. Note, that the optimized design solution for the common mode choke causes equal losses than the DUT although the number of turns N increases and the wire diameter is nearly the same. The reason is the reduction in height h which causes the total length of the winding l_W to remain almost at the same value. So the total resistance of the wire does not change significantly. The outer and inner radius of the core are increased but the difference $r_O - r_I$ is reduced so the effective cooling area A is not much affected.

TABLE III
PARAMETERS OF THE OPTIMIZATION RESULT AND DEVIATION TO DUT

Parameter	Value	Deviation
$r_{O,o}$	42 mm	9.5 mm
$r_{I,o}$	32 mm	13 mm
h	15 mm	-10 mm
N	21	10
d_W	2.8 mm	-0.03 mm
L	1.89 mH	0.29 mH
P	2.18 W	0.01 W
A	133.12 cm²	2.91 cm² / ≈ 2 %
V	55.3 cm³	-20 cm³ / − 36 %

The most significant enhancement is the volume reduction of the common mode choke. It can be reduced by 36 % without any constraint violation. This optimization result is assumed to be quiet accurate because the calculation models are calibrated by measurements. Although the volume enhancement is quiet impressive the design method is not verified yet by a practical realization because toroidal cores with optimized parameters of table III are not part of a standard portfolio of core manufacturers. Such cores need to be fabricated individually.

V. CONCLUSION

A design method for common mode chokes with toroidal cores for line filter applications based on a multi-objective optimization approach is presented considering common mode and differential mode noise simultaneously. The method allows to elaborate the Pareto-Front as a compromise between losses

978-1-4673-9551-9/16 $31.00 © 2016 IEEE

and volume of the common mode choke. The depiction of the Pareto-Points characterizes the influence of the priority parameter which can be chosen according to the designers choice. All models for the characterization of geometrical, electrical and thermal behavior are derived and partly validated by measurements of an existing common mode choke. It is briefly shown that the realization of effective differential and common mode inductance within one electrical component is possible but largely constrained by the chosen relative permeability of the core material. The performance of the design procedure is demonstrated exemplarily by an optimization of an already designed common mode choke for a single-phase application where the total volume was reduced by 36 % without degrading choke losses.

ACKNOWLEDGMENT

This research and development project is funded by the German Federal Ministry of Education and Research (BMBF) with the Leading-Edge Cluster Intelligent Technical Systems OstWestfalenLippe (its OWL) and managed by the Project Management Agency Karlsruhe (PTKA).

REFERENCES

[1] S. Busquets-Monge, J.-C. Crebier, S. Ragon, E. Hertz, D. Boroyevich, Z. Gurdal, M. Arpilliere, and D. Lindner, "Design of a boost power factor correction converter using optimization techniques," *Power Electronics, IEEE Transactions on*, vol. 19, no. 6, pp. 1388–1396, Nov 2004.

[2] A. Nagel and R. De Doncker, "Systematic design of emi-filters for power converters," in *Industry Applications Conference, 2000. Conference Record of the 2000 IEEE*, vol. 4, Oct 2000, pp. 2523–2525 vol.4.

[3] J. Muhlethaler, M. Schweizer, R. Blattmann, J. Kolar, and A. Ecklebe, "Optimal design of lcl harmonic filters for three-phase pfc rectifiers," *Power Electronics, IEEE Transactions on*, vol. 28, no. 7, pp. 3114–3125, July 2013.

[4] J. Muhlethaler, H. Uemura, and J. Kolar, "Optimal design of emi filters for single-phase boost pfc circuits," in *IECON 2012 - 38th Annual Conference on IEEE Industrial Electronics Society*, Oct 2012, pp. 632–638.

[5] L. Zhao, R. Chen, and J. van Wyk, "An integrated common mode and differential mode transmission line rf-emi filter," in *Power Electronics Specialists Conference, 2004. PESC 04. 2004 IEEE 35th Annual*, vol. 6, June 2004, pp. 4522–4526 Vol.6.

[6] *VDI Heat atlas*, 2nd ed. Springer-Verlag Berlin Heidelberg, 2010.

[7] F. Giezendanner, J. Biela, J. Kolar, and S. Zudrell-Koch, "Emi noise prediction for electronic ballasts," in *Power Electronics Specialists Conference, 2008. PESC 2008. IEEE*, June 2008, pp. 4392–4398.

[8] S. Wang, P. Kong, and F. Lee, "Common mode noise reduction for boost converters using general balance technique," *Power Electronics, IEEE Transactions on*, vol. 22, no. 4, pp. 1410–1416, July 2007.

[9] F.-Y. Shih, Y.-T. Chen, Y.-P. Wu, and Y.-T. Chen, "A procedure for designing emi filters for ac line applications," *Power Electronics, IEEE Transactions on*, vol. 11, no. 1, pp. 170–181, Jan 1996.

[10] K. Stille, C. Romaus, and J. Bocker, "Online capable optimized planning of power split in a hybrid energy storage system," in *EUROCON, 2013 IEEE*, July 2013, pp. 1158–1163.

[11] K. V. Price, R. M. Storn, and J. A. Lampinen, *Differential Evolution A Practical Approach to Global Optimization*, ser. Natural Computing Series, G. Rozenberg, T. Bäck, A. E. Eiben, J. N. Kok, and H. P. Spaink, Eds. Berlin, Germany: Springer-Verlag, 2005. [Online]. Available: http://www.springer.com/west/home/computer/foundations?SGWID=4-156-22-32104365-0&teaserId=68063&CENTER_ID=69103

[12] *Information technology equipment - Radio disturbance characteristics - Limits and methods of measurement (CISPR 22:2008, modified); German version EN 55022:2010*, EN55022 Std., 2010.

[13] *Residual current-operated protective devices (RCDs) for household and similar use - Electromagnetic compatibility (IEC 61543:1995 + A2:2005)*, IEC 61543/A2 Std., 2005.

Electromagnetic Noise Coupling and Mitigation for Fast Response On-die Temperature Sensing in High Power Modules

Chengcheng Yao, Pengzhi Yang, Mingzhi Leng, He Li, Lixing Fu, Jin Wang
Center for High Performance Power Electronics (CHPPE)
The Ohio State University
Columbus, Ohio - USA
Email: wang.1248@osu.edu

Ke Zou and Chingchi Chen
Ford Motor Company
Dearborn, Michigan - USA
Email: cchen4@ford.com

Abstract—A heavy low pass filter is usually applied to an on-die temperature sensor's output to filter out the strong noises during switching transients. Aiming at high bandwidth junction temperature sensing, this paper evaluates the electromagnetic noise coupling of an on-die temperature sensing diode in a high power module. Challenges of fast response on-die temperature sensing are reviewed first. Noise coupling mechanisms are analyzed under different grounding configurations and operating conditions. Based on the analysis, a method is proposed to estimate the parasitic capacitance between the sensor and the power device. Special attention has been paid to achieve a high bandwidth and noise immunity test setup. With that, the sensor coupled noises are evaluated in experiments and compared with the model. Noise propagation impedance compensation is applied and verified by experiment. Considerations of sensing circuit design are also discussed. A 100 kHz low pass filter is used to deal with the residual noises during switching transients. With the designed sensing circuitry, short circuit tests are conducted to demonstrate sensor's dynamic response.

Keywords—junction temperature; on-die temperature sensing; fast response; electromagnetic noise coupling; high power module

I. INTRODUCTION

The junction temperature information of a power semiconductor device is beneficial to its thermal management, protection and aging monitoring. Three types of methods are mainly used in power devices temperature sensing: optical methods, electrical methods and physically contacting methods [1]. Optical methods using infrared camera (IR) can directly obtain a temperature map of the power device, but fast response measurement (in tens of micro-second range) is difficult. Electrical methods using thermo-sensitive electrical parameters (TSEPs) of the device itself can achieve fast response. However, extensive calibration is usually needed. Some methods even need to interrupt the converter operation when the measurement is taking [2], [3]. One of the commonly used contact methods is to mount a negative temperature coefficient thermistor (NTC) on the ceramic substrate, however, large distance to the semiconductor device

and large thermal mass of the base plate introduce steady-state error and limit sensor's dynamic response [4]. Another contact method that becomes increasingly popular is to integrate a temperature sensing diode onto the semiconductor chip, which can save space and improve both the static and dynamic performance of temperature sensing. Nevertheless, a sensing diode's output signal is vulnerable to noises due to its low voltage amplitude, low sensitivity and close distance to the power device. A heavy low pass filter is usually applied to attenuate the coupled noise, however, by doing so the temperature sensing's bandwidth is sacrificed.

This paper evaluates the dynamic performance of a diode based on-die temperature sensor, focusing on the electromagnetic interference (EMI) between the sensing diode and the power device during switching transients. The paper starts with a review of challenges in fast response on-die temperature sensing. Then, the noise coupling mechanisms under different grounding configurations and operating conditions are analyzed. The sensor coupled noises during the switching transient are investigated and compared with the proposed model. Considerations in sensing circuit design are then discussed and verified with experiment. At the end, the sensor's dynamic response in short circuit tests are shown.

II. CHALLENGES OF FAST RESPONSE ON-DIE TEMPERATURE SENSING

As shown in Fig. 1, the target is a half-bridge Insulated-gate Bipolar Transistor (IGBT) module with two IGBTs and Free-wheeling Diodes (FWDs) in parallel. Each IGBT chip has a string of diodes fabricated in the polysilicon, which is on the surface of the IGBT chip's emitter side. Chip temperature can be detected by measuring the forward voltage drop when it is forward biased. Like most silicon diodes the forward voltage drop decreases with the increasing junction temperature. To alleviate noise problem, a string of diodes are used to provide a high enough sensing voltage. However, using the temperature sensing diode during continuous switching operation is still challenging due to the following reasons.

978-1-4673-9551-9/16 $31.00 © 2016 IEEE

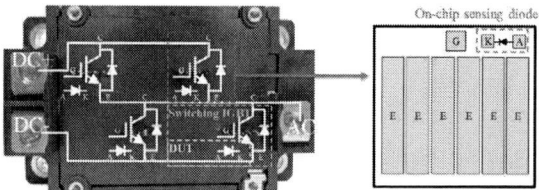

Fig. 1. An IGBT module with on-die temperature sensing diodes.

A. Sensing diodes' low forward voltage and sensitivity

Fig. 2 shows a typical V_F versus T_j static characteristic of a temperature sensing diode, with a 500 uA forward bias. The result shows that the diode has a very linear resolution of $V_F /T_j = 5.8$ mV/ °C within the target operating temperature range. However, its low signal amplitude and sensitivity make it vulnerable to noises, which pose a challenge to temperature sensing during switching transients.

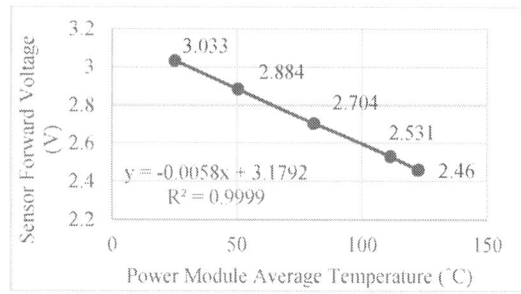

Fig. 2. Static response of the on-die temperature sensing diode.

B. Close distance to the high power switching device

Integrating sensors onto the chip can significant decrease both temperature sensing's steady state error and response time. However, the close distance to noise sources also means easy for noise propagation. For instance, parasitic capacitances between the two devices are likely to be larger compared with the NTC solution, as shown in Fig. 3. Displacement current incurred during the dv/dt switching transient may affect the sensor's output signal, which will be even more critical in WBG devices. Moreover, these capacitance value are difficult to measure directly as they are all interconnected with the device parasitic capacitances.

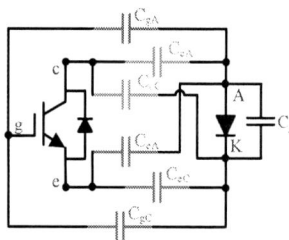

Fig. 3. Parasitic capacitances between the sensing diode and the power device.

C. Reqirements of high bandwidth and noise immunity sensing circuit

If the sensing circuitry is not designed carefully, noises coupled to it can easily mask sensing diode's signal. Applying heavy low pass filtering blindly can attenuate the noises, however, it becomes difficult to distinguish the noises caused by the sensor and the sensing circuit. More importantly, the sensor's bandwidth is sacrificed. Thus, careful EMI control on the measurement system was implemented according to [5]. Two one GHz bandwidth passive probes are used. The probes zero test and open test results under 200 V, 200 A double pulse test (DPT) are shown in Fig. 4 and Fig. 5. In the probes zero test, two probes were shorted to ground, in which conductive common-mode noises and inductive coupled noises were evaluated. In the probes open test, the tips of two probes were floating, in which capacitive coupled noises were evaluated.

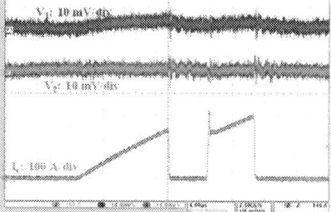

Fig. 4. Two probes zero test result.

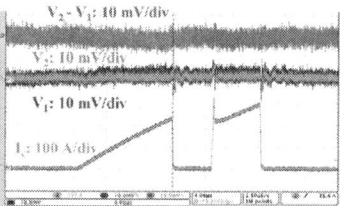

Fig. 5. Two probes open test result.

III. NOISE COUPLING MECHANISMS OF THE SENSING DIODE

The sensor coupled noises can be evaluated by the sensor unbiased test, in which the sensor was not biased and two probes were connected to its two electrodes. DPT and short circuit tests with two grounding configurations were used to investigate the dominate noise coupling mechanisms and to estimate parasitic capacitances, as shown in Fig. 6. The lower right side IGBT was controlled and the other three IGBTs were remained off during the tests, as shown in Fig. 1. The sensing diode on the switching IGBT was chosen as the device under test (DUT).

Three possible noise coupling mechanisms are discussed. Two of them become dominant in two different grounding configurations and operating conditions. Fig. 7 is schematic of sensor noise coupling paths when the sensing circuit is grounded at the IGBT power emitter. . Due to the symmetry, only diode anode side measurement unit and equivalent noise sources are included in the schematic.

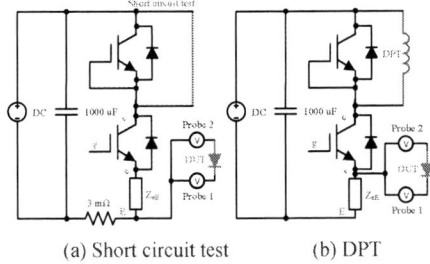

(a) Short circuit test   (b) DPT

Fig. 6. Test setups for sensor coupled noises evaluation.

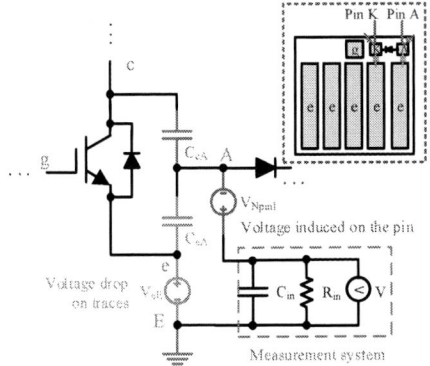

Fig. 7. On-die temperature sensor noise coupling mechanisms when grounded at the IGBT power emitter.

A. Inductive coupled noise on the sensor lead frames

The time-varying magnetic fields generated during the switching transient can induce noises on the measurement system, DUT and interconnections. The inductive coupling in the test setup has been confirmed to be negligible, as shown in Fig. 4. However, the lead frames of the sensor could have pick up some magnetic fields. The induced voltage on V_1 and V_2 would be close to di_c/dt. Moreover, if the voltages induced on the two pins are slightly different, $V_2 - V_1$ will not be zero and will have the same shape as V_1 and V_2. However, since the lead frames are relative short and small, the induced voltages are not likely to become the dominant noises. Experiment shows that extending the lead frame length does not affect the noises.

B. Capacitive coupled noise from the IGBT emitter

When only the capacitive coupling is considered, Fig. 7 can be simplified into Fig. 8. V_{eE} represents the voltage drop on the impedance Z_{eE} in Fig. 6. It is formed by the stray inductance and resistance between the kelvin and power emitter of the IGBT, which can be significant when large current or large di_c/dt occurs.

As the sensing diode is fabricated on the surface of the IGBT's emitter side, which makes C_{cA} negligible compared with C_{eA}. When V_{ce} is small, the measured V_{AE} is likely to share similar waveform as V_{eE}. A parametric study on C_{cA} and C_{cK} was conducted, as shown in Fig. 9. V_{ce} waveform was

obtained from a 50 V short circuit test. When C_{cA} and C_{cK} are small, V_{AE} and V_{KE} exhibit similar waveform as V_{eE}. However, when they become large enough, V_{AE} and V_{KE} will be influenced by V_{ce}. Luckily, these two capacitances are usually in the tens of fF range so V_{cE}'s impact can be neglected.

Moreover, since measurement system's input impedance C_{in} is already known, C_{eA} can be estimated based on the noise divider ratio, V_{AE}/V_{eE} and V_{KE}/V_{eE}. To achieve this, a short circuit test with strong V_{eE} and weak V_{ce} can be used to estimate C_{eA} and C_{eK}. These conditions can be met by a low voltage short circuit test with large turn-on and turn-off gate resistances.

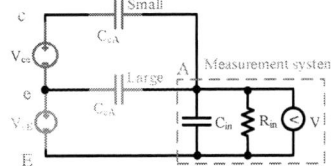

Fig. 8. On-die temperature sensor capacitive coupled noise when grounded at the IGBT power emitter.

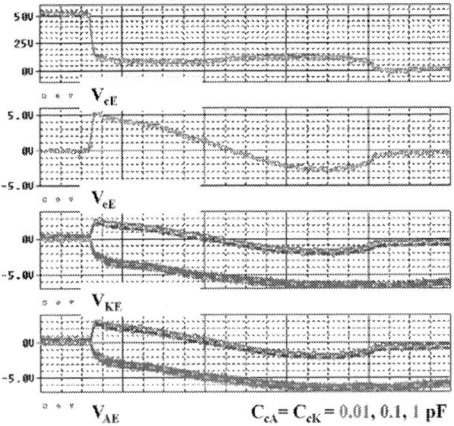

Fig. 9. C_{cA} and C_{cK}'s impact on coupled noises with grounding at power emitter

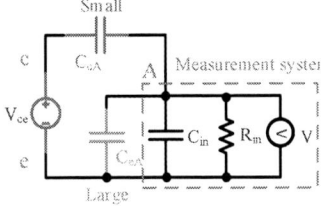

Fig. 10. On-die temperature sensor noise coupling mechanisms when grounded at the IGBT kelvin emitter.

C. Capacitive coupled noise from the IGBT collector

If the measurement system is grounded at the kelvin emitter, the dominate noise source is no longer V_{eE}, which is no longer in the loop. V_{Ae} becomes the voltage drop on the new capacitive divider's lower arm. V_{ce} becomes the new noise source, as shown in Fig. 10. Thus, V_{Ae} is expected to be a scaled down version of V_{ce}. Moreover, a high voltage DPT with small gate resistances can be used to estimate C_{cA} once C_{eA} is known.

IV. SIMULATION AND EXPERIMENTAL VERIFICATION OF NOISE COUPLING MECHANISMS

Experiments have been conducted in both test setups in Fig. 6 to verify the previous analysis. Fig. 11 is a DPT result at 400 V, 200 A with the grounding at the kelvin emitter. Fig. 13 is a short circuit test result at 50 V, 1.4 kA with grounding at the power emitter. Ac coupling are used for the sensor measurements. Both results show that: 1) strong common-mode (CM) noises present on both the anode and cathode, 2) the CM noises are unbalanced and can be converted into differential-mode (DM) noise, 3) when biased, the sensor's output signal variation would be small compared to the CM noises. All of these match with the previous analysis.

The noise signatures are different in the two tests, which feature differential grounding and operating conditions. In the DPT result, the CM noise voltages and 'V_F' are closely related to V_{ce}, which proves the analysis in section III, part C. On the other hand, in the short circuit test, the CM noise voltages are similar to V_{eE}, which proves the analysis in section III, part B.

In addition, based on the measured noise voltages, the parasitic capacitances can be estimated as in Table 1. It proves that C_{cA} and C_{cK} are much smaller than C_{eA} and C_{eK}, which are in the tens of fF range. The capacitive dividers are unbalanced, which lead to CM to DM transformation. As a result, "V_F" is not zero during the unbiased test. Then, the estimated parasitics capacitances were used for the developed equivalent noise propagation model. With the measured V_{ce} and V_{eE}, simulation results well match the experimental results in both waveform shapes and amplitudes, as shown in Fig. 12 and Fig. 14.

Table 1. Estimated parasitic capacitances

Parasitic caps	C_{cA}	C_{eA}	C_{cK}	C_{eK}	C_{cin}
Value (pF)	0.043	3.50	0.028	3.00	3.45

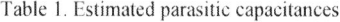

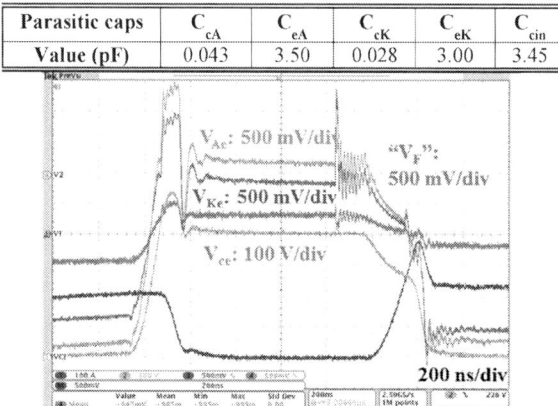

Fig. 11. Sensor noise in a DPT with grounding at kelvin e.

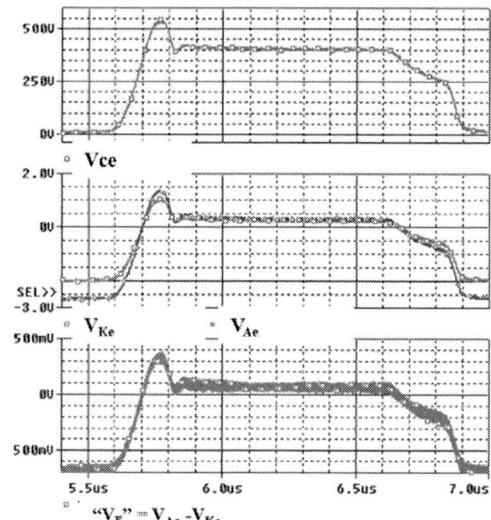

Fig. 12. Simulation of sensor noise in a DPT with grounding at kelvin e.

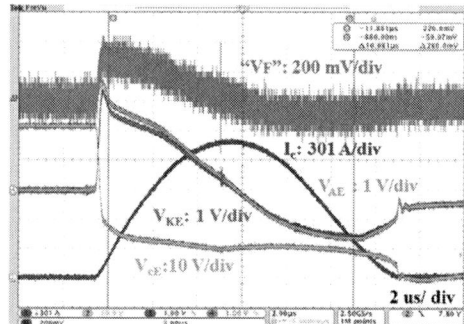

Fig. 13. Sensor noise in a short circuit test with grounding at E.

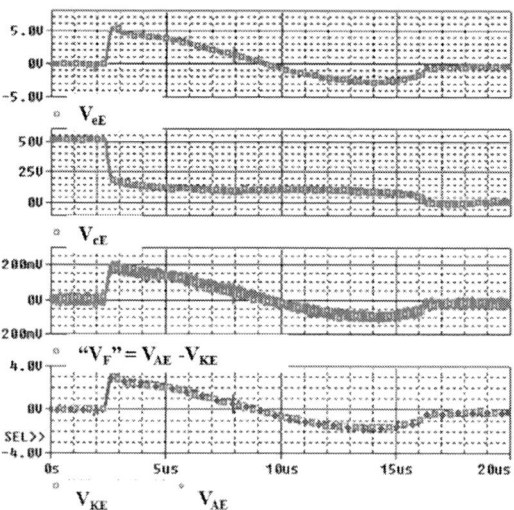

Fig. 14. Simulation of sensor noise in a short circuit test with grounding at E.

Another approach to verify the analysis is to add an external capacitor to modify the capacitive divider ratio. Compared with Fig. 13, "V_F" flipped the polarity in Fig. 15, in which an 3.2 pF capacitor is in parallel with C_{eK}. In contrast, when a large enough capacitor was added in parallel with C_{eA}, the V_{KE} became almost the same as V_{eE}, which is the noise source itself. Meanwhile, V_2 and V_1 signals maintained the same waveform shapes during the two verification tests. Therefore, the noise coupling mechanisms have been verified.

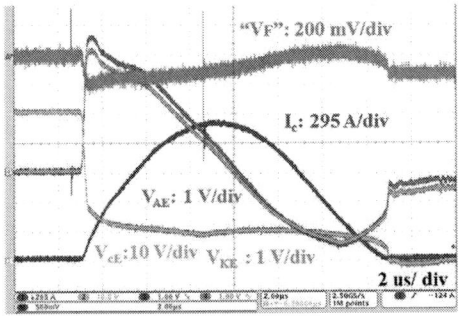

Fig. 15. 3.2 pF external capacitor in parallel with C_{eK}.

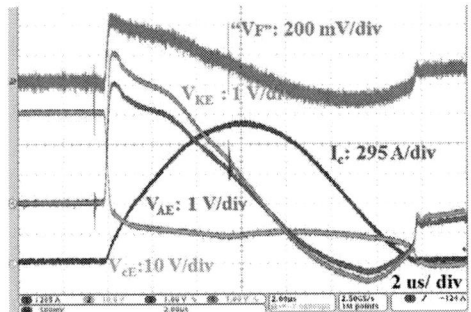

Fig. 16. 560 pF external capacitor in parallel with C_{eA}.

V. CONSIDERATIONS ON SENSING CIRCUIT DESIGN

In real implementation, the sensor's biasing and sensing circuitry are usually referred to the kelvin emitter. Thus, only kelvin emitter grounding configuration will be discussed for the sensing circuit design.

A. Requirement on high CM noise rejection at high frequency

The CM noises presented on the sensor significantly reduce the Signal Noise Ratio (SNR) and could also transform CM noise into DM noise if any impedance unbalance exists. A fully symmetric sensing circuit is preferred to achieve a high common-mode rejection ratio (CMRR). One way is to use an instrumentation amplifier (in-amp) with high CMRR, as shown in Fig. 18. However, during switching transients, the equivalent noise source (V_{ce}) frequency is typically in the MHz or even higher range. Nevertheless, a typical in-amp's

CMRR drops significantly from dc to high frequency. Fig. 17 shows the CMRR test result when both in-amp's inputs were tied to the sensor's anode. At the turn-off transient, Vce has a 2 MHz equivalent frequency and the in-amp used has a CMRR below 10 dB, which explains the 50 mV peak noise voltage.

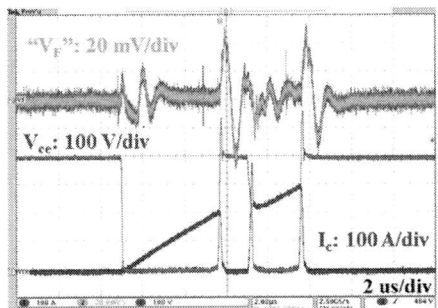

Fig. 17. An in-amp's CMRR test with V_{Ae} as the noise source.

B. Impedance unbalance compensation

The unbalanced CM noise voltage maybe even more critical due to the CM to DM transformation. This unbalance can be compensated by adding an external capacitor, which is estimated to be 2 pF in parallel with C_{eA}. The sensor noise with compensation is shown in Fig. 19. Compared with Fig. 11, the CM voltages difference is reduced significantly.

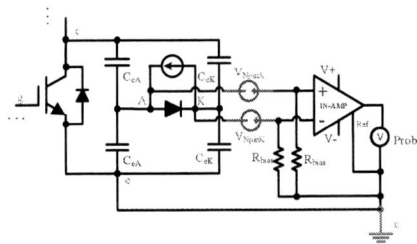

Fig. 18. Sensor dynamic measurement with IN-AMP circuit.

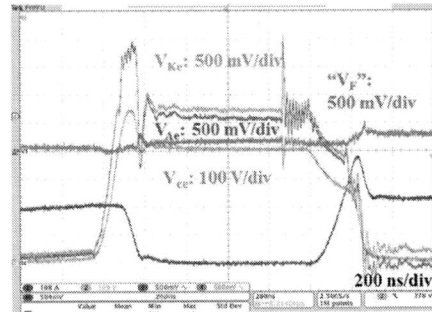

Fig. 19. Sensor noise in a DPT with grounding at kelvin e (with impedance unbalance compensation).

C. Minimum filter design using noises in DPT

To forward bias the sensing diode, a current source fed by a battery is used to minimize noises. In the DPT, the junction

978-1-4673-9551-9/16 $31.00 © 2016 IEEE

temperature barely change due to the small losses. Therefore, it can be used for filter design to deal with the residual noises. A sensor biased test result in a 400 V, 200 A DPT is shown in Fig. 20. To achieve a 10 us time constant, a 2nd order Butterworth filter with 100 kHz cutoff frequency is applied, as shown in Fig. 21. The filter is capable of attenuating most the noises during switching transients.

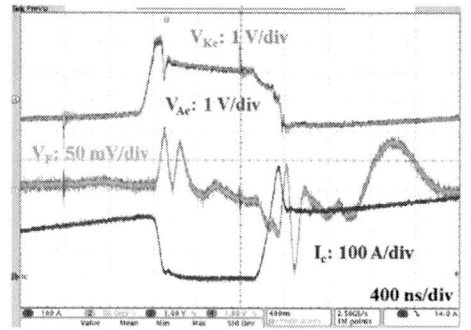

Fig. 20. Sensor biased test in a 400 V, 200 A DPT

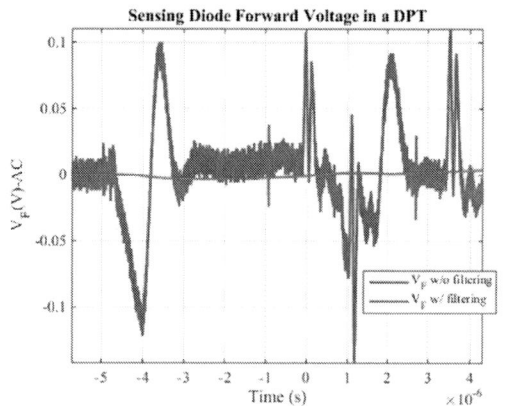

Fig. 21. Sensor noises with a 100 kHz filter.

VI. EVALUATION OF SENSOR DYNAMIC RESPONSE

Short circuit tests can be used to heat up the device so that the temperature sensor's dynamic response can be evaluated. A set of short circuits at 100 V, 1.5 kA with different pulse widths were conducted. Sensor's thermal responses are shown in Fig. 22, which indicates a thermal time constant of around 2 milliseconds. Therefore, the designed filter with 100 kHz cutoff frequency is effective in filtering out the switching noises while maintaining the sensor's bandwidth. Fig. 23 shows the filtered response comparison with different filter cutoff frequencies, which verifies the filter design.

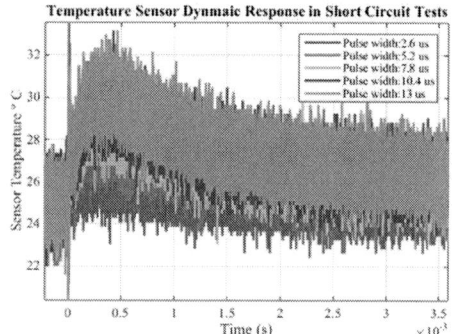

Fig. 22. Sensor dynamic response in short circuit conditions

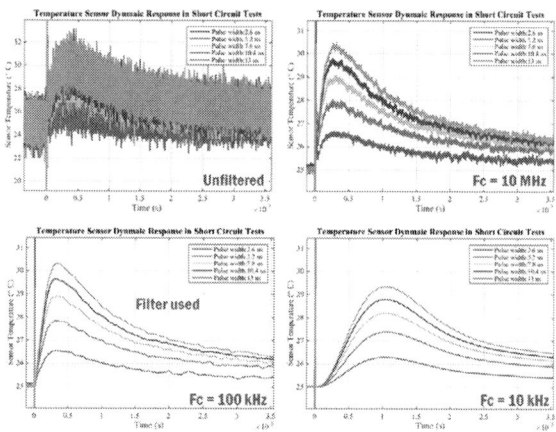

Fig. 23. Sensor signals with different filter cutoff frequencies

VII. CONCLUSIONS

A heavy low pass filter is usually added at the on-die temperature sensor's output to attenuate the strong noises during switching transients. To increase the bandwidth of on-die temperature sensing, this paper investigated the electromagnetic noise coupling between the sensing diode and the power device in high power modules. The challenges of fast response on-die temperature sensing are discussed: 1) sensing diodes' low forward voltage and sensitivity, 2) close distance to the high power switching devices, 3) requirements on high bandwidth and noise immunity sensing circuit.

The dominate noise coupling mechanisms are analyzed under two different grounding configurations and operating conditions. The proposed models can be used for parasitic capacitances estimation. With a high bandwidth and noise immunity test setup, the unbiased sensor coupled noises are evaluated. The results show that the capacitive coupled noise from the emitter dominates when the sensing circuit is grounded at the power emitter. In contrast, the capacitive coupled noise from the collector dominates when the sensing circuit is grounded at the kelvin emitter. Unbalance in parasitic capacitance is also found thus CM to DM transformation is expected. Based on proposed model and estimated parasitic capacitances, simulation and experiment has been conducted to verify the analysis in different aspects.

Sensing circuit design is critical in achieving high SNR. In particular, 1) high CMRR at high frequency is needed for CM noise attenuation, 2) impedance unbalance compensation can be applied to minimize the CM to DM noise transformation, 3) sensor biased test noise in a DPT can be used for filter design. A filter with 10 us time constant is designed to attenuate the residual noises. The designed sensing circuitry is applied to evaluate sensor's dynamic response in short circuit conditions, which demonstrates a 2 milliseconds thermal time constant. The designed filter does not affect sensor's bandwidth.

REFERENCES

[1]   Yvan Avenas, Laurent Dupont, and Zoubir Khati "Temperature Measurement of Power Semiconductor Devices by Thermo-Sensitive Electrical Parameters—A Review" *IEEE Transactions on Power Electronics, VOL. 27, NO. 6, June 2012*

[2]   Z. Xu, F. Xu, Fred. Wang, "Junction Temperature Measurement of IGBTs Using Short-Circuit Current as a Temperature-Sensitive Electrical Parameter for Converter Prototype Evaluation," *Industrial Electronics, IEEE Transactions on* , vol.62, no.6, pp.3419,3429, June 2015

[3]   R. Schmidt and U. Scheuermann, "Using the chip as a temperature sensor—The influence of steep lateral temperature gradients on the Vce(T)-measurement," in *Proc. 13th Eur. Conf. Power Electron. Appl.*, 2009, pp. 1–9.

[4]   Motto, E., et al. "MAXISS: A New Servo Duty IPM With On-Chip Temperature Sensing." *Powerflex,[Online]*.

[5]   C. Yao, M. Leng, H. Li, L. Fu, K. Zou, CC. Chen, F. Luo, and J. Wang, "Electromagnetic noise coupling and mitigation in dynamic tests of high power switching devices," in *Energy Conversion Congress and Exposition (ECCE)*, 2015 IEEE, pp.6610-6615, 20-24 Sept. 2015

Ultra-Low Inductance Vertical Phase Leg Design with EMI Noise Propagation Control for Enhancement Mode GaN Transistors

Xuning Zhang, Zhiyu Shen, Nidhi Haryani, Dushan Boroyevich and Rolando Burgos

Center for Power Electronics Systems
The Bradley Department of Electrical and Computer Engineering
Virginia Polytechnic Institute and State University
Blacksburg, VA 24061 USA
Xuning45@vt.edu

Abstract— **This paper presents an improved phase leg power loop design for enhance mode lateral structure Gallium Nitride (GaN) transistors. Static characterization results of a 650V/30A GaN transistor are presented to determine the design parameters of the gate driver circuits. The control of Common Mode (CM) noise current propagation is considered during the gate driver design by optimizing the power distribution and grounding structure of the gate driver and digital control circuits. By differentiating the propagation path impedance of digital control circuits and their power supply circuits, conductive CM noise can propagate through power supply path to protect the digital control circuits. In order to reduce current commutation loop inductance within the GaN phase leg, an improved power loop design with vertical structure is proposed for lateral structure GaN transistors which can significantly reduce power loop inductance compared with conventional lateral power loop design. The design is verified through experiments on a phase leg prototype which prove the performance of the proposed phase leg on the overvoltage reduction during current transition along with less cross-coupling between power loop and gate loop compared with conventional lateral power loop design. A full bridge voltage source inverter is implemented with the designed phase leg and tested with EMI noise measurement that verifies the effectiveness of the CM propagation path control.**

Keywords—GaN HMET, Characterization, Gate Driver, EMI; CM noise propagation.

I. INTRODUCTION

Wide band-gap power switches outperform silicon counterparts in terms of switching speed and on-resistance, and therefore provide the potentials to improve the power density of various converters by shrinking the size of passive components and improving the power conversion efficiency. [1~6] For 600V rated power devices, gallium nitride switches have shown significant advanced performance compared with Si and even SiC devices. In this voltage level, the cascode structure has been dominant, however, it has issues with lower reliability and high common source inductance [7]. Recently, the enhancement mode lateral structure GaN HEMTs become commercially available.[8,9] In lateral structures, current flows at near surface in the devices which enable the usage of Si substrates instead of GaN free standing

substrates to reduce the cost of GaN devices. Moreover, the lateral HEMTs have a low parasitic capacitance which means that these devices have both lower conduction losses and low switching losses, providing the potentials for ultra-high switching frequency in power conversion. However, fast switching also generates high dv/dt and di/dt in the system that requires advanced technologies of packaging and circuit design for noise control such as advanced isolation for gate driver circuits and improved power loop layout design to achieve proper EMI noise control and small parasitic inductance for both power loop and gate loop [10]. In order to extract the full benefits from these the enhanced-mode lateral GaN devices, this paper presents an improved phase leg power loop design with vertical power loop structure and CM noise current propagation control for a 650V/30A enhancement mode GaN switch (GS66508) from GaN systems. The static characterization results are presented that verifies the better performance of GaN switches compared with Si MOSFETs. Based on the static characterization results, a gate drive circuit design is presented considering the CM noise current propagation control by differentiating the propagation path impedance of digital control circuits and their power supply circuits to use the power supply path to bypass more conductive CM noise and protect the digital control circuits. Moreover, a vertical power loop layout is proposed to minimize the current commutation loop inductance. The design is verified through experiments on a phase leg prototype which proves the effectiveness of the proposed phase leg on the overvoltage reduction during current transition along with less cross-coupling between power loop and gate loop compared with conventional lateral power loop design. Finally, a full bridge voltage source inverter is designed and tested based on the proposed phase leg with time domain and frequency domain measurement that verifies the effectiveness of CM noise propagation control.

II. STATIC CHARACTERIZATION FOR GATE DRIVER DESIGN

In order to verify the better performance of GaN switches compared with Si MOSFETs and get parameters for gate

978-1-4673-9551-9/16 $31.00 © 2016 IEEE

driver circuit design, a static characterization test is conducted for a 650V/30A enhancement mode GaN switches with top side cooling(GS66508P) from GaN systems with curve tracer. Figure 1(a) shows the static characterization results of device I-V characterization under different gate voltages, which indicates that the channels is fully enhanced at gate voltage above 7 V which is set to be the driving voltage of the gate driver circuit considering that the maximum gate voltage rating is 10V from the device datasheet. The test results also indicate that the switch can work at reverse conduction mode with almost the same I-V characteristics with normal conduction mode, therefore there is no need for anti-parallel diodes. From the test results, it is clear that negative gate voltage will increase the on state voltage drop in the reverse conduction mode which will increase the conduction loss significantly, therefore, negative gate voltage is not implemented in the gate driver design. Figure 1(b) shows the static characterization results of device I-V characterization under 150 °C junction temperature, where the R_{ds_on} increases from 55 mΩ to around 130 mΩ which needs to be considered for higher current system design. Figure 1(c) shows the capacitance of the devices changes with drain-source voltage, where both the input and output capacitances of the devices are much lower than those of the silicon devices, which indicate that GaN switches have much faster gate charge and lower capacitor charge loss during switching. Therefore GaN switches is very suitable for ultra-fast switching applications. However the fast switching speed also exaggerate electromagnetic interference (EMI) in the system, the high dv/dt and di/dt due to fast switching will increase the level of both conductive and radiative EMI noise in the system and interfere with other circuits like the gate driver IC or the digital control IC.

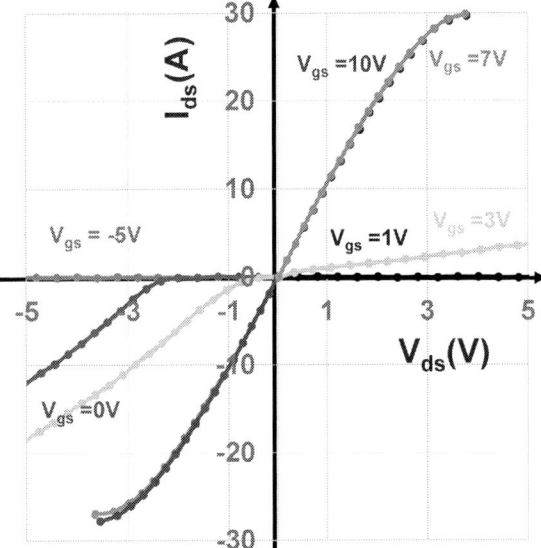

Figure 1(b): I-V characteristics at 150 °C

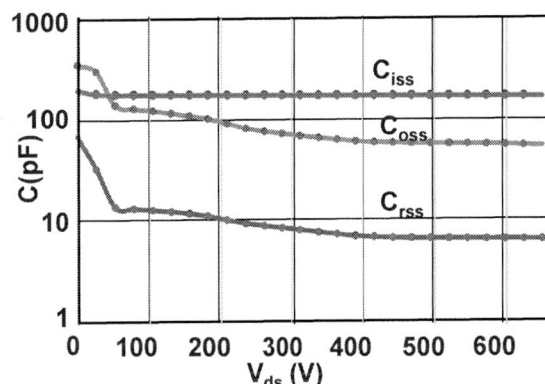

Figure 1 (c): Capacitance characteristics at 20 °C

In order to reduce the conductive and radiative EMI noise and improve electromagnetic compatibility (EMC) of the system, these circuits require advanced layout design to achieve proper EMI noise control and small parasitic inductances for both power loop and gate loops to avoid voltage overshoot during switching. Based on the static characterization results, a half-bridge phase leg structure with no anti-parallel diodes as shown in figure 2 is studied. The phase leg is designed with power electronics building block (PEBB) concept for easy scalability with self-noise-containment and optimized gate and drive loop design.

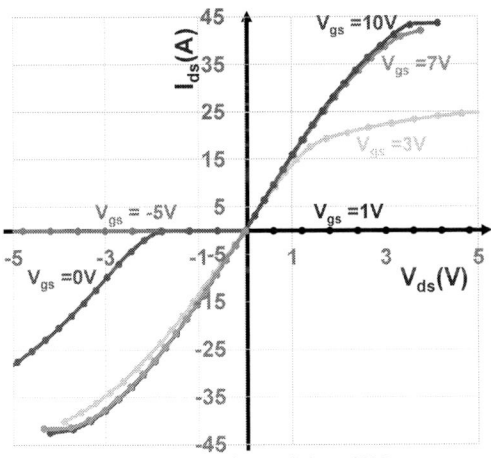

Figure 1(a): I-V characteristics at 20°C

978-1-4673-9551-9/16 $31.00 © 2016 IEEE

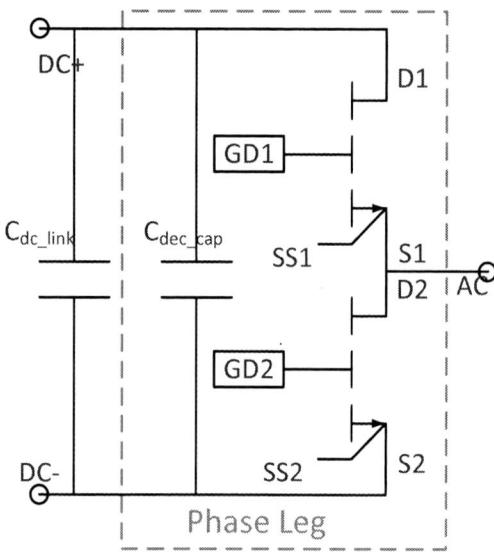

Figure 2: Phase leg configuration under consideration

III. GATE DRIVER CIRCUIT DESIGN WITH CM NOISE PROPAGATION CONTROL

Isolation is necessary to block the high frequency noise generated by the high dv/dt in the power stage to protect digital control circuits. Especially for GaN devices with ultra-fast switching speed, where the voltage slew rate can be as high as hundreds of volts per nanosecond and the harmonic frequency related with the turning-on and turning-off of the devices may be up to several hundreds of mega-hertz, these high dv/dt can generate high frequency EMI noise that propagates through the parasitic capacitance to the whole system including the power stage and control circuits and bring electromagnetic compatibility (EMC) issues to the system. Therefore, EMI noise needs to be well controlled to avoid its interference with the digital control circuits, especially the propagation of the common mode EMI noise current. With this consideration the gate driver circuit is designed with digital isolators SI8410 from Siliconlabs to provide high isolation between logic control circuits and power circuits, a commercial gate driver chip IXDN609 from IXYS is implemented as the gate driver and current booster. A 5.2kVdc rated isolated power supply MEJ2S1509SC from Murata is implemented to provide isolation between gate driver power supply and the power circuits. Two non-isolated linear regulators are used to provide the +7V for gate driver and +5V for digital isolator. No negative voltage is used for the driving signals to avoid excessive conduction loss increase. Figure 3 shows the gate driver circuit design.

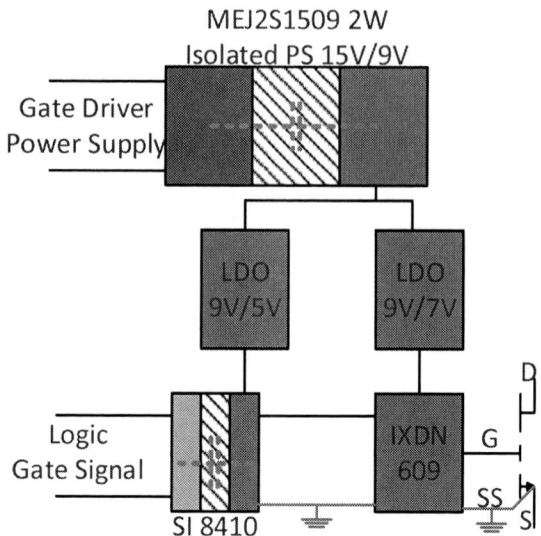

Figure 3 Gate driver circuit design

The high switching speed of GaN devices will generate high dv/dt that can induce high CM current that propagates from power loop to gate loop. The main propagation path is through stray capacitance of the isolated power supplies and digital isolators as shown in Figure 3. In order to improve the electromagnetic compatibility of the system, CM noise propagation is controlled by differentiating the propagation path impedance of digital control circuits and their power supply circuits. The block diagram is shown in Figure 4, where digital isolators with ultra-low isolation capacitance are selected for both top and bottom devices to provide good isolation for high frequency EMI noise. The power supply of the digital control circuit is also selected with ultra-low isolation capacitance to create a high impedance path to reduce CM noise current through the digital control circuits. Meanwhile the gate driver power supply is selected with relatively higher isolation capacitance to create a lower impedance path for CM noise. In addition, CM chokes are added to both signal and power supply path to maintain higher impedance at high frequency. The selection of choke impedance also need to ensure that the digital control circuits have higher noise propagation path impedance than its power supply. In this configuration, the digital control circuit will have ultra-high propagation path impedance and the gate drive power supply will provide a bypassing path with relatively low propagation path impedance in parallel with the digital control circuits which can effectively reduce CM noise current through the digital control circuits. With the proposed design, gate drive power supply circuits will sustain high CM current, however, the power supply control circuits should have higher susceptibility and noise immunity considering its relatively smaller dimension and higher circuit integration level.

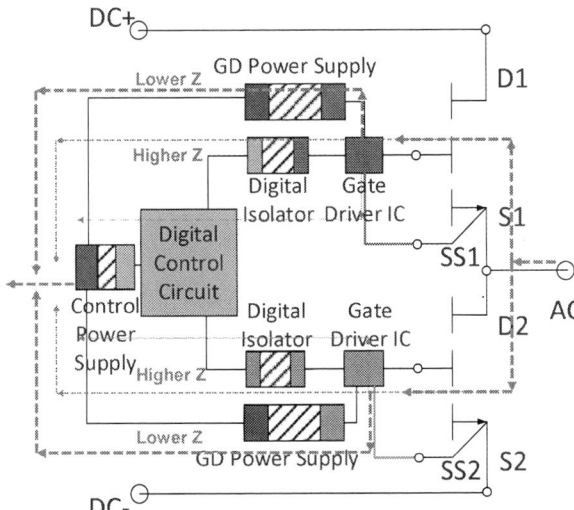

Figure 4: Gate driver circuit design
with CM noise propagation control

IV. VERTICAL POWER LOOP DESIGN FOR LOOP INDUCTANCE MINIMIZATION

The fast switching of GaN devices also generates high di/dt in the power loop, if the parasitic loop inductance is too high, the overvoltage during switching will also be high and it will increase the voltage stress on the devices which degrade the maximum bus voltage and prevent further increase of the switching speed. Therefore it is very important to design the power stage layout with low loop inductance. [8] Figure 5 shows the reference layout recommended by the devices vendors [12] and Figure 5(b) shows the power loops and gate loops in the reference design. The current needs to propagate through the two lateral located devices which creates a relatively high current loop with higher parasitic loop inductance. Moreover, the power loop and gate loops are paralleled which will increase the near field coupling between the power loop and the gate loops. Moreover, the DC input and AC output are coupled together which increases the interaction between input and output EMI noise and make it harder for EMI filter design.

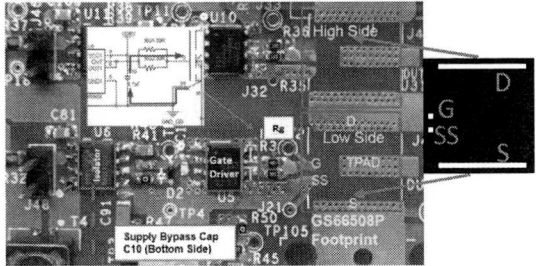

Figure 5(a) Reference layout

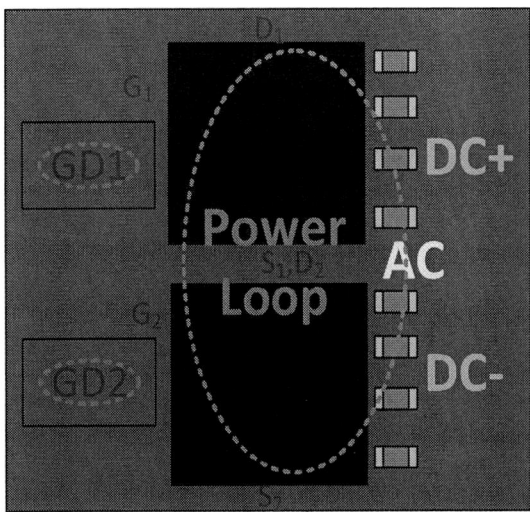

Figure 5(b) Reference design power loop

To solve these issues, a vertical power loop design is proposed and shown in Figure 6 where GaN devices are mounted on both sides of the PCB board and the gate driver circuits are also separated to different sides of the boards to minimize noise coupling. Figure 6(b) (c) shows the current loop and gate loop in the proposed design, where the power loop is folded to increase mutual coupling between the current through the two devices and the decoupling capacitors are mounted near the devices as close as possible to reduce the length of the power loop, therefore the current commutation loop inductance can be reduced significantly compared with the reference design. For lateral GaN devices, the current conducts along the devices shown as in Fig. 6(b) which will increase the mutual coupling within the current commutation path, therefore the proposed vertical layout is more suitable for lateral devices compared with vertical devices. Table I shows the parasitic loop inductance estimation through Q3D extraction which proves that the power loop inductance is reduced 10 times compared with the reference layout design. Figure 6(c) also indicates that the gate loop is perpendicular to the current loop which can also reduce the near field coupling due to the high di/dt in the current commutation loop. Moreover, the DC input and AC output terminals are separated in the proposed layout, which reduces the interaction between input and output noise. The thermal design is more challenging in the proposed layout since the devices are overlapping each other and the heat has to dissipate along the PCB board. However, the heat dissipation can be improved by implementing the power loop with direct bonded copper (DBC) or using the newly-released top-cooled devices.

978-1-4673-9551-9/16 $31.00 © 2016 IEEE

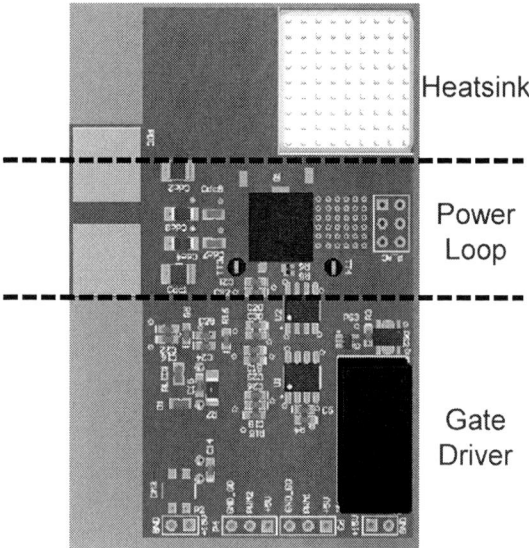

Figure 6(a): Proposed phase leg design

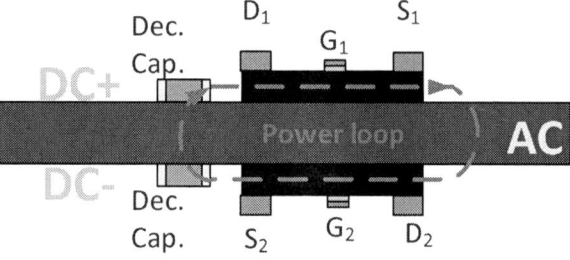

Figure 6(b): Proposed design power loop (side view)

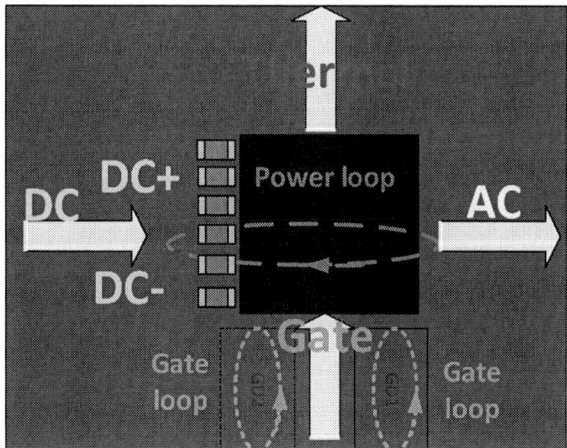

Figure 6(c): Proposed design power loop (top view)

Table I Parasitic inductance comparison

	Reference Lateral Layout	Proposed Vertical Layout
Power Loop	28.7 nH	3.1 nH
Gate Loop	0.4 nH	0.2 nH

V. EXPERIMENTAL VERIFICATION

Two phase legs are implemented following the reference layout (Fig.5 (a)) and the proposed design layout (Fig. 6(a)) respectively to compare the performance of different layouts. A standard double pulse test is conducted for both phase legs with the same gate resistance (R_g=10Ω) to verify the effectiveness of the proposed phase leg design. Figure 7 shows the test results of the bottom device at 400V dc link voltage and 32A inductor peak current with the proposed phase leg design. Table II shows the comparison of the peak device voltage during turning-off and peak gate voltage during turning-on. For reference design, the peak voltage is already near 600V with 32A switching current which prevent further increase of the device switching speed or the dc link voltage, however, the proposed phase leg layout can reduce voltage overshoot during turning-off by 100V, therefore, the device can switch faster with no issues, which verifies the effectiveness of the proposed layout design for loop inductance reduction.

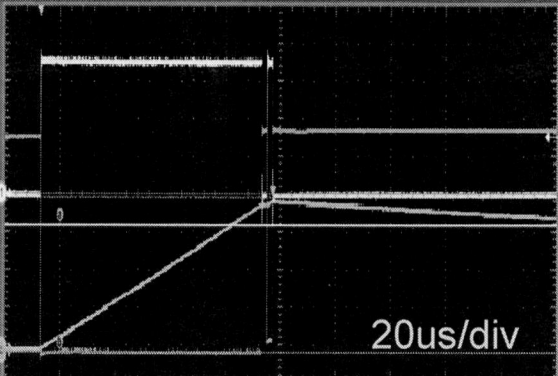

Figure 7(a): Double pulse test results
(yellow: Gate voltage: Vgs (2V/div) ; blue: Inductor current :IL(8A/div); green: Drain-source voltage: Vds (70V/div))

978-1-4673-9551-9/16 $31.00 © 2016 IEEE   1565

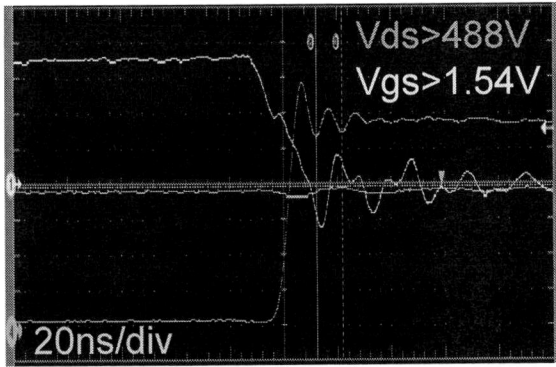

Figure 7(b):Zoomed turning-off results

(yellow: Gate voltage: Vgs (2V/div) ; blue: Inductor current :IL(8A/div); green: Drain-source voltage: Vds (70V/div))

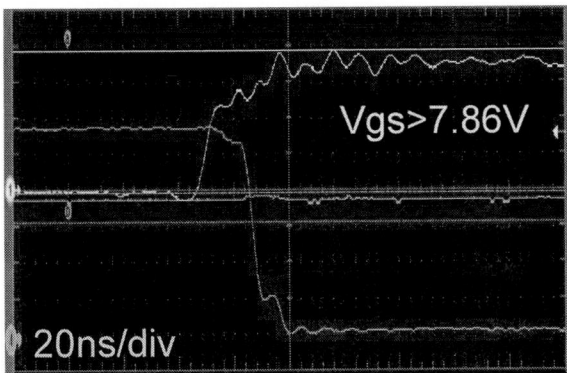

Figure 7(c): Zoomed turning-on results

(yellow: Gate voltage: Vgs (2V/div) ; blue: Inductor current :IL(8A/div); green: Drain-source voltage: Vds (70V/div))

Table II Double pulse test comparison

	Vds_max	Vgs_max
Reference Design	588 V	8.59V
Proposed Design	488 V	7.88V

The phase legs are designed and implemented as power electronics building blocks (PEBB), two phase legs are assembled to a custom designed dc link baseboard with dc bulk capacitors and distributed dc link capacitors to form a full bridge voltage source inverter (VSI) for continuous testing. Figure 8(a) shows the implemented GaN based full bridge voltage source inverter and figure 8(b) shows the structure of the test circuit with RL load. The test conditions are: dc link voltage V_{dc}= 270 V; switching frequency fs = 100 kHz, fundamental frequency f1= 400 Hz; Load inductance L_L= 500 uH; Load resistor R_L= 16ohm; modulation index M=0.7, output power Pout= 500W; modulation method is triangular unipolar modulation. Figure 9 shows the time domain test results of the system. The system continuously runs for over 20 minutes with no fault or protection triggered which verify the validity of the electromagnetic compatibility (EMC) design of the system.

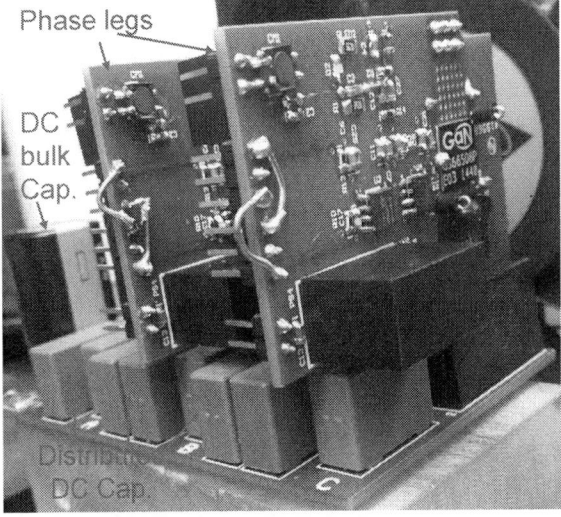

Figure 8(a) Implemented GaN based full bridge VSI

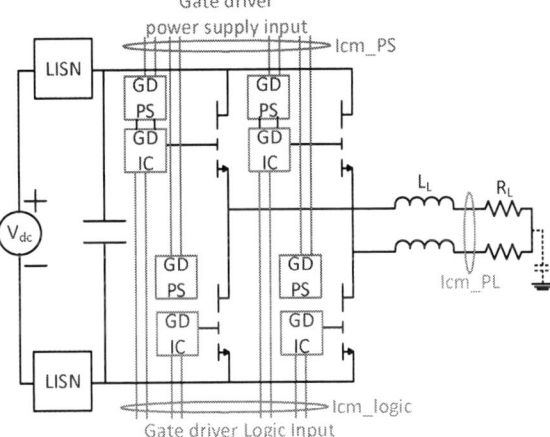

Figure 8 (b) Structure of test circuit and EMI noise measurement points

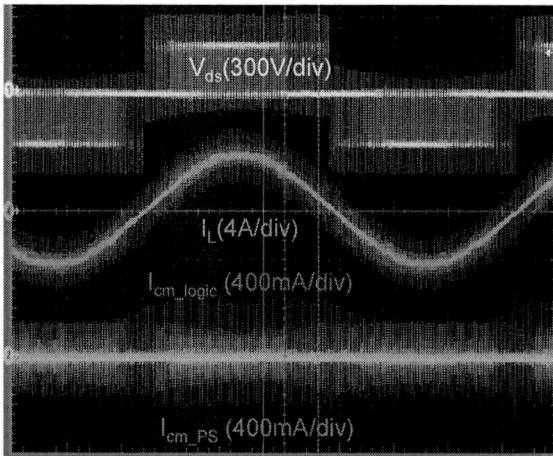

(a) Test results in time domain (200us/div)

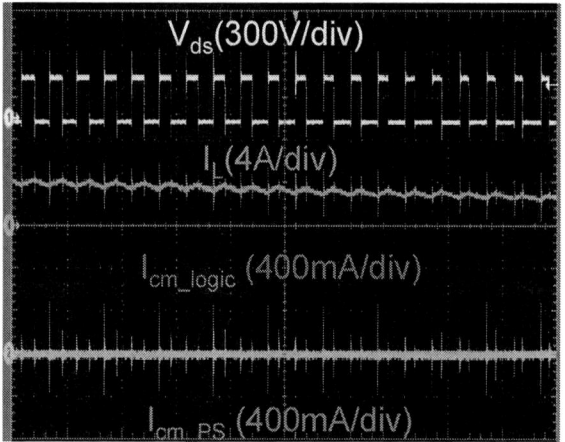

(b) Zoomed Results for CM current noise (10us/div)

Figure 9 Time domain test results
(Drain-source voltage Vds(300V/div): yellow; load current IL(4A/div):green;
gate driver logic path CM current Icm_logic(400mA/div):blue; gate driver
power supply path CM current Icm_PS (400mA/div:pink;)

EMI noise measurement is conducted per DO160 standard with LISN connected at DC side as shown in Figure 8(b). Output CM and DM noise current is measured and CM noise current through gate drive signal path and gate drive power supply path is also measured to verify the design of CM noise propagation control. The measuring points are marked in Figure 8(b). The time domain noise current is shown in Figure 9 and Figure 10 shows the frequency domain noise spectrum between 150 kHz and 30 MHz. It is clear the CM noise that propagates through gate driver signal path and gate driver power supply path have the same profile of the power line CM noise. With the designed gate driver circuit configuration, the gate driver power supply path have higher CM noise current than the gate driver signal propagation

path, which verifies the effectiveness of controlling CM noise propagation by controlling the propagation path impedance to use power supply path to bypass CM noise and protect signal path with digital control circuits.

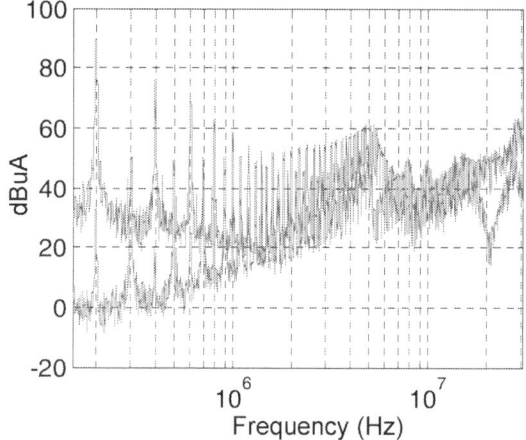

(a) Power circuit EMI noise current measurement results
(DM noise: pink; CM noise: red)

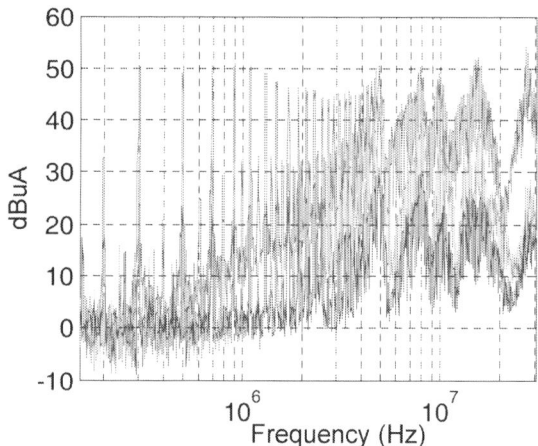

(b) CM noise current comparison (Gate driver logic path: blue; Gate drive
power path: green; Power Line : red)

Figure 10 EMI noise measurement results in frequency domain.

VI. CONCLUSIONS

This paper presents an improved phase leg design for lateral enhance mode GaN switches with vertical power loop structure and CM noise current propagation control. Static characterization results are presented which verifies the better performance of GaN switches comparing with silicon MOSFET. However, the fast switching speed of GaN devices also exxagrate EMI issues to gate driver circuit design. A gate driver circuit is designed based on the characterization results with the consideration of the CM noise current propagation control by differentiating the propagation path impedance of digital control circuits and their power supply circuits. Moreover, a vertical power loop layout is proposed to minimize the current commutation loop inductance. The design is verified through experiments on a double pulse test on a half bridge phase leg, which proves the effectiveness of the proposed phase leg on the overvoltage reduction during current transition along with less cross-coupling between power loop and gate loop compared with conventional lateral power loop design. Finally, a full bridge voltage source inverter designed and tested with EMI noise measurement based on the proposed phase leg that verifies the effectiveness of the CM noise propagation control.

REFERENCES

[1] M. Kasper, D. Bortis, and J. W. Kolar, "Classification and Comparative Evaluation of PV Panel-Integrated DC-DC Converter Concepts," IEEE Trans. Power Electron., vol. 29, no. 5, pp. 2511-2526, 2014.

[2] M. Rodriguez, M. Roberg, A. Zai, E. Alarcon, Z. Popovic, and D. Maksimovic, "Resonant Pulse-Shaping Power Supply for Radar Transmitters," IEEE Trans. Power Electron., vol. 29, no. 2, pp. 707-718, 2014.

[3] Z. Xuan, Y. Chengcheng, L. Cong, F. Lixing, G. Feng, and W. Jin, "A Wide Bandgap Device-Based Isolated Quasi-Switched-Capacitor DC/DC Converter," IEEE Trans. Power Electron., vol. 29, no. 5, pp. 2500-2510, 2014.

[4] R. Mitova, R. Ghosh, U. Mhaskar, D. Klikic, W. Miao-Xin, and A. Dentella, "Investigations of 600-V GaN HEMT and GaN Diode for Power Converter Applications," IEEE Trans. Power Electron., vol. 29, no. 5, pp. 2441-2452, 2014.

[5] T. Uesugi and T. Kachi, "Which are the future GaN power devices for automotive applications, lateral structures or vertical structures?" in CS Mantech Tech. Dig., 2011, pp. 1–4.

[6] Kaminski, N., "State of the art and the future of wide band-gap devices," Power Electronics and Applications, 2009. EPE '09. 13th European Conference on , vol., no., pp.1,9, 8-10 Sept. 2009

[7] Zhengyang Liu; Xiucheng Huang; Lee, F.C.; Qiang Li, "Package Parasitic Inductance Extraction and Simulation Model Development for the High-Voltage Cascode GaN HEMT," Power Electronics, IEEE Transactions on , vol.29, no.4, pp.1977,1985, April 2014

[8] Application note,"How to Drive GaN Enhancement Mode Power Switching Transistors", GaN Systems inc., 2014.

[9] T. McDonald, "GaN Based Power Technology Stimulates Revolution in Conversion Electronics", Bodo's Power Systems (www.bodospower.com), April 2009

[10] Delaine, J.; Jeannin, P.; Frey, D.; Guepratte, K., "High frequency DC-DC converter using GaN device," Applied Power Electronics Conference and Exposition (APEC), 2012, pp.1754,1761, 5-9 Feb. 2012

[11] Boroyevich, Dushan; Zhang, Xuning; Bishinoi, Hemant; Burgos, Rolando; Mattavelli, Paolo; Wang, Fred, "Conducted EMI and Systems Integration," in Integrated Power Systems (CIPS), 2014 8th International Conference on , pp.1-14, 25-27 Feb. 2014

[12] Application note- "How to Drive GaN Enhancement Mode Power Switching Transistors", GaN Systems inc., 2014.

[13] Preliminary datasheet- "GS66508P-E03- 650V enhancement mode GaN transistor", GaN Systems inc., 2014.

[14] Zhang, Xuning; Haryani, Nidhi; Shen, Zhiyu; Burgos, Rolando; Boroyevich, Dushan. "Ultra-Low Inductance Phase Leg Design for GaN-Based Three-Phase Motor Drive Systems,"The 3rd IEEE Workshop on Wide Bandgap Power Devices and Applications WiPDA 2015, Nov. 2015

[15] Zhang, Xuning; Boroyevich, Dushan; Burgos, Rolando, "On discussion of mixed mode noise in H-bridge converters," in Energy Conversion Congress and Exposition (ECCE), 2015 IEEE , vol., no., pp.255-262, 20-24 Sept. 2015

Decoupling of Interaction Between WBG Converter and Motor Load for Switching Performance Improvement

Zheyu Zhang, Fred Wang, Leon M. Tolbert, Benjamin J. Blalock, and Daniel J. Costinett

Department of Electrical Engineering and Computer Science
The University of Tennessee, Knoxville, TN, USA
zzhang31@vols.utk.edu

Abstract— **High speed switching of WBG devices causes their switching behavior to be highly susceptible to the parasitics in the circuit, including inductive loads. An inductive load consisting of a motor and power cable significantly worsens the switching speed and losses of SiC MOSFETs in a PWM inverter. This paper focuses on the motor plus power cable based inductive load, and aims at mitigating its negative influence during the switching transient. An auxiliary filter is designed and inserted between the converter and inductive load so that the parasitics of the load will not be "seen" from the converter side during the switching transient. Test results with Cree 1200-V/20-A SiC MOSFETs show that the proposed auxiliary inductor enables the switching performance with a practical inductive load (e.g., motor plus cable based inductive load) to exhibit behavior close to that when the optimally-designed double pulse test load inductor is employed.**

I. INTRODUCTION

High switching-speed capability of WBG devices leads to low switching loss and enables high switching frequency. However, high di/dt and dv/dt during the fast switching transient worsens the electromagnetic environment of loads. In the meantime, fast switching makes the switching performance of WBG devices significantly susceptible to the loads' parasitics. In the end, the interaction between converter and load due to fast switching WBG devices challenges the performance and reliability of the whole system [1, 2].

First, voltage pulses with fast rise time generated by PWM switching of power devices can cause serious non-uniform voltage distribution in motor windings, and voltage doubling affect at motor terminals for motor loads with long power cables. The phenomenon is detrimental to motor insulation and requires a dedicated filter or special motor to mitigate. This is a well-known issue with Si device based PWM drives, and becomes more severe when WBG devices are utilized. Moreover, high dv/dt induced by WBG devices can cause larger bearing current in motor loads, detrimental to motor reliability. Several researchers have taken the initiatives of exploring dv/dt related issues for WBG power electronics converters [3, 4].

On the other hand, high speed switching of WBG devices will be affected by parasitics of loads [5, 6]. Fig. 1(a) depicts the impedance of a 7.5-kW induction motor plus 2-meter power cable with the frequency range from 10 kHz to 100 MHz. The motor load is no longer inductive in the switching-related frequency range which is determined by the switching speed and typically at several MHz to tens of MHz for WBG devices considering switching intervals of tens of nanoseconds [7]. The load and cable parasitic impedances worsen the WBG devices' switching performance. Due to parasitics of the inductive load in Fig. 1(a), as compared to the switching waveform exhibited by using an optimally-designed inductor load in a double pulse test (DPT), the tested switching time of SiC MOSFETs increases up to 42% during turn-on, and doubles during turn-off; an additional 32% of energy loss is dissipated during the switching transient [8].

For the higher power rating induction motor with longer power cable, the associated impedance at high frequency is even lower [9]. Thus, the motor load selected in [8] indicates a conservative impact of the induction motor on the switching performance. Accordingly, it is critical to understand and mitigate the adverse effect of the inductive load on switching behavior so as to achieve optimal switching behavior of WBG devices in actual converters for practical applications.

This paper focuses on the motor plus power cable based inductive load, and aims at mitigating its negative influence during the switching transient. First, the high frequency inductive load modeling is presented, as the fundamental knowledge for developing the subsequently proposed solution. Second, a basic concept of decoupling the interaction between the inductive load and power device in the voltage source converter during the switching transient is presented. Based on this concept, an auxiliary filter is designed and inserted between the converter and inductive load to reshape the high frequency impedance of the inductive load and mitigate its adverse influence. Finally, a double pulse tester with Cree 1200-V/20-A SiC MOSFETs is established to demonstrate the validity and effectiveness of this proposed approach.

978-1-4673-9551-9/16 $31.00 © 2016 IEEE

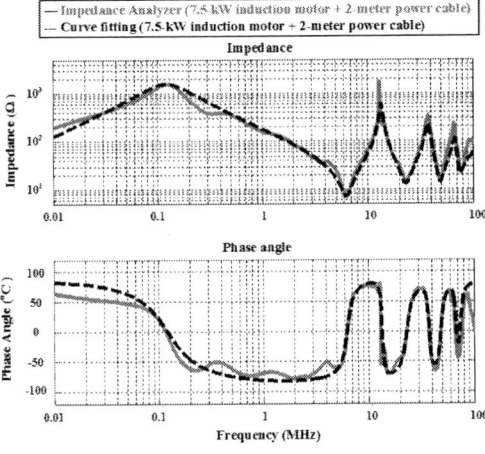

(a) Impedance comparison.

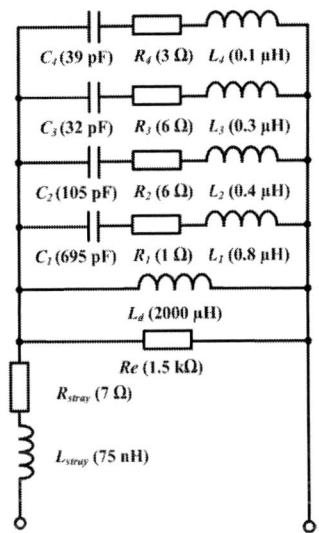

(b) High frequency behavior model.

Fig. 1. High frequency impedance and circuit model of motor plus cable based inductive load.

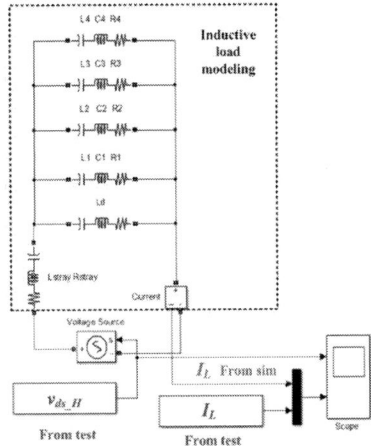

(a) Evaluation platform built by Matlab/Simulink.

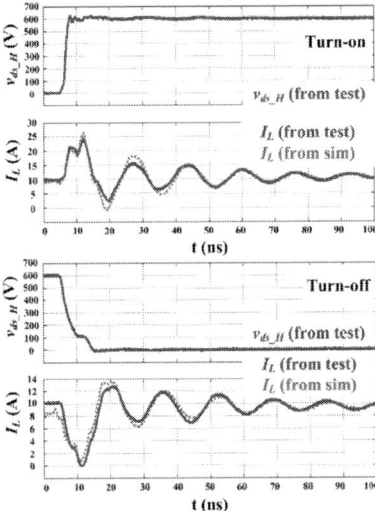

(b) Experiment versus model based simulation.

Fig. 2. Accuracy verification of the high frequency modeling in time domain.

II. INDUCTIVE LOAD HIGH FREQUENCY MODELING

According to the high frequency impedance of the inductive load consisting of motor plus cable measured by an impedance analyzer, a circuit model is derived, as shown in Fig. 1(b) [10-12]. Several *LRC* series resonant networks are employed in a paralleled structure to simulate the high frequency behavior of the inductive load under investigation. As can be observed in Fig. 1(a), the fitted curves of the inductive load impedance based on the circuit model can represent the critical characteristics of the motor plus cable impedances in frequency domain.

Based on the behavior model proposed above, a simulation circuit is built in Matlab/Simulink to verify the accuracy of this high frequency model in time domain. In Fig. 2(a), the voltage across the inductive load (i.e., drain-source voltage of the upper switch v_{ds_H}) measured by the double pulse test is added at terminals of the inductive load's circuit model. Then, the current flowing through the inductive load in the simulation is monitored and compared with the tested inductive current. The switching waveform comparison in Fig. 2 (b) shows that the simulated inductive current I_L is almost identical to that based on test results during both turn-on and turn-off trainsets. In total, the derived high frequency circuit model is accurate in both frequency and time domains and can be used for the following investigation.

III. BASIC CONCEPT FOR MITIGATION OF ADVERSE EFFECT OF INDUCTIVE LOAD

According to the circuit model in Fig. 1(b), the large inductance L_d can be considered as a current source

978-1-4673-9551-9/16 $31.00 © 2016 IEEE

during the switching transient. It is the *LRC* series resonant networks that affect the switching behavior. Taking the turn-on transient as an example, as can be observed in Fig. 3(a), during the *dv/dt* transient, there will be resonant current excited per each *LRC* series resonant branch (i.e., I_1 to I_4). This additional current flows into the device and increases the channel current, resulting in the slower turn-on speed with larger switching losses. On the other hand, during the turn-off transient in Fig. 3(b), the resonant current induced by a *LRC* series resonant branch decreases the equivalent inductive current, leading to a longer turn-off time for charging/discharging of the device' output capacitance. Hence, the turn-off speed decreases as well.

To mitigate the impact of the inductive load's parasitics on the switching performance, resonant current suppression during the switching transient is critical. There are two basic concepts: 1) minimization of the resonant currents for the *LRC* branches with respect to the high resonant frequency (e.g., f_2 to f_4 in Fig. 4(a)), 2) regulation of the resonant period of the low resonant frequency based *LRC* branch (e.g., f_1 in Fig. 4(a)) to be much longer than the switching commutation time so that there is almost no response for this series resonant network to the switching voltage excitation during the switching transient.

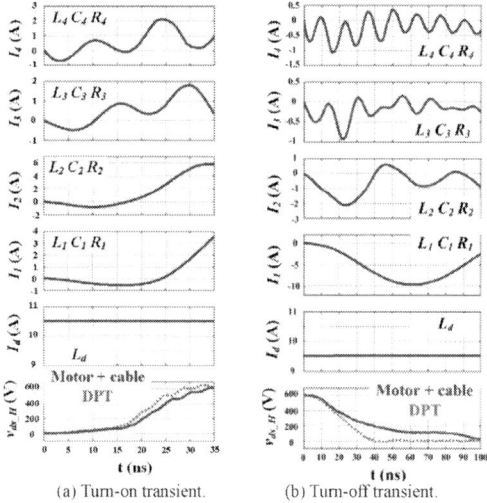

Fig. 3. Impact of parasitics of inductive load on the switching behavior.

To realize these two concepts, an auxiliary inductor with small EPC is introduced in series with the existing inductive load to modify the high frequency impedance of the inductive load. As can be found in Fig. 4(a), first, this inductor has an excellent inductive characteristic at high frequency. Hence, the *LRC* branches with respect to the resonant frequency of f_2 to f_4 will disappear. Second, this inductor is capable of tuning the resonant frequency of the $L_1R_1C_1$ branch from f_1 to f_{ring}, enabling its corresponding resonant period T_{ring} to be much longer than the switching commutation time. Thus, the resonant current I_1 remains small during the switching

transient (see Fig. 4(b)).

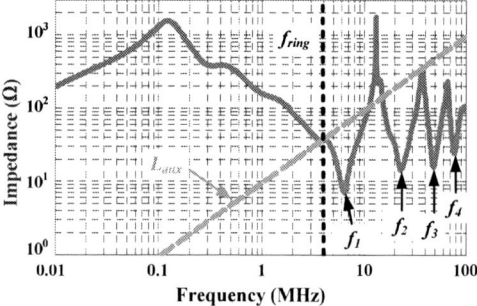

(a) High frequency impedance comparison: inductive load versus auxiliary inductor.

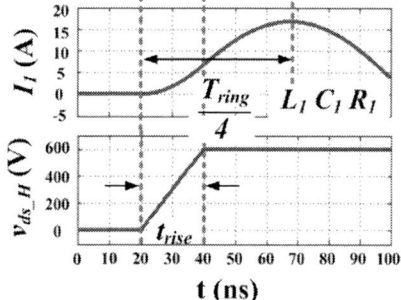

(b) Resonant current I_1 during the switching commutation time.

Fig. 4. Basic concept for mitigation of the adverse effect due to inductive load.

IV. DESIGN CRITERIA OF THE AUXILIARY FILTER

According to the basic idea discussed above, the inductance for the auxiliary inductor L_{aux} is determined by f_{ring}, which is directly related to the switching voltage commutation time. However, the switching voltage commutation time changes as the operating condition varies. Therefore, it is critical to select a switching voltage commutation time such that the switching performance will not be affected by the inductive load under most significant operating conditions.

As can be observed in Fig. 5, the switching voltage commutation time tested by the double pulse tester greatly depends on the operating current, and the longest switching voltage commutation time occurs during the turn-off transient under the light inductive load. If f_{ring} is determined based on the turn-off switching voltage commutation time at the extremely light load, the required L_{aux} will be fairly large, but the benefit that can obtained is limited since the turn-off switching loss at the light load is always almost zero [13, 14]. Thus, it is wise to select a switching voltage commutation time longer than all of the turn-on commutation time and part of the turn-off commutation time so that except the turn-off time at the light load, the switching time under the rest of the operating conditions together with the switching losses under all of the operating points will not be affected by the parasitics of the inductive load. In the following case study, f_{ring} is determined by the switching voltage commutation time of 34 ns during the turn-off transient under the operating condition of 600-V/10-A (see Fig. 5).

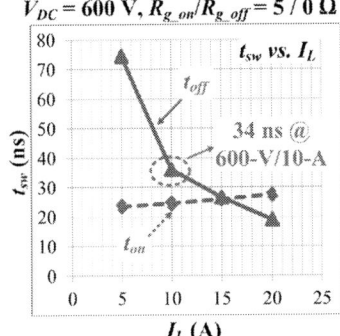

Fig. 5. Voltage commutation time dependence on load current measured by double pulse test.

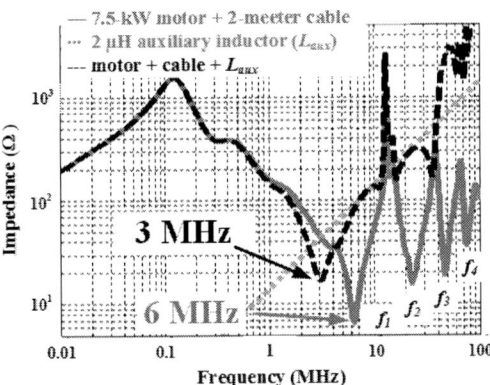

Fig. 6. High frequency impedance comparison: inductive load versus auxiliary inductor.

Based on the aforementioned design concept, f_{ring} should satisfy,

$$\frac{1}{f_{ring}} \gg t_{v,worst} \qquad (1)$$

where $t_{v,worst}$ refers to the switching voltage commutation time at the selected worst operating conditions (e.g., 34 ns at 600-V/10-V in this case study). f_{ring} of 3 MHz is selected so that its corresponding resonant period T_{ring} is 10 times longer than $t_{v,worst}$. Combining with the high frequency impedance of the inductive load, a 2 µH auxiliary inductor is needed. Fig. 6 illustrates that after insertion of a 2 µH auxiliary inductor, high resonant frequencies, such as f_2, f_3 and f_4 existing in the original motor plus cable load is suppressed. Moreover, the resonant frequency of f_1 is shifted from 6 MHz to 3 MHz. In total, a 2 µH auxiliary inductor achieves all the design requirements described above. Considering the three-phase conversion system, the auxiliary inductor per phase is 1.4 µH.

In addition to the inductance selection, physical design of the auxiliary inductor is important because it

affects the loss dissipated by this auxiliary inductor. The design objective is to minimize the extra loss so that combining with the switching loss reduction due to the proposed auxiliary inductor, the total power loss of the converter can decrease. In this case study, considering the relatively small inductance, an air core inductor is employed. The design methodology is described as follows.

(a) Typical air core inductor.

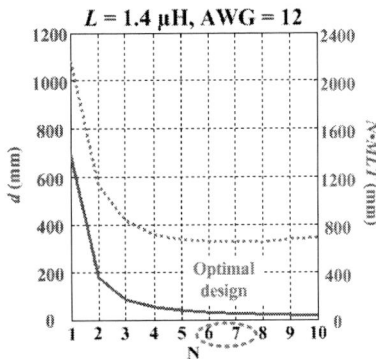

(b) Design optimization of air core inductor: d dependence on N and corresponding $N \times MLT$.

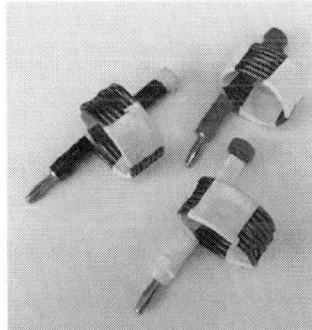

(c) Three air core inductors designed as auxiliary inductors.

Fig. 7. Design of auxiliary inductors.

First, the size of the magnet wire is selected based on the maximum operating current. In this case study, based on 11 Arms current per each phase in the three-phase voltage source inverter together with 4 A/mm² maximum current density for copper wires, 12 AWG magnet wire is selected.

Second, diameter d and number of turns N of the air core inductor are designed to achieve the required inductance with the shortest length of the copper wire

978-1-4673-9551-9/16 $31.00 © 2016 IEEE

(i.e., smallest $N \times MLT$ (mean-length-per turn)) for the minimization of power loss dissipated by this auxiliary inductor. Fig. 7(a) shows a typical air core inductor, and its inductance is given by

$$L = (25.4 \times d^2 \times N^2)/(18d + 40l) \qquad (2)$$

where L is inductance in µH, d is the coil diameter in mm, N is number of turns, l is coil length in mm, which can be determined by N and diameter of the copper wire [15]. According to (2), the relationship between N and d to achieve a 1.4 µH air core inductor is illustrated in Fig. 7(b). To minimize the length of the copper wire (i.e., $N \times MLT$), there will be an optimal design as a function of N and d. In this case study with 12 AWG copper wire, the smallest $N \times MLT$ is achieved when N of 6 or 7 is selected. In the end, based on the parameters listed in Table I, three air core inductors are designed and fabricated, as shown in Fig. 7(c).

TABLE I. PARAMETERS OF THE DESIGNED AIR CORE INDUCTOR

Max. current density	RMS current per phase	AWG	N	d
4.0 A/mm²	11 A	12	7	30 mm

V. EXPERIMENTAL VERIFICATION

To evaluate the effectiveness of the proposed auxiliary inductor on the switching performance improvement, the comparison experiments with Cree 2nd generation 1200-V/20-A SiC MOSFETs are carried out under three different inductive loads listed in Table II, including double pulse test based optimally-designed load inductor, 7.5-kW induction motor plus 2-meter power cable based inductive load, and induction motor plus power cable with proposed auxiliary inductor.

TABLE II. THREE COMPARISON GROUPS WITH DIFFERENT INDUCTIVE LOADS

1st Group	Double pulse test based load inductor (DPT)
2nd Group	7.5-kW induction motor plus 2-meter power cable (IM-PC)
3rd Group	Induction motor plus power cable with auxiliary inductor (IM-PC-L_{aux})

Fig. 8 shows the comparison switching waveforms with three different inductive loads under the operating condition of 600-V/10-A with 5-Ω gate resistance during the turn-on transient and 0-Ω gate resistance during the turn-off transient. As can be observed from the overall switching waveforms during both turn-on and turn-off transients, there is a 6.0 MHz frequency ringing in the switching current with the motor plus cable based inductive load (see dashed red waveforms in Fig. 8). After inserting 1.4 µH air core inductor per each phase, the ringing frequency is varied from 6.0 MHz to 3.0 MHz (see dotted black waveforms in Fig. 8), which agrees with the aforementioned design.

Also, as predicted, during the turn-on transient, the auxiliary inductor improves the switching performance as compared to that with induction motor plus power cable (IM-PC) based inductive load: dv/dt increases

from 40 V/ns to 43 V/ns, and the turn-on switching loss reduces from 166 µJ to 158 µJ. Additionally, thanks to the auxiliary inductor, its corresponding switching behavior is almost identical to that by using the optimally-designed DPT based inductor load.

Similarly, during the turn-off transient, the auxiliary inductor allows the dv/dt with IM-PC based inductive load to increase from 5.6 V/ns to 14.0 V/ns, which almost approaches to the dv/dt of 16.8 V/ns by using the DPT inductor. Furthermore, the total switching losses of motor plus cable inductive load with the auxiliary inductor decreases from 224 µJ to 204 µJ, which is nearly the same as 200 µJ, the total switching losses achieved by DPT inductor.

In summary, during the switching transient, the proposed auxiliary inductor successfully decouples the interaction between the inductive load and power device, enabling the switching performance with the practical inductive load (e.g., motor plus cable based inductive load in this case study) to achieve the similar fast switching behavior comparable to when the optimally-designed DPT inductor is employed.

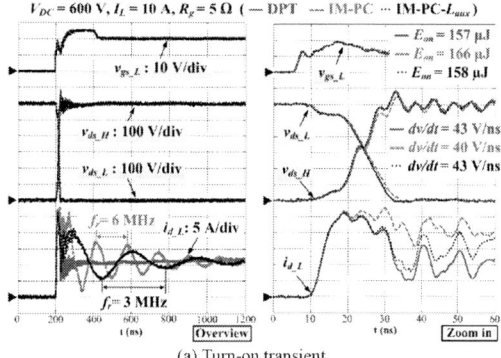

(a) Turn-on transient.

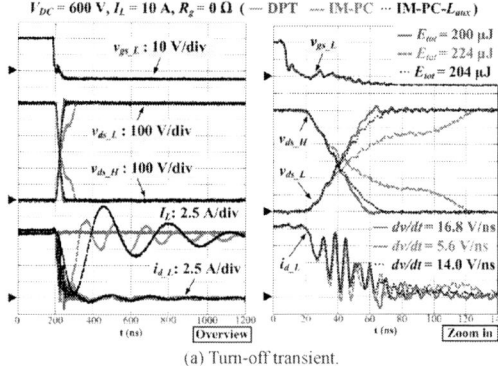

(a) Turn-off transient.

Fig. 8. Switching waveform comparisons among different inductive loads.

Additionally, based on the design consideration of the auxiliary inductor, to minimize the required inductance, there will be a penalty for the turn-off time at the light load (e.g., < 10 A in this case study). As can be observed in Fig. 9, under the operating current of 5-A, the turn-off time with the motor plus cable based inductive load becomes even longer when the auxiliary inductor is applied. This is because that unlike the short switching voltage commutation time illustrated in Fig. 4(b), the relatively long turn-off commutation time at the light operating current cannot avoid the impact of the resonant current induced by the parasitics of the inductive load. Also, the proposed auxiliary inductor reduces the resonant frequency from 6.0 MHz to 3.0 MHz, which means that the duration of the resonant current with this lower resonant frequency becomes longer, and then the turn-off time by using the auxiliary inductor turns to be longer as well. However, the tested turn-off switching loss at the light load current stays nearly constant, and theoretically equals to the energy stored in the junction capacitance of the power device and its antiparallel diode. Fig. 9 illustrates that although the turn-off time difference with different inductive loads are significant; their turn-off switching losses are almost identical.

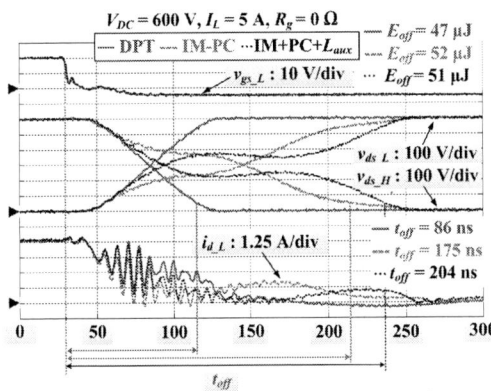

Fig. 9. Turn-off switching performance at light load.

Fig. 10 gives more switching data comparison among different inductive loads under different operating currents. Fig. 10(b) shows that, thanks to the auxiliary inductor, the total switching loss with motor plus cable becomes lower, and is almost identical to the switching losses tested by using the DPT inductor. Similar behavior can be found in terms of turn-on switching time in Fig. 10 (c). In Fig. 10 (d), except for the light load (i.e., < 10 A in this case study), the turn-off time with motor plus cable based inductive load after the insertion of the auxiliary inductor decreases, and becomes identical to that when the DPT inductor is employed.

Fig. 11 illustrates the comparison test results dependence on the junction temperature. Note that labels of comparison groups is the same as that defined in Fig. 10(a). As predicted, the experimental data demonstrate the effectiveness of the proposed auxiliary inductor on the switching performance improvement: lower total switching loss together with shorter turn-on/off switching times are achieved as compared to the data tested based on motor plus cable based inductive load without using the auxiliary inductor. Also, the proposed auxiliary inductor enables the similar fast switching behavior comparable to when the

optimally-designed DPT inductor is employed among wide operating temperature range.

In total, the proposed auxiliary inductor is capable of mitigating the interaction between the inductive load and power device among a wide operating range: except the turn-off time at the light load, the switching time under the rest of the operating conditions together with the switching losses under all of the operating points will not be affected by the parasitics of the inductive load.

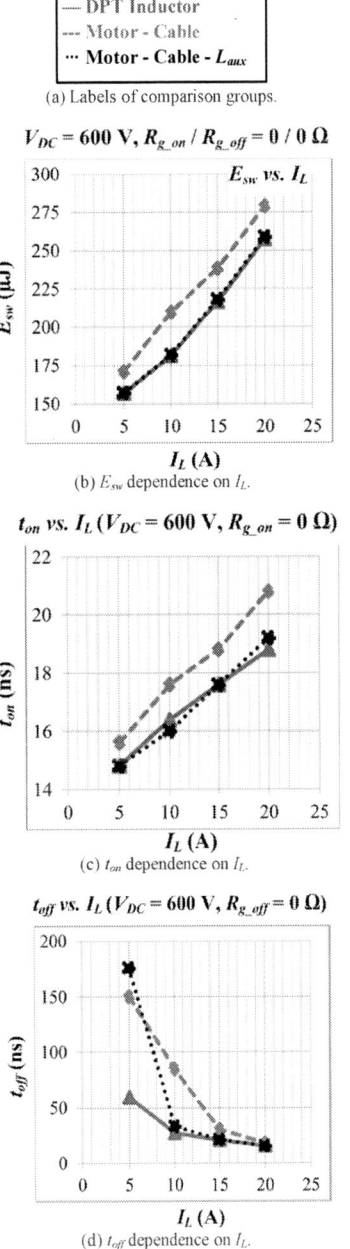

Fig. 10. Comparison test results dependence on load inductor with 600-V dc bus voltage and 0-Ω gate resistance.

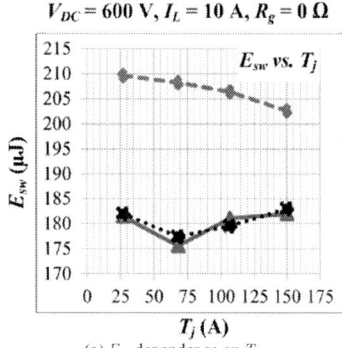

$V_{DC} = 600$ V, $I_L = 10$ A, $R_g = 0$ Ω

(a) E_{sw} dependence on T_j.

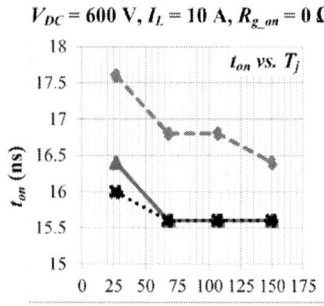

$V_{DC} = 600$ V, $I_L = 10$ A, $R_{g\_on} = 0$ Ω

(b) t_{on} dependence on T_j.

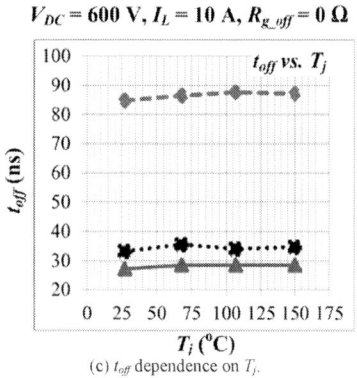

$V_{DC} = 600$ V, $I_L = 10$ A, $R_{g\_off} = 0$ Ω

(c) t_{off} dependence on T_j.

Fig. 11. Comparison test results dependence on junction temperature under 600-V/ 10-A with 0-Ω gate resistance (Labels of comparison groups is the same as that defined in Fig. 10(a)).

In addition to suppress the adverse impact of an inductive load on the switching behavior, the auxiliary inductor is able to mitigate the dv/dt and over-voltage across the motor terminals. Fig. 12 illustrates that under the operating condition of 600-V/10-A, the auxiliary inductor decreases the motor terminal over-voltage from 1049 V to 1018 V together with the dv/dt reduction up to 44%. The reason causing the mitigation of over-voltage and dv/dt is because the ringing frequency of the voltage across motor terminals becomes lower when the auxiliary inductor is employed. More test data under different operating currents are shown in Fig. 13. Thus, the proposed auxiliary inductor is beneficial to the insulation of an induction motor.

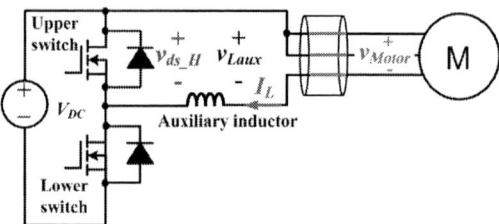

(a) Equivalent circuit and measurement terminals.

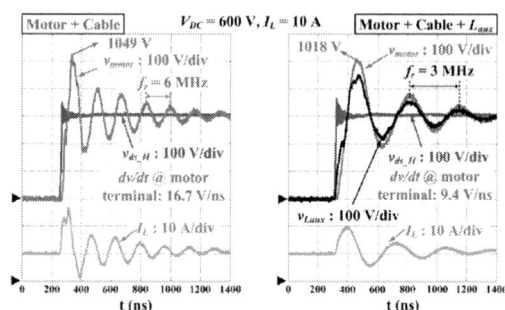

(b) Experimental waveform when lower switch turns on.

Fig. 12. Impact of auxiliary inductor on motor terminal voltage: v_{ds_H} is the drain-source voltage of the upper switch; v_{motor} is motor terminal voltage; v_{Laux} is converter terminal voltage (after L_{aux}).

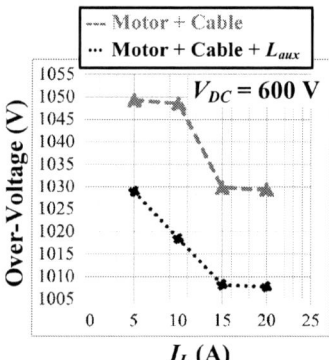

(a) Motor terminal over-voltage dependence on I_L.

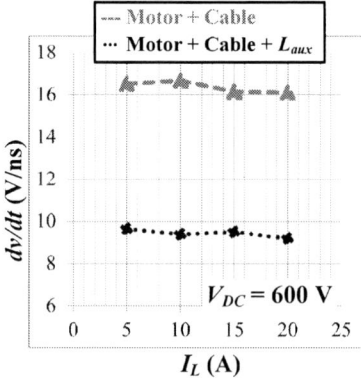

(b) Motor terminal dv/dt dependence on I_L.

Fig. 13. Motor terminal over-voltage and its dv/dt comparison dependence on load currents.

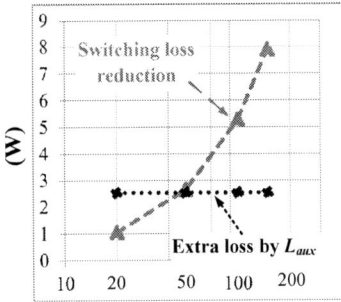

Switching frequency (kHz)

Fig. 14. Power loss comparison: switching loss reduction versus extra loss by L_{aux}.

To make a fair judgment of the benefits gained from the auxiliary inductor, the additional power loss dissipated by this auxiliary inductor should be taken into account. As shown in Fig. 14, considering the skin effect, the extra power loss induced by L_{aux} is lower than the switching loss reduction when the switching frequency is above 50 kHz. Therefore, for high switching frequency applications, the proposed approach is able to improve the overall efficiency of the power conversion system.

V. CONCLUSION

A dedicated auxiliary inductor is designed and inserted between the converter and inductive load so that the parasitics of the load will not be "seen" from the converter side during the switching transient. The test results with Cree 2nd generation 1200-V/20-A SiC MOSFETs verify that the proposed auxiliary inductor is capable of mitigating the interaction between the inductive load and power device under different operating conditions: except the turn-off time at the light load, the switching time under the rest of the operating conditions together with the switching losses under all of the operating points will not be affected by the parasitics of the inductive load. Moreover, the proposed auxiliary filter is capable of mitigating motor terminal over-voltage and dv/dt. Furthermore, it is noted that the extra loss dissipated by the auxiliary inductor is lower than the switching loss reduction for the high switching frequency applications (> 50 kHz in this case study), leading to an overall efficiency improvement in the power conversion system.

ACKNOWLEDGEMENT

The authors would like to thank II-VI Foundation for its support of this research work. This work made use of the Engineering Research Center Shared Facilities supported by the Engineering Research Center Program of the National Science Foundation and DOE under NSF Award Number EEC-1041877 and the CURENT Industry Partnership Program.

REFERENCES

[1] Z. Zhang, F. Wang, L. M. Tolbert, B. J. Blalock, and D. J. Costinett, "Understanding the limitations and impact factors of wide bandgap devices' high switching-speed capability in a voltage source converter," in *IEEE Workshop on Wide Bandgap Power Devices and Applications (WiPDA)*, 2014, pp. 7-12.

[2] F. Wang, Z. Zhang, T. Ericsen, R. Raju, R. Burgos, and D. Boroyevich, "Adances in power conversion and drives for shipboard systems," *Proceedings of the IEEE*, 2015, accepted.

[3] M. J. Scott, J. Brockman, H. Boxue, F. Lixing, X. Longya, W. Jin, and R. Darbali Zamora, "Reflected wave phenomenon in motor drive systems using wide bandgap devices," in *IEEE Workshop on Wide Bandgap Power Devices and Applications (WiPDA)*, 2014, pp. 164-168.

[4] N. Oswald, P. Anthony, N. McNeill, and B. H. Stark, "An Experimental Investigation of the Tradeoff between Switching Losses and EMI Generation With Hard-Switched All-Si, Si-SiC, and All-SiC Device Combinations," *IEEE Transactions on Power Electronics*, vol. 29, pp. 2393-2407, 2014.

[5] E. Matheson, A. von Jouanne, and A. Wallace, "Evaluation of inverter and cable losses in adjustable speed drive applications with long motor leads," in *International Conference Electric Machines and Drives*, 1999, pp. 159-161.

[6] S. Walder and X. Yuan, "Effect of load parasitics on the losses and ringing in high switching speed SiC MOSFET based power converters," in *IEEE Energy Conversion Congress and Exposition (ECCE)*, 2015, pp. 6161-6168.

[7] Z. Zhang, B. Guo, F. Wang, L. M. Tolbert, B. J. Blalock, Z. Liang, and P. Ning, "Methodology for switching characterization evaluation of wide band-gap devices in a phase-leg configuration," in *IEEE Applied Power Electronics Conference and Exposition (APEC)*, 2014, pp. 2534-2541.

[8] Z. Zhang, F. Wang, L. M. Tolbert, B. J. Blalock, and D. J. Costinett, "Evaluation of Switching Performance of SiC Devices in PWM Inverter-Fed Induction Motor Drives," *IEEE Transactions on Power Electronics*, vol. 30, pp. 5701-5711, 2015.

[9] B. Mirafzal, G. L. Skibinski, R. M. Tallam, D. W. Schlegel, and R. A. Lukaszewski, "Universal Induction Motor Model With Low-to-High Frequency-Response Characteristics," *IEEE Transactions on Industry Applications*, vol. 43, pp. 1233-1246, 2007.

[10] M. Schinkel, S. Weber, S. Guttowski, W. John, and H. Reichl, "Efficient HF modeling and model parameterization of induction machines for time and frequency domain simulations," in *IEEE Applied Power Electronics Conference and Exposition (APEC)*, 2006, pp. 1181-1186.

[11] W. Liwei, C. N. m. Ho, F. Canales, and J. Jatskevich, "High-Frequency Modeling of the Long-Cable-Fed Induction Motor Drive System Using TLM Approach for Predicting Overvoltage Transients," *IEEE Transactions on Power Electronics*, vol. 25, pp. 2653-2664, 2010.

[12] A. Boglietti and E. Carpaneto, "Induction motor high frequency model," in *IEEE Industry Applications Conference*, 1999, pp. 1551-1558.

[13] Z. Zhang, F. Wang, D. J. Costinett, L. M. Tolbert, B. J. Blalock, and L. Haifeng, "Dead-time optimization of SiC devices for voltage source converter," in *IEEE Applied Power Electronics Conference and Exposition (APEC)*, 2015, pp. 1145-1152.

[14] X. Li, L. Zhang, S. G. Y. Lei, A. Huang, and B. Zhang, "Understanding Switching Losses in SiC MOSFET: Toward Lossless Switching," in *IEEE Workshop on Wide Bandgap Power Devices and Applications (WiPDA)*, 2015.

[15] Air Core Inductor Inductance Calculator [Online]. Available: http://www.daycounter.com/Calculators/Air-Core-Inductor-Calculator.phtml

978-1-4673-9551-9/16 $31.00 © 2016 IEEE

Control and Characterization of Electromagnetic Emissions in Wide Band Gap Based Converter Modules for Ungrounded Grid-Forming Applications

Robert Cuzner, Rasoul Hosseini
Department of Electrical
Engineering and Computer Science
University of Wisconsin-Milwaukee
Milwaukee, USA
robcuzner@ieee.org

Andrew Lemmon
Department of Electrical and
Computer Engineering
University of Alabama
Tuscaloosa, USA
Andrew.n.lemmon@ieee.org

James Gafford, Michael Mazzola
Department of Electrical and
Computer Engineering
Mississippi State University
Starkville, USA
gafford@CAVS.MsState.Edu

Abstract— Electromagnetic emissions of a 1.2kV, 120A SiC-based half-bridge switching at 100kHz that includes grounding paths and that can be extended to a 100kVA inverter and, eventually, systems of paralleled and cascaded inverters suitable for shipboard and solar farm applications is studied. This switching pole forms the basis of a test platform specifically designed to discover sensitivities to resonant paths so that design guidelines for peripheral structures and EMI mitigating components can be developed. Ungrounded grid-forming inverters are considered in this work because such systems present a worst case scenario when it comes to the effects of resonances through grounding paths being excited by "near-RF" frequencies associated with Wide Band Gap implementations.

Keywords— Wide bandgap power electronic systems; EMI testing

I. INTRODUCTION

WBG device based power converters present tremendous opportunities for increased power density and ease of plug and play solutions in grid-forming applications such as ship service inverters and solar farms, but the implications of higher switching frequency and high di/dt and dv/dt on both the packaging and installation into the system application must be well understood. These types of grid-forming inverter applications require high power quality and low conducted and radiated Electromagnetic Interference (EMI) as determined by compliance with stringent standards. WGB device based converters can enable future markets if power density, efficiency and affordability can be increased well beyond the capabilities of present silicon (Si) power semiconductor technology. However, implementations will not be successful if design decisions are made, including the application of WBG devices and selection of switching frequency, exclusively on the basis of other system-level metrics such as electrical efficiency and power density, without regard to EMI implications, which have recently surfaced as a prominent challenge [1]-[19]. A root cause of this increased EMI is excitation of resonances in parasitic impedances in the multi-chip packaging structure and gate drive [4]-[9],[13]-[14],[16]-

[19], in the grounding paths through heat sink [3], [6], [8], [12], [15], [17], chassis and external loads and within the EMI mitigating components themselves [1]. These phenomena are a direct result of the both the higher switching frequency capability of WBG devices and the higher signal edge rates, both of which have resulted in an exansion of generated spectral content into the "near-RF" domain where such parasitic effects are more likely to be excited [20], [22]. At the same time, these parasitic effects are difficult to predict. This has resulted in a surge in literature addressing parameter extraction [22]-[24], behavioral and "gray" box models [15], [25], [26] and the integration of measurement methods previously relegated to the RF designer's domain into power electronics [27] such as the use of Scattering Parameters (S-Parameters) and Time Domain Reflectometry (TDR).

So far, some success has been reported in mitigating these EMI effects by minimizing and damping the Differential Mode (DM) parasitic circuit around the devices [3]-[8], [10], [14], [17], [22] through gate drive design [28], [29], proper cable shielding design [11] and passive filter design [1]. However, the promise of highly compatible plug and play power electronics with higher power density and efficiency is still unproven—particularly as the system impacts outside of the power electronic components themselves are considered, along with the interactions between multiple cascaded and paralleled power converters connected into a system. Moving outside of packaging around the multi-chip modules themselves, the main consideration of system compatibility is in the ground paths. Some consideration has been given to both characterizing and mitigating ground current—or Common Mode (CM) current—paths through novel approaches, such as segregating heat sinks [2], but with some limitations and for fairly low power systems. So far, for SiC MOSFET based inverters, CM currents at frequencies beyond 1MHz have been identified as the largest contributors to EMI [15]. These CM currents flow through parasitic impedances which are difficult to characterize in most cases for practical converters. In addition, increased switching frequency and dv/dt of WBG converters can excite

978-1-4673-9551-9/16 $31.00 © 2016 IEEE

ground impedance resonances, further complicating this analysis.

The purpose of this paper is to characterize the impact of the cabinet grounding structure on conducted emissions for a 1.2kV, 120A half-bridge SiC MOSFET module based inverter. Custom-designed Line Stabilization Networks (LISNs) are developed that can withstand the stresses imposed by unfiltered DM and CM voltage pulses from a switching power converter operating up to power levels of 100 kVA. A three half-bridge converter suitable for DC to AC and DC to DC conversion is constructed around 1.2kV, 120A SiC MOSFET modules from Cree [30]. Tests are performed progressively with this set-up by operating one half-bridge in a pseudo-continuous manner with a Clamped-Inductive-Load (CIL) [21], [22]. The gate drive and bus layout was designed according to [22], which demonstrated switching frequencies up to 1MHz. DM resonant effects are understood and controlled and carried over into this work so that no significant resonances are excited. The main purpose of this paper is to characterize the behavior of this system with the heat sink mounted on a ground plane and in the EMI test environment specified by [34]. Since resonances at the switching edge are controlled at the source as in [22], the very best possible situation is presented for studying the effects of resonant paths beyond the inverter pole itself and for building up a compatible grid-forming inverter. Conducted emissions are characterized against specified limits in the EMI test environment.

II. IMPACT OF GROUND IMPEDANCE RESONANCE IN UNGROUNDED SYSTEMS

Distribution systems in many applications are increasingly being fed by power electronic converters in order to introduce Distributed Energy Resources (DERs). Many such systems have a DC microgrid for the purpose of interfacing multiple DERs and energy resources which must also feed into a conventional AC distribution system through DC to AC conversion. Such a system is shown in Fig. 1. The interface between the DC microgrid and the AC distribution system can be scaled through the use of multiple inverter sub-modules in parallel. In some applications paralleled non-isolated DC to DC converter sub-modules might also be connected upstream of the inverter sub-modules to provide voltage bucking or boosting or to provide a means for feeding the inverter from more than one DC bus [35]. An inverter sub-module is shown in Fig. 2 and a buck converter sub-module with three-interleaved half-bridges is shown in Fig. 3. In order to save cost, space, and to maintain efficiency, the sub-modules may not include transformer isolation (typically used to break CM current paths). Instead, CM inductors within the inverter modules may be used to limit any CM circulating current between the modules in addition to internal controls that control the low frequency CM current output. The main concern for limiting the CM circulating current is between the inverter/converter sub-modules. However, resonances in the grounding paths between inverter/converter sub-module chassis may also be a concern if the CM mode inductors saturate and the resonant frequency of a grounding path in series with the sub-module(s) can be excited by the inverter/converter CM voltages. Saturation may occur if excessive external capacitance to ground is introduced into the AC distribution system, i.e. for the purpose of EMI mitigation at individual power electronic loads on the AC distribution bus. Fig. 1 shows this total line to ground capacitance combined at a single Point of Common Coupling PCC as C_{og} and multiple

ground impedance paths, i.e. from the inverter sub-module chassis (Z_{cg1}, Z_{cg2}, Z_{cg3}) and lumped impedances to ground in the DC and AC systems (Z_{ig}, Z_{og}) to some common ground point. Each inverter sub-module produces CM voltage through power semiconductor switching which couples to the external system through deliberate paths at the output EMI filter and parasitic paths from the devices to the heat sink and from the heat sink to chassis (Z_{sc}) as shown in Fig. 2. In ungrounded systems, these parasitic path impedances are not easily determined but knowledge of them may be essential in order to model CM induced behaviors. Normally, the primary consequence of switching through these parasitic paths is an increase in EMI. However, other side-effects are possible. For example, Fig. 4 shows an extreme condition where saturation of the CM inductors in each sub-module has shifted a system resonance to the switching frequency. The result is serious DM distortion and voltage spikes at the PCC where combined capacitance to ground effects interact with the rest of the system to produce resonant behavior in the CM path that is also reflected into the DM path. The scenario of Fig. 4 is uncommon for Si IGBT based systems which have switching frequencies in the range of 10kHz. However, as switching frequencies increase into the range of 50-100kHz in order to reduce the size of inverter sub-module filters, the potential for exciting these resonances will likely increase.

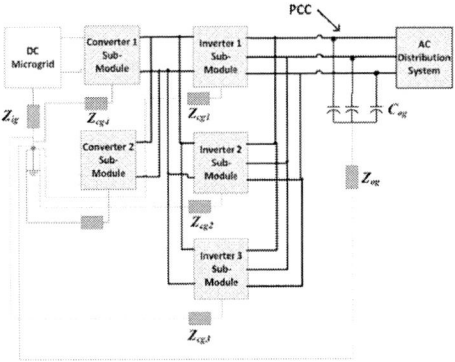

Fig. 1 Interface between DC microgrid and AC distribution system

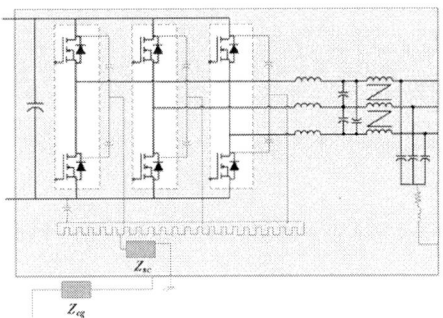

Fig. 2 Inverter sub-module

978-1-4673-9551-9/16 $31.00 © 2016 IEEE

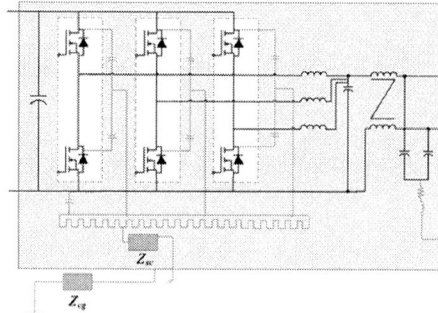

Fig. 3 Buck converter sub-module

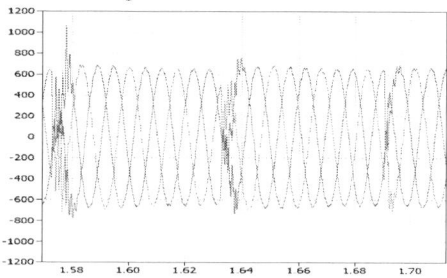

Fig. 4 Simulated voltage distorion at PCC caused by resonances in the ground path in the system of Fig. 1

In a shipboard system, each inverter cabinet is mounted to a conducting hull, so ground currents induced by inverter switching are more tightly coupled to the ground path throughout the ship than in typical land-based systems. The importance of understanding these ground paths and modeling them in order to ensure inter-compatibility of a large number switching power converters in a confined space—as is expected to be the case with future Navy ships [31]—has been recognized recently [32], [33]. Also, the present approach for ensuring system compatibility is for power electronic sources and loads to be individually tested for compliance with an Electromagnetic Compatibility (EMC) standard [34]. With the movement towards WBG devices and the possibility of higher switching frequencies and attendant high signal edge rates, it will be important to develop a strong understanding of the interplay of CM behavior in such systems. This understanding must start with the sources of CM behavior and CM impedance paths at the individual inverter sub-modules, and then progress into the larger system.

III. CIRCUIT LAYOUT AND EMI TEST FIXTURE DESIGN

A generic three half-bridge converter was designed and fabricated as part of this work to serve as a platform for characterizing the flow of CM currents in ungrounded systems. This converter was designed for an input of 800Vdc and intended AC output of 450Vrms, 60Hz, 100kVA or DC 300-700Vdc, 100-200kW in configurations of Fig. 2 and Fig. 3, respectively. Every effort was taken to incorporate best practices from prior work by the authors [5], [7], [22], [23], [29] to mitigate parasitic inductance and capacitance within the bridge structure—including minimization of gate drive and bus layout interconnecting and cross-coupling inductances.

Fig. 5 shows the single-phase CIL configuration of the converter with input and output LISNs connected. The DC link capacitors and three half-bridge power modules of the power switching portion of the inverter are all affixed to a common printed circuit board (PCB) which constitutes all DC bus interconnections. The design of this PCB is a multi-layer interleaved design. The interleaved positive and negative dc bus planes result in an ultra-low inductance high current connection between the DC link capacitors and the power modules. Each half-bridge power module is controlled by an independent gate driver card which consists of two isolated gate driver channels for driving the high-side and low-side devices, respectively. The gate driver design is comprised of a high-bandwidth magnetically isolated signal interface, a commercially available gate drive integrated circuit, and a high-current BJT totem pole output stage. With respect to the source of each driven device, the gate driver is capable of producing a bipolar output. Per the module manufacturer's recommended operating conditions, the turn-off bias voltage is -5VDC and the turn-on bias voltage is +20VDC. The BJT totem pole output stage is capable of very high slew rates which are adjustable via a series gate resistor; in this work, the gate resistor value used was 5 Ω.

One of the two load inductor elements used for the CIL test set-up is shown in Fig. 6. Three toroidal high flux powder cores were used and the inductor was wound sparsely with litz wire to minimize inter-winding capacitance and AC resistance. In addition, an impedance analyzer was used to verify that a desired inductance of >100µH was achieved for a series combination of two of the inductor elements shown in Fig. 6. An inductance vs. frequency sweep for the resulting inductor bank is shown in Fig. 7. Linear inductance was achieved up to 3.6MHz.

The Line Stabilization Networks (LISNs) were designed to withstand the unfiltered switching pulses of a three phase inverter operating with an input bus voltage as high as 1kV. A switching simulation was used in order to account for DM and CM level switching harmonics and a 100% de-rating was included to account for additional contributions associated with high dv/dt switching. LISN inductors were designed to withstand the rated currents, minimize inter-winding capacitance, and ensure linear inductance to at least 10MHz, according to best practices. This was done by winding air core inductors on PVC tubes and carefully measuring inductor impedance versus frequency in order to ensure that the desired characteristics were achieved. A high frequency range of the LISN inductor will block multi-MHz frequency excitation of the CIL inductor and de-couple the EMI test stand from the effects of an upstream switching power supply. With LISNs connected to the input and output of the converter, the CM currents in the system can be measured. The LISNs form the common mode voltages that reflect onto the half-bridge DC input legs in the CIL configuration. It is therefore very important to ensure that resonant effects in the LISNs do not couple back into the EMI test set-up. LISN inductors wound both with stranded copper wire and litz wire were constructed. It was determined that, from a linear inductance vs. frequency range perspective, the use of litz wire resulted in only a marginal improvement. Magnitude impedance vs. frequency sweeps are shown for the two types of LISN inductors in Fig. 8 and Fig. 9. Table I shows a comparison of the parallel RLC equivalent-circuit values for all of the inductors built for the test stand.

Finally, the entire system was mounted on a copper ground plane as shown in Fig. 10. The main purpose for the initial testing documented in this paper is to understand the impact of the ground path from the power semiconductors mounted on a heat sink or cold plate with respect to the larger system. In this case, the ground reference is considered to be the chassis of power converter into which an inverter or converter sub-module is mounted. The SiC power MOSFETs are cooled by a cold plate as shown in Fig. 10. This set-up allowed the adjustment of the mechanical and electrical connection between the cold plate and ground plane by adding spacers as shown in Fig. 11. These spacers limit the galvanic paths through which current may flow and increase the distance between the cold plate and the ground plane but do not galvanically isolate the cold plate from the ground plane.

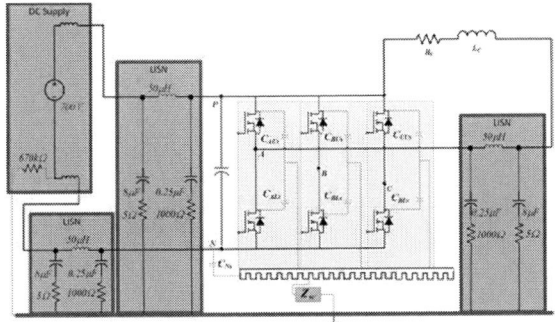

Fig. 5 CIL Test Circuit with input and output LISNs

Fig. 6 Load inductor for CIL test set-up

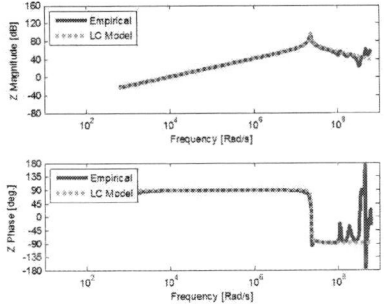

Fig. 7 Load inductor frequency sweep

TABLE I.   INDUCTOR MEASUREMENTS

Component Description	Inductor Parameters			
	L (μH)	R_p ($k\Omega$)	C_p (pF)	f_o (MHz)
Load Inductor	115.6	55.1	17.2	3.6
LISN Inductor w/ stranded wire	56.1	137	3.9	10.7
LISN Inductor w/ litz wire	49.1	66.8	3.2	11.5

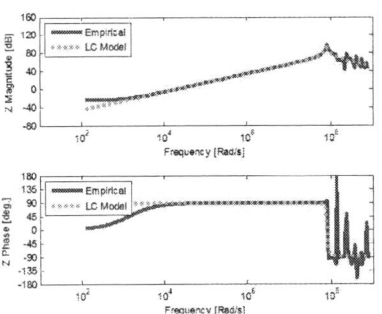

Fig. 8 LISN inductor frequency sweep, stranded wire version

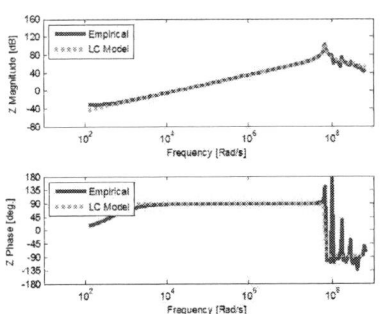

Fig. 9 LISN inductor frequency sweep, litz wire version

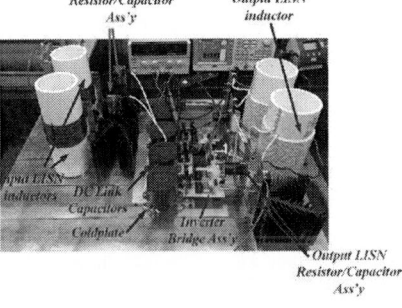

Fig. 10 Hardware Set-Up showing ground plane

Fig. 11 Implementation of spacers between coldplate and ground plane

IV. EQUIVALENT COMMON MODE QUANTITIES

The single-phase CIL configuration described in the previous section provides a good starting point for characterizing the CM behavior of the system and isolating contributors to CM behavior. Previous work has identified the following contributors to CM emissions [2], [26]:

978-1-4673-9551-9/16 $31.00 © 2016 IEEE

(1.) Application of CM voltage (or level switching) at the switching frequency and its harmonics to CM impedances external to the system
(2.) Switching of voltages across parasitic capacitances to ground within the half-bridge converter structure
(3.) CM current coupling paths between two or more switching half-bridges

All of these effects except for (3.) can be demonstrated by CIL testing a single half-bridge in the configuration of Fig. 5. Fig. 12 shows the CIL test set-up with an equivalent circuit for one of the three half-bridge inverters. The parasitic capacitances from device drains to the cold plate are shown in Fig. 5 as C_{AUs}, C_{BUs}, and C_{CUs} for the upper devices and C_{ALs}, C_{BLs}, and C_{CLs} for the lower devices. Reference [17] calculates this parasitic capacitance to be 90pF for the CAS120M12BM2 devices. In the CIL configuration used here, the upper devices are gated off and their total capacitance to sink becomes:

$$C_{Us} = C_{AUs} + C_{BUs} + C_{CUs} \qquad (1)$$

Only one of the lower devices is being actively switched but this capacitance to ground, C_{ALs}, becomes the main source of contribution (2.) to CM currents produced by the half-bridge structure at the output through the pole current, I_{Ao}. The influence of contributor (1.) is limited to the CM voltage produced by the switching of the lower device and the resultant common mode voltage at the output. Due to test infrastructure limitations, only four signals could be simultaneously measured in this test setup, so not all of the signals required to characterize the CM behavior could be concurrently measured. However, between multiple test runs and the application of a few simplifying assumptions, a good picture of the CM behavior of this system can be obtained. The output CM voltage was approximated as follows:

$$V_{CMo} \cong V_{AN} + V_{NLISN} \qquad (2)$$

In the experimental set-up of Fig. 12 only V_{PLISN} was measured. Therefore, an estimate of V_{NLISN} is given by:

$$V_{NLISN} \cong V_{PLISN} - V_{PN} \qquad (2)$$

where V_{PN} is the differential mode voltage from **P** to **N**. The data showed a very low value of ripple during the off state voltages of V_{AN}, so V_{PN} was considered simply as a dc offset. V_{CMo} is approximated from the measured data as:

$$V_{CMo} \cong V_{ALISN} + V_{PLISN} - \langle V_{PN} \rangle \qquad (3)$$

Similarly, the input CM voltage, V_{CMi}, was estimated from the following expression:

$$V_{CMi} \cong \frac{2 \cdot V_{PLISN} - V_{PN}}{2} \qquad (4)$$

Note that all of these approximations neglect the effects of voltage drop across the LISN capacitors. At frequencies >300kHz, it can be shown that these contributions are negligible. More importantly, the approximated CM voltages provide some accurate insight into the CM current behaviors.

In Fig. 12, the input CM current, I_{CMi}, was directly measured and the output CM current, I_{CMo}, was derived from the following

$$I_{CMo} = I_{DMo} - I_{Ao} \qquad (5)$$

I_{CMo} represents the current flowing through the output LISN into the ground plane. I_{CMi} was measured directly and reflects the influence of the output CM voltage and impedances

reflecting back into the system. All of the CM impedances in the system are from the LISNs, with the exception of the parasitic impedances to the ground plane internal to the half-bridge modules.

V. EXPERIMENTAL RESULTS

The electrical signals characterizing CM behavior of the system were measured with the spacing between the cold plate and the ground plane at three positions: (1.) No spacers, i.e. the cold plate directly mounted to the ground plane; (2.) 0.185 inch spacers between the cold plate and the ground plane; and (3.) 0.75 inch spacers between the cold plate and the ground plane. For each position, data was captured during two runs. Fig. 13 and Fig. 14 show the results for direct cold plate mounting to the ground plate. Fig. 15 and Fig. 16 show results for 0.185 inch spacing and Fig. 17 and Fig. 18 show results for 0.75 inch spacing. There are a few important points to note. First of all, I_{CMi} in Fig. 13 has a negative DC offset. This is due to the fact that the source of CM noise in the system, V_{CMo}, has a DC offset and effectively charges and discharges the RC circuits in the input LISNs as CM voltage level shifts.

Voltage across the lower switching device, V_{AN}, and the switching across the internal capacitance C_{ALs} are the main sources of CM output current, I_{CMo}, in Fig. 14. Comparing Fig. 14 with Fig. 16 and Fig. 18, the amplitude of I_{CMo} is significantly reduced once spacing is introduced between the cold plate and the ground plane—which effectively increases the impedance path in series with C_{ALs}. This result is significant because it indicates that single pole dv/dt induced EMI impacts on an external system may be essentially eliminated in practical implementations, leaving only the CM voltage as the main source of EMI. Of course, these dv/dt effects will still have a significant impact on system self-compatibility. The main takeaway from the results of Fig. 13-Fig. 18 is that the CM current circulating through the ground path is significantly reduced as the spacing between the cold plate and the ground plane is increased.

Finally, Fig. 19 and Fig. 20 show the conducted EMI measured according to [34] at the input and the output of the CIL test stand; and how these results are affected by the change in spacing between the cold plate and the ground plane.

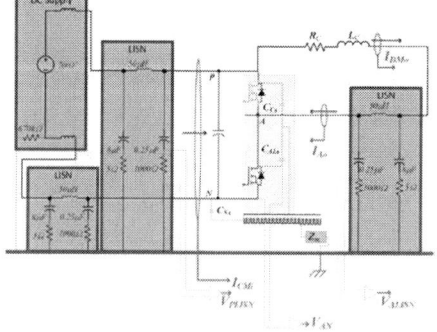

Fig. 12 CIL Test Set-up with equivalent half bridge circuit

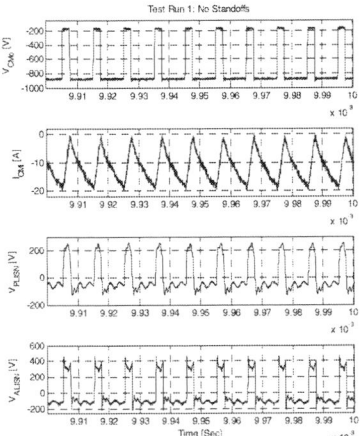

Fig. 13 Experimental and derived quantities from data run #1, direct mounting of cold plate to ground plane

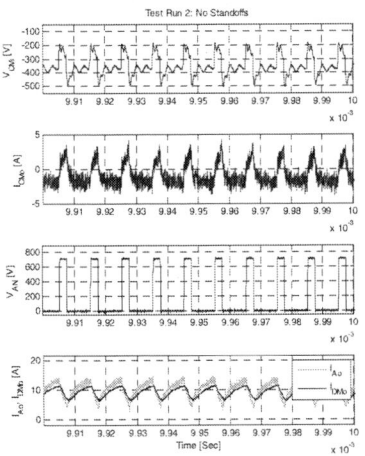

Fig. 14 Experimental and derived quantities from data run #2, direct mounting of cold plate to ground plane

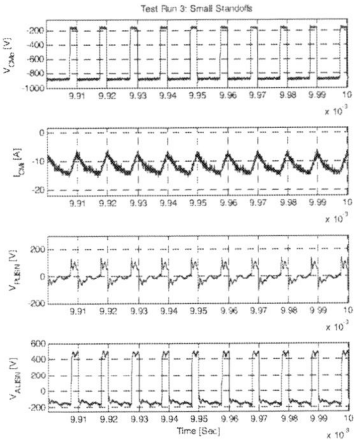

Fig. 15 Experimental and derived quantities from data run #3, 0.185 inch spacers between the cold plate and the ground plane

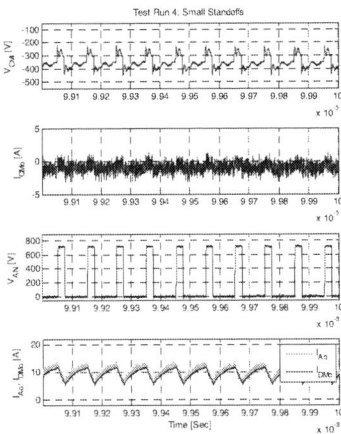

Fig. 16 Experimental and derived quantities from data run #4, 0.185 inch spacers between the cold plate and the ground plane

Specifically, Fig. 19 and Fig. 20 show voltages measured across the LISN resistors on the input and output or V_{PLISN} and V_{ALISN}, respectively. These results show that the difference between cold plate mounting spacing accounts for 3-10dB across the entire frequency range. The impact is greatest on the input LISN and represents a 10dB reduction if the cold plate is spaced anywhere from 0.185 inches to 0.75 inches. Although the emissions are not greatly affected by this change, there appears to be some EMI benefit to increasing the spacing between the cold plate and ground plane. In addition, it is notable that the amount of ground current is reduced significantly by increasing this spacing. Since this is the current that must be handled by CM output inductors used to bring a system into compliance, it is reasonable to expect that leveraging this technique may lead to a reduction in the size of the required CM inductors to meet the conducted EMI requirements of [34].

VI. CONCLUSIONS

This work has demonstrated the effects of the ground impedance path on the CM currents and the conducted emissions for a three half-bridge circuit configured with a CIL load on one phase. Results show that induced CM current through the ground path can be controlled through the heat sink interface to the rest of the system and that this has some effect on EMI performance. Additional efforts to further characterize the EMI performance of this system and to describe the influence of additional parameters of sensitivity are underway, and will be reported in a future paper.

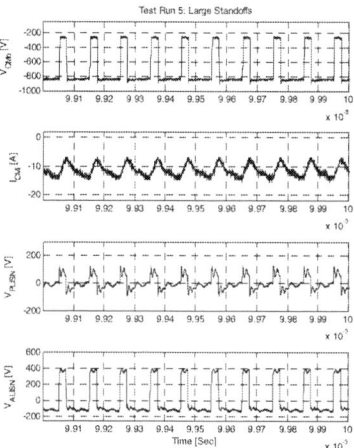

Fig. 17 Experimental and derived quantities from data run #5, 0.75 inch spacers between the cold plate and the ground plane

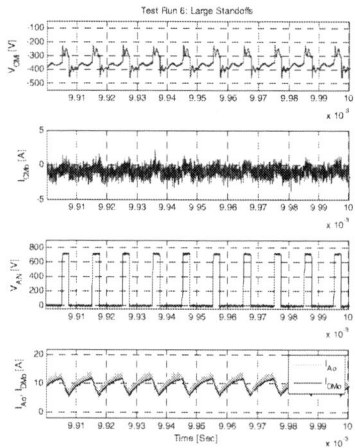

Fig. 18 Experimental and derived quantities from data run #6, 0.75 inch spacers between the cold plate and the ground plane

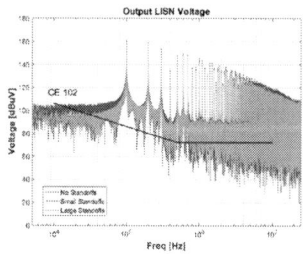

Fig. 19 Conducted emissions at the output of the CIL test stand

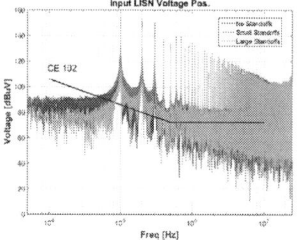

Fig. 20 Conducted emissions at the input of the CIL test stand

REFERENCES

[1] Xun Gong; Josifovic, I.; Ferreira, J.A., "Comprehensive CM filter design to suppress conducted EMI for SiC-JFET motor drives," Power Electronics and ECCE Asia (ICPE & ECCE), 2011 IEEE 8th International Conference on , vol., no., pp.720,727, May 30 2011-June 3 2011

[2] Xun Gong; Ferreira, J.A., "Investigation of Conducted EMI in SiC JFET Inverters Using Separated Heat Sinks," in Industrial Electronics, IEEE Transactions on , vol.61, no.1, pp.115-125, Jan. 2014

[3] Josifovic, I.; Popovic-Gerber, J.; Ferreira, J.A., "Improving SiC JFET Switching Behavior Under Influence of Circuit Parasitics," in Power Electronics, IEEE Transactions on , vol.27, no.8, pp.3843-3854, Aug. 2012

[4] Hazra, S.; Madhusoodhanan, S.; Bhattacharya, S.; Moghaddam, G.K.; Hatua, K., "Design considerations and performance evaluation of 1200 V, 100 a SiC MOSFET based converter for high power density application," in Energy Conversion Congress and Exposition (ECCE), 2013 IEEE , pp.4278-4285, 15-19 Sept. 2013

[5] Lemmon, A.; Mazzola, M.; Gafford, J.; Parker, C., "Stability Considerations for Silicon Carbide Field-Effect Transistors," in Power Electronics, IEEE Transactions on , vol.28, no.10, pp.4453-4459, Oct. 2013

[6] X. Gong and J. A. Ferreira, "Comparison and Reduction of Conducted EMI in SiC JFET and Si IGBT-Based Motor Drives," IEEE Trans. Power Electron., vol. 29, no. 4, pp. 1757–1767, Apr. 2014.

[7] Lemmon, A.; Mazzola, M.; Gafford, J.; Parker, C., "Instability in Half-Bridge Circuits Switched With Wide Band-Gap Transistors," in Power Electronics, IEEE Transactions on , vol.29, no.5, pp.2380-2392, May 2014

[8] N. Oswald, P. Anthony, N. Mcneill, and B. H. Stark, "An Experimental Investigation of the Tradeoff between Switching Losses and EMI Generation With Device Combinations," IEEE Trans. Power Electron., vol. 29, no. 5, pp. 2393–2407, 2014

[9] Mantooth, H.A.; Glover, M.D.; Shepherd, P., "Wide Bandgap Technologies and Their Implications on Miniaturizing Power Electronic Systems," in Emerging and Selected Topics in Power Electronics, IEEE Journal of , vol.2, no.3, pp.374-385, Sept. 2014

[10] Yuchen Yang; Zhengyang Liu; Lee, F.C.; Qiang Li, "Analysis and filter design of differential mode EMI noise for GaN-based interleaved MHz critical mode PFC converter," in Energy Conversion Congress and Exposition (ECCE), 2014 IEEE , vol., no., pp.4784-4789, 14-18 Sept. 2014

[11] Scott, M.J.; Brockman, J.; Boxue Hu; Lixing Fu; Longya Xu; Jin Wang; Darbali Zamora, R., "Reflected wave phenomenon in motor drive systems using wide bandgap devices," in Wide Bandgap Power Devices and Applications (WiPDA), 2014 IEEE Workshop on , vol., no., pp.164-168, 13-15 Oct. 2014

[12] Jiangbiao He; Tiefu Zhao; Xin Jing; Demerdash, N.A.O., "Application of wide bandgap devices in renewable energy systems — Benefits and challenges," in Renewable Energy Research and Application (ICRERA), 2014 International Conference on , pp.749-754, 19-22 Oct. 2014

[13] Nayak, P.; Krishna, M.V.; Vasudevakrishna, K.; Hatua, K., "Study of the effects of parasitic inductances and device capacitances on 1200 V, 35 A SiC MOSFET based voltage source inverter design," in Power Electronics, Drives and Energy Systems (PEDES), 2014 IEEE International Conference on , vol., no., pp.1-6, 16-19 Dec. 2014

[14] Scott, M.J.; Lixing Fu; Chengcheng Yao; Xuan Zhang; Longya Xu; Jin Wang; Zamora, R.D., "Design considerations for wide bandgap based motor drive systems," in Electric Vehicle Conference (IEVC), 2014 IEEE International , vol., no., pp.1-6, 17-19 Dec. 2014

[15] Bingyao Sun; Burgos, R., "Assessment of switching frequency impact on the prediction capability of common-mode EMI emissions of sic power converters using unterminated behavioral models," Applied Power Electronics Conference and Exposition (APEC), 2015 IEEE, pp.1153-1160, 15-19 March 2015

[16] Di Han; Sarlioglu, B., "Study of the switching performance and EMI signature of SiC MOSFETs under the influence of parasitic inductance in an automotive DC-DC converter," in Transportation Electrification Conference and Expo (ITEC), 2015 IEEE , vol., no., pp.1-8, 14-17 June 2015

[17] Chen, Hao; Divan, Deepak, "High speed switching issues of high power rated silicon-carbide devices and the mitigation methods," in Energy Conversion Congress and Exposition (ECCE), 2015 IEEE , vol., no., pp.2254-2260, 20-24 Sept. 2015

[18] Song, Xiaoqing; Huang, Alex Q.; Lee, Mengjia; Wang, Gangyao, "A dynamic measurement method for parasitic capacitances of high voltage SiC MOSFETs," in Energy Conversion Congress and Exposition (ECCE), 2015 IEEE , vol., no., pp.935-941, 20-24 Sept. 2015

[19] Zare, Firuz; Kumar, Dinesh; Lungeanu, Marian; Andreas, Aupke, "Electromagnetic interference issues of power, electronics systems with wide band gap, semiconductor devices," in Energy Conversion Congress and Exposition (ECCE), 2015 IEEE , vol., no., pp.5946-5951, 20-24 Sept. 2015

[20] N. Oswald, B. H. Stark, D. Holliday, C. Hargis, and B. Drury, "Analysis of Shaped Pulse Transitions in Power Electronic Switching Waveforms for Reduced EMI Generation," IEEE Trans. Ind. Appl., vol. 47, no. 5, pp. 2154–2165, Sep. 2011.

[21] Zheyu Zhang; Ben Guo; Wang, F.; Tolbert, L.M.; Blalock, B.J.; Zhenxian Liang; Puqi Ning, "Methodology for switching characterization evaluation of wide band-gap devices in a phase-leg configuration," in Applied Power Electronics Conference and Exposition (APEC), 2014 Twenty-Ninth Annual IEEE , pp.2534-2541, 16-20 March 2014

[22] A. Lemmon, R. Graves, and J. Gafford, "Evaluation of 1.2 kV, 100A SiC Modules for High-Frequency, High-Temperature Applications," Applied Power Electronics Conference and Exposition (APEC), 2015 IEEE , vol., no., pp.789-793, 15-19 March 2015

[23] A. Lemmon, R. Graves, "Parameter extraction for Silicon Carbide modules," IEEE International Workshop on Integrated Power Packaging (IWIPP), 3-6 May 2015

[24] A. Dutta, S. Ang, "Electromagnetic interference simulations of power electronic modules," IEEE International Workshop on Integrated Power Packaging (IWIPP), 3-6 May 2015

[25] Bishnoi, H.; Mattavelli, P.; Burgos, R.; Boroyevich, D., "EMI Behavioral Models of DC-Fed Three-Phase Motor Drive Systems," in Power Electronics, IEEE Transactions on , vol.29, no.9, pp.4633-4645, Sept. 2014

[26] Sun, Bingyao; Burgos, Rolando; Zhang, Xuning; Boroyevich, Dushan, "Differential-mode EMI emission prediction of SiC-based power converters using a mixed-mode unterminated behavioral model," in Energy Conversion Congress and Exposition (ECCE), 2015 IEEE , vol., no., pp.4367-4374, 20-24 Sept. 2015

[27] N. Bondarenko, L. Zhai, B. Xu, G. Li, T. Makharashvili, D. Loken, P. Berger, T. P. Van Doren, D. G. Beetner, "A Measurement-Based Model of the Electromagnetic Emissions From a Power Inverter", IEEE Trans. Power Electron., vol. 30, no. 10, pp. 5522–5531, Oct. 2015

[28] Anthony, P.; McNeill, N.; Holliday, D., "High-speed resonant gate driver with controlled peak gate voltage for silicon carbide MOSFETs," in Energy Conversion Congress and Exposition (ECCE), 2012 IEEE , vol., no., pp.2961-2968, 15-20 Sept. 2012

[29] Shahverdi, Masood; Mazzola, Michael; Schrader, Robin; Lemmon, Andrew; Parker, Christopher; Gafford, James, "Active gate drive solutions for improving SiC JFET switching dynamics," in Applied Power Electronics Conference and Exposition (APEC), 2013 Twenty-Eighth Annual IEEE, pp.2739-2743, 17-21 March 2013

[30] "1.2kV, 13 mΩ All-Silicon Carbide Half-Bridge Module," CAS120M12BM2 datasheet, Cree Inc., Durham, NC, 2014.

[31] Cuzner, R. M., "Power electronics packaging challenges for future warship applications," in Integrated Power Packaging (IWIPP), 2015 IEEE International Workshop on , pp.5-8, 3-6 May 2015

[32] Rahmani, Maryam; Mazzola, Michael, "Modeling of common mode currents on electric ship hull using scattering parameters," Electric Ship Technologies Symposium (ESTS), 2015 IEEE , pp.156-160, 21-24 June 2015

[33] Graber, L.; Mohebali, B.; Bosworth, M.; Steurer, M.; Card, A.; Rahmani, M.; Mazzola, M., "How scattering parameters can benefit the development of all-electric ships," in Electric Ship Technologies Symposium (ESTS), 2015 IEEE , pp.353-357, 21-24 June 2015

[34] "DOD Interface Standard, Requirements for the control of electromagnetic interference characteristics of subsystems andequipment, Mil-Std-461F, 10 Dec 2007

[35] Hegner, H.; Desai, B., "Integrated fight through power," Power Engineering Society Summer Meeting, 2002 IEEE , vol.1, no., pp.336,339 vol.1, 25-25 July 2002

A Practical Switching Time Model for Synchronous Buck Converters

Yuan Rao†, Surinder P. Singh‡, and Taisuke Kazama†
Texas Instruments
DC Solutions (DCS) Business Unit†, WEBENCH® Design Center‡
12500 TI Blvd, Dallas, TX 75243, USA
Email: yrao1@ti.com

Abstract — This paper presents a practical physics-based switching time (turn-on and turn-off time) model of MOSFET in synchronous buck converters. The turn-on and turn-off time of the MOSFET switching element has a strong impact of the operating efficiency of the converter. The proposed model can be implemented in predictive, practical efficiency model of DC/DC converters, like WEBENCH® Power Designer. Unlike the more complex models that exist in literature, the new model needs fewer parameters while still being physics-based. The proposed model simplifies the calculation through linear approximation of the voltage and current waveforms of the high-side FET. Different from other state-of-the-art models, it provides the detailed calculation of each parameter, allowing quick estimation of the switching losses. The model is useful in practical applications where the designer needs to predict the switching performance of the buck converter with the minimum available information of the internal circuit. Experimental results are described in detail to support the accuracy of the analytical model presented.

Keywords— buck converter, switching time model, switching loss turn-on time, turn-off time, turn-on loss, turn-off loss, rise time, fall time, MOSFET loss, efficiency model

I. INTRODUCTION

Power electronics is at the heart of the modern electronic revolution. Power designers are faced with enormous challenges in the design process: they have to select the appropriate topology, the power device from a large set of possibilities, and the best passives and switching elements like MOSFETs from a large set of vendors. They also have to make engineering design decisions to make intelligent trade-off between solution size, cost and thermal performance with efficiency [1-4]. Hence, accurate, predictive, computationally-efficient modeling of efficiency is at the heart of modern power design.

The most challenging part of computing efficiency is modeling the semiconductor switching element. It is widely acknowledged that modeling conduction losses is easier; modeling of switching losses is the real challenge [5]. The modeling of the turn-on and turn-off times, also referred to as rise- and fall-times, of the MOSFET continues to be a topic of keen interest in academia and industry [5-16].

Synchronous buck converters are widely used in low-voltage, high-current applications. Its switching frequency has been increasing for the last few years to achieve small footprints. With the increasing frequencies, the switching loss

associated with the turn-on and turn-off of the MOSFET are the dominant loss mechanism. This paper proposes a practical, physics-based, computationally efficient model of the turn-on and turn-off loss of a MOSFET.

There are three basic types among previously reported switching loss models [5-17]: physical model, behavioral model and analytical model.

The physical model uses finite-element or finite-different Technology CAD (TCAD) simulations of the MOSFET and parasitic using physical structure parameters like doping density, oxide thickness, and layout [5] [9] [12]. While these models are the most sophisticated and have the best matches with the experimental data, the simulation times are extremely long. Reference [5], for instance, reports that it took one month of CPU time to simulate a single efficiency curve of just 18 points. Authors of [5] report simulation time of two days for just one switching cycle. The physical models are impractical for power design.

Behavioral models use circuit simulation engines like SPICE or PSpice along with SPICE models of MOSFETs, for example BSIM [18] or PSP [19] model, in the detailed simulations to extract losses [13] [14] [15] [16]. While they are faster than the physical models, the run time is still unacceptably long. This is because the switching phenomenon in a switching regulator occurs at very small time scales as compared to the overall switching cycle. To resolve both time scales, very small time steps are needed, resulting in long simulation times. Furthermore, the accuracy of the power loss depends critically on the accuracy of MOSFET model employed. Even when these models capture the physics, they require extensive effort to calibrate against bench data. Despite these efforts, these models do not provide insight into the physical mechanisms readily [11] [12].

The third model type, analytical model, or mathematical model, is based on equations derived from equivalent circuits of the power converters [5,6,8-11]. These models are computationally efficient and therefor well suited for data processing and even hand analysis. They are the preferred candidates for regular efficiency modeling in power design CAD tools. These analytical models, unlike the physical and behavioral models, provide clear insight in the variables and mechanisms of power loss in converters. The accuracy of analytical models, however, depends on the assumption and estimation used to simplify the equations. Fewer

978-1-4673-9551-9/16 $31.00 © 2016 IEEE

approximations yield more predictive, but complex, difficult to calibrate, often unwieldy models. Excessive simplifications may render the model too simplistic and inapplicable outside the narrow range of assumptions. The challenge for the model developer is to develop the model with the right level of sophistication and making it simple, practical and computationally efficient. The model should have few parameters and should be easy to calibrate. Additionally, the model should provide insights for the power designer. Success and failure of the model depends on these subtle effects.

In this paper an analytical physics-based switching time model is proposed. It captures the turn-on and turn-off switching losses using the physics of the MOSFET in switch-mode power supply. The non-linear MOSFET capacitances are accounted for in the model. The dependence of turn-on and turn-off time on the input voltage and output current is brought out explicitly. The model needs a minimal set of parameters—two parameters for turn-on and two for turn-off times. These parameters can be determined either by simple bench measurements or estimated by the Fab data on the MOSFET. For commercial MOSFETs, they can be derived from the datasheet parameters. This makes the model accurate, practical and CPU time efficient and is suitable for power supply design tools [1].

This paper is organized as follows. In Section II the model equations are developed and the final model is proposed in Section II.C. Model verification against bench measurements for several devices from Texas Instruments is shown in Section III. Conclusions are drawn in Section 0.

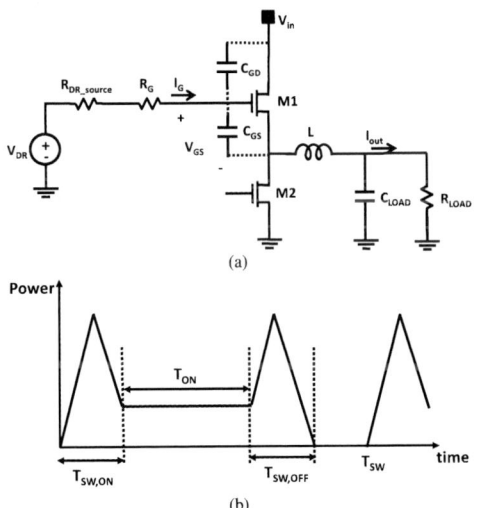

Fig. 1. A typcial synchronous buck converter and its switching power loss.

II. PROPOSED SWITCHING TIME MODEL

A. Switching Time in Switching Loss Model

In a typical synchronous buck converter shown in Fig. 1 (a), the switching loss induced by the High-side (HS)

MOSFET M1 dominates the overall switching loss of the buck converter. The Low-side (LS) MOSFET M2 acts as a synchronous rectifier (SR) and has near zero switching loss. There are two parts of the HS MOSFET switching loss: turn-on switching loss and turn-off switching loss. As shown in Fig. 1 (b), when HS MOSFET turns on, its drain-source voltage (V_{DS}) drops from the input voltage (V_{in}) to zero. Meanwhile, the current (I_D) flows through HS MOSFET, storing the input energy to the inductor. Switching loss of the power MOSFET occurs when both the drain current and drain voltage are present. So the overlapping of the V_{DS} decrease and I_D increase contributes to the turn-on switching loss. Likewise, when HS MOSFET turns off, the overlapping of the V_{DS} increase and I_D decrease causes the turn-off switching loss. The overall switching loss P_{SW} can therefore be expressed as

$$P_{SW} = P_{SW,ON} + P_{SW,OFF} \qquad (1)$$

where the turn-on switching loss $P_{SW,ON}$ and turn-off switching loss $P_{SW,OFF}$ are determined by the input voltage V_{in}, switching frequency F_{SW}, output dc current I_{out}, output current ripple $I_{\Delta ripple}$, turn-on time $T_{SW,ON}$, turn-off time $T_{SW,OFF}$,

$$P_{SW,ON} = \tfrac{1}{2} V_{in} F_{SW} \big(I_{out} - I_{\Delta ripple} \big) T_{SW,ON} \qquad (2)$$

$$P_{SW,OFF} = \tfrac{1}{2} V_{in} F_{SW} \big(I_{out} + I_{\Delta ripple} \big) T_{SW,OFF} \qquad (3)$$

When the converter output capacitor (C_{LOAD}) is large enough, the output current ripple $I_{\Delta ripple}$ is negligible compared with the output dc current I_{out}. The switching loss equation can therefore be simplified as

$$P_{SW} = \tfrac{1}{2} V_{in} f_{SW} I_{out} \big(T_{SW,ON} + T_{SW,OFF} \big) \qquad (4)$$

It's easy to get the input voltage, switching frequency and output dc current information from the datasheet. However, a good estimation of switching time ($T_{SW,ON}$ and $T_{SW,OFF}$) is complicated and challenging, due to the non-linear relationship between the gate charge and gate capacitances. To solve this, a new simplified switching time model is proposed as described below.

B. Proposed Switching Time Model

1) Turn On Time Model

Fig. 2 gives the MOSFET turn on sequence. There are four regions *t0, t1, t2* and *t3* as described below.

- *Time period t0:* V_{GS} begins to rise, charging C_{GS} and C_{GD}. Because C_{GS} is much larger than C_{GD}, most drive current is flowing into C_{GS} rather than C_{GD}. The drain current has not started to flow, so V_{DS} keeps unchanged.
- *Time Period t1:* I_D begins to flow. But V_{DS} still keeps unchanged because LS FET is still on. I_D reaches the

maximum at the end of this period. Meanwhile the drain current keeps charging C_{GS}.

- *Time Period t2 (Plateau region):* The gate is at the steady voltage V_{PL} to carry the entire load current and LS FET is turned off. The drain voltage begins to fall, while I_{DS} keeps constant. The driver current is diverted to discharge C_{GD} to help V_{DS} drop.
- *Time Period t3:* By applying higher V_{GS}, the HS FET channel is fully conducting. The final voltage V_{DR} determines the R_{DSON} of HS FET. The driver current keeps charging C_{GS} and C_{GD}. However the drain current is still constant and V_{DS} is slightly dropped due to reduced R_{DSON}.

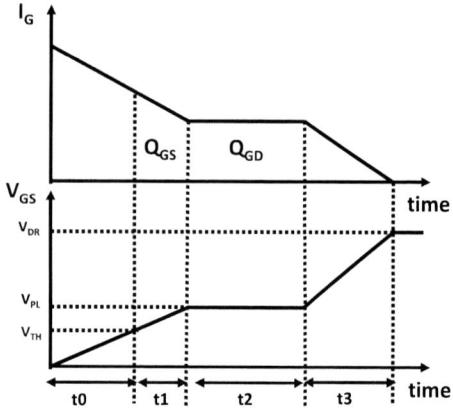

Fig. 2. MOSFET turn on sequence

Above all, turn on time is determined by how long it takes the driver to move the gate charge of Q_{GS} and Q_{GD}, so the turn on time is expressed as

$$T_{SW,ON} = t_1 + t_2 = \frac{Q_{GS}}{I_{DRV,t1}} + \frac{Q_{GD}}{I_{DRV,t2}}$$

where

$$I_{DRV,t1} = \frac{V_{DR} - 0.5(V_{TH} + V_{PL})}{R_{DR\_source} + R_G}$$

$$I_{DRV,t2} = \frac{V_{DR} - V_{PL}}{R_{DR\_source} + R_G}$$

Here we assume linear, and use the middle point $0.5(V_{TH} + V_{PL})$ as an approximate of V_{GS} during t1. For t1, MOSFET starts carrying current (I_{DS} is increasing) and the device is in linear region. The gate current charges both C_{GS} and C_{GD}, where the gate voltage increases from V_{TH} to V_{PL}.

$$Q_{GS} = (C_{GS} + C_{GD})(V_{PL} - V_{TH})$$

During t2, MOSFET is in Miller Plateau region, where I_{DS} is constant and equal to I_{out}. The drain-source voltage V_{DS} drops from V_{in} to zero as the driver discharges C_{GD}.

$$Q_{GD} = C_{GD}V_{in}$$

Then substitute everything into Equation (5).

$$T_{SW,ON} = (\frac{(V_{PL} - V_{TH})(C_{GS} + C_{GD})}{V_{DR} - 0.5(V_{TH} + V_{PL})}$$
$$+ \frac{V_{in}C_{GD}}{V_{DR} - V_{PL}})(R_{DR\_source} + R_G)$$

If $Q_{GS2} \ll Q_{GD}$ or $t_2 \gg t_1$

$$T_{SW,ON} \cong \frac{V_{in}C_{GD}}{V_{DR} - V_{PL}}(R_{DR\_source} + R_G) \tag{5}$$

2) Turn Off Time Model

Similarly, Fig. 3 gives the MOSFET turn off sequence.

- *Time period t0:* V_{GS} begins to decrease, discharging C_{GS} and C_{GD}. The drain current has not changed, and V_{DS} slightly increases due to higher R_{DSON}.
- *Time Period t1 (Plateau region):* V_{DS} increases from I_D*R_{DSON} to the final voltage for turn off and is then clamped. I_D reaches maximum at the end of t1. The driver keeps charging C_{GD}, as V_{GS} keeps constant for this time period.
- *Time Period t2:* V_{GS} drops from V_{PL} to V_{TH}. Most of the gate current is flowing from C_{GS} because C_{GD} is virtually fully charged in *t1*. The drain current I_D begins to fall and reaches zero at end of t2, while V_{DS} keeps clamped.
- *Time Period t3:* V_{GS} is further reduced to be less than V_{TH}, as C_{GS} is further discharged. HS FET is finally off with zero gate current at the end of *t3*.

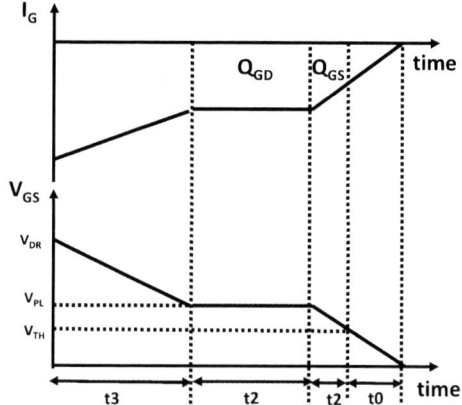

Fig. 3. MOSFET turn off sequence

Above all, when V_{GS} begins to decrease, discharging C_{GS} and C_{GD}, and turn off time can be derived as

$$T_{SW,OFF} = t_1 + t_2 = \frac{Q_{GD}}{I_{DRV,t3}} + \frac{Q_{GS2}}{I_{DRV,t4}}$$

where

$$I_{DRV,t1} = \frac{V_{PL}}{R_{DR\_sink} + R_G} \quad I_{DRV,t2} = \frac{0.5(V_{TH} + V_{PL})}{R_{DR\_sink} + R_G}$$

$$T_{SW,OFF} = (\frac{(V_{PL}-V_{TH})(C_{GS} + C_{GD})}{0.5(V_{TH} + V_{PL})} + \frac{V_{in}C_{GD}}{V_{PL}})(R_{DR\_sink} + R_G)$$

If $Q_{GS2} \ll Q_{GD}$ or $t_1 \gg t_2$

$$T_{SW,OFF} \cong \frac{V_{in}C_{GD}}{V_{PL}}(R_{DR\_sink} + R_G) \qquad (6)$$

3) Parameter Calculation

Equation (5) and Equation (6) give the simplified switching on/off time model. To make hand calculation possible, Table I provides the calculation of the parameters.

TABLE I. CALCULATION OF SWITCHING MODEL PARAMETERS

Parameter	Descripltion	Calculation
V_{PL}	Miller Plateau voltage	$V_{PL} = V_{TH} + \sqrt{\frac{I_{out}}{K_{prime}}}$
$R_{DR\_sink,}$ R_{DR_source}	Driver sinking resistance Driver sourcing resistance	Data from Design
R_G	Gate resistance	Data from Design (or can be ignored)
V_{DR}	Driver voltage	When $V_{in}<V_{CLAMP}$, $V_{DR}=V_{in}$ When $V_{in}>=V_{CLAMP}$, $V_{DR}= V_{CLAMP}$
C_{GS}	Gate-source capacitance	Datasheet (can be treated as a constant)
C_{GD_AVG}	Gate-drain capacitance	$\cong 2C_{GD@SPEC}\sqrt{\frac{V_{DS@SPEC}}{V_{in}}}$ [20]

In TABLE I, the miller voltage is derived from K_{prime}. It is because when converter is switching, HS FET works in linear region as a voltage-controlled resistor.

$$R_{on} \cong \frac{1}{\mu_n C_{ox} \frac{W}{L}(V_{GS} - V_T)} = \frac{1}{2K_{prime}(V_{GS} - V_T)}$$

and therefore

$$K_{prime} = \frac{1}{2R_{on}(V_{GS}-V_T)} \qquad (7)$$

For synchronous buck converter datasheet such as [21] [22] [23] where R_{on} and its corresponding V_{GS} are available, K_{prime} can be derived by Equation (7).

Modeling non-linear capacitance is more complicated. For external PowerFET [24], the datasheet normally provides C_{GS}, C_{GD} under certain V_{DS}, where C_{GS} is almost a constant. Average C_{GD} is a non-linear function of variable V_{DS}, as below

$$C_{GD}(V_{DS}) = C_{GD@SPEC} \times \sqrt{V_{DS@SPEC}} \times \frac{1}{\sqrt{V_{DS}}}$$

$$Q_{GD} = \int_{V_{in}}^0 C_{GD}dV_{DS} = C_{GD@SPEC} \times \sqrt{V_{DS@SPEC}}$$
$$\times \int_{V_{in}}^0 \frac{1}{\sqrt{V_{DS}}}dv$$

So

$$C_{GD\_AVG} = \frac{Q_{GD}}{V_{in}} = 2C_{GD@SPEC}\sqrt{\frac{V_{DS@SPEC}}{V_{in}}}$$

Otherwise, for internal FET, the gate charge curve (V_{GS} vs. Q_g) in designer's PDK and the LDMOS Size (W/L) are required to get the decent estimation of C_{GS} and C_{GD}, which is out of the scope of this paper.

Note that datasheet parameters may lead to contradictory results [10], so care must be taken to interpret them.

C. Model Extraction

The proposed model can be further extracted to estimate the impact of V_{in} and I_{out} on the switching time. By substituting variable equations in Table I into Equation (5) and Equation (6) separately, the switching time model is expressed as

$$T_{SW,ON} \cong \frac{2V_{in}C_{GD@SPEC}\sqrt{\frac{V_{DS@SPEC}}{V_{in}}}}{V_{DR} - V_{TH} - \sqrt{\frac{I_{out}}{K_{prime}}}}(R_{DR\_source} + R_G)$$

For a given converter design, $C_{GD@SPEC}$, $\sqrt{V_{DS@SPEC}}$, K_{prime}, $R_{DR_{source}}$ and R_G can be taken as a constant and extracted out. The on-time model is now simplified as

$$T_{SW,ON} \cong \frac{K_{ON1}\sqrt{V_{in}}}{K_{ON2}-\sqrt{I_{out}}} \qquad (8)$$

where

$$K_{ON1} = 2C_{GD@SPEC}\sqrt{V_{DS@SPEC}}\sqrt{K_{prime}}(R_{DR_{source}} + R_G)$$

$$K_{ON2} = (V_{DR} - V_{TH})\sqrt{K_{prime}}$$

Similarly, for off-time model, it can be extracted as

$$T_{SW,OFF} \cong \frac{K_{OFF1}\sqrt{V_{in}}}{K_{OFF2}+\sqrt{I_{out}}} \qquad (9)$$

where

$$K_{OFF1} = 2C_{GD@SPEC}\sqrt{V_{DS@SPEC}}\sqrt{K_{prime}}(R_{DR_{sink}} + R_G)$$

978-1-4673-9551-9/16 $31.00 © 2016 IEEE

$$K_{OFF2} = V_{TH}\sqrt{K_{prime}}$$

Above all, Equation (8) and Equation (9) predict switching on and switching off time as a function of V_{in} and I_{out}. This estimation significantly improves the switching loss accuracy for wide V_{in} and I_{out} devices. The relationship is demonstrated in next section with both measurement and simulation results.

III. MODEL VERIFICATION AND DISCUSSION

The proposed switching time model was validated against laboratory measurements. Turn-on and turn-off time measured on several buck DC-DC converters made by Texas Instruments, Inc. was compared against the proposed model. The functional dependence of the model on input voltage an output current was validated by these measurements. We also demonstrate that the use of the proposed model on modeling efficiency and show that this results in better matches to bench data.

A. Switching Time matching

1) Switching time versus Input voltage

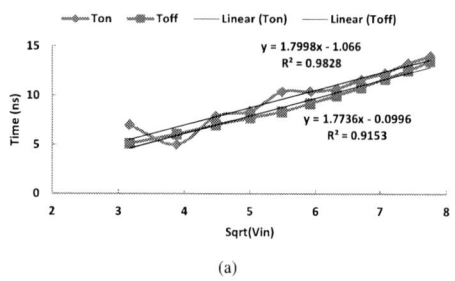

(a)

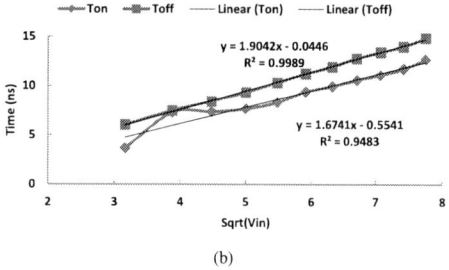

(b)

Fig. 4 Measurement and model result of switching time at different input voltage for TPS54560 device.

The switching time of TPS54560 [23], a commercial 60-V input, 5-A, step-down DC-DC Converter, was measured in the lab at different input voltage range and output current range. Fig. 4 demonstrates the model accuracy by plotting switching time versus square root of input voltage, V_{in}, using the method of linear trend line fitting, where y stands for the linear expression, and R^2 describes the goodness of fit (the closer R^2 is to unity, the better the fit). It is evident that the quality of the fit is excellent, which experimentally proves that turn-on and turn-off times are proportional to the square root of input voltage.

2) Switching time vs. Output current

The measured turn-on and turn-off times for TPS54560A are plotted as a function of output current in Fig. 5. The ratio of square root of input voltage to switching time, $\sqrt{V_{in}}/T_{SW,ON}$, is plotted versus square root of output current, $\sqrt{I_{out}}$ and fit by a linear trend line. The goodness of the linear fit in both cases demonstrates the validity of the model. The dependence on output current is weak, but is linear and in alignment with the model.

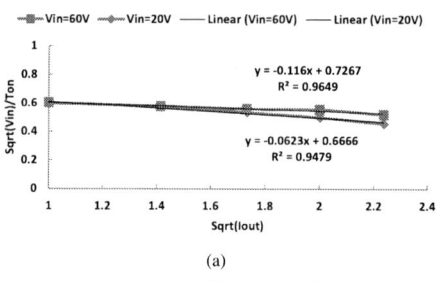

(a)

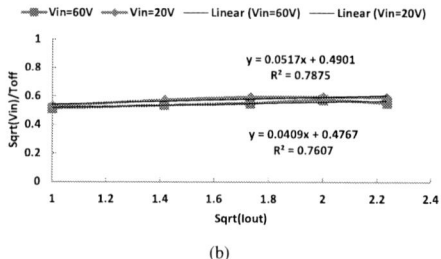

(b)

Fig. 5. Measurement and model result of switching time at different output current for TPS54560.

B. Efficiency Matching

We implemented the proposed switching time model in a standard efficiency model [17] and compared it against the original model of constant turn-on and turn-off times. Here we choose TPS54335A device [21] efficiency operating at a high 1.2MHz because the switching loss plays a bigger role at high frequency. As can be seen in Fig. 6, the new model fits the measured efficiency much more closely over the wide range of input voltage and output current. This demonstrates the predictive nature of the switching time model.

IV. CONCLUSION

A practical and simplified switching time model is proposed based on classic piece-wise-linear model. The model uses datasheet values and circuit parameters to calculate the switching time, as well as switching loss with decent accuracy. To improve the discrepancy of conventional switching time models which use constant turn-on and turn-off at all conditions, the proposed model provides clear trend of the switching time versus input voltage and output current. The experimental measurement verifies that the model provides a good approximation with the measurement, and can be very practical and helpful predicting the switching performance of synchronous buck converters.

TPS54335A 1.2MHz (Vin=12V)

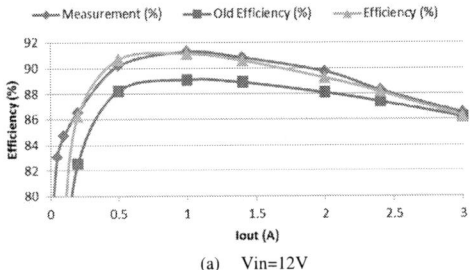

(a) Vin=12V

TPS54335A 1.2MHz (Vin=24V)

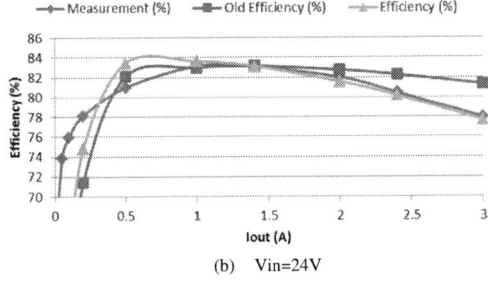

(b) Vin=24V

Fig. 6. Measured efficiency data from TPS54335A compared with efficiency model with the proposed switching time model versus old constant switching times.

REFERENCES

[1] Texas Instruments, Inc., "WEBENCH Power Designer tools," [Online]. Available: http://www.ti.com/webench. [Accessed 10 November 2015].

[2] J. Perry, "Ease power supply design with design tools," 2011 [Online]. Available: http://www.ti.com/lit/ml/snvp001/snvp001.pdf. [Accessed 10 November 2015].

[3] J. Perry, "MOSFET selection for switching power supply systems," 3 April 2012. [Online]. Available: http://powerelectronics.com/power-electronics-systems/mosfet-selection-switching-power-supply-systems. [Accessed 10 November 2015].

[4] J. Perry, "Optimizing LED Lighting Systems for Efficiency, Size and Cost," January 2011. [Online]. Available: http://powerelectronics.com/sitefiles/powerelectronics.com/files/archiv e/powerelectronics.com/images/OptimizingLEDLightingSyste. [Accessed 10 November 2015].

[5] Y. Ren, M. Xu, J. Zhou and F. C. Lee, "Analytical loss model of power MOSFET," *IEEE Trans. Power Electron.*, vol. 21, no. 2, pp. 310-319, 2005.

[6] M. Rodriguez, A. Rodriguez and P. F. Miaja, "A complete analytical switching loss model for power MOSFETs in low voltage converters," in *13th European Conference on Power Electronics and Applications*, 2009.

[7] J. Brown, "Power MOSFET Basics: understanding gate charge and using it to assess switching performance," December 2004. [Online]. Available: http://www.vishay.com/docs/73217/73217.pdf. [Accessed 10 November 2015].

[8] W. Eberle, Z. Zhang, Y. Liu and P. Sen, "A practical switching loss

model for buck boltage regulators," *IEEE Transactions on Power Electronics*, vol. 24, no. 3, pp. 700-713, 2009.

[9] J. Shen, Y. Xiong, X. Cheng, Y. Fu and P. Kumar, "'Power MOSFET switching loss analysis: a new insight," in *IEEE Industry Applications Conference*, 2006.

[10] J. Brown, "Modeling the Switching Performance of a MOSFET in the High Side of a Non-Isolated Buck Conveter," *IEEE Trans. Power Electron.*, vol. 21, no. 1, pp. 3-10, 2006.

[11] M. Rodriguez, A. Rodriguez., P. Miaja, D. G. Lamar and J. S. Zuniga, "An Insight into the Switching Process of Power MOSFETs: An Improved Analytical Loss Model," *IEEE Trans. Power Electron.*, vol. 25, no. 6, pp. 1626-1640, 2010.

[12] Y. Xiong, S. Sun, H. Jia, P. Shea and Z. J. Shen, "New physical insights on power MOSFET switching losses," *IEEE Trans. Power Electron.*, vol. 24, no. 2, pp. 525-531, 2009.

[13] U. Drofenik and J. W. Kolar, "A general scheme for calculating switching-and conduction-losses of power semiconductors in numerical circuit simulations of power electronic systems," in *Proceedings of the 2005 International Power Electronics Conference*, Niigata, Japan, 2005.

[14] O. Al-Naseem, R. Erickson and P. Carlin, "Prediction of switching loss variations by averaged switch modeling," in *Applied Power Electronics Conference and Exposition, Fifteenth Annual IEEE*, New Orleans, LA, 2000.

[15] C. Peng and C. J. Wang, "An analysis of buck converter efficiency in PWM/PFM mode with simulink," *Energy and Power Engineering*, vol. 5, no. 3, pp. 64-69, 2013.

[16] J. V. Gragger, A. Haumer and M. Einhorn, "Averaged Model of a Buck Converter for Efficiency Analysis," *Eneineering Letters*, vol. 18, no. 1, p. 49, 2010.

[17] D. Jauregui, B. Wang and R. Chen, "Power loss calculation with common source inductance consideration for synchronous buck converters," July 2011. [Online]. Available: http://www.ti.com/lit/an/slpa009a/slpa009a.pdf. [Accessed November 2015].

[18] B. J. Sheu, D. L. Scharfetter, P. K. Ko and M. C. Jeng, "BSIM: Berkeley short-channel IGFET model for MOS transistors," *IEEE J. Solid-State Circuits*, vol. 22, no. 4, pp. 558-566, 1987.

[19] G. Gildenblat, Ed., Compact modeling: principles, techniques and applications, Springer Science & Business Media.

[20] A. Pathak, "MOSFET/IGBT drivers theory and applications," [Online]. Available: www.ixys.com/Documents/AppNotes/IXAN0010.pdf. [Accessed 10 November 2015].

[21] Texas Instruments, Inc., "TPS54335A: 4.5V to 28V Input, 3A, Synchronous, Step-Down Converter with Eco-mode™," [Online]. Available: http://www.ti.com/product/tps54335A. [Accessed 10 November 2015].

[22] Texas Instruments, Inc., "TPS54336A: 4.5V to 28V Input, 3A, Synchronous Step-Down SWIFT™ Converter with Eco-mode™," [Online]. Available: http://www.ti.com/product/tps54336A. [Accessed 10 November 2015].

[23] Texas Instruments, Inc., "TPS54560: 60 V Input, 5 A, Step-Down DC-DC Converter with Eco-Mode™," [Online]. Available: http://www.ti.com/product/tps54560. [Accessed 10 November 2015].

[24] Texas Instruments, Inc., "CSD18504A: 40V, N-Channel NexFET Power MOSFET™," [Online]. Available: http://www.ti.com/product/CSD18504Q5A. [Accessed 10 November 2015].

Off-line identification of digitally controlled power converters using an analog frequency response analyzer

Marco Meola
Digital Power Management, ZMDI
Munich, GERMANY
marco.meola@zmdi.com

Anthony Kelly
Altera Corporation
Limerick, IRELAND
akelly@altera.com

Abstract— **The use of a digital architecture in PWM controllers for point-of-load (POL) applications, together with system identification techniques, allows the development of fully automated routines for *in-situ* system performance optimization where controller parameters are specifically tailored to the application. In this context, this paper proposes a method for performing parametric system identification of digitally controlled power converters using a conventional analog frequency response analyzer (FRA). Nonlinearities intrinsic to the digital loop are taken into account, thus leading to accurate estimation of converter parameters. The proposed method has been verified on a digitally controlled POL with V_{in}=12V, V_{out}=1.2V, I_{out}=10A and f_{sw}=400kHz for various bulk capacitor scenarios.**

Keywords— system identification, frequency response, quantization, digital control, PWM controller, DC-DC converter

I. INTRODUCTION

The use of a digital architecture in pulse-width-modulation (PWM) controllers for point of load (POL) applications allows access to a further level of system performance optimization where controller parameters are configured and fine-tuned directly *in situ* to achieve a power system specifically tailored to the application.

System identification techniques, i.e. techniques able to identify the characteristics of the power stage by means of experimental measurements, have been intensively investigated for integration into digital PWM controllers [1]-[7] with the purpose of developing a fully automated compensator design procedure [8]-[9]. Different system identification methods have been proposed to obtain more accurate solutions with less expensive hardware. Depending on the type of the injected stimuli (pseudo-random noise [2]-[3], sinusoid [7], pulse [3]-[5], step [6]) and the injection point (duty cycle [1]-[6], reference voltage [7]), different techniques can be used to derive the frequency response of the power plant (correlation analysis [1]-[2], Fourier analysis [7]) using parametric [4] and non-parametric [1]-[3] approaches.

Despite the capabilities of the proposed system identification techniques, the following drawbacks make the

use of embedded frequency response measurement methods problematic in PWM controllers: 1) perturbations are injected in the closed-loop system while in steady state, thus causing jitter on the PWM signal and unwanted noise on the output voltage; 2) the computational complexity of the system identification process is high; 3) the identification phase requires a long learning phase.

Although on-line system identification, i.e. performed continuously in the application, has disadvantages, system identification during the product development phase still remains a very powerful tool. Off-line system identification may be used to identify the converter transfer function by estimating the converter parameters so that the digital controller can be fine-tuned to accommodate deviations from the modelled converter dynamics to the measured ones. In such a scenario when system identification is embedded directly in the PWM controller, the computational resources used for this operation may be used only occasionally in the whole life cycle of the system. Also, the automatic system optimization routines associated with it would not be applicable to PWM controllers that do not embed hardware for system identification, thus limiting their usage. It is important, therefore, to investigate system identification methods that do not rely on any hardware support and that can be performed with instruments that are commonly used in the product development phase. In this context, this paper investigates the identification and estimation of the parameters of a converter by means of the open-loop frequency response measurements obtained from an analog frequency response analyzer (FRA). Although traditionally frequency response measurements are a non-parametric system identification technique, this paper demonstrates how to estimate the power plant model parameters from the data. Once converter parameters are estimated, an off-line automated procedure can be used to scale the original controller as proposed in [10] to obtain the desired closed-loop dynamics.

The proposed identification method accounts for nonlinear behaviors intrinsic to the digital control loop, does not require any embedded hardware in the digital PWM controller and therefore can be applied to any PWM controller. The use of a parametric identification technique makes the system identification procedure robust against measurement noise and guarantees the identification process is completed successfully upon an appropriate selection of the power stage model.

978-1-4673-9551-9/16 $31.00 © 2016 IEEE

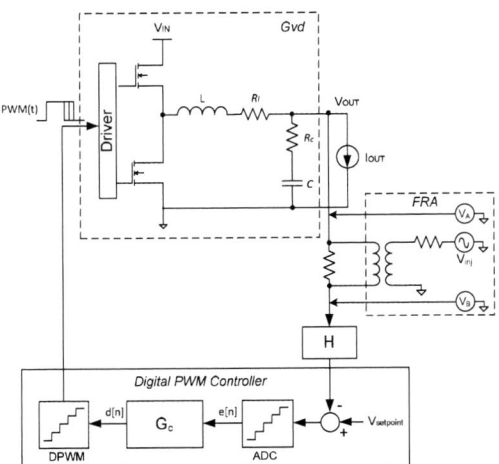

Figure 1 Block diagram of a digital POL converter including connection to the analog frequency response analyzer (FRA) for voltage injection open-loop measurements.

The proposed estimation method is also well suited to be integrated in a graphical user interface (GUI) with the intent of implementing an automatic off-line routine for fine-tuning a compensator based on [10].

The proposed method has been verified on a digitally controlled POL with the parameters V_{in}=12V, V_{out}=1.2V, I_{out}=10A and f_{sw}=400kHz for different bulk capacitor scenarios.

II. PARAMETRIC IDENTIFICATION OF CONVERTERS USING ANALOG FREQUENCY RESPONSE ANALYZER

Without loss of generality, let us consider a buck converter as in Figure 1 where the frequency response of the compensated open-loop transfer function $T(s)$ can be measured experimentally by a common analog FRA.

For ease of explanation, let us model the control loop in the continuous time domain neglecting the delay of the digital loop for the reason explained in the next section. The open-loop gain $T(s)$ can be expressed as

$$T(s) = G_{vd}(s) \cdot G_{ADC}(s) \cdot G_{dpwm}(s) \cdot G_c(s) \qquad (1)$$

where G_{vd} is the control-to-output transfer function of the converter to be identified; G_{ADC} and G_{dpwm} are the transfer functions of the ADC and DPWM respectively; and G_c is the controller transfer function in the s-domain.

$T(s)$ is measured by injecting a sinusoidal perturbation signal V_{inj} of frequency ω_i within a selected frequency range and measuring the ratio V_A/V_B (Figure 1). The converter control-to-output transfer function (G_{vd}) can be inferred from (1) by observing that the controller G_c is known and the ADC and DPWM behave as an approximately constant gain in the frequencies of interest of the control loop such that any variation is attributed to a variation in G_{vd}.

Let us denote the measured open-loop frequency response by $T_{meas}(j\omega_i)$ $i=\{1,...n\}$. Depending on the accuracy of the models used to predict converter dynamics $T_{meas}(j\omega_i)$ will differ from the expected $T(j\omega_i)$ in (1). The aim of the identification process is to search for an estimated converter transfer function $\hat{G}_{vd}(s)$ such that $T_{meas}(j\omega_i) = \hat{T}(j\omega_i)$ for $i=\{1,...n\}$, where $\hat{T}(j\omega_i)$ is defined as

$$\hat{T}(j\omega_i) = \hat{G}_{vd}(j\omega_i) \cdot G_{ADC}(j\omega_i) \cdot G_{dpwm}(j\omega_i) \cdot G_c(j\omega_i),$$
$$i=\{1,...n\}$$

Methods to perform parametric identification of transfer function are well suited for this purpose. In fact, taking advantage of the knowledge of the structure of the converter plant, parameters of the control-to-output transfer function can be estimated directly in the frequency domain by means of a common machine-learning algorithm.

III. ESTIMATION OF CONVERTER PARAMETERS

As we are fitting a parametric model, let us start with a converter model of the following form:

$$G_{vd}(s) = \frac{\left(\frac{s}{\omega_{esr}}+1\right)V_{in}}{\left(\frac{s^2}{\omega_0^2}+\frac{2D_0 s}{\omega_0}+1\right)} \qquad (2)$$

where V_{in} is the input voltage; ω_o is the LC resonant frequency; D_o is the damping factor; and ω_{esr} is the ESR zero of the output capacitor. The set of converter parameters to be identified is $P = \{\hat{V}_{in}, \hat{\omega}_o, \hat{D}_o, \hat{\omega}_{esr}\}$, where ^ indicates the estimated value of the specific parameter. The converter transfer function in (2) can then be tuned with the values of P:

$$G_{vd}(s) \rightarrow \hat{G}_{vd}(s) = \hat{G}_{vd}(s, P)$$

In this way, the identification process of the converter control-to-output transfer function reduces to a search for a set of parameters P that results in the best fit of the frequency response data $T_{meas}(j\omega_i)$ $i=\{1,...n\}$.

A logarithmic method as in [11] is used in this paper. This has the advantage that low-gain regions are not neglected in the search process, thus making the proposed approach well suited for frequency response data over a wide dynamic range. Additionally, for a minimum phase transfer function, as in the case of a buck converter, the logarithmic frequency response magnitude data is sufficient. This latter consideration simplifies the identification procedure because the use of the phase data may require a more complex multi-frequency model of the power plant to provide an accurate parameters estimation. When only magnitude data is used in the identification process, the cost function to be minimized by the search algorithm has the form:

$$J(P) = \frac{1}{2}\sum_{i=1}^{n_d}\left|\left|\ln T(j\omega_i, P)\right| - k_{dB}D(j\omega_i)\right|^2 \ln\left[\frac{\omega_{i+1}}{\omega_{i-1}}\right] \qquad (3)$$

The parameters of the set P are tuned using a stochastic gradient descendent method (SGD). Due to the different ranges of the elements of P, the generic parameters $p \in P$ can be normalized with respect to its range. The normalized parameter $p_n \in P$ is $p_n = p/(p_{max} - p_{min})$ where p_{max} and p_{min} are the maximum and minimum values defined for p. A variable step size based on *backtracking line search* [12] has been used to guarantee convergence of the SGD algorithm.

It is important to note that $J(P)$ in (3) is a *convex function*. By definition, $J(P)$ has then a minimum and this minimum is a *global minimum* for the minimization problem. Therefore when the model of the parametric power plant (2) is of the same order as the experimental plant to be identified, it is possible to find a set of parameters P_{opt} that fits the experimental plant and this set of parameters is *unique*.

IV. MEASUREMENT ERRORS IN THE MEASUREMENT OF THE OPEN LOOP FREQUENCY RESPONSE

The ADC and DPWM blocks employed in digital PWM controllers exhibit nonlinear behaviors for input signals with amplitude comparable to their quantization levels. Such nonlinear behavior affects the accuracy of the open-loop measurements taken with an analog FRA. The ADC and DPWM can be modelled as quantizers, where the effect of quantization results in the introduction of high-order harmonics visible on the spectrum of the regulated output voltage. The high-order harmonics within the controller bandwidth are fed back in the loop, causing a dependency between the amplitude of the injected signal and the measured gain. Figure 2 shows the block diagram of the open-loop frequency response measurement performed using the FRA. The signals of interest are the inputs of the quantizer blocks; i.e., the error signal e and the duty cycle d. As the resolution of the DPWM is commonly designed to be very high or at least higher than the ADC resolution to avoid limit cycles [13], the quantization of the ADC is the dominant nonlinear behavior observed in the frequency response measurements of the digital control loop.

When a sinusoid V_{inj} of a given frequency ω_i is injected, the error signal e seen at the input of the ADC is

$$E(j\omega_i) = -\frac{H(j\omega_i)}{1 + H(j\omega_i)K_{adc}G_c(j\omega_i)K_{dpwm}G_{vd}(j\omega_i)} V_{inj}(j\omega_i) \quad (4)$$

From (4) it can be seen that the error signal is a function of open-loop frequency response. Assuming the amplitude of the injected sinusoid V_{inj} to be constant over the whole range of injected frequency ω_i the amplitude of the error signal becomes very small at frequencies where the open-loop gain is very high.

The quantization of error signal introduced by the ADC results in the generation of high-order harmonics that are visible on the output voltage spectrum (Figure 3) and that can be seen as

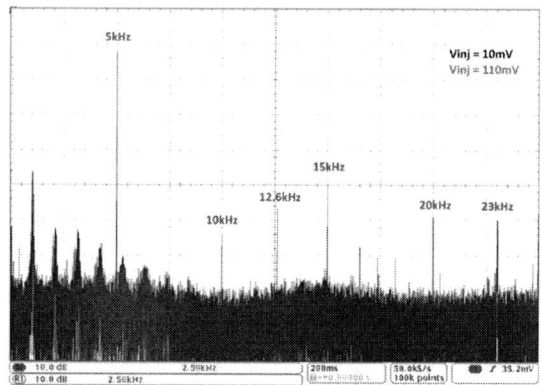

Figure 3 Spectrum of the output voltage for a 5kHz injected sinusoid with amplitude 10mV (black trace) and 110mV (red trace).

an h_{oh} term added to the output of the ADC for modelling purposes. Especially in the case where a low frequency sinusoid is injected, high-order harmonics within the controller bandwidth are fed back into the loop. This results in a measured gain lower than expected and a distortion of the phase curve. A similar behavior, together with a modelling procedure, has been presented in [14].

Figure 4 shows the measured open-loop frequency response of a digitally controlled buck converter for different amplitudes V_{inj} of the injected signal. As V_{inj} increases, the error signal e reaches more bins of the ADC thus resulting in higher gain. It is important to notice that nonlinear effects such as saturations in the control loop signal path result in a reduction in the gain of the open-loop frequency response *only*, thus not affecting the phase behavior. This effect can be seen in Figure 4 in the frequency region above the control bandwidth for injected amplitude V_{inj} higher than 130mV. Saturation effects on the open-loop frequency response of the converter can be modelled using first-order sinusoidal input-describing functions as in [13]. First-order sinusoidal input-describing functions analyze the gain of a nonlinear system at the frequency ω_i of the injected sinusoid and neglect the effect that higher order harmonics have on it. Therefore, such an approach is not suitable to analyze the gain/phase behavior shown in Figure 4. A higher-order sinusoidal input describing function as proposed in [14] should be used instead.

The nonlinear behavior illustrated in Figure 4 affects the accuracy of the estimated converter parameters because minimization of the cost function (3) would be achieved by

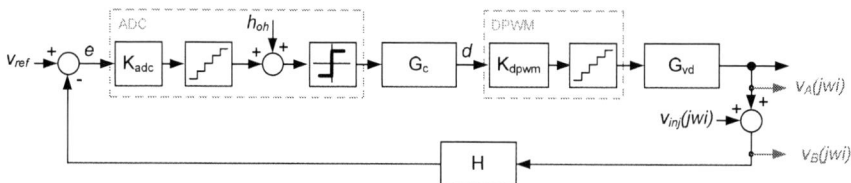

Figure 2: Block diagram of the digital control loop including the ADC and DPWM as quantizers. The injected signal V_{inj} of the FRA can be seen as a disturbance injected into the loop. Such disturbance propagates through the loop and by measuring the ratio V_A/V_B the gain and phase of the open loop frequency response at ω_i is obtained.

978-1-4673-9551-9/16 $31.00 © 2016 IEEE

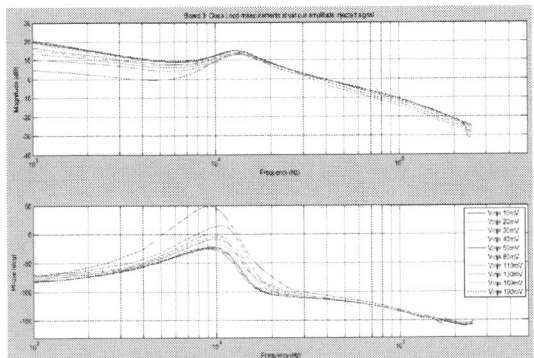

Figure 4 Comparison of open loop frequency measurements for different amplitudes of the injected sinusoidal signal

uniformly distributing the error terms $\big|\|\ln T(j\omega_i, P)\| - k_{dB}D(j\omega_i)\big|^2$ over the whole range of frequencies used in the measurement.

In order to compensate for the measurement errors described above, controller asymptotes can be used. Controller transfer function G_c can be decomposed in its proportional, integral and derivative asymptotes, which describe controller behavior in low, middle and high frequency regions of the open-loop frequency response respectively.

$$G_c(s) = K_p + \frac{K_i}{s} + \frac{K_d s}{\left(\frac{s}{\omega_d} + 1\right)}$$

Starting from the designed values K_p, K_i, K_d, and ω_d, the set of estimated parameters P of the SGD algorithm can be extended to also include controller parameters, resulting in the new parameter set $P = \{\hat{V}_{in}, \hat{\omega}_o, \hat{D}_o, \hat{\omega}_{esr}, \hat{K}_p, \hat{K}_i, \hat{K}_d, \hat{\omega}_d\}$. As the parameters of the converter power stage will be searched to match the measured open-loop frequency response, controller parameters will be tweaked to compensate for gain errors occurring with different magnitudes at different frequency ranges.

V. EXPERIMENTAL RESULTS

The proposed method to estimate converter parameters using frequency response measurements obtained from an analog FRA has been applied to a digital POL converter using two different output capacitor scenarios. Converter parameters are V_{in}=12V, V_{out}=1.2V, I_{out}=10A, f_{sw}=400kHz, L=0.33µH, R_l=0.2mΩ. Setup 1 uses 8x100µF ceramic capacitors with ESR=3.12µΩ while setup 2 uses 8x100µF ceramic capacitors with ESR=3.12µΩ and 2x470µF tantalum capacitors with ESR=8mΩ. The digital PWM controller has an ADC resolution of 1.5mV and a DPWM resolution of 163ps. Open-loop frequency measurements have been taken for both setups using the N4L frequency response analyzer.

Table 1 shows a comparison of the estimated converter parameters using the full set of parameters ($P = \{\hat{V}_{in}, \hat{\omega}_o, \hat{D}_o, \hat{\omega}_{esr}, \hat{K}_p, \hat{K}_i, \hat{K}_d, \hat{\omega}_d\}$) and the limited set of parameters ($P = \{\hat{V}_{in}, \hat{\omega}_o, \hat{D}_o, \hat{\omega}_{esr}\}$) for the estimation. When controller asymptotes are not used to compensate for gain errors, then the DC gain of the power stage is lower than the expected DC gain (nominal value). This gain error affects the accuracy of the estimation algorithm, which in the process of minimizing the cost function (3), pushes the location of the ESR zero to a very high frequency. However, when the full set of parameters is used in the estimation algorithm, the controller asymptotes tend to compensate for gain errors occurring over frequency thus leading to more accurate results. Table 2 shows the resulting correction factor applied to the original controller parameters K_p, K_i, K_d, and ω_d to allow correct parameter estimation of the converter power stage. To compensate the low frequency gain/phase distortion, the gain of the integral asymptote K_i is reduced and the proportional asymptote K_p is adjusted such that the zero formed with the integral asymptote is moved to lower frequencies to fit the phase boost seen in the frequency region of the converter's natural frequency ω_o. According to the amplitude of the injected signal, K_d and ω_d may be adjusted to compensate the gain attenuation resulting from saturation of the ADC input window, as explained in section IV.

Figure 5 and Figure 6 show the open-loop frequency responses obtained for the estimated set of parameters $P = \{\hat{V}_{in}, \hat{\omega}_o, \hat{D}_o, \hat{\omega}_{esr}\}$ with their corresponding frequency response measurements. Figure 7 and Figure 8 show the same results obtained for the estimated set of parameters $P = \{\hat{V}_{in}, \hat{\omega}_o, \hat{D}_o, \hat{\omega}_{esr}, \hat{K}_p, \hat{K}_i, \hat{K}_d, \hat{\omega}_d\}$. The gain plots in Figure 7 and Figure 8 match the experimental results quite well over the whole range of measured frequencies.

	Board 1			Board 2		
	Nominal	Limited set P	Full set P	Nominal	Limited set P	Full set P
G_{vdo}	12	6.70	11.44	12	6.99	11.25
ω_o [krad/s]	65.66	97.85	84.40	41.75	54.92	53.47
D_o	0.26	0.22	0.25	0.08	0.48	0.38
ω_{esr} [krad/s]	$3.52*10^3$	$26.05*10^3$	$3.92*10^3$	265.96	481.38	157.08

Table 1: Comparison between estimated power stage parameters obtained using the full and limited set of parameters P for the estimation. G_{vdo} is the DC gain of the power stage; ω_o is its natural frequency; and Do is its damping factor. ω_{esr} is the location of the ESR zero of the output capacitance.

	Board 1	Board 2
$\hat{K}_p/K_p$	0.567	0.4275
$\hat{K}_i/K_i$	0.519	0.7181
$\hat{K}_d/K_d$	0.962	0.4911
$\hat{\omega}_d/\omega_d$	0.383	0.2873

Table 2: Correction factors applied to original controller parameters to account for measurement errors at different frequency ranges.

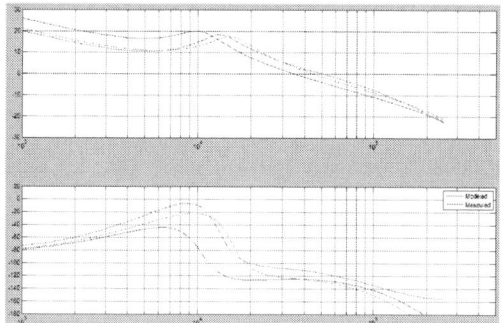

Figure 5: Comparison of the open-loop frequency response obtained for the estimated converter parameters for setup 1 using the limited parameter set $P = \{\widehat{V}_{in}, \widehat{\omega}_o, \widehat{D}_o, \widehat{\omega}_{esr}\}$. *Blue:* nominal curve. *Red:* measured curve. *Green:* estimated curve.

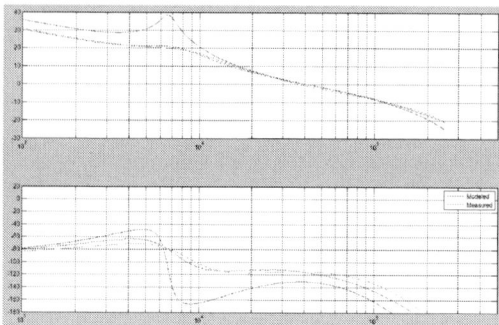

Figure 6: Comparison of the open-loop frequency response obtained for the estimated converter parameters for setup 2 using the limited parameter set $P = \{\widehat{V}_{in}, \widehat{\omega}_o, \widehat{D}_o, \widehat{\omega}_{esr}\}$. *Blue:* nominal curve. *Red:* measured curve. *Green:* estimated curve.

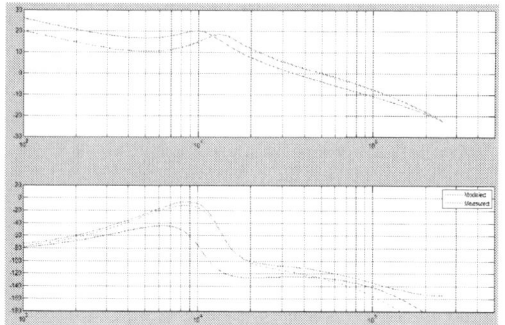

Figure 7 Comparison of the open-loop frequency response obtained for the estimated converter parameters for setup 1 using the full parameter set $P = \{\widehat{V}_{in}, \widehat{\omega}_o, \widehat{D}_o, \widehat{\omega}_{esr}, \widehat{R}_p, \widehat{R}_l, \widehat{R}_d, \widehat{\omega}_d\}$. *Blue:* nominal curve. *Red:* measured curve. *Green:* estimated curve.

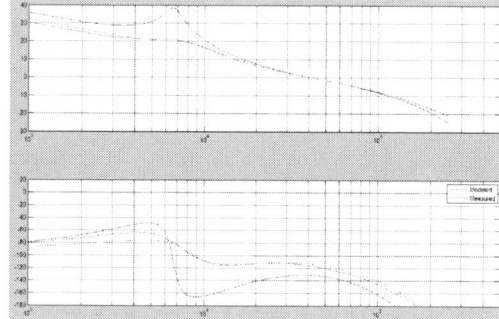

Figure 8: Comparison of the open-loop frequency response obtained for the estimated converter parameters for setup 2 using the full parameter set $P = \{\widehat{V}_{in}, \widehat{\omega}_o, \widehat{D}_o, \widehat{\omega}_{esr}, \widehat{R}_p, \widehat{R}_l, \widehat{R}_d, \widehat{\omega}_d\}$. *Blue:* nominal curve. *Red:* measured curve. *Green:* estimated curve.

As expected, open-loop gain data are sufficient to correctly match both gain and phase measurements with the open-loop data obtained using the estimated power stage. Additional phase delay at high frequencies is due to (i) inaccurate modelling of controller loop delay and (ii) the use of a small signal average model neglecting the effect of the PWM modulator in the open-loop measurements as described in [15]. Measurement of the control-to-output transfer function has been performed to validate the estimated converter parameters. In order to run the measurement using the same FRA without opening the control loop, the following procedure has been used. Once the converter is operating in steady state, controller parameter K_i has been set to 0. The converter is still stable because the phase lag of -90° associated with the integral asymptote is removed and the zero originated by the proportional and derivative asymptotes still boosts the phase of the converter at the frequency ω_o. Afterwards ω_d has been moved to very high frequencies and K_d is set to 0. Due to the high output current, the converter damping factor D_o is such that, together with the ESR of the output capacitors, it prevents the phase reaching -180°.

Eventually K_p has been set to its minimum digital value, thus achieving a compensator with a unity gain.

Measurement results are shown in Figure 9. It can be seen that the identified converter frequency response matches the measured data quite closely up to the converter's natural frequency ω_o.

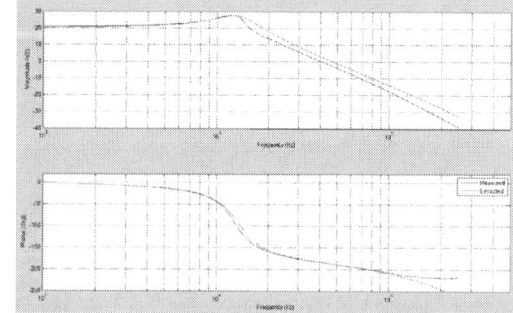

Figure 9 Comparison between measured (blue) and identified (magenta) converter control-to-output frequency response for setup 1.

For frequencies above ω_b, the injected sinusoid saturates the ADC output thus resulting in an attenuation of the gain curve. As saturations do not affect the phase behavior, it can be seen that the phase of the identified converter power stage follows the measured phase data quite closely over the whole range of measured frequencies, thus verifying the correctness of the identified converter parameters.

VI. CONCLUSIONS

A method to perform parametric system identification of digitally controlled power converters using a conventional analog frequency response analyzer has been presented. The effectiveness of the proposed method has been verified for a digital point-of-load converter for two different output capacitance scenarios. Controller asymptotes have been used to account for gain errors due to the intrinsic nonlinear behavior of the digital control loop, thus making the proposed method robust against measurement errors.

The proposed method can be applied to any digital PWM controller since it does not require any specific hardware to be embedded in the controller. Estimated converter parameters can be used to fine-tune the digital controller as in [10] to obtain the desired closed-loop performance. The proposed SGD algorithm can be easily integrated in a GUI with the controller fine-tuning procedure, thus resulting in a fully automated setup for system performance optimization of digital PWM controllers.

REFERENCES

[1] B. Miao, R. Zane, and D. Maksimovic, "System identification of power converters with digital control through cross-correlation methods," *IEEE Trans. Power Electron.*, vol. 20, no. 5, pp. 1093–1099, Sep. 2005.

[2] M. Shirazi, J. Morroni, A. Dolgov, R. Zane, and D. Maksimovic, "Integration of frequency response measurement capabilities in digital controllers for DC–DC converters," *IEEE Trans. Power Electron.*, vol. 23, no. 5, pp. 2524–2535, Sep. 2008.

[3] M. Shirazi, R. Zane, and D. Maksimovic, "An autotuning digital controller for DC–DC power converters based on online frequency-response measurement," *IEEE Trans. Power Electron.*, vol. 24, no. 1, pp. 2578–2588, Nov. 2009.

[4] L. Corradini, P. Mattavelli, W. Stefanutti, and S. Saggini, "Simplified model reference-based autotuning for digitally controlled SMPS," *IEEE Trans. Power Electron.*, vol. 23, no. 4, pp. 1956–1963, Nov. 2008.

[5] A. Costabeber, P. Mattavelli, S. Saggini, and A. Bianco, "Digital autotuning of DC-DC converters based on model reference impulse response," in *Proc. 25th IEEE Appl. Power Electron. Conf. Expo (APEC'10)*, pp. 1287–1294, 2010.

[6] M.M. Peretz and S. Ben-Yaakov, "Time domain identification of pulse width modulated converters," *IEEE Power Electron. IET*, vol. 5, no. 2, pp. 166–172, Jan. 2012.

[7] N. Kong, A. Davoudi, M. Hagen, E. Oettinger, M. Xu, D. S. Ha, and F. C. Lee, "Automated system identification of digitally controlled multi-phase DC-DC converters," in *Proc. 24th IEEE Appl. Power Electron. Conf. Expo (APEC'09)*, pp. 259–263, 2009.

[8] B. Miao, R. Zane, and D. Maksimovic, "Automated digital controller design for switching converters," *IEEE 36th Annu. Power Electron. Spec. Conf (PESC'05)*, pp. 2729–2735, 2005.

[9] A. Davoudi, N. Kong, M. Hagen, M. Muegel, and P.L. Chapman, "A general framework for automated tuning of digital controllers in multiphase dc-dc converters," in *Proc. 24th IEEE Appl. Power Electron. Conf. Expo (APEC'09)*, pp. 626–630, 2009.

[10] M. Meola, A. Cinti, and A. Kelly, "Controller scalability methods for digital point of load converters," in *Proc. 30th IEEE Appl. Power Electron. Conf. Expo (APEC'15)*, pp. 430–436, 2015.

[11] M.D. Sidman, F.E. Deangelis, and G.C. Verghese, "Parametric system identification on logarithmic frequency response data," *IEEE Trans. Automatic Control*, vol. 36, no. 9, pp. 1065–1070, 1991.

[12] L. Armijo, "Minimization of functions having Lipschitz continuous first partial derivatives," *Pacific J. Math*, vol. 16, no. 1, pp. 1–3, 1966.

[13] H. Peng, A. Prodić, E. Alarcon, and D. Maksimović, "Modeling of quantization effects in digitally controlled DC-DC converters," *IEEE Trans. Power Electron.*, vol. 22, no. 1, pp. 208–215, 2007.

[14] P. Nuij, M. Steinbuch, and O. Bosgra, "Measuring the higher order sinusoidal input describing functions of a non-linear plant operating in feedback," *Control Engineering Practice*, vol. 16, no. 1, pp. 101–113, 2008.

[15] Y. Qiu, M. Xu, K. Yao, J. Sun, and F.C. Lee, "Multifrequency small-signal model for buck and multiphase buck converters," *IEEE Trans. Power Electron.*, vol. 21, no. 5, pp. 1185–1192, 2006.

Extended Wide-Load Range Model for Multi-Level Dc-Dc Converters and a Practical Dual-Mode Digital Controller

Nenad Vukadinović, Aleksandar Prodić
Laboratory for Power Management and Integrated SMPS
ECE Department, University of Toronto,
Toronto, ON, Canada
{vukadinovic, prodic}
@ece.utoronto.ca

Brett A. Miwa, Cory B. Arnold, Michael W. Baker
Maxim Integrated
San Jose, CA, USA
{brett.miwa, cory.arnold,
michael.baker}
@maximintegrated.com

Abstract— **This paper addresses limitations of previous models of multi-level flying capacitor (ML-FC) dc-dc converters, in terms of not being able to predict instability in the regulation of the flying capacitor voltage under non-negligible inductor current ripple conditions, usually existing under light to medium load operating conditions. Assuming small-ripple approximation, linear ac equivalent circuit is derived for a general N-level FC dc-dc buck converter and, through a geometrical analysis, limitations of that model are explained. Also, related stability problems are addressed. Then, an extended mathematical model that takes into account the ripple component of the inductor current is derived and used in the design of a practical dual-mode digital controller, which for non-negligible ripple operating conditions changes its mode of operation. Validity of the model and the functionality of the introduced controller are verified both through simulations and experimental verifications. Performance of this controller is tested with a wide input 15 V/3 A, 250 kHz, 45 W, three-level experimental prototype.**

Keywords: Low-power dc-dc converter, flying capacitor, small-signal modelling, dual-mode digital controller.

I. INTRODUCTION

Multi-level flying capacitor (ML-FC) converters, introduced by Meynard [1]-[3], whose 3-level buck version is shown in Fig.1, are potentially a very attractive alternative to the conventional buck converter in space constrained low power applications [4],[5]. This is because in those applications cost-effective implementation and high power processing efficiency are of key importance. Examples include numerous dc-dc converters used in virtually all electronic devices today, processing power from a fraction of watt to tens of watts. In a ML-FC converter both the voltage stress across the power switches and the swing of the switching node voltage (v_{SW}), labeled in Fig. 1, are reduced by N-1 times, compared to the conventional two-level implementation. Here, N defines "the converter level", i.e. the number of different voltages that can be obtained at the switching node. These two advantages result in a reduction of semiconductor related losses and allow for a significant reduction of the inductance value [4],[5]. The advantages of multi-level converters over the conventional two-level solutions have already been proven and utilized in

numerous high-power applications [3]. However, in the targeted low-power dc-dc applications conventional two-level converters are almost exclusively used. Among the main reasons why the adoption of the ML-FC converters has been slowed down, are more complex dynamic behavior, which in some cases is not fully understood, and lack of hardware-efficient control solutions for them.

The control of ML-FC converter requires simultaneous regulation of both the output and flying capacitor voltages, which in some cases is a challenging task [6]. For an ideal converter, the control of flying capacitor voltages is not required [7],[8], due to its natural balancing. However, in practice mismatches in gate-drive signals or, on-resistances of the switches, or in resistances of the charging and discharging flying capacitor paths, will cause imbalance of the flying capacitor voltages. Usually, a properly designed output voltage controller is able to regulate the output voltage, even under unbalanced

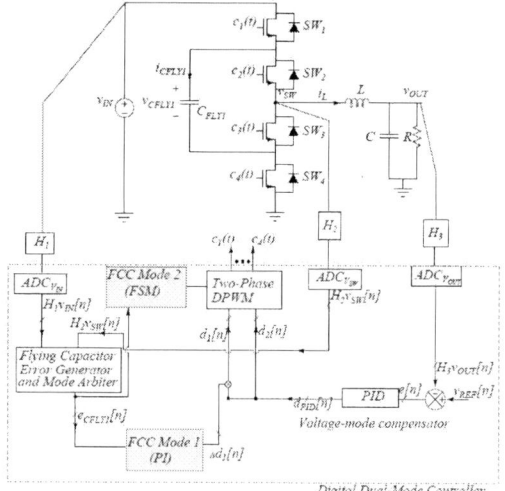

Fig. 1. Three-level buck converter with digital dual-mode flying capacitor balancing controller;

This work of Laboratory for Power Management and Integrated SMPS is supported by Maxim Integrated.

conditions of the flying capacitor voltages, but several other implications occur in the circuit. Those include increased inductor current and output voltage ripples, presence of low-frequency harmonics, and larger voltage stress of the switches. All of these can lead to the increased power losses of the converter, reliability problems, and possible unwanted circuit behavior. Therefore, understanding the dynamics of the flying capacitor voltage behavior and its control are of a great importance for designing the multi-level flying capacitor converters.

The main goal of this paper is to show limitations of small-signal average modeling of multi-level buck converters. In particular, it addresses the problems related to the regulation of the flying capacitor voltages for the cases when the inductor current ripple is not negligible. Also, the paper shows a practical digital dual-mode controller that deals with the previously mentioned limitations providing simultaneous regulation of the output and flying capacitors voltages.

Paper is organized as follows. In the following section, conventional a small-signal modeling of a general N-level FC converter is presented. Also, the limitations of this approach in the presence of a non-negligible current ripple are presented and the source of the flying capacitor voltage instability is described. Section III presents a practical dual-mode digital controller that provides control of flying capacitor voltage over full range of the loads. In Section IV experimental results obtained with the introduced controller regulating a discrete three-level prototype are presented. The results verify the aforementioned instability of the flying capacitor voltage and demonstrate the effectiveness of the introduced balancing method.

II. CIRCUIT MODELING AND LIMITATIONS OF THE CONVENTIONAL MODELING APPROACH

The balancing of the flying capacitor and modeling of the multi-level converters have been addressed in [6]-[8]. These models were developed under an assumption that the inductor current has a small-ripple which can be neglected. The models show that when the charging and discharging times of the flying capacitors are equal the capacitor balance is automatically maintained. In this section, it will be shown that for the cases when the inductor current is not negligible stability problems

related to regulation of the flying capacitor can occur and that the conventional compensators designed based on the small-ripple approximation falls apart in those cases.

Analysis given in [6], based on assumption that the inductors behave as currents sources and resulting in ideal square-wave waveforms, shows that a natural balancing of the flying capacitor voltages is achieved as long as the same duty ratio is applied to all upper switches of the converter, for example to switches SW_1 to SW_2 for the converter of Fig. 1. The same work also gives a model explaining dynamics of the flying capacitor voltage under the assumed conditions. For the targeted low power applications, where limited computational power is available and fairly simple PI and/or PID compensators are used, direct utilization of the developed relatively complex model is quite challenging. A simple flying capacitor voltage controller for low-power 3-level buck converters has been presented in [9],[10]. That controller utilizes the assumption that whenever the upper switches are turned on (SW_1) for a longer time than the switches (SW_2) in the lower branch of a converter the flying capacitor (C_{FLY1}) is charged and vice versa. In this section it will be shown that the above presented assumptions are only valid when the inductor behaves as a constant current source and that, in the cases, when the inductor current ripple is taken into account, stability problems can occur.

To demonstrate this problem a small-ripple approximation (SRA) based average model of the N-level converter is derived and its deficiencies explained through a geometrical analysis. Then, simulation results confirming stability problems of previous control methods and standard modeling approaches are described. Finally, an extended average model that takes into account the inductor current ripple effect and, therefore, gives foundation for avoidance of the previously mentioned stability problems, is derived.

A. Conventional Small-Signal Averaged ac Model

Using the well-known tools [11], small-signal equivalent circuit of N-level buck converter for continuous conduction mode is obtained as shown in Fig. 2. From the equivalent circuit corresponding equation is given:

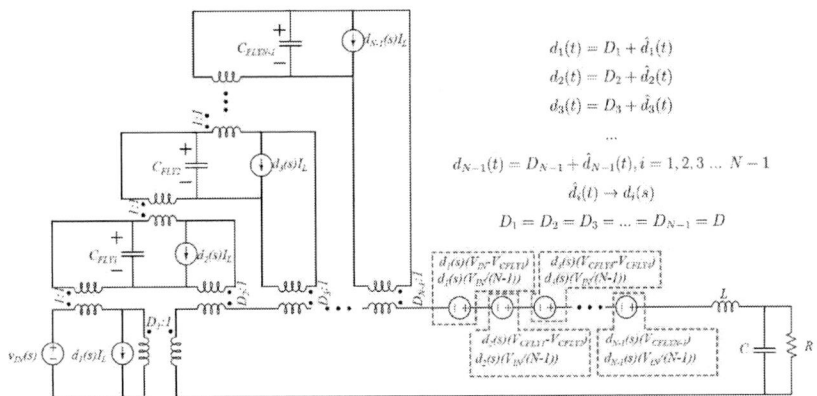

$$d_1(t) = D_1 + \hat{d}_1(t)$$
$$d_2(t) = D_2 + \hat{d}_2(t)$$
$$d_3(t) = D_3 + \hat{d}_3(t)$$
$$...$$
$$d_{N-1}(t) = D_{N-1} + \hat{d}_{N-1}(t), i = 1, 2, 3 ... N - 1$$
$$\hat{d}_i(t) \to d_i(s)$$
$$D_1 = D_2 = D_3 = ... = D_{N-1} = D$$

Fig. 2. Averaged small-signal ac model for ML-FC buck converter;

978-1-4673-9551-9/16 $31.00 © 2016 IEEE   1598

$$\begin{bmatrix} i_{C_{FLY1}} \\ i_{C_{FLY2}} \\ \vdots \\ i_{C_{FLYN-1}} \end{bmatrix} = \begin{bmatrix} +I_L \\ 0 \\ \vdots \\ 0 \end{bmatrix} \cdot d_1(s) + \begin{bmatrix} -I_L \\ +I_L \\ \vdots \\ 0 \end{bmatrix} \cdot d_2(s) + \cdots + \begin{bmatrix} 0 \\ 0 \\ \vdots \\ -I_L \end{bmatrix} \cdot d_{N-1}(s), \quad (1)$$

where $i_{C_{FLY,i}}(s)$, presents the current of flying capacitor $C_{FLY,i}$, $d_i(s)$ small-signal components of duty ratios corresponding to the switches SW_i, for $i = 1,2,3,...N-1$ and I_L dc value of output inductor current. It can be seen that the SRA based model predicts that, for the case when all duty ratios are equal, the flying capacitors are balanced, i.e. the net current going into flying capacitors is zero. This prediction coincides with the previously presented modeling approaches [6]-[10]. Equation (1) can also be related to the simple control approach used in [8], where a 3-level converter is controlled through a difference between $d_1(t)$ and $d_2(t)$ values. The control is performed such that by providing $d_1(t)$ larger than $d_2(t)$ the flying capacitor is charged and for the opposite case it is discharged. However, the assumption of charging and discharging capacitors falls apart when a non-negligible current ripple is present. To demonstrate this problem, simulations of a 3-level converter for $d_1(t) > d_2(t)$ for the case when an inductor is used and when the inductor is replaced with an ideal dc current source, like in previous approaches, are shown in Fig. 3.

The results show that, although the average model predicts charging of the flying capacitor, in the presence of the inductor current ripple, the capacitor actually gets discharged. This deficiency of the model can cause serious stability problems,

since the ripple effectively introduces a positive feedback into the system. The following subsection describes a physical origin of this phenomenon and gives an extended mathematical model that incorporates this effect, providing foundation for stable controller design.

B. The Effect of Current Ripple and Extended Mathematical Model

Here, on the example of a three-level buck converter the effect of the inductor current ripple on the flying capacitor voltage causing potential instability of the system is descried. To explain the problem, geometrical analysis based on the capacitor charge calculation is utilized. It will be shown that although, this effect is especially emphasized at low output currents, when the inductor current ripple is comparable to the average output current, the problems can also occur under medium load conditions, where the converter is relatively far away from the discontinuous conduction mode (DCM). Fig. 4 shows the key waveforms of a three-level buck converter for a case when $v_{CFLY} \neq v_{IN}/2$, causing different rising slopes of the inductor current during the first and the second half of a switching cycle. To find the expression for the charging and discharging currents of the capacitor in this case, realistic current waveforms are taken into account and the capacitor charging and discharging currents are averaged over a half of the switching period T_S, as shown in the following equations:

$$\langle i_{CFLY1}(t)\rangle_{(T_S/2)I} = \langle i_L(t)\rangle_{(T_S/2)I} \cdot d_2(t)$$
$$+ \frac{\langle v_{CFLY1}(t)\rangle_{(T_S/2)II} d_1(t) d_2(t)\big(d_2(t) - d_1(t)\big)}{2f_S L}, \quad (2)$$

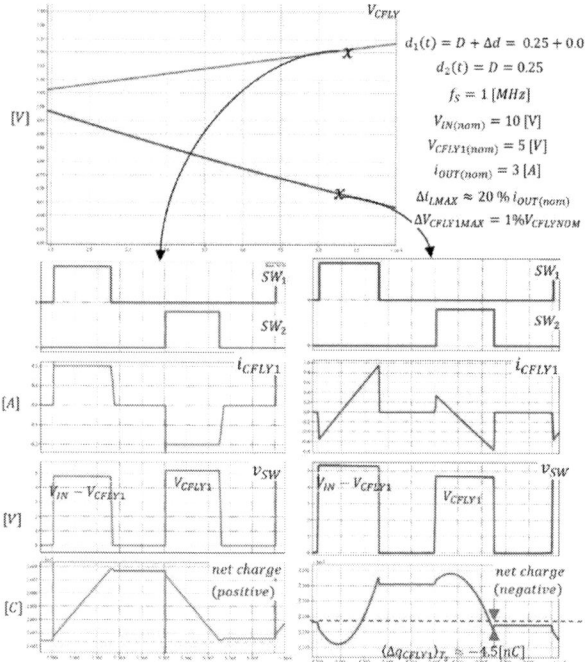

$$d_1(t) = D + \Delta d = 0.25 + 0.01$$
$$d_2(t) = D = 0.25$$
$$f_S = 1 \, [MHz]$$
$$V_{IN(nom)} = 10 \, [V]$$
$$V_{CFLY1(nom)} = 5 \, [V]$$
$$i_{OUT(nom)} = 3 \, [A]$$
$$\Delta i_{LMAX} \approx 20\% \, i_{OUT(nom)}$$
$$\Delta V_{CFLY1MAX} = 1\% V_{CFLYNOM}$$

Fig. 3. Simulation waveforms for different duty cycles, $d_1(t) > d_2(t)$ under light load operating condition $i_{OUT} = 7\% \, i_{OUT(max)}$. Top Waveform – Flying capacitor voltages: Blue line shows a case when ideal current source is used instead of the output inductor (ideal SRA); Red line shows a realistic case when inductor is used; Bottom figures show switching waveforms over one cycle;

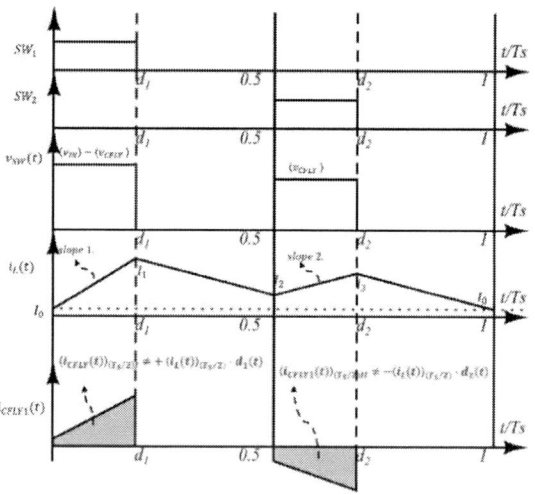

Fig. 4. Key switching waveforms of the three-level buck converter for the case when $d_1 \neq d_2$ and $V_{CFLY1} \neq V_{IN}/2$

$$\langle i_{CFLY1}(t)\rangle_{(T_S/2)II} = \; -\langle i_L(t)\rangle_{(T_S/2)II}\cdot d_2(t)$$
$$-\frac{\langle v_{CFLY1}(t)\rangle_{(\frac{T_S}{2})II}d_1(t)d_2(t)\big(d_2(t)-d_1(t)\big)}{2f_S L}$$
$$+\frac{\langle v_{IN}(t)\rangle_{\frac{T_S}{2}}d_1(t)d_2(t)\big(d_2(t)-d_1(t)\big)}{2f_S L}, \qquad (3)$$

$$\langle i_{CFLY1}(t)\rangle_{T_S} = \frac{\Delta\langle q\rangle_{T_S}}{T_S} = \langle i_{CFLY1}(t)\rangle_{(T_S/2)I} + \langle i_{CFLY1}(t)\rangle_{(T_S/2)II}, \qquad (4)$$

where, $\langle i_{CFLY1}(t)\rangle_{(T_S/2)I}$ stands for the average charging current in the first half of the switching cycle and $\langle i_{CFLY1}(t)\rangle_{(T_S/2)II}$ for average discharging current in the second half of the switching cycle, while $\Delta\langle q\rangle_{T_S}$ represents the overall net charge over the full switching cycle. It should be noted that the average inductor currents $\langle i_L(t)\rangle_{(T_S/2)I}$ and $\langle i_L(t)\rangle_{(T_S/2)II}$ are not the same. To demonstrate that, in Fig. 5 the flying capacitor charge during one switching cycle is plotted as a function of the difference in duty cycles for various load conditions for the example shown in Fig. 3. In this example inductor current ripple has 20% of the nominal current value as shown in Fig. 3. Also, converter operates in forced continuous conduction mode for load currents lower than the inductor current ripple. The results show that, for some operating conditions, discharging of the flying capacitor occurs even when $d_1(t) > d_2(t)$. This effect is also confirmed with zoomed-in waveforms of Fig. 3, which shows that the total net charge over a switching cycle is negative, even though applied difference in the duty ratios is positive.

III. Dual-Mode Controller

The previous analysis shows that potential stability problems exist in some conventionally used controllers for multi-level converters. The analysis also indicates that the finding boundary between stable and unstable region of operation is a complex task requiring extensive calculations. This task becomes even more complex as the number of levels in a converter increases. To solve this issue, here a simple control strategy is introduced, in a form of a dual-mode digital controller, and it is described on an example of a three-level buck converter. This controller

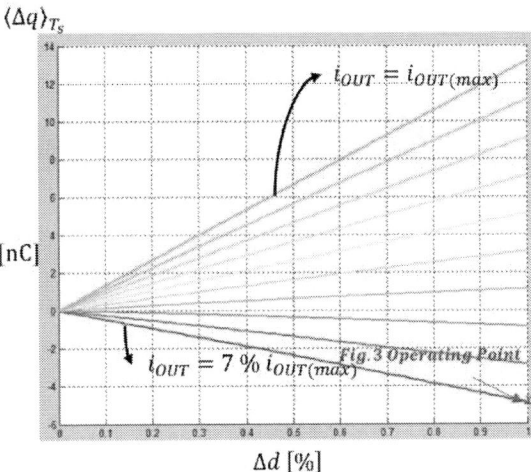

Fig. 5. Analytical dependence of average net charge (Eq. (4)) on $\Delta d = d_1(t) - d_2(t)$, for different load currents where $d_1(t) = 0.25 + \Delta D$, $d_2(t) = 0.25$.

mitigates aforementioned problem of the sign changing of the flying capacitor average net charge and it can be extended to the converters with a larger number of levels. The controller has two modes of operation. In the first mode, for the cases when the inductor current ripple has a smaller effect on net charge and cannot cause instability, it operates as a conventional compensator developed around the small-signal model derived in Section II. For the cases when the ripple cannot be neglected it operates in a by-pass mode that does not rely on the difference between the two duty ratio values. To eliminate the need for complex calculation of the stability boundary, the switching between the modes is performed based on the detection of a possible instability.

The operation of the controller can be explained with the help of Fig. 1, where the complete system is shown, and Fig. 6 showing detailed block diagram of the controller. In the first mode, *Flying Capacitor Control (FCC) Mode* 1, which is used when the ripple can be neglected, adjustment of $\Delta d_1[n]$

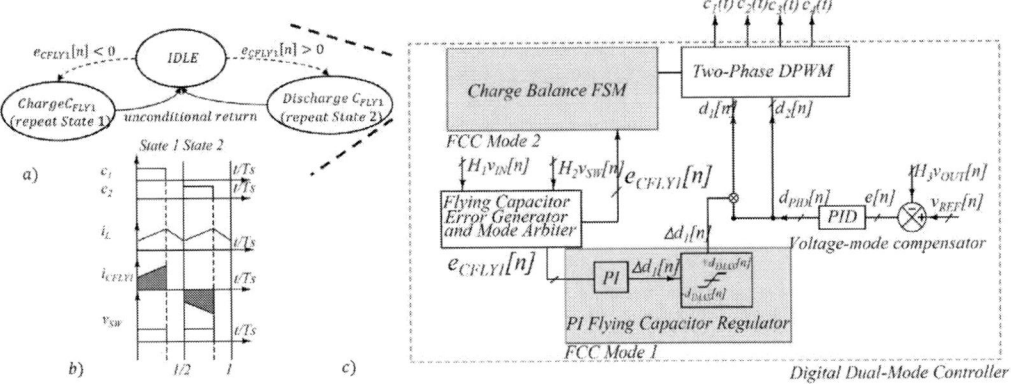

Fig. 6. a) Finite State Machine of the *FCC Mode* 2; b) Basic switching waveforms of the three-level converter; c) Block diagram of the digital dual-mode controller;

978-1-4673-9551-9/16 $31.00 © 2016 IEEE

associated with duty cycle $d_1[n]$, is obtained from PI regulator block. In this mode, a $d_1[n]$ greater than $d_2[n]$ results in an increase of the flying capacitor net charge. From Fig. 6, it can be noticed that, in this mode, only one of the duty ratios, $d_1[n] = d_{PID}[n] + \Delta d_1[n]$, is being controlled by the flying capacitor voltage loop while the other one, $d_2[n] = d_{PID}[n]$ is directly provided by the PID. This allows for an easy extension of the control method for converters with a larger number of levels and flying capacitors. Also, to avoid interference of two loops and ensure tight regulation of the output voltage, the flying capacitor inner loop is made slower than the output voltage loop. In a case of a sudden load transient the new duty ratio for the output voltage loop, $d_{PID}[n]$, is first calculated and, then, relative duty ratio, $\Delta d_1[n]$, is provided by the flying capacitor loop.

The second mode, named *Flying Capacitor Control (FCC) Mode 2*, is utilized when the ripple is not negligible. The controller turns on *FCC Mode 2* when instability of *FCC Mode 1* is detected. The instability is easily detected, since the flying capacitor loop is slow. Operation in *FCC Mode 2* is similar to that of the controller introduced in [12], where control action, instead of relying on fine adjustments of duty cycle Δd_1 as in *FCC Mode 1*, is based on repetition of different circuit states.

The operation of the finite-state machine (FSM) which controls the flying capacitor voltage in this mode and the corresponding switching states are shown in Fig. 6. In the *IDLE* state (Fig. 6 a), the converter operates with regular switching pattern, State 1 -> State2, and in this FSM state, measurement of the flying capacitor voltage is done. Based on the measurement the next FSM state is defined. For instance, when a lower voltage of the flying capacitor C_{FLY1} (Fig. 1) is detected, the FSM enters the state *Charge C_{FLY1}* (dashed line Fig. 6 a)) in which only State 1 of the converter is repeated Fig. 6 b). After that FSM unconditionally returns to *IDLE* (solid line Fig. 6 a)), evaluation of the next stage starts again.

IV. EXPERIMENTAL RESULTS

In order to demonstrate previously mentioned stability problems in the flying capacitor regulation and verify performance of the aforementioned controller, a low-power three-level buck converter, $V_{IN} = 15$ V, 10 mA $< I_{OUT} < 3$ A, $f_S = 250$ kHz (resulting in the inductor current ripple $f_{ripple} = 500$ kHz), has been built, based on the diagrams of Figs. 1 and 6. The digital controller is implemented with an FPGA based system. Nominal voltage of the flying capacitor under light load should be $V_{IN}/2 = 7.5$ V, while for higher output currents this voltage is lower, due to non-zero on-resistance of the switches and reduced input voltage due to source resistance.

Fig. 7 shows waveforms of heavy to light load transient, $I_{OUT} = 1.5$ A $\rightarrow$ 200 mA of the converter. The top waveform of the Fig. 7 shows the conventional case, when the control of the flying capacitor voltage is done only based on the adjustment of duty cycle (only *FCC Mode 1* employed for both loads) and the bottom waveform shows a case when the introduced dual-mode flying capacitor balancing controller is used (*FCC 1 Mode* used under heavy load and *FCC Mode 2* for the light load).

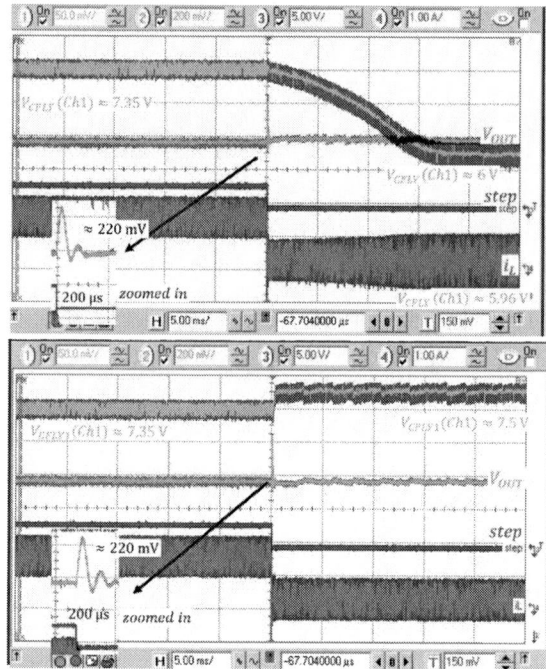

Fig. 7. Comparison of the heavy to light transient of 3-level buck converter for $V_{OUT} = 3.3$ [V], $I_{OUT} = 1.5$ [A] $\rightarrow$ 0.2 [A]. a) With Conventional Flying Capacitor Controller (top) and b) With *Dual-Mode* Flying Capacitor Controller (bottom). For both waveforms: Time scale: 2 [ms]/div; Ch1. Differential Probe 50x10 [mV]/div: V_{CFLY}, flying capacitor voltage (Fig 1.); Ch2. 200 [mV]/div: v_{OUT}, output voltage (Fig. 1.); Ch3. 5 [V]/div: – Load-step signal Ch4: 1[A]/div: i_L output inductor current (Fig. 1);

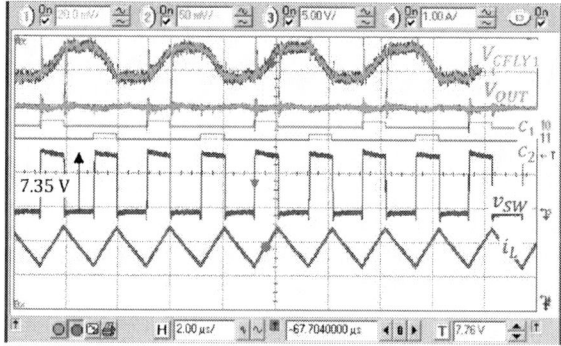

Fig. 8. Steady state operation after transient of 3-level buck converter for $V_{OUT} = 3.3$ [V], $I_{OUT} = 1.5$ [A] (heavy load). With *Dual-Mode* Flying Capacitor Controller. For both waveforms: Time scale: 2 [μs]/div; Ch1. Differential Probe 20x10 [mV]/div: V_{CFLY1}, flying capacitor voltage (Fig 1.); Ch2. 50 [mV]/div: v_{OUT}, output voltage (Fig 1.); Ch3. 5 [V]/div: – switching node voltage. Ch4: 1 [A]/div: i_L output inductor current (Fig. 1), Digital probes c_1, c_2 digital control signals of switches SW_1 and SW_2

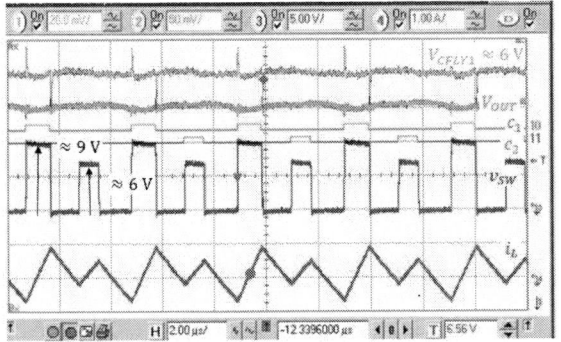 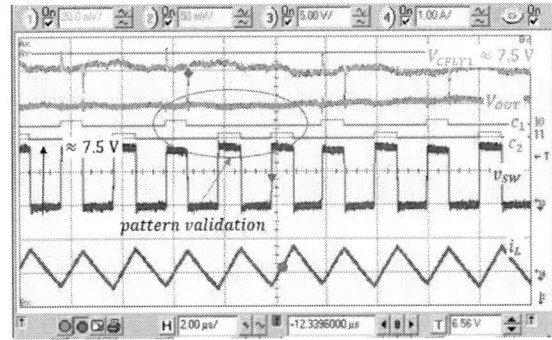

Fig. 9. Comparison of the steady state operation after transient of 3-level buck converter for $V_{OUT} = 3.3$ [V], $I_{OUT} = 0.2$ [A] (light load). a) With Conventional Flying Capacitor Controller (left) and b) With *Dual-Mode* Flying Capacitor Controller (right). For both waveforms: Time scale: 2 [μs]/div; Ch1. Differential Probe 20x10 [mV]/div: V_{CFLY1} flying capacitor voltage (Fig 1.); Ch2. 50 [mV]/div: v_{OUT} output voltage (Fig 1.); Ch3. 5 [V]/div: – switching node voltage Ch4: 1 [A]/div: i_L output inductor current (Fig. 1), Digital probes c_1, c_2 digital control signals of switches SW_1 and SW_2

Zoomed-in waveforms of Fig. 8 show that at the heavy load *FCC Mode 1* provides tight regulation of the flying capacitor voltage. Fig. 9 shows zoomed-in version of the waveforms for light load operation of the converters. Waveforms on the left side show a case when the regulation of the flying capacitor voltage is attempted with FCC *Mode 1* controller, as previously mentioned, designed without taking into account the ripple influence. Imbalance of the flying capacitor voltage in this case can be clearly seen from the switching node voltage, with two different high voltage values $V_{IN} - V_{CFLY1} \neq V_{IN}/2$ and $V_{CFLY1} \neq V_{IN}/2$. As a result, this imbalance increases the inductor current ripple. Right side of the Fig. 9 shows a case when the proposed controller, employing *FCC Mode 2* is used. It is shown that the proposed controller effectively balances the flying capacitor voltage.

V. CONCLUSION

In this paper a limitation of existing modeling approaches of multi-level converters, which do not take into account inductor current ripple, and associated stability problems are explained. Physical background of the problem is described through a geometrical analysis of the flying capacitor charge in the presence of a significant current ripple. An extended mathematical model that takes into account the inductor current ripple has been developed. Accordingly, a practical digital controller that does not suffer from the related stability issues and can be used in low-power applications has been developed. The effectiveness of the model and the controller have been experimentally verified.

REFERENCES

[1] T. A. Meynard, H. Foch, "Multi-level conversion: high voltage choppers and voltage-source inverters," *in Proc. IEEE PESC '92*, vol. 1. pp.397-403 July 1992.

[2] T. A. Meynard, H. Foch, "Multilevel converters and derived topologies for high power conversion," *in Proc. 1995 IEEE 21st Int. Conf. Industrial Electronics, Control, and Instrumentation*, pp. 21–26. Nov. 1995.

[3] T. A. Meynard, H. Foch, P. Thomas, J. Courault, R. Jakob and M. Nahrstaedt, "Multicell converters: basic concepts and industry applications," *IEEE Trans on Industrial*

[4] R. Y. Lei, W.C. Liu and R.C.N. Pilawa-Podgurski, "An Analytical Method to Evaluate Flying Capacitor Multilevel Converters and Hybrid Switched-Capacitor Converters for Large Voltage," *IEEE Workshop on Control and Modeling for Power Electronics (COMPEL)* July 2015. (In Press)

[5] N. Vukadinović, A. Prodić, B. A. Miwa, C. B. Arnold, M. W. Baker., "Volume and Efficiency Comparison Between Multi-level Dc-Dc Converters and Buck Converter for Low-Power Mobile Applications" *18th International Symposium on Power Electronics Ee 2015*, Novi Sad, Serbia, October 2015

[6] T. A. Meynard, M. Fadel; Aouda, N., "Modeling of multilevel converters," *Industrial Electronics, IEEE Transactions on*, vol.44, no.3, pp.356, 364, Jun 1997

[7] R. H. Wilkinson, T. A. Meynard, H. du Toit Mouton, "Natural Balance of Multicell Converters: The Two-Cell Case," *Power Electronics, IEEE Transactions on*, vol.21, no.6, pp.1649, 1657, Nov. 2006.

[8] R. H. Wilkinson, T. A. Meynard, H. du Toit Mouton, "Natural Balance of Multicell Converters: The General Case," *Power Electronics, IEEE Transactions on*, vol.21, no.6, pp.1658, 1666, Nov. 2006

[9] V. Yousefzadeh, E. Alarcon, D. Maksimovic, "Three-level buck converter for envelope tracking in RF power amplifiers," *Applied Power Electronics Conference and Exposition, 2005. APEC 2005. Twentieth Annual IEEE*, vol.3, no., pp.1588, 1594 Vol. 3, 6-10 March 2005

[10] G. V. Piqué, E. Alarcón,: CMOS Integrated Switching Power Converters: A Structured Design Approach. Springer May 20, 2011.

[11] R. W. Erickson and D. Maksimović, Fundamentals of Power Electronics. New York: Springer-Verlag, 2001.

[12] A. Radić, A. Prodić, "Buck Converter with Merged Active Charge-Controlled Capacitive Attenuation," *IEEE Transactions on Power Electronics*, March 2012, Vol.27, Issue. 3, pp. 1049-1054.

Burst Mode Control and Switched-Capacitor Converters Losses

Michael Evzelman and Regan Zane

UPEL Power Electronics Laboratory
ECE Department, Utah State University
North Logan, UT, 84341.
Email: evzelman@gmail.com; Website: http://power.usu.edu/

Abstract—**The feasibility of applying burst mode control in regulation of classical switched-capacitor converters is reevaluated. The results show that contrary to switched inductor based converters, this simple and easy to implement control has no efficiency advantage in terms of conduction losses over frequency modulation control when used to regulate classical switched-capacitor based converters. It is expected, however, that the burst mode could be viable for very low power applications in reducing biasing and housekeeping associated losses if reduced during the burst off time.**

Keywords— Switched Capacitor; Burst; Control; Losses.

I. INTRODUCTION

Switched Capacitor Converter (SCC) circuits are getting more and more attention during the last decade [1-8]. Despite the fact that many works in the field present an excellent analysis, modeling solutions and insights into the operation of SCC [9-20], there are still misconceptions related to the SCC basic operation. One of them, which is the subject of this work, is that the efficiency of voltage regulated SCC could be improved by running it in a burst, or on-off control mode.

It appears that the concept stems from the sibling family, the switched inductor converters (SIC), inductor based switch mode converters. Burst mode control in SIC is used to overcome the low efficiency at lower power levels where switching losses dominate [21-22]. Turning the converter on for short periods to operate at nominal power level, where the efficiencies in switched inductor converter are generally much higher, and then turning it off running the load from an energy buffer, effectively increases the overall system efficiency. Unfortunately, due to the fundamental difference in principal of operation between the SCC and SIC families, burst mode approach doesn't yield the same efficiency increase in classical switched capacitor based converters.

II. THEORETICAL CONSIDERATIONS

Figure 1a shows a generic model of SCC as accepted by many in the field [12-20]. Where the open circuit converter voltage is represented by V_T, the losses associated with converter operation are represented by the equivalent resistance R_e, which is in series with the target voltage and with the load resistance R_l. Burst mode operation could be described as an on/off signal, which on one hand disconnects the load from the converter, while on the other hand puts the converter into an off mode (Fig. 1b). Off mode is a reduced power, standby mode of the converter, where it is fully ready but not actively supplying the load and the load is fed by the energy buffer/output capacitor C_o. A common method to generate the on/off signal is based on a comparator, with two level hysteresis [23]. Alternatively time based modulation, or sigma delta modulation techniques could also be used [24, 25]. Time based modulation is used in this work to simplify the analysis, but the result holds for other modulation methods as well. Adequate burst period is determined, and the converter is turned on or off for the duration of burst duty cycle, D_{Burst}, which is based on the sensing signal arriving from ADC. This technique features constant burst period T_{Burst}, and constant frequency (Fig. 1c).

Based on Fig. 1c, burst mode could be summarized as an operation of the converter at two different power levels: P_1 during D_{Burst}, and P_2 during $(1-D_{Burst})$. The average power is

$$P_{avg} = D_{Burst} \cdot P_1 + (1 - D_{Burst}) \cdot P_2 \big|_{P_2=0} = D_{Burst} \cdot P_1 \ (1).$$

The quantity of interest in this case is the power loss on the equivalent resistor R_e, P_{1Loss} in a burst mode operation, which based on Fig. 1a can be expressed as

$$P_{1Loss} = D_{Burst} \cdot (I_o^2 \cdot R_e). \tag{2}$$

From (2) the model of Fig. 1a is refined to include two phase (as described in Fig. 1c) burst mode operation, so that the full equivalent resistance R_{e_Burst} takes the form of

$$R_{e\_Burst} = \frac{R_e}{D_{Burst}}, \tag{3}$$

and P_{1Loss} could be expressed as

$$P_{1Loss} = I_o^2 \cdot R_{e\_Burst}. \tag{4}$$

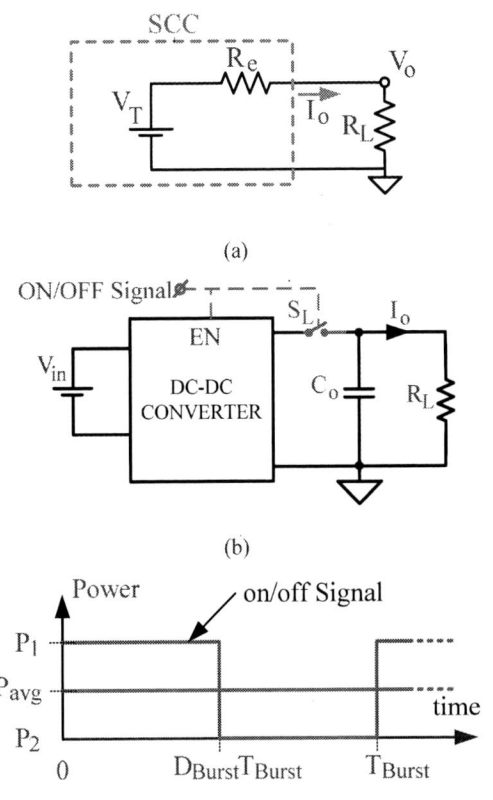

(a)

(b)

(c)

Fig. 1. (a) Generic Model of SCC; (b) Block diagram of a burst mode control implementation; (c) Burst Mode operation power transfer profile, and on/off signal.

Substituting R_{e_Burst} in place of R_e in the model of Fig. 1a, results in a more general model, capable to predict SCC behavior in burst mode. For the classical case of no burst mode, D_{Burst} will be equal to unity.

SCC regulation is carried out by adjusting the equivalent resistance R_e (Fig. 1a) such that the voltage drop generated across R_e due to the output current I_o satisfies the output voltage regulation objective

$$V_o = V_T - R_e \cdot I_o. \tag{5}$$

It becomes apparent from (3), (4) and Fig. 1a that introducing a new control variable doesn't change the fundamental regulation mechanism to achieve the goal (5). The equivalent resistance still needs to be adjusted to ensure proper voltage drop between V_T and V_o, either by applying a Burst mode control, or any of the classic control modes such as frequency, duty cycle, or any other modulation. This equivalent resistance adjustment procedure, generally referred as regulation, will inevitably carry out efficiency penalties.

III. SIMULATION

To evaluate the theoretical predictions presented in the previous section, a practical case of a lithium ion battery step up regulator from 3.3 V to 5.0 V is examined. The doubler SCC topology suitable to carry the regulation task was constructed in PSIM (Fig. 2a). Frequency modulation control was selected as a classical regulation strategy to compare performance to the burst mode control. Optimal operation point was selected for the Burst mode control, residing in the Partial Charge (PC) region as developed in [20]. Simulation was based on the practical system parameters as summarized in Table I. Figure 2b shows simulation results for output voltage regulated by the doubler SCC operated in burst mode and in frequency control mode. Straight lines are the averages of both control circuits. The values of cycle-by-cycle simulation equivalent resistances were calculated to be practically the same: for the burst mode control $R_{e\_Burst} = 17.092\ \Omega$, and for the frequency control $R_{e\_Frequency} = 17.093\ \Omega$. Model calculated values coincide with cycle-by-cycle simulation values for both cases, rendering the extension of the model to include burst mode as an excellent fit. Simulated efficiencies are summarized in Table II.

TABLE I. EXPERIMENTAL AND SIMULATION SYSTEM PARAMETERS

Parameter	Vin	$R\_S1/S2$	C_fly	R_ESR	C_o	R_L	$f\ s\ H$	$f\ s\ L$
Value	3.3 V	0.9 Ω	5.7 μF	65 mΩ	65 μF	52 Ω	122 kHz	21 kHz

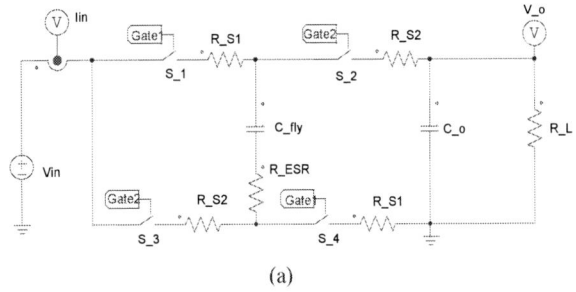

(a)

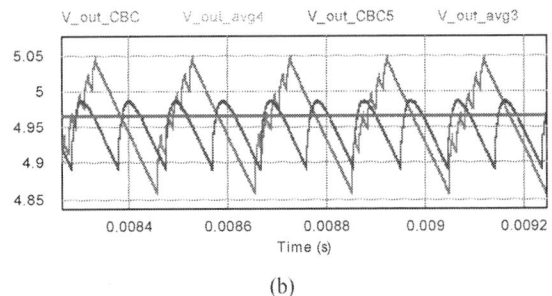

(b)

Fig. 2. (a) Simulation Circuit; (b) Simulation results of Doubler SCC - Output voltages: Red Trace (V_out_CBC) – Burst mode, Blue Trace (V_out_CBC5) – Frequency mode, Pink Trace (V_out_avg3) – Burst mode calculation based on average voltage. Green Trace (V_out_avg4) – Low frequency mode calculated based on average model (coincides with pink trace).

978-1-4673-9551-9/16 $31.00 © 2016 IEEE

IV. EXPERIMENTAL VALIDATION

To validate the theoretical predictions and simulation results a laboratory prototype based on a commercial SCC IC SP6660 manufactured by Sipex, was built (Fig. 3). Prototype measured parameters are summarized in Table I. Control task was carried out using a low power TI MSP430F2013 microcontroller. Input and output voltages were measured using BK Precision (Model: 2831E) 4½, and currents were measured using BK Precision (Model: 5492B) 5½ digit multimeters. Input, output voltages and currents along with equivalent resistance, power and efficiency calculations, and output ripple measurements are summarized in Table II. Experimental waveforms are shown in Fig. 4a for burst mode control and in Fig. 4b for frequency modulation control mode. Note that the driving signal is twice the switching frequency due to the internal design of the commercial IC used. The output ripple in Fig. 4a for burst mode control is significantly higher than that of frequency mode control in Fig. 4b, while the efficiency according to Table II is maintained to be the same.

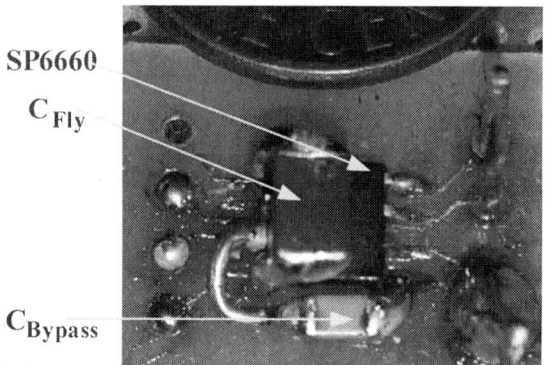

Fig. 3. Experimental prototype (Magnification x3, Output capacitor and load resistor are connected on the other side of the board).

V. DISCUSSION

As demonstrated in Section II, the basic principle of SCC operation limits the practical advantage of burst mode control. The same holds true even if faster turn on/off switches are used to reduce the conduction loss during switch transition times as highlighted in [26], and for a common remedy used in SICs, replacing the switches with lower R_{ds_on} [27-28]. Both methods will fail to improve the efficiency of regulated SCC. The results in Table II confirm that there is no efficiency gain for operating SCC in its optimal range and turning it off for periods of time in a burst mode control, versus a frequency modulation control and operation in a less attractive operation mode. Auxiliary quiescent power losses, such as microcontroller power, ADC/DAC reference circuits, operational amplifiers, gate drive biasing and level shifting, increase the losses of practical circuits. In very low power applications where quiescent losses may dominate, it can be advantageous to operate in the burst mode provided that the control circuit is able to enable and disable the biasing circuits efficiently within the burst mode off-time.

SCC circuits are particularly attractive candidates for on-chip converter implementation. In on chip environment the parasitic capacitances of the converter, such as the bottom plate capacitance, become a significant source of leakage contributing to the losses of the SCC [29-30]. Another parasitic capacitance is switch output capacitance, which is charged and discharged every cycle, resulting in additional loss. Gate drive losses are another source of losses associated with switching frequency. Further investigation is required to address the potential of switching loss reduction in SCC circuits by applying different regulation methods.

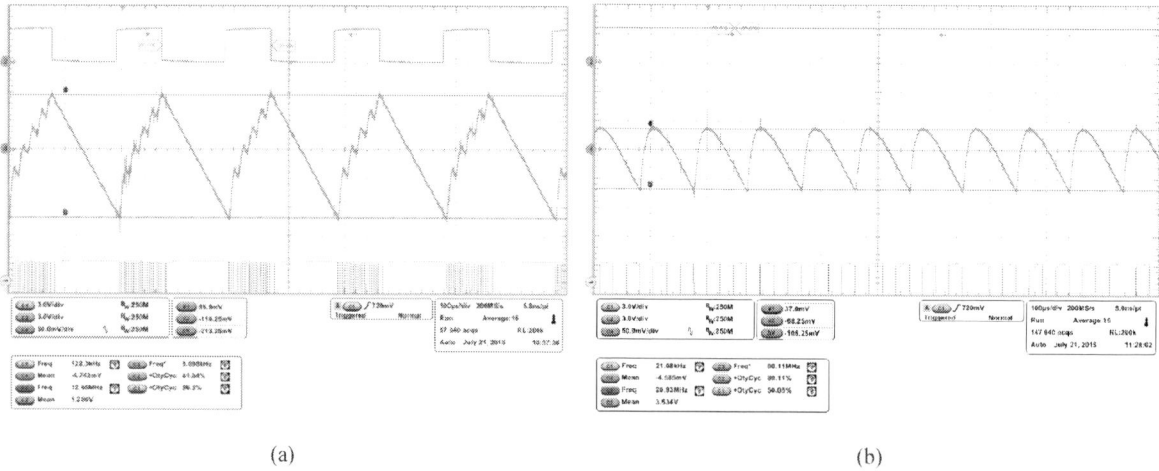

(a)   (b)

Fig. 4. Experimental results (Timebase 100µs/Div): (a) Burst mode control. (b) Frequency modulation control. Upper trace – Burst modulation signal; Middle trace – Output voltage AC coupling (50mV/Div); Bottom trace – Driving Signal (the driving signal is double the switching frequency due to the commercial IC internal design).

TABLE II.   Experimental and Simulation Results

Parameter	Simulation		Experimental	
	Burst mode	Frequency mode	Burst mode	Frequency mode
Base Frequemncy [kHz]	122	21	122.4	21.1
Vin [V]	3.3	3.3	3.304	3.302
Vout [V]	4.959	4.949	4.962	5.004
Vpk-pk$_{Ripple}$ [mV]	185	92	213	105
Iin [mA]	189.7	189.2	191	193
Iout [mA]	95.17	95.17	95.42	96.23
Req [Ω]	17.092*	17.093*	~17.25	~16.63
Pin [W]	0.626	0.624	0.632	0.637
Pout [W]	0.473	0.471	0.4735	0.4815
Efficiency η [%]	**75.56**	**75.48**	**74.92**	**75.59**

VI.   Conclusions

Burst mode control strategy applied to switched capacitor converters is reevaluated. The dependence of the equivalent resistance on the burst mode duty cycle is derived and an extension to the basic model to include the burst mode control parameter is developed. Theoretical predictions were validated by simulation and experiment, using a practical scenario of a lithium ion battery voltage doubling converter, boosting and regulating voltage from 3.3 V to 5.0 V at power levels of approximately 500mW.

The results reaffirm that burst mode control has no efficiency advantage in terms of conduction losses over classical frequency modulation control when applied to regulate switched capacitor based converters. Burst mode control could be favorable in some applications requiring for example cheaper control implementation. However, the same amount of conduction losses, with potentially higher ripple should be expected when compared to the classical control modes like frequency and duty cycle modulation strategies. It is expected that in some circumstances burst mode control may be viable in reducing the switching or quiescent control losses.

References

[1] A. Ioinovici, "Switched-capacitor power electronics circuits," *IEEE Circuits Syst. Mag.*, vol. 1, no. 3, pp. 37-42, 2001.

[2] O. Keiser, P.K. Steimer, and J.W. Kolar, "High power resonant Switched-Capacitor step-down converter," *IEEE Power Electron. Spec. Conf., PESC 2008*, pp. 2772-2777, 15-19 Jun. 2008.

[3] D. Cao and F.Z. Peng, "Multiphase Multilevel Modular DC–DC Converter for High-Current High-Gain TEG Application," *IEEE Trans. on Ind. Appl.*, vol. 47, no. 3, pp. 1400-1408, 2011.

[4] Yuang-Shung Lee; Yi-Pin Ko; Ming-Wang Cheng; Li-Jen Liu, "Multiphase Zero-Current Switching Bidirectional Converters and

Battery Energy Storage Application," *IEEE Transactions on Power Electronics*, vol. 28, no. 8, pp. 3806-3815, Aug. 2013.

[5] Y. Yuanmao, K.W.E. Cheng, and Y.P.B. Yeung, "Zero-Current Switching Switched-Capacitor Zero-Voltage-Gap Automatic Equalization System for Series Battery String," *IEEE Trans. on Power Electron.*, vol. 27, no. 7, pp. 3234-3242, Jul. 2012.

[6] B. Van Tassell, S. Yang, C. Le, L. Huang, S. Liu, P. Chando, A. Byro, X. Liu, D.L. Gerber, E.S. Leland, S. Sanders, P.R. Kinget, I. Kymissis, D. Steingart, S. O'Brien, "Metacapacitors: Printed Thin Film, Flexible Capacitors for Power Conversion Applications," *IEEE Transactions on Power Electronics*, vol. PP, no. 99, pp. 1-1.

[7] A. Cervera, M. Mordechai Peretz, "Resonant Switched-Capacitor Voltage Regulator With Ideal Transient Response," *IEEE Transactions on Power Electronics*, vol. 30, no. 9, pp. 4943-4951, Sept. 2015.

[8] Lim Seungbum, J. Ranson, D.M. Otten, D.J. Perreault, "Two-Stage Power Conversion Architecture Suitable for Wide Range Input Voltage," *IEEE Transactions on Power Electronics*, vol. 30, no. 2, pp. 805-816, Feb. 2015.

[9] C. K. Tse, S. C. Wong and M. H. L. Chow, "On Lossless Switched-Capacitor Power Converters", *IEEE Trans. on Power Electron.*, vol. 10, no. 3, pp. 285-261, 1995.

[10] P. Favrat, P. Deval, and M.J. Declercq, "A high-efficiency CMOS voltage doubler," *IEEE J. of Solid-State Circuits*, vol. 33, no. 3, pp. 410-416, Mar. 1998.

[11] Fan Zhang, Lei Du, Fang Zheng Peng, and Zhaoming Qian, "A New Design Method for High-Power High-Efficiency Switched-Capacitor DC–DC Converters," *IEEE Trans. Power Electron.*, vol. 23, no.2, pp. 832-840, 2008.

[12] Y. P. B. Yeung, K. W. E. Cheng, S. L. Ho, K. K. Law, and D. Sutanto, "Unified analysis of switched-capacitor resonant converters," *IEEE Transactions on Industrial Electronics*, vol. 51, no. 4, pp. 864-873, Aug. 2004.

[13] J. W. Kimball and P. T. Krein, "Analysis and Design of Switched Capacitor Converters", *IEEE Appl. Power Electron. Conf., APEC- 2005*, vol. 3, 1473 -1477.

[14] B. Arntzen and D. Maksimovic, "Switched-Capacitor DC/DC Converters with Resonant Gate Drive" *IEEE Trans. on Power Electron.*, vol. 13, no. 5, pp. 892-902, Sept. 1998.

[15] M. S. Makowski and D. Maksimovic, "Performance Limits of Switched-Capacitor DC-DC Converters", *IEEE Power Electron. Spec. Conf., PESC-1995*, vol. 2, pp. 1215 -1221.

[16] M. D. Seeman and S. R. Sanders, "Analysis and Optimization of Switched Capacitor DC-DC Converters", *IEEE Trans. on Power Electron.*, vol. 23, no. 2, pp. 841-851, Mar. 2008.

[17] I. Oota, N. Hara, and F. Ueno, "A General Method for Deriving Output Resistances of Serial Fixed Type Switched-Capacitor Power supplies", *IEEE Int. Symp. on Circuits Syst., ISCAS-2000*, vol. 3, pp. 503-506.

[18] J.M. Henry and J.W. Kimball, "Practical Performance Analysis of Complex Switched-Capacitor Converters," *IEEE Trans. on Power Electron.*, vol. 26, no.1, pp.127-136, Jan. 2011.

[19] J.M. Henry and J.W. Kimball, "Switched-Capacitor Converter State Model Generator," *IEEE Trans. on Power Electron.*, vol. 27, no. 5, pp. 2415-2425, May 2012.

[20] M. Evzelman, S. Ben-Yaakov, "Average-Current-Based Conduction Losses Model of Switched Capacitor Converters," *IEEE Transactions on Power Electronics*, vol. 28, no. 7, pp. 3341-3352, July 2013.

[21] G. Oggier, M. Ordonez, "High Efficiency DAB Converter Using Switching Sequences and Burst-Mode," *IEEE Transactions on Power Electronics*, vol. PP, no. 99, pp. 1-1.

[22] Shuze Zhao; Jiale Xu; O. Trescases, "Burst-Mode Resonant LLC Converter for an LED Luminaire With Integrated Visible Light Communication for Smart Buildings," IEEE *Transactions on Power Electronics*, vol. 29, no. 8, pp. 4392-4402, Aug. 2014.

[23] D. Vasic, Yuan-Ping Liu, F. Costa, D. Schwander, "Piezoelectric transformer-based DC/DC converter with improved burst-mode control," *2013 IEEE Energy Conversion Congress and Exposition*, pp. 140-146, 2013.

[24] M.M. Peretz, S. Ben-Yaakov, "Digital Control of Resonant Converters: Enhancing Frequency Resolution by Dithering," *APEC 2009 Twenty-Fourth Annual IEEE Applied Power Electronics Conference and Exposition, 2009*, pp. 1202-1207, 15-19 Feb. 2009.

[25] T.A.D. Riley, M.A. Copeland, T.A. Kwasniewski, "Delta-sigma modulation in fractional-N frequency synthesis," *IEEE Journal of Solid-State Circuits*, vol. 28, no. 5, pp. 553-559, May 1993.

[26] M. Evzelman and S. Ben-Yaakov, "The effect of switching transitions on switched capacitor converters losses," in *IEEE 27th Convention of Electrical & Electronics Engineers in Israel (IEEEI)*, pp. 1-5, 14-17 Nov. 2012.

[27] S. Ben-Yaakov, "On the Influence of Switch Resistances on Switched-Capacitor Converter Losses," in *IEEE Transactions on Industrial Electronics*, vol.59, no.1, pp.638-640, Jan. 2012.

[28] M. Evzelman and S. Ben-Yaakov, "Optimal switch resistances in Switched Capacitor Converters," in *IEEE 26th Convention of Electrical and Electronics Engineers in Israel (IEEEI)*, pp.436-439, 17-20 Nov. 2010.

[29] T. Tanzawa, "On Two-Phase Switched-Capacitor Multipliers With Minimum Circuit Area," in *IEEE Transactions on Circuits and Systems I*, vol. 57, no. 10, pp. 2602-2608, Oct. 2010.

[30] T. Tanzawa, "A comprehensive optimization methodology for designing charge pump voltage multipliers," in *IEEE International Symposium on Circuits and Systems (ISCAS)*, pp. 1358-1361, 24-27 May 2015.

Equivalent Circuit Modeling of LLC Resonant Converter

Shuilin Tian, Fred C. Lee and Qiang Li

Center for Power Electronic Systems, Virginia Tech, Blacksburg, VA, 24061

sltian87@vt.edu

Abstract—LLC resonant converter is widely used in industry. However, up to now, no simple and accurate small-signal equivalent circuit model is available. This paper proposes an equivalent circuit model of LLC resonant converter. The simple equivalent circuit model is derived based on modification and simplification of extended describing function method. The model can well predicts the small-signal behaviors observed in LLC resonant converter, whenever switching frequency is below, close to or above the resonant frequency. For the first time, analytical expressions for control to output voltage, input to output voltage, input impedance and output impedance are provided to aid close loop feedback design. Simplis simulation and experimental results are presented to prove the accuracy of the model.

Keywords— LLC resonnant converter, small-signal model, equivlaent circuit model, extended describing function.

I. INTRODUCTION

Generally speaking, DC-DC converters can be divided into two categories: pulse-width-modulation (PWM) converters and resonant converters. As most of the applications involve a regulated voltage output, therefore a feedback loop is incorporated into the control system to stabilize the output voltage. For optimal design purpose, small-signal equivalent circuit models are indispensable. For PWM converters, equivalent circuit models are available to engineers for most of the control methods: voltage mode control (power stage three-terminal switch model) [1][2]; equivalent circuit model of current mode control including inductor current sideband [3]-[6]; equivalent circuit model of V^2 control including both inductor current sideband and capacitor voltage sideband [7] – [13]; Therefore, for PWM converters, feedback design is straightforward with the help of all these research efforts.

However, the scenario is different for resonant converters. Series resonant converter (SRC) is the simplest resonant converter. The state-space averaging method breaks down for resonant converters. The reason is that for resonant converters, some of the state variables do not have dc components but contain strong switching frequency component and its harmonics, whereas the dc components are the dominant parts of the state variables for PWM converters. Due to the strong oscillatory nature of resonant states, the switching frequency interacts with the natural resonant frequency. This results in an interesting phenomenon which is often referred to as the beat frequency dynamics where the high-frequency response is determined by a pair of double pole located at the beat frequency [14]. The most successful model to deal with resonant converter is based on extended describing function

method [15] and recently a decent simple third-order equivalent circuit model is developed to explain all the small-signal properties in SRC [16] [17].

LLC resonant converter, shown as Fig. 1, is the most popular resonant converter for front-end DC-DC converters used in distributed power systems in telecom, computer and network applications [18]-[19].Besides, they are also widely adopted in other applications, such as LCD, LED and plasma display in TV and flat panels; iron implanter arc power supply; solar array simulator in photovoltaic application; fuel cell applications and so on. Available small-signal models published in literatures include extended describing function method [20]-[23], approach based on communication theory [24], sampled-data modeling approach [25], analysis based on Simplis simulation [26] or bench measurement results [27]. All of the above models use numerical solutions instead of analytical solutions. As a result, no simple equivalent circuit model is available and no analytical expressions of transfer functions are presented.

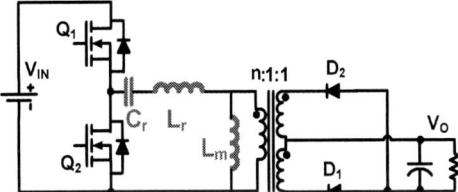

Fig. 1. Schematic of half-bridge LLC resonant converter (LLC)

This paper tries to propose a simple equivalent circuit model as a good design tool for LLC resonant converter. The simple equivalent circuit model is derived based on modification and simplification of extended describing function method, following similar modeling strategy on SRC resonant converter [16][17]. The following two main objectives will be achieved: (1) a simple third-order equivalent circuit model of LLC resonant converter is proposed. The equivalent circuit model can well predict beat frequency dynamics and provide fruitful physical insights to understand the small-signal behavior. (2) For the first time, analytical expressions for all transfer functions are provided to help engineers to design the feedback control appropriately.

II. EQUIVALENT CIRCUIT MODEL OF LLC RESONANT CONVERTER

For LLC resonant converter, one major advantage is its ZVS capability for zero to full load range. Generally speaking, ZVS

978-1-4673-9551-9/16 $31.00 © 2016 IEEE

is preferred for applications using MOSFET. For SRC, ZVS can only be achieved when switching frequency is above the resonant frequency. However, for LLC, due to the effect of Lm, ZVS can also be achieved when switching frequency is below the series resonant frequency Fo. From resonance point of view, the resonant tank is different when comparing $F_s \geq F_o$ with $F_s \leq F_o$: For $F_s \geq F_o$, only Lr resonates with Cr and Lm is clamped by the output voltage. For $F_s < F_o$, there is some time period that L_m also participates in resonance. Due to different resonant behavior, the small-signal models are developed for each case, as follows:

A. Equivalent circuit model of LLC for $F_s \geq F_o$.

As shown in steady-state waveform in Fig. 2, in this case, LLC behaves like SRC. The magnetizing inductor, L_m, is either clamped by Vo or –Vo and never participates in resonance.

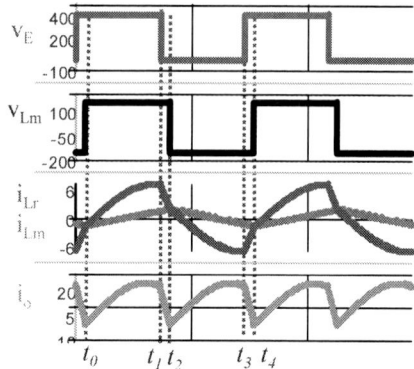

Fig. 2 Steady-state waveforms for Fs=1.4Fo.

The model of inverter, rectifier, resonant inductor and resonant capacitor can be derived following the same methodology as SRC shown in [16][17]. The small-signal model of LLC resonant converter for Fs ≥ Fo is shown as Fig. 3. Note that since half-bridge inverter is used in LLC resonant converter, compared with the full-bridge inverter model shown in [16][17], the magnitude of the input current source and output voltage source is reduced to half.

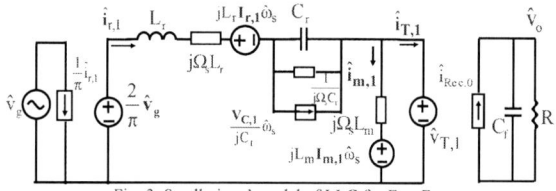

Fig. 3. Small-signal model of LLC for Fs ≥ Fo.

Similar as SRC [16][17], the resonant capacitor behaves like an equivalent inductor with respect to modulation frequency. Therefore, the resonant tank can be simplified as shown in Fig. 4.

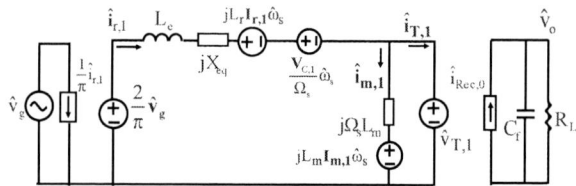

Fig. 4. Small-signal model of LLC for Fs ≥ Fo with simplified resonant capacitor branch.

The expression of equivalent inductor L_e and impedance X_{eq} is shown as (1):

$$L_e = L_r + \frac{1}{C_r \Omega_s^2} = L_r(1 + \frac{\Omega_o^2}{\Omega_s^2})$$

$$X_{eq} = \Omega_s L_r - \frac{1}{\Omega_s C_r}$$

(1)

The small-signal model shown in Fig. 4 has complex terms and can not be used for simulation. Following the similar methodology as SRC [16][17], the complex terms can be eliminated by separating the resonant tank into sine part and cosine part, shown as Fig. 5.

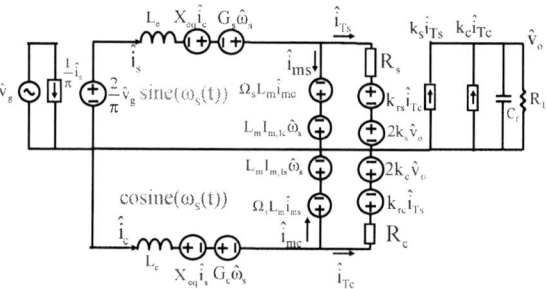

Fig. 5. Separated third-order equivalent circuit model of LLC for $F_s \geq F_o$.

Similar as SRC [16][17], the superposition theorem is applied to derive a non-coupled equivalent circuit model. The non-coupled equivalent circuit is shown as Fig. 6.

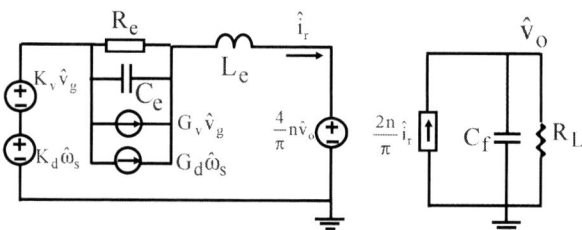

Fig. 6 Non-coupled third-order equivalent circuit model of LLC for $F_s \geq F_o$.

Le, Re, Ce are used to represent beat frequency dynamics and their expressions are same as SRC. The expressions of Kv, Gv, Kd and Gd are different from SRC and will be affected by the design of L_m. All the expressions are shown as (2).

Due to page limitations, some detailed derivation process is omitted here and the readers can refer to [28] for thorough details.

$$L_e = L_r(1 + \frac{\Omega_o^2}{\Omega_s^2}), C_e = \frac{1}{L_e(\Omega_s - \Omega_o)^2}, R_e = \frac{L_e|X_{eq}||\Omega_s - \Omega_o|}{R_{eq}}$$

$$G_d = \frac{2V_g}{\pi} \frac{L_n}{\omega_o} \frac{1}{R_e} \left(\frac{\frac{1}{\omega_n}(\frac{1}{\omega_n^2} - \omega_n^2)(\frac{\pi^2}{8}QL_n)^2 - (L_n + 1 - \frac{1}{\omega_n^2})(\frac{2}{\omega_n^2})}{\left[\sqrt{(L_n + 1 - \frac{1}{\omega_n^2})^2 + ((\frac{1}{\omega_n} - \omega_n)\frac{\pi^2}{8}QL_n)^2} \right]^3} + \frac{2}{L_n^2} \right), \quad (2)$$

$$K_d = -\frac{4V_g}{\pi} \frac{1}{\omega_o L_n}, G_v = \frac{1}{\pi} \frac{X_{eq}}{\sqrt{X_{eq}^2 + R_{eq}^2}},$$

$$K_v = \frac{4}{\pi^2} \frac{V_g L_n \omega_n}{R_{eq}} \frac{L_n + 1 - \frac{1}{\omega_n^2}}{\sqrt{(L_n + 1 - \frac{1}{\omega_n^2})^2 + ((\frac{1}{\omega_n} - \omega_n)\frac{\pi^2}{8}QL_n)^2}},$$

$$R_{eq} = \frac{8}{\pi^2} n^2 R_L, X_{eq} = \Omega_s L_r - \frac{1}{\Omega_s C_r}, L_n = \frac{L_m}{L_r}$$

B. Equivalent circuit model of LLC for Fs < Fo

Fig.7 shows steady-state waveforms when $F_s=0.8F_o$ and Fig. 8 shows the operating modes at different time periods. Obviously, in this case, the resonant tank changes at different time periods: in time periods [t0, t1] and [t2, t3], the magnetizing inductor, Lm, is either clamped by Vo or –Vo and never participates in resonance. This is exactly the same condition as previous case when Fs ≥ Fo; in time periods [t1,t2] and [t3,t4], the magnetizing inductor Lm participates in resonance and the resonant tank is comprised of Cr and Lr in series with Lm. In the meanwhile, the output load is decoupled from the resonant tank. As the resonant tank is changing within one switching period, the LLC essentially belongs to multi-resonant structure when operating in the region Fs< Fo.

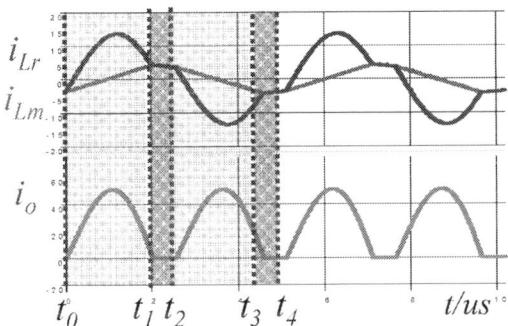

Fig. 7. Steady-state waveforms for Fs=0.8Fo.

In time periods [t0, t1] and [t2, t3], tank and load are coupled and the relation of Lm, rectifier and output load are same as Fs ≥ Fo case, as shown in Fig. 9 (a). In time periods [t1,t2] and [t3,t4], tank and load are decoupled. In this case, Lr is in series with Lm and the sum of the two inductances resonates with Cr, as shown in Fig. 9(b).

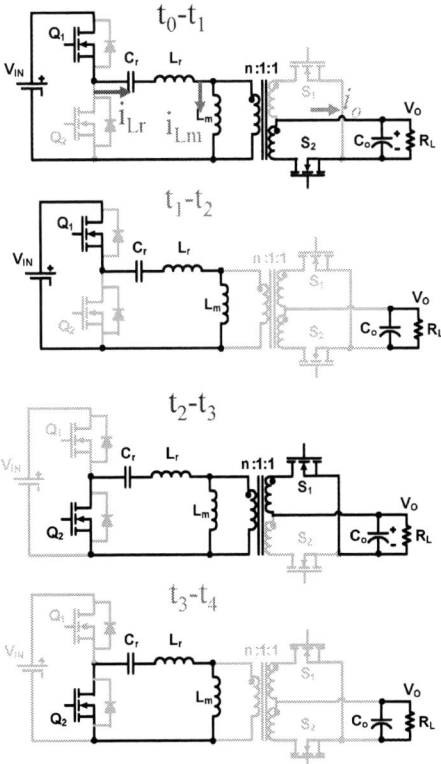

Fig. 8. Operating modes of LLC for Fs < Fo.

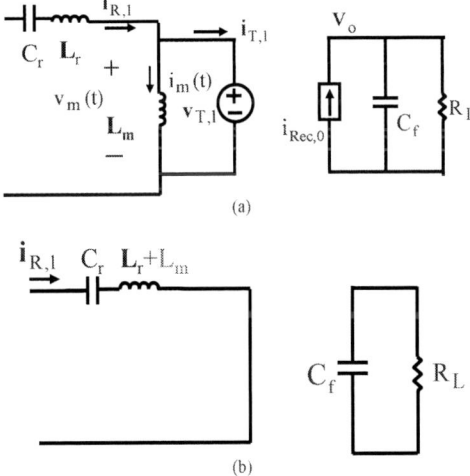

Fig. 9. The relation of Lm, rectifier and output load for Fs < Fo. (a) In time periods [t0, t1] and [t2, t3] (b) In time periods [t1,t2] and [t3,t4].

The length of time periods [t0, t1] and [t2, t3] is the resonant period To and the length of time periods [t1,t2] and [t3,t4] is Ts-To. For the whole switching period, the modulation model can be derived by combining the modulation

978-1-4673-9551-9/16 $31.00 © 2016 IEEE   1610

model of Fig. 9(a) and Fig. 9(b), with the ratio of To/Ts and (Ts-To)/Ts, respectively. The large signal modulation model is shown as Fig.10.

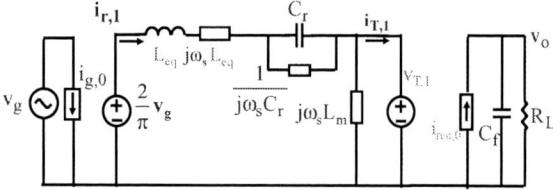

Fig. 10. Modulation model of LLC for Fs < Fo.

The equivalent resonant inductor, Leq is related with resonant inductor and magnetizing inductor. The expression of Leq is shown as (3):

$$L_{eq} = L_r + L_m \frac{\Omega_o - \Omega_s}{\Omega_o} \tag{3}$$

The model of the rectifier should be modified, as the voltage across the magnetizing inductor is a quasi-square wave instead of a square wave, the modified large-signal rectifier model is shown as Fig. 11 and the expressions are shown in (4) for detailed derivation, please refer to [28].

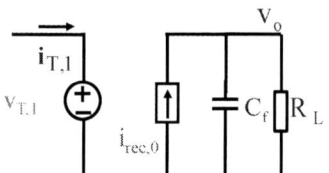

Fig. 11. Modulation model of the rectifier of LLC for Fs < Fo.

$$v_{T,1} = \frac{4}{\pi} n v_o \sin(\frac{1}{2}\frac{\Omega_s}{\Omega_o}\pi) \frac{i_{T,1}}{\|i_{T,1}\|} \tag{4}$$

$$i_{rec,0} = \frac{2}{\pi} n \|i_{T,1}\|$$

Following the similar methodology as previous case, the small-signal model with complex impedance (similar as Fig.3 and Fig. 4), separated small-signal model (similar as Fig. 5) and the non-coupled equivalent circuit model (similar as Fig. 6) can all be derived for this case. The non-coupled equivalent circuit is shown as Fig. 12. The expressions of Le`, Kv, and Kd are shown as (5).

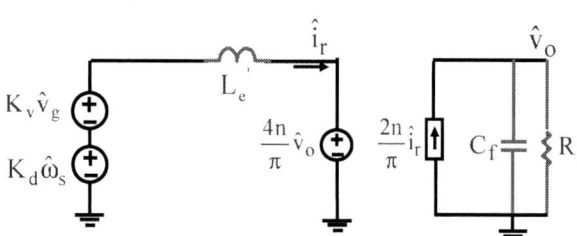

Fig. 12. Non-coupled equivalent circuit model of LLC for Fs< Fo.

$$L_e^{'} = L_r(1+\frac{1}{\omega_n^2}) + L_m(1-\omega_n)$$

$$K_d = \frac{2V_g}{\pi}\frac{L_n}{\omega_o} \left[\frac{\left[(\frac{1}{\omega_n^2}-\omega_n^2)\left(\frac{\pi^2}{8}Q_rL_n\right) - \left(L_n+1-\frac{1}{\omega_n^2}\right)\left(\frac{2}{\omega_n^2}\right)\right]\frac{1}{\omega_n}\frac{1}{\sin(\frac{\pi}{2}\omega_n)} +}{\sqrt{\left[\left(L_n+1-\frac{1}{\omega_n^2}\right)^2 + \left(\frac{1}{\omega_n}-\omega_n\right)\frac{\pi^2}{8}Q_rL_n\right)^2\right]}} \frac{-\frac{\pi}{2}\frac{\cos(\frac{\pi}{2}\omega_n)}{\sin^2(\frac{\pi}{2}\omega_n)}}{\sqrt{\left(L_n+1-\frac{1}{\omega_n^2}\right)^2 + \left(\left(\frac{1}{\omega_n}-\omega_n\right)\frac{\pi^2}{8}Q_rL_n\right)^2}} \right] \tag{5}$$

$$K_v = \frac{2}{\pi}\frac{L_n}{\sin(\omega_n\pi/2)}\frac{1}{\sqrt{\left(L_n+1-\frac{1}{\omega_n^2}\right)^2 + \left(\left(\frac{1}{\omega_n}-\omega_n\right)\frac{\pi^2}{8}Q_rL_n\frac{1}{\sin(\omega_n\pi/2)}\right)^2}}$$

In this case, there is no beat frequency dynamics and the circuit is essentially second-order.

C. Unified equivalent circuit model of LLC.

The small-signal models shown in Fig. 6 and Fig. 12 can be combined to obtain a model which is valid both for Fs ≥ Fo and Fs<Fo. The unified equivalent circuit model is shown in Fig. 13. The expressions of Le, Re, Ce, are shown as (6). The expressions of Kv, Gv, Kd and Gd are shown in (2) and (5).

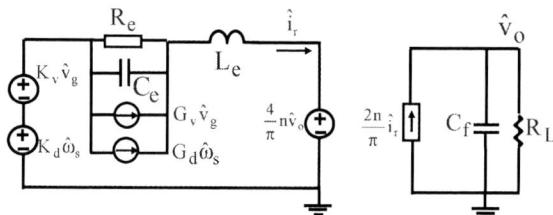

Fig. 13. Unified equivalent circuit model of LLC.

$$L_e = \begin{cases} (1+\frac{1}{\omega_n^2})L_r & \text{for } \omega_n \geq 1 \\ (1+\frac{1}{\omega_n^2})L_r + (1-\omega_n)L_m & \text{for } \omega_n < 1 \end{cases}$$

$$R_e = \begin{cases} \frac{L_e|X_{eq}|\|\Omega_s - \Omega_o\|}{R_{eq}}, & \text{for } \omega_n \geq 1 \\ 0, & \text{for } \omega_n \leq 1 \end{cases} \tag{6}$$

$$C_e = \frac{1}{L_e(\Omega_s - \Omega_o)^2}$$

III. DISCUSSION AND PREDICTIONS OF PROPOSED EQUIVALENT CIRCUIT MODEL

A. Steady-state voltage conversion ratio

The steady-state voltage conversion ratio can be derived as in (7) and plotted in Fig. 14. When Fs≥ Fo, the DC gain is smaller than 1 and shows the characteristic of SRC; When Fs<Fo, the DC Gain is greater than 1 and it shows the characteristic of Parallel Resonant Converter (PRC). As shown in Section IV, since the mode change is considered when Fs<Fo, the voltage conversion ratio derived in this paper is more accurate than traditional fundamental analysis [18].

$$M = \frac{1}{\sin(\alpha/2)} \left\| \frac{j\omega_n L_n}{j\omega_n \left(L_n + 1 - \frac{1}{\omega_n^2}\right) + \frac{\pi^2}{8} Q \frac{1}{\sin(\alpha/2)}(1 - \omega_n^2)) L_n \omega_n} \right\| \quad (7)$$

$\alpha = \omega_n \pi \text{ for } F_s < F_o$

$\alpha = \pi \text{ for } F_s \geq F_o$

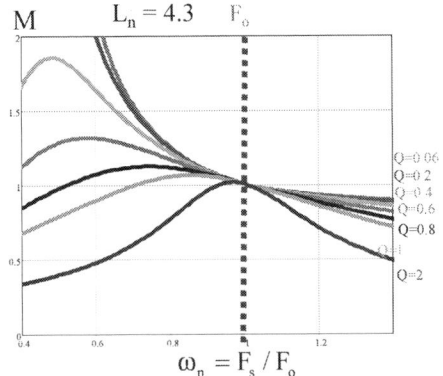

Fig.14. Voltage conversion ratio of LLC.

B. DC gain and beat frequency dynamics

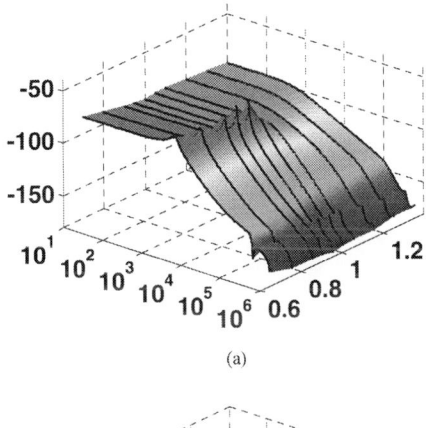

(a)

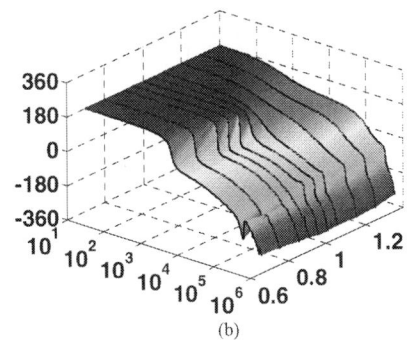

(b)

Fig. 15. 3D Bode plot of control-to-output voltage transfer function (a) Gain plot. (b) Phase plot.

With the help of Fig. 13, the 3D-plot of control to output voltage transfer function can be plotted as Fig. 15. The DC gain and poles can be explained clearly using the proposed non-coupled equivalent circuit model shown in Fig. 13.

The DC gain of the control-to-output voltage transfer function can be derived easily. The DC gain can be related with the slope of the voltage conversion ratio curve by the relations shown in (8):

$$G_{DC} = \frac{\partial V_o}{\partial \Omega_s} = \frac{V_g}{2n\Omega_o} \frac{\partial M}{\partial \Omega_s} \quad (8)$$

The beat frequency dynamic performance of the circuit can also be well explained by the equivalent circuit. The components Re, Ce and Le represent beat frequency dynamics. The equivalent inductor Le is probably resonant with the equivalent output capacitor or the equivalent capacitor Ce, depending on the value of Re.

When switching frequency Fs is much larger than resonant frequency Fo, Re is large. Le is resonant with Ce, which forms the beat frequency double pole. The output capacitor and load resistor forms a single pole on the load side. In this case, the double pole position its quality factor can be easily derived from Fig. 13 as in (9). Note that the position and quality factor of beat frequency double pole is same as SRC. This is reasonable as when Fs>>Fo, LLC behaves like SRC and Lm does not participates in resonance, therefore the effect of Lm on beat frequency double pole is very little.

$$\omega_p = |\Omega_s - \Omega_o|, Q_p = \frac{\left| \Omega_s L_r - \frac{1}{\Omega_s C_r} \right|}{R_{eq}} \quad (9)$$

When switching frequency Fs is larger than Fo but close to Fo, Re is small. The double pole caused by Le and Ce will be damped out and split, one moves to high frequency and the other one moves to low frequency. This low frequency pole will combine with low pass filter pole and forms a double pole. In other words, Le resonates with equivalent output capacitor Cf and load resistance determines the damping factor of this double pole. In this case, the double pole position and its quality factor are shown as (10).

$$\omega_p = \sqrt{\frac{1}{L_e \frac{\pi^2}{8n^2} C_f}}, Q_p = \frac{8n}{\pi^2} R_L \sqrt{\frac{C_f}{L_e}} \quad (10)$$

$$L_e = (1 + \frac{1}{\omega_n^2}) L_r$$

When switching frequency Fs is below Fo, in this case, no beat frequency double pole exists. Le resonates with equivalent output capacitor Cf and load resistance determines the damping factor of this double pole. However, in this case, Le is modified by Lm according to (6) as there is a certain time period that Lm also participated in resonance. As a result, the double pole moves to a lower frequency with reduced quality factor. In this case, the double pole position and its quality factor are shown as (11):

$$\omega_p = \sqrt{\frac{1}{L_e \frac{\pi^2}{8n^2} C_f}}, Q_p = \frac{8n}{\pi^2} R_L \sqrt{\frac{C_f}{L_e}}$$

$$L_e = L_r(1 + \frac{1}{\omega_n^2}) + L_m(1 - \omega_n)$$

(11)

The following illustration example is provided to explain the beat frequency dynamics. When Fs=1.4Fo, Re=6.5Ω is large, Le resonates with Ce. Fbeat=100kHz, Qbeat=0.6. The 2D Bode plot is shown as the red curve in Fig. 16, there is a beat frequency double pole at 100kHz and a single pole at low frequency caused by the output filter. When Fs=Fo, Re=0Ω is very small. The beat frequency double pole is split. Le resonates with output capacitor Cf. Shown as the blue curve in Fig.16, the double pole formed by Le and equivalent output capacitor, instead of beat frequency double pole, is observed. In this circumstance, the circuit is second order instead of third-order. When Fs=0.8Fo, the double pole reduces to around 3kHz due to the increase of the equivalent resonant inductance. There is no beat frequency double pole in region Fs < Fo.

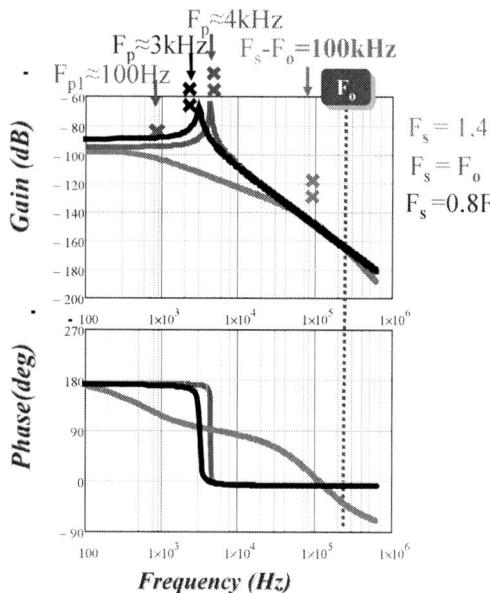

Fig. 16. Bode plots of control-to-output voltage transfer functions for LLC.

C. Analytical Expressions of Transfer Functions

The analytical transfer functions are provided in Table 1 and Table 2 for easy reference. These transfer functions are very helpful in designing the outer feedback compensator. For $F_s \geq F_o$, the transfer functions are generally third-order and can be reduced to second order when F_s is close to F_o. For $F_s < F_o$, the transfer functions are second-order. Due to space limitation, here only control to output voltage transfer functions are listed, for other transfer functions, please refer to [28].

Table I. Analytical transfer functions of LLC resonant converter for Fs ≥ Fo

$$\frac{\hat{v}_o(s)}{\hat{\omega}_s(s)} = G_{DC} \frac{X_{eq}^2 + R_{eq}^2}{(s^2 L_e^2 + sL_e R_{eq} + X_{eq}^2)(1 + R_L C_f s) + R_{eq}(sL_e + R_{eq})}$$

$$G_{DC} = \frac{V_g}{2n} \frac{L_n}{\omega_o \omega_n} \frac{(\frac{1}{\omega_n^2} - \omega_n^2)\left(\frac{\pi^2}{8} Q_t L_n\right)^2 - \left(L_n + 1 - \frac{1}{\omega_n^2}\right)\left(\frac{2}{\omega_n^2}\right)}{\left[\sqrt{\left(L_n + 1 - \frac{1}{\omega_n^2}\right)^2 + \left((\frac{1}{\omega_n} - \omega_n)\frac{\pi^2}{8} Q_t L_n\right)^2}\right]^3}$$

(12)

$$L_e = (1 + \frac{\Omega_o^2}{\Omega_s^2})L_r, R_{eq} = \frac{8}{\pi^2} n^2 R_L,$$

$$X_{eq} = \Omega_s L_r - \frac{1}{\Omega_s C_r}, Q = \frac{\sqrt{L_r/C_r}}{n^2 \cdot R_L}, L_n = \frac{L_m}{L_r}$$

Table II. Analytical transfer functions of LLC resonant converter for Fs < Fo

$$\frac{\hat{v}_o(s)}{\hat{\omega}_s(s)} = G_{DC} \frac{1}{1 + \frac{s}{Q_p \omega_p} + \left(\frac{s^2}{\omega_p^2}\right)}$$

$$G_{DC} = \frac{V_g}{2n} \frac{L_n}{\omega_o} \frac{\left[\left[(\frac{1}{\omega_n^2} - \omega_n^2)\left(\frac{\pi^2}{8} Q_t L_n\right)^2 - \left(L_n + 1 - \frac{1}{\omega_n^2}\right)\left(\frac{2}{\omega_n^2}\right)\right]\frac{1}{\omega_n} \cdot \frac{1}{\sin(\frac{\pi}{2}\omega_n)}}{\left[\sqrt{\left(L_n + 1 - \frac{1}{\omega_n^2}\right)^2 + \left((\frac{1}{\omega_n} - \omega_n)\frac{\pi^2}{8} Q_t L_n\right)^2}\right]^3} + \frac{-\frac{\pi}{2}\frac{\cos(\frac{\pi}{2}\omega_n)}{\sin^2(\frac{\pi}{2}\omega_n)}}{\sqrt{\left(L_n + 1 - \frac{1}{\omega_n^2}\right)^2 + \left((\frac{1}{\omega_n} - \omega_n)\frac{\pi^2}{8} Q_t L_n\right)^2}}\right]$$

(13)

$$Q_p = \frac{8n}{\pi^2} R_L \sqrt{\frac{C_f}{L_e}}, \omega_p = \sqrt{\frac{1}{L_r \frac{\pi^2}{8n^2} C_f}}, L_e = L_r(1 + \frac{1}{\omega_n^2}) + L_m(1 - \omega_n)$$

IV. SIMULATION AND EXPERIMENTAL VERIFICATIONS

The SIMPLIS simulation tool is used to verify the voltage conversion ratio and small-signal analysis. Circuit parameters are shown as follows: Vg=400V, Lr=14uH, Cr=30nF, Fo=250kHz, Cf=660μF, n=4, RL=2.3Ω, the corresponding Q=0.6. At resonant frequency, the output voltage is around 48V with 1kW full power.

Fig. 17 shows the comparison of the voltage conversion ratio using analytical equation (7) and simulation results. The analytical solutions have good accuracy when Fs ≥ 0.8 Fo. However, as fundamental approximation is still used in this approach, at low switching frequency, the gain curve starts to lose accuracy as there is more harmonics in the resonant variables.

Fig. 18 shows Simplis verification of control to output voltage transfer function for Fs=1.4Fo, Fs=Fo and Fs=0.8Fo. In all three cases, the model match very well with simulation results. As previous analysis shows, when Fs=1.4Fo, there is still a beat frequency double pole; when Fs=Fo, the beat frequency double pole splits and a new double pole formed by Le and equivalent output capacitor Cf shows up; when Fs=0.8Fo, the double pole moves to a little lower frequency as equivalent resonant inductor is increased to include the effect of magnetizing inductor.

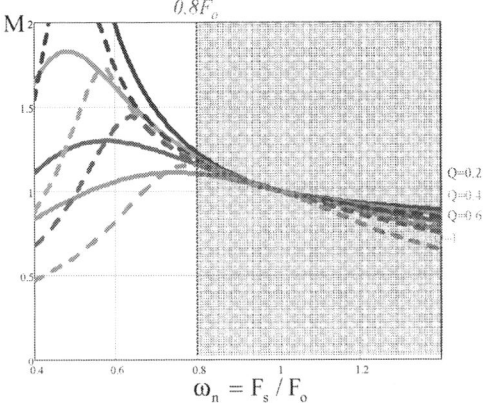

Fig. 17. Simplis verification of voltage conversion ratio for LLC converter.

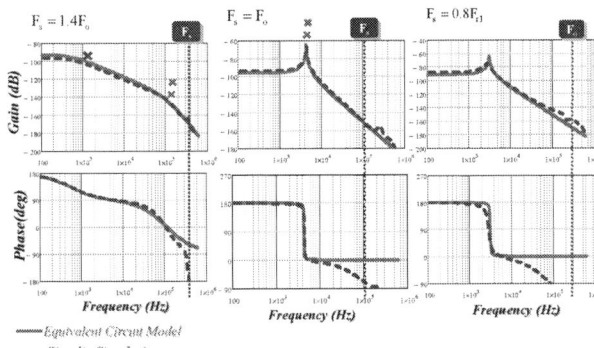

— Equivalent Circuit Model
— · Simplis Simulation

Fig. 18. Simplis verification of control-to-output transfer function for F_s=1.4F_o, Fs=Fo and Fs=0.8Fo

To verify other transfer functions, Fig. 19 shows the comparison of all the transfer functions between the equivalent circuit model and simulation results for Fs=1.2Fo. All the transfer functions match very well. For more verifications, please refer to [28].

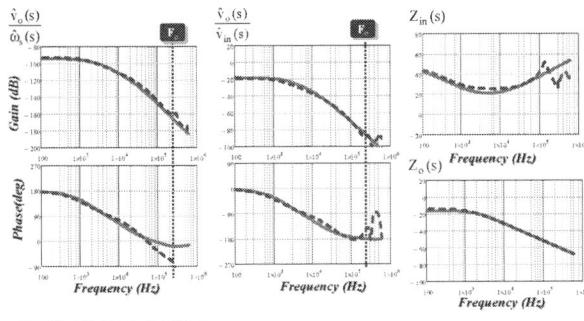

— Simplified Equivalent Circuit
— · Simplis Simulation

Fig. 19. Simplis verification of small-signal equivalent circuit model for F_s=1.2F_o.

The experimental verification of control to output voltage is shown as Fig. 20. The circuit parameters of the hardware are as follows: Vin=50V, Lr = 360nH, Lm = 2.1uH, Cr = 110nF, Cf = 50uF, n=5:1, RL=1Ω, the corresponding Q=0.07, fo=800kHz, fs=1.1MHz. From Fig. 20, the small-signal model matches very well with the experimental data.

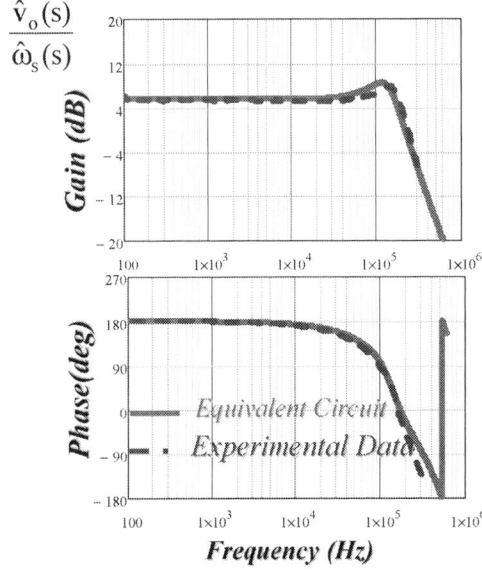

Fig.20. Experimental verification of control-to-output transfer function.

V. SUMMARY

This paper proposes an equivalent circuit model for LLC resonant converter. When Fs ≥ Fo, Lm is clamped by the output voltage and LLC behaves very similar as SRC. As a result, the dynamic behavior is similar as SRC: when switching frequency is larger than resonant frequency, the beat frequency double pole show up and the circuit is third-order; when switching frequency is close to resonant frequency, beat frequency double pole disappear and a new double pole formed by equivalent inductor Le and equivalent output capacitor Cf show up. The circuit reduces to second order.   When Fs<Fo, Lm participates in resonance and the circuit is essentially a multi-resonant structure. An approximated model is proposed where the equivalent resonant inductor is modified to include the effect of Lm. As a result, the double pole will move to a little lower frequency. For the first time, analytical solutions are provided for all the transfer functions which are very helpful for feedback design. Simulation and experimental results verify that the equivalent circuit model can well predict the dynamic behavior when switching frequency is below, close to or above resonant frequency.

Acknowledgment

This work was supported by PMC consortium in Center for Power Electronics System, Virginia Tech. In particular, the authors would like to express thanks to Delta Corporation for providing experimental data.

978-1-4673-9551-9/16 $31.00 © 2016 IEEE

REFERENCES

[1] R. Tymerski, V. Vorperian, F. C. Lee and W. T. Baumann, "Nonlinear Modeling of the PWM Switch," IEEE Trans. Power Electron., vol.4, no. 2, pp. 225-233, 1989.

[2] V. Vorperian, "Simplified analysis of PWM converters using model of PWM switch. Continuous conduction mode," IEEE Trans. Aerosp., Vol. 26, pp. 490 –496, May 1990.

[3] J. Li, F. C. Lee, "New modeling approach and equivalent circuit representation for current mode control" IEEE Trans. Power Electron., vol. 25, no. 5, pp.1218–1230, May. 2010.

[4] Y. Yan, F. C. Lee, and P.Mattavelli, "Unified three-terminal switch model for current mode controls," IEEE Trans. Power Electron., vol. 27, no. 9, pp. 4060–4070, Sep. 2012.

[5] S. Tian, F. C. Lee, J. Li, Q. Li and P. Liu, "Equivalent Circuit Model of Constant On-time Current Mode Control With External Ramp Compensation," in Proc. IEEE ECCE, 2014, pp.3747—3754.

[6] S. Tian, F. C. Lee, J. Li, Q. Li and P. Liu, "Three-terminal Switch Model of Constant On-time Current Mode with External Ramp Compensation," IEEE Trans. Power Electron., Vol. PP, No. 99. Accepted and to be published.

[7] J. Li and F. C. Lee, "Modeling of V^2 current-mode control," IEEE Tran. Circuits and Systems, Part I, Vol. 57, No 9. Sep.2009, pp. 2552-2563.

[8] S. Tian, K. Cheng, F. C. Lee and P. Mattavelli, "Small-signal model analysis and design of constant on-time V^2 control for low-ESR caps with external ramp compensation," in Proc. IEEE ECCE, 2011, pp. 2944-2951.

[9] S. Tian, F. C. Lee, P. Mattavelli, K. Cheng and Y. Yan, "Small-signal Analysis and Optimal Design of External Ramp for Constant On-Time V^2 Control with Multilayer Ceramic Caps," IEEE Trans. Power Electron., vol.29, no.8, pp.4450-4460, Aug. 2014.

[10] S. Tian, F. C. Lee, P. Mattavelli and Y. Yan, "Small-signal analysis and design of constant frequency V^2 peak control," in Proc. IEEE APEC, 2013.pp. 1717 - 1724.

[11] S. Tian, F. C. Lee, P. Mattavelli and Y. Yan, "Small-signal Analysis and Optimal Design of Constant Frequency V^2 Control," IEEE Trans. Power Electron., vol.30, no.3, pp. 1724-1733, Mar. 2015.

[12] S. Tian, F. C. Lee, Q. Li and Y. Yan, "Unified equivalent circuit model of V^2 control," in Proc. IEEE APEC, 2014, pp. 1016 - 1023.

[13] S. Tian, F. C. Lee, Q. Li and Y. Yan, "Unified Equivalent Circuit Model and Optimal Design of V^2 Controlled Buck Converters," IEEE Trans. Power Electron., vol. 31, no. 2, pp. 1734-1744, Feb. 2016.

[14] V. Vorperian, "Approximate small-signal analysis of the series and the parallel resonant converters," IEEE Trans. Power Electron., vol. 4, no. 1,

pp. 15–24, Jan. 1989.

[15] E. X. Yang, F. C. Lee and M. M. Jovanovic, "Small-signal modeling of power electronic circuits by extended describing function concept," in Proc. Virginia Power Electronics Center Seminar, 1991, pp. 167-178.

[16] S. Tian, F. C. Lee, Q. Li and B. Li, " Small-signal equivalent model of series resonant converter," in Proc. IEEE ECCE, 2015, pp. 172-179.

[17] S. Tian, F. C. Lee, and Q. Li, "A Simplified Equivalent Circuit Model of Series Resonant Converter," IEEE Trans. Power Electron., vol. PP, Issue:99, 2015, accepted and to be published.

[18] B. Yang, F.C. Lee, A.J. Zhang and G. Huang, "LLC Resonant Converter For Front End DC-DC Conversion," in Proc. IEEE APEC, 2002, vol. 2, pp. 1108-1112.

[19] (2007, May.) STMicroelectronics. AN 2321 application note: Reference design high performance, L6599-based HB-LLC adapter with PFC for laptop computers. Avalable:
http://www.mouser.com/catalog/specsheets/EVAL6599-90W.pdf

[20] C. Chang, E. Chang, C. Cheng, H. Cheng, and S. Lin, "Small Signal Modeling of LLC Resonant Converters Based on Extended Describing Function," in Proc. Inter. Sympo. On. Comp. Cons. and Cont. (IS3C), 2012, pp. 365-368.

[21] C. Buccella, C. Cecati, H. Latafat, P. Pepe and K. Razi, "Linearization of LLC Resonant Converter Model Based on Extended Describing Function Concept," in Proc. IEEE Intern. Workshop on Intelligent Energy System (IWIES), 2013, pp. 131-136.

[22] Z. Zahid, J. Lai, X. Huang, S. Madiwale and J. Hou, "Damping Impact on Dynamic Analysis of LLC Resonant Converter," in Proc. IEEE APEC, 2014, pp. 2835-2841.

[23] C. Buccella, C. Cecati, H. Latafat, P. Pepe, and K. Razi, "Observer-Based Control of LLC DC/DC Resonant Converter Using Extended Describing Functions," IEEE Trans. on Power Electronics, vol. 30, no. 10, pp. 5881-5891, Oct. 2015.

[24] B. Cheng, F. Musavi and W. Dunford, "Novel Small Signal Modeling and Control of an LLC Resonant Converter," in Proc. IEEE APEC, 2014, pp.2828-2834.

[25] J. Stahl, T. Hieke, C.oeder and T. Duerbaum, "Small-signal Analysis of the Resonant LLC Converter," in Proc. IEEE ECCE-Asia, 2013, pp. 25-30.

[26] B. Yang and F. C. Lee, "Small-signal analysis of LLC Resonant Converter," in Proc. Center for Power Electronics System Seminar, 2003, pp. 1-6.

[27] H. Huang, "Feedback Loop Design of an LLC Resonant Power Converter," Application Report, Texas Instruments, Nov, 2010.

[28] S. Tian, "Equivalent Circuit Model of High Frequency PWM and Resonant Converters," Ph.D. dissertation, Dept. Elect. Comput. Eng., Virginia Tech, Blacksburg, VA, USA, Aug. 2015.

Small Signal Modeling of the Hysteretic Modulator with a Current Ripple Synthesizer

Yi Huang, Chun Cheung
Intersil Corporation
440 U.S. Highway 22 East, Suite 100
Bridgewater, NJ 08807

Abstract— **In this paper, the synthetic ripple modulator for synchronous buck converter is modeled by the describing function approach. A detailed circuit configuration of the PWM modulator is introduced, including the hysteretic band control and the ripple synthesizer. The transfer function of the modulator is derived. By comparing the Bode plots for the open-loop hysteretic modulator and the closed-loop system, a good match is obtained between the analytical model and the SIMPLIS simulation results. The proposed model can be used in the design of the current ripple synthesizer to achieve fast transient response, and in the prediction of the system stability.**

I. INTRODUCTION

As the development of modern microprocessors, such as CPU and GPU progress, the demand for fast transient response from voltage regulator modules (VRM) has rapidly increased. Conventional voltage and current mode control cannot keep up with the high slew rate of the current demands from the loads. Therefore, researchers and circuit designers have been devoted to the development of novel control architectures to meet the challenge brought from the microprocessor design. Among the novel control techniques, free-running converters, or ripple regulators, have gained much attention, due to its intrinsic fast response [1]- [3]. As shown in Fig. 1, the ripple modulator generates a PWM signal based on the process of the sensed ripple information of the output voltage.

Although multiple controller ICs for the ripple regulators are readily available on the market, the bottleneck for the promotion of the regulators has been their modeling. The attempt in modeling ripple regulators by the conventional averaging technique always yielded some degree of error. Therefore, this bottleneck increases the system design difficulty in predicting system stability and dynamic characteristics.

Over the past few years, a significant amount of effort in academia and industry has been devoted to the modeling of ripple regulators. The constant On-time ripple regulator was successfully modeled in [4] and has been extended in [5] using the describing function (DF) approach. The DF approach is a powerful mathematical tool for predicting the dynamic characteristics of the control systems. Previously, it has been employed in the analysis of a general PWM feedback system [6], predicting phase shift [7] and phase lag [8] of the PWM modulators. As it is mentioned in [4], the DF method is an effective solution for modeling the ripple regulators due to its time domain basis.

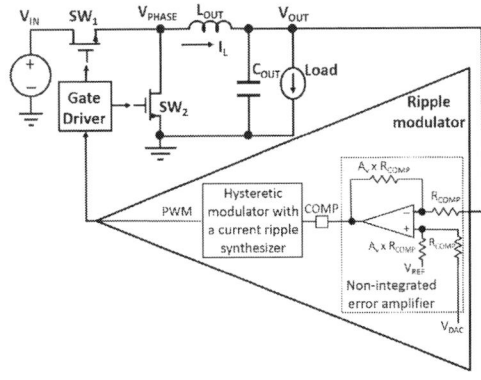

Fig. 1. Schematic of a synchronous buck ripple regulator, consisting of non-integrated error amplifier compensation and hysteretic modulator with a current ripple synthesizer.

Although the DF approach has been used in modeling different control schemes: constant On-time and Off-time voltage mode control [9], peak current mode control [4], variable frequency current mode control [4], [5], average current mode control [10], the constant and variable frequency V^2 control [11]- [13], resonant tanks in series resonant converters [14], [15], and LLC resonant converters [16], this approach has not been reported in modeling hysteretic controlled ripple regulators. The existing hysteretic control modeling in [17]- [19] is still based on the averaging technique and not applicable to ripple regulators. The purpose of this paper is to apply the DF approach to the ripple-based control of hysteretic buck regulators, with an emphasis on a special case with the current ripple synthesizer. The proposed model can also be extended to general hysteretic PWM modulators.

This paper is organized as follows: a detailed circuit configuration of the variable frequency PWM modulator, with the current ripple synthesizer, is discussed in Section II, together with the derivation of its transfer function. The analysis of a closed loop buck converter is presented in Section III. Model verification is shown in section IV, and the conclusions are given in Section V.

II. MODELING OF HYSTERETIC MODULATOR WITH A CURRENT RIPPLE SYNTHESIZER

The concept of the synthetic ripple modulator [20], [21] has gone through a fast evolution over the past few years. State-

978-1-4673-9551-9/16 $31.00 © 2016 IEEE

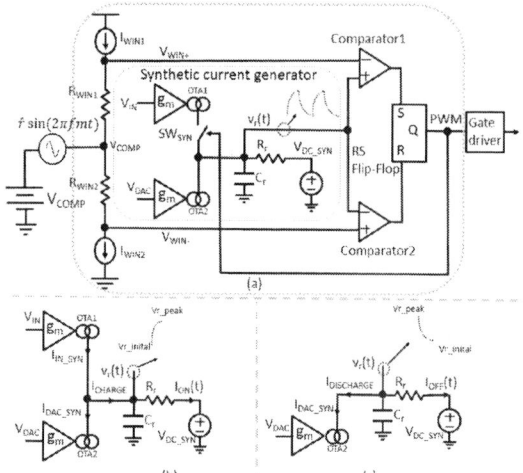

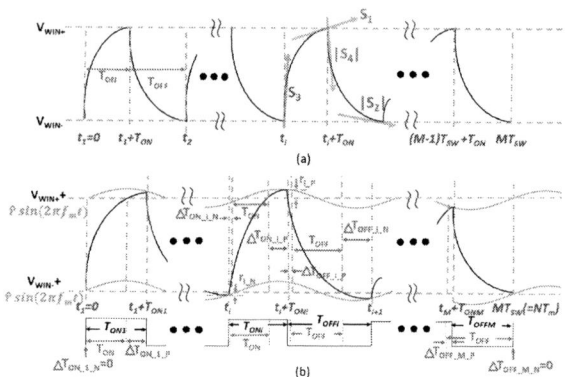

Fig. 3. The transient waveform of $v_r(t)$ and PWM: (a) steady state operation (b) under perturbation.

from $v_r(t)$ to GND, while the current going through R_r is defined as $I_{OFF}(= (v_r(t) - V_{DC\_SYN})/R_r)$.

Within a complete switching cycle, the transient expressions of this artificial current ripple $V_r(t)$ can be described as:

$$
\begin{cases}
C_r \dfrac{dv_r(t)}{dt} = g_m V_{IN} - g_m V_{DAC} - \dfrac{v_r(t) - V_{DC\_SYN}}{R_r}, \\
\qquad\qquad\qquad (iT_{SW} \le t \le iT_{SW} + T_{ON}) \\
C_r \dfrac{dv_r(t)}{dt} = -g_m V_{DAC} - \dfrac{v_r(t) - V_{DC\_SYN}}{R_r}, \\
\qquad\qquad\qquad (iT_{SW} + T_{ON} \le t \le (i+1)T_{SW})
\end{cases}
\tag{1}
$$

where i can be either 0 or any positive integer, and T_{SW} is the switching cycle of the PWM.

The differential equations in (1) can thus be solved as:

$$
\begin{cases}
v_r(t) = A - B e^{\frac{-t}{R_r C_r}}, & (iT_{SW} \le t \le iT_{SW} + T_{ON}) \\
v_r(t) = C + D e^{\frac{-t}{R_r C_r}}, & (iT_{SW} + T_{ON} \le t \le (i+1)T_{SW})
\end{cases}
\tag{2}
$$

where

$$
\begin{aligned}
A &= V_{DC\_SYN} + R_r g_m (V_{IN} - V_{DAC}), \\
B &= R_r g_m V_{IN} \left[1 - \frac{\left(1 - e^{\frac{-V_{DAC}}{V_{IN} R_r C_r f_{SW}}}\right) e^{\frac{-1}{R_r C_r f_{SW}}}}{e^{\frac{-V_{DAC}}{V_{IN} R_r C_r f_{SW}}} \left(1 - e^{\frac{-1}{R_r C_r f_{SW}}}\right)} \right], \\
C &= V_{DC\_SYN} - R_r g_m V_{DAC}, \\
D &= R_r g_m V_{IN} \frac{1 - e^{\frac{-V_{DAC}}{V_{IN} R_r C_r f_{SW}}}}{e^{\frac{-V_{DAC}}{V_{IN} R_r C_r f_{SW}}} \left(1 - e^{\frac{-1}{R_r C_r f_{SW}}}\right)}.
\end{aligned}
$$

Based on (2), the four slopes at the switching decision points, shown in Fig. 3-a, can be derived. By differentiating $v_r(t)$ at the specific time point, the absolute values of the slopes are expressed in (3).

B. Description of modulation law under perturbation

As shown in Fig. 2-a, an open loop test can be setup to derive the modulation law under perturbation: initially break the $COMP$ node from the compensation network; then place a DC voltage source V_{COMP} onto this node in order to maintain the steady state operation of the modulator; inject a

Fig. 2. Hysteretic modulator with current ripple synthesizer: (a) schematic of the open loop test (b) steady state operation during PWM On-time of the current ripple synthesizer (c) steady state operation during PWM Off-time of the current ripple synthesizer.

of-the-art technology has combined it with non-integrated error amplifier compensation technology [22], [23] to provide ultra-fast transient response [24]- [26]. The current ripple synthesizer discussed in this text is a simplified version of the topologies discussed in [27].

A. Steady state operation

In Fig. 1, a non-integrated error amplifier inverts and amplifies the ripple information of the output voltage of the buck converter. The output node of the error amplifier v_{COMP} is then feed into the hysteretic modulator, to generate the PWM signal of the system. The internal circuit configuration of the hysteretic modulator is shown within the dashed line in Fig. 2-a. Two identical current sources I_{WIN1} and I_{WIN2} go through resistors R_{WIN1} and R_{WIN2} to form the hysteresis window, with V_{COMP} as its central level. This window is compared with the synthetic current ripple $v_r(t)$ via two hysteretic comparators ($Comparator1$ and $Comparator2$), to generate the PWM carrier signal via a RS Flip-Flop.

The steady state operation of the current ripple synthesizer is interpreted here: during the On-time of the PWM signal, SW_{SYN} in Fig. 2-a conducts. The input voltage of the buck converter V_{IN} goes through a transconductance amplifier ($OTA1$) to generate a current I_{IN_SYN}, while a sink current I_{DAC_SYN} is formed by another DC voltage V_{DAC} via a second transconductance amplifier ($OTA2$). As it is shown in Fig. 2-b, a charging current $I_{CHARGE}(= I_{IN\_SYN} - I_{DAC\_SYN})$ is going towards the capacitor C_r. C_r is connected to a DC source voltage V_{DC_SYN} via a series resistor R_r. The current going through R_r is defined as $I_{ON}(= (v_r(t) - V_{DC\_SYN})/R_r)$. During the Off-time of the PWM signal, SW_{SYN} in Fig. 2-a is turned off. Similarly, as shown in Fig. 2-c, $I_{DISCHARGE}(= I_{DAC\_SYN})$ sinks current

small modulation sinusoidal signal $\hat{r}\sin(2\pi f_m t)$ in superposition with V_{COMP}. The AC perturbation signal at V_{COMP} is replicated at the DC levels of the window boundaries (V_{WIN+} and V_{WIN-}) simultaneously, as shown in the blue traces Fig. 3-b. The modulation law can be derived based on the perturbation analysis at two window boundaries seperately.

As shown in Fig. 3-b, in the ith switching cycle, the perturbed On-time and Off-time of the PWM signal are denoted as T_{ON_i} and T_{OFF_i}. Since this is a variable frequency modulation scheme, the changes in T_{ON_i} and T_{OFF_i} contribute to the change of time duration of the ith switching cycle. As shown in Fig. 3-b, the relationship between the perturbed signal and the steady state operation can be expressed as:

$$\begin{cases} T_{ON\_i} = T_{ON} + \Delta T_{ON\_i\_N} + \Delta T_{ON\_i\_P}, \\ T_{OFF\_i} = T_{OFF} + \Delta T_{OFF\_i\_P} + \Delta T_{OFF\_i\_N}, \end{cases} \quad (4)$$

where $\Delta T_{ON\_i\_N}$ and $\Delta T_{OFF\_i\_N}$ are caused by the perturbation at the V_{WIN-}, while $\Delta T_{ON\_i\_P}$ and $\Delta T_{OFF\_i\_P}$ come from the perturbation at the V_{WIN+}.

To start the analysis of the perturbed waveforms, several assumptions must be made:

- The peak amplitude $\hat{r}$ of the modulation signal $\hat{r}\sin(2\pi f_m t)$ is much smaller than the steady state hysteresis window size. Therefore, T_{ON} and T_{OFF} are much greater than the perturbed time intervals.
- The values of all four slopes in Fig. 3-a are assumed to be constant within a small neighboring region of their corresponding switching decision time points. Therefore, for each cycle under perturbation, the time interval between the DC window boundaries is still T_{ON} and T_{OFF}.
- The initial phase shift θ_0 between the modulation signal and the first switching cycle is assumed to be zero. In other words, $\hat{r}\sin(2\pi f_m t + \theta_0) = \hat{r}\sin(2\pi f_m t)$ when the perturbed signal is injected at $t_1 = 0$.
- The modulation frequency and the switching frequency are assumed to be commensurable: $M f_m = N f_{SW}$, while M and N are two positive integers [4], [9].

To derive the general expressions of the switching decision points t_i and $t_i + T_{ON\_i}$, the performance of the first two switching cycles can analyzed as the starting point:

- For the first switching cycle, $\Delta T_{ON\_1\_N} = 0$ due to the zero initial phase shift (Fig. 3-b). The duration of the entire cycle is $T_{ON} + \Delta T_{ON\_1\_P} + \Delta T_{OFF\_1\_P} + T_{OFF} + \Delta T_{OFF\_1\_N}$.
- The duration of the first cycle is also the starting point t_2 of the second switching cycle, where $t_2 = T_{SW} + \Delta T_{ON\_1\_P} + \Delta T_{OFF\_1\_P} + \Delta T_{OFF\_1\_N}$. The entire time interval of the second cycle is $T_{ON} + \Delta T_{ON\_2\_N} + \Delta T_{ON\_2\_P} + \Delta T_{OFF\_2\_P} + T_{OFF} + \Delta T_{OFF\_2\_N}$.

Based on the analysis above, one can conclude that

$$t_i = (i-1)T_{SW} + \sum_{k=1}^{i-1} \Delta T_{ON\_k\_N} + \sum_{k=1}^{i-1} \Delta T_{ON\_k\_P} \quad (5)$$
$$+ \sum_{k=1}^{i-1} \Delta T_{OFF\_k\_P} + \sum_{k=1}^{i-1} \Delta T_{OFF\_k\_N},$$

and

$$t_i + T_{ON\_i} = (i-1)T_{SW} + T_{ON} + \sum_{k=1}^{i} \Delta T_{ON\_k\_N} \quad (6)$$
$$+ \sum_{k=1}^{i} \Delta T_{ON\_k\_P} + \sum_{k=1}^{i-1} \Delta T_{OFF\_k\_P} + \sum_{k=1}^{i-1} \Delta T_{OFF\_k\_N}.$$

In Fig. 4, the small neighboring regions of the switching decision points at the ith cycle in Fig. 3-b are expanded at: the start of the switching cycle t_i (Fig. 4-c), the end of the On-time $t_i+T_{ON\_i}$ (Fig. 4-b), and the end of this cycle t_{i+1}, which is also the start of the following cycle (Fig. 4-d).

1) $\Delta T_{ON\_i\_N}$: As shown in Fig. 4-c, the amplitude of the perturbed sinusoidal wave is r_{i_N}, and the perturbed On-time duration near this window boundary is $\Delta T_{ON\_i\_N}$.

The perturbation amplitude r_{i_N} can be expressed as

$$r_{i\_N} = \hat{r}\sin(2\pi f_m t_i). \quad (7)$$

$$S_1 = \frac{dv_r(t)}{dt}\Big|_{t=iT_{SW}+T_{ON-}} = \frac{g_m V_{IN}}{C_r}\left[e^{\frac{-V_{DAC}}{V_{IN}R_r C_r f_{SW}}} - \frac{\left(1-e^{\frac{-V_{DAC}}{V_{IN}R_r C_r f_{SW}}}\right)e^{\frac{-1}{R_r C_r f_{SW}}}}{1-e^{\frac{-1}{R_r C_r f_{SW}}}} \right]$$

$$S_2 = \left| \frac{dv_r(t)}{dt}\Big|_{t=(i+1)T_{SW-}} \right| = \left| \frac{-g_m V_{IN}}{C_r} \frac{\left(1-e^{\frac{-V_{DAC}}{V_{IN}R_r C_r f_{SW}}}\right)e^{\frac{-1}{R_r C_r f_{SW}}}}{e^{\frac{-V_{DAC}}{V_{IN}R_r C_r f_{SW}}}\left(1-e^{\frac{-1}{R_r C_r f_{SW}}}\right)} \right|$$

$$S_3 = \frac{dv_r(t)}{dt}\Big|_{t=iT_{SW+}} = \frac{g_m V_{IN}}{C_r}\left[1 - \frac{\left(1-e^{\frac{-V_{DAC}}{V_{IN}R_r C_r f_{SW}}}\right)e^{\frac{-1}{R_r C_r f_{SW}}}}{e^{\frac{-V_{DAC}}{V_{IN}R_r C_r f_{SW}}}\left(1-e^{\frac{-1}{R_r C_r f_{SW}}}\right)} \right]$$

$$S_4 = \left| \frac{dv_r(t)}{dt}\Big|_{t=iT_{SW}+T_{ON+}} \right| = \left| \frac{-g_m V_{IN}}{C_r} \frac{\left(1-e^{\frac{-V_{DAC}}{V_{IN}R_r C_r f_{SW}}}\right)}{\left(1-e^{\frac{-1}{R_r C_r f_{SW}}}\right)} \right|$$

$$(3)$$

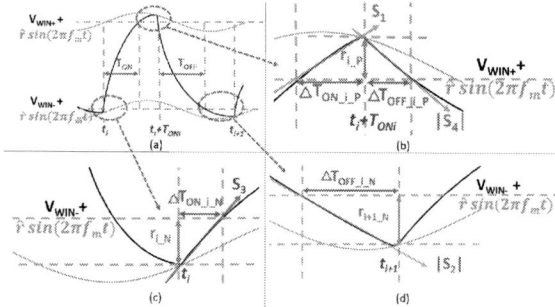

Fig. 4. Expanded view of ith switching cycle of the perturbed $V_r(t)$ waveform: (a) overview (b) near the falling edge of the PWM signal (c) near the rising edge of the PWM signal (d) near the end of one switching cycle.

Given the assumption that the slope S_3 does not change from t_i to $t_i + \Delta T_{ON\_i\_N}$,

$$|\Delta T_{ON\_i\_N}||S_3| = |r_{i\_N}|. \tag{8}$$

The signs of both sides of (8) can be determined by looking into Fig. 4-c without losing generality: r_{i_N} is below the DC level V_{WIN-} and thus is negative. The fact that r_{i_N} is below V_{WIN-} leads to a positive extension of the PWM On-time at this boundary. Since S_3 is the absolute value of the slope, it is always positive. Equation (8) can be rewritten as

$$\Delta T_{ON\_i\_N} S_3 = -r_{i\_N}. \tag{9}$$

As a result,

$$\Delta T_{ON\_i\_N} = -\frac{r_{i\_N}}{S_3} = -\frac{\hat{r}\sin(2\pi f_m t_i)}{S_3}. \tag{10}$$

2) $\Delta T_{ON\_i\_P}$: In Fig. 4-b, the perturbation at the upper window boundary leads to a second portion of the perturbed On-time, $\Delta T_{ON\_i\_P}$. Following a similar procedure in analyzing $\Delta T_{ON\_i\_N}$,

$$|\Delta T_{ON\_i\_P}||S_1| = |r_{i\_P}| = |\hat{r}\sin(2\pi f_m(t_i + T_{ON\_i}))|. \tag{11}$$

Due to the fact that a positive r_{i_P} extends the On-time near the upper window boundary, (11) can be rewritten as

$$\Delta T_{ON\_i\_P} S_1 = r_{i\_P}, \tag{12}$$

hence,

$$\Delta T_{ON\_i\_P} = \frac{r_{i\_P}}{S_1} = \frac{\hat{r}\sin(2\pi f_m(t_i + T_{ON\_i}))}{S_1}. \tag{13}$$

3) $\Delta T_{OFF\_i\_P}$: In Fig. 4-b, since the perturbation at $t_i + T_{ON\_i}$ causes the change at both On-time and Off-time, it is straightforward to express

$$\Delta T_{OFF\_i\_P} = \frac{r_{i\_P}}{S_4} = \frac{\hat{r}\sin(2\pi f_m(t_i + T_{ON\_i}))}{S_4}. \tag{14}$$

4) $\Delta T_{OFF\_i\_N}$: Similar to the analysis for $\Delta T_{ON\_i\_N}$, in Fig. 4-d, $\Delta T_{OFF\_i\_N}$ can be expressed as:

$$\Delta T_{OFF\_i\_N} = -\frac{r_{i+1\_N}}{S_2} = -\frac{\hat{r}\sin(2\pi f_m t_{i+1})}{S_2}. \tag{15}$$

C. Fourier analysis

By implementing the modulation law, each switching decision point has been obtained from (10), (13), (14) and (15). Fourier analysis can be performed to derive the transfer function from v_{COMP} to perturbed duty cycle. According to [4], the duty cycle at the ith switching cycle can be described as:

$$d(t)|_{t_i \leq t \leq t_i + T_{ON\_i}} = u(t - t_i) - u(t - t_i - T_{ON\_i}), \tag{16}$$

and

$$d(t)|_{0 \leq t \leq MT_{SW}} = \sum_{i=1}^{M}\left[u(t - t_i) - u(t - t_i - T_{ON\_i})\right], \tag{17}$$

where $u(t)$ is the step function.

Applying Fourier analysis to (17), the Fourier coefficient at perturbation frequency f_m can be written as:

$$
\begin{aligned}
C_m(d) &= \frac{j2\pi f_m}{N\pi}\sum_{i=1}^{M}\int_{t_i}^{t_i + T_{ON\_i}} d(t)e^{-j2\pi f_m t}dt \\
&= \frac{1}{N\pi}\sum_{i=1}^{M}\left(e^{-j2\pi f_m t_i} - e^{-j2\pi f_m(t_i + T_{ON\_i})}\right).
\end{aligned} \tag{18}
$$

By inserting (5) and (6) in (18), $C_m(d)$ can be expressed as (18), where

$$
\begin{cases}
\Delta T_{ON\_k} = \Delta T_{ON\_k\_N} + \Delta T_{ON\_k\_P}, \\
\Delta T_{OFF\_k} = \Delta T_{OFF\_k\_N} + \Delta T_{OFF\_k\_P}.
\end{cases} \tag{19}
$$

In (18), since both ΔT_{ON_k} and ΔT_{OFF_k} are much less than 1, the Taylor series expansion can be applied to the terms related to both of them:

$$
\begin{aligned}
&e^{-j2\pi f_m\left(\sum_{k=1}^{i-1}\Delta T_{ON\_k} + \sum_{k=1}^{i-1}\Delta T_{OFF\_k}\right)} \\
&\approx 1 - j2\pi f_m\underbrace{\left(\sum_{k=1}^{i-1}\Delta T_{ON\_k} + \sum_{k=1}^{i-1}\Delta T_{OFF\_k}\right)}_{\Delta t_A} \\
&e^{-j2\pi f_m\left(\sum_{k=1}^{i}\Delta T_{ON\_k} + \sum_{k=1}^{i-1}\Delta T_{OFF\_k}\right)} \\
&\approx 1 - j2\pi f_m\underbrace{\left(\sum_{k=1}^{i}\Delta T_{ON\_k} + \sum_{k=1}^{i-1}\Delta T_{OFF\_k}\right)}_{\Delta t_B}.
\end{aligned} \tag{20}
$$

$$C_m(d) = \frac{1}{N\pi}\sum_{i=1}^{M}\left(e^{-j2\pi f_m\left((i-1)T_{SW} + \sum_{k=1}^{i-1}\Delta T_{ON\_k} + \sum_{k=1}^{i-1}\Delta T_{OFF\_k}\right)} - e^{-j2\pi f_m\left((i-1)T_{SW} + T_{ON} + \sum_{k=1}^{i}\Delta T_{ON\_k} + \sum_{k=1}^{i-1}\Delta T_{OFF\_k}\right)}\right) \tag{18}$$

With the aid of (20), (18) can be further implied to:

$$
\begin{aligned}
C_m(d) \quad &= \underbrace{\frac{1}{N\pi} \sum_{i=1}^{M} e^{-j2\pi f_m(i-1)T_{SW}}}_{=0} \\
&\quad -\underbrace{\frac{1}{N\pi} \sum_{i=1}^{M} e^{-j2\pi f_m((i-1)T_{SW}+T_{ON})}}_{=0} \\
&\quad -\underbrace{\frac{j2\pi f_m}{N\pi} \sum_{i=1}^{M} e^{-j2\pi f_m(i-1)T_{SW}} \Delta t_A}_{C_{m\_A}} \\
&\quad +\underbrace{\frac{j2\pi f_m}{N\pi} \sum_{i=1}^{M} e^{-j2\pi f_m((i-1)T_{SW}+T_{ON})} \Delta t_B}_{C_{m\_B}},
\end{aligned}
\tag{21}
$$

whose first two terms are equal to zero (see Appendix A). Now Δt_A and Δt_B in (21) need be derived to calculate C_{m_A} and C_{m_B}.

By the observation of Δt_A and Δt_B in (20), three different terms need to be derived separately: $\sum_{k=1}^{i-1} \Delta T_{ON\_k}$, $\sum_{k=1}^{i} \Delta T_{ON\_k}$ and $\sum_{k=1}^{i-1} \Delta T_{OFF\_k}$. Once these three terms are available, one can proceed to calculate C_{m_A} and C_{m_B}.

Starting from the analysis of $\sum_{k=1}^{i-1} \Delta T_{ON\_k}$, and considering (10), (13) and (19),

$$
\begin{aligned}
\sum_{k=1}^{i-1} \Delta T_{ON\_k} &= \sum_{k=1}^{i-1} \left(\Delta T_{ON\_k\_N} + \Delta T_{ON\_k\_P} \right) \\
&= \sum_{k=1}^{i-1} \left(-\frac{\hat{r}\sin(2\pi f_m t_k)}{S_3} + \frac{\hat{r}\sin\left(2\pi f_m(t_k + T_{ON\_k})\right)}{S_1} \right).
\end{aligned}
\tag{22}
$$

By making a local assumption that $t_k \approx (k-1)T_{SW}$ and $T_{ON\_k} \approx T_{ON}$, (22) can be simplified and expressed as the form of the exponential function in (23), according to Euler's formula. Similarly, the expressions for $\sum_{k=1}^{i} \Delta T_{ON\_k}$ and $\sum_{k=1}^{i-1} \Delta T_{OFF\_k}$ can be obtained in (24) and (25).

With the expressions in (23), (24) and (25), Δt_A and Δt_B in (20) can be described. By applying the useful geometric series given in the Appendix B, Δt_A and Δt_B are expressed as (26) and (27) for the following derivation of C_{m_A} and C_{m_B} in (21).

Given that C_{m_A} is the summation of the product of $e^{-j2\pi f_m(i-1)T_{SW}}$ and Δt_A, the results of the vector analysis listed in Appendix A can be employed to derive its expression. This applies to C_{m_B} as well. With the aid of the expressions of C_{m_A} and C_{m_B} in (28) and (29), the Fourier coefficient in (21) is expressed in (30).

Considering that $M f_M = N f_{SW}$, the transfer function from control to perturbed duty cycle of this hysteretic modulator can be calculated from (30) and described in the s-domain in (31). Equation (31) indicates that the dynamic characteristic of this hysteretic modulator is a nonlinear function of system

$$
\sum_{k=1}^{i-1} \Delta T_{ON\_k} \approx \hat{r} \sum_{k=1}^{i-1} \frac{-e^{j2\pi f_m(k-1)T_{SW}} + e^{-j2\pi f_m(k-1)T_{SW}}}{2jS_3} + \hat{r} \sum_{k=1}^{i-1} \frac{e^{j2\pi f_m((k-1)T_{SW}+T_{ON})} - e^{-j2\pi f_m((k-1)T_{SW}+T_{ON})}}{2jS_1}
\tag{23}
$$

$$
\sum_{k=1}^{i} \Delta T_{ON\_k} \approx \hat{r} \sum_{k=1}^{i} \frac{-e^{j2\pi f_m(k-1)T_{SW}} + e^{-j2\pi f_m(k-1)T_{SW}}}{2jS_3} + \hat{r} \sum_{k=1}^{i} \frac{e^{j2\pi f_m((k-1)T_{SW}+T_{ON})} - e^{-j2\pi f_m((k-1)T_{SW}+T_{ON})}}{2jS_1}
\tag{24}
$$

$$
\sum_{k=1}^{i-1} \Delta T_{OFF\_k} \approx \hat{r} \sum_{k=1}^{i-1} \frac{e^{j2\pi f_m((k-1)T_{SW}+T_{ON})} - e^{-j2\pi f_m((k-1)T_{SW}+T_{ON})}}{2jS_4} - \hat{r} \sum_{k=1}^{i-1} \frac{e^{j2\pi f_m k T_{SW}} - e^{-j2\pi f_m k T_{SW}}}{2jS_2}
\tag{25}
$$

$$
\begin{aligned}
\Delta t_A = \frac{\hat{r}}{2j} \Bigg(&-\frac{1}{S_3}\frac{1-e^{j2\pi f_m(i-1)T_{SW}}}{1-e^{j2\pi f_m T_{SW}}} + \frac{1}{S_3}\frac{1-e^{-j2\pi f_m(i-1)T_{SW}}}{1-e^{-j2\pi f_m T_{SW}}} + \frac{e^{j2\pi f_m T_{ON}}}{S_1}\frac{1-e^{j2\pi f_m(i-1)T_{SW}}}{1-e^{j2\pi f_m T_{SW}}} \\
&-\frac{e^{-j2\pi f_m T_{ON}}}{S_1}\frac{1-e^{-j2\pi f_m(i-1)T_{SW}}}{1-e^{-j2\pi f_m T_{SW}}} + \frac{e^{j2\pi f_m T_{ON}}}{S_4}\frac{1-e^{j2\pi f_m(i-1)T_{SW}}}{1-e^{j2\pi f_m T_{SW}}} - \frac{e^{-j2\pi f_m T_{ON}}}{S_4}\frac{1-e^{-j2\pi f_m(i-1)T_{SW}}}{1-e^{-j2\pi f_m T_{SW}}} \\
&-\frac{e^{j2\pi f_m T_{SW}}}{S_2}\frac{1-e^{j2\pi f_m(i-1)T_{SW}}}{1-e^{j2\pi f_m T_{SW}}} + \frac{e^{-j2\pi f_m T_{SW}}}{S_2}\frac{1-e^{-j2\pi f_m(i-1)T_{SW}}}{1-e^{-j2\pi f_m T_{SW}}} \Bigg)
\end{aligned}
\tag{26}
$$

$$
\begin{aligned}
\Delta t_B = \frac{\hat{r}}{2j} \Bigg(&-\frac{1}{S_3}\frac{1-e^{j2\pi f_m i T_{SW}}}{1-e^{j2\pi f_m T_{SW}}} + \frac{1}{S_3}\frac{1-e^{-j2\pi f_m i T_{SW}}}{1-e^{-j2\pi f_m T_{SW}}} + \frac{e^{j2\pi f_m T_{ON}}}{S_1}\frac{1-e^{j2\pi f_m i T_{SW}}}{1-e^{j2\pi f_m T_{SW}}} \\
&-\frac{e^{-j2\pi f_m T_{ON}}}{S_1}\frac{1-e^{-j2\pi f_m i T_{SW}}}{1-e^{-j2\pi f_m T_{SW}}} + \frac{e^{j2\pi f_m T_{ON}}}{S_4}\frac{1-e^{j2\pi f_m(i-1)T_{SW}}}{1-e^{j2\pi f_m T_{SW}}} - \frac{e^{-j2\pi f_m T_{ON}}}{S_4}\frac{1-e^{-j2\pi f_m(i-1)T_{SW}}}{1-e^{-j2\pi f_m T_{SW}}} \\
&-\frac{e^{j2\pi f_m T_{SW}}}{S_2}\frac{1-e^{j2\pi f_m(i-1)T_{SW}}}{1-e^{j2\pi f_m T_{SW}}} + \frac{e^{-j2\pi f_m T_{SW}}}{S_2}\frac{1-e^{-j2\pi f_m(i-1)T_{SW}}}{1-e^{-j2\pi f_m T_{SW}}} \Bigg)
\end{aligned}
\tag{27}
$$

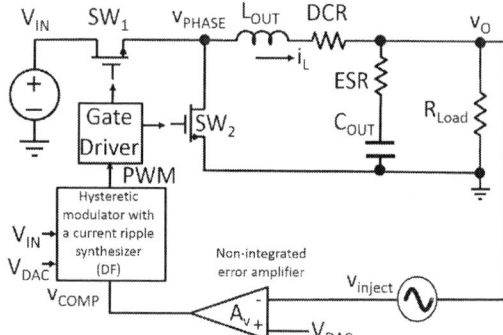

Fig. 5. The equivalent closed loop model.

configurations V_{IN}, V_{DAC} and f_{SW}, as well as the parameters of the current ripple synthesizer: gm, C_r and R_r.

III. CLOSED LOOP ANALYSIS

In the closed loop operation, the feedback output ripple goes through the non-integrated error amplifier. Although the amplifier itself has a limited bandwdith, it is generally much higher than the frequency range of interest for the loop analysis of switching regulators. As a result, it is fair to conclude that this non-integrated error amplifier will not attenuate the high frequency components of its input signal. Therefore, it can be modeled as a constant gain $-A_v$, where, the negative sign represents a 180° phase shift between its input and output.

Considering the inverted and amplified output voltage ripple is superpositioned with the DC voltage at $COMP$ node, the same transient signal is shown in the upper and lower window boundaries in the closed loop operation. If one tries to accurately model the inverted and amplified ripple information at v_{COMP}, then apply the describing function approach in the perturbed analysis, the modeling will be considerably complicated. Consequently, in the following closed loop analysis, an assumption is made that the inverted and amplified output voltage ripple is equivalent to a DC offset at $COMP$ node.

Based upon this assumption, the transfer function from v_{COMP} to the perturbed duty cycle, derived in (31), is ready to be utilized in the closed loop analysis.

The equivalent circuit model is shown in Fig. 5: break the loop at v_O; the injected signal v_{inject} enters the non-integrated error amplifier and generates the output signal v_{comp}; v_{comp} enters the hysteretic modulator, with the current ripple synthesizer and the power stage, to generate the perturbed switching signal v_{phase}; and v_{phase} generates the perturbed output voltage v_O via the LC low pass filter.

Based on the signal flow in the loop of Fig. 5, the entire loop gain transfer function can be expressed as

$$T(s) = T_1(s)T_2(s)T_3(s) = \frac{v_{comp}(s)}{v_{inject}(s)} \frac{v_{phase}(s)}{v_{comp}(s)} \frac{v_O(s)}{v_{phase}(s)}. \tag{32}$$

In (32), by revisiting DF(s) as the transfer function from $v_{COMP}(s)$ to the perturbed duty cycle and defining the equivalent impedance of the output capacitor and the load resistor $Z_{CAP\_Load}(s)$ as:

$$Z_{CAP\_Load}(s) = \frac{1}{\frac{1}{C_{OUT} \cdot s} + ESR} // R_{Load}, \tag{33}$$

the three transfer functions in (32) can be summarized as:

$$T_1(s) = -A_v, \tag{34}$$

$$T_2(s) = \frac{v_{phase}(s)}{v_{comp}(s)} = DF(s)V_{IN},$$

$$T_3(s) = \frac{v_O(s)}{v_{phase}(s)} = \frac{Z_{CAP\_Load}(s)}{L_{OUT} \cdot s + DCR + Z_{CAP\_Load}(s)}.$$

In conclusion, the loop gain transfer function can be expressed as

$$T(s) = \frac{-A_v DF(s) V_{IN} Z_{CAP\_Load}(s)}{L_{OUT} \cdot s + DCR + Z_{CAP\_Load}(s)}. \tag{35}$$

IV. MODEL VERIFICATION

To verify the proposed open loop model in (31) and the closed loop model in (35), two cases with different configurations are deliberately chosen, such as, the duty cycle and the type of the output capacitors, to cover a broad range of the

$$C_{m\_A} = \frac{\hat{r}M}{2j} \left[\frac{1}{S_3 \left(1 - e^{j2\pi f_m T_{SW}}\right)} - \frac{e^{j2\pi f_m T_{ON}}}{S_1 \left(1 - e^{j2\pi f_m T_{SW}}\right)} - \frac{e^{j2\pi f_m T_{ON}}}{S_4 \left(1 - e^{j2\pi f_m T_{SW}}\right)} + \frac{e^{j2\pi f_m T_{SW}}}{S_2 \left(1 - e^{j2\pi f_m T_{SW}}\right)} \right] \tag{28}$$

$$C_{m\_B} = \frac{\hat{r}M}{2j} \left[\frac{e^{j2\pi f_m T_{SW}}}{S_3 \left(1 - e^{j2\pi f_m T_{SW}}\right)} - \frac{e^{j2\pi f_m T_{ON}} e^{j2\pi f_m T_{SW}}}{S_1 \left(1 - e^{j2\pi f_m T_{SW}}\right)} - \frac{e^{j2\pi f_m T_{ON}}}{S_4 \left(1 - e^{j2\pi f_m T_{SW}}\right)} + \frac{e^{j2\pi f_m T_{SW}}}{S_2 \left(1 - e^{j2\pi f_m T_{SW}}\right)} \right] \tag{29}$$

$$C_m(d) = \frac{M f_m \hat{r}}{N \left(1 - e^{j2\pi f_m T_{SW}}\right)} \left(\frac{-1 + e^{j2\pi f_m (T_{SW} - T_{ON})}}{S_3} + \frac{-e^{j2\pi f_m T_{SW}} + e^{j2\pi f_m (T_{SW} - T_{ON})}}{S_2} + \frac{-e^{j2\pi f_m T_{SW}} + e^{j2\pi f_m T_{ON}}}{S_1} + \frac{-1 + e^{j2\pi f_m T_{ON}}}{S_4} \right) \tag{30}$$

$$DF(s) = \frac{\hat{d}}{\hat{v}_{COMP}} = \frac{C_m(d)}{\hat{r}} = \frac{f_{SW}}{1 - e^{sT_{SW}}} \left(\frac{-1 + e^{s(T_{SW} - T_{ON})}}{S_3} + \frac{-e^{sT_{SW}} + e^{s(T_{SW} - T_{ON})}}{S_2} + \frac{-e^{sT_{SW}} + e^{sT_{ON}}}{S_1} + \frac{-1 + e^{sT_{ON}}}{S_4} \right) \tag{31}$$

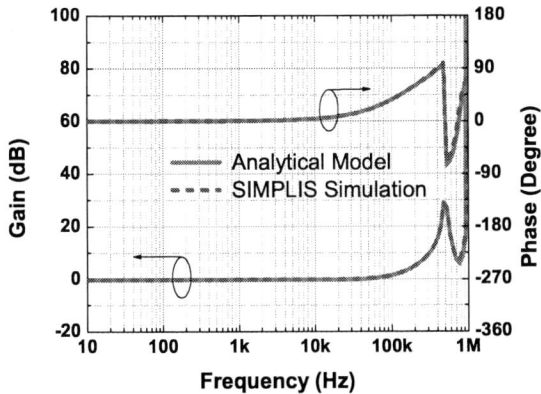

Fig. 6. Case 1: open loop analytical model DF(s) vs. SIMPLIS Simulation.

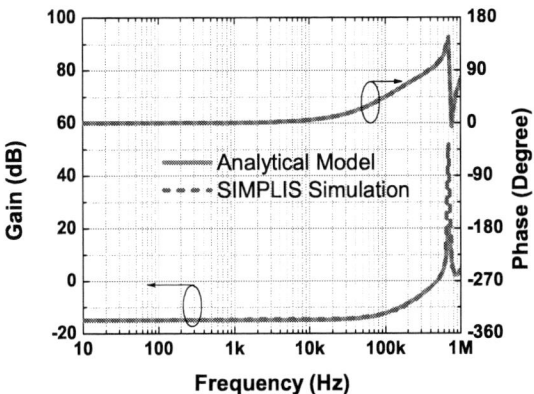

Fig. 7. Case 2: open loop analytical model DF(s) vs. SIMPLIS Simulation.

applications. The preliminary data of the analytical model is obtained from MathCAD, while the simulation results come from the SIMPLIS simulator. The parameters for both cases are summarized in Table I.

TABLE I
SYSTEM CONFIGURATIONS OF TWO CASES FOR MODEL VERIFICATION

Parameters	Case 1	Case 2
V_{IN} (V)	5	12
V_{DAC} (V)	3.3	1
Duty cycle (%)	66	8.3
f_{SW} (kHz)	500	700
R_{Load} (Ω)	0.33	0.033
A_v	14	49
V_{DC_SYN} (V)	2	2
g_m (μS)	1.25	1.25
R_r $(k\Omega)$	200	400
C_r (pF)	5	3.5
L_{OUT} (nH)	470	230
DCR $(m\Omega)$	0.32	0.3
Type of the output capacitor	Bulk	MLCC
Equivalent C_{OUT} (μF)	470×2	22×30
Equivalent ESR $(m\Omega)$	$\frac{5}{2}$	$\frac{4}{30}$

For the open loop modulator, the analytical model and the SIMPLIS simulation results are plotted in Fig. 6 and Fig. 7. The well matched plots validate the proposed model of the control scheme in Fig. 2-a. By observation of Fig. 6 and Fig. 7, the hysteretic modulator with the current ripple synthesizer, can be approximated as a block containing a single zero within the switching frequency. The location of this zero depends on the duty cycle and all four slopes derived in (3).

For the closed loop system of the case *1*, as shown in Fig. 8, the analytical model matches with the SIMPLIS simulation results up to $f_{SW}/2$. Some discrepancy can be observed by comparison of the phase plot between $f_{SW}/2$ to f_{SW}. For the case *2*, as shown in Fig. 9, the discrepancy in phase plot can be found from around the loop bandwidth range. The gain plots of both cases show a good match up to $f_{SW}/2$. The discrepancies shown in the phase plots may be

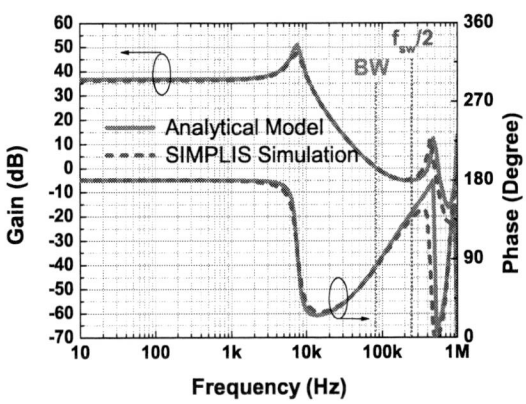

Fig. 8. Case 1: closed loop model vs. SIMPLIS Simulation.

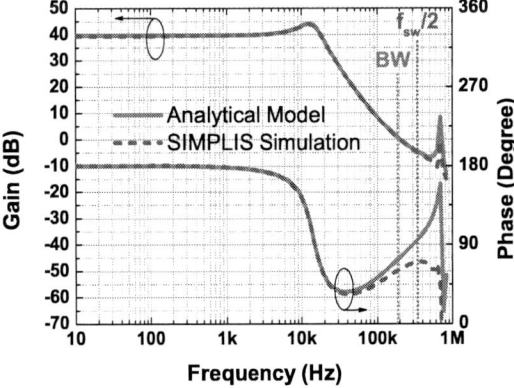

Fig. 9. Case 2: closed loop model vs. SIMPLIS Simulation.

978-1-4673-9551-9/16 $31.00 © 2016 IEEE

attributed from the assumption proposed in the last section: the inverted and amplified output voltage ripple is assumed to be an equivalent DC offset at $COMP$ node in the closed loop operation, thus the transfer function from v_{COMP} to the perturbed duty cycle in (31) is used in the loop gain transfer function in (35). The analysis of a similar scenario of V^2 control can be found in [13]: for V^2 control, since the output voltage is a direct feedback without any low-pass filter, v_{COMP} contains a certain amount of switching frequency ripple from v_O. Therefore, in the closed loop operation, the high frequency phase response of the PWM modulator is impacted by this switching frequency component, and the sidebands effect associated with it in the frequency spectrum. As shown in [13], the sidebands effect causes additional phase delay in high frequency range. This could be the reason why the simulation models shown in Fig. 8 and Fig. 9 have more phase delays than the analytical model in high frequency range. A further investigation will be performed to improve the accuracy of the closed loop model, as well as to derive a simplified model to predict the stability and improve the dynamic performance.

V. Conclusion

In this paper, the hysteretic modulator with the current ripple synthesizer is modeled using the describing function method. Based on the analysis of steady state operation of the current ripple synthesizer and the hysteresis window, the modulation law and the transfer function of the modulator are derived. Both open loop and closed loop models are validated by comparing the SIMPLIS simulation results to the MathCAD calculation. This model can be used in the system design by predicting system stability and dynamic characteristics.

Appendix

A. Vector analysis

Since $M f_M = N f_{SW}$, the following equations can be obtained by vector analysis:

$$\sum_{i=1}^{M} e^{-j2\pi f_m(i-1)T_{SW}} = 0$$
$$\sum_{i=1}^{M} e^{-j2\pi K f_m(i-1)T_{SW}} = 0,$$

where K is an integer, such as $\pm1, \pm2$, etc.

B. Useful geometric series

$$\sum_{k=1}^{i-1} e^{j2\pi f_m(k-1)T_{SW}} = \frac{1-e^{j2\pi f_m(i-1)T_{SW}}}{1-e^{j2\pi f_m T_{SW}}}$$
$$\sum_{k=1}^{i-1} e^{-j2\pi f_m(k-1)T_{SW}} = \frac{1-e^{-j2\pi f_m(i-1)T_{SW}}}{1-e^{-j2\pi f_m T_{SW}}}$$
$$\sum_{k=1}^{i-1} e^{j2\pi f_m k T_{SW}} = e^{j2\pi f_m T_{SW}} \frac{1-e^{j2\pi f_m(i-1)T_{SW}}}{1-e^{j2\pi f_m T_{SW}}}$$
$$\sum_{k=1}^{i-1} e^{-j2\pi f_m k T_{SW}} = e^{-j2\pi f_m T_{SW}} \frac{1-e^{-j2\pi f_m(i-1)T_{SW}}}{1-e^{-j2\pi f_m T_{SW}}}$$
$$\sum_{k=1}^{i} e^{j2\pi f_m(k-1)T_{SW}} = \frac{1-e^{j2\pi f_m i T_{SW}}}{1-e^{j2\pi f_m T_{SW}}}$$
$$\sum_{k=1}^{i} e^{-j2\pi f_m(k-1)T_{SW}} = \frac{1-e^{-j2\pi f_m i T_{SW}}}{1-e^{-j2\pi f_m T_{SW}}}$$

References

[1] J. Sun, "Characterization and performance comparison of ripple-based control for voltage regulator modules," *IEEE Trans. on Power Electronics*, vol. 21, no. 2, pp. 346–353, 2006.

[2] R. Redl and J. Sun, "Ripple-based control of switching regulatorsAn overview," *IEEE Trans. on Power Electronics*, vol. 24, no. 12, pp. 2669–2680, 2009.

[3] B. P. Schweitzer and A. B. Rosenstein, "Free running-switching mode power regulator: Analysis and design," *IEEE Trans. on Aerospace and Electronic Systems*, vol. 4, no. 2, pp. 1171–1180, 1964.

[4] J. Li and F. C. Lee, "Modeling of current-mode control," *IEEE Trans. on Circuits and Systems I*, vol. 57, no. 9, pp. 2552–2563, 2010.

[5] S. Tian et al., "Equivalent circuit model of constant on-time current mode control with external ramp compensation," in *IEEE Energy Conversion Congress and Exposition (ECCE)*, 2014, pp. 3747–3754.

[6] W. R. Kolk, "A Study of Pulse Width Modulation in Feedback Control Systems," Ph.D. dissertation, University of Connecticut, 1972.

[7] D. Mitchell, "Pulsewidth modulator phase shift," *IEEE Trans. on Aerospace and Electronic Systems*, vol. 3, no. AES-16, pp. 272–278, 1980.

[8] R. Middlebrook, "Predicting modulator phase lag in PWM converter feedback loops," in *Proc. Powercon 8*, 1981, pp. H4: 1–6.

[9] J. Sun, "Small-signal modeling of variable-frequency pulsewidth modulators," *IEEE Trans. on Aerospace and Electronic Systems*, vol. 38, no. 3, pp. 1104–1108, 2002.

[10] F. Yu, F. C. Lee, and P. Mattavelli, "A small signal model for average current mode control based on describing function approach," in *IEEE Energy Conversion Congress and Exposition*, 2011, pp. 405–412.

[11] S. Tian, F. C. Lee, P. Mattavelli, K.-Y. Cheng, and Y. Yan, "Small-Signal Analysis and Optimal Design of External Ramp for Constant On-Time V^2 Control With Multilayer Ceramic Caps," *IEEE Trans. on Power Electronics*, vol. 29, no. 8, pp. 4450–4460, 2014.

[12] S. Tian, F. C. Lee, P. Mattavelli, and Y. Yan, "Small-Signal Analysis and Optimal Design of Constant Frequency Control," *IEEE Trans. on Power Electronics*, vol. 30, no. 3, pp. 1724–1733, 2015.

[13] S. Tian et al., "Unified Equivalent Circuit Model and Optimal Design of V2 Controlled Buck Converters," *IEEE Trans. on Power Electronics*, vol. 31, no. 2, pp. 1734–1744, 2016.

[14] E. X. Yang, F. C. Lee, and M. M. Jovanovic, "Small-signal modeling of series and parallel resonant converters," in *7th Annual IEEE Applied Power Electronics Conference and Exposition*, 1992, pp. 785–792.

[15] S. Tian et al., "Small-signal equivalent model of series resonant converter," in *IEEE Energy Conversion Congress and Exposition (ECCE)*, 2015, pp. 172–179.

[16] S. Tian, "Equivalent Circuit Model of High Frequency PWM and Resonant Converters," Ph.D. dissertation, Virginia Tech., 2015.

[17] T. Suntio, J. Lempinen, K. Hynynen, and P. Silventoinen, "Analysis and small-signal modeling of self-oscillating converters with applied switching delay," in *7th Annual IEEE Applied Power Electronics Conference and Exposition*, vol. 1, 2002, pp. 395–401.

[18] T. Nabeshima et al., "Analysis and design considerations of a buck converter with a hysteretic PWM controller," in *35th Annual IEEE Power Electronics Specialists Conference*, vol. 2, 2004, pp. 1711–1716.

[19] J. Park and B. Cho, "Small signal modeling of hysteretic current mode control using the PWM switch model," in *IEEE Workshops on Computers in Power Electronics*, 2006, pp. 225–230.

[20] K. D. Ngo, S. K. Mishra, and M. Walters, "Synthetic-ripple modulator for synchronous buck converter," *IEEE Power Electronics Letters*, vol. 3, no. 4, pp. 148–151, 2005.

[21] R. S. Philbrick, E. Chen, G. Luff, and R. J. Parikh, "System and method of equivalent series inductance cancellation," Nov. 22 2013, US Patent App. 14/087,638.

[22] R. S. Philbrick, M. B. Harris, and S. P. Laur, "Switching regulator with balanced control configuration with filtering and referencing to eliminate compensation," Apr. 10 2012, US Patent 8,154,268.

[23] ——, "Switching regulator with balanced control configuration with filtering and referencing to eliminate compensation," Jan. 1 2013, US Patent 8,344,717.

[24] "ISL68200: Single-Phase R4 Hybrid Digital PWM Controller with Integrated Driver, PMBus and PFM," Intersil Corporation, Data Sheet, 2016 [online]. Available: http//www.intersil.com.

[25] "ISL95872: Buck PWM Controller with Internal Compensation and External Reference Tracking," Intersil Corporation, Data Sheet, 2012 [online]. Available: http//www.intersil.com.

[26] "ISL95901: Wide V_{IN} Dual Integrated Buck Regulator With 4A/4A Continuous Output Current and LDOs," Intersil Corporation, Data Sheet, 2012 [online]. Available: http//www.intersil.com.

[27] S. P. Laur and R. S. Philbrick, "System and method for determining output voltage level information from phase voltage for switched mode regulator controllers," Oct. 30 2012, US Patent 8,299,764.

978-1-4673-9551-9/16 $31.00 © 2016 IEEE

A Black-box Modeling Approach for DC Nanogrids

A. Francés, R. Asensi, O. García, R. Prieto and J. Uceda

Centro de Electrónica Industrial (CEI)

Universidad Politécnica de Madrid

Madrid, Spain

Email: airan.frances@upm.es

Abstract—The smart grid concept is increasingly becoming popular within the academia and industry. The integration of electronic power converters as an enabler for the massive deployment of distributed renewable energy sources, along with the inclusion of control, monitoring and automation systems in the grid, has drawn the attention of many researchers. Furthermore, a transition towards dc distribution is currently under investigation due to its more suitable interface with most of the modern loads and sources, which offers benefits in terms of size, cost and reliability of the whole system.

This paper proposes a black-box polytopic modeling approach as a tool for the system-level design of dc based nanogrids. This strategy allows both small and large-signal analysis of power distribution systems even when commercial off-the-shelf converters have to be integrated. In addition, the main characteristics of the dc bus signaling control, i.e. droop control, changes in power converter control mode and disconnection of loads, have been incorporated in the modeling structure.

I. INTRODUCTION

In current power systems, electricity is mainly transported and distributed in a 50/60 Hz ac network. Although it provides many advantages, there are some drawbacks that can be avoided using electronic power distribution systems, which can use ac, dc or a hybrid ac/dc topology. For example, as generation, transport and consumption are strongly coupled, slow dynamics are necessary in order to guarantee stability [1], [2].

Another drawback is that power flows depend on the system (lines and transformers) impedances. To avoid some of these problems in transportation HVDC links have been implemented. In distribution, on the other hand, the transition is not straightforward due to political, economic and technical issues. Among the advantages of dc distribution stand out the reduction of ac/dc and dc/ac power conversions, the fact that the power transmission capability of the lines is not limited by the reactance and the elimination of problems with reactive power and frequency synchronization.

The practical implementation of dc nanogrids requires not only power converters and electrical switchgear for the new applications, but also the development of electrical models to understand the behavior of the system and to control its parts both in static and in dynamic scenarios. Besides, in practice these systems comprise the integration of several commercial off-the-shelf electronic equipments whose manufacturers provide very limited information about. Therefore, there is a need for simulation tools to overcome the lack of details about the power converters behavior.

In this context, black-box models have been proposed with two different approaches: Wiener-Hammerstein structure and small-signal LTI (linear time invariant) models. In [3] a hybrid Wiener-Hammerstein structure is obtained from the data provided by the datasheets and the transient response of the converter. It considers a nonlinear static block that defines the steady state characteristics of the converter and linear dynamic networks to account for the transient behavior. The second method was proposed in [4], [5] where small-signal G-parameters models are used in order to simulate power converters. The LTI models are obtained from the frequency response of the converters under specific tests. A combination was presented in [6], where a Hammerstein structure is implemented, using G-parameters models as the linear dynamic networks. These structures have been proved to be very useful for both commercial converter modeling and system-level integration analysis, however their transient performance is restricted to converters with constant dynamic behavior. To account for highly nonlinear behaviors both in static and dynamic conditions, in [7] a polytopic structure is proposed, where the idea is to integrate the small-signal models obtained at different operating points by means of suitable weighting functions. Finally, [8] makes an effort to simplify the models according to the converters behavior. A mix among the previous structures is proposed with a method to obtain the LTI models from its transient response to avoid the use of expensive equipment.

This paper continues the analysis of polytopic models, focusing on the particularities of dc bus signaling control for smart dc microgrids. Modifications are proposed to improve the accuracy of highly nonlinear converter models during sharp steps and to include droop control to existing models. Besides, modeling of power converter which change their control mode and disconnection of loads are studied.

II. TERMINAL CHARACTERIZATION OF ELECTRONIC POWER CONVERTERS

Electronic system-level design usually entails the integration of a huge number of elements. These basic units can be represented as N-port networks whose variables are the current and voltage of each port. Any linear circuit can be represented as an N-port network provided that it does not contain an independent source and satisfies the port condition. The idea behind the terminal characterization is to build a model in which these variables are related by means of transfer functions that have information about the internal dynamics of the system. In this way, the designer sees the subsystem as a black box with a certain behavior, ready to be integrated into the next level of abstraction.

978-1-4673-9551-9/16 $31.00 © 2016 IEEE

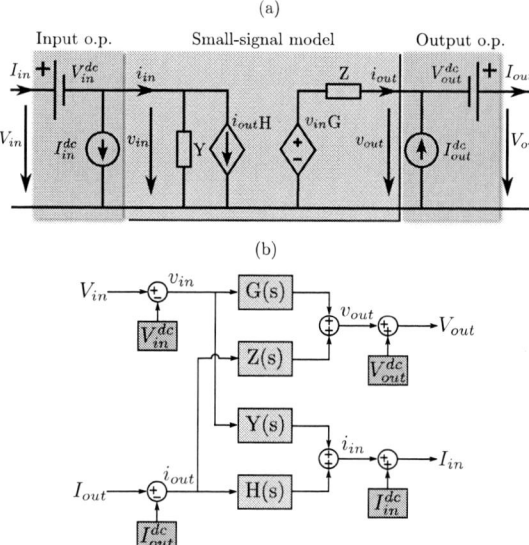

Figure 1: G-parameters model. (a) Equivalent electrical circuit, (b) Block diagram notation.

This paper focuses on the two-port terminal modeling of dc/dc power converters as they are the key elements when it comes to designing a dc based power distribution system. This model can be derived in different ways, depending on the definition of the variables as inputs or outputs. This choice determines the designation of the transfer functions as impedance (Z), admittance (Y), hybrid (H) or inverse hybrid (G) parameters. Although they are equivalent, an adequate choice of parameters results in the development of simpler structures.

The most widely used structure for dc/dc converter models is the inverse hybrid parameters (G-parameters) model (Fig. 1), since it fits with the common representation of the converters using Norton and Thèvenin equivalent circuits to represent input and output, respectively.

Thus, the transfer functions receive the subsequent physical interpretation (1).

Audio-susceptibility   Input admittance

$$G(s) = \left.\frac{\tilde{v}_{out}}{\tilde{v}_{in}}\right|_{\tilde{i}_{out}=0} \qquad Y(s) = \left.\frac{\tilde{i}_{in}}{\tilde{v}_{in}}\right|_{\tilde{i}_{out}=0} \quad (1)$$

Output impedance   Back current gain

$$Z(s) = -\left.\frac{\tilde{v}_{out}}{\tilde{i}_{out}}\right|_{\tilde{v}_{in}=0} \qquad H(s) = \left.\frac{\tilde{i}_{in}}{\tilde{i}_{out}}\right|_{\tilde{v}_{in}=0}$$

A. Un-terminated model

Following a black-box approach, these parameters can be obtained linearizing the converter around an operating point by means of tests performed at its terminals. This method can be performed either in frequency [4], [5] or time domain [8]. In this work frequency domain has been used due to its suitability

to concentrate the information about fast and slow dynamics of the converter.

The procedure to define the parameters in frequency domain implies performing two tests, one to each output variable (Fig. 2). In the G-parameter model input current and output voltage are measured while the converter is at the selected operating point. During the measurement, one of the input variables is perturbed while the other is kept constant.

In order to perform the measurements, it is necessary to introduce two external elements: a source and a load. However, the external impedances of these devices will interact with the power converter and, therefore, they will adulterate the G-parameters transfer functions. A model which is not affected by these interactions is known as un-terminated model.

Source and load interactions with power converters have been analyzed mathematically by means of G-parameters models in [9], [10] as a tool for stability analysis of electronic power systems. Afterwards, in [4], [5] this mathematical analysis of the interactions was applied to uncouple the un-terminated two-port model of the power converters from the source and load dynamics.

B. Power converter G-parameters model performance

Some experiments have been performed in order to illustrate the G-parameters model capabilities and limitations. To achieve this goal an unregulated synchronous buck converter has been designed. For the sake of validity, the converter has been implemented and tested using PSIM, whereas the identification process has been performed employing MATLAB. Finally, the resulting model is put together in PSIM so their behaviors can be compared.

Fig. 3 depicts both the switching model and the obtained G-parameters model behaviors under the same input voltage and load steps. Additionally, the switching converter inductor current is represented to discern between CCM (continuous conduction mode) and DCM (discontinuous conduction mode). It can be appreciated how the G-parameters model is able to predict the switching model behavior under substantial variations as long as it remains in CCM: input voltage, from

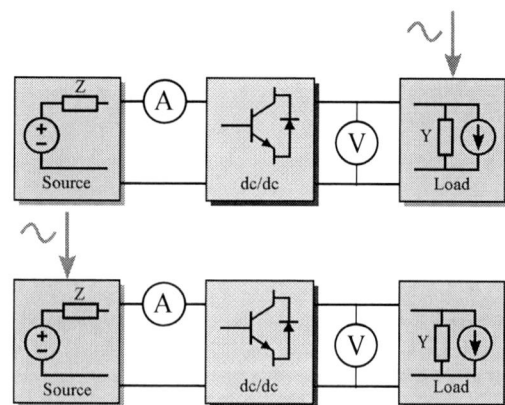

Figure 2: Scheme of the test for terminal characterization of converters in frequency domain.

978-1-4673-9551-9/16 $31.00 © 2016 IEEE   1625

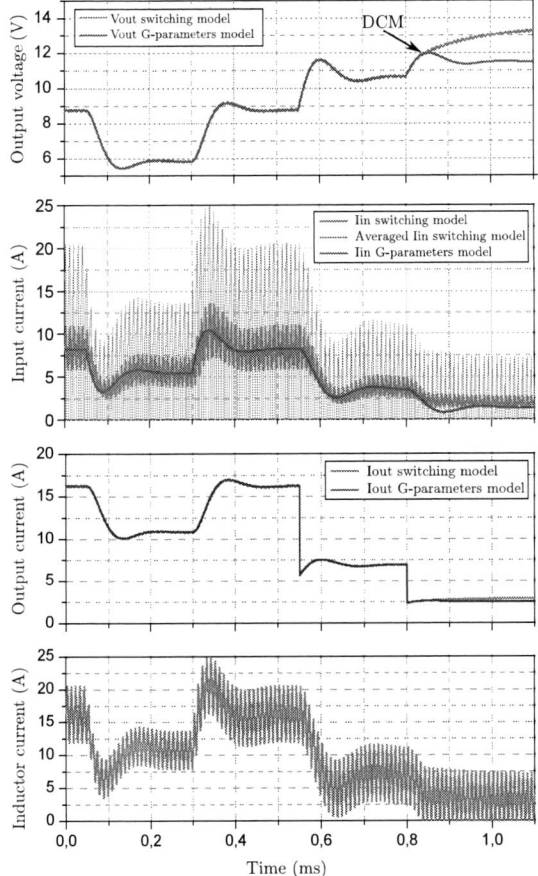

Figure 3: G-parameters performance for an unregulated synchronous buck converter in CCM and DCM.

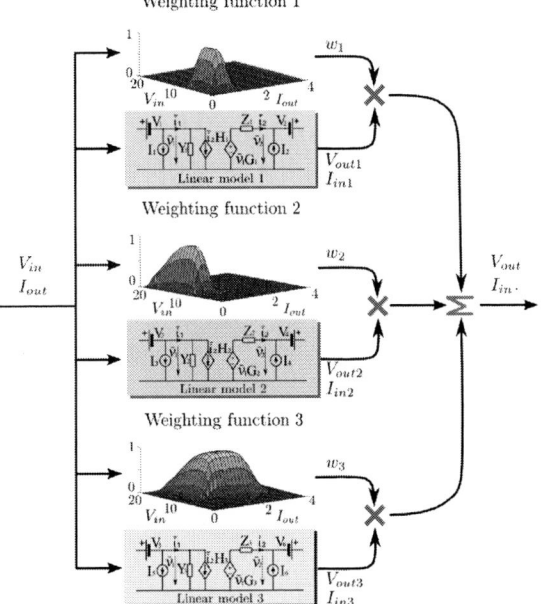

Figure 4: Two-dimensional polytopic model scheme.

$Vin = 24$ V to $Vin = 16$ V in $t = 0.05$ ms and back in $t = 0.3$ ms, and load, from $R = 0.54$ Ω to $R = 1.54$ Ω in $t = 0.55$ ms. However, after the second load step, from $R = 1.54$ Ω to $R = 4.54$ Ω in $t = 0.8$ ms, the converter enters in DCM operation and visible differences arise between the models.

This test exposes how the G-parameters model precisely reproduce power converters with linear behaviors, whereas it fails when nonlinear phenomena appear.

III. POLYTOPIC MODELING OF ELECTRONIC POWER CONVERTERS

DC power distribution systems usually present several characteristics where the nonlinear effects become significant. In order to overcome this problem, the natural upgrade of the linear two-port model calls for the iteration of the linearization process applied to different operating points. Afterwards, this set of linear parameters can be integrated into a single system. This nonlinear structure, developed in a piece-wise linear fashion, has been labeled as polytopic model. This approach has been utilized in a wide range of applications in different fields such as robotics [11], aeronautics [12] or mechanics [13]. In fuzzy logic applications this procedure is known as Takagi-Sugeno model [14], which has been applied broadly in neural networks [15]. The first applications of polytopic models in power electronic based distribution systems were [16] and [17] in order to numerically obtain Lyapunov function candidates. Thereafter, the development of black-box polytopic models of power converters was presented in [7]. This paper proposes the extension of its usage to dc nanogrids.

A. Model development

The development of polytopic models faces three essential choices: variables, number of linear models and type of scheduling functions (Fig. 4). The set of relevant operating points can be generated varying current, voltage or other variables like temperature. However the addition of each variable poses an exponential increase in the number of linear models. Therefore a trade-off between accuracy and complexity is to be made. Likewise, the number of considered variables will drastically restrain the separation between evaluated operating points for a given level of complexity. From a computational perspective the number of linear models entails further calculations and hence an increased simulation time.

Mathematically the polytopic structure is defined as follows:

$$\dot{x} = \sum_{i=1}^{n} \omega_i(\alpha, \beta, ...)(A_i x + B_i u)$$
$$\dot{y} = \sum_{i=1}^{n} \omega_i(\alpha, \beta, ...)(C_i x + D_i u)$$

(2)

where α and β are the considered variables. The weighting or scheduling functions $\omega_i(\alpha, \beta, ...)$ must always comply with the restrictions (3) to avoid distortion of the model.

$$0 \leq \omega_i(\alpha, \beta, ...) \leq 1$$
$$\sum_{i=1}^{n} \omega_i(\alpha, \beta, ...) = 1 \qquad (3)$$

A wide range of weighting or scheduling functions are suitable for this purpose, from the straightforward triangular function used in Takagi-Sugeno models [14] to more elaborated nonlinear alternatives as double sigmoid functions. In this paper the latter has been used due to its flexibility, thus defining as:

$$\omega_i(\alpha, \beta, ...) = \left(\frac{1}{1 + e^{v_{\alpha i}(\alpha)}} - \frac{1}{1 + e^{v_{\alpha i+1}(\alpha)}} \right) \cdot$$
$$\cdot \left(\frac{1}{1 + e^{v_{\beta i}(\beta)}} - \frac{1}{1 + e^{v_{\beta i+1}(\beta)}} \right) \cdot \qquad (4)$$
$$...$$

with

$$v_{xi}(x) = -m_{xi}(x - c_{xi}) \qquad (5)$$

where m_{xi} and c_{xi} are the slope and the center of the i-th sigmoid of the variable x, respectively. This definition enables the development of complex multidimensional functions, while weighting function constrains are assured. Furthermore, these two new parameters offer two variables that can be adjusted to improve the transition among small-signal models. Finally, these parameters are suitable to be optimized for a given objective function, e.g. minimum error.

B. Power converter polytopic model performance

The capabilities of polytopic modeling will be presented adding a voltage regulator to the previous synchronous buck converter. In order to illustrate the nonlinear behavior of this system, the variation of the frequency response functions is depicted in Fig. 5, where the output current varies from $Iout = 0.5$ A to $Iout = 6$ A. This portion has been selected aiming to consider the critical transition from DCM to CCM.

The first comparison represents a slow load step. In order to capture the transition from DCM to CCM, which as presented before is known to occur around $Iout = 4$ A when $Vin = 24$ V, the output current is varied from $Iout = 1.5$ A to $Iout = 5.5$ A, whereas the input voltage remains constant. In Fig. 6 the polytopic model, along with the linear G-parameters models implemented, is compared with the switching model. Notice that the linear model around $Iout = 4$ A has been avoided because the small signal model of an edge point does not provide meaningful information, instead small signal models around $Iout = 3.5$ A and $Iout = 4.5$ A have been included. The different linear models show relatively large differences in performance, e.g. the output voltage variation is 10 times higher in the case of the small signal model around $Iout = 1$ A than the model around $Iout = 6$ A. This comparison shows how, with a slow load step, the polytopic model adjusts the

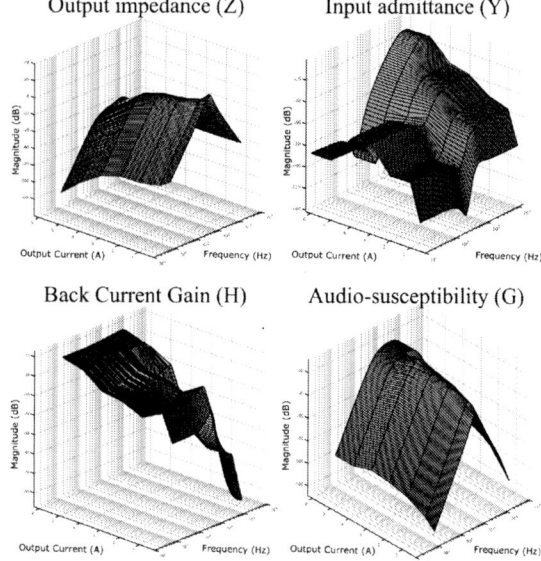

Figure 5: Regulated synchronous buck converter frequency response variation with output current.

weights of the different linear models according to the output current variation, resulting in a reasonably good approximation of the transition from DCM to CCM of the switching model.

Nevertheless, in dc distribution systems sharp load steps are not unusual. In case the polytopic model uses the output current as an input parameter, the sudden load step will cause an abrupt change between the linear models involved. In Fig. 7 this situation is represented. In this case, the steady state is still well represented, whereas noticeable differences are found in the transient behaviors. This can be explained by the dynamic of the regulator, which is ignored by the polytopic model. In order to overcome this problem, the input variables can be filtered before they are used in the scheduling functions (Fig. 8). The low pass filter will be tunned by analyzing the transition among linear models of the converter. The improvement in the transient behavior is also depicted in Fig. 7, where the light blue signal represents the polytopic model using the filtered input signals for the weighting functions. Also the filtered signals which are used as inputs for the scheduling functions are depicted in gray.

IV. DC NANOGRID BLACK-BOX MODEL

DC power distribution systems usually include a high level control system which can manage power flow or parallel connection of power sources, among other functionalities. One of the most popular control mode in dc nanogrids is dc bus signaling [2], [18]. It is a decentralized control which allows the desired Plug & Play capability. This strategy implements droop control [19] for the parallel connection of power sources to the bus, using the droop slope as a variable to control the power driven by each power converter. Likewise, the power converters use the bus voltage level as a communication

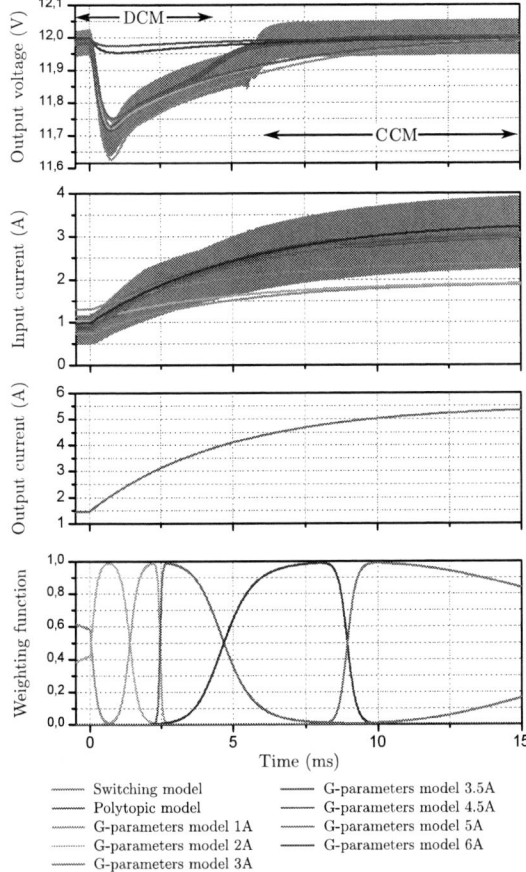

Figure 6: Polytopic model performance for a regulated synchronous buck converter with a slow step from DCM to CCM.

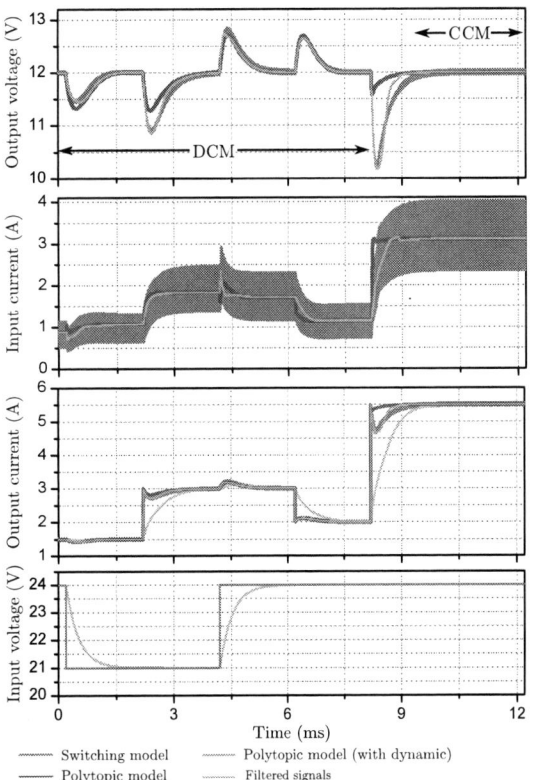

Figure 7: Polytopic model performance for a synchronous buck converter with sharp steps.

system, so they may change their operation mode according to the state of the system.

A. DC bus signaling control model

This strategy has two main characteristics: droop control and changes in the power converters control mode. The droop control causes a decrease in the output voltage which is proportional to the output current. As shown in Fig. 9 this effect can be included in the black-box model, subtracting the product $I_{out} \cdot K_{droop}$ from the dc output voltage level. Furthermore, K_{droop} can be modified freely, allowing both testing and designing high level control strategies. Finally, as the dc value of the output voltage has been modified, the dc value of the input current should be adjusted accordingly to keep the power balance. Assuming the efficiency does not change, which is usually realistic considering the maximum output voltage variation is often limited to 5%, the input current can be modified as follows:

$$I_{in}^2 = I_{in}^1 \left(\frac{V_{out} - I_{out} \cdot K_{droop}}{V_{out}} \right) \qquad (6)$$

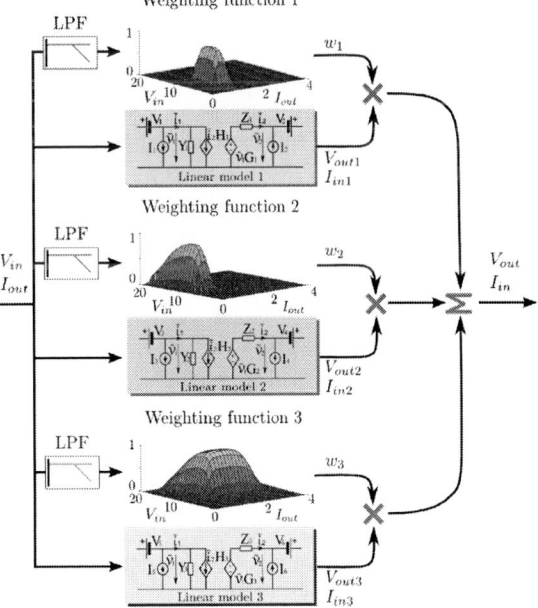

Figure 8: Polytopic model scheme with the dynamic modification.

As for changes in the power converter control mode, the linear model of the converter working in constant current mode must be obtained and, instead of using weighting functions for the transition, perform an abrupt change to this linear model when the proper conditions are met.

In order to test these modeling approaches, a full-bridge switching model has been designed using a double control loop, the inner is a fast current control, whereas the outer is a slow voltage control loop. The change of control mode is studied by means of a current limitation, imposing a 6 A reference to the inner loop when the outer loop sets a reference higher than 6.2 A.

The condition used in the polytopic model for switching to the current limited linear model is the same, output current higher than 6.2 A, however to go back to the voltage controlled models two conditions are included: the output current must be less than 6 A and the output voltage of the current limited model must be higher than the output voltage of the voltage controlled models. The latter considers the transient when the output voltage recovers its controlled value. Indeed, while the output voltage is lower than the reference, the output current of the voltage controlled linear models will be higher than 6.2 A and an oscillation between current limited and voltage controlled models would arise without the mentioned condition. Finally, notice that the switching converter measures the inductor current, while the polytopic model uses the output current, hence in this case it is particularly important to use the dynamic limitation presented before (Fig. 8) for the input of the scheduling functions.

The performance of the black-box model under these phenomena is presented in Fig. 10. The droop control behavior is represented with a sharp load step, where the output current changes from $I_{out} = 1.5$ A to $I_{out} = 5.5$ A, whereas the change in the control mode occurs when a second load step attempts to rise the output current beyond 6 A. Notice how the delay in the transition from voltage to current control has been well described owing to the dynamic limitation, which is also depicted in gray along with the output current. Finally, both the transient and the steady state after the change of control of the black-box model are in good agreement with the switching model. One of the G-parameters models has been included to compare changes in transient and steady state behavior from small-signal to polytopic model.

B. DC nanogrid simulation

A dc nanogrid switching model has been developed aiming to validate the application of black-box models for its system-level design. A voltage droop control has been implemented in the sources that share the bus. The studied nanogrid (Fig. 11) consists of a 24 V bus supplied by two isolated dc/dc converters: a full-bridge is connected to 380 V, whereas a flyback is connected to 250 V, which includes an inner current loop with a limited reference of 4 A and an outer output voltage loop with droop control. Two variable loads are supplied from the bus through regulated point-of-load converters. Finally, they are integrated by means of several line impedances.

The proposed model takes advantage of the relation between voltage and current, established by the load impedance, to

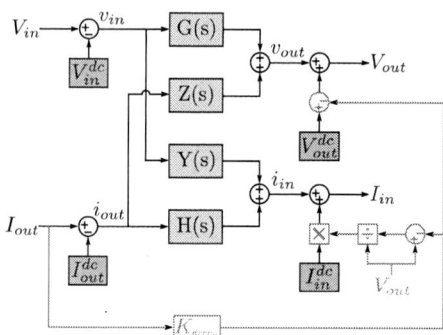

Figure 9: Droop control implementation in the G-parameters structure.

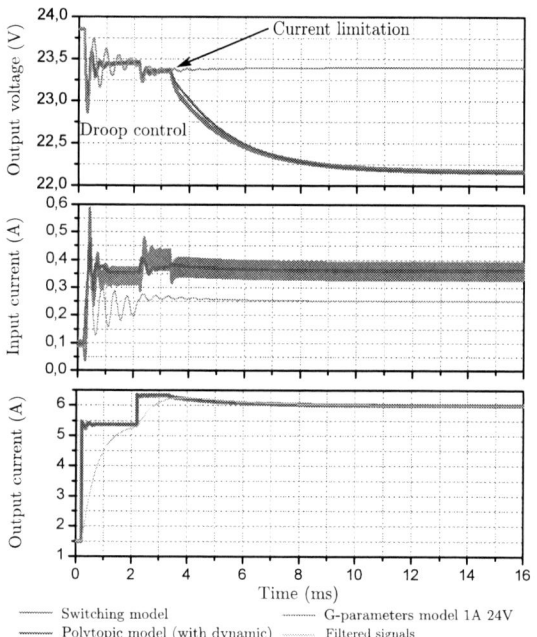

Figure 10: Polytopic model performance for a full-bridge converter using droop control and current limitation.

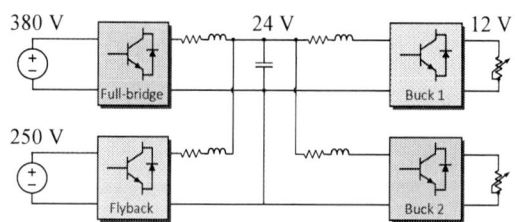

Figure 11: Studied dc nanogrid scheme.

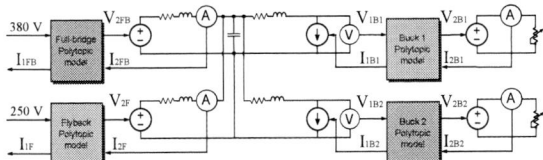

Figure 12: DC nanogrid polytopic model scheme.

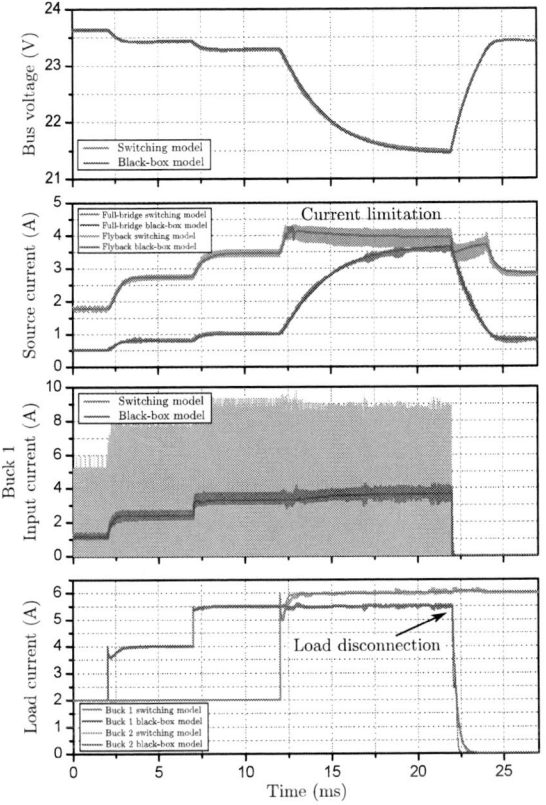

Figure 13: Polytopic model performance for a dc nanogrid with droop control, current limitation and load disconnection.

reduce the inputs of the model to those actually needed in real conditions. In Fig. 12 the scheme of the black-box structure is presented, where only two inputs are necessary to simulate the nanogrid.

The black-box models have been developed according to the methodology previously detailed. Afterwards, these models were integrated in the nanogrid and both responses were compared under different conditions. In Fig. 13 the comparison between switching model and black-box model signals is presented.

The results demonstrate the capability of the black-box polytopic models to reproduce the dynamic behavior of the converters. Additionally, the change of the control mode in the flyback, from voltage control to current control, is well reproduced. Furthermore, a load disconnection that make the flyback

converter work with droop control again has been included. Notice how this transition causes a peculiar phenomenon in the output current of the flyback model due to the bus voltage recovery, which is in good agreement in both models. Finally, the interactions between converters are also accurately reflected by this structure.

V. CONCLUSION

Lately, researches on smart grids have drawn attention to electronic power converters integration into the electric grid and the possibility of a transition to dc distribution systems. This paper presents a black-box modeling approach for the simulation of dc nanogrids, which cover the particular events involved in the dc bus signaling control. For that purpose, the linear G-parameters model and the nonlinear polytopic model have been detailed.

Regarding individual power converters, performance comparison between the G-parameters model and polytopic model has been presented. Strong nonlinearities such as the transition from DCM to CCM of a synchronous buck converter has been analyzed both with slow and sharp steps, proposing a dynamic modification in the polytopic model in order to take into account the regulator transient response. Concerning dc bus signaling control, droop control and changes in the operation mode of power converters have been considered.

Finally, a dc based nanogrid black-box polytopic model has been developed by the combination of different types of power converters. It includes parallel connection of sources by means of droop control, changes in the control mode of the power converters and disconnection of loads. The variables of the model have been related such that exclusively the actual inputs of the system are needed to simulate the power distribution network. The potential of this approach has been demonstrated presenting that not only large-signal behavior of the regulated converters can be accurately reproduced, but also the interactions among them.

ACKNOWLEDGMENT

This work was supported by the Spanish Ministry of Economy and Competitiveness under the project EVA-ANRI with reference DPI2013-47176-C2-1-R.

REFERENCES

[1] T. Dragicevic, J. Vasquez, J. Guerrero, and D. Skrlec, "Advanced lvdc electrical power architectures and microgrids: A step toward a new generation of power distribution networks." *Electrification Magazine, IEEE*, vol. 2, no. 1, pp. 54–65, March 2014.

[2] D. Boroyevich, I. Cvetkovic, D. Dong, R. Burgos, F. Wang, and F. Lee, "Future electronic power distribution systems a contemplative view," in *Optimization of Electrical and Electronic Equipment (OPTIM), 2010 12th International Conference on*, May 2010, pp. 1369–1380.

[3] J. Oliver, R. Prieto, J. Cobos, O. Garcia, and P. Alou, "Hybrid wiener-hammerstein structure for grey-box modeling of dc-dc converters," in *Applied Power Electronics Conference and Exposition, 2009. APEC 2009. Twenty-Fourth Annual IEEE*, Feb 2009, pp. 280–285.

[4] L. Arnedo, D. Boroyevich, R. Burgos, and F. Wang, "Un-terminated frequency response measurements and model order reduction for black-box terminal characterization models," in *Applied Power Electronics Conference and Exposition, 2008. APEC 2008. Twenty-Third Annual IEEE*, Feb 2008, pp. 1054–1060.

[5] I. Cvetkovic, D. Boroyevich, P. Mattavelli, F. Lee, and D. Dong, "Un-terminated, low-frequency terminal behavioral model of dc-dc converters," in *Applied Power Electronics Conference and Exposition (APEC), 2011 Twenty-Sixth Annual IEEE*, March 2011, pp. 1873–1880.

978-1-4673-9551-9/16 $31.00 © 2016 IEEE

[6] ——, "Non-linear, hybrid terminal behavioral modeling of a dc-based nanogrid system," in *Applied Power Electronics Conference and Exposition (APEC), 2011 Twenty-Sixth Annual IEEE*, March 2011, pp. 1251–1258.

[7] L. Arnedo, D. Boroyevich, R. Burgos, and F. Wang, "Polytopic black-box modeling of dc-dc converters," in *Power Electronics Specialists Conference, 2008. PESC 2008. IEEE*, June 2008, pp. 1015–1021.

[8] V. Valdivia, A. Barrado, A. Roldan, C. Fernandez, and P. Zumel, "Black-box modeling of dc-dc converters based on transient response analysis and parametric identification methods," in *Applied Power Electronics Conference and Exposition (APEC), 2010 Twenty-Fifth Annual IEEE*, Feb 2010, pp. 1131–1138.

[9] T. Suntio, D. Gadoura, and K. Zenger, "Input filter interactions in current-mode controlled converters-a unified analysis approach," in *IECON 02 [Industrial Electronics Society, IEEE 2002 28th Annual Conference of the]*, vol. 2, Nov 2002, pp. 1179–1184.

[10] M. Hankaniemi, M. Karppanen, T. Suntio, A. Altowati, and K. Zenger, "Source-reflected load interactions in a regulated converter," in *IEEE Industrial Electronics, IECON 2006 - 32nd Annual Conference on*, Nov 2006, pp. 2893–2898.

[11] J. Chen, R. Li, and C. Cao, "Convex polytopic modeling for flexible joints industrial robot using tp-model transformation," in *Information and Automation (ICIA), 2014 IEEE International Conference on*, July 2014, pp. 1046–1050.

[12] Y. Huang, C. Sun, C. Qian, J. Zhang, and L. Wang, "Polytopic lpv modeling and gain-scheduled switching control for a flexible air-breathing hypersonic vehicle," *Systems Engineering and Electronics, Journal of*, vol. 24, no. 1, pp. 118–127, Feb 2013.

[13] P. Grof, Z. Petres, and J. Gyeviki, "Polytopic model reconstruction of a pneumatic positioning system," in *Applied Computational Intelligence and Informatics, 2009. SACI '09. 5th International Symposium on*, May 2009, pp. 61–66.

[14] T. Takagi and M. Sugeno, "Fuzzy identification of systems and its applications to modeling and control," *Systems, Man and Cybernetics, IEEE Transactions on*, vol. SMC-15, no. 1, pp. 116–132, Jan 1985.

[15] F.-J. Lin, M.-S. Huang, Y.-C. Hung, C.-H. Kuan, S.-L. Wang, and Y.-D. Lee, "Takagi-sugeno-kang type probabilistic fuzzy neural network control for grid-connected lifepo4 battery storage system," *Power Electronics, IET*, vol. 6, no. 6, pp. 1029–1040, July 2013.

[16] S. Sudhoff, S. Glover, S. Zak, S. Pekarek, E. Zivi, D. Delisle, and D. Clayton, "Stability Analysis Methodologies for DC Power Distribution Systems," in *Thirteenth International Ship Control Systems Symposium (SCSS)*, 2003, pp. 1–10.

[17] S. F. Glover, "Modeling and stability analysis of a power electronics based systems," Ph.D. dissertation, Dept. Elect. Eng., Purdue Univ., Lafayette, IN, 2003.

[18] J. Guerrero, J. Vasquez, and R. Teodorescu, "Hierarchical control of droop-controlled dc and ac microgrids 2014; a general approach towards standardization," in *Industrial Electronics, 2009. IECON '09. 35th Annual Conference of IEEE*, Nov 2009, pp. 4305–4310.

[19] Y. Ito, Y. Zhongqing, and H. Akagi, "Dc microgrid based distribution power generation system," in *Power Electronics and Motion Control Conference, 2004. IPEMC 2004. The 4th International*, vol. 3, Aug 2004, pp. 1740–1745.

Design and Evaluation of Isolated Gate Driver Power Supply for Medium Voltage Converter Applications

Krishna Mainali, Sachin Madhusoodhanan, Awneesh Tripathi, Kasunaidu Vechalapu, Ankan De, Subhashish Bhattacharya

FREEDM Systems Center, Department of Electrical and Computer Engineering
North Carolina State University
Raleigh, NC, USA
Email: {kmainal}, {sachin}, {aktripat}, {kvechal}, {ade}, {shatta4}@ncsu.edu

Abstract—The commercial gate drivers are available upto 6.5 kV IGBTs. With the advances in the SiC, power devices rated beyond 10 kV are being researched. These devices will have use on medium voltage power converters. Commercial gate drivers rated for such high voltages are not available. These power devices have very high dv/dts (30-100 kV/μs) at switching transitions. Such high dv/dts bring in challenges in the gate driver design. The isolation stage of the gate power supply needs to have very low coupling capacitance to limit the high frequency circulating currents from reaching the gate driver control circuits. Also, the isolation stage has to be designed with insulation several times higher than the peak system voltage level. In this paper, design, development and evaluation of the gate power supply for medium voltage level applications have been investigated. Several isolation transformer designs have been investigated and optimum design, with very low coupling capacitance ≈ 0.5 pF, has been identified and used in the gate driver design. Experimental characterization of the transformer has been done. The performance of the gate driver power supply has been evaluated in several MV power converters, using 10 kV SiC MOSFETs.

I. Introduction

In the recent years 4H-SiC based high voltage power devices have been of interest for efficient power conversion in medium voltage (MV) levels. The SiC are viable alternative to Si due to much higher critical electric field, better thermal conductivity and higher temperature capability [1]. In the recent years, SiC power devices rated for more than 10 kV have been developed [2]-[4]. With these high voltage devices, simpler two-level and three-level based power converters are feasible for MV power conversion applications.

One of the challenges in using these high voltage SiC devices is their very high turn-on and turn-off dv/dts at the switching instants. The very high dv/dt can give rise to high frequency circulating current through the gate driver circuit, which may also disrupt the gate drive control signals. These high frequency currents will find the paths through the inter-winding coupling capacitances of the isolation stage of the gate driver power supplies. The high frequency high dv/dt stress on the isolation stage will also reduce the life time of its insulation. Commercial gate drivers are available for IGBTs rated up to 4-6.5 kV. These commercial gate drivers' isolation transformers have significant coupling capacitance in the range of 8-14 pF [5]. In the presence of high dv/dt, such high coupling capacitance gives rise to large high frequency circulating in the circuit. The gate driver for high voltage 15 kV SiC IGBT reported in [6] has the coupling capacitance in the range of 3.4 pF. Gate driver for 10kV/100A SiC MOSFET has been reported in [7]. The net effective coupling capacitance of the isolation stage is 5 pF. With optimum design approach, these capacitances may be further reduced. Contactless air core transformer based isolated power supply for gate driver have been proposed in [8], [9]. The power conversion efficiency of the isolation stage is 80-85%. A double galvanic isolation transformer has been proposed in [10] to achieve high voltage isolation for gate driver power and also to isolate the gate drive signal.

The detail gate driver power supply design has been presented in this paper, specifically the isolation stage transformer design is analyzed and method to reduce the coupling capacitance has been investigated. Several possible isolation transformers of different soft magnetic materials and winding techniques are evaluated through electrostatic finite element methods (FEM) and the optimum design is selected. Considering the intended application for MV power converters, the gate drivers' isolation stage is designed for very high the dc insulation (>30 kV). Coupling capacitance as low as ≈ 0.5 pF has been achieved. The complete gate driver power supply design is presented and supported by measurement results. The functionality of the gate driver power supplies on MV power converters are validated by using them on switching low side and high side devices.

In this paper, first a short introduction to the characteristics of MV SiC devices is presented in Section II. In Section III, the complete gate driver power supply design is discussed. Isolation transformers based on different magnetic materials and winding techniques are investigated in Section IV. The isolation transformers measured parameters are evaluated in Section V. One of the isolation transformers is put in the MV converter tests and the experimental results demonstrating the functionality of the gate driver power supply have been presented.

978-1-4673-9551-9/16 $31.00 © 2016 IEEE

II. HIGH VOLTAGE SiC DEVICE CHARACTERISTICS

The switching characteristics of power devices are captured on a standard clamped inductive Double Pulse Test (DPT) circuit. The clamped inductive turn-on and turn-off voltage transitions of SiC 15 kV, 20 A IGBT with dc-link voltage of 11 kV are shown in Fig.1 [6], [11]. The IGBT turn-on current is 5A with gate resistance of 200 Ω and turn-off current is 10A with gate resistance of 10 Ω. From this switching characteristic, the high dv/dts (3.3 kV/μs to 110 kV/μs) at the switching instants are noticeable. Even with such high turn-on resistance, the dv/dts are drastically high.

Fig. 2 shows the switching characteristics of the 10 kV SiC MOSFET switching at 10 kV and 10A current and gate resistance of 15 Ω [12]. Unlike the IGBT, the SiC MOSFET

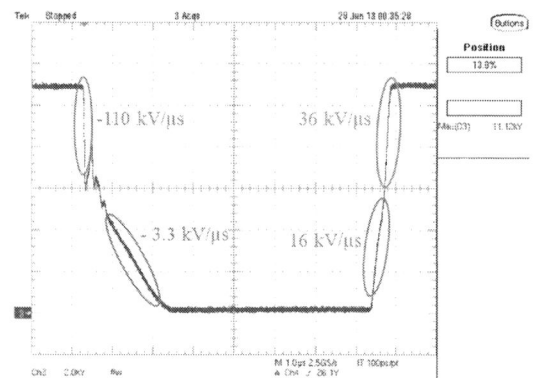

Fig.1. Turn-on and turn-off switching voltage transitions at 11 KV, for 15 kV SiC N-IGBT [6]

has single slope transition and the dv/dts are comparatively lower. However, the dv/dts are in the range of 20-42 kV/ μs, which are considerably high.

The dv/dts produced by the high voltage SiC IGBTs and MOSFETs are significantly higher than those produced by lower voltage silicon devices. Such high dv/dts cause displacement current through the parasitic coupling capacitance of the gate power dc-dc isolation barrier. For a dv/dt of 100 kV/ μs and transformer coupling capacitance of 5 pF, a peak of 500 mA current will be flowing through the coupling capacitor of the transformer. This current will find the return path through the control circuitry back to the power circuit. The voltage spikes introduced on the interconnecting impedances by such high frequency circulating currents are likely to disrupt the operation of the controller and thus the power converter also. The isolation stage of the gate driver should be designed with minimal coupling capacitance.

III. GATE DRIVER POWER SUPPLY DESIGN

In power electronic converters, the PWM signals and the associated power supplies driving the gate of the power devices' whose emitters/sources are at switching nodes have to be floating with respect to the system ground. Such gate drives are often referred to as high side drivers. At high voltages, the PWM signals are isolated by using the optical fiber to

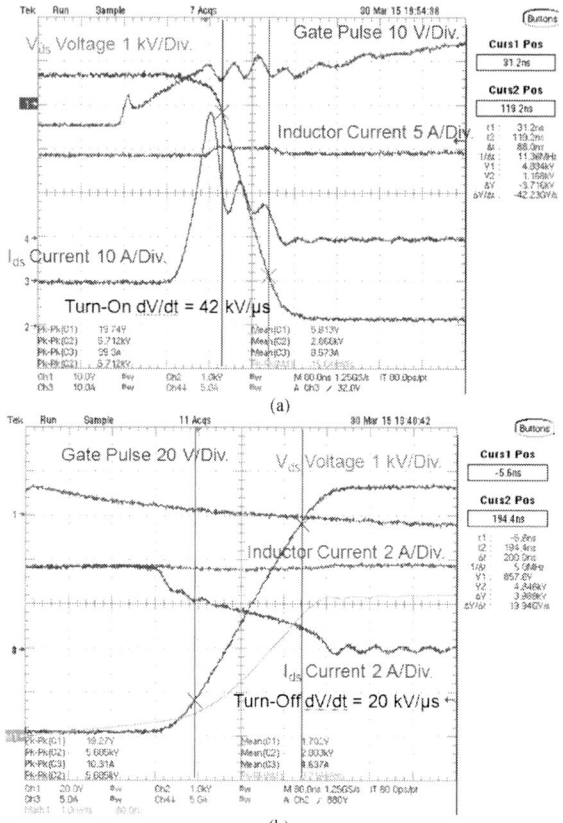

(a)

(b)

Fig.2. Switching voltage transitions at 6 KV, for 10 kV SiC MOSFET (a) Turn-on (b) Turn-off [12]

communicate the gating signals. The gate drive power supplies are derived using an isolated dc-dc converter. The isolation stage of the dc-dc converter has to be designed with low coupling capacitance and also it should not show significant degradation due to high voltage and high frequency stress.

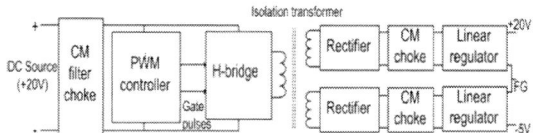

Fig.3. Schematic of isolated gate driver power supply (FG: floating ground)

The high voltage SiC power devices typically require +20 V and -5 V gate voltages to turn-on and turn-off respectively. A negative gate drive is required to overcome the possibility of gate turn-on due to the effect of collector-gate 'Miller' capacitance on device during turn-off. An isolated dc-dc converter is used for generating these voltages. The schematic of the dc-dc power supply is shown in Fig. 3. A push-pull PWM controller LM5030, in open-loop, has been used to generate the complementary pulses to drive the MOSFET H-bridge BD6231F-E2. A 50 kHz switching frequency has been selected for the H-bridge. The H-bridge provides the bipolar 50 kHz square wave to the primary of the isolation transformer.

TABLE I.   GATE DRIVER SPECIFICATIONS

Parameter	Value
Turn-on Voltage	18 V -> 20V
Turn-off Voltage	-5 V
Supply Input Voltage	18-20 V
Switching Frequency	Up to 20 kHz
Turn-on Gate Resistance	10-100 Ω
Turn-off Gate Resistance	10-33 Ω
Isolation Voltage	Up to 20 kV
dv/dt capability	> 50 kV/μs
Isolation Transformer Coupling Capacitance	< 5 pF (1- 100 MHz)

The isolation transformer has two secondary windings; the winding ratios have been selected so as to generate +20 V and -5 V upon rectification. To maintain the high voltage isolation of the transformer, no voltage feedback control has been used. However, the output voltages are regulated using linear regulators at the output stage of the gate drive power supply. In Fig. 3, FG corresponding to the floating ground which may be connected to the floating emitter/source of power device the gate drive power supply is connected to. Common mode chokes are used both at the input and output of the gate driver power supply to limit the circulating common mode currents.

The fundamental component of this power supply is the isolation stage; the design of the isolation stage is discussed in next section. Very low ESL capacitors have been used both on primary and secondaries of the power supply. The specifications of the gate driver for high voltage SiC IGBT/MOSFET is summarized in the Table I.

IV. DESIGN OF ISOLATION TRANSFORMER

The isolation stage of the dc-dc converter must be designed to have robust insulation with sufficient creepage. It is recommended to have much higher insulation than actually required in the system. The reason for the over insulation design is that the insulation degradation over time is faster when they are subjected to high voltage high frequency stress as compared with just DC.

The selection of the magnetic core material, size and the winding design plays significant role in the determination of the coupling capacitance. The magnetic core has to be selected such that it can house the required number of the turns and also give high magnetizing inductance. The core has to be big enough to have enough clearance between the primary and secondary windings. The two secondary windings need not be separated as there is no isolation requirement between them. The insulations between the windings and the core also have to be sufficient to achieve the required isolation. As the voltages of each individual winding are low, a 24 AWG varnished

TABLE II.   CORE PARAMETERS

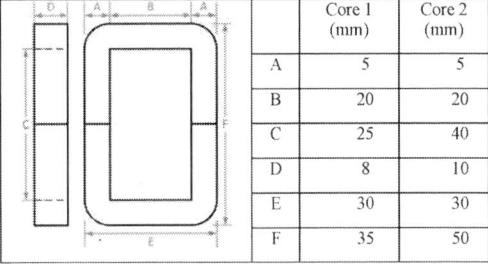

	Core 1 (mm)	Core 2 (mm)
A	5	5
B	20	20
C	25	40
D	8	10
E	30	30
F	35	50

copper wire with insulation rated for 1 kV is used. Three layers of nomex paper of thickness 0.38 mm have been used on each winding for insulation from the core. Also, five layers of kapton tape have been used to hold the nomex paper, which also contributes to the insulation. This insulation, by design, gives more than 30kV isolation between primary and secondary windings. Various available magnetic cores, shapes and material, are selected for the transformer design. Keeping the similar windings (ten turns on primary, eleven and five turns on secondaries) and the insulations, electrostatic FEM simulations are done on different cores to find the primary to secondary coupling capacitances for each case.

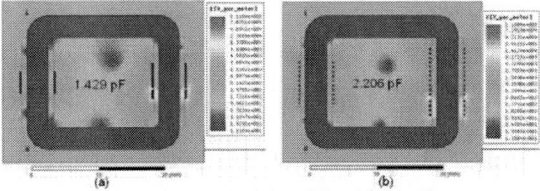

Fig.4. Electrostatic FEM simulation of Core 1 for coupling capacitance (a) compact winding (b) distributed winding

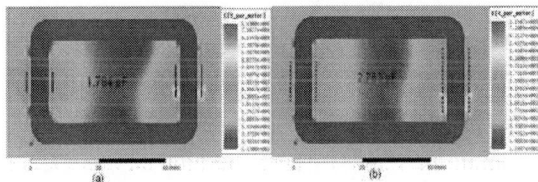

Fig.5. Electrostatic FEM simulation of Core 2 for coupling capacitance (a) compact winding (b) distributed winding

Two amorphous based cut C-cores are selected for the transformer design. The dimensions of these cores are shown in Table II. The core 2 has slightly bigger cross sectional area than core 1. The core 2 is also longer than the core 2. Two separate winding techniques are tried on these cores. In first winding approach, the varnished wire is compactly arranged in both the primary and secondary windings. In the second winding approach, the windings are slightly distributed along the core. This pattern of windings is achieved when a PVC insulated wires are used in winding the transformer. Fig. 4 and 5 show the FEM simulation results. In both the cases, the coupling capacitances have increased when the windings are distributed. In core 1, the coupling capacitance has increased from 1.4 pF to 2.2 pF; while for core 2 it increased from 1.78

978-1-4673-9551-9/16 $31.00 © 2016 IEEE   1634

pF to 2.75 pF. The core 2 has slightly higher coupling capacitance as its core area is bigger than that of core 1. The capacitance between the core and the windings will be more in the case of core 2, resulting in higher coupling capacitance.

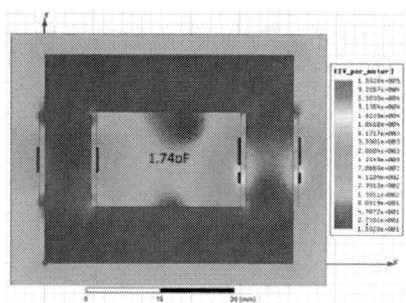

Fig.6. Electrostatic FEM simulation of ferrite core UR 42/25/12 for coupling capacitance

Fig. 6 shows the FEM simulation result for coupling capacitance when ferrite UR 42/25/12 core from Magnetics is used. The estimated coupling capacitance between the primary and secondary windings is 1.74 pF. As seen from earlier case the distributed winding will give higher coupling capacitance thus in this and subsequent cases, only the compact windings are considered.

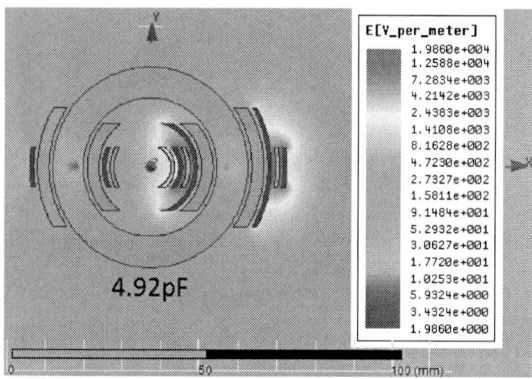

Fig.7. Electrostatic FEM simulation of ferrite toroidal core ZW44916TC for coupling capacitance

Toroidal core ZW44916TC from Magnetics is analyzed next. The compact windings with ten turns on primary, eleven and five turns on secondaries are considered. The estimated coupling capacitance between the primary and secondaries is

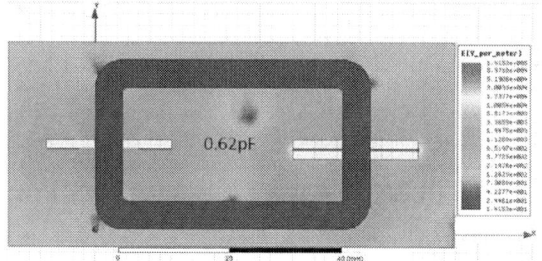

Fig.8. Electrostatic FEM simulation of amorphous core for coupling capacitance with PCB windings

4.92 pF, shown in Fig. 7. Besides, the coupling capacitance, in toroidal cores the edges of the windings are much closer than compared with rectangular cores. Thus, attention has to the creepage between the primary and secondary windings have to be made.

As discussed earlier, the parasitic capacitance of a winding depends on the winding area on the core. The wider the winding area, the higher the winding capacitance which will lead to higher coupling capacitance. Thus to reduce the winding area, the windings are made on the printed circuit board (PCB). Also, the insulation is also provided on the FR4 PCb board. A 3mm gap is left all around the core. Fig.8 shows estimated coupling capacitance between the primary and secondary windings, which is 0.62 pF. In this case an amorphous core 2, parameters on Table II, has been used. This in fact verifies that the area between the winding and the core has to be kept to minimal to get the best minimal coupling capacitance.

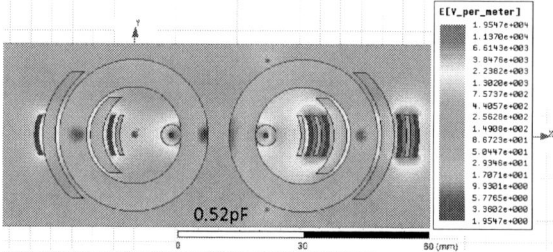

Fig.9. Electrostatic FEM simulation of double galvanic isolation transformer based on ferrite toroid ZW43610TC

Double galvanic isolation requires two magnetic cores. However, it helps to increase the isolation level. Also, as the parasitic capacitances of two isolation stages will be in series the net effective coupling capacitance will be reduced. Fig. 9 shows the FEM simulations result for coupling capacitance for double galvanic isolation when ferrite toroids ZW43610TC are used. The first transformer has ten turns on primary and single turn on secondary. The second transformer has single turn on primary. Its secondaries have eleven and five turns. The estimated coupling capacitance is 0.52 pF. Thus with the double galvanic isolation stage it is possible to achieve a very small coupling capacitance.

FEM analysis on nano-crystalline core, with dimension of core 1 in Table II, has also been done. The coupling capacitance for this case is similar to the one shown in Fig. 4. Amorphous or nano-crystalline cores have high permeability and high flux saturation level. However, at switching frequency of 50 kHz, the amorphous cores' permeability is relatively low resulting in low magnetizing impedance.

Comparing FEM based estimates of the coupling capacitances; it shows that winding to core capacitance dominates in contributing the net coupling capacitance as compared to direct winding to winding capacitance. Thus, increasing the distance between the primary and secondary windings alone will not help to reduce the coupling capacitor.

V. EXPERIMENTAL VALIDATION

In this section, the measurement results on above designed isolation transformers will be presented. The designed gate driver power supply will be put in the switching tests, driving both the low and high side high voltage devices.

a) Measurement on isolation transformers

The hardware prototypes of the transformers designed in Section IV are made. All transformers are made with 24 AWG varnished wire. The transformers have ten turns on primary, eleven and five turns on secondaries. The insulations used on all transformers are also similar. Fig. 10 shows the photos of the isolation transformers.

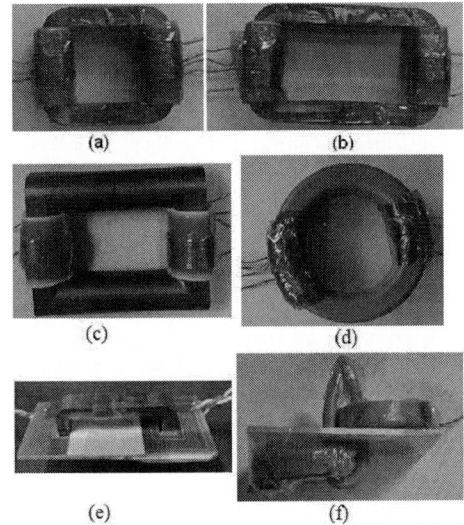

Fig.10. Isolation transformer prototypes (a) Core 1 (Table II) (b) Core 2(Table II) (c) UR 42/25/12 based transformer (d) ZW44916TC based transformer (e) Transformer with PCB winding (f) double galvanic

The impedance analyzer HP4294A has been used to measure the magnetizing inductance and parasitic coupling capacitances of all the transformers shown in Fig. 10. The measured coupling capacitances of these transformers in the range of 10 kHz to 110 MHz are shown in Fig. 11. The measured capacitances are relatively close to the values estimated through FEM simulations presented in Section IV. As predicted through the FEM estimation, the double galvanic isolation transformer has the least coupling capacitance. The values of the measured coupling capacitances at 50 MHz and magnetizing inductances at 50 kHz are tabulated in Table III.

The magnetizing inductances have been measured at 50 kHz which is the isolated power supply's switching frequency. The amorphous core's measured magnetizing inductance is comparatively small. Such low magnetizing impedance is likely to overload the H-bridge, BD6231F-E2. Thus, for the gate driver power supply evaluation a nano-cyrstalline based transformer has been selected. The dimensions of this core are same as that of amorphous core 1, shown in Table II. The measured magnetizing inductance and coupling capacitance of

the nano-crystalline based transformer are also shown in Table III. The measured coupling capacitance in 10 kHz- 110 MHz is shown in Fig. 12.

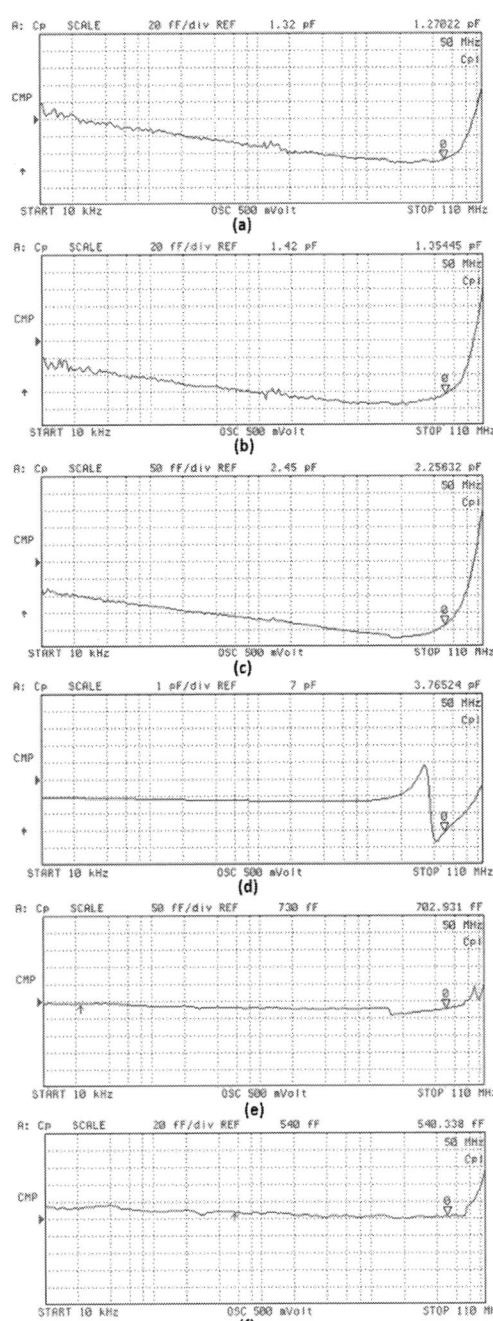

Fig.11. Measured coupling capacitances of (a) Core 1 (Table II) (b) Core 2(Table II) (c) UR 42/25/12 based transfromer (d) ZW44916TC based transformer (e) Transformer with PCB winding (f) double galvanic isolation transformer

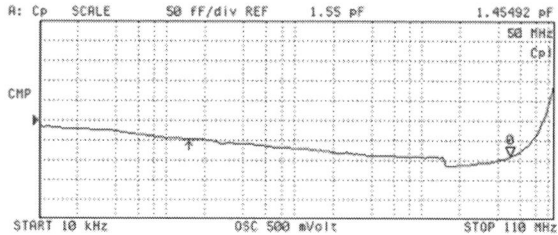

Fig.12. Measured coupling capacitances of nano-crystalline core of dimension core 1 in Table II

TABLE III.   TRANSFORMERS' MEASURED PARAMETERS

Transformer	Magnetizing inductance (μH) @50kHz	Coupling capacitance (pF) @50MHz
Amorphous core 1	63	1.27
Amorphous core 2	85	1.35
Ferrite core UR 42/25/12	152	2.2
Ferrite core ZW44916TC	503	3.7
PCB winding	302	0.7
Double galvanic isolation	217	0.54
Nano-crystalline core 1	1050	1.45

b) Switching tests on gate driver

The isolated gate driver power supplies the regulated +20 V and -5 V to the gate driver board. The gate driver board is similar to the one presented in [10]. The PWM signal are transmitted from the controller through the fiber optical cable; thus isolating the controller from the power converter. Fig. 13 shows the hardware prototype of the gate drive power supply, the top board, with nano-crsytalline based transformer. The input voltage of +20V is given to the primary side of the converter. The isolated and regulated +20 V and -5 V are given to the gate driver board.

Fig. 14 shows the test waveforms captured from the gate driver board. The CH1 and CH2 show the regulated +20 V and -5V respectively. CH3 shows the gate signal received from the controller. The CH4 shows the gate voltage applied to the power device. This verifies the basic functioning of the gate

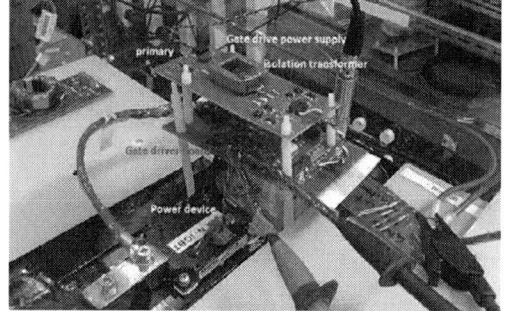

Fig.13. Gate driver power supply (top PCB board) and gate driver hardware prototype (bottom PCB board)

driver board and the gate power supplies.

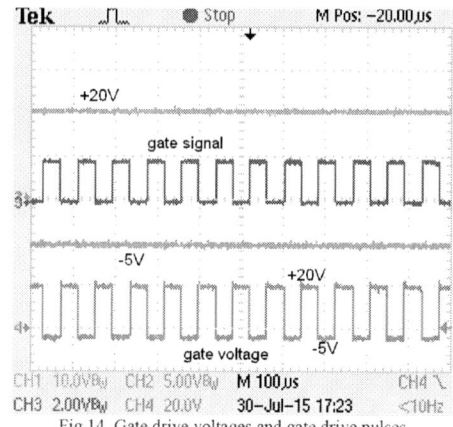

Fig.14. Gate drive voltages and gate drive pulses

To qualify the gate driver power supplies for medium voltage applications, the test circuits shown in Fig. 15 are used. In these test circuits the switches S1 and S2 are 10 kV/10A SiC MOSFETs. The diodes D1 and D2 are 20 kV SiC JBS diodes. Test circuit shown in Fig. 15 (a) is simple double pulse test circuit. The capacitor is charged using a high voltage low current source. The capacitor stores the energy sufficient for two switching pulses. The 10 kV SiC MOSFET on the low side are switched at 6 kV voltage and 10 A purely inductive load. Fig. 16 and 17 show the switching waveforms for the 10 kV SiC MOSFET. The functionality verification of the gate driver power supply and the gate driver board has done through the expected operation of the circuit.

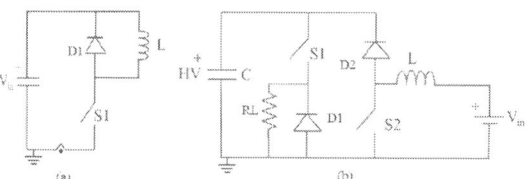

Fig.15. Gate drive test circuits: (a) double pulse test (b) boost and buck converter

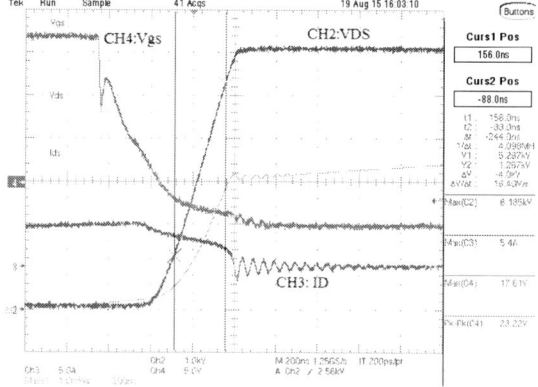

Fig.16. Turn- off transition at 6kV, 10A with Rgoff of 10Ω

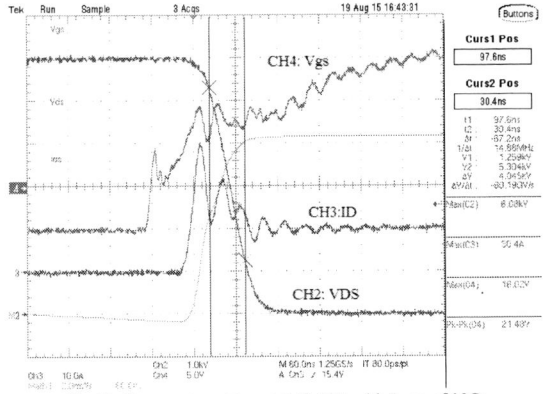

Fig.17. Turn- on transition at 6kV, 10A with Rgon of 10Ω

The test circuit shown in Fig. 15(b) is used for continuous switching tests on the gate driver. This converter has a boost converter followed by a buck converter. The continuous switching tests are done first without the buck converter. The input to the boost converter is given through a regulated laboratory dc source. The output of the boost converter is regulated at 5 kV and supplying a resistive load of 4.3 kW. The boost converter is switched at 5 kHz. The experimental test is run for thirty minutes. Fig. 18 shows the test waveforms captured at the end of the experiment. The gate power supply and the gate driver showed reliable operation on driving a low side device.

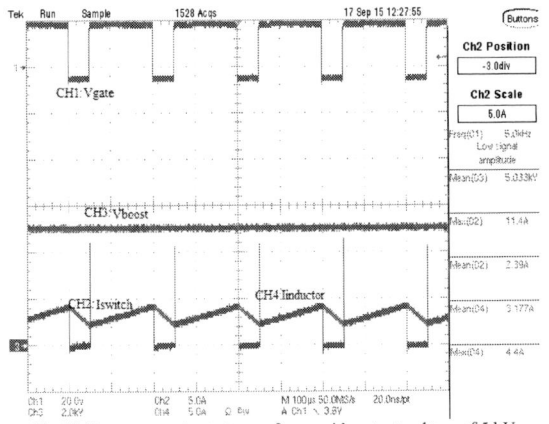

Fig.18. Boost converter test waveforms with output voltage of 5 kV

The functionality of the gate driver power supply and gate driver on switching high side power device are evaluated by using the gate driver on the buck converter. After completion of the boost test, the buck converter is operated to load the boost converter. The continuous switching tests are done at 5 kV dc bus voltages. Both the boost and buck converters are switched at 10 kHz switching frequency. Fig. 19 shows the test waveforms captured at the end of thirty minutes continuous switching. In this case the gate driver on the buck converter is

subjected to very high dv/dt, typically in the range of 15-70 kV/ μs. The gate driver power supply and the gate driver performed reliably without any sign of degradation.

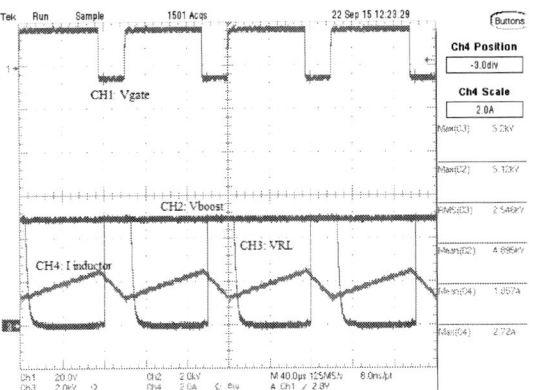

Fig.19. Boost and buck converter test waveforms with dc link voltage of 5 kV boost duty ratio of 75% and buck duty ratio of 25%

Thus, the experimental tests conducted showed that the gate driver power supplies are suitable for the medium voltage SiC devices. The evaluation of other presented isolation transformers are in progress.

VI. CONCLUSIONS

In this paper, the design and experimental validation of gate driver power supplies with very low coupling capacitance for SiC power devices used in MV have been presented. These high voltage power devices are finding more and more applications in MV applications. However, commercial gate drivers are not yet available. Several transformer design has been evaluated and optimum design with inter-winding coupling capacitance $\approx 0.5pF$ has been achieved. It has been found that the area of the winding with the core has to be minimized to reduce the coupling capacitance. Also, common mode chokes have been added on both sides of the isolation stage to further reduce the common mode circulating currents. The experiment results have been shown to verify the coupling capacitance of the isolation stage. An isolated gate driver power supply using nano-crystalline core, with coupling capacitance of 1.45 pF, has been built. The functionality of the gate driver has been verified. The evaluation and validation of the gate driver power supplies with different converter test topologies has been presented. The evaluation of other designed isolation transformers are in progress, which will be reported in future articles.

ACKNOWLEDGMENT

The information, data, or work presented herein was funded in part by the Office of Energy Efficiency and Renewable Energy (EERE), U.S. Department of Energy, under Award Number DE-EE0006521 with North Carolina State University, PowerAmerica Institute. This work made use of FREEDM ERC shared facilities supported by NSF under award no. EEC-0812121.

DISCLAIMER

The information, data, or work presented herein was funded in part by an agency of the United States Government. Neither the United States Government nor any agency thereof, nor any of their employees, makes any warranty, express or implied, or assumes any legal liability or responsibility for the accuracy, completeness, or usefulness of any information, apparatus, product, or process disclosed, or represents that its use would not infringe privately owned rights. Reference herein to any specific commercial product, process, or service by trade name, trademark, manufacturer, or otherwise does not necessarily constitute or imply its endorsement, recommendation, or favoring by the United States Government or any agency thereof. The views and opinions of authors expressed herein do not necessarily state or reflect those of the United States Government or any agency thereof.

REFERENCES

[1] J. Palmour, et al, "Silicon Carbide Power MOSFETs: Breakthrough Performance from 900 V Up to 15 kV," in *IEEE International Symposium on Power Semiconductor Devices and ICs* (ISPSD) 2014, pp. 79-82.

[2] M. K. Das, et al., "10 kV, 120 A SiC half H-bridge power MOSFET modules suitable for high frequency, medium voltage applications," in Proc. Energy Conversion Congress and Exposition (ECCE), 2011, pp. 2689-2692.

[3] J. Palmour, et al, "Silicon Carbide Power MOSFETs: Breakthrough Performance from 900 V Up to 15 kV," in IEEE International Symposium on Power Semiconductor Devices and ICs (ISPSD) 2014, pp. 79-82.

[4] S. Ryu, et al, "Ultra high voltage IGBTs in 4H-SiC," in IEEE Workshop on Wide Bandgap Power Devices and Applications (WiPDA), 2013, pp. 36-39.

[5] http://igbt-driver.com/sites/default/files/product_document/data_sheet/ISO3116I_2.pdf

[6] A. Kadavelugu and S. Bhattacharya, "Design considerations and development of gate driver for 15 kV SiC IGBT," in Proc. IEEE Applied Power Electronics Conference and Exposition (APEC), 2014, pp. 1497-1501.

[7] D.W. Berning, T. H. Duong, J. M. Ortiz-Rodriguez, A. Rivera-Lopez and A. R. Hefner Jr. , "High-voltage isolated gate driver circuit for 10kV, 100A SiC MOSFET/JBS Power Module," in Proc. IEEE Industrial Applications Society Annual Meeting, 2008, pp.1-7.

[8] C. Marxgut, J. Biela, J. W. Kolar, R. Steiner and P. K. Steimer, "DC-DC converter for gate power supplies with an optimal air transformer," in Proc. IEEE Applied Power Electronics Conference and Exposition (APEC), 2010, pp. 865-1870.

[9] R. Steiner and P. K. Steimer, F. Krismer and J. W. Kolar, , "Contactless energy transmission for isolated 100W gate driver supply of a medium voltage converter," in Proc. IEEE Industrial Electronics Conference (IECON), 2009, pp. 302-307.

[10] S. Brehaut and F. Costa., "Gate driving of high power IGBT through a double galvanic insulation transformer," in Proc. IEEE Industrial Electronics Conference (IECON), 2006, pp. 2505-2510.

[11] Kadavelugu, et al, "Characterization of 15 kV SiC n-IGBT and its Application Considerations for High Power Converters," in Proc. IEEE Energy Conversion Congress and Exposition (ECCE) 2013, pp. 2528-2535.

[12] S. Madhusoodhanan et al., "Solid State Transformer and MV Grid Tie Applications Enabled by 15 kV SiC IGBTs and 10 kV SiC MOSFETs based Multilevel Converters," IEEE Trans. Ind. Applicat., Vol. 51, no. 4, pp. 3343-3360, Mar. 2015.

General-Purpose Clocked Gate Driver (CGD) IC with Programmable 63-Level Drivability to Reduce Ic Overshoot and Switching Loss of Various Power Transistors

Koutaro Miyazaki[*], Seiya Abe[†], Masanori Tsukuda[†], Ichiro Omura[†], Keiji Wada[‡], Makoto Takamiya[*] and Takayasu Sakurai[*]

[*]University of Tokyo, Tokyo, Japan
[†]Kyushu Institute of Technology, Fukuoka, Japan
[‡]Tokyo Metropolitan University, Tokyo, Japan
Email: koutaro@iis.u-tokyo.ac.jp

Abstract—A general-purpose clocked gate driver (CGD) IC to generate an arbitrary gate waveform is proposed to provide a universal platform for fine-grained gate waveform optimization handling various power transistors. The fabricated IC with 0.18μm BCD process has 63 PMOS and 63 NMOS driver transistors on a chip whose activation patterns are controlled by 6-bit digital signals and 25-MHz clock (= 40-ns time step control). In the 500-V switching measurements, the proposed CGD reduces the I_C overshoot by 25% and 41% and the energy loss by 38% and 55% for Si-IGBT and SiC-MOSFET, respectively.

Keywords—Gate driver; IGBT; SiC; Clock; Programmable

I. INTRODUCTION

A gate driver is a key technology for the switching of power devices to minimize the switching loss and the current overshoot. Conventional gate drivers, however, have two problems: (1) customized design to each power transistor (e.g. Si-IGBT, SiC-MOSFET) increases the development cost and turnaround time (TAT), and (2) limited programmability (e.g. 2 gate resistances [1-5], multi voltage levels [6-9], and 9 output resistances of the segmented gate drivers [10]) prevents a precise gate waveform optimization for the low noise and the low loss of the power transistors. To solve the problems, a general-purpose clocked gate driver (CGD) IC is proposed to provide a universal platform for fine-grained gate waveform optimization handling various power transistors including Si-IGBT and SiC-MOSFET, thereby reducing the development cost and TAT for the gate drivers. The proposed CGD IC enables a fine-grained programmability of 63-level drivability and 40-ns step timing control, which reduces I_C overshoot and turn-on energy loss for Si-IGBT and SiC-MOSFET in the 500-V switching measurement. The programmability of CGD is the finest compared with the previous gate drivers [1-10].

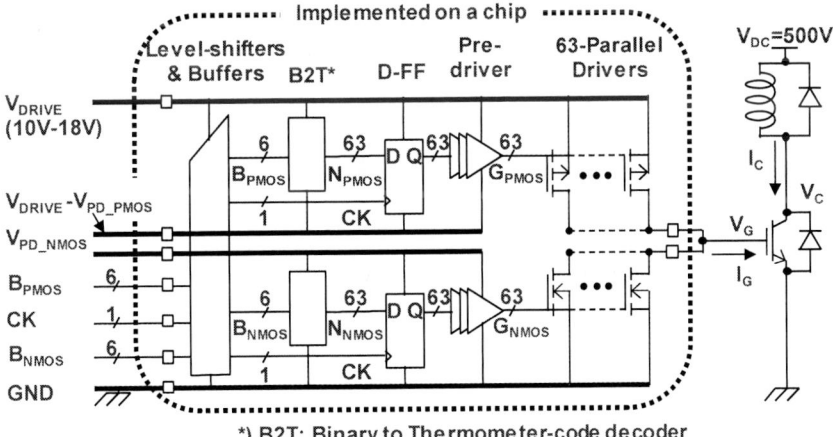

Fig. 1 Schematic diagram of general-purpose clocked gate driver (CGD) IC.

II. System Implementation

The schematic diagram of the implemented general-purpose CGD IC is shown in Fig. 1. CGD IC is developed for the switching of power devices at $V_{DC} = 500V$. In order to realize programmable 63-level drivability, 63 parallel drivers are connected to the gate of the power device and a 6-bit binary control signal, B_{PMOS} (B_{NMOS}), is applied to specify the number of activated PMOS (NMOS) driver transistors, N_{PMOS} (N_{NMOS}). A pair of 6-bit signals (B_{PMOS} and B_{NMOS}) are latched by the clock (CK) and activate the final 63 PMOS (NMOS) transistors. CK frequency is 25MHz and 40-ns time step control of the drivability is achieved. The power supply voltage (V_{DRIVE}) of CGD IC is 10V–18V, and V_{DRIVE} of 15V is used in the following measurements. The voltage swing of input digital signals (B_{PMOS}, B_{NMOS}, and CK) is 5V, and the swing is increased to V_{DRIVE} by level-shifters. By adjusting the pre-driver voltage swing, V_{PD_PMOS} and V_{PD_NMOS}, from 1.2V to 5V, the output drivability of a single driver MOS transistor can be tuned from 3mA to 80mA. The peak drivability is 63 times of the single driver, which corresponds to the maximum peak current of the gate current (I_G) from 0.19A (= 3mA x 63) to 5A (= 80mA x 63). V_{PD_PMOS} and V_{PD_NMOS} of 1.8V is used in the following measurements.

The binary-coded input is indispensable since 63 x 2 input pins are too many to handle. The binary signals, however, may cause glitch problems in the gate voltage (V_G). The glitch will break down the power transistors. For example, when the binary input changes from 011111 (31) to 100000 (32), there is a possibility that the state goes from 011111 (31) to 111111 (63) to 100000 (32) causing a few nano-seconds glitch at the pre-driver, if there are variability of devices and interconnection designs which make the most significant bit change faster than the other bits. This is the cause of the glitch problems. To prevent this problem, a small-sized binary to thermometer-code decoder in Fig. 2 is employed.

Fig. 3 shows operation waveforms for 63 PMOS transistors to pull up V_G in CGD IC. The operation for 63 NMOS transistors to pull down V_G is similar. An arbitrary I_G waveform is generated by applying a control bit pattern (B_{PMOS} (B_{NMOS})) in each clock cycle with 40-ns step and digitally specifying time and current pairs of t_i and I_{Gi} (i=1,2,3...n).

III. Modeling of Gate Driver

In this chapter, the modeling of the 63 parallel drivers in Fig. 1 is discussed. In the previous the segmented gate drivers [10], the transistors in the segmented gate drivers (Fig. 4(a)) were modeled as a resistor (Fig. 4(b)). In this paper, it is proposed that the transistors in the segmented gate drivers should be modeled as current-source (Fig. 4(c)) instead of the resistor (Fig. 4(b)). Fig. 5 shows the SPICE simulated pull-up and pull-down waveforms of V_G with two models in Fig. 4. The capacitance in Fig. 4 is 22nF emulating the gate capacitance of the power devices. V_{PD_PMOS} and V_{PD_NMOS} are 5V and 1.8V in Figs. 5 (a) and (b), respectively. Compared with the resistor model, the current-source model is in good agreement with the driver with transistors. Therefore, the gate driver behaves like the constant-current driver (Fig. 4(c)) rather than the resistor (Fig. 4(b)) because of the high output resistance of MOS transistors in a saturation region.

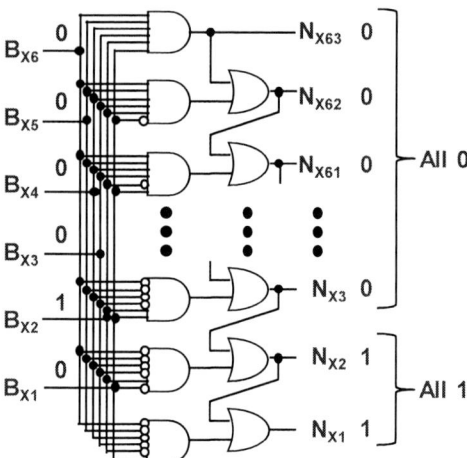

Fig. 2 Binary to thermometer-code decoder.

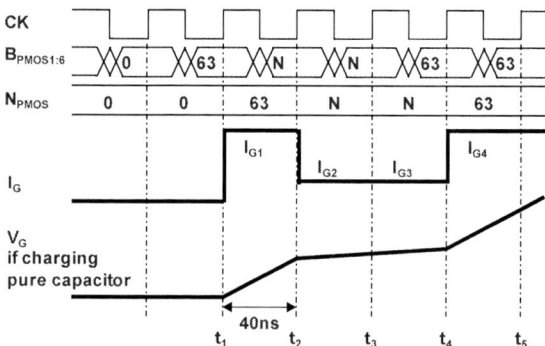

Fig. 3 Operation waveforms for 63 PMOS transistors to pull up V_G in CGD IC.

IV. Measurement Results

The proposed general-purpose CGD IC is fabricated with 40V, 0.18μm BCD process. Fig. 6 shows a die photo of CGD IC. The core size is 2300μm by 730μm. Fig. 7 shows photos of PCB. The 2.5-mm square CGD IC is placed on the top side of PCB. Si-IGBT and SiC diodes are placed on the reverse side of PCB.

Turn-on characteristics are measured with a double-pulse setup shown in Fig. 1 with SiC diodes (C4D10120D, 1200V, 18A) at $V_{DC} = 500V$. To demonstrate the versatility of the proposed general-purpose CGD IC, both Si-IGBT (IRG7PH46UPbF, 1200V, 75A) and SiC-MOSFET (SCH2080KE, 1200V, 40A) are driven by CGD IC. Although in Fig. 1, an IGBT symbol is used for a power device, the IGBT is replaced by SiC-MOSFET when SiC-MOSFET is under test. Notations such as I_C and V_C are used even for the SiC-MOSFET device just for simplicity.

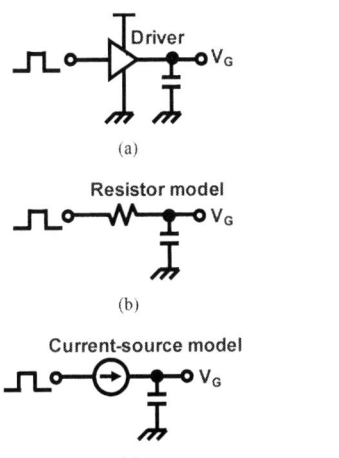

(a)

Resistor model

(b)

Current-source model

(c)

Fig. 4 Modeling of gate drivers. (a) Original gate driver. (b) Conventional resistor model. (c) Proposed current-source model.

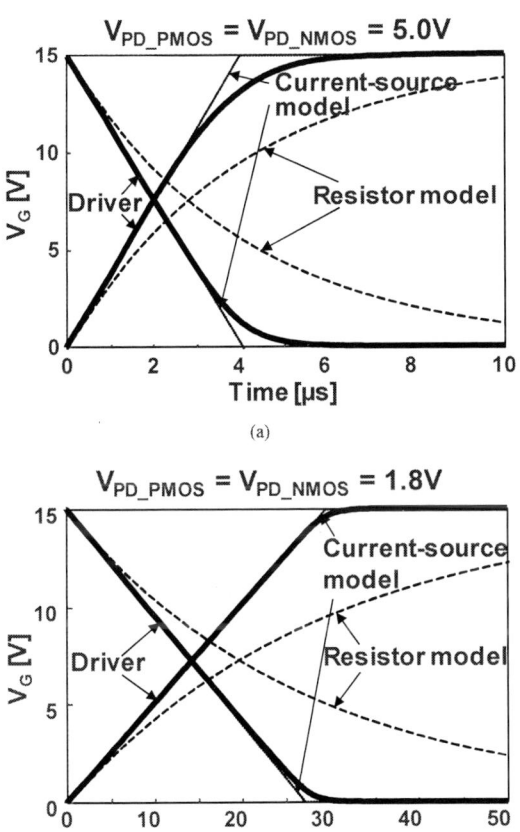

(a)

(b)

Fig. 5 SPICE simulated pull-up and pull-down waveforms of V_G with two models in Fig. 4. V_{PD_PMOS} and V_{PD_NMOS} are 5V and 1.8V in (a) and (b), respectively.

To show the advantage of the proposed CGD IC with programmable 63-level drivability, three types of gate waveforms shown in Fig. 8 are compared. Fig. 8 (a) shows a conventional "no active gate drive" [11]. To show the trade-off between the turn-on energy loss and I_C overshoot, I_G to pull-up V_G is varied by N_{PMOS} in the measurement. Fig. 8 (b) shows a conventional "9-level active gate drive" emulating the 9-level segmented gate drivers [10]. This waveform is based on [5, 7, 9-10, 12]. At the turn-on, N_{PMOS} changes from 0 to m and keeps m for t_1. Then, N_{PMOS} changes from m to 9 level of i (i = 2 to 58 with 7 increments in between) and keeps i for t_2. Finally, N_{PMOS} changes from i to m. Fig. 8 (c) shows the proposed "63-level active gate drive". Fig. 8 (c) is the same as Fig. 8 (b) except for i. In Fig. 8 (c), i is from 0 to 63 with 1 increment in

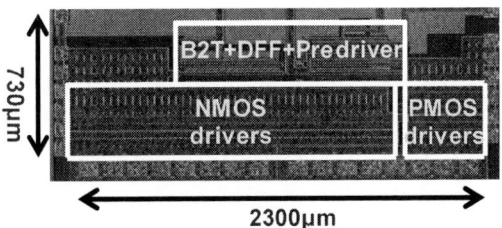

Fig. 6 Die photo of CGD IC.

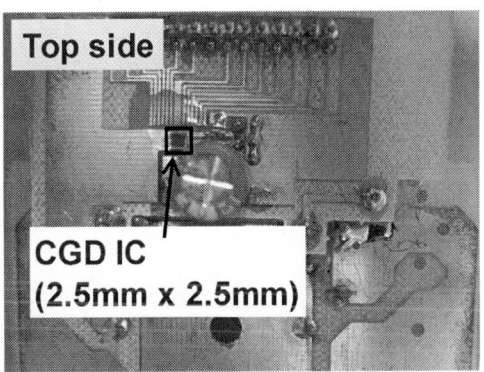

Fig. 7 Photos of PCB.

978-1-4673-9551-9/16 $31.00 © 2016 IEEE

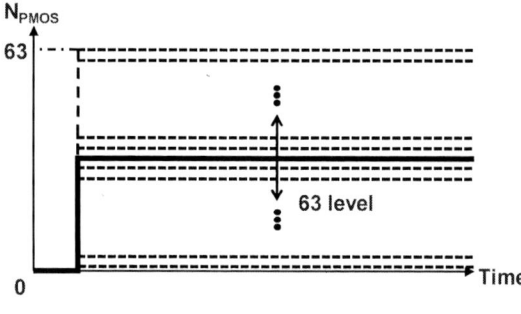

(a)

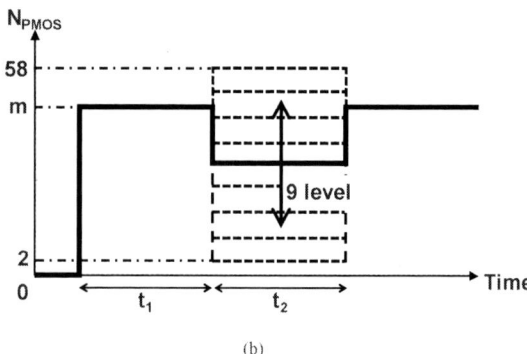

(b)

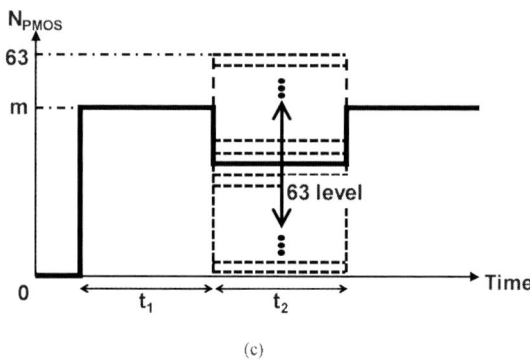

(c)

Fig. 8 Three types of gate waveforms. (a) No active gate drive. (b) 9-level active gate drive. (c) Proposed 63-level active gate drive.

Table I Parameters used in measurements for Si-IGBT and SiC-MOSFET

	Si-IGBT	SiC-MOSFET
m	31	63
t_1	160ns	40ns
t_2	160ns	80ns

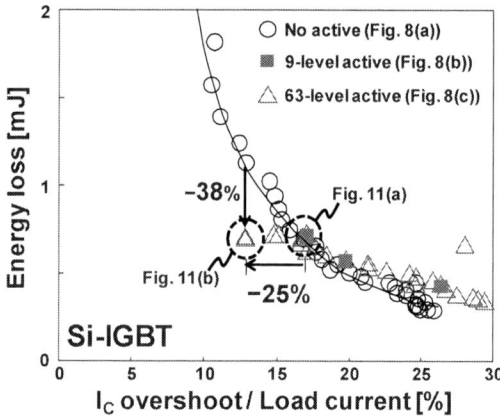

Fig. 9 Measured energy loss vs. I_C overshoot in turn-on characteristics at 500-V switching for Si-IGBT.

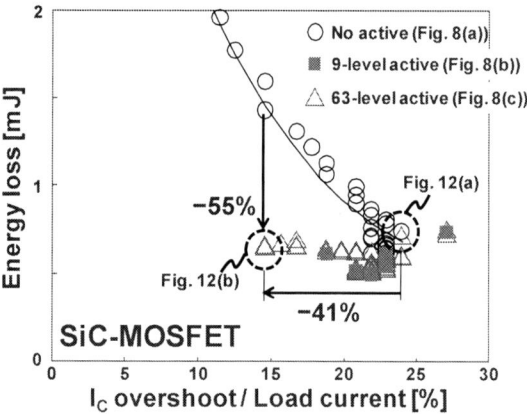

Fig. 10 Measured energy loss vs. I_C overshoot in turn-on characteristics at 500-V switching for SiC-MOSFET.

between. Table I shows m, t_1, and t_2 in the measurements for Si-IGBT and SiC-MOSFET, respectively.

Figs. 9 and 10 show measured energy loss versus I_C overshoot in turn-on characteristics at 500-V switching with the three gate waveforms shown in Fig. 8 for Si-IGBT and SiC-MOSFET, respectively. In the no active gate drive, the trade-off between the turn-on energy loss and I_C overshoot is observed. By using 63-level active gate drive, however, the loss-overshoot trade-off can be optimized more compared with cases of 9-level active gate drive [10] and no active gate drive. The proposed 63-level active gate drive reduces the measured energy loss at the same I_C overshoot by 38% (Fig. 9) and 55%

(Fig. 10) for Si-IGBT and SiC-MOSFET, respectively. Similarly, the proposed 63-level active gate drive reduces the measured I_C overshoot at the same energy loss by 25% (Fig. 9) and 41% (Fig. 10) for Si-IGBT and SiC-MOSFET, respectively. The corresponding measured waveforms of N_{PMOS}, V_G, V_C, and I_C for Si-IGBT and SiC-MOSFET are shown in Figs. 11 and 12, respectively. The 25% and 41% reduction of I_C overshoot are clearly shown in Figs. 11 and 12, respectively.

Table II shows a comparison of the proposed CGD IC with previous gate drivers. This work achieved the 40-ns step timing

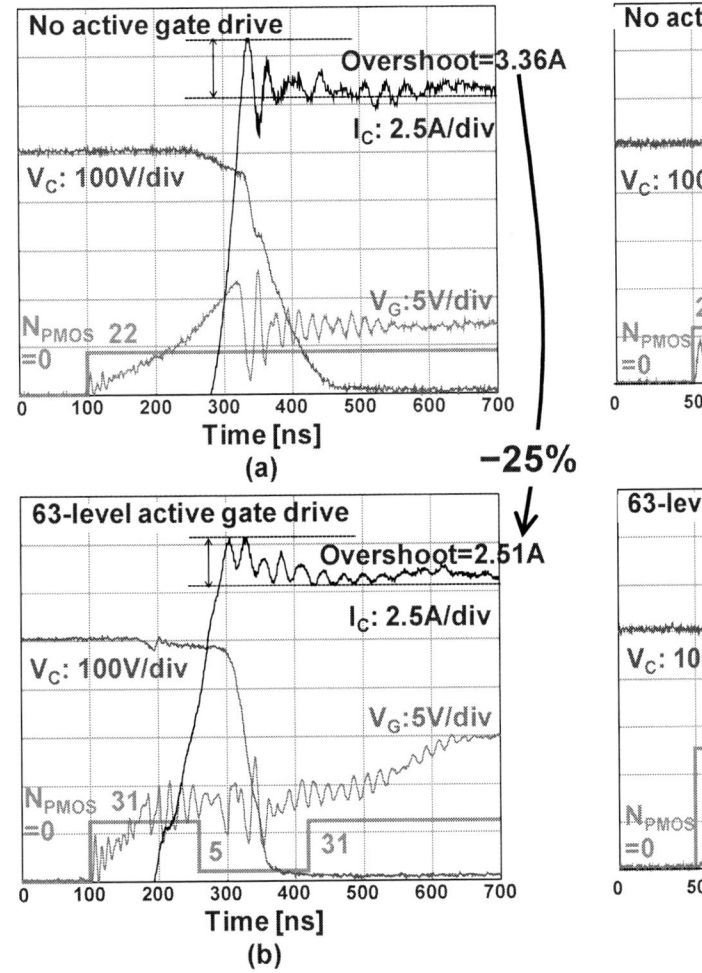

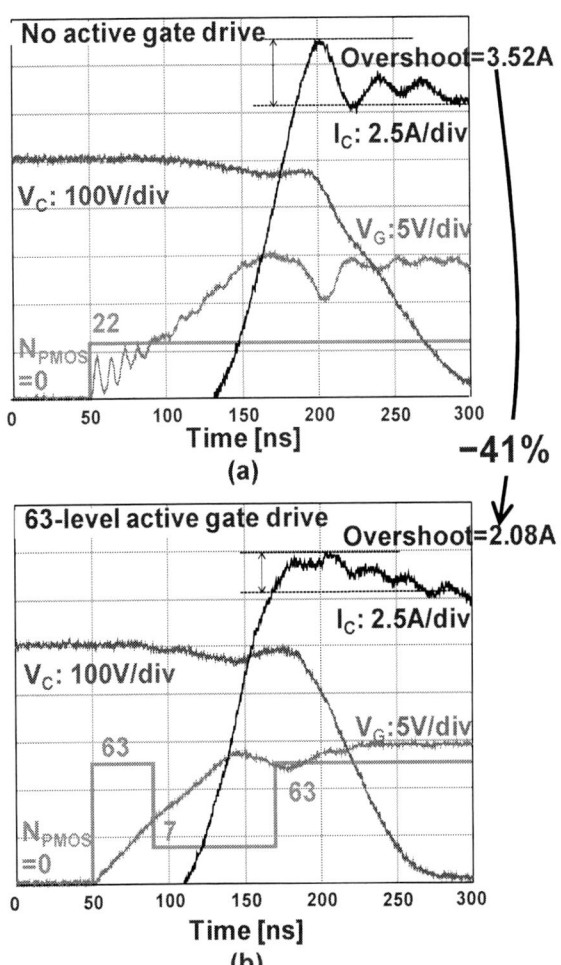

Fig. 11 Measured waveforms for Si-IGBT corresponding to Fig. 9. (a) No active gate drive. (b) Proposed 63-level active gate drive.

Fig. 12 Measured waveforms for SiC-MOSFET corresponding to Fig. 10. (a) No active gate drive. (b) Proposed 63-level active gate drive.

control and 63-level drivability, thereby enabling the gate waveform optimization for both Si-IGBT and SiC-MOSFET. The time programmability is achieved for the first time and the 63-level drivability is the largest number of the drivability levels.

V. CONCLUSIONS

The general-purpose CGD IC to generate an arbitrary gate waveform is the universal platform for fine-grained gate waveform optimization handling various power transistors. The 40-ns step timing programmability is achieved for the first time and the 63-level drivability is the largest number of the drivability levels in the previously published gate drivers. In the 500-V switching measurements, the proposed CGD

Table II Comparison with previous gate drivers

	[1]	[2]	[6]	[10]	This work
Implementation	PCB	PCB	PCB	IC	IC
Target power device	Si-IGBT	Si-IGBT	SiC-MOSFET	Si-IGBT	Si-IGBT & SiC-MOSFET
Time programmability	NA	NA	NA	NA	40-ns step
Number of drivability levels	2	2	4	9	63
How to change drivability	R_G	R_G	Drive voltage	Driver size	N_{PMOS}, N_{NMOS} V_{PD_PMOS}, V_{PD_NMOS}
Gate current	NA	NA	NA	NA	3mA~5A

978-1-4673-9551-9/16 $31.00 © 2016 IEEE

reduces the I_C overshoot by 25% and 41% and the energy loss by 38% and 55% for Si-IGBT and SiC-MOSFET, respectively.

REFERENCES

[1] Z. Wang, X. Shi, L. M. Tolbert, F. Wang, and B. J. Blalock, "A di/dt feedback-based active gate driver for smart switching and fast overcurrent protection of IGBT modules," IEEE Trans. on Power Electronics, Vol. 29, No. 7, pp 3720-3732, Jul. 2014.

[2] M. Sasaki, H. Nishio, and W. T. Ng, "Dynamic gate resistance control for current balancing in parallel connected IGBTs," IEEE Applied Power Electronics Conf. and Expo., pp. 244-249, Mar. 2013.

[3] N. Teerawanich and M. Johnson, "Design optimization of quasi-active gate control for series-connected power devices," IEEE Trans. on Power Electronics, Vol. 29, No. 6, pp. 2705-2714, Jun. 2014.

[4] Y. Miki, M. Mukunori, T. Matsuyoshi, M. Tsukuda, and I. Omura, "High speed turn-on gate driving for 4.5kV IEGT without increase in PIN diode recovery current," IEEE Int. Symp. on Power Semiconductor Devices and ICs, pp. 347-350, May. 2013.

[5] V. John, B. S. Suh, and T. A. Lipo, "High-performance active gate drive for high-power IGBT's," IEEE Trans. on Industry Application, Vol. 35, No. 5, Sep. 1999.

[6] Z. Zhang, F. Wang, L. M. Tolbert, B. J. Blalock, and D. J. Costinett, "Active gate driver for fast switching and cross-talk suppression of SiC devices in a phase-leg configuration," IEEE Applied Power Electronics Conf and Expo., pp.774-781, Mar. 2015.

[7] Y. Lobsiger and J. W. Kolar, "Closed-loop di/dt and dv/dt IGBT gate driver," IEEE Trans. on Power Electronics, Vol. 30, No. 6, pp. 3402-3417, Jun. 2015.

[8] Z. Dong, Z. Zhang, X. Ren, X. Ruan, and Y. F. Liu, "A gate drive circuit with mid-level voltage for GaN transistor in 7-MHz isolated resonant converter," IEEE Applied Power Electronics Conf. and Expo., pp. 731-736, Mar. 2015.

[9] N. Idir, R. Bausière, and J. J. Franchaud, "Active gate voltage control of turn-on di/dt and turn-off dv/dt in insulated gate transistors," IEEE Trans. on Power Electronics, Vol. 21, No. 4, Jul. 2006.

[10] A. Shorten, W. T. Ng, M. Sasaki, T. Kawashima, and H. Nishio, "A segmented gate driver IC for the reduction of IGBT collector current over-shoot at turn-on," IEEE Int. Symp. on Power Semiconductor Devices and ICs, pp. 73-76, Mar. 2013.

[11] S. Azzopardi, A. Kawamura, and H. Iwamoto, "Switching performances of 1200V conventional planar and trench punch-through IGBTs for clamped inductive load under extensve measurements," Int. Power Electronics and Motion Control Conf., Vol. 1, pp. 64-69, Aug. 2000.

[12] I. Baraia, J. A. Barrena, G. Abad, J. M. Canales, and U. Iraola, "An experimentally verified active gate control method for the series connection of IGBT/diodes," IEEE Trans. on Power Electronics, Vol. 27. No. 2, pp. 1025-1038, Feb. 2012.

An Integrated SiC CMOS Gate Driver

Matthew Barlow, Shamim Ahmed,
H. Alan Mantooth

Dept. of Electrical Engineering
University of Arkansas
Fayetteville, Arkansas, USA
mbarlow@uark.edu

A. Matt Francis

Ozark Integrated Circuits, Inc.
Fayetteville, Arkansas, USA
francis@ozarkic.com

Abstract – **In this work, the first reported integrated silicon carbide (SiC) CMOS gate driver is presented. The gate driver is designed in a 15V, 1.2 µm silicon carbide CMOS process, and simulated from 25 ℃ to 300 ℃. The gate drivers are packaged and tested over a wide temperature range. Measured peak output current exceeds the 4 A source, 8 A sink current design goals at room temperature. The measured gate driver rise and fall times are 74.2 ns and 36.8 ns, respectively, while driving a SiC power MOSFET at a package temperature above 500 ℃. Switching operation is demonstrated while at temperature.**

Keywords— high-temperature electronics; driver circuits; silicon carbide

I. INTRODUCTION

Wide bandgap power devices such as gallium nitride (GaN) and silicon carbide (SiC) are pushing the operational boundaries of power electronics [1]. Harsh environments such as car engine compartments, jet engines, down-hole oil drilling, and space exploration have high temperature environments that prohibit conventional silicon devices from operating *in situ* [2]. Silicon carbide power MOSFETs have shown high temperature performance that cannot be matched by conventional silicon power devices. In order to drive these devices effectively a gate driver is required to be located close to the power MOSFET, which subjects the gate driver to the same high temperature environment [3].

Previous work in high temperature gate drivers focused on high temperature silicon-on-insulator (SOI) processes with upper temperature limits of 225 ℃ [4], [5]. Recently, a 2.0 µm NFET-only process was used to design gate drivers that pushed the high temperature capabilities past 400 ℃ [6], [7]. However, due to the limitations imposed by NFET circuit design, the system is difficult to implement with multiple supply voltages and exhibits high quiescent current. Work with a SiC NPN process [8] has yielded a high temperature driver capable of 500 ℃ operation, though with limited voltage and current support needed to drive a SiC power MOSFET. Instead, this work targets SiC BJT devices with lower active drive voltages near 3-4 V.

This research was funded by the National Science Foundation Grant #IIP1237816. Any opinions, findings, and conclusions or recommendations expressed in this material are those of the author(s) and do not necessarily reflect the views of the National Science Foundation.

II. DESIGN

A gate driver was designed with an intended use in high temperature environments above 400 ℃ and up to a 500 kHz switching frequency. Drive strength properties were selected to be comparable to silicon 4 A gate drivers. The gate driver was designed for easy integration into a module with a power MOSFET to minimize parasitic inductance from the gate driver to the MOSFET.

To achieve this goal, a silicon carbide IC process was selected. The process is a 1.2 µm minimum channel length CMOS P-well process, with two polysilicon layers and a high-temperature top metal layer [9]. The nominal power supply voltage rating for the process is 15 V, resulting in a sufficiently high output voltage to drive SiC power MOSFETs directly without needing an LDMOS-type output stage. Device measurements on earlier foundry material was the basis for detailed transistor models. These models allowed simulation at discrete temperatures between 25 ℃ and 300 ℃, as well as projected process corners and aging effects.

The initial design step was to identify an output transistor size that allowed for maximum system performance. Due to the lack of extra metal routing layers, it became clear that a single uniform transistor structure was inadvisable due to the parasitic metal and polysilicon conductor resistances in an output device capable of sinking or sourcing multiple amperes. To reduce the effort in iterating the layout design, a system was developed to parametrically generate output transistor layouts. The generated layouts were then validated using design rule checking (DRC) and layout versus schematic (LVS) checks, and the parasitic layout resistances and capacitances were then extracted. Transistor array aspect ratios and conductor widths were swept parametrically, and simulated over temperature and process variation to identify optimal transistor array parameters. The transistor dimensions were chosen to maintain performance over the entire temperature range.

The basic topology for the gate driver is shown in Fig. 1. The design consists of a CMOS control logic block responsible for reducing switching overlap, followed by a set of CMOS inverters for drive strength gain to drive the pull-up PFET and the pull-down NFET. The pull-up and pull-down transistors are then connected off-chip to minimize losses due to on-chip routing. Additionally, the split output transistors also allows for the flexibility to selectively change the pull-up or pull-down drive strength with the addition of gate resistors. In order to achieve an acceptable performance at the output, the output transistors were divided into sixteen slices. The digital logic responsible for control and dead-time generation use a PFET to NFET drive ratio of 5:1 [10]. A four-stage buffer follows the control logic, with the transistor sizes listed in Table 1. The pull-up and pull-down transistors were designed to source 4 A peak, and sink 8 A peak over the simulated temperature range. The design was simulated from 25 ℃ to 300 ℃ to verify functionality and performance, and then the layout was designed. The final dimensions of the gate driver IC are 4.5 mm by 5.0 mm, or 22.5 mm², and a die micrograph is shown in Fig. 2.

TABLE 1. TRANSISTOR SIZES

Stage	NFET Dimensions (μm / μm)	PFET Dimensions (μm / μm)
1x Digital Logic	4 / 1.2	20 / 1.2
S1	12 / 1.2	60 / 1.2
S2	40 / 1.2	200 / 1.2
S3	200 / 1.2	1,000 / 1.2
S4	780 / 1.2	3,900 / 1.2
Output Slice	36,000 / 1.2	48,000 / 1.2
Total Output	288,000 / 1.2	384,000 / 1.2

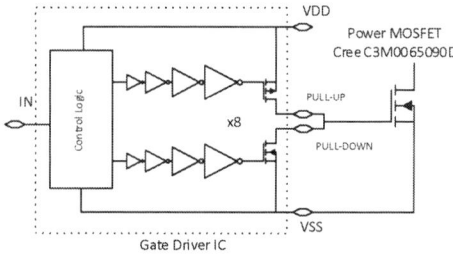

Fig. 1. Gate driver block diagram showing buffering and power device under test.

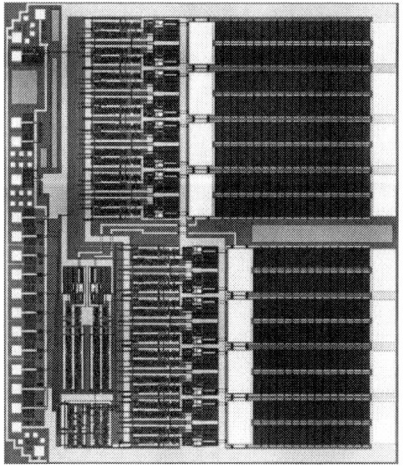

Fig. 2. Die micrograph of the silicon carbide CMOS gate driver. Dimensions are 5.0 mm x 4.5 mm.

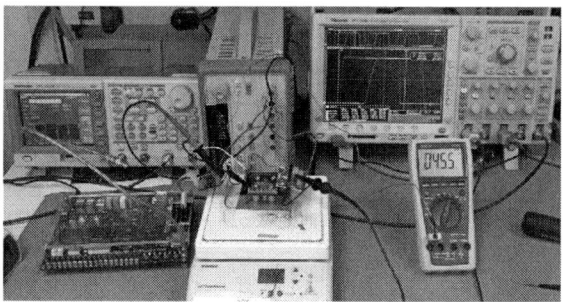

Fig. 3. High temperature test setup in operation driving a power MOSFET. Measured temperature at the bottom of the gate driver package is 455 °C, and the input frequency is 1 MHz.

III. TESTING

High temperature testing beyond standard temperatures provides a constant stream of materials challenges to achieve a reliable result. Performance related goals of minimizing wire and trace distances are at odds with needs to physically isolate the hottest portions of the test setup. A high temperature test jig (Fig. 3) was used to selectively apply high temperatures directly to the package of the gate driver IC without needing to build a complete setup (load, passives, etc.) rated for the upper temperature limits. The gate driver package is suspended by the package pins over an opening in the daughterboard PCB. A copper "hot finger" passes through the opening in the daughterboard PCB to transfer heat from the hot plate directly to the bottom of the gate driver package. A K-type thermocouple is placed in a groove in the "hot finger" to measure the temperature at the bottom of the gate driver package. This configuration was then used to test the gate driver performance over temperature.

With this system, the temperature of the bottom of the package is measured by contact of a K-type thermocouple instead of measuring the die temperature directly. Thermocouple temperature measurements were compared to die temperatures using a platinum RTD. The RTD was epoxied in the die cavity of an identical ceramic package with the same high-temperature conductive epoxy used for die attach. The RTD leads were epoxied to the package pad ring across multiple pins, and the package was then attached to a daughterboard PCB, to replicate the mounting and thermal conditions as closely as possible. The RTD resistance was measured using a four-wire Kelvin connection to eliminate sources of measurement error. The RTD temperature measured less than 10 ℃ lower than the thermocouple over the duration of the test, up to a temperature of 470 ℃. These measurements indicate that the actual die temperature is well predicted by the thermocouple measurement on the bottom of the package.

The finished gate driver IC was packaged in a ceramic leaded chip carrier package. A high temperature, conductive epoxy was used to attach the die to the package, and 1 mil gold bond wires were used to connect the IC pads to the package lead frame. The package lead frame was then soldered to the daughterboard made of Rogers 4350 using a high melting point solder. A second, larger daughterboard was placed on top of the first to provide inputs and outputs to the gate driver IC, as well as to provide power supply decoupling and the power MOSFET for testing. The 900 V Cree C3M0065090D was selected as the target MOSFET as it shows favorable $R_{DS(on)}$ at the 15 V output of the gate driver.

Gate drive current is a measure of the peak current that can be switched by the gate driver. To measure gate drive current, a 1 µF ceramic capacitor is placed from the output to ground. A 0.2 Ω resistor is used as a current shunt to measure the current flowing through the output of the device, and an oscilloscope is used to measure the voltage drop across the resistor. At room temperature, the peak source current was 6 A, and the peak sink current was 12 A.

Gate driver pull-up and pull-down resistance was measured using a DC current source to source or sink a 100 mA current on the gate driver output. The measured output voltage and current were used to evaluate the capability of the gate driver to resist gate charging currents coupled through the parasitic C_{GD} (gate-to-drain Miller capacitance) path. The pull-up and pull-down strength stays steady to 200 ℃, where the resistance increases slowly to 500 ℃. The calculated resistance is shown in Fig. 6.

The propagation delay as well as the rise and fall times of the gate driver while driving the 900 V Cree C3M0065090D power MOSFET over temperature are given in Figs. 4 and 5, respectively. As can be seen in Fig. 4, the propagation delay decreases in the range of 25 ℃ to 200 ℃, and then remains relatively constant above 200 ℃. The falling edge propagation delay decreases by 30 %, while the rising edge propagation delay decreases by only 25%. The rise and fall times shown in Fig. 5 show a more complicated dynamic, with the rise time increasing over temperature, while the fall time has a local minima at 200 ℃. Above 200 ℃, both rise and fall times increase at a slow rate with temperature. Representative waveforms under normal operation are shown in Fig. 7.

As previously mentioned, several aspects of the driver's performance show an inflection point around 200 ℃. The performance degrades a bit if the temperature is increased further. In SiC MOSFETs, the presence of interface states is very common. Additionally, the SiC-SiO$_2$ interface is not very smooth. As a result, both surface roughness and interface trapped charge induced coulomb scattering make the mobility of SiC MOSFETs very low at room temperature. As the temperature increases, the probability of carriers being trapped becomes smaller, thus the mobility becomes higher. The increase in mobility and threshold voltage improve the MOSFET's performance until temperature reaches 200 ℃. Beyond 200 ℃, almost all the interface states are empty. Also, phonon scattering becomes higher due to increasing lattice vibrations. As a result, mobility starts to fall and the devices' gain decreases. This is why the gate driver performance slowly degrades after 200 ℃.

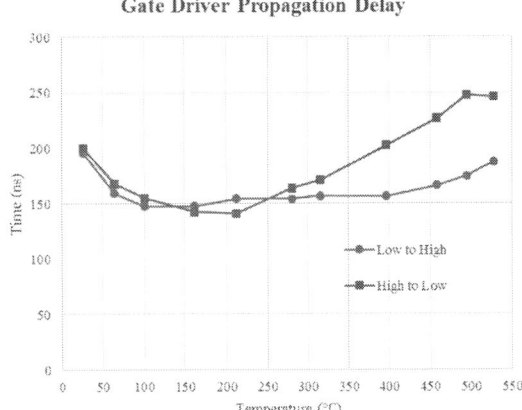

Fig. 4. Gate driver propagation delay while driving a C3M0065090D.

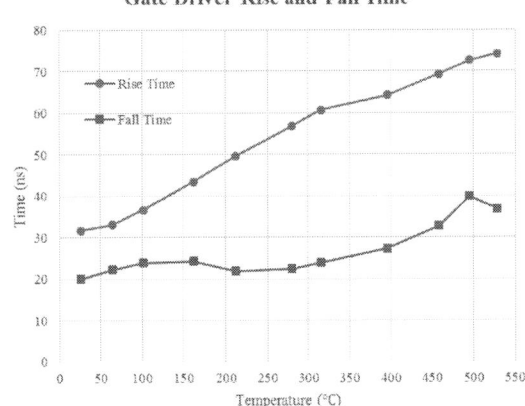

Fig. 5. Gate driver rise and fall times while driving a C3M0065090D.

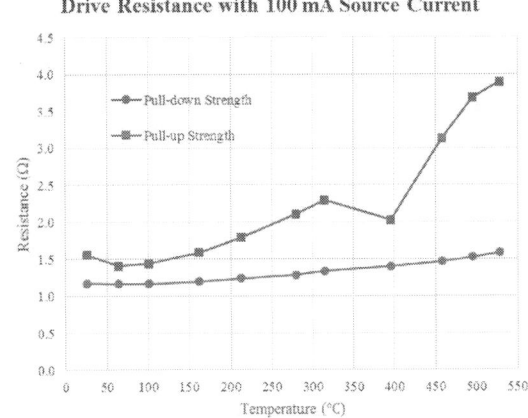

Fig. 6. Gate driver output resistance over temperature.

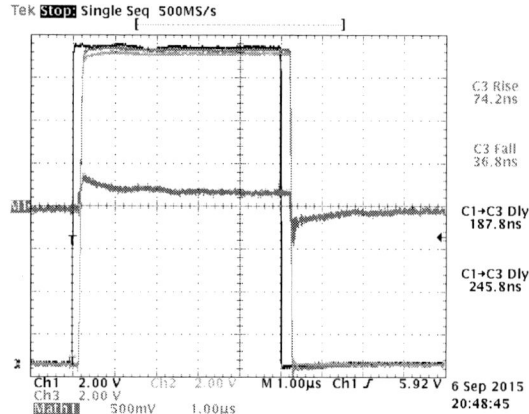

Fig. 7. Gate driver in operation driving a C3M0065090D at 528 ℃.

IV. CONCLUSION AND FUTURE WORK

This work demonstrates a gate driver IC capable of driving a SiC power MOSFET above 500 ℃ without significant degradation in performance over temperature. High drive strength and fast switching are shown, which are necessary for driving SiC power MOSFETs. This work consists of a single IC, with minimal external component requirements, high switching frequency, fast output slew rate, low quiescent current, and only a single gate driver supply voltage, allowing for easy integration into power modules.

Additional work will focus on integrating the gate driver IC into a power module with SiC power MOSFETs. Future gate driver IC revisions may include adjustments to device sizes to compensate for lessons learned.

V. ACKNOWLEDGEMENTS

The authors would like to thank Ewan P. Ramsay, Robin F. Thompson, Robert A. R. Young, and Jennifer D. Walls from Raytheon UK for their help and support during the design and fabrication of this work. In addition, this work wouldn't be possible without the support of the many members at the NSF – Partners for Innovation Building Innovation Capacity project. Finally, these IC wouldn't be more than die on a wafer without the packaging support of Dr. Michael Glover and Michael Steger at the University of Arkansas' High Density Electronics Center.

VI. REFERENCES

[1] H. A. Mantooth, M. D. Glover, and P. Shepherd, "Wide Bandgap Technologies and Their Implications on Miniaturizing Power Electronic Systems," *IEEE J. Emerg. Sel. Top. Power Electron.*, vol. 2, no. 3, pp. 374–385, Sep. 2014.

[2] P. G. Neudeck, R. S. Okojie, and L.-Y. Chen, "High-temperature electronics - a role for wide bandgap semiconductors?," *Proc. IEEE*, vol. 90, no. 6, pp. 1065–1076, Jun. 2002.

[3] A. Lostetter, J. Hornberger, B. McPherson, B. Reese, R. Shaw, M. Schupbach, B. Rowden, A. Mantooth, J. Balda, T. Otsuka, K. Okumura, and M. Miura, "High-temperature silicon carbide and silicon on insulator based integrated power modules," in *IEEE Vehicle Power and Propulsion Conference, 2009. VPPC '09*, 2009, pp. 1032–1035.

[4] J. Valle-Mayorga, C. P. Gutshall, K. M. Phan, I. Escorcia-Carranza, H. A. Mantooth, B. Reese, M. Schupbach, and A. Lostetter, "High-Temperature Silicon-on-Insulator Gate Driver for SiC-FET Power Modules," *IEEE Trans. Power Electron.*, vol. 27, no. 11, pp. 4417–4424, Nov. 2012.

[5] R. L. Greenwell, B. M. McCue, L. M. Tolbert, B. J. Blalock, and S. K. Islam, "High-temperature SOI-based gate driver IC for WBG power switches," in *2013 Twenty-Eighth Annual IEEE Applied Power Electronics Conference and Exposition (APEC)*, 2013, pp. 1768–1775.

[6] R. R. Lamichhane, N. Ericsson, S. Frank, C. Britton, L. Marlino, A. Mantooth, M. Francis, P. Shepherd, M. Glover, S. Perez, T. McNutt, B. Whitaker, and Z. Cole, "A wide bandgap silicon carbide (SiC) gate driver for high-temperature and high-voltage applications," in *2014 IEEE 26th International Symposium on Power Semiconductor Devices IC's (ISPSD)*, 2014, pp. 414–417.

[7] N. Ericson, S. Frank, C. Britton, L. Marlino, S.-H. Ryu, D. Grider, A. Mantooth, M. Francis, R. Lamichhane, M. Mudholkar, P. Shepherd, M. Glover, J. Valle-Mayorga, T. McNutt, A. Barkley, B. Whitaker, Z. Cole, B. Passmore, and A. Lostetter, "A 4H Silicon Carbide Gate Buffer for Integrated Power Systems," *IEEE Trans. Power Electron.*, vol. 29, no. 2, pp. 539–542, Feb. 2014.

[8] S. Kargarrazi, L. Lanni, A. Rusu, and C.-M. Zetterling, "A monolithic SiC drive circuit for SiC Power BJTs," in *2015 IEEE 27th International Symposium on Power Semiconductor Devices IC's (ISPSD)*, 2015, pp. 285–288.

[9] D. T. Clark, E. P. Ramsay, A. E. Murphy, D. A. Smith, R. Thompson, R. A. R. Young, J. D. Cormack, C. Zhu, S. Finney, and J. Fletcher, "High temperature silicon carbide CMOS integrated circuits," in *Materials Science Forum*, 2011, vol. 679, pp. 726–729.

[10] A. M. Francis, Moudy, T., Holmes, J. A., and Mantooth, H. A., "Towards Standard Component Parts in SiC-CMOS," in *Proceedings of IEEE Aerospace Conference 2015*, Big Sky, Mt., 2015.

Digital active gate drives using sequential optimization

Daniel J. Rogers
Department of Engineering Science
University of Oxford
Oxford, UK

Boris Murmann
Department of Electrical Engineering
Stanford University
Stanford, CA, USA

Abstract—This paper presents a digital Active Gate Drive (AGD) methodology for power semiconductor devices. The inherent latency limitation of digital signal processing systems is addressed by a sequential optimization procedure that uses voltage and current signals recorded at the previous switching edge to develop an optimized gate drive waveform for the next switching edge. Experimental results using a half-bridge circuit operating at 180 V/100 A show that the proposed scheme is capable of minimizing switching losses whilst constraining the overvoltage peaks occurring at turn-on and turn-off to as low as 200 V and 300 V respectively.

Keywords—converters, inverters, power semiconductor devices, optimzation, digital control

I. INTRODUCTION

Active Gate Drive (AGD) systems dynamically alter the shape of the gate drive waveform applied to a power semiconductor device during each turn-on and turn-off event. This synthesized waveform provides fine-grained control over the switching behavior of the power transistor, allowing particular EMI or device stress objectives to be met across the whole range of operating points by suitable adjustment of wave shape. The most popular method reported in the literature is the 'direct-analogue' closed-loop control providing control over turn-on and turn-off emitter di/dt (typically using voltage sensing across the emitter inductance) and collector-emitter dv/dt [1]–[4]. In these works, particularly [1], the direct-analogue method has been experimentally verified with great success, demonstrating a wide di/dt and dv/dt control range using relatively small amounts of AGD circuity. Other methods have also been presented, including the control of di/dt by dynamic selection of different gate drive resistors [5] or use of wave shaping circuits [6], as well as the addition of wider functionality by introducing over-current protection alongside controlled switching [7]. Analogue AGD circuits have also been proposed and tested for SiC devices [8], which present a particularly challenging control problem due to their very rapid switching characteristics.

This paper proposes a new control methodology for an AGD for power semiconductor devices. The method is based almost entirely in the digital domain, which offers great flexibility when choosing control objectives, but differs significantly from true closed loop schemes presented previously. Whilst in theory it should be possible to digitally implement schemes such as that proposed in [1] by suitably translating the continuous-time transfer function of the controller to a discrete-time representation running on a high-speed FPGA or DSP, in reality this would prove very challenging. Performing adequately low-latency high-bandwidth signal acquisition of device current and voltage waveforms is one major limiting factor: High-performance ADCs currently available on the market are capable of sampling at over 1 GS/s but have latencies of 10 clock cycles or more (particularly for resolutions >8 bits), meaning that acquired data is delayed by a minimum of 10 ns. If the control-loop processing delay is a further (minimal) 15 ns and the DAC output delay a further 5 ns, the total input-to-output latency of the closed-loop system will be approximately 30 ns. This is significant when compared to typical power device switching times, which can contain dynamics occurring over 50 ns, even for large (>600 V, >100 A) IGBT devices. Therefore it appears technically very challenging to construct 'direct-digital' closed-loop AGD systems using commonly available cost-effective hardware at the present time.

II. A SEQUENTIAL OPIMISATION APPROACH

Due to the limitations described above, this paper does not pursue direct-digital closed loop control, but instead proposes an optimization-based pseudo-closed-loop control which operates from one switching cycle to the next (rather than on one switching 'edge' only). This method uses the signals recorded at previous switching edges to generate a new gate drive waveform for a future switching edge, providing much longer periods of time to absorb signal acquisition and processing latencies. This process would ideally be repeated from edge-to-edge, i.e. data recorded for the previous switching edge is used at the next switching edge. As IGBT switching frequencies are typically less than 100 kHz, this is feasible with relatively modest ADC, DAC and DSP (or FPGA) hardware. A timing diagram illustrating the organization and operation of the 'digital sequential optimization' active gate drive (DSO-AGD) concept is shown in Fig. 1.

978-1-4673-9551-9/16 $31.00 © 2016 IEEE

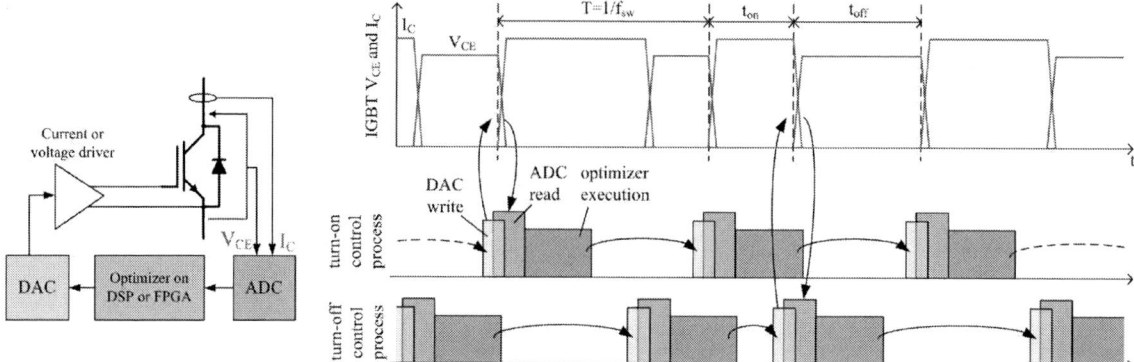

Fig. 1. Basic system and timing diagram showing operation of the digital sequential optimisation active gate drive (DSO-AGD) scheme from cycle to cycle. DAC reads/ADC writes in the turn-on and turn-off control processes cannot overlap (because they use the same hardware) and optimizer execution must complete before the next edge. These constraints dictate the minimum IGBT on and off times, unless cycles are skipped.

Here, each cluster of DAC-optimizer-ADC blocks in the timing diagram is effectively a single evaluation of the optimizer objective function:

1. The DAC applies a trial decision vector in the form of a gate drive waveform.
2. The ADC records the resulting device voltage and current waveforms.
3. The optimizer uses the voltage and current waveforms to evaluate the objective function and constraints at the trial decision vector, e.g. by identifying peak voltages, calculating total switching energy loss etc.
4. The optimizer evaluates the fitness of the solution and generates a new trial decision vector ready for the next switch edge.

An underlying assumption is that the operating condition of the circuit does not change significantly from one cycle to the next, such that a gate drive waveform constructed on the basis signals recorded on the previous switching edge will be valid for the next switching edge. This will be the case assuming the switching frequency is many times the frequency of the modulating signal, which is true in many grid interfaces or machine drive applications operating at 50 Hz or 60 Hz, for example. In fact, if the modulating signal is sufficiently slow compared to the switching frequency, it may be feasible to perform the evaluation only every N switching periods, allowing a longer period NT for processing to complete. This could significantly reducing processing hardware costs.

III. EXPERIMENTAL SETUP

The experimental setup is built around an IGBT silicon half-bridge power module (Fig. 2) operated from a DC bus voltage of 180 V. The lower IGBT is controlled via a custom-built galvanically-isolated signal generator and high-bandwidth voltage amplifier capable of driving +/-3 A. The upper IGBT is held in the off state by shorting the gate and emitter connections. The upper diode forms a freewheel current path when the lower IGBT is off. A current smoothing inductor and resistive load bank form the load. Lower IGBT

collector-emitter voltage and collector current and upper diode voltage and current are recorded using off-the-shelf probes and a four-channel digital oscilloscope. The signal generator triggers the oscilloscope so that the applied gate voltage and resulting device voltage and current waveforms are accurately and repeatably synchronized in time. MATLAB, running on a standard desktop PC, is used to acquire signals from the oscilloscope, process data (e.g. to calculate switching energy from voltage and current waveforms) and perform the optimization process.

As a whole, this apparatus provides a platform for developing optimization algorithms and objective functions, and to demonstrate that optimization of gate drive waveforms is possible. However, it does not provide the capability to perform edge-to-edge optimization as proposed in Fig. 1 because of the slow PC-signal generator-oscilloscope-PC loop time (the main limitation being the roughly 150 ms taken to trigger the oscilloscope and then to transfer data to the PC). As a result, although the IGBT switching frequency is 20 kHz, only roughly one in 3000 switching edges is used to perform an objective function evaluation. This system can still be used to demonstrate the proposed DSO-AGD scheme on the condition the system remains in 'steady state' over the length of the optimization cycle (about 5-11 seconds). This means the load current must be held constant during any one optimization run (this is achieved by maintaining a fixed duty cycle).

Fig. 3 shows an example set of waveforms resulting from the operation of the apparatus with the fastest possible switching edges, i.e. with the DSO-AGD system inactive and the gate voltage reference for the voltage amplifier set to a simple square wave. Device power dissipation is calculated by multiplying device voltage and current. Switching energy may then be calculated by integrating the power waveform over the period of the switching edge.

A. Software

The optimization algorithm used to generate the results presented in this paper is the built-in MATLAB

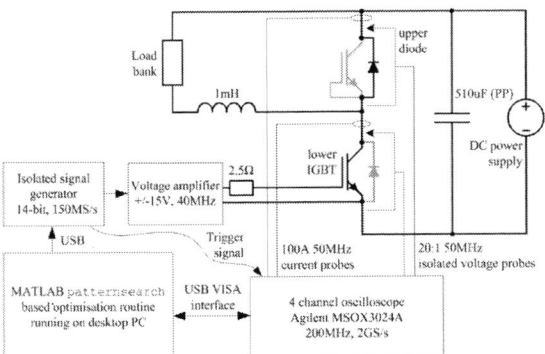

Fig. 2. System diagram of the experimental DSO-AGD apparatus. The time required for one objective function evaluation is approximately 150ms. Half-bridge IGBT module is Fuji 2MBI300U4N-120-50.

`patternsearch` function [9]. This algorithm is a gradient-free method and is particularly suited for use with objective functions where finite-difference approximations may not be reliable. During the development of the experimental system, finite difference calculations for the objective function were observed to be unreliable due to their sensitivity to measurement noise. In the results presented here, between 30 and 70 objective function evaluations (i.e. switching edges) were required to find a solution[1].

B. The gate drive waveform

The gate drive waveform for one switching period is constructed from the six regions labelled in Fig. 4:

1. A turn-on pre-charge period designed to bring the IGBT gate-emitter voltage to just below the threshold voltage (approximately 7 V for the module used here). As the gate driver is a voltage amplifier feeding the capacitive gate through a resistor, the gate voltage follows an RC exponential.
2. A turn-on optimization period which is manipulated by the `patternsearch` function. This region is defined by a number of samples which form the decision vector of the optimization problem.
3. A hold-on period which ensures the IGBT is held in the fully-on state until turn-off.
4. A turn-off pre-charge period designed to bring the IGBT gate-emitter voltage to just above the threshold voltage.
5. A turn-off optimization period defined by a different decision vector manipulated by the `patternsearch` function.
6. A hold-off period which ensures the IGBT is held in the fully-off state until turn-on.

The gate voltage reference in the turn-on and turn-off optimization regions is a piecewise-linear waveform defined

---

[1] `patternsearch` parameters: `TolMesh = 10⁻⁴`, `TolFun = 10⁻³`, `TolCon = 10⁻³`, `InitialMeshSize = 10⁻¹`, all other parameters as defaults. When successful, the algorithm normally terminates after 10-30 iterations when the mesh size tolerance is reached.

by the decision vector. In this paper, the turn-on decision vector has length 2 and the turn-off vector has length 3. These choices were found to give a good tradeoff between the quality of the result and the complexity of the optimization problem. Other important parameters are given in Table 1. Interestingly, this gate waveform 'design' is quite similar to that reported in some early work on AGD systems [10] which used timing of consecutive gate drive pulses to control the switching process.

Note that turn-on and turn-off is guaranteed to complete in the periods illustrated in Fig. 4 because the length of the optimization period is fixed and the hold-on and hold-off periods will eventually force the IGBT into the desired state. This means that the optimizer cannot, for example, create arbitrarily slow switching edges in an attempt to reduce overvoltage peaks.

TABLE I.   IMPORTANT GATE WAVEFORM PARAMETERS

Parameter	Value	Description
V_{pon}	6.5 V	Turn-on pre-charge voltage
t_{pon}	300 ns	Turn-on pre-charge length
t_{xon}	290 ns	Turn-on optimizer linear piece length
V_{poff}	8.0 V	Turn-off pre-charge voltage
t_{poff}	300 ns	Turn-off pre-charge length
t_{xoff}	320 ns	Turn-off optimizer linear piece length

C. Objective function formulation

The optimization objective used throughout this paper is the minimization of the total switching energy lost in the turn-on and turn-off periods. The optimization is also subject to a variable constraint (V_{lim}) on the peak overvoltages that occur across the diode during IGBT turn-on and across the IGBT during IGBT turn-off. The turn-on overvoltage is caused by the rapid decay of the freewheeling diode's reverse recovery current during diode turn-off which causes a voltage rise across the stray DC link inductance. The turn-off overvoltage is caused by the rapid rate-of-change of collector current causing a similar rise (these overvoltages are clearly observed in Fig. 3). A 'softer' IGBT turn-on or turn-off can be produced by an appropriate gate waveform shape in order to reduce these effects dramatically.

The turn-on optimization problem may is defined as

$$\mathbf{x}^* = \arg \min E_{on}(\mathbf{x})$$
subject to: $\qquad$ (1)
$$V_{dpk}(\mathbf{x}) \le V_{lim} \quad \text{and} \quad V_{gmin} \le \mathbf{x}_{(1..N_{on})} \le V_{gmax}$$

where the turn-on switching energy is

$$E_{on} = \int_{t_1}^{t_2} V_q(\mathbf{x})I_q(\mathbf{x}) + V_d(\mathbf{x})I_d(\mathbf{x}) . dt .\qquad (2)$$

The subscripts g, q and d respectively indicate lower IGBT gate, lower IGBT and upper diode voltages or currents. The turn-on process occurs in the period $[t_1, t_2]$, i.e. it corresponds to the time range shown in the left plot of Fig. 4. V_{dpk} is the peak upper diode voltage during turn on. V_{gmin} and V_{gmax} are the gate voltage limits imposed by the gate drive circuit and/or the IGBT (typically +/-15 V).

978-1-4673-9551-9/16 $31.00 © 2016 IEEE   1652

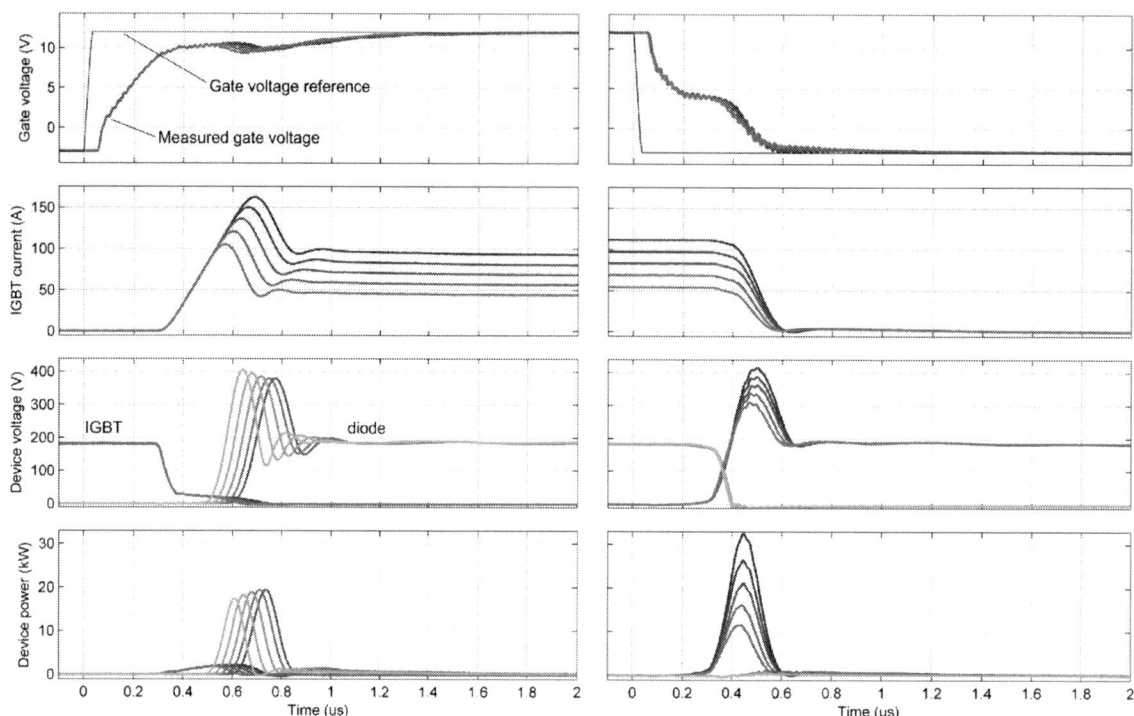

Fig. 3. Device waveforms for switch-on (left) and switch-off (right) as the load current is varied. Darker lines indicate higher load current. The gate drive waveform is set to a square wave in (which produces the fastest possible turn-on and turn-off edges) in order to simulate a system with no AGD control.

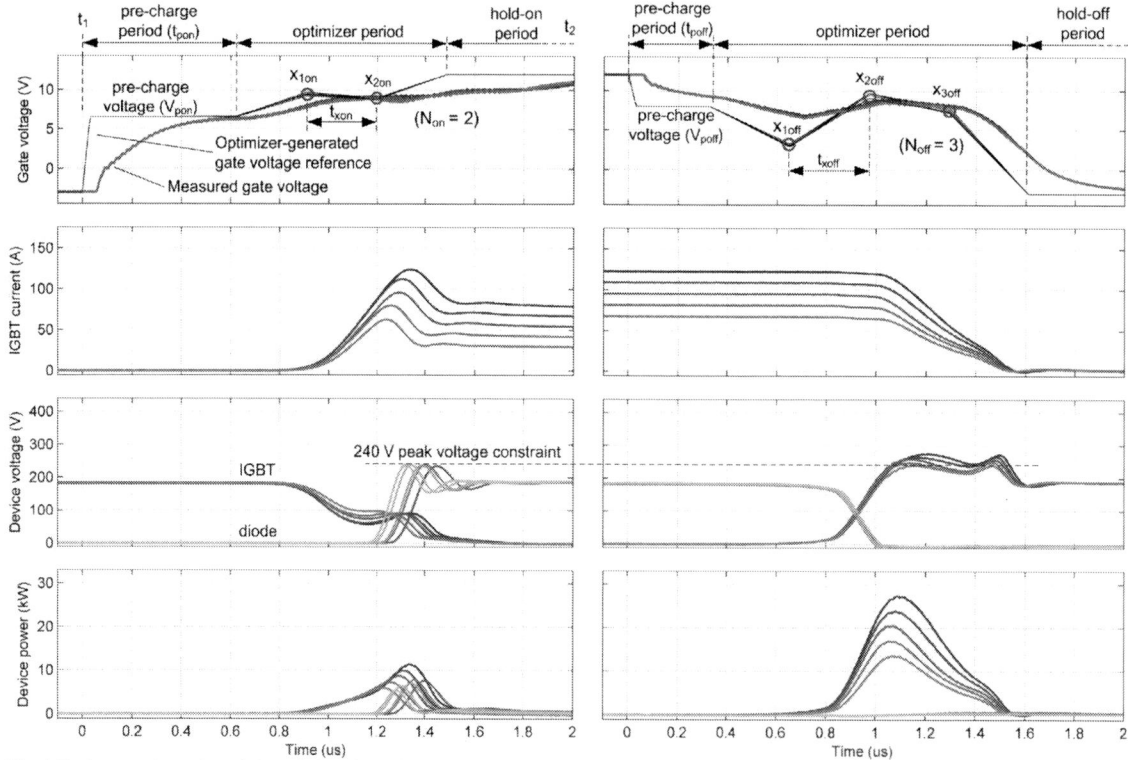

Fig. 4. Device waveforms for switch-on (left) and switch-off (right) after the DSO-AGD process has terminated. Darker lines indicate higher load current. The optimizer peak voltage constraint is fixed at 240V. Important optimizer parameters are labelled in the gate voltage plot (see Table 1).

except that turn-off energy and lower IGBT peak voltage are used in place of E_{on} and V_{dpk} respectively.

Practically, the introduction of a peak voltage constraint is interesting because it provides a way for the power electronic system designer to limit the voltage stresses applied to the power module to within a well-defined envelope. This could allow a different approach to the physical layout of the circuit and power module, selection of power devices, DC link capacitors and other components that would otherwise dictate the peak voltages occurring in the circuit. The magnitude of the reverse recovery and turn-off voltage peaks in power electronic systems can require the use of power devices rated very substantially above the nominal DC link voltage. Therefore, the application of the DSO-AGD system might allow the designer to select a lower voltage rated module than otherwise possible, resulting in lower conduction losses. This would help compensate for the increase in switching losses inevitably produce by the slower switching speed imposed by an AGD system when attempting to limit the overvoltage peaks.

IV. EXPERIMENTAL RESULTS

In Fig. 3, the gate voltage amplifier drives the lower IGBT with a simple square wave. It therefore represents operation without AGD and therefore switching speed, switching loss and peak voltages are dictated purely by the properties of the power module IGBT and diodes. It serves as the reference case with which to compare the DSO-AGD operation of Fig. 4 and Fig. 5. These figures show the optimal solution ($\mathbf{x}^*$) found by the `patternsearch` algorithm, i.e. they show the final

gate drive waveform and *not* the many trial evaluations of the objective function leading up to the termination of the algorithm.

The IGBT gate is capacitive in nature and is charged by the gate driver voltage amplifier via a low value resistor. The measured gate voltage does not therefore directly follow the gate voltage reference, but instead appears as a low-pass filtered version of the reference. Due to the Miller effect, during periods of high collector-emitter dv/dt the measured gate voltage is held almost constant and so the gate amplifier behaves as a current driver. The gate voltage plots of Figures 4 and 5 show that small changes (~500 mV) in the gate voltage reference are sufficient to produce the required control over the turn-on and turn-off processes, i.e. that the IGBT is sensitive to small variations in the applied gate current during the switching transitions.

A. Performance with varying load current

Fig. 4 shows the behavior of the DSO-AGD system as the load current is varied with V_{lim} fixed at 240 V. It is clear that the peak voltage limit is obeyed during turn-on. However, during turn-off the limit is violated in all cases except at the lowest load current. This indicates that controlling the di/dt of the IGBT current during turn-on is substantially more challenging than during turn-on. Increasing the length of the turn-off decision vector and/or tweaking the t_{xoff} parameter did not produce a noticeable improvement in performance. It is posited that the gate voltage amplifier used in the experimental setup is unable to drive sufficient current to fully control the IGBT during this period.

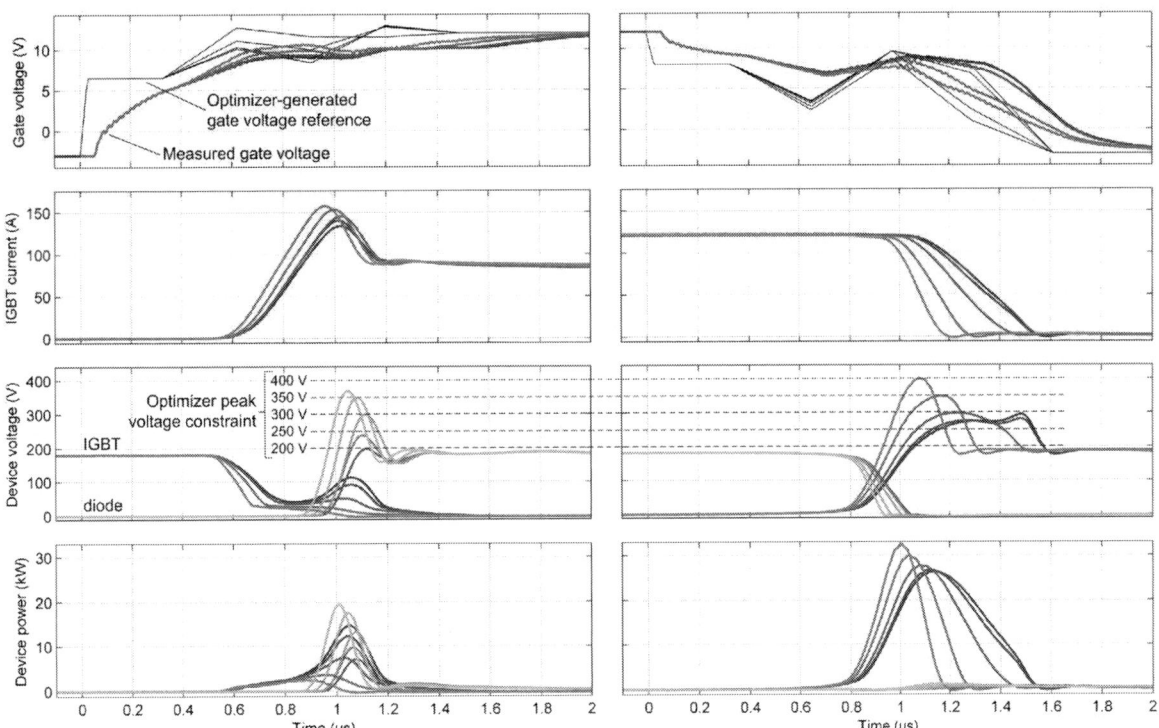

Fig. 5. Device waveforms for switch-on (left) and switch-off (right) after the DSO-AGD process has terminated. Darker lines indicate lower peak voltage (V_{lim}) constraints. Average load current is fixed at 100A.

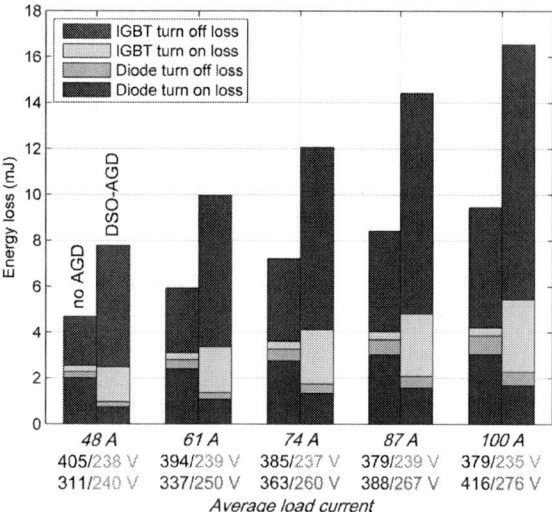

Fig. 6. Comparison of switching losses between the no-AGD case (i.e. fastest switching speed, Fig. 3) and the peak voltage DSO-AGD case (Fig. 4) as the load current is varied. Figures in green indicate an active V_{lim} constraint, figures in red indicate a violated constraint.

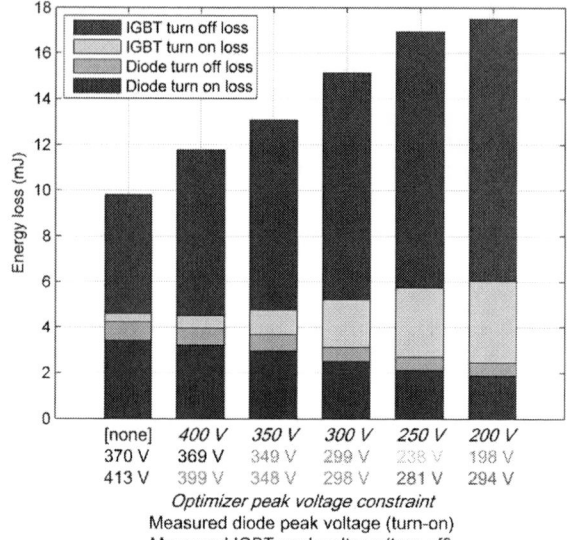

Fig. 7. Switching losses in the DSO-AGD case when the peak voltage constraint is varied (Fig. 5, average load current 100 A). Figures in green indicate an active V_{lim} constraint, figures in red indicate a violated constraint. The orange figure indicates an inactive constraint.

Fig. 6 provides a graphical comparison of the switching losses occurring in the IGBT and diode, along with values for the overvoltage peaks. As expected, overall switching losses increase as the overvoltage limit is decreased (this can also be seen by the increase in area under the power curves in Fi. 4). Energy losses are redistributed from the diode to the IGBT with total diode losses in the DSO-AGD case about one-half of those in the no-AGD case.

B. Performance with varying peak overvoltage constraint

Fig. 6 shows the behavior of the DSO-AGD system as V_{lim} is varied at a fixed average load current (100 A). The peak voltage limit is obeyed for turn-on and turn-off for 400, 350 and 300 V (note that the peak turn-on voltage does not reach 400 V even when the gate reference is square and so this constraint is not active when $V_{lim} = 400$ V). At lower limit voltages, the limit is obeyed at turn-on but not during turn-off, again indicating that control of the turn-off overvoltage is more challenging. One unusual result is evident at turn-on when the limit is set to 250 V. Here, the optimizer converges on a solution which produces a peak voltage significantly below the limit (238 V), i.e. the constraint is obeyed but not active. It is likely that a lower switching loss could be found if the peak voltage was allowed to reach the limit (constraint active) and so this should be considered only a local minimum of the problem. This highlights the challenge of ensuring that the algorithm reliably converges to the true global minimum.

Fig. 7 summaries the effect on switching losses as the voltage limit constraint is varied. Again, as expected, a lower

voltage limit increases total losses significantly whilst diode losses are reduced due to the lower peak voltage stress.

C. Overvoltages duing the optimization process

It is important to note that during the optimization process occurring in the experimental system, some trial function evaluations (i.e. trial gate waveforms) were observed to cause the overvoltage limit to be exceeded. This is caused by the `patternsearch` testing a point in the decision vector space that causes too-rapid changes in the collector current and hence large voltage rises across the DC bus inductance. Clearly, if a low voltage power module was specified under the expectation that the voltage limit would never be exceeded, the circuit could be damaged. Further work is therefore required to develop an optimization algorithm purpose-designed for DSO-AGD systems. Such an algorithm should avoid violating the constraint function during all evaluations of a trial decision vector. Effectively, this means designing an algorithm tailored to the specific properties of the problem (in contrast to the generic `patternsearch` algorithm) and possibly further development of the constraint function to avoid operation outside a specific set of acceptable gate drive waveforms.

D. Alternative objective functions

The objective function proposed here is only one possible formulation in the sense that any other quantity that can be extracted from the available current and voltage measurements could be substituted (or added), e.g. a constraint could be created to limit the rate of change of collector-emitter voltage and/or collector current as way to reduce EMI emissions above a certain frequency. One intriguing possibility is to attempt to synthesize optimal gate drive waveforms in order to obtain Gaussian shape switching edges resulting in minimum EMI emission [11].

V. CONCLUSION

This paper has presented the concept and initial experimental validation of a new AGD methodology. The unique feature is the use of a sequential online optimization approach implemented in the digital domain to generate optimal gate drive waveforms in a 'pseudo-closed-loop' manner. The objective and constraint functions were formulated to minimize switching losses and limit peak overvoltages respectively. This choice of objective function demonstrates the flexibility of the DSO-AGD concept: it allows the whole switching process to be analyzed and then acted upon *after* it has occurred. Information that is calculated in a post-processing step (such as switching loss) can be used to inform the generation of the next gate waveform. This is not possible with other closed-loop AGD systems which can only act on instantaneous quantities (such as rate-of-change of collector current etc.) This illustrates the potential of digital AGD approaches to tackle complex and time-varying control objectives which would be very difficult to achieve using an analogue AGD system.

REFERENCES

[1] Y. Lobsiger and J. W. Kolar, "Closed-Loop d d and d d IGBT Gate Driver," *IEEE Trans. Power Electron.*, vol. 30, no. 6, pp. 3402–3417, Jun. 2015.

[2] L. Chen and F. Z. Peng, "Closed-Loop Gate Drive for High Power IGBTs," in *Twenty-Fourth Annual IEEE Applied Power Electronics Conference and Exposition, 2009. APEC 2009*, 2009, pp. 1331–1337.

[3] S. Park and T. M. Jahns, "Flexible dv/dt and di/dt control method for insulated gate power switches," *IEEE Trans. Ind. Appl.*, vol. 39, no. 3, pp. 657–664, May 2003.

[4] Z. Wang, X. Shi, L. M. Tolbert, F. Wang, and B. J. Blalock, "A di/dt Feedback-Based Active Gate Driver for Smart Switching and Fast Overcurrent Protection of IGBT Modules," *IEEE Trans. Power Electron.*, vol. 29, no. 7, pp. 3720–3732, Jul. 2014.

[5] G. Chen, Y. Wang, X. Cai, and S. Igarashi, "Adaptive digital drive for high power and voltage IGBT in multi-MW wind power converter," in *Power Electronics and Motion Control Conference (IPEMC), 2012 7th International*, 2012, vol. 2, pp. 1452–1456.

[6] P. J. Grbovic, "An IGBT Gate Driver for Feed-Forward Control of Turn-on Losses and Reverse Recovery Current," *IEEE Trans. Power Electron.*, vol. 23, no. 2, pp. 643–652, Mar. 2008.

[7] L. Chen, B. Ge, and F. Z. Peng, "Modeling and analysis of closed-loop gate drive," in *2010 Twenty-Fifth Annual IEEE Applied Power Electronics Conference and Exposition (APEC)*, 2010, pp. 1124–1130.

[8] M. Shahverdi, M. Mazzola, R. Schrader, A. Lemmon, C. Parker, and J. Gafford, "Active gate drive solutions for improving SiC JFET switching dynamics," in *2013 Twenty-Eighth Annual IEEE Applied Power Electronics Conference and Exposition (APEC)*, 2013, pp. 2739–2743.

[9] "Find minimum of function using pattern search - MATLAB patternsearch - MathWorks United Kingdom." [Online]. Available: http://uk.mathworks.com/help/gads/patternsearch.html. [Accessed: 23-Jul-2015].

[10] V. John, B.-S. Suh, and T. A. Lipo, "High-performance active gate drive for high-power IGBT's," *IEEE Trans. Ind. Appl.*, vol. 35, no. 5, pp. 1108–1117, Sep. 1999.

[11] N. Patin and M. L. Vinals, "Toward an optimal Heisenberg's closed-loop gate drive for Power MOSFETs," in *IECON 2012 - 38th Annual Conference on IEEE Industrial Electronics Society*, 2012, pp. 828–833.

One Adaptive Turn-off Method for PFC Converter with Voltage Spike Limitation

Qunfang Wu, Qin Wang, Lan Xiao, Jialin Xu and Hongxu Li

Jiangsu Key Laboratory of New Energy Generation and Power Conversion
Nanjing University of Aeronautics & Astronautics
Nanjing, 211106, P. R. China
Email: wangqin@nuaa.edu.cn, qfwu55@aliyun.com

Abstract—The voltage source driver is widely used in a conventional drive circuit of PFC converter, where the fixed parameters of drive circuit couldn't optimize the switching loss of power switches. In view of this problem, one adaptive turn-off method of power switch for PFC converter is proposed, in which the turn-off current could be regulated adaptively with the input voltage changed periodically. The design principle is in reference to the voltage spike of the power switch under the peak input voltage, and making sure the voltage spike wouldn't be higher than that in other time. In this paper, firstly, the proposed drive method is presented. Then, the switch turn-off process is discussed. Finally, the adaptive turn-off drive circuit is designed and the experimental results verify the theory.

Keywords—adaptive turn-off; PFC converter; voltage spike limitation

I. INTRODUCTION

The gate-driver has a great influence on the switching speed, switching loss, current/ voltage variations (di/dt and du/dt), the efficiency and electromagnetic interference (EMI) of the converters [1]-[2]. In AC-DC applications, the power factor correction (PFC) technique is used widely, where most commonly utilize the conventional voltage source drive (VSD), as shown in Fig.1. However, the VSD has a big defect that all the gate-driver energy dissipates through the drive resistance and the fixed-drive-parameters couldn't optimize the switching loss of the power switches.

In order to improve the performance of VSDs, current-source drivers (CSDs) are proposed [3]-[5], which could reduce the switching loss by generating the constant drive current to charge and discharge the gate capacitance to accelerate the turn-on/turn-off speed. For this technology, the switching loss is reduced. But the gate drive loss needed to be compromised in design of the CSDs. To optimize this issue, the adaptive drive current inherently depending on the switching current of the switches for a boost PFC converter is built[6]-[7]. Nevertheless, they ignore the problem that the stronger drive current means faster switching speed, also means it results in higher current/voltage spike during the switching transient, which leads to the problem of EMI. In addition, in spite of the current/voltage variations di/dt and du/dt could be controlled in [8]-[9] and the current/voltage spike can be controlled, but they lose the merits of the CSD.

In view of these problems, one adaptive turn-off scheme

for the PFC converter is proposed in this paper, where the turn-off current can be regulated adaptively with the changed input voltage. The turn-off switching loss could be reduced, and simultaneously, the voltage spike can be limited to a decent value. The drive principle, design consideration and the drive circuit are discussed at length. Finally, the proposed drive scheme is tested and compared to the conventional drive method. The experimental results are presented to verify the validity of the proposed drive strategy.

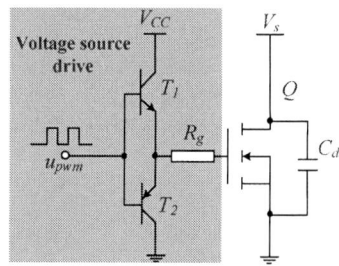

Fig. 1. The conventional voltage source driver.

II. PROPOSED DRIVE STRATEGY

A. Proposed turn-off strategy

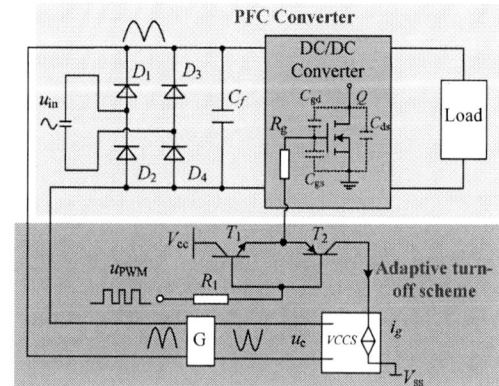

Fig. 2. The proposed turn-off scheme.

Fig. 2 shows the proposed turn-off scheme. A voltage-controlled current source (VCCS) is in series with the transistor T_2 in the turn-off power loop of the high frequency PFC converter. The regulator G adjusts the uncontrolled rectifier output voltage u_f reversely and outputs the control signal u_c. Then, by the VCCS, the signal u_c determines the turn-off current i_g. Therefore, i_g will change with the variation of u_f. Thus, the adaptive turn-off scheme for the power switch Q can be realized.

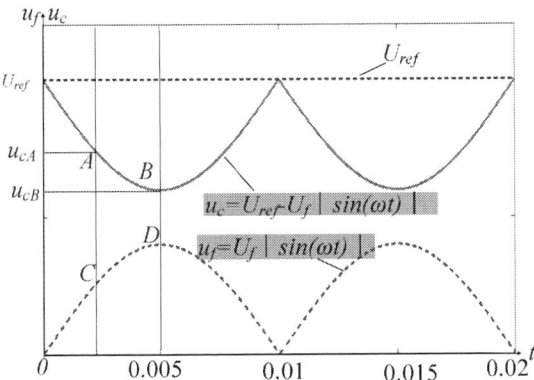

Fig. 3. Relation curves of control signal u_c and u_f

It is worth noticing that the faster turn-off speed is, the lower power loss will be. However, the faster turn-off speed must result in higher voltage spike. Therefore, the principle of the proposed drive-off method is that the turn-off current i_g is regulated by u_f adaptively and simultaneously, the voltage spike couldn't be higher than that under the peak input voltage. The signals regulated by regulator G are shown in Fig.3. U_{ref} is a constant voltage. u_f can be expressed as $U_f | \sin(\omega t) |$, where U_f is the amplitude of u_f. Thus, the signal u_c can be derived as $U_{ref} - U_f | \sin(\omega t) |$. From Fig.3, it can be obtained that the lower input voltage is, the higher voltage u_c and turn-off current i_g will be. For example, C and D are two points on the curve u_f in a half line cycle. Correspondingly, u_{cA}, u_{cB} are the control voltages of the VCCS. Because of $u_{fD} > u_{fC}$, the control voltage can be designed as $u_{cB} < u_{cA}$, which means $i_{gA} > i_{gB}$.

B. Turn-off Process

The switching loss and the voltage spike resulted from turn-off transient are two key factors, which should be considered deeply when designing the drive circuit. So it is necessary to discuss the switching characteristics of the switch Q (MOSFET) with a constant current driving. In this section, a typical boost PFC converter is employed to clarify the MOSFET turn-off process. The equivalent switching circuit model of the boost converter and qualitative waveforms of the turn-off transient are given in Fig.4 (a) and (b), respectively. In Fig.4 (a), the gate-source capacitance C_{gs}, gate-drain capacitance C_{gd}, and drain-source capacitance C_{ds} of the MOSFET are considered. The freewheeling diode D is modeled by a diode in parallel with its parasitic capacitance C_D. L_s is the stray inductance in the power loop. R_g is gate

drive resistance, which consists of the internal gate drive resistor and the external gate drive resistor.

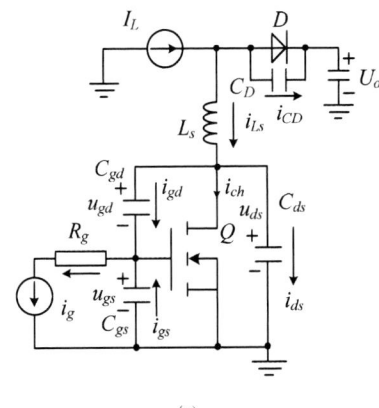

(a)

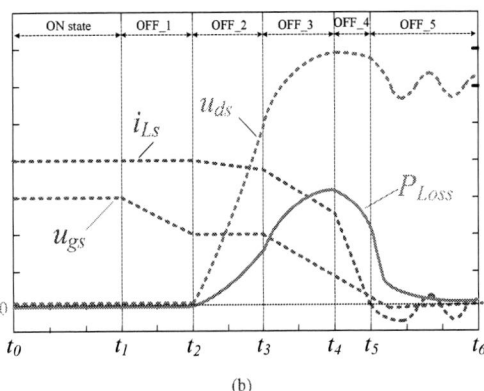

(b)

Fig. 4. (a) Equivalent switching circuit model of the boost converter. (b) Qualitative waveforms of the turn-off process.

To simplify the analysis of the turn-off process, the following conditions are assumed.

1) $C_{iss} = C_{gd} + C_{gs}$ and $C_{oss} = C_{gd} + C_{ds}$ are represented the input and output capacitances of the MOSFET, respectively.

2) In all the analytical calculations, the MOSFET is considered to be a resistance, an open circuit or a dependent current source whose behavior is described as $i_{ch} = g_{fs}(u_{gs} - V_{TH})$. g_{fs} and V_{TH} are the MOSFET transconductance and the threshold voltage, respectively.

3) The inductor L can be modeled as a constant current source I_L during the turn-off process.

4) The output voltage and the turn-off current are unchangeable in one switching cycle.

The equivalent circuit mode of the turn-off process is shown in Fig. 5, which can be divided into four modes and they are thoroughly analyzed in the following section.

Mode 1 [Fig5 (a). OFF_1: t_1-t_2] At t_1, the PWM signal u_{PWM} is set as zero. The turn-off current i_g begins to discharge the input capacitor C_{iss} and the voltage u_{gs} begins to fall. But the MOSFET are also working at the ohmic region and the channel current i_{ch}, u_{ds} remain unchanged in this mode.

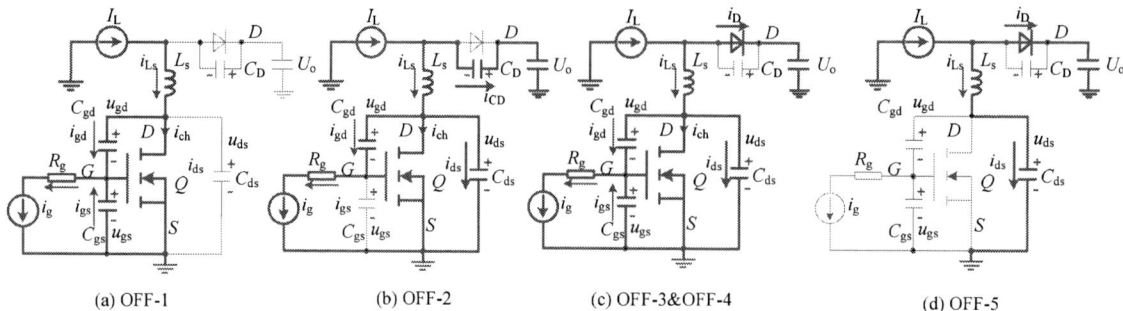

(a) OFF-1   (b) OFF-2   (c) OFF-3&OFF-4   (d) OFF-5

Fig. 5. Equivalent circuits of the turn-off process.

Mode 2 [Fig5 (b), OFF_2: t_2-t_3] At t_2, when the gate-source voltage u_{gs} falls down to V_{TH}+I_L/g_{fs}, this mode starts and the MOSFET begins to operate in saturation region. The channel current i_{ch} decreases suddenly and the excess current I_L-i_{ch}-i_{CD} charges the capacitors C_{gd} and C_{ds}. Thus the voltage u_{ds} starts to increase until it reaches U_o and the current i_{Ls} begins to decrease from I_L. In this interval, the parasitic capacitor C_D of the freewheeling diode discharges from U_o to zero.

Mode 3 [Fig5 (c), OFF_3- OFF_4: t_3-t_5] At t_3, the voltage of C_D drops to zero and the freewheeling diode ceases to block the voltage. Then the inductor current I_L starts to divert from the MOSFET to freewheeling diode. Because the stray inductor L_s impedes a sudden change in the drain current, the drain-source voltage u_{ds} continues increasing to the peak voltage. To get the peak voltage, the analytical method in [10] is adopted here. This mode can be subdivided into two stages and the specific demonstration is as follows.

I) *Current falling period (I)* [t_3-t_4] In this period, assuming the gate voltage keeps constant and as well as the channel current. The circuit condition can be simplified by the independent current and voltage source in the circuit and a rearrangement of the sources, which is a simple parallel resonant circuit as shown in Fig. 6(a). It is clear that the inductor L_s will resonate with capacitor C_{ds}, thereby, u_{ds} can be expressed from Fig. 6(a).

$$
\begin{cases}
u_{ds}(t) = U_o + (I_L - I_{ch\infty})L_s \dfrac{1}{\sqrt{L_s C_{ds}}} \sin[\dfrac{1}{\sqrt{L_s C_{ds}}}(t-t_3)] \\
I_{ch\infty} = 2(\dfrac{I_L}{g_{fs}} + V_{TH}) - \dfrac{i_g(C_D + C_{oss})}{g_{fs}C_{gd}}
\end{cases}
\tag{1}
$$

Where $I_{ch\infty}$ is the approximation value of the channel current at t_3, which is derived based on [10].

This stage ends when i_{Ls} arrives at the channel current $I_{ch\infty}$ at t_4, the voltage u_{ds} has reached its peak value u_{peak} and it can be given by.

$$
u_{peak} = u_{ds}(t_4) = U_o + (I_L - I_{ch\infty})L_s \frac{1}{\sqrt{L_s C_{ds}}} \sin[\frac{1}{\sqrt{L_s C_{ds}}}(t_4 - t_3)]
\tag{2}
$$

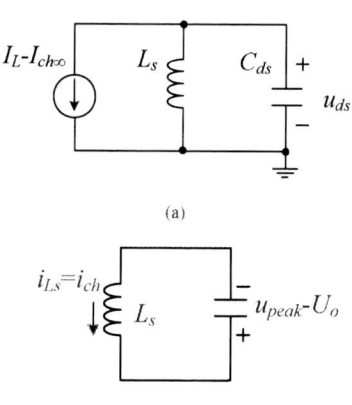

(a)

(b)

Fig. 6. The rearrangement equivalent circuits during mode3. (a) for current falling period (I), (b) for current falling period (II).

II) *Current falling period (II)* [t_4-t_5] During this period, the voltage u_{peak}-U_o effects on L_s. The drain current i_{Ls} will be forced to continue falling to zero. As the current difference i_{Ls}-i_{ch} charging the capacitor C_{ds}, however, there will not be a large amount of current to charge C_{ds} because the value of i_{Ls} is very close to that of i_{ch}. Hence, the voltage u_{ds} can be taken into account being equal to u_{peak} approximately. To simplify this period, the channel current i_{ch} is assumed to be controlled by the voltage U_o. The equivalent circuit is shown in Fig. 6 (b) and i_{ch} can be approximated as.

$$
i_{ch}(t) = I_{ch\infty} - \frac{u_{peak} - U_o}{L_s}(t - t_4)
\tag{3}
$$

At t_5, the channel current i_{ch} decreases to zero and u_{gs} decrease to V_{TH}, this stage ends.

Mode 4 [Fig5 (d), OFF_5: t_5-t_6] After the channel current i_{ch} decreases entirely to zero, u_{gs} reduces from V_{TH} to zero and the MOSFET works in the cutoff region. At this moment, a resonate circuit formed by L_s and C_{oss} begins to oscillate, however, the stray resistance R_s will damp the oscillation until the voltage u_{ds} reaches its steady-state value U_o and this mode ends.

III. DESIGN CONSIDERATION AND DRIVE CIRCUIT

A. Design Consideration

From (1), it is easy to find that the voltage peak value is determined by on U_o, I_L, i_g, L_s and the inherent parameters of MOSFET and freewheeling diode (g_{fs}, V_{TH}, C_{oss}, C_{ds}, C_{gs}, C_D i.e.). An example of boost PFC converter setup is used to demonstrate the design of the turn-off current i_g. The relational parameters are listed in table I. Combing the drive principle shown in the section II-A with the parameters, the relationship among u_{peak}, i_{Ls} and i_g can be drawn as shown in Fig.7. If the drain-source voltage peak value and the stray inductance remain constant, the turn-off current i_g will decrease with the inductor current i_{Ls} rising. Therefore, in practice, if u_{peak} is limited to 550V, from Fig.7, the turn-off current i_g could be approximately designed as:

$$\begin{cases} i_g = I_{ref} - Ki_{Ls} \\ i_{Ls} = I_L \left| \sin(\omega t) \right| \end{cases} \quad (4)$$

Where, I_{ref} and K are two constants, I_L is the amplitude value of i_{Ls}.

Table I The relational parameters

Input voltage	180~265VAC	Inductor L	1mH
Rated power	1.5 kW	Stray inductor L_s	1.5μH
Output voltage	380VDC	Input capacitor C_f	1μF
Switching frequency		50KHz	
MOSFET(Q): SPW20N60C3; Diode(D): RHRG1560_F085			
C_{oss}(Q)	160 pf	C_{iss}(Q)	2400 pf
C_{rss}(Q)	7 pf	g_{fs}(Q)	17.5
V_{TH}(Q)	3V	C_D(D)	30 pF

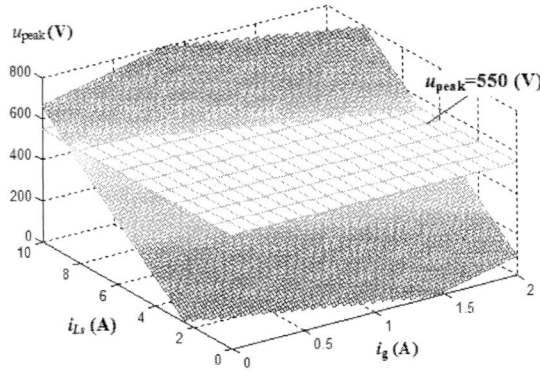

Fig. 7. The relationship of u_{peak}, i_g and i_{Ls}.

B. The Drive Circuit

From Fig.7, the maximum turn-off current i_g is less than 2.0 A. In real circuit, a bipolar transistor can be adopted as the VCCS to implement this function, which is not limited to this kind of method. Fig.8 shows the characteristic of bipolar transistor. The collector current i_g is equal to βi_b, where i_b is the base current. Therefore, combing with the design consideration and (4), the drive circuit can be designed as shown in Fig.9, where a IC LM358 is used to achieve subtraction operation. If the zener diode D_{01} is 5.1V, the magnification of Q_{02} is β and $R_{06}=R_{07}$, then the current i_g can be obtained by:

$$i_g = \frac{\beta}{R_{08}} \left(10.2 - \frac{R_{04}}{R_{03} + R_{04}} u_{in} \left| \sin \omega t \right| \right) \quad (5)$$

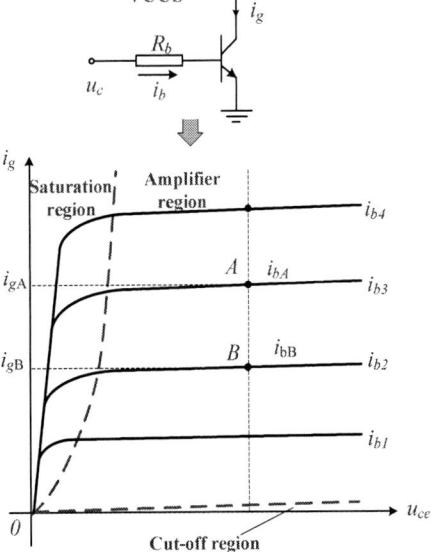

Fig. 8. The VCCS and the working characteristic of bipolar transistor.

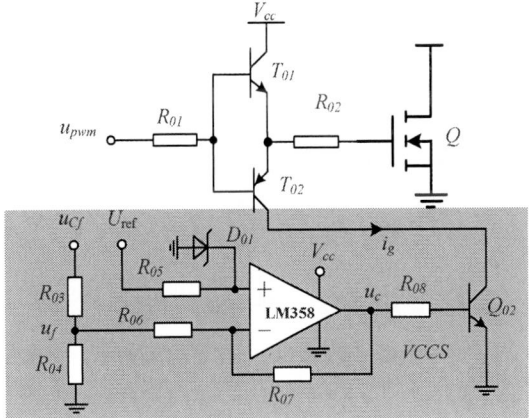

Fig. 9. The drive circuit.

IV. EXPERIMENTAL RESULTS

To verify the proposed adaptive drive turn-off scheme, one Boost PFC converter prototype has been built. The specifications are listed in Table I; the control IC is UC2818. It is noticed that the smaller the stray inductance, the better the performance of the converter because voltage spike during the turn-off period could be lower. But in this prototype, to reflect the principle of the proposed adaptive turn-off scheme, the stray inductance is set about 1.5μH. According to the analysis in Section-III-A, if the voltage spike are limited to 550V, the turn-off current i_g can be set within 0.5~1.5A, which can be obtained from Fig. 7. Corresponding to drive circuit shown in Fig.9, Q_{02} can choose 2SC2655, and the turn-off current i_g can achieve by regulating R_{03}, R_{04} and R_{08}. However, in real experimental process, the i_g was regulated within 1~1.5A adaptive variation to attain a better performance. Much of this discrepancy is to do with the precision of turn-off switching model.

Fig 10 shows the waveforms of input voltage u_{in}, input current i_{in}, the gate-source voltage u_{gs} and the control signal u_c. It is observed that the input current is sinusoidal and follows the input voltage perfectly, thus, the function of power factor correction is realized. Moreover, the control signal u_c for adjusting the turn off current is able to follow the output voltage of the uncontrollable rectifier.

Fig.11 illustrates the comparison of the typical waveforms of i_{Ls}, u_{ds} and u_{gs} during turn-off process with the VSD and the proposed method at the peak input voltage. It can be seen that both the voltage spikes are limited to the 550V and the turn-off time is approximately 130ns. Fig.12 shows that at half peak input voltage. It can clearly see that the turn-off time decreases to 90ns and the voltage spike would not exceed 550V with the adaptive driving method. Noticeably, the turn-off time is not changed with the VSD scheme. Therefore, with the proposed turn-off method, the overlap region of the drain-current and the drain-source voltage will be smaller. So the switching loss could be reduced, and the efficiency of converter can be higher.

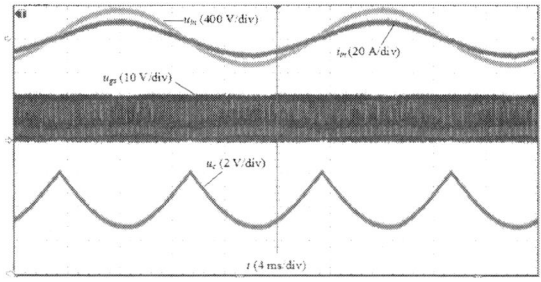

Fig. 10. The input voltage u_{in}, input current i_{in}, gate-source voltage u_{gs} and control signal u_c.

Fig. 13 shows the efficiency comparison of the proposed adaptive driver and the conventional VSD driver. It is shown that the efficiency using the proposed driver rises by more 1.3% than that with the conventional driving method.

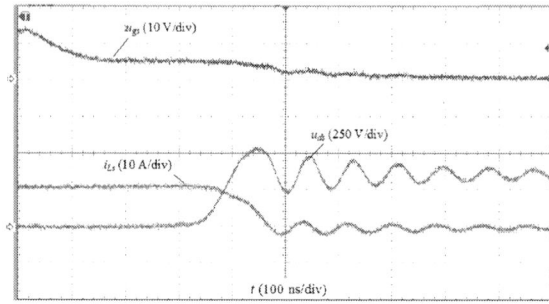

(a) u_{gs}, u_{ds} and i_{Ls} with the conventional driver.

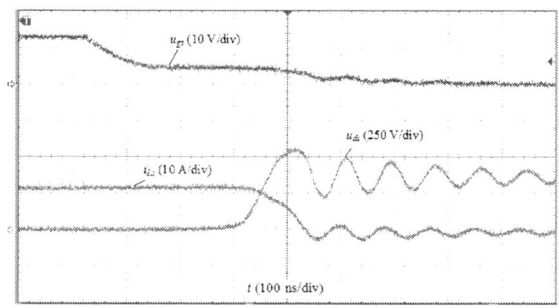

(b) u_{gs}, u_{ds} and i_{Ls} with the proposed driver.

Fig. 11. The waveforms comparison with two different drivers at peak input voltage.

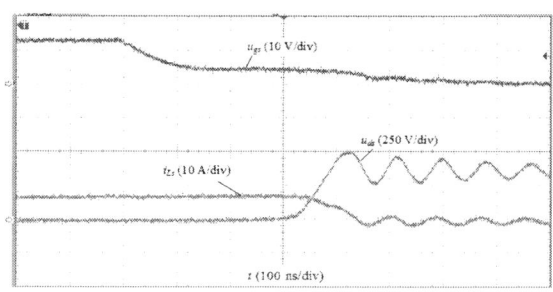

(a) u_{gs}, u_{ds} and i_{Ls} with the conventional driver.

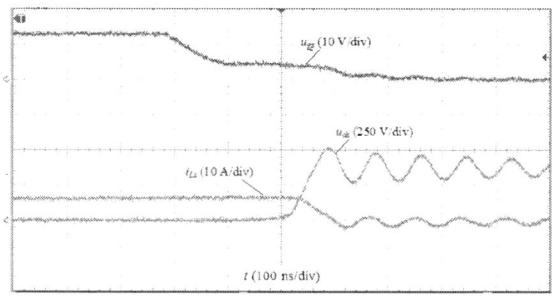

(b) u_{gs}, u_{ds} and i_{Ls} with the proposed driver.

Fig. 12. The waveforms comparison with two different drivers at half peak input voltage.

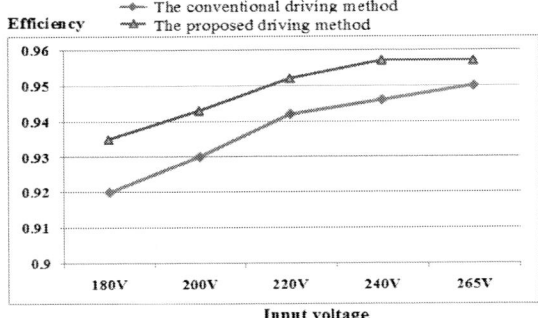

Fig. 13. The efficiency comparison

V. CONCLUSION

This paper presents an adaptive turn-off scheme applied in PFC converter, which could limit the voltage spike of the power switch when the turn-off speed is regulated adaptively with the change of input voltage. Firstly, the working principle of the proposed scheme is analyzed. Then, the voltage spike could be obtained from the analysis of the turn-off process and the drive circuit is designed. Finally, the prototype with a boost PFC converter has been established to verify the proposed turn-off scheme. The efficiency using the proposed driver rises by more 1.3% than that with the VSD driver.

ACKNOWLEDGMENT

This work was supported by the National Nature Science of China under Award 51377082, Funding of Jiangsu Innovation Program for Graduate Education KYLX15_0275 and the Fundamental Research Funds for the Central Universities.

REFERENCES

[1] R. J. E. Hueting, E.A. Hijzen, A. Heringa, A.W.Ludikhuize, M. A. A. int' Zandt, "Gate-Drain Charge Analysis for Switching in Power Trench MOSFETs", *IEEE Trans. on Electron. Devices*, Vol. 51, No. 8, pp. 1323-1330, Aug. 2004.

[2] N. Idir, R. Bausière, J. J. Franchaud, "Active gate voltage control of turn-on di/dt and turn-off dv/dt in insulated gate transisitors," *IEEE Trans. Power Electron.*, vol. 21,no. 4, pp. 849–855, Jul. 2006.

[3] W. Eberle, Z. Zhang, Y. F. Liu, P. C. Sen, "A current source gate driver achieving switching loss savings and gate energy recovery at 1-MHz," *IEEE Trans. Power Electron.*, vol. 25, no. 6, pp. 1439–1443, Jun. 2010.

[4] Z. Zhang, J. Fu, Y. F. Liu, P. C. Sen, "Discontinuous current source drivers for high frequency power MOSFETs," *IEEE Trans. Power Electron.*, vol. 25, no. 7, pp. 1863–1876, Jul. 2010.

[5] Z. Zhang, W. Eberle, P. Lin, Y. F. Liu, P. C. Sen, "A 1-MHz high efficiency 12-V buck voltage regulator with a new current-source gate driver," *IEEE Trans Power Electron.*, vol. 23, no. 6, pp. 2817–2827, Nov. 2008.

[6] Z. Zhang, P. Xu, Y. F. Liu, "Adaptive continuous current source drivers for 1-MHz boost PFC converters," *IEEE Trans. Power Electron.*, vol. 28,no. 5, pp. 2457–2467, May 2013.

[7] Z. Zhang, C. Xu, Y. F. Liu, "A digital adaptive discontinuous current source driver for high-frequency interleaved boost PFC converters," *IEEE Trans. Power Electron.*, vol. 29, no. 3, pp. 1298–1310, Mar. 2014.

[8] S. Park, T. M. Jahns, "Flexible dv/dt and di/dt control method for insulated gate power switches," *IEEE Trans. Ind. Appl.*, vol. 39, no. 3, pp. 657–664, May 2003.

[9] H. Riazmontazer, S. K. Mazumder, "Optically switched-drive-based unified independent dv/dt and di/dt control for turn-off transition of power MOSFETs ," *IEEE Trans. Power Electron.*, vol. 30,no. 4, pp. 2338–2449, Apr. 2015.

[10] M. Rodr´ıguez, A. Rodr´ıguez, P. F. Miaja, D. G. Lamaret, "An insight into the switching process of power MOSFETs: an improved analytical losses model," *IEEE Trans. Power Electron.*, vol. 25,no. 6, pp. 1626–1640, Jun. 2010.

A digital implementation for PWM Phase-Frequency synchronization in SMPS systems

Luca Bizjak, Emanuele Bodano, Ante Gotovac, Sergii Tkachov

Infineon Technologies
Villach, Austria
luca.bizjak@infineon.com, emanuele.bodano@infineon.com, ante.gotovac@infineon.com, sergii.tkachov@infineon.com

Abstract—This paper describes a novel concept of PWM phase and frequency synchronization. This approach applies for digitally controlled switched mode power supplies (SMPS) which need to regulate the switching frequency externally. One of the well-known advantages of having a system with a controllable switching behavior is the mitigation of the Electromagnetic Interference (EMI), which leads to the system cost reduction as a smaller input filter is required. The proposed solution overcomes analog and digital state of the art drawbacks such as limitation of the synchronization frequency range, susceptibility to noise and glitches, undesired overshoots/undershoots on the output voltage and the tuning of the IC digital clock which leads to an over-constrained design. In order to confirm the synchronization performance the proposed solution is implemented as part of an SMPS controller integrated in a 130nm BCD (Bipolar-CMOS-DMOS) technology. The output voltage over/under-shoot caused by an abrupt frequency synchronization change has been analyzed and measured. The frequency and phase settling time have been characterized with respect to various configurations. The robustness against synchronization frequency and phase noise has been measured for different noise levels.

Keywords—synchronization; PWM; SMPS; frequency; phase; digital

I. INTRODUCTION

In the field of multi-SMPS systems a well-known topic is the synchronization of the frequency and phase of the PWM signals controlling the power stages. In Fig. 1 an application with more than one SMPS has been presented, one is used as a master and the others are slaves. The slaves PWM signals are synchronized towards the one of the master SMPS. The phase offset of the slave devices is selected by the system designer to achieve good system performance.

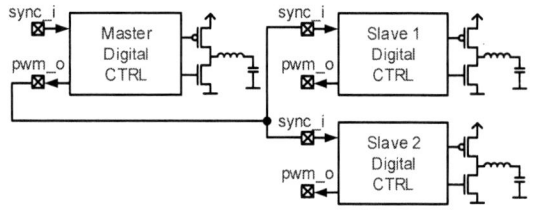

Fig. 1. Multi SMPS system with synchronization

In [1, 2, 3, 4, 5] the benefits (i.e. EMI mitigation, system cost reduction) of having a synchronization between different PWM based SMPS ICs on a single printed circuit board (PCB) are presented. Synchronization is also important in Radio Frequency systems in order to place the radiated harmonics into an allowed frequency band. Furthermore the PWM synchronization is relevant for multi-phase SMPS systems where the load balance is strictly dependent on the frequency and phase synchronization of the different branches. As last highlight of the PWM synchronization importance we would like to mention the possibility to introduce a smaller stress on the system supply, leading to a reduction of the bill of material (BoM) for what concerns the choice of the SMPS input capacitors [3, 5].

Several analog solutions to the problem can be found in recent state-of-the-art [6, 7], which present some drawbacks inherent to the analog implementation. One is the limitation to get the synchronization only for lower frequencies with respect to the IC's default one. Secondly the synchronization robustness can suffer from spurious glitches on the synchronization signal since this is not filtered before passing it to the analog synchronizer and the system can therefore lock to a higher frequency or generate noise on the SMPS switching frequency. Finally, the transition from the free running to the synchronized PWM and vice versa is normally done in an abrupt way leading to undershoots/overshoots on the regulated voltage as it can be seen in [6] Fig. 21 and Fig. 22.

In [4] a digital solution to the synchronization problem is described. The proposed digital approach overcomes most of the weaknesses of the analog implementation however, since the phase and frequency synchronization of the PWM signal is achieved by digitally tuning the SMPS main oscillator, the clock frequency of the entire SMPS is changed (i.e. increased) and therefore an over constrained design with larger area and a higher power consumption has to be taken into account.

In this paper we present a digital solution for achieving the phase-frequency synchronization of the SMPS switching frequency. The proposed approach avoids both weaknesses of the state of the art analog and digital solutions keeping the system clock frequency constant while the reset value (OSF, over-sampling factor) of the DPWM counter [8, 9] is adapted as shown in Fig 2.

978-1-4673-9551-9/16 $31.00 © 2016 IEEE

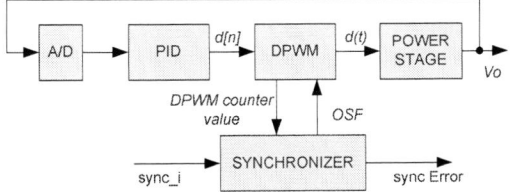

Fig. 2. Overview of SMPS with synchronizer

This paper proceeds as follows: in section II we present the conceptual description and implementation of the proposed solution. The dimensioning of the synchronization system is analyzed in section III. The experimental results obtained from the fabricated prototype chip are evaluated in section IV. Conclusions are addressed in section V.

II. PHASE-FREQUENCY SYNCHRONIZAZION CONCEPT AND IMPLEMENTATION

A Graphical representation of the complete PWM phase and frequency synchronization concept is given in the Fig. 3. The upper part of the diagram is defined to achieve the frequency synchronization and the lower part the phase synchronization.

The frequency of the train of pulses on the input signal "sync_i" is determined measuring the period between two subsequent synchronization signal rising edges with the counter CNT. CNT is clocked with the system clock clk. In order to avoid abrupt frequency changes the sampled period OS_{IN} is processed by a first order digital Low Pass Filter (AVG) before passing it to the subsequent logic.

The phase synchronization consists of the phase error generation, the phase error processing and the phase reconstruction. In order to detect the phase error, the current phase of the synchronization input signal has to be detected and compared to the desired phase at which the PWM signal has to lock. This can be achieved by sampling the current DPWM counter value on the rising edge of the synchronization pulse and subtracting it to the target phase. The resolution of the calculated phase error is limited to the digital clock frequency. In order to get the loop stable the phase error (Ph_{ERROR}) needs to be clamped to +/- OSF/2, which is equivalent to a phase error of +/- 180 deg.

In order to stabilize the phase correction loop the phase error needs to be filtered by a PI (Proportional/Integral) filter. This implementation mimics the functionality of the loop filter within a PLL system [10]. The total phase error correction (Fperr) is calculated as the sum of the proportional (FPph) and integral (FIph) error components from which the calculated phase noise (FNph) gets subtracted. Here, FNph represents the "decimal" part of the calculation that is truncated and fed back into the sum node in order to increase the resolution of the phase reconstruction. The Fperr is clamped in order to avoid too big phase compensation which can abruptly modify the final OSF calculation.

The total correction of the switching frequency is defined as the sum of the averaged frequency of the synchronization input signal (Fsa) and the negated value of the calculated total phase error (Fperr). The resulting value is provided to the DPWM counter for the generation of the PWM signal in the next switching cycle.

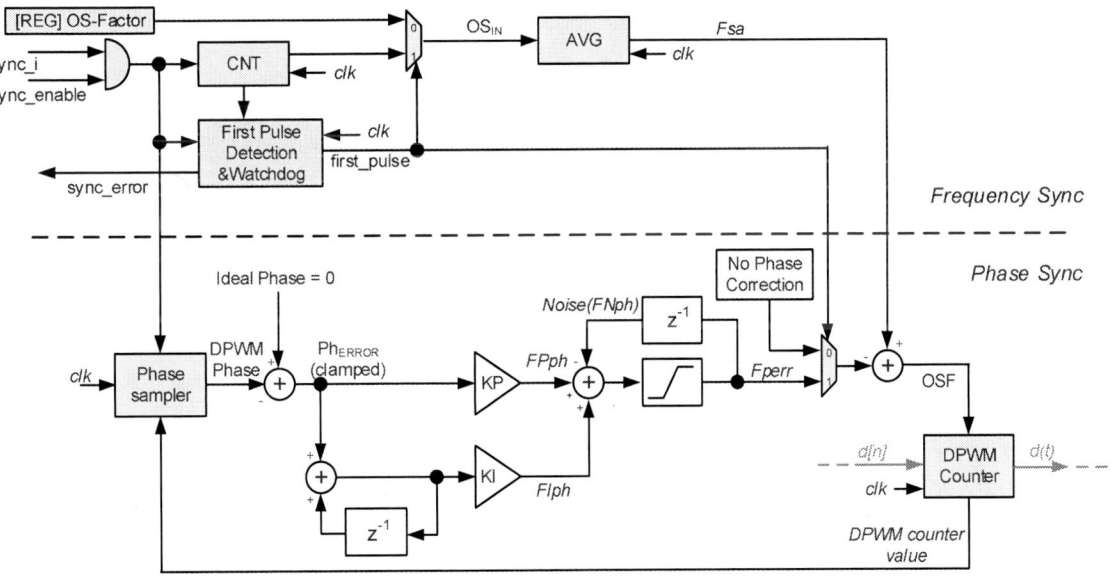

Fig. 3. Frequency and Phase synchronization concept

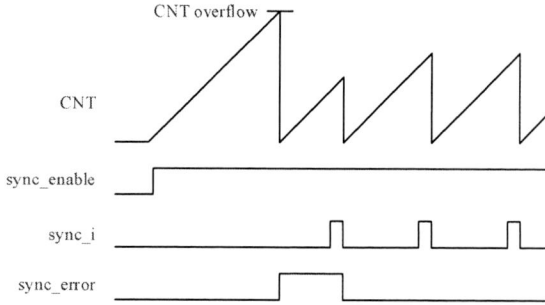

Fig. 4. Synchronization error management for first pulse recognition

In addition to the phase and frequency synchronization the concept presents also the synchronization error management feature to avoid spurious glitches on the sync input or synchronizing to not allowed frequencies (i.e. too low frequencies which will degrade the SMPS controller performances).

If a synch pulse is not received (see Fig. 4) within the first full counter cycle (CNT in Fig. 3) then the sync_error is asserted and kept high until the first sync pulse is coming or the synchronization mechanism is disabled (sync_enable = 0). The converter works with the default switching frequency (i.e. with the default OSF for the DPWM counter) until the first pulse on the sync input has been detected (first_pulse = 1).

Another watchdog feature, shown in Fig. 5, performs a continuous min/max check on the frequency of sync_i. For this purpose a min and max limits are defined for the counter CNT. The min limit has the purpose to skip spurious or unwanted fast pulses that could set the switching frequency at an undesired high value. The max limit is set to bind the minimum switching frequency allowed by the DCDC converter.

III. SYNCHRONIZATION SYSTEM DIMENSIONING

In this section the mathematical analysis of the frequency and the phase synchronization dimensioning is addressed.

The block diagram of the frequency synchronization path in connection with the SMPS loop is depicted in Fig. 6 where the latter is composed by an analog to digital converter A/D, a

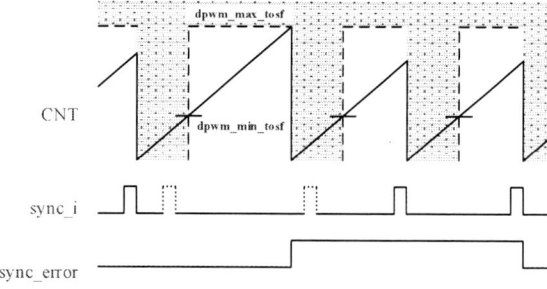

Fig. 5. Switching frequency watchdog

digital compensator PID, the DPWM and the power stage with its driver. In order to dimension the frequency synchronization path, the effect of a sync_i frequency variation to the converter output voltage Vo(t) is analyzed. The sync_i frequency is measured by the counter CNT (see Fig. 3) and its outcome OS_{IN} can be expressed as follows:

$$OS_{IN} = \frac{fclk}{fsync\_i} \tag{1}$$

Following Fig. 6, OS_{IN} is processed by the low pass filter AVG to generate Fsa. Since in this analysis the Phase synchronization loop effect is not considered than Fperr is assumed to be equal to zero and thus OSF = Fsa. The AVG filter is a first order low pass filter, and its transfer function [11] is given by:

$$T_{lpf}(z) = 2^{(\beta-\alpha)} \frac{z}{z - 1 + 2^{(\beta-\alpha)}} \tag{2}$$

where β is a constant and α is a configuration variable of the filter. The filter output OSF is the initial value for the DPWM counter. When the DPWM counter value is higher than the duty cycle d[n], determined by the SMPS control loop, the driver on/off control signal d(t) is set high, vice versa is set low. The relation between d(t) and OSF is therefore given by the formula:

$$d(t) = \frac{t_{ON}}{T_{SW}} = 1 - \frac{t_{OFF}}{T_{SW}} = 1 - \frac{d[n]/f_{clk}}{OSF/f_{clk}} = 1 - \frac{d[n]}{OSF} \tag{3}$$

It's thus clear that an OSF perturbation causes a variation in the d(t) and the OSF to d(t) transfer function can be calculated as follows:

$$T_{DPWM,OS} = \frac{\partial d(t)}{\partial OSF} = \frac{d[n]}{OSF^2} = \frac{1-D}{OSF} \tag{4}$$

where D=Vo/Vin [2].

The d(t) to output voltage Vo(t) transfer function [2] is equal to:

$$G_{vd}(s) = V_{in} \frac{1}{s^2 LC + s\dfrac{L}{R_{load}} + 1} \tag{5}$$

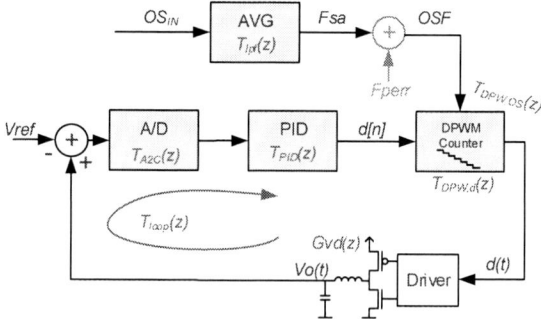

Fig. 6. Frequency synchronization transfer function diagram

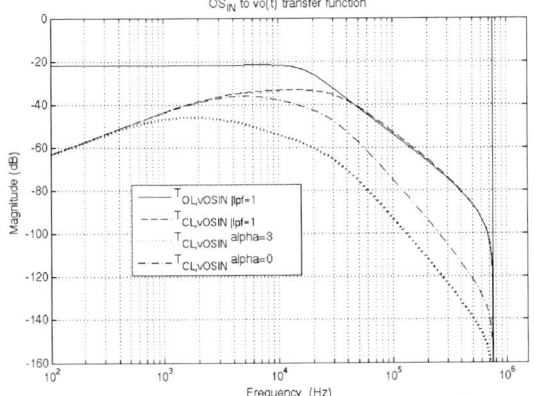

Fig. 7. Magnitude Bode plot of open and closed loop OS_{IN} to Vo(t) transfer function with AVG low pass filter enabled $\alpha=3$, $\alpha=0$ and with AVG filter disabled. In (8) the design parametrs are: LSB=10mV, P=53, I=0.1, D=3.3

In Fig. 7 are reported the OS_{IN} to Vo(t) transfer function in four different configuration for a buck with: Vin = 14 V, Vout = 5.8 V, Iout = 0.5 A, L = 4.7 µH, C = 22 µF, f_{sync_i} = 1.5 MHz. If we assume the buck loop to be settled, to open the control loop and we do not consider the AVG filter effect $T_{lpf}=1$, then the OS_{IN} to Vo(t) Open Loop Transfer Function can be calculated using (4) and (5):

$$T_{OL,vOS_{IN}}(s)\Big|_{T_{lpf}=1} = T_{DPWM,OS} \cdot G_{vd}(s) \qquad (6)$$

In Fig. 7 the solid line shows the $T_{OL,vOSIN}\big|_{T_{lpf}=1}$ that has a DC gain of -22 dB. We can observe that the high frequencies are attenuated by the buck output filter.

The buck negative control loop helps to mitigate the disturbances of the OSF perturbation to the output. The buck Open Loop Transfer Function can be calculated:

$$T_{loop}(z) = T_{A2D}(z) \cdot T_{PID}(z) \cdot T_{DPWM,d} \cdot G_{vd}(z) \qquad (7)$$

where the design is:

$$\begin{cases} T_{A2D}(z) = 1/LSB \\ T_{PID}(z) = P + I\dfrac{z}{z-1} + D\dfrac{z-1}{z} \\ T_{DPWM,d} = -\dfrac{1}{OSF} \end{cases} \qquad (8)$$

The transfer function OS_{IN} to Vo(t) in Closed Loop is:

$$T_{CL,vOS_{IN}}(z)\Big|_{T_{lpf}=1} = \frac{T_{OL,vOS_{IN}}(z)\Big|_{T_{lpf}=1}}{1 - T_{loop}(z)} \qquad (9)$$

The closed loop attenuates the low frequency reducing the maximum gain to -33 dB (dashed line in Fig.7). Considering a Δf_{sync_i} abrupt variation, this produces a ΔOS_{IN} and the output voltage perturbation can be estimated by the following coarse equation:

$$\Delta Vo = \Delta OS_{IN} \cdot 10^{MaximimGain/20} \qquad (10)$$

In order to decrease further the gain the filter AVG is introduced. The Closed Loop Transfer Function becomes:

$$T_{CL,vOS_{IN}}(z) = \frac{T_{lpf}(z) \cdot T_{DPWM,OS} \cdot G_{vd}(s)}{1 - T_{loop}(z)} \qquad (11)$$

In order to be effective the filter cut frequency must be lower than the bandwidth of the SMPS output filter. In Fig. 7 the dotted line and the dash-dot line represent respectively the system behaviour with the parameter α equal to 0 and equal to 3 (with β = 5). The case with α = 0 has a maximum gain = -45 dB. By the coarse estimation (10) it can be estimated that a Δf_{sync_i} perturbation will be further attenuated and this result will be discussed and compared with measurements in section IV.

The phase synchronization branch, shown in Fig. 3, is a closed loop system that mimics the functionality of the loop filter within a PLL system [10]. In order to guarantee the loop stability we calculate the transfer function of the DPWM and the PI filter. Since the system is discretized by the clock clk, if OSF changes by one then the DPWM counter phase Φ changes by one clk period. Therefore the new phase can be written as follows:

$$\Phi[n] = \Phi[n-1] + \Delta OSF[n] \qquad (12)$$

Considering the phase synchronization branch similar to a PLL, the DPWM acts as a Digitally Controlled Oscillator (DCO) and its frequency behavior is equal to an integrator. Moving from the discrete time to the z domain (12) can be rewritten as follows:

$$\Phi(z) = \Phi(z) \cdot z^{-1} + \Delta OSF(z) \qquad (13)$$

$$T_{DPWM,\Phi}(z) = \frac{\Delta\Phi(z)}{\Delta OSF(z)} = \frac{1}{1 - z^{-1}} \qquad (14)$$

The filter architecture is a PI and its transfer function equation is:

$$T_{PI,\Phi}(z) = 2^{-1-KP} + \frac{2^{-8-KI}}{z-1} \qquad (15)$$

where KP and KI are filter configuration factors. Then the phase synchronization Open Loop Transfer Function is:

$$T_{phase}(z) = T_{DPWM,\Phi}(z) \cdot T_{PI,\Phi}(z) \qquad (16)$$

The Bode diagram varying the KI and KP factors is reported in Fig. 8.

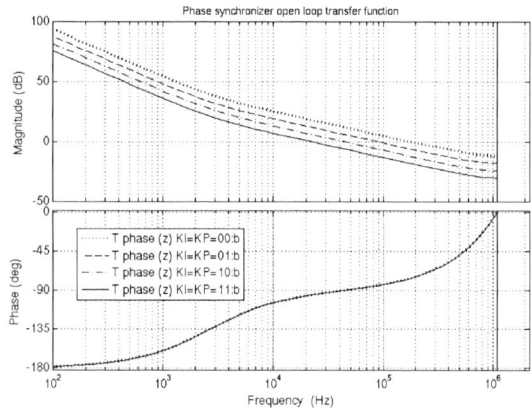

Fig. 8. Phase Synchronizer open loop transfer function varying KI and KP

The phase synchronization loop phase margin is always >> 45deg, so the stability of the loop is guaranteed. However it has to be considered that this is a discrete and sampled system and thus the models are valid only for frequencies much lower than the half of the regulator switching frequency. A deeper analysis is out of the scope of this paper.

IV. EXPERIMENTAL RESULTS

The synchronizer has been integrated in a buck regulator with the following specification: Vin = 14 V, Vo = 5.8 V, Iout max = 1.5 A, L = 4.7 μH and C = 22 μF. Furthermore in the presented implementation a 12-bit resolution DPWM counter has been selected to generate a programmable switching frequency in the range of 400 kHz - 2.5 MHz. The system clock frequency is fclk = 160 MHz. The area of the digital controller part is ca. 0.5mm^2 and the synchronization logic is ca. 0.05mm^2. Several measurements have been performed where the sync_i signal has been generated by Keysight 33522B and the measurement is performed by Lecroy HDO8108 oscilloscope.

The improvement brought by the introduction of the synchronizer described in this paper is highlighted in Fig. 9. This measurement shows the output voltage response to a sync_i frequency abrupt change from 2 MHz down to 1.5 MHz. Without the synchronizer an overshoot of about 700 mV is measured, while it is reduced to less than 100 mV when the synchronizer is enabled. This result shows a good agreement with the coarse estimation computed in (10). The estimated overshoot value considering T_{lpf} = 1 for a fsync_i change from 1.5 MHz to 2 MHz, which results from (1) in a ΔOS_{IN} = 26, is equal to ΔVo = 582 mV. On the other hand, taking in consideration the AVG filter with the variable α = 0 has a maximum gain = -45 dB and from (10) the estimation of the overshoot results equal to ΔVo = 130 mV. Measurements are thus a confirmation that the coarse calculation (10) gives a pessimistic value but it provides a good insight of the system performance.

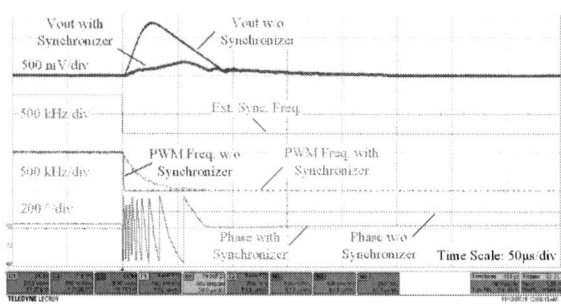

Fig. 9. Output voltage overshoot due to a sync_i frequency step from 2 MHz down to 1.5 MHz with (α = 00:b, KI = 11:b, KP = 11:b) and without synchronizer enabled

The transient response to a 500 kHz frequency step variation on the sync input is shown in Fig. 10 and Fig. 11 with 3 different AVG filter settings.

It can be observed that after an abrupt change of the synchronization input frequency, the output PWM signal locks as expected at the synchronization frequency and phase. The PWM signal changes its frequency smoothly to limit system disturbances (undershoot/overshoot).

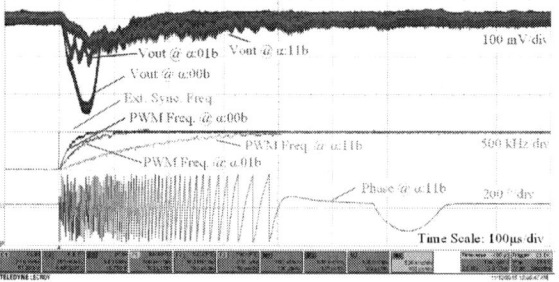

Fig. 10. Output voltage undershoot, frequency and phase synchronization response to a sync_i frequency step from 1.5 MHz to 2 MHz, α = 00, 01, 11:b, KI = 11:b, KP = 11:b. Phase is plotted only for the case α = 11:b

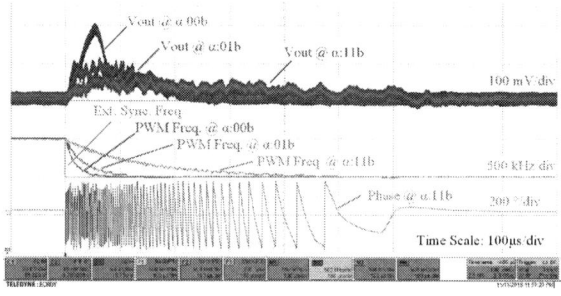

Fig. 11. Output voltage overshoot, frequency and phase synchronization response to a sync_i frequency step from 2 MHz down to 1.5 MHz, α = 00, 01, 11:b, KI = 11:b, KP = 11:b. Phase is plotted only for the case α = 11:b

Furthermore the convergence of the phase difference between the sync and the PWM signal is visible. As expected and shown in table I the under/overshoot is larger for larger AVG bandwidth (higher values of α).

TABLE I.   UNDER/OVERSHOOT DEPENDENCY TO AVG FILTER COEFFICIENT α, WITH $KI = 11$:B, $KP = 11$:B, SYNC FREQUENCY STEP BETWEEN 2 MHZ AND 1.5 MHZ.

α	Frequency Settling Time for undershoot [us]	Vo under shoot [mV]	Frequency Settling Time for overshoot[us]	Vo over shoot [mV]
00	163	236	210	184
01	174	112	230	86
10	300	84	350	72
11	715	62	761	54

Table II shows frequency and phase settling time for various KP and KI factors for a fixed α where Table III shows frequency and phase settling time for all possible AVG configuration factor α and fixed KP and KI factors. From Table II it can be observed that the variation of the configuration of the PI filter has a minor impact on the frequency settling time respect to the phase settling time. From Table III it can be seen that the factor α has a major impact on the frequency settling time doubling for every configuration step. The phase settling is also increasing since this can be achieved only after the frequency has locked. The definite settling time demonstrates that the phase loop is for all the tested condition stable.

TABLE II.   DEPENDENCY OF FREQUENCY AND PHASE SETTLING TIME FROM KP, KI WITH A FIXED AVG CONFIGURATION FACTOR $\alpha = 10$:B

KI	KP	Frequency Settling Time [us]	Phase Settling time [us]
00	00	160	200
00	01	172	299
00	10	194	342
00	11	234	362
10	10	211	368
11	00	151	233
11	01	194	298
11	10	231	307
11	11	294	336

TABLE III.   DEPENDENCY OF FREQUENCY AND PHASE SETTLING TIME FROM AVG CONFIGURATION FACTOR α WITH $KP = 00$:B, $KI = 11$:B

α	Frequency Settling Time [us]	Phase Settling time [us]
00	39	61
01	76	109
10	151	233
11	299	460

In order to prove the robustness of the synchronization module against frequency noise on the sync input further measurements have been performed. The sync input signal is generated with a constant frequency and super-imposing a disturbance on the frequency. Its variation is defined by the instrument as a percentage of the nominal frequency. In Table IV the rejection level to different input synch frequency noise

is summarized. The performance shows a rejection level lower than -15dB even with 30% noise applied.

In Table V the noise on the phase is considered. Since the PWM phase is measured at the Power Stage switching node, the SMPS control loop presents a phase noise which leads to a sdev offset of 11.5 deg. The robustness to phase noise is demonstrated, as highlighted in column D), up to a 10% level.

TABLE IV.   REJECTION TO THE NOISE ON THE FREQUENCY OF THE SYNCHRONIZATION INPUT SIGNAL. A) LEVEL SET AT THE SYNC WAVEFORM GENERATOR. B) AND C) STANDARD DEVIATION MEASURED AT THE OSCILLOSCOPE, D) 20LOG ((Δ PWM NOISE)/(Δ SYNCH NOISE))

A) Noise level [%]	B) Synch Noise [KHz]	C) PWM Noise [kHz]	D) Noise Rejection [dB]
0	0,628	11,2	-
5	15,87	12,1	-24,58
10	31,8	14,8	-18,75
20	63,4	21,4	-15,78
30	95,1	26,2	-15,98

TABLE V.   REJECTION TO THE NOISE ON THE PHASE OF THE SYNCHRONIZATION INPUT SIGNAL. COLUMN A) IS THE LEVEL SET AT THE SYNCH WAVEFORM GENERATOR. B) AND C) ARE SDEV VALUES MEASURED AT OSCILLOSCOPE. D) SHOWS THE PWM PHASE NOISE TO WHICH THE PHASE OFFSET IS SUBTRACTED.

A) Noise level [% (deg)]	B) Sync Phase Noise [deg]	C) PWM Phase Noise [deg]	D) PWM Phase Noise – offset [deg]
0	1,36	11,5	0
1 (3,6)	1,59	11,5	0
2 (7,2)	2,08	11,5	0
5 (18)	4,06	11,5	0
10 (36)	7,7	12,6	1,1

In Fig. 12 is shown the case when the sync_i signal is abruptly removed and the error management reacts setting the switching frequency equal to the minimum allowed (i.e. set to 1 MHz), and asserting the Sync. Error output. Once the sync_i is reactivated, the Sync. Error is set back low, and the PWM frequency synchronizes to the fsync value (i.e. set to 2MHz).

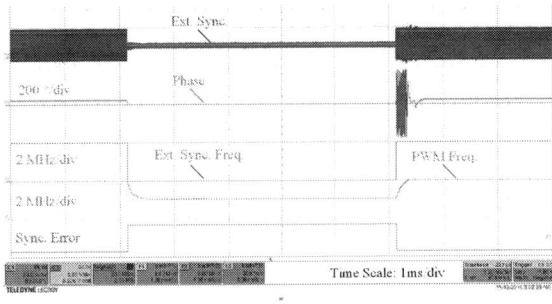

Fig. 12. Synchronization error management in case the sync signal drops

V.   CONCLUSIONS

A digital module for frequency and phase synchronization of a PWM signal to an external synchronization signal has been presented. A novel concept has been described analyzing the realization of both the frequency and the phase synchronization to the input signal.

The presented solution overcomes analog and digital limitations of similar concepts as described in [4, 6, 7], by keeping the system clock frequency constant while the reset value (OSF) of the DPWM counter is adapted. Furthermore the introduction of a digital AVG to smooth the sync_i frequency changes and the adoption of a PI filter grant the phase and frequency synchronization without major perturbations to the regulated SMPS output voltage.

The Synchronizer has been implemented in an integrated buck converted and measurements show that the circuit improves the output voltage (under) overshoot caused by an abrupt input frequency change by a factor bigger than 10.

The frequency and phase settling time have been characterized in respect to the filters configurations (α, KI and KP).

The presented experimental results show that the proposed concept is robust against frequency noise and it shows a frequency phase noise rejection performance lower than -15dB.

Finally the synchronization error management has been introduced and a special case has been measured. In this way it has been shown how the synchronization algorithm can gain robustness if the synchronization signal shows an unexpected behavior.

REFERENCES

[1] M. Dunn, (2014, May 5) *Avoid problems with multiple DC-DC converters* [Online]. Available: http://www.edn.com

[2] Erickson, R.W. and D. Maksimovic, Fundamentals of Power Electronics. 2nd ed. *Colorado Power Electronics Center Department of Electrical and Computer Engineering University of Colorado, Boulder, CO 80309, Kluwe Academic Publishers*

[3] P. Zumel 1, O. Garcia, J. A. Cobos, J. Uceda, "EMI Reduction by Interleaving of Power Converters", *Applied Power Electronics Conference and Exposition (APEC), 2004 Nineteenth Annual IEEE. (1295894)*

[4] Eamon O' Malley, Karl Rinne, Anthony Kelly, Basil Almukhtar, Paul Kelleher "Digital control scheme for robust clock tuning and PWM phase synchronization in digitally controlled multi-POL applications", *Applied Power Electronics Conference and Exposition (APEC), 2010 Twenty-Fifth Annual IEEE. (05433497)*

[5] T. Hegarty, (2008, January 28), *Reducing buck converter input capacitance through multi-phasing and clock synchronization* [Online] Available: http://www.edn.com

[6] Texas Instruments "LM27341-Q1 Datasheet", 2013, Available from, http://www.ti.com

[7] Linear Technology "LTC3833 Datasheet", 2010, Available from http://www.linear.com

[8] Eamon O' Malley, Karl Rinne, "A Programmable Digital Pulse Width Modulator Providing Versatile Pulse Patterns and Supporting Switching Frequencies Beyond 15 MHz", Applied Power Electronics Conference and Exposition (APEC), Nineteenth Annual IEEE (1295787)

[9] Albert M. Wu, Jinwen Xiao, Dejan Markovic, Seth R. Sanders, "Digital PWM control: application in voltage regulation modules", Power Electronics Specialists Conference, 1999. PESC 99. 30th Annual IEEE. (788984)

[10] B. Razavi, RF Microelectronics. Upper Saddle River, NJ: Prentice-Hall, 2000.

[11] J.G.Proakis, D.G. Manolakis, Digital Signal Processing. Prentice Hall; 4 edition, 2006.

A High Accuracy and High Bandwidth Current Sense Circuit for Digitally Controlled DC-DC Buck Converters

David Stack,
Anthony Kelly,
Power R&D, Altera Corp.
Limerick, IRELAND
{david.stack, anthony.kelly}@altera.com

Thomas Conway
Electronics and Computer Engineering Dept.,
University of Limerick, IRELAND
thomas.conway@ul.ie

Abstract— **This paper presents a high accuracy and high bandwidth current sense circuit for digitally controlled DC-DC buck converters. The circuit uses lossless inductor DCR current sensing and a sigma delta modulator ADC to sense the average current in the inductor to high accuracy. These accurate but low bandwidth measurements are unsuitable for use in feedback control loops as their acquisition delay would limit the achievable loop bandwidth. To solve this issue a Kalman Filter estimates the inductor current at the switching frequency of the converter. The output of the estimator is shown through experimental results to achieve 0.3A static current accuracy when used in a buck converter switching at 1MHz and using a power efficient inductor with a DCR of only 165μOhms.**

Keywords—Current sensing, analog to digital conversion, digital control, point of load, DC-DC converter

I. INTRODUCTION

Current sensing circuits are an important part of digital controllers in DC-DC converters. In single phase controllers they provide over-current protection, diagnostic information on the POL (point of load) and control the switch over between CCM (continuous conduction mode) and DCM (discontinuous conduction mode) in multi-mode operation. Additionally in multi-phase systems current sense circuits are used in making phase add and drop decisions, for current share between POLs and implementing active voltage position control.

Three broad measurement methods are used in current sense circuits, sense-fets, sensorless observer circuits and voltage drop measurement techniques. Of these methods, sense-fet's [1] - [3] require the power fet and the sensing fet to match accurately and hence are limited by needing the digital power controller to be integrated onto the same IC as the power-stage or in multi-chip solutions require the use of a power-stage with integrated sense-fets. Sensorless observer circuits [4] - [6] synthesise the inductor current from known variables such as input voltage, output voltage and duty cycle but suffer from limited accuracy which can only be improved with tuning circuits. While in voltage drop methods a voltage

proportional to the current and a known resistance is sensed. These circuits are also difficult to implement as the sensing resistance is typically small for efficiency reasons and hence a small voltage must be sensed.

Lossless inductor DCR current sensing [7] is the most popular current sense approach to be used with digital power controllers [8] - [10]. As shown in Fig. 1 this sensing scheme uses the DC resistance of the inductor and a matched filter to extract the current information. A disadvantage of this current sensing approach is that the voltage to be sensed, $V_{sense}(t) = R_L * I_L(t)$, is proportional to the inductor's DC resistance, R_L, while R_L is usually made small to minimize the power loss in the inductor. For example in high current (80A) point-of-load (POL) modules with a low output voltage used in FPGA or DSP applications the DCR can typically be 160μOhms and the current to voltage conversion is only 160μV/A. In such modules with digital current sharing and active voltage position control loops, the current sense ADC acts as a quantizer in the current feedback loop and the output current needs to be sensed to high resolution to avoid steady state limit cycling [11]. In addition this current needs to be sensed at approximately the switching frequency of the converter to minimize the effect of filter phase lag while maintaining reasonable bandwidth in the control loops. The combined requirements of sensing a small voltage with high resolution and high bandwidth makes the design of the current sense circuit a formidable challenge.

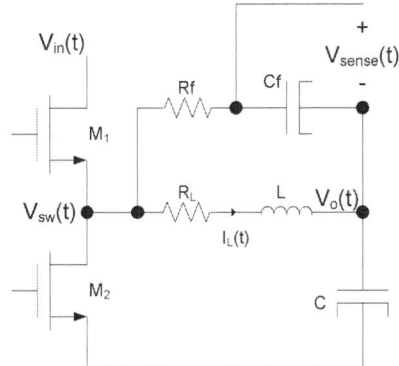

Fig. 1   Inductor DCR current sensing

This work of Altera Power R&D group is supported by the Irish Research Council (IRC).

To sense such a small voltage with high bandwidth and resolution the signal is usually amplified by an instrumentation amplifier before being digitized [12], [13]. Instrumentation amplifiers that achieve low offset and high bandwidth use chopper and auto-zero offset stabilization schemes and are difficult to implement [14]. Alternatively in systems where the current sense information is used for diagnostic purposes only, sigma delta modulator ADCs are used [15]. These ADCs employ over sampling and noise shaping techniques to achieve high resolution but typically the bandwidth that can be achieved is much lower than the switching frequency of the converter.

The focus of this paper, as depicted in Fig. 2, is to use a sigma delta modulator ADC with 2 outputs, $I_{LADC1}[n]$ and $I_{LADC2}[n]$, that have been decimated to different data rates to digitize the current sense signal. $I_{LADC1}[n]$ is a high resolution but low bandwidth signal that accurately represents the average inductor current but contains no high frequency information. In contrast $I_{LADC2}[n]$ is only decimated to the switching frequency of the converter and hence contains higher frequency information but is also corrupted by an unacceptably high level of quantisation noise. The inductor current samples, $I_{LADC1}[n]$ and $I_{LADC2}[n]$, are then combined with a KF (Kalman filter) using a compact inductor current model to predict the current at the switching frequency of the converter. This method allows the load current to be sensed to high static accuracy and resolution using standard sigma delta modulator ADC techniques. It also greatly reduces the bandwidth requirements of the current sense input path as the high frequency samples of the inductor current are estimated by the KF.

The remainder of this paper is organised as follows : Section II describes the average inductor current model as used in the KF. Section III shows how the average inductor current model, SDM ADC and KF are combined during static and transient operation of the power converter. Section IV and V present how the KF is implemented and section VI describes measurement results to demonstrate the circuits performance. Finally conclusions are listed in section VII.

II. Proposed Inductor Current Model

The discrete time model for average inductor current that is to be used in the KF is derived for the buck converter circuit shown in Fig. 1. In this diagram Vin(t) and Vo(t) denote the buck converters input and output voltages, and L and R_L are the inductance and its dc resistance. The inductor current waveform for the buck converter is shown in Fig. 3, in which $I_{LV}[n+1]$ denotes the sampled valley current at time $[n+1]Ts$, $I_{LV}[n]$ is the previous sampled valley current, $I_{LAVG}[n]$ is the average inductor current for the nth period, d[n] is the duty cycle for the nth period, d'[n] = 1- d[n], T_{sw} is the converter's switching period, m1 is the slope of the inductor current while the high side FET is switched on and m2 is the slope of the inductor current while the low side FET is switched on. From Fig. 3 we can deduce that the inductor valley and average current dynamics can be calculated as

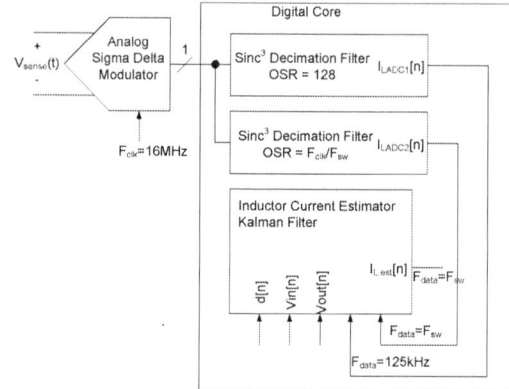

Fig. 2   Estimator working with SDM ADC

$$I_{LV}[n+1] = I_{LV}[n] + m_1 d[n]T_{sw} + m_2 d'[n]T_{sw} \qquad (1)$$

$$I_{LAVG}[n] = I_{LV}[n] + \frac{T_{sw}}{2}[m_1 d[n](2 - d[n]) + m_2 d'[n]^2] \qquad (2)$$

The usual expressions for the inductor current slopes [16] are

$$m_1 = \frac{V_{in} - V_o}{L} \qquad (3)$$

$$m_2 = -\frac{V_o}{L} \qquad (4)$$

Combining (1) and (2) with (3) and (4) yields

$$I_{LAVG}[n+1] = I_{LAVG}[n] - U_{DLAVG}[n-1] + U_{DLV}[n-1] + U_{DLAVG}[n] \qquad (5)$$

where we have defined the deterministic input terms for the nth period as follows

$$U_{DLV}[n] = \frac{T_{sw}}{L}d[n]V_{in} - \frac{T_{sw}}{L}V_o \qquad (6)$$

$$U_{DLAVG}[n] = \frac{T_{sw}}{2L}d[n](2 - d[n])V_{in} - \frac{T_{sw}}{2L}V_o \qquad (7)$$

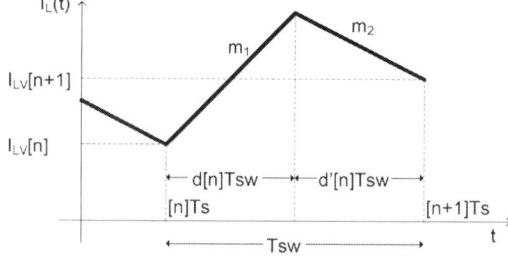

Fig. 3   Inductor Current Waveform

978-1-4673-9551-9/16 $31.00 © 2016 IEEE

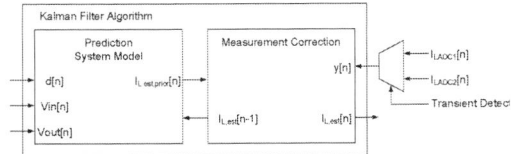

Fig. 4   Kalman Filter recursive algorithm and SDM ADC Outputs

The average inductor current model given in (5) is inaccurate as it neglects non-idealities such as parasitic resistances of the components and non overlap times of the gate driver. These model errors will be dealt with in section IV as the KF using the model is developed.

III. COMBINING KALMAN FILTER, INDUCTOR CURRENT MODEL AND SDM ADC

The basic recursive KF [17] algorithm working with the average inductor current model and SDM ADC outputs is depicted in Fig. 4. The KF algorithm consists of a prediction step and measurement correction step. In the prediction step the system model and prior knowledge about the inductor current are used to generate an a priori estimate of the average inductor current, $I_{L,est,prior}[n]$. In the measurement correction step this estimate is then corrected by the inductor current measurement to form an a posteriori estimate, $I_{L,est}[n]$. Hence to use a KF to estimate the average inductor current the system model given in (5) needs to be combined with a measurement model that represents the sigma delta modulator output to allow the predicted states be compared against the measurements and be corrected. To make this comparison during steady state converter operation the estimated value for $I_{LAVG}[n]$ can be corrected with the measured value $I_{LADC1}[n]$. A transient detect circuit decides if the converter is in a static operating region and controls the mux that allows the measurement update of the KF based on $I_{LADC1}[n]$ to take place. However if the converter is found to be in a transient operating region, as shown for a transient current event in the simulation plot in Fig. 5, $I_{LADC1}[n]$ is no longer equal to the average current over a switching cycle, $I_{LAVG}[n]$, as its bandwidth is too low to track the rapidly changing current. During these times the measurement update is based on $I_{LADC2}[n]$ and the assumed sensor measurement white noise is increased to take care of the extra quantization noise.

IV. AUGMENTED KALMAN FILTER

The average inductor current model given in (5) is inaccurate as no attempt has been made to model non-idealities of the components and the gate driver. These simplifications have been made to minimize the calculations in the KF based on the values of non-idealities that are not very well known. However model errors can cause divergence [18] in the KF and an augmented KF is developed where the model error state, ε_k, is used to represent model error at time t_k. To also simplify the deterministic input calculations only quantities relating to the present switching cycle are used in prediction calculations. From these considerations the system model as

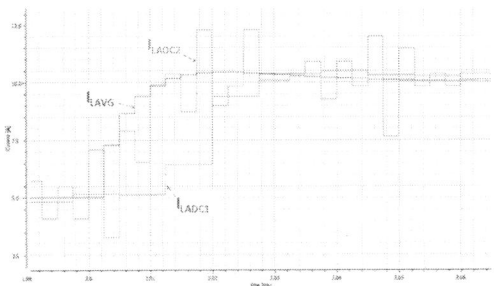

Fig. 5   Inductor Current and SDM ADC Outputs during transient current event

used in the KF can be derived from (5) and is written as (subscript k is subsequently taken to mean a variable at time t_k).

$$\begin{bmatrix} I_{LAVG,k} \\ \varepsilon_k \end{bmatrix} = \begin{bmatrix} 1 & 1 \\ 0 & 1 \end{bmatrix} \begin{bmatrix} I_{LAVG,k-1} \\ \varepsilon_{k-1} \end{bmatrix} + \begin{bmatrix} U_{DLAVG,k} \\ 0 \end{bmatrix} + v_k \tag{8}$$

The measurement model depends on the operating region as discussed in section III and is found as

$$y_k = \begin{cases} I_{LADC1,k} + w_{1,k} & static\ operation \\ I_{LADC2,k} + w_{2,k} & transient\ operation \end{cases} \tag{9}$$

In (8) and (9) v_k represents the model noise and $w_{1,k}$ and $w_{2,k}$ represents the SDM ADC noise for outputs $I_{LADC1,k}$ and $I_{LADC2,k}$. The EKF is then developed in the usual manner [17], [18] based on (8) and (9) and will be explained in the following section.

V. AUGMENTED KALMAN FILTER DEVELOPMENT

To develop the KF the system model (8) is written in compact form, with state vector $x_k = [I_{LAVG,k}, \varepsilon_k]^T$, input vector $u_k = [U_{DLAVG,k}, 0]^T$ and measurement vector y_k

$$\begin{aligned} x_k &= f(x_{k-1}, u_k) \\ y_k &= h(x_k, u_k) \end{aligned} \tag{10}$$

The system function, $f(x_{k-1}, u_k)$ and measurement function, $h(x_k, u_k)$ are defined as

$$f(x_k, u_k) = \begin{bmatrix} 1 & 1 \\ 1 & 0 \end{bmatrix} x_k + u_k \tag{11}$$

$$h(x_k, u_k) = [1 \quad 0] x_k \tag{12}$$

This allows the derivative matrices A_k and H_k to be calculated

$$A_k = \frac{d}{dx} f(x_{k-1}, u_k) = \begin{bmatrix} 1 & 1 \\ 1 & 0 \end{bmatrix} \tag{13}$$

$$H_k = \frac{d}{dx} h(x_k, u_k) = [1 \quad 0] \tag{14}$$

The final variables needed to realize the KF are the initial covariance matrix, P_0, the covariance matrix, R_V, associated with model noise V_k and covariance R_{W1} and R_{W2} associated with SDM ADC noise $W_{1,k}$ and $W_{2,k}$ respectively.

THE KF algorithm which includes prediction and correction stages is then composed of a prediction stage

$$\hat{x}_k^- = f(\hat{x}_{k-1}, u_k) \tag{15}$$
$$\hat{y}_k = h(\hat{x}_k^-) \tag{16}$$
$$P_k^- = A_k P_{k-1} A_k^T + R_V \tag{17}$$

where $\hat{x}_k^-$ denotes the a priori estimate of x_k, $\hat{y}_k$ is the estimated output based on $\hat{x}_k^-$, and P_k^- is defined as the a priori error covariance matrix.

The correction stage is defined as

$$L_k = P_k^- H_k^T \left[H_k^T P_k^- H_k + R_{Wi} \right]^{-1} \tag{18}$$
$$P_k = [I - L_k H_k] P_k^- \tag{19}$$
$$y_{err,k} = y_k - \hat{y}_k \tag{20}$$
$$\hat{x}_k = \hat{x}_k^- + L_k y_{err,k} \tag{21}$$

where L_k is the Kalman Gain, R_{Wi} is the SDM ADC noise covariance which is chosen as R_{W1} during static operation and R_{w2} during transient operation as explained in section III and IV. P_k denotes the a posteriori error covariance matrix, I is the 2x2 identity matrix, $y_{err,k}$ is the estimation error and $\hat{x}_k$ is the updated estimate.

VI. EXPERIMENTAL PLATFORM AND RESULTS

An experimental system was built for the proposed current sense circuit having the following parameters: $V_{in} = 12V$, $V_o = 1.2V$, $F_{sw} = 1MHz$, $L = 330nH$, $C = 2700uF$ and $R_L = 165u\Omega$. The digital controller and KF have been implemented on a FPGA. The digital controller was designed for bandwidth $F_c = 15kHz$ and phase margin $PM = 65°$. The WE744309033 [19] has been used as the power inductor and the AMC1305L25 [20] second order SDM has been used with a sampling rate of 16MHz to implement the ADC. This ADC achieves an equivalent input noise of 0.15Arms at the 125kHz data rate on the $I_{LADC1}[n]$ output and 10Arms equivalent input noise at the 1MHz data rate on $I_{LADC2}[n]$ for the inductor with $R_L = 165u\Omega$. This sets the SDM ADC error covariance for $I_{LADC1}[n]$ and $I_{LADC2}[n]$ as

$$R_{W1} = 0.15^2 \tag{22}$$
$$R_{W2} = 10^2 \tag{23}$$

The remaining error covariance matrices P_0 and R_V have been set based on the observed model accuracy as

$$P_0 = \begin{bmatrix} 2^2 & 0 \\ 0 & 2^2 \end{bmatrix} \tag{24}$$
$$R_V = \begin{bmatrix} 2^2 & 0 \\ 0 & 2^2 \end{bmatrix} \tag{25}$$

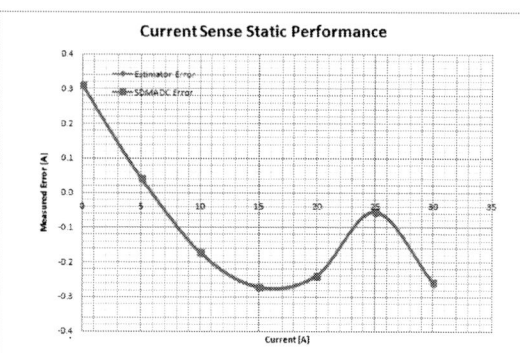

Fig. 6   Estimated Inductor Current Error

A. Static Current Sense Accuracy

The accuracy of the current estimator output is tested by varying the output load current from 0 to 30A and monitoring the estimated current. The data is shown in Fig. 6. The current sense error is an excellent 0.3A when used with the extremely small DCR of $165u\Omega$. The equivalent input noise of the estimated current is measured as 0.15Arms and the peak to peak noise of the SDM ADC and estimator output is shown for over 3000 consecutive output samples in Fig. 7 for a load current of 25A.

B. Augmented Kalman Filter Stability

The stability of the augmented kalman filter is tested for a transient load step. Fig. 8 shows that the state variables $I_{LEST}[n]$ and $\varepsilon[n]$ settle quickly after the transient load event is finished and indicate that the system model of the process is robust. In addition Fig. 9 shows the step response of $P11^-$ and P11, the a priori estimation error and a posteriori estimation error for I_{LAVG}. Clearly both error terms settle quickly to values less than P_0 once the transient current step is finished.

C. Transient Current Sense Performance

Fig. 10 demonstrates the current sense performance during a

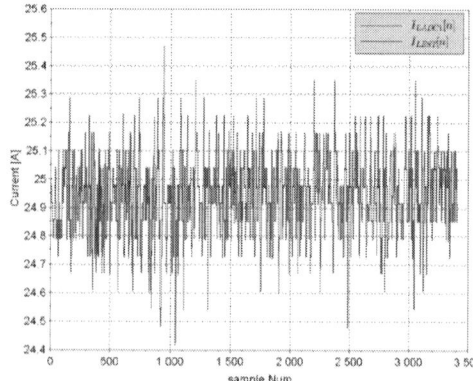

Fig. 7   Estimated Inductor Current Noise

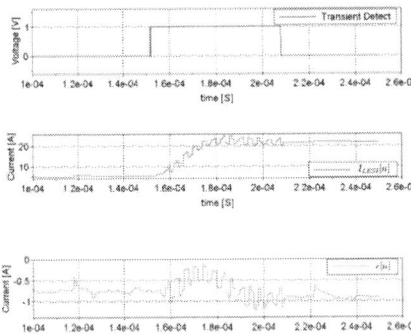

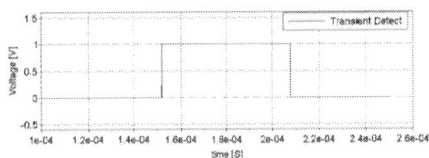

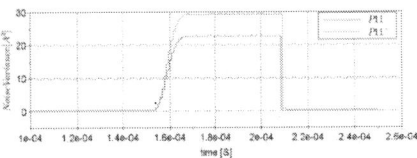

Fig. 8 Step Response of State variables

Fig. 9 Step response of Estimation Errors

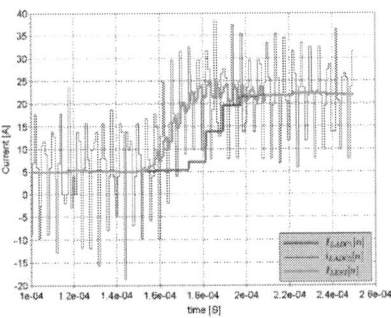

Fig. 10 Transient Current Sense Performance

current step. As expected $I_{LADC1}[n]$ reacts slowly to the current transient but has low noise while $I_{LADC2}[n]$ has a much higher bandwidth but is corrupted by a large noise signal. The estimator output, $I_{LEST}[n]$, can be seen to have the same latency as $I_{LADC2}[n]$ but with much improved noise performance. When the transient is over it can also be seen that $I_{LEST}[n]$ is compared to $I_{LADC1}[n]$ and its static noise performance and accuracy improves as discussed previously.

VII. CONCLUSIONS

In this paper a high accuracy and high bandwidth current sense circuit has been introduced. The circuit combines existing SDM ADC techniques with a kalman filter to achieve a current sense circuit suitable for integration with digital power controllers. The proposed circuit can sense extremely small current sense signals to high resolution and work with SMPS operating at high switching rates. The estimator has been implemented on an FPGA and experimental results show correct operation with a buck converter switching at 1MHz and DCR value of *165uΩ*.

REFERENCES

[1] M. Du and H. Lee, "An integrated speed- and accuracy-enhanced CMOS current sensor with dynamically biased shunt feedback for current-mode buck regulators," *IEEE Trans. Circuits Syst. I Regul. Pap.*, vol. 57, no. 10, pp. 2804–2814, 2010.

[2] S. Rao, S. Member, Q. Khan, S. Bang, and D. Swank, "A 1 . 2-A Buck-Boost LED Driver With On-Chip," vol. 46, no. 12, pp. 1–12, 2011.

[3] V. Michal, "Absolute Value ,1% Linear and Lossless Converters With Integrated Power Stage," *IEEE J. Solid-State Circuits*, vol. 49, no. 5, pp. 1256–1270, 2014.

[4] P. Midya, P. T. Krein, and M. F. Greuel, "Sensorless Current Mode Control - An Observer-Based Technique for Dc-Dc Converters," *Trans. Power Electron.*, vol. 16, no. 4, pp. 522–526, 2001.

[5] A. Kelly and K. Rinne, "Sensorless current-mode control of a digital dead-beat DC-DC converter," *Ninet. Annu. IEEE Appl. Power Electron. Conf. Expo. 2004. APEC '04.*, vol. 3, pp. 1790–1795.

[6] Z. Luki, S. M. Ahsanuzzaman, A. Prodi, and Z. Zhao, "Self-tuning sensorless digital current-mode controller with accurate current sharing for multi-phase DC-DC converters," *Conf. Proc. - IEEE Appl. Power Electron. Conf. Expo. - APEC*, pp. 264–268, 2009.

[7] E. Dallago, M. Passoni and G. Sassone, 'Lossless current sensing in low-voltage high-current DC/DC modular supplies', IEEE Transactions on Industrial Electronics, vol. 47, no. 6, pp. 1249-1252, 2000.

[8] Zentrum Mikroelektronik Dresden AG, ZSPM1000 Data Sheet, June 2015 [Online]. Available: www.zmdi.com

[9] Intersil Corp., ZL8801 Data Sheet, June 2015 [Online]. Available: www.intersil.com

[10] Texas Instruments Inc., TPS40428 Data Sheet, June 2015 [Online]. Available: www.ti.com

[11] A. Peterchev and S. Sanders, 'Quantization resolution and limit cycling in digitally controlled PWM converters', IEEE Transactions on Power Electronics, vol. 18, no. 1, pp. 301-308, 2003.

[12] T. Liu, H. Yeom, B. Vermeire, P. Adell and B. Bakkaloglu, 'A Digitally Controlled DC-DC Buck Converter with Lossless Load-Current Sensing and BIST Functionality', in Solid-State Circuits Conference Digest of Technical Papers (ISSCC), San Francisco, CA, 2011, pp. 388 - 390.

[13] H. Forghani-zadeh and G. Rincon-Mora, 'An Accurate, Continuous, and Lossless Self-Learning CMOS Current-Sensing Scheme for Inductor-Based DC-DC Converters', IEEE J. Solid-State Circuits, vol. 42, no. 3, pp. 665-679, 2007.

[14] J. F. Witte, J. H. Huijsing, and K. A. A. Makinwa, "A current-feedback instrumentation amplifier with 5uV offset for bidirectional high-side current-sensing," IEEE J. Solid-State Circuits, vol. 43, no. 12, pp. 2769–2775, 2008.

[15] Linear Technol. Corp., LTC2974 Data Sheet, June 2015 [Online]. Available: www.linear.com

[16] R. Erickson and D. Maksimović, Fundamentals of power electronics. Norwell, Mass.: Kluwer Academic, 2001.

[17] R. Brown, P. Hwang and R. Brown, Introduction to random signals and applied Kalman filtering. New York: J. Wiley, 1992.

[18] M. Grewal and A. Andrews, Kalman filtering: Theory and Practice Using Matlb. New York: Wiley, 2008.

[19] Wurth Elektronik Group, WE744309033 Data Sheet, June 2015 [Online]. Available: www.we-online.com

[20] Texas Instruments Inc., AMC1305L25 Data Sheet, June 2015 [Online]. Available: www.ti.com

Modular Multilevel Dual Active Bridge DC-DC Converter with ZVS and Fast DC Fault Recovery for Battery Energy Storage Systems

Yuxiang Shi, Rui Li, Hui Li
Center for Advanced Power System
Florida State University
Tallahassee, FL, USA
hli@caps.fsu.edu

Abstract—This paper proposes a novel modular multilevel dual-active-bridge (MMDAB) dc-dc converter for spilt-battery energy storage system (BESS) in medium-voltage dc (MVDC) grid application. The proposed converter can be derived from conventional DAB converter and therefore exhibits favorite characteristics of galvanic isolation, inherent zero-voltage-switching (ZVS) condition and small passive components. In addition, compared to cascaded DAB converter, the proposed converter has limited dc fault current rising and fast dc fault recovery due to the leakage inductance and the kept cell energy during the fault. A new modulation method with minimal passive components requirement and low *dv/dt* has been proposed, and the corresponding control system is also developed to realize the power flow control and balancing control. A case study of 500 kW, 20 kHz MMDAB-based BESS is performed in simulation to validate the performance.

Keywords—modular multilevel converter; dual active bridge; battery energy storage system; ZVS; dc fault

I. INTRODUCTION

Due to the low voltage of battery pack (< 1000 V), split-battery energy storage system (sBESS) based on modular multilevel dc-dc converters (MMDC) presents significant advantages in MVDC systems, in terms of modularity, easy battery management system (BMS) design, high efficiency and redundancy [1]-[7]. However, the non-isolated characteristic limits their applications. Recently, isolated MMDC (iMMDC) consisting of two MMC converters cascaded through transformer have been reported for multi-terminal HVDC or HVDC to MVDC applications [8]-[14]. However, they are not for energy storage application. Cascaded dual-active-bridge (DAB) dc-dc converter with sBESS has been reported in [15], nonetheless the dc-link capacitors are discharged under fault condition, resulting in extra fault current and long recovery time.

In this paper, we propose a MMDAB converter for sBESS application, see Fig. 1. A cascaded multilevel converter (CMC) is connected to a quasi-two-level converter (Q2LC) [13] through a high frequency transformer, split-battery units are integrated into each submodule of low-voltage-side (LVS) CMC, while the dc terminal of high-voltage-side (HVS) Q2LC is connected to the MVDC bus. Half-bridge cells are implemented for submodules in both side. Our analysis reveals

that the proposed MMDAB converter can be derived from conventional DAB converter, therefore it inherits the advantages of DAB converter. A similar topology using MMC instead of Q2LC in HVS has been published for sBESS in [16], however its operation principle and control is based on MMC instead of DAB, making it different from that of the proposed MMDAB converter. Compared to the topology using full-bridge cells in [16], the proposed MMDAB converter employs less semiconductor devices, fewer magnetic components, and smaller cell capacitors; moreover, the inherent ZVS condition for all the switches enables high-frequency operation. Unlike cascaded DAB converter, the cell capacitor energy in MMDAB converter can be kept under dc fault, therefore the system can be fast recovered when fault is cleared. In addition, the fault current rising is limited by the leakage inductance since there is no dc fault loop exists in the converter.

Based on quasi-two-level (Q2L) modulation [13] and two-level (2L) modulation [17], a new modulation method is proposed with small cell capacitance requirements and low *dv/dt*. A simple control system is also developed accordingly to realize power and balancing control. Simulation results are provided for verification.

II. OPERATION PRINCIPLE OF PROPOSED MMDAB sBESS

A. Derivation of MMDAB Converter from DAB Converter

As mentioned in the introduction, the MMDAB converter can be derived from the DAB converter. Fig. 2 shows evolution of MMDAB converters from traditional DAB converter. Assuming the DAB converter is connected to dc sources, the H-bridge can be split into two half-bridge cells connected to identical dc sources as shown in Fig. 2 (a). Considering the split half-bridge cell with dc source as a submodule, a cascaded half-bridge dc-dc structure shown in Fig. 2 (b) can be formed by cascading multiple half-bridge submodules. By applying the same transformation to multiport DAB converter with dual output in Fig. 2 (c), the MMDAB prototype in Fig. 2 (d) can be obtained where an output dc terminal is formed between common nodes of positive and negative arms.

978-1-4673-9551-9/16 $31.00 © 2016 IEEE

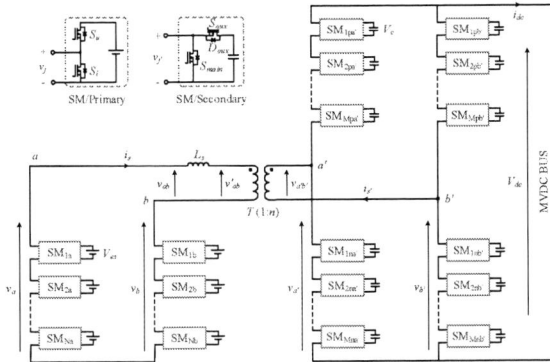

Fig. 1. Proposed MMDAB based sBESS

B. Operation Principle of MMDAB Converter

The operation principle of proposed MMDAB converter is similar to the conventional DAB converter where the power flow is control by the phase shift angle between the LVS side and HVS side. As a multilevel converter with no dc-link in LVS, modulation methods for conventional DAB converter, like conventional phase shift (CPS) modulation and dual phase shift (DPS) modulation, have to be modified.

In Q2L modulation, the carrier of submodules in an arm are phase-shifted with such a small angle that the arms just function as single switches but with reduced dv/dt [13], [14]. Since the capacitor inside submodule operates as voltage clamper and is not engaged in main power transfer, Q2L modulation is helpful for the capacitance reduction but not applicable for interfacing energy storage device (ESD). The 2L modulation in [17] operates the iMMDC as an "electronic dc tap changer", with each submodule as an energy buffer for main power transfer, making it suitable for ESD employment.

To interface battery and reduce capacitor together, a modified multilevel phase shift modulation integrating the features of Q2L and 2L modulations is proposed for the MMDAB based sBESS. The key waveforms under discharging mode are illustrated in Fig. 3 (a). The two LVS phase are complimentarily operated with pulse width modulation for balancing purpose. With half-bridge submodules, the output voltage of LVS phase can varies from 0 to NV_{es}, where N is the submodule number per phase and V_{es} is the voltage of battery units. Modulation index for phase a as example is defined as

$$M_a = \begin{bmatrix} m_1 & m_2 \\ 2D\pi & 2(1-D)\pi \end{bmatrix} \quad m_2 < m_1 \le N \quad (1)$$

$$\text{active} \qquad \text{inactive}$$

where m_1 and m_2 is the numbers of submodules actively connected in the phase a during the phase angel $2D\pi$ (active state) and $2(1-D)\pi$ (inactive state), respectively. With the modulation, the battery units of $N-m_1$ inactively connected submodules during the active state will not be discharged, while m_2 actively connected battery units during the inactive state will be charged by the ac current. This mechanism helps to realize the charge balancing of battery units. For the HVS

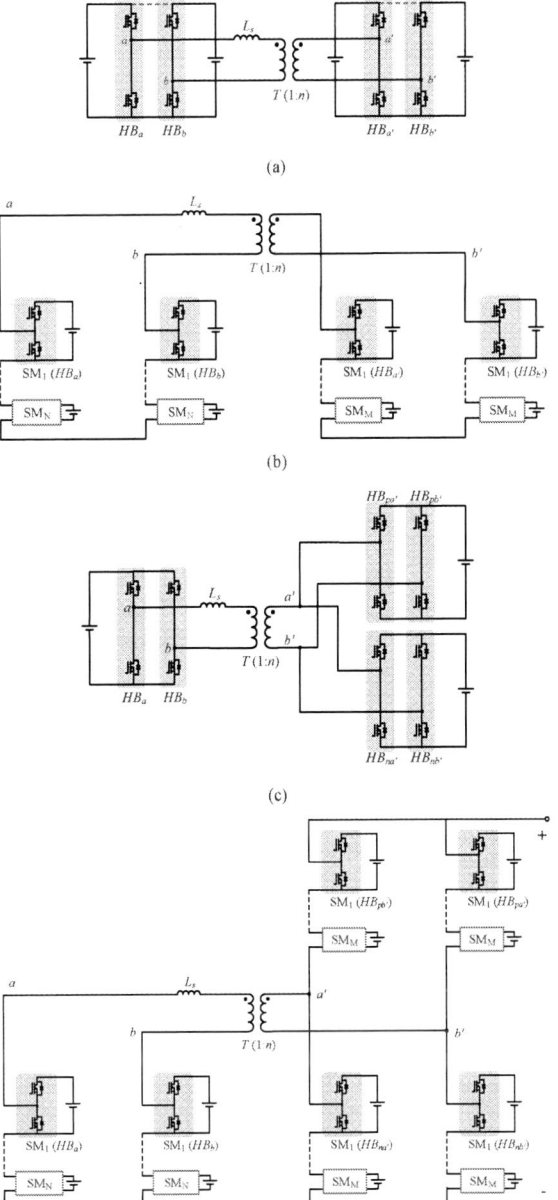

Fig. 2. Topology derivation of MMDAB converter: (a) single phase DAB with split dc source, (b) Cascaded half-bridge dc-dc structure, (c) DAB converter with dual output, and (d) MMDAB converter

side, the two phase are 180° shifted, and each phase operating as two-level switch with duty cycle of 0.5. The corresponding modulation index M_s is defined as

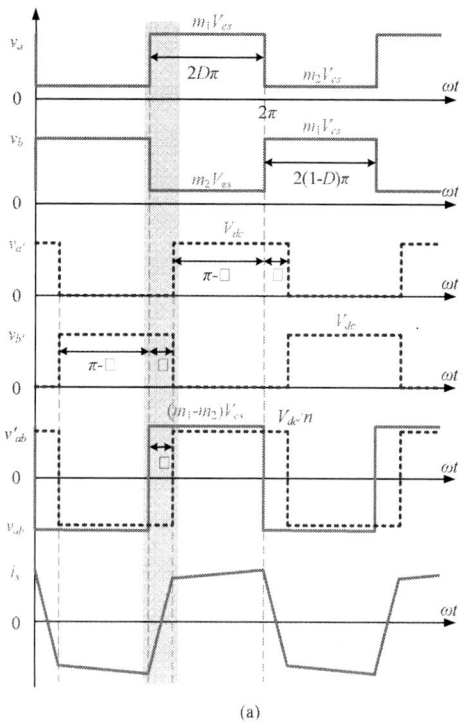 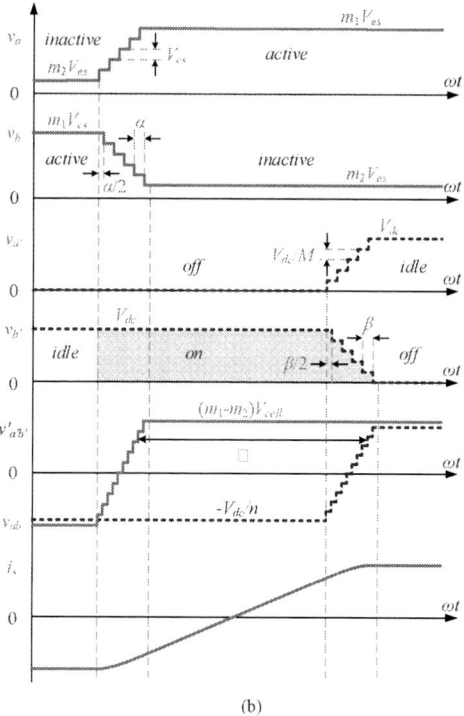

Fig. 3. Key waveforms of MMDAB converter under discharging mode: (a) voltage and current waveforms, and (b) zoom view of the shaded phase range in (a).

$$M_s = \begin{bmatrix} - & M & 0 \\ \pi - \phi & \phi & \pi \end{bmatrix} \tag{2}$$
$$\text{idle} \quad \text{on} \quad \text{off}$$

where M is submodule numbers per arm for HVS side and $\square$ is the phase shift angle. Beside conventionally on state and off state, an idle state is employed, of which neither the main switch nor auxiliary switch is active. With commutation current, the cell capacitors are first charged during the $\pi - \square$ idle state through auxiliary diodes, and then they are reset during $\square$ on state through the auxiliary switches. In such a way, ZVS turn-on of the main and auxiliary switches are achieved; moreover, the cell capacitor energy can be maintained when dc fault occurs, since they only get discharged during the small period of on state.

To achieve multilevel operation, trapezoidal modulation with shifted complimentary switching (SCS) sequence is adopted [14]. The SCS sequence is illustrated in Fig. 3 (b), a detail view of the shaded phase range in Fig. 3 (a). For the LVS side, the submodules in the same arm are switched on and off in sequence with small phase step α; while the switching sequences of the two phase are shifted by $\alpha/2$. As a result, the ac voltage v_{ab} presents in staircase pattern with reduced dv/dt. The dwell time corresponding to α must account for the switching time of the submodule and ensure an acceptable dv/dt. Meanwhile, the entire switching commutation in a phase should be finished before the ac current change the polarity to maintain ZVS condition for all the switches. Normally, small dwell time is preferred. Similarly, SCS sequence with phase step β is applied for the phases in HVS side. The positive and negative arms can be commutated either in complementary or non-complementary switching sequence [14]. Since each arm has to block the bus voltage V_{dc}, the nominal voltage for auxiliary capacitors is given as V_{dc}/M. And the capacitance required can be much smaller than that in the MMC as they are only engaged in power transfer during the commutating process.

III. SYSTEM CONTROL

A simple control system diagram for the proposed MMDAB based sBESS are illustrated in Fig. 4, mainly including voltage balancing control of HVS capacitors, state-of-charge (SOC) balancing control of battery units, power flow control of the converter and dc fault management.

A. Voltage Balancing

Since the phases in HVS Q2LC operates in two-level mode, all the arm voltage are automatically reset to V_{dc} during the on state. However, the cell voltage in an arm can be varied due to Q2L modulation. To balance the voltage, sorting method is adopted which adapts the switching order of cells inside arms.

B. SOC Balancing

The SOC balancing control are provided in two level: cell balancing control and phase balancing control.

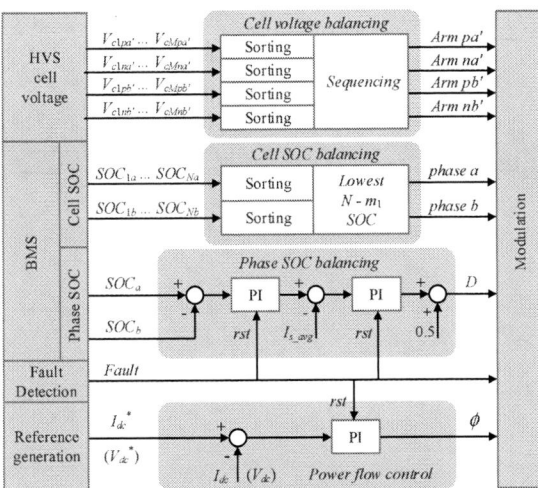

Fig. 4. Control block diagram of the sBESS

TABLE I. CIRCUIT PARAMETERS OF THE MMDAB BASED sBESS

Items	Descriptions	Value
P_N	Rated power	500 kW
V_{dc}	DC bus voltage	6 kV
V_{es}	LiFePO4 battery voltage	650-850 V
V_{esN}	Rated battery voltage	750 V
Q_{es}	Nominal battery-unit capacity	200 Ah
C_{es}	Battery-paralleled capacitor	100 μF
C_{cell}	HVS cell capacitor	10 μF
S_{LVS}	LVS half bridges	CAS300M12BM2
S_{main}	HVS main switches	CAS120M12BM2
S_{aux}	HVS auxiliary switches	C2M0040120D
N	LVS module number per arm	5
M	HVS module number per arm	8
n	Transformer turns ratio	2
f_s	Switching frequency	20 kHz
L_s	Leakage inductor	25 μH
$[m_1\ m_2]$	Modulation index	[4 0]

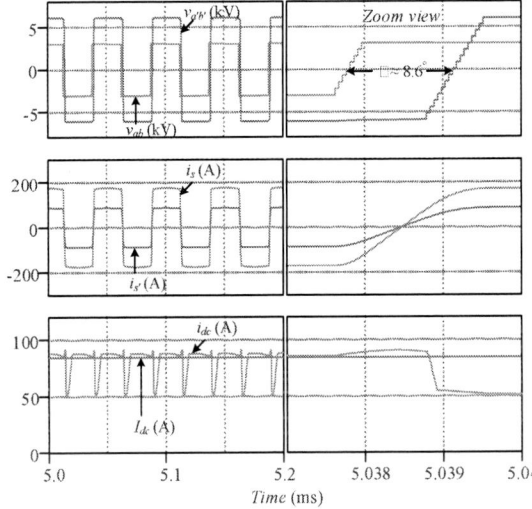

Fig. 5. Key waveforms of the MMDAB converter at rated power

The cell SOC balancing control is aiming to balance the SOC of battery units in the same phase. The SOC data of each battery unit for the battery management system (BMS) is collected, and sorting algorithm is used to determine N-m_1 submodules with lowest SOC in each phase. With the proposed modulation, these selected submodules will not get discharged in the active state of the phase which will balance the cell SOCs. To enhance the cell balancing capability, the selected submodules can be also charged in the inactive state of the phase, i.e. $m_2 = N$-m_1. However, this requires more submodules for the same ac voltage and will reduce the system efficiency due to reverse power in the batteries. Therefore, $m_2 = 0$ is preferred and used in this paper.

The phase balancing control ensures the average SOC of the two phase are balanced and is realized through duty cycle control with dual loop compensator. The phase SOC difference is compensated through PI controller in the outer loop, and the output is used as the reference for the inner loop to regulate the dc circulating current which serves the phase SOC balancing.

C. Power Flow Control

The power flow management of the MMDAB is obtained by controlling the dc-link current in HVS through phase shift angle □. Active power command form remote power dispatch center, and charging or discharging profile from BMS are send to the reference generation block which generates dc-link current reference accordingly. The MMDAB can be also controlled in voltage source mode to support the MVDC bus when required.

D. DC Fault Management

When dc fault occurs, the dc current will rise rapidly. For the MMDAB converter, as the cell capacitors are not connected during idle state, no dc fault loop exists and the fault current will flow through the transformer. Thereby the fault rising rate is determined by the leakage inductor. By sensing the ac current, fault signal is generated when the current is over the threshold, which will reset the controller and trigger

protection. With undischarged cell capacitors, the system can be fast recovered once the fault is cleared and the dc bus is restored.

IV. SIMULATION RESULTS

A 500 kW model of proposed MMDAB based sBESS connected to 6 kV MVDC bus was built in MATLAB/Simulink with PLECS. Table I summarizes the main parameters of the designed sBESS system. The nominal voltage of submodules is selected to be 750 V, so that SiC MOSFET devices can be implemented to fully take advantage of ZVS characteristic of the MMDAB converter. An ESL of

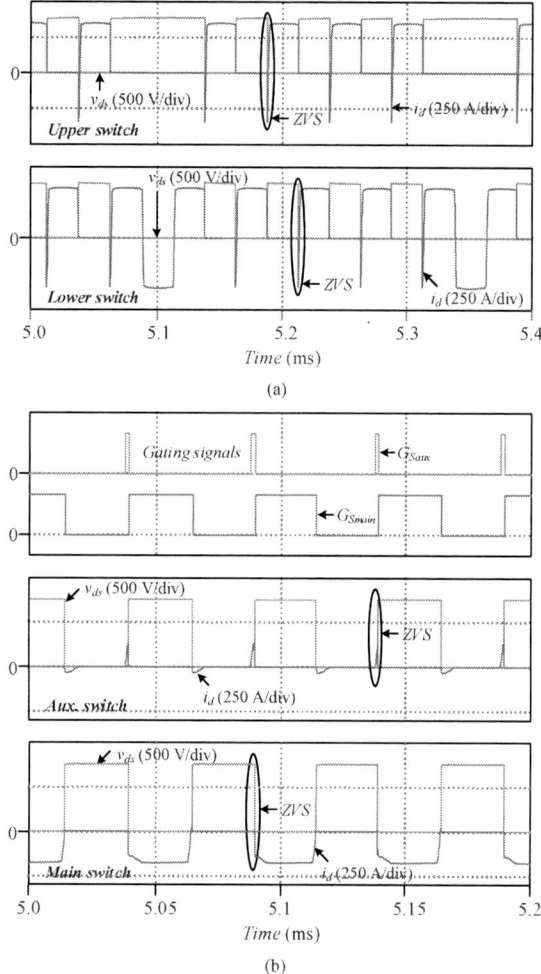

(a)

(b)

Fig. 6. Switching waveforms in submodules: (a) LVS, and (b) HVS

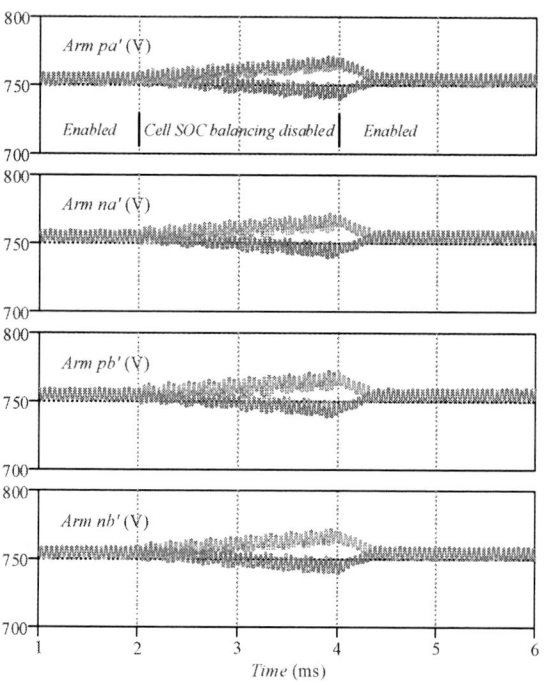

Fig.7. Cell voltage balancing in HVS arms

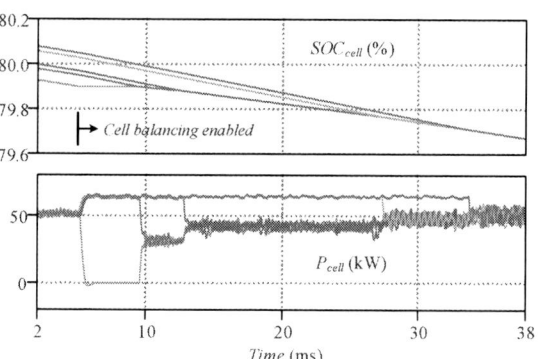

Fig. 8. Cell SOC balancing in LVS arms

100 nH and ESR of 58 mΩ per submodule are considered for the HVS arms based on the datasheets of the selected devices. LiFePO4 battery model in [18] is adopted and switching frequency is set as 20 kHz. Simulation results under discharging mode in Fig. 5 - 9 are provided to verify the performance of the proposed sBESS.

As shown in Fig. 5, the key waveforms of the MMDAB converter are very similar to that of the conventional DAB converter, while the staircase voltage waveforms in the zoomed view illustrates the multilevel operation under Q2L modulation. Considering the fast switching of SiC device, dwell time for the successive switching event is set to 100 ns for both sides in the simulation, i.e. $\alpha = \beta = 100$ ns.

Fig. 6 depicts the voltage and current waveforms in the submodules of MMDAB converter. As seen, ZVS is achieved for all the switches in LVS and HVS. For balancing purpose, one submodule in LVS is not actively connected in each

switching cycle. Accordingly, no switching will occur in such switching cycle for that submodule, as shown in Fig. 6 (a). In Fig. 6 (b), the current flow through the auxiliary switch or diode is much smaller compared to the main switch since it is not engaged in the main power transfer. Thereby, a much lower current rating device can be selected to for the auxiliary switch.

The cell voltage balancing performance in HVS arms are shown in Fig. 7. In the beginning, sorting algorithm is executed every switching cycle and the capacitor voltage are well regulated around 750 V, with a voltage ripple less than ±1%. At $t = 2$ ms, sorting algorithm is disabled and the submodules are switched in and out in a fixed rotating sequence instead. As expected, the cell voltage begin to diverge. At $t = 4$ ms, the

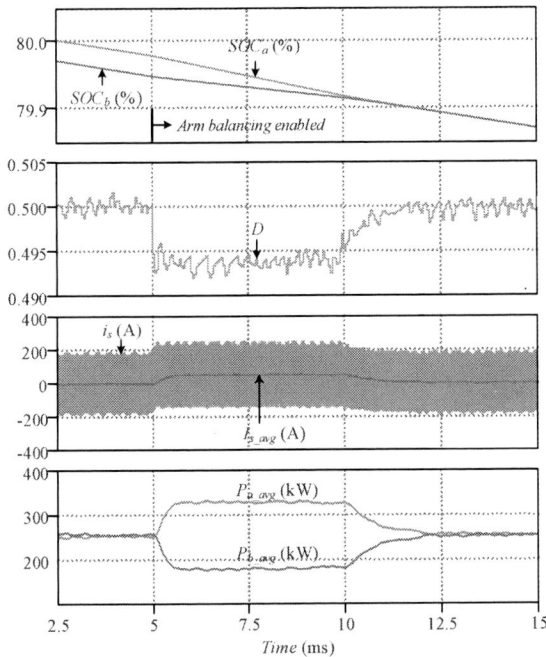

Fig. 9. Arms SOC balancing with an accelerating factor of 1000

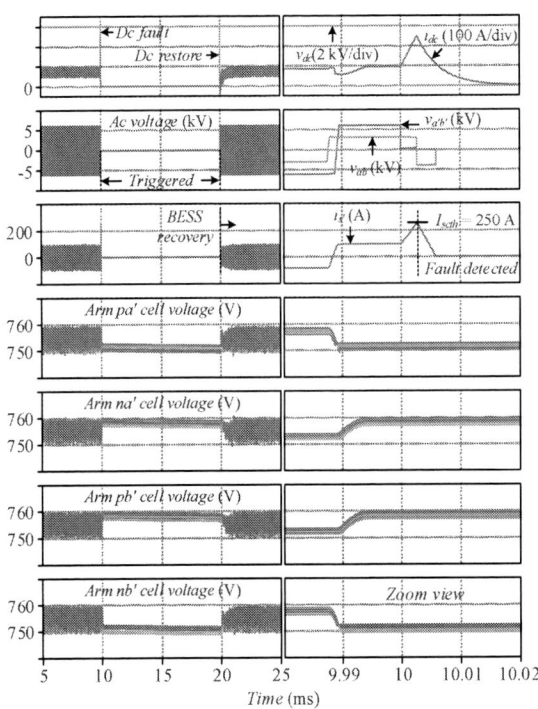

Fig. 10. System operation under dc fault

balancing control is re-enabled and the cell voltage converges to 750 V quickly.

Fig. 8 illustrates the cell SOC and output power of phase *a* in LVS during the balancing process. It should be noted that the SOC balancing is too slow to be directly simulated with circuit model, hence an accelerating factor of 1000 is applied in all the SOC calculation, namely 1 ms in the simulation represents 1 s in the real case. In the beginning, each submodule in the arm outputs the same power and the cell SOCs differs from each other. At $t = 5$ ms, the cell balancing control is enabled. The power of the cell with lowest SOC drops to zero and the SOC stays constant, while the power of other cells increases. Then at $t = 9.5$ ms, the SOC become the same for two cells with the lowest SOC. Then this two cell begin to output 1/2 power of the rest cells, thus their SOC drop slower than others. When the 3 cells share the same lowest SOC, a similar process starts until all the cells have the same SOC.

Fig. 9 reveals the averaged SOC balancing between the two phases. In the beginning, the phases have different averaged SOCs. The dc component of LVS current is regulated to be zero with duty cycle around 0.5. The averaged power of the two phases are the same, about 250 kW. To balancing the SOCs, phase balancing control is enabled at $t = 5$ ms, and the duty cycle slightly decreases to 0.494, resulting a small dc voltage applied in the LVS circuit. This dc voltage excites a dc current which redistributes the power in the two phases and balances the SOCs. At steady state, all the dc voltage will all lands on the ESR in LVS. Since the ESR is usually small, a very small duty cycle variation is enough to inject the balancing current. In the simulation, the balancing current is limited to 50 A considering the device rating. When the two SOCs getting close, the duty cycle is regulated slowly towards 0.5, and the averaged SOCs are finally balanced at around 12.5 ms.

The converter operation under dc fault is shown in Fig. 10. At $t = 10$ ms, dc fault occurs and the dc current begins to increase. At the same time, the ac current increases due to the drop of dc bus. When the ac current rise over threshold of 250 A, dc fault is detected which triggers the converter. The fault signal will also reset the controllers for next recovery. As can be seen, the cell capacitor voltage is maintained during the fault, enabling fast recovery. At $t = 20$ ms, the fault is cleared and the dc bus is restored. Simultaneously, the BESS relaunches and is fully recovered in about 2 ms.

V. CONCLUSION

In this paper, MMDAB dc-dc converter is proposed for BESS application in MVDC systems. Similar to voltage-fed DAB converter, the proposed converter provides galvanic isolation and inherent ZVS operation. The operating principle of MMDAB converter was analyzed, and a simple and effective control system was developed for the BESS accordingly. With the advantages of MMDAB converter and the proposed modulation method, high efficiency and high power density can be achieved for BESS. In addition, the MMDAB converter can limit the dc fault current rising rate and has fast recovery capability. The simulation results have confirmed the superior performance of MMDAB converter for

BESS in MVDC systems. By adding another two arms to form a dc terminal in LVS, this MMDAB converter can be easily extended for other high-voltage dc-dc power conversion applications.

REFERENCES

[1] H. Akagi, "Classification, terminology, and application of the modular multilevel cascade converter (MMCC)," *IEEE Trans. Power Electron.*, vol. 26, no. 11, pp.3119-3130, 2011.

[2] M. Vasiladiotis and A. Rufer, "Analysis and control of modular multilevel converters with integrated battery energy storage," *IEEE Trans. Power Electron.*, vol. 30, no. 1, pp.163-175, Jan. 2015.

[3] M. Quraan, i. Yeo, P. Tricoli, "Design and Control of Modular Multilevel Converters for Battery Electric Vehicles," *IEEE Trans. Power Electron.*, vol. 31, no. 1, pp.507-517, Jan. 2016.

[4] T. Soong, P.W. Lehn, "Evaluation of Emerging Modular Multilevel Converters for BESS Applications," *IEEE Trans. Power Del.*, vol. 29, no. 5, pp. 2086-2094, Aug. 2014.

[5] D. Montesinos-Miracle, M. Massot-Campos, J. Bergas-Jane, S. Galceran-Arellano, and A. Rufer, "Design and control of a modular multilevel DC/DC converter for regenerative applications," *IEEE Trans. Power Electron.*, vol. 28, no. 8, pp. 3970-3979, Aug. 2013.

[6] J. A. Ferreira, "The multilevel modular DC converter," *IEEE Trans. Power Electron.*, vol. 28, no. 10, pp. 4460-4465, Oct. 2013.

[7] G.J. Kish, M. Ranjram, P.W. Lehn, "A Modular Multilevel DC/DC Converter With Fault Blocking Capability for HVDC Interconnects," *IEEE Trans. Power Electron.*, vol. 30, no. 1, pp. 148-162, Dec. 2013.

[8] C. Oates, "A methodology for developing 'Chainlink' converters," in *Proc. 13th Eur. Conf. Power Electron. Appl.*, 2009, pp.1-10.

[9] T. Luth, M. Merlin, T. Green, F. Hassan and C. Barker, "High frequency operation of a DC/AC/DC system for HVDC applications," *IEEE Trans. Power Electron.*, vol. 29, no. 8, pp.4107-4115, Aug. 2014.

[10] R. Xie, Y. Shi, and H. Li "Modular Multilevel DAB (M2DAB) Converter for Shipboard MVDC System with Fault Protection and Ride-Through Capability," in *Proc. Electric Ship Technology Symposium*, Jun. 2015, pp. 427-432.

[11] S. Kenzelmann, A. Rufer, M. Vasiladiotis, D. Dujic, F. Canales, and Y. R. de Novaes, "A versatile DC–DC converter for energy collection and distribution using the modular multilevel converter," in *Proc. 14th Eur. Conf. Power Electron. Appl.*, Aug. 30/Sep. 1, 2011, pp. 1-10.

[12] S. Kenzelmann, D. Dujic, F. Canales, Y. R. de Novaes, and A. Rufer, "Modular DC/DC converter: Comparison of modulation methods," in *Proc. Int. Power Electron. and Motion Control Conf.*, 2012, LS2a.1-1 - LS2a.1-7.

[13] I. Gowaid, G. Adam, A.M. Massoud, S. Ahmed, D. Holliday, and B. Williams, "Quasi Two-Level Operation of Modular Multilevel Converter for Use in a High-Power DC Transformer With DC Fault Isolation Capability," *IEEE Trans. Power Electron.*, vol. 30, no. 1, pp.108-123, Jan. 2015.

[14] I. Gowaid, G. Adam, S. Ahmed, D. Holliday, and B. Williams, "Analysis and Design of a Modular Multilevel Converter With Trapezoidal Modulation for Medium and High Voltage dc-dc Transformers," *IEEE Trans. Power Electron.*, vol. 30, no. 10, pp. 5439-5457, Oct. 2015.

[15] M. Bragard, N. Soltau, S. Thomas, and R. W. De Doncker, "The balance of renewable sources and user demands in grids: Power electronics for modular battery energy storage systems," *IEEE Trans. Power Electron.*, vol. 25, no. 12, pp. 3049–3056, Dec. 2010.

[16] R. Mo, R. Li, and H. Li "Isolated Modular Multilevel (IMM) DC/DC Converter with Energy Storage and Active Filter Function for Shipboard MVDC System Applications," in *Proc. Electric Ship Technology Symposium*, Jun. 2015, pp. 113-117.

[17] S. Kenzelmann, A. Rufer, M. Vasiladiotis, D. Dujic, F. Canales, and Y. R. de Novaes, "Isolated DC/DC Structure Based on Modular Multilevel Converter," *IEEE Trans. Power Electron.*, vol. 30, no. 1, pp.89-98, Jan. 2015.

[18] M. Michalczuk, B. Ufnalski, L. M. Grzesiak, P. Rumniak, "Power converter-based electrochemical battery emulator," *Electrical Review*, R. 90 NR 7/2014, pp. 18-22, Jul. 2014.

An Analytical Framework to Design A Dynamic Frequency Control Scheme for Microgrids Using Energy Storage

Ajit A. Renjit
Student Member, IEEE
The Ohio State University
Electrical and Computer Engineering
Columbus, OH 43210, USA
renjit.1@osu.edu

Feng Guo and Ratnesh Sharma
Member, IEEE
NEC Laboratories America, Inc.
Energy Management Department
Cupertino, CA 95014, USA
fguo@nec-labs.com

Abstract— **Microgrids with increased penetration of renewables have significant frequency excursions to power generation changes. A straight forward approach to solve this problem is by enhancing the inertia of the system using energy storage (ES). They provide Dynamic Frequency Control (DFC) support to the interconnected Distributed Energy Resources (DERs) in the microgrid. However, the amount of inertial support required from the ES varies based on the type of interconnected DERs, the amount of load change and many other factors. This paper proposes an analytical framework for calculating the inertia required from the ES systems during a generation change. A case study to corroborate the proposed framework for the DFC scheme has also been studied.**

Keywords— *DER, ES, DFC, Virtual Synchronous Generator (VSG), Distributed power generation, energy conversion, microgrids*

I. INTRODUCTION

Recently, the increasing demand for improved power quality and reliability along with the distributed nature of renewable energy resources has enabled distributed generation to be a desirable alternative to the traditional centralized generation. Since most of the distributed energy resources (DERs) have an inverter as their utility interface to the grid they bring down the system inertia significantly causing the grid to be more prone to frequency instability [1]-[3]. Moreover, the high proliferation of renewables further compromises the system due to its intermittent nature and also being operated as non-dispatchable resources [4]. Under such circumstances, an islanded microgrid with renewables presents one of the greatest technical challenges for frequency regulation as it is more sensitive to generation-load imbalances [2], [4]. Such disturbances on an islanded microgrid can result in sizeable frequency excursions for system disturbances, increasing the risk of under frequency load shedding (UFLS) and cascading outages.

In order to improve the reliability of islanded microgrids, it is essential to tackle the inertia issue. Several solutions have been proposed in the past, such as extra rotating masses added to power systems via the connection of generators dispatched in synchronous compensation mode [5]. The recent advancements in automotive industry have resulted in significant improvements in battery chemistry technologies [6]. Therefore, another option to enhance the reliability of islanded microgrids is by complementing the system with energy storage and by that adding inertia to the system virtually [7], [8]. In [9] the authors have proposed a virtual inertia based control algorithm for energy storage that provides Dynamic Frequency Control (DFC) support in the French island of Guadeloupe which has a significant penetration of renewables. Here, the control algorithm was designed in such a way that it deploys the energy storage completely at the instant when the rate of change of frequency crosses a threshold value during a large generation change.

Another popular technique to provide DFC is by enabling the energy storage to emulate the behavior of traditional synchronous generators, popularly known as Virtual Synchronous Generators (VSG) [10]-[13]. Unlike the traditional generators, the VSG has the advantage of modifying its inertia based on the type of disturbance and also on the frequency control system of every DER connected in the microgrid. However, one of the key challenges in developing such a scheme is in choosing the amount of inertia that needs to be programmed in its controller under different operating conditions. In [14]-[15], the idea of adoptive inertia in a VSG was proposed, which allows the selection of moment of inertia real-time to enhance the fast response of the VSG in tracking the steady-state frequency. A self-tuning VSG has been proposed in [16] which makes use of an optimization algorithm to minimize the frequency deviations (amplitude and rate of change) and the power flow through the energy storage. Although, these literatures provide promising results, the analytical framework behind the choice of the moment of inertia has not been reported so far. The objective of this work is to develop an analytical framework for the DFC scheme in energy storage systems that could be used to calculate the amount of virtual inertia to be programmed in the controller based on the type of disturbance and also on the frequency control system of the microgrid.

978-1-4673-9551-9/16 $31.00 © 2016 IEEE

II. DYNAMIC FREQUENCY CONTROL

A. Traditional Power Systems with Inertia

Fig. 1 shows the block diagram of a traditional steam turbine with its governor, prime-mover and rotating mass.

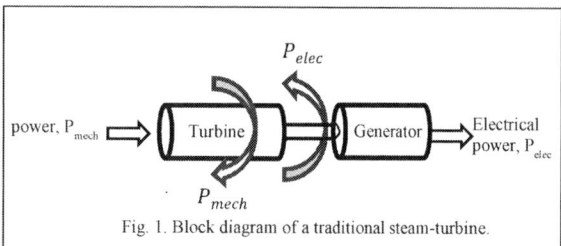

Fig. 1. Block diagram of a traditional steam-turbine.

During a large disturbance, an unbalance between the mechanical power input and the electrical load demand happens which needs to be corrected by 'frequency regulation'. There are several levels of frequency regulation techniques that operate hierarchically to perform the balancing act. It could be classified into 4 levels that are usually present in large interconnected systems: 1. Inertial response (initial few seconds) 2.Governor response (i.e. Primary Frequency Control – in the time frame of few seconds) 3. Automatic Generation Control (i.e. Secondary Frequency Control – in the time frame of few minutes) 4. Tertiary control

Among these techniques, the inertial response is uncontrollable and primarily influences the transient behavior of the generator immediately after the disturbance. It could be better explained using the electromechanical energy conversion dynamic equation in the turbine

$$\begin{pmatrix} Mech.\,Energy\,Input \\ from\,Engine \end{pmatrix} = \begin{pmatrix} Elec.\,Energy \\ Output \end{pmatrix} + \begin{pmatrix} Energy \\ Dissipated \end{pmatrix} + \begin{pmatrix} Increase\,in\,Kinetic \\ Energy, \Delta KE \end{pmatrix}$$

In terms of power quantities, this dynamic equation during transient conditions can be defined as

$$\frac{\Delta KE}{\Delta t} = P_{mech} - P_{elec} - P_{loss} \qquad (1)$$

or

$$J\omega \frac{\Delta\omega}{\Delta t} = P_{mech} - P_{elec} - P_{loss} \qquad (2)$$

$$\frac{\Delta\omega}{\Delta t} = \frac{P_{mech} - P_{elec} - P_{loss}}{J\omega} \qquad (3)$$

where P_{mech} is the mechanical input power, P_{elec} is the electrical output power, and P_{loss} accounts for the power dissipated as losses, J represents the moment of inertia and ω represents the angular frequency of the generator. The stored kinetic energy ($J\omega \frac{\Delta\omega}{\Delta t}$) in the system compensates for the difference between the electrical and mechanical power in (2). The amount of change in speed, $\Delta\omega$ and the rate at which the speed changes, $\frac{\Delta\omega}{\Delta t}$ are inversely proportional to the inertia, J of the prime-mover. From (3) it could be understood that the inertial response plays a major role in the initial frequency decline/increase after a disturbance. Along with the governor response, it dictates the maximum frequency deviation, $\Delta\omega_{min}$.

B. Microgrids with Virtual Inertia

With the increased penetration of renewables in the power system, the system inertia has been steadily declining. This causes large frequency transients, which makes the power system more prone to instability during disturbances. Islanded microgrids are especially more sensitive to these disturbances as they have small-sized DERs with low inertia. The addition of renewables in islanded systems further worsens the case.

In order to improve the reliability of microgrids where the main grid supply is unavailable, the DERs are supplemented with Energy Storage (ES). Nowadays, ES can be operated in such a way that it satisfies both short- and long-term energy requirements in microgrid applications [17]. It can be interfaced to the microgrid bus using a power electronics module comprising both dc/dc converter and dc/ac inverter [18], [19]. In this paper, the energy storage units are used to provide the necessary Dynamic frequency Control (DFC) support in an islanded microgrid with non-inertial renewable energy resources.

The idea of DFC is to control the energy storage to emulate the behavior of traditional synchronous generators thereby providing inertial support to the microgrid. Fig. 2 shows the control architecture of the DFC scheme. It should be noted that the well-known swing equation of traditional generators is used to develop DFC. As presented in (4) & (5) the ES either injects or absorbs power after a disturbance based on the amount of inertia programmed in its controller.

$$P_k^* - P_k = M_k \left(\frac{d\omega_k^*}{dt} \right) + D_k (\omega_k^* - w_{nom}) \qquad (4)$$

$$P_k^* - P_k = J_k \omega_k^* \left(\frac{d\omega_k^*}{dt} \right) + D_k (\omega_k^* - w_{nom}) \qquad (5)$$

This is exactly similar to the traditional generators with stored kinetic energy in its rotating shaft. However, one notable feature of DFC is that the amount of inertia is controllable unlike the traditional rotating generators. This gives us the additional degree of freedom to control the amount of frequency deviation, $\Delta\omega_k$ based on the severity of the disturbance and the inertial support provided by the other interconnected DERs.

Fig. 3 shows the control block diagram of the ES-based DFC that has been implemented in the outer loop of the inverter controls. One of the main challenges in implementing the DFC in the ES is to calculate the most appropriate value of angular momentum, 'M_k' to be incorporated in the controller to maintain the system frequency within acceptable limits. The next section presents a detailed analysis to calculate 'M_k' for the islanded microgrid.

III. DFC SETTINGS CALCULATION USING REDUCED ORDER AND STATE SPACE MODELS

A. Control Strategy

For a rotating generator in the standalone mode of operation, the amount of frequency deviation during a disturbance could be explained using (3). The generator's shaft angular momentum, M_k, the amount of load change and the governor response are the key reasons that determine the

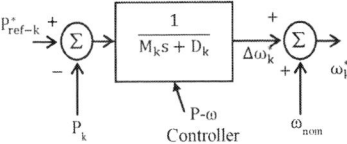

Fig. 2. Block diagram of the DFC scheme.

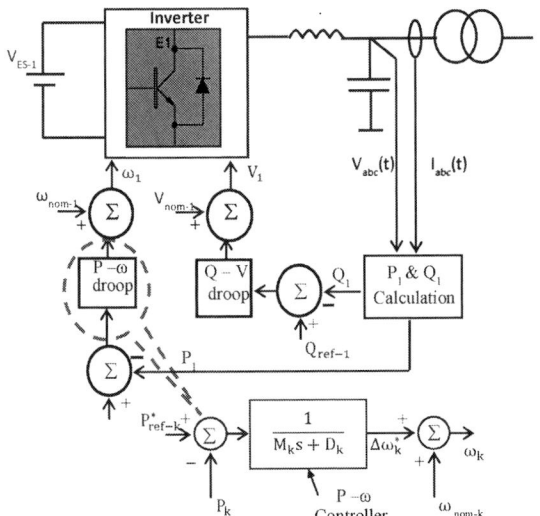

Fig. 3. Block diagram of the DFC scheme.

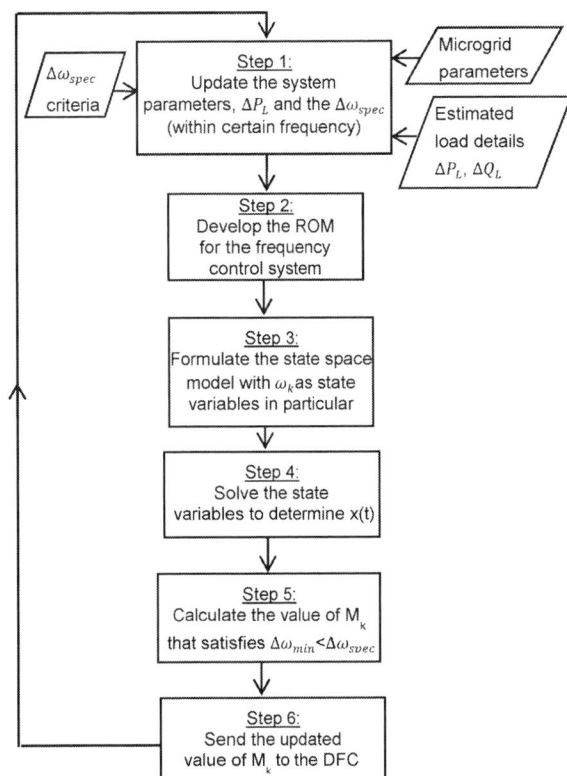

Fig. 4. Flowchart representing the control strategy. Mechanical

steady-state power sharing between the diesel-generator and the ES is determined by the parameter D_k.

B. System Description

In this paper, the system chosen to study the effects of the DFC during a power generation change is a PV-diesel-ES hybrid power system shown in Fig. 5. The system comprises of a PV generator equipped with a Maximum Power Point Tracking (MPPT) control [20], a diesel-generator set, an ES and AC load. Besides, it is assumed that the microgrid is always disconnected from the main grid and independently operated as a stand-alone system. Further information about the DERs and their ratings are mentioned in Table I.

Predominantly, the renewable energy-based DERs like PV and wind operate as non-dispatchable sources using a grid-following converter [21]. In such cases, these converters cannot be operated in the islanded mode without a grid-forming converter in parallel. When operating in its maximum power as a non-dispatchable source, the grid-following DERs use an MPPT controller in its outer loop to regulate its real power output. Therefore, the PV generator (DER-3) is modeled as a conventional grid-following current sourced inverter. It always operates at its maximum power at unity power factor by injecting a prescribed current in phase with the 3-phase voltage vector. This schematic of the model developed in MATLAB®/Simulink™ is shown in Fig. 6.

amount of frequency deviation from the nominal. However, in a microgrid the scenario is completely different. Microgrids comprises of different types of DERs (DERs with inverter as their utility interface against DERs that have synchronous generator as their utility interface) with different frequency control schemes to regulate frequency. Moreover, microgrids may also feature dispatchable and non-dispatchable DERs (renewables that operate at their MPPT) which also influence the system frequency in a different way.

In this paper, a novel control strategy to calculate the controller parameters of the DFC is proposed as shown in Fig.4. The idea is to use the different parameters of the microgrid to develop a Reduced Order Model (ROM) of the frequency control system. Using this model, the state space equations of the system are derived with the frequency of the DERs, $\Delta\omega_1, \Delta\omega_2$ as mandatory state variables. The dynamic equations of the state variables are then used to determine the value of angular momentum, M_k that reduces the maximum frequency deviation, $\Delta\omega_{min}$ within the acceptable limits specified, $\Delta\omega_{spec}$.

The energy storage unit is programmed to be a grid-supporting unit by enabling real and reactive power droop controllers in its outer loop. The amount of droop required for

978-1-4673-9551-9/16 $31.00 © 2016 IEEE

Sl. No	Type of DER	% of the total rating	Actual rating (kW)
1	Synchronous generator	62.5%	200kW
2	PV generator	25%	80kW
3	Energy storage	12.5%	40kW

TABLE I.   SYSTEM CONFIGURATION

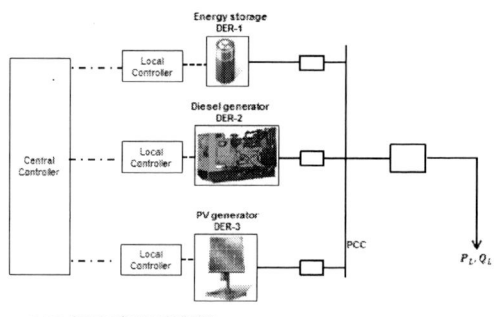

Fig. 5. System chosen for studying the effects of DFC.

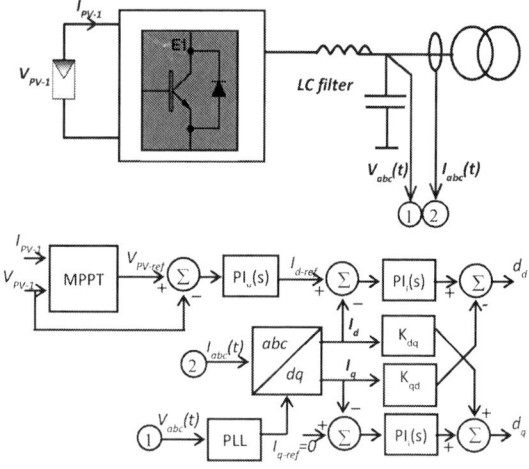

Fig. 6. Schematic of the grid-following PV generator with the MPPTcontrols.

The diesel generator (DER-2) used in the system is modeled as a voltage source on the lines of the author's earlier work in [22], [23] is shown in Fig. 7. This ensures that the PV generator can operate in parallel with other DERs as it requires a grid-forming converter or a synchronous generator to set the voltage amplitude and frequency in the islanded microgrid. It is understood from earlier literature that the diesel generator model has a number of subsystems to represent its physical behavior. Typically, it has a diesel engine coupled to the synchronous generator through a mechanical shaft. The engine

torque is controlled automatically using a speed governor for changes in the genset speed. The synchronous generator's field circuit is self-excited through a brushless exciter. A digital voltage regulator is used to control the excitation system which in turn regulates the terminal voltage. In order to enable the DERs to be grid-supporting, real and reactive power droop controllers are programmed in its outer loop. The schematic of the detailed model developed MATLAB®/Simulink™ is shown in Fig. 7.

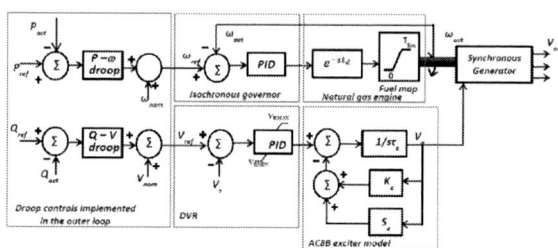

Fig. 7. Block diagram of the diesel generator model.

The AC microgrid frequency and voltage are determined by the steady state real and reactive power shared by the two grid-supporting units i.e. diesel generator and the energy storage. The grid-following PV generator follows the frequency and magnitude of the voltage vector set by the grid-supporting DERs in the microgrid. A synchronous reference frame phase-locked loop (SRF-PLL) is used for this purpose which synchronizes the *d-axis* with the grid voltage vector as shown in Fig. 6.

C. Reduced Order and State Space Models

A single line diagram of the system used to study the effect of DFC is shown in Fig. 8. As it is difficult to calculate the frequency deviation of this system during a disturbance, a simplified reduced order model representative of the interconnected DERs in Fig. 8 has been presented in Fig. 9. Since the PV generator is modeled as a constant power generator which could track the system frequency at any instant using its PLL, it could be represented as a negative load as shown in Fig. 8 & 9.

In the energy storage, DER-1 branch, P_{ref-1} is the reference power to the energy storage system. ΔP_L denotes the amount of load change that causes frequency disturbance in the system. P_{PV} is the PV output power. $P_{tie}(s)$ is the power flowing in the tie line between the DERs. The DFC generates the frequency compensation signal, which gets added to the nominal frequency, ω_{nom-1} to generate the actual frequency signal ω_1.

In the diesel generator, DER-2 branch, P_{ref-2} acts as the reference power to the droop controller of the diesel generator. $1/D_2$ is the droop gain which generates the frequency compensation signal that gets added to the nominal frequency of the diesel generator, ω_{nom-2} to produce the reference speed to the PID governor. The difference between this input and the actual speed of the diesel generator, ω_2 is the frequency error ω_{e2}. The output of the PI controller, T_{pm2} is the reference torque which is the input to the engine delay block. The engine delay

978-1-4673-9551-9/16 $31.00 © 2016 IEEE

block then generates the actual mechanical torque of the diesel generator T_{m2}. $T_{tie}(s)$ is the equivalent electrical torque due to the tie line power flow. The mechanical coupling block is then used to model the actual inertia of the synchronous generator coupled to the diesel engine.

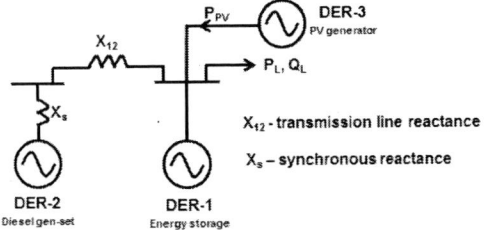

Fig. 8. Single line diagram of the AC microgrid considered for simulation.

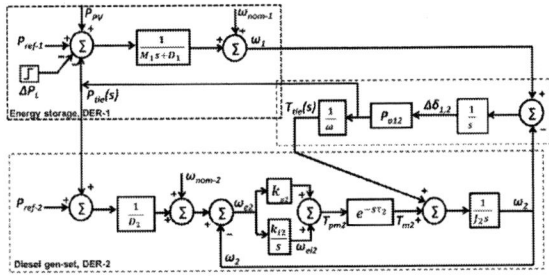

Fig. 9. Block diagram of the Reduced Order Model (ROM) used to derive the state space equations.

In the tie-line power calculation loop, the difference between the frequency of the DERs, ω_1 and ω_2 is sent to a integrator, which provides the phase angle difference, $\Delta\delta_{1,2}$ between the DERs. By multiplying the phase angle difference with the parameter, $P_{o1,2}$, the tie line power $P_{tie}(s)$ can be calculated. $T_{tie}(s)$ can be obtained by multiplying $P_{tie}(s)$ with the reciprocal of the system frequency.

$$P_{o1,2} = \frac{V_1 * V_2}{X_{tie}}$$
$$X_{tie} = X_{12} + X_s$$

During transient power sharing $X_s = X'_d$. X'_d is the generator's d-axis transient reactance. The system differential equations of Fig. 9 are

$$\Delta\delta'_{21} = \Delta\omega_1 - \Delta\omega_2 \qquad (6)$$

$$\Delta\omega'_1 = \frac{-(\Delta P_L + \Delta\delta_{21} * P_{o21} + D_1 * \Delta\omega_1)}{M_1} \qquad (7)$$

$$\Delta\omega'_2 = \frac{\Delta T_{m2} - \Delta T_{e2}}{J_2} \qquad (8)$$

$$\Delta T'_{m2} = \frac{\Delta T_{pm2} - \Delta T_{m2}}{\tau_2} \qquad (9)$$

$$\Delta\omega'_{ei2} = \Delta\omega_{e2} * -k_{i2} \qquad (10)$$

Representing the system differential equations in the form
$$\dot{x}(t) = A\,x(t) + B\,u(t);$$

$$y(t) = C\,x(t) + D\,u(t);\ x(0) = x_0;$$

$$x(t) = [\Delta\delta_{12}\ \ \Delta\omega_1\ \ \Delta\omega_2\ \ \Delta T_{m2}\ \ \Delta\omega_{e2}];\ u(t) = [\Delta P_L];\ (11)$$

$$A = \begin{bmatrix} 0 & 1 & -1 & 0 & 0 \\ -\frac{P_{o12}}{M_1} & -\frac{D_1}{M_1} & 0 & 0 & 0 \\ \frac{P_{o12}}{\omega*J_2} & 0 & 0 & \frac{1}{J_2} & 0 \\ \frac{k_{p2}*P_{o12}}{D_2*\tau_2} & 0 & -\frac{k_{p2}}{\tau_2} & -\frac{1}{\tau_2} & \frac{1}{\tau_2} \\ \frac{k_{i2}*P_{o12}}{D_2} & 0 & -k_{i2} & 0 & 0 \end{bmatrix};$$

$$B = \begin{bmatrix} 0 \\ -\frac{1}{M_1} \\ 0 \\ 0 \\ 0 \end{bmatrix};$$

$$C = \begin{bmatrix} 1 & 0 & 0 & 0 & 0 \\ 0 & 1 & 0 & 0 & 0 \\ 0 & 0 & 1 & 0 & 0 \\ 0 & 0 & 0 & 1 & 0 \\ 0 & 0 & 0 & 0 & 1 \end{bmatrix};\ D = \begin{bmatrix} 0 \\ 0 \\ 0 \\ 0 \\ 0 \end{bmatrix}.$$

IV. SYSTEM SIMULATION FOR TRANSIENT RESPONSE

A. Reduced Order Model Validation

Time-domain simulations were carried out in Matlab®/Simulink™ to validate the performance of the reduced order model against the actual simulink model. Although there are slight discrepancies in the results, the maximum frequency deviation is closely matching using both the models. Table II lists the amount of frequency deviation in both the reduced order and the actual simulink model when the system is subjected to different load changes. This confirms that the reduced order model can be used to calculate the minimum frequency deviation during a disturbance. It further helps to calculate the appropriate value of 'M_k' that has to be programmed in the DFC scheme to limit the frequency deviation within acceptable limits.

TABLE II.   PERFORMANCE ANALYSIS OF THE PROPOSED FRAMEWORK FOR DIFFERENT TEST CASES

Sl. No	ΔP_L(kW)	M_1 (kg.m²/sec)	$\Delta\omega_1, \Delta\omega_2$ (Simulink model)	$\Delta\omega_1, \Delta\omega_2$ (Reduced order model)
1	50	5	2.3, 2.0	2.3, 1.95
2	100	5	4.5, 4.1	4.4, 4.0

978-1-4673-9551-9/16 $31.00 © 2016 IEEE

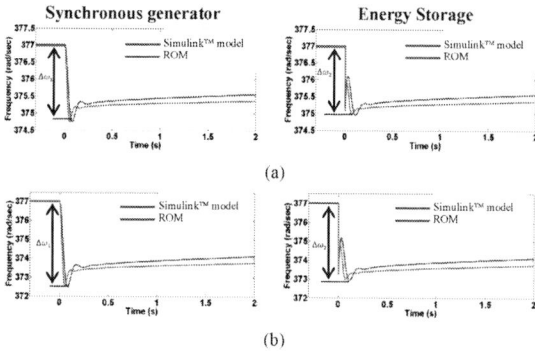

Synchronous generator   **Energy Storage**

(a)

(b)

Fig. 10. Response of the Simulink™ and Reduced order Model (ROM) of the DERs for (a) 50kW and (b) 100kW step load transients.

B. Case Study

The performance of the proposed control strategy to determine the DFC parameters is explained using a case study in this section. The system used for this study is shown in Fig. 5 and the parameters of the system are listed in Table. III. The

dynamic behavior of the microgrid for a large load change without the DFC implemented in the ES is presented in the left column of Fig. 11. It is understandable that the system has to shed a significant amount of the connected load as the grid frequency has exceeded the under-frequency relay setpoint.

TABLE III.   TEST CONDITIONS OF THE MICROGRID

Sl. No	Parameter	Value
1	ω/P droop gain – DER-1 & DER-3	2*pi (rad/sec/p.u)
2	V/Q droop gain- DER-1 & DER-3	0.1 (V/p.u)
3	DER-2 maximum power @ 25°C and 1000 W/m2	40kW
5	Load change, ΔP_L	140-280kW
6	Under-frequency relay setpoint	±2Hz/ ±12.57rad/sec
7	Calculated angular momentum, M_1 to be programmed in the DFC	495 J.s/rad.^{2}

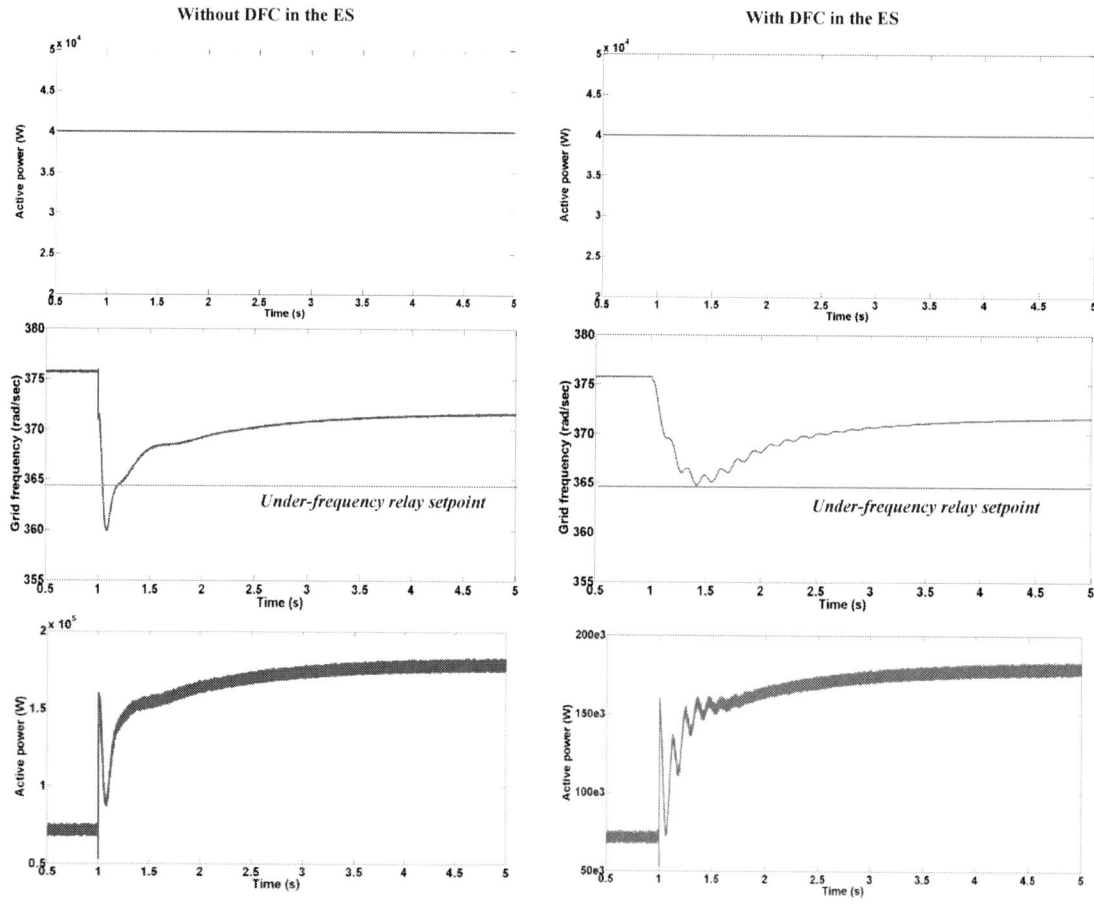

Without DFC in the ES   **With DFC in the ES**

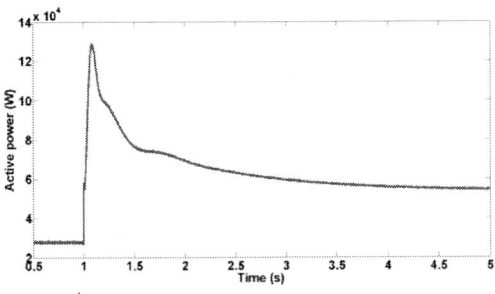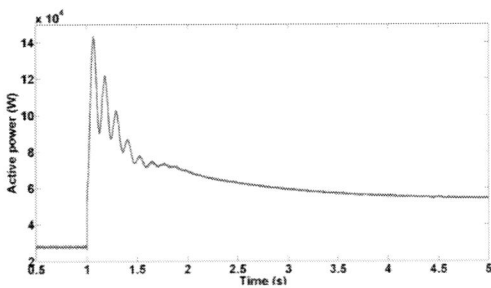

Fig. 11. Response of the DERs with and without the DFC scheme in the ES.

Now the same parameters of the microgrid and the amount of load change, ΔP_L are substituted in the solved frequency ($\Delta\omega_1$ and $\Delta\omega_2$) equations (11) obtained from the state-space model. The appropriate value of M_1 that reduces the frequency deviation to stay within the limits set by the under-frequency relay is then programmed in the DFC. The results obtained for the same load change is presented in the right column of Fig. 11. The DFC has enabled the ES to increase its dynamic power share, thereby lowering the grid frequency deviation from its nominal value.

V. CONCLUSION

The power system must maintain a balance between the amount of power generated and consumed. Disturbances in the grid could cause unbalances which are generally corrected using different frequency regulation techniques. DFC is a recent technique used in energy storage systems to provide frequency regulation especially when they are operated in a weak grid with poor inertial support. The amount of inertia required to provide frequency regulation varies based on the kind of interconnected DERs and also the amount of load change. A control strategy to effectively implement DFC in ES to maintain the frequency deviation within acceptable limits has been proposed in this paper. The idea of ROMs to depict the frequency control system of a microgrid and to calculate its frequency deviations for load changes has been validated. Finally, a case study to corroborate the proposed control strategy for the DFC scheme has been presented.

REFERENCES

[1] A. Azmy and I. Erlich, "Impact of distributed generation on the stability of electrical power system," in *Proc. IEEE Power Eng. Soc. Gen. Meeting*, 2005, pp. 1056–1063.

[2] F. Katiraei, M. R. Iravani, and P. W. Lehn, "Micro-grid autonomous operation during and subsequent to islanding process," *IEEE Trans. Power Del.*, vol. 20, no. 1, pp. 248–257, Jan. 2005.

[3] L. Meegahapola and D. Flynn, "Impact on transient and frequency stability for a power system at very high wind penetration," in *Proc. IEEE Power Eng. Soc. Gen. Meeting*, 2010, pp. 1–8.

[4] A. Mullane, G. Bryans, and M. O'Malley, "Kinetic energy and frequency response comparison for renewable generation systems," in *Proc. Int. Conf. Future Power Systems*, Nov. 16-18, 2005, p. 6

[5] J. Bomer, K. Burges, C. Nabe, and M. Poller, All Island TSO Faciliation of Renewables Study—Final Report for Work Package 3 Ecofys, Germany, Tech. Rep. PEGEDE083532, Jun. 2010.

[6] S. Vazquez, S. Lukic, E. Galvan, L. Franquelo, and J. Carrasco, "Energy storage systems for transport and grid applications," *IEEE Trans. Ind. Electron.*, vol. 57, no.12, pp. 3881-3895, Dec. 2010.

[7] M. P. N. vanWesenbeeck, S.W.H. deHann, P.Varela, and K.Visscher, "Grid tied converter with virtual kinetic storage," in *Proc. IEEE Power Tech Conf.*, Bucharest, Romania, Jul. 2009.

[8] J. Driesen and K. Visscher, "Virtual synchronous generators," in *Proc. IEEE PES General Meeting*, Pittsburgh, PA, Jul. 2008.

[9] G. Delille, B. Francois, and G. Malarange, "Dynamic frequency control support by energy storage to reduce the impact of wind and solar generation on isolated power system's inertia," *IEEE Trans. Sustain. Energy*, vol. 3, no. 4, pp. 931–939, Oct. 2012.

[10] K. Visscher and S. W. H. De Haan, "Virtual synchronous machines (VSG's) for frequency stabilisation in future grids with a significant share of decentralized generation," in *Proc. IET-CIRED Seminar Smart-Grids Distrib.*, 2008, pp. 1–4.

[11] H. P. Beck and R. Hesse, "Virtual synchronous machine," in *Proc. IEEE EPQU Conf.*, 2007, pp. 1–6.

[12] Q.-C. Zhong and G.Weiss, "Synchronverters: Inverters thatmimic synchronous generators," *IEEE Trans. Ind. Electron.*, vol. 58, no. 4, pp.1259–1267, Apr. 2011.

[13] M. Torres and L. A. C. Lopes, "Virtual synchronous generator control in autonomous wind-diesel power systems," in *Proc. IEEE-EPEC Conf.*, 2009, pp. 1–6.

[14] T. Shintai, Y. Miura and T. Ise "Oscillation Damping of a Distributed Generator Using a Virtual Synchronous Generator," *IEEE Trans. Power Delivery*, vol. 29, no. 2, pp. 668-676, Apr. 2014.

[15] J. Alipoor, Y. Miura and T. Ise, "Power System Stabilization Using Virtual Synchronous Generator With Alternating Moment of Inertia," *IEEE Journal of Emerging and Selected Topics in Power Electronics*, vol. 3, no. 2, pp. 451-458, June 2015.

[16] M. Torres, L.A.C Lopes, and L.A Moran, and J.R Espinoza, "Self-Tuning Virtual Synchronous Machine: A Control Strategy for Energy Storage Systems to Support Dynamic Frequency Control," *IEEE Trans. Energy Conversion*, vol. 29, no. 4, pp. 833-840, Dec. 2014.

[17] Q. Fu, A. Hamidi, A. Nasiri, V. Bhavaraju, S. B. Krstic and P. Theisen, "The role of energy storage in a microgrid concept: Examining the opportunities and promise of microgrids," *IEEE Electrification Magazine*, vol. 1, no. 2, pp. 21--29, Dec 2013.

[18] B. Kroposki, C. Pink, R. DeBlasio, H. Thomas, M. Simões, and P. K. Sen, "Benefits of Power Electronic Interfaces for Distributed Energy Systems," *IEEE Trans. Energy Conversion*, vol. 25, no. 3, pp. 901-908, Sept. 2010.

[19] E. Alegria, T. Brown, E. Minear, and R. Lasseter, "CERTS Microgrid Demonstration With Large-Scale Energy Storage and Renewable Generation," *IEEE Trans. Smart Grid*, vol. 5, no. 2, pp. 937-943, Mar. 2014.

[20] E. Figueres, G. Garcera, J. Sandia, F. Gonzalez-Espin and J.C. Rubio, "Sensitivity Study of the Dynamics of Three-Phase Photovoltaic Inverters With an LCL Grid Filter," *IEEE Trans. Industrial Electronics*, vol. 56, no. 3, pp. 706-717, March 2009.

[21] J. Rocabert, A. Luna, F. Blaabjerg, and P. Rodriguez, "Control of power converters in AC microgrids," *IEEE Trans. on Power Electronics*, vol. 27, pp. 4734-4749, 2012.

[22] A. A. Renjit, M. S. Illindala and D. A. Klapp, "Graphical and Analytical Methods for Stalling Analysis of Engine Generator Sets," *IEEE Trans. Industry Applications*, vol. 50, no. 5, pp. 2967-2975, Sept.-Oct.2014.

[23] A. A. Renjit, M. S. Illindala, R. H. Lasseter, M. J. Erickson and D. Klapp, "Modeling and control of a natural gas generator set in the CERTS microgrid," *IEEE Energy Conversion Congress and Exposition (ECCE)*, pp. 1640-1646, Sept. 2013.

Comparative Evaluation of LiFePO$_4$ Cell SOC Estimation Performance with ECM Structure and Noise Model/Data Rejection in the EKF for Transportation Application

Hyun-jun Lee
Dpt. Electrical Engineering
University of Soongsil
Seoul, Korea
vowgood@naver.com

Joung-hu Park
Dpt. Electrical Engineering
University of Soongsil
Seoul, Korea
wait4u@ssu.ac.kr

Jounghoon Kim
Dpt. Electrical Engineering
University of chosun
Gwangju, Korea
qwzxas@hanmail.net

Abstract—This study presents a comparison of several methods to improve state-of-charge (SOC) estimation performance using the extended Kalman Filter (EKF) algorithm for a commercial a lithium iron phosphate (LiFePO$_4$) cell. Firstly, this work attempts to show the comparison of SOC performance according to the number of RC-ladder. Secondly, this work shows a comparison of SOC estimation with and without minor loop to overcome the difference between charging open-circuit voltage (OCV) and discharging OCV. The SOC performance with and without noise model and data rejection in the EKF is finally compared.

Keywords—LiFePO$_4$; SOC; EKF; Minor loop; Noise model

I.   INTRODUCTION

As the interest in EV (electric vehicle) and ESS (energy storage system) is becomes higher, researches on battery have been actively conducted as the key component of the system have also been actively conducted. Because of a higher energy density and convenience of high voltage application development, the requirement especially for a lithium secondary battery cell has been abruptly increased [1]-[11]. In particular, a lithium iron phosphate (LiFePO$_4$) cell can have an ability of increased the charge and discharge frequency two times over in comparison with that of the existing lithium secondary battery. Furthermore, in terms of safety, this LiFePO$_4$ battery is superior of the existing lithium second battery that has the problem of the explosion and ignition. While LiFePO$_4$ battery has many advantages, it is very expensive. Price of battery is a half of transportation application. In order to use this battery safety and long, BMS (battery management system) is necessarily required. Then, SOC (state of charge) is key factor of the BMS. The biggest factor in the aging of the battery is over-charge and discharge. So, knowing the exact SOC of the battery is very important to prevent the over-charge and discharge.

There are several SOC estimation methods. The most basic method of the SOC estimation is ampere-hour counting [12] using a definition of the SOC. But, because of the simple integral algorithm, it is not possible to overcome the initial value of the error and the accumulated error according to the

internal state changes. Therefore, it is difficult to use in a application requiring high accuracy such as EV and HEV. The neural network [13] is a method to find a function related to input and output on the basis of the huge quantities of data. It does not reflect the actual physical model of the battery in terms of accuracy less than other methods. In particular electric-powered transportation, if the input and output waveforms of the electric current changed rapidly, accuracy of the SOC estimation is reduced.

EKF (extended kalman filter) [14] is a method based on battery ECM (equivalent circuit model). If the correct ECM based, EKF method has a high accuracy and great strength for the error correction of the internal state. But, because of the non-linear characteristics of the battery, accurate modeling is difficult. So, the model error caused unavoidable, it should be compensated by additional techniques such as noise model and data rejection techniques.

Modeling the LiFePO$_4$ cell is difficult due to it has a hysteresis characteristics. To overcome this weakness, this paper introduces several methods to improve SOC estimation based on the EKF. Firstly, this work clearly shows an improvement of SOC performance according to the number of increased RC-ladder in ECM, namely parameterization. Secondly, this work designs several minor loops to overcome the differences between charging OCV and discharging OCV curve. Finally, this work introduces some noise models and data rejection methods to the SOC estimation to reduce the modeling error caused by high current and RC-ladder's dynamic in the EKF. Consequently, this proposed work makes an effort to provide an improved SOC estimation of a LiFePO$_4$ cell.

II.   EQUIVALENT CIRCUIT MODEL (ECM)

Fig. 1 is a charging and discharging OCV curve of the 14 Ah pouch type LiFePO$_4$ cell, namely major loop. The charging OCV is higher than that of the discharging curve caused by the hysteresis effect. So, the equivalent circuit model (ECM) of LiFePO$_4$ cell is divided into the charging and discharging part, as shown in Fig. 2.

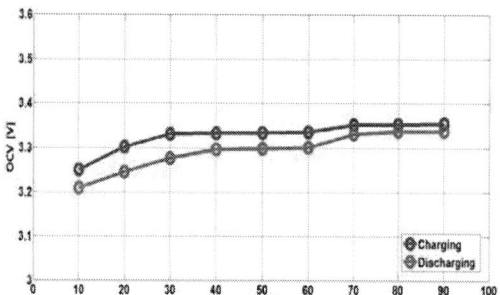

Fig. 1. Charging/Discharging OCV of the LiFePO₄

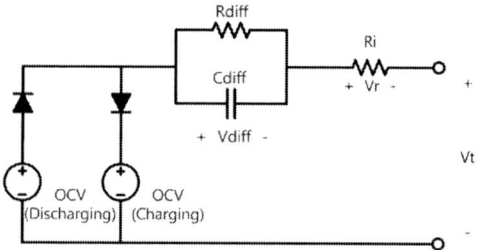

Fig. 2. ECM of the LiFePO₄ cell with one RC-ladder

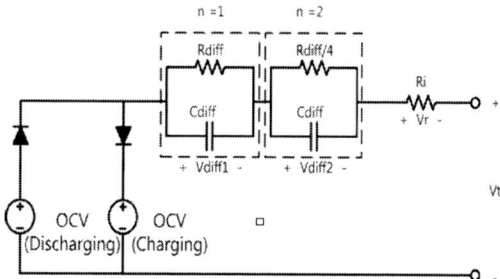

Fig. 3. ECM of the LiFePO₄ cell with two RC-ladder

The diffusion region is expressed as an infinite number of RC-ladder, the parallel connection of R_{diff} and C_{diff}. The resistance R_{diff} becomes small, inversely proportional to the n^2, where n is number of the RC-ladder. According to the decreases in R_{diff}, effect of the capacitance C_{diff} is decreased, caused by the effect of the time constant in RC-ladder. Therefore, the RC-ladder can be simplified, which does not significantly affect the accuracy of the SOC estimation. For the proposed works, ECM consists of single and double RC-ladder, as shown in Fig. 2 and 3.

III. MINOR HYSTERESIS LOOP OF OCV

$$OCV_n(SOC_n) = OCV_{n-1} + \int_{SOC_{n-1}}^{SOC_n} e^{a_1(SOC_n - t)} \cdot u_1 dt \quad (1)$$

$$OCV_n(SOC_n) = OCV_{n-1} - \int_{SOC_{n-1}}^{SOC_n} e^{a_1(SOC_n - t)} \cdot u_1 dt \quad (2)$$

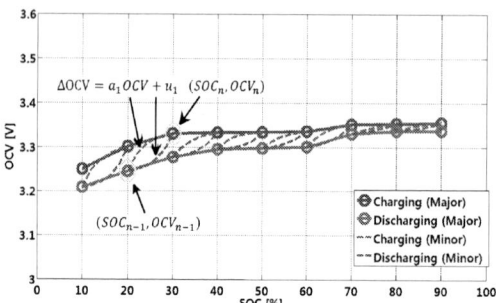

Fig. 4. Proposed major loop and minor loop of the LiFePO₄ cell

Charging and discharging OCV difference is very large model error in the EKF algorithm that SOC estimation based on the model. In particular, charging and discharging OCV often iterate in short period, as the battery cell repeat the charging and discharging actions in a small SOC range in transportation application, such as HEV(Hybrid Electric Vehicle) and EV. And, there is a charging and discharging OCV difference caused by a hysteresis characteristic in a small SOC range. To overcome the OCV difference of the LiFePO₄ cell between charging and discharging, the minor loop of the OCV trajectory can be additionally applied to the fundamental major loop [15]. Thus, this paper proposes to apply the multiple minor hysteresis loops by SOC 10% interval, as illustrated in Fig. 4. Design equations are the first-order differential equation form representing the exponential function. Equation (1) and (2) are minor loop of the charging and discharging direction. The value of a_1 and u_1 are selected to represent the best accuracy through the multiple experiments.

IV. EXTENDED KALMAN FILTER (EKF)

$$\begin{bmatrix} SOC_k \\ V_{Diff,k} \end{bmatrix} = \begin{bmatrix} 1 & 0 \\ 0 & 1 - \frac{\Delta t}{R_{Diff}C_{Diff}} \end{bmatrix} \begin{bmatrix} SOC_{k-1} \\ V_{Diff,k-1} \end{bmatrix} + \begin{bmatrix} -\frac{\Delta t}{C_n} \\ \frac{\Delta t}{C_{Diff}} \end{bmatrix} i_{k-1} \quad (3)$$

$$\begin{bmatrix} SOC_k \\ V_{Diff1,k} \\ V_{Diff2,k} \end{bmatrix} =$$

$$\begin{bmatrix} 1 & 0 & 0 \\ 0 & 1 - \frac{\Delta t}{R_{Diff}C_{diff}} & 0 \\ 0 & 0 & 1 - \frac{4\Delta t}{R_{Diff}C_{diff}} \end{bmatrix} \begin{bmatrix} SOC_{k-1} \\ V_{Diff1,k-1} \\ V_{Diff2,k-1} \end{bmatrix} + \begin{bmatrix} -\frac{\Delta t}{C_n} \\ \frac{2\Delta t}{C_{Diff}} \\ \frac{2\Delta t}{C_{Diff}} \end{bmatrix} i_{k-1} \quad (4)$$

$$V_k = h_k(OCV, V_{Diff}) - R_i i_k = OCV - V_{Diff} - R_i i_k \quad (5)$$

$$V_k = h_k(OCV, V_{Diff1}, V_{Diff2}) - R_i i_k = OCV - V_{Diff1} - V_{Diff2} - R_i i_k \quad (6)$$

$$If \; R_k = 0, K_k = P_k H_k^T [H_k P_k H_k^T + R_k]^{-1} = H_k^{-1} \quad (7)$$

$$If \; R_k = \infty, K_k = P_k H_k^T [H_k P_k H_k^T + R_k]^{-1} = 0 \quad (8)$$

978-1-4673-9551-9/16 $31.00 © 2016 IEEE

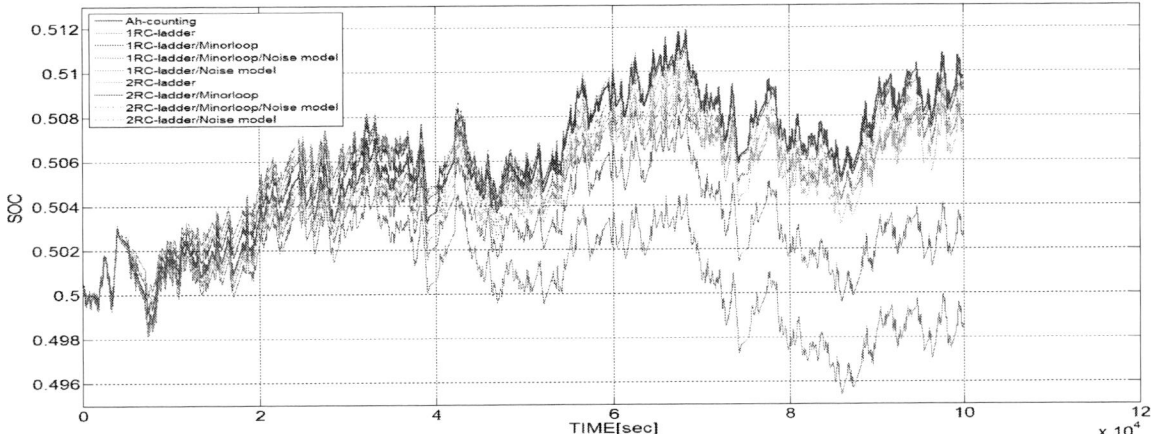

Fig. 5. Comparison of SOC estimation result

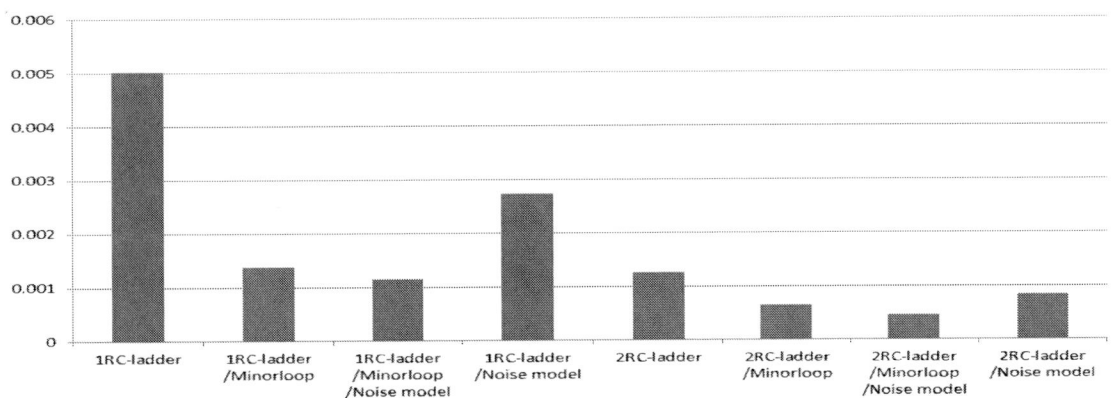

Fig. 6. Comparison of average error

The EKF algorithm is divided by state equation and measurement equation. State equation consists of ampere-hour counting equation and equation of the capacity of the RC-ladder, as shown in (3) and (4). Equation (3) and (4) are valid when the number of RC-ladders is one and two. Measurement equation indicates the terminal voltage of battery based on ECM. Equation (5) and (6) present the measurement equations, the same as the state equation. Kalman gain K_k determines how to use the measurement equation. And, K_k is determined by measurement error variance R_k, as shown in (7) and (8). R_k presents the accuracy of the measurement equation. If the value of R_k is 0, the internal state is completely determined by measurement equation. On the other hand, if the value of R_k is very large, the internal state is determined by the state equation. P_k and H_k indicate the error variance.

V. NOISE MODEL AND DATA REJECTION

TABLE I. NOISE MODEL AND DATA REJECTION IN THE EKF

Measurement noise model by battery current
$R_{k-1} = R_k$, reliable current ($|i| < 8A$)
$R_{k+1} = R_k[1+G_i(|i|-8A)]$, unreliable current ($|i| > 8A$)
$G_i = 4A^{-1}$
Measurement noise model by dynamic of RC-ladder
$R_{k+1} = R_k[1+G_{step}(step\_time)]$
$G_{step} = 0.1sec^{-1}$
Data rejection technique
$R_k = \infty$, reject time ($\Delta I > 10A$)
reject time = 100ms

EKF algorithm has a more complex calculation structure than the other algorithms. In particular, if the EKF algorithm

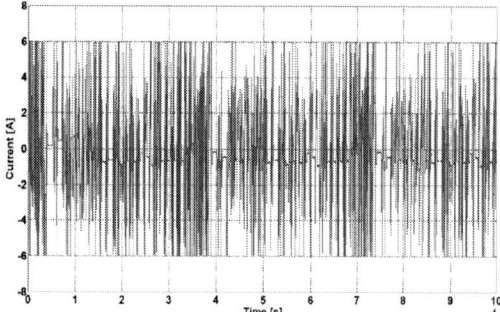

Fig. 7. Scale-down charging and discharging current profile for HEV

has a large number of internal states, calculation time becomes longer. So, EKF algorithm uses a simplified ECM structure. And, measurement model error occurs in the ECM structure itself inevitably. Computation time and model error has a trade-off relationship. Additional technique is required to compensate for model error, such as noise model and data rejection technique. In case the error which is big in the measurement equation happens, it is the general idea of data rejection not to use for the internal state estimation. And it has the measurement noise model to express this measurement noise as the specific model which the constant is not. There are several noise models in the battery model that can be measured at the battery ECM, such as extreme SOC, high current, high temperature and dynamic of RC-ladder. In this study, we used a measurement noise model in variable high current as well as dynamic of RC-ladder. And, data rejection techniques were used according to the amount of current change, as shown in Table.I. The measurement noise model and data rejection technique is controlled by the measurement noise covariance R_k, similar to the (7) and (8). Measurement noise model is defined by the battery current is used to compensate for changes in the resistance, when a large charging or discharging current is imposed. Measurement noise model by dynamic of RC-ladder is used to compensate the simplified RC-ladders. Because of the simplified RC-ladder, model error occurs when a dynamic of current is very extreme. The data rejection technique is used to compensate the model error when the change of current is very high.

VI. Verification

This work compares the SOC estimation performances of a LiFePO$_4$ cell using an EKF algorithm depending on the number of RC-ladder, with and without the minor loops, with and without a noise model and data rejection technique For the experiment, a charging and discharging voltage data is extracted from applying a scale-down current profile for HEV to a 14 Ah voltage range 2 ~ 3.65 V pouch type LiFePO$_4$ cell. Fig. 5 shows a result of SOC estimation of the LiFePO$_4$ cell. In order to analyze the results numerically, the average error was calculated to compare with ampere-hour counting, as shown in Fig. 6. Analysis of the result, it can be confirmed that the SOC estimation accuracy significantly increase when the number of RC-ladder increases. Also, accuracy of the SOC estimation increase gradually when using minor loops, compared with one

that does not use a minor loop. On the other hand, the effect of the noise model and data rejection is slightly smaller than that of the other methods. The reason is that the absolute model error is small because of the scale-down current profile, as shown in Fig. 7. (maximum current is 6 A $\approx$ 1/2.3 C). In real transportation, impact of the noise model and data rejection technique is more significant than that of the scale-down current profile.

VII. Conclusion

This paper compares some of the SOC estimation result of LiFePO$_4$ cell using EKF algorithms and additional techniques. Consequently, parametrization, application of minor loops, and a measurement noise model method are effective for SOC estimation of LiFePO$_4$ cells. In particular, it is confirmed that the exact ECM structure is the most important factor, such as parametrization and application of minor loops. However, these methods complicate the algorithm realization. Because, elaboration of the battery ECM is proportionate to calculation amount of the EKF algorithm. Thus, these methods should be suitably selected and used with consideration of the battery characteristics.

References

[1] H. J. Lee, J. H. Park, J. H. Kim, "SOC estimation performance comparison based on the equivalent circuit model using an EKF in commercial LiCoO2 and LiFePO4 cells," EVS 28, May, 2015.

[2] T. Katrasnik, "Analytical method to evaluate fuel consumption of hybrid electric vehicles at balanced energy content of the electric storage devices," Applied Energy, vol. 87, No. 11, pp. 3330-3339, Nov, 2010.

[3] L. Wang, Y. Cheng, J. Zou, "Battery available power prediction of hybrid electric vehicle based on improved Dynamic Matrix Control algorithms," Journal of Power Sources, vol. 261 pp. 337-347, Sep, 2014.

[4] W. Waag, C. Fleischer, D. U. Sauer, "Critical review of the methods for monitoring of lithium-ion batteries in electric and hybrid vehicles," Journal of Power Sources, vol. 258, pp. 321-339, Jul, 2014.

[5] E. H. Miliani, "Leakage current and commutation losses reduction in electric drives for Hybrid Electric Vehicle," Journal of Power Sources, vol. 255, pp. 266-273, Jun, 2014.

[6] Z. Chen, C. C. Mi, R. Xiong, J. Xu, C. You, "Energy management of a power-split plug-in hybrid electric vehicle based on genetic algorithm and quadratic programming," Journal of Power Sources, vol. 248, pp. 416-426, Feb, 2014.

[7] J. L. Torres, R. Gonzalez, A. Gimenez, J. Lopez, "Energy management strategy for plug-in hybrid electric vehicles. A comparative study," Applied Energy, vol. 113, pp. 816-824, Jan, 2014.

[8] X. Wu, B. Cao, X. Li, J. Xu, X. Ren, "Component sizing optimization of plug-in hybrid electric vehicles," Applied Energy, vol. 88, no. 3, pp. 799-804, Mar, 2011.

[9] J. Han, Y. Park, D. Kum, "Optimal adaptation of equivalent factor of equivalent consumption minimization strategy for fuel cell hybrid electric vehicles under active state inequality constraints," Journal of Power Sources, vol. 267, pp. 491-502, Dec, 2014.

[10] L. Xu, J. Li, M. Ouyang, J. Hua, G. Yang, "Multi-mode control strategy for fuel cell electric vehicles regarding fuel economy and durability," International Journal of Hydrogen Energy, vol. 39, no. 5, pp. 2374-2389, Feb, 2014.

[11] L. Xu, M. Ouyang, J. Li, F. Yang, L. Lu, J. Hua, "Optimal sizing of plug-in fuel cell electric vehicles using models of vehicle performance and system cost," Applied Energy, vol. 103, pp. 477-487, Mar, 2013.

[12] S. Piller, M. Perrin, A. Jossen, "Methods for state-of-charge determination and their applications," Journal of Power Sources, vol. 96, no.1, pp. 113-120, Jun, 2001.

[13] C. Zhang, J. Jiang, W. Zhang, S. M. Sharkh, "Estimation of State of Charge of Lithium-Ion Batteries Used in HEV Using Robust Extended Kalman Filtering," Energies, vol. 5 no. 4, pp. 1098-1115, 2012.

[14] G. L. Plett, "Extended Kalman filtering for battery management systems of LiPB-based HEV battery packs: Part 1. Background," Journal of Power Sources, vol. 134, no. 2, pp. 252-261, Aug, 2004.

[15] H. Zhang, M. Y. Chow, "On-line PHEV Battery Hysteresis Effect Dynamics Modeling," IECON 2010 - 36th Annual Conference on IEEE Industrial Electronics Society, pp. 1844-1849, Nov,. 2010

A Power Sharing Scheme for Series Connected Offshore Wind Turbines in a Medium Voltage DC Collection Grid

Michael T. Daniel, *Student Member, IEEE*, and Prasad N. Enjeti, *Fellow, IEEE*
Department of Electrical and Computer Engineering
Texas A&M University
College Station, TX, USA

Abstract— This work introduces a new method for connecting offshore wind turbine generators (WTGs) in series when interfaced to a medium voltage DC (MVDC) collection grid. Limiting the series stacking to pairs of WTGs and introducing a power sharing converter (PSC) stage between them allows series connected WTGs to operate at independent power levels in a voltage sourced collection grid. This allows for reduced collection grid losses compared to previously proposed series connected current sourced approaches when operating below nominal power. Simulation results demonstrate system operation during a 1:2 power unbalance between series connected WTGs. A 100 W, 100 V_{dc} three-phase lab-scale prototype input rectifer stage demonstrates proper WTG current shaping with >0.97 displacement power factor magnitude, as well as balanced DC link voltages.

Keywords—Offshore Wind Farm, Wind Turbine Generator (WTG), MVDC, High Frequency Transformer (HFT), Resonant Soft Switching, Power Sharing Converter (PSC)

I. INTRODUCTION

Around the world people are continuing to attain higher standards of living, resulting in a continually growing global demand for energy[1]. To satisfy this demand within the contemporary context of global climate change, numerous countries are now aggressively developing their offshore wind energy resources[2]-[4].

Offshore wind resources are now being developed further out to sea than ever[5]. While Germany's Global Tech I farm, currently operating approximately 100 km from shore[6], is one of the most distant, there are plans for wind farms to be constructed nearly twice that distance from shore[7]. Due to their large distance from shore, many of these modern wind farms rely on offshore HVDC stations to transmit their energy to shore[8].

Additionally, the ≥5 MW WTGs used in modern offshore farms require nearly 1 km of inter-turbine spacing[9], and 10 MW WTGs currently in development [10] may require >1 km of inter-turbine spacing.

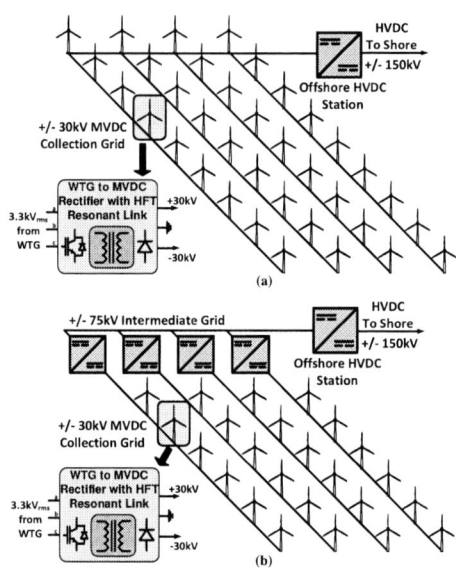

Fig. 1. (a) Single-stage and (b) two-stage WTG interface to offshore wind farm with +/-30 kV MVDC collection grid. The proposed interface approach is suitable for both (a) and (b), although for the sake of simplicity (a) is assumed for this work. © IEEE 2015.

Due to the large inter-turbine spacing and presence of an offshore HVDC converter, many researchers are now investigating MVDC collection grids for aggregating energy within the farm, including converter architectures for interfacing WTGs to such an MVDC grid[11]-[24], as summarized in Fig. 1 [24]. As part of this research thrust there has been significant work in the area of series connected WTGs[25]-[32]. However, these series connected approaches require many WTGs to share the same DC current. This requires each WTG to modulate its DC output voltage to inject power into the current sourced DC line running through it, similar to a classic line-commutated HVDC converter. This is not desirable as collection grid conduction losses are constant, even when little power is being generated. A power sharing strategy similar to [33]-[36] can be introduced to allow series connected WTGs to operate at independent power levels.

978-1-4673-9551-9/16 $31.00 © 2016 IEEE

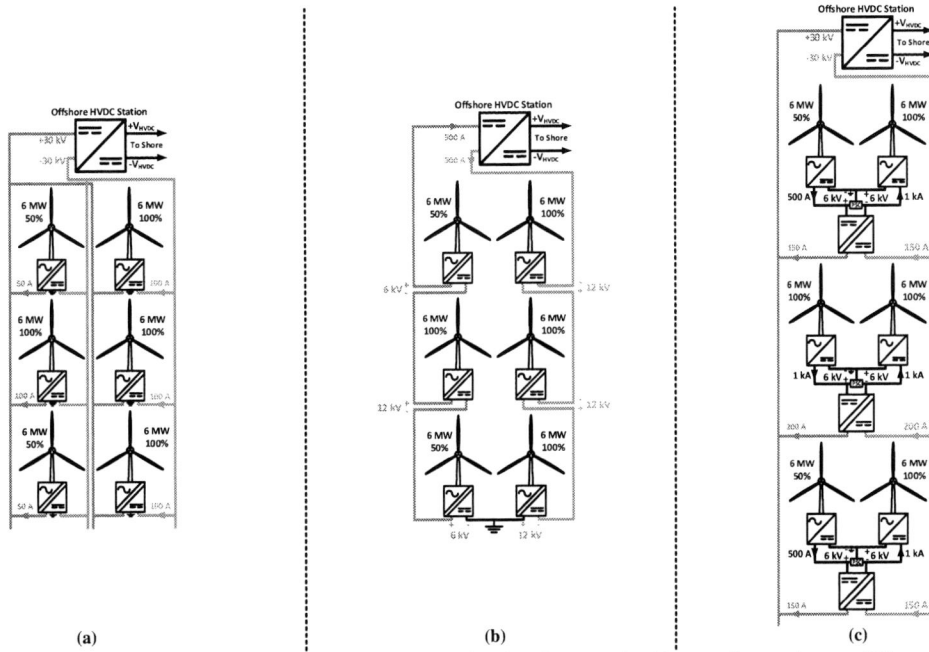

Fig. 2. Example wind farm with six 6 MW WTGs using (a) parallel connection (b) series connection (c) proposed connection; case (b) is current sourced. Note the same pattern of different WTG operating points, indicated as a percentage of rated power, is enforced in all three example architectures.

II. PROPOSED APPROACH

This work proposes the WTG-to-MVDC grid connection shown in Fig. 2c. Compared to the traditional connections of [11]-[31], illustrated via Figs. 2a and 2b, the proposed approach has the following advantages:

1) Reduced common mode voltage stress at individual WTG converters.

2) Lower voltage step-up required from WTG to MVDC collection grid due to series connected WTGs.

3) PSC stage allows series connected WTGs to operate at independent power levels.

4) Isolated DC/DC for each series pair eliminates need to balance MVDC voltage among many WTGs.

5) Reduced collection grid losses when operating at less than 100% power compared to traditional series connection.

Observing Fig. 2c, the proposed approach utilizes a PSC stage to allow two series connected WTGs to source different currents, and hence different amounts of power, while providing a higher pole-to-pole voltage to the input of the isolated DC/DC converter. PSC operation with the top two series connected WTGs from Fig. 2c is shown in detail in Fig. 3. In steady state both top and bottom pairs of PSC switches maintain 50% duty cycle, regardless of the state of power unbalance between the WTGs, as zero average voltage must appear across L_{PSC} in steady state. When the state of

unbalance changes, the switches temporarily increase or decrease their duty cycles to adjust the current in L_{PSC}. The PSC is important as individual WTG output powers can vary greatly over a relatively small range of wind speed since WTG power scales with the cube of wind speed[37]. Each WTG of the series connected pair from Fig. 3 is shown in detail in Fig. 4. The AC/DC stage provides input current shaping capability for power factor correction, and the inverter stage of the isolated DC/DC features soft switching at the resonant frequency provided by L_{res} and C_{res}. The currents indicated in Figs. 2, 3, and 4 are DC, although other components are present due to rectifier and inverter switching action.

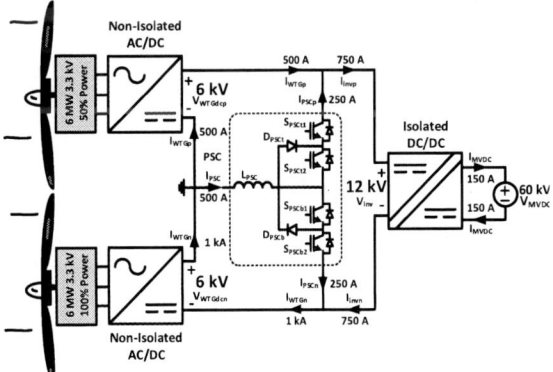

Fig. 3. Details of power sharing converter and resulting DC components of converter currents.

978-1-4673-9551-9/16 $31.00 © 2016 IEEE

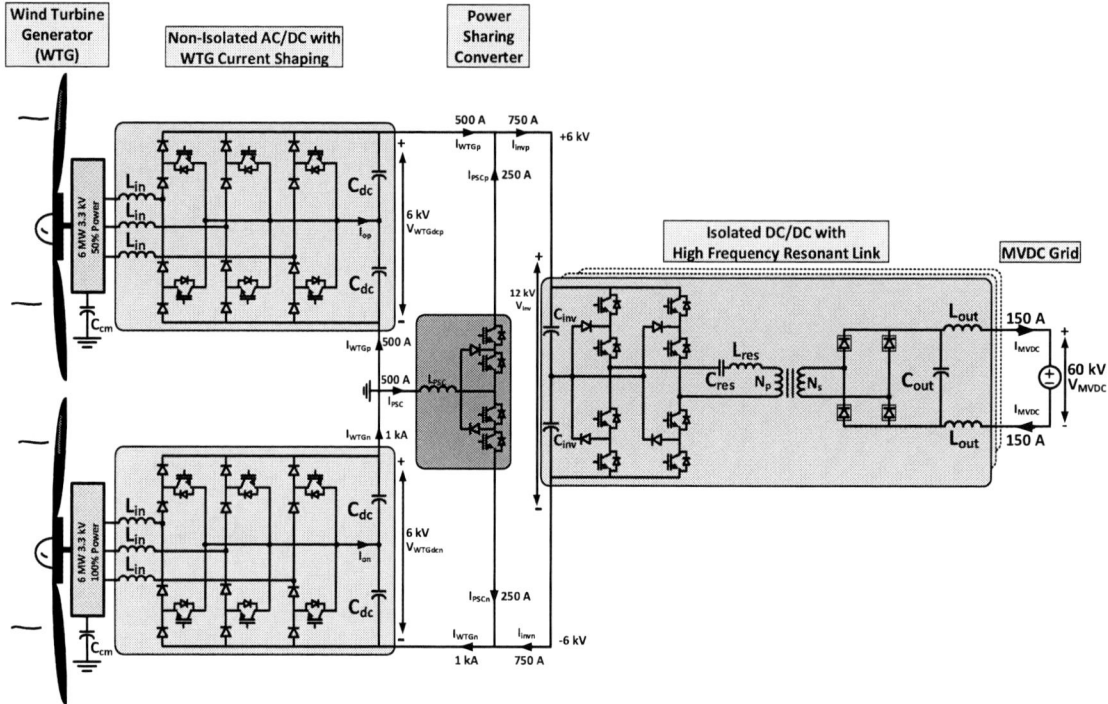

Fig. 4. Detail of series connected WTG pair. WTG neutral is ungrounded, hence C_{cm} is included. Dashed outlines indicate that multiple parallel isolated DC/DC modules may be used since a single state-of-the-art HFT is likely not capable of processing the total available power from both WTGs.

III. DESIGN AND CONTROL

The detailed design of components such as C_{dc}, L_{res}, C_{res}, C_{out}, and L_{out}, as well as the detailed discussion on the control strategies for the input three-phase rectifier and isolated DC/DC stages are presented in [24] and [38]. For the purposes of this work it is sufficient to say that the input rectifier control operates in the manner of a PWM rectifier, maintaining the DC link voltage and sinusoidal input current by adjusting the phase and amplitude of the fundamental-frequency voltage appearing at each phase terminal, while the HF inverter in the isolated DC/DC stage is controlled as a phase-shifted inverter, maintaining the desired output current to the MVDC grid by adjusting the relative phase of square wave switching in each inverter leg.

The key additional component that must be designed for this work is the power sharing inductor L_{PSC}. As the PSC is essentially functioning as a 1:2 boost converter between the 6 kV DC link of the pole "n" WTG and the 12 kV input to the inverter, L_{PSC} can be designed using the familiar boost inductor design equation with a fixed duty cycle of 50% which yields (1). The inductor ripple ΔI_{PSC} is specified to be <1% for this work.

$$L_{PSC} = \frac{V_{WTGdc}}{2\Delta I_{PSC} f_{PSC}} \qquad (1)$$

Control of the PSC is achieved by sensing the DC values of both I_{WTGp} and I_{WTGn} and calculating the error between the average difference between these currents and I_{PSC}. This error is routed through a proportional-integral (PI) controller, and the resulting signal is used to adjust the relative duty cycle of the high- and low-side PSC IGBTs. This control scheme is summarized in Fig. 5.

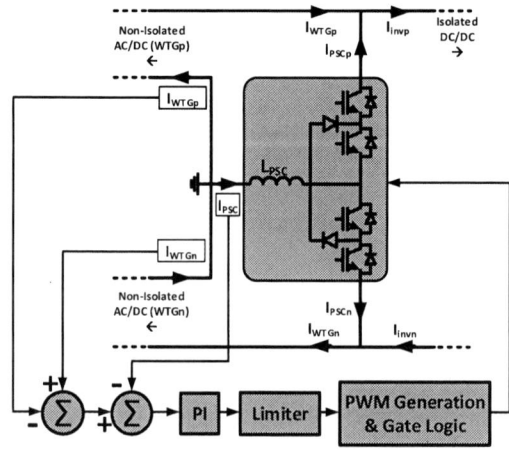

Fig. 5. PSC control scheme. All indicated currents are DC values.

Note that the neutral-point-clamped (NPC) structure featured in both the PSC and inverter stage is utilized for voltage sharing purposes between series connected IGBTs. Freewheeling zero-voltage states using NPC diodes are not applied, although they may be used in a future iteration of this work. Therefore, using the nomenclature of Fig. 3 we can say that S_{PSCt1} and S_{PSCt2} maintain identical switching states, as do S_{PSCb1} and S_{PSCb2}. This yields the steady state relationships between PSC DC currents described by (2)-(4).

$$I_{PSC} = I_{WTGn} - I_{WTGp} \qquad (2)$$

$$I_{PSCn} = I_{PSCp} = \frac{I_{PSC}}{2} \qquad (3)$$

$$I_{invp} = I_{invn} \qquad (4)$$

By considering that in power sharing mode at least one pair of PSC switches must be closed to provide a path for I_{PSC}, and that equal and opposite voltage polarity is applied to L_{PSC} when each pair of switches is alternately closed and opened, it is obvious that in steady state the duty cycle of each switch pair must be 50% to apply zero net volt-seconds to L_{PSC} and maintain constant I_{PSC}.

IV. SIMULATION RESULTS

For simulation purposes all inductors were considered to have an ωL:R ratio of 200:1. In addition, capacitor ESRs were accounted for by introducing a series resistor across which 0.1% of each capacitor voltage was dropped. IGBT and diode on-state voltages of 3.1 V [39] and 1.7 V [40], respectively, were included. C_{cm} was taken to be 100 nF[41],[42]. Detailed parameters of the simulated system are outlined in Table I, and results of simulating the unbalanced WTG power case described by Figs. 2-4 are summarized in Figs. 6-9.

TABLE I.   WTG-TO-MVDC INTERFACE SIMULATION PARAMETERS

Parameter	Value
P_{WTG}	6 MW
V_{WTG}	3.3 kV$_{LL,rms}$
L_{WTG}	2.40 mH
f_{WTG}	60 Hz
Displacement PF$_{WTG}$	> 0.95
L_{in}	10 μH
f_{rect}	1.5 kHz
V_{WTGdc}	6 kV
C_{dc}	10 mF
L_{PSC}	65 mH
f_{PSC}	15 kHz
C_{inv}	10 mF
f_{inv}	10 kHz
N_p:N_s	1:7
V_{MVDC}	± 30 kV (60 kV)
C_{out}	500 μF
L_{out}	500 mH

A 6 MW permanent magnet synchronous generator (PMSG) having a nominal terminal voltage of 3.3 kV$_{LL,rms}$ and nominal frequency of 60 Hz was assumed. According to [37] the WTG generator voltage and frequency scale linearly with wind speed for such a PMSG as that assumed in this approach. Therefore, since WTG power scales with the cube of wind speed the WTG current must scale with the square of wind speed. As such, a wind speed of 79.4% of nominal is required for operation at 50% power. The WTG voltage and frequency is also 79.4% of nominal at 50% power, while the current is 63% of nominal. The current of the 50% power turbine is shown in the top plot of Fig. 6. Both DC link voltages are controlled to be 6 kV, while DC link current is pulsating with average values of 500 A and 1 kA, as shown in Fig. 7. The PSC is able to redirect the difference in DC WTG current through L_{PSC} and on to the isolated DC/DC stage, shown in Fig. 8. In Fig. 9 the DC/DC stage converts pulsating input current to smooth DC current for the MVDC grid.

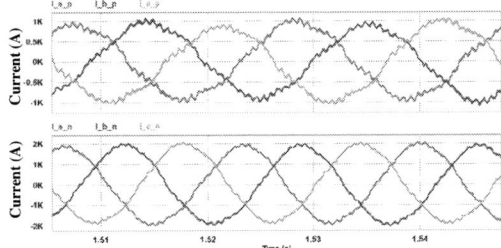

Fig. 6. One can observe that pole "p" generator currents (top) are much less than pole "n" generator currents (bottom) since it is operating at 50% power. Pole "p" currents are 48 Hz, or 79.4% of nominal frequency.

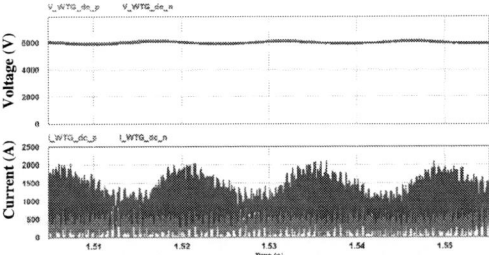

Fig. 7. Rectifier output DC link voltages (top) and currents (bottom). Average value of pole "p" DC link current is approximately 500 A, while average of pole "n" current is approximately 1 kA. 3 MW are sourced by pole "p" WTG and 6 MW are source by pole "n" WTG.

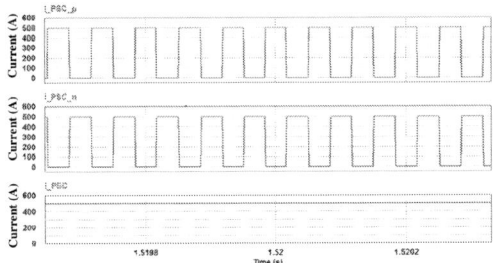

Fig. 8. Current in pole "p" PSC IGBTs (top), pole "n" PSC IGBTs (middle), and PSC inductor (bottom). Steady state duty cycle is 50%, therefore average current is each switch is half I_{PSC}. I_{PSC} is the difference between the average of "p" and "n" DC link currents from Fig. 7.

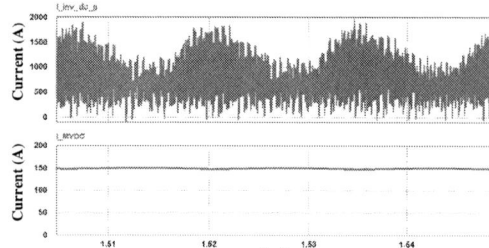

Fig. 9. Current into isolated DC/DC stage (top) and current injected into MVDC grid (bottom). Average value of $I_{inv\_dc\_p}$ is 750 A, and MVDC grid voltage is fixed by HVDC converter station at 60 kV pole-to-pole, hence 9 MW are injected to MVDC grid.

V. EXPERIMENTAL RESULTS

A 100 W, 100 V_{dc} three-phase lab-scale prototype was constructed and tested to validate the AC/DC stage control and PFC capability with the large synchronous reactance associated with such PMSGs[43],[44]. The WTG voltage was 42 $V_{LL,rms}$ and the synchronous inductance was designed to be 4.45 mH per phase; detailed experimental parameters are outlined in Table II.

It can be observed in Fig. 10 that the switching rectifier is able to properly shape the WTG current. This is done by temporarily closing the IGBT switches of each phase, thereby creating a low-impedance path around the DC link that allows the internal WTG voltage to boost the current flowing in the WTG synchronous reactance. When the switches are opened, this boosted current in L_{WTG} forces the outer rectifier diodes to conduct, clamping the WTG terminal to either the positive or negative pole of the DC link, depending on current direction.

While this creates a nearly sinusoidal WTG current waveform, the WTG terminal voltage takes the form of a pulse-width-modulated square wave, as illustrated in Figs. 11 and 12. The fundamental component of this PWM square wave voltage can be controlled in both phase and amplitude to control both the active and reactive power extracted, respectively, from the WTG as discussed in [38].

TABLE II. WTG-TO-MVDC INTERFACE EXPERIMENTAL PARAMETERS

Parameter	Value
V_{WTG}	42$V_{LL,rms}$
L_{WTG}	4.45 mH
f_{WTG}	60 Hz
Displacement PF$_{WTG}$	> 0.97
f_{rect}	1.5 kHz
V_{dc}	100 V
C_{dc}	2.20 mF
P_{out}	100 W
IGBT	BSM 150GB 60 DLC
Diode	IRK D196-16

Observing Fig. 13 it can be seen that only a portion of the shaped input current conducts to the DC link via rectifier diodes. Since the input current portions that conduct to the DC link have been boosted and are now decaying to the DC link, they have high peak values and conduct for a relatively

short time. In many cases the instantaneous duty cycle of rectifier IGBTs can be as high as 75%, meaning that energy must be delivered from the WTG to the DC link in only 25% of the available conducting time.

The result is that energy is transferred to the DC link in brief high-current bursts, leading to a high peak-to-average ratio of DC link current in order to sustain the required average current using only brief conducting periods. A large diode and IGBT current de-rating is therefore required, and stray inductance must be reduced such that voltages induced by the high di/dt do not cause failure of switching devices.

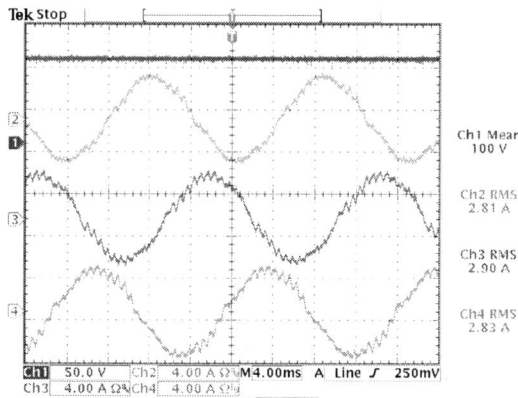

Fig. 10. Three phase rectifier current from WTG. Rectifier switching action shapes WTG current and boosts WTG voltage from approximately 60 $V_{LL,peak}$ to the 100 V_{dc} bus.

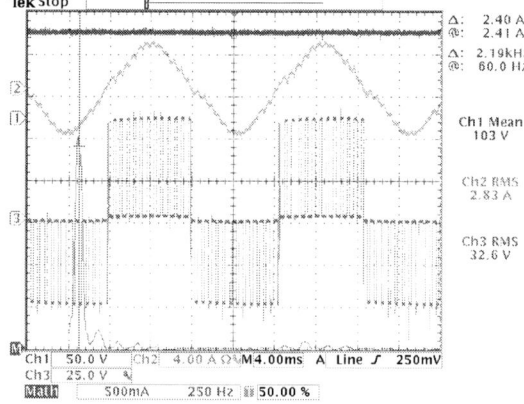

Fig. 11. Phase "a" detail showing current and switching voltage at rectifier terminals to be in phase. Note switching voltage is measured from phase to ground. Also note that the FFT of WTG current shows a strong fundamental component and negligible 5th and 7th harmonics that would normally be present when using a passive 6-pulse rectifier.

Comparing Fig. 13 to Fig. 7 the same characteristic current can be observed pulsing from zero to the peak WTG inductor current: zero to 2 kA for WTG$_n$ in Fig. 7, and zero to 4.86 A in Fig. 13. This waveform is much clearer on the shorter time scale, and lower current scale, of Fig. 13, where the individual pulses of current can be observed.

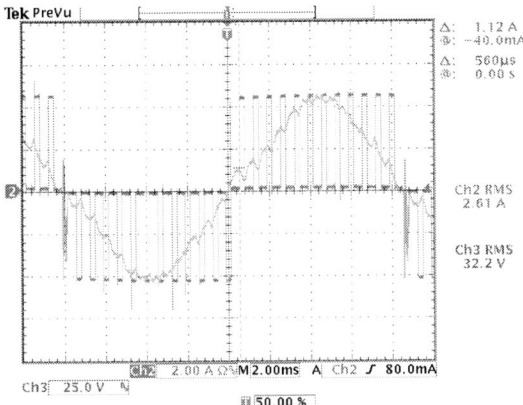

Fig. 12. Here it is observed that fundamental phase voltage and current zero crossings coincide within 0.56 ms of each other, yielding a displacement power factor magnitude of >0.97 at 60 Hz.

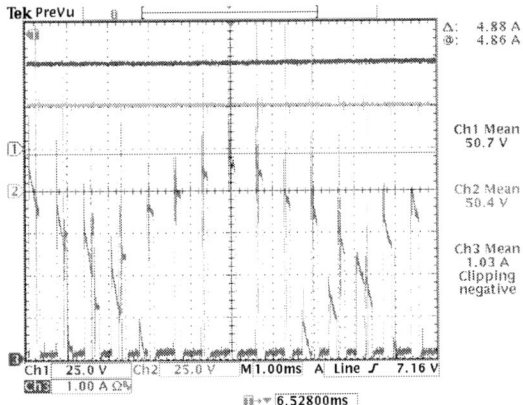

Fig. 13. Ch1 & Ch2 show that DC link voltage is balanced between top and bottom DC link capacitors, and as expected the DC link current has a high peak-to-average ratio due to the rectifier switching action which allows decaying portions of input current to conduct to the DC link via diodes but circulates boosting portions of input current through the DC link neutral point via IGBTs.

VI. CONCLUSION

This work has introduced a new method for connecting offshore WTGs in series when interfaced to an MVDC collection grid. Limiting the series stacking to pairs of WTGs and introducing a power sharing converter stage between them allows series connected WTGs to operate at independent power levels in a voltage sourced collection grid. This allows for reduced collection grid losses compared to previously proposed series connected current sourced approaches when operating below nominal power. Simulation results demonstrate system operation during a 1:2 power unbalance between series connected WTGs. A 100 W, 100 V_{dc} three-phase lab-scale prototype input rectifer stage demonstrates proper WTG current shaping with >0.97 displacement power factor magnitude, as well as balanced DC link voltages.

REFERENCES

[1] "Annual Energy Outlook 2015", EIA, US Department of Energy, 2015 [Online]. Available: http://www.eia.gov/forecasts/aeo/

[2] Li Rennian; Wang Xin, "Status and challenges for offshore wind energy," Materials for Renewable Energy & Environment (ICMREE), 2011 International Conference on , vol.1, no., pp.601,605, 20-22 May 2011

[3] Higgins, P.; Foley, A.M., "Review of offshore wind power development in the United Kingdom," Environment and Electrical Engineering (EEEIC), 2013 12th International Conference on , vol., no., pp.589,593, 5-8 May 2013

[4] Kling, W.L.; Gibescu, M.; Ummels, B.C.; Hendriks, R.L., "Implementation of wind power in the Dutch power system," Power and Energy Society General Meeting - Conversion and Delivery of Electrical Energy in the 21st Century, 2008 IEEE , vol., no., pp.1,6, 20-24 July 2008

[5] "The European Offshore Wind Industry – Key Trends and Statistics 2013", European Wind Energy Association, January 2014 [Online]. Available: http://www.ewea.org/fileadmin/files/library/publications/statistics/European_offshore_statistics_2013.pdf

[6] "Pioneering performance in the North Sea: The offshore wind farm Global Tech I", Global Tech I Offshore Wind GmbH, 2012, [Online]. Available:http://www.globaltechone.de/en/about/

[7] "Nord-Ost Passat I", 4C Offshore, 2015, [Online]. Available: http://www.4coffshore.com/windfarms/nord-ost-passat-i-germany-de1p.html

[8] Das, D.; Pan, J.; Bala, S., "HVDC Light for large offshore wind farm integration," Power Electronics and Machines in Wind Applications (PEMWA), 2012 IEEE , vol., no., pp.1,7, 16-18 July 2012

[9] A. Westerhellweg, B. Cañadillas, F. Kinder, T. Neumann,"Detailed Analysis of Offshore Wakes Based on Two Years Data of alpha ventus and Comparison with CFD Simulations", DEWI Wilhelmshaven, DEWI Magazine, No. 42, pp. 65-70, Feb. 2013.

[10] "Sea Titan 10 MW Wind Turbine: Maximum Power per Tower for Ofsfhore Environment", Windtec Solutions, American Superconductor Corp., 2012 [Online]. Available: http://www.amsc.com/documents/seatitan-10-mw-wind-turbine-data-sheet/

[11] Chen, W.; Huang, A. Q.; Li, C.; Wang, G.; Gu, W., "Analysis and Comparison of Medium Voltage High Power DC/DC Converters for Offshore Wind Energy Systems," Power Electronics, IEEE Transactions on , vol.28, no.4, pp.2014,2023, April 2013

[12] Daniel, M.T.; Krishnamoorthy, H.S.; Enjeti, P.N., "A new wind turbine interface to MVDC grid with high frequency isolation and input current shaping," Industrial Electronics Society, IECON 2014 - 40th Annual Conference of the IEEE , vol., no., pp.1924,1930, Oct. 29 2014-Nov. 1 2014

[13] Lam, J.; Jain, P.K., "A high efficient medium voltage step-up DC/DC converter with zero voltage switching (ZVS) and low voltage stress for offshore wind energy systems," Power Electronics and Applications (EPE'14-ECCE Europe), 2014 16th European Conference on , vol., no., pp.1,10, 26-28 Aug. 2014

[14] Steimer, P.K.; Apeldoorn, O., "Medium voltage power conversion technology for efficient windpark power collection grids," Power Electronics for Distributed Generation Systems (PEDG), 2010 2nd IEEE International Symposium on , vol., no., pp.12,18, 16-18 June 2010

[15] Meyer, C.; Hoing, M.; Peterson, A.; De Doncker, R.W., "Control and Design of DC Grids for Offshore Wind Farms," Industry Applications, IEEE Transactions on , vol.43, no.6, pp.1475,1482, Nov.-dec. 2007

[16] Prasai, A.; Jung-Sik Yim; Divan, D.; Bendre, A.; Seung-Ki Sul, "A New Architecture for Offshore Wind Farms," Power Electronics, IEEE Transactions on , vol.23, no.3, pp.1198,1204, May 2008

[17] Carmeli, M.S.; Castelli-Dezza, F.; Marchegiani, G.; Mauri, M.; Rosati, D., "Design and analysis of a Medium Voltage DC wind farm with a transformer-less wind turbine generator," Electrical Machines (ICEM), 2010 XIX International Conference on , vol., no., pp.1,6, 6-8 Sept. 2010

[18] Ho-Sung Kim; Ju-Won Baek; Myung-Hyu Ryu; Jong-Hyun Kim; Jee-Hoon Jung, "The High-Efficiency Isolated AC–DC Converter Using the

Three-Phase Interleaved LLC Resonant Converter Employing the Y-Connected Rectifier," Power Electronics, IEEE Transactions on , vol.29, no.8, pp.4017,4028, Aug. 2014

[19] Parastar, A.; Jul-Ki Seok, "High power step-up modular resonant DC/DC converter for offshore wind energy systems," Energy Conversion Congress and Exposition (ECCE), 2014 IEEE , vol., no., pp.3341,3348, 14-18 Sept. 2014

[20] Jovcic, D., "Step-up DC-DC converter for megawatt size applications," Power Electronics, IET , vol.2, no.6, pp.675,685, Nov. 2009

[21] Jun Li; Bhattacharya, S.; Huang, A.Q., "A New Nine-Level Active NPC (ANPC) Converter for Grid Connection of Large Wind Turbines for Distributed Generation," Power Electronics, IEEE Transactions on , vol.26, no.3, pp.961,972, March 2011

[22] Lam, J.; Jain, P.K., "Single-stage three-phase AC/DC step-up medium voltage resonant converter for offshore wind power systems," Energy Conversion Congress and Exposition (ECCE), 2014 IEEE , vol., no., pp.4612,4619, 14-18 Sept. 2014

[23] Fujin Deng; Zhe Chen, "Control of Improved Full-Bridge Three-Level DC/DC Converter for Wind Turbines in a DC Grid," Power Electronics, IEEE Transactions on , vol.28, no.1, pp.314,324, Jan. 2013

[24] Daniel, M.T.; Krishnamoorthy, H.S.; Enjeti, P.N., "A New Wind Turbine Interface to MVdc Collection Grid With High-Frequency Isolation and Input Current Shaping," in Emerging and Selected Topics in Power Electronics, IEEE Journal of , vol.3, no.4, pp.967-976, Dec. 2015

[25] Gang Shi; Zhibing Wang; Miao Zhu; Xu Cai; Liangzhong Yao, "Variable speed control of series-connected DC wind turbines based on generalized dynamic model," Renewable Power Generation Conference (RPG 2013), 2nd IET , vol., no., pp.1,6, 9-11 Sept. 2013

[26] Takemura, A.; Tatsuta, F.; Yokoyama, H.; Nishikata, S., "Studies on field current control method for constant tip speed ratios of series connected wind turbine generators in a wind farm," Electrical Machines and Systems (ICEMS), 2012 15th International Conference on , vol., no., pp.1,5, 21-24 Oct. 2012

[27] Holtsmark, N.; Bahirat, H.J.; Molinas, M.; Mork, B.A.; Hoidalen, H.K., "An All-DC Offshore Wind Farm With Series-Connected Turbines: An Alternative to the Classical Parallel AC Model?," Industrial Electronics, IEEE Transactions on , vol.60, no.6, pp.2420,2428, June 2013

[28] Nishikata, S.; Tatsuta, F., "Dynamic control of series connected wind turbine generating system," PowerTech, 2011 IEEE Trondheim , vol., no., pp.1,6, 19-23 June 2011

[29] Garces, A.; Barrera-Cardenas, R.; Molinas, M., "Optimal control for an HVDC system with series connected offshore wind turbines," Energy Conversion Congress and Exposition (ECCE), 2013 IEEE , vol., no., pp.3918,3925, 15-19 Sept. 2013

[30] Gjerde, S.; Undeland, T.M., "Dynamic performance of the modular series connected converter in a 100 kVdc- transformerless offshore wind turbine," OCEANS - Bergen, 2013 MTS/IEEE , vol., no., pp.1,9, 10-14 June 2013

[31] Tatsuta, F.; Nishikata, S., "Basic investigations on a wind power plant consisting of series-connected wind turbine generators," Electrical Machines and Systems (ICEMS), 2014 17th International Conference on , vol., no., pp.1345,1349, 22-25 Oct. 2014

[32] Veilleux, E.; Lehn, P.W., "Interconnection of Direct-Drive Wind Turbines Using a Series-Connected DC Grid," Sustainable Energy, IEEE Transactions on , vol.5, no.1, pp.139,147, Jan. 2014

[33] Hawke, J.T.; Krishnamoorthy, H.S.; Enjeti, P.N., "A Family of New Multiport Power-Sharing Converter Topologies for Large Grid-Connected Fuel Cells," Emerging and Selected Topics in Power Electronics, IEEE Journal of , vol.2, no.4, pp.962,971, Dec. 2014

[34] Krishnamoorthy, H.S.; Essakiappan, S.; Enjeti, P.N.; Balog, R.S.; Ahmed, S., "A new multilevel converter for Megawatt scale solar photovoltaic utility integration," Applied Power Electronics Conference and Exposition (APEC), 2012 Twenty-Seventh Annual IEEE , vol., no., pp.1431,1438, 5-9 Feb. 2012

[35] McClurg, J.; Pilawa-Podgurski, R.C.N.; Shenoy, P.S., "A series-stacked architecture for high-efficiency data center power delivery," Energy Conversion Congress and Exposition (ECCE), 2014 IEEE , vol., no., pp.170,177, 14-18 Sept. 2014

[36] Shenoy, P. S.; Krein, P. T., "Differential Power Processing for DC Systems," Power Electronics, IEEE Transactions on , vol.28, no.4, pp.1795,1806, April 2013

[37] Polinder, H.; Bang, D.; Van Rooij, R. P J O M; McDonald, A.S.; Mueller, M.A., "10 MW Wind Turbine Direct-Drive Generator Design with Pitch or Active Speed Stall Control," Electric Machines & Drives Conference, 2007. IEMDC '07. IEEE International , vol.2, no., pp.1390,1395, 3-5 May 2007

[38] Daniel, M.T.; Krishnamoorthy, H.S.; Enjeti, P.N., "An improved offshore wind turbine to MVDC grid interface using high frequency resonant isolation and input power factor control," in Power and Energy Conference at Illinois (PECI), 2015 IEEE , vol., no., pp.1-8, 20-21 Feb. 2015

[39] "5SNA 1500E3303005 HiPak IGBT Module", ABB Switzerland Ltd., Feb 2014 [Online]. Available: http://www08.abb.com/global/scot/scot256.nsf/veritydisplay/d740233e8 18310bc83257ca9002c95b1/$file/5SNA%201500E330305%205SYA% 201407-07%2002-2014.pdf

[40] "5SDF 20L4521 Fast Recovery Diode", ABB Switzerland Ltd., Sept 2014 [Online]. Available: http://www08.abb.com/global/scot/scot256.nsf/veritydisplay/c085c2d71 2d4b4c283257d70002cd749/$file/5SDF%2020L4521_5SYA1186-01%20Sep%2014.pdf

[41] Rendusara, D.A.; Cengelci, E.; Enjeti, P.N.; Stefanovic, Victor R.; Gray, J.W., "Analysis of common mode voltage-"neutral shift" in medium voltage PWM adjustable speed drive systems," Power Electronics, IEEE Transactions , vol.15, no.6, pp.1124,1133, Nov 2000

[42] Palma, L.; Harfman Todorovic, M.; Enjeti, P.N., "Analysis of Common-Mode Voltage in Utility-Interactive Fuel Cell Power Conditioners," Industrial Electronics, IEEE Transactions on , vol.56, no.1, pp.20,27, Jan. 2009

[43] Reigstad, T., "Direct Driven Permanent Magnet Synchronous Generators with Diode Rectifiers for Use in Offshore Wind Turbines", NTNU, Jun. 2007 [Online]. Available: http://www.diva-portal.org/smash/get/diva2:347491/FULLTEXT01.pdf

[44] Rucker, J.E.; Kirtley, J.L.; McCoy, T.J., "Design and analysis of a permanent magnet generator for naval applications," Electric Ship Technologies Symposium, 2005 IEEE , vol., no., pp.451,458, 25-27 July 2005

Fault Ride-Through Performance Evaluation of an Interleaved Grid-Connected Converter Employing Low Switching Frequency

Lorand Bede[1], Ghanshyamsihn Gohil[1], Mihai Ciobotaru[2], Tamas Kerekes[1], Remus Teodorescu[1], Vassilios G Agelidis[2]

[1]Department of Energy Technology, Aalborg University, Aalborg, Denmark
[2]School of Electrical Engineering, University of New South Wales, Sydney, Australia
Email: lbe@et.aau.dk

Abstract—This paper presents the performance evaluation under grid voltage sags of a two parallel-interleaved grid-connected converters employing low switching frequency. A current controller based on the synchronous reference frame including voltage feed-forward is used. The dominant frequency components from the direct- and quadrature-axis voltage V_d and V_q are eliminated by using an adaptive moving average filter. The dynamic performance of the converter during voltage sags is improved using an adaptive approach which minimizes the filter delay during the faults. The presented results conclude the performance of the converter system.

Keywords—feed forward filtering, low switching, grid connected, grid control, LVRT, parallel interleaved,

I. INTRODUCTION

In the last decades the percentage of the renewable energy sources contributing to the energy production has increased significantly [1].To ensure stable grid operation more and more stringent grid codes are introduced, one of the most stringent one being the German BDEW standard [2]. Traditionally, in order to comply with the grid codes, a Voltage Source Converter (VSC) is employed [1]. In order to maintain low switching losses, the switching frequency (f_{sw}) of the VSCs has to be limited to a few kHz in the case of MW size power electronic converters. Furthermore, high order filters (i.e. *LCL*) are usually employed at the output of the converter[1], to be able to comply with the stringent standard requirements.

In general, the power converters of large-scale renewable energy power plants are often connected to the medium voltage network to reduce transmission losses. Since the operating voltage for these converters is limited, a step up transformer is required when connecting to medium voltage line. Furthermore, the leakage inductance of this transformer can be used as the grid side inductor of an *LCL* filter [3].

Multiple power converters can be connected in parallel to increase the power handling capability of these converter systems [4]. To reduce the filtering requirements, carrier interleaving can be employed [5]. By interleaving the carriers, the first order switching harmonic is removed from the total current [6]. On the other hand, carrier interleaving causes an additional circulating current to flow between the modules of the converters connected in parallel [7]. To suppress this circulating current, multiple solutions are available [8, 9]. In

this paper a coupled inductor (CI) is added between the corresponding phases of the converters to limit the circulating current[10, 11].

There are different methods on how to control the converter current injected into the grid [12, 13]. The most common method is to use a Proportional Integral (PI) controller in the synchronous reference frame. It has been proven that, if the direct and quadrature axis voltage (V_d and V_q) is fed-forward to the output of the PI controller, the performance of the PI controller is enhanced during grid voltage transients [14]. A Phased-Locked Loop (PLL) is used to obtain the angle of the grid voltage which is measured at the Point of Common Coupling (PCC) [15, 16]. There are two connection points where the grid voltage can be measured depending on whether a step up transformer is used or not. One solution is to measure the voltage on the medium voltage side of the transformer. This solution is not preferred since the medium voltage sensors are expensive. The preferred solution is to measure the voltage on the low voltage side of the transformer. By doing so, the voltage will be measured across the capacitor of the output filter, in the case when *LCL* filters are used. However, the voltage across the capacitor is distorted and it contains multiple harmonics owing to the modulation technique. On the other hand, V_d and V_q should not contain any harmonics, otherwise the harmonic performance of the system might be compromised. A filter can be used to filter V_d and V_q accordingly. Nevertheless, the filter will introduce delays, which could affect the converter dynamics during voltage sags.

This paper evaluates the fault ride-through capability of a grid-connected converter employing low switching frequency. An adaptive Moving Average Filter (MAF) is used, which eliminates the dominant frequency components from V_d and V_q. This filter has the ability to minimize its delay during voltage sags, thus helping the dynamic performance of the converter system.

II. SYSTEM DESCRIPTION

A. Converter system

A system consisting of two 2-level parallel interleaved converters using full bridge topology and IGBTs was considered in this paper, as presented in Fig. 1.

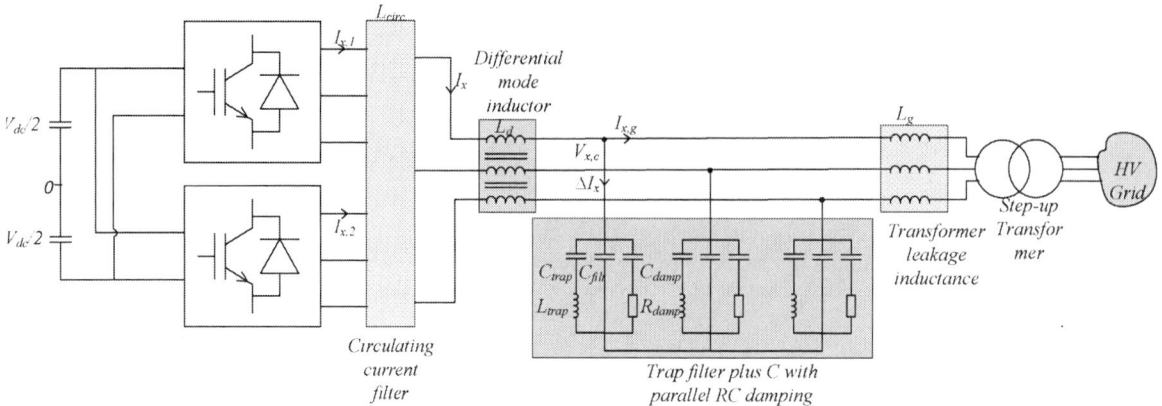

Fig. 1Schematic of the investigated converter system

Furthermore, three CIs were used (one for each phase) to limit the circulating currents caused by the interleaving.

Due to the interleaving, the major high frequency component in the total current (I_x, where x= [a,b,c]) is at the double of the switching frequency. A trap filter is designed to eliminate this frequency component from the grid current ($I_{x,g}$)[17]. Along with the trap filter, an *LCL* filter with passive damping is used. The leakage inductance of the step-up transformer, which is used to be able to connect to the medium voltage network, acts as the grid side inductor of the *LCL* filter.

B. Control system

The current control uses PI controllers in synchronous reference frame. A three-phase PLL provides the voltage phase angle used to obtain the direct and quadrature axis components of the grid currents(I_d, I_q) and PCC voltage (V_d, V_q), respectively. During the voltage sags, it is critical to have an accurate feed-forward value of V_d and V_q to help the converter dynamics. In the case when no feed-forward is used or an incorrect value is fed-forward, the current drawn from the converter will increase rapidly and the converter may trip. The rapid increase in current is due to the low controller bandwidth as a result of low switching frequency. On the other hand, some of the grid standards require the converter to stay connected during grid voltage sags and support the network. Therefore, the tripping of the converter in this situation should be avoided.

The structure of the control system is presented in Fig. 2. The three phase voltages are sampled at the capacitor of the filter, and they are fed to the PLL block. The PLL block calculates the grid angle (θ) and the direct and quadrature components of the voltage (V_d and V_q).

These values are used as inputs to the filter, which can be either a Moving Average Filter (MAF) or an Adaptive MAF (AMAF). The filtered values (V_{d_filt} and V_{q_filt}) are fed forward to the output of the current controller for the corresponding phase.

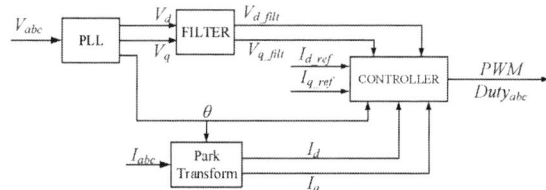

Fig. 2Software structure of the three-phase converter system.

C. Moving Average Filter

An MAFis designed to filter the major harmonic components from V_d and V_q, which are fed-forward to the controller. Thus, there is a need for two filters, one to filter the V_d and one to filter the V_q.

The filter cut-off frequency is tuned to remove the 300 Hz and its upper order harmonics caused by the modulation. The equation of this filter [16] is represented by

$$h[n] = \frac{1}{N}\sum_{k=0}^{N-1}\delta[n-k] \tag{1}$$

where h is the output vector of the filter, δ is the input vector of the filter, and N is the number of samples used for averaging. The value of N can be determined as follows:

$$N = floor(\frac{1}{f_{cut} \cdot T_s}) \tag{2}$$

where f_{cut} is the frequency to be filtered, and T_s is the sampling period. Since the double update technique is used here, the value of T_s is $1/(2*f_{sw})$.

The equation of the MAF is as follows:

$$\frac{1}{N}\sum_{k=0}^{N-1}V_{x\_sampled}[n-k] \tag{3}$$

where x represents the d and the q axis.

The drawback of this solution is that during transient operations, such as voltage sags, the filter introduces a delay which will affect the transient response of the system.

D. Adaptive Moving Average Filter

In order to overcome the shortcomings of the MAF, this paper proposes an Adaptive MAF (AMAF) which enhances the transient performance of the system, and in the same time has the same performance in steady state as the MAF.

A simple algorithm is used to detect voltage sags based on previous samples and to adapt the feed-forward value to the new conditions, as shown by(4). The algorithm compares the actual measured value with the previously measured one and if the difference between the two samples is bigger than a predefined value the algorithm interprets it as a fault.

$$V_{x\_filt} = \begin{cases} \dfrac{1}{N} \displaystyle\sum_{k=0}^{N-1} V_{x\_sampled}\left[n-k\right] & if \ no \ fault \\ \\ V_{x\_sampled} & if \ fault \end{cases} \qquad (4)$$

To ensure a fast tracking of the V_d and V_q, the algorithm sets the feed-forward value to the actual V_d and V_q (without any filtering) during the voltage sag.

III. RESULTS

A. Experimental results

To test the performance of the MAF and the AMAF an experimental setup consisting of 2 parallel interleaved converters has been built. The total power of the system is 10KW and the control system has been implemented on a Texas Instruments TMS320F28346 floating point Micro Controller Unit (MCU). The parameters of the filter components along with other system parameters are presented in TABLE I. The switching frequency used is 2.55 kHz. Due to the limited switching frequency, the bandwidth of the controller is also limited. In this system the interleaving technique is used, hence the output of each converter is connected to 3 CIs, in order to limit the circulating current between the converters. In this case the voltage to which the synchronization is made is measured at the filter capacitance. This is done because in the high power setup the MV voltages would be too expensive to measure. The drawback of this measurement is that the measured voltage will be distorted and will contain a lot of harmonics. Since the paper wants to comply with the BDEW standard, a trap filter is employed, which is tuned to eliminate the second carrier harmonics from the grid current.

The following three cases have been investigated in order to determine the performance of the system under a symmetrical grid fault, when the grid voltage is dropped to 25% of its nominal value: 1) when no filtering was applied on the V_d and V_q values, 2) When the MAF was used and 3) When the AMAF was used. The comparison criteria has been the current Total Harmonic Distortion (THD) in steady state and the current overshoot during the transient operation. Furthermore, to show the effectiveness of the filter at the specific harmonic components the FFT analysis is also included up to 1Khz.

For all three cases the measured currents and voltages are shown. For a better understanding, the direct and quadrature

TABLE I System Parameters

Parameter	Value
Power	10KW
Nominal Frequency	50 Hz
Switching Frequency	2.55 KHz
Nominal Voltage	230 Vrms
DC link Voltage(V_{DC})	650V
Differential mode inductor(L_d)	2.2 mH
Trap capacitance (C_{trap})	4.4 µF
Trap inductance (L_{trap})	232 µH
Damping Capacitance (C_{damp})	2.2 µF
Damping Resistance(R_{damp})	30 Ω
Filter Capacitance (C_f)	2.2 µF
Grid side inductance (L_g)	3.6 mH

axis component of both the currents and voltages are also presented. To be able to make a fair comparison, for all three cases the grid faults happened at the same instance (at the same grid angle). This was possible by using a grid emulator from California Instruments.

1) No filtering of the feed forward terms

In this case the measured V_d and V_q components of the grid voltage have been fed forward to the output of the I_d and I_q current controllers.

The experimental results showing the grid currents and the grid voltages are shown by Fig. 3 and Fig. 4, respectively. As it is visible from the grid currents (Fig. 3), during steady state in the currents there are dominant harmonics such as the 5th and the 7th. Because of this the THD value of the currents is ~ 5%. Moreover, during the voltage sag the currents start to oscillate and this oscillation is damped only after a fundamental period. The capacitor voltages corresponding to the grid currents shown previously are depicted by Fig. 4. It has to be noted that the capacitor voltages contain double switching harmonics because of the high order filter.

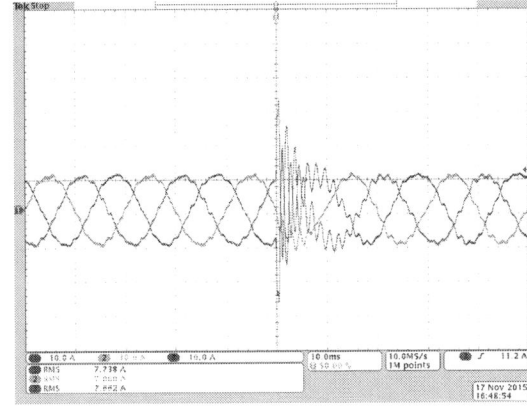

Fig. 3 Grid currents when no filtering is applied for V_d, Ch1(*blue*) grid current Phase A (I_{Ag}),Ch2(*cyan*) grid current Phase B (I_{Bg}),Ch3(*magenta*) grid current Phase C (I_{Cg})

978-1-4673-9551-9/16 $31.00 © 2016 IEEE

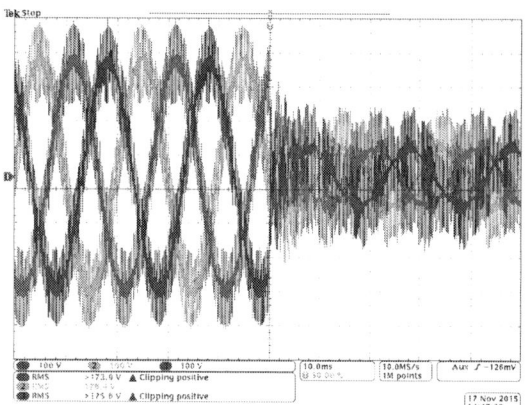

Fig. 4 Voltage on the filter capacitor with no feed filtering on the V_d;Ch1 (*blue*) Capacitor voltage Phase A ($V_{A,c}$),Ch2(*cyan*) Capacitor voltage Phase B ($V_{B,c}$),Ch3(*magenta*) Capacitor voltage Phase C ($V_{C,c}$)

2) MAF of the feed forward terms

In this case the value of V_d and V_q are fed to the controller after they have been filtered by the MAF. The system in steady state and during transient operation is depicted by the current and voltage waveform on Fig. 5 and Fig. 6, respectively.

In this case the THD value of current during steady state operation has been decreased from ~5% to ~3%. Moreover, unlike in the previous case, during the voltage sag the current is free from high frequency resonances. On the other hand, because of the filtering the overshoot is higher by 25%.

3) AMAF of the feed forward terms

The purpose of this filtering is to have the steady state behavior of the MAF and in the same time have fast transient response. During steady state the AMAF acts as a MAF, and because of this the current THD is only ~3%. However, when the algorithm detects that a fault happened, the filtering is suspended and the feed forward values are the measured ones

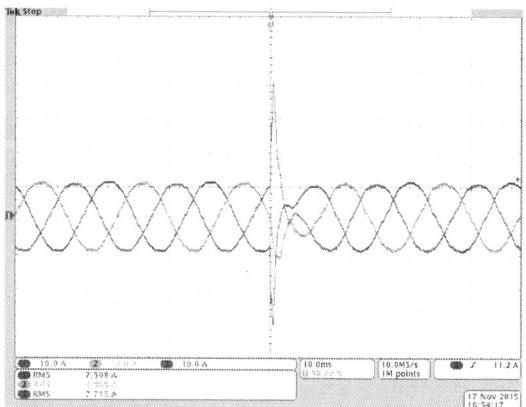

Fig. 5 Grid currents when the MAF is applied for V_d;Ch1(*blue*) grid current Phase A ($I_{A,g}$),Ch2(*cyan*) grid current Phase B ($I_{B,g}$),Ch3(*magenta*) grid current Phase C ($I_{C,g}$)

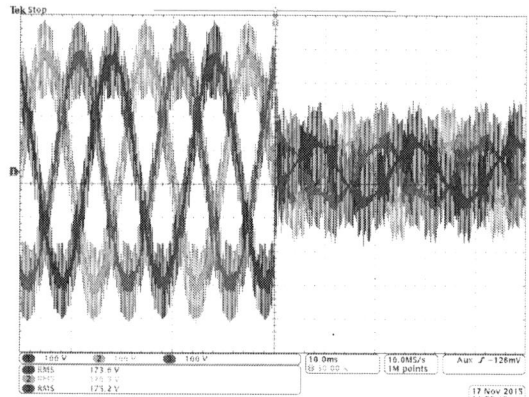

Fig. 6 Voltage on the filter capacitor when the Vd was filtered with the MAF;Ch1 (*blue*) Capacitor voltage Phase A ($V_{A,c}$),Ch2(*cyan*) Capacitor voltage Phase B ($V_{B,c}$),Ch3(*magenta*) Capacitor voltage Phase C ($V_{C,c}$)

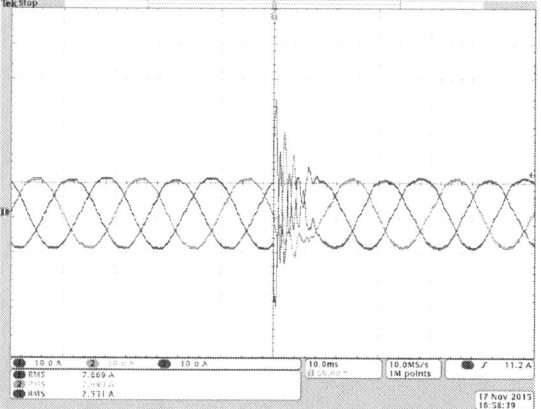

Fig. 7 Grid currents when the AMAF is applied for V_d;Ch1(*blue*) grid current Phase A ($I_{A,g}$),Ch2(*cyan*) grid current Phase B ($I_{B,g}$),Ch3(*magenta*) grid current Phase C ($I_{C,g}$)

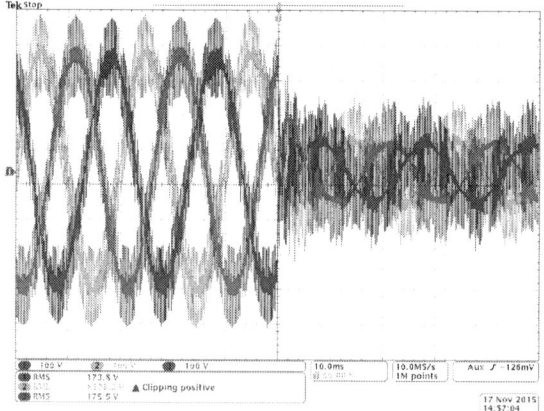

Fig. 8 Voltage on the filter capacitor when the Vd was filtered with the AMAF;Ch1 (*blue*) Capacitor voltage Phase A ($V_{A,c}$),Ch2(*cyan*) Capacitor voltage Phase B ($V_{B,c}$),Ch3(*magenta*) Capacitor voltage Phase C ($V_{C,c}$)

without any filtering. This can be seen from the current and voltage values in Fig. 7 and Fig. 8, respectively. During transient the same performance is observable as in the case where no filtering has been used, while in steady state the current THD is the same as in the case of the MAF.

IV. DISCUSSION

For a better understanding, the direct and quadrature axis voltages (V_d and V_q) and currents (I_d and I_q) were obtained and are depicted by Fig. 10 and Fig. 11, respectively.

A. Steady state operation

In order to quantify the effectiveness of the MAF, the THD levels of the currents have been compared. In the case when there was no filtering applied (Fig. 3) the THD value of the current was ~5%, while in the case when the MAF was applied (Fig. 5) the current THD decreased to ~3%. The same THD value (~3%) was also obtained for the AMAF case. Furthermore, on Fig. 9 the Fast Fourier Transform (FFT) of the grid currents during steady state operation are shown, for all 3 above mentioned cases. For the MAF and AMAF cases the dominant harmonic components are reduced significantly compared to the non-filtered case. As mentioned before, the AMAF filter works as the MAF during steady-state operations as it is also visible from Fig. 9.

B. Transient operation

During the fault, when there is no filtering applied on the V_d and V_q the grid current starts to oscillate. This phenomenon is also visible from the direct and quadrature axis current plots (Fig. 10, blue lines). However, in the case of the MAF, the current oscillation is removed, but the overshoot of the currents is higher (Fig. 7). From Fig. 10 (red lines) it is visible that the overshoot of the I_d is higher than in the non-filtered case. Moreover, the transient response is also slower. As for the I_q, the oscillation is also removed from this current, but the overshoot is much smaller compared to the non-filtered one.

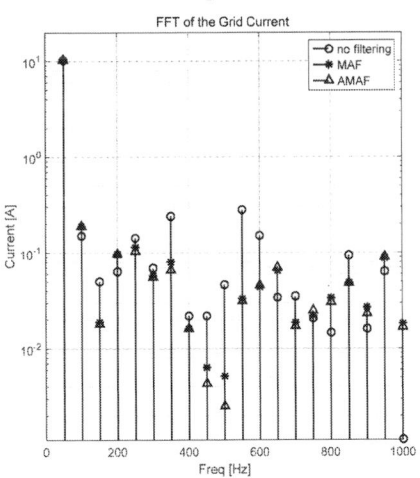

Fig. 9 FFT of the grid current for the 3 investigated cases (with no filtering, MAF and AMAF)

without any filtering. This can be seen from the current and voltage values in Fig. 7 and Fig. 8, respectively. During transient the same performance is observable as in the case where no filtering has been used, while in steady state the current THD is the same as in the case of the MAF.

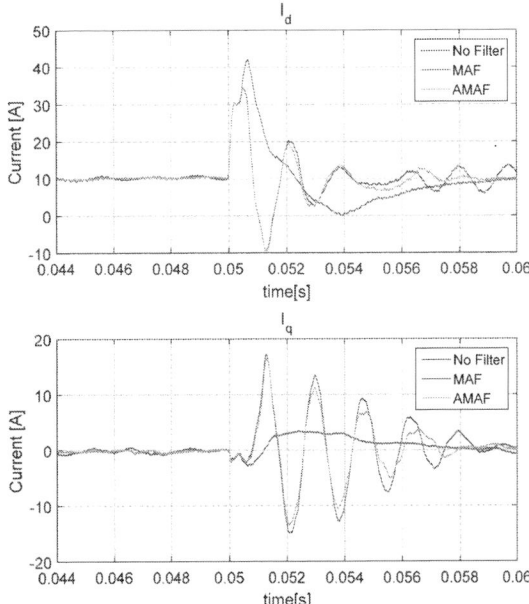

Fig. 10 I_d(top) and I_q(bottom) currents for the 3 different cases; *Blue*—no filter; *Red*—MAF; *Orange*—AMAF

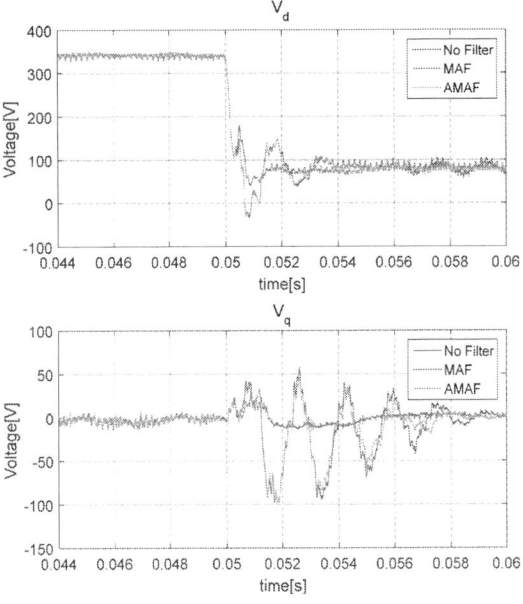

Fig. 11 V_d(top) and V_q(bottom) parts of the capacitor voltages; *Blue*—no filter; *Red*—MAF; *Orange*—AMAF

V. CONCLUSIONS

This paper uses a MAF in order to eliminate the harmonics from the grid current, which are caused by modulation and by the feed-forward of the grid voltages, which can contain also harmonics. This is done by filtering the V_d and V_q values which are used for the feed-forward part of the current controllers. The shortcoming of this method is that during transients the response of the control system will be increased along with the current overshoot. The overshoot of the grid current has been increased by 25%. To overcome this limitation the paper proposes an AMAF, which has the same performance as the MAF during steady state operation, but during transient the filtering is suspended and a faster response can be achieved (compared to the MAF). Moreover, the overshoot is also decreased, hence the proposed method helps the control system during transients to be able to remain connected.

An FFT analysis of the grid current has also been presented for the 3 cases, where the individual harmonics are presented up until 1 kHz. It is visible from the graph that the dominant frequency components in the grid current are decreased significantly in the case when either of the filtering has been applied.

A setup consisting of two parallel interleaved converters has been built, and the methods were tested on this system. The experimental results reflect the performance of the filtering method both in steady state and during transient operations.

REFERENCES

[1] F. Blaabjerg, M. Liserre, and M. Ke, "Power Electronics Converters for Wind Turbine Systems," *Industry Applications, IEEE Transactions on,* vol. 48, pp. 708-719, 2012.

[2] B. B. d. Energieund and Wasserwirtschaft e.V., "Technical guidline: Generating plants connected to the medium-voltage network," ed, 2008.

[3] G. Gohil, L. Bede, R. Teodorescu, T. Kerekes, and F. Blaabjerg, "Line Filter Design of Parallel Interleaved VSCs for High Power Wind Energy Conversion System," *Power Electronics, IEEE Transactions on,* vol. PP, pp. 1-1, 2015.

[4] B. Andresen and J. Birk, "A high power density converter system for the Gamesa G10x 4.5 MW wind turbine," in *Power Electronics and Applications, 2007 European Conference on,* 2007, pp. 1-8.

[5] J. S. S. Prasad and G. Narayanan, "Minimization of Grid Current Distortion in Parallel-Connected Converters Through Carrier Interleaving," *Industrial Electronics, IEEE Transactions on,* vol. 61, pp. 76-91, 2014.

[6] S. K. T. Miller, T. Beechner, and S. Jian, "A Comprehensive Study of Harmonic Cancellation Effects in Interleaved Three-Phase VSCs," in *Power Electronics Specialists Conference, 2007. PESC 2007. IEEE,* 2007, pp. 29-35.

[7] R. Maheshwari, G. Gohil, L. Bede, and S. Munk-Nielsen, "Analysis and modelling of circulating current in two parallel-connected inverters," *Power Electronics, IET,* vol. 8, pp. 1273-1283, 2015.

[8] G. Gohil, L. Bede, R. Teodorescu, T. Kerekes, and F. Blaabjerg, "An Integrated Inductor For Parallel Interleaved Three-Phase Voltage Source Converters," *Power Electronics, IEEE Transactions on,* vol. PP, pp. 1-1, 2015.

[9] G. Gohil, L. Bede, R. Teodorescu, T. Kerekes, and F. Blaabjerg, "An Integrated Inductor for Parallel Interleaved VSCs and PWM Schemes for Flux Minimization," *Industrial Electronics, IEEE Transactions on,* vol. PP, pp. 1-1, 2015.

[10] G. J. Capella, J. Pou, S. Ceballos, J. Zaragoza, and V. G. Agelidis, "Current-Balancing Technique for Interleaved Voltage Source Inverters With Magnetically Coupled Legs Connected in Parallel," *Industrial Electronics, IEEE Transactions on,* vol. 62, pp. 1335-1344, 2015.

[11] F. Forest, E. Laboure, T. A. Meynard, and V. Smet, "Design and Comparison of Inductors and Intercell Transformers for Filtering of PWM Inverter Output," *Power Electronics, IEEE Transactions on,* vol. 24, pp. 812-821, 2009.

[12] A. Timbus, M. Liserre, R. Teodorescu, P. Rodriguez, and F. Blaabjerg, "Evaluation of Current Controllers for Distributed Power Generation Systems," *Power Electronics, IEEE Transactions on,* vol. 24, pp. 654-664, 2009.

[13] L. Bede, G. Gohil, T. Kerekes, M. Ciobotaru, R. Teodorescu, and V. G. Agelidis, "Comparison between grid side and inverter side current control for parallel interleaved grid connected converters," in *Power Electronics and Applications (EPE'15 ECCE-Europe), 2015 17th European Conference on,* 2015, pp. 1-10.

[14] F. Blaabjerg, R. Teodorescu, M. Liserre, and A. V. Timbus, "Overview of Control and Grid Synchronization for Distributed Power Generation Systems," *Industrial Electronics, IEEE Transactions on,* vol. 53, pp. 1398-1409, 2006.

[15] P. Rodriguez, A. Luna, M. Ciobotaru, R. Teodorescu, and F. Blaabjerg, "Advanced Grid Synchronization System for Power Converters under Unbalanced and Distorted Operating Conditions," in *IEEE Industrial Electronics, IECON 2006 - 32nd Annual Conference on,* 2006, pp. 5173-5178.

[16] S. Golestan, M. Ramezani, J. M. Guerrero, F. D. Freijedo, and M. Monfared, "Moving Average Filter Based Phase-Locked Loops: Performance Analysis and Design Guidelines," *Power Electronics, IEEE Transactions on,* vol. 29, pp. 2750-2763, 2014.

[17] G. Gohil, L. Bede, R. Teodorescu, T. Kerekes, and F. Blaabjerg, "Design of the trap filter for the high power converters with parallel interleaved VSCs," in *Industrial Electronics Society, IECON 2014 - 40th Annual Conference of the IEEE,* 2014, pp. 2030-2036.

Analysis of Two Charging Modes of Battery Energy Storage System for a Stand-Alone Microgrid

Jongmin Jo and Hanju Cha

Dept. of Electrical Engineering, Chungnam National University, Daejeon, Korea

Abstract— In this paper, two charging modes of battery energy storage system (BESS) for a stand-alone microgrid are analyzed. The stand-alone microgrid system consists of 50kW BESS, 50kW diesel generator (DG) and controllable loads, where BESS is composed of 115kWh battery bank and 50kW DC-AC inverter. The operation modes of the stand-alone microgrid system are divided into four modes, and BESS is connected to DG to charge an insufficient SOC (State of Charge) of battery bank through two charging modes. Charging mode I is that BESS performs constant voltage constant frequency (CVCF) control as main source and DG operates in active power control. On contrary, charging mode II is that DG performs CVCF control as main source and BESS is charged by constant current constant voltage (CC-CV) method. The operation of BESS is similar to grid-connected characteristic in the charging mode II, where BESS is charged from DG for increase of SOC. PR voltage + P current control in the stationary reference frame for charging mode I and PI + R control in the synchronous reference frame is applied for charging mode II. Stability of two charging modes is analyzed by using root locus in the discrete time z-domain. Demonstration site is constructed and performance of the proposed two control methods is verified through experiment, where THD of charging current is 1% in charging mode I and is 2.2% in charging mode II.

I. INTRODUCTION

A stand-alone microgrid system have been rapidly grown in the whole world market. In general, a stand-alone microgrid system is based on Battery Energy Storage System (BESS) whose main role is to supply the sinusoidal voltage and current having lower harmonics to the variety of loads [1]. However, if the State of Charge (SOC) of battery bank is in an insufficient state, BESS is not possible to act as a continuous supplier. Therefore, another important point is to charge the battery bank. To maintain a lifetime and increase an efficiency of the battery bank, BESS is supplied to sinusoidal current by adopting the stable and robust current control from either renewable energy sources or Diesel Generator (DG) in case of the stand-alone microgrid system [2]. In stand-alone microgrid system, there are several operation modes and these modes are changed depending on SOC of the battery bank or external environmental conditions under connected renewable sources. In case that the BESS is connected to DG, the roles of BESS

and DG can be changed according to the operation modes when charging of BESS. Therefore, BESS requires the stable and robust control methods whether performs either Constant Voltage Constant Frequency (CVCF) or grid-connected mode. In the controller point of view, Proportional Resonant (PR) control is widely used to supply stable voltage and current and it is possible to make the steady-state error to zero in stationary reference frame [3]. The performance and principle of PR control is the same as PI control in the synchronous reference frame. To attenuate the switching frequency ripple, LCL filters have been widely employed. Usage of LCL filters is preferred due to the many advantages such as reduction of the switching frequency ripple, decrease of overall filter size and weight compared with L filter [4]. However, one of the most important problem of LCL filter can introduce a resonance that make the control system unstable considerably. Moreover, the possible wide range of grid impedance value challenges the stability and the effectiveness of the LCL-filter-based current control [5]. For solution of these problems, many methods are proposed as passive damping or active damping which are lead network, multi-loop using converter or grid current feedback with filter capacitor voltage or current, or notch filtered methods [6]-[8].

This paper analyzes two charging modes of battery energy storage system for a stand-alone microgrid. To charge SOC of the battery bank, the control methods of two charging modes are proposed and stability of control methods is analyzed by using root locus in z-domain. 50kW-class demonstration site for a stand-alone microgrid is constructed, and performance and stability of two control methods for BESS are verified through experiments.

II. PROPOSED TWO CHARGING MODES

A. Configuration of a stand-alone microgrid

Figure 1 shows configuration of a stand-alone microgrid that consist of 50kW BESS, 50kW diesel generator and controllable loads. In BESS, DC link is connected to 115kWh Lead-Acid battery bank with 48 200AH/12V batteries and LCL filter is adopted to attenuate switching frequency ripples of current. The grid inductances L_2 includes the leakage

978-1-4673-9551-9/16 $31.00 © 2016 IEEE

inductance of isolation transformer and synchronous inductance of diesel generator. System parameters are listed in Table I.

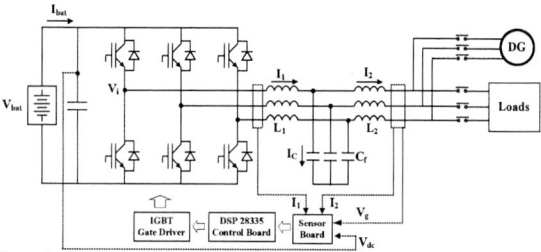

Figure 1: The configuration of a stand-alone microgrid

Table I System parameters

Parameters	Symbol	Values
Rated power	P	50kW
Rated voltage	V_g	$380V_{rms}$
Switching frequency	f_s	10kHz
Inverter side inductance	L_1	1.04mH
Grid side inductance	L_2	7.89mH
Filter capacitor	C_f	32uF

B. The operation modes of the stand-alone microgrid

Figure 2 shows the operation mode of the stand-alone microgrid, which are divided into four modes. In normal mode, BESS performs CVCF and DG stops. In charge mode, BESS is controlled by CVCF and DG performs active power control. It corresponds to charging mode I when charging from DG. On the contrary, in manual mode, DG performs CVCF control and BESS is operated by CC-CV control. It corresponds to charging mode II. When SOC of battery bank is insufficient, BESS is normally charged through DG in either charging mode I or charging mode II depending on operating modes.

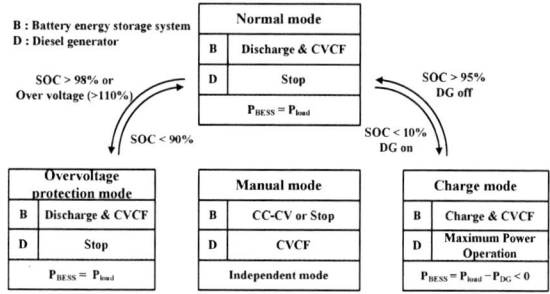

Figure 2: The operation mode of the stand-alone microgrid

C. Control in two charging modes

Figure 3 shows control methods of the two charging modes in the stand-alone microgrid. Figure 3 (a) shows control in charging mode I that includes an outer PR voltage feedback

loop for line to line voltage and an inner P current feedback loop. In the outer voltage loop, PR controller regulates fundamental positive/negative sequence voltage as well as selective harmonics compensation for three, fifth, seventh. Damping of the resonance of LCL filter is possible to by controlling capacitor current which are calculated by the difference of grid and inverter currents. Figure 3 (b) shows control in charging mode II where the outer loop regulates dc link voltage to maintain battery bank voltage and the inner current loop maintains unity power factor. Current control is based on PI controller in the synchronous reference frame and R controller is adopted to compensate lower harmonics. The control methods must guarantee the performance and the stability of the whole system.

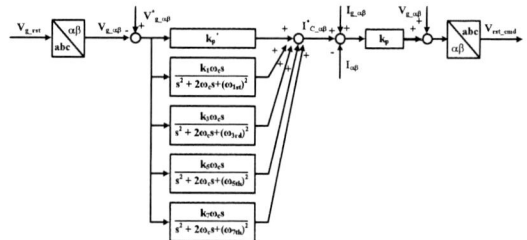

(a) Charging mode I

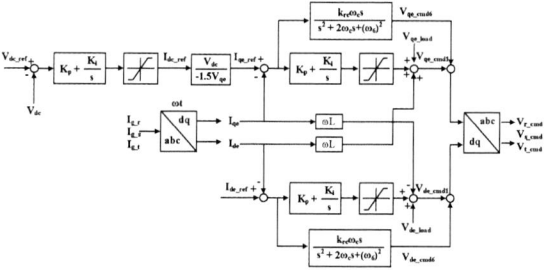

(b) Charging mode II

Figure 3: Control of two charging methods

III. ANALYSIS IN THE STABILITY OF TWO CHARGING MODES

A. Stability of the charging mode I

In the stand-alone microgrid, the way to charge battery bank of BESS is to connection with DG either charging mode I or charging mode II. Figure 4 shows continuous time modeling of voltage and current control loops for charging mode I that includes an outer PR voltage feedback loop for line to line voltage and an inner P current feedback loop. (1) is to the transfer function of PR controller for fundamental voltage control. An inner current feedback loop controls the filter capacitor current. It means that active damping of multi-loop methods is applied to attenuate resonance influence of LCL filters as well as improve stability of voltage control. In charging mode I, the role and the stability of the voltage control are very important since BESS operates CVCF control as main source. In this paper, system delay is taken the computational and PWM modulator delay in terms of $1.5T_s$ delay as shown (2). For converting discrete time modeling from continuous time modeling, PR controller and LCL filters are transformed by tustin and ZOH methods, respectively. System delay is represented to one step delay.

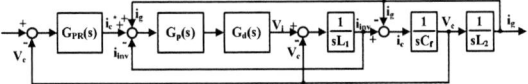

Figure 4: Continuous time modeling of current control for charging mode I

$$G_{PR}(s) = k_p + \frac{k_i \omega_c s}{(s^2 + 2\omega_c s + \omega_0^2)} \tag{1}$$

$$G_d(s) = e^{-1.5T_s * s} \tag{2}$$

Figure 5 shows the stability of controller for charging mode I using root locus in z-domain in accordance with increase of PR voltage controller proportional gain k_p. In root locus of charging mode I, all poles are located in unit circle at selected gain k_p. The overall control system is marginally stable because the location of a conjugate pair of resonance poles is near to unit circle line. However, the stability of the system is guaranteed through the robust and stable multi-loop control method at selected gain k_p.

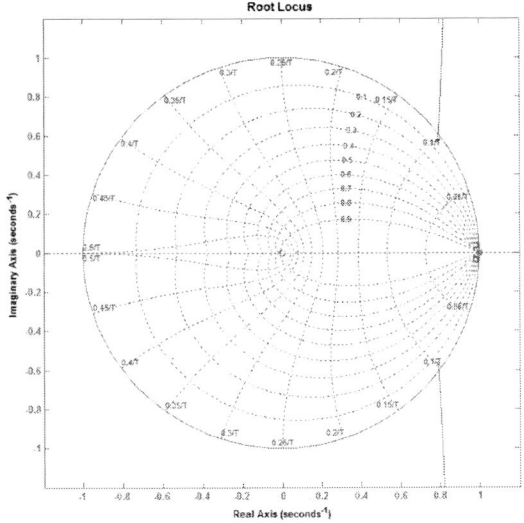

Figure 5: The stability of charging mode I using root locus in z-domain

B. Stability of the charging mode II

Figure 6 shows continuous time modeling of current control loop for charging mode II. Each block represents PI controller, system delay and LCL filter transfer functions, respectively. (3) is to the frequency transfer functions of PI controller. (4) is to the frequency transfer function of grid current feedback loop with LCL filters. In Figure 6, d-q decoupling terms and voltage feedback is not taken into consideration. For converting discrete time modeling from continuous time modeling, PI controller is transformed by tustin method.

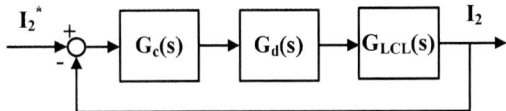

Figure 6: Continuous time modeling of current control for charging mode II

$$G_c(s) = k_p \left(\frac{T_i s + 1}{T_i s} \right) \tag{3}$$

$$G_{LCL}(s) = \frac{I_2(s)}{V_i(s)} = \frac{1}{L_1 s} \frac{z_{LC}^2}{(s^2 + \omega_{res}^2)} \tag{4}$$

Current control with LCL filters can introduce a resonance problem that make the control system unstable considerably. In charging mode II, passive damping method that passive resistors are connected to filter capacitors in series is adopted to attenuate resonance influence. Therefore, Transfer function of (4) is replaced with (5). The small passive resistors values as 0.3Ω is selected in order to minimize the losses caused by passive damping resistors and that is small enough to ignore. For discrete modeling, (5) is transformed by ZOH method .

$$G_{LCL}(s) = \frac{I_2(s)}{V_i(s)} = \frac{1}{L_1 s} \frac{s \frac{R_d}{L_2} + z_{LC}^2}{(s^2 + sR_d C_f \omega_{res}^2 + \omega_{res}^2)} \tag{5}$$

Figure 7 shows the stability of controller for charging mode I using root locus in z-domain in accordance with increase of PI current controller proportional gain k_p. It shows root locus when grid side inductance is to 7.89mH that synchronous inductance of diesel generator is summed and the resonance frequency is to 943Hz. In this case, all poles are located in unit circle at selected gain k_p. The overall control system is marginally stable because a conjugate pair of resonance poles get closed to unit circle line. However, the robust current control guarantees the stability of the system at selected gain k_p through attenuation of resonance influence by using passive damping method. A conjugate pair of resonance poles can closely move to origin by increasing passive resistor values.

978-1-4673-9551-9/16 $31.00 © 2016 IEEE

This results in the stability improvement of current control. But, it leads to the growth of power losses.

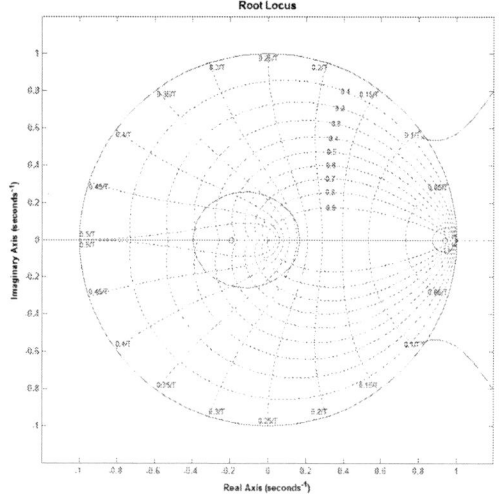

Figure 7: The stability of charging mode II using root locus in z-domain

IV. EXPERIMENTAL RESULTS

Figure 8 shows the hardware configuration of the stand-alone microgrid that consists of 50kW BESS and 50kW diesel generator. DC link is connected to 115kWh battery bank that is composed of Lead-Acid battery with 48 200AH/12V. To verify stability of control between BESS and diesel generator, controllable loads is neglected.

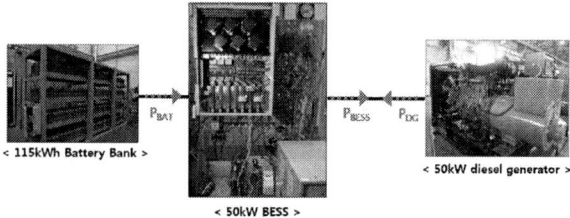

Figure 8: Hardware configuration of Battery Energy Storage System and Diesel Generator

A. Charging mode I

Figure 9 shows experimental results of charging mode I between BESS and diesel generator. Figure 9 (a) shows that initial charge is performed as the stable state from 0kW to 40kW as well and Figure 9 (b) shows line to line voltage by CVCF of BESS and three-phase current in 40kW steady-state. PR + P controller of BESS performs very well since it makes the steady-state error to zero and compensates low frequency harmonics perfectly. Resonance has no influence by using multi-loop method for active damping. THD of current is to

1%. Likewise, the performance and the stability of BESS is satisfactory as shown in Figure 9. The experimental results are equal to the stability analysis of charging mode I using root locus in z-domain.

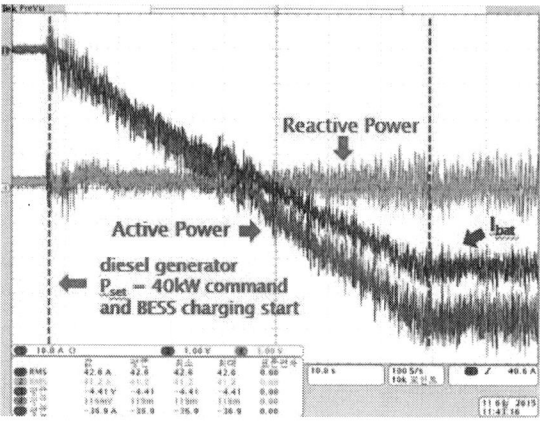

(a) BESS 40kW charging

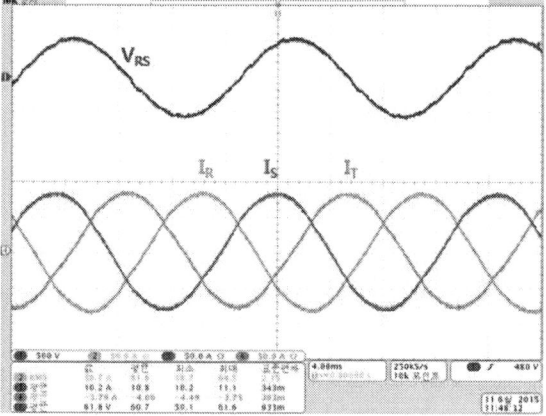

(b) Voltage and current in 40kW

Figure 9: Experimental results of charging mode I

B. Charging mode II

Figure 10 shows the experimental results of charging mode II between BESS and diesel generator. Figure 10 (a) shows that initial charge is performed as the stable state from 0kW to 40kW without resonance problem. Figure 10 (b) shows the controller DA output about CC-CV operation. Figure 10 (c) shows three-phase current in 40kW steady-state and BESS is charged from the stable three-phase current by PI + R current control. THD of current is to 2.2%. The performance of the current control is satisfactory as shown in Figure 10. The experimental results are equal to the stability analysis of charging mode II using root locus in z-domain.

978-1-4673-9551-9/16 $31.00 © 2016 IEEE   1711

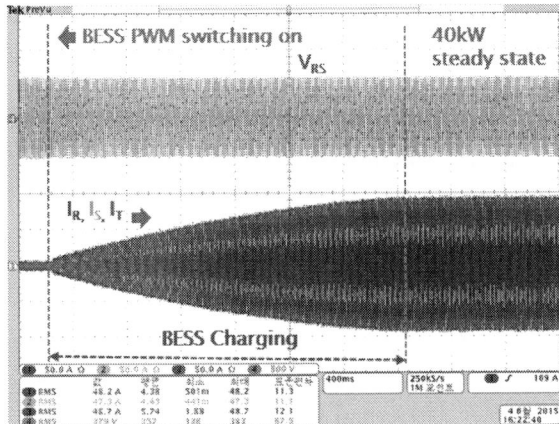

(a) BESS 40kW charging

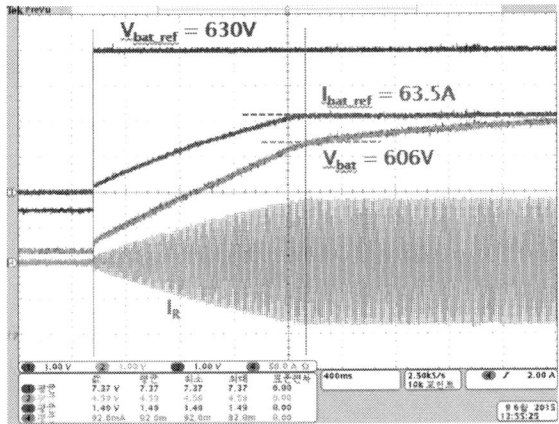

(b) CC-CV operation

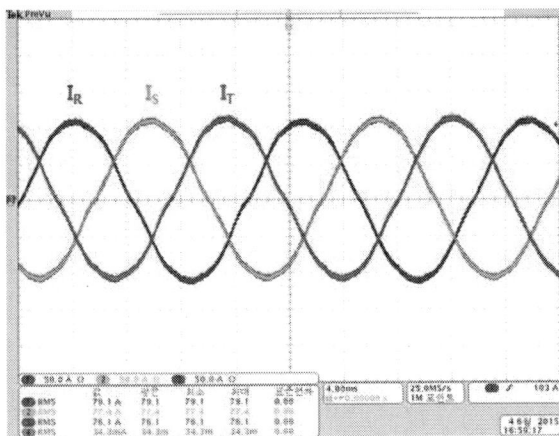

(c) Phase current in 40kW

Figure 10: Experimental results of charging mode II

V. CONCLUSION

In this paper, two charging modes of battery energy storage system for a stand-alone microgrid have been analyzed. BESS has been connected to DG to charge an insufficient SOC of battery bank through two charging modes. Charging mode I is that BESS performs CVCF control, which has been controlled by PR voltage + P current control, and DG operates active power control. In charging mode I, multi-loop method for active damping has been adopted to damp resonance. On contrary, charging mode II is that DG performs CVCF control and BESS is charged by CC-CV method, which has been controlled by PI + R current control. In this case, Passive damping method has been applied to attenuate resonance influence. The stability of two charging modes have been analyzed by using root locus in the discrete time z-domain and two charging modes have been drawn a stable conclusion that all poles have been located in unit circle at selected gain k_p. Demonstration site has been constructed and performance of the proposed two control methods have been verified through experiment, where THD of charging current is 1% in charging mode I and is 2.2% in charging mode II.

REFERENCES

[1] J. Philip, B. Singh and S. Mishra, "Design and operation for a standalone DG-SPV-BES microgrid system", Power India International Conference (PIICON), 2014 6th IEEE, pp.1-6, 5-7 Dec. 2014.

[2] Bo Zhao, Xuesong Zhang, Jian Chen and Caisheng Wang, "Operation Optimization of Standalone Microgrids Considering Lifetime Characteristics of Battery Energy Storage System", IEEE Trans. Sustainable Energy, vol. 4, no. 4, pp. 934-943, Jan 2013.

[3] P. C. Loh, M. J. Newman, D. N. Newman and D. G. Holmes, "A comparative analysis of multiloop voltage regulation strategies for single and three-phase UPS systems", IEEE Trans. Power Electronics, vol. 18, no. 5, pp. 1176-1185, Sep 2003.

[4] S. G. Parker, B. P. McGrath and D. G. Holmes, "Regions of Active Damping Control for LCL Filters", IEEE Trans. Industry Applications, vol. 50, no. 1, pp. 424-432, Jan. 2014.

[5] R. Pene-Alzola, M. Liserre, F. Blaabjerg and M. Ordonez, "LCL-Filter Design for Robust Active Damping in Grid-Connected Converters", IEEE Trans. Industrial Informatics, vol. 10, no. 4, pp. 2192-2203, Nov. 2014.

[6] R. Pene-Alzola, M. Liserre, F. Blaabjerg, M. Ordonez and T. Kerekes, "A Self-commissioning Notch Filter for Active Damping in a Three-Phase LCL-Filter-Based Grid-Tie Converter", IEEE Trans. Power Electronics, vol. 29, no. 12, pp. 6754-6761, Dec 2014.

[7] J. Dannehl, C. Wessels and F. W. Fuchs, "Limitations of Voltage-Oriented PI Current Control of Grid-Connected PWM Rectifiers With LCL Filters", IEEE Trans. Industrial Electronics, vol. 56, no. 2, pp. 380-388, Feb 2009.

[8] M. Liserre, R. Teodorescu and F. Blaabjerg, "Stability of Photovoltaic Wind Turbine Grid-Connected Inverters for a Large Set of Grid Impedance Values", IEEE Trans. Power Electronics, vol. 21, no. 1, pp. 263-272, Jan 2006.

Proposition and Experimental Verification of a Bi-Directional Isolated DC/DC Converter for Battery Charger-Discharger of Electric Vehicle

Ryota Kondo, Yusuke Higaki and Masaki Yamada
Advanced Technology R&D Center
Mitsubishi Electric Corporation
Amagasaki, Hyogo, Japan
Kondo.Ryota@db.MitsubishiElectric.co.jp

Abstract— This paper proposes a bi-directional isolated DC/DC converter for the battery charger and discharger of electric vehicles controlled by a new phase-shift topology. The proposed DC/DC converter consists of two full-bridge inverters, an isolated transformer, and boost reactors. The converter provides bi-directional transmission, buck-boost conversion and zero-voltage switching, adjusting the phase-shift of two full bridge inverters continuously. A 400 V - 3.5 kW experimental system provides stable bi-directional buck-boost conversion and seamless transition between charging and discharging modes. A maximum efficiency of 93.3 % was achieved for this experimental system.

Keywords— *Bi-directional isolated DC/DC converter, Phase-shift control, Battery, Electric vehicle(EV)*

I. INTRODUCTION

In recent years, smart houses are attracting much attention for saving energy. A smart house is composed of a grid and renewable energy sources (photovoltaic, battery, fuel cell and so on) , and realizes optimum power management. Also an emergency power feed from the renewable energy sources to the house in an unpredictable disaster [1]. Fig. 1 shows the system configuration of a smart house [2]. Distributed energy sources consist of photovoltaic (PV) and battery fitted to an electric vehicle (EV battery). A power conditioner for EV battery (EV-PCS) is connected between the EV battery and the grid and compensates for the different power between the generating power from the PV and electricity consumption in the grid. The EV-PCS is integrated with a bi-directional isolated DC/DC converter that is a key device of the system. This converter provides the 3 functions below.

- Galvanic isolation between the EV battery and grid.

- Bi-directional operation for charging and discharging the EV battery.

- The buck mode or boost mode operation corresponding to different battery voltages.

In particular, this converter needs to change the charging mode and the discharging mode quickly and seamlessly to compensate for the unpredictable generating power in the PV or a steep consumption power in the grid.

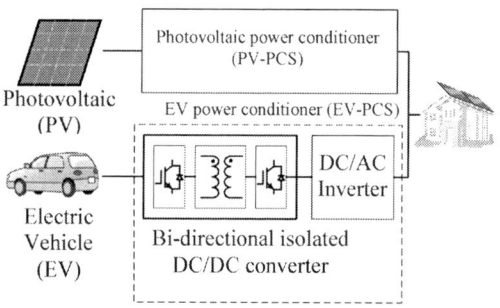

Fig. 1. System configuration of PV power conditioner and EV power conditioner linked by AC power line.

The DAB (dual-active-bridge) type converter was proposed for a bi-directional isolated DC/DC converter [3]-[4]. Since then, the 2-level type DAB converter and 3-level converter have been mainly studied. Compared to the 2-level DAB converter [5]-[9], the 3-level type converter can lower the voltage applied to the current high frequency reactors connected in the primary and secondary circuits of the isolation transformer [10]. However, the 3-level type DAB converter requires three control variables consisting of two variables of the phase-shift amount for each of the two full bridge inverters, and one variable of the phase-shift amount between two full bridge inverters, resulting in a complicated control system [11].

To resolve this issue, we have now developed a 3-level type DAB converter having only one control variable. The proposed driving method synchronizes one leg of each full bridge inverter connected to both sides of the isolation transformer, and controls the phase shift amount of either one of the two full bridge inverters seamlessly for a stable buck-boost/bi-directional power transfer and seamless adjustment of charging and discharging power.

This paper introduces the circuit configuration of the proposed bi-directional isolated DC/DC converter and the specifications of the phase shift control. Then, the steady-state charge and discharge characteristics, seamless switching

978-1-4673-9551-9/16 $31.00 © 2016 IEEE

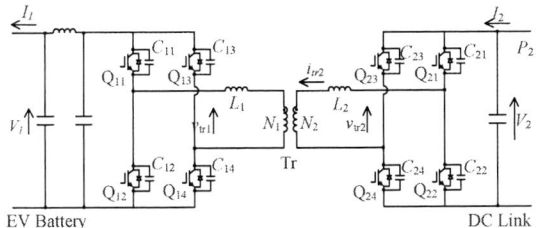

Fig. 2. Circuit structure of the proposed bi-directional DC/DC converter.

characteristics between charging and discharging modes, and the efficiency characteristics of the rated operation, are presented based on the test results obtained from the 400 V - 3.5 kW experimental system.

II. CIRCUIT DESCRIPTION

A. Circuit Configuration

Fig. 2 depicts a circuit diagram of the proposed bi-directional isolated DC/DC converter. As shown in Fig. 2, the proposed bi-directional isolated DC/DC converter consists of two full bridge inverters, one isolation transformer and high frequency reactors (AC reactor) connected to both sides of the isolation transformer. The side in the EV battery defines the primary side, and the side in the DC Link defines the secondary side. The current flow from the secondary side to the primary side defines the charging mode, and the opposite current flow (from primary side to secondary side) defines the discharging mode. Semiconductors (Q_{11}-Q_{24}) which configure both full bridge inverters have loss-less snubber capacitors (C_{11}-C_{24}). In the charging mode, L_1 and L_2 act as Zero Voltage switching [12] with C_{21}-C_{24} at the secondary side inverter, also L_1 and L_2 act as a boost reactor at the primary side inverter. In the discharging mode, L_1 and L_2 act as a boost reactor at the secondary side inverter, also L_1 and L_2 act as Zero Voltage switching with C_{11}-C_{14} at the primary side inverter.

B. Operating theory

The proposed bi-directional isolated DC/DC converter is operated by the phase shift control in the primary and secondary side. Fig. 3 depicts the switching pattern specifications at the gate of switching semiconductors (Q_{11}-Q_{24}) in controlling the phase shift amount of the primary side inverter. Fig. 4 depicts the switching pattern specifications at the gate of the switching semiconductors (Q_{11}-Q_{24}) in controlling the phase shift amount of the secondary side inverter. As seen in Fig. 3 and Fig. 4, all semiconductors are driven at a 50% duty cycle, the switching pattern of Q_{11} and Q_{12} is synchronized to that of Q_{21} and Q_{22}, and also the dead time ($\theta_{td} T$) is set to the switching timing of each leg. The gating phases for Q_{11} and Q_{12} are fixed and those for Q_{13} and Q_{14} are shifted by θ_1 in the primary side and the gating phase for Q_{21} and Q_{22} are fixed and those for Q_{23} and Q_{24} are shifted by θ_2 in the secondary side. Here θ_1 and θ_2 are dimensionless because of being standardized by T. Fig. 5 depicts the

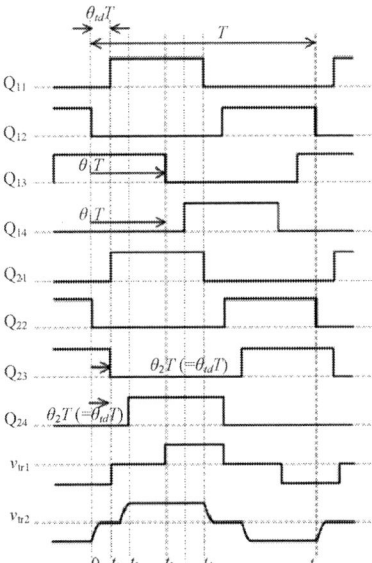

Fig. 3. Switching patterns when $\theta_2 = \theta_{td}$ and $\theta_1 > \theta_{td}$.

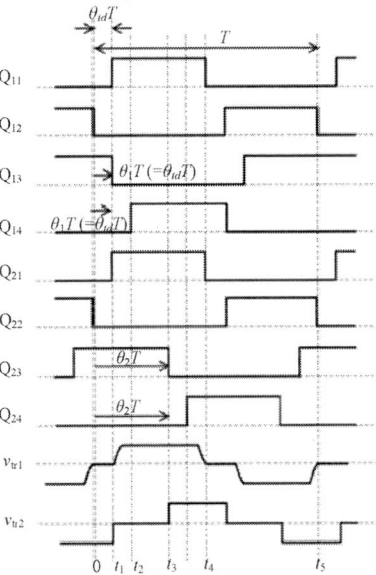

Fig. 4. Switching patterns when $\theta_1 = \theta_{td}$ and $\theta_2 > \theta_{td}$.

conceptual diagram between a phase shift specification and the charging/discharging power of the bi-directional DC/DC converter at the three conditions (a-c). The proposed control method controls the phase shift amount of either one of the two full bridge inverters seamlessly and fixes that of other one to the initial value θ_{td}. The switching condition of θ_1 or θ_2 between charging mode and discharging mode is varied by the

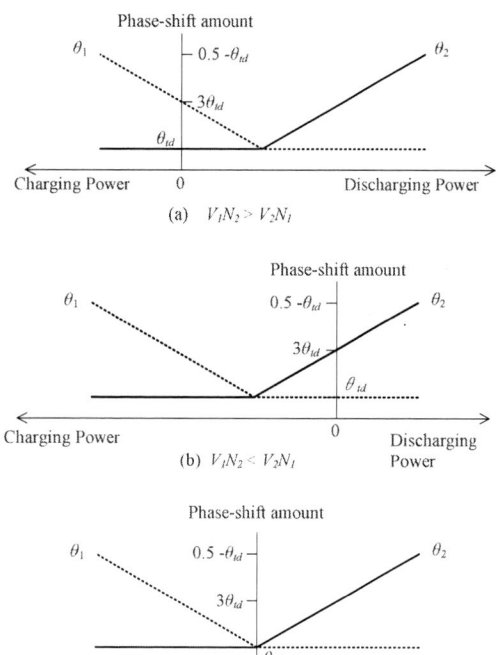

(a) $V_1N_2 > V_2N_1$

(b) $V_1N_2 < V_2N_1$

(c) $V_1N_2 = V_2N_1$

Fig. 5. The conceptual diagram between a phase shift specification and charging - discharging power of the bi-directional DC/DC converter.

relationship between V_1N_2 and V_2N_1. In the condition $V_1N_2 > V_2N_1$ shown in Fig. 5(a), the boundary condition between charging mode and discharging mode is $\theta_1 = 3\theta_{td}$. Adjusting θ_1 to over $3\theta_{td}$ seamlessly (θ_2 is fixed to θ_{td}) makes the charging power increase. By contrast, adjusting θ_1 to under $3\theta_{td}$ and θ_2 to over θ_{td} seamlessly (θ_1 is fixed to θ_{td}) makes the discharging power increase. In this condition, the charging operation is defined as a boost-charging mode and discharging operation is defined as a buck-discharging mode. In the condition $V_1N_2 < V_2N_1$ shown in Fig. 5(b), the boundary condition is $\theta_2 = 3\theta_{td}$. Adjusting θ_2 to over $3\theta_{td}$ seamlessly (θ_1 is fixed to θ_{td}) makes the discharging power increase. By contrast, adjusting θ_2 to under $3\theta_{td}$ and θ_1 to over θ_{td} seamlessly (θ_2 is fixed to θ_{td}) makes the charging power increase. In this condition, the charging operation is defined as a buck-charging mode, and the discharging operation is defined as a boost-discharging mode. In the condition $V_1N_2 = V_2N_1$ shown in Fig. 5(c), the boundary condition is $\theta_1 = \theta_2 = 3\theta_{td}$. Adjusting θ_1 to over θ_{td} seamlessly (θ_2 is fixed to θ_{td}) makes the charging power increase. By contrast, adjusting θ_2 to over θ_{td} seamlessly makes the discharging power increase.

The operating theory in the condition $V_1N_2 > V_2N_1$ shown in Fig. 5(a) can be explained by Fig. 6 and Fig. 7. Fig. 6 shows the current flow in boost-charging mode by adjusting θ_1. Terms (a) $t_2 - t_3$ and (b) $t_3 - t_4$ are defined in Fig. 3. In the terms $t_2 - t_3$ as seen in Fig. 6(a), Q_{21} and Q_{24} are turned on. The v_{tr2} was injected to the isolated transformer and i_{tr1} flows from

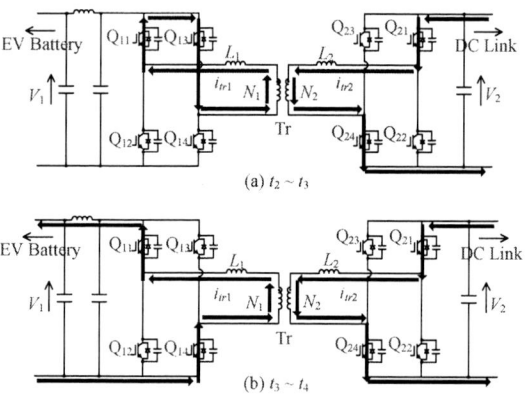

Fig. 6. Operation in boost charging mode.

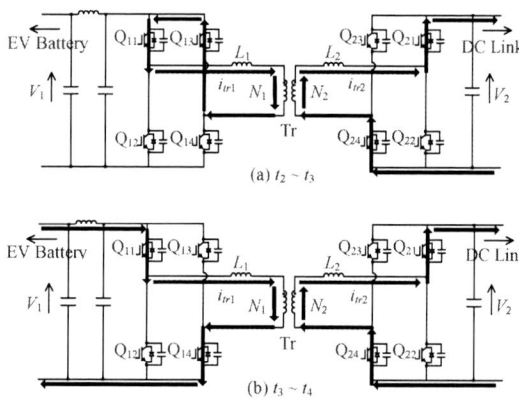

Fig. 7. Operation in buck discharging mode.

the DC Link to the isolated transformer. However, i_{tr2} is free-wheeled in the primary side and L_1, L_2 are excited because Q_{11} and Q_{13} are turned on. In the terms $t_3 - t_4$ as seen in Fig. 6(b), Q_{11} and Q_{14} are turned on and i_{tr2} flows to EV battery through Q_{11} and Q_{14}. This operating method, as seen in Fig. 6, realizes a boost operation in adjusting the term ratio of $t_2 - t_3$ and $t_3 - t_4$ to control the excited power of L_1 and L_2. It should be noted that term $0 - t_2$ can be ignored because term $0 - t_2$ is a dead time and doesn't have any influence on boost operation. The operation principle in term $t_4 - t_5$ is same as that in $0 - t_4$ according to the symmetry. Fig. 7 shows the current flow in buck-discharging mode by adjusting θ_1. The terms (a) $t_2 - t_3$ and (b) $t_3 - t_4$ are defined in Fig. 3. In the terms $t_2 - t_3$ as seen in Fig. 7(a), Q_{11} and Q_{13} are turned on and there is no power transmitted from the EV battery. The i_{tr1} is free-wheeled in the primary side. In the terms $t_3 - t_4$ as seen in Fig. 7(b), Q_{11} and Q_{14} are turned on and i_{tr1} is flowed from the EV battery to the DC Link through Q_{11} and Q_{14}. This operating method, as seen in Fig. 7, realizes a buck-operation because of adjusting the discharge ratio from the EV battery. It should also be noted that term $0 - t_2$ can be ignored because term $0 - t_2$ is a dead time and doesn't have any influence on buck operation. The

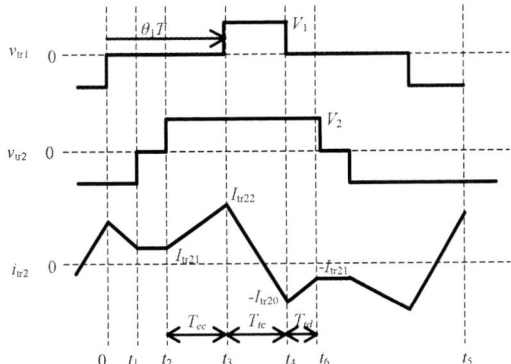

Fig. 8. Idealizing operating waveforms for analysis. ($V_1 N_2 > V_2 N_1$)

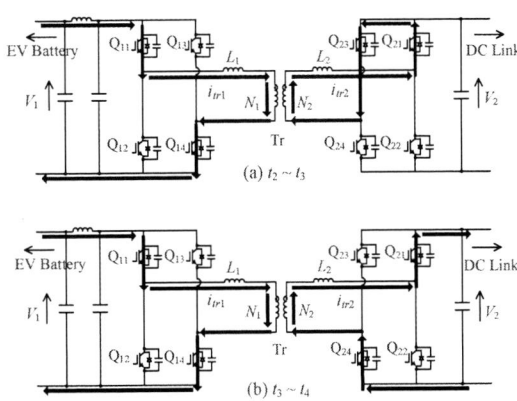

Fig. 9. Operation in boost discharging mode.

operation principle in term t_4 - t_5 is same as that in 0 - t_4 according to the symmetry.

Fig. 8 depicts the idealized operation waveforms of v_{tr1}, v_{tr2}, and i_{tr2} in the condition $V_1 N_2 > V_2 N_1$. The operating condition of v_{tr1}, v_{tr2}, and i_{tr2} as seen in Fig. 8 is limited to the right load condition so that the polarity of i_{tr2} is switched and reactive power is generated to model the transition operating between charging mode and discharging mode. The following theoretical analysis proves seamless transition of charging/discharging power to the seamless adjustment of θ_1. In Fig. 8, the excited term (or free-wheeling term)T_{ec}, the transmitted term T_{tc}, and the dead time T_{td} can be expressed as follows:

$$T_{ec} = (\theta_1 - \theta_2 - \theta_{td})T \qquad (1)$$

$$T_{tc} = (1 - 2\theta_1)\frac{T}{2} \qquad (2)$$

$$T_{td} = \theta_{td}T \qquad (3)$$

Then, i_{tr2} in excited term T_{ec} can be expressed as,

$$\{L_2 + (N_2 / N_1)^2 L_1\}\frac{(I_{tr22} - I_{tr21})}{(\theta_1 - \theta_2 - \theta_{td})T} = V_2 \qquad (4)$$

Then, i_{tr2} in transmitted term T_{tc} can be expressed as,

$$\{L_2 + (N_2 / N_1)^2 L_1\}\frac{(I_{tr22} + I_{tr20})}{(1/2 - \theta_1)T} = V_1\left(\frac{N_2}{N_1}\right) - V_2 \qquad (5)$$

Then, i_{tr2} in dead time T_{td} can be expressed as,

$$\{L_2 + (N_2 / N_1)^2 L_1\}\frac{(I_{tr20} - I_{tr21})}{\theta_{td}T} = V_2 \qquad (6)$$

The charging/discharging power P_2 can be calculated by summing the power in T_{ec}, the power in T_{tc}, and the power in T_{td}. Finally, considering switching frequency f_{sw}, P_2 can be expressed as,

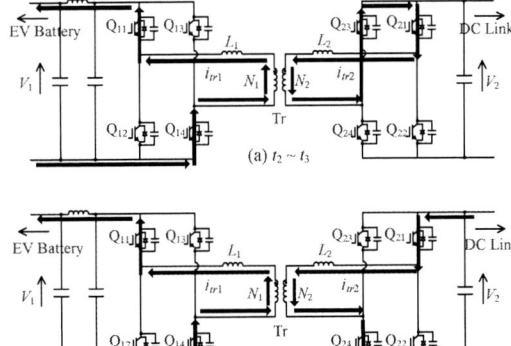

Fig. 10. Operation in buck charging mode.

$$P_2 = \begin{bmatrix} V_2 \cdot \dfrac{I_{tr21} + I_{tr22}}{2} \cdot (\theta_1 - \theta_2 - \theta_{td})T - V_2 \cdot \dfrac{I_{tr20} + I_{tr21}}{2} \cdot \theta_{td}T \\[2mm] + \dfrac{1}{2}\{L_2 + (N_2 / N_1)^2 L_1\}\dfrac{V_2 \cdot I_{tr22}^2}{\{V_1(N_2 / N_1) - V_2\}} \\[2mm] - \dfrac{1}{2}\{L_2 + (N_2 / N_1)^2 L_1\}\dfrac{V_2 \cdot I_{tr20}^2}{\{V_1(N_2 / N_1) - V_2\}} \end{bmatrix} 2f_{sw}$$

$$= \frac{V_1(N_2 / N_1) \cdot V_2 \cdot (\theta_1 - 3\theta_{td})\left(\dfrac{1}{2} - \theta_1\right)T^2}{\{L_2 + (N_2 / N_1)^2 L_1\}}f_{sw} \qquad (7)$$

From equation (7), the polarity of P_2 becomes positive in the condition $\theta_1 > 3\theta_{td}$. Symmetrically the polarity of P_2 becomes negative in the condition $\theta_1 < 3\theta_{td}$. The charging and discharging modes seamlessly transfer at the boundary condition of $\theta_1 = 3\theta_t$.

The operating theory in the condition $V_1 N_2 < V_2 N_1$ shown in Fig. 5(b) are explained by Fig. 9 and Fig. 10. Fig. 9 shows the current flow in boost-discharging mode by adjusting θ_2. Terms (a) t_2 - t_3 and (b) t_3 - t_4 are defined in Fig. 4. In the terms t_2 - t_3 as seen in Fig. 9(a), Q_{11} and Q_{14} are turned on. The v_{tr1} was injected to the isolated transformer and i_{tr2} flows from the EV battery to the isolated transformer. However, i_{tr1} is free-wheeled in the secondary side and L_1, L_2 are excited because Q_{21} and Q_{23} are turned on. In the terms t_3 - t_4 as seen in Fig. 9(b), Q_{21} and Q_{24} are turned on and i_{tr2} flows to the DC Link through Q_{21} and Q_{24}. This operating method, as seen in Fig. 9, realizes a boost operation in adjusting the term ratio of t_2 - t_3 and t_3 - t_4 to control the excited power of L_1 and L_2. Where it is noted that term 0 - t_2 can be ignored because term 0 - t_2 is a dead time and doesn't have any influence on boost operation. The operation principle in term t_4 - t_5 is the same as that in 0 - t_4 according to the symmetry. Fig. 10 shows the current flow in buck-charging mode by adjusting θ_2. The terms (a) t_2 - t_3 and (b) t_3 - t_4 are defined in Fig. 4. In the terms t_2 - t_3 as seen in Fig. 10(a), Q_{21} and Q_{23} are turned on and there is no power transmitted from the DC Link. The i_{tr2} is free-wheeled in the secondary side. In the terms t_3 - t_4 as seen in Fig. 10(b), Q_{21} and Q_{24} are turned on and i_{tr2} is flowed from the DC Link to the EV battery through Q_{21} and Q_{24}. This operating method, as seen in Fig. 10, realizes a buck operation because of adjusting the discharge ratio from the DC Link. It is also noted that term 0 - t_2 can be ignored because it is a dead time and doesn't have any influence on the buck operation. The operation principle in term t_4 - t_5 is the same as that in 0 - t_4 according to the symmetry.

Fig. 11 depicts idealizing operation waveforms of v_{tr1}, v_{tr2}, and i_{tr2} in the condition $V_1 N_2 < V_2 N_1$. The operating condition of v_{tr1}, v_{tr2}, and i_{tr2} as seen in Fig. 11 is also limited to the right load condition as well as in condition $V_1 N_2 > V_2 N_1$. The following theoretical analysis proves the seamless transition of charging/discharging power to the seamless adjustment of θ_2. In Fig. 11, the excited term (or free-wheeling term) T'_{ec}, the transmitted term T'_{tc}, and the dead time T'_{td} can be expressed as,

$$T'_{ec} = (\theta_2 - \theta_1 - \theta_{td})T \tag{8}$$

$$T'_{tc} = (1 - 2\theta_2)\frac{T}{2} \tag{9}$$

$$T'_{td} = \theta_{td}T \tag{10}$$

Then, i_{tr2} in excited term T'_{ec} can be expressed as,

$$\{L_2 + (N_2 / N_1)^2 L_1\}\frac{(I'_{tr22} - I'_{tr21})}{(\theta_2 - \theta_1 - \theta_{td})T} = V_1\left(\frac{N_2}{N_1}\right) \tag{11}$$

Then, i_{tr2} in transmitted term T'_{tc} can be expressed as,

$$\{L_2 + (N_2 / N_1)^2 L_1\}\frac{(I'_{tr20} + I'_{tr22})}{(1/2 - \theta_2)T} = V_2 - V_1\left(\frac{N_2}{N_1}\right) \tag{12}$$

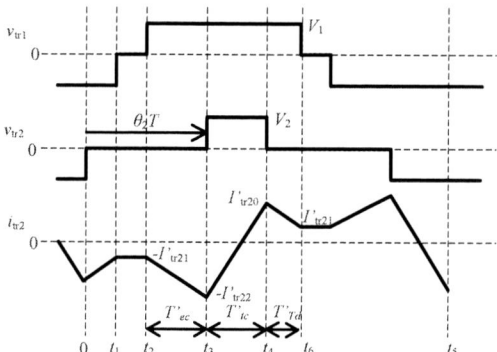

Fig. 11. Idealizing operating waveforms for analysis. ($V_1 N_2 < V_2 N_1$)

Then, i_{tr2} in dead time T'_{td} can be expressed as,

$$\{L_2 + (N_2 / N_1)^2 L_1\}\frac{(I'_{tr20} - I'_{tr21})}{\theta_{td}T} = V_1\left(\frac{N_2}{N_1}\right) \tag{13}$$

The charging/discharging power P_2 can be calculated by summing the power in T'_{ec}, the power in T'_{tc}, and the power in T'_{td}. Finally, considering the switching frequency f_{sw}, P_2 can be expressed as,

$$P_2 = \begin{bmatrix} \frac{1}{2}\{L_2 + (N_2/N_1)^2 L_1\}\frac{V_2 \cdot I'^2_{tr20}}{\{V_2 - V_1(N_2/N_1)\}} \\ -\frac{1}{2}\{L_2 + (N_2/N_1)^2 L_1\}\frac{V_2 \cdot I'^2_{tr22}}{\{V_2 - V_1(N_2/N_1)\}} \end{bmatrix} 2f_{sw}$$

$$= -\frac{V_1(N_2/N_1)\cdot V_2 \cdot \left(\theta_2 - 3\theta_{td}\right)\left(\frac{1}{2} - \theta_2\right)T^2}{\{L_2 + (N_2/N_1)^2 L_1\}}f_{SW} \tag{14}$$

From equation (14), the polarity of P_2 becomes positive in the condition $\theta_2 > 3\theta_{td}$. Symmetrically the polarity of P_2 becomes negative in the condition $\theta_2 < 3\theta_{td}$. The charging and discharging modes seamlessly transfer at the boundary condition of $\theta_2 = 3\theta_{td}$. From equations (7) and (14), it is verified that the charge and discharge amount can be seamlessly controlled by the seamless adjustment of one variable of the phase-shift amount.

In the condition $V_1 N_2 = V_2 N_1$, the charging and discharging modes seamlessly transfer at the boundary condition of $\theta_1 = \theta_2 = \theta_{td}$. When θ_1 and θ_2 are θ_{td}, L_1 and L_2 cannot be excited because the excited voltage and the excited terms are 0. Thus, adjusting θ_1 makes boost-charging mode and adjusting θ_2 makes boost-discharging mode.

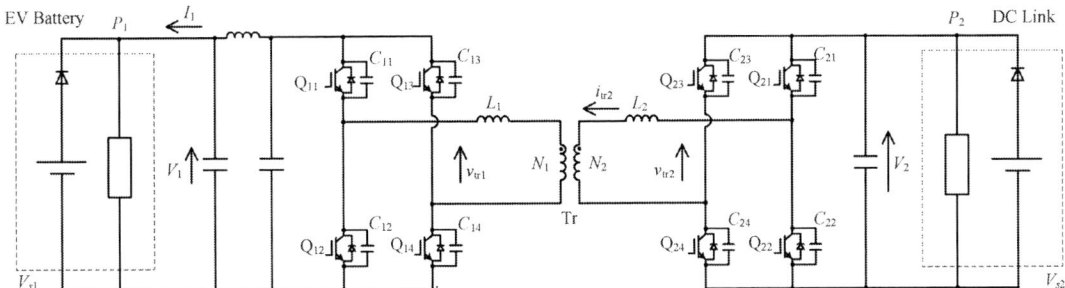

Fig. 12. The 400 V- 3.5 kW experimental circuit configuration of the bi-directional DC/DC converter.

TABLE I. EXPERIMENTAL SPECIFICATION OF THE BI-DIRECTIONAL DC/DC CONVERTER

EV Battery voltage	V_1	290 - 400V
EV Battery current	I_1	12 A
Power rating	P_1	3.5 kW
DC-Link voltage	V_2	380 V
Switching frequency	f_{sw}	20 kHz
Transformer turn ratio	$N_1 : N_2$	32 : 35
Boost and soft switching inductor	L_1, L_2	30 μH
Snubber capacitor	$C_{11}, C_{12},$ C_{13}, C_{14}	12000 pF
Snubber capacitor	$C_{21}, C_{22},$ C_{23}, C_{24}	8200 pF

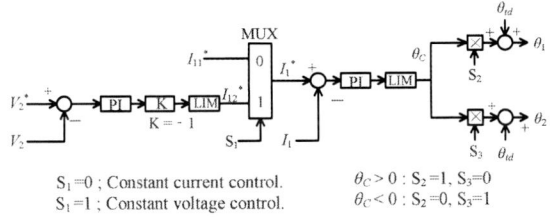

S_1=0 ; Constant current control.
S_1=1 ; Constant voltage control.

$\theta_C > 0$: S_2=1, S_3=0
$\theta_C < 0$: S_2=0, S_3=1

Fig. 13. Control block diagrams of the bi-directional DC/DC converter.

III. EXPERIMENTAL VERIFICATION

A 400 V - 3.5 kW experimental system was designed and manufactured to verify the charging and discharging mode of the operation. Fig. 12 shows the 400 V - 3.5 kW experimental circuit configuration. Table I shows the details of the experimental specifications. The DC Link voltage V_2 was designed at 380 V to correspond the operating voltage of DC/AC inverter as seen in Fig. 1 that converts the commercial voltage, 200 V, to the DC Link Voltage V_2. The rated charging/discharging current is 12 A and the rated charging/discharging power is 3.5 kW. The polarity of charging mode that current flows from the secondary side to the primary side is defined as a positive polarity. The DC voltage sources V_{s1} and V_{s2} make it possible to transmit for bi-direction by using a DC voltage, diode connected in series to the DC voltage source and a register connected in parallel to the DC voltage source and diode. The inductance of L_1 and L_2 is designed at 30 μH to satisfy the maximum output power in the maximum condition of each one of the phase shift amounts, θ_1 or θ_2. The capacitances of C_{11}-C_{24} are designed to satisfy the resonance condition with L_1 and L_2 by the topology of zero voltage switching[12]. Fig. 15 depicts a control block diagram of the proposed bi-directional DC/DC converter. The fully digital controller used in this experiment consists mainly of a digital signal processor (DSP). The proposed bi-directional DC/DC converter has a constant current control and a constant voltage control. The multiplexer (MUX) selects one of either control. The constant current control regulates I_1 to a constant value I_1^* and the constant voltage control regulates V_2 to a constant value V_2^*. In the constant current control, I_{11}^* is inputted to MUX as a external signal from EV. This external signal I_{11}^* from EV controls the charging system on the EV side for example the start/stop of the system or preventing overcharging according to the state of charge (SOC) in the EV battery. In the constant voltage control, the PI control unit outputs current reference I_{12}^* from the deviation value of voltage reference value V_2^* and detected voltage value V_2. MUX selects either I_{11}^* or I_{12}^* as a current reference according to S_1. Then, the PI control unit outputs the phase shift reference θ_C from the deviation of current reference I_1^* and the detected current value I_1. Here θ_C is a dimensionless value like θ_1 and θ_2. If θ_C is a positive value, the phase shift amount θ_1 is calculated by adding the θ_C to θ_{td} setting S_2 to 1 and the phase shift amount θ_2 is fixed to θ_{td} setting S_3 to 0. If θ_C is a negative value, the phase shift amount θ_1 is fixed to θ_{td} setting S_2 to 0 and the phase shift amount θ_2 is calculated by adding the θ_C to θ_{td} setting S_3 to 1.

Fig. 14 and Fig. 15 represent the steady-state characteristics of buck-charging mode and boost-discharging mode respectively under the conditions of V_2 = 380 V, V_1 = 290 V and a rated power of 3.5 kW. Fig. 16 and Fig. 17 represents the steady-state characteristics of boost-charging mode and buck-discharging mode respectively under the conditions of V_2 = 380V, V_1 = 400V and a rated power of 3.5 kW. As seen in Fig. 14 showing the actually measured waveforms of buck-charging mode, the output term of v_{tr2} is restricted. This means that secondary side inverter restricts the transmitting power from the DC Link by adjusting θ_2. Thus I_1

978-1-4673-9551-9/16 $31.00 © 2016 IEEE   1718

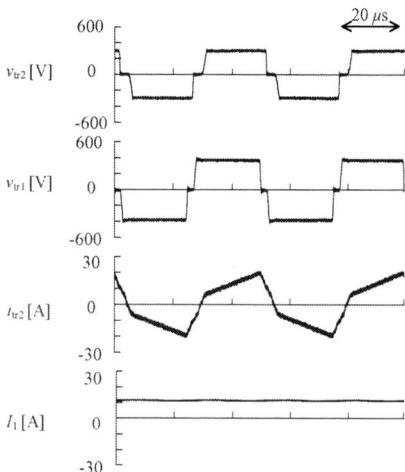

Fig. 14. Experimental waveforms in buck charging mode.
($V_2 = 380$ V, $V_1 = 290$ V, $I_1 = 12$ A)

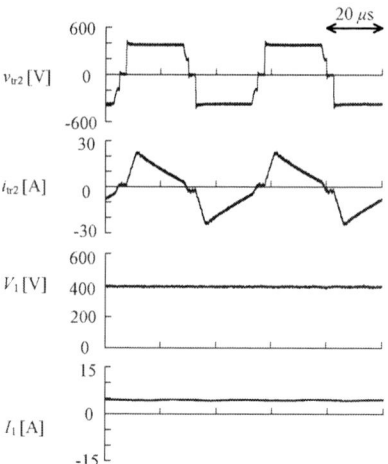

Fig. 16. Experimental waveforms in boost charging mode.
($V_2 = 380$ V, $V_1 = 400$ V, $I_1 = 8.7$ A)

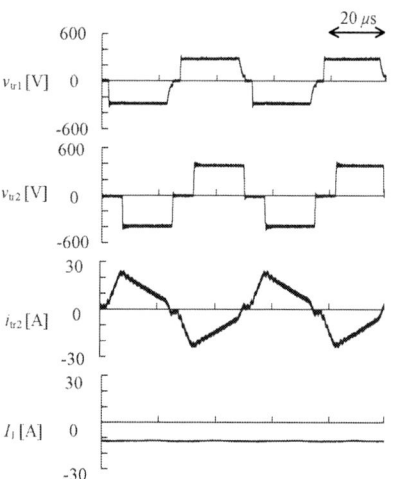

Fig. 15. Experimental waveforms in boost discharging mode.
($V_2 = 380$ V, $V_1 = 290$ V, $I_1 = -12$ A)

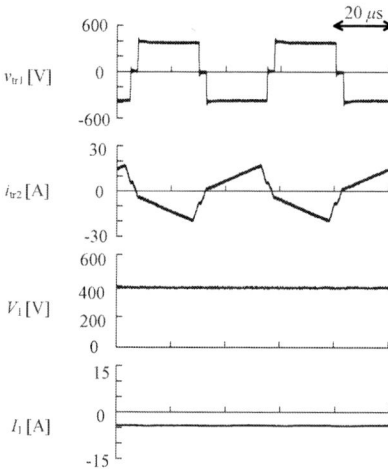

Fig. 17. Experimental waveforms in buck discharging mode.
($V_2 = 380$ V, $V_1 = 400$ V, $I_1 = -8.7$ A)

is controlled to 12A constantly and the proper operation of the buck-charging mode is verified. As seen in Fig. 15 showing the actually measured waveforms of boost-discharging mode, the output term of v_{tr2} is restricted and i_{tr2} is increased for exciting L_1 and L_2 in this term. After that, v_{tr2} outputs V_2 and the i_{tr2} is decreased by the different voltages of v_{tr1} and v_{tr2}. This result means adjusting θ_2 excites L_1 and L_2 and transmits the excited power to the DC Link. Thus I_1 is controlled to -12 A constantly and the proper operation of the boost-discharging mode is verified. As seen in Fig. 16 showing the actually measured waveforms of boost-charging mode, i_{tr2} is increased for exciting L_1 and L_2 as well as boost-discharging mode and I_1 is controlled to 8.7 A constantly. Thus the proper operation of the boost-charging mode is verified. As seen in Fig. 16 showing the actually measured waveforms of buck-discharging mode, the output term of v_{tr1} is restricted as well

as the buck-charging mode and I_1 is controlled to -8.7 A constantly. Thus the proper operation of the buck-discharging mode is verified.

Fig. 18 represents the experimental waveforms at the transition from charging mode to discharging mode under the constant current control. The operational verification conditions are prepared such that the primary side current reference I_{11}^* is changed stepwise from +10 A to -10 A while V_2 is fixed to 380 V and V_1 is fixed to 350 V. As seen in Fig. 18 showing the actually measured waveforms, I_1 is seamlessly switched from +10 A to -10 A within 14.9 ms. Fig. 19 represents the experimental waveforms at the load changed condition under the constant current control. The operational verification conditions are prepared such that the secondary side load is changed stepwise from the rated 3.5 kW to no-load 0 kW while V_2 is fixed to 380 V and V_1 is fixed to 350 V.

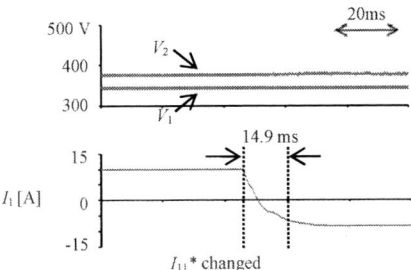

Fig. 18. Experimental waveforms at the transition from charging mode to discharging mode under the constant current control.

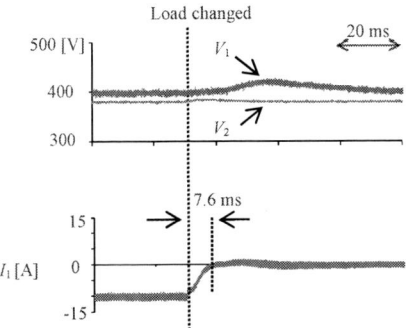

Fig. 19. Experimental waveforms at the load changed condition under the constant voltage control.

As seen in Fig. 19 showing the actually measured waveforms, I_1 is seamlessly switched from $+ 10$ A to $- 0$ A within 7.6 ms and V_2 is controlled to 380 V constantly.

Fig. 20 depicts the efficiency measured at the rated power of 3.5 kW in the charging mode and discharging mode. The efficiency of the charging mode and discharging mode is equal at $V_1 = 350$ V. This $V_1 = 350$ V and $V_2 = 380$ V is a voltage balancing condition considering the turn ratio of the isolated transformer ($N_1 : N_2 = 32 : 35$). The maximum efficiency in the charging mode is 93.3 % at $V_1 = 320$ V and that in the discharging mode is 93.3 % at $V_1 = 380$ V.

IV. CONCLUSION

This paper has presented a new driving method and operational specifications for a bi-directional isolated DC/DC converter for charging and discharging EV batteries. The proposed DC/DC converter consists of two full bridge inverters, an isolation transformer and boost reactors. The proposed driving method synchronizes one leg of each full bridge inverter connected to both sides of the isolation transformer, and controls the phase shift amount of either one of the two full bridge inverters seamlessly for a stable buck-boost/bi-directional power transfer and seamless adjustment of charging and discharging power. A 400 V - 3.5 kW experimental system provides stable bi-directional buck-boost conversion, seamless transition between charging and discharging modes and seamless transition at steep load

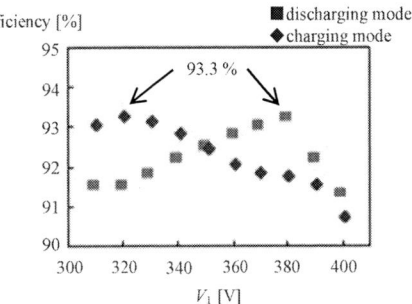

Fig. 20. The measured efficiency at the rated power of 3.5 kW.

changes. A maximum efficiency of 93.3 % was achieved for this experimental system.

REFERENCES

[1] Ogami K, Tanaka K, Uchida K, Yona A, Senjyu T and Funabashi T, "Optimum operation planning of controllable loads in smart house", Power Electronics and Drive Systems, 2012 IEEE 4th International Conference on Cloud Computing Technology and Science, pp. 141-146, 2012

[2] Kato T, Kuboyama H, Izumi K and Okuda T, "Development of Power Conditioning System for Electric Vehicles," 2015 National Convention Record, IEE Japan, No. 4, pp. 240-241 (2015) (in Japanese)

[3] Rik W. A. A. De Doncker, "A Three - Phase Soft - Switched High - Power – Density DC - DC Converter for High-Power Applications," IEEE Transactions on Industry Applications, Vol. 27, No. 1, pp. 63-73, 1991

[4] Mustansir H. Kheraluwala and Randal W. Gascoigne, "Performance Characterization of a High-Power Dual Active Bridge DC - DC Converter," IEEE transactions on Industry Applications, Vol. 28, No. 6, pp. 1294-1301, 1992

[5] H.L. Chan, K.W.E. Cheng, and D. Sutanto, "A Novel Square-Wave Converter with Bidirectional Power Flow," Power Electronics and Drive Systems, Vol. 2, pp. 966-971, 1999

[6] F. Z. Peng, H. Li, G.-J. Su, and J. S. Lawler, "A new ZVS bi – directional DC - DC converter for fuel cell and battery application," IEEE Transactions on Power Electronics, Vol. 19, No. 1, pp. 54-65, Jan 2004

[7] S. Inoue and H. Akagi, "A bidirectional DC-DC converter for an energy storage system with galvanic isolation," IEEE Transactions on Power Electronics, Vol. 22, No. 6, pp. 2299-2306, Nov 2007

[8] H. Zhou, and A. M. Khambadkone, "Hybrid modulation for dual-active-bridge bidirectional converter with extended power range for ultracapacitor application," IEEE Transactions on Industry Applications, Vol. 45, No. 4, pp. 1434-1442, July/Aug 2009

[9] S. Inoue and H. Akagi, "A bidirectional isolated DC-DC converter as a core circuit of the next-generation medium-voltage power conversion system," IEEE Transactions on Power Electronics, Vol. 22, No. 2, pp. 535-542, Mar 2007

[10] K. Wu, C. W. Silva, and W. G. Dunford, "Stability analysis of isolated bidirectional dual active full-bridge DC-DC converter with triple phase-shift control," IEEE Transactions on Power Electronics, Vol. 27, No. 4, pp. 2007-2017, Apr 2012

[11] B. Zhao, Q. Song, W. Liu and Y. Sun, "Overview of Dual-Active-Bridge Isolated Bidirectional DC-DC Converter for High-Frequency-Link Power-Conversion System," IEEE Transactions on Power Electronics, Vol. 29, No. 8, pp. 4091-4106, Aug 2014

[12] G.J.Torvetjonn, A. Petterteig, and T. M. Undeland, "Analysis and measurements on a PWM DC-DC converter with lossless snubbers", in Conference Record of the 1993 IEEE Industry Applications Society Annual Meeting, vol. 2, pp. 1057-1064, Oct 1993.

A CLLC-Compensated High Power and Large Air-Gap Capacitive Power Transfer System for Electric Vehicle Charging Applications

Fei Lu[1,3], *Student Member, IEEE*, Hua Zhang[2,3], *Student Member, IEEE*, Heath Hofmann[1], *Senior Member, IEEE*, and Chris Mi[3,*], *Fellow, IEEE*

[1]University of Michigan-Ann Arbor, Ann Arbor, MI, 48105, USA
[2]Northestern Polytechnical University, Xi'an, 710072, China
[3]San Diego State University, San Diego, CA, 92182, USA
[*]Email: cmi@sdsu.edu

Abstract — This paper proposes a CLLC-compensated capacitive power transfer system for electric vehicle charging applications. Four metal plates are utilized to form two capacitors to transfer power through an air-gap distance of 150 mm. The CLLC compensation circuit topology is used on both the primary and secondary sides to resonate with the power transmitting capacitors. The resonance provides high voltage on the plates to increase the system power level. A comparison to the previously proposed LCLC circuit topology is also presented, which shows that the resonance inductance can be reduced. The circuit model of the coupling plates is provided and the capacitance variation with differing misalignments and air-gap distances is also analyzed. A 2.9kW input power CPT system with a CLLC compensation circuit is designed and implemented. The experimental prototype operates at 1MHz, and its dc to dc efficiency is 89.3% at 2.57 kW output power and a 150 mm air gap distance, which validates the effectiveness of the proposed CLLC compensation circuit. In the future, the system will be optimized to increase its efficiency.

Keywords — capacitive power transfer (CPT), electric field, CLLC compensation, high frequency resonant circuit, electric vehicle charging, misalignment ability

I. INTRODUCTION

Capacitive power transfer (CPT) technology is becoming a practical solution for high power and high efficiency wireless power transfer [1]. Compared to the inductive power transfer (IPT) system, the CPT system has two advantages. First, the CPT system is not sensitive to the presence of metal material nearby, as it does not generate extra heat in the metal due to the eddy current loss. Second, the CPT system utilizes metal plates to transfer power, instead of coils made of Litz-wire [2, 3], so the system cost is reduced.

The current CPT system can be classified by the compensation network topology working with the coupling capacitors. The most commonly used designs consist of one or two inductors resonating with the capacitors in series [4-6].

The advantage of these designs is their simplicity. However, they require a large coupling capacitor value in the nF range [7], which limits the transfer distance of the CPT system. Also, the series resonant circuit is sensitive to the variation of the capacitance due to the large misalignment [8]. A high frequency class *E* or class *D* inverter can also be used in the CPT system to increase the switching frequency by up to 10s of MHz. The advantage of doing so is that the coupling capacitor can be reduced to 10's of pF range, and the transferred distance is therefore increased. The drawback is that the efficiency of the high-frequency inverter is limited, usually around 85% [9]. To reduce the switching frequency, a PWM switch-mode converter is also a good candidate for the CPT system [10, 11]. It usually necessitates much larger capacitance value in the 10s of nF range, and it works in hard-switching mode, which lowers the system efficiency.

Compared with the previous CPT systems, the double-sided LCLC compensated system is proposed in [12] to achieve high power transfer though an air-gap distance of 150 mm. The compensation circuit provides high voltage (up to kV level) on the plates to increase the power density of the CPT system. The system is also not sensitive to the air-gap variation and misalignment conditions. The key aspect of the design of the LCLC network is to use a large capacitor connected in parallel with the coupling capacitor to increase the equivalent

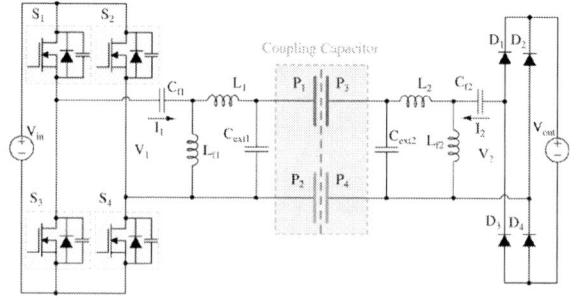

Fig. 1. Circuit Topology of CLLC-compensated CPT System

capacitance. In this way, the resonant inductance is reduced. However, in reference [12], the inductance is still larger than 200µH, which is difficult to make in practice. In this paper, the CLLC compensation topology is proposed to reduce the required inductance value, as shown in Fig. 1. In the experiment, the system can also achieve high power transfer with 89.3% efficiency. In this paper, the inter-coupling between the plates is considered as well, in order to make the modeling of the coupling capacitors more accurate.

This paper will be arranged in six sections. Section one introduces the concept of a CLLC topology. Section two analyze its working principle. Section three designs the capacitive coupler. Section four presents a 2.9kW design example. Section five validates the design by experiments. The last section gives the conclusion.

II. CLLC COMPENSATED TOPOLOGY

The circuit topology of a CLLC-compensated CPT system is shown in Fig. 1. Two pairs of metal plates are used to form two capacitors in a loop to transfer power through a capacitive coupling. The compensation circuit resonates with the coupling capacitors to generate high voltage on the plates in order to achieve power transfer. On the primary side, a full-bridge inverter is used to provide ac excitation to the resonant tank. On the secondary side, a full-bridge diode rectifier is utilized to provide dc voltage to the load.

A. Circuit Model of a Capacitive Coupler

The circuit model of the coupling plates should be derived to design the compensation circuit parameters. In reference [12], the two pairs of plates are arranged 500 mm away from each other, and the cross-coupling between the two pairs is neglected. In this design, the couplers are placed closer together to make the system more compact. Therefore, the inter-coupling should be considered and modeled. The six coupling capacitors between the plates and the equivalent circuit model with four capacitors are shown in Fig. 2. The simplification process will be provided. If a current source I_{P1} is applied at the plates P_1 and P_2, the nodal current equation is expressed as,

$$\begin{cases} (C_{12}+C_{13}+C_{14}) \cdot V_{P1} - C_{13} \cdot V_{P3} - C_{14} \cdot V_{P4} = I_{P1}/(j\omega_0) \\ -C_{13} \cdot V_{P1} + (C_{13}+C_{23}+C_{34}) \cdot V_{P3} - C_{34} \cdot V_{P4} = 0 \\ -C_{14} \cdot V_{P1} - C_{34} \cdot V_{P3} + (C_{14}+C_{24}+C_{34}) \cdot V_{P4} = 0 \end{cases} \quad (1)$$

where V_{P1}, V_{P2}, V_{P3}, and V_{P4} are the voltage on each plate ($V_{P2}=0$ is set to be the reference node), $\omega_0=2\pi f_{sw}$, f_{sw} is the switching frequency, and I_{P1} is the fundamental external input current flowing into P_1. In order to simplify the circuit model,

the relationship between the plate voltage is derived.

$$\begin{cases} V_{P3} = \dfrac{C_{13}C_{14}+C_{13}C_{24}+C_{13}C_{34}+C_{14}C_{34}}{C_{34}(C_{13}+C_{14}+C_{23}+C_{24})+(C_{13}+C_{23})(C_{14}+C_{24})} \cdot V_{P1} \\ V_{P4} = \dfrac{C_{13}C_{14}+C_{14}C_{23}+C_{13}C_{34}+C_{14}C_{34}}{C_{34}(C_{13}+C_{14}+C_{23}+C_{24})+(C_{13}+C_{23})(C_{14}+C_{24})} \cdot V_{P1} \end{cases} \quad (2)$$

Considering (2) and the first equation in (1), the equivalent input capacitance, $C_{in}=I_1/(j\omega_0 V_{P1})$, seen from the P_1 and P_2 side is expressed as,

$$\begin{aligned} C_{in} = C_{12} &+ \frac{C_{34}(C_{13}+C_{14})(C_{23}+C24)}{C_{34}(C_{13}+C_{14}+C_{23}+C_{24})+(C_{13}+C_{23})(C_{14}+C_{24})} \\ &+ \frac{C_{13}C_{23}(C_{14}+C_{24})+C_{14}C_{24}(C_{13}+C_{23})}{C_{34}(C_{13}+C_{14}+C_{23}+C_{24})+(C_{13}+C_{23})(C_{14}+C_{24})} \end{aligned} \quad (3)$$

For the plates, the voltage between P_1 and P_2 is treated as the input, and the voltage between P_3 and P_4 is treated as the output. The transfer function between the two voltages can be defined as $H=(V_{P3}-V_{P4})/V_{P1}$. Considering (2), the transfer function H is expressed as,

$$H = \frac{C_{13}C_{24}-C_{14}C_{23}}{C_{34}(C_{13}+C_{14}+C_{23}+C_{24})+(C_{13}+C_{23})(C_{14}+C_{24})} \quad (4)$$

The plates structure can be designed to be symmetric between the primary and secondary sides. For the equivalent model in Fig. 2, there exists $C_{s1}=C_{s2}$, and $C_{int1}=C_{int2}$. The input capacitance and transfer function are expressed as,

$$\begin{cases} C_{in} = C_{int1} + \dfrac{C_s \cdot C_{int2}}{C_s + C_{int2}} \\ H = \dfrac{C_s}{C_s + C_{int2}} \\ Cs = \dfrac{C_{s1} \cdot C_{s2}}{C_{s1} + C_{s2}} \end{cases} \quad (5)$$

The equivalent capacitors can be expressed as,

$$\begin{cases} C_{int1} = C_{in} \cdot \dfrac{1}{1+H} \\ Cs = C_{in} \cdot \dfrac{H}{1-H^2} \end{cases} \quad (6)$$

Therefore, using (3), (4), and (6), the equivalent capacitor model can be derived for any given plates' dimensions.

B. Circuit Working Principle

Replace the four plates in Fig. 1 with the equivalent circuit model in Fig. 2. The circuit topology is re-drawn in Fig. 3. The CLLC topology is used to work with the plates. The external

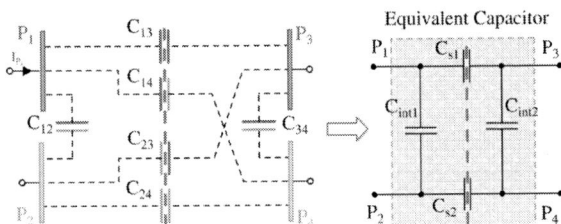

Fig. 2. Circuit Model of the Coupling Plates

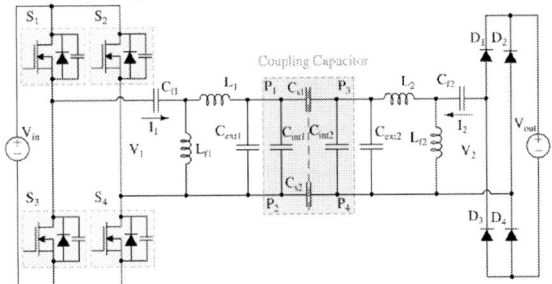

Fig. 3. Circuit Topology of a CPT System with the Simplified Capacitor Model

978-1-4673-9551-9/16 $31.00 © 2016 IEEE

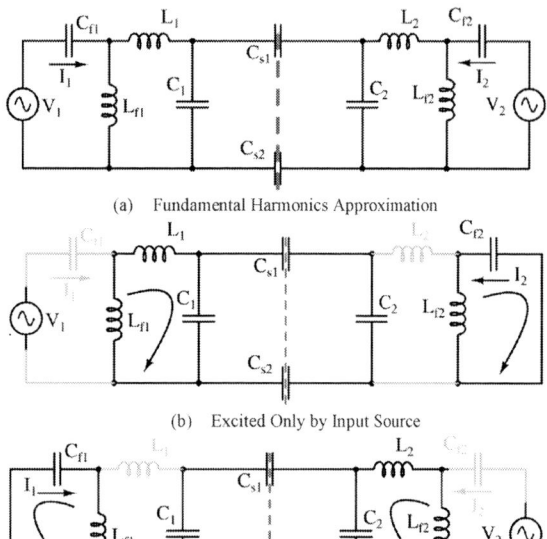

(a) Fundamental Harmonics Approximation

(b) Excited Only by Input Source

(c) Excited Only by Output Source

Fig. 4. Circuit Working Principle Analysis

capacitors C_{ext1} and C_{ext2} are connected in parallel with the internal capacitors C_{int1} and C_{int2}. The total compensation capacitance can be defined as $C=C_1=C_2=C_{ext1}+C_{int1}=C_{ext2}+C_{int2}$. The compensation parameters are also designed to be symmetric, then $L_f=L_{f1}=L_{f2}$, $C_f=C_{f1}=C_{f2}$. The fundamental harmonics approximation (FHA) can be used to analyze the working principle of the circuit at the resonant frequency. The input and output square wave sources are treated as sinusoidal source, as shown in Fig.4 (a).The superposition theorem is used to analyze the two sources separately.

Fig 4 (b) shows the circuit is excited only by the input source. The two resonances can be expressed as,

$$\begin{cases} L_{f2} = \dfrac{1}{\omega_0^2 \cdot C_{f2}} \\ L_1 = \dfrac{1}{\omega_0^2 \cdot C_{in}} - L_{f1} \end{cases} \tag{7}$$

The output current on C_{f2} can be expressed as,

$$I_2 = \frac{L_{f1}+L_1}{L_{f1}} \cdot \frac{C_s}{C_s+C_2} \cdot \frac{V_1}{j\omega_0 L_{f2}} \tag{8}$$

Similarly, Fig 4 (c) shows the circuit is excited only by the output source. The two resonances can be expressed as,

$$\begin{cases} L_{f1} = \dfrac{1}{\omega_0^2 \cdot C_{f1}} \\ L_2 = \dfrac{1}{\omega_0^2 \cdot C_{in}} - L_{f2} \end{cases} \tag{9}$$

Also, the input current on C_{f1} can be expressed as,

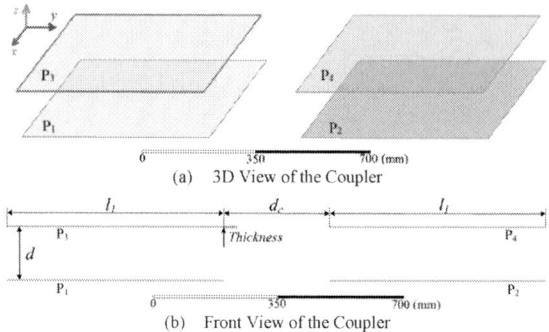

(a) 3D View of the Coupler

(b) Front View of the Coupler

Fig. 5. Dimensions of the Capacitive Coupler

$$I_1 = \frac{L_{f2}+L_2}{L_{f2}} \cdot \frac{C_s}{C_s+C_1} \cdot \frac{V_2}{j\omega_0 L_{f1}} \tag{10}$$

Since a full-bridge diode rectifier is used on the secondary side, the output voltage and current are in phase with each other. Considering equations (7)-(10), the output power can be expressed as,

$$P_{out} = |V_2| \cdot |-I_2| = \frac{\omega_0 \cdot C_s \cdot C_{f1} C_{f2}}{C_1 C_2 + C_1 C_s + C_2 C_s} \cdot |V_1| \cdot |V_2| \tag{11}$$

Compared to reference [12], the output power of the CLLC compensated system is the same as that of the LCLC system. The inductances of L_1 and L_2 can be reduced to make it easier to implement them.

III. Capacitive Coupler Design

The dimensions of the capacitive coupler are shown in Fig. 5. The area of the plate determines the coupling capacitance. Each plate has a square shape, and the length l_1 is 24 in (610mm). The two pairs are separated, and the distance between them, d_c, is 300mm. The air gap, d, is 150 mm. The thickness of the plates does not relate to the coupling

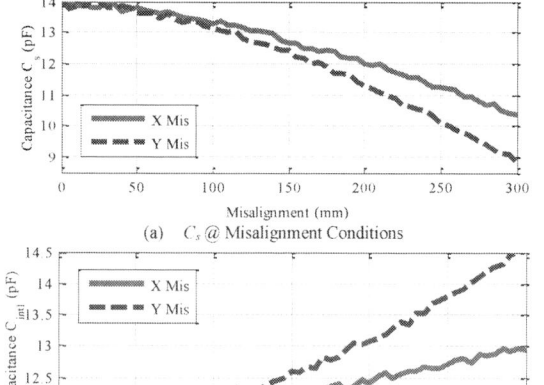

(a) C_s @ Misalignment Conditions

(b) C_1 @ Misalignment Conditions

Fig. 6. Capacitance Value at Different Misalignment Conditions

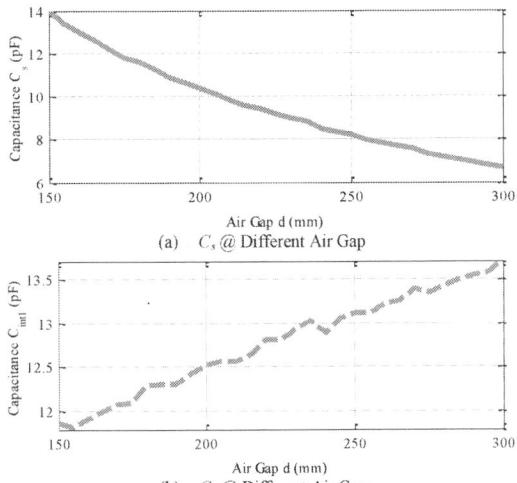

(a) C_s @ Different Air Gap

(b) C_1 @ Different Air Gap

Fig. 7. Capacitance Value at Different Air Gap Conditions

capacitance and it is set to be 2mm.

Finite element analysis (FEA) by Maxwell is used to determine the capacitance matrix that contains the coupling capacitance between each pair of plates, as shown in Fig 2. Based on the FEA results, the equivalent capacitances, C_{int1}, C_{int2}, C_{s1}, and C_{s2} can be calculated using equations (3)-(6). The misalignment ability is also an important design specification. The X, Y, and Z directions are indicated in Fig. 5(a). When there is X and Y misalignment, the variation of the coupling capacitance is as shown in Fig 6. It shows that C_s decreases with the increasing misalignment. Since equation (11) shows that the system output power is proportional to C_s, it means the system power will decrease with the misalignment. For the other capacitor, C_{int1}, which is used to resonate with L_1, its value increases with the misalignment. However, its variation is relatively small and its influence on the system power can be neglected. When comparing the X and Y direction misalignments, Fig. 6 also shows that the capacitances are more sensitive to Y direction misalignment.

The variation of air gap distance is also studied, and the capacitances are shown in Fig. 7. Fig. 7 shows that when the air gap distance increases from 150 mm to 300 mm, the coupling capacitance, C_s reduces by one half, which means that the system power will also reduce by half. The other capacitance, C_{int1} is not sensitive to the air gap variation, hence it only changes by about 15%.

IV. A 2.9 kW CPT SYSTEM DESIGN

Table I. System Specifications and Circuit Parameters

Parameter	Design Value	Parameter	Design Value
V_{in}	400 V	V_{out}	450 V
l_1	610 mm	d	150 mm
f_{sw}	1 MHz	C_s	14.0 pF
L_{f1}	11.76 μH	L_{f2}	11.76 μH
C_{f1}	2.15 nF	C_{f2}	2.15 nF
L_1	164.0 μH	L_2	165.8 μH
C_1	130 pF	C_2	130 pF

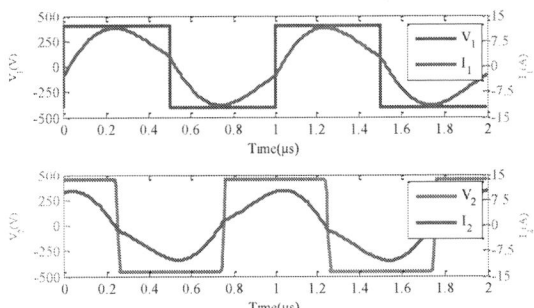

Fig. 8. Simulated Input and Output Voltage and Current Waveforms

After the coupler structure and compensation circuit topology have been designed in the previous sections, a 2.9 kW input power CLLC-compensated CPT system is designed according to the power requirement in equation (11). The parameter values are calculated using equations (7) and (9). All the system specifications and circuit parameter values are shown in Table I.

The input dc voltage is 400 V, and the output dc voltage is 450V to represent the battery pack on the vehicle side. Since the system power is proportional to the switching frequency, the frequency is set to be 1MHz to increase the output power. Compared to in reference [12], compensation inductor L_1 is decreased from 231 μH to 164μH, which is easier to make in practice. It also needs to be emphasized that inductor L_2 is designed to be larger than L_1 to provide soft-switching condition to the input side inverter.

With all the parameters given in Table I, the system performance is simulated in the software LTspice. All the input and output voltage and current waveforms are shown in Fig. 8. Fig. 8 shows that the input voltage and current are in phase, which means there is no reactive power injected into the resonant tank, and the system efficiency is maintained. The cut-off input current at the switching transient is about 3A, which can help to achieve soft-switching for the input side inverter. On the secondary side, the voltage and current are also in phase for the diode rectifier.

V. EXPERIMENT

The experiment prototype is constructed as shown in Fig 9. Since the switching frequency is 1 MHz, AWG 46 Litz-wire is

Fig 9. A 2.9 kW Input Power CLLC Compensated CPT System

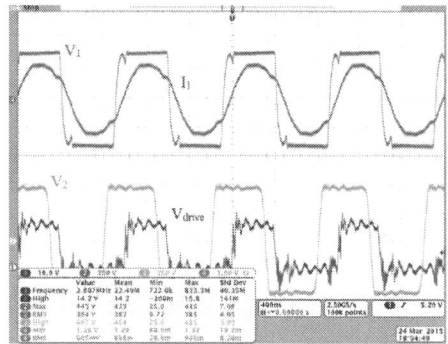

(a) Input and Output Waveforms

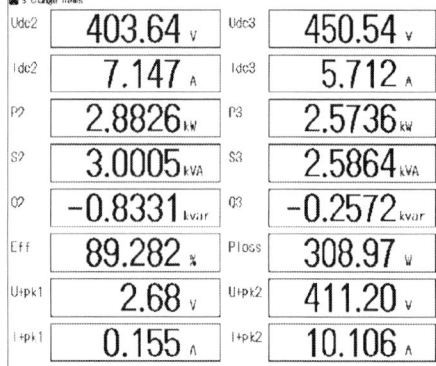

Udc2	403.64 v	Udc3	450.54 v
Idc2	7.147 A	Idc3	5.712 A
P2	2.8826 kW	P3	2.5736 kW
S2	3.0005 kVA	S3	2.5864 kVA
Q2	−0.8331 kvar	Q3	−0.2572 kvar
Eff	89.282 %	Ploss	308.97 W
U+pk1	2.68 v	U+pk2	411.20 v
I+pk1	0.155 A	I+pk2	10.106 A

(b) Maximum Output Power

Fig. 10. Experiment Results of the Prototype

used to make all the inductors. The skin depth of copper is 65 μm at 1MHz, and the diameter of AWG 46 wire is 40 μm. Therefore, the skin effect can be neglected. The capacitors use high-power, high-frequency thin-film capacitors from KEMET to reduce the power losses. The prototype's dissipation factor is 0.18% at 1MHz. For the semiconductor devices, Silicon carbide (SiC) MOSFETs (C2M0080120D) from CREE are used in the input side inverter, and SiC diodes (IDW30G65C) from Infineon are utilized in the output rectifier.

After the experiment is conducted, the experimental waveforms and output power are as shown in Fig. 10. The waveforms are similar to the simulation results in Fig 8. Fig 10 (a) shows that the input voltage and current are almost in phase with each other, and the cut-off current at the switching transient is about 3A to help to achieve soft-switching for the MOSFET. There is noise on the driver signal, but its amplitude

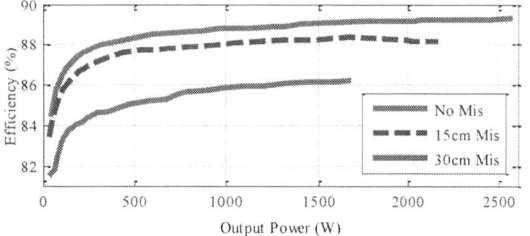

Fig. 11. System output Power and Efficiency

does not exceed the threshold voltage for the MOSFET, thus the operation is safe.

Fig. 10(b) shows that the maximum output power is 2.57kW with 89.3% efficiency, which is comparable to the experimental results in reference [12]. The misalignment in the X and Y direction of the system is also tested. The efficiency at different power levels and misalignment conditions is shown in Fig 11. Fig. 11 shows that the system can maintain 1.7kW output at a 30 cm misalignment in the X direction.

VI. CONCLUSIONS AND FUTURE WORKS

This paper proposes a CLLC compensated circuit topology for capacitive power transfer. Compared to the LCLC topology, this design can reduce the required inductance value of L_1 and L_2. The circuit model and the simplification method of the capacitive coupler are also provided for system parameter design. The working principle of the CLLC compensated system is presented. Then, a 2.9 kW input power system is designed and constructed as an example to validate the proposed topology. The system's efficiency reaches 89.3% at the highest power, and it has good misalignment ability, which can help with the application of CPT technology. Future work will focus on system optimization to improve the total efficiency.

REFERENCES

[1] J. Dai and D. Ludois, "A Survey of Wireless Power Transfer and a Critical Comparison of Inductive and Capacitive Coupling for Small Gap Applications", *IEEE Trans. Power Electron.*, vol. PP, pp. 1-14, 2015.

[2] S. Li, W. Li, J. Deng, T.D. Nguyen, C.C. Mi, "A Double-Sided LCC Compensation Network and Its Tuning Method for Wireless Power Transfer," *IEEE Transactions on Vehicle Technology*, pp. 1-12, 2014.

[3] J. Deng, F. Lu, C. Mi, S. Li, "Development of a High Efficiency Primary Side Controlled 7kW Wireless Power Charger," in *IEEE 2014 International Electric Vehicle Conference (IEVC)*, pp. 1-6.

[4] K. Wang, S. Sanders, "Contactless USB – A Capacitive Power and Bidirectional Data Transfer System," in *IEEE 2014 Applied Power Electronics Conference*, pp. 1342-1347.

[5] D.C. Ludois, M.J. Erickson, J.K. Reed, "Aerodynamic Fluid Bearing for Translational and Rotating Capacitors in Noncontact Capacitive Power Transfer Systems," *IEEE Trans. Ind. Appl.*, vol. 50, pp. 1025-1033, 2014.

[6] C. Liu, A.P. Hu, G.A. Covic, N.C. Nari, "Comparative Study of CCPT Systems with Two Different Inductor Tuning Position," *IEEE Trans. Power Electron.*, vol. 27, pp. 294-306, 2012.

[7] D. Shmilovitz, S. Ozeri, M. Ehsani, "A Resonant LED Driver with Capacitive Power Transfer," in *IEEE 2014 Applied Power Electronics Conference*, pp. 1384-1387.

[8] C. Liu, A.P. Hu, B. Wang, N.C. Nair, "A Capacitively Coupled Contactless Matrix Charging Platform with Soft Switched Transformer Control," *IEEE Trans. Indus. Electron.*, vol. 60, pp. 249-260, 2013.

[9] L. Huang, A.P. Hu, A. Swwain, X. Dai, "Comparison of Two High Frequency Converters for Capacitive Power Transfer," *Proc. IEEE Energy Convers. Congr. Expo.*, pp. 5437-5443, 2014.

[10] J. Dai, D.C. Ludios, "Single Active Switch Power Electronics for Kilowatt Scale Capacitive Power Transfer," *IEEE Joun. of Emerg. And Selec. in Power Elect.*, vol. 3, pp. 315-323, 2015.

[11] J. Dai, D.C. Ludios, "Wireless Electric Vehicle Charging via Capacitive Power Transfer Through a Conformal Bumper," in *IEEE 2015 Applied Power Electronics Conference*, pp. 3307-3313.

[12] F. Lu, H. Zhang, H. Hofmann C.C. Mi, "A Double-Sided LCLC-compensated Capacitive Power Transfer for Electric Vehicle Charging," IEEE Transaction on Power Electronics, pp. 1-4, 2015.

978-1-4673-9551-9/16 $31.00 © 2016 IEEE

A Large Air-Gap Capacitive Power Transfer System with a 4-Plate Capacitive Coupler Structure for Electric Vehicle Charging Applications

Hua Zhang[1,3], *Student Member, IEEE*, Fei Lu[2,3], *Student Member, IEEE*, Heath Hofmann[2], *Senior Member, IEEE*, Weiguo Liu[1], *Senior Member, IEEE*, and Chris Mi[3,*], *Fellow, IEEE*

[1]Northwestern Polytechnical University, Xi'an, 710072, China
[2]University of Michigan-Ann Arbor, Ann Arbor, MI, 48105, USA
[3]San Diego State University, San Diego, CA, 92182, USA
*Email: cmi@sdsu.edu

Abstract—This paper proposes a compact 4-plate capacitive coupler structure for large-air-gap capacitive power transfer (CPT). An LCL-compensated circuit topology is applied at both the primary and secondary sides to resonate with the power transmitting plates. The resonance boosts the voltage on the plates to kV levels to increase the power capacity of the CPT system. On each side, the two plates are aligned vertically with a small distance between them to provide a large coupling capacitance, which can eliminate the need for an external capacitor. An equivalent circuit model of the 4-plate coupler is presented to simplify the design process of the system parameters. Finite Element Analysis (FEA) is used to determine the coupler capacitance. The leakage electric field is also simulated at high power to demonstrate the potential safety issues. A 2.2kW input power CPT system is designed and implemented, which can achieve a dc-dc efficiency of 85.87% at a 150 mm air-gap distance. This study demonstrates that the CPT system is also a solution to electric vehicle charging applications.

Keywords—wireless power transfer (WPT), large air-gap, capacitive power transfer (CPT), coupler structure, 4-Plate structure, electric vehicle (EV) charging

I. INTRODUCTION

Inductive power transfer (IPT) and capacitive power transfer (CPT) are two typical wireless power transfer technologies. The IPT system utilizes magnetic fields to transfer power and has been widely applied in the charging of mobile devices [1] and electric vehicles [2]. It consists of power transmitting coils and the corresponding compensation network. An IPT system can be classified by the topology of the compensation circuit, such as series-series, series-parallel, parallel-series, and parallel-parallel [3]. In addition to these circuits, the LCC compensation topology is proposed to achieve better system performance [4]. The efficiency from the dc source to dc load has reached 96% at 7 kW output power for the LCC compensated IPT system [5], which is already comparable to that of the wire connected system.

However, the leakage magnetic field of the IPT system can induce eddy current losses in the nearby conductive material, which lowers the system efficiency. The material may also heat up to cause a dangerous fire hazard in practical applications. Also, since the IPT system switches at a high frequency, it requires the use of Litz-wire to build up the coils [6], which increases the total cost of the system.

Unlike the IPT system, the CPT system uses metal plates to establish an electric field and transfer power [7]. Electric fields do not produce significant power losses in the nearby conductive material, so the CPT system is safer to apply. Since the coils are replaced by metal plates and the aluminum sheet is much cheaper than Litz-wire, the system cost can be dramatically reduced. However, most of the existing CPT technology focuses on low power applications and very short distances (e.g., 1 mm), which makes the coupling capacitances between plates in the range of several nF [8, 9]. In addition, the performance of the CPT system is limited by the compensation circuit topology. The series compensated structure is mostly used in CPT systems, where only a single inductor is connected in series with the plates to provide resonance. Although the system power can be increased to kW levels, the system is sensitive to the capacitance value variations [10-12].

For electric vehicle charging applications, an LCLC-compensated topology has been proposed in reference [13]. It can boost the voltage on the metal plates to kV level and dramatically increase the transferred power. As a result, the system power has reached 2.4kW with 90.8% efficiency through an air-gap distance of 150 mm. In reference [13], two pairs of aluminum plates are arranged and separated by 500 mm to reduce the cross-coupling between the plates. This structure is very simple and is convenient for extracting the circuit model of the plates; however, it takes up more space than is necessary and reduces the power density of the CPT system. When there is misalignment, the cross-coupling cannot be avoided, and the compensation circuit cannot work properly. In the worst case scenario, when the plates are rotated by 90°,

978-1-4673-9551-9/16 $31.00 © 2016 IEEE

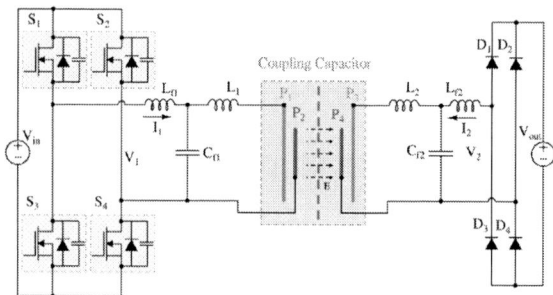

Fig. 1. Circuit Topology of LCL-compensated CPT System

the output power will be reduced to zero. Therefore, it is necessary to improve the coupling structure.

In this paper, a more compact plate structure is proposed, in which the four metal plates are vertically arranged. It can maintain the coupling when there is horizontal and rotatory misalignment. This structure has been mentioned in reference [14] and [15]. However, previously, the cross-coupling between each pair of plates has not been considered, the accurate equivalent circuit has not been derived, and the compensation topology for the coupler has not been studied. In this paper, all three of these questions will be solved in detail. First, the coupling between each pair of plates is considered. Second, the equivalent circuit model of the plates is presented and a LCL compensated circuit is used to resonate with the plates. Also, a 2.2 kW input power prototype is designed and constructed to validate the compact structure.

This paper will be arranged in six sections. Section one introduces the compact plate structure. Section two discusses the LCL compensation topology for the coupler. Section three presents the coupler structure and derives its equivalent circuit. Section four designs a 2.2kW CPT system with the coupler. Section five validates the system by experiment. The last section gives the conclusion.

II. LCL COMPENSATED TOPOLOGY

The circuit topology of the LCL compensated CPT system is shown in Fig. 1. MOSFETs S_1-S_4 form the input inverter to provide the AC excitation to the resonant circuit. Diodes D_1-D_4 form a full-bridge rectifier to serve dc current to the output load. The LCL network is used to resonate with the plates and generate high voltage to transfer power. To analyze the working principle, the equivalent model of the plates needs to be derived first.

The equivalent capacitors of the coupling plates are shown

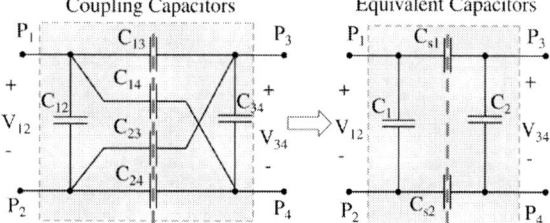

Fig. 2. Circuit Model of the Coupling Capacitors

in Fig. 2. There is capacitive coupling between each pair of the plates to form a capacitor in the circuit model. Considering all the couplings, there are six capacitors needed to reperesent the coupler. In the circuit model in reference [13], where the LCLC compensation topology is adopted, the six capacitor model can be simplified as a four capacitor model, as shown in Fig. 2. Different from [13], capacitors C_1 and C_2 are not externally connected discrete components; instead, they are integrated in the plate model. Therefore, the total number of components is reduced and the system's structure is simplified. The relationship between the component values in Fig. 2 will be discussed in the next section.

Using the equivalent circuit model in Fig. 2, the circuit topology can be simplified as in Fig. 3 (a), where the fundamental harmonics approximation (FHA) is applied. The power loss of each component is also neglected to simplify the analysis. Since inductor L_f and capacitor C_f form a low-pass filter, there is no high-order current injected into the resonant circuit. Therefore, the square wave voltage at the input and output can be approximated as a sinusoidal source with acceptable error. Due to the linearity of the circuit model, the superposition theorem can be used to analyze the two sources separately, and then add the two cases together.

Fig. 3 (b) shows the circuit excited only by the input voltage source, V_1. There are two parallel resonances highlighted in the circuit. L_{f2} and C_{f2} resonate on the secondary side and act as an infinite impedance, so there is no current flowing through L_2. The resonance can be expressed as follows:

$$L_{f2} \cdot C_{f2} = 1/\omega_r^2 \qquad (1)$$

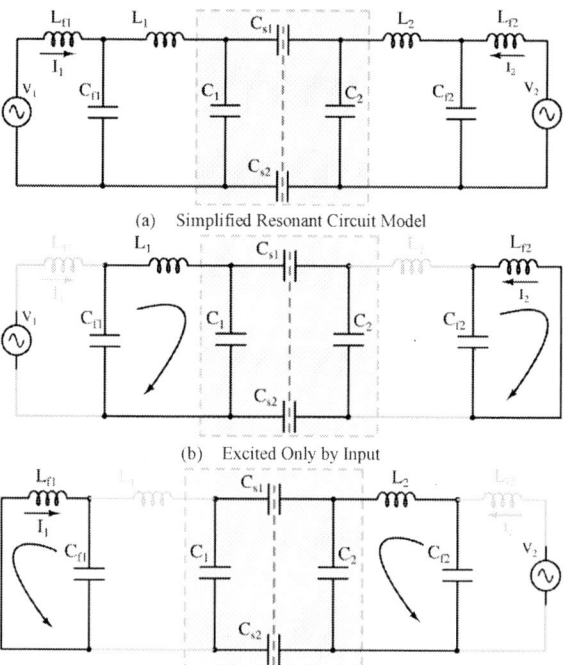

(a)   Simplified Resonant Circuit Model

(b)   Excited Only by Input

(c)   Excited Only by Output

Fig. 3. Fundamental Harmonics Approximation of the Circuit Topology

where, ω_r is the resonant angular frequency. The other resonance includes L_1, C_{f1}, C_1, C_2, C_{s1}, and C_{s2}, which also act as an infinite impedance, meaning the input current flowing through L_{f1} is zero. It also means the input current does not rely on the input voltage. The parameter relationship can be expressed as,

$$
\begin{cases}
C_s = C_{s1} \cdot C_{s2} / (C_{s1} + C_{s2}) \\
C_{in,pri} = C_1 + C_s \cdot C_2 / (C_s + C_2) \\
L_1 = 1/(\omega_r^2 C_{f1}) + 1/(\omega_r^2 C_{in,pri})
\end{cases}
\tag{2}
$$

where, C_s is defined as the equivalent coupling capacitance, $C_{in,pri}$ is the input capacitance of the coupler seen from the primary side, and L_1 is the external resonant inductance.

Fig. 3 (c) shows the circuit excited only by the output voltage source V_2. Similar to in the previous case, there are two parallel resonances in the circuit. L_{f1} and C_{f1} form one resonance, and there is no current flowing through L_1. The resonance is expressed as follows.

$$
L_{f1} \cdot C_{f1} = 1/\omega_r^2
\tag{3}
$$

The other resonance includes L_2, C_{f2}, C_1, C_2, C_{s1} and C_{s2}, and it acts as an infinite impedance. There is no current flowing through L_{f2}, so the output current does not depend on the output voltage. The resonant inductance is calculated as,

$$
\begin{cases}
C_{in,sec} = C_2 + C_s \cdot C_1 / (C_s + C_1) \\
L_2 = 1/(\omega_r^2 C_{f2}) + 1/(\omega_r^2 C_{in,sec})
\end{cases}
\tag{4}
$$

Where, $C_{in,sec}$ is the input capacitance of the coupler seen from the secondary side. From the above analysis, the LCL compensated system behaves as a current source for both the input and output. Therefore, the output current on the inductor L_{f2} can be calculated as follows.

$$
I_{Lf2} = -I_2 = \frac{C_{f1}}{C_{in,pri}} \cdot \frac{C_s}{C_s + C_2} \cdot \frac{-V_1}{j\omega_r L_{f2}} = \frac{j\omega_r C_s \cdot C_{f1} C_{f2} \cdot V_1}{C_1 C_2 + C_1 C_s + C_2 C_s}
\tag{5}
$$

The above shows that output current I_{Lf2} leads the input voltage by 90°. Similarly, Fig. 3 (c) shows that input current I_1 lags the output voltage by 90°. Since a full-bridge diode rectifier is used on the secondary side, the output voltage is in phase with the output current. Therefore, the input voltage and current are in phase, which means the unity power factor is achieved. Based on the current in (5), the system output power can be expressed as,

$$
P_{out} = |V_2| \cdot |-I_2| = \frac{\omega_r C_s \cdot C_{f1} C_{f2}}{C_1 C_2 + C_1 C_s + C_2 C_s} \cdot \frac{2\sqrt{2}}{\pi} V_{in} \cdot \frac{2\sqrt{2}}{\pi} V_{out}
\tag{6}
$$

where, the input and output dc sources are used to represent the RMS value of voltages V_1 and V_2. It shows that the system power is proportional to coupling capacitor C_s and voltages V_{in} and V_{out}.

III. CAPACITIVE COUPLER DESIGN

Four aluminum plates are used to form the capacitive coupler. The plates are vertically arranged, and the primary side is designed to be symmetric to the secondary side, as shown in Fig. 4. The two outer plates, P_1 and P_3 are larger than the inner plates, P_2 and P_4. All the plates are designed to be

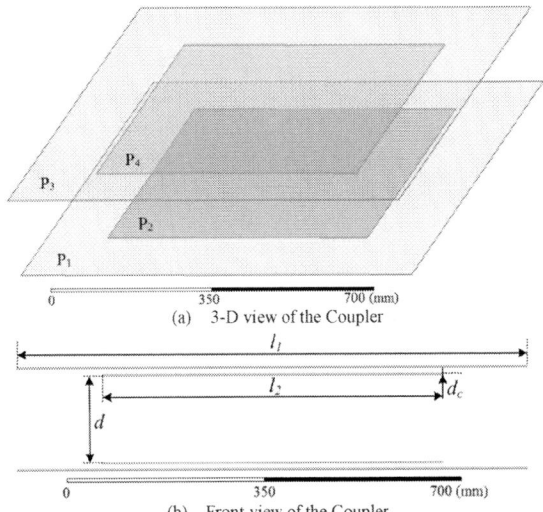

(a)   3-D view of the Coupler

(b)   Front view of the Coupler

Fig. 4. The Dimensions of the Coupling Plates

square to simplify the structure design. The outer plate length is l_1, the inner plate length is l_2, the distance between P_1 and P_2 is d_c, and the distance between P_2 and P_4 is d, which is 150 mm. The value of d_c is usually much smaller than d to provide a larger compensation capacitance. The thickness of the aluminum plate does not affect the capacitances significantly and it is set to be 2mm. There is capacitance between each pair of plates, and the Maxwell FEA software is used to simulate all the capacitance values for circuit calculation.

The circuit model of the plates is shown in Fig. 2. Because of the symmetric structure of the coupler, $C_{12}=C_{34}$ and $C_{14}=C_{23}$. The Extra Element Theorem (EET) [16] is applied to simplify the circuit model, and capacitor C_{34} can be treated as the extra element. The input capacitance $C_{in,pri}$ from the primary side can be expressed as,

$$
\begin{cases}
C_{in,pri} = C_{12} + C_{in,inf} \cdot \dfrac{1 + C_{34} / C_{e,inf}}{1 + C_{34} / C_{e,0}} \\
C_{in,inf} = \dfrac{C_{13} \cdot C_{23}}{C_{13} + C_{23}} + \dfrac{C_{14} \cdot C_{24}}{C_{14} + C_{24}} \\
C_{e,inf} = \dfrac{C_{13} \cdot C_{14}}{C_{13} + C_{14}} + \dfrac{C_{23} \cdot C_{24}}{C_{23} + C_{24}} \\
C_{e,0} = \dfrac{(C_{13} + C_{23}) \cdot (C_{14} + C_{24})}{C_{13} + C_{23} + C_{14} + C_{24}}
\end{cases}
\tag{7}
$$

Where $C_{in,inf}$ is the input capacitance with C_{34} removed, $C_{e,0}$ is the capacitance seen by C_{34} with C_{12} short-circuited, and $C_{e,inf}$ is the capacitance seen by C_{34} with C_{12} removed.

The transfer function H, relating V_{34} to V_{12} is also an important parameter. It can be expressed as in (8). H_{inf} is the

$$
\begin{cases}
H = \dfrac{H_{inf}}{1 + C_{34} / C_d} \\
H_{inf} = \dfrac{C_{13}}{C_{13} + C_{23}} - \dfrac{C_{14}}{C_{14} + C_{24}} \\
C_d = \dfrac{(C_{13} + C_{23}) \cdot (C_{14} + C_{24})}{C_{13} + C_{23} + C_{14} + C_{24}}
\end{cases}
\tag{8}
$$

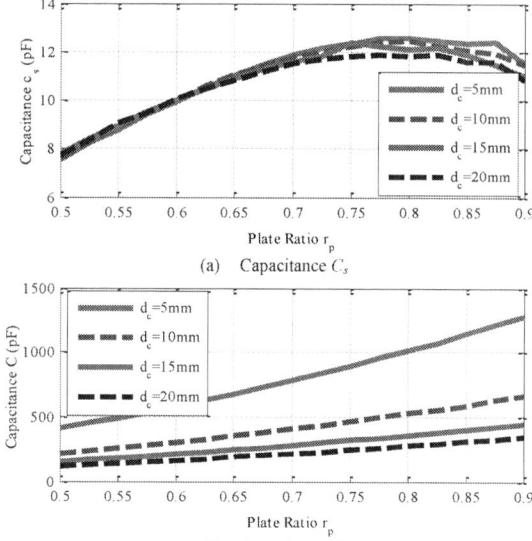

(a) Capacitance C_s

(b) Capacitance C

Fig. 5. Capacitances at Different r_p and d_c when l_1=914 mm

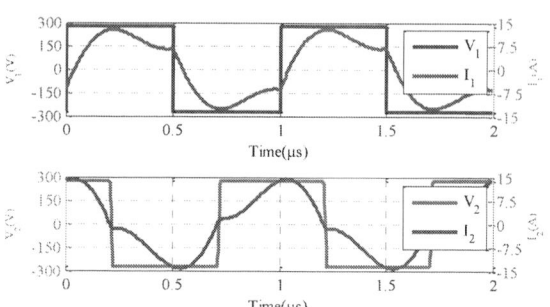

Fig. 6. Simulated Input and Output Voltage and Current Waveforms

transfer function with C_{34} removed, and C_d is the capacitance across terminals 3 and 4 with C_{34} removed and C_{12} short-circuited, which is the same as $C_{e,0}$ in (7).

According to (7) and (8), the equivalent capacitances in Fig. 2 can be expressed as in (9). Therefore, for a given coupler size, the FEA analysis can simulate the capacitance matrix, and equations (7), (8), and (9) can be used to simplify the circuit.

$$\begin{cases} C = C_1 = C_2 = C_{in,pri} \cdot \dfrac{1}{1+H} \\ C_s = C_{in,pri} \cdot \dfrac{H}{1-H^2} \end{cases} \quad (9)$$

In Fig. 4, the outer plate length l_1 is set to be 36 inches (910 mm). The ratio between the inner and outer plates' lengths is defined as $r_p=l_2/l_1$. The capacitances C_s and C_1, as a function of r_p and d_c, are shown in Fig 5. It indicates that capacitance C_s is not sensitive to plate distance d_c, while C is sensitive to both d_c and r_p. In this system, the outer and inner plates' lengths are selected to be 36 in (914 mm) and 24 in (610 mm), respectively, and the plate ratio r_p is therefore 0.667. The distance between the plates can be varied to regulate the capacitance C, which resonates with external inductor L_1. In this design, d_c is set to be 10 mm.

Table I. System Specifications and Circuit Parameter Values

Parameter	Design Value	Parameter	Design Value
V_{in}	270 V	V_{out}	270 V
l_1	914 mm	l_2	610 mm
d_c	10 mm	r_p	0.667
f_{sw}	1 MHz	C_s	11.3 pF
L_{f1}	2.76 μH	L_{f2}	2.76 μH
C_{f1}	9.19 nF	C_{f2}	9.19 nF
L_1	65.3 μH	L_2	65.9 μH
C_1	390 pF	C_2	390 pF

IV. A 2.2KW CPT SYSTEM DESIGN

After the coupler dimensions are determined in the previous section, a 2.2 kW input power system can then be designed according to equations (1)-(4) and (6). The input and output dc voltage is 270 V, and the system specifications and circuit parameters are shown in Table I. It needs to be emphasized that L_1 is smaller than L_2 to generate a soft-switching condition in the switches.

LTspice software is used to simulate the system performance with the parameters in Table I. The input and output voltages and current waveforms are shown in Fig. 6. It shows that the input voltage and current are nearly in phase with each other, which indicates that most of the power injected into the resonant tank is active power. Therefore, the system efficiency is maintained. Fig. 6 also shows that the cut-off current of the switch at the turn-on transient is 3 A, which is sufficient for zero-voltage turn on.

The issue of safety is also important in practical applications. The simulation shows that the RMS voltage between plates P_1 and P_3 is 3.8 kV and the RMS voltage between P_1 and P_2 is 5.1 kV. The breakdown voltage of air is about 3 kV/mm and the distance between P_1 and P_2 is 10mm, so there is no concern with arcing.

V. EXPERIMENT

With the parameters in Table I, a 2.2 kW input power CPT system is constructed, as shown in Fig. 7. Since the system is switching at 1MHz, AWG 46 Litz-wire is used to mitigate power losses induced by the skin and proximity effect. The Litz-wire is used to make the four inductors, $L_{1,2}$ and $L_{f1,2}$. The

Fig. 7. A 2.2 kW Input Power Prototype of CPT System

978-1-4673-9551-9/16 $31.00 © 2016 IEEE

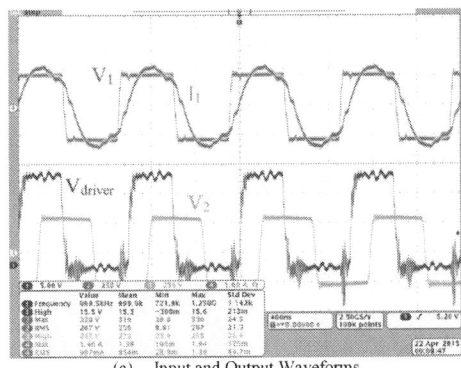

(a)   Input and Output Waveforms

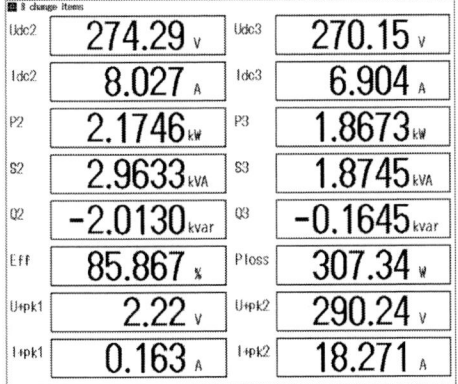

(b)   Maximum Output Power

Fig. 8. Experiment Results of the Prototype

capacitors, $C_{f1,2}$, are high-power, high-frequency thin-film capacitors from KEMET with a dissipation factor of 0.18% at 1MHz. The inverter utilizes silicon carbide (SiC) MOSFETs C2M0080120D from CREE, and the rectifier uses SiC diode IDW30G65C from Infineon. A ceramic spacer is used between the plates as an insulator.

The input and output waveforms are shown in Fig. 8(a). They are similar to the simulation results in Fig. 6. The input current and voltage are in phase with each other. There is also noise on the driver signal at high power; however, its amplitude does not exceed the threshold voltage. At the nominal power point, the system working status is as shown in Fig. 8(b). It shows that the system efficiency reaches 85.87% at 1.87 kW output power. Fig 9 shows the system performance at different output powers. The system can maintain an efficiency higher than 85% when the power is higher than 600W.

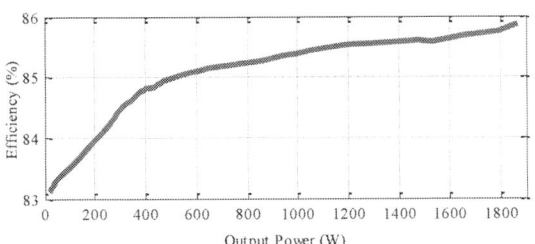

Fig. 9. System output Power and Efficiency

VI. CONCLUSIONS AND FUTURE WORK

This paper proposes a compact coupler structure for capacitive power transfer and the corresponding LCL compensation circuit topology. The equivalent circuit model of the coupler is presented, and Maxwell FEA simulation results help design the parameters of the coupler. A coupler with 914 mm × 914 mm dimensions is designed and implemented for a 2.2 kW input power system. The CPT system prototype is also constructed to verify the design, and its efficiency reaches 85.87% at 1.87 kW output. In future work, the radiated EMI will be studied more thoroughly and the system efficiency will be improved.

REFERENCES

[1] Q. Li, Y.C. Liang, "An Inductive Power Transfer System with a High-Q Resonant Tank for Mobile Device Charging," *IEEE Transactions on Power Electronics*, vol. 30, pp. 6203-6212, 2015.

[2] F. Lu, H. Hofmann, J. Deng, C.C. Mi, "Output Power and Efficiency Sensitivity to Circuit Parameter Variations in Double-sided LCC-compensated Wireless Power Transfer System," in *IEEE 2015 Applied Power Electronics Conference (APEC)*, pp. 507-601.

[3] J. Kim, D. Kim, Y. Park, "Analysis of Capacitive Impedance Matching Networks for Simultaneous Wireless Power Transfer to Multiple Devices," *IEEE Transactions on Industrial Electronics*, vol. 62, pp. 2807-2813, 2014.

[4] S. Li, W. Li, J. Deng, T.D. Nguyen, C.C. Mi, "A Double-Sided LCC Compensation Network and Its Tuning Method for Wireless Power Transfer," *IEEE Transactions on Vehicle Technology*, pp. 1-12, 2014.

[5] J. Deng, F. Lu, C. Mi, S. Li, "Development of a High Efficiency Primary Side Controlled 7kW Wireless Power Charger," in *IEEE 2014 International Electric Vehicle Conference (IEVC)*, pp. 1-6.

[6] J. Deng, F. Lu, W. Li, S. Li, "ZVS Double-sided LCC Compensated Resonant Inverter with Magnetic Integration for Electric Vehicle Wireless Chargers," in *IEEE 2015 Applied Power Electronics Conference (APEC)*, pp. 1131-1136.

[7] D. Shmilovitz, S. Ozeri, M.M. Ehsani, "A Resonant LED Driver with Capacitive Power Transfer," in *IEEE 2014 Applied Power Electronics Conference (APEC)*, pp. 1384-1387.

[8] C. Liu, A.P. Hu, X. Dai, "A contactless power transfer system with capacitively coupled matrix pad," *IEEE 2011 Energy Conversion Congress and Expo. (ECCE)*, pp. 3488-3494.

[9] M. Kline, I. Izyumin, B. Boser, S. Sanders, "Capacitive Power Transfer for Contactless Charging," in *IEEE 2011 Applied Power Electronics Conference*, pp. 1398-1404.

[10] J. Dai and D. Ludois, "A Survey of Wireless Power Transfer and a Critical Comparison of Inductive and Capacitive Coupling for Small Gap Applications", *IEEE Trans. Power Electron.*, vol. PP, pp. 1-14, 2015.

[11] J. Dai and D. Ludois, "Wireless Electric Vehicle Charging via Capacitive Power Transfer through a Conformal Bumper," in *IEEE 2015 Applied Power Electronics Conference*, pp. 3307-3313.

[12] J. Dai, D.C. Ludios, "Single Active Switch Power Electronics for Kilowatt Scale Capacitive Power Transfer," *IEEE Joun. of Emerg. And Selec. in Power Elect.*, vol. 3, pp. 315-323, 2015.

[13] F. Lu, H. Zhang, H. Hofmann C.C. Mi, "A Double-Sided LCLC-compensated Capacitive Power Transfer for Electric Vehicle Charging," *IEEE Transaction on Power Electronics*, pp. 1-4, 2015.

[14] Shinji Goma, "Capacitive Coupling Powers Transmission," [online]: *http://www.murata.com/~/media/webrenewal/about/newsroom/tech/power/wptm-ta1291.ashx*

[15] T. Komaru, H. Akita, "Positional Characteristics of Capacitive Power Transfer as a Resonance Coupling System," *Proc. IEEE Wireless Power Transfer Conf.* pp. 218-221, 2013

[16] R.D. Middlebrook , "Null double Injection and the Extra Element Theorem," *IEEE Transactions on Education*, vol. 32, no. 3, pp. 167-180, Aug. 1989.

978-1-4673-9551-9/16 $31.00 © 2016 IEEE

Dynamic Wireless Power Transfer System for Electric Vehicles to Simplify Ground Facilities - Power Control and Efficiency Maximization on the Secondary Side -

Katsuhiro Hata, Takehiro Imura, and Yoichi Hori
The University of Tokyo
5–1–5, Kashiwanoha, Kashiwa, Chiba, 277–8561, Japan
Phone: +81-4-7136-3881, Fax: 81-4-7136-3881
Email: hata@hflab.k.u-tokyo.ac.jp, imura@hori.k.u-toyko.ac.jp, hori@k.u-tokyo.ac.jp

Abstract—A dynamic wireless power transfer (WPT) system for electric vehicles can extend their cruising distance and reduce the size of their energy storage system. Power control and efficiency maximization of WPT are preferable to be controlled on the secondary side because ground facilities of the dynamic charging system have to be simplified. Although previous research has proposed a secondary-side simultaneous control of the maximum efficiency and the desired power, the battery charging current cannot be controlled directly. In this paper, a novel secondary-side control method for power control and efficiency maximization is proposed. The battery charging power is controlled by the DC-DC converter and the transmitting efficiency is maximized by Half Active Rectifier. These control strategies and the controller design are proposed based on the WPT circuit analysis and the power converter model. The effectiveness of the proposed method is verified by simulation and experiment.

Keywords—*Electric vehicle, Dynamic wireless power transfer, Efficiency maximization, Power control, Secondary-side control*

I. INTRODUCTION

Electric vehicles (EVs) have gathered attention for their highly environmental performance. Additionally, their electric motors can achieve a high performance in motion control because of a faster torque response over internal combustion engines [1]. However, their limited mileage per charge, which are caused by a low energy density of their energy storage system, imposes the need for a frequent and complicated charging on users.

Wireless power transfer (WPT) can mitigate complicated charging operations and endure the frequent charging. Additionally, a dynamic WPT system can provide electricity to EVs in motion. As a result, the cruising distance can be extended and the size of the energy storage system can be reduced [2–5]. However, ground facilities of the dynamic WPT system, which are composed of power source, high-frequency inverters, transmitters, and so on, are applied to rugged roadways over long distances. Consequently, a feasible control strategy for the dynamic charging system is different from a stationary charging system. In order to simplify ground facilities, a secondary-side control is preferable to a primary-side control [6] or a dual-side control [7]. Therefore, this paper focuses on the secondary-side control without signal communication.

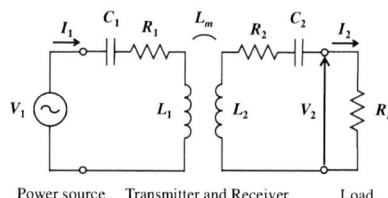

(a) Equivalent circuit of magnetic resonant coupling.

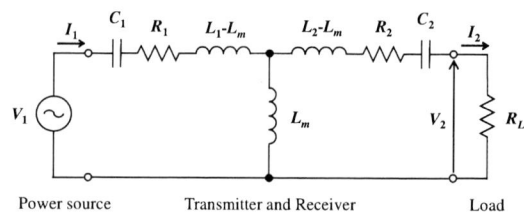

(b) T–type equivalent circuit.

Fig. 1. Equivalent circuit of wireless power transfer system.

Previous research on the secondary-side control has proposed maximum efficiency control [8], [9] and power control [10]. Additionally, efficiency and power can be controlled simultaneously using two power converters [11]. This control method uses a DC-DC converter and Half Active Rectifier (HAR), which is a role as an AC-DC converter. The transmitting efficiency is maximized by the DC-DC converter and the transmitting power is controlled by the HAR. However, this method cannot control battery charging power directly.

This paper proposes a novel secondary-side control method for power control and efficiency maximization. The proposed method directly controls the battery charging current using the DC-DC converter. Because the DC link voltage becomes unstable [12], the proposed method stabilizes the DC link voltage using the HAR and maximize the transmitting efficiency by determining the reference value of the DC link voltage.

Fig. 2.   Transmitter and receiver coils.

TABLE I.   SPECIFICATIONS OF COILS.

	Primary side	Secondary side
Resistance R_1, R_2	1.24 Ω	1.23 Ω
Inductance L_1, L_2	615 μH	615 μH
Capacitance C_1, C_2	4000 pF	4000 pF
Resonant frequency f_1, f_2	101 kHz	101 kHz
Outer diameter	440 mm	
Number of turns	50 turns	
Coil gap	300 mm	
Mutual inductance L_m	37.8 μH	
Coupling coefficient k	0.0615	

II.   WIRELESS POWER TRANSFER VIA MAGNETIC RESONANCE COUPLING

A. Characteristics at resonance frequency

This paper uses WPT via magnetic resonance coupling [13], which is compensated by a series-series (SS) circuit topology. Fig. 1 shows an equivalent circuit of the WPT system [14]. The transmitter and receiver coils are connected to the resonance capacitors in series. They are characterized by the self-inductances L_1, L_2, the series-resonance capacitances C_1, C_2, and the internal resistances R_1, R_2, respectively. L_m is the mutual inductance between the transmitter and the receiver. V_1 is the RMS voltage of the power source and its angular frequency ω_0 is the same as the resonance angular frequency of the transmitter and the receiver, which are expressed as follows:

$$\omega_0 = \frac{1}{\sqrt{L_1 C_1}} = \frac{1}{\sqrt{L_2 C_2}}. \quad (1)$$

The transmitter and the receiver that used in this study are shown in Fig. 2 and their specifications are described in TABLE. I.

When the load resistance is R_L, the transmitting efficiency η and the transmitting power P can be analyzed by the circuit equation and they are obtained as follows [15]:

$$\eta = \frac{(\omega_0 L_m)^2 R_L}{(R_2 + R_L)\{R_1 R_2 + R_1 R_L + (\omega_0 L_m)^2\}} \quad (2)$$

$$P = \frac{(\omega_0 L_m)^2 R_L}{\{R_1 R_2 + R_1 R_L + (\omega_0 L_m)^2\}^2} V_1^2. \quad (3)$$

When the amplitude of V_1 equals to 100 V, Fig. 3 shows the load resistance R_L versus the transmitting efficiency η and the charging power P.

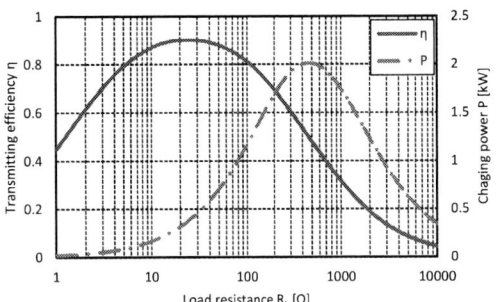

Fig. 3.   Load resistance vs. transmitting efficiency and charging power.

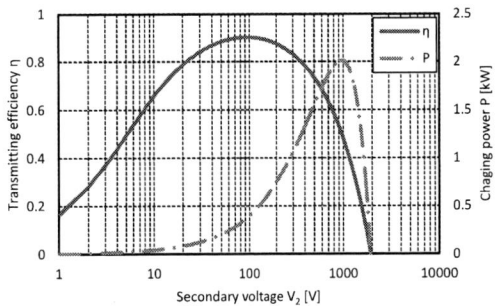

Fig. 4.   Secondary voltage vs. transmitting efficiency and charging power.

B. Maximization of transmitting efficiency

In order to maximize the transmitting efficiency η, the load resistance R_L should be optimized as follows [15]:

$$R_{L\eta\max} = \sqrt{R_2 \left\{ \frac{(\omega_0 L_m)^2}{R_1} + R_2 \right\}}. \quad (4)$$

In a dynamic WPT system for EVs, the mutual inductance L_m changes depending on the motion of the vehicle. Therefore, R_L has to be controlled according to L_m.

As a method to control R_L, Secondary voltage control methods have been proposed [8], [9]. In order to obtain the optimal secondary voltage $V_{2\eta\max}$, the secondary voltage V_2 versus the transmitting efficiency η and the charging power P are depicted in Fig. 4. For efficiency maximization, $V_{2\eta\max}$ value is determined as follows [8]:

$$V_{2\eta\max} = \sqrt{\frac{R_2}{R_1}} \frac{\omega_0 L_m}{\sqrt{R_1 R_2 + (\omega_0 L_m)^2} + \sqrt{R_1 R_2}} V_1. \quad (5)$$

In this study, the amplitude of the primary voltage V_1 is fixed to simplify ground facilities. Then, if R_1 is assumed to be constant and given, L_m can be estimated from the secondary side [9]. Therefore, efficiency maximization using secondary voltage control can be achieved based on secondary-side information, which are the secondary voltage V_2 and the secondary current I_2.

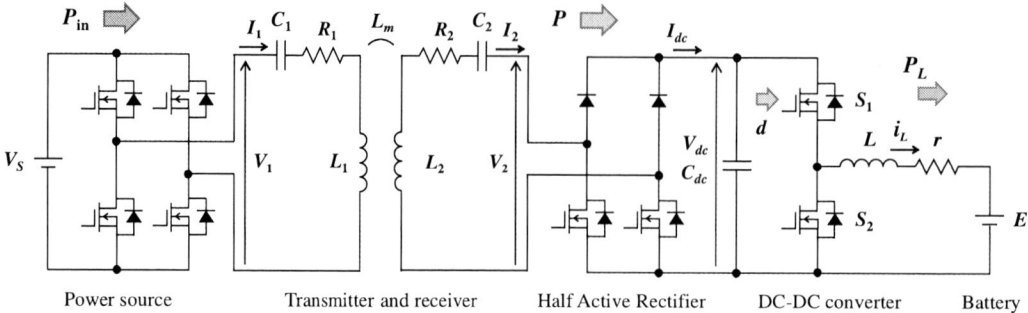

Fig. 5. Circuit diagram of the wireless power transfer system.

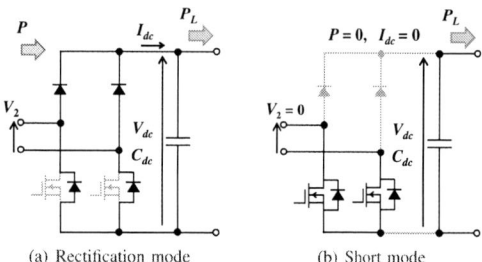

(a) Rectification mode　　(b) Short mode

Fig. 6. Operation modes of Half Active Rectifier.

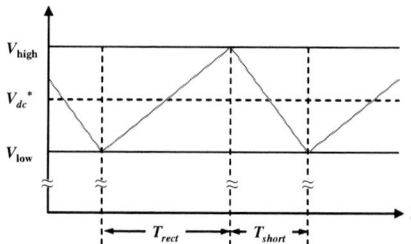

Fig. 7. Waveform of the DC link voltage.

C. System configuration

Previous research on maximum efficiency control has proposed using a diode rectifier and a DC-DC converter on the secondary side [8], [9]. However, this control cannot achieve the desired charging power because the secondary voltage V_2 is controlled for efficiency maximization. As a result, the charging power P is determined by $V_{2\eta\max}$, which is given by eq. (5).

In this paper, a secondary-side simultaneous control method of efficiency maximization and power control is proposed. Fig. 5 shows the circuit diagram of the WPT system for the simultaneous control. The diode rectifier is replaced with the HAR, which maximize the transmitting efficiency η. Then, the DC link voltage V_{dc} is regulated by the HAR to achieve eq. (5). Additionally, the battery charging current I_L is controlled by the DC-DC converter. These control strategies are described further below.

III. EFFICIENCY MAXIMIZATION BY HALF ACTIVE RECTIFIER

A. DC link voltage control

The HAR is operated by two modes, which are shown in Fig. 6. In the rectification mode, the HAR is operated as the diode rectifier. If the charging power P is larger than the load power P_L, V_{dc} is increased. On the other hand, the short mode is worked by turning on lower arm MOSFETs. Then, P is cut-off and P_L is supplied by the DC link capacitor. As a

result, V_{dc} is decreased in the short mode. Therefore, V_{dc} can be controlled by switching between the rectification mode and the short mode. In this paper, V_{dc} is controlled using hysteresis comparator [16].

The upper bound V_{high} and the lower bound V_{low} are defined as follows:

$$V_{\mathrm{high}} = V_{dc}^* + \Delta V \tag{6}$$
$$V_{\mathrm{low}} = V_{dc}^* - \Delta V, \tag{7}$$

where V_{dc}^* is the reference value of V_{dc} and ΔV is the hysteresis band. If V_{dc} becomes smaller than V_{low}, the HAR is operated in the rectification mode. Additionally, when V_{dc} becomes larger than V_{high}, the HAR switches to the short mode. As shown in Fig. 7, V_{dc} is kept within the desired range.

B. Efficiency maximization

In order to achieve the maximum efficiency, V_{dc}^* has to be equal to $V_{dc\eta\max}$, which is given as follows [8]:

$$V_{dc\eta\max} = \sqrt{\frac{R_2}{R_1}} \frac{\omega_0 L_m}{\sqrt{R_1 R_2 + (\omega_0 L_m)^2} + \sqrt{R_1 R_2}} V_S. \tag{8}$$

Then, the transmitting efficiency η can be maximized during the rectification mode. Meanwhile, losses in the short mode is small compared to losses in the rectification mode. This is because the secondary voltage V_2 is nearly equal to zero and the input power P_{in} is drastically decreased in the short mode. In this paper, losses during the short mode are assumed to be negligible to losses during the rectification mode.

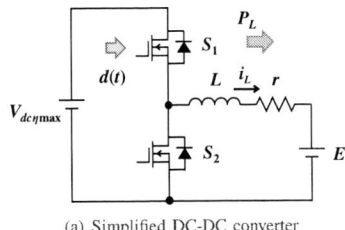

(a) Simplified DC-DC converter

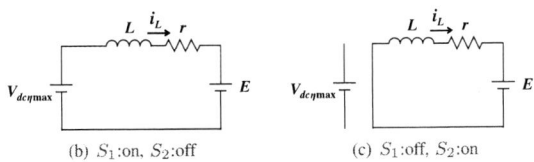

(b) S_1:on, S_2:off   (c) S_1:off, S_2:on

Fig. 8.　Circuit diagram of the DC-DC converter.

IV. POWER CONTROL BY THE DC-DC CONVERTER

A. Circuit configuration

If the DC link voltage V_{dc} is regulated by the HAR, the circuit diagram of the DC-DC converter can be indicated as Fig. 8 (a). In this study, $V_{dc\eta max}$ is used as the nominal value of the DC link voltage to simplify the DC-DC converter model. E is the battery voltage, L is the inductance of the reactor coil and r is the internal resistance of the reactor coil and the battery. In order to achieve power control, the load current i_L has to be controlled in battery charging.

B. Modeling of the DC-DC converter

This paper assumes that the DC-DC converter is operated in the continues conduction mode because the MOSFETs of the DC-DC converter are alternatively turned on and off. Therefore, the operation modes are expressed in Fig. 8 (b) and Fig. 8 (c).

The plant model of the DC-DC converter is obtained by the state space averaging method. From the circuit equation, the state equation of Fig.8 (b) is described as follows:

$$\frac{d}{dt}i_L(t) = -\frac{r}{L}i_L(t) - \frac{1}{L}E + \frac{1}{L}V_{dc\eta max}. \tag{9}$$

Also, the state equation of Fig.8 (c) is expressed as follows:

$$\frac{d}{dt}i_L(t) = -\frac{r}{L}i_L(t) - \frac{1}{L}E. \tag{10}$$

When $d(t)$ is defined as the duty cycle of the upper side MOSFET S_1, the state space model of the DC-DC converter is obtained as follows:

$$\frac{d}{dt}i_L(t) = -\frac{r}{L}i_L(t) - \frac{1}{L}E + \frac{V_{dc\eta max}}{L}d(t). \tag{11}$$

As the DC-DC converter is a non-linear system, it is linearized around an equilibrium point to apply the linear control theory on the controller design. By defining I_L and

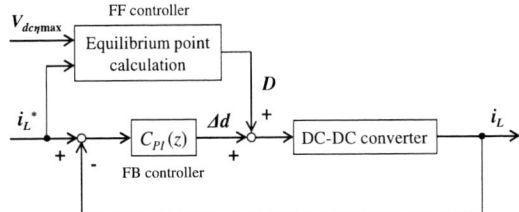

Fig. 9.　Block diagram of load current control.

D as the equilibrium point, $i_L(t)$ and $d(t)$ are expressed as follows:

$$i_L(t) = I_L + \Delta i_L(t) \tag{12}$$

$$d(t) = D + \Delta d(t), \tag{13}$$

where $\Delta i_L(t)$ and $\Delta d(t)$ are the microscopic fluctuations around the equilibrium point. By substituting eq. (12) and eq. (13) in eq. (11), the linearized DC-DC converter model is obtained as follows:

$$\frac{d}{dt}\Delta i_L(t) = -\frac{r}{L}\Delta i_L(t) + \frac{V_{dc\eta max}}{L}\Delta d(t). \tag{14}$$

Therefore, the transfer function from $\Delta d(s)$ to $\Delta i_L(s)$ is given as follows:

$$\Delta P_i(s) = \frac{\Delta i_L(s)}{\Delta d(s)} = \frac{V_{dc\eta max}}{Ls + r}. \tag{15}$$

C. Controller design

Fig. 9 shows the block diagram of the load current control. The feedforward controller is the same as the equilibrium point calculation, which is given by the constraint equation of the DC-DC converter. Assuming $i_L{}^*$ is the reference value of i_L, the equilibrium point I_L and D are obtained as follows:

$$I_L = i_L{}^* \tag{16}$$

$$D = \frac{E + rI_L}{V_{dc\eta max}}. \tag{17}$$

The feedback controller is designed by the pole placement method. As the plant model of the DC-DC converter is expressed by the first-order transfer function, we use a PI controller $C_{PI}(s)$, which is described as follow:

$$C_{PI}(s) = \frac{sK_P + K_I}{s}. \tag{18}$$

If closed loop poles are expressed by a multiple root ω_c, the gains are obtained as follows:

$$K_P = \frac{2L\omega_c - r}{V_{dc\eta max}} \tag{19}$$

$$K_I = \frac{L\omega_c{}^2}{V_{dc\eta max}}. \tag{20}$$

In order to implement the discretized controller $C_{PI}(z)$, $C_{PI}(s)$ is redesigned by Tustin transform.

978-1-4673-9551-9/16 $31.00 © 2016 IEEE

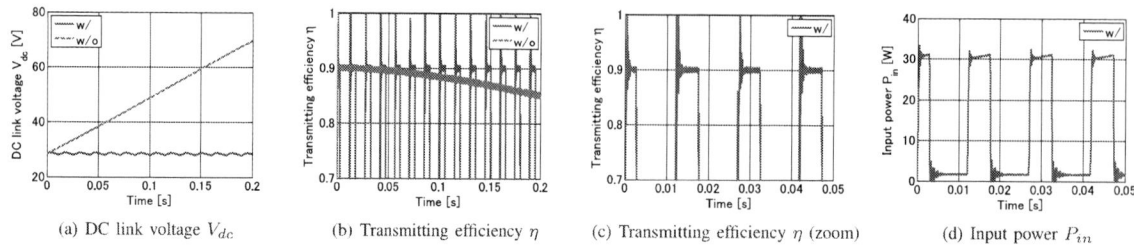

(a) DC link voltage V_{dc} (b) Transmitting efficiency η (c) Transmitting efficiency η (zoom)   (d) Input power P_{in}

Fig. 10. Simulation results of efficiency maximization by Half Active Rectifier

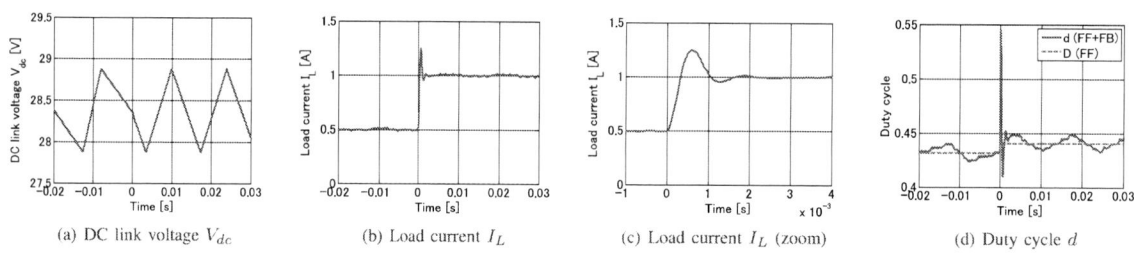

(a) DC link voltage V_{dc} (b) Load current I_L (c) Load current I_L (zoom)   (d) Duty cycle d

Fig. 11. Simulation results of power control by the DC-DC converter with Half Active Rectifier.

TABLE II.   SIMULATION AND EXPERIMENTAL CONDITIONS.

Parameter	Value
Power source voltage V_S	30 V
Operating frequency f_0	101 kHz
DC link voltage reference $V_{dc}{}^*$	28.38 V
Hysteresis band ΔV	0.5 V
Battery voltage E	12 V
Reactor resistance r	0.5 Ω
Reactor inductance L	1000 μH
DC link capacitance C_{dc}	3300 μF
Carrier frequency f_c	20 kHz

Fig. 12.   Half Active Rectifier and DC-DC converter.

V. SIMULATION

Simulations are performed using MATLAB Simlink Sim-PowerSystems. The circuit configuration is shown in Fig. 5. Simulation conditions are described in TABLE II. The inverter supplies the transmitter with a square wave voltage.

A. Efficiency maximization by Half Active Rectifier

In order to verify the effectiveness of maximum efficiency control by the HAR, this simulation replaced the DC-DC converter with a constant power load, which is modeled using controlled current source [12], independent of the power control performance by the DC-DC converter. The load power P_L was set to 10 W. In case of without control, the HAR was operated in the rectification mode at all times.

Fig. 10 shows simulations results of efficiency maximization by the HAR. From Fig. 10 (a) and (b) without control, the DC link voltage V_{dc} is unstable and departs from the reference voltage $V_{dc\eta max}$, which maximize the transmitting efficiency η. On the other hand, the HAR control using the hysteresis comparator can stabilize V_{dc} within the desired range and

maximize η during the rectification mode as shown in Fig. 10 (c). Additionally, Fig. 10 (d) shows that the input power P_{in} is reduced during the short mode. Therefore, it is verified that efficiency maximization by the HAR is effective.

B. Power control by the DC-DC converter with Half Active Rectifier.

In this simulation, V_{dc} was regulated by the HAR and the load current i_L was controlled by the DC-DC converter. The closed loop poles of the load current control were placed at -3000 rad/s (multiple root). The load current reference $i_L{}^*$ was changed from 0.5 A to 1 A at $t = 0$ s.

Simulation results of power control by the DC-DC converter with the HAR are shown in Fig. 11. The HAR can regulate V_{dc} within the desired range as shown in Fig. 11 (a). The step response of i_L is shown in Fig. 11 (b) and (c). The proposed control achieves the fast response without steady-state errors. Fig. 11 (d) shows the duty cycle of the DC-DC converter. The feedforward controller updates the equilibrium

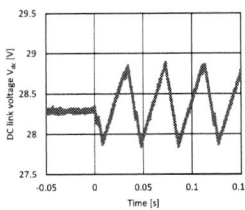 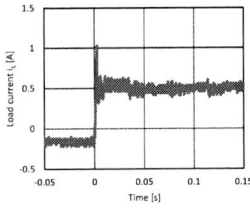 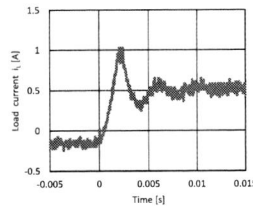 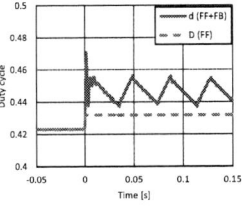

(a) DC link voltage V_{dc} (b) Load current i_L (c) Load current i_L (zoom)   (d) Duty cycle d

Fig. 13. Experimental results of power control by the DC-DC converter with Half Active Rectifier.

point properly and the feedback controller compensates for the error of V_{dc} from the nominal voltage $V_{dc\eta max}$. From these results, the effectiveness of the proposed method is verified.

VI. EXPERIMENT

The experiment was demonstrated using the experimental equipment. The HAR and the DC-DC converter are shown in Fig. 12. Experimental conditions are indicated in TABLE II. The closed loop poles of the load current control were placed at -300 rad/s (multiple root). The load current reference $i_L{}^*$ was changed from 0 A to 0.5 A at $t = 0$ s. The feedback controller was worked from $t = 0$ s.

Experimental results are shown in Fig. 13. The DC link voltage V_{dc} keeps within the desired range as shown in Fig. 13 (a). As a result, the HAR with the hysteresis comparator are able to work properly. Fig. 13 (b) and (c) shows the load current i_L. Although the steady-state error of i_L occurs before $t = 0$ s due to the modeling error of the DC-DC converter, it is suppressed by the feedback controller after $t = 0$ s. Additionally, the feedback controller also compensates for the parameter error of V_{dc} as shown in Fig. 13(d). Therefore, the proposed method can achieve efficiency maximization using the HAR and power control by the DC-DC converter simultaneously.

VII. CONCLUSION

This paper proposed a simultaneous control method of efficiency maximization by HAR and power control by a DC-DC converter. The HAR can regulate the DC link voltage using a hysteresis comparator and it can maximize the transmitting efficiency based on the WPT circuit analysis. The DC-DC converter was modeled under the HAR control and a load current feedback controller was designed. The effectiveness of the proposed method is verified by simulation and experiment.

Future works are to propose efficiency maximization considering losses during the short mode of the HAR and to implement a high power prototype for EV applications.

REFERENCES

[1] Y. Hori, "Future vehicle driven by electricity and control-research on four-wheel-motored "UOT electric march II"," *IEEE Transactions on Industrial Electronics*, vol. 51, no. 5, pp. 954–962, Oct. 2004.

[2] S. Chopra and P. Bauer, "Driving range extension of EV with on-road contactless power transfer—a case study," *IEEE Transactions on Industrial Electronics*, vol. 60, no. 1, pp. 329–338, Jul. 2013.

[3] J. Shin, S. Shin, Y. Kim, S. Ahn, S. Lee, G. Jung, S. Jeon, and D. Cho, "Design and implementation of shaped magnetic-resonance-based wireless power transfer system for roadway-powered moving electric vehicles," *IEEE Transactions on Industrial Electronics*, vol. 61, no. 3, pp. 1179–1192, Mar. 2014.

[4] K. Lee, Z. Pantic, and S. M. Lukic, "Reflexive Field Containment in Dynamic Inductive Power Transfer Systems," *IEEE Transactions on Industrial Electronics*, vol. 29, no. 9, pp. 4592–4602, Sep. 2014.

[5] L. Chen, G. R. Nagendra, J. T. Boys, and G. A. Covic, "Double-coupled systems for IPT roadway applications," *IEEE Journal of Emerging and Selected Topics in Power Electronics*, vol. 3, no.1, pp. 37–49, Mar. 2015.

[6] J. M. Miller, O. C. Onar, and M. Chinthavali, "Primary-side power flow control of wireless power transfer for electric vehicle charging," *IEEE Journal of Emerging and Selected Topics in Power Electronics*, vol. 3, no.1, pp. 147–162, Mar. 2015.

[7] H. H. Wu, A. Gilchrist, K. D. Sealy, and D. Bronson, "A high efficiency 5 kW inductive charger for EVs using dual side control," *IEEE Transactions on Industrial Informatics*, vol. 8, no. 3, pp. 585–595, Aug. 2012.

[8] M. Kato, T. Imura, and Y. Hori, "Study on maximize efficiency by secondary side control using DC-DC converter in wireless power transfer via magnetic resonant coupling," in *Proc. The 27th International Electric Vehicle Symposium and Exhibition (EVS)*, 2013, pp. 1–5.

[9] D. Kobayashi, T. Imura, and Y. Hori, "Real-time coupling coefficient estimation and maximum efficiency control on dynamic wireless power transfer for electric vehicles," in *Proc. IEEE PELS Workshop on Emerging Technologies: Wireless Power*, 2015, pp. 1–6.

[10] S. Li and C.C. Mi, "Wireless power transfer for electric vehicle applications," IEEE Journal of Emerging and Selected Topics in Power Electronics, vol. 3, no.1, pp. 4–17, Mar. 2015.

[11] G. Lovison, M. Sato, T. Imura, and Y. Hori, "Secondary-side-only simultaneous power and efficiency control for two converters in wireless power transfer system," in *41st Annual Conference of the IEEE Industrial Electronics Society (IECON)*, 2015, pp. 4824–4829.

[12] D. Gunji, T. Imura, H. Fujimoto, "Stability analysis of constant power load and load voltage control method for wireless in-wheel motor," in *Proc. The 9th International Conference on Power Electronics - ECCE Asia (ICPE)*, 2015, pp. 1–6.

[13] A. Kurs, A. Karalis, R. Moffatt, J. D. Joannopoulos, P. Fisher, and M. Soljacic, "Wireless power transfer via strongly coupled magnetic resonance," *Science Express on 7 June 2007*, vol. 317, no. 5834, pp. 83–86, Jun. 2007.

[14] T. Imura and Y. Hori, "Maximizing air gap and efficiency of magnetic resonant coupling for wireless power transfer using equivalent circuit and Neumann formula," *IEEE Transactions on Industrial Electronics*, vol. 58, no. 10, pp. 4746–4752, Oct. 2011.

[15] M. Kato, T. Imura, and Y. Hori, "New characteristics analysis considering transmission distance and load variation in wireless power transfer via magnetic resonant coupling," in *IEEE 34th International Telecommunications Energy Conference (INTELEC)*, 2012, pp. 1–5.

[16] D. Gunji, T. Imura, and H. Fujimoto, "Basic study of transmitting power control method without signal communication for wireless in-wheel motor via magnetic resonance coupling," in *Proc. The IEEE/IES International Conference on Mechatronics (ICM)*, 2015, pp. 313–318.

Uniform-Gain Frequency Tracking of Wireless EV Charging for Improving Alignment Flexibility

Yabiao Gao[1], Antonio Ginart[1,2], Kathleen Blair Farley[3], Zion Tsz Ho Tse[1]
[1]College of Engineering, The University of Georgia, Athens, GA 30602, USA
[2]Sonnenbatterie GmbH, Norcross, GA 30093, USA
[3]Southern Company Services, Inc. Birmingham, AL, 35291, USA
ygao@uga.edu, ziontse@uga.edu

Abstract—Wireless power transfer is a promising alternative option for electric vehicle charging due to its non-contact operation. However, the magnetic coupling variation caused by misaligned coils limits its practical application. The output voltage can significantly drop due to the coupling change, lowering the power transfer capability. A uniform-gain frequency tracking control is proposed to keep the output voltage stable within a large misalignment. The uniform-gain control is achieved through voltage gain and impedance analysis across the frequency domain. Experimental results demonstrate that the voltage variation of uniform gain control is within 3.3% across the misalignment range of 200mm, while it is 57.2% for the same misalignment range under fixed frequency control.

Keywords—*Wireless power transfer; inductive charging; wireless charging; electric vehicles*

I. INTRODUCTION

Transportation electrification has been a popular trend due to the increase in greenhouse gas emission, environmental pollutants, and fossil-fuel price fluctuation [1, 2]. Despite the increasing number of plug-in hybrid electric vehicles (PHEVs) and pure electric vehicles (EVs) hitting the road over the past decade, PHEVs and EVs still need to address range anxiety, high battery cost, and the inconvenience of charging before their widespread acceptance [1, 3-5]. Wireless power transfer (WPT) has been a growing research area for several years with the potential to overcome the inconvenience of EV charging [3, 6]. A typical WPT system mainly consists of a high-frequency power inverter, primary and secondary coils, compensation capacitors, rectifiers, and DC-DC converters.

Magnetic coupling variations can lower the efficiency and power delivery of wireless EV charging[4, 7]. Advanced coupler designs have been proposed, such as bipolar pads and Double-D or Double-D Quadrature pads, to improve the coupling when misalignment occurs [1-3, 8, 9]. This paper aims to solve the misalignment issue from the control side and presents a uniform-gain frequency tracking method to compensate for the voltage and efficiency drop caused by misalignment.

Adaptive frequency control was shown to increase the WPT efficiency during misalignment [1, 8, 10, 11]. However, the WPT output voltage can vary under a variable frequency and a DC-DC converter is applied to maintain a stable output for battery charging. The uniform-gain frequency tracking

proposed in this paper takes both output voltage and efficiency into account. The proposed frequency tracking control is based on voltage gain and impedance analysis across the frequency domain, and the uniform voltage gain control is achieved through phase angle feedback at the primary side. The proposed control method is simpler than other solutions because it lowers design requirements of the DC-DC converter on the secondary side and also requires no communication between the primary and secondary sides.

II. WPT ANALYSIS AND CONTROL LOOP DESIGN

A. WPT analysis

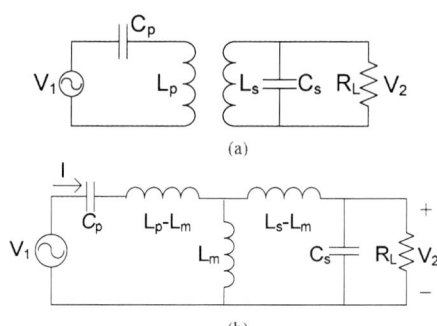

Fig. 1. (a) SP topology for WPT; (b) Equivalent circuit of SP topology

Fig. 1(a) is a simplified series-parallel circuit diagram, and Fig. 1(b) is the corresponding equivalent circuit. C_p is the series tuning capacitance of the primary side, and C_s is the tuning capacitance of the secondary side. L_m is the mutual inductance. The battery's equivalent resistance, R_L, can be calculated using the delivered power and voltage across the battery. For a desired 1.4 kW/120 V battery charging condition, R_L is 20 Ω. The coupling coefficient is defined as

$$k = L_m \big/ \sqrt{L_p L_s}$$

The circuit impedance is given by

$$Z(\omega) = \frac{1}{j\omega C_p} + j\omega(L_p - L_m) + j\omega L_m \,/\!/\, \left(j\omega(L_s - L_m) + \frac{1}{j\omega C_s} \,/\!/\, R_L \right)$$

978-1-4673-9551-9/16 $31.00 © 2016 IEEE   1737

The voltage gain G is the ratio of output voltage V_2 over input voltage V_1 as shown in Fig. 1 and is determined by

$$G = \left| \frac{\omega^2 L_m C_p R_L}{\omega^4 L_p L_s C_p C_s R_L (1-k^2) - \omega^2 R_L (L_p C_p + L_s C_s) + R_L - j(\omega^3 L_p L_s C_p (1-k^2) - \omega L_s)} \right|$$

Fig. 2(a, b, c) shows the simulation results of impedance, phase angle, and voltage gain against frequency for multiple coupling cases. The total impedance on the right side of f_0 (Fig. 2) is inductive in this frequency domain where the input voltage leads the input current, which also realizes a zero-voltage switching (ZVS) operation of the inverter. Therefore, the frequency range marked in Fig. 2(c) is considered as the uniform gain control area. For each curve of the phase angle in Fig. 2(b), since one zero-crossing point exists, which indicates the resonant frequency point, the switching frequency for a fixed voltage gain can be shifted to some value higher than the resonant frequency through detecting the phase angle.

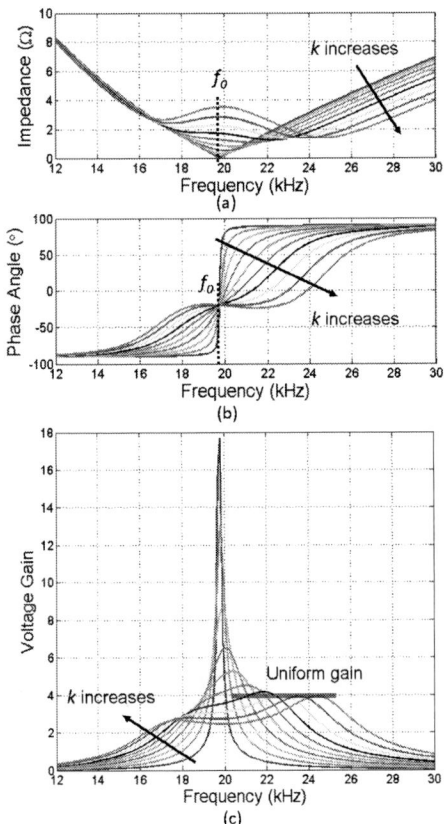

Fig. 2. (a) Impedance magnitude; (b) Impedance phase characteristics; (c) Voltage gain in frequency domain. The simulation conditions were L_p=65 μH, L_s = 65 μH, C_p = 1 μF, C_s = 1 μF, R_L = 20 Ω, and L_m = 2.5, 5, 7.5, 10, 12.5, 15, 17.5, 20, 22.5, and 25 μH.

B. Control loop design

Fig. 3 shows the flowchart of uniform-gain control. Firstly, the WPT system searches for the resonant frequency for a specific alignment condition. The resonant frequency can be different each time a driver parks an EV. As shown in Fig. 3, the resonant frequency is located through frequency shifting (Δfp) and phase angle comparison between current and previous angles ($\theta_{current}$ and $\theta_{previous}$). p is a direction flag that determines whether to shift the frequency to larger or smaller frequencies. The resonant frequency is set when the current phase angle ($\theta_{current}$) is lower than t, which is close to zero. Secondly, the tuning program increases the frequency step by step (Δf) while measuring the load phase angle ($\theta_{uniform}$) to determine the uniform gain frequency.

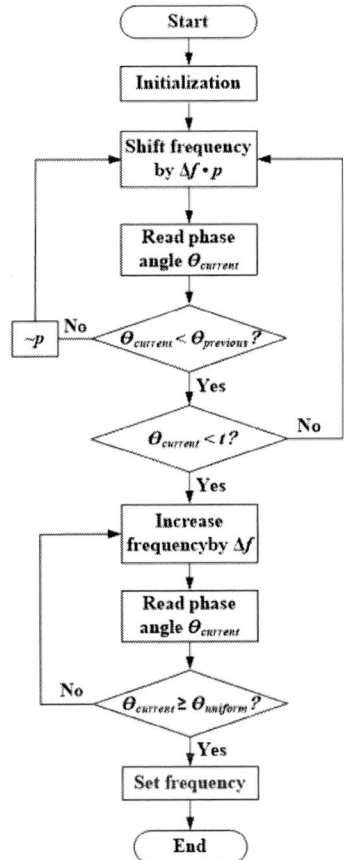

Fig. 3. Flowchart of control loop design for uniform-gain frequency tracking.

III. EXPERIMENAL VALIDATION

A. Experimental setup

The alignment conditions in the experiment were set by a 3-axis motorized platform. The coils shown in Fig.4a are 12-turn, and 56cm in diameter. Fig.4b shows the primary electronics including a full-bridge inverter built by four IGBTs, a DSP controller for frequency tracking, a phase-angle measurement module, and a resistor-capacitor-diode (RCD) snubber circuit. The phase-angle between output voltage and current is measured by an exclusive-OR gate (XOR) and the phase delay equals the pulse width of output waveform of the XOR [7]. A RC low-pass filter was used to convert the pulse signal into a DC voltage for the DSP controller to acquire the phase angle by reading its ADC (analog-to-digital converter) channel values. A current transformer is employed to measure the circulating current. Since it's difficult for a commercialized hall-effect voltage sensor to measure the inverter output voltage at such a high frequency, a voltage probe with a potentiometer is directly connected to the inverter output. The relays in Fig. 4b are to isolate the power side and control electronics in case the high power might destroy the phase angle measurement module. The air gap was set at 100mm in the experiment. The experiment was conducted under fixed frequency control and uniform-gain frequency tracking to compare the two approaches in terms of output voltage, efficiency and misalignment.

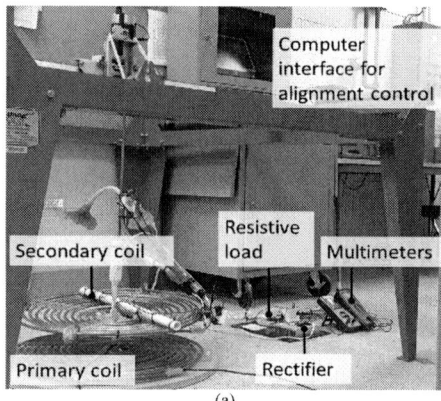

(a)

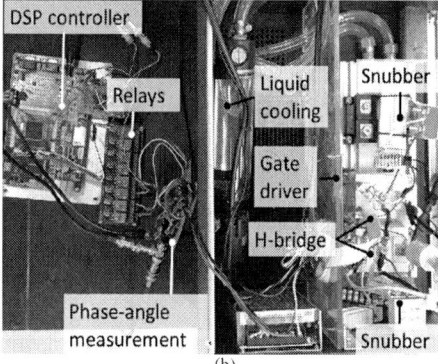

(b)

Fig. 4. (a) Experimental platform; (b) Power inverter and control electronics.

B. Experimental comparison with misalignment

Fig.5 shows the efficiency and gain curves under fixed frequency and uniform gain control. The experiment was conducted under a DC input voltage of 20V and 40V separately. The uniform gain control is better than fixed frequency control in both output voltage and efficiency (Fig.5), especially when the misalignment is bigger than 75mm. The voltage variation of uniform gain control is within 3.3% across the misalignment range of up to 200mm, while it is 57.2% for the same misalignment range under fixed frequency control. The efficiency drop of uniform gain control is 27.2% when the

misalignment changes from 0 to 200m and the DC input is 40V, which is much lower than that of fixed frequency control for the same experimental conditions (57.8%).

While the voltage gain under fixed frequency control began to vary significantly at a misalignment of >100mm, the gain (G $=V_b/V_1$, V_b is the secondary output after rectifier) was maintained at about 3.04 across the misalignment range up to 200mm, which is quite close to the simulation result. The theoretical peak-peak voltage gain of 4.0 (Fig. 2(c)) equals 3.27 when taking the rectification into account. Note that the DC output voltage V_b after the rectifier is the rectifier input RMS voltage V_{rms} multiplied by a constant value: $V_2 = (\sqrt{3}/2)V_b$ [12].

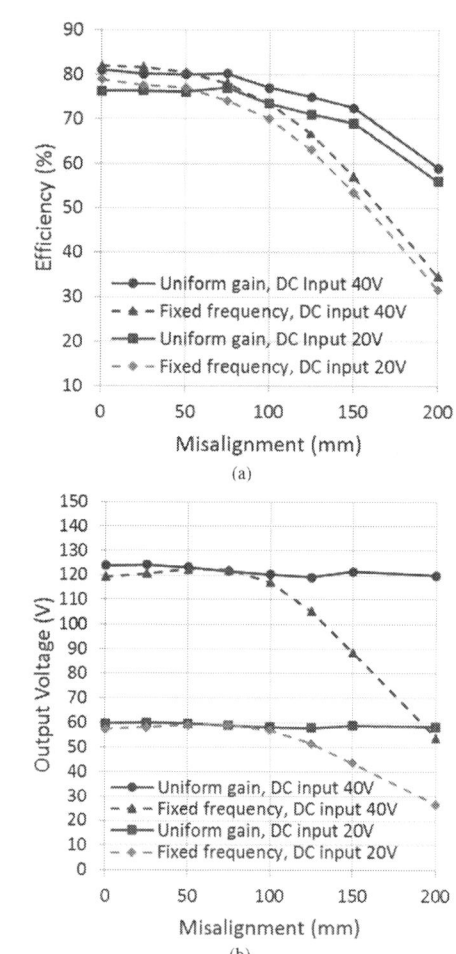

Fig. 5. (a) Efficiency at fixed frequency and uniform gain control; (b) Output voltage at fixed frequency and uniform gain control.

C. Frequency tracking at a large misalignment

Fig. 6 shows the circulating current of the primary and the gate drive signals when the misalignment is 200mm. The two scope images were recorded under a DC input voltage of 40V.

The phase angle between the two signals in Fig. 6(a) is about 56° for fixed frequency control, while the phase delay is around 29.3° under uniform-gain operation (Fig. 6(b)). The gate switching signal is seen to be consistent with the inverter's output voltage. The measured mutual inductance between the primary and secondary is 9.69μH when the misalignment is 200 mm and the air gap is 100mm. Fig. 2 shows the theoretical phase angle is around 30.5° for this particular mutual inductance, meaning that it coincides well with the experimental value which is about 29.3° as shown in Fig. 6(b). According to Fig. 5 and 6, the RMS current of fixed frequency control is 13.4% smaller than that of uniform gain tracking control, while a 76.4% higher efficiency was achieved by using uniform-gain frequency tracking instead of the fixed frequency control.

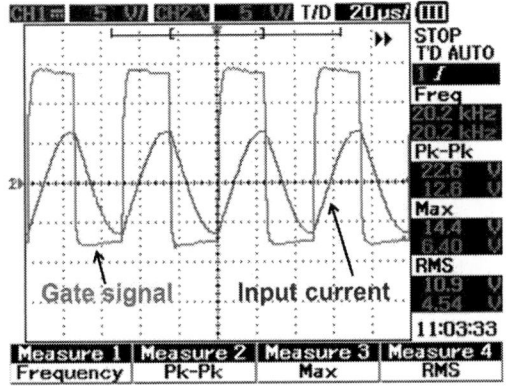

(a)

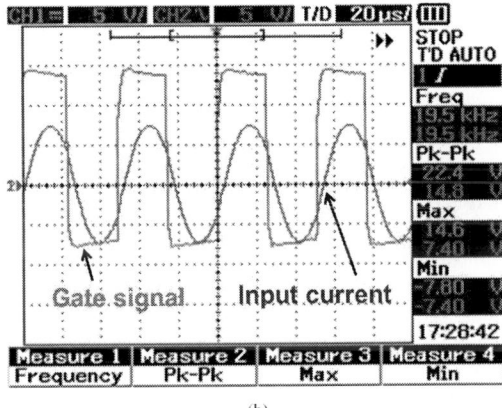

(b)

Fig. 6. Gate signals and the coil circulating current under an extreme condition (misalignment = 200mm, air gap = 100mm) (a) Fixed frequency control (20.2kHz). Sinusoidal current waveform: Max = 32A, RMS = 22.7A, (b) Uniform gain frequency tracking (19.5kHz). Sinusoidal current waveform: Max = 37A, RMS = 26.2 A. The voltage-current ratio of the current sensor is 0.2 V/A.

IV. CONCLUSIONS AND FUTURE WORK

A uniform-gain frequency tracking control for wireless charging can generate a stable voltage to overcome the output voltage drop and efficiency issues resulting from misalignment in EV applications. Simulation and experimental validation were conducted, showing that the uniform voltage gain control offers certain advantages over the traditional fixed frequency control. The uniform gain control allows for operation under greater misalignment and worse magnetic couplings. Therefore, it can counteract the drawback of a common circular coupler, which is that the coupling coefficient drops relative quickly. In addition, since the uniform-gain frequency tracking generates a stable secondary output voltage, it can lower the design requirements of a DC-DC converter on the secondary side. Potentially, a battery could be directly charged with no need of the DC-DC converter using the uniform gain control plus duty-cycle regulation, which could reduce the amount of electronic elements involved in WPT systems. Future work would be focused on the duty-cycle control for wireless power delivery.

REFERENCES

[1] G. A. Covic and J. T. Boys, "Modern trends in inductive power transfer for transportation applications," *Emerging and Selected Topics in Power Electronics, IEEE Journal of,* vol. 1, pp. 28-41, 2013.

[2] T. M. Fisher, K. B. Farley, Y. Gao, H. Bai, and Z. T. H. Tse, "Electric vehicle wireless charging technology: a state-of-the-art review of magnetic coupling systems," *Wireless Power Transfer,* vol. 1, pp. 87-96, 2014.

[3] J. Deng, W. Li, T.-D. Nguyen, S. Li, and C. Mi, "Compact and Efficient Bipolar Pads for Wireless Power Chargers: Design and Analysis," *Power Electronics, IEEE Transactions on,* vol. 20, pp. 6130 - 6140, 2015.

[4] Y. Gao, K. B. Farley, A. Ginart, and Z. T. H. Tse, "Safety and efficiency of the wireless charging of electric vehicles," *Proceedings of the Institution of Mechanical Engineers, Part D: Journal of Automobile Engineering.* p. 0954407015603863, 2015.

[5] Y. Gao, K. B. Farley, and Z. T. H. Tse, "Investigating safety issues related to Electric Vehicle wireless charging technology," in *Transportation Electrification Conference and Expo (ITEC), 2014 IEEE,* 2014, pp. 1-4.

[6] C. Duan, C. Jiang, A. Taylor, and K. H. Bai, "Design of a zero-voltage-switching large-air-gap wireless charger with low electric stress for electric vehicles," *IET Power Electronics,* vol. 6, pp. 1742-1750, 2013.

[7] Y. Gao, K. B. Farley, and Z. T. H. Tse, "A Uniform Voltage Gain Control for Alignment Robustness in Wireless EV Charging," *Energies,* vol. 8, pp. 8355-8370, 2015.

[8] N. Liu and T. G. Habetler, "Design of a Universal Inductive Charger for Multiple Electric Vehicle Models," *Power Electronics, IEEE Transactions on,* vol. 30, pp. 6378 - 6390, 2015.

[9] G. R. Nagendra, G. Covic, and J. T. Boys, "Determining the physical size of inductive couplers for IPT EV systems," *Emerging and Selected Topics in Power Electronics, IEEE Journal of,* vol. 2, pp. 571-583, 2014.

[10] A. P. Sample, D. A. Meyer, and J. R. Smith, "Analysis, experimental results, and range adaptation of magnetically coupled resonators for wireless power transfer," *Industrial Electronics, IEEE Transactions on,* vol. 58, pp. 544-554, 2011.

[11] D. Kar, P. Nayak, S. Bhuyan, and S. Panda, "Automatic frequency tuning wireless charging system for enhancement of efficiency," *Electronics Letters,* vol. 50, pp. 1868-1870, 2014.

[12] J. Miller and A. Daga, "Elements of Wireless Power Transfer Essential to High Power Charging of Heavy Duty Vehicles," *Transportation Electrification, IEEE Transactions on,* vol. 1, pp. 26 - 39, 2015.

978-1-4673-9551-9/16 $31.00 © 2016 IEEE

Design and Optimization of a Multi-Coil System for Inductive Charging with Small Air Gap

Christopher Joffe, Andreas Roßkopf, Stefan Ehrlich, Christian Dobmeier, Martin März
Fraunhofer Institute for Integrated Systems and Device Technology
Schottkystraße 10, 91058 Erlangen, Germany
Email: christopher.joffe@iisb.fraunhofer.de
Telephone: +49 9131 761-296, Fax -212

Abstract—This paper focuses on the magnetic design of a multi-coil system for an inductive power transfer (IPT) system for charging electric vehicles with a small air gap between transmitter and receiver coils. Thus, external stray fields can be reduced and the expense of material decreases compared to systems using a large air gap. The magnetic design of the multi-coil system is optimized for the given package space of the vehicle by using a new coupled numeric approach and considering additional coil geometries as well as alternative winding structures. The result of this design process presents the basis to build up a prototype system, including an additional secondary side DC/DC converter. Therewith, an efficiency between 92% and 96% at an output power of up to 3.5 kW is achieved.

Index Terms—inductive power transfer, design algorithm, resonant converter, series-series compensation, DC/DC-converter, litz wire, FEM

I. INTRODUCTION

Wireless charging using inductive power transfer is a key technology for the prevalence of electric vehicles. In this regard the power transfer across a large air gap ($\Delta z > 100$ mm) is comprehensively published in a variety of system approaches [1]–[4]. However, this large mechanical distance leads to numerous problems, such as a large package volume and a high effort to limit the electric and magnetic field exposure [4], [5]. Hence, the minimization of the air gap in order to decrease the coil size and magnetic stray fields seems a promising solution. This is feasible with a mechanical adaption [6] or by utilizing the license plate [7]. Nevertheless, these systems still have a strongly restricted positioning tolerance due to their coil design and arrangement. Consequently, further examination of coil designs and arrangements with a small air gap need further examination. In literature a variety of published work deals with different coil geometries [3], [7]–[13], but none of those examine a coil system transferring several kilo watts across a comparably small air gap and a large lateral displacement in the range of more than half a coil radius. In regard of the the power electronics, the design as well as the fundamental function and operation of a series-series compensated resonant converter is discussed extensively [14]–[17] and is not object of this paper. Furthermore, methodologies for the entire IPT system design are published [12], [18]. However, these do not consider essential parameters such as the frequency dependent winding resistance.

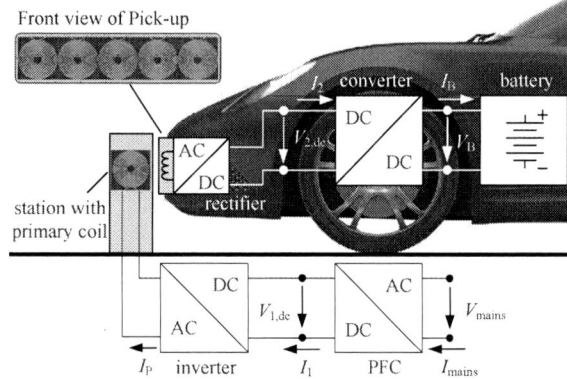

Fig. 1. Schematic of the inductive charging system topology.

In this paper the main objective of the proposed concept is the minimization of the air gap between transmitter and receiver coils due to an autonomous parking system as shown in Fig. 1. Thus, values smaller than 20 mm are feasible. Therefore, a design and optimization process is performed. For the magnetic design of the multi-coil system, it is important to specify the mechanical and electrical constraints. After the definition of installation space and expectable air gap, a coil geometry is chosen on basis of finite elements method (FEM) simulation. Subsequently, the achievable coupling factor and realization of the required inductance is carried out. Further, the frequency dependent winding losses are investigated by two different numeric approaches. Finally, these design aspects are considered to build up a prototype system to verify the simulation results. This IPT system consists of one primary and a set of five secondary circular coils with an outer coil radius of 55 mm. These secondary coils are aligned in a single pick-up housing.

II. SYSTEM DESCRIPTION AND DESIGN ALGORITHM

From a mechanical point of view the multi-coil system is mounted in front of the car. Thus, small air gaps ($\Delta z = 10$-20 mm) depending on the housing could be achieved by autonomous parking. Therefore, the receiver coils are adapted

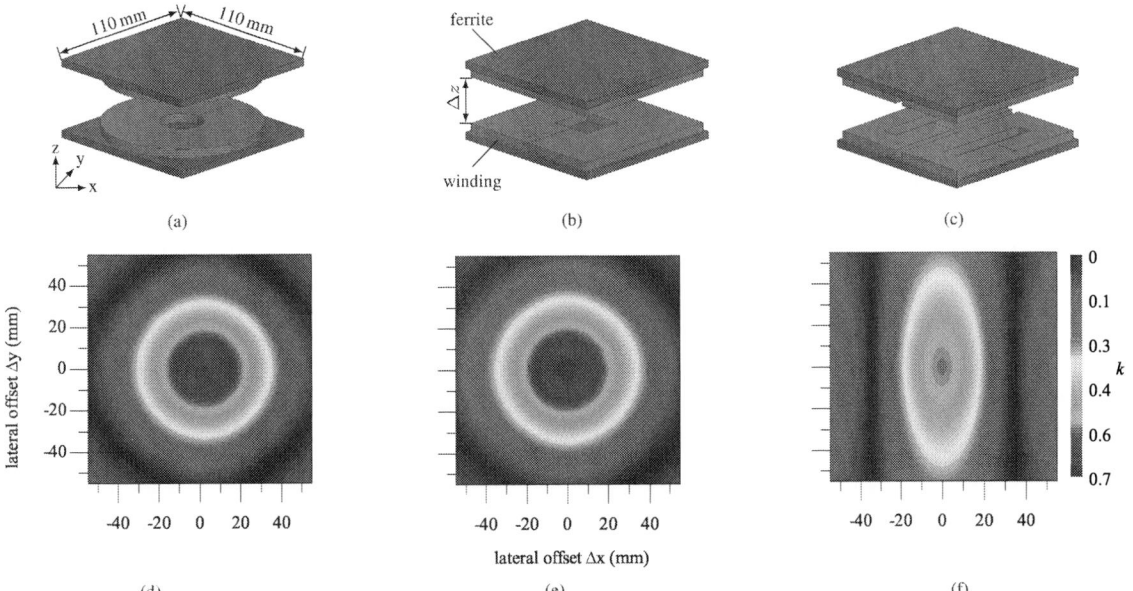

Fig. 2. Simulated coupling map of coupling factor k for lateral displacement in x-y plane with an air gap Δz of 12 mm for (a) a circular coil $A_{k>0.3} = 3560$ mm^2, (b) a square coil $A_{k>0.3} = 4400$ mm^2 and (c) a segmented square coil $A_{k>0.3} = 3050$ mm^2. $A_{k>0.3}$ is the area of the 110 mm × 110 mm examined coupling map with a coupling factor greater than 0.3.

to the form factor of a standardized European license plate (520 mm × 110 mm). Due to parking tolerances and loading of the vehicle a displacement between the coils in vertical (y-axis) and lateral (x-axis) direction occurs.

The electric input and output parameters are given by the peripheral infrastructure (see Fig. 1), namely on the primary side by the mains connection (e.g. $V_{mains} = 230$ Vac with $P_{in} = 3.6$ kW in Europe) and on the secondary side by the traction battery of the vehicle (e.g. $V_B = 320$ - 400 Vdc). The system efficiency for mains to battery shall be >90%. Furthermore, the switching frequency is chosen in the range of 145 kHz as stated in the draft standard [19].

As described above the construction space for a single secondary transfer coil is defined to 110 mm × 110 mm. In a first step, a suitable coil geometry is carried out by a 3D FEM simulation in order to achieve best coupling performance within a specified displacement. In this simulation, the coil is not discretized with separate winding to minimize the computational effort. However, manufacturing aspects of the coils such as the bending radius of litz wire windings are also taken into account. After the coil geometry and the coupling range is determined a suitable converter topology is chosen [1], [14], [17]. Subsequently, the first iteration of power electronics design for a series - series compensated converter is performed to calculate the required self-inductance [20]. Thereby, the main objective is the transfer capability of the specified power within the given boundary conditions. Additional design criteria such as the prevention of "over coupling" [21], [22] within the load and coupling range are

also considered. Next, the maximum current in the primary winding is determined under the proviso of the maximum power that can be dissipated as heat in the coil system. The maximum power dissipation, thus the thermal limits can be either roughly calculated with approximation formulae [23], measured by maximum current rating of the winding or simulated with FEM. On basis of the maximum current and the maximum power dissipation an initial copper cross section is chosen to design and optimize the winding setup and realize the required inductance. In this regard, different approaches on basis of 2D or 3D FEM simulations are possible. In this paper on the one hand a 2D FEM simulation coupled with an analytic loss calculation [24] and on the other hand a 3D FEM simulation coupled with a numeric loss calculation [25] are used. The 2D FEM simulation approach is feasible for parameter sweeps. Therewith, results for parametric optimizations for rotational symmetric geometries can be achieved relatively fast compared to the 3D approach. However, the 2D rotational symmetric FEM simulation is not feasible for more complex 3D geometries such as non-symmetric coils and their lateral displacement. Therefore, 3D FEM combined with the partial element equivalent circuit method (PEEC) provides a valuable solution. Nevertheless, the computational effort for the loss calculation with the 3D FEM approach is significant higher.

III. MAGNETIC DESIGN

A. Coil Geometry

In a first step a coil geometry is determined and optimized. Therefore, the three most common coil geometries [3], [7]–

[13] are compared by 3D FEM simulation of the magnetic coupling characteristics: a circular coil (Fig. 2(a)), a square coil (Fig. 2(b)) and a segmented coil (Fig. 2(c)). Thereby, the segmented coil consists of two parallel connected rectangular coils as shown in Fig. 2(c). The outer dimension of each coil is set to 110 mm × 110 mm. Furthermore, the coils are not resolved as separate windings in the first simulation step due to simplifications. Instead, they are implemented as single winded full copper geometry which is in terms of coupling comparable to a discrete winded coil. Each coil geometry has a closed ferrite structure, in order to achieve maximum magnetic coupling and self-inductance which is proven in section III-C. The good agreement between measurement and simulation of coupling is proven in [20]. Therefore, the presented coupling factors are entirely simulated. The coupling characteristics shown in Fig. 2(a) - 2(c) are achieved by moving the secondary coil over the x-y plane with a constant air gap of 12 mm while the primary coil is fixed. The resulting coupling maps for the three different coil shapes are shown in Fig. 2(d) - 2(f). A minimum coupling limit of $k = 0.3$ is chosen in the power electronics design process in order to limit the primary reactive current I_p to 20 A. This value of k is used to evaluate the coupling performance of the coils for the considered area of displacement. Obviously, the examined circular and square coil achieve quiet similar areas with a higher coupling factor compared to the segmented coils. A detailed comparison between the circular and the square coil examines a 10%

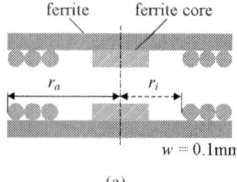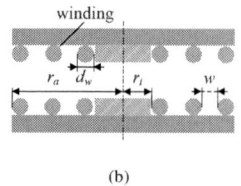

Fig. 4. Schematic of the evaluated winding setups with an example of $N = 3$. (a) *winding scheme I* with a decreasing r_i and (b) *winding scheme II* with a decreasing winding distance w for increasing N.

larger area with $k > 0.3$ for the square coil. In contrast, the segmented coil has a 25% lower coupling performance compared to the square coil. Next, the coupling performance of the three different coil geometries as a function of the air gap is examined by means of the 3D FEM simulation. Thereby, the coils have no offset in x- or y-direction. It is observed that the coupling factor of the circular and square coil are relatively similar whereas the segmented winding shows a 15%-40% lower coupling performance.

Accordingly, a square form would be the preferable coil shape. However, a perfectly square coil shape is not feasible with the use of litz wire due to the limited bending radius of the winding. Thus, the shape of the square coil would converge to a rounded shape and the differences of the coupling performance decreases. Consequently, a circular coil is chosen for the application.

B. Multi-Coil System

So far, the coupling performance of single identical primary and secondary coils were examined. However, this configuration might be not sufficient for every parking situation. Following, the improvement of the positioning tolerance of the chosen circular coil in the x-y-plane is investigated. Therefore, the center of the primary coil is moved to the outer edge of the secondary coil (*Pos. 2*) as shown in Fig. 3(b). At this position, the outer radius $r_{a,p}$ of the primary coil is increased in order to improve the magnetic coupling. Additionally, the impact of a second receiver coil on the magnetic coupling is analyzed. The results of the 3D simulation are plotted in Fig. 3(a). It can be clearly seen that a coupling factor greater than 0.3 is only achievable with a primary coil radius $r_{a,p} = 100$ mm and an additional secondary winding. Furthermore, the magnetic coupling in dependence of the primary coil radius for the center position (*Pos. 1*) is shown in Fig. 3(a). It indicates that k has a maximum at $r_{a,p} = 65$ mm for the specified air gap of $z = 12$ mm. Nevertheless, for the following researches both coils, the primary and secondary, have a similar size ($r_a = 55$ mm).

C. Winding Setup

For the chosen circular coil geometry, the influence of the winding arrangement and additional ferrite on the coupling factor is investigated with 2D and 3D FEM simulations. In this section the winding setup will be determined on basis of

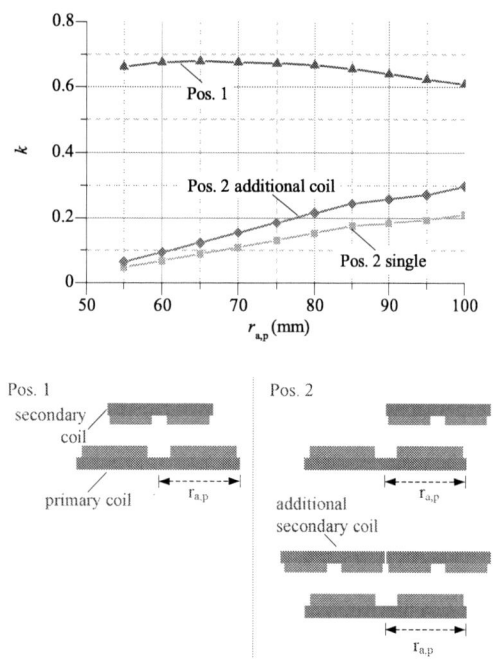

Fig. 3. k vs. $r_{a,p}$ for nonlateral displacement ($x = 0$) for *Pos. 1*, maximum lateral displacement ($x = 55$ mm) for *Pos. 2* and an additional secondary coil at *Pos. 2*.

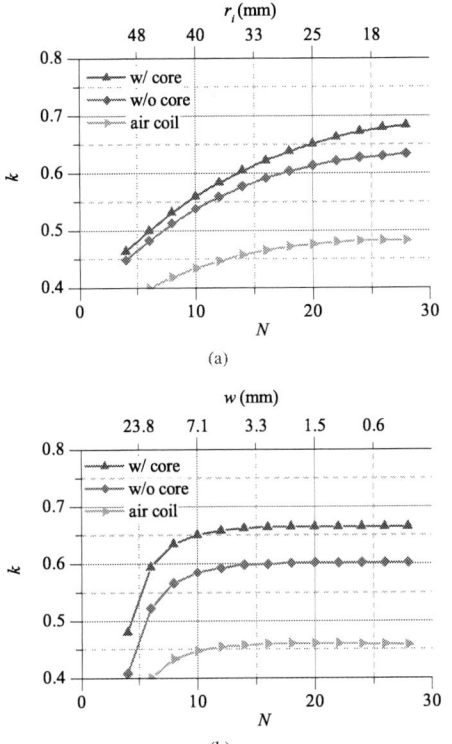

Fig. 5. (a) k as a function of the number of windings for decreasing r_i for *winding scheme I*. (b) k plotted against the number of windings and the winding spacing w for *winding scheme II* for a double layer winding (simulation).

the boundary conditions which are carried out by a power electronic design published in [20]. These are in detail the maximum current rating ($I_{p,max} = 20$ A) within a given coupling factor range ($k = 0.3 - 0.65$) and a required self-inductance ($L \approx 100 \mu$H). First, the required inductance is realized by means of FEM simulation. The modeling of discrete windings in a 2D FEM simulation has a much lower computational effort compared to a 3D simulation in case of parameter sweeps. In this regard, a 2D FEM simulation approach is feasible due to rotational symmetry of the circular coil. In order to perform the 2D FEM simulation with discrete windings an initial winding radius d_w of 2.9 mm is chosen. Furthermore, there are two winding arrangements considered: *winding scheme I* with a variable r_i (see Fig. 4(a)) and *winding scheme II*, with a fixed inner coil radius r_i and equidistant spacing w with an increasing number of windings (see Fig. 4(b)). Additionally, the impact of the winding scheme on coupling is considered.

As mentioned previously and shown in Fig. 4(a), the outer coil radius r_a is fixed to 55 mm for *winding scheme I* and the number of turns is increased to a maximum of 14 ($r_{i,min} = 13$ mm), due to limited bending radius of the winding. To examine differences in self-inductance and magnetic coupling the parameterized coils are simulated as air coils,

as coils with a ferrite back plane (w/o core) and with an additional ferrite core (w/ core) in the free winding space in the center of the coil (see Fig. 4(a)). Subsequently, 2D FEM simulations for the different winding schemes were performed. As a result, the required value of self-inductance $L = 100 \mu$H is only achievable with a double layer coil, a ferrite back plane and an additional ferrite core in the center.

Next, the dependency of the winding schemes on coupling is examined for the double layer coil for three different configurations mentioned above. Regarding *winding scheme I*, Fig. 5(a) illustrates that the coupling factor decreases with an increasing r_i for all configurations, whereas an additional ferrite in the center of the coils improves the coupling up to 15 %. However, the improvement of k for smaller r_i diminishes and k converges. This effect is more distinct for the *winding scheme II* shown in Fig. 5(b). In this case, k converges for less than three windings. Moreover, the placement of windings in the coil centre leads to increasing proximity losses due to higher flux densities in the center. Therefore, the coil quality factor Q decreases.

D. Winding Resistance and Quality Factor

Besides the self-inductance and the magnetic coupling of the coil system, additional design criteria apply such as the afore-mentioned current rating of the primary and secondary coil. In order to reduce frequency dependent winding losses, caused by skin and proximity effects, litz wires are used. The frequency dependent ohmic losses R_{ac} of a litz wire winding are separated in skin and proximity losses. Thereby, the skin losses comprise the inner skin and proximity effects caused by the internal magnetic field. The winding losses caused by external magnetic fields are external proximity losses [26]. However, these litz wires have a complex geometric structure which makes accurate simulation or calculation of losses challenging for a certain coil configuration.

The chosen semi-analytic approach combines the analytic calculation of ohmic losses in litz wires [24], [27] with field distribution calculated in a 2D FEM simulation. In a first step, the coil is modeled with discrete windings (see Fig. 4(b)) and the external magnetic field is simulated. Subsequently, in a post processing step the external magnetic field is assigned to each winding. Finally, the winding losses P_{cu} in a litz wire caused by internal and external fields are calculated analytically. The core losses P_{fe} are taken into account separately in a 2D FEM simulation for the given operation point.

To verify the validity of the above discussed approach R_{ac} is measured for an exemplary single air coil using *winding scheme I*, with $w = 0.5$ mm, $r_a = 55$ mm and $N = 12$ and a 800 x 0.071 mm Litz wire. Furthermore, the non-ideal behavior of litz wire is considered in calculation in terms of additional parameters discussed in [28]. The measurement results and simulation shown in Fig. 6(a) corresponds very well. Therefore, the 2D semi-analytic approach is chosen for further optimization and design steps in order to determine the needed litz wire. In this regard, the ac resistance factor $F_R = R_{ac}/R_{dc}$ is introduced to compare certain litz wires.

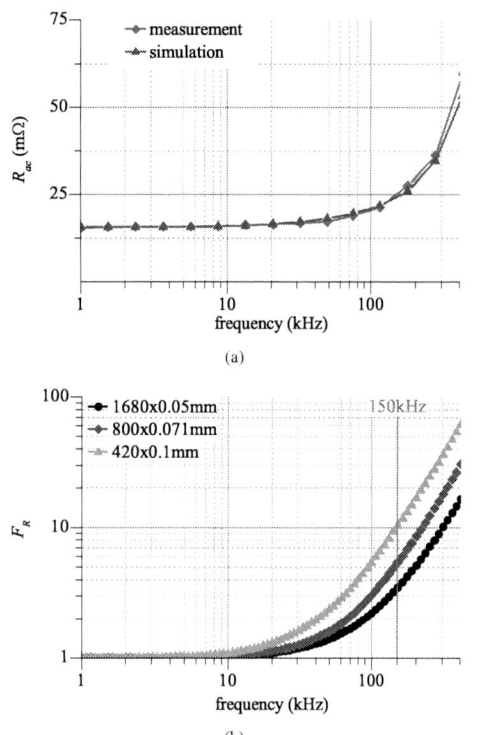

(a)

(b)

Fig. 6. (a) calculation and measurement of an exemplary spiral spiral air coil with 800×0.071mm, $w=0.5$mm, $r_a = 55$ mm and $N=12$. (b) simulated F_R for 420×0.1 mm, 800×0.071 mm, 1680×0.05 mm.

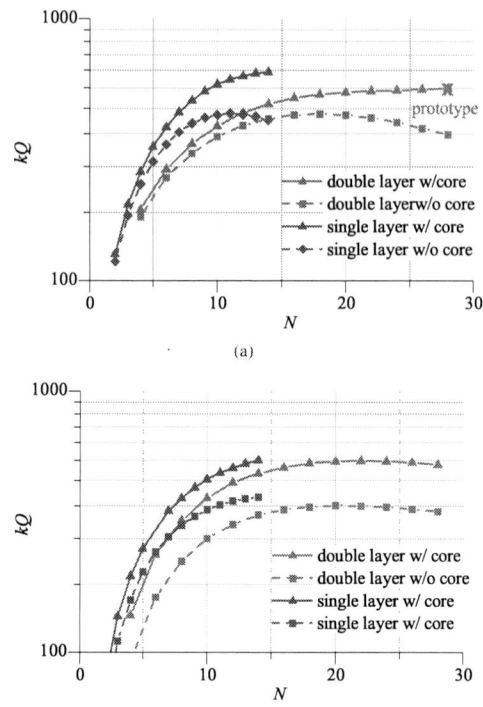

(a)

(b)

Fig. 7. FOM of kQ as a function of the winding number for (a) *winding scheme I* and (b) *winding scheme II* with a 1680×0.05 mm litz wire (simulation).

On basis of the inductance considerations a coil system with a double layer winding structure using *winding scheme I* with ferrite back and ferrite core ($N = 28$, $r_a = 55$mm) is simulated. Thereby, the winding radius d_w is set to 2.9 mm and three different litz wires with equivalent copper cross section are chosen. The results are shown in Fig. 6(b). It can be clearly seen that the number of strands have a strong impact on F_R. The F_R of the litz wire with a strand diameter of 0.1 mm is almost three times higher compared to the litz wire using a 0.05 mm strand diameter. The boundary conditions for the choice of litz wire are given by the thermal limits of the coils, thus the maximum power dissipation for the winding:

$$P_{\mathrm{cu}} = I_{\max}^2 \, F_R \, R_{\mathrm{dc}} \qquad (1)$$

Consequently, a minimum F_R applies in order to meet the thermal limits. With this approach modeling a rectangular litz wire is not feasible. For the application however a rectangular litz wire might be useful in order to increase the copper fill factor and decrease the winding losses additionally.

To obtain an IPT system with optimized performance at the mentioned operating points, the magnetic parameters formerly discussed have to be optimized. Therefore, Waffenschmidt et al. states that the coupling factor k and the coil quality factor $Q = \omega L / R_{\mathrm{ac}}$ are a figure of merit (FOM) to optimize

inductive power transfer systems. In order to minimize losses the product kQ has to be maximized. The FOM shown in Fig. 7 are simulation results for the *winding scheme I* and *winding scheme II* with a 1680×0.05 mm litz wire. The results in Fig. 7(b) show no distinct maximum, whereas the configuration in Fig. 7(a) exhibits a slight maximum of kQ for the double layer (DL) and single layer (SL) without core. This confirms the statement of increasing proximity losses due to field shaping in the coil center.

The semi-analytic approach for the calculation of winding losses is not feasible for complex 3D geometries. However, the lateral displacement of the proposed coils leads to a 3D problem. Therefore, the coil system was simulated with a full 3D coupled numeric approach that combines FEM simulations and PEEC method [25], [29], [30]. Thus, the simulation of winding losses for lateral misalignment and more complex 3D geometries such as the lateral displacement of the transfer coils as shown in Fig. 8 is possible. The effect of the lateral displacement, with regard to the modified field distribution, on F_R is depicted in Fig. 8. Thereby, a 800×0.071 mm litz wire is simulated due to the high computational effort for litz wires with a high number of strands. It can be seen that the ac resistance changes only slightly with the lateral displacement, although the coupling factor decreases drastically from 0.7 at

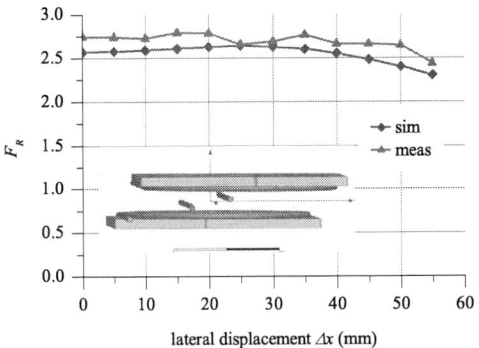

Fig. 8. resistance factor F_R as a function of the lateral displacement for a 800×0.071 mm litz wire at 145 kHz simulated wit 3D FEM and discrete windings (see schematic) and measured.

$x = 0$ to 0.04 at $x = 55$ mm. Hence, the lateral displacement has a minor impact on the resistance factor compared to the coupling factor.

IV. SYSTEM VERIFICATION

Considering the preliminary magnetic design process, a prototype, capable of transferring up to 3.5 kW across an air gap of $\Delta z = 12$ mm, is set up. The coil system consists of one primary and five secondary, structurally identical, double layer circular coils ($L_1 = L_2 = 100\mu H$). Each coil has 28 windings with a ferrite back plane and a ferrite core in the center of the core. In order to increase the copper cross section for the given construction space a rectangular litz wire with 2880×0.05 mm strands and a dimension of 2.9 mm $\times$3.6 mm is used.

TABLE I
SPECIFICATION OF THE PROTOTYPE SYSTEM.

Symbol	Quantity	Value
P_{in}	power rating	3.6 kW
$V_{1,\text{dc}}$	primary intermediate DC link voltage	400 V
$V_{2,\text{dc}}$	secondary intermediate DC link voltage	450 V
V_B	battery voltage	320-400 V

Fig. 9(a) shows the test setup with the mechanical integration behind a nonmetallic license plate and the peripheral power electronics. The measurements were performed using an additional DC/DC buck converter, with an impedance control scheme, to adapt the wide battery voltage range to the resonant converter. Consequently, system efficiencies from 92% to 96% ($V_{1,\text{dc}}$ to V_B) for the defined coupling and battery voltage range are achieved and are shown in Fig. 9.

V. CONCLUSION

The magnetic design of a multi-coil inductive charging system for an electric vehicle, with a small air gap between transmitter and receiver coil is shown. 2D and 3D FEM simulation approaches coupled with numeric and analytical calculations are performed to optimize the coil geometry,

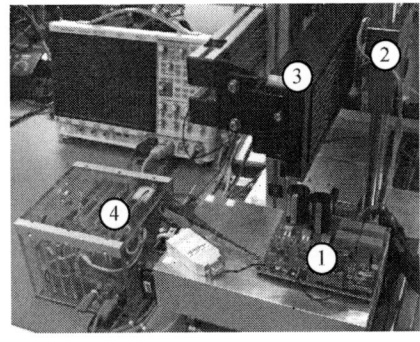

(a)

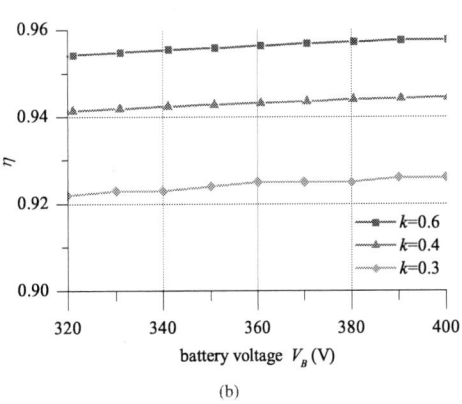

(b)

Fig. 9. (a) Setup of the evaluated IPT system (1-primary inverter, 2-primary coil, 3-secondary integrated license plate coil system with rectifier, 4-secondary dc/dc converter). (b) Measured efficiency from $V_{1,\text{dc}}$ to V_B with an output power of 3.5 kW at $k = 0.4 \ldots 0.6$ and reduction to 3 kW at $k = 0.3$.

winding arrangement and inductance. The result of this design process is the basis to build up a prototype system, achieving a system efficiency between 92% and 96% at an output power of up to 3.5 kW over a broad range of coil displacement. Furthermore, a design algorithm is presented which examines the coupling between the electrical network and the magnetic design. This iterative process is verified and discussed on the basis of simulations and measurements. Additionally, the design process of the litz wire using 3D coupled numeric approach is discussed.

ACKNOWLEDGMENT

The research leading to these results has received funding as part of the Energy Campus Nuremberg (EnCN) which is financed by the State of Bavaria as part of the program Bavaria on the move. Moreover, this contribution was supported by the Bavarian Ministry of Economic Affairs and Media, Energy and Technology as part of the Bavarian project: Leistungszentrum Elektroniksysteme (LZE). The authors would like to thank Dr. S. Ditze, Cluster of Excellence Engineering of Advanced Materials (EAM) for her valuable scientific input.

978-1-4673-9551-9/16 $31.00 © 2016 IEEE

REFERENCES

[1] C. Rui, Z. Cong, Z. U. Zahid, E. Faraci, Y. Wengsong, S. L. Jih, M. Senesky, D. Anderson, and G. Lisi, "Analysis and parameters optimization of a contactless IPT system for EV charger," in *Applied Power Electronics Conference and Exposition (APEC), 2014 Twenty-Ninth Annual IEEE*, 2014, pp. 1654–1661. [Online]. Available: http://ieeexplore.ieee.org/stamp/stamp.jsp?arnumber=6803528

[2] J. Sallán, J. Villa, A. Llombart, J. Sanz, and Fco, "Optimal Design of ICPT Systems Applied to Electric Vehicle Battery Charge," *Industrial Electronics, IEEE Transactions on*, vol. 56, no. 6, pp. 2140–2149, 2009.

[3] M. Budhia, G. Covic, and J. Boys, "Design and Optimization of Circular Magnetic Structures for Lumped Inductive Power Transfer Systems," *IEEE Transactions on Power Electronics*, vol. 26, no. 11, pp. 3096–3108, 2011.

[4] M. Ibrahim, L. Pichon, L. Bernard, A. Razek, J. Houivet, and O. Cayol, "Advanced Modeling of a 2-kW Series–Series Resonating Inductive Charger for Real Electric Vehicle," *Vehicular Technology, IEEE Transactions on*, vol. 64, no. 2, pp. 421–430, 2015.

[5] K. Hongseok, S. Chiuk, K. Jonghoon, D. H. Jung, S. Eunseok, K. Sukjin, and K. Jiseong, "Design of magnetic shielding for reduction of magnetic near field from wireless power transfer system for electric vehicle," in *Electromagnetic Compatibility (EMC Europe), 2014 International Symposium on*, 2014, pp. 53–58. [Online]. Available: http://ieeexplore.ieee.org/stamp/stamp.jsp?arnumber=6930876

[6] (2015) PRIMOVE for e-cars. [Online]. Available: http://primove.bombardier.com/applications/e-car.html

[7] Daniel Kürschner, F. Turki, Chris Yotta, Svenn Thamm, and Christian Rathge, "Comparison of Planar and Solenoid Coil Arrangements for Inductive EV-Charging Application," in *PCIM Europe*, 2013, pp. 385–391.

[8] W. X. Zhong, X. Liu, and S. Hui, "A Novel Single-Layer Winding Array and Receiver Coil Structure for Contactless Battery Charging Systems With Free-Positioning and Localized Charging Features," *Industrial Electronics, IEEE Transactions on*, vol. 58, no. 9, pp. 4136–4144, 2011.

[9] M. Budhia, G. Covic, and J. Boys, "A new IPT magnetic coupler for electric vehicle charging systems," in *IECON 2010 - 36th Annual Conference on IEEE Industrial Electronics Society*, 2010, pp. 2487–2492.

[10] G. R. Nagendra, G. A. Covic, and J. T. Boys, "Determining the Physical Size of Inductive Couplers for IPT EV Systems," *Emerging and Selected Topics in Power Electronics, IEEE Journal of*, vol. 2, no. 3, pp. 571–583, 2014.

[11] R. Bosshard, J. Muehlethaler, J. W. Kolar, and I. Stevanovic, "Optimized magnetic design for inductive power transfer coils," in *Applied Power Electronics Conference*, 2013.

[12] Y. H. Chang, J. E. James, and G. A. Covic, "Design Considerations for Variable Coupling Lumped Coil Systems," *Power Electronics, IEEE Transactions on*, vol. 30, no. 2, pp. 680–689, 2015.

[13] A. Zaheer, D. Kacprzak, and G. Covic, "A bipolar receiver pad in a lumped IPT system for electric vehicle charging applications," in *Energy Conversion Congress and Exposition (ECCE), 2012 IEEE*, 2012, pp. 283–290. [Online]. Available: http://ieeexplore.ieee.org/stamp/stamp.jsp?arnumber=6342811

[14] Z. Wei, C. W. Siu, C. K. Tse, and C. Qianhong, "Analysis and Comparison of Secondary Series- and Parallel-Compensated Inductive Power Transfer Systems Operating for Optimal Efficiency and Load-Independent Voltage-Transfer Ratio," *Power Electronics, IEEE Transactions on*, vol. 29, no. 6, pp. 2979–2990, 2014.

[15] S. Ditze, "Steady-state analysis of the bidirectional CLLLC resonant converter in time domain," in *Telecommunications Energy Conference (INTELEC), 2014 IEEE 36th International*, 2014, pp. 1–9. [Online]. Available: http://ieeexplore.ieee.org/stamp/stamp.jsp?arnumber=6972179

[16] S. Ditze, T. Heckel, and M. Maerz, "Influence of the Junction Capacitance of the Secondary Rectifier Diodes on Output Characteristics in Multi-Resonant Converters," in *Applied Power Electronics Conference*, 2016.

[17] B. J. Gyu and B. Cho, "An energy transmission system for an artificial heart using leakage inductance compensation of transcutaneous transformer," *IEEE Transactions on Power Electronics*, vol. 13, no. 6, pp. 1013–1022, 1998.

[18] D. Kurschner, C. Rathge, and U. Jumar, "Design Methodology for High Efficient Inductive Power Transfer Systems With High Coil Positioning Flexibility," *Industrial Electronics, IEEE Transactions on*, vol. 60, no. 1, pp. 372–381, 2013.

[19] IEC, "Draft: IEC 61980-1: Electric vehicle wireless power transfer systems (WPT) - Part 1: General requirements," 2013.

[20] C. Joffe, S. Ditze, and A. Rosskopf, "A novel positioning tolerant inductive power transfer system," in *Electric Drives Production Conference (EDPC), 2013 3rd International*, 2013, pp. 1–7. [Online]. Available: http://ieeexplore.ieee.org/stamp/stamp.jsp?arnumber=6689747

[21] C. Bowick, J. Blyler, and C. J. Ajluni, *RF circuit design: [a bestseller now thoroughly revised ; two new chapters on RF front-end design and RF design tools ; perfect for the practical, hard-working RF professional]*, 2nd ed. Amsterdam: Elsevier/Newnes, 2008.

[22] S. W. Chwei, G. A. Covic, and O. H. Stielau, "Power transfer capability and bifurcation phenomena of loosely coupled inductive power transfer systems," *IEEE Transactions on Industrial Electronics*, vol. 51, no. 1, pp. 148–157, 2004.

[23] *VDI heat atlas*, 2nd ed., ser. Springer reference. Heidelberg: Springer, 2010.

[24] Rossmanith and Albach, "Fast and precise calculation of winding losses in P-, RM- and ETD-ferrite core inductors," in *2nd International Conference on Automotive Power Electronics - APE 2007*, 2007.

[25] A. Rosskopf, C. Joffe, E. Baer, and C. Bonse, "Calculation of power losses in litz wire systems by coupling FEM and PEEC method," *Power Electronics, IEEE Transactions on*, vol. PP, no. 99, p. 1, 2015.

[26] C. Sullivan, "Optimal choice for number of strands in a litz-wire transformer winding," *IEEE Transactions on Power Electronics*, vol. 14, no. 2, pp. 283–291, 1999.

[27] M. Albach, "Two-dimensional calculation of winding losses in transformers," in *Power Electronics Specialists Conference, 2000. PESC 00. 2000 IEEE 31st Annual*, vol. 3, 2000, pp. 1639 –1644 vol.3.

[28] H. Rossmanith, M. Doebroenti, M. Albach, and D. Exner, "Measurement and Characterization of High Frequency Losses in Nonideal Litz Wires," *IEEE Transactions on Power Electronics*, vol. 26, no. 11, pp. 3386–3394, 2011.

[29] A. Rosskopf, C. Joffe, and E. Bar, "Calculation of ohmic losses in litz wires by coupling analytical and numerical methods," in *Electric Drives Production Conference (EDPC), 2014 4th International*, 2014, pp. 1–6. [Online]. Available: http://ieeexplore.ieee.org/stamp/stamp.jsp?arnumber=6984423

[30] R. Y. Zhang, J. K. White, J. G. Kassakian, and C. R. Sullivan, "Realistic litz wire characterization using fast numerical simulations," in *Applied Power Electronics Conference and Exposition (APEC), 2014 Twenty-Ninth Annual IEEE*, 2014, pp. 738–745. [Online]. Available: http://ieeexplore.ieee.org/stamp/stamp.jsp?arnumber=6803390

Core Design for Better Misalignment Tolerance and Higher Range of Wireless Charging for HEV

Mostak Mohammad
Student Member, IEEE
Department of Electrical and Computer
Engineering, University of Akron,
Akron, OH, USA
Email: mm251@zips.uakron.edu

Dr. Sangshin Kwak
Member, IEEE
School of Electrical and Electronics
Engineering,Chung-ang University,
Seoul, Korea
Email: sskwak@cau.ac.kr

Dr. Seundeog Choi
Member, IEEE
Department of Electrical and Computer
Engineering, University of Akron,
Akron, OH, USA
Email: schoi@uakron.edu

Abstract—An optimized core design for wireless power transfer system for hybrid vehicle application is proposed in this paper. One of the main challenge of wireless power in hybrid vehicle charging application is the limited distance range and significant efficiency drop under small misalignments between the transmitter (Tx) and the receiver (Rx) coil. Appropriately designed transmitter and receiver coil with core can make the system more tolerant against misalignment. For a planar spiral coil, an optimum core is designed here, using minimum amount of core and higher range of wireless power is achieved with better misalignment tolerance. The proposed model is simulated in Finite Element Analysis (FEA) and verified with experimental data of an implemented system of 3kW power for different gap and misalignment. Implementing the proposed optimized core for the system, the maximum efficiency is achieved much higher compared to the system without core.

I. INTRODUCTION

BOTH the low and high power wireless power transfer systems are being extensively researched for last ten years. In this system, as the distance between the transmitter and the receiver increases, the wireless power transfer become challenging due to lower magnetic coupling between the Tx and Rx. Specially, the high power medium range (several hundred millimeters) wireless power system has the challenges of high switching loss in inverter circuit and resistive loss in Tx and Rx coil [1]. The loss even increases as the switching frequency, the distance or misalignment increases between Tx and Rx; hence the efficiency decreases significantly [2]. As it takes six to twenty hours for Level 1 (120 VAC at 8-15A) & level 2 (240 VAC at 40A) charging system to fully charge the HEV battery, stability and high efficiency is essential and such drop of efficiency due to misalignment is not practically tolerable. The inductive wireless charging is getting more matured for HEV charging and prediction indicates that, this market will expand to $4.6 billion by 2019.

In commonly used contact-less charging application, transmitter is placed on the floor and the receiver is mounted at the bottom of the vehicle as shown in Fig. 1(a). As the vehicle is parked, it is usual that there will be certain level of misalignment between that transmitter and receiver. The application of wireless power in Hybrid Electric Vehicle (HEV) is more challenging because of this unpredictable car parking, which causes mainly the horizontal misalignment as shown in Fig. 1(b). For a typical car parking on a flat parking

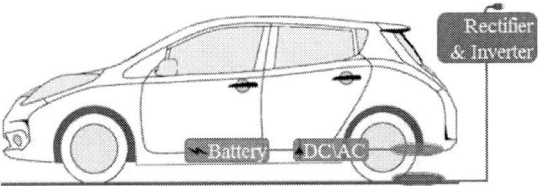

(a) Basic components of wireless HEV charger.

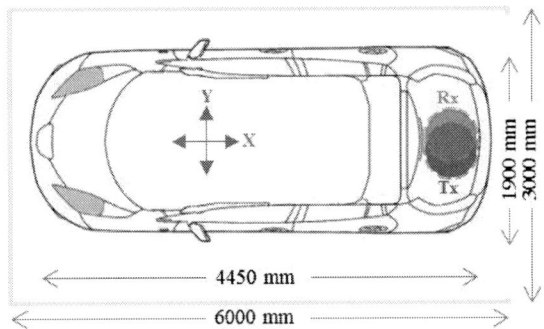

(b) Typical misalignment of wireless charging in car parking.

Fig. 1. Wireless charging (a) basic components and (b) typical misalignment.

area, the horizontal misalignment is more likely to happen than angular or vertical misalignment, therefore, the system requires to be designed such that it become more tolerant to the horizontal misalignment. Strong coupling between the high quality Tx and Rx coils can enhance the range of operation. In [3], the core is found to significantly enhance the inductance and quality factor for micro-scale low power wireless charger.

A properly designed core can enhance these misalignment tolerances of this high power wireless charging within certain limit of misalignment. Recently, wireless charger with planar coil and ferrite core has been investigated in its simplest form [4], [5]. But it doesn't provide the optimum structure and configuration of the core specially considering the application of HEV. To make the system more misalignment insensitive, array of inductor is proposed in [13], which although increases the misalignment tolerance, but it significantly reduces the range of vertical gap between Tx and Rx, therefore not feasible

978-1-4673-9551-9/16 $31.00 © 2016 IEEE

for HEV application where the air-gap is comparatively large. The application of cylindrical core for helical transmitter coil is shown to improve the flux linkage capability and transmission efficiency in [7]. Again, the cylindrical coil increases the height of the system and it uses too much core, and therefore, not suitable for HEV charging application. For HEV charging, the Tx and Rx have to be highly efficient as well as comfortably small and of minimum weight. Therefore, planer coil is best suitable with planer core and shield. Specially, core with the receiver side will add additional weight, which needs a serious consideration of optimization. Again, with introduction of core, significant core loss would incur, especially at high operating frequency and high current. Therefore, an optimization approach for core design is critically important considering core loss, size and weight. In this paper, a core optimization approach is proposed maximize the magnetic coupling and efficiency considering all those constrains and compared with a system without core as well as several other core configuration.

II. POWER TRANSFER EFFICIENCY

Inductive power transfer (IPT) and magnetic resonance wireless power transfer are the two most popular near field wireless power transfer methods. Inductive power transfer is highly efficient for limited distances and strong magnetic coupling, and therefore, it has been adopted for this high power automotive application.

The efficiency of the power transfer is given as follows [1], which shows the dependency of the efficiency on mutual inductance M_0 and resonance frequency ω_0,

$$\eta = \frac{Z_L}{(R_s + Z_L)(1 + \frac{R_p(R_s + Z_L)}{(\omega_0 M_0)^2})} \quad (1)$$

where, R_p and R_s are parasitic resistance of Tx and Rx, Z_L is the load impedance, ω_0 is the resonant frequency and M_0 is the mutual inductance between Tx and Rx in the air without core. The high frequency rectified output is fed to the battery charging circuit and Z_L is the equivalent load impedance of the battery. Although, the battery impedance changes gradually during charging [8], but for the simplicity of analysis, the load impedance is considered constant here. For higher efficiency, the parasitic resistance R_p and R_s needs to be minimized, and $\omega_0 M$ needs to be maximized for high quality of the coil. But, the switching frequency of the inverter circuit, is limited by switching loss in semiconductor switches (MOSFET, IGBT, etc.), and conductor loss due to ac resistance of high frequency skin and proximity effect. And mutual inductance M_0 is limited by gap and misalignment between Tx and Rx.

For a the proposed wireless charging system with fixed the primary and secondary side parasitic resistance and fixed load impedance at 50Ω, the relation between mutual inductance and efficiency as given in equation (1), for this investigating system without core is plotted in Fig. 2, which shows that, for a specific resonant frequency, the efficiency can be increased by increasing the mutual inductance. Therefore, main focus in this paper is given on increasing mutual inductance M between transmitter and receiver by guiding flux using an optimally designed high permeability ferrite core. The effect of high permeability core on planer spiral inductor is investigated in [10], which showed a significant increase of its inductance. For

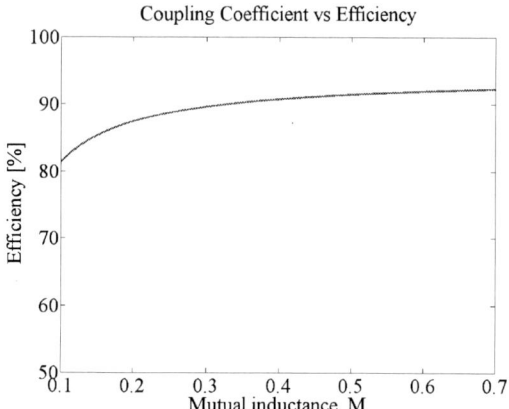

Fig. 2. Efficiency vs coupling coefficient.

the two coil wireless power transfer system proposed here, as the core partially provide the low reluctance path for a selected portion of area, the effective increase of the mutual inductance between the Tx and Rx can be given by,

$$M = \acute{\mu}_r \times M_0 \quad (2)$$

where, α is effective relative permeability with core, and it depends on the core material and geometry.

The relation between the mutual inductance M and self-inductances L_p, L_s represented by coupling coefficient. The coupling coefficient of two magnetically coupled system, which are transmitter and receiver for wireless power transfer system, are given as follows.

$$k = \frac{\acute{\mu}_0 M_0}{\sqrt{L_p L_s}} \quad (3)$$

For high efficiency, a minimum stable coupling is required, but the coupling factor between these air core coils are much lower than the traditional iron core transformer and it is also highly susceptible to fluctuate under misalignment. Therefore, for effective power transfer, the coils are to be strongly tuned at the same resonance frequency while maintaining a minimum coupling factor.

The mutual inductance M is dependent on the core size, shape, distance and the effective permeability of the total flux path. Fig. 3 shows the magnetically coupled transmitter and receiver under arbitrary horizontal misalignment Δ. Using Neumann Formula the mutual inductance between the transmitter and the receiver can be expressed as follows [9].

$$M = \frac{\acute{\mu}_r \mu_0}{4\pi} \oint_{Tx} \oint_{Rx} \frac{\vec{dl}_{Tx} \vec{dl}_{Rx}}{r} \quad (4)$$

For this system M is calculated analytically for different gap and misalignment using (4), and also simulated it in FEA analysis and validated experimentally.

A proposed system diagram of the complete wireless charging system is presented in Fig. 4 , which shows that the

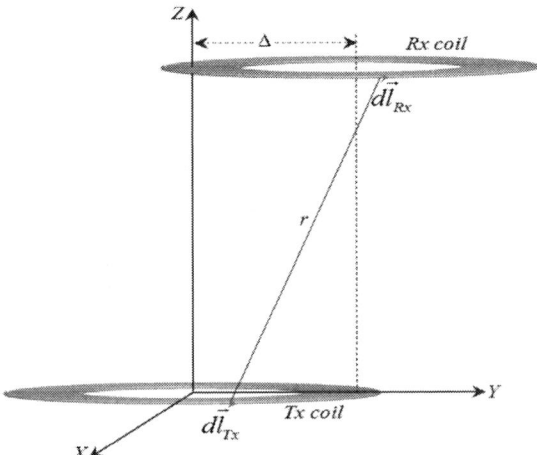

Fig. 3.   Simplified structure of two magnetically coupled coils.

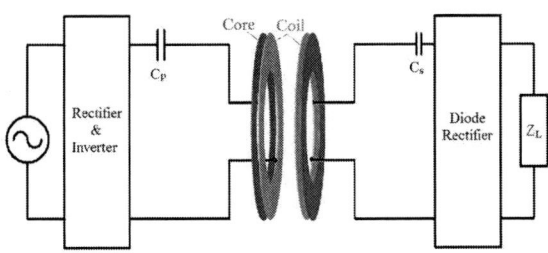

Fig. 4.   Proposed system of wireless power transfer with core

two electrically isolated magnetically coupled coils are tuned to same resonance frequency ω_0 by series capacitor C_p and C_s. The resonance frequency of this series LC circuit is given by,

$$\omega_0 = \frac{1}{\sqrt{L_p C_p}} = \frac{1}{\sqrt{L_s C'_s}} \tag{5}$$

The quality factor, Q of a coil is another important parameter that signifies the efficiency of the power transfer, which is given by,

$$Q = \frac{\omega_0 L_p}{R_p} = \frac{\omega_0 L_s}{R_s} \tag{6}$$

where, L_p and L_s are the self- inductance of Tx and Rx as indicated in the equivalent circuit of the proposed system is shown in Fig. 5. With higher Q coils, the power can be effectively transferred over a larger gap and misalignment. Under the constrain of size and wire resistivity, the coil turn number, the inductance is practically limited. As the operating frequency, $f = \omega/2\pi$ is limited within several hundred kHz due switching loss of the driver circuit, only the high permeability, low loss flux path can increase the inductance and the quality factor of planer spiral coil without increasing number of turn or switching frequency.

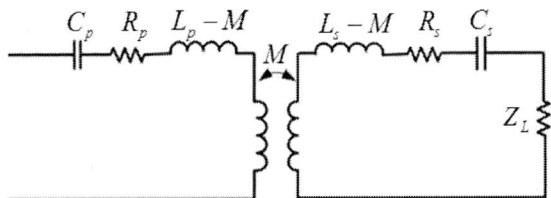

Fig. 5.   Equivalent circuit diagram of WPT system.

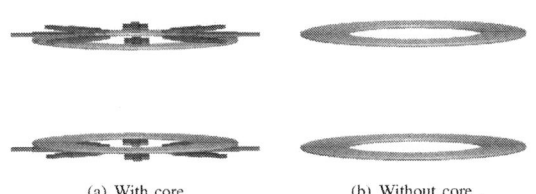

(a) With core   (b) Without core

Fig. 6.   Magnetic circuit with and without core.

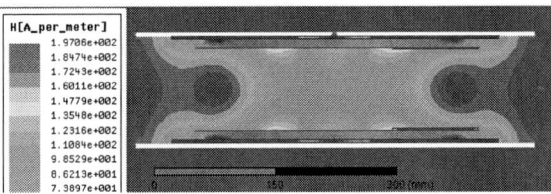

Fig. 7.   Magnetic circuit with and without core.

The magnetic circuit diagram of the total system with and without core is shown in Fig. 6 that indicate the low reluctance flux path through the high permeability ferrite core and the magnetic field distribution of with core is shown in Fig. 7, which proves that the flux is effectively passing through this new path. The detail effect the core and the proposed concept of increasing inductance and coupling is simulated and tested to verify in the next sections.

III.   CORE DESIGN

The core in transmitter and receiver increases coupling factor k, and quality factor Q at the resonance frequency ω_0. Hence, the optimized core increases the range of wireless power transfer efficiency and the misalignment tolerance. But, as the wireless power transfer operate at high frequency, the incorporation of core material will also introduces core losses. Therefore, the dimension and geometry of the core needs to be determined to minimize the core loss. Weight is another critical factor for vehicle charging application, especially for the receiver, as it is directly mounted at the bottom of the vehicle. As ferrite is a heavy material, if included in the design of the core, the total amount of material used in the receiver will be required to minimized. Therefore, an objective function Φ is proposed in (7) to reduce the core loss and weight, while maximizing the magnetic coupling and efficiency under certain constrain of weight and efficiency.

$$(\Phi)^2 = c_1 \left(\frac{P_e}{P_{e,ref}} \right)^2 + c_2 \left(\frac{P_h}{P_{h,ref}} \right)^2 \\ + c_3 \left(\frac{W_{core}}{W_{core,ref}} \right)^2 + c_4 \left(\frac{k_{ref}}{k} \right)^2 \tag{7}$$

where,
Φ= Objective Function
P_e= eddy current loss
$P_{e,ref}$= reference eddy current loss
P_h= hysteresis loss in core
$P_{h,ref}$= reference hysteresis loss in core
W_{core} = total core weight
$W_{core,ref}$ = reference total core weight
k = magnetic coupling factor
k_{ref} = reference magnetic coupling factor

and, w_1, w_2, w_3, w_4 are the weight factor for eddy current loss, hysteresis loss, total weight and magnetic coupling factor correspondingly and all the reference values are taken from initially optimized system. The function has been defined to maximize the coupling coefficient while minimizing the core losses and weight under proposed constrains.

The ferrite material has high resistivity which reduces the eddy current loss and high flux density at high frequency increases the hysteresis loss. Therefore, the hysteresis loss is more significant than eddy current loss for ferrite materials. Therefore, in the objective function, higher weight is given to hysteresis loss than eddy current loss for ferrite core.

The core losses are dependent on the properties of core material, core geometry, magnetic flux density in the core and the operating frequency. Again, different core configurations show different effect on impedance characteristics of the system.

A. Core Geometry Analysis

The core geometry and pattern affect effective relative permeability, equivalent resistance, capacitance and mutual coupling factor. Hence, the core geometry needs to select such that it gives minimum reluctance along flux path and minimum saturation in the core. The gradual increase of each outer turn for planer spiral coil produces a symmetric flux distribution with maximum flux density around the mid-turns as shown for a planer core in Fig. 9. Therefore, the core geometry is selected with maximum core volume in mid-turn are of the spiral coil. Different core is analyzed to validate this analytical finding through simulation as shown in Fig. 8.

B. Core Flux Density and Loss Distribution

Under the given constrain of core shape, the core can be consist of a single core block, or small blocks of core electrically insulated from one another as shown in the Fig. 8(c). The larger the core size, the eddy current path in the core increases, therefore, eddy current loss increases. This eddy current can be reduced by increasing core resistance and segregation of a large core block into small core blocks and electrically isolate each other. Again, for mechanical stability,

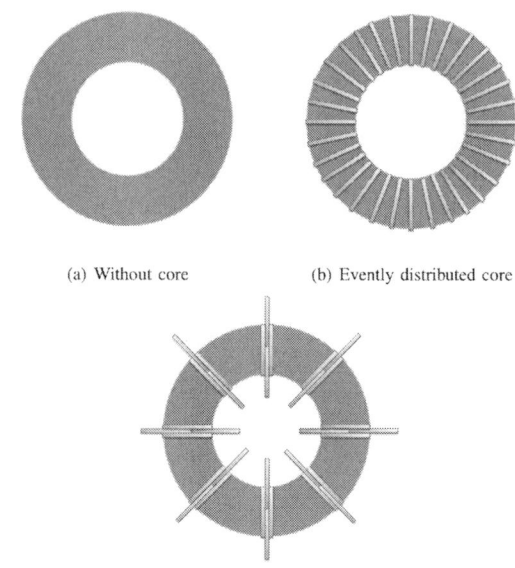

(a) Without core   (b) Evenly distributed core

(c) Optimally core configuartion

Fig. 8. Core structure: (a) optimally configured core and (b) without core and (c) evenly distributed core.

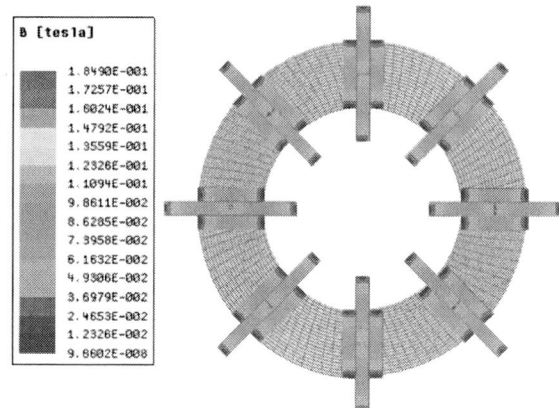

Fig. 9. Flux density distribution on the receiver core surface

each small core block is considered to have a minimum size. The flux density distribution of the optimized planer core is shown in Fig. 9, which show that the flux density is higher at the center region of the core, so is the hysteresis loss.

C. Inductance and Quality factor

The introduction of core increases the inductance, L of the coil compared to only coil. The the quality factor ($Q = \omega L/R$) of a coil at a certain resonance frequency depends on the inductance as well as the equivalent resistance. The core increases the inductance as well as the equivalent resistance R, hence the increase in quality factor is dependent on the relative increase of L and R and both are dependent on the

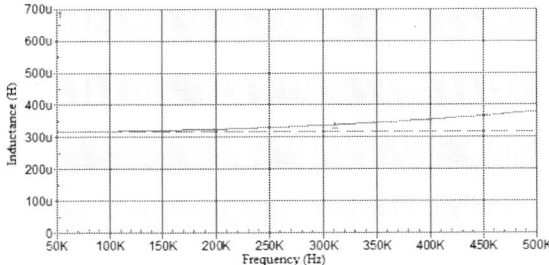

Fig. 10. Frequency dependency of effective inductance.

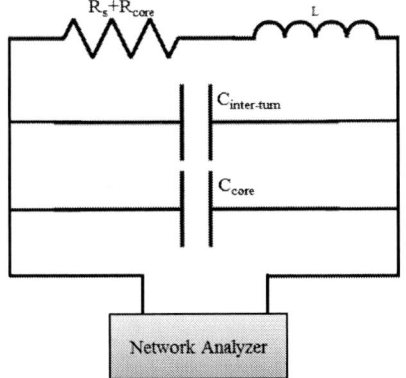

Fig. 11. Core characterization circuit using Network Analyzer BODE 1000.

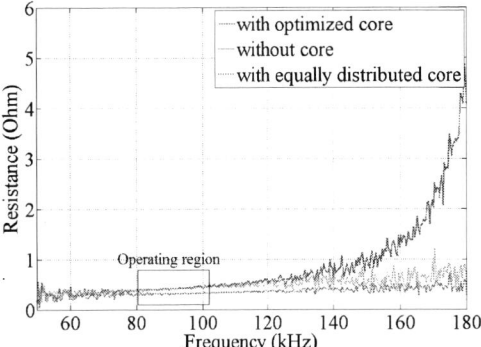

Fig. 12. Frequency dependency of parasitic resistance under different core configurations.

core configuration. The core equivalent resistance for different core configuration is shown in section D.

Both for mutual and leakage flux, of a one side distributed high permeability core provides a partial lower reluctance flux path. Therefore, both the mutual and self-inductance increases. Still, for mid-range wireless power transfer system, this covers only a small portion of the flux path, compared to the air-gap in between. Therefore, the effective inductance of the transmitter can be given by,

$$L_p(f) = \acute{\mu}_r L_{p0} \qquad (8)$$

where, L_{p0} is the inductance of the coil in air without core. Again, $\acute{\mu}_r$ depends on frequency, core material permeability and relative geometry of coil and core [11]. Therefore, the effective inductance is depended on frequency as well which is shown in Fig. 10.

D. Core Impedance

The transmitter and receiver characteristics can be evaluated like the way we do for power inductor [11], where equivalent parasitic resistance can be characterized as equivalent series resistance and total inter-turn parasitic capacitance as equivalent parallel capacitance. The core introduce additional resistance due to core loss in the equivalent circuit which can be found from the frequency response of the system. The setup for the frequency response measurement is shown in the Fig. 11, where $C_{inter-turn}$ is the inter-turn capacitance and C_{core} is the equivalent capacitance between core and coil.

The value of both C_{core} and $C_{inter-turn}$ can be measured from self resonating frequency of the Tx and Rx coil without matching capacitor. The equivalent series resistance also can be measured from impedance measurement from the Network Analyzer. Comparison of parasitic resistance without core and with two other core configuration is shown in Fig. 12. It shows that, without core the parasitic resistance is lowest and it increases monotonously with frequency until it reaches its self resonating frequency and without core, the self-resonance frequency is approximately 850kHz, which is much higher than other configurations, and it indicates that it has very low inter-turn capacitance $C_{inter-turn}$. Two type of core configurations shows comparatively higher equivalent resistance and lower self-resonance frequency. Therefore, it can be concluded that the core has introduced additional series resistance R_{core} and parallel capacitance C_{core}, which are also dependent of core configuration with same amount of core. Fig. 12 also shows that our optimally designed core can be used only if the operating frequency ω_0 is lower than 200kHz.

Optimizing the core by the objective function given in (7) for the proposed transmitter and receiver, the dimension of each core block is chosen 100mm x 20mm x 3mm and total 32 bars are used with each coil. The optimized core configuration is shown in Fig. 13.

IV. SIMULATION AND FEA ANALYSIS

For the proposed system, two spiral planar coils are designed as transmitter and receiver. Each coil, made of copper litz wire to reduce the skin effect, has twenty turns with 300 mm inner diameter and 500 mm outer diameter. The self-inductance of both the Tx and Rx coils without core is 226uH and the coupling coefficient, which also indicate the mutual inductance as shown in (3), is plotted in Fig. 14 for different gap and misalignment.The design models with ferrite magnetic core are shown in Figure 4 (a) and (b). The proposed systems are simulated in ANSYS Maxwell for magnetostatic and eddy current analysis to calculate the coupling coefficient,core loss and copper loss. The mutual inductance and the coupling coefficient are simulated for different gap and misalignment, and the results are plotted in Fig. 14. It shows that, introducing core, the magnetic coupling between the transmitter and the

978-1-4673-9551-9/16 $31.00 © 2016 IEEE

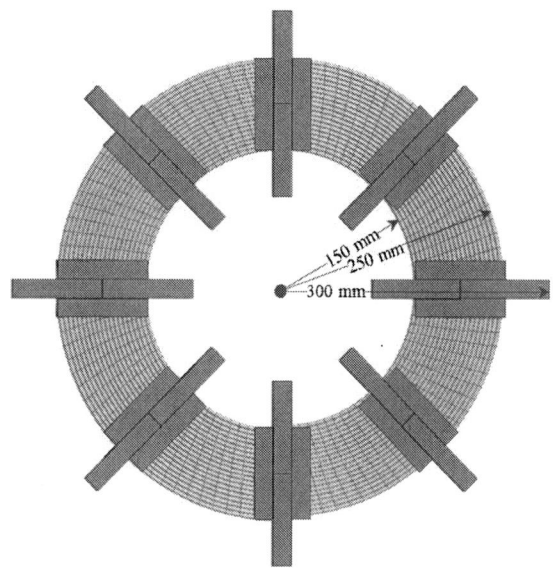

Fig. 13. FEA model of optimized core configuration.

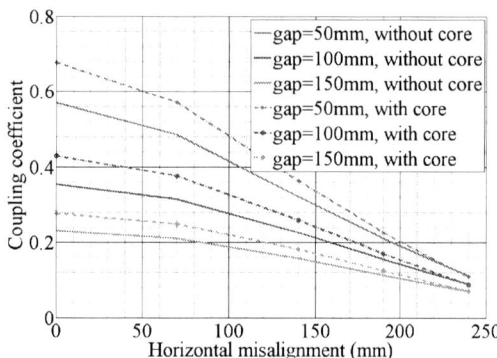

Fig. 14. Misalignment vs coupling coefficient.

receiver has been significantly increased for a wide range of misalignment.

The transmission efficiency of the system is calculated through S-parameter of a two port network as shown in Fig. 15. The S-parameters can be expressed by the (9). Here, S_{11} and S_{21} are the two most important S-parameters that signify the reflected and transmitted power respectively [12].

$$\begin{vmatrix} S_{11} & S_{12} \\ S_{21} & S_{22} \end{vmatrix} = \begin{vmatrix} b_1/a_1 & b_1/a_2 \\ b_2/a_1 & b_2/a_2 \end{vmatrix} \qquad (9)$$

The power transmission efficiency from transmitter to receiver can be given by (10).

$$\eta = S_{21}^2 \times 100\% \qquad (10)$$

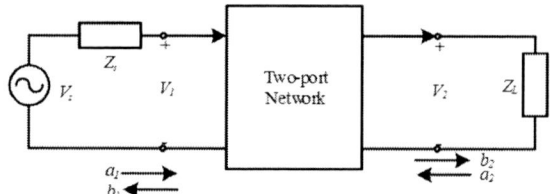

Fig. 15. Two port network representation of tuned Transmitter-Receiver.

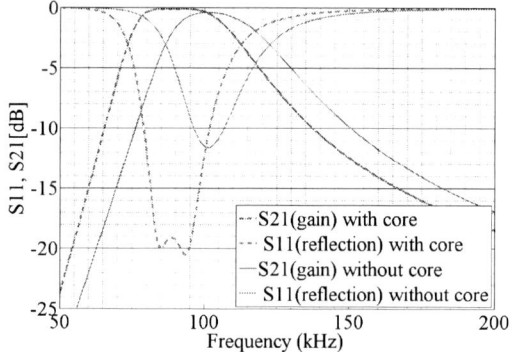

Fig. 16. S-parameter: gain (S_{21}) and reflection (S_{11}).

TABLE I. SYSTEM PARAMETERS

Parameter	Transmitter	Receiver
Number of turns	20	20
Inner Radius (mm)	150	151
Outer radius (mm)	254	252
Wire type	AWG38x800	AWG38x800
Wire diameter (mm)	4.1	4.1
Core material	ferrite	ferrite
Self-inductance without core (uH)	246.3	259.8
Self-inductance with core (uH)	331.6	341.2
Compensating series capacitor (uF)	0.01	0.01
Parasitic resistance (Ω)	0.21	0.23

The measured efficiency and S-parameters are compared with and without core for different separation as shown in Fig. 21 and Fig. 16.

V. EXPERIMENTAL SETUP AND RESULTS

The proposed system is implemented and tested for different core configurations as shown in Fig. 17(a), 17(b) and 17(c). The transmission efficiency of the system is measured in two port network configuration using network analyzer BODE 100, as shown in Fig. 15. The resonance frequency of the system with and without core is measured as 88 kHz and 101 kHz. All other system design parameter is shown in Table I.

Fig. 20 shows that, the transmission efficiency has been increased with core. The inductance L_p and L_s are also increased with core and therefore as given in (5), the resonance frequency has been decreased.

The coupling factor for coil pair with no core and optimized core configuration is given in Fig. 19.

978-1-4673-9551-9/16 $31.00 © 2016 IEEE

(a) Optimally configured core

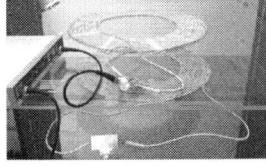

(b) Setup without core

(c) Evently distributed core

Fig. 17. Experimental setup for (a) optimally configured core and (b) without core and (c) evenly distributed core.

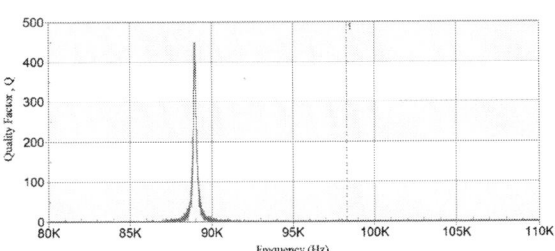

Fig. 18. Q factor for different core configuration.

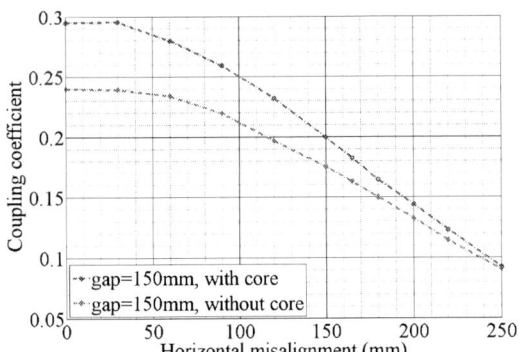

Fig. 19. Horizontal misalignment vs coupling coefficient.

The measured S-parameter is plotted in Fig. 16, which shows that, at resonance frequency, the transmission gain S_{21} is increased significantly including the optimized core. The measured coupling coefficient is given in Fig. 19, which

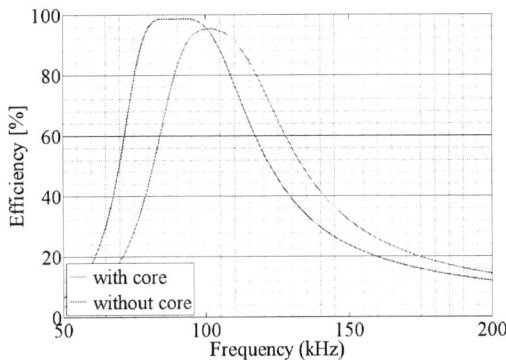

Fig. 20. Frequency vs transmission efficiency.

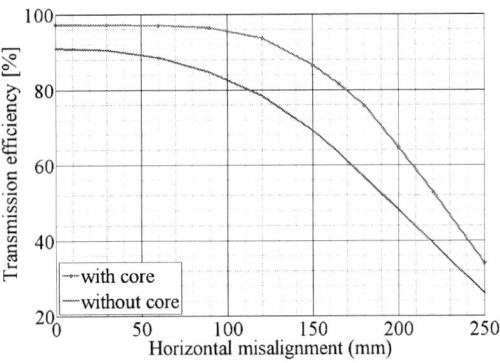

Fig. 21. Horizontal misalignment vs transmission efficiency.

shows a significant increase with core. Finally, the transmission efficiency against horizontal misalignment in shown in Fig. 21 for 150mm vertical gap between the transmitter and the receiver. It shows that, at the horizontally aligned position, the efficiency is increased by 6.3% with incorporation of core.

VI. CONCLUSION

The core design for inductive wireless power is a critical factor for flexible and efficient power charging of HEV. An optimized core design considering loss, weight and efficiency can increase the distance range of charging and provide more spatial freedom and flexibility. This paper focused on designing a efficient and less misalignment sensitive wireless charging system using ferrite core. It is found that ferrite core increases coupling coefficient, self and mutual inductance of the transmitter and receiver. The different configuration of ferrite core also increases the parasitic resistance and capacitance of the transmitter and receiver. The transmission efficiency was approximately same for 120 mm horizontal misalignment for the system with core. In contrast, for the core-less system, the transmission efficiency gradually dropped after 35 mm horizontal misalignment. The maximum transmission efficiency of the investigated system without core is found 90.9% for 150mm gap. Implementing the proposed core for the same system, for the same gap, the maximum efficiency is achieved 97.3%, which shows 6.3% efficiency improvement..

ACKNOWLEDGMENT

This work was partially supported by the UA NSF I-Corps Sites Program (F14-012) funded by National Science Foundation (NSF), USA and National Research Foundation of Korea (NRF) grant funded by the Korea government (MSIP) (2014R1A2A2A01006684).

REFERENCES

[1] J. Garnica, J. Casanova, and J. Lin, "High efficiency midrange wireless power transfer system," in *IEEE MIT-S Int. Microwave System Digest*, pp. 73-76, 2011.

[2] M. Pinuela, D. C. Yates, S. Lucyszyn and P. D. Mitcheson,, "Maximizing DC-to-Load Efficiency for Inductive Power Transfer," *IEEE Trans. Power Electron.*, vol. 28, no. 5, pp. 2437-2447, May 2013.

[3] Xuming Sun, Yang Zheng, Zhongliang Li, Xiuhan Li and Haixia Zhang, "Stacked flexible parylene-based 3D inductors with Ni80Fe20 core for wireless power transmission system," in *Micro Electro Mechanical Systems (MEMS)*, pp. 849-852, Jan 2013.

[4] T. Diekhans and R. W. De Doncker, "A Dual-Side Controlled Inductive Power Transfer System Optimized for Large Coupling Factor Variations and Partial Load," *IEEE Trans. Power Electron.*, vol. 30, no. 11, pp. 6320-6328, Nov. 2015.

[5] S. Santalunai, C. Thongsopa and T. Thosdeekoraphat, "An increasing the power transmission efficiency of flat spiral coils by using ferrite materials for wireless power transfer applications," in *Electrical Engineering/Electronics, Computer, Telecommunications and Information Technology (ECTI-CON)*, May 2014.

[6] H. R. Ahn, M. S. Kim and Y. J. Kim, "Inductor array for minimising transfer efficiency decrease of wireless power transmission components at misalignment," in *Electronics Letters*, vol. 50, no. 5, pp. 393-394, Feb. 2014.

[7] Weidong Ding and Xu Wang, "Magnetically coupled resonant using Mn-Zn ferrite for wireless power transfer," in *Electronic Packaging Technology (ICEPT)*, pp. 1561-1564, Aug. 2014.

[8] D.I. Stroe, M. Swierczynski, A.I. Stan, V. Knap, R. Teodorescu and S.J. Andreasen, "Diagnosis of lithium-ion batteries state-of-health based on electrochemical impedance spectroscopy technique," in *Energy Conversion Congress and Exposition (ECCE), 2014 IEEE*, pp. 4576-4582, Sept. 2014.

[9] T. Imura and Y. Hori, "Maximizing Air Gap and Efficiency of Magnetic Resonant Coupling for Wireless Power Transfer Using Equivalent Circuit and Neumann Formula," *IEEE Trans. Ind. Electron.*, vol. 58, no. 10, pp. 4746-4752, Oct. 2011.

[10] A. Balakrishnan, W. D. Palmer, W. T. Joines and T. G. Wilson, "The inductance of planar structures," in *Applied Power Electronics Conference and Exposition*, pp. 912-921, Mar 1993.

[11] Qin Yu, T. W. Holmes and K. Naishadham, "RF equivalent circuit modeling of ferrite-core inductors and characterization of core materials," *IEEE Trans. Electromagnetic Compatibility*, vol. 44, no. 1, pp. 258-262, Feb 2002.

[12] H. C. Jing and Yuanxun Ethan Wang, "Capacity Performance of an Inductively Coupled Near Field Communication System," in *Antennas and Propagation Society International Symposium, 2008. IEEE*, July 2008.

[13] H. R. Ahn, M. S. Kim and Y. J. Kim, "Inductor array for minimising transfer efficiency decrease of wireless power transmission components at misalignment," in *Electronics Letters*, vol. 50, no. 5, pp. 393-394, Feb. 2014.

978-1-4673-9551-9/16 $31.00 © 2016 IEEE

A 25 kW Industrial Prototype Wireless Electric Vehicle Charger

Mariusz Bojarski[1], Erdem Asa[1,3], Kerim Colak[2], Dariusz Czarkowski[3]

[1]Hevo Power Inc., New York, USA
[2]Istanbul Ulasim A.S., Istanbul, Turkey
[3]New York University, Polytechnic School of Engineering, New York, USA

Abstract— **A 25 kW wireless charger design for electric vehicles is presented in this paper. The wireless charger consist of three phase power factor corrector (PFC), three phase resonant inverter, primary and secondary coils with series resonant compensation, and a rectifier. In the proposed design, PFC provides a constant voltage DC Bus and the whole output regulation is done by the resonant inverter. It simplifies the rectifier structure to the simple full-bridge topology. The design is verified with experiments at the output power from 0 W to 25 kW. The measured system efficiency was up to 91%.**

Keywords—multi-phase, resonant, inverter, frequency, control, power factor, wireless, EV, charger

I. INTRODUCTION

Efficient wireless transfer systems have been recently receiving a growing attention for their numerous potential applications such as electric vehicle charging, medical implant devices, smartphone charging platforms [1]-[3]. With a proper magnetic and power electronic circuit design, the overall efficiency of the system can be increased by minimizing conduction and switching losses [4]-[7].

The wireless charging process of EVs can be carried out when vehicles are on roads, called dynamic charging; or parked, refer to stationary charging. An example of stationary wireless EV charging system is shown in Fig. 1. The charger consists basically of two parts: 1) an ac/dc converter stage with a rectifier and a power factor correction (PFC), regulates the input current to meet the requirements for the power factor (PF) and total harmonic distortion (THD), and delivers a nominal dc voltage as an input voltage source for the second stage, 2) wireless power transfer (WPT) stage with an inverter in the primary side and rectifier in the secondary side, the dc output voltage connected the battery is provided by the second stage with a high frequency air-gap contactless transformer and a proper impedance matching system. The power flow is regulated using a suitable switch mode controller to charge the battery with a constant current (CC) and constant voltage (CV) operations [8]-[9].

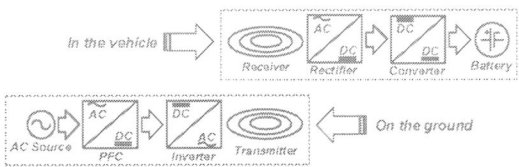

Fig. 1. Standard EV wireless power transfer system.

The recently published studies on stationary wireless energy transfer deal with optimization and design concerns of the system resonant and magnetic coil parameters [10]-[11]. However, there has been little investigation on high power applications and their effect on the wireless system performance [12]-[16]. Korea Rail Research Institute has shown a great progress achieving up to 1 MW output power for the dynamic charging of trains [15]. The achieved system efficiency was, however, up to 82.7%. For the commercial applications, electric vehicle charging products are available in the marketplace. Wireless Advance Vehicle Electrification (WAVE) Company establishes an inductive power transfer up to 50 kW at around 90% efficiency with 20 cm clearance [16].

This paper presents a 25 kW wireless EV charger which consists of three-phase resonant converter and three phase power factor corrector. To verify the proposed such a structure, the experimental analysis is performed with a maximum efficiency of 91%. A related circuit and control system are set up and theoretical analysis of the converter is discussed more detailed in the following chapters.

II. CIRCUIT DESCRIPTION

The proposed wireless power transfer system with a three-phase PFC boost converter and three-phase resonant converter are shown in Fig. 2. It consists of an ac input voltage source V_S, three phase PFC boost converter, three switching legs, three intercell transformers ICT, a wireless power link, impedance matching auto-transformer, full bridge current driven rectifier, and a battery. Each intercell transformer has two windings, which are connected as shown in the figure.

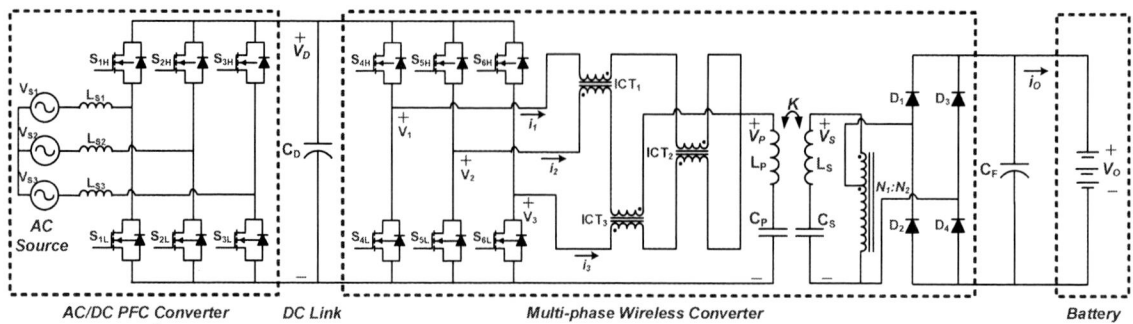

Fig. 2. The proposed high power wireless EV battery charging modeling.

The wireless power link consist of two identical coils L_P, L_S and two series resonant capacitors C_P, C_S. These coils are coupled with each other with the coupling factor K. The wireless power link with series compensation on both sides can be described in Fig. 3 [17].

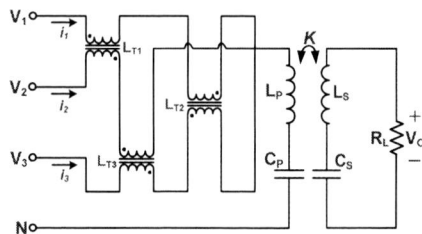

Fig. 3. The proposed circuit diagram of wireless power link with series compensations on both sides.

The switching legs and the dc input voltage V_D form square-wave voltage sources in the wireless link. Since the resonant currents i_1, i_2, and i_3 at the switching leg outputs are sinusoidal, only the power of the fundamental component of each input voltage source is transferred to the output. Therefore, the square wave voltage sources can be replaced by sinusoidal voltage sources V_1, V_2, and V_3 which represent the fundamental components. The secondary side transformer and rectifier are replaced by the equivalent resistance R_l connected in series with wireless transformer. The intercell transformer in the primary side provides equal current sharing between paralleled inverter legs. The fundamental components of the voltage sources in the circuit are

$$v_1 = V_m sin(\omega t - \phi_1) \tag{1}$$

$$v_2 = V_m sin(\omega t) \tag{2}$$

$$v_3 = V_m sin(\omega t + \phi_2) \tag{3}$$

where the amplitude V_m is given by

$$V_m = \frac{2}{\pi} V_D \tag{4}$$

and ϕ_1 and ϕ_2 are the phase shifts in the range from 0 to π. The voltages at the inputs of the intercell transformers are expressed in the complex domain by

$$V_1 = V_m e^{-j\phi_1} \tag{5}$$

$$V_2 = V_m \tag{6}$$

$$V_3 = V_m e^{j\phi_2} \tag{7}$$

Considering intercell transformers ICT_1 through ICT_3, the magnetizing inductance at primary sides and secondary sides are L_{T1P} through L_{T3P} and L_{T1S} through L_{T3S}, respectively. It is assumed that all of the windings have the same magnetizing inductance and leakage inductance. Then, the winding impedance of each intercell transformer is equivalent to the difference of the magnetizing inductance $L_{T,mag}$, the leakage inductance $L_{T,leak}$. An each phase current of the phase-controlled inverter flows through two windings and the circuit can be simplified as shown in Fig. 4.

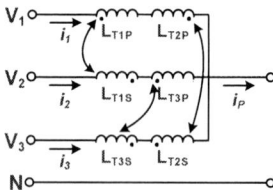

Fig. 4. Equivalent circuit of the intercell transformer.

Three voltage sources with the intercell transformers are replaced by a single voltage source V_P as

$$V_P = \frac{V_m[e^{-j\phi_1} + 1 + e^{j\phi_2}]}{3} \tag{8}$$

978-1-4673-9551-9/16 $31.00 © 2016 IEEE

An expression for the current of each phase can be derived based on the system circuit analysis, these currents can be described as

$$I_1 = \frac{1}{3}\left(I_P + \frac{2V_1 - V_2 - V_3}{2j\omega L_{T,leak} - j\omega L_{T,mag}}\right) \tag{9}$$

$$I_2 = \frac{1}{3}\left(I_P + \frac{2V_2 - V_1 - V_3}{2j\omega L_{T,leak} - j\omega L_{T,mag}}\right) \tag{10}$$

$$I_2 = \frac{1}{3}\left(I_P + \frac{2V_3 - V_1 - V_2}{2j\omega L_{T,leak} - j\omega L_{T,mag}}\right) \tag{11}$$

The obtained simplified equivalent circuit, the magnetizing inductance of the intercell transformer cannot be neglected here as it is affecting the output voltage of the equivalent voltage source. The inductor of intercell transformer can be represented with the combined leakage inductance as

$$L_T = \frac{2L_{T,leak} - 4L_{T,mag}}{3} \tag{12}$$

The equivalent circuit of wireless power link is defined by using described functions in Fig. 5.

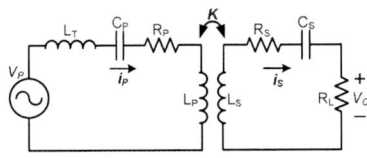

Fig. 5.   The equivalent circuit of wireless power link.

The wireless power link can be represented as a typical circuit model of a transformer with leakage and magnetizing inductances. Both wireless coils L_P and L_S are assumed to be identical and equal to L. Then, the simplified model can be equivalently represented by the circuit in Fig. 6.

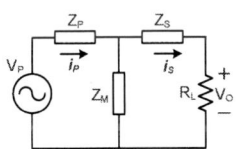

Fig. 6.   The simplified wireless model.

In this model, the equivalent impedances of Z_P, Z_S, and Z_M can be expressed as

$$Z_P = j\omega L_T + \frac{1}{j\omega C_P} + R_P + j\omega L_L \tag{13}$$

$$Z_M = j\omega L_M \tag{14}$$

$$Z_S = R_S + \frac{1}{j\omega C_S} + j\omega L_L \tag{15}$$

where $L_M = K\sqrt{L_P L_S} = KL$ and $L_L = L - L_M = (1 - K)L$. The dc to dc voltage transfer function of the converter can be given as

$$M_V = \left|\frac{V_O}{V_P}\right| = \left|\frac{Z_P}{R_L} + \frac{Z_S(Z_P + Z_M)}{R_L Z_M} + \frac{Z_P + Z_M}{Z_M}\right|^{-1} \tag{16}$$

III.   THE SYSTEM CONTROL INFRASTRUCTURE

Combining the two traditional modulation, phase shift modulation and frequency control, phase-frequency hybrid control strategy is proposed in this work. The principle of operation is to utilize the phase shift for regulating the system output and the frequency to operate as close to the resonant frequency as possible. The proposed hybrid control loops as shown in Fig. 7.

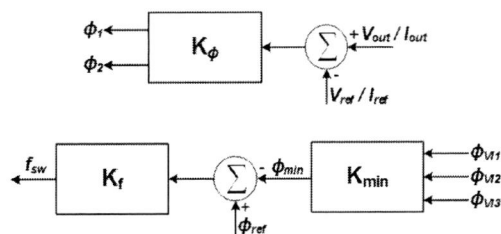

Fig. 7.   The proposed hybrid control block diagram.

The first loop is adjusting control phase shifts ϕ_1 and ϕ_2 according to the reference signal V_{ref}/I_{ref} and the output measurement V_{out}/I_{out}. The second loop is maintaining minimum phase shift ϕ_{min} between voltage and current ϕ_{VI1}, ϕ_{VI2}, ϕ_{VI3} at the switching leg outputs by adjusting the operating frequency f_{sw}. There is an important constrain on the operating frequency. It must assure that the load for all switching legs is inductive with a minimum phase angle between voltage and current. An inductive load is a necessary but not sufficient condition for the zero voltage turn on. Therefore, the energy stored in the parasitic capacitance of the switching leg is investigated. This capacitance C_{oss} is equal to the sum of the output capacitances of the transistors in the switching leg. The charge stored in C_{oss} is

$$Q_{oss} = C_{oss}V_D \tag{17}$$

With the determined necessary charge Q_{oss}, the minimum phase shift between the voltage and the current at the particular switching leg output can be obtained from the following equation as

$$\int_0^{\frac{\phi}{\omega}} I_{Pm} \sin \omega t \, dt \geq Q_{oss} \tag{18}$$

which leads to

$$\phi \geq \arccos\left(1 - \frac{\omega Q_{oss}}{I_{Pm}}\right) \qquad (19)$$

where I_{Pm} is the magnitude of resonant tank current value at the minimum load. This condition is the necessary and sufficient condition to obtain a zero voltage turn on. The proposed hybrid control algorithm is implemented in FPGA and DSP microprocessor. The designed control block diagram of FPGA and DSP microprocessor is demonstrated in Fig. 8.

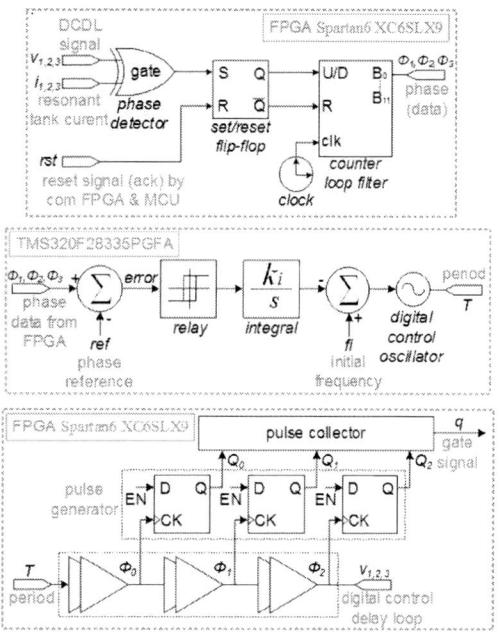

Fig. 8.   The implementation of control algorithm blocks.

The blocks phase detector (PD), loop filter (LF), and digital control delay loop (DCDL) are implemented in FPGA Spartan6 XC6SLX9 and feedback controller is managed by microprocessor TMS320F28335PGFA. The block of phase detector finds the phase error between reference and output signals. Then, its output is filtered by counter loop filter and applied to the microprocessor. The proposed hybrid control strategy evaluates the tuning frequency and phase shift function based on reference and output signal and send the signal to DCDL block in FPGA. The evaluated signals are driven by gate driver that is driven from FPGA ports.

IV.   EXPERIMENTAL VERIFICATION

A 25 kW experimental prototype of the proposed inverter was built and tested. A summary of parameters and component values of the prototype is presented in Table I.

TABLE I

Symbol	Parameter	Values
V_D	PFC Output Voltage	750 V
V_O	Output Voltage	0-375 V
I_O	Output Current	66.67 A
P_O	Rated Power	25 kW
C_P, C_S	Resonant Capacitor	62 nF
L_P, L_S	Coil Self Inductance	60 µH
$L_{T.mag}$	Intercell Magnetizing Inductance	20 µH
$L_{T.leak}$	Intercell Leakage Inductance	1.5 µH
K	Coupling Factor	0.2
D	Coils Distance	21 cm
C_F	Filter Capacitance	170 µF
N_1/N_2	Transformer Turn Ratio	22:9
sw	Main Switch	C2M0025120D
D	Rectifier Diode	DSEI2X101-06A
f_{sw}	Operating Frequency	85-88 kHz
t_{dead}	Dead Time	300 ns

The K is coupling factor between primary and secondary coil and D is a distance between those coils. Both coils are identical and tuned with series resonant capacitors to the resonant frequency $f_o = 1/\sqrt{C_P(L_L + L_T)}$. They are coreless and has dimension of 60 by 90 cm. N_1 and N_2 are number of turns of the impedance matching auto-transformer. In the result, the wireless power link sees two times lower impedance. Six C2M0025120D MOSFETs were used for the inverter prototype. 1 nF capacitors were added at the output of each switching leg of the inverter to alleviate high $dv=dt$ at turn-off. The inverter with wireless power link was loaded with a Class D full bridge current driven rectifier with resistive load. The rectifier was built using four DSEI2X101-06A diodes. The intercell transformers (ICTs) were built using 0077101A7 cores from Magnetics. The resonant capacitor was constructed by paralleling 62 pieces of 1 nF capacitors rated for 2 kV. The designed experimental prototype converter picture is given in Fig. 9. The controller is also assembled in the prototype.

Fig. 9.   The designed experimental prototype.

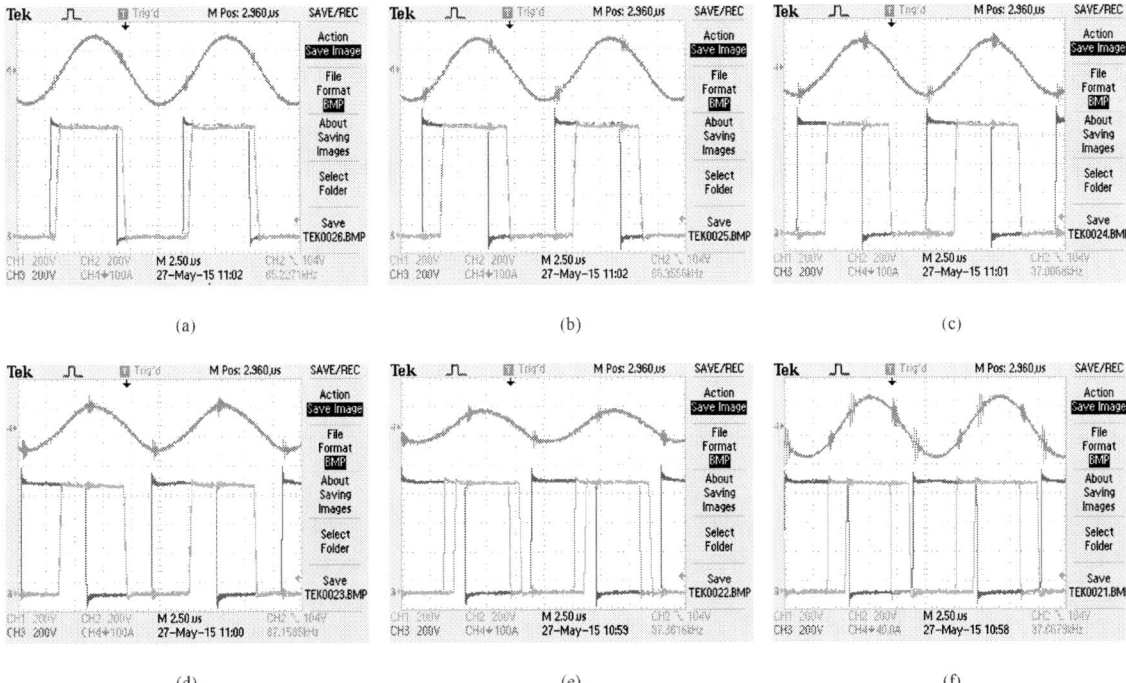

Fig. 10. Measured waveforms of phase voltages and output current for output power of a) 25 kW, b) 20 kW, c) 15 kW, d) 10 kW, e) 5 kW, f) 3 kW.

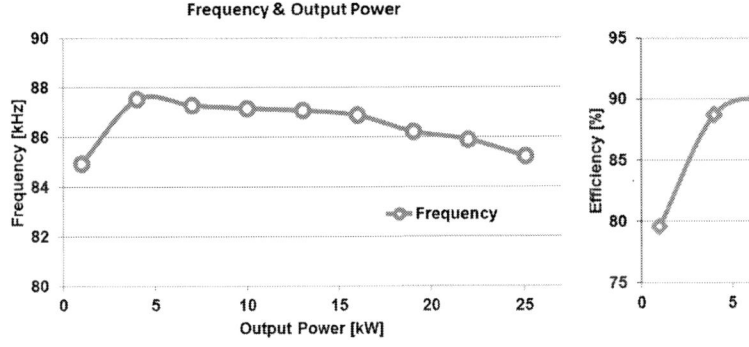

Fig. 11. Inverter frequency variation with the proposed control method at the different output power values.

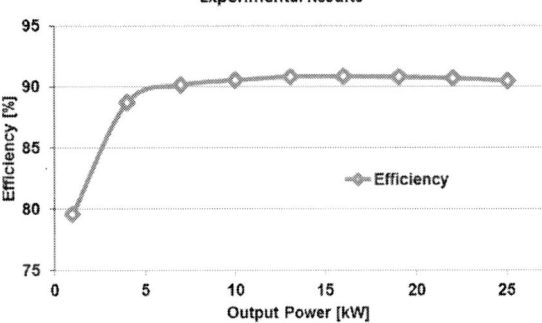

Fig. 12. System efficiency from AC input to DC battery and inverter operating frequency with the proposed control strategy.

The inverter was tested with a resistive load 5 Ω. The phase-frequency hybrid control strategy is implemented for the experimental results. The obtained experimental results are presented as the following waveforms. Phase voltage and resonant circuit current waveforms are shown in Fig. 10 for the hybrid control and 5 Ω load resistance. Fig. 10 (a) shows waveforms recorded at the high output power of 25 kW. Fig. 10 (c) shows waveforms for a medium output power of 15 kW. It can be seen that two phases are in phase and regulation is performed by the third one. Fig. 10 (e) illustrates operation at 5

kW of output power. At this power level, phase shift between two phases is fixed to 120° and regulation is done by the third one. It is important that the operating frequency does not change much while the output power is regulated. It can be also seen that voltage waveform slopes are smooth and there are almost no oscillations after switching which means that all phases are working at ZVS conditions regardless of the output power level.

The system performance analysis is shown using the proposed control strategy with the following waveforms. The graph shown in Fig. 11 presents frequency of the inverter for hybrid control. The AC to DC system efficiency for the proposed control is shown in Fig. 12. It clearly illustrates that the hybrid control allows for keeping the operating frequency within limited range in the whole regulation range. The hybrid control also does not have low-power operation limitations. As opposed to the frequency control which requires very high frequencies at low power, the proposed hybrid control strategy regulates from no power to full load in a narrow frequency range.

V. CONCLUSIONS

In this paper, a 25 kW wireless energy transfer system is presented. An industry prototype was built and tested to verify the proposed concepts. Experimental results showed that the proposed phase-frequency hybrid control strategy performs well. Firstly, the proposed concept provides full-range regulation from zero to full power without losing ZVS conditions. Secondly, it keeps the operating frequency range narrow, in particular, it can operate within the frequency range suggested by SAE J2954 for wireless charging electric vehicles. And thirdly, the presented inverter has high efficiency in a wide range of operation parameters.

REFERENCES

[1] P. S. Riehl, A. Satyamoorthy, H. Akram, Y. C. Yen, J. C. Yang, B. Juan, C. M. Lee, F. C. Lin, V. Muratov, W. Plumb, P. F. Tustin, "Wireless Power Systems for Mobile Devices Supporting Inductive and Resonant Operating Modes," IEEE Transactions on Microwave Theory and Techniques, vol.63, no.3, pp.780-790, Mar. 2015.

[2] E. Asa, K. Colak, M. Bojarski, D. Czarkowski, "A Novel Multi-Level Phase-Controlled Resonant Inverter with Common Mode Capacitor for Wireless EV Chargers," IEEE International Conference on Transportation Electrification (ITEC), pp.1-6, Jun. 2015

[3] O. Knecht, R. Bosshard, J. Kolar, "High Efficiency Transcutaneous Energy Transfer for Implantable Mechanical Heart Support Systems," IEEE Transactions on Power Electronics, vol.30, no.11, pp.6221-6236, Nov. 2015.

[4] K. Colak, E. Asa, M. Bojarski, D. Czarkowski, O. C. Onar, "A Novel Phase Shift Control of Semi-Bridgeless Active Rectifier for Wireless Power Transfer" IEEE Transaction on Power Electronics, vol.30, no.11, pp.6288-6297, Nov. 2015.

[5] M. Nalbant, "Wireless Power Transmitter Having Low Noise and High Efficiency, and Related Methods," U.S. Patent, 2014/0132077 A1 May 15, 2014.

[6] M. Bojarski, E. Asa, D. Czarkowski, "Effect of Wireless Power Link Load Resistance on the Efficiency of the Energy Transfer," IEEE International Electric Vehicle Conference (IEVC), pp.1-7, Dec. 2014.

[7] E. Asa, K. Colak, M. Bojarski, D. Czarkowski, "A Novel Phase Control of Semi Bridgeless Active Rectifier for Wireless Power Transfer Applications," IEEE Applied Power Electronics Conference and Exposition (APEC), pp.3225-3231, Mar. 2015.

[8] M. Bojarski, E. Asa, M. T. Outeiro, D. Czarkowski, "Control and Analysis of Multi-level Type Multi-phase Resonant Converter for Wireless EV Charging," 41th Annual Conference of IEEE Industrial Electronics Society (IECON), pp.1-6, Nov. 2015.

[9] K. Colak, M. Bojarski, E. Asa, D. Czarkowski, "A Constant Resistance Analysis and Control of Cascaded Buck and Boost Converter for Wireless EV Chargers," IEEE Applied Power Electronics Conference and Exposition (APEC), pp.3157-3161, Mar. 2015.

[10] J. A. S. Gonzalez, K. M. Elbaggari, K. K. Afridi, D. J. Perreault, "Design of Class E Resonant Rectifiers and Diode Evaluation for VHF Power Conversion," IEEE Transactions on Power Electronics, vol.30, no.9, pp.4960-4972, Sep. 2015.

[11] D.W. Ferreira, R. V. Sabariego, L. Lebensztajn, L. Krahenbuhl, F. Morel, C. Vollaire, "Homogenization Methods in Simulations of Transcutaneous Energy Transmitters," IEEE Transactions on Magnetics, vol.50, no.2, pp.1017-1020, Feb. 2014.

[12] M. Bojarski, E. Asa, D. Czarkowski, "Three Phase Resonant Inverter for Wireless Power Transfer," IEEE Wireless Power Transfer Conference (WPTC), pp.1-4, May 2015.

[13] J. F. Sanz, J. L. Villa, J. Sallan, J. M. Perie, L. G. Duarte, "UNPLUGGED Project: Development of a 50 kW Inductive Electric Vehicle Battery Charge System," EVS27 Electric Vehicle Symposium and Exhibition, pp.1-7, Nov. 2013.

[14] I. Fujita, T. Yamanaka, Y. Kaneko, S. Abe, T. Yasuda, "A 10kW Transformer with A Novel Cooling Structure of A Contactless Power Transfer System for Electric Vehicles," IEEE Energy Conversion Congress and Exposition, pp.3643-3650, Sep. 2013.

[15] J. H. Kim, B. S. Lee, J. H. Lee, S. H. Lee, C. B. Park, S. M. Jung, S. G. Lee, K. P. Yi, J. Baek, "Development of 1-MW Inductive Power Transfer System for a High-Speed Train," IEEE Transactions on Industrial Electronics, vol.62, no.10, pp.6242-6250, Oct. 2015.

[16] M. Jurjevich, "Large-Scale, Commercial Wireless Inductive Power Transfer (WPT) for Fixed Route Bus Rapid Transportation", IEEE Transportation and Electrification Newsletter, Oct. 2014.

[17] M. Bojarski, D. Czarkowski, F. de Leon, Q. Deng, M. K. Kazimierczuk, and H. Sekiya, "Multiphase resonant inverters with common resonant circuit," in Proc. IEEE International Symposium on Circuits and Systems (ISCAS), pp. 2445-2448, Jun. 2014.

Full-Bridge Series Resonant Multi-Inverter Featuring new 900-V SiC Devices for Improved Induction Heating Appliances

Mario Pérez-Tarragona, Héctor Sarnago, Óscar Lucía, and José M. Burdío
Department of Electronic Engineering and Communications. Universidad de Zaragoza.
María de Luna, 1. 50018 Zaragoza, España.
E-mail: maperta@unizar.es

Abstract—**Induction heating is a key technology due to its advantages in terms of performance and efficiency. Recent developments in domestic induction heating including flexible cooking surfaces with multiple output power converters. Besides, advances in wide bandgap devices allow new design possibilities such as cost-effective and high efficiency power converters. In this paper, an improved full-bridge series resonant multi-inverter with SiC devices is presented. The proposed converter and the modulation strategies to obtain proper output power control are detailed. Finally, a two-load experimental prototype is built, showing the experimental results and the advantages of the converter.**

Keywords—Electronic Power, Induction Heating (IH), Resonant Power Conversion, Home Appliances, SiC Devices.

I. INTRODUCTION

Domestic Induction Heating (IH) is currently the heating technology of choice due to its advantages in terms of efficiency, fast and accurate heating, cleanness, and safety [1]. Usually, IH cookers are composed of three or four coils according to the appliance model. Each coil is fed by an inverter, and shared electromagnetic compatibility filter and rectifier [2, 3]. Nevertheless, current IH cooker market is moving towards to flexible cooking surfaces [4] (Fig. 1). Those are multi-coil structures which achieve better benefits because they give more freedom to select pot size, position and shape. This configuration implies a high manufacturing cost due to high number of coils and large number of electronic components. In order to reduce the cost, multi-inverter topologies applied to multiple-inductive load systems have been proposed in previous works [5-10].

Moreover, recent technological advances in wide bandgap power devices, such as silicon carbide (SiC) switching devices [11, 12], and their commercial availability, are enabling the design of higher efficiency and increased power density power electronic converters [12, 13] since they feature lower impedance and faster switching times than Si devices.

Fig. 1. Flexible cooking surface.

The aim of this paper is therefore to propose an improved full-bridge series-resonant multi-inverter taking advantage of the recently developed 900-V SiC devices. This fact implies lower switching, and on-state losses, further increasing efficiency, and enabling both reduced cooling requirements and smaller coils, since these devices allow higher operation frequencies. Besides, several modulation strategies are proposed to achieve an effective power distribution in every load.

The remainder of this paper is organized as follows. Section II presents the full bridge series resonant multi-inverter topology. Section III details the modulation strategies allow controlling output power accurately. In Section IV the implemented prototype and the main experimental results are detailed. Finally, Section V summarizes the main conclusions of this paper.

II. FULL BRIDGE SERIES RESONANT MULTI-INVERTER

A. Topology

The proposed full-bridge series resonant multi-inverter (FB-SRMI) topology is shown in Fig. 2. It features two main blocks, a Common Inverter Block (CIB), based on full-bridge topology [14] with SiC semiconductor devices, and a Resonant Induction Load Block (RILB) which are activated with specific semiconductor devices. The load is composed of the coil electrical equivalent, $R_{eq,i}$ and $L_{eq,i}$, and its associated resonant capacitor, $C_{r,i}$. Besides, the bus capacitor C_s ensures

978-1-4673-9551-9/16 $31.00 © 2016 IEEE

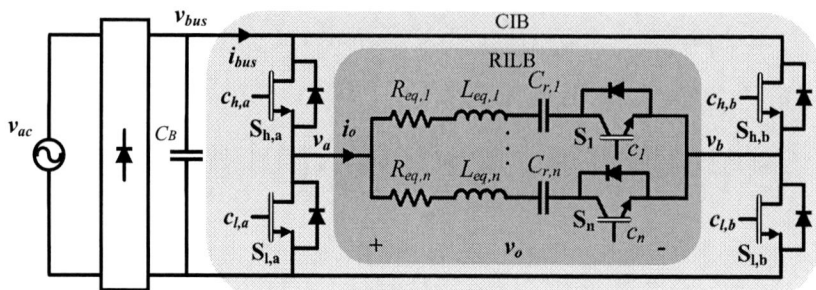

Fig. 2.   Full-bridge series resonant multi-inverter (FB-SRMI).

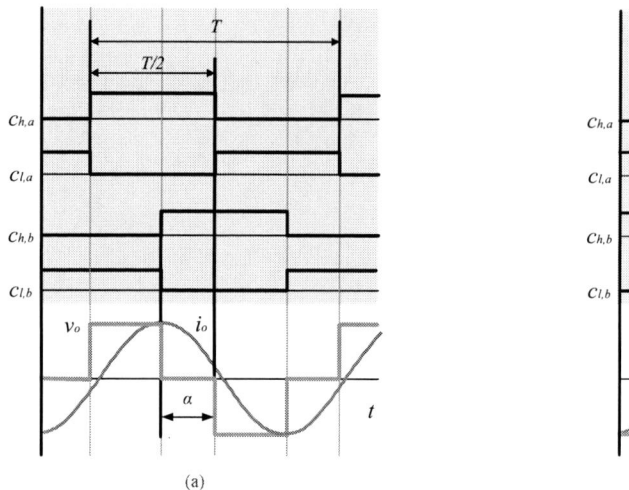

Fig. 3.   Phase Shift, PS, (a) and Asymmetrical Duty Cycle control, ADC, (b).

the smooth operation of the converter. Finally, a diode bridge rectifies the mains alternate voltage.

On the one hand, this topology works as a usual full-bridge with two main half-bridge branches, a and b, each composed of two commutation devices, $S_{h,x}$ and $S_{l,x}$. In this way, the voltage in the branch middle point, v_x, can be zero or v_{bus}.

On the other hand, every load is effectively activated or deactivated with a specific commutation device with a parallel diode, S_i. It is remarkable that the transistor have to be commuted when diode is driving current in order to avoid high voltage among its terminals, since the component can be destroyed.

The main benefit of this configuration is that a single high performance SiC inverter is used, leading to a cost-effective implementation. Since the proposed inverter enables higher switching frequencies, smaller coils and resonant capacitors are necessary. Moreover, in comparison with half-bridge, the full-bridge topology doubles the load voltage, reducing current

trough the coils and devices and, consequently, enabling lower on-state losses.

B. Analysis

The applied output voltage of the proposed converter, i.e. $v_o = v_a - v_b$, can be defined according to the activated common devices as

$$
v_o = \begin{cases} v_{bus}, & (c_{h,a}, c_{l,b}) \\ -v_{bus}, & (c_{l,a}, c_{h,b}) \\ 0, & (c_{h,a}, c_{h,b}) \\ 0, & (c_{l,a}, c_{l,b}) \end{cases}. \tag{1}
$$

Consequently, taking into account the first harmonic in a square wave control, the output voltage results

$$
V_{o,rms} = \frac{4V_{bus,rms}}{\pi}. \tag{2}
$$

978-1-4673-9551-9/16 $31.00 © 2016 IEEE   1763

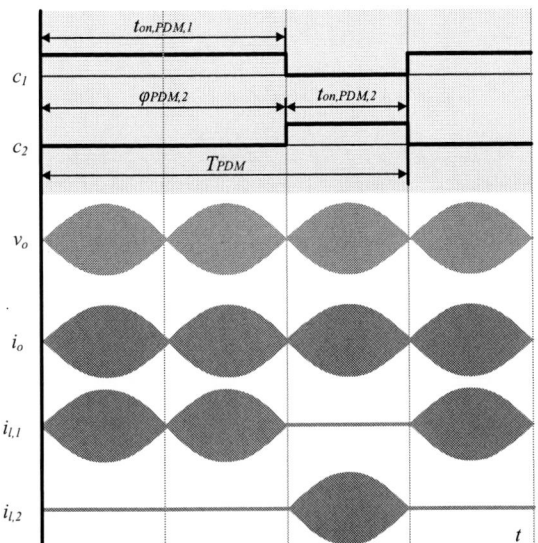

Fig. 4.   Low Frequency Pulse Density Modulation with two loads.

III.   MODULATION STRATEGIES

The FB-SRMI enables several modulations with increased degrees of freedom to allow controlling the output power supplied to each IH load. CIB modulations allow controlling the overall output power whereas RILB modulations allow controlling the output power at each load.

On the one hand, CIB devices can be modulated using both Phase Shift (PS), with the phase angle, α, and Asymmetrical Duty Cycle (ADC) control, with the duty cycle, D, which are shown in Fig. 3. These strategies have the advantage that the modulation frequency is constant, improving electromagnetic interference behavior, and avoiding acoustic noise.

On the other hand, RILB devices are modulated using mains synchronized Low Frequency Pulse Density Modulation (LF-PDM) [15], which is shown in Fig. 4. The specific activation ratio is defined as

$$m = \frac{t_{on,PDM,i}}{T_{PDM}}, \qquad (5)$$

where $t_{on,PDM,i}$ is the i load activation time, and T_{PDM} is the common term. Besides, the specific device phase ($\varphi_{PDM,i}$) allows distributing power uniformly to comply with EMC regulations. This strategy activates or deactivates transistors when bus voltage is zero, avoiding high voltage, and decreasing stress in devices. Therefore, every temporal parameter is a multiple of mains frequency half-cycle. The power can be modifies using PS or ADC modulation according to (6) or (7) respectively. In Fig. 5, the normalized power, $P_{i,n} = P_i / P_{max,i}$, using these modulation strategies are shown.

In this way, the power delivered to the i load at the resonant frequency, i.e.

$$f_o = \frac{1}{\left(2\pi\sqrt{L_{eq,i}C_{r,i}}\right)}, \qquad (3)$$

can be calculated as

$$P_{max,i} = \frac{8V_{bus,rms}^2}{\pi^2 R_{eq,i}}. \qquad (4)$$

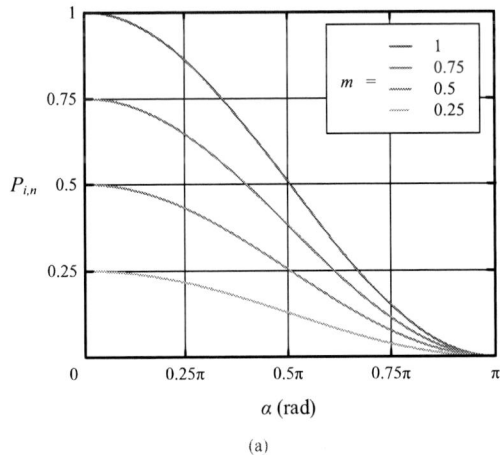

(a)

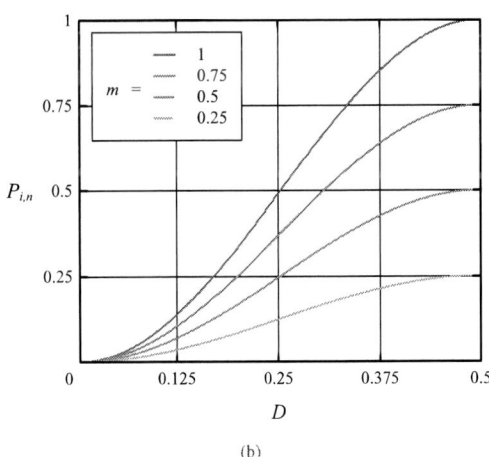

(b)

Fig. 5.   Power maps with Phase Shift, PS, (a) and Asymmetrical Duty Cycle control, ADC, (b).

Fig. 6. CIB implementation.

$$P_i = m \cos^2\left(\frac{\alpha}{2}\right) P_{\max,i}. \qquad (6)$$

$$P_i = m \sin^2\left(\pi D\right) P_{\max,i}. \qquad (7)$$

IV. IMPLEMENTATION AND EXPERIMENTAL RESULTS

In order to prove the feasibility and performance of the proposed converter, an experimental prototype has been designed and built. This section summarizes the main implementation and experimental results.

A. Experimental prototype

In order to test the feasibility of the proposed topology, and its modulation strategies, an experimental prototype for induction heating applications has been designed, and implemented (Fig. 6). The selected common devices are the recently developed 900-V C3M0065090D SiC devices from Cree to significantly improve its performance, whereas the selected specific devices are the 1600-V H30R1602 Si devices from Infineon in order to avoid overvoltage breaks. The prototype can feed two 700 W loads, which are composed of a 6.8 nF resonant capacitor and the induction load defined with its equivalent series parameters: 60 Ω and 200 μH. The converter is controlled by an FPGA-based versatile digital control architecture [16] which generates the PWM control signal for every power device. The modulation parameter can be modified from a PC since the FPGA is synchronized with a PC via the USB connector and serial protocol. Finally, the stage is completed with a zero crossing detector that enables measuring the mains zero-cross signal and performing the mains synchronized LF-PDM strategy. The next lines summarizes the mains experimental results obtained with the described prototype.

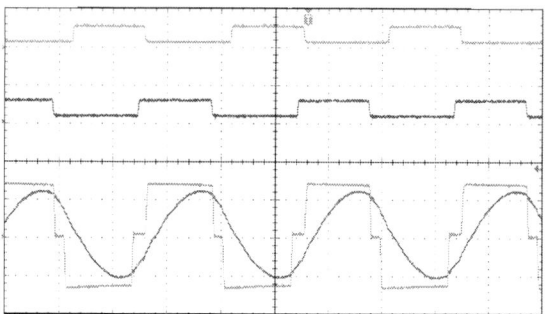

Fig. 7. Main waveforms of the BIC using PS modulation with $\alpha = \pi/6$: $c_{l,a}$ (50 V/div, blue), $c_{l,b}$ (50 V/div, dark blue), v_o (250 V/div, green), i_o (4 A/div, purple. Time: 2 μs/div.

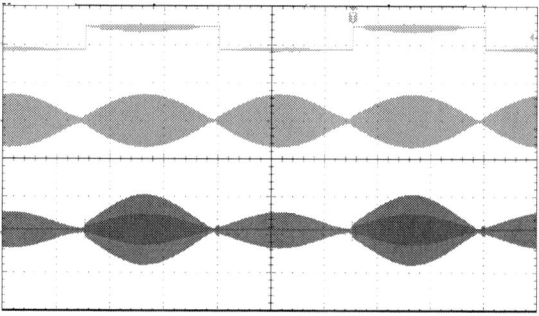

Fig. 8. Current through the coils with one pulsed load: c_2 (50 V/div, blue), v_o (500 V/div, green), i_o (10 A/div, purple), i_{L2} (10 A/div, dark blue). Time: 4 ms/div.

B. Experimental results

The experimental measurements of the proposed converter working at 170 kHz, i.e. close to the resonant frequency, are shown below. More in detail, Fig. 7 shows the main waveforms of the CIB using PS modulation with $\alpha = \pi/6$ and delivering 600 W to one load, whereas Fig. 8 shows LF-PDM modulation of the RILB feeding two loads, one of them activates only 50% of TPDM, delivering 700 W, and 350 W.

V. CONCLUSIONS

In this paper, a series-resonant full-bridge multi-inverter with SiC devices has been presented. The topology enables a reduced component count due to an only inverter is necessary to feed several loads. Using SiC devices, switching, and on-state losses are reduced, increasing the converter efficiency, besides, the full-bridge topology allows reducing the current trough devices, and coils. Furthermore, switching frequency is higher than with Si devices, enabling smaller inductors, and capacitors. A prototype able to feed two 700 W loads has been designed, and implemented obtaining satisfactory results. As a conclusion, a cost-effective, and high-efficiency multi-inverter that improves domestic induction heating application has been proposed.

ACKNOWLEDGMENT

This work was partly supported by the Spanish MINECO under Project TEC2013-42937-R, Project CSD2009-00046, and Project RTC-2014-1847-6, by the DGA-FSE, by the University of Zaragoza under Project JIUZ-2014-TEC-08, and by the BSH Home Appliances Group.

REFERENCES

[1] O. Lucía, P. Maussion, E. Dede, and J. M. Burdío, "Induction heating technology and its applications: Past developments, current technology, and future challenges," *IEEE Transactions on Industrial Electronics*, vol. 61, pp. 2509-2520, May 2014.

[2] F. P. Dawson and P. Jain, "A comparison of load commutated inverter systems for induction heating and melting applications," *Power Electronics, IEEE Transactions on*, vol. 6, pp. 430-441, 1991.

[3] H. W. Koertzen, J. D. Van Wyk, and J. A. Ferreira, "Design of the half-bridge, series resonant converter for induction cooking," in *Power Electronics Specialists Conference, 1995. PESC '95 Record., 26th Annual IEEE*, 1995, pp. 729-735 vol.2.

[4] O. Lucía, J. Acero, C. Carretero, and J. M. Burdío, "Induction heating appliances: Towards more flexible cooking surfaces," *IEEE Industrial Electronics Magazine*, vol. 7, pp. 35-47, September 2013.

[5] O. Lucía, J. M. Burdío, L. A. Barragán, J. Acero, and I. Millán, "Series-resonant multiinverter for multiple induction heaters," *IEEE Transactions on Power Electronics*, vol. 24, pp. 2860-2868, November 2010.

[6] O. Lucía, J. M. Burdío, L. A. Barragán, J. Acero, and C. Carretero, "Series resonant multi-inverter with discontinuous-mode control for improved light-load operation," *IEEE Transactions on Industrial Electronics*, vol. 58, pp. 5163-5171, November 2011.

[7] C. Bernal, J. M. Burdío, I. Garde, S. Llorente, O. Lucía, I. Millán, *et al.*, "Kochfeld mit einer Mehrzahl von Heizelementen," German Patent Patent, 2009.

[8] H. Pham, H. Fujita, K. Ozaki, and N. Uchida, "Phase angle control of high-frequency resonant currents in a multiple inverter system for zone-control induction heating," *IEEE Transactions on Power Electronics*, vol. 26, pp. 3357-3366, 2011.

[9] F. Forest, S. Faucher, J.-Y. Gaspard, D. Montloup, J.-J. Huselstein, and C. Joubert, "Frequency-synchronized resonant converters for the supply of multiwindings coils in induction cooking appliances," *IEEE Transactions on Industrial Electronics*, vol. 54, pp. 441-452, February 2007.

[10] F. Forest, E. Labouré, F. Costa, and J.-Y. Gaspard, "Principle of a multi-load/single converter system for low power induction heating," *IEEE Transactions on Industrial Electronics*, vol. 15, pp. 223-230, March 2000.

[11] H. Sarnago, O. Lucía, and J. M. Burdío, "A comparative evaluation of SiC power devices for high performance domestic induction heating," *IEEE Transactions on Industrial Electronics*, vol. 62, pp. 4795-4804, 2015.

[12] H. Sarnago, O. Lucía, A. Mediano, and J. M. Burdío, "Improved operation of SiC-BJT-based series resonant inverter with optimized base drive circuit," *IEEE Transactions on Power Electronics*, vol. 29, pp. 5097-5101, March 2014.

[13] J. Jordan, V. Esteve, E. Sanchis-Kilders, E. J. Dede, E. Maset, J. B. Ejea, *et al.*, "A Comparative performance study of a 1200 V Si and SiC MOSFET intrinsic diode on an induction heating inverter," *IEEE Transactions on Power Electronics*, vol. 29, pp. 2550-2562, May 2014.

[14] H. Sarnago, O. Lucía, A. Mediano, and J. M. Burdío, "Analytical model of the half-bridge series resonant inverter for improved power conversion efficiency and performance," *IEEE Transactions on Power Electronics*, vol. 30, pp. 4128-4143, August 2015.

[15] O. Lucía, J. M. Burdío, I. Millán, J. Acero, and D. Puyal, "Load-adaptive control algorithm of half-bridge series resonant inverter for domestic induction heating," *IEEE Transactions on Industrial Electronics*, vol. 56, pp. 3106-3116, August 2009.

[16] O. Jiménez, O. Lucía, L. A. Barragán, D. Navarro, J. I. Artigas, and I. Urriza, "FPGA-based test-bench for resonant inverter load characterization," *IEEE Transactions on Industrial Informatics*, vol. 9, pp. 1645-1654, August 2013.

A Novel Phase Control of Single Switch Active Rectifier for Inductive Power Transfer Applications

Kerim Colak[1], Erdem Asa[2,3], Dariusz Czarkowski[3]
[1]Istanbul Ulasim A.S., Istanbul, Turkey
[2]Hevo Power Inc., New York, USA
[3]New York University, Polytechnic School of Engineering, New York, USA

Abstract— In this study, a novel phase controlled of single switch active rectifier (S-SAR) is investigated for applications in wireless energy transfer. The proposed rectifier structure is based on the connection of controllable single switch thorough full bridge rectifier and regulation by a phase-shifted PWM signal. Theoretical and simulation results show that the performance of the proposed active rectifier topology is appropriate for resonant converters and secondary side controlled structure requiring power control at the secondary side such as contactless energy transfer systems. A 1 kW laboratory prototype system is designed to validate the proposed topology in terms of power and efficiency. Experimental results are provided for various loads using 8 inches air gap coreless transformer, which has dimension 2.5 by 2.5 feet, with a 120 V input and a maximum efficiency of 93 %.

Keywords— *active rectifier, single switch, phase shift, secondary control, wireless,*

I. Introduction

An efficient inductive power transfer (IPT) system has been received recently a growing attention from both academia and industry for its numerous potential applications [1]-[5]. The overall system efficiency can be improved by decreasing conduction and switching losses with a proper impedance matching and power circuit design [6]-[7]. A conventional inductive energy transfer system is demonstrated in Fig. 1. The system consists of two main stages: the transmitter and receiver platforms as seen in the figure. The first stage role is to deliver energy to the second stage with the impedance matching network. The dc output voltage is provided to the load by the second stage with the impedance matching network, a high-frequency rectifier, and a non-isolated dc/dc converter.

The impedance matching network is important to improve the system performance and is usually used at both primary and secondary sides. However, due to the load resistance or coupling coefficient variation, the contactless system overall impedance matching diverges from the designed characteristic, which means that the designed impedance matching maybe not

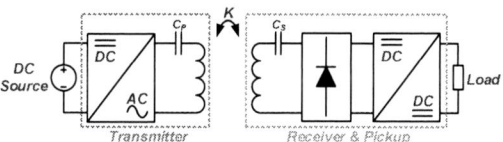

Fig. 1. A diagram of a pickup wireless energy transfer system.

working effectively causing a decrease in the system efficiency as compared to the designed performance values.

Various primary and secondary side controller synthesis, circuit topologies, and compensation strategies have been presented in the literature [8]-[17]. A switched compensator type between LC parallel and LC series resonant topologies is proposed and its change over algorithm is demonstrated in [8]. A common mode multi-phase half-wave semi-synchronous rectifier is investigated for applications in wireless energy transfer systems in [9]-[10]. In [11]-[12], researchers have studied reflected power to the transmitter side for low-middle power inductive power transfer applications by using cascaded boost and buck or cascaded buck and boost converter. Primary side steady-state load identification method of inductive power transfer system has been introduced based on switching capacitors in [13]. A novel phase shift control of semi-bridgeless active rectifier topology is proposed in [14]-[15]. In order to increase the coupling from primary to secondary pads, intermediate couplers have been shown for IPT systems in [16]. In [17], optimal resonant load transformation is analyzed for high power applications.

In this paper, a novel phase shift controlled of single switch active rectifier is proposed for the wireless power transfer applications. In this topology it is possible to control the output voltage without changing any primary side control parameters such as frequency, pulse-width modulation, etc. In the proposed secondary side active rectifier topology S-SAR, a controllable single switch is connected thorough the full bridge rectifier. The controllable switch is driven by a phase-shifted signal to obtain a higher voltage gain without changing the

978-1-4673-9551-9/16 $31.00 © 2016 IEEE

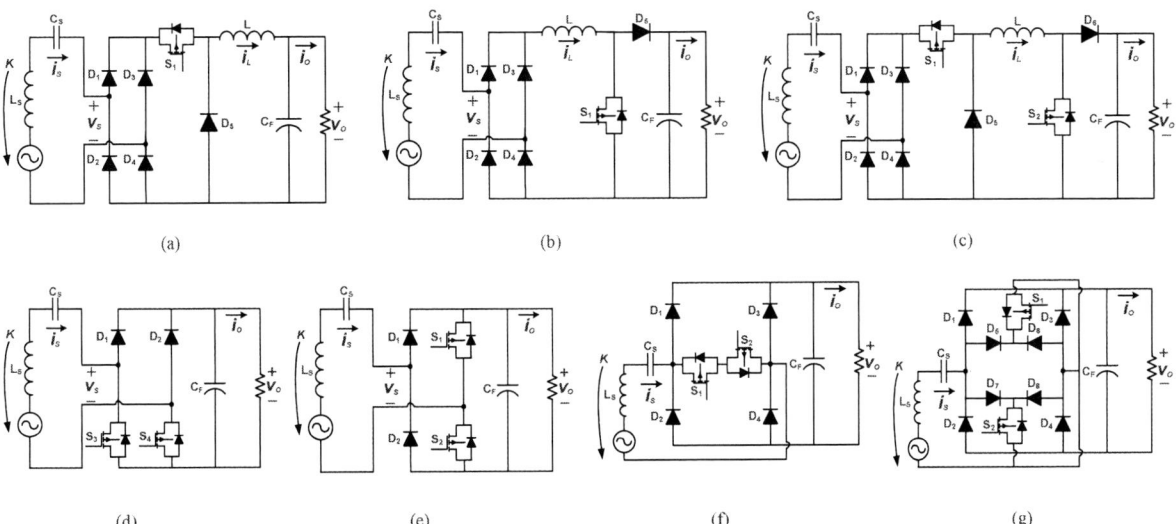

Fig. 2. Conventional single phase pulse width modulation (PWM) rectifiers a) buck, b) boost, and c) cascaded buck-boost converters [12]; d)-e) semi-bridgeless active rectifier topologies [14]-[15], f)-g) the proposed new type four quarant active rectifier topologies.

operating frequency. This novel approach brings an easy and cost effective solution considering the existed methods in the secondary side. EMI problems can be reduced with this new control approach that provides transmitter side control at the constant frequency. The converter model controllability is analyzed and transfer function of the converter is extracted. The system performance is confirmed with experimental results at 20 cm (8 inch) air gap wireless transformer and resonant converter, 1 kW output load with a maximum efficiency of 93 % in the laboratory conditions.

II. SINGLE SWITCH TOPOLOGY DERIVATION AND SYSTEM CIRCUIT ANALYSIS OF THE PROPOSED CONVERTER

A single phase buck, boost, and cascaded buck-boost type pulse width modulation (PWM) rectifiers, widely employed in commercially available in IPT, are shown in Fig. 2(a)-(c). These converter systems have many advantages that they employ only one or two controllable switch, fast switched diode, and a single inductor. Also, they can be managed with relatively less complicated control strategies, where simple digital integrated circuits can be engaged. However, the current in the inductance is carried out to the all semiconductor devices in each switch transition, causing high switching and conduction losses. Comparison among these type traditional topologies, semi-bridgeless active rectifier topologies, have only two switches conducting one of their current conduction states, were recommended as seen in Fig. 2(d)-(e). The converter concepts presented in Fig. 2(f)-(g) are proposed for IPT receiver side technologies that present a similar operating principle with the semi-bridgeless topologies.

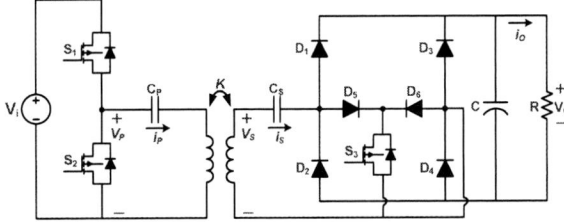

Fig. 3. The proposed single switch phase controlled of S-SAR.

The main difference can be found that four quadrant active switches are implemented for the bidirectional power flow. Proposed four quadrant active rectifier technologies can be evaluated with a single semiconductor switch. Hence, a single switch active rectifier (S-SAR) circuit topology is proposed for the wireless power transfer as shown in Fig. 3. It comprises a half bridge resonant inverter, a wireless transformer, and a single switch active rectifier in the secondary side. The active rectifier is consisted of one transistor and two diodes connected opposite side thorough a full bridge rectifier. The phase shift angle of the secondary side transistor regulates the output voltage and power in the load.

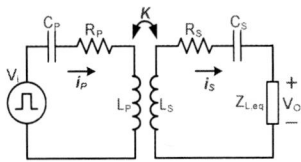

Fig. 4. The equivalent circuit model of the proposed wireless converter.

In order to perform the circuit analysis, the wireless power link can be represented as two coupled inductors and two resonant capacitors connected in series as shown in Fig. 4. In this model, input voltage source is V_i, load impedance is $Z_{L,eq}$ where $Z_{L,eq} = \{jX_{L,eq} + R_{L,eq}\}$, two coupled inductors are L_P and L_S with equivalent series resistances R_S and R_P. K is a coupling factor between the two coils and C_P and C_S are resonant capacitors. The two coupled inductors can be equivalently modeled as a transformer with proper leakage and magnetizing inductances.

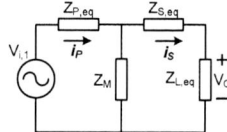

Fig. 5. The simplified model of wireless power link.

To simplify analysis, both coils L_P and L_S are assumed to be identical and equal to L. Then, the model can be equivalently represented by the circuit in Fig. 5. In this model $V_{i,1}$ is a fundamental component of voltage source V_i, the equivalent impedances of $Z_{P,eq}$, $Z_{S,eq}$, and Z_M can be expressed as

$$Z_{P,eq} = R_P + \frac{1}{j\omega C_P} + j\omega L_L \tag{1}$$

$$Z_M = j\omega L_M \tag{2}$$

$$Z_{S,eq} = R_S + \frac{1}{j\omega C_S} + j\omega L_L \tag{3}$$

where L_M and L_L can be described by the following equations as

$$L_M = K\sqrt{L_P L_S} = KL \tag{4}$$

$$L_L = L - L_M = (1-K)L \tag{5}$$

When the secondary switches are in ZVS region, the reflected equivalent impedance can be found by using first harmonic approximation (FHA) as

$$Z_{L,eq} = \frac{R}{2\pi^2}[3 - cos(\beta - \alpha)]\sqrt{10 - 6cos(\beta - \alpha)}e^{j(\phi-\beta)} \tag{6}$$

where β is the conduction angle of the half wave rectifier, ϕ is phase angle of receiver side voltage. The equivalent resistance value can obtained by taking real portion of (3), as

$$R_{L,eq} = Re\{Z_{L,eq}\} \tag{7}$$

and taking imaginary portion of the equivalent impedance, the corresponding reactance value is defined as

$$X_{L,eq} = Im\{Z_{L,eq}\} \tag{8}$$

With the help of the above described equations, the voltage transfer function of the system can be calculated as

$$M_V = \frac{2}{\sqrt{10 - 6cos(\beta - \alpha)}} \left| \frac{AZ_{L,eq}}{[B^2 + 2AB + CZ_{L,eq}]} \right| \tag{9}$$

where A, B, and C are described as follows

$$A = \frac{j\omega_N K Z_O}{(1-K)} \tag{10}$$

$$B = \frac{(1-\omega_N^2)Z_O}{j\omega_N} \tag{11}$$

$$C = \frac{(1-K-\omega_N^2)Z_O}{(1-K)j\omega_N} \tag{12}$$

Also, the normalized parameters of the system Z_O, Q, and ω_N are given with the following equations

$$Z_O = \sqrt{\frac{L_L}{C_P}} \tag{13}$$

$$Q = \frac{R_{L,eq}}{Z_O} \tag{14}$$

$$\omega_N = \frac{\omega_{sw}}{\omega_R} \tag{15}$$

$$\omega_R = \frac{1}{\sqrt{L_L C_P}} \tag{16}$$

where characteristic impedance Z_O affects the operating frequency range of the system. The quality factor Q is defined as the ratio of the total average stored energy and the dissipated energy per cycle. The normalized frequency ω_N depends on the switching ω_{sw} and resonant frequency ω_R.

III. ANALYSIS OF THE PROPOSED CONVERTER

The operating waveforms and switch state transitions are presented in Fig. 6 and Fig. 7 to show the behavior of the proposed converter. To simplify the circuit analysis, secondary side rectifier diodes and switches are ideal, the output capacitor is assumed to be large enough for a constant dc output, and filter losses are neglected. The proposed circuit is examined under these conditions in the following operation modes:

Mode 1 [t0<t<t1]

During this interval, the rectifier diodes D_2, D_3, D_5, and D_6 are off-state, and switch S_3 is turned off as demonstrated in Fig. 6.(a). The current i_S flows in positive direction through diodes D_1 and D_4 and the energy is transferred from the transformer secondary side and the resonant capacitor C_S to the filter capacitor C and the load resistance R.

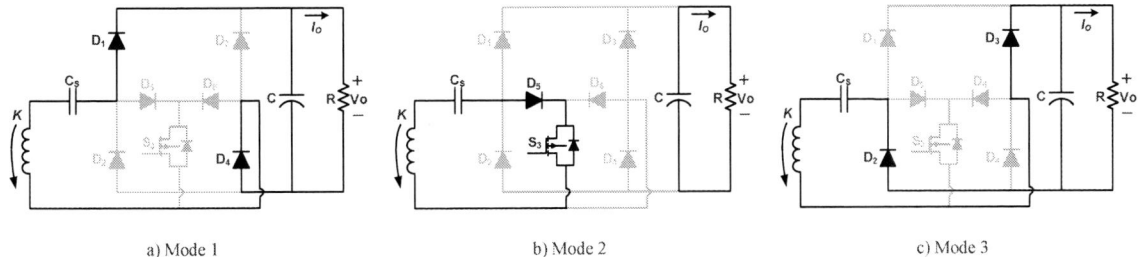

a) Mode 1   b) Mode 2   c) Mode 3

Fig. 6. Mode analysis, current path and switching transition.

Mode 2 [t1<t<t2]

In this mode, switch S_3 is turn-on and diode D_5 is on-state as shown in Fig. 6.(b). The current i_S is shorted from the switch and circulates through resonant capacitor C_S, the current waveform is depicted in Fig. 7. The filter capacitor C discharge into the load resistance R in this mode.

Mode 3 [t2<t<t3]

The resonant current decreases to the negative in this interval and reaches the minimum value within this mode is completed as plotted in Fig. 7. The switch S_3 is off, while diodes D_2 and D_3 are in the on state. The current flows from wireless transformer to diodes D_2, D_3 as shown in Fig. 6.(c).

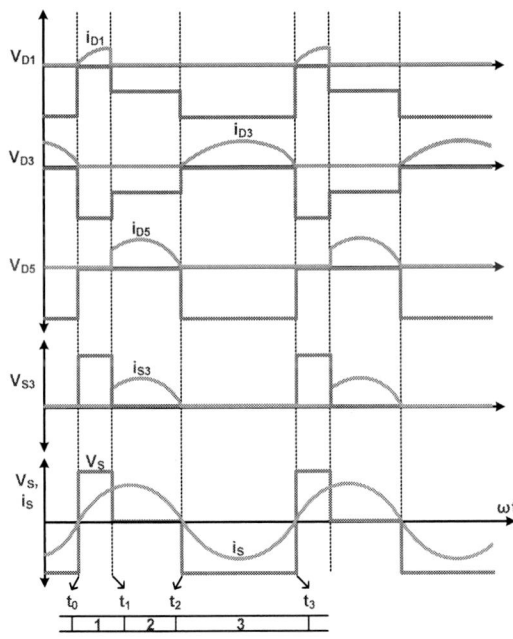

Fig. 7. Voltage and current waveforms in the switch and diodes.

IV. EXPERIMENTAL RESULTS

In order to verify the proposed converter idea and show the operating principle, a laboratory prototype of the proposed wireless system is designed for 1 kW, 120 V input voltage considering 10Ω and 20Ω output load conditions. The coreless transformer is tested with 8 inch distances between coils which results in 0.3 coupling factor. A summary of parameters and component values of the prototype is presented in Table I.

TABLE I

Symbol	PARAMETER	Value
V_t	dc input voltage	120 V
P_O	maximum output power	1 kW
C_P	primary side capacitor	40 nF
C_S	secondary side capacitor	40 nF
L_P	primary coil self-inductances	25 μH
L_P	secondary coil self-inductances	25 μH
d	square coil dimension	2.5 x 2.5 feet
n	coil turn number	4
f_{sw}	operating frequency	150 kHz
C	filter capacitor	10 μF

The characteristic waveforms of the proposed converter described above are given in the following records. Fig. 8 shows the output voltage characteristics with phase shift angle at 10 Ω and 20 Ω. When the switching frequency is constant at 150 kHz, the phase shift angle is swept to obtain the output voltage waveform. As seen in the figure, the characteristic of output voltage basically changes in two regions when the phase shift angle is lower than zero in the first region ($\alpha<0$) and greater than zero ($\alpha>0$) in the second region. The wide output voltage range can be achieved in these two regions.

The experimental results at 30, 60, 90, 120, 150, and 180 degrees for $\alpha>0$ at 10 Ω and $\alpha<0$ at 20 Ω are given in Fig. 9. The phase-shift angle controls the output voltage by changing the transformer secondary side voltage as shown. This variation is provided by the equivalent resistance as given in theoretical analysis in equations (6), (7), and (8).

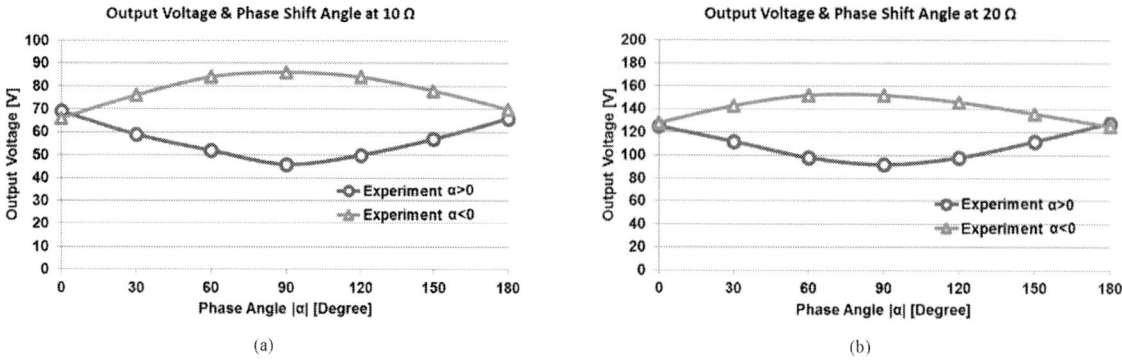

(a)   (b)

Fig. 8. The output voltage characteristics with phase shift angle for a) 10 Ω and b) 20 Ω.

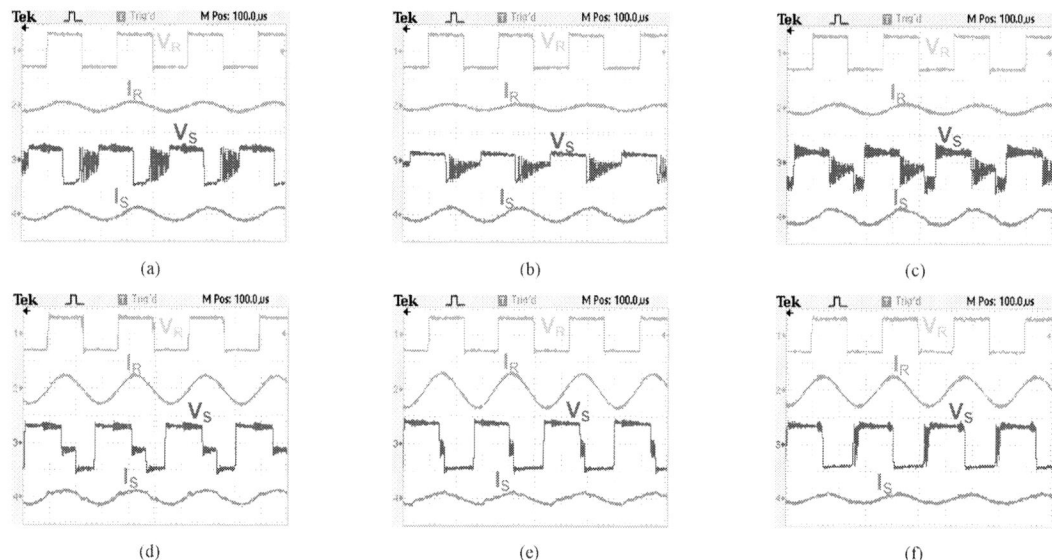

Fig. 9. Primary resonant tank voltage (V_R), current (I_R) and secondary side voltage (V_S), current (I_S): a, b, c for $\alpha > 0$ at 10 Ω and V_R *(100V/div)*, I_R *(5A/div)*, V_S *(100V/div)*, I_S *(5A/div)*, d, e, f for $\alpha < 0$ at 20 Ω and V_R *(100V/div)*, I_R *(5A/div)*, V_S *(200V/div)*, I_S *(5A/div)*.

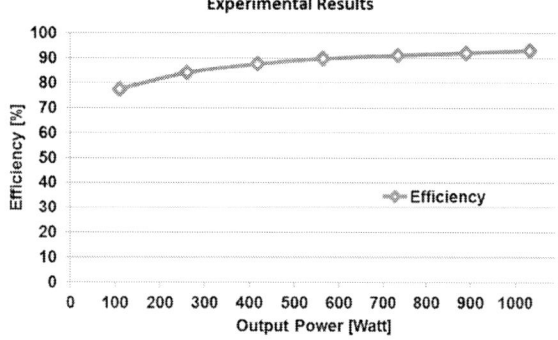

Fig. 10. Efficiency characteristics with the output power.

The experimental result of efficiency with output power is shown in Fig. 10. As it can be seen from the figure, a wide output power range can be reached with high efficiency.

V. CONCLUSIONS

In this study, a new phase control of single switch active rectifier (S-SAR) is presented for wireless power transfer applications. Analytic equations of the converter are explained by describing the switch state conditions. Voltage and current waveforms are displayed in all operating modes. It is shown that the proposed converter rectifier topology provides wide output voltage range with phase-shift angle control. The other important result is that the designed converter does not require any frequency tuning, even the constant frequency can provide wide output voltage range by phase-shift control. Maximum efficiency of 93 % is observed at the full power 1 kW and obtained at the zero current switching in the secondary side.

REFERENCES

[1] E. Asa, K. Colak, D. Czarkowski, "Analysis of Cascaded Multi-Output-Port Converter for Wireless Plug-in Hybrid on-Board EV Chargers," IEEE Applied Power Electronics Conference and Exposition (APEC), pp.1-5, Mar. 2016.

[2] M. Nalbant, "Wireless Power Transmitter Having Low Noise and High Efficiency, and Related Methods," U.S. Patent, 2014/0132077 A1 May 15, 2014.

[3] E. Asa, K. Colak, M. Bojarski, D. Czarkowski, "A Novel Multi-Level Phase-Controlled Resonant Inverter with Common Mode Capacitor for Wireless EV Chargers," IEEE Transportation Electrification Conference and Expo (ITEC), pp.1-6, Jun. 2015.

[4] M. J. Chabalko, A. P. Sample, "Three-Dimensional Charging via Multimode Resonant Cavity Enabled Wireless Power Transfer," IEEE Transactions on Power Electronics, vol.30, no.11, pp.6163-6173, Nov. 2015.

[5] M. Bojarski, E. Asa, D. Czarkowski, "A 25 kW Industrial Prototype Wireless Electric Vehicle Charger," IEEE Applied Power Electronics Conference and Exposition (APEC), pp.1-6, Mar. 2016.

[6] K. Colak, E. Asa, D. Czarkowski, H. Komurcugil, "A Novel Multi-level Bi-directional DC/DC Converter for Inductive Power Transfer Applications," 41th Annual Conference of IEEE Industrial Electronics Society (IECON), pp.3827-3831, Nov. 2015.

[7] E. Asa, K. Colak, D. Czarkowski, B. Tamyurek, "An Efficiency Analysis of Bi-directional DC/DC Converter for Wireless Energy Transfer Applications," IEEE Energy Conversion Congress and Exposition (ECCE), pp.594-598, Sep. 2015.

[8] H. Matsumoto, Y. Neba, H. Asahara, "Switched Compensator for Contactless Power Transfer Systems," IEEE Transactions on Power Electronics, vol.30, no.11, pp.6120-6129, Nov. 2015.

[9] K. Colak, E. Asa, M. Bojarski, D. Czarkowski, "A Novel Common Mode Multi-phase Half-wave Semi-synchronous Rectifier for Inductive Power Transfer Applications," IEEE Transportation Electrification Conference and Expo (ITEC), pp.1-6, Jun. 2015.

[10] E. Asa, K. Colak, D. Czarkowski, "Analysis of Multi-Output Half-Wave Semi-Synchronous Rectifier with a Uniform Magnetic Field Transmitter," IEEE Applied Power Electronics Conference and Exposition (APEC), pp.1-5, Mar. 2016.

[11] D. Ahn and S. Hong, "Wireless Power Transfer Resonance Coupling Amplification by Load-Modulation Switching Controller," IEEE Transactions on Industrial Electronics, vol.62, no.2, pp.898-909, Feb. 2015.

[12] K. Colak, M. Bojarski, E. Asa, D. Czarkowski, "A Constant Resistance Analysis and Control of Cascaded Buck and Boost Converter for Wireless EV Chargers," IEEE Applied Power Electronics Conference and Exposition (APEC), pp.3157-3161, Mar. 2015.

[13] Y. G. Su, H. Y. Zhang, Z. H. Wang, A. P. Hu, L. Chen, Y. Sun, "Steady-State Load Identification Method of Inductive Power Transfer System Based on Switching Capacitors," IEEE Transactions on Power Electronics, vol.30, no.11, pp.6349-6355, Nov. 2015.

[14] E. Asa, K. Colak, M. Bojarski, D. Czarkowski, "A Novel Phase Control of Semi Bridgeless Active Rectifier for Wireless Power Transfer Applications," IEEE Applied Power Electronics Conference and Exposition (APEC), pp.3225-3231, Mar. 2015.

[15] K. Colak, E. Asa, M. Bojarski, D. Czarkowski, O. C. Onar, "A Novel Phase Shift Control of Semi-Bridgeless Active Rectifier for Wireless Power Transfer" IEEE Transaction on Power Electronics, vol.30, no.11, pp.6288-6297, Nov. 2015.

[16] A. Kamineni, G. A. Covic, J. T. Boys, "Analysis of Coplanar Intermediate Coil Structures in Inductive Power Transfer Systems," IEEE Transactions on Power Electronics, vol.30, no.11, pp.6141-6154, Nov. 2015.

[17] M. Bojarski, E. Asa, D. Czarkowski, "Effect of Wireless Power Link Load Resistance on the Efficiency of the Energy Transfer," IEEE International Electric Vehicle Conference (IEVC), Dec. 2014.

978-1-4673-9551-9/16 $31.00 © 2016 IEEE

Optimal Shaped Dipole-Coil Design and Experimental Verification of Inductive Power Transfer System for Home Applications

Duy T. Nguyen, Eun S. Lee, Byeung G. Choi, and Chun T. Rim
Dept. of Nuclear and Quantum Engineering
KAIST
Daejeon, Korea
Email: {danielnguyen, eunsoo86, choibk09, ctrim}@kaist.ac.kr

Abstract—1m-off long distance and high efficiency inductive power transfer system (IPTS) with optimal shaped dipole-coil structure is proposed for home appliance charging applications. Conductive reflectors for transmitter (Tx) and receiver (Rx) dipole coils are investigated to improve power efficiency and to mitigate electromagnetic field for human safety. By adopting the Tx reflector behind the Tx core, the exposure level of magnetic flux density can be reduced by 94% in average verified by a finite-element method simulation. The optimal switching frequency of 200 kHz was experimentally found for maximum power efficiency, meeting an international guideline of Power Matters Alliance (PMA). It was experimentally verified that 4.2% of power efficiency reduction for the Rx reflector and 7.8% of the power efficiency improvement for the Tx reflector were observed. A prototype of the proposed IPTS for home appliances has achieved 83.1% of high efficiency with 150W of output power transfer.

Keywords—wireless TV; wireless power transfer system; inductive power transfer system; optimal shaped dipole-coil

I. INTRODUCTION

Wireless power transfer system (WPTS) has been widely used in electric vehicles, medical devices, defense systems, lightings, home appliances, and mobile devices due to the convenience of providing power without wires and battery problem. The WPTS can be typically classified to coupled magnetic resonance system (CMRS) and inductive power transfer system (IPTS) [1]-[7]. The CMRS was firstly introduced, which can deliver up to 2.2 m at 60 W wireless power [1]. This system adopts four large self-resonant coils and operates over several MHz with a high quality factor of 2,000, which results in large voltage stress of resonant capacitors, volumetrically bulky shape, and hyper-sensitive characteristic to surroundings such as temperature, humidity, and human proximity. On the other hands, the IPTS is known to be used for 5m-off 209W and 7m-off 11W of wireless power delivery by adopting dipole coil resonance system (DCRS) [5]-[7]. The DCRS has a low quality factor of 100, and reduces bulky and volumetric coil structures for several meters of long distance

wireless power transfer. Therefore, it can be said that the DCRS is a good candidate to be highly applicable for home appliance charging applications such as Television (TV), microwave, and cooker, which require high power capability and long distance with high power efficiency. Previous researches on WPTS for home appliances such as laptop or TV have demerits, i.e., low power efficiency, short transfer distance with multi-resonant coils and volumetrically bulky coils, which are impractical for commercialization [2], [8]-[9].

In this paper, 1m-off long distance IPTS using long slim dipole coils structure of transmitter (Tx) and receiver (Rx), which can be applicable for home appliances, is proposed, as shown in Fig. 1. The effects of Tx and Rx reflectors for high efficiency are investigated by simulation and experiment. Optimum design of the proposed IPTS is derived and experimentally verified to achieve maximum power efficiency under given Tx and Rx core structures and materials and to provide a design guide for a long-distance WPTS of home appliance charging applications.

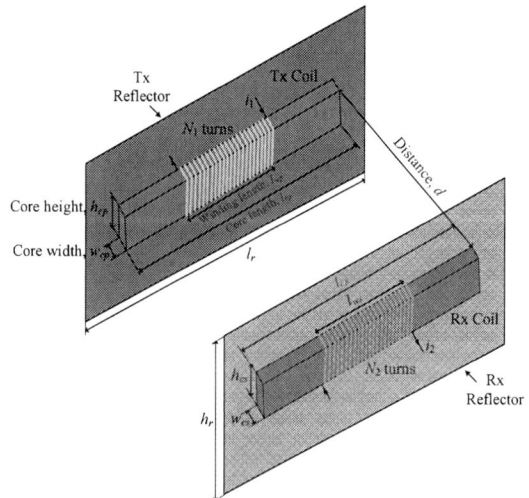

Fig. 1. Overall configuration of the proposed IPTS with the Tx and Rx reflectors.

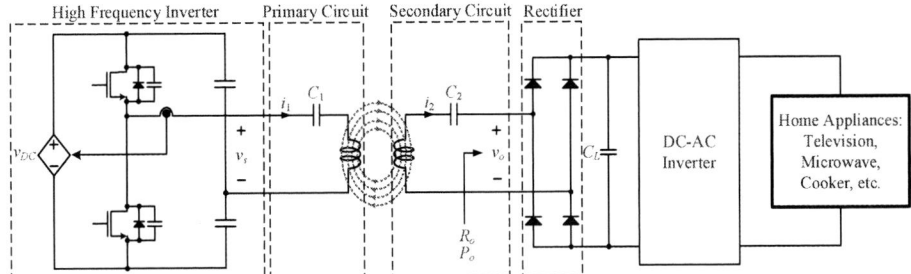

Fig. 2. Overall system configuration of the proposed IPTS for home appliance charging applications.

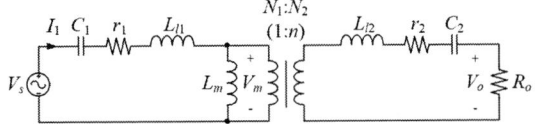

Fig. 3. An equivalent circuit of the proposed IPTS, considering only the fundamental switching frequency.

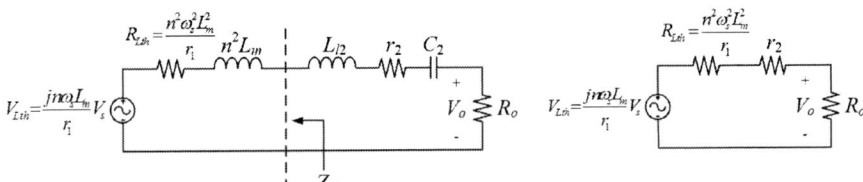

Fig. 4. Simplified circuits from the output side viewpoint.

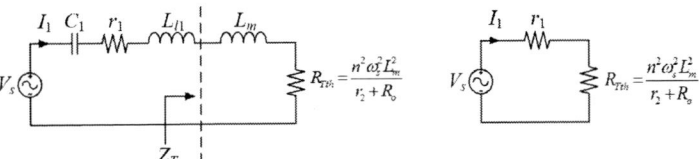

Fig. 5. Simplified circuits from the source side viewpoint.

II. STATIC ANALYSIS AND DESIGN OF THE PROPOSED IPTS

A. Equivalent Circuit Modeling of the Proposed IPTS

The overall system configuration of the proposed IPTS is shown in Fig. 2, which is composed of a high frequency inverter for generating high frequency AC current, a primary circuit, a secondary circuit, a rectifier and a DC-AC inverter for supplying AC voltage from DC voltage for home appliance load. Considering only fundamental component of the output voltage of the inverter v_s, the equivalent circuit of the proposed IPTS is shown in Fig. 3, where L_{l1}, L_{l2} are leakage inductances of Tx and Rx coils, respectively, and L_m is magnetizing inductance between Tx and Rx coils with turn-ratio n. r_1 and r_2 include equivalent series resistances (ESRs) of capacitors C_1 and C_2, core and copper losses of Tx and Rx coils, respectively. For simplicity of analysis, home appliance load is replaced by resistor R_o. To guarantee zero voltage switching operation of the inverter, the switching frequency of the inverter was selected slightly higher than the resonant frequency of primary circuit composed of C_1 and L_1 [10]. The resonant frequency of secondary circuit composed of C_2 and L_2 is tuned exactly to the switching frequency as follows:

$$\omega_s = 1.05\omega_{r1} = \frac{1.05}{\sqrt{L_1 C_1}}, \quad \omega_s = \omega_{r2} = \frac{1}{\sqrt{L_2 C_2}} \quad (1a)$$

$$\because L_1 = L_m + L_{l1}, \ L_2 = n^2 L_m + L_{l2} \ . \quad (1b)$$

By applying Thevenin circuit theorem and converting the equivalent circuit in Fig. 3 to output side and source side viewpoints combining with resonant conditions (1), the power efficiency η can be derived as follows [4], [7]:

$$\eta \equiv \frac{P_o}{P_{in}} = \frac{n^2 \omega_s^2 L_m^2 R_o}{r_1 \left(r_2 + R_o\right)^2 + n^2 \omega_s^2 L_m^2 \left(r_2 + R_o\right)} \ . \quad (2)$$

The equivalent resistances of Tx coil r_1 and Rx coil r_2, and the magnetizing inductance L_m can be modeled as follows [11]-

[12], where k_1, k_2, and L_{mo} are assumed to be constant in this paper for simplicity of analysis.

$$r_1 = g_1(N_1) \cong k_1 N_1^2, \quad r_2 = g_2(N_2) \cong k_2 N_2^2 \quad (3a)$$

$$L_m = g_3(N_1) = N_1^2 L_{mo} \quad (3b)$$

From (3), the maximum power efficiency can be achieved at the optimal number of secondary turns N_{2m} (5) under given conditions of the Tx and Rx core structures as follows [4]:

$$\eta_{max} = \eta \big|_{N_2 = N_{2m}} = \frac{\omega_s^2 L_{mo}^2}{2k_1 k_2 + \omega_s^2 L_{mo}^2 + 2\sqrt{k_1^2 k_2^2 + k_1 k_2 \omega_s^2 L_{mo}^2}} \quad (4)$$

where N_{2m} can be found as follows:

$$\frac{\partial \eta}{\partial N_2}\bigg|_{N_2 = N_{2m}} = 0 \quad \Rightarrow \quad N_{2m} = \sqrt[4]{\frac{k_1 R_o^2}{k_1 k_2^2 + k_2 \omega_s^2 L_{mo}^2}} \quad (5)$$

As one of experimental condition that is $f_s = 200$ kHz, $k_1 = 0.003$, $k_2 = 0.002$, $R_o = 20\ \Omega$, and $L_{mo} = 2.2 \times 10^{-8}$ in this paper, the optimal number of secondary turns $N_{2m} = 30$ turns was determined from (5), and the maximum power efficiency of 83.8 % for $N_2 = 30$ turns was obtained from (4).

B. Reflector Effect on the IPTS

To increase the power efficiency and to block the magnetic flux from the back side of the Tx and Rx dipole coils for human safety, the reflectors having high conductivity with almost unity relative permeability such as aluminum or copper, are considered behind the Tx and Rx dipole coils, as shown in Fig. 1. By adopting the proposed reflectors, the magnetic flux generated by eddy current in the Tx reflector mitigates the magnetic flux inside the Tx core generated from the Tx coil; hence, the magnetic flux inside the Tx core can be reduced to improve the core loss in the Tx core, which is the majority of total losses, although the magnetic flux inside the Rx core is reduced.

Unfortunately, it is almost impossible to analyze this reflector effect by numerical methods or explicit equivalent magnetic circuit modeling. Nevertheless, it is noteworthy to verify the reflector effect by simulation and experiment so that the qualitative characteristics are roughly investigated according to the physical dimensions of Tx and Rx reflectors.

III. SIMULATION VERIFICATION FOR REFLECTOR EFFECTS

To investigate the effect of each Tx and Rx reflector, the Tx and Rx reflectors are placed in back side of Tx and Rx coils, respectively, as shown in Fig. 6. As one of example, an aluminum plate having 2 mm thickness is adopted for the Tx

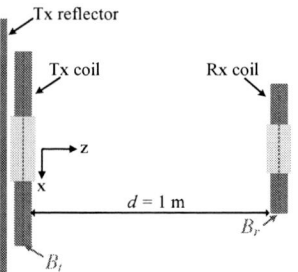

(a) Tx reflector case.

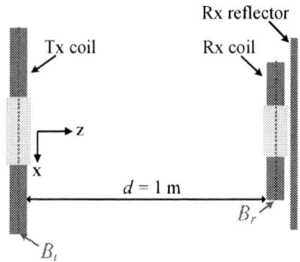

(b) Rx reflector case.

Fig. 6. Simulation configurations for the Tx and Rx reflectors.

TABLE I
PHYSICAL DIMENSIONS OF THE PROPOSED IPTS

Parameters	Values	Parameters	Values
l_{cp}	150 cm	l_{cs}	90 cm
l_{wp}	50 cm	l_{ws}	30 cm
h_{cp}	10 cm	h_{cs}	14 cm
w_{cp}	4 cm	w_{cs}	3 cm

and Rx reflectors. A finite-element method simulation tool (Ansoft Maxwell simulation) is implemented to evaluate the magnetic flux density inside the Tx and Rx coils B_t and B_r in Fig. 6, according to the length and the height of the Tx and Rx reflectors. For simulation configuration, the Tx and Rx coil structures in Fig. 1 are established based on one of physical dimensions summarized in Table I, and primary ampere-turns $N_1 I_1$ is set to be 90.

A. Rx Reflector

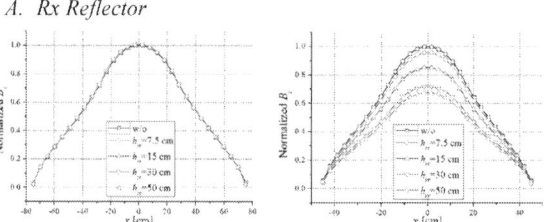

(a) The normalized B_t (b) The normalized B_r

Fig. 7. The normalized B_t and B_r for different h_{rr} when l_{rr}=120 cm.

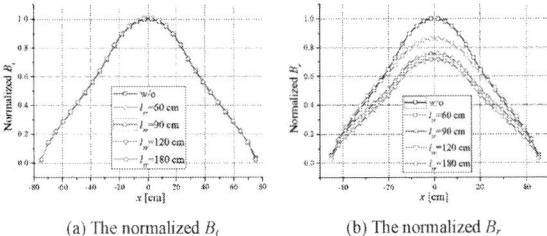

(a) The normalized B_t (b) The normalized B_r

Fig. 8. The normalized B_t and B_r for different l_{rr} when h_{rr}=30 cm.

When the length and height of the Rx reflector h_{rr} and l_{rr} are changed, the simulation results of the magnetic flux density of Tx and Rx core B_t and B_r are shown in Figs. 7-8. As a result, B_t is always constant regardless of physical dimension of the Rx reflector, whereas B_r decreases when h_{rr} and l_{rr} increases. Because the lower B_t and larger B_r are desirable by the Rx reflector, it is concluded that the Rx reflector is not suitable to improve the power efficiency.

B. Tx Reflector

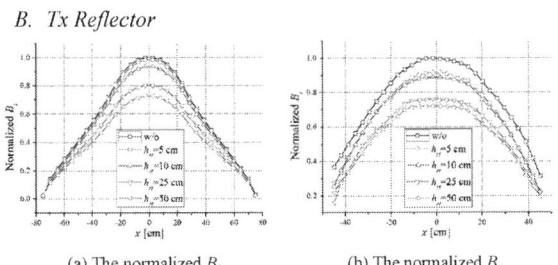

(a) The normalized B_t (b) The normalized B_r

Fig. 9. The normalized B_t and B_r for different h_{rt} when l_{rt} = 250 cm.

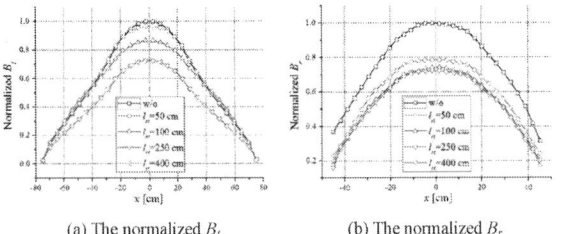

(a) The normalized B_t (b) The normalized B_r

Fig. 10. The normalized B_t and B_r for different l_{rt} when h_{rt} = 50 cm.

In the same way, the simulation results of B_t and B_r are shown in Figs. 9-10 for various sizes of the length and height of the Tx reflector. Contrary to the Rx reflector, both B_t and B_r decrease when both h_{rt} and l_{rt} increase. Accordingly, although the core loss of the Tx core is reduced, the induced voltage of Rx coil is also reduced. Therefore, it is necessary to find the optimum sizes of h_{rt} and l_{rt}, considering the reduction ratio of the magnetic flux density when the Tx reflector is adopted for B_t and B_r, as follows:

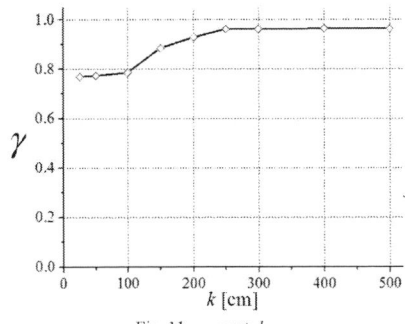

Fig. 11. γ w.r.t. k.

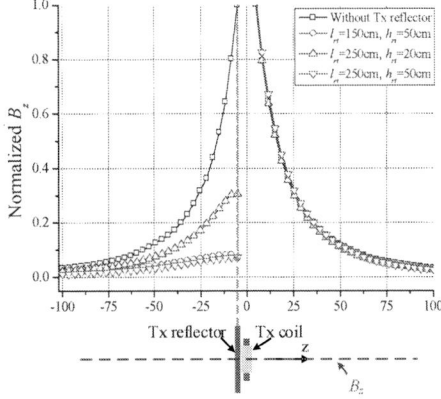

Fig. 12. Simulation results of the magnetic shielding effect.

$$\alpha \equiv \frac{\left.\left|\widetilde{B}_t\right|\right|_{l_{rt}=k}}{\left.\left|\widetilde{B}_t\right|\right|_{w/o}} \tag{6a}$$

$$\beta \equiv \frac{\left.\left|\widetilde{B}_r\right|\right|_{l_{rt}=k}}{\left.\left|\widetilde{B}_r\right|\right|_{w/o}} , \tag{6b}$$

where $\left.\left|\widetilde{B}_t\right|\right|_{l_{rt}=k}$ and $\left.\left|\widetilde{B}_t\right|\right|_{w/o}$ in (6a) are the average values of the magnetic flux density in the Tx core for $l_{rt} = k$ and without the Tx reflector cases, respectively, and $\left.\left|\widetilde{B}_r\right|\right|_{l_{rt}=k}$ and $\left.\left|\widetilde{B}_r\right|\right|_{w/o}$ in (6b) are the average values of the magnetic flux density in the Rx core for $l_{rt} = k$ and without the Tx reflector cases, respectively.

From (6), the ratio of β and α, i.e., γ ($\equiv \beta/\alpha$), is investigated w.r.t. k, as shown in Fig. 11, where the higher γ is better in this case. Consequently, $k = 250$ cm in Fig. 11 is adopted throughout this paper, considering the saturation point that γ is no longer increased when k increases.

978-1-4673-9551-9/16 $31.00 © 2016 IEEE   1776

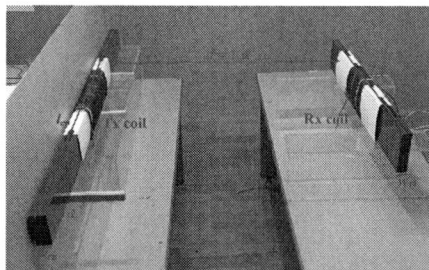

(a) Side view.

Fig. 15. A prototype of half-bridge inverter.

(b) Front view.

Fig. 13. An experimental kit of the proposed IPTS including the Tx reflector.

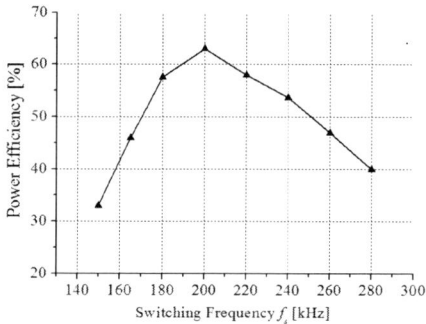

Fig. 16. Experimental results of the power efficiency w.r.t f_s.

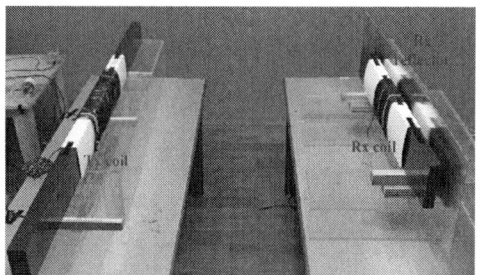

(a) Side view.

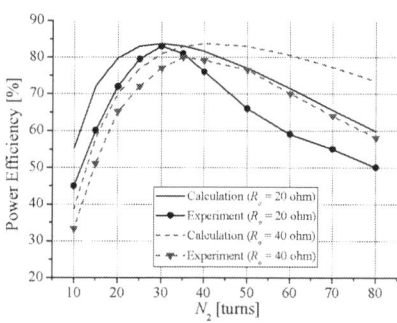

Fig. 17. Calculation and experimental results of the power efficiency w.r.t N_2.

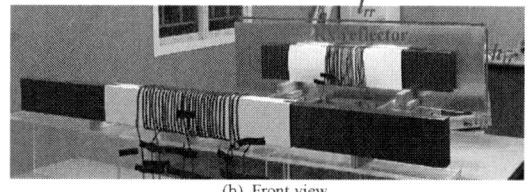

(b) Front view.

Fig. 14. An experimental kit of the proposed IPTS including the Rx reflector.

C. Magnetic Shielding

It is important for all the magnetic flux to be blocked toward the human being, which is around the Tx core. As shown in Fig. 12, the effect of the magnetic flux shielding by the Tx reflector is evaluated for three different sizes of it. It is found that the magnetic flux density behind Tx reflector is significantly reduced when the Tx reflector is adopted. Especially, the longer height of Tx reflector, i.e., $h_{rt} = 50$ cm, has much impact to reduce the magnetic flux density. On the other hand, the over 150 cm length of the Tx reflector, which is the same length with the Tx core, has little influence for magnetic shielding effect. As a result, the exposure level of magnetic flux density was reduced by 94% in average for $l_{rt} = 250$ cm and $h_{rt} = 50$ cm of the Tx reflector.

IV. EXPERIMENTAL VERIFICATION

The experimental kits of the proposed dipole-dipole IPTS adopting the Tx and Rx reflectors are shown in Figs. 13-14. The physical dimensions of the Tx and Rx coils are summarized in Table I, and the number of primary turns N_1 was selected as 30 turns. An aluminum plate as the reflector was adopted in Tx core, as identified from simulation results: l_{rt} = 250 cm and $h_{rt} = 50$ cm with 2 mm of thickness was selected in this paper. Mn-Zn-type PL-F2 ferrite core made by Samhwa Inc. was conducted to build the Tx and Rx cores, and power efficiency considering power loss of the half-bridge inverter in Fig. 15 was measured when the output power P_o is 150 W.

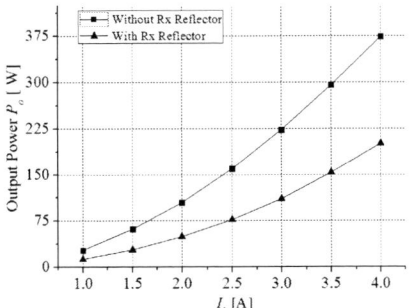

(a) Output power for with and without the Rx reflector cases. ($R_o = 20\Omega$)

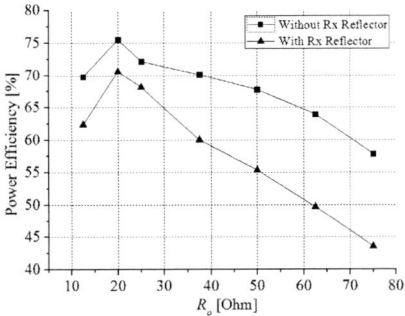

(b) Power efficiency for with and without the Rx reflector cases.

Fig. 18. Experimental results of output power and power efficiency for the Rx reflector.

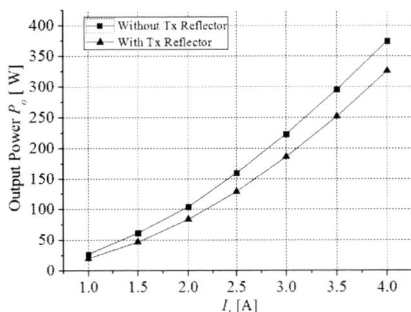

(a) Output power for with and without the Tx reflector cases. ($R_o = 20\Omega$)

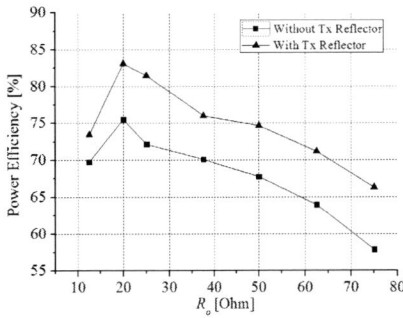

(b) Power efficiency for with and without the Tx reflector cases.

Fig. 19. Experimental results of output power and power efficiency for the Tx reflector.

A. The Optimal Switching Frequency and Secondary Turns for Maximum Power Efficiency

The various switching frequencies to find maximum power efficiency under given Tx and Rx core sizes and material were tested, as shown in Fig. 16: optimal switching frequency was experimentally selected as 200 kHz in this paper. In Fig. 16, when $f_s < 200$ kHz, it requires higher current in Tx coil for the output power of 150 W, which causes high core and copper losses in Tx coil. On the other hand, core losses of the Tx and Rx coils become significant when $f_s > 200$ kHz because core loss is proportional to f_s by the Steinmetz equation [6].

As identified from (5), experimental results of the power efficiency for different secondary turns N_2 are shown in Fig. 17, where the experimental results are good agreement with the proposed design analysis. The discrepancy mainly comes from the modeling of core loss in Tx and Rx coils, and assumption that k_1 and k_2 are constant in (3).

B. Reflector Effect on the IPTS

Experimental results of output power and power efficiency for Tx and Rx reflectors are shown in Figs. 18-19. The Tx reflector having a size of 250 cm × 50 cm × 2 mm and the Rx reflector having a size of 120 cm × 30 cm × 2 mm are set up in the back sides of 1 cm distance from the Tx and Rx coils, respectively. As identified from simulation results, the Rx reflector degrades the power efficiency in Fig. 18, whereas the Tx reflector improves the power efficiency in Fig. 19: 4.2% of power efficiency reduction for the Rx reflector and 7.8% of the power efficiency improvement for the Tx reflector were observed at $R_o = 20\Omega$. As a result, 83.1% of the power efficiency in Fig. 19 was achieved by the proposed IPTS with the Tx reflector.

V. CONCLUSIONS

1m-off long distance IPTS with dipole coil structures for 150W output power transfer has been demonstrated to achieve high power efficiency. The optimum secondary turns can be appropriately selected by the proposed design procedure for maximum power efficiency. The characteristics of the Tx and Rx reflectors are specifically investigated by simulation and experiments. As a result, the Rx reflector degrades the power efficiency, whereas the Tx reflector improves the power efficiency. It is expected that the proposed IPTS with the Tx reflector will be suitable for home appliance charging applications.

REFERENCES

[1] A. Kurs, A. Karalis, R. Moffatt, J. D. Joannopoulos, P. Fisher, and M. Soljacic, "Wireless power transfer via strongly coupled magnetic resonance," *Science*, vol. 317, no. 5834, pp. 83-86, Jun. 2007.

[2] A. P. Sample, D. T. Meyer, and J. R. Smith, "Analysis, experimental results, and range adaption of magnetically coupled resonator for wireless power transfer," *IEEE Trans. Ind. Electronics*, vol. 58, pp. 544-554, Feb. 2011.

[3] E. Lee, J. Huh, X. V. Thai, S, Choi, and C. Rim, "Impedance transformers for compact and robust coupled magnetic resonance systems," in *2013 ECCE conf.*, 2013, pp. 2239-2244.

[4] B. Choi, E. Lee, J. Huh, and C. Rim, "Lumped impedance transformers for compact and robust coupled magnetic resonance systems," *IEEE Trans. Power Electron.*, vol. PP, no. 99, pp. 1, Jan. 2015.

[5] C. Park, S. Lee, and C. Rim, "5m-off-long-distance inductive power transfer system using optimum shaped dipole coils," *in 2012 IPEMC conf.*, 2012, pp. 137-1142.

[6] C. Park, S. Lee, G. Cho, and C. Rim, "Innovative 5-m-off-distance inductive power transfer systems with optimally shaped dipole coils," *IEEE Trans. Power Electron.*, vol. 30, no. 2, pp. 817-827, Nov. 2014.

[7] B. Choi, E. Lee, J. Kim, and C. Rim, "7m-off-long-distance extremely loosely coupled inductive power transfer system using dipole coils," in *2014 ECCE conf.*, 2014, pp. 858-863.

[8] J. W. Kim, H. C. Son, D. H. Kim, and Y. J. Park, "Optimal design of a wireless power transfer system with multiple self-resonators for an LED TV," *IEEE Trans. on Consumer Electron.*, vol. 58, no. 3, pp. 775-780, Sep. 2012.

[9] Hongseok Kim et al., "Suppression of leakage magnetic field from a wireless power transfer system using ferrimagnetic material and metallic shielding," in *2012 IEEE International Symposium on Electromagnetic Compatibility (EMC)*, 2012, pp. 640-645.

[10] J. Huh, S. W. Lee, W. Y. Lee, G. H. Cho, and C. T. Rim, "Narrow-Width Inductive Power Transfer System for Online Electrical Vehicles," *IEEE Trans. Power Electron.*, vol. 26, no. 12, pp. 3666-3679, Dec. 2011.

[11] P. L. Dowell, "Effects of eddy currents in transformer windings," in *Proceedings of the Institution of electrical Engineers*, 1966, pp. 1387-1394.

[12] B. Fincan, O. Ustun, "A study on solutions for wireless energy transfer limitations," in *7th IET International Conference on Power Electronics, Machines and Drives (PEMD 2014)*, 2014, pp. 1-6.

A Novel Time-Sharing Current-Fed ZCS High Frequency Inverter-applied Resonant DC-DC Converter for Inductive Power Transfer

Kyohei Konishi, Tomokazu Mishima
Mechatronics Engineering Division
Graduate School of Maritime Science, Kobe University
Kobe, Hyogo 658–0022, Japan
Email: 143w511w@stu.kobe-u.ac.jp, mishima@maritime.kobe-u.ac.jp

Mutsuo Nakaoka
University of Malaya
Kuala Lumpur 50603, Malaysia

Abstract—This paper presents a novel prototype of a time-sharing frequency doubler principle-based current-fed zero current soft-switching (ZCS) high frequency resonant (HF-R) inverter for inductive power transfer (IPT) systems. The newly-proposed ZCS HF-R inverter is suitable for producing a higher frequency resonant current with switching power loss reduction by using a middle-class switching frequency insulated-gate-bipolar-power transistor (IGBT). In order to continuously regulate the output power, resonant current phasor control (RCPC) is newly applied. The performances of the newly-proposed IPT resonant power converter are demonstrated by experiment, after which the feasibility of the circuit topology and control method is discussed from a practical point of view.

Keywords—*high frequency-resonant (HF-R) inverter, time-sharing principle, zero current soft-switching (ZCS), Si-IGBT/SiC-SBD hybrid module, inductive power transfer (IPT), primary-side series/secondary-side parallel (SP) compensation, resonant current phasor control (RCPC).*

I. Introduction

IPT systems have drawing much attention in the wide variety areas of industrial and automotive electric power applications such as battery chargers in Automated Guided Vehicle (AVG) and Electric Vehicles these days[1]–[3]. A high frequency inverter being able to generate around 100 kHz and its multiple-frequency resonant current is strongly demanded in those applications. This makes a switching power loss outstanding, then switching surges and ringing results in an electromagnetic noise.

As an emerging solution for this technical issue, high efficiency IPT power converter using the wide band gap (WBG) power devices such as Silicon Carbide (SiC) and Gallium Nitride (GaN) transistors with the high-speed transition and low conduction loss have been developing in the relevant researches. However, the low reliability and ruggedness as well as high cost still remain obstacles for applications of the WBG power devices together with their gate driver into the practical level[4]–[7].

A general IPT system is depicted in Fig.1. In order to realize the cost-effective and practical power converter circuit scheme suitable for IPT, a novel resonant dc-dc power converter assisted with a time-sharing principle-based current-fed ZCS HF-R inverter is proposed in this research. The HF-R inverter with the self-commutated power device IGBT has been originally presented by the authors of this paper for induction heating applications[8]–[10]. In the HF-R inverter applied herein, a high frequency current, e.g. 100 kHz, which is twice or multiple times as much as the switching frequency, can be produced effectively even with middle speed-range (20 kHz–50 kHz) Si-IGBT which features a good balance between voltage and current ratings suitable for kW-class power processing. Thereby, the switching loss can be minimized while supplying a higher frequency resonant current into a sending coil. In addition, the current ripple of the dc source side can be reduced effectively owing to the multi-phase current-fed inverter topology. The power devices can be protected by the DCLs from over-current in case of a short load, thereby the HF-R inverter can operate near its load resonant frequency.

The rest of this paper is organized as follows. The circuit topology and operating principle are explained with favorable features in Section II. The theory of output power control method, RCPC, which is adopted in the proposed HF-R inverter is described in Section III, revealing the corresponding equivalent circuit models. The experimental verifications of validity for adopting RCPC on the proposed circuit and the effectiveness applying SiC-SBD which have both high breakdown voltage and low forward voltage to anti-parallel diode to improve the power conversion efficiency are demonstrated in Section IV.

II. Circuit Configuration and Operations

A schematic diagram of the proposed IPT power converter is illustrated in Fig.2. Additionally, its operating waveforms, operating mode transitions and equivalent circuits during the one switching cycle are depicted in Fig.3 and Fig.4, respectively.

The proposed HF-R inverter consists of the two-phase Class-E ZCS inverters U_1 and U_2 which share a sending coil in series with the power factor tuned capacitor C_s. The two-phase Class-E ZCS inverters are composed of DCLs L_{d1}, L_{d2} ($L_{d1} = L_{d2}$), the resonant capacitors C_1, C_2 ($C_n = C_1 = C_2$), and the reversely conducting (RC)-type IGBTs Q_1, Q_2. Those resonant-link capacitors contribute for blocking dc currents from L_{d1} and L_{d2} flowing into the sending coil. At the secondary side, a receiving coil is connected with a parallel

978-1-4673-9551-9/16 $31.00 © 2016 IEEE

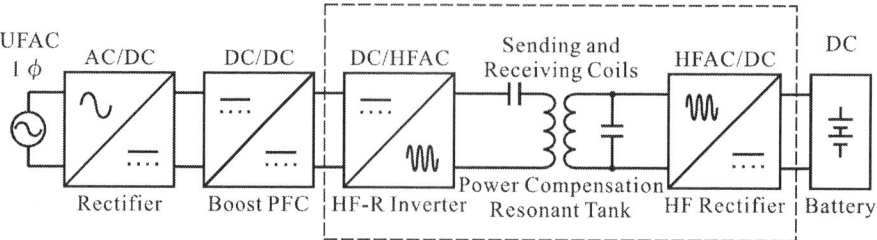

Fig. 1. Power conversion process of battery charger IPT system.

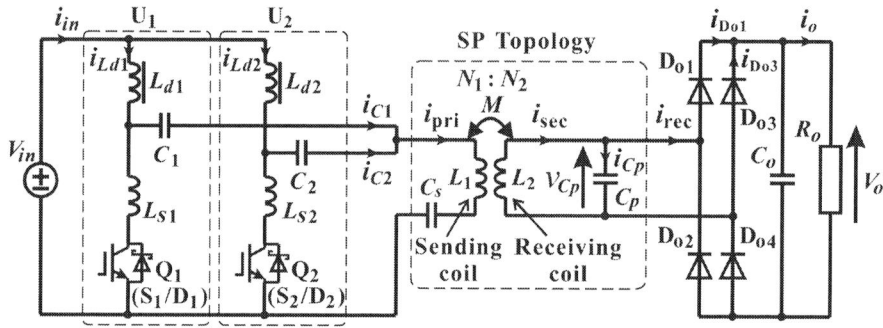

Fig. 2. A proposed resonant DC-DC converter using a time-sharing two-phase class-E ZCS inverter.

resonant capacitor C_p. A smoothing capacitor C_o and a DC load are connected with them through a full wave rectifier composed of rectifier diodes D_{o1}-D_{o4}.

The two active switches Q_1, Q_2 are driven by the inter-leaved gate clocking. Then, the two-phase class-E ZCS invert-ers U_1 and U_2 generate the two resonant link-capacitor cur-rents i_{C_1} and i_{C_2}, respectively. The synthesized link-capacitor currents i_{C_1} and i_{C_2} produce the sending coil current i_{pri} by the time sharing manner. A single switching action yields one-cycle resonant full-wave current the device-conducting pathway of which is divided into the active switch and anti-parallel diode. Therefore, SiC-SBD is suitable for this IPT converter since it has both high breakdown voltage and low forward voltage. The reverse recovery current can be reduced regardless of the type of anti-parallel diode by the ZCS-assisted inductors L_{s1}, L_{s2} ($L_s = L_{s1} = L_{s2}$) in effect.

The ZCS operation of Q_1 and Q_2 can be achieved by the effect of L_{s1} and L_{s2}. The small inductors attenuate the high di/dt turn-on behavior which is inherent with a conventional load resonant inverter operating with a capacitive load network [11]–[13]. In addition, the gate signals for S_1, S_2 are removed while conducting the anti-parallel diodes D_1, D_2 of Q_1, Q_2, and then ZCS & ZVS turn-off commutation can be attained. Thus, switching power losses, especially the turn-off power losses due to the tail current of the bipolar power devices can be completely eliminated.

The series resonant frequency f_r of the series resonant network that consists of leakage inductance of the sending and receiving coils, link capacitor and ZCS-assisted inductor

in each phase can be defined by

$$f_r = \frac{1}{2\pi\sqrt{L_c C_r}} \tag{1}$$

$$L_c = L_s + \frac{M}{L_2}, \quad C_r = \frac{C_n \cdot C_s}{C_n + C_s}. \tag{2}$$

In order to perform the ZCS turn-on of active switches and ZCS turn-off of parallel diodes, the switching frequency f_s is selected as

$$f_s = \frac{f_o}{2} > \frac{f_r}{2}, \tag{3}$$

where f_o represents the sending coil current frequency. As a result, a higher frequency resonant current can be produced efficiently to the sending coil while suppressing the switching frequency. In addition, all the diodes of the secondary-side rectifier are also commutated in ZCS due to the primary-side series and secondary-side parallel (SP) compensation resonant tanks. Adoption of the SP compensation can be justified by considering the voltage boost function of the current-fed ZCS HF-R inverter [14]. The parallel resonant frequency f_p can be defined by

$$f_p = \frac{1}{a'} \cdot \frac{1}{2\pi\sqrt{(1-k^2)L_1 C_p}} \tag{4}$$

$$a' = \frac{M}{L_2} = k\sqrt{\frac{L_1}{L_2}}, \quad k = \frac{M}{\sqrt{L_1 L_2}}, \tag{5}$$

where M denotes the mutual inductances of sending and receiving coils, a' and k represents the winding turns ratio and the coupling coefficient of the ideal transformer.

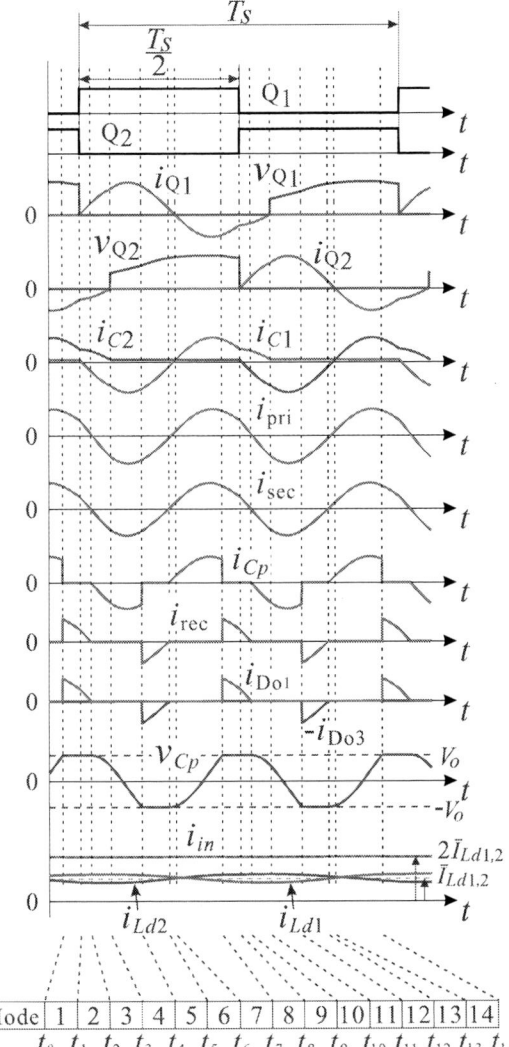

Fig. 3. Key voltage and current waveforms.

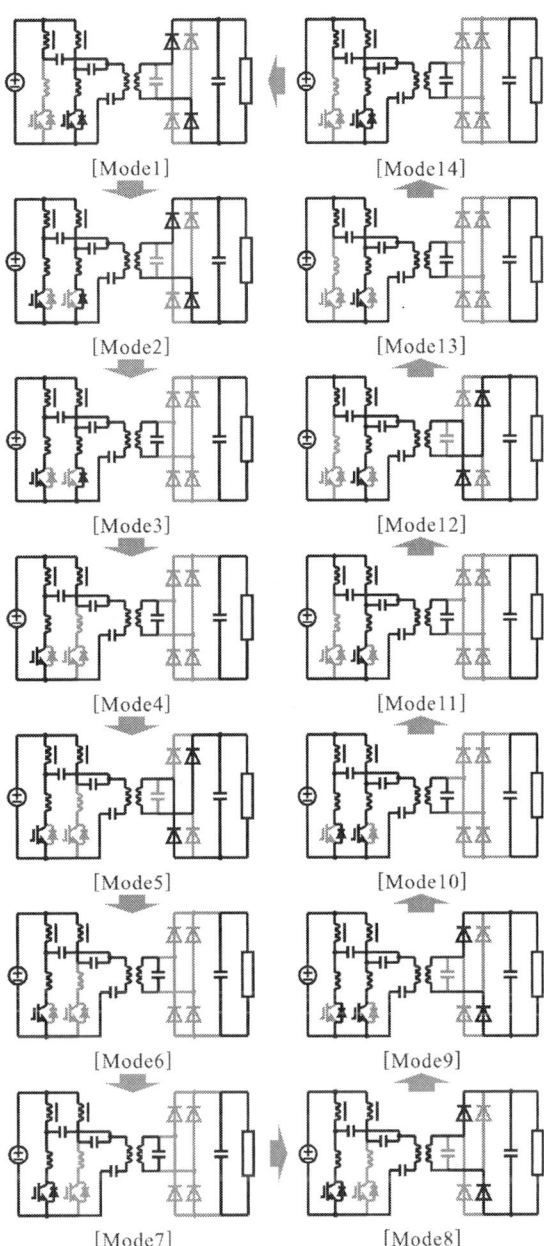

Fig. 4. Operating mode transitions and equivalent circuits during the one switching-cycle.

The one switching cycle can be divided into the fourteen sub-mode intervals with ZCS commutation of the active switches and diodes as follows:

[Mode 1 ($t_0 \leq t < t_1$): Q_2 ON-state and D_{o1} & D_{o4} conducting] The series resonant current circulates in the primary-side sending coil with the anti-parallel diode D_2 of Q_2 conducting. The rectifier diodes D_{o1} and D_{o4} are forward-biased, then power is fed from the receiving coil to the load.

[Mode 2 ($t_1 \leq t < t_2$): Q_1 ZCS turn-on and Q_2 ZCS & ZVS turn-off] The gate signal is supplied to S_1 at $t = t_1$. Accordingly, the current through Q_1 rises gradually due to the effect of L_{s1}, whereby ZCS turn-on transition stars. At the same time, the conduction current through S_2 naturally decays due to the effect of L_{s2}.

[Mode 3 ($t_2 \leq t < t_3$): secondary-side diodes reversely biased and parallel compensation circuit resonates] The rectifier input current i_{rec} naturally decays to zero at $t = t_2$, then the secondary-side diodes D_{o1} and D_{o4} are reversely biased with the output voltage V_o. Accordingly, the parallel compensation circuit of the receiving coil self-inductance L_s and C_p starts to resonate.

[Mode 4 ($t_3 \leq t < t_4$): **single-loop series resonance in the primary-side inverter]** The current through D_2 decreases to zero with a certain slope at $t = t_3$. Thus, the reverse recover current of the anti-parallel diode can be mitigated effectively.

[Mode 5 ($t_4 \leq t < t_5$): **secondary-side rectifier diodes forward-biased]** The voltage across C_p builds up in the negative polarity, and reaches to $-V_o$ at $t = t_4$. Accordingly, the rectifier diodes D_{o2} and D_{o3} are forward-biased and the power is fed to the load via the receiving coil. During this interval, the parallel capacitor voltage v_{cp} are clamped to $-V_o$.

[Mode 6 ($t_5 \leq t < t_6$): **secondary-side rectifier diodes reversely-biased and parallel compensation circuit resonates]** The rectifier input current i_{rec} naturally decays to zero at $t = t_5$, then the secondary-side diodes D_{o2} and D_{o3} are reversely biased with the output voltage V_o. Accordingly, the parallel compensation circuit of the receiving coil self-inductance L_2 and C_p stars to resonate similarly to Mode 3.

[Mode 7 ($t_6 \leq t < t_7$): **anti-parallel diode forward-biased in Q_1, and primary-/secondary-winding currents reverse direction.** The current through Q_1 decreases gradually to zero with a certain slope at $t = t_6$, and its anti-parallel diode is forward-biased. Thus, the positive half-cycle of the first full-wave appears in the sending-coil current i_{pri}. This sub-mode transition continues until v_{cp} in the secondary-side grows up to V_o at $t = t_7$.

The operation from sub-mode 8 to 14 corresponds with the second full-wave interval of i_{pri}, thereby the mode transitions are symmetrical to those of Sub-mode 1 to 7 as mentioned above.

III. POWER CONTROL METHOD

A. Resonant Current Phasor Control

In order to carry out the output power control, RCPC of the two-phase resonant capacitor currents is applied under the constant switching frequency condition[15].

The phase difference between the gate signals of Q_1 and Q_2 is reduced from $180°$ as shown in Fig.5(a),(b) on the basis of the RCPC principle. This results in continuously adjusting the RMS value of the HF-R inverter current, thereby the output power control is achievable under the high frequency condition. In this case, the phase difference angle of the gate signal ϕ_s is defined as

$$\phi_s = \frac{t_{\phi_s}}{T_s} \times 360 \; [\text{deg}]. \tag{6}$$

The output power can be controlled only by adjusting ϕ_s, whereby simplification of the power circuit and control system can be attained as compared to other power regulation schemes such as pulse amplitude modulation (PAM).

The simulated power control characteristics by RCPC of the proposed IPT converter is depicted in Fig.6 with the same parameters of prototype experiment as after mentioned. The output power is continuously controlled from rating power to minimum power by RCPC. Additionally, ZCS turn-on and ZCS & ZVS turn-off transitions are ensured while thorough the selected power control range.

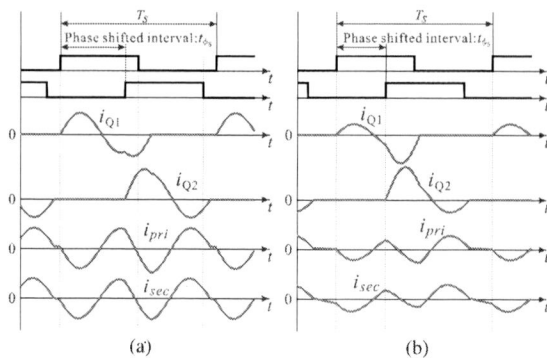

Fig. 5.   RCPC-based pulse patterns and key current waveforms: (a) $\phi_s = 150°$, and (b) $\phi_s = 115°$.

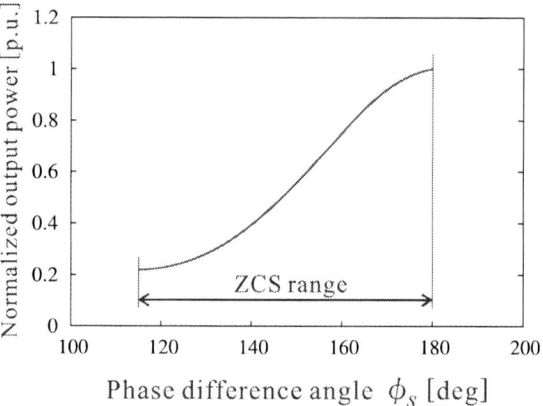

Fig. 6.   Simulated output power versus phase difference angle curves.

It can be seen from Fig.5 that the double frequency component of the sending and receiving coil currents are reduced in accordance with ϕ_s, and a single-frequency operation of the switching frequency component is dominant in the smaller area of ϕ_s. Therefore, in order to keep the double frequency current, the maximum value of ϕ_s is $180°$, and the control range of the set value to the single-frequency operation appears.

The harmonic analysis of sending and receiving coil currents is demonstrated by simulation to determine the control range of ϕ_s. Sending and receiving coil current waveforms and Fast Fourier Transform (FFT) harmonic analysis results are depicted in Fig.7 (a)–(e). The double frequency component keeps the high profile as compared to the switching frequency component in the range of $\phi_s = 115°$–$180°$. In contrast to that, the relationship between the switching frequency of the receiving coil current i_{sec} exceeds the double frequency components at $\phi_s = 110°$. Additionally, the switching frequency component completely overwhelms the double frequency component both in i_{pri}, i_{sec} at $\phi_s = 95°$. This implies that there is no generating the double frequency component for the load resonant current under than $\phi_s = 95°$. This simulation analysis justifies the phase angle range is set to $\phi_s = 115°$–$180°$ in the experimental verification as mentioned in the next section.

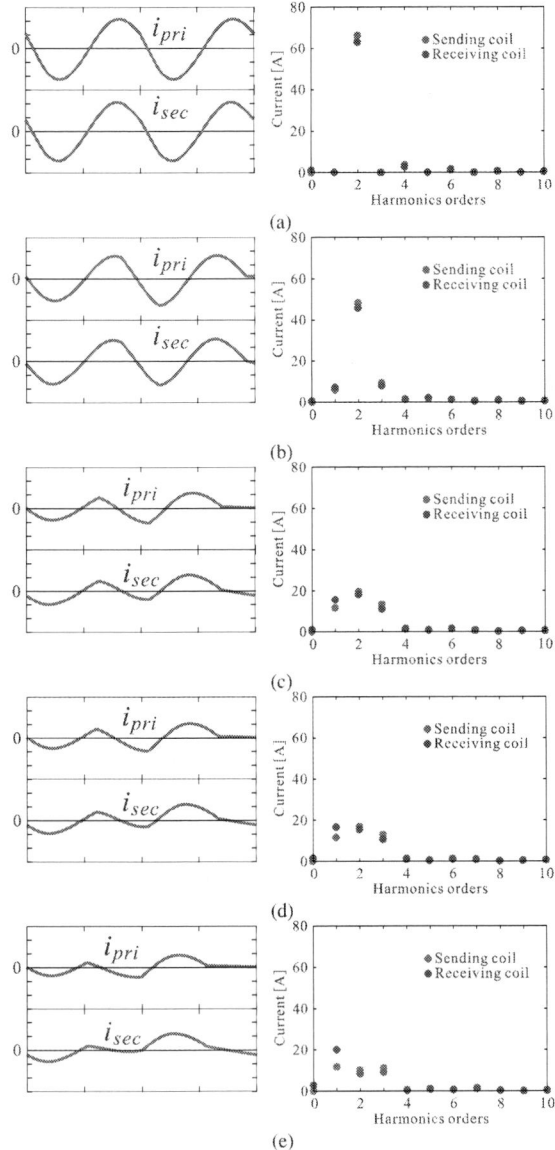

Fig. 7. Simulated sending and receiving coils current waveforms with FFT analysis under the various output power conditions: (a) $\phi_s = 180°$, (b) $\phi_s = 150°$, (c) $\phi_s = 115°$, (d) $\phi_s = 110°$, and (e) $\phi_s = 95°$ (30 A/div, 5 μs/div).

IV. EXPERIMENTAL RESULTS AND DISCUSSIONS

A. Experimental Set-Up and Specification

The practical feasibility of the proposed IPT power converter is investigated in experiment based on a 2.0 kW-100 kHz laboratory prototype.

The exterior appearance of the prototype and experimental set-up is depicted in Fig. 8. The circuit parameters of the prototype including the sending and receiving coils are indi-

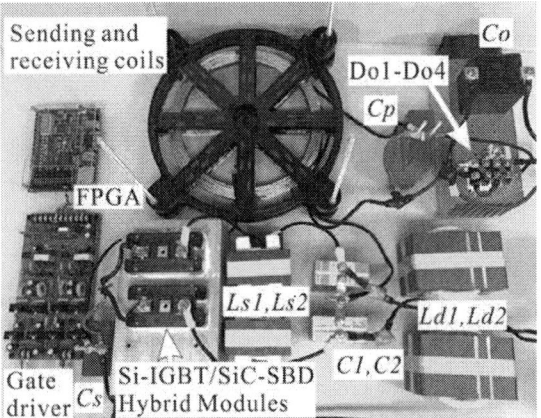

Fig. 8. Exterior appearance of the IPT power converter prototype.

TABLE I. EXPERIMENTAL CIRCUIT PARAMETERS

Item	Symbol	Value[unit]
Output power rating	P_o	2 [kW]
Input voltage	V_{in}	170 [V]
Output voltage	V_o	192 [V]
DCLs	L_{d1}, L_{d2}	500 [μH]
ZCS-assisted inductors	L_{s1}, L_{s2}	7 [μH]
Resonant capacitors	C_1, C_2	300 [nF]
Power factor tuned capacitor	C_s	600 [nF]
Parallel resonant capacitor	C_p	390 [nF]
Load resistor	R_o	18 [Ω]
Switching frequency	f_s	50 [kHz]
HF-R inverter output frequency	f_o	100 [kHz]
Series resonant frequency	f_r	86 [kHz]
Parallel resonant frequency	f_p	99 [kHz]
Sending and receiving coils		
Self inductance of sending coil	L_1	22.4 [μH]
Self inductance of receiving coil	L_2	22.4 [μH]
Mutual inductances	M	16.5 [μH]
Coupling coefficient of L_1 and L_2	k	0.74
* Winding turns of L_1 and L_2	N_1/N_2	6/6 [turn]
* Air gap length of L_1 and L_2	g	10 [mm]

* Active switches : Mitsubishi CMH100DY-24NFH, 1200[V], 100[A]
* Rectifier diodes : IXYS DSEI 2x31-10B, 600[V], 2×30[A]

cated in TABLE I. It should be noted here that the switching and resonant frequencies are based on the self-inductances L_1, L_2 of the sending and receiving coils that are measured in their gap-length $g = 10$ mm. Once setting the sending coil frequency f_o to 100 kHz, the switching frequency can be determined to 50 kHz based on the time-sharing principle. The loaded resonant frequency f_r is selected to 86 kHz in the prototype by considering with Eq.(3).

B. Steady-State Performances and Characteristics

The switching operations and effectiveness of the RCPC-based output power regulation with the time sharing principle are demonstrated in experiment of the prototype circuit.

The observed switching waveforms of Q_1, Q_2, sending and receiving coils as well as the rectifier diode are depicted in Figs.9-10 for the rated output power at $\phi_s = 180°$. The ZCS

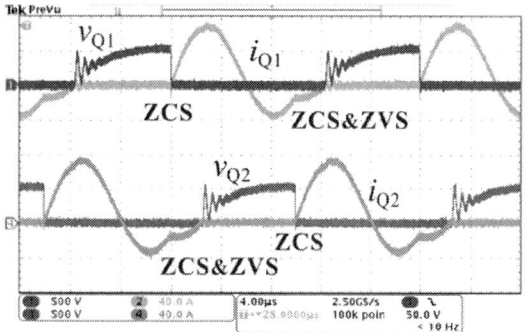

Fig. 9. Observed switching voltage and current waveforms at $\phi_s = 180°$, $P_o = 2.0\,\text{kW}$ (500 V/div, 40 A/div, 4 μs/div).

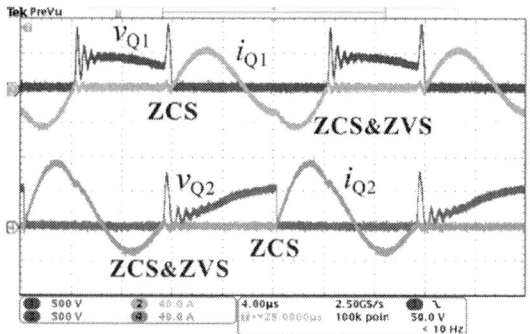

Fig. 12. Observed switching voltage and current waveforms at $\phi_s = 150°$, $P_o = 1.15\,\text{kW}$ (500 V/div, 40 A/div, 4 μs/div).

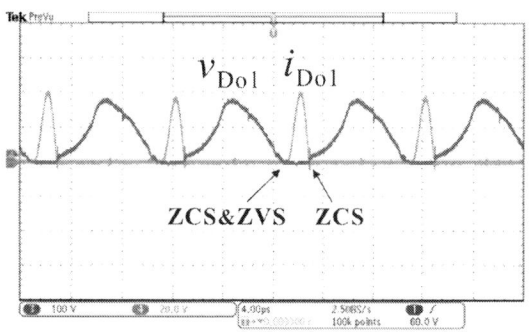

Fig. 10. Observed voltage and current of rectifier a diode at $\phi_s = 180°$, $P_o = 2.0\,\text{kW}$ (100 V/div, 20 A/div, 4 μs/div).

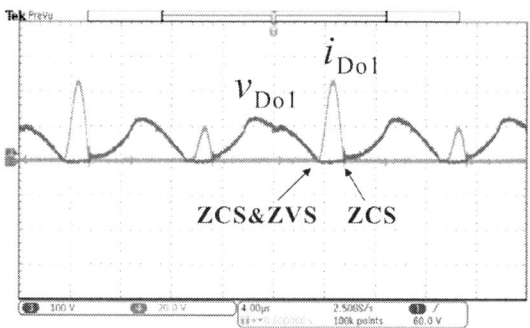

Fig. 13. Observed voltage and current of rectifier a diode at $\phi_s = 150°$, $P_o = 1.15\,\text{kW}$ (100 V/div, 20 A/div, 4 μs/div).

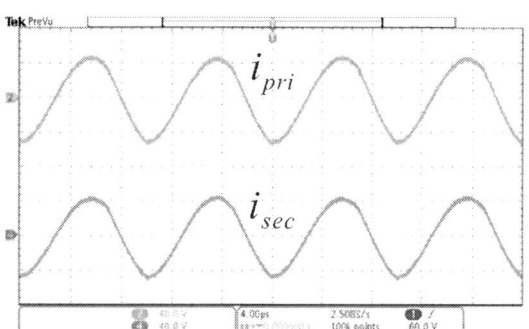

Fig. 11. Observed current waveforms of sending and receiving coils at $\phi_s = 180°$, $P_o = 2.0\,\text{kW}$ (40 A/div, 4 μs/div).

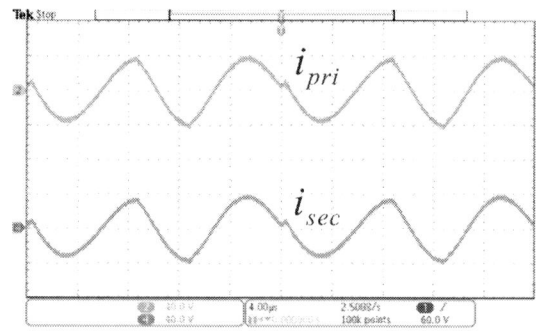

Fig. 14. Observed current waveforms of sending and receiving coils at $\phi_s = 150°$, $P_o = 1.15\,\text{kW}$ (40 A/div, 4 μs/div).

turn-on and turn-off commutations can be confirmed for Q_1 and Q_2 in Fig.9. In addition, the ZCS turn-on and turn-off are also achieved in the secondary-side rectifier diode as show in Fig.10 due to the parallel resonant tank of the secondary-side power stage. Fig.11 demonstrates the resonant currents i_{pri}, i_{sec} in phase, which proves the effective power transfer is actually achievable between the sending and receiving coils in the proposed IPT power converter.

The ZCS commutations and power transfer can also be observed at $\phi_s = 150°$ as depicted in Figs.12-13. The sending and receiving coil currents get in the boundary condition of continuous and discontinuous conduction modes as depicted in Fig.14. This implies the switching frequency component begins to appear in the sending and receiving coils.

The ZCS commutations of the active switches and diodes still maintain at $\phi_s = 115°$ in Figs.15 and 16. However, the

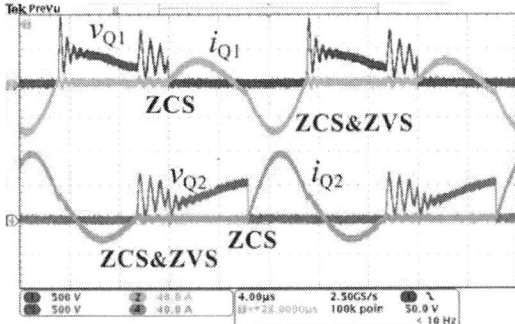

Fig. 15. Observed switching voltage and current waveforms at $\phi_s = 115°$, $P_o = 440\,\text{W}$ (500 V/div, 40 A/div, 4 μs/div).

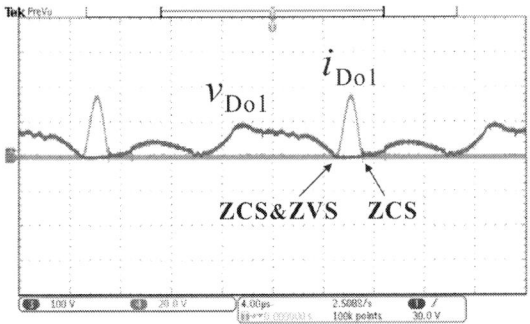

Fig. 16. Observed voltage and current of rectifier a diode at $\phi_s = 115°$, $P_o = 440\,\text{W}$ (100 V/div, 20 A/div, 4 μs/div).

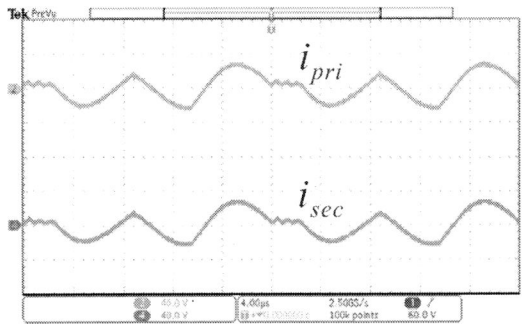

Fig. 17. Observed current waveforms of sending and receiving coils at $\phi_s = 115°$, $P_o = 440\,\text{W}$ (40 A/div, 4 μs/div).

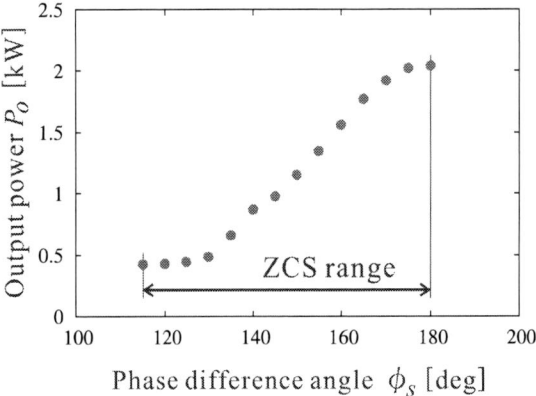

Fig. 18. Experimental output power versus phase difference angle curves.

switching waveforms are discontinuous conduction mode, and the half of the full-wave are canceled out each other between i_{Q_1} and i_{Q_2}. As a result, the sending and receiving currents, i_{pri} and i_{sec} contain harmonics while keeping the double frequency principle.

At this phase difference angle, switching currents i_{Q_1}, i_{Q_2} include DC component and harmonic components. As a result, i_{Q_1}, i_{Q_2} become distorted waves including except for a double frequency components.

It should be remarked herein that the high frequency magnetizing current, which corresponds to the circulating current between the sending and receiving coils, can be minimized regardless of ϕ_s condition owing to effect of the SP compensation circuit.

The steady-stage characteristics of output power and phase difference angle are indicated together with the ZCS range in Fig. 18. The output power is related continuously from the ratted output power 2.0 kW to minimum 440 W at $\phi_s = 115°$, thus the validity of RCPC-based high frequency power regulation is actually demonstrated for the IPT power converter.

The actual power conversion efficiencies of the IPT power converter prototype are provided in Fig. 19. The three stages of power conversion efficiencies are evaluated; from the dc input to the inverter output $\eta_{HF,i}$, the high frequency output

(rectifier input) $\eta_{HF,o}$, and the dc output $\eta_{DC,o}$, respectively. The maximum total dc-dc conversion efficiency 79.0 % is recorded at $P_o = 2.0\,\text{kW}$, while 91.0 % achieves in the dc-HFAC conversion stage. This analysis reveals the copper losses of the sending and receiving coils account for a large portion of the total power loss under the condition of rated power. In contrast to that, the power loss of the HF-R inverter accounts for a larger part of the total power loss in the low power setting. The switching power losses of all the power devices can be minimized in the whole power range due to the effect of the wide-range ZCS commutations deriving from the proposed circuit topology.

C. Power Loss Analysis

In order to confirm the effect of the Si-IGBT / SiC-SBD hybrid power module, the power loss analysis of the inverter is indicated in Fig. 20 as compared with the case of using Si-IGBT / Si-PNDiode (PND) power module CM100DU-24NFH (1200 V-100 A). The indexes of the power loss breakdown are as follows: active switches S_1, S_2 conduction losses P_{S_1}, P_{S_2}, anti-parallel diodes D_1, D_2 conduction losses P_{D_1}, P_{D_2}, DCLs L_{d1}, L_{d2} total power loss P_{L_d}, ZCS-assisted inductors L_{s1}, L_{s2} total power loss P_{L_s}, and primary-side capacitors total power loss P_c.

978-1-4673-9551-9/16 $31.00 © 2016 IEEE

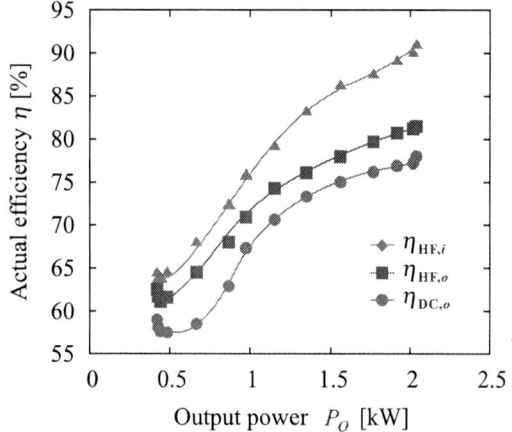

Fig. 19. Actual efficiencies of inverter output, high frequency output and DC output stages.

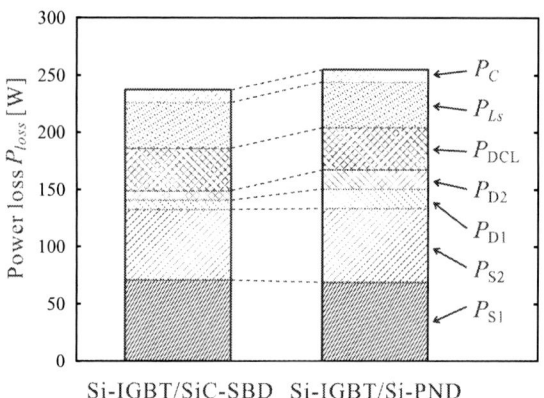

Fig. 20. Measured power loss analysis.

From this analysis, it can be confirmed that the power loss of the reverse conduction to anti-parallel diode is approximately halved by applying a SiC-SBD. As a result, the power conversion efficiencies from the DC input to HF-R inverter with the Si-IGBT/SiC-SBD power module improves to 91% from the result of 90% with the Si-IGBT/Si-PND power module. In order to attain higher efficiency, S_1 and S_2 conduction losses need to be reduced.

V. CONCLUSIONS

A novel time-sharing current-fed ZCS high frequency resonant inverter-applied resonant dc-dc converter for inductive power transfer has been proposed and evaluated in this research. It has been clarified by experimental results that high frequency current generation with ZCS commutations can be achieved under the time sharing principle, which is suitable for bipolar switching power devices. Furthermore, it has been revealed that the effectiveness of the output power control by the resonant current phasor control, and the power loss can be reducing effect by applying a SiC-SBD to anti-parallel diode

of the two-phase class-E ZCS inverter in the proposed IPT power converter.

The detailed analysis on the steady-state characteristics under the wider gap of the sending and receiving coils will be the future challenge for the proposed IPT power converter.

REFERENCES

[1] G.A. Covic, and J.T. Boys, "Inductive power transfer," *Proc. The IEEE*, vol.101. no.6, pp.1276-1289, Jun. 2013.

[2] S.Y.R. Hui, W. Zhong, and C.K. Lee, "A critical review of recent progress in mid-range wireless power transfer," *IEEE Trans. Power Electron.*, vol.29, No.9, pp.4500-4511, Sep. 2014.

[3] K. Yan, Q. Chen, J. Hou, X. Ren, and X. Ruan, "Self-oscillating contactless resonant converter with phase detection contactless current transformer," *IEEE Trans. Power Electron.*, vol.29, No.8, pp.4438-4449, Aug. 2014.

[4] B. Yuan, X. Yang, D. Li, Y. Pei, J. Duan, and J. Zhai, "A current-fed multiresonant converter with low circulating energy and zero-current switching for high step-up power conversion.", *IEEE Trans. Power. Electron.*, vol.26, no.6, pp.1613-1619, Jun. 2011.

[5] C. Zheng, R. Chen, E. Faraci, Z.U. Zahid, M. Senesky, D. Anderson, J. Lai, W. Yu, and C. Lin, "High efficiency contactless power transfer system for electric vehicle battery charging." *Proc. 2013-ECCE*, pp.3243-3249, Sep. 2013.

[6] I. Fujita, T. Yamanaka, Y. Kaneko, S. Abe, and T. Yasuda, "A 10kW transformer with a novel cooling structure of a contactless power transfer system for electric vehicles." *Proc. 2013-ECCE*, pp.3643-3650, Sep. 2013.

[7] R. Haldi, and K. Schenk, "A 3.5kW wireless charger for electric vehicles with ultra high efficiency." *Proc. 2014-ECCE*, pp.668-674, Sep. 2014.

[8] K. Konishi, T. Mishima, and M. Nakaoka, "Operation analysis of a time-sharing multiplex-controlled current-source ZCS high-frequency inverter for induction heating applications", *proc. IEEJ JIASC*, pp.[4-156]-[4-524], Aug. 2014.

[9] K. Konishi, T. Mishima, and M. Nakaoka, "A time-sharing multiplex-controlled current-fed ZCS high-frequency resonant inverter for aluminum utensil induction heating", *The Papers of Joint Technical Meeting on Semiconductor Power Converter and Motor Drive, IEE Japan*, SPC-14-143, HCA-14-51, VT-14-38, pp.1-6, Dec. 2014.

[10] T. Mishima, K. Konishi, and M. Nakaoka, "Current-source ZCS high-frequency resonant inverter based on time-sharing frequency doubler principle and its induction heating applications." *Proc. IEEE Power Electronics and Drive Systems Conf.* (PEDS) 2015, pp.598-603, Jun. 2015.

[11] J.G. Kassakian, "A new current mode sine wave inverter." *IEEE Trans. Ind. Appl.*, vol. IA-18, no.3, pp.273-278, May/Jun. 1982.

[12] N. Sanajit, and A. Jangwanitlert, "A series-resonant half-bridge inverter for induction-iron appliances." *Proc. 2011 IEEE Power Electronics and Drive Systems Conf.* (PEDS), pp.46-50, Dec. 2011.

[13] R.W. Erickson and D. Maksimović, "Fundamentals of power electronics – second edition," Kluwer Academic Publications, ISBN:0-7923-7270-0, 2004.

[14] J. Hou, Q. Chen, K. Yan, X. Ren, S. Wong, and Chi.K Tse, "Analysis and control of S/SP compensation contactless resonant converter with constant voltage gain." *Proc. 2013-ECCE*, pp.2552-2558, Sep. 2013.

[15] T. Mishima, C. Takami, and M. Nakaoka, "A new current phasor-controlled ZVS twin half-bridge high-frequency resonant inverter for induction heating," *IEEE Trans. Ind. Electron.*, vol.61, No.5, pp.2531–2545, May. 2014.

Optimization of Coils for Magnetically Coupled Resonant Wireless Power Transfer System based on Maximum Output Power

Dan Jiang[1], Yong Yang[1], Fuxin Liu[1], Xinbo Ruan[1], Xuling Chen[2]

Jiangsu Key Laboratory of New Energy Generation and Power Conversion[1], College of Mechanical and Electrical Engineering[2]
Nanjing University of Aeronautics and Astronautics
Nanjing, Jiangsu Province, China
xiaobai_bingo@nuaa.edu.cn

Abstract—**Magnetically coupled resonant (MCR) wireless power transfer (WPT) technology is efficient and practical for mid-range wireless energy transmission. For the design of MCR WPT system, it is necessary to seek optimal transmission performance under different applications. This paper presented an equivalent analytical model for such system to incorporate spatial misalignments between the transmitting coils and receiving coils. According to the obtained model, the relationship among the output power, spatial misalignments and coil parameters were analyzed in detail. To achieve maximum output power and high stability of power transfer in a specific misalignments range, a normalization method was introduced, providing critical insight into the optimal design of coils. Relative design considerations and optimization procedures were further stated. Experiments had also been carried out to evaluate the accuracy of theoretical analysis and confirm the validity of the proposed method.**

I. INTRODUCTION

Wireless power transfer (WPT) is an emerging technology to be used in many areas, such as electric vehicles, consumer electronics and implantable devices, due to its superiority on convenience of being cordless, safety during power charging and ability to operate in hostile environments[1-6]. Magnetically coupled inductive (MCI) and magnetically coupled resonant (MCR) are two major techniques of wireless power transfer based on near field[7]. Some MCI WPT cases have exhibited strong power transfer capability of several kilowatts with energy efficiency higher than 90%. However, once the distance increases, the efficiency will drop significantly, this is why it is usually performed in short-range applications[8]. MCR is another wireless power transfer technique suitable for applications with mid-range (10-300cm). In 2007, scientists in Massachusetts Institute of Technology made a breakthrough in such advanced field, successfully lighting up a 60 watts bulb over a distance above 2m[9]. Follow-up studies have shown that MCR WPT provides higher efficiency in longer transmission distance compared to MCI WPT[10]. Therefore, MCR is considered to be one of the most potential techniques for mid-range WPT applications at present.

In MCR WPT system, the transmitting coils and receiving coils are designed to resonate at the operating frequency, establishing an efficient energy channel for power transfer. For the design of such system, it is crucial to seek optimal transmission performance under different specifications. Previous work mainly focused on how to improve transmission capability for longer distance when the transmitter and receiver are perfectly aligned along their axis. In [11], optimum design of the two-coil coupler is discussed and the impact of coil parameters on the transmission efficiency is emphatically analyzed. In [12], low coupling coefficient is thought to be the dominant factor that results in difficulties to achieve high efficiency or large amount of power transfer, so it proposes a method to improve the coupling coefficient by optimizing the structure of coils. Deficiently, neither [11] nor [12] takes spatial misalignments into account. Actually, in some practical situations, spatial scales of coils including lateral misalignment and angular misalignment may vary randomly. The varying spatial scales will cause fluctuations in output power and transmission efficiency, leading to instability of the overall WPT system. So it is critical to predict the misalignment tolerance and specify the geometric boundaries of steady operation. In [13], a novel power transfer function is introduced which combines coil spatial misalignments in deducing the relationship among the transmission efficiency and several key parameters of the system. In [14], the lateral misalignment of coils is investigated and the high efficiency range is identified by using a 3-D physical simulation model. Unfortunately, none of these researches develops a systematic design procedure to guide the selection of optimal configurations for a desired output power.

This paper proposed a normalization method to explore an optimal way of realizing maximum and stable power transfer in a specific misalignments range. The equivalent analytical model of MCR WPT system was put forward to involve spatial misalignments of coils. According to the obtained model, expressions of the output power and mutual inductance in arbitrary spatial scales were further derived. By normalizing mutual inductance into a function only related to coil turns and solving the differential equation of output power, the steady operation range and optimum value of coil turns can be achieved. Finally, a prototype was built and tested to validate the theoretical analysis.

Sponsored by grants from the National Natural Science Foundation of China(51505223), the Natural Science Foundation of Jiangsu Province, China(BK20151471), the Lite-on Research Program, and Jiangsu province university outstanding science and technology innovation team project.

978-1-4673-9551-9/16 $31.00 © 2016 IEEE

II. MODELING AND ANALYSIS OF TRANSMISSION PERFORMANCE

A. MCR WPT System

To enhance the capacity of power transfer, improve the system efficiency, and lower the voltage and current stress of source and load, resonances need to be ensured in both the transmitting side and receiving side[15]. To meet such requirement, resonant capacitors should be introduced for power compensation. They can be provided not only by external capacitors but also by parasitic capacitors of the coils. Here we add external capacitors in the circuit for modulating the operating frequency. Coil parasitic capacitors whose values are so small that can be ignored in analysis.

Fig. 1 shows one kind of topology configuration of MCR WPT system. In Fig. 1, the resonant tank is driven by a half-bridge inverter with dc input voltage. The output voltage is rectified by a voltage-doubler rectifier. L_S and L_D are resonant inductors. C_S and C_D are resonant capacitors. M is the mutual inductance. d is the transmission distance. R_L is the load resistance. R_S and R_D are parasitic resistances of the coils.

Fig. 2 depicts key voltage and current waveforms of the system. The dashed lines represent the fundamental harmonic of the square-wave voltage. v_{gs1} and v_{gs2} are the gate-drive voltages of power switches. v_{ds1} and v_{ds2} are the drain-to-source voltages of power switches. i_{rs} and i_{rd} are the resonant currents of transmitting coils and receiving coils respectively. i_{dr1} and i_{dr2} are currents flowing through the rectifier diodes. From Fig. 2, it can be known that two power switches are turned on or off alternately with a small dead time for soft switching. During the dead time, the parasitic capacitor of the power switch is discharged by the lagged resonant current in case of inductive impedance condition. Before the power switch is turned-on, the drain-to-source voltage is clamped to zero, resulting in no turned-on switching loss. Thus the power switches can realize zero-voltage-switching (ZVS). Besides, the currents of rectifier diodes are naturally decreased to zero when they are turned off, which indicates that the rectifier diodes can operate with zero-current-switching (ZCS). Both ZVS and ZCS operation are beneficial for high power efficiency and low EMI noise, especially in high frequency applications.

Generally, in the WPT system, quality factors of the resonant coils (mostly made of litz wire) are very high. The resonant currents are almost sinusoidal waveforms. So fundamental harmonic analysis (FHA) can be adopted for the steady-state analysis, which only considers the fundamental component and neglects the dc component and higher harmonics. The equivalent circuit of the system based on FHA is shown in Fig. 3, in which phasor $\dot{U}_S$ is the fundamental harmonic of the square-wave voltage applied on the transmitting coils, phasors $\dot{I}_S$ and $\dot{I}_D$ are the fundamental harmonics of the resonant currents, and R_W is the equivalent load resistance.

B. Output Power and Transmission Efficiency

In order to simplify the analysis, the following assumptions are made:
1) Resonant inductor $L_S = L_D$, resonant capacitor $C_S = C_D$;
2) Parasitic resistance of coils $R_S = R_D$, and they are assumed to be constant.

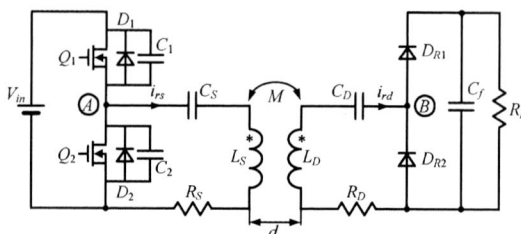

Fig.1 Topology configuration of MCR WPT system

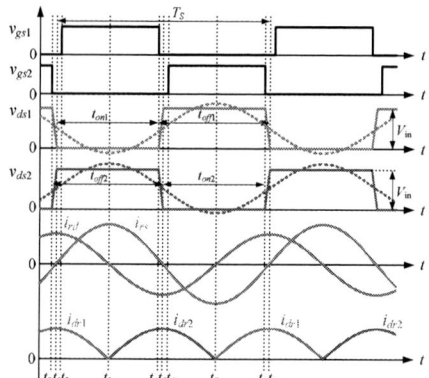

Fig.2 Key waveforms of MCR WPT system

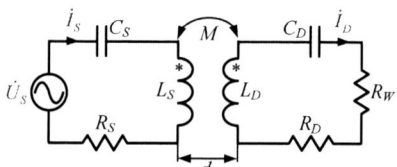

Fig.3 Equivalent circuit of MCR WPT system

Due to the half-bridge configuration, the norm of $\dot{U}_S$ can be calculated as

$$U_S = \left\| \dot{U}_S \right\| = \frac{\sqrt{2}}{\pi} V_{in} \tag{1}$$

The relationship between the equivalent load resistance R_W and the real load resistance R_L is

$$R_W = \frac{2}{\pi^2} R_L \tag{2}$$

According to the equivalent circuit shown in Fig. 3 and Kirchhoff voltage law (KVL), we can obtain

$$\begin{bmatrix} \dot{U}_S \\ 0 \end{bmatrix} = \begin{bmatrix} R_S + jX_S & -j\omega M \\ -j\omega M & R_D + R_W + jX_D \end{bmatrix} \cdot \begin{bmatrix} \dot{I}_S \\ \dot{I}_D \end{bmatrix} \tag{3}$$

where X_S and X_D are equivalent reactances of the transmitter and the receiver respectively,

$$X_S = \omega L_S - \frac{1}{\omega C_S}, \quad X_D = \omega L_D - \frac{1}{\omega C_D} \tag{4}$$

The resonant frequency of MCR WPT system is defined as

978-1-4673-9551-9/16 $31.00 © 2016 IEEE

$$f_r = 1 \Big/ \left(2\pi \sqrt{L_r C_r} \right) \tag{5}$$

When the transmitting coils and receiving coils are both designed to resonate at the operating frequency, we can get $X_S = X_D = 0$. Then, the output power of the system can be solved,

$$P_O = \frac{U_S^{\,2} \cdot \left(\omega M \right)^2 R_W}{\left[R_S \left(R_D + R_W \right) + \left(\omega M \right)^2 \right]^2} \tag{6}$$

Assuming that the radiation loss and power losses are negligible, the transmission efficiency of the system can be expressed as

$$\eta = \frac{\left(\omega M \right)^2 R_W}{R_S \left(R_D + R_W \right)^2 + \left(\omega M \right)^2 \left(R_D + R_W \right)} \tag{7}$$

Equations (6) and (7) show that P_O and η are both the function of the mutual inductance, which not only depends on coil parameters but also is significantly affected by the spatial misalignments of coils.

C. Mutual Inductance with Varying Spatial Scales

The varying spatial scales of transmitting and receiving coils will alter the mutual inductance, further affect the output power. To describe the relationship among the output power, spatial misalignments and coil parameters directly, it is important to obtain the mutual inductance formula in arbitrary spatial scales. This can be realized by solving the double integral in Neumann's formula[16],

$$M = \frac{\mu_0 N_S N_D}{4\pi} \oint \oint \frac{\mathbf{dl_S} \cdot \mathbf{dl_D}}{r_{SD}} \tag{8}$$

where N_S, N_D, $\mathbf{l_S}$, $\mathbf{l_D}$, $\mathbf{dl_S}$ and $\mathbf{dl_D}$ are respectively the coil turns, the length of each turn and the infinitesimal of $\mathbf{l}$ of the transmitting coils and receiving coils. r_{SD} is the distance between $\mathbf{dl_S}$ and $\mathbf{dl_D}$. μ_0 is the magnetic permeability of vacuum. Fig. 4 gives the coil configuration with general misalignments where both the lateral displacement and angular tilt are included. For clarity, we only show the case of one-turn coil in the diagram.

The line elements $\mathbf{dl_S}$ and $\mathbf{dl_D}$, and the distance r_{SD} can be computed by geometrical methods,

$$\mathbf{dl_S} = r_S(-\sin\theta \mathbf{x} + \cos\theta \mathbf{y})d\theta \tag{9}$$

$$\mathbf{dl_D} = r_D(-\sin\phi\cos\alpha \mathbf{x} + \cos\phi \mathbf{y} + \sin\phi\sin\alpha \mathbf{z})d\phi \tag{10}$$

$$
\begin{aligned}
r_{SD} = \big[& r_S^{\,2} + r_D^{\,2} + d^2 + \Delta^2 + 2\Delta r_D \cos\phi\cos\alpha - 2\Delta r_S \cos\theta \\
& -2r_S r_D(\cos\theta\cos\phi\cos\alpha + \sin\theta\sin\phi) - 2r_D d\cos\phi\sin\alpha \big]^{\frac{1}{2}}
\end{aligned}
\tag{11}
$$

where r_S and r_D are coil radiuses. Substituting (9)-(11) into (8) yields,

$$M = \frac{\mu_0 N_S N_D r_S r_D}{4\pi} \oint d\phi \oint \frac{\sin\theta\sin\phi\cos\alpha + \cos\theta\cos\phi}{r_{SD}} d\theta \tag{12}$$

Further substituting (12) into (6) and (7), the relationship between the transmission performance and the spatial misalignments can be revealed apparently. As a result, the

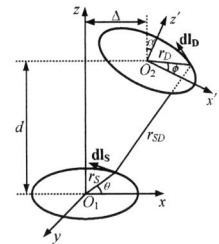

Fig.4 Coil configuration with general misalignments

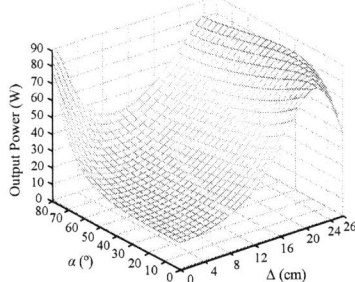

Fig.5 A 3-D view of output power with general misalignments

output power as the function of lateral and angular misalignment is plotted in Fig. 5. The specifications are listed as follows: V_{in}=24V, f_s=200kHz, R_L=10Ω, r_S=r_D=11cm, N_S=N_D=14, d=24cm.

III. PARAMETER OPTIMIZATION OF COILS

As indicated above, the transmission performance of MCR WPT system is sensitive to coil spatial misalignments. From Fig. 5, it can be observed that output power varies in an unpredictable manner when receiving coils are laterally or angularly misaligned from transmitting coils. So it is desirable to optimize the system to improve transmission stability. This paper focused on the optimal design of coil parameters by using a normalization method. According to different specifications, it can be classified into three cases: 1) both the transmitting coils and receiving coils are flexible to change; 2) when the dimension of transmitting coils is mechanically fixed, only receiving coils are allowed to be optimized; 3) when the redesign of receiving coils is forbidden, only transmitting coils can be modified. As analysis processes of case 2 and case 3 are similar, for simplicity, we just take case 1 and case 2 as examples for discussion.

A. Unified Optimization of Two Coils

Assuming the transmitting coils and receiving coils are identical helical coils employing the same way of winding with same parameters: $N_S = N_D = N$, $r_S = r_D = r$, $R_S = R_D = R$, then (12) can be simplified as

$$M_1 = \frac{\mu_0 N^2 r^2}{4\pi} \oint d\phi \oint \frac{\sin\theta\sin\phi\cos\alpha + \cos\theta\cos\phi}{r_{SD}} d\theta \tag{13}$$

Based on (13), there are two variables (coil radius and coil turns) can be optimized. For specific applications, coil radius is always dimensionally determined, so we preset it as a constant and normalize the mutual inductance into a function only

related to coil turns,

$$M_1 = x_1(\Delta, \alpha) \cdot N^2 \qquad (14)$$

where $x_1(\Delta, \alpha)$ is a variable coefficient depending on the lateral misalignment Δ and angular misalignment α.

Fig. 6 illustrates the mutual inductance curves against the square of coil turns with various spatial misalignments. From Fig. 6, it can be seen that the mutual inductance almost increases linearly following the growth of the square of coil turns. The slope remains almost unchanged within the misalignments range of 0~10cm and 0~50°. The average value of the slope (0~10cm, 0~50°) is calculated to be 1.20×10^{-8}. We use this calculated result to replace x_1(0~10cm, 0~50°) in further analysis. Here, the specifications are: r=11cm, d=24cm.

From (6), it is known that P_O is also related to the parasitic resistances of coils, which are mainly composed of two parts: the dc component and the ac component,

$$R = R_{dc} + R_{ac} = \left(\frac{2}{\sigma n r_a^2} + \frac{1}{n r_a}\sqrt{\frac{\omega \mu_0}{2\sigma}} \right) \cdot N \cdot r \qquad (15)$$

where σ, n and r_a are respectively the conductivity of coil material, the number of strands and single-strand radius of the wire. ω is the operating angular frequency. As σ, n, r_a and ω are a set of given parameters, following the same normalization method above, (15) can be expressed as

$$R = x_2 N \qquad (16)$$

Substituting (14) and (16) into (6) yields,

$$P_O = \frac{1}{\left(a_1 N^2 + a_2 + a_3/N \right)^2} \qquad (17)$$

where

$$a_1 = \frac{\omega}{U_s \sqrt{R_W}} \cdot x_1(\Delta, \alpha) \qquad (18)$$

$$a_2 = \frac{1}{U_s \omega \sqrt{R_W}} \cdot \frac{x_2^2}{x_1(\Delta, \alpha)} \qquad (19)$$

$$a_3 = \frac{\sqrt{R_W}}{U_s \omega} \cdot \frac{x_2}{x_1(\Delta, \alpha)} \qquad (20)$$

By differentiating P_O with respect to N and equating the differential function to zero, the optimum value of N for maximum output power can be obtained as

$$N_{OPT} = \sqrt[3]{\frac{R_W}{2\omega^2} \cdot \frac{x_2}{x_1(\Delta, \alpha)^2}} \qquad (21)$$

When (21) is satisfied, the maximum output power can easily be solved,

$$P_{O\_MAX} = \frac{U_s^2}{\left(\sqrt[6]{\frac{R_W}{16\omega^2} \cdot \frac{x_2^4}{x_1(\Delta, \alpha)^2}} + \frac{1}{\omega\sqrt{R_W}} \cdot \frac{x_2^2}{x_1(\Delta, \alpha)} + \sqrt[6]{\frac{4R_W}{\omega^2} \cdot \frac{x_2^4}{x_1(\Delta, \alpha)^2}} \right)^2} \qquad (22)$$

Substituting the value of x_1(0~10cm, 0~50°) and other defined parameters into (21) and (22), we can get: $N_{OPT}\approx7$,

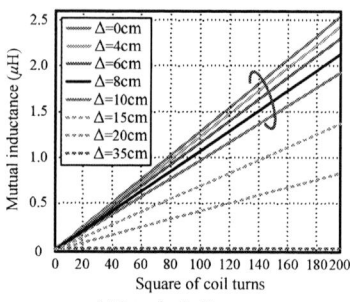

(a) lateral misalignment

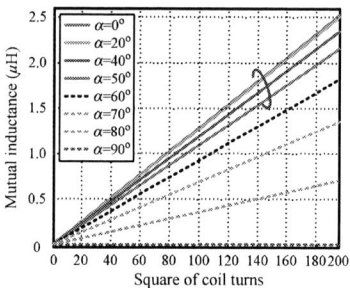

(b) angular misalignment

Fig.6 Mutual inductance with spatial misalignments in case 1

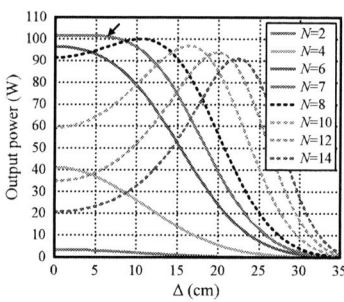

(a) lateral misalignment

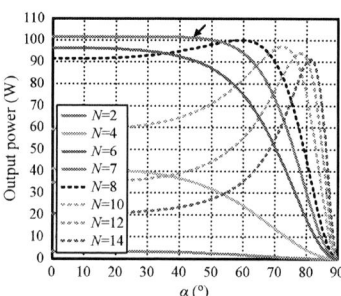

(b) angular misalignment

Fig.7 Output power with spatial misalignments in case 1

$P_{O\_MAX}\approx101.53$W. To verify the calculation results, Fig. 7 depicts the output power curves against spatial misalignments under different coil turns according to (6). By comparison, it is observed that the output power reaches maximum value and remains almost constant within the misalignments range of 0~10cm and 0~50° when $N=N_{OPT}$.

B. Optimization of Receiving Coils

Sometimes, the dimension of transmitting coils is mechanically fixed which means neither the radius r_S nor coil turns N_S of transmitter can be changed. To realize perfect transmission performance, modifying receiving coils must be a terrific option. There are also two variables (the radius r_D and coil turns N_D of receiver) alternative for optimization. Similarly, presetting the radius of receiving coils r_D and normalizing the mutual inductance into a function only related to coil turns N_D, we can get

$$M_2 = y_1(\Delta, \alpha) \cdot N_D \qquad (23)$$

where $y_1(\Delta, \alpha)$ is a variable coefficient depending on the lateral misalignment Δ and angular misalignment α.

Fig. 8 illustrates the mutual inductance curves against the coil turns of receiver. From Fig. 8, it can be concluded that the mutual inductance almost increases linearly following the growth of the coil turns of receiver. The slope remains almost constant within the misalignments range of 0~10cm and 0~50° and its average value is calculated to be 1.70×10^{-8}. We use this calculated value to replace $y_1(0\text{~}10\text{cm}, 0\text{~}50°)$. Here, the specifications are: $r_S = r_D = 11\text{cm}$, $N_S = 14$, $d = 24\text{cm}$.

The parasitic resistance of receiving coils is

$$R_D = R_{D\_dc} + R_{D\_ac} = \left(\frac{2}{\sigma n r_a^2} + \frac{1}{n r_a} \sqrt{\frac{\omega \mu_0}{2\sigma}} \right) \cdot N_D \cdot r_D \qquad (24)$$

Normalizing R_D into a function only related to coil turns N_D, then (24) can be expressed as

$$R_D = y_2 N_D \qquad (25)$$

Substituting (23) and (25) into (6) yields,

$$P_O = \frac{1}{\left(b_1 N_D + b_2 + b_3 / N_D \right)^2} \qquad (26)$$

where

$$b_1 = \frac{\omega}{U_s \sqrt{R_W}} \cdot y_1(\Delta, \alpha) \qquad (27)$$

$$b_2 = \frac{R_S}{U_s \omega \sqrt{R_W}} \cdot \frac{y_2}{y_1(\Delta, \alpha)} \qquad (28)$$

$$b_3 = \frac{\sqrt{R_W} R_S}{U_s \omega} \cdot \frac{1}{y_1(\Delta, \alpha)} \qquad (29)$$

By differentiating P_O with respect to N_D and equating the differential function to zero, the optimum value of N_D for maximum output power can be obtained as

$$N_{D\_OPT} = \frac{\sqrt{R_S R_W}}{\omega} \cdot \frac{1}{y_1(\Delta, \alpha)} \qquad (30)$$

The maximum output power is

$$P_{O\_MAX} = \frac{U_S^2}{\left(2\sqrt{R_S} + \frac{R_S}{\omega \sqrt{R_W}} \cdot \frac{y_2}{y_1(\Delta, \alpha)} \right)^2} \qquad (31)$$

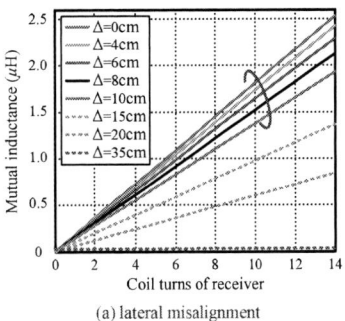

(a) lateral misalignment

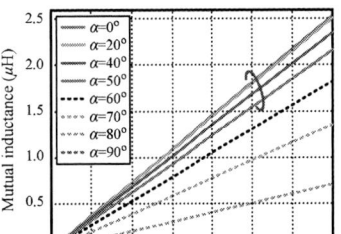

(b) angular misalignment

Fig.8 Mutual inductance with spatial misalignments in case 2

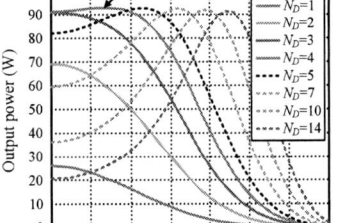

(a) lateral misalignment

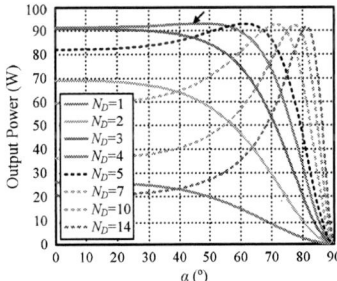

(b) angular misalignment

Fig.9 Output power with spatial misalignments in case 2

Substituting the value of $y_1(0\text{~}10\text{cm}, 0\text{~}50°)$ and other defined parameters into (30) and (31), we can get: $N_{D\_OPT} \approx 4$, $P_{O\_MAX} \approx 92.44\text{W}$. Likewise, Fig. 9 depicts the output power curves against the spatial misalignments under different coil turns according to (6). By comparison, it is observed that the output power reaches maximum value and remains almost constant within the misalignments range of 0~10cm and 0~50° when $N_D=N_{D\_OPT}$.

IV. EXPERIMENTAL VERIFICATIONS

In order to validate the theoretical analysis, a prototype had been built and tested in the laboratory, as shown in Fig. 10. The main specifications and coil parameters before and after optimization are listed in Table I.

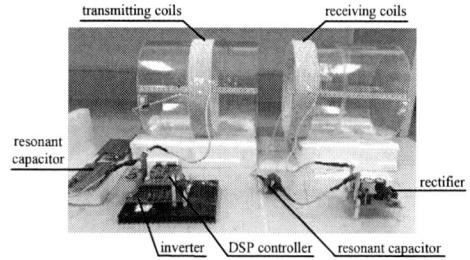

Fig.10 Prototype of MCR WPT system

TABLE I. PARAMETERS OF THE PROTOTYPE

Parameter		Symbol	Value		
			Case 1	Case 2	Case 3
Transmitter	Radius	r_S	11cm	11cm	11cm
	Turns	N_S	7	14	4
Receiver	Radius	r_D	11cm	11cm	11cm
	Turns	N_D	7	4	14
Input voltage		V_{in}	24V		
Operating frequency		f_s	200kHz		
Transmission distance		d	24cm		
Load resistance		R_L	10Ω		

Experimental measured data in three optimization cases are plotted in Fig.11-13 by solid lines, along with the theoretical curves, given by (6) in dashed lines. Compared with the measured data, it is observed that experimental results are well in agreement with the theoretical analysis. Besides, as seen in Fig.11-13, after optimization, maximum and nearly stable output power can be achieved within the misalignments range of 0~10cm and 0~50°, proving the normalization method valid for the optimal design of coils.

Fig. 14 shows the key experimental waveforms with both lateral and angular misalignment when $\Delta=10$cm and $\alpha=30°$. As shown in Fig. 14, before the switch turns on, the drain current i_D is negative, which indicates that the parasitic diode of the switch has conducted. In the meantime, the drain-to-source voltage v_{ds} is clamped to zero. So ZVS can be realized. Besides, resonant currents are good sinusoidal waveforms and the phase difference between them is about 90°. Obviously, the experimental results are well in agreement with the theoretical analysis, verifying the accuracy of the analytical model.

V. CONCLUSIONS

This paper focused on the optimal design of coil parameters to realize maximum and stable power transfer in a specific misalignments range. To facilitate analysis of the transmission

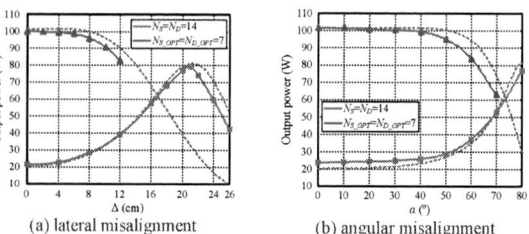

(a) lateral misalignment   (b) angular misalignment

Fig. 11 Measured output power with spatial misalignments in case 1

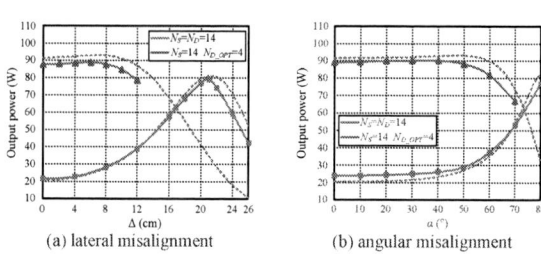

(a) lateral misalignment   (b) angular misalignment

Fig. 12 Measured output power with spatial misalignments in case 2

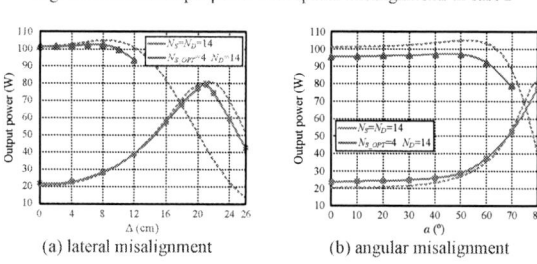

(a) lateral misalignment   (b) angular misalignment

Fig. 13 Measured output power with spatial misalignments in case 3

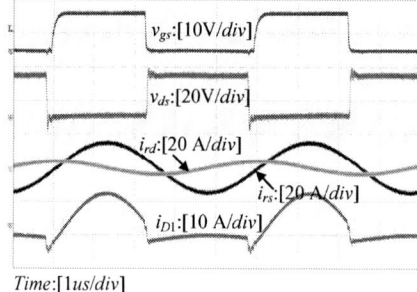

Time:[1*us/div*]

Fig. 14 Experimental waveforms when $\Delta=10$cm and $\alpha=30°$

performance in arbitrary spatial scales, an equivalent model of MCR WPT system and a normalization method were introduced. By normalizing the mutual inductance formula and solving the output power differential equation, steady operation range and optimum value of coil turns was easily obtained. Research results show that after the optimal design of coil parameters, maximum and nearly stable output power can be achieved within the misalignments range of 0~10cm and 0~50°. Beyond this range, the output power will decrease sharply with the increasing lateral or angular misalignment. Finally, a prototype had been built and tested to validate the theoretical analysis. Experimental results well prove the effectiveness of the proposed method.

ACKNOWLEDGMENT

This work was financially supported by the National Natural Science Foundation of China(51505223), the Natural Science Foundation of Jiangsu Province, China(BK20151471), the Lite-on Research Program, and Jiangsu province university outstanding science and technology innovation team project.

REFERENCES

[1] C. K. Lee, W. X. Zhang, and S. Y. R. Hui, "Recent progress in mid-range wireless power transfer," IEEE Energy Conversion Congress and Exposition (ECCE), 2012, pp. 3819-3824.

[2] N. Shinohara, "Wireless power transmission progress for electric vehicle in Japan," IEEE Radio and Wireless Symposium (RWS), 2013, pp. 109-111.

[3] S. Y. R. Hui and W. W. C. Ho, "A new generation of universal contactless battery charging platform for portable consumer electronic equipment," IEEE Transactions on Power Electronics, vol. 20, no. 3, pp. 620-627, May 2005.

[4] H. G. Lim, Y. H. Yoon, C. W. Lee, I. Y. Park, B. S. Song, and J. H. Cho, "Implementation of a transcutaneous charger for fully implantable middle ear hearing device," IEEE Conference of Engineering in Medicine and Biology Society, 2005, pp. 6813-6816.

[5] T. Hosotani and I. Awai, Cho, "A novel analysis of ZVS wireless power transfer system using coupled resonators," IEEE Microwave Workshop Series (IMWS), 2012, pp. 235-238.

[6] M. F. Fu, C. B. Ma , and X. E. Zhu, "A cascaded boost-buck converter for high-efficiency wireless power transfer systems," IEEE Transactions on Industrial Informatics, vol. 10, no. 3, pp. 1972-1980, August 2014.

[7] S. Y. R. Hui, W. X. Zhong, and C. K. Lee, "A critical review on recent progress in mid-range wireless power transfer," IEEE Transactions on Power Electronics, vol. 29, no. 9, pp. 4500-4511, September 2014.

[8] M.G. Egan, D.L. O'Sullivan, J.G. Hayes, M.J. Willers, and C.P. Henze, "Power-factor-corrected single-stage inductive charger for electric vehicle batteries," IEEE Transactions on Industrial Electronics, vol. 54, no. 2, pp. 1217-1226, April 2007.

[9] A. Kurs, A. Karalis, R. Moffatt, J. D. Joannopoulos, P. Fisher, and M. Soljacic, "Wireless power transfer via strongly coupled magnetic resonances," Science, vol. 317, no. 5834, pp. 83-86, July 2007.

[10] X. Zhang and J. Chae, "Working distance comparison of inductive and electromagnetic couplings for wireless and passive underwater monitoring system of rinsing process in semiconductor facilities," IEEE Sensors Journal, vol. 11, no. 11, pp. 2932-2939, May 2011.

[11] L. Gao, W. S. Hu, X. W. Xie, Q. J. Deng, Z. D. Wu, H. Zhou, and Y. Jiang, "Optimum design of coil for wireless energy transmission system based on resonant coupling," IEEE International Conference on Control and Automation (ICCA), 2013, pp. 190-195.

[12] H. C. Li, K. P. Wang, L. Huang, J. Li, and X. Yang, "Coil structure optimization method for improving coupling coefficient of wireless power transfer," IEEE Applied Power Electronics Conference (APEC), 2015, pp. 2518-2521.

[13] J. H. Wang, J. G. Li, S. L. Ho, W. N. Fu, Y. Li, H. L. Yu, and M. G. Sun, "Analytical design study of a novel Witricity charger with lateral and angular misalignments for efficient wireless energy transmission," IEEE Transactions on Magnetics, vol. 47, no. 10, pp. 2616-2619, October 2011.

[14] Z. G. Dang, and J. A. A. Qahouq, "Modeling and investigation of magnetic resonance coupled wireless power transfer system with lateral misalignment," IEEE Applied Power Electronics Conference (APEC), 2014, pp. 1317-1322.

[15] H.C. Li, J. Li, K.P. Wang, W.J. Chen, and X. Yang, "A maximum efficiency point tracking control scheme for wireless power transfer systems using magnetic resonant coupling," IEEE Transactions on Power Electronics, vol. 30, no. 7, pp. 3998-4008, August 2014.

[16] S. Ramo, J. R. Whinnery, and T. Van Duzer, "Fields and Waves in Communication Electronics," New York: Wiley, 1965.

Online Regulation of Receiver-side Power and Estimation of Mutual Inductance in Wireless Inductive Link Based on Transmitter-Side Electrical Information

Jeff Po Wa CHOW*
*Centre for Smart Energy Conversion and Utilization Research, City University of Hong Kong, Hong Kong
Email: pwchow2-c@my.cityu.edu.hk

Henry Shu Hung CHUNG*‡
‡ Research Institute of Electronic Automation, Shanghai Maritime University, Shanghai, China
Email: eeshc@cityu.edu.hk

Chun Sing CHENG*
Email: cscheng9-c@my.cityu.edu.hk

Abstract — **It is well-known that the power transfer efficiency and the power transmitted over a wireless inductive link are significantly affected by the strength of the magnetic coupling and the spatial displacement between the transmitting and receiving coils. Misalignment between the transmitting and receiving coils is practically unavoidable. In order to control and regulate the receiver-side power, on-the-spot measurement of electrical quantities and establishment of communication link between the transmitter and receiver are typically required. This paper will present an investigation into the use of the transmitter-side electrical information to estimate the mutual inductance and regulate the power consumption of the receiver side. The nonlinear characteristics of the diode-bridge rectifier are taken into account in the mathematical formulations. The proposed technique is successfully implemented on a 4W wireless-powered LED driver prototype. Experimental results reveal that the LED power can be regulated within ±25% spatial misalignment over the operating zone. The estimated mutual inductance is also found to be in close agreement with the theoretical predictions.**

Keywords—Wireless inductive link, wireless power transfer, mutual inductance, coil misalignments, load detection.

I. INTRODUCTION

As illustrated in Fig. 1, a wireless inductive link consists of an excitation source, input matching circuit, transmitting and receiving coils, output matching and rectifying circuit and load (end-use device). The centers of the transmitting and receiving coils are typically placed in coaxially aligned position and the electric energy is delivered from the transmitting coil to the receiving coil through alternating magnetic fields. The key characteristic of the wireless inductive link is that the transmitter and receiver have no physical contact but are coupled by magnetic field. In principle, the receiver has spatial freedom to be displayed freely from the transmitter. However, the power transfer efficiency and the transmitted power are significantly affected by the strength of the magnetic coupling

and the spatial displacement between the coils [1]. In practice, inevitable spatial misalignment between the two coils leads to the variation of the strength of the magnetic coupling. In order to regulate the receiver-side power, on-the-spot measurement of the electrical quantities and establishment of communication link between the transmitter and receiver are sometimes applied [2]-[6]. The load modulation scheme, as depicted in [7]-[9], is often applied to perform both power and signal transfer in the wireless inductive link.

Many algorithms [10]-[11] have been proposed to estimate the mutual inductance online and regulate the load side power based on the transmitter-side information. Therefore, no output measurement and controller is embedded into the receiver-side and hence, the circuit complexity in the receiver is simplified. However, the load is typically assumed to be linear.

This paper gives another perspective of determining the mutual inductance between the two coils, the receiver-side electrical information and the load power, obtained by sensing the transmitter-side input voltage and input current. The nonlinear load characteristic of the rectifying circuit on the receiver is taken into consideration in the time-domain analysis.

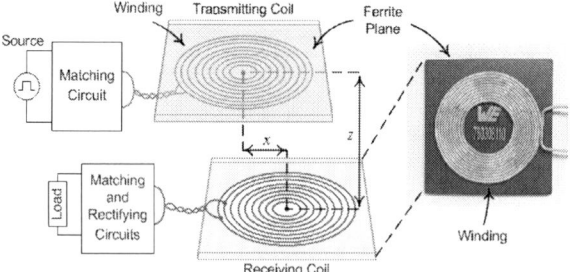

Fig. 1 Configuration of Wireless Inductive Link.

978-1-4673-9551-9/16 $31.00 © 2016 IEEE

II. MATHEMATICAL MODEL

A. Modeling and Parameter Identifications

Fig. 1 shows the simplified structure of a wireless inductive link system. The windings are placed on the ferrite plates. The equivalent circuit of the entire system is shown in Fig. 2, in which a transformer model is used to represent the magnetic coupling between the transmitting and receiving coils. The excitation source v_{in}, which is typically the output of a switching network, consists of square pulses. Assume that the mark-space ratio of v_{in} equals unity. i_1 and i_2 are the input and output currents, respectively. The voltage v_o is rectified by a diode bridge circuit. V_L and I_L are the average load voltage and load current.

The load-side diode-capacitor rectifying circuit leads to nonlinear behaviors on the wireless inductive link [12]-[14]. For example, the current i_2 is non-sinusoidal. Such property will give challenges to methods for linear systems, such as the frequency-domain analysis [10]. Instead, a time-domain analysis technique that takes such nonlinear behaviors into account is discussed. The system in Fig. 2 can be described by the following equations,

$$v_{in}(\omega t)-v_{L1}(\omega t)-v_{C1}(\omega t)-i_1(\omega t)\,r_1+\omega M\frac{d\,i_2(\omega t)}{d\,\omega t}=0 \quad (1)$$

$$v_o(\omega t)+v_{L2}(\omega t)+v_{C2}(\omega t)+i_2(\omega t)\,r_2-\omega M\frac{d\,i_1(\omega t)}{d\,\omega t}=0 \quad (2)$$

$$i_1(\omega t)=\omega\,C_1\frac{d\,v_{C1}(\omega t)}{d\,\omega t} \quad (3)$$

$$i_2(\omega t)=\omega\,C_2\frac{d\,v_{C2}(\omega t)}{d\,\omega t} \quad (4)$$

$$v_{L1}(\omega t)=\omega^2\,L_1\,C_1\frac{d^2\,v_{C1}(\omega t)}{d\,\omega t^2} \quad (5)$$

$$v_{L2}(\omega t)=\omega^2\,L_2\,C_2\frac{d^2\,v_{C2}(\omega t)}{d\,\omega t^2} \quad (6)$$

B. Time-Domain Equations

Fig. 4 shows the simulated waveforms (in red color) of v_{in}, i_1, v_{C1}, i_2, v_{C2}, v_o under the aligned position with $x = 0$mm and $z = 4$mm. The forward voltage drop of the diodes is assumed to be zero. The amplitude of v_{in} is V_{in}. The V_{C20} and V_L are the amplitudes of v_{C2} and v_o, respectively. The angle α in Fig. 4 is determined by the phase difference between the fundamental components of i_1 and v_{in}. Mathematically,

$$\alpha=\frac{\pi}{2}+(\angle i_{1,fund}-\angle v_{in,fund}) \quad (7)$$

As the wireless inductive link is excited at the resonant frequency f, i_1 is dominated by the fundamental component,

which is depicted in green color in Fig. 4(b). The time-domain equations of i_1, i_2, v_{L1}, v_{L2} and v_{C1} as labeled in Fig. 2 can be derived by solving v_{C2} at different time intervals. A non-homogenous differential equation for v_{C2} can be formulated as

$$\frac{d^4v_{C2}}{d\,\omega t^4}+a_3\frac{d^3v_{C2}}{d\omega t^3}+a_2\frac{d^2v_{C2}}{d\omega t^2}+a_1\frac{dv_{C2}}{d\omega t}+a_0 v_{C2}=-a_0 v_o \quad (8)$$

The coefficients a_3, a_2, a_1, and a_0 are given in Appendix A.

As all waveforms in Fig. 4 are odd functions and have rotational symmetry with respect to $\omega t = \pi$, the time-domain analysis is conducted for $\omega t \in [0,\ \pi]$. Two time-domain functions are defined for $0\le\omega t<\alpha$ and $\alpha\le\omega t<\pi$, respectively. Thus, the solutions of (8) can be expressed as

$$v_{C2}=\begin{cases} K_1e^{\lambda_1\omega t}+K_2e^{\lambda_2\omega t}+K_3e^{\lambda_3\omega t}+K_4e^{\lambda_4\omega t}-v_o, & \text{if } 0\le\omega t<\alpha \\[6pt] K_5e^{\lambda_1\omega t}+K_6e^{\lambda_2\omega t}+K_7e^{\lambda_3\omega t}+K_8e^{\lambda_4\omega t}-v_o, & \text{if } \alpha\le\omega t<\pi \end{cases}$$
$$(9)$$

By using (1)-(6) and (9),

$$i_1=\begin{cases} \begin{aligned} &\frac{1}{2\omega M}[K_1\frac{G_{L1}}{\lambda_1}(2e^{\lambda_1\omega t}-e^{\lambda_1\alpha}-1)+K_2\frac{G_{L2}}{\lambda_2}(2e^{\lambda_2\omega t}-e^{\lambda_2\alpha}-1)\\ &+K_3\frac{G_{L3}}{\lambda_3}(2e^{\lambda_3\omega t}-e^{\lambda_3\alpha}-1)+K_4\frac{G_{L4}}{\lambda_4}(2e^{\lambda_4\omega t}-e^{\lambda_4\alpha}-1)\\ &+K_5\frac{G_{L1}}{\lambda_1}(e^{\lambda_1\alpha}-e^{\lambda_1\pi})+K_6\frac{G_{L2}}{\lambda_2}(e^{\lambda_2\alpha}-e^{\lambda_2\pi})\\ &+K_7\frac{G_{L3}}{\lambda_3}(e^{\lambda_3\alpha}-e^{\lambda_3\pi})+K_8\frac{G_{L4}}{\lambda_4}(e^{\lambda_4\alpha}-e^{\lambda_4\pi})]\quad\text{if }0\le\omega t<\alpha \end{aligned}\\[16pt] \begin{aligned} &\frac{1}{2\omega M}[K_1\frac{G_{L1}}{\lambda_1}(1-e^{\lambda_1\alpha})+K_2\frac{G_{L2}}{\lambda_2}(1-e^{\lambda_2\alpha})\\ &+K_3\frac{G_{L3}}{\lambda_3}(1-e^{\lambda_3\alpha})+K_4\frac{G_{L4}}{\lambda_4}(1-e^{\lambda_4\alpha})\\ &+K_5\frac{G_{L1}}{\lambda_1}(2e^{\lambda_1\omega t}+e^{\lambda_1\pi}+e^{\lambda_1\alpha})+K_6\frac{G_{L2}}{\lambda_2}(2e^{\lambda_2\omega t}+e^{\lambda_2\pi}+e^{\lambda_2\alpha})\\ &+K_7\frac{G_{L3}}{\lambda_3}(2e^{\lambda_3\omega t}+e^{\lambda_3\pi}+e^{\lambda_3\alpha})+K_8\frac{G_{L4}}{\lambda_4}(2e^{\lambda_4\omega t}+e^{\lambda_4\pi}+e^{\lambda_4\alpha})]\\ &\hspace{5cm}\text{if }\alpha\le\omega t<\pi \end{aligned} \end{cases}$$
$$(10)$$

where K_1 to K_8 are all constants, λ_1 to λ_4 are the roots of the general solutions of (8) and the parameters G_{L1} to G_{L4} are given in Appendix B.

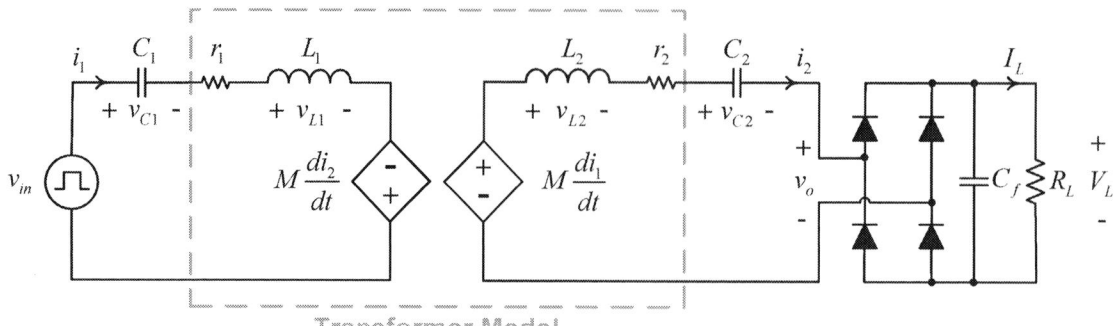

Fig. 2. Equivalent Circuit of the Wireless Inductive Link.

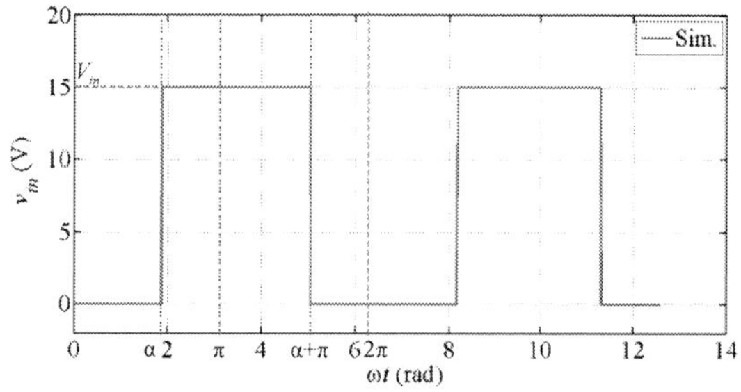

(a) v_{in} (Simulated Waveform).

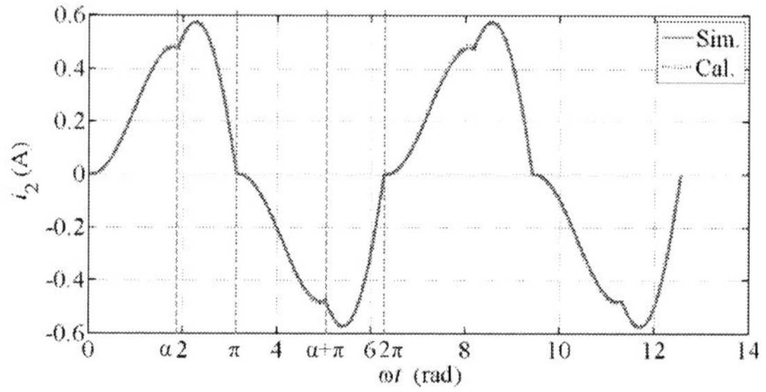

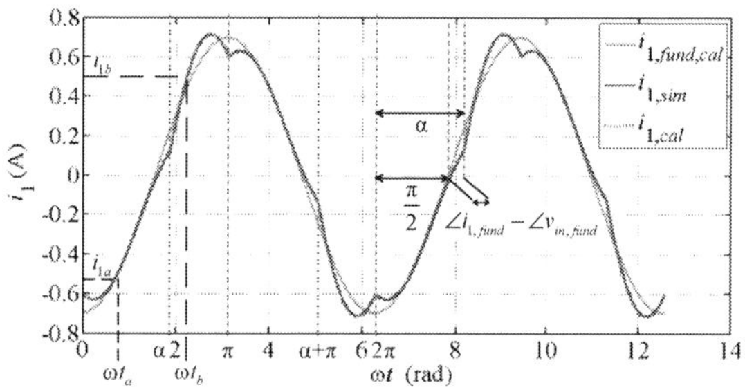

(b) i_1.

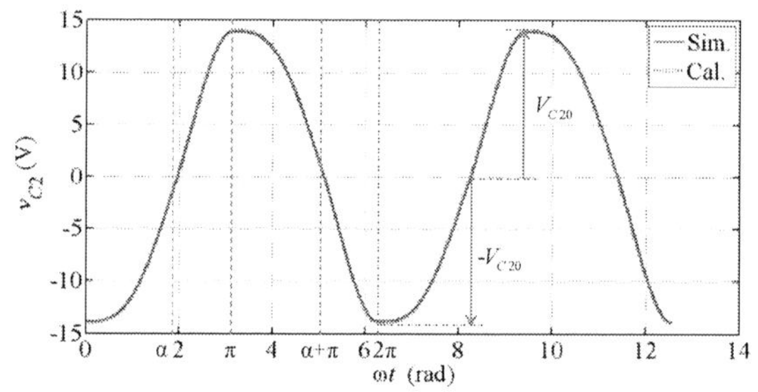

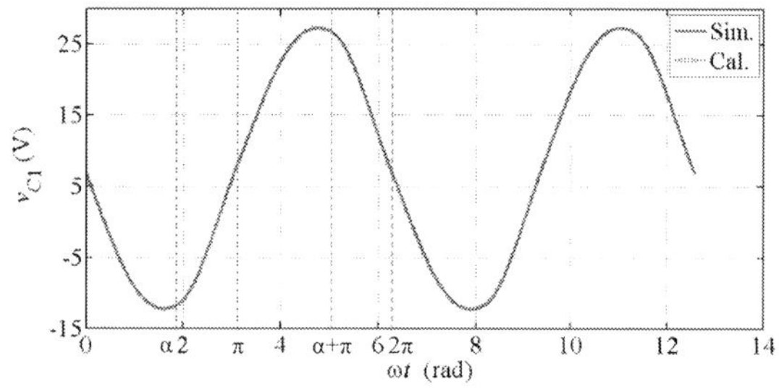

(c) v_{C1}.

(d) i_2.

(e) v_{C2}.

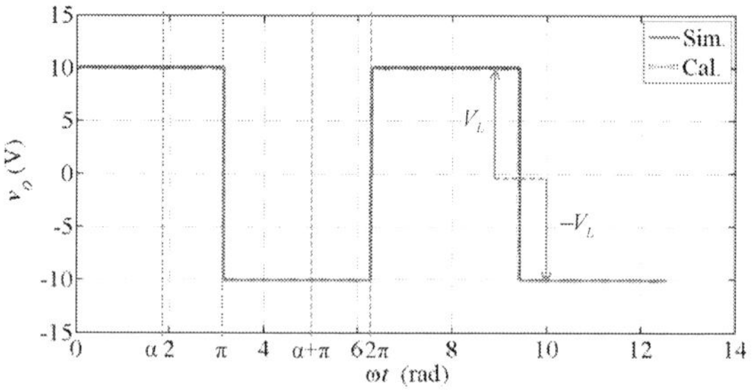

(f) v_o.

Fig. 4 Ideal waveforms at the aligned position ($x = 0$mm).

978-1-4673-9551-9/16 $31.00 © 2016 IEEE   1797

C. Average Load Voltage and Current

The average load voltage V_L, average load current I_L, and average load power P_L are determined. Based on the waveforms given in Fig. 4, the initial conditions of v_{in}, i_1, v_{C1}, i_2, v_{C2}, v_o, and $v_{L1} - v_{in} - \omega M \dfrac{di_2}{d\,\omega t}$ are tabulated. The initial conditions are substituted into (1)-(10) to give (11)-(20) below. By solving (11)-(20) for $K_1 \sim K_8$, V_{C20}, and V_L, I_L and P_L can be shown to be

$$I_L = \frac{2\,\omega\,C_2}{\pi}\,V_{C20} \qquad (21)$$

$$P_L = V_L\,I_L \qquad (22)$$

III. PROCEDURE OF ESTIMATING MUTUAL INDUCTANCE AND LOAD POWER

A method of using the electrical information on the transmitter side, including v_{in} and i_1, to estimate the mutual inductance M and load power is discussed. M varies between the minimum value M_{min} and the maximum value M_{max}. With the help of Fig. 5, an estimation procedure is described as follows,

1. Measure the values of L_1, L_2, r_1, r_2, C_1, and C_2.
2. Sample v_{in} and i_1 over one cycle.
3. Apply Goertzel algorithm to calculate the fundamental components of v_{in} and i_1 to estimate α with (7).
4. Choose the initial values for $\omega t_a \in [0, \alpha]$ and $\omega t_b \in [\alpha, \pi]$, as illustrated in Fig. 4(b).
5. Extract $i_{1a} = i_1(\omega t_a)$ and $i_{1b} = i_1(\omega t_b)$ from the measured waveform of i_1 conducted in Step 2).
6. Let the value of M be M_{min}.
7. Calculate $K_1 \sim K_8$, V_{C20}, and V_L by solving (11)-(20).
8. Estimate the input current $i_{1,cal}(\omega t)$ with (10) over one half cycle for $\omega t \in [0, \pi]$.
9. Calculate the mean squared error (MSE) between the actual input current i_1 measured in Step 2) and $i_{1,cal}$ for a half cycle.
10. If MSE < MSE$_{min}$, MSE$_{min}$ is updated with the new MSE and $V_{C20,cal}$, $V_{L,cal}$, and M_{cal} are equal to V_{C20}, V_L, and M calculated in Step 7), respectively.
11. Increase the value M by ΔM and repeat the procedure from Step 7) until $M \geq M_{max}$.
12. When $M \geq M_{max}$, V_{C20}, V_L, and M are equal to $V_{C20,cal}$, $V_{L,cal}$, and M_{cal}, respectively, and calculate average load current I_L and load power P_L by (21) and (22), respectively.

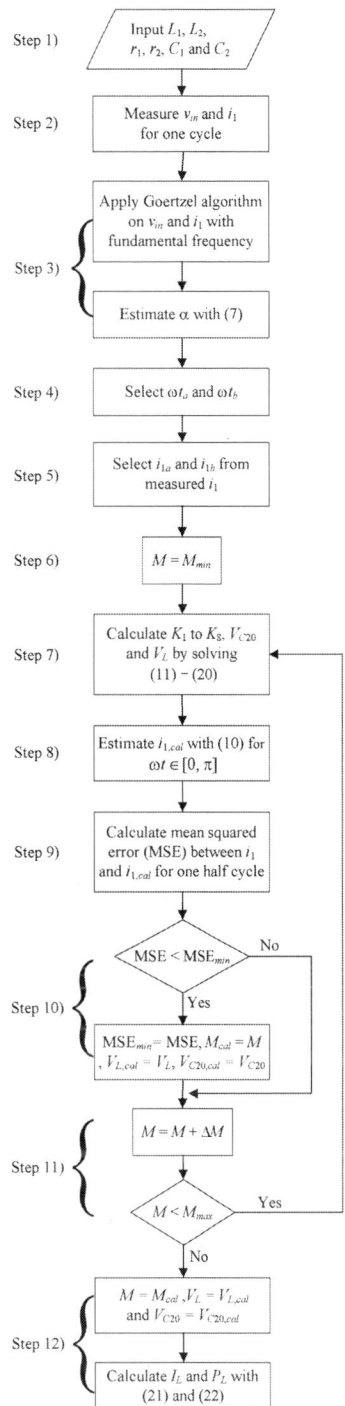

Fig. 5 Flowchart of the Proposed Estimation Algorithm.

$$K_1\lambda_1 + K_2\lambda_2 + K_3\lambda_3 + K_4\lambda_4 = 0 \tag{11}$$

$$K_1 + K_2 + K_3 + K_4 + V_{C20} - V_L = 0 \tag{12}$$

$$K_5 e^{\lambda_1 \pi} + K_6 e^{\lambda_2 \pi} + K_7 e^{\lambda_3 \pi} + K_8 e^{\lambda_4 \pi} - V_{C20} - V_L = 0 \tag{13}$$

$$K_5 \lambda_1 e^{\lambda_1 \pi} + K_6 \lambda_2 e^{\lambda_2 \pi} + K_7 \lambda_3 e^{\lambda_3 \pi} + K_8 \lambda_4 e^{\lambda_4 \pi} = 0 \tag{14}$$

$$(K_1 - K_5)e^{\lambda_1 \alpha} + (K_2 - K_6)e^{\lambda_2 \alpha} + (K_3 - K_7)e^{\lambda_3 \alpha} + (K_4 - K_8)e^{\lambda_4 \alpha} = 0 \tag{15}$$

$$(K_1 - K_5)\lambda_1 e^{\lambda_1 \alpha} + (K_2 - K_6)\lambda_2 e^{\lambda_2 \alpha} + (K_3 - K_7)\lambda_3 e^{\lambda_3 \alpha} + (K_4 - K_8)\lambda_4 e^{\lambda_4 \alpha} = 0 \tag{16}$$

$$K_1 G_{V1} + K_2 G_{V2} + K_3 G_{V3} + K_4 G_{V4} + K_5 G_{V1} e^{\lambda_1 \pi} + K_6 G_{V2} e^{\lambda_2 \pi} + K_7 G_{V3} e^{\lambda_3 \pi} + K_8 G_{V4} e^{\lambda_4 \pi} = 0 \tag{17}$$

$$(K_1 - K_5)G_{V1} e^{\lambda_1 \alpha} + (K_2 - K_6)G_{V2} e^{\lambda_2 \alpha} + (K_3 - K_7)G_{V3} e^{\lambda_3 \alpha} + (K_4 - K_8)G_{V4} e^{\lambda_4 \alpha} = -V_{in} \tag{18}$$

$$\frac{1}{2\omega M}[K_1 \frac{G_{L1}}{\lambda_1}(1 - e^{\lambda_1 \alpha}) + K_2 \frac{G_{L2}}{\lambda_2}(1 - e^{\lambda_2 \alpha}) + K_3 \frac{G_{L3}}{\lambda_3}(1 - e^{\lambda_1 \alpha}) + K_4 \frac{G_{L4}}{\lambda_4}(1 - e^{\lambda_4 \alpha}) + K_5 \frac{G_{L1}}{\lambda_1}(2e^{\lambda_1 \omega t_b} + e^{\lambda_1 \pi} + e^{\lambda_1 \alpha})$$

$$+ K_6 \frac{G_{L2}}{\lambda_2}(2e^{\lambda_2 \omega t_b} + e^{\lambda_2 \pi} + e^{\lambda_2 \alpha}) + K_7 \frac{G_{L3}}{\lambda_3}(2e^{\lambda_3 \omega t_b} + e^{\lambda_3 \pi} + e^{\lambda_3 \alpha}) + K_8 \frac{G_{L4}}{\lambda_4}(2e^{\lambda_4 \omega t_b} + e^{\lambda_4 \pi} + e^{\lambda_4 \alpha})] = i_{1b} \tag{19}$$

$$\frac{1}{2\omega M}[K_1 \frac{G_{L1}}{\lambda_1}(2e^{\lambda_1 \omega t_a} - e^{\lambda_1 \alpha} - 1) + K_2 \frac{G_{L2}}{\lambda_2}(2e^{\lambda_2 \omega t_a} - e^{\lambda_2 \alpha} - 1) + K_3 \frac{G_{L3}}{\lambda_3}(2e^{\lambda_3 \omega t_a} - e^{\lambda_3 \alpha} - 1) + K_4 \frac{G_{L4}}{\lambda_4}(2e^{\lambda_4 \omega t_a} - e^{\lambda_4 \alpha} - 1)$$

$$+ K_5 \frac{G_{L1}}{\lambda_1}(e^{\lambda_1 \alpha} - e^{\lambda_1 \pi}) + K_6 \frac{G_{L2}}{\lambda_2}(e^{\lambda_2 \alpha} - e^{\lambda_2 \pi}) + K_7 \frac{G_{L3}}{\lambda_3}(e^{\lambda_3 \alpha} - e^{\lambda_3 \pi}) + K_8 \frac{G_{L4}}{\lambda_4}(e^{\lambda_4 \alpha} - e^{\lambda_4 \pi})] = i_{1a} \tag{20}$$

where the parameters G_{L1} to G_{L4} and G_{V1} to G_{V4} are given in Appendix B.

IV. EXPERIMENTAL VERIFICATION

In order to study the online regulation of the receiver-side power and estimation of the mutual inductance in the wireless inductive link, a wireless-powered LED driver is built and the schematic is shown in Fig. 7. The driver is composed of two power conversion stages. The first stage is a buck DC-DC converter, which is used to regulate the DC-link voltage V_{DC}. The second stage is a half-bridge series-resonant-parallel-loaded inverter. The switching frequency is 150kHz. The duty cycle of each switch is slightly less than 0.5. A controller board, Red Pitaya V1.0, is used to perform three major functions: 1) Sample the input voltage v_{in} and input current i_1 simultaneously with the sampling rate of 125M samples per second, 2) Estimate the average load power P_L and mutual inductance M by the iterative algorithm, depicted in Sec. III. 3) Update the duty cycle of the switch in the buck converter to regulate the load power P_L.

Fig. 8 shows the experimental test bed. The component values of the wireless inductive link are given in Table I. The receiving coil and the LED are allowed to move freely within the operating zone, which is defined on a Cartesian coordinate system as in Fig. 8(b). The dimension of the operating zone is 73mm x 73mm. The center of the receiving coil is moved within a region of 10mm x 10mm. Thus, the maximum misalignment is around ±14mm (i.e., $\pm 10\sqrt{2}$) between the coils, which is equivalent to approximately ±25% of the coil diameter.

Table I - Component Values and Parameters in Conducting Simulations and Experiments

L_1 (μH)	L_2 (μH)	r_1 (Ω)	r_2 (Ω)	C_1 (nF)	C_2 (nF)	C_f (μF)	V_s (V)
30.69	30.32	0.21	0.15	37.32	37.82	31.55	24

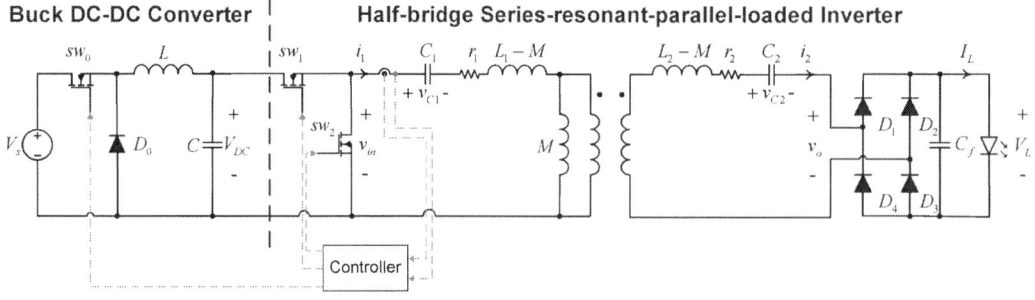

Fig. 7 Schematic of the Wireless-Powered LED Driver

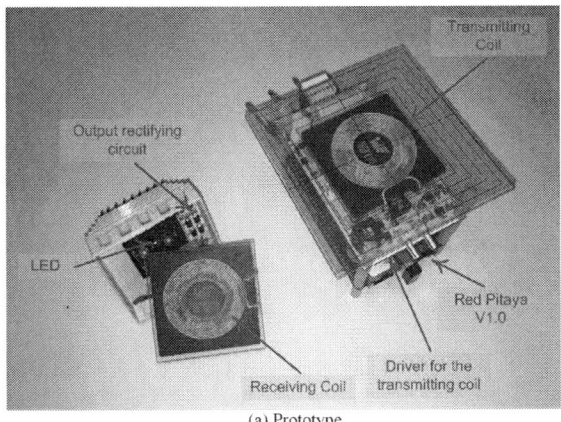

(a) Prototype.

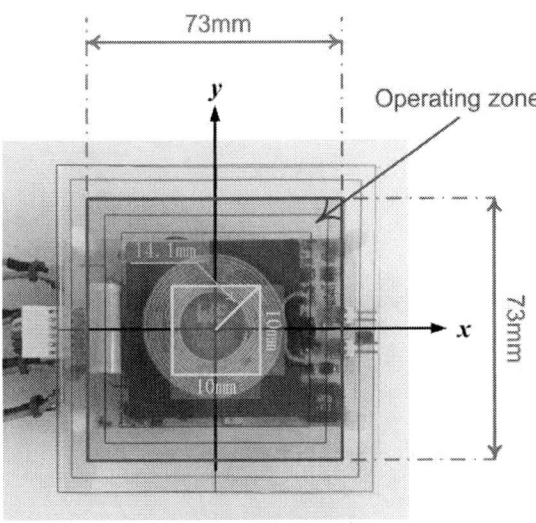

(b) Top View of the Transmitter.
Fig. 8 Experimental Setup.

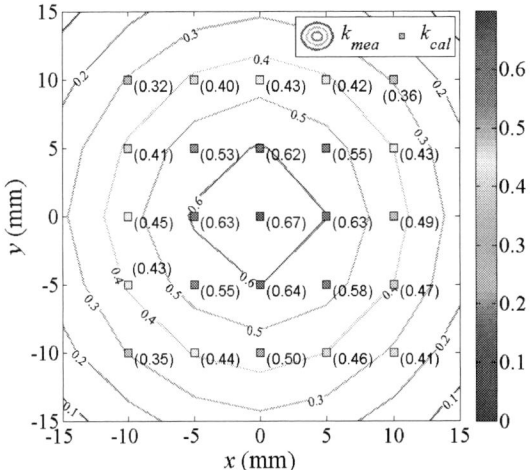

(a) Measured and Calculated Coupling Coefficient at Different Positions.

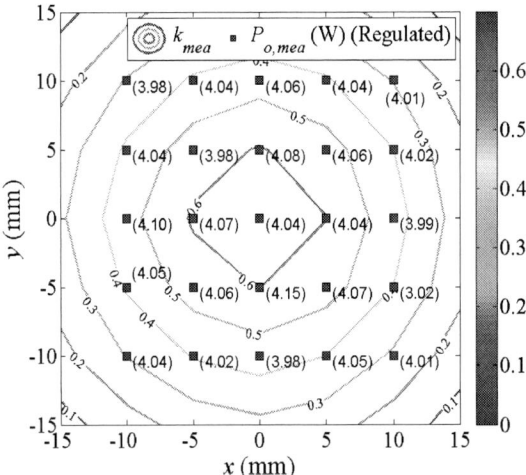

(b) Measured Coupling Coefficient and Output Power at Different Positions.
Fig. 9 Online Estimation of the Coupling Coefficient and Regulation of the Output Power.

Furthermore, after solving the equations from (11) to (20), the mutual inductance and the receiver-side power are estimated by the controller within a reasonable computational time. For the sake of comparison, the mutual inductance between the two coils is represented by a coupling coefficient. In Fig. 9(a), the measured coupling coefficients k_{mea} are depicted by a two-dimensional contour graph and the online calculated coupling coefficients k_{cal} are labeled by square dots on the same diagram. k_{mea} varies between 0.3 and 0.67 within the operating zone (±10mm, ±10mm). The values of k_{mea} and k_{cal} are in close agreement. The reference for the receiver-side power P_L is set at 4W. The measured average output power $P_{o,mea}$ at different locations is sampled and calculated by using an oscilloscope. Fig. 9(b) shows the results, revealing that the output power can be regulated with the maximum error of 4%.

V. CONCLUSIONS

An estimation algorithm, which is based on using the transmitter-side electrical information to estimate the mutual inductance of the wireless inductive link and regulate the receiver-side power, has been proposed. The nonlinear load characteristics of the diode-bridge rectifier have been taken into account. The proposed algorithm has been successfully demonstrated on a setup for regulating the power to an LED lighting load. The estimated results and experimental measurements are in close agreement.

ACKNOWLEDGMENT

The authors would like to thank RS Components for the support of Red Pitaya V1.0 in conducting the measurement and analysis.

This work was supported by a grant from the Research Grants Council of the Hong Kong Special Administrative Region, China, through Project CityU 112613 and also under the program for professor of special appointment (Eastern Scholar) at Shanghai Institutions of Higher learning.

APPENDIX

A. Expressions of the parameters in (8)

$$a_0 = \frac{1}{\omega^2 C_1 C_2 (\omega^2 L_1 L_2 - \omega^2 M^2)} \tag{A1}$$

$$a_1 = \frac{r_1 / \omega C_2 + r_2 / \omega C_1}{\omega^2 L_1 L_2 - \omega^2 M^2} \tag{A2}$$

$$a_2 = \frac{L_1 / C_2 + r_1 r_2 + L_2 / C_1}{\omega^2 L_1 L_2 - \omega^2 M^2} \tag{A3}$$

$$a_3 = \frac{r_1 \omega L_2 + r_2 \omega L_1}{\omega^2 L_1 L_2 - \omega^2 M^2} \tag{A4}$$

B. Expressions of the parameters in (10)-(20)

$$G_{L1} = \omega^2 L_2 C_2 \lambda_1^2 + r_2 \omega C_2 \lambda_1 + 1 \tag{B1}$$

$$G_{L2} = \omega^2 L_2 C_2 \lambda_2^2 + r_2 \omega C_2 \lambda_2 + 1 \tag{B2}$$

$$G_{L3} = \omega^2 L_2 C_2 \lambda_3^2 + r_2 \omega C_2 \lambda_3 + 1 \tag{B3}$$

$$G_{L4} = \omega^2 L_2 C_2 \lambda_4^2 + r_2 \omega C_2 \lambda_4 + 1 \tag{B4}$$

$$G_{r1} = \frac{C_2}{M} [(\omega^2 L_1 L_2 - \omega^2 M^2)\lambda_1^2 + r_2 \omega L_1 \lambda_1 + \frac{L_1}{C_2}] \tag{B5}$$

$$G_{r2} = \frac{C_2}{M} [(\omega^2 L_1 L_2 - \omega^2 M^2)\lambda_2^2 + r_2 \omega L_1 \lambda_2 + \frac{L_1}{C_2}] \tag{B6}$$

$$G_{r3} = \frac{C_2}{M} [(\omega^2 L_1 L_2 - \omega^2 M^2)\lambda_3^2 + r_2 \omega L_1 \lambda_3 + \frac{L_1}{C_2}] \tag{B7}$$

$$G_{r4} = \frac{C_2}{M} [(\omega^2 L_1 L_2 - \omega^2 M^2)\lambda_4^2 + r_2 \omega L_1 \lambda_4 + \frac{L_1}{C_2}] \tag{B8}$$

REFERENCES

[1] W. H. Ko, S. P. Liang, and C. D. Fung, "Design of Radio-Frequency Powered Coils for Implant Instruments," *Medical & Biological Engineering & Computing*, vol.15, no.6, pp. 634-640, Nov. 1977.

[2] H. L. Li, A. P. Hu, G. A. Covic and C. Tang, "A New Primary Power Regulation Method for Contactless Power Transfer," in *Proc. International Conference on Industrial Technology (ICIT)*, Gippsland, VIC, February, 10-13, 2009, pp. 1-5.

[3] J. M. Miller, O. C. Onar and M. Chinthavali, "Primary-Side Power Flow Control of Wireless Power Transfer for Electricc Vehicle Charging," *IEEE Journal of Emerging and Selected Topics in Power Electron.*, vol. 3, no. 1, pp. 142-163, March, 2015.

[4] N. Y. Kim, K. Y. Kim, J. Choi, and C. W. Kim, "Adaptive frequency with power-level tracking system for efficient magnetic resonance wireless power transfer," *IEEE Electron. Lett.*, vol. 48, no. 8, pp. 452–454, April, 12, 2012.

[5] G. Simard, M. Sawan and D. Massicotte, "High-Speed OQPSK and Efficient Power Transfer Through Inductive Link for Biomedical Implants," *IEEE Trans. Biomed. Circuits Syst.*, vol. 4, no. 3, pp. 192-200, June, 2010.

[6] G. Wang, P. Wang, Y. Tsang and W. Liu, "Analysis of Dual Band Power and Data Telemetry for Biomedical Implants," *IEEE Trans. Biomed. Circuits Syst.*, vol. 6, no. 3, pp. 208-215, June, 2012.

[7] E. Waffenschmidt, "Wireless power for mobile devices," in *Proc. IEEE Telecommunications Energy Conference (INTELEC), 33rd International*, Amsterdam, Netherlands, October, 9-13, 2011, pp. 1–9.

[8] I. Nam, R. Dougal, and E. Santi, "Novel control approach to achieving efficient wireless battery charging for portable electronic devices," in *Proc. Energy Conversion Congress and Exposition (ECCE)*, Raleigh, NC, USA, September, 15-20, 2012, pp. 2482-2491.

[9] V. Boscaino, F. Pellitteri, R. L. Rosa, and G. Capponi, "Wireless battery chargers for portable applications: design and test of a high-efficiency power receiver," *IET Power Electron.*, vol. 6, pp. 20–29, January, 20, 2013.

[10] J. Chow and H. Chung, "Use of Primary-Side Information to perform Online Estimation of the Secondary-Side Information and Mutual Inductacne in Wireless Inductive Link," in *Proc. Applied Power Electronics Conference and Exposition (APEC), 2015 IEEE*, Charlotte, NC, USA, March, 15-19, 2015, pp. 2648-2655.

[11] D. J. Thrimawithana and U. K. Madawala, "A Primary Side Controller for Inductive Power Transfer Systems," in *Proc. IEEE International Conference on Industrial Technology (ICIT)*, Vi a del Mar, March, 14-17, 2010, pp. 661-666.

[12] C. Florian, F. Mastri and R. P. Paganelli, "Theoretical and Numerical Design of a Wireless Power Transmission Link With GaN-Based Transmitter and Adaptive Receiver," *IEEE Trans. Microwave Theory and Techniques*, vol. 62, no. 4, pp. 931-946, April 2014.

[13] P. P. Mercier and A. P. Chandrakasan, "Rapid Wireless Capacitor Charging Using a Multi-Tapped Inductive-Coupled Secondary Coil," *IEEE Trans. Circuits and Systems-I*, vol. 60, no. 9, pp. 2263-2272, September 2013.

[14] S. L. Ho, X. Zhang and W. N. Fu, "Extension of Time-Demain Finite Element Method to Nonlinear Frequency-Sweeping Problems," *IEEE Trans. Magnetics*, vol. 49, no. 5, pp. 1781–1784, May 2013.

Dynamic Period Switching of PRS-PWM with Run-Length Limiting Technique for Spurious and Ripple Reduction in Fast Response Wireless Power Transmission

Takahiro Moroto, Toru Kawajiri, and Hiroki Ishikuro
Department of Electronics and Electrical Engineering, Keio University
Yokohama, Japan
E-mail:moroto@iskr.elec.keio.ac.jp

Abstract— **In this paper, we propose a control method in a wireless power transmission that targets small battery-less devices. Dynamic period switching of PRS-PWM (Pseudo Random Sequence – Pulse Width Modulation) with run length limiting technique has been developed to break through the tradeoff between transient response speed, rectified voltage ripple, and spurious emission. Developed wireless power transmission system has achieved ripple reduction by 44%, response speed of 70µs, and spurious reduction by 15.6dB.**

Keywords— *Wireless power transfer, PRS-PWM, spurious, fast response*

I. INTRODUCTION

Recently, wireless power transmission technology begins to be applied in various fields, and many kind of new applications are expected to be realized in the near future. This study aimed small battery-less application such as artificial retina (Fig.1) [1,2]. In medical implantable devices, such as the artificial retina, power delivery to the device is a difficult challenge. In the case of using wire, there are problems such as the risk of infection and failure of metal contact. In the case of using a battery, the device size becomes large and battery replacement requires surgery, which imposes a strain to patients.

Wireless power transmission solves the above mentioned problems. However, in the wireless power transmission to battery-less devices, excess power transmission degrades the efficiency due to heat generation and sometimes becomes harmful for human body. In several hundred kHz band, frequency modulation (FM) [3] or pulse width modulation (PWM) [4,5] are usually used to control the transmitting power. However, several hundred kHz band is not suited for fast power control because the switching frequency is not enough high. If MHz band is used, the speed of power control can be improved. However, fast power control brings spurious emission which becomes interference to the other electronic systems. And the regulation of the spurious emission is stringently restricted in MHz band (Fig.2) [6]. Therefore, FM and PWM are difficult to be used for power control.

Fast transmitting power control by using 6.78MHz or 13.56 MHz switching frequency [7-10] have proposed to use ΔΣ

modulation or PRS-PWM to reduce spurious emission. However, there were tradeoff between response time, ripple, and spurious emission. To break through the tradeoff, dynamic period switching of PRS-PWM (Pseudo Random Sequence – Pulse Width Modulation) with run length limiting technique is proposed and studied in this work.

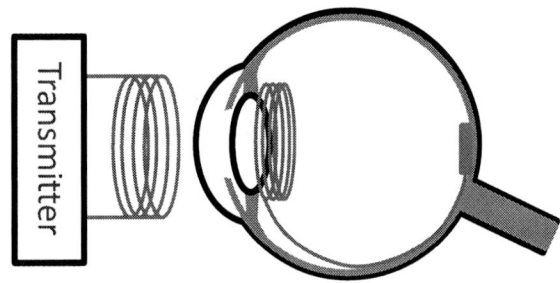

Fig.1. Concept view of wireless power delivery system for artificial retina

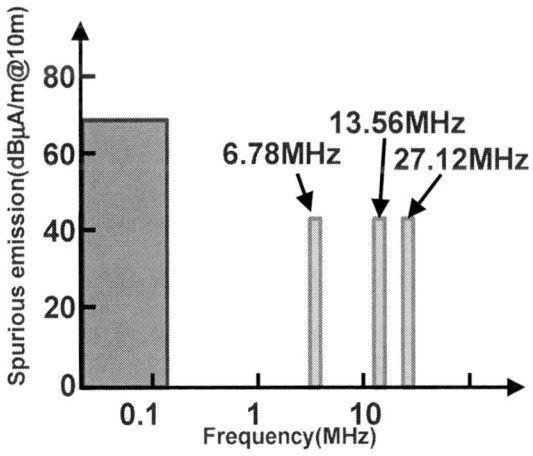

Fig.2. ISM Band frequency

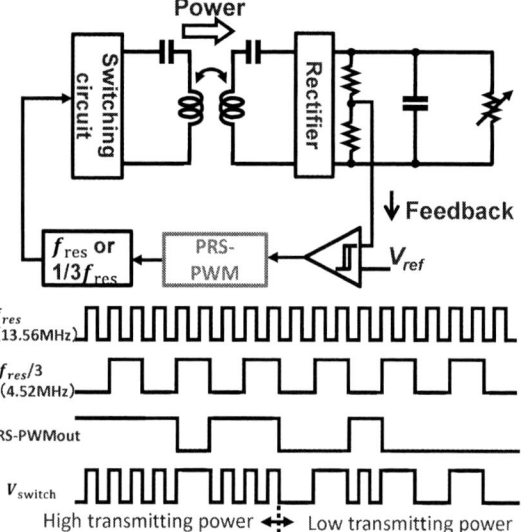

Fig.3. Block diagram of the developed wireless power transmission system

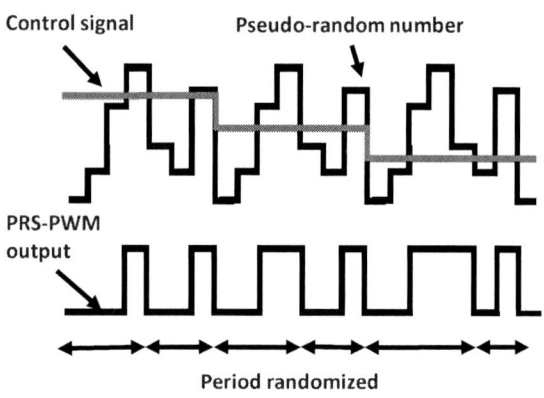

Fig.4. Waveform of PRS-PWM

II. SYSTEM OVERVIEW

Fig.3 shows the block diagram of the developed wireless power transmission system. The output voltage is compared with target voltage by a comparator, and the transmitting power is controlled by changing the switching frequency between resonant frequency of LC tank (13.56MHz) and one third of the resonant frequency (4.52MHz) [9]. If the switching clock frequency is one third of the resonant frequency, the third harmonic component of the clock signal is mainly transmitted. Since the amplitude of third order harmonics is 1/3 of the fundamental component, transmitting power can be reduced to 1/9. By changing the ratio between the resonant frequency (13.56MHz) and one third of the

TABLE.I. Comparison between long and short period of PRS-PWM

Period of random number	Suprious emission	Response speed	Ripple
Short	✕	○	○
Long	○	✕	✕

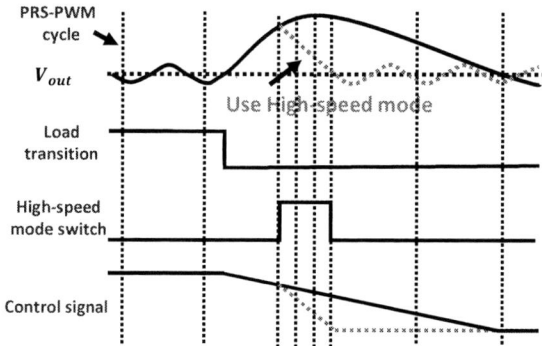

Fig.5. Dynamic period switching to improve the transient response

resonant frequency (4.52MHz), transmitting power can be continuously controlled. However, if the frequency is controlled directly by the output of the comparator, spurious emission at frequencies out of ISM band becomes high due to periodicity of the control signal.

The spurious components can be spread by using PRS-PWM [9,10] (Fig.4). In a usual PWM, control signal is compared with the value of the triangular wave and PWM signal is generated. As a result, the PWM signal contains a strong frequency component at triangular wave frequency, which becomes spurious component in the wireless power transfer system. By contrast, in PRS-PWM, the control signal is compared with the value of the PRS. Therefore, the output of PRS-PWM is randomized and spurious component can be spread.

If the period of the PRS-PWM becomes long, the PRS signal become to be more randomized and the spurious components can be suppressed more effectively. However, in the long period PRS-PWM, the PRS signal is consecutively takes large value or small value for some duration. As a result, the PRS-PWM consecutively outputs "High" or "Low" for long duration in one PRS period, which can be seen that the run-length is long. This brings the rectified voltage ripple. The other disadvantage of the long period PRS-PWM is response speed degradation (Table 1). To enjoy both the advantage of short period and long period PRS-PWM, the proposed system dynamically switches the period of PRS-PWM by monitoring the sequence of the comparator output.

A. High-speed response mode

Fig.5 shows the timing chart of PRS-PWM period switching technique. When the output of the comparator keeps same polarity for certain clock cycles, it is decided that the system is in the transient state and short period PRS-PWM is selected. After the system becomes steady state, the controller is switched to long period PRS-PWM. Fig.6 shows the implemented block diagram of dynamic period switching PRS-PWM controller. When the controller switches between the long period PRS-PWM and short period PRS-PWM, discontinuous change may be arise in control signal. To solve the issue, the counter increment value is also changed depending on the period of PRS-PWM. As a result, fast response is realized when the load is changed and spurious can be suppressed at steady state.

B. Ripple suppression

Fig.6 and Fig.7 show block diagram and timing chart of ripple suppression system. As the period of the PRS-PWM is longer, the PRS signal consecutively takes large value or small value for some duration. During that time, the controller tries to increase or decrease the transmitting power, which brings large ripple. To solve this issue, the sum of PRS signal in three

cycles is monitored and switch the control signal from PRS-PWM to direct comparator output when the controller detects the consecutive large value or small value. As a result, the spurious increases a little due to it is short time, and the ripple is suppressed greatly.

III. MESUREMENT RESULT

Fig.8 shows the wireless power transmitter and receiver modules and silicon chips. The size of the coil is 2cm×2cm, and transmitter and rectifier are integrated into two silicon chips. The distance between the transmitter coil and receiver coil is set at 3mm. The distance of the magnetic field probe and the secondary coil is 3cm. The controller is implemented into FPGA. In this work, the feedback signal was sent to the transmitter by wireline.

A. Spurious emission

Fig.9 shows measurement results of the spurious emission. The magnetic strength measured by the magnetic probe at distance of 3cm were converted to magnetic distance at 10m. Conventional bang-bang control technique and proposed PRS-PWM technique were measured for comparison. The transmitting power is changed between 0.04W and 0.56W. At high or low transmitting power, spurious emission level of both the control techniques are almost same, because the frequency switching between resonant frequency (13.56MHz) and one third (4.52MHz) of resonant frequency is not so often. However, it can be seen that spurious is suppressed when the transmitting power is middle. The spurious emission level can be suppressed by 15.6dB at maximum.

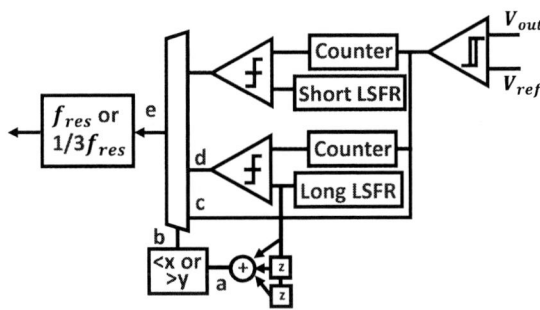

Fig.6. Block diagram of ripple suppression controller

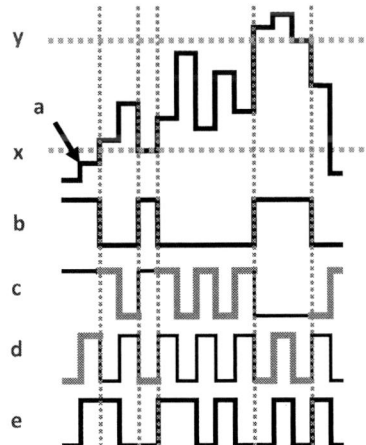

Fig.7. Timing chart in the ripple suppression controller

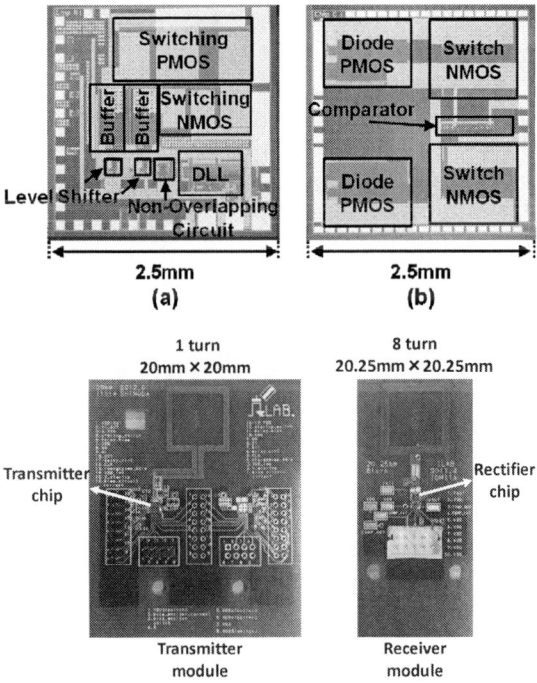

Fig.8. Wireless power transceiver modules and silicon chips

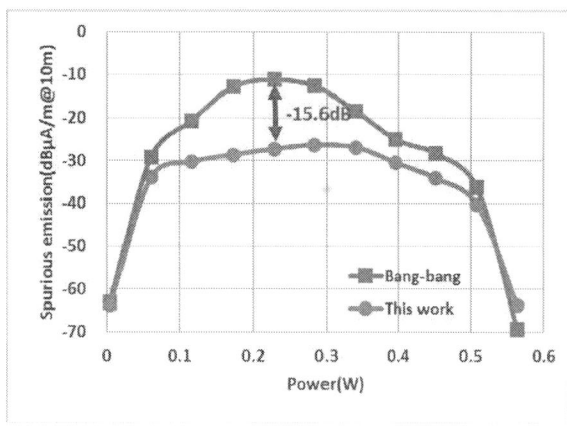

Fig.9. Power vs. spurious emission

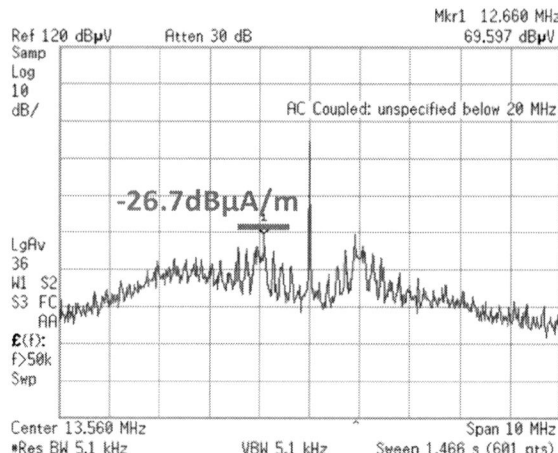

Fig.12. Spurious of propose PRS-PWM (Dynamic period switching between 5bit and 10bit)

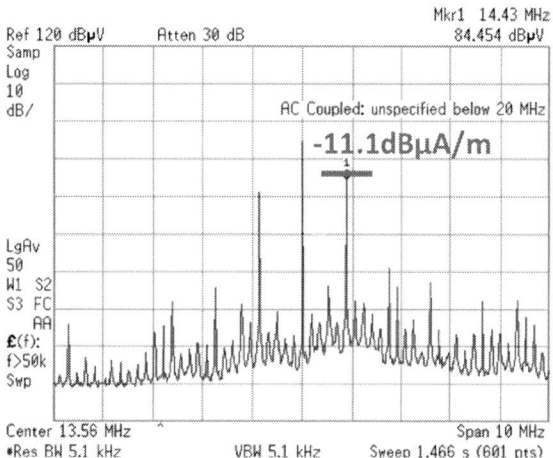

Fig.10. Spurious of Bang-bang control

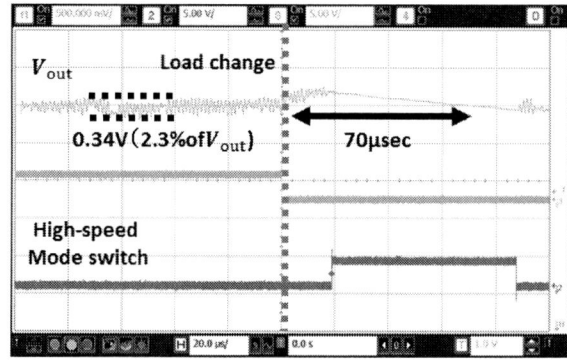

Fig.13. Waveform of transient response

Fig.10, 11 and 12 show measurement results of the spurious of Bang-bang control and PRS-PWM with short period and proposed dynamic period switching PRS-PWM technique. Load resistance in the receiver side is 1kΩ. It can be seen that spurious is suppressed when long period PRS-PWM is used. It reduced the spurious level by 15.6dB.

B. Response speed

Fig.13 shows the measured response speed in proposed PRS-PWM when the load is changed from 0.33W to 0.067W. Output voltage is regulated at 15V. It was achieved that the response speed is 70µsec.

Fig.14 shows the measured results of the response speed and spurious emission. By the proposed dynamic period switching technique, the response speed can be improved while keeping the spurious level. It can be seen that the response speed is reduced by 70µsec when the spurious emission is kept at same level.

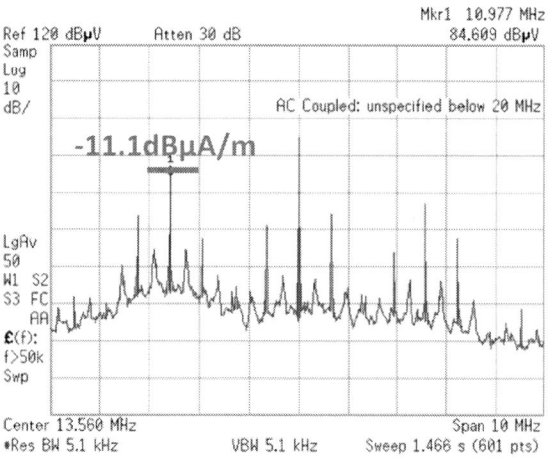

Fig.11. Spurious of normal PRS-PWM (5bit)

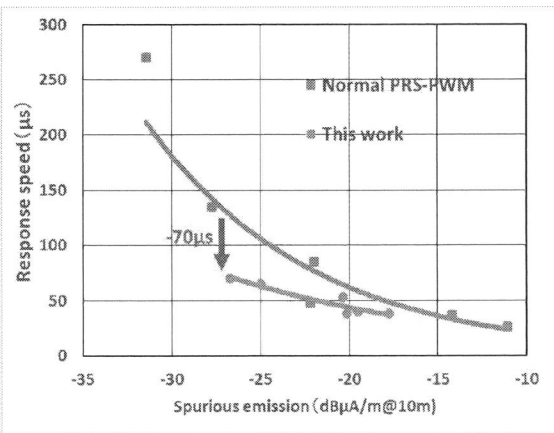

Fig.14.   Response speed vs. spurious emission

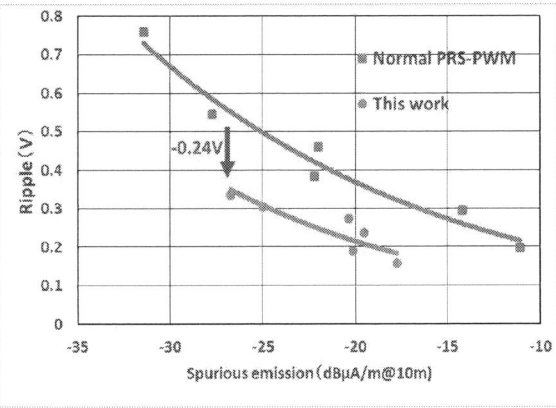

Fig.15.   Ripple vs. spurious emission

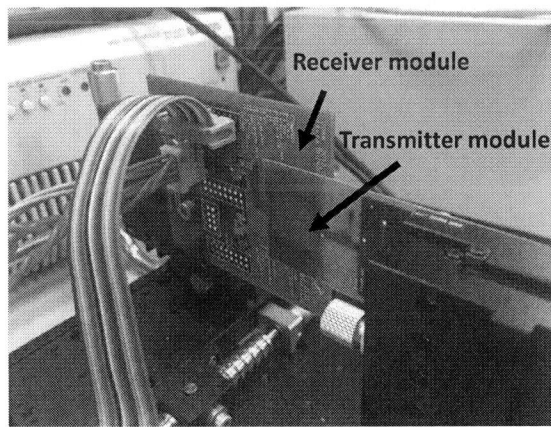

Fig.16.   Photo of the experiment

C. Ripple

Fig.15 shows the measurement results of ripple and spurious emission. The measured ripple is the maximum value of the peak-to-peak voltage when the power consumption by the receiver side load is 0.25W. The ripple is suppressed in proposed PRS-PWM even at longer period than that of a normal PRS-PWM. It was achieved that the ripple is 0.34V in 10bit. It can be seen that the ripple is reduced by 0.24V (-44%) maximally when the spurious emission is kept at same level.

IV.   CONCLUSION

Dynamic period switching of PRS-PWM with run length limiting technique has been proposed to improve the response speed while reducing the rectified voltage ripple and spurious emission level. The developed wireless power transmitting system has achieved the response speed of 70μsec, ripple reduction by 44%, and the spurious suppression by 15.6dB.

REFERENCES

[1] C. Brendler, N. P.Aryan, V. Rieger, A. Rothermel, "Wireless power delivery for a biomedical retinal prosthesis" Electronics, Circuits, and Systems (ICECS), Dec.2013, pp.465-468.

[2] David C. Ng, E. Skafidas, "Coupling invariant inductive link for wireless power delivery to a retinal prosthesis" Engineering in Medicine and Biology Society (EMBC), Jul.2013, pp.3250-3253

[3] Ping Si et al., "A Frequency Control Method for Regulating Wireless Power to Implantable Devices" Transaction on Biomedical Circuits and Systems(TBioCAS), March 2008, vol. 2, pp.22-29

[4] Jun-Young Lee, Byung-Moon Han, "A Bidirectional Wireless Power Transfer EV Charger Using Self-Resonant PWM" Transactions on Power Electronics, April 2015, pp1784-1787

[5] Chen, X.W, Wu, Z.H, Zhao, M.J, Li, B, "An adaptive feedback for implantable wireless power transmission system" International Conference of Electron Devices and Solid-State Circuits (EDSSC), June 2013, pp1-2

[6] K. Finkenzeller, Translated by R. Waddington, "RFID HANDBOOK Second Edition," Munich/FRG, original German language published by Carl Hanser Verlag, The Atrium/Southern Gate/Chichester/West Sussex PO19 8SQ/England, John Wiley & Sons Ltd, pp. 161-181, 2003.

[7] Y. Hasegawa, K. Tomita, S. Ishihara, R. Honma, and H. Ishikuro, "Single-Inductor-Dual-Output wireless power receiver with synchronous Pseudo-Random-Sequence PWM switched rectifiers" Solid-State Circuits Conf. (A-SSCC), Nov. 2013, pp. 261-264.

[8] K. Kusaka, J.-I. Itoh, "Proposal of Switched-mode Matching Circuit in power supply for wireless power transfer using magnetic resonance coupling" Applied Power Electronics Conference and Exposition (APEC), Feb.2012, pp.653-660

[9] R. Shinoda, K. Tomita, Y. Hasegawa, H. Ishikuro, "Voltage-boosting wireless power delivery system with fast load tracker by ΔΣ-modulated sub-harmonic resonant switching" IEEE ISSCC, Feb. 2012, pp. 288-290.

[10] S. Bhuvanasundaram, Y. Vagapov, S. Lupin, R. Chakirov, "A random PWM technique for Z-source inverter based on pseudorandom binary sequence bits" Electrical and Electronic Engineering Conference (EIConRusNW), Feb. 2015, pp. 275 - 278

978-1-4673-9551-9/16 $31.00 © 2016 IEEE   1806

A Flyback AC/DC Converter Using Power Semiconductor Filter for Input Power Factor Correction

Chung-pui TUNG*, *Student Member, IEEE* and Henry Shu-hung CHUNG**, *Senior Member, IEEE*

Centre for Smart Energy Conversion and Utilization Research
City University of Hong Kong, Kowloon Tong, Hong Kong
*cptung2-c@my.cityu.edu.hk, **eeshc@cityu.edu.hk

Abstract - An application of an input current harmonic filtering technology, namely power semiconductor filter (PSF), for conventional flyback AC/DC converter will be presented. By connecting a series-pass device (SPD) in series with the input of the flyback converter, the input current of the entire system has the waveform profiled to be in phase with the supply voltage and its magnitude adjusted to perform output regulation. To minimize the loss of the SPD, its operating point is kept at the boundary between the active and saturation region through controlling the duty cycle of the switch in the flyback converter. A 100W, 85-265Vac / 36Vdc flyback AC-DC converter prototype has been built. A comprehensive study into the performance characteristics, additional loss due to the SPD, and thermal distribution of the system will be given. Results reveal that the input power factor can be kept above 0.995 over the operating range.

Keywords - AC/DC converter, rectifier, power factor corrector, harmonic filtering, PFC

I.   INTRODUCTION

Flyback AC/DC converter has been a popular power conversion topology for low-power applications, such as LED drivers and notebook chargers. By modulating the duty cycle of the switching device in the flyback converter, the low-frequency component of the input current is profiled to be in the same phase and wave shape as the supply voltage, so that high input power factor can be assured. To eliminate high-frequency input harmonics, which is caused by the pulsating input current of the flyback converter, an *LC*-based input filter is used [1]. Due to the interaction between the filter and the current pulses, the input current is typically non-sinusoidal [2-3]. The frequency response of the passive components also determines the filtering bandwidth [3]. Nevertheless, the physical size of the input filter will be one of the limiting factors in advancing the power density of the overall power conversion stage [4].

An investigation into the potential of replacing the passive filter with a series-pass device (SPD) to perform input current filtering in a DC/DC converter is given in [5]. The entire system is controlled by using two controllers to profile the waveform of the input current, minimize the loss of the SPD,

and regulate the output voltage. Apart from being amenable to monolithic integration of the filter part, the technology can also avoid possible acoustic resonance caused by the inductive components. An extension of such technology for flyback AC/DC converter will be presented in this paper. Detailed modeling, design, and analysis are given.

II.   BRIEF DESCRIPTION OF THE OPERATING PRINCIPLE

Fig. 1 shows the system architecture. The SPD, which can be a BJT or MOSFET, is connected in series with the input of the conventional flyback converter. Its operating point is maintained at the boundary between the active and saturation region by two control blocks. They are "i-control" and "v-control" blocks. The "i-control" block is used to 1) profile the input current i_{in} to be in the same phase and wave shape as the rectified supply voltage $|v_s|$, and 2) adjust the magnitude of i_{in} to regulate the output voltage v_o. Its input i_{con} is derived from an error amplifier EA, which compares v_o with the output reference $V_{o,ref}$. It generates a control signal i_b to control the operating point of the SPD. The "v-control" block is used to regulate the voltage across the SPD, v_T, around its saturation voltage. It senses v_T and compares it with a voltage reference $V_{T,ref}$ to generate the gate signal for the switch in the flyback converter. The mechanism is similar to a hysteresis control for regulating v_T. Typically, $V_{T,ref}$ is less than 0.8V, making the voltage drop across the SPD comparable with the voltage drop of a diode. As derived in [5], the overall efficiency of the entire system, η, is

$$\eta = 1 - \frac{V_{T,ref}}{V_s} \qquad (1)$$

where V_s is the rms value of the supply voltage.

As $V_{T,ref} \ll V_s$ in high-voltage applications, the conversion efficiency can be maintained at the high value. With the above mentioned mechanism, the input power factor can be controlled to be near unity.

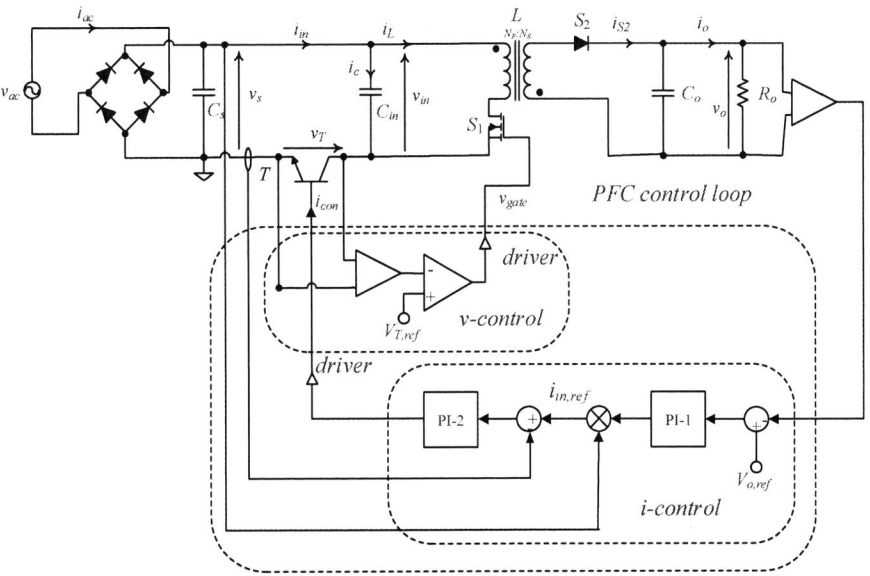

Fig. 1. Architecture of the flyback AC/DC converter with PSF.

Fig. 2(a) shows the theoretical waveforms of the input current i_{in}, flyback transformer current i_L, input capacitor current i_C, voltage across the SPD v_T, gate signal v_{gate} to the switch S_1 in the flyback converter. Fig. 2(b) shows the details of the waveforms in two switching cycles. While the input current is ripple free, the input current of the flyback converter is pulsating (which is equal to i_L). Thus, the input capacitor

C_{in} absorbs the high-frequency harmonics generated by the flyback converter.

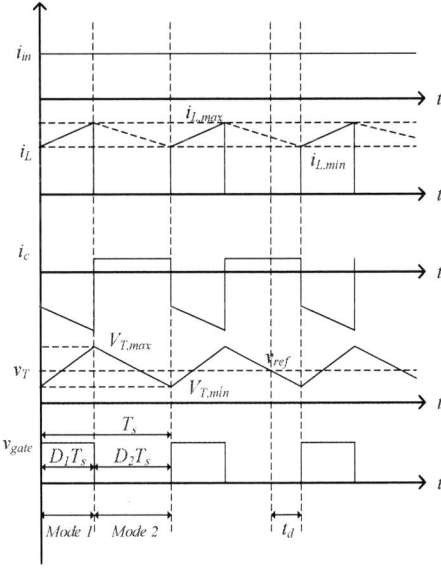

(a) Key waveforms.

(b) Enlarged waveform in Fig. 2(a).

Fig. 2. Key Waveforms

978-1-4673-9551-9/16 $31.00 © 2016 IEEE   1808

III. STATIC CHARACTERISTICS

For the sake of simplicity in the analysis, the following assumptions have been made as follows,

1. All components are ideal.
2. The SPD is able to maintain the input current i_{in} constant over a switching cycle.

Fig. 2(b) shows the operation of the converter within several switching cycle. The proposed converter is operated between boundary conduction mode (BCM) and continuous conduction mode (CCM). Consider the operation in the n-th switching cycles.

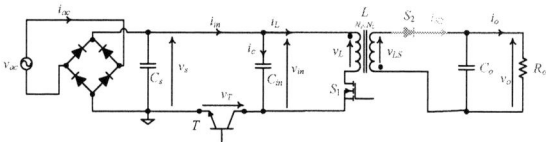

Fig. 3. Operation Mode 1 of the proposed converter

Mode 1: S_1 is on and S_2 is off.

Fig 3 shows the converter operation under Mode 1.

$$i_L(t) = I_L(0) + \frac{V_s - v_T(t)}{L} D_1 T_s \quad (2)$$

$$\left[\Delta i_L(t)\right]_{Mode1} = \frac{V_s - V_{T,min}}{L} D_1 T_s \quad (3)$$

where v_T is treated as the minimum voltage of T, $v_T(0) = V_{T,min}$ since the start of duty cycle implies the voltage across the SPD T, v_T changes from decrease to increase. And, $\left[\Delta i_L(t)\right]_{Mode1}$ is the change of inductor current i_L within the time interval of S_1 turns on and S_2 turns off.

$$i_C(t) = I_{in} - i_L(t) \quad (4)$$

Substituting (2) into (4),

$$i_C(t) = I_{in} - I_L(0) - \frac{V_s - v_T(t)}{L} D_1 T_s \quad (5)$$

$$v_{in}(t) = \frac{1}{C_{in}} \int i_C(t) dt + V_{in}(0) \quad (6)$$

Substitute (5) into (6),

$$v_{in}(t) = \frac{1}{C_{in}} \int \left[I_{in} - I_L - \frac{V_s - v_T(t)}{L} D_1 T_s \right] dt + V_{in}(0) \quad (7)$$

$$v_L(t) = L \frac{di_L(t)}{dt}$$
$$= L \frac{\Delta i_L(t)}{\Delta t} \quad (8)$$
$$= L \frac{V_{in}(t)}{D_1 T_s}$$

$$v_T(t) = V_s - v_{in}(t) \quad (9)$$

where T_s is the switching period, I_{in} and V_s are the average values of the input current i_{in} and rectified source voltage v_s, respectively. $I_L(0)$ is the initial current of inductor before Mode 1.

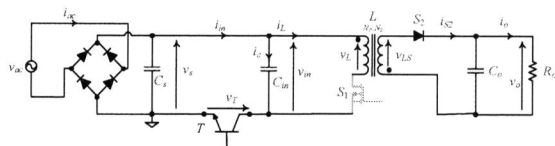

Fig. 4. Operation Mode 2 of the proposed converter

Mode 2: S_1 is off and S_2 is on.

Fig 4 shows the converter operation under Mode 2.

$$i_L(t) = 0 \quad (10)$$

$$i_C(t) = I_{in} \quad (11)$$

$$v_L(t) = -v_{LS}(t) \frac{N_P}{N_S}$$
$$= -v_o(t) \frac{N_P}{N_S} \quad (12)$$
$$= L \frac{di_L(t)}{dt}$$

$$\frac{di_L(t)}{dt} = \frac{\Delta i_L(t)}{\Delta t}$$
$$= \frac{\Delta i_L(t)}{(1 - D_1) T_s} \quad (13)$$
$$= \frac{-v_o(t)}{L} \frac{N_P}{N_S}$$

$$[\Delta i_L(t)]_{Mode2} = \frac{-v_o(t)(1-D_1)T_s}{L}\frac{N_P}{N_S} \quad (14)$$

$$i_{S_2}(t) = \left\{ i_L(D_1T_s) + [\Delta i_L(t)]_{Mode2} \right\}\frac{N_S}{N_P} \quad (15)$$

$$[\Delta i_L(t)]_{Mode1} + [\Delta i_L(t)]_{Mode2} = 0 \quad (16)$$

Substitute (3) and (14) into (16),

$$\frac{V_s - V_{T,\min}}{L}D_1T_s + \frac{-V_o(1-D_1)T_s}{L}\frac{N_P}{N_S} = 0 \quad (17)$$

$$\begin{aligned}
v_{in}(t) &= v_{in}(D_1T_s) + \frac{(1-D_1)T_sI_{in}}{C_{in}} \\
&= V_s - V_{T,\max} + \frac{(1-D_1)T_sI_{in}}{C_{in}}
\end{aligned} \quad (18)$$

By simplifying (17), the duty cycle D_1 of the corresponding switching cycle can be obtained.

$$D_1 = \frac{V_oN_P}{(V_s - V_{T,\min})N_S + V_oN_P} \quad (19)$$

where V_o is the average value of output voltage v_o. $[\Delta i_L(t)]_{Mode2}$ is the change of inductor current i_L within the time interval, where S_1 is off and S_2 is on. i_{S_2} is the current throw through the output diode S_2.

Consider the whole switching cycle when the converter is in CCM,

$$\begin{aligned}
I_{L,\max} &= I_L + \frac{\Delta I_L}{2} \\
&= I_L + \frac{[\Delta i_L(t)]_{Mode1}}{2}
\end{aligned} \quad (20)$$

$$\begin{aligned}
I_{L,\min} &= I_L - \frac{\Delta I_L}{2} \\
&= I_L - \frac{[\Delta i_L(t)]_{Mode1}}{2}
\end{aligned} \quad (21)$$

$$I_L = \frac{(V_s - V_{T,\min})I_{in}N_S}{(1-D_1)V_oN_P} \quad (22)$$

By using (21), if the converter is operating in BCM, the minimum inductor current $I_{L,\min}$ is 0A. Therefore, the minimum inductance for non-discontinuous conduction mode can be found.

$$I_{L,\min} = 0 = \frac{V_{in}I_{in}N_S}{(1-D_1)V_oN_P} - \frac{V_s - V_{T,\max}}{2L_c}D_1T_s \quad (23)$$

$$L_c = \frac{(V_s - V_{T,\max})(1-D_1)D_1T_sV_o}{2(V_s - V_{T,\min})I_{in}}\frac{N_P}{N_S} \quad (24)$$

where I_L is the average inductor current i_L within the time interval of Mode 1, L_c is the minimum inductance of the inductor for the converter to operate in the boundary between BCM and CCM.

Table I shows the key equations for the PSF flyback converter.

TABLE I. KEY EQUATIONS FOR THE PSF FLYBACK CONVERTER

Parameter	CCM
D_1	$\dfrac{V_oN_P}{(V_s - V_{T,\min})N_S + V_oN_P}$
V_o	$\dfrac{V_{in}D_1N_S}{(1-D_1)N_P}$
$I_{L,\max}$	$\dfrac{(V_s - V_{T,\min})I_{in}N_S}{(1-D_1)V_oN_P} + \dfrac{V_s - V_{T,\max}}{2L}D_1T_s$
$I_{L,\min}$	$\dfrac{(V_s - V_{T,\min})I_{in}N_S}{(1-D_1)V_oN_P} - \dfrac{V_s - V_{T,\max}}{2L}D_1T_s$
L_c	$\dfrac{(V_s - V_{T,\max})(1-D_1)D_1T_sV_o}{2(V_s - V_{T,\min})I_{in}}\dfrac{N_P}{N_S}$
V_{in}	$V_s - V_T$
Δv_T	$\dfrac{V_s - V_{T,\min}}{C_{in}L}(D_1T_s)^2$
$v_{T,\max}$	$V_{T,ref} + \Delta v_T - \dfrac{I_{in}}{C_{in}}t_d$
$v_{T,\min}$	$V_{T,ref} - \dfrac{I_{in}}{C_{in}}t_d$

IV. EXPERIMENTAL VERIFICATION

A 100W, 85-265Vac / 36Vdc flyback AC-DC converter prototype has been built. The photo is shown in Fig. 5. The model number and values of the components are tabulated in Table II. Digital control has been implemented for controlling the proposed topology. In addition, soft-start scheme for powering up has also been integrated into the control algorithm.

Fig. 5.   100W Prototype of the proposed flyback

TABLE II.   COMPONENT VALUES

Component	Value
C_{in}	1μF
S_1	STP8N65M5
T	FZT853
L	750811351
S_2	C3D20060
Controller	STM32F334

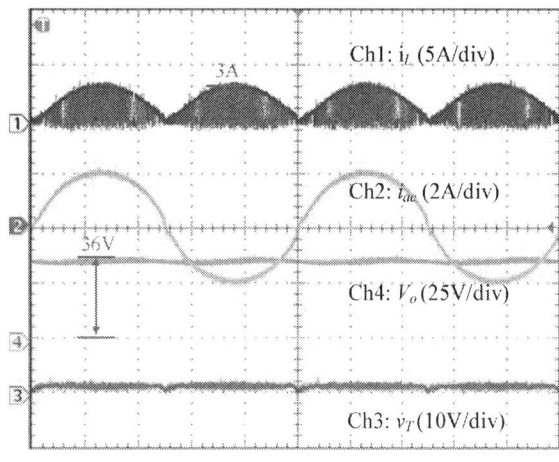

(a) Supply voltage = 85Vac. (Time base = 4ms/div)

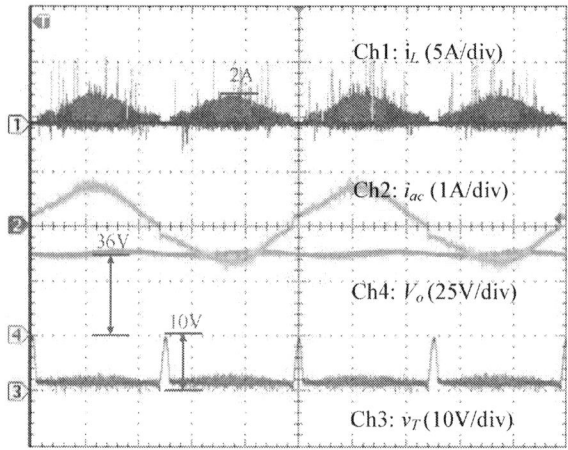

(b) Supply voltage = 265Vac. (Time base = 4ms/div)

Fig. 6.   Steady state waveforms operating at 100W.

To demonstrate the performance of the PSF flyback converter, the experimental results are shown as follow. Figs. 6 (a) and 6 (b) show the waveforms of i_L, i_{ac}, v_o, and v_T of the PFC under full-load operation with the supply voltage of 85Vac and 265Vac. respectively. Those steady state measurements shows the PSF that is able to provide a sin-like input current by filtering out current harmonics. Figs. 7 (a)-(c) show the output load step response of the converter between 50% load to 100% under low line, 85Vac, and high line, 265Vac, respectively. Moreover, soft-start scheme has been applied to controller of the converter. Figs. 8 (a)-(b) show the power up waveforms of the converter with slight overshoot under low line and high line respectively.

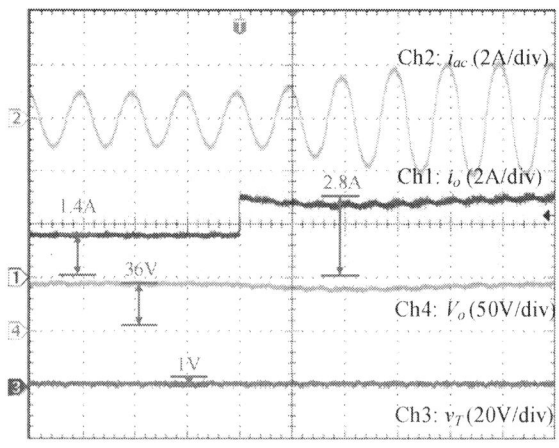

(a) 50% to 100% loading under 85Vac. (Time base = 20ms/div)

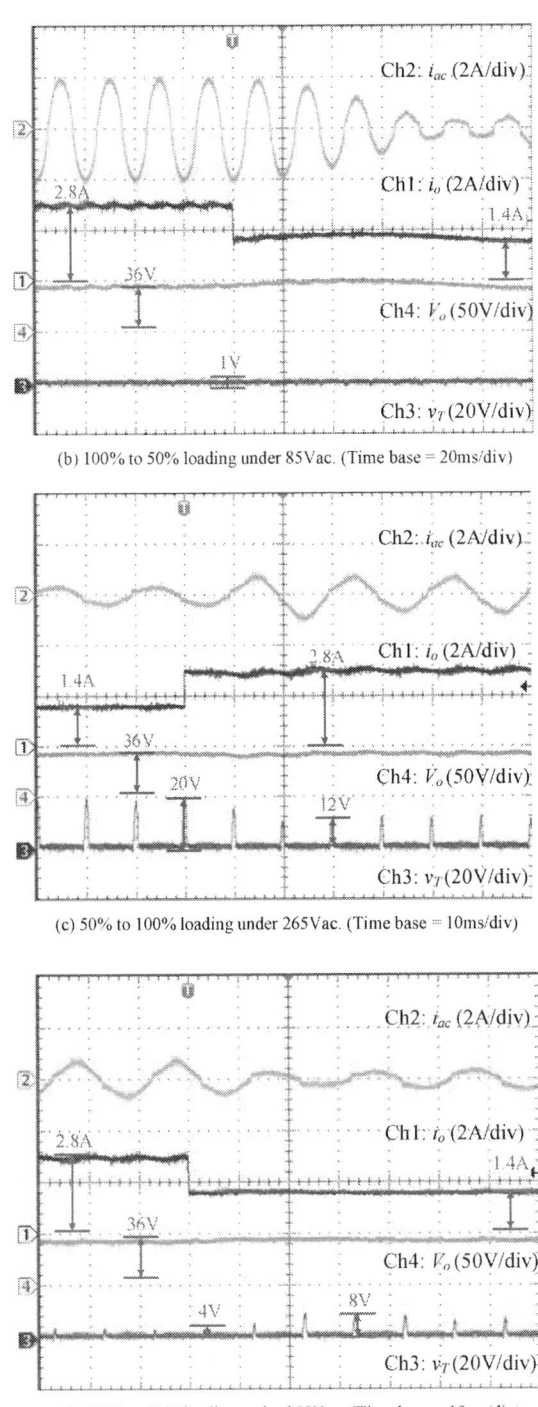

(b) 100% to 50% loading under 85Vac. (Time base = 20ms/div)

(c) 50% to 100% loading under 265Vac. (Time base = 10ms/div)

(d) 100% to 50% loading under 265Vac. (Time base = 10ms/div)

Fig. 7.   Transient responses under Different Supply Voltages.

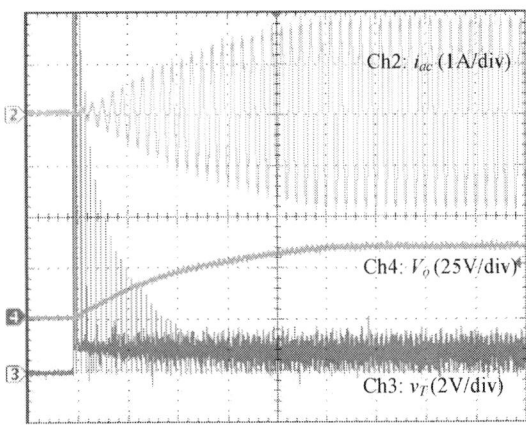

(a) Startup under 85Vac. (Time base = 100ms/div)

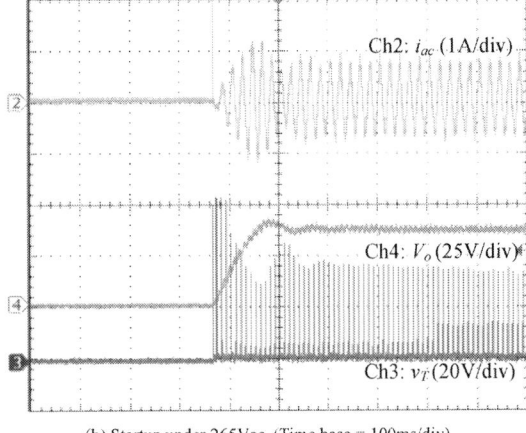

(b) Startup under 265Vac. (Time base = 100ms/div)

Fig. 8.   Startup waveform of soft-start scheme under high and low line voltage respectively

Comparing i_L with i_{ac}, the PSF can effectively eliminate the high-frequency harmonics generated by the flyback converter and gives a near sinusoidal input current waveform under both supply voltages. Figs. 9 (a)-(d) show the measured power factor and THD of the input current. The power factor is above 0.995 and the THD of the input current $\mathrm{THD_i}$ is around 7%.

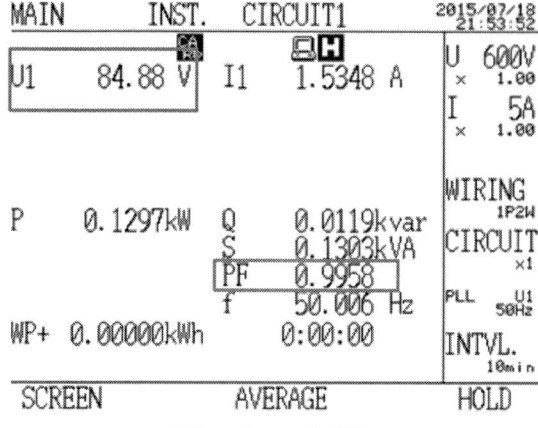

(a) Power factor under 85Vac

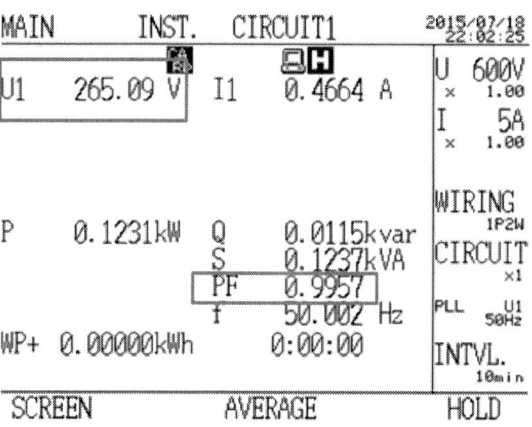

(b) Power factor under 265Vac

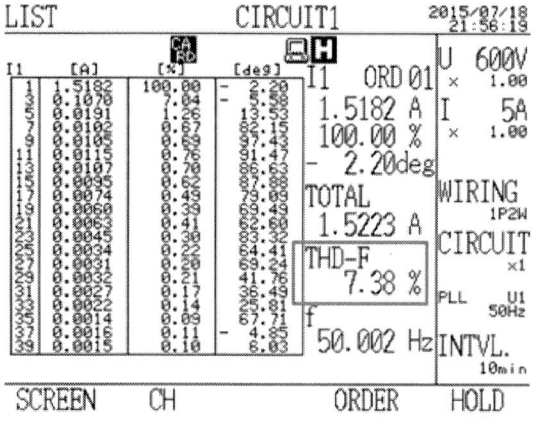

(b) Total harmonic distortion under 85Vac

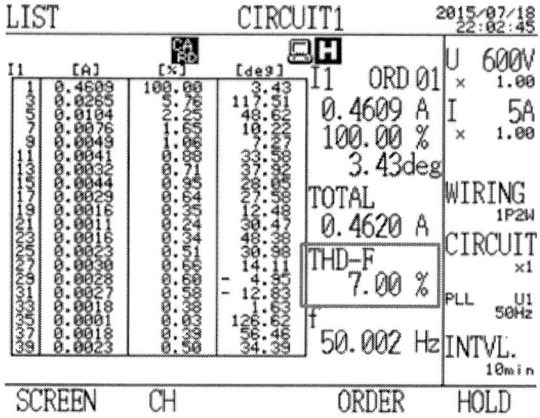

(d) Total harmonic distortion under 265Vac

Fig. 9. Power factor and THD of line current i_{ac} of proposed Flyback converter under 85Vac and 265Vac

Fig. 10. Temperature distribution.

Fig. 10 shows temperature distribution of the components. The major power lossy components include the switching devices S_1 and diodes S_2 in the flyback converter, and the input diode bridge. The SPD does not have significant power loss, confirming the merit of the PSF that the power loss on the SPD T is small. Fig. 11 shows the efficiency curves with and without PSF. The introduced PSF only causes an additional 1 to 2% power loss while comparing with a conventional 100W AC/DC flyback converter without using any input filter.

Table III, additionally, shows the efficiency difference of the same converter design between using PSF and using LC-based EMI filter under full loading. The efficiency of two approaches under 85V and 265V input voltage is nearly the same, and only around ±2% of difference between 110V and 220V. The efficiency difference is defined as the efficiency of using PSF substrates by the one's using LC-based filter. The result illustrates that the efficiency of using PSF is compatible with the one which using LC-based filter. Finally, Fig. 12

shows the measured power factor and THD of the input current under different line voltages. The results reveal that it is easy to achieve nearly unity input power factor by using PSF for power factor correction.

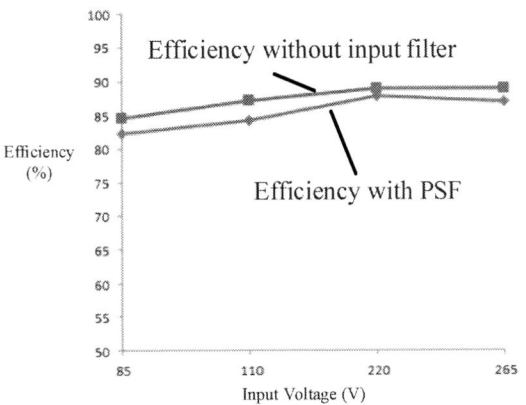

Fig. 11. Efficiency curves of using PSF and without any input filter under different supply voltages in full load.

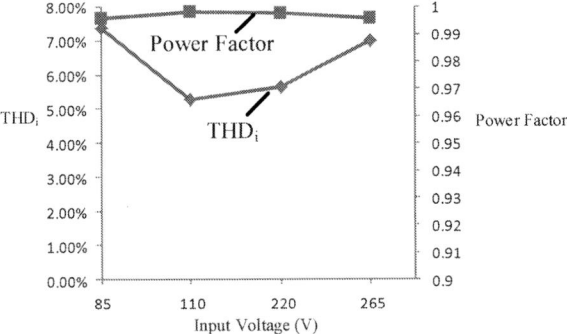

Fig. 12. The measured power factor and THD of the input current under different line voltages

TABLE III. EFFICIENCY DIFFERENCE OF THE SAME CONVERTER DESIGN BETWEEN USING PSF AND USING LC-BASED EMI FILTER UNDER FULL LOADING

Input Voltage	85V	110V	220V	265V
Efficiency Difference	0.129%	-1.94%	1.564%	0.2%

V. CONCLUSION

An investigation into the use of the previously-proposed input filtering technology, namely power semiconductor filter, for conventional flyback AC/DC converter will be presented. Experimental results of a 100W prototype allowing universal supply voltage reveal that the PSF can effectively eliminate the harmonics generated by the flyback converter and shape the input waveform in order to achieve a high input power factor. Time domain circuit analysis of the PSF for the flyback AC/DC converter is also given.

VI. ACKNOWLEDGMENT

This work was supported by a grant from the City University of Hong Kong through Project #7004231.

REFERENCES

[1] H. Patel, "Flyback Power Supply EMI Signature and Suppression Techniques," in Proceeding of Joint International Conference on Power System Technology and Power System Technology and IEEE Power India Conference, New Delhi, 2008, pp. 1-6, Oct. 12-15, 2008.

[2] Iftikhar, M.U.; Sadarnac, D.; Karimi, C., "Conducted EMI Suppression and Stability Issues in Switch-mode DC-DC Converters," in *Multitopic Conference, 2006. INMIC '06. IEEE* , vol., no., pp.389-394, 23-24 Dec. 2006

[3] Kotny, J.L.; Duquesne, T.; Idir, N., "EMI Filter design using high frequency models of the passive components," in *Signal Propagation on Interconnects (SPI), 2011 15th IEEE Workshop on* , vol., no., pp.143-146, 8-11 May 2011

[4] B. Touré, J. Schanen, L. Gerbaud, T. Meynard, J. Roudet, and R. Ruelland, "EMC Modeling of Drives for Aircraft Applications: Modeling Process, EMI Filter Optimization, and Technological Choice," *IEEE Trans. Power Electron.*, vol. 28, no.3, pp. 1145-1156, Mar. 2013.

[5] W. Fan, K. Yuen, and H. Chung, "Power Semiconductor Filter: Use of a Series-Pass Device in Switching Converters for Filtering Input Current Harmonics," *IEEE Trans. Power Electron.*, (Early access).

978-1-4673-9551-9/16 $31.00 © 2016 IEEE

Reducing the Variation Range of the Switching Frequency for CRM Boost PFC Converter by Injecting 3rd Harmonic into the Input Current

Yi Wang, Kai Yao
School of Automation
Nanjing University of Science and Technology
Nanjing 210094, China

Abstract—CRM Boost PFC converter features zero-current turn-on for the switch and no reverse recovery in diode. However, its switching frequency varies over a wide range, which leads to the efficiency deterioration and the EMI filter design complexity. In this paper, a minimum switching frequency variation range control (MFVC) strategy is proposed by injecting the optimal 3rd harmonic into the input current. Compared to that with the traditional control (TC), the proposed MFVC strategy achieves a minimum variation range of the switching frequency, an output voltage ripple reduction and a higher efficiency. A 60W input prototype has been fabricated and tested in the laboratory. The experimental results are given to verify the effectiveness of the proposed method.

Index Terms—*CRM Boost PFC, switching frequency, minimum switching frequency variation range control*

I. INTRODUCTION

Power factor correction (PFC) converters have been widely used in ac-dc power conversions to achieve high power factor (PF) and low harmonic distortion. There are different topologies for implementing PFC techniques, among which, boost converter is the representing topology. Depending on the inductor current to be continuous or not, the boost PFC converter may be designed to operate in three modes: continuous current mode (CCM), critical current mode (CRM) and discontinuous current mode (DCM).

When boost PFC converter operates in CCM, the inductor current ripple is very small, leading to low root-mean-square (RMS) currents on the inductor and switch, and low EMI. However, the switch always operates at hard switching, and the diode suffers reverse recovery. CCM boost PFC converter is mainly used in high and medium power applications. DCM boost PFC converter has such advantages as zero-current turning on of switch, no reverse recovery of diode and constant switching frequency. However, the PF is not so high, especially at high input voltage. DCM boost PFC converter is widely used in medium and low power applications.

Same as DCM approach, CRM boost PFC converter features zero-current turning on for the switch, no reverse

recovery of diode and high PF. But it operates in variable switching frequency, resulting in difficulties in the design of the inductor and EMI filter. CRM boost PFC converter is mainly used in medium and low power applications.

Ref. [10]-[13] propose variable on-time control strategies of CRM boost PFC converter, in order to decrease distortion of the input current waveform and suppress total harmonic distortion (THD). Interleaved DCM boost PFC converter is discussed in [14] and [15]. When multiple converters are paralleled or interleaved, the output power can be extended and the input and output current ripples can be reduced, leading to a smaller EMI filter. However, the circuit is complicated.

The objective of this paper is to propose a minimum switching frequency variation range control (MFVC) scheme for the CRM boost PFC converter over the whole input voltage range of 176-264 VAC. Section II analyzes the input current and the range of switching frequency of CRM boost PFC converter in detail. In section III, the optimal 3rd harmonic for minimum variation range of switching frequency is derived and the implementation circuit for MFVC is presented. In section IV, the comparison between the proposed MFVC and the traditional control (TC) is made in terms of the variation range of switching frequency, the output voltage ripple and the design of inductor. The analytical results show that the MFVC can reduce the variation range of switching frequency and output voltage ripple. A 60W prototype has been built and tested, and the experimental results are presented in Section V.

II. OPERATION PRINCIPLE OF CRM BOOST PFC CONVERTER

Fig. 1 shows the main circuit of a Boost PFC converter. Fig. 2 shows the inductor and switch current waveform in a switching cycle when the converter operates at CRM. Fig. 3 shows the peak and RMS current waveforms of the inductor in a half line cycle when the converter operates at CRM.

The input voltage V_{in} and the rectified voltage v_g can be defined as

$$v_{in} = V_m \sin \omega t \qquad (1)$$

$$v_g = V_m \cdot |\sin \omega t| \qquad (2)$$

where V_m is the amplitude of the input voltage, ωt is the angular frequency of the input voltage.

This work was supported by the national natural science foundation of China (51307085).

Kai Yao is the corresponding author. E-mail:13813980876@163.com.

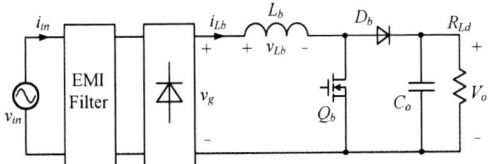

Fig. 1. Main circuit of Boost PFC converter.

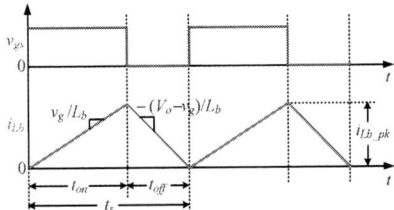

Fig. 2. Inductor current waveform in switching cycles.

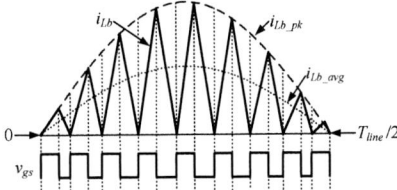

Fig. 3. The peak and average current waveforms of the inductor in a half line.

In a switching cycle, the inductor peak current $i_{\text{Lb\_pk}}$ is

$$i_{Lb\_pk} = \frac{v_g}{L_b} t_{on} = \frac{V_m |\sin \omega t|}{L_b} t_{on} \quad (3)$$

where t_{on} is the on-time value, L_b is the inductance value.

In each switching cycle, the inductor has the volt-second balance, so the off-time value is

$$t_{off} = \frac{i_{Lb\_pk}}{(V_o - v_g)/L_b} = \frac{V_m |\sin \omega t|}{V_o - V_m |\sin \omega t|} t_{on} \quad (4)$$

where V_o is the amplitude of the output voltage.

From (3), the average inductor current in a switching cycle can be derived as

$$i_{Lb\_av} = \frac{V_m |\sin \omega t|}{2 L_b} t_{on} \quad (5)$$

From (5) and Fig. 3, the input current i_{in} is

$$i_{in} = \frac{V_m \sin \omega t}{2 L_b} t_{on} \quad (6)$$

According to (6), it is known that if t_{on} is constant in a switching cycle when the average input current is sine waveform, the PF is unity. From (4), it can be seen that t_{off} varies with the input voltage, so the frequency is variable in a switching cycle.

Assuming the efficiency of the converter is 100%, namely

$$P_o = P_{in} = \frac{1}{\pi} \int_0^\pi v_{in} \cdot i_{in} \, d\omega t = \frac{V_m^2}{4 L_b} t_{on} \quad (7)$$

Then the on time is

$$t_{on} = \frac{4 L_b P_o}{V_m^2} \quad (8)$$

where $2P_o/V_m$ is the amplitude of the fundamental component.

Substituting (8) into (5) and (6), $i_{\text{Lb\_av}}$ and i_{in} can be calculated as

$$i_{Lb\_av} = \frac{2 P_o}{V_m} |\sin \omega t| \quad (9)$$

$$i_{in} = \frac{2 P_o}{V_m} \sin \omega t \quad (10)$$

From (4) and (8), off-time value can be derived as

$$t_{off} = \frac{4 L_b P_o}{V_m^2} \frac{V_m |\sin \omega t|}{V_o - V_m |\sin \omega t|} \quad (11)$$

From (8) and (11), the switching frequency is

$$f_s = \frac{1}{t_{on} + t_{off}} = \frac{V_m^2}{4 L_b P_o} \frac{V_o - V_m |\sin \omega t|}{V_o} \quad (12)$$

According (12), it can be seen that the minimum switching frequency occurs at the zero crossings of the input voltage and the maximum value at the peak, i.e.,

$$f_{s\_max} = f_{s\_0} = \frac{V_m^2}{4 L_b P_o} \quad (13)$$

$$f_{s\_min} = f_{s\_\pi/2} = \frac{V_m^2}{4 L_b P_o} \left(1 - \frac{V_m}{V_o} \right) \quad (14)$$

The variation range of the switching frequency can be defined as the ratio of f_{s_max} and f_{s_min}. That is

$$\frac{f_{s\_max}}{f_{s\_min}} = \frac{1}{1 - \frac{V_m}{V_o}} \quad (15)$$

Setting the minimum switching frequency, the maximum value of the inductor can be figured out from (14), i.e.,

$$L_b = \frac{V_m^2}{4 f_{s\_min} P_o} \left(1 - \frac{V_m}{V_o} \right) \quad (16)$$

III. THE OPTIMAL 3RD HARMONIC FOR MINIMUM VARIATION RANGE OF SWITCHING FREQUENCY

When the input current contains the 3rd harmonic with an initial phase of 0, it can be expressed as

$$i_{in} = I_1 \left(\sin \omega t + I_3^* \sin 3\omega t \right) = \frac{2 P_o}{V_m} \left(\sin \omega t + I_3^* \sin 3\omega t \right) \quad (17)$$

where I_1 is the amplitude of the fundamental component, I_3^* is the normalized amplitude of the 3rd harmonic with the base of fundamental component.

From (6) and (17), t_{on} can be calculated as

$$t_{on} = \frac{4L_b P_o}{V_m^2}\left[1 + I_3^*\left(3 - 4\sin^2 \omega t\right)\right] \quad (18)$$

According to (4) and (18), f_s can be derived as

$$f_s = \frac{1}{t_{on} + t_{off}} = \frac{V_m^2}{4L_b P_o}\frac{1 - \dfrac{V_m}{V_o}\sin\omega t}{\left[1 + I_3^*\cdot\left(3 - 4\sin^2 \omega t\right)\right]} \quad (19)$$

From (19), the switching frequency at the zero crossings and the peak of input voltage are

$$f_{s\_0} = \frac{V_m^2}{4L_b P_o\left(1 + 3I_3^*\right)} \quad (20)$$

$$f_{s\_\pi/2} = \frac{V_m^2}{4L_b P_o}\frac{1 - V_m/V_o}{1 - I_3^*} \quad (21)$$

A. Derivation of the Optimal 3rd harmonic

The derivative of (19) with respect to ωt is

$$f_s' = -\frac{V_m^2}{4L_b P_o\left(1 + 3I_3^*\right)}\frac{\cos\omega t\left(\dfrac{V_m}{V_o}\dfrac{4I_3^*}{1 + 3I_3^*}\sin^2 \omega t - 2\dfrac{4I_3^*}{1 + 3I_3^*}\sin\omega t + \dfrac{V_m}{V_o}\right)}{\left(1 - \dfrac{4I_3^*}{1 + 3I_3^*}\sin^2 \omega t\right)^2} \quad (22)$$

Let $f_s' = 0$, and ωt can be derived as

$$\omega t = \arcsin\left(\frac{V_o}{V_m} - \sqrt{\frac{V_o^2}{V_m^2} - \frac{1 + 3I_3^*}{4I_3^*}}\right) \quad (23(a))$$

$$\omega t = \pi - \arcsin\left(\frac{V_o}{V_m} - \sqrt{\frac{V_o^2}{V_m^2} - \frac{1 + 3I_3^*}{4I_3^*}}\right) \quad (23(b))$$

$$\omega t = \pi/2 \quad (23(c))$$

When $I_3^* < \dfrac{V_m}{8V_o - 7V_m}$, i.e., $\dfrac{V_o}{V_m} - \sqrt{\dfrac{V_o^2}{V_m^2} - \dfrac{1 + 3I_3^*}{4I_3^*}} > 1$, 23(a) and 23(b) are imaginary roots. During $[0, \pi/2]$, $f_s' \le 0$, the switching frequency decreases with ωt increasing. So the maximum and minimum value occur at $\omega t=0$ and $\omega t=\pi/2$ respectively. In the same way, when $I_3^* \ge \dfrac{V_m}{8V_o - 7V_m}$, i.e., $\dfrac{V_o}{V_m} - \sqrt{\dfrac{V_o^2}{V_m^2} - \dfrac{1 + 3I_3^*}{4I_3^*}} \le 1$, the switching frequency decreases with ωt increasing during $[0, \arcsin\left(\dfrac{V_o}{V_m} - \sqrt{\dfrac{V_o^2}{V_m^2} - \dfrac{1 + 3I_3^*}{4I_3^*}}\right)]$ and increases with ωt increasing during $[\arcsin\left(\dfrac{V_o}{V_m} - \sqrt{\dfrac{V_o^2}{V_m^2} - \dfrac{1 + 3I_3^*}{4I_3^*}}\right), \pi/2]$.

Substitution of (23(a)) into (19) leads to

$$f_{s\_\arcsin\left(\frac{V_o}{V_m} - \sqrt{\frac{V_o^2}{V_m^2} - \frac{1+3I_3^*}{4I_3^*}}\right)} = \frac{V_m^2}{4L_b P_o}\frac{\dfrac{V_m}{V_o}\sqrt{\dfrac{V_o^2}{V_m^2} - \dfrac{1 + 3I_3^*}{4I_3^*}}}{\left(1 + 3I_3^*\right)\left[2 - \dfrac{8I_3^*}{1 + 3I_3^*}\left(\dfrac{V_o^2}{V_m^2} - \dfrac{V_o}{V_m}\sqrt{\dfrac{V_o^2}{V_m^2} - \dfrac{1 + 3I_3^*}{4I_3^*}}\right)\right]} \quad (24)$$

According to (20) and (21), it can be seen that the switching frequency at $\omega t=0$ is higher than at $\omega t=\pi/2$ when $I_3^* \le \dfrac{V_m}{4V_o - 3V_m}$. On the contrary, when $I_3^* \ge \dfrac{V_m}{4V_o - 3V_m}$ the switching frequency at $\omega t=\pi/2$ is higher than at $\omega t=0$.

Based on the above analysis, dividing the value range of I_3^* into three parts, the ratio of the maximum and minimum switching frequency can be expressed as

$$\frac{f_{s\_max}}{f_{s\_min}} = \begin{cases} \dfrac{f_{s\_0}}{f_{s\_\pi/2}} = \dfrac{1 - I_3^*}{\left(1 + 3I_3^*\right)\left(1 - \dfrac{V_m}{V_o}\right)} & 0 \le I_3^* \le \dfrac{V_m}{8V_o - 7V_m} \\[3ex] \dfrac{f_{s\_0}}{f_{s\_\arcsin\left(\frac{V_o}{V_m}\sqrt{\frac{V_o^2}{V_m^2}\frac{1+3I_3^*}{4I_3^*}}\right)}} = \dfrac{\left[2 - \dfrac{8I_3^*}{1+3I_3^*}\left(\dfrac{V_o^2}{V_m^2} - \dfrac{V_o}{V_m}\sqrt{\dfrac{V_o^2}{V_m^2} - \dfrac{1+3I_3^*}{4I_3^*}}\right)\right]}{\dfrac{V_m}{V_o}\sqrt{\dfrac{V_o^2}{V_m^2} - \dfrac{1+3I_3^*}{4I_3^*}}} & \dfrac{V_m}{8V_o - 7V_m} \le I_3^* \le \dfrac{V_m}{4V_o - 3V_m} \\[3ex] \dfrac{f_{s\_\pi/2}}{f_{s\_\arcsin\left(\frac{V_o}{V_m}\sqrt{\frac{V_o^2}{V_m^2}\frac{1+3I_3^*}{4I_3^*}}\right)}} = \dfrac{\dfrac{1 - \dfrac{V_m}{V_o}}{1 - I_3^*}}{\dfrac{\dfrac{V_m}{V_o}\sqrt{\dfrac{V_o^2}{V_m^2} - \dfrac{1+3I_3^*}{4I_3^*}}}{\left(1+3I_3^*\right)\left[2 - \dfrac{8I_3^*}{1+3I_3^*}\left(\dfrac{V_o^2}{V_m^2} - \dfrac{V_o}{V_m}\sqrt{\dfrac{V_o^2}{V_m^2} - \dfrac{1+3I_3^*}{4I_3^*}}\right)\right]}} & \dfrac{V_m}{4V_o - 3V_m} < I_3^* < 1 \end{cases} \quad (25)$$

So when the output voltage is 400V and the input voltage changes from 176V to 264V, the value of V_m/V_o varies between 0.62 and 0.93. According to (25), the range of switching frequency that depends on V_m/V_o and I_3^* is plotted in Fig. 4. It can be seen that, there must be a corresponding optimal I_3^* at any input voltage which leads to a minimum range of the switching frequency.

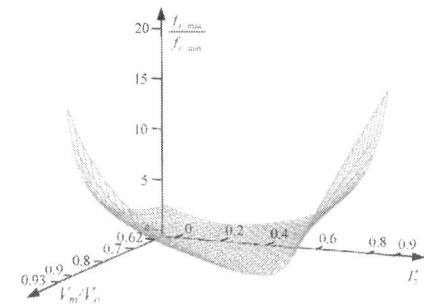

Fig. 4. Surface of variation range of the switching frequency as the function of V_m/V_o and I_3^*.

The derivative of (25) with respect to I_3^* leads to

$$\frac{d\dfrac{f_{s\_max}}{f_{s\_min}}}{dI_3^*} < 0 \quad 0 \leq I_3^* < \frac{V_m}{4V_o - 3V_m}$$

$$\frac{d\dfrac{f_{s\_max}}{f_{s\_min}}}{dI_3^*} = 0 \quad I_3^* = \frac{V_m}{4V_o - 3V_m} \qquad (26)$$

$$\frac{d\dfrac{f_{s\_max}}{f_{s\_min}}}{dI_3^*} > 0 \quad \frac{V_m}{4V_o - 3V_m} < I_3^* \leq 1$$

Therefore the optimal I_3^* which enables the minimum variation range of the switching frequency is

$$I_{3\_optimal}^* = \frac{\dfrac{V_m}{V_o}}{4 - \dfrac{3V_m}{V_o}} = \frac{V_m}{4V_o - 3V_m} \qquad (27)$$

Substituting (27) into (17) and (18), the input current and on time can be rewritten respectively as

$$i_{in} = \frac{8P_o\left(V_o - V_m\sin^2\omega t\right)\sin\omega t}{V_m\left(4V_o - 3V_m\right)} \qquad (28)$$

$$t_{on} = \frac{4L_b P_o V_o\left(V_o - V_m\sin^2\omega t\right)}{V_m^2\left(4V_o - 3V_m\right)} \qquad (29)$$

According to the standard requirement of IEC61000-3-2, Class D, the ratio of 3rd harmonic content and input power should meet

$$\frac{I_3/\sqrt{2}}{P_{in}} = \frac{I_3^* I_1/\sqrt{2}}{\left(V_m/\sqrt{2}\right)\cdot\left(I_1/\sqrt{2}\right)} \leq 3.4\cdot 10^{-3} \qquad (30)$$

That is

$$I_3^* \leq I_{3\_limit}^* = 3.4\cdot 10^{-3}\cdot\left(V_m/\sqrt{2}\right) \qquad (31)$$

where $I_{3\_limit}^*$ is the harmonic upper-limit value.

According to (27) and (31), $I_{3\_optimal}^*$ and $I_{3\_limit}^*$ can be plotted in Fig. 5. As seen, at any input voltage between 176V and 264V, the optimal 3rd harmonic content is always lower than the harmonic limit of IEC61000-3-2, Class D.

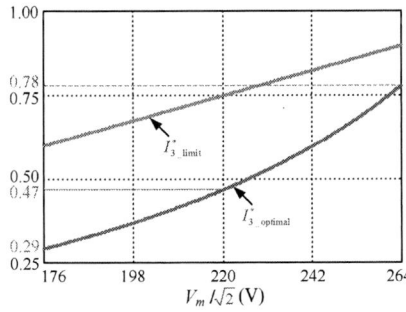

Fig. 5. $I_{3\_optimal}^*$ and $I_{3\_limit}^*$ over the whole input voltage range.

As seen in Fig. 5, the optimal I_3^* are 0.29, 0.47 and 0.78 respectively, when the input voltage are 176, 220 and 264 VAC. On the basis of convertor design index (which will be given at section IV), assigning the inductance value L_b=0.64 mH with TC, f_s with different 3rd harmonic content can be plotted in Fig. 6. It can be seen that, with I_3^* increasing, f_s gradually decreases at ωt=0 and increases at ωt=π/2, the minimum f_s still occurs at ωt=π/2. With I_3^* further increasing, the minimum f_s will occur between period [0, π/2] and [π/2, π], and the variation range of switching frequency gradually decreases until $I_3^* = I_{3\_optimal}^*$. After $I_3^* > I_{3\_optimal}^*$, f_s at ωt=0 is higher than at ωt=π/2 and the variation range of switching frequency begins increasing.

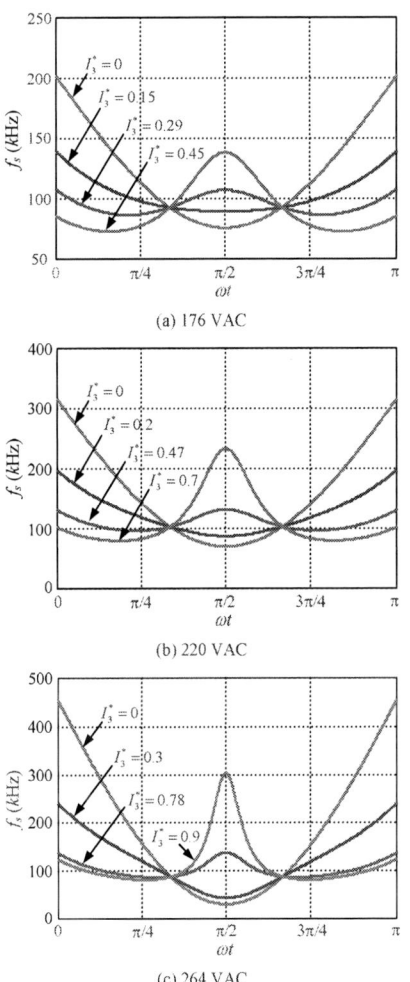

Fig. 6. Switching frequency with different 3rd harmonic content.

B. Implementation of the MFVC Circuit.

The implementation circuit for MFVC is shown in Fig. 7. The rectified input voltage is sensed through a voltage divider composed of R_1 and R_2, and $v_{gf} = k_{vg}V_m|\sin \omega t|$, where k_{vg} is the voltage sensor gain. R_3, D_1, C_1 and R_4 are the circuit to obtain the peak value of the rectified input voltage, i.e., $v_B = v_C = k_{vg} \cdot V_m$. Please be noted that here R_3 is used to limit the charging current of C_1, and it is very small compared to R_4 The output voltage is sensed through a voltage divider composed of R_5 and R_6, and let $R_5/R_6 = R_1/R_2$, so $v_D = k_{vg}V_o$. Multiply v_A by v_D, then we get $v_E = k_{vg}V_m\sin^2 \omega t$. Letting $R_7 = R_8$ and $R_9 = R_{10}$, then $v_F = k_{vg}(V_o - V_m\sin^2 \omega t)$. And multiplying v_A by v_D, then $v_G = k_{vg}^2 (V_o - V_m\sin^2 \omega t)V_m\sin \omega t$. The output voltage is regulated through the error amplifier, and the sensed output voltage through a voltage divider composed of R_{11} and R_{13} compares with the reference voltage V_{og}, R_{12} and C_2 form the compensation network. v_G and v_{EA} are sent to the multiplier, then v_H is sent to the comparator and compared with v_{Rt} which determined by inductor current. The output of the comparator and ZCD determines the on-time value, which varies as expressed in (29).

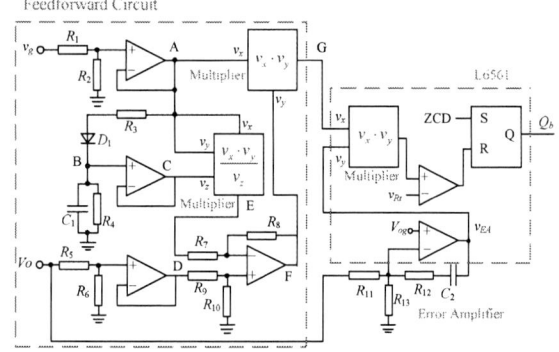

Fig. 7. Implementation circuit for MFVC.

IV. PERFORMANCE COMPARISON

A. The design of the Inductor

Substitution of (27) into (24) leads to

$$f_{s\_min} = \frac{V_m^2}{32L_bP_o} \frac{\left(4 - \frac{3V_m}{V_o}\right)}{\sqrt{\frac{V_o}{V_m}}\left(\sqrt{\frac{V_o}{V_m}} - \sqrt{\frac{V_o}{V_m} - 1}\right)} \quad (32)$$

From (32) it can be figured out that

$$L_b = \frac{V_m^2}{32f_{s\_min}P_o} \frac{\left(4 - \frac{3V_m}{V_o}\right)}{\sqrt{\frac{V_o}{V_m}}\left(\sqrt{\frac{V_o}{V_m}} - \sqrt{\frac{V_o}{V_m} - 1}\right)} \quad (33)$$

According to (16) and (32), the inductance value with TC and MFVC are depicted in Fig. 8. It can be seen that the critical inductance with TC and MFVC are $L_{b1} = 1.28mH$ and $L_{b2} = 3.66mH$ respectively.

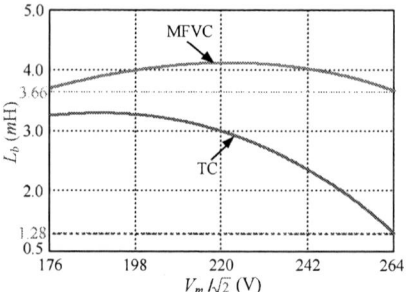

Fig. 8. Critical inductors over the input voltage range.

The inductor peak current with TC and MFVC are

$$i_{Lb1\_pk} = \frac{4P_o}{V_m}|\sin \omega t| \quad (34)$$

$$i_{Lb2\_pk} = \frac{4P_o}{V_m}\left(\sin \omega t + \frac{V_m}{4V_o - 3V_m}\sin 3\omega t\right) \quad (35)$$

From (34) and (35), the peak value of the inductor current with two control scheme can be obtained, which are plotted in Fig. 9. As seen, the peak value is reduced with MFVC at 176 and 220 VAC.

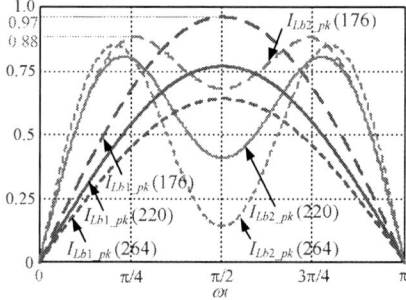

Fig. 9. The peak value of the inductor current with two control scheme.

In a line cycle the RMS value of the inductor current, the diode current and the switching tube current are

$$I_{Lb\_rms} = \frac{2\sqrt{2}}{\sqrt{3}}\frac{P_o}{V_m}\sqrt{1 + I_3^{*2}} \quad (36)$$

$$I_{Qb\_rms} = \sqrt{\frac{16P_o^2}{3\pi V_m^2}\int_0^\pi \left(\sin \omega t + I_3^*\sin 3\omega t\right)^2\left(1 - \frac{V_m\sin \omega t}{V_o}\right)d\omega t} \quad (37)$$

$$I_{Db\_rms} = \sqrt{\frac{16P_o^2}{3\pi V_m V_o}\int_0^\pi \left(\sin \omega t + I_3^*\sin 3\omega t\right)^2\sin \omega t d\omega t} \quad (38)$$

Substituting $I_3^* = 0$ and (27) into (36)-(38), I_{Lb1_rms}, I_{Qb1_rms}, I_{Db1_rms} and I_{Lb2_rms}, I_{Qb2_rms}, I_{Db2_rms} can be obtained, which are plotted in Fig. 10. As seen, I_{Lb2_rms}, I_{Qb2_rms}, I_{Db2_rms} are slightly higher than I_{Lb1_rms}, I_{Qb1_rms}, I_{Db1_rms}.

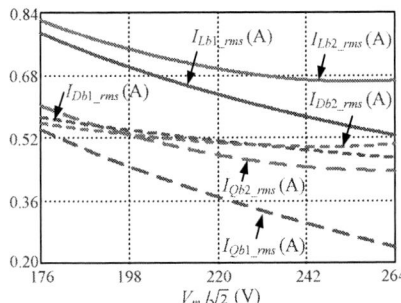

Fig. 10. The RMS value of the current with two control scheme.

B. Reduction of the Variation Range of Switching Frequency

Substitution of (27) into (17), f_s with MFVC is derived as

$$f_s = \frac{V_m^2}{4 L_b P_o} \cdot \frac{1 - \frac{V_m}{V_o}\sin\omega t}{\left[1 + \frac{V_m}{4 V_o - 3 V_m}\cdot(3 - 4\sin^2\omega t)\right]} \quad (39)$$

Substituting $L_{b1}=1.28mH$ into (12) and $L_{b2}=3.66mH$ into (39), the switching frequency with TC and MFVC are depicted in Fig. 11.and Fig. 12.

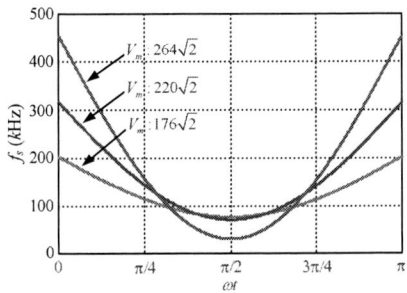

Fig. 11. Switching frequency with TC.

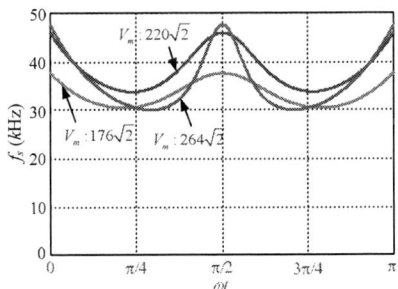

Fig. 12. Switching frequency with MFVC.

Substituting (27) into (25) leads to

$$\frac{f_{s\_max}}{f_{s\_min}} = 2\left(\frac{V_o}{V_m} - \sqrt{\frac{V_o^2}{V_m^2} - \frac{V_o}{V_m}}\right) \quad (40)$$

According to (15) and (40), the variation range of switching frequency with TC and MFVC are depicted in Fig. 13. It can be seen the variation range of switching frequency is reduced greatly especially at high input voltage. When the input voltage is 264 VAC, the ratio is reduced from 15 to 1.2.

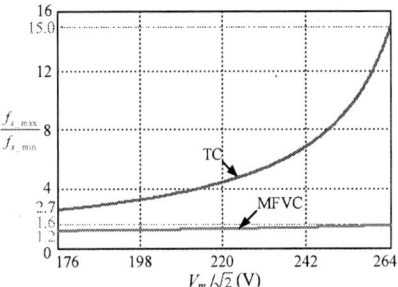

Fig. 13. Variation range of switching frequency with TC and MFVC

C. Reduction of the Output Voltage Ripple

When the CRM Boost PFC converter is operated with TC, from (1) and (5), the normalized instantaneous input power can be derived as

$$p_{in\_1}^*(t) = \frac{v_{in}(t)\cdot i_{in}(t)}{P_o} = 2\sin^2\omega t \quad (41)$$

From (1) and (28), the normalized instantaneous input power with MFVC is derived as

$$p_{in\_2}^*(t) = 2\sin\omega t\left(\sin\omega t + \frac{V_m}{4V_o - 3V_m}\sin 3\omega t\right) \quad (42)$$

From (41) and (42), the normalized instantaneous input power in a half line cycle with two control scheme are plotted in Fig. 14.

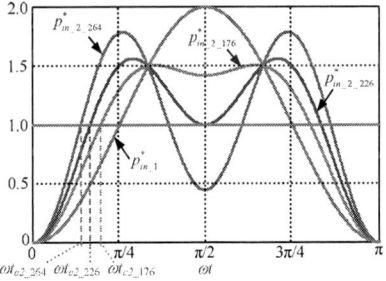

Fig. 14. Normalized instantaneous input power in a half line cycle with TC and MFVCS.

According to Fig. 14, When $p_{in}^*>1$, the storage capacitor C_o is charged, and when $p_{in}^*<1$, C_o is discharged. The normalized energy discharging C_o (which equals the charged energy) in a half line cycle with TC and MFVD are

$$\Delta E_1^* = \left\{ 2 \int_0^{t_{c1}} \left[1 - p_{in,1}^*(t) \right] \cdot dt \right\} \Big/ \left(T_{line}/2 \right) \qquad (43)$$

$$\Delta E_2^* = \left\{ 2 \int_0^{t_{c2}} \left[1 - p_{in,2}^*(t) \right] \cdot dt \right\} \Big/ \left(T_{line}/2 \right) \qquad (44)$$

where t_{c1} and t_{c2} are the time instants when p_{in}^* crosses 1 with TC and MFVC, respectively.

ΔE_1^* and ΔE_2^* can be also expressed as:

$$\Delta E_1^* \approx \frac{\frac{1}{2}C_o\left(V_o + \frac{\Delta V_{o1}}{2}\right)^2 - \frac{1}{2}C_o\left(V_o - \frac{\Delta V_{o1}}{2}\right)}{P_o T_{line}/2} = \frac{2C_o V_o \cdot \Delta V_{o1}}{P_o T_{line}} \qquad (45)$$

$$\Delta E_2^* \approx \frac{\frac{1}{2}C_o\left(V_o + \frac{\Delta V_{o2}}{2}\right)^2 - \frac{1}{2}C_o\left(V_o - \frac{\Delta V_{o2}}{2}\right)}{P_o T_{line}/2} = \frac{2C_o V_o \cdot \Delta V_{o2}}{P_o T_{line}} \qquad (46)$$

where ΔV_{o1} and ΔV_{o2} are the output voltage ripple with TC and MFVC, respectively.

From (43)-(46), ΔV_{o1} and ΔV_{o2} are derived as

$$\Delta V_{o1} = 2P_o \int_0^{t_{c1}} \left[1 - p_{in,1}^*(t) \right] dt \Big/ C_o V_o \qquad (47)$$

$$\Delta V_{o2} = 2P_o \int_0^{t_{c2}} \left[1 - p_{in,2}^*(t) \right] dt \Big/ C_o V_o \qquad (48)$$

According to (41), (42), (47), (48) and the specifications of the converter which will be given in Section V, the curves of output voltage ripple can be plotted in Fig. 15. As seen, over the whole range of the input voltage, with TC, the output voltage ripple keeps at 7.02V. With MFVC, the output voltage ripple decreases from 5.33V to 3.89V. The output voltage ripple with MFVC is reduced greatly compared to that with TC.

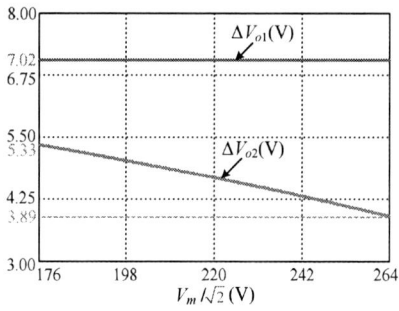

Fig. 15. Output voltage ripple with TC and MFVC.

V. EXPERIMENTAL VERIFICATION

In order to verify the validity of the proposed control scheme, a prototype has been built and tested in the lab. The specifications of the prototype are as follows:

- input voltage: $v_{in} = 176 \sim 264$ VAC / 50 Hz;

- output voltage: $V_o = 400$ VDC;

- output power: $P_o = 60$ W;

- minimum of switching frequency: $f_{s\_min} = 30$ kHz.

- Boost inductor: 1.28 mH (TC), 3.66 mH (MFVC);

- output filter capacitor: $C_o = 68$ μF;

- control IC: L6561.

Fig. 16(a) and Fig. 17(a) show the experimental waveforms of the rectified input voltage, cubic harmonic (which is used to generate 3rd harmonic), output voltage ripple and inductor current with TC and MFVC at 198 VAC input.

Figs. 16(b)-(c) and Figs.17(b)-(d) show the experimental waveforms of the inductor current in switching cycles with TC and MFVC at 198 VAC input.

Combining with Fig. 16(a) and Fig. 17(a), it can be seen that compared to TC, the output voltage ripple is reduced with MFVC. And the waveform of input current with MFVC has been distorted.

From Figs. 16(b)-(c) and Figs.17(b)-(d), it can be seen that the variation range of switching frequency is reduced greatly with MFVC.

The experimental results agree well with the analysis.

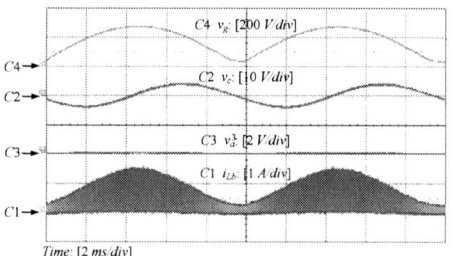

(a) Experimental waveforms in frequency cycle

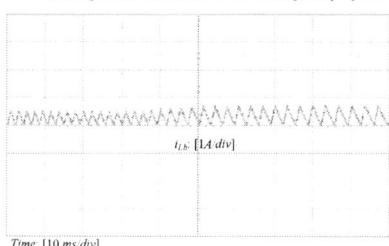

(b) The inductor current in switching cycles at $\omega t = 0$

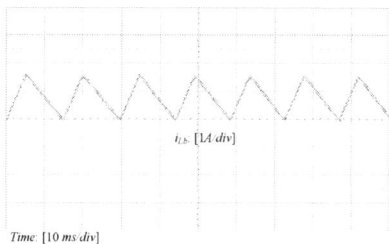

(c) The inductor current in switching cycles at $\omega t = \pi/2$

Fig. 16. Eperimental waveforms with TC when $v_{in}=198$V.

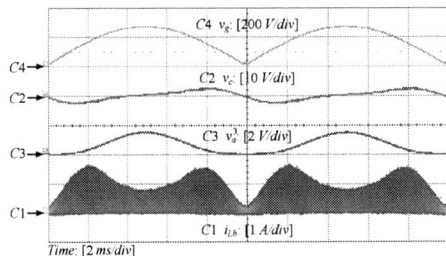

(a) Experimental waveforms in frequency cycle

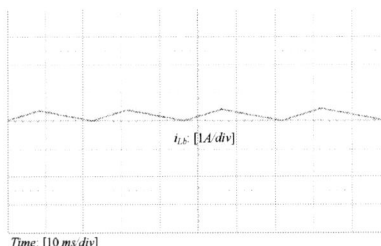

(b) The inductor current in switching cycles at $\omega t = 0$

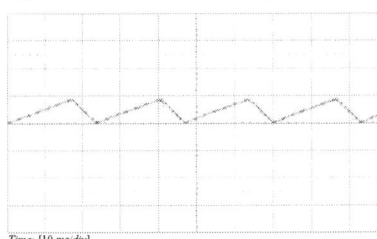

(c) The inductor current in switching cycles at $\omega t = \pi/2$

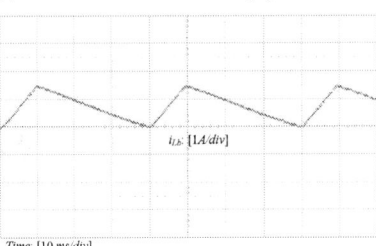

(d) The inductor current in switching cycles at $\omega t = \arcsin\left(\dfrac{V_o}{V_m} - \sqrt{\dfrac{V_o^2}{V_m^2} - \dfrac{1+3I_3^*}{4I_3^*}}\right)$

Fig. 17. Eperimental waveforms with MFVC when v_{in}=198V.

VI. CONCLUSIONS

The paper proposes a minimum switching frequency variation range control for CRM Boost PFC converter. Compared to that with traditional control,

1). The switching frequency range is reduced, which is helpful for the design of the inductor and EMI filter;

2). The output voltage ripple can be greatly reduced;

3). The input current harmonics still meet the IEC61000-3-2 Class D standard.

REFERENCES

[1] Z. Yang and P. C. Sen, "Recent developments in high power factor switch-mode converters," in *Proc. IEEE CCECE*, pp. 477-480, 1998.

[2] N. Genc, I. Iskender, and M. A. Celik, "Application of interleaved bridgeless boost PFC converter without current sensing," in *Proc. IEEE PEOCO*, pp. 1-6, 2014.

[3] H. J. Chen, S. Y. Lee, Y. M. Chen, Y. L. Chen, and K. H. Liu, "A stepping on-time adjustment method for interleaving three-channel critical mode boost PFC converter," in *Proc. IEEE ECCE*, pp. 749-754, 2013.

[4] Y. M. Liu and L. K. Chang, "Single-stage soft-switching ac-dc converter with input-current shaping for universal line applications," *IEEE Trans. Ind. Electron.*, vol. 56, no. 2, pp. 467-479, 2009.

[5] H. Ma, Y. Ji, and Y. Xu, "Design and analysis of single-stage power factor correction converter with a feedback winding," *IEEE Trans. Power Electron.*, vol. 25, no. 6, pp. 1460-1470, 2010.

[6] L. S. Fan and A. M. Khambadkone, "A Simple Digital DCM Control Scheme for Boost PFC Operating in Both CCM and DCM," *IEEE Trans. Ind. Electron.*, vol. 47, no. 4, pp. 1802-1812, 2011.

[7] J. S. Lai and D. Chen, "Design consideration for power factor correction boost converter operating at the boundary of continuous conduction mode and discontinuous conduction mode," in Proc. IEEE APEC, pp. 267-273, 1993.

[8] L. Hang, M. Zhang, L. M. Tolbert, and Z. Lu, "Digitized feed forward compensation method for high-power-density three-phase Vienna PFC converter," IEEE Trans. Ind. Electron., vol. 60, no. 4, pp. 1512–1519,Apr. 2013.

[9] Doron Shmilovitz, "On the Definition of Total Harmonic Distortion and Its Effect on Measurement Interpretation," IEEE Transactions on Power Delivery, vol. 20, no.1, pp. 526-528, January 2005.

[10] J. W. Kim, S.-M. Choi, and K. T. Kim, "Variable On-Time Control of the Critical Conduction Mode Boost Power Factor Correction Converter to Improve Zero-crossing Distortion," *Power Electronics and Drives Systems, PEDS 2005*, pp. 1542-1546, 2005.

[11] J. S. Tsai, Y. T. Chen, Y. T. Chen, C. L. Ni, C. Y. Chen, and k. H. Chen, "Perturbation On-Time (POT) Technique in Power Factor Correction (PFC) Controller for Low Total Harmonic Distortion and High Power Factor," *IEEE Trans. Power Electron.*, vol. 28, no. 1, pp. 199-212, Jan. 2013.

[12] Y. P. Su, C. L. Ni, C. Y. Chen, Y. T. Chen, J. C. Tsai, and K. H. Chen, "Boundary Conduction Mode Controlled Power Factor Corrector With Line Voltage Recovery and Total Harmonic Distortion Improvement Techniques," IEEE Trans. Power Electron., vol. 61, no. 7, pp.3220-3231, Jul. 2014.

[13] S. H. Tang, D. Chen, C.-S. Huang, C. Y. Liu, and H. Liu, "A new on-time adjustment scheme for the reduction of input current distortion of critical mode power factor correction boost converters," in Proc. IEEE Int. Power Electron. Conf., pp. 1717–1724, Jun. 2010.

[14] T. Ishii and Y. Mizutani, "Power factor correction using interleaving technique for critical mode switching converters," in Proc. IEEE Power Electron.Spec. Conf., pp. 905–910, May 1998.

[15] B. Lu, "A novel control method for interleaved transition mode PFC," in Proc. IEEE Appl. Power Electron. Conf., pp. 697–701, Feb. 2008.

A sustained increase of input current distortion in active input current shapers to eliminate electrolytic capacitor for designing ac to dc HB-LED drivers for retrofit lamps applications

D. G. Lamar, M. Arias, A. Rodriguez, and J. Sebastian
Universidad de Oviedo. Grupo de Sistemas Electrónicos de Alimentación (SEA)
Edificio Departamental nº 3. Campus Universitario de Viesques.
33204 Gijón. SPAIN
gonzalezdiego@uniovi.es

A. Fernandez[1] and J. A. Villarejo[2]
[1]ESA - European Space Agency
Noordwijk – THE NETHERLANDS
arturo.fernandez@esa.int
[2]Universidad P de Cartagena
Departamento de Tecnologia Electronica
Cartagena, SPAIN
jose.villarejo@upct.es

Abstract—Nowadays, the solid-sate lighting technology evolution has changed traditional solutions in lighting. High-Brightness Light–Emitting Diodes (HB-LEDs) have become very attractive light sources due to their excellent characteristics: high efficiency, high life-time and low maintenance. It is evident that HB-LED drivers must be durable and efficient to achieve these advantages. Moreover, for replacing incandescent bulbs, the ac-dc HB-LED driver must be low cost and comply with international regulations (i.e. injection of low frequency harmonics into the mains). Traditionally, authors have focused its efforts on increasing efficiency. All these solutions obviate the elimination of traditional electrolytic capacitor of ac to dc converters, highlighting that this is the price to pay for a very low-cost solution. This paper presents a new proposal to design a simple and low-cost ac to dc HB-LED driver for retrofit lamps without electrolytic capacitor. The proposed solution comes from a very well-known technique used in the past: Active Input Current Shapers (AICS), but in this case without electrolytic capacitor. If the electrolytic capacitor of an AICS is removed, then low frequency ripple arises in its intermediate dc bus, increasing the distortion of the line input which already has appreciable distortion. However, the increase of distortion is very slight. Also, the low frequency ripple is not transferred to the output due to the high output dynamic response of AICS, avoiding flickering. This paper presents a theoretical analysis that guarantees a trade of between compliance with international regulations and the use of other capacitor technologies different from the electrolytic one. Finally, a 24W experimental prototype has been built and tested in order to validate the theoretical results presented in this digest.

Keywords—Ac to dc power conversion, harmonic distortion, LEDs, lighting, power factor, switched mode power supplies.

I. INTRODUCTION

It is known that High-Brightness Light–Emitting Diodes (HB-LEDs) are a fast emergent technology, considered as the true alternative to many mature technologies (i.e. incandescent

bulbs, compact fluorescent lamps, etc.) due to its high efficiency, low maintenance and durability. To perform these advantages HB-LED drivers must be both durable and efficient.

Since HB-LEDs are diodes, the default method for driving them is controlling the dc forward current trough this semiconductor device. If the primary energy source is the ac line, then some type of ac to dc converter must be placed between the line and the HB-LEDs. Also, it is known that the low-frequency harmonic content of the line current must comply with specific regulations (IEC 61000-3-2 [3-5] and ENERGY STAR® program [6]). Traditionally, these regulations establish a very strict harmonic content for lighting (e.g. IEC 61000-3-2, Class C), so that only sinusoidal line waveforms is able to comply with the aforementioned regulations. Therefore, the only practical method to comply with these regulations is to use active high Power Factor (PF) converters. These converters are known as Power Factor Correctors (PFCs), which are expensive and complex solutions. However, for power levels lower than 25 W the compliance with IEC 61000-3-2 regulation becomes more relaxed due to the fact that luminaries must comply with this regulation in Class D [6]. Hence, new solutions can arise.

A possible application for substituting incandescent bulbs lamps is to use two strings of around 10 HB-LEDs of 1 W in parallel connected to the output of an ac to dc driver to produce the same luminance flux as the one produced by a 100 W incandescent bulb. These configurations supply output voltages around of 20 V and power levels lower than 24 W. The most extended solution adopted is to use a flyback converter operating in Discontinuous Conduction Mode (DCM) with switching frequencies below 100 kHz in order to obtain efficiencies around 82 %. Traditionally, authors have focused their efforts on increasing the efficiency of the ac-to-dc driver in spite of increasing its cost and complexity. Some examples are solutions based on an asymmetrical half bridge flyback converter [7], two

---

This work was supported by the Spanish Ministry of Education and Science under Project MINECO-13-DPI2013-47176- C2-2-R, by Government of the Principality of Asturias under the Project FC-15-GRUPIN14-143 and by European Regional Development Fund (ERDF) grants.

stage resonant buck converter [8] or tapped-inductor buck converter [9-10]. However, all these proposals exhibit a main drawback: the use of an electrolytic capacitor to reduce the low frequency ripple of the output current.

This paper presents a low-cost ac to dc HB-LED driver conceived from a well-known concept: the Active Input Current Shapers (AICS). The proposal of this solution comes from the latest regulations modifications for low power lighting equipment (i.e. IEC 61000-3-2, Class D) which are now more relaxed than previous ones (i.e. IEC 61000-3-2, Class C), and therefore, a sinusoidal input current is not needed. In AISCs, if the electrolytic capacitor that stabilize intermediate bus is substituted by other technology, then some low frequency ripple arises in the intermediate bus, increasing the distortion of the line current. However, this added distortion is slight in comparison to traditional distortion of AICS, validating the proposed idea without electrolytic capacitor. Moreover, due to its fast dynamic response, the low frequency ripple is not transferred to the output. Finally, an ac to dc HB-LED driver for retrofit lamps applications without electrolytic capacitor is achieved.

II. REVIEWING ACTIVE INTUT CURRENT SHAPER (AICS)

A. Basic concepts about AISC

The concept of the AICS is very well known in the design of ac to dc Switching Mode Power Supplies (SMPS), [13-17]. This solution is based on conventional dc to dc converters with a slight modification: an additional output, obtained from the converter transformer (Fig. 1a), is connected between the diode bridge and the bulk capacitor (C_B).

This output is called "delayed output" in [11] and it was proposed into the context of two fully regulate outputs in dc to dc converters [12]. It seems similar to a conventional forward output. However, an extra inductor (L_D) is placed between one terminal of the secondary side transformer and the diode D_1 (Fig. 1a). With this extra inductor and with L working in Continuous Conduction Mode (CCM, i.e. L>> L_D), the Thévenin equivalent circuit of the "delayed output" becomes a voltage source (V_S, see Fig. 1b) a loss-free resistor (R_{LF}, see Fig. 1b). This "delayed output" recycles some amount of energy redirecting it to the input in order to shape line input current. By properly choosing the values of these two elements (i.e. V_S and R_L), the AISC can achieve both high efficiency and limited low-frequency harmonic content of input current.

The current in a half cycle of input voltage can be easily deduced from the AICS behavior. The input rectifier start to conduct when input voltage (i.e. $v_g(t)=V_{gp}\cdot|\sin(\omega_L t)|$) reaches ($V_S-V_C$). Thus, the expression of the rectified input current can be expressed as:

$$i_{gdc}(t) = \frac{V_{gp}\cdot|\sin(\omega_L t)|-V_C+V_S}{R_{LF}}, \quad (1)$$

where V_C is the voltage of intermediate bus, ω_L and V_{gp} are the angular frequency and the peak value of input voltage respectively. It should be noted that this expression is only valid for the interval [$(\pi-\phi_C)/2$, $(\pi+\phi_C)/2$], being ϕ_C the conduction angle (see Fig. 1). By equaling (1) to zero, the expression of the conduction angle can be easily calculated:

$$\phi_C = 2\cos^{-1}\left(\frac{V_C-V_S}{V_{gp}}\right). \quad (2)$$

Therefore, the line input current is defined by (1) into the [$(\pi-\phi_C)/2$, $(\pi+\phi_C)/2$] interval and by zero outside of this interval of positive semi-cycle of the line input voltage. Also, in the same way for the negative semi-cycle of the line input voltage (see Fig. 1). Moreover, it is important to say that the higher ϕ_C the greater amount of energy recycled to the input, and therefore, the lower efficiency.

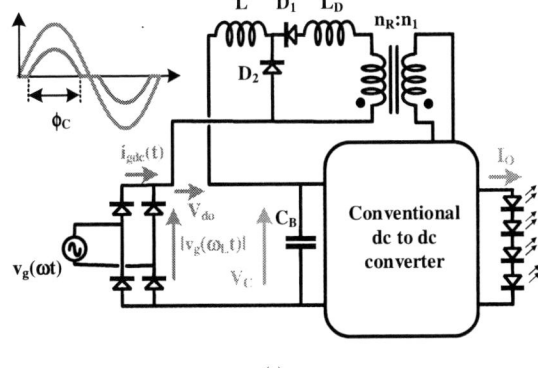

(a)

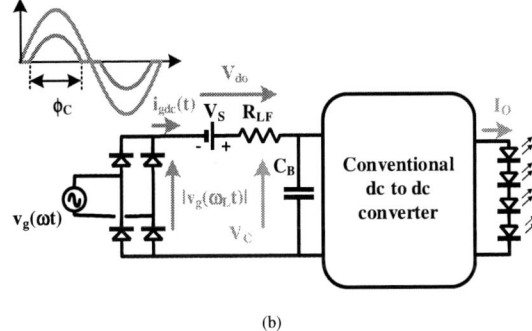

(b)

Fig. 1. a) AISC solution. b) Equivalent circuit of AISC.

From the expression of the input voltage, (1) and (2), the average input power will be:

$$P_g = \frac{1}{\pi}\int_{(\phi_C-\pi)/2}^{(\phi_C+\pi)/2}\left[i_{gdc}(t)\cdot V_{gp}\cdot|\sin(\omega_L t)|\right]dt =$$
$$= \frac{V_{gp}^2}{2\pi R_{LF}}\left(\phi_C - \sin(\phi_C)\right). \quad (3)$$

The rectified input current can be rewritten as a function of the average input power, conduction angle and peak value of the input voltage by using (1), (2) and (3):

$$i_{gdc}(t) = \frac{2\pi P_g}{V_{gp}}\left(\frac{|\sin(\omega_L t)|-\cos\left(\frac{\phi_C}{2}\right)}{\phi_C-\sin(\phi_C)}\right). \quad (4)$$

Also, from (4), it is easy to obtain the minimum ϕ_C value which complies with international regulations for a given input voltage (i.e. the minimum ϕ_C which introduce higher efficiency). Table I shows these minimum values (i.e. ϕ_{Cmin}) which are the same for the american and european mains supply. Some of these values have been previously calculated in [15, 17]. As you can see in Table I, the more restrictive regulation the higher value of ϕ_{Cmin}. At this point, the input current of the AICS can be represented. Figure 2 shows the normalized input current for several optimized designs that both comply international regulations at nominal input voltage and maximize efficiency. All Fig. 2 designs have been performed by following [15, 17] design procedure.

TABLE I. MINIMUN VALUE OF ϕ_C WHICH COMPLIES WITH INTERNATIONAL REGULATIONS

	ϕ_{Cmin} (°)
EN 61000-3-2 Class C regulations	140.49
EN 61000-3-2 Class D regulations	63.12
ENERGY STAR® residential applications	103.87
ENERGY STAR® for commercial applications	55.59

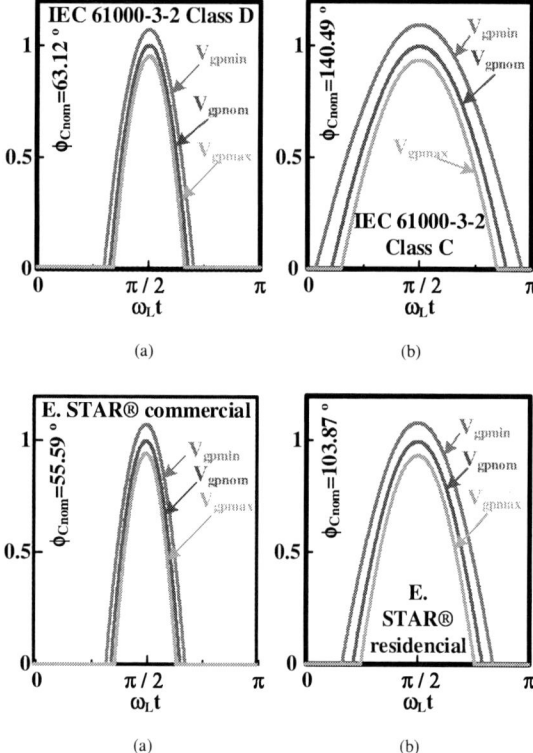

(a)

(b)

(a)

(b)

Fig. 2. Normalized input current for different optimized designs at different peak values of $v_g(t)$: a) Class D european design. b) Class C european design. c) ENERGY STAR® american desing for comercial applications. d) ENERGY STAR® american desing for residential applications.

B. Inplementation of the voltage source and the LFR with the forward "delayed output"

From the analysis of the forward "delayed output" presented in [11], V_S and R_{LF} can be calculated. Figure 3 shows the equivalent circuit of the "delayed output". As you can see, it is

a forward output, but with an additional inductor L_D in series with the rectifier diode D_1. Due to the action of this inductor, there is a delay between the turn-off of D_2 in comparison to traditional forward output. In fact D_2 stops conducting later because L_D must be charged until $i_L(t)$ (i.e. $i_{LD}(t)$ reaches $i_L(t)$) by the action of the voltage reflected to the secondary side of the transformer of the forward "delayed output" (see Fig.3b).

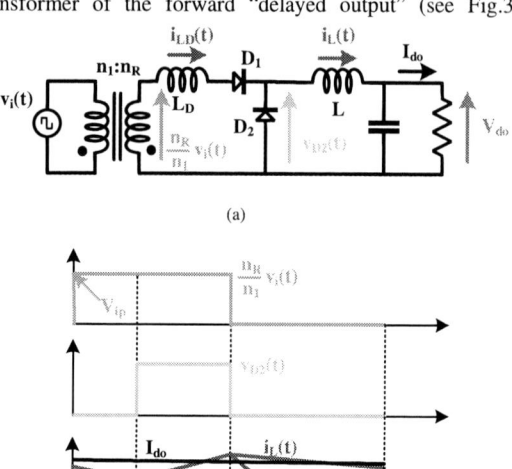

(a)

(b)

Fig. 3. a) "Delayed ouput". b) Main waveforms.

From Fig. 3b the delay time can be deduced applying Faraday´s law to the "delayed output":

$$t_d = \frac{i_L(t=t_d)}{\frac{n_R}{n_1}v_i(t)}, \tag{5}$$

where $v_i(t)n_R/n_1$ is the voltage reflected in the secondary side of the forward "delayed output", n_R/n_1 being the turns ratio of the transformer.

The effective duty cycle at the input of the output LC filter can be deduced from Fig. 3b:

$$d' = d - t_d \cdot f_s. \tag{6}$$

where d is the duty cycle and $f_S = 1/T_S$ the switching frequency, T_S being the switching period.

Assuming that there is no ripple through inductor L because the forward "delayed output" operates in CCM (i.e. L>>L_D), the output voltage of the "delayed output" is:

$$V_{do} = \frac{n_R}{n_1} \cdot V_{ip} \cdot d - L_D \cdot f_s \cdot I_{od}, \tag{6}$$

where V_{ip} is the peak value of $v_i(t)$ and I_{od} is the output current of the "delayed output". From Fig. 1, we can deduce that the forward "delayed output" becomes a real source voltage. Therefore equation (6) can be rewritten as follows:

$$V_{do} = V_S - R_{LF} \cdot I_{od}, \qquad (7)$$

where:

$$V_S = \frac{n_R}{n_1} \cdot V_{ip} \cdot d, \qquad (8)$$

$$R_{LF} = L_D \cdot f_s. \qquad (9)$$

It should be noted that no energy is dissipated in the R_{LF} if all components are ideal. Finally, it is important to say that L_D energy is transferred to the primary side of the transformer, in this case to the equivalent voltage source $v_i(t)$.

C. Using a flyback converter to design the AISC

Figure 4 shows the implementation of an AISC in a flyback converter (it will be equal on a member of the flyback's family of dc to dc converters, that is, SEPIC, Cuk and Zeta). First, Fig. 4a defines the basic implementation. Second, two modifications of this implementation are shown in Fig. 4b and Fig. 4c, where the transformer becomes an autotransformer. Finally, Fig. 4d shows a particularization on Fig. 4c solution. This is an easy implementation of proposed idea, ideal for low-cost solutions. This implementation only introduce two extra inductors and two extra diodes in comparison to traditional flyback topology. The price to pay is the loss of a degree of freedom in the design of the AISC, because the autotransformer disappears (i.e. $n_R=n_1$).

By using a flyback topology in order to implement the AICS, the input voltage of the dc to dc converter becomes V_C. Taking into account CCM operation, the following equation can be written:

$$V_O = \frac{n_2}{n_1} \cdot V_C \cdot \frac{d}{1-d} \;, \qquad (10)$$

where n_2 is the number of turns of the secondary side of the transformer. Moreover equation (8) becomes:

$$V_S = \frac{n_R}{n_1} \cdot V_C \cdot d. \qquad (11)$$

As (11) shows, V_S depend on V_C, the duty cycle and the turn ratio of the "delayed output". In fact a properly choice of n_R/n_1 allow us to freely set V_S. Also V_C and V_S are related by the fact that the output voltage of the AISC must be kept constant by the action of the feedback loop. A new equation must be deduced by using (2), (10) and (11):

$$V_C - \frac{n_R}{n_1} \cdot V_C \cdot \frac{V_O}{\frac{n_2}{n_1} \cdot V_C + V_O} = V_{gp} \cdot cos\left(\frac{\phi_C}{2}\right). \qquad (12)$$

From (3) and (12), the evolution of V_C as a function of the design parameters (i.e. the conduction angle for nominal conditions and full load, ϕ_{Cnom}, and the duty cycle for minimum peak value of the input voltage, d_{max}) can be calculated. V_C could be represented versus input power for different V_{gp} values. Figure 5 shows the voltage on the intermediate bus for different optimized designs following [15, 17] design procedure. (the same as Fig. 2 designs). [15,17] optimized design procedures is focused on minimizing VC value and the amount of recycled energy, keeping compliance with international regulations at nominal input voltage and full load. By an adequate choice of n_R/n_1 the voltage drop across the series connection of V_S and R_{LF} could be zero at minimum input voltage V_{gpmin} and full load (P_{gmax}). In this conditions V_C (i.e V_{Cmin}) becomes equal to V_{gpmin}.

Although V_C is minimized, it is not maintained constant for different operation conditions (i.e. P_g and V_{gp} variations),

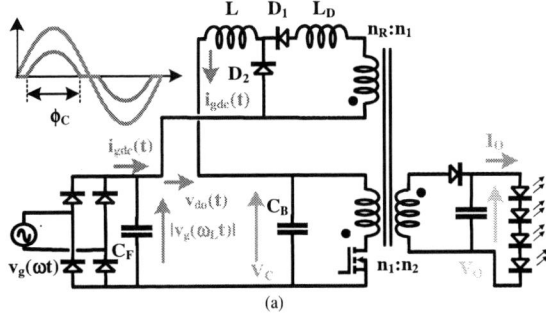

(a)

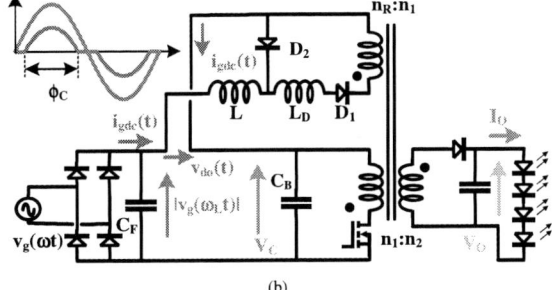

(b)

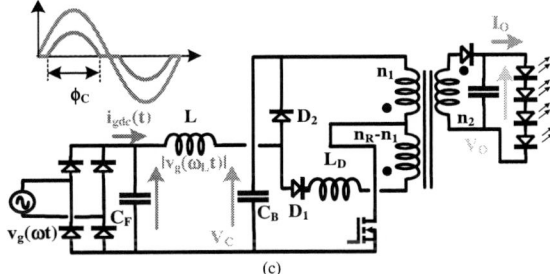

(c)

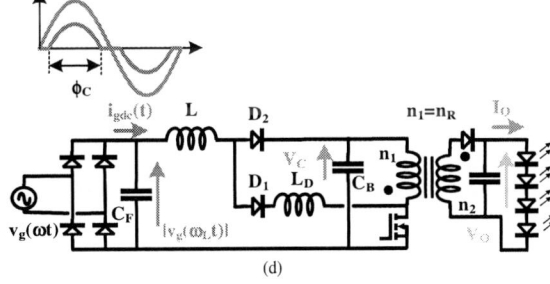

(d)

Fig. 4. Implementation of AISC based on flyback converter. a) Basic scene. b) After moving L, L_D, D_1 and D_2. c) Using an extra tap instead of "delayed output". d) Using no extra tap ($n_1=n_R$).

at least if the flyback converter is operating at constant switching frequency, as Fig. 5 shows. This is the price to pay due to the simplicity of this solution in comparison to two-stage solution, where the voltage across the intermediate bus is controlled. Finally, it is important to say that in Fig. 4d structure, V_C limiting cannot be achieved due to the fact that n_R/n_1 is set a priori.

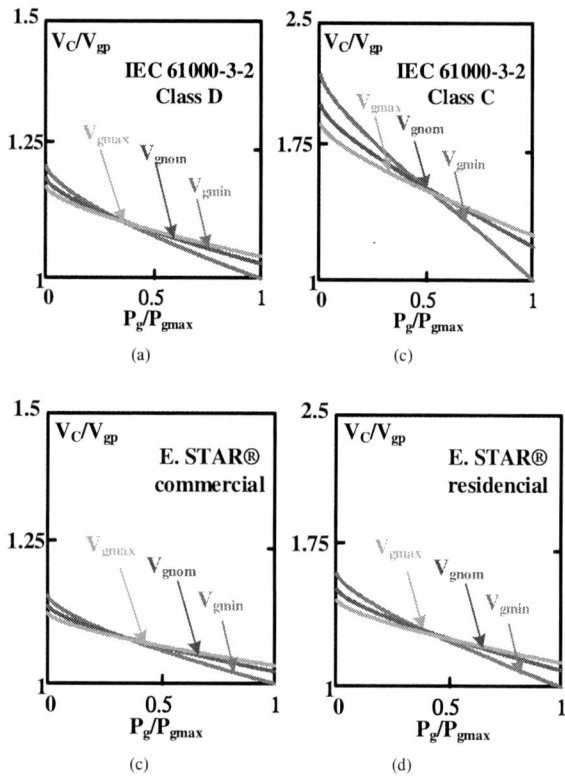

Fig. 5. Normalized voltage of the intermediate bus versus normalized power at different peak values of the input voltage for differerent optimized designs. a) Class D european design. b) Class C european design. c) ENERGY STAR® american desing for comercial applications. d) ENERGY STAR® american desing for residential applications.

III. EXPERIMENTAL RESULTS: INCREASING LOW FREQUENCY VOLTAGE RIPPLE OF AISC IN THE INTERMEDIATE BUS TO ELIMINATE ELECTROLITHYC CAPACITOR

A prototype of the proposed ac-dc HB-LED driver based on an AISC was design to widely comply with regulations (i.e. IEC 61000-3-2, Class D and ENERGY STAR® program requirements for commercial applications), built and tested. A design has been performed following [15, 17] for next specifications: $\phi_{Cnom}=70°$, $P_g=24$ W, $V_O=19$ V, $f_S=110$ kHz, american design (i.e. $90\sqrt{2}<V_{gp}<130\sqrt{2}$ and 60 Hz), CCM operation of the "delayed output" (i.e. L=1,8 mH) and $d_{max}=0.6$. The circuit has been performed according to the scheme given in Fig. 6a, where $R_{LF}=43.45$ (i.e. $L_D=0.39$ mH), $n_S=n_1$ and $n_2/n_1=0.1$. The prototype was controlled using a commercial IC as is shown in Fig. 6b (UC2825 by Texas Instruments). Finally, the converter output is connected to a matrix of two strings in parallel of 6 HB-LEDs each one. Table II summarize all main components.

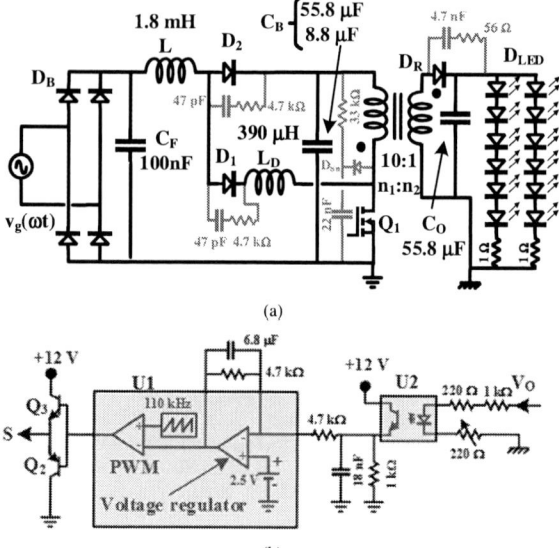

Fig. 6. Experimental prototype based on flyback converter. a) Power stage. b) Control stage.

TABLE II.   COMPONENTS OF HE EXPERIMENTAL PROTOTYPE

Fig. 6 reference	Value
D_1	BYP08P140
D_2	HFADBTB
D_B	3KBP04M
D_R	8TQ100
D_{Sn}	MUR4100
D_{LED}	LXK2PW14T00 (Luxeon)
Q_1	FQPF8N80C
Q_2 and Q_3	BD140 and BD139
U_1	UC3825
U_2	MCT2

A. AISC without low frequency ripple in the intermediate bus ($C_B=55.8\ \mu F$)

The prototype was tested until both the prototype temperature and the HB-LEDs temperature stabilized at the aforementioned specifications. The final operating temperature was reached after 45 min of operation. Figure 7 shows the line input current, voltage in the intermediate bus, input voltage and output voltage of the AISC. As expected, the experimental results of $i_g(t)$ match with the theoretical ones. Also, the voltage of intermediate bus is around 200 V, being the expected one. In this implementation V_C cannot be controlled, however, this is the price to pay for using an implementation as simple as the one of proposed (i.e. $n_1=n_R$).

B. AISC with low frequency ripple in the intermediate bus ($CB=8.8\ \mu F$)

In this second test, the electrolytic capacitor of the intermediate bus ($C_B=47\ \mu F$) has been removed and only the ceramic capacitor remains ($C_B=4$ x $2.2\ \mu F$). As a consequence of this, some low frequency ripple arises at the voltage of the intermediate bus (see $v_C(t)$ in Fig. 8) which increase the traditional distortion of the AICS line input current. This added distortion is slight in comparison to traditional distortion of AICS (see $i_g(t)$ in Fig. 8). Moreover, this slightly increase of the input current distortion can be explicitly checked in comparison to the first test (Fig. 9). As you can see, compliance with IEC 61000-3-2 Class D international regulations is achieved too. Table III shows also compliance with ENERGY STAR® program requirements for commercial applications and the slight increase of THD and slight decrease of PF.

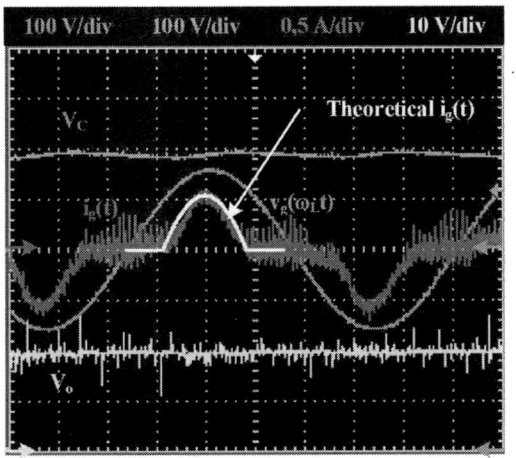

Fig. 7. Line input current ($i_G(t)$), voltage of the intermediate bus (V_C), line input voltage ($v_g(t)$) and output voltage (V_O) of the AISC without low frequency ripple in the intermediate bus.

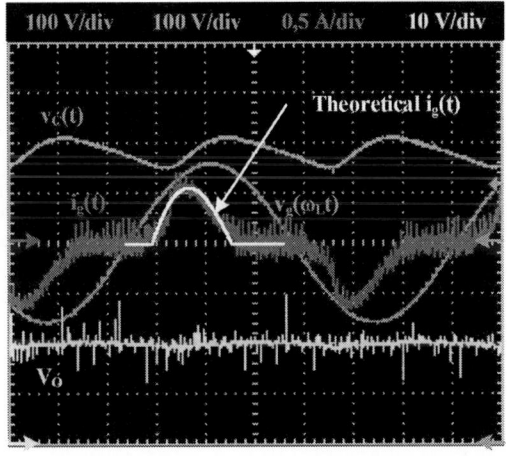

Fig. 8. Line input current ($i_g(t)$), voltage of the intermediate bus ($v_c(t)$), line input voltage ($v_g(t)$) and output voltage (V_O) of the AISC with low frequency ripple in the intermediate bus.

Now the question is how the low frequency ripple of the intermediate bus is reflected at the output of the AICS. The answer is shown in Fig. 10. As you can see, the low frequency ripple of the output voltage (V_O) and output current (I_O) is very low, because the action of the output voltage feedback loop (Fig. 6b), which have been designed to eliminate this ripple.

In order to validate the absence of flickering, [18] considerations have been followed. To limit the biological effects and detection of flicker in general illumination, the Modulation (%) should be kept within the shaded region defined by [18]. Modulation (%) must be calculated by assuming perfect ac power line conditions, being:

$$Modulation\ (\%)_C = 100 \cdot \frac{(L_{max} - L_{min})}{(L_{max} + L_{min})},\quad (13)$$

where L_{max} and L_{min} correspond to the maximum and minimum luminance of each harmonic of the ac component of the output current, respectively. In this test a proportionality between luminance and ac component of output current can been assumed. Results of this analysis are shown in Fig. 11. As you can see, all ac harmonic content is within the shaded region.

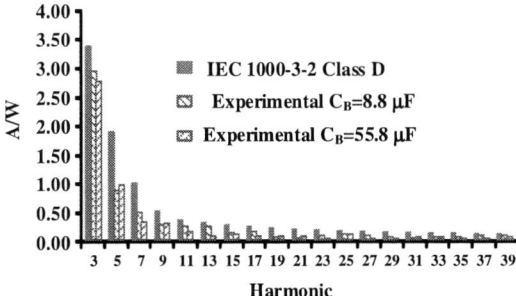

Fig. 9. Experimental harmonic content with and without electrolytic capacitor.

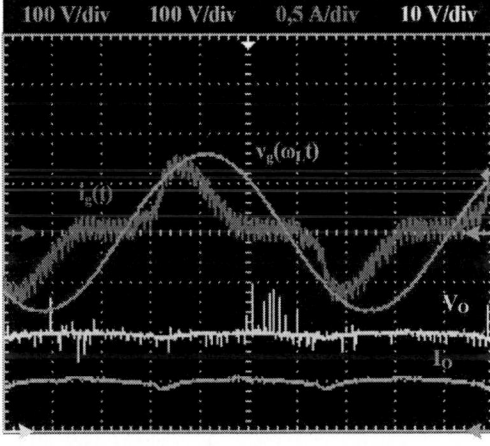

Fig. 10. Line input current ($i_g(t)$), input linevoltage ($v_g(t)$), output voltage (V_O) and output current (I_O) of the AISC with low frequency ripple in the intermediate bus.

Finally, the efficiency measured in both prototypes are the same, 82 %. This efficiency is lower than other proposed topologies for replacing incandescent bulb lamps [7-10]. However, this is the price to pay for eliminating the electrolytic capacitor.

978-1-4673-9551-9/16 $31.00 © 2016 IEEE

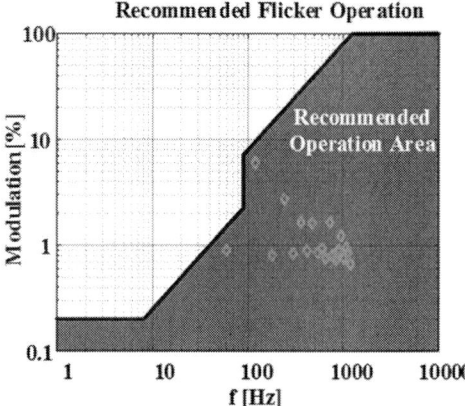

Fig. 11. Modulation (%) of the ouput current of proposed desing into recommended operation area defined in [18].

TABLE III. PF AND THD OF BOTH TEST

TEST	PF	THD(%)
AISC without low frequency ripple in the intermediate bus (CB=55.8 µF)	0.871	56.2
AISC with low frequency ripple in the intermediate bus (CB=8.8 µF)	0.818	62.9

IV. ANALYSIS OF THE AISC WITH LOW FREQUENCY VOLTAGE RIPPLE IN THE INTERMEDIATE BUS

At this point, it is obvious that a theoretical analysis of the AISC solution with low frequency ripple in the intermediate bus must be done. This analysis must be focused on the distortion of the line input current in order to validate the experimental results presented in the second test of the previous section.

If some ripple arises in the intermediate bus of AISC due to the substitution of the electrolytic capacitor by other technology, the constant voltage V_C becomes $v_C(t)$:

$$v_C(t) = V_{Cdc} - V_{Cac}\sin(2\omega_L t) = V_{Cdc}(1 - k\sin(2\omega_L t)), \quad (13)$$

where V_{Cdc} and V_{Cac} are the dc component and ac component of the voltage across the intermediate bus, and k is the value of the relative ripple of $v_C(t)$. It is important to say that only the component of twice the line frequency of $v_C(t)$ has been taking into account in the sake of simplicity.

The study will be carried out for a flyback converter operating in CCM (or a member of the flyback's family of dc to dc converters, that is, SEPIC, Cuk AND Zeta). Equation (13) and a modification of equation (10) (i.e. changing d by d(t) and V_C by $v_C(t)$), can be used in order to calculate the duty ratio:

$$d(t) = \frac{V_O}{\frac{n_2}{n_1}(V_{Cdc}(1 - k\sin(2\omega_L t)))}. \quad (14)$$

Now the duty cycle varies with the line frequency due to the action of the output voltage feedback loop, which is designed to keep either the $i_O(t)$ or $v_O(t)$ constant. This output voltage feedback loop of the vAISC can be designed with very fast dynamic response in order to eliminate the low frequency ripple, which comes from the input of the flyback dc to dc converter (i.e.

intermediate bus of the AICS). This characteristic of the AICS [14-17] is the key to not transfer the low frequency ripple of $v_C(t)$ to the output, and to enable that the removal of the electrolytic capacitor does not involve flickering at the output. However, this variation of the duty cycle plus the low frequency ripple of $v_C(t)$ has consequences on V_S (which becomes $v_S(t)$ in this analysis). From a modification of (11) (i.e. changing d by d(t), V_C by $v_C(t)$ and V_S by $v_S(t)$), (13) and (14), the expression of $v_S(t)$ can be deduced:

$$v_S(t) = V_o \frac{\frac{n_R}{n_1}}{\frac{n_2}{n_1}} \frac{(1 - k\sin(2\omega_L t))}{(1 - k\sin(2\omega_L t)) + \frac{V_o}{\frac{n_2}{n_1}V_{Cdc}}}. \quad (15)$$

As (15) shows, now $v_S(t)$ is not a constant voltage source, and therefore, the line input current will be not sinusoidal during the conduction of the diodes of the rectifier bridge. Using a modification of (3) (i.e., V_C being $v_C(t)$ and V_S being $v_S(t)$), (13) and (15), the line input current will be:

$$i_{gdc}(t) = \frac{1}{R_{LF}} \left[V_{gp}|\sin(\omega_L t)| + V_o \frac{\frac{n_R}{n_1}}{\frac{n_2}{n_1}} \frac{(1 - k\sin(2\omega_L t))}{(1 - k\sin(2\omega_L t)) + \frac{V_o}{\frac{n_2}{n_1}V_{Cdc}}} - V_{Cdc}(1 - k\sin(2\omega_L t)) \right]. \quad (16)$$

It should be noted that this expression is only valid for the interval where $v_g(t)$ is greater than $v_C(t)-v_S(t)$. This interval can be calculated by equaling to zero (16):

$$V_{gp}|\sin(\omega_L t_i)| + V_o \frac{n_S}{n} \frac{(1 - k\sin(2\omega_L t_i))}{(1 - k\sin(2\omega_L t_i)) + \frac{V_o}{nV_{Cdc}}} - V_{Cdc}(1 - k\sin(2\omega_L t_i)) = 0; \quad i = 1,2, \quad (17)$$

where the conduction angle becomes:

$$\phi_c = 2\pi \frac{t_2 - t_1}{T} \quad (18)$$

As you can deduct from (16), average input current of the AISC with ripple in the intermediate bus is non-sinusoidal during the interval [t_1, t_2]

Finally, the expression of R_{LF} can be deduced from the input power by using (17). In sake of simplicity R_{LF} has been considered constant in the theoretical analysis:

$$R_{LF} = \frac{1}{P_g \frac{T}{2}} \int_{t_1}^{t_2} i_{gdc}(t) V_{gp}|\sin(\omega_L t)| dt =$$

$$\frac{1}{P_g \frac{T}{2}} \int_{t_1}^{t_2} \left[V_{gp}|\sin(\omega_L t)| + V_o \frac{n_S}{n} \frac{(1 - k\sin(2\omega_L t))}{(1 - k\sin(2\omega_L t)) + \frac{V_o}{nV_{Cdc}}} - V_{Cdc}(1 - k\sin(2\omega_L t)) \right] \cdot V_{gp}|\sin(\omega_L t)| dt. \quad (19)$$

At this point, the line input current of the AISC can be theoretically calculated for a given specifications. Figure 12, shows the normalized input current for the same optimized design presented in Fig. 2, but now introducing some ripple on $v_C(t)$. From Fig. 12 waveforms analysis, we can conclude that the distortion of the line input current due to the low frequency ripple in the intermediate bus is negligible in comparison to the distortion naturally generated by the AISC operation.

Finally, with the theoretical model presented in this section the line input current of experimental results test B (i.e. k=0.2) can be calculated. As you can see in Fig. 10, the experimental results match with theoretical ones, validating the proposed model.

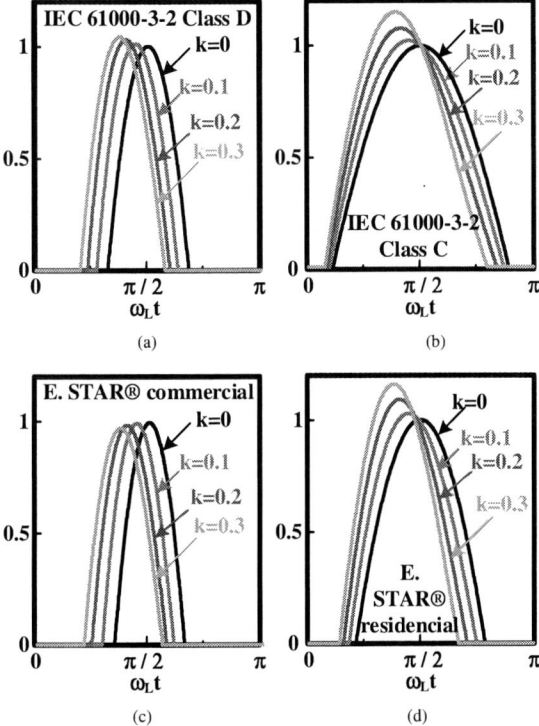

Fig. 12. Normalized input current for different optimized designs at different peak values of $v_g(t)$ and k values: a) Class D european design. b) Class C european design. c) ENERGY STAR® american desing for comercial applications. d) ENERGY STAR® american desing for residential applications.

V. CONCLUSIONS

This paper presents an ac to dc HB-LED driver with no electrolytic capacitor, based on the AICS solution. By removing the electrolytic capacitor, some low frequency ripple arises in the intermediate bus of the AICS. As a consequence of this, the distortion of the line current is increased. However, as theoretical and experimental results show, this increase in distortion is slight in comparison to the one of a standard AISC, and compliance with international regulations (i.e IEC 61000-3-2 Class D) is achieved. Moreover, no low frequency ripple is translated to the output due to the fast dynamic response of AICS, and therefore, no flickering performance is obtained in the ac to dc one-stage topology without electrolytic capacitor. However, the proposed solution presents two main drawbacks: slightly lower efficiency in comparison to other solutions and no wide input voltage range performance. This is the price to pay for a very low-cost solution without electrolytic capacitor.

REFERENCES

[1] Azevedo, I.L.; Morgan, M.G.; Morgan, F, "The Transition to Solid-State Lighting," Proceedings of the IEEE, vol.97, no.3, pp.481-510, March 2009.

[2] Shur, M.S.; Zukauskas, R. "Solid-State Lighting: Toward Superior Illumination," Proceedings of the IEEE, vol.93, no.10, pp.1691-1703, Oct. 2005.

[3] Electromagnetic compatibility (EMC)-part 3: Limits-section 2: Limits for harmonic current emissions (equipment input current<16A per phase), IEC1000-3-2 Document, 1995.

[4] [4] Draft of the proposed CLC Common Modification to IEC 61000-3-2 Document, 2006.

[5] Draft of the proposed CLC Common Modification to IEC 61000-3-2/A2 Document, 2010.

[6] Revised ENERGY STAR Program Requirements for Solid-State Lighting Luminaires: Eligibility Criteria - Version 1.1, December 2008.

[7] Buso, S.; Spiazzi, G.; Sichirollo, F., "Study of the Asymmetrical Half-Bridge Flyback Converter as an Effective Line-Fed Solid-State Lamp Driver," Industrial Electronics, IEEE Transactions on , vol.61, no.12, pp.6730,6738, Dec. 2014

[8] X. Qu, S.-C. Wong, and C. K. Tse, "Resonance-assisted buck converter for offline driving of power LED replacement lamps," IEEE Trans. Power Electron., vol. 26, no. 2, pp. 532–540, Feb. 2011.

[9] Lamar, D.G.; Fernandez, M.; Arias, M.; Hernando, M.M.; Sebastian, J., "Tapped-Inductor Buck HB-LED AC–DC Driver Operating in Boundary Conduction Mode for Replacing Incandescent Bulb Lamps," Power Electronics, IEEE Transactions on , vol.27, no.10, pp.4329,4337, Oct. 2012.

[10] Lamar, D.G.; Arias, M.; Hernando, M.M.; Sebastian, J., "Using the Loss-Free Resistor Concept to Design a Simple AC–DC HB-LED Driver for Retrofit Lamp Applications," Industry Applications, IEEE Transactions on , vol.51, no.3, pp.2300,2311, May-June 2015.

[11] Sebastian, J.; Uceda, J., "Two different types of fully regulated two-output dc‑to‑dc converters with one switch". Second International Conference on Power Electronics and Variable Speed Drives, Birmingham (Reino Unido), noviembre 1986, pp. 172-176.

[12] Sebastian, J.; Uceda, J., "An alternative method for controlling two-output DC-to-DC converters using saturable core inductor," in Power Electronics, IEEE Transactions on , vol.10, no.4, pp.419-426, Jul 1995.

[13] Huber, L.; Jovanovic, M.M., "Single-stage, single-switch, isolated power supply technique with input-current shaping and fast output-voltage regulation for universal input-voltage-range applications," Applied Power Electronics Conference and Exposition, 1997. APEC '97 Conference Proceedings 1997., Twelfth Annual , vol.1, no., pp.272,280 vol.1, 23-27 Feb 1997.

[14] Huber, L.; Jovanovic, M.M., "Design optimization of single-stage single-switch input-current shapers," Power Electronics, IEEE Transactions on , vol.15, no.1, pp.174,184, Jan 2000.

[15] Sebastian, J.; Hernando, M.M.; Fernandez, A.; Villegas, P.J.; Diaz, J., "Input current shaper based on the series connection of a voltage source and a loss-free resistor," Industry Applications, IEEE Transactions on , vol.37, no.2, pp.583,591, Mar/Apr 2001.

[16] Yimin Jiang; Lee, F.C.; Hua, G.; Tang, W., "A novel single-phase power factor correction scheme," in Applied Power Electronics Conference and Exposition, 1993. APEC '93. Conference Proceedings 1993., Eighth Annual , vol., no., pp.287-292, 7-11 Mar 1993.

[17] Villarejo, J.A.; Sebastian, J.; Soto, F.; de Jodar, E., "Optimizing the Design of Single-Stage Power-Factor Correctors," Industrial Electronics, IEEE Transactions on , vol.54, no.3, pp.1472,1482, June 2007.

[18] IEEE Recommended Practices for Modulating Current in High-Brightness LEDs for Mitigating Health Risks to Viewers," in IEEE Std 1789-2015 , vol., no., pp.1-80, June 5 2015.

Reduced Current Stress Bridgeless Cuk PFC Converter with New Voltage Multiplier Circuit

Yi-Hung Liao

Department of Electrical Engineering
National Penghu University of Science and Technology
Penghu 880, Taiwan, R.O.C.
Email: yhlmliao@gmail.com

Abstract—In this paper, a bridgeless Cuk derived AC/DC converter is proposed. The present circuit topology separates the input PFC current and output voltage-regulation current, which reduces the switch current stresses and improves system thermal management compared to the conventional bridge and bridgeless Cuk topologies. In addition, the proposed bridgeless topology has semi-soft switching function of all active switches to reduce converter switching losses. Therefore, the switch conduction losses and switching power losses all can be decreased due to the reduced current stress and semi-soft switching function. To understand the proposed Cuk derived converter, the circuit operation is explained and the steady-state behavior is analyzed. Finally a prototype system with DSP TMS320F28335 controller is implemented. Some simulation and experimental results are offered to verify the validity of the proposed bridgeless Cuk derived PFC converter.

Keywords—bridgeless; Cuk derived; AC/DC converter; power factor correction; semi-soft switching

I. INTRODUCTION

With the development of computer, communication, consumer electronic production and electric vehicle, the power supplies and/or power chargers with active power correction techniques are becoming necessary to meet harmonic regulation and standards, such as IEC 61000-3-2 [1]. So far, most power factor correction circuits are implemented by the boost type converters due to its low cost and high performance in terms of simplicity, efficiency and unity power factor ability. However, for universal input voltage, the boost converter suffers from lower efficiency at low input voltage, which led to the development of bridgeless topology [2]-[7]. A system review [8] of performance comparison with the conventional PFC boost rectifier was drawn attention. The bridgeless boost rectifier also has the same major practical drawbacks as the conventional one, including that the startup inrush current is high, the dc output voltage is always higher than the peak input voltage, and input-output isolation cannot easily be implemented [9]-[12]. Furthermore, the boost converter operating in discontinuous current mode can provide some merits such as reduced diode reversed-recovery losses, soft turn-on of the main switch, simple control and inherent PFC ability. Nevertheless, the DCM operation required a high quality boost inductor due to high peak ripple currents and voltages. To overcome this, a robust input filter must be

This research is sponsored by the Ministry of Science and Technology of R.O.C. under Grant MOST 103-2221-E-346 -004 -MY2.

adopted to suppress the high-frequency pulsating input current, which results in the weight and cost of the rectifier.

To improve the drawbacks of a bridgeless PFC boost rectifier, the step up/step down AC-DC rectifiers [10]-[12] for Sepic and Cuk converter were proposed. However, the topologies all suffer from high current stresses of active switches due to the switch currents flowing together coming from input PFC current and output load current such as bridge [8] and bridgeless [10] Cuk derived AC/DC converters shown in Fig. 1(a) and Fig. 1(b), respectively, which results in increasing conduction losses and switch current stresses so as to decrease the power rating and lifetime of the power converter.

In this paper, a new single phase ac-dc Cuk derived PFC bridgeless rectifier with voltage multiplier is introduced. The bridgeless topology with reduced switch current stress can increase the power rating and lifetime of the power converter. Although the four switches are utilized, the active switches all have simi-soft switching function, which implies there only exists one-half switching power losses of the four switches. In addition, the voltage gain also can be extended without extreme duty cycle operation which makes the proposed topology suitable for universal line voltage applications.

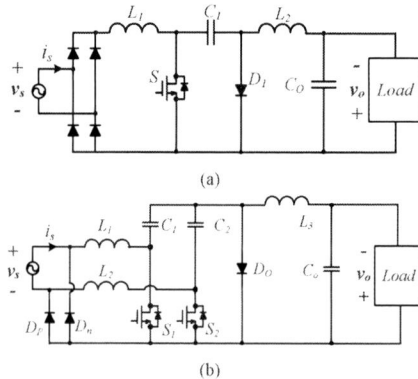

(a)

(b)

Fig. 1 Conventional single-phase (a) bridge [8] and (b) bridgeless Cuk derived AC/DC converters [10].

II. CIRCIT OPERATION

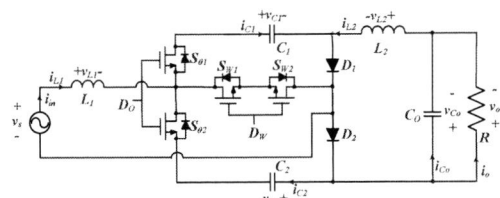

Fig. 2 Proposed bridgeless single-phase Cuk derived PFC topology.

The proposed bridgeless Cuk derived PFC rectifier with voltage multiplier cell is shown in Fig. 2. Where S_{W1} and S_{W2} is the PFC switch module and controlled by the PFC-controlled duty D_w, and So1 and So2 is the output voltage switch module and controlled by the output-voltage-controlled duty D_o. Since the proposed circuit consists of two symmetrical configurations, the circuit is only analyzed in the positive half line cycle. First, assuming that the L_1 inductor is operating in CCM and the L_2 is operating in DCM, then the key waveforms of proposed bridgeless converter during one switching period T_s in a positive half-line period are shown in Fig. 3 and the circuit operation can be divided into four distinct operating modes as shown in Fig. 4(a)-(d), which can be described as follows.

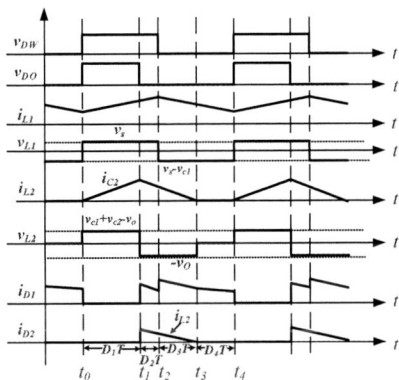

Fig. 3 The key waveforms of proposed bridgeless converter during positive half-line period.

1) **State 1** ($t_0 < t < t_1$): In this stage, as shown in Fig. 4(a), the PFC switch module and output-voltage switch module are turned on. Inductor current i_{L1} is increasing to store energy. Capacitors C_1 and C_2 release their energy to the inductor L_2 and output. The state and output equations can be expressed as follows.

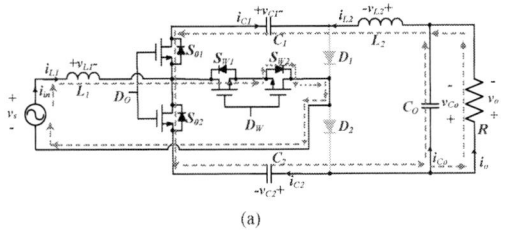

(a)

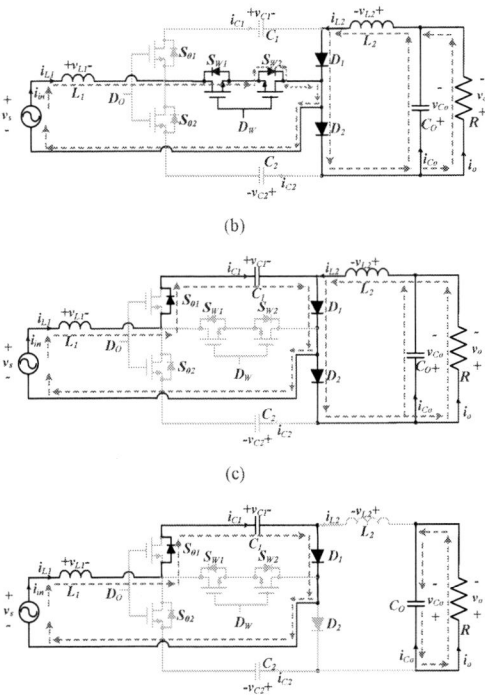

(b)

(c)

(d)

Fig. 4 Operation circuits of the proposed bridgeless Cuk derived converter under (a)State 1 (b)State 2 (c)State 3 (d)State 4 during positive half-line period.

$$\frac{dx}{dt} = A_1 x + B_1 u \qquad (1)$$

$$y = C_1 x \qquad (2)$$

Where the state and output vectors are defined as

$$x = \begin{bmatrix} i_{L1} & i_{L2} & v_{C1} & v_{C2} & v_{CO} \end{bmatrix}^T, \quad y = \begin{bmatrix} i_{in} & v_o \end{bmatrix}^T$$

$$A_1 = \begin{bmatrix} 0 & 0 & 0 & 0 & 0 \\ 0 & 0 & \frac{1}{L_2} & \frac{1}{L_2} & -\frac{1}{L_2} \\ 0 & -\frac{1}{C_1} & 0 & 0 & 0 \\ 0 & -\frac{1}{C_2} & 0 & 0 & 0 \\ 0 & \frac{1}{C_o} & 0 & 0 & -\frac{1}{RC_o} \end{bmatrix}, \quad B_1 = \begin{bmatrix} \frac{1}{L_1} \\ 0 \\ 0 \\ 0 \\ 0 \end{bmatrix},$$

$$C_1 = \begin{bmatrix} 1 & 0 & 0 & 0 & 0 \\ 0 & 0 & 0 & 0 & 1 \end{bmatrix}, \quad u = v_s$$

2) **State 2** ($t_1 < t < t_2$): In this stage, as shown in Fig. 4(b), the PFC switch module remains on and output-voltage switch module is turned off. Inductance current i_{L1} continues increasing to store energy. The inductor L_2 releases its current to output load. The state and output equations can be described as follows.

$$\frac{dx}{dt} = A_2 x + B_2 u \qquad (3)$$

$$y = C_2 x \qquad (4)$$

978-1-4673-9551-9/16 $31.00 © 2016 IEEE   1832

Where

$$A_2 = \begin{bmatrix} 0 & 0 & 0 & 0 & 0 \\ 0 & 0 & 0 & 0 & -\frac{1}{L_2} \\ 0 & 0 & 0 & 0 & 0 \\ 0 & 0 & 0 & 0 & 0 \\ 0 & \frac{1}{C_o} & 0 & 0 & -\frac{1}{RC_o} \end{bmatrix}, \; B_2 = \begin{bmatrix} \frac{1}{L_1} \\ 0 \\ 0 \\ 0 \\ 0 \end{bmatrix}, \; C_2 = \begin{bmatrix} 1 & 0 & 0 & 0 & 0 \\ 0 & 0 & 0 & 0 & 1 \end{bmatrix}$$

3) **State 3** ($t_2 < t < t_3$): In this stage, as shown in Fig. 4(c), the PFC switch module turns off and output-voltage switch module remains off. The inductor L_1 releases its current to the capacitor C_1. The inductor L_2 continues to release its current to output load. The state and output equations can be expressed as follows.

$$\frac{dx}{dt} = A_3 x + B_3 u \tag{5}$$

$$\frac{dx}{dt} = A_3 x + B_3 u \tag{6}$$

Where

$$A_3 = \begin{bmatrix} 0 & 0 & -\frac{1}{L_1} & 0 & 0 \\ 0 & 0 & 0 & 0 & -\frac{1}{L_2} \\ \frac{1}{C_1} & 0 & 0 & 0 & 0 \\ 0 & 0 & 0 & 0 & 0 \\ 0 & \frac{1}{C_o} & 0 & 0 & -\frac{1}{RC_o} \end{bmatrix}, \; B_3 = \begin{bmatrix} \frac{1}{L_1} \\ 0 \\ 0 \\ 0 \\ 0 \end{bmatrix}, \; C_3 = \begin{bmatrix} 1 & 0 & 0 & 0 & 0 \\ 0 & 0 & 0 & 0 & 1 \end{bmatrix}$$

4) **State 4** ($t_3 < t < t_4$): In this stage, as shown in Fig. 4(d), the PFC switch module remains off and output-voltage switch module remains off. The inductor L_1 continues to release its current to the capacitor C_1. The inductor current i_{L2} had dropped to zero. The output voltage is supplied from the output capacitor C_o. The state and output equations can be expressed as follows.

$$\frac{dx}{dt} = A_4 x + B_4 u \tag{7}$$

$$y = C_4 x \tag{8}$$

Where

$$A_4 = \begin{bmatrix} 0 & 0 & -\frac{1}{L_1} & 0 & 0 \\ 0 & 0 & 0 & 0 & 0 \\ \frac{1}{C_1} & 0 & 0 & 0 & 0 \\ 0 & 0 & 0 & 0 & 0 \\ 0 & 0 & 0 & 0 & -\frac{1}{RC_o} \end{bmatrix}, \; B_4 = \begin{bmatrix} \frac{1}{L_1} \\ 0 \\ 0 \\ 0 \\ 0 \end{bmatrix}, \; C_4 = \begin{bmatrix} 1 & 0 & 0 & 0 & 0 \\ 0 & 0 & 0 & 0 & 1 \end{bmatrix}$$

III. STEADY-STATE ANALYSIS

The analysis of voltage gain and switch current stress of the proposed converter operated in steady state is discussed in this section. Furthermore, the comparison of conventional Cuk converter [8], bridgeless Cuk converter [10], and the proposed converter are also discussed to verify the validity of proposed converter.

A. Voltage gain

Consider the converter is operating in one switching period during positive half-line period. By utilizing state-space averaged technique and voltage second balance theory in the inductor L_1 during the D_w duty period, one can obtain the relationship as follows:

$$v_{C1} = \frac{v_s}{(1-D_W)} \tag{9}$$

Similarly, for the inductor L_2 during the D_o duty period, the voltage relationship can be obtained

$$v_O = (v_{C1} + v_{C2})D_0 \tag{10}$$

Next, Consider the converter is operating in one switching period during the negative half-line period, and one can obtain the following equations:

$$v_{C2} = \frac{-v_s}{(1-D_W)} \tag{11}$$

$$v_O = (v_{C1} + v_{C2})D_0 \tag{12}$$

According to the equations (9)-(12), the voltage gain of proposed converter can be derived as follows:

$$v_o = D_0(v_{c1} + v_{c2}) = D_0 \left(\frac{|v_s|}{1-D_W} + \frac{|v_s|}{1-D_W} \right) = \frac{2D_0}{1-D_W} |v_s| \tag{13}$$

For convenient comparison, assume that $D_O = D_W$. Fig. 5 shows the comparisons of voltage gain as a function of duty ratio for the three converters. As can be seen from Fig. 5, the proposed converter has twice the voltage gain compared with the conventional bridge [8] and bridgeless [10] AC-DC Cuk derived converters.

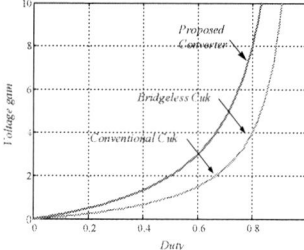

Fig. 5 Comparisons of voltage gain for the conventional Cuk [8], bridgeless Cuk [10],and the proposed converter.

B. Current stresses of active switches

According to the circuit operation in section II, the current stress of PFC switch module and OV switch module is i_{in} or i_o. Compared with the conventional Cuk [8] and bridgeless Cuk [10], consider the converter is operated at the same input voltage V_s and output load I_o. The switch current stress of the conventional Cuk and bridgeless Cuk is $i_{in} + i_o$ as shown in Fig. 6 for illustration. Although the proposed converter needs four switches, the proposed converter has lower switch current stress and lower cost of each switch. Therefore, it can be utilized for higher input current application compared to the conventional Cuk and bridgeless Cuk.

Table I is provided to summarize comparisons of the switch/diode/inductance number, current stress, voltage gain for the conventional bridge [8], bridgeless [10], and the proposed Cuk derived rectifiers.

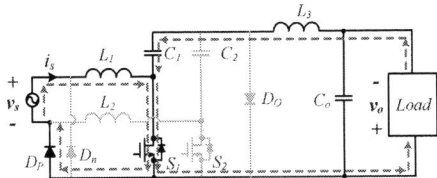

Fig. 6 The illustration of switch current stress of conventional bridgeless Cuk converter [10].

TABLE I. COMPARISONS OF CUK DERIVED AC/DC CONVERTERS

	Conventional Cuk [8]	Bridgeless Cuk [10]	Proposed
Slow diode	4	2	0
Fast diode	1	1	2
Inductance	2	3	2
Switch count	1 switches (Hard switching)	2 switches (Hard switching)	4 switches (All switches with semi-soft switching)
Switch current stress	$i_{in}+i_o$	$i_{bi}+i_o$	i_{in} or i_o
Voltage gain	$v_O = \dfrac{D}{1-D} \cdot \lvert v_{\sin} \rvert$	$v_O = \dfrac{D}{1-D} \cdot \lvert v_{\sin} \rvert$	$v_o = \dfrac{2D_0}{1-D_W} \lvert v_{\sin} \rvert$
Bridgeless	No	Yes	Yes

IV. SIMULATION AND EXPERIMENTAL RESULTS

To verify the validity of the proposed bridgeless Cuk derived rectifier, some simulation results are executed and then a prototype system is constructed to facilitate the theoretical results as verification. The input voltage is AC grid with $110V_{rms}$ and 60Hz fundamental frequency. The controlled output voltage is 250V and the output load is 250Ω. The assigned output power ratings is 250W. Fig. 7 shows the simulation results with the proposed feedforward control scheme. From Fig. 7, with the proposed feedforward control scheme, the waveform of input line current is smooth and closer to sinusoidal waveform. Fig. 8 shows the experimental results with the proposed feedforward control scheme. From Fig. 8, one can see that the input current is sinusoidal and in phase with the input voltage. Fig. 9 shows the measured harmonic distortion. One can find that the measured harmonic currents meet the IEC 61000-3-2 harmonics standards. The power factor is 0.99 at 250W output power. When the output power is 250W, the measured efficiency is about 92.5%.

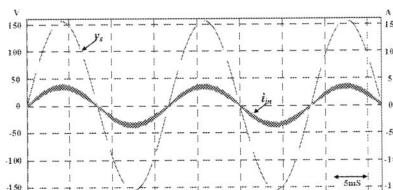

Fig. 7 Simulation results of the input voltage v_s and current i_{in} with proposed feedforward control scheme.

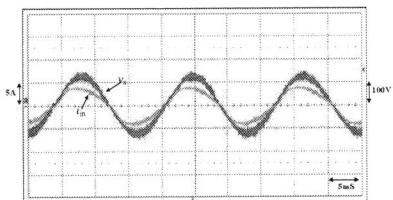

Fig. 8 Experimental results of the input voltage v_s and current i_{in} with proposed feedforward control scheme.

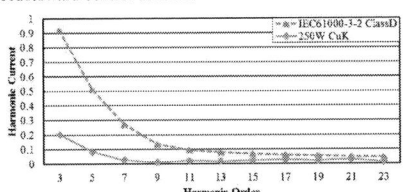

Fig. 9 The measured harmonic distribution of the input current i_{in} compared with IEC standard.

REFERENCES

[1] European power supply manufactures association "Harmonic Current Emissions guidelines to the standard EN 61000-3-2", Nov. 2010 edition.

[2] G. Moschopoulos and P. Jain, "A novel single-phase soft-switched rectifier with unity power factor and minimal component count," IEEE Trans. Ind. Electron., vol. 51, no. 3, pp. 566–576, Jun. 2004.

[3] W.-Y. Choi, J.-M. Kwon, E.-H. Kim, J.-J. Lee, and B.-H. Kwon,"Bridgeless boost rectifier with low conduction losses and reduced diode reverse-recovery problems," IEEE Trans. Ind. Electron., vol. 54, no. 2, pp. 769–780, Apr. 2007.

[4] J. C. Liu, C. K. Tse, N. K. Poon, B. M. Pong, and Y. M. Lai, "A PFC voltage regulator with low input current distortion derived from a rectifierless topology," IEEE Trans. Power Electron., vol. 21, no. 4, pp. 906–911,Jul. 2006.

[5] P. Kong, S. Wang, and F. C. Lee, "Common mode EMI noise suppression for bridgeless PFC converters," IEEE Trans. Power Electron., vol. 23, no. 1, pp. 291–297, Jan. 2008.

[6] W.-Y. Choi, J.-M. Kwon, and B.-H. Kwon, "Bridgeless dual-boost rectifier with reduced diode reverse-recovery problems for power-factor correction,"IET Power Electron., vol. 1, no. 2, pp. 194–202, Jun. 2008.

[7] Y. Jang, M. M. Jovanovic, and D. L. Dillman, "Bridgeless PFC boost rectifier with optimized magnetic utilization," in Proc. IEEE Appl. Power Electron. Conf. Expo., 2008, pp. 1017–1021.

[8] D. S. L. Simonetti, J. Sebastian, and J. Uceda, "The discontinuous conduction mode Sepic and Cuk power factor preregulators: Analysis and design, "IEEE Trans. Ind. Electron., vol. 44, no. 5, pp. 630–637, Oct. 1997.

[9] L. Huber, Y. Jang, and M. M. Jovanovic, "Performance evaluation of bridgeless PFC boost rectifiers," IEEE Trans. Power Electron., vol. 23, no. 3, pp. 1381–1390, May 2008.

[10] A. A. Fardoun, E. H. Ismail, A. J. Sabzali, M. A. Al-Saffar, "New Efficient Bridgeless Cuk Rectifiers for PFC Applications," Power Electronics, IEEE Transactions on , vol.27, no.7, pp.3292-3301, July 2012.

[11] A. M. Al Gabri, A. A. Fardoun, E. H. Ismail, "Bridgeless PFC-Modified SEPIC Rectifier With Extended Gain for Universal Input Voltage Applications," Power Electronics, IEEE Transactions on, vol.30, no.8, pp.4272-4282, Aug. 2015.

[12] A. J. Sabzali, E. H. Ismail, M. A. Al-Saffar, A. A. Fardoun, "New Bridgeless DCM Sepic and Cuk PFC Rectifiers With Low Conduction and Switching Losses," Industry Applications, IEEE Transactions on , vol.47, no.2, pp.873,881, March-April 2011.

Implementation of Multi-level Bridgeless PFC Rectifiers for Mid-Power Single Phase Applications

Trong Tue Vu
Icergi Ltd., Dublin, Ireland
Email: ttrongvu@icergi.com

George Young
Icergi Ltd., Dublin, Ireland
Email: georgeyoung@icergi.com

Abstract— **Multi-level approaches to AC/DC power conversion have been successfully deployed in high-voltage high-power industrial applications, and are recently considered for mid-power ranges due to their capability of reducing component sizes and improving efficiency. The main challenge for mid-power deployment is the complexity of sensing circuits and control algorithms used to maintain the capacitor voltage balance, which typically results in expensive and impractical designs. Although several studies have been carried out to address the issue, they have not simplified the system complexity and fully exploited the advantages of multi-level converters. This paper presents a new four-level topology for PFC rectifiers, and demonstrates its performance and practicality through a 200W prototype. The theoretical works and experimental results are also provided to confirm the feasibility of the solution.**

Keywords— **PFC; multi-cell conversion; bridgeless rectifiers; voltage balancing**

I. Introduction

In offline switched mode power supplies, the conversion from alternative current (AC) to direct current (DC) was conventionally performed by means of a half-wave or full-wave rectifier followed by a storage capacitor. The combination of these components forms nonlinear impedance seen from the mains, which distorts the current drawn from the power supply input and, in turn, pollutes the AC mains with rich harmonic content. Since only the fundamental component contributes to real power flowing into the power supply, the presence of high frequency and high magnitude harmonics is not desirable, and, according to IEC 61000-3-2 [1], should be limited for applications with input power above 75W in general, and lighting applications with a lower threshold in particular. Therefore, power factor correction (PFC) becomes a universal requirement in such contexts.

Driven by market demand for smaller and greener power supplies, there is considerable attraction in being able to reduce the size of the PFC choke and EMI filter which can account for 30% plus of the volume utilization. Two well-known solutions are to either interleave several boost converters [2] or rely on multi-cell topologies [3] – [5]. Although they both allow inductor volume reduction through current ripple cancelation, multi-cell power conversion is much more advantageous in terms of efficiency, EMI filters and device stresses, and, as a result, is mainly considered in the development of a new PFC approach.

Multilevel conversion techniques have been widely adopted in high voltage and high power DC-to-AC inverters and DC-to-DC choppers, where a single switching device either is unable to handle a significant voltage swing, for example over 5kV, or cannot provide adequate performance in terms of speed, efficiency and electro-magnetic interference (EMI). The key purposes of multilevel approaches are to use multiple switches for evenly sharing the voltage stress, which allows the deployment of lower voltage rating but faster switching devices.

Recently, several studies were dedicated to development of universal PFC rectifiers using multilevel approaches [2] – [4]; however, they either borrow the same setup aimed for high voltage and high power applications [2], limit themselves to three levels only [3, 4], or require a dedicated downstream stage to balance capacitor voltages [5]. There is no PFC study that fully takes advantages of multilevel approaches, i.e. deployment of low voltage MOSFETs for applications with universal input voltage, and simplifies the control complexity associated with inherent voltage balancing issues. Therefore, this paper focuses on developments of a new four-level bridgeless PFC topology that is able to overcome all these cited limitations.

II. Genesis of the New Multi-level Bridgeless PFC Rectifier

Minimizing conduction losses for the PFC rectifier is desirable in the context of power supplies, and this topic has been intensively discussed in various studies [6] – [11]. Among the proposed solutions, the totem pole bridgeless topology is the most attractive one due to its simplicity and low conduction losses [6], [9] – [11]. Despite all these features, the totem pole bridgeless approach is not particularly welcomed in practice. The main setbacks come from poor reverse-recovery performance of the body diodes of the high voltage MOSFETs operating in continuous conduction mode (CCM), and the complication in control design and gate drive circuitry.

Given the fact that the reverse recovery issues are more relevant to high voltage MOSFETs than to lower voltage rating ones, the hard switching challenges associated with the totem pole arrangement can be elegantly addressed by exploiting multilevel conversion techniques [3]. In particular, the two-switching-device totem pole leg is replaced by a four level flying capacitor multilevel converter stage as illustrated

978-1-4673-9551-9/16 $31.00 © 2016 IEEE

in Fig. 1. Flying capacitors are chosen instead of other multi-cell topologies because (a) a least number of switches is required for implementation, (b) voltage balance comes naturally for a given modulation scheme, and (c) output harmonic spectrum is theoretically reduced by 4 times as compared to a conventional totem pole approach.

In addition to the use of multiple switches, two thyristors T_1 and T_2 as shown in Fig. 1 are also used in the position of rectifier diodes, allowing ease of inrush current management and startup. In order to prevent inrush current from going through the inductor L and MOSFET string on startup, two bypass diodes D_1 and D_2 are provided to allow an alternating path to the bulk capacitor C_{bulk}.

The inductor L, MOSFET string, and flying capacitors C_{lv} and C_{lh} form a 4-level synchronous boost converter. However, unlike existing multi-level topologies, the flying capacitor voltages v_{hv} and v_{lv} can be initialized and forced to return to their balanced levels during transient responses by a passive clamped circuit enclosed by the dashed line in Fig. 1.

The proposed circuit as shown in Fig. 1 inherits the operating principle of the original totem pole arrangement. Particularly, in the positive half line-cycle, where T_2 is conducted and T_1 is off, the converter operates in a similar fashion as synchronous boost converters with MOSFETs Q_4, Q_5, and Q_6 operating as boost switches while the reverse conduction characteristics of MOSFETs Q_1, Q_2, and Q_3 serve as rectifiers feeding the bulk capacitor C_{bulk}. The operation in the opposing half line-cycle is also similar to synchronous boost converters; however, those MOSFETs swap their roles and the inductor current flows in an inverse direction.

The interactions between the multi-level boost converter and the diode clamped circuitry are quite complex; therefore, for ease of interpretation, the operation of each sub circuit is presented separately in Sections III and IV.

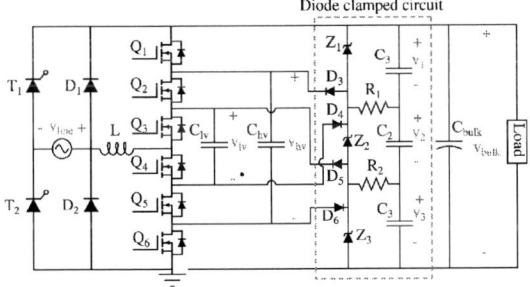

Fig. 1: Proposed 4-level bridgeless totem pole boost rectifier with balance enforcement circuitry

III. FOUR-LEVEL SYNCHRONOUS BOOST CONVERTER

A. Operating principle

Due to the symmetry of the proposed rectifier, only positive half-line cycles are considered here for simplicity, and the thyristor T_2 is assumed to be conducted while T_1, D_1 and D_2 are off. Without the balance enforcement mechanism, the circuit as

shown in Fig. 1 can be simplified to a four-level boost converter which is illustrated in Fig. 2.

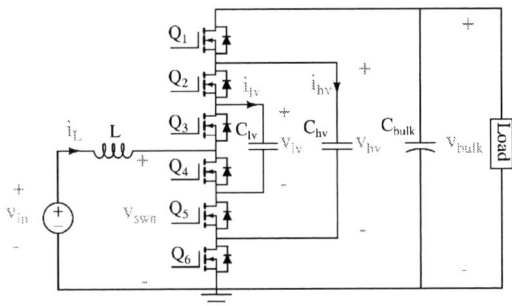

Fig. 2: Four-level flying capacitor boost converter

The operation of six MOSFETs in the string can be summarized as follow. Firstly, the switching devices are split into 3 pairs of switches including (Q_1, Q_6), (Q_2, Q_5), and (Q_3, Q_4). Each arrangement of paired switches is operated in a complementary manner, i.e. when one switch of the pair is on, the other must be off and vice versa. The control signals for the three top switches Q_1, Q_2 and Q_3 share the same duty ratio D and the switching frequency f_{pwm}, but are different in phase by an angle of 120 degrees while Q_6, Q_5 and Q_4 are driven by negating the driving pulses of Q_1, Q_2 and Q_3, respectively.

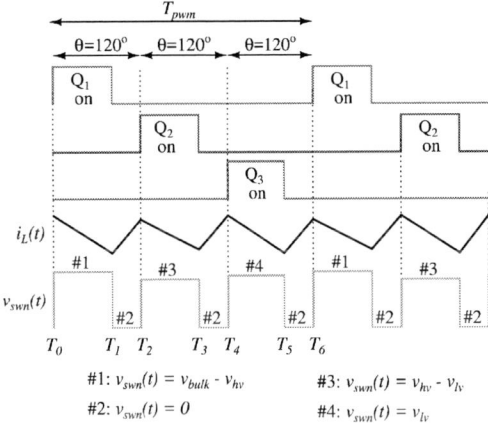

Fig. 3: Operational waveforms of a four-level synchronous boost converter with $0 < D \leq 1/3$

Since a phase shift of 120 degrees is equivalent to a time shift of one third of the switching period, the four-level boost converter changes its switching sequence every time D crossing the 1/3 and 2/3 points, which is illustrated in Fig. 3, Fig. 4 and Fig. 5, respectively. In particular, each number going with a hash symbol in those figures represents a unique combination of the states of Q_1, Q_2 and Q_3, and a unique relationship between the switched node voltage $v_{swn}(t)$, the flying capacitor voltages and the output voltage. If we consider the duty ratio range of (0 1/3] as shown in Fig. 3 as an example, the switching sequence within a switching cycle is #1, #2, #3, #2, #4, #2, and this order is preserved over the next cycles.

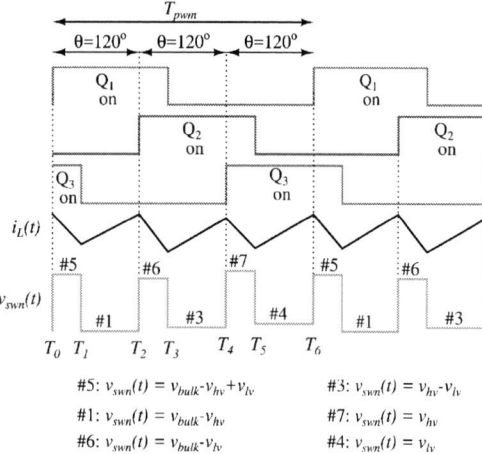

Fig. 4: Operational waveform of a four-level synchronous boost converter with $1/3 \le D \le 2/3$

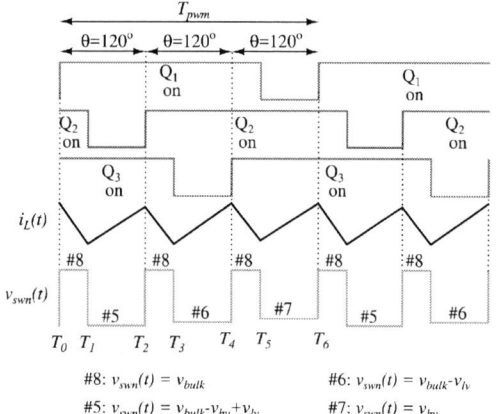

Fig. 5: Operational waveforms of a four-level synchronous boost converter with $2/3 \le D < 1$

Deriving the steady state responses for three operating modes of the multi-level boost converter can be carried out in a similar manner; therefore, for simplicity, Sections III.B and III.C focus on the analysis for D between 0 and 1/3 only, and summarize the final results for other cases.

B. DC analysis

The converter is assumed to operate in continuous conduction mode (CCM), and have negligible voltage ripples across the capacitors. The inductor current and switched node voltage waveforms are illustrated in Fig. 3: Operational waveforms of a four-level synchronous boost converter with 0

$< D \le 1/3$. Applying the voltage-second balance principle to the inductor gives

$$\sum_{i=0}^{5} v_L[i]\Delta T_i = \sum_{i=0}^{5} (v_{in}[i] - v_{swn}[i])\Delta T_i = 0 \quad (1)$$

where the index i denotes the time interval from the time instant T_i to T_{i+1} and $\Delta T_i = T_{i+1} - T_i$. Substituting the expression of $v_{swn}(t)$ as given in Fig. 3 into (1) gives

$$\sum_{i=0}^{5} v_{line}[i]\Delta T_i - D v_{bulk} = 0 \quad (2)$$

In most scenarios, the input voltage changes very slowly and can be considered as constant over a switching cycle. This assumption allows simplification of (2) into

$$\frac{V_{bulk}}{V_{in}} = \frac{1}{D} \quad (3)$$

The net charge accumulated by two flying capacitors over one switching cycle should be also zero at steady state, which means

$$\sum_{i=0}^{5} i_{lv}[i]\Delta T_i = 0,$$
$$\sum_{i=0}^{5} i_{hv}[i]\Delta T_i = 0. \quad (4)$$

One can verify from Fig. 2 that the capacitors C_{lv} and C_{hv} gets charged and discharged by the inductor current i_L for the same amount of time within each switching cycle. This suggests that the left-hand sides of (4) will be always zero if the assumption of negligible inductor current ripples is applied. Therefore, one needs a better approximation of i_L so as to get more useful information from these equations. One simple solution here is to take account of increments in the inductor current during each subinterval of the switching cycle. Particularly, if the voltage across the inductor during the time interval $[T_i, T_{i+1}]$ is $v_L[i]$, the inductor current during such an interval can be approximated by

$$i_L[i] = I_L + \frac{v_L[i]\Delta T_i}{2L} \quad (5)$$

where I_L is the averaged inductor current without ripples. Following such a principle allows us to transform (4) into

$$V_{bulk} - 2V_{hv} + V_{lv} = 0,$$
$$V_{hv} - 2V_{lv} = 0. \quad (6)$$

Solving (6) for V_{lv} and V_{hv} gives

$$V_{lv} = \frac{1}{3}V_{bulk},$$
$$V_{hv} = \frac{2}{3}V_{bulk}. \quad (7)$$

where V_{bulk} denotes the DC value of the bulk capacitor voltage. One can easily confirm that (3) and (7) are valid for all operating modes of the converter, i.e. any value of the duty ratio within the range $(0, 1)$. Given the voltages across the flying capacitor, the operating voltage of all MOSFETs can be approximated by $V_{bulk}/3$ which is technically reduced by a factor of 3 as compared to that of a conventional totem pole arrangement. If the nominal voltage of the bulk capacitor is assumed to be 400V, the nominal operating voltages of switching devices are 133.3V. This allows the usage of 200V MOSFETs which switch much faster and more efficiently than 600V counterparts; hence, conventional switching operation can be exploited here with minimum loss penalty. Another benefit of low voltage MOSFETs is the capability to switch at high frequencies which allows material reduction in the size of the PFC choke.

C. Capacitor voltage and inductor current ripples

Knowing the voltage and current ripples is important in the context of component selection as well as performance comparison. The simplest way is to examine the voltage across the inductor and the currents going through the flying capacitors at steady state. In particular, the DC values of v_{lv} and v_{hv} as described in (7) imply that the switched node voltage $v_{swn}(t)$ as highlighted in Fig. 2 can be well approximated by a periodic square wave having a frequency of three times the switching frequency. Knowing $v_{swn}(t)$ allows derivation of the peak-to-peak current ripples as given by

$$\Delta I_L = \left(\frac{1}{3} - D\right)\frac{DV_{bulk}T_{pwm}}{L}. \qquad (8)$$

Similarly, the ripples of the flying capacitor voltages can be derived by modelling the inductor as a constant current source having a magnitude of

$$I_L = \frac{P_{out}}{DV_{bulk}}, \qquad (9)$$

where P_{out} denotes the load power. Given the switching sequence and timing chain in Fig. 3, the peak-to-peak capacitor voltage ripples can be simply obtained via

$$\Delta V_{lv} = \frac{D^2 T_{pwm} P_{out}}{C_{lv} V_{bulk}},$$

$$\Delta V_{hv} = \frac{D^2 T_{pwm} P_{out}}{C_{hv} V_{bulk}}. \qquad (10)$$

For other operating modes of the converter, steady state and ripple analyses can be performed in a similar fashion as discussed in this section. The final results associated with each duty ratio range are collected in TABLE I. As compared to a conventional boost converter, one can see a great similarity in the input-to-output gain but significant differences in the inductor current ripple. Specifically, ΔI_L is suppressed not only at two extreme values of the duty ratio but also at mode transition points, i.e. $D = 1/3$ and $2/3$.

TABLE I: DC GAIN, VOLTAGE AND CURRENT RIPPLES OF THE PROPOSED FOUR-LEVEL TOTEM POLE RECTIFIER

	$0 < D \leq \frac{1}{3}$	$\frac{1}{3} \leq D \leq \frac{2}{3}$	$\frac{2}{3} \leq D \leq 1$
DC gain V_{bulk}/V_{in}	$\dfrac{1}{D}$	$\dfrac{1}{D}$	$\dfrac{1}{D}$
DC flying capacitor voltage V_{lv}	$V_{bulk}/3$	$V_{bulk}/3$	$V_{bulk}/3$
DC flying capacitor voltage V_{hv}	$2V_{bulk}/3$	$2V_{bulk}/3$	$2V_{bulk}/3$
Inductor current ripple ΔI_L (peak-to-peak)	$\left(\dfrac{1}{3} - D\right)\dfrac{V_{in}T_{pwm}}{L}$	$\left(D - \dfrac{1}{3}\right)\left(\dfrac{2}{3D} - 1\right)\dfrac{V_{in}T_{pwm}}{L}$	$\left(1 - \dfrac{2}{3D}\right)(1 - D)\dfrac{V_{in}T_{pwm}}{L}$
Flying capacitor voltage ripple ΔV_{lv} (peak-to-peak)	$\dfrac{DP_{out}T_{pwm}}{V_{in}C_{lv}}$	$\dfrac{D}{3}\dfrac{P_{out}T_{pwm}}{V_{in}C_{lv}}$	$(1 - D)\dfrac{P_{out}T_{pwm}}{V_{in}C_{lv}}$
Flying capacitor voltage ripple ΔV_{hv} (peak-to-peak)	$\dfrac{DP_{out}T_{pwm}}{V_{in}C_{hv}}$	$\dfrac{DP_{out}T_{pwm}}{V_{in}C_{hv}}$	$\dfrac{DP_{out}T_{pwm}}{V_{in}C_{hv}}$

If the PFC output is assumed to be unchanged at steady state, the voltage and current ripples as formulated in TABLE I can be described as functions of D, which allows theoretical derivation of maximums of these quantities, which is summarized in TABLE II.

The results as presented in TABLE II show that the maximum current ripple in the proposed four-level boost converter is one ninth of that of the conventional one. In other words, for the same current ripple, the four-level approach requires nine times less inductance than the conventional boost converter, or equivalently nine time less stored energy/volume for the same power rating.

TABLE II: MAXIMAL VOLTAGE AND CURRENT RIPPLES

ΔI_L is maximal at $D = \frac{1}{6}$, $\frac{1}{2}$, and $\frac{5}{6}$	$\dfrac{1}{36}\dfrac{V_{out}T_{sw}}{L}$
ΔV_{lv} is maximal at P_{out_max} and $V_{in} \le \frac{V_{bulk}}{3}$	$\dfrac{P_{out\_max}T_{pwm}}{V_{out}C_{lv}}$
ΔV_{hv} is maximal at P_{out_max} and $V_{in} \le \frac{V_{bulk}}{3}$	$\dfrac{P_{out\_max}T_{pwm}}{V_{out}C_{hv}}$

IV. FLYING CAPACITOR INITIALIZATION AND VOLTAGE BALANCING

The diode clamped circuit as illustrated in Fig. 1 is proposed to address to two real challenges including capacitor initialization and voltage balancing associated with implementation of a four-level boost converter. In particular, if Z_1, Z_2 and Z_3 are 160V Zener diodes, turning Q1 and Q2 on while keeping other MOSFETs off will activate Z_2 and Z_3, which in turn charges C_{lv} and C_{hv} to $v_{bulk} - 320V$ and $v_{bulk} - 160V$, respectively. Therefore, the worst-case voltage stress on the switching devices is 160V which is safe for 200W rating MOSFETs. Two pairs of diodes (D$_3$, D$_4$) and (D$_5$, D$_6$), current limit resistors R$_1$ and R$_2$, and a capacitor array (C$_1$,C$_2$,C$_3$) are provided to supply additional charge to C_{lv} and C_{hv} during the condition where $v_{lv} + v_{hv} \le v_{bulk}$. Although the balance enforcement circuit has no effect when $v_{lv} + v_{hv} > v_{bulk}$, it actually helps to damp/interrupt all disturbances occurring in the flying capacitor voltages due to continual variations in the duty ratio over the line cycle. In order to demonstrate the effectiveness of balance enforcement, a simulation is set up in LTSpice for two operating conditions of the proposed converter, one with balance enforcement and one without such a functionality. The input current is programmed to follow the shape of the input voltage and provide power of 200W to the output. Simulated results for the two test conditions are plotted in Fig. 6 and Fig. 7, respectively.

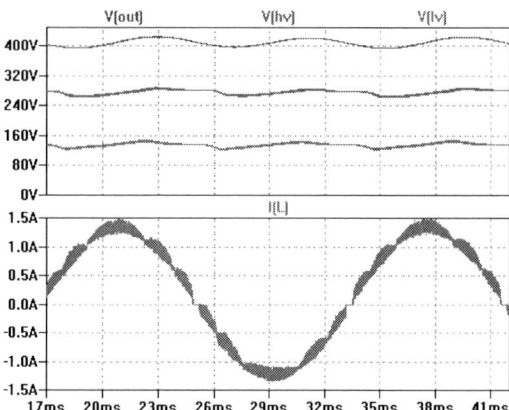

Fig. 6: Capacitor voltage waveforms with balance enforcement. The values of the damping resistance and capacitor string used in the simulation are R$_1$ = R$_2$ = 68Ω and C$_1$=C$_2$=C$_3$ = 100nF.

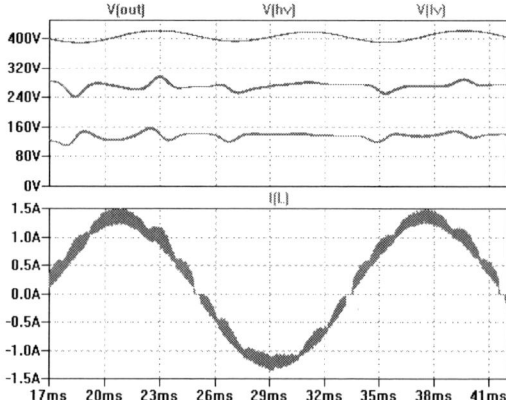

Fig. 7: Capacitor voltage waveforms without balance enforcement

Comparing the data between these figures shows that without balance enforcement, the flying capacitor voltages tend to deviate from the balanced points when the duty ratio command varies, which seems to be unavoidable in PFC applications. One may argue that disturbances in v_{lv} and v_{hv} seems benign, and do not have much effect on the converter operation except slightly distortion in the current waveform as can be seen in Fig. 7. The main concern here is unbalanced voltage stresses on switching devices, which typically gets worse during start-up and step-load transitions Therefore, the role of the clamped circuit is obviously critical in this context.

Although the clamped circuit looks quite complex, it is composed of low-power devices only and, as a consequence, neither take up much real estate from the PCB design nor contribute much to the build of material (BOM) costs.

V. DESIGN AND IMPLEMENTATION

The input data for the design of a four-level converter prototype are listed in TABLE III.

TABLE III: CONVERTER DESIGN SPECIFICATIONS

Input voltage, V_{line}	$85V_{rms}$ - $265V_{rms}$
Bulk capacitor voltage, V_{bulk}	$175V_{DC}$ - $400V_{DC}$
Maximum pk-pk capacitor voltage ripples, ΔV_{lv_max}, ΔV_{hv_max}	10V
Switching frequency, f_{pwm}	150kHz
Maximum output power, P_{out_max}	200W
Efficiency, η	98%
Hold-up time, t_{holdup}	20ms

A. Power stage design

1) Boost inductor selection:

Inductance values are generally determined in an iterative fashion so as to ensure converter operating condition, i.e. continuous or discontinuous conduction mode, component stresses, switching losses and many others. For simplicity, maximal peak current and current ripple are used as a basis for inductor value determination here. If the peak-to-peak current ripple is assumed to be negligible, the maximal peak inductor current can be approximated by

$$I_{pk\_max} = \frac{\sqrt{2}P_{out\_max}}{\eta V_{line\_min}} = \frac{\sqrt{2}*200}{0.98*85} = 3.39A. \quad (11)$$

Choosing the maximal current ripple of about 5% of I_{pk_max} yields $\Delta I_{L\_max} = 169.5mA$. Given the expression of ΔI_L in TABLE II, one can obtain the minimum inductor value via

$$L_{min} = \frac{V_{bulk}T_{pwm}}{36\Delta I_{L\_max}} = \frac{400*6.7\bullet10^{-6}}{36*0.1695} = 437\mu H. \quad (12)$$

Notice that a 5% ripple specification is actually desirable in the context of EMI filter, core losses and control stability, but is rarely adopted in design of conventional 200W PFC rectifiers as doing this requires a significant large magnetics with an inductance value of as high as 4mH.

2) Flying capacitors determination

The flying capacitors can be calculated from the voltage ripple requirement via

$$C_{hv\_min} = \frac{P_{out\_max}T_{pwm}}{V_{bulk}\Delta V_{hv\_max}} = \frac{200*6.7\bullet10^{-6}}{400*10} = 333nF,$$

$$C_{lv\_min} = \frac{P_{out\_max}T_{pwm}}{V_{bulk}\Delta V_{hv\_max}} = \frac{200*6.7\bullet10^{-6}}{400*10} = 333nF. \quad (13)$$

Considering a margin of 20% for voltage roll off gives $C_{lv} = C_{hv} = 1.2*333nF = 400nF$.

3) Output capacitor design

Generally, three factors including output voltage ripples, output current ripples, and hold-up requirement are involved in the design of the bulk capacitor. However, in most cases, the hold-up time t_{holdup} is the main factor influencing the size and the value of the bulk capacitor.

Due to the constraints on the input voltage range of conventional downstream stages, the PFC output typically operates between around 300V and 400V, which implies that only 56.25% of the energy storage in the bulk capacitor is recycled for hold-up. For a better use of output capacitance, a stacked multi-phase DC/DC converter with a wide input range [12] is deployed after the PFC stage. The output voltage now can be drained to a value as low as 175V, which suggests minimal bulk capacitance of

$$C_{bulk\_min} = \frac{2t_{holdup}P_{out\_max}}{V_{bulk\_nom}^2 - V_{bulk\_min}^2}$$
$$= \frac{2*0.02*200}{400^2 - 175^2} = 61.8\mu F. \quad (14)$$

In this case, a $68\mu F$ capacitor can satisfy the minimal capacitance condition with a margin of 10%.

B. Implementation

The design procedure as discussed in Section V.A suggests the following components: $C_{bulk} = 68uF – 450V$ rating- electrolytic, $C_{lv} = C_{hv} = 400nF$ - 450V rating - ceramic, $C_1 = C_2 = C_3 = 100nF$ - 250V rating- ceramic, and $R_1 = R_2 = 68\Omega$. The PFC choke is implemented by a standard toroid having an outer diameter of 18mm, and an A_L value of 89 nanohenries per turn squared which can provide inductance of 461uH with 72 turns. A roll-off of 60% in the inductance is to be expected at low-line voltage; however, the inductance is over-designed in (12) so it is safe to operate with such configuration The main 200V switches are BSZ900N20NS3G with Drain-Source on-resistance R_{DSon} of $90m\Omega$, and controlled by ARM Cortex M0 (STM32F051) through proprietary isolated gate-drive circuitry

VI. EXPERIMENTAL RESULTS

Operational waveforms at low line and high line input voltages are illustrated in Figures 3 and 4, respectively. Both experiments are configured to provide a quasi-constant output voltage of 400V. The experimental results show that the converter operates in a stable manner with no sign of voltage imbalance, which confirms the feasibility of the deployment of 200V MOSFETs here. Thanks to the multi-level structure, the inductor current has minimum ripples even no EMI input filter is present during the measurement.

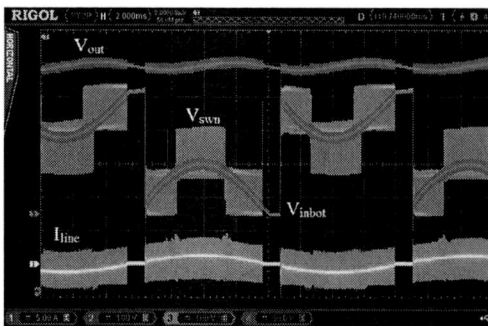

Fig. 8: Operational waveforms of the four-level PFC rectifier at low line input voltage and P_{out} = 100W: (CH1) I_{line} -Input current, (CH2) V_{swn} - Switched node voltage, (CH3) V_{inbot} - Input voltage referred to the bottom, (CH4) V_{bulk} – Bulk capacitor voltage

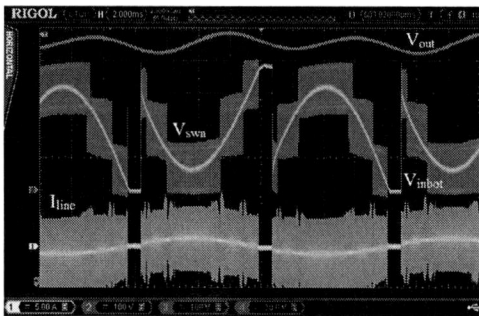

Fig. 9: Operational waveforms of the four-level PFC rectifier at high line input voltage and P_{out} = 200W: (CH1) I_{line} - Input current, (CH2) V_{inbot} - Input voltage, (CH3) V_{swn} - Switched node voltage, referred to the bottom, (CH4) V_{bulk} – Bulk capacitor voltage

VII. Conclusions

This paper presents a commercially-deployed implementation of universal four-level PFC rectifiers using 200V MOSFETs, allowing nine times reduction in the PFC choke, four times reduction in the complete product volume, and materially improvements in the rectifier efficiency and power density. The study also confirms that balancing multi-level flying capacitor converters can be practically and effectively achieved by a passive clamping network, although it is found that cost-effective and low-complexity properties are lost when extending the design to six-level converters or higher. The works also include insight into the theoretical analyses of the four-level synchronous boost converter as well as the balance enforcement mechanism, and discuss the design of the power stage in details.

Some of the technologies and implementation details presented in this paper may be subject of patent applications.

References

[1] Limits for Harmonic Current Emissions (Equipment Input Current < 16A per Phase), IEC Std. 61000-3-2.

[2] L. Balogh and R. Redl, "Power-factor correction with interleaved boost converter in continuous-inductor-current mode", in *Proc. IEEE Applied Power Electron. Conf.*, Mar. 1993, pp. 168-174

[3] T. A. Meynard, and H. Foch, "Multi-level conversion high voltage coppers and voltage-source inverters", 23rd *Ann. IEEE Power Electron. Specialists Conf.*, 1992

[4] F. Forest, T. A. Meynard, S. Faucher, F. Richardeau, J. J. Huselstein and C. Joubert, "Using the multilevel imbricated cells topologies in the design of low-power power-factor-corrector converters", *IEEE Trans. Industrial Electronics*, Vol. 52, No. 1, Feb. 2005, pp. 151-161

[5] B. Mahdavikhan, and A. Prodic, "Low-volume PFC rectifier based on nonsymmetric multilevel boost converter", *IEEE Trans. Power Electron.*, Vol. 30, No. 3, Mar. 2015, pp. 1356-1372

[6] J. C. Salmon, "Circuit topologies for PWM boost rectifiers operated from 1-phase and 3-phase ac supplies and using either single or split dc rail voltage output," in *Proc. IEEE Applied Power Electron. Conf.*, Mar. 1995, pp. 437-479

[7] A. F. Souznad and I. Barbi, "High power factor rectifier with reduced conduction and commutation losses", in *Proc. Int. Telecommunication Energy Conf.*, Jun. 1999

[8] W. Y. Choi, J. M. Kwon, E. H. Kim, J. J. Lee, and B. H. Kwon, "Bridgeless boost rectifier with low conduction losses and reduced diode reverse-recovery problems", *IEEE. Trans. Power Electron.*, Vol 54, No. 2, Apr. 2007, pp. 769-780

[9] Q. Li, M. A. E. Andersen, and O. C. Thomsen, "Conduction losses and common mode EMI analysis on bridgeless power factor correction," *Int. Conf. Power Electron. and Drive System*, Nov. 2009.

[10] B. Su, J. Zhang, and Z. Lu, "Totem-pole boost bridgeless PFC rectifier with simple zero-current detection and full range ZVS operating at the boundary of DCM/CCM," *IEEE Trans. Power Electron.*, vol. 26, no. 2, Feb. 2011.

[11] L. Zhou, and Y. Wu, "99% efficiency true-bridgeless totem-pole PFC based on GaN HEMTs", *Power Conversion Intelligent Motion* (PCIM), May 2013.

[12] T. T. Vu, G. Young, "Stack multiphase asymmetrical half-bridge topology offering advance performance and efficiency", *IEEE Applied Power Electron. Conf*, Mar., 2016, accepted for publication

[13] Erickson, R. W. and Maksimovic, D. *Fundamental of power electronics*, 2nd Edition, Kluwer Academic Publishers, 2001

US Mains Stacked Very High Frequency Self-oscillating Resonant Power Converter with Unified Rectifier

Jeppe A. Pedersen*, Mickey P. Madsen*, Jakob D. Mønster*,
Thomas Andersen*, Arnold Knott*, Michael A. E. Andersen*
*Technical University of Denmark
Richard Petersens Plads, building 325
2800 Kongens Lyngby
Denmark
Email: {jarpe, akn, ma}@elektro.dtu.dk

Abstract—This paper describes a Very High Frequency (VHF) converter made with three Class-E inverters and a single Class-DE rectifier. The converter is designed for the US mains (120 V, 60 Hz) and can deliver 9 W to a 60 V LED. The converter has a switching frequency of 37 MHz and achieves an efficiency of 89.4%. With VHF converters the power density can be improved and the converter described in this paper has a power density of 2.14 W/cm3. The power factor (PF) requirements of mains connected equepment is fulfilled with a power factor of 0.96.

I. INTRODUCTION

In traditional Switch Mode Power Supplies (SMPS) the passive components occupies most of the volume. Increasing the switching frequency reduces the physical size of the power supply since the passive components scale inversely with frequency. The switching frequency of hard switching converters is typically limited to a few megahertz due to switching losses and limited selection of commercial available gate drivers operating at higher frequencies. In resonant converters using soft switching the switching losses can be neglected and is therefore not limiting the frequency. Therefore when increasing the switching frequency to the Very High Frequency (VHF) range (30 MHz - 300 MHz) resonant converter topologies are used [1]–[5]. Most VHF converters consist of an inverter and a rectifier. The VHF converter described in this paper consists of three Class-E inverters [6], [7], and one Class-DE rectifier described in [8]–[12]. One of the challenges with the Class-E resonant inverter when using a high input voltage is the peak voltage across the MOSFET which theoretically is $V_{peak} \approx 3.6 \cdot V_{in}$ when the duty cycle is 50% [8]. If a single Class-E inverter were to operate from the US mains (120 V, 60 Hz) the MOSFET should be at least 650 V devise to withstand the voltage peak. As there do not exist 650 V devises with good characteristics, three inverters is connected in series to divide the input voltage as in [13]. This reduces the peak voltage across the MOSFET in each Class-E inverter to one third. To decrease the input voltage of the three inverters further, the load is placed in series with the input of the inverters as described in [14]. The three inverters

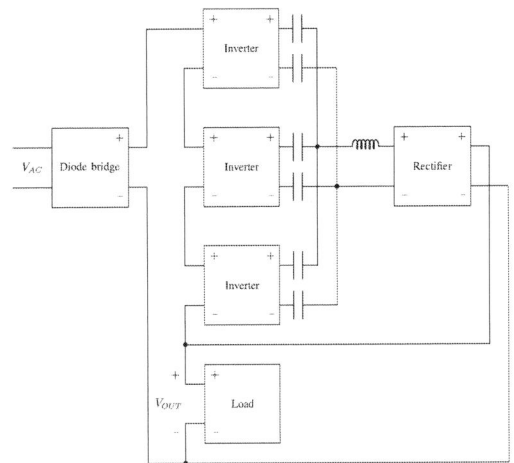

Fig. 1. Configuration of the converter, each box has input on the left and output on the right side. The resonant tank is shown between the inverters and the rectifier.

share a single rectifier and has a combined resonant tank made with six capacitors and one inductor. The configuration of the inverters, rectifier and load is shown in Fig. 1. Each inverter is implemented with the self-oscillating gate drive described in [15] and used in [16], [17].

The schematic of the converter is seen in Fig. 2 where the dotted line shows the separation between the inverter, resonant tank and the rectifier. In this configuration each inverter's output is galvanic isolated from the rectifier by the two resonant capacitors, this is necessary because of the series connection of the inverters and the load. The converter is however not galvanicly isolated since the load is directly connected to the input diode bridge as seen in Fig. 1.

The converter described in this paper is designed to run from the US-mains delivering 9 W to a 60 V LED. The full specifications are shown in Table I. The LED market

978-1-4673-9551-9/16 $31.00 © 2016 IEEE

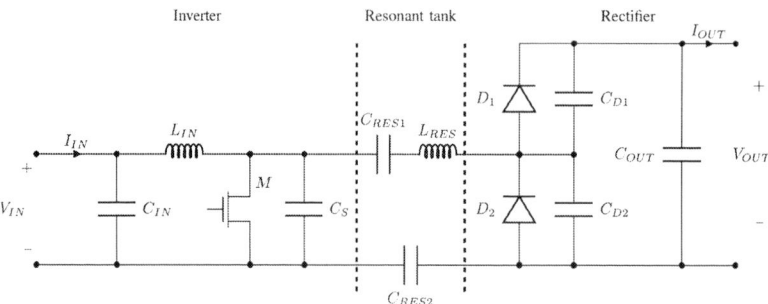

Fig. 2. Schematic of the converter topologi used in this paper, the dotted line shows seperation between the Class-E inverter, Resonant tank and the Class-DE rectifier. In the final converter three inverters are combined to a single rectifier.

TABLE I
SPECIFICATIONS

$V_{IN,RMS}$	120 V
V_{OUT}	60 V
P_{OUT}	9 W
f_s	37 MHz

is growing rapidly both with replacement bulbs and new LED fixtures. In this market a small size, high efficiency, and increased regulations regarding power factor (PF) is a challenge. This paper shows how this can be achieved by use of VHF converters.

II. THEORY

The converter consists of three Class-E inverters and a single Class-DE rectifier which is connected as shown in Fig. 1. When designing a resonant converter the rectifier is usually designed first. Then the inverter is designed to match the input impedance of the rectifier. The load is a 60 V LED and due to the high output voltage the Class-DE rectifier topology is chosen. The Class-DE rectifier has the advantage that the peak voltage across the diode is equal to the output voltage, which makes is suitable for high voltage applications. The design of a Class-DE rectifier is described in [8], [11]. The capacitances C_{D1} and C_{D2} is found by:

$$R_L = \frac{V_{out}}{I_{out}} = \frac{2 \cdot \pi}{\omega \cdot (C_{D1} + C_{D2})} \quad (1)$$

The input impedance of the Class-DE rectifier with $D = 0.25$ is:

$$Z_{IN} = \frac{1}{2 \cdot \pi} \cdot R_L \quad (2)$$

The design for the Class-E inverter is described in [16]. Each inverter is designed to work with a third of the rectifiers input impedance. The resonant inductor is designed as three separate inductors and then combined into one inductor. The resonant inductor can be combined since the output of the three inverters is in parallel. The value of the combined resonant inductor is equal to a parallel connection of the three separately calculated resonant inductors. With a combined inductor a

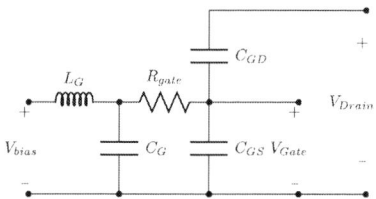

Fig. 3. Schematic of the gate drive with MOSFET parasitics.

higher Q can be achieved. This increases the efficiency and reduces component count. Another advantage of using a single resonant inductor is that the switching frequency of the three inverters is forced to synchronize.

The gate drive used in the inverters is shown in Fig. 3. This is a simple passive gate drive described in [15] where C_{GD}, C_{GS} and R_{gate} is the parasitic components of the MOSFET. The gate drive creates a phase shift of $180°$ from drain to gate and maintains a sinusoidal gate voltage. The bias voltage is kept around the threshold voltage of the MOSFET to achieve 50% duty cycle.

III. DESIGN AND SIMULATIONS

The converter is designed to deliver 9 W from US mains (120 V at 60 Hz) with a 60 V LED as load. The converter is designed with a switching frequency of 37 MHz this frequency is chosen to place the second harmonic below the FM band (87.5 - 108 MHz) and the third harmonic above. The power balance design of the converter is described in this section. One of the three Class-E inverter is simulated together with a Class-DE rectifier designed for a single stage as a DC-DC converter as shown in Fig. 2. To speed-up the simulation process. The 37 Mhz converter is evaluated at different DC input voltages to equivalent the quasi stationary behaiviuor of the slow 60 Hz AC mains input. LT-Spice is used to simulate the converter with a simple model of the MOSFET including the parasitic capacitances and gate resistor.

When designing a resonant inverter it is important that the converter can operate in the desired voltage range. The converter is designed to deliver 2.5 W at $V_{IN} = 30$ which is

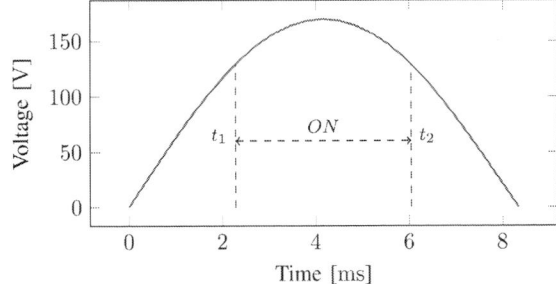

Fig. 4. The input voltage after the rectification with markings for when the full converter are running.

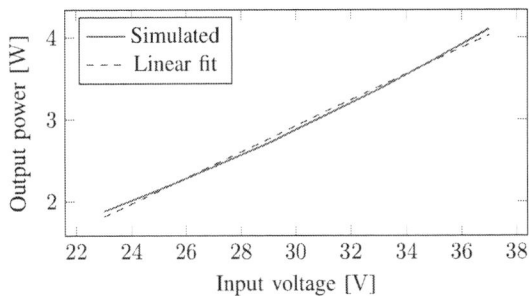

Fig. 5. Simulated output power vs. input voltage.

a reasonable starting point. As each inverter will deliver less than a third of the total power to the load. The first simulation is used to find the minimum stable voltage. It shows that the converter start at 20 V where the gate drive has enough gain to maintain the resonating gate voltage. In case all three inverters are not perfectly sharing the input voltage the minimum turn on voltage is raised to 23 V. The maximum voltage can be found by (3).

$$V_{inv,peak} = \frac{V_{IN,peak} - V_{out}}{3} \qquad (3)$$

When $V_{IN,peak} = V_{IN,RMS} \cdot \sqrt{2} = 120V \cdot \sqrt{2} \approx 170V$ the maximum voltage for each inverter is 37 V.

With the minimum and maximum voltage for each inverter found and assuming the output voltage is a constant 60 V the converters ON period can be defined. The ON period of the rectified 60 Hz period is shown in Fig. 4.

When the operating voltage and the ON period is found the average input current $I_{IN,AVG}$ in the ON period of the total converter can be found from (4).

$$P_{IN} = \frac{T_{ON}}{T_{Period}} \cdot \int_{t_1}^{t_2} V_{IN,peak} \cdot sin(t) \cdot I_{IN,avg} \, dt \qquad (4)$$

The input power is $P_{IN} = \frac{P_{OUT}}{0.9} = 10\,W$ assuming an efficiency of 90%. The integral is found from t_1 to t_2 to integrate over the total ON perriod T_{ON}. By solving (4) for the average input current the result is $I_{IN,avg} \approx 100\,mA$.

When connecting the load in series with the inverters the output power can be split up into two different parts. The $P_{INcurrent}$ is the part coming from the input current that runs through the inputs of the inverters. The P_{REC} is the part coming from the rectifier.

$$P_{OUT} = P_{INcurrent} + P_{REC}. \qquad (5)$$

$P_{INcurrent}$ is found to be $P_{INcurrent} = I_{IN,avg} \cdot V_{Load} \cdot \frac{t_{ON}}{t_{Period}} = 2.71\,W$. P_{REC} is the total power from the rectifier and therefore each inverter need to deliver a third of this $\frac{P_{REC}}{3} \approx 2.1\,W$. The single converter is now simulated at

different input voltages where the output power is measured, this is done to avoid long simulations of the half sine at 60 Hz. The result of the simulations is shown in Fig. 5.

By making a linear aproximation of the measurements shown in Fig. 5 a relationship between the input voltage and the output is defined (6).

$$Pout = 0.16 \cdot V_{Inv} - 1.8 \qquad (6)$$

This can then be used to see how much power each converter delivers to the load over one period. The voltage across each inverter follows the form described in (7).

$$V_{Inv} = \frac{V_{IN,peak} \cdot sin(t) - V_{OUT}}{3} \qquad (7)$$

When combining (6) and (7) the output power can be integrated over an ON period which is used to verify the output power of a single inverter stage is reaching 2.1 W.

$$P_{out} = \frac{t_{ON}}{t_{Period}} \cdot \int_{t_1}^{t_2} 0.16 \cdot \frac{V_{IN,peak} \cdot sin(t) - V_{OUT}}{3} - 1.8 \, dt \qquad (8)$$

The single converter is delevering 2.13 W in a full period. The components used in a single converer stage is described in Table II. The component values will change in the resonant tank and the rectifier when all three inverters shares one single rectifier.

TABLE II
COMPONENT VALUES FOR A SIGLE INVERTER AND RECTIFIER.

Component	Value
C_{IN}	1 µF
L_{IN}	1 µH
C_{S1}	18 pF
C_{RES1}	39 pF
C_{RES2}	4.7 nF
L_{RES}	1.24 µH
C_{D1}	7 pF
C_{D2}	7 pF
C_{OUT}	1 µF
L_G	82 nH
C_G	39 pF

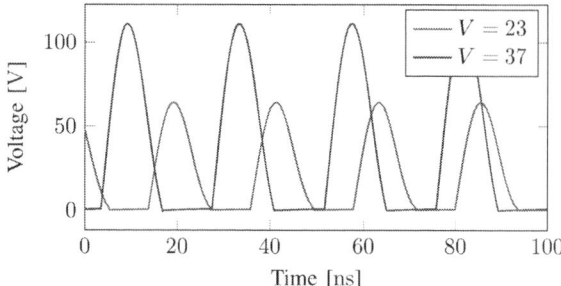

Fig. 6. Drain waveforms from simulation of the inverter at the mimimum and maximum input voltages.

To verify the behavior of the converter within the full voltage range simulations are performed. The converter is operating in the full voltage range. The drain waveforms from simulation at minimum and maximum input voltage are shown in Fig. 6, the converter is ZVS at both input voltage. The self-oscilating gatedrive causes the switching frequency to change depending on the input voltage. The frequency change comes from the nonlinear parasitic capacitances of the MOSFETs. The change in frequency are relativly small is in not a problem regarding the FM-band.

IV. EXPERIMENTAL RESULTS

The converter is implemented with a single rectifier to reduce component count. Because the rectifier is connected in parallel with all three inverter the input impedance should be three times higher that for a single inverter stage. By looking at (1) that means that the two capacitances is getting 3 times as large. The three resonant tank inductors can also be combined into a single resonant inductor. A single resonant inductor will only be a third on the value and can have a higher Q which means higher efficiency. The component list is shown in Table III. The inductors used as input inductor's (L_{IN}) is the 1812CS series from Coilcraft and the gate inductor's (L_G) is the 1008CS series. The commen resonant inductor is chosen to be a 2929SQ air inductor because this have a high Q at 37 MHz. The MOSFET used in the inverters is IRF5802, which is a 150 V silicon MOSFET from International Rectifier. The diodes used in the rectifier is the 60 V schottky diode PMEG6010AED from NXP. The input bridge rectifier is the MBS2 which is rated for 200 V.

A picture of the converter is shown in Fig. 7. It measures 20.2mm x 38.2mm. The PCB is devided into two box volumes one is to the left of the large resonant inductor and one with the large resonant inductor. The distance from the left side of the PCB to the resonant inductor is 22.7 mm and the height in this part is 2.7 mm, the part with the resonant inductor is 9.3 mm in height. The converter is achieving an efficiency of 89.4% and a power density of 2.14 $\frac{W}{cm^3}$. The temperature of the converter in steady state is shown in Fig. 8 were the warmest spot is the MOSFETs in the middle of the PCB reaching $T_{max} = 58\ ^{\circ}C$. The temperature test was performed

TABLE III
COMPONENT VALUES FOR THE IMPLEMENTED VHF CONVERTER.

Component	Value	PCS
C_{IN}	1 µF	3
L_{IN}	1 µH	3
C_{S1}	18 pF	3
C_{RES1}	39 pF	3
C_{RES2}	4.7 nF	3
L_{RES}	430 nH	1
C_{D1}	21 pF	1
C_{D2}	21 pF	1
C_{OUT}	1 µF	1
L_G	82 nH	3
C_G	39 pF	3

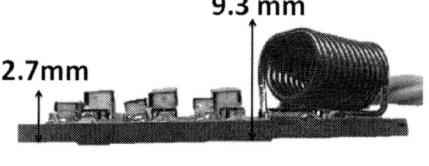

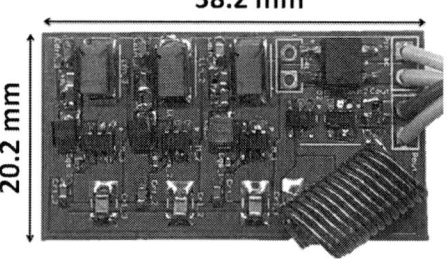

Fig. 7. The converter with three inverters and one rectifier.

with the PCB placed horisontal in a vise with 21 °C ambient temperature.

The converter is designed to run from the US mains and high PF is often a requrement for offline converters. This means that the converter starts once each inverter's input voltage reaches 23 V on the input and runs until the input voltage becomes too low for the self-oscillating gate drive to operate. This converter

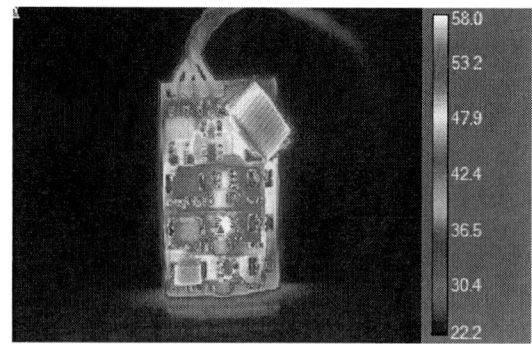

Fig. 8. Image of the converter in steady state. The temperatures are shown in °C.

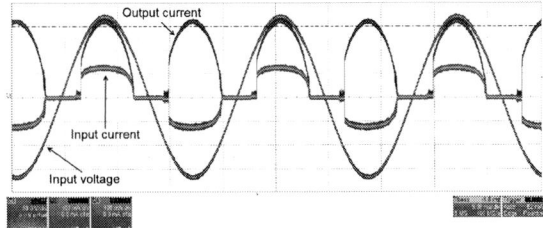

Fig. 9. Measurements of the input voltage (yellow), input current (green) and output current through the LED (blue).

type works as a voltage controlled current source. The input voltage and input/output currents are shown in Fig. 9. It is evident that the input current is in phase with the voltage which is needed to achieve a high PF. The converter reaches a PF = 0.96, which is high for an offline LED driver in this power range. The converter achieves a high PF, this makes it suitable as a power factor correction (PFC) converter if prober contols is implemented. The converter shuts down twice every 60 Hz period which results in a 120Hz flicker at the output. To reduce this low frequency flicker an energy storage is needed. This could be a large capacitor at the output, however this would reduce the power density.

V. CONCLUSION

This paper describes a VHF converter driving a 60 V LED from the US mains. The converter consists of three Class-E inverters and one Class-DE rectifier. The converter has a switching frequency of 37 MHz and an efficiency of 89.4%. Increasing the swithcing frequency to the VHF range reduces the size of the passive components and thereby the size of the converter. This size reduction results in a high power density of $2.14 \frac{W}{cm^3}$. The input current is in phase and proportional to the input voltage which results in a PF of 0.96.

REFERENCES

[1] T. Andersen, S. Christensen, A. Knott, and M. A. E. Andersen, "A VHF class E dc-dc converter with self-oscillating gate driver," in *Applied Power Electronics Conference and Exposition (APEC), 2011 Twenty-Sixth Annual IEEE*, 2011, pp. 885–891.

[2] M. Kovacevic, A. Knott, and M. Andersen, "A VHF interleaved self-oscillating resonant sepic converter with phase-shift burst-mode control," in *Applied Power Electronics Conference and Exposition (APEC), 2014 Twenty-Ninth Annual IEEE*, March 2014, pp. 1402–1408.

[3] M. Madsen, A. Knott, and M. Andersen, "Very high frequency half bridge dc/dc converter," in *Applied Power Electronics Conference and Exposition (APEC), 2014 Twenty-Ninth Annual IEEE*, March 2014, pp. 1409–1414.

[4] M. K. Song, M. Dehghanpour, J. Sankman, and D. Ma, "A VHF-level fully integrated multi-phase switching converter using bond-wire inductors, on-chip decoupling capacitors and dll phase synchronization," in *Applied Power Electronics Conference and Exposition (APEC), 2014 Twenty-Ninth Annual IEEE*, March 2014, pp. 1422–1425.

[5] W. Cai, Z. Zhang, X. Ren, and Y.-F. Liu, "A 30-MHz isolated push-pull VHF resonant converter," in *Applied Power Electronics Conference and Exposition (APEC), 2014 Twenty-Ninth Annual IEEE*, March 2014, pp. 1456–1460.

[6] R. Redl, B. Molnar, and N. Sokal, "Class E resonant regulated dc/dc power converters: Analysis of operations, and experimental results at 1.5 MHz," *Power Electronics, IEEE Transactions on*, vol. PE-1, no. 2, pp. 111–120, April 1986.

[7] M. Kazimierczuk and J. Jozwik, "Resonant dc/dc converter with class-E inverter and class-E rectifier," *Industrial Electronics, IEEE Transactions on*, vol. 36, no. 4, pp. 468–478, 1989.

[8] D. Kazimierczuk, Marian K. Czarkowski, *Resonant Power Converters*, 2nd ed. Wiley, 2011.

[9] K. Fukui and H. Koizumi, "Analysis of half-wave class DE low dv/dt rectifier at any duty ratio," *Power Electronics, IEEE Transactions on*, vol. 29, no. 1, pp. 234–245, Jan 2014.

[10] D. Hamill, "Class DE inverters and rectifiers for dc-dc conversion," in *Power Electronics Specialists Conference, 1996. PESC '96 Record., 27th Annual IEEE*, vol. 1, 1996, pp. 854–860 vol.1.

[11] K. Fukui and H. Koizumi, "Half-wave class DE low dv/dt rectifier," in *Circuits and Systems (APCCAS), 2012 IEEE Asia Pacific Conference on*, Dec 2012, pp. 69–72.

[12] L. Raymond, W. Liang, and J. Rivas, "Performance evaluation of diodes in 27.12 MHz class-D resonant rectifiers under high voltage and high slew rate conditions," in *Control and Modeling for Power Electronics (COMPEL), 2014 IEEE 15th Workshop on*, June 2014, pp. 1–9.

[13] M. Kovacevic, A. Knott, and M. Andersen, "VHF series-input parallel-output interleaved self-oscillating resonant sepic converter," in *Energy Conversion Congress and Exposition (ECCE), 2013 IEEE*, Sept 2013, pp. 2052–2056.

[14] M. Madsen, M. Kovacevic, J. Mønster, J. Pedersen, A. Knott, and M. Andersen, "Input-output rearrangement of isolated converters," in *Power and Energy Conference at Illinois (PECI), 2015 IEEE*, Feb 2015, pp. 1–6.

[15] M. Madsen, J. Pedersen, A. Knott, and M. Andersen, "Self-oscillating resonant gate drive for resonant inverters and rectifiers composed solely of passive components," in *Applied Power Electronics Conference and Exposition (APEC), 2014 Twenty-Ninth Annual IEEE*, March 2014, pp. 2029–2035.

[16] J. Pedersen, M. Madsen, A. Knott, and M. Andersen, "Self-oscillating galvanic isolated bidirectional very high frequency dc-dc converter," in *Applied Power Electronics Conference and Exposition (APEC), 2015 IEEE*, March 2015, pp. 1974–1978.

[17] J. Mønster, M. Madsen, J. Pedersen, and A. Knott, "Investigation, development and verification of printed circuit board embedded air-core solenoid transformers," in *Applied Power Electronics Conference and Exposition (APEC), 2015 IEEE*, March 2015, pp. 133–139.

Digital-Based Interleaving Control for GaN-based MHz CRM Totem-pole PFC

Zhengyang Liu, Zhengrong Huang, Fred C. Lee, and Qiang Li
Center for Power Electronics Systems
The Bradley Department of Electrical and Computer Engineering
Virginia Polytechnic Institute and State University
Blacksburg, Virginia, USA
LZY@vt.edu

Abstract — **In this paper, the performance of different interleaving control methods for gallium-nitride (GaN) devices based MHz critical conduction mode (CRM) totem-pole power factor correction (PFC) circuit is compared. Both closed-loop interleaving and open-loop interleaving are good for low frequency CRM PFC; but for MHz very high frequency CRM PFC with microcontroller (MCU) implementation, open-loop interleaving outperforms closed-loop interleaving with small and non-amplified phase error. After software optimization, the phase error of open-loop interleaving is smaller than 3 degree at 1MHz, when the control is implemented by 60MHz low cost MCU. Significant ripple cancellation effect and differential-mode (DM) filter size reduction is achieved with good interleaving. For a 1.2kW MHz totem-pole PFC, the DM filter size is reduced to one quarter compared to the counterpart of a 100kHz PFC. Last but not least, the stability of open-loop interleaving is also analyzed indicating that the MHz CRM totem-pole PFC with voltage-mode control, open-loop interleaving, and turn-on instant synchronization can maintain critical mode operation with better stability compared to low frequency CRM PFC.**

Keywords—Interleaving, GaN, MHz, totem-pole PFC, MCU, DM filter

I. INTRODUCTION

The totem-pole bridgeless power-factor-correction (PFC) circuit is becoming popular attributing to the emerging high voltage gallium-nitride (GaN) devices [1,2]. According to paper [3-5], soft switching operation is important to achieve MHz very high frequency operation for 600V GaN devices. Critical conduction mode (CRM) is a most simple way to achieve soft switching; and when applied to boost-type PFC circuit, it is easy to achieve good power factor with CRM operation.

A 1.2kW 1-3MHz GaN-based CRM totem-pole PFC was built with close to 99% peak efficiency and more than 200W/in³ power density [6,7]. Figure 1 shows the circuit diagram of the two-phase interleaved totem-pole PFC with cascode GaN devices. In addition, the MHz impact of PFC is not limited to itself, but has even more significant impact on the input filter design. According to paper [8], when the

switching frequency is above 400kHz, the higher switching frequency, the smaller filter size. Then in paper [6,9,10], it is demonstrated that from 100kHz to 1MHz, the DM filter is simplified from 2 stage to 1 stage and the volume is reduced by 50%. It also claims that if two-phase PFC interleaved with 180 degree phase shift, then another 50% volume reduction is expected.

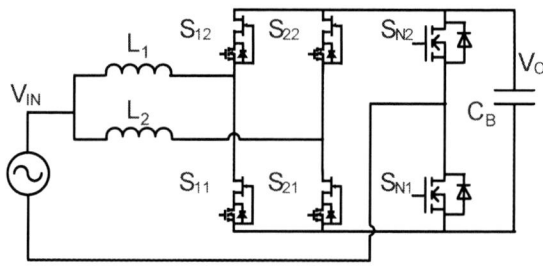

Figure 1. Circuit diagram of two-phase interleaved totem-pole PFC with cascode GaN devices

The challenge of interleaving control for CRM PFC is that, the nature of circuit is variable frequency operation. For a given input and load condition, the frequency varies 3-5 times in a half line cycle. For different input or load conditions, the frequency range varies as well. According to literature, generally there are two categories of interleaving control methods proposed for the variable frequency CRM PFC, closed-loop interleaving [11-15] and open-loop interleaving [16-21]. For low frequency, both methods work well to maintain the phase error to a minimal value. However, when the frequency is pushed 10 times higher to multi-MHz, the interleaving control becomes a new challenge.

In this paper, the impact of interleaving control on MHz CRM totem-pole PFC and DM filter is introduced in Section II. Then the performance comparison between closed-loop interleaving and open-loop interleaving for MHz totem-pole PFC is discussed in Section III. The optimization and experimental results of open-loop interleaving is presented in Section IV. Finally, the stability analysis of open-loop interleaving is elaborated in Section V.

II. Impact of Interleaving Control on MHz Totem-Pole PFC and its DM Filter

CRM PFC suffers from large current ripple. For a single phase CRM PFC, the input current ripple is always more than two times higher than its average current. Multi-phase interleaving techniques are widely used so that the total input current ripple is reduced, which is beneficial to have lower conduction loss, longer capacitor lifetime, and most importantly, smaller input filter size.

For a two phase interleaved CRM PFC, when the phase shift is 180 degree, the first order components in the input current is totally canceled while only second order and higher orders components exist. With the proposed DM noise model for CRM PFC [22], the DM noise spectrum is predicted for the dual-phase MHz totem-pole PFC (Figure 2). It is clearly shown that the ripple cancellation effect is very sensitive to phase error. Even just a few degree phase error leads to quick increase of the first order noise components, while the critical value is 4.6 degree and 1 degree for 1-stage DM filter and 2-stage DM filter respectively. So in order to have full benefits on the DM filter, the phase error should be kept smaller than the critical value. Previously it is not a big issue to maintain small phase error in low frequency range like 70kHz or 130kHz PFC, however, when it goes to multi-MHz very high frequency, it is really a challenge to make phase error smaller than the critical value because only a few nano-second delay error leads to significantly large phase error.

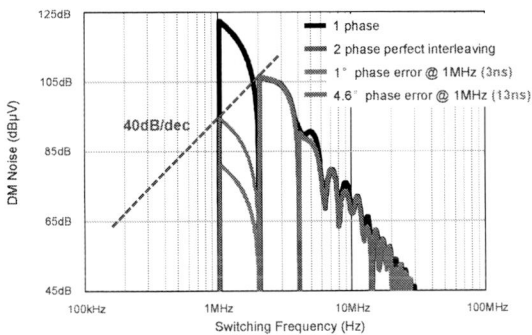

Figure 2. Predicted DM noise spectrum for 1.2kW MHz totem-pole PFC

III. Performance Comparison between Closed-Loop Interleaving and Open-Loop Interleaving for MHz CRM Totem-Pole PFC

According to literatures, previously proposed different interleaving control methods are divided into two categories, closed-loop interleaving and open-loop interleaving. In this paper, master-slave relationship between two phases, and voltage-mode control are applied to both cases. Particularly the turn-on instant synchronization is used in open-loop interleaving analysis.

For the system control implementation, no commercial controller supporting MHz CRM PFC operation is available at the time of this research conducted; discrete components based analog control was tried at the first time, which cannot

offer good enough control accuracy for interleaving; MCU based control is considered as a viable alternative to achieve good performance and reasonable cost.

60MHz MCU is used in the hardware demonstration. When the interleaving control is implemented by MCU, there are two limitations. The first limitation is the interleaving control cannot be done cycle by cycle. As shown in Figure 3, different control functions are executed in series sequence so that the total control cycle takes 240 system clock cycles to complete. This is equivalent to 4us with a 60MHz MCU so that the control cycle is much longer than the switching cycle. As a result, the interleaving control is performed once in each 4us.

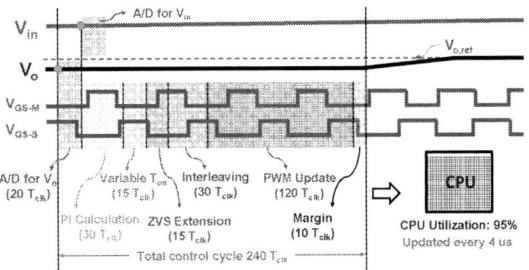

Figure 3. MCU control implementation sequence

The second limitation is that all control signals are synchronized to MCU system clock. Then the switching period and delay time are all integer compared to the clock cycle. So even the nature of CRM PFC is continuously smooth changing frequency but when implemented by MCU, the actual frequency is discrete.

When the two limitations combined together, phase error oscillation is observed with the closed-loop interleaving method. Figure 4 shows the simulation waveform. There is significant oscillation in the half line cycle input current which indicates the interleaving is not accurate. The zoom-in waveform further illustrates how the phase error is amplified and keeps oscillating. Due to this phenomenon, there is up to 24-degree phase error at 1MHz which is far larger than the critical value and made the closed-loop interleaving unacceptable. Figure 5 is the experimental waveform with closed-loop interleaving. The blue and the red waveforms are inductor current of each phase while the first waveform is the total input current after interleaving. Very similar to the simulation waveform, the measured total input current has unstable ripple randomly occurred as well. So with 24 degree phase error, the first order component is becoming dominant and its magnitude is even higher than the second order component as indicated on Figure 2.

On the contrary, the open loop interleaving has small and non-amplified phase error. Although two limitations still exist, the phase error of open-loop interleaving is able to be kept smaller than one clock cycle. Detailed analysis of phase error adjustment process of open-loop interleaving with MCU implementation is shown on Figure 6. Finally, Figure 7 offers a side by side comparison which demonstrates that open-loop

978-1-4673-9551-9/16 $31.00 © 2016 IEEE

interleaving outperforms the closed-loop interleaving with small and stable phase error.

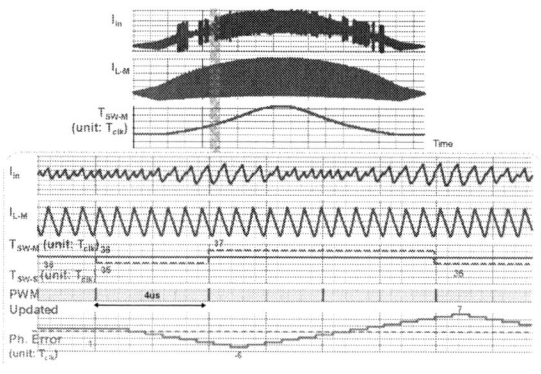

Figure 4. Simulated closed-loop interleaving of MHz CRM totem-pole PFC

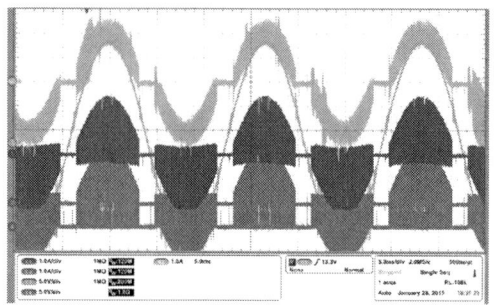

Figure 5. Simulated closed-loop interleaving of MHz CRM totem-pole PFC

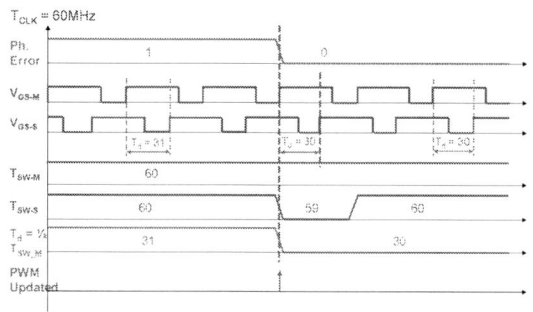

Figure 6. Phase error adjustment with open-loop interleaving implemented by 60MHz MCU

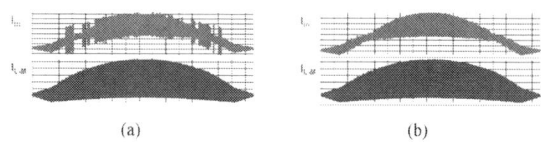

Figure 7. Interleaving performance comparison of (a) closed-loop interleaving and (b) open-loop interleaving

IV. OPTIMIZATION OF OPEN-LOOP INTERLEAVING

Due to the discrete frequency, there is always one clock cycle phase error existing in open-loop interleaving. With 60MHz, this is equivalent to 6 degree phase error at 1MHz which is still larger than the critical value. Besides increasing the CPU speed of MCU, software improvement is possible.

Figure 8 shows the improvement method. The idea is that the total delay time is divided into two separate parts. The basic delay time is still an integer of the system clock. At the same time, it is also possible to give another extra delay time which positions the PWM single edge with one half clock cycle resolution. Then the edge is synchronized to system clock when the master phase switching cycle is an even number; and the edge is placed in the middle of two adjacent system clock to provide one half clock cycle delay when the master phase switching cycle is an odd number. Ideally, when two parts delay time combined, the total delay time exactly equals to one half of the master phase switching period no matter it is an even number or an odd number.

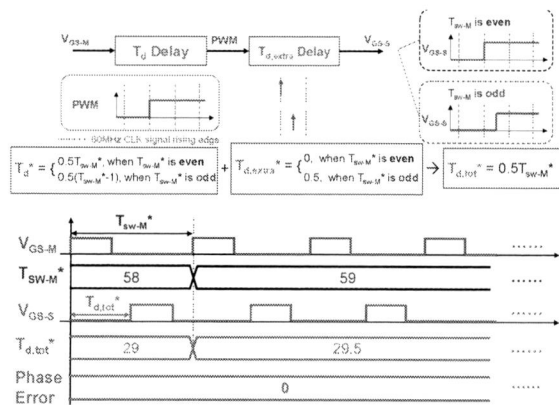

Figure 8. Software optimization of open-loop interleaving

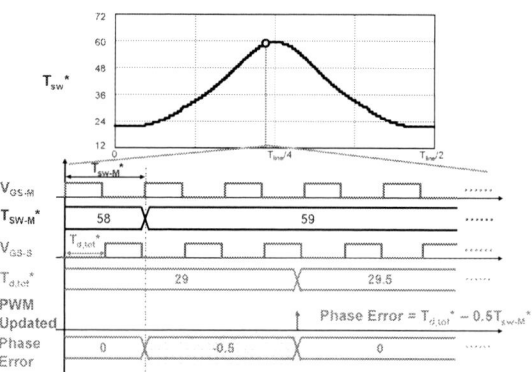

Figure 9. Phase error worst case of open-loop interleaving after optimization

Figure 9 further shows the phase error worst case of open-loop interleaving after the software optimization. It occurs in the switching frequency abrupt change point. Again, as the 4us control cycle is larger than the switching cycle, so between the switching frequency change and next control cycle adjustment there is one half clock cycle phase error. It equals to 3 degree phase error and it is smaller than the 4.6 degree critical phase error value if one stage DM filter is used.

Figure 10 shows the measured DM noise spectrum comparison before and after optimization. So with optimization method, the 2MHz noise component became the dominant factor for 1-stage DM filter design in Figure 10(b). Figure 11 is the experimental waveform shows good interleaving and ripple cancellation effect in different instant of a line cycle.

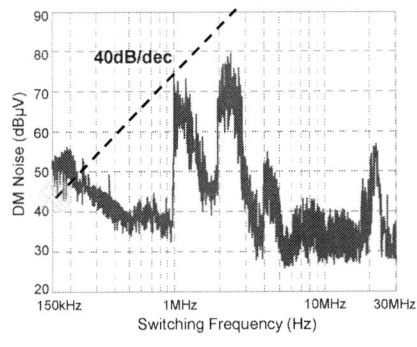

(a)

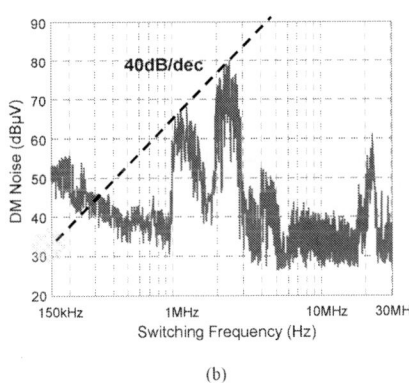

(b)

Figure 10. Measured DM noise spectrum of open-loop interleaving (a) before optimization and (b) after optimization

Although the 60MHz low cost MCU is used but the scope of the paper is not limited to one specific MCU. Several representative samples are selected for performance comparison and shown in Table I. As a simple function of CPU speed, the corresponding phase errors with closed-loop interleaving and open-loop interleaving are calculated respectively. The benefits offered by open-loop interleaving is clear through this comparison.

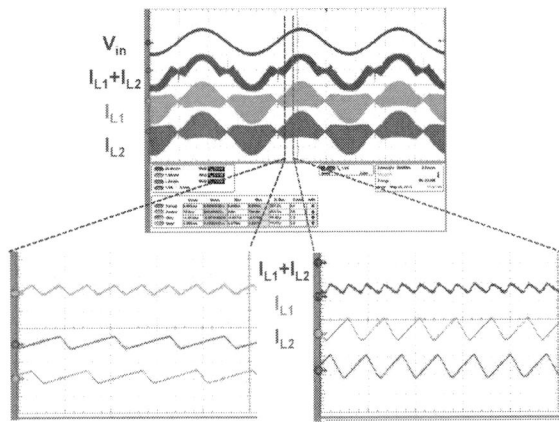

Figure 11. Experimental waveform of open-loop interleaving

TABLE I. INTERLEAVING PERFORMANCE WITH DIFFERENT MCU

	Series	CPU Speed	Closed-loop interleaving (4×T_CLK)			Open-loop interleaving (0.5×T_CLK)		
			Phase error	2-stage (1°)	1-stage (5°)	Phase error	2-stage (1°)	1-stage (5°)
Piccolo	F28027	60 MHz	24°	✗	✗	3°	✗	✓
	F28069	90 MHz	16°	✗	✗	2°	✗	✓
	F28075	120 MHz	12°	✗	✗	1.5°	✗	✓
Delfino	F28335	150 MHz	9.6°	✗	✗	1.2°	✗	✓
	F2837xS	200 MHz	7.2°	✗	✗	0.9°	✓	✓
	C28346	300 MHz	4.8°	✗	✓	0.6°	✓	✓

V. STABILITY ANALYSIS OF OPEN-LOOP INTERLEAVING AT MHZ HIGH FREQUENCY

According to paper [21], there are four scenarios of open-loop interleaving implementation, current mode with turn-on instant synchronization, current mode with turn-off instant synchronization, voltage mode with turn-on instant synchronization, and voltage mode with turn-off instant synchronization. Current mode with turn-on instant synchronization is the only stable case for both duty cycle smaller or larger than 0.5 and for positive and negative delay time perturbation. In the analysis of paper [21], there is an assumption that the resonant time is negligible compared to the switching period. This assumption is reasonable for low frequency CRM PFC. However, for MHz CRM PFC the resonant time to achieve ZVS occupies up to 10% of total switching period; or in other words, there is significant duty cycle loss so that the CRM mode at MHz is more like discontinuous current mode (DCM) instead of ideally boundary mode.

Due to this reason, the MHz totem-pole PFC with voltage mode control and turn-on instant synchronization open loop interleaving method is becoming more stable compared to low frequency case. Figure 12 shows the simulation waveform. Delay time perturbation, which in this case equals to 10%

switching period as an example, is applied to both 70kHz CRM boost and 1MHz CRM boost with the same input and output condition (V_{in}=100V and V_o=400V). After the perturbation, in the 70kHz case the inductor current of slave phase immediately goes to continuous current mode (CCM) with a very long settle down time (tens of switching period), while in the 1MHz case CCM operation just happens in one or two switching cycles and the current settle down to steady state quickly. In Figure 12, both red dash lines are inductor current of slave phase without perturbation. They can be treated as a reference of steady state operation. In addition, it is also verified that for both duty cycle smaller or larger than 0.5, similar conclusions can be made as well.

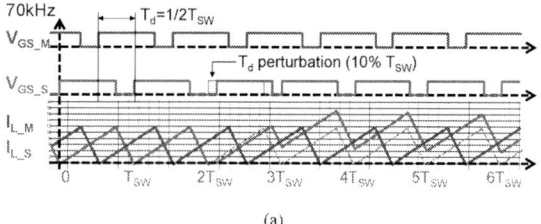

(a)

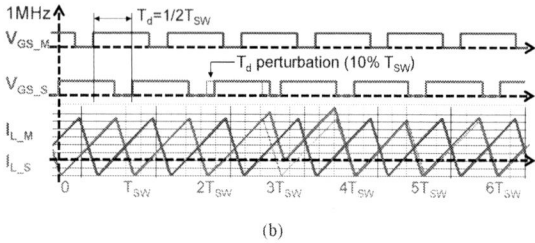

(b)

Figure 12. Open-loop interleaving under delay time perturbation of (a) 70kHz CRM PFC and (b) 1MHz CRM PFC

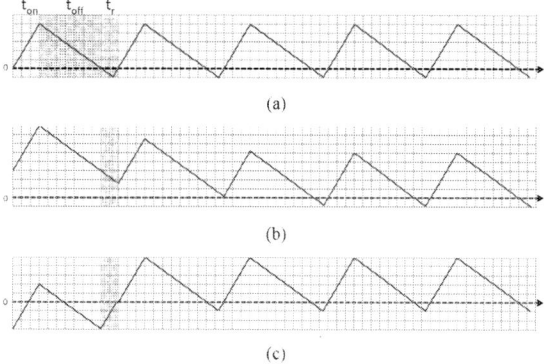

(a)

(b)

(c)

Figure 13. Stability mechanism of (a) CRM, (b) CCM, and (c) QSW

The theory to explain this phenomenon is volt-second unbalance. Figure 13 (a) shows the inductor current in critical mode operation and steady state with triangular shape approximation. Then Figure 13 (b) and (c) verifies that no matter the circuit goes to CCM or quasi-square-waveform

mode (QSW) after perturbation, the inductor current gradually settles down to CRM steady state. For CCM case, the original resonant time is becoming off time so that the average inductor current decreases each cycle, while for QSW case the original resonant time is becoming on time so that the average inductor current increases each cycle. Eventually, the unstable cases settle down very quickly due to large duty cycle loss at MHz very high frequency.

For the other two control implementations, which is current mode with turn-off synchronization and voltage mode with turn-off synchronization, the unstable phenomenon is sub-harmonic oscillation. They are still considered as unstable at MHz.

VI. CONCLUSIONS

Accurate interleaving control is very critical to effectively reduce input current ripple and to achieve expected EMI filter size reduction for the MHz CRM totem-pole PFC. With MCU implementation, open-loop interleaving has better performance than closed-loop interleaving. Less than 3 degree phase error is accomplished with open-loop interleaving and 60MHz MCU. Furthermore, 1-stage filter is less sensitive to phase error thus it is more suitable for MHz CRM totem-pole PFC.

ACKNOWLEDGMENT

The information, data, or work presented herein was funded in part by the Office of Energy Efficiency and Renewable Energy (EERE), U.S. Department of Energy, under Award Number DE-EE0006521 with North Carolina State University, PowerAmerica Institute.

This work is supported in part by the Power Management Consortium (PMC) in CPES, Virginia Tech.

This work was conducted with the use of GaN device samples donated in kind by Transphorm of the CPES Industry Consortium Program.

This work was conducted with the use of SIMPLIS donated in kind by Simplis Technologies of the CPES Industry Consortium Program.

DISCLAIMER

The information, data, or work presented herein was funded in part by an agency of the United States Government. Neither the United States Government nor any agency thereof, nor any of their employees, makes any warranty, express or implied, or assumes any legal liability or responsibility for the accuracy, completeness, or usefulness of any information, apparatus, product, or process disclosed, or represents that its use would not infringe privately owned rights. Reference herein to any specific commercial product, process, or service by trade name, trademark, manufacturer, or otherwise does not necessarily constitute or imply its endorsement, recommendation, or favoring by the United States Government or any agency thereof. The views and opinions of authors expressed herein do not necessarily state or reflect those of the United States Government or any agency thereof.

REFERENCES

[1] Liang Zhou, Yi-Feng Wu, Umesh Mishra, "True Bridgeless Totem-pole PFC based on GaN HEMTs", PCIM Europe 2013, 14-16 May, 2013, pp.1017-1022.

[2] Zhou, Liang; Wu, Yifeng; Honea, Jim; Wang, Zhan, High-efficiency True Bridgeless Totem Pole PFC based on GaN HEMT Design Challenges and Cost-effective Solution, in PCIM Europe 2015; International Exhibition and Conference for Power Electronics, Intelligent Motion, Renewable Energy and Energy Management; Proceedings of , vol., no., pp.1-8, 19-20 May 2015

[3] Xiucheng Huang; Zhengyang Liu; Qiang Li; Lee, F.C., "Evaluation and Application of 600 V GaN HEMT in Cascode Structure," Power Electronics, IEEE Transactions on , vol.29, no.5, pp.2453,2461, May 2014

[4] Xiucheng Huang; Qiang Li; Zhengyang Liu; Lee, F.C., "Analytical Loss Model of High Voltage GaN HEMT in Cascode Configuration," Power Electronics, IEEE Transactions on , vol.29, no.5, pp.2208,2219, May 2014

[5] Zhengyang Liu; Xiucheng Huang; Lee, F.C.; Qiang Li, "Package Parasitic Inductance Extraction and Simulation Model Development for the High-Voltage Cascode GaN HEMT," Power Electronics, IEEE Transactions on , vol.29, no.4, pp.1977,1985, April 2014

[6] Liu, Zhengyang; Lee, Fred C.; Li, Qiang; Yang, Yuchen, "Design of GaN-based MHz totem-pole PFC rectifier," in Energy Conversion Congress and Exposition (ECCE), 2015 IEEE , vol., no., pp.682-688, 20-24 Sept. 2015

[7] Yang, Yuchen; Mu, Mingkai; Liu, Zhengyang; Lee, Fred C.; Li, Qiang, "Common mode EMI reduction technique for interleaved MHz critical mode PFC converter with coupled inductor," in Energy Conversion Congress and Exposition (ECCE), 2015 IEEE , vol., no., pp.233-239, 20-24 Sept. 2015

[8] Bing Lu; Wei Dong; Wang, S.; Lee, F.C., "High frequency investigation of single-switch CCM power factor correction converter," Applied Power Electronics Conference and Exposition, 2004. APEC '04. Nineteenth Annual IEEE , vol.3, no., pp.1481,1487 Vol.3, 2004

[9] Yuchen Yang; Zhengyang Liu; Lee, F.C.; Qiang Li, "Analysis and filter design of differential mode EMI noise for GaN-based interleaved MHz critical mode PFC converter," Energy Conversion Congress and Exposition (ECCE), 2014 IEEE , vol., no., pp.4784,4789, 14-18 Sept. 2014

[10] Zhengyang Liu; Xiucheng Huang; Mingkai Mu; Yuchen Yang; Lee, F.C.; Qiang Li, "Design and evaluation of GaN-based dual-phase interleaved MHz critical mode PFC converter," Energy Conversion Congress and Exposition (ECCE), 2014 IEEE , vol., no., pp.611,616, 14-18 Sept. 2014

[11] M. S. Elmore, "Input current ripple cancellation in synchronized, parallel connected critically continuous boost converters," in Proc. IEEE Appl. Power Electron. Conf. (APEC), Mar. 1996, pp. 152–158.

[12] M. S. Elmore and K. A. Wallace. "Zero voltage switching supplies connected in parallel." U.S. Patent 5 793 191, Aug. 11, 1998.

[13] X. Xu and A. Huang, "A novel closed loop interleaving strategy of multiphase critical mode boost PFC converters," in Proc. IEEE Appl. Power Electron. Conf. (APEC), Feb. 2008, pp. 1033–1038.

[14] Huber, L.; Irving, B.T.; Jovanovic, M.M., "Review and Stability Analysis of PLL-Based Interleaving Control of DCM/CCM Boundary Boost PFC Converters," Power Electronics, IEEE Transactions on , vol.24, no.8, pp.1992,1999, Aug. 2009

[15] Xiliang Chen, "Research on Totem-Pole Bridgeless PFC Converter", MS Thesis, Zhejiang University, 03.24.2014

[16] Kolar, J.W.; Kamath, G.R.; Mohan, N.; Zach, Franz C., "Self-adjusting input current ripple cancellation of coupled parallel connected hysteresis-controlled boost power factor correctors," Power Electronics Specialists Conference, 1995. PESC '95 Record., 26th Annual IEEE , vol.1, no., pp.164,173 vol.1, 18-22 Jun 1995

[17] T. Ishii and Y. Mizutani, "Power factor correction using interleaving technique for critical mode switching converters," in Proc. IEEE Power Electronics Specialists Conf. (PESC), May 1998, pp. 905–910.

[18] T. Ishii and Y. Mizutani, "Variable Frequency Switching of Synchronized Interleaved Switching Converters," U.S. Patent 5 905 369, May 18, 1999.

[19] B. T. Irving, Y. Jang, and M. M. Jovanovic´, "A comparative study of soft-switched CCM boost rectifiers and interleaved variable-frequency DCM boost rectifier," in Proc. IEEE Applied Power Electronics Conf. (APEC), Feb. 2000, pp. 171–177.

[20] J. Zhang, J. Shao, F. C. Lee, and M. M. Jovanovic´, "Evaluation of input current in the critical mode boost PFC converter for distributed power systems," in Proc. IEEE Applied Power Electronics Conf. (APEC), Feb. 2001, pp. 130–136.

[21] L. Huber, B. T. Irving, and M. M. Jovanovic, "Open-Loop Control Methods for Interleaved DCM/CCM Boundary Boost PFC Converters," IEEE Trans. On Power Electron., vol. 23, no. 4, pp. 1649-1655, Jul. 2008.

[22] Zijian Wang, Shuo Wang, Pengju Kong, Fred C. Lee, "DM EMI Noise Prediction for Constant On-Time, Critical Mode Power Factor Correction Converters", IEEE Transactions on Power Electronics, vol. 27, issue 7, pp. 3150-3157, July 2012.

A Novel AC-to-DC Adaptor with Ultra-High Power Density and Efficiency

Yan-Cun Li, Fred C. Lee, Qiang Li, Xiucheng Huang, and Zhengyang Liu
Center for Power Electronics Systems, the Bradley Department of Electrical and Computer Engineering
Virginia Polytechnic Institute and State University, Blacksburg, VA, 24061, USA

Abstract— **This paper proposes a novel ac-to-dc adaptor circuit with ultra-high power density and efficiency, which is based on a two-stage topology, consists of a bridgeless boost converter followed by a LLC converter. The bridgeless boost converter is in charge of regulating output voltage and minimizing bulk capacitor, while the LLC converter is operated as a dc transformer (DCX) and plays a role of providing isolated voltage step-down, maximizing circuit efficiency and simplifying the control circuit. Furthermore, in order to achieve ultra-high power density and high efficiency, the proposed circuit applies gallium nitride (GaN) devices and planar transformer, while operates the circuit to around 1MHz to shrink both the sizes of EMI filter and magnetic components. By applying the proposed circuit, both the features of high performance and compact volume could be achieved; up to 44.22W/in³ power density and 94% efficiency are obtained from a 65W ac-to-dc adaptor prototype.**

Keywords— *Gallium nitride (GaN); Bridgeless Boost Converter; LLC dc transformer (DCX); ac-to-dc adaptor.*

VI. INTRODUCTION

In the power electronics industries, the ac-to-dc adaptor is one of largest markets. The ac-to-dc adaptor acquires the ac line voltage (power) from wall plugs while converts it into a low dc voltage (power) to supply various electronic products. In recent years, since the markets of portable electronic products, such as laptop computers, tablets, smart phones, and etc. are flourishing, the ac-to-dc adaptor industries becomes more prosperity. Furthermore, shrinking size is one of the important demands in portable electronic products, while it leads the developments on power density enhancement and size/ profile reduction in ac-to-dc adaptors. For the ac-to-dc notebook adaptors in the market, the power density of an ac-to-dc adaptor is usually above 5-6W/in³, while the average efficiency is above 85% [1-6]. Review the better 65W notebook adaptors in the markets, it can be seen that Delta Electronics provides a 65W adaptor with high power density and low profile, the power density is about 11W/in³ (97.6×50.6×19.2 mm); Apple Inc. produces a compact 60W adaptor with 7W/in³ power density (74×74×29mm); FSP-Powerland Technology Inc. fabricates a highest power density 65W adaptor in the world, the power density is about 13W/in³(69x49x24 mm). It is looked forward to further increase the power density and shrink the size. However, the power density is limited by the safety regulation on the surface temperature of adaptor case. By the way, the surface temperature depends on the power loss (efficiency) and surface area (size) of the adaptor.

Besides the demand on power density enhancement, the power efficiency improvement is another focused in the adaptor markets, which is also one of the main trends in power electronics industries. In addition, due to the energy is gradually exhausted, the whole world has put more and more emphasis on the energy utilization and exploitation. On the other hand, because of the huge usage amount of the ac-to-dc adaptors, it creates an enormous opportunity to saving the energy by reducing the power consumption and enhancing the efficiency. [1-2, 5-7]

In recent years, although whole world engineers and researchers have paid more attention on developing the higher power density and efficiency ac-to-dc adaptors, and led the industry moving forward significantly. However, consumers always expect to have a more compact, smaller size, and higher efficiency ac-to-dc adaptor, and are never satisfied. Therefore, this paper focuses on further enhancing the power density and efficiency. Following the description of the proposed circuit is the insight into the operating principle, the design equations and an example application of a novel ac-to-dc adaptor with ultra-high power density and efficiency. Finally, the experimental results verify the accuracy of theoretical analyses under a sample design of a 65W ac-to-dc adaptor.

VII. REVIEW OF CONVENTIOAL ADAPTOR CIRCUIT

Fig. 2 shows a typical ac-to-dc adaptor circuit, which consists of a diode-bridge rectifier followed by a bulk dc bus capacitor and a flyback converter with critical conduction mode (CRM) operation. Both the efficiency and power density are limited in such adaptor circuit, due to the significant power loss on diode-bridge rectifier and the huge dc bus capacitor [1-4, 7-8]. In general, the conventional adaptor circuit is usually operated in critical conduction mode with switching frequency less than 200kHz to maintain the well efficiency and EMI performance, while such low switching frequency causes a bulky EMI filter. Moreover, a bulky bus capacitor with about 120μF capacitance is usually applied for a 65W design [7-10], which also occupies a lot of space. The power density of the conventional ac-to-dc adaptor is limited accordingly.

On the other hand, the diode-bridge rectifier consumes a considerable conduction loss and reduces the circuit efficiency, especially in low line operation. In addition, the efficiency is remarkably encumbered with the leakage inductance when applying flyback couple inductor. The snubber circuit is necessary to dissipate the leakage energy, which causes an extra

978-1-4673-9551-9/16 $31.00 © 2016 IEEE

power loss [7, 11-12]. For the conventional adaptor circuit, only the limited power density and efficiency can be achieved.

VIII. THE PROPOSED SYSTEM STRUCTURE AND OPERATION PRINCIPLES

In order to heighten the power density and efficiency greatly, this paper proposed a two-stage topology composed of a bridgeless boost converter followed by a LLC converter, which is shown in Fig. 3. The bridgeless boost converter converts the time-varying line voltage to a stable dc bus voltage, V_{BUS}, and deposit the energy on the bulk dc bus capacitor, C_{BUS}, while it also regulates both the dc bus and output voltage, V_O; the LLC converter is operated as a dc transformer (DCX) to step-down V_{BUS} to the desired V_O, while it also provides the electrical isolation in order to comply with the safety regulation.

Furthermore, the bridgeless boost converter step-up the line voltage to a higher dc bus voltage. The energy stored in a capacitor can be expressed by the following equation.

$$E = \frac{1}{2} \times C \times V^2 \tag{1}$$

where E is the stored energy; C is the capacitance; V is the voltage imposed on the capacitor.

Since the energy stored in a capacitor is in proportional to the square voltage across the capacitor, it can be seen that the energy density can be significantly enhanced by increasing the voltage imposed on the capacitor. Hence, the bus capacitance can be significantly reduced by imposing a higher bus voltage. In addition, due to the diode-bridge rectifier is replaced by the bridgeless boost converter, the conduction loss of rectifier can be remarkably reduced. To take all mentioned advantages, the bridgeless boost converter is chosen for the first stage of the proposed ac-to-dc adaptor circuit. Moreover, by applying a harmonics injection control on bridgeless boost converter can further reduce the required bus capacitance [13-14], while the power density can be enhanced accordingly.

On the other hand, the LLC resonant converter an attractive candidate for the DCX, which operates the LLC resonant converter at resonant frequency. In this way, the LLC resonant converter features zero-voltage switching (ZVS) on the primary-side switches while zero-current switching (ZCS) for secondary-side rectifiers [15-17]. Since ZVS and ZCS operations can extremely reduce the switching losses of devices, the operation frequency can be greatly raised. Furthermore, by applying gallium nitride (GaN) devices, above MHz operation frequency can be easily achieved, such that both the passive component and EMI filter sizes can be further shrunk [18-20]. Accordingly, the proposed circuit is characteristic of both high efficiency and high power density through cascading the bridgeless boost converter with the LLC DCX.

According to the operation principles described above, both the characteristics of the bridgeless boost converter and the LLC DCX are unaltered, while they can be regarded as two independent conversion stages, and thus the detail of each converter operation can be explained separately.

A. Bridgeless boost converter with harmonics injection

In order to enhance the high power density, the bridgeless boost converter is operated to above 1MHz, which is about 5 to 10 times of conventional design. In literatures [20-22], it shows that the EMI filter size can be reduced over 50%, it is one of the main reasons running bridgeless boost converter to above 1MHz.

Moreover, ZVS technique is required to eliminate the significant switching-on loss during the high-frequency operation. The CRM operation of bridgeless boost converter causes a resonance between the inductor and device junction capacitors, while such resonance brings the natural ZVS or valley-switching of devices. For boost-type CRM converter, the natural ZVS can only be achieved during the input voltage is lower than one half of the bus voltage. When the input voltage is higher than one half of the bus voltage, the device drain-source voltage can't be resonated to zero, and a switching-on loss occurs at next turn-on instant accordingly. Such non-ZVS loss is in proportional to switching frequency (f_s), and it would be much significant and dominant the circuit efficiency when the switching frequency is pushed to MHz levels. For sake of avoiding this issue, ZVS extension strategy is applied. In this way, the bridgeless boost converter is no longer operated in CRM and changed to quasi-square-wave (QSW) mode, which applies a purposely delay to store enough energy in the inductor for achieving ZVS, while the switching losses of the switches are minimized accordingly [20-22].

On the other hand, in order to further shrink the bus capacitor size and enhance the power density, the harmonics injection technique is applied. Literatures [13-14] have discussed reducing the bus capacitance by injecting harmonic currents on PFC for LED lightings, the higher orders or larger harmonic currents injection the smaller bus capacitance. In addition, since this paper focus on a 65W adaptor design, it is unnecessary to achieve high power factor and low total harmonic distortion (THD), and thus larger or higher orders harmonic currents can be injected to further shrink bus capacitance.

Detail mathematical derivation and explain of harmonic current injection are follows.

The bridgeless boost converter is supplied by the line voltage source.

$$v_{ac}(t) = V_m \sin(\omega t) \tag{2}$$

where V_m and ω are the amplitude and angular frequency of the line voltage source, respectively.

To consider the harmonic currents, the input line current can be expressed as

$$i_{ac}(t) = \sum_{n=1}^{k} I_{mn} \sin(n \cdot \omega t) \tag{3}$$

where I_{mn} is the amplitude of the n-order harmonic current, n is the harmonic order, and k is a constant to express the highest harmonic order.

The instantaneous input power is the product of input $v_{ac}(t)$ and $i_{ac}(t)$, which is

$$p_{acn}(t) = V_m \sin(\omega t) \sum_{n=1}^{k} I_{mn} \sin(n \cdot \omega t) \tag{4}$$

According to equation (4), the instantaneous input power with harmonic currents injection can be calculated. For example, the instantaneous input power with injection of third-order harmonic current can be described as

$$p_{ac\_1+3}(t)=V_m\left[I_{m1}\frac{1-\cos(2\omega t)}{2}+I_{m3}\frac{\cos(2\omega t)-\cos(4\omega t)}{2}\right] \quad (5)$$

Assume I_{m1} is equal to I_{m3}, then equation (5) would become

$$p_{ac\_1+3}(t)=V_m I_{m1}\left[\frac{1-\cos(4\omega t)}{2}\right] \quad (6)$$

To compare equation (5) with (6), it can be seen that the second order component is eliminated when I_{m1} is equal to I_{m3}. If both the third-order and fifth-order harmonic current are injected, and make all the order current amplitudes are the same, and thus the instantaneous input power would be

$$p_{ac\_1+3+5}(t)=V_m I_{m1}\left[\frac{1-\cos(6\omega t)}{2}\right] \quad (7)$$

According to equations (6) and (7), if all the order current amplitudes are identical, the instantaneous input power can be redefined as

$$p_{ac\_n}(t)=V_m I_{m1}\left\{\frac{1-\cos[(n+1)\omega t]}{2}\right\} \quad (8)$$

Form equation (8), it can be seen that when injected the n order harmonic current, all the harmonic components less than n+1 order would be eliminated. According to this point, the harmonic components of instantaneous input power can be pushed to higher order, i.e. the higher frequency. Based on the low-pass filter concept, the required bus capacitance can be reduced when it is applied at higher frequency. Fig. 4 shows the theoretical waveforms, which is based on a given condition listed on the figure. In Fig. 4, it can be observed that during harmonic current is injected, the bus voltage ripple is reduced, while the input peak current increases. Fig. 5 illustrates the impacts of the required bus capacitance and efficiency during the harmonic current injection. According to Fig. 5(a), it can be seen that the bus capacitance can be further shrunk during inject higher order harmonic current. However, there is a penalty when injecting the harmonic currents, the input current (inductor current) increasing, which also results in larger conduction and switching losses in the circuit. The impact of circuit efficiency during the harmonic current injection is illustrated in Fig. 5(b).

Based on Fig. 5(a) and Fig. 5(b), a trade-off between bus capacitance and circuit efficiency can be made. It can be observed when the 3rd order harmonic is injected, the bus capacitance can be reduced above 50% and the efficiency drops about 1%. If the 5th harmonic current is further injected, the efficiency still drops 1% but the capacitance reduction is less than 30%; if the 7th order harmonic current is further injected, the efficiency still drops 1% but the capacitance reduction is about 25%. The benefits of harmonic current injection become smaller when injecting the higher order harmonic current. Therefore, in this design, to inject 3rd harmonic current and achieve 96.4% efficiency with 33uF bus capacitance is selected.

B. LLC DCX

The LLC DCX stage is also operated at resonant frequency, in this design, 1MHz of resonant frequency is selected. In order to enhance the circuit efficiency, the synchronous rectifier (SR) technique is applied to the secondary-side rectifier. However, there are several SR topologies for LLC DCX, such as center-tapped SR, full-bridge SR, and voltage-doubler SR. Every SR topology has different characteristics that resulting in its own advantages and drawbacks. Since this research focus on a 65W adaptor application, the output current is relative small, and thus the conduction losses are less. Therefore, it needs to be paid more attention on magnetic core loss.

The magnetic core loss is dominant by voltage-second across the transformer, in order to reduce the magnetic core loss, the voltage-doubler SR is chosen to scale down the voltage-second. The steady-state operation of can be explained by four modes within one switching cycle.

Fig. 7 and Fig. 8 show the equivalent circuit and theoretical waveforms of the proposed LLC DCX. According to Fig. 7(a) and Fig. 7(b), it can be seen that the voltage across transformer secondary-side is equal to one half of output voltage, and thus the voltage-second imposed on the transformer can reduced by half. In this way, the magnetic core loss can be shrunk accordingly. Also, from Fig. 8, it can be observed that both V_{DS5} and V_{DS6} drop during the dead time (t_d), and reach zero before Q_5 and Q_6 are switched on, i.e. ZVS can be achieved on these two devices. On the other hand, it also can be seen that the secondary-side current reaches zero before SR devices, Q_7 and Q_8, are turned off. Since there is no current flows through the SR devices during they are turned off, and thus ZCS operation is accomplished.

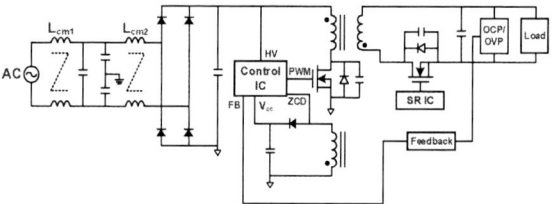

Fig. 2. The conventional adaptor circuit.

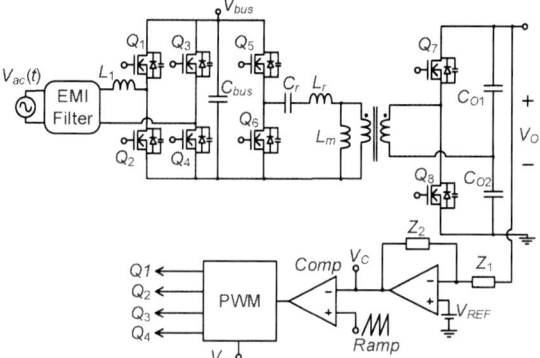

Fig. 3. The proposed adaptor circuit.

978-1-4673-9551-9/16 $31.00 © 2016 IEEE

$$v_{ac}=90\text{V},\ V_{BUS}=400\text{V},\ P_o=65\text{W},\ C_{BUS}=65\mu\text{F}$$

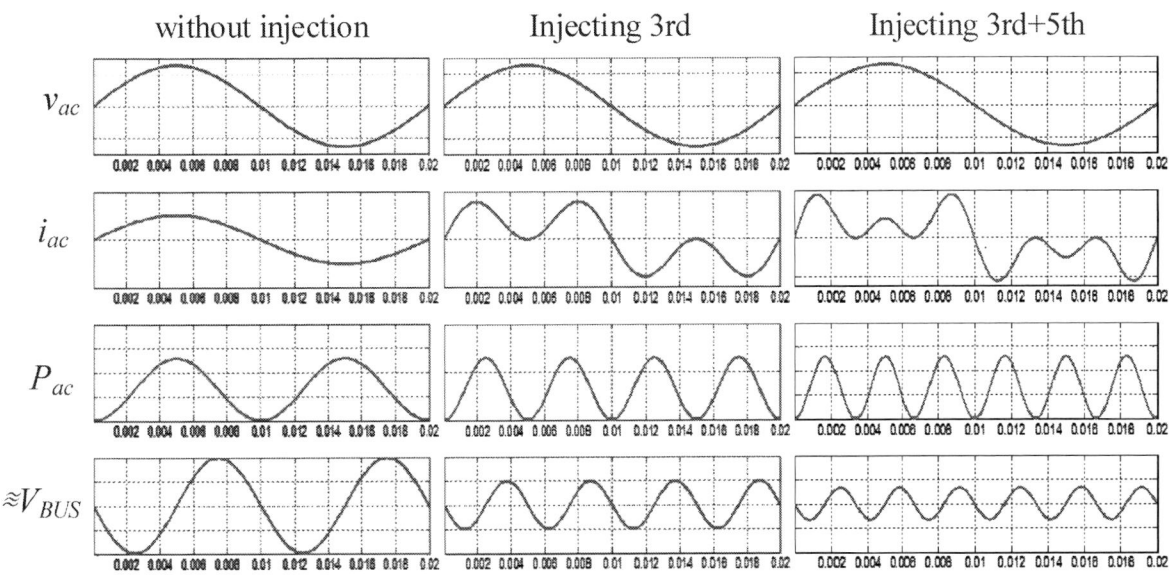

v_{ac}: 100V/div, i_{ac}: 2A/div, P_{ac}, 50W/div, ΔV_{BUS}: 2V/div

Fig. 4. Theoretical waveforms during harmonic injection.

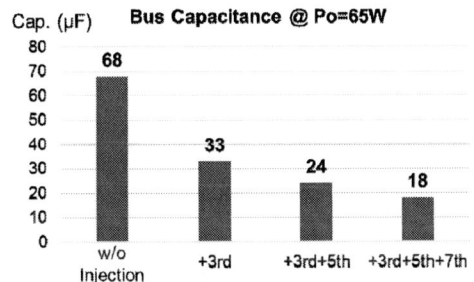

(a) The impact of the required bus capacitance

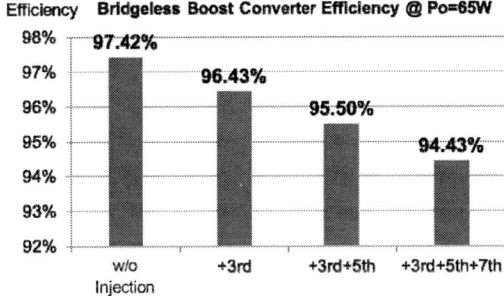

(b) The impact of the efficiency

Fig. 5. The impact of the required bus capacitance and efficiency during harmonic injection.

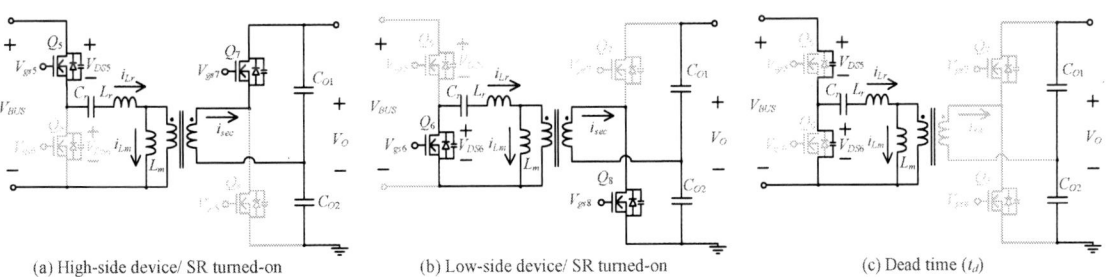

(a) High-side device/ SR turned-on   (b) Low-side device/ SR turned-on   (c) Dead time (t_d)

Fig. 6. The equivalent circuit of LLC DCX.

978-1-4673-9551-9/16 $31.00 © 2016 IEEE   1856

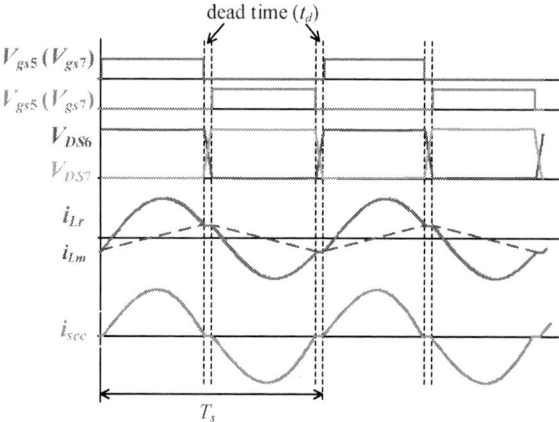

Fig. 7. Theoretical waveforms of LLC DCX with voltage-doubler SR.

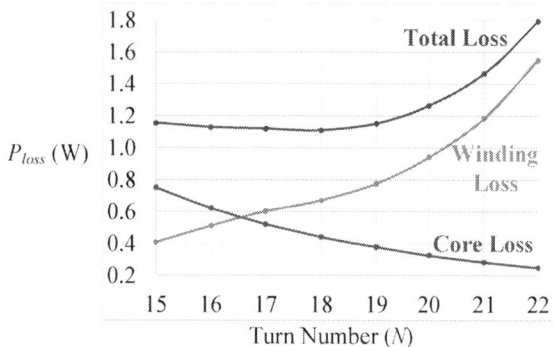

Fig. 8. The inductor losses vs. turn number (N).

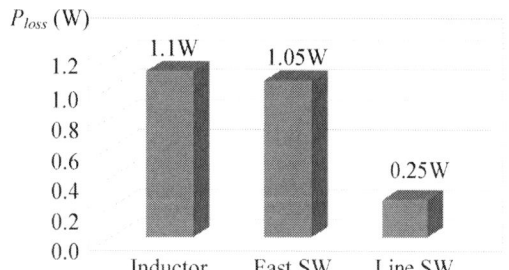

Fig. 9. Loss breakdown of bridgeless boost converter.

IV. DESIGN CONSIDERATIONS

According to the operation principles described above, both the characteristics of the bridgeless boost converter and the LLC DCX are unaltered, even though they are cascaded. Therefore, the proposed ac-to-dc adaptor circuit can be regarded as two independent conversion stages, the bridgeless boost converter and the LLC DCX.

A. Design consideration of bridgeless boost converter with harmonic current injection

Besides harmonic current injection, the basic design consideration of bridgeless boost converter is discussed in literatures [21-22]. As mentioned in previous section, the input current (inductor current) is increased when injecting the harmonic current, while it can be seen the input current (inductor current) waveform is also distorted in accordance with Fig. 4. Therefore, for the inductor design, the worst case is no longer at the peak line voltage after injecting harmonic currents. General speaking, the worst case of inductor design is under low line voltage and the largest inductor current. Since above 1MHz switching frequency is desired, the boost inductance can be designed accordingly. In this 65W ac-to-dc adaptor design, the required boost inductance is 12 μH.

Based on the desired boost inductance, and following the design methods in literatures [22-24], the inductor loss can be estimated during different turn number (N), which is shown in Fig. 8. According to Fig. 8, the minimum loss point is 1.11W under $N=18$. In addition, the loss breakdown of bridgeless boost converter is shown in Fig. 9, it also shows the total loss of bridgeless boost converter with third order harmonic current injection is about 2.4W, while the circuit efficiency is about 96.43%.

B. Design consideration of LLC DCX

To achieve high power density and high efficiency, the LLC DCX is designed at 1MHz. The detail design procedures are as follows whereas the circuit specifications are listed in Table 1.

Step 1: Determine transformer turn ratio (N_t)

The transformer turn ratio of LLC DCX with voltage-doubler SR can be calculated as

$$N_t = \frac{N_1}{N_2} = \frac{V_{BUS}}{V_O} = \frac{400}{20} = 20 \qquad (9)$$

Step 2: Optimize t_d and magnetizing inductance (L_m)

Since the magnetizing inductor current (i_{Lm}) charges/ discharges the device junction capacitors to achieve ZVS operation of primary-side devices (Q_5 and Q_6) during dead time, while the larger i_{Lm} can fully charge/ discharge the device junction capacitors faster, i.e. ZVS operation can be achieved in a smaller dead time (t_d). However, the larger i_{Lm} causes larger circulating energy, which leads higher power loss and impedes the circuit efficiency. Also, under a given operating point listed in Table 1, i_{Lm} is directly determined by L_m. Therefore, in order to achieve high efficiency, it is very important to design L_m and t_d carefully.

Fig. 10 and Fig. 11 illustrate the rms current variations of primary-side and secondary-side under different t_d. At the same time, the total power loss varies with t_d can also be estimated through these currents, while the optimal t_d can be selected in according with the minimum loss point, the relationship between total power loss and t_d is shown in Fig. 12. 150ns of t_d and 67μH of L_m are selected to achieve 97.5% circuit efficiency.

TABLE I. CIRCUIT SPECIFICATIONS AND PARAMETERS

Input line voltage, v_{ac}	90V- 265V
Output power, P_o	65W
Output voltage, V_o	20V
Output voltage ripple, ΔV_o (%)	±1%
Boost inductance, L_1	12μH
Bus capacitance, C_{BUS}	33μF
Magnetizing Inductance, L_m	95μH
Leakage Inductance, $L_{lk}(L_r)$	1.5μH
Turns-ratio, N_t	20
Resonant capacitance, C_r	8.3nF
Output capacitance, C_O	176μF
Fast devices (GaN), Q_1,Q_2,Q_5, and Q_6	PGA26E19BV
Line devices, Q_3 and Q_4	STL57N65M5
SR devices (GaN), Q_7 and Q_8	EPC2023

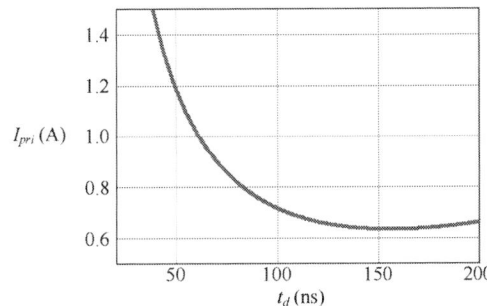

Fig. 10. The primary-side rms current vs. t_d.

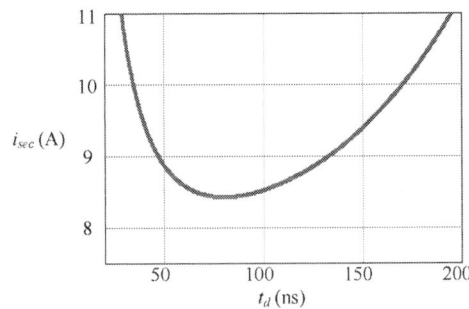

Fig. 11. The secondary-side rms current vs. t_d.

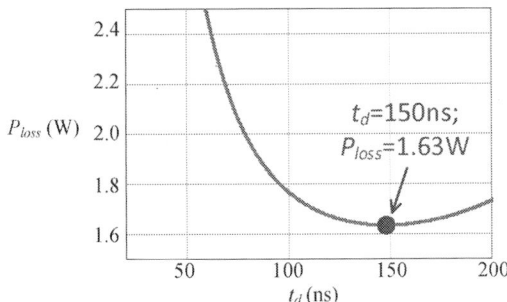

Fig. 12. The total power loss of LLC DCX vs. t_d.

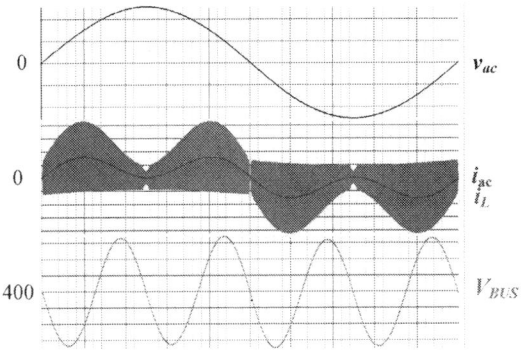

i_{in}:500mA/div; i_L: 1A/div; V_{BUS}: 1V/div; time: 2ms/div @ v_{ac}=90V

Fig. 13. The simulation waveforms of bridgeless boost converter.

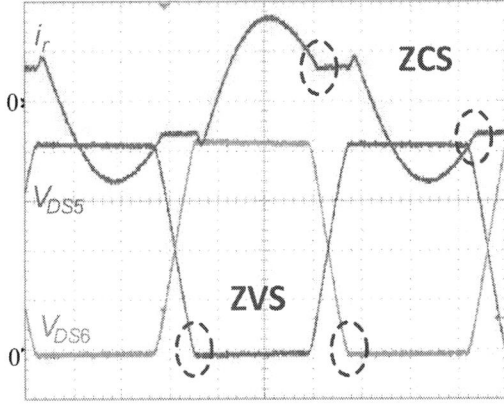

i_r:500mA/div; V_{DS}: 100V/div; time: 200ns/div

Fig. 14. The experimental waveforms of LLC DCX.

V. SIMULATION AND EXPERIMENTAL RESULTS

The simulation waveforms of bridgeless boost converter with third-order harmonic current injection are shown in Fig. 13. It can be seen that the bus voltage ripple can fulfill the desired specification, less than ±1%, when injecting third order harmonic current with 33μF bus capacitance, which coincides the mathematical analysis above.

Fig. 14 shows the experimental waveforms of LLC DCX, while it can be seen that both ZVS and ZCS operations are accomplished, and thus 97.5% of circuit efficiency can be achieved.

The prototype is shown in Fig. 15, the dimension is 1.47 in³ (66×36.5×10 mm³). The primary GaN devices are PGA26E19BV from Panasonics Inc for both bridgeless boost converter and LLC DCX, the line frequency devices of bridgeless boost converter are STL57N65M5, and the secondary-side SR devices are EPC2023. By increasing the switching frequency above MHz with GaN devices, the transformer, EMI filter and output filter size are significantly reduced, while about 44.22W/in³ (exclude case) power density can be achieved. On the other hand, the total efficiency of the proposed ac-to-dc adaptor circuit is about 94%.

Fig. 15. The prototype of 65W ac-to-dc adaptor.

VI. CONCLUSIONS

This paper presents the design consideration of cascading MHz bridgeless boost converter with LLC DCX for ac-to-dc adapter application. A prototype circuit verifies the theoretical analysis and system design feasibility. Based on the proposed design procedures, much higher power density and higher efficiency can be accomplished compared with state of art products. Also, the planar transformer is applied for LLC DCX, which features high power density, controllable parasitics, and easy to achieve automation manufacture. Future work will focus on further improvements of power density, light load efficiency, and EMI performance.

ACKNOWLEDGMENT

The information, data, or work presented herein was funded in part by the Office of Energy Efficiency and Renewable Energy (EERE), U.S. Department of Energy, under Award Number DE-EE0006521 with North Carolina State University, PowerAmerica Institute.

This work was supported in part by the Power Management Consortium in CPES, Virginia Tech.

This work was conducted with the use of GaN device samples by Panasonic Inc. and EPC of the CPES Industry Consortium Program.

DISCLAIMER

The information, data, or work presented herein was funded in part by an agency of the United States Government. Neither the United States Government nor any agency thereof, nor any of their employees, makes any warranty, express or implied, or assumes any legal liability or responsibility for the accuracy, completeness, or usefulness of any information, apparatus, product, or process disclosed, or represents that its use would not infringe privately owned rights. Reference herein to any specific commercial product, process, or service by trade name, trademark, manufacturer, or otherwise does not necessarily constitute or imply its endorsement, recommendation, or favoring by the United States Government or any agency thereof. The views and opinions of authors expressed herein do not necessarily state or reflect those of the United States Government or any agency thereof.

REFERENCES

[1] Y. Panov and M.M. Jovanovic, "Performance Evaluation of 70-W Two-Stage Adaptors for Notebook Computers," in *IEEE APEC*, 1999, pp. 1059-1065.

[2] D.-Y. Kim, C.-E. Kim, and G.-W. Moon, "High-Efficiency Slim Adapter with Low-Profile Transformer Structure," *IEEE Trans. on Ind. Electron.*, vol. 59, no. 9, pp. 3445-3449, Sept. 2012.

[3] J.-B. Baek, J.-T. Kim, and B.-H. Cho, "Low-profile AC/DC Converter for Laptop Adaptor," in *IEEE APEC*, 2011, pp. 60-64.

[4] B.T. Bucheru, M. A. Davila, and I. D. Jitaru, "Increasing power density of adapters by using DC link chopper," in *IEEE APEC*, 2012, pp. 796-801.

[5] Y.-K. Lo, S.-C. Yen, and C.-Y. Lin, "A High-Efficiency AC-to-DC Adaptor with a Low Standby Power Consumption," *IEEE Trans. on Ind. Electron.*, vol. 55, no. 2, pp. 963-965, Feb. 2010.

[6] External power adapters and chargers report-2015, IHS Technology.

[7] X. Huang, W. Du, F. C. Lee, and Q. Li, "A novel driving scheme for synchronous rectifier in MHz CRM flyback converter with GaN devices, " in *IEEE ECCE*, 2015, pp. 5089-5095.

[8] L. Huber, and M. Jovanovic, "Evaluation of flyback topologies for notebook acldc adapterlcharger applications", in proc. *High Frequency Power Conversion Con.*, May 1995, pp. 284-294.

[9] 65W (130W Surge) Flyback Power Supply for Laptop Adapter Apps w/85-265VAC Input Reference Design-2014, Texas Instruments Inc.

[10] 19V- 65W quasi-resonant flyback adapter using L6566B and TSM1014, Application Note-2010, STMicroelectronics.

[11] Robert Watson, Fred C. Lee, and Guichao Hua, "Utilization of an Active-Clamp Circuit to achieve Soft Switching in Flyback Converters," *IEEE Trans. on Power. Electron.*, vol.11, pp. 162-169, January 1996.

[12] Y. C. Li and C. L. Chen, "A Novel Single-Stage High-Power-Factor AC- to-DC LED Driving Circuit With Leakage Inductance Energy Recycling," *IEEE Trans. Ind. Electron.*, vol. 59, no. 2, pp. 793-802, Feb. 2012.

[13] L. Gu, X. Ruan, M. Xu, and K. Yao, "Means of Eliminating Electrolytic Capacitor in AC/DC Power Supplies for LED Lightings," *IEEE Trans. on Power. Electron.*, vol. 24, no. 5, pp. 1399-1408, May 2009.

[14] B. Wang, X. Ruan, K. Yao, and M. Xu, "A Method of Reducing the Peak-to-Average Ratio of LED Current for Electrolytic Capacitor-Less AC–DC Drivers," *IEEE Trans. on Power. Electron.*, vol. 25, no. 3, pp. 592-601, Mar. 2010.

[15] D. Huang, D. Gilham, W. Feng, P. Kong, D. Fu, and F. C. Lee, "High power density high efficiency dc/dc converter," in *IEEE ECCE*, 2011, pp. 1392-1399.

[16] W. Feng, D. Huang, P. Mattavelli, D. Fu, and F. C. Lee, "Digital implementation of driving scheme for synchronous rectification in LLC resonant converter," in *IEEE ECCE*, 2010, pp. 256-263.

[17] R. Ren, S. Liu, J. Wang, and F. Zhang, "High Frequency LLC DC-Transformer based on GaN devices and the dead time optimization," in *IEEE ECCE.*, 2014, pp. 462-467.

[18] Y. Yang, D. Huang, F. C. Lee, and Q. Li, "Analysis and reduction of common mode EMI noise for resonant converters," in *IEEE APEC*, 2014, pp. 566-571.

[19] D. Fu, S. Wang, P. Kong, F. C. Lee, and D. Huang, "Novel Techniques to Suppress the Common-Mode EMI Noise Caused by Transformer Parasitic Capacitances in DC–DC Converters," *IEEE Trans. Ind. Electron.*, vol. 60, no. 11, pp. 4968-4977, Nov. 2013.

[20] Y. Yang, Z. Liu, F. C. Lee, and Q. Li, "Analysis and filter design of differential mode EMI noise for GaN-based interleaved MHz critical mode PFC converter," in *IEEE ECCE.*, 2014, pp. 4784-4789.

[21] Z. Liu, X. Huang, M. Mu, Y. Yang, F. C. Lee, and Q. Li, "Design and evaluation of GaN-based dual-phase interleaved MHz critical mode PFC converter," in *IEEE ECCE.*, 2014, pp. 611-616.

[22] Z. Liu, F. C. Lee, Q. Li, and Y. Yang, "Design of GaN-based MHz totem-pole PFC rectifier," in *IEEE ECCE.*, 2015, pp. 682-688.

[23] M. Mu and F. C. Lee, "Comparison and optimization of high frequency inductors for critical model GaN converter operating at 1MHz," in *IEEE PEAC*, 2014, pp. 1363-1368.

[24] M. Mu, F. Zheng, Q. Li, and F. C. Lee, "Finite element analysis of inductor core loss under DC bias condition," *IEEE Trans. on Power. Electron.*, vol. 28, no. 9, pp. 4414-4421, Mar. 2013.

A Single-stage Single-phase Isolated AC-DC Converter based on LLC Resonant Unit and T-type Three-level Unit for Battery Charging Applications

Yikai Gao, *Student Member, IEEE*, Wen Cai, *Student Member, IEEE* and Fan Yi, *Student Member, IEEE*
Department of Electrical Engineering
The University of Texas at Dallas
Richardson, USA
Email: {yikai.gao, wen.cai, fan.yi}@utdallas.edu

Abstract—This paper proposes a four-switch single-phase single-stage isolated AC-DC converter with high efficiency for battery charging applications. By combining one LLC resonant unit and one T-type three-level unit with switch multiplexing technique, a single-stage single-phase isolated LLC resonant (SSPLLC) converter is proposed to achieve AC-DC conversion. In comparison with conventional two-stage isolated AC-DC topologies, the proposed topology works in discontinuous conduction mode (DCM) with small inductor requirement. It can improve both efficiency and power density because of soft-switching capability and wide operating range. The mode analysis, steady-state operation performance and control method are discussed sequentially. Experimental results using a 250W prototype are presented to verify the feasibility and superior performance of the proposed single-stage AC-DC converter.

Keywords—*AC-DC converter; battery charging; LLC resonant converter; switch multiplexing; soft switching; rectifier*

I. INTRODUCTION

Storage devices, like Li-ion battery and lead-acid battery, to name a couple, have been used widely in numerous applications [1]. They can allow for uninterrupted power supply, removing power randomness caused by renewable sources and improving efficiency and power density in hybrid electric vehicles [2]. As for batteries, high grade battery charger is an integral component under various conditions [3]. A typical topology for battery charging is two-stage isolated AC-DC-DC converter (active power factor correction and phase-shift full-bridge converter). However, in two-stage power converter, eight switches are needed to achieve the requirement of isolation and AC grid input. In consideration of drive circuits and control part, it is difficult to decrease the system cost and size using this topology. Therefore, new topology especially single-stage isolated AC-DC converter is worth further studying for efficiency and power density enhancement over a wide range of voltage application.

Recently, in the area of single-phase isolated rectifier based on LLC unit, there have been several topologies proposed for performance improvement. [4] proposes a single-stage AC-DC driver with coupled inductor for LED applications. Its feasibility has been verified including low total harmonic distortion (THD) and high power factor. However, this topology is complicated and three inductors are necessary. Similar topology is used for multi-channel LED driving in [5]. In addition, [6] proposes and analyzes single-stage half-bridge converter with only two switches and several passive components. Even though high power factor is achieved, the THD of the input current can't be restricted. Furthermore, additional switch or auxiliary circuit is necessary in them. It is still an important issue to achieve high efficiency charging over a wide output power range.

In this paper, by combining LLC resonant unit and T-type three-level unit using switch multiplexing, a new single-stage single-phase AC-DC converter is proposed with high efficiency. This topology maintains the advantages of two basic units. Compared with two-stage AC-DC converter, this topology can achieve soft-switching for all the switches with efficiency improvement. In addition, compared with traditional LLC resonant converter with fixed DC link voltage, the DC link voltage in this topology can be regulated in accordance to the operating situation. This can help to expand the soft-switching region of the LLC branch and to maintain high efficiency. This is useful for battery charging system because the batter is a varying load. When the state of charge (SOC) is low, high-power charging is desirable for time saving. While when the SOC rises, it is better to transfer to low-power charging to extend the lifetime. An experimental prototype is built to verify the feasibility of the proposed topology. Its performance including total harmonic distortion (THD) and efficiency is also tested based on experimental waveforms.

II. DERIVATION AND OPERATING PRINCIPLE OF SSPLLC CONVERTER

A. Topology derivation

As stated in the introduction, single-stage isolated AC-DC converter is desirable to replace two-stage converters for efficiency improvement. By combining the resonant unit and three-level unit using switch multiplexing technique, their

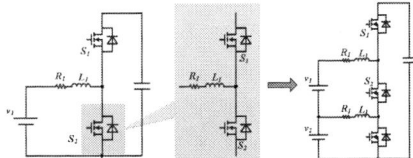

Fig. 1. Basic concept of switch multiplexing technique.

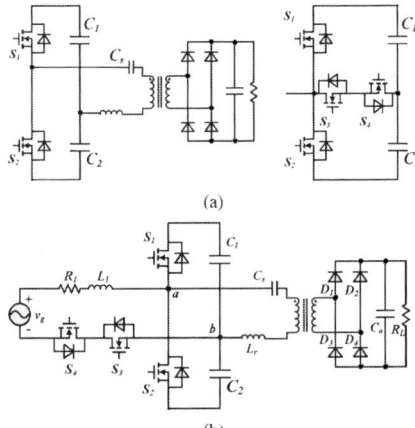

(a)

(b)

Fig. 2. Proposed topology: (a) resonant unit and T-type three-level unit; (b) four-switch isolated single-stage single-phase LLC AC-DC converter.

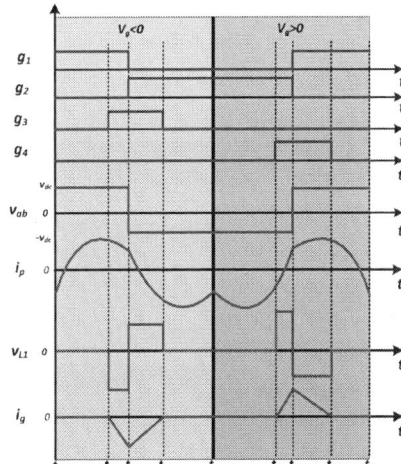

Fig. 3. Switching waveform of the proposed topology.

advantages are maintained, but the switch number is decreased. Consequently, the efficiency and power density of the resulting topology might be further improved. The basic idea of switch multiplexing has been proposed in [7] which is shown in Fig. 1. [8] also introduced the corresponding modulation strategy for the derived topology. Similarly, with the two units in Fig. 2(a), the single-stage single-phase isolated LLC resonant (SSPLLC) topology is derived and shown in Fig. 2(b). There are only four switches necessary in the proposed topology. For this topology, modulation strategy needs to be modified as well.

B. Mode analysis

In order to analyze the topology and verify its performance, operating principle based on switching modes is to be discussed. In order to make sure that the resonant branch work with high efficiency, the switches S_1 and S_2 should turn on and off alternatively. Their duty cycles are the same (50%) if dead time is ignored. On the other hand, S_3 and S_4 consist of a bi-directional switch. Therefore, during one switching period, only one switch is chopping and the other is kept off. However, their states would change when the grid voltage inverts. Fig. 3 shows the basic waveform for two cases ($v_g<0$ and $v_g>0$). The drive signals for S_1 and S_2 are always the same. Meanwhile, when $v_g<0$, S_4 is always turned off. Conversely, S_3 is off under the condition. The voltage and current waveforms are also illustrated in Fig. 3. The time period t_0-t_4 corresponds to the condition and the other is for $v_g>0$. When $v_g<0$, the switch S_3 is turned on before the instant

at which S_1 is turned off. After that, S_3 is turned off when/after the inductor current i_{L1} goes to zero. The same modulation method can be used for S_4 when (i.e. it should be turned on before S_2 is turned off.).

Based on the drive signals shown in Fig. 3, it can be found that there are 8 modes. They are divided and listed in Figure 4. The capacitor voltage is assumed constant and higher than the amplitude of grid voltage so as to simplify the theoretical analysis. In addition, the switching frequency is close to the resonant frequency of the used LLC circuit for efficiency optimization.

The four modes when $v_g<0$ are listed and explained as follows (the dotted line means that the semiconductors are cut off):

Mode 1 (t_0-t_1): Under this condition (Figure 4(a)), switches S_2, S_3, S_4 are turned off and only S_1 is on. Hence, there is no current through the AC port ($i_{L1}<0$). Meanwhile, $v_{ab}=v_{dc}$, the primary current of the transformer starts oscillating positively. During this period, diodes D_1 and D_4 are on. D_2 and D_3 are off automatically.

Mode 2 (t_1-t_2): At t_1, switch S_3 is turned on. The other switches are kept unchanged. The transformer current keeps oscillating continuously. The voltage of inductor L_1 is v_g - v_{dc}. Because $v_g<0$, the current i_L increases negatively as Figure 4 shows. Considering that there is no current through this branch before t_1 and the current increment is limited by inductor L_1, the switching loss of switch S_3 is very small.

Mode 3 (t_2-t_3): The states of switches S_3 and S_4 are not changed. At t_2, S_1 is turned off and S_2 is turned on. The voltage of LLC branch is $v_{ab}= -V_{dc}$, so the current oscillates naturally to the negative value. At the same time, the inductor voltage v_{L1} equals to v_g+V_{dc}. Since $Amp(v_g)<V_{dc}$, the current would increase from the minimum value. Figure 4 shows this process as well.

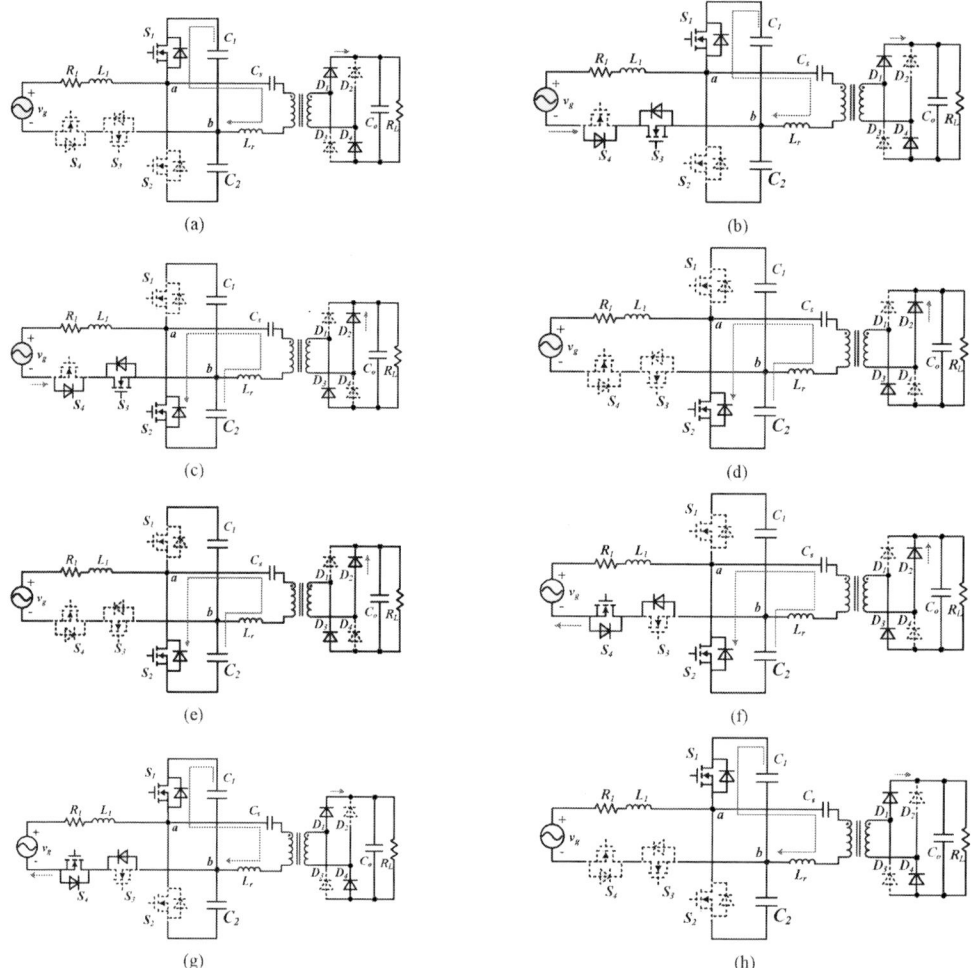

Fig. 4. Theoretical modes analysis: (a) Mode 1 (t0-t1); (b) Mode 2 (t1-t2); (c) Mode 3 (t2-t3); (d) Mode 4 (t3-t4); (e) Mode 5 (t4-t5); (f) Mode 6 (t5-t6); (g) Mode 7 (t6-t7); (h) Mode 8 (t7-t8).

Mode 4 (t_3-t_4): When the inductor current i_L increases to 0, the current diode in parallel with S_4 would cut off automatically. After that, the current through utility grid remains zero. It can be found that the switch S_3 is turned off without any current, so zero-current switching is achieved as long as S_3 is turned off after the current becomes zero. Correspondingly, the other four modes when v_g>0 are listed and explained as follows:

Mode 5 (t_4-t_5): This mode is similar with mode 4. Switch S_1, S_3 and S_4 are off and S_2 is turned on. The transformer current through the primary winding is oscillating to the negative direction because $v_{ab} = -V_{dc}$. During this period, there is no current through the utility grid.

Mode 6 (t_5-t_6): S_4 are turned on at t_5. The voltage of the inductor is $v_{L1} = v_g + V_{dc}$. Because v_g>0, the current increases linearly. Since the switches S_1 and S_2 don't change modes, the current through LLC branch keeps oscillating. The current through switch S_4 is limited by the inductor, so the switching loss is small, as well.

Mode 7 (t_6-t_7): At t_6, S_1 is turned on and S_2 is turned off with certain dead time. The current through LLC branch oscillates to positive direction in consideration of that $v_{ab} = V_{dc}$. On the other hand, the inductor voltage is $v_{L1} = v_g - V_{dc}$<0. The current i_{L1} goes down from the maximum value to zero.

Mode 8 (t_7-t_8): This mode is similar with mode 1. The current through LLC branch keeps oscillating until zero. There is no current through the utility grid.

C. Soft-switching characteristic

From the mode analysis, soft-switching characteristic should be analyzed for efficiency estimation. It can be seen that the two switches S_3 and S_4 can almost achieve zero-switching. When $v_g<0$, switch S_4 is always off, so there is no switching loss. Meanwhile, the freewheeling diode with S_4 is cut off with almost zero current. At the same time, S_3 is turned on and off with nearly zero current as well. Therefore, it can achieve zero current switching (ZCS) and the switching loss is small. Similarly, the switching loss can be estimated as near zero when $v_g>0$ for these two switches. In terms of the other two switches S_1 and S_2, their switching process can be analyzed as well. When $v_g<0$, both mode 1 and mode 4 in Fig. 3 reveals that the current through the leg is negative when S_1 is turned on and S_2 is turned off. Hence, zero voltage switching (ZVS) can be done for S_2. On the other hand, when S_1 is turned off and S_2 is turned on, the current through the leg is i_p-i_{L1} which is positive. Therefore, the switch S_1 can be turned off with zero voltage as well. For the case $v_g>0$, the situation can be analyzed as well to show that zero voltage switching is also achieved.

In comparison with conventional two-stage AC-DC converter, the derived circuit has only four switches with less conduction loss. At the same time, the above theoretical analysis demonstrates that the circuit can achieve soft-switching during most switching process. In addition, unlike traditional LLC resonant converter, the DC link voltage in this topology is adjustable. The soft-switching region can be expanded in accordance to the load and the input voltage. Hence, its efficiency is supposed to be better than two-stage AC-DC converter.

III. Steady-State Circuit Analysis

Circuit modeling is necessary to design closed-loop control so as to make sure the input current is sinusoidal and output voltage constant. However, as introduced in Section 2, the inductor current through the grid is discontinuous and the switching frequency is supposed to change as an input variable for the LLC converter. These two features make circuit modeling difficult. In this section, steady-state waveform is analyzed first. After that, circuit model is

deduced and control strategy is designed in detail.

For the utility grid, the discontinuous current can be calculated based on the time when the switch is turned on. Assume the switching period is T_s, the time between t_1 and t_2 is t_r, the capacitor voltage is v_{dc}, the transformer ratio is n and the output voltage is v_{out}. The relationship between the voltage through the inductor from t_1 to t_2 and inductor current is given by:

$$L_1 \frac{di_{L1}}{dt} = v_g - V_{dc} \qquad (1)$$

The maximum current can be obtained as (2).

$$I_{max} = \frac{|v_g| + V_{dc}}{L_1} t_r \qquad (2)$$

During the time between t_2 and t_3, the inductor voltage is $v_g + V_{dc}$, the time for the current decreasing from I_{max} to 0 can be calculated as (3).

$$t_f = \frac{V_{dc} + |v_g|}{V_{dc} - |v_g|} t_r \qquad (3)$$

Hence, the average current through the utility grid during one switching period is:

$$I_{ave} = \frac{\dfrac{|v_g| + V_{dc}}{2L_1} t_r (t_r + t_f)}{T_s} = \frac{V_{dc}}{L_1 T_s} \frac{|v_g| + V_{dc}}{V_{dc} - |v_g|} t_r^2 \qquad (4)$$

Therefore, if the utility grid voltage vg and the time t_r are known, the input power from utility grid would be (5).

$$P_{in} = \int_0^{T_u} v_g i_{ave} dt = \frac{V_{dc}}{L_1 T_s} \int_0^{T_s} \frac{|v_g(t)| + V_{dc}}{V_{dc} - |v_g(t)|} v_g(t) t_r(t)^2 dt \qquad (5)$$

It is worth pointing out that the relationship between the average current and the peak current is not linear which can be written as (6). If both grid voltage and average current are sinusoidal, the waveform of peak current using specific

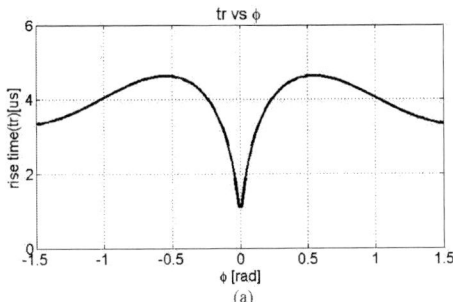
(a)

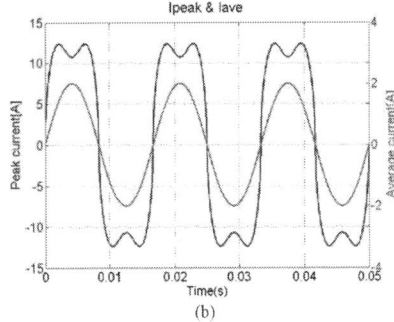
(b)

Fig. 5. Relationship between peak current, duty cycle and phase angel: (a) grid peak current vs time; (b) duty cycle vs phase angel.

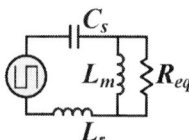

Fig. 6. Equivalent circuit of LLC resonant converter.

hardware parameters is shown in Fig. 5(b). This represents the envelop curve of the inductor current.

$$i_{max} = sign(i_{ave})\sqrt{\frac{v_g|T_s\left(v_{dc}^2 - v_g^2\right)}{L_1 V_{dc}}} \quad (6)$$

According to the literature, the basic equivalent circuit for LLC branch is shown in Fig. 6. The output voltage is $v_{eq} = nv_{out}$ and the equivalent resistor is estimated as $R_{eq} = 8n^2 R_L/\pi^2$.

Based on the equivalent circuit, the relationship between the output voltage and the input voltage can be obtained and represented in (7).

$$\frac{v_{out}}{v_{dc}} = \frac{1}{n}\frac{1}{\dfrac{L_r}{L_m}\left(\dfrac{\omega_s^2}{\omega_m^2}-1\right)\dfrac{\omega_r^2}{\omega_s^2} + j\dfrac{\pi^2}{8}Q\left(\dfrac{\omega_s}{\omega_r}-\dfrac{\omega_r}{\omega_s}\right)} \quad (7)$$

Where, ω_s represents the switching frequency $\omega_s = 2\pi/T_s$, $\omega_r = 1/\sqrt{L_r C_s}$, $\omega_m = 1/\sqrt{(L_r + L_m)C_s}$ and $Q = \sqrt{L_s C_s}/n^2 R_L$. For the steady-state condition, the output voltage is:

$$V_{out} = \frac{V_{dc}}{n\sqrt{\dfrac{L_r^2}{L_m^2}\left(\dfrac{\omega_s^2}{\omega_m^2}-1\right)^2\dfrac{\omega_r^4}{\omega_s^4} + \dfrac{\pi^4}{64}Q^2\left(\dfrac{\omega_s}{\omega_r}-\dfrac{\omega_r}{\omega_s}\right)^2}} \quad (8)$$

If one assumes the input power equals to the output power, the capacitor voltage can be obtained by solving (9).

$$\frac{V_{dc}}{L_1 T_s}\int_0^{T_s}\frac{|v_g(t)|+V_{dc}}{V_{dc}-|v_g(t)|}v_g(t)t_r(t)^3\,dt = \frac{V_*^2}{n^2\left[\dfrac{L_r^2}{L_m^2}\left(\dfrac{\omega_s^2}{\omega_m^2}-1\right)^2\dfrac{\omega_r^4}{\omega_s^4}+\dfrac{\pi^4}{64}Q^2\left(\dfrac{\omega_s}{\omega_r}-\dfrac{\omega_r}{\omega_s}\right)^2\right]R_L} \quad (9)$$

With the capacitor voltage, all the other steady-state variables can be obtained correspondingly. It is worth pointing out that the relationship between the average inductor current and the start time t_r is not linear as shown in

(4). If the current is supposed to be controlled as sinusoidal waveform and in phase with utility grid voltage, the start time t_r should be regulated in real time.

IV. HYBRID CONTROL METHOD WITH PREDICTOR

For the proposed AC-DC converter, the input current should be controlled in order to restrain the THD. Meanwhile, the output voltage should be controlled in order to control the output power. Besides, the dc bus voltage needs be limited within a certain range. Two control loops are designed for input current and output voltage control respectively.

A. Input Current and Output Voltage Control

There are three variables to be controlled in this topology. First, the capacitors' voltage should be controlled as constant to act as DC link and decouple the input and output. Second, the input current from utility grid is supposed to be controlled as sinusoidal waveform with low THD and it should be in phase with grid voltage to maintain unity power factor. Third, the output voltage needs be controlled at a constant value in consideration of the load requirement. Referring to the control strategy of two-stage AC-DC converter, the designed control scheme has been displayed in Fig. 7.

B. DC Bus Voltage Control

Different from control of input current and output voltage, the DC bus voltage would be affected by the low-frequency voltage ripple. However, this voltage ripple is because of the instantaneous power being unbalanced and the DC bus capacitor compensating such power difference. In consideration of the fact that the frequency of voltage ripple on DC bus is constant, it is proposed to use Proportional-Integrator-Resonant (PIR) control instead of Proportional-Integrator (PI) control for DC bus voltage control. Detailed analysis has been analyzed in [9]. Compared with conventional PI controller, PIR controller can suppress the ripple propagation and achieve better steady-state performance since the voltage ripple frequency is fixed. [10] has analyzed this control method. Based on the equivalent model and PIR control method, the closed loop bode diagram has been shown in Fig. 8.

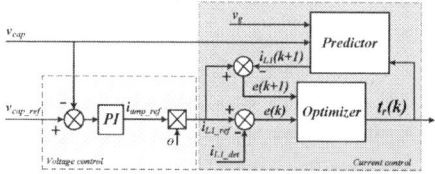

Fig. 7. Control scheme of the dc bus voltage and input current in SSPLLC converter.

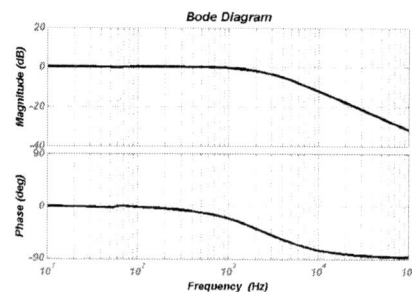

Fig. 8. Bode diagram of DC bus voltage closed-loop control system.

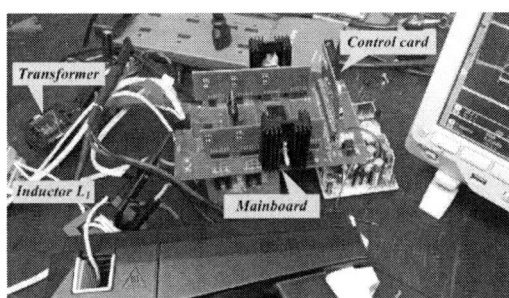

Fig. 9. 250W prototype based on proposed isolated AC-DC topology.

TABLE I. HARDWARE PARAMETERS OF FOUR-SWITCH AC-DC CONVERTER IN FIGURE 2(B)

Parameters	Value
V_{in}	90-120V AC
V_{dclink}	300-400V DC
V_{out}	24.0V
Inductor (L_1)	20μH, 0.01Ω
DC link capacitors (C_1 & C_2)	100μF
Output capacitor (C_o)	220μF
Switching frequency	150-250kHz
Switches	C2M0080120D

V. EXPERIMENTAL RESULT

In order to verify the feasibility and superiority of the proposed SSPLLC converter, a 250 W experimental prototype is built. Fig. 9 shows the picture of the overall system. The hardware parameters have been listed in Table 1. The switching frequency region is set at [150, 250]-kHz. In consideration of the grid voltage, the DC link voltage is selected at [300, 400] *V*. Here, SiC MOSFET C2M0080120D-ND is used for the requirement of high switching frequency. The varying-frequency modulation and multi-loop control method are implemented via a 32-bit ARM Coretex MCU STM32F405 from STMicroelectronics.

The grid current waveform has been shown in Fig. 10. Even though the current is discontinuous, its envelope curve is in phase with grid voltage which is shown with large-periodic waveform in Fig. 10(a). Unlike the conventional DCM PFC, the envelope curve is not sinusoidal. This is because the voltage drop between the inductor is $-v_{dc}-|v_g|$ and $v_{dc}-|v_g|$. When the grid voltage is high, the current rate of increase becomes high but the rate of decrease is low. The grid current in Fig. 10(a) shows that the average current is sinusoidal. Its THD is 4.9% at 100 W output which is less than 5%. Besides, the switching-periodic inductor current waveform and switching voltage of the switch S_3 are shown in Fig. 10(b). This figure indicates that the current through MOSFET is almost zero during switching process. Hence, its switching loss is very small.

The voltage and current waveforms through LLC branch have been displayed in Fig. 10(c). The inductor current and series capacitor voltage are oscillating simultaneously. From the switch voltage waveform, it can be observed that zero voltage switching is achieved. It is similar with the waveform of conventional LLC converter. The DC link voltage and output voltage waveforms are shown in Fig.

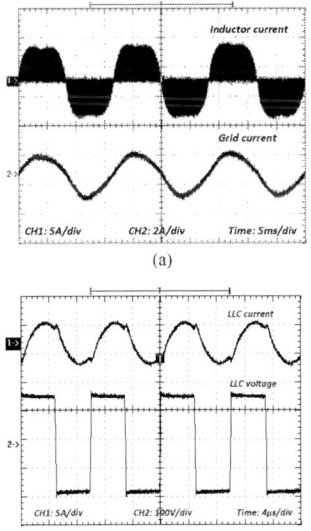

(a)

(c)

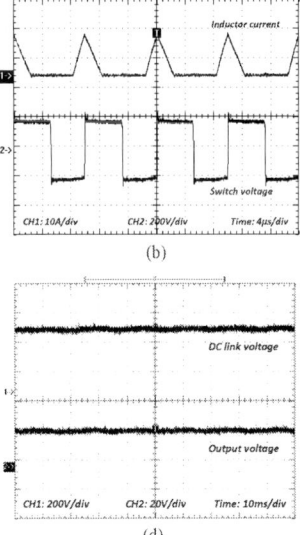

(b)

(d)

Fig. 10. Grid current in experiment: (a) Large periodic waveform; (b) switching periodic voltage and current; (c) switching periodic voltage and current through LLC branch; (d) DC link voltage and output voltage waveforms

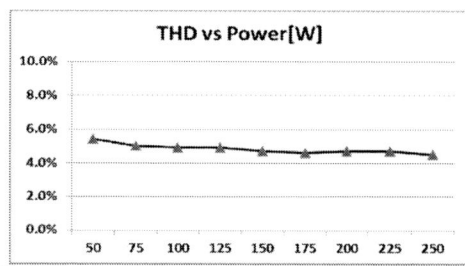

Fig. 11. THD vs power [W].

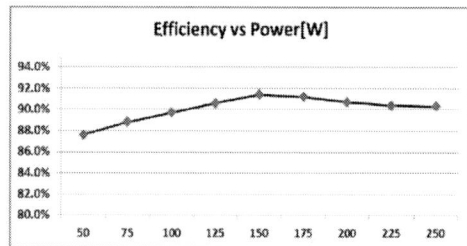

Fig. 12. THD vs power [W].

10(d).

The THD of the grid current vs output power is tested and listed in Fig. 11. This figure indicates that the current THD can be controlled to less than 5.0% when the output power is larger than 50W which is in compliance with IEC61000-3-2 class D regulations. The experimental efficiency of the proposed circuit has been shown in Fig. 12. The maximum efficiency is 91.1% which occurs at 150W. In addition, the efficiency keeps within the range [87.6%, 91.1%] when the output power changes from 50W to 250W. This efficiency curve is optimized by varying DC link voltage instead of maintaining a fixed voltage.

VI. CONCLUSION

A new single-stage single-phase isolated AC-DC converter is derived by combining a LLC unit and a three-level unit. Only four switches and four diodes are necessary in this topology. The benefit of this topology is that the DC link voltage is flexible which can help to expand the soft-switching region of the half bridge connected to LLC branch. Meanwhile, both the input current and output voltage can be controlled with start time and frequency as input variables.

Experimental results verify the topology's validity and reveal that the circuit can work with more than 87.6% efficiency and 91.1% maximum efficiency in the power range [50, 250]W.

ACKNOWLEDGMENT

This work is done in the Renewable Energy and Vehicular Technology (REVT) Laboratory, the University of Texas at Dallas. The authors would like to thank the founding director Dr. Babak Fahimi for his support.

REFERENCES

[1] J. Wei and B. Fahimi, "Multiport Power Electronic Interface-Concept, Modeling, and Design," *Power Electronics, IEEE Transactions on Power Electronics*, vol. 26, pp. 1890-1900, 2011.

[2] Y. Xibo, F. Wang, R. Burgos, L. Yongdong, and D. Boroyevich, "Dc-link voltage control of full power converter for wind generator operating in weak grid systems," in *Applied Power Electronics Conference and Exposition, 2008. APEC 2008. Twenty-Third Annual IEEE*, 2008, pp. 761-767.

[3] P. Shamsi and B. Fahimi, "Dynamic Behavior of Multiport Power Electronic Interface Under Source/Load Disturbances," *Industrial Electronics, IEEE Transactions on*, vol. 60, pp. 4500-4511, 2013.

[4] C. Chun-An, C. Chien-Hsuan, C. Tsung-Yuan, and Y. Fu-Li, "Design and Implementation of a Single-Stage Driver for Supplying an LED Street-Lighting Module With Power Factor Corrections," *Power Electronics, IEEE Transactions on*, vol. 30, pp. 956-966, 2015.

[5] C. Chien-Hsuan, C. Chun-An, M. Jinno, and C. Hung-Liang, "An interleaved single-stage LLC resonant converter used for multi-channel LED driving," in *Power Electronics Conference (IPEC-Hiroshima 2014 - ECCE-ASIA), 2014 International*, 2014, pp. 3333-3340.

[6] S. Kyoung-Wook and K. Bong-Hwan, "A novel single-stage half-bridge AC-DC converter with high power factor," *Industrial Electronics, IEEE Transactions on*, vol. 48, pp. 1219-1225, 2001.

[7] C. Woo-Young and Y. Joo-Seung, "A Bridgeless Single-Stage Half-Bridge AC/DC Converter," *Power Electronics, IEEE Transactions on*, vol. 26, pp. 3884-3895, 2011.

[8] M. Marchesoni and C. Vacca, "New DC–DC Converter for Energy Storage System Interfacing in Fuel Cell Hybrid Electric Vehicles," *Power Electronics, IEEE Transactions on*, vol. 22, pp. 301-308, 2007.

[9] X. Chen, S. Duan, W. Cai, M. Ma, C. Chen, and Y. Chen, "Low frequency ripple propagation analysis in LLC resonant converter base on signal modulation-demodulation theory," in *Industrial Electronics (ISIE), 2013 IEEE International Symposium on*, 2013, pp. 1-5.

[10] C. Xi, M. Mengyin, D. Shanxu, C. Wen, and C. Changsong, "Low frequency ripple propagation analysis in LLC resonant converter based on signal modulation-demodulation theory and reduction based on PIR control strategy," in *Energy Conversion Congress and Exposition (ECCE), 2013 IEEE*, 2013, pp. 5390-5394.

DC-DC Power Converter Controller for SOC Balancing of Paralleled Battery System

Jaber A. Abu Qahouq, *Senior Member, IEEE*, Lin Zhang, *Student Member, IEEE*,
Yuan Cao, *Student Member, IEEE*, and Bharat Balasubramanian

The University of Alabama
Department of Electrical and Computer Engineering
Tuscaloosa, Alabama 35487, USA

Abstract— This paper presents a DC-DC power converter control scheme and system architecture for batteries which are connected in parallel in order to maintain State-Of-Charge (SOC) balancing between batteries without the need for additional circuitries and their associated controllers. When the battery cells or battery packs are connected in parallel, it is desired to maintain SOC balancing during both charging mode and discharging mode. Using conventional balancing circuits is energy inefficient and/or might be complicated/not suitable. This paper addresses this by presenting a controller that is able to maintain a real-time natural charge balance between the in parallel connected batteries while maintaining output voltage regulation at the same time.

Index Terms — Battery; DC-DC; State of Charge; Control; Power; Converter; Electronic System; Power Management.

I. INTRODUCTION

Battery systems are increasingly becoming important in many applications such as plug-in hybrid electric vehicles (PHEVs), electric vehicles (EVs), smart grid, laptops, and other portable and mobile applications [1-9].

Lithium-Ion (Li-Ion) batteries or battery packs can be connected in series to yield higher voltages and can be connected in parallel to yield higher current capability. When connected in series, batteries carry same value of current and therefore they discharge and charge at the same rate if they are symmetric and equally healthy. On the other hand, when connected in parallel batteries are at higher risk of carrying different current values and therefore charge or discharge at different rates due to any asymmetry between batteries or the connections/wiring or layout. This can cause one or more Li-Ion battery to be charged or discharged at much faster rate which can cause catastrophic failures and/or not full utilization of system capacity. In the case of in parallel connection, unlike in the case of in series connection, common cell balancing schemes [6-9] might be ineffective or not suitable.

This paper presents a control scheme for in parallel connected battery system architecture where desired SOC balance is naturally maintained in real-time. Section II gives an overview of batteries' connection schemes for higher current capability. Section III describes the SOC power converter control scheme for in parallel-connected batteries. Section IV presents preliminary experimental results and Section V gives the conclusion.

II. AN OVERVIEW OF BATTERIES CONNECTION SCHEMES FOR HIGHER CAPACITY/CURRENT

Figure 1 diagram illustrates a conventional in parallel connection scheme (Scheme 1) for Li-Ion batteries in order to be able to supply higher current (higher capacity) to a load through a DC-DC power converter with voltage/current regulation. Fig. 2 shows an alternative scheme (Scheme 2) to achieve higher capacity and/or higher current to a load where each battery has its own power converter and the power converters of the batteries are connected in parallel rather than directly connecting the batteries terminals themselves in parallel.

For Scheme 1, it is required that the batteries have same voltage, and for optimal and safe operation the batteries should be of the same type and characteristics. Mismatches, for examples due to different internal and/or branch resistances and/or different charge/discharge characteristics, can result in issues such as power loss and heating. Mismatches could be due to asymmetric aging, asymmetric temperature rise, or asymmetric connections. A shorted battery could cause excessive heating and a fire risk. This scheme does not provide a mechanism to balance or control the State-Of-Charge (SOC) of the batteries. It also does not allow for the control of energy usage from each battery individually.

On the other hand, in the case of Scheme 2 the voltages of the batteries can be different and the batteries can have mismatched characteristics. Each battery is interfaced with the output/load through its own power converter and the outputs of the power coveters are connected in parallel in order to provide higher current to the load. Current sharing control [10] can be used to balance the current between the power converters.

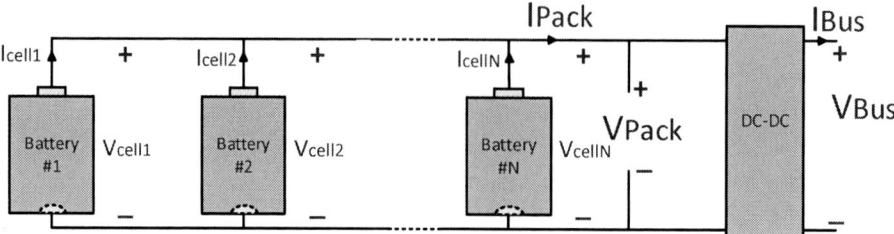

Fig. 1. Illustration diagram for how Li-Ion batteries can conventionally be connected in parallel (Scheme 1).

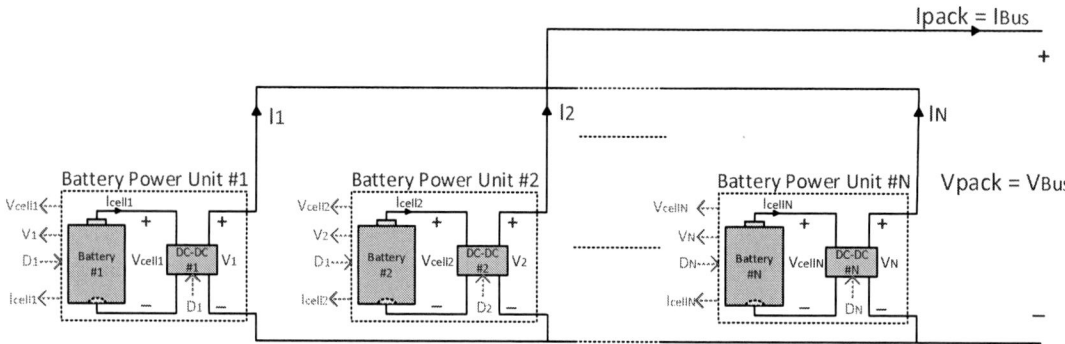

Fig. 2. Illustration diagram for an alternative scheme to realize in parallel connection with Li-Ion batteries for higher capacity and/or load current (Scheme 2).

III. POWER CONVERTER SOC CONTROL SCHEME FOR IN PARALLEL-CONNECTED BATTERIES

This section presents a controller for SOC balancing or regulation between batteries for Scheme 2 of Fig. 2 while regulating the output bus voltage to the load (Vbus = VPack) at the same time.

Figure 3 shows the basic control block diagram of the SOC balancing controller presented in this paper during the discharge operation mode. A single voltage-mode control loop is used and is realized by a single transfer function Gv(s) or Gv(z) which receives the difference between the measured Vbus and the reference value Vbus-ref as an input and outputs the duty cycle Dt. Several duty cycle values (D1 through DN) are then generated by multiplying Dt by SOC coefficients (αI1 through αIN) generated by an SOC control loop. Gv(s) or Gv(z) can for example be a PI (Proportional-Integral) or PID (Proportional-Integral-Derivative) controller compensators/filters.

The SOC control loop generates the SOC coefficients (αI1 through αIN) that are a function of the SOC of each battery. For each battery, this loop is realized by a transfer function GI-SOC (which can for example be a PI or PID compensator) which has its input from the difference between the measured SOCr value for each battery and the reference SOCI-ref value (SOCI-ref − SOCr) and generates an output that has αIr SOC coefficient after being subtracted from number one (or a unit or a fixed number), where r = 1, 2... N.

For N Battery Power Units (BPUs), the SOC reference value is generated by Equation (1) if it is desired that all batteries have same SOC value during operation.

$$SOC_{I-ref} = (SOC_1 + SOC_2 + \cdots + SOC_N)/N \qquad (1)$$

Therefore, Dt value generated by the voltage-mode closed-loop ensures maintaining a regulated output/bus voltage while the SOC loops ensure maintaining SOC balance by altering the final duty cycle for each converter based on Dt and the αI1 through αIN values.

The BPUs share the same output voltage as given by Equation (2).

$$V_{bus} = V_{Pack} = V_1 = V_2 = \cdots = V_N \qquad (2)$$

When each of the power converters is a DC-DC boost power converter:

$$\frac{I_{cellr}}{I_r} = \frac{V_{bus}}{V_{cellr}} = \frac{1}{1 - \alpha_r D} \qquad (3)$$

$$I_{cellr} = \frac{1}{1 - \alpha_r D} I_r \qquad (4)$$

Where r = 1, 2… N. Ir is the output current of the rth power converter. I$cellr$ and V$cellr$ are the current and voltage of the rth battery, respectively.

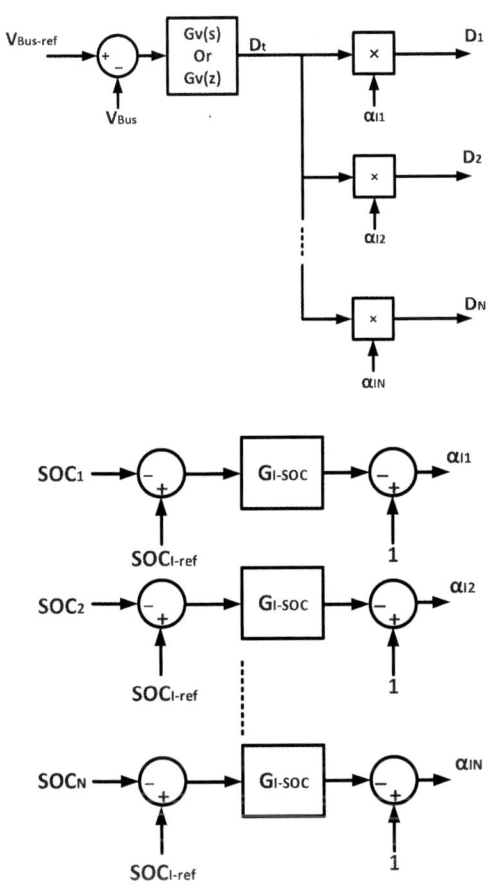

Fig. 3. Main controller block diagram for discharging mode.

IV. PRELIMINARY RESULTS

A proof of concept experimental prototype is built in the laboratory to test the charge balancing controller. It has two DC-DC boost power converters which are connected in parallel (two-phase DC-DC power converters) as illustrated in Fig. 4. The input for each boost converter is from a 2.6 Ah 3.7 V$_{nominal}$ Li-Ion battery and the outputs of the converters are connected/shared. The total capacity for the two batteries is 2.6 + 2.6 = 5.2 Ah. The controller is implemented using TMS320F28335 microcontroller.

Figure 5 shows the SOC curves for each battery during the first 60 minutes of discharge operation. At the start of the discharge period, the SOC1 of the first battery is intentionally made lower than the SOC2 of the second battery by 7% (0.93) in order to evaluate the SOC balancing ability/operation of the controller. The discharge rate used is 0.5 C (2.6 A discharge current). It can be observed how the controller is able to balance the SOC values and keep them balanced during the discharge operation.

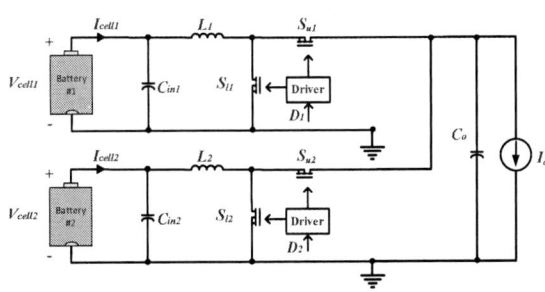

Fig. 4. Illustration of power stage block diagram with batteries.

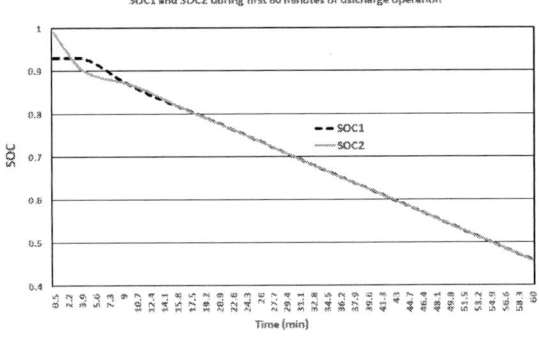

Fig. 5. Preliminary experimental prototype results for charge balancing during discharging operation.

V. CONCLUSION

The paper presents a controller for DC-DC power converters (connected in parallel at the output) which is able to achieve SOC balancing between the batteries at the input of each converter while regulating the output voltage at the same time. It eliminates the need for additional charge balancing circuits and controllers. Since each converter has a battery at its input and the converter outputs are connected, this configuration is equivalent to connecting the batteries in parallel but with the ability to control each battery's discharge and charge rate individually, which is a characteristic that the presented controller utilizes for SOC balancing. The paper presents preliminary results for discharging operation. Future work includes but is not limited to developing the controller for both discharging and charging operations and developing additional related optimization techniques for State-of-Charge, State-of-Health, and efficiency.

ACKNOWLEDGEMENT

This work is supported in part by The Center for Advanced Vehicle Technologies (CAVT) at The University of Alabama – Tuscaloosa (UA) and in part by the Alabama NASA EPSCoR Research Infrastructure Development grant no. NNX13AB09A. Any opinions, findings, and conclusions or recommendations expressed in this material are those of the author(s) and do not necessarily reflect the views of these entities.

REFERENCES

[1] W. Huang and J. Abu Qahouq, "Energy Sharing Control Scheme for State-of-Charge Balancing of Distributed Battery Energy Storage System," IEEE Transaction on Industrial Electronics, vol. 62, no. 5, pp. 2764-2776, May 2015.

[2] W. Huang and J. Abu Qahouq, "Distributed battery energy storage system architecture with energy sharing control for charge balancing," The 2014 IEEE Applied Power Electronics Conference and Exposition, APEC'2014, Pages: 1126 - 1130, March 2014.

[3] M. Yilmaz and P. T. Krein. "Review of Battery Charger Topologies, Charging Power Levels, and Infrastructure for Plug-In Electric and Hybrid Vehicles," IEEE Transactions on Power Electronics, vol. 28, no. 5, pp. 2151 - 2169, August 2012.

[4] David Velasco de la Fuente, Cesar L. Trujillo Rodriguez, and Gabriel Garcera, "Photovoltaic Power System with Battery Backup with Grid-Connection and Islanded Operation Capabilities," IEEE Transactions on Industrial Electronics, vol. 60, no. 4, pp. 1571-1581, April 2013.

[5] W. Huang and J. A. Abu Qahouq, "An Online Battery Impedance Measurement Method Using DC–DC Power Converter Control." Industrial Electronics, IEEE Transactions on, vol. 61, no. 11, pp. 5987-5995, Jun. 2014.

[6] J. Cao, N. Schofield, and A. Emadi, "Battery balancing methods: A comprehensive review," in Proc. IEEE Veh. Power Propulsion Conf., pp. 1–6, Sep. 2008.

[7] Y. Yuanmao, K. W. E. Cheng, and Y. P. B. Yeung, "Zero-current switching switched-capacitor zero-voltage-gap automatic equalization system for series battery strings," IEEE Trans. Power Electron., vol. 27, no. 7, pp. 3234–3242, Jul. 2012.

[8] C.-H. Kim, M.-Y. Kim, H.-S. Park, and G.-W. Moon, "A modularized two stage charge equalizer with cell selection switches for series-connected lithium-ion battery string in an HEV," IEEE Trans. Power Electron., vol. 27, no. 8, pp. 3764–3774, Aug. 2012.

[9] M. Udo and K. Tanaka, "Double-switch single-transformer cell voltage equalizer using a half-bridge inverter and a voltage multiplier for series connected supercapacitors," IEEE Trans. Veh. Technol., vol. 61, no. 9, pp. 3920–3930, Nov. 2012.

[10] J. Abu Qahouq, "Analysis and Design of N-Phase Current Sharing Auto-Tuning Controller," IEEE Transactions on Power Electronics, Vol. 25, No. 6, Pages: 1641 - 1651, June 2010.

Ultra-Step-Up DC-DC Converter with Integrated Autotransformer and Coupled Inductor

Yam P. Siwakoti, *Member, IEEE*, Frede Blaabjerg, *Fellow, IEEE*, Poh Chiang Loh

Department of Energy Technology,

Aalborg University, Pontoppidanstræde 101, 9220 Aalborg, DENMARK.

yas@et.aau.dk, fbl@et.aau.dk, pcl@et.aau.dk

Abstract—This paper introduces a new single-switch non-isolated dc-dc converter with very high voltage transfer ratio and reduced semiconductor voltage stress. The converter utilizes an integrated autotransformer and a coupled inductor on the same core to achieve a high step-up voltage gain without extreme duty cycle. Further, an integrated passive regenerative circuit recycles the leakage energy of the coupled magnetics and transfer the leakage energy to the load, which helps to avoid high voltage spikes across the switch. This feature along with low stress on the switching device enables the designer to use a low voltage and low R_{DS-on} MOSFET, which reduces the cost, and also the conduction and turn on losses of the switch. The principle of operation and theoretical analysis supported by key simulation and experimental waveforms are presented in details.

Index Terms—dc-dc power conversion, coupled inductor, autotransformer, and distributed generation.

INTRODUCTION

A dc-dc converter with high voltage gain and efficiency is required in many applications such as in front-end converter for intermittent renewable and distributed power generations including solar PV, fuel cells and power systems based on battery banks and supercapacitors [1], [2]. In addition, hybrid electric vehicle, high voltage Light Emitting Diode (LED) lamps, Uninterruptible Power Supply (UPS) also require high efficient dc-dc converter to convert the low and varying voltage of the battery (12 – 48 V) and fuel cell (25 – 50 V) to standard dc bus voltage of 380 – 400 V [3]. A conventional boost converter is not sufficient to boost the input voltage to 10 – 20 times due to high losses and associated stress on the switching components. Further, higher stress on the switch demands a higher voltage switch, which has higher R_{DS-on} and corresponding high conduction losses. Various high step-up dc-dc converter topologies are presented in the literature using different voltage boosting techniques such as multilevel, interleaved, cascaded or using voltage multiplier cells, switched capacitor and coupled inductors [4]-[11]. However, the numbers of components to design multilevel, interleaved, cascaded including voltage multiplier cells and switched capacitor are large, especially

switch and associated driving circuitry. In addition, the voltage stress on the switch is high and suffers high conduction losses [8]. The attractive option among them is to use coupled inductor or transformer in the converter circuit, which reduces the number of components and complexity in the circuit. However, the efficiency of the converter degrades with load current and frequency due to the loss associated with leakage inductances. The use of an active clamp circuit can effectively recycle the leakage energy and yet reduce the main switch voltage stress, but at the expense of circuit complexity and loss in the extra circuitry. Moreover, these converters require a high voltage rated switch [10].

The aforementioned and related drawbacks associated with existing high boost converter encourage to investigate a high step-up converter topology, which enables to use a lower voltage rated, lower R_{DS-on} MOSFET in order to provide a better efficiency by reducing both conduction and switching losses. To address this issue, a very high voltage gain dc-dc converter is proposed in this digest using integrated autotransformer and coupled inductor for the voltage boosting without extreme duty ratio. The leakage energy of the converter is also recycled to the load, which increases the efficiency and avoids lossy snubber or active-clamp circuit. Further, this new topology utilizes only eight components with integrated single core magnetics. Therefore, this new topology opens a new horizon in the area of Switch Mode Power Supply (SMPS) to design a compact and high efficient dc-dc converter. A mathematical derivation of the proposed converter is documented in Section II, which is validated by experimental results in Section III and findings drawn are finally concluded in Section IV.

PROPOSED TOPOLOGY AND PRINCIPLE OF OPERATION

The proposed converter consists of an autotransformer and a coupled inductor wound on the same core and three diodes and the same number of capacitors as shown in Fig. 1 (a). The advantage of the proposed topology are: (1) very high voltage gain, which is particularly suitable for (a) low voltage output Fuel Cell (25 – 50 V) to stabilize the output voltage to 400 VDC and, (b) high voltage Light Emitting Diode (LED) lamps (which require 100 – 600 V for a series/parallel string of LED from a battery input of 12 – 24 V), (2) low voltage stresses on switch (Q), (3) no voltage spikes across the switch (diode D_1 and capacitor C_2 forms a regenerative snubber for the switch Q), and (3) only eight components are required to design the converter. The winding design is simplified by winding the autotransformer and coupled inductor on the same core that reduces the space and the component count.

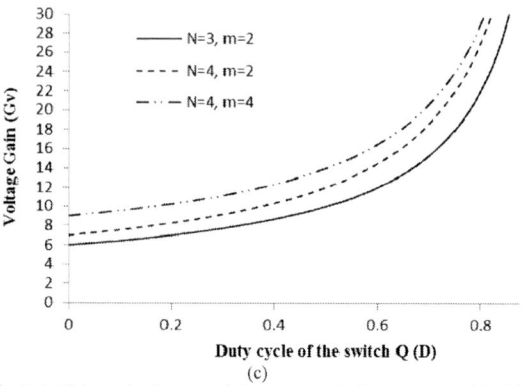

(c)

Fig. 1. (a) Schematic of proposed ultra-step-up dc-dc converter and its (b) steady-state waveforms in Continuous Conduction Mode (CCM) operation, and (c) theoretical voltage transfer characteristics.

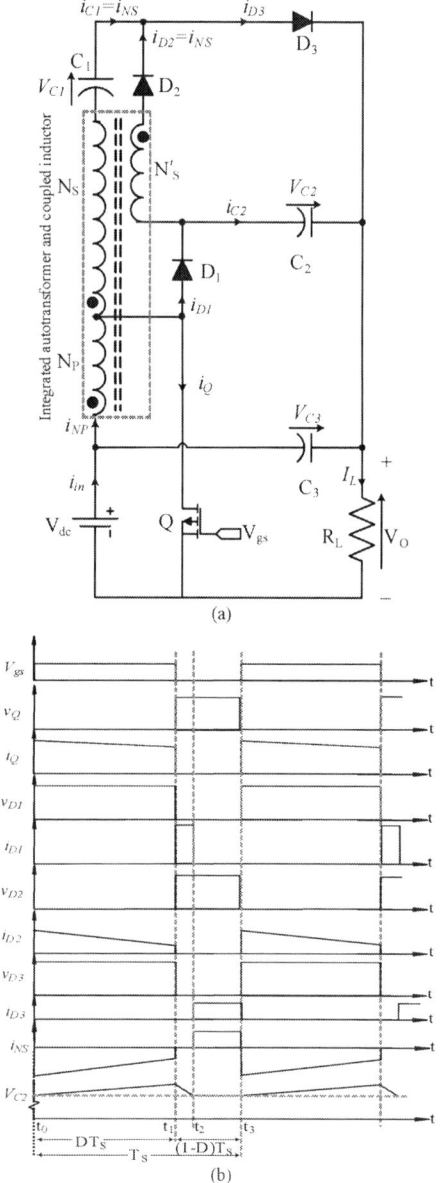

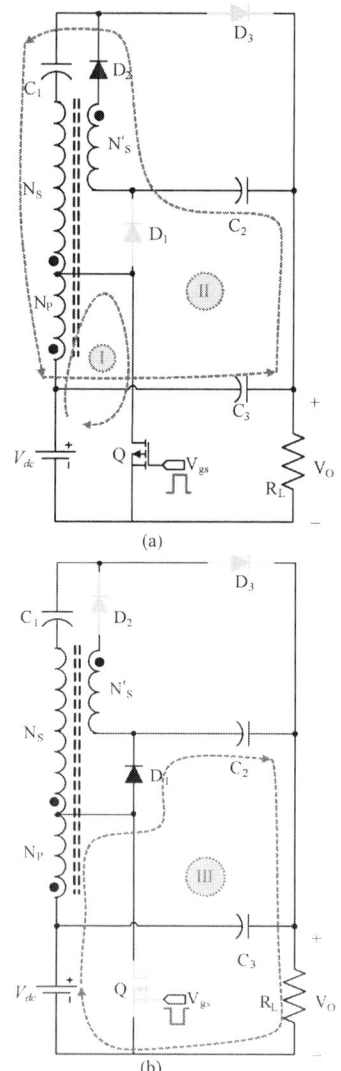

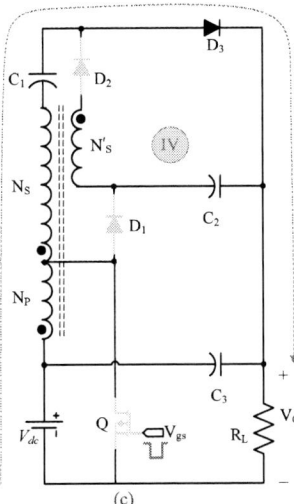

(c)

Fig. 2. Equivalent circuit of the proposed converter during different operating modes: (a) Mode 1 $[t_0 - t_1]$, (b) Mode 2 $[t_1 - t_2]$, and (c) Mode 3 $[t_2 - t_3]$.

The steady-state waveforms of the proposed converter operating in Continuous Conduction Mode (CCM) are depicted in Fig. 1 (b). The converter has three operating mode in a switching period. The equivalent circuit of the converter in each operating mode is shown in Fig. 2 (a)-(c). Each mode is described briefly as follows:

Mode 1 ($[t_0 - t_1]$): During this interval, the switch Q turns-on and current flows through the primary winding, secondary winding of the autotransformer and on the coupled inductor through diode D_2 and charge capacitor C_1 and C_2. Diode D_1 and D_3 are reversed biased. There are two closed loops for the current to flow as shown in Fig. 2(a). The relevant circuit expressions for this mode can then be written like in (1) and (2), where v_L is the voltage across the primary inductor winding (N_P).

$$\frac{N_S}{N_P} v_L + V_O - V_{C2} + \frac{N'_S}{N_P} v_L - V_{C1} = 0$$

$$\Rightarrow v_L = (V_{C1} + V_{C2} - V_O)/((N_S + N'_S)/N_P) \quad (1)$$

$$\text{and } v_L = V_{dc} \quad (2)$$

From (1) and (2) $\quad V_{C1} + V_{C2} = \frac{N_S + N'_S}{N_P} V_{dc} + V_O \quad (3)$

Mode 2 ($[t_1 - t_2]$): The switch is turned-off during this interval. Diode D_2 is reverse biased, whilst diode D_3 continue its reverse biased until the end of this mode. The load current flows through diode D_1 and primary winding of an autotransformer, and discharge the capacitor C_2. The circuit expression for this mode can then be written as (4)

$$V_{dc} - v_L + V_{C2} - V_O = 0$$

$$\Rightarrow v_L = V_{dc} + V_{C2} - V_O \quad (4)$$

Mode 3 ($[t_2 - t_3]$): The diode D_1 remains on in the previous mode until capacitor C_2 is discharged to V_{C2}. This reverse biasing the diode D_1 and the load current path changes to loop IV as shown in Fig. 2 (c). The circuit expression for this mode can then be written as (5)

$$V_{dc} - v_L - \frac{N_S}{N_P} v_L + V_{C1} - V_O = 0$$

$$\Rightarrow v_L = -\frac{V_{C1} - V_O - V_{dc}}{\frac{N_S + N_P}{N_P}} \quad (5)$$

Applying volt-sec balancing to (2) and (4) further results in (6), where D represents duty cycle of Q.

$$\int_0^{DT_S} v_L dt + \int_{DT_S}^{T_S} v_L dt = 0$$

$$\Rightarrow DV_{dc} + (1 - D)(V_{dc} + V_{C2} - V_O)$$

$$\Rightarrow V_{C2} = (V_O(1 - D) - V_{dc})/(1 - D) \quad (6)$$

Applying volt-sec balancing to (1) and (5) further results in (7).

$$\int_0^{DT_S} v_L dt + \int_{DT_S}^{T_S} v_L dt = 0$$

$$\frac{(V_{C1} + V_{C2} - V_O)}{\frac{N_S + N'_S}{N_P}} D - \frac{(V_{C1} - V_O - V_{dc})}{\frac{N_S + N_P}{N_P}} (1 - D) = 0 \quad (7)$$

The voltage gain (G_v) and hence the voltage transfer characteristics equation of the converter is found by solving (3) and (7) as

$$G_v = \frac{V_O}{V_{dc}} = \left[\frac{1 + N}{1 - D} + m \right] \quad (8)$$

where, $N = (N_P + N_S)/N_P$ is the autotransformer voltage transfer constant and $m = N'_S/N_P$ is the turns ratio of the coupled inductor. The voltage transfer characteristic of the proposed converter is shown in Fig. 1 (c) with varying N, m and D.

TABLE I.

COMPARISON OF THE PROPOSED CONVERTER WITH SOME OF THE HIGH STEP-UP DC-DC CONVERTER.

Ref.	Voltage Gain $\left(G_v = \frac{V_O}{V_{dc}}\right)$	Switch Stress	No. of components
Proposed	$\left[\frac{1 + \frac{N_1 + N_2}{N_1}}{1 - D} + \frac{N_3}{N_1}\right]$	$\frac{V_{dc}}{1 - D}$	1 coupled inductor, 3 diodes, 4 capacitors, 1 switch
[7]	$1 + \frac{N_3}{N_1} + \frac{1}{1 - D} + \frac{D}{1 - D}\frac{N_2}{N_1}$	$\frac{V_{dc}}{1 - D}$	1 coupled inductor, 3 diodes, 4 capacitors, 1 switch
[8]	$\left[\frac{2 - D + \frac{N_2}{N_1}}{1 - D} + \frac{N_3}{N_1}\right]$	$\frac{V_O}{1 + \frac{N_2 + N_3}{N_1}D + N_2(1 - D)}$	1 coupled inductor, 4 diodes, 4 capacitors, 1 switch
[9]	$\frac{3 + D}{1 - D}$	$\frac{2V_{dc}}{3 + D}$	2 inductors, 5 diodes, 4 capacitors, 1 switch

978-1-4673-9551-9/16 $31.00 © 2016 IEEE

Table I summarises the voltage conversion ratio and the switch voltage stress for the proposed converter and the other single switch high step-up converter topologies introduced in [7]-[9]. Fig. 3 shows that the proposed converter has higher conversion ratio, lower switch voltage stress and can be realized with a small number of components compared to other higher boost topologies (for comparison with other converter topologies $N_P = N_1$, $N_S = N_2$ and $N'_S = N_3$).

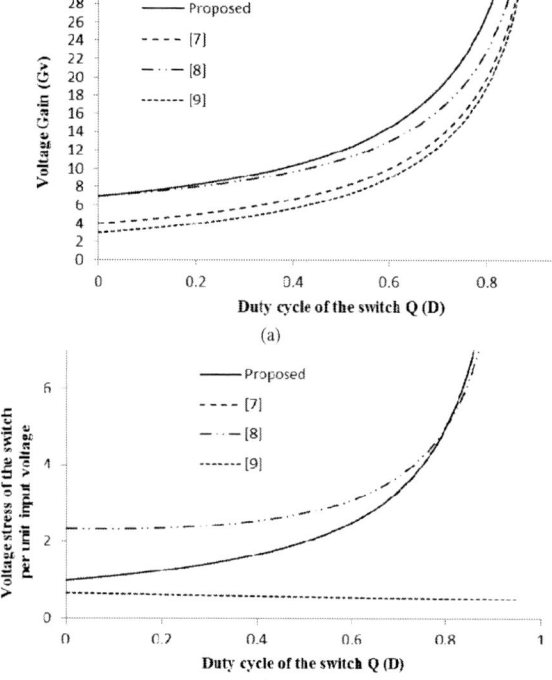

Fig. 3. Comparison of (a) voltage gain and (b) switch voltage stress of the proposed topology with different three winding coupled inductor based high voltage dc-dc converters for $(N_1 : N_2 : N_3 = 1 : 3 : 2)$ over a wide range of switch duty cycle D.

SIMULATION AND EXPERIMENTAL RESULTS

Simulations are carried out in Matlab-Simulink with PLECS toolboxes included to verify the performance of the proposed converter. The converter was simulated with $N = 4$, $m = 2$ ($N_P : N_S : N'_S = 1 : 3 : 2$), $D = 0.65$ and $f_s = 24\ kHz$. With these conditions, the output voltage is boosted to $V_O = 392$ V for $V_{dc} = 25\ V$, which is consistent with (8) as shown in the fourth trace of Fig. 4 (a). The output voltage and load current are constant and stable as shown in the fourth and fifth traces of Fig. 4 (a). The drain source voltage of the switch are around 70 V as shown in third trace of Fig. 4 (b), which helps to select low voltage and a low $R_{DS\text{-}on}$ switch. Other simulated waveforms are also noted to be in agreement with the theoretical values derived in Section II. The performances expected from the converter are thus smoothly verified.

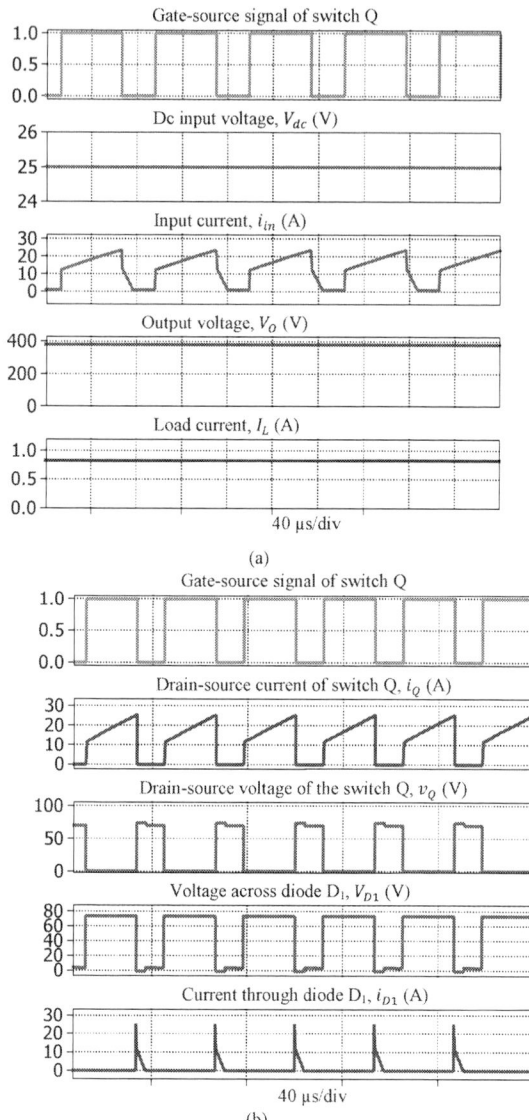

Fig. 4. Simulated waveforms of the proposed converter at $N = 4$, $m = 2$, $D = 0.65$, $V_{dc} = 25$ V and $f_s = 24\ kHz$ conditions showing (a) input-output voltage and current waveforms, and (b) switch Q and diode D_1 voltage and current waveforms.

A 300-W experimental prototype as shown in Fig. 5 has been built in the laboratory to confirm the theory and to validate the simulation results. Same conditions are used as of the simulations with a low voltage and low $R_{DS\text{-}on}$ switch (PSMN5R6 (100 V, 100 A, 5.6 mΩ)), diode D_1 (43CTQ100 (100 V, 40 A)), Diode D_2 and D_3 (STTH20R04FP (400 V, 20 A, $t_{rr} = 18$ ns)), capacitor C_1-C_3 (220 µF, 400 V). Some experimental waveforms showing the input, output voltage and current of the converter along with major waveforms of the switch and diodes are shown in Fig. 6. The corresponding boosted output voltage is the same as of the theoretical and

978-1-4673-9551-9/16 $31.00 © 2016 IEEE

simulated value as read from the fourth experimental trace shown in Fig. 6(c). The blocking voltage across the switch Q is also noted to clamp at around 90 V in Fig. 6 (a), which quantitatively is only 25% of the output voltage. The voltage stress across diodes and capacitors are also minimal and corresponds to the theoretical and simulation results as shown in Fig. 6 (b) - (c). This helps to choose lower-voltage switch and diodes with lower $R_{DS\text{-}on}$ to reduce the conduction losses and to improve the efficiency.

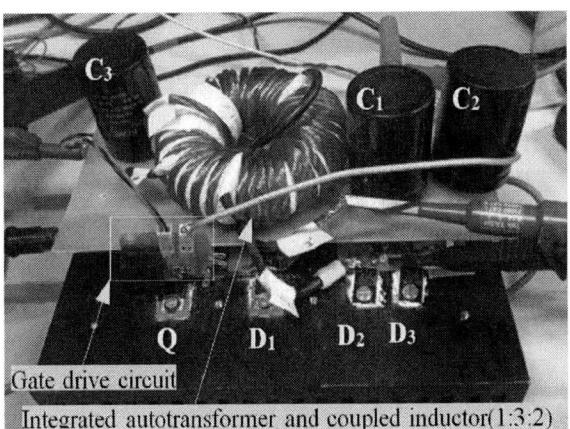

Fig. 5. A 300-W experimental prototype of the proposed ultra-step-up dc-dc converter.

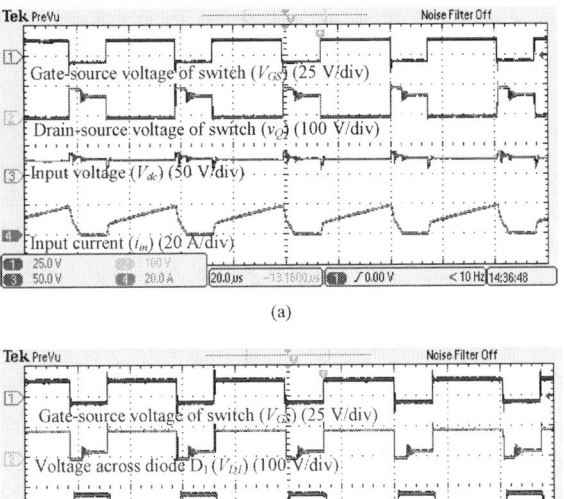

(a)

(b)

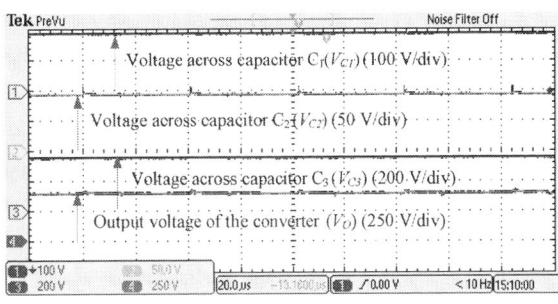

(c)

Fig. 6. Some experimental waveforms of the converter at $N = 4$, $m = 2$, $D = 0.65$, $V_{dc} = 25$ V and $f_s = 24\ kHz$.

Fig. 7 (a) shows the measured efficiency of the converter over a wide range of voltage gain. The input voltage was varied from 24 V to 80 V, whilst keeping the output voltage constant at 400 V. The highest efficiency measured was 98.3% at a voltage gain of 5. It decreases with the voltage gain due to the increase in input current and magnitude of the ripple content. The efficiency plot of the converter over a wide load range (50 W – 300 W) is shown in Fig. 7 (b) at a constant voltage gain of 14.2. The highest efficiency measured was 95.5%. Thus, the proposed converter is suitable for power conversion system, such as distributed generation, electric vehicles and high voltage LED lamps.

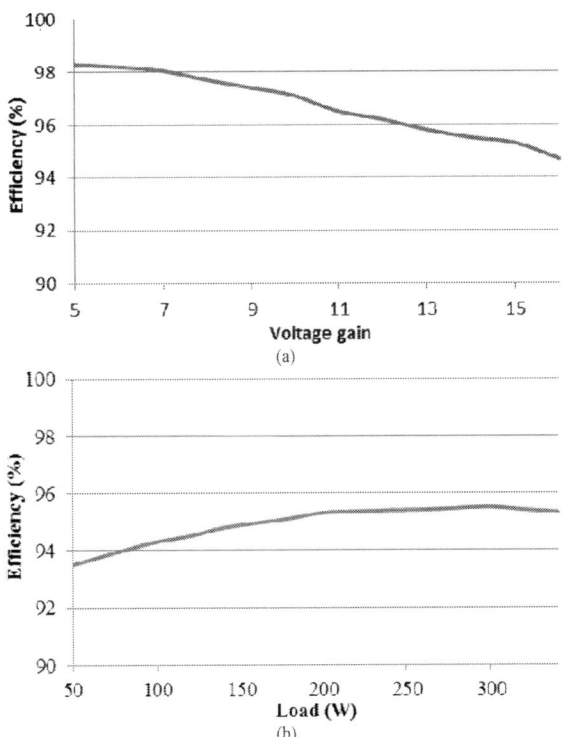

Fig. 7. Measured efficiency of the proposed converter at (a) different voltage gain (V_{dc} varies from 80 V – 24 V for $V_O = 400$ V) and (b) different load power (W) and by keeping constant voltage gain of 14.2.

CONCLUSION

A high step-up single–switch dc-dc converter with high voltage gain and reduced switch voltage stress has been proposed in this paper using an integrated autotransformer and coupled inductor. The leakage energy is recycled to the load and the voltage spike across the switch is alleviated, which helps to choose a low voltage and low R_{DS-on} switch to improve the efficiency. Simulation and experimental results have verified the principle of operation and confirmed the presented theoretical analysis. A full load efficiency of 95.5% at a gain of 14.2 was achieved. Further, the proposed converter stands out to be a competitive alternative for a practical application where a high voltage gain is demanded, such as for a fuel cell and high voltage Light Emitting Diode (LED) lamps.

REFERENCES

[1] F. Blaabjerg, R. Teodorescu, Z. Chen, and M. Liserre, "Power converters and control of renewable energy systems," in *Proc. ICPE 2004*, Pusan, Korea, Oct. 2004.

[2] F. Blaabjerg, Z. Chen and S. B. Kjaer, "Power Electronics as Efficient Interface in Dispersed Power Generation Systems," *IEEE Trans. Power Electron.*, vol. 19, no. 5, pp. 1184-1194, Sep. 2004.

[3] Q. Zhao and F. C. Lee, "High-Efficiency, High Step-Up DC-DC Converters," *IEEE Trans. Power. Electron.*, vol. 18, no. 1, pp. 65-73, Jan. 2003.

[4] A. Tomaszuk and A. Krupa, "High Efficiency High Step.Up DC-DC Converters-A Review," *Bulletin of the Polish Academy of Sciences*, vol. 59, no. 4, 2011.

[5] W. Li and X. He, "Review of Nonisolated High-Step-Up DC-DC Converters in Photovoltaic Grid-Connected Applications," *IEEE Trans. Ind. Electron.*, vol. 58, no. 4, pp. 1239-1250, Apr. 2011.

[6] J. Dawidziuk, "Review and Comparison of High Efficiency High Power Boost DC-DC Converters for Photovoltaic Applications," *Bulletin of the Polish Academy of Sciences*, vol. 59, no. 4, 2011.

[7] R. J. Wai, C. Y. Lin, R. Y. Duan, and Y. R. Chang, "High-Efficiency DC-DC Converter With High Voltage Gain and Reduced Switch Stress," *IEEE Trans. Ind. Electron.*, vol. 54, no. 1, pp. 354-364, Feb. 2007.

[8] K. C. Tseng, J. T. Lin, and C. C. Huang, "High Step-Up Converter With Three-Winding Coupled Inductor for Fuel Cell Energy Source Applications," *IEEE Trans. Power Electron.*, vol. 30, no. 2, pp. 574-581, Feb. 2015.

[9] A. A. Fardoun, and E. H. Ismail, "Ultra Step-Up DC-DC Converter With Reduced Switch Stress," *IEEE Trans. Ind. Electron.*, vol. 46, no. 5, pp. 2025-2034, Sep./Oct. 2010.

[10] Q. Zhao, "Performance Improvement of Power Conversion by Utilizing Coupled Inductors," A PhD thesis from Virginia Polytechnic Institute and State University, Blacksburg, Virginia, USA, Feb. 2003.

[11] Y. P. Siwakoti, F. Z. Peng, F. Blaabjerg, P. C. Loh and G. E. Town, "Impedance Source Network for Electric Power Conversion — Part I: A Topological Review" *IEEE Trans. on Power Electron.*, vol. 30, no. 2, pp. 699-716, Feb. 2015.

Optimal Dynamic Phase Add/Drop Mechanism in Multiphase DC-DC Buck Converters

Anandha Ruban T T[†], Preetam Tadeparthy[†], Sankaran Aniruddhan[*]
Vikram Gakhar[†] and Muthusubramanian Venkateswaran[†]
[*]Dept. of Electrical Engineering, Indian Institute of Technology Madras, Chennai, India
Email: ani@ee.iitm.ac.in
[†]Texas Instruments (India) Pvt. Ltd., Bangalore, India
Email: {anandtt, preetam, vgakhar, muthusub}@ti.com

Abstract—**In multiphase buck converters, phases are dynamically added or dropped based on the output load, to improve efficiency. This is typically achieved by programming a lookup table containing the range of threshold currents for each phase, independent of input voltage and switching frequency. In this work, an algorithm is proposed to recalibrate the threshold currents for operation at optimum number of phases across changes in input voltage and switching frequency. To evaluate the algorithm, a 6-phase synchronous buck converter is tested in forced CCM (Continuous Conduction Mode) over several input voltages and switching frequencies. A maximum efficiency improvement of 1.6 % is demonstrated with the proposed algorithm, when compared with the conventional scheme.**

I. INTRODUCTION

A multiphase buck converter may be viewed as a combination of multiple units of single-phase buck converters connected in parallel. The total load current is shared among the various phases, resulting in a reduction of resistive losses and increase in switching losses. Operating at higher number of phases is efficient for heavy loads, while operating at lower number of phases is efficient for light loads. Therefore, for a given load current, there exists an optimum number of phases at which the converter should be operated.

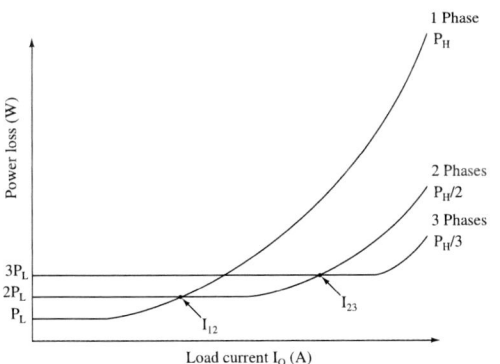

Fig. 1. Representative model of power loss curves.

Fig. 1 shows a representative model of power loss curves for various number of phases. The point of intersection of adjacent curves indicates the optimum threshold current for switching between phases. In general, these threshold current

values depend on the specific operating settings of the converter. Use of a fixed set of threshold currents will result in sub-optimal operation. Some techniques have been previously proposed to generate a look-up table of these threshold currents [1]–[6]. However, none of these are able to adapt the table values without manual intervention.

In this work, the analysis for power loss of an N-phase converter in [2] is extended to enable automatic adaptation of the threshold current lookup table. Based on this analysis, an algorithm is proposed to operate the converter optimally across input voltage and switching frequency changes. Finally, the proposed algorithm is tested on a 6-phase buck converter with different combinations of V_{IN} (10 V, 16 V) and F_S (400 kHz, 600 kHz).

II. ANALYSIS OF POWER LOSS

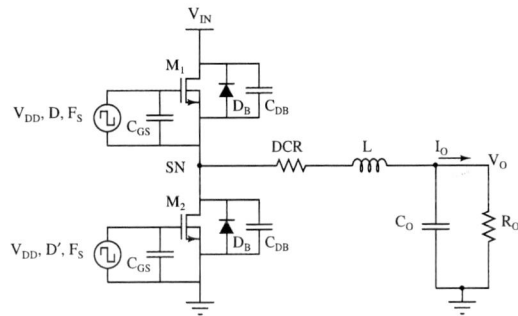

Fig. 2. Practical buck converter.

Fig. 2 shows the schematic of a practical buck converter with all the lossy elements in place. There are basically two types of losses.

1) Resistive losses: Losses due to ON-resistance of MOSFETs and DC resistance of Inductors (represented by "P_{RL}").

2) Switching losses: Losses due to body diode dead-time ("P_{DT}"), drain-source voltage and current overlap ("P_{OL}"), reverse recovery ("P_{RR}"), gate drive ("P_{GD}") and output capacitance ("P_{OC}"). Switching losses are frequency dependent losses. The total power loss in an N-phase buck converter can therefore be written as [2]

$$P(I_O, N) = P_{RL} + P_{DT} + P_{OL} + P_{RR} + P_{GD} + P_{OC} \quad (1)$$

978-1-4673-9551-9/16 $31.00 © 2016 IEEE

To show dependencies on input voltage, switching frequency and number of phases, this can be expressed in the following way

$$P(I_O, N) = \left(\frac{I_O{}^2}{N} + N\frac{\Delta I^2}{12}\right)(DR_{DS}(M_1) + D'R_{DS}(M_2)$$
$$+ DCR) + I_O\left(V_F T_{DT} + \frac{1}{2}(V_{IN} + V_F)(T_R + T_F)\right)F_S$$
$$+ N(V_{IN}Q_{RR} + V_{DD}Q_G(M_1, M_2)$$
$$+ \frac{1}{2}V_{IN}{}^2 C_{DB}(M_1, M_2))F_S \quad (2)$$

Eqn. 2 can be rewritten as

$$P(I_O, N) = \frac{I_O{}^2}{N}k_2 + I_O k_1 + N k_0$$
$$= N\left(\left(\frac{I_O}{N}\right)^2 k_2 + \frac{I_O}{N}k_1 + k_0\right) \quad (3)$$

The threshold current to switch between N and $N + 1$ phases, denoted by $I_{N,N+1}$, may therefore be obtained by equating the power loss of these two phases.

$$P(I_O, N) = P(I_O, N + 1) \quad (4)$$

Thus, the threshold current is given by

$$I_{N,N+1} = \sqrt{N(N + 1)}\sqrt{\frac{k_0}{k_2}} \quad (5)$$

Eqn. 5 demonstrates that the threshold currents depend on k_0 and k_2, which are direct functions of V_{IN} and F_S. Therefore, the contents of the threshold current lookup table need to adapt to changes in these two quantities. Techniques described previously in [2], [4]–[6] use the relation in Eqn. 5. However, it is very difficult to precisely determine the loss parameters in Eqn. 2 individually. Hence, Eqn. 5 cannot be directly used to recalibrate the thresholds.

In this paper, the power loss for N phases is expressed in terms of the power loss for a single phase, to automatically determine the threshold currents. The power loss of a multiphase converter operating with N phases is N times the power loss of the converter when operating with a single phase. Each phase of the multiphase converter supplies a load current of $\frac{I_O}{N}$. Consequently, the relationship between the power loss with N active phases and that with a single active phase may be expressed in the following way

$$P(I_O, N) = NP\left(\frac{I_O}{N}, 1\right) \quad (6)$$

Unlike Eqn. 5, it is possible to use the relation between the power loss of multiple phases and a single phase directly, given by Eqn. 6, to determine optimal thresholds. Since all phases are nominally identical, the loss parameters will change for all phases uniformly.

In order to verify the accuracy of this relationship, the loss of a practical 6-phase buck converter with $150\,nH$ per-phase inductance is recorded with $2\,A$ steps in the load current, for each number of phases individually. The power loss curves of all phases other than the single-phase one are remapped onto an equivalent single-phase power loss curve using Eqn. 6, and plotted in Fig. 3 for comparison.

As expected, Fig. 3 shows that all remapped power loss curves overlap to a large extent, with some small differences.

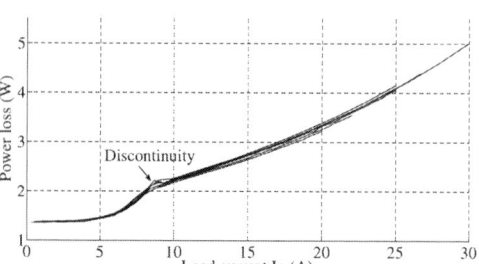

Fig. 3. Power loss curves at $V_{IN} : 16\,V$, $F_S : 600\,kHz$ and $V_O : 1.7\,V$.

To better understand the differences between the curves, a second-order polynomial fit was applied to all curves. While the coefficient of the square term k_2 and the constant term k_0 remain the same, the coefficient of the linear term k_1 differs for all curves. From Eqn. 2 it is clear that k_1 is a function of the dead time, and rise and fall times, which are expected to vary. Eqn. 5 shows that the threshold currents are independent of k_1. Hence, minor variations in k_1 will not change the threshold currents significantly and it is still possible to predict the threshold currents accurately.

It can be seen from Fig. 3 that there is a discontinuity in the slope of the power loss curve at $I_O \approx \frac{\Delta I}{2}$. In forced CCM mode, when I_O is less than $\frac{\Delta I}{2}$ and M_2 is ON, the inductor current will ramp down to a negative value. During dead time, this negative current flows into the body diode of M_1 instead of M_2. When M_1 turns ON, the current switches from its body diode to its own channel. Hence, there is no reverse recovery loss. Overlap losses are still present, which however depend only on the fall times of M_1 and M_2, instead of both rise and fall times of M_1. In practice, the fall time is much smaller than the rise time, leading to reduced overlap losses.

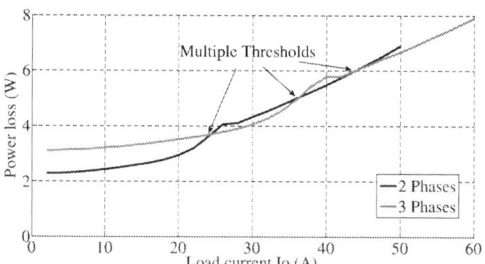

Fig. 4. Power loss curves of adjacent phases at $V_{IN} : 16\,V$, $F_S : 400\,kHz$.

Since the power loss mechanisms are different for I_O less than and greater than $\frac{\Delta I}{2}$, two different functions define the power loss for single-phase. In DCM mode, this discontinuity will be steeper as the reduction in loss is higher, when I_O is less than $\frac{\Delta I}{2}$. It is, in fact, possible to have more than one intersection point, leading to multiple thresholds. This is demonstrated in Fig. 4, where the power loss curves of the converter with 2 and 3 active phases intersect at 3 points. Techniques described in [1]–[6] do not consider the possibility of multiple thresholds, but this is covered in this work by the proposed algorithm described in the next section.

III. PROPOSED ALGORITHM

From experimental data, it is clear that the relation between a single phase and multiple phases given by Eqn. 6 is still valid and the power loss curve for N phases can be derived using the single-phase power loss curve. Therefore, the power loss curve for single-phase operation needs to be recorded first. It is also necessary to measure and correct any other losses such as PCB losses, which are independent of N and I_O. As an example, the constant loss for a 6-phase converter is shown in Fig. 5.

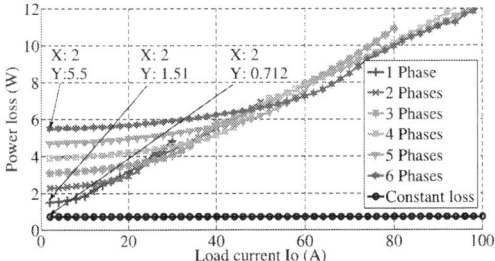

Fig. 5. Constant loss at $V_{IN} : 16\,V$, $F_S : 400\,kHz$ and $V_O : 1.7\,V$.

At light loads the loss that depends on I_O is eliminated. Once the average addition of loss per-phase at light load is determined, the loss that depends on N can also be identified. Therefore, the constant loss for a 6-phase converter may be written as

$$C = P(0,1) - \frac{(P(0,6) - P(0,1))}{5} \quad (7)$$

The inductor ripple current may be obtained from the following relation

$$\Delta I = \frac{DD'V_{IN}}{LF_S} \quad (8)$$

From Fig. 3, it is clear that there is a discontinuity at $\frac{\Delta I}{2}$. A cubic fit is required to represent the region of the curve where I_O is less than $\frac{\Delta I}{2}$, while a square fit is sufficient to represent the rest of the curve. Since the power loss curves of higher phases are derived from that of the single-phase, the fitting function needs to be accurate and sufficient number of samples are needed for best fit. For the 6-phase converter, four samples are sufficient for each region. Thus, the range of load currents is divided into two regions and four sub-regions as shown in Fig. 6. The threshold current to switch from 1 to 2 phases is then increased to the maximum per-phase current limit (30A for the 6-phase converter in this work). This is done to force the converter to operate in single-phase mode for the full range of per-phase load currents. Since the load current value cannot be forced during real operating conditions, the power loss is recorded whenever I_O falls into one of these regions. Once the samples from all the regions are obtained, cubic and square polynomial functions are used to fit the regions where I_O is less than and greater than $\frac{\Delta I}{2}$ respectively. The constant loss calculated at no load (Eqn. 7) is then subtracted from the polynomial function obtained after curve-fitting, resulting in the single-phase power loss curve. The power loss curves of

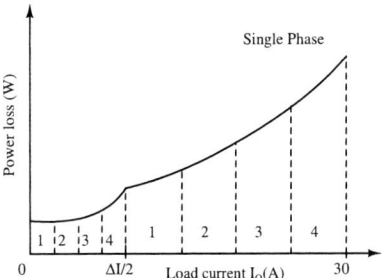

Fig. 6. Tracing of single-phase power loss curve.

other phases are derived from the relation expressed in Eqn. 6, and are used to predict the optimum number of phases for a given load current. This procedure is repeated whenever the input voltage and switching frequency settings are changed to recalibrate the threshold current table. At light loads, the switching activities may be distributed across all phases to reduce the ageing mismatches among them [7].

IV. EXPERIMENTAL RESULTS

In order to test the algorithm, a 6-phase buck converter is tested at different settings of input voltage and switching frequency. Fig. 7 shows the setup used to test the algorithm. All measuring devices are connected to a computer through a General Purpose Interface Bus (GPIB) interface. The 6-phase controller, which controls the number of active phases, is connected to the computer through a Power Management Bus (PMB).

The ripple amplitude and constant loss are first determined for each setting. Next, a single phase is activated and power loss is recorded as $(V_{IN}I_{IN} + V_{DD}I_{DD} - V_OI_O)$ for all the 8 load current values in each region. Finally, the power loss curves for other phases are derived from the measurement for a single phase and a table of optimum number of phases for all current values is generated. The efficiency curve is recorded for the entire range of load currents while choosing the phases based on the lookup table generated above. The efficiency curves for a fixed lookup table with pre-programmed thresholds are also measured. Both these curves are compared to study the improvement in efficiency.

Fig. 8 shows a comparison between two efficiency curves. The first is when the converter is operated at $V_{IN} : 16\,V$, $F_S : 600\,kHz$, while the threshold currents are pre-programmed for $V_{IN} : 10\,V$ and $F_S : 400\,kHz$. The second curve is generated based on the proposed algorithm based on adapted threshold currents. A peak efficiency improvement of $1.6\,\%$ is observed at a load current of $18\,A$.

An efficiency comparison is performed for different settings of V_{IN} and F_S. The difference in percentage efficiency between the proposed algorithm and inbuilt fixed threshold current lookup table for each setting is shown in Fig. 9. It is evident that as the V_{IN} and F_S settings deviate from the pre-programmed fixed threshold current lookup table (measured for $V_{IN} : 10\,V$ and $F_S : 400\,kHz$ in this case), the gain in efficiency is higher. Fig. 9(a) shows very little difference

978-1-4673-9551-9/16 $31.00 © 2016 IEEE

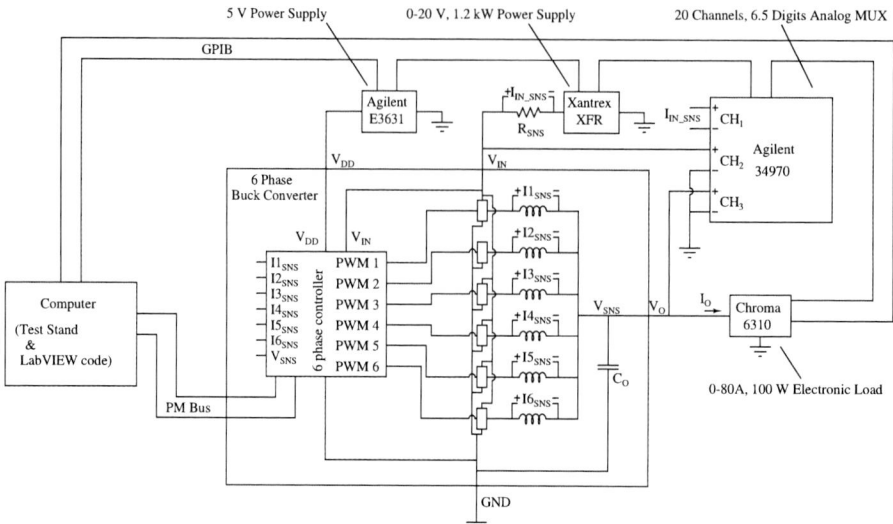

Fig. 7. Measurement Setup.

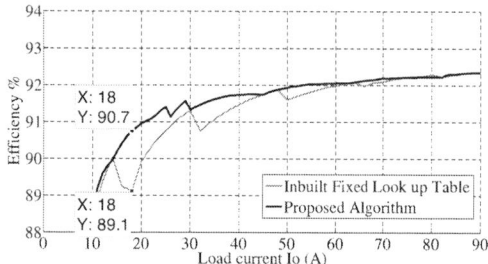

Fig. 8. Efficiency at $V_{IN} : 16\,V$, $F_S : 600\,kHz$ and $V_O : 1.7\,V$.

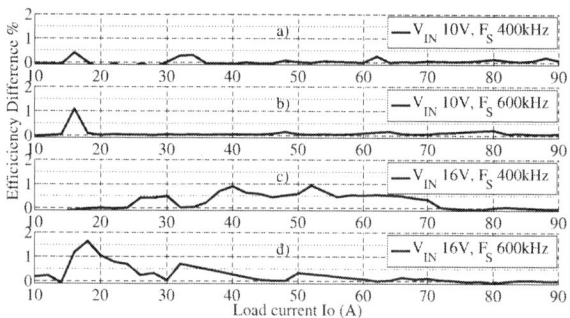

Fig. 9. Efficiency difference at different settings but with same $V_O : 1.7\,V$.

in efficiency between the proposed algorithm and the fixed lookup table. However, a considerable difference in efficiency is observed when the input voltage and switching frequency are changed (Fig. 9(c) and 9(d) respectively). The converter may therefore be operated more optimally when using the proposed algorithm.

V. CONCLUSION

The optimum threshold currents to switch phases do not remain the same when the input voltage and switching frequency are changed. There is a need to recalibrate the system to optimize the number of phases at a given load current. An algorithm based on the relationship between single-phase and multiphase power loss curves is proposed, and tested on a 6-phase converter. Peak efficiency improvements of $1.6\,\%(\approx 0.5\,W)$ at $18\,A$ (Fig. 9(d)) and $1\,\%(\approx 0.9\,W)$ at $52\,A$ (Fig. 9(c)) are observed by employing the proposed algorithm.

REFERENCES

[1] X. Huijie, B. Jinying, W. Chunsheng, and X. Honghua, "Design and implementation of a pv dc/dc converter with high efficiency at low output power," in *Power System Technology (POWERCON), 2010 International Conference on*, Oct 2010, pp. 1–6.

[2] W. Qiu, C. Cheung, S. Xiao, and G. Miller, "Power loss analyses for dynamic phase number control in multiphase voltage regulators," in *Applied Power Electronics Conference and Exposition, 2009. APEC 2009. Twenty-Fourth Annual IEEE*, Feb 2009, pp. 102–108.

[3] P. Zumel, C. Fernnndez, A. de Castro, and O. Garcia, "Efficiency improvement in multiphase converter by changing dynamically the number of phases," in *Power Electronics Specialists Conference, 2006. PESC '06. 37th IEEE*, June 2006, pp. 1–6.

[4] P. Bartal and I. Nagy, "Game theoretic approach for achieving optimum overall efficiency in dc/dc converters," *Industrial Electronics, IEEE Transactions on*, vol. 61, no. 7, pp. 3202–3209, July 2014.

[5] S. Waffler and J. Kolar, "Efficiency optimization of an automotive multi-phase bi-directional dc-dc converter," in *Power Electronics and Motion Control Conference, 2009. IPEMC '09. IEEE 6th International*, May 2009, pp. 566–572.

[6] J.-T. Su and C.-W. Liu, "A novel phase-shedding control scheme for improved light load efficiency of multiphase interleaved dc-dc converters," *Power Electronics, IEEE Transactions on*, vol. 28, no. 10, pp. 4742–4752, Oct 2013.

[7] Y. Ahn, I. Jeon, and J. Roh, "A multiphase buck converter with a rotating phase-shedding scheme for efficient light-load control," *Solid-State Circuits, IEEE Journal of*, vol. 49, no. 11, pp. 2673–2683, Nov 2014.

A Universal Self-Calibrating Dynamic Voltage and Frequency Scaling (DVFS) Scheme with Thermal Compensation for Energy Savings in FPGAs

Shuze Zhao[1], Ibrahim Ahmed[1], Carl Lamoureux[1], Ashraf Lotfi[2], Vaughn Betz[1] and Olivier Trescases[1]

[1]University of Toronto, Toronto, ON, Canada

10 King's College Road, Toronto, ON, M5S 3G4, Canada

[2]Altera Corp., Hampton, NJ, 08827, USA

Email: szhao@ece.utoronto.ca

Abstract—**Field Programmable Gate Arrays (FPGAs) are widely used in telecom, medical, military and cloud computing applications. Unlike in microprocessors, the routing and critical path delay of FPGAs is user dependent. The design tool suggests a maximum operating frequency based on the worst-case timing analysis of the critical paths at a fixed nominal voltage, which usually means there is significant voltage or frequency margin in a typical chip. This paper presents a universal offline self-calibration scheme, which automatically finds the FPGA frequency and core voltage operating limit at different self-imposed temperatures by monitoring design-specific critical paths. These operating points are stored in a calibration table and used to dynamically adjust the frequency and core voltage according to the FPGA temperature when the application circuit is running. The self-calibration process is demonstrated on an Altera Cyclone IV 65-nm FPGA with a digitally controlled dc-dc converter, leading to 40% power savings in a typical digital filter application.**

I. INTRODUCTION

Field Programmable Gate Arrays (FPGAs) can outperform microprocessors and Digital Signal Processors (DSPs) in many applications, thanks to their ability to implement massively parallel algorithms [1]–[3]. Since FPGAs can be reprogrammed to accommodate evolving standards, they eliminate the custom manufacturing and resulting high Non-Recurring Engineering (NRE) costs and development time of Application-Specific Digital ICs (ASICs). Thus FPGAs are widely used in telecom, medical, military and cloud computing applications. However, the flexibility of FPGAs comes at a significant cost; they typically consume ten times the dynamic power of an ASIC performing the same task [4], making power reduction techniques crucial for FPGAs. Dynamic Voltage and Frequency Scaling (DVFS) has been widely deployed in microprocessor applications over the past decade [5]–[11]. The fact that an FPGA can be programmed to perform any digital function gives rise to some unique challenges in designing a DVFS control system. Unlike microprocessors, the speed-limiting paths of a specific FPGA IC are unknown at manufacturing time; hence mimicking the critical path and setting the minimum core voltage for the DVFS control system is a major challenge.

Currently, FPGA designers operate each IC at its rated nominal voltage, and must choose a clock frequency at or below the limit predicted by the Computer-Aided Design (CAD) tool's timing analysis. This timing analysis is extremely conservative, using worst-case models for process corners, on-chip voltage drop, temperature and aging. In the vast majority of chips and systems, however, the supply voltage can be reduced significantly below nominal in order to obtain energy savings. Operating the IC at a lower voltage also reduces the impact of aging effects such as Bias-Threshold Instability (BTI), and improves the chip lifetime [12], [13].

A. Prior Work

In [14], a DVS scheme with a Complex Programmable Logic Device (CPLD) load is presented. A CPLD is similar to an FPGA but has a non-volatile configuration memory. This scheme does not monitor the logic error due to the timing violation. In [15], a Logic Delay Measurement Circuit (LDMC) is used to determine the voltage at which the application circuit has a timing failure, and adjusts the supply voltage accordingly. It assumes that the critical path in the application circuit can be exercised by randomly generated inputs during calibration, which is not valid in modern FPGA applications. Approaches in [14] and [15] also rely on a non-valid assumption that the VCO/LDMC delay value perfectly tracks the delay variation in the application circuit critical paths with temperature and aging.

In [16]–[18], online timing slack measurement is achieved by using a phase-shifted clock and one shadow register for each critical path to determine timing headroom in a circuit during operation. This approach has several notable shortcomings:

1) the timing slack measurement is dependent on the input data, which cannot be controlled during normal operation,
2) the technique is limited to FPGA components where a second capture register can be added at the end of a critical path, which is not feasible for important 'hard' blocks such as the on-chip RAM,

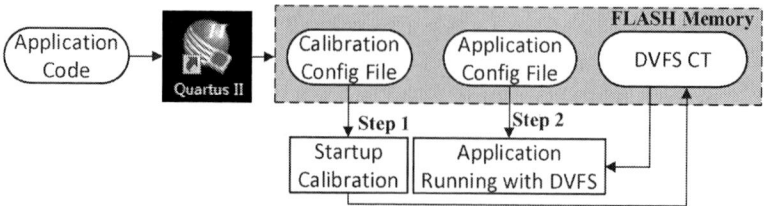

Fig. 1. Two step DVFS scheme.

3) the scheme requires extra logic elements (LEs) and clock resources, increasing circuit power and reducing the usable capacity of the FPGA.

In addition, the past works [15]–[18] do not employ a high-frequency digital dc-dc converter to generate the variable core supply voltage. Many important practical issues, such as the converter response time and quantization issues, are therefore ignored.

In this work, a new offline universal self-calibration scheme that requires close interaction with the digital dc-dc converter is proposed to automatically characterize the exact relationship between the maximum operating frequency for each core voltage and temperature corner. This information is saved in a calibration table that is used during normal operation for DVFS.

II. Self-Calibration Concept with Two-Step Configuration

The proposed universal self-calibration process is intended to run on a system production line, or regularly during each power-up sequence of a system. It is therefore important that this process (1) be reasonably fast and (2) require the minimum possible FPGA resource overhead. The self-calibration method has three steps and requires the FPGA to be programmed twice, as shown in Fig. 1:

1) The user's design is automatically analyzed by the augmented CAD tool to extract the logic paths having the most critical timing. A design-specific self-calibration configuration file is then created. The critical paths used in self-calibration are *exact* replicas of the critical paths in the application; they are placed and routed using the identical resources (routing wires, LEs, etc.). All inputs along the critical path are set to non-controlling values to guarantee that the path is synthesized. These non-controlling values are selected to mimic the worst-case rising and falling pattern reported by the tool. The CAD automation ensures that no additional designer effort is required.

2) The FPGA is programmed once with the self-calibration configuration file, as shown in Fig. 3(a). The on-chip configuration contains:

- the design-specific critical paths with error checking circuit,
- flip-flop chain based logic blocks configured as programmable heaters for temperature control,
- a temperature sensing circuit,
- a frequency synthesizer,

- a digital dc-dc controller,
- a calibration controller.

Each of the critical paths is exercised by toggling the source register and checking that the sink register captures the correct value. And for each critical path, a fast path (a buffer or an inverter) that behaves as the critical path is synthesized to identify what the correct value is. The output of this fast path and the critical path is compared together. The fast path is designed such that it does not fail the timing at the maximum applied frequency.

Heater cells are distributed across the entire chip, and are used to create different die temperature conditions. The FPGA proceeds to run the calibration, using self-heating and automatic timing error checking, and populates the DVFS calibration table (CT). An ideal calibration table is demonstrated in Fig. 2. The detailed calibration scheme is described in the following session.

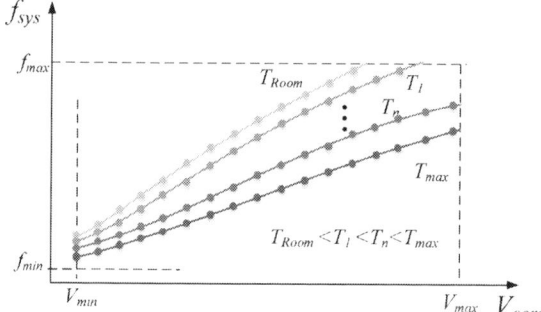

Fig. 2. An ideal CT as the result of the self-calibration process.

3) Finally, when the self-calibration is complete, the FPGA is automatically programmed a second time with the user's regular configuration file, as well as the DVFS control system which relies on the extracted calibration table, as shown in Fig. 3(b). Based on the clock frequency requirements and chip temperature, the DVFS control core refers to the calibration table and set the according core voltage, V_{core}.

III. System Level Architecture

The system architecture is shown in Fig. 4 and includes a 65-nm CMOS Cyclone IV FPGA (EP4CE115F29C7N). The two-phase Buck converter has an input voltage of 5 V and regulates the FPGA core voltage, V_{core}, between 0.85 - 1.35 V. The main phase, which delivers the majority of

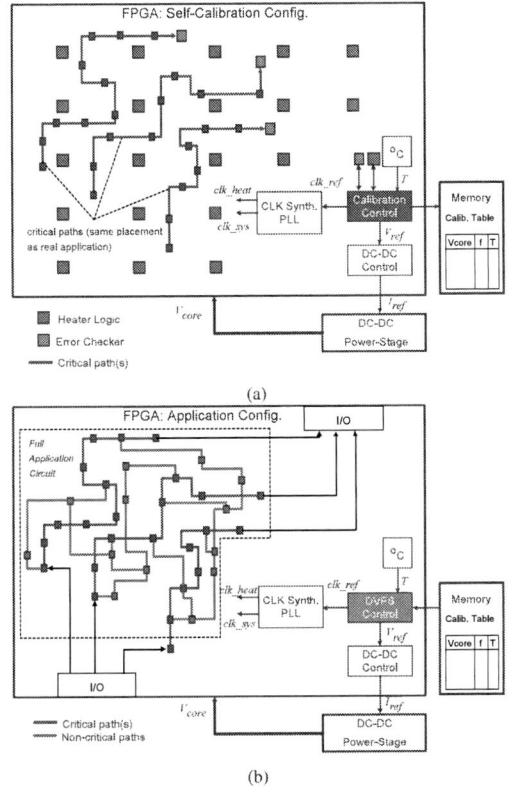

(a)

(b)

Fig. 3. FPGA with (a) self-calibration configuration and (b) application configuration for DVFS system.

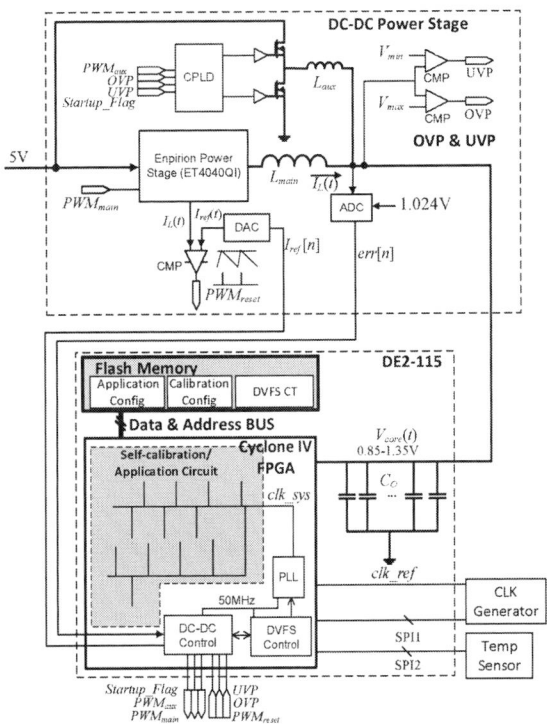

Fig. 4. Detailed system-level implementation.

the FPGA power, is implemented using an Enpirion power module, ET4040QI. The main phase is rated at 10 W and operates in digital peak current mode control, where the peak current command, $I_{ref}[n]$, is generated within the FPGA and is converted to an analog reference, $I_{ref}(t)$, using a high-speed DAC. The outer voltage loop is also implemented on the FPGA, based on the sampled voltage error signal, $err[n]$. The controller is carefully optimized to operate down to the minimum FPGA core voltage, V_{min}. While the latest-generation Altera FPGAs include both on-chip temperature and core voltage sensing, these are implemented off-chip in this initial phase of the project.

The auxiliary phase, which has a lower power rating of 3 W, is controlled by a non-volatile CPLD to assist with the startup process when the main-phase controller in the FPGA is not powered. The auxiliary phase can also be used to improve the dynamic response, similar to [19]. The CPLD can be removed in future implementations, where the startup control can be integrated into the ET4040QI for example.

In an ideal application scenario, the self-calibration/application configurations and the CT would be stored in the on-board non-volatile memory. To simplify the process, the FPGA is manually programmed with different configurations.

The fully automated calibration process is shown in Fig. 5 and can be explained as follows. The heater blocks are first enabled to cause the die temperature to ramp up. Each heater cell is programmable and consists of $N_{heater} = 8$ chains with 88 flip-flops per chain switching at 100 MHz. With heater cells enabled and $V_{core} = 1.2$ V, the FPGA package reaches 85 °C. At every integer temperature value, CT entries are obtained and stored in the on-board Flash Memory. During each sweep, the dc-dc controller drops V_{core} to $V_{min} = 0.832$ V and starts to increases the clock frequency, f_{sys}, from the lowest operating frequency. The increasing clock frequency, f_{sys}, is applied to the critical paths until a logic error is detected ($err\_flag = 1$, when $f_{sys} = f_{max}$) by the error-checking blocks. Once an error is detected, V_{core} is increased by $\Delta V = 16$ mV until $V_{core} = V_{max} = 1.328$ V at the end of the sweep. Since a higher voltage always allows for a higher frequency, the frequency range only needs to be swept once with this method. In between sweeps, V_{core} is set to 1.2 V.

The duration of the full calibration process is limited by the system's thermal time response, which is considerably longer than the dc-dc converter dynamics. For each temperature value, one sweep of frequency and voltage takes less than 100 ms, while the entire temperature sweep takes approximately 2 minutes. The calibration time can be greatly reduced by optimizing the heater design, the voltage range and other calibration parameters depending on application needs.

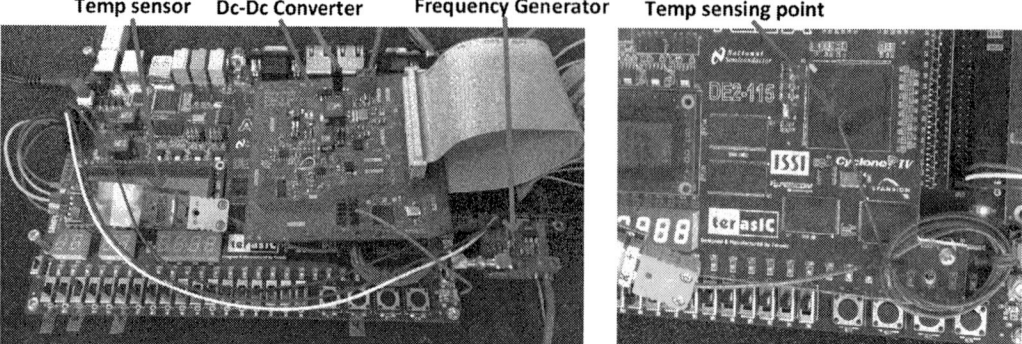

Fig. 6. Experimental setup.

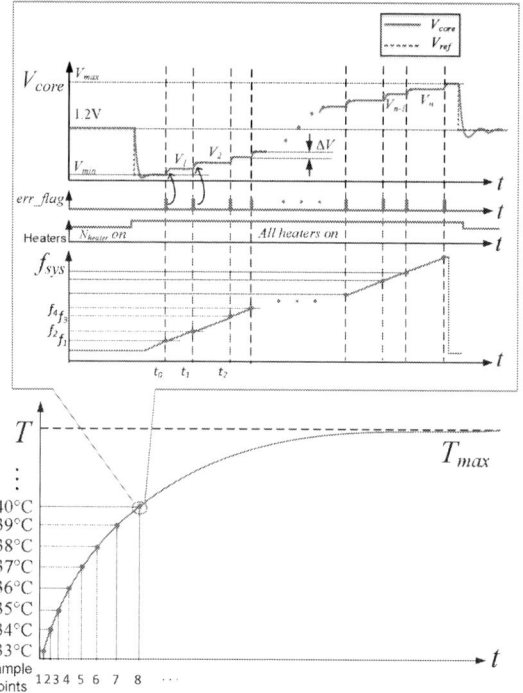

Fig. 5. Ideal waveform for the entire self-calibration process and detailed waveform at each temperature point.

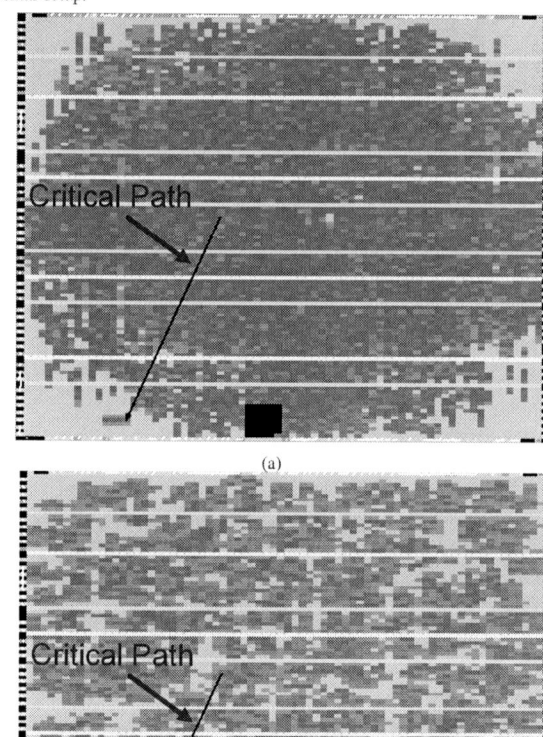

Fig. 7. The FPGA Chip Planner view of (a) self-calibration configuration and (b) application configuration.

IV. Experimental Results

The automated self-calibration process was demonstrated using a common application, a digital FIR filter design. The package temperature ranges from 30 °C - 85 °C for V_{core} from 0.832 V - 1.328 V. Fig. 6 shows the experiment setup. The power stage supplying V_{core} on the DE2-115 is disconnected, and the customized dc-dc converter is mounted on top of the FPGA, while its output is connected to the decoupling capacitors on the DE2-115 board through vias with a short path to supply V_{core}. The frequency generator feeds in the input clock, clk_ref, through an SMA connector with the frequency of $f_{sys}/4$. The on-chip PLL boosts the frequency for 4 times to f_{sys} as the system clock, clk_sys. The testing point of the thermocouple is fixed on the FPGA package, as shown in Fig. 6.

Altera Quartus II Chip Planner provides the visualized FPGA on-chip configuration and shows exactly which logic elements are used by the circuit. The on-chip configuration of the self-calibration and the application are shown in Fig. 7(a) and Fig. 7(b), respectively. Each blue box represents a Logic

Array Block (LAB) consisting of a number of logic elements. The darkness of the LAB represents the relative number of LEs used in the LAB. The LABs comprising the application's most critical path are shown in red. The black arrow connecting the two red areas represents part of the critical path routing. This arrow is only a representation of the connection and does not reflect the actual routing. As shown in Fig. 7, the two on-chip configurations are significantly different, however the critical path resources are identical, as in Fig. 3. Only the most critical path is monitored in this experiment to verify the concept, but in a real application, there might be multiple critical pathes which have a similar delay. All these near critical pathes should be monitored to ensure the safe operation of the device.

The entire calibration process is shown in Fig. 8(a): each voltage spike (as noted by "$\star$") corresponds to one full sweep of f_{max} versus V_{core} at the given temperature. One such sweep is shown in Fig. 8(b), which reveals the converter dynamics.

During heating, V_{core} is held at 1.2 V and then ramped down slowly to 832 mV when the target temperature is reached. All the heater circuits are turned on at this point to simulate the worst-case internal voltage drop in the application. f_{sys} is increased until the failing indicator, err_flag, goes high at which time the frequency is stored with the corresponding core voltage V_{core} in the CT, $V_{ref}[n]$ is then increased and the process repeats.

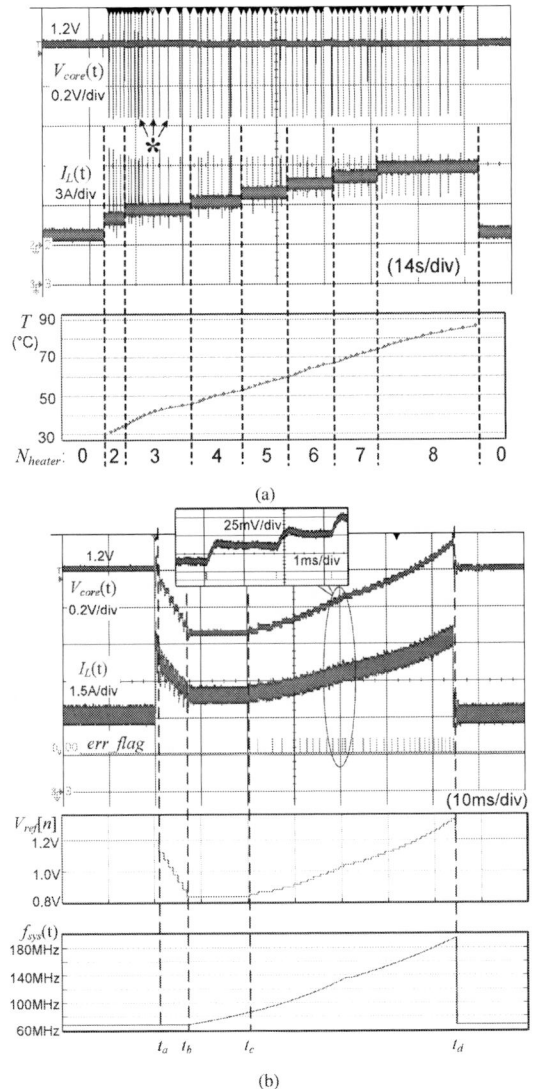

Fig. 8. (a) An entire self-calibration process from 30°C to 85°C. N_{heater} is dynamically controlled to achieve an approximately linear rise in temperature, T. (b) One sweep of frequency, f_{sys}, and core voltage, V_{core}, at 45°C.

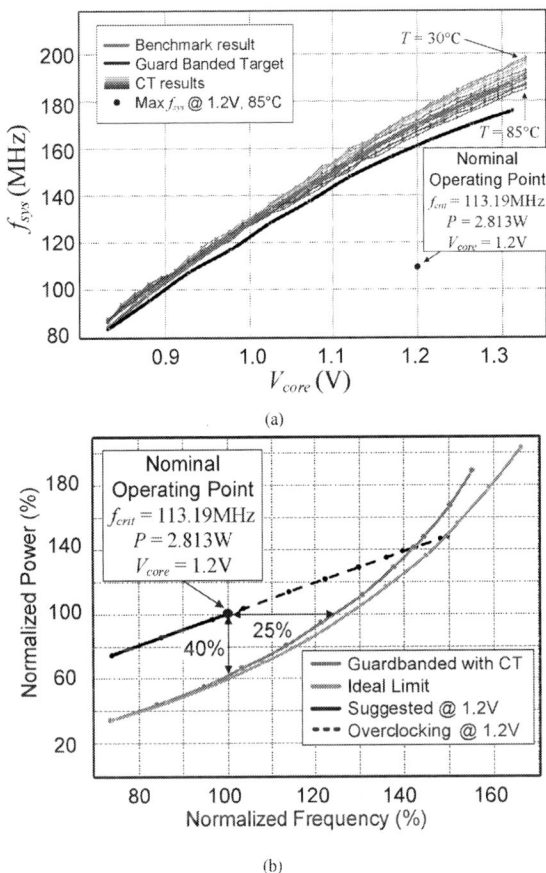

(a)

(b)

Fig. 9: (a) Extracted calibration data and guardband target. (b) Power saving comparison between the proposed method and ideal limit.

The stored CT data of f_{max} versus V_{core} versus T is plotted in Fig. 9(a), with one curve per 5 °C temperature increment. The effect of temperature is more noticeable at higher V_{core}. For example, at V_{core}= 1.3 V, f_{max} drops by 6.25% over a temperature range of ΔT = 55 °C. In order to check the accuracy of the CT table data, which is generated from the calibration configuration (ie: Fig. 3(a)), the maximum clock frequency of the full FIR application (ie:

978-1-4673-9551-9/16 $31.00 © 2016 IEEE

in Fig. 3(b)) was independently checked using an exhaustive random data generator and error checking. The result is shown as the Benchmark result in Fig. 9(a) and matches very well with the CT results. The thick black curve is a conservative 5%-guardbanded operating target, based on the CT data for the purpose of power consumption comparison. The measured power consumption of the benchmark is shown in Fig. 9(b), with constant voltage (black line) and DVFS based on the guard-banded CT-table (red line). The frequency axis is normalized to the maximum value specified by the timing analysis of the CAD tool, f_{crit} (ie: the best available data for designers currently). The green line corresponds to the minimum possible DVFS operation power without guard-band.

Several key points can be drawn from the data: (1) even with V_{core} fixed at 1.2 V, the benchmark circuit can operate up to 50% above f_{crit} (dashed line). This shows that the CAD tool timing is necessarily conservative as expected, since it must account for worst-case temperature and process variations; (2) using DVFS enables 40% power savings at f_{crit} in this application; (3) for the same power consumption, DVFS enables a 25% increase in the clock frequency.

V. Conclusion

While DVFS is highly successful in microprocessors, it remains elusive in FPGAs mainly due to the fundamental challenge of a user-dependent critical paths. The proposed self-calibration technique can be universally applied to any user design, has a very low resource overhead and guarantees no logic errors during operation. The technique allows FPGA designers to safely operate each FPGA at its optimal performance point, reaching power savings on the order of 40%. This procedure is fast enough to be applied automatically at board burn-in/test time, or possibly even at each board power-up.

VI. Acknowledgement

This work was supported by Altera Corporation, the Ontario Centres of Excellence, the Natural Sciences and Engineering Research Council of Canada, the Canadian Foundation for Innovation and the Ontario Research Fund.

References

[1] B. Sukhwani, H. Min, M. Thoennes, P. Dube, B. Brezzo, S. Asaad, and D. Dillenberger, "Database analytics: A reconfigurable-computing approach," *Micro, IEEE*, vol. 34, no. 1, pp. 19–29, Jan 2014.

[2] M. Lavasani, H. Angepat, and D. Chiou, "An fpga-based in-line accelerator for memcached," *Computer Architecture Letters*, vol. 13, no. 2, pp. 57–60, July 2014.

[3] A. Putnam, A. Caulfield, E. Chung, D. Chiou, K. Constantinides, J. Demme, H. Esmaeilzadeh, J. Fowers, G. Gopal, J. Gray, M. Haselman, S. Hauck, S. Heil, A. Hormati, J.-Y. Kim, S. Lanka, J. Larus, E. Peterson, S. Pope, A. Smith, J. Thong, P. Xiao, and D. Burger, "A reconfigurable fabric for accelerating large-scale datacenter services," *Micro, IEEE*, vol. 35, no. 3, pp. 10–22, May 2015.

[4] I. Kuon and J. Rose, "Measuring the gap between fpgas and asics," *Computer-Aided Design of Integrated Circuits and Systems, IEEE Transactions on*, vol. 26, no. 2, pp. 203–215, Feb 2007.

[5] A. Kahng, S. Kang, R. Kumar, and J. Sartori, "Enhancing the efficiency of energy-constrained dvfs designs," *Very Large Scale Integration (VLSI) Systems, IEEE Transactions on*, vol. 21, no. 10, pp. 1769–1782, Oct 2013.

[6] M. Yadav, M. Casu, and M. Zamboni, "Laura-noc: Local automatic rate adjustment in network-on-chips with a simple dvfs," *Circuits and Systems II: Express Briefs, IEEE Transactions on*, vol. 60, no. 10, pp. 647–651, Oct 2013.

[7] M. Gerards, J. Hurink, and J. Kuper, "On the interplay between global dvfs and scheduling tasks with precedence constraints," *Computers, IEEE Transactions on*, vol. 64, no. 6, pp. 1742–1754, June 2015.

[8] M. Al-Mothafar and K. Hammad, "Small-signal modelling of peak current-mode controlled buck-derived circuits," *Electric Power Applications, IEE Proceedings -*, vol. 146, no. 6, pp. 607–619, Nov 1999.

[9] B. Das and H. Onodera, "Frequency-independent warning detection sequential for dynamic voltage and frequency scaling in asics," *Very Large Scale Integration (VLSI) Systems, IEEE Transactions on*, vol. 22, no. 12, pp. 2535–2548, Dec 2014.

[10] T. Lin, K.-S. Chong, J. Chang, and B.-H. Gwee, "An ultra-low power asynchronous-logic in-situ self-adaptive v_{DD} system for wireless sensor networks," *Solid-State Circuits, IEEE Journal of*, vol. 48, no. 2, pp. 573–586, Feb 2013.

[11] Y.-W. Ma, J.-L. Chen, C.-H. Chou, and S.-K. Lu, "A power saving mechanism for multimedia streaming services in cloud computing," *Systems Journal, IEEE*, vol. 8, no. 1, pp. 219–224, March 2014.

[12] E. Mintarno, J. Skaf, R. Zheng, J. Velamala, Y. Cao, S. Boyd, R. Dutton, and S. Mitra, "Self-tuning for maximized lifetime energy-efficiency in the presence of circuit aging," *Computer-Aided Design of Integrated Circuits and Systems, IEEE Transactions on*, vol. 30, no. 5, pp. 760–773, May 2011.

[13] N. Gong, J. Wang, S. Jiang, and R. Sridhar, "Tm-rf: Aging-aware power-efficient register file design for modern microprocessors," *Very Large Scale Integration (VLSI) Systems, IEEE Transactions on*, vol. 23, no. 7, pp. 1196–1209, July 2015.

[14] O. Trescases and J. Ng, "Variable output, soft-switching dc/dc converter for vlsi dynamic voltage scaling power supply applications," in *Power Electronics Specialists Conference, 2004. PESC 04. 2004 IEEE 35th Annual*, vol. 6, June 2004, pp. 4149–4155 Vol.6.

[15] C. Chow, L. Tsui, P. Leong, W. Luk, and S. Wilton, "Dynamic voltage scaling for commercial fpgas," in *Field-Programmable Technology, 2005. Proceedings. 2005 IEEE International Conference on*, Dec 2005, pp. 173–180.

[16] J. Levine, E. Stott, G. Constantinides, and P. Cheung, "Online measurement of timing in circuits: For health monitoring and dynamic voltage & frequency scaling," in *Field-Programmable Custom Computing Machines (FCCM), 2012 IEEE 20th Annual International Symposium on*, April 2012, pp. 109–116.

[17] J. M. Levine, E. Stott, G. Constantinides, and P. Y. Cheung, "Smi: Slack measurement insertion for online timing monitoring in fpgas," *Field Programmable Logic and Applications (FPL), 2013 23rd International Conference on*, pp. 1–4, Sept 2013.

[18] J. M. Levine, E. Stott, and P. Y. Cheung, "Dynamic voltage & frequency scaling with online slack measurement," in *Proceedings of the 2014 ACM/SIGDA International Symposium on Field-programmable Gate Arrays*, ser. FPGA '14. New York, NY, USA: ACM, 2014, pp. 65–74. [Online]. Available: http://doi.acm.org/10.1145/2554688.2554784

[19] Y. Wen and O. Trescases, "Dc-dc converter with digital adaptive slope control in auxiliary phase for optimal transient response and improved efficiency," *Power Electronics, IEEE Transactions on*, vol. 27, no. 7, pp. 3396–3409, July 2012.

Morphing Switched-Capacitor Step-Down DC–DC Converters with Variable Conversion Ratio

Song Xiong*, Ying Huang*, Siew-Chong Tan*, and Shu-Yuen (Ron) Hui*[†]

Email: sxiong@eee.hku.hk,sctan@eee.hku.hk.

*Department of Electrical and Electronic Engineering, The University of Hong Kong, Hong Kong

[†]Department of Electrical and Electronic Engineering, Imperial College London, U.K.

Abstract—High-voltage-gain and wide-input-range DC–DC converters are widely used in various electronics and industrial products such as portable devices, telecommunication, automotive, and aerospace systems. The two-stage converter is a widely adopted architecture for such applications, and it is proven to have a higher efficiency as compared with that of the single-stage converter. This paper presents a modular-cell-based morphing switched-capacitor (SC) converter for application as a front-end converter of the two-stage converter. The conversion ratio of this converter is flexible and can be freely extended by increasing more SC modules. The varying conversion ratio is achieved through the morphing of the converter's structure corresponding to the amplitude of the input voltage. This converter is light and compact, and is highly efficient over a very wide range of input voltage and load conditions. Experimental results show that the efficiency of a single SC module is higher than 98%.

I. Introduction

High-voltage-gain, wide input range DC–DC converters are widely applied in various electronics and industrial products such as portable devices, telecommunication systems, automotive systems, and aerospace systems, where the common power bus voltage (e.g. 12 V, 24 V, 48 V) is much higher than the voltage of the loads required (e.g. 1 V–1.8 V for point-of-loads [1], [2], 0.4 V for processor's supply voltage by 2026 [3]). On the other hand, a battery backup sub-system is typically required in some of these system. The output voltage of a battery can vary widely according to its state-of-charge, e.g. 18 V–58 V for 42 V automotive system [4], [5], 36 V–72 V in telecommunication applications [6]. Therefore, the design of a high-voltage-gain, wide-input-range converter is important.

A conventional solution applied to high-voltage-gain conversion is through the use of single-stage non-isolated converters, e.g. the buck converter. This, however, will require the buck converter to work at an extremely low duty ratio, i.e., $D = \frac{V_o}{V_{in}}$, which leads to many issues. First, operating the converter at a very low duty cycle gives a low efficiency [7], [8]. Second, the extremely low duty ratio also limits the dynamic respond as the room of reducing the duty ratio is limited [5]. Third, there is a limit to the operating input voltage range of the converter [9]. This is similar for other single-stage converters, such as the buck-boost, Ćuk, Sepic, and Zeta converters.

Alternatively, transformer-based converters, such as LLC resonant converter, and dual-active-bridge (DAB) converter, etc. [10]–[12], have also been used for the high-voltage-gain conversion application. For these converters, the transformer is the core component used for achieving the high-voltage conversion. For low power applications where galvanic isolation is not mandatary, the use of transformer will be excessive and costly as compared to non-isolated solutions.

The two-stage converter, which is found capable of having an overall higher efficiency than that of a single-stage DC–DC converter in high-voltage-gain applications [7], [8], [13]–[19], is the trend moving forward and has been applied in industry. In the two-stage architecture, the first-stage converter is typically used for the stepping down of the input voltage to an intermediate bus voltage efficiently, and the second-stage converter performs both the stepping down of the input voltage and the regulation of the output voltage with high efficiency and tight regulation. Note that the first-stage front-end converter can output either a regulated or an unregulated intermediate bus voltage without affecting the performance of the overall converter.

In this paper, an unregulated variable-conversion-ratio morphing switched-capacitor (SC) converter is proposed as the front-end converter of the two-stage converter. This morphing SC converter is composed of many highly-efficient modules of SC cells, of which the variable conversion ratio is achieved through the morphing control of the SC converter. The principle is to morph the operating number of modular SC cells according to the amplitude of the input voltage of the front-end converter to achieve the highest possible conversion efficiency. This variable-conversion-ratio morphing SC converter has many advantages.

(a) The conversion ratio of this converter is adapted to the input voltage to give a specified range of the output voltage. For conventional SC converter used for performing only voltage transformation, the conversion ratio is typically fixed [20]–[32].

(b) Each SC module is optimally design to achieve a very high efficiency. The efficiency is higher than 98%.

(c) The converter is easily extendable with the additional cascade of more SC modules.

(d) It is highly compact with low weight and small size. SC converters are composed of switches and capacitors, and do not contain any magnetic element. It can be easily fabricated as an integrated-circuit (IC) chip [33]–[35].

(e) The modular design leads to many advantages, such as availability, maintainability, flexible system structure, and layout [2].

978-1-4673-9551-9/16 $31.00 © 2016 IEEE

II. PROPOSED MORPHING SC CONVERTER

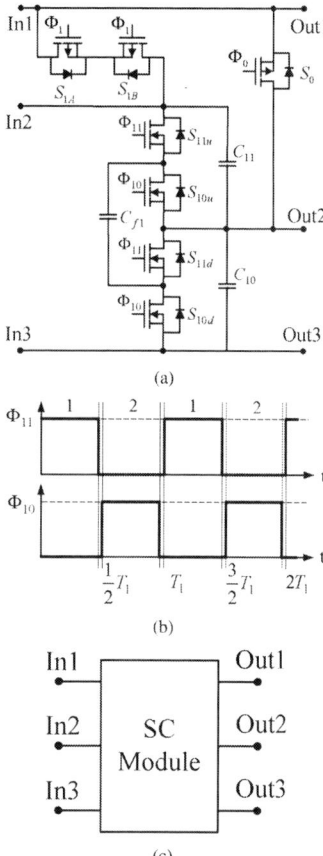

(a)

(b)

(c)

Fig. 1. (a) Topology, (b) timing diagram and (c) block diagram of the SC cell.

Fig. 1(a) shows the topology of a module of the SC cell of the proposed morphing SC converter. It contains three input and three output ports. The power source is connected to ports In1 (+ve) and In3 (-ve), and the load is connected to ports Out2 (+ve) and Out3 (-ve). Ports In2 and Out1 are extension ports for the connection of an additional SC module to the morphing SC converter. The SC cell has two operation modes, which are shown in Fig. 2.

In Mode 1, the voltage conversion ratio is 1, which is shown in Fig. 2(a). In this mode, only switch S_0 is turned on, and all other switches are off. The output is directly connected to the input via switch S_0.

In Mode 2, the voltage conversion ratio is 0.5. In this mode, switch S_0 is off, switches S_{1A} and S_{1B} are ON, and switches S_{11u}, S_{11d}, S_{10u} and S_{10d} are operated with the timing diagram shown in Fig. 1(b).

The proposed morphing SC converter is composed of N number of SC cells connected in cascade as shown in Fig. 3. The three input ports (In1, In2, and In3) of k-th cell is connected to the three output ports (Out1, Out2, and Out3) of the $(k-1)$-th cell, where $k = 2 \cdots N$. The power source

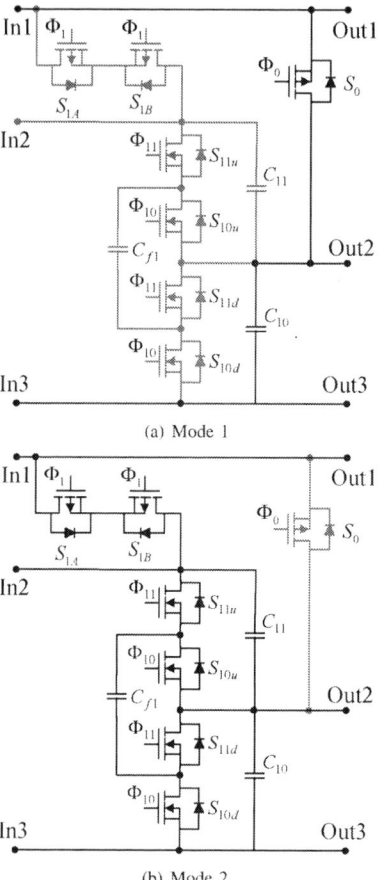

(a) Mode 1

(b) Mode 2

Fig. 2. Operation modes of basic SC cell at conversion ratio (a) 1 and (b) 0.5.

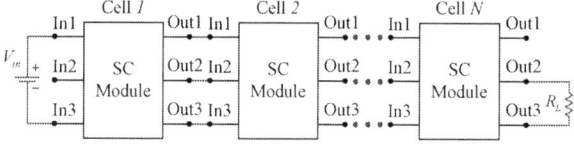

Fig. 3. Proposed morphing SC converter with N SC cells.

is connected to the input ports In1 and In3 of the first cell, and the load is connected to the output ports Out2 and Out3 of the N-th cell. This N-module morphing SC converter has $N + 1$ conversion ratios, which are 0.5^j, where $j = 0 \cdots N$.

III. OPERATION OF THE TWO-MODULE MORPHING SC CONVERTER

A two-module morphing SC converter as shown in Fig. 4(a) is hereon used for the detailed discussion of the operation and properties of the morphing SC converter. Fig. 4(b) shows the pair of complementary PWM signals for cell 1 and cell 2. The control signal required for driving the two-module morphing SC converter is summarized as shown in Table I. By adopting this control, the three operation modes of the two-module morphing SC converter gives the conversion ratios of 1, 0.5, and 0.25, respectively.

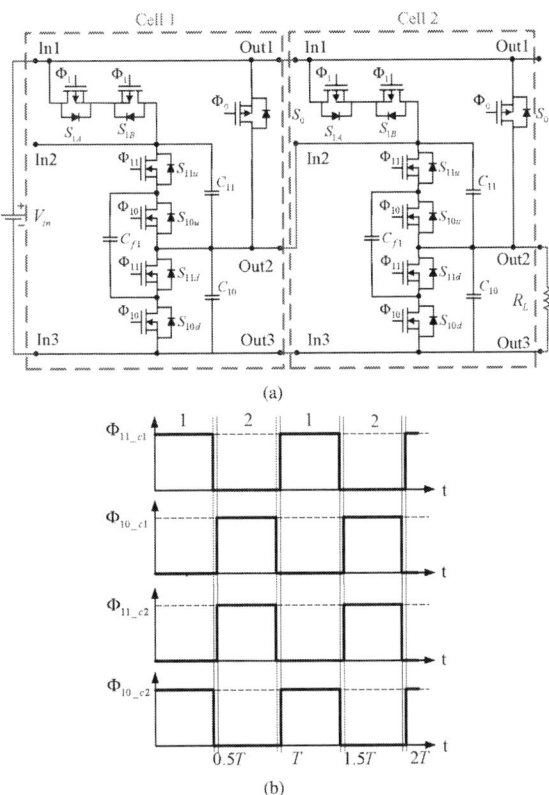

(a)

(b)

Fig. 4. (a) Topology of two-module morphing SC converter and its (b) control PWM signals.

A. Three Operation Modes of the Two-Module Morphing SC Converter

(1) The first mode (Mode 1) is shown in Fig. 5(a), of which the conversion ratio is 1. Here, only S_0 of cell 2 is turned on, while all other switches are off. As only S_0 of cell 2 is working, the only conduction loss is on S_0. Therefore, the converter is of high efficiency.

(2) In Mode 2 as shown in Fig. 5(b), the conversion ratio is 0.5. Here, cell 1 is shut down. Switch S_0 of cell 2 is off, while switches S_{1A} and S_{1B} of cell 2 are turned on. The switches S_{11u}, S_{11d}, S_{10u} and S_{10d} of cell 2 are driven by the control signal of Φ_{11_c2}, and Φ_{10_c2} given in Fig. 4(b).

(3) In Mode 3 as shown in Fig. 5(c), the conversion ratio is 0.25. Here, both cell 1 and cell 2 are in operation. Switch S_0 of cell 1 and switches S_{1A}, S_{1B} and S_0 of cell 2 are off. The switches S_{1A} and S_{1B} of cell 1 are turned on. Switches S_{11u}, S_{11d}, S_{10u} and S_{10d} of cell 1 are driven by the timing diagram of Φ_{11_c1}, and Φ_{10_c1} given in Fig. 4(b). The switches S_{11u}, S_{11d}, S_{10u} and S_{10d} of cell 2 are driven by control signals Φ_{11_c2}, and Φ_{10_c2} which are complementary to that of cell 1, and are shown in Fig. 4(b).

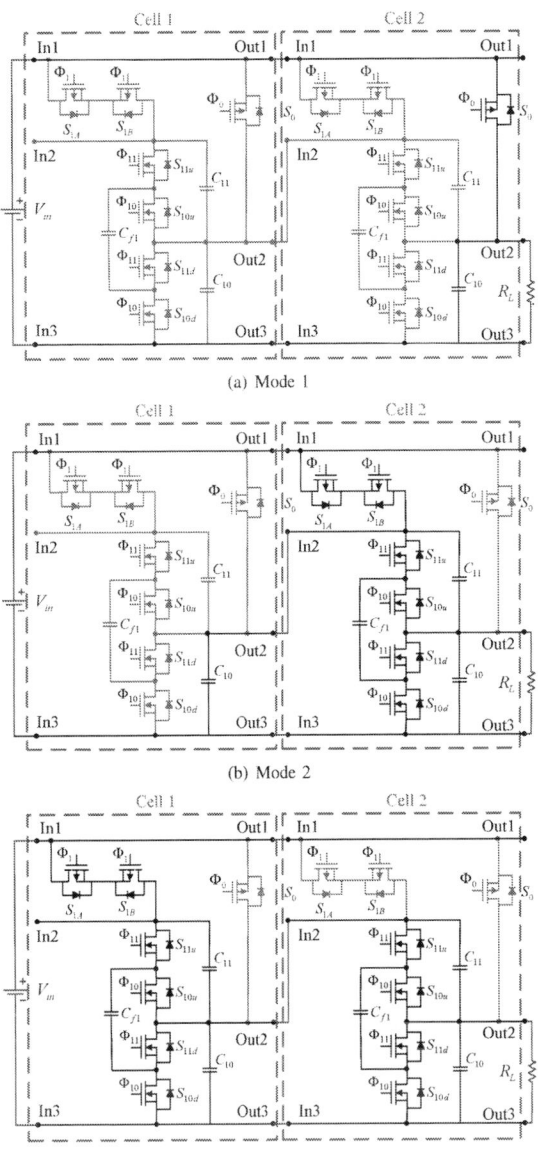

(a) Mode 1

(b) Mode 2

(c) Mode 3

Fig. 5. Three operation modes of two-module morphing SC converter. (a) Mode 1, (b) Mode 2 and (c) Mode 3.

B. Control Methodology

The mode selection control diagram of the two-module morphing SC converter is shown in Fig. 6(a), and Fig. 6(b) shows the converter's operating mode at different input voltages. The two-module morphing SC converter has four mode transition voltages. When the input voltage of the two-module morphing SC converter is increased to $V_{act,12}$, the converter is changed from Mode 1 ($M_{1increase}$) to Mode 2 ($M_{2increase}$), and when the converter's input voltage is increased to $V_{act,23}$, it is changed from Mode 2 ($M_{2increase}$) to Mode 3 ($M_{3increase}$). When the converter's input voltage is decreased to $V_{act,32}$,

TABLE I
CONTROL SIGNAL OF TWO-MODULE MORPHING SC CONVERTER.

Mode	Cell 1				Cell 2			
	S_0	S_{1A}, S_{1B}	S_{11u}, S_{11d}	S_{10u}, S_{10d}	S_0	S_{1A}, S_{1B}	S_{11u}, S_{11d}	S_{10u}, S_{10d}
Mode 1	OFF	OFF	OFF	OFF	ON	OFF	OFF	OFF
Mode 2	OFF	OFF	OFF	OFF	OFF	ON	OFF	OFF
Mode 3	OFF	ON	Φ_{11_c1}	Φ_{10_c1}	OFF	OFF	Φ_{11_c2}	Φ_{10_c2}

the two-module morphing SC converter is changed from Mode 3 ($M_{3decrease}$) to Mode 2 ($M_{2decrease}$), and when the converter's input voltage is decreased to $V_{act,21}$, it is changed from Mode 2 ($M_{2decrease}$) to Mode 1 ($M_{1decrease}$). By adopting this control approach, the output voltage will be in a small range as shown in Fig. 6(c).

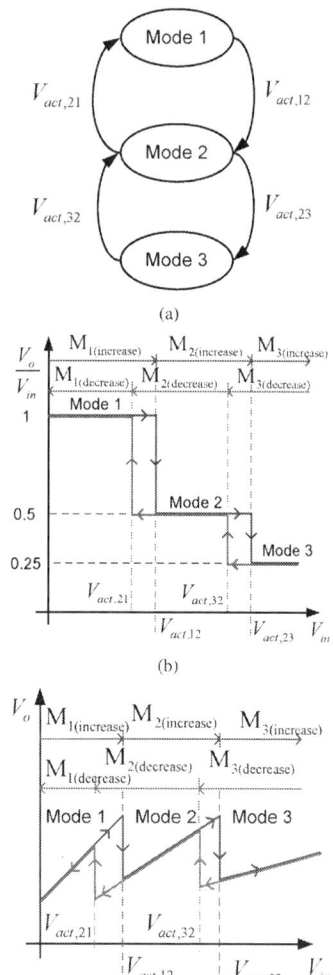

(a)

(b)

(c)

Fig. 6. Mode morphing diagram of the two-module morphing SC converter. (a) Mode selection diagram, (b) operation mode at different input voltage, and (c) the output voltage versus the input voltage plot.

IV. EXPERIMENT RESULTS

A prototype is built to verify the performance of the proposed morphing SC converter. The components used are shown in Table II. The control parameters of the two-module morphing SC converter are shown in Table III. A single module of the SC cell is shown in Fig. 7(a). The power stage of a single module SC cell is shown in Fig. 7(a). The volumetric breakdown is shown in Fig. 7(b).

TABLE II
COMPONENTS USED IN THE MODULE OF SINGLE SC CELL.

Component	Part no.
C_{10}	$2 \times 47\mu F$ (two GEM32ER61C476K paralleled)
C_{11}	$2 \times 47\mu F$ (two GEM32ER61C476K paralleled)
C_{f1}	$6 \times 47\mu F$ (six GEM32ER61C476K paralleled)
S_0	SIS443DN
S_{1A}, S_{1B}	SIS443DN
S_{11u}, S_{11d}	FDMC7660
S_{10u}, S_{10d}	FDMC7660

TABLE III
CONTROL PARAMETERS OF TWO-MODULE MORPHING SC CONVERTER.

Component	Part no.
V_{in}	6 V–30 V
V_o	3.5 V–8.5 V
f_s	100 kHz
$V_{act,12}$	8.28 V
$V_{act,23}$	17.06 V
$V_{act,32}$	15.25 V
$V_{act,21}$	5.9 V

Fig. 7(c) shows the plot of the output voltage of the two-module morphing SC converter with an input voltage of 6 V–30 V at 25 W load. The output voltage is converted to a relatively smaller voltage range of 3.5 V–8.5 V.

The efficiency of the two-module morphing SC converter in the three operating modes are shown in Figs. 8(a)-8(c). The measurement is based on a constant output power of 25 W with a varying input voltage. Fig. 8(a) shows the efficiency curve of Mode 1 for different input voltages. The efficiency is higher than 99% when the input voltage is higher than 8 V. Fig. 8(b) is the efficiency curve of the converter in Mode 2 for different input voltages and switching frequencies. The results show that the efficiency of the SC converter working at 50 kHz switching frequency is close to 98% when the input voltage is lower than 18 V. Fig. 8(c) is the efficiency curve of the SC converter in Mode 3 for different input voltages. The highest efficiency in this mode is also higher than 94.5% when the input voltage is higher than 20 V, while the highest efficiency is close to 96% at 100 kHz switching frequency.

The morphing SC converter is applied to the two-stage configuration where the second stage is a buck converter using

978-1-4673-9551-9/16 $31.00 © 2016 IEEE

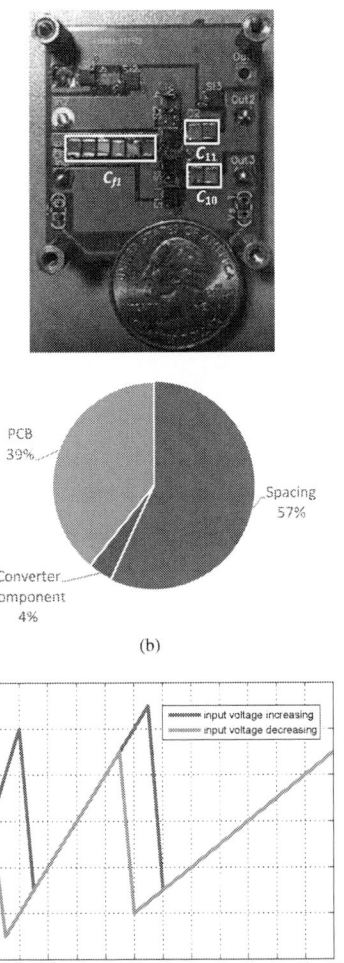

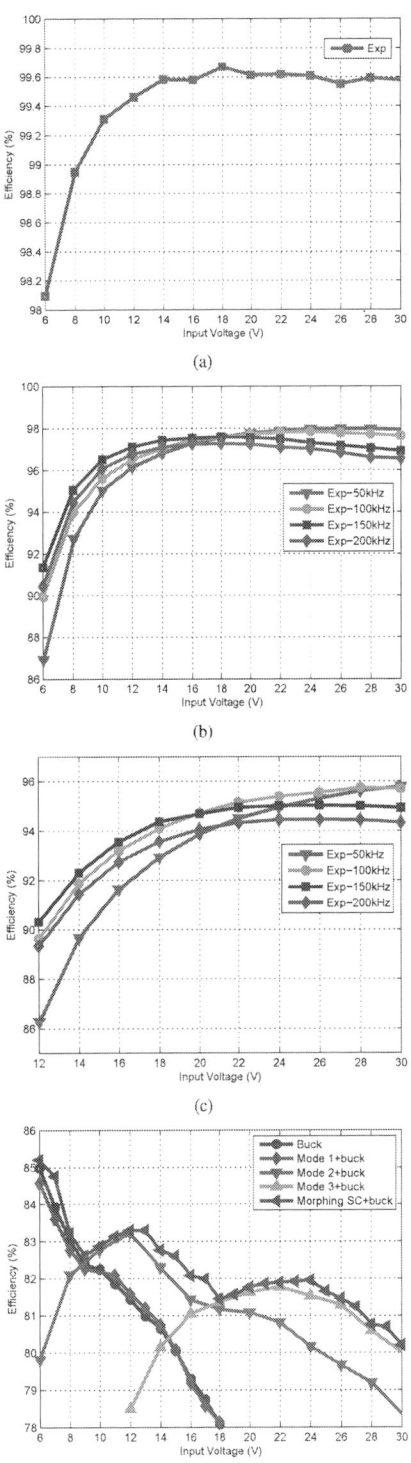

Fig. 7. (a) Power stage of a single module of the SC cell, (b) power stage volume breakdown without considering driver circuit, and (c) the output voltage of the two-module morphing SC converter at 25 W load and 100 kHz switching frequency.

the commercial IC IR3820. The efficiency of this two-stage converter is shown in Fig. 8(d). The two-stage converter has a high efficiency over the entire range of the input voltage between 6 V and 30 V. The efficiency is higher than 80% over the entire range, which is higher than the buck converter itself. Moreover, the applicable input voltage range is also extended with the incorporation of the SC converter (the actual maximum application voltage of IC IR3820 is 21 V).

V. CONCLUSIONS

A morphing SC converter, which is of a high voltage gain and that is suitable for wide-input-range application, is proposed. This morphing SC converter is adaptive to the change in the magnitude of the input voltage and the converter's conversion ratio is variable through the morphing of the structure.

Fig. 8. The measured efficiency curves of the two-stage morphing SC in (a) Mode 1, (b) Mode 2, and (c) Mode 3 with 25 W constant power output. (d) The efficiency curve of the morphing SC-buck converter at 15 W output power in different modes.

The morphing SC converter is easily extendable through the cascade of extra SC cells. It is experimentally demonstrated that a high efficiency, small volume, and light weight morphing SC converter is achievable. When the morphing SC converter is applied as the front-end converter in a two-stage converter with the buck converter being the second-stage converter, the overall efficiency of the two-stage converter is improved as compared with that of the single buck converter itself. Moreover, the morphing SC converter can also extend the input voltage range of the buck converter.

ACKNOWLEDGMENT

This work is fully supported by the Hong Kong Research Grant Council under GRF project 17207314.

REFERENCES

[1] K. K. Leong, G. Deboy, K. Krischan, and A. Muetze, "A single stage 54 V to 1.8 V multi-phase cascade buck voltage regulator module," in *IEEE Appl. Power Electron. Conf. and Exposit. (APEC)*, pp. 1966–1973, Mar. 2015.

[2] S. G. Luo and I. Batarseh, "A review of distributed power systems part I: DC distributed power system," *IEEE Magaz. of Aerosp. and Electron. Syst.*, vol. 20, no. 8, pp. 5–16, Aug. 2005.

[3] C. Schaef and J. T. Stauth, "A multilevel VR implementation and MIMO control scheme for vertically stacked microprocessor cores," in *IEEE Appl. Power Electron. Conf. and Exposit. (APEC)*, pp. 2090–2096, Mar. 2015.

[4] Application Note, "Choosing the right DC–DC converter for automotive applications," http://www.maximintegrated.com/cn/app-notes/index.mvp/id/1845.

[5] R. Guo, Z. G. Liang, and A. Q. Huang, "A family of multimodes charge pump based DC–DC converter with high efficiency over wide input and output range," *IEEE Trans. on Power Electron.*, vol. 27, no. 11, pp. 4788–4798, Nov. 2012.

[6] L. Q. Chen, H. F. Wu, P. Xu, H. B. Hu, and C. G. Wan "A high step-down non-isolated bus converter with partial power conversion based on synchronous LLC resonant converter," in *IEEE Appl. Power Electron. Conf. and Exposit. (APEC)*, pp. 1950–1955, Mar. 2015.

[7] J. Sun, M. Xu, Y. Ying, and F. C. Lee, "High power density high efficiency system two-stage power architecture for laptop computers," in *IEEE Power Electron. Spec. Conf. Rec. (PESC)*, pp. 4008–4015, Jun. 2006.

[8] M. Xu, J. Sun, and F. C. Lee, "Voltage divider and its application in the two-stage power architecture," in *IEEE Appl. Power Electron. Conf. and Exposit. (APEC)*, vol. 2, pp. 499–505, Mar. 2006.

[9] Datasheet of IR3820, http://www.irf.com/product-info/datasheets/data/ir3820am.pdf

[10] E. S. Kim, J. H. Park, J. S. Joo, S. M. Lee, K. Kim, and Y. S. Kong, "Bidirectional DC–DC converter using secondary LLC resonant tank," in *IEEE Appl. Power Electron. Conf. and Exposit. (APEC)*, pp. 2104–2108, Mar. 2015.

[11] D. Patil, A. K. Rathore, and D. Srinivasan, "A non-isolated bidirectional soft switching current fed LCL resonant DC–DC converter to interface energy storage in DC microgrid," in *IEEE Appl. Power Electron. Conf. and Exposit. (APEC)*, pp. 709–716, Mar. 2015.

[12] D. Doncker, D. M. Divan, and M. H. Kheraluwala, "A three-phase soft-switched high-power-density DC/DC converter for high-power applications," *IEEE Trans. on Indust. Applic.*, vol. 27, no. 1, pp. 63–73, Jan./Feb. 1991.

[13] R. D. Middlebrook, "Transformerless DC-to-DC converters with large conversion ratios," *IEEE Trans. on Power Electron.*, vol. 3 no. 4, pp. 484–488, Oct. 1988.

[14] J. Wei, P. Xu, H. P. Wu, F. C. Lee, K. Yao, and M. Ye, "Comparison of three topology candidates for 12 V VRM," in *IEEE Appl. Power Electron. Conf. and Exposit.(APEC)*, pp. 245–251, Mar. 2001.

[15] B. Axelrod, Y. Berkovich, and A. Ioinovici, "Switched-capacitor/switched-inductor structures for getting transformerless hybrid DC–DC PWM converters," *IEEE Tran. on Cir. and Sys.*, vol. 55 no. 2, pp. 687–696, Mar. 2008.

[16] Y. H. Chang, "Variable-conversion-ratio switched-capacitor-voltage-multiplier/divider DC–DC converter," *IEEE Trans. on Cir. and Sys.–I: Regular Papers*, vol. 58, pp. 1944–1957, Aug. 2011.

[17] S. Xiong, S. C. Tan, and S. C. Wong, "Analysis and design of a high-voltage-gain hybrid switched-capacitor buck converter" *IEEE Trans. on Cir. Sys. I, Regul. Pap.*, vol. 59, no. 5, pp. 1132–1141, May. 2012.

[18] R. C. N. Pilawa-Podgurski, D. M. Giuliano, and D. J. Perreault, "Merged two-stage power converter with soft charging switched-capacitor stage in 180 nm CMOS," *IEEE Jour. of Solid-state. Cir.*, vol. 47, no. 7, pp. 1557–1567, July 2012.

[19] S. Xiong, S. C. Wong, S. C. Tan, and C.K. Tse, "A family of exponential step-down switched-capacitor converters and their applications in two-stage converters," *IEEE Trans. on Power Electron.*, vol. 29, no. 4, pp. 1870–1880, Apr. 2014.

[20] S. V. Cheong, H. Chung, and A. Ioinovici, "Inductorless DC–DC converter with high power density," *IEEE Trans. on Ind. Electron.*, vol. 41, no. 2, pp. 208–215, Apr. 1994.

[21] M. S. Makowski and D. Maksimovic, "Performance limits of switched-capacitor DC–DC converters," in *IEEE Power Electron. Special. Conf. (PESC)*, vol. 2, pp. 1215–1221, Jun. 1995.

[22] A. Ioinovici, "Switched-capacitor power electronics circuits," *IEEE Circuits Syst. Mag.*, vol. 41, no. 2, pp. 37–42, Sept. 2001.

[23] J. W. Kimball and P. T. Krein, "Analysis and design of switched capacitor converters," in *IEEE Appl. Power Electron. Conf. and Exposit. (APEC)*, vol. 3, pp. 1473–1477, Mar. 2005.

[24] F. Zhang, L. Du, F. Z. Peng, and Z. M. Qian, "A new design method for high-power high-efficiency switched-capacitor DC–DC converters," *IEEE Trans. on Power Electron.*, vol. 23, no. 2, pp. 832–840, Mar. 2008.

[25] V. W. Ng, M. D. Seeman, and S. R. Sanders, "High-efficiency, 12V-TO-1.5V DC–DC converter realized with switched-capacitor architecture," in *Symp. on VLSI Circ.*, pp. 168–169, Jun. 2009.

[26] S. C. Tan, S. Kiratipongvoot, S. Bronstein, A. Ioinovici, Y. M. Lai, and C. K. Tse, "Interleaved switched-capacitor converters with adaptive control," in *IEEE Energy Convers. Congr. Exposit. (ECCE)*, pp. 2725–2732, Sep. 2010.

[27] D. Cao, X. H. Yu, X. Lu, W. Qian, and F. Z. Peng, "A double-wing multilevel modular capacitor-clamped DC–DC converter with reduced capacitor voltage stress," in *IEEE Energ. Convers. Congr. and Exposit. (ECCE)*, pp. 545–552, Sept. 2011.

[28] W. Qian, D. Cao, J. G. Cintron-Rivera, M. Gebben, D. Wey, and F. Z. Peng, "A switched-capacitor DC–DC converter with high voltage gain and reduced component rating and count," *IEEE Trans. on Ind. Electron.*, vol. 48, no. 4, pp. 1397–1406, July-Aug. 2012.

[29] K. Zou, M. J. Scott, and J. Wang, "A switched-capacitor voltage tripler with automatic interleaving capability," *IEEE Trans. on Power Electron.*, vol. 27, no. 6, pp. 2857–2868, Jun. 2012.

[30] M. Evzelman and S. Ben-Yaakov, "Average-current-based conduction losses model of switched capacitor converters," *IEEE Trans. on Power Electron.*, vol. 28, no. 7, pp. 3341–3352, Jul. 2013.

[31] D. Cao, S. Jiang, and F. Z. Peng, "Optimal design of a multilevel modular capacitor-clamped DC–DC converter," *IEEE Trans. on Power Electron.*, vol. 28, no. 8, pp. 3816–3826, Aug. 2013.

[32] B. Wu, S. Li, K. Smedley, and S. Singer, "A family of two-switch boosting switched-capacitor converters," *IEEE Trans. on Power Electron.*, vol. 30, no. 10, pp. 5413–5424, Oct. 2015.

[33] T. Santa, M. Auer, C. Sandner, and C. Lindholm, "Switched capacitor DC–DC converter in 65nm CMOS technology with a peak efficiency of 97%," in *IEEE Inter. Symp. on Cir. and Sys. (ISCAS)*, pp. 1351–1354, May 2011.

[34] T. M. Andersen, F. Krismer, J. W. Kolar, T. Toifl, C. Menolfi, L. Kull, T. Morf, M. Kossel, M. Brandli, P. Buchmann, and P. A. Francese, "A $4.6W/mm^2$ power density 86% efficiency on-chip switched capacitor DC–DC converter in 32 nm SOI CMOS," in *IEEE Appl. Power Electron. Conf. and Exposit. (APEC)*, pp. 692–699, Mar. 2013.

[35] T. M. Andersen, F. Krismer, J. W. Kolar, T. Toifl, C. Menolfi, L. Kull, T. Morf, M. Kossel, M. Brandli, P. Buchmann, and P. A. Francese, "A deep trench capacitor based 2:1 and 3:2 reconfigurable on-chip switched capacitor DC-DC converter in 32 nm SOI CMOS," in *IEEE Appl. Power Electron. Conf. and Exposit. (APEC)*, pp. 1448–1455, Mar. 2014.

Compact Modular Switched-Capacitor DC/DC Converters with Exponential Voltage Gain

Ying Huang*, Song Xiong*, Siew-Chong Tan*, and Shu-Yuen (Ron) Hui*†

Email: yhuang@eee.hku.hk sxiong@eee.hku.hk sctan@eee.hku.hk

*Department of Electrical and Electronic Engineering, The University of Hong Kong, Hong Kong

†Department of Electrical and Electronic Engineering, Imperial College London, U.K.

Abstract—A compact modular switched-capacitor DC/DC converter with exponential voltage gain, high efficiency, light weight, and bidirectional power flow, is proposed in this paper. The proposed converter is suitable for applications in high temperature environments since it does not contain magnetic element nor temperature-sensitive capacitor. With an output voltage that is 2^N (N is the cell number) times the low-side voltage, the proposed converter has a considerably low component count and suffers from the lowest overall capacitor voltage stress as compared with other existing switched-capacitor converters. Besides, the proposed converter adopts a modular structure and simple control scheme, which enables it to be easily extended, through a repeated cascade of the same module, to achieve a higher voltage conversion. Moreover, since the load is charged via two complementary paths in an interleaved operation, the output voltage ripple is small, which is beneficial for high efficiency. Experimental results of an $8\times$-gain prototype at 5 V input and 70 W output power, are provided to validate the performance of the proposed converter. The achievable efficiency is up to 95.78% (including the driver's loss).

I. INTRODUCTION

High-voltage-gain DC/DC step-up converters are an important class of converters that are adopted to boost a low-level DC voltage input to a high-level DC voltage output. It is widely used in renewable and industrial applications as front-end converters in clean energy systems (e.g. photovoltaic (PV) panels [1], thermoelectric generators (TEG) [2], and reverse electrodialysis (RED) stacks [3]), and telecommunication systems [4]. For examples, the output voltage of TEG reported in [5] is close to 5 V and that of the RED [6] is generally ranged from a few volts to a few tens of volts. In recent years, there is an increasing demand for this type of DC/DC converters which should be light and compact, and that can simultaneously achieve high voltage gain and high efficiency in high temperature environments [7].

Existing converters that can achieve the boost operation can be classified into three categories: transformer-based converters, coupled-inductor-based converters, and non-isolated converters. Transformer-based converters and coupled-inductor based converters can easily achieve a high voltage ratio through the use of large transformer's or inductor's turns ratio [2], [8]–[10]. However, large turns ratio necessitates the use of large magnetic component, which increases the size, weight, and cost, and lowers the converters' efficiency. Moreover, at a high temperature, the permeability of magnetics decreases significantly as temperature increases, leading to a high current

ripple and power loss, which further increases its temperature [1]. Such drawbacks make these converters less appropriate for some applications, e.g. high-temperature applications. For non-isolated converters, the most conventional converter for step-up conversion is the boost converter. However, a boost converter can only achieve limited voltage conversion gain (< 5) with an extremely high duty ratio and the efficiency achievable is low [1], [11], [12]. Therefore it cannot meet the requirement of high-voltage-gain conversion. With consideration to the requirements posed above (light, compact, high-voltage-gain, high efficiency and available for high temperature working environments), the switched-capacitor (SC) converters, which do not contain any magnetic element or temperature-sensitive capacitor (e.g. electrolytic capacitor), will be a suitable solution.

Several topologies of SC converters that can achieve a high-voltage gain have been reported in the literature. This includes the series-parallel (SP) SC converters [13], [14], Fibonacci (FIB) SC converters [15], [16], 2^N (N is an integer) SC converters [17]–[19], and the multilevel modular capacitor clamped DC/DC converters (MMCCC) [20]. For both the SP SC converter and MMCCC, the voltage gain increases linearly with the increasing number of SC cells (i.e., for N cells, under ideal condition, the gain is $N + 1$). Therefore, a large number of cells are required to achieve a high voltage gain. Moreover, for step-up applications, a high-voltage capacitor is required to be in parallel with the load to provide energy to the load for half a period. A nX SC converter, which combines the modular structure and simple control of the MMCCC, has been proposed in [1] to further improve the conversion ratio with a reduced component count (i.e., for N cells, under ideal condition, the gain is $2N$). However, the achievable component count as compared with FIB SC and 2^N SC converters is still unsatisfactory for high-voltage-gain applications. On the other hand, the major disadvantage of the FIB and 2^N SC converters is that their power switches have to handle a much larger voltage stress, and therefore the use of high-voltage MOSFETS (more costly and has a higher $R_{DS_{on}}$) is required in these converters. Moreover, a high-voltage capacitor is required in FIB SC converter in step-up applications to hold the output voltage.

In this paper, a compact modular SC DC/DC converter with exponential voltage gain, high efficiency, light weight, and bidirectional power flow, is proposed. Compared with

978-1-4673-9551-9/16 $31.00 © 2016 IEEE

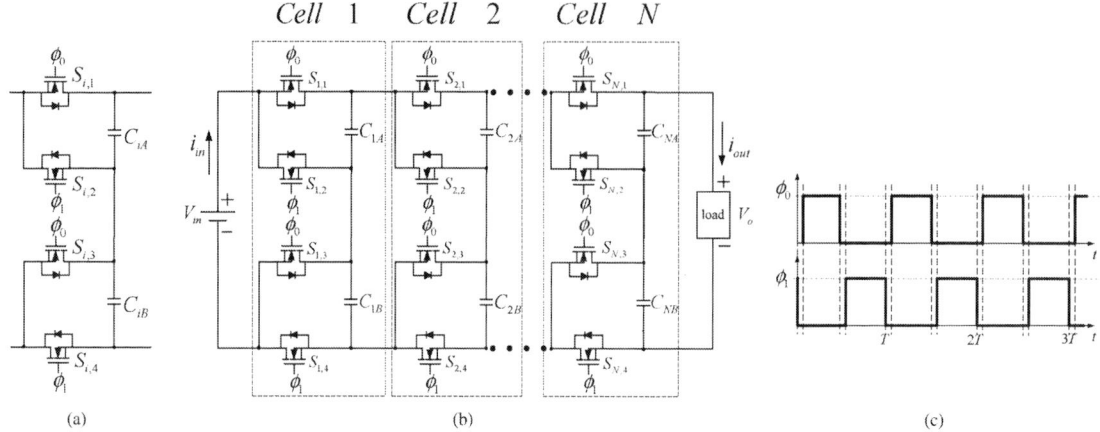

Fig. 1. (a) The basic SC cell, (b) the topology, and (c) the control timing diagram of the proposed SC converter.

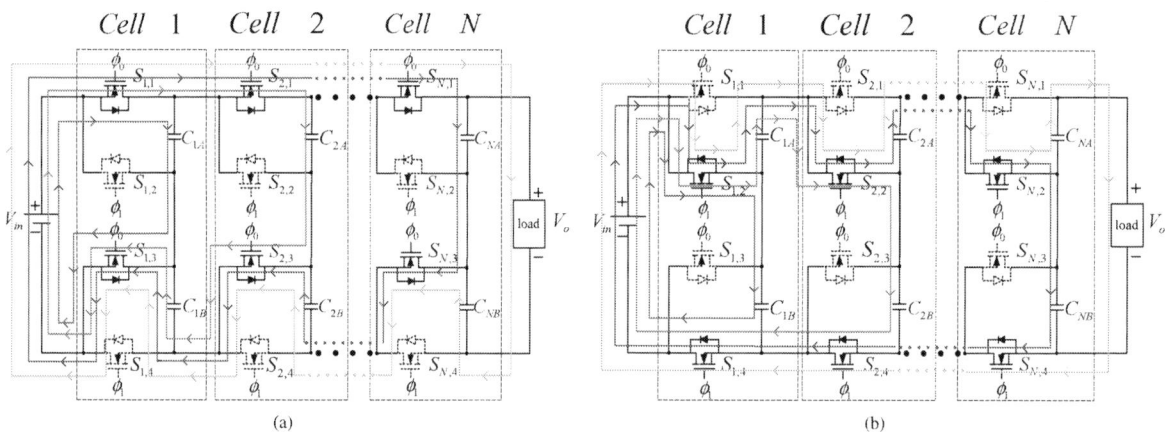

Fig. 2. The two operating states of the proposed SC converter: (a) State I and (b) State II.

the SC converters described above (SP SC, FIB SC, 2^N SC, nX SC converters, and MMCCC), the proposed converter has a relatively lower component count and capacitor voltage stress. A lower switch number will lead to a simpler driver circuit. The switch voltage stress is naturally clamped by the capacitors in the converter, which means that the proposed converter does not suffer the same issue of high-voltage switch voltage stress as that by the SP, FIB and 2^N SC converters. Besides, with only two operating states, the control for this converter circuit is simple, requiring only two complementary gate signals. For some SC converters, e.g. flying capacitor multilevel dc-dc converter (FCMDC) [21], the permitted charging/discharging time is decreased with an increased voltage ratio as the number of subintervals required is equal to its conversion ratio. Consequently, at high switching frequency and high voltage ratio, the on-time for the switches in FCMDC will be comparable to its transition time (rise or fall time), which deteriorates the converter's efficiency. However, with the proposed converter, this is not the case.

The charging/discharging time will be kept constant even for high voltage ratio. Thus, the proposed SC converter can be operated at high frequency to achieve size reduction.

II. THE PROPOSED SC CONVERTER

A. Converter Topology

Fig. 1(a) shows the basic cell of the proposed SC converter. The circuit structure of the proposed compact modular 2^N-time SC converter is shown in Fig. 1(b), which consists of a cascade of N number of cells (N is a positive integer). The switches $S_{i,1}$ and $S_{i,3}$ ($i = 1, 2, \cdots, N$) are driven by a PWM signal ϕ_0 while switches $S_{i,2}$ and $S_{i,4}$ ($i = 1, 2, \cdots, N$) are driven by a complementary PWM signal ϕ_1 just as shown in Fig. 1(c). The operation of the converter is based on the full charging operation of all the capacitors [24].

B. Operation of the Proposed SC Converter

Fig. 2 shows the two operating states of the proposed converter. During State I, only switches $S_{i,1}$ and $S_{i,3}$ ($i =$

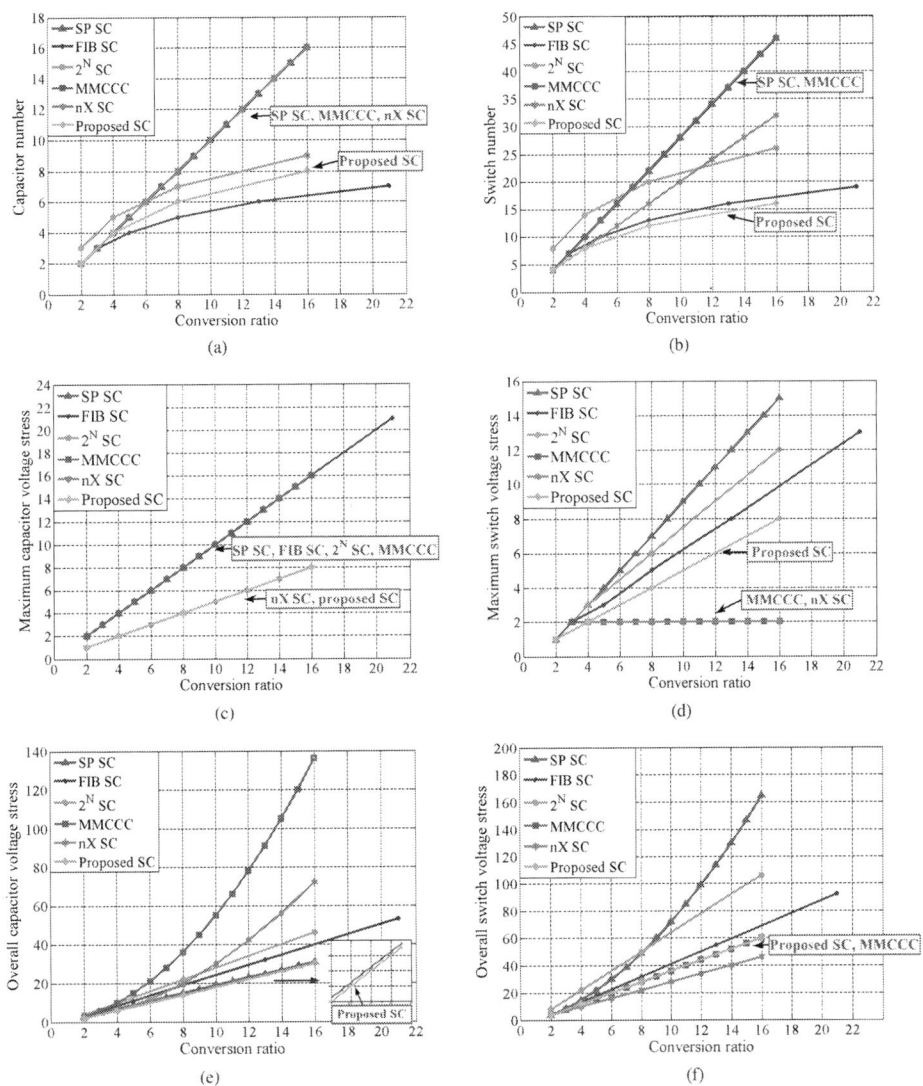

Fig. 3. A comparison of (a) capacitor number, (b) switch number, (c) maximum capacitor voltage stress, (d) maximum switch voltage stress, (e) overall capacitor voltage stress, and (f) overall switch voltage stress among SP SC, FIB SC, 2^N SC, nX SC converters, and MMCCC with the proposed SC converter in step-up applications.

TABLE I
COMPARISON AMONG VARIOUS SC CONVERTERS IN STEP-UP APPLICATIONS.

	SP SC	FIB SC	2^N SC	MMCCC	nX SC	Proposed SC
Load capacitor	Yes (large)	Yes (large)	Yes (small)	Yes (large)	No	No

TABLE II
PARAMETERS OF THE $8\times$-GAIN SC CONVERTER PROTOTYPE.

$S_{1,1}, S_{1,2}, S_{1,3}, S_{1,4}$	BSC010NE2LS1	$S_{2,1}, S_{2,2}, S_{2,3}, S_{2,4}$	BSC010NE2LS1
$S_{3,1}, S_{3,2}, S_{3,3}, S_{3,4}$	BSC014N04LS	C_{1A}, C_{1B}	32 paralleled C3216X5R1H106K160AB
C_{2A}, C_{2B}	21 paralleled C3216X5R1H106K160AB	C_{3A}, C_{3B}	9 paralleled C5750X7S2A106M230KB
Filter capacitors at the input port		A 10 μF multilayer ceramic capacitor and a 1 μF ceramic capacitor	

978-1-4673-9551-9/16 $31.00 © 2016 IEEE

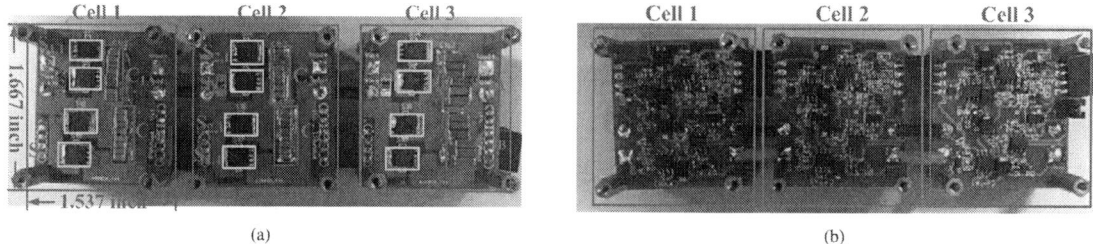

Fig. 4.   Photograph of the 8×-gain prototype, (a) top side (power circuit), (b) bottom side (driver circuit).

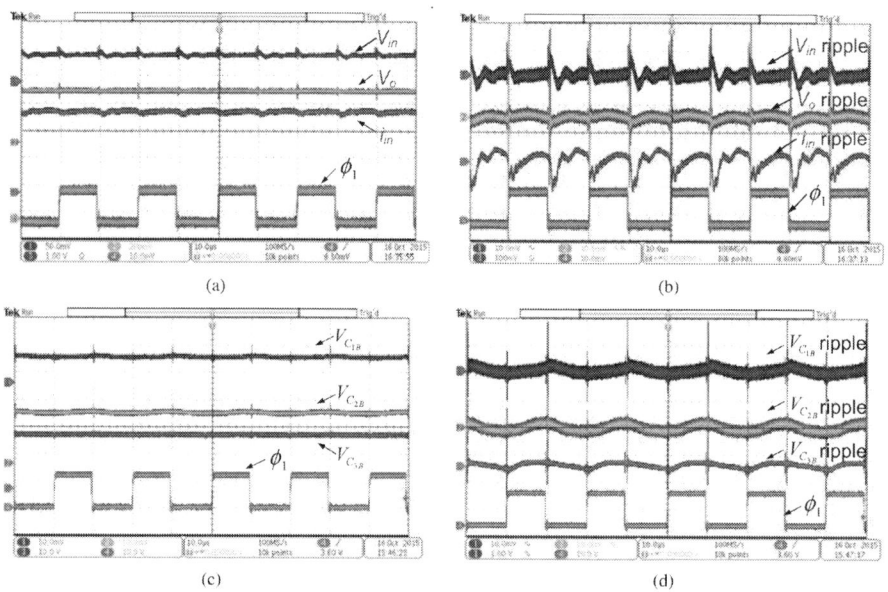

Fig. 5.   Key waveforms of the 8×-gain SC converter operating at 50 kHz switching frequency, 5 V input voltage and 70 W output power, (a) input and output waveforms [V_{in} (5 V/div); V_o (20 V/div); i_{in} (5 A/div); ϕ_1 (10 V/div)], and (b) their ripple waveforms [V_{in} (1 V/div); V_o (1 V/div) (filter bandwidth 250 MHz); i_{in} (0.5 A/div); ϕ_1 (10 V/div)], (c) capacitor voltage waveforms [$V_{C_{1B}}$ (5 V/div); $V_{C_{2B}}$ (5 V/div); $V_{C_{3B}}$ (10 V/div); ϕ_1 (10 V/div)], and (d) their ripple waveforms [$V_{C_{1B}}$ (1 V/div); $V_{C_{2B}}$ (1 V/div); $V_{C_{3B}}$ (1 V/div); ϕ_1 (10 V/div)].

$1, 2, \cdots, N)$ are turned on. The capacitor C_{1A} is directly charged from the source. Concurrently, the input source is in series with C_{1B} to charge capacitor C_{2A}; in series with both C_{1B} and C_{2B} to charge C_{3A}; $\cdots$; and in series with $C_{1B}, C_{2B}, ..., C_{(N-1)B}$ to charge capacitor C_{NA}. Hence, in the steady state,

$$\begin{cases} V_{C_{1A}} = V_{in}, & \text{(1a)} \\ V_{C_{iA}} = V_{in} + \sum_{k=1}^{i-1} V_{C_{kB}}, & \text{(1b)} \end{cases}$$

for $i = 2, 3, \cdots, N$. $V_{C_{iB}}$ is the voltage of capacitor C_{iB}, $V_{C_{iA}}$ is the voltage of capacitor C_{iA}, and N is the total number of SC basic cells used in the converter.

During State II, switches $S_{i,2}$ and $S_{i,4}$ ($i = 1, 2, \cdots, N$) are turned on, while the other switches are off. The capacitor C_{1B} is directly charged by the input source V_{in}. Concurrently, the input source is in series with C_{1A} to charge capacitor C_{2B};

in series with both C_{1A} and C_{2A} to charge C_{3B}; $\cdots$; and in series with $C_{1A}, C_{2A}, ..., C_{(N-1)A}$ to charge capacitor C_{NB}. Hence, in the steady state,

$$\begin{cases} V_{C_{1B}} = V_{in}, & \text{(2a)} \\ V_{C_{iB}} = V_{in} + \sum_{k=1}^{i-1} V_{C_{kA}}, & \text{(2b)} \end{cases}$$

for $i = 2, 3, \cdots, N$.

Combining (1a), (1b), (2a) and (2b), the voltages of the capacitor C_{iA} and C_{iB} are

$$V_{C_{iA}} = V_{C_{iB}} = 2^{(i-1)} \cdot V_{in}, \quad \text{(3)}$$

for $i = 1, 2, \cdots, N$.

Thus, for the proposed converter with N cells, the output voltage is

$$V_o = V_{C_{NA}} + V_{C_{NB}} = 2^N \cdot V_{in}. \quad \text{(4)}$$

C. Comparison with Other Existing SC Converters

Fig. 3 shows a comparison of the proposed SC converter with the SP SC, FIB SC, 2^N SC, nX SC converters, and MMCCC in terms of the capacitor number, switch number, maximum capacitor voltage stress, maximum switch voltage stress, overall capacitor voltage stress and overall switch voltage stress in step-up applications. The maximum capacitor voltage stress is defined as the ratio of the largest capacitor voltage to the input voltage. The maximum switch voltage stress is defined as the ratio of the largest switch voltage stress to the input voltage. The overall capacitor voltage stress is the ratio of the sum of the capacitor voltage stress to the input voltage. The overall switch voltage stress is the ratio of the sum of the switch voltage stress to the input voltage. It should be noted that the SP SC, FIB SC converters, and MMCCC have the same maximum capacitor voltage stress. It is because of this that a large-volume capacitor is required to be connected in parallel with the load to supply the load with the energy in the converters' step-up applications for half a switching period. For the 2^N SC converter, although there are two complementary charge paths to provide energy to the load, a load capacitor is also required because of the dead time between the switching operations. Thus, the largest capacitor voltage stress of the SP SC, FIB SC, 2^N SC converters, and MMCCC are equal and relative to the output voltage. However, for the proposed converter, an extra load capacitor is not needed. The comparison of load capacitor requirement among the various converters is shown in Table I. The term "large" and "small" signify respectively the size of the required capacitor. It is clear that the proposed converter has the lowest switch number while still having the lowest overall capacitor voltage stress. A low switch number signifies a simpler driver circuit which contributes to a smaller converter size. Besides, its maximum and overall switch voltage stress is also low as compared to the FIB, 2^N, and SP SC converters, which means that switches with a lower on-state resistance that can potential result in a high converter's efficiency (can be adopted). In addition, to achieve the same conversion ratio, the proposed converter requires only slightly more capacitors as compared to the FIB SC converter (e.g., one capacitor more when conversion ratio is 8).

III. EXPERIMENTAL RESULTS

An $8\times$-gain 70 W prototype with input voltage of 5 V and output voltage of 40 V has been developed for practical verification. The prototype consists of three series cells as shown in Fig. 4. The area of each cell is small at 2.56 inch^2 (1.537 inch long and 1.667 inch width). No heat dissipation device is required.

The parameters of the prototype are shown in Table II. Considering operation safety, the rated voltage of the switches are chosen to be twice of the voltage stress. For the switches in cell 1 and cell 2, the drain-to-source voltage stress are 5 V and 10 V respectively, and the rated voltage of the selected switches are 25 V. For the switches in cell 3, the drain-to-source voltage stress is 20 V, and the rated voltage of the

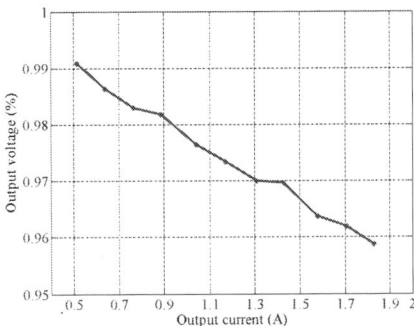

Fig. 6.   Open-loop output voltage versus output current at 50 kHz switching frequency and 5 V input voltage (the baseline voltage is 40 V at no load).

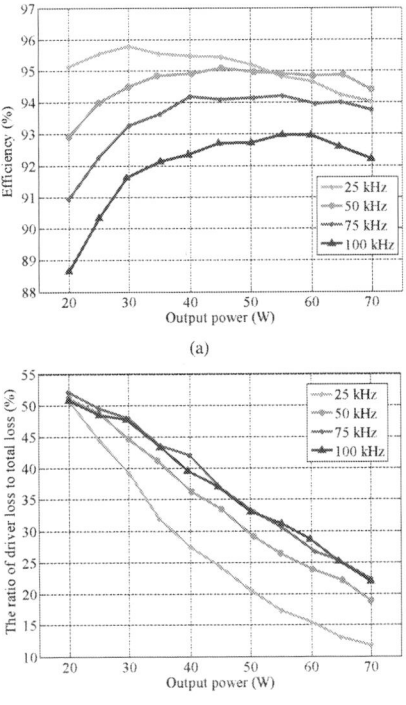

(a)

(b)

Fig. 7.   (a) Measured efficiency versus output power plot of the converter (including the driver's loss) and (b) the ratio of driver's power loss to the total power loss at different switching frequencies.

selected switches are 40 V. Based on the capacitor DC-bias characteristic parameter, the effective capacitance value for C_{1A} and C_{1B} is about 293 μF, for C_{2A} and C_{2B} is about 146 μF, and for C_{3A} and C_{3B} is about 73 μF.

Fig. 5 shows the key experimental waveforms of the converter operation at 5 V input voltage, 70 W output power and 50 kHz switching frequency. The peak-to-peak ripple of the output voltage is about 0.6 V, which is 1.56% of the output voltage. The capacitor voltage waveforms are shown in Figs. 5(c) and 5(d). The converter output voltage ripple is

978-1-4673-9551-9/16 $31.00 © 2016 IEEE

reduced as the converter works in complementary way, and the low voltage ripple contributes to potential high efficiency.

The experimental results in Fig. 7 are measured using a PM6000 power analyzer. Fig. 6 shows the load regulation of the proposed converter with open-loop control. The experiment is performed under 5 V input voltage and 50 kHz switching frequency. The baseline voltage is 40 V, which is the output voltage of the 8×-gain prototype under no load. The experimental results show that the voltage drop is less than 5% for the entire load range at 50 kHz switching frequency.

Fig. 7(a) shows the plots of measured efficiency of the proposed converter as a function of the output power at various switching frequencies. The efficiency of the converter (including the driver's loss) at 25 kHz is up to 95.78%. Fig. 7(b) shows the plots of the ratio of driver's power loss to the total loss over the full load range. It is worth noting that the driver's loss is severe, and is up to 52.5% of total loss at 75 kHz switching frequency. The efficiency of the proposed converter can be further improved if the driver's loss can be further reduced.

IV. CONCLUSIONS

This paper presents an exponential gain bidirectional switched-capacitor (SC) converter which is compact, modular, light and of high efficiency. Compared with other existing SC converters, it has a relatively low component count, overall switch voltage stress and the lowest overall capacitor voltage stress. Besides, no load capacitor is required for this proposed converter in step-up applications. An 8×-gain prototype at 5 V input, 70 W output power with three cascaded cells has been successfully tested to verify the proposed converter performance and each cell area is small at 2.56 inch^2. The prototype's output voltage drop is less than 5% at full load, 50 kHz switching frequency. Moreover, the prototype's efficiency is up to 95.78% (including the driver's loss) at 25 kHz switching frequency. As the driver's loss is up to 52.5% loss at 75 kHz switching frequency, the efficiency of the prototype can be further improved if the driver's loss can be further reduced.

ACKNOWLEDGEMENT

This work is fully supported by the Hong Kong Research Grant Council under GRF project 17207314.

REFERENCES

[1] F. Z. Peng, M. I. Gebben, and B. Ge, "A compact nX DC–DC converter for photovoltaic power systems," in *IEEE Energ. Conv. Cong. and Expo. (ECCE)*, pp. 4780–4784, Sep. 2013.

[2] I. Laird and D. D. Lu, "High step-up DC/DC topology and MPPT algorithm for use with a thermoelectric generator," *IEEE Trans. on Power Electron.*, vol. 28, no. 7, pp. 3147–3157, Jul. 2013.

[3] J. W. Post, J. Veerman, H. V. M. Hamelers, G. J. W. Euverink, S. J. Metz, K. Nymeijer, and C. J. N. Buisman, "Salinity-gradient power: evaluation of pressure-retarded osmosis and reverse electrodialysis," *Jour. of Membr. Sci.*, vol. 288, no. 1–2, pp. 218–230, Feb. 2007.

[4] I. Barbi and R. Gules, "Isolated DC–DC converters with high-output voltage for TWTA telecommunication applications," *IEEE Trans. on Power Electron.*, vol. 18, no. 4, pp. 975–984, Jul. 2003.

[5] S. Lineykin and S. Ben-Yaakov, "Modeling and analysis of thermoelectric modules," *IEEE Trans. on Ind. Electron.*, vol. 43, no. 2, pp. 505–512, Mar.–Apr. 2007.

[6] J. Veerman, R. M. de Jong, M. Saakes, S. J. Metz, and G. J. Harmsen, "Reverse electrodialysis: comparison of six commercial membrane pairs on the thermodynamic efficiency and power density," *Jour. of Membr. Sci.*, vol. 343, no. 1–2, pp. 7–15, Nov. 2009.

[7] D. Cao, W. Qian, and F. Z. Peng, "A high voltage gain multilevel modular switched-capacitor DC–DC converter," in *IEEE Energ. Convers. Congr. and Exposit. (ECCE)*, pp. 5749–5756, Sep. 2014.

[8] W. Li, Y. Zhao, Y. Deng, and X. He, "Interleaved converter with voltage multiplier cell for high step-up and high-efficiency conversion," *IEEE Trans. on Power Electron.*, vol. 25, no. 9, pp. 2397–2408, Sep. 2010.

[9] B. Yuan, X. Yang, D. Li, Y. Pei, J. Duan, and J. Zhai, "A current-fed multiresonant converter with low circulating energy and zero-current switching for high step-up power conversion," *IEEE Trans. on Power Electron.*, vol. 26, no. 6, pp. 1613–1619, Jun. 2011.

[10] V. T. Liu and L. J. Zhang, "Design of high efficiency boost-forward-flyback converters with high voltage gain," in *IEEE Int. Conf. on Contr. & Autom.*, pp. 1061–1066, Jun. 2014.

[11] R. W. Erickson and D. Maksimovic, *Fundamentals of Power Electronics*, 2nd ed., Norwell, MA: Kluwer, 2001.

[12] S. Choi, V. G. Agelidis, J. Yang, D. Coutellier, and P. Marabeas, "Analysis, design and experimental results of a floating-output interleaved-input boost-derived DC–DC high-gain transformer-less converter," *IET Power Electron.*, vol. 4, no. 1, pp. 168–180, Jan. 2011.

[13] M. Mihara, Y. Terada, and M. Yamada, "Negative heap pump for low voltage operation flash memory," in *Symp. VLSI Circ. Dig. Tech. Pap.*, pp. 76–77, Jun. 1996.

[14] J. S. Brugler, "Theoretical performance of voltage multiplier circuits," *IEEE Jour. of Solid-state Circ.*, vol. 6, no. 3, pp. 132–135, Jun. 1971.

[15] F. Ueno, T. Inoue, I. Oota, and I. Harada, "Emergency power supply for small computer systems," in *IEEE Int. Symp. Circ. Syst.*, pp. 1065–1068, Jun. 1991.

[16] A. Cabrini, L. Gobbi, and G. Torelli, "Voltage gain analysis of integrated Fibonacci-like charge pumps for low power applications," *IEEE Trans. on Circ. and Syst. II: Exp. Briefs*, vol. 54, no. 11, pp. 929–933, Nov. 2007.

[17] L. Gobbi, A. Cabrini, and G. Torelli, "A discussion on exponential-gain charge pump," in *Eur. Conf. Circ. Theory Des.*, pp. 615–618, Aug. 2007.

[18] L. K. Chang and C. H. Hu, "High efficiency MOS charge pumps based on exponential-gain structure with pumping gain increase circuits," *IEEE Trans. on Power Electron.*, vol. 21, no. 3, pp. 826–831, May 2006.

[19] C. Y. Tsui, H. Shao, W. H. Ki, and F. Su, "Ultra-low voltage power management and computation methodology for energy harvesting applications," in *Symp. on VLSI Circ.*, pp. 316–319, Jun. 2005.

[20] F. H. Khan and L. M. Tolbert, "A multilevel modular capacitor clamped DC–DC converter," *IEEE Trans. on Ind. Electron.*, vol. 43, no. 6, pp. 1628–1638, Nov.–Dec. 2007.

[21] F. Z. Peng and F. Zhang, "A novel compact DC/DC converter for 42 V systems," *Power Electron. in Trans.*, pp. 143-148, Oct. 2002.

[22] M. D. Seeman and S. R. Sanders, "Analysis and optimization of switched-capacitor DC–DC Converters," *IEEE Trans. on Power Electron.*, vol. 23, no. 2, pp. 841–851, Mar. 2008.

[23] M. S. Makowski and D. Maksimovic, "Performance limits of switched-capacitor dc–dc converters," *IEEE Power Electron. Spec. Conf.*, pp. 1215-1221, Jun. 1995.

[24] C. K. Cheung, S. C. Tan, C. K. Tse, and A. Ioinovici, "On energy efficiency of switched-capacitor converters," *IEEE Trans. on Power Electron.*, vol. 28, no. 2, pp. 862–876, Feb. 2013.

Study and Implementation of a High Step-Up Voltage DC-DC Converter Using Coupled-Inductor and Cascode Techniques

Tsorng-Juu Liang, Yung-Ting Huang, Jian-Hsing Lee, and Lo Pang-Yen Ting

Green Energy Electronics Research Center (GREERC)/ Advanced Optoelectronic Technology Center (AOTC)
Department of Electrical Engineering, National Cheng Kung University, Tainan, Taiwan
Email: tjliang@mail.ncku.edu.tw

Abstract—A novel high efficiency high step-up DC-DC converter is proposed in this paper. The proposed converter can achieve high voltage ratio with appropriate duty cycle by using the coupled-inductor and cascode technique. In order to achieve high efficiency, a capacitor is used to recycle the leakage energy of the coupled-inductor and reduce the voltage stress on the switch. Therefore, the low-voltage rating MOSFET with low conduction resistance can be used. The operational principles and steady-state analysis of the proposed converter are discussed in detail. Finally, a prototype circuit with input voltage 24 V, output voltage 200 V and output power 250 W is implemented to verify the performances of the proposed converter. The experimental results reveals that the highest efficiency of the proposed converter is 94.6%, the full load efficiency is 92.3%, and the 10% load efficiency is 93.8%.

Keywords—high step-up, cascode technology, coupled-inductor.

I. INTRODUCTION

Many nonisolated topologies were proposed to obtain high step-up voltage gain in the past decade. These nonisolated converters can be used with coupled inductor technique, voltage-lifttechnique[1]–[3], switched capacitor techniques[4]–[5], cascode technique[6]–[8],and cascade technique [9]–[11] to obtain high voltage gain with the appropriate duty ratio.But However, the leakage inductor of the coupled inductor occurs a voltage spike on the main switch and affects the conversion efficiency. For this reason, the converters using the coupled-inductor technique with an active clamp circuit have also been proposed. But the active clamp circuit will add extra cost. High step-up gain can be achieved by the use of the switched capacitor technique. However, the main switch will suffer high transient current, and the conduction loss is high. Another method for achieving high step-up gain is the use of the voltage-lift technique . However, it has the same drawback.

This paper proposes a high efficiency, high step-up voltage gain converter,which uses the coupling inductor and cascode techniques. The system configuration of the proposed nonisolated dc–dc topology is shown in Fig. 1. The transformer includes magnetizing inductance L_m and leakage inductances L_{kp} and L_{ks}. This converter consists of a DC input voltage V_{dc}, one power switch, one coupled inductor, three diodes and three capacitors. The boost converter is adopted to generate a stable voltage V_{c1} and supply the energy for the V_{c3} when the S_1 turn on.

Additionally, the diode D_1 is tuned on when switch S_1 is in turn off period, and the energy stored in the leakage inductance is recycled into C_1 the voltage across switch S_1 is clamped at a low voltage level. Since switch S_1 has a low voltage rating and low conducting resistance $r_{ds}(on)$, the proposed converter has high efficiency. The cascode technique is applied to plus the voltage V_{c2} and V_{c3}, obtain a higher output voltage V_o. Furthermore, the turn ratio of the coupled inductor is adjusted to achievea high step-up voltage gain. Finally, a prototype circuit with 24-V input voltage, and 400-V output voltage, and 200-W output power is implemented in the laboratory to verify the performance.

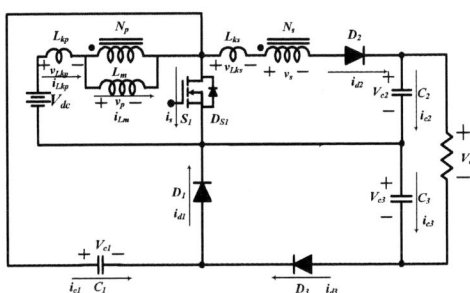

Fig. 1. Circuit configuration of the proposed converter

The features of the proposed nonisolated converter are as follows.

1) High step-up voltage gain.
2) Energy of leakage inductor L_{kp} is released to C_1 thus, increasing the efficiency.
3) Voltage stress of MOSFET is clamped by C_1, so this circuit can use low voltage rated switch.
4) High voltage step-up converter is achieved at appropriate duty cycle.

II. OPERATIONAL PRINCIPLE

In order to simplify the analysis of the proposed converter,the following are assumed over one switching period.

1) Capacitors C_1, C_2, and C_3 are large enough; thus,V_{c1}, and V_{c3} are regarded as constant values.
2) Active switches and all diodes are regarded as ideal.

978-1-4673-9551-9/16 $31.00 © 2016 IEEE

3) Turn ratio of the coupled inductor $n = N_s / N_p$

4) Parasitic inductors, capacitors, and resistors of circuit traces are ignored.

The proposed converter operating in continuous conduction mode (CCM) and discontinuous conduction mode (DCM) are analysed as follows.

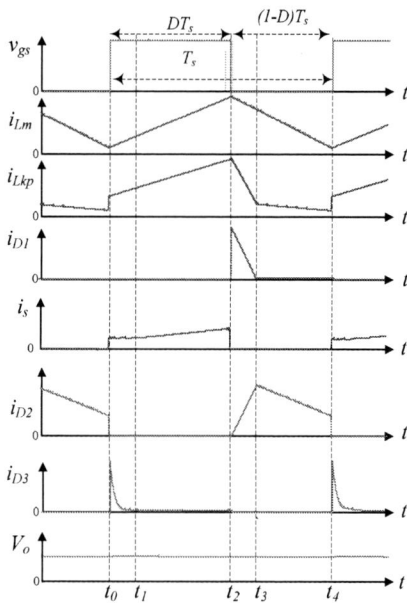

Fig. 2. The key waveforms of the proposed converter operated in (CCM).

A. Continuous Conduction Mode

Based on the above assumption, the operating principle of CCM is divided into four modes during each switching period. Fig. 2 illustrates some typical key waveforms under CCM operation in one switching period. The operating modes are described as follows:

Mode I [$t_0 \sim t_1$] : At $t = t_0$, S_1 is turned on, D_3 is turned on, D_1 and D_2 are turned off. Fig.3(a) shows the equivalent circuit of the proposed converter in this mode. Since the capacitor C_1 and C_3 are in parallel connection and C_1 is discharged to C_3, the current i_{D3} decreases quickly. The energy stored in C_3 continues to release to the output. Because the magnetising inductor L_m and leakage inductor L_{kp} are charged by V_{dc}, i_{Lm} and i_{Lkp} are increased linearly. The mode ends when $V_{C1} = V_{C3}$ at $t = t_1$.

Mode II [$t_1 \sim t_2$] : At $t = t_1$, S_1 is turned on, D_3 is turned on, D_1 and D_2 are turned off. Fig. 3b presents the equivalent circuit of this mode. In this time interval, the inductances L_{kp} and L_m are charged by the input voltage V_{dc}. The currents, i_{Lm} and i_{Lkp}, are equally increased linearly with a slope of V_{dc} / L_m ($L_m \gg L_{kp}$). The load energy is supplied by output capacitors C_2 and C_3. This mode is ended when the

switch S_1 is turned off at $t = t_2$.

Mode III [$t_2 \sim t_3$] : At $t = t_2$, S_1 is turned off, D_1 and D_2 are turned on, D_3 is turned off. Fig.3(c) illustrates the equivalent circuit of the proposed converter in this mode. The energy stored in L_m and L_{kp} are released to capacitors C_1 and C_2 when D_1 and D_2 start to conduct in this mode. Thus, the leakage inductance energy can be recycled to C_1, and the voltage stress of the switch can be limited by V_{C1}. This mode is ended when $i_{Lkp} = i_{D2}$ at $t = t_3$.

Mode IV [$t_3 \sim t_4$] : In this mode, S_1 is still turned off, D_1 and D_2 are turned on, D_3 is turned off. The equivalent circuit of the proposed converter in this mode is shown in Fig. 3(d). The current i_{Lk} and i_{D2} are equally decreased. The load energy is supplied by the output capacitors C_1 and C_2. This mode is ended when the switch S_1 is turned on at $t = t_4$.

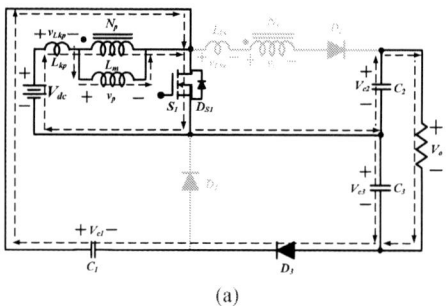

(a)

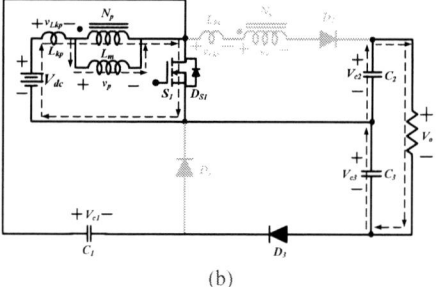

(b)

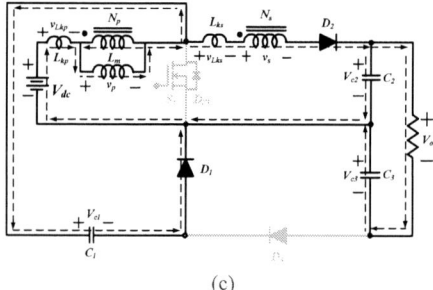

(c)

978-1-4673-9551-9/16 $31.00 © 2016 IEEE

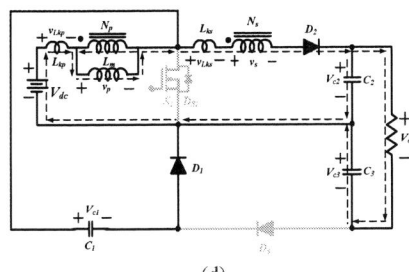

(d)

Fig.3. Equivalent circuits of the proposed c for CCM Operation. (a) Mode I. (b) Mode II. (c) Mode III. (d) Mode IV.

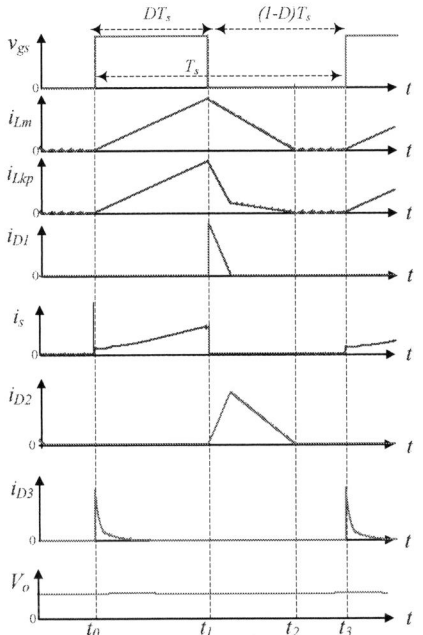

Fig. 4. The key waveforms of the proposed converter operated in (DCM).

B. Discontinuous Conduction Mode

The key waveforms of the proposed converter at discontinuous condition mode (DCM) operation are shown in Fig. 4. To simplify the analysis for DCM operation, the leakage inductances L_{kp} and L_{ks} of the transformer are neglected. Equivalent circuits of the proposed converter operated in DCM are shown in Fig. 5. There are three operating modes in DCM.

Mode I [$t_0 \sim t_1$] : At $t = t_0$, S_1 is turned on, D_3 is turned on, D_1 and D_2 are turned off. The equivalent circuit of the proposed converter is show in the Fig. 5(a). In this interval, the inductor L_m is charged by input voltage V_{dc}. The current i_{Lm} is increasing linearly. The voltage V_{C1} is equal to V_{C3}. The load energy is supplied by by output capacitor C_2 and C_3. This mode ends when switch S_1 is turned off at $t = t_1$.

Mode II [$t_1 \sim t_2$] : At $t = t_1$, S_1 is turned off, D_1 and D_2 are turned on, D_3 is turned off. The equivalent circuit of the proposed converter is show in the Fig. 5(b). In the beginning of this interval, the energy store in L_{kp} is released to C_1 through the D_1. When i_{Lkp} decreases to zero, the current i_{Lm} is released to C_2 through the coupled incuctor. This mode ends when magnetising current i_{Lm} reaches to zero at $t = t_2$.

Mode III [$t_2 \sim t_3$] : At $t = t_2$, i_{Lm} is equal to zero. S_1 is still turned off and D_1, D_2 and D_3 are also turned off. Fig. 5(c) shows the quivalent circuit of the proposed converter in this mode. The energy stored in output capacitor C_2 and C_3 are released to the load during this interval. This mode ends when the switch S_1 is turned on at $t=t_3$.

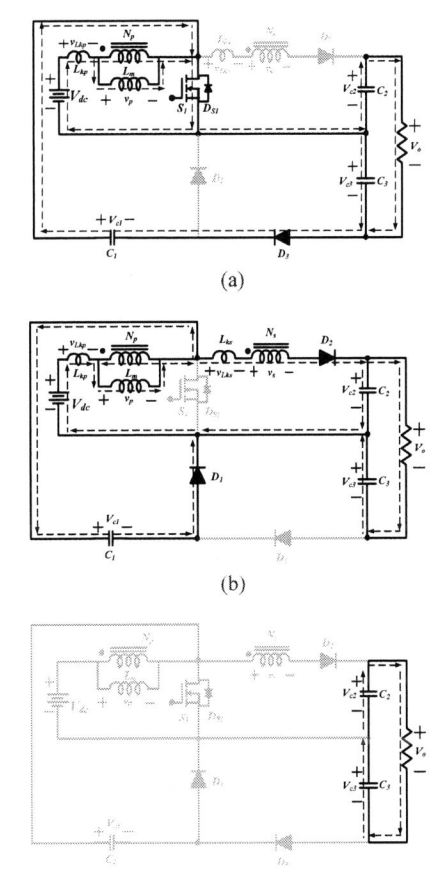

Fig.5. Equivalent circuits of the proposed c for CCM Operation. (a) Mode I. (b) Mode II. (c) Mode III.

III. STEADY- STATE ANALYSIS OF THE PROPOSED CONVERTER

When the proposed converter operated in CCM, the leakage inductance L_{k1} is further smaller than L_m. Therefore, L_{k1} can be neglected in this analysis.

A. CCM Operation

While S_1 on , primary magnetizing inductance voltage:

$$v_p = V_{Lm} = V_{dc} \tag{1}$$

$$V_{c1} = V_{c3} \tag{2}$$

While S_1 off, primary magnetizing inductance voltage:

$$v_s = V_{c1} - V_{c2} \tag{3}$$

$$v_p = V_{dc} - V_{c1} \tag{4}$$

$$v_p = V_{dc} - V_{c1} = \frac{N_p}{N_s}[V_{c1} - V_{c2}] \tag{5}$$

According to the Volt-Second balance law on N_p and N_s of the coupled inductor:

$$\frac{1}{T}\int_0^T v_{Lm}dt = 0 \tag{6}$$

$$V_{dc}DT_s = [V_{c1} - V_{dc}](1-D)T_s \tag{7}$$

$$V_{c1} = \frac{1}{(1-D)}V_{dc} = V_{c3} \tag{8}$$

From (5) and (8) V_{C2} is derived as

$$V_{dc} - V_{c1} = \frac{N_p}{N_s}[V_{c1} - V_{c2}] \tag{9}$$

$$V_{c2} = \frac{1+nD}{1-D}V_{dc} \tag{10}$$

From (8) and (10) V_{C2}, V_{C3} and V_o are derived as

$$V_o = V_{c2} + V_{c3} = \frac{2+nD}{1-D}V_{dc} \tag{11}$$

Equation(9) indicates that the proposed converter accomplishes a high voltage gain by utilising the cascode technique and increasing the turns ratio of the coupled inductor. Fig. 6 plots the ideal voltage gain of the proposed conveerter against the duty ratio, under various turns ratios(n=2 and n=3). These family curves can be used to determine the duty ratio of the converter and the turns ratio of the coupled inductor. Fig. 6 illustrates the voltage gain against duty ratio of the proposed converter and the other coupled inductor-based converters under a fixed turns ratio of coupled inductor, n=2. The proposed converter achieved the highest voltage gain among these counterparts. However, the components of the proposed converter are more than these counterparts.

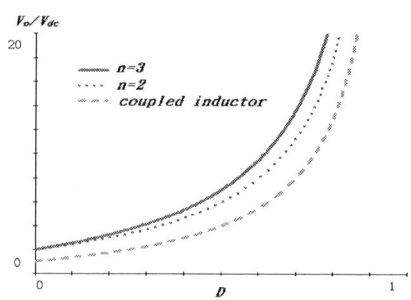

Fig.6. Voltage gain comparison of traditional boost converter and proposed converter at CCM operation.

According to (7) and (8), the voltage stresses on the switch and the diodes are as follows

$$V_{DS} = V_{c1} = \frac{1}{(1-D)}V_{dc} \tag{12}$$

$$V_{D1} = V_{c1} = \frac{1}{(1-D)}V_{dc} \tag{13}$$

$$V_{D2} = V_{c2} + nV_{dc} = \frac{1+n}{(1-D)}V_{dc} \tag{14}$$

$$V_{D3} = V_{c3} = \frac{1}{(1-D)}V_{dc} \tag{15}$$

B . DCM Operation

By using the voltage-second balance principle on N_p and N_s of the coupled inductor, the following equations are derived as:

$$\int_0^{DT}v_{dc}dt + \int_{DT}^{(D+D_L)T}(-v_{c1})dt = 0 \tag{16}$$

$$\int_0^{DT}v_{dc}dt + \int_{DT}^{(D+D_L)T}\frac{1}{n}[v_{c1} - v_{c2}]dt = 0 \tag{17}$$

From (14) and (15), V_{C1}, V_{C2} and V_o are given by

$$V_{c1} = \frac{D+D_L}{D_L}V_{dc} \tag{18}$$

$$V_{c2} = \frac{nD+D+D_L}{D_L}V_{dc} \tag{19}$$

$$V_o = V_{c1} + V_{c2} = [\frac{nD+2D}{D_L} + 2]V_{dc} \tag{20}$$

D_L can be derived from (18)

$$D_L = \frac{(2D+nD)V_{dc}}{V_o - 2V_{dc}} \tag{21}$$

If the proposed converter is operated in boundary-condition mode(BCM), the average value of i_{co} is computed as

$$\frac{V_{dc}}{L_m}DT_s = \frac{2(1+n)}{D_L}\cdot\frac{V_o}{R} \tag{22}$$

The boundary normalized magnetizing inductance time constant is defined as

$$\tau = \frac{L_m}{RT_s} = \frac{(2+n)V_{dc}^2 \cdot D^2}{2 \cdot V_o \cdot (1+n)[V_o - 2V_{dc}]} \quad (23)$$

If the proposed converter is operated in the boundary conduction mode (BCM), then the voltage gain of CCM operation is equal to the voltage gain of DCM operation. From (9) and (29), the boundary normalised inductor time constant τ_{LB} can be derived as:

$$\tau_{LB} = \frac{D(D^2 - 2D + 1)}{2(2n + nD + n^2 + 2)} \quad (24)$$

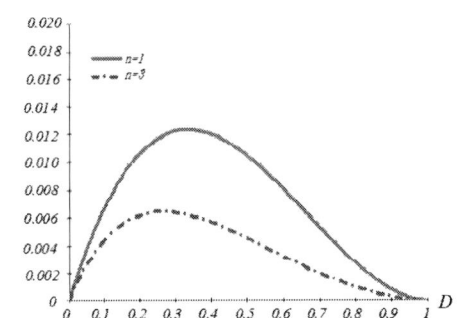

Figure 7   Boundary normalised inductor time constant against duty ratio.

IV. EXPERIMENTAL RESULTS

The laboratory prototype sample is implemented to demonstratemthe practicability of the proposed converter. The system specifications and components are as follows:
1) input dc voltage V_{dc}: 24 V;
2) output dc voltage V_o: 200 V;
3) maximum output power: 250 W;
4) operating frequency: 50 kHz;
5) diodes D_1, D_3: STPS61H100C Schottky diode, and D_2: DSEP30-06A Schottky diode;
6) switches S_1: IXFH160N15T2;
7) transformer: ETD-39, core PC-40, $Np : Ns = 1 : 3$, $Lm = 22.52\ \mu H$, and $L_k = 0.18\ \mu H$;
8) boosting capacitors C_1:100-μF/100V;
9) output capacitor C_2:100-μF/100VandC_3:220-μF/200V;

Figure 8 to figure 10 are the main waveforms at light load(I_o = 0.25 A), half load(I_o = 0.625 A) and full load(I_o = 1.25 A), respectively. As shown in figure 8(b), when the secondary current i_{D2} decreases to zero, the load receives the energy from the output capacitor. The proposed converter is operated in DCM. As shown in figure 8(c), the coupled inductor L_m and the output capacitor of the switch are resonant when the switch S_1 is at the off state. Figure 9(b) shows that the switch is turned on when the current i_{D2} decreases to zero. Therefore, the proposed converter is

operated in BCM. As shown in figure 9(b), the waveform of the current i_{D2} shows that the proposed converter is operated in CCM. According to figure 10(a), the current i_{ds} is the sum of the magnetizing current and the charging current from the capacitor C_1 to C_3. The turn-on transient of the current i_{ds} is caused by the resonance of leakage inductance and the capacitance. The current i_{D3} flowing through the resistance R_{ds} increases as the power increases, which causes higher conduction loss. According to figure 10(c), the voltage stress of the switch is lower than 100 V. Therefore, the switch can be chosen with a small on-resistance so that the conduction loss is reduced to increase the efficiency. The experimental results can verify the feasibility of the proposed converter.

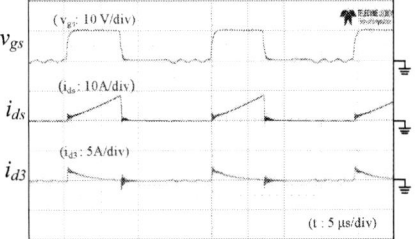

Fig. 8(a) The waveform of v_{gs}, i_{ds} and i_{d3} at P_o=50W

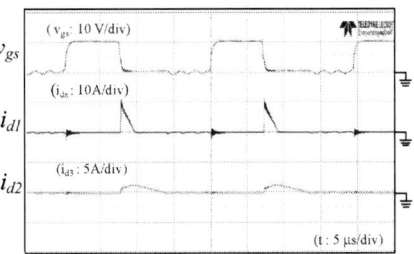

Fig. 8(b) The waveform of v_{gs}, i_{d1} and i_{d2} at P_o=50W

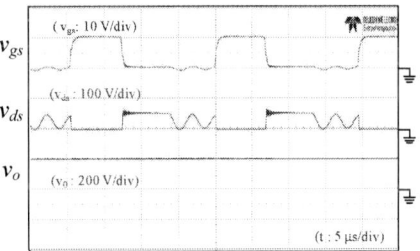

Fig. 8(c) The waveform of v_{gs}, v_{ds} and v_o at P_o=50W

Fig. 9(a) The waveform of v_{gs}, i_{ds} and i_{d3} at P_o=125W

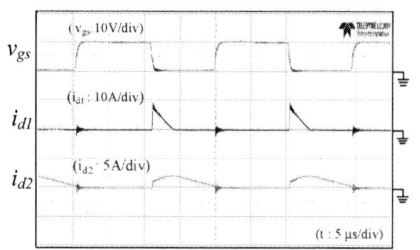

Fig. 9(b) The waveform of v_{gs}, i_{d1} and i_{d2} at P_o=125W

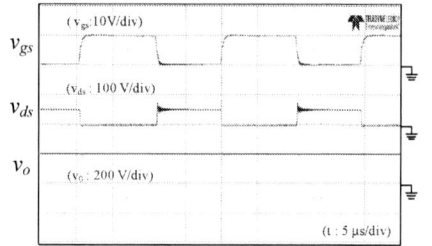

Fig. 9(c) The waveform of v_{gs}, v_{ds} and v_o at P_o=125W

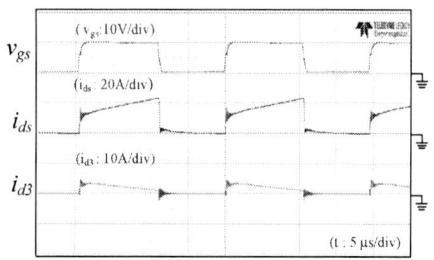

Fig. 10(a) The waveform of v_{gs}, i_{ds} and i_{d3} at P_o=250W

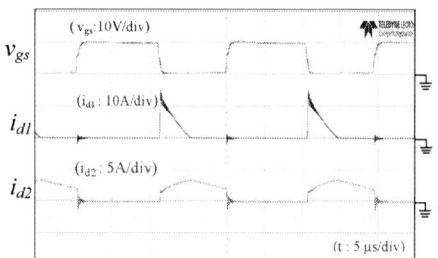

Fig. 10(b) The waveform of v_{gs}, i_{d1} and i_{d2} at P_o=250W

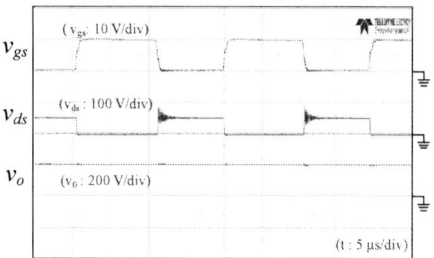

Fig. 10(c) The waveform of v_{gs}, v_{ds} and v_o at P_o=250W

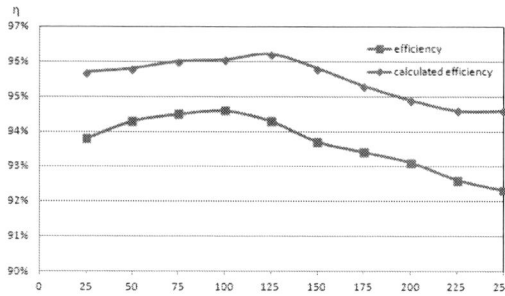

Fig. 11. Experimental conversion efficiency of proposed is converter.

Fig. 11 shows the experimental conversion efficiency of the proposed converter. The maximum efficiency is 94.6%, and efficiency is 92.3% at full load. High current on the primary side causes higher conduction loss and resistance R_T of transformer loss under high-power condition. Thus, the efficiency is decreased at high power. Hence, the efficiency decays when the output power increases.

V. CONCLUSION

This paper presents a step-up DC–DC converter which uses Cascode technique, adjusts the turns ratio of the coupled inductor and achieve high step-up voltage gain. The proposed converter is high efficiency because it recycles the leakage energy of the coupled inductor. Moreover, the voltage across the switch is clamped at the lower voltage level, so the converter can use low voltage rating switch to improve efficiency. Finally, a prototype converter is implemented to verify the performance of the proposed converter. The CCM and DCM operating principles and steady-state of voltage gain have been analyzed in detail. Finally, a prototype of the proposed converter with an input voltage of 24V, an output voltage of 200V, an output power of 250W, a maximum efficiency of 94.6%, and switch voltage spikes lower than 100 V has been achieved in the laboratory.

ACKNOWLEDGEMENT

The authors gratefully acknowledge financial support from the Ministry of Science and Technology, Taiwan under project No. 104-2221-E-006-085 and No. 103-2221-E-006-105-MY3.

REFERENCES

[1] F. L. Luo "Seven Self-lift DC-DC Converters, Voltage Lift Technique," *IEEE proc . on Electric Power Applications*, vol 148, no 4, pp. 329 - 338, Jul. 2001.

[2] L. S. Yang, T. J. Liang and J. F. Chen "Transformerless DC–DC Converters With High Step-Up Voltage Gain," *IEEE proc. on Electric Power Applications*, vol 56, no 1, pp.3144 - 3152, Aug. 2009.

[3] P. H. Kuo, T. J. Liang, K. C. Tseng, J. F. Cheng, and S. M. Cheng, "An Isolated High Step-Up Forward/Flyback Active-Clamp Converter with Output Voltage Lift," *Energy Conversion Congress and Exposition (ECCE)*, vol. 152, no. 2, pp. 542-548, Sept. 2010.

[4] A. Ioinovici, "Switched-capacitor Power Electronics Circuits," *IEEE Trans. On Circuits and System*, vol. 1, no. 4, pp. 37-42, 2001.

[5] Y. P. Hsieh, J. F. Chen, T. J. Liang and L. S. Yang, "Novel High Step-Up DC–DC Converter With Coupled-Inductorand . Switched-Capacitor Techniques," *IEEE Trans. on Industrial Electronics*, vol. 59, pp. 998-1007, Feb. 2012.

[6] B. Axelrod, Y. Berkovich, and A. Ioinovici, "Switched Coupled-inductor cell for DC-DC Converters with Very Large Conversion Ratio," *IEEE Industrial Electronics*, IECON, pp. 2366 - 2371, Nov. 2006.

[7] L. S. Yang, T. J. Liang, H. C. Lee and J. F. Chen, "Novel High Step-Up DC–DC Converter With Coupled-Inductor and Voltage-Doubler Circuits," *IEEE Trans. on Industrial Electronics*, vol. 58, pp. 4196-4206, Sept 2011.

[8] F. L. Luo and H. Ye, "Positive Output Cascade Boost Converters," *IEE Proc. on Electric Power Applications*, vol. 151, no. 5, pp. 590-606, Sep. 2004.

[9] Q. Zhao and F. C. Lee, "High-efficiency, High Step-up DC-DC Converters," *IEEE Trans. on Power Electronics*, vol. 18, no. 2, pp. 65-73, Jan. 2003.

[10] T. F. Wu, T. S. Lai, J. C. Hung, and Y. M. Chen, "Boost Converter with Coupled Inductors and Buck-boost Type of Active Clamp," *IEEE Trans. on Industrial Electronics*, vol. 55, no. 1, pp.154-162, Jan. 2008.

[11] X. Wu, J. Zhang, X. Ye, and Z. Qian, "A Family of Non-Isolated ZVS DC–DC Converter Based on A New Active Clamp Cell," *In proc. IEE IECON* , 2005.

20 mV Input, 4.2 V Output Boost Converter with Methodology of Maximum Output Power for Thermoelectric Energy Harvesting

Taichi Ogawa[1], Takeshi Ueno[1], Takayuki Miyazaki[2] and Tetsuro Itakura[1]

Corporate Research & Development Center, Toshiba Corporation, Kawasaki, Japan[1],
Center for Semiconductor Research & Development, Toshiba Corporation, Kawasaki, Japan[2]
Email: taichi.ogawa@toshiba.co.jp

Abstract— **This paper presents a low-input-voltage boost converter for thermoelectric energy harvesting. In environments with 1-2 K thermal difference, such as in the case of a body-wearable application, TEG generates several micro watts and tens of mV to the boost converter. For the low-input-voltage operation, the power consumption of the control circuit of the proposed boost converter is reduced by using a duty-cycle bandgap reference voltage circuit. In order to maximizing the output power, the input voltage of the boost converter is conventionally set to half of an open voltage of a thermoelectric generator (TEG). In this setting, however, conduction losses such as those of an inductor and a power switch are not considered. The conventional boost converter cannot output the maximum output power. We propose a methodology of the maximum output power considering the conduction losses in the boost converter. The boost converter was implemented in a 0.13 μm CMOS process. The output voltage can be boosted from 20 mV input which is the open voltage of TEG of 1.5 Ω source resistance. The measurement results show that the output power is increased by approximately 10% using the proposed methodology for the open voltage between 20 mV and 100 mV.**

Keywords—DC-DC converter; Thermolelectric generetor; Energy harvest; Maximum power point; Low-input voltage; Low power

I. INTRODUCTION

For sensor ICs operating for a long time without maintenance, development of a boost converter connected to a thermoelectric generator (TEG) and a secondary battery such as a Li-ion battery is underway [1-3]. In environments with 1-2 K thermal difference, such as in the case of a body-wearable application, TEG generates several micro watts and tens of mV to the boost converter. Since the output power of the TEG is very small, power consumption in a control circuit of the boost converter should be reduced. Furthermore, the impedances of the boost converter and the TEG should be matched so that maximum output power at the boost converter is achieved. With the conventional methodology, the input voltage of the boost converter is set to half of an open voltage of the TEG [2,4-5]. In this setting, however, conduction losses such as those of an inductor and a power switch are not considered. The conventional boost converter cannot output the maximum output power. We propose a methodology to achieve the

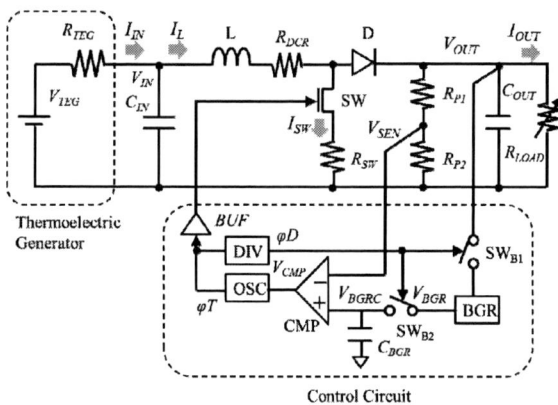

Fig. 1. Proposed circuit.

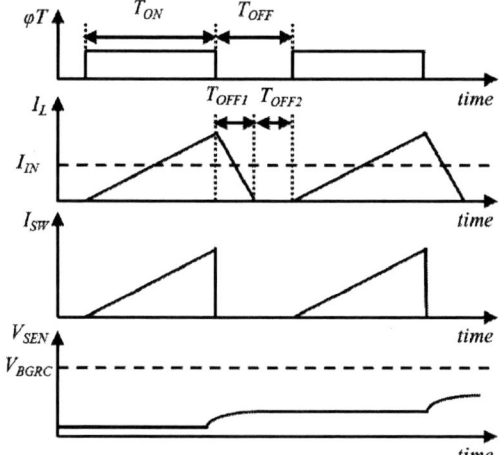

Fig. 2. Waveforms of power switch's clock, inductor current, power switch's current and output-sensing voltage.

maximum output power where the conduction losses in the boost converter are taken into account.

II. METHODOLOGY OF MAXIMUM OUTPUT POWER FOR BOOST CONVERTER FROM THERMOELECTRIC GENERATOR

Fig.1 shows the proposed boost converter and the TEG. The TEG is modeled with a voltage source, V_{TEG}, and a source resistance, R_{TEG}. An input voltage, V_{IN}, is applied to the proposed boost converter from the TEG and an output power, P_{OUT}, is provided into an output capacitor, C_{OUT}, and a load, R_{OUT} by the proposed converter. Then an output voltage of the proposed converter is V_{OUT}. A control circuit turns a power switch, SW, on and off until a divided output voltage, V_{SEN}, reaches a reference voltage, V_{BGRC}.

Fig.2 shows the waveforms of inductor current, I_L, SW's current, I_{SW}, V_{SEN}, V_{BGRC}, and a timing signal, φT, of a driving voltage to SW. φT consists of fixed T_{ON} and T_{OFF} by an oscillator, OSC, of the control circuit. During T_{ON}, SW is on, and I_L increases. During T_{OFF}, SW is off, and I_L decreases. During T_{OFF1}, I_L is supplied to the output through a diode, D. During T_{OFF2}, I_L is 0A. Since the TEG outputs several mV, φT is set to satisfy $T_{ON} \gg T_{OFF}$ and to operate in discontinuous current mode (DCM). Considering the small V_{IN} and the parasitic resistances of L, SW and wiring, the maximum output power methodology is explained.

P_{OUT} is given by the difference between P_{IN} and $(P_{LOSS}+P_{CIR})$. Here, proposed boost converter input power, P_{IN}, which is the TEG output power, is obtained by V_{IN} and a RMS value of I_{IN}. The RMS value of I_{IN} equals to an average current of I_{IN} by a capacitor, C_{IN}. P_{LOSS} is the power consumption at R_{DCR}, R_{SW} and D. Since $T_{ON} \gg T_{OFF1}$, the RMS value of current through D is much small. The power consumption at D can be neglected. Moreover the RMS value of I_L, I_{LRMS}, can be approximately equal to the RMS value of I_{SW}. P_{CIR} is the fixed power consumption of the control circuit.

P_{OUT} is given as

$$P_{OUT} = P_{IN} - P_{LOSS} - P_{CIR}$$
$$= V_{IN}I_{IN} - I_{LRMS}{}^2 R_{DCR} - I_{SWRMS}{}^2 R_{SW} - I_{DRMS}V_D - P_{CIR}$$
$$\approx V_{IN}I_{IN} - I_{LRMS}{}^2(R_{DCR}+R_{SW}) - P_{CIR} \qquad .(1)$$

V_D is a diode forward voltage of D. I_{DRMS} is a RMS value of D's current.

The average of I_{IN} is equal to the average of I_L, I_{LAVG}. Note that the RMS value of I_L is not equal to the average of I_L. I_{IN} is given as

$$I_{IN} = I_{LAVG} = \frac{1}{T_{ON}+T_{OFF}} \int_0^{T_{ON}-T_{OFF}} I_L(t)\,dt \approx \frac{1}{T_{ON}} \int_0^{T_{ON}} \frac{V_{IN}}{L} t\,dt = \frac{T_{ON}V_{IN}}{2L} \qquad (2)$$

By contrast, I_{LRMS} is shown as

$$I_{LRMS} = \sqrt{\frac{1}{T_{ON}+T_{OFF}} \int_0^{T_{ON}+T_{OFF}} \left(\frac{V_{IN}}{L}t\right)^2 dt} \approx \frac{T_{ON}V_{IN}}{\sqrt{3}\,L} \qquad (3)$$

V_{IN} is given as

$$V_{IN} = V_{TEG} - R_{TEG}I_{IN} \qquad (4)$$

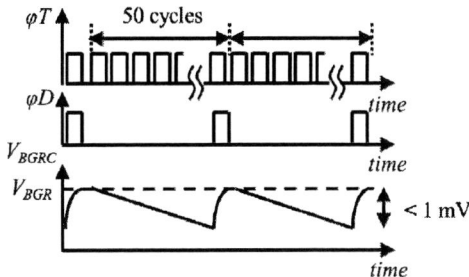

Fig. 3. Waveforms of power switch's input clock, reference-voltage's clock and capacitor voltage.

From (1)-(4), P_{OUT} is shown as

$$P_{OUT} = \frac{V_{IN}(V_{TEG}-V_{IN})}{R_{TEG}} - \frac{4(V_{TEG}-V_{IN})^2(R_{DCR}+R_{SW})}{3R_{TEG}{}^2} - P_{CIR} \cdot (5)$$

From (5), V_{INOPT}, which is V_{IN} at the maximum output power, P_{OUTMAX}, is calculated as from $\partial P_{OUT}/\partial V_{IN}=0$,

$$V_{INOPT} = \frac{8(R_{DCR}+R_{SW})+3R_{TEG}}{8(R_{DCR}+R_{SW})+6R_{TEG}} V_{TEG} \qquad (6)$$

By ignoring $(R_{DCR}+R_{SW})$ in (6), the conventional methodology sets $V_{IN}=V_{TEG}/2$. However, this input voltage is not optimal because $(R_{DCR}+R_{SW}) \ll R_{TEG}$ does not always hold.

Furthermore, from (1)-(4), (6). P_{OUTMAX} is also obtained as

$$P_{OUTMAX} = \frac{2LT_{ON}-\frac{4}{3}(R_{DCR}+R_{SW})T_{ON}{}^2}{(R_{TEG}T_{ON}+2L)^2} V_{TEG}{}^2 - P_{CIR} .(7)$$

From (7), T_{ONOPT}, that is T_{ON} at P_{OUTMAX}, is shown as from $\partial P_{OUT}/\partial T_{ON}=0$,

$$T_{ONOPT} = \frac{6L}{8(R_{DCR}+R_{SW})+3R_{TEG}} \qquad (8)$$

Since T_{ONOPT} does not depend on V_{TEG}, T_{ONOPT} can be fixed. In the proposed boost converter, T_{ON} can be set to T_{ONOPT} at OSC to obtain the maximum output power.

III. ARCHITECTURE OF CONTROL CIRCUIT

The control circuit of the proposed boost converter consists of OSC, a comparator, CMP, a bandgap reference circuit, BGR, a divider, DIV, a buffer amp, BUF, and switches SW_{B1} and SW_{B2}. Power supply of the control circuit is provided by the output voltage, V_{OUT}, of the boost converter. Fig.3 shows the waveforms of φT, φD and the voltage, V_{BGRC} of the capacitor, C_{BGR}, where φD controls the on and off of BGR.OSC consists of current-starved inverter chain and outputs φT to BUF and DIV. DIV outputs φD, so that the on and off time of SW_{B1} and SW_{B2} are set to 1:49. BGR outputs the reference voltage, V_{BGR} during SW_{B1} on. For low power consumption of the control circuit, BGR is driven at duty-cycle operation by φD. Since

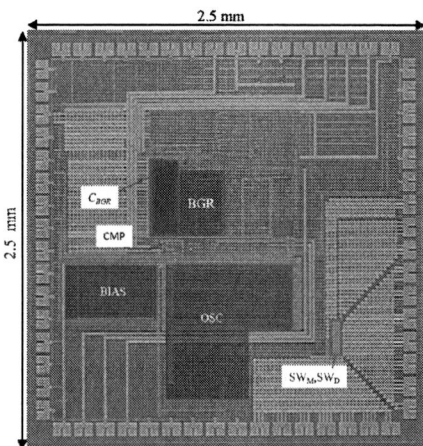

Fig. 4. Chip photo.

Table 1. Parameters of the proposed circuit.

L	47 μH
R_{TEG}	1.5 Ω
R_{DCR}	0.6 Ω
R_{SW}	0.29 Ω
C_{IN}	100 μF
C_{BGR}	7.5 pF
$R_{P1}+R_{P2}$	4.2 MΩ
F_{SW}	25 kHz

Table 2. Power consumption of the control circuit.

BUF	4.1 μW
OSC	5.7 μW
BIAS	3 μW
CMP	2 μW
$R_{P1}+R_{P2}$	1 μW
BGR+DIV	1.3 μW
total	17.1 μW

OSC is shared between the boost converter and BGR, a new dedicated oscillator for BGR is unnecessary. During the on time, BGR refreshes V_{BGRC} to V_{BGR}. During the off time, V_{BGR} is kept by C_{BGR}. V_{SEN} is V_{OUT} divided by R_{P1} and R_{P2}. CMP compares V_{SEN} and V_{BGRC}. When $V_{BGRC} < V_{SEN}$, OSC is enabled and outputs φT to BUF. During $V_{BGRC} > V_{SEN}$, OSC is disabled and output φT and φD are set to "Low" and "Hi" respectively. Since the proposed circuit only operates at DCM, a compensator is unnecessary.

IV. MEASUREMENT RESULTS

The proposed boost converter is implemented in 0.13 μm CMOS process. Fig.4 shows the chip photo. Table 1 shows the

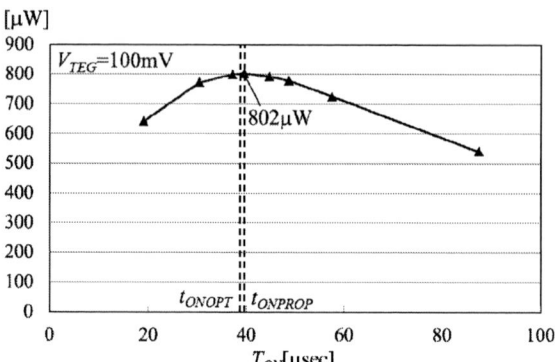

Fig. 5. Measurement results of P_{OUT} versus T_{ON}.

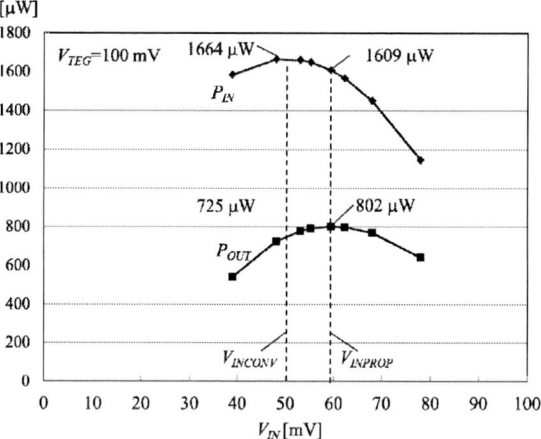

Fig. 6. Measurement result of P_{IN} and P_{OUT} versus V_{IN}.

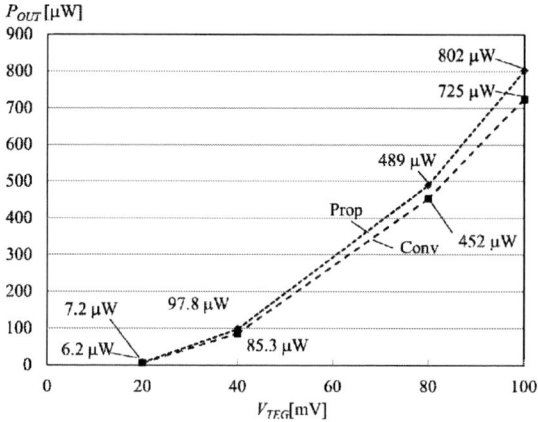

Fig. 7. Measurement results of P_{OUT} versus V_{TEG}.

parameters of the proposed circuit. The input of the boost converter is connected to R_{TEG} and V_{TEG} that simulate a TEG. R_{TEG} is set to 1.5 Ω. V_{OUT} is set to 4.2 V.

978-1-4673-9551-9/16 $31.00 © 2016 IEEE   1909

Table 3. Comparison with state-of-the-art low-input-voltage boost converters.

	This work	[2]	[3]	[4]	[5]
Min. V_{TEG}	20 mV	70 mV	20 mV	25 mV	20 mV
V_{OUT}	4.2 V	3 V-5.8 V	2.35 V-5 V	2.4-5 V	1.1 V
R_{TEG}	1.5 Ω	8 Ω	1 Ω-10 Ω	5 Ω	2 Ω
η@V_{TEG}=0.1 V	49 % (end-to-end)	65 % (just boost converter)	27 % (just boost converter)	59 % (end-to-end)	75 % (end-to-end)

Table 2 shows the measurement results of the power consumptions of each block in the control circuit. The power consumption of BGR decreases from 32.3 μW at steady operation to 1.3 μW at duty-cycle operation. V_{BGRC} is 1.203 V and its variation by the duty-cycle operation is less than 1mV. The total power P_{CIR} is 17 μW. Fig. 5 shows P_{OUT} versus T_{ON} at V_{TEG}=100 mV. For maximizing P_{OUT}, the proposed theoretical value of T_{ONOPT} is 38.6 μsec from eq. (8). Then, in proposed boost converter, T_{ON} can be set T_{ONPROP} 39.5 μsec, which is the closest to T_{ONOPT} by setting the OSC. The maximum output power of 802 μW is obtained at T_{ONPROP}. Fig. 6 shows the P_{IN}, P_{OUT} versus V_{IN} at V_{TEG}=100 mV by same measurement result of Fig. 5. In the conventional methodology where V_{IN} is set to V_{INCONV} (=50 mV=V_{TEG}/2), the input power, P_{IN}, is maximized. Then, P_{OUT} at V_{INCONV} is 725 μW. VIN, that a voltage of T_{ONPROP}, is 61.9 mV. V_{IN} of obtaining the maximum output power is not V_{INCONV} but V_{INPROV}. Fig.7 shows P_{OUT} using the proposed methodology and the conventional methodology versus V_{TEG}. The improvement of the output power can be increased by around 10 % from V_{TEG}=20 mV to 100 mV.

Table 3 shows the comparison with state-of-the-art low-input, high-output boost converters. The end-to-end efficiency is the output power divided by the maximum input power from the TEG. The proposed boost converter circuit can be operated at the lowest open voltage of the TEG. The proposed boost converter can output higher voltage than [5], and can be operated at higher efficiency than [3].

V. CONCLUSIONS

We presented a low-input-voltage boost converter for thermoelectric energy harvesting. In order to achieving the maximum output power, the proposed methodology, which includes the effect of resistance components at the boost converter, is used. Furthermore, for low-input-voltage operation, the power consumption of the BGR in the control circuit is decreased by duty-cycle operation using the oscillator of the boost converter. The output power is increased by 10 % compared with the conventional methodology.

REFERENCES

[1] S.Lineykin and S. Ben-Yaakov, "Modeling and Analysis of Thermoelectric Modules," IEEE Transactions on Industry Applications, vol.43, no.2, pp.505-512, 2007.

[2] J. Kim and C. Kim, "A DC–DC Boost Converter with Variation-Tolerant MPPT Technique and Efficient ZCS Circuit for Thermoelectric Energy Harvesting Applications," IEEE Transactions on Power Electronics, vol.28, no.8, pp.3827-3833, 2013.

[3] Linear Technology, LTC 3108 datasheet, 2010, http://cds.linear.com/docs/en/datasheet/3108fc. pdf.

[4] Y. K. Ramadass and A. P. Chandrakasan, "A Battery-Less Thermoelectric Energy Harvesting Interface Circuit With 35 mV Startup Voltage," IEEE Journal of Solid-State Circuits, vol.46, no.1, pp.333-341, 2011.

[5] A. Shrivastava, N. E. Roberts, O. U. Khan, D. D. Wentzloff and B. H. Calhoun, "A 10 mV-Input Boost Converter With Inductor Peak Current Control and Zero Detection for Thermoelectric and Solar Energy Harvesting With 220 mV Cold-Start and - 14.5 dBm, 915 MHz RF Kick-Start," IEEE Journal of Solid-State Circuits, Aug. 2015

Clarification of Relationship between Current Ripple and Power Density in Bidirectional DC-DC Converter

Hoai Nam Le, Koji Orikawa, and Jun-ichi Itoh
Department of Electrical engineering
Nagaoka University of Technology, NUT
Niigata, Japan
lehoainam@stn.nagaokaut.ac.jp

Abstract— This paper clarifies the relationship between the current ripple and the power density in bidirectional DC-DC converters. In the conventional power density design method, in order to obtain the pareto-front curve of the power density and the efficiency, the current ripple is designed as constant value, whereas the switching frequency is varied. As a result, the possibilities of higher power density or higher efficiency at different current ripple are not considered. Therefore, in this paper, the current ripple is also varied in order to evaluate all the designable power density. Specifically, a design flow chart is introduced to show step-by-step how to express all the losses and the volume of the converter as functions of the current ripple. Several 1-kW prototypes are constructed in order to confirm the validity of the design flow chart. By varying the current ripple, the highest power density of 10.1 kW/dm³ with the efficiency of 98.55% is achieved at the current ripple of 60%. Furthermore, the maximum error between the calculated and experimental power density and efficiency are 19.5% and 0.22 pt. respectively.

Keywords—pareto-front curve; power density; current ripple

I.   INTRODUCTION

In recent years, non-isolated bidirectional DC-DC converters are widely applied to battery systems in Hybrid Electric Vehicles (HEVs), or Uninterruptible Power Supplies. This DC-DC converter is required to have small size and high efficiency. The minimization of converters provides not only the material cost reduction but also the easy implementation for applications which requires space-saving power system. Because the passive components such as inductors and capacitors account for the majority of the volume of DC-DC converters, the miniaturization of these passive components leads to compact DC-DC converter. Many minimization methods have been proposed such as high frequency switching, or resonant circuit [1]-[2]. However, higher switching frequency leads to the increase of switching loss, noise level, and also requires high-speed controller [3]-[4].

The boost inductor is one of the components which contributes mainly to the converter volume. By reducing the inductance, the volume of the boost inductor can be reduced. Instead of increasing the switching frequency, the inductance can be reduced by increasing the current ripple with interleaved topology [5]. In the interleaved topology, the total current ripple can still be maintained at small value despite that the current ripple in each phase is increased. This avoids the

complexity of the input filter design in the applications where the current ripple is strictly limited [6]. However, many challenges arise when the current ripple is increased such as the high flux density ripple, which increases the core loss. Besides, the high current ripple increases the effective current for the same average current. This leads to higher conduction loss in the semiconductor devices, the boost inductor, and higher current rating for the capacitors, which result in larger volume of the heatsink and the capacitors. Therefore, the increase of the current ripple introduces the tradeoff relationship among the volume of the boost inductor, the power conversion efficiency and the volume of other components in the circuit. In order to design the converter at the high power density with the acceptable power conversion efficiency, the relationship between the current ripple and the power density is necessary to be clarified. According to the knowledge of the author, this relationship has not been researched thoroughly yet.

In this paper, the relationship between the current ripple and the power density of the bidirectional DC-DC converter is clarified by the pareto-front curve of the power density and the efficiency. Specifically, a flow chart is introduced to show step-by-step how to obtain the pareto-front curve when the current ripple is varied. First, the semiconductor devices' losses are calculated in order to design the heatsink volume. Second, the method is explained to design the capacitor volume with the condition that both the current ripple and the voltage ripple are satisfied. One of the most difficult challenges is the design of the inductor. When the inductance is reduced by increasing the current ripple, the winding loss decreases due to the decrease of the winding turns. However, the core loss increases due to the high flux density ripple. This results in another tradeoff relationship between the winding loss and the core loss in the inductor design step. Therefore, the method to achieve the smallest volume of the inductor with the lowest loss is explained. Finally, several 1-kW prototypes are constructed in order to verify the validity and the effectiveness of the proposed power density design flow chart.

II.   CIRCUIT TOPOLOGY

Fig. 1 shows the circuit of the non-isolated bidirectional DC-DC converter. In order to achieve the high power density with the acceptable power conversion efficiency, the analysis of the relationship between losses and volume is conducted. In this circuit, four main kinds of loss occur; the conduction loss and the switching loss due to the on-resistance and the non-ideal switching of the semiconductor devices (P_{cond}, P_{sw}), the core loss and the winding loss of the boost inductor (P_{core}, P_{wind}). On the other hand, there are three factors which contribute to the volume; the heatsink of the semiconductor devices Vol_H, the boost inductor Vol_L and the capacitor Vol_C. By varying both the switching frequency and the current ripple, every possibilities of the designable power density can be evaluated. This paper limits to the one-phase DC-DC converter. However, the optimal phase for the interleaved topology can be calculated simply with the proposed power density design method.

III. FLOWCHART OF POWER DENSITY DESIGN METHOD

Fig. 2 shows the flowchart to design the power density when both the switching frequency and the current ripple are varied. The main contribution of the paper to the conventional power density design method is to consider the current ripple as a variable. First, the range of the switching frequency f_{sw} and the current ripple ΔI are selected.

Fig. 3 shows the definition of the boost inductor current ripple ΔI and the output voltage ripple ΔV_{p-p}. In this paper, the switching frequency f_{sw} is kept constant at 50 kHz, whereas the current ripple ΔI is varied in 10%~200%, because the design of the power density by varying the switching frequency f_{sw} at the constant current ripple ΔI is considered as the conventional design method. It should be noted that, when the current ripple is larger than 100%, the DC-DC converter can be operated in either Continuous Current Mode (CCM) or Discontinuous Current Mode (DCM). Second, at certain (f_{sw},ΔI), the variables such as the inductance L, the capacitance C, and the maximum current I_{max} are calculated. Then, the heatsink volume Vol_H is calculated from the semiconductor device losses P_{cond} and P_{sw}. After that, the effective value of the capacitor current I_{C_rms} is calculated. The capacitor volume is designed in the condition that both the capacitance C and the allowable effective current are satisfied.

Next, from the inductance L and the maximum current I_{max}, the required storage energy in the boost inductor is calculated in order to select the core size. Because the high power density is preferred, the minimum designable core size is selected. With certain core size, the evaluation range of the air gap is selected. Next, at certain air gap, the winding turns N is calculated and decided whether this design can be implemented. In case that all the windings can be fit into the window area of the core, the inductor volume Vol_L, and the inductor loss P_{core}, P_{wind} are calculated. After that, the length of the air gap is increased in order to find another combination of P_{core} and P_{wind}. In case that all the windings cannot be fit into the window area of the core, the current ripple and the switching frequency are varied in order to evaluate the losses and the volumes at the next combination of (f_{sw},ΔI).

A. Design of heatsink

The volume of the heatsink changes according to the loss from the semiconductor devices. The estimation of the heatsink volume is conducted based on the Cooling System Performance Index (CSPI) [7]. CSPI is defined as the inverse factor of the thermal resistance per unit volume. Furthermore,

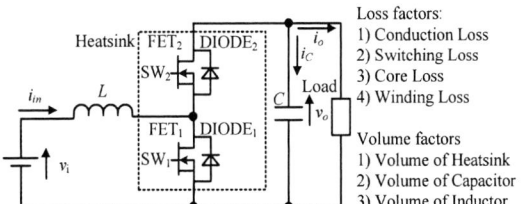

Fig. 1. Non-isolated bidirectional DC-DC with simple buck-boost topology. The losses and volume is calculated in order to design the power density and the power conversion efficiency.

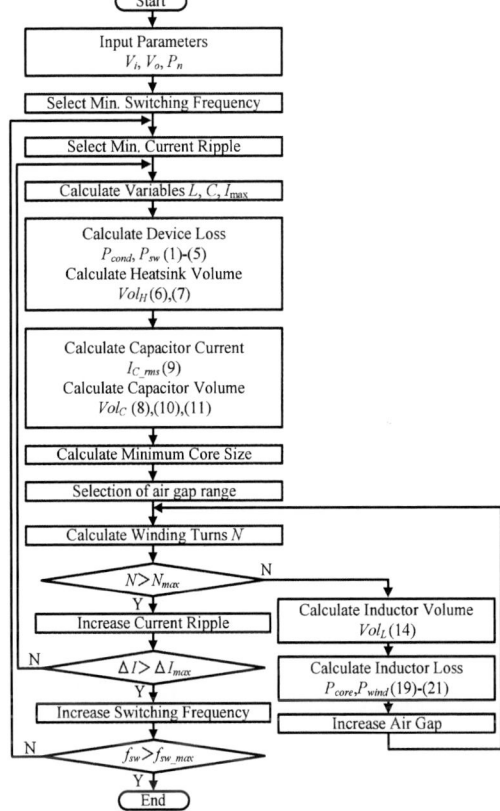

Fig. 2. Flowchart to design power density when both switching frequency and current ripple are varied. By this method, every possibilities of the designable power density can be evaluated.

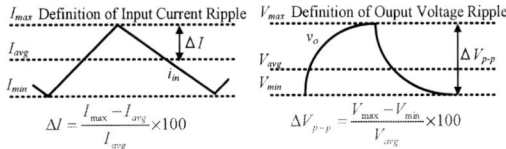

Fig. 3. Definition of boost inductor current ripple and output voltage ripple. By increasing the current ripple, the inductance of the boost inductor is decreased. This leads to the size reduction of the boost inductor.

the higher CSPI becomes, the more effective the heat is dissipated through the heatsink.

First, the conduction loss P_{cond} and the switching loss P_{sw} of the semiconductor devices are derived as follows. In this paper, MOSFET is chosen as the semiconductor devices. The conduction loss P_{cond} is calculated by the effective current flowing through MOSFET and the on-resistance of MOSFET,

$$P_{cond} = \frac{1}{T_{sw}} \int_0^{T_{sw}} i_{in}^2 R_{on} dt = R_{on}\left(I_{avg}^2 + \frac{\Delta I^2}{3}\right) \tag{1}$$

Where, T_{sw} is the switching period, R_{on} is the on-resistance of MOSFET, and i_{in}, I_{avg}, ΔI are the instantaneous value, the average value and the ripple of the boost inductor current respectively as shown in Fig. 3.

Fig. 4 shows the pattern of the switching losses when the current ripple is varied. When the current ripple is smaller than 100%, the DC-DC converter is operated in only CCM. However, when the current ripple is larger than 100%, the DC-DC converter is operated in either CCM or DCM. The turn-on loss P_{sw_on} is calculated by,

$$P_{sw\_on} = \frac{1}{T_{sw}} \int_0^{t_r} v_{ds} i_{ds} dt = f_{sw} K_{sw} V_o t_r (I_{avg} - \Delta I) \tag{2}$$

Where, v_{ds} is the voltage applied to MOSFET, i_{ds} is the current flowing through MOSFET, K_{sw} is the coefficient derived from the switching waveform [9], V_o is the output voltage, and t_r is the rise time of the MOSFET current i_{ds}. Similarly, the turn-off loss P_{sw_off} is calculated by,

$$P_{sw\_off} = \frac{1}{T_{sw}} \int_0^{t_f} v_{ds} i_{ds} dt = f_{sw} K_{sw} V_o t_f (I_{avg} + \Delta I) \tag{3}$$

Where, t_f is the fall time of the MOSFET current i_{ds}. On the other hand, the reverse recovery loss in the parasitic diode $P_{sw\_recovery}$ is calculated by [8],

$$P_{sw\_recovery} = f_{sw} V_o \left[\frac{I_{rrm}}{k_{Ifall}}|I_{avg} - \Delta I| + Q_{rr}\right] \tag{4}$$

Where, I_{rrm} is the peak value of the reverse recovery current, k_{Ifall} is the current decrease rate, and Q_{rr} is the reverse recovery charge. The total switching loss of MOSFET changes corresponding to the current ripple, and is classified as shown in Fig. 4,

$$\Delta I < 100\%(CCM) : P_{sw} = P_{sw1\_on} + P_{sw1\_off} + P_{sw1\_recovery}$$
$$\Delta I \geq 100\%(CCM) : P_{sw} = P_{sw1\_off} + P_{sw2\_off} \tag{5}$$
$$\Delta I \geq 100\%(DCM) : P_{sw} = P_{sw1\_off}$$

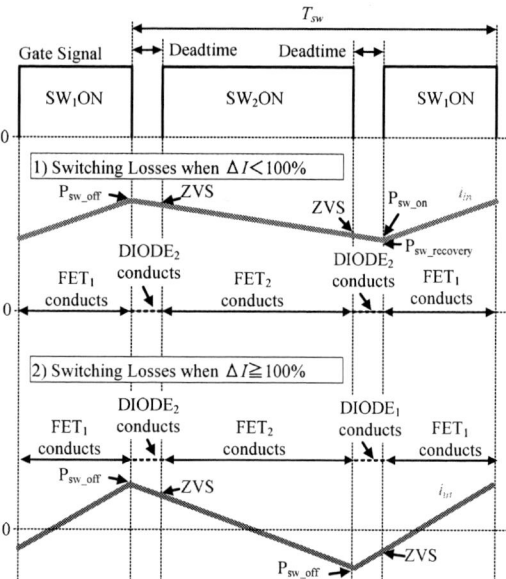

a) Switching losses in CCM.

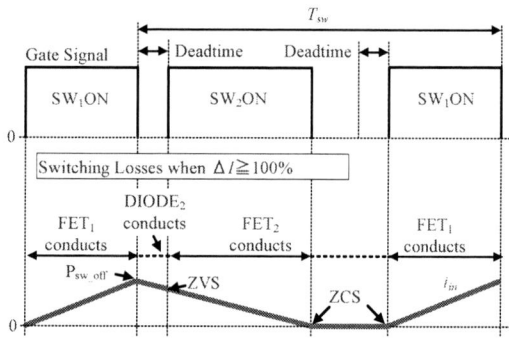

b) Switching losses in DCM.

Fig. 4. Pattern of switching losses when current ripple is varied. When the current ripple is smaller than 100%, the reverse recovery, which results in high switching loss, occurs in the parasitic diode. When the current ripple is larger than 100%, the DC-DC can be operated in either CCM or DCM.

Fig. 5 shows the relationship between the current ripple and the MOSFET loss. When the current ripple is increased, the effective current also increases. This leads to the increase in the conduction loss. On the other hand, when the current ripple is smaller than 100%, the reverse recovery, which results in high switching loss, occurs in the parasitic diode. Furthermore, when the current ripple is larger than 100%, by applying DCM, not only the switching loss but also the conduction loss are reduced due to the decrease in the effective current. However, the control method for DCM becomes complex due to the nonlinearity in DCM [10].

From (1)-(5), the required thermal resistance between the heatsink and the ambience $R_{th(s-a)}$ is calculated by [11].

$$R_{th(s-a)} = \frac{T_j - T_a}{P_{cond} + P_{sw}} - R_{th(j-s)} \quad (6)$$

Where, T_j is the junction temperature of the MOSFET, T_a is the ambient temperature, and $R_{th(j-s)}$ is the thermal resistance between the junction and the heatsink. Next, the volume of the heatsink Vol_H is estimated by [7],

$$Vol_H = \frac{1}{R_{th(s-a)} CSPI} \quad (7)$$

Where, CSPI is the coefficient which depends on the cooling methods such as natural cooling or water cooling. CSPI can be extracted from the datasheet of the heatsink or can be obtained in optimal heatsink design [7].

B. Design of capacitor

The volume of the capacitor depends on the capacitance C, the allowable effective current, and the type of the implemented capacitors. There are three main types of the capacitor; the electrolytic capacitor, the film capacitor, and the ceramic capacitor. The film capacitor and the ceramic capacitor are evaluated to have longer life time than the electrolytic capacitor in high temperature environment such as HEVs. Moreover, with the same required capacitance, the ceramic capacitor volume is smaller than the film capacitor volume due to the high dielectric coefficient of the ceramic capacitor. Therefore, in this paper, the ceramic capacitor is applied. First, the capacitance C is calculated as a function of the desired peak-to-peak voltage ripple ΔV_{p-p},

$$C = \frac{I_{avg}}{0.5 \Delta V_{p-p}} \frac{1}{f_{sw}} D_1(1 - D_1) \quad (8)$$

Where, D_1 is the duty ratio of SW_1. Then, the effective current I_{C_rms} flowing to the capacitor is calculated by,

$$I_{C\_rms} = \sqrt{\frac{1}{T_{sw}} \int_0^{T_{sw}} i_C^2 dt}$$
$$= \sqrt{D_1 \left[\frac{(I_{max} - I_{min})^2}{3} + (I_{max} - I_o)(I_{min} - I_o) \right] + (1 - D_1)I_o^2} \quad (9)$$

Where i_C is the instantaneous value of the capacitor current, and I_o is the average output current as shown in Fig. 1.

Fig. 6 shows the relationship between the capacitance change rate due to DC bias voltage from Murata Manufacturer. When the DC bias voltage is applied to the ceramic capacitor, the effective capacitance decreases [12]. Therefore, when applying the ceramic capacitor, the capacitor is necessary to be designed based on the DC bias voltage characteristic in the datasheet. The actual designed capacitance C_{design} is calculated by the capacitance change rate due to DC bias voltage K_{DCbias} and the required capacitance C,

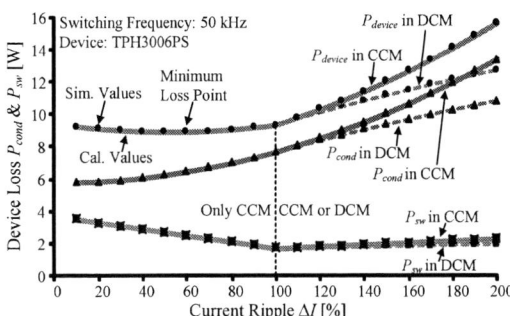

Fig. 5. Relationship between current ripple and MOSFET loss. When the current ripple is increased, the conduction loss increases. On the other hand, when the current ripple is set as 100%, the DC-DC converter is operated in Boundary Conduction Mode. In this mode, the switching loss reaches the smallest value. When the current ripple is larger than 100%, by applying DCM, not only the switching loss but also the conduction loss are reduced due to the decrease in the effective current [10].

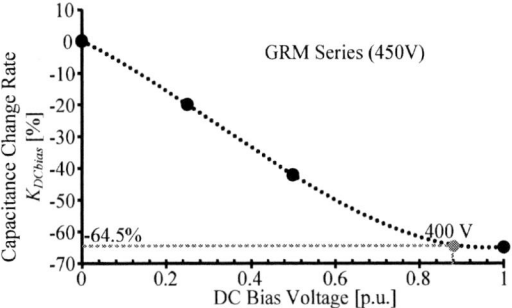

Fig. 6. Relationship between capacitance change rate due to DC bias voltage from Murata Manufacturer. When the DC bias voltage is applied to the ceramic capacitor, the effective capacitance decreases [12]. Therefore, when applying the ceramic capacitor, the capacitor is necessary to be designed based on the DC bias voltage characteristic in the datasheet.

$$C_{design} = C \frac{100}{100 + K_{DCbias}} \quad (10)$$

Fig. 7 shows the ratio between the capacitance and the allowable effective current from the datasheet [12]. It is necessary to make the capacitor current below the allowable effective current. The method to satisfy both the allowable effective current and the actual designed capacitance is to connect capacitors in parallel. However, this might increase the capacitor volume. Therefore, the procedure to achieve the smallest designable capacitor volume is applied. First, the dashed line which shows the designable capacitors in Fig. 7, is calculated by,

$$\frac{I_{rms\_data}}{C_{data}} \geq \frac{I_{C\_rms}}{C_{design}} \quad (11)$$

Where, I_{rms_data} and C_{data} are the allowable effective current and the capacitance for each series of capacitor in the datasheet respectively. From Fig. 7, it is understood that, by connecting the capacitors with high ratio between the allowable effective current and the capacitance, the allowable effective current is simply satisfied. Then, from the designable capacitors, the capacitor which achieves the smallest volume is selected.

Fig. 8 shows the relationship between the capacitor volume and the actual designed capacitance. It is understood that, the capacitor with the highest ratio between the allowable effective current and the capacitance does not always achieve the smallest volume. By this procedure of capacitor design, the smallest volume of capacitor is achieved with the satisfaction of both the allowable effective current and the actual designed capacitance.

C. Design of boost inductor

1) Estimation of inductor volume

In this section the design of the boost inductor to achieve the smallest volume and lowest loss is explained. First, the inductance L is calculated from the boost inductor current ripple ΔI,

$$ L = \frac{1}{f_{sw}\Delta I} \frac{V_i(V_o - V_i)}{2V_o} \tag{12} $$

Where, V_i is the input voltage. The inductor volume Vol_L is calculated by Area Product [13].

$$ Vol_L = K_V \cdot \left(\frac{LI_{max}^2}{K_u B_{max} J} \right)^{\frac{3}{4}} \tag{13} $$

Where, K_V is the coefficient which depends on the shape of the core [13], K_u is the window utilization factor, B_{max} is the maximum flux density of the core, and J is the current density. By substituting (12) into (13), the equation of the inductor volume is rewritten as the function of the current ripple ΔI and the switching frequency f_{sw},

$$ Vol_L = K_V \cdot \left[\frac{V_i(V_o - V_i)I_{avg}}{2K_u B_{max} J V_o} \frac{1}{f_{sw}} \left(\sqrt{\Delta I} + \frac{1}{\sqrt{\Delta I}} \right) \right]^{\frac{3}{4}} \tag{14} $$

It is clearly understood from (14) that, the inductor volume depends on both the switching frequency f_{sw} and the current ripple ΔI. Therefore, in case only the switching frequency is varied, the smallest designable inductor volume might be unconsidered.

Fig. 9 shows the relationship between the current ripple and the volume of the DC-DC converter. Because the heatsink volume depends on the MOSFET loss as shown in (6), and (7), the dependence of the heatsink volume on the current ripple is similar to the relationship between the MOSFET loss and the current ripple as shown in Fig. 5. Besides, the capacitor volume is relatively small because the ceramic capacitor with the high dielectric coefficient is applied. On the other hand, the inductor volume reaches the smallest value when the current ripple is designed at 100%. When the current ripple is increased over 100%, the required storage energy in the inductor can be no

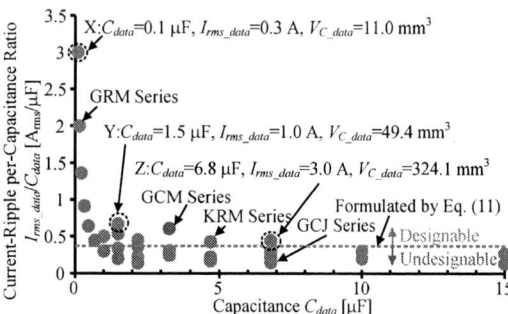

Fig. 7. Relationship between allowable current ripple and capacitance from Murata Manufacturer [12]. The higher the ratio between the allowable effective current and the capacitance becomes, the more simply the allowable effective current is satisfied.

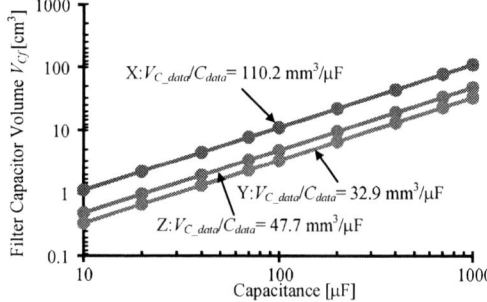

Fig. 8. Relationship between capacitor volume and capacitance. From the designable capacitors shown in Fig. 7, the capacitor which achieves the smallest volume is selected. It should be noted that the capacitor with the highest ratio between the allowable effective current and the capacitance, does not always achieve the smallest volume.

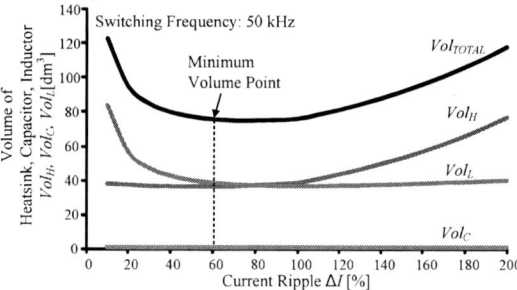

Fig. 9. Relationship between current ripple and converter volume. By varying the current ripple, the smallest volume of the converter is achieved at current ripple of 60%.

longer decreased due to the increase of the current peak I_{max} as shown in (13) and (14). In order to achieve the highest power density, the smallest core size which satisfies (15) is selected [13].

$$ A_e A_w \geq \frac{LI_{max}^2}{JB_{max}K_u} \tag{15} $$

Where A_e, and A_w are the cross-sectional area, and the window area of the core.

2) Calculation of winding turns

With selected core, the air gap is varied in order to achieve the smallest inductor loss. This section explained the method to calculate the required winding turns at certain air gap l_g. The winding turns N is calculated by,

$$N = \sqrt{L(R_g + R_c)} \qquad (16)$$

Where, R_g and R_c are the reluctance of the gap and the core. In the case that the air gap is applied to avoid the flux saturation, the fringing flux near the gap is necessary to be consider in order to calculate the winding turns accurately. Specifically, the conventional and improved equations to calculate the reluctance of the gap are as follows respectively,

$$R_{g\_conv} = \frac{1}{A_e} \frac{l_g}{\mu_o} \qquad (17)$$

$$R_{g\_impr} = \frac{\sigma_x \sigma_y}{A_e} \frac{l_g}{\mu_o} \qquad (18)$$

Where, μ_o is the relative permeability of the core material, and σ_x, σ_y are the fringing factors considering the fringing effects in the x-direction and y-direction respectively. The procedure to calculate the fringing factor σ_x and σ_y with certain air gap shape has been explained thoroughly in [14].

3) Calculation of winding loss

The method to calculate the winding loss considering the skin effect is explained in this section. First, the effective cross-sectional area of the winding wire S_{wind} which considers the skin effect is calculated by,

$$S_{wind} = 4\pi\delta \frac{\left[r - \delta\left(1 - e^{-\frac{r}{\delta}}\right) \right]^2}{r - \frac{\delta}{2}\left(1 - e^{-\frac{r}{2\delta}}\right)} \qquad (19)$$

Where, r is the radius of the winding wire, and δ is the skin depth at certain switching frequency [13]. Then, the winding loss P_{wind} is calculated by the effective value of the boost inductor current and the resistance of the winding wire R_{wind},

$$P_{wind} = \frac{1}{T_{sw}} \int_0^{T_{sw}} i_{in}^2 R_{wind} dt = \frac{1}{\sigma} \frac{l_{wind}}{S_{wind}} \left(I_{avg}^2 + \frac{\Delta I^2}{3} \right) \qquad (20)$$

Where, l_{wind} is the length of the winding wire, and σ is the electrical conductivity of the winding material. In this paper, because the switching frequency is limited up to 50 kHz, the proximity effect is considered to be small enough to be negligible.

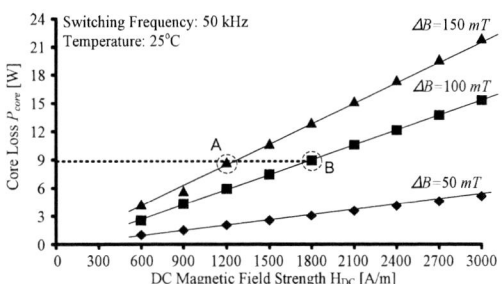

Fig. 10. LOSS MAP under DC bias conditions (ferrite PC40). By increasing the winding turn, the flux density ripple is reduced. However, because the premagnetization H_{DC} is increased, the core loss reduction is not effective.

4) Calculation of core loss

When the current ripple is increased, the flux density ripple becomes high, which leads to the high core loss. The core loss P_{core} is calculated by the improved Generalized Steinmetz Equation (iGSE),

$$P_{core} = V_e k_i' f_{sw}^{\ \alpha} \Delta B^{\beta-\alpha} \left[\left| \frac{\Delta B}{D_1} \right|^\alpha D_1 + \left| \frac{\Delta B}{D_2} \right|^\alpha D_2 \right] \qquad (21)$$

where V_e is the effective core volume, ΔB is the flux density ripple and D_2 is the duty ratio of SW$_2$, k_i', α, β are the iGSE parameters. The calculation of the core loss is conducted as following steps:

i) The LOSS MAP, which is the measured core losses of the triangular flux waveforms at different DC bias conditions, is obtained [15]-[16]. In this paper, the LOSS MAP is obtained at 3 different values of ΔB.

ii) The iGSE parameters k_i', α, β are derived from the LOSS MAP in order to be able to calculate the core loss in various conditions [17].

Fig. 10 shows the LOSS MAP of PC40 (Ferrite) at 25°C. It is understood from Fig. 10 that, in case that the impact of DC bias to core losses cannot be negligible as in ferrite PC40, the reduction of ΔB is not always a good solution to reduce the core loss. Specifically, by increasing the winding turns, ΔB can be reduced from point A to point B; however, this increases the premagnetization H_{DC}, which increases the core loss. As the result, the increase in the winding turns cannot reduce the core loss but only increases the winding loss.

IV. EXPERIMENTAL RESULTS

Table 1 shows the experimental condition and the parameters of the semiconductor devices and the inductor.

Fig. 11 shows one of the prototypes which are constructed in order to verify the validity of the proposed power density design method. In this paper, both the highest power density of 10.1 kW/dm^3 and the highest efficiency of 98.55% are achieved at the current ripple of 60%. By the waveform of the input current, it is confirmed that the core do not saturate at rated power.

Table 1. Experimental and Design Parameters.

Parameters of Circuit			Parameters of Capacitor		
Input Voltage	V_{in}	240 V	Output Voltage Ripple	ΔV_{P-P}	5 V
Output Voltage	V_{out}	400 V	Cera. Capacitor GR355DD72W474KW01L		
Rated Power	P_n	1 kW	Parameter of Inductors (PC40)		
Rated Input Current	I_{in}	4.17 A	Saturated Flux Density	B_{max}	0.39 T
Switching Frequency	f_{sw}	50 kHz	Volume factor	K_V	35.8
Parameters of Switch (TPH3006PS)			Current Density	J	6 A/mm²
On-Resistance	R_{on}	0.33 Ω	Space factor	K_u	0.5
Rated Voltage	V_{DS}	500 V	Gap Length	l_{g1}	0.75 mm
Rated Current	I_{DS}	12 A		l_{g2}	1.25 mm
Reverse Current	I_{rrm}	11 A		l_{g3}	1.5 mm
Reverse Recovery Charge	Q_{rr}	54 nC	Turn Numbers	N_1	33
				N_2	42
Parameters of Heat Sink				N_3	60
Junction Temperature	T_j	80 °C	Parameters of Prototype		
Ambient Temperature	T_a	25 °C	Length	L	59 mm
Junction-to-Case Thermal Resistance	$R_{th(j-s)}$	1.55 °C/W	Width	W	42 mm
			Height	H	40 mm
Cooling System Performance Index	CSPI	4	Power Density	ρ	10.1 kW/dm³
			Maximum Efficiency	η	98.55 %

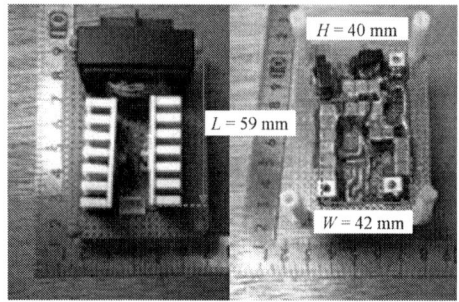

(a) Prototype Layout.

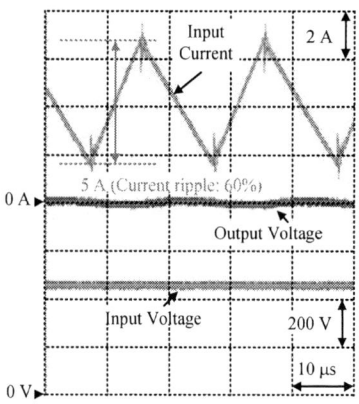

(b) Operation Waveform.

Fig. 11. 1-kW non-isolated DC-DC converter with power density of 10.1 kW/dm³ and efficiency of 98.55%.

Fig. 12 shows the calculated and experimental inductance when the fringing flux is unconsidered/considered. When the air gap increases from 0.5 mm to 1.25 mm, the error between the calculated and experimental inductance increases from 34% to 79%. Furthermore, when the fringing flux is considered, the error is reduced from 79% to 6%.

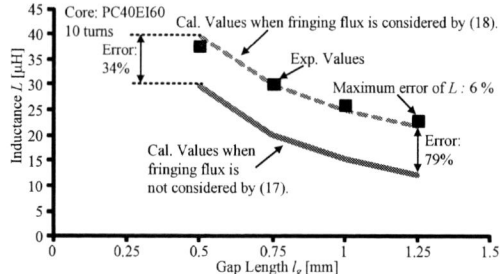

(a) Inductance with 10 turns.

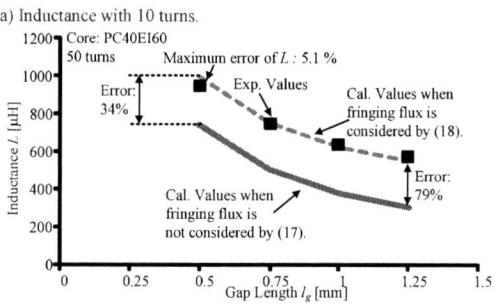

(b) Inductance with 50 turns.

Fig. 12. Inductance design with/without consideration of fringing flux around gap. By considering the fringing effect, the desired inductance is designed accurately. The winding turn design step is crucial to the inductor loss design.

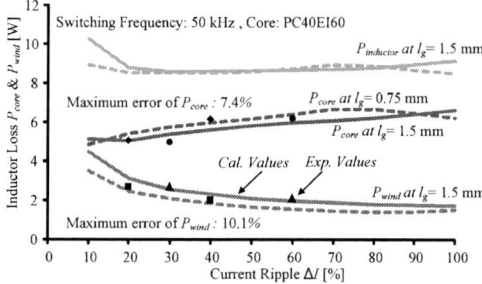

Fig. 13. Relationship between current ripple and inductor loss. The trade-off relationship between the winding loss and the core loss requires the optimal design of both the current ripple and the air gap.

Fig. 13 shows the relationship between the inductor loss and the current ripple. By increasing the air gap, the winding turns can be increased without making the core become saturated. However, the core loss does not always decrease by increasing the winding turns as explained above. The calculation of the winding loss by (20) still results with error of above 10% due to the inconsideration of the proximity effect.

Fig. 14 shows the experimental and calculated total loss. The experimental results agree to the calculated results with the error smaller than 10%. For simplicity, the air gap is designed at constant value (1.25 mm). However, as mentioned above, the air gap should be varied in order to find the lowest inductor loss. By calculating accurately the total loss, the efficiency can be designed simply.

978-1-4673-9551-9/16 $31.00 © 2016 IEEE

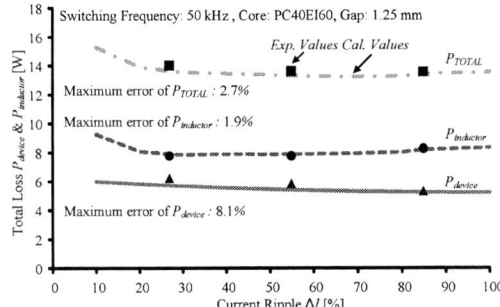

Fig. 14. Experimental and calculated results of total loss. By calculating the total loss accurately, the efficiency can be designed simply.

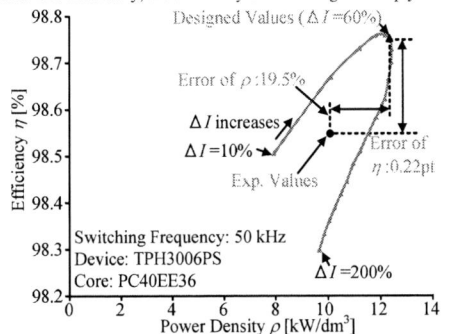

Fig. 15. η-ρ pareto front with current ripple as variable. The proposed power density design method enables to evaluate every possibilities of the designable power density.

Fig. 15 shows the pareto-front curves of the proposed design with the varied current ripple. The maximum errors in the power density and the efficiency are 19.5% and 0.22 pt., respectively. The error of the power density occurs because the air space between the devices is not considered in the proposed power density design method, whereas the errors of the efficiency occurs due to the inconsideration of the proximity effect and the fringing flux loss. However, the experimental power density without the air space almost agrees with the calculated value. This result confirms the validity of the proposed design. Furthermore, in this consideration, the highest power density can be reached with the highest efficiency. However, the parameters such as the switching frequency and the core material should be varied in order to find the optimum point of the power density or the efficiency.

V. CONCLUSION

In this paper, the relationship between the current ripple and the power density of the bidirectional DC-DC converter was clarified. The flowchart which was used to draw every possibilities of the designable power density in the pareto-front curve was introduced. Furthermore, the method to calculate the ceramic capacitor with condition that both the effective capacitor current and the capacitance are satisfied was explained. Besides, the method to design the smallest inductor with the lowest loss was also explained. Several 1-kW prototypes were constructed in order to verify the validity of

the proposed power design method. The highest power density of 10.1 kW/dm^3 was achieved with the efficiency of 98.55% at the current ripple at 60%.

In the future, the consideration of the losses from the proximity effect and the fringing flux will be conducted in case that those losses cannot be neglected. Furthermore, the input filter will also be considered in order to satisfy the regulations of the noises.

REFERENCES

[1] R. C. N. Pilawa-Podgurski, A. D. Sagneri, J. M. Rivas, D. I. Anderson, D. J. Perreault, "Very-High-Frequency Resonant Boost Converters", Transactions on Power Electronics, Vol. 24, No. 6, pp. 1654-1665, 2009.

[2] Ch. Huang, P. K. T. Mok: "An 84.7% Efficiency 100-MHz Package Bondwire-Based Fully Integrated Buck Converter With Precise DCM Operation and Enhanced Light-Load Efficiency", Journal of Solid-State Circuits, Vol. 48, No. 11, pp. 2895-2607, 2013.

[3] P. Haaf, J. Harper, "Understanding Diode Reverse Recovery and its Effect on Switching Losses", Fairchild Semiconductor Europe, Farchild Power Seminar 2007.

[4] J. Itoh, T. Araki, K. Orikawa, "Experimental Verification of EMC Filter Used for PWM Inverter with Wide Band-Gap Devices", IEEJ Journal of Industry Applications, Vol. 4, No. 3, pp. 212-219, 2015.

[5] J. Zhang, J. Sh. Lai, R. Y. Kim, W. Yu, "High-Power Density Design of a Soft-Switching High-Power Bidirectional dc–dc Converter", Transactions on Power Electronics, Vol. 22, No. 4, pp. 1145-1153, 2007.

[6] L. Ni, D. J. Patterson, J. L. Hudgins, "High Power Current Sensorless Bidirectional 16-Phase Interleaved DC-DC Converter for Hybrid Vehicle Application", Transactions on Power Electronics, Vol. 27, No. 3, pp. 1141-1151, 2012.

[7] U. Drofenik, G. Laimer, J. W. Kolar, "Theoretical Converter Power Density Limits for Forced Convection Cooling", Proceedings of the Inter. Pow. Conv. Intel. Motion (PCIM) Europe 2005 Conf., Ger., June 7 - 9, pp. 608 – 619, 2005.

[8] Y. Kashihara, J. Itoh, "Power Losses of Multilevel Converters in Terms of the Number of the Output Voltage Levels", The 2014 International Power Electronics Conference, No. 20A4-4, pp. 1943-1949, 2014.

[9] Y. Louvrier, P. Barrade, A. Rufer, "Weight and efficiency optimization strategy of an interleaved DC-DC converter for a solar aircraft", in Proc. 13th Eur. Conf. Power Elec. Application, (EPE), Sep. 2009, pp. 1–10.

[10] H. N. Le, K. Orikawa, J. Itoh, "Efficiency Improvement at Light Load in Bidirectional DC-DC Converter by Utilizing Discontinuous Current Mode", 17th Conference on Power Electronics and Applications, EPE'15-ECCE Europe, No. DS1b-Topic 3-0484, 2015.

[11] Kandarp I. Pandya, W. McDaniel, "A Simplified Method of Generating Thermal Models for Power MOSFETs", Semiconductor Thermal Measurement and Management, 2002.

[12] Catalog of Multilayer Ceramic Capacitors, Murata Manufracturing Corporation, 2014-2015.

[13] Wm. T. Mclyman, "Transformer and inductor design handbook", Marcel Dekker Inc., 2004.

[14] J. Muhlethaler, J. W. Kolar, A. Ecklebe, "A Novel Approach for 3D Air Gap Reluctance Calculations", 2011 IEEE 8th International Conference on Power Electronics and ECCE Asia (ICPE & ECCE), pp. 446 – 452.

[15] Mingkai Mu, Qiang Li, David Gilham, Fred C. Lee, Khai D. T. Ngo, "New Core Loss Measurement Method for High Frequency Magnetic Materials", 2010 IEEE Energy Conversion Congress and Exposition (ECCE), pp. 4384 – 4389.

[16] C. A. Baguley, U. K. Madawala, and B. Carsten, "A new technique for measuring ferrite core loss under DC bias conditions," IEEE Transactions on Magnetics, vol. 44, no. 11, pp. 4127–4130, 2008.

[17] J. Muhlethaler, J. Biela, J. W. Kolar, A. Ecklebe, "Core Losses under DC Bias Condition based on Steinmetz Parameters", 2010 International Power Electronics Conference (IPEC), pp. 2430 – 2437.

Grid-Voltage Feedforward Based Control for Grid-Connected *LCL*-Filtered Inverter with High Robustness and Low Grid Current Distortion in Weak Grid

Jinming Xu, Qiang Qian, Shaojun Xie, and Binfeng Zhang
College of Automation Engineering
Nanjing University of Aeronautics and Astronautics
Nanjing, China
xjinming01@163.com

Abstract—**For grid-connected *LCL*-filtered inverters, recent applications often observe a weak grid at the point of common coupling (PCC) with non-negligible grid impedance. In this case, the previous control methods would perform poor or even cause the inverter protection. Especially when the feedforward of PCC voltage is designed to suppress grid current harmonics, phase and gain margins will be largely reduced and rejections of grid current harmonics will be aggravated if a weak grid is connected. In fact, the grid voltage feedforward in the weak grid introduces a positive feedback loop related to the grid impedance, and then the unexpected feedback loop makes the inverter dynamic badly influenced by the grid impedance at low frequencies. Therefore, this study proposes to feed forward only the fundamental grid voltage through a second-order generalized integrator (SOGI) and to design the harmonic resonant controller with adaptability to large grid impedance, in order to avoid the undesired dynamic interactions at low frequencies. With the proposed design and control, the grid-connected *LCL*-filtered inverter is capable of realizing high robustness and low grid current distortion in the weak grid. Comparative experiments verify the effectiveness of the proposed control method.**

Keywords—*grid-connected inverter; LCL filter; grid impedance; voltage feedforward; stability; SOGI*

I. INTRODUCTION

Grid-connected inverters are widely used. Besides, for the purpose of suppressing switching-frequency harmonics in the grid-connected inverter, an *LCL* filter is widely used [1]-[3], but the *LCL* resonance is not beneficial for the current control performance. In last decade, several current control methods capable of *LCL* resonance damping have been well-established. The existing active damping (AD) methods include the filter-based AD in [4]-[5], capacitor current feedback AD in [6]-[11], capacitor voltage feedback AD in [12]-[13] and grid current feedback AD in [14]. On the premise that the *LCL* resonance has been solved, the grid voltage feedforward or resonant controller is used to suppress low-order current harmonics caused by inverter non-ideal effect and grid voltage harmonics

[7], [15]-[16]. That is, previous studies prove that the current quality issue for grid-connected *LCL*-filtered inverter has been well solved by active damping and harmonic compensators in the case that the grid impedance is not considered.

However, as long as many power electronic devices are linked to the grid at the point of common coupling (PCC) and the proportion of distributed generations grows with years, the grid at PCC always shows some complicated features such as the voltage sag and harmonics, the grid impedance and so on. Especially, the grid impedance at PCC varies largely. In [17]-[22], current control performances were studied in the case of a weak grid with large inductive impedance at low frequencies. In [17]-[18], the impedance-based criterion told that instability was aroused by large inductive impedance. Studies in [19] found that if the PCC grid-voltage feedforward was used, many current harmonics were aroused. Besides, in [19]-[21] where a harmonic resonant or predictive controller was used, many grid current harmonics were produced in the presence of large grid impedance. For stability improvement, in [22], an adaptive grid voltage feedforward algorithm based on the on-line impedance estimation was proposed, but, the impedance estimation caused adverse impact on the grid current quality and increased the implementation cost unfortunately. In summary, the inverter performance with the typical harmonic compensation methods in the weak grid case is not satisfactory.

Under the above technical backgrounds, this study aims to provide a solution for the problem of grid-voltage feedforward and resonant controller in weak grid case in order to maintain high robustness and low grid current distortion. In Section II, the parameters and control of the grid-connected *LCL*-filtered inverter are briefly introduced. Section III analyzes the control performance in the case of a weak grid with varied impedance. Section IV gives the proposal and design of the current control with modified grid-voltage feedforward and resonant controller. In Section V, the proposed control is compared with the typical control through experiments. Finally, Section VI concludes the whole paper.

978-1-4673-9551-9/16 $31.00 © 2016 IEEE

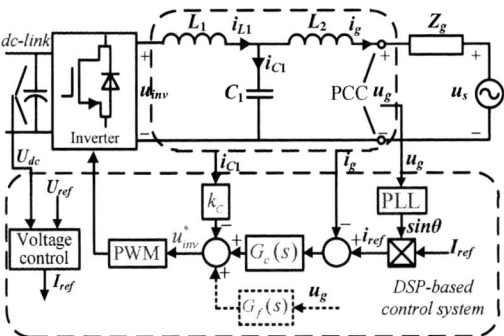

Fig. 1. Grid-connected *LCL*-filtered inverter with typical dual-loop current control and grid voltage feedforward

II. SYSTEM DESCRIPTION AND CONTROL

Fig. 1 gives the structure of a typical grid-connected *LCL*-filtered inverter with the dual-loop current control using the capacitor current and grid current feedbacks. U_{dc} is the dc-link voltage, and U_{ref} is the reference for dc-link voltage control. I_{ref} is the amplitude reference for grid current control. A phase-lock-loop (PLL) is used to monitor the phase of grid voltage u_g. Through the PLL, the sinusoidal reference i_{ref} is obtained. The inverter output voltage is denoted as u_{inv}. L_1 is the inverter-side inductance, C_1 is the capacitance, L_2 is the grid-side inductance, Z_g represents the grid impedance and u_s is the ideal grid voltage. Besides, k_C is a proportional factor, i_{C1} is the capacitor current, i_{L1} is the inverter-side current and i_g is the grid current. $G_c(s)$ is the current controller and $G_f(s)$ is the PCC voltage feedforward. With the control in Fig. 1, the PWM reference u_{inv}^* is obtained.

Without the grid-voltage feedforward (i.e., $G_f(s)=0$), it is obtained:

$$\left[(i_{ref}-i_g)\cdot G_c(s)-i_{C1}\cdot k_C\right]\cdot k_{\text{PWM}}=u_{inv}^*\cdot k_{\text{PWM}}=u_{inv} \\ =i_{L1}\cdot L_1 s+i_g\cdot L_2 s+u_g \quad (1)$$

where k_{PWM} represents the transfer function from u_{inv}^* to u_{inv}.

With grid-voltage feedforward (i.e., $G_f(s)\neq 0$), it is obtained:

$$\left[(i_{ref}-i_g)\cdot G_c(s)-i_{C1}\cdot k_C+u_g\cdot G_f(s)\right]\cdot k_{\text{PWM}} \\ =i_{L1}\cdot L_1 s+i_g\cdot L_2 s+u_g \quad (2)$$

Unlike that in (1), the current controller $G_c(s)$ in (2) can be seen only in charge of outputting the reference for the voltage across the two inductors when the grid-voltage feedforward

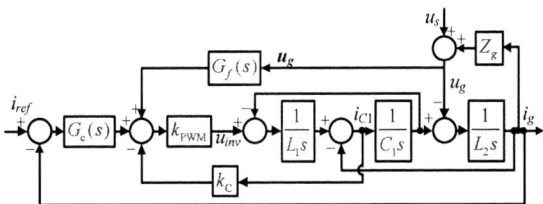

Fig. 2. Typical current control structure in the weak grid case

factor $G_f(s)$ equals $1/|k_{\text{PWM}}|$ (In this study, $|k_{\text{PWM}}|$ is treated as 1 for simplification). Thus, the implementation of grid-voltage feedforward owns two merits: 1) fast response during the start-up procedure and the grid voltage sag because of the reduction of the burden of current controller; 2) good steady-state performance with high power factor and low distortion because the terms related to u_g at both sides of (2) cancel each other out approximately. Because of the above merits, the grid-voltage feedforward has been widely applied in both single-phase and three-phase inverters.

III. PROBLEM OF GRID-VOLTAGE FEEDFORWARD IN WEAK GRID WITH GRID IMPEDANCE VARIATION

In the practical application where the grid impedance is considered, the voltage used for feedforward is actually the PCC voltage, i.e., $u_g=i_g\cdot Z_g+u_s$. Therefore, in weak grid case, the voltage feedforward brings in an additional positive feedback of i_g with the feedback factor Z_g [22]. The control diagram in the weak grid case is shown in Fig. 2. As a consequence, the open-loop transfer function from i_{ref} to i_g with high inductive grid impedance ($Z_g=L_g\cdot s$) is expressed as (see the bottom of this page)

In (3), an extra term related to the feedforward factor $G_f(s)$ is found. Fig. 3 presents the open-loop Bode plots without and with that extra term for L_g=4 mH (i.e., $3.3\times(L_1+L_2)$). Note that $G_c(s)$ and k_C are designed according to [19]. A PI controller is used here, i.e., $G_c(s)=k_p+k_i/s$. The system parameters are given in Table I. With the use of Fig. 3, the impact of the weak-grid-voltage feedforward can be observed: phase and gain margins are largely reduced. This phenomenon indicates that the *LCL*-filtered inverter may still work stably in the case of small grid impedance, but many grid current harmonics exist due to the insufficient margins; however, if L_g increases much larger, the large amount of grid current harmonics will easily trigger the protection of inverter [22].

It should be mentioned that the delay existed in the digital control system would affect the stability, as studied in several literatures. Luckily, several approaches in [8]-[11] and [23] can be used to eliminate the adverse effect of digital control.

$$G_{I_{ref}\_o}^{i_g}(s)=\frac{k_{\text{PWM}}G_c(s)}{L_1(L_2+L_g)C_1 s^3+k_C k_{\text{PWM}}(L_2+L_g)C_1 s^2+(L_1+L_2+L_g)s-L_g s\cdot k_{\text{PWM}}G_f(s)} \quad (3)$$

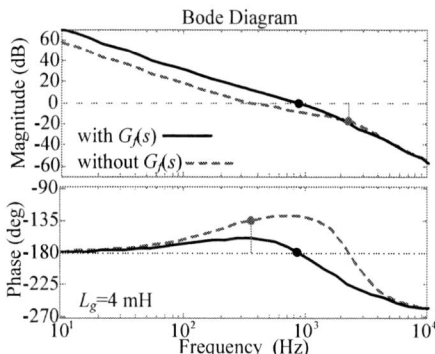

Fig. 3. Open-loop Bode plots with and without $G_f(s)$

TABLE I
SYSTEM PARAMETERS

Parameters	Symbols	Values
Grid voltage	U_g	220 V
Grid frequency	f_0	50 Hz
Rated power	P	5 kW
Switching frequencies	f_s	15 kHz
Inverter-side inductance	L_1	0.75 mH
Grid-side inductance	L_2	0.45 mH
Filter capacitance	C_1	6.8 μF
PI controller	k_p, k_i	9, 11000
iC1 feedback factor	k_C	10.6

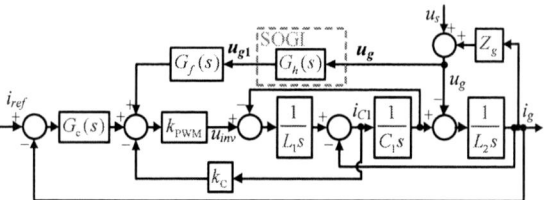

Fig. 4. Improved grid-voltage feedforward based control

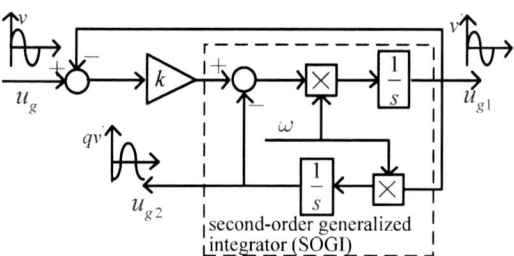

Fig. 5. Block of the used SOGI model

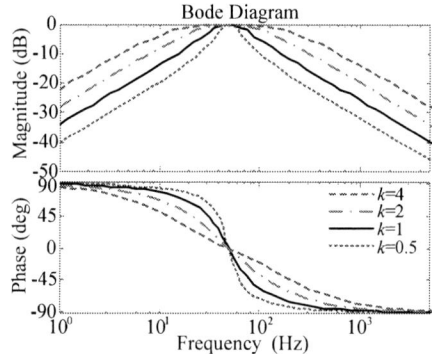

Fig. 6. Bode plots of SOGI with different k

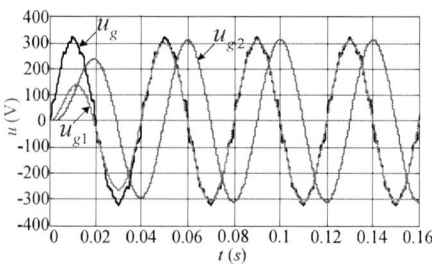

Fig. 7. Typical waveforms of SOGI (k=1) under a distorted PCC voltage

IV. PROPOSAL AND DESIGN OF IMPROVED GRID-VOLTAGE FEEDFORWARD BASED CONTROL

It is concluded from the above that the control performance degradation is produced by the improper voltage feedforward. With the control in Fig. 2, the voltage feedforward injects all the information of u_g containing the dynamic of Z_g into the main control loop so that unexpected dynamic interactions are yielded. Seen from Fig. 3, the dynamic interactions mainly lie at low frequencies. Thus, this study proposes to add a digital filter before $G_f(s)$ in Figs. 1 and 2 to filter out the harmonics information from u_g in order to solve the unexpected dynamic interactions, as shown in Fig. 4. Then, the resonant controller can be used to suppress low-order grid current harmonics if properly designed.

A. Design of digital filter $G_h(s)$

An SOGI is adopted to subtract the voltage harmonics from u_g in Fig. 5. The transfer function from u_g to u_{g1} is (ω is the fundamental angular frequency):

$$G_h(s) = \frac{k\omega s}{s^2 + k\omega s + \omega^2} \qquad (4)$$

Fig. 6 shows the Bode plots of (4) with different k. If k is small, harmonics in u_g are highly attenuated, but the bandwidth is reduced and the transient response would be a little slower. Therefore, a trade-off design is required. Fig. 7 provides some

waveforms of SOGI under a distorted PCC voltage. It is clear that the output of SOGI, i.e., u_{g1}, does not contain low-order harmonics. Thus, the harmonic information in PCC voltage can be eliminated. However, the major demerit of using SOGI is

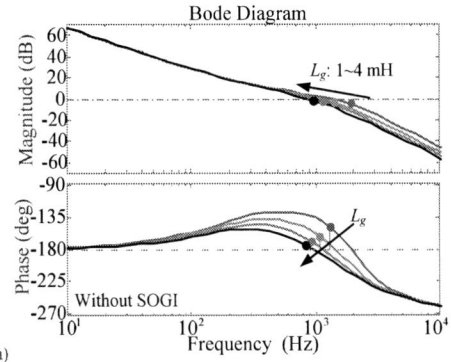
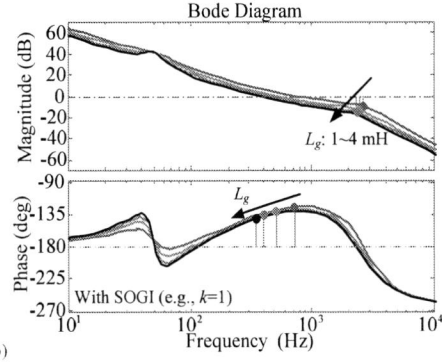

(a)　　　　　　　　　　　　　　　(b)

Fig. 8. Open-loop Bode plots (a) without and (b) with the proposed grid-voltage feedforward

also seen in Fig. 7. The voltage signal u_{g1} takes about one grid cycle to track the fundamental amplitude of u_g accurately. Thus, the start-up transient performance may be unsatisfactory (i.e., the current overshoot) if all the algorithms including the current control, PLL and SOGI are enabled at the same time. Thus, in Section V, a simple approach will be provided to improve the start-up transient.

With the improved control in Fig. 4, the open-loop transfer function from i_{ref} to i_g is (see the bottom of this page)

In order to show the effect of SOGI on the stability, open-loop Bode plots with and without the SOGI-based feedforward for different L_g are given in Fig. 8. Note that $G_c(s)$ is the same PI controller as that used in Fig. 3. Seen from Fig. 8(a), without SOGI, the phase response is seriously affected by the weak-grid feedforward. On the contrary, because the SOGI filters out almost all the harmonic information from u_g, the open-loop phase response is decoupled from the weak-grid feedforward. Therefore, both the gain margin (GM) and phase margin (PM) are improved, as shown in Fig. 8(b). Even if L_g increases to 4 mH, the PM of 38.4° is much higher than that in Fig. 8(a). The inverter with the proposed grid-voltage feedforward is more robust than that with the typical grid-voltage feedforward.

B. Design of resonant controller

On the basis of improving the stability margins in the weak grid case, a resonant controller can be used to further suppress low-frequency grid current harmonics. The resonant controller is expressed as:

$$G_c(s) = G_{PR}(s) = k_p + \sum_{n=1,3,5}^{n_{max}} \frac{k_r s}{s^2 + \omega_c s + (n\omega_0)^2} \quad (6)$$

where k_p is the proportional gain (its value is the same with that

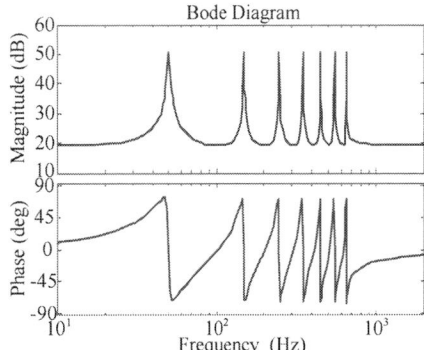

Fig. 9. Magnitude and phase characteristics of $1^{st} \sim 13^{th}$ PR controller

of the PI controller), ω_0 equals $2\pi f_0$ and f_0 is the fundamental frequency, ω_c represents the bandwidth of the resonant control, and k_r/ω_c is the resonant gain at $n\omega_0$. Note that n_{max} denotes the maximum order of current harmonics to be suppressed. Among these parameters, k_r and ω_c can follow the existing design like that in [15]. The value of ω_c is designed considering the slight variation of grid fundamental frequency in practice. Following [15], ω_c is selected as 6. Besides, k_r is selected as 2000 here. Unlike k_r and ω_c, n_{max} has some relations with the stability in the weak grid case, which should be seriously considered.

Before determining the maximum value of n_{max}, the basic feature of PR is investigated. Fig. 9 shows the Bode plot of (6) with $n_{max}=13$. High peaks as well as sharp phase variations at n-order frequencies are observed. If the magnitude and phase responses in Fig. 9 are added into those in Fig. 8(b), frequent crossings with 0dB and −180° may occur. Seen from Fig. 8(b), in the weak grid case, the increase of grid impedance decreases the open-loop 0dB-crossing frequency. If the value of $n_{max}f_0$

$$G_{i_{ref}\_o\_improved}^{i_g}(s) = \frac{k_{PWM} G_c(s)}{L_1(L_2 + L_g)C_1 s^3 + k_C k_{PWM}(L_2 + L_g)C_1 s^2 + (L_1 + L_2 + L_g)s - L_g s \cdot k_{PWM} G_f(s) G_h(s)} \quad (5)$$

978-1-4673-9551-9/16 $31.00 © 2016 IEEE　　　　1922

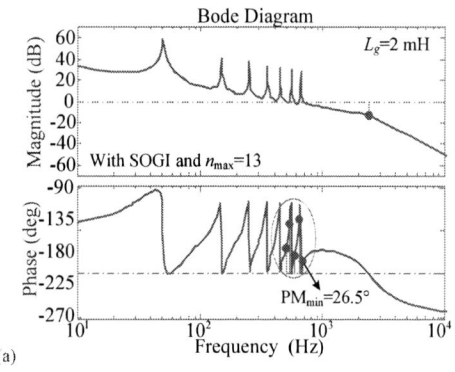

(a)

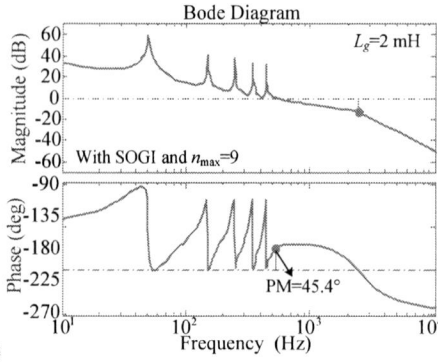

(b)

Fig. 10. Open-loop Bode plots with the proposed control when L_g is 2 mH and n_{max} equals (a) 13 and (b) 9 separately

exceeds the 0dB-crossing frequency, additional crossings with 0dB or $-180°$ will be yielded. As a consequence, the minimum stability margin is likely to be reduced. Thus, $n_{max}f_0$ should be lower than the 0dB-crossing frequency in order not to affect the stability margin.

First, the 0dB-crossing frequency without the resonant part of (6) is calculated. At low frequencies, (5) can be simplified by letting C_1 and $G_h(s)$ equal 0. That is, (5) is expressed as:

$$G_{i_{ref}\_o\_improved}^{i_g}(s) \approx \frac{k_{PWM}G_c(s)}{(L_1 + L_2 + L_g)s} \qquad (7)$$

Then, the 0dB-crossing frequency in Hz approximates:

$$f_{cr\_Lg} \approx \frac{|k_{PWM}|k_p}{2\pi(L_1 + L_2 + L_g)} \qquad (8)$$

Then, to avoid the frequent crossings with 0dB and $-180°$, the value of n_{max} is recommended to be:

$$n_{max} < \frac{f_{cr\_Lg}}{f_0} \qquad (9)$$

Depending on the maximum grid impedance in a practical application, the value of n_{max} is different in order to ensure that the stability in the weak grid case is hardly affected by PR. For instance, if the maximum L_g is 2 mH, n_{max} equals 9 according to (9). Fig. 10 shows the open-loop Bode plots of (5) when L_g is 2mH and n_{max} equals 9 and 13 separately. In Fig. 10(a), the magnitude curve has multiple crossings with 0dB-line so that the minimum PM (26.5°) occurs at about 13th frequency. It is mentioned that the phase will also have multiple crossings with $-180°$ once the value of n_{max} is further increased. However, if (9) is satisfied, the above frequent crossings are avoided so that the PM and GM are still satisfactory in the weak grid case, as indicated in Fig. 10(b).

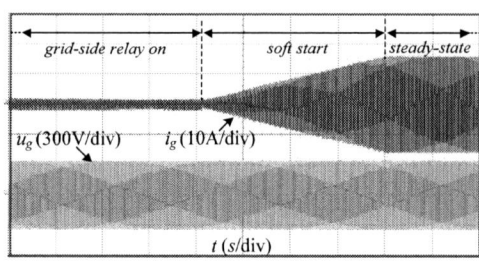

Fig. 11. Grid current and PCC voltage waveforms during start-up

V. VERIFICATIONS

A grid-connected *LCL*-filtered inverter has been built in the lab. The *LCL* parameters are the same with Table I. The signal sampling, current reference tracking, voltage feedforward and PLL are all implemented in DSP TMS320F28335 from TI. In the study, the PLL is also achieved with the use of SOGI [24]. In this section, test results with the proposed and typical control for several values of the inductive grid impedance (i.e., 1.45 and 2.66 mH) will be provided.

As mentioned in Section IV-A, the demerit of using SOGI-based grid-voltage feedforward is that the generation of u_{g1} takes a relatively long time. Then, the current controller bears the burden of generating almost all the PWM reference during the start-up so that the current overshoot may occur. To avoid this inconvenience, here, a simple way is adopted. During the start-up, the SOGI-based PLL is enabled at first. After the fundamental phase of the grid voltage is obtained, the current tracking algorithm and the feedforward which uses u_{g1} as input are then activated at the zero-crossing instant determined by the PLL. Besides, for the sake of soft start-up, the magnitude of grid current reference is increased gradually till it reaches the desired value. Fig. 11 shows the measured start-up grid current and PCC voltage waveforms.

To verify the improvement of the proposed control over the typical control, selected waveforms with different values of L_g are presented in Figs. 12 and 13. With the typical grid-voltage feedforward based control, the grid current has many current

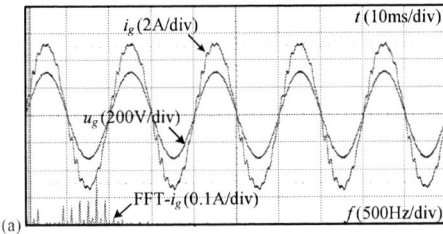 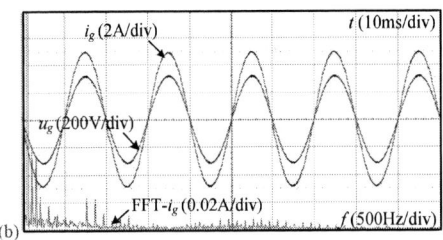

Fig. 12. Experimental waveforms with 1.45mH-grid impedance with (a) the typical control and (b) the proposed control

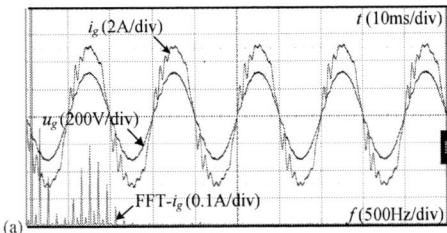 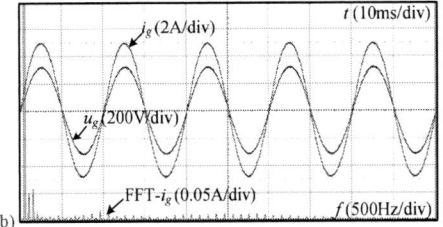

Fig. 13. Experimental waveforms with 2.66mH-grid impedance with (a) the typical control and (b) the proposed control

harmonics, and the magnitudes of grid current harmonics grow with the increase of grid impedance. On the contrary, with the proposed control, the grid current is always synchronous with the PCC voltage and the grid current distortion is much lower than that with the typical control.

VI. CONCLUSIONS

In the weak grid case, the voltage feedforward aggravates the current control performance because the inverter current control dynamics at low frequencies are affected by the grid impedance through the additional positive feedback loop. Thus, in order to avoid the unexpected dynamic interactions, this study proposes to feed forward only the fundamental voltage at PCC through an SOGI. It is proved that the stability margins with the SOGI-based voltage feedforward are improved. Then, the harmonic resonant controller is used to suppress low-order grid current harmonics. In order to achieve a high robustness, the design of resonant controller has been carefully dealt with. The maximum order of grid current harmonics pending to be suppressed is determined by the bandwidth and grid impedance value. Generally speaking, the proposed method is simpler than the existing adaptive control with on-line impedance estimation. With the proposed design and control, the *LCL*-filtered inverter can maintain high robustness and low grid current distortion in the weak grid case.

ACKNOWLEDGMENT

This work was supported in part by the National Natural Science Foundation of China (No. 51477077) and the Funding of Jiangsu Innovation Program for Graduate Education (No.

CXZZ12_0153, Fundamental Research Funds for the Central Universities).

REFERENCES

[1] A. Reznik, M.G. Simões, A. Al-Durra, and S.M. Muyeen, "*LCL* filter design and performance analysis for grid-interconnected systems," *IEEE Trans. Ind. Appl.*, vol. 50, no. 2, pp.1225-1232, Mar./Apr. 2014.

[2] J. Muhlethaler, M. Schweizer, R. Blattmann, J.W. Kolar, and A. Ecklebe, "Optimal design of *LCL* harmonic filters for three-phase PFC rectifiers," *IEEE Trans. Power Electron.*, vol. 28, no. 7, pp.3114-3125, July 2013.

[3] X. Zheng, L. Xiao, Y. Lei, and Z. Wang, "Optimisation of *LCL* filter based on closed-loop total harmonic distortion calculation model of the grid-connected inverter," *IET Power Electron.*, vol. 8, no. 6, pp. 860-868, June 2015.

[4] J. Dannehl, M. Liserre and F. W. Fuchs, "Filter-based active damping of voltage source converters with *LCL* filter," *IEEE Trans. Ind. Electron.*, vol. 58, no. 8, pp. 3623–3633, Aug. 2011.

[5] W. Yao, Y. Yang, X. Zhang, and F. Blaabjerg, "Digital notch filter based active damping for *LCL* filters," *in Proc. IEEE APEC*, 2015, pp. 2399-2406.

[6] Y. Tang, P. C. Loh, P. Wang, F. H. Choo, F. Gao and F. Blaabjerg, "Generalized design of high performance shunt active power filter with output *LCL* filter," *IEEE Trans. Ind. Electron.*, vol. 59, no. 3, pp. 1443-1452, Mar. 2012.

[7] J. Xu, T. Tang, and S. Xie, "Research on low-order current harmonics rejections for grid-connected *LCL*-filtered inverters," *IET Power Electron.*, vol. 7, no. 5, pp.1227-1234, May 2014.

[8] Z. Zou, Z. Wang, and M. Cheng, "Modeling, analysis, and design of multifunction grid-interfaced inverters with output *LCL* filter," *IEEE Trans. Power Electron.*, vol. 29, no. 7, pp. 3830–3839, July 2014.

[9] D. Pan, X. Ruan, C. Bao, W. Li, and X. Wang, "Capacitor-current-feedback active damping with reduced computation delay for improving robustness of *LCL*-type grid-connected inverter," *IEEE Trans. Power Electron.*, vol. 29, no. 7, pp. 3414-3427, July 2014.

[10] J. Xu, S. Xie, J. Kan, and B. Zhang, "Research on stability of grid-connected *LCL*-filtered inverter with capacitor current feedback active damping control," *in Proc. ICPE-ECCE Asia*, 2015, pp. 682-687.

[11] D. Yang, X. Ruan, and H. Wu, "A real-time computation method with dual sampling modes to improve the current control performances of the *LCL*-type grid-connected inverter," *IEEE Trans. Ind. Electron.*, vol. 62, no. 7, pp. 4563-4572, July 2015.

[12] J. Dannehl, F.W. Fuchs, S. Hansen, and P.B. Thøgersen, "Investigation of active damping approaches for PI-based current control of grid-connected pulse width modulation converters with *LCL* filters," *IEEE Trans. Ind. Appl.*, vol. 46, no. 4, pp. 1509–1517, July/Aug. 2010.

[13] M. Malinowski and S. Bernet, "A simple voltage sensorless active damping strategy for three-phase PWM converters with an *LCL* filter," *IEEE Trans. Ind. Electron.*, vol. 55, no. 4, pp. 1876–1880, Apr. 2008.

[14] J. Xu, S. Xie, and T. Tang, "Active damping-based control for grid-connected *LCL*-filtered inverter with injected grid current feedback only," *IEEE Trans. Ind. Electron.*, vol.61, no.9, pp.4746–4758, Sept. 2014.

[15] G. Shen, X. Zhu, J. Zhang, and D. Xu, "A new feedback method for PR current control of *LCL*-filter-based grid-connected inverter," *IEEE Trans. Ind. Electron.*, vol. 57, no. 6, pp. 2033-2041, June 2010.

[16] M. Xue, Y. Zhang, Y. Kang, Y. Yi, S. Li, and F. Liu, "Full feedforward of grid voltage for discrete state feedback controlled grid-connected inverter with *LCL* filter," *IEEE Trans. Power Electron.*, vol. 27, no. 10, pp. 4234–4247, Oct. 2012.

[17] J. Sun, "Impedance-based stability criterion for grid-connected inverters," *IEEE Trans. Power Electron.*, vol. 26, no. 11, pp. 3075-3078, Nov. 2011.

[18] D. Yang, X. Ruan, and H. Wu, "Using virtual impedance network to improve the control performances of *LCL*-type grid-connected inverter under the weak grid condition," *in Proc. IEEE APEC*, 2014, pp.1233-1239.

[19] J. Xu, S. Xie, and T. Tang, "Evaluations of current control in weak grid case for grid-connected *LCL*-filtered inverter," *IET Power Electron.*, vol. 6, no. 2, pp. 227-234, Feb. 2013.

[20] Liserre, M., Teodorescu, R., and Blaabjerg, F, "Stability of photovoltaic and wind turbine grid-connected inverters for a large set of grid impedance values," *IEEE Trans. Power Electron.*, vol. 21, no. 1, pp. 263–272, Jan. 2006.

[21] YA-RI. Mohamed, "Suppression of low- and high-frequency instabilities and grid-induced disturbances in distributed generation inverters," *IEEE Trans. Power Electron.*, vol. 26, no. 12, pp. 3790–803, Dec. 2011.

[22] J. Xu, S. Xie, and T. Tang, "Improved control strategy with grid-voltage feedforward for *LCL*-filter-based inverter connected to weak grid," *IET Power Electron.*, vol. 7, no. 10, pp. 2660-2671, Oct. 2014.

[23] X. Li, X. Wu, Y. Geng, X. Yuan, C. Xia, and X. Zhang, "Wide damping region for *LCL*-type grid-connected inverter with an improved capacitor-current-feedback method", *IEEE Trans. Power Electron.*, vol. 30, no. 9, pp. 5247–5259, Sep. 2015.

[24] M. Ciobotaru, R.Teodorescu, and F. Blaabjerg, "A new single-phase PLL structure based on second order generalized integrator," *in Proc. IEEE PESC*, 2006, pp.1-6.

Evaluation of PV Frequency-Watt Function for Fast Frequency Reserves

J. Neely, *Member IEEE*, J. Johnson, *Member IEEE*, J. Delhotal, S. Gonzalez, M. Lave, *Member IEEE*,

Sandia National Laboratories
Albuquerque, NM, USA
jneely@sandia.gov

Abstract—Increasing the penetration of distributed renewable sources, including photovoltaic (PV) sources, poses technical challenges for grid management. The grid has been optimized over decades to rely upon large centralized power plants with well-established feedback controls, but now non-dispatchable, renewable sources are displacing these controllable generators. By programming autonomous functionality into distributed energy resources—in particular, PV inverters—the aggregated PV resources can act collectively to mitigate grid disturbances. This paper focuses on the problem of frequency regulation. Specifically, the use of existing IEC 61850-90-7 grid support functions to improve grid frequency response using a frequency-watt function was investigated. The proposed approach dampens frequency disturbances associated with variable irradiance conditions as well as contingency events without incorporating expensive energy storage systems or supplemental generation, but it does require some curtailment of power to enable headroom for control action. Thus, this study includes a determination of the trade-offs between reduced energy delivery and dynamic performance. This paper includes simulation results for an island grid and hardware results for a testbed that includes a load, a 225 kW diesel generator, and a 24 kW inverter.

Keywords—photovoltaics, IEC advanced functions, advanced inverters, frequency regulation, frequency-watt, PV curtailment

I. INTRODUCTION

With plans to replace conventional power generation with renewable power, utilities fear large frequency and voltage excursions resulting from variability of renewable power generation, reduced mechanical inertia (for power electronic-coupled systems), and reduced grid support, i.e. voltage and frequency regulation. Some island power systems have realized high renewable penetration in a short time frame, and technical issues encountered therein are regarded by many as predictive for larger power systems in the U.S. In these island grids, PV systems are often required to also install expensive smoothing batteries [1] to mitigate the effect of PV fluctuations on the grid. The requirements are driven by concerns that the variability of renewable generators will overpower frequency regulating reserves provided by traditional generators, resulting in large frequency swings. These strict, expensive requirements could be lightened if the entire system were stabilized (rather than smoothing each plant individually). One method involves the implementation of grid support functions in grid-connected

photovoltaic sources [2]. In this paper, the proposed method is studied for the power grid on the island of Lanai.

A. Lanai Power System

Lanai is a 140.5 square mile Hawaiian island with approximately 3,200 residents in 1,150 households, living mostly in Lanai city. Fig. 1 presents an overview of the power system which includes a 10.4 MW diesel power plant, three 12.47 kV distribution circuits, the 1.2 MW La Ola PV power plant [3], and a 6 MW peak load.

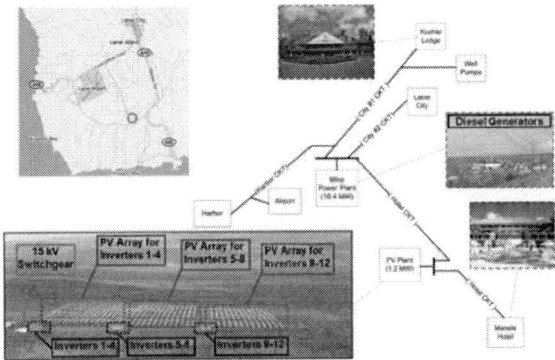

Fig. 1. Overview of the Lanai power system; reproduced from [1]

A previous study of the 1.2 MW PV plant in Lanai, HI demonstrated that real power curtailment alone was an effective tool to reduce PV ramp rates and assist with frequency stability [3]. Unfortunately, when performing this control strategy, much of the DC power is 'left on the array.' In this work, the Lanai power system is modeled and simulated assuming PV resources are configured to implement frequency-watt control. Evaluations of techno-economic performance are determined over a full day using a simplified model implemented in MATLAB [4]; and fault performance is evaluated over short time frames using a high-fidelity model developed in General Electric's *Positive Sequence Load Flow* (GE's PSLF). Several scenarios are considered with different PV penetrations and grid-support function settings. In addition, a testbed was assembled using select hardware at the *Distributed Energy Technology Laboratory* (DETL), including a diesel generator and a PV inverter with frequency-watt capability. This testbed was used to evaluate the frequency-watt function's ability to mitigate frequency deviations on a small power system with large steps in load.

This work was supported by the U.S. Department of Energy SunShot program under Award Number 29092. Sandia National Laboratories is a multi-program laboratory managed and operated by Sandia National Laboratories, a wholly owned subsidiary of Lockheed Martin Corporation, for the U.S. Department of Energy's National Nuclear Security Administration under contract DE-AC04-94AL85000.

978-1-4673-9551-9/16 $31.00 © 2016 IEEE

B. Frequency-Watt Control

Figure 2 shows an example of the *International Electrotechnical Commission* (IEC) *Technical Report* (TR) 61850-90-7 [5] FW22 frequency-watt curve with a deadband which curtails the output power at nominal grid frequency, allows generation to increase at grid frequencies below F_2 (if the power is available), and decreases generation when the grid frequency rises above F_3.

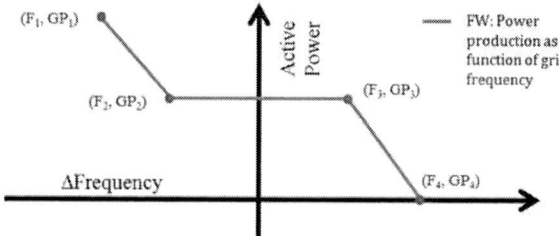

Fig. 2. Frequency-Watt control curve [5]

The next section provides details of the simulation models developed for this study in MATLAB and PSLF. Section III presents simulation results that evaluate the trade-offs of energy loss and frequency deviation. In particular, use of frequency-watt control is compared to that of fixed curtailment, and it is shown to provide a superior trade-off in frequency response versus energy loss. Section IV presents the results of hardware experiments intended to demonstrate and evaluate the application of frequency-watt to a power system that mimics an island grid. Finally, Section V provides conclusions and cites future work.

II. SIMULATION MODELS FOR THE LANAI POWER SYSTEM

The Lanai power system was evaluated in simulation using irradiance data collected at the 1.2 MW La Ola PV power plant located there. Specifically, irradiance data was used in concert with the Sandia-developed Wavelet Variability Model [6]-[9] to compute available PV power. Four irradiance profiles ranging from low to high variability were identified for the study. To evaluate the power system dynamics, a simplified MATLAB model was developed to efficiently simulate the day-long effects of PV variability and frequency-watt function parameters on frequency response and renewable energy production. In addition, a model was developed using PSLF to evaluate grid response using high-fidelity component models of the island grid. Both models were developed for three PV penetration levels (20%, 70% and 120%), and simulations were conducted using irradiance data from three different times of day (early morning, late morning, and midday). In this work, penetration level was defined as installed PV power divided by peak load. Assuming a 6 MW peak load, these PV penetration levels correspond to 1.2 MW, 4.2 MW and 7.2 MW installed capacity respectively. Both MATLAB and PSLF models were validated against PV power and frequency data collected at the 1.2 MW La Ola PV plant in Lanai for the 20% case.

A. Irradiance Data and the Wavelet Variability Model

Available PV power output was computed based on irradiance measured using a single LI-COR® silicon pyranometer at the 1.2 MW La Ola PV plant on Lanai [10]. Irradiance data was available at 1-second resolution. This irradiance point sensor was scaled to plant-average irradiance which accounts for the spatial smoothing across the array area for the PV plant. This allowed for simulation of any size PV plant; using measured power output from the La Ola plant would only be representative of the 1.2 MW case. In particular, the Wavelet Variability Model (WVM) [6] was used to smooth the irradiance at each simulation PV array location. The WVM applied different smoothing based on the amount of PV, the PV density, and the daily cloud speed (see Fig. 3). The PV density was assumed to be 40 W/m^2 for utility-scale PV plants (such as La Ola) and 5 W/ m^2 for distributed PV (i.e., rooftop). The daily cloud speed was assumed to be 10 m/s for Hawaii based on [11].

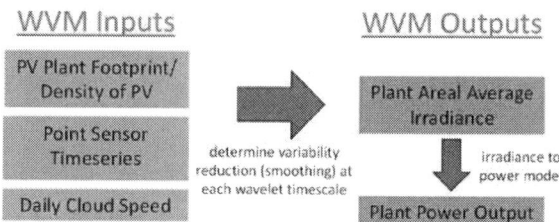

Fig. 3. Inputs and outputs for the wavelet variability model (WVM)

The output of the WVM was plant area average irradiance, which was converted into an available AC power value using the Sandia Array Performance Model (irradiance to DC) [12] and the Sandia Inverter Model (DC to AC) [13]. This procedure has been validated at other PV plants [8]-[9] and is expected to have similar accuracy at the Lanai location.

B. Irradiance Profiles

Irradiance profiles from four days were used for this study to cover a wide range of irradiance conditions and capture the system dynamics during the worst-case ramp rates:

- December 7th, 2010 – Least Variable
- September 3rd, 2010 – Average Variability
- April 23rd, 2011 – Largest Single Ramps
- November 4th, 2010 – Most Variable Total Day

Irradiance profiles for these four days are displayed in Fig. 4.

C. MATLAB Dynamic Model

To evaluate the effect of the frequency-watt function on power system performance over a full solar day (sunrise to sunset), a simplified model was developed in MATLAB. At the Miko power plant on Lanai, 2015 operations typically include two 2.8 MW diesel generators in a master-slave configuration providing isochronous control with additional 1.2 MW droop control diesel generators connected as needed to meet the island load.

In this model, the generators were aggregated and represented as one spinning mass, the governor controls were aggregated, and the PV power was summed. Fig. 5 shows the integrated Miko power plant model.

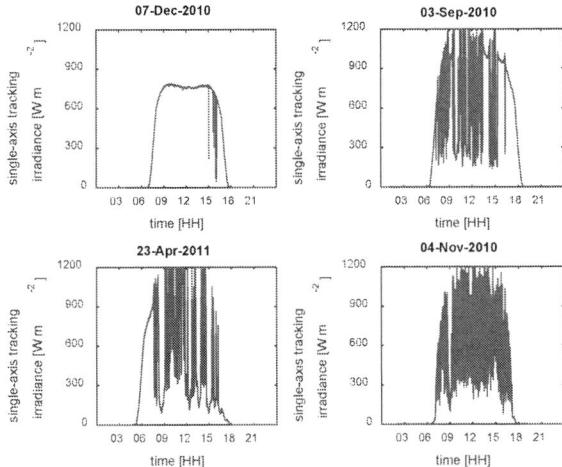

Fig. 4. Shows Irradiance profiles collected from the La Ola PV power plant showing (top left) little variability on Dec. 7th, 2010; (top right) typical variability seen on Sept. 3rd, 2010; (bottom left) April 23, 2011, largest single ramps; and (bottom right) day with highest total variability, Nov. 4th, 2010

With isochronous control, combustion time constant and rotor velocity, the system model is third order. The system dynamics were computed using four-step integration with fixed time step of 2 ms. Since the available PV power was computed using the WVM with 1-second resolution, intermediate values were determined at the smaller time step using linear interpolation. The available PV power and system frequency were then used to compute the PV power applied to the system using the frequency-watt function definition.

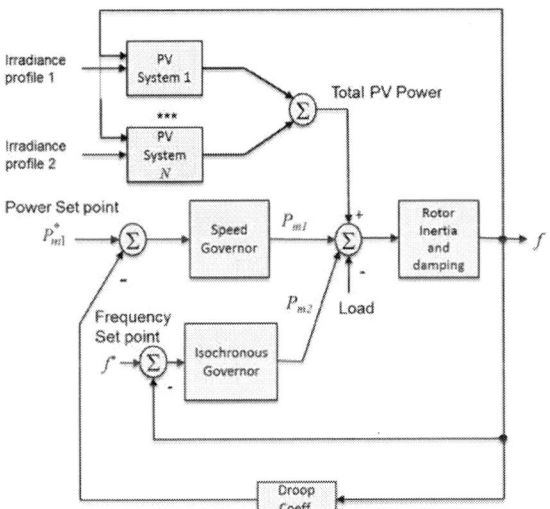

Fig. 5. Overview of the simplified power plant model which includes a mixture of droop and isochronous speed control as well as PV power from multiple locations

D. PSLF Model

GE's PSLF platform was used to evaluate the grid response to two grid event scenarios: generator outage and line-to-ground fault on a distribution line feeding a well pump. The PSLF model of the Lanai electricity grid includes the transmission system, substation transformers, diesel generators, and PV generators, such that high-fidelity transient system dynamics could be studied on short time scales (i.e. several minutes). To account for the time-varying PV power availability, PV power values were pre-computed using WVM and read into the 240 second PSLF simulations. Unlike the MATLAB model—wherein PV power was summed—in the PSLF model, individual PV plants and aggregated rooftop solar was simulated with inverter controller dynamics and connected to the appropriate bus in the circuit. The PSLF model included 49 buses for the 20% penetration case, 56 buses for the 70% penetration case and 59 buses for the 120% case. The model was based on previous work at Sandia National Laboratories [14].

Three pre-calculated irradiance profiles were created for the PSLF simulations to represent early morning (irradiance varies from 176 to 493 W/m²), late-morning (irradiance varies from 292 to 1,033 W/m²), and midday (irradiance varies from 1,012 to 1,147 W/m²). The available PV power profiles were computed using WVM variability, which considers the density of installed PV panels, cloud speed and whether the panels are fixed or tracking. In the case of multiple PV plants, irradiance data was time-shifted based on geographical distance and 10 m/s cloud speed. For the 120% penetration case, three locations were considered: La Ola power plant with concentrated PV collection with single axis trackers, Lanai city with distributed PV collection using fixed plane rooftop solar, and the Manele Hotel with concentrated PV collection with single axis trackers. The available normalized PV power was computed for the nine cases (three locations and three times of day) and shown in Fig. 6.

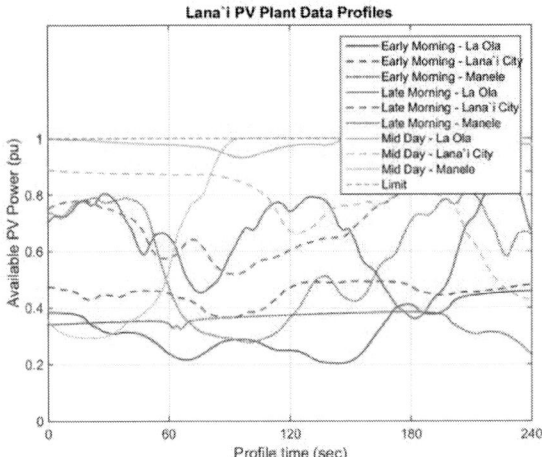

Fig. 6. Available PV power profiles computed for PSLF simulation for 120% PV penetration case; power values given in per unit (pu)

Finally, it is noted that the frequency-watt control would not be instantaneous. In practice, changes in inverter output power would lag behind measured frequency changes. Herein, the PSLF inverter models were assumed to have a first-order response with 100 msec time constant for both increasing and decreasing output power.

E. Model Validation

The MATLAB and PSLF grid frequency simulations were validated using Lanai frequency data collected at the La Ola PV power plant for the 20% penetration case. The OSIsoft PI server data [15] contains time-synchronized PV power and frequency data. A six minute period of high PV variability was replayed in both the simplified MATLAB model and the PSLF system model to generate grid frequency data. A comparison of the grid frequency from the PI data and the two models is shown in Fig. 7.

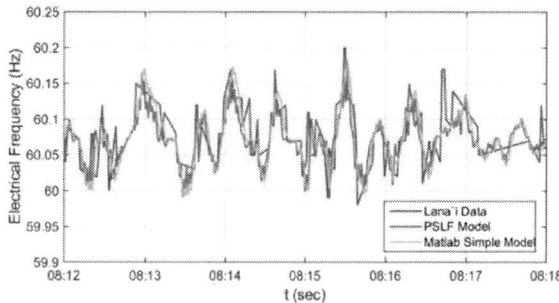

Fig. 7. Comparison of the Lanai frequency data to the MATLAB and PSLF simulations.

On inspection, there is good correlation between the MATLAB and PSLF model outputs and the measured frequency data. The simulation data was down sampled to correspond to the measured data, and the agreement was evaluated using the Pearson's correlation coefficient, computed over the six minute window, where $-1<\rho<1$ with $\rho=1$ being the best match. The Pearson's correlation coefficient relating the PSLF simulation result to La Ola data was $\rho=0.724$, relating the MATLAB result to La Ola data was $\rho=0.782$, and relating the two models was $\rho=0.939$. Thus, the models agree well with one another and moderately well with the data. It is noted that the load was constant in both models but was likely varying in the physical system, which would contribute to frequency variation.

III. SIMULATION RESULTS

The effect of FW function parameters was evaluated in simulation using both the MATLAB and PSLF models. The MATLAB simulation example is presented first.

A. MATLAB Simulations

To illustrate performance of the PV system under high solar variability conditions and high penetration, the system is evaluated using the irradiance data from 4 Nov, 2010 with 70% PV penetration, and the system response was simulated using the FW curves defined in Fig. 8. Fig. 9 shows the output power, ramp rates, and frequency plotted over time. The ramp rate and grid frequency range performance metrics compared to lost PV power generation were obtained, as seen in Fig. 10.

The energy lost is defined as the percent difference between the PV energy output from a given curtailment scenario and the PV energy available with no curtailment. This percentage would represent the cost of implementing the frequency-watt function, depending on the price of electricity. However, as PV penetrations increase, PV systems may be curtailed anyway, so there would no longer be an economic penalty to enacting this control scheme.

The ramp rates were defined using the difference between samples of the output power, spaced 1-second apart. The frequency performance was determined by the difference between the maximum and minimum frequency experienced during the simulation of a day. It is noted in Figure 10b that a 15% capacity curtailment results in the 99% ramp rate dropping from 79.6 kW/sec to 78.6 kW/sec and the frequency range drops from 1208 mHz to just 1039 mHz: almost negligible benefit. However, with FW22 set 2, the 99% ramp rates drop to 50.0 kW/sec and frequency range is 739 mHz for a 2% loss of PV energy.

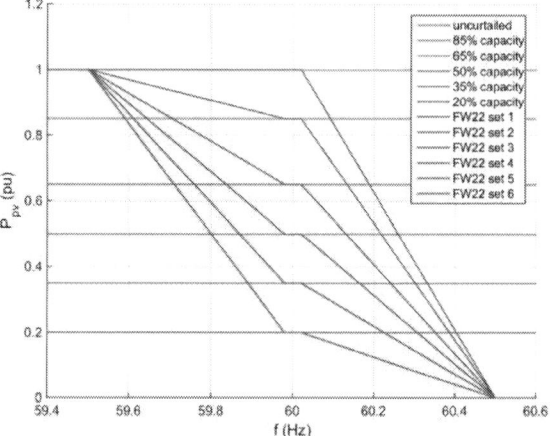

Fig. 8. Family of curves showing capacity curtailment and FW function defintions used in MATLAB study

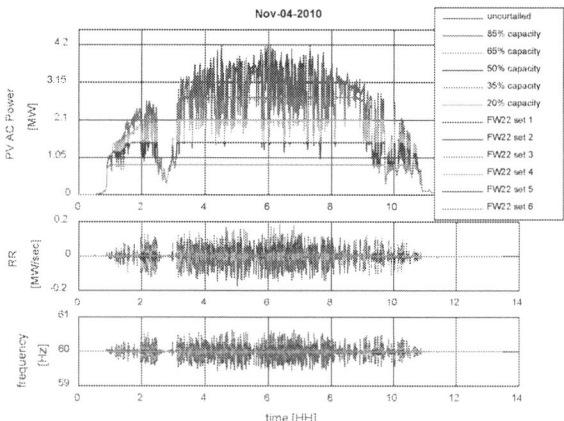

Fig. 9. Time domain plot of MATLAB simulation for 70% penetration case showing PV AC Power, Ramp Rates (RRs) and system frequency for variety of curtailment values and FW function definitions

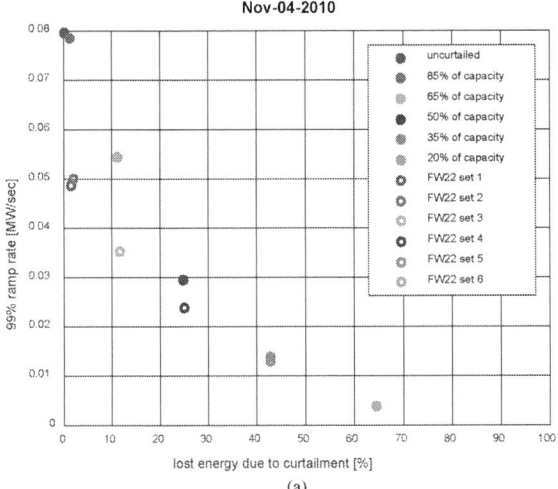

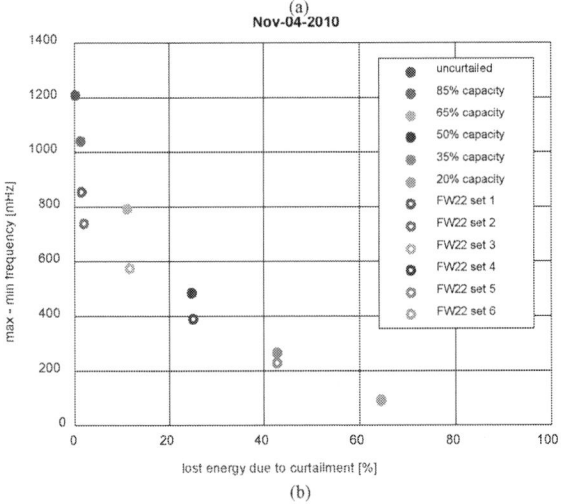

Fig. 10. Plot of (a) ramp rates vs energy loss and (b) frequency range vs energy loss for November 4th, 2010 dataset assuming 70% penetration

Similar benefits are seen for the April 23rd, 2011 dataset. Frequency range results are shown in Fig. 11, assuming again a 70% PV penetration. Therein, the base case results in 1566 mHz of range. When a simple 15% curtailment is applied, this range drops to 1415 mHz and results in 3.2% less energy. With a 35% curtailment, the range drops to 1166 mHz but costs 13.4% of the PV energy. The trade-off improves by applying the FW control. FW set 1 results in 1011 mHz range with 1% loss of energy, set 2 results in 902 mHz with 3.7% loss of energy, and set 3 results in 776 mHz with 13.6% loss.

The September 3rd, 2010 dataset is evaluated in Fig. 12 assuming 120% penetration (7.2 MW installed PV). In this case, a minimum curtailment is necessary since the installed capacity is greater than load. Therein, large benefits are demonstrated for the use of the FW function. For 35% curtailment, the system frequency deviates 1976 mHz and

results in 13% energy loss, but the FW set 3 limits the frequency range to 1271 mHz with just 14.2% energy loss.

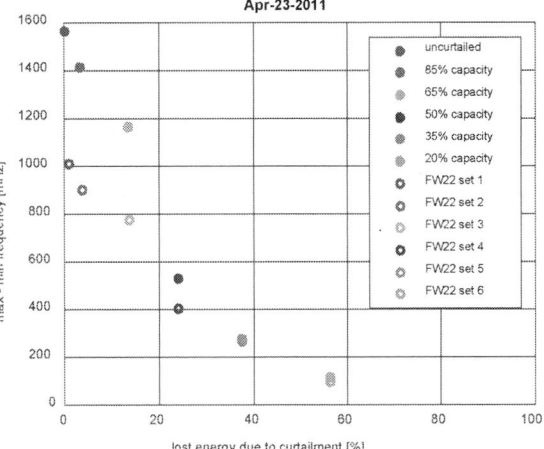

Fig. 11. Plot of frequency range vs energy loss for April 23rd, 2011 dataset assuming 70% penetration

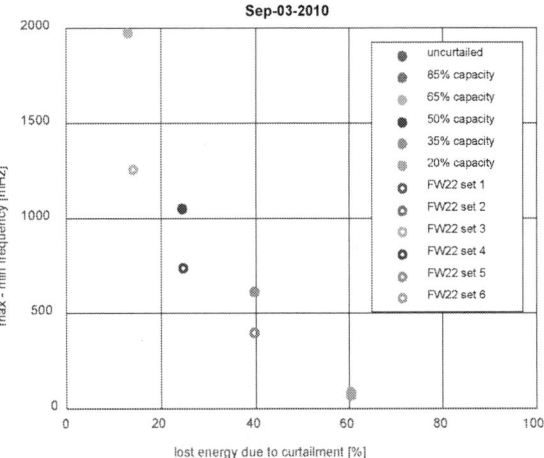

Fig. 12. Plot of frequency range vs energy loss for September 3rd, 2010 dataset assuming 120% penetration

As a final example, the December 7th, 2010 case assuming 20% PV penetration is shown in Fig. 13. In this low penetration clear-sky case, the frequency range is very small, and neither curtailment nor FW function implementation provides a benefit that is worth the loss of energy.

B. PSLF Simulations

To illustrate the benefits of the FW implementation during a contingency event, two examples are shown below comparing the FW function to PV active power curtailment. Both scenarios consider a 70% PV penetration scenario with late morning irradiance profile. The FW curve was defined with $GP_2 = GP_3 = 0.6$, 59.95 to 60.05 Hz deadband, and a droop of 80% of nameplate power/Hz; the curtailment was defined at

0.6 for all frequencies. Fig. 14a shows the simulated electrical frequency at the Miko power plant following loss of a 1.2 MW droop-controlled generator at t = 120 sec, and Fig. 14b shows the total PV power.

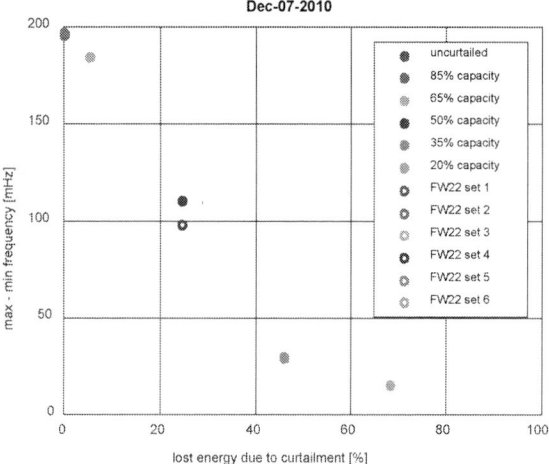

Fig. 13. Plot of frequency range vs energy loss for December 7th, 2010 dataset assuming 20% penetration

The FW and curtailment scenarios exhibit similar response to PV variation from t = 0-120 sec since they have the same limit at nominal frequency and frequency variation is primarily within the deadband of the FW function. However, following the loss of the 1.2 MW generator, a sizable drop in frequency is observed. Comparing the two responses, an improvement is seen in the frequency response with the FW implementation. The frequency drops to 59.32 Hz with simple curtailment, but the nadir is 59.52 Hz with the FW implementation. In addition, the generator speed rises to 60.11 Hz around t = 205 sec due to PV variation in the curtailment case, but is limited to 60.05 Hz in the FW implementation. The improvement is due to the FW control action. In particular, PV power rises by almost 223 kW (to full available power) in the FW case to compensate for the decrease in frequency following loss of generation; see Fig. 14b. The control action is best depicted, however, in the *power-frequency* phase plane together with the FW and curtailment limits; see Fig. 15. As the frequency falls below the deadband of the FW curve, the FW limit increases and PV power rises, helping the system to return to nominal frequency.

Fig. 16 shows results for the simulated line fault. The line-fault causes the generators to speed up. As with the previous example, the control action of the FW function mitigates the frequency deviation, resulting in a max frequency of f_{max_f} = 60.41 Hz compared to 60.70 Hz for the curtailment case. The resultant dynamic response results in a slightly lower minimum frequency of f_{min_f} = 59.80 Hz compared to 59.94 Hz for the curtailment case, however.

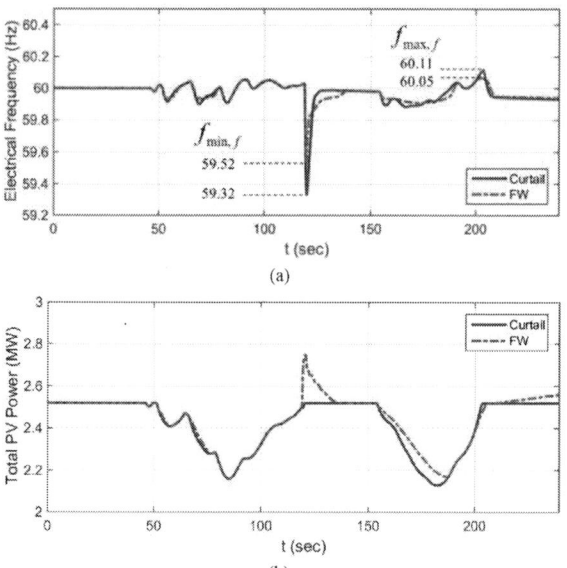

Fig. 14. Impact of FW and curtailment for simulated loss of 1.2 MW generator at t=120 sec showing (a) electrical frequency at the power plant and (b) total PV power

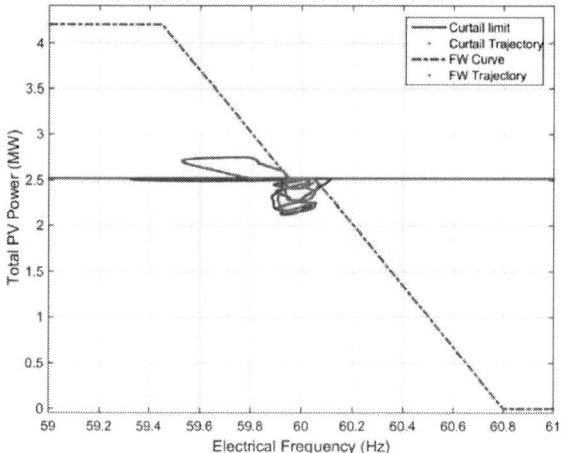

Fig. 15. FW implementation versus fixed curtailment, showing response in the PV power as generator speed varies

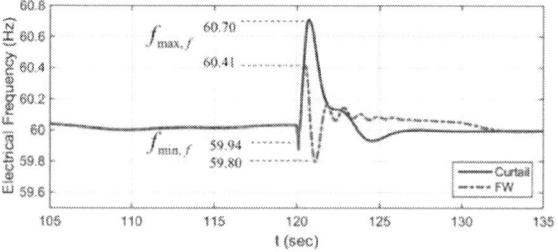

Fig. 16. Impact of FW and curtailment for simulated line fault at t=120 sec showing close-up of electrical frequency

IV. EXPERIMENTAL RESULTS

To demonstrate the ability of frequency-watt implementation to mitigate frequency (generator speed) deviations, experiments were performed at DETL. These experiment configurations included a 225 kW diesel generator, a PV array simulator, a 24 kW inverter with frequency-watt functionality, a 50 kW resistive load, and a 25 kW resistive load. Four scenarios were considered: (a) the inverter operating with 100% of capacity with FW enabled, (b) the inverter operating with 100% of capacity with FW disabled, (c) the inverter operating at 50% of capacity with FW enabled, and (d) the inverter disconnected. Fig. 17 shows the configuration of the experimental setup, and Fig. 18 shows the FW function settings used in the inverter for scenarios (a) and (c). Therein, F_3 = 60.5 Hz and F_4 = 62.0 Hz.

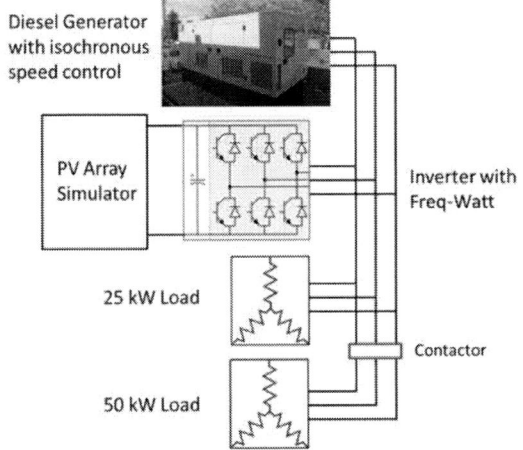

Fig. 17. Experimental test setup including diesel generator, PV inverter equiped with frequency-watt capability and loads.

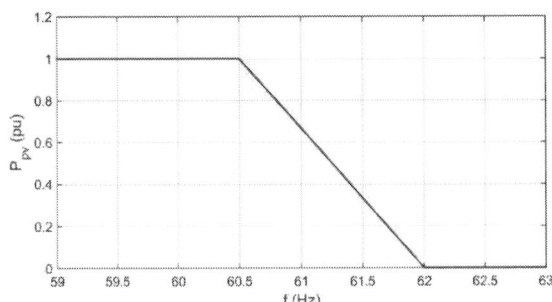

Fig. 18. FW function settings used for hardware experiments

Under all cases, the DC power available to the inverter was set to the DC input rating of the inverter using a PV array simulator. In case (c) the inverter is commanded to operate at 50% of rated active power. Fig. 19 shows the frequency response when the load is stepped from 75 kW to 25 kW by opening a contactor. In each case, generator speed increases rapidly after the load shed, but the peak frequency is lessened and then the return to nominal speed is improved in the cases with frequency-watt capability enabled. Only the case of

increasing grid frequency was tested because of the limits in the inverter FW settings. Specifically, in this particular inverter, it was not possible to set up the FW parameters to allow power increase when $f < F_2$ because it was designed for the German grid code requirement VDE-AR-N 4105 [16].

To quantitatively compare the frequency performance of each scenario, a figure of merit was chosen to represent overall frequency deviation with time: $f_e = \int_{t_0}^{t_s}(f(\tau) - f^*)^2 d\tau$, where f_e is the integral square frequency error, expressed in Hz^2·s, or simply Hz; f is the system frequency; f^* is the scheduled frequency; t_0 is the time that the step change in the load occurs; and t_s is the time at which the frequency transient settles. The error is calculated over a 5-second time period. Another figure of merit is the total PV power generation before the event. Results are shown in Table 1. Therein, it is noted that the first case demonstrates the best frequency response and the greatest PV power penetration for this set of experiments. The same PV penetration occurs in the second case, but f_e is greater. The third case has a similar f_e as the first, but the PV penetration is half. The final case has slightly lower f_e than the second but includes no PV power.

Table 1: Performance comparison for a 75 kW to 25 kW stepped load

Scenario	Integral Squared Frequency Error f_e (Hz)	Initial PV Power (kW)
Inverter at 100% capacity with FW enabled	0.50993	24.0
Inverter at 100% capacity with FW disabled	0.75181	24.0
Inverter at 50% capacity with FW enabled	0.51738	12.0
Diesel generator with no inverter	0.61614	0.0

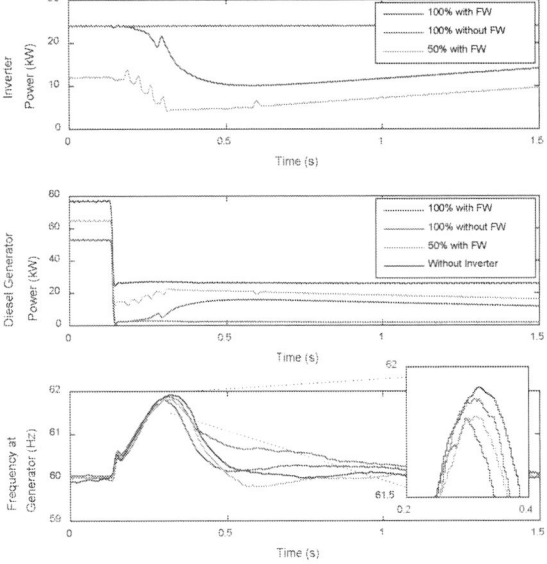

Fig. 19. Response to a 75 kW to 25 kW step change in load

This demonstration was intended to represent frequency variations that occur with 100-200 msec rise-times on a small electrical grid (much smaller than the Lanai power system). The frequency-watt control does exhibit a small lag in responding to a positive frequency (power reduction) swing, which is evident in Fig. 19, but the control is still effective for this experiment. However, it is noted that the power is slow to rise after the frequency returns to nominal. The exact cause of this is unknown but may be an intentional ramp limit. This slow response time could reduce the effectiveness of the control to compensate for drops in frequency.

Frequency events in the continental US would experience slower frequency changes than what is modeled and demonstrated here; thus, it is expected that FW control could provide effective frequency support in a larger grid. However, to be most effective, work is needed to ensure that the inverter response time is sufficiently small for both increasing and decreasing power output (in response to falling and rising frequency respectively), and that FW parameters can be adjusted to allow for power increase when $f<F_2$.

V.   CONCLUSIONS AND FUTURE WORK

In this work, use of the frequency-watt function to mitigate frequency deviations in a power system with high penetration of PV power is investigated. Using models of an island grid that have been validated against field-data, the approach is demonstrated in simulation, and the trade-off between renewable energy production and frequency deviation is quantified. The simulations show that using frequency-watt functionality and fixed curtailment both improve the frequency deviations on the grid during normal operation and during fault transients, but the frequency-watt functions provide a greater improvement for a given curtailment level. Furthermore, feasibility of implementing the frequency-watt function in an electrical system experiencing a loss of load fault was demonstrated using a hardware testbed.

In the future, the research team plans to investigate the optimal FW settings to provide frequency regulation and contingency reserve capabilities while minimizing the loss of PV power from curtailment. The team also plans to develop specifications and recommendations for limits on inverter response time.

REFERENCES

[1] V. Gevorgian, S. Booth, Review of PREPA Technical Requirements for Interconnecting Wind and Solar Generation, NREL technical report, NREL/TP-5D00-57089, Nov. 2013.

[2] Hawai'i Natural Energy Institute, Summary Review of Advanced Inverter Technologies for Residential PV Systems, Sept. 2014.

[3] J. Johnson, B. Schenkman, J. Quiroz, and A. Ellis, "Initial operating experience of the La Ola 1.2 MW photovoltaic system," Sandia National Laboratories Technical Report SAND2011-8848, 2011.

[4] The Mathworks, Inc., MATLAB the language of technical computing, 3 Apple Hill Drive, Natick, MA 01760-2098 USA.

[5] IEC Technical Report 61850-90-7, "Communication networks and systems for power utility automation–Part 90-7: Object models for power converters in distributed energy resources (DER) systems," Edition 1.0, Feb 2013.

[6] M. Lave, J. Kleissl, and J. S. Stein, "A Wavelet-Based Variability Model (WVM) for Solar PV Power Plants," Sustainable Energy, IEEE Transactions on, pp. 1-9, 2012.

[7] M. Lave and J. Kleissl, "Cloud speed impact on solar variability scaling – Application to the wavelet variability model," Solar Energy, vol. 91, pp. 11-21, 2013.

[8] M. Lave and J. Kleissl, "Testing a wavelet-based variability model (WVM) for solar PV power plants," in IEEE Power and Energy Society General Meeting, 2012.

[9] M. Lave, J. Kleissl, and J. S. Stein, "A Wavelet-Based Variability Model (WVM) for Solar PV Power Plants," IEEE Transactions on Sustainable Energy, 2013.

[10] S. Kuszmaul, A. Ellis, J.S. Stein, L. Johnson, Lanai High-Density Irradiance Sensor Network for characterizing solar resource variability of MW-scale PV system, 35th IEEE Photovoltaic Specialists Conference (PVSC), Honolulu, HI, 2010.

[11] L. M. Hinkelman, "Differences between along-wind and cross-wind solar irradiance variability on small spatial scales," Solar Energy, vol. 88, pp. 192-203, 2013.

[12] D. King, W. Boyson, and J. Kratochvil, "Photovoltaic Array Performance Model," SAND2004-3535, 2004.

[13] D. King, S. Gonzalez, G. Galbraith, and W. Boyson, "Performance Model for Grid-Connected Photovoltaic Inverters," SAND2007-5036, Sandia National Laboratories, 2007.

[14] KEMA, Inc., Lanai PV Interconnect Requirements Study: System Impact Study, June 5, 2008, Interconnection Customer: SunPower, KEMA, Inc., Raleigh, NC, USA, 26707.

[15] OSIsoft, The PI System - the industry standard in enterprise infrastructure for management of real-time data and events, visited 22 Nov, 2015, URL: https://www.osisoft.com/software-support/what-is-pi/What_Is_PI.aspx

[16] VDE Reference "VDE-AR-N 4105 Power generation systems connected to the low voltage distribution network -Technical minimum requirements for the connection to and parallel operation with low voltage distribution networks", Aug. 2008.

A Systematic Design Method and Verification for a Zero-ripple Interface for PV/Battery-to-Grid Applications

Suvankar Biswas, Ned Mohan and William Robbins

Department of Electrical and Computer Engineering
University of Minnesota
Minneapolis, MN
{biswa029, mohan, robbins}@umn.edu

Abstract—A systematic method of designing a zero-ripple Ćuk converter for PV/Battery-to-grid applications is presented in this paper. The integrated magnetic core design uses an intuitive flux-reluctance model to arrive at the Area Product for this kind of structure. Unlike the earlier designs for this converter, it provides a completely analytical approach to design this converter for a range of specifications. The target application is grid interface of PV or battery. The validity of the proposed method is confirmed using finite element analyses (both 2-D and 3-D), circuit simulations in pspice as well as preliminary experimental validation.

I. INTRODUCTION

DC-DC converters for interfacing PV panels to grid-tied inverters have been discussed in the recent literature. The boost converter is usually the topology of choice [1]–[3]. However,the problem of EMI (electromagnetic interference) in residential microgrids, as alluded to, in [1], [2] can be addressed better by the Ćuk converter. The Ćuk converter can soft-start, on account of its voltage conversion ratio [4]. Also, being a current-fed converter, it can interface with PV. The control of the Ćuk converter is a little involved, but our target application is not switched-mode power supplies, where load transients are frequent. The idea of this integrated magnetic converter is not novel in itself. An analytical condition on the inductance matrix can be derived which shows that the zero-ripple operation is theoretically possible, but the design of this magnetic structure is somewhat complex. There are a few notable solutions to date [5]–[7], but are mostly semi-analytical. For example, [5] has deficiencies in picking the area-product simply because the formulation does not have all known parameters. In [6], the authors define a "leakage parameter", and provide a list of that parameter for a number of EE-cores. However, the leakage does not depend just on the core geometry but also the winding geometry. Lack of a systematic method determining the power converter requirements into a viable magnetic design has prevented the Ćuk converter from being used.

This paper places emphasis on the magnetic design process in the following way. Section II talks about the general requirements of a PV-to-grid DC-DC converter, and the inductance matrix requirements for zero-ripple. Section III describes the flux-reluctance model for the chosen magnetic structure and an approximate zero-ripple condition. This uses the "leakage parameter" defined earlier, but does not require explicit knowledge of this quanity. The Area-Product [8], [9] is then derived using these approximate models in order to provide a systematic approximate design for this converter. The FEM and simulation results are discussed that fine tune the design of this converter in Section IV. Section V presents conclusion and future work.

II. DESCRIPTION OF THE CONVERTER

The converter schematic is shown in Fig. 1. It is designed such that the operable duty ratio $D \in (0.3, 0.75)$. The upper limit is due to influence of non-idealities on the voltage conversion ratio. The lower limit can be attributed to high peak flux density in the core, which is discussed later. The converter is designed to operate at a DC bus voltage of 340V at output with a solar panel which has same specifications as the SunModule ® Plus SW 270 mono (270 W) [10]. As such, this converter can be used for standalone residential applications among others.

For a four-winding coupled inductor, we have the following set of equations:

$$
\begin{bmatrix} v_1 \\ v_2 \\ v_3 \\ v_4 \end{bmatrix} = \begin{bmatrix} L_{11} & L_{12} & L_{13} & L_{14} \\ L_{12} & L_{22} & L_{23} & L_{24} \\ L_{13} & L_{23} & L_{33} & L_{34} \\ L_{14} & L_{24} & L_{34} & L_{44} \end{bmatrix} \times \frac{d}{dt} \begin{bmatrix} i_1 \\ i_2 \\ i_3 \\ i_4 \end{bmatrix} \quad (1)
$$

The voltages v_1, v_2, v_3, v_4 are proportional to each other. For zero-ripple in i_1 and i_2, their time derivatives must be zero at all times. This gives:

$$ nL_{14} = L_{24}, L_{23} = L_{34}, L_{24} = L_{44} \quad (2) $$
$$ L_{13} = L_{33}, L_{14} = L_{34}, nL_{13} = L_{23} \quad (3) $$

However, this provides very little insight into the magnetic design process, although it specifies the end-goal. What we need is a core structure and a method to do an approximate

978-1-4673-9551-9/16 $31.00 © 2016 IEEE

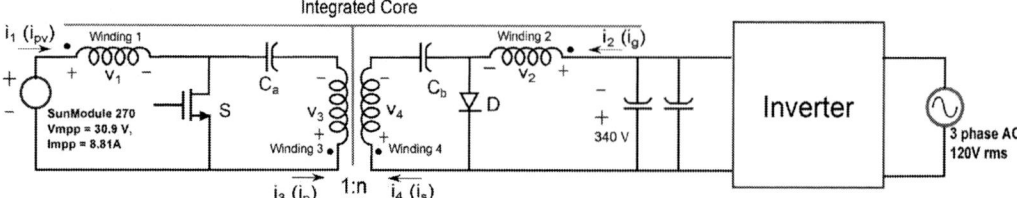

Fig. 1. Complete System with Specifications

design which gives us the winding turns and conductor dimensions for each winding. This is explained in the next section. This reduces the problem to simply a matter of picking the correct air-gap at which the relationships in (2) and (3) will hold true, which is done by FEM. Please note that henceforth, capital letters denote the dc values of a current or voltage.

III. PRINCIPLE OF OPERATION AND AREA PRODUCT

A. Magnetizing Inductance

The basic idea for zero-ripple is to shift the total interface winding ripple (windings 1 & 2) to the magnetizing inductance of the isolation transfomer (Windings 3 & 4 in Fig.1) as explained by Ćuk in [11]. The other quantities are as labeled in Fig.1. If the peak-to-peak ripple in magnetizing current is Δi_M, then let us impose the requirement

$$\Delta i_M = f_r(i_{1,rms} + ni_{2,rms}) \tag{4}$$

where

$$i_{1,rms} = I_{pv}; i_{2,rms} = I_g \tag{5}$$

since i_1 and i_2 are zero-ripple quantities. For a Ćuk converter (ideally),

$$I_g \approx \frac{I_{pv}(1-D)}{nD}; \Delta i_M = \frac{V_{pv}D}{L_p f_s} \tag{6}$$

where D = duty-ratio of S, f_s = switching frequency, L_p = isolation transformer magnetizing inductance.

From (4),(5) and (6) we have

$$\Delta i_M = \frac{f_r I_{pv}}{D} \tag{7}$$

From (6) and (7), then

$$L_p = \frac{D^2 Z_{pv}}{f_r f_s} \tag{8}$$

Since the converter will most likely operate at the maximum power point of the PV panel for maximum efficiency, it seems reasonable to pick

$$L_p = \frac{D_{max}^2 Z_{MPP}}{f_r f_s} \tag{9}$$

where Z_{MPP} = PV source impedance at maximum power point. Now usually for a PV panel, the stable region is to the right of the MPP (maximum power point). If the grid voltage is asssumed to be constant throughout, then the maximum duty ratio of a Ćuk converter corresponds to the MPP.

B. Core Structure, Zero-ripple and Peak Flux Density

An EE-core with a spacer air gap is chosen for symmetry reasons and tunability as explained in [5].The core structure and the corresponding intuitive flux-reluctance model is shown in Fig.2. The reluctances are defined in terms of R.

$$R = \frac{2x}{\mu_0 A_c} \tag{10}$$

where $\frac{x}{2}$ = spacer airgap.

$R/2$ and $R/4$ represent the reluctances due to air-gap, while R_l, R_{l1} and R_{l2} represent those due to leakage. The zero-ripple condition is derived in [12]. It is given by:

$$\frac{N_{pv}}{N_p} = f = 2 + \frac{x}{l} \tag{11}$$

where l = leakage parameter [5], N_{pv} = No. of turns of winding 1, N_p = No. of primary turns (winding 3). Also

$$N_s(\text{secondary turns(winding 4)}) = nNp \tag{12}$$

$$N_g(\text{winding 2}) = nN_{pv} \tag{13}$$

If we define the following:

$$N_1 i_1 = N_{pv} i_{pv} \tag{14}$$

$$Ni = N_p i_p + N_s i_s \tag{15}$$

$$N_1 i_2 = N_g i_g \tag{16}$$

Solving the circuit of Fig.2 yields the following equations:

$$\phi_1 = \frac{1.5 N_1 i_1 - 0.5 N_1 i_2 + Ni}{R} \tag{17}$$

$$\phi_2 = \frac{1.5 N_1 i_2 - 0.5 N_1 i_1 + Ni}{R} \tag{18}$$

$$\phi = \frac{N_1(i_1 + i_2) + 2Ni}{R} \tag{19}$$

Peak flux densities are needed to in order to find the correct winding area A_c. To find it, we make use of the following facts:

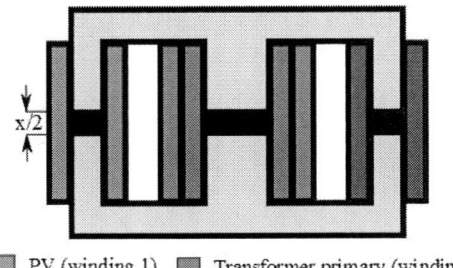

Center leg cross section area = A_c

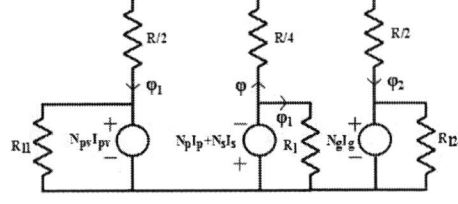

PV (winding 1) ▨ Transformer primary (winding 3)
Transformer secondary (winding 4) ▨ Grid (winding 2)

Fig. 2. Core Structure and Flux Reluctance Model

- i_1 and i_2 are purely dc.
- The ideal converter approximation, $\frac{V_g I_g}{V_{pv} I_{pv}} \approx 1$.

Using the above two, and the first equation in (6), we can show that at quasi-steady state the peak flux densities corresponding to ϕ_1, ϕ_2, ϕ are

$$\hat{B}_2 = \frac{N_{pv} I_{pv} f_{\phi_2}(D) + 2 * max(N_p i_p + N_s i_s)}{RA_c} \quad (20)$$

$$\hat{B}_1 = \frac{N_{pv} I_{pv} f_{\phi_1}(D) + 2 * max(N_p i_p + N_s i_s)}{RA_c} \quad (21)$$

$$\hat{B} = \frac{N_{pv} I_{pv} f_{\phi}(D) + 2 * max(N_p i_p + N_s i_s)}{RA_c} \quad (22)$$

where f_{ϕ_2}, f_{ϕ_1} and f_ϕ are defined as follows.

$$f_{\phi_2}(D) = 2\frac{1-D}{D}(2 - \frac{0.5}{D}) \quad (23)$$

$$f_{\phi_1}(D) = 2(2 - \frac{0.5}{D}) \quad (24)$$

$$f_\phi(D) = \frac{1}{D} \quad (25)$$

f_{ϕ_2}, f_{ϕ_1} and f_ϕ are plotted in Fig.3(a). Inspecting it tells us that $B_{peak} = \hat{B}(D = 0.3)$ for the given range of duty ratio. We also see why its a good choice to limit D_{min} to 0.3 to prevent probable core saturation.

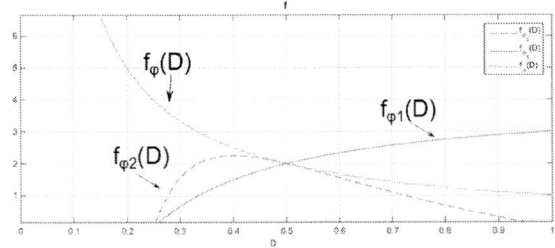

Fig. 3. Determination of max. flux density

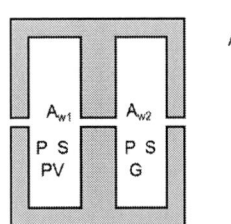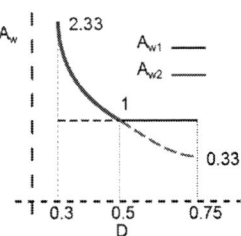

Fig. 4. Window Area Determination

C. Window Area

Using the template of Fig.4, and equations (6),(11)-(13), the window areas can be expressed in terms of

$$A_{w1} = \frac{fN_p I_{pv} + N_p i_{p,rms} + nN_p i_{s,rms}}{k_{Cu} J_{rms}} \quad (26)$$

$$A_{w2} = \frac{\frac{1-D}{D} fN_p I_{pv} + N_p i_{p,rms} + nN_p i_{s,rms}}{k_{Cu} J_{rms}} \quad (27)$$

Defining normalized versions of (25) and (26) as

$$a_{w1} = \frac{k_{Cu} J_{rms} A_{w1} - N_p i_{p,rms} - nN_p i_{s,rms}}{fN_p I_{pv}} \quad (28)$$

$$a_{w2} = \frac{k_{Cu} J_{rms} A_{w2} - N_p i_{p,rms} - nN_p i_{s,rms}}{fN_p I_{pv}} \quad (29)$$

These are plotted in Fig.4 as a function of D.

It is seen that selecting a_{w2} (henceforth A_{w2}) at $D = 0.3$ takes care of the entire design space (since in the actual physical core, both window areas are identical, and this will be the worst case design).

The final expression is :

$$max(A_w) = A_{w2,D=0.3} = \left(\frac{N_g I_g + N_p I_p + N_s I_s}{k_{Cu} J_{rms}}\right)_{D=0.3} \quad (30)$$

978-1-4673-9551-9/16 $31.00 © 2016 IEEE   1936

It is seen that the worst cases occur at $D = 0.3$ for both peak flux density and window area. Hence it would be a natural choice to define the area product at this point.

D. Area Product

1) $D \in (0.3, 0.5)$: From (22) & (5),

$$N_p = \frac{\hat{B} A_c R}{f I_{pv} + n f I_g + max(i_p + n i_s)} \qquad (31)$$

From (29) & (5),

$$N_p = \frac{A_{w2} k_{Cu} J_{rms}}{n f I_g + i_{p,rms} + i_{s,rms}} \qquad (32)$$

2) $D \in (0.5, 0.75)$: From (22) & (5),

$$N_p = \frac{\hat{B} A_c R}{2 f I_{pv}(2 - \frac{0.5}{D}) + max(i_p + n i_s)} \qquad (33)$$

From (29) & (5),

$$N_p = \frac{A_{w1} k_{Cu} J_{rms}}{f I_{pv} + i_{p,rms} + i_{s,rms}} \qquad (34)$$

The total core reluctance seen by the windings 3 & 4 can be evaluated to $R/2$. Hence the primary magnetizing inductance

$$L_p = \frac{2 N_p^2}{R} \qquad (35)$$

From (31)-(35), we can deduce the area product:

$$A_p(\text{Area Product}) = A_c A_w = \frac{L_p I \hat{I}}{2 \hat{B} k_{Cu} J_{rms}} \qquad (36)$$

where

$$\hat{I} = \begin{cases} f I_{pv} + n f I_g + max(i_p + n i_s) & \text{if } D \in (0.3, 0.5) \\ 2 f I_{pv}(2 - \frac{0.5}{D}) + max(i_p + n i_s) & \text{if } D \in (0.5, 0.75). \end{cases} \qquad (37)$$

and

$$I = \begin{cases} n f I_g + i_{p,rms} + i_{s,rms} & \text{if } D \in (0.3, 0.5) \\ f I_{pv} + i_{p,rms} + i_{s,rms} & \text{if } D \in (0.5, 0.75). \end{cases} \qquad (38)$$

The waveforms for $i_{p,rms}$ and $i_{s,rms}$ are shown in Fig.5. Then we can deduce:

$$i_{p,rms} = \sqrt{T_1 + T_2 + T_3 + T_4} \quad \text{where} \qquad (39)$$

$$T_1 = (1 - D) I_{pv}^2 \qquad (40)$$

$$T_2 = D(0.5 \Delta i_M - \frac{(1 - D) I_{pv}}{D})^2 \qquad (41)$$

$$T_3 = \Delta i_M (0.5 \Delta i_M - \frac{(1 - D) I_{pv}}{D}) \frac{2D - D^2}{1 - D} \qquad (42)$$

$$T_4 = \Delta i_M^2 \frac{3D - D^2 + D^3}{3(1 - D)^2} \qquad (43)$$

$$i_{s,rms} = \sqrt{T_5 + T_6 + T_7 + T_8} \quad \text{where} \qquad (44)$$

$$T_5 = \frac{(1 - D)^2 I_{pv}^2}{D n^2} \qquad (45)$$

$$T_6 = (1 - D)(\frac{I_{pv}}{n} + \frac{\Delta i_M}{2n})^2 \qquad (46)$$

$$T_7 = \Delta i_M(\frac{I_{pv}}{n} + \frac{\Delta i_M}{2n}) \frac{1 - D^2}{nD} \qquad (47)$$

$$T_8 = \Delta i_M^2 \frac{1 - D^3}{3 n^2 D^2} \qquad (48)$$

It can be seen that A_p is an extremely complex function of D, provided the parameters of the PV panel are provided. An iterative method, composed of completely deterministic steps is discussed in the next section.

IV. CONVERTER SPECIFICATIONS AND RESULTS

The specifications for the SW 270 mono panel were :

$f_r = 0.6$	$Z_{MPP} = 3.5\Omega$	$f_S = 100kHz$
$I_{pv} = 9.44A$	$J_{rms} = 4A/mm^2$	$B_{sat} = 0.47T$
$k_{Cu} = 0.4$	$\mu_r = 1790$	$f = 2.25$

The material used is the 3C94 power ferrite available from Ferroxcube. It is designed for use up to 300 kHz. The choice of factor f was due to guidelines given in [4]. With these constants, the area-product is plotted as a function of duty-ratio (D') in Fig.6.

In Fig.5, $A_{p1} = f_1(f_{\phi_1}(D), A_{w1})$ and $A_{p2} = f_2(f_\phi(D), A_{w2})$. The worst case design for a particular duty ratio is highlighted in black for $D \in (0.3, 0.75)$. Clearly, the worst case for the entire span is $D = 0.3$, which confirms our previous hypotheses in sections (III)B & (III)C.

The entire iterative design algorithm is outlined in the flowchart of Fig.6. Once an appropriate core is picked for the worst case design and N_p calculated, the air-gap is computed using an accurate expression of inductance. This is necessary because the inductance varies very rapidly near zero air-gap

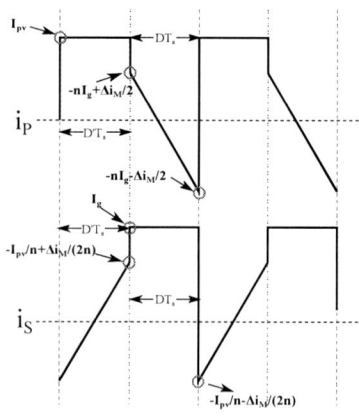

Fig. 5. Transformer Primary and Secondary Current Waveforms

and the approximate equation (35) is no longer adequate. However, we cannot use the actual equation in picking the core because the core was not yet known to us. The validity of the design is therefore verified by the condition $g < \frac{10 l_m}{\mu_r}$. Additionally, the ratio N_{pv}/N_p needs to be fairly accurate, i.e., 2.25, hence the minimum integer value for N_p needs to be 4. If N_p needs to be increased from the intial value N_{p0}, there is the additional problem that the windings will no longer fit according to (26)-(27). Hence a custom version of the original core with the same area product, but with a more skewed aspect ratio is needed as shown in the flowchart.

For windings 1 & 2, since they have very low ripple, skin effect is considered negligible and so round conductors are used. The winding sizes were calculated using the following equations:

$$d_{pv}(\text{Winding 1 diameter}) = \sqrt{\frac{4 I_{pv}}{\pi . J_{rms}}} \quad (49)$$

$$d_g(\text{Winding 2 diameter}) = \sqrt{\frac{4 I_g}{\pi . J_{rms}}} \quad (50)$$

Foil conductors are used for windings 3 & 4. The skin depth for Copper at 100 kHz is $\delta = 0.2mm$ and the layer porosity factor is chosen to be $\eta = 0.9$. Then the maximum number of turns per layer are:

$$nl_{p,max} = \left\lfloor \eta \sqrt{\frac{4}{\pi}} \frac{l_w}{d_p} \right\rfloor \quad (d_p = \sqrt{\frac{4 i_{p,rms}}{\pi J_{rms}}}) \quad (51)$$

$$nl_{s,max} = \left\lfloor \eta \sqrt{\frac{4}{\pi}} \frac{l_w}{d_s} \right\rfloor \quad (d_p = \sqrt{\frac{4 i_{s,rms}}{\pi J_{rms}}}) \quad (52)$$

Here d_p and d_s are the diameters if primary and secondary windings were built with round conductors. l_w is the window height of the selected core, with zero air-gap. Turns per layer are given by:

$$nl_p = \frac{\eta l_w}{\left\lfloor \frac{\pi d_p^2}{4\delta} \right\rfloor} \quad ; \quad nl_s = \frac{\eta l_w}{\left\lfloor \frac{\pi d_s^2}{4\delta} \right\rfloor} \quad (53)$$

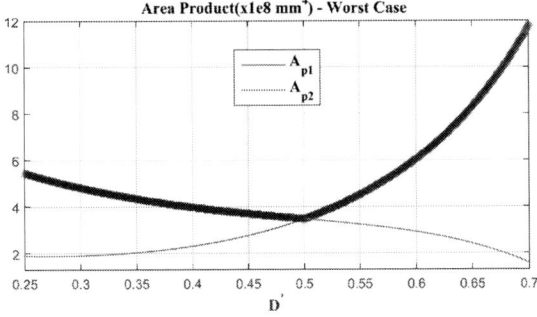

Fig. 6.   Worst Case Design

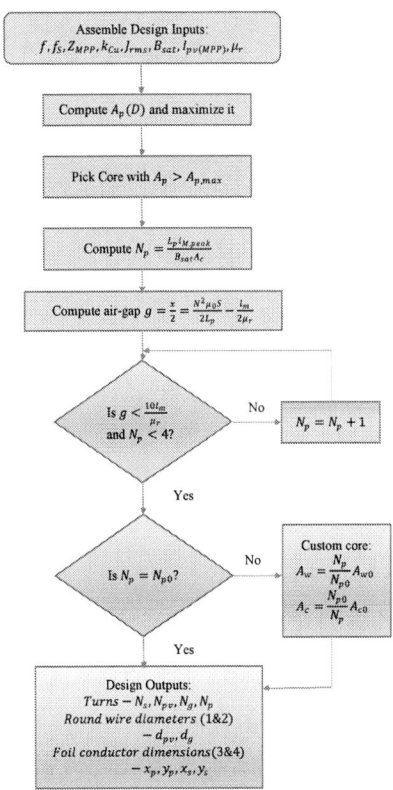

Fig. 7.   Design Flowchart

Foil widths:

$$y_p = \frac{\eta l_w}{\lfloor nl_p \rfloor} \quad ; \quad y_s = \frac{\eta l_w}{\lfloor nl_s \rfloor} \quad (54)$$

Foil thicknesses:

$$x_p = \frac{\pi d_p^2}{4 y_p} \quad ; \quad x_s = \frac{\pi d_s^2}{4 y_s} \quad (55)$$

For the given solar panel, the design outputs were obtained with a first pass of the algorithm. The design outputs are:

$N_p = 8$	$N_s = 40$	$N_{pv} = 18$
$N_g = 90$	$x_p = 0.195mm$	$x_s = 0.189mm$
$y_p = 11.1mm$	$y_s = 3.7mm$	$d_g = 1.2mm$
$d_{pv} = 2.6mm$		

The core chosen was the standard Ferroxcube E55/28/21.

Finite Element Modelling : The results of magnetostatic 2-D FEA (Finite-Element Analysis) on this core using ANSYS Maxwell are shown in Fig. 8. It is seen that the peak flux densities occur at the junctions of the center-limb, as mentioned

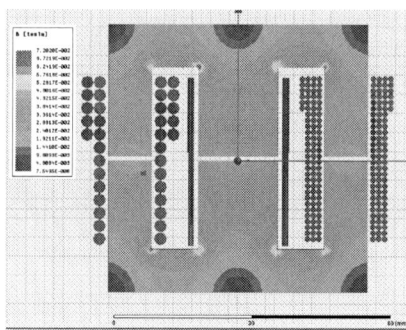

Fig. 8.   2D core structure showing flux densities

in [12]. The results of parametric sweeps across the air-gap are shown in figures 9 & 10.

The 2D FEA is approximate in the sense that the current return paths are infinite and the inductance values are extracted per metre in the z-direction and then multiplied with z-dimension of the core. However, the computational time is an order of magnitude less than its 3D counterpart. According to this analysis, the optimum air-gap is approximately 1.8mm, where the equations (2) & (3) hold true.

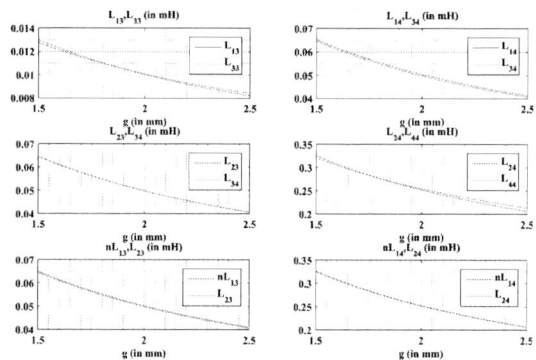

Fig. 9.   plots of the relevant quantities in (2) & (3) with 2D FEA

The 3D FEA is performed using symmetry cuts to reduce computation time. The optimum air-gap in this case is 0.9mm, and also $N_p = 7$ and $N_s = 37$ in order to get the curves to match. The coupling coefficients extracted are the following:

$k_{14} = 0.50957$	$k_{12} = 0.34138$	$k_{34} = 0.99705$
$k_{23} = 0.51044$	$k_{13} = 0.50738$	$k_{24} = 0.51247$

The self-inductances at this air-gap are as follows: $L_{11} = 48.82\mu H$, $L_{22} = 1.23 mH$, $L_{33} = 12.68\mu H$, $L_{44} = 319\mu H$.

Circuit Simulations: The circuit shown in Figure 11 is simulated in Cadence Pspice with the coupling coefficient extracted above. The FET (Field-Effect-Transistor) used is IRF540, while the diode is HFA50PA60C.

The simulation results in PSpice using the coupling coefficients extracted by the FEA in Fig.612 show that the currents

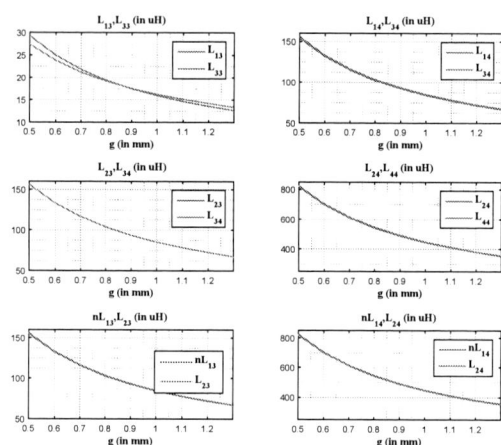

Fig. 10.   plots of the relevant quantities in (2) & (3) with 3D FEA

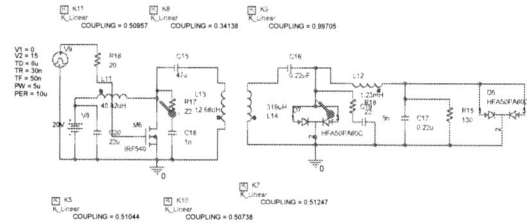

Fig. 11.   Pspice Circuit

i_{pv} and i_g are low-ripple. The peak-to-peak ripple percentages are 2 and 3 respectively.

V.   CONCLUSION AND FUTURE WORK

In this paper, a general method to design a zero-ripple Ćuk converter for a PV/Battery to grid application has been presented. Criteria for core saturation and converter non-idealities have been accounted into formulating the area-product, as well as designing for an entire range of duty-ratio as opposed to the earlier designs. The results of FEA and Pspice simulations have proved the method to be accurate and hardware ready. Hardware design should be fairly straightforward from this point. Future work entails generalizing the design for bidirectional power flow, as well as for battery-tied grid system. An investigation of the efficiency of the converter, and a design method based on optimizing the copper/core loss of the converter including proximity effects can also be researched. This will utilize the K_g approach [8], [13].

REFERENCES

[1] S. Kjaer, J. Pedersen, and F. Blaabjerg, "A review of single-phase grid-connected inverters for photovoltaic modules," *Industry Applications, IEEE Transactions on*, vol. 41, no. 5, pp. 1292–1306, Sept 2005.

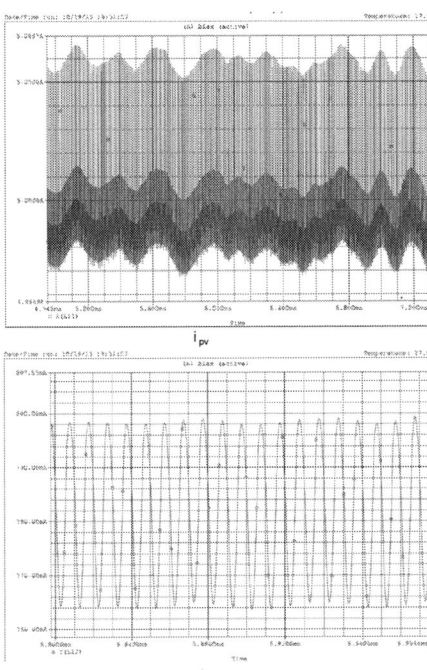

i_{pv}

i_g

Fig. 12.   Simulated Terminal Currents

[2]   S. Jain and V. Agarwal, "A single-stage grid connected inverter topology for solar pv systems with maximum power point tracking," *Power Electronics, IEEE Transactions on*, vol. 22, no. 5, pp. 1928–1940, Sept 2007.

[3]   F. Nejabatkhah, S. Danyali, S. Hosseini, M. Sabahi, and S. Niapour, "Modeling and control of a new three-input dc-dc boost converter for hybrid pv/fc/battery power system," *IEEE Transactions on Power Electronics*, vol. 27, no. 5, pp. 2309–2324, May 2012.

[4]   R. D. Middlebrook and S. Ćuk, *Advances in Switched-Mode Power Conversion*.   Pasadena, CA: TeslaCo, 1983, vol. I and II.

[5]   P. Jose, "Novel zvs bidirectional Ćuk converter with h_{∞} average-current control for dual voltage systems in automobiles," Ph.D. dissertation, University of Minnesota, Minneapolis, MN, 2004.

[6]   Z. Zhang, "Coupled inductor magnetics in power electronics," Ph.D. dissertation, California Institute of Technology, Pasadena, CA, 1987.

[7]   S. Biswas, S. Dhople, and N. Mohan, "A three-port bidirectional dc-dc converter with zero-ripple terminal currents for pv/microgrid applications," in *Industrial Electronics Society, IECON 2013 - 39th Annual Conference of the IEEE*, Nov 2013, pp. 340–345.

[8]   C. W.T.McLyman, *Transformer and Inductor Design Handbook*.   Marcel Dekker Inc., 1988.

[9]   N.Mohan, W. Robbins, and T.Undeland, *Power Electronics : Converters, Applications and Design*.   Wiley Academic Publishers, 2002.

[10]   *Sunmodule Plus SW 270 mono Solar Panel Datasheet*, Solarworld GmBH, 2014.

[11]   S. Cuk, "A new zero-ripple switching dc-to-dc converter and integrated magnetics," in *Power Electronics Specialists Conference, 1980. PESC. IEEE*, June 1980, pp. 12–32.

[12]   S. Biswas, S. Dhople, and N. Mohan, "Zero-ripple analysis methods for three-port bidirectional integrated magnetic Ćuk converters," in *Industrial Electronics Society, IECON 2014 - 40th Annual Conference of the IEEE*, Oct 2014, pp. 1889–1895.

[13]   R.W.Erickson and D.Maksimovic, *Fundamentals of Power Electronics*. Springer-Verlag GmBH, 2004.

978-1-4673-9551-9/16 $31.00 © 2016 IEEE

Grid-Voltage-Feedforward Active Damping for Grid-Connected Inverter with *LCL* Filter

Minghui Lu, Xiongfei Wang, Frede Blaabjerg
Department of Energy Technology
Aalborg University, Aalborg, 9220, Denmark
E-mail: {mil, xwa, fbl}@et.aau.dk

S.M. Muyeen, Ahmed Al-Durra, Siyu Leng
Department of Electrical Engineering
The Petroleum Institute, P.O. Box 2533, Abu Dhabi, U.A.E.
E-mail: {smmuyeen, aaldurra, sleng}@pi.ac.ae

Abstract—For the grid-connected voltage source inverters, the feedforward scheme of grid voltage is commonly adopted to mitigate the current distortion caused by grid background voltages harmonics. This paper investigates the grid-voltage-feedforward active damping for grid connected inverter with *LCL* filter. It reveals that proportional feedforward control can not only fulfill the mitigation of grid disturbance, but also offer damping effects on the *LCL* filter resonance. Digital delays are intrinsic to digital controlled inverters; with these delays, the feedforward control can be equivalent to a damping resistor and a reactance paralleled with filter capacitor. The damping performance in different frequency regions are discussed through Bode diagrams. Compared to other widely used active damping strategies, no extra sensor is needed because the Point of Common Coupling (PCC) voltage is sampled for Phase Locked Loop (PLL). Simulation and experiment results are provided for verifying the theoretical analyses.

Keywords— *Digital time delay, feedforward, grid-connected inverter, LCL filter, resonance damping*

I. INTRODUCTION

The Pulse Width Modulation (PWM) inverters are widely applied as interface for connecting renewable energy sources with the utility grid [1], [2]. In order to smooth the current injected into the grid, output power filters are usually placed between the inverter and the grid to obtain high quality grid current. Compared with the inductive *L* filter, the *LCL* filter has demonstrated greater high-frequency harmonic attenuation ability even with a smaller total reactive element used [3], [4]. However, *LCL* filter may bring resonance problems, which may lead to instability. Active Damping (AD) schemes are commonly adopted to damp the resonance peak of *LCL* filter [3]-[5]. Compared with the passive damping, AD schemes have the advantages of lower losses and better flexibility. The filter-based AD [3] and the AD methods based on extra feedback of filter components variables, such as voltage and current, are proposed in recent literature [4], [5]. However, more complex algorithms are required and more sensors need to be installed in the system.

Recently, the stable operation of grid-connected inverter without any additional damping under single current loop control strategy has been proved to be possible [6]-[8]. The simple and effective single loop control strategies with typical

linear Proportional Integral (PI) or Proportional Resonant (PR) controllers are promising options for industrial applications. In fact, the essential theoretical foundations of these work are the critical role of digital time delays in stabilizing the grid connected inverters with *LCL* filter. The impact of digital delay on *LCL* filtered grid connected inverter stability and active damping have been discussed in recent literatures [9]-[11]. Parker *et al.* [6] demonstrates that one sixth of the sampling frequency is the critical *LCL* filter resonance frequency and concludes that if the resonance frequency is equal to $f_s/6$, the system will be always unstable even with capacitor-current-feedback active damping. Zou *et al.* [9] comprehensively studies relationship between the time delay and stability of single-loop controlled grid-connected inverters that employ inverter current feedback or grid current feedback. For the sake of less sensors, [10] regards grid-current-feedback as an alternative AD scheme.

Feedforward control is a simple and effective method to mitigate the grid disturbance and improve the system dynamic performance [12]-[16]. Full feedforward techniques are proposed in [12], however, digital delays are hastily ignored and high order derivative elements will make the system sensitive to noise. [13] adopts both capacitor-current-feedback active damping and PCC voltage feedforward control to improve the system adaptability to weak grid condition. [14] finds that feedforward control with digital delay taken into account improves system stability for *LCL* filtered inverter when the inverter side current is controlled. On the contrary, [15] claims that the feedforward voltage will result in a grid current positive feedback, which will essentially degrade the system stability and control performance. Therefore, it is necessary to have a more comprehensive discussion on the grid voltage feedforward control.

In view of the above issues, detailed modeling and damping effects of the PCC voltage feedforward control has been explored in this paper, beginning with an overall description of configuration and control of the grid-connected inverter in § II, the delay mechanism is explicitly explained. § III then presents the feedforward control by the equivalent impedance paralleled with filter capacitor; § IV continues with damping performance through Bode diagram; § V analyzes system stability against

---

This work was supported in part by European Research Council (ERC) under the European Union's Seventh Framework Program (FP/2007-2013)/ERC Grant Agreement [321149-Harmony].

978-1-4673-9551-9/16 $31.00 © 2016 IEEE

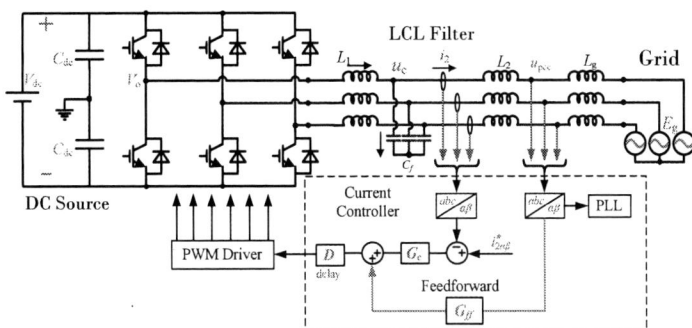

Fig. 1. Schematic configuration of *LCL*-filtered grid inverter.

the grid impedance variation; § VI then finalizes the paper by showing simulation and experimental results.

II. MODELING OF GRID-CONNECTED INVERTER

Fig. 1 shows the topology structure of a three-phase voltage source inverter with *LCL* filter. L_1 is the inverter-side inductor, C_f is the filter capacitor, and L_2 is the grid-side inductor. Zero resistance is considered in the filter for emulating a worst-case scenario. The fundamental target for this system is to regulate the grid current i_2 to manage the power injected into the grid. Typically, the grid impedance at the PCC consists of an inductor series with resistor. In this paper, the grid resistance is also neglected for the non-damping situation; so the grid impedance is treated as a pure inductance.

Modern grid-connected inverter controllers usually are implemented by digital microprocessors, which inevitably require algorithm computation time after sampling the measured variables. Time delays in digital controller include one-sample delay and PWM update delay [17]. The measured variable is sampled at the beginning of control period, but the updated reference value only is employed and compared against a triangular carrier during next control period. Intrinsically, this process introduces one period delay z^{-1}; On the other hand, sampling and hold (ZOH) process will bring 0.5 control period delay due to volt-second balancing. Thus, an overall 1.5 control period T_s delays should be taken into account in the control loop. The digital delays are represented by a block D in Fig. 1, which can be written as

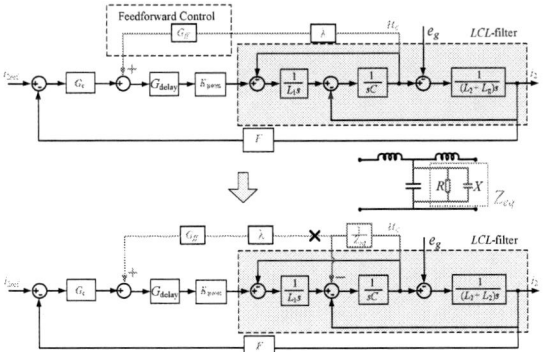

Fig. 2. Equivalent virtual impedance of the GVFF scheme.

$$D = e^{-1.5sT_s} \tag{1}$$

Fig. 2 presents the diagram of the grid current control. The inverter is replaced by gain K_{pwm}, the current controller G_c is a typical linear PI or PR controller, F is the grid current feedback coefficient. The open loop gain can be expressed as

$$T_{LCL}(s) = \frac{F \cdot G_c \cdot K_{pwm} \cdot e^{-1.5T_ss}}{(L_1+L_2+L_g)s} \cdot \frac{\omega_{res}^2}{s^2+\omega_{res}^2} = T_L(s) \cdot \frac{\omega_{res}^2}{s^2+\omega_{res}^2} \tag{2}$$

where $T_L(s)$ is the open loop gain of the L filtered inverter. It can be seen that a resonant part is undesirably introduced, which may make the system oscillation, even unstable. Hence, damping strategies are required in the system.

III. DAMPING CHARACTERISTIC OF GRID VOLTAGE FEEDFORWARD CONTROL

Assume the studied inverter is three-phase symmetrical, thus it can be analyzed as single phase configuration. Due to the existence of grid impedance, the PCC voltage u_{pcc} contains not only the information of ideal grid voltage u_g but also that of capacitor voltage u_c, as shown in (3). It indicates that filter components variable (u_c), which may contribute active damping effect to the filter resonance, is feedforwarded. Obviously, this behavior will lead to non-negligible impacts on system stability, inherent damping or stability deterioration.

$$u_{pcc} = \lambda \cdot u_c + (1 - \lambda) \cdot e_g \tag{3}$$

For the PCC voltage feedforward, $\lambda = L_g / (L_2 + L_g)$; for the capacitor voltage feedforward, $\lambda = 1$. As shown in Fig. 2, the feedforward term of u_c is moved from the output of $G_c(s)$ to the output of $1/sL_1$, the virtual impedance Z_{eq} connected in parallel with the filter capacitor is obtained as

$$Z_{eq}(s) = -\frac{L_1}{\lambda G_{ff}K_{pwm}} \cdot s \cdot e^{1.5sT_s} \tag{4}$$

Substitute $s = j\omega$ to the expression above, the virtual impedance $Z_{eq}(\omega)$ in ω-domain is presented as

$$Z_{eq}(s = j\omega) = \frac{L_1}{\lambda G_{ff}K_{pwm}} \cdot \left(\frac{\omega}{\sin(1.5T_s\omega)} \parallel \frac{\omega}{j\cos(1.5T_s\omega)}\right)$$
$$= R_d \parallel X_d \tag{5}$$

978-1-4673-9551-9/16 $31.00 © 2016 IEEE

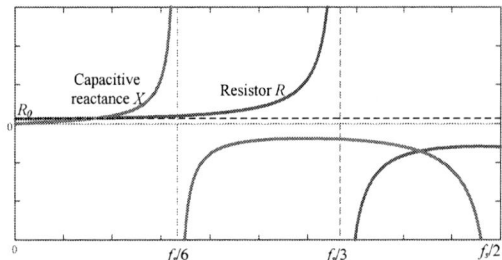

Fig. 3. Curves of resistor R_d and reactance X_d.

where the sign $\parallel$ means impedance parallel operation.

The virtual impedance Z_{eq} can be represented by a virtual resistor R_d paralleled with capacitive reactance X_d. It can be seen that Z_{eq} is closely associated with the value λ. For the PCC voltage feedforward, if the grid is stiff enough, $L_g = 0$, $\lambda = 0$, then Z_{eq} approaches to infinity, and the effect on system stability can be neglected; however, if the coefficient $\lambda \neq 0$, then Z_{eq} is equal to a specific value, the stability characteristics change. For the filter capacitor voltage feedforward, $\lambda = 1$, hence Z_{eq} will always influence the stability, regardless of the stiffness of the power grid. The resistor part R_d may damp the resonance of the LCL filter in certain frequency region, and the X_d part shifts the resonance frequency. The expressions of R_d and X_d are given as

$$R_d = R_0 \cdot \frac{\omega}{\sin(1.5T_s\omega)}, X_d = R_0 \cdot \frac{1.5\omega T_s}{\cos(1.5\omega T_s)} \quad (6)$$

where R_0 is the limit value when the frequency ω is approaching to zero, the expression is written as

$$\lim_{\omega \to 0} R_d = R_0 \cdot \lim_{\omega \to 0} \frac{1.5\omega T_s}{\sin(1.5\omega T_s)} = R_0 = \frac{L_1}{\lambda G_{ff} K_{pwm} 1.5 T_s} \quad (7)$$

Equation (7) gives the relationship between the damping resistor limit value and system parameters. Fig. 3 shows the curve of virtual resistor R_d and reactance X_d, where both the value of R_d and X_d are frequency dependent.

IV. DAMPING PERFORMANCE

According to the analysis above, the feedforward control offers damping effect on the resonance peak of LCL filter in specific region. This section discusses the damping performance in detail. From the equation (7), the damping resistor R_d and reactance X_d closely related to the index λ and G_{ff}. If the index $\lambda = 0$, it means that the grid inductance $L_g = 0$ and damping resistor $R_d = \infty$ and reactance $X_d = \infty$, there will be no damping effect on the resonance. For a specific $\lambda \neq 0$, the value of feedforward coefficient G_{ff} influences the damping effect. The open loop gain $T'_{LCL}(s)$ can be expressed as

$$T'_{LCL}(s) = \frac{F \cdot G_c \cdot K_{pwm} \cdot e^{-1.5T_s s}}{(L_1 + L_2 + L_g)s} \cdot \frac{\omega_{res}^2}{s^2 + \omega_{res}^2 \cdot (1 - \beta \cdot \lambda \cdot G_{ff} \cdot K_{pwm} \cdot e^{-1.5T_s s})},$$

$$\beta = \frac{L_2 + L_g}{L_1 + L_2 + L_g} \quad (8)$$

where $\omega_{res} = \sqrt{\frac{L_1 + L_2 + L_g}{L_1(L_2 + L_g)C_f}}$.

The Bode diagram of $T'_{LCL}(s)$ with $F = 1$, $G_c = 1$ are shown in Fig. 4. In the region $(0, f_s/6)$, X_d is capacitive reactance and yields a lower resonance frequency, the damping resistor R_d damps the resonance, which is beneficial to system stability, as shown in Fig. 4 (a). In the region $(f_s/6, f_s/3)$, X_d is inductive reactance, yielding a higher resonance frequency, the damping resistor R_d damps the resonance, as shown in Fig. 4 (b). Note that the resonance frequency may step over $f_s/3$ if a specific G_{ff} is selected. For the cases above, the R_d is positive and damps the resonance peak. In the region $(f_s/3, f_s/2)$, X_d is inductive reactance and yields a higher resonance frequency, however, the R_d is a negative value, as shown in Fig. 4 (c), which may worsen the system stability.

V. STABILITY ANALYSIS

The effect of GVFF contains positive and negative varied resistor R_d and reactance X_d that bring a challenge in the stability analysis. This section is dedicated to address this issue.

Typically, modeling systems in the continuous time domain results in clearer and lower order transfer functions, so the analysis will be much simpler. However, when the digital delays are taken into account, the accuracy of these models in s-domain is only guaranteed in low frequency range since first or higher order *Padé* approximation methods are usually employed to replace the digital delays. Therefore, the analysis in this paper will be carried out in the z-domain. The loop gain $T(z)$ is used to analyze and evaluate the system stability.

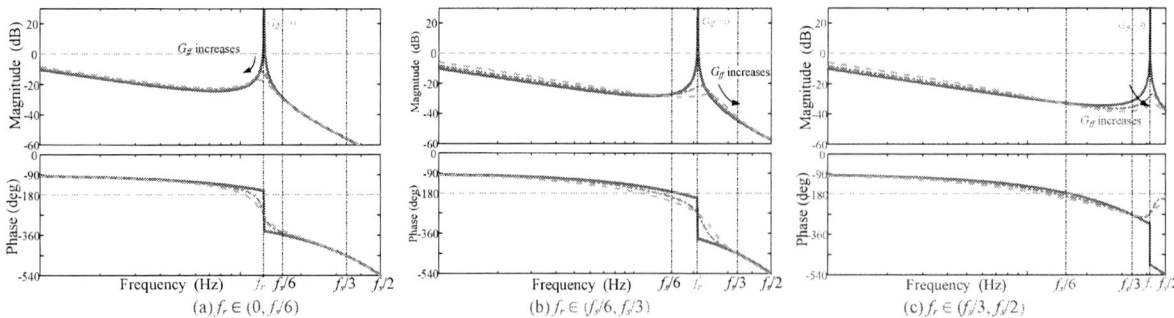

Fig. 4. Bode diagrams of the loop gain $T'_{LCL}(s)$ in different frequency regions.

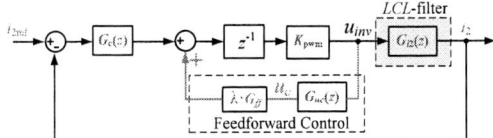

Fig. 5. Control block diagram in z-domain.

Fig. 5 presents the control structure of grid-connected inverter with GVFF in z-domain, $G_c(z)$ is the discrete representation of controller, G_{ff} is the grid voltage feedforward coefficient, $G_{i2}(z)$ and $G_{uc}(z)$ are the discrete expressions of the $G_{i2}(s)$ and $G_{uc}(s)$, respectively.

$$G_{i2}(z) = \frac{\omega_r T_s[z^2 - 2z \cdot \cos(\omega_r T_s) + 1] - \sin(\omega_r T_s)(z-1)^2}{\omega_r(L_1 + L_2 + L_g)(z-1)[z^2 - 2z \cdot \cos(\omega_r T_s) + 1]} \quad (9)$$

$$G_{uc}(z) = \frac{1 - \cos(\omega_r T_s)}{L_1 C_f \omega_r^2} \cdot \frac{z+1}{[z^2 - 2z \cdot \cos(\omega_r T_s) + 1]} \quad (10)$$

The loop gain $T(z)$ can be derived as

$$T(z) = \frac{G_c(z) \cdot K_{pwm}}{\omega_r(L_1 + L_2 + L_g)} \cdot \frac{\omega_r T_s[z^2 - 2z \cdot \cos(\omega_r T_s) + 1] - \sin(\omega_r T_s)(z-1)^2}{(z-1)[z(z^2 - 2z\cos(\omega_r T_s) + 1) - k_a(z+1)(1 - \cos(\omega_r T_s))]} \quad (11)$$

where $k_a = \frac{\lambda G_{ff} K_{pwm}}{L_1 C \omega_r^2}$.

Fig. 6 shows the root locus with different feedforward coefficient G_{ff}, it is obvious that the grid voltage feedforward control has great impact on the system stability. As shown in the diagram, when the G_{ff} is equal to zero, the system is unstable because the grid impedance makes the resonance frequency lower than critical frequency. When the feedforward control is added, the poles are located in the unit circle if proper control gain is selected. The stable region is much wider than the case without feedforward control. However, if G_{ff} is too large, the poles will go outside the unit circle again. The loop gain expression $T(z)$ in z-domain is shown in (8), the partial factor from the denominator of $T(z)$ is:

$$\text{Den}(z) = z(z^2 - 2z\cos\omega_r T_s + 1) - \frac{\lambda G_{ff} K_{pwm}}{L_1 C \omega_r^2}(z+1)(1 - \cos\omega_r T_s) \quad (12)$$

Define the coefficient k_a as:

$$k_a = \frac{\lambda G_{ff} K_{pwm}}{L_1 C \omega_r^2} \quad (13)$$

The polynomial Den(z) can be rewritten as:

$$\text{Den}(z) = a_0 z^3 + a_1 z^2 + a_2 z^1 + a_3 z^0 \quad (14)$$

where the coefficients are written as: $a_0 = 1$, $a_1 = -2\cos\omega_r T_s$, $a_2 = 1 - k_a + k_a \cdot \cos\omega_r T_s$, $a_3 = k_a \cdot \cos\omega_r T_s - k_a$.

The Jury's criterion is adopted to analyze the distribution of the poles [18], [19]. The stable boundary for feedforward coefficient is given as

$$0 < G_{ff} < \frac{L_1 + L_2 + L_g}{L_g \cdot K_{pwm}} \cdot \frac{1 + 2\cos(\omega_{res} T_s)}{1 - \cos(\omega_{res} T_s)} \quad (15)$$

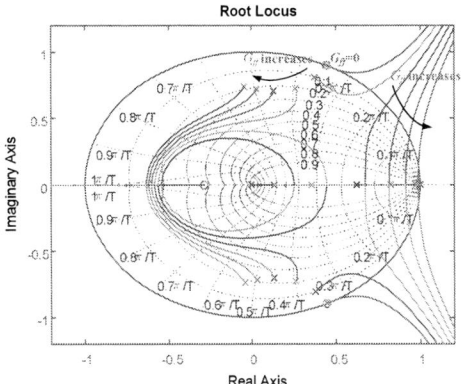

Fig. 6. Root locus with different feedforward coefficient.

VI. SIMULATION RESULTS

In this section, simulation results are presented to validate the theoretical study of the previous sections. The simulations are performed with Matlab/Simulink, the inverter and control

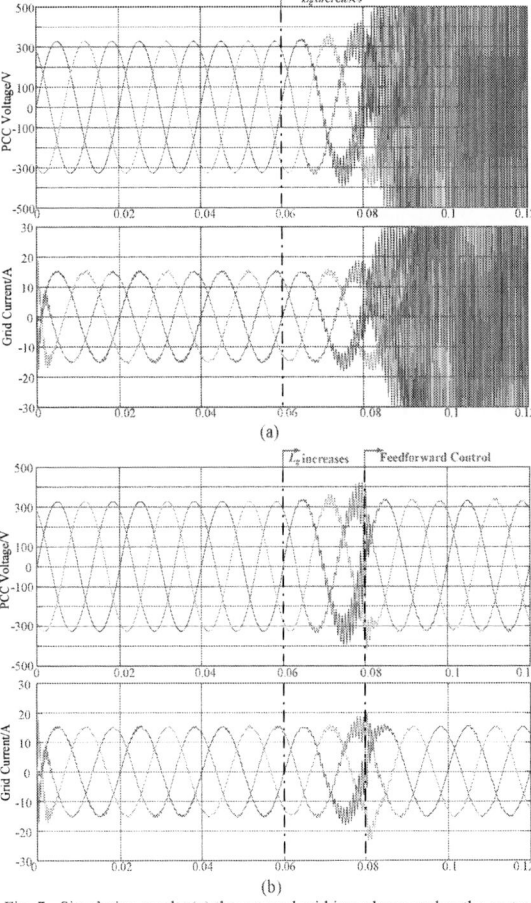

Fig. 7. Simulation results (a) the external grid impedance makes the system unstable; (b) the GVFF scheme return system to stable.

TABLE I
SYSTEM PARAMETERS

CIRCUIT PARAMETER		Value
Converter Side Inductor	L_1	3.2mH
Grid Side Inductor	L_2	1mH
Filter Capacitor	C_f	6.6uF
Filter Resonance	f_r	2.24kHz
Grid Inductor	L_g	1mH
Switching Frequency and Sampling Frequency	$f_{sw} = f_s$	10kHz

strategy as in Fig. 1 are implemented. System parameters are listed in the Table I, it can be seen that the *LCL* filter resonance frequency is higher than critical frequency, i.e. $f_r > f_s/6$, the inverter can work stably without any additional damping strategy. The external inductive grid impedance makes the resonance frequency lower than $f_s/6$, then the system will be unstable.

As shown in Fig. 7 (a), when t $\in$(0,0.06s), the grid impedance L_g is zero; at t =0.06s, L_g increases, it is obvious that the system will unstable if L_g exceeds a specific value. To verify the damping effect and stability improvement of feedforward control; at t=0.08s, the feedforward control is employed, the system will be stable again, as shown in Fig. 7 (b). From the simulation results, high robustness against grid impedance variation is obtained. It demonstrates that feedforward exhibits damping effect on the filter resonance.

VII. EXPERIMENTAL RESULTS

A 3-kW inverter with *LCL* filter prototype is built in the laboratory. The controller is implemented in dSPACE system – MicroLabBox. Since the real power grid has uncertain grid impedance, a 15 kW programmable grid emulator is used in the experiment to offer selectable grid voltage, and the grid impedance is replaced by an external inductor which makes the system more realistic.

Fig. 8 shows the experimental waveforms with the Table I parameter, of which filter resonance frequency belongs to region 1. It should be noted that the critical frequency is equal to one-sixth of the sampling frequency for the delay of 1.5 times the sampling period considered in the paper [7]. The

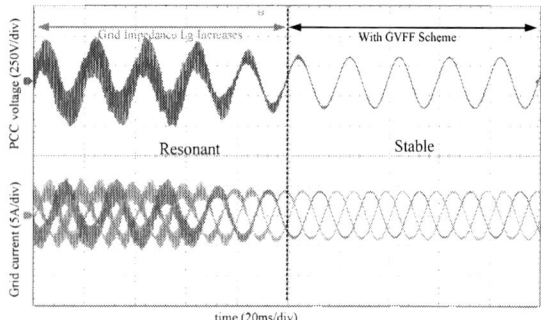

Fig. 8. Root locus with different feedforward coefficient.

selected resonance frequency is placed above the critical frequency, and the external grid impedance is then added to make the resonance frequency lower than critical frequency, leading to an unstable case for single grid current control loop. As shown in Fig. 8, the system becomes undesirably oscillatory when the grid impedance increases, then the system returns to be stable if the single proportional GVFF is enabled, the feedforward coefficient $G_{ff} = 1/K_{pwm}$. This is in agreement with the root locus shown in Fig. 6, where the resonant poles track inside the unit circle and, thus, describe a system that is damped and stable. It demonstrates the active damping effects of the feedforward scheme of grid voltage on the *LCL* filter resonance.

VIII. CONCLUSION

This paper investigates the grid-voltage-feedforward active damping for grid connected inverter with *LCL* filter. It reveals that proportional feedforward control can not only fulfill the mitigation of grid disturbance, but also offer damping effects on the *LCL* filter resonance. The damping performance in different frequency regions are discussed through bode diagram. Moreover, the numerical zone of the feedforward coefficient is given. Compared to other widely used active damping strategies, no extra sensor is needed. Simulation and experiment results are provided to validate the theoretical analysis.

REFERENCES

[1] S. M. Muyeen, R. Takahashi, T. Murata, and J. Tamura, "A variable speed wind turbine control strategy to meet wind farm grid code requirements," IEEE Trans. Power Syst., vol. 25, no. 1, pp. 331–340, Feb. 2010.

[2] Y. W. Li, F. Blaabjerg, D. M. Vilathgamuwa, and P. C. Loh, "Design and comparison of high performance stationary-frame controllers for DVR implementation," IEEE Trans. Power Electron., vol. 22, no. 2, pp. 602– 612, Mar. 2007.

[3] J. Dannehl, M. Liserre, and F. W. Fuchs, "Filter-based active damping of voltage source converters with LCL filter," IEEE Trans. Ind. Electron., vol. 58, no. 8, pp. 3623–3633, Aug. 2011.

[4] E. Twining and D. G. Holmes, "Grid current regulation of a three-phase voltage source inverter with an *LCL* input filter," IEEE Trans. Power Electron., vol. 18, no. 3, pp. 888–895, May 2003.

[5] D. Pan, X. Ruan, C. Bao, W. Li, X. Wang, "Capacitor-Current-Feedback active damping with reduced computation delay for improving robustness of LCL-type grid-connected inverter," IEEE Trans. Power Electron., vol. 29, no. 7, pp. 3414-3426, July. 2014.

[6] M. Lu, X. Wang, P. C. Loh and F. Blaabjerg, "An analysis method for harmonic resonance and stability of multi-paralleled LCL-filtered inverters," in Proc. 5th Int. IEEE Power Electron. Distrib. Gen. Syst., Aachen, Germany, Jun. 2015, pp. 1–6.

[7] S.G. Parker, B. P. McGrath, D.G. Holmes, "Regions of Active Damping Control for LCL Filters," IEEE Trans. Ind. Appl., vol. 50, no. 1, pp. 424-432, Jan./Feb. 2014.

[8] J. Yin, S. Duan, B. Liu, "Stability analysis of grid-connected inverter with LCL filter adopting a digital single-loop controller with inherent damping characteristic," IEEE Trans. Ind. Inf., vol. 9, no. 2, pp. 1104-1112, May. 2013.

[9] C. Zou, B. Liu, S. Duan, R. Li, "Influence of delay on system stability and delay optimization of grid-connected inverters with LCL filter," IEEE Trans. Ind. Inf., vol. 10, no. 3, pp. 1175-1784, Aug. 2014.

[10] X. Wang, F. Blaabjerg, P.C. Loh, "Grid-current-feedback active damping for LCL resonance in grid-connected voltage source converters," IEEE Trans. Power Electron., vol. PP, no. 99, pp. 1-1, Mar. 2015.

[11] M. Lu, X. Wang, P. C. Loh and F. Blaabjerg, "Interaction and aggregated modeling of multiple paralleled inverters with LCL filter," *in Proc. IEEE Energy Convers. Congr. Expo.*, 2015, pp. 1954-1959.

[12] W. Xuehua, R. Xinbo, L. Shangwei, and C. K. Tse, "Full Feedforward of Grid Voltage for Grid-Connected Inverter With LCL Filter to Suppress Current Distortion Due to Grid Voltage Harmonics," IEEE Trans. Power Electron., vol. 25, pp. 3119-3127, 2010.

[13] J. Xu, S. Xie, T. Tang, "Improved control strategy with grid voltage feedforward for lcl filter based inverter connected to weak grid," IET Power Electron., vol. 7, no. 10, pp. 2660-2671, 2014.

[14] C. Zou, B. Liu, S. Duan, and R. Li, "A feedfoward scheme to improve system stability in grid-connected inverter with LCL filter," in Proc. IEEE Energy Convers. Congr. Expo., 2013, pp. 4476-4480.

[15] J. Wang, Y. Song, A. Monti, "A study of feedforward control on stability of grid-parallel inverter with various grid impedance," in Proc. IEEE 5th International Symposium on Power Electronics for Distributed Generation Systems (PEDG), pp. 1-8, June 2014.

[16] X. Zhang, F. Wang, W. Cao, and Y. Ma, "Influence of voltage feed-forward control on smallsignal stability of grid-tied inverters," in *Proc. of APEC*, pp. 1216-1221, Mar. 2015.

[17] D. G. Holmes, T. A. Lipo, B. P. McGrath, and W. Y. Kong, "Optimized Design of Stationary Frame Three Phase AC Current Regulators, " *IEEE Trans. Ind. Inf.*, vol. 24, no. 11, pp. 2417-2426, Nov. 2009.

[18] E. I. Jury, Theory and Application of the Z-Transform Method. New York: Wiley, 1964.

[19] E. Jury, "A simplified stability criterion for linear discrete systems," *Proc. IRE*, vol. 50, no. 6, pp. 1493–1500, 1962.

A High Power Density Single-Phase Inverter Using Stacked Switched Capacitor Energy Buffer

Colin McHugh, Sreyam Sinha, Jeffrey Meyer, Saad Pervaiz, Jie Lu, Fan Zhang, Hua Chen, Hyeokjin Kim,
Usama Anwar, Ashish Kumar, Alihossein Sepahvand, Scott Jensen, Beomseok Choi, Daniel Seltzer, Robert Erickson,
Dragan Maksimovic, Khurram K. Afridi
University of Colorado Boulder, Boulder, CO, 80309, USA
Email: colin.mchugh@colorado.edu

Abstract— This paper presents a high power density 2 kW single-phase inverter, with greater than 50 W/in³ power density and 90% line-cycle average efficiency. This performance is achieved through innovations in twice-line-frequency (120 Hz) energy buffering and high frequency dc-ac power conversion. The energy buffering function is performed using an advanced implementation of the recently proposed stacked switched capacitor (SSC) energy buffer architecture, and the dc-ac power conversion is performed using a soft-switching SiC-FET based converter, with a digital implementation of variable frequency constant peak current control.

Keywords—dc-ac power conversion, energy buffer, switched capacitor, twice line frequency, constant peak current variable frequency, digital current control.

I. INTRODUCTION

Single-phase, high power factor inverters inherently require energy storage to buffer the difference between their instantaneous input and output power. For a variety of applications including single-phase photovoltaic (PV) inverters and uninterruptible power supplies (UPS), the input power drawn by the inverter is required to be constant, with small ripple, while the output power pulsates at twice the line frequency, warranting the need for an intermediate energy storage element. Typically, electrolytic capacitors are utilized as the energy storage element on the dc side to buffer the twice line frequency energy. In conventional inverter architectures, these capacitors are directly connected across the input dc bus. A limitation on the dc bus voltage ripple imposes constraints on the energy utilization of the buffering capacitor. In order to have acceptable voltage ripple on the dc bus, a large capacitor is needed, resulting in poor energy utilization and a substantial fraction of the total volume being utilized for energy buffering. In order to increase the energy utilization of the buffering capacitor and achieve high power densities, an alternative energy buffering solution is required. Several such solutions have been presented in the literature, including the use of bi-directional dc-dc converters, [1],[2], energy buffers incorporated in the power stage design, [3]-[5], and switched capacitor energy buffers [6]-[14]. Among these, the stacked switched capacitor (SSC) energy buffer architecture has been shown to achieve a substantial reduction in the capacitor volume by effective utilization of the stored energy, enabling high power densities [8]-[14].

Further enhancements in inverter power density require a compact and efficient dc-ac power conversion stage. Traditional PWM based dc-ac converters operate in continuous conduction

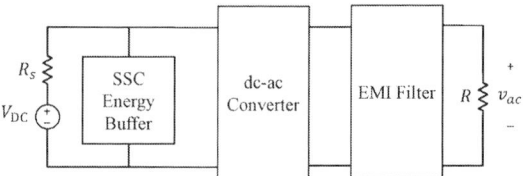

Figure 1: Architecture of the proposed single phase inverter.

mode (CCM) in high power applications, and in discontinuous/boundary conduction mode (DCM/BCM) for mid and low power designs [15]-[21]. Operating in DCM/BCM has several advantages, including smaller filtering inductors and soft-switching capability. However, conventional DCM/BCM control techniques suffer from reduced efficiencies at low power levels, resulting in low weighted efficiencies. As a solution, a variable-frequency constant peak current control strategy was recently proposed in [22], with high efficiencies demonstrated across a wide range of operating power levels.

This paper presents a single-phase, high efficiency, high power density 2 kW inverter. For twice-line-frequency energy buffering, this inverter utilizes an advanced variant of the stacked switched capacitor energy buffer architecture. For the dc-ac conversion stage, a high frequency, SiC-MOSFET based solution using variable frequency constant peak current control is utilized. A novel digital implementation of this control scheme is introduced as part of this work.

The remainder of this paper is organized as follows: section II of this paper presents the proposed inverter architecture. Section III details the design optimization of the SSC energy buffer and the dc-ac converter. Experimental results are presented in Section IV, while Section V concludes the paper.

II. PROPOSED ARCHITECTURE

The architecture of the proposed high power density, single-phase inverter is shown in Fig. 1. It comprises a stacked switched capacitor (SSC) energy buffer across the dc input bus, and a dc-ac converter utilizing a digital implementation of variable frequency constant peak current control. The following subsections discuss the operational principles of the SSC energy buffer and the dc-ac converter.

A. SSC Energy Buffer

A stacked switched capacitor (SSC) energy buffer comprises two series connected blocks of switches and

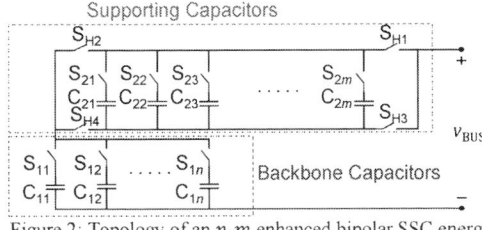

Figure 2: Topology of an n-m enhanced bipolar SSC energy buffer.

capacitors, referred to as the backbone and supporting blocks. The basic operational principle of an SSC energy buffer is that the voltage across the individual blocks is allowed to vary across a wide range, but the variation across one block compensates for the variation across the other block, such that the total bus voltage across the buffer, which is the sum of the individual block voltages, remains within the desired ripple specifications. By allowing a larger voltage swing within the individual blocks, the SSC energy buffer utilizes the stored energy effectively, resulting in significantly reduced volume for the buffer capacitors.

An important metric for evaluating the performance of an energy buffer is its energy buffering ratio (EBR), which is defined as the ratio of the total energy that can be extracted from the buffer within a charge/discharge cycle to the rated energy capacity of the buffer [8]-[14]. A higher EBR means that a buffer utilizes a larger portion of its total energy capacity. For a given amount of energy to be buffered, this results in a lower energy storage requirement for the buffer, which directly translates into reduced passive volume.

Two primary categories of SSC energy buffers have been previously presented in the literature: the unipolar and bipolar variants. In a unipolar design, the supporting block capacitors can only connect in series with the backbone block capacitor.

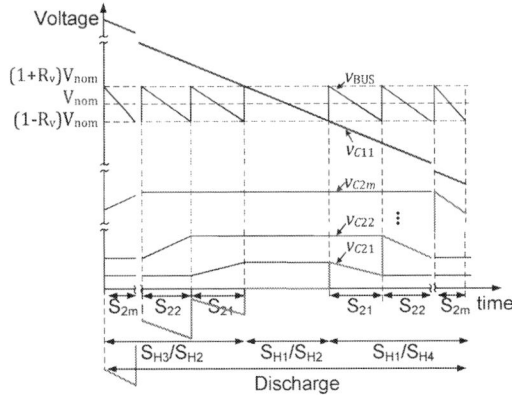

Figure 3: Switch states, individual capacitor voltages and resulting bus voltage over a discharge cycle for an n-m enhanced bipolar SSC energy buffer with capacitors of unequal capacitance values and a bus voltage ripple ratio of R_v.

On the other hand, in a bipolar design the supporting block capacitors can switch between a series and an anti-series connection with the backbone block capacitor(s). Each variant has an enhanced version, which can be further optimized for EBR by the appropriate choice of capacitance ratios. The enhanced unipolar design consists of 1 backbone capacitor, m supporting capacitors, and $m+1$ switches, while the enhanced bipolar design consists of n backbone capacitors, m supporting capacitors, and $n+m+4$ switches.

Figure 2 shows the topology of an n-m enhanced bipolar SSC energy buffer. Before the energy buffer starts normal operation, all capacitors are precharged to pre-determined voltage levels. In standard operation, one supporting capacitor is always connected in series or anti-series with one backbone capacitor. In enhanced operation, two of the H-bridge switches, in series, can bypass all supporting capacitors, providing additional states. At the start of the charging cycle, all capacitors are in their lowest energy state. At this time, the supporting capacitor with the highest voltage level (C_{2m}) is connected in series with one of the backbone capacitors (C_{11}) and the two capacitors are charged together. When the bus voltage reaches its maximum allowed value, switch S_{2m} is turned off and $S_{2(m-1)}$ is turned on to connect the next supporting capacitor ($C_{2(m-1)}$) in series with the backbone capacitor and these two capacitors are then charged in series. This process is continued until all the supporting capacitors are charged. Next, the H-bridge switches are flipped and the same process is continued, but this time the supporting capacitors are connected in anti-series with the backbone capacitor. This allows the supporting capacitors to be discharged to their original voltages while the backbone capacitor is charged. The backbone capacitor reaches its peak value, which is higher than the bus voltage, when all the supporting capacitors have been discharged. This charging process is repeated for the remaining backbone capacitors. Once all the backbone capacitors have been fully charged, the discharge cycle begins, and proceeds in simply the reverse manner. The switch states and voltage waveforms of individual capacitors during a discharge cycle for an n-m enhanced bipolar SSC energy buffer with capacitors of equal capacitance values, a nominal bus voltage of V_{nom}, a nominal-to-peak ripple voltage of $R_v V_{nom}$ and the resultant bus voltage V_{BUS} are shown in Fig. 3. A detailed description of the operation of an n-m enhanced bipolar SSC energy buffer can be found in [8], [10], [13].

B. dc-ac Converter

The topology of the dc-ac converter is shown in Fig. 4. This converter is based on an H-bridge configuration, where one of

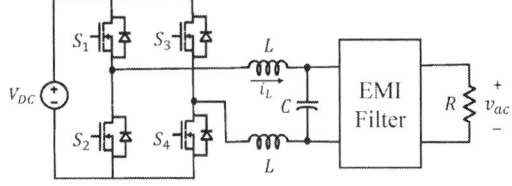

Figure 4: dc-ac converter architecture.

978-1-4673-9551-9/16 $31.00 © 2016 IEEE

the legs of the bridge switches at high frequency to shape the output current, while the other functions as an unfolder switched at line frequency. The roles of the two legs alternate over each half-line cycle. The control strategy for the high frequency switching is based on the variable frequency constant peak current control scheme introduced in [22].

Typical inductor current and average output current reference of the dc-ac converter operated with this control scheme are shown in Fig. 5(a), while the switching frequency profile across a half line cycle is shown in Fig. 5(b). In this control scheme, the converter is operated in DCM/BCM. At each required average power level, the peak current of the inductor is controlled to an optimal constant value, given by [22]:

$$I_{pk,ref,optimal} = 2\,I_{out,pk}. \tag{1}$$

Here $I_{out,pk}$ is the peak of the desired average output current. This optimal choice of inductor peak current results in improved efficiency over varying average power levels. A constant maximum peak inductor current also greatly simplifies EMI filter design. As may be observed from Fig. 5(b), the switching frequency in this control scheme scales with required instantaneous output power, keeping frequency-dependent losses to a minimum. In particular, as opposed to conventional schemes, the switching frequency in the proposed scheme reduces at low instantaneous power levels (near the line zero-crossings), resulting in much lower switching losses. Referring to Fig. 5(a), an additional point of note is that the peak current is scaled down near the line voltage zero-crossings, so as to maintain a specified minimum switching frequency.

As means of further illustration of this control strategy, consider a typical inductor current waveform in a switching

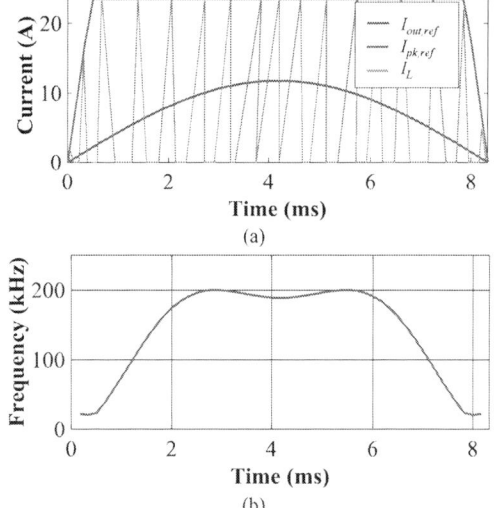

Figure 5: (a) Inductor current and average output current profile and (b) Variation of switching frequency over half line cycle.

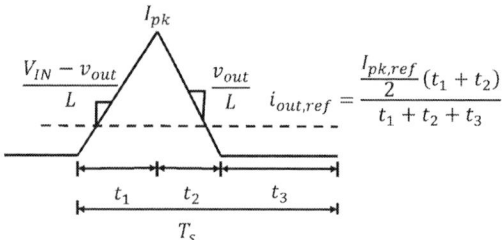

Figure 6: Illustration of the relationship between the average and peak current references in a typical switching cycle.

cycle within the line period, as shown in Fig. 6. The relationship between the average current reference (which varies across the line period, but is considered constant within a switching cycle), and the peak current reference (which remains constant across the line period for a given average power requirement), can be expressed as:

$$\left(\frac{I_{pk,ref}}{2} - i_{out,ref}\right)(t_1 + t_2) + \left(-i_{out,ref}\right)t_3 = 0, \tag{2}$$

where $I_{pk,ref}$ is the constant peak current reference, $i_{out,ref}$ is the average output current reference for the switching period under consideration, and t_1, t_2 and t_3 are the subintervals of the switching cycle, as defined in Fig. 6. These subintervals, and hence the switching period, T_s, were generated using an analog integrator to obtain the desired average current in [22].

In this work, we present a digital implementation of the control scheme described above. This digital control scheme is implemented as follows: two comparators are used, one to compare the inductor current to its peak reference value, and the other to detect the current zero crossing. At the beginning of the switching cycle, the high-side transistor of one of the inverter legs is turned on, and a counter with zero initial value is activated. The inductor current rises, and the counter increments in steps equal to $\left(\frac{I_{pk,ref}}{2} - i_{out,ref}\right)$ over each clock cycle, until the sensed inductor current equals its peak reference value. At this point, the high-side transistor is turned off, and the low-side transistor is turned on after an appropriate dead time. The inductor current now begins to ramp down. The counter continues to increment in the same steps as before, until the inductor current reaches zero. The low-side transistor is now turned off, and the counter begins to decrement in steps equal to $i_{out,ref}$. When the count falls to zero, the next switching period is initiated by turning the high-side transistor on again. In effect, this cycle-by-cycle digital control strategy implements the control objective specified in (2), resulting in the desired average current being obtained. This implementation of the variable frequency constant peak current control scheme is illustrated via a flowchart in Fig. 7.

III. SYSTEM DESIGN

The following subsections describe the methodologies used to design the SSC energy buffer and the dc-ac converter, based upon the specifications provided in Table I.

Table I: System Design Specifications

dc Input Voltage	400 V
ac Output Voltage	240 Vrms
Output Power	2 kW
dc Bus Voltage Ripple Ratio	1.5 %

A. SSC Energy Buffer

As mentioned earlier, SSC energy buffers can be categorized into unipolar and bipolar variants. The choice between a unipolar and a bipolar SSC energy buffer represents a tradeoff between switch complexity and energy buffering ratio. While unipolar designs offer simpler switch structures, bipolar designs have the potential of achieving higher energy buffering ratios. Previous work on the optimization of unipolar and bipolar SSC energy buffers concentrated on maximizing the energy buffering ratio of a given topology (for instance, a 1-2 enhanced bipolar SSC energy buffer) across a range of voltage ripple ratios.

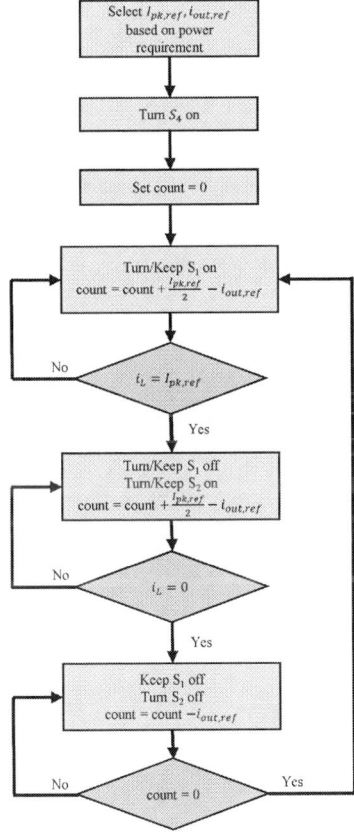

Figure 7: Flowchart demonstrating the digital implementation of the variable frequency constant peak current control scheme of the dc-ac converter.

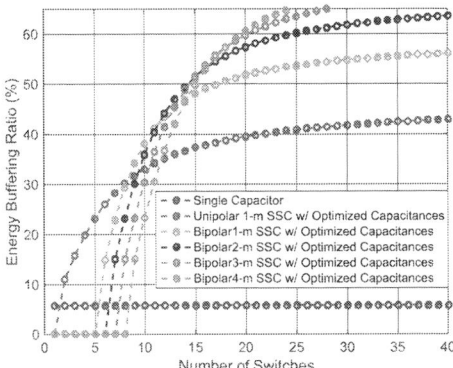

Figure 8: Energy buffering ratio (EBR) vs. number of switches for multiple SSC energy buffer topologies.

For a typical inverter application, the dc bus voltage ripple ratio is constrained by the given specifications. With this in mind, we present a new optimization approach that evaluates a range of SSC energy buffer topologies for a fixed voltage ripple ratio. The objective of this optimization is to minimize the total volume of the buffer, while maintaining a realizable switch complexity. A reasonable estimate of the passive volume of an SSC energy buffer is provided by the energy buffering ratio (EBR). As such, the EBR of a range of SSC buffer topologies is evaluated as a function of their switch count. An example of this evaluation is presented in Fig. 8. It may be observed that the curves in Fig. 8 are monotonic, and do not have optima for any SSC buffer topology and switch count. Therefore, regardless of topology, the passive volume of the buffer can always be reduced by using a higher switch count. However, increasing the switch count results in an increase in the volume of transistors and associated gate drive circuitry, thereby negatively impacting the total system volume. In order to minimize the total system volume, the volume of active

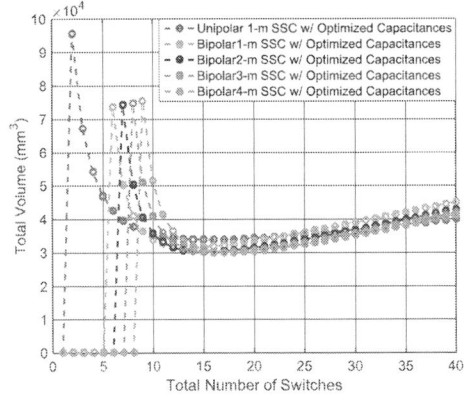

Figure 9: Total volume including transistors and gate drive circuitry, for multiple SSC energy buffer topologies, vs. number of switches.

circuitry (transistors and gate drives) needs to be taken into account in the optimization procedure. Furthermore, to accurately map the EBR of buffer topologies to actual capacitor volume, the energy densities of commercially available capacitors must be considered. A study was performed on market available ceramic capacitors to identify the highest energy density products across voltage ratings, and hence determine the efficient frontier of capacitive energy density. With this data in hand, the total system volume is then evaluated as a function of switch count. This optimization approach was implemented for the inverter specifications of Table I. The results, for the same buffer topologies as in Fig. 8, are shown in Fig. 9. With active volume and capacitor energy density data included, the curves in Fig. 9 can be observed to contain minima; for any SSC buffer topology, there exists a switch count that minimizes total volume. A decision on the optimum SSC energy buffer topology can now be made. For the given system specifications, it can be seen from Fig. 9 that a 3-9 enhanced bipolar SSC energy buffer has the lowest total volume, and was selected for this work. The capacitance values used in the buffer are provided in Table II.

For control of the 3-9 enhanced bipolar SSC energy buffer, a hysteretic controller was implemented. This controller monitors the dc bus voltage and compares it to high and low voltage threshold values, which are determined by the ripple specifications. When the bus voltage crosses either threshold, the current supporting capacitor is switched out of the circuit, and the next supporting capacitor is switched in. The switching sequence is decided using a state machine.

B. dc-ac Converter

The power stage inductor for the dc-ac converter is designed taking into consideration the peak instantaneous current across its operating power range, as well as the limits of switching frequency. For maintaining the maximum required power of 2 kW at 240 V_{rms} output voltage, the maximum required output peak current is calculated, which then translates to an optimal peak inductor current ($I_{pk,ref,optimal}$) using (1). The upper limit on switching frequency is determined by the allowable

frequency-dependent losses (including inductor core losses, turn-off switching losses etc.), while the lower frequency limit is set by the human audible range. With these limits, the inductor is designed for a peak current of 23.6 A, and a switching frequency in the range of 20 kHz to 200 kHz. The minimum inductance value that meets these limits is given by:

$$L_{min} = \left. \frac{8 I_{out,pk} V_{BUS}^2}{27 I_{pk,ref}^2 V_{out,pk} f_{s,max}} \right|_{P_{out,avg}=P_{out,max}}. \quad (3)$$

Here, $I_{out,pk}$ and $I_{pk,ref}$ are as defined earlier, $V_{out,pk}$ is the peak of the output voltage, V_{BUS} is the dc input voltage of the dc-ac converter, $f_{s,max}$ is the upper limit on switching frequency, $P_{out,avg}$ is the average output power, and $P_{out,max}$ is the maximum required average output power. The inductor is designed for this minimum inductance value with a view to minimize losses. Given this inductance, $L = L_{min}$, the fraction of the line period (α) for which the peak current is scaled down so as to maintain a minimum switching frequency, as described earlier and defined in Fig. 5(a), can be obtained from:

$$L = \left. \frac{2 I_{out,pk} V_{out,pk} \sin^2(2\pi\alpha)(V_{BUS} - V_{out,pk}\sin(2\pi\alpha))}{I_{pk,ref}^2 V_{BUS} f_{s,min}} \right|_{P_{out,avg}=P_{out,max}}, \quad (4)$$

where $f_{s,min}$ is the lower limit on switching frequency.

IV. EXPERIMENTAL RESULTS

To validate the design methodologies outlined in the previous section, a prototype 2 kW inverter was designed, built and tested. The volume of the system excluding the thermal management components is 27 in³, while the total volume is 40 in³. A photograph of the prototyped system is shown in Fig. 10. An inductance value of 30 µH is selected for the dc-ac converter based on the design method presented in the previous section. Figure 11 shows the SSC energy buffer waveforms with a 500 W resistive load at 400 V nominal bus voltage with 1.5% voltage ripple. As designed, the bus voltage ripple is maintained within ripple specifications throughout the line cycle. The voltage across the supporting block is also shown in the figure. It can be seen that the voltage across the supporting capacitor block increases at first and then decreases for the portion of the line cycle where the H-bridge is set up in series, and the voltage remains at zero for the portion of the line cycle where the H-bridge is in anti-series, i.e., when the other terminal of the supporting capacitors constitutes the intermediate node.

Table II: Capacitance values and voltage ratings for 3-9 enhanced bipolar SSC energy buffer

Capacitor	Required Linear Capacitance (µF)	Selected Ceramic Capacitance (µF)	Required Voltage Rating (V)	Selected Voltage Rating (V)
C_{11}	24	80	553	630
C_{12}	24	80	553	630
C_{13}	24	80	553	630
C_{21}	146	1474	14	16
C_{22}	146	2200	26	35
C_{23}	132	940	37	50
C_{24}	105	317	49	100
C_{25}	88	416	60	100
C_{26}	78	406	70	100
C_{27}	70	505	80	100
C_{28}	64	601	90	100
C_{29}	59	101	100	250

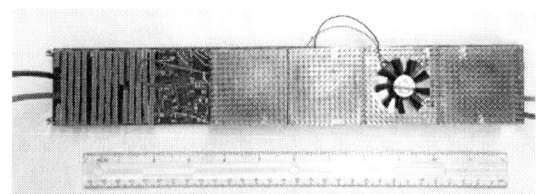

Figure 10: A photograph of the prototype 2 kW inverter with the heat sinks connected. A standard ruler is also shown for comparison.

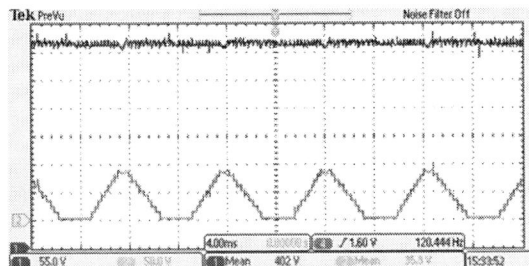

Figure 11: Demonstration of SSC Energy Buffer working at 400 V nominal bus voltage with 1.5% ripple with 500 W resistive load.

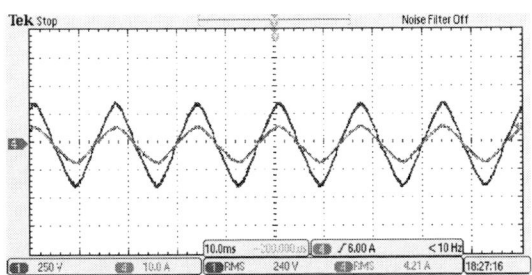

Figure 12: Output voltage and output current waveforms of the inverter working at 1 kW resistive load.

The output voltage and current waveforms of the dc-ac converter are shown in Fig. 12 with a 1 kW resistive load. The efficiency of the dc-ac converter has been measured across a range of output power levels, from 200 W to 1.4 kW. A plot of the measured efficiency as a function of output power is shown in Fig. 13. The dc-ac converter maintains an efficiency of greater than 90% for power levels greater than 600 W.

V. CONCLUSIONS

This paper presents a 50 W/in³ power density, 2 kW single-phase inverter with 90% line-cycle average efficiency. This inverter features innovations in twice-line-frequency (120 Hz) energy buffering and high frequency dc-ac power conversion. The energy buffering stage utilizes an advanced implementation of the recently proposed stacked switched capacitor (SSC) energy buffer architecture, while the dc-ac power conversion is performed using a soft-switching SiC-FET based converter, with digitally implemented variable frequency constant peak current control.

ACKNOWLEDGMENT

The authors would like to thank Google Inc. for financially supporting this work.

REFERENCES

[1] A.C. Kyritsis, N. Papanikolaou and E. Tatakis, "Enhanced Current Pulsation Smoothing Parallel Active Filter for Single Stage Grid Connected AC-PV Modules," *Proceedings of the International Power Electronics and Motion Control Conference (EPE-PEMC)*, pp.1287-1292, Poznan, Poland, 2008.

[2] H. Wang and H. Chung, "A Novel Concept to Reduce the DC-Link Capacitor in PFC Front-End Power Conversion Systems," *Proceedings of*

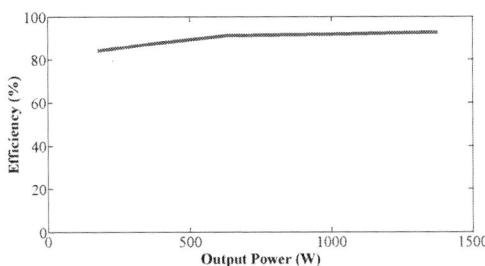

Figure 13: Measured efficiency of the dc-ac converter.

the *IEEE Applied Power Electronics Conference (APEC)*, pp. 1192-1197, Orlando, FL, 2012.

[3] P.T. Krein and R.S. Balog, "Cost-Effective Hundred-Year Life for Single-Phase Inverters and Rectifiers in Solar and LED Lighting Applications Based on Minimum Capacitance Requirement and a Ripple Power Port," *Proceedings of the IEEE Applied Power Electronics Conference (APEC)*, pp. 620-625, Washington, DC, Feb. 2009.

[4] B.J. Pierquet and D.J. Perreault, "Single-Phase Photovoltaic Inverter Topology with Series-Connected Power Buffer," *Proceedings of the IEEE Energy Conversion Congress and Exposition (ECCE)*, pp. 2811-2818, Atlanta, GA, Sept. 2010.

[5] M. Chen, KK. Afridi, D.J. Perreault, "A Multilevel Energy Buffer and Voltage Modulator for Grid-Interfaced Micro-Inverters", *Proceedings of the IEEE Energy Conversion Congress and Exposition (ECCE)*, Denver, CO, Sep. 2013.

[6] S. Sugimoto, S. Ogawa, H. Katsukawa, H. Mizutani and M. Okamura, "A Study of Series-Parallel Changeover Circuit of a Capacitor Bank for an Energy Storage System Utilizing Electric Double Layer Capacitors," *Electrical Engineering in Japan*, vol. 145, no. 3, pp. 33-42, Nov. 2003.

[7] X. Fang, N. Kutkut, J. Shen and I. Batarseh, "Ultracapacitor Shift Topologies with High Energy Utilization and Low Voltage Ripple," *Proceedings of the International Telecommunications Energy Conference (INTELEC)*, Orlando, FL, Jun. 2010.

[8] M. Chen, K.K. Afridi and D.J. Perreault, "Stacked Switched Capacitor Energy Buffer Architecture," *IEEE Transactions on Power Electronics*, vol. 28, no. 11, pp. 5183-5195, Nov. 2013.

[9] A.H. Chang, J.J. Colley and S.B. Leeb, "A Systems Approach to Photovoltaic Energy Extraction," *Proceedings of the IEEE Applied Power Electronics Conference (APEC)*, pp. 59-70, Orlando, FL, Feb. 2012.

[10] K.K. Afridi, M. Chen and D.J. Perreault, "Enhanced Bipolar Stacked Switched Capacitor Energy Buffers," *IEEE Transactions on Industry Applications*, vol. 50, no. 2, pp. 1141-1149, Mar./Apr. 2014.

[11] Y. Ni, S. Pervaiz, M. Chen and K.K. Afridi, "Energy Density Enhancement of Unipolar SSC Energy Buffers through Capacitance Ratio Optimization," *IEEE Workshop on Control and Modelling for Power Electronics (COMPEL)*, Santander, Spain, June, 2014.

[12] M. Chen, Y. Ni, C. Serrano, B. Montgomery, D.J. Perreault, K.K. Afridi, "An Electrolytic-Free Offline LED Driver with a Ceramic-Capacitor-Based Compact SSC Energy Buffer," *Proceedings of the IEEE Energy Conversion Congress and Exposition (ECCE)*, Pittsburgh, PA, Sep. 2014.

[13] S. Pervaiz, Y. Ni, KK. Afridi, "Improved capacitance ratio optimization methodology for stacked switched capacitor energy buffers", *Proceedings of the IEEE Applied Power Electronics Conference (APEC)*, Charlotte, SC, Mar. 2015.

[14] S. Pervaiz, A. Kumar, K.K. Afridi, "A Compact Electrolytic-Free Two-Stage Universal Input Offline LED Driver", *Proceedings of the IEEE Applied Power Electronics Conference (APEC)*, Long Beach, CA, Mar. 2016.

[15] R. Erickson and A. Rogers, "A microinverter for building-integrated photovoltaics," *Proceedings of the IEEE Applied Power Electronics Conference (APEC)*, Feb. 2009, pp. 911–917.

[16] Y. Bo, L. Wuhua, G. Yunjie, C. Wenfeng, and H. Xiangning, "Improved transformerless inverter with common-mode leakage current elimination

978-1-4673-9551-9/16 $31.00 © 2016 IEEE

for a photovoltaic grid-connected power system," *IEEE Transactions on Power Electronics*, vol. 27, no. 2, pp. 752–762, Feb. 2012.

[17] B. Chen, B. Gu, L. Zhang, Z. U. Zahid, J. S. Lai, Z. Liao, and R. Hao, "A high efficiency MOSFET transformerless inverter for non-isolated microinverter applications," *IEEE Transactions on Power Electronic*, 2015.

[18] R. Gonzalez, J. Lopez, P. Sanchis, and L. Marroyo, "Transformerless inverter for single-phase photovoltaic systems," *IEEE Transactions on Power Electronics*, vol. 22, no. 2, pp. 693–697, Mar. 2007.

[19] A. Amirahmadi, U. Somani, C. Lin, N. Kutkut, and I. Batarseh, "Variable boundary dual mode current modulation scheme for three-phase microinverter," *Proceedings of the IEEE Applied Power Electronics Conference (APEC)*, 2014, pp. 650–654.

[20] A. Amirahmadi, H. Haibing, A. Grishina, Z. Q. Zhang, C. Lin, U. Somani, and I. Batarseh, "Hybrid ZVS BCM current controlled three-phase microinverter," *IEEE Transactions on Power Electronics*, vol. 29, no. 4, pp. 2124,2134, Apr. 2014.

[21] Q. Zhang, H. Hu, D. Zhang, X. Fang, Z. J. Shen, and I. Batarseh, "A controlled-type ZVS technique without auxiliary components for the low power dc/ac inverter," *IEEE Transactions on Power Electronics*, vol. 28, no. 7, pp. 3287–3296, Jul. 2013.

[22] Y. Levron and R.W. Erickson, "High Weighted Efficiency in Single-Phase Solar Inverters by a Variable-Frequency Peak Current Controller", *IEEE Transactions on Power Electronics*, vol. 31, no. 1, pp. 248-257, Jan. 2016.

A Novel Single-stage Dual-Active Bridge based Isolated DC-AC Converter

Shiladri Chakraborty, *Student Member, IEEE* and Souvik Chattopadhyay, *Member, IEEE*
Department of Electrical Engineering
Indian Institute of Technology Kharagpur
Kharagpur - 721302, India
Email: shiladri@ee.iitkgp.ernet.in

Abstract—**This paper presents a single-stage, isolated DC-AC converter topology based on the dual active bridge concept. The proposed topology has the advantages of low device count and high efficiency due to the single-stage power processing structure and zero-voltage-switching (ZVS) operation of the MOSFETs. Double-line-frequency energy buffering can be achieved via the high voltage side dc bus capacitor, resulting in reduction of decoupling capacitance requirement. Thus it can be considered as a suitable candidate for high-efficiency, isolated, single-phase DC-AC converter applications such as PV microinverter. Basic principle of operation of the circuit is described followed by an explanation of the control strategy. Numerical simulation results are presented to validate the approach. Practical evaluation of converter performance is done on a 200 W experimental prototype.**

I. Introduction

Recently, some single-stage, high-frequency ac link based topologies suited for isolated photovoltaic (PV) microinverter applications have been proposed in literature [1]- [3]. Their basic architecture consists of a high-frequency resonant inverter, a high-frequency transformer, and a half-bridge or full-bridge cycloconverter (with four-quadrant switches). Operationally, these circuits function like the series resonant dc-ac dual active bridge (DAB) converter [4], wherein a high-frequency quasi-square wave generated by the full-bridge inverter is fed to the series resonant tank, which creates a high frequency ac current waveform with its amplitude modulated at line frequency. The matrix converter (with four-quadrant switches) following the transformer is used to down-convert the high-frequency ac current. Power flow is regulated by phase-shift control and zero-voltage-switching (ZVS) operation of the devices is possible by keeping the switching frequency above the tank resonant frequency. Despite these advantages, these topologies suffer from high device count (12), which is the same as that required in a typical two-stage solution [5].

Another distinct disadvantage of existing single-stage high-frequency link based dc-ac solutions is the lack of an inherent power decoupling mehanism [6]. Such a power decoupling arrangement is invariably needed in all single-phase dc-ac (or indeed ac to dc) conversions to balance the mismatch between the steady dc power and the pulsating ac power. This entails the use of life-limiting electrolytic capacitors (placed directly across the low voltage PV input) in the topologies of [1]- [3]. It is possible to integrate the power decoupling arrangement in these topologies by using auxiliary circuitry, as is done in [7], but this increases the device count further.

This paper presents a novel low-device count, isolated, single-stage, dual active bridge-based solution for PV microinverter type of applications. The proposed topolgy has the following specific advantages.

1) It uses only eight devices, which to the knowledge of the authors, is the least among all existing isolated, single-stage, DAB-based solutions or other non-DAB but dual-bridge-based solutions [6].

2) Being a DAB-derived structure, the proposed topology shares the same benefits as the classical DAB converter - that is inherent ZVS turn-on operation of devices and absence of turn-off voltage spikes across them.

3) Power decoupling can be affected via the high voltage (secondary) side dc bus, thereby resulting in considerable reduction in decoupling capacitance requirement.

4) Possibility of interfacing a battery on the PV side makes the proposed topology a multiport solution with integrated energy storage capability [8]- [9].

The remaining part of the paper is organised as follows. Section II introduces the topology followed by an explanation of its working principle. Details of two possible control approaches are discussed in section III. Finally relevant simulation results and some preliminary experimental results are presented in sections IV and V respectively to validate the approach.

II. Proposed Topology

A. Topology Description

The proposed topology is shown in Fig. 1. It consists of two full-bridges ($S_1 - S_4$ and $S_5 - S_8$) with floating dc buses, one on either side of a high frequency transformer. The filter components (L_{f1}, L_{f2}, C_{f1}, L_{f3}, L_{f4}, C_{f2}), configured as shown, are used to integrate the dc and ac ports with the respective bridges and transformer windings. Though L_{f1}, C_{f1} and L_{f4}, C_{f2} are not explicitly required for functioning of the circuit, they ensure that no high-frequency component is present in the current drawn from or delivered to the respective ports. C_{b1}, C_{b2} are small series capacitors, which prevent low-frequency current injection into the transformer. The external leakage inductor L_{lk} connected in series with either of the transformer windings serves as a high-frequency energy storage and delivery element, as in a conventional DAB converter.

978-1-4673-9551-9/16 $31.00 © 2016 IEEE

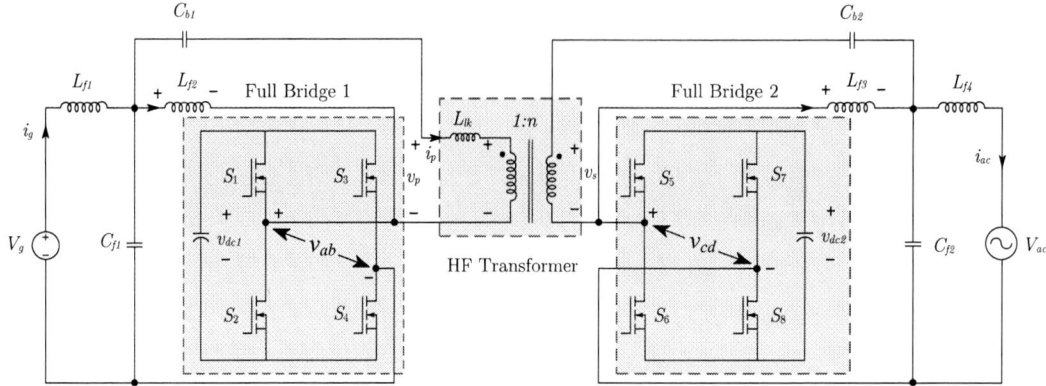

Fig. 1: Schematic of the proposed topology.

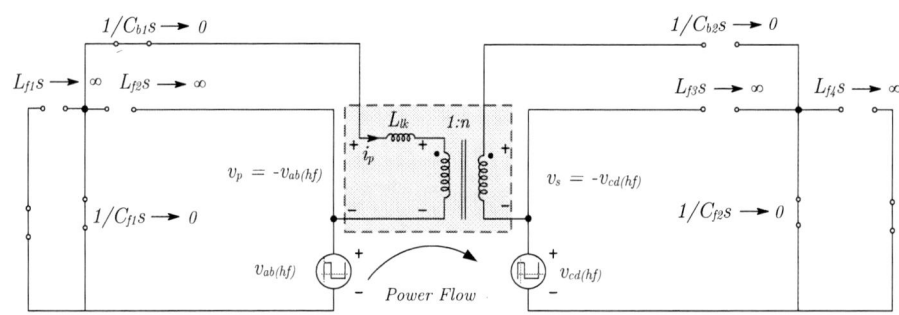

(a) High frequency (hf) equivalent circuit

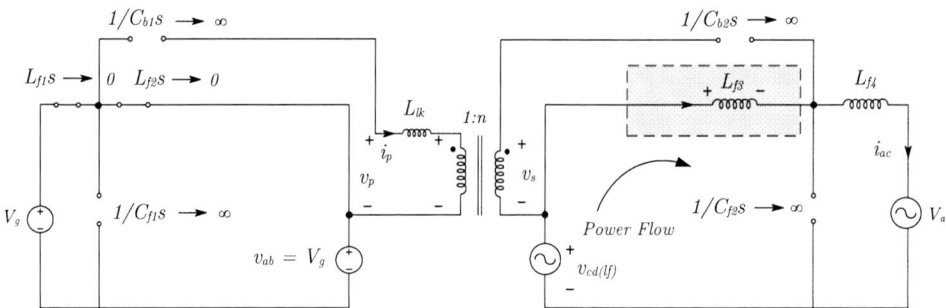

(b) Low frequency (lf) equivalent circuit

Fig. 2: Explaining operation of the circuit. Power flow can be conceptually thought of as occurring in two different frequency regimes. In the hf (switching frequency) circuit, power flow is as in a classical DAB converter. In the lf (line frequency) circuit, power flow is between two low-frequency phase-shifted ac sources separated by an inductor.

B. Working Principle

The working principle of the circuit can be better understood by analysing the high frequency (switching frequency) and low frequency (line frequency) equivalent circuits separately, as shown in Fig. 2. At high frequency, the inductive and capacitive impedances being high and low respectively, the filter inductors and capacitors are represented by an equivalent open and an equivalent short respectively in the high frequency equivalent circuit (Fig. 2a). Moreover, the low-frequency sources i.e the dc source and the line frequency ac source also get shorted. Following similar arguments, the filter capacitors are open and the filter inductors shorted in the low frequency equivalent circuit (Fig. 2b).

Key waveforms explaining circuit operation are shown in Fig. 4. Diagonal devices of each full-bridge are driven by the same switching signal. The switching signal on the primary side is generated by comparing a triangular carrier tri_p with a constant modulating signal, which results in a constant (non-0.5) duty-ratio D_1. On the secondary side, the switching signal is synthesized by comparing another triangular carrier tri_s with a sinusoidally varying modulating signal $d_2(t)$ given by

$$d_2(t) = \frac{1}{2} + m_a \sin \omega t \ , \qquad (1)$$

where $0 < m_a < 0.5$ refers to the amplitude modulation index of sine-PWM modulation and ω is the angular grid frequency.

In general, both modules generate a rectangular voltage waveform, which has a dc component (averaged over the switching period T_s). For the dc side module, this dc component is constant and must be equal to the dc voltage V_g in order to ensure volt-second balance of the filter inductor L_{f2}. From this condition, the average voltage of the primary-side dc bus can be obtained as

$$v_{dc1} = \frac{V_g}{2D_1 - 1} \ , \qquad (2)$$

which shows that at steady-state, the duty-ratio D_1 is always greater than 0.5 and v_{dc1} is greater than V_g. In this context, it should be mentioned that in an ideal circuit, the blocking capacitor C_{b1} is redundant. However, in presence of practical circuit non-idealities such as series resistance of L_{f2}, C_{b1} serves to block the dc component of the non-zero-mean voltage that would otherwise be impressed across the transformer primary. This prevents dc current injection into the primary, thereby reducing core losses and minimising the chance of transformer saturation [10].

On the secondary side, the synthesized voltage (v_{cd}) has an average varying sinusoidally over the line cycle given by

$$v_{cd(LF)} = v_{dc2}(2d_2(t) - 1) = 2v_{dc2}m_a \sin \omega t \ . \qquad (3)$$

For power-flow to occur in the low frequency equivalent circuit, $v_{cd(LF)}$ leads the grid voltage V_{ac}, and the (small) difference of these low frequency voltages is applied across the filter inductor L_{f3}. Capacitor C_{b2} serves the crucial purpose of blocking this low frequency voltage, thereby preventing it from directly appearing across the transformer's secondary terminals. Thus, in the hf equivalent circuit, a zero-mean rectangular voltage waveform, corresponding to the high-frequency component $v_{cd}(hf)$ of the synthesized voltage v_{cd} gets applied on the secondary side.

As discussed earlier, volt-second balance of L_{f2} and the presence of C_{b1} ensures a zero-mean rectangular voltage also gets applied on the primary side. Hence, following the principle of the classical DAB converter, power flow in the high frequency equivalent circuit is possible by simple control of the phase-shift between these two impressed voltage waveforms. Noting that both these waveforms are asymmetric in nature, the direction of power flow is determined by the phase-relationship between the fundamental components. Hence for power to flow from the primary side to the secondary side, the fundamental of the primary voltage should lead that of the secondary, which can be ensured my making the primary-side carrier tri_p lead the secondary side-carrier tri_s by an angle δ_{ps}, as shown in Fig. 4.

C. Battery Integration

Another key aspect of the proposed topology is that an energy-storage element like a battery can be easily integrated with it, as Fig. 3 illustrates. The main energy source, like a PV module is now connected to the primary side dc bus, with the associated LC filter ensuring that constant current is drawn from it. It should be mentioned that for this scheme to work successfully, V_{pv} should be greater than V_{bat}, which follows from (2). This condition is most likely to be satisfied in typical applications, where a low voltage (12-18 V) battery is used in conjunction with a standard commercial PV module (MPP voltage of 25 -45 V).

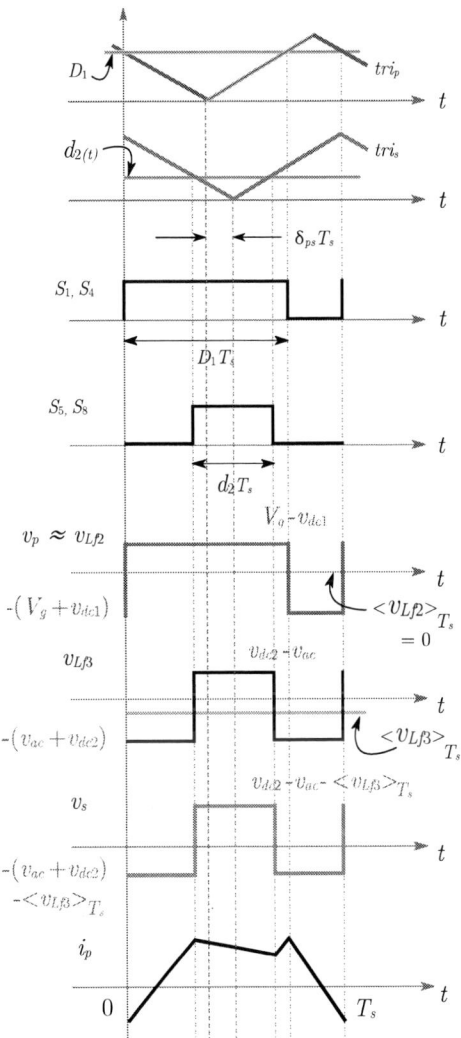

Fig. 4: Key waveforms over a high-frequency switching period. For power flow to take place from primary to secondary, the primary-side carrier tri_p should lead the secondary-side carrier tri_s.

978-1-4673-9551-9/16 $31.00 © 2016 IEEE

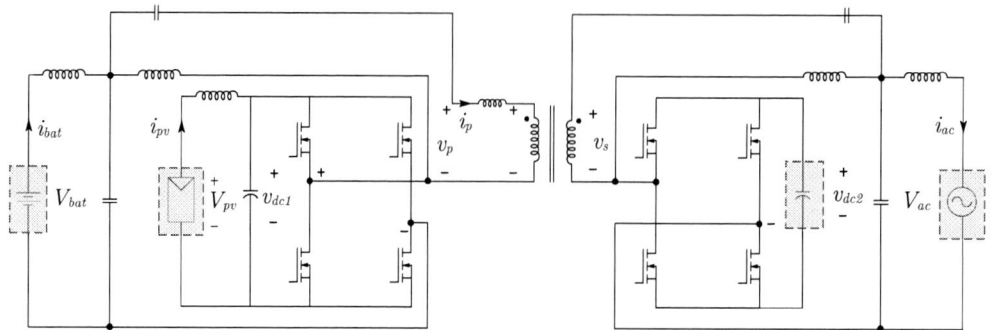

Fig. 3: Schematic illustrating battery-integration in the proposed topology.

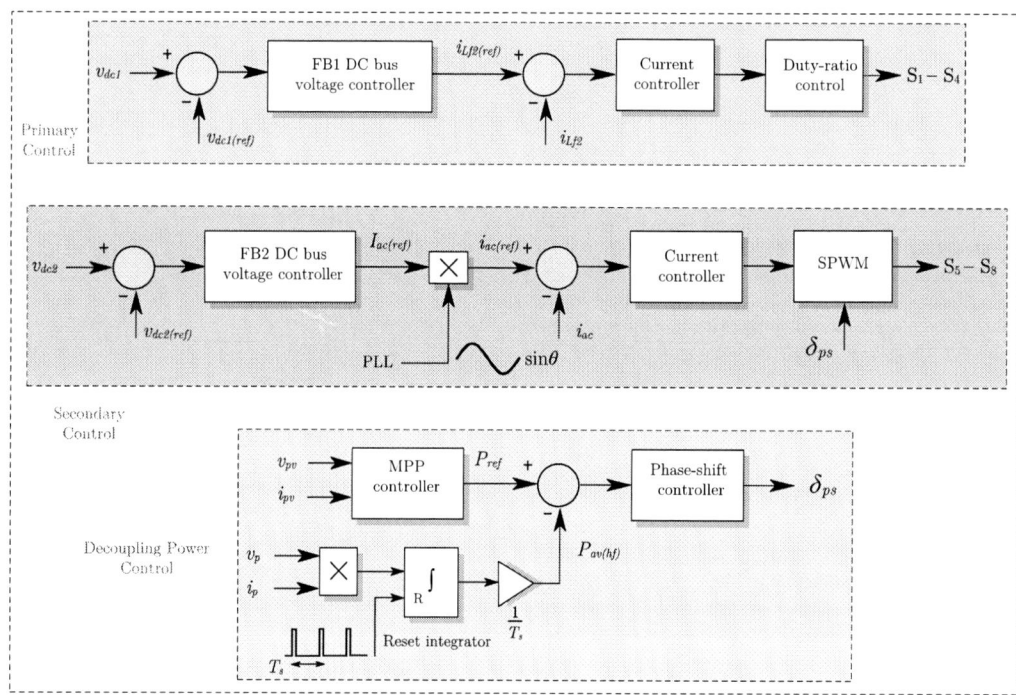

Fig. 5: The overall control architecture for power decoupling control (without battery-integrated operation). For battery-integrated operation, v_{dc1} control is absent and reference $i_{Lf2(ref)}$ for current controller is directly fed.

III. CONTROL SCHEME

The overall control structure for operation without battery-integration is depicted in Fig. 5. On both primary and secondary sides, an outer voltage control loop cascaded with an inner current loop controls the average value of the dc bus voltages. On the dc side, the current controller output is compared with a triangular carrier (tri_p) to generate the switching signals, as shown in Fig. 4. On the ac side, using the amplitude of the ac current reference signal ($I_{ac}(ref)$) and the reference signal from the phase locked loop (PLL), the instantaneous ac reference current signal is synthesized.

The inner current controller tracks this sinusoidal reference by controlling the switching signals of S_5-S_6, using standard bipolar sinusoidal pulse-width modulation (SPWM), as discussed earlier. It should be noted that, for battery-integrated operation, dc bus voltage control on dc side is absent and the reference $i_{Lf2(ref)}$ for current-mode control is explicitly obtained from the charging algorithm of the battery.

There could be two approaches to control power flow in the proposed converter. The simplest method is to give the v_p waveform a constant phase lead with respect to v_s and vary this phase-shift for varying load conditions. Though simple, this approach would result in double line-frequency power pulsations being transmitted to the PV port, resulting in high decoupling capacitance requirement as in [2] and [3]. A more advanced control strategy, shown in Fig. 5, involves varying the phase-shift δ_{ps} dynamically over the line-cycle to keep the high-frequency average power $P_{av(hf)}$ through the transformer constant at the intended reference value P_{ref} (obtained from say the MPP reference generator.) Computation of the high-frequency cycle-by-cycle average is done by passing the instantaneous product of v_p and i_p through a reset integrator. This would result in power decoupling being achieved exclusively through the ac side module.

IV. SIMULATION RESULTS AND DISCUSSIONS

A 310 W model of the proposed converter with 40 V input and 50 Hz, 110 V rms, 4 A peak output, working under closed-loop power decoupling control has been developed in the PLECS simulation platform. The switching frequency is 50 KHz. Values of the various passive elements are listed in Tab. I. Relevant simulation results are shown in Fig. 6 - 11.

TABLE I: Parameter values for the simulation model

L_{f1}, L_{f2}	C_{f1}	C_{b1}	C_{dc1}	L_{lk}
200 μH	20 μF	1 μF	20 μF	25 μH

Turns-ratio (n)	L_{f3}, L_{f4}	C_{f2}	C_{b2}	C_{dc2}
1:4	3 mH	4 μF	1 μF	10 μF

Fig. 6 shows the voltage v_{ab} synthesized by the full bridge module on the dc side. It can be seen that the synthesized waveform has a dc component equal to the dc voltage V_g. Fig. 7 shows the voltage and current waveforms of the transformer, from which it can be observed that the primary (dc side) voltage leads the secondary one, implying that power flows towards the ac side.

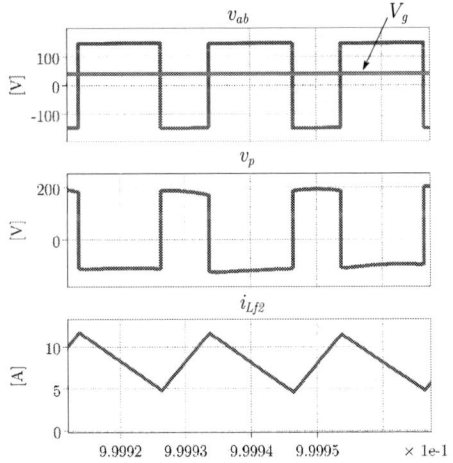

Fig. 6: The full-bridge module output voltage on the dc side (v_{ab}), transformer primary voltage (v_p) and the filter inductor current (i_{Lf1}).

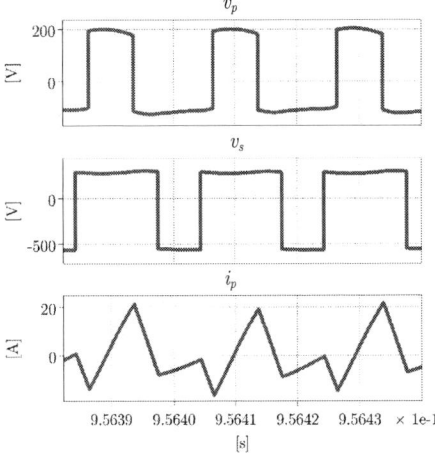

Fig. 7: Transformer voltage and current waveforms.

The full-bridge voltage controllers regulate the average values of the dc and ac side bus at 150 V and 500 V, as Fig. 8 shows. The double line-frequency ripple on the capacitor voltage (v_{dc2}) of the ac side module can be clearly observed. Absence of any such low frequency ripple on the dc side module's capacitor voltage (v_{dc1}) indicates that power decoupling is affected solely through the secondary side module. The peak-peak magnitude of the low-frequency ripple on v_{dc2} can be seen to be around 200 V, which matches very closely with the theoretically predicted value [6] of 197 V, obtained from (4) by substituting the chosen values of 310 W, 10 μF and 500 V for P_{dc}, C_{dc} and V_{dc} respectively.

$$C_{dc} = \frac{P_{dc}}{\omega_0 V_{dc} \Delta v_{dc}} \ . \tag{4}$$

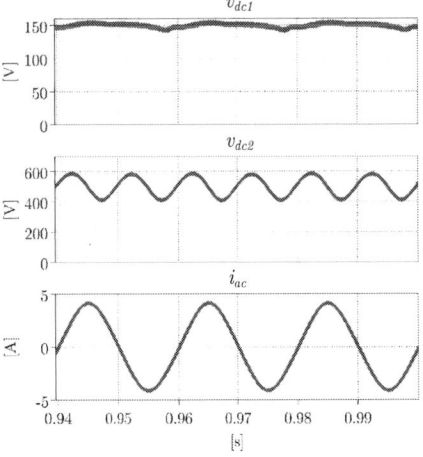

Fig. 8: DC bus voltages of the modules and the current injected into the grid (i_{ac}).

978-1-4673-9551-9/16 $31.00 © 2016 IEEE

The effectiveness of the power decoupling control is further illustrated through Fig. 9, which shows that the (switching period) average of the instantaneous power delivered through the primary terminals of the transformer is almost maintained constant. Dynamic variation of the phase-shift δ_{ps} over the the ac line cycle, which makes this possible, is also shown. Fig. 9 also confirms injection of high-quality sinusoidal current into the grid at unity power factor.

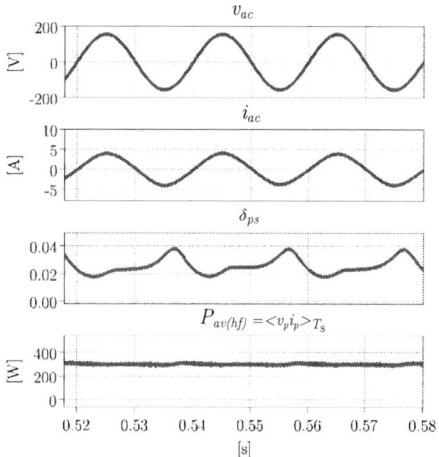

Fig. 9: Illustrating closed-loop variation of phase-shift D_ϕ to keep the (switching period) average power $P_{av(hf)}$ through the transformer constant.

Fig. 10 illustrates the working of the battery-integrated circuit, in which a 12 V battery is interfaced with a 40 V PV module. Operation in both the charging and discharging mode of the battery are shown for a constant load power requirement of 310 W (4 A peak).

Waveforms exploring occurrence of ZVS operation in the primary and secondary side devices are shown in Fig. 11. Specifically, the current through the switches at the turn-on-instant is of interest. If this current is in the negative direction, the drain-source capacitor of the MOSFETs has a possibility of being fully discharged resulting in ZVS turn-on. As can be seen, at the chosen instant of the ac line cycle, ZVS operation of all primary devices and one diagonal pair of secondary devices occur. In general, ZVS operation of the devices occur over a range of the ac line cycle.

Results of comparison of the device ratings, ZVS operation and decoupling capacitance requirement of the proposed topolgy with some other existing topologies are listed in Tab. II. Operating conditions and parameters are same as before, except that the average values of the dc bus voltages are regulated at 100 V and 250 V respectively. It can be seen that at the considered operating point, the net switch utilization factor [11] (defined as $P/\Sigma V_i I_i$) of the proposed topology is marginally lower compared to the other topologies, which is indicative of a slightly higher cumulative VA rating of the devices. However, the proposed topology still has the advantages of reduced device count, moderate rms current ratings and low decoupling capacitance.

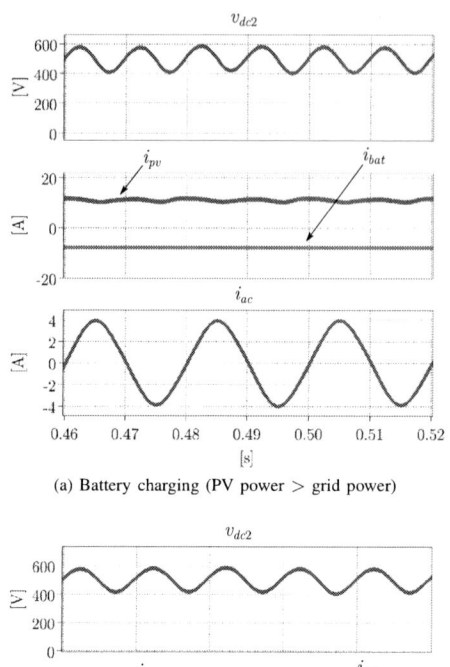

(a) Battery charging (PV power > grid power)

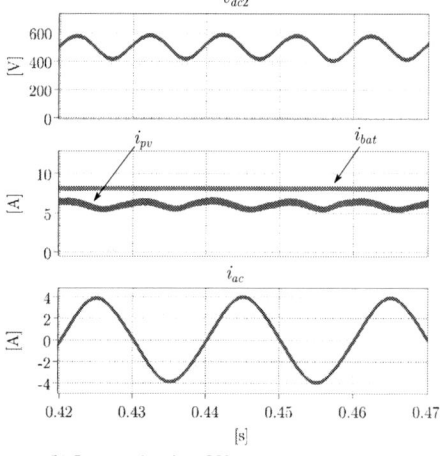

(b) Battery charging (PV power < grid power)

Fig. 10: Illustrating operation with battery-integration.

V. EXPERIMENTAL RESULTS

A 50 KHz, 50 V input, 110 V (rms), 4 A (rms) output proof-of-concept prototype of the converter has been built and tested in the open-loop standalone mode. Specifications of the components used in the prototype are listed in Tab. III. Each full-bridge module is realized using separate daughter boards, which comprise the devices, heat sinks and dc bus capacitor along with gate-drivers and a dc bus voltage sensor. The daughter boards are interfaced with a motherboard which holds all the filter components, the blocking capacitors, an external leakage inductor, the high frequency transformer and associated sensing circuitry. Gate pulses are given to the converter using a Xilinx XC3S400 FPGA-based control card.

Preliminary experimental results in open-loop condition, with a constant phase-shift applied between primary and secondary, are shown in Fig. 12 - 15. Fig. 12 shows the dc bus voltages on both sides and the synthesized ac waveform.

TABLE II: Comparison of device ratings, ZVS range and buffer capacitance of the proposed topology with existing topologies. $V_g = 40V$, $V_{ac} = 110$ V (rms), $P = 310$ W, $f_s = 50$ KHz for all cases. FR denotes ZVS operation over full range of the ac line cycle, PR denotes partial range ZVS operation.

	Primary-side switches			Secondary-side switches			i_p (RMS)	Switch utilization	C_{dcp}
	RMS current (I_i)	Peak voltage (V_i)	ZVS	RMS current (I_i)	Peak voltage (V_i)	ZVS		$P/\Sigma V_i I_i$	
Proposed topology	10.8 A (S_1, S_4)	100 V	FR	2.8 A (S_6, S_7)	300 V	PR	10.5 A	0.043	40 μF
($n = 4$)	9.3 A (S_2, S_3)	100 V	PR	2.5 A (S_5, S_8)	300 V	PR			
Reference [2]	15.5 A	40 V	PR	2.64 A	155 V	FR	22 A	0.053	12 mF
($n = 5.88$)									
Two-stage DAB [5]	6.5 A	40 V	FR	1.6 A	300 V	FR	9 A	0.058	40 μF
($n = 4$)	6.2 A	40 V	FR	1.55 A	300 V	FR			
				2 A	300 V	PR			

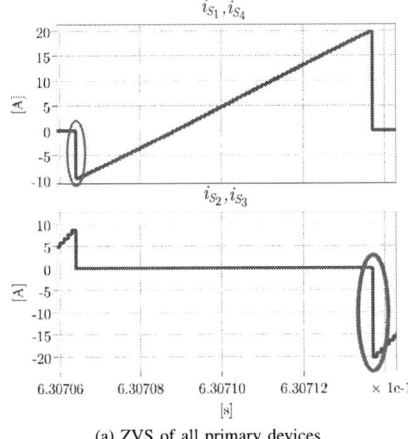

(a) ZVS of all primary devices

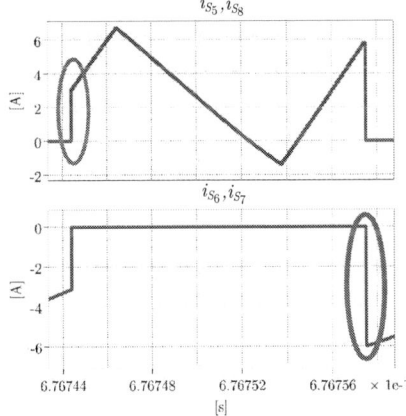

(b) ZVS of one diagonal pair of secondary devices

Fig. 11: Illustrating ZVS operation of devices at a particular instant of the ac line cycle. Transitions encircled with green denote ZVS turn-on.

TABLE III: Components used in the hardware prototype

Component	Value	Specification/Part no.
$S_1 - S_4$	200 V, 10.7 mΩ	IPP110N20NAXK
$S_5 - S_8$	600 V, 185 mΩ	24N60CFD
C_{dc1}, C_{dc2}	20 μF, 500 V	MKP1848C62050JP4
L_{lk} (external)	22 μH	-
Turns-ratio	1 : 3	-
L_{f2}	5 mH, 4 A (peak)	-
L_{f3}	5 mH, 4 A (peak)	-
C_{f2}, C_{b1}, C_{b2}	2.2 μF, 850 V	B32656S8225K561

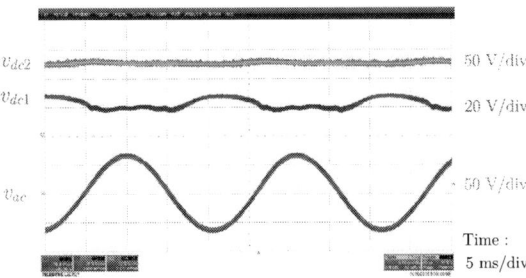

Fig. 12: DC bus voltages (v_{dc1} - dc side and v_{dc2} - ac side) and output ac voltage.

High-frequency transformer voltage and current waveforms for two different points of the ac line cycle are shown in Fig. 13. From the secondary voltage waveform, both amplitude modulation (due to variation of v_{ac}) and pulse-width modulation (due to SPWM control) can be observed. The dc side filter inductor current i_{Lf2} and the synthesized voltage v_{ab} are shown in Fig. 14. Primary and secondary devices can be seen to be undergoing ZVS for the particular switching period shown in Fig. 15.

978-1-4673-9551-9/16 $31.00 © 2016 IEEE

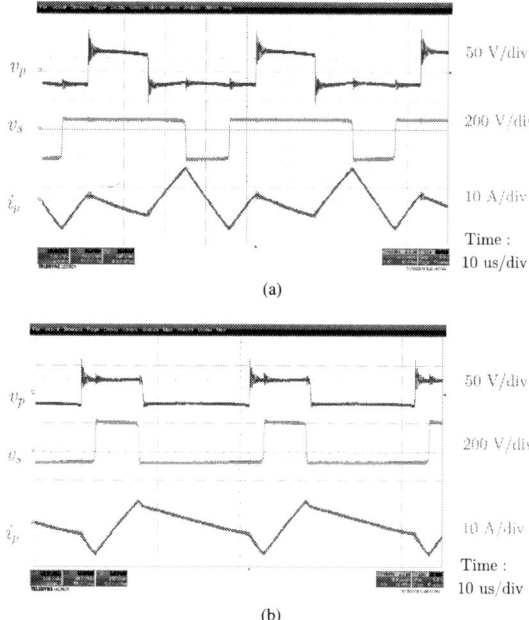

(a)

(b)

Fig. 13: Transformer voltage and current waveforms at two different portions of the line cycle.

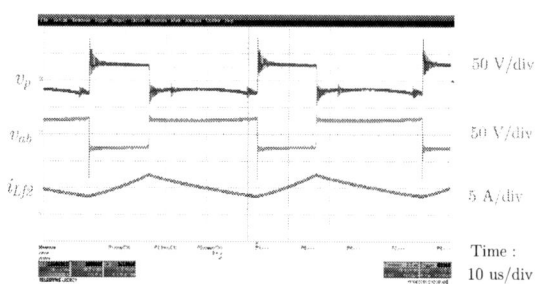

Fig. 14: Transformer primary voltage (v_p), voltage synthesized by primary full-bridge (v_{ab}) and current through filter inductor L_{f2}.

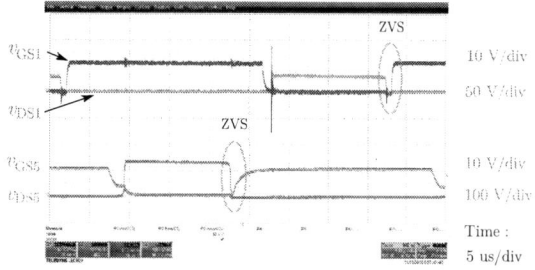

Fig. 15: Illustrating ZVS turn-on of devices (S_1 - primary side, S_5 - secondary side).

VI. CONCLUSION

In this paper, a novel architecture of a single-stage, isolated dual active bridge-based DC-AC converter has been proposed. The topology uses only eight devices, which is the least compared to existing single-stage or two-stage DAB-based solutions or other non-DAB but dual-bridge-based solutions. ZVS operation of devices and the single-stage power processing result in low overall losses. Also, by varying the phase-shift, it is possible to achieve power decoupling exclusively on the ac-side module resulting in low decoupling capacitance requirement. Integrated energy storage is also possible, which makes the proposed converter a multiport converter. All these advantages and the possibility of modular operation and inter-facing different types of sources, make the proposed solution an excellent candidate for a wide variety of applications.

REFERENCES

[1] D. R. Nayanasiri, D. M. Vilathgamuwa and D. L. Maskell, "Half-Wave Cycloconverter-Based Photovoltaic Microinverter Topology With Phase-Shift Power Modulation," *IEEE Trans. Power Electronics*, 28(6), pp. 2700-2710, Nov. 2013.

[2] H. Krishnaswami, "Photovoltaic Microinverter using Single-stage Iso-lated High-frequency link Series Resonant Topology," *IEEE Energy Conversion and Congress Exposition (ECCE) 2011*, pp. 495-500, Sep. 2011.

[3] A. Trubitsyn, J. Pierquet, A. K. Hayman, G. E. Gamache, C. R. Sullivan and D. J. Perreault, "High-Efficiency Inverter for Photovoltaic Applica-tions," *IEEE Energy Conversion and Congress Exposition (ECCE) 2010*, pp. 2803-2810, Sep. 2010.

[4] M. H. Kheraluwala, R.W. Gascoigne, D.M. Divan and E. D. Baumann, "Performance characterization of a high-power dual active bridge dc-to-dc converter," *IEEE Trans. Ind. Applications*, 28(6), pp. 1294-1301, Nov./Dec. 1992.

[5] H. Quin and J. W. Kimball, "Closed-Loop Control of DC-DC Dual-Active-Bridge Converters Driving Single-Phase Inverters," *IEEE Trans. Power Electronics*, 29(2), pp. 1006-1017, Feb. 2014.

[6] Soeren Baekhoej Kjaer, John K. Pedersen and Frede Blaabjerg, "A Review of Single-Phase Grid-Connected Inverters for Photovoltaic Modules," *IEEE Trans. Ind. Applications*, vol 41(5), pp. 1292-1306, September-October 2005.

[7] M. Chen, K. K. Afridi and D. J. Perreault , "A Multilevel Energy Buffer and Voltage Modulator for Grid-Interfaced Micro-Inverters," *IEEE Energy Conversion and Congress Exposition (ECCE) 2012*, pp. 3070-3080, Sep. 2012.

[8] S. Poshtkouhi, M. Fard, H. Hussein, L. M. Dos Santos, O. Trescases, M. Varlan and T. Lipan, "A dual-active-bridge based bi-directional micro-inverter with integrated short-term Li-Ion ultra-capacitor storage and active power smoothing for modular PV systems," *Applied Power Electronics Conference and Exposition (APEC) 2014*, pp.643-649, March 2014.

[9] V. R. Chowdhury, M. A. Chaaban, S. Essakiappan, M. Manjrekar and Y. Xue, "A grid connected PV micro-inverter with optimized battery storage utilization," *IEEE Energy Conversion and Congress Exposition (ECCE) 2015*, pp.3455-3461, Sep. 2015.

[10] G. Ortiz, L. Fassler, J. W. Kolar and O. Apeldoorn, "Flux Balancing of Isolation Transformers and Application of "The Magnetic Ear" for Closed-Loop Volt-Second Compensation," *IEEE Trans. Power Electron-ics*, 29(8), pp. 4078-4090, March 2014.

[11] R. W. Erickson and D. Maksimovic, " Fundamentals of Power Elec-tronics," (2nd ed) Springer International Edition.

Ultra-low Ripple Inverters for Distributed Generation Applications

Ang Shen (shena@mst.edu), Pourya Shamsi (shamsip@mst.edu), Mehdi Ferdowsi (ferdowsi@mst.edu)
Department of Electrical and Computer Engineering
Missouri University of Science and Technology
Rolla, Missouri 65409 USA

Abstract—This paper investigates a Zero First-order Ripple (ZFR) inverter topology for integration with distributed energy resources. The proposed ZFR topology is free from first order triangular ripples and will demonstrate minimal sinusoidal ripples. Therefore, ZFR inverters can offer low harmonic distortions utilizing small coupled inductors. Additionally, elimination of electrolyte capacitors in ZFR topologies improves the reliability and life expectancy of the overall system. After the introduction of the ZFR topology, the proposed inverter is modeled and the steady state values of state variables are derived. Lastly, experimental results are provided to evaluate the proposed inverter.

I. INTRODUCTION

Massive integration of distributed energy resources has resulted in a demand for miniaturized inverter topologies with minimal requirements for filtering elements. Switching converters rely on the switching between multiple modes of operation to generate a desired output. The majority of power electronic converters utilize semiconductor switches to generate a voltage pattern to be filtered by inductors and capacitors. Although this conventional method of designing switching converters is simple and requires a low number of components, it suffers from large ripples. To reduce the ripples, bulky inductors and capacitors are required. This problem suggests utilization of ripple-canceling techniques to cancel ripples instead of filtering them.

Another challenge with conventional switching converters is the utilization of electrolyte capacitors. These capacitors are the first cause of failure in switching converters. An average life span of an industrial grade electrolyte capacitor is 5 to 7 years. However, the remaining circuit components including the semiconductor switches offer more than 30 years of life expectancy. Hence, by using electrolyte capacitors in a solar energy system, the expected life span of the system is reduced to less than 10 years which can jeopardize the return-on-investment analysis. Hence, inverter topologies with minimal filtering capacitors are of interest.

Ultra low ripple converters which are often known as zero ripple converters utilize coupled inductors to compensate for the switching ripples [1]. Using this technique, such converters offer Zero First-order Ripples (ZFR). In some ZFR topologies, only current ports are tuned for canceling ripples [2], [3]. However, using the same coupled inductor structure, both current and voltage ripples can be canceled [4]. [5]–[7] have

investigated applications of ripple cancellation for non-isolated and isolated buck topologies. [8] has investigated ZFR boost topologies with their corresponding damping circuitry. Also, ZFR boost has been integrated with power factor correctors (PFC) in [9]–[11]. Recently, a multi-port ZFR Ćuk for photovoltaic energy harvesting was introduced in [12], [13], and a ZFR asymmetric bridge for motor drive applications was investigated in [14]. The damping circuitry is a critical design aspect of high-order converters which has been studied in [15]–[17]. Although there are various research on dc-dc ZFR converters, no significant research has explored applications of ZFR topologies in dc-ac applications.

In this paper, ZFR techniques have been extended to dc-ac converters for developing an ultra low ripple inverter with a miniaturized footprint. First, a ZFR inverter topology is introduced. Next, a model of the proposed ZFR inverter is developed. Lastly, experimental results are provided to evaluate the proposed ZFR inverter.

II. ZFR INVERTERS

In this section, single and three phase ZFR inverters are introduced. ZFR inverter is designed by extending the traditional inverter leg topology using voltage and current ripple cancellation tanks. In this methods, a coupled inductor is introduced to create voltage and current ripples with similar amplitudes and opposite signs to cancel the ripples induced by the switching function. The schematic of the proposed ZFR inverter is shown in Figure 1. Unlike traditional inverters, the proposed ZFR inverter does not have a filtering inductor on the phase outputs and the main inductors are the magnetizing inductance of the phase transformers (i.e. L_{3a}, L_{3b}, and L_{3c}) which are drawn using a dashed line. In practice, these inductors are formed within the transformers and no external inductors are required.

Theoretically, this converter does not have any input/output current/voltage ripples. However, in a practical implementation, exact matching of inductive elements cannot be achieved. Hence, C_{4a}, C_{4b}, and C_{4c} are added to filter the high frequency ripples on the phase outputs. Selection of these capacitors depends on the accuracy of the design of the coupled inductor. We suggest a value in the range of 0.1 to 1 μF. The dc bus capacitor C_1 has a similar situation. Ideally, this capacitor is not needed and is included to compensate for the non-ideal coupling of T_{p_1} and T_{p_3} where p is an arbitrary phase.

978-1-4673-9551-9/16 $31.00 © 2016 IEEE

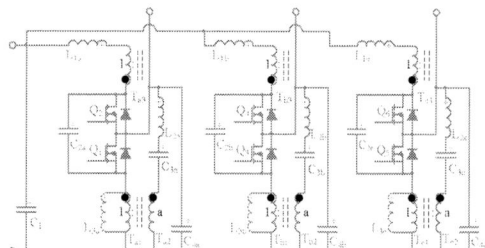

Fig. 1. The proposed 3-phase ZFR inverter topology.

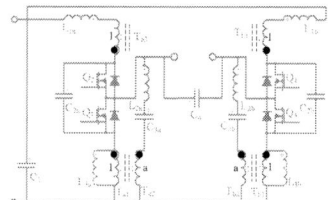

Fig. 2. Single phase ZFR inverter with two ZFR legs.

A ZFR single phase converter has a similar structure. Instead of three ZFR legs, one can utilize two to form a single phase converter as shown in Figure 2. In another approach, one can utilize one ZFR leg and one standard leg to eliminate the extra components as shown in Figure 3. In this case, the PWM should be applied to the ZFR leg while the standard leg is only for switching between the positive and negative half-waves. In a single phase ZFR inverter, C_1 cannot be eliminated as this capacitor will provide the single phase 120Hz power ripple.

To study the behavior of the proposed ZFR converter, the two modes of operation for the phase a of this converter are illustrated in Figures 4a and 4b. Benefits of a ZFR converter vanish during discontinuous conduction modes and hence, such conditions are not studied here.

In mode boost A as shown in Figure 4a, Q_1 is active and power flows from the output to charge the magnetizing inductor of L_{3a}. During this mode, magnetizing current of L_{3a} is increasing. To compensate for the rise in the input current, the coupling of $T_{a_1} : T_{a_2}$ is utilized to sample the voltage of L_{3a} and create a voltage of $av_{L_{3a}}$ across T_{a_2}.

In the steady state operation of the converter, C_{3a} is charge

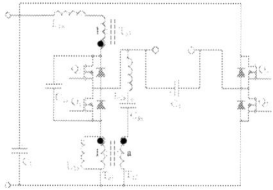

Fig. 3. Single phase ZFR inverter with one ZFR leg.

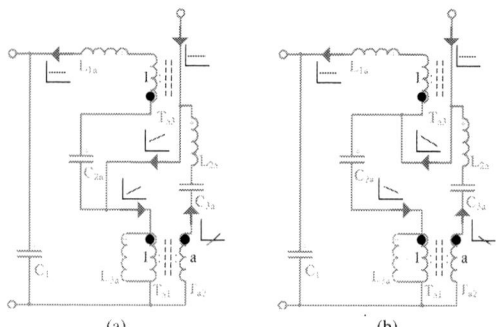

Fig. 4. Switching modes of the converter under CCM with the phase current flowing into the converter, (a) switching mode A, (b) switching mode B.

to the momentarily phase a voltage (i.e. $v_a(t) = v_{C_{3a}}(t)$). To achieve this, a sufficiently small value of C_{3a} is required. On the other hand, if C_{3a} is too small, it will not maintain a fixed voltage during one switching cycle. Selection of C_{3a} depends on the output ac voltage, rated power as well as the switching frequency. To design a 1kW ZFR leg, we have selected a 3.9 μF capacitor.

Considering a fixed voltage for C_{3a} during one switching period, the induced voltage by the transformer will directly result in a current rise within T_{a_2} as $L_{2a}di_{L_{2a}}/dt = v_a - av_a - v_{C_{3a}} \simeq -av_a$. Comparing this equation with $L_{3a}di_{L_{3a}}/dt = v_a$ and $i_{T_{a_1}} = -ai_{L_{2a}}$, we derive that if $L_{2a} = a(1-a)L_{3a}$, the rise in the current of L_{3a} is canceled by the ripple of L_{2a}. Hence, the effective input ripple is almost zero (validity of these equations is with the assumption that $dv_{C_{3a}}/dt \simeq 0$). Based on this equation, there is a limit in selection of a as $0 < a < 1$. In practice, $a = 0.25$ is a good performing selection. Higher values of a result in a larger voltage on L_{2a} and demands a larger inductor. Selection of L_{3a} is achieved by setting a maximum ripple limit on $i_{L_{3a}}$. This limit defines the range of continuous conduction mode for the proposed ZFR inverter.

On the dc side, C_{2a} is being discharged into the dc bus capacitor C_1 at a constant rate of $i_{dc} = i_{L_{1a}}$. In the steady state operation of the converter C_{2a} has the same voltage as the dc bus capacitor. Hence, when the Q_1 is active, variations of the current of L_{1a} is defined by $L_{1a}di_{L_{1a}}/dt = v_a + v_{C_{2a}} - v_{T_{a_3}} - v_{dc}$. The transformer ratio of $1 : 1$ between T_{a_3} and T_{a_1} induces the input voltage to T_{a_3} which cancels the effect of the input voltage. Also, $v_{C_{2a}} \simeq v_{dc}$ by neglecting the ripples on the C_{2a}. To achieve this, we have used a 10μF capacitor. Therefore, the current of L_{1a} is almost constant. Hence, by properly designing the converter with respect to the switching period, the claim of *almost zero ripple* is valid. In conclusion, the larger C_{2x} and C_{3x} are selected, a better zero ripple behavior is achieved but at a higher cost of manufacturing and larger real estate requirements.

In the second mode of operation, Q_2 is active as is illustrated in Figure 4b. In this mode $v_{L_{3a}} = v_a - v_{C_{2a}} < 0$. This

978-1-4673-9551-9/16 $31.00 © 2016 IEEE

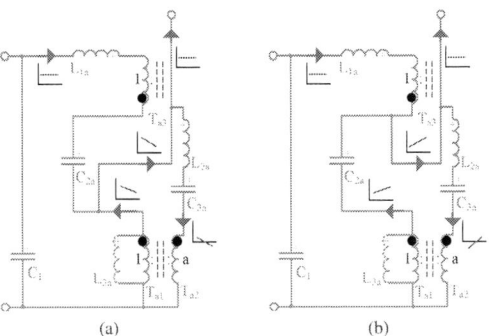

Fig. 5. Switching modes of the converter under CCM with the phase current flowing outwards, (a) switching mode A, (b) switching mode B.

Fig. 6. Experimental setup of three ZFR inverter legs.

will discharge the ripple induced in the magnetizing inductor during the first mode. The equations proving the almost zero ripples are still valid and hence, the converter will operate with a negligible input or output current ripples.

Figures 4a and 4b illustrated the operation modes of the ZFR inverter leg in a boosting current pattern (i.e. current flowing from phase a to the ZFR inverter leg). The bucking mode where the current flows from the dc bus to phase a has a similar analysis. The two modes for the buck ZFR inverter leg are illustrated in Figures 5a and 5b.

III. MODELING

First, the average model of one inverter leg is derived for a fixed duty cycle of $d_a = (1 - d'_a)$ where d_a is the duty cycle of Q_2. Assuming that C_1 is a voltage source (i.e. $v_{C_{1a}} = v_{dc}$, we have

$$C_{2a}\dot{v}_{C_{2a}} = -d'_a i_{L_{1a}} - d_a i_{T_{a1}} + d_a i_{L_{3a}} \tag{1a}$$

$$C_{3a}\dot{v}_{C_{3a}} = i_{L_{2a}} \tag{1b}$$

$$C_{4a}\dot{v}_{C_{4a}} = i_{T_{a1}} - (i_{L_{1a}} + i_{L_{2a}} + i_{L_{3a}}) - i_a \tag{1c}$$

$$L_{1a}\dot{i}_{L_{1a}} = v_{C_{2a}} - v_{dc} \tag{1d}$$

$$L_{2a}\dot{i}_{L_{2a}} = (1 - a)v_{C_{4a}} - v_{C_{3a}} + ad_a v_{C_{2a}} \tag{1e}$$

$$L_{3a}\dot{i}_{L_{3a}} = v_{C_{4a}} - d_a v_{C_{2a}} \tag{1f}$$

where i_a is the inward phase current. If the transformer was ideal, $i_{T_{a1}}$ would be equal to $i_{L_{1a}} + a i_{L_{2a}}$. However, a practical transformer depends on th variations of flux, $d\phi/dt$, which is zero for dc frequencies. Therefore, no practical transformer can have a coupling of 1:1 for dc frequencies. Hence, one should implement a practical transformer as a dynamical system with a transfer function of $s/(s + 2\pi f_p)$ where f_p defines the pass band of the transformer. In this method, we will introduce a zero at the dc frequency and a pole at $s = -2\pi f_p$ to cancel this zero for higher frequencies. This approach will model a practical transformer that does not pass any dc signals but can pass ac signals with a gain 1 (assuming that the frequency of interest, f, is far higher than the location

of the pole $10 f_p \leq f$). To achieve this, a dummy state variable of $i_{T_{a1}}$ is introduced as

$$\dot{i}_{T_{a1}} = -2\pi f_p i_{T_{a1}} + (v_{C_{2a}} - v_{dc})/L_{1a}$$
$$+ a((1 - a)v_{C_{4a}} - v_{C_{3a}} + ad_a v_{C_{2a}})/L_{2a} \tag{2}$$

which can emulate the behavior of a practical transformer. Using this model, the dc components of the average model of the system for a duty cycle of d_a can be calculated as

$$v_{C_{2a}} = v_{dc} \tag{3a}$$

$$v_{C_{3a}} = d_a v_{dc} \tag{3b}$$

$$v_{C_{4a}} = d_a v_{dc} \tag{3c}$$

$$i_{L_{1a}} = i_a \tag{3d}$$

$$i_{L_{2a}} = 0 \tag{3e}$$

$$i_{L_{3a}} = i_a(1 - d_a)/(d_a) \tag{3f}$$

which suggests that the converter should not get close to $d_a = 0$. In the normal sinusoidal applications where $d_a = 0.5 + \kappa \cos(\omega t)$, there should be no problems as long as $\kappa < 0.5$. However, if one is interested in using zero vectors to get additional benefits such as 33% reduction in the switching losses by maintaining one phase in on or off states for one third of each cycle, then large magnetizing currents will be observed in L_{3a}. Hence, those methods should not be applied for a ZFR converter. A ZFR converter has to maintain a $d_a > 0$.

Based on this model, we observe that the ZFR converter acts similar to a standard inverter leg. Hence, $v_a/v_{dc} = d_a$. Therefore, as long as $d_a > 0$, all of the existing Pulse Width Modulation (PWM) methods available for traditional inverters can be applied to a ZFR inverter.

IV. EXPERIMENTAL RESULTS

The experimental setup consists of a three-phase ZFR inverter constructed using three individual ZFR inverter legs. Our designed ZFR inverter has a switching frequency of 200kHz. Figure 6 illustrates the proposed ZFR inverter leg design with a rating of 1kW. Figure 7 illustrates the voltage waveforms of the converter at a fixed dc bus voltage of 100V and a fixed duty

978-1-4673-9551-9/16 $31.00 © 2016 IEEE

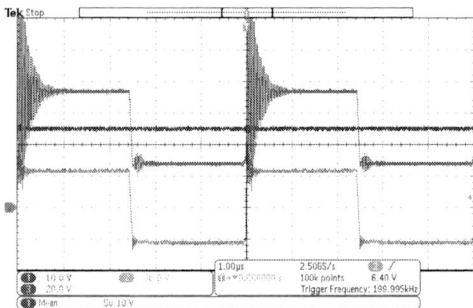

Fig. 7. Measured signals under a fixed duty cycle operation.

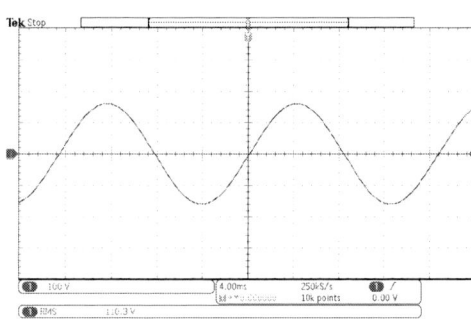

(a)

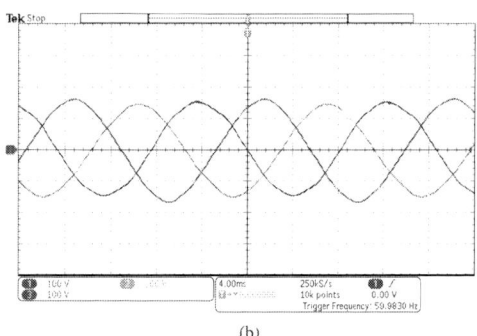

(b)

Fig. 8. Output voltage waveform (a) single phase, (b) three phase.

cycle of 50%. In this scenario, the phase voltage is shown with the blue line (probe 1), the drain voltage of Q_2 is shown as pink (probe 3), and the source voltage of Q_1 is shown as cyan (probe 2). It can be observed that the phase voltage has no ripples. However, the internal switching components observe large variations of voltage. In this figure, one can notice that the switching function induces a train of pulses on L_{3a} which is the source of Q_1. From this figure, it can be observed that the mean of this voltage is zero which was expected. The positive amplitude is $v_{C_{4a}} = v_a$ and the negative amplitude is $v_{C_{4a}} - v_{C_{2a}} = v_{dc} - v_a$. On the drain of Q_2, we have T_{a_3}. The voltage on this pin is $v_{C_{2a}} + v_{T_{a_1}}$ which is then added by $-v_{T_{a_3}}$ to form the voltage across L_{1a}. If a coupling of 1:1 is achieve between these two windings then $v_{L_{1a}} = 0$

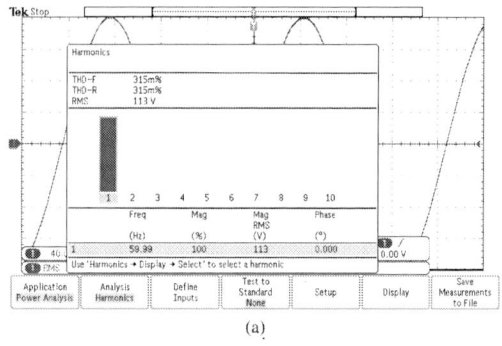

(a)

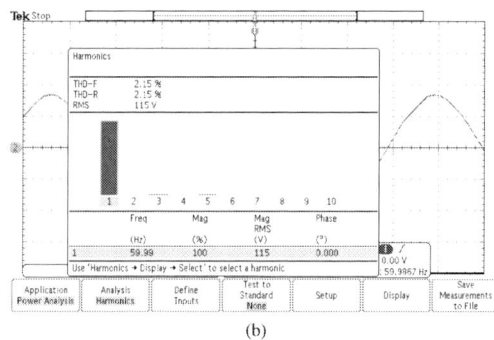

(b)

Fig. 9. Harmonic distortion analysis (a) voltage THD of a single phase ZFR inverter, (b) THD for a three phase case.

as $v_{C_{2a}} \simeq v_{dc}$ which results into almost zero input current ripples. In Figure 7, the ringings on probe 3 are not occurring in the actual circuit. These ringings are due to the oscillations caused by our differential voltage probe and high dv/dt of the circuit.

In the next scenario, sinusoidal duty cycles are applied to a single phase ZFR inverter and the load output voltage is illustrated in Figure 8a. For this example, the duty cycles are generated using a simple sinusoidal PWM as $d = 0.5 \pm \kappa cos(\omega t)$ for legs a and b. It can be observed that the output has no visible ripples. Similarly, Figure 8b illustrates the output waveform for a three phase ZFR inverter. It can be observed that the Total Harmonic Distortion (THD) is visibly higher compared to Figure 8a. One reason for this effect is the utilization of a vector based PWM generation as $d_a = 0.5 + \sqrt{2/3}\kappa \cos(\omega t)$, $d_b = 0.5 - \sqrt{1/6}\kappa \cos(\omega t) - \sqrt{1/2}\kappa \sin(\omega t)$, and $d_c = 0.5 - \sqrt{1/6}\kappa \cos(\omega t) + \sqrt{1/2}\kappa \sin(\omega t)$ which can increase the quantization error in an embedded digital implantation considering the fact that the sinusoidal functions were calculated using Taylor expansion. In addition, the three phase ZFR inverter requires switches with higher breaking voltage. Hence, the switches used for the three phase converter are different from the single phase case and the dead-time was increased from 50ns in the single phase to 300ns in the three phase inverter. In the three phase system, voltages are measured with respect to the neutral point of the load. In our scenario, three 15Ω resistors where used in a wye connection as the main load. However, the resistors are not exactly balanced as it can be observed from Figure 8b. The THD of the voltage signal is calculated using the power analysis module of a Tektronix oscilloscope as shown in Figure 9a. It can be observed that the THD of the single phase inverter is close to zero. Ideally, if there were no numerical errors in the calculation of sinusoidal functions, dead-times were zero, and the cancellation within the converter were fully achieved, we would get a zero THD. Based on the factors mentioned above, the THD of the three phase inverter is higher. Figure 9b illustrates the THD of the three phase inverter. Harmonics 3 and 5 can be observed on this signal which are in particular magnified due to the phase to neutral connection of the probes and the slight unbalanced loading.

V. CONCLUSIONS

This paper proposed a new inverter topology that offers ultra-low ripples using small coupled inductors. This ZFR topology utilizes a coupled inductor to cancel the current and voltage ripples generated by the switching function. In this paper, first a ZFR inverter was introduced. Then a simple average model of the inverter was developed. Lastly, experimental results were provided to evaluate the proposed topology.

REFERENCES

[1] R. P. Martinelli and C. Ashley, "Coupled inductor boost converter with input and output ripple cancellation," in *Applied Power Electronics Conference and Exposition, 1991. APEC'91. Conference Proceedings, 1991., Sixth Annual.* IEEE, 1991, pp. 567–572.

[2] D. C. Hamill and P. T. Krein, "A zero ripple technique applicable to any dc converter," in *Power Electronics Specialists Conference*, 1999, pp. 1165–1171.

[3] S. R. Hui, S. N. Li, X. H. Tao, W. Chen, and W. Ng, "A novel passive offline led driver with long lifetime," *Power Electronics, IEEE Transactions on*, vol. 25, no. 10, pp. 2665–2672, 2010.

[4] Y. Gu, D. Zhang, and Z. Zhao, "Input/output current ripple cancellation and rhp zero elimination in a boost converter using an integrated magnetic technique," *Power Electronics, IEEE Transactions on*, vol. 30, no. 2, pp. 747–756, 2015.

[5] M. J. Schutten, R. L. Steigerwald, J. Sabaté *et al.*, "Ripple current cancellation circuit," in *Applied Power Electronics Conference and Exposition, 2003. APEC'03. Eighteenth Annual IEEE*, vol. 1. IEEE, 2003, pp. 464–470.

[6] D. Diaz, O. Garcia, J. Oliver, P. Alou, Z. Pavlovic, J. Cobos *et al.*, "The ripple cancellation technique applied to a synchronous buck converter to achieve a very high bandwidth and very high efficiency envelope amplifier," *Power Electronics, IEEE Transactions on*, vol. 29, no. 6, pp. 2892–2902, 2014.

[7] M. Jacobs, "Ripple reduction for switch-mode power conversion," Oct. 26 2010, uS Patent 7,821,799. [Online]. Available: http://www.google.com/patents/US7821799

[8] Z. Lu, H. Chen, Z. Qian, and T. Green, "An improved topology of boost converter with ripple free input current," in *Applied Power Electronics Conference and Exposition, 2000. APEC 2000. Fifteenth Annual IEEE*, vol. 1. IEEE, 2000, pp. 528–532.

[9] D. Garinto, "A new zero-ripple boost converter with separate inductors for power factor correction," in *Power Electronics Specialists Conference, 2007. PESC 2007. IEEE.* IEEE, 2007, pp. 1309–1313.

[10] E. Chou, F. Chen, C. Adragna, and B. Lu, "Ripple steering ac-dc converters to minimize input filter," in *Energy Conversion Congress and Exposition, 2009. ECCE 2009. IEEE*, Sept 2009, pp. 1325–1330.

[11] V. Tarateeraseth and W. Khan-ngern, "Reducing the electromagnetic interference for single-stage single-switch ac/dc power factor correction by ripple-steering technique," in *Power Electronics and Drive Systems, 2003. PEDS 2003. The Fifth International Conference on*, vol. 1. IEEE, 2003, pp. 332–337.

[12] S. Biswas, S. Dhople, and N. Mohan, "Zero-ripple analysis methods for three-port bidirectional integrated magnetic ćuk converters," in *Industrial Electronics Society, IECON 2014-40th Annual Conference of the IEEE.* IEEE, 2014, pp. 1889–1895.

[13] ——, "A three-port bidirectional dc-dc converter with zero-ripple terminal currents for pv/microgrid applications," in *Industrial Electronics Society, IECON 2013-39th Annual Conference of the IEEE.* IEEE, 2013, pp. 340–345.

[14] P. Shamsi, "Near zero-ripple switched reluctance drives," in *IEEE Transportation Electrification Conference.* IEEE, 2016, pp. 1–5.

[15] P. Jia, T. Q. Zheng, and Y. Li, "Parameter design of damping networks for the superbuck converter," *Power Electronics, IEEE Transactions on*, vol. 28, no. 8, pp. 3845–3859, 2013.

[16] M. U. Iftikhar, D. Sadarnac, and C. Karimi, "Input filter damping design for control loop stability of dc-dc converters," in *Industrial Electronics, 2007. ISIE 2007. IEEE International Symposium on.* IEEE, 2007, pp. 353–358.

[17] D. Diaz, D. Meneses, J. Á. Oliver, Ó. García, P. Alou, and J. A. Cobos, "Dynamic analysis of a boost converter with ripple cancellation network by model-reduction techniques," *Power Electronics, IEEE Transactions on*, vol. 24, no. 12, pp. 2769–2775, 2009.

A 15 kV SiC MOSFET Gate Drive with Power over Fiber based Isolated Power Supply and Comprehensive Protection Functions

Xuan Zhang, He Li, John A. Brothers, and Jin Wang
Center for High Performance Power Electronics
The Ohio State University
Columbus, Ohio, USA
zhang.1973@osu.edu

Lixing Fu
Texas Instrument Inc.
Warwick, Rhode Island, USA
l-fu@ti.com

Mico Perales and John Wu
MH GoPower Co., Ltd.
Kaohsiung City, Taiwan
mico.perales@mhgopower.com

Abstract—**This paper presents a 15 kV SiC MOSFET gate drive circuit, which features high common-mode (CM) noise immunity, small size, light weight, and robust yet flexible protection functions. To enhance the gate-drive power reliability, a power over fiber (PoF) based isolated power supply is designed to replace the traditional design based on isolation transformer. It delivers the gate-drive power by laser light via optical fiber over a long distance (>1 m), so a high isolation voltage (>20 kV) is achieved, and the circuit size and weight are reduced. More importantly, it eliminates the parasitic CM capacitance coupling the power stage and control stage, and thus eradicates the control signal distortion caused by high *dv/dt* in switching transients of the high-voltage SiC devices. In addition, the gate drive circuit design integrates comprehensive protection functions, including the over-current protection, under/over-voltage lockout, gate-clamping, soft turn-off, and fault report. The over-current protection responds within 400 ns. Experimental results from a 15 kV double pulse tester are presented to validate the design.**

Keywords— common-mode noise immunity; gate drive; high voltage SiC device; isolated power supply; laser power; over current protection; protection functions; power over fiber

I. INTRODUCTION

In recent years, there has been a significantly growing interest in the high-voltage (HV) Silicon Carbide (SiC) based semiconductor power devices and their applications. To date, 10 kV and 15 kV SiC MOSFETs [1-3], 13 kV p-i-n diode [4], 15 kV SiC IGBTs [5-13], and 16 kV n-channel IGBT [4] have been developed and demonstrated. 10 kV SiC MOSFET power modules have already been applied in power electronics building blocks [14-16]. Applications of these HV SiC power devices include medium voltage drives and circuit breakers, power converters for renewable energy integration, STATCOM, active power filters, FACTS devices, and solid-state transformers [1-18].

The gate drive circuit of the HV SiC power devices is critical to ensure fast and reliable switching operation. It is comprised of four major subsystems: 1) an isolated power supply, 2) an isolated signal transfer circuit, 3) the driving circuit, and 4) protection circuits against faults such as over-current/short-circuit, over/under-voltage, over temperature, and etc.

The isolated power supply of the gate drive circuit must be able to sustain high isolation voltage (>15 kV) and achieve high common-mode (CM) *dv/dt* (>110 kV/µs) rejection [19-20]. In traditional designs, the isolation is implemented with transformers [3], [9], [20-24]. To achieve the high isolation voltage, the transformer requires a large magnetic core to separate the primary and secondary windings and create sufficient creepage distance. This results in a large and heavy gate drive circuit. Moreover, there is inevitable CM parasitic capacitance coupling the windings of the transformer, which makes the gate drive circuit sensitive to high CM *dv/dt*, causing distortion in the gate-drive input signals, cross talk between gate drives, and malfunction in upstream control circuits [19-20]. As an alternative solution for the isolated gate-drive power supply, wireless power transfer using separated coils has been studied in [25-27]. However, this solution is also sensitive to EMI, and it requires large space for isolation and shielding.

To achieve reliable yet compact design of the gate drive circuit, in this paper a power over fiber (PoF) based isolated power supply is proposed. It delivers sufficient gate-drive power by laser light via optical fiber over a long distance (>1 m), so a >20-kV isolation voltage is achieved. It eliminates the parasitic CM capacitance coupling the power stage and control stage, and achieves significant size and weight reduction of the gate drive circuit. Although the laser emitter, which can be placed far away from the main power stage, may offset the overall system volume reduction, the main power stage layout can be optimized with smaller gate drive circuits.

To ensure safe operation, the gate drive circuit must integrate fast and reliable protection functions against faults such as over-current/short-circuit. Compared to Si devices, SiC devices tend to have lower short circuit withstand capability, and the high *dv/dt* and *di/dt* noises also make fault detection more challenging [28-30]. Although over-current protection schemes for 1.2 kV SiC MOSFETs have been discussed in

978-1-4673-9551-9/16 $31.00 © 2016 IEEE

[30], the over-current protection for the HV SiC devices is more challenging because of much higher isolation voltage and higher dv/dt [1], [10], [19]. This paper presents the design of comprehensive protection functions for the 15 kV SiC MOSFET, including over-current protection, gate-drive power under/over-voltage lockout, gate-clamping, and soft turn-off.

II. DESIGN OF THE ISOLATED GATE-DRIVE POWER SUPPLY

A. Power over Fiber based Laser Receiver

Power over fiber (PoF) is a technology in which a fiber optic cable carries optical power, which is used as an energy source rather than, or as well as, carrying data [31-32]. This allows a device to be remotely powered, while providing electrical isolation between the device and the power supply. *MH GoPower Co., Ltd.* offers a PoF laser receiver YCH-0.5-6V, which is shown in Fig. 1. The major specifications of this laser receiver are as follows:

- Optimized for 915 nm, 940 nm, or 975 nm laser sources.
- Conversion efficiency at 1 W input: Typical ~24%.
- Max electrical output power: ~0.5 W.

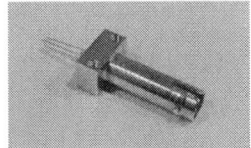

Fig. 1.   The 0.5 W PoF laser receiver YCH-0.5-6V from *MH GoPower Co., Ltd.*

Fig. 2 shows the output I-V curve and P-V curve of the PoF laser receiver YCH-0.5-6V at 1 W input optical power.

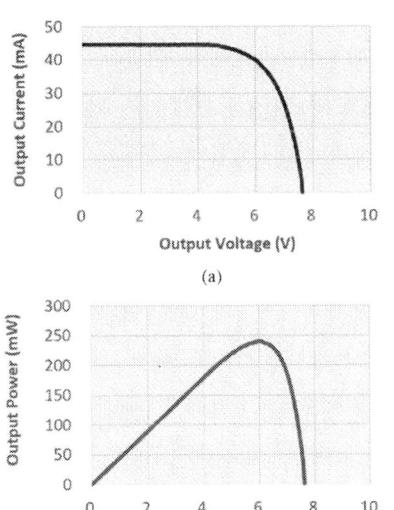

Fig. 2.   The output (a) I-V curve and (b) P-V curve of the 0.5 W PoF laser receiver YCH-0.5-6V at 1 W input optical power.

The maximum electrical power supplied by this PoF laser receiver is 0.5 W. It is low but sufficient for driving the 15 kV, 10 A SiC MOSFET switching at 10 kHz, given the measured device input capacitance is around 4.24 nF.

B. Transformer-based and PoF-based Isolated Gate-drive Power Supply

To demonstrate the CM noise immunity improvement and circuit size reduction of the PoF-based isolated gate-drive power supply, two designs of the isolated gate-drive power supply are presented, including a traditional transformer-based design and a proposed PoF-based design. The schematics of the two designs are shown in Fig. 3.

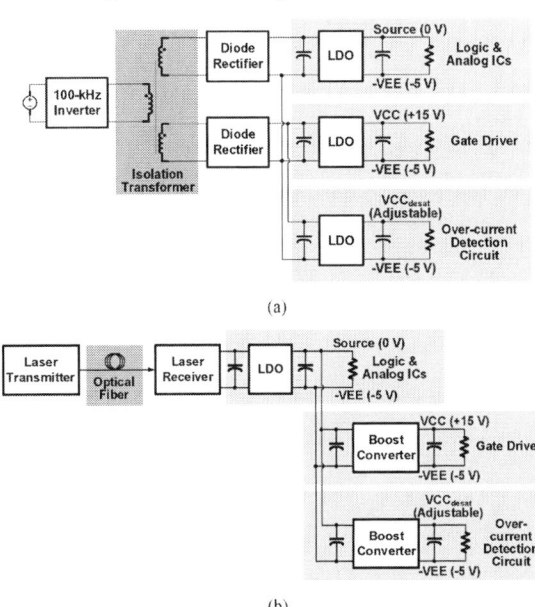

Fig. 3.   Schematics of the two designs of the isolated gate-drive power supply, including (a) a transformer-based design, and (b) a PoF-based design.

It can be seen from Fig. 2 that the laser receiver is indeed a photovoltaic cell, which features a maximum power point in operation. Thus, there exists a minimum input laser irradiance that matches the maximum output power of the laser receiver with the required gate-drive power, and the operation at such a sweet spot minimizes the gate-drive power loss. However, power converters built upon the 15 kV SiC devices are typically operated at 5~10 kHz because of the device thermal limitation [1-16], so the gate-drive power loss is negligible compared to the main power stage loss. Therefore, for simplification of the design, the matching between the input laser light irradiance and required gate-drive power is not implemented.

Based on the two different designs of the isolated gate-drive power supply presented in Fig. 3, two prototypes of the gate drive circuit for the 15 kV SiC MOSFET are built and presented side by side for comparison, as shown in Fig. 4.

978-1-4673-9551-9/16 $31.00 © 2016 IEEE

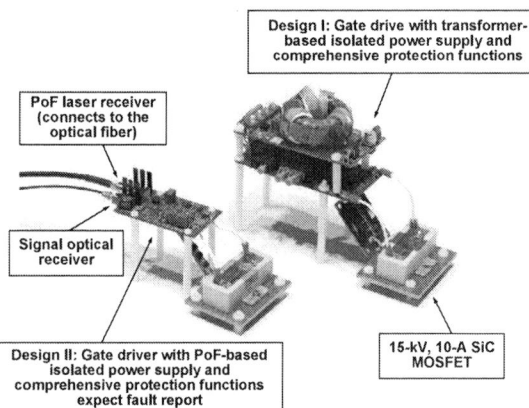

Fig. 4. Two prototypes of the gate drive circuit for the 15 kV SiC MOSFET. They are with two different designs of the isolated gate-drive power supply, including a transformer-based design and a PoF-based design.

The transformer-based isolated gate-drive power supply suffers inevitable parasitic CM capacitance coupling the isolated transformer windings. The measured results of this capacitance are shown in Tab. I. In the PoF-based design, this parasitic CM capacitance is eliminated, as a result of the long-distance (>1 m) laser transmission via fiber optics.

TABLE I. MEASURED PARASITIC CM CAPACITANCE COUPLING THE ISOLATED TRANSFORMER WINDINGS

Frequency	1 kHz	10 kHz	50 kHz	100 kHz
CM capacitance	3.5 pF	3.47 pF	3.48 pF	3.49 pF

Compared to the transformer-based isolated power supply, the PoF-based design has the following advantages:

- Higher insulation voltage (>20 kV).
- Reduced size and weight.
- Higher CM transient rejection (>200 kV/μs) because of the elimination of parasitic CM capacitance.

III. DESIGN OF COMPREHENSIVE PROTECTION FUNCTIONS

The gate drive design integrates comprehensive protection functions, including the over-current protection, under/over-voltage lockout, gate-clamping, soft turn-off, fault report, and protection mode selection to allow either single or multiple faults. Fig. 5 shows the schematic for the protection circuits, where the under/over-voltage lockout function is not included as it is implemented with voltage monitoring ICs.

A. Over-current Detection

For the over-current protection design, the "desaturation detection" method used for IGBTs-based power circuits is employed. This method is generally applicable for HV MOSFETs given that their output characteristics are similar to IGBTs in the active region. It is also already demoed in the over-current protection for 1.2 kV SiC MOSFETs [30]. For the 15 kV SiC MOSFETs, the challenges to implement this method have been the high voltage bias and high dv/dt on the device drain terminal.

To sense the drain-to-source voltage drop across the switch in the ON state and block the high drain voltage bias in the OFF state, five 3.3 kV SiC Schottky diodes (D_{ss1}-D_{ss5}) are connected in series to the switch's drain terminal. This is not only to achieve sufficient reverse-bias blocking voltage (>15 kV), but also to greatly reduce the equivalent sensing diode junction capacitance. In the presence of high dv/dt at the switch's drain terminal, lower junction capacitance of the sensing diodes causes less displacement current flowing from the switch drain terminal to the sensing analog circuits, and thus reduces the risk of false trigger of the protection.

However, the series connection of the five sensing diodes causes an equivalent diode forward voltage five times higher, which results in a large offset in the sensed voltage and thus lower the sensing resolution. In the design, a high-speed, 5 V analog comparator is chosen to achieve fast response time. To match the comparator's input voltage range, the sensed voltage is scaled down with a resistive voltage divider (R_{div1} and R_{div2}). Also, the voltage level (VCC_{desat}) of the PWM buffer output needs to be adjusted accordingly.

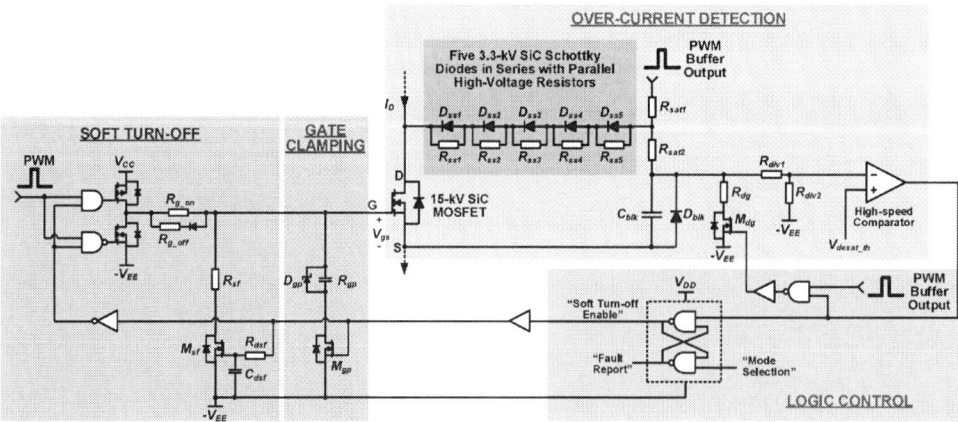

Fig. 5. Schematics of the protection functions including over-current protection, gate-clamping, soft turn-off, fault report, and protection mode selection to allow either single or multiple faults.

978-1-4673-9551-9/16 $31.00 © 2016 IEEE

Another problem of the series connection of the sensing diodes is the non-uniform voltage distribution across those diodes when reverse-biased, which would cause breakdown failure of those diodes. To solve this problem, HV rated resistors (R_{ss1}-R_{ss5}) are added in parallel with each sensing diode to balance the voltage distribution.

The over-current protection for the 15 kV SiC MOSFET is currently designed to be triggered in the device triode region rather than the saturation region. This is because the behaviors of 15 kV SiC MOSFETs in the saturation region are not yet experimentally investigated due to the limited testing samples. However, the over-current protection in the device triode region allows the triggering at a much lower device current, which enables a safer device operation.

B. Over-current Protection Response Time

The response time of the over-current protection circuit relies heavily on the blanking time delay, which is required to bypass the device turn-on transient current to avoid false triggering. As shown in [1] and also in the section V of this paper, the normal turn-on transient current of the 15 kV SiC MOSFET could surge to a very high value during the blanking time. The blanking time is set by the blanking capacitance (C_{blk}) and oscillation damping resistor (R_{sat1}). The selection of these parameters are discussed in [30]. Another resistor R_{sat2} increases the blanking time when the over-current protection is triggered when the sensing diodes D_{ss1}-D_{ss5} are reverse biased. To prevent false triggering of the protection during the device turn-off transient, an auxiliary switch M_{dg} and a current limiting resistor R_{dg} discharges the C_{blk} after the device is turned off. The designed over-current protection total response time under hard-switching fault (HSL) is presented in Tab. II.

TABLE II.   OVER-CURRENT FAULT RESPONSE TIME

Detection Delay	Comparator Delay	Logic Control Delay	Total Delay
330 ns	4.5 ns	45 ns	379.5 ns

IV.   DESIGN OF A 15 KV DOUBLE PULSE TESTER

A 15 kV double pulse tester is built to verify the designed gate drive circuit for the 15 kV SiC MOSFET. The schematic of the double pulse tester is shown in Fig. 6, and the prototype and setup are shown in Fig. 7.

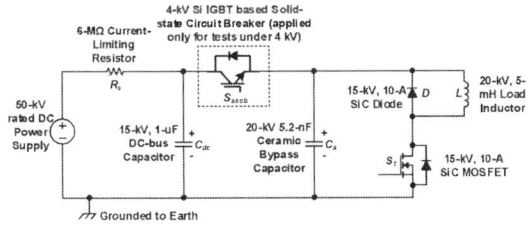

Fig. 6.   Schematic of the 15 kV double pulse tester.

The gate drive circuit over-current protection is debugged in the tests with dc-bus voltage applied under 4 kV. During the

debugging test, a designed 4 kV Si IGBT based solid-state circuit breaker is applied in the test connections for additional short-circuit protection, as shown in Fig. 6.

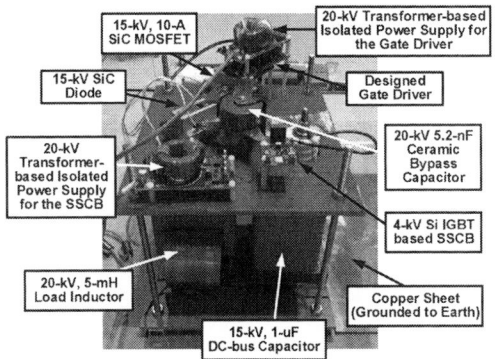

Fig. 7.   Prototype of the 15 kV double pulse tester.

The design of the 20-kV, 5-mH inductor is presented in [21]. To enhance the insulation capability, the inductor winding cables are upgraded to the 20 kV rated cable ETN 2022 from *Wiremax*. This cable is also used to make connections between the power stage components of the 15 kV double pulse tester.

V.   EXPERIMENTAL VERIFICATIONS

A.   PoF-based Gate Drive Circuit Test

To verify the driving power capability, the designed PoF-based gate drive circuit is tested to drive a 12 nF load capacitor, at a 0.5 duty ratio and a 10 kHz switching frequency. The experimental waveforms are shown in Fig. 8. The 12 nF load capacitance is 3 times larger than the input capacitance (4.24 nF) of the 15 kV SiC MOSFET, so it is verified that the designed PoF-based gate drive circuit can supply sufficient driving power.

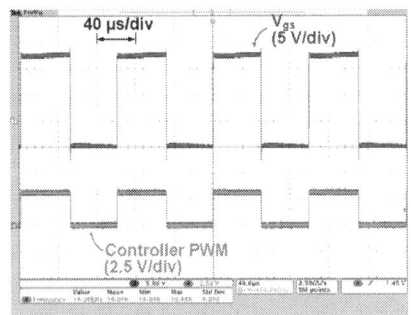

Fig. 8.   The experimental results of the designed PoF-based gate drive circuit driving a 12 nF load capacitor, at a 0.5 duty ratio and a 10 kHz switching frequency.

B. Protection Functions Test

To verify the designed comprehensive protection functions, a single pulse test of the 15 kV SiC MOSFET is conducted on the designed 15 kV double pulse tester. The test setup is shown in Fig. 6 and Fig. 7. Before the test, the insulation between critical electrical points of the setup is verified, with 8 kV dc voltage applied for 10 minutes. Insulation tests at higher voltage will be continued in future work. In the single pulse test, the turn-on gate resistance is 66 Ω, and the turn-off gate resistance is 10 Ω. Tab. III shows the measurement equipment.

TABLE III.   MEASUREMENT EQUIPMENT SPECIFICATIONS

Equipment	P. N.	Bandwidth	Measurement
Oscilloscope	DPO4054B	500 MHz	
20 kV single-ended probe	P6015A	75 MHz	V_{ds}
300 V passive probe	TPP1000	1 GHz	V_{gs}
30 A current probe	TCP0030A	120 MHz	I_d

Fig. 9 shows the experimental waveforms before and after triggering the over-current protection, which is set at 9.6 A. It can be seen that the designed over-current protection, gate clamping, and soft turn-off are all verified. The device drain current (I_d) after turn-off shows a tail which damps out slowly. Similar test result can be seen in [1]. This could be caused by the resonance between the load inductor and the device output capacitance, which needs further study to confirm.

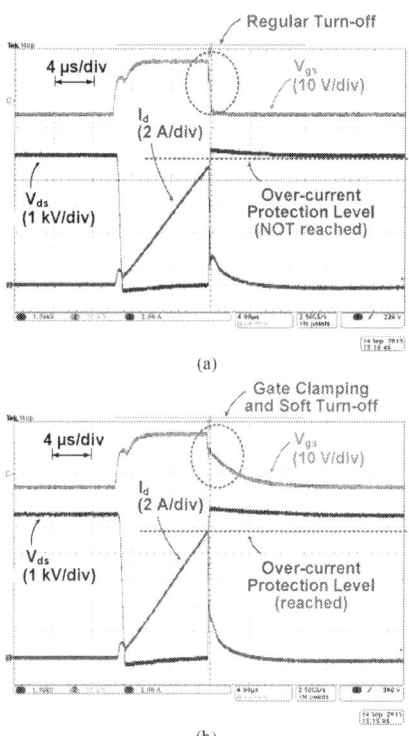

(a)

(b)

Fig. 9.   The experimental results (a) before and (b) after triggering the designed protection functions including the over-current protection, gate clamping, and soft turn-off.

C. Double Pulse Test of the 15 kV SiC MOSFET

After the protection functions are verified, the experiments with gate drive circuit are continued to a preliminary 6 kV double pulse test of the 15 kV SiC MOSFETs. The test setup is shown in Fig. 6 and Fig. 7. In the test, in order to increase the device switching speed, the turn-on gate resistance is reduced to 10 Ω, and the turn-off gate resistance is reduced to 2 Ω. The measurement equipment are kept the same as specified in Tab. III.

The experimental waveforms are shown in Fig. 10. It can be seen that the device dv/dt only reached about 30 kV/μs, which is much less than the results presented in [1]. This is because: 1) the tested I_d is low, which slows down the charge transfer between the MOSFET and the diode in switching transients, 2) the applied gate voltage (+15 V) is low, which limits the device transconductance, 3) the applied gate resistance is still relatively large, and 4) the main power loop has excessive stray inductance because of the large loop length and non-optimized loop layout.

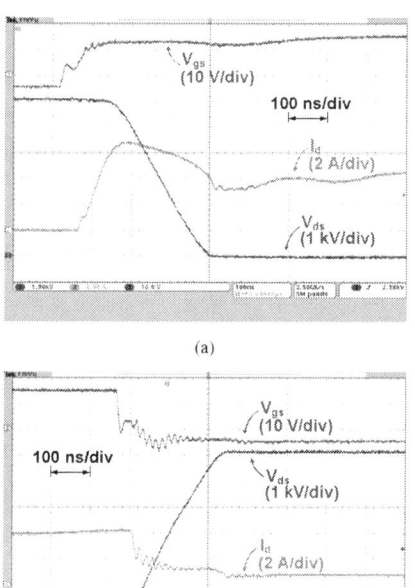

(a)

(b)

Fig. 10.   Experimental waveforms of the preliminary 6 kV double pulse test of the 15 kV SiC MOSFET in the (a) turn-on and (b) turn-off events, under the test conditions R_{g_on} = 10 Ω, R_{g_off} = 2 Ω, and V_{gs} = +15 V/-5 V.

VI.   CONCLUSIONS

This paper propose a PoF-based gate drive circuit for the 15 kV SiC MOSFET. Compared to the traditional transformer-based isolated gate-drive power supply, the PoF-based design features higher insulation voltage (>20 kV), reduced circuit volume and weight, and higher CM transient rejection (>200 kV/μs) due to the elimination of parasitic CM capacitance.

The PoF-based gate drive circuit has sufficient driving power for the 15 kV SiC MOSFET, which is experimentally verified.

The gate-drive circuit design also integrates comprehensive protection functions, including the over-current protection, under/over-voltage lockout, gate-clamping, soft turn-off, and fault report. The over-current protection for the 15 kV SiC MOSFET is currently designed to be triggered in the device triode region. It enable the protection at a low device current. The designed over-current protection total response time under HSL fault is within 400 ns.

A 15 kV double pulse tester is built to verify the designed gate drive circuit. Experimental results of a preliminary 6 kV double pulse test of the 15 kV SiC MOSFET are presented. Further study will be continued to raise the switching speed and test the designed gate drive under harsher dv/dt conditions.

ACKNOWLEDGMENT

The authors would like to thank *Wolfspeed* for providing the 15 kV SiC device samples and support.

REFERENCES

[1] V. Pala, E. V. Brunt, L. Cheng, M. O'Loughlin, J. Richmond, A. Burk, S. T. Allen, D. Grider, J. W. Palmour, and C. J. Scozzie, "10 kV and 15 kV silicon carbide power MOSFETs for next-generation energy conversion and transmission systems," *Proc. IEEE ECCE 2014*, pp. 449-454, Sept. 2014.

[2] M. K. Das, C. Capell, D. E. Grider, R. Raju, M. Schutten, J. Nasadoski, S. Leslie, J. Ostop, and A. Hefner, "10 kV, 120 A SiC half H-bridge power MOSFET modules suitable for high frequency, medium voltage applications," *Proc. IEEE ECCE* 2011, pp. 2689-2692, Sept. 2011.

[3] S. Madhusoodhanan, A. Tripathi, D. Patel, K. Mainali, A. Kadavelugu, S. Hazra, S. Bhattacharya, and K. Hatua, "Solid-State Transformer and MV Grid Tie Applications Enabled by 15 kV SiC IGBTs and 10 kV SiC MOSFETs Based Multilevel Converters," *IEEE Transactions on Industry Applications*, vol. 51, no. 4, pp. 3343-3360, July-Aug. 2015.

[4] F. Fukuda, D. Okamoto, M. Okamoto, T. Deguchi, T. Mizushima, K. Takenaka, H. Fujisawa, S. Harada, Y. Tanaka, Y. Yonezawa, T. Kato, S. Katakami, M. Arai, M. Takei, S. Matsunaga, K. Takao, T. Shinohe, T. Izumi, T. Hayashi, S. Ogata, K. Asano, H. Okumura, and T. Kimoto, "Development of Ultrahigh-Voltage SiC Devices," *IEEE Transactions on Electron Devices*, , vol. 62, no. 2, pp. 396-404, Feb. 2015.

[5] A. Kadavelugu, S. Bhattacharya, S.-H. Ryu, E. V. Brunt, D. Grider, A. Agarwal, and S. Leslie, "Characterization of 15 kV SiC n-IGBT and its Application Considerations for High Power Converters," *Proc. IEEE ECCE 2013*, pp. 2528-2535, Sept. 2013.

[6] D. Rothmund, G. Ortiz, T. Guillod, and J. W. Kolar, "10 kV SiC-based isolated DC-DC converter for medium voltage-connected Solid-State Transformers," in *Proc. IEEE APEC 2015*, pp. 1096-1103, Mar. 2015.

[7] A. Kadavelugu, K. Mainali, D. Patel, S. Madhusoodhanan, A. Tripathi, K. Hatua, S. Bhattacharya, S.-H. Ryu, D. Grider, and S. Leslie, "Medium voltage power converter design and demonstration using 15 kV SiC N-IGBTs," *Proc. IEEE APEC 2015*, pp. 1396-1403, Mar. 2015.

[8] A. K. Tripathi, K. Mainali, D. C. Patel, A. Kadavelugu, S. Hazra, S. Bhattacharya, and K. Hatua, "Design Considerations of a 15-kV SiC IGBT-Based Medium-Voltage High-Frequency Isolated DC–DC Converter," *IEEE Transactions on Industry Applications*, vol. 51, no. 4, pp. 3284-3294, July-Aug. 2015.

[9] K. Mainali, A. Tripathi, S. Madhusoodhanan, A. Kadavelugu, D. Patel, S. Hazra, K. Hatua, and S. Bhattacharya, "A Transformerless Intelligent Power Substation: A three-phase SST enabled by a 15-kV SiC IGBT," *IEEE Power Electronics Magazine*, vol. 2, no. 3, pp. 31-43, Sept. 2015.

[10] S. Madhusoodhanan, K. Mainali, A. Tripathi, D. Patel, A. Kadavelugu, S. Bhattacharya, and K. Hatua, "Performance Evaluation of 15 kV SiC IGBT based Medium Voltage Grid Connected Three-Phase Three-Level NPC Converter," *Proc. IEEE ECCE 2015* (to be published).

[11] K. Vechalapu, A. Tripathi, K. Mainali, B. J. Baliga, and S. Bhattacharya, "Soft Switching Characterization of 15 kV SiC n- IGBT and Performance Evaluation for High Power Converter Applications," *Proc. IEEE ECCE 2015* (to be published).

[12] A. Tripathi, K. Mainali, S. Madhusoodhanan, D. Patel, A. Kadavelugu, S. Hazra, S. Bhattacharya, and K. Hatua, "MVDC Microgrids enabled by 15kV SiC IGBT based Flexible Three Phase Dual Active Bridge Isolated DC-DC Converter," *Proc. IEEE ECCE 2015* (to be published).

[13] S. Madhusoodhanan, A. Tripathi, K. Mainali, A. Kadavelugu, D. Patel, S. Bhattacharya, and K. Hatua, "Three-Phase 4.16 kV Medium Voltage Grid Tied AC-DC Converter based on 15 kV/40 A SiC IGBTs," *Proc. IEEE ECCE 2015* (to be published).

[14] D. Grider, M. Das, R. Raju, M. Schutten, S. Leslie, J. Ostop, A. Hefner, "10 kV/120 A SiC DMOSFET half H-bridge power modules for 1 MVA solid state power substation," *IEEE ESTS*, pp. 131-134, 2011.

[15] C. DiMarino, I. Cvetkovic, Z. Shen, R. Burgos, and D. Boroyevich, "10 kV, 120 a SiC MOSFET modules for a power electronics building block (PEBB)," *Proc. 2014 IEEE Workshop on Wide Bandgap Power Devices and Applications (WiPDA)*, pp. 55-58, Oct. 2014.

[16] I. Cvetkovic, Z. Shen, M. Jaksic, C. DiMarino, F. Chen, D. Boroyevich, and R. Burgos, "Modular scalable medium-voltage impedance measurement unit using 10 kV SiC MOSFET PEBBs," *Proc. 2015 IEEE Electric Ship Technologies Symposium (ESTS)*, pp. 326-331, June 2015.

[17] C. Peng, A. Q. Huang, and X. Song. "Current commutation in a medium voltage hybrid DC circuit breaker using 15 kV vacuum switch and SiC devices." in *Proc. IEEE APEC 2015*, pp. 2244-2250, Mar. 2015.

[18] C. Peng, I. Husain, A. Q. Huang, B. Lequesne, and R. Briggs, "Design and Experimental Investigations of a Medium Voltage Ultra-Fast Mechanical Switch for Hybrid AC and DC Circuit Breakers", *Proc. IEEE ECCE 2015*, Sep. 2015.

[19] K. Vechalapu, S. Bhattacharya, E. V. Brunt, S.-H. Ryu, D. Grider, and J. W. Palmour, "Comparative Evaluation of 15 kV SiC MOSFET and 15 kV SiC IGBT for Medium Voltage Converter under Same dv/dt Conditions," *Proc. IEEE ECCE 2015* (to be published).

[20] A. Kadavelugu, and S. Bhattacharya, "Design considerations and development of gate driver for 15 kV SiC IGBT," *Proc. IEEE APEC 2014*, pp. 1494-1501, Mar. 2014.

[21] L. Fu, X. Zhang, H. Li, X. Lu, and J. Wang, "The development of a high-voltage power device evaluation platform", in *Proc. IEE WiPDA 2014*, pp. 13-17, Oct. 2014.

[22] L. Fu, H. Li, X. Lu, and J. Wang, "Overview and evaluation methodologies of high voltage power devices", in *Proc. IEEE IPMHVC 2014*, pp. 546-549, Jun. 2015.

[23] CONCEPT, "2015 Product Catalog," [Online], Available: https://igbt-driver.power.com/products/product-documents/2015-product-catalog/

[24] M. N. Nguyen, et al., "Compact, intelligent, digitally controlled IGBT gate drivers for a PEBB-based ILC Marx modulator," IPAC'10, pp. 1-3, May 2010.

[25] K. Kusaka, K. Orikawa, J.-I. Itoh, K. Morita, and K. Hirao, "Isolation system with wireless power transfer for multiple gate driver supplies of a medium voltage inverter," *Proc. IPEC-Hiroshima 2014 - ECCE-ASIA 2014*, pp. 191-198, May 2014.

[26] K. Kusaka, M. Kato, K. Orikawa, J.-i. Itoh, I. Hasegawa, K. Morita, and T. Kondo, "Galvanic Isolation System for Multiple Gate Drivers with Inductive Power Transfer Drive of Three-phase Inverter," *Proc. IEEE ECCE 2015* (to be published).

[27] S. Brehaut and F. Costa, "Gate driving of high power IGBT by wireless transmission," *Proc. CES/IEEE 5th International Power Electronics and Motion Control Conference, 2006 (IPEMC 2006)*, pp. 1-5, Aug. 2006.

[28] D. Othman, M. Berkani, S. Lefebvre, A. Ibrahim, Z. Khatir, and A. Bouzourene, "Comparison study on performances and robustness between SiC MOSFET & JFET devices–Abilities for aeronautics application," *Proc. Eur. Symp. Reliab. Electron Devices, Failure Phys. Anal.*, vol. 52, no. 9, pp. 1859–1964, Sep. 2012.

[29] M. K. Dasa, S. Haney, J. Richmond, A. Olmedo, J. Zhang, and Z. Ring, "SiC MOSFET reliability update," *Mater. Sci. Forum*, vol. 717–720, pp. 1073–1076, May 2012.

[30] Z. Wang, X. Shi, Y. Xue, L. M. Tolbert, F. Wang, and B. J. Blalock, "Design and Performance Evaluation of Overcurrent Protection Schemes for Silicon Carbide (SiC) Power MOSFETs," *IEEE Transactions on Industrial Electronics*, vol. 61, no. 10, pp. 5570-5581, Oct. 2014.

[31] Fraunhofer ISE, "Power-by-Light," [Online], Available: https://www.ise.fraunhofer.de/en/business-areas/iii-v-and-concentrator-photovoltaics/research-topics/power-by-light

[32] J.-G. Werthen, "Powering Next Generation Networks by Laser Light over Fiber," in *Proc. Conference on Optical Fiber communication /National Fiber Optic Engineers Conference 2008. (OFC/NFOEC 2008)*, pp.1-3, Feb. 2008.

A 15-kV Class Intelligent Universal Transformer for Utility Applications

Jih-Sheng Lai[1], Wei-Han Lai[2], Seung-Ryul Moon[1], Lanhua Zhang[1], Arindam Maitra[3]

[1]Virginia Polytechnic Institute and State University, Blacksburg, VA
[2]Enertronics, Inc., Blacksburg, VA
[3]Electric Power Research Institute, Charlotte, NC

ABSTRACT-This paper discloses the development of 15-kV class intelligent universal transformer (IUT) to show system level design, circuit topology, and prototype test results. The complete system is split into two stages: (1) high-voltage ac to low-voltage dc and (2) low-voltage dc to low-voltage ac. With the adoption of silicon carbide (SiC) devices, the high-voltage front-end ac to low voltage dc conversion stage achieves 98.4% efficiency, and the complete power stage can be naturally cooled. With the adoption of auxiliary resonant soft-switching inverter, the second stage achieves 99.2% efficiency. Overall the system has been demonstrated at 97.5% efficiency without forced-air cooling. Extended 8-hour testing was conducted to ensure long-term operation reliability. Overall efficiency and voltage regulation were compared with that of a conventional transformer for justification of IUT adoption.

Keywords: IUT, Intelligent Universal Transformer, Medium voltage, SST, Solid State Transformer

I. INTRODUCTION

The solid-state transformer (SST) concept was initiated for size and weight reduction [1] and drew some interest from utility industry [2]-[7]. The intelligent universal transformer (IUT) is to incorporate smart functions into an SST for utility applications [8]-[10]. Major IUT features that are superior to the conventional transformers were identified as follows:

- Instantaneous voltage regulation under load dynamics and transients
- Maintain unity input power factor under reactive load condition
- Maintain clean sinusoidal input current under harmonic distorted nonlinear load condition
- Protection against unbalanced load conditions
- Protection against overload and output short-circuit
- Voltage sag compensation
- Outage compensation
- Capacitor switching protection

These features have been mostly proven with hardware implementation [11]-[12]. The purpose of this paper is to show some performances of a recently developed IUT in medium voltage (MV) level, mainly efficiency and voltage regulation. The results are also compared with a conventional copper-iron based transformer for considerations of further design improvement and identification of niche applications.

With the most popular MV class in United States at 15-kV, this development aims at unidirectional 8-kV single-phase ac input and 240-V single-phase ac output. To operate SST at this level, one can use devices with high enough voltage blocking capability or relatively low voltage but cascading devices or converters to high voltage levels [13]. Direct ac-to-ac SST was also considered [14]. Voltage blocking capability of silicon devices has been limited to 6.5 kV for pulse-with-modulated (PWM) operation [15]-[16]. These relatively high-voltage devices, however, are too slow to allow high frequency operation, and thus suffering from low efficiency and bulky passive component size. High-voltage silicon carbide devices with 10-kV blocking capability are under development and evaluation for medium voltage SST applications [17]-[23]. Their voltage slew rate (dv/dt) induced noise through common mode coupling can impact the circuit design and possibly require substantial switching speed reduction to allow reliable operation. At this point, none of the SiC 10-kV level devices have been reported for 15-kV class SST implementation.

The IUT presented in this paper adopts the design with series-cascading multiple converter modules using relatively low voltage (1.2-kV level) device. The entire development was dated back in 2003 and followed with hardware demonstration at 2.4-kV ac input voltage level [24]. Recently, the work was extended to electric vehicle fast chargers [25] and 7.2-kV and 8-kV single-phase ac input voltage level. Early stage work was considered conceptual development without actual implementation for utility voltage level. Now with availability of advanced power semiconductor devices such as silicon carbide (SiC) devices, the 15-kV class IUT can be realized with efficiency comparable to the conventional transformers. The prototype has been going through two generations. This paper reveals key test results of the second generation 15-kV class IUT.

In utility environment, the transformer needs to have 100% over-load capability for up to 30-minute operation, therefore, the component sizing for the newly developed 25-kW IUT was for 50-kW. The most concern in overload operation is the input and output inductor saturation. Another concern is the thermal handling capability. The system needs to be ultrahigh efficient to avoid excessive heat. The prototype described in this paper was tested at

978-1-4673-9551-9/16 $31.00 © 2016 IEEE

200% load for individual modules to ensure the overload capability. At the rated 25-kW, the overall IUT measured efficiency was 97.5%, and the unit was able to operated 8-hour continuously under full-load testing without forced-air cooling. The complete prototype has been designed, built, and tested with input voltage level at 7.2-kV rms nominal and 8-kV rms maximum to meet the utility voltage regulation requirement.

II. System Architecture and Circuit Configuration

The complete IUT consists of an active front-end (AFE) ac-dc converter, an isolated dc-dc converter, and a dc-ac inverter. Since the AFE converter directly interfaces high voltages, it is the most critical stage, and the combination of the AFE and isolated dc-dc stage has been the design focus. However, the performance of dc-ac stage should not be neglected.

The proposed combined stage with integration of front-end ac-dc and isolated dc-dc was designed in a modular configuration that allows simple stacking for the increased voltage level. In this module, the AFE is a 3-level ac-dc boost converter, and the isolation is provided by an LLC soft-switched converter. The output of the module is an isolated 400-V dc which supplies to the dc-ac stage to produce 240-V ac output. A hybrid switch based soft-switching inverter [26]-[28] is adopted for high efficiency design.

Fig. 1 shows the block diagram of the 8-kV input IUT. A 3-level active-front-end (AFE) boost converter and a high-frequency isolated dc-dc converter are combined together as a module. The module inputs are connected in series, and the outputs are connected in parallel. Such an input series and output parallel structure tends to suffer from balancing problem under no-load condition. To solve this problem, two main approaches were taken. The first critical approach is to construct high frequency transformers uniformly with consistent leakage and magnetizing inductances. The second approach is to supply the auxiliary power by the 400-V output bus. With well-uniformly constructed high frequency transformers, the input voltage sharing can be balanced with a small loading coming from the auxiliary power that supplies gate drives, sensors, control, and optical fiber interface circuits, which sums to about 100-W for the complete IUT.

Fig. 2 shows the individual power module circuit for the AFE and dc-dc converters. The AFE circuit is a 3-level boost converter with two SiC MOSFET devices in parallel to lower the temperature under natural convection condition. Interleaving operation for 10 modules is employed to reduce the switching current ripple. The dc link mid-bus voltage 700 V is guaranteed by balanced transformer construction that ties the secondary in parallel and the primary in series.

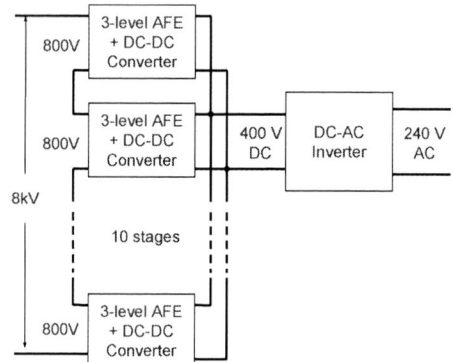

Fig. 1. Ten-module configuration for the Gen-2 IUT showing voltage level under 7.2-kV input condition.

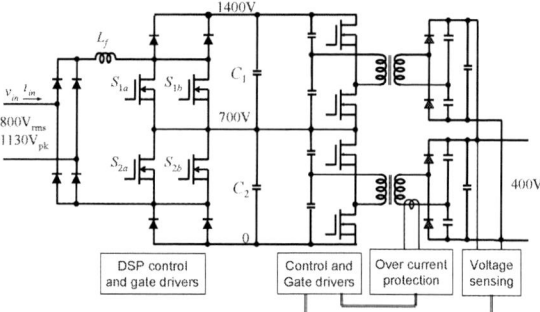

Fig. 2. individual power module circuit for the AFE and dc-dc converters.

The dc-dc converter is a half-bridge LLC based converter. The switches of the input side half-bridge circuit are the same SiC MOSFET used in the AFE stage. High-frequency rated polypropylene film capacitors are used as the resonant capacitor. The secondary diode bridge nominal voltage output is 400-V dc, which feeds the next stage inverter to produce 240-V ac. Voltage sensing signal feeds back to the digital signal processor (DSP) controller for feedforward control of the output ac voltage.

A tertiary transformer turn is used for current sensing and over-current protection. A high-voltage isolated auxiliary power supply is needed to provide gate drive and control voltages for individual power stage. These auxiliary power supplies are critical and need to have minimum capacitive coupling between input and output because they can couple the common mode voltage in between different power modules.

The power module of the combined AFE and LLC dc-dc stage utilize silicon carbide (SiC) power MOSFET rated 1200V/80mΩ and SiC diode rated 1200V/20A. The use of SiC MOSFET as compared to Si devices for dc-dc converter allows significant reduction on the magnetizing current requirement for zero-voltage switching, thus avoiding excessive core loss under light-load condition. The dc-dc converter output diode is 600-V, 60-A fast recovery diode.

978-1-4673-9551-9/16 $31.00 © 2016 IEEE

The most critical design consideration is the dielectric insulation between primary and secondary. All the transformers were tested with a 20-kV high-potential tester. Similarly, all the auxiliary power supplies for gate drivers were also tested with 20-kV insulation level. These gate drive power supplies are rated at 6 W each and are developed in house with resonant power conversion to ensure 90% efficiency.

Fig. 3 shows photograph of the entire 8-kV input AFE and LLC dc-dc stage. The overall size of this stage is 6-ft height, 3-ft wide, and 1-ft depth. A set of 8-kV to 220-V low-frequency (LF) transformers located on the right bottom corner are used to provide initial auxiliary power supply requirements. The 10 modules are split in top and bottom halves with five modules each. High-voltage film capacitors are located between AFE and dc-dc converters, and high-frequency high-voltage insulated transformers are installed between the front-end power board and the secondary rectifier boards. The combined AFE and dc-dc power stage was designed with different levels of protection. The gate drive circuit has built-in short circuit protection. The individual dc-dc converter has a third winding to feedback over current protection signal. The dc bus voltage of each module has an over-voltage protection that can shut down individual modules locally and instantly while sending fault signal to DSP for system level shut down. All the fault signals produced from individual modules are fed back to DSP through optical fibers.

In order to maximize the efficiency, the LLC isolated dc-dc stage and the dc-ac inverter stage are operating under 100% duty cycle. In other words, they are all under open-loop operation. The AFE converter, however, is closed-loop controlled that takes signals from ac input i_{in} and dc output V_{dc} for power factor correction control while regulating dc bus voltage to obtained 240 V ac at the inverter output. The AFE current loop adopts the repetitive controller to eliminate all harmonic contents. The dc bus voltage adopts the conventional PI controller for dc bus voltage regulation. The dc bus voltage reference varies with the load condition, which is scaled by the input current magnitude.

A DSP control board is developed in house using TI TMS320F28335 to interface with a computer programmable logic device (CPLD) to produce 10 sets of interleaving signals for each boost converter. The outputs of CPLD are connected to optical fiber transmitters which are then feeding signals to individual stages. Consider the spacing and creepage requirement, all the modules need to have sufficient separation and are spread across the entire cabinet. Thus conventional signal wire with opto-coupler isolation will not work due to the transmission distance and high-voltage isolation requirement. Fig. 4 illustrates the basic control concept of the entire IUT. The DSP controller that controls the AFE AC-DC converter is responsible for input power factor correction as well as output voltage regulation.

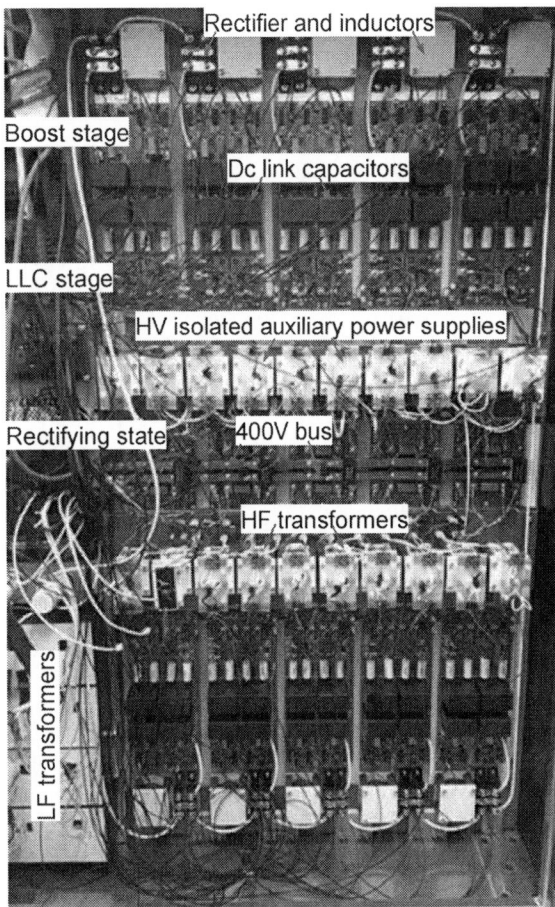

Fig. 3. Photograph of 8-kV AFE and dc-dc stage.

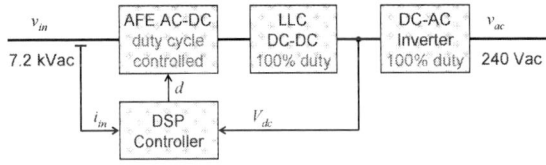

Fig. 4. Basic control block diagram.

III. HARDWARE IMPLEMENTATION AND EXPERIMENTAL RESULTS

A. Test Equipment

The converter that consists of AFE and dc-dc stages is integrated with the 25-kW soft-switching inverter shown in Fig. 5(a) for complete system performance evaluation. A 240V/8kV commercial transformer shown in Fig. 5(b) is used to produce 8 kV source for the system level testing. A high-precision potential transforer (PT) is used for scope measurement.

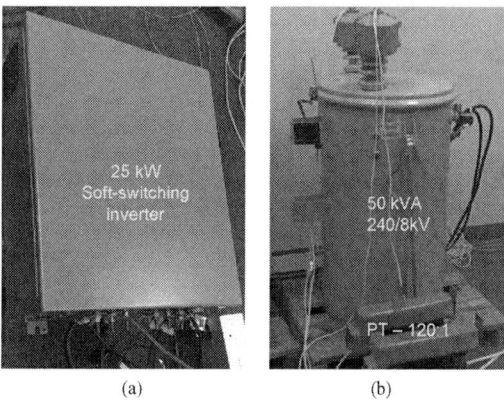

(a)   (b)

Fig. 5. (a) photograph of the 25-kW soft-switching inverter; and (b) Photograph of a 50-kW commercial transformer serving as the source.

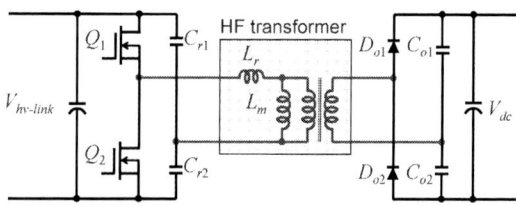

Fig. 6. Half-bridge LLC resonant converter circuit diagram.

B. Half-Bridge LLC Converter

Fig. 6 shows circuit diagram of the half-bridge resonant converter. Key components are listed as follows:

- Switch Q_1, Q_2: C2M0080120D SiC MOSFET
- Resonant capacitor C_{r1}, C_{r2}: 0.15 µF polypropylene
- Transformer: E6527 core, 3C95 ferrite, L_m = 840 µH, L_r = 10.5 µH
- Output diode D_{o1}, D_{o2}: 60EPF06
- Output capacitor C_{o1}, C_{o2}: 6.6 µF polypropylene

The output side is also a half-bridge type rectifying circuit that reduces the number of diodes to half as compared to the full-bridge rectifier. Since the output diodes D_{o1} and D_{o2} are turned off at zero current, they do not need to be ultrafast reverse recovery, and thus low conduction voltage drop diodes can be used to reduce the conduction losses. The input of the half-bridge converter is the high-voltage dc link voltage, $V_{hv\text{-}link}$, which has a nominal value of 700V, and the output voltage V_{dc} is regulated at 400-V nominal. The leakage inductance L_r is adjusted to the same value for all 20 transformers.

Fig. 7 shows the resonant circuit operation of the dc-dc converter under no-load condition, which indicates a low magnetizing current with peak less than 1 A. The primary side voltage peaks around 750 V, and the operating frequency is 93 kHz. Note that the commercial capacitor values vary and tend to be lower than the marking. The switching frequency needs to be adjusted accordingly for efficiency optimization.

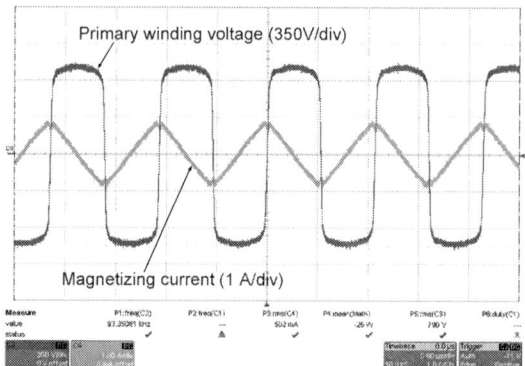

Fig. 7. Transformer primary voltage and current waveforms showing magnetizing current of <1 A at 93 kHz switching.

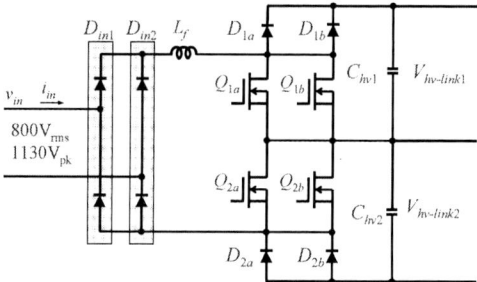

Fig. 8. Rectifier AFE converter stage circuit diagram

C. AFE Converter Stage

Fig. 8 shows circuit diagram of the input side rectifier and AFE converter circuit. Key components are listed as follows:

- Switches Q_{1a}, Q_{1b}, Q_{2a}, Q_{2b}: C2M0080120D
- Diodes D_{1a}, D_{1b}, D_{2a}, D_{2b}: C2D20120D SiC diode
- Capacitor $C_{hv\text{-}link1}$, $C_{hv\text{-}link2}$: 40 µF×2 polypropylene
- Input diode module D_{in1}, D_{in2}: MDD95-22N1B
- Input inductor L_f: KoolMu core, 500 µH

Before connecting to the ac source, the power stage was tested with dc source to check the efficiency. Fig. 9 plots the efficiency evaluation results with 800-V dc input and 400 V dc output. The individual LLC converter as shown in Fig. 6 circuit was tested first. The second test is to combine the AFE circuit shown in Fig. 8 and two LLC converters connected in input series and output parallel configuration.

The individual LLC converter peaks near 99% efficiency, while the combined AFE and LLC converter module achieves 98% peak efficiency. The AFE converter efficiency drops significantly under light load (<10%) condition. The LLC converter, on the other hand, maintains high efficiency at light load because its magnetizing current is relatively small, which translates low magnetizing circuit or no-load loss. The test also allows observation of dc link voltage $V_{hv\text{-}link}$ balancing condition. If the two LLC transformers are not well matched, the two $V_{hv\text{-}link}$ will show unbalance under light load condition.

978-1-4673-9551-9/16 $31.00 © 2016 IEEE

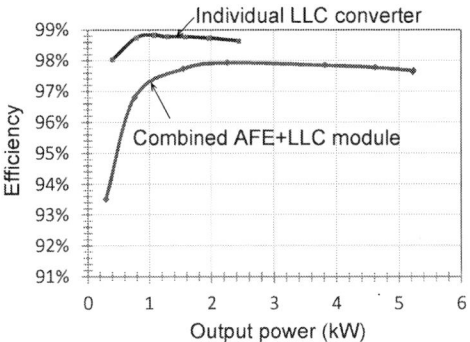

Fig. 9. Measured efficiency of individual power module running under dc-dc condition.

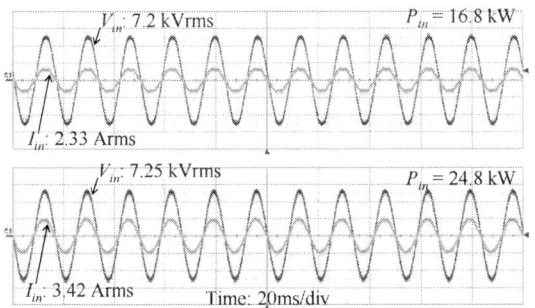

Fig. 10. Input voltage and current under partial and full-load conditions.

Power factor correction function was implemented with repetitive control to ensure low harmonic distortion [29]-[30]. Fig. 10 shows the tested voltage and current waveforms under partial- and full-load conditions. The voltage and current for partial load, or 2/3 load condition are 7.2 kV rms and 2.33 A rms, and for full load are 7.25 kV rms and 3.42 A rms. Both conditions have total harmonic distortion (THD) well below 5%. Note that the supply side voltage is directly fed by the utility transformer output, and its magnitude changes continuously. The current, however, is well controlled to track the voltage.

D. DC-AC Inverter Stage

Fig. 11 shows one phase leg of the auxiliary coupled magnetic clamped soft-switching inverter circuit diagram used in the 25-kW IUT. The main switches Q_1, M_1, Q_2, and M_2 are hybridized switch that consists of two IGBTs rated 600-V, 200-A and two super-junction MOSFETs rated 650-V, 19-mΩ in parallel. These main switches are packaged in one module with a total of 500-A continuous current capability. The auxiliary circuits are also packaged in a module consisting of IGBT devices Q_{x1} and Q_{x2} rated 150-A and all the auxiliary diodes including D_{x1}, D_{x2}, D_{x5}, D_{x6}, D_{x3a}, D_{x3b}, D_{x4a}, and D_{x4b} that are all 600-V rated.

With the inverter switching frequency at 20 kHz, the output LC filter cut-off frequency is designed at 3.5 kHz. The entire filter consists of four series connected inductors, each rated 400 µH, and a filter capacitor rated 200 µF.

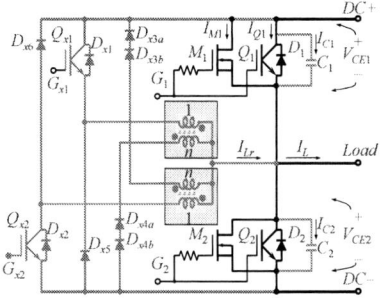

Fig. 11. Circuit diagram of auxiliary coupled-magnetic clamped soft-switching inverter.

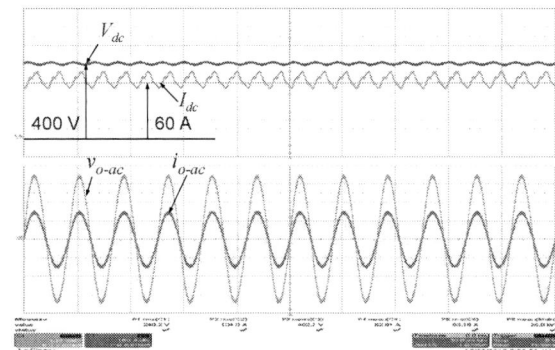

Fig. 12. Tested waveforms of dc bus and output voltage and current under full-load condition

The auxiliary switches turn on before the main switch so the load current can be diverted to the resonant circuit. The resonant operation ensures the device voltage down to zero before they are turned on, so the main devices are all switched at zero-voltage condition. After the main switch turns on, the resonant current can be reset to zero, and thus the auxiliary switch can be turned off at zero current condition. Detailed operation and performance of a hybrid-switch based inverter can be referred to [28].

Fig. 12 depicts the experimental waveforms of dc bus voltage V_{dc} = 403 V, current I_{dc} = 62.8 A, output voltage v_{o-ac} = 240 V rma, and current i_{o-ac} = 104 A under full-load condition. With dc bus consisting of four 450-V, 8.2 mF rated ELX-451LGN822MFK0M, the dc bus current still presents some a 17-A peak-to-peak ripple at 120 Hz. The dc bus voltage, however, is relatively flat, and the output ac waveform is also quite clean. The THD under resistive load condition is less than 2.5%.

Fig. 13 shows the overall IUT efficiency with separation of AC-DC-DC front stage and DC-AC inverter stage. Note that the entire front stage efficiency is higher than that of the individual module because the inductor current ripple is reduced through interleaved operation. The peak AFE stage efficiency of 98.4% occurs at the rated power condition. The inverter efficiency reaches 99% when the load is about 50%. Overall system efficiency at the nominal load is about 97.5%.

978-1-4673-9551-9/16 $31.00 © 2016 IEEE

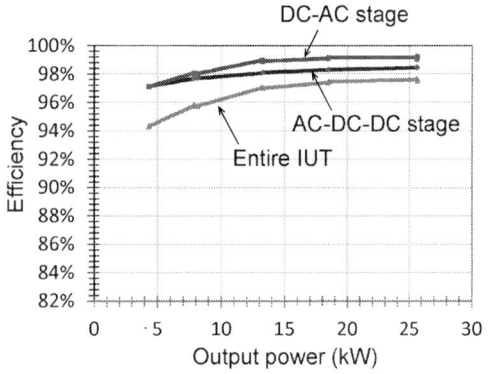

Fig. 13. Measured IUT system level efficiency.

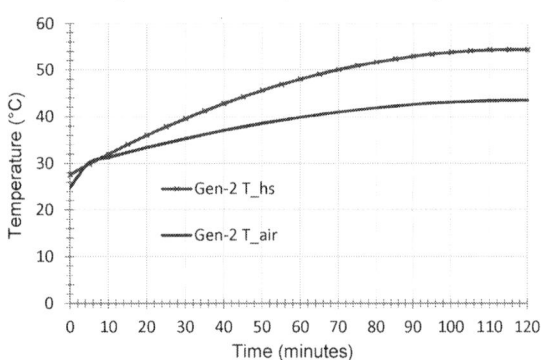

Fig. 14. Recorded temperature of IUT heat sink and inside air during long-hour test.

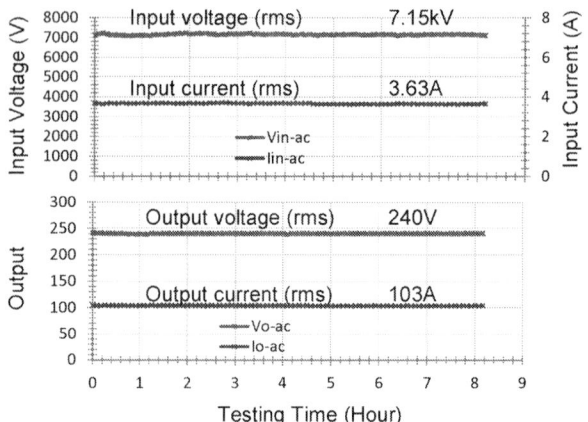

Fig. 15. Recorded input and output voltages and currents during long-hour testing.

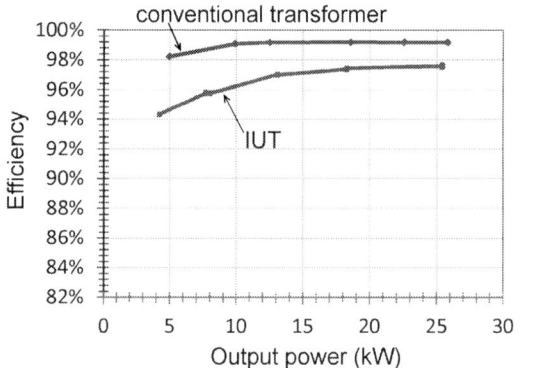

Fig. 16. Efficiency comparison between conventional transformer and IUT.

E. Long-Hour Testing Results

The power stage containing AFE and LLC converters was tested for a long period. Temperatures of heat sink and inside cabinet air were first checked over a two-hour period to ensure that they are not excessively high. Fig. 14 plots the temperature profile that indicates the temperature is about to reach the steady state at about 2 hours. Note that the entire IUT power stage was enclosed without forced-air cooling. The heat sink is cooled by the nature convection. Its temperature stabilizes at about 54 °C, or 26 °C temperature rise. Similarly, the cabinet inside is not cooled by any air circulation, so its temperature rise is noticeable. The steady-state cabinet inside air temperature is about 43 °C, or 17 °C temperature rise.

After confident about the thermal condition, the entire IUT was tested for an 8-hour period. Fig. 15 plots the input voltage, input current, output voltage and output current. All curves maintain nearly constant throughout the entire 8-hour testing. The input voltage starts at 7.2 kV but varies around 7.15 kV throughout the test due to the utility voltage fluctuation. However, the output voltage remains constant at 240 V after temperature is stabilized.

IV. PERFORMANCE COMPARISON WITH A CONVENTIONAL TRANSFORMER

A commercially available 50-kVA transformer was tested for the performance comparison, mainly on efficiency and voltage regulation. Fig. 16 shows efficiency comparison between the conventional transformer and IUT. The conventional transformer has an efficiency of 98% at its 10% load, or 5 kVA and maintains 99.1% efficiency from its 20% load and above. The IUT efficiency is 94.4% at its 20% load, or 5 kW and reaches 97.5% at the rated 25 kW condition. Such an IUT efficiency profile is not very attractive as compared with the conventional transformer. However, the functionality of IUT is not just in voltage conversion, but also in many other aspects that have been mentioned and verified in [12].

Voltage regulation is considered an important benefit that IUT can provide to reduce the labor on tap changing operation. Two types of voltage regulations including load regulation and line regulation were tested and compared.

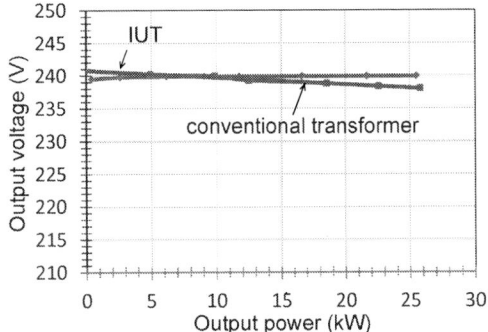

Fig. 17. Load regulation comparison.

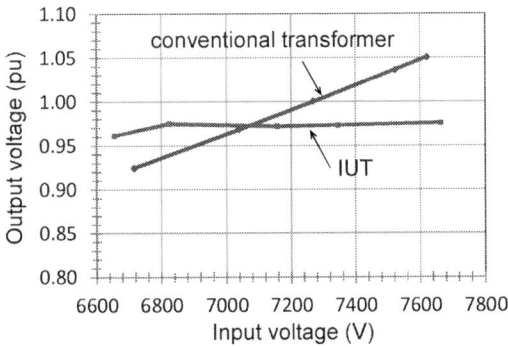

Fig. 18. Line regulation comparison.

Fig. 17 compares the load regulation between a commercially available 50-kVA conventional transformer and the 25-kW IUT prototype. The conventional transformer has a 2.2% voltage regulation between no load and full load conditions. On the other hand, the IUT output is nearly flat under all load conditions.

Fig. 18 compares the line regulation between the 50-kVA conventional transformer and the IUT prototype. The tap changing position typically varies ±10%, and the actual operation is regulating at ±5% range. In this test, the input line voltage runs from 6.7 kV to 7.6 kV. The conventional transformer output voltage linearly follows the input voltage, but the IUT maintains its output voltage nearly constant.

In principle, the IUT output can be regulated very well under all different source and load conditions. In actual implementation, however, the IUT output voltage can vary due to the temperature condition of the voltage sensor and the voltage loop control method. Such variations can be minimized with proper compensation methods. As such, the voltage regulation is perhaps the most noticeable benefit that an SST can offer because it allows elimination of conventional tap-changing effort.

V.   CONCLUSION

A 15-kV class single-phase IUT has been successfully developed and tested for basic performance verification. This paper describes its system architecture, circuit topology, control strategy, and design considerations. Extensive testing was conducted to characterize the performance and to ensure reliable operation. Individual power stage was tested with 100% overload. The complete system was tested at the full load for more than 8 hours. The peak efficiency of the entire MV ac to LV ac system reaches 97.5% under the rated load condition. The steady-state heat sink temperature stays below 55°C with natural convection cooling.

Key benefit of IUT in voltage conversion is the voltage regulation against both load and source. Test results indicate that the IUT output voltage can be well regulated under different load and line voltage variations. The conventional transformer output voltage varies with both load and source variations, and thus additional effort of tap-changing is needed.

In terms of efficiency evaluation results, the three individual power stages including AFE converter, LLC converter, and soft-switching inverter, all achieved around 99% peak efficiency, which already excels the state-of-the-art power conversion efficiency, but the complete system efficiency is shown inferior to the conventional iron-copper based transformer. Although further efficiency improvement is possible, identifying niche applications for IUT is necessary for the industry to adopt such a new technology.

The power quality advantage and intelligent functions identified in previous papers [11]-[12] were considered the major advantages for IUT to penetrate into the utility industry. Another key advantage of the IUT is the availability of dc bus voltage. Test results in reference [25] clearly indicate that using medium voltage input and dc bus output for electric vehicle fast charging system can significantly improve the energy efficiency. This dc bus voltage availability allows many other initiatives such as dc micro grid and dc distribution system in data centers.

Other application areas and industries that were not emphasized in this paper are medium voltage power conversions for marines and railways where size and weight reduction is critical for payload reduction and system level efficiency improvement.

ACKNOWLEDGMENT

The authors would like to thank the financial support from Tennessee Valley Authority, Southern Company, and Northeast Utility, the test equipment and facility provided by Howard Industry, and the initial development effort by former Virginia Tech graduates including Hide Miwa, Younghoon Cho, and Chien-Liang Chen.

REFERENCES

[1] J. L. Brooks, Solid State Transformer Concept Development, *Report of Naval Material Command, Civil Engineering Laboratory*, Naval Construction Battalion Center, Port Hueneme, CA, 1980.

[2] EPRI Report, TR-105067, *Proof of the Principle of the Solid-State Transformer – The AC/AC Switch-mode Regulator, Research Project 8001-13*, San Jose State University, San Jose, CA, August 1995.

[3] M. Kang, P. Enjeti, and I. Pitel, "Analysis and design of electronic transformers for electric power distribution system," *IEEE Trans. Power Electron.*, vol. 14, no. 6, pp. 1133–1141, 1999.

[4] S. D. Sudhoff, "Solid State Transformer," US Patent No. 5,943,229, August 24, 1999.

[5] E. R. Ronan, Jr., S. D. Sudhoff, S. F. Glover, D. L. Galloway, "Application of Power Electronics to the Distribution Transformer," in Conf. Rec. of APEC, Feb. 2000, New Orleans, LA, pp. 861 – 867.

[6] E. Ronan, S. Sudhoff, S. Glover, and D. Galloway, "A power electronic-based distribution transformer," *IEEE Trans. Power Del.*, vol. 17, no. 2, pp. 537–543, Apr. 2002.

[7] E. C. Aeloiza, P. N. Enjeti, L.A. Moran, I. Pitel, "Next Generation Distribution *Transformer: To Address Power Quality for Critical Loads,"* in *Proc. of Power Electronics Specialists Conference*, June 2003, Acapulco, Mexico, pp. 1266 – 1271

[8] J.-S. Lai, "Designing the Next Generation Distribution Transformers: New Power Electronic-Based Hybrid and Solid-State Design Approaches," in *Proc. of IASTED Power and Energy Systems Conference*, Palm Spring, CA, Feb. 2003, pp. 262 – 267.

[9] J.-S. Lai, A. Mansood, A. Maitra, and F. Goodman, "Multifunction hybrid intelligent universal transformer," U.S. Patent #6,954,366, Oct. 11, 2005.

[10] J.-S. Lai, A. Mansood, A. Maitra, and F. Goodman, "Multilevel converter based intelligent universal transformer," U.S. Patent #7,050,311, May 2006.

[11] J.-S. Lai, A. Maitra, A. Mansoor, and F. Goodman, "Multilevel Intelligent Universal Transformer for Medium Voltage Applications," in *Conf. Rec. of IEEE IAS Annual Meeting*, Hong Kong, Oct. 2005, pp. 1893-1899.

[12] J.S. Lai, A. Maitra, and F. Goodman, "Performance of a Distributed Intelligent Universal Transformer under Source and Load Disturbances," in *Conf. Rec. of IEEE IAS Annual Meeting*, Tampa, FL, Oct. 2006, pp. 719 – 725.

[13] X. She, A.Q. Huang, R. Burgos, "Review of solid-state transformer technologies and their application in power distribution systems," *IEEE Journal of Emerging and Selected Topics in Power Electronics*, vol. 1, no. 3, pp. 186–198, September, 2013.

[14] A. De and S. Bhattacharya, "Design, Analysis and Implementation of Discontinuous Mode Dyna-C AC/AC Converter for Solid State Transformer Applications," in *Proc. of IEEE Energy Conversion Congress and Exposition, Sept. 2015*, pp. 5030-5037.

[15] J .S. Lai, A. Hefner, A. Maitra, and F. Goodman, "Characterization of a Multilevel HV-IGBT Module for Utility Distribution Applications," in *Conf. Rec. of IEEE IAS Annual Meeting*, Tampa, FL, Oct. 2006, pp. 743 – 753.

[16] J.-S. Lai, "Characterization of HV-IGBT for High-Power Inverter Applications," in Conf. Rec. of IEEE IAS Annual Meeting, Hong Kong, Oct. 2005, pp. 377-382.

[17] F. Wang; G. Wang; A. Huang, W. Yu, X, Ni, "Design and operation of A 3.6kV high performance solid state transformer based on 13kV SiC MOSFET and JBS diode," in *Proc. of IEEE Energy Conversion Congress and Exposition, Sept. 2014*, pp. 4553-4560.

[18] A. Kadavelugu, K. Mainali, D. Patel, S. Madhusoodhanan, A. Tripathi, K. Hatua, S. Bhattacharya, S.-H Ryu, and D. Grider, "Medium Voltage Power Converter Design and Demonstration Using 15 kV SiC N-IGBTs," in *Proc. of IEEE Applied Power Electronics Conf. and Exposition*, Charlotte, NC, March 2015, pp. 1396 – 1403.

[19] S. Madhusoodhanan, K. Mainali, A. Tripathi, D. Patel, A. Kadavelugu, S. Bhattacharya, K. Hatua, "Performance Evaluation of 15 kV SiC IGBT based Medium Voltage Grid Connected Three-Phase Three-Level NPC Converter," in *Proc. of IEEE Energy Conversion Congress and Exposition, Sept. 2015*, pp. 3710-3717.

[20] S. Madhusoodhanan, K. Mainali, A. Tripathi, D. Patel, A. Kadavelugu, S. Bhattacharya, K. Hatua, "Three-Phase 4.16 kV Medium Voltage Grid Tied AC-DC Converter Based on 15 kV/40 A SiC IGBTs," in *Proc. of IEEE Energy Conversion Congress and Exposition, Sept. 2015*, pp. 6675-6682.

[21] D. Rothmund, G. Ortiz, Th. Guillod, and J. W. Kolar, "10kV SiC-Based Isolated DC-DC Converter for Medium Voltage-Connected Solid-State Transformers," in *Proc. of IEEE Applied Power Electronics Conf. and Exposition*, Charlotte, NC, March 2015, pp. 1096 – 1103.

[22] T. Zhao, G. Wang, S. Bhattacharya, A.Q. Huang, "Voltage and power balance control for a cascaded H-bridge converter-based solid-state transformer," *IEEE Trans. on Power Electronics*, vol. 28, no. 4, pp. 1523–1532, April, 2013.

[23] S. Xu, Y. Xunwei, F. Wang, A.Q. Huang, "Design and demonstration of 3.6 kV-120 V/10 kVA solid-state transformer for smart grid application," *IEEE Trans. on Power Electronics*, vol. 29, no. 8, pp. 3982–3988, August, 2014.

[24] EPRI Report, 1020087, *4.16KV 25KVA Field Prototype Development of Intelligent Universal Transformer*, Dec. 2010.

[25] EPRI Report, 1022009, *Field Prototype Development of Intelligent Universal Transformer*, Dec. 2011.

[26] W. Yu, J.-S. Lai, and S.-Y. Park, "An Improved Zero-Voltage-Switching Inverter Using Two Coupled Magnetics in One Resonant Pole," *IEEE Trans. on Power Electronics*, pp. 952-961, April 2010.

[27] P. Sun, J.-S. Lai, C. Liu, W. Yu, "A 55kW Three-Phase Inverter Based on Hybrid-Switch Soft-Switching Modules for High Temperature Hybrid Electric Vehicle Drives Application," *IEEE Trans. on Industry Applications*, May/June 2012, pp. 962–969.

[28] J.-S. Lai, W. Yu, P. Sun, S. Leslie, B. Arnet, C. Smith and A. Cogan, "Hybrid Switch Based Soft-Switching Inverter for Ultrahigh Efficiency Traction Motor Drives," *IEEE Trans. on IAS*, pp. 1966 - 1973, May/June 2014.

[29] Y. Cho, H. Mok, and J.-S. Lai, "Analysis of the Admittance Component for Digitally Controlled Single-phase PFC Converter," *Journal of Power Electronics*, vol. 13, no. 4, pp. 600-608, July 2013.

[30] Y. Cho and J.-S. Lai, "Digital Plug-In Repetitive Controller for Single-Phase Bridgeless PFC Converters," *IEEE Trans. on Power Electronics*, Jan. 2013, pp. 165 - 175.

Modulation Technique for Common Mode Voltage Reduction in a Matrix Converter Drive Operating with High Voltage Transfer Ratio

Varsha Padhee
Rockwell Automation
Mequon, Wisconsin 53092, USA
Email: padhe001@umn.edu

Ashish Kumar Sahoo
University of Minnesota, Twin Cities
Minneapolis, Minnesota 55455–0170, USA
Email: saho0007@umn.edu

Ned Mohan
University of Minnesota, Twin Cities
Minneapolis, Minnesota 55455–0170
Email: mohan@umn.edu

Abstract—This paper discusses a modified indirect space vector pulse-width modulation (ISVPWM) technique for common mode voltage (CMV) reduction in a matrix converter (MC). The ISVPWM technique offers immense flexibility in making an intelligent choice of space vectors to not only achieve the objective of variable speed operation of the drive but also target other advanced features like reduced CMV and improved output voltage distortion. A smart use of these space vectors can result in the generation of different levels in the output voltage and this capability has been utilized to address an improvement in its distortion. The fundamentals of appropriate choice of space vectors in accordance to a new space vector diagram and its corresponding switching sequence have been developed and described in detail. The effectiveness of the algorithm is substantiated by simulations in MATLAB/Simulink and experiments on a laboratory prototype.

Keywords—*Matrix converter (MC), Space vector pulse-width modulation (SVPWM), Common mode voltage (CMV)*

I. INTRODUCTION

Matrix converters (MC) are a modern all-silicon solution to AC-AC power conversion [1]. They target at being an economical substitute to conventional back-to-back connected voltage source converters for industrial adjustable speed drives [2]. MC's gained limelight over the past two decades because of certain indisputable advantages like single stage power conversion, absence of a bulky DC-link capacitor, open loop input power factor correction, regenerative capability and high power density which gives them a major edge over back-to-back converters [3] [4]. This can directly translate to improved lifetime, reliability and compactness of the system. The Matrix converter can be implemented via two different topologies

- Direct Matrix Converter (DMC): the output load is connected to the grid through an array of 9 bidirectional switches forming a matrix Fig. 1(a).

- Indirect Matrix Converter (IMC): a virtual DC-link is formed from the grid voltages using a current source rectifier (CSR) followed by an inverter which interfaces it with the load Fig. 1(b).

In any drive topology, an important problem is the switching common mode voltage and high dv/dt at machine terminals resulting in generation of high frequency bearing and ground currents. These are major reasons for motor winding

failures, electromagnetic interference (EMI) and shaft voltage buildup which reduces the machines lifetime [5] [6]. These can be suppressed using passive chokes and filters at the expense of making the system bulkier and expensive. Thus software based modulation strategies have been investigated to reduce the aforementioned disadvantages of volume and cost.

Space vector pulse width modulation (SVPWM) strategies for CMV reduction in current source rectifier based drives [7], [8] and voltage source inverter based drives [9], [10] is well established in literature. Modulation strategies for CMV reduction in a matrix converter drive also exists. Most discussion is for the indirect matrix converter (IMC) topology as shown in Fig. 1(b), formed by a current source rectifier and a voltage source inverter connected by a virtual DC-link [11]–[14]. The conventional direct matrix converter (DMC) as shown in Fig. 1(a) generates a 42% lower peak CMV in [15] by suitable choice of space vectors. In [16], synchronously rotating vectors are used to drive the converter with zero CMV. However, both the above techniques limit the operation of the MC to a low voltage transfer ratio of below 0.5. The concept of completely eliminating CMV with an improved maximum voltage transfer ratio of 1.5 has been discussed in [17] for an open-end winding AC machine driven by two DMCs. But this requires two DMCs and hence double the number of semiconductor switches. A similar concept of completely eliminating CMV while operating the MC over its full range of modulation has been discussed in [18] [19] for dual MCs. An alternative SVM implementation for DMCs, based on the concept of replacing traditional zero vectors with rotating vectors has been proposed in [20]. A method to appropriately choose zero vectors for SVM to reduce the harmonic distortion in the output voltage and peak CMV is proposed in [21] for an IMC and in [22] for a DMC.

In SVPWM, a particular state of switches of the matrix converter corresponds to a space vector. This further is indicative of a particular instantaneous value of CMV and level of output voltage in the converter. It can therefore be concluded that, a wise choice of these switching vectors can reduce the peak value of CMV and improve the distortion in output voltage while still maintaining the fundamental sinusoidal component of current and voltage. Most drives operate close to the rated speed and hence run the converter at high voltage transfer ratios. A new modulation technique for this range of operation is addressed in this paper. Novel space vector

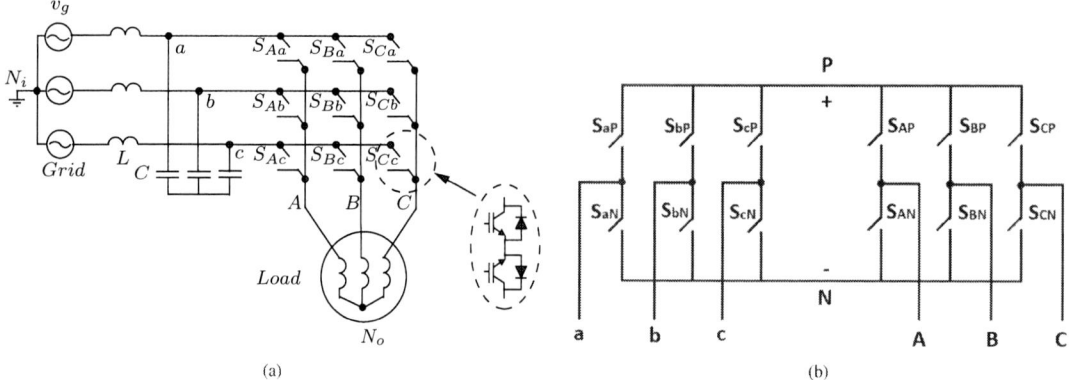

Fig. 1: (a) Direct matrix converter (DMC) topology, (b) Indirect matrix converter (IMC) topology

diagrams, expressions for duty cycles and switching sequence are explained in detail. Apart from reducing the peak value of CMV by a factor of $\sqrt{3}$ as compared to the conventional method, the proposed method comes with additional advantages of improved output voltage distortion, lower core losses in the rotating machine, without an increase in the net number of switching transitions. The paper is organized in the following manner: The conventional and proposed modulation techniques for the control of the DMC is discussed in Section II and III respectively. Section IV and V deal with the simulation and experimental results which validate the effectiveness of the proposed algorithm. Lastly the paper concludes in Section VI.

II. ANALYSIS : CONVENTIONAL SVM TECHNIQUE AND CMV

The direct matrix converter is the most conventional topology for single stage AC-AC power conversion as shown in Fig. 1(a), unlike the IMC topology which is a two stage AC-AC power converter with a fictitious DC-link. The modulation technique is essentially a control strategy which must ensure that the converter formulates sinusoidal voltages and currents of desired magnitude and frequency. This fact translates to the operation of the DMC at different modulation indices. This objective can be achieved through indirect SVPWM, where for a sufficiently small time interval, the reference voltage and current vector can be generated by the usage of stationary space vectors. The two major aspects of SVPWM are selection of switching vectors and computation of corresponding time intervals. This section presents the theory behind the conventional space vector modulation (C-SVM) technique and expression for net common mode voltage.

A. Conventional SVM Technique

In a DMC the three output phases A, B, C are connected to the three input phase a, b, c through a matrix of nine bidirectional switches. The switching function S_{Ba} for example can take two values of 0 or 1. "$S_{Ba}=1$ (ON)" implies that the output phase B is connected to the input phase a and "$S_{Ba}=0$ (OFF)" implies that B and a are disconnected. This naming convention can be similarly extended to the other

eight switches. A switching configuration (SC) of a DMC can be defined as $[x_A, x_B, x_C]$ where x_Y ($Y \epsilon A, B, C$) refers to the input phase to which the corresponding output phase is connected.

Indirect space vector modulation is used to synthesize the reference output voltage and input current space vectors by use of appropriate switching states based on the fictitious IMC topology as shown in Fig. 1(b). The switching signals are then mapped to the DMC topology. There are independent space vector diagrams for the formation of the current reference vector $\overline{I_{in}}$ as in Fig. 2(a) and voltage reference vector $\overline{V_o}$ as in Fig. 2(b). Each diagram comprises of six active vectors. The six active current vectors are I_1-I_6 while the six active voltage vectors are V_1-V_6. A combination of one active current vector and one active voltage vector results in the formation of different switching configurations of the DMC as summarized in Table. I. For example the current vector I_{ab} and the voltage vector V_{100} results in the SC of '+1' which translates to a switching state of $[a, b, b]$. This further implies that the output phases A, B, C are connected to the input phases a, b, b respectively through the switches with switching functions $S_{Aa} = 1$, $S_{Bb} = 1$ and $S_{Cb} = 1$ while the other switches are in the 0 state. The switching states resulting from the SCs: ± 1 - ± 9 replicates itself for the SCs $\pm 1'$ - $\pm 9'$. Apart from these states, there are three zero vectors 0_a, 0_b and 0_c where all the output phases are connected to the same input phase. There are thus 21 unique states for the DMC which are used for output voltage generation. There also exists six synchronous states where each output phase is connected to a different input phase, but this results in lower output voltage generation. Table. II summarizes the switching states used and its sequence of implementation for a current reference lying in sector Si_x and a voltage reference lying in sector Sv_y where $x, y \epsilon (1, 2, 3, 4, 5, 6)$.

In C-SVM, the reference vectors $\overline{I_{in}}$ and $\overline{V_o}$ are generated by the usage of two active vectors and one zero vector whose duty ratios are given by (1) and (2) respectively. The switching states this fact translates to, for a DMC is summarized in Table. III for a current and voltage reference, both lying in Sector 1. This can be extended to the current and voltage reference lying

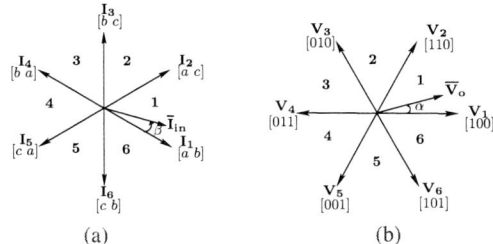

Fig. 2: Conventional SVM: (a) Space vectors for $\bar{I}_{in}$. (b) Space vectors for $\bar{V}_o$

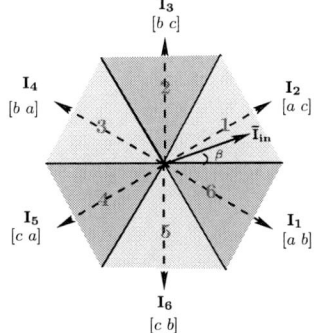

Fig. 3: Modified space vectors of CSR in proposed SVM technique

SC No.	I_x	V_x	A	B	C	V_x	I_x	SC No.
+1	I_{ab}	[1,0,0]	a	b	b	[0,1,1]	I_{ba}	+1'
+2	I_{ab}	[1,1,0]	a	a	b	[0,0,1]	I_{ba}	+2'
+3	I_{ab}	[0,1,0]	b	a	b	[1,0,1]	I_{ba}	+3'
-1	I_{ab}	[0,1,1]	b	a	a	[1,0,0]	I_{ba}	-1'
-2	I_{ab}	[0,0,1]	b	b	a	[1,0,0]	I_{ba}	-2'
-3	I_{ab}	[1,0,1]	a	b	a	[0,1,0]	I_{ba}	-3'
+4	I_{ac}	[1,0,0]	a	c	c	[0,1,1]	I_{ca}	+4'
+5	I_{ac}	[1,1,0]	a	a	c	[0,0,1]	I_{ca}	+5'
+6	I_{ac}	[0,1,0]	c	a	c	[1,0,1]	I_{ca}	+6'
-4	I_{ac}	[0,1,1]	c	a	a	[1,0,0]	I_{ca}	-4'
-5	I_{ac}	[0,0,1]	c	c	a	[1,1,0]	I_{ca}	-5'
-6	I_{ac}	[1,0,1]	a	c	a	[0,1,0]	I_{ca}	-6'
+7	I_{bc}	[1,0,0]	b	c	c	[0,1,1]	I_{cb}	+7'
+8	I_{bc}	[1,1,0]	b	b	c	[0,0,1]	I_{cb}	+8'
+9	I_{bc}	[0,1,0]	c	b	c	[1,0,1]	I_{cb}	+9'
-7	I_{bc}	[0,1,1]	c	b	b	[1,0,0]	I_{cb}	-7'
-8	I_{bc}	[0,0,1]	c	c	b	[1,1,0]	I_{cb}	-8'
-9	I_{bc}	[1,0,1]	b	c	b	[0,1,0]	I_{cb}	-9'
0_a	-	-	a	a	a	-	-	
0_b	-	-	b	b	b	-	-	
0_c	-	-	c	c	c	-	-	

TABLE I: Summary of Switching Configurations and Space Vectors

TABLE II: Summary of Switching States - Conventional SVM

	Si_1	Si_2	Si_3	Si_4	Si_5	Si_6
Sv_1	0_b +1 +2 0_a +5 +4 0_c	0_a +5 +4 0_c +7 +8 0_b	0_c +7 +8 0_b -2' -1' 0_a	0_b -2' -1' 0_a -4' -5' 0_c	0_a -4' -5' 0_c -8' -7' 0_b	0_c -8' -7' 0_b +1 +2 0_a
Sv_2	0_b +3 +2 0_a +5 +6 0_c	0_a +5 +6 0_c +9 +8 0_b	0_c +9 +8 0_b -2' -3' 0_a	0_b -2' -3' 0_a -6' -5' 0_c	0_a -6' -5' 0_c -8' -9' 0_b	0_c -8' -9' 0_b +3 +2 0_a
Sv_3	0_b +3 -1 0_a -4 +6 0_c	0_a -4 +6 0_c +9 -7 0_b	0_c +9 -7 0_b +1' -3' 0_a	0_b +1' -3' 0_a -6' +4' 0_c	0_a -6' +4' 0_c +7' -9' 0_b	0_c +7' -9' 0_b +3 -1 0_a
Sv_4	0_b -2 -1 0_a -4 -5 0_c	0_a -4 -5 0_c -8 -7 0_b	0_c -8 -7 0_b +1' +2' 0_a	0_b +1' +2' 0_a +5' +4' 0_c	0_a +5' +4' 0_c +7' +8' 0_b	0_c +7' +8' 0_b -2 -1 0_a
Sv_5	0_b -2 -3 0_a -6 -5 0_c	0_a -6 -5 0_c -8 -9 0_b	0_c -8 -9 0_b +3' +2' 0_a	0_b +3' +2' 0_a +5' +6' 0_c	0_a +5' +6' 0_c +9' +8' 0_b	0_c +9' +8' 0_b -2 -3 0_a
Sv_6	0_b +1 -3 0_a -6 +4 0_c	0_a -6 +4 0_c +7 -8 0_b	0_c +7 -9 0_b +3' -1' 0_a	0_b +3' -1' 0_a -4' +6' 0_c	0_a -4' +6' 0_c +9' -7' 0_b	0_c +9' -7' 0_b +1 -3 0_a

TABLE III: Switching States & Space Vectors for $\bar{I}_{in}$ and $\bar{V}_o$ in Sector 1 for C-SVM

	I_{ab}				I_{ac}				I_z
	$V_0[000]$	$V_1[100]$	$V_2[110]$	$V_7[111]$	$V_7[111]$	$V_2[110]$	$V_1[100]$	$V_0[000]$	-
	$dI_1\frac{dV_z}{2}$	dI_1dV_1	dI_1dV_2	$dI_1\frac{dV_z}{2}$	$dI_2\frac{dV_z}{2}$	dI_2dV_2	dI_2dV_1	$dI_2\frac{dV_z}{2}$	dI_z
A	b	a	a	a	a	a	a	c	c
B	b	b	a	a	a	a	c	c	c
C	b	b	b	a	a	c	c	c	c
SC	0_b	+1	+2	0_a	0_a	+5	+4	0_c	0_c

in any of the six corresponding sectors giving rise to 36 such possibilities. The duty ratios for the nine switches of the DMC can be formulated from this table. For example, the duty ratio of the switch with switching function S_{Ca} can be defined as the sum of dI_1dV_z and dI_2dV_z Here m_I ($= I_{in}/I_{DC}$) and m_V ($= V_o/V_{DC}$) are the modulation indices, while β and α are the corresponding angles between the first vector and reference vector in Fig. 2. Here I_{in} and V_o are peak input current and peak output voltage to be synthesized and I_{DC}, V_{DC} are corresponding average DC-link quantities in Fig. 1(b).

$$dI_1 = m_I \sin\left(\frac{\pi}{3} - \beta\right)$$
$$dI_2 = m_I \sin\beta$$
$$dI_z = 1 - dI_1 - dI_2 \tag{1}$$

$$dV_1 = \sqrt{3}m_V \sin\left(\frac{\pi}{3} - \alpha\right)$$
$$dV_2 = \sqrt{3}m_V \sin\alpha$$
$$dV_z = 1 - dV_1 - dV_2 \tag{2}$$

B. Common Mode Voltage Analysis for DMC

The common mode voltage (CMV) in a DMC can be defined as the voltage difference between the load neutral (N_o) and the grid neutral (N_i) of Fig. 1(a). The CMV can further be expressed in terms of output phase voltages referred to the grid neutral (ground) as in (3). It can therefore be justified that the CMV depends on the switching states of the DMC and the instantaneous values of grid voltage and a variation in either parameter can result in change in value of CMV at that instant.

$$V_{CMV} = \frac{1}{3}\left(v_{AN_i} + v_{BN_i} + v_{CN_i}\right) \tag{3}$$

III. PROPOSED SVM TECHNIQUE

This section describes the proposed space vector modulation technique (P-SVM) which targets the operation of the DMC at a higher voltage transfer ratio. The basic point of difference between the C-SVM and P-SVM is the modified current space vector diagram as shown in Fig. 3 and the usage of current vectors for reference formation. It can be noticed that the sectors of current space vector diagram have been displaced in the anti-clockwise direction by an angle of $\pi/6$ as compared to the conventional diagram in Fig. 2. On the other hand, the voltage space vector diagram is retained as in C-SVM. Unlike the conventional method, the proposed method uses three adjacent active current space vectors for the formation of the current reference. This results in the generation of an output voltage with lower distortion. This method eliminates the zero vectors but retains the usage of same active voltage vectors as C-SVM, but with normalized duty ratios. Elimination of zero vectors reduces the peak value of net CMV by 42%.

Table. V summarizes the range of operation, vector diagram for current reference generation in sector 1 and corresponding duty cycles of current and voltage vectors where $m\ (= 1.5m_I m_V)$ is the modulation index of the DMC. It should be noted that for operation of the DMC at the lower modulation index range, the method described in [16] using synchronously rotating vectors should be used because it completely eliminates CMV.

TABLE IV: Summary of P-SVM

m Index Range	Vector Diagram	Duty Ratios of Current Vectors	Duty Ratios of Voltage Vectors
$0.577 < m < 0.866$		$dI_1 = 1 - m_I \sin\left(\frac{\pi}{6} + \beta\right)$ $dI_2 = \sqrt{3}m_I \sin\left(\frac{\pi}{3} + \beta\right) - 1$ $dI_3 = 1 - m_I \cos\beta \quad (3)$	$dV_1' = dV_1/(dV_1 + dV_2)$ $dV_2' = dV_2/(dV_1 + dV_2) \quad (4)$

TABLE V: Summary of Switching States in P-SVM

	Si_1	Si_2	Si_3	Si_4	Si_5	Si_6
Sv_1	+1 +2 +5 +4 +7 +8	+5 +4 +7 +8 -2' -1'	+7 +8 -2' -1' -4' -5'	-2' -1' -4' -5' -8' -7'	-4' -5' -8' -7' +1 +2	-8' -7' +1 +2 +5 +4
Sv_2	+3 +2 +5 +6 +9 +8	+5 +6 +9 +8 -2' -3'	+9 +8 -2' -3' -6' -5'	-2' -3' -6' -5' -8' -9'	-6' -5' -8' -9' +3 +2	-8' -9' +3 +2 +5 +6
Sv_3	+3 -1 -4 +6 +9 -7	-4 +6 +9 -7 +1' -3'	+9 -7 +1' -3' -6' +4'	+1' -3' -6' +4' +7' -9'	-6' +4' +7' -9' +3 -1	+7' -9' +3 -1 -4 +6
Sv_4	-2 -1 -4 -5 -8 -7	-4 -5 -8 -7 +1' +2'	-8 -7 +1' +2' +5' +4'	+1' +2' +5' +4' +7' +8'	+5' +4' +7' +8' -2 -1	+7' +8' -2 -1 -4 -5
Sv_5	-2 -3 -6 -5 -8 -9	-6 -5 -8 -9 +3' +2'	-8 -9 +3' +2' +5' +6'	+3' +2' +5' +6' +9' +8'	+5' +6' +9' +8' -2 -3	+9' +8' -2 -3 -6 -5
Sv_6	+1 -3 -6 +4 +7 -9	-6 +4 +7 -9 +3' -1'	+7 -9 +3' -1' -4' +6'	+3' -1' -4' +6' +9' -7'	-4' +6' +9' -7' +1 -3	+9' -7' +1 -3 -6 +4

TABLE VI: Switching States & Space Vectors for $\overline{I}_{in}$ and $\overline{V}_o$ in Sector 1 for P-SVM

	I_{ab}		I_{ac}		I_{bc}	
	$V_1[100]$	$V_2[110]$	$V_2[110]$	$V_1[100]$	$V_1[100]$	$V_2[110]$
	dI_1dV_1	dI_1dV_2	dI_2dV_2	dI_2dV_1	dI_3dV_1	dI_3dV_2
A	a	a	a	a	b	b
B	b	a	a	c	c	b
C	b	b	c	c	c	c
SC	+1	+2	+5	+4	+7	+8

To be more specific, in order to generate a current reference in Sector 1, adjacent active current vectors I_{ab}, I_{ac} and I_{bc} are used. Similarly for generation of the voltage reference in Sector 1, $V_{[100]}$ and $V_{[110]}$ are used. Knowledge of the individual current and voltage space vectors, gives us information about the switching states of the DMC as obtained from Table. I. Table. V summarizes the switching states and switching sequence for the six sectors. It can be noticed that none of the cells have the zero vectors with SC of 0_a, 0_b or 0_c. This method eliminates its usage and achieves the functionality of the conventional modulation technique by the usage of only active space vectors. The switching sequence here ensures minimum number of switching transitions. The duty ratios of the nine switches of the DMC can be found out in the same fashion as described in Section. II. For P-SVM, this is summarized for a current and voltage reference vector in sector 1 in Table. VI. It can be observed how a combination of an active voltage and active current vector in ISVPWM can be translated to a switching configuration (SC) for the DMC. The SC follows the same duty cycle as formulated by ISVPWM. For example, I_{ab} and $V_1[100]$ results in the SC of '+1'. Furthermore the duty cycle of a switch with switching configuration S_{Cc} (say) can be given by the sum of dI_2dV_2, dI_2dV_1, dI_3dV_1 and dI_3dV_2. Here the denotation of the parameters stay the same as defined in the previous section for C-SVM.

IV. SIMULATION RESULTS

The proposed modulation technique has been validated through simulations in MATLAB/Simulink. For the purpose of comparison, the conventional technique is also simulated. The parameters are: grid voltage at $100V_{LL_{RMS}}$, switching frequency of $5kHz$, output frequency of $30Hz$, per phase balanced load of $R = 5.4\ \Omega$ and $L = 22\ mH$ and a modulation index of $m = 0.7$.

The simulation results for line-line output voltage, load currents and net CMV for C-SVM are given in Fig. 4(a)

and P-SVM in Fig. 4(b). It can be observed that both the methods of modulation result in the same output currents for the same load. The proposed method therefore achieves the basic purpose of SVPWM of forming currents and voltage of a desired magnitude. Fig. 4(c) presents a comparison of THD of the output line-neutral voltage of the converter. The dotted blue line and red line corresponds to THD values in p.u. for C-SVM and P-SVM respectively. Here 1 p.u = 100% THD. It can be seen that the THD is much lower with proposed modulation technique. There is also a very distinct difference in the shape of the output voltage waveform. This fact accounts to an improvement in its degree of distortion. There is a 25% reduction in percentage of THD at $m = 0.7$. It can be noticed that the peak values of CMV has been decreased by 42%.

V. EXPERIMENTAL RESULTS

Both the P-SVM and C-SVM algorithms of modulation have been tested on a laboratory prototype of the DMC. Integrated power IGBT module APTGF90TA60PG from Microsemi and gate driver 6SD106EI from CONCEPT are used. Control signals for SVM are generated from a FPGA (Xilinx XC3S500E). The block diagram of the experimental setup is provided in Fig. 5. Three grid voltages and three load currents are sent to the FPGA through a sensor board made up using LEM sensors. The grid voltages sensed are used for space vector modulation and the load currents sensed are used for safe 4-step commutation of the matrix converter. A diode bridge based clamp circuit is used for protection. The FPGA generates the 18 PWM signals for the switches on the MC.

The hardware based results for the same operational parameters as given in the previous section are presented in Fig. 6. The waveforms in Fig. 6(a) and Fig. 6(b) correspond to operation of the MC with C-SVM, while Fig. 6(c) and Fig. 6(d) correspond to the MC's operation with P-SVM. Fig. 6(a) and Fig. 6(c) presents the grid voltage, grid current, output line-neutral voltage and net CMV waveforms. The input line filter was designed to provide sufficient attenuation of

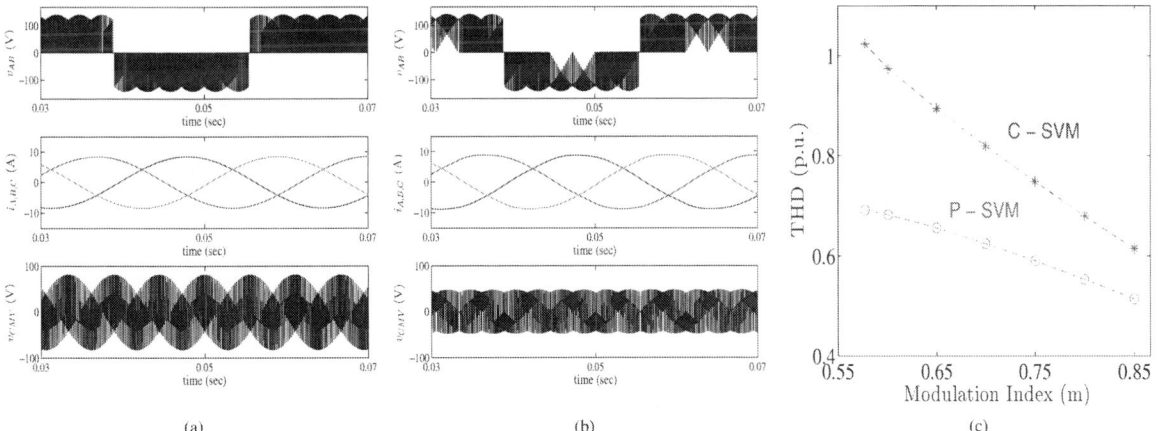

(a)   (b)   (c)

Fig. 4: Simulation results for DMC at $m = 0.7$ with (a) C-SVM and (b) P-SVM. From top to bottom - line-line output voltage, load currents and net CMV. (c) Output voltage THD (p.u) comparison

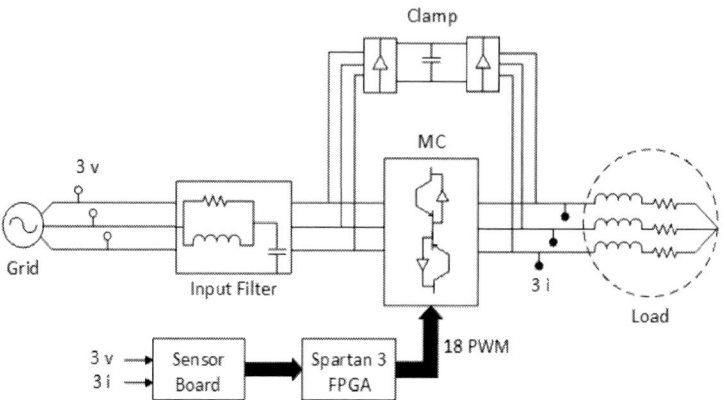

Fig. 5: Block Diagram of Experimental Setup

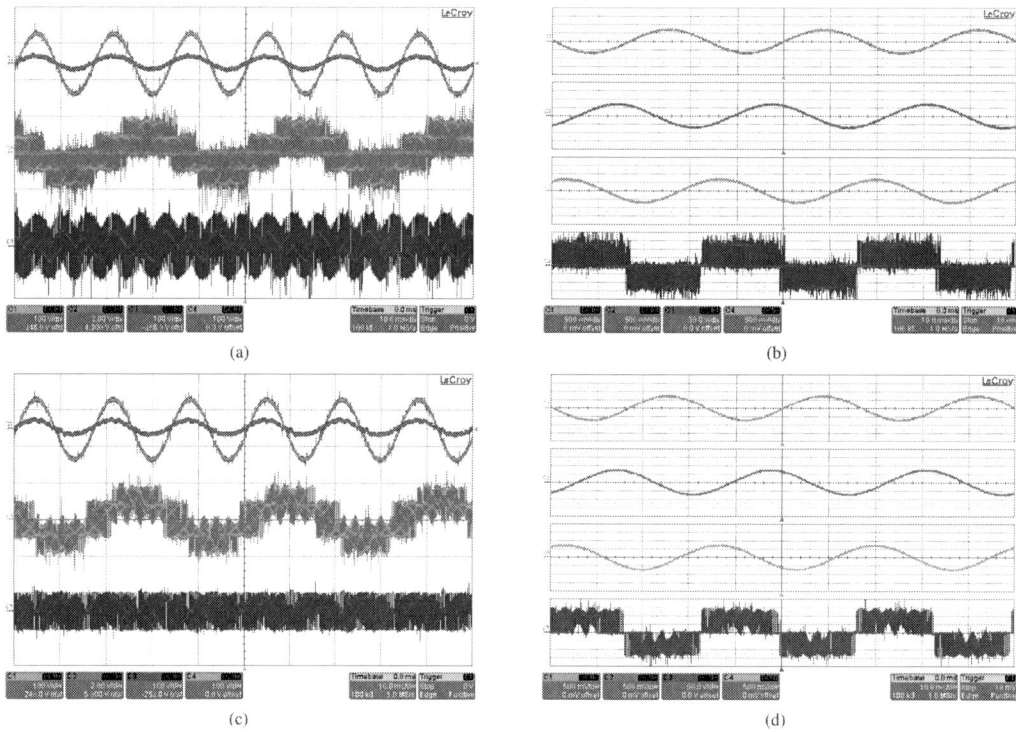

Fig. 6: Experimental results for DMC at $m = 0.7$ with C-SVM (top) and P-SVM (bottom). From top to bottom (for (a) & (c)) - grid current (20A/div) and grid voltage (100V/div), output line-neutral voltage (100V/div) and net CMV (100V/div). From top to bottom (for (b) & (d)) three phase load currents (5A/div) and output line-line voltage (50V/div)

injected current harmonics and ensure that the grid voltage and grid current were in phase with each other [23]. The difference in shape of the output voltage waveform which has been accounted for in the previous section, has been validated experimentally. The peak value of CMV has been decreased in P-SVM by a factor of $\sqrt{3}$ as compared to C-SVM.

Fig. 6(b) and Fig. 6(d) presents the three phase load currents and the output line-line voltage waveforms. These results are consistent with the simulation based results discussed in the previous section. Fig. 7 presents a comparison of RMS value of net CMV with C-SVM (in blue) and P-SVM (in red). The dotted continuous line and the solid star shaped points in these

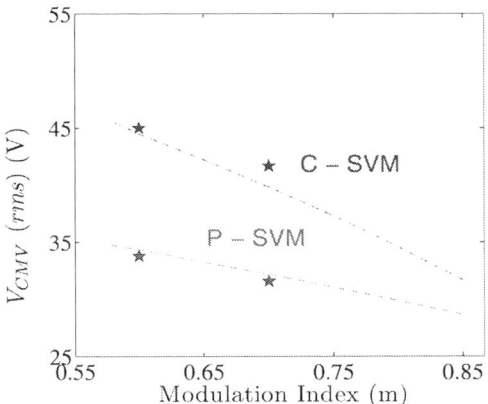

Fig. 7: Comparison of RMS of net CMV

plots corresponds to simulation results and experimental results respectively. There is an improvement in both the parameters of comparison.

VI. CONCLUSION

A new SVPWM technique for control of a matrix converter is proposed in this paper. This results in reduction of the peak value of CMV by a factor of $\sqrt{3}$ and also lowers output voltage distortion. The method targets the operation of the drive for a higher modulation index range ($0.577 \leq m \leq 0.866$). Existing SVM method which exclusively targets the operation at the lower modulation index range and completely eliminates CMV should be used in conjunction with the proposed method described in this paper to gain access to the full operational range of the DMC. The proposed technique has been validated through simulation and experimental results.

ACKNOWLEDGMENT

The authors would like to thank ONR (Office of Naval Research), Grant No. - N00014-13-1-0511 for their support.

REFERENCES

[1] A. Alesina and M. Venturini, "Analysis and design of optimum-amplitude nine-switch direct ac-ac converters," *Power Electronics, IEEE Transactions on*, vol. 4, no. 1, pp. 101–112, Jan 1989.

[2] P. Wheeler, J. Rodriguez, J. Clare, L. Empringham, and A. Weinstein, "Matrix converters: a technology review," *Industrial Electronics, IEEE Transactions on*, vol. 49, no. 2, pp. 276–288, Apr 2002.

[3] T. Friedli, J. Kolar, J. Rodriguez, and P. Wheeler, "Comparative evaluation of three-phase ac-ac matrix converter and voltage dc-link back-to-back converter systems," *Industrial Electronics, IEEE Transactions on*, vol. 59, no. 12, pp. 4487–4510, Dec 2012.

[4] A. Sahoo, K. Basu, and N. Mohan, "Comparison of filter components of back-to-back and matrix converter by analytical estimation of ripple quantities," in *Industrial Electronics Society, IECON 2013 - 39th Annual Conference of the IEEE*, Nov 2013, pp. 4831–4837.

[5] S. Chen, T. Lipo, and D. Fitzgerald, "Source of induction motor bearing currents caused by pwm inverters," *Energy Conversion, IEEE Transactions on*, vol. 11, no. 1, pp. 25–32, Mar 1996.

[6] J. Erdman, R. Kerkman, D. Schlegel, and G. Skibinski, "Effect of pwm inverters on ac motor bearing currents and shaft voltages," *Industry Applications, IEEE Transactions on*, vol. 32, no. 2, pp. 250–259, Mar 1996.

[7] J. Shang, Y. W. Li, N. Zargari, and Z. Cheng, "Pwm strategies for common-mode voltage reduction in current source drives," *Power Electronics, IEEE Transactions on*, vol. 29, no. 10, pp. 5431–5445, Oct 2014.

[8] N. Zhu, B. Wu, D. Xu, N. Zargari, and M. Kazerani, "Common-mode voltage reduction methods for medium-voltage current source inverter-fed drives," in *Energy Conversion Congress and Exposition (ECCE), 2011 IEEE*, Sept 2011, pp. 3136–3143.

[9] L. Haifeng, S. Shuang, Y. Junjie, L. Yituo, W. Lixun, and J. Shijun, "A nonzero vector pwm method to reduce common-mode voltage," in *Applied Power Electronics Conference and Exposition (APEC), 2012 Twenty-Seventh Annual IEEE*, Feb 2012, pp. 1604–1608.

[10] A. Hava and E. Un, "A high-performance pwm algorithm for common-mode voltage reduction in three-phase voltage source inverters," *Power Electronics, IEEE Transactions on*, vol. 26, no. 7, pp. 1998–2008, July 2011.

[11] T. Nguyen and H.-H. Lee, "Modulation strategies to reduce common-mode voltage for indirect matrix converters," *Industrial Electronics, IEEE Transactions on*, vol. 59, no. 1, pp. 129–140, Jan 2012.

[12] R. Pena, R. Cardenas, E. Reyes, J. Clare, and P. Wheeler, "Control of a doubly fed induction generator via an indirect matrix converter with changing dc voltage," *Industrial Electronics, IEEE Transactions on*, vol. 58, no. 10, pp. 4664–4674, Oct 2011.

[13] T. Nguyen and H.-H. Lee, "A new svm method for an indirect matrix converter with common-mode voltage reduction," *Industrial Informatics, IEEE Transactions on*, vol. 10, no. 1, pp. 61–72, Feb 2014.

[14] V. Padhee, A. K. Sahoo, and N. Mohan, "Svpwm technique with varying dc-link voltage for common mode voltage reduction in an indirect matrix converter," in *Energy Conversion Congress and Exposition (ECCE), 2015 IEEE*, Sept 2015, pp. 875–881.

[15] H.-H. Lee and H. Nguyen, "An effective direct-svm method for matrix converters operating with low-voltage transfer ratio," *Power Electronics, IEEE Transactions on*, vol. 28, no. 2, pp. 920–929, Feb 2013.

[16] H.-N. Nguyen and H.-H. Lee, "An enhanced svm method to drive matrix converters for zero common-mode voltage," *Power Electronics, IEEE Transactions on*, vol. 30, no. 4, pp. 1788–1792, April 2015.

[17] R. Gupta, A. Somani, K. Mohapatra, and N. Mohan, "Space vector pwm for a direct matrix converter based open-end winding ac drives with enhanced capabilities," in *Applied Power Electronics Conference and Exposition (APEC), 2010 Twenty-Fifth Annual IEEE*, Feb 2010, pp. 901–908.

[18] R. Baranwal, K. Basu, and N. Mohan, "Carrier-based implementation of svpwm for dual two-level vsi and dual matrix converter with zero common-mode voltage," *Power Electronics, IEEE Transactions on*, vol. 30, no. 3, pp. 1471–1487, March 2015.

[19] R. Baranwal, K. Basu, and N. Mohan, "An alternative carrier based implementation of space vector pwm for dual matrix converter drive with common mode voltage elimination," in *Industrial Electronics Society, IECON 2014 - 40th Annual Conference of the IEEE*, Oct 2014, pp. 1208–1213.

[20] J. Espina, C. Ortega, L. de Lillo, L. Empringham, J. Balcells, and A. Arias, "Reduction of output common mode voltage using a novel svm implementation in matrix converters for improved motor lifetime," *Industrial Electronics, IEEE Transactions on*, vol. 61, no. 11, pp. 5903–5911, Nov 2014.

[21] H. J. Cha and P. Enjeti, "An approach to reduce common-mode voltage in matrix converter," *Industry Applications, IEEE Transactions on*, vol. 39, no. 4, pp. 1151–1159, July 2003.

[22] C. Ortega, A. Arias, C. Caruana, and M. Apap, "Common mode voltage in dtc drives using matrix converters," in *Electronics, Circuits and Systems, 2008. ICECS 2008. 15th IEEE International Conference on*, Aug 2008, pp. 738–741.

[23] A. Sahoo, K. Basu, and N. Mohan, "Systematic input filter design of matrix converter by analytical estimation of rms current ripple," *Industrial Electronics, IEEE Transactions on*, vol. 62, no. 1, pp. 132–143, Jan 2015.

Soft-Switched Discontinuous Pulse-Width Pulse-Density Modulation Scheme

Arash Rahnamaee, Alireza Mojab, Hossein Riazmontazer, Sudip K. Mazumder, and Milos Zefran

Department of Electrical and Computer Engineering
University of Illinois, Chicago, IL: 60607
a.rahnamaee@gmail.com

Abstract— **This paper presents a soft-switched discontinuous pulse-width pulse-density modulation (PWPDM) scheme to decrease the switching losses of the capacitor-less high-frequency pulsating dc-link (HFPDCL) inverters. The proposed modulation scheme employs both pulse-width modulation (PWM) and pulse-density modulation (PDM) schemes to increase the overall efficiency of the inverter and generate high-quality output line-frequency (LF) Sine waveforms. In order to increase the efficiency of the converter, it decreases the switching requirement of the converter and it also provides soft-switching condition for the output inverter. The performance of the proposed PWPDM scheme is verified using am experimental 2 kW HFPDCL inverter.**

I. INTRODUCTION

The HFPDCL inverters [1]-[8] are compact, small foot-printed, modular, and reliable since they do not employ large dc-link capacitors and LF transformers. HFPDCL inverters have high-power density due to the elimination of the bulky dc-link capacitors and using of high-frequency (HF) transformers to provide galvanic isolation. They do not need any electrolytic capacitors, which results in a higher reliability than the fixed dc-link HFL inverters. However, they have three power conversion stages which require us to address the switching losses of the converter. The power conversion stages of the HFPDCL inverters are not decoupled since they do not employ any dc-link capacitors. As a result, they need a modulation scheme to synchronize the operation of the conversion stages and reduce the switching losses of the converter. It also decreases the switching losses and requirement of the HFPDCL inverter.

Fig. 1 shows the schematic of a HFPDCL inverter that consists of three conversion stages: 1) a dc/pulsating-dc converter that comprises two full-bridge converters and HF transformers, 2) an ac/pulsating-dc converter, and 3) a pulsating-dc/ac converter that generates the desired LF output waveforms. Therefore, this converter consists of three conversion stages. As a result, the switching losses of the converter need to be addresses and mitigated. Auxiliary circuits can be used to provide soft-switching condition for the converter to mitigate the switching losses of the switches.

However, auxiliary circuits require additional components and they increase the complicity of the topology. They impact the modularity of the converter as well. Therefore, modulation-based soft-switching schemes are better since they do not require any additional components. The presented PWPDM scheme is a modulation-based soft-switching scheme for the HFPDCL inverters to decrease the switching requirement and losses of the converter.

Modulation scheme are to decrease the total harmonic distortion (THD) of the output waveforms [9]-[13] or decrease the switching losses of the inverter [14]-[24]. Discontinuous modulation schemes reduce the switching requirement of the inverters by clamping one or two legs of the converter to the dc rail for a fraction of the line period. Therefore, the switching requirement can be decreased by 33% [14]-[19] or up to 66% [20]-[24]. The switching requirement can be decreased up to 33% without modulating the pulses on the pulsating-dc link (PDCL) waveform. PWPDM scheme clamps two predetermined legs of the pulsating-dc/ac converter to the dc bus. It also provides ZVS condition for the remaining leg of the pulsating-dc/ac converter. Therefore, it practically mitigates the switching losses of the pulsating-dc/ac converter. The presented PWPDM scheme takes advantage of the pulsating nature of the PDCL waveform. It employs PDM [25]-[29] and pulse-width modulation (PWM) schemes. It is only possible due to the pulsating nature of the PDCL waveform. PDM schemes decrease the switching requirement of the inverters by varying the frequency operation of the converter. It synthesizes the output sine waveforms by varying the density of the pulses instead of varying the width of the pulses.

II. PULSE-WIDTH PULSE-DENSITY MODULATION

The frond-end dc/ac converter synthesizes the PDCL waveform using PWM scheme. It synthesizes a series of pulses on the PDCL waveform representing the maximum output reference signal. The generated PDCL voltage waveform is shows in Fig. 2. The maximum reference signal changes every 60° of the output line cycle. Therefore, the dc/ac converter generates a stream of pulses based on the maximum reference signal during each operating sector [30].

978-1-4673-9551-9/16 $31.00 © 2016 IEEE

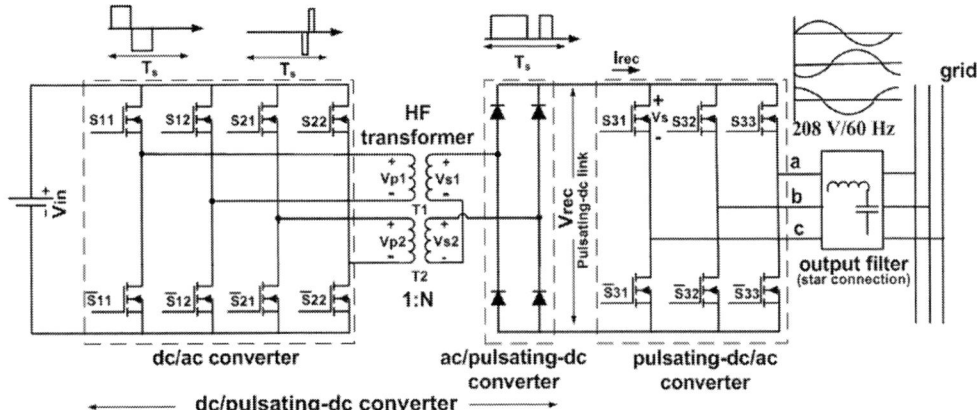

Fig. 1: Schematic of a three-phase HFPDCL inverter.

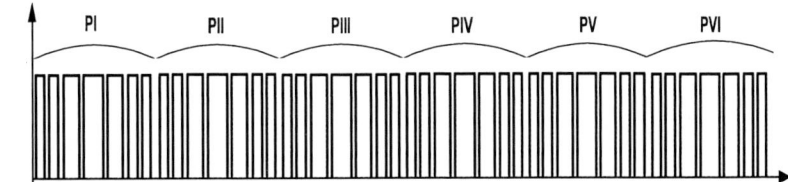

Fig. 2: The PDCL waveform when the dc/ac converter generates a stream of pulses based on the maximum reference signal.

Then, two predetermined legs of the pulsating-dc/ac converter are clamped to the dc-rail. The zero states are used to provide ZVS condition for the pulsating-dc/ac converter. Therefore, the zero-states need to be placed and adjusted on the PDCL waveform to turn on/off the switches of the pulsating-dc/ac converter during the generated zero states. The zero-states are generated on the PDCL waveform by the front-end dc/ac converter. Fig. 3 shows the generated PDCL waveform using the maximum reference signal as the modulation signal of the dc/ac converter. Two legs of the pulsating-dc/ac converter are clamped to the dc bus. The third leg turns on/off during the zero states of the PDCL waveform. The pulses are placed in the middle of the switching cycles to turn on/off the switches of the pulsating-dc/ac converter at the beginning and end of each switching cycle. It makes the modulation of the dc/ac and the pulsating-dc/ac converter symmetric [30].

Fig. 4 shows the synthesized signals of the pulsating-dc/ac converter over a line cycle. It shows that the generated output pulses of the inverter are a mix of PDM and PWM schemes. Therefore, two predetermined legs of the pulsating-dc/ac converter are clamped to dc bus. And the remaining legs, is modulated by PDM scheme, which operates under ZVS condition. The operation frequency of the remaining leg is varying and its duty cycle is always 50% [30].

In summary, the PWDM scheme uses the advantages of both PWM and PDM schemes. The PWM of the PDCL waveform decreases the THD of the output sine waveforms. This is possible due to pulsating nature of the PDCL waveform in contrast to the fixed-dc link inverter.

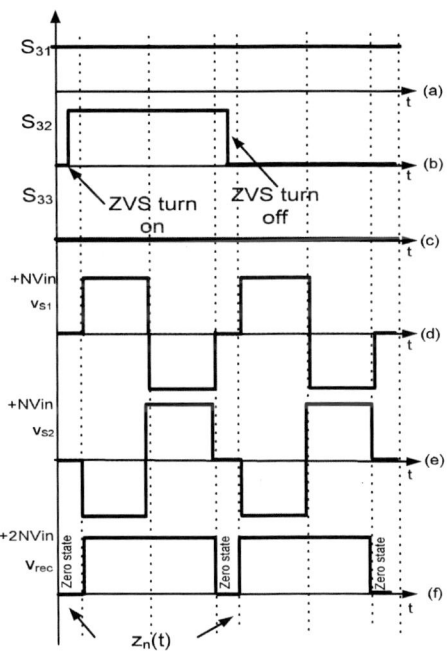

Fig. 3: Pulse placement, the generation of the PDCL waveform and ZVS operation of the pulsating-dc/ac converter: (a)-(c) the synthesized gate signals of S31, S32, and S33, (e)-(f) the secondary voltages of the HF transformer, and (g) the generated dc-link waveform.

978-1-4673-9551-9/16 $31.00 © 2016 IEEE

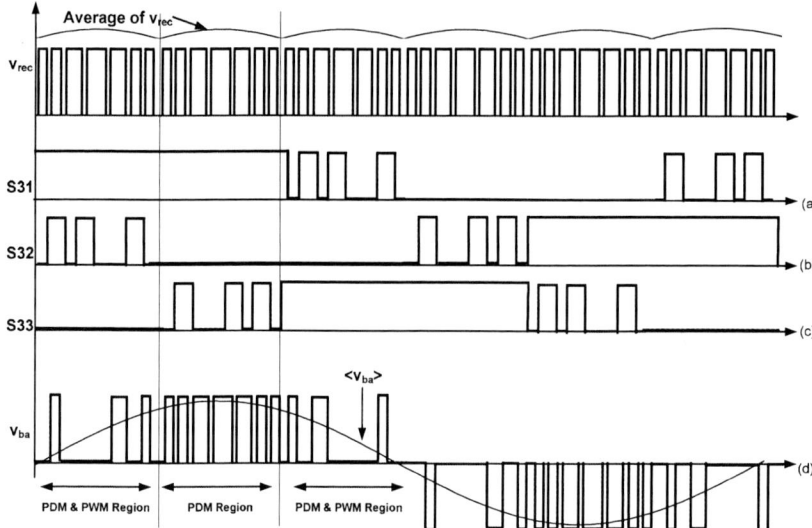

Fig. 4: The switching signals of the pulsating-dc/ac converter using PWPDM scheme. a) Gate signal of S_{31}, b) Gate signal of S_{32}, c) Gate signal of S_{33}, and d) The achieved output voltage waveform v_{ba}.

Each line cycle is divided into six operation sectors: PI-PVI. Sectors PI and PIV, Sectors PII and PV, and PIII and PVI are complementary. Therefore, gate signals of S_{31}, S_{32}, and S_{33} are inverted in Sectors PI and PIV, Sectors PII and PV, and Sectors PIII and PVI. As a result, PWPDM scheme divides in three distinctive operating sectors. For example during sectors PI and PIV, the first leg and the third leg of the pulsating-dc/ac converter are clamped to the dc bus in PI and PVI. Fig. 5 shows the controller for Sectors PI and PIV. Variable v_C is the upper and lower limits of the hysteresis band.

III. Experimental Results

The performance of the presented PWPDM scheme is evaluated by the implemented modular and compact HFPDCL inverter. The specifications of the implemented HFPDCL inverter are summarized in Table I. Fig. 6 shows the implemented capacitor-less HFPDCL inverter to verify the performance of the proposed PWPDM scheme. The soft-switching operation of a leg of the pulsating-dc/ac converter using PWPDM is shown in Fig. 7. It shows that switches of the pulsating-dc/ac converter turn on and turn off under zero voltage switching condition.

TABLE I. The Specifications and Main Components of the Implemented HFLDC Inverter.

Frequency	Input voltage	Output phase-to-phase voltage	Output power	Transformer turns ratio
20 kHz	200 V	208 V	2 kW	1:3.2
Name	**Components**			
Switches	IFS75S12N3T4, 1200 V, 75 A			
HF transformers	PQ 5050 Ferrite Core			

Fig. 6: The implemented HFPDCL inverter.

The PWM signals of the high-side switches of the converter are shown in Figs. 8 and 9. They show that two legs of the pulsating-dc/ac converter clamps to the dc rail and the remaining leg is modulated using PDM. This dramatically reduces the switching requirement of the pulsating dc/ac converter. The remaining leg of the pulsating-dc/ac converter operates under ZVS condition. Therefore, the switching losses of the pulsating-dc/ac converter are practically mitigated. The output voltage waveforms of the inverter before output filters are shown in Figs. 10 and 11. They also

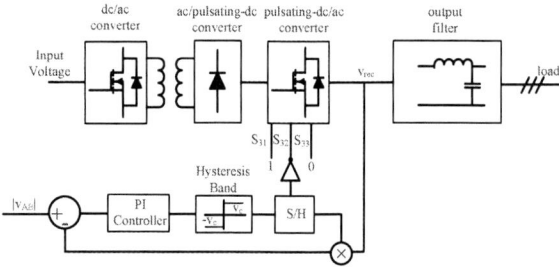

Fig. 5: The PWPDM modulator of the pulsating-dc/ac converter during PI and PIV.

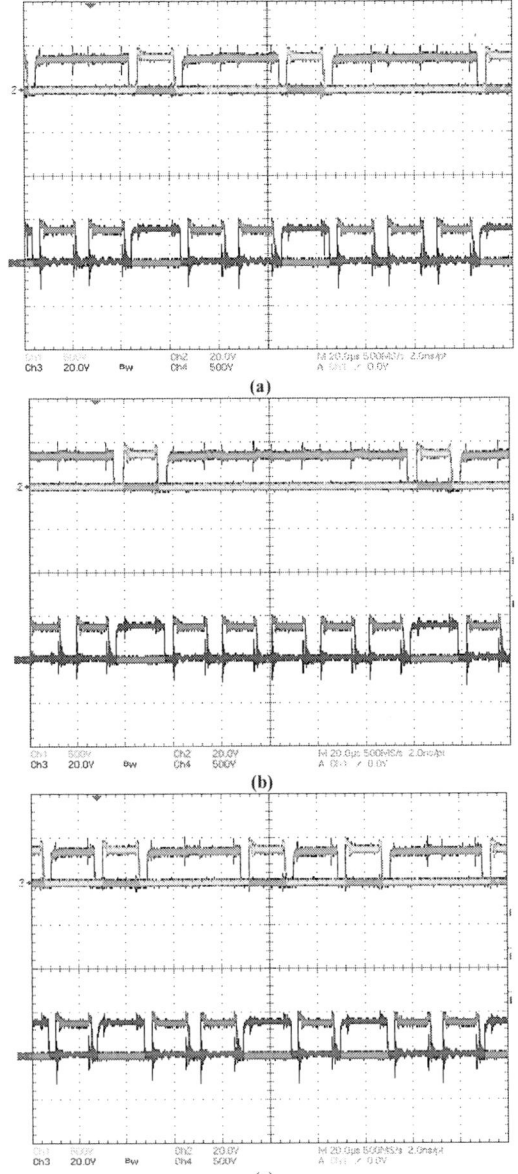

Fig. 7: Soft-switching operation of the pulsating-dc/ac converter using PWPDM: (a) S31 & S ⎯31,(b) S32, S⎯32, and (c) S33, S ⎯33.

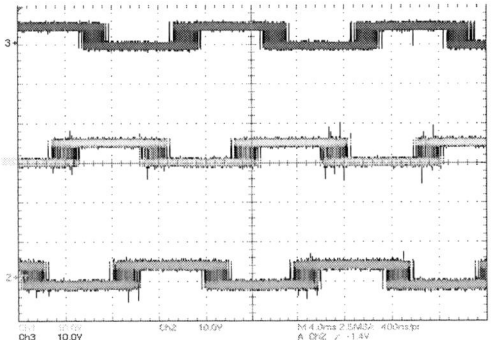

Fig. 8: PWM signals of the high-side switches of the pulsating-dc/ac converter.

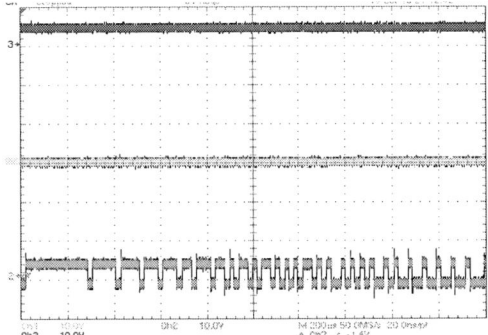

Fig. 9: Discontinuous and PDM operation of the switches of pulsating-dc/ac converter.

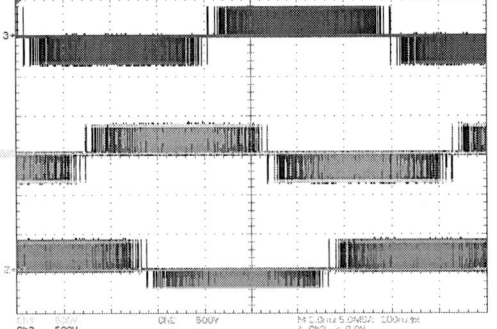

Fig. 10: Output waveforms of the HFPDCL inverter before filters over a line cycle.

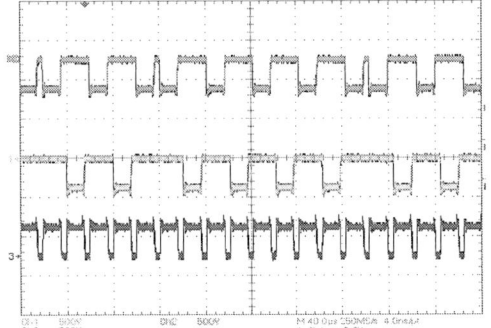

Fig. 11: The output voltage phase to phase waveforms before output filters.

show the PDM operation of the pulsating-dc/ac converter and the modulation of the output pulses using both PWM and PDM schemes. The front-end dc/ac converter indirectly participates in the PWM modulation of the output waveforms.

The resulting output sine waveforms of the HFPDCL inverter using PWPDM scheme is shown in Fig. 12, which shows high-quality output waveforms. Fig. 13 shows the overall efficiency of the HFPDCL inverter using the proposed PWPDM scheme. Fig. 14 shows the output THD of the HFPDCL inverter using PWPDM scheme. The THD is well below 5% for all range of output power.

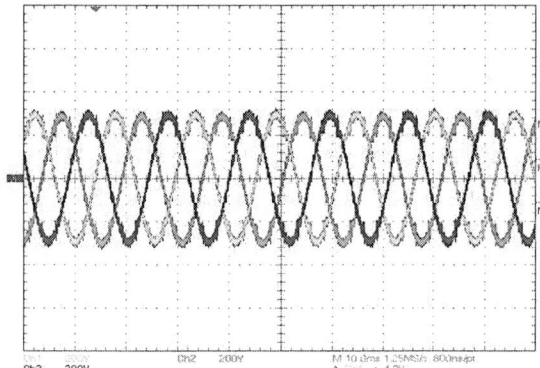

Fig. 12: The output sine waveforms of the HFPDCL inverter using PWPDM scheme.

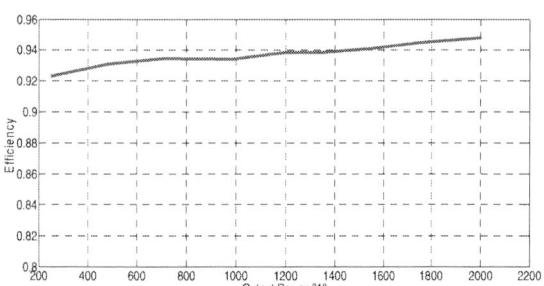

Fig. 13: The overall efficiency of the HFPDCL inverter using PWPDM scheme.

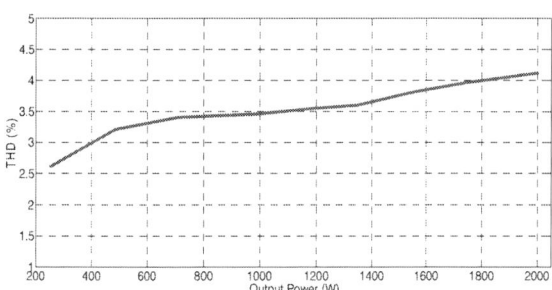

Fig. 14: The output THD of the HFPDCL inverter using PWPDM scheme.

The modulation scheme of the front-end dc/ac converter is PWM, and it operates at fixed frequency. It generates and synthesizes the pulses of the PDCL waveform. Therefore, its operating frequency is fixed and independent from the modulation index and load. The modulation index changes the width of the pulses on the PDCL waveform generated by the front-end dc/ac converter. However, it does not affect the operating frequency of the front-end dc/ac converter. The maximum operating frequency of the pulsating-dc/ac converter is limited to the frequency of the PDCL waveform, which is 20 kHz for the implemented converter. The minimum frequency of the pulsating-dc/ac converter can be set by the controller, which is 3 kHz for the implemented converter. Therefore, the minimum and maximum operating frequency of the converter is set by the controller and it is independent from the modulation index and load.

IV. CONCLUSIONS

A discontinuous pulse-width pulse-density modulation is proposed to improve the performance of the dc-link capacitor-less high-frequency pulsating dc-link inverters. It decreases the switching losses and requirement of the HFPDCL inverter. It provides zero-voltage switching condition for the pulsating-dc/ac converter. It also decreases the switching requirement of the converter. A compact and modular HFPDCL inverter is implemented to verify the performance of the proposed modulation scheme. The THD of the output waveforms of the inverter is well below 5% and the overall efficiency of the inverter reached to 95%.

V. REFERENCES

[1] J. Zhang, F. Huang; Bin Yan, and D. Chen, "Unidirectional buck DC-DC converter mode photovoltaic grid-connected inverters with high frequency link," in *Future Energy Electronics Conference (IFEEC)*, Nov. 2013, pp.376-381.

[2] D. Chen and L. Li, "Novel static inverters with high frequency pulse DC link," *IEEE Trans. On Power Electron.*, vol.19, no.4, pp.971-978, Jul. 2004.

[3] J. Pereda and J. Dixon "High-frequency link: A solution for using only one DC source in asymmetric cascaded multilevel inverters", *IEEE Trans. Ind. Electron.*, vol. 58, no. 9, pp. 3884 -3892 2011.

[4] H. Fujimoto, K. Kuroki, T. Kagotani, and H. Kidoguchi, "Photovoltaic inverter with a novel cycloconverter for interconnection to a utility line," in *IEEE IAS Annu. Meeting*, 1995, pp. 2461–2467.

[5] E. Koutroulis, J. Chatzakis, K. Kalaitzakis, and N. C. Voulgaris, "Bidirectional, sinusoidal, high-frequency inverter design," in *Inst. Electr. Eng. Electr. Power Appl.*, vol. 148, no. 4, pp. 315–321, Jul. 2001.

[6] S.K. Mazumder and R. Huang, "Multiphase converter apparatus and method", USPTO Patent# 7,768,800 B2, awarded on Aug 3, 2010.

[7] S.K. Mazumder, R.K. Burra, and K. Acharya, "A ripple-mitigating and energy-efficient fuel cell power-conditioning system", *IEEE Trans. On Power Electron.*, vol. 22, no. 4, pp. 1437-1452, 2007.

[8] S.K. Mazumder, "High-frequency inverters: From photovoltaic, wind, and fuel-cell based renewable- and alternative-energy DER/DG systems to battery-based energy-storage applications", Book Chapter in Power electronics handbook, Editor M.H. Rashid, Academic Press, Burlington, Massachusetts, accepted, 2010.

[9] K.J. Pratheesh, G. Jagadanand, and R. Ramchand, "An improved space vector PWM method for a three-level inverter with reduced THD," in *9th Int. Conf. Compatibility and Power Electronics* (CPE), Jun. 2015, pp.167-172.

[10] L.H.S.C Barreto, G.A.L Henn, P.P. Praca, R.N.A.L Silva, D.S. Oliveira, and E.R.C da Silva, "Carrier-based PWM modulation for THD and losses reduction on multilevel inverters," in *27th annu. Appl. Power Electron. Conf. and Expo. (APEC)*, Feb. 2012, pp.2436-2441.

[11] V. Agarwal, A. Kumar, R. Singh, and T.J. Robin, "Modified PWM schemes for Cyclo-inverters," in Int. *Power Eng. Conf., Dec. 2007* , vol., no., pp.655-660.

[12] M. Aleenejad, P. Moamaei, H. Mahmoudi, and R.Ahmadi, "Unbalanced Selective Harmonic Elimination for fault-tolerant operation of three phase multilevel Cascaded H-bridge inverters," in *Power Electron. Conf. and Expo. (APEC)*, Mar. 2015, vol., no., pp.1589-1594.

[13] M. Aleenejad, P. Moamaei, H. Mahmoudi, and R.Ahmadi, "A New Fault-Tolerant Strategy Based on a Modified Selective Harmonic Technique for Three Phase Multilevel Converters," in *IEEE Trans. Power Electron.*, vol.PP, no.99, pp.1-1, 2015.

[14] S. Ogasawara, H. Akagi, and A. Nabae, "A novel PWM scheme of voltage source inverter based on space vector theory," *in Rec. European Power Electron Conf.*, 1989, pp. 1197–1202.

[15] J. W. Kolar, H. Ertl, and F. C. Zach, "Influence of the modulation method on the conduction and switching losses of a PWM converter system", *IEEE Trans. Ind. Applicat.*, vol. 27, pp. 1063-1075, 1991.

[16] H. W. Van Der Broeck, "Analysis of the harmonics in voltage fed inverter drives caused by PWM schemes with discontinuous switching operation", in *Conf. Rec. European Power Electronics Conf.*, 1991, pp. 261 -266, 1991.

[17] K. Taniguchi, Y. Ogino, and H. Irie, "PWM technique for power MOSFET inverter", *IEEE Trans. Power Electron.*, vol. 3, pp. 328 -334, 1988.

[18] A.M. Hava, R.J. Kerkman, and T.A. Lipo "A high performance generalized discontinuous PWM algorithm", *IEEE Trans. on Ind. Appl.*, vol. 1 A, no. 5, pp.1059 -1071, 1998.

[19] K. W. Leen and F. C. Lai, "Operation principles of bi-directional full-bridge dc/pulsating-dc converter with unified soft-switching scheme and soft-starting capability,"in *Appl. Power Electron. Conf. and Expo. (APEC)*, 2000, pp.111-118.

[20] R. Huang and S. K. Mazumder, "A soft-switching scheme for an isolated dc/pulsating-dc converter with pulsating dc output for a three-phase high-frequency-link PWM Converter," *IEEE Trans. Power Electron.*, vol. 24, no. 10, pp. 2276-2288, 2009.

[21] R. Huang and S.K. Mazumder, "A soft switching scheme for multiphase dc/pulsating-dc converter for three-phase high-frequency-link PWM inverter," *IEEE Trans. on Power Electron.*, vol. 25, no. 7, pp. 1761-1774, 2010.

[22] S. K. Mazumder, "A novel hybrid modulation scheme for an isolated high-frequency-link fuel cell inverter," *IEEE Power and Energy Society General Meeting*, 2008, pp.1-7.

[23] A. Rahnamaee, S. K. Mazumder, and A. Tajfar, "A novel modulation-based zero-voltage and zero-current switched high-frequency-link inverter for renewable-energy systems," in *IEEE Power Electron. Specialist Conf.*, 2011, pp. 784-790.

[24] A. Rahnamaee and S. K. Mazumder, "A Soft-Switched Hybrid-Modulation Scheme for a Capacitor-Less Three-Phase Pulsating-DC-Link Inverter," *IEEE Trans. Power Electron.*, vol. 29, no. 8, pp. 3893–3906, Aug. 2014.

[25] Y. Nakata and J. Itoh, "Pulse density modulation control using space vector modulation for a single-phase to three-phase indirect matrix converter," in IEEE Energy Conversion Congr. and Exp.(ECCE), 2012, pp. 1753–1759.

[26] Y. Nakata and J. Itoh, "An experimental verification and analysis of a single-phase to three-phase matrix converter using PDM control method for high-frequency applications," in *9th IEEE Int. Conf. on Power Electron and Drive Syst.*, 2011, pp. 1084–1089.

[27] B. Jacob and M. R. Baiju, "Spread spectrum modulation scheme for two-level inverter using vector quantised space vector-based pulse density modulation," *IET Electr. Power Appl.*, vol. 5, no. 7, p. 589, 2011.

[28] A. Sandali, A. Cheriti, and P. Sicard, "Design Considerations for PDM Ac/ac Converter Implementation," in *Power Electron. Conf. and Expo. (APEC)*, 2007, pp. 1678–1683.

[29] J. Essadaoui, P. Sicard, E. Ngandui, and A. Cheriti, "Power inverter control for induction heating by pulse density modulation with improved power factor," in *Canadian Conf. on Elect. and Comput.Eng.. Toward a Caring and Humane Technology*, 2003, vol. 1, pp. 515–520.

[30] A. Rahnamaee, "Soft-Switched Hybrid-Modulation for an Isolated DC-Link-Capacitor-Less Pulsating-DC-Link Inverter," Ph.D. dissertation, Dept. Elect. and Comput.Eng., Univ of Illinois., Chicago, IL, 2015.

A Novel Flux Estimator Based on SOGI with FLL for Induction Machine Drives

Rende Zhao
Electrical Engineering Department
China University of Petroleum(Eastern China)
Qingdao China
zhaorende@126.com

Zhen Xin, Poh Chiang Loh, Frede Blaabjerg
Energy Department
Aalborg University
Aalborg Denmark
zxi@et.aau.dk, pcl@et.aau.dk, fbl@et.aau.dk,

Abstract— **It is very important to estimate flux accurately in implementing high-performance control of AC motors. Theoretical analysis has been made to illustrate the performance of the pure-integration-based and the Low-Pass Filter (LPF) based flux estimators. A novel flux estimator based on Second-Order General Integrator (SOGI) with Frequency Locked-Loop (FLL) is investigated in this paper for induction machine drives. A single SOGI instead of pure integrator or LPF is used to integrate the back electromotive force (EMF). It can solve the problems of the integration saturation and the dc drift caused by the initial conditions with no need for the magnitude and phase compensation. Because the dc and harmonic components are inversely proportional to the speed in the estimated flux, the performance of the single SOGI-based estimator become worse at low speed. A multiple SOGI-based flux estimator is the proposed to solve the problem. It can deeply attenuate the dc and harmonic components and then it has an excellent performance in a wide speed range. Theoretical study, simulation and experimental results validate the effectiveness of the proposed estimator.**

Keywords—Flux estimation, induction motor drives, second order generalized integrator, frequency locked loop

I. INTRODUCTION

Flux estimation is an important task in implementing high-performance control of AC motors. The voltage model is more attractive with no need for the speed sensors and the flux can be simply obtained by integrating the back ElectroMotive Forces (EMFs). However, it is well known that pure integrators suffer from two major problems. The first problem is dc drift, because a dc component generated by measurement devices, circuits and A/D conversion processes is inevitable. The dc component, no matter how small it is, can drive the integrator into saturation. The second problem is caused by the initial conditions, which lead to the appearance of a dc offset at the output of the integrator, which actually does not exist.

So a pure integrator is usually replaced by LPF[1],[2]. However, a LPF will produce errors in both the magnitude and the phase of the estimated flux, especially when the motor runs at a frequency lower than the cut-off frequency [2]. This problem has motivated diverse researches on the techniques of eliminating the magnitude and phase errors inherent in the LPF. The techniques can be classified into two kinds according to whether the cutoff frequency is programmable or not. The estimation of the fluxes using a fixed cut-off LPF is easy to implement. But practically, it is difficult to select an

appropriate cut-off frequency because the trade-offs must be made among the magnitude and phase errors, the damping of the dc drifts and the dynamic response [1]. In [2] and [3] a first-order programmable LPFs with the magnitude and phase compensation are used for flux estimation. The cut-off frequency of the LPFs, the parameters of the magnitude and phase compensation are functions of the synchronous angular frequency, which must be estimated. The accuracy of the synchronous angular frequency estimation is an important factor of the whole flux estimation. It is estimated with a Phase Locked Loop (PLL) [2] or as a function of the back-EMFs and fluxes [3]. B. K. Bose proposed a programmable cascaded LPF [4] which consists of n identical first-order LPFs with angular compensation. Yu Wang proposed two improved flux estimation method with programmable cascade LPFs. The first one contains a three-stage cascade LPF and a high-pass filter [5]. The second one contains a five-stage cascade LPF and a high-pass filter [6]. Each filter in the two methods is programmable according to the synchronous angular frequency.

As mentioned previously, both the pure integrator and the LPF have application problems, when they are used for flux estimation. To solve these problems, a new methodology for flux estimation based on SOGI with FLL is proposed. The effectiveness of the proposed method is proved by theoretical analysis and experiments.

II. PROBLEM OF FLUX ESTIMATION

In the stationary reference frame, the stator and rotor fluxes estimation based on the voltage model can be expressed as:

$$\psi_s = \int (u_s - R_s i_s) dt = \int e_s dt \tag{1}$$

$$\psi_r = \frac{L_r}{L_m} \left[\int (u_s - R_s i_s) dt - \sigma L_s i_s \right] = \frac{L_r}{L_m} \left[\int e_s dt - \sigma L_s i_s \right] \tag{2}$$

where R_s is the stator resistance. L_s is the stator inductance. $\sigma = (L_s L_r - L_m^2)/L_m^2$ is the leakage flux coefficient. L_m and L_r are the magnetizing and rotor inductances respectively. e_s is the back-EMF vector of the stator winding. $\psi_s = \psi_{s\alpha} + j\psi_{s\beta}$, $\psi_r = \psi_{r\alpha} + j\psi_{r\beta}$, $u_s = u_{s\alpha} + ju_{s\beta}$, $i_s = i_{s\alpha} + ji_{s\beta}$ are vectors of the stator flux, rotor flux, stator voltage and stator current respectively. It can be seen that the

---

This work was supported by Shandong Provincial Natural Science Foundation, China (ZR2014EEM012).

978-1-4673-9551-9/16 $31.00 © 2016 IEEE

estimation of $\psi_{s\alpha}$ is the same as $\psi_{s\beta}$. And the rotor flux is similar to the stator flux from (1) and (2). When discussing the estimation of $\psi_{s\alpha}$ for example, the same results are applicable to the estimation of $\psi_{s\beta}$ and ψ_r. $\psi_{s\alpha}$ can be expressed as

$$\psi_{s\alpha} = \int (u_{s\alpha} - R_s i_{s\alpha})dt = \int e_{s\alpha}dt \qquad (3)$$

A. Practical back-EMF signal

In practice $e_{s\alpha}$ contains not only the fundamental component but also dc and harmonic components. The dc component can be generated by voltage and current sensors, operational amplifier, voltage bias circuit and A/D conversion [7],[8]. The harmonics are caused by nonlinearities of the inverter and the motor, such as dead time and motor cogging effects[7]. So $e_{s\alpha}$ is described by

$$e_{s\alpha} = A_0 + A_1 \sin(\omega_1 t + \varphi_{01}) + \sum A_h \sin(\omega_h t + \varphi_{0h}) \qquad (4)$$

where A_0 is the dc component. $A_1 \sin(\omega_1 t + \varphi_{01})$ is the fundamental component, and $A_h \sin(\omega_h t + \varphi_{0h})$ represents the h-th harmonic component. The fundamental frequency and magnitude will decrease when the speed decreases. But the magnitude of the harmonics will rise when the speed decreases because the number of the dead-time error pulses increases in a fundamental period. So the problem caused by dc drift and harmonics are more serious when the speed is lower. The Laplace transform of $e_{s\alpha}$ is

$$E_{s\alpha}(s) = L\left[e_{s\alpha}\right] = \frac{A_0}{s} + A_1 \frac{\omega_1 \cos\varphi_{01} + s \sin\varphi_{01}}{s^2 + \omega_1^2} $$
$$+ \sum A_h \frac{\omega_h \cos\varphi_{0h} + s \sin\varphi_{0h}}{s^2 + \omega_h^2} \qquad (5)$$

B. Analysis of Flux Estimator Based on Pure Integrator

According to (3), the flux estimator based on a pure integrator can be given as

$$\Psi_{s\alpha\_I}(s) = \frac{1}{s} E_{s\alpha}(s) \qquad (6)$$

where $\Psi_{s\alpha\_I}(s)$ is the Laplace transform of flux $\psi_{s\alpha\_I}$ gained by the pure integrator. From (5) and (6), $\psi_{s\alpha\_I}$ is given by the inverse Laplace transform of $\Psi_{s\alpha\_I}(s)$ as

$$\psi_{s\alpha\_I} = L^{-1}\left[\Psi_{s\alpha\_I}(s)\right]$$
$$= A_0 t + \frac{A_1 \cos\varphi_{01}}{\omega_1} + \sum \frac{A_h \cos\varphi_{0h}}{\omega_h} \qquad (7)$$
$$+ \frac{A_1}{\omega_1} \sin(\omega_1 t + \varphi_{01} - \frac{\pi}{2}) + \sum \frac{A_h}{\omega_h} \sin(\omega_h t + \varphi_{0h} - \frac{\pi}{2})$$

Equation (7) is a time-domain expression of the flux estimated by a pure integrator. The integration of dc

component increases with time. So no matter how small the dc offset is, it can finally drive the pure integrator into saturation. New dc drift occurs caused by initial conditions of the fundamental and harmonic components. The drift, representing a constant dc flux in a motor, does not exist during normal motor operation. It will lead to the eccentric of the flux circle and it will affect the control.

The magnitude of the fundamental flux is the magnitude of the fundamental back-EMF divided by the fundamental frequency. The phase of the flux lags behind the back-EMF for 90 degrees. The harmonic components of the flux is similar to the fundamental one.

Pure integrator shows good harmonic attenuation characteristic, when the frequency is high. Harmonics are not a problem at a high speed and can usually be ignored. But A_h will become larger and ω_h will become smaller when the speed decreases, which means that the harmonic must not be ignored at low speed.

Among the five terms of (7), only the fundamental flux is needed. But it is mixed with the other four terms. There have been significant efforts on how to reduce or remove the effects of the four terms. Elimination of the saturation caused by the dc offset and the dc drift caused by initial conditions is the primary issue to be solved.

C. Analysis of Flux Estimator Based on LPF

To solve the problem of the saturation and the dc drift, a first-order LPF is used to replace pure integrator. Then the flux estimator can be expressed as

$$\Psi_{s\alpha\_F}(s) = \frac{1}{s + \omega_c} E_{s\alpha}(s) \qquad (8)$$

where $\Psi_{s\alpha\_F}(s)$ is the Laplace transform of flux $\psi_{s\alpha\_F}$ gained by LPF. ω_c is the cutoff frequency of the LPF. ω_c is usually much less than ω_1. From (5) and (8), $\psi_{s\alpha\_F}$ is given by the inverse Laplace transform of $\Psi_{s\alpha\_F}(s)$ as

$$\psi_{s\alpha\_F} = L^{-1}\left[\Psi_{s\alpha\_F}(s)\right]$$
$$= \frac{A_0}{\omega_c} - \frac{A_0}{\omega_c} e^{-\omega_c t}$$
$$+ \frac{A_1 \cos(\varphi_{01} + \theta_1)}{\sqrt{\omega_1^2 + \omega_c^2}} e^{-\omega_c t} + \sum \frac{A_h \cos(\varphi_{0h} + \theta_h)}{\sqrt{\omega_h^2 + \omega_c^2}} e^{-\omega_c t} \qquad (9)$$
$$+ \frac{A_1}{\sqrt{\omega_1^2 + \omega_c^2}} \sin(\omega_1 t + \varphi_{01} - \frac{\pi}{2} + \theta_1)$$
$$+ \sum \frac{A_h}{\sqrt{\omega_h^2 + \omega_c^2}} \sin(\omega_h t + \varphi_{0h} - \frac{\pi}{2} + \theta_h)$$

where $\theta_1 = \tan^{-1}\dfrac{\omega_c}{\omega_1}$ is the phase error of the estimated fundamental flux. $\theta_h = \tan^{-1}\dfrac{\omega_c}{\omega_h}$ is the phase error of the estimated h-th harmonic flux.

Comparing (9) with (7), the flowing differences can be found.

Firstly, the saturation caused by integrating the dc offset becomes a steady-state dc offset and a dynamic component decays with time. The steady-state dc offset is inversely proportional to ω_c. The dc drift caused by initial conditions decays exponentially with time and disappear in the steady-state. Secondly, the magnitude and phase errors occur in the fundamental component. So the magnitude and phase compensation must be taken. Thirdly, the harmonic components still exist in the LPF estimator and their magnitudes and phases have a little difference from the pure integrator.

The flux estimator based on LPF solves the saturation and dc drift problems of the pure integrator. But it brings the amplitude and phase errors to the fundamental flux simultaneously as well as the harmonic problems at low speed still exist.

III. PROPOSED FLUX ESTIMATOR

A. Second Order Generalized Integrator(SOGI)

SOGI is widely used to obtain the phase of grid voltage [9-12]. It is usually used to extract sinusoidal signal and its quadrature component in order to obtain the phase and magnitude of the signal. The traditional structure of a SOGI is shown in Fig. 1.

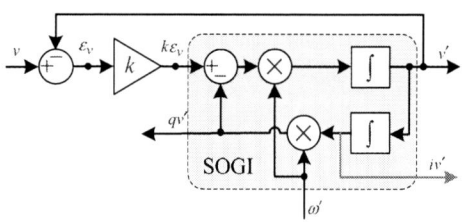

Fig. 1. Traditional structure of SOGI.

In Fig. 1 ω' and k are the tuning frequency and the damping factor. If the input signal v is a sinusoidal signal, v' is the observation of v. ε_v is the error between v and v'. qv' is the quadrature component of v'. So the phase and magnitude of the input sinusoidal signal can easily be calculated from the two in-quadrature outputs. There is an intermediate variable iv' in Fig.1. Its phase is 90-degree lagged behind the input and its magnitude is that of the input divided by its frequency. So iv' is the integration of the input sinusoidal signal.

B. Analysis of Flux Estimator Based on a Single SOGI

A flux estimator is obtained by modifying the tradition structure of SOGI to obtain the integration of the input, as shown in Fig. 2.

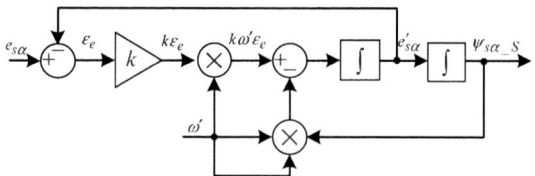

Fig. 2. Flux estimator based on a single SOGI.

According to Fig. 2, the flux estimator can be expressed as

$$\Psi_{s\alpha\_S}(s) = \frac{k\omega'}{s^2 + k\omega's + \omega'^2} E_{s\alpha}(s) = S(s)E_{s\alpha}(s) \quad (10)$$

$$S(s) = \frac{\Psi_{s\alpha\_F}(s)}{E_{s\alpha}(s)} = \frac{k\omega'}{s^2 + k\omega's + \omega'^2} \quad (11)$$

where $\Psi_{s\alpha\_S}(s)$ is the Laplace transform of the flux $\psi_{s\alpha\_S}$ obtained by the single SOGI-based estimator. $S(s)$ is the transfer function of the estimator. From (5) and (10), $\psi_{s\alpha\_S}$ is given by inverse Laplace transform of $\Psi_{s\alpha\_S}(s)$.

$$
\begin{aligned}
\psi_{s\alpha\_S} &= L^{-1}\left[\Psi_{s\alpha\_S}(s)\right] \\
&= \frac{A_0 k}{\omega'} + f_0(t)e^{-\frac{1}{2}k\omega't} + f_1(t)e^{-\frac{1}{2}k\omega't} + \sum f_h(t)e^{-\frac{1}{2}k\omega't} \\
&+ \frac{A_1}{\omega_1}\frac{k\omega'\omega_1}{\sqrt{(\omega'^2 - \omega_1^2)^2 + (k\omega'\omega_1)^2}}\sin(\omega_1 t + \varphi_{01} - \frac{\pi}{2} + \gamma_1) \\
&+ \sum \frac{A_h}{\omega_h}\frac{k\omega'\omega_h}{\sqrt{(\omega'^2 - \omega_h^2)^2 + (k\omega'\omega_h)^2}}\sin(\omega_h t + \varphi_{0h} - \frac{\pi}{2} + \gamma_h)
\end{aligned}
\quad (12)
$$

where $f_0(t)$, $f_1(t)$, $f_h(t)$ are the transient components of the dc, the fundamental initial condition and the harmonic initial conditions respectively. They all decay exponentially and will be zero in steady state. The detailed expressions of $f_0(t)$, $f_1(t)$ and $f_h(t)$ are given in Appendix. γ_1 and γ_h are expressed as

$$\gamma_1 = \arctan\frac{(\omega'^2 - \omega_1^2)}{k\omega'\omega_1} \quad (13)$$

$$\gamma_h = \arctan\frac{(\omega'^2 - \omega_h^2)}{k\omega'\omega_h} \quad (14)$$

In steady state, if the fundamental frequency is precisely observed, which means $\omega' = \omega_1$, then (12) is rewritten as

$$
\begin{aligned}
\psi_{s\alpha\_S} &\approx \frac{A_0 k}{\omega_1} + \frac{A_1}{\omega_1}\sin(\omega_1 t + \varphi_{01} - \frac{\pi}{2}) \\
&+ \sum \frac{kA_h}{h\omega_h}\sin(\omega_h t + \varphi_{0h} - \frac{\pi}{2} + \gamma_h)
\end{aligned}
\quad (15)
$$

Comparing (15) with (7) and (9) , the flowing conclusion can be drawn.

Firstly, the single SOGI-based flux estimator can avoid the saturation caused by the dc component. But there is still a dc offset inversely proportional to ω_l . Dc drift caused by the initial conditions is eliminated in steady state. Secondly, there are no magnitude and phase errors in the fundamental component on the basis of the precise observation of the fundamental frequency. So the magnitude and phase compensations are not needed, which are excellent characteristics. Harmonic components are still existing. The magnitude of the h-th harmonic component in the pure integrator is inversely proportional to ω_h . The proportion in LPF is almost equal. This proportion in the single SOGI-based estimator is $h\omega_h/k$, where k is usually $\sqrt{2}$ and $h = 5, 7, 11, 13 \cdots$. So the harmonics are attenuated more severely by the single SOGI-based estimator than by the pure integrator and the LPF. But the attenuation also declines with the speed.

Bode diagrams of the pure integrator and the single SOGI-based estimator with $\omega' = 1$ Hz, 10 Hz, 50 Hz are shown in Fig. 3. At the frequency ω', the single SOGI-based estimator has the same magnitude and phase characteristic as the pure integrator. The dc offset attenuation of the single SOGI-based estimator declines with ω'. ω' is equal to the fundamental frequency ω_l in steady state, and ω_l reflects the motor speed. The harmonic components is attenuated more by the single SOGI-based estimator than by the pure integrator.

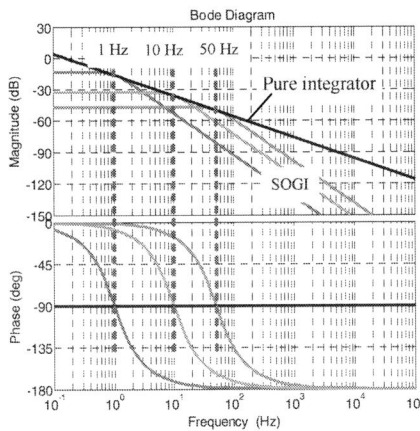

Fig. 3. Bode diagram of the single SOGI-based estimator and the pure integrator.

The single SOGI-based flux estimator shows good performance, but there are still problems such as the fundamental frequency observation, the dc offset and harmonics. Precise observation of the fundamental frequency is the key to the estimator. Ref. [11] and [12] develop a frequency locked loop to observe the frequency of the grid. It can be modified to obtain the fundamental frequency of the back-EMF.

C. Single SOGI-Based Flux Estimator with FLL

According to [11] and [12], a flux estimator based on the single SOGI with FLL is shown by Fig. 4. Different from the traditional FLL, the integration component is used instead of the quadrature component to observe the frequency. So the gain normalization part is changed to $\dfrac{k\omega'^2}{e_{s\alpha1}'^2 + \omega'^2 \psi_{s\alpha\_S}}$.

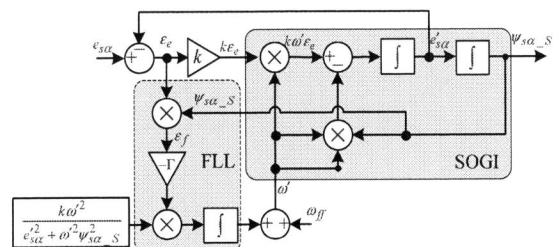

Fig. 4. The single SOGI-based flux estimator with FLL.

The signal ε_f collects information about error in frequency observation and it is suitable to act as the control signal of the FLL [11]. From Fig. 4 ε_f is given by

$$\varepsilon_f = \omega' \bullet \varepsilon_e \bullet \psi_{s\alpha\_S} = \frac{\dot{\omega}'}{-\gamma} \tag{16}$$

where $\dot{\omega}'$ is the differential of ω', and γ is the gain of the FLL. γ is positive to shift the SOGI resonance frequency ω' until matching the input frequency. γ varies with the magnitude, frequency of the input signal and k, which is the gain of flux estimation loop. In order to get a normalized gain $-\Gamma$ in the FLL, γ can be written as

$$\gamma = \frac{k\omega'^2}{e_{s\alpha1}'^2 + \omega'^2 \psi_{s\alpha\_S}} \Gamma \tag{17}$$

According to Fig. 4 and (16) the dc and harmonic components in $\psi_{s\alpha\_S}$ will influence the signal ε_f, then the accuracy of the frequency observation and the flux estimation will be affected especially at low speed. The dc and harmonic components in the flux estimation must be further attenuated to get precise frequency and fluxes.

D. Multiple SOGI-Based Flux Estimator with FLL

A multiple SOGI-based flux estimator is proposed to further attenuate the dc and harmonic components by adding the corresponding SOGI modules to the single SOGI-based flux estimator as shown in Fig. 5. The whole diagram of the estimator is shown in Fig. 5(a). The estimator contains three parts: the fundamental flux estimation, the dc and harmonics attenuation and the fundamental frequency observation. A normalized structure of the SOGI is shown in Fig.5(b). As the traditional structure of the SOGI is only suitable for ac signals, the so-called normalized structure means that it is suitable for not only ac signals but also dc signals. The dc component can be separated from the input signal by substituting 0 for h in

Fig. 5(b). The fundamental and harmonic components of the back-EMF can be separated by substituting a proper number for h. As only the fundamental flux is wanted, the SOGI modules of dc and harmonic output the estimated back-EMF only.

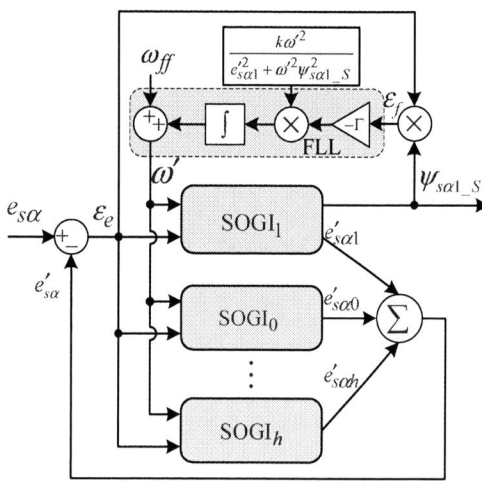

(a)

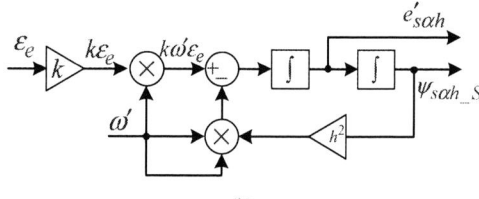

(b)

Fig. 5. The multiple SOGI-based flux estimator. (a) The whole diagram, (b) normalized structure of SOGI.

According Fig. 5(b), the transfer function between the estimated back-EMF and the system error is given by

$$B_h(s) = \frac{e'_{s\alpha h}}{\varepsilon_e}(s) = \frac{k\omega' s}{s^2 + h^2 \omega'^2} \quad (18)$$

The transfer function between the system error and the input back-EMF is given by

$$N(s) = \frac{\varepsilon_e}{e_{s\alpha}}(s) = \frac{1}{1 + B_0(s) + B_1(s) + \cdots}$$
$$= \frac{1}{1 + \sum_{h=0,1,\cdots} B_h(s)} \quad (19)$$

So the transfer function between the estimated h-th EMF and the input back-EMF is given by

$$M_h(s) = \frac{e'_{s\alpha h}}{e_{s\alpha}}(s) = \frac{B_h(s)}{1 + \sum_{h=0,1,\cdots} B_h(s)} \quad (20)$$

The transfer function between the estimated fundamental flux and the input back-EMF is given by

$$S_1(s) = \frac{\psi_{s\alpha 1\_S}}{e_{s\alpha}}(s) = \frac{1}{s} \frac{B_1(s)}{1 + \sum_{h=0,1,\cdots} B_h(s)} \quad (21)$$

According to (11) and (21), the Bode diagrams as shown in Fig. 6 can be plotted. The fundamental frequency is set to 0.1Hz, which means $\omega' = 0.2\pi$ rad/s. The 5th and 7th harmonics are shown as examples. From the diagrams, the following observations are made.

Firstly, the flux estimator $S_1(s)$ performs as a pure integrator at the fundamental frequency. The magnitude gain of the pure integrator is calculated as $20\log(1/0.2\pi) = 4.0364$. The phase of the pure integrator is 90 degrees delayed. The magnitude and phase responses of $S_1(s)$ in the Bode diagram are identical to them. Secondly, there is a notch for the dc component in the magnitude response of $S_1(s)$. So the dc component is eliminated in steady state. As there is a positive magnitude gain in $S(s)$, the dc component will be enlarged in a single SOGI-based flux estimator. Thirdly, there are notch characteristics at the 5th and 7th harmonic frequency in the magnitude response of $S_1(s)$. The harmonic components are suppressed deeply in the multiple SOGI-based estimator.

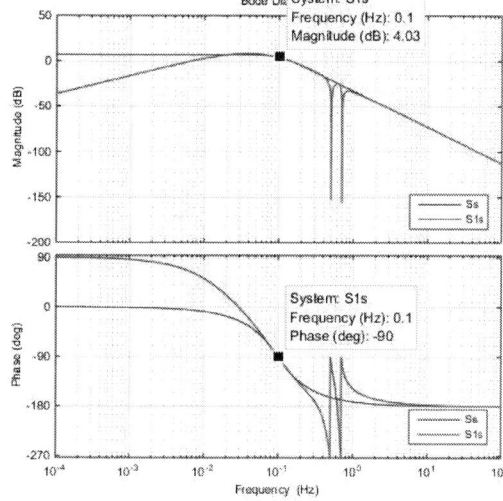

Fig.6. Bode diagram of the flux estimator based on single SOGI and multiple SOGIs.

IV. SIMULATION RESULTS

The conditions of different motor speed are considered for demonstrating the excellent performance of the proposed flux estimator by simulation.

Firstly the fundamental frequency is set to 50 Hz as being the high speed condition. The back-EMF is given by

$$e_{s\alpha} = \frac{311*3}{1000} + 311\sin(100\pi t) + \frac{311*2}{100}\sin(5*100\pi t) \quad (22)$$

where the magnitude of the fundamental component is set to 311V according to the normal phase voltage of the motor. The initial phase angle is set to 0 as the worst initial condition. The dc component is set to 0.3% of the normal phase voltage. And the 5th harmonic is set to 2% of the normal phase voltage. The observed fundamental frequencies and errors of the estimated fluxes based on the single SOGI and the multiple SOGIs with FLL are shown in Fig. 7(a) and Fig. 7(b). Although the two estimators both show good performance, the multiple SOGI-based estimator performs better than the single SOGI-based one.

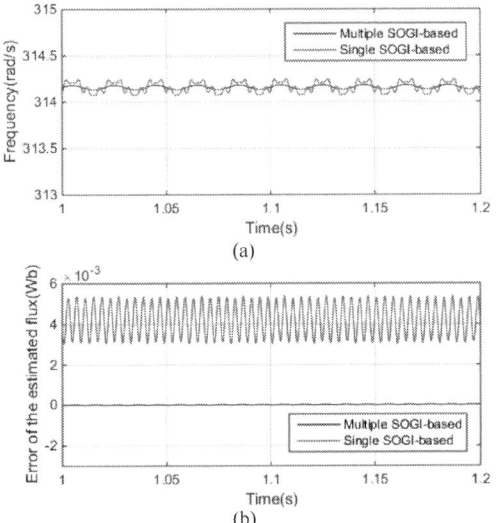

(a)

(b)

Fig. 7. Simulation results at high speed (50 Hz) (a) the observed fundamental frequencies, (b) the errors of estimated fluxes.

Next, the fundamental frequency is set to 2 Hz as the low speed condition. The back-EMF is given by

$$e_{s\alpha} = \frac{311*3}{1000} + \frac{311}{25}\sin(\frac{100}{25}\pi t) + \frac{311*2}{100}\sin(5*\frac{100}{25}\pi t) \quad (23)$$

where the magnitude of the fundamental component is changed proportionally to the fundamental frequency. The initial phase angle is set to 0 as the worst initial condition. The dc component and the 5th harmonic remain unchanged. The observed fundamental frequencies, the estimated fluxes and the errors of estimated fluxes based on the single SOGI and the multiple SOGIs with FLL are shown in Fig. 8 (a), (b) and (c). The multiple SOGI-based estimator still shows excellent performance but the performance of the single SOGI-based estimator become worse at low speed.

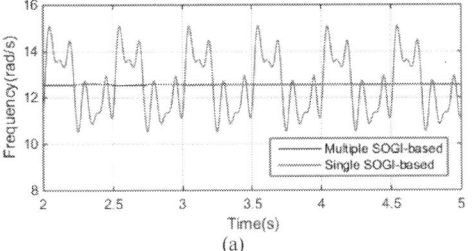

(a)

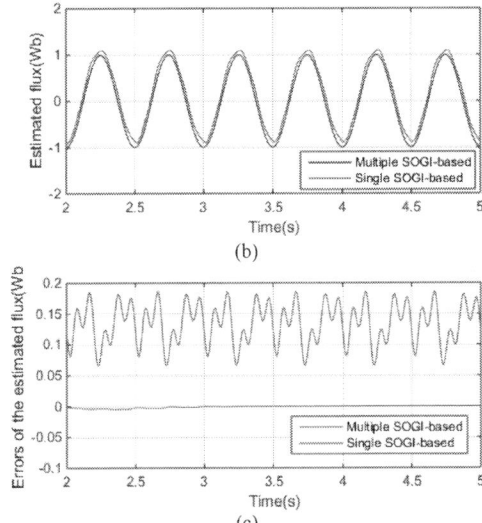

(b)

(c)

Fig.8. Simulation results at low speed (2 Hz)(a) the observed fundamental frequencies, (b) the estimated fluxes, (c) the errors of the estimated fluxes.

The multiple SOGI-based flux estimator shows excellent performance at both high and low speed. The single SOGI-based flux estimator shows good performance at high speed but poor performance at low speed as the dc and harmonic components are inversely proportional to the fundamental frequency according to (12).

V. EXPERIMENTAL RESULTS

The performance of the proposed flux estimator has been examined by the experimental setup shown in Fig. 9. The setup contains an 11 kW induction motor (IM) and a permanent magnet synchronous generator (PMSG), which is mechanically coupled with the IM through a gearbox (7.22:1) to produce load torque. The rated parameters of the IM and PMSG are listed in Table I and Table II respectively. An incremental encoder is used to obtain the speed of the IM.

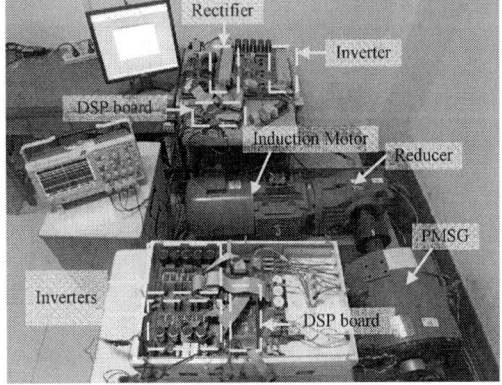

Fig. 9. Experimental setup to test the flux estimation methods.

TABLE I

Parameters of induction motor for test

Rated power [kW]	11	Rated speed [r/min]	1480
Rated voltage [V]	380	Rated current [A]	24.8
Stator resistance [Ω]	0.308	Rotor resistance [Ω]	0.365
Stator inductance [mH]	98.5	Pole pairs	2
Mutual inductance [mH]	95		

TABLE II
Parameter of PMSG for test

Rated power [kW]	15	Rated speed [r/min]	150
Rated voltage [V]	260	Rated current [A]	24
d-axis inductance[mH]	6.5	Pole pairs	10
q-axis inductance[mH]	13.5	Stator resistance[Ω]	0.362

The whole control scheme of the IM is shown in Fig. 10 and Texas Instruments TMS320F2812 DSP is adopted to execute the whole control algorithm. The switching frequency of the inverter is 19.2 kHz and the dead time is set to be $1.5\,\mu s$.

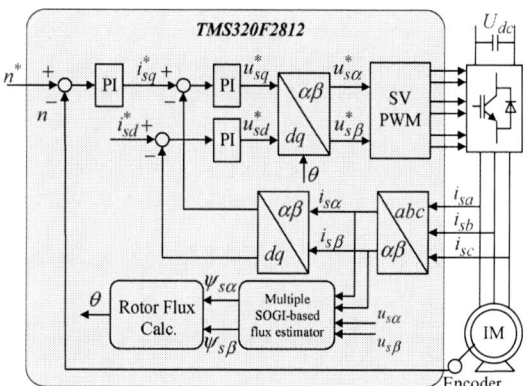

Fig.10. Control scheme of the IM based on the proposed flux estimator.

Steady state and dynamic experiments have been carried out and some of them are shown in Fig. 11-Fig. 13.

The three-dimensional flux-circle trajectories of the flux estimator based on a pure integrator, a LPF and the proposed multiple SOGIs with FLL, at the speed of 1200 r/min, are shown in Fig. 11. It can be seen the flux circle is eccentric gradually in the pure integrator, because of the integration of the dc component in the back-EMF. The LPF-based and the multiple SOGI-based flux estimators both solve this problem. There are less noises in the multiple SOGI-based estimator compared to the LPF-based estimator, because the former has a better high frequency attenuation characteristics than the latter.

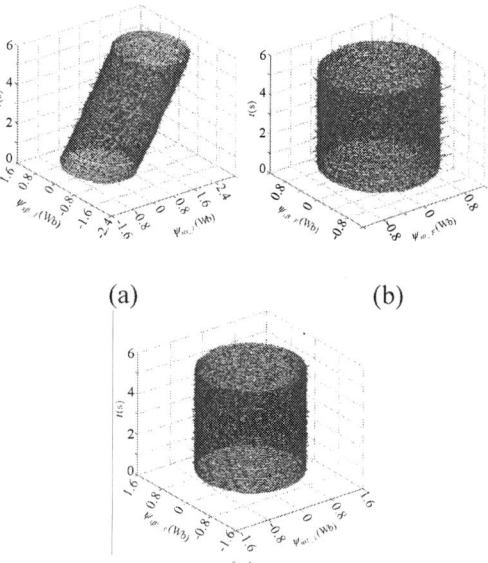

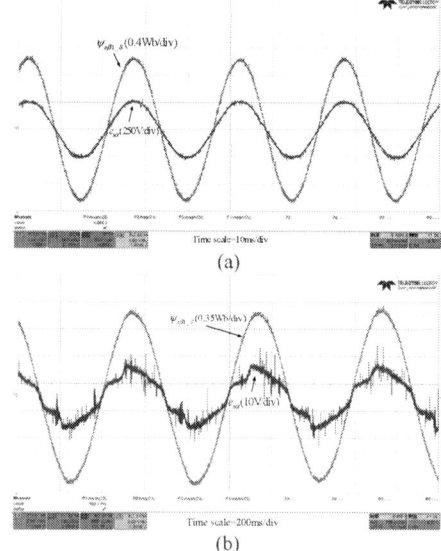

Fig. 11. Three-dimensional flux-circle trajectories (a) pure integrator, (b) LPF, (c) Multiple SOGI-based estimator.

The fluxes estimated by the multiple SOGI-based flux estimator at high speed and at low speed are shown in Fig. 12. Theoretically $e_{s\alpha}$ has the same phase with $\psi_{s\beta1\_S}$, for $e_{s\alpha}$ and $\psi_{s\beta1\_S}$ both lag behind $e_{s\beta}$ with 90 degrees. The waveform of the back-EMF is distorted at low speed, but the estimated flux is still sinusoidal. Its magnitude remains unchanged and the phase is still identical with $e_{s\alpha}$. The results reveal a perfect steady-state performance of the proposed estimator.

Fig.12. The back-EMF and the estimated flux. (a) 40 Hz, (b) 1.69 Hz.

The dynamic performance of the proposed estimator is shown in Fig. 13, when the motor speed changes from 500 r/min to 1200 r/min. The observed fundamental frequency changes in line with the speed as expected. The phase of the estimated flux can follow the change quickly. The results show a perfect dynamic performance of the multiple SOGI-based flux estimator.

both high and low speed. The excellent performance of the proposed flux estimator is verified by simulations and experiment results.

REFERENCES

[1] J. Hu, B. Wu, "New Integration Algorithms for Estimating Motor Flux over a Wide Speed Range", *IEEE Trans. Power Electron.*, vol. 13, no. 5, pp. 969-977, Sep. 1998.

[2] Mihai Comanescu, Longya Xu, "An Improved Flux Observer Based on PLL Frequency Estimation for Sensorless Vector Control of Induction Motor", *IEEE Trans. Power Electron.*, vol. 53, no. 1, pp. 51-56, Feb. 2006.

[3] Myoung-Ho Shin, Dong-Seok Hyun, Soon-Bong Cho and Song- Yul Choe, "An Improved Stator Flux Estimation for Speed Sensorless Stator Flux Orientation Control of Induction Motors", *IEEE Trans. Power Electron.*, vol.15, no.2, pp.312- 318, Mar. 2000.

[4] Bimal K. Bose, Nitin R. Patel, "A Programmable Cascaded Low-Pass Filter-Based Flux Synthesis for a Stator Flux-Oriented Vector-Controlled Induction Motor Drive", *IEEE Trans. Power Electron.*, vol.44, no.1, pp.140-143, Feb. 1997

[5] Yu Wang and Zhiquan Deng, "Improved Stator Flux Estimation Method for Direct Torque Linear Control of Parallel Hybrid Excitation Switched-Flux Generator", *IEEE Trans. Energy Conversion*, vol.27, no.3, pp.747-756, Sep. 2012.

[6] Yu Wang and Zhiquan Deng, "An Integration Algorithm for Stator Flux Estimation of a Direct-Torque-Controlled Electrical Excitation Switched-Flux Generator", *IEEE Trans. Energy Conversion*, vol.27, no.2, pp.411-420, Sep. 2012.

[7] K. Cho, and J. Seok, "Pure-Integration-Based Flux Acquisition With Drift and Residual Error Compensation at a Low Stator Frequency," *IEEE Trans. Ind. Applicat.*, Vol. 45, pp.1276-1285, Jul./Aug. 2009

[8] M. Karimi-Ghartemani, S. A. Khajehoddin,P. K. Jain, A. Bakhshai, and M. Mojiri, "Addressing DC Component in PLL and Notch Filter Algorithms," *IEEE Trans. Power Electron.*, Vol. 27, pp.78-86, Jan. 2012

[9] M. Ciobotaru, R. Teodorescu, and F. Blaabjerg, "A new single-phase PLL structure based on second order generalized integrator," in *Proc. IEEE Power Electron. Spec. Conf.* (PESC'06), Jun. 2006, pp. 1–6

[10] P. Rodriguez, R. Teodorescu, I. Candela, A. V. Timbus, and F. Blaabjerg, "New positive-sequence voltage detector for grid synchronization of power converters under faulty grid conditions," in Proc. PESC,2006, pp. 1–7.

[11] P. Rodriguez, A. Luna, M. Ciobotaru, R. Teodorescu, and F. Blaabjerg, "Advanced grid synchronization system for power converters under unbalanced and distorted operating conditions," in Proc. 32nd IEEE IECON, Paris, France, Nov. 6–10, 2006, pp. 5173–5178

[12] P. Rodríguez, A. Luna, I. Candela, R. Mujal, R. Teodorescu and F. Blaabjerg, "Multiresonant Frequency-Locked Loop for Grid Synchronization of Power Converters Under Distorted Grid Conditions", *IEEE Trans. Ind. Electron.*, Vol. 58, No. 1, pp.127-138, Jan. 2011.

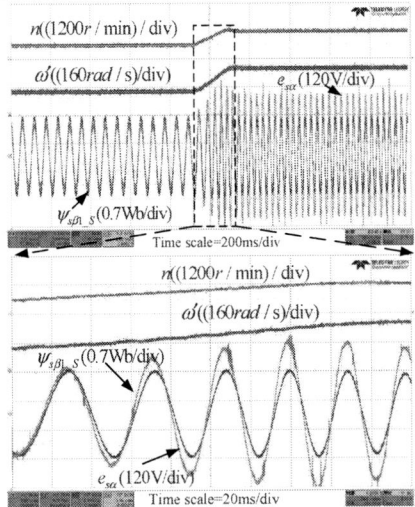

Fig. 13. Dynamic performance of the multiple SOGI-based estimator during the speed chang.

VI. CONCLUSION

In this paper a novel flux estimator based on SOGI using FLL is proposed in order to obtain precise fluxes at a wide speed range. Detailed theoretical analysis has been made to demonstrate the performance of the flux estimator based on the pure integrator, the LPF, the single SOGI and the multiple SOGIs with FLL. The SOGI-based flux with FLL can solve the problems caused by the initial conditions and the saturation caused by integrating the dc component, and no compensation for the phase and magnitude is needed. At high speed, the single SOGI-based flux estimator performs well, but the accuracy of the flux estimation is greatly influenced by the dc and harmonic components at low speed. The multiple SOGI-based flux estimator shows a perfect performance at

Appendix

$$f_0(t) = -\frac{A_0 k^2 \sinh\left(\frac{1}{2}t\omega'\sqrt{k^2-4}\right)}{\omega'\sqrt{k^2-4}} - \frac{A_0 k \cosh\left(\frac{1}{2}t\omega'\sqrt{k^2-4}\right)}{\omega'} \tag{A-1}$$

$$f_h(t) = A_h k \sqrt{\omega'^2(k^2-4)} \frac{\left[-k\omega'(\omega'^2+\omega_h^2)\sin\varphi_{0h}+\omega_h(k^2\omega'^2-2\omega'^2+2\omega_h^2)\cos\varphi_{0h}\right]\sinh\left(\frac{1}{2}t\sqrt{\omega'^2(k^2-4)}\right)}{\omega'(k^2-4)\left[(\omega'^2-\omega_h^2)^2+(k\omega'\omega_h)^2\right]} +$$

$$A_1 k\omega' \frac{\cosh\left(\frac{1}{2}t\sqrt{\omega'^2(k^2-4)}\right)\cos(\varphi_{01}+\gamma_1)}{\sqrt{(\omega'^2-\omega_h^2)^2+(k\omega'\omega_h)^2}} \tag{A-2}$$

Substituting 1 for h in (A-2), $f_1(t)$ can be obtained.

978-1-4673-9551-9/16 $31.00 © 2016 IEEE

Performance Characterization of Random Pulse Width Modulation Algorithms in Industrial and Commercial Adjustable Speed Drives

Kevin Lee	Guangtong Shen	Wenxi Yao	Zhengyu Lu
kevinlee@eaton.com	gt@purdue.edu	ywxi@zju.edu.cn	eeluzy@cee.zju.edu.cn
Eaton Corporation	Purdue University	Zhejiang University	Zhejiang University

Abstract – **One of the problems of the pulse width modulation (PWM) controlled AC machines is the acoustic noise that could become unacceptable when used in sensitive environments. Random PWM (RPWM) in industrial and commercial adjustable speed drives (ASD) results in the partial transfer of power from the discrete spectrum (narrowband noise) of the output voltage to the continuous spectrum (wideband noise), with advantageous effects on reducing acoustic noise in the motor drive system. In this paper, the theoretical power spectrum analysis as the basis for RPWM is presented. Five state-of-the-art RPWM strategies, their voltage, current and acoustic noise spectra characteristics are quantitatively evaluated. The PWM schemes and theoretical analysis are validated through a 2.2kW, 380V, 50Hz ASD, induction machine (IM) experimental setup. The results provide valuable data for practicing engineering community to choose the best option in real world applications.**

I. INTRODUCTION

PWM based ASDs are the driving force in industrial and commercial applications to meet ever growing demand for energy efficiency, feature rich functionalities and lower total cost of ownership. The harmonic voltage components from ASDs can produce a high pitch sound in the motor with mechanical resonance and vibration. Higher audible noise can cause lower acceptance particularly in buildings such as hospitals, schools, office complex, etc. The acoustic noise in ASD driven motor systems can be categorized in three parts: (a). Mechanical noise due to surface accidents, eccentricities, too accentuated or insufficient axial or radial displacement, shaft displacement, magnetics core vibration, cooling fans [1]. (b). Aero-dynamical noise constituted by air turbulences, siren effect and cavity resonance. These two noise types are practically independent from the electrical supply of the motor. Their frequencies in a Fourier analysis are typically below 2kHz in industrial applications. (c). Harmonics rich electromagnetic noise which is produced by excitations of each harmonic in the ASD output voltage spectrum to the motor it's driving [2-5].

Different PWM approaches are used in various ASD applications: lower switching frequency gives acoustic noise with low losses, higher switching frequency increases the losses in the inverter, but can minimize the acoustic problem. Significant improvement in the acoustic and electromagnetic noise in RPWM converters has been reported in many publications for keeping an average low switching frequency with low audible noise profile simultaneously. There are two categories of RPWM: one has a fixed switching frequency, the other is based on the variable switching frequency. An advanced RPWM technique is presented in [6] for improving

acoustic and EMI characteristics. The theoretical analysis based on a lead-lag RPWM is treated in [7]. The techniques based on adjusting the duration of the zero-vectors or adjusting the three pulse positions in a switching period are introduced in [8-9]. [10-11] address an intelligent RPWM varying ±1 kHz range while avoiding motor mechanical resonance frequency. The RPWM mathematical formulation based on power spectrum, instead of harmonic Fourier Series spectrum is investigated in dc/dc converters and dc/ac inverters [12-13]. The current sampling strategy and control performance effect under RPWM are explored in [14]. Formulas are given in [15] to help select the random PWM switching frequencies. A fixed switching frequency with adjusted duration of zero vectors or the three-pulse positions is proposed in [16]. Switching based on Markov chains in used in [17] and it is argued that RPWM could be beneficial for any power converter operation. In [18], each of the three phase pulses is placed randomly in each switching interval, such that at higher modulation index, the degree of freedom in RPWM still exists. A variable delay RPWM suitable for high volume, low cost automotive and uninterruptable power supplies (UPSs) applications is demonstrated in [19]. [20] proposes a scheme possessing the hybrid characteristics of the random pulse position PWM and the random carrier frequency PWM. A systematic approach to select the dithering frequency range for a typical ac drive is explained in [21]. For the removal of spectral power at frequencies harmful for the system, the method in [22] produces spectral nulls at selected frequency components. A two-phase separately pulse position pulse width modulation scheme is described in [23], which can obtain the randomization effect over the entire modulation index range.

However, a unified approach to quantitatively compare the line-to-line voltage, phase current and power spectra of dominating RPWM methods are not readily available. In this paper, theoretical and simulation analysis of state-of-the-art RPWM techniques including random distribution of zero voltage vector (RZD-PWM), random pulse center displacement (RCD-PWM), separately randomized pulse position PWM (SRP-PWM), random switching frequency PWM (RSF-PWM), variable delay random PWM (VDR-PWM) are studied in detail. The characteristics of their discrete frequency spectrum distribution, continuous frequency domain power density, current ripple harmonics, computation burden in digital signal processors (DSP) implementation, furthermore, the effects of PWM modulation index are presented. The results are validated in a 2.2kW, 380V, 50Hz experimental system.

978-1-4673-9551-9/16 $31.00 © 2016 IEEE

II. POWER SPECTRUM ANALYSIS OF RANDOM PWM

In order to effectively judge various randomized PWM switching implementations, we need to choose a definite and quantifiable analytical framework. This can not only make evaluation and verification of different schemes possible, but will also point out capabilities and limitations of different RPWM methods.

The Fourier transform is a display of the frequency components of a deterministic signal, but the Fourier transform expression of a random signal is itself a random function. Instead, power spectrum is often used as a benchmark to compare system performances because it is deterministic [12]. The power spectrum of a static stochastic process is deterministic so that the frequency characteristic of the process can be analyzed in a deterministic expression. In addition to theoretical derivation, the power spectrum can be estimated by enough samples of the random signal.

A. Power Spectrum Expression of a Random PWM Signal

A time domain PWM signal $q(t)$ is a switching function defined in Eq. (1), where $u_k(t-\xi_k)$ is a random pulse of the k-th cycle, ξ_k is the time at which the k-th cycle starts. Fig. 1 (a) represents $q(t)$, and Fig. 1 (b) is the waveform of $u_k(t-\xi_k)$, where T_k is the k-th cycle duration, α_k is on-state duration, ε_k is the delay within this cycle.

$$q(t) = \sum_{k=-\infty}^{\infty} u_k(t-\xi_k) \tag{1}$$

According to Wiener-Khintchine principle, the power spectrum of a static stochastic process $q(t)$ is the Fourier transform of the its autocorrelation, which is noted as $S_q(f)$ and expressed as in Eq. (2), where $U(f)$ is the Fourier transform of $u_k(t)$, $E\{.\}$ represents the expectation operation and $T = E\{T_k\}$ [12].

$$S_q(f) = \frac{1}{T} E\left\{|U(f)|^2\right\} - \frac{1}{T}\left|E\{U(f)\}\right|^2$$
$$+ \frac{1}{T^2}\left|E\{U(f)\}\right|^2 \sum_{k=-\infty}^{\infty} \delta\left(f - \frac{k}{T}\right) \tag{2}$$

The first two terms in Eq. (2) denote continuous power spectrum, whereas the third term is the discrete power spectrum at frequency $f=k/T$.

For the traditional space vector PWM and sinusoidal PWM, $E\{.\}$ can be omitted since they are deterministic signals. Thus the first two terms are dropped off with only discrete power spectrum at multiples of the switching frequency. The intensity of a discrete spectrum is propositional to the power at f.

For PWM with random pulse positions (RZD-PWM, RCD-PWM, and SRP-PWM), all three terms exist, which means the signal contains continuous spectrum and discrete spectrum at multiples of switching frequency. For PWM with random switching frequencies in a continuous range (RSF-PWM), the discrete term does not exist [24]. If the switching frequencies alternate randomly among a finite set of a few frequencies, the discrete power spectrums will appear at the common multiples of the set [15]. For variable delay random PWM (VDR-PWM), discrete power spectrums exist at multiples of the fixed sampling frequency [19].

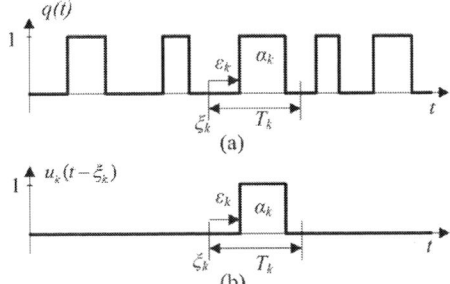

Fig. 1. (a). The switching waveform of $q(t)$; (b). the pulse of $u_k(t-\xi_k)$ at the k-th cycle of $q(t)$.

B. Estimation of Power Spectrum

The Power Spectrum Expressions of different random PWM techniques can be derived with complex calculations [12][15]. However, the power spectrum of a random PWM signal can be estimated and plotted from enough signal sample sequence.

For a sample sequence $\{x_n\}$, n= 0...N-1, if the Fourier transform is $X_N(\omega)$, the Power Spectrum can be obtained by Eq. (3).

$$P_N(\omega) = \frac{1}{N}\left|X_N(\omega)\right|^2 \tag{3}$$

The square of the magnitude of $X_N(\omega)$ in Eq. (3) gives the energy density at ω. After dividing the sequence length N, it becomes the power density. Therefore, the estimated power spectrum has the same shape with the energy spectrum $\left|X_N(\omega)\right|^2$. The discrete spectrum originally depicted with δ function in Eq. (2) has limited value in the estimation.

Such estimation is called periodogram method in Matlab, where the power spectrum is shown in log-scale. PSD (Power spectrum density) = $10\log_{10}(P_k)$, where P_k is the discrete $P_N(\omega)$.

III. SYSTEM DESCRIPTIONS OF RANDOMNIZED PWM SWITCHING STRATEGIES

A. Random Distribution of Zero Voltage Vector (RZD) Scheme

The standard SVPWM switching pattern in symmetry is illustrated in Fig. 2(a). Suppose that d_a, d_b, and d_c are the duty ratios of phases A, B, C.

Fig. 2(b) illustrates the case of RZD-PWM. In a three-phase, three-wire system, the duration of the zero voltage vector does not alter the line voltages. This fact is utilized in the random distribution of the zero voltage vector, where the proportion between the time durations for the two zero-vector states 111 and 000 are randomized in a switching cycle. All pulses are center-aligned as in a standard SVPWM. The random variable of this scheme is the time duration of zero states 111, which can be expressed as:

$$R_{rzd} = r \cdot (1 - d_{max}) + (1-r) \cdot d_{min} \tag{4}$$

where r is the random number which range from 0~1, d_{max} is the maximum duty ratio of three phases, and d_{min} is the minimum duty ratio of three phases.

978-1-4673-9551-9/16 $31.00 © 2016 IEEE

$$R_{rsf} = r \cdot (T_{s\,max} - T_{s\,min}) + T_{s\,min} \qquad (7)$$

where T_{smax} and T_{smin} is the allowed maxim and minimum PWM cycle.

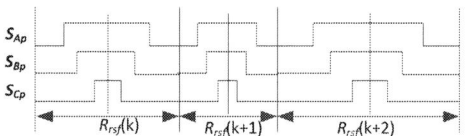

Figure 3. RSFPWM scheme.

E. Schemes Variable Delay Random PWM (VDR-PWM) Scheme

The problem of variable sample time is solved in VDR-PWM, which preserves both the regular sampling of the classic SVPWM method and the harmonic mitigation of the RSF-PWM technique. The VDR-PWM involves two independent strategies: one for fast determination of the switching pattern, subsequently called arithmetic pulse-width modulation (APWM), and another for randomization of switching periods. Fig. 4 illustrates the principle of VDR-PWM method. There are two random variables in this scheme and can be obtained with recursion equation:

$$\begin{aligned} R_{vdr1} &= R_{vdr2} \\ R_{vdr2} &= r \cdot (T_{s\,max} - T_{s\,min}) + T_{s\,min} - T_{sample} \end{aligned} \qquad (8)$$

where T_{sample} is the fixed sample cycle.

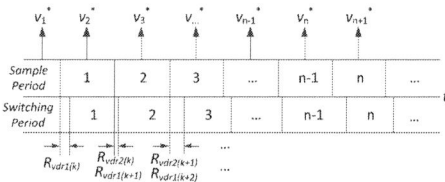

Figure 4. Configuration of VDR-PWM Scheme.

F. Unified implementation

The above five different RPWM methods are implemented with a unified configuration diagram in Fig. 5 where the PWM period (PRD), set and reset time are controlled separately by a microcontroller unit (MCU).

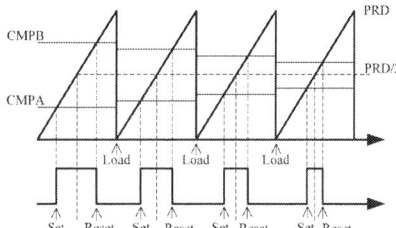

Figure 5. Realization of various PWM implementation.

Table I describes the configuration of different RPWM schemes, where d is the original duty cycle, T_s is the PWM cycle or center PWM cycle in RSF and VDR schemes, f_{cnt} is the frequency of PWM counter, subscript

---

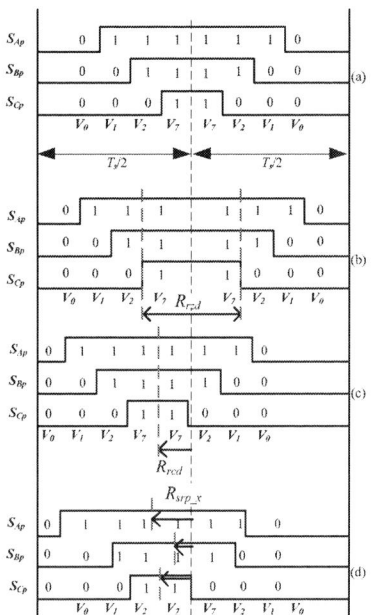

Figure 2. Switching patterns of (a). Classic SVPWM; (b). RZD-PWM; (c). RCD-PWM, and (d). SRP-PWMs, all with fixed switching frequency.

B. Random Pulse Center Displacement(RCD) Schemes

Fig. 2(c) depicts the RCD-PWM where the pulses are mutually center-aligned, but the common pulse center is displaced apart from the period middle randomly. The freedom of displacement is limited when the maximum duty ratio is high. Therefore the shifted time between the center of duty ratio and the half time of PWM cycle is the random variable, expressed as:

$$R_{rcd} = r \cdot (1 - d_{max}) / 2 \qquad (5)$$

C. Schemes Separately Randomized Pulse Position PWM (SRP-PWM) Scheme

In this scheme, the pulse position of each phase is separately placed. Even when the duty ratio is high, there is still enough displacement freedom of low duty ratio pulses. Fig. 2(d) represents a typical SRP-PWM method. There are three random variables in this scheme which can be configured separately as:

$$R_{srp\_x} = r \cdot (1 - d_{max\_x}) / 2, \ \mathrm{x = a,b,c} \qquad (6)$$

where x denotes the different phase.

D. Schemes Random Switching Frequency (RSF) PWM Scheme

As shown in Fig. 3, the most common RPWM technique is referred as random switching frequency RSF-PWM, where the switching and sampling cycles of the digital modulator coincide. This is inconvenient, because the modulator forms only a part of a larger control scheme designed with a specific bandwidth, dependent on the sampling frequency. The random variable is the PWM cycle in this scheme:

'x' denotes the different phase. It shows that the implementations of these RPWMs are not very complex. Relatively speaking, the SPR-PWM and VDR-PWM need a bit more computation burden than others.

TABLE I. COMPARISON OF VARIOUS RPWM IMPLEMENTATIONS (x = a,b,c)

PWM Type	PRD	CMPA	CMPB
SVM	$T_s f_{cnt}$	$(PRD-d_x)/2$	$(PRD+d_x)/2$
RZD	$T_s f_{cnt}$	$(PRD-d_x-R_{rzd})/2$	$(PRD+d_x+R_{rzd})/2$
RCD	$T_s f_{cnt}$	$(PRD-d_x)/2-R_{rcd}$	$(PRD+d_x)/2-R_{rcd}$
SRP	$T_s f_{cnt}$	$(PRD-d_x)/2-R_{srp\_x}$	$(PRD+d_x)/2-R_{srp\_x}$
RSF	$(T_s+R_{rsf})f_{cnt}$	$(PRD-R_{rsf}*d_x)/2$	$(PRD+R_{rsf}*d_x)/2$
VDR	$(T_s+R_{vdr2}-R_{vdr1})f_{cnt}$	$(PRD-d_x)/2$	$(PRD+d_x)/2$

IV. SIMULATION RESULTS AND ANALYSIS

A time-domain Matlab/Simulink simulation platform including power electronics, IM circuits, and control systems is developed to evaluate and quantify the performances of the five RPWM methods.

As a benchmark, where the IM speed has low modulation index at 10Hz, the power spectrum density of the output current using SVPWM is plotted in Fig. 6. The harmonics at discrete multiples of switching frequency are significant causing disturbing acoustic noises. Its line-to-line voltage THD is plotted in Fig. 7 with a value reaching 231.7%. Fig. 8 shows the results of VDR-PWM with a varying switching frequency between 3.57 and 8.33kHz at 40Hz IM speed with higher modulation index. The power spectrum density distribution is more continuous and the audible noise is significantly improved. The corresponding line-to-line voltage THD is shown in Fig. 9 with a value reduced to 48.34%.

TABLE II. INDUCTION MACHINE AND INVERTER PARAMETERS

Parameter	Value
Rated Power	2200W
Line Voltage	380Vrms
Rated Current	5.1Arms
Rated Frequency	50Hz
Stator Resistance R_s	3Ω
Rotor Resistance R_r	3Ω
Magnetizing Inductance L_m	220mH
Stator Leakage Inductance L_{ls}	9mH
Rotor Leakage Inductance L_{lr}	9mH
Moment of Inertia J	0.0067kg·m²
Pole Pairs P	2
Base PWM Frequency	5kHz
dc Bus Voltage	540V
Dead Time	3us

In addition to the power spectrum density evaluation, the IM output line-line voltage and phase current THDs are quantified as well for all scenarios described in the paper. Table III demonstrates that in all cases, RSF-PWM has the best overall THD performance with the switching frequency greater than the sampling frequency.

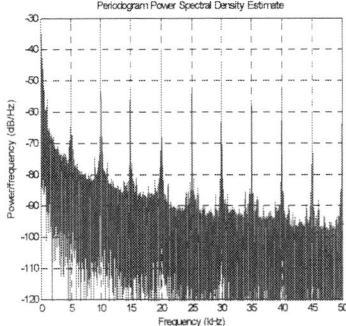

Figure 6. The output current (10Hz) power spectrum density vs. frequency under classic SVPWM method.

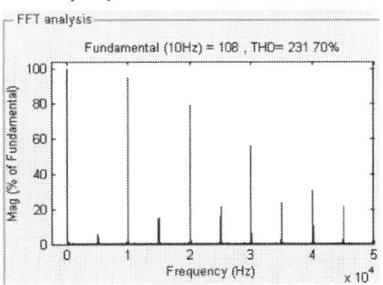

Figure 7. The output line-to-line voltage THD (10Hz) under classic SVPWM method.

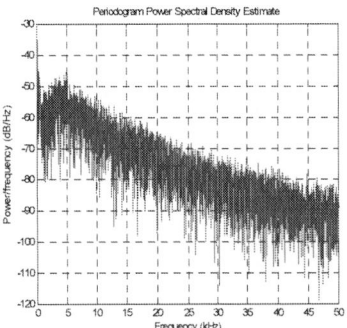

Figure 8. The output current (40Hz) power spectrum density vs. frequency under VDR-PWM method.

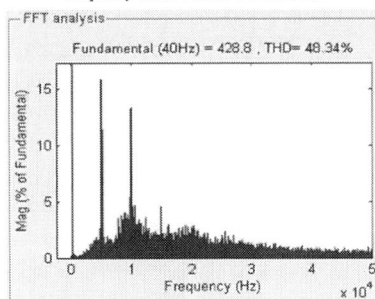

Figure 9. The output line-to-line voltage THD (40Hz) under VDR-PWM method.

In low modulation index at 10Hz, all RPWM methods have drastic improvement in THD performance as compared to the classical SVPWM method. In high modulation index, even though the line-to-line voltage THD is better from RPWM methods, the output current THDs are similar among the classic SVPWM, RZD, RCD, and SRP PWMs. VDR-PWM has better output current waveform quality among all RPWM methods at 40Hz.

TABLE III. SUMMARY OF IM LINE-LINE VOLTAGE AND PHASE CURRENT UNDER VARIOUS RPWMs.

Method	THD% (10Hz)	THD% (10Hz)	THD% (40Hz)	THD% (40Hz)
SVPWM	231.7	3.05	77.4	4.47
RZD	120.5	1.76	65.7	4.41
RCD	121.0	1.82	65.2	4.35
SRP	119.3	1.87	51.8	4.34
RSF	115.3	1.47	38.5	2.33
VDR	128.0	1.69	48.3	3.73

V. EXPERIMENTAL VERIFICATION

The methods and results of the five RPWM schemes are validated through the experiments on a platform shown in Fig. 10 including a three-phase full bridge inverter, an induction motor and control systems. The system parameters are the same as in the simulation platform, shown in Table II.

For all tests, the motor is controlled with open loop V/Hz method and the current is adjusted to the same value (at the amplitude of about 5A). To test high and low modulation index conditions, the command frequency is 40Hz and 10Hz respectively. The base switching frequency is 5kHz and the frequency range for random switching frequency strategy is chosen to be 5kHz to 7kHz.

Classic SVPWM is deterministic and its line voltage and phase current is measured and analyzed as a benchmark. Only discrete power spectrum exists in voltage and current signals of the classic SVPWM, as discussed in Section II.

A. FFT of Voltages and Currents of Random PWM

The line voltage and phase current in each random PWM experiment are recorded with an oscilloscope. The data was imported to Matlab and analyzed with FFT. The FFT results of all random PWM methods are given in Table IV.

B. Power Spectrum of Voltages and Currents of RPWM

Then the sampled voltages and currents of each random PWM, as well as the classical SVPWM are used to estimate the power spectrum of the signals, which are shown in Fig. 11 and Fig. 12. The upper and lower plots I each case are power spectrum for line voltage and phase current.

C. Comparison of the Random PWM Strategies

To suppress acoustic noise, the voltage and current power spectrum should be as continuous as possible. Discrete power spectrum brings high pitch acoustic noise. The effect of RPWM is to spread the voltage signal power to the whole frequency range instead of gathering at some discrete frequencies, thus the electromagnetic noise spectrum is also spread out.

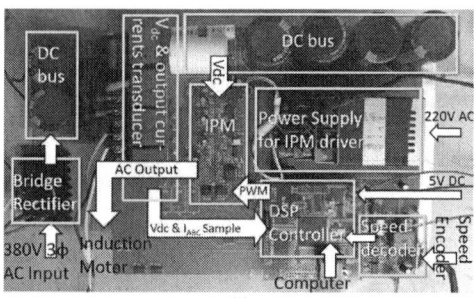

(a)

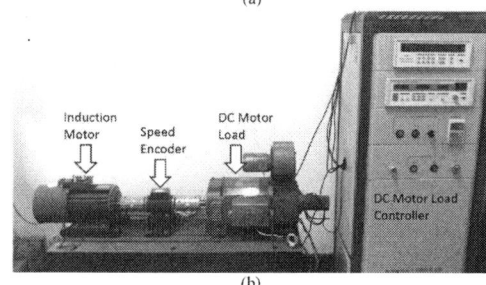

(b)

Figure 10. Experimental Setup. (a) The power and control board of the tested inverter. (b) The experimental induction motor and dc motor load.

1) Classical SVPWM

The fundamental voltage of classical SVPWM at 40Hz is about 4 times larger than at 10Hz. So harmonics percentage is lower at 40Hz and the voltage THD is lower than at 10Hz in Table IV. But the current THD at 40Hz is larger than at 10Hz, which means noise at high speed will be louder.

TABLE IV FFT RESULTS OF DIFFERENT RANDOM PWMs

	Random PWM Method	Line Voltage		Phase Current	
		Fundamental Magnitude	*THD*	*Fundamental Magnitude*	*THD*
10Hz	SVPWM	110.5	190.70%	4.885	4.20%
	RZD	110.4	114.96%	4.890	3.89%
	RCD	110.3	116.18%	4.912	3.79%
	SRP-PWM	109.5	115.06%	4.897	4.06%
	RSF	109.6	108.69%	4.837	3.61%
	VDR-PWM	109.7	109.30%	4.845	3.86%
40Hz	SVPWM	427.4	76.29%	4.807	6.92%
	RZD	428.6	67.09%	4.830	6.98%
	RCD	428.4	66.81%	4.834	6.89%
	SRP-PWM	424.5	56.57%	4.803	7.16%
	RSF	424.2	42.39%	4.795	4.75%
	VDR-PWM	422.9	43.94%	4.667	5.40%

In Fig. 11 (a) and Fig. 12 (a), continuous power spectrums are under -20dB, showing that almost all the signal power is at discrete frequencies. That's why high frequency noise of classic SVPWM is highly pronounced.

2) RZD

At 10Hz, RZD-PWM reduced harmonics of both voltage and current compared to the classical SVPWM, thus the THDs of RZD-PWM are smaller. The power spectrum is distributed evenly in Fig. 11 (b) but has discrete spectrum at multiples of switching frequency.

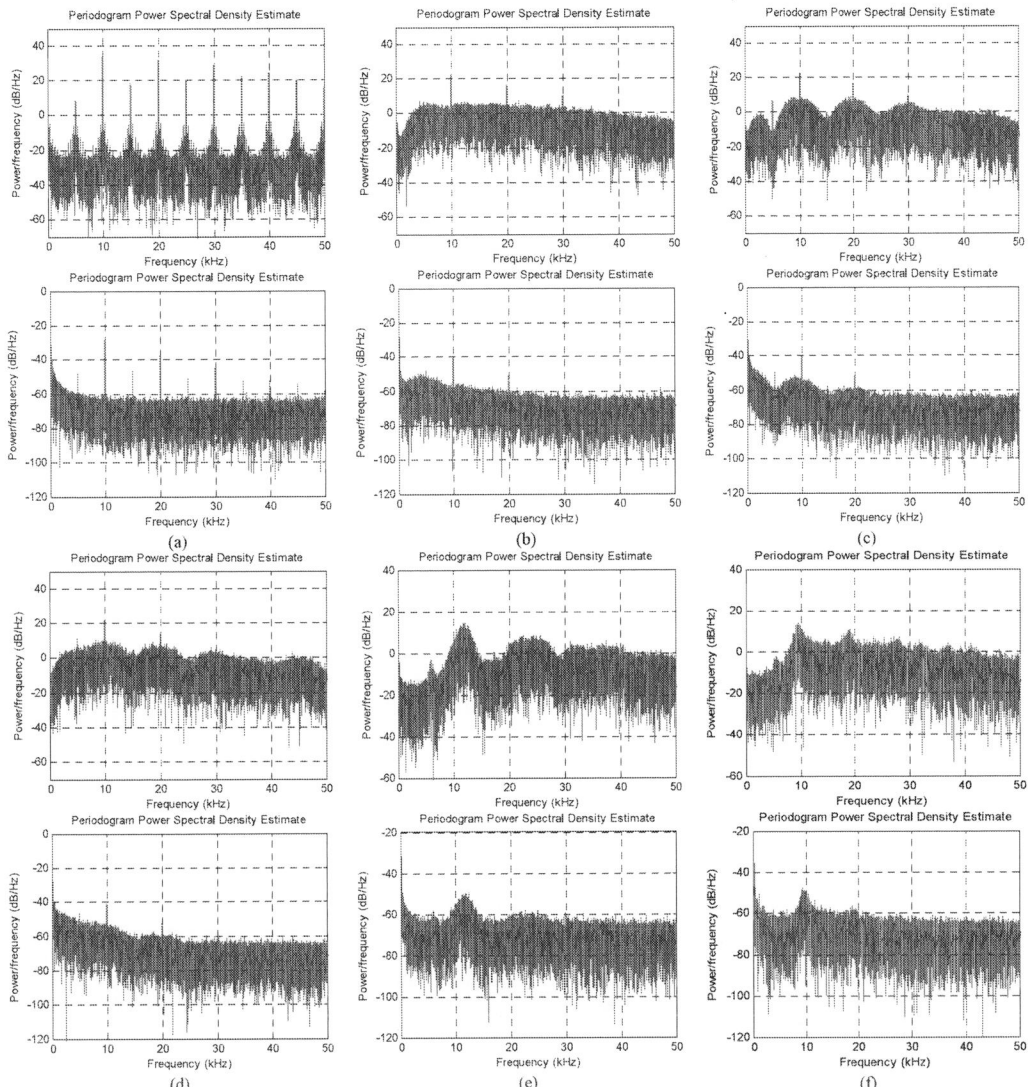

Figure 11. Power spectrum estimation using periodogram method of line voltage (upper window in each case) and phase current (lower window in each case) at fundamental frequency of 10Hz and base PWM frequency of 5kHz. (a). classic SVPWM; (b). RZD-PWM; (c). RCD-PWM; (d). SRP-PWM; (e) RSF-PWM (Frequency Range: 5~7kHz); (f) VDR-PWM (Frequency Range:3.57~8.33kHz).

At 40Hz in Fig 12 (b), the performance of RZD PWM is approaching to that of the classic SVPWM with signal power focusing on multiples of switching frequency because zero-vector at high modulation is short and has little random distribution range.

3) RCD

RCD in Fig.11 (c) spreads power in continuous spectrum clusters with discrete spectrum among them. In Fig. 12 (c), its effect is reduced when modulation index is high. Current THD of RCD is relatively small since the pulses are still symmetric.

4) SRP-PWM

Power spectrum is distributed more evenly than RCD at 10 Hz and 40Hz as shown in Fig. 11 (d) and Fig. 12 (d). But due to asymmetric pulse distribution of PWM, the current THD is larger. SRP-PWM keeps random freedom at high modulation index so that the voltage THD at 40Hz

is much smaller than that of RZD-PWM and RCD-PWM. The continuity of power spectrum of SRP-PWM is the best among random position PWM strategies.

5) RSF

The experiment of RSF-PWM changes switching frequency between 5 kHz to 7 kHz randomly and the sampling frequency is kept at 5 kHz.

Compared to the three random position PWM strategies above, RSF-PWM has stronger ability to spread power spectrum. No discrete spectrum appears in Fig. 11 (e) and Fig.12 (e), but there is some power clustering around 6kHz, which led to a little acoustic noise.

At high modulation index, RSF still has the same random freedom to distribute power spectrum continuously. In addition, the voltage and current THDs are reduced greatly even at high modulation index in Table IV.

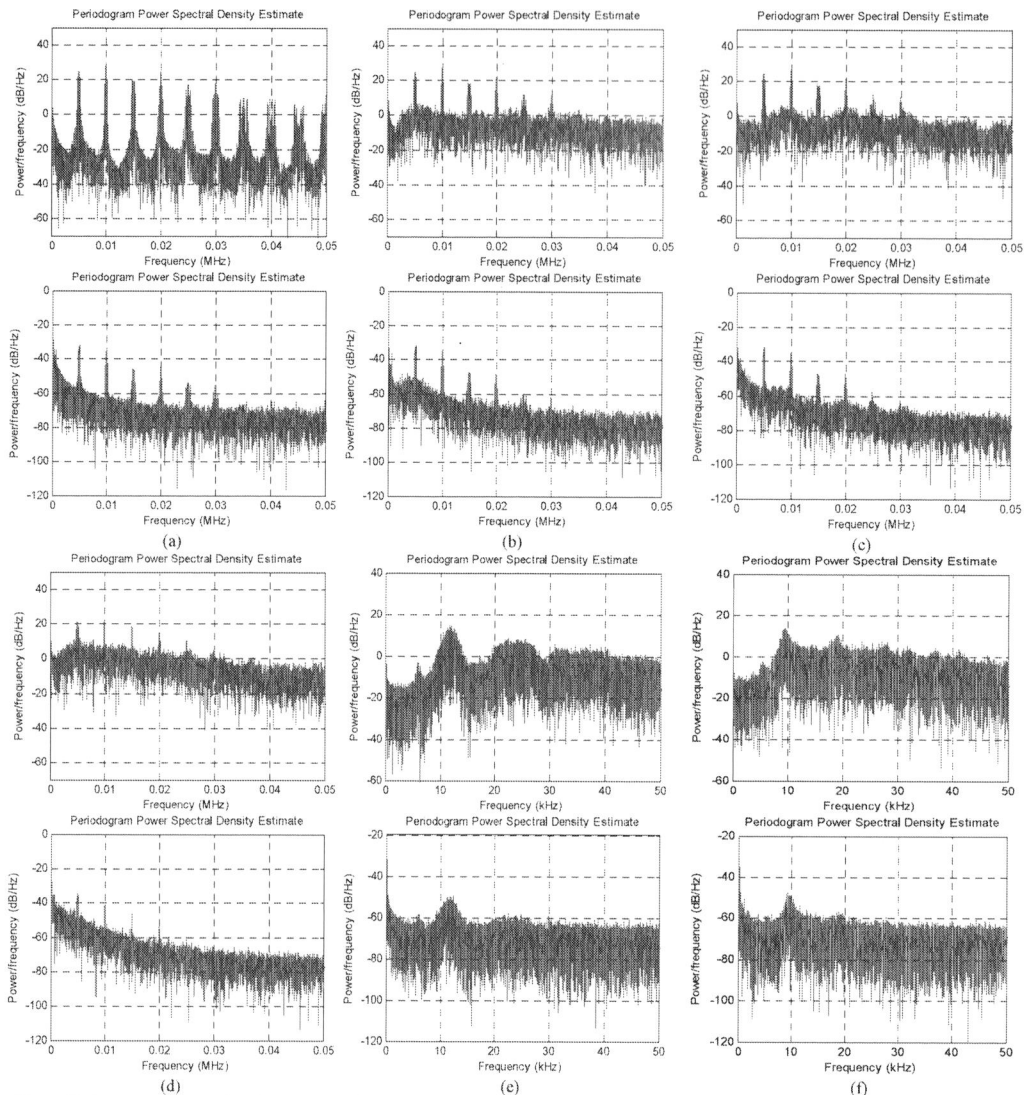

Figure 12. Power spectrum estimation using periodogram method of line voltage (upper window in each case) and phase current (lower window in each case) at fundamental frequency of 40Hz and base PWM frequency of 5kHz. (a). classic SVPWM; (b). RZD; (c). RCD; (d). SRP-PWM; (e) RSF (Frequency Range: 5~7kHz); (f) VDR-PWM (Frequency Range:3.57~8.33kHz).

6) VDR-PWM

The power spectrum is discrete at the control loop frequency (current sampling frequency). The power spectrum can be expanded to a wide range at high or low modulation index. However, the current sampling accuracy is lowered because of a fixed sampling frequency while the PWM frequency is changing.

VI. CONCLUSIONS

In this paper, the performance characterization of five state-of-the-art random PWM modulation algorithms is quantitatively evaluated in industrial and commercial ASDs. Power spectrum theory is described for analyzing the frequency components of RPWM signal. The additional comparative results taking into account of sampling effects, current ripple harmonics, influence of PWM modulation index, computation burden are evaluated. Table V is a comparative summary on various RPWM methods.

TABLE V. COMPARATIVE SUMMARY OF RPWMs.

Method	Audible Noise (dB)	IM Ripple Current	Sampling Error	CPU Burden
SVPWM	High	Low	Low	Low
RZD	Medium	Medium	Low	Low
RCD	Medium	Medium	Medium	Medium
SRP	Medium	Medium	High	High
RSF	Lower	Low	High	Medium
VDR	Medium	Medium	High	High

SVPWM is efficient overall with an exception of a high pitch audible noise. RZD-PWM solves the problem the best with low sampling error and CPU burden. RCD-PWM has similar performance as RZD-PWM, but more costly in its implementation. SRP-PWM can maintain randomness better at higher PWM index, but it turns out more costly than RCD-PWM, this method is not recommended for adoption. RSF-PWM has good performance at both low and high speed or modulation index, provided that the switching frequency is larger than its sampling rate. In addition to a lower tone mixture, RSF-PWM has significantly lower current THDs as compared to other RPWMs. VDR-PWM is comparable to SRP PWM in Table V. Experimental tests of a 2.2kW, 380V, 50Hz ASD and IM system have been conducted for verifying the findings in theory and simulation studies. Based on the results, it is recommended that RZD-PWM and RSF-PWM are suitable for achieving the performance and implementation cost parity. Users can select the appropriate RPWM strategy for their applications more effectively.

REFERENCES

[1]. S. L. Nau, H. G. G. Mello, "Acoustic noise in induction motors: causes and solutions." *Record of Conference Papers in Industry Applications Society 47th Annual Petroleum and Chemical Industry Conference*, pp. 253-263, 2000.

[2]. R. J. M. Belmans, L. D'Hondt, A. J. Vandenput, and W. Geysen, "Analysis of the audible noise of three-phase squirrel-cage induction motors supplied by inverters." *IEEE Transactions on Industry Applications*, vol. IA-23, issue 5, pp. 842-847, 1987.

[3]. A. K. Wallace, R. Spee, and L. G. Martin, "Current harmonics and acoustic noise in ac adjustable-speed drives." *IEEE Transactions on Industry Applications*, vol. 26, issue 2, pp. 267-273, 1990.

[4]. W. C. Lo, C. C. Chan, Z. Q. Zhu, X. Lie, D. Howe, and K. T. Chau, "Acoustic noise radiated by PWM-controlled induction machine drives." *IEEE Transactions on Industrial Electronics*, vol. 47, issue 4, pp. 880-889, 2000.

[5]. A. C. Binojkumar, B. Saritha, and G. Narayanan, "Acoustic noise characterization of space-vector modulated induction motor drives - an experimental approach." *IEEE Transactions on Industrial Electronics*, vol. 62, issue 6, pp. 3362-3371, June 2015.

[6]. S. Legowski, and A.M. Trzynadlowski, "Advanced random pulse width modulation technique for voltage-controlled inverter drive systems." *Conference Proceedings of Sixth Annual Applied Power Electronics Conference and Exposition*, pp. 100-106, 1991.

[7]. R. L. Kirlin, S. Kwok, S. Legowski, and A.M. Trzynadlowski, "Power spectra of a PWM inverter with randomized pulse position." *IEEE Transactions on Power Electronics*, vol. 9, no. 5, pp. 463-472, September 1994.

[8]. A. M. Trzynadlowski, R. L. Kirlin and S. F. Legowski, "Space vector PWM technique with minimum switching losses and a variable pulse rate." *IEEE Transactions on Industrial Electronics*, vol. 44, no. 2, pp. 173-181, April 1997.

[9]. A. M. Trzynadlowski, F. Blaabjerg, J. K. Pedersen, R. L. Kirlin, and S. Legowski, "Random pulse width modulation techniques for converter-fed drive systems - a review." *IEEE Transactions on Industry Applications*, vol. 30, no. 5, pp. 1166-1175, September/October 1994.

[10]. J. K. Pedersen, F. Blaabjerg and P.S. Frederiksen. "Reduction of acoustic noise emission in AC-machines by intelligent distributed random modulation". *Fifth European Conference on Power Electronics and Applications*, vol. 4, pp. 369-375, 1993.

[11]. F. Blaabjerg, J. K. Pedersen, E. Ritchie, and P. Nielsen, "Determination of mechanical resonances in induction motors by random modulation and acoustic measurement." *IEEE Transactions on Industry Applications*, vol. 31, no. 4, pp. 823-829, July/August, 1995.

[12]. A. M. Stankovic, G. C. Verghese and D. J. Perreault, "Analysis and synthesis of randomized modulation schemes for power converters." *IEEE Transactions on Power Electronics*, vol. 10, no. 6, pp. 680-693, November 1995.

[13]. A. M. Stankovic, "Random pulse modulation with applications to power electronic converters." *Ph.D. dissertation*, Massachusetts Institute of Technology, 1993.

[14]. C. B. Jacobina, A. M. N. Lima, E. R. C. da Silva, and A. M. Trzynadlowski, "Current control for induction motor drives using random PWM." *IEEE Transactions on Industrial Electronics*, vol. 45, no. 5, pp. 704-712, 1998.

[15]. R. L. Kirlin, M. M. Bech and A.M. Trzynadlowski, "Power spectral density analysis of randomly switched pulse width modulation for DC/AC converters." Proceedings of the Tenth IEEE .Workshop on Statistical Signal and Array Processing, pp. 373-377, 2000.

[16]. M. M. Bech, F. Blaabjerg and J. K. Pedersen, "Random modulation techniques with fixed switching frequency for three-phase power converters." *IEEE Transactions on Power Electronics*, vol. 15, no. 4, pp. 753-761, July 2000.

[17]. Stankovic, X., M. A. and H. Lev-Ari, "Randomized modulation in power electronic converters." *Proceedings of the IEEE Journals & Magazines*, vol. 90, issue 5, pp. 782-799, 2002.

[18]. S. H. Na, Y. G. Jung, Y. C. Lim, and S. H. Yang, "Reduction of audible switching noise in induction motor drives using random position space vector PWM." *IEE Proceedings in Electric Power Applications*, vol. 149, issue 3, pp. 195-200, 2002.

[19]. A. M. Trzynadlowski, K. Borisov, Y. Li and L. Qin, "A novel random PWM technique with low computational overhead and constant sampling frequency for high-volume, low-cost applications." *IEEE Transactions on Power Electronics*, vol. 20, no. 1, pp. 116-122, January 2005.

[20]. K. S. Kim, Y. G. Jung, and Y. C. Lim, "A new hybrid random PWM scheme." *IEEE Transactions on Power Electronics*, vol. 24, no. 1, pp. 192-200, January 2009.

[21]. A. A. Fardoun, and E. H. Ismail, "Reduction of EMI in ac drives through dithering within limited switching frequency range." *IEEE Transactions on Power Electronics*, vol. 24, no. 3, pp. 804-811, March 2009.

[22]. R. L. Kirlin, C. Lascu and A. M. Trzynadlowski, "Shaping the noise spectrum in power electronic converters." *IEEE Transactions on Industrial Electronics*, vol. 58, no. 7, pp. 2780-2788, July 2011.

[23]. Y. C. Lim, Y. G. Jung, S. Y. Oh, and J. G. Kim, "A two-phase separately randomized pulse position PWM (SRP-PWM) scheme with low switching noise characteristics over the entire modulation index." *IEEE Transactions on Power Electronics*, vol. 27, no. 1, pp. 362-369, January 2012.

[24]. Kirlin, R.L., M.M. Bech and A.M. Trzynadlowski, Analysis of power and power spectral density in PWM inverters with randomized switching frequency. IEEE Transactions on Industrial Electronics, 2002. 49(2): p. 486-499

Stability Analysis and Controller Synthesis for Digital Single-Loop Voltage-Controlled Inverters

Xiongfei Wang, Poh Chiang Loh, Frede Blaabjerg
Department of Energy Technology, Aalborg University, Aalborg, Denmark
xwa@et.aau.dk, pcl@et.aau.dk, fbl@et.aau.dk

Abstract—This paper analyzes first the stability of single-loop digital voltage control scheme for the LC-filtered voltage source inverters. It turns out that the phase lag, caused by the time delay of digital control system and by the use of integral controller, can stabilize the voltage loop without damping of LC-filter resonance. The stability regions are then identified with alternative voltage controller synthesized. For further widening the stability region, an active damping approach is proposed and co-designed with the voltage controller in the discrete *z*-domain. Simulations and experimental results of both 50-Hz and 400-Hz systems validate the theoretical analyses and the performance of the approach.

Keywords—time delay, stability, voltage-controlled inverters, LC-filter resonance

I. INTRODUCTION

Applications of Voltage Source Inverters (VSIs) have been growing constantly over the last few years. Voltage-controlled VSIs with LC-filters are widely found in uninterruptible power supplies [1], grid-interactive distributed generations [2], ground power units for airplanes [3], and other high-performance ac voltage sources, e.g. grid emulators [4] and power amplifiers in hardware-in-the-loop tests [5]. A stable ac output voltage with high waveform quality is generally required in these systems.

A large number of ac voltage control strategies have been developed, which can, in general, be classified into two groups, i.e. double-loop voltage-current control [6]-[8], and single-loop voltage control [9]-[11]. In the double-loop control methods, an inner control loop based on the feedback of the filter inductor- or capacitor-current is used for damping of LC-filter resonance, while the ac voltage is controlled in the outer loop [6]. Such a double-loop control structure usually works well for the VSIs with a high pulse-ratio, which is namely the ratio of switching to ac fundamental frequency [10]. However, it hardly fulfills the performance requirements for high-power inverters with a low pulse-ratio, e.g. the 400-Hz ground power units [11].

Single-loop voltage control has thus been preferred for low-pulse-ratio VSIs [9]-[11]. To mitigate the influence of LC-filter resonance, the direct pole placement method based on discrete *z*-domain model of the inverter has been discussed in [9], [10]. It exhibits a better dynamic performance than the double-loop control schemes, but it does also make the voltage control more sensitive to the parameter variation of LC-filter. To overcome the limit, a single-loop Resonant (R) control scheme is recently reported in [11]. Instead of the Proportional + Integral (PI) or P

This work was supported by European Research Council (ERC) under the European Union's Seventh Framework Program (FP/2007-2013)/ERC Grant Agreement [321149-Harmony].

+ Resonant (PR) controllers, only the R controller is adopted in [11] for stable operation. Multiple R controllers have also been employed for high waveform quality control in the presence of nonlinear loads. Yet, the stabilizing mechanism underlying this approach is not clearly discussed, and its stability region is not identified either.

This paper presents an in-depth analysis on the stability of digital single-loop voltage-controlled inverters. Considering the P controller first, it is shown that the digital computation and modulation delays lead to a critical frequency for the LC-filter resonance, above which the system can be stabilized without any resonance damping, and meanwhile the proportional gain should be kept lower than unity for the stable operation, which results in a non-trivial steady-state error. Then, the use of R or I controller is analyzed. It turns out that the phase lag of the R or I controller can further reduce the critical frequency to a lower value. This equivalently widens the stability region of voltage control loop. Moreover, an active damping control based on the voltage feedback loop is developed, which further widens the stability region with better transient response. Simulations and experimental tests for both 50-Hz and 400-Hz inverters are carried out. The results confirm the theoretical analyses and the performance of the proposed approach.

II. VOLTAGE-CONTROLLED VSIS

A. System Description

Fig. 1 shows a simplified one-line diagram of a three-phase voltage-controlled VSI with an LC-filter, where the ac voltage across the filter capacitor is controlled by the voltage controller $G_v(z)$. The dc-link voltage V_{dc} is assumed constant, which thus allows to modeling the ac voltage control loop as a linear time-invariant system in the stationary frame [12], and consequently the voltage control loop also applies to single-phase VSIs. The main circuit parameters of the VSI are provided in Table I.

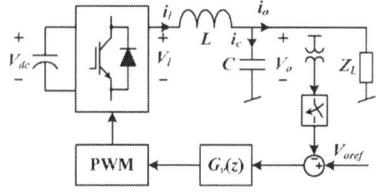

Fig. 1. Simplified one-line diagram of a three-phase voltage-controlled VSI with an LC-filter.

TABLE I.   MAIN CIRCUIT PARAMETERS

Symbol	Electrical Constant	Value
f_1	AC output fundametnal frequency	50/400 Hz
f_{sw}	Switching frequency	5/10 kHz
f_s	Sampling frequency of control system	5/10 kHz
V_{dc}	DC-link voltage	730 V
L	Filter inductance	1.5 mH
C	Filter capacitance	5/10 μF
Z_L	Resistive load	236 Ω

B. Digital Single-Loop Voltage Control

Fig. 2 depicts the block diagram of the digital single-loop voltage control system. The Pulse Width Modulation (PWM) is represented as a Zero-Order Hold (ZOH) block to account for the time delay effect of the uniformly sampled PWM [13]. One sampling period delay (z^{-1}) is also included to model the delay effect of the computation and the update of duty cycle [14]. Thus, the total time delay in the system can be modeled as $G_d(s)$ in the continuous s-domain, which is given by

$$G_d(s) = e^{-1.5T_s s} \tag{1}$$

where $T_s = 1/f_s$, f_s is the sampling frequency of the inverter. The voltage control loop in the s-domain can also be represented as shown in Fig. 3. $G_p(s)$ and $Z_o(s)$ depict the dynamics of the LC-filter plant, which are the transfer functions from the inverter voltage and load current to the output voltage.

$$G_p(s) = \left.\frac{V_o}{V_{pwm}}\right|_{i_o=0} = \frac{1}{LCs^2 + 1} \tag{2}$$

$$Z_o(s) = -\left.\frac{V_o}{i_o}\right|_{V_{pwm}=0} = \frac{Ls}{LCs^2 + 1} \tag{3}$$

The voltage controller $G_v(s)$ in [11] only uses the R controller.

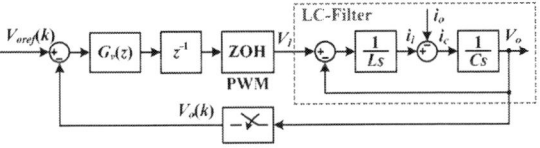

Fig. 2. Block diagram of single-loop digital voltage control system.

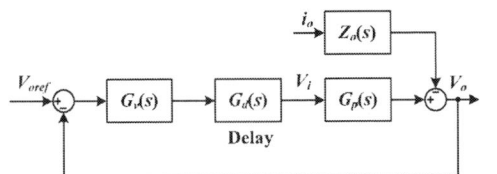

Fig. 3. Single-loop voltage control system in the continuous s-domain.

III.   STABILITY ANALYSIS AND CONTROLLER SYNTHESIS

This section derives first the critical frequency of the LC-filter resonance considering the effect of time delay $G_d(s)$. The use of R controller is then analyzed with the stability region identified and the alternative voltage controllers synthesized.

A. P Control

The open-loop gain of voltage control loop can be given by

$$T(s) = G_v(s)G_d(s)G_p(s). \tag{4}$$

Considering the P controller for $G_v(s)$, the phase and magnitude of $T(j\omega)$ are given by

$$\angle T(j\omega) = \begin{cases} -1.5T_s\omega, & \omega < \omega_r \\ -1.5T_s\omega - \pi, & \omega \geq \omega_r \end{cases} \quad \omega_r = \frac{1}{\sqrt{LC}} \tag{5}$$

$$|T(j\omega)| = \frac{k_p}{1 - LC\omega^2} \tag{6}$$

where ω_r is the resonance frequency of the LC-filter, k_p is the proportional gain. Based on the Nyquist stability criterion, the following two conditions can be formulated for stable system:

- The frequency at which the phase of $T(j\omega)$ is $-\pi$, i.e. the phase crossover frequency, ω_p, is below ω_r. A critical value of ω_r, i.e. $\omega_c = \omega_s/3$, can thus be derived from (5), above which the control loop has a non-negative Phase Margin (PM).

- The magnitude of $T(j\omega)$ is not above 0dB at the phase crossover frequency for the loop to have a non-negative Gain Margin (GM), which leads to $k_p < 1$ based on (6).

Fig. 4 plots the bode diagrams of $T(s)$ for $f_s = 5$ kHz. It is seen that the system is stable (GM >0, PM>0) for $C = 5$ μF, $k_p = 0.1$, owing to $\omega_r > \omega_s/3$. In contrast, the system is unstable for the other two cases, due to the GM<0 or PM<0. Moreover, the condition of $k_p < 1$ results in a non-trivial voltage tracking error at the steady-state, though the system is stable when $\omega_r > \omega_s/3$.

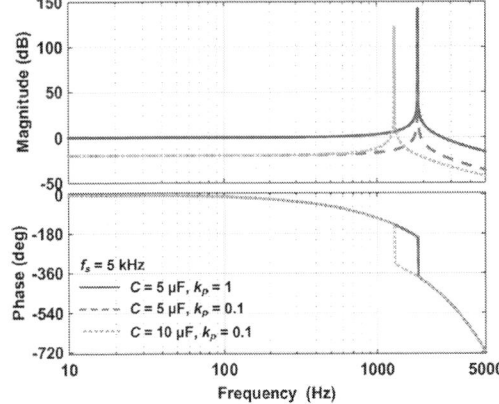

Fig. 4. Bode diagrams of $T(s)$ with P controller at $f_s = 5$ kHz.

B. R Control

For zero steady-state tracking error, the R controller needs to be used in the stationary frame with $G_v(s)$, whose basic form is given by

$$G_v(\text{s}) = k_i R(\text{s}), \quad R(\text{s}) = \frac{s}{s^2 + \omega_1^2} \tag{7}$$

where $\omega_1 = 2\pi f_1$ is the ac output fundamental frequency. Using R controller for single-loop voltage control has been reported in [11], yet its stabilization mechanism is not revealed.

Fig. 5 shows the bode diagrams of $T(s)$ with $R(s)$ only used for $G_v(s)$. Two cases with different sampling and fundamental frequencies are plotted. It is worth noting that the R controller provides two superior features with the voltage loop: 1) a phase lag of $\pi/2$ above the fundamental frequency, and 2) reducing the $|T(j\omega)|$ below 0 dB with the exceptions at the frequencies ω_1 and ω_r. The critical frequency ω_c is thus reduced as $\omega_s/6$, as derived in (8), and the stability region is widened than that with the P controller. Comparing Figs. 4 and 5(a), it is clear that the unstable cases with the P controller are stabilized by using the R controller.

$$-1.5T_s\omega_c - \frac{\pi}{2} = -\pi \tag{8}$$

It also can be seen from Fig. 5 (b) that the case of $C = 10 \mu F$ is unstable at $f_s = 10$ kHz, due to $\omega_r < \omega_s/6$. Hence, the phase lag of the R controller plays a critical role in stabilizing the voltage loop. Using PR controller as usual can only reduce the phase lag of R controller and leads to a narrowed stability region.

Besides the phase lag of the basic R controller given in (7), the discretization of R controller may bring in additional phase lag, affecting the stability region. Fig. 6 gives a comparison of frequency responses for the R controller discretized with three different methods, including the Tustin transformation with the prewarping at ω_1, the two-integrator-based scheme with the forward and backward Euler integrators [15], and the Zero-Order Hold (ZOH) transformation [11]. It is noted that the first method has no additional phase lag, while the latter two cause an additional $0.5T_s$ time delay, which consequently leads to a critical frequency $\omega_c = \omega_s/6$ with $2T_s$ time delay. Fig. 7 shows the bode diagrams of $T(z)$ at $f_s = 10$ kHz yet with two different

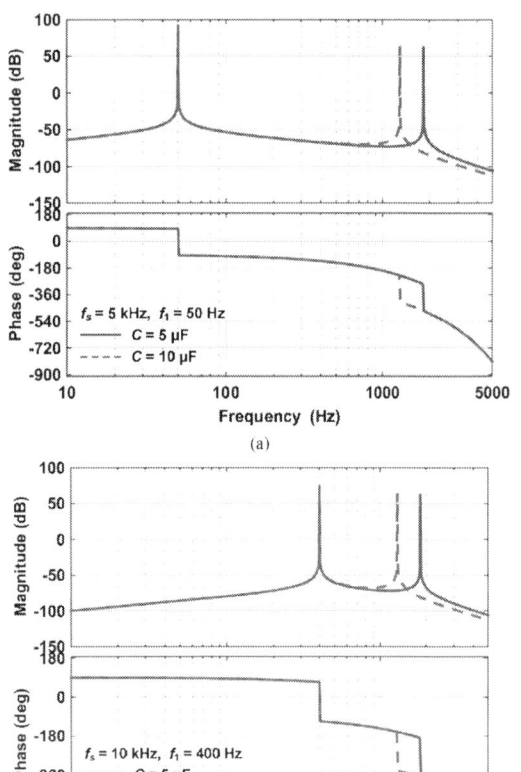

(a)

(b)

Fig. 5. Bode diagrams of $T(s)$ with the unity-gain-R controller, i.e. R(s). (a) $f_s = 5$ kHz, $f_1 = 50$ Hz. (b) $f_s = 10$ kHz, $f_1 = 400$ Hz.

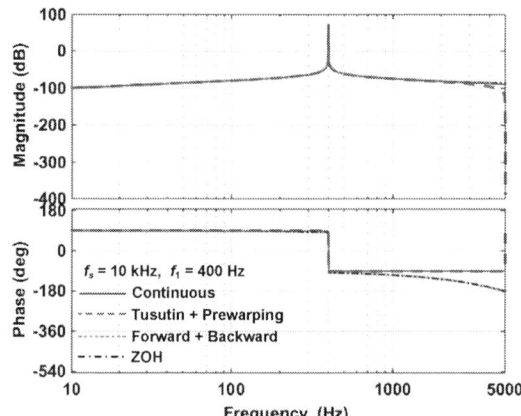

Fig. 6. Frequency responses of $R(s)$ with different discretization methods.

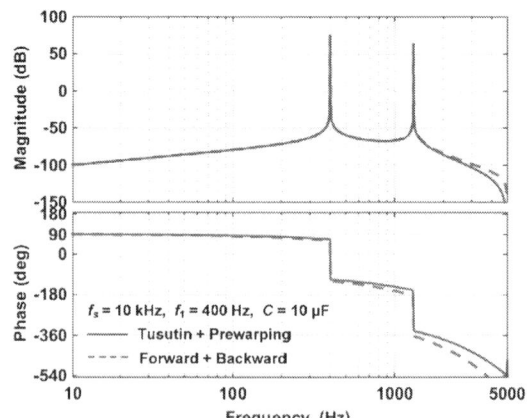

Fig. 7. Bode diagrams of $T(z)$ with two different discretized R controller $R(z)$.

discretized R controllers, where $T(z)$ can be derived from Fig. 2 with the ZOH-discretized $G_p(s)$. It can be seen that the unstable case in Fig. 5 (b) is stabilized by using the two-integrator-based R controller. This also explains why the system in [11] is kept stable even if the LC-filter resonance frequency is below $f_s/6$.

C. Proposed I+R (IR) control

As an alternative to the use of R controller only, the basic I controller can also be used to obtain the phase lag of $\pi/2$, which together with the PR controller, yields a I+R (IR) controller for $G_v(s)$, as shown in Fig. 8. In the approach, the R controller is used to trim the zero steady-state error yet has little effect on the phase of $T(s)$ at the phase crossover frequency, ω_p [16]. The gain of R controller can thus be designed as [17]

$$k_i \approx \frac{k_p}{10} \qquad (9)$$

Then, the system stability will mainly be determined by the P controller gain, k_p, and the discretized I controller.

Fig. 9 shows the frequency responses of $T(s)$ with the basic I controller with unity gain, i.e. $1/s$, at $f_s = 10$ kHz, $f_1 = 400$ Hz. Compared to Fig. 5(b), the same stabilizing effect as that using the R control can be observed. Fig. 10 plots then the frequency responses of discretized I controllers. It is clear that the Tustin method has no additional phase lag, similarly to Tustin with prewarping in Fig. 6. Yet, the forward Euler causes $0.5T_s$ delay, and the backward Euler reduces $0.5T_s$ delay. As a consequence, the critical frequency ω_c is shifted to $\omega_s/8$ ($2T_s$ delay) and $\omega_s/4$ (T_s delay), respectively.

Instead of the PR controller shown in Fig. 8, the Vector PI (VPI) controller, as expressed in (9) [18], can also be used with

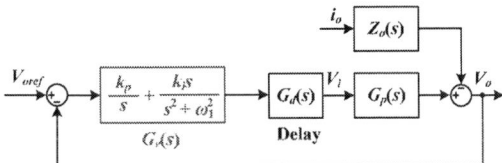

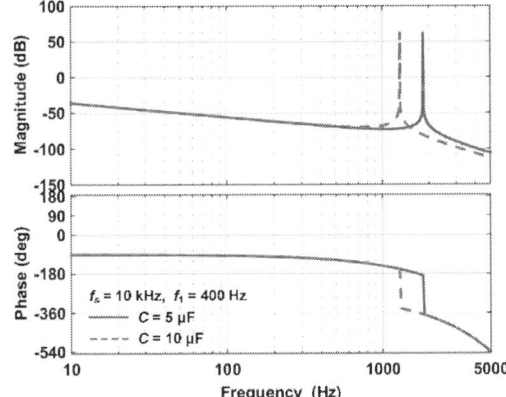

Fig. 8. Single-loop voltage control diagram with the proposed IR controller.

Fig. 9. Bode diagrams of $T(s)$ with the basic I controller with unity gain.

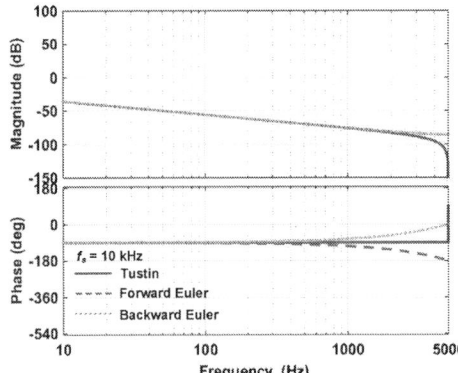

Fig. 10. Frequency responses of the different discretized I controllers.

the I controller. It then yields a voltage controller in the form of (10), whose phase lag will be the same as the R controller used in Fig. 5 if k_i is designed to have little effect on the phase of $T(s)$, and the system stability will be mainly dependent on k_p.

$$G_{VPI}(s) = \frac{k_p s^2 + k_i s}{s^2 + \omega_1^2} \qquad (10)$$

$$G_v(s) = G_{VPI}(s) \cdot \frac{1}{s} = \frac{k_p s + k_i}{s^2 + \omega_1^2} \qquad (11)$$

IV. VOLTAGE-FEEDBACK ACTIVE DAMPING

In this section, an active damping scheme is developed with the voltage feedback loop, which can further widen the system stability region and improve the transient response.

A. Operation Principle

Fig. 11 illustrates the block diagram of the proposed control scheme, where a damping controller, $G_a(s)$, based on a negated low-pass filter is integrated into the voltage feedback loop. The voltage controller $G_v(s)$ can be either the IR controller given in Fig. 8 or the R controller in (7).

The closed-loop response of Fig. 11 can thus be derived as

$$V_o = \frac{G_v(s)G_d(s)G_p(s)}{1 + [G_v(s) + G_a(s)]G_d(s)G_p(s)} V_{oref}$$
$$- \frac{Z_o(s)}{1 + [G_v(s) + G_a(s)]G_d(s)G_p(s)} i_o \qquad (12)$$

where the open-loop gain $T_{ad}(s)$ with $G_a(s)$ is given by

$$T_{ad}(s) = [G_v(s) + G_a(s)]G_d(s)G_p(s) . \qquad (13)$$

As $G_v(s)$ can be simplified as the term k/s for stability analysis, where k is k_i for the R controller in (7) and is k_p for the IR controller in Fig. 8 and (10), $T_{ad}(s)$ can be equivalent to (14). It is shown that $G_a(s)$, together with $G_v(s)$, essentially synthesizes a lead-lag filter in cascade with the I controller.

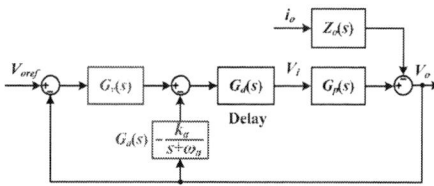

Fig. 11. Block diagram of the proposed active damping scheme.

$$T_{ad}(s) = \left[\frac{k}{s} - \frac{k_a}{s+\omega_a}\right] G_d(s)G_p(s)$$

$$= \frac{(k-k_a)s+k\omega_a}{s+\omega_a} \cdot \frac{1}{s} \cdot G_d(s)G_p(s)$$

(14)

Further on, the phase of the lead-lag filter can be derived as

$$\angle\left\{(k-k_a)\frac{j\omega+\dfrac{\omega_a}{1-k_a/k}}{j\omega+\omega_a}\right\} = \begin{cases} [-\dfrac{\pi}{2},\ 0], & k_a < k \\[2mm] [-\dfrac{\pi}{2},\ 0], & k_a = k \\[2mm] [-\pi,\ 0], & k_a > k \end{cases}$$

(15)

It is clear that the lead-lag filter introduces additional phase lag, which stabilizes the control system. Moreover, in the case with $k_a > k$ leads to a wider range of phase lag, which consequently widens the system stability region. Yet, it also adds a right half-plane zero into the open-loop gain, and thus results in the non-minimum phase characteristic of the control system.

B. Co-Design of Active Damping and Voltage Controller

For digital implementation, $G_a(s)$ is discretized by using the Tustin method, while for $G_v(s)$, the R controller is discretized by the Tustin with prewarping and the I controller is discretized with the Tustin method, in order not to bring in additional time delay. The sampling frequency is $f_s = 10$ kHz, as in this case C

$= 10$ μF leads to $\omega_r < \omega_s/6$, which necessitates the use of $G_a(s)$.

From (13) to (15), it is seen that three controller parameters mainly affect the system stability: k_a, ω_a, and k. A co-design procedure for these parameters is formulated below by using the root contours in the discrete z-domain. The overall rule is to shift the root loci for equating the natural frequencies of two conjugate pole pairs [19].

1) The root locus of $T(z)$ without $G_a(z)$ is plotted first for showing the effect of varying k on the closed-loop poles of the system.

2) A few typical k values are then chosen based on the root locus of $T(z)$, and the root contours of $T_{ad}(z)$ with the changes of k_a and ω_a are plotted. Each root locus shows the movement of closed-loop poles along with the increase of k_a and a given k, and a sweep of ω_a from $0.5\omega_s$ to $2\omega_s$ at the step of $0.5\omega_s$ is performed to identify the range of controller parameters.

3) The parameters can finally be chosen by sweeping ω_a at a smaller step and slightly tuning k value.

For illustration, Fig. 12 plots the root contours to design the IR controller and $G_a(z)$ for $C = 10$ μF, where k is k_p, and k_i can be designed by (9). The controller parameters apply to both 50-Hz and 400-Hz VSIs. Three values of k are chosen for plotting the root contours with the change of k_a and ω_a. The dashed line is the root locus for $T(z)$, which shows that the system is unstable without $G_a(z)$. The solid lines are the root loci with the different ω_a values. It is clear that two conjugate pole pairs will appear if the parameters are not properly chosen, and their natural frequencies may affect the system response.

From Figs. 12(a) and 12(c), it is seen that simply increasing ω_a hardly equate the two conjugate pole pairs, and the tuning of k_p is needed. By taking a further look at Fig. 12(b), it is found that the possible range for ω_a is $1.5\omega_s < \omega_a < 2\omega_s$. With a sweep at a smaller step, $0.1\omega_s$ for example, the parameters for single conjugate pole pair can be found by the root contours.

Fig. 13 shows the root contours for the case of $C = 5$ μF. It can be seen from the root locus for $T(z)$ (dashed line) that the system is initially stable without damping, which matches with

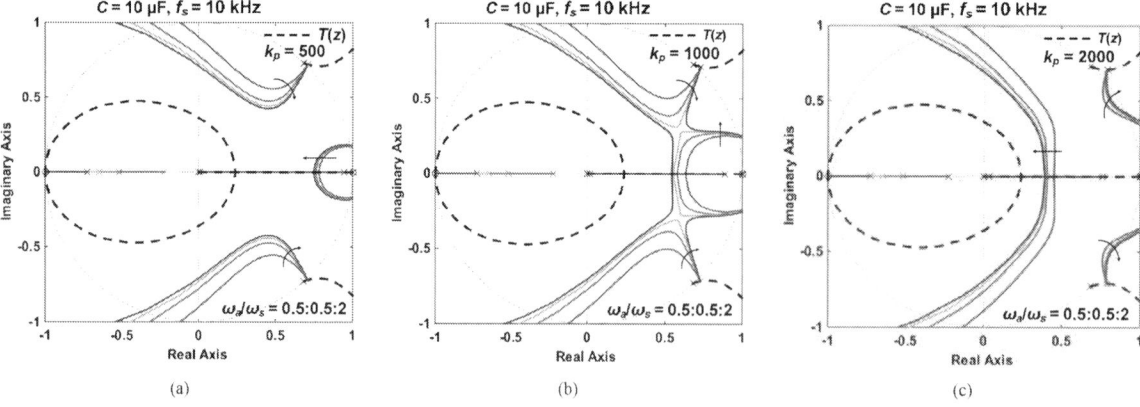

Fig. 12. Root contours to design the IR controller and $G_a(z)$ for $C = 10$ μF. (a) $k_p = 500$. (b) $k_p = 1000$. (c) $k_p = 2000$.

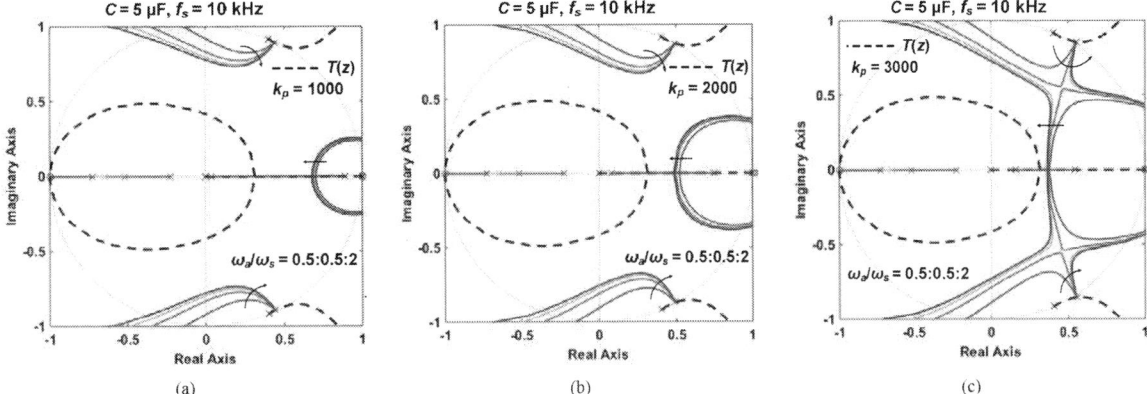

Fig. 13. Root contours to design the IR controller and $G_a(z)$ for $C = 5\ \mu\text{F}$. (a) $k_p = 1000$. (b) $k_p = 2000$. (c) $k_p = 3000$.

the frequency response analysis in Fig. 9. Similarly to Fig. 12, the solid lines depict the movement of closed-loop poles along with the varying k_a and ω_a. Figs. 13(a) and 13(b) clearly show that a cross tuning of k_p and ω_a is required to equate the natural frequencies of two conjugate pole pairs. Fig. 13(c) shows that ω_a should be chosen slightly higher than ω_s, and k_a value can be found at the intersection point. The same design procedure also applies to the design of the R controller in (7) and $G_a(z)$.

V. SIMULATIONS AND EXPERIMENTAL RESULTS

For validations, the time-domain simulations with Simulink and PLECS and the experimental measurements are performed. The Equivalent Series Resistance (ESR = 0.8 Ω) is considered for the filter inductor and the case of $C = 10\ \mu\text{F}$ is evaluated. Table II provides the co-designed controller parameters. Both 50-Hz and 400-Hz system are tested. All the voltages are line-to-line voltage, as a three-phase three-wire system is built in this work.

A. Simulations Results

Fig. 14 shows the simulated ac output voltage with the R controller in (7) only. To see the effect of time delay on system stability, two discretization methods, i.e. the Tustin method with prewraping at the fundamental frequency and the two-integrator-based (feedforward + backward Euler integrals) discretization, are switched at the time instant of 0.4s. It is clear that the system becomes unstable when using Tustin method with prewraping. This matches with the stability region analysis in Fig. 7, where additional time delay ($0.5T_s$) caused by the two-integrator-based discretization method widens the

TABLE II.   CONTROLLER PARAMETERS

Symbol	Electrical Constant	Value
k_{i1}	Gain for using R controller only	1200
k_p	I controller gain of the IR controller	1800
k_{i2}	R contorller gain of the IR controller	180
k_a	Gain of $G_a(s)$	46200
ω_a	Cut-off frequency of $G_a(s)$	$1.71\omega_s$

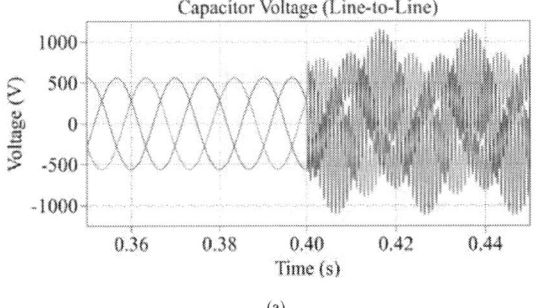

(a)

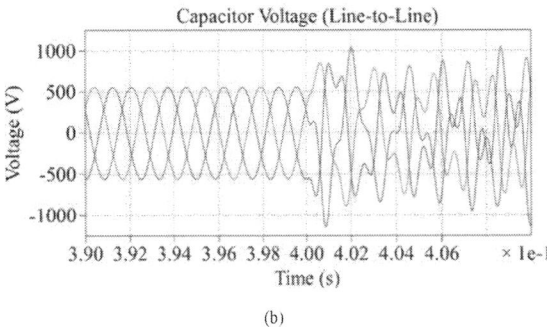

(b)

Fig. 14. Simulated ac output voltage for using the R controller only, where the R controller discretized by two-integrator (feedforward + backward Euler integrals) is changed at 0.4s to the R controller discretized by the Tustin with prewarping at the ac fundamental frequency. (a) 50-Hz system. (b) 400-Hz system.

system stability region.

Fig. 15 then shows the simulated ac output voltage with the IR controller. The IR controller is designed initially unstable, as the LC-filter resonance frequency is lower than the critical frequency in this case. Hence, the active damping controller $G_a(s)$ is adopted for system stabilization. For 50-Hz system, the damping controller is activated at the time instant 0.4s, whereas it is disabled at the same time instant in 400-Hz system. The effectiveness of the damping controller can be clearly seen.

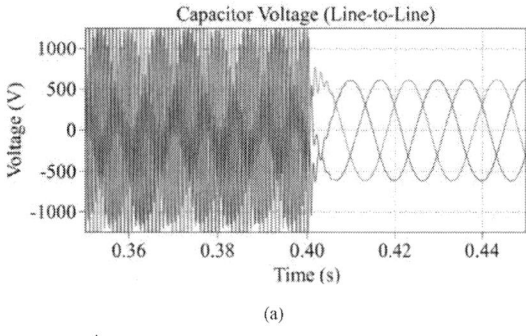

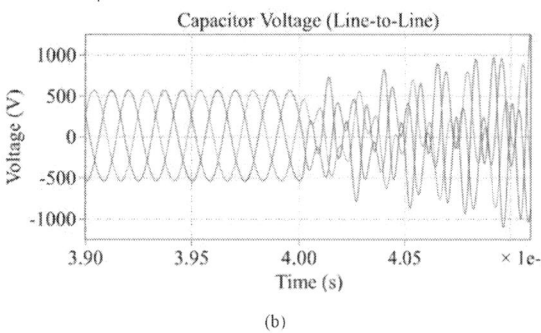

Fig. 15. Simulated ac output voltage with the IR controller, where the damping controller is enabled at 0.4s in (a) 50-Hz system, and is deactivated at 0.4s in (b) 400-Hz system.

B. Experimental Results

In experiments, the DS1007 dSPACE system is used for the digital controller implementation, which integrates the DS2004 high-speed 16-bit A/D sampling board and the DS5101 digital waveform output board for switching pulses generation. The single-update duty cycle mode with the sampling frequency of 10 kHz is used with the controller. A frequency converter with a constant dc-link voltage supply is used as the inverter.

To further verify the simulation results, experimental tests on the same cases as simulation studies are demonstrated in Figs. 16 and 17. First, the influence of discretization method of R controller is evaluated in Fig. 16. It shows that the system becomes unstable when the R controller is discretized by the Tustin method with prewarping, which agrees with the analysis in Fig. 7 and simulation results in Fig. 14. This also confirms the time-delay-dependent stability characteristic of the digital single-loop voltage control, which has not been discussed yet. Moreover, comparing Figs. 16(b) and 14(b), a closer match with the simulation result can be seen in the 400-Hz system, while the less resonance is obtained in the experimental 50-Hz system, as shown in Fig. 16(a). This may be due to the parasitic damping of the test setup at the LC-filter resonance frequency.

Fig. 17 then shows the measured voltage waveforms with the developed IR controller and damping controller. Compared to Fig. 15, the performance of damping controller is evaluated in the same way as the simulation study. The control loop with the IR controller only theoretically cannot be designed stable, since the LC-filter resonance frequency is below the critical

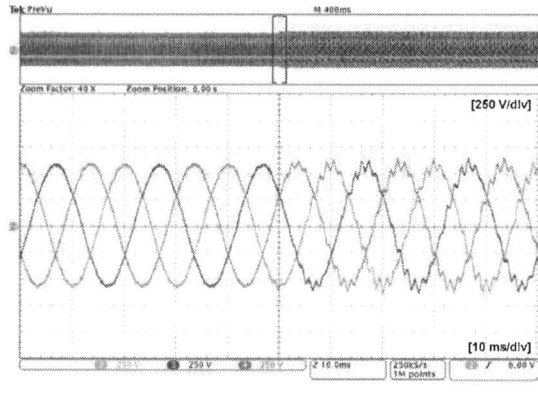

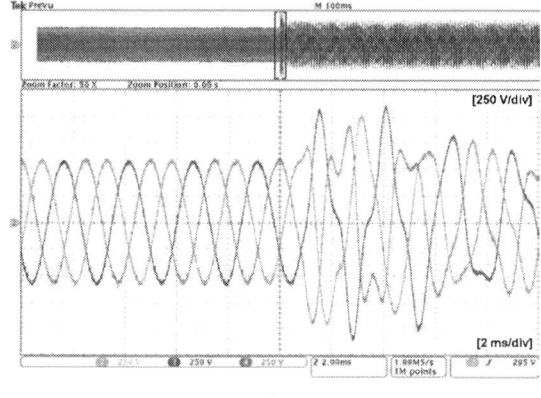

Fig. 16. Measured ac output voltage waveforms with the R controller only, where the discretization of R controller is changed from two-integrator-based structure to the Tustin method with prewarping. (a) 50-Hz system. (b) 400-Hz system.

frequency determined by the digital computation and PWM delay, and the implementation of the I controller by using the Tustin method does not introduce additional time delay. With the damping controller enabled in 50-Hz system, it is clear that the system is stabilized. However, due to the design of low R controller gain, the slow variation of the voltage magnitude can be introduced, as shown by envelope of the voltage waveform in the above zoom-region. Similarly to Fig. 15 (b), the 400-Hz system in the experimental tests begins to run with the damping controller, and then deactivate it to see its stabilizing effect. A correlation with Fig. 15 (b) can be observed.

VI. CONCLUSIONS

This paper has discussed digital single-loop voltage control for LC-filtered VSIs. Simulations and experimental tests have confirmed that the phase lag of the control loop, caused by the time delay or the use of integral controller, plays a critical role in the stability analysis. This is similar to the current control of LCL-filtered VSIs [20]. A feedback-type damping scheme based on the main feedback loop has also been discussed. It has been found that the damping controller, together with the main

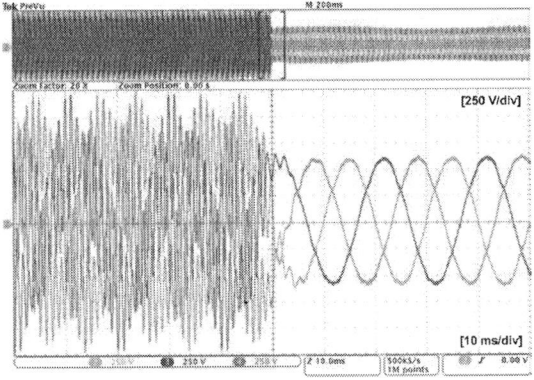

(a)

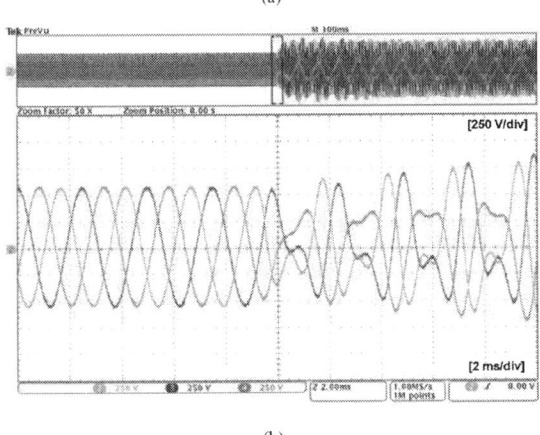

(b)

Fig. 17. Measured ac output voltage waveforms with the IR controller and the active damping controller, where the damping controller is activated in (a) 50-Hz system, and is disabled in (b) 400-Hz system.

controller, leads to an equivalent lead-lag filter added into the control loop. The damping performance has been validated.

REFERENCES

[1] M. J. Ryan, W. E. Brumsickle, and R. D. Lorenz, "Control topology options for single-phase UPS inverters," *IEEE Trans. Ind. Appl.*, vol. 33, no. 2, pp. 493–501, Mar./Apr. 1997.

[2] X. Wang, J. M. Guerrero, Z. Chen, and F. Blaabjerg, "Distributed energy resources in grid interactive ac microgrids," in *Proc. IEEE PEDG 2010*, 2010, pp. 806-812.

[3] L. Mihalache, "DSP control of 400 Hz inverters for aircraft applications," in *Proc. IEEE IAS'02*, 2002, pp. 1564–1571.

[4] N. R. Averous, M. Stieneker, and R. W. De Doncker, "Grid emulator requirements for a multi-megawatt wind turbine test-bench," in *Proc. IEEE PEDS 2015*, 2015, pp. 419-426.

[5] M. Steurer, C. S. Edrington, M. Sloderbeck, W. Ren, and J. Langston, "A megawatt-scale power hardware-in-the-loop simulation setup for motor drives," *IEEE Trans. Ind. Electron.*, vol. 57, no. 4, pp. 1254–1260, Apr. 2010.

[6] X. Wang, F. Blaabjerg, Z. Chen, and W. Wu, "Resonance analysis in parallel voltage-controlled distributed generation inverters," in *Proc. IEEE APEC 2013*, 2013, pp. 2977-2983.

[7] X. Wang, Y. W. Li, F. Blaabjerg, and P. C. Loh, "Virtual-impedance-based control for voltage-source and current-source converters," *IEEE Trans. Power Electron.*, vol. 30, no. 12, pp. 7019-7037, Dec. 2014.

[8] P. C. Loh and D. G. Holmes, "Analysis of multiloop control strategies for LC/CL/LCL-filtered voltage-source and current-source inverters," *IEEE Trans. Ind. Appl.*, vol. 41, no. 2, pp. 644-654, Mar./Apr. 2005.

[9] R. Turner, S. Walton, and R. Duke, "Robust high-performance inverter control using discrete direct-design pole placement," *IEEE Trans. Ind. Electron.*, vol. 58, no. 1, pp. 348–357, Jan. 2011.

[10] U. B. Jensen, F. Blaabjerg, and J. K. Pedersen, "A new control method for 400-Hz ground power units for airplanes," *IEEE Trans. Ind. Appl.*, vol. 36, no. 1, pp. 180–187, Jan./Feb. 2000.

[11] Z. Li, Y. Li, P. Wang, H. Zhu, C. Liu, and F. Gao, "Single-loop digital control of high-power 400-Hz ground power unit for airplanes," *IEEE Trans. Ind. Electron.*, vol. 57, no. 2, pp. 532–543, Feb. 2010.

[12] S. Hiti, D. Boroyevich, and C. Cuadros, "Small-signal modeling and control of three-phase PWM converters," in *Proc. IEEE IAS 1994*, 1994, pp. 1143-1150.

[13] D. M. Van de Sype, K. D. Gusseme, A. P. Van den Bossche, and J. A. Melkebeek, "Small-signal laplace-domain analysis of uniformly-sampled pulse-width modulators," in *Proc. IEEE PESC 2004*, 2004, pp. 4292-4298.

[14] S. Buso and P. Mattavelli, *Digital Control in Power Electronics*, San Francisco, CA: Morgan & Claypool Publ., 2006.

[15] A. Yepes, F. Freijedo, J. Gandoy, O. Lopez, J. Malvar, and P. Comesana, "Effects of discretization methods on the performance of resonant controllers," *IEEE Trans. Power Electron.*, vol. 25, no. 7, pp. 1692-1712, Jul. 2010.

[16] X. Wang, F. Blaabjerg, and P. C. Loh, "Virtual RC damping of LCL-filtered voltage source converters with extended selective harmonic compensation," *IEEE Trans. Power Electron.*, vol. 30, no. 9, pp. 4726-4737, Sept. 2015.

[17] D. G. Holmes, T. A. Lipo, B. P. McGrath, and W. Y. Kong, "Optimized design of stationary frame three phase ac current regulators," *IEEE Trans. Power Electron.*, vol. 24, no. 11, pp. 2417-2426, Nov. 2009.

[18] C. Lascu, L. Asiminoaei, I. Boldea, and F. Blaabjerg, "High performance current controller for selective harmonic compensation in active power filters," *IEEE Trans. Power Electron.*, vol. 22, no. 5, pp. 1826-1835, Sept. 2007.

[19] X. Wang, F. Blaabjerg, and P. C. Loh, "High-performance feedback-type active damping of LCL-filtered voltage source converters," in *Proc. IEEE ECCE 2015*, 2015, pp. 2629-2636.

[20] X. Wang, F. Blaabjerg, and P. C. Loh, "Grid-current-feedback active damping for LCL resonance in grid-connected voltage-source converters," *IEEE Trans. Power Electron.*, vol. 31, no. 1, pp. 213-223, Jan. 2016.

High Efficiency, Hybrid Selective Harmonic Elimination Phase-Shift PWM Technique for Cascaded H-Bridge Inverters to Improve Dynamic Response and Operate in Complete Normal Modulation Indices

Amirhossein Moeini, Zhao Hui, Shuo Wang

Power Electronics and Electrical Power Research Laboratory, Electrical and Computer Engineering,
University of Florida
Gainesville, Florida, USA
ahm1367@ufl.edu, zhaohui@ufl.edu, shuo.wang@ece.ufl.edu

Abstract- Selective Harmonic Elimination PWM (SHE-PWM) method has slow dynamic response, due to using the Fourier transform before the Look-Up Tables (LUTs) in practices. Also, in some cases, the SHE-PWM cannot find any solutions in normal modulation ranges. On the other hand, Phase-Shift PWM (PSPWM) does not need Fourier transform in practice to operate, but it has lower efficiency than SHE-PWM. Furthermore, the PSPWM can work properly in all of the normal modulation indices. Using the combination of these two methods can achieve high efficiency and fast dynamic response at the same time. Moreover, the inverter can work in all of the normal modulation indices. To reach these goals, a hybrid Selective Harmonic Elimination PWM with Phase-Shift Pulse Width Modulation (HSHE-PSPWM) is introduced by applying SHE-PWM in steady state condition and applying PSPWM in dynamic conditions. Also, for some modulation indices that the SHE-PWM does not have any solutions, the PSPWM has been applied to the Cascaded H-Bridge (CHB) inverter. A controller is also proposed to switch between the SHE-PWM and PSPWM. Also, a RLC low pass filter is designed to attenuate high order harmonics of the proposed HSHE-PSPWM and it will be shown that the proposed technique can fully meet the IEEE-519 2014 in steady state and transient conditions. Finally, to verify the effectiveness and applicability of the proposed method, simulation and experimental results are obtained from a 7-level Cascaded H-Bridge inverter.

Keywords— Hybrid Selective Harmonic Elimination Phase-Shift PWM; Multilevel Inverter; High power applications; Dynamic Performance;

I. INTRODUCTION

High power multilevel inverters have taken a lot of attentions nowadays, due to their advantages in comparison to conventional two-level inverters [1-2] such as lower dv/dt on each switch, lower THD, higher efficiency, and lower EMI noise [2].

In high power applications, increasing the efficiency of the inverter is so important [3]. Moreover, the efficiency of the inverter depends on the switching losses of the solid state devices. The switching losses can be controlled in the power switches by using the low frequency modulation techniques such as Selective Harmonic Elimination PWM (SHE-PWM)

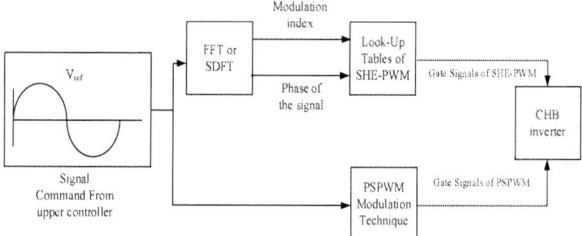

Fig. 1. The comparison of the SHE-PWM and PSPWM techniques in dynamic conditions.

and Selective Harmonic Mitigation-PWM (SHM-PWM) [4-8].

Also, having the high dynamic performance in the power electronic inverters is critical in power grid applications [9-11] such as STATCOM [12] or in Direct Torque Control of the Induction fed Motor [13].

The dynamic performance of the inverter depends on the time delays on upper controller performance and time delays on the operation of modulation technique.

As it is shown in the Fig. 1, with the same sine wave command signal from the upper controller, the SHE-PWM or SHM-PWM need a (FFT or SDFT) and Phase Lock Loop (PLL) blocks to find the modulation index and phase of the reference signal. The time delay of these blocks reduce the dynamic performance of the SHE-PWM. For this reason, in [14], due to using the SDFT block before the LUTs, the dynamic response of the inverter had time delay more than 1 cycle [14].

In this paper, a hybrid SHE-PWM and PSPWM method is proposed to improve the dynamic performance of the SHE-PWM method. In the proposed method, in the steady state condition, SHE-PWM method which has the low switching frequency is used to achieve high efficiency on the inverter. Also, the PSPWM method which does not need Fourier transform is employed on the dynamic conditions. Moreover, a simple controller will be presented to switch between the SHE-PWM and PSPWM techniques in the dynamic conditions. The proposed HSHE-PSPWM can also work in all of the normal modulation indices, even though the SHE-PWM does not have any solutions for some ranges of the modulation

This work was supported by National Science Foundation under award number ECCS-1540118.

indices. For higher order harmonics of the SHE-PWM and PSPWM also a RLC filter is used. The designed RLC filter can meet the requirements of IEEE-519 2014 for higher order harmonics which cannot meet by the modulation technique. Finally, in order to keep the efficiency of the inverter as high as possible, the lowest amount of switching frequency of the carrier which can eliminate the low order harmonics up to 19th in PSPWM technique has been applied to the 7-level CHB inverter.

II. PROPOSED HYBRID SHE-PSPWM TECHNIQUE

The Selective Harmonic Elimination has two main challenges when it is used for high dynamic performance [14-18].

1) As it can see in the Fig. 1, in comparison to the PSPWM, to implement the SHE-PWM, the Fast Fourier Transform (FFT) or Sliding Discrete Fourier Transform (SDFT) must be used before the look-up tables (LUTs), because the upper controller command is the sine wave, but the look-up tables need modulation index to operate. The FFT and SDFT blocks need time to find the modulation indices. As a result, it can significantely reduce the dynamic performance of the SHE-PWM.

2) The obtained solutions in the SHE-PWM cannot completely cover all of the required normal modulation indices. In other words, in SHE-PWM technique, for some modulation indices, there is not any solutions on LUTs to be applied to the CHB inverter. So, it is impossible to implement SHE-PWM in all of the modulation indices.

To solve these two challenges, a Hybrid Selective Harmonic Elimination PWM and Phase-Shift PWM has been implemented in this paper.

In Fig. 2, the general circuit model of the N-cell CHB inverter which is connected to the RL loads has been illustrated. The DC link voltages of the proposed technique can be connected to the grid voltage by using active or passive rectifiers [19].

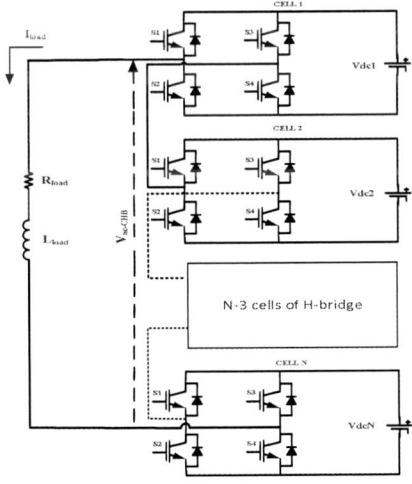

Fig. 2. The general single-phase overview of CHB multilevel inverter.

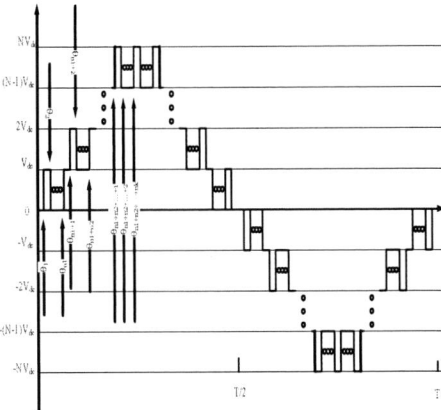

Fig. 3. The general predefined waveform of the SHE-PWM technique.

A. The Selective Harmonic Elimination technique

The SHE-PWM method tries to employ the Fourier series equations in order to eliminate the low order harmonics of the inverter [20, 21].

In the Fig. 3, the general predefined waveform of selective harmonic elimination technique for CHB converter which consists of N-cell has been shown. In high power applications, it is desirable to have low number of switching transitions in each level to reduce the switching losses of each switch. The general equations of the SHE-PWM technique based on the predefined waveform in Fig. 3 can be written as,

$$V_{ac-CHB} = \sum_{m=1}^{\infty} \frac{4V_{dc}}{\pi m}(a_m \sin(m\omega t) + b_m Cos(m\omega t));$$

$$
\begin{aligned}
a_m &= (\cos(m\theta_1) - \cos(m\theta_2) + ... + \cos(m\theta_{n1}) + \cos(m\theta_{n1+1}) \\
&- \cos(m\theta_{n1+2}) + ... + \cos(m\theta_{n1+n2}) + ... + \cos(m\theta_{n1+n2+...+1}) \\
&- \cos(m\theta_{n1+n2+...+2}) + ... + \cos(m\theta_{n1+n2+...+nk}))
\end{aligned}
\tag{1}
$$

$$
\begin{aligned}
b_m &= (-\sin(m\theta_1) + \sin(m\theta_2) - ... - \sin(m\theta_{n1}) - \sin(m\theta_{n1+1}) \\
&+ \sin(m\theta_{n1+2}) - ... - \sin(m\theta_{n1+n2}) - ... - \sin(m\theta_{n1+n2+...+1}) \\
&+ \sin(m\theta_{n1+n2+...+2}) - ... - \sin(m\theta_{n1+n2+...+nk}))
\end{aligned}
$$

where, *m* is the order of each harmonic. *θ* is the switching angle that should be applied by the converter. The general modulation index of the equation (1) can be written as,

$$
M_a = \sqrt{
\begin{aligned}
&(Cos\theta_1 - Cos\theta_2 + ... + Cos\theta_{n1} + Cos\theta_{n1+1} - Cos\theta_{n1+2} \\
&+ ... + Cos\theta_{n1+n2} + ... + Cos\theta_{n1+n2+...+1} - Cos\theta_{n1+n2+...+2} \\
&+ Cos\theta_{n1+n2+...+nk})^2 + (-Sin\theta_1 + Sin\theta_2 - ... - Sin\theta_{n1} - \\
&Sin\theta_{n1+1} + Sin\theta_{n1+2} - ... - Sin\theta_{n1+n2} - ... - Sin\theta_{n1+n2+...+1} \\
&+ Sin\theta_{n1+n2+...+2} - ... - Sin\theta_{n1+n2+...+nk})^2
\end{aligned}
}
\tag{2}
$$

In order to have voltage waveform similar to the Fig. 3, the switching angles should satisfy the following constraints.

$$0 < \theta_1 < \theta_2 < ... < \theta_{n1} < \theta_{n1+1} < \theta_{n1+2} < ... < \theta_{n1+n2+...+1}$$

$$< \theta_{n1+n2+...+2} < \theta_{n1+n2+...+nk} < \frac{\pi}{2}
\tag{3}$$

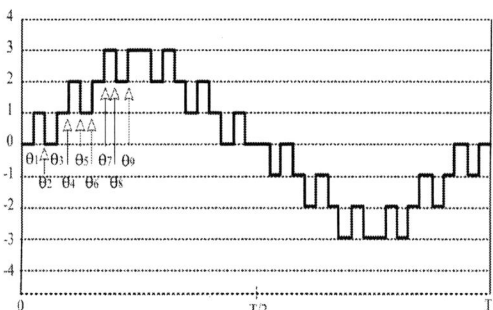

Fig. 4. The predefined waveform of the 7-level SHE-PWM.

In the equation (1), the magnitude and phase of each harmonic depends on a_m and b_m components. However, the symmetry in the predefined waveform can eliminate some terms from the output voltage. As a result, due to quarter wave symmetry in the predefined waveform in the Fig. 3, all of the b_m are equal to zero. Moreover, the even order harmonics in the a_m also are equal to zero. So, the odd harmonics only need to be controlled in the SHE-PWM technique.

In this paper, 9-switching transitions in each quarter of the period has been employed in the 7-level CHB inverter, as it can see in the Fig. 4. Each cell of the inverter has only 3-switching transitions in each quarter of the period, so the switching frequency of each switch in the proposed SHE-PWM method is 180 Hz. The Fourier series equations of the predefined waveform can be written as,

$$M_a = (Cos\theta_1 - Cos\theta_2 + Cos\theta_3 + Cos\theta_4$$
$$- Cos\theta_5 + Cos\theta_6 + Cos\theta_7 - Cos\theta_8 + Cos\theta_9)$$
$$a_m = \frac{4V_{dc}}{m\pi}(Cosm\theta_1 - Cosm\theta_2 + Cosm\theta_3 + Cosm\theta_4 \quad (4)$$
$$- Cosm\theta_5 + Cosm\theta_6 + Cosm\theta_7 - Cosm\theta_8 + Cosm\theta_9)$$
$$Form = 3,5,7,9,11,13,15,17$$
$$0 < \theta_1 < \theta_2 < \theta_3 < \theta_4 < \theta_5 < \theta_6 < \theta_7$$
$$< \theta_8 < \theta_9 < \frac{\pi}{2}$$

In the equation (4), M_a is the modulation index and the θ_1-θ_9 are the switching angles of the SHE-PWM approach. V_{dc} is the DC link voltage of the CHB inverter.
To find the solutions of the SHE-PWM approach, the Particle Swarm Optimization technique has been used to solve the equation (4) in the wide range of modulation indices. The objective function which is used in PSO technique is written below,

$$OF = \sqrt{\frac{(M_a - (\cos(\theta_1) - \cos(\theta_2) + ... + \cos(\theta_9)))^2}{+ (a_3)^2 + (a_5)^2 + ... + (a_{17})^2}} \quad (5)$$

where, in the equation (5), each low order harmonic of the output voltage up to 17th must be equal to zero, when the modulation index is equal to certain value. As it is shown in the Fig. 5, the obtained solutions of the PSO technique has been obtained in wide range of modulation indices (1.67, 2).

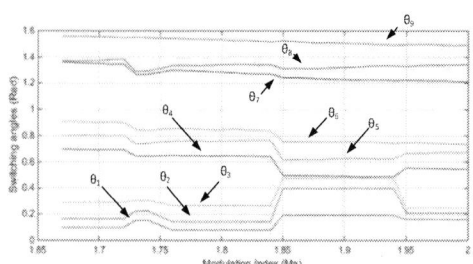

Fig. 5. The obtained solutions of the SHE-PWM.

However, the obtained solutions cannot cover the normal range of modulation indices between 0 to 1.67 and 2 to 3. Consequently, when the upper controller in the Fig. 1 applies the modulation index between these two regions, the SHE-PWM cannot eliminate the low order harmonics of the voltage up to 17th.

B. The Phase-Shift Pulse Width Modulation Technique

The Phase-Shift PWM technique has been mentioned in many papers due to its simplicity, high dynamic performance and acceptable THD and harmonic spectra [2]. In Fig. 6, the structure of the PSPWM technique has been shown. As it is shown, to generate the gate signals for 4 switches in each H-bridge, the reference signal must be compared with the carrier signal ($V_{carrier}$) and ($-V_{carrier}$) in bipolar dual-edge Phase-Shift PWM technique [23, 24].

By considering the reference voltage that is commanded by the upper controller to the PSPWM technique as,

$$V_{ref}(t) = R_1 Cos(\omega_m t + \theta_m) \quad (6)$$

here, the equation of the PSPWM can be written as [26],

$$V_{a-CHB} = M_a V_{dc} Cos(\omega_m t + \theta_m)$$
$$+ \sum_{b=1}^{\infty} \sum_{a=-\infty}^{\infty} \frac{2V_{dc}}{Nb\pi} J_a(bM_a\pi) Sin((2Nb + a + 2)\frac{\pi}{2}) Cos[2Nb(\omega_{carrier} t + \theta_{carrier}) \quad (7)$$
$$+ a(\omega_m t + \theta_m)]$$

where, in the above equation, V_{dc} is the DC link voltages, M_a is the modulation index that can be obtained by,

$$M_a = \frac{NR_1}{R_{tri}} \quad (8)$$

where, R_{tri} is the magnitude of the triangular signal of the carrier, N is the number of CHB cells in the inverter, J_a in equation (7) is the ath order Bessel function of the first kind which can be obtained by the following integral,

$$J_a(z) = \frac{j^{-a}}{\pi} \int_0^{2\pi} e^{jzCos\theta} e^{ja\theta} d\theta$$

In equation (7), the $\omega_{carrier}$ is the carrier frequency of the each cell of the CHB inverter. So, when the 2N carriers are used for three-cell of the CHB inverter, the carrier frequency of the output voltage of the inverter (ω_s) can be obtained by $\omega_s = 2 \times N \times (2\pi F_{carrier})$. $F_{carrier}$ is the carrier frequency of the PSPWM technique for each cell. ω_m is the frequency of the reference waveform. The ratio between the ω_s and ω_m can be written as,

$$K_f = \frac{\omega_s}{\omega_m} \quad (9)$$

978-1-4673-9551-9/16 $31.00 © 2016 IEEE

If there is N number of cells in each phase of the CHB inverter, it would be better to choose the phase difference (θ_{p-h}) for the carrier signals of each cell by the following equation to have lower harmonic distortion [26],

$$\theta_{p-h} = \pi(h-1)/N \qquad (10)$$

where, h is the number of each cell of the CHB inverter which can be between 1 to N. So, for the three-cell CHB inverter the 0, $\pi/3$, and $2\pi/3$ are suitable phases for the triangular carriers of the CHB cells.

In the equation (7), b and a are the baseband and sideband order harmonics of the PSPWM technique.

The harmonic spectra of the PSPWM technique in equation (7) has harmonics in the basebands (bf_s) and sidebands of the baseband harmonics (bf_s+af_m, and bf_s-af_m). In order to obtain the lowest number of switching transitions in the PSPWM technique which can meet the harmonic spectra requirements of the SHE-PWM technique (Eliminating the low order harmonics of the output waveform up to the 17th order harmonic) the sidebands of the first baseband harmonic (b=1) should not have high magnitude for harmonics lower than 19th. In other words, the considerable sideband order harmonic for first baseband must be at 19th order harmonic. To reach this goal, the considerable sideband harmonics of the first baseband (b=1) which is at the carrier frequency should be analyzed. According to the Carson bandwidth rule [25], the bandwidth of the PSPWM modulation signal in (7) can be approximated by the following equation,

$$BW_{PWM}^{b} \approx 2(b\pi M_a + 2)f_m \qquad (11)$$

In the above equation, the bandwidth (BW_{PWM}^{1}) can be used to find the highest ($f_s+BW_{PWM}^{1}$) and lowest ($f_s-BW_{PWM}^{1}$) considerable sideband harmonics of the PSPWM technique.

So, by considering the highest modulation index (M_a=3) (worst case condition), and the first baseband harmonic (b=1), the considerable bandwidth of the PSPWM technique from the equation (11) must be at $22.85f_m \approx 23f_m$. Furthermore, to ensure that the sideband harmonics of the PSPWM can eliminate up to 19th, the switching frequency of the carrier should be chosen based on following equation,

K_f = {The highest order of harmonic which should be eliminated by PSPWM + BW_{PWM}^{1} } $\qquad (12)$

Thus, the K_f = 40 in the proposed HSHE-PSPWM has been chosen. So, the switching frequency of the carrier has been selected by $f_{carrier}$=400Hz. This is the lowest limit of the switching frequency of the PSPWM that can eliminate harmonics up to 17th. Both of the modulation techniques generate switching transitions according to the command signal and applied it to the Modulation Selector block.

C. The Proposed HSHE-PSPWM Technique

The proposed controller for the HSHE-PSPWM method has been shown in the Fig. 7 and 8. As it is shown, the reference signal has been assigned by the upper controller and applied to both SHE-PWM and PSPWM techniques.

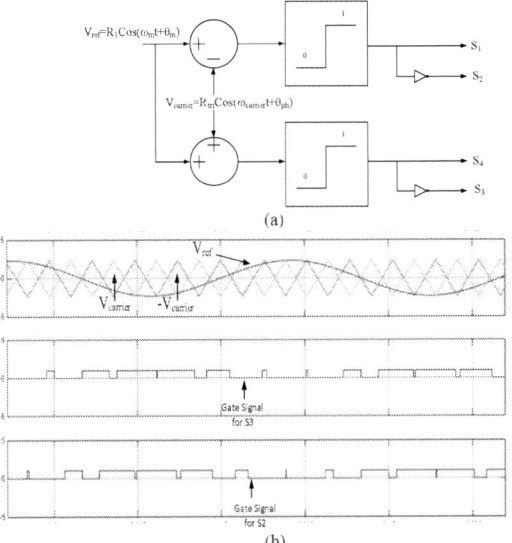

Fig. 6. (a) The modulation procedure for each cell of the CHB in the PSPWM modulation technique (b) The carrier signals and gate signals for one cell of the CHB inverter.

However, in SHE-PWM technique, the magnitude of the signal command (M_a) and the phase of the signal ($\omega_m t$) should be applied. As a result, the FFT or SDFT blocks for obtaining the magnitude of the signal and PLL for estimating the phase of the command signal should be used before the LUTs. However, these blocks significantly reduce the dynamic performance of the SHE-PWM technique. To solve this issue, in dynamic conditions, the PSPWM technique will be employed, since it can work properly without any FFT or SDFT blocks. In order to switch properly between these two modulation techniques, the modulation selector block which is shown in Fig. 7 has been used in this paper.

From the equation (1, 4), the voltage of the SHE-PWM technique (V_{SHE}) can be calculated by,

$$V_{SHE} = \frac{4M_a V_{dc}}{\pi} Sin(\omega_m t) \qquad (13)$$

So, by applying the modulation index and phase of the command signal, the output voltage of the SHE-PWM can be estimated. Moreover, due to having the time delay for the FFT or SDFT blocks, the V_{SHE} can clearly show the time delay of the Fourier block. As a result, this command has been compared to the reference command of the upper controller (V^*) in the Modulation Selector block, based on the flowchart of the Fig. 7, to switch between the SHE-PWM and PSPWM techniques in the dynamic conditions. In the Fig. 8, the flowchart of the Modulation Selector block of the proposed HSHE-PSPWM has been depicted.

In this method, first, the calculated modulation index (M_{SHE}) has been checked to be in the available solution region of the Fig. 5.

978-1-4673-9551-9/16 $31.00 © 2016 IEEE

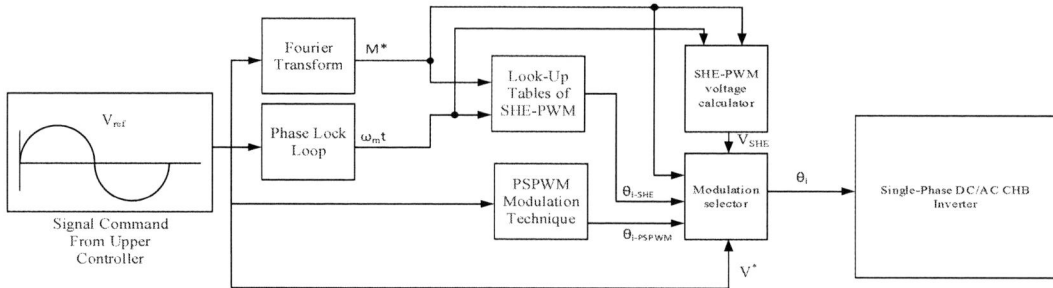

Fig. 7. The proposed HSHE-PSPWM technique block diagram.

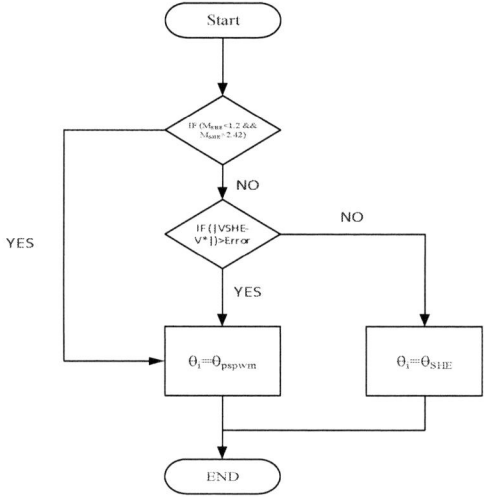

Fig. 8. The modulation selector flowchart of the controller block diagram of the proposed HSHE-PSPWM technique.

If the M_{SHE} is not in the available solution region, the PSPWM switching transitions will be applied to the CHB inverter. Otherwise, the difference between the reference modulation index (V_{SHE}) and the reference voltage of the upper controller (V^*) are checked to be lower than an error.

If the absolute value of the ($V_{SHE}-V^*$) is higher than the error, the PSPWM must be used, otherwise the SHE-PWM switching angles must be applied to the CHB inverter to have high efficiency.

D. Filter Design for the proposed HSHE-PSPWM

According to the Fig. 9, the RLC filter has been used in this paper to meet the requirement of the IEEE-519 2014 [27]. In the Fig. 9, the transfer function of the RLC filter can be written as,

$$\frac{V_{out}}{V_{ac-CHB}} = \frac{\frac{1}{LC}}{s^2 + s\frac{1}{CR} + \frac{1}{LC}} \qquad (14)$$

To design the filter, worst case conditions for the higher order harmonics in the SHE-PWM and PSPWM should be considered. Because the low order harmonics of the SHE-PWM and PSPWM can be eliminated, the highest magnitude

of the high order harmonic should be considered for designing the filter. As it can see in the Fig. 10, the highest magnitude of the 19th order harmonic is at $M_a = 1.9$ for SHE-PWM technique. So, the magnitude of this harmonic must be used to design the filter in SHE-PWM technique.

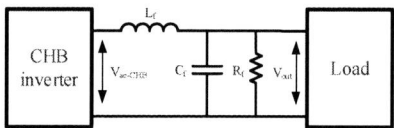

Fig. 9. The configuration of the CHB inverter that is connected to load by the RLC filter.

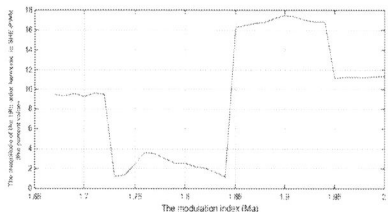

Fig. 10. the magnitude of the SHE-PWM for 19th order harmonic.

The highest magnitude of the PSPWM technique in the equation (7) depends on the Bessel function $J_a(bM_a\pi)$. Also, in equation (7), it is obvious that,

$$Sin((2Nb+a+2)\frac{\pi}{2}) = 0 , \text{ when } a \text{ is even} \qquad (15)$$

so only the odd value of the a should be considered. Thus, from the Fig. 11, the highest magnitude of the harmonics of the PSPWM has been occurred at ($a=1$).

To obtain the x that has the highest value for $J_1(x)$, from [29],

$$J_1(x) = \frac{Sinx}{x^2} - \frac{Cosx}{x} \qquad (16)$$

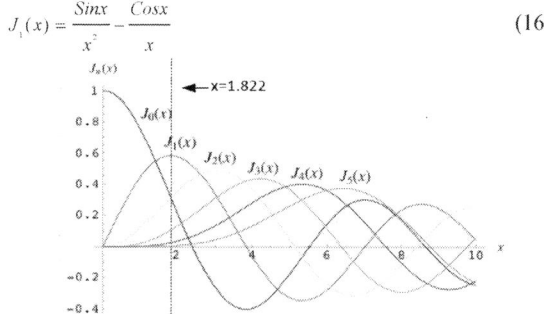

Fig. 11. The first kind of Bessel function waveforms up to 5th order.

to find the maximum value of the $J_1(x)$ near 2, the derivative of the equation (16) should become equal to zero or,

$$\frac{d(J_1(x))}{dx} = 0 \qquad (17)$$

Thus, the equation can be written as,

$$\frac{Cos(x)(x + x^2) + Sin(x)(x^2 - 2)}{x^3} = 0 \qquad (18)$$

In the equation (18), when x is equal to 1.822 the $J_1(x)$ has highest value. By considering the x=1.822, the M_a=0.58 has the highest value of the harmonic magnitude.

By knowing the worst case condition for both of the techniques, the corner frequency of the RLC filter should be designed based on the worst case condition of these two techniques. As it can see, in the Fig. 12 and 13, the red line is the 40dB/dec attenuation of the RLC filter. So, in order to attenuate the magnitude of the current of the PSPWM technique at switching frequency equal to 400Hz, the corner frequency of the RLC filter should be designed at 250Hz. So, based on this corner frequency the RLC filter can be designed. It is worth to note that the corner frequency of the PSPWPM technique depends on the switching frequency. As it can see in Fig. 12, For the SHE-PWM waveform also the harmonics can be attenuated with the same corner frequency.

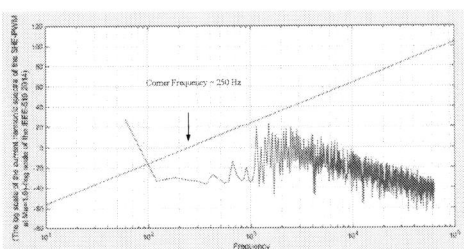

Fig. 12 The Harmonic Spectra of the current of the SHE-PWM at M_a=1.9 (worst case condition in SHE-PWM).

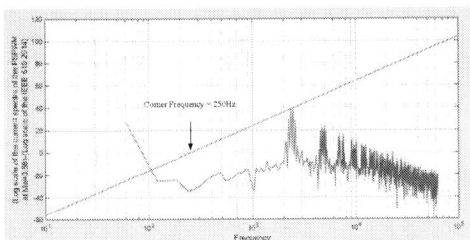

Fig. 13. The harmonic Spectra of the load current of the PSPWM at M_a = 0.58 (worst case condition in PSPWM).

III. SIMULATION RESULTS

To verify the effectiveness and superiority of the proposed HSHE-PSPWM method in comparison to the conventional SHE-PWM technique, simulation results are provided in SIMULINK MATLAB environment. The circuit parameters of the simulation and experiment are presented in the Table I. In the Fig. 14, first, the converter works in the steady state condition and the modulation index is equal to 1.7. Then, at t=0.305s, the command of modulation index has been changed to

1.9. So, the PSPWM technique has been applied to each cell of the CHB inverter according to Fig. 6(a).

TABLE I. THE CIRCUIT PARAMETERS OF THE PROPOSED HSHE-PSPWM

Parameters	Symbol	Value
Number of Cells	N	3
DC link voltage	V_{dc}	50V
Switching Frequency of the SHE-PWM	Fsw	180Hz
Switching Frequency of the PSPWM	Fsw	400Hz
Resistance of the load	R_l	100Ω
Inductance of the load	L_l	0.01H
Inductance of the filter	L_f	0.007H
Capacitor of the filter	C_f	60uF
Resistance of the filter	R_f	10Ω

Finally, at t=0.328s, because the error between the V_{SHE} and V^* become less than (5V), the SHE-PWM switching transitions which has better efficiency than the PSPWM technique has been applied to the CHB inverter.

In the Fig. 14 (a), the time domain response of the proposed HSHE-PSPWM has been shown. In comparison to the conventional SHE-PWM (Fig. 14 (b)), the high switching frequency (PSPWM) which does not need FFT blocks can track the reference voltage much better than the conventional SHE-PWM. Also, the PSPWM is only applied around 1 cycle, so the switching frequency of the CHB inverter does not increase too much. In Fig. 14 (c), to better compare the conventional SHE-PWM and the proposed HSHE-PSPWM, the magnitude of the voltage of the CHB inverter for both of the techniques has been obtained by the Fourier block of the MATLAB SIMULINK. As it can see in this figure, the proposed HSHE-PSPWM can better track the reference voltage of the upper controller. But, the conventional SHE-PWM need to spend at least one cycle to track the reference voltage. Moreover, due to switching modulation technique between a high frequency modulation technique (PSPWM) and the low frequency modulation technique (SHE-PWM) at t=0.328s in the Fig. 14 (c), the Fourier block of the MATLAB generates some oscillations. So, this oscillation does not occur in the voltage and current of the system.

The current harmonic spectra of the SHE-PWM at M_a=1.9 has been shown in the Fig. 15 (a). As it can see, the requirement of the IEEE 519-2014 can be met by the RLC filter. Also, for the PSPWM technique at M_a = 0.58 has been shown in Fig. 15 (b). Also, in the PSPWM technique, the current harmonic spectra of has been met by using the RLC filter.

IV. EXPERIMENTAL RESULTS

In this section, the experimental result of the proposed HSHE-PSPWM method has been presented. As it can see in Fig. 16, for the hardware prototype, the proposed method have acceptable dynamic performance, when the modulation index has been changed between 1.7 and 1.9. In Fig. 16 (c), the magnitude of the reference voltage of the upper controller with the voltages of the CHB inverter in the conventional SHE-PWM and proposed HSHE-PSPWM has been compared.

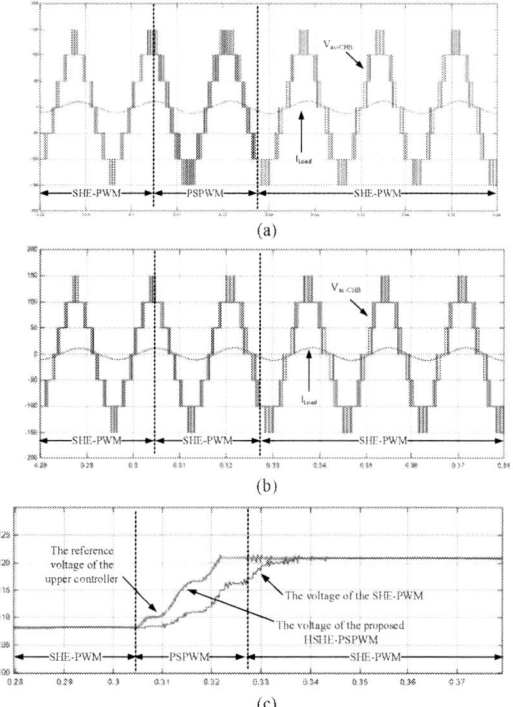

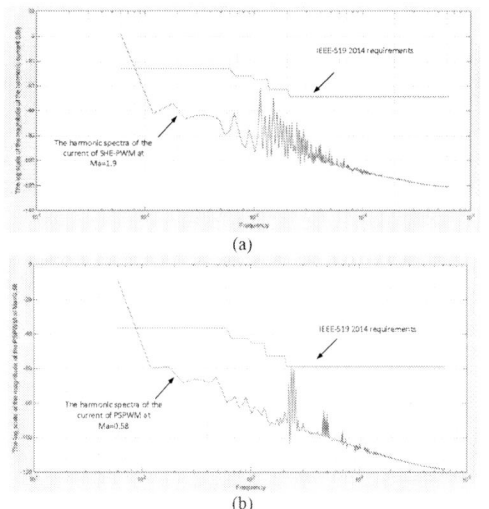

Fig. 14. (a) The Dynamic performance of the proposed HSHE-PSPWM. (b) The dynamic performance of the conventional SHE-PWM. (c) Comparison between dynamic performance of the proposed HSHE-PSPWM and conventional SHE methods in the magnitude of the fundamental harmonic.

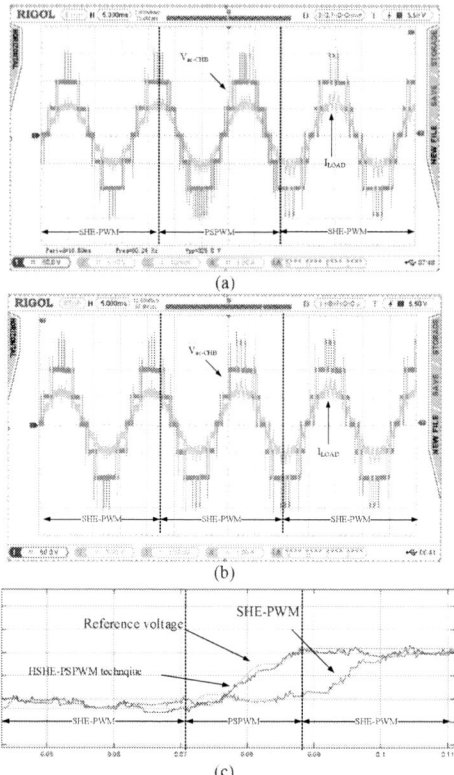

Fig. 16. The experimental result of the proposed HSHE-PSPWM technique in dynamic condition (a) The voltage and current of the proposed HSHE-PWM (b) The voltage and current of the conventional SHE-PWM. (c) The comparison of the magnitude of the proposed HSHE-PWM and SHE-PWM in dynamic condition.

Fig. 15, The log scale of the current harmonic spectra of the SHE-PWM and PSPWM, (a) SHE-PWM at Ma=1.9, (b) PSPWM at Ma=0.58.

As it is shown in Fig. 16 (c), similar to Fig. 14 (c), the magnitude of the HSHE-PSPWM can better track the reference voltage than the SHE-PWM technique, due to not using any FFT or SDFT block to select the switching transitions.

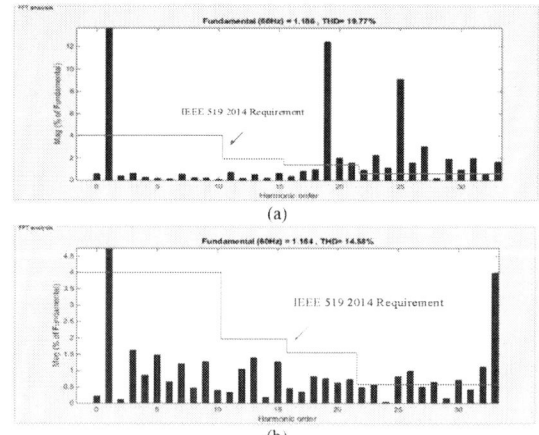

Fig. 17. (a) The current harmonic spectra of the SHE-PWM technique in the M_a=1.9, (b) The current harmonic spectra of the PSPWM technique in transition condition.

In Fig. 17 (a), the proposed PSPWM technique has lower low order harmonics than SHE-PWM during transition. So it has no problems to meet the harmonic standards with the existing passive harmonic filter.

V. CONCLUSIONS AND FUTURE WORK

In this paper, a new hybrid SHE-PSPWM modulation technique has been proposed to improve the dynamic performance of the high power applications in low frequency modulation technique. The proposed method can significantly improve the dynamic performance of the CHB inverter as it was shown in the simulation and experimental results. Moreover, the PSPWM method which is employed in this paper has the low number of switching frequency to increase the efficiency of the converter. It is worth to note that the proposed method is so simple to implement and also can provide high performance in the dynamic condition. The implementation of the proposed method in the grid connected converters is the future work in this area.

REFERENCES

[1] Khomfoi, S., Tolbert, L. M.: "Multilevel Power Converters," Power Electronics Handbook, The University of Tennessee, Department of Electrical and Computer Engineering.

[2] Rodriguez, J., Lai, J. S., Peng, F. Z.: "Multilevel inverters: a survey of topologies, controls, and applications," *Industrial Electronics, IEEE Transactions on*, Aug 2002 , vol. 49, no. 4, pp. 724- 738.

[3] Watson, A. J., Wheeler, P. W., Clare, J. C., "A Complete Harmonic Elimination Approach to DC Link Voltage Balancing for a Cascaded Multilevel Rectifier," *Industrial Electronics, IEEE Transactions on*, Dec. 2007, vol. 54, no. 6, pp. 2946-2953.

[4] Marzoughi, A., Imaneini, H., and Moeini, A.:'An optimal selective harmonic mitigation technique for high power converters' International Journal of Electrical Power & Energy Systems, July.2013, vol. 49, pp.34-39.

[5] Fei, W., Ruan, X., Wu, B.: "A Generalized Formulation of Quarter-Wave Symmetry SHE-PWM Problems for Multilevel Inverters," *Power Electronics, IEEE Transactions on*, July 2009, vol. 24, no.7, pp. 1758-1766.

[6] Moeini, A.; Iman-Eini, H.; Bakhshizadeh, M., "Selective harmonic mitigation-pulse-width modulation technique with variable DC-link voltages in single and three-phase cascaded H-bridge inverters," in *Power Electronics, IET*, vol.7, no.4, pp.924-932, April 2014.

[7] Moeini, A.; Iman-Eini, H.; Marzoughi, A., "DC link voltage balancing approach for cascaded H-bridge active rectifier based on selective harmonic elimination-pulse width modulation," in *Power Electronics, IET*, vol.8, no.4, pp.583-590, 4 2015.

[8] Moeini, A.; Marzoughi, A.; Iman-Eini, H.; Farhangi, S., "A modified control strategy for cascaded H-bridge rectifiers based on the low frequency SHE-PWM," in *Environment and Electrical Engineering (EEEIC), 2013 12th International Conference on*, vol., no., pp.501-506, 5-8 May 2013.

[9] Zheng Wang; Bin Wu; Dewei Xu; Zargari, N.R., "Hybrid PWM for High-Power Current-Source-Inverter-Fed Drives With Low Switching Frequency," in *Power Electronics, IEEE Transactions on*, vol.26, no.6, pp.1754-1764, June 2011.

[10] Silva, C.; Oyarzun, J., "High Dinamic Control of a PWM Rectifier using Harmonic Elimination," in *IEEE Industrial Electronics, IECON 2006 - 32nd Annual Conference on*, vol., no., pp.2569-2574, 6-10 Nov. 2006.

[11] Barreto, L.H.S.C.; Henn, G.A.L.; Praca, P.P.; Silva, R.N.A.L.; Oliveira, D.S.; da Silva, E.R.C., "Carrier-based PWM modulation for THD and losses reduction on multilevel inverters," in *Applied Power Electronics Conference and Exposition (APEC), 2012 Twenty-Seventh Annual IEEE*, vol., no., pp.2436-2441, 5-9 Feb. 2012.

[12] Fang Zheng Peng; Jih-Sheng Lai, "Dynamic performance and control of a static VAr generator using cascade multilevel inverters," in *Industry Applications Conference, 1996. Thirty-First IAS Annual Meeting, IAS '96., Conference Record of the 1996 IEEE*, vol.2, no., pp.1009-1015 vol.2, 6-10 Oct 1996.

[13] Jie Zhang, "High performance control of a three-level IGBT inverter fed AC drive," in *Industry Applications Conference, 1995. Thirtieth IAS Annual Meeting, IAS '95., Conference Record of the 1995 IEEE*, vol.1, no., pp.22-28 vol.1, 8-12 Oct 1995

[14] Hua Zhou; Yun Wei Li; Zargari, N.R.; Zhongyaun Cheng; Ruoshui Ni; Ye Zhang, "Selective Harmonic Compensation (SHC) PWM for Grid-Interfacing High-Power Converters," in *Power Electronics, IEEE Transactions on*, vol.29, no.3, pp.1118-1127, March 2014.

[15] Yu Liu; Hoon Hong; Huang, A.Q., "Real-Time Calculation of Switching Angles Minimizing THD for Multilevel Inverters With Step Modulation," in *Industrial Electronics, IEEE Transactions on*, vol.56, no.2, pp.285-293, Feb. 2009

[16] Aggrawal, H.; Leon, J.I.; Franquelo, L.G.; Kouro, S.; Garg, P.; Rodriguez, J., "Model predictive control based selective harmonic mitigation technique for multilevel cascaded H-bridge converters," in *IECON 2011 - 37th Annual Conference on IEEE Industrial Electronics Society*, vol., no., pp.4427-4432, 7-10 Nov. 2011.

[17] Kouro, S.; La Rocca, B.; Cortes, P.; Alepuz, S.; Bin Wu; Rodriguez, J., "Predictive control based selective harmonic elimination with low switching frequency for multilevel converters," in *Energy Conversion Congress and Exposition, 2009. ECCE 2009. IEEE*, vol., no., pp.3130-3136, 20-24 Sept. 2009.

[18] Haghi, H.V.; Bina, M.T., "Complete harmonic-domain modeling and performance evaluation of an optimal-PWM-modulated STATCOM in a realistic distribution network," in *Nonsinusoidal Currents and Compensation, 2008. ISNCC 2008. International School on*, vol., no., pp.1-7, 10-13 June 2008.

[19] Singh, Bhim; Singh, B.N.; Chandra, A.; Al-Haddad, K.; Pandey, A.; Kothari, D.P., "A review of three-phase improved power quality AC-DC converters," in *Industrial Electronics, IEEE Transactions on*, vol.51, no.3, pp.641-660, June 2004.

[20] Aleenejad, M.; Mahmoudi, H.; Moamaei, P.; Ahmadi, R., "A New Fault-Tolerant Strategy Based on a Modified Selective Harmonic Technique for Three Phase Multilevel Converters," in *Power Electronics, IEEE Transactions on*, vol.PP, no.99, pp.1-1.

[21] Aleenejad, M.; Ahmadi, R.; Moamaei, P., "A modified selective harmonic elimination method for fault-tolerant operation of multilevel cascaded H-bridge inverters," in *Power and Energy Conference at Illinois (PECI), 2014*, vol., no., pp.1-5, Feb. 28 2014-March 1 2014.

[22] Eberhart, Russ C., and James Kennedy. "A new optimizer using particle swarm theory." *Proceedings of the sixth international symposium on micro machine and human science*. Vol. 1. 1995.

[23] Cai, X. J., Wu, Z. X., Li, Q. F., & Wang, S. X. (2015). Phase-Shifted Carrier Pulse Width Modulation Based on Particle Swarm Optimization for Cascaded H-bridge Multilevel Inverters with Unequal DC Voltages. *Energies*, 8(9), 9670-9687.

[24] Odavic, M., Sumner, M., Zanchetta, P., & Clare, J. C. (2010). A theoretical analysis of the harmonic content of PWM waveforms for multiple-frequency modulators. *Power Electronics, IEEE Transactions on*,(1), 131-141.

[25] B. Carlson, Communication Systems, McGraw-Hill International baseband error are plotted, together with their associated spectra. Editions, 3rd ed, 1986.

[26] Liu, Xudan, Andreas Lindemann, and Hadi Amiri. "A theoretical and experimental analysis of n+ 1 and 2n+ 1 phase-shifted carrier-based pwm strategies in modular multilevel converters." *PCIM Europe 2014; International Exhibition and Conference for Power Electronics, Intelligent Motion Renewable Energy and Energy Management; Proceedings of*. VDE, 2014.

[27] IEEE Recommended Practice and Requirements for Harmonic Control in Electric Power Systems," in *IEEE Std 519-2014 (Revision of IEEE Std 519-1992)*, vol., no., pp.1-29, June 11 2014.

[28] Abramowitz, Milton; Stegun, Irene A., eds. (December 1972) [1964]. "Chapter 9". *Handbook of Mathematical Functions with Formulas, Graphs, and Mathematical Tables*. Applied Mathematics Series 55 (10 ed.). New York, USA.

Implementation and Experimental Validation of Efficiency Improvement in PMSM Drives through Switching Frequency Reduction

Parag Kshirsagar
United Technologies Research Center
411 Silver Ln,
East Hartford, CT-06108
kshirsp@utrc.utc.com , parag@vt.edu

R.Krishnan, *Fellow IEEE*
Professor Emeritus
ECE Dept., Virginia Tech
Blacksburg,VA-24060
kramu@vt.edu

Abstract−Energy efficiency improvement in variable speed motor drives is of widespread interest because of the rising cost of energy. Permanent magnet synchronous motor (PMSM) drives are especially favored in variable speed applications due to their high efficiency of operation. This research focuses on improving efficiency of the PMSM drive system by reducing the switching frequency of the inverter driving the motor. At reduced switching frequencies, programmed pulse width modulation methods that results in elimination or reduction of the low ordered sidebands of the currents are considered in this work. The study showcases new implementation aspects of programmed PWM and closed loop current control at reduced switching frequencies necessary to operate the PMSM drive system at higher efficiency. The experimental measurements validate the total system efficiency improvement at reduced switching frequencies.

I. INTRODUCTION

The PMSM drive system efficiency can be improved by optimizing the losses in the machine or inverter or both. As the motor resistive losses are proportional to the magnitude of the square of the current and the core losses are a function of the frequency and flux density, then for a given the stator current and frequency, the machine losses can be considered constant. Then the remaining dominant losses are in the inverter in the form of its conduction and switching losses. The former loss is proportional to the magnitude of the current while the latter is a function of the current and switching frequency of the inverter. Given the current in the inverter, then the losses that can be minimized are dependent on the switching frequency. Therefore, the approach considered in this research is that given an optimally designed permanent magnet machine and its control algorithm, the only option to improve the system efficiency is by reducing the inverter switching losses by minimizing the switching frequency. At reduced switching frequencies, the small impedance offered by PMSM to the inverter output voltage causes higher current waveform distortion and higher losses in the motor. To reduce this distortion, programmed pulse width modulators (PWM) are preferred over conventional carrier based modulators [1].

Fig 1(a) shows the three-phase PMSM drive system and the stator current trajectories in stationary reference frames considering space vector modulation at 10 kHz and synchronous optimal programmed PWM at 675 Hz are shown in Fig 1(b) and (c), respectively. Both the methods yield the same average torque in the motor but yield different system efficiencies. Accordingly, the focus of this paper is the efficiency improvement in PMSM drive system at reduced switching frequencies, its implementation and experimental validation.

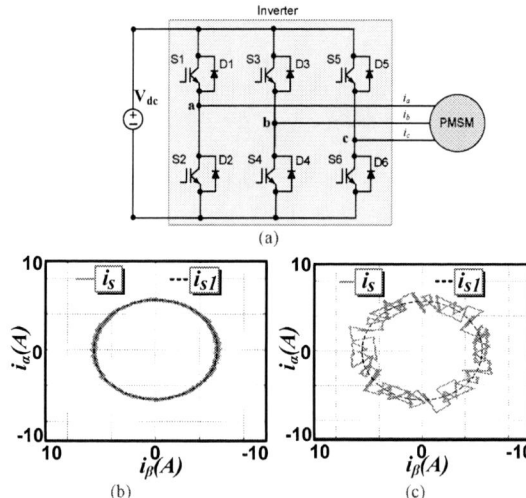

Fig.1 (a) Three phase PMSM drive system, experimental result of PMSM stator current trajectory in stationary reference considering (b) space vector modulation at 10 kHz and (c) synchronous optimal programmed PWM at 675 Hz.

Research on reduced switching frequency operation of permanent magnet motors remains yet to be explored in depth. In comparison to induction machines, PMSMs have significant advantages in power density, and motor and drive system efficiencies. The power density is enhanced in the PMSM due to PM-field excitation which reduces the ampere-turn requirement on the stator side and additionally having no windings on the rotor side resulting in elimination of resistive losses on it. The permanent magnets contribute to a reduction of stator and elimination of rotor resistive losses as well as harmonic core losses [2-4]. In [5], it was shown that in induction machines operating at reduced switching frequencies and asynchronous PWM (800Hz), the rotor harmonic losses were up to 50 percent of the total machine losses. Since PMSMs do not have rotor windings, the corresponding total harmonic losses are lower allowing further reduction of switching frequency.

State of the art three phase variable speed motor drives have switching frequencies in the range of 2 kHz to 20 kHz depending on their power level. High switching frequencies (above 10 kHz) are used in applications where precise position control or low acoustic noise is desired. For applications where such requirements are not stringent, the switching frequency can be significantly reduced. The operation of motor drives with low switching frequency was a highly researched topic in

978-1-4673-9551-9/16 $31.00 © 2016 IEEE

the 1960's and the 70's because of use of slow devices such as thyristors and bipolar transistors [6-8] that dominated the field of variable speed drives at that time. In order to minimize the motor harmonic losses at low switching frequencies, programmed pulse width modulation (PWM) methods were resorted to [8]. The optimized PWM methods were mainly developed for reduction of losses in the stator as well as in the rotor of the induction machines [9]. After the advent of faster switching devices such as MOSFETs and IGBTs, programmed PWM methods are currently limited to medium voltage high power drives [10] and mainly for induction machines. A preliminary analysis of efficiency improvement in PMSM drives considering reduced switching frequency operation of the inverter is presented in section II.

At higher switching frequencies, carrier based PWM is commonly used. It is implemented by comparing a sinusoidal reference signal with a high frequency triangle (carrier) signal [11], 12]. The high frequency carrier signal is generated using an up-down counter which is readily available in microcontrollers for motion control. The programmed PWM, on the other hand, is implemented by storing offline calculated commutation angles in lookup tables and then recalled for PWM generation using dedicated digital logic circuit [8], [13], [14]. Therefore the microcontrollers that have digital hardware catered for carrier based PWM and cannot be easily modified for programmed PWM generation. Due to the complexity of implementation and hardware requirements, programmed PWM is less favored in low voltage drives [11]. Hence a simple method for implementation of programmed PWM on state of the art DSP is desired and discussed in section III.

Current control affects in general not the system efficiency but only the damping and possible instability. Accordingly, for stable motor performance, simple current controller implementation is essential. The existing literature focuses on complex variable [15] and model predictive control [16], [17] of induction motor drives operating low switching frequency. These methods require detailed model of the inverter and motor to pre-calculate the voltage vector that has to be applied to the motor terminals. It is known that induction machine inherently has stable open loop response while PMSM can be operated on open loop only for a short duration of time with oscillatory speeds and chaotic torque generation. Hence it can be inferred that during low switching frequency operation, when the inverter control bandwidth is low, induction machines can have more stable operation while the capacity for the stable operation of PMSM is extremely limited both in its torque and speed responses as well as shrunken torque capabilities. The current literature only covers low switching frequency operation of induction motors, while the challenges for similar current regulation methods for PMSM have yet to appear. These aspects are discussed in section IV.

Finally, with respect to PMSM drive system efficiency improvement at reduced switching frequencies, the literature is very limited. In [18], total system losses in PMSM drive were calculated by varying the switching frequency from 16 kHz to 450 Hz using asynchronous carrier based PWM and also six-step modulation. The analysis mainly focused on the re-design

of PMSM for higher inductance and did not include programmed PWM methods. The experimental validation of benefits of reduced switching frequency incorporating programmed PWM methods for permanent magnet motors driven by two-level inverter has not been reported. Hence section V addresses the same and then conclusions are presented in section VI.

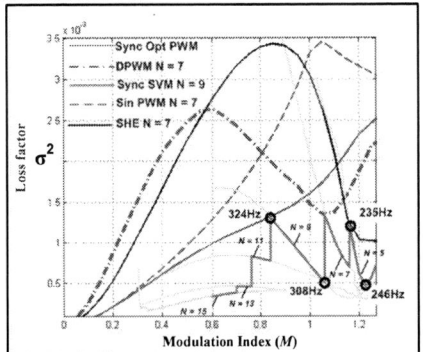

Fig 2. Simulation results: loss factor comparison for various PWM modulation strategies.

II. LOSS ANALYSIS OF PMSM DRIVES FOR REDUCED SWITCHING FREQUENCY OPERATION

Operation of PMSM with various PWM methods at low switching frequency for improving the system efficiency requires a careful analysis of the motor harmonic losses. The motor harmonic losses are proportional to the mean square magnitude of the harmonic current. The root mean square amplitude of the motor harmonic current I_σ is defined as [1],

$$I_\sigma = \frac{1}{\omega_1 L_s}\,\sigma, \qquad \sigma = \sqrt{\frac{1}{2}\sum_{n=3,\cdots}^{\infty}\left(\frac{V_n}{n}\right)^2} \qquad (1)$$

where ω_1 is the fundamental frequency and L_s is the motor inductance, V_n is the harmonic amplitude of the inverter voltage. To estimate the harmonic distortion factor σ, the amplitude of the harmonic voltages (V_n) is calculated using fast Fourier transform of the output switching voltage waveform. Note that the square of harmonic distortion factor σ^2 is also known as loss factor to gauge the harmonic losses in the motor [19] The loss factor σ^2 varies with modulation strategies and is a function of the modulation index M. In order to reduce the magnitude of loss factor it is not uncommon to have a PWM that utilizes different switching to fundamental frequency ratios N [19].The value of N is reduced for higher modulation indices mainly to reduce the switching losses.

Fig.2 shows the relationship between the loss factor and modulation index for different PWM methods considering reduced switching frequencies. The PWMs being compared are: (a) synchronous optimal PWM with the number of switching pulses varied from $N = 15$ to 5, (b) discontinuous PWM (DPWM) with $N = 7$, (c) synchronous SVM with $N = 9$, (d) sine PWM with $N = 7$and (e) selective harmonic elimination (SHE) PWM with $N = 7$. It is seen that synchronous optimal modulator with $N = 15$ through 5 has lower loss factor than all other modulators. It is achieved by combining different pulse numbers to achieve lower loss factor for the given modulation range. For this analysis, the PWM patterns are computed offline considering steady state operation only.

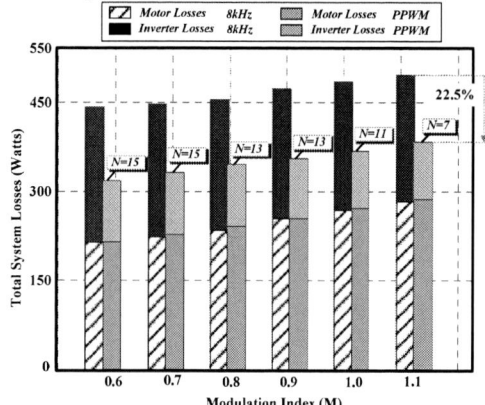

Fig.3. An 8kW PMSM drive loss comparison considering SVM at 8 kHz and programmed PWM with maximum switching frequency of 1kHz. Motor maximum electrical fundamental frequency is 133 Hz, load torque is 38 Nm.

Given the programmed PWM method results in Fig 2, losses in an 8 kW PMSM drive system are estimated and presented in Fig 3. The system parameters are given in the Appendix. In the analysis, two modulators are being compared and they are (a) space vector modulation (SVM) at switching frequency of 8 kHz and (b) programmed PWM with maximum switching frequency of 1 kHz. Fig 3 shows that in the given modulation range, the total drive system losses are reduced by 22.5% considering reduced switching frequency operation. This leads to a total system efficiency improvement of around 1%. The loss distribution between the inverter and motor shows that the increase in PMSM harmonic losses is not significant in comparison to the inverter loss reduction at reduced switching frequencies.

Since the main focus of this work is on implementation and experimental validation of the efficiency improvement, the following section will discuss the carrier based implementation of programmed PWM.

III. IMPLEMENTATION OF PROGRAMMED PWM FOR REDUCED SWITCHING FREQUENCY OPERATION

In comparison to carrier based modulators, programmed PWM does not require comparing the sine wave and triangle waveform. Instead, the pre-calculated commutation angles are compared directly with the rotor position to synthesize the switching (commutation) events. Typical digital hardware required for such an implementation comprises of timers, counters, comparators and memory with lookup tables [6]. While such digital logic can be implemented on a field programmable gate array (FPGA), implementing the same on digital signal processor dedicated for motion control application poses challenges due to limitation of the counter

size and clock frequency. Therefore a novel implementation method of programmed PWM using carrier signal that is catered for motion control DSP platforms is proposed. Fig 4(a) shows step-by-step sequence of programmed PWM generation using three commutation angles for half of the fundamental cycle. The commutation angles are given as α_1, α_2 and α_3. The commutation angles are computed offline as a function of modulation index M assuming a steady state operation using a solver ('Matlab- fminsearch'). As the PWM waveform exhibit quarter wave symmetry, the angles above $\pi/2$ are given as $\pi - \alpha_1, \pi - \alpha_2$ and $\pi - \alpha_3$. The six commutation angles are then compared with the rotor position θ_r to generate the switching pattern S_a. It should be noted that when $\theta_r < \pi/2$, the commutation angles with odd indices generate rising pulse edges while those with even indices generate falling pulse edges whereas when $\theta_r > \pi/2$, the commutation angles with odd integers generate falling pulse edges while those with even integers generate rising pulse edges. The switching pattern S_a

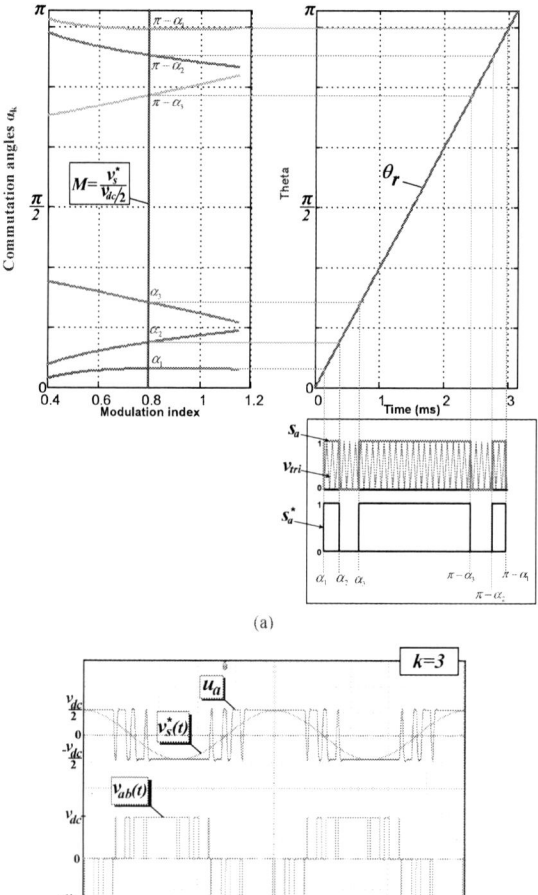

Fig.4. (a) Carrier based programmed PWM (a) implementation method and (b) its experimental results showing inverter phase and line-line voltage waveform for three commutation angles.

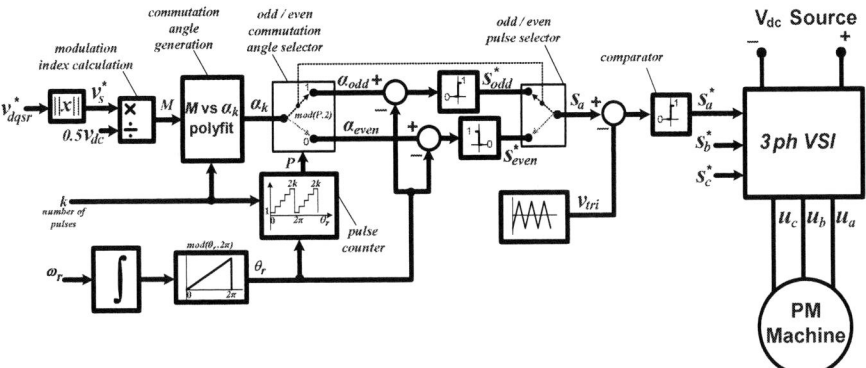

Fig.5 Carrier based programmed PWM implementation block diagram.

is compared with a high frequency carrier signal v_{tri} to generate the desired output PWM voltage waveform S_a^*.

Fig 4 (b) shows the experimental results of the inverter phase and the line-line voltage waveform using the aforementioned steps. It is seen that for the three commutation angles ($k = 3$), seven pulses are generated in the phase voltage waveform. In this illustration, the switching frequency is 210 Hz, while the frequency of the carrier signal is 15 kHz. The frequency of the carrier signal should be high enough to reduce the output voltage error of the modulator.

Fig 5 shows the block diagram of the programmed PWM implementation method. The inputs to the PWM generator are the reference voltages v_{dqsr}^* in rotor reference frame, number of switching pulses k, and rotor position θ_r. From the reference voltages v_{dqsr}^*, the scalar magnitude of the voltage vector v_s^* is calculated as $v_s^* = sqrt(v_{qsr}^2 + v_{dsr}^2)$. The modulation index is calculated based on the DC link voltage as $M = v_s^*/(v_{dc}/2)$. The instantaneous value of commutation angles α_k corresponding to the reference modulation index are estimated from the polynomial functions. The polynomial functions are determined offline and used instead of the look-up tables. The angles are then compared with the rotor angle to synthesize the odd switching pulses S_{odd}^* and even switching pulses S_{even}^*. These pulses are selected based on the pulse counter P and combined to produce the desired switching pattern S_a. The modulating waveform S_a is compared with a high frequency carrier signal resulting in PWM signal S_a^*. The switching signals for phase B and C are generated similarly by offsetting the rotor position by $-2\pi/3$ and $2\pi/3$, respectively.

The above approach can be considered as asynchronous sampling of the discontinuous PWM signal S_a by the carrier signal. The phase error φ between the reference and output PWM is determined by the carrier frequency. Assuming first order time delay approximation, the relationship between the maximum switching frequency f_{sw} and carrier frequency f_{tri} is given by from [20] as,

$$f_{tri} = \frac{f_{sw}}{tan(\varphi)} \qquad (2)$$

From (2), it is seen that if a very small phase error is desired, the denominator approaches to zero resulting in a very high carrier frequency. For example if the phase error desired

is $\varphi = 0.05$ rad with maximum switching frequency of 1 kHz, then the required carrier frequency is 20 kHz. Although a higher carrier frequency is desired, the lowest carrier frequency is determined by the influence of phase error on output current waveform distortion. If the phase error is too large, then the asynchronous sampling of the reference voltage will result in output voltage of smaller magnitude and large phase delays that may result in unstable system operation.

Fig 6 (a) and (b) show the motor current trajectory and time domain response using the proposed carrier based PWM implementation. The switching frequency is 1 kHz while the carrier frequency is 20 kHz and 50 kHz for the cases illustrated in Fig 6 (a) and (b), respectively. As seen in Fig 6 (a), when the phase error φ is large (0.05 rad), the three-phase currents are unbalanced. The stator current trajectory has offset from the origin. The current amplitude is also smaller due to lower voltage applied to the motor. When the phase error is reduced to 0.02 rad by increasing the carrier frequency to 50 kHz, the current unbalance is removed as seen in Fig 6(b). Note that the stator current trajectory also becomes symmetric around the origin.

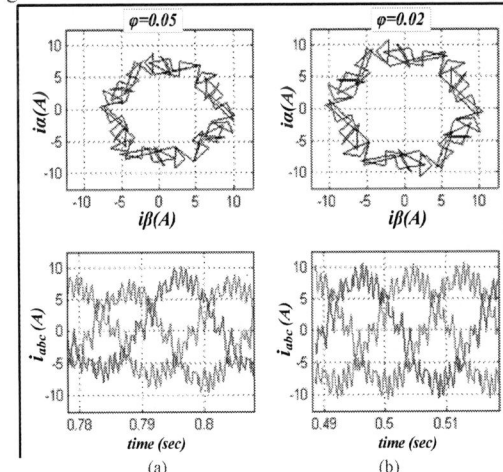

Fig 6 (a) Motor current trajectory and time domain response for phase errors at 1 kHz switching frequency applying programmed modulation technique (a) $\varphi = 0.05$ rad and (b) $\varphi = 0.02$ rad

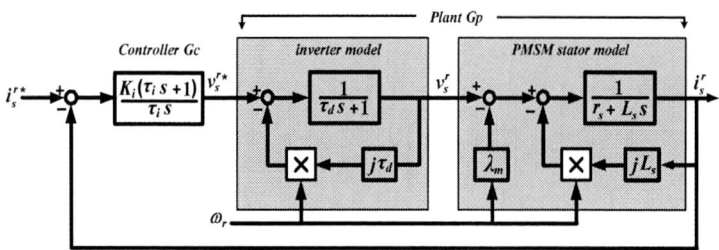

Fig 7. Space phasor block diagram implementation of current regulator

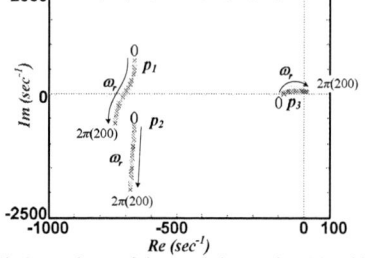

Fig 8. Root locus of characteristic equation $C(s)$ with sampling frequency of 2kHz

Thus selection of carrier frequency is critical in the proposed implementation method. The next section will discuss the current control method at reduced switching frequency.

III. PMSM CURRENT CONTROL AT REDUCED SWITCHING FREQUENCY

Current control affects in general not the system efficiency but only the damping and possible instability. Accordingly, for stable motor performance, simple current controller implementation is essential. In the following sections limitation of conventional PI (proportional and integral gain) current regulator design at low switching will be shown and then a simple approach of controlling current with feedback oversampling is provided.

A. PI current regulator

Fig 7 shows a space phasor based current control model of a PMSM drive system. The transfer functions of the current controller G_c and the plant G_p, incorporating inverter time delays and cross coupling effect between the windings, are given as,

$$G_c = \frac{K_i(1 + s\tau_i)}{s\tau_i},$$
$$G_p = \frac{K_p}{(1 + s\tau_r + j\omega_r\tau_r)(1 + s\tau_d + j\omega_r\tau_d)} \quad (3)$$

where the plant gain $K_p = 1/r_s$, the PMSM stator time constant $\tau_r = L_s/r_s$, τ_d is sampling and microprocessor delay [21], with $\tau_d = 1.5/(2f_{sw})$ and f_{sw} is the inverter switching frequency, τ_i is the current controller time constant and K_i is the proportional gain.

When the ratio of switching to fundamental frequency N is small, the sampling frequency is small. In that case the cross coupling between the direct and quadrature axes due to the inverter delay τ_d must be considered. Such an inverter model is given in [15] and the space phasor model of PMSM is described in [4].

The current regulator is designed by applying pole zero cancellation by setting $\tau_i = \tau_r$, and comparing the characteristic equation with a second order system to have a damping ratio of 0.707 [22]. Then the corresponding PI regulator parameters are given as,

$$K_i = \frac{\tau_r}{2K_p\tau_d}, \tau_i = \tau_r \quad (4)$$

The aforementioned PI current regulator design method ignores the influence of inverter delay related cross coupling terms with the assumption that the sampling time delay $\tau_d \ll \tau_r$. This is true at higher switching frequencies where even in presence of the inverter voltage cross coupling between the d and q axes, the current controller is able to force the stator current to track the reference value. However, when $\omega_r \neq 0$ and the value of τ_d is close to τ_r, the cross coupling effect between the d and q axes becomes evident. It is explained using the closed current loop transfer function using the parameters in equation (4) as,

$$G_{cl} = \frac{\frac{\tau_r}{2\tau_d}(1 + s\tau_r)}{s\tau_r(1 + s\tau_r + j\omega_r\tau_r)(1 + s\tau_d + j\omega_r\tau_d) + \frac{\tau_r}{2\tau_d}(1 + s\tau_r)} \quad (5)$$

The equation (5) is of third order and has three roots p_1, p_2 and p_3. The root locus of the characteristic equation $C(s) = (1 + p_1)(1 + p_2)(1 + p_3)$ is derived from (5) and plotted on the complex plane in Fig 8. In this illustration, the sampling frequency is 2 kHz and the motor electrical frequency is varied from zero to 200 Hz. The roots p_1 and p_2 are referred as single complex poles for $\omega_r \neq 0$, whereas for $\omega_r = 0$, they form a complex conjugate pair [15]. The intercept on real axis of the poles p_1 and p_2 is mainly determined by the inverter time delay. The root p_3 is closer to the imaginary axis and approaches the imaginary axis when the rotor speed approaches 200 Hz. In Fig 8, the complex pole p_3 is compensated by the classical current regulator at zero speed because the pole has only real part at this operating point. However, at higher speeds due to presence of imaginary part, the pole remains uncompensated and moves towards imaginary axis due to rotor speed variation. Consequently, the system will exhibit poor dynamic performance and even unstable operation. The above PMSM closed-loop current control model has similar characteristics to that of the induction motor model presented in [15].

B. PI current regulator with feedback oversampling

One approach of designing the current regulators with stable operation at low switching frequency is by increasing the sampling frequency of the feedback currents in order to reduce the sampling delay [23]. In this approach, the new reference voltages are computed after every sampling instant and the most updated value is applied at the next switching cycle thus

978-1-4673-9551-9/16 $31.00 © 2016 IEEE

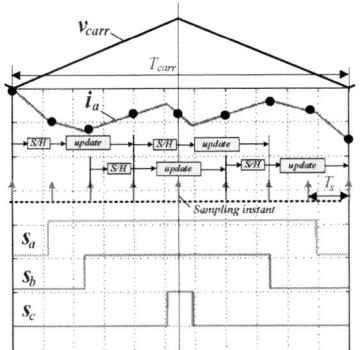

Fig 9.From top to bottom-carrier signal (v_{carr}), instantaneous phase current i_a, and sampling instants with over sampling at 8 times per carrier period and duty cycle update 4 times per carrier period, three-phase switch voltages S_{abc}

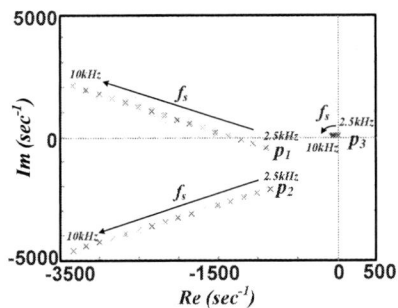

Fig 10.Root locus of characteristic equation $C(s)$ for various sampling frequencies. Fundamental frequency is 200 Hz.

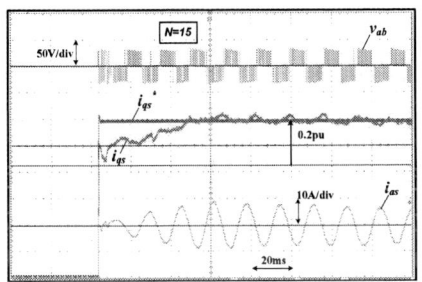

Fig 11Experimental results: transient response of current controller with programmed PWM (N = 15) and sampling frequency of 15 kHz.

improving the current controller bandwidth. An illustration of inverter sampling at 8 times the carrier period with update rate of four times the switching period is shown in Fig 9. Based on the illustration, for a switching frequency is 1 kHz, the sampling frequency is 8 kHz, and the update frequency is 4 kHz. With this approach, the sampling delay is reduced four times and the computation delay by half.

Fig 10 shows the root locus of the characteristic equation $C(s) = 1 + G_c G_p$ at various sampling frequencies with fundamental frequency of 200 Hz. It is seen that when the sampling frequency is increased from 2.5 kHz to 10 kHz, the pole p_3 moves toward left half of complex plane while the complex poles, p_1 and p_2 move further left. Using this approach, the assumption $\tau_d \ll \tau_r$ holds and the current

controller can be designed using the method described in part A of this section. Although the system is stable, the dynamic response becomes slower at higher rotor frequencies as the PWM update frequency is still small. The response is mainly determined by the location of the complex pole p_3. Depending on the application, the complex pole p_3 can be left uncompensated. However, for faster dynamic response it must be compensated by a complex zero in which case the current regulator takes the form of complex current controller as discussed in [15]. Most of the variable speed drive applications for fan or pump type applications do not require fast dynamic response. In such applications, the speed loop bandwidth requirement is low, and therefore the current controller bandwidth is also low. In such case, the gains of the current controller are given by

$$K_i = \frac{\tau_r}{2\tau_{dn}}, \tau_{dn} = \frac{2}{f_{sw}}, f_{sw} = Nf_1, \tau_i = \tau_r \qquad (5)$$

where τ_{dn} is the inverter PWM delay which is inversely proportional to the switching frequency f_{sw}. The switching frequency for programmed PWM is given by the number of pulses times the fundamental frequency f_1. The integral time constant of the controller is set to be the same as the stator time constant for pole zero cancellation. The current controller bandwidth can be increased by increasing the proportional gain K_i. Accordingly, Fig 11 shows the experimental results of the current controller with feedback oversampling for programmed PWM with switching frequency of 319Hz and sampling frequency of 15 kHz. From the transient response, it is seen that the feedback current i_{qs} tracks the reference current i_{qs}^* within 40 msec. The motor stator time constant τ_r is 12.5 msec. This result validates the stable operation of the current controller using the proposed approach.

IV. PMSM DRIVE SYSTEM EFFICIENCY EVALUATION AT REDUCED SWITCHING FREQUENCIES

In this section experimental validation of efficiency improvement in PMSM drives at reduced switching frequency is presented.

Figure 12 (a-c) shows the experimental results of an 8 pole PMSM drive operating with SVM at 10 kHz, discontinuous PWM [19] at 1.67 kHz and synchronous optimal programmed PWM at 675 Hz, respectively. The drive parameters are listed in Appendix. It is noted that the three methods generate the same average load torque of 8 Nm, however at reduced switching frequencies the average input DC current required by discontinuous PWM and synchronous optimal programmed PWM is lower than SVM. Thus the total power consumption of the motor and drive system is reduced.

The above test is repeated at different operating points and the input and output powers are measured to validate efficiency improvement. The load torque is kept constant at 8.0 Nm and the rotor frequency is varied from 25 Hz to 45 Hz proportionally varying shaft output power from 320 W up to 570W. The total system efficiency is the ratio of the motor shaft output power to the DC input power. Based on the input and output power measurements, percentage losses in the inverter and the motor can be analyzed at various operating

978-1-4673-9551-9/16 $31.00 © 2016 IEEE

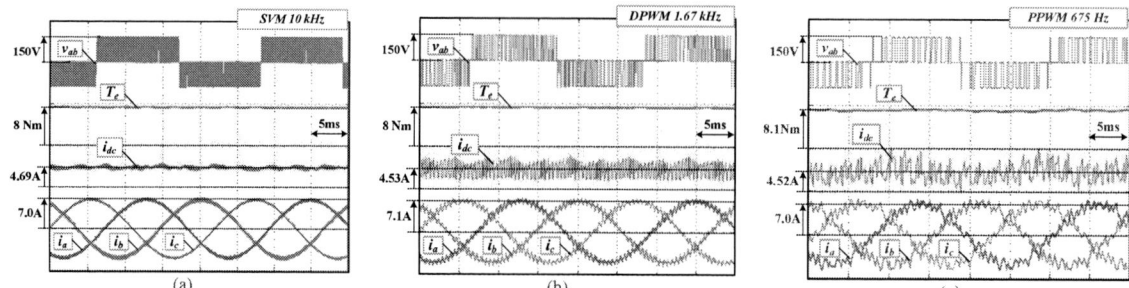

Figure 12. (a) Three-phase PMSM drive system Comparison of motor line-line voltage v_{qb}, phase current i_a and back EMF voltage e_a using (b) space vector PWM at 10 kHz and (c) synchronous optimal PWM at 675Hz. Fundamental frequency = 45 Hz, Modulation index M = 0.9 and reference current $i_{qs}*$ = 7A.

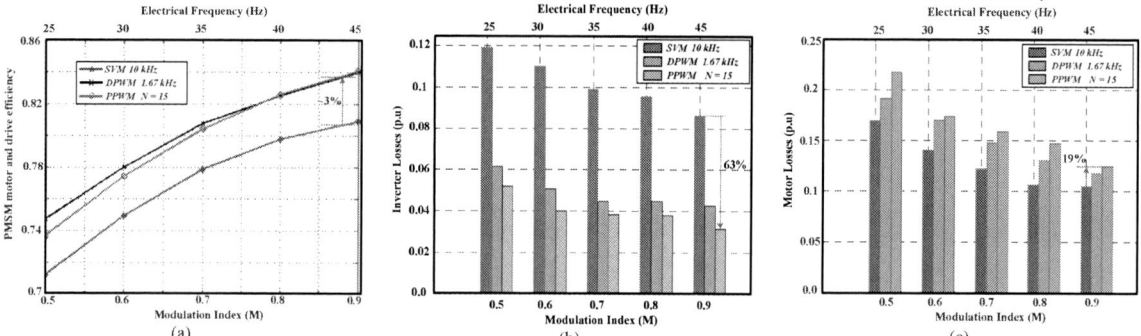

Figure13 Comparison of PMSM drive system performance at different switching frequencies and modulation indices (a) total system efficiency (b) inverter losses (c) motor losses. Losses normalized to total system input power for each operating point.

points. If the drive input power is measured as P_d, motor input power as P_m and the shaft output power as P_o then the normalized motor and inverter losses denoted as P_{loss_m} and P_{loss_i} respectively are given as,

$$P_{loss\_m} = \frac{P_m - P_o}{P_d} (p.u),$$
$$P_{loss\_i} = \frac{P_d - P_m}{P_d} (p.u)$$

(4)

Figure 13 (a) shows that in the given PMSM drive system, applying DPWM and PPWM schemes results in 3% higher efficiency than SVM at 10 kHz at the rated load condition. This result validates the proposed methodology of improving total system efficiency of PMSM drives by reducing the switching frequency of the inverter. It should be noted that DPWM, at modulation index below 0.8, has higher efficiency than PPWM, while PPWM has marginally higher efficiency than DPWM above $M = 0.8$. The low efficiency of PPWM at lower modulation indices is due to the higher harmonic content in the stator currents. The difference in efficiencies between PPWM and DPWM becomes noticeable as the power level and loss in the semiconductor devices increases.

Fig 13(b) shows the per unit losses in the inverter. It is seen that in all three modulation methods, the inverter losses decrease as the modulation index increases. This trend is due to the reduction of the root mean square amplitude of the current at higher modulation indices. Since the conduction loss of the inverter is proportional to the motor rms current, the normalized losses of the inverter demonstrate similar trend.

Considering the operating point at 45 Hz and modulation index of 0.9, it is seen that both DPWM and programmed PWM methods operating at reduced switching frequencies result in more than 63% inverter loss reduction in comparison to SVM operating at 10 kHz. In the given setup and operating condition, programmed PWM with N=15 has lowest inverter losses amongst the three modulation methods.

Fig 13(c) shows the normalized losses of motor which are higher than that of inverter. The motor harmonic losses have similar trend as that of the inverter for the given speed range. Since the motor losses are directly proportional to the harmonic current distortion, programmed PWM and DPWM schemes show higher losses than SVM at 10 kHz. Considering the operating point at 45 Hz and modulation index of 0.9, it is seen that programmed PMW method results in 19% higher motor losses when compared to SVM at 10 kHz. At this operating point, the net loss reduction of the PMSM drive system is 23% resulting in drive system efficiency improvement.

The experimental results validate that the total system efficiency of the PMSM drive can be improved by significantly reducing the switching frequency of the inverter driving it.

V CONCLUSIONS

The main focus of this research is on the analysis, implementation and experimental validation of efficiency improvement in PMSM drive system operating at low switching frequency with the following salient contributions:

1. For low switching frequency operation, programmed PWM (PPWM) is advantageous. Implementation of PPWM on a present state of art digital signal processor has limitations as they are customized for carrier based modulators. Therefore a novel method of implementing programmed PWM using carrier signal is proposed and is experimentally demonstrated on state of art digital signal processor.

2. At low switching frequencies, the sampling rate and PWM update rate are low. This results in unstable operation of the motor. An alternative current controller approach with feedback signal oversampling is applied in conjunction with programmed PWM. The analysis and experimental results show that the method can achieve stable operation with acceptable dynamic performance.

3. PMSM drive system efficiency is experimentally evaluated using three modulation methods (a) Space Vector Modulation (SVM) at high switching frequency (10kHz), (b)Discontinuous PWM (DPWM) at 1.67kHz and (c) Programmed PWM at 675 Hz. The experimental results show that the total system losses can be reduced by reducing the switching frequency. Since motor harmonic losses do not increase significantly at reduced switching frequency, the total system efficiency is improved. Programmed PWM method is advantageous at higher modulation indices due to its lower harmonic distortion content.

4. This efficiency improvement is achieved at no additional cost of control hardware or machine redesign.

Appendix

Table A.1 PMSM drive parameters

Parameter	Value
r_s	0.37 ohm
L_s	4.7 mH
λ_m	0.2 Vp/rad/s
Poles	8
Speed	3000 rpm
Torque	42 Nm
Power	8 kW
DC voltage	640
Motor voltage	305
Magnet type	NdFeB
Lamination Steel	M800
IGBT rating (for Fig 2)	1200V, 30A
IGBT rating (for Fig 13)	1200V, 200A

REFERENCES

[1] G.S. Buja, and G.B. Indri, "Optimal Pulsewidth Modulation for Feeding AC Motors." *IEEE Trans. Ind. Appl.*, vol. IA-13, no. 1 Jan 1977: 38 –44.

[2] G.R. Slemon and X. Liu, "Core losses in permanent magnet motors," *IEEE Trans. Magnetics*, vol. 26, no. 5, pp. 1653-1655, Sept.1990.

[3] M. Melfi, S. Evon, and R. McElveen, "Induction versus permanent magnet motors," *IEEE Ind. Appl. Mag.*, vol. 15, no. 6, pp. 28 –35, Dec. 2009.

[4] R. Krishnan, *Permanent Magnet Synchronous and Brushless DC Motor Drives*, CRC Press, 2010.

[5] E. Dlala and A. Arkkio, "A General Model for Investigating the Effects of the Frequency Converter on the Magnetic Iron

Losses of a Squirrel-Cage Induction Motor," *Magnetics, IEEE Transactions on*, vol. 45, no. 9, pp. 3303 –3315, Sep. 2009.

[6] P.H.Nayak, and R.G. Hoft, "Optimizing the PWM Waveform of a Thyristor Inverter." *IEEE Trans. Ind. Appl.*, no. 5 (October 1975), pp. 526–530.

[7] D. Murphy and F.G. Turnbull, *Power Electronic Control of AC Motors*, Pergamon Press, Oxford, 1988

[8] P.N.Enjeti, P.D. Ziogas, and J.F. Lindsay, "Programmed PWM Techniques to Eliminate Harmonics - A Critical Evaluation." *IEEE Industry Applications Society Annu. Mtg., 1988. Conference Record of the 1988*, vol.1, pp. 418 –430, 1988.

[9] F.C. Zach, and H. Ertl, "Efficiency Optimal Control for AC Drives with PWM Inverters." *IEEE Trans. Ind. Appl.*, IA-21, no. 4 (July 1985), pp. 987 –1000.

[10] J. Holtz and X. Qi, "Optimal Control of Medium-Voltage Drives—An overview," *IEEE Trans. Ind. Electron.*, vol. 60, no. 22, pp. 5472–5481, Dec 2013.

[11] S.R. Bowes, and D. Holliday, "Optimal regular-sampled PWM inverter control techniques," *IEEE Trans. Ind. Electron.*, vol. 54, no. 3, pp. 1547–1559, 2007.

[12] J. Sun, and H. Grotstollen, "Optimized Space Vector Modulation and Regular Sampled PWM: A Reexamination," *Conf. Proc. IEEE IAS'96*, vol. 2, pp. 956-963, 1996.

[13] S.R. Bowes, and A. Midoun, "Microprocessor implementation of new optimal PWM switching strategies," *IEE Proceedings Electric Power Applications*, B, vol. 135, no. 5, pp. 269–280, 2002.

[14] J.R. Wells, B.M. Nee, P.L. Chapman, and P.T. Krein, "Selective Harmonic Control: A General Problem Formulation and Selected Solutions." *IEEE Trans. Power Electron.*, vol. 20, no. 6 (2005), pp. 1337–1345.

[15] J. Holtz, J. Quan, G. Schmittt, J. Pontt, J. Rodriguez, P. Newman, and H. Miranda, "Design of Fast and Robust Current Regulators for High Power Drives Based on Complex State Variables." *Industry Applications Conference, 2003. Conf. Rec 38th IAS Annu. Mtg.*, vol.3, pp. 1997 – 2004, 2003

[16] J. Holtz and S. Stadtfeld, "A predictive controller for the stator current vector of AC machines fed from a switched voltage source," Proc. *Int. Power Electronics Conf. (IPEC)*, Tokyo, 1983, pp. 1665–1675.

[17] T. Geyer and G. Papafotiou, "Model predictive control in power electronics: A hybrid systems approach," Proc. *4th IEEE Conf. Decision and Control and European Control Conf.*, Seville, Spain, 2005, pp. 5606–5611.

[18] J. Itoh, T. Ogura, "Evaluation of Total Loss for an Inverter and Motor by Applying Modulation Strategies", *EPE-PEMC 2010*, S12-21 – S12-28

[19] J. Holtz, "Pulsewidth Modulation for Electronic Power Conversion." *Proc. IEEE 82*, no. 8 (August 1994): 1194 –1214.

[20] A. Hambley, "*Electrical Engineering: Principles And Applications*", 4th Edition, Pearson Education, Inc, 2008

[21] T. Ohmae, T. Matsuda, K. Kamiyama, and M. Tachikawa, "A Microprocessor-Controlled High-Accuracy Wide-Range Speed Regulator for Motor Drives," *IEEE Trans. Ind. Electron.*, vol. IE-29, no. 3, pp. 207 –211, Aug. 1982.

[22] V. Blasko, V. Kaura, W. Niewiadomski, "Sampling of Discontinuous Voltage and Current Signals in Electrical Drives:A System Approach," *IEEE Trans. Ind. Appl.*, Vol 34, no. 5,1998, pp. 1123 –1130.

[23] Bocker, J, Buchholz, O. "Can oversampling improve the dynamics of PWM controls?", (ICIT), IEEE International Conference on Industrial Technology, pp 1818 – 1824, Feb. 2013.

Sensorless Speed Control of Symmetrical Triple-Star Nine-Phase Interior Permanent Magnet Machines

Olorunfemi Ojo and Mehdi Ramezani
Department of Electrical and Computer Engineering/Centre for Energy Systems Research
Tennessee Technological University, Cookeville, TN 38505, USA

Abstract— This paper presents a sensor-less method to control triple-star nine-phase Interior Permanent Magnet (IPM) machines based on a decoupled machine model. The decoupled model removes the control complexity of the drive caused by the coupling inductance terms between different sets of three-phase windings of the triple-star machine. The rotor position is estimated using high frequency voltage injections into the stator fifth sequence circuit of the machine which is non torque producing. The control strategy which seeks to minimize stator copper loss (equivalently, the peak phase current) is set forth and tested by simulations using the full order coupled model of the machine. It is further validated with experimental results.

Keywords— multi-phase machines, triple-star connections, loss minimization, decoupled model of triple star machines, sensor-less drive; position detection; high frequency voltage injection.

I. INTRODUCTION

To further increase the reliability and fault tolerance of multi-phase machines, the stator windings have been connected as double, triple, quadruple and higher three phase machines with separate and unconnected neutral (star) points [1-4]. For the nine-phase IPM machines, instead of one single neutral point for all nine-phases of the stator phases, they are connected as three isolated three-phase machines with three isolated star (neutral) points [5]. In this configuration if one or two of the three phase sets experience(s) a fault, it or they can be disenabled while the remaining set(s) support a reduced load. In this paper the minimum stator copper loss vector control strategy of a triple star IPM machine is set forth based on a decoupled model which eliminates potential control complexity due to the coupling of the three stator winding sets. The rotor position and speed are estimated from the processing of stator currents resulting from the non-torque generating high frequency voltage injections in the fifth sequence circuit of the machine. Using the full order coupled model of the machine, accounting for all sequences, the simulation results of the proposed position sensor-less vector control are presented. Experimental results are also included to validate the machine model and the control strategy.

II. THE MODEL OF THE MACHINE

The stator windings of the triple-star IPM machine is schematically shown in Figure 1. The phase 'a' windings of the three sets of three phase connections are phase shifted by an angle of 40 degrees from each adjacent other. The voltage equations of the machine in the natural and rotor reference

$$V_{abci} = r_s i_{abci} + p\lambda_{abci}, \, i = 1,2,3 \qquad (1)$$

Figure 1. The phase connections of the machine.

frame variables are given in equation (1-2), where quantities with subscripts 1, 2, 3 refer to the individual three machine sets. The zero sequence equations of the machine set are decoupled from their qd components and from each other. There are coupling terms between the qd axis inductances of the three, three phase stator winding sets. These coupling terms are the non-diagonal terms of the inductance matrix in equation (2) which complicates the design of the vector controllers. To remove these coupling terms, a new reference frame transformation which is a combination of the rotor reference frame transformation and a second transformation is determined to diagonalize the impedance matrix in equation (2) [6-7]. The inductance matrix is separated into two components comprising of a diagonal matrix of the leakage inductances and inductance matrix equation L_{qd} (3). The diagonalization of equation (3) is possible since it has distinct eigen-values [8]. For this machine, the matrix P is presented in equation 4. The new decoupling transformation matrix for the model is given in equation 5. where,' $T(\theta_r)$ 'is the rotor reference frame transformation matrix. The machine equations in decoupled reference frame are presented in equation (6) where the inductance matrix is diagonal and there is no couplings

$$
\begin{bmatrix} V_{q1} \\ V_{d1} \\ V_{o1} \\ V_{q2} \\ V_{d2} \\ V_{o2} \\ V_{q3} \\ V_{d3} \\ V_{o3} \end{bmatrix} = r_{sc}\begin{bmatrix} i_{q1} \\ i_{d1} \\ i_{o1} \\ i_{q2} \\ i_{d2} \\ i_{o2} \\ i_{q3} \\ i_{d3} \\ i_{o3} \end{bmatrix} + \omega_r \left(\begin{bmatrix} 0 & 1 & 0 & 0 & 0 & 0 & 0 & 0 & 0 \\ -1 & 0 & 0 & 0 & 0 & 0 & 0 & 0 & 0 \\ 0 & 0 & 0 & 0 & 0 & 0 & 0 & 0 & 0 \\ 0 & 0 & 0 & 0 & 1 & 0 & 0 & 0 & 0 \\ 0 & 0 & 0 & -1 & 0 & 0 & 0 & 0 & 0 \\ 0 & 0 & 0 & 0 & 0 & 0 & 0 & 0 & 0 \\ 0 & 0 & 0 & 0 & 0 & 0 & 0 & 1 & 0 \\ 0 & 0 & 0 & 0 & 0 & 0 & -1 & 0 & 0 \\ 0 & 0 & 0 & 0 & 0 & 0 & 0 & 0 & 0 \end{bmatrix}\begin{bmatrix} L_{q1q1}+L_{ls} & 0 & 0 & L_{q1q2} & 0 & 0 & L_{q1q3} & 0 & 0 \\ 0 & L_{d1d1}+L_{ls} & 0 & 0 & L_{d1d2} & 0 & 0 & L_{d1d3} & 0 \\ 0 & 0 & L_{ls} & 0 & 0 & 0 & 0 & 0 & 0 \\ L_{q2q1} & 0 & 0 & L_{q2q2}+L_{ls} & 0 & 0 & L_{q2q3} & 0 & 0 \\ 0 & L_{d2d1} & 0 & 0 & L_{d2d2}+L_{ls} & 0 & 0 & L_{d2d3} & 0 \\ 0 & 0 & 0 & 0 & 0 & L_{ls} & 0 & 0 & 0 \\ L_{q3q1} & 0 & 0 & L_{q3q2} & 0 & 0 & L_{q3q3}+L_{ls} & 0 & 0 \\ 0 & L_{d3d1} & 0 & 0 & L_{d3d2} & 0 & 0 & L_{d3d3}+L_{ls} & 0 \\ 0 & 0 & 0 & 0 & 0 & 0 & 0 & 0 & L_{ls} \end{bmatrix}\begin{bmatrix} i_{q1} \\ i_{d1} \\ i_{o1} \\ i_{q2} \\ i_{d2} \\ i_{o2} \\ i_{q3} \\ i_{d3} \\ i_{o3} \end{bmatrix} + \begin{bmatrix} 0 \\ \lambda_{pmd1} \\ 0 \\ 0 \\ \lambda_{pmd1} \\ 0 \\ 0 \\ \lambda_{pmd1} \\ 0 \end{bmatrix} \right)
$$

$$
+ p\left(\begin{bmatrix} L_{q1q1}+L_{ls} & 0 & 0 & L_{q1q2} & 0 & 0 & L_{q1q3} & 0 & 0 \\ 0 & L_{d1d1}+L_{ls} & 0 & 0 & L_{d1d2} & 0 & 0 & L_{d1d3} & 0 \\ 0 & 0 & L_{ls} & 0 & 0 & 0 & 0 & 0 & 0 \\ L_{q2q1} & 0 & 0 & L_{q2q2}+L_{ls} & 0 & 0 & L_{q2q3} & 0 & 0 \\ 0 & L_{d2d1} & 0 & 0 & L_{d2d2}+L_{ls} & 0 & 0 & L_{d2d3} & 0 \\ 0 & 0 & 0 & 0 & 0 & L_{ls} & 0 & 0 & 0 \\ L_{q3q1} & 0 & 0 & L_{q3q2} & 0 & 0 & L_{q3q3}+L_{ls} & 0 & 0 \\ 0 & L_{d3d1} & 0 & 0 & L_{d3d2} & 0 & 0 & L_{d3d3}+L_{ls} & 0 \\ 0 & 0 & 0 & 0 & 0 & 0 & 0 & 0 & L_{ls} \end{bmatrix}\begin{bmatrix} i_{q1} \\ i_{d1} \\ i_{o1} \\ i_{q2} \\ i_{d2} \\ i_{o2} \\ i_{q3} \\ i_{d3} \\ i_{o3} \end{bmatrix} + \begin{bmatrix} 0 \\ \lambda_{pmd1} \\ 0 \\ 0 \\ \lambda_{pmd1} \\ 0 \\ 0 \\ \lambda_{pmd1} \\ 0 \end{bmatrix} \right)
\tag{2}
$$

$$
L_{qd} = \begin{bmatrix} L_{q1q1} & 0 & L_{q1q2} & 0 & L_{q1q3} & 0 \\ 0 & L_{d1d1} & 0 & L_{d1d2} & 0 & L_{d1d3} \\ L_{q2q1} & 0 & L_{q2q2} & 0 & L_{q2q3} & 0 \\ 0 & L_{d2d1} & 0 & L_{d2d2} & 0 & L_{d2d3} \\ L_{q1q3} & 0 & L_{q3q2} & 0 & L_{q3q3} & 0 \\ 0 & L_{d1d3} & 0 & L_{d3d2} & 0 & L_{d3d3} \end{bmatrix}
\tag{3}
$$

between the models of the three sets of machines. It is observed that in the new reference frame, the axis currents and voltages are composites of the original axes motor currents and voltages in equation (2). However, individual axis currents of the original machine set can be recovered by using the inverse transformation of the new diagonalizing/decoupling matrix. The determined decoupling model which eliminates the complexities arising from the axis coupling inductances of the original rotor reference frame model, simplifies the design and implementation of the vector controlled triple-star IPM motor drive undertaken in the next section.

$$
\begin{aligned}
P &= \begin{bmatrix} V_1 & V_2 & V_3 & V_4 & V_5 & V_6 \end{bmatrix} \\
&= \begin{bmatrix} 1 & 1 & -1 & 0 & 0 & 0 \\ 0 & 0 & 0 & -1 & 1 & 1 \\ -2 & 1 & 0 & 0 & 0 & 0 \\ 0 & 0 & 0 & 0 & 1 & -2 \\ 1 & 1 & 1 & 0 & 0 & 0 \\ 0 & 0 & 0 & 1 & 1 & 1 \end{bmatrix}
\end{aligned}
\tag{4}
$$

III. MINIMUM COPPER LOSS VECTOR CONTROL

The vector control drive is designed based on the minimum copper loss strategy while meeting the load torque requirements. The problem is cast in terms of optimizing the stator copper loss subject to the constraint of the load torque using the Lagrange multipliers [9]. The total stator copper loss is defined in the equation 7. By defining 'λ' as the Lagrange multiplier, the Lagrange function is defined as equation 8. The conditions for the minimum stator copper loss subject to the constraint of the load torque is obtained by solving the Lagrange equations and derivatives. The Lagrange derivations are presented in equations 9 to 15. The relationship between the q1n and d1n currents at the optimum point is obtained from the equations 9 and 10 as equation 16. The speed controller loop also can be designed based on the mechanical dynamic equation of the rotor according to equation 17. Also using the equations 16 and 17 the references for the q1n axis current that achieve minimum loss operation are given equation 18. In equation 18 the q1n axis current can be calculated from the load torque and the rotor speed error (δ_{ω_r}). The reference current of q1n axis can be substituted in the equation 16 to obtain the d1n axis current reference. After obtaining the reference values for q1n and d1n axis currents the current regulators can be designed. The current regulators are designed based on the dynamic equations of the q1n and d1n voltages, presented in equation 6. The 'i_{dq1n}' controllers are designed as equations 19 and 20.

978-1-4673-9551-9/16 $31.00 © 2016 IEEE

$$T_n(\theta_r) = P^{-1}T(\theta_r) =$$

$$\begin{bmatrix} 0 & -\frac{1}{3} & 0 & -\frac{1}{3} & 0 & -\frac{1}{3} \\ -\frac{1}{3} & 0 & -\frac{1}{3} & 0 & -\frac{1}{3} & 0 \\ -\frac{1}{2} & 0 & 0 & 0 & \frac{1}{2} & 0 \\ 0 & -\frac{1}{2} & 0 & 0 & 0 & \frac{1}{2} \\ \frac{1}{6} & 0 & -\frac{1}{3} & 0 & \frac{1}{6} & 0 \\ 0 & \frac{1}{6} & 0 & -\frac{1}{3} & 0 & \frac{1}{6} \end{bmatrix} \times \frac{2}{3} \begin{bmatrix} C(\theta_r+\alpha_1) & C(\theta_r+\alpha_1-\gamma) & C(\theta_r+\alpha_1+\gamma) & 0 & 0 & 0 & 0 & 0 & 0 \\ S(\theta_r+\alpha_1) & S(\theta_r+\alpha_1-\gamma) & S(\theta_r+\alpha_1+\gamma) & 0 & 0 & 0 & 0 & 0 & 0 \\ 0 & 0 & 0 & C(\theta_r+\alpha_2) & C(\theta_r+\alpha_2-\gamma) & C(\theta_r+\alpha_2+\gamma) & 0 & 0 & 0 \\ 0 & 0 & 0 & S(\theta_r+\alpha_2) & S(\theta_r+\alpha_2-\gamma) & S(\theta_r+\alpha_2+\gamma) & 0 & 0 & 0 \\ 0 & 0 & 0 & 0 & 0 & 0 & C(\theta_r+\alpha_3) & C(\theta_r+\alpha_3-\gamma) & C(\theta_r+\alpha_3+\gamma) \\ 0 & 0 & 0 & 0 & 0 & 0 & S(\theta_r+\alpha_3) & S(\theta_r+\alpha_3-\gamma) & S(\theta_r+\alpha_3+\gamma) \end{bmatrix} =$$

$$\frac{2}{3}\begin{pmatrix} -\frac{S(\theta_r+\alpha_1)}{3} & -\frac{S(\theta_r+\alpha_1-\gamma)}{3} & -\frac{S(\theta_r+\alpha_1+\gamma)}{3} & -\frac{S(\theta_r+\alpha_2)}{3} & -\frac{S(\theta_r+\alpha_2-\gamma)}{3} & -\frac{S(\theta_r+\alpha_2+\gamma)}{3} & -\frac{S(\theta_r+\alpha_3-\beta)}{3} & -\frac{S(\theta_r+\alpha_3-\gamma)}{3} & -\frac{S(\theta_r+\alpha_3+\gamma)}{3} \\ -\frac{C(\theta_r+\alpha_1)}{3} & -\frac{C(\theta_r+\alpha_1-\gamma)}{3} & -\frac{C(\theta_r+\alpha_1+\gamma)}{3} & -\frac{C(\theta_r+\alpha_2)}{3} & -\frac{C(\theta_r+\alpha_2-\gamma)}{3} & -\frac{C(\theta_r+\alpha_2+\gamma)}{3} & -\frac{C(\theta_r+\alpha_3)}{3} & -\frac{C(\theta_r+\alpha_3-\gamma)}{3} & -\frac{C(\theta_r+\alpha_3+\gamma)}{3} \\ -\frac{C(\theta_r+\alpha_1)}{2} & -\frac{C(\theta_r+\alpha_1-\gamma)}{2} & -\frac{C(\theta_r+\alpha_1+\gamma)}{2} & 0 & 0 & 0 & \frac{C(\theta_r+\alpha_3)}{2} & \frac{C(\theta_r+\alpha_3-\gamma)}{2} & \frac{C(\theta_r+\alpha_3+\gamma)}{2} \\ -\frac{S(\theta_r+\alpha_1)}{2} & -\frac{S(\theta_r+\alpha_1-\gamma)}{2} & -\frac{S(\theta_r+\alpha_1+\gamma)}{2} & 0 & 0 & 0 & \frac{S(\theta_r+\alpha_3)}{2} & \frac{S(\theta_r+\alpha_3-\gamma)}{2} & \frac{S(\theta_r+\alpha_3+\gamma)}{2} \\ \frac{C(\theta_r+\alpha_1)}{6} & \frac{C(\theta_r+\alpha_1-\gamma)}{6} & \frac{C(\theta_r+\alpha_1+\gamma)}{6} & -\frac{C(\theta_r+\alpha_2)}{3} & -\frac{C(\theta_r+\alpha_2-\gamma)}{3} & -\frac{C(\theta_r+\alpha_2+\gamma)}{3} & \frac{C(\theta_r+\alpha_3)}{6} & \frac{C(\theta_r+\alpha_3-\gamma)}{6} & \frac{C(\theta_r+\alpha_3+\gamma)}{6} \\ \frac{S(\theta_r+\alpha_1)}{6} & \frac{S(\theta_r+\alpha_1-\gamma)}{6} & \frac{S(\theta_r+\alpha_1+\gamma)}{6} & -\frac{S(\theta_r+\alpha_2)}{3} & -\frac{S(\theta_r+\alpha_2-\gamma)}{3} & -\frac{S(\theta_r+\alpha_2+\gamma)}{3} & \frac{S(\theta_r+\alpha_3)}{6} & \frac{S(\theta_r+\alpha_3-\gamma)}{6} & \frac{S(\theta_r+\alpha_3+\gamma)}{6} \end{pmatrix},$$

(5)

$$\alpha_1 = \frac{2\pi}{9}, \alpha_2 = 0, \alpha_1 = -\frac{2\pi}{9}, \gamma = \frac{2\pi}{3}$$

$$\begin{bmatrix} V_{q1n} \\ V_{d1n} \\ V_{q2n} \\ V_{d2n} \\ V_{q3n} \\ V_{d3n} \end{bmatrix} = r_{sn}\begin{bmatrix} i_{q1n} \\ i_{d1n} \\ i_{q2n} \\ i_{d2n} \\ i_{q3n} \\ i_{d3n} \end{bmatrix} + \omega_{rn}\left(\begin{bmatrix} L_{ls}+3L_{md} & 0 & 0 & 0 & 0 & 0 \\ 0 & L_{ls}+3L_{mq} & 0 & 0 & 0 & 0 \\ 0 & 0 & L_{ls} & 0 & 0 & 0 \\ 0 & 0 & 0 & L_{ls} & 0 & 0 \\ 0 & 0 & 0 & 0 & L_{ls} & 0 \\ 0 & 0 & 0 & 0 & 0 & L_{ls} \end{bmatrix}\begin{bmatrix} i_{q1n} \\ i_{d1n} \\ i_{q2n} \\ i_{d2n} \\ i_{q3n} \\ i_{d3n} \end{bmatrix} + \begin{bmatrix} 0 \\ -\lambda_{pm} \\ 0 \\ 0 \\ 0 \\ 0 \end{bmatrix}\right)$$

(6)

$$+ \begin{bmatrix} L_{ls}+3L_{md} & 0 & 0 & 0 & 0 & 0 \\ 0 & L_{ls}+3L_{mq} & 0 & 0 & 0 & 0 \\ 0 & 0 & L_{ls} & 0 & 0 & 0 \\ 0 & 0 & 0 & L_{ls} & 0 & 0 \\ 0 & 0 & 0 & 0 & L_{ls} & 0 \\ 0 & 0 & 0 & 0 & 0 & L_{ls} \end{bmatrix}\begin{bmatrix} pi_{q1n} \\ pi_{d1n} \\ pi_{q2n} \\ pi_{d2n} \\ pi_{q3n} \\ pi_{d3n} \end{bmatrix}, T_e = \frac{9}{2}\left(\frac{P}{2}\right)\left((L_{md}-L_{mq})i_{d1n}i_{q1n}+\lambda_{pm}i_{q1n}\right)$$

$$P_{loss} = \frac{3}{2}\sum_{i=1}^{3}r_s\left(i^2_{qin}+i^2_{din}\right) =$$

$$\frac{3}{2}r_s\left(i^2_{q1n}+i^2_{d1n}+i^2_{q2n}+i^2_{d2n}+i^2_{q3n}+i^2_{d3n}\right)$$

(7)

$$L = \frac{3}{2}r_s\left(i^2_{q1n}+i^2_{d1n}+i^2_{q2n}+i^2_{d2n}+i^2_{q3n}+i^2_{d3n}\right)+$$

$$\lambda\left(T_L - \frac{3P}{4}\left(3(L_{md}-L_{mq})i_{d1n}i_{q1n}+\lambda_{pm}i_{q1n}\right)\right)$$

(8)

$$\frac{dL}{di_{q1n}} = 2r_s\left(i_{q1n}\right) - \lambda\frac{3P}{4}\left(3(L_{md}-L_{mq})i_{d1n}+\lambda_{pm}\right) = 0 \Rightarrow$$

$$\lambda = \frac{8r_s i_{q1n}}{3P\left(3(L_{md}-L_{mq})i_{d1n}+\lambda_{pm}\right)}$$

(9)

$$\frac{dL}{di_{d1n}} = 2r_s\left(i_{d1n}\right) - \lambda\frac{3P}{4}\left(3(L_{md}-L_{mq})i_{q1n}\right) = 0 \Rightarrow$$

$$\lambda = \frac{8r_s i_{d1n}}{9P\left(L_{md}-L_{mq}\right)i_{q1n}}$$

(10)

$$\frac{dL}{di_\lambda} = T_L - \frac{3P}{4}\big(3(L_{md} - L_{mq})i_{d1n}i_{q1n} + \lambda_{pm}i_{q1n}\big) = 0 \quad (11)$$

$$\frac{dL}{di_{q2n}} = 2r_s(i_{q2n}) = 0 \Rightarrow i_{q2n} = 0 \quad (12)$$

$$\frac{dL}{di_{d2n}} = 2r_s(i_{d2n}) = 0 \Rightarrow i_{d2n} = 0 \quad (13)$$

$$\frac{dL}{di_{q3n}} = 2r_s(i_{q3n}) = 0 \Rightarrow i_{q3n} = 0 \quad (14)$$

$$\frac{dL}{di_{d3n}} = 2r_s(i_{d3n}) = 0 \Rightarrow i_{d3n} = 0 \quad (15)$$

$$i^2_{d1n} + \frac{\lambda_{pm}i_{d1n}}{3(L_{md} - L_{mq})} - i^2_{q1n} = 0 \Rightarrow$$
$$i_{d1n} = -\frac{\lambda_{pm}}{6(L_{md} - L_{mq})} \pm \sqrt{\left(\frac{\lambda_{pm}}{6(L_{md} - L_{mq})}\right)^2 + i^2_{q1n}} \quad (16)$$

$$T_e = J\left(\frac{2}{P}\right)p\omega_r + T_L \Rightarrow T_e - T_L =$$
$$J\left(\frac{2}{P}\right)p\omega_r = K_{\omega r}(S)(\omega_r^* - \omega_r) = \delta_{\omega_r} \quad (17)$$

$$9(L_{md} - L_{mq})^2 i^4_{q1n} + \frac{4\lambda_{pm}i_{q1n}(\delta_{\omega_r} - T_L)}{3P} = \left(\frac{4(\delta_{\omega_r} - T_L)}{3P}\right)^2 \quad (18)$$

$$r_s i_{d1n} + p(L_{ls} + 3L_{mq})i_{d1n} = \delta_{d1n} =$$
$$V_{d1n} - \omega_r\big((L_{ls} + 3L_{md})i_{q1n}\big) =$$
$$\left(K_{Pq1} + \frac{K_{Iq1}}{p}\right)(i_{d1n}^* - i_{d1n}) \quad (19)$$

$$r_s i_{q1n} + p(L_{ls} + 3L_{md})i_{q1n} = \delta_{q1n} =$$
$$V_{q1n} + \omega_r\big(-(L_{ls} + 3L_{mq})i_{d1n} + \lambda_{pm}\big) =$$
$$\left(K_{Pq1} + \frac{K_{Iq1}}{p}\right)(i_{q1n}^* - i_{q1n}) \quad (20)$$

The 'i_{dq2n}' controllers are also designed as:

$$r_s i_{d2n} + pL_{ls}i_{d2n} = V_{d2n} + \omega_r L_{ls}i_{q2n} =$$
$$\delta_{d2n} = \left(K_{pd2} + \frac{K_{Id2}}{p}\right)(i_{d2n}^* - i_{d2n}) \quad (21)$$

$$r_s i_{q2n} + pL_{ls}i_{q2n} = V_{q2n} - \omega_r L_{ls}i_{d2n}$$
$$= \delta_{q2n} = \left(K_{pq2} + \frac{K_{Iq2}}{p}\right)(i_{q2n}^* - i_{q2n}) \quad (22)$$

The 'q3n' and 'd3n' axis also have the same controller structure as 'q2n' and 'd2n', therefore they are not repeated here. Based on equations 19 to 22 the transfer functions of the current regulators of the different axis can be derived as:

$$\frac{i_{d1n}}{i_{d1n}^*} = \frac{\dfrac{(pK_{Pd1} + K_{Id1})}{(L_{ls} + 3L_{mq})}}{p^2 + p\dfrac{(r_s + K_{Pd1})}{(L_{ls} + 3L_{mq})} + \dfrac{K_{Id1}}{(L_{ls} + 3L_{mq})}} \quad (23)$$

$$\frac{i_{q1n}}{i_{q1n}^*} = \frac{\dfrac{(pK_{Pq1} + K_{Iq1})}{(L_{ls} + 3L_{md})}}{p^2 + p\dfrac{(r_s + K_{Pq1})}{(L_{ls} + 3L_{md})} + \dfrac{K_{Iq1}}{(L_{ls} + 3L_{md})}} \quad (24)$$

$$\frac{i_{d2n}}{i_{d2n}^*} = \frac{\dfrac{(pK_{Pd2} + K_{Id2})}{L_{ls}}}{p^2 + p\dfrac{(r_s + K_{Pd2})}{L_{ls}} + \dfrac{K_{Id2}}{L_{ls}}} \quad (25)$$

$$\frac{i_{q2n}}{i_{q2n}^*} = \frac{\dfrac{(pK_{Pq2} + K_{Iq2})}{L_{ls}}}{p^2 + p\dfrac{(r_s + K_{Pq2})}{L_{ls}} + \dfrac{K_{Iq2}}{L_{ls}}} \quad (26)$$

By setting the denominator of the transfer functions equal to the second order Butterworth polynomial which is presented in equation 27, the controller's coefficients are given Table I [10].

$$p^2 + \sqrt{2}p\omega_o + \omega_o^2 = \sqrt{2}p200 + (200)^2 \quad (27)$$

Table-I. The machine parameters and controllers coefficients.

Machine Parameters		Controllers Coefficients	
r_s	0.01 (Ω)	K_{Pd1}, K_{Id1}	2.14, 312
L_{ls}	0.01 (mH)	K_{Pq1}, K_{Iq1}	0.915, 100
L_{md}	0.016 (H)	$K_{Pd2,3}, K_{Id2,3}$	5.62, 400
L_{mq}	0.033 (H)	$K_{Pq2,3}, K_{Iq2,3}$	5.62, 400

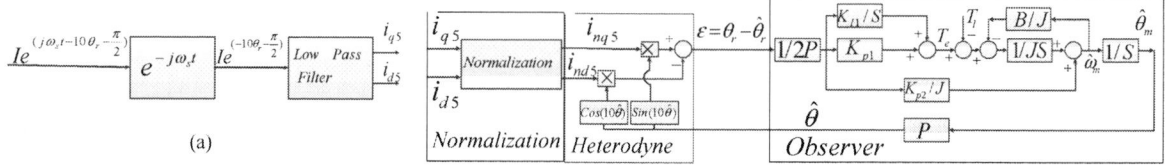

(b)

Figure 2. (a) The block diagram used to extract signal for observer, (b) The Luenberger observer.

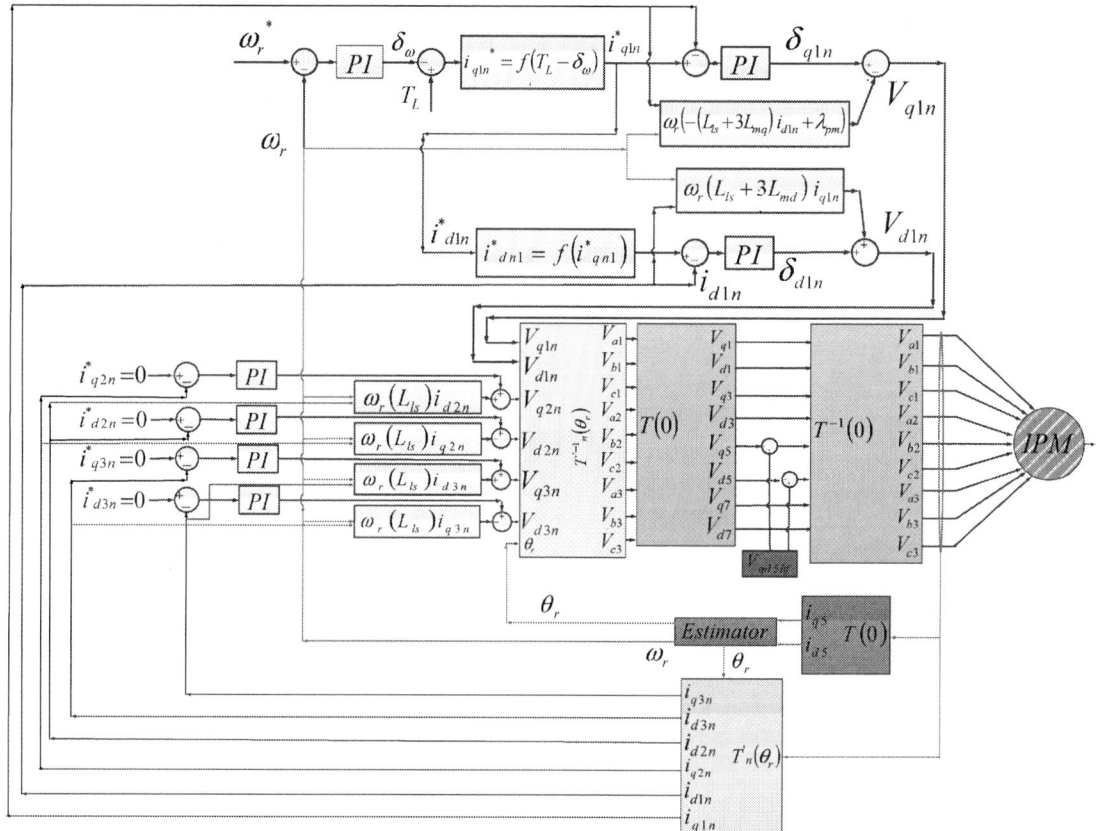

Figure 3. The controller of the symmetrical machine.

IV. THE ROTOR POSITION OBSERVER

To implement the speed and current controllers the rotor position and speed are required. These are estimated using currents due to the high frequency voltages into the stator windings and and a Luenberger observer. The high frequency voltage set is injected to the fifth sequence circuit of the machine model in the stationary reference frame. To design the signal processing method, the general form of the stator winding currents due to the high frequency voltage is determined. The expressions of the inductances of the fifth sequence of the stationary reference frame are given below.

$$L_{q5} = \frac{a_3}{2}\left(\frac{N_5^2}{25}\right)Cos(10\theta_r) \tag{28}$$

$$L_{d5} = \frac{a_3}{2}\left(\frac{N_5^2}{25}\right)Cos(10\theta_r + \frac{\pi}{2}) \tag{29}$$

A high frequency voltage set in the form of equation 30 is injected into the fifth sequence of the nine-phase IPM in the stationary reference frame to generate rotor angle dependent current components in the stator windings.

$$\begin{bmatrix} V_{q5} \\ V_{d5} \end{bmatrix} = A\begin{bmatrix} Cos(\omega_s t) \\ Cos(\omega_s t + \frac{\pi}{2}) \end{bmatrix} \tag{30}$$

Using the qd inductances in equations 28 and 29, the qd stationary reference frame voltage equation for fifth sequence circuit of the machine in stationary reference frame can be expressed as:

$$\begin{bmatrix} V_{q5} \\ V_{d5} \end{bmatrix} = A \begin{bmatrix} Cos\,(\omega_s t) \\ Cos\,(\omega_s t - \dfrac{\pi}{2}) \end{bmatrix} = r_s \begin{bmatrix} i_{q5} \\ i_{d5} \end{bmatrix} + p \begin{bmatrix} \lambda_{q5} \\ \lambda_{d5} \end{bmatrix} = r_s \begin{bmatrix} i_{q5} \\ i_{d5} \end{bmatrix} -$$

$$p \begin{bmatrix} \dfrac{a_3}{2}\left(\dfrac{N_5^{\,2}}{25}\right) Cos\,(10\,\theta_r) & 0 \\ 0 & \dfrac{a_3}{2}\left(\dfrac{N_5^{\,2}}{25}\right) Sin\,(10\,\theta_r) \end{bmatrix} \begin{bmatrix} i_{q5} \\ i_{d5} \end{bmatrix} \tag{31}$$

Using the harmonic balance technique, the currents of the fifth sequence circuit of the machine in the stationary reference frame can are obtained as [11]:

$$I = I_1 e^{j(\omega_s t - 10\theta_r - \frac{\pi}{2})} + I_2 e^{j(\omega_s t + 10\theta_r - \frac{\pi}{2})}, \; \theta_r = \omega_r t \tag{32}$$

where:

$$I_1 = I_2 = \frac{100\,A}{\omega_s a_3 N_5^{\,2}} \tag{33}$$

The currents have two frequencies, one is related to the frequency of the injected high frequency voltage and the second component related to the rotor angle. To extract the rotor position information, equation 33, is first multiplied by $e^{-j(\omega_s t)}$ resulting in :

$$i_{qd5} = \frac{100Ae^{j(10\theta_r - \frac{\pi}{2})}}{(\omega_s)a_3 N_5^{\,2}} + \frac{100Ae^{j(-10\theta_r - \frac{\pi}{2})}}{(\omega_s)a_5 N_5^{\,2}} \tag{34}$$

A low pass filter removes any remaining of the high frequency components after heterodyning. The outputs of the filters have variable magnitudes which vary with the rotor speed. Using equation 35, the currents are normalized to bring their magnitudes to unity.

$$i_{nq5} = \frac{i_{q5}}{\sqrt{i_{q5}^{\,2} + i_{d5}^{\,2}}}, \; i_{nd5} = \frac{i_{d5}}{\sqrt{i_{q5}^{\,2} + i_{d5}^{\,2}}} \tag{35}$$

The described current synthesis and conditioning are illustrated in Figure 2(a). The currents of the equation 34 are used as the input variables to the Luenberger observer shown in Figure 2(b). The schematic diagram of the position and speed sensor-less vector control scheme is shown in Figure 3. This controller has 6 loops for controlling the currents of the different axis of the decoupled model. The rotor speed and the position are estimated using the observer that operates on the fifth sequence of the machine currents in the stationary reference frame. After generating the voltages of the decoupled axis they are transformed to the natural variables (using the transformation matrix of equation 5 to obtain the three sets of the three phase voltages for each machine set. Then the voltages are transformed to the stationary reference frame. After adding the high frequency voltages, they are transformed back to the natural variables to obtain the phase voltages to be applied to the stator windings of the machine. The stator currents are all

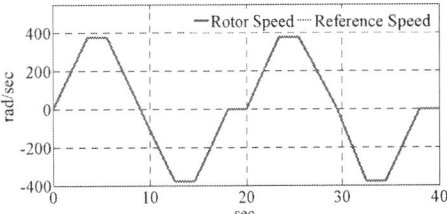

Figure 4. The reference and rotor speed.

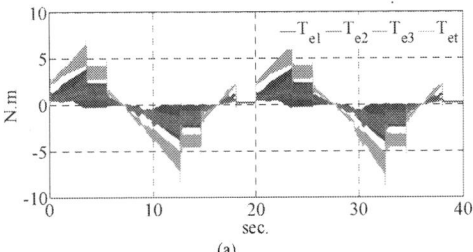

(a)

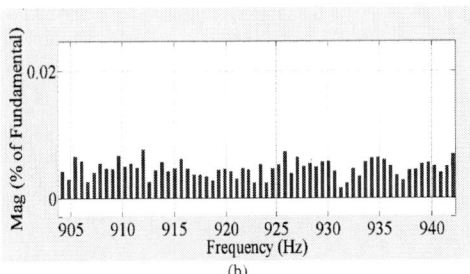

(b)

Figure 5: (a) The electromagnetic torque of each machine and the total, (b) The zoomed view of the spectrum of the steady-state electromagnetic torque around the frequency of 920 (Hz) .

measured by current sensors, transformed to the stationary reference frame to extract the fifth sequence current for the position estimator. Also these measured currents are transformed to the decoupled reference frame to obtain the feedback currents for the implementation of the controllers.

V.   SIMULATION OF THE SENSORLESS SYMMETRICAL TRIPLE-STAR IPM DRIVE

In this section the speed and position sensor-less vector controller is simulated with the full order coupled model of the machine [12]. In this simulation the motor controller receives a trapezoidal speed reference as the input command. The speed reference starts from the zero and goes to the positive rated speed and from the positive rated speed it goes to the negative rated speed and returns to zero. Figure 4 shows the tracking of the reference speed. Figure 5(a) shows the electromagnetic torque generated by each machine and the total electromagnetic torque. The spectrum of the torque is shown in Figure 5 (b) where it is observed that at the frequency of 920 Hz (the frequency of the injected voltages) the machine produces a non-significant torque. Figure 6 shows the reference and the feedback currents of $qd1n$ axis. Figure 7 shows the currents of the $q2n$ and $d2n$ axes with zero as reference currents. The $q3n$ and $d3n$ axes have the same current references and similar responses and are not repeated here. Figure 8 shows the fifth sequence currents after heterodyning and normalizing along with the estimated position and rotor angle around the zero speed intervals of rotor speed. The observer accurately tracks the rotor angle during negative, positive and zero rotor speeds.

VI. EXPERIMENTAL RESULTS

The proposed rotor angle and speed estimation method have been implemented on a 2hp, nine-phase IPM using a DSP-FPGA controller which controls the machine via three sets of three-phase voltage source inverter. The model of the observer has been discretized and implemented in the DSP (TMS320C6713 DSK). The current sensors are connected to the controller with nine Analogue/Digital converters with the sampling rate of 50 kHz. After generating the voltages inside the DSP they are sent to the FPGA which generates the PWM pulses for the inverter. The currents due to the high frequency voltages are processed using the presented method and the rotor position is estimated. Figure 9 shows the experimental result for

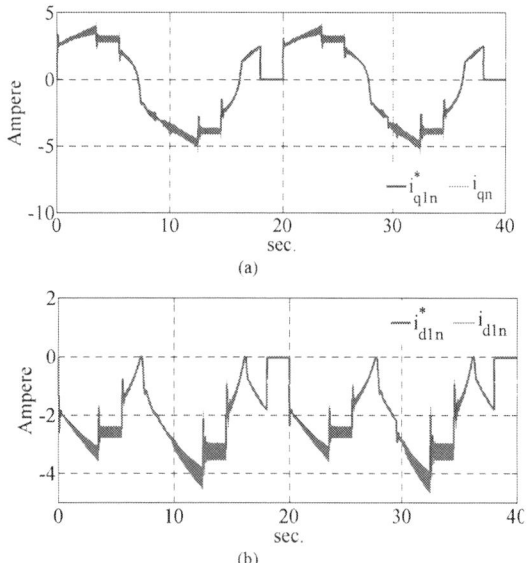

Figure 6. (a) The q1n reference and feedback currents, (b) The d1n reference and feedback currents.

the trapezoidal speed reference spanning negative, positive and zero speeds, demonstrating satisfactory tracking. Figure 10 shows the reference and measured currents of the q1n and d1n axis of the decoupled reference frame with acceptable correlations. These currents are generated by the current regulator loops shown in Figure 3. The normalized currents of the fifth sequence are shown in the Figure 11 along with the estimated and measured rotor angle around the zero speed

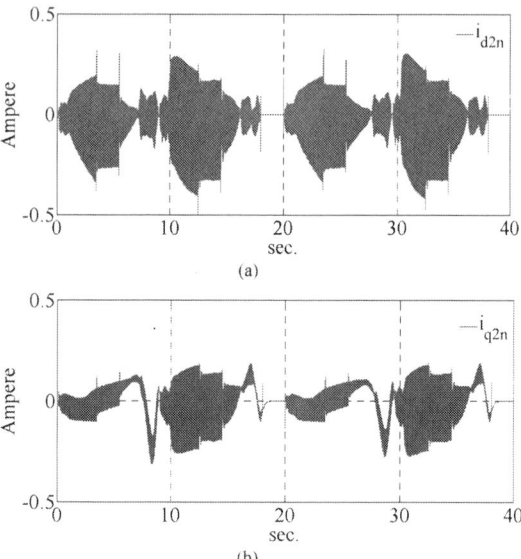

Figure 7. (a) The currents of the zero axis for, (a) d2n, (b) q2n

intervals. From this figure, it is clear that the estimated angle accurately tracks the rotor angle.

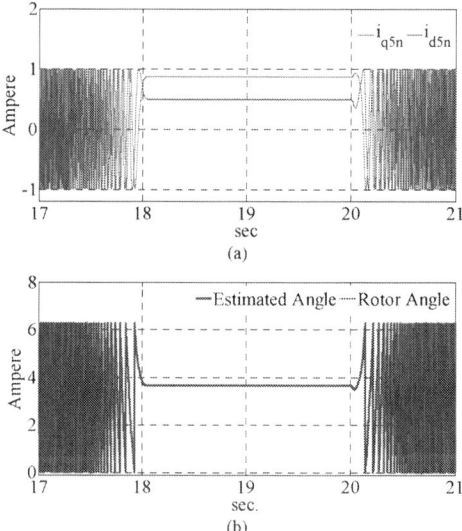

Figure 8. (a) The normalized axis currents of the fifth sequence, (b) The actual and estimated angle during the zero speed intervals.

VII. CONCLUSION

In this paper a controller for a triple-star nine-phase IPM is presented based on a decoupled model. The proposed model made possible by a diagonalizing reference frame transformation simplifies the controller design and hardware implementation by eliminating coupling q-d axis inductances between the three sets of three phase connections and within each of them. The control strategy is based on

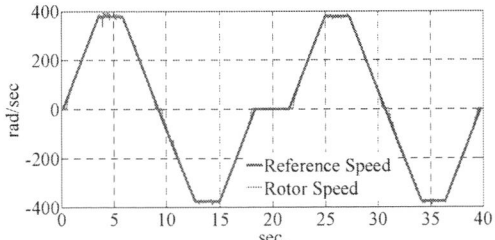

Figure 9. The reference and rotor speed.

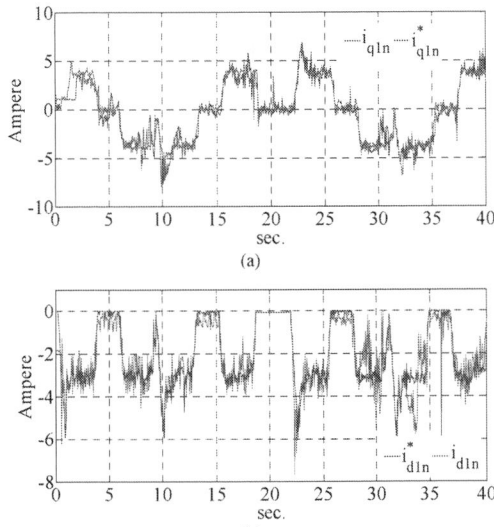

Figure 10. (a) The reference and actual q1n axis current, (b) The reference and actual d1n axis current.

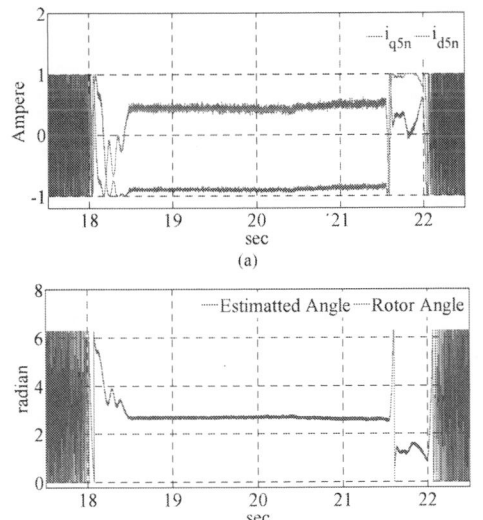

Figure 11. (a) The normalized currents of the fifth sequence, (b) the actual and estimated angle during the zero crossing.

minimizing the stator copper loss in the presence of the load torque, in which the torque producing currents are determine using the multiplier Lagrange constrained optimization methodology. The rotor position and speed needed for the drive is estimated using high currents generated by impressing defined non-torque producing high frequency voltages in the fifth sequence circuit of the machine. The processing of the high frequency currents are set forth and with a Luenberger observer, the rotor position and speed are estimated. The performance of the proposed position and speed sensor-less vector control under minimum stator copper loss strategy is validated by simulation results based on the full order coupled model of the machine. Confirmatory experimental results are presented, demonstrating the utility of the scheme in all speed regimes, including close to zero speed operations.

VIII. REFERENCES

[1] E. A. Klingshirn, "High phase order induction motors- Part I – Description and theoretical considerations," IEEE Transactions on Power Apparatus and Systems, vol. PAS-102, no. 1, pp. 47-53, January 1983.

[2] J. Karttunen, S. Kalilio, P. Peltoniemi, P. Silventoinen and O. Pyrhonen, "Decoupled vector control scheme for dual three-phase permanent magnet synchronous machines," IEEE Transactions on Industrial Electronics, vol. 61, no. 5, pp. 2185-2196, May 2104.

[3] A.Tessarolo and C.Bassi, "Stator harmonic currents in VSI-Fed synchronous motors with multiple three-phase armature winding," IEEE Transactions on Energy Conversion, vol. 25, no. 4, pp. 974-982, 2010.

[4] A.Tessarolo, G.Zocco and C. Tonello, "Design and testing of a 45MW, 100Hz quadruple-star synchronous motor for a liquefied natural gas turbo-compressor drive," IEEE Transactions on Industry Applications, vol. 47, no. 3, pp. 1210-1219, 2011.

[5] E. Prieto-Araujo, A. Junyent-Ferre, D. Lavernia-Ferrer and O. Gomis-Bellmunt, " Decentralized control of nine-phase permanent magnet generator for offshore wind turbines," IEEE Transactions on Energy Conversion, vol. 30, no. 3, pp. 1103-1112, September 2015.

[6] S. Kallio, M. Andriollo, A. Tortella and J. Karttunen, "Decoupled d-q model of double –star interior permanent synchronous machines," IEEE Transactions on Industrial Electronics, vol. 60, no. 6, pp. 2486-2494, June 2013.

[7] G. J. Retter, **Matrix and Space-Phasor Theory of Electrical Machines**, Akademiai Kiado, Budapest 1987.

[8] Chi-Tsong Chen, **Linear System Theory and Design**, Oxford University Press, London, 1984.

[9] R. Fletcher, **Practical Methods of Optimization**, John Wiley and Sons, New York, 1987.

[10] B. Friedland, **Control System Design: An Introduction to State-Space Methods**,McGraw-Hill Inc, New York, 1986.

[11] A. H. Nayfeh and B. Balachandran, **Applied Nonlinear Dynamics, Analytical, Computational and Experimental Methods**, John Wiley and Sons, New York, 1995.

[12] M. Ramezani and Olorunfemi Ojo, "A new sensor-less position estimation method for a nine-phase interior permanent magnet machine using a high frequency injection in a non-torque generating circuit." Conference Proceedings of the 30th Annual IEEE Applied Power Electronic Conference and Exposition, pp. 1548-1555, March 2015. November/December. 2001.

978-1-4673-9551-9/16 $31.00 © 2016 IEEE   2042

Mitigation of Common-mode Noise in Wide Band Gap Device Based Motor Drives

Sneha Narasimhan*, Saurabh Tewari†, Eric Severson‡, Rohit Baranwal‡, and Ned Mohan‡, *Life Fellow, IEEE*
Email: {naras024, tewari, sever212, baran065, mohan}@umn.edu
*Rockwell Automation, Rockwell Automation, Mequon, Wisconsin, 53092
†MTS Systems Corporation, Eden Prairie, MN 55344
‡Department of Electrical and Computer Engineering, University of Minnesota, Minneapolis, MN 55455

Abstract—MOSFETs built using wide band gap (WBG) materials offer numerous benefits to power electronic circuits. These benefits are quite apparent in applications requiring breakdown voltages ≥ 600 V, where Silicon IGBTs are typically used due to their combination of high breakdown voltage and low conduction losses. Compared to Silicon IGBTs, WBG MOSFETs offer very short turn-ON and turn-OFF times, which reduce switching losses and enable significantly higher switching frequencies. This paper explores the application of WBG MOSFETs to motor drives, where higher switching frequencies reduce motor losses and torque ripple and allow higher control bandwidth, thus enabling greater output frequencies needed to operate motors at higher speeds. Specifically, two-level voltage source inverters utilizing Silicon Carbide (SiC) MOSFETs are constructed to operate a 1 HP induction motor. Experimental results are presented which show that the short turn-ON and turn-OFF transients as well as high switching frequencies lead to increased shaft voltage and conducted ground currents. Mitigation techniques are implemented and evaluated, including clamp-on ferrites and an open-end winding drive implementation. The shaft voltage and ground currents are found to be best suppressed in an open-end winding drive utilizing clamp-on ferrites.

Index Terms—common-mode voltage, electromagnetic interference, open-end drive, silicon carbide, gallium nitride;

I. INTRODUCTION

Several manufacturers have recently commercialized Power MOSFETs utilizing Silicon Carbide (SiC) and Gallium Nitride (GaN) WBG materials. These products have attracted considerable attention for use in power electronic circuits due to a combination of large breakdown voltages and low conduction losses with short switching times. Previously, designers often relied on Silicon IGBTs, that have significantly slower switching speeds, to obtain acceptable conduction losses in motor drive applications where DC bus voltages over 600 V are common. At a similar ON-state voltage drop as a Silicon IGBT, WBG MOSFETs allow drive operation at much higher switching frequencies [1]. The increased switching frequency reduces harmonic losses and torque ripple in the machine. Higher switching frequencies also allow higher bandwidth control of the power electronic inverter that results in better precision and enables higher output fundamental frequencies necessary for high-speed machines e.g. [2] and machines with a greater numbers of poles. Higher speed operation and a greater number of poles, both, improve the motor power density [3].

It is well known that pulse width modulated (PWM) power electronic motor drives cause shaft voltage buildup and ground currents in motors [4]–[12]. The shaft voltage can result in sparks due to discharge events, which pose a major risk in combustible environments [10], [13]. The ground currents cause electromagnetic interference (EMI) and include bearing currents, which have been shown to significantly reduce bearing lifetime [4]–[12]. The cause of both of these effects is the switching common-mode voltage applied at the motor terminals interacting with parasitic capacitances in the machine. The transition speed of the common-mode voltages (dv/dt corresponding to the switching speeds of the MOSFETs or IGBTs used) and how often these transitions take place (corresponding to the switching frequency) play critical roles in the extent to which the shaft voltage and ground currents are a problem. To obtain the maximum benefit from WBG MOSFETs, motor drives using these devices will be designed with both greater switching speeds and switching frequencies, and are therefore expected to suffer from greater shaft voltage and ground currents compared to drives using the traditional Silicon IGBTs.

Several active and passive techniques have been developed for Si IGBT motor drives to mitigate shaft voltage and ground current, e.g. [7], [12], [14]–[27], but little research exists on their application to WBG device based drives. This paper will explore two of these techniques: additional common-mode impedance and an open-end winding drive configuration, for a SiC MOSFET based motor drive. It is found that the increased switching speeds and switching frequency mean that a smaller common-mode impedance can be used to noticeably reduce shaft voltage and ground currents. The lowest values of shaft voltage and ground currents are obtained when common-mode impedance is added in an open-end winding drive configuration. Experimental results are presented to demonstrate these points.

II. COMMON-MODE NOISE

The source of common-mode voltage in PWM drives and the conduction paths in the electric motor are now explained for the conventional two-level voltage source inverter (VSI).

A. Switching common-mode voltage from a VSI drive

The standard two-level VSI is shown in Fig. 1. In conventional space vector pulse width modulation (SVPWM) of

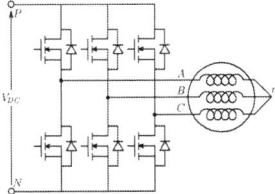

Fig. 1. Conventional two-level VSI drive

the two-level inverter, the available active vectors adjacent to the desired output voltage vector, and the available zero vectors are alternately applied to generate the desired output voltage in an average sense. Since each space vector has a different common-mode voltage value, a switched common-mode voltage is generated at the inverter output. The output common-mode voltage with respect to the DC bus negative terminal is defined below.

$$v_{\mathrm{com}} = \frac{v_{AN} + v_{BN} + v_{CN}}{3} \qquad (1)$$

An example of the common-mode voltage applied over one switching interval is shown in Fig. 2. The rate of change of the common-mode voltage is determined by the speed at which the semiconductor switches turn-ON and turn-OFF. The common-mode voltage consists of low frequency triplen harmonics of the inverter fundamental output frequency and high frequency harmonics of the inverter switching frequency. While the low frequency common-mode voltage is included by design to improve the DC bus utilization, the high frequency components are undesirable byproducts of the pulse width modulation.

B. Common-mode paths

The simplified construction of an induction machine equipped with ball bearings is shown in Fig. 3(a). The windings of the machine are coupled to the stator frame and the shaft through parasitic capacitances as illustrated in Fig. 3(b). The shaft of the machine is supported on ball bearings, which ride upon a non-conductive lubricant film when the machine is rotating near rated speed. The lubricant film appears as a parasitic capacitance between the shaft and the stator frame, and the current taking this path travels through the bearings. The stator frame is typically grounded and the DC power rails of the VSI are coupled to ground, forming a potential

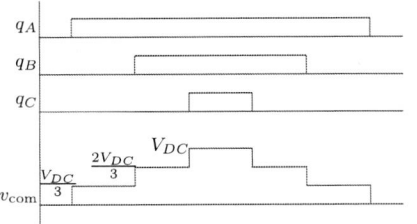

Fig. 2. Example of varying CMV in a single VSI over a switching period.

conduction path for common-mode currents to flow. This path is illustrated in Fig. 3(c).

All common-mode currents are undesirable in that they cause losses without producing any torque and may interfere with other equipment. The common-mode currents that travel through the bearings are particularly troubling as they have been shown to cause premature bearing failures. In addition to the capacitive current path through the bearing there is also a resistive path that forms when the shaft voltage exceeds the lubricant breakdown voltage. When this occurs, a sudden, localized spike of high current density flows through the bearings in what is called an electrical discharge machining (EDM) current [28]. This EDM current is especially harmful to the bearing since it results in pitting and fluting. The degree to which EDM currents are a problem depend upon 1) the machine design: the lubricant's dielectric strength and film thickness determining the breakdown voltage; and 2) the common-mode voltage output of the inverter coupling to the shaft through the machine capacitances, determining the magnitude of shaft voltage.

Fig. 3(c) shows that the common-mode paths are through distributed LC branches. At the resonant frequencies of these LC networks, the common-mode source (the inverter) would see little impedance resulting in a high common-mode current, were the common-mode voltage to contain components at frequencies close to the resonant frequency. These resonances typically occur at high frequencies, in the multi-MHz range, which means that the higher switching frequencies and shorter switch transition times characteristic of inverters utilizing WBG devices are expected to cause greater common-mode currents.

Because the occurrence of EDM currents depends on the instantaneous lubricant thickness and the shaft voltage, they occur at seemingly random intervals, unlike the capacitive common-mode currents which occur during the common-mode voltage transitions. Therefore, to evaluate the degree to which common-mode EMI is a problem, this paper considers both, shaft voltage and conducted ground currents.

III. COMMON-MODE NOISE MITIGATION TECHNIQUES

The two approaches for common-mode noise mitigation considered in this paper are now explained: 1) the use of dual-VSIs with an open-end winding motor to nearly eliminate switching common-mode voltages; and 2) the use of common-mode impedance to suppress currents from flowing in response to common-mode voltage.

A. Open-end winding drive

An open-end winding drive has a power electronic converter connected to either end of an electric motor. In this paper, a dual two-level inverter drive is used for the open-end winding motor. A circuit diagram of this motor drive is shown in Fig. 4. Both the converters are powered by single DC bus with voltage V_{DC}. The converter labeled 'Positive end converter' is connected to terminals A, B and C, and the converter labeled 'Negative end converter' is connected to terminals A', B' and C'. The common-mode voltages $v_{\mathrm{com,p}}$ and

978-1-4673-9551-9/16 $31.00 © 2016 IEEE

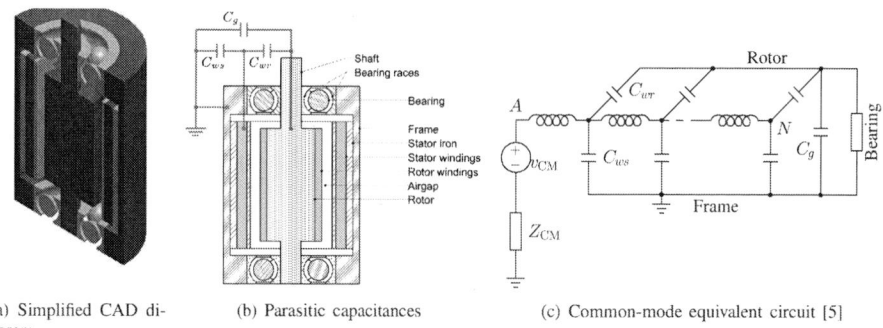

(a) Simplified CAD di-agram   (b) Parasitic capacitances   (c) Common-mode equivalent circuit [5]

Fig. 3. Common-mode current path of an electric motor: (a) CAD cross-sectional diagram of an electric motor with ball bearings, (b) and (c) circuit diagrams.

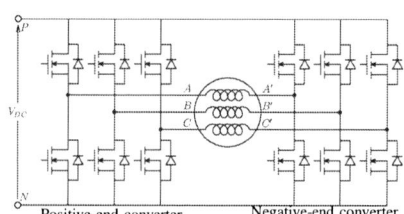

Fig. 4. Dual two-level VSI based open-end winding drive.

$v_{\mathrm{com,n}}$ at positive and negative end machine terminals are defined in equation (2). The difference of these voltages $v_{\mathrm{cm,diff}}$ leads to low frequency circulating currents which cause extra losses, while their sum $v_{\mathrm{cm,sum}}$ leads to the previously described high frequency common-mode ground currents and shaft voltages that plague conventional VSI motor drives. If the converter is operated in a manner that keeps the CMV constant at the positive and negative end machine terminals, i.e. $v_{\mathrm{com,p}} = v_{\mathrm{com,n}} = \mathrm{const.}$, these problems can theoretically be eliminated.

$$v_{\mathrm{com,p}} = \frac{v_{AN} + v_{BN} + v_{CN}}{3}$$
$$v_{\mathrm{com,n}} = \frac{v_{A'N} + v_{B'N} + v_{C'N}}{3} \qquad (2)$$

The space vectors of positive and negative end converters in Fig. 4 are termed as $\mathbf{U}$ and $\mathbf{W}$ respectively. Their definition is given in equation (3) and they are shown in Fig. 5. The vectors $\mathbf{U_1}$, $\mathbf{U_3}$, $\mathbf{U_5}$, $\mathbf{W_1}$, $\mathbf{W_3}$ and $\mathbf{W_5}$ have a CMV of $\frac{V_{DC}}{3}$; the vectors $\mathbf{U_2}$, $\mathbf{U_4}$, $\mathbf{U_6}$, $\mathbf{W_2}$, $\mathbf{W_4}$ and $\mathbf{W_6}$ have a CMV of $\frac{2V_{DC}}{3}$. If only one of these two sets of vectors are

used for PWM of the dual-VSI drive, then the positive and negative CMV are held equal and at constant values. Then, $v_{\mathrm{cm,diff}}$ is zero, while $v_{\mathrm{cm,sum}}$ is a constant value.

$$\mathbf{U} = v_{AN} + v_{BN}e^{j\frac{2\pi}{3}} + v_{CN}e^{-j\frac{2\pi}{3}}$$
$$\mathbf{W} = -(v_{A'N} + v_{B'N}e^{j\frac{2\pi}{3}} + v_{C'N}e^{-j\frac{2\pi}{3}}) \qquad (3)$$

The voltage vectors $\mathbf{U_1}$, $\mathbf{U_3}$, $\mathbf{U_5}$, $\mathbf{W_1}$, $\mathbf{W_3}$ and $\mathbf{W_5}$ can be combined to give six resultant active space vectors labeled $\mathbf{V_1}$ through $\mathbf{V_6}$, as shown in Fig. 6. For example, $\mathbf{U_1}$ and $\mathbf{W_5}$ add up to give $\mathbf{V_2}$. In addition, three zero vectors are also obtained by applying $\mathbf{U_1}$ and $\mathbf{W_1}$ together, or $\mathbf{U_3}$ and $\mathbf{W_3}$ together, or $\mathbf{U_5}$ and $\mathbf{W_5}$ together. Hence, we have a set of active and zero vectors for the dual-VSI, which consists only of vectors that have equal CMV at the positive and negative ends. In Fig. 6, the output voltage vector $\overline{\mathbf{V}}_\mathbf{o}$ can be in any of the six sectors labeled 1 through 6 and rotates at the speed of motor drive's output frequency ω_o.

It should be noted that the PWM control of a dual-VSI drive need not involve the standard space vector approach of sector determination logic based on the output voltage vector using the $\alpha\beta$ transformation. In this paper, a carrier-based approach has been used to operate the dual-VSI drive, as explained in [29], [30].

In the discussion regarding the dual-VSI drive so far, the non-idealities of switching devices have not been considered.

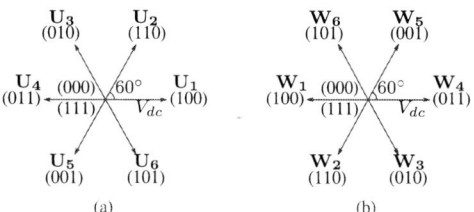

Fig. 5. (a) Positive end space vectors and (b) Negative end space vectors.

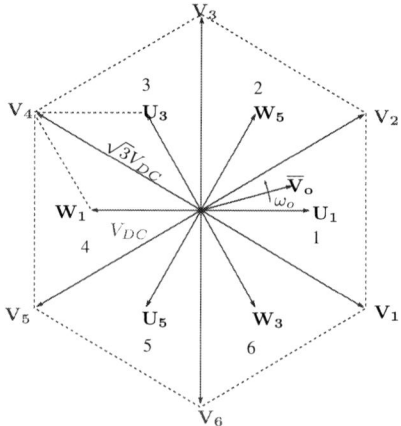

Fig. 6. Dual-VSI vectors

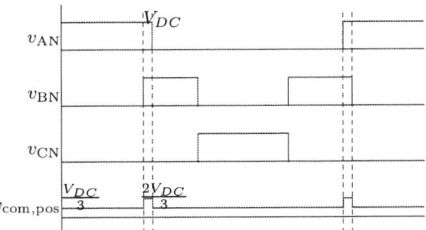

Fig. 7. CMV glitch in dual-VSI due to dead band (illustrated for positive end converter)

In a real power electronic converter, the devices do not turn ON and OFF instantaneously. To account for this, dead-time is introduced in each phase leg between when one switch is turned OFF and the other is turned ON so that the DC link is not short circuited. During the dead-time intervals, the output space vectors at the motor terminals are not necessarily the vectors intended by the modulation; the output vectors are instead determined by the motor currents' direction and the freewheeling diodes that carry these currents during the dead-time. This behavior leads to fluctuations in the CMV [21] and is illustrated in Fig. 7. These CMV fluctuations cause the dual-VSI drive to experience common-mode ground currents and, since each VSI experiences this phenomenon separately, it also leads to low frequency circulating currents within the drive. As discussed in [21], increasing the dead-time t_d for a fixed switching frequency increases the magnitude of ground currents and circulating currents, because it increases the amount of time that the unintentional space vectors are applied. Since WBG switches are able to turn ON and OFF very rapidly, inverters using these devices are able to minimize the dead-time, which will be shown later to significantly reduce the common-mode currents.

B. Common-mode chokes

An alternative method to suppress common-mode currents is with the use of common-mode chokes. These chokes can be in the form of a clamp-on ferrite that encircles the connecting cable of all three phases or as a continuous core upon which the connecting cable of all three phases is wound. These devices are standard components that can be purchased through electronic vendors.

Circuit connection diagrams for the use of the common-mode ferrites in both single VSI motor drives and dual-VSI motor drives are shown in Fig. 8. The common-mode ferrite choke acts like a lossy common-mode transformer, where any common-mode current is transformed to a secondary that consists of an RL branch. This RL branch appears in series in the equivalent common-mode circuit of the motor drive, but does not appear in the differential-mode circuit. Common-mode ferrites are typically most effective for high frequency components, which suggests that these devices could be highly effective for inverters utilizing WBG switches.

In a dual-VSI drive, the CMV is already reduced compared to a single VSI and the residual CMV is due to unintentional space vectors applied during the short dead-time. This means

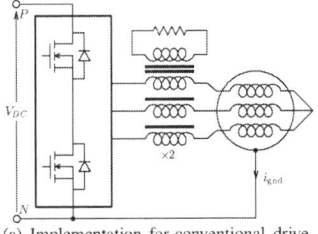

(a) Implementation for conventional drive

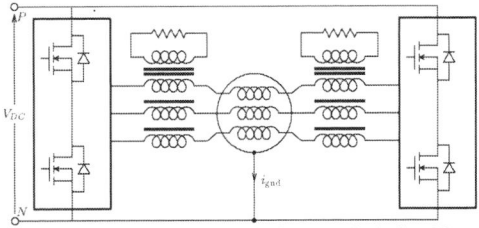

(b) Implementation for dual two-level open-end winding drive

Fig. 8. Drive circuit diagrams with common-mode chokes.

that the switching CMV of a dual-VSI is predominately at higher frequencies than in the single VSI and will be more effectively mitigated by a common-mode ferrite.

IV. EXPERIMENTAL SETUP AND OPERATING POINTS

The experimental setup consists of SiC Power MOSFET based half-bridge circuits and a 1 hp AC induction motor (ACIM). Three half-bridge 'phase legs' are mounted on a set of DC rails to form the voltage source inverter (VSI). Two identical VSIs are used to realize the dual-VSI open-end winding drive. The ACIM is modified such that it can be used either as a wye-connected, or as an open-end winding machine. The ACIM is coupled to a 1 hp DC motor that serves as the mechanical load. The phase legs utilize SCH2080KE SiC MOSFETs (ROHM Semiconductor); further information about the phase leg circuit design and performance can be found in [31]. The circuit diagrams of the drive configurations are illustrated in Fig. 1 and 4, and a photograph of the experimental setup is shown in Fig. 9. The drive configurations for results taken with clamp-on ferrites (common-mode impedance mitigation technique) are depicted in Fig. 8. The primary operating points used to compare the mitigation techniques are shown in Table I.

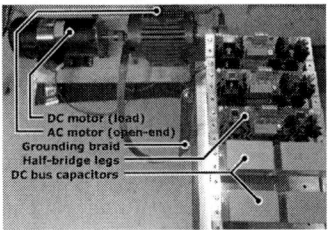

Fig. 9. Picture of the experimental setup showing the VSIs and the ACIM coupled to a DC motor load.

978-1-4673-9551-9/16 $31.00 © 2016 IEEE

Symmetric SVPWM [32] is used to drive the machine in the wye-connected configuration (Fig. 1). As previously mentioned, the open-end winding configuration of Fig. 4 uses carrier-based PWM for dual-VSI drives [29].

The shaft voltage is measured using a non-isolated probe and a brush that is electrically isolated from the motor body. A grounding braid connects the motor body to the DC bus negative rail that is also connected to the earth ground. The motor set is otherwise isolated from the earth ground. A high-bandwidth current probe ($400\,\text{Hz} - 200\,\text{MHz}$) is employed to measure the high-frequency common-mode current entering the motor terminals. Another current probe suitable for lower frequencies ($1\,\text{Hz} - 20\,\text{MHz}$) is used to measure any circulating currents [21] that may flow in the open-end winding configuration.

The output of the high-bandwidth current probe is the total common-mode current. For a conventional VSI drive, this current is $\sum i_{\text{CM}} = i_A + i_B + i_C$ where A, B, C are the load terminals. For an open-end winding drive, this current is $\sum i_{\text{CM}} = i_A + i_B + i_C + i_{A'} + i_{B'} + i_{C'}$ where the load terminals $A, \ldots, A', \ldots$ are marked in Fig. 4.

In both configurations, the current $\sum i_{\text{CM}}$ must flow through the grounding braid, into the earth ground. Therefore $\sum i_{\text{CM}}$ is synonymous with the ground current. To distinguish the measurements made with the high-bandwidth probe and the lower bandwidth probe in the open-end winding drive, identifiers HF and LF have been used.

V. RESULTS

A. Results from a two-level VSI

As described in Section II-A, a two-level VSI generates a switching common-mode voltage that causes a shaft-voltage to develop and a common-mode current, including the destructive bearing current, to flow. The experimental results of the shaft voltage and the common-mode current that flows into the ground, and their spectra for a two-level SVPWM VSI driving the ACIM (wye-connected) are shown in Fig. 10. Results are shown at the operating points OP1 and OP2.

The magnitude of the common-mode current is larger at OP2 (Figs. 10(c) and 10(f)) which is consistent with the literature: higher PWM frequencies will exacerbate the effects of the inevitable common-mode voltage generated by a typical PWM VSI drive.

In addition to the higher PWM frequencies offered by WBG devices, the higher switching speeds also affect the common-mode currents as discussed earlier. To investigate this, the MOSFET switching speed in the two-level VSI drive of Fig. 1 was intentionally reduced by increasing the gate resistance from 5Ω to 33Ω and adding 10 nF of capacitance between the

TABLE I
OPERATING POINTS USED FOR EXPERIMENTS.

	Switching frequency	Dead-time	Output fundamental	Output voltage
OP1	20 kHz	1 μs	30 Hz	90 V (ll, rms)
OP2	100 kHz	100 ns	30 Hz	90 V (ll, rms)

gate and the source. The results are shown in Fig. 13, where it is seen that the turn-ON time increases from 25.1 ns to 85 ns. This decreased the observed common-mode current magnitude by approximately 22%, validating the expectation of short switching intervals contributing to common-mode currents.

The dead-time interval, however, was not found to affect the common-mode currents when it was increased from 100 ns at OP2 to 500 ns. This is also consistent with conclusions found in the literature that the primary cause of common-mode currents is the fast changing common-mode voltage of the intentionally applied space vectors, and not due to unintentional space vectors applied during the dead-time interval (unlike the discussion at the end of Section III-A).

Of course, the reduction in the common-mode current should not be interpreted as an argument in favor of lower PWM frequencies and slower switching speeds — slowing the switching speeds of WBG devices intentionally limits their performance and requires longer dead-time intervals, thus introducing departure from the expected modulation performance. To fully exploit the features offered by the wide band gap devices and avail the advantages discussed in the Introduction, the PWM frequency (switching frequency) and the switching speeds must be increased. At the same time, the effects of the inverter generated common-mode voltage must be curtailed. The results provided in the next subsection show that the open-end winding drive discussed earlier succeeds in reducing the shaft voltage and the ground current while operating at high switching frequencies and very short transition times.

B. Open-end winding drive for common-mode elimination

As described in Section III-A, dual-VSI-based open-end winding (dual-VSI OE Wdg.) drives can eliminate the high-frequency common-mode components in the output voltage by using a smaller set of space vectors, and also improve the DC bus utilization. The shaft voltage and the common-mode current in the open-end winding drive of Fig. 4 for operating points OP1 and OP2 are provided in Fig. 11. A comparison with Fig. 10 reveals a significant reduction in the shaft voltage and the common-mode current, as expected.

As explained in Section III-A, the only significant switching common-mode voltage present is due to the unintentional space vectors applied during the dead-time intervals. To illustrate the strong influence this has, results are presented in Fig. 12 when the dead-time interval of OP2 has been increased from 100 ns to 500 ns. In comparing Fig. 12 to Fig. 11(e) and (f), it is readily apparent that both the shaft voltage and the common-mode current have increased significantly.

In addition to poorer high-frequency common-mode performance, increasing the dead-time also results in low-frequency circulating currents in an OE Wdg. drive. These undesirable circulating currents cause additional losses, but are an unavoidable consequence of the necessary dead-time. Since the faster switching speeds of WBG devices permit very short dead-time intervals, open-end winding drives stand to gain even more from these devices. Observed waveforms of the circulating currents are included in Fig. 11. This topic has been studied

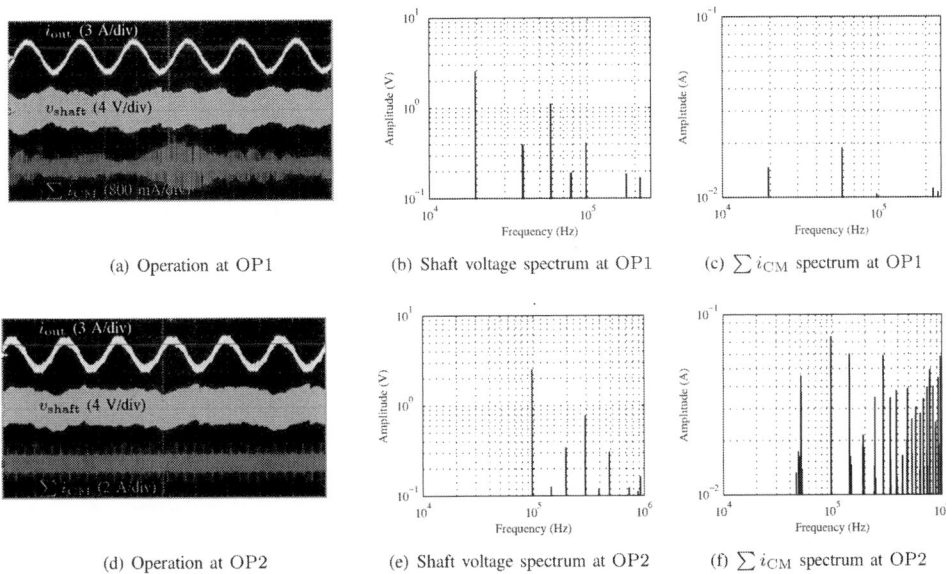

(a) Operation at OP1   (b) Shaft voltage spectrum at OP1   (c) $\sum i_{\mathrm{CM}}$ spectrum at OP1

(d) Operation at OP2   (e) Shaft voltage spectrum at OP2   (f) $\sum i_{\mathrm{CM}}$ spectrum at OP2

Fig. 10. Shaft voltage and common-mode current in the SiC two-level VSI drive of Fig. 1: (a), (b), (c) at OP1, and (d), (e), (f) at OP2

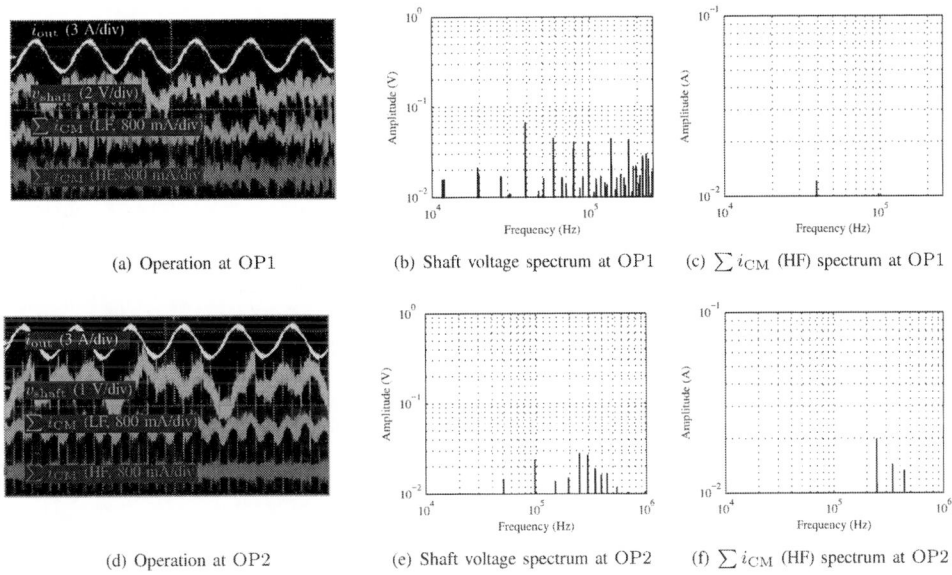

(a) Operation at OP1   (b) Shaft voltage spectrum at OP1   (c) $\sum i_{\mathrm{CM}}$ (HF) spectrum at OP1

(d) Operation at OP2   (e) Shaft voltage spectrum at OP2   (f) $\sum i_{\mathrm{CM}}$ (HF) spectrum at OP2

Fig. 11. Shaft voltage and common-mode current in the SiC dual-VSI OE Wdg. drive of Fig. 4: (a), (b), (c) at OP1, and (d), (e), (f) at OP2

extensively for Silicon based dual-VSI OE Wdg. drives in [21]. The applicability of these conclusions for a WBG based OE Wdg. drive is being investigated as a future research topic.

C. Common-mode chokes for common-mode elimination

As described in Section III-B, common-mode impedance can be used as a passive common-mode current mitigation technique. To investigate this, clamp-on ferrites were added to both the conventional and dual-VSI OE Wdg. drives. In the

case of the dual-VSI drive, the clamp-on ferrites also suppress the circulating currents. Results are presented in Fig. 14, which can be compared to Fig. 10 and 11. Significant reductions in the shaft voltage and common-mode currents are readily apparent as compared to the results without ferrites (compare Figs. 10(d) and 14(a); Figs. 11(d) and 14(b)). Most remarkably, for the open-end winding drive, the high frequency ground currents are nearly eliminated. Note that the same number of ferrite chokes were used in both experimental setups (conventional VSI and dual-VSI).

978-1-4673-9551-9/16 $31.00 © 2016 IEEE   2048

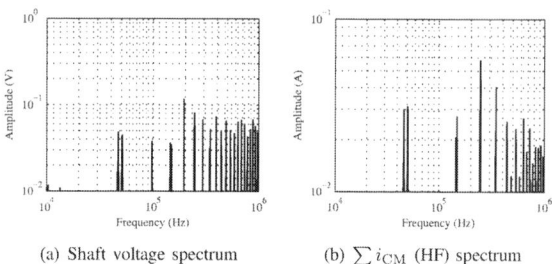

(a) Shaft voltage spectrum   (b) $\sum i_{CM}$ (HF) spectrum

Fig. 12. Shaft voltage and common-mode current in the SiC dual-VSI OE Wdg. drive when the dead-time of OP2 has been increased from 100 ns to 500 ns.

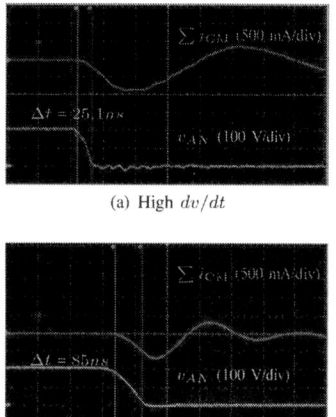

(a) High dv/dt

(b) Low dv/dt

Fig. 13. Common-mode current waveforms illustrating the impact of MOS-FET turn-on, turn-off times.

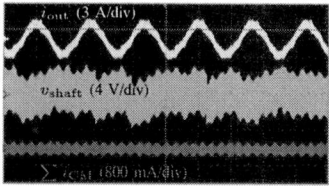

(a) Conventional with clamp-on ferrites

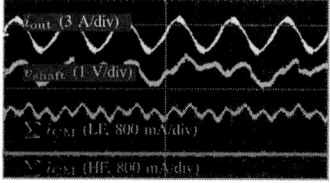

(b) OE Wdg. with clamp-on ferrites

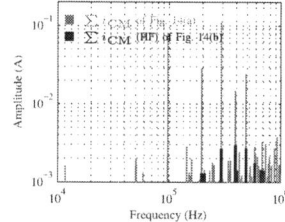

(c) Spectra of HF common-mode currents

Fig. 14. A comparison of the sum of the common-mode currents in SiC-based (a) conventional VSI and (b) dual-VSI open-end winding drive with ferrites clamped on the motor cables. Motor is operating at OP2.

VI. DISCUSSION

Two common-mode noise mitigation techniques have been explored for motor drives: dual-VSI OE Wdg. drives and clamp-on ferrite cores. Both techniques taken alone were demonstrated to provide noticeable reductions to the amount of common-mode noise observed as compared to the conventional VSI drive. When combined (dual-VSI drives with clamp-on ferrites), the common-mode ground currents were nearly eliminated. However, both techniques introduce additional hardware and associated complexities which must be considered when determining the true value of each solution.

Dual-VSI drives have twice the number of switches and, for the same semiconductor device, have twice the conduction losses and similar switching losses (if the improved voltage transfer ratio is not utilized). However, if a Silicon-based conventional VSI drive is replaced by a WBG-based dual-VSI drive, the switching losses can be reduced greatly; thus enabling PWM frequencies impractical with Silicon devices and reducing the machine losses. Furthermore, the high maximum junction temperature reduces thermal management requirements. Overall, using WBG devices can reduce the size of a dual-VSI drive (compared to a Silicon-based conventional drive) and the total system losses. Therefore, WBG (SiC)

device-based dual-VSI drives are considered to be a viable option for preventing common-mode noise.

Mitigation using ferrite cores requires additional space and may impose constraints on the type of cabling used for the drive. The authors were very intrigued by the success obtained when combining the clamp-on ferrites with a dual-VSI drive. No special design of the ferrite cores was conducted to select these components, and the near-elimination of ground currents suggests that these ferrites can be used as a field measure if additional attenuation of the common-mode currents is required. A thorough design of a common-mode filter may therefore not be necessary if a dual-VSI drive is used. Attenuation of the peak common-mode current using ferrites was also observed in the conventional VSI drive, although the overall reduction was not as dramatic as the OE Wdg. drive.

VII. CONCLUSION

Notwithstanding the significant enhancements, drives availing the advantages of WBG devices are expected to have more pronounced common-mode related issues. This paper has experimentally demonstrated increases in common-mode noise due to the higher switching frequencies and shorter switch turn-ON and turn-OFF times associated with WBG devices. However, because the effect of WBG devices is to shift the noise spectrum to higher frequencies, clamp-on ferrites can be effectively used to reduce the common-mode current and eliminate current spikes — a potentially simpler and field-executable solution compared to a dedicated common-mode filter. Alternatively, a dual-VSI OE Wdg. drive can instead be used to achieve significant reduction of the common-mode currents. It was shown that the combination of a dual-VSI drive with ferrite cores was the most effective approach, resulting in a near-elimination of common-mode currents. Due to the low losses in WBG devices, combined with high temperature ratings, a dual-VSI drive can be constructed at a small volume and weight investment toward greatly extended bearing lifetime and reduced EMI.

While both mitigation techniques come at the expense of additional components, the problems associated with increasing levels of common-mode noise may prove far more costly for many applications.

VIII. ACKNOWLEDGMENT

This research was supported by the Office of Naval Research (ONR) under grant number N00014-13-1-0511.

REFERENCES

[1] N. Mohan and T. M. Undeland, *Power electronics: converters, applications, and design.* John Wiley & Sons, 2007.

[2] E. Severson, S. Gandikota, and N. Mohan, "Practical implementation of dual purpose no voltage drives for bearingless motors," *Industry Applications, IEEE Transactions on*, vol. PP, no. 99, pp. 1–1, 2015.

[3] D. Han, Y. Li, and B. Sarlioglu, "Analysis of sic based power electronic inverters for high speed machines," in *Applied Power Electronics Conference and Exposition (APEC), 2015 IEEE*, March 2015, pp. 304–310.

[4] J. Erdman, R. Kerkman, D. Schlegel, and G. Skibinski, "Effect of pwm inverters on ac motor bearing currents and shaft voltages," *Industry Applications, IEEE Transactions on*, vol. 32, no. 2, pp. 250–259, Mar 1996.

[5] S. Chen, T. Lipo, and D. Fitzgerald, "Source of induction motor bearing currents caused by pwm inverters," *Energy Conversion, IEEE Transactions on*, vol. 11, no. 1, pp. 25–32, Mar 1996.

[6] ——, "Modeling of motor bearing currents in pwm inverter drives," *Industry Applications, IEEE Transactions on*, vol. 32, no. 6, pp. 1365–1370, Nov 1996.

[7] D. Busse, J. Erdman, R. Kerkman, D. Schlegel, and G. Skibinski, "System electrical parameters and their effects on bearing currents," *Industry Applications, IEEE Transactions on*, vol. 33, no. 2, pp. 577–584, Mar 1997.

[8] S. Chen and T. Lipo, "Circulating type motor bearing current in inverter drives," *Industry Applications Magazine, IEEE*, vol. 4, no. 1, pp. 32–38, Jan 1998.

[9] ——, "Bearing currents and shaft voltages of an induction motor under hard- and soft-switching inverter excitation," *Industry Applications, IEEE Transactions on*, vol. 34, no. 5, pp. 1042–1048, Sep 1998.

[10] F. Wang, "Motor shaft voltages and bearing currents and their reduction in multilevel medium-voltage pwm voltage-source-inverter drive applications," *Industry Applications, IEEE Transactions on*, vol. 36, no. 5, pp. 1336–1341, Sep 2000.

[11] S. Bell, T. Cookson, S. Cope, R. Epperly, A. Fischer, D. Schlegel, and G. Skibinski, "Experience with variable-frequency drives and motor bearing reliability," *Industry Applications, IEEE Transactions on*, vol. 37, no. 5, pp. 1438–1446, Sep 2001.

[12] A. Muetze and C. Sullivan, "Simplified design of common-mode chokes for reduction of motor ground currents in inverter drives," *Industry Applications, IEEE Transactions on*, vol. 47, no. 6, pp. 2570–2577, Nov 2011.

[13] M. Melfi, F. Ladonne, and D. Ankele, "Can a shaft brush be safely applied on a motor in a class i hazardous location?" *Industry Applications, IEEE Transactions on*, vol. PP, no. 99, pp. 1–1, 2015.

[14] S. Ogasawara, H. Ayano, and H. Akagi, "An active circuit for cancellation of common-mode voltage generated by a pwm inverter," *Power Electronics, IEEE Transactions on*, vol. 13, no. 5, pp. 835–841, Sep 1998.

[15] H. Zhang, A. von Jouanne, S. Dai, A. Wallace, and F. Wang, "Multilevel inverter modulation schemes to eliminate common-mode voltages," *Industry Applications, IEEE Transactions on*, vol. 36, no. 6, pp. 1645–1653, Nov 2000.

[16] M. Baiju, K. Mohapatra, R. Kanchan, and K. Gopakumar, "A dual two-level inverter scheme with common mode voltage elimination for an induction motor drive," *Power Electronics, IEEE Transactions on*, vol. 19, no. 3, pp. 794–805, May 2004.

[17] A. von Jauanne and H. Zhang, "A dual-bridge inverter approach to eliminating common-mode voltages and bearing and leakage currents," *Power Electronics, IEEE Transactions on*, vol. 14, no. 1, pp. 43–48, Jan 1999.

[18] H.-J. Kim, H.-D. Lee, and S.-K. Sul, "A new pwm strategy for common-mode voltage reduction in neutral-point-clamped inverter-fed ac motor drives," *Industry Applications, IEEE Transactions on*, vol. 37, no. 6, pp. 1840–1845, Nov 2001.

[19] A. Hava and E. Un, "A high-performance pwm algorithm for common-mode voltage reduction in three-phase voltage source inverters," *Power Electronics, IEEE Transactions on*, vol. 26, no. 7, pp. 1998–2008, July 2011.

[20] A. Sandulescu, F. Meinguet, X. Kestelyn, E. Semail, and A. Bruyere, "Flux-weakening operation of open-end winding drive integrating a cost-effective high-power charger," *Electrical Systems in Transportation, IET*, vol. 3, no. 1, pp. 10–21, March 2013.

[21] A. Somani, R. Gupta, K. Mohapatra, and N. Mohan, "On the causes of circulating currents in pwm drives with open-end winding ac machines," *Industrial Electronics, IEEE Transactions on*, vol. 60, no. 9, pp. 3670–3678, Sept 2013.

[22] S. Tewari, R. Gupta, A. Somani, and N. Mohan, "A new sinusoidal input-output three-phase full-bridge direct power converter," in *Industrial Electronics Society, IECON 2013 - 39th Annual Conference of the IEEE*, Nov 2013, pp. 4824–4830.

[23] M. Di Piazza, G. Tine, and G. Vitale, "An improved active common-mode voltage compensation device for induction motor drives," *Industrial Electronics, IEEE Transactions on*, vol. 55, no. 4, pp. 1823–1834, April 2008.

[24] V. Somasekhar, S. Srinivas, and K. Kumar, "Effect of zero-vector placement in a dual-inverter fed open-end winding induction-motor drive with a decoupled space-vector pwm strategy," *Industrial Electronics, IEEE Transactions on*, vol. 55, no. 6, pp. 2497–2505, June 2008.

[25] P. Pairodamonchai, S. Suwankawin, and S. Sangwongwanich, "Design and implementation of a hybrid output emi filter for high-frequency common-mode voltage compensation in pwm inverters," *Industry Applications, IEEE Transactions on*, vol. 45, no. 5, pp. 1647–1659, Sept 2009.

[26] B. Zhu, Y. Jia, U. Prasanna, K. Rajashekara, and H. Kubo, "An input switched multilevel inverter for open-end winding induction motor drive," in *Power Electronics Conference (IPEC-Hiroshima 2014 - ECCE-ASIA), 2014 International*, May 2014, pp. 1594–1600.

[27] J. Kalaiselvi and S. Srinivas, "Bearing currents and shaft voltage reduction in dual-inverter-fed open-end winding induction motor with reduced cmv pwm methods," *Industrial Electronics, IEEE Transactions on*, vol. 62, no. 1, pp. 144–152, Jan 2015.

[28] D. Busse, J. Erdman, R. Kerkman, D. Schlegel, and G. Skibinski, "The effects of pwm voltage source inverters on the mechanical performance of rolling bearings," *Industry Applications, IEEE Transactions on*, vol. 33, no. 2, pp. 567–576, Mar 1997.

[29] R. Baranwal, K. Basu, and N. Mohan, "Carrier-based implementation of svpwm for dual two-level vsi and dual matrix converter with zero common-mode voltage," *Power Electronics, IEEE Transactions on*, vol. 30, no. 3, pp. 1471–1487, March 2015.

[30] ——, "Dual two level inverter carrier svpwm with zero common mode voltage," in *Power Electronics, Drives and Energy Systems (PEDES), 2012 IEEE International Conference on*, Dec 2012, pp. 1–6.

[31] L. Ravi, E. Severson, S. Tewari, and N. Mohan, "Circuit-level characterization and loss modeling of sic-based power electronic converters," in *Industrial Electronics Society, IECON 2014 - 40th Annual Conference of the IEEE*, Oct 2014, pp. 1291–1297.

[32] D. G. Holmes and T. A. Lipo, *Pulse width modulation for power converters: principles and practice.* John Wiley & Sons, 2003, vol. 18.

A High-efficient Driving Isolated Drive-by-Microwave Half-Bridge Gate Driver for a GaN Inverter

Shuichi Nagai, Yasufumi Kawai, Osamu Tabata, Songbaek Choe, Noboru Negoro, and Tetsuzo Ueda
Green Autonomous Technology Development Center
Automotive & Industrial Systems Company, Panasonic Corporation
Moriguchi, Osaka 570-8501, Japan
Email: nagai.shuichi@jp.panasonic.com

Abstract—A compact isolated Drive-by-Microwave (DBM) half-bridge gate driver is newly developed, which can drive GaN-GITs with its constant and low power consumption of 0.9 W even a high switching frequency up to 1.0 MHz due to the gate power time sharing by using the 2.4GHz microwave wireless power transmission. The fabricated GaN isolated DBM half-bridge gate driver provides enough output gate power for GaN-GIT's driving under 140 ℃ (Ta). Moreover, the fabricated isolated power source free GaN-GIT inverter module with the isolated DBM half-bridge gate drivers successfully demonstrated a 3-phase motor drive with high power conversion efficiency by covering from its low output.

Keywords—isolated gate driver; GaN; inverter; half-bridge

I. INTRODUCTION

A power system using emerging GaN power devices [1,2] is very promising for a compact system with small inductors and a small heat sink because these operate at high frequency under high temperature. a high speed level shift, very fast propagation delay, and large common mode suppression are required in order to drive a GaN power device with a high speed,. Thus, an isolated gate driver is best suitable for such a high speed power device because it satisfies these demands. Herein, since the conventional isolated gate driver with a photo-coupler has the problem of its short life time by high temperature operation, several digital isolators without an optical device have been developed [3-6]. In particular, the DBM (Drive-by-Microwave) isolated gate driver [5-7] is very attractive because it doesn't need an isolated power source by providing isolated gate signal and isolated gate power in together though wireless power transmission. However, the power consumption of conventional isolated gate drivers increases at higher frequency operation in accordance with the turn-on number of times, without exception. Therefore, there is large demand for such a high speed driver for GaN power device, which keeps its low power consumption even in high switching operation and can operate at high temperature.

In this paper, we describe the new GaN isolated DBM half-bridge gate driver with the self-turn-off signal output RF rectifiers, which has low power consumption regardless its operation frequency due to gate power time-sharing. A 3-phase

This work is partially supported by the New Energy and Industrial Technology Development Organization (NEDO), Japan.

motor drive by the fabricated GaN-GIT inverter module with the fabricated isolated DBM half-bridge gate driver was performed.

II. AN ISOLATED DBM HALF-BRIDGE GATE DRIVER

A. DBM Gate Drive System

Fig. 1 shows the system block diagram of the new isolated DBM half-bridge gate driver, which consists of a DBM transmitter, isolation electro-magnetic resonant couplers, and a DBM receiver. This half bridge driver can control two GaN power devices for a half bridge circuit by two PWM signal inputs. The PWM signals are converted to the 2.4 GHz enveloped modulated signals by DBM transmitter, which has a 2.4GHz Oscillator and a double balanced RF switching mixer. After going through the half-wing isolated butterfly couplers

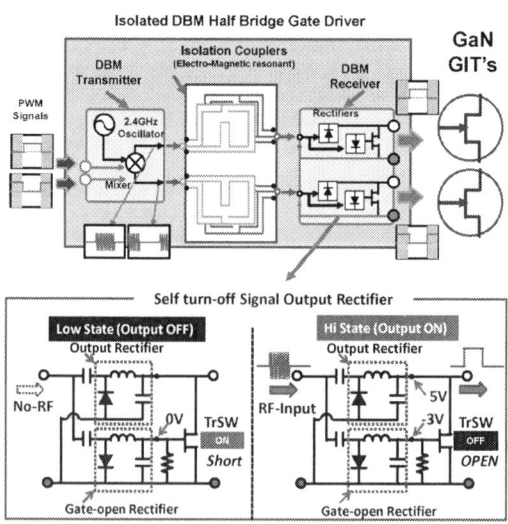

Fig.1 System block diagram of the isolated DBM half-bridge gate driver with the self-turn-off signal output rectifier. Circuit operation of the self-turn-off signal output rectifier in the bottom.

for DC isolation, the 2.4GHz modulated signals are converted to isolated gate signals by the DBM receiver with RF rectifiers in order to drive the GaN power devices. Although the conventional DBM gate driver internally needs two microwave signals for turn-on and turn-off of one power device, the new DBM half-bridge gate driver requires only one microwave signal for one power device thanks to the newly installed self-turn-off signal-output rectifier in the DBM receiver.

The circuit operations of the self-turn-off signal output rectifier in the DBM receiver are illustrated in fig.1. At first, the output port of the self-turn-off signal-output rectifier is shorted for Low-state in the case of no-RF signal input, because the TrSW is a D-mode (normally-off) transistor and the gate voltage of TrSW at the gate-open rectifier is 0V. This very low impedance of the gate port at off state is very useful to prevent a gate glitch by external noise. On the other hand, when the RF signal is input in the self-turn-off signal-output rectifier, the output gate power is created by the output rectifier for Hi-state. At the same time, the output port is opened by the off state of the TrSW because the input RF signal generates the negative gate voltage for the TrSW off by the gate-open rectifier. Due to this new rectifier circuit, the gate power time-sharing for the Hi and Low side drive by one 2.4GHz oscillator is realized for an efficient gate drive.

B. Fabricated DBM Half-Bridge Gate Driver

The internal photo in the packaged DBM half-bridge gate driver is shown in fig.2. Since this DBM half-bridge gate driver

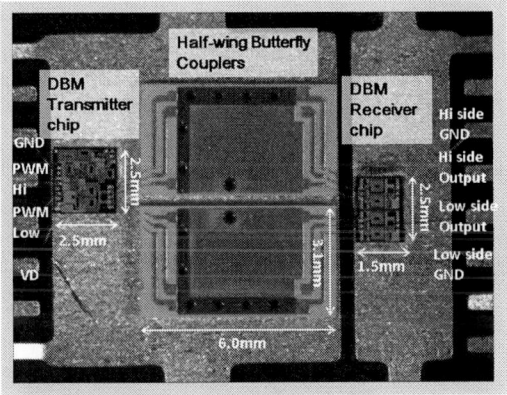

Fig.2 Internal photo in the packaged DBM half-bridge driver.

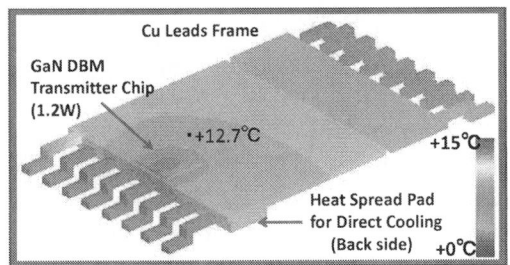

Fig.3 Thermal simulation of the DBM half-bridge gate driver package.

can provide an isolated gate power and gate signal by mixing, the special thermal management in the package is employed. This package has the heat spread pad underneath the DBM transmitter chip, which is exposed for direct cooling. From the thermal simulation of the package in fig.3, the rising temperature by the 2.5W heat can be suppressed at +13.0℃ increasing thanks to the internal heat spread pad.

The DBM transmitter and receiver chip were fabricated with the D-mode GaN-HFETs on Si substrates. The fabricated single transistor with gate length for 0.8μm exhibits a Ft of 15GHz and Fmax of 55GHz, respectively. The output power of 21.0dBm at 2.4GHz is obtained from the fabricated GaN DBM transmitter chip. The output voltage of the fabricated GaN DBM receiver chip is shown in fig.4 as a function of 2.4GHz signal input power. The output voltage from the DBM receiver has a threshold at 16dBm RF input power because of the GaN transistor of the TrSW.

The isolated half-wing butterfly coupler [7] was fabricated with a printed circuits board. The measured isolation voltage of the fabricated isolation coupler was over 5.0 kV because of its 0.28μm layer thickness. The small signal S-parameters of the fabricated half-wing butterfly coupler are shown in fig.5. As seen from fig.5, the insertion loss (S21) of the fabricated coupler is very low as 1.0dB with very wide bandwidth of

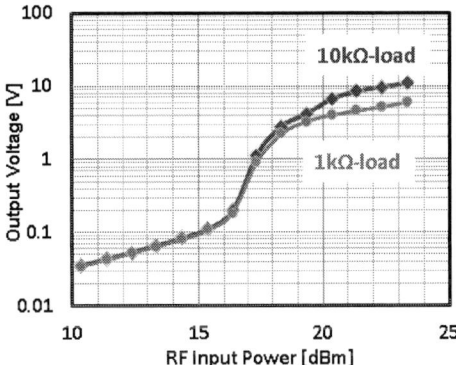

Fig.4 Output voltage of the fabricated GaN DBM receiver chip as a function of 2.4GHz input power.

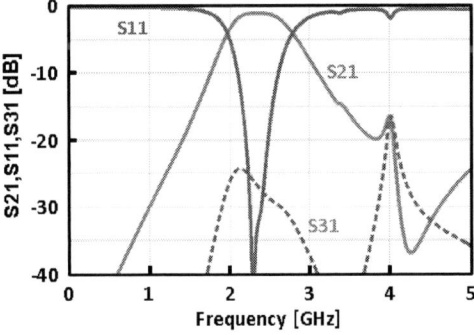

Fig.5 Small signal S-parameters of the fabricated half-wing butterfly couplers.

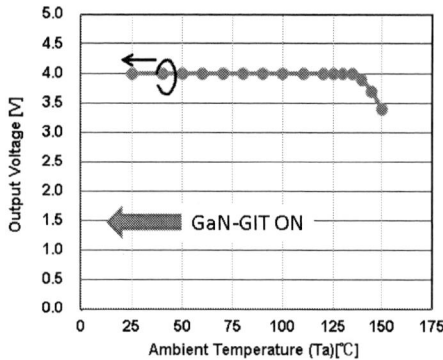

Fig.6 Temperature dependency of the output voltage from the fabricated DBM half bridge gate driver

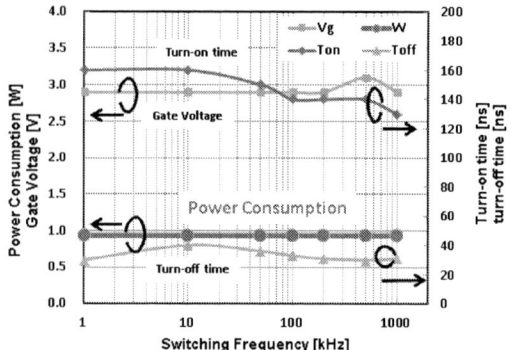

Fig.7 Switching frequency dependency of the fabricated DBM half bridge gate driver.

1.0GHz because this coupler is based on the microwave electro-magnetic resonant coupling [8]. And, it also shows that the cross talk (S31) between two ports is low as a -25dB.

C. Chircactistics of the DBM Half-bridge Gate Driver

The temperature dependency of the output voltage from the fabricated DBM half-bridge gate driver is shown in fig.6. This driver constantly provides about 10mA gate current while Ga-GIT is ON-state. It is the best for a gate driver for a GaN-GIT because the GaN-GIT need gate current to keep its ON due to gate current injection device. In order to prevent the gate port damage, the output voltage from the driver is regulated to 4.0V by a crippling diode at the output port. This fabricated gate driver provides enough output gate voltage for GaN-GIT ON up to 150℃ without an additional heat-sink. Consequently, it is noted that this driver can be closed to a hot power device in order to drive it with a high speed by eliminating the parasitic inductance between the driver and the power device.

The 20V switching with 100Ω resistance load by a 15A GaN-GIT [9] (PGA26C09DV) that is driven by the fabricated DBM half-bridge gate driver is performed as shown in fig7. The very fast turn on time was achieved because the ON resistance of the TrSW at the DBM receiver chip is very small as around 1.5Ω. However, the turn on time is a little slow due to the low gate peak current from the driver. Even at the 1.0MHz switching, the power consumption of the fabricated DBM half-bridge gate driver is constant as 0.9 W (Vdd:12V, Idd:75mA). Moreover, the propagation delay of this gate driver is very fast as 7.2ns, although it includes the input/output lines delays on an evaluation board. These results imply that this isolated DBM half-bridge gate driver is very suitable for a high frequency and a high temperature GaN power system with a small heat sink.

III. A GaN-GIT INVERTER MODULE WITH THE DBM HALF-BRIDGE GATE DRIVERS

As shown in fig.8, we fabricated a compact 3-phase GaN-GITs inverter module with the isolated DBM half-bridge gate drivers. This does not need bulky and large isolated voltage sources, any more. The package size of the DBM half-bridge

gate driver is compact as a 1.0cm x 1.4cm. The total power consumption for the gate drivers is only 2.7W. Also, because this driver is specified for a GaN power device, such as gate resistors or speed-up capacitors are not necessarily. The breakdown voltage of the 15A GaN-GIT is over 600V, then this invertor module hands about 5.0kW power.

The power conversion efficiency of the fabricated GaN inverter module is shown in fig.9, when it drives a 5.0kW-class motor with a 10 kHz carrier frequency. The fabricated GaN inverter module demonstrated very high conversion efficiency over 94% even the low output power of 100W as compared with that of the 30A-IGBT [10](STGP30NC60W) inverter module with photo-coupler base isolated gate drivers. Because the GaN-GIT has no offset voltage (such as collector emitter threshold voltage in an IGBT), the GaN-GIT exhibits a low ON resistance at the low applied voltage (VDS). The conversion efficiency at large output from the fabricated GaN inverter might be limited by the Ron of the GaN-GIT (RDSON: 71mΩ).

IV. CONCLUSIONS

We developed the compact isolated Drive-by-Microwave half-bridge gate driver with a gate power time-sharing for a

Fig.8 Isolated voltage source free GaN-GIT inverter module with the DBM half bridge gate drivers

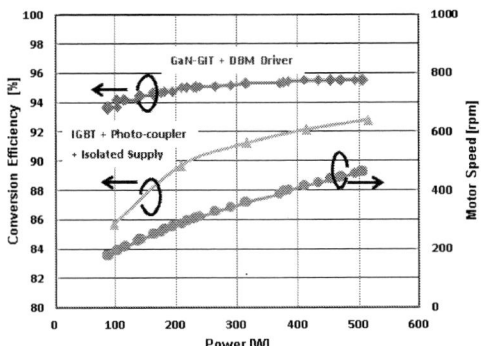

Fig.9 Power conversion efficiency at a motor drive by the fabricated GaN-GIT inverter module.

GaN power device. This does not need isolated voltage sources because this driver provide isolated gate power and gate signal all together by 2.4GHz microwave wireless power transmission. The newly installed self-turn-off signal output rectifier in the DBM receiver contributes the constant power consumption as 0.9 W even a high switching frequency up to 1.0 MHz or more. The fabricated GaN isolated DBM half-bridge gate driver provides enough output gate power for the GaN-GIT's diving up to under 140 ℃ (Ta). Moreover, the fabricated isolated power source free GaN-GIT inverter module with the isolated DBM half-bridge gate drivers successfully demonstrated a 3-phase motor drive with high power conversion efficiency over 94% by covering from its low output.

ACKNOWLEDGMENT

The authors would thank Mr. Y. Uemoto, Dr. Manabu Yanagihara, Mr. Masahiro Hikita at Panasonic Semiconductor Solutions Co., Ltd. This work is partially supported by the New Energy and Industrial Technology Development Organization (NEDO), Japan.

REFERENCES

[1] Y. Uemoto, M. Hikita, H. Ueno, H. Matsuo, H. Ishida, M. Yanagihara, T. Ueda, T. Tanaka, D. Ueda, "Gate Injection Transistor (GIT) – A Normally-Off AlGaN/GaN Power Transistor Using Conductivity Modulation," IEEE Transaction on Electron Device, vol. 54, no. 12, pp. 3393-3399, 2007.

[2] T. Ueda, M. Ishida, T.Tanaka, and D.Ueda, "GaN transistors on Si for switching and high-frequency applications," Japanese Journal of Applied Physics, Vol.53, 100214, 2014

[3] S. Ma, T. Zhao, and B. Chen, "4A Isolated Half-Bridge Gate Driver with 4.5V to 18V Output Drive Voltage," in IEEE Applied Power Electronics Conference and Exposition, 2014, pp. 1490-1493.

[4] H. Shinoda and T. Terada, "Insulated Signal Transmission System using Planar Resonant Coupling Technology for High Voltage IGBT Gate Driver," in IEEE Energy Conversion Congress and Exposition, pp.265-270.

[5] S. Nagai, Y.Kawai, O.Tabata, H.Fujiwara, N.Otsuka, D.Ueda, N.Negoro,and M.Ishida," A Compact Drive-by-Microwave Gate Driver with Coupler Integrated in a Package" in IEEE Applied Power Electronics Conference and Exposition, 2014, pp. 1464-1464.

[6] Y. Kawai, S.Nagai, O.Tabata, H.Fujiwara, N.Negoro, H.Ueno, M.Ishida, N.Otsuka, "An Isolated DC Power Supply Free Compact GaN Inverter Module," in IEEE International Conference on Power Electronics and Drive Systems, 2015, pp. 84-88.

[7] S.Nagai, Y.Kawai, O. Tabata, H. Fujiwara, N. Negoro, M. Ishida, N. Otsuka, "A Drive-by-Microwave isolated gate driver with gate current charge for IGBTs," Power Electronics and Applications (EPE'14-ECCE Europe), pp.1-6, Aug 2014.

[8] Y. Okuyama, J. Ao, I. Awai, and Y. Ohno, "Wireless Inter-Chip Signal Transmission by Electromagnetic Coupling of Open-Ring Resonators" Japanese Journal of Applied Physics, 48, 04C025, 2009.

[9] http://www.semicon.panasonic.co.jp/en/products/powerics/ganpower/.

[10] http://www.st.com/web/jp/catalog/sense_power/FM100/CL826/SC68/PF 122657.

978-1-4673-9551-9/16 $31.00 © 2016 IEEE

Sensing Gallium Nitride HEMT Junction Temperature Using Gate Drive Output Transient Properties

He Niu
University of Wisconsin-Madison, WEMPEC
Madison, WI, 53705
hniu@wisc.edu

Robert D. Lorenz
University of Wisconsin-Madison, WEMPEC
Madison, WI, 53705
rdlorenz@wisc.edu

Abstract—**The gate drive output transient properties of conventional power semiconductors (Silicon MOSFET, Silicon-Carbide MOSFET, and Silicon IGBT) have been studied for online junction temperature (T_j) sensing. This method utilizes the semiconductor intrinsic T_j-dependent characteristics and the gate drive output dynamics. In this paper, the method is applied to Gallium Nitride (GaN) high electron mobility transistors (HEMTs). To demonstrate the GaN HEMT T_j sensing, two passivated die HEMTs are implemented in a half-bridge and driven by two integrated push-pull type gate drives. The gate drive turn-on current transient is used for GaN HEMT T_j sensing. The "gate drive-HEMT" switching properties are modeled to explain the T_j-dependent gate drive output dynamics. Experimental results are compared with LTSpice simulations.**

Keywords—*Gallium Nitride HEMT; junction temperature; online sensing; gate drive*

I. INTRODUCTION

A. Conventional T_j Sensing Methods

Switching power semiconductor real-time T_j sensing is essential for power converter switching quality monitoring, package strain balancing, and lifetime optimization [1]. With high demands on sensing bandwidth and sensing accuracy, contact-based temperature sensors (such as direct bonded copper-mounted temperature detectors) are not feasible for GaN HEMTs device with passivated die packages.

Modern power semiconductor fabrication allows the "diode-on-die technology" to be used for T_j estimation [2]. By implementing a small p-n junction on the die of IGBT or MOSFET, the temperature dependence of the p-n junction voltage drop can be used for MOSFET/IGBT T_j estimation. However, applying the "diode-on-die technology" to GaN HEMTs has the following limitations. Firstly, considering the limited number of stringed diodes used for T_j sensing, the sensing consistency control is challenging. As a result, the performance of the sensors fabricated in the same batch has

variance that cannot be ignored [3]. Secondly, the on-chip p-n junction T_j sensor is usually fabricated on a small area on the power semiconductor chip (as shown in figure 8 in [4] and figure 14 in [5]). The temperature on a small area of the chip is not an accurate estimation of the overall junction temperature of the semiconductor chip.

B. T_j Sensing Using Device Intrinsic Properties

Power MOSFET on-state resistance $R_{DS\text{-}on}$ is one of the most popular device intrinsic parameters that are often used for online T_j sensing [6]. Power MOSFET $R_{DS\text{-}on}$ can be digitally calculated online with (1.1) by measuring the MOSFET on-state voltage $V_{DS\text{-}on}$ and on-state current $I_{DS\text{-}on}$ [7]. The online calculated $R_{DS\text{-}on}$ can be correlated to T_j based on the power MOSFET "$R_{DS\text{-}on} - T_j$" relationship. The value of $R_{DS\text{-}on}$ can vary from a few milli-ohms to a few hundred milli-ohms, according to different GaN HEMTs from EPC Co. [8]. Thus, the online $R_{DS\text{-}on}$ measurement based on (1.1) can require very high resolution voltage sensing with galvanic isolation.

$$R_{DS\text{-}on} = \frac{V_{DS\text{-}on}}{I_{DS\text{-}on}} \qquad (1.1)$$

Si MOSFET switching behavior in a converter configuration can be described as an under-damped LCR circuit under step excitation (also referred to as a "voltage source-$R_{DS\text{-}on}$-L-C" resonant model). The time constant of the load voltage ringing decay (or the DC bus current ringing decay, in some topologies) can be used to estimate $R_{DS\text{-}on}$ and hereby the corresponding T_j, [9]. Based on the same methodology in [9], [10] utilizes the "gate drive-$R_{DS\text{-}on}$-L-C" resonant model to estimate Si MOSFET T_j. However, high speed online frequency domain analysis and high speed analog to digital conversion (ADC) are required for both [9] and [10].

C. Research Opportunities

[11], [12], and [13] proposed a T_j sensing method based on the dynamic interaction between a gate drive and a switching power semiconductor. The method was applied to different types of switching power semiconductors (Si

978-1-4673-9551-9/16 $31.00 © 2016 IEEE

MOSFET, SiC MOSFET, and Si IGBT) and different types of gate drives (push-pull gate drive, current mirror gate drive, and inductor based gate drive).

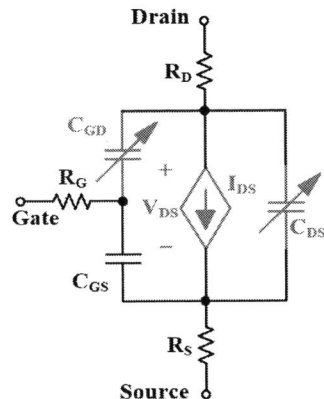

Fig. 1 GaN T$_j$ dependent model from [14]

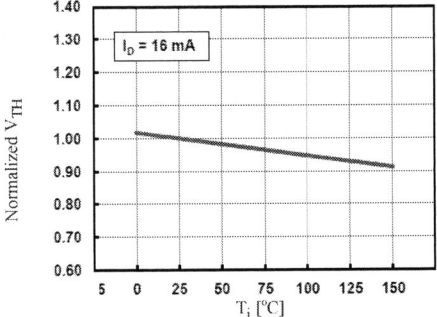

Fig. 2 EPC2030 "V$_{TH}$ – T$_j$" dependency, normalized

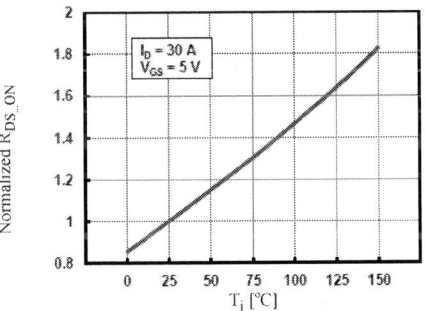

Fig. 3 EPC2030 "R$_{DS\_ON}$ – T$_j$" dependency, normalized

This research extends the aforementioned method (based on the dynamic interaction between gate drive and switching power semiconductor) to GaN HEMTs online T$_j$ sensing. This paper presents the T$_j$ dependent modeling of GaN HEMTs, a push-push based integrated HEMT gate drive, and a half-bridge test circuit system. The gate drive turn-on output current (i$_{G\_ON}$) is used for online T$_j$ sensing. The "i$_{G\_ON}$ - T$_j$" relationship is experimentally evaluated, and compared with LTSpice simulation. A peak detection method is proposed to

extract the T$_j$ of GaN HEMTs online. The implementation details and the control of the peak detection are investigated.

II. GAN HEMT TEMPERATURE DEPENDENT MODELING

A GaN HEMT temperature dependent model is developed in [14] as shown in Fig. 1, where the temperature-dependent components are marked in red. A voltage-dependent current source I$_{DS}$ is used to characterize the device's forward and reverse conduction properties. Two voltage-dependent intrinsic capacitances, C$_{GD}$ and C$_{DS}$, as well as a voltage-independent intrinsic capacitance, C$_{GS}$, are used to model the device's switching transient properties. Three constant resistors, R$_G$, R$_D$, and R$_S$, are used to represent the device's intrinsic parasitics. In this research, EPC2030 GaN HEMT is used for evaluation, and the temperature dependency of the device gate-source threshold voltage V$_{TH}$ (as shown in Fig. 2) is characterized by C$_{GD}$, and the temperature dependency of the on-state resistance R$_{DS-on}$ (as shown in Fig. 3) is characterized by I$_{DS}$. Simulation and experimental results used in [14] indicate that load dependency, gate voltage dependency, and T$_j$ dependency are properly characterized in the proposed GaN HEMT model.

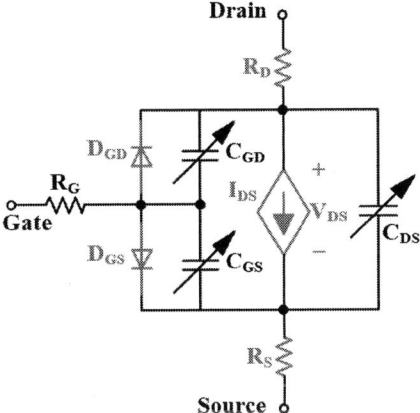

Fig. 4 GaN T$_j$ dependent model from [15]

The simulation model for some GaN HEMT products from EPC Co. are not fully temperature characterized. In the LTSpice sub-circuit model for EPC2010 [15], V$_{TH}$ is set to a constant value of 2.2V, and R$_G$ is set to a constant value of 0.6Ω. Based on the LTSpice simulation model, an equivalent circuit model can be created as shown in Fig. 4, where the temperature dependent components are marked in red. In the HEMT model in Fig. 4, two temperature-sensitive diodes, D$_{GD}$ and D$_{GS}$, are added to characterize the gate-source over-voltage properties.

To introduce the temperature dependency of V$_{TH}$ (as discussed in [14]) and R$_G$ (as discussed in [16]) to the LTSpice simulation model, V$_{TH}$ and R$_G$ are re-characterized in (2.1) and (2.2), respectively.

$$\text{Normalized } V_{TH} = K_{T1}\,(T_j - 300K) + V_{TH0} \quad (2.1)$$

$$R_G = K_{T2}\,(T_j - 300K) + R_{G0} \quad (2.2)$$

978-1-4673-9551-9/16 $31.00 © 2016 IEEE

The intrinsic capacitance of GaN HEMTs are well understood as bias dependent [17]. However, the temperature dependent properties of the GaN HEMTs intrinsic capacitance [18] are not always fully characterized by the device model. In this research, the temperature dependency of the device intrinsic capacitance are characterized by (2.3) and (2.4).

$$C_{GS} = K_{T3} (T_j - 300K) + C_{GS0} \qquad (2.3)$$

$$C_{GD} = K_{T4} (T_j - 300K) + C_{GD0} \qquad (2.4)$$

Based on the "V_{TH} - T_j" dependency and the "semiconductor resistance – T_j" dependency of EPC2030 GaN HEMT [19], the device switching transient can be fully characterized with $K_{T1} = 6.7 \times 10^{-3}K^{-1}$, $V_{TH0} = 1$, $K_{T2} = 3.7 \times 10^{-3}\Omega/K$, $R_{G0} = 0.4\Omega$. The LTSpice model in [19] covers (2.1), but (2.2), (2.3), and (2.4) are not considered. A modified model with (2.1) to (2.4) included is shown in Fig. 5, where the T_j dependencies of V_{TH} and R_G are characterized by the components drawn in green.

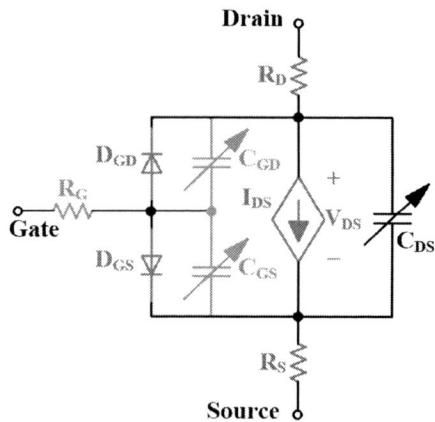

Fig. 5 Modified GaN HEMT T_j dependent model

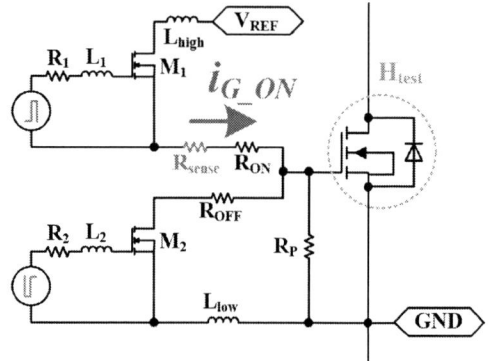

Fig. 6 Push-pull type gate drive equivalent circuit

III. Gate Drive and Evaluation Circuit Modeling

In this research, classic push-pull type gate drives are used as shown in Fig. 6 with parasitics and GaN HEMT. The

pull-up transistor M_1 and the push-down transistor M_2 are integrated in the Texas Instrument LM5114 driver. R_{sense}, R_{ON}, R_{OFF}, and R_P are implemented on the printed circuit board (PCB). The values of off-chip resistors are listed in Table I. R_{sense} is used for the gate drive turn-on transient output current sensing. To turn on the HEMTs, V_{REF} is regulated to 5V (referred to GND).

TABLE I. Values of Off-chip Resistors

Parameter	Value [Ω]
R_{sense}	360
R_{ON}	36
R_{OFF}	1.5
R_P	1800

To demonstrate the GaN HEMT T_j sensing, a half-bridge inverter is designed as shown in Fig. 7, including the DC bus voltage source, two GaN HEMTs, two integrated HEMT gate drives, and a resistive load. The two HEMTs will be switched using the same duty ratios, but with 180° phase difference. The switching time line for the half bridge inverter is illustrated in Fig. 8. By synchronously adjusting the high/low side duty ratios (D_{high} and D_{low} as defined in (3.1) and (3.2)) of the two HEMTs, T_j can be manipulated while other circuit conditions remain the same.

$$D_{high} = \frac{t_1}{2T} \qquad (3.1)$$

$$D_{low} = \frac{t_2}{2T} \qquad (3.2)$$

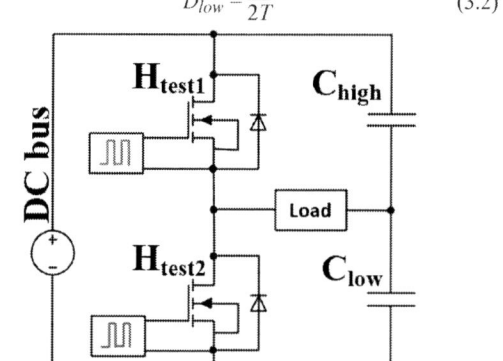

Fig. 7 Half bridge inverter

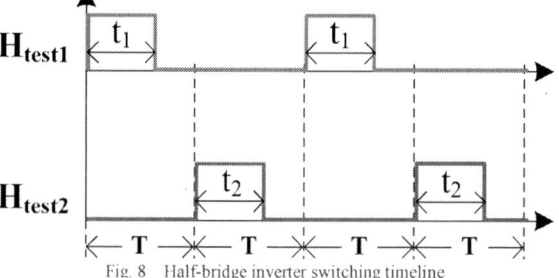

Fig. 8 Half-bridge inverter switching timeline

IV. SIMULATION

The simulation parameters are selected based on the prototype test setup as discussed in section V. The semiconductor models and parameters are based on the sub-circuit and datasheet from the semiconductor manufacturer as discussed in section II and III.

By setting T_j to 31°C, 41°C, and 51°C, the system dynamic model is simulated under different junction temperatures. The simulation results for the gate drive turn-on current, i_{G_ON}, are shown in Fig. 9. The comparison between simulation results and experimental results will be included in section V.

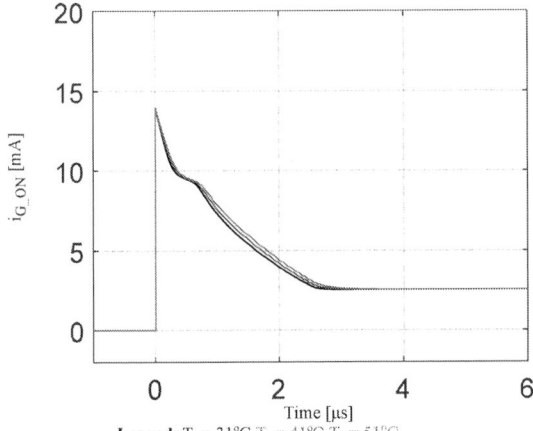

Legend: $T_j = 31°C$ $T_j = 41°C$ $T_j = 51°C$
Fig. 9 Simulated i_{G_ON} waveforms under different T_j

V. EXPERIMENTAL RESULTS

The half-bridge inverter prototype with two EPC2030 GaN HEMTs is shown in Fig. 10. A zoomed-in picture for the GaN HEMTs are shown in Fig. 11. The "Target HEMT" in Fig. 11 is painted with black coating to realize constant surface emissivity. The DC bus voltage is set to 38V (EPC2030 GaN HEMT is rated at 40V), a 3Ω resistive load is applied to the inverter, and the PWM frequency is 3.6kHz.

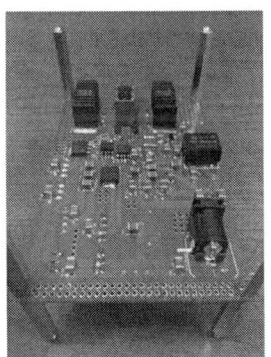
top side

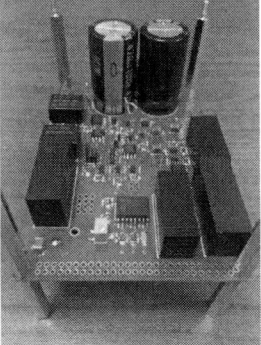
bottom side

Fig. 10 Half-bridge inverter prototype

Fig. 11 Half-bridge inverter zoomed-in picture with GaN HEMTs

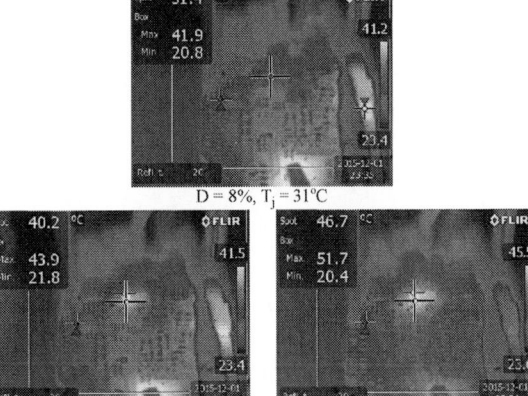

D = 8%, T_j = 31°C

D = 23%, T_j = 41°C   D = 38%, T_j = 51°C

Fig. 12 Thermal images captured by Flir camera under different D

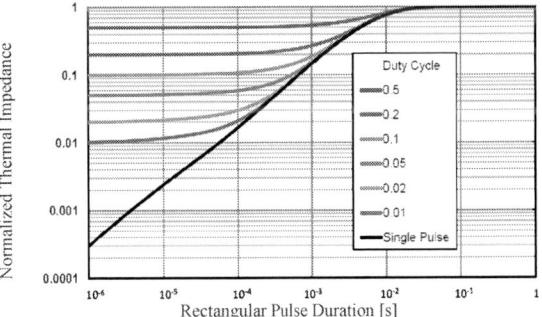

Fig. 13 EPC2030 device junction to case thermal impedance

In this research, D_{high} and D_{low} are set to the same value D ($D = D_{high} = D_{low}$), with 180° phase delay between D_{high} and D_{low}. By setting D equal to 8%, 23%, and 38%, the steady-state case temperature of EPC2030 GaN HEMT is measured to be 31°C, 41°C, and 51°C, respectively. Fig. 13 illustrates the thermal images captured by the Flir thermal camera under different test conditions. Based on the device package thermal impedance as described in Fig. 13, [19], the T_j values are calibrated to be 31°C, 41°C, and 51°C, respectively (The thermal impedance between the junction and the case is modest).

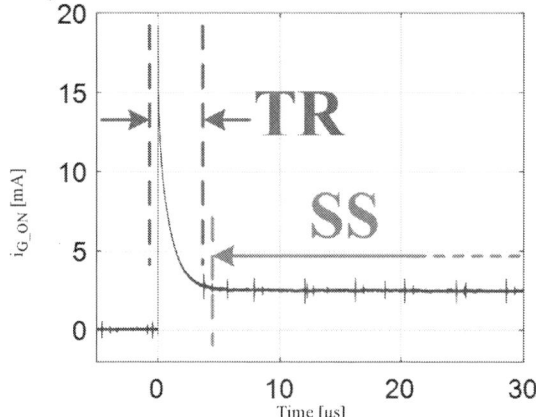

Fig. 14 Experimentally measured i_{G_ON} waveform, overview

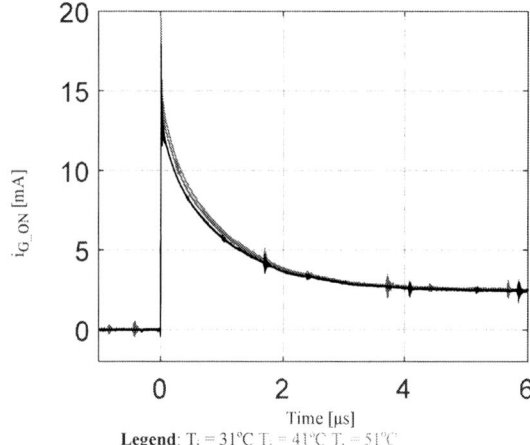

Legend: $T_j = 31°C$ $T_j = 41°C$ $T_j = 51°C$

Fig. 15 Experimentally measured i_{G_ON} waveforms under different T_j

The gate drive turn-on current, i_{G_ON} is calculated by measuring the voltage drop across R_{sense}. An overview waveform of i_{G_ON} is shown in Fig. 14, where the first part of the waveform "TR" is the GaN HEMT turn-on transient gate current, and the second transient "SS" is the steady state gate drive output current bypassed by R_p (as shown in Fig. 6). A zoom-in plot for "TR" under different T_j (31°C, 41°C, and 51°C) is shown in Fig. 15. Based on the simulation (Fig. 9) and experimental results (Fig. 15), the GaN HEMT turn-on transient gate current is sensitive to T_j. With fixed DC bus voltage, higher T_j will result in larger gate charge (Q_G). Specifically, in the experimental results, the peak value of i_{G_ON} is positively correlated to T_j. Thus, a peak detection

method is feasible for the online extraction of T_j.

The T_j dependency of the GaN HEMT gate switching properties has been investigated in [18] [20] [21]. Descriptions for the related symbols are summarized in Table II.

TABLE II. LIST OF SYMBLS IN (5.1) TO (5.4)

symbol	Description
n	Charge density
D	Density of states
V_T	Thermal voltage
E_f	Position of Fermi level
E_0	Position of the first energy level
E_1	Position of the second energy level
V_g	Gate potential (gate voltage)
V	Local quasi-Fermi potential
q	Electron charge
d	Layer thickness
γ_0	Parameters determined from experiment
ε	Material permittivity
W	Channel width
L	Channel length
n_D	Drain charge carrier density
n_S	Source charge carrier density

In the GaN HEMT structure, the charge density accumulated in the potential well can be calculated by (5.1).

$$n = DV_T\left[\ln\left(e^{(E_f - E_0)/V_H} + 1\right) + \ln\left(e^{(E_f - E_1)/V_H} + 1\right)\right]$$
(5.1)

A simple charge control dynamic model can be derived from (5.1), as shown in (5.2). This model is valid in all operation regions. It effectively relates the applied voltage and the charge carrier concentration.

$$V_{g0} - V = \frac{qdn}{\varepsilon} + \gamma_0\, n^{2/3} + V_T ln\left(\frac{n}{DV_T}\right)$$
(5.2)

The gate charge can be calculated by integrating the charge density along the channel over the gate area, as described in (5.3).

$$Q_G = W\int_0^L q\,n(x)\,dx$$
(5.3)

By substituting (5.1) and (5.2) into (5.3), and simplifying the expression, the gate charge can be calculated with (5.4).

$$Q_G = W\,L\,q\,\frac{\frac{qd}{3\varepsilon}(n_D^3 - n_S^3) + \frac{1}{4}\gamma_0(n_D^{8/3} - n_S^{8/3}) + \frac{1}{2}V_T(n_D^2 - n_S^2)}{\frac{qd}{2\varepsilon}(n_D^2 - n_S^2) + \frac{2}{5}\gamma_0(n_D^{5/3} - n_S^{5/3}) + V_T(n_D - n_S)}$$
(5.4)

In (5.4), V_T, n_D, and n_S are T_j sensitive. As a result, the gate charge value of the device is a function of T_j. In this research, the gate charge can be regarded as the integration of the transient gate current waveform "TR" in Fig. 14. Thus, the

978-1-4673-9551-9/16 $31.00 © 2016 IEEE

"i_{G_ON} – T_j" relationship can be explained as the T_j dependencies of V_T, n_D, and n_S.

In validation, n_D and n_S can be experimentally measured, and V_{TH} can be calculated by (5.5).

$$V_T = \frac{kT}{q} \qquad (5.5)$$

VI. EXTRACTING THE TEMPERATURE DEPENDENCY ONLINE

As can be seen from the experimentally measured i_{G_ON} waveform in Fig. 15, the peak value of the GaN HEMT turn-on transient gate current is explicitly T_j sensitive. In this research, an analog peak detector is implemented to measure the peak value of i_{G_ON}. After the peak detection period, the measured i_{G_ON} peak will be latched for the analog to digital conversion (ADC).

The implementation of the analog peak detector is shown in Fig. 16, where S_1 is used to activate peak detection, and S_2 is used to reset the measured peak value. The time table for the peak detector circuit is shown in Fig. 17. The control combinations for the peak detector are listed in Table III.

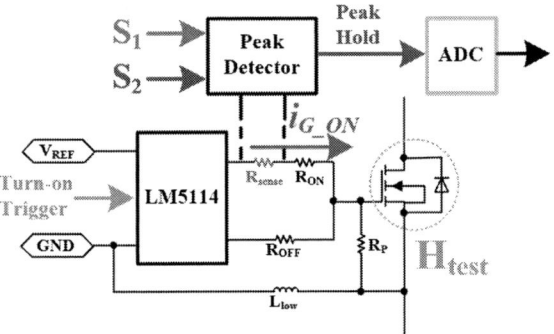

Fig. 16 Peak detector implemented in the GaN HEMT gate drive

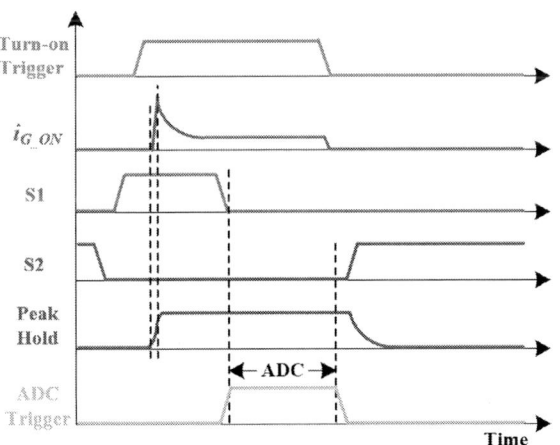

Fig. 17 Peak detector control time table

The analog peak detection can be realized with different electrical circuitries. In this research, by using a feedback op-amp circuit to charge a capacitor, pulse peak detection is

configured as shown in Fig. 18. The input signal (i_{G_ON}, in Volt unit) will be tracked by "Op Amp1" with unity closed-loop gain. If the tracked "i_{G_ON}" is larger than the voltage across the 470pF capacitor, the capacitor will be charged. If the tracked "i_{G_ON}" is less than the voltage across the 470pF capacitor, the diode will block the capacitor charging. Note that the voltage stored in the 470pF capacitor is always negative (referred to the ground in Fig. 18). In this circuit, the 470pF capacitor is used to hold the sensed peak value; S_1 is used to activate peak detection; and S_2 is used to reset the capacitor voltage.

TABLE III. CONTROL SIGNAL COMBINATIONS

	S_1 high	S_1 low
S_2 high	invalid	peak detection OFF peak detection reset ADC OFF
S_2 low	peak detection ON ADC OFF	peak detection OFF ADC ON

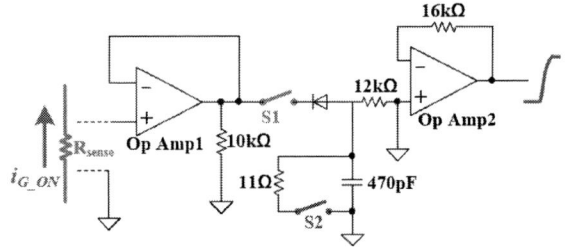

Fig. 18 Circuit implementation of the pulse peak detector

The experimentally measured peak detector output waveforms (the purple "analog out" signal in Fig. 16 and Fig. 17) under different T_j are shown in Fig. 19. It can be seen that the output voltage levels of the peak detector are sensitive to the T_j of the GaN HEMT.

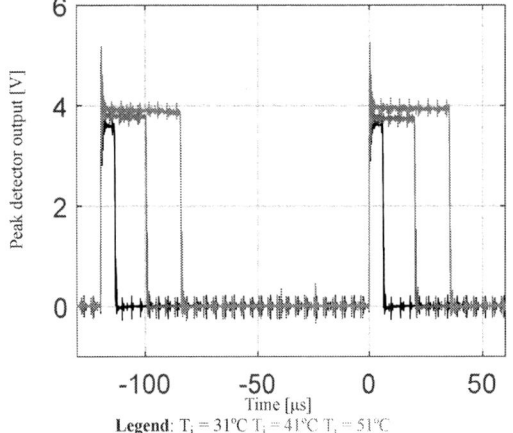

Legend: $T_j = 31°C$ $T_j = 41°C$ $T_j = 51°C$

Fig. 19 The output voltage of the peak detector under different T_j

About 260kHz of transient noise can be observed from the peak detector output, as can be seen in Fig 19. This transient noise is the result of switching transient from the power supply of the gate drive and the peak detector. To improve the output reliability of the peak detector, very stable power supplies or an output filter is desired in the "gate drive – peak detection" system.

The "peak vs. T_j" relationship derived experimentally with calibration is shown in Fig. 20 (The peak values are measured by the peak detector as shown in Fig 12. T_j is measured by the Flir thermal camera, as shown in Fig. 13. The calibration is achieved based on the device thermal impedance, as shown in Fig. 13.). It can be seen from Fig. 20 that the measured peak is positively correlated to the T_j of the GaN HEMT.

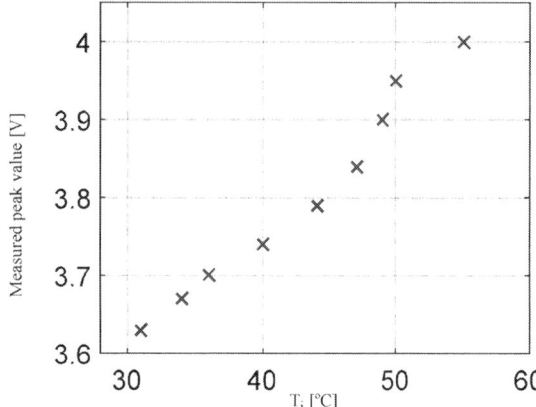

Fig. 20 Experimentally measured "peak vs. T_j" relationship

The "peak vs. T_j" should be measured offline first and stored in the controller of the system. By comparing the off-line measured relationship with the online measured peak detector output value, the T_j of the GaN HEMTs can be derived in real time.

VII. THE EFFECT OF OFF-STATE VOLTAGE

With the half-bridge inverter setup evaluated in this research, during the HEMT turn-on transient, the voltage stress across the device drain and source (V_{DS}) decreases from half of the DC bus voltage to almost zero. The switching transient properties of GaN HEMTs are proved to be sensitive to the bias voltage (for both on state and off state) [17]. As a result, varying the DC bus voltage of the half-bridge inverter can affect the switching transient properties of the GaN HEMTs.

To include the effect of the V_{DS} bias voltage in the proposed T_j sensing method, the 2-dimensional "peak vs. T_j" relationship (as shown in Fig. 20) is extended to a 3-dimensional "peak vs. T_j vs. DC bus" relationship, as illustrated in Fig. 21. The "peak vs. T_j vs. DC bus" relationship is experimentally derived based on the method discussed in section V and VI, with the DC bus varying from 22V to 36V, as illustrated in Fig. 21.

It can be seen from Fig. 21 that the measured peak is positively correlated to the DC bus voltage of the half-bridge inverter. In other power circuitry, if the on-state and off-state V_{DS} can vary during online operation, a 3-dimentional "peak vs. T_j vs. DC bus" relationship is required for the proposed online T_j sensing method.

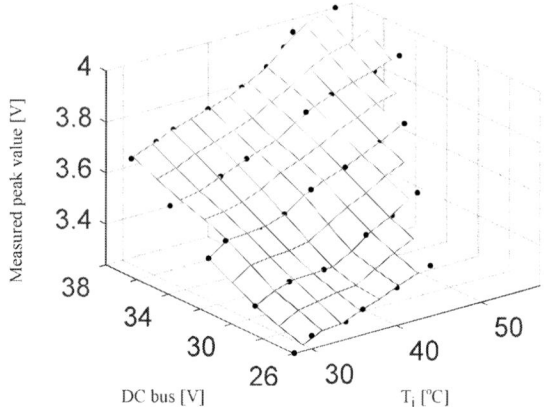

Fig. 21 Experimentally measured "peak vs. T_j vs. DC bus" relationship

VIII. CONCLUSIONS AND CONTRIBUTIONS

This paper proposed a method for GaN HEMT junction temperature online sensing. The key contributions made in this paper are concluded as follows:

- Developed a "gate drive – GaN HEMT" transient model to analyze the GaN HEMT turn-on dynamics.
- Theoretically and experimentally investigated the T_j dependency of the gate drive turn-on transient output current for GaN HEMT.
- Established a modified GaN HEMT dynamic model with fully characterized T_j dependency from both the gate side and the power side.
- Proposed the method for the real time measurement of the peak value of the gate drive turn-on transient output current for GaN HEMT.
- Provided the method of deriving and calibrating the "peak vs. T_j" relationship.
- Studied the effect of drain-to-source bias voltage on the proposed T_j sensing method..

ACKNOWLEDGMENT

The authors wish to acknowledge the motivation provided by the Wisconsin Electric Machines and Power Electronics Consortium (WEMPEC) of the University of Wisconsin-Madison.

REFERENCES

[1] H. Niu, R.D. Lorenz, "Evaluating Different Implementations of Online Junction Temperature Sensing for Switching Power Semiconductors", *Energy Conversion Congress and Exposition (ECCE)*, Montreal, Canada, 2015, pp. 5696-5703.

[2] R.C.S. Freire, S. Daher, G.S. Deep, "A highly linear single p-n junction temperature sensor", *IEEE Trans. Instrum. Meas.*, vol. 43, pp. 127-132, Apr., 1994.

[3] T. Kajiwara *et al.*, "New intelligent power multi-chips modules with junction temperature detecting function", *Industry Application Conference*, St. Louis, MO, 1993, pp. 1085-1090.

[4] E.R. Motto, J.F. Donlon, "IGBT module with user accessible on-chip current and temperature sensors", *Applied Power Electronics Conf. and Exposition (APEC)*, St. Louis, MO, 1993, pp. 1085-1090.

[5] M. Berthou, P. Godignon, J. Millan, "Monolithically Integrated Temperature Sensor in Silicon Carbide Power MOSFETs", *IEEE Trans. Power Electron.*, vol. 29, pp. 4970-4977, Sep., 2014.

[6] J.L. Barnette, "Lifetime Synchronization of Parallel DC/DC Converters", M.S. Thesis, North Carolina A&D Univ., Greensboro, NC, 2006.

[7] A. Koenig, T. Plum, P. Fidier, R.W. De Doncker, "On-line Junction Temperature Measurement of CoolMOS Devices", *Intl. Conf. on Power Electronics and Drive Systems*, Bangkok, Thailand, 2007, pp. 90-95.

[8] Efficient Power Conversion Co., Ltd., "eGaN FETs", Jul., 2015, [Online]. Available: http://epc-co.com/epc/Products/eGaNFETs.aspx

[9] M.L. Walters, "Circuit Modeling Methodology for Isolated, High Bandwidth Junction Temperature Estimation", *Intl. Power Electrn. Conf.*, Singapore, 2010.

[10] H. Niu, R.D. Lorenz, "Sensing Power MOSFET Junction Temperature Using Circuit Output Current Ringing Decay", *IEEE Trans. Ind. Appl.*, vol. 51, pp. 1763-1773, Jul., 2014.

[11] H. Niu, R.D. Lorenz, "Sensing Power MOSFET Junction Temperature Using Gate Drive Turn-on Current Transient Properties", *Energy Conversion Congress and Exposition (ECCE)*, Pittsburgh, PA, 2014, pp. 2909-2916.

[12] H. Niu, R.D. Lorenz, "Sensing IGBT Junction Temperature Using Gate Drive Output Transient Properties", *Applied Power Electronics Conf. and Expo (APEC)*, Charlotte, NC, 2015, pp. 2492-2499.

[13] H. Niu, R.D. Lorenz, "The Effect of Gate Drive Topology on Online Silicon Carbide MOSFET Junction Temperature Sensing", *Energy Conversion Congress and Exposition (ECCE)*, Montreal, Canada, 2015, pp. 7015-7022.

[14] K. Peng, E. Santi, "Characterization and Modeling of a Gallium Nitride Power HEMT", *Energy Conversion Congress and Exposition (ECCE)*, Pittsburgh, PA, 2014, pp. 113-120.

[15] Efficient Power Conversion Co., Ltd., "EPC2010 – Enhancement Mode Power Transistor", Feb., 2013, [Online]. Available: http://epc-co.com/epc/Products/eGaNFETs/EPC2010.aspx

[16] N. Baker, S. Munk-Nielsen, M. Liserre, F. Iannuzzo, "Online junction temperature measurement via internal gate resistance during turn-on", *European Conf. on Power Electronics and Application (EPE)*, Lappeenranta, Finland, 2014, pp. 1-10.

[17] A. Zhang, *et al.*, "Analytical Modeling of Capacitances for GaN HEMTs, Including Parasitic Components", *IEEE Trans. Electron Devices*, vol. 61, pp. 755-761, Mar., 2014.

[18] J. Lee, K.J. Webb, "A Temperature-dependent Nonlinear Analytic Model for AlGaN-GaN HEMTs on SiC", *IEEE Trans. Microwave Theory and Techniques*, vol. 52, pp. 2-9, Jan., 2004.

[19] Efficient Power Conversion Co., Ltd., "EPC2030 – Enhancement Mode Power Transistor", May, 2015, [Online]. Available: http://epc-co.com/epc/Products/eGaNFETs/EPC2030.aspx

[20] N.V. Drozdovski, R.H. Caverly, M.J. Quinn, "Large-signal modeling of microwave gallium nitride-based HFETs", *Asia-Pacific Microwave Conference (APMC)*, Taipei, Taiwan, 2001, pp. 248-251.

[21] F.M. Yigletu, *et. al*, "Compact physical models for gate charge and gate capacitances of AlGaN/GaN HEMTs", *Intl. Conf. Simulation of Semiconductor Processes and Devices (SISPAD)*, Glasgow, UK, 2013, pp. 268-271.

Design and Application of a 1200V Ultra-fast Integrated Silicon Carbide MOSFET Module

Suxuan Guo, Liqi Zhang, Yang Lei, Xuan Li, Wensong Yu, Alex Q. Huang

FREEDM Systems Center
North Carolina State University, Raleigh
sguo6@ncsu.edu

Abstract — With the commercial introduction of wide bandgap power devices such as Silicon Carbide (SiC) and Gallium Nitride (GaN) in the last few years, the high power and high frequency power electronics applications have gained more attention. The fast switching speed and high temperature features of SiC MOSFET break the limit of the traditional silicon MOSFET. However, the EMI problem under high dI/dt and dV/dt is an unneglectable problem. The overshoot and oscillation on drain-source voltage and gating signal could cause breakdown of the switches. This paper proposes a 1200V integrated SiC MOSFET module. With the ultra-fast gate driver integrated with the SiC MOSFET, the parasitic inductance and capacitance could be reduced dramatically, which accordingly suppress the EMI problem caused by the parasitic parameters. Thus zero gate resistance could be adopted in the module to further increase the switching speed. The switching performance of the integrated SiC module is shown better than the discrete package device. The switching loss of the SiC MOSFET module is measured by the inverter level measurement and composition method. Zero switching loss could be achieved when the drain current is lower than a critical value. The module has been tested at 1.5MHz and 3.38MHz switching frequency to prove its high speed capability. For isolated topology applications, the impact of high frequency on the power density and efficiency is discussed in this paper.

Keywords—SiC integrated module, zero switching loss, ultra-high frequency application

I. INTRODUCTION

In the last few years, the demands for high temperature and high power density converters have become more and more evident due to the rapid development of electric vehicles and electric aircrafts. In high voltage (BV>600V) Silicon (Si) based converters, its internal parasitic parameters, tail losses and large reverse recovery charge limit the operation switching frequency to less than one hundred kilohertz and less than 20 kHz in the case of Si IGBT[1]. The narrow bandgap of Si limits the device peak temperature to be below 125 °C in most cases. The emerging technology of wide bandgap devices provides an very attractive alternatives to Si devices. Gallium Nitride (GaN) devices have been developed and widely used in under 600V and Silicon Carbide (SiC) devices up to 1700V are commercially available with research devices exceeding 15 kV. Many researches have been done to compare the performance of Si and wide bandgap devices in various applications [2-4]. The SiC MOSFET die size are 50% smaller than the Si device with the similar rating, which means the parasitic capacitance of the die is much smaller as shown in Fig. 1 [5]. Due to the smaller parasitic parameters, the switching speed of the SiC MOSFET could be high accordingly. The junction temperature could be above 150 °C

for SiC MOSFET, which not only reduces the volume of the thermal management system, but also increase the reliability of the system. However, the high switching speed of the SiC MOSFET does not necessarily lead to high switching frequency. The internal parasitic in the package and the external parasitic induced by the circuit board are inevitable. With fast switching speed, these parasitic inductance and capacitance could cause the overshoot and oscillation on the drain-source voltage and gate source voltage and the breakdown of the device. In order to solve the EMI problem, the parasitic should be reduced. Multi-chip SiC MOSFET modules such as H bridge and three-phase modules are fabricated to reduce the parasitic of the power switching loop [6-8]. This will suppress the overshoot on the drain source voltage to some extent. However, the power switching loop size is still relatively large. Meanwhile, without optimizing the gate loop, the switching speed could not be increased either and the gate breakdown hazard is possible by the overshoot and oscillation of the gate voltage.

In this paper, an integrated SiC MOSFET half bridge module has been developed and fabricated. A fast driver chip is closely integrated with the SiC MOSFET to reduce the parasitic of the gate loop. The power switching loop is carefully designed and optimized to reduce the parasitic of the power switching loop. Comparing with the discrete TO-247 device with the same die, the integrated SiC MOSFET module has much better switching performance with minimum overshoot and oscillation. With the suppression of EMI problem, zero gate resistance could be used to further increase the switching speed to 96 kV/us at 800V, 10A test. Thanks to the high switching speed, the turn off loss could be reduced to zero when the drain current is lower than a critical value.

This paper is organized as following. The switching transient and the influence of the parasitic parameters are introduced in section I. The configuration and the layout of the integrated module are included in section II. Double pulse test and continuous test are used to verify the switching performance and compared with discrete TO-247 device. The

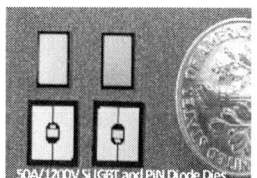 

Figure. 1 Size comparison between Si vs. SiC

turn off loss is measured in inverter-level measurements. The module is operated at 3.38MHz in a half bridge inverter to prove its zero turn off loss. Finally, the design of a 4kW LLC resonant converter and transformer are discussed in this paper.

II. SWITCHING TRANSIENT ANALYSIS AND PARASITIC PARAMETERS INFLUENCES

A. Switching Transient Analysis

Normally, the switching loss is measured by the double pulse test. The integral of the drain current times the drain-source voltage is calculated during the switching transient as the switching loss. However, with an insightful understanding of the turn on and turn off transients, it is the channel current that generates the actual power loss and this current is not equal to the measured drain current. Therefore conventional measurement using the drain current could not reflect the true switching loss [9].

The turn-off simulation using a finite element simulator is shown in Fig. 2 which is able to capture the 'channel current'. When the switch turns off, the drain current flows through the channel, and charges the parasitic capacitors C_{gd} and C_{ds}. From $t_0 \sim t_1$, the gate-source capacitor C_{gs} discharge. The channel current starts to drop at t_0. The difference of the drain current and channel current discharges the C_{gs}. From $t_1 \sim t_2$, the gate current continues the discharge of the C_{gs} and the gate voltage decreases. At t_2, the gate voltage reaches the threshold voltage, the drain-source output capacitor $C_{ds} + C_{gd}$ starts to charge. Due to the large capacitance at low V_{ds} voltage, there is little V_{ds} voltage rise while the gate-source capacitor C_{gs} still discharges quickly, the channel current drops quickly to zero and the channel is cut off at t_3. From $t_3 \sim t_4$, the drain current charges $C_{ds} + C_{gd}$. Until the drain-source voltage V_{ds} reaches bus voltage at t_4, the turn off switching transient ends. From

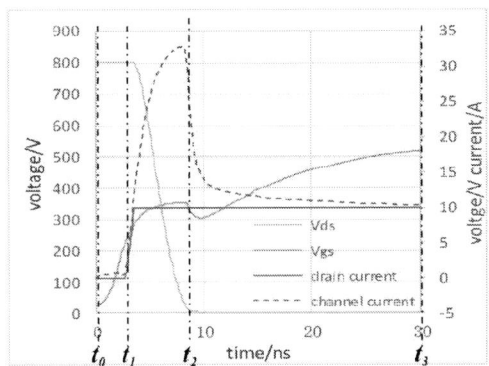

Figure. 3 Turn on transient, V=800V, Iload=10A

the turn off switching transient, only the period $t_2 \sim t_3$ generates the switching loss. If this period is short enough, which is made possible by the fast switching speed and small drain current, zero turn off switching loss could be realized. The part of drain current charges C_{ds} and C_{gd} (primarily from t_3 to t_4) is stored as E_{oss} which is not an actual loss.

During the turn on process, from $t_0 \sim t_1$, the gate voltage is charged from -4V to threshold voltage, the channel turns on at t_1. The gate voltage keeps charging to plateau voltage, energy stored in the energy stored in C_{ds} and C_{gd} start to release. The discharge current of C_{ds} and C_{gd} flows through the channel, which makes the channel current bigger than the measured drain current. The plateau gate voltage is charged to a high level to support the discharge current. The channel current includes the drain current and the C_{oss} discharging current. At t_2, C_{ds} is fully discharged, the drain-source voltage drops to zero, the current through the channel starts to decrease, C_{gs} keeps charging and C_{gd} keeps discharging from $t_2 \sim t_3$. At t_3, C_{gs} is charged to 20V, the turn on transient ends. In the turn on process, the period $t_1 \sim t_3$ generates the switching loss. The current flows through the channel is larger than the drain current.

According to the above analysis, the switching loss measured by the drain current is not precise. The turn on switching loss is larger than the measured value, and the turn off loss is actually smaller. The turn on switching speed is determined by the driver capability. If utilizing soft switching during turn on transient, the high turn on loss could be avoided. During the turn off transient, if the overlapping of channel current and drain-source voltage is small enough, which could be realized by the fast switching driver, the turn off loss could also be minimized or even zero if the load current is lower than a critical value. Thus, the method of increasing gate drive switching speed should be explored.

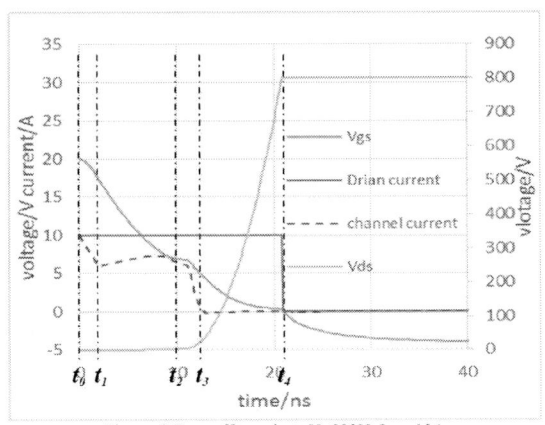

Figure. 2 Turn off transient, V=800V, Iload=10A

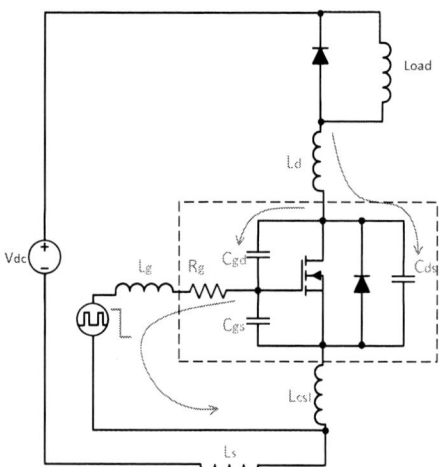

Figure 4. Parasitics parameters in MOSFET

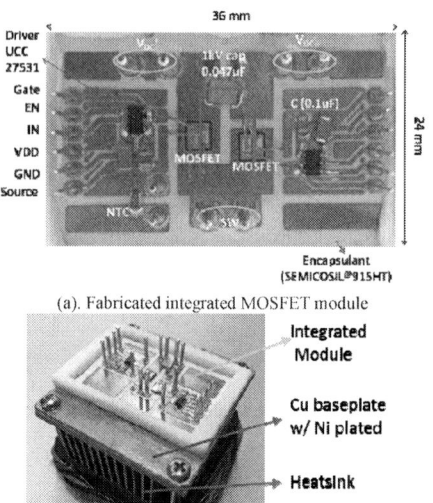

(a). Fabricated integrated MOSFET module

Figure 5. (b) Integrated module with cooling system

B. Parasitic Parameters Influences during Switching Transient

The parasitic inductance and capacitance are inevitable in a power electronic circuit and they can slow down the switching speed. The chip size of the device is proportional to the parasitic capacitance of the MOSFET. Thus the parasitic capacitance of SiC device is much smaller than the Si device. The wire bonds and leads for connection induce the parasitic inductance. From the external circuit, the PCB trace and layout could both influence the external parasitic inductance and capacitance of the MOSFET. The parasitic inductance is proportional to the area of the loop. The layout of the module and PCB circuit should both be designed to minimize the loop area.

As shown in Fig. 4, in the power switching loop, the parasitic inductance includes the L_s, L_d and common source inductance L_{cm}. The gate loop parasitic inductance includes L_g and L_{cm}. When the switch turns off, the drain current charges C_{ds} and miller capacitor C_{gd}. The high dI/dt during the turn off transient cause the spike voltage on L_s and L_d, which causes the overshoot on drain-source voltage. Meanwhile, L_s and L_d resonant with C_{ds}, the drain-source voltage has oscillation. Similar process occurs in the gate switching loop. During the turn off transient, C_{gs} discharges and resonant with L_g, the current has voltage drop on the gate resistance R_g. If the voltage drop is too large, fault trigger will happen and breakdown the device. The common source inductance L_{cm} couples the noise from the power switching loop to the gate loop. To avoid the overshoot and oscillation problems, the gate resistance should be used to limit the switching speed to a lower level. Thus, the switching loss will be increased accordingly. By reducing the parasitic parameters in the loop, the switching speed could be increased, the switching loss could be minimized. When the drain current is less than a critical value, the switch could naturally achieve zero turn off loss. With the soft switching during turn on transient, the MOSFET switch will not have any switching loss. Of course there is still losses in the gate driver.

III. INTEGRATED MODULE CONFIGURATION AND TESTS

The integrated half bridge SiC MOSFET module is consisted of two 1200V 31.6A SiC MOSFET bare dies CPM2-1200-0080B from Cree and two 5A gate drivers UCC27531 from Texas Instrument, as shown in Fig. 5. Zero external gate resistor is used in the module while the internal gate resistance of the SiC MOSFET is 4.6•. 47nF decoupling capacitor is soldered between the high voltage dc terminals to decrease the drain loop parasitic inductance. Direct bonded copper (DBC) substrate is used here to improve the thermal impedance from junction to the baseplate. The bottom size of the DBC is soldered to a nickel plated copper baseplate. The dimension of the module is 36mm × 24mm. A 45mm×45mm×25mm heatsink and 1.68W fan are used to cool the module, which is attached firmly with the baseplate.

In order to reduce the parasitic inductance for the gate loop, the minimum loop area should be designed, so the driver should be placed as near as possible to the MOSFET die. However, there is a thermal coupling between the driver and the MOSFET die. The temperature of the MOSFET die will be above 100 ˚C during the operation, consistent with the 150 ˚C peak junction temperature of SiC MOSFET device. Strong thermal coupling will cause the drive temperature to rise. Meanwhile, the driver loss during the operation could add additional heat to the driver chip. The total driver loss is:

$$P_{loss\_dr} = Q_g \cdot \Delta V_{gs} \cdot f_{sw} \qquad (1)$$

978-1-4673-9551-9/16 $31.00 © 2016 IEEE

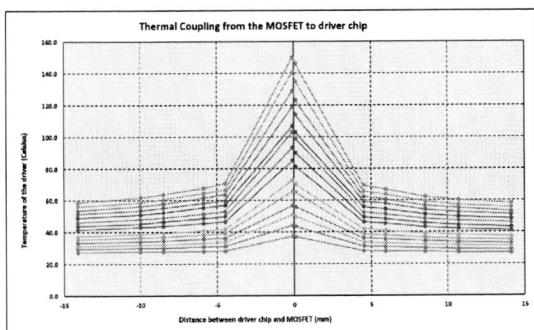

Figure 6. Thermal coupling from MOSFET to driver

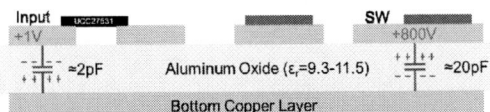

Figure 7. Parasitic capacitance of integrated module

A portion of the loss will be in the driver chip and a portion will be in the MOSFET internal resistance Rg. The higher the switching frequency is, the larger the temperature rising of the driver chip. The driver temperature is the sum up of the self-heating and the thermal coupling from the MOSFET die. Thus, the placement of the driver chip should be designed to minimize thermal coupling from the MOSFET die.

Multiple NTCs are soldered on a plain DBC with the MOSFET to test the thermal coupling effect when the module the cooled by the integrated cooling fan. The distance between the NTC and the MOSFET die varies from 4.5mm to 14.1mm. There are two MOSFET dies on the DBC, thus the NTCs are symmetrical. A current source is used to inject power from the DC terminals of the module, the body diodes of the MOSFET are self-heated. During the experiment, fourteen sets of data are recorded as shown in Fig. 6. The midpoint of the module is at 0mm location. The temperatures of NTCs are recorded to test the thermal coupling from the MOSFET die. According to the results, when the MOSFET junction temperature is 150 °C, the NTC which is 5.9mm away from the MOSFET is 66.8 °C, while at 8.5mm away from the MOSFET the temperature is 62.1 °C. With 85 °C lower than the MOSFET, even a silicon driver is in a safe region even if the MOSFET is at 150 °C. Thus, the distance of the driver chip to MOSFET is chosen in the range of 5.9mm and 8.5mm.

Another challenge is the parasitic capacitance in the integrated module. As the driver chip and the MOSFET share the same bottom copper layer, parasitic capacitance exists between the top copper layer and the bottom copper layer of the DBC. Current loop is formed by the parasitic capacitance and the bottom copper. Fig. 7 shows the side view of the integrated module. During the continuous operation, the half bridge SW node has high dV/dt, which is coupled from the bottom layer to the input gating signal of the gate driver. The parasitic capacitance between the switching point and the bottom copper layer is about 20pF. When the switching speed

is fast, the noise current caused by the parasitic capacitance is

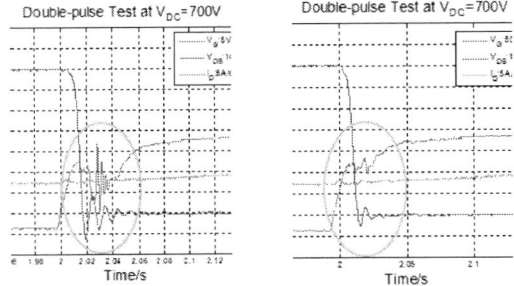

Figure 8. Gate signal (a) without capacitor   (b) with capacitor

unneglectable. The input capacitance of the driver chip could be discharged if the noise current is too large. The gate voltage could be unstable especially during the turn on transient. The solution to this problem is to add a small 10pF capacitor to the input side of the driver, which helps to keep the gate voltage stable. The gating signals with and without the capacitance during turn on transient are shown in Fig. 8. With the external 10pF capacitor, the gating signal is stable during the fast turn on transient.

With the integrated fast driver chip, the gate loop area of the integrated module is much smaller than the conventional TO247 design. The power switching loop is then optimized in the module to have the minimum L_d and L_s. In order to eliminate the common source inductance, Kelvin connection is utilized in this module. The gate loop and the power switching loop are separated as shown in the layout. The discrete TO-247 packaged device C2M0080120D with the same die is chosen for comparison. The power switching loop area of the integrated module is 42mm², comparing to 200mm² loop area of the TO-247 device. The gate loop area is 18mm², comparing to 100mm² loop area of the TO-247 device. The parasitic inductance of gate loop is much smaller in the integrated module. The double pulse tests are carried out for TO-247 device and the integrated device, as shown in Fig. 9. The test is under 800V and 10A, the switching performance of the

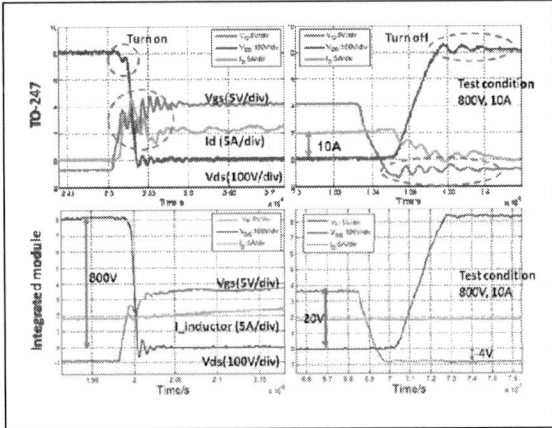

Figure 9. Continuous test comparison of integrated module and TO-247

integrated module is much better than TO-247. The dV/dt of the integrated module reaches 96kV/us during the turn on transient and 39kV/us during the turn off transient. The gate-source voltage and drain-source voltage are clean comparing with TO-247, the overshoot and oscillation are minimized.

Due to highly integrated drain loop, the device current could not be measured in the double pulse tester. On the other hand, the measurement using the double pulse test is not precise enough as explained early. To compare the switching loss of the integrated module and the TO-247 device, half bridge inverter with inductive load is built to operate at 510kHz under 800V, 20A peak current. By using the same source and inductive load, the two setups are operating in identical condition. The difference between the input power of the two setups indicates the difference of the power loss on MOSFET. In order to measure dc input current, a RLC filter is used which is connected to the input side. The waveforms of the two setups are shown in Fig. 10. According to the calculation, the input power of the integrated module is 4.78W less than the TO-247 device, which corresponds to 9.4uJ lower turn off loss.

IV. SWITCHING LOSS ESTIMATION AND CONTINOUS TEST

The double pulse test measurement is traditionally used to measure the switching loss. However, according to the analysis of the switching transient, the double pulse test is not accurate for the turn on and turn off loss. High bandwidth current probes are needed for the drain current measurement, which is normally coaxial shunt resistor or Rogowski coil [10-11]. For the turn on switching transient, the turn on time is less than

10ns, the current probe with enough bandwidth is hard to find. On the other hand, the decoupling capacitor is integrated in the module, the drain current could only be measured from the wire bonds, which is not feasible. Thus, the inverter-level measurement method is used in this paper. Compared with the double pulse test, the inverter-level measurement method is based on the continues operation power loss.

The thermal performance of the integrated module is measured at first. A current source is connected to the dc terminals of the module, body diodes of MOSFETs are self-heated to generate a constant power loss and they are the only heat source in the test. The junction temperature and heatsink temperature is measured in this test as shown in Fig. 11. The thermal resistance of the cooling system θ_{hs_a} is calculated as 0.6672 °C/W since:

$$P_{dis} = \frac{T_{heat\sin k} - T_{ambient}}{\theta_{hs\_a}} \quad (2)$$

As the module encapsulate is thermally insulated, the heat generated in the module could only be dissipated down to the heatsink. The power dissipated by the heatsink is therefore assumed to be the same with the power loss in the module. Then the half bridge inverter is tested at high switching frequency with inductive load. The MOSFET is naturally soft switching when turn on, so there is only turn off loss measured. The power loss in the module is composited as:

$$P_{dis} = P_{cond} + P_{sw} + P_{dt} + P_{driver} \quad (3)$$

The conduction loss and deadtime loss could be calculated as:

$$P_{cond} = R_{on} \cdot I_{ds\_rms}^2 \quad (4)$$

$$P_{dt} = 2 \cdot \Delta t \cdot I_{pk} \cdot V_F \cdot f_{sw} \quad (5)$$

The turn off switching loss could be calculated as:

$$E_{off} = \frac{P_{sw}}{2 \cdot f_{sw}} \quad (6)$$

As the R_{on} is related to the drain current and the junction temperature, the junction temperature need to be measured too.

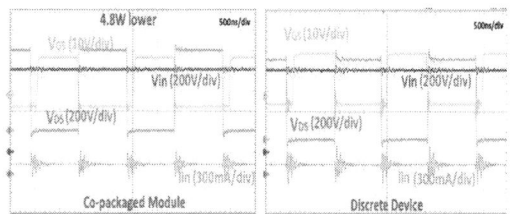

Figure 10. Continuous test comparison between TO-247 device and integrated module

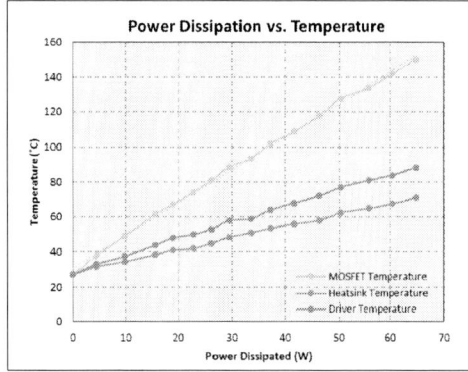

Figure 11. Power dissipation vs. temperature

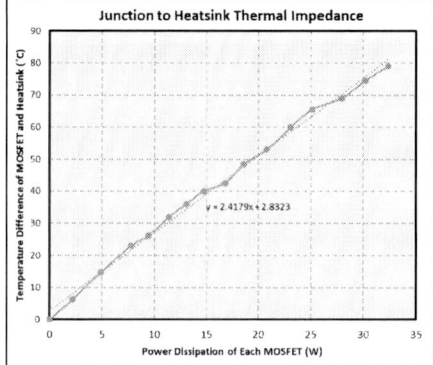

Figure 12. Junction to heatsink thermal impedance

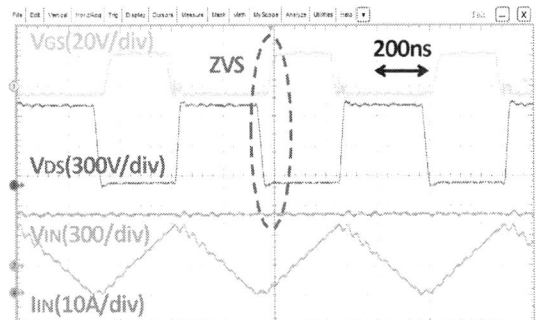

Figure 13. Waveforms of 1.5MHz synchronous boost converter

which is impossible during the operation. Thus, the junction temperature is estimated by the thermal impedance from junction to heatsink. The thermal impedance θ_{j_hs} is 2.42 °C/W as shown in Fig. 12. Thus the junction temperature could be estimated by:

$$T_j = T_{hs} + \theta_{j\_hs} \cdot P_{mos} \qquad (7)$$

According to the junction temperature and drain current, the R_{on} is calibrated from the datasheet. Megahertz continuous tests are then carried out to verify the low switching loss of the module. Firstly, a 1.5MHz synchronous boost converter is built. The input voltage is 400V, with 50% duty cycle, the voltage stress on the MOSFETs is 800V. The load is 800 Ω resistor and the dc link capacitance is 54uF. The boost inductor is 5.4uH, which provides 24.6A peak to peak current to the MOSFET. The experimental results are shown in Fig. 13. The deadtime is 62ns. Zero voltage switching is realized for both of the MOSFETs, thus only turn off loss exists. The heatsink temperature is 40 °C, the junction temperature could be estimated to be 61.8 °C. The rms MOSFET current is 7.1A. According to the datasheet of the MOSFET, R_{on} is 85mΩ. The power dissipated in the module is calculated to be 18.02W. The driver loss with 1.5MHz is 3.54W in the module. Thus, the power loss of the MOSFETs are 14.48W. By deducting the deadtime loss and conduction loss, the turn off loss is calculated as 2.84uJ. This loss is very low and can be considered near zero turn-off loss.

Then a 3.38MHz continuous test in half bridge inverter is carried out to further prove the zero switching loss when the drain current is relatively low. The half bridge inverter is operated at 800V and 4.6A peak to peak current. The experiment results are shown in Fig. 14. With the fast switching speed and high switching frequency, the drain-source voltage is still without overshoot and oscillation. The heatsink temperature is 35.1°C, even lower than the 1.5MHz test as the drain current is lower. The increasing of switching frequency does not increase the power loss, which proves the conduction loss is the dominant part for the module, not the switching loss. The junction temperature is estimated to be 48 °C. According to the calculation, the turn off switching loss is 0uJ, which verifies that in this condition, the fast switching speed could naturally achieve zero switching loss without causing any EMI problem.

V. HIGH FREQUENCY TRANSFORMER DEAIGN AND LIMITATION

From our analysis, the SiC MOSFET could achieve zero switching loss if ZVS turn-on is also utilized in the circuit. Therefore the semiconductor limitation of the switching frequency is way above 1 MHz for the 1200V SiC MOSFET. This enables many applications including isolated high voltage DC/DC converters. In this case, the LLC resonant converter is very suitable as it has naturally soft switching feature [12-14]. With the switching frequency equals to the resonant frequency, zero voltage switching could be achieved for turn on. The turn off current is very low and therefore the turn-off loss is also zero. By utilizing the integrated module, the switching loss for the converter is close to zero and the conduction loss is dominant for the MOSFET, which is not depending on the switching frequency. By increasing the switching frequency, the volume of the heatsink for MOSFET remains the same, but the volume of the transformer could be reduced significantly. Thus, for the optimization of the LLC resonant convert, magnetic components are the major part which need to be optimized. The transformer for a 4kW half bridge LLC resonant converter is designed as an example. The Steinmetz equation is conventionally used to evaluate the core loss:

$$P_{cv} = k \cdot f_s^{\alpha} \cdot B_{pk}^{\beta} \qquad (8)$$

P_{cv} is the core loss per unit volume, f_s is the switching frequency in Hz, B is the peak flux density. k, α and β are the coefficients relating to materials. The peak flux density is

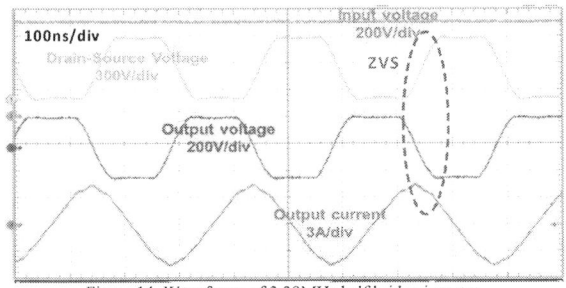

Figure 14. Waveforms of 3.38MHz half bridge inverter

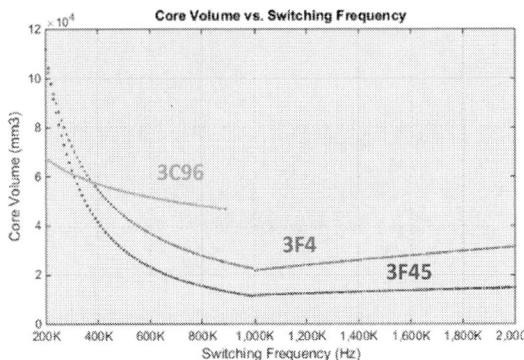

Figure 15. Core volume vs. switching frequency

determined by:

$$B_{pk} = \frac{D \cdot V_{in}}{2 \cdot N_p \cdot A_e \cdot f_s} \qquad (9)$$

An assumption is made that the cross section area of the transformer relates to the volume according to different core shapes [14]:

$$V_e = y \cdot A_e^C \qquad (10)$$

Thus, bringing (9) and (10) to (8), the volume of the transformer is relating to the switching frequency:

$$V_e = (\frac{P_{core}}{k})^{\frac{c}{c-\beta}} \cdot y^{\frac{\beta}{\beta-c}} \cdot f_s^{\frac{c(\beta-\alpha)}{c-\beta}} \cdot (\frac{V_{pri}}{4N_p})^{\frac{\beta c}{\beta-c}} \qquad (11)$$

Three ferrite core materials 3C96, 3F4 and 3F45 are analyzed to show the impact of high frequency to the volume of the transformer. For a given fixed core loss of 6W and an input voltage of 200V, the volume of the transformer vs. switching frequency is shown as Fig. 15. When the switching frequency is lower than 1MHz, the volume of the transformer could be reduce with higher switching frequency. With the switching frequency higher than 1MHz, the volume of the transformer is not significantly depending on the switching frequency. According to Fig. 15, 3F4 could be chosen for the high frequency transformer around 1MHz. A matrix transformer consisting of two planar E38/8/25 cores can be used for an input voltage of 800V half bridge design so that the primary side of the transformer is connected in series, the secondary side is connected in parallel. The output voltage is 400V, thus the turns ratio is 1:2 for each transformer. The current of the primary side is determined by the magnetizing inductance:

$$I_{rms\_p} = \frac{V_o}{8nR_L}\sqrt{\frac{2n^4 R_L^2 T_{sw}^2}{L_m^2} + 8\pi^2} \qquad (12)$$

The larger the magnetizing inductance is, the smaller the rms current is. However, in order to realize soft switching, the magnetizing current should be enough to charge the parasitic

capacitance of the MOSFET and turn on the body diodes before the gate signal. Thus, the maximum magnetizing inductance is:

$$L_m \leq \frac{T_{sw} t_{dead}}{16 C_{eff}} \qquad (13)$$

Thus, the minimum rms current is determined by the deadtime. The winding loss of the transformer is:

$$P_{wire\_p} = I_{rms\_p}^2 \cdot F\_R \cdot N_p \cdot R_{dc\_p}$$
$$P_{wire\_s} = I_{rms\_s}^2 \cdot F\_R \cdot N_s \cdot R_{dc\_s} \qquad (14)$$

F_R is the AC-to-DC resistance ratio given by the Dowells equation, which is given by:

$$F\_R = X \frac{e^{2X} - e^{-2X} + 2\sin(2X)}{e^{2X} + e^{-2X} - 2\cos(2X)} + 2X \frac{p^2-1}{3} \frac{e^X - e^{-X} - 2\sin(2X)}{e^X + e^{-X} + 2\cos(2X)} \quad (15)$$

$$X = \frac{\sqrt{\pi} d_{wire}}{\delta_{wire}} \qquad (16)$$

p is the layer of the litz wire, δ_{wire} is the skin depth of the litz wire, d_{wire} is the diameter of the litz wire. In this paper, 175 strands of #46 litz wire is chosen for the primary side and secondary side. Two wires are winded in parallel for primary side, single wire is winded for secondary side. The core loss is:

$$P_{core} = V_e \cdot k \cdot f_s^\alpha \cdot (\frac{D \cdot V_{in}}{2 \cdot N_p \cdot A_e \cdot f_s})^\beta \qquad (17)$$

Thus, the transformer loss is:

$$P_{xfmr} = P_{core} + P_{wire\_p} + P_{wire\_s} \qquad (18)$$

With the switching frequency changing from 200kHz to 1MHz, and the winding of the primary side changes from 1 to 15, the transformer loss is shown in Fig. 16. Extracting the minimum transformer loss for each switching frequency, the minimum transformer loss and corresponding primary

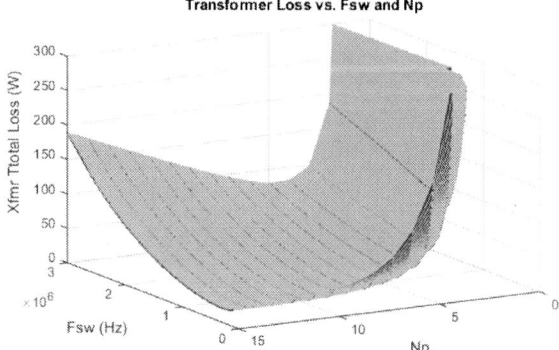

Figure 16. Transformer loss vs. frequency and primary windings

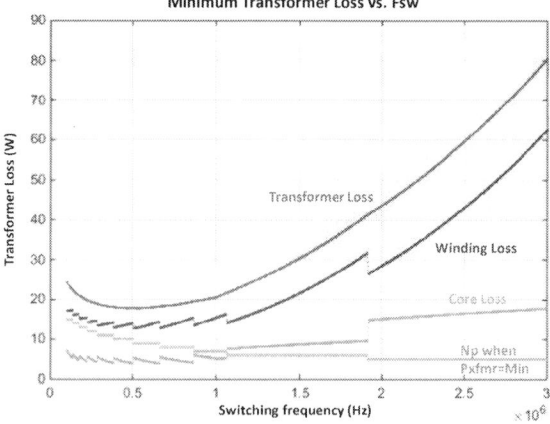

Figure 17. Minimum transformer loss vs. switching frequency

windings are shown in Fig 17. The winding loss is the dominant loss for transformer when the switching frequency is above 1MHz for this design. From the results, the minimum transformer loss is 17.9W, which is achieved at 470kHz, when primary winding is 10. The transformer loss is only 0.44% of the output power. This design example clearly shows that the frequency in an actual converter is now determined by the magnetic design, not by the semiconductor device.

VI. CONCLUSION

In order to improve the switching speed of the SiC MOSFET and utilize its maximum capability, the parasitic parameters which are the major limitation should be reduce. By integrating the ultra-fast driver chip with the SiC MOSFET, the parasitic inductance of both the power switching loop and the gate loop could be reduced significantly. Zero gate resistance is used to push the switching speed to above 96 kV/us. Double pulse tests and continuous test verify the better switching performance and smaller switching loss of the proposed module. Inverter-level measurement method is used here to calculate the turn off loss. 3.38MHz continuous test is carried out to verify no turn off loss when the drain current is below a critical value. Utilization of the integrated module in resonant converter is discussed, the transformer design of 4kW LLC resonant converter is analyzed in this paper. Better materials and design concepts will be needed for the transformer to operate at multimegahertz switching frequency.

ACKNOWLEDGMENT

The author would like to thank TI for the helpful discussion on the module design and the support of the project.

REFERENCES

[1] Alex Q. Huang. "New Unipolar Switching Power Device Figures of Merit." IEEE Electron Device Letters 25, no. 5 (May 2004): 298–301.

[2] Xiaoqing Song, Alex Q. Huang, Xijun Ni, Liqi Zhang, "Comparative Evaluation of 6kV Si and SiC Power Devices for Medium Voltage Power Electronics Applications", WiPDA 2015

[3] J. Biela, M. Schweizer, S. Waffler, B. Wrzecionko, and J.W. Kolar "SiC vs. Si - Evaluation of Potentials for Performance Improvement of Power Electronics Converter Systems by SiC Power Semiconductors," IEEE Trans. Industrial Electronics, vol. 58, no. 7, pp. 2872-2882, Jul. 2011.

[4] Fei Xue, Ruiyang Yu, Wensong Yu, Alex Huang, "Distributed Energy Storage Device Based On A Novel Bidirectional DC-DC Converter With 650V GaN Transistors", In the 6th International Symposium on Power Electronics for Distributed Generation Systems (PEDG), 2015

[5] Liang, Zhenxian, F. Wang, and L. Tolbert. "Development of Packaging Technologies for Advanced SiC Power Modules." In *2014 IEEE Workshop on Wide Bandgap Power Devices and Applications (WiPDA)*, 42–47, 2014.

[6] Z. Chen, Y. Yao, D. Boroyevich, K. Ngo, P. Mattavelli and K. Rajashekara, "A 1200-V, 60-A SiC MOSFET MultichipPhase-Leg Module for High-Temperature, High-Frequency Applications" IEEE Trans. Power Electronics, vol. 29, no. 5, pp. 2307-2320, May 2014.

[7] Cree, CAS100H12AM1 1200 V, 100 A SiC half-bridge module datasheet, available online at http://www.cree.com/.

[8] Powerex, QJD1210010 1200 V, 100 A split dual SiC MOSFET module datasheet, available online at http://www.pwrx.com/.

[9] Xuan Li, Liqi Zhang, Suxuan Guo, Yang Lei, Alex Q. Huang and Bo, Zhang "Understanding Switching Losses in SiC MOSFET: Towards Lossless Switching", to appear in WIPDA 2015.

[10] Cougo, B., H. Schneider, and T. Meynard. "Accurate Switching Energy Estimation of Wide Bandgap Devices Used in Converters for Aircraft Applications." In *2013 15th European Conference on Power Electronics and Applications (EPE)*, 1–10, 2013.

[11] Cougo, B., H. Schneider, and T. Meynard. "High Current Ripple for Power Density and Efficiency Improvement in Wide Bandgap Transistor-Based Buck Converters." *IEEE Transactions on Power Electronics* 30, no. 8 (August 2015): 4489–4504.

[12] Kai Tan; Ruiyang Yu; Suxuan Guo; Huang, A.Q., "Optimal design methodology of bidirectional LLC resonant DC/DC converter for solid state transformer application," Industrial Electronics Society, IECON 2014 - 40th Annual Conference of the IEEE , vol., no., pp.1657,1664, Oct. 29, 2014 - Nov.1, 2014.

[13] Yu, Ruiyang, Godwin Kwun Yuan Ho, Bryan Man Hay Pong, B.W.-K. Ling, and J. Lam. "Computer-Aided Design and Optimization of High-Efficiency LLC Series Resonant Converter." *IEEE Transactions on Power Electronics* 27, no. 7 (July 2012): 3243–56.

[14] Reusch, D., and F.C. Lee. "High Frequency Bus Converter with Low Loss Integrated Matrix Transformer." In *2012 Twenty-Seventh Annual IEEE Applied Power Electronics Conference and Exposition (APEC)*, 1392–97, 2012.

Active Gate Charge Control Strategy for Series-Connected IGBTs

Fan Zhang, Xu Yang, Yu Ren, Ying Chen
School of Electrical Engineering
Xi'an Jiaotong University
Xi'an, China
zhangfan2014@yahoo.com

Ruifeng Gou

Xi'an XD Power Systems Co., LTD
Xi'an, China
gourf@xdps.com.cn

Abstract—**Insulated gate bipolar transistors (IGBTs) are usually connected in series to satisfy the requirements of high-power and high-voltage in power electronics applications. However, due to the parameter deviations of the series-connected IGBTs, it is difficult to ensure an equal voltage sharing between them during both transient and steady-state operations. This paper proposed a novel active gate drive (AGD) which operates basing on the active gate charge control strategy. The proposed active gate drive was able to achieve both minimized switching loss and proper voltage sharing between the series-connected IGBTs. The performance of the proposed active gate drive and active gate control method have been validated by experimental results, and promising results have been obtained.**

Keywords—insulated gate bipolar transistors; series-connected; active gate drive

I. INTRODUCTION

Power electronic converters, which based mainly on power semiconductor devices, have been widely used in different applications. Due to the relatively narrow band gap of the silicon (Si) material, the commercial available Si-IGBTs have a maximum voltage-blocking capability of 6.5kV. Clearly, for high-power and high-voltage applications, the voltage rating of a single IGBT is not adequate for the requirements. One promising solution for that is having several devices connected in series to provide the expected voltage rating. Nevertheless, the series association of the IGBTs is not easy due to the unequal voltage sharing between them during both steady-state and transient operations.

Methods have been applied for minimizing the voltage imbalance of series-connected IGBTs. Generally, those methods can be grouped into three categories [1]: passive snubber circuits, active clamping circuits, and active gate control methods. The most widely used passive snubber circuits often implement a RC or RCD snubber in parallel with the IGBT for transient voltage sharing. By slowing down the voltage gradient during switching transient of the series-connected devices, the voltage unbalance can be minimized at the cost of costly snubber circuits and massive power loss [2] [3]. Active clamping circuits aim at clamping the V_{ce} at a reference voltage level. The IGBT is slightly turned on if the voltage across it exceeds the reference voltage. This method actually provides an overvoltage protection circuit and the voltage balance cannot be guaranteed. Besides, the extra loss generated due to the implementing of the clamping circuit make it unattractive in high-frequency applications [4] [5]. In the active gate control methods, the way to charge the gate terminal is controlled to get the desired wave-shape of the V_{ce}. But they also show drawbacks like uncontrolled power loss or close-loop instability [6] ~ [9]. Thus, a new method which could overcome the drawbacks of previous works and maintain good voltage balancing performance need to be put forward for series-connected IGBTs.

In this paper, a novel active gate drive is proposed. By adjusting the switching time of each IGBT, The switching loss can be effectively decreased and the voltage balancing of the series-connected IGBTs is also achieved. The active gate drive is described in section II. In Section III, the implementation method of the AGD is discussed, and the effectiveness of the AGD has been verified by experimental results in Section IV.

II. PROPOSED ACTIVE GATE DRIVE

A. Switching Transients Description

Gate drive of today's IGBTs are in most cases voltage source based and with fixed gate resistors. Typical turn-on/off waveforms of a voltage source driven IGBT under a clamped inductive load is sketched in Fig. 1(a). The turn-on characteristics show the turn-on delay time before the rising of the collector current (T_1), the current rising period (T_2) and the voltage falling period (T_3). And the turn-off transients include the turn-off delay time before the collector voltage rising (T_4), the collector voltage rising period (T_5), the current falling period and the tail current period (T_6). More detailed description of the switching transients is available in [10] [11] and [12]. Previous works have already illustrated that the value of the gate resistor impact the IGBT switching performance significantly. A small value of the gate resistor can lead to small turn-on and turn-off delay time and high voltage changing rate, but also big current overshoot (I_{os}) during turn-on transient and big voltage overshoot (V_{os}) during turn-off transient. Consequently, the switching loss of the IGBT can be minimized at the price of relatively high device stress. On the contrary, if a large value of the gate resistor is applied, the device stress can be suppressed, but the switching loss will increase to an undesirable level. Such contradiction can be effectively resolved if the gate drive is able

---

This work was supported by the National Key Basic Research Program of China (973 Program) under project 2015CB251001 and 2015CB251004

to actively accelerate specified switching periods, which is shown in Fig. 1(b).

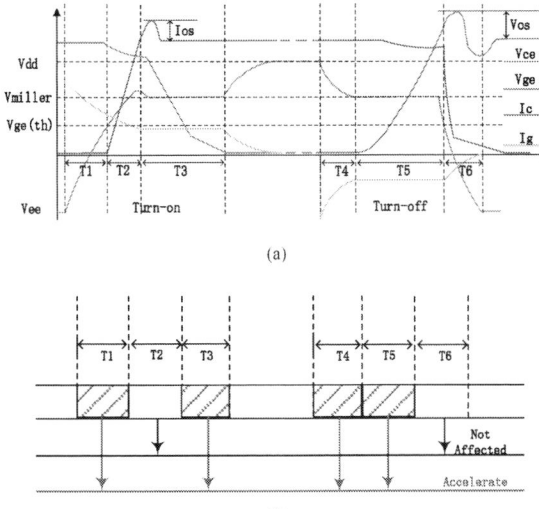

(a)

(b)

Fig. 1. (a) Switching waveforms of a voltage source driven IGBT. (b) Illustration of the switching period acceleration method.

B. Design of the Active Gate Drive

An active gate drive is proposed to both achieve the desired switching performance of a single IGBT and synchronize the voltage of series-connected devices. As shown in Fig. 2, a voltage source based conventional gate driver (CGD) is used as the main drive of the IGBT, and a complementary current source (CCS) is added to accelerate the switching of the IGBT. A local controller is applied to each IGBT to sample the gate-emitter voltage (V_{ge}) and collector-emitter voltage (V_{ce}) of the device, to provide drive signals to the CGD, and to control the CCS through a DA converter. Moreover, the load current I_{load} is sampled by the global controller. The global controller, which communicates with local controllers through the optocouplers,

is implemented to synchronize the gate drive signal of each IGBT, and provide the time reference to the local controller according to the sampled load current I_{load}. With the help of the CCS, each local controller is able to adjust the switching time of its corresponding IGBT by controlling the current amplitude and acting period of the CCS.

III. IMPLEMENTATION OF THE PROPOSED AGD

A. Implementation of the CCS

To improve the performance of the IGBT, a CCS is implemented to accelerate specified switching periods. When turn-on, T_1 and T_3 intervals need to be accelerated to reduce the delay time and voltage falling time, consequently the switching loss can be minimized. However, the current rising period T_2 should remain unaffected for not magnifying the reverse recovery current amplitude (I_{os}). Equation (1) and (2) show how the gate charge injected into the gate is calculated when the IGBT is driven by CGD.

$$Q_{T_1} = \int_0^{T_1} \frac{V_{dd} - V_{ge}(t)}{R_g} dt \quad (1)$$

$$Q_{T_3} = \int_{T_2}^{T_2+T_3} \frac{V_{dd} - V_{miller}}{R_g} dt \quad (2)$$

where Q_{T_1} is the gate charge within the T_1 interval, Q_{T_3} is the gate charge within the T_3 interval, V_{dd} is the positive gate drive voltage, R_g is the gate drive resistor value and V_{miller} is the miller platform voltage, .

Assume that the expected turn-on switching period is t_1, t_2 and t_3 if AGD is applied ($t_1 < T_1, t_2 = T_1, t_3 < T_3$), the equivalent circuit in t_1 and t_3 are shown in Fig. 3(a) and Fig. 3(b). In order to achieve the expected switching period, the CCS current within t_1 ($I_{gc(t_1)}$) can be calculated with (1) and (3), and

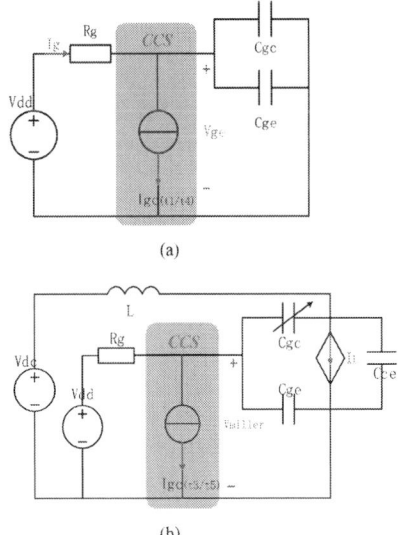

(a)

(b)

Fig. 3. Equivalent circuit of the IGBT. (a) T1 and T4 interval. (b) T3 and T5 interval

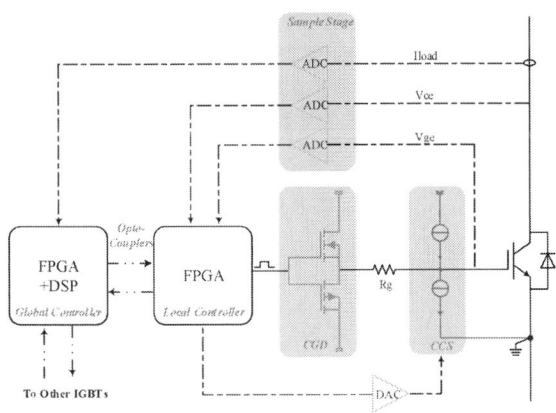

Fig. 2. Schematic diagram of the proposed active gate drive.

978-1-4673-9551-9/16 $31.00 © 2016 IEEE   2072

CCS current within t_3 interval ($I_{gc(t_3)}$) can be derived with (2) and (4).

$$\begin{cases} Q_{t_1} = Q_{T_1} = \int_0^{t_1} \frac{V_{dd}-V_{ge}(t)}{R_g} + I_{gc(t_1)}dt \\ V_{ge}(t) = V_{dd} - R_g[I_{gc(t_1)} + C_{iss}\frac{dV_{ge}(t)}{dt}] \end{cases} \quad (3)$$

$$Q_{t_3} = Q_{T_3} = \int_{t_2}^{t_2+t_3} \frac{V_{dd}-V_{miller}}{R_g} + I_{gc(t_3)} \, dt \quad (4)$$

where $I_{gc(t_1)}$ is the CCS current during t_1 interval, $I_{gc(t_3)}$ is the CCS current during t_3 interval and C_{iss} is the input capacitance of the IGBT

When turn-off, T_4 and T_5 intervals also need to be accelerated to achieve proper turn-off delay time and minimized turn-off loss. Within T_4 and T_5. The CCS will act similarly with that in the T_1 and T_3 intervals respectively since their equivalent circuit are actually the same. Therefore, the corresponding current amplitude of the CCS ($I_{gc(t_4)}$ and $I_{gc(t_5)}$) can be derived in the same method that $I_{gc(t_1)}$ and $I_{gc(t_3)}$ use.

B. Implementation of the Voltage Balancing Principle

Since the AGD has got the ability to adjust the duration of different switching state (T_1, T_3, T_4 and T_5), for IGBTs connected in series, voltage balancing will be achieved if the time references (t_1, t_3, t_4 and t_5) for all devices are given the same by the global controller.

The active gate drive will perform in an adaptive manner. The local controller will sample the switching variables (V_{ce} and V_{ge}) of the IGBT during the current switching cycle. The sampled data are collected and proceeded in the local controller. Before next cycle occurs, the global controller would give the same switching time reference value to all the local controllers. Basing on the sampled data of the current cycle and the time reference value given, each local controller will decide how and when the CCS works to achieve the best voltage balancing of the IGBT series.

IV. EXPERIMENT VERIFICATIOM

A. Turn-on & Turn off Performance Comparison

In order to validate the effectiveness of the proposed active gate drive, a gate drive circuit has been developed in which the functionalities are described earlier. The device under test is an Infineon 3.3kV/1.5kA IGBT/Diode module (FZ1500R33HE3), and the circuit used is a typical double pulse test circuit with a DC bus voltage of 1kV and a load inductance of 2.4mH.

Turn-on switching waveforms comparison of a single IGBT (bus voltage: 1000V, load current: 300A, gate resistance: 5Ω) are sketched in Fig. 4. It has shown that, with the conventional gate drive, the turn-on delay is around 1.6us, while it is reduced to around 1us when the AGD is implemented. The miller plateau period is reduced from 4us to 2.1us, thus the voltage falling period is decreased. Besides, the turn-on current overshoot is not magnified due to that the current rising period is not accelerated by the CCS, thus the current stress of the IGBT is not enlarged and the turn-on loss is also suppressed effectively. Similarly, as sketched in Fig. 5, the turn-off switching characteristics are also

improved by the AGD. The turn-off delay time of the IGBT is decreased by 1us, and the turn-off voltage rising period is also accelerated. However, since the current rising period is not affected by the CCS, the voltage overshoot during turn-off remains unaffected.

More detailed IGBT switching characteristics (Turn-on loss, Turn-on peak current, Turn-off loss and Turn-off peak voltage)

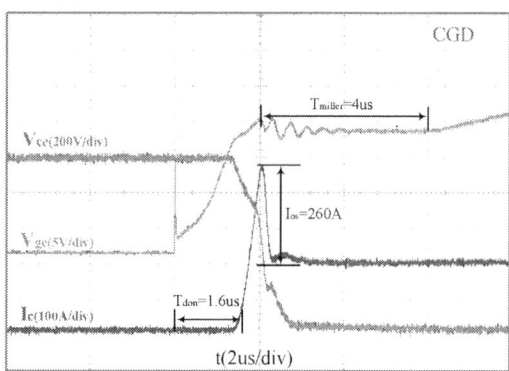

(a)

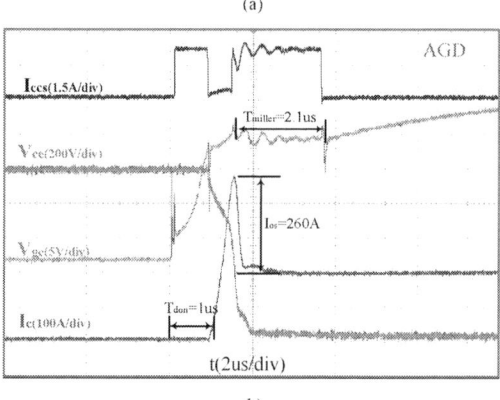

(b)

Fig. 4. Turn-on switching waveforms comparison. (a) CGD. (b) AGD

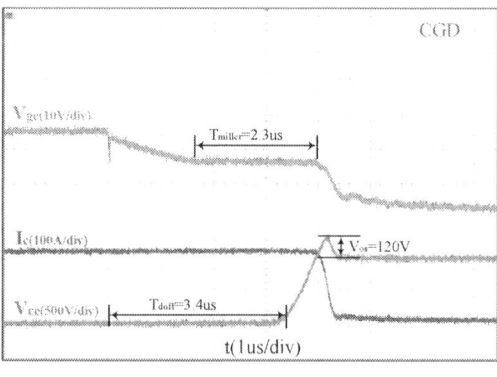

(a)

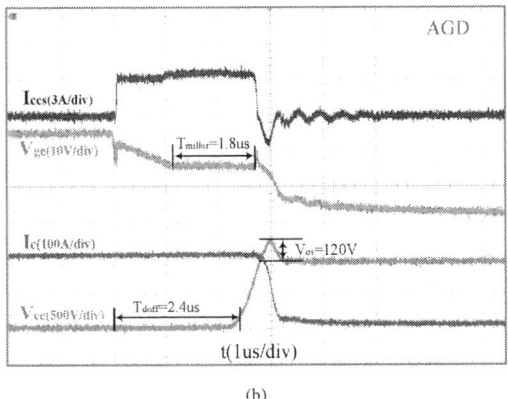

(b)

Fig. 5. Turn-off switching waveforms comparison. (a) CGD. (b) AGD

as functions of different gate resistance and different load current under 1000V bus voltage for both CGD and AGD are shown in Fig. 6.

As turn-on gate resistance increased, the turn-on delay time and the turn-on loss increase accordingly, while the turn-on current overshoot caused by the reverse recovery current of the free-wheeling diode decreases. The turn-on loss decreased by 20-50% and when different gate resistance and load current are applied. The same condition turns out when it comes to the turn-off transient. It can be observed that, in comparison of CGD, the turn-off loss of the IGBT is reduced by 10%~25% with AGD when different gate resistance and load current are combined. Besides, turn-off peak voltage remain unaffected when comparing with CGD.

B. Verification of the Voltage Balancing Principle

Fig. 7 shows the turn-off switching waveforms comparison of two series-connected IGBTs under 2000V bus voltage and 430A load current. In Fig. 7(a), when conventional gate drive is used to drive the series-connected IGBTs, different turn-off delay time of CGD leads to collector voltage deviation between two IGBTs during both the voltage rising period and the tail current period.

After AGD is applied in Fig. 7(b), both turn-off transients of two IGBTs are accelerated and the turn-off delay time have been synchronized by injecting different CCS current into the gate. Besides, the voltage rising period has also been accelerated to minimize the switching loss. Voltage spikes in the gate-emitter voltage (V_{ge}) waveforms indicates the current injecting period of the CCS. It has shown that the CCS starts extracting current from the gate as soon as the IGBT starts to turn-off, and right before the current falling period begins, the CCS is deactivated, the voltage overshoot will be suppressed therefore. It can be also observed that the voltage deviation of two IGBTs during the tail current period is minimized. This phenomenon can be attributed to the fact that the IGBT's tail current amplitude declines as their turn-off voltage reach higher. Therefore, the voltage unbalance during the tail current period can be resolved automatically if the transient turn-off voltage of the IGBTs are precisely synchronized.

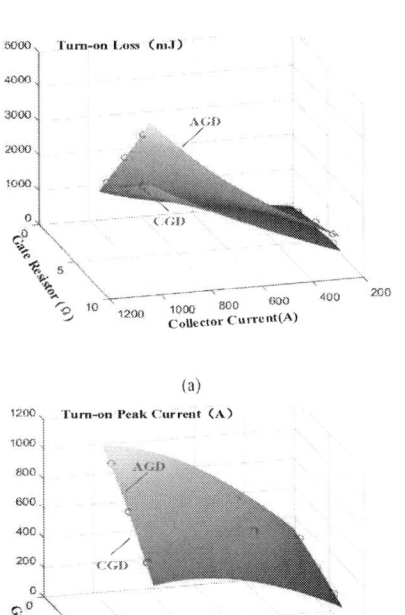

(a)

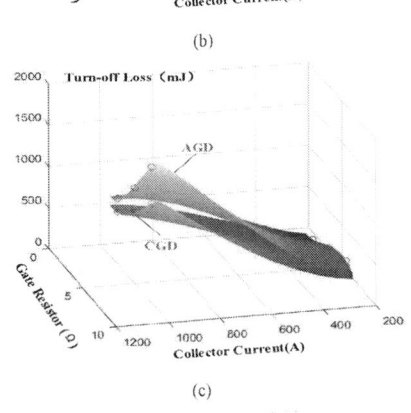

(b)

(c)

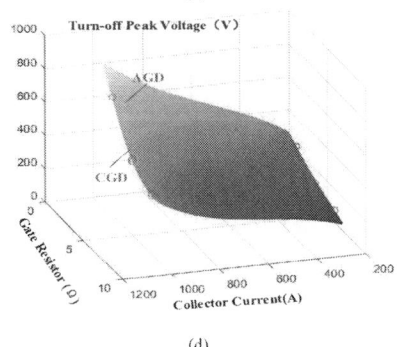

(d)

Fig. 6. Comparison of switching related characteristics under different gate resistor and load current between CGD and AGD. (a) Turn-on loss. (b) Turn-on peak current. (c) Turn-off loss. (d) Turn-off peak voltage.

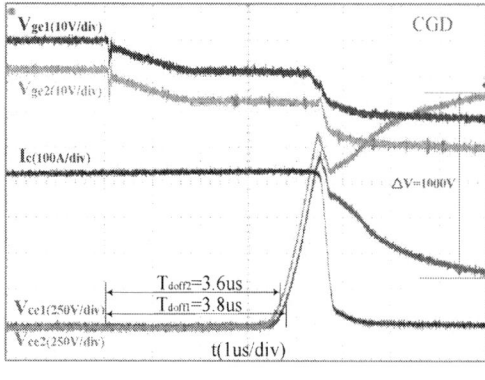

(a)

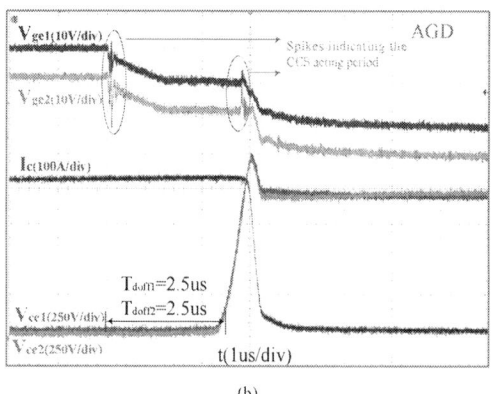

(b)

Fig. 7. Turn-off switching waveforms comparison of two series-connected IGBTs.

(a) CGD. (b) AGD.

V.　CONCLUSIONS

In this paper, an active gate drive was proposed for both switching loss reduction and voltage balancing of series-connected IGBTs. The schematic diagram of the active gate drive is sketched and the control method is discussed. A double-pulse test circuit for two series-connected IGBTs is established. The performance of the proposed active gate drive has been verified by experiments and promising results have been obtained.

Compared to the conventional gate drive, the proposed method has the following advantages:

- The switching speed can be adaptively adjusted according to the working condition of the present cycle and the reference of the following cycle. As a result, the switching stress of a single IGBT is effectively suppressed, and the switching loss is also minimized.

- By adding a complementary current source into the gate drive, the driving capability of the conventional gate drive is enlarged by injecting extra gate charge current during turn-on and extracting extra gate charge current during turn-off.

- When controlling the IGBTs connected in series, the time reference of all the devices are given the same. Therefore, the IGBTs will follow the time reference and switch at the same instance. Their voltage balancing will be assured accordingly.

- By utilizing the correlation between the turn off voltage and the tail current amplitude, voltage difference between different IGBTs at the tail current period can also be effectively reduced through the gate terminal.

Future works will focus on the reliability investigation of the active gate drive.

ACKNOWLEDGMENT

This work was supported by the National Key Basic Research Program of China (973 Program) under project 2015CB251001 and 2015CB251004.

REFERENCES

[1]　R. Withanage and N. Shammas, "Series connection of insulated gate bipolar transistors (IGBTs)," IEEE Trans. Power Electron, Vol. 27, No. 4, pp. 2204-2212, 2012.

[2]　J. F. Chen, J. N. Lin, and T. H. Ai, "The techniques of the serial and paralleled IGBTs," in Proc. IEEE IECON 22nd Int. Conf., 1996, vol. 5–10, pp. 999–1004.

[3]　R. Roesner, J. Holtz, and R. Kennel, "Cellular driver/snubber scheme for series connection of IGCTs," in Proc. IEEE 32nd Annu. Power Electron. Spec. Conf., 2001, pp. 637–641.

[4]　J. Saiz, M. Mermet, D. Frey, P. O. Jeannin, J. L. Schanen, and P. Muszicki, "Optimisation and integration of an active clamping circuit for IGBT series association," in Proc. Conf. Record 36th Ind. Appl. Conf., Sep. 30/Oct. 4, 2001, vol. 2, pp. 1046–1051.

[5]　T. Lu, Z. M. Zhao, S. Q. Ji, H. L. Yu, and L. Q. Yuan, "Parameter design of voltage balancing circuit for series connected HV-IGBTs," in Proc. IEEE IPEMC, Harbin, China, 2012, pp. 1502–1507.

[6]　A. Piazzesi and L. Meysenc, "Series connection of 3.3kV IGBTs with active voltage balancing," in Proc. IEEE 35th Annu. Power Electron. Spec. Conf., 2004, pp. 893–898.

[7]　C. Abbate, G. Busatto, and F. Iannuzzo, "High-voltage, high-performance switch using series-connected IGBTs," IEEE Trans. Power Electron., vol. 25, no. 9, pp. 2450–2459, Sep. 2010.

[8]　P. R. Palmer and A. N. Githiari, "The series connection of IGBTs with optimised voltage sharing in the switching transient," in Proc. IEEE Power Electron. Spec. Conf., Jun. 1995, pp. 44–49.

[9]　P. J. Grbovic, "An IGBT gate driver for feed-forward control of switch-on losses and reverse recovery current," IEEE Trans. Power Electron., vol. 23, no. 2, pp. 643–652, Mar. 2008.

[10]　F. Calmon, J. P. Chante, E. Reymond and A. Senes, "Analysis of the IGBT dv/dt in hard switching mode," in Proc. EPE'95, 1995, pp.1.234–1.239.

[11]　F. Zhang, Y. Ren, M. F. Tian and X. Yang, "A novel active gate drive for HV-IGBTs using feed-forward gate charge control strategy," 2015 IEEE Energy Conversion Congress and Exposition (ECCE), Sep. 2015, pp. 7009-7014.

[12]　V. Venketashan, M. Eshaghi, R. Borras, and S. Deuty, "IGBT turn-off characteristics explained through measurements and device simulation," in Proc. IEEE APEC'97, 1997, pp. 175-178

978-1-4673-9551-9/16 $31.00 © 2016 IEEE

A MV Intelligent Gate Driver for 15kV SiC IGBT and 10kV SiC MOSFET

Awneesh Tripathi, Krishna Mainali, Sachin Madhusoodhanan
Akshat Yadav, Kasunaidu Vechalapu, Subhashish Bhattacharya
Department of Electrical and Computer Engineering
North Carolina State University, Raleigh, NC 27606
{aktripat}, {kmainal}, {sbhatta4}@ncsu.edu

Abstract—This paper presents an Intelligent Medium-voltage Gate Driver (IMGD) for 15 kV SiC IGBT and 10 kV SiC MOSFET devices. The high voltage-magnitude and high dv/dt(> 30 kV/µs) of these MV SiC devices, pose design challenge in form of isolation and EMI. This problem is solved by development of a < 1 pF isolation capacitance power-supply. But due to applied high stress, smaller short-circuit withstand time and the criticality of the application, these devices need to be monitored, well protected, active gate-driven and safely shut-down. This paper presents an EMI hardened IMGD built around a CPLD, sensing and optical interfacing unit. It provides advanced gate-driving, added protection and optically isolated state-monitoring features. The device operating conditions such as module temperature and $V_{ds}(on)$ can be data-logged. They can be used for diagnosis/prognosis purposes such as to predict failure and safely shut-down the system. This paper describes the functionality of different building blocks. The 15 kV SiC IGBT has higher second switching slope above its punch-through level which is moderated without increasing losses by using digitally controlled active gate-driving. The shoot-through protection time can be reduced below withstand time by advanced gate driving. Soft turn-on and over-current triggered gate-voltage reduction helps reducing blanking time and quick turn-off reduces the protection response time. In this paper, the IMGD is high side tested at 5 kV with device state monitoring on. The active gate-driving is tested at 6 kV.

I. INTRODUCTION

The recent developments in SiC MOSFET and SiC IGBT device technology have paved way for simple non-series & non-cascaded medium-voltage (MV) converter topologies [1], [2]. These devices can operate at higher voltages and can switch at higher frequencies compared to their Si counterparts [3]. These characteristics enable applications such as high-speed MV drives, MV dc micro-grids and compact grid-connected converters for renewable energy [4], [5]. The challenges for a 15 kV SiC IGBT & 10 kV SiC MOSFET Gate-Driver (GD) design are the high-voltage isolation and high dv/dt (30 kV/µs) EMI immunity [6], [7]. A suitable such basic MV GD with < 1 pF isolation capacitance power-supply is presented previously in [8]. The driving and desat function is supported by a 2.5 Amp AVAGO driver with integrated (Vce) desat detection and fault status feedback [8]. It has recommended nominal blanking time of 2.8 µs and extra 2 µs fall time [9]. The typical protection time provided by this driver is 5 µs. The mentioned devices are still in research and

development stage and costly. The short-circuit (SC) protection of these devices has currently not been implemented as it is very challenging due to their characteristics [8]. Higher blocking capability for same buffer thickness and better thermal conductivity of these wide band-gap SiC devices have enabled high power density of the chip. But the reduction in chip size has also reduced short-circuit withstand time (SCWT).

These devices with small SCWT, cannot be protected by above AVAGO driver. A proper matching has to be done between protection time and SCWT. Reducing blanking time results in spurious trips. The functionality of the proposed Intelligent MV Gate Driver (IMGD) is shown in Table I. This IMGD provides sensing, advance gate-driving, protection and communication functions on top of the basic gate-driving. Advanced gate-driving using the CPLD has to be applied for improving SC protection. The 10 kV SiC MOSFET and the 15 kV SiC IGBT do not have sharp saturation curve for SC protection [8], [10]. In this paper, an active saturation control of SiC device is attempted. An over-current triggered device saturation is controlled using gate-voltage magnitude using a shunt connected BJT. This helps reduce blanking time and increasing SCWT. Active gate-driving is implemented for 15 kV SiC IGBT using CPLD controlled switching circuit to moderate the above punch-through dv/dt without slowing the device with gate-resistance.

Due to criticality of application and high stress on the device, it is important to acquire operating data from the device under test for diagnosis or predicting failure beforehand. Some of these functions have been previously implemented for low-voltage Si devices [11]–[13]. However this EMI hardened IMGD, based on a CPLD digital logic, provides optically isolated data-logging under given high dv/dt and isolation voltage. Even in the harsh environment, it measures near-chip module temperature (T_{mod}), drain on-voltage ($V_{ds}(on)$) & device current (I_d). The fault signal and measured data are optically sent to the control side FPGA based IMGD interface board for decoding, data-logging and diagnosis [14], [15]. The purpose of the intelligence addition is generally optimization in terms of the switching loss and the EMI. It also helps for matched series/parallel operation of the devices for scalability [14], [15]. The CPLD provides flexibility in configuring the signals and protections based on custom application. Together

978-1-4673-9551-9/16 $31.00 © 2016 IEEE

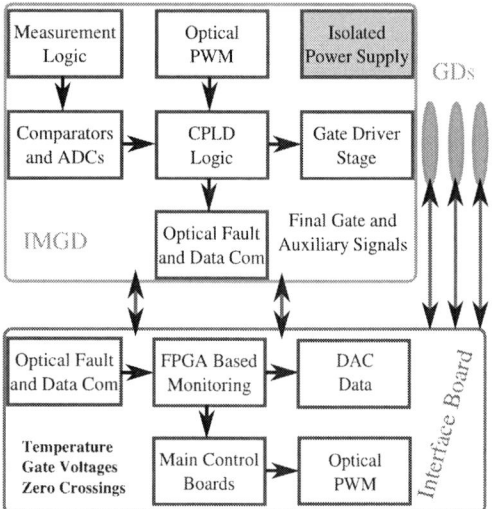

Fig. 1: Block Diagram of Intelligent MV Gate Driver

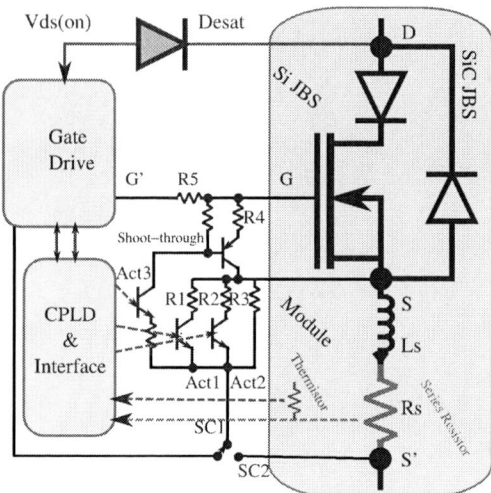

Fig. 2: Active Gating & Protection Circuit

TABLE I. Intelligent MV Gate Driver Functions

Parameters/Functions	Value[Units]
Isolation Voltage	20 kV
Isolation Capacitance	< 1.5 pF
Drive Voltages	+20/ − 5 V
Optical Communication	$V_{ds(on)}, I_{ds}, T$
Shoot-through Protection	V_{gs} Controlled
Protections	Local OT, OC, ST
Active Gating	Clock Timed

with a local clock, it helps in timing the events such as advanced gating and protections. It interfaces ADCs with the optical signals.

This paper is organized in following sections. Section II presents the basic block diagram. Section III presents the advanced gate-driving feature. Section IV presents sensing circuits for I_d, $V_{ds}(on)$ and T_{mod}. Section V presents the populated IMGD and interface board. Section VI presents the digital implementation and programming. Section VII presents the experimental results. The study is finally concluded in section VIII.

II. BLOCK DIAGRAM

Fig. 1 shows the block diagram of IMGD. It mainly consists of the sensing & measurements, comparators, ADCs, communication, active-gating and the driver stages. A 44 pin ALTERA CPLD EPM3032A is used as the local brain. The simple 6 pin, 40 MHz serial 8 bit, 3 MSPS ADCs are used to convert the critical measurements. The PWM and fault signals are routed through CPLD. The local Over-Temperature (OT) and Over-Current (OC) fault signals are generated using ultra-fast comparators. This board has two optical receivers and four optical transmitters which are configured by CPLD as PWM, clock, fault and serial ADC signals. These IMGDs communicate with an interface board which can do the data-logging and independent abnormal condition controls. The FPGA at the

interface board can implement the communication decoding and other digital functions such as status display. The main control boards based on DSP or FPGA, implement the actual control algorithm for the converter systems such as Grid-tie Front End Converter [4] and DC-DC Dual Active Bridge [5]. They also have separate provisions for over-voltage and OC protections of the converters. They need feedback sensors such as voltage and current sensors for closed-loop control. The PWM cables are optical. The feedback sensors must have very small coupling capacitance and good isolation level for EMI resistant functioning.

III. ACTIVE GATING AND SHOOT-THROUGH PROTECTION CIRCUITS

A. Active Gate Driving

As shown in Fig. 2, a pair of resistance paths are switched between S and GD ground SC1 to create a dynamically changing effective gate resistance during device switching. The sequence and delay of these switching is digitally controlled by the CPLD in steps of local 100 MHz clock. The advantage is mostly for SiC IGBTs above punch-through voltage switching [7]. The steeper dv/dt can now be controlled without increasing the gate-resistance and the losses. This feature can also be used with MOSFET/IGBT to reduce the spike current during turn-on so that desat blanking time could be reduced by avoiding spurious fault detections. Since the AVAGO driver has 300 ns delay between input PWM and output pulse, the active BJTs timings are relative to the output pulse by feeding it back to the CPLD [9].

B. Shoot-through or SC protection by controlled saturation

The present AVAGO driver is retained on this IMGD for driving stage similar to the previous 15 kV SiC IGBT GD [8]. Fig. 3a and Fig. 3b show the SiC MOSFET and IGBT forward characteristics. It can be seen that both devices have limited saturation current for V_{gs} in range of 6 to 8 Volts. Fig. 4 shows the timing diagram of AVAGO driver. After the fault

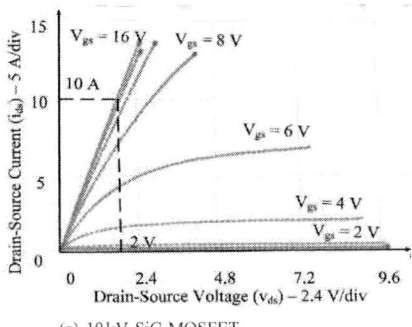

(a) 10 kV SiC MOSFET

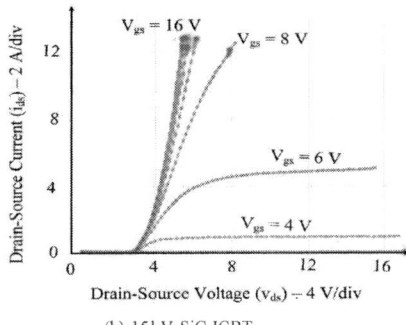

(b) 15 kV SiC IGBT

Fig. 3: Forward Characteristics of the MV SiC Devices

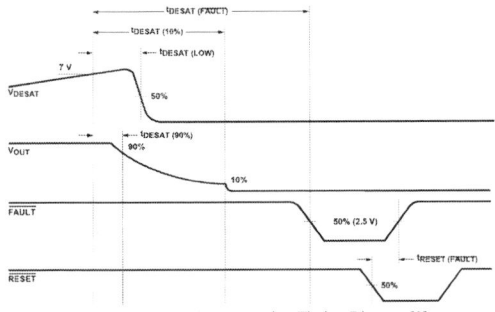

Fig. 4: AVAGO Driver Protection Timing Diagram [9]

is detected, it takes $t_{desat}(90\%V_{gs}) = 300$ns to start dropping the V_{gs}. The $t_{desat}(10\%V_{gs})$ depends on R_g and C_{gs}, and can be typically in order of $2\,\mu$s. A blanking time of $t_{blank}=2.8\,\mu$s is recommended to avoid dv/dt noise. So the minimum protection time is $t_{desat}(10\%V_{gs}) + t_{blank}$ which is typically $5\,\mu$s. The shoot-through solution for SiC 10 kV MOSFET and 15 kV IGBT is difficult as it does not allow this magnitude time duration before it fails. Therefore this typical time has to be shortened: 1. By turning device quickly off during desat=7V detection using very low off gate-resistance in order of 1 or $2\,\Omega$, 2. Minimizing t_{blank} but at the cost of increased noise pick-up. The second active pulse terminal can be used to reduce gate resistance during desat trigger but it is internal signal and its difficult to change gate-resistance externally without this trigger information. Fig. 2, a gate-voltage level reduction using a shunt switch circuit during short/overload is used to protect the device by forcefully saturating it. A regulator cannot achieve this as it cannot discharge 20 V quickly, so a switching topology is used to change voltage quickly [15]. The OC level for triggering the shunt circuit is kept 2-3 times peak nominal current and is enabled using a digital blanking time to avoid spurious trigger which would result in abnormal gate-driving. At the start of the gate-pulse, the comparator is disabled for specific time duration and then used to latch this OC fault. With reduced gate-voltage during OC the MOSFET is more likely to saturate [10]. This enlarges the SCWT of the device and also makes the desat detection fast together with the soft turn-on using active gating described above. The AVAGO

driver can now generate a desat fault signal in case of the real SC equivalent to smaller blanking time. Another provision for SC protection is to dynamically use device S and $\overline{S}$ terminals with overlapping switches such that during SC, gate voltage is naturally lowered by sense resistor to protect the device. Both of these methods are tried in this GD board as marked in Fig. 2.

IV. CIRCUIT IMPLEMENTATIONS

The SiC MOSFET and IGBT modules are supposed to have inbuilt shunt resistor and a thermistor temperature sensor as shown in Fig. 2. They provide foundation for the IMGD main functions. Fig. 1 block diagram components are explained in following points.

- A thermistor signal amplifier is on board for signal conditioning. The INA330 is used for this purpose and the resistor values are designed using (1) and configured for temperature measurement up to 80°C currently [16].

$$R_m^{Min/Max} = \frac{R_G}{\frac{R_G}{R_1} \pm 1.65} \qquad (1)$$

- Two current sense differential amplifiers are placed for device and gate current signals. They are based on fixed gain (G=20) current sense amplifier ADM4073 [17].
- Differential amplifier for on-state drain voltage measurement is also included.
- Ultra-fast high temperature EMI hardened comparators & analog to digital converters (ADC) convert the measured signals into digital domain for outside optical communication. The used comparator LMV7219 has 7ns delay with internal hysteresis [18]. The used ADC AD7278 has high speed serial interface with 3 MSPS throughput [19]. Fig. 5 shows the timing diagram for ADC interfacing. The clock-speed for nominal throughput is 36 MHz. Due to limitation of optical bandwidth of 10 MHz, a lower word data rate of 250 kSPS is implemented. The spike data such as gate-current does not need to be converted to 8 bit, rather it can be sensed with comparators for few selected di/dt slopes.
- Optical Links: Individual IMGDs of a converter need to send fault signals to main control board. All the IMGDs to

978-1-4673-9551-9/16 $31.00 © 2016 IEEE

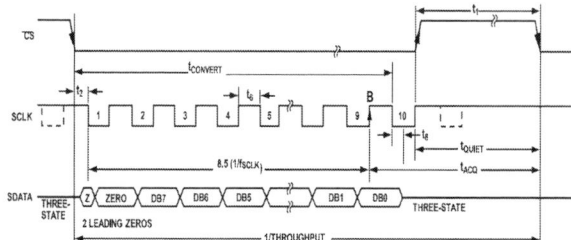

Fig. 5: ADC Timing Diagram [9]

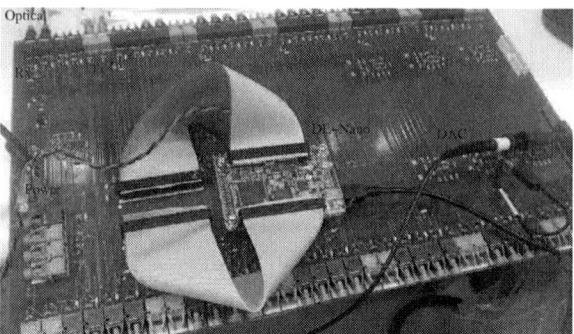

Fig. 7: IMGD Interface Circuit Board

Fig. 6: Intelligent Gate Driver Circuit Board

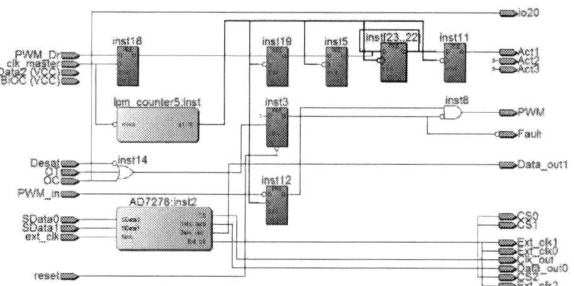

Fig. 8: Altera Quartus II RTL View of the CPLD Implementation on IMGD

and data-logging. Based on any abnormality or any fault signal from any particular IMGD, all the IMGDs can be turned off together for safety.

VI. Digital Implementations

The motivation for using CPLD is that it is digital and the in-circuit programmable chip once programmed, works as a hard wired custom logic. There is no loading time delay as in case of a FPGA with an EEPROM. The digital processing is more noise resistant compared to the analog processing for active gate-drive, smart protection and optical-communication functions. Considering the harsh EMI environment for the GD, a CPLD EPM3032A with 34 I/O, 5 ns pin to pin delay, 32 macro-cells and 600 gates size is selected to implement the required digital functions. There is individual on-board clock 100 MHz for timing the active-gating and desat protection signals. This allows 10 ns resolution in the timing. For the active-gating, the driver output PWM is synchronized with the local clock using a register and other cascaded registers

interface communication including PWM and fault signal, is via optical fiber cables to provide necessary isolation. Important ADC and comparator information also needs to be sent as digital signal using simple protocols. The optical cable uses Avago HFBR-1528 transmitter and HFBR-2528 receiver with 10 MBd signaling rate. This signaling rate is sufficient for sending important information to the interface board.

V. IMGD and Interface Boards

Fig. 6 shows the populated IMGD board with all mentioned functions. This is the first version board with extra components than the requirement. In the final version this will be optimized. It has different analog and digital functional sections which are separated from one another for better EMI performance. Fig. 7 shows the populated interface board which communicates with the IMGDs of a MV converter. It has on-board DACs and FPGA board for implementing monitoring and data-logging function. Each IMGD has 4 Tx and 2 Rx used respectively as Fault, DT0, DT1, CS, CK and PWM. The interface has the opposite type optical Tx/Rx for all signals and 12 IMGDs. The interface FPGA transmits a common clock for IMGD ADC operation and communication. The IMGD generates CS signal for framing of serial data and each bit is read synchronous to the transmitted clock of interface. The critical measurements can be converted by DACs for display

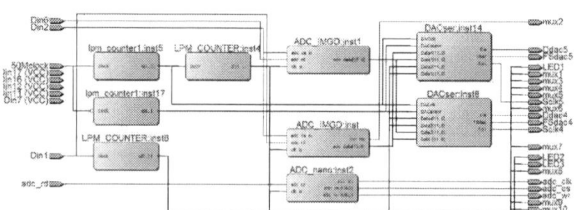

Fig. 9: Altera Quartus II RTL View of the Interface Implementation

978-1-4673-9551-9/16 $31.00 © 2016 IEEE

TABLE II: IMGD CPLD Utilization

Parameters	Value[Units]
Family	MAX3000A
Device	EPM3032ATI44-10
Timing Models	Final
Total macrocells	27/32(84%)
Total pins	30/34(88%)

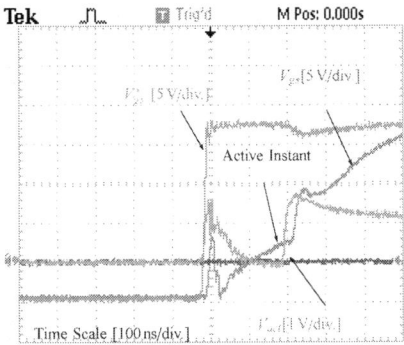

Fig. 10: Active Pulse Generation Test

Fig. 11: Double-Pulse Test Setup

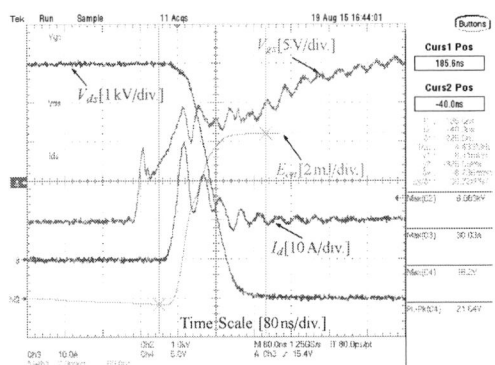

Fig. 12: Passive Gating Double-pulse Test

are fed with appropriate divided clocks to generate delay. The interface generated optically received clock is used for ADC operation and communication as there is no need for synchronization and more than 10MBd speed. The CPLD implements five main functions a.) ADC protocol, b.) Oring and latching the local protections b.) Optical comm. to interface board, c.) Active gate-drive by switching transistors with series resistance, d.) Smart protection by shunt transistor-series resistance switching in case of over-current. Fig. 8 shows the RTL view of the simple implementation of latched fault protected PWM gate-driving with active-gating, data-logging and OT, OC protections. It takes sensor digital inputs, optical PWM, local & external data clocks, desat, driver output and generates digital data, serial data-frame CS, common-fault and PWM for desired purpose. Table II shows the device utilization for the CPLD with mentioned implementation. The selected CPLD is optimum in terms of complexity and resources. Fig. 9 shows the RTL view of the simple implementation of FPGA based interface board. It is configured for this basic test to read two data lines of an IMGD and direct it to on-board two DACs for logging. It provides PWM and communication clock to this IMGD as well.

VII. Experimental Results

The designed gate-driver is tested on a 10 kV SiC MOSFET and 15 kV SiC IGBT. The gate voltage control, temperature measurement, active gate resistance and current sensing features are tested. The digitally controlled active gating pulse w.r.t. the Driver output PWM is shown in Fig. 10. The waveform shown is BJT base voltage w.r.t. control ground. The V_{gs} is the gate-voltage measured at the IMGD output pins. The connecting wire inductance causes a positive spike during turn-on transition but it does not appear on the device termination pins. Fig. 11 shows the Double-pulse (DP) test hardware setup photograph for validating the IMGD for its driving and active gating capability. Fig. 12 shows DP switching test of 10kV SiC MOSFET at 6kV and 10A with 10Ω R_g. A turn-on spike can be seen shooting to 30A. Fig. 13 shows the result under same conditions for testing the active-gating functionality. The starting R_g is 50Ω. It can be seen that the active circuit with 4.7Ω R_g is switched on after certain delay. This delay can be tuned such as to reshape the V_{ds}. This will be helpful in soft turn-on and also in moderating 15 kV SiC IGBT dv/dt as

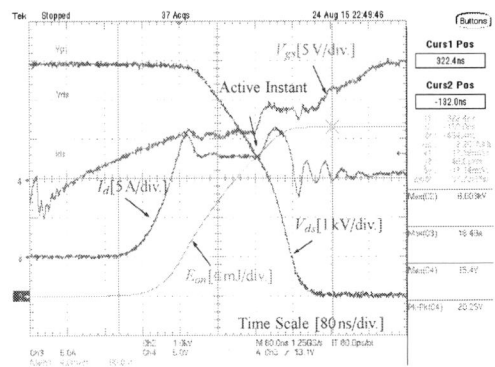

Fig. 13: Active Gating Double-pulse Test

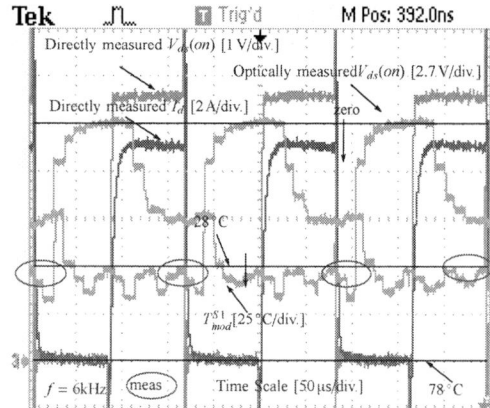

Fig. 14. Monitoring Test with IMGD using Genesic Desat Diode at 60V

Fig. 15: Boost-Buck Setup

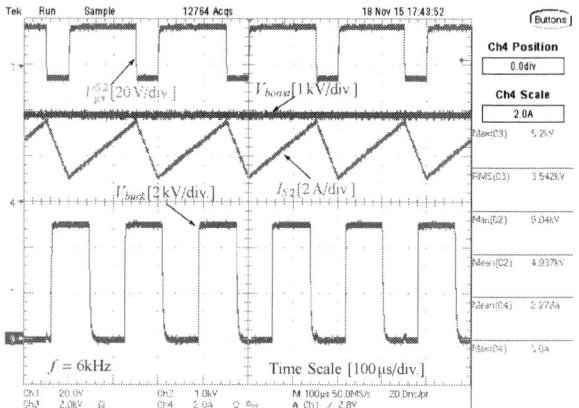

Fig. 16: 5 kV Boost-Buck GD Qualification Results

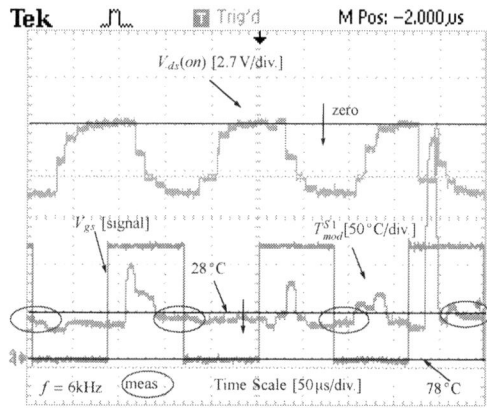

Fig 17. 5 kV Boost-Buck GD Qualification Interface side Results

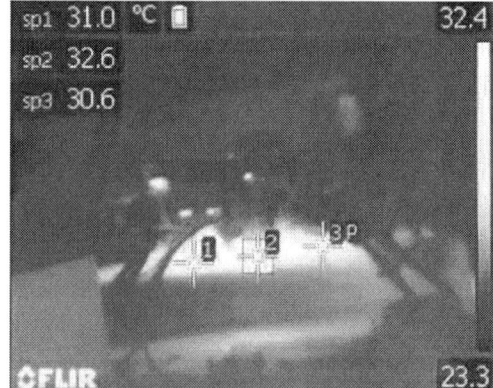

Fig. 18: Boost-Buck Test Thermal Photograph

discussed earlier.

Fig. 14 shows the LV test for verifying optical measurements from the IMGD and the desat function using Genesic 8 kV desat diode. This is a 50% duty test on a resistor load in series with the IGBT. This desat setting allows upto 9 A current for this IGBT. The desat-sensing, $V_{ds}(on)$, T_{mod} and I_d measurements are verified in this setup. The $V_{ds}(on)$ signal is shown in negative polarity for current DAC setting but it can be scaled as per requirement. Fig. 15 shows the boost-buck hardware setup photograph for validating the IMGD with its basic functionality. 15 kV SiC IGBT is selected for this test as it applies harsher dv/dt on the IMGD above punch-through level. Fig. 16 shows the results for 5 kV operation for the IMGD at MV potential. The boost input is 1.25 kV and output is 5 kV. The boost duty is 25%. The buck converter IGBT switch is now at higher potential so that its IMGD gets 5 kV pulsating stress. The buck duty is 50% feeding a resistive load for simplicity. The IMGD input power supply is at ground reference and the output which powers the IMGD, is at 5 kV stress. Fig. 17 shows the optically transmitted $V_{ds}(on)$ and T_{mod}^{S1} measurements coming from the IMGD out of high voltage setup. The T_{mod}^{S1} signal gets distorted near device voltage transition due to change in the ground level and EMI. Normally it is not required to be logged so fast, therefore it can be measured only in the encircled zone and averaged for logging. Fig. 18 shows the thermal photograph of the setup in Fig. 15 after 30 min thermal run at 5 kV and 3 kW power in boost-buck operation.

VIII. CONCLUSIONS

A MV intelligent gate-driver has been presented in this paper which is not only capable of active gate driving and extended shoot-through protection for SiC 10 kV MOSFET and 15 kV IGBT but also able to fetch real-time operating data to the outside control for diagnosis and prognosis purposes. The $V_{ds}(on)$ can be very useful to determine any aging issue. Some of the features of this gate-driver such as temperature, current, $V_{ds}(on)$ sense in high side 5 kV converter and 6 kV double-pulse active gating have been presented in this paper. The shoot-through test has not been completed.

IX. ACKNOWLEDGMENT

The information, data, or work presented herein was funded in part by the Office of Energy Efficiency and Renewable Energy (EERE), U.S. Department of Energy, under Award Number DE-EE0006521 with North Carolina State University, PowerAmerica Institute. This work made use of FREEDM ERC shared facilities supported by NSF under award no. EEC-0812121.

X. DISCLAIMER

The information, data, or work presented herein was funded in part by an agency of the United States Government. Neither the United States Government nor any agency thereof, nor any of their employees, makes any warranty, express or implied, or assumes any legal liability or responsibility for the accuracy, completeness, or usefulness of any information,apparatus, product, or process disclosed, or represents that its use would not infringe privately owned rights. Reference herein to any specific commercial product, process, or service by trade name, trademark, manufacturer, or otherwise does not necessarily constitute or imply its endorsement, recommendation, or favoring by the United States Government or any agency thereof. The views and opinions of authors expressed herein do not necessarily state or reflect those of the United States Government or any agency thereof.

REFERENCES

[1] W. van der Merwe and T. Mouton, "Solid-state transformer topology selection," in *IEEE International Conference on Industrial Technology (ICIT)*, 2009, pp. 1–6.

[2] K. Hatua, S. Dutta, A.Tripathi, S. Baek , G. Karimi, S. Bhattacharya, "Transformer less Intelligent Power Substation design with 15kV SiC IGBT for grid interconnection", *IEEE Energy Conversion Congress and Exposition (ECCE)*, pp. 4225-4232, Sept. 2011.

[3] A. Kadavelugu, S. Bhattacharya, S.-H. Ryu, E. Van Brunt, D. Grider, S. Leslie, "Experimental switching frequency limits of 15 kV SiC N-IGBT module," in *Power Electronics Conference (IPEC-Hiroshima 2014-ECCE-ASIA), 2014 International.* IEEE, 2014, pp. 3726–3733.

[4] S. Madhusoodhanan, A. Tripathi, D. Patel, K. Mainali, A. Kadavelugu, S. Hazra, S. Bhattacharya, and K. Hatua, "Solid State Transformer and MV grid tie applications enabled by 15 kV SiC IGBTs and 10 kV SiC MOSFETs based multilevel converters," *Power Electronics Conference (IPEC-Hiroshima ECCE-ASIA)*, pp. 16261633, 2014.

[5] A. Tripathi, K. Mainali, D. Patel, A. Kadavelugu, S. Hazra, S. Bhattacharya, K. Hatua, "Design Considerations of a 15-kV SiC IGBT-Based Medium-Voltage High-Frequency Isolated DC-DC Converter," *IEEE Transactions on Industry Applications*, vol. 51, no. 4, pp. 3284-3294, July 2015.

[6] A. Kadavelugu, S. Bhattacharya, S. Ryu, E. V. Brunt, D. Grider, A. Agarwal, S. Leslie, "Characterization of 15 kV SiC n-IGBT and its Application Considerations for High Power Converters", *Energy Conversion Congress and Exposition (ECCE)*, pp. 2528-2535, 2013.

[7] A. Kadavelugu, S. Bhattacharya, S. Leslie, Sei-Hyung Ryu, D. Grider, and K. Hatua. "Understanding dv/dt of 15 kV SiC N-IGBT and Its Control Using Active Gate Driver." *IEEE Energy Conversion Congress and Exposition (ECCE)*, 221320, 2014.

[8] A. Kadavelugu, S. Bhattacharya, "Design Considerations and Development of Gate Driver for 15 kV SiC IGBT," *Applied Power Electronics Conference and Exposition (APEC)*, pp. 1494-1501, 2014.

[9] "Avago HCPL-316J Datasheet" [Online]. Available: http://www.avagotech.com/docs/AV02-0717EN

[10] M.K. Das, C. Capell, D.E. Grider, R. Raju, M. Schutten, J. Nasadoski, S. Leslie, J. Ostop, A. Hefner, "10 kV, 120 A SiC half H-bridge power MOSFET modules suitable for high frequency, medium voltage applications", *IEEE Energy Conversion Congress and Exposition (ECCE)*, pp. 2689 - 2692, 2011.

[11] M. Zdanowski, J. Rabkowski, and R. Barlik, "Design issues of the high-frequency interleaved DC/DC boost converter with Silicon Carbide MOSFETs," in *2014 16th European Conference on Power Electronics and Applications (EPE'14-ECCE Europe)*, Aug. 2014, pp. 1–10.

[12] T. Kosan, J. Molnar, L. Streit, L. Polacek, and Z. Peroutka, "Complete design of down-scale prototype of mining machine converter based on four-level voltage-source converter with flying capacitors," in *Power Electronics and Motion Control Conference (EPE/PEMC), 2012 15th International*, Sept. 2012, pp. DS2b.4–1–DS2b.4–6.

[13] Z. Wang, X. Shi, L. Tolbert, F. Wang, and B. Blalock, "A di/dt Feedback-Based Active Gate Driver for Smart Switching and Fast Overcurrent Protection of IGBT Modules," *IEEE Transactions on Power Electronics*, vol. 29, no. 7, pp. 3720–3732, July 2014.

[14] Y. Lobsiger and J. Kolar, "Closed-Loop d d and d d IGBT Gate Driver," *IEEE Transactions on Power Electronics*, vol. 30, no. 6, pp. 3402–3417, June 2015.

[15] D. Bortis, J. Biela, J.W. Kolar, "Active Gate Control for Current Balancing of Parallel-Connected IGBT Modules in Solid-State Modulators", *IEEE Trans. Plasma Sci.*, vol. 36, no. 5, pp. 26322637, Oct 2008.

[16] "Texas Instruments INA330 Datasheet" [Online]. Available: http://www.ti.com/lit/ds/sbos260/sbos260.pdf

[17] "Analog Devices ADM4073 Datasheet" [Online]. Available: http://www.farnell.com/datasheets/690988.pdf

[18] "Texas Instruments LMV7219 Datasheet" [Online]. Available: http://www.ti.com/lit/ds/symlink/lmv7219.pdf

[19] "Analog Devices AD7278 Datasheet" [Online]. Available: http://www.analog.com/en/products/analog-to-digital-converters/ad-converters/ad7278.html#product-overview

Linear Temperature Sensors in High-Voltage GaN-HEMT Power Devices

Richard Reiner, Patrick Waltereit, Beatrix Weiss, Matthias Wespel, Dirk Meder, Michael Mikulla, Rüdiger Quay, and Oliver Ambacher

Fraunhofer Institute for Applied Solid State Physics (IAF),
Tullastrasse 72, 79108 Freiburg, Germany, Phone: +49 761 5159-552,
Email: richard.reiner@iaf.fraunhofer.de

Abstract— This work presents a high-voltage GaN-based power HEMT with a highly-linear, monolithically-integrated temperature sensor. The principle is shown and compared to other concepts. The sensor is fabricated by using a interconnect metallization without additional process steps. The performance of the sensor as well as of the power device is characterized. The 600 V power device achieves an on-state resistance of $R_{ON} = 55\ m\Omega$ at a corresponding drain current $I_D = 30$ A and an advanced dynamic performance with a low gate charge of 20 nC.

Keywords—thermal; overload; thermistor; HFET; PTC.

I. INTRODUCTION

Gallium nitride (GaN) power devices allow the operation at high off-state voltages, high on-state currents, and high junction temperatures, with fast switching edges and high converter efficiencies, as shown e.g. in [1-3]. Furthermore GaN can be grown on inexpensive large-area Si-substrates. These capabilities permit the reduction of weight, volume, and costs of power electronic circuits in a wide range of applications such as automotive, aviation, and energy conversion. On the one hand, the capability of GaN-HEMTs to work at high operation temperatures reduces the effort for cooling. On the other hand, the device performance and reliability suffer at high temperatures [4]. A monitoring of the precise in-chip temperature enables an optimum control of the operation adapted to the matters of the application.

In previous works the temperature dependence of the two-dimensional electron gas (2DEG) has been used as resistive sensor, e.g. in [5] and [6], however its temperature dependence is non-linear. A lookup table or a more complex model has to be used for the signal interpretation. In [7] a proportional-to-absolute-temperature voltage source has been used to sense the temperature of an integrated circuit, but this concept is not suitable for large area temperature sensing on power HEMTs. In [8] a temperature sensor is processed above an ungated GaN-HEMT channel. The structure is suitable for physical characterization of 2DEG, but cannot be implemented in high-voltage, fast switching GaN-HEMT devices.

This work presents a high-voltage GaN-on-Si HEMT with a highly-linear, monolithically-integrated temperature sensor for power electronic applications.

This work was supported partly by the German Federal Ministry for Environment. (BMU) through grant "GaNPV" FKZ: 0325529 and by the German Federal Ministry of Education and Research (BMBF) through grant "ZuGaNG" FKZ: 16ES0076K.

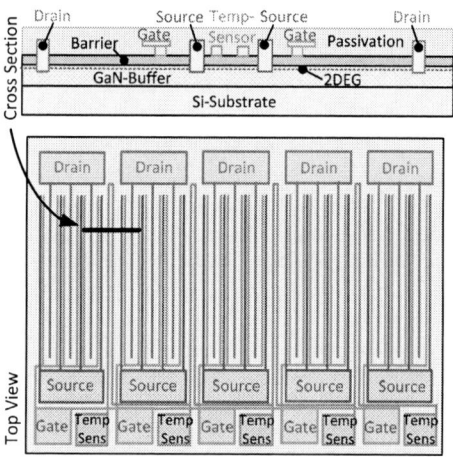

Fig. 1. Functional principle and design of a GaN HEMT with integrated temperature sensor. The design is shown as cross section and as the top view.

II. DESIGN AND FABRICATION

The power device is fabricated in a high-voltage AlGaN/GaN-on-Si technology, which is published in [9]. The schematic layout of the chip is illustrated in Fig. 1. The chip has 14 separated comb cells, and each unit has a gate width of 24 mm. Thus the $4\times4\ mm^2$ chip has a total gate width of $W = 337$ mm. The temperature sensor is processed as a thin-film gold metallization chain between two passivation layers in the gap between the HEMT cells and is realized using a interconnect metallization of the HEMT process. The temperature sensor loops can be accessed separately at each cell or end to end to detect the total chip temperature. The HEMT cells are connected in parallel by bond wires on common drain- and source-pad-areas on a laser structured Cu / AlN / Cu power board, which is shown in Fig. 2. The measurement circuit is shown in Fig. 3. A constant sense current of 10 µA is applied and flow through the sensor turns. The temperature signal is measured as voltage directly at the temperature sensor pads. Therefore two separated bond wires are connected with both outer sensor pads to achieve an accurate four-point measurement.

Fig. 2. Test assembly for the characterization of the integrated temperature sensor. Layout of a GaN-HEMT with integrated temperature sensor. The HEMT has a total gate width of $W = 337$ mm and a total area of 4×4 mm².

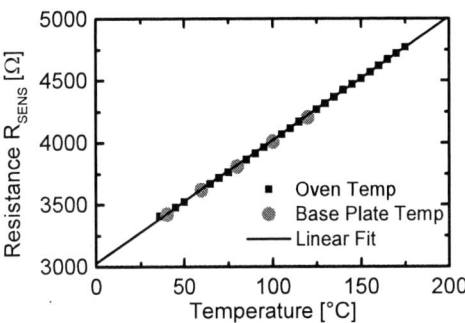

Fig. 4. Characterization of the integrated temperature sensor. Sensor resistance is measured as a function of the temperature. The measurements were made on a hot plate as well as in an oven.

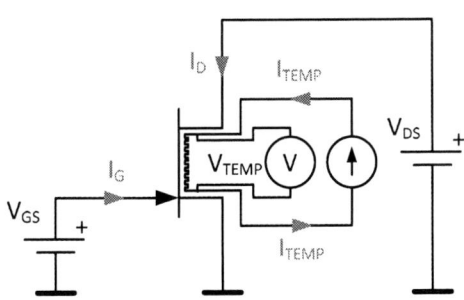

Fig. 3. Test circuit of the GaN HEMT with integrated temperature sensor. The sensor signal is measured by a four point measurement circuit. A constant current of $I_{TEMP} = 10$ µA is applied to generated a sensor signal which is detected by a voltmeter.

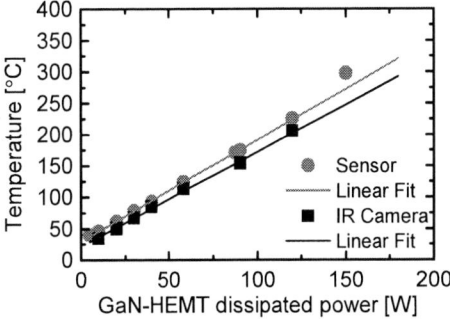

Fig. 5. Sensor temperature as function of the dissipated power of the GaN-HEMT. The assembly is mounted on a cooled heat sink with a base plate temperture of 25 °C.

III. CHARACTERIZATION OF THE TEMPERATURE SENSOR

The temperature-dependent resistance R_{SENS} of the sensor was characterized in an oven as well as on a hot plate. Both methods show well coincident results, which are shown in Fig. 4. The resistance of the gold metallization increases linearly with temperature in the operation range of the power devices. The sensor increase it resistance R_{SENS} starting from 3400 Ω at 35 °C up to 4800 Ω at 175 °C. Thus a linear approximation can be assumed and expressed by:

$$T_{SENS} = 0.1 \frac{°C}{\Omega} \cdot R_{SENS} - 305 \, °C \cdot \quad (1)$$

In a following experiment the sensor values have been used to characterize the in-chip temperature as a function of the dissipated GaN-HEMT power, which is shown in Fig. 5. Different operational points have been used to generate heat in the GaN-HEMT chips. The assembly was mounted on a base plate held at a constant temperature of $T_{BASE} = 25$ °C. By using the sensor values the thermal resistance of the assembly is found to be $R_{TH} = (T_{CHIP} - T_{BASE})/P = 1.6$ K/W.

In Fig. 5 the sensor in-chip temperature is correlated to the chip surface temperature, which was measured using an infrared (IR) camera. As expected, the surface temperature is slightly cooler compared to the in-chip temperature because the in-chips sensor turns are close to the transistor channels, which act as heat sources. Whereas the chip surface temperature is cooled by the ambient air.

The transient thermal behavior of the assembly is characterized in Fig. 6. Therefore the sensor signal response of a power step has been measured. The tested device (shown in Fig. 2) was assembled on a heat sink without active cooling at room temperature. The power step was realized applying a current step from 0 to $I_D = 5$ A. The measurements have been made in the on-state with a gate-source voltage of $V_{GS} = 0$ V. Under these conditions a power of around $P = 1.4$ W is dissipated in the chips. As step response the in-chip temperature raises from 21 °C to 30 °C within few seconds. This slowly increase is caused by the thermal capacities of the assembly. However the temperature sensor itself reacts within microseconds to the power step and starts to increase the temperature signal, because of the physical proximity between sensor turns and heat source.

978-1-4673-9551-9/16 $31.00 © 2016 IEEE

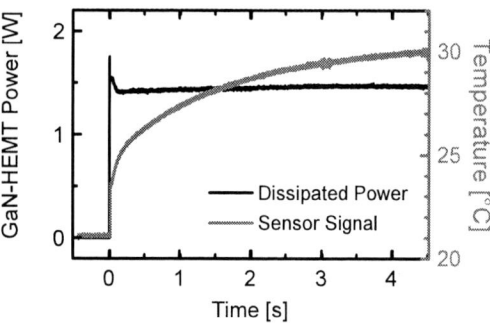

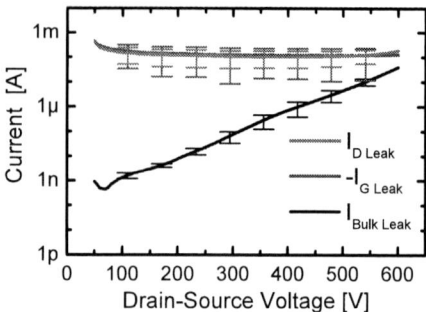

Fig. 6. Power step response of the assembly measured by the temperature sensor. The power step is generated by a current step of $I_D = 5$ A at drain. The assembly is mounted on a heat sinks at room temperture.

Fig. 8. Off-state characterization of Schottky leakage and vertical buffer leakage. Measurements were performed using a high voltage on-wafer prober setup.

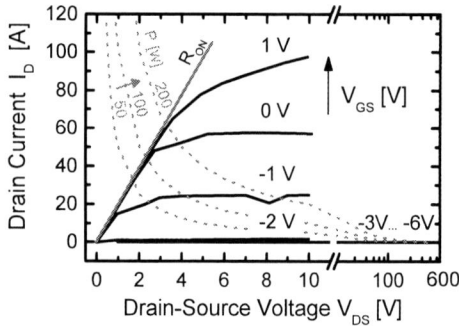

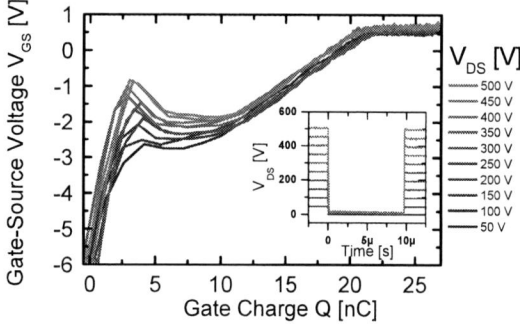

Fig. 7. Output characteristics of the GaN-HEMT. The high-current values are characterized in a pulse setup with $t_{PLS} = 100$ µs.

Fig. 9. Gate charge measurement in a pulse setup. The pulses are measured up to $V_{DS} = 500$ V in 50 V steps.

IV. CHARACTERIZATION OF THE POWER DEVICE

The large-area GaN-HEMT device has a total gate width of $W = 337$ mm and achieves a high power performance, which was characterized in this work. Fig. 7 shows the output characteristic of the HEMT. In the on-state a maximum drain current of up to $I_{D\,MAX} = 100$ A has been achieved in four-point measurements with a pulse time of $t_{PLS} = 100$ µs. An on-state resistance below $R_{ON} = 55$ mΩ was shown at a corresponding current of $I_D = 30A$. Under this condition a power of around 50 W is dissipated in the device, as shown by the power hyperbola plotted in Fig. 7.

The off-state was characterized in semi-automatically on-wafer measurements for all devices on the 4-inch wafer and the results are illustrated in Fig. 8. In the off-state the lateral leakage currents at the drain $I_{D\,Leak}$ and gate $-I_{G\,Leak}$ are below 3 µA/mm. Furthermore the vertical bulk isolation is measured and show as buffer leakage current $I_{Bulk\,Leak}$ in Fig. 8. The vertical bulk current rises up to a leakage of $I_{Bulk\,Leak} < 100$ µA at a corresponding voltage of $V_{DS} = 600$ V. The devices on wafer yield high uniformity regarding leakage currents. 27 working devices have measured up to an off-state voltage of 600 V out of 37 wafer cells.

Because of the wide bandgap properties GaN-HEMTs are predestined for efficient and fast switching voltage converters. High switching frequencies are desirable, because the energy storage elements in converters, as inductors and capacitor, can be designed in smaller size. Thus volume, weight and costs are reduced. Fast switching transistors require low gate charge quantities to switch the HEMT from the off-state into the on-state and back. This dynamic charge process at the gate was recorded during turn-on with a pulse setup. The curve is shown in Fig. 9. The gate charge is found by calculating the integral of the gate current $Q_G = \int I_G(t)\, dt$. At drain the off-state voltage is increased in 50 V steps up to $V_{DS} = 500$ V. Gate charges is for all drain voltages in the range of $Q_G = 20$ nC. The product of the on-state resistance and gate-charge $R_{ON} \times Q_G$ is known to be a figure-of-merit for efficient, fast switching power devices. Thus the device in this works achieves a value of $R_{ON} \times Q_G = 1.1$ Ω·nC. A theoretical limit of $R_{ON} \times Q_G$ as well as a comparison to other state of the art devices in literature can be found in [10].

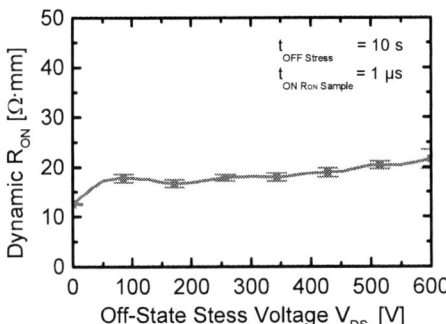

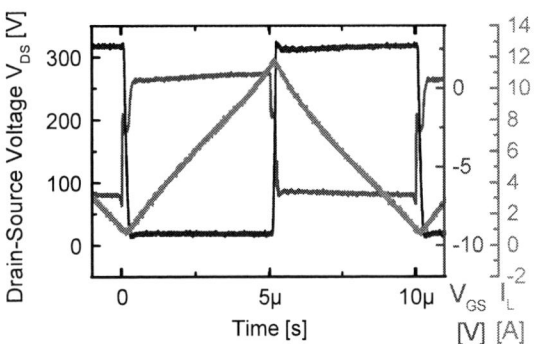

Fig. 10. Dynamic on-state resitance measurement on test devices with same technology and same intrinsic layout. Three HEMT devices with a total gate width of $W = 1$ mm have been performed in a Auriga™ pulse setup.

Fig. 11. Boost converter swiching operation of the GaN HEMT at 100 kHz with a storage inductance of $L = 66$ µH and an ohmic load of 100 Ω.

Another important dynamic characteristic of GaN-HEMTs is the response to time dependent off-state stress voltages. High electric fields in the off-state can charge defects in the HEMT structure. These fixed charges constrain the electron flow through the channel in the on-state, which lead to increased dynamic on-resistance. The time constant of the defect detrapping is in the range of microseconds. A suitable layout and technology eliminated this undesired trapping effects [11,12]. The remaining weak increase of dynamic on-state resistance is characterized and shown in Fig. 10 for the used technology and the used intrinsic HEMT-layout at stress voltages up to $V_{DS} = 600$ V with an off-state stress time of $t_{OFF\ Stress} = 10$ s. The dynamic on-state resistance is measured $t_{ON} = 1$ µs after turn-on by using an Auriga™ pulse setup.

Furthermore the pulse behavior of the GaN-chip is tested under in-circuit conditions. The device was used as power switch in a boost converter demonstrator. Fig. 11 shows the transient voltages at drain and gate as well as the current though the storage inductance with a value of $L = 66$ µH. The device operates at a switching frequency of $f_{SW} = 100$ kHz and a duty cycle of 0.5 in continuous conduction mode. The voltage across the switch node achieves $V_{DS} = 325$ V and the inductor yields a peak current of $I_L = 12$ A in this boost converter application. The DC output features 3.1 A of current and 1 kW of power across the 100 Ω ohmic load.

CONCLUSIONS

This work demonstrates the functionality of an integrated temperature sensor in a 600 V-class high performance GaN-on-Si power HEMT. The sensor is fabricated by using an ordinary interconnect metallization of the HEMT technology without needing any additional process steps. It achieves a linear behavior in the operation range of the power device and it is able to monitor the in-chip temperature at the time. The fabricated high-performance device can be used in applications for a temperature-controlled operation, for characterization of the assembly technology, for reliability prediction, and for overload-, and error-mode detection. Thus the integration of such a linear sensor increases the functionality of GaN-based power devices without extra cost.

REFERENCES

[1] E. A. Jones, F. Wang, B. Ozpineci, "Application-based review of GaN HFETs," IEEE Workshop on Wide Bandgap Power Devices and Applications (WiPDA), pp. 24-29, 13-15 Oct. 2014.

[2] W. S. Tan, M. J. Uren, P. W. Fry, P. A. Houston, R. S. Balmer, and T. Martin, "High temperature performance of AlGaN/GaN HEMTs on Si substrates," Solid-State Electr., vol. 50, no.3, pp. 511-513, Mar. 2006.

[3] D. Donovala, M. Floroviča, D. Gregušováb, J. Kováča, and P. Kordoš, "High-temperature performance of AlGaN/GaN HFETs and MOSHFETs," Microelectrnics Reliability, vol. 48, iss. 10, pp. 1669-1672, Mar. 2006.

[4] M. Dammann, H. Czap, J. Rüster, M. Baeumler, F. Gütle, P. Waltereit, F. Benkhelifa, R. Reiner, M. Cäsar, H. Konstanzer, S. Müller, R. Quay, M. Mikulla, and O. Ambacher, "Reverse bias stress test of GaN HEMTs for high-voltage switching applications," In Proc. of IEEE International Integrated Reliability Workshop (IRW), pp. 105-108, Oct. 2012.

[5] A. H. Zahmani, A. Nishijima, Y. Morimoto, H. Wang, J.-F. Li, and A. Sandhu, "Temperature dependence of the resistance of AlGaN/GaN heterostructures and their applications as temperature sensors," Japanese Journal of Applied Physics, vol. 49, no. 4S, Apr. 2010.

[6] J. Roberts, G. Klowak, Di Chen, A. Mizan, "Drive and protection methods for very high current lateral GaN power transistors," in IEEE Applied Power Electronics Conference and Exposition (APEC), pp. 3128-3131, Mar. 2015.

[7] A. M. H. Kwan, Y. Guan, X. Liu, and K. J. Chen,, "A highly linear integrated temperature sensor on a GaN smart power IC platform," Transactions on Electron Devices, vol. 61, no. 8, Aug. 2014.

[8] O. Arenas, E.Al Alam, A. Thevenot, Y. Cordier, A. Jaouad, V. Aimez, H. Maher, R. Ares, and F. Boone, "Integration of micro resistance thermometer detectors in AlGaN/GaN devices," IEEE Journal of the Electron Devices Society, vol. 2, no. 6, pp. 145-148. Oct. 2014.

[9] P. Waltereit, R. Reiner, H. Czap, D. Peschel, S. Müller, R. Quay, M. Mikulla, O. Ambacher, "GaN-based high voltage transistors for efficient power switching", Physica Status Solidi, vol. 10, no. 5, pp. 831-834, Feb. 2013.

[10] R. Reiner, P. Waltereit, B. Weiss, M. Wespel, R. Quay, M. Schlechtweg, M. Mikulla, and O. Ambacher, "Integrated reverse-diodes for GaN-HEMT structures," in Proc. of the 27th IEEE International Symposium on Power Semiconductor Devices & IC's (ISPSD), pp. 45-48. Jun. 2015.

[11] M. Wespel, M. Dammann, V. Polyakov, R. Reiner, P. Waltereit, B. Weiss, R. Quay, M. Mikulla, O. Ambacher, "High-voltage stress time-dependent dispersion effects in AlGaN/GaN HEMTs," in International Reliability Physics Symposium, pp. 19-23, Apr. 2015.

[12] M. Wespel, V. M. Polyakov, M. Dammann, R. Reiner, P. Waltereit, R. Quay, M. Mikulla, and O. Ambacher "Trapping effects at the drain edge in 600 V GaN-on-Si HEMTs", IEEE Transactions on Electron Devices, in press

An Innovative Power Module with Power-System-In-Inductor Structure

Laili Wang and Doug Malcolm
Sumida Technologies Inc.
Sumida Corporation
Kingston, Canada
laili_wang@us.sumida.com

Yan-Fei Liu Fellow IEEE
Department of Electrical and Computer Engineering
Queen's University
Kingston, Canada
yanfei.liu@queensu.ca

Abstract— This paper presents an integrated power module with the features of high power density and high efficiency. A multi-functional integrated magnetic component is designed. The component has the roles of both the filter inductor and the case of the whole power module. With the assist of finite element analysis (FEA), design and optimization of the proposed inductor are demonstrated. It has higher inductance value than the inductor used in conventional designs. It has bigger coil winding and larger surface area, leading to lower DCR and better thermal performances than plastic packaging. Benefiting from these advantages, the power module constructed based on the inductor can achieve higher efficiency and lower temperature than those based on traditional plastic packaging solutions. Design of the inductor is demonstrated through the combination of analytical and simulation methods. An inductor prototype is built and used in an integrated power module to do experimental tests. Loss breakdown of the power module is executed to show the loss of the magnetic component. The proposed power module shows better performances than the plastic packaging solution.

Keywords—power-system-in-inductor packaging; magnetic perfomance; inductor design; thermal;

I. INTRODUCTION

Power modules have been widely used in telecom equipment, computer servers, and consumer electronics to provide an integrated solution of power management. In a small package, the power modules should integrate all the functions of their counterparts designed with discrete components so that the system design engineers do not need to spend too much time on the power supply solution, saving time for products development. Besides, integrated power module has better performance than the solution of discrete components in terms of reliability and space saving. Non-isolated step-down power modules are the most popular power modules used in today's industry. They are composed of an integrated buck regulator (or controller with two Mosfets), an inductor, some input and output capacitors, and some auxiliary components.

Figure 1 shows the schematic of a step-down power module. There are quite a few literatures discussing about the

package of high density power modules [1-15]. Embedding magnetics and capacitors in the printed circuit boards is demonstrated in [1], however, core loss of the embedded magnetic material is significant. Low temperature co-fired ceramic (LTCC) technology becomes very popular recently in research and can achieve very high power density [2-4, 8-15]. Micro-inductor and micro-transformer based on thin film technology is another key technology used for power modules [5-7]. The disadvantage of this technology is that it has higher Direct Current Resistance (DCR) and causes more loss than conventional inductors. In industry, there are two typical packaging solutions for making the power modules. One is based on PCB substrate. The regulator die, chip inductor and other parts are soldered or wire-bonded on the substrate, and the whole substrate is packaged using injection molding. Manufacturers, such as Linear uses this technology. The other one is based on lead frame. In this way, all the components are connected to the lead frame through wire-bonding, and the whole module is then packaged with injection molding. TI, Micrel, and Intersil use this technology to package their modules. No matter which technology is employed to package the module, the chip inductor is a necessary part in a power module. And to leave some space for plastic material, the chip inductor should be smaller than the mold. It generally takes up about 1/4~1/3 of the whole volume of the module. This paper proposes a new packaging technology called power-system-in-inductor (PSI²). A customized magnetic component is designed, acting as both the case of the whole power module and the filter inductor of the converter. Section II of the paper presents the structure of the power module; Section III will demonstrate design of the customized inductor based on the combination of analytical and simulation methods. Section IV will build a prototype and do comparison experiments with plastic packaging solution. Section V concludes the paper.

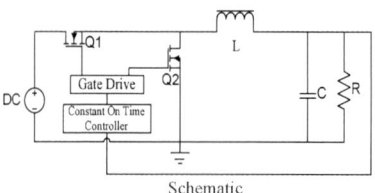

Schematic

Figure 1. Schematic of a step-down power module.

978-1-4673-9551-9/16 $31.00 © 2016 IEEE

II. PRINCIPLE OF STRUCTURE

This section will introduce the structure of the proposed power module. A structure based on traditional packaging technology will also be introduced with the purpose of comparison. Power modules in the market are generally packaged with plastic material by injection molding process. Figure 2 shows the structure of the traditional plastic packaging power module. It has a regulator (in the monolithic version) or several active devices (including a controller and two discrete Mosfets in the multiple chips version), an inductor, and some auxiliary components mounted on the substrate together, then the whole substrate is packaged with plastic material. Figure 3 shows the structure of the proposed power module. The proposed structure has a regulator and some auxiliary components, and an inductor mounted on the PCB board. In contrast to conventional plastic packaged power modules, the inductor also acts as a package case of the whole converter. The magnetic core has a cavity in one half and an embedded coil in the other half. The buck regulator and the auxiliary components are embedded underneath the cavity. Thermally conductive glue is used to attach the top of the regulator to the ceiling of the cavity with the purpose of transferring heat from the regulator to the magnetic core. The embedded coil forms an inductor together with the magnetic core. Therefore, the magnetic core has functions of both power module package and filter inductor in the converter. Benefiting from this package technology, the proposed converter has two advantages over the power modules based on conventional plastic package technology. Firstly, without the necessity of leaving enough room for plastic packaging, the inductor size

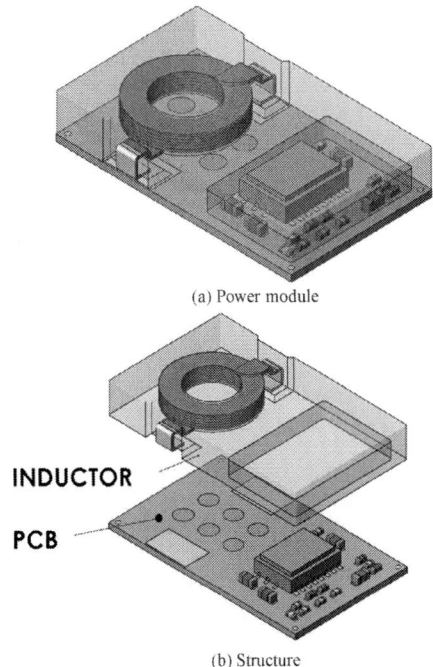

(a) Power module

(b) Structure

Figure 3. Proposed magnetic packaging.

could be as big as the package size, which means both the winding and magnetic core will be bigger than those of the inductors used in plastic packing. Compared with a conventional plastic packaged power module with a small metal composite inductor inside, the proposed magnetic packaged power module has lower DCR, higher inductance value since the volume of the inductor is larger. Secondly, in the power module, the heat sources are the semiconductor devices and the inductor. In a plastic packing power module, the heat sources are packaged by the plastic surface (generally 1mm thick), which increases the thermal resistance from the heat sources to ambient. Moreover, the proposed power module has the package composed of magnetic material with higher thermal conductivity than plastic material. It means the magnetic core has better thermal co-efficient and thus lower temperature, which will in turn improve the performance of the whole power module.

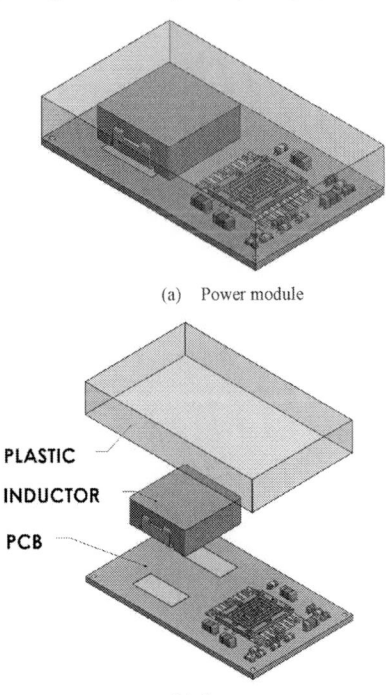

(a) Power module

(b) Struture

Figure 2. Conventionl plastic packing.

III. INDUCTOR DESIGN AND ANALYSIS

There are quite a few literature [9-10, 13-18] discussing about the model of inductance calculation and design. Most of them are for inductors on the semiconductor or PCB, which are not suitable for inductor design in this paper. Since the magnetic core has a customized shape, and there's no existing models for calculating the inductance value, finite element analysis (FEA) is used to simulate the inductance value accurately and design the power module. Benefiting from the proposed structure, the proposed inductor has the same width as the module itself. Therefore, the proposed inductor size is 9mm×9mm×2.8mm. However, the metal powder composite inductor designed for the plastic packaging has to be smaller than the module to leave enough room for plastic packaging. Generally, the thickness of the plastic layer is around 1mm for thick power modules. Thus the maximum inductor size is 6mm×6mm×1.8mm by considering the extended soldering points on the inductor leads. This Section analyzes the relationship between the inductance L, coil radius R, coil width w, and wire thickness t based on 15mm×9mm×2.8mm package size chosen for prototype demonstration. After a rough calculation to obtain the number of turns, a 3D FEA simulation is used to optimize the inductance and resistance by sweeping parameters above, based on which further analysis and optimization are executed.

Before the simulation, a rough calculation is executed to obtain the possible number of turns based on the initial winding dimension. Figure 4 shows the top view of the winding embedded in the magnetic core. A_s and A_e show the areas of magnetic materials through which the flux lines go. To guarantee the mechanical stress of the iron powder at the inductor terminal (the coil has to be bended), the iron power thickness from the winding coil to the edge of the magnetic core should be large enough, which might results in A_s being larger than A_e. However, they are assumed equal to simplify the design process of the magnetic path when calculating the number of turns. The inductance value calculating equation is expressed in

$$L = \frac{\mu_r \mu_0 N^2 A_e}{l} \tag{1}$$

Where A_e is the effective area for flux density; l is the length of magnetic path; N is the number of turns, μ_r is the relative permeability of the magnetic material. Both A_e and l highly depend on the radius R, width w and thickness t of the coil. The effective area A_e can be calculated by (2)

$$A_e = \pi \cdot r^2 \tag{2}$$

The estimated magnetic path length can be calculated through (3)

$$l = 2h + 2w + r \tag{3}$$

Where h is the total thickness of the winding, and it can be expressed by the height of the inductor H and thickness of magnetic material above or below the winding e in (4)

$$h = H - 2e \tag{4}$$

Substitute (2), (3) and (4) in (1), yielding (5)

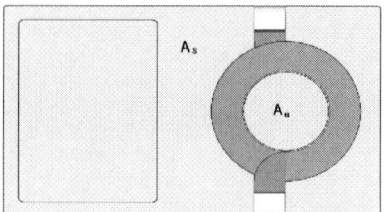

Figure 4. Dimension of the winding.

$$N = \sqrt{\frac{L(2(H - 2e) + 2w + r)}{2\mu_r \mu_0 \pi r^2}} \tag{5}$$

In this design, the total height of the inductor H is 2.4mm; thickness of magnetic materials above and below the winding e is set to be 0.65 mm in turn number calculation; the relative permeability of the magnetic material μ_r is 10 and the proposed inductance value is 1μH. Number of turns is calculated by substituting these parameters into (5) and sweeping internal radius r and conductor width w. TABLE 1 shows the calculated results. The number of turns increases with the decrease of the internal radius r and increases with the width of the coil w. For the proposed inductor in this paper, the number of turns should include a half turn since it has two terminals leading out from two sides. Therefore, the calculated number of turns should be modified to number of turns in a practical design as shown in TABLE 2. Inductance value calculated based on the TABLE 2 is still close to 1μH although there should be some variations. Based on the calculation and modification, the number of turns can be 3.5, 4.5, 5.5, 6.5, 7.5, 8.5, 9.5 and 10.5 depending on the internal radius and coil width. Even with a simulation tool, it is still very time-consuming to simulate every case of turns. Some of the cases can be easily excluded because of their higher DCR. The equation of DCR calculation is shown in (6):

$$R_{DC} = \rho \frac{2\pi N \cdot (r + \frac{w}{2})}{(\frac{H - 2e}{[N]} - s) \cdot w} \tag{6}$$

Where N is the number of turns listed in TABLE 2; ρ is the resistivity of the copper; s is the thickness of the insulation layer; $[N]$ is the celling integral of the calculated turns listed in TABLE 1.

TABLE 1 CAUCULATED NUMBER OF TURNS FOR THE PROPOSED INDUCTOR

w (mm) / r (mm)	1.00	1.20	1.40	1.60	1.80	2.00	2.20	2.40	2.60	2.80	3.00
1.0	8.1	8.4	8.7	9.0	9.3	9.6	9.8	10.1	10.3	10.6	10.8
1.25	6.6	6.9	7.1	7.3	7.6	7.8	8.0	8.2	8.4	8.6	8.8
1.50	5.7	5.9	6.1	6.2	6.4	6.6	6.8	6.9	7.1	7.2	7.4
1.75	5.0	5.1	5.3	5.4	5.6	5.7	5.9	6.0	6.2	6.3	6.4
2.00	4.4	4.6	4.7	4.8	5.0	5.1	5.2	5.3	5.5	5.6	5.7
2.25	4.0	4.1	4.3	4.4	4.5	4.6	4.7	4.8	4.9	5.0	5.1
2.50	3.7	3.8	3.9	4.0	4.1	4.2	4.3	4.4	4.5	4.6	4.7
2.75	3.4	3.5	3.6	3.7	3.8	3.9	4.0	4.0	4.1	4.2	4.3
3.00	3.2	3.3	3.4	3.4	3.5	3.6	3.7	3.8	3.8	3.9	4.0

TABLE 2 Practical Design Number of Turens for The Proposed Inductor

w (mm) / r (mm)	1.00	1.20	1.40	1.60	1.80	2.00	2.20	2.40	2.60	2.80	3.00
1.0	8.5	8.5	8.5	9.5	9.5	9.5	9.5	10.5	10.5	10.5	10.5
1.25	6.5	6.5	7.5	7.5	7.5	7.5	8.5	8.5	8.5	8.5	8.5
1.50	5.5	5.5	6.5	6.5	6.5	6.5	6.5	6.5	7.5	7.5	7.5
1.75	5.5	5.5	5.5	5.5	5.5	5.5	5.5	6.5	6.5	6.5	6.5
2.00	4.5	4.5	4.5	4.5	5.5	5.5	5.5	5.5	5.5	5.5	5.5
2.25	4.5	4.5	4.5	4.5	4.5	4.5	4.5	4.5	4.5	5.5	5.5
2.50	3.5	3.5	3.5	4.5	4.5	4.5	4.5	4.5	4.5	4.5	4.5
2.75	3.5	3.5	3.5	3.5	3.5	3.5	4.5	4.5	4.5	4.5	4.5
3.00	3.5	3.5	3.5	3.5	3.5	3.5	3.5	3.5	3.5	3.5	4.5

TABLE 3 Calculated DCR

w (mm) / r (mm)	1.00	1.20	1.40	1.60	1.80	2.00	2.20	2.40	2.60	2.80	3.00
1.0	46.3	41.1	37.5	72.3	67.8	64.2	61.3	215.8	208.2	201.8	196.2
1.25	17.5	15.5	23.9	22.0	20.5	19.3	33.0	31.5	30.3	29.2	28.3
1.50	11.7	10.3	15.8	14.4	13.4	12.5	11.8	11.3	18.5	17.8	17.2
1.75	13.2	11.5	10.3	9.4	8.6	8.1	7.6	12.3	11.8	11.3	10.9
2.00	8.3	7.2	6.4	5.8	9.5	8.8	8.3	7.8	7.5	7.1	6.9
2.25	9.1	7.9	7.0	6.3	5.8	5.4	5.1	4.8	4.5	7.7	7.3
2.50	5.2	4.5	4.0	6.9	6.3	5.8	5.4	5.1	4.9	4.6	4.4
2.75	5.7	4.9	4.3	3.9	3.5	3.3	5.8	5.5	5.2	4.9	4.7
3.00	6.1	5.2	4.6	4.1	3.8	3.5	3.2	3.1	2.9	2.7	5.0

TABLE 3 lists the calculated DCR values obtained through (6). The DCR value increases with the reduction of internal radius r and the increase of the number of turns. When $r=1.0$, 1.5mm, the DCR are much higher than other cases, so they are excluded from the candidates. The left solutions have the number of turns 3.5, 4.5, 5.5, 6.5. Those four winding solutions will be put into simulation to do further analysis.

The number of turns is obtained through (5) by assuming that the total inductance value is 1µH. However, it is only used to calculate the rough number of turns. The inductor designed with the parameters in the calculation might not get the assumed inductance value. This is because (5), which uses the calculated inductance value of closed magnetic path with constrained regular effective area A_e such as E I ferrite magnetic cores, can introduce some error in inductance calculation. The proposed inductor does not have constant effective area A_e along the magnetic path. The equivalent length of magnetic path could also contribute some error. A FEA simulation for each number of turns is executed to further get the right parameters for designing the inductance value. Figure 5 and Figure 6 show the simulated inductance value and DCR for different number of turns. For all the cases, the inductance value has the trend of increase with the increase of radius while it has the trend of decrease with the increase of the coil width. The outer diameter expressed in (7) is limited by the width of the magnetic core 9mm which means $r + w$ should be less than 4.5mm in the simulation. Therefore, in Figure 5 and Figure 6, the maximum radius varies for different values of coil width.

$$d = 2(w + r) \qquad (7)$$

In practice, more restrictions are added to leave some margin at both sides of terminals. DCR of the inductor has the same trend as inductance. It increases with increase of the radius r and reduces with increases of coil width w. Therefore, the design of such an inductor is essentially choosing the right parameters to reduce DCR while achieving the expected inductance value. For the 3.5 turn design, DCR is very small since it has less number of turns and thicker copper for each turn, but the inductance value is not high enough to meet the requirement of 1µH specification. It can be excluded from the comparison, then only 4.5 turn, 5.5 turn, 6.5 turn are further considered as candidates. For the 4.5 turn design, the maximum inductance value can be higher than 1µH, it is at about $r=2.75$mm, $w=1$mm. There are some points, which are not shown in the curves, reaching 1µH point. For example, $r=2.85$, $w=1.1$mm, they are very similar to the point $r=2.75$mm, $w=1$mm. For 5.5 turn design, there are three combinations. $r=1.8$mm, $w=1$mm; $r=2$mm, $w=1.4$mm; $r=2.25$, $w=1.8$. For 6.5 turn design, there are five candidates. $r= 1.2$mm, $w=1$mm; $r=1.3$mm, $w=1.4$mm; $r=1.4$mm, $w=1.8$mm; $r=1.45$mm, $w=2.2$mm; $r=1.5$mm, $w=2.6$mm. For different groups of parameters, the DCR shown in Figure 6 is different.

Since the design of an inductor is essentially finding the trade off point of the inductance and DCR under the certain volume condition, it is hard to determine through either the inductance value curves or the DCR value curves. A new

TABLE 4 Simulation Results for Selected Candidates

No.	1	2	3	4	5	6	7	8	9
Turns	4.5	5.5	5.5	5.5	6.5	6.5	6.5	6.5	6.5
Radius (mm)	2.75	1.8	2	2.25	1.2	1.3	1.4	1.45	1.5
Width (mm)	1	1	1.4	1.8	1	1.4	1.8	2.2	2.6
r + w (mm)	3.75	2.8	3.4	4.05	2.2	2.7	3.2	3.65	4.1
DCR (mΩ)	10.5	10.8	9.28	8.4	15	12.5	11.3	9.52	9.5
L/DCR (µH/mΩ)	0.095	0.088	0.107	0.12	0.068	0.080	0.093	0.098	0.103

978-1-4673-9551-9/16 $31.00 © 2016 IEEE

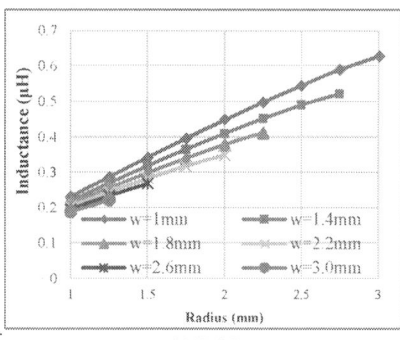

(a) N=3.5

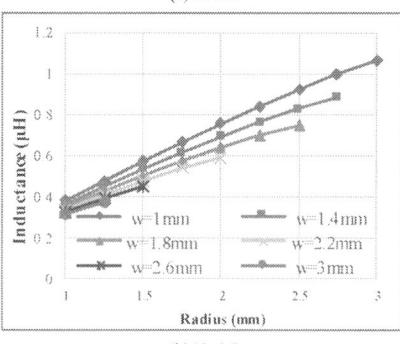

(b) N=4.5

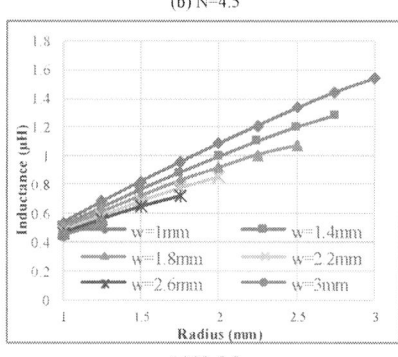

(c) N=5.5

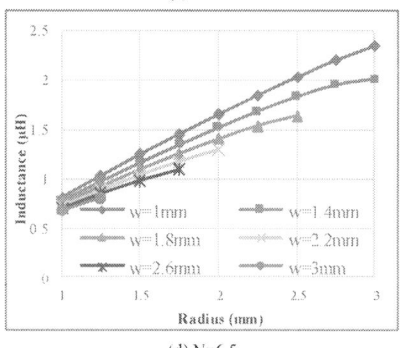

(d) N=6.5

Figure 5. Inductance vs Radius with coil width as a parameter.

parameter *L/R* is defined in this paper as a new criterial to choose and optimize the design. It should be noted that this parameter is different from the quality factor $\omega L/R$ which also reflects the AC performance of the magnetics. It is manly used to show the low frequency characteristics of an inductor. Figure 7 shows the curves of L/R for number of turns 4.5, 5.5, 6.5. It can be observed that the *L/R* is higher for N=5.5 than the other two number of turns for the same radius and coil and *L/R* of the selected candidates.

In TABLE 4, No. 4 (N=5.5, *r*=2.25mm, *w*=1.8mm) has the lowest DCR and the highest L/R. However, the *r + w* value is 4.05 which is close to half width of the module. In the real practice of manufacture, by considering the mechanical stress of the magnetic material when bending the coil terminals, there should be 3mm margin (1.5mm on both sides), which means $r + w \leq 3\text{mm}$. With this restriction applied to the candidates, three options are left. They are No. 2 (N=5.5, *r*=1.8mm, *w*=1mm), No. 5 (N=6.5, *r*=1.2mm, *w*=1mm) and No. 6 (N=6.5, *r*=1.3mm, *w*=1.4mm). No.2 has the highest L/R value and it can be selected as a final design. Since the *r + w* value for No. 2 is 2.8mm, *w* can be further increased by 0.2mm (*r + w* is not larger than 3mm) which will not lead to significant decrease of the inductance value. And prototypes will be made based on this parameters (N=5.5, *r*=1.8mm, *w*=1.2mm). A new simulation is executed based on the selected parameters. The inductance value at no load is 1.1μH, it drops to 1.01μH which still satisfies the specification. A field distribution of flux density at 6A current excitation in the designed inductor is shown in Figure 8. The maximum flux density happens around the winding, which is 0.24mT. The saturation flux density of the magnetic materials is about 1T. Therefore, the inductor still has the potential to be used in higher output current application, such as 12V input, 5V/8A output power module.

IV. PROTOTYPE AND EXPERMENT

In this section, inductor prototype will be made with the parameters obtained in the last section. Figure 9(a) and Figure 9(b) show the top view and bottom view of the inductor. Two leads were extended from the winding and bended at the bottom of the inductor for soldering on the small PCB board

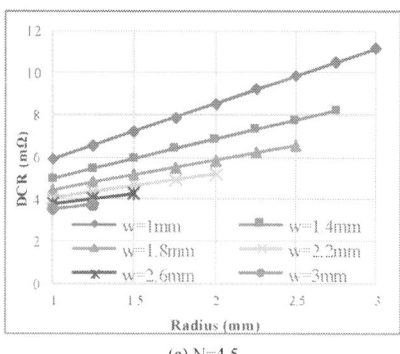

(a) N=4.5

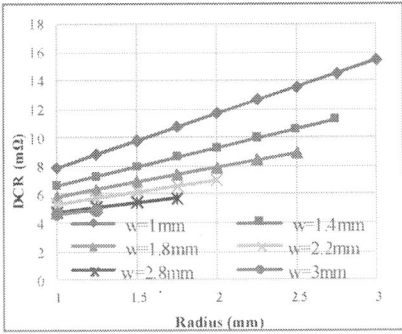

(b) N=5.5

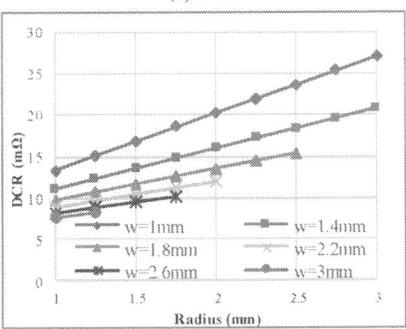

(c) N=6.5

Figure 6. DCR vs Radius with coil width as a parameter.

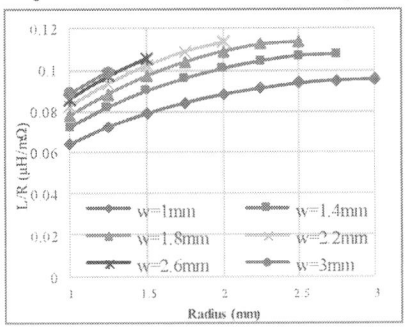

(a) N=4.5

(b) N=5.5

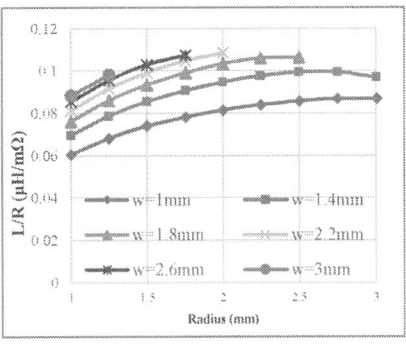

(c) N=6.5

Figure 7. L/R vs Radius with coil width as a parameter.

which is 0.4mm thick. Discrete components, such as regulator and resistors in the power circuit, are soldered on the area of PCB board underneath the cavity of the inductor. Red glue is used at the bottom of the magnetic core where winding is embedded and the ceiling of the cavity where the regulator chip is located to strengthen the adhesive force during the assembling process. Figure 9(c) shows the bottom view of the assembled power module. Before assembling the inductor on the PCB, inductance vs DC bias is measured using a Wayne Kerr 3260B LCR meter. With the purpose of comparison, it is also simulated through FEA simulation. Figure 10 shows the test and simulation results. They correlate with each other very well. Figure 11 shows the resistance versus switching frequency by measurement and simulation. The power module is then soldered on an evaluation board to do loss analysis and performances test. Loss characteristic of the magnetic material is firstly teste to obtain the parameters of Stenmez equations used in FEA simulation and calculate the core loss. Figure 12 shows the power loss the magnetic materials versus flux density with frequency as a parameter. With the simulated core loss, loss breakdown of the power module is executed. Figure 13 shows the loss of the inductor and the switches.

Two comparison experimental tests will be done with the assembled power module to verify its performance. The first one is to show the advantage of the proposed packaging technology over the conventional plastic packing technology by using the same PCB and circuit but different inductors and packaging. The second one is to compare the performances of the proposed power module with the state-of-art plastic packaging power modules in industry.

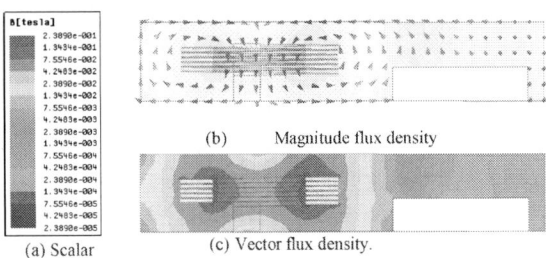

(a) Scalar

(b)   Magnitude flux density

(c) Vector flux density.

Figure 8. Simulated flux distribution.

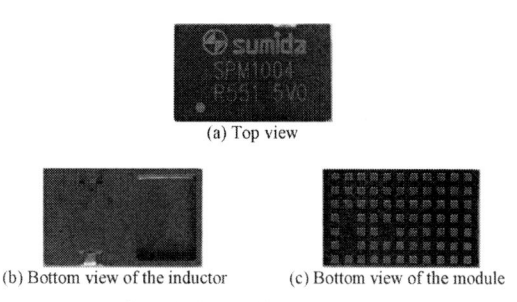

(a) Top view

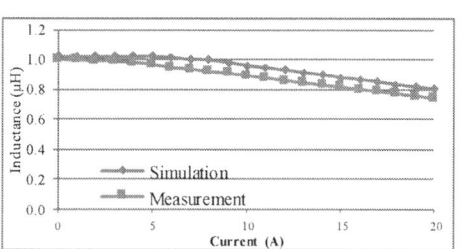

(b) Bottom view of the inductor　　(c) Bottom view of the module

Figure 9. Picture of the power module.

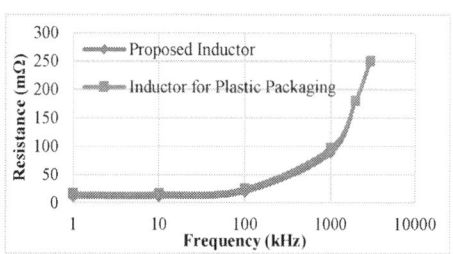

Figure 10. Inductance value of the proposed inductor via DC bias.

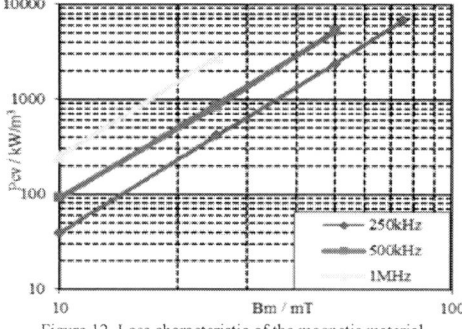

Figure 11. Resistance vs frequency.

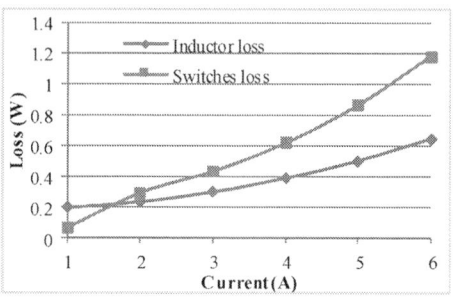

Figure 12. Loss characteristic of the magnetic material.

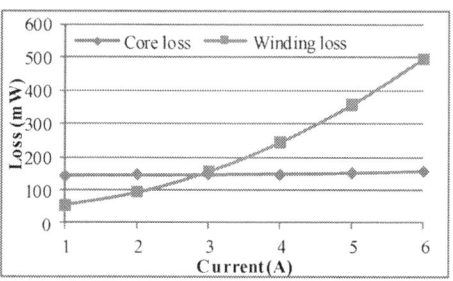

(a) Inductor loss and switches loss.

(b) Winding loss and core loss.

Figure 13. Loss breakdown of the power module.

As described in Section I, one of the advantages of the proposed power module is that it is able to have larger volume of inductor than that in conventional plastic packaging which has to leave enough margin for injecting plastic material during packaging process. For inductors made with the same magnetic material, larger volume means higher inductance value or lower DCR. Figure 14 shows the picture of the selected metal powder composite inductor used for plastic packaged module. The volume of the inductor is 6mm×6mm×2.2mm. It has 0.9μH inductance value and 17mΩ DCR. Figure 15 shows its inductance value versus DC bias. The prototype of the proposed inductor has 1μH inductance value and 11mΩ DCR. The inductor for plastic packaging has much smaller saturation current than the proposed inductor because it has smaller volume and thinner wire. Both inductors are soldered on the same PCB board to test their efficiency performances under the same load conditions. Figure 16 shows the small board with inductor soldered on it. Figure 17 shows the efficiency comparison results. It can be observed that the PCB board with the proposed inductor has about 2% higher efficiency across the whole load range since it has lower DCR and better performances.

Figure 14. Selected metal powder composite inductor for plastic packaging.

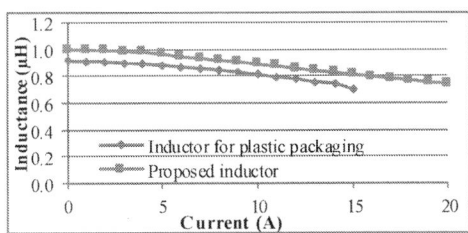

Figure 15. Inductance of the proposed inductor and the selected inductor for plastic packaging.

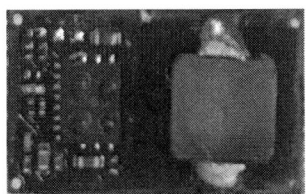

Figure 16. PCB of the power module with selected metal powder composite inductor.

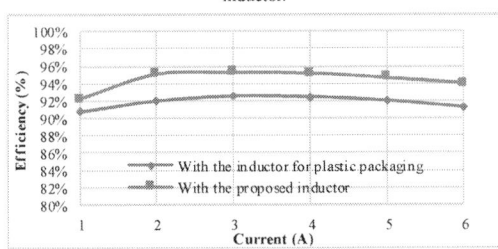

Figure 17. Efficiency curves of the power modules with different inductors.

V. CONCLUSION

The special magnetic component proposed in this paper has functions of both inductor and the packaging. A PSI^2 structured power module is designed to show the benefit of the magnetic component. Compared with plastic packaging solution, the proposed design has larger inductor volume, and thus higher inductance value. Experimental test shows that the designed power module has a 95.2% high peak efficiency and high saturation current.

REFERENCES

[1] E. Waffenschmidt; B. Ackermann; and J. A. Ferreira, "Design method and material technologies for passives in printed circuit board embedded circuits," *IEEE Trans. Power Electron.*, vol. 20, no. 3, pp. 576–584, May 2005.

[2] M. H. Lim; J. D. van Wyk; and F. C. Lee, "Hybrid integration of a low voltage, high-current power supply buck converter with an LTCC substrate inductor," *IEEE Trans. Power Electron.*, vol. 25, no. 9, pp. 2287–2298, Sep. 2010.

[3] Qiang Li; Yan Dong; F.C.Lee; Gilham, D., "High-Density Low-Profile Coupled Inductor Design for Integrated Point-of-Load Converters," in *Power Electronics, IEEE Transactions on*, vol.28, no.1, pp.547-554, Jan. 2013.

[4] Laili Wang; Yunqing Pei; Xu Yang; Zhaoan Wang, "Design of Ultrathin LTCC Coupled Inductors for Compact DC/DC Converters," in *Power Electronics, IEEE Transactions on*, vol.26, no.9, pp.2528-2541, Sept. 2011.

[5] Mino, M.; Yachi, T.; Tago, Akio; Yanagisawa, Keiichi; Sakakibara, K., "A new planar microtransformer for use in micro-switching converters," in *Magnetics, IEEE Transactions on*, vol.28, no.4, pp.1969-1973, Jul 1992.

[6] Yun, Eui-Jung; Jung, Myunghee; Chae Il Cheon; Hyoung Gin Nam, "Microfabrication and characteristics of low-power high-performance magnetic thin-film transformers," in *Magnetics, IEEE Transactions on*, vol.40, no.1, pp.65-70, Jan. 2004.

[7] Kowase, I.; Sato, T.; Yamasawa, K.; Miura, Y., "A planar inductor using Mn-Zn ferrite/polyimide composite thick film for low-Voltage and large-current DC-DC converter," in *Magnetics, IEEE Transactions on*, vol.41, no.10, pp.3991-3993, Oct. 2005.

[8] Laili Wang; Yunqing Pei; Xu Yang; Yang Qin; Zhaoan Wang, "Improving Light and Intermediate Load Efficiencies of Buck Converters With Planar Nonlinear Inductors and Variable On Time Control," in *Power Electronics, IEEE Transactions on*, vol.27, no.1, pp.342-353, Jan. 2012.

[9] Lim, M.H.F.; van Wyk, J.D.; Zhenxian Liang, "Internal Geometry Variation of LTCC Inductors to Improve Light-Load Efficiency of DC-DC Converters," in *Components and Packaging Technologies, IEEE Transactions on*, vol.32, no.1, pp.3-11, March 2009.

[10] Lim, M.H.; van Wyk, J.D.; Lee, F.C.; Ngo, K.D.T., "A Class of Ceramic-Based Chip Inductors for Hybrid Integration in Power Supplies," in *Power Electronics, IEEE Transactions on*, vol.23, no.3, pp.1556-1564, May 2008.

[11] Hahn, R.; Krumbholz, S.; Reichl, H., "Low profile power inductors based on ferromagnetic LTCC technology," in *Electronic Components and Technology Conference, 2006. Proceedings. 56th*, vol., no., pp. 528–533.

[12] Mingkai Mu; Yipeng Su; Qiang Li; Lee, F.C., "Magnetic characterization of low temperature co-fired ceramic (LTCC) ferrite materials for high frequency power converters," in *Energy Conversion Congress and Exposition (ECCE), 2011 IEEE*, vol., no., pp.2133-2138, 17-22 Sept. 2011.

[13] Yipeng Su; Qiang Li; Mingkai Mu; Gilham, D.; Reusch, D.; Lee, F.C., "Low profile LTCC inductor substrate for multi-MHz integrated POL converter," in *Applied Power Electronics Conference and Exposition (APEC), 2012 Twenty-Seventh Annual IEEE*, vol., no., pp.1331-1337, 5-9 Feb. 2012.

[14] Qiang Li; Lee, F.C., "High Inductance Density Low-Profile Inductor Structure for Integrated Point-of-Load Converter," in *Applied Power Electronics Conference and Exposition, 2009. APEC 2009. Twenty-Fourth Annual IEEE*, vol., no., pp.1011-1017, 15-19 Feb. 2009.

[15] Yipeng Su; Qiang Li; Lee, F.C., "Design and Evaluation of a High-Frequency LTCC Inductor Substrate for a Three-Dimensional Integrated DC/DC Converter," in *Power Electronics, IEEE Transactions on*, vol.28, no.9, pp.4354-4364, Sept. 2013.

[16] Lopera, J.M.; Prieto, M.J.; Pernia, A.M.; Nuno, F., "A multiwinding modeling method for high frequency transformers and inductors," in *Power Electronics, IEEE Transactions on*, vol.18, no.3, pp.896-906, May 2003.

[17] Jizheng Qiu; Sullivan, C.R., "Design and Fabrication of VHF Tapped Power Inductors Using Nanogranular Magnetic Films," in *Power Electronics, IEEE Transactions on*, vol.27, no.12, pp.4965-4975, Dec. 2012.

[18] Jian Lu; Hongwei Jia; Xuexin Wang; Padmanabhan, K.; Hurley, W.G.; Shen, Z.J., "Modeling, Design, and Characterization of Multiturn Bondwire Inductors With Ferrite Epoxy Glob Cores for Power Supply System-on-Chip or System-in-Package Applications," in *Power Electronics, IEEE Transactions on*, vol.25, no.8, pp.2010-2017, Aug. 2010.

978-1-4673-9551-9/16 $31.00 © 2016 IEEE

Thermal Analysis of a Magnetic Packaged Power Module

Laili Wang, Doug Malcolm
Sumida Technologies Inc.
Sumida Corporation
Kingston, Canada
laili_wang@us.sumida.com

Wenbo Liu, Yan-Fei Liu *Fellow IEEE*
Department of Electrical and Computer Engineering
Queen's University,
Kingston, Canada
liu.wenbo@queensu.ca, yanfei.liu@queensu.ca

Abstract—Power density of converters have been dramatically increased through the innovations of packaging and integration technologies. Meanwhile, it also imposes more challenges on the thermal performances. An integrated power module packaged with magnetic component is proposed to improve both electrical and thermal performances. This paper presents the thermal analysis of the proposed power module. The magnetic component acts as both the filter inductor in the converter and the package of the power module. Benefiting from this package technology, the inductor can be designed with a bigger winding of lower resistance, thus generating less heat. The magnetic material has better thermal conductivity than plastic material used in conventional plastic packaged power modules; therefore, the power module has better thermal performance. Simulation is executed to show thermal effect of winding configurations. A thermal evaluation board is built to compare thermal performances of the proposed power module and two other commercial products. The proposed power module has 11°C lower than the other part with the same size.

Keywords—electrical packaging; thermal resistance; analytical model; finite element method

I. INTRODUCTION

Thermal performance is a significant issue in high power density converters from small power devices used as a power supply of a mother board to high power equipment used in wind turbine [1-10]. It is related to the efficiency, the materials of the components, and the package of the whole converter. In industry, it is a trend to increase the power density of the converters in electronic system with the purpose of reducing the total volume or leaving more space to integrate more functions. Different kinds of packaging and integration technology have been proposed to realize this purpose [2]. However, with the increased power density, the thermal issue becomes even more critical since either the total power is increased or the size of the converter is reduced. In some cases, extra measurements should be taken to dissipate the heat [4], which might impose adversary effect on increasing the power density. Besides, they can result in the increase of the total cost, too.

The integrated non-isolated power modules are the most popular DC/DC converters used in industry. They are widely used in telecom equipment, network servers, FPGAs, and other electronic devices. The input voltage rails can be 5V, 12V or even higher, the output voltage varies from 0.6V~5.5V. The output can be from hundreds of the milliamp to tens of amp depending on the applications. They generally consist of a regulator (or controller with discrete Mosfets), a metal composite inductor, and some auxiliary components. As a necessary part in the buck converter, the metal composite inductor takes about 1/5~1/3 of the whole volume. The substrate of the integrated power modules are made of PCB or lead frame, taking about 1/10 of the whole volume. They are packaged by plastic with all the components inside. The left space of the power modules are filled with plastic material, resulting in inefficient overall utilization of space and higher thermal resistance from the heat sources to the ambient from the top side. Generally, to guarantee enough high power density, heat sink is not applicable for these converters, then thermal issue becomes a big challenge since all the components are completely integrated in one block and heat is hard to be dissipated from the top side. However, more effort should be done by reducing the loss and reducing the thermal resistance from the heat sources to the ambient on the top side.

A new idea is proposed to package the converter with customized magnetic component which is not only the inductor but also the case of the power module. With this package and integration technology, the thermal performance of the integrated power module can be significantly improved. Section II of the paper illustrates the structure and thermal model of the power module; Section III will analyze the thermal effect of the winding loss and do a comparison with the plastic solution; Section IV shows a prototype and evaluates its performance; Section V concludes the paper.

II. STRUCTURE AND THERMAL MODEL

In this section, the detailed structure and material characteristics of the power module are presented based on which analytical thermal model is further developed. Figure 1(a) shows the structure of the magnetic core and the DC/DC converter. The converter is composed of two parts. One part is a magnetic core on the top; the other part is PCB substrate at the bottom. The terminal of the windings are soldered on the PCB substrate while some thermal conductive glue is used to provide thermal conductive path from the top of the regulator to the ceiling of the cavity. Since general magnetic material has

978-1-4673-9551-9/16 $31.00 © 2016 IEEE

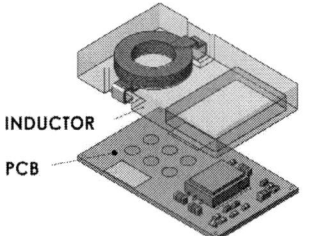

(a) Structure of the power module.

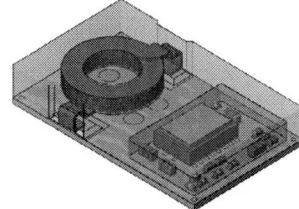

(b) Heat sources of the power module.

Fig. 1. Magnetic packaged power module

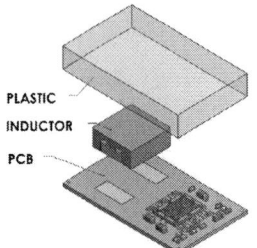

(a) Structure of the power module.

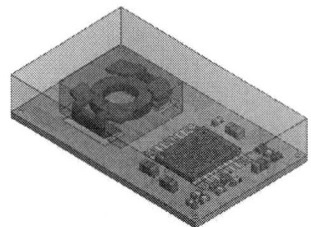

(b) Heat sources of the power module.

Fig. 2. Palstic packaged power module

better thermal conductivity, compared with plastic packaging material used in conventional power module with a small metal composite inductor inside, the proposed magnetic packaged power module has a higher thermal conductive case, which will in turn improve the performance of the whole power module. In the proposed module, there are three major components of loss: IC loss, winding loss, and core loss. The heat generated by the loss power has two major paths to dissipate to the ambient. The first one is PCB board underneath; the second one the magnetic core on the top. Figure 1(b) shows the heat sources and the major paths for dissipating the heat to ambient. A standard four-layer board with 80×60mm size is used to dissipate the heat from the power module to the PCB board.

With the purpose of comparison, a module with the same package size but based on conventional plastic packaging is also built. Figure 2 shows the structure and distribution of the heat sources in the plastic packaged power module. Since enough space has to be left for injecting epoxy materials, the inductor in the module can't be the same size as the magnetic packaged power module in which the inductor has the same width of the module. Therefore, the inductor in plastic packaged power module is subject to having higher DCR or lower inductance value. The heat sources in plastic packaged power module are the IC die and the inductor. Compared with the proposed magnetic packaged power module, the conventional plastic power module also has two paths to dissipate the heat. One is PCB substrate at the bottom; the other one is the epoxy on the top. Very thick epoxy is injected on the top of the die as well as the inductor which make the heat dissipation less efficient than the magnetic packaged power module. The other heat dissipation path has nearly the same thermal resistance as in the magnetic packaged power module. Table I shows the definition of thermal resistance in both Figure 1 and Figure 2. To calculate the thermal resistance in both models, thermal characteristics of different materials are

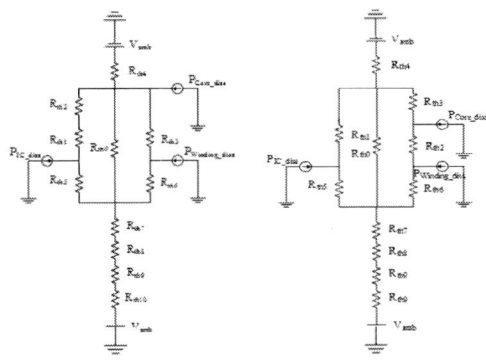

(a) Integrated power module   (b) plastic packaged module

Fig. 3. thermal model of the two power modules

listed in TABLE II. With above parameters, thermal models of the both power modules can be obtained. Figure 3 shows the thermal model of the two power modules. Further loss thermal analysis will be done based on it. Thermal resistance value in the thermal models are obtained through FEA simulation. Both the proposed magnetic packaged power module and the plastic package power module have complicated internal structure. Components such as the magnetic case and the inductor winding do not have regular shapes. Therefore, it is very difficult to calculate the thermal resistance through analytical equations. However, with the help of the simulation, we can easily get the thermal resistance in an easier way. 3D simulation modules are built to calculate the thermal resistance. Thermal conductivity parameters used in

978-1-4673-9551-9/16 $31.00 © 2016 IEEE

TABLE I.   A. THERMAL RESISTANCE IN MAGNETIC PACKAGED POWER MODULE

Designator	Thermal resistance	Value (K/W)
R_{th1}	IC plastic thermal resistance, from the IC die to its plastic case	148
R_{th2}	Red glue thermal resistance, from the IC case to the magnetic case	2
R_{th3}	Winding insulation layer thermal resistance, from the copper to the magnetic case	3.5
R_{th4}	Magnetic case thermal resistance, from the magnetic case to the ambient	170
R_{th5}	Joints thermal resistance of the IC, from the IC die to the substrate of the module	1.2
R_{th6}	Joints thermal resistance of the inductor, from the winding to the substrate of the module	14.5
R_{th7}	Substrate thermal resistance, from the top of the substrate to the bottom of the substrate	1
R_{th8}	Joints thermal resistance of the module, from the bottom of the module to the Test board	0.5
R_{th9}	Thermal board resistance, from the top of the thermal board to ambient	2

TABLE I.   B. THERMAL RESISTANCE IN PLASTIC PACKAGED POWER MODULE

Designator	Thermal resistance	Value (K/W)
R_{th1}	Thermal resistance of IC top plastic, from the IC die to the top surface of plastic case	244
R_{th2}	Winding insulation layer thermal resistance, from the copper to the magnetic case	17.6
R_{th3}	Thermal resistance from the inductor to the top surface of plastic case	223
R_{th4}	Thermal resistance of the plastic case, from the plastic case to the ambient	170
R_{th5}	Joints thermal resistance of the IC, from the IC die to the substrate of the module	4.2
R_{th6}	Joints thermal resistance of the inductor, from the winding to the substrate of the module	12
R_{th7}	Substrate thermal resistance, from the top of the substrate to the bottom of the substrate	1
R_{th8}	Joints thermal resistance of the module, from the bottom of the module to the Test board	0.5
R_{th9}	Thermal board resistance, from the top of the thermal board to ambient	2

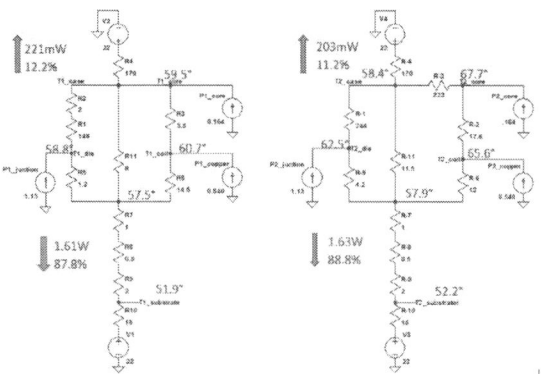

(a) Magnetic packaged module   (b) Plastic packaged module

Fig. 4.   Simulation results of thermal models

TABLE II.   THERMAL CONDUCTIVITY OF THE MATERIALS

Materials	Thermal conductivity(W/K.m)
FR4	0.35
Epoxy	0.6
Copper	400
Red glue	1
Magnetic core	3
Solder	62

the simulation are listed in TABLE II. In the simulation, any component can be assigned as a heat source. Suppose heat flowing through a neighbor component of the source is P_{diss}, and temperature differences of the external surface and internal surface of the component is obtained in the simulation. The thermal resistance of the component can be calculated by (1). In this way, all the thermal resistance are obtained, and they are added in TABLE I, too.

$$R_{th} = \frac{\Delta T}{P_{diss}} \qquad (1)$$

The thermal module is then put into a Spice software to do simulation, Figure 4 shows the temperature of different layers and the heat dissipation through each way. It can be seen that proposed power module has lower junction temperature and case temperature benefiting from its package technology.

III.   LOSS AND THERMAL ANALYSIS

In this section, loss of the converter is firstly analyzed and the thermal effect is simulated through the FEA simulation. Loss of the inductor is mainly determined by the parameters of the winding. Therefore, the winding parameters will be swept to simulate its effect on thermal performances of the whole converter. The analytical thermal models have shown the proposed inductor has smaller thermal resistance from heat sources to the ambient than that of power module packaged based on conventional plastic packaging technology. Besides this benefit, the proposed inductor also has larger inductor size than the inductor used in conventional plastic packaging technology. Therefore, the inductor can have much smaller copper loss. Details about how to design the inductor are discussed in another paper. This paper only addresses the effect of loss and thermal performances of the inductor by the parameters. The inductance value L, DCR with coil radius R, coil width w, and wire thickness t as parameters are analyzed by FEA simulation. Figure 5 shows the inductance value versus radius and width of the inductor. Figure 6 shows the DCR versus radius and width of the inductor. According to figure 5, higher inductance value could be obtained with smaller width and larger radius winding. However, as shown in Figure 6, the DCR will also increase with this kind of parameter configurations which will result in more copper loss and higher temperature. From the perspective of thermal performance, there should be a tradeoff between inductance value and DCR. Therefore, L/R is defined as a new parameter

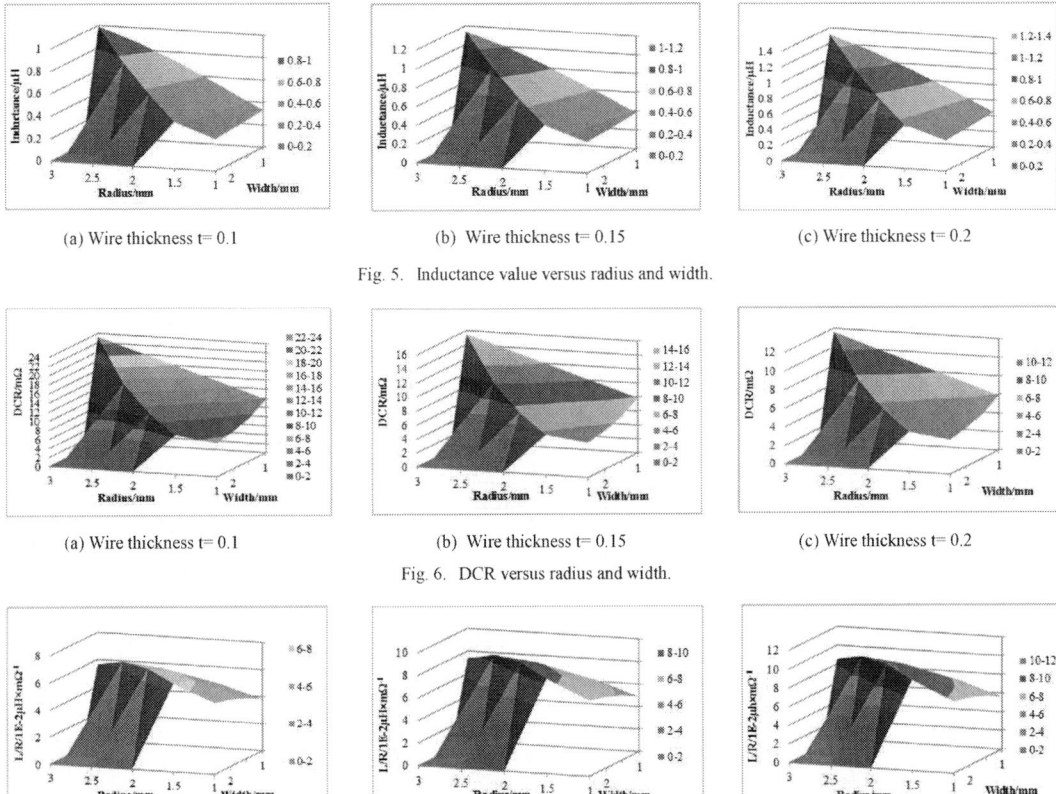

(a) Wire thickness t= 0.1   (b) Wire thickness t= 0.15   (c) Wire thickness t= 0.2

Fig. 5. Inductance value versus radius and width.

(a) Wire thickness t= 0.1   (b) Wire thickness t= 0.15   (c) Wire thickness t= 0.2

Fig. 6. DCR versus radius and width.

(a) Wire thickness t= 0.1   (b) Wire thickness t=0.15   (c) Wire thickness t=0.2

Fig. 7. L/R versus radius and width.

to select the right radius, height and width to simulate the thermal effect and temperature. Figure 7 shows the L/R value with radius and width as parameters.

It can be seen that the reasonable designs for different wire thickness are around (R=1.5~2.5mm, w=1~2mm). So several groups of the parameters are picked out from these configurations to simulate the thermal effect of winding parameters. They are show in Table III. Thermal performance of the proposed power module is simulated based on the losses of different winding configurations. The results are shown in Figure 8. It can be seen the winding parameter of the inductor also plays an important role on the thermal performances of the whole converter. With different parameters, the DCR differentiates a lot, and thus the winding loss, which can result in 4°C different on the case.

Although Group 3 in Table III has only 5.8mΩ DCR value, the winding can't be used in the inductor since the total thickness of the winding is too thick, leaving very thin magnetic material on top and bottom. Also, it is not practical from the perspective of the manufacturing. In the following analysis, Group 2 will be chosen to do comparison. As discussed in last section, benefiting from the larger space for the inductor winding in the proposed module, either higher

inductance or lower DCR can be achieved. In other words, the inductor used in plastic packaging will have either lower inductance or higher DCR value since its volume is smaller than the proposed inductor for leaving enough space to do plastic packaging. In this paper, a commercial inductor with 5×5×2.2mm size, 1μH inductance value is selected to do the comparison. Its DCR value is 17mΩ. The selected inductor is soldered on the same substrate of the proposed power module and then packaged with plastic material to do comparison. Another power module with same DCR value but different package is also simulated to see the effect of packaging itself on thermal. Both the power modules are simulated to show their thermal performances in the software.

TABLE III.   LOSS VS INDUCTOR CONFIGURATION

Groups	Thickness (mm)	Width (mm)	Radius (mm)	DCR (mΩ)	Loss (mW)
Group 1	0.1	1	2.5	17.6	752
Group 2	0.15	1	1.8	11.7	500
Group 3	0.2	1	1.5	5.8	248

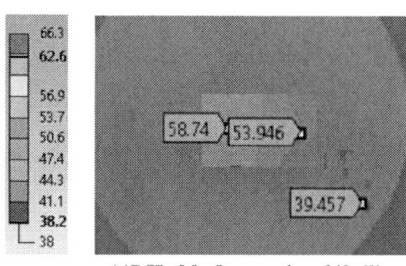 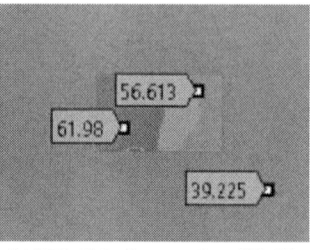 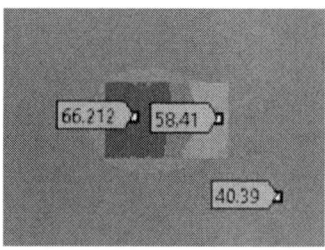

(a)DCR=5.8mΩ, copper loss=248mW;   (b) DCR=11.7mΩ, copper loss=500mW;   (c) DCR=17.6mΩ, copper loss=752mW.

Fig. 8. Thermal simulation results of magnetic packaged power modules with different winding configuration. (a) *t*=0.2mm, *R*=2mm, *w*=1.5mm; (b) *t*=0.15mm, *R*=1.5mm, *w*=1.2mm; (c) *t*=0.1mm, *R*=1.5mm, *w*=1mm.

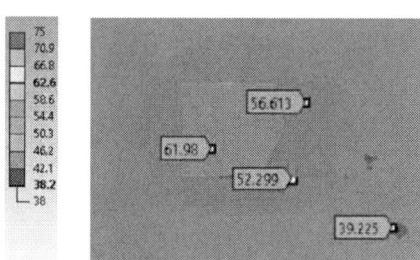 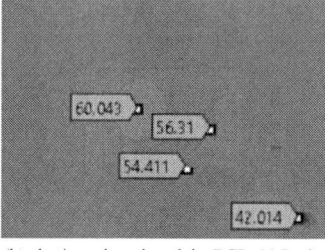 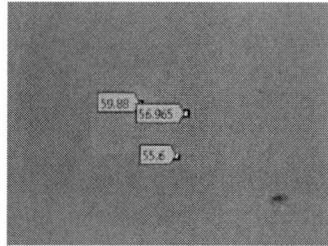

(a) Magnetic packaged module, DCR=11.7mΩ;   (b) plastic packaged module, DCR=11.7mΩ;   (c) plastic packaged module, DCR=17.6mΩ

Fig. 9. Case thermal plots for magnetic and plastic packaged modules.

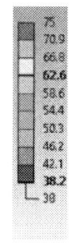 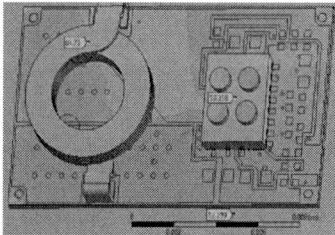 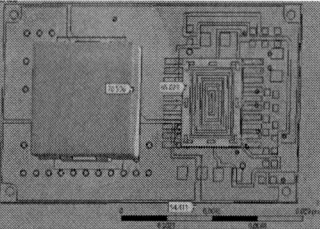 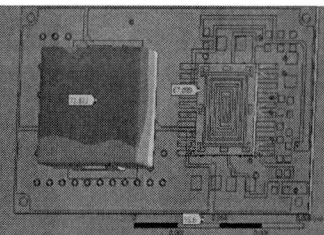

(a) Magnetic packaged module, DCR=11.7mΩ;   (b) plastic packaged module, DCR=11.7mΩ;   (c) plastic packaged module, DCR=17.6mΩ

Fig. 10. Junction thermal plots for magnetic and plastic packaged modules.

A 3D FEA simulation is used to calculate the power loss, based on which further thermal analysis is executed. A test is set up to measure the thermal conductivity coefficient of the magnetic core. The result is 3W/m-k, which is 5 times of the value for plastic packaging material 0.6W/m-k. With the thermal conductivity coefficient, a thermal simulation is also executed to show the thermal effect of inductor loss determined by different dimension configurations. In the thermal simulation, the module is soldered on an 80×60mm four layer PCB board to help dissipate the heat. Figure 9 and 10 shows the case and junction thermal plots of the power module at full load (6A). According to the results, junction temperature of magnetic package can be 11°C lower (61.7°C compared with 72.8°C). By considering several factors, such as the physical size limitation, power loss, saturation current, the specification of the inductor is chosen to be w=1.2mm, t=0.15mm, R=1.75mm for making prototypes.

IV. PROTOTYPE AND TEST

In this section, a prototype of the proposed power module is made to test its performances. The power module is composed of an inductor on the top and the PCB substrate at the bottom. A regulator together with some auxiliary components is soldered on the PCB (underneath the inductor cavity). Figure 11 shows the picture of the inductor, and the winding used for making the inductor. A 5.5 turn winding is used in the inductor. The winding is formed with 1.2×0.15mm planar copper wire with the purpose of reducing the eddy current loss. The whole inductor has the size of 15×9×2.4mm and the winding is embedded on one side of the inductor, leaving a 6×5×1mm cavity on the other side to put the components underneath. Figure 12 shows the schematic of the power module, top and bottom views of the assemble power module. The power module is then soldered on an evaluation board to do further

978-1-4673-9551-9/16 $31.00 © 2016 IEEE   2099

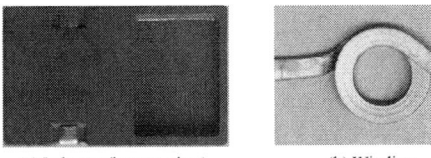

(a) Inductor (bottom view).   (b) Winding.

Fig. 11. Picture of the inductor and the winding

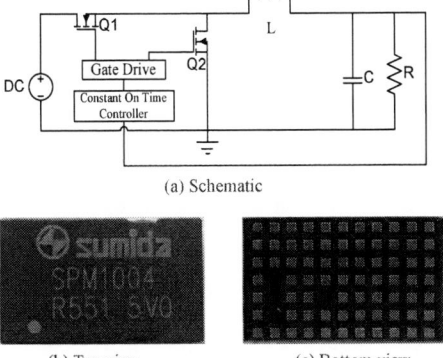

(a) Schematic

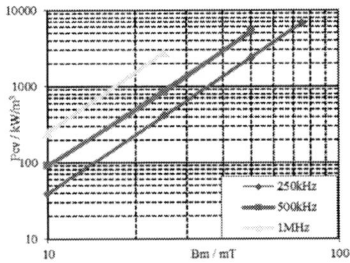

(b) Top view.   (c) Bottom view.

Fig. 12. Schematic and picture of the power module.

Fig. 13. Loss character of magnetic material

measurement and analysis. The power module operates under the constant on time control with the purpose of high light load efficiency. To do loss break down of the converter, the loss character of the magnetic material is measured, it is shown in Figure 13. Core loss is then simulated with FEA software by extracting the parameters based on the loss character curves. The core loss is much smaller than the winding loss since the major output current component is DC. Based on the inductor loss analysis, Total loss breakdown is executed to show loss distribution of the power module. The results are shown in Figure 14.

Benefiting from the higher thermal conductivity of the iron powder and high efficiency regulator, the magnetic packaged converter has better thermal performance than its plastic packaging counterparts. A thermal evaluation board is made to evaluate the prototype and the other two plastic packaging power modules with the same input and output specifications. Figure 15 shows the comparison of efficiency curves between the proposed power module and plastic packaged power module with same DCR and substrate. For the two plastic packaged power modules, #3 has the same size as the prototype; the other one has bigger size of 15×15×2.8mm. Figure 16 shows the comparison of efficiency curves between the 3 power modules. They are soldered on the thermal evaluation board with the same copper area as the prototype is. All the three power modules have the 12V input, 5V/6A output. Figure 17 shows the picture of the thermal evaluation board and the thermal image when the output power is 30W (5V/6A). It can be observed that the prototype has the same temperature with the bigger size power module (#2), and 11 °C less than the power module with the same size (#1). Without the outer plastic package layer, the magnetic packaged power module has bigger inductor volume and less thermal resistance from the heat sources to ambient. To show this benefit, loss analysis of three power modules is executed and the corresponding contributions to temperature rise are calculated. Table IV shows the results. The prototype has the same temperature rise with #2 although it has smaller volume. The prototype has less temperature rise than #3 because it has less inductor loss and better heat dissipating condition.

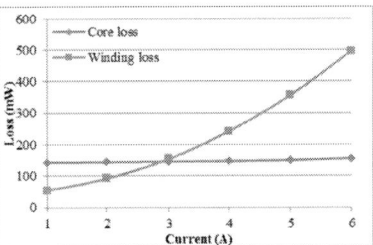

Fig. 14. Total loss break down

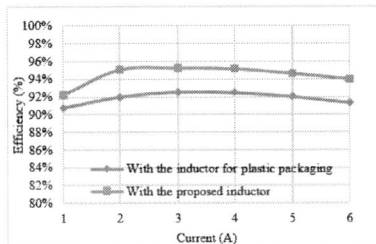

Fig. 15. Efficiency curves for proposed power modules and plastic packaged power module with same DCR and substrate

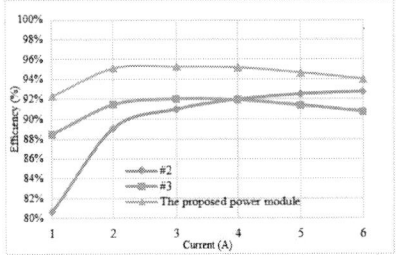

Fig. 16. Efficiency curves for three different power modules

(a) Thermal board

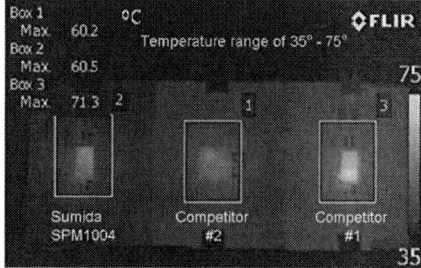

(b) Thermal imagine

Fig. 17. Thermal board picture and the thermal image.

TABLE IV.   INDUCTOR LOSS AND ITS CONTRIBUTION TO
TEMPURETURE RISE

	Prototype (SPM1004)	#2	#3
Inductor loss (W)	0.82	0.81	1.02
Temperature rise (℃)	15.1	15.3	19.6

V. CONCLUSION

This paper analyses the thermal performance of the magnetic packaged power module. Benefiting from the high thermal conductivity of the magnetic material, the power module has less thermal resistance from the heat source to ambient, thus better thermal performance. Compared with the plastic packaging, the magnetic packaged power module has less case temperature and higher efficiency. Detailed analysis on how much the magnetic package can enhance the thermal performance has also been presented.

REFERENCES

[1] Ke Ma; Bahman, A.S.; Beczkowski, S.; Blaabjerg, F., "Complete Loss and Thermal Model of Power Semiconductors Including Device Rating Information," in Power Electronics, IEEE Transactions on , vol.30, no.5, pp.2556-2569, May 2015.

[2] Shuojie She; Wenli Zhang; Xiucheng Huang; Weijing Du; Zhengyang Liu; Lee, F.C.; Qiang Li, "Thermal analysis and improvement of cascode GaN HEMT in stack-die structure," in Energy Conversion Congress and Exposition (ECCE), 2014 IEEE , vol., no., pp.5709-5715, 14-18 Sept. 2014

[3] Isidori, A.; Rossi, F.M.; Blaabjerg, F.; Ma, K., "Thermal Loading and Reliability of 10-MW Multilevel Wind Power Converter at Different Wind Roughness Classes," in Industry Applications, IEEE Transactions on , vol.50, no.1, pp.484-494, Jan.-Feb. 2014.

[4] Josifovic, I.; Popović-Gerber, J.; Ferreira, J.A., "Thermal management of compact SMT multilayer power converters," in Energy Conversion Congress and Exposition (ECCE), 2011 IEEE , vol., no., pp.52-59, 17-22 Sept. 2011

[5] Gautam, D.; Wager, D.; Musavi, F.; Edington, M.; Eberle, W.; Dunford, W.G., "A review of thermal management in power converters with thermal vias," in Applied Power Electronics Conference and Exposition (APEC), 2013 Twenty-Eighth Annual IEEE , vol., no., pp.627-632, 17-21 March 2013

[6] Senturk, O.S.; Helle, L.; Munk-Nielsen, S.; Rodriguez, P.; Teodorescu, R., "Converter Structure-Based Power Loss and Static Thermal Modeling of The Press-Pack IGBT Three-Level ANPC VSC Applied to Multi-MW Wind Turbines," in Industry Applications, IEEE Transactions on , vol.47, no.6, pp.2505-2515, Nov.-Dec. 2011.

[7] Seung-Yo Lee; Pfaelzer, A.G.; van Wyk, J.D., "Thermal analysis for improved packaging of 4-channel 42 V/14 V DC/DC converter," in Industry Applications Conference, 2004. 39th IAS Annual Meeting. Conference Record of the 2004 IEEE , vol.4, no., pp.2330-2336 vol.4, 3-7 Oct. 2004.

[8] Gerstenmaier, Y.C.; Castellazzi, A.; Wachutka, G.K.M., "Electrothermal simulation of multichip-modules with novel transient thermal model and time-dependent boundary conditions," in Power Electronics, IEEE Transactions on , vol.21, no.1, pp.45-55, Jan. 2006

[9] Cheng, M.-C.; Feixia Yu, "Heat flow models for silicon-on-insulator structures," in Electron Devices for Microwave and Optoelectronic Applications, 2003. EDMO 2003. The 11th IEEE International Symposium on , vol., no., pp.244-249, 17-18 Nov. 2003.

[10] DeVoe, J.; Ortega, A., "An investigation of board level effects on compact thermal models of electronic chip packages," in Semiconductor Thermal Measurement and Management, 2002. Eighteenth Annual IEEE Symposium , vol., no., pp.8-14, 12-14 March 2002.

Analysis of A Low-Inductance Packaging Layout for Full-SiC Power Module Embedding Split Damping

Yu Ren, Xu Yang, Fan Zhang, Linlin Tan, Xiangjun Zeng
Power Electronics and Renewable Energy Research Center
Xi'an Jiaotong University
Xi'an, P.R. China
Email: ryrenyu@stu.xjtu.edu.cn

Abstract—A novel low-inductance packaging layout for Full-SiC (Silicon Carbide) MOSFET (Metal Oxide Semiconductor Field Effect Transistor) Module with split damping capacitors embedding inside is proposed in this paper. The 3-Demision model of the optimized layout was built and analyzed with ANSYS Q3D. The total self-inductance of the single commutation loop is only 6.2nH. Additionally, 55V voltage spike over the semiconductor on the current rate of 6717A per microsecondand remarkable current sharing characteristic are observed by LTspice simulation based on the synchronous buck circuit. Two comparative 1.2KV, 40A prototypes based on regular wire-bond structure were fabricated to verify the low inductance layout design. Experimental result of double pulse test demonstrates the validity of the design. The ultra-fast switching speed can be achieved by using the proposed layout design.

Keywords—*silicon carbide MOSFET; power module; low-inductance packaging; layout; damping capacitor*

I. INTRODUCTION

High power modules are widely used not only in industrial field but also in HEV (Hybrid Electric Vehicle), EV (Electric Vehicle) and renewable energy area such as Photovoltaic Grid inverter and wind power [1]. These power module must have high efficiency and high reliable features because they play a core role in power electronics equipment. Meanwhile, the application of Si (Silicon) device is more and more close to the limitation of Si which cannot be exceeded. A promising way is to use wide band gap semiconductor device like SiC device in high power application.

SiC has ten times the dielectric breakdown field strength, three times the bandgap, and three times the thermal conductivity compared to Si [2], [3]. These properties make SiC device attractive in modern power electronics module manufacture which can far outstrip the performance of their Si counterparts [4].

Stray inductance is a major concern in design and layout of packages and power stages with both high switching speed and high power handling requirements [5-6]. The reasons are that stray inductance can cause a voltage spike among the semiconductor device due to high current rate, increase switching losses and cause electromagnetic interference and compatibility EMI and EMC problems. These problems must be handled more carefully in SiC MOSFET power module design since

This work was supported by the National Key Basic Research Program of China (973 Program) under project 2015CB251001 and 2015CB251004.

the higher switching frequency potential. For this reason, conventional power module structure is entirely unsuitable for the SiC MOSFET. Planer interconnection type packaging structures are proposed for the low parasitic inductances and high temperature application [7]-[8]. In those packaging structure, power device dies are directly bonded to DBCs or PCBs with solder or bumps instead of the wire bonding technology for realizing the extremely short interconnection length. Meanwhile, wire-bond is still the most popular and mature connecting technical. Therefore, it is imperative to innovate a low inductance layout using wire bonding technical which suit to the characteristics of SiC devices.

Researchers propose a method to decrease the influence of stray inductance which is to fabricate two capacitors near DC terminals in literature [9]. In this study, a newly designed Full-SiC layout fabricating several split damping capacitors inside the module is proposed. The parasitic parameters including resistor, self-inductance and mutual inductance of the layout were extracted by Ansys Q3D extractor. The results of LTspice simulation combination of SiC MOSFET bare die spice model and parasitic parameters of the layout show better performance both on voltage spike among semiconductor device and uniformity of the current distribution between parallel chips. Double pulse test of two comparative prototypes shows the validation of the design and the high frequency application potential of the proposed layout.

II. STRUCTURE OF NEWLY-DESIGNED LAYOUT

SiC MOSFET chips are usually connected in parallel to achieve high current capability since the current level of single SiC MOSFET chip is really low to use in high power system. To simplify the parasitic parameters model, each switch contains two SiC MOSFET chips in this study. The proposed layout design called Design-A is shown in Fig. 1 (a). To separate contributions of well-designed layout and split damping capacitors inside the module to the small voltage spike, Design-B shown in Fig. 1(b) was also modeled.

In this design, the power SiC MOSFETs and SBDs are commercially available manufactured by CREE in the form of bare SiC dies, both rated at 20 A and 1200 V, the DC-Link capacitors can operate at a temperature range from -55°C to 125°C.The substrate layouts of the module are illustrated in Fig. 1. As seen, the semiconductor devices are distributed on one piece of DBC substrates since the model is an experimental model only rating at 1.2KV, 40A. The electrical interconnection is achieved by 0.35 millimeters in diameter bonding aluminum (Al) wires on top of the dies and soldering the dies on Cu traces of a direct bond copper (DBC) substrate; electrical insulation is provided by the aluminum nitride (ALN) ceramic slice inside with high thermal conductivity which conducive to the high temperature performance of SiC device. The thickness of ALN ceramic is 0.635 millimeters; the thickness of DBC on both side of ceramic is 0.3 millimeters.

The layout of multiple SiC dies on DBC substrate and their interconnections has been optimized to reduce stray inductances. Specifically, DC copper traces of this design are put together against AC copper trace locates in the middle of DC copper traces in most commercial phase-leg modules. Fig.2 show that optimized layout can largely reduce the commutation loop area and thus reduce loop stray inductance. The voltage overshoot across the switch during the turn-off transient is proportional to the loop inductance and the rate of current change di/dt:

$$V_s = L_{Loop} \cdot \frac{d_i}{d_t} \qquad (1)$$

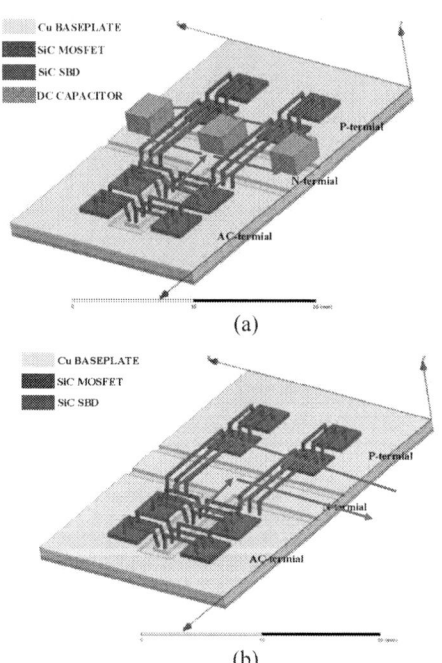

(a)

(b)

Fig. 1. 3-Dimension view of novel Full-SiC power module layouts: (a) Design A with DC capacitor inside the module; and (b) Design B without DC-damping capacitors inside the module

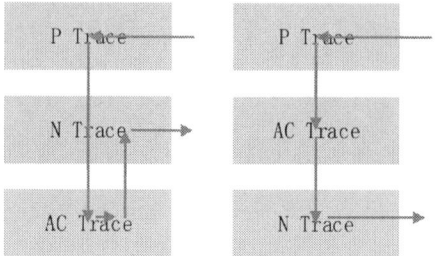

Fig. 2. Commutation loop of (a) proposed layout and (b) conventional layout.

Therefore, this design can directly reduce the voltage spike on the switch when the switch pair is in hard switching conditions. Another consideration to put two DC traces together is to create space putting split damping capacitors beside the switch pair. These capacitors have a function of "closing" the commutation loop shown in Figure 1 with red lines which lead to even smaller loop inductance.

III. EXTRACTION OF PARASITICAL PARAMETERS

For the accurate parasitic parameters extraction, the real layout of both Design-A and Design-B are modeled in Ansys Q3D extractor electromagnetic FEA tool shown in Fig. 1. By providing real dimensions, material properties (resistivity of conductors and permittivity of insulators) and boundary conditions (the conductors and current paths), the software can extract the structural impedances of arbitrary geometry [12]. In reality, stray inductance is only defined for closed loops, and the stray inductance critically depends on the chosen closed path. However, it is difficult to figure out the exact closed loop in the layout analysis. To solving this problem, we divide the layout include copper on the substrate and bonding wire into several small part, and each small part of the loop can be represented by resistors, self-inductance and mutual inductance, thus equivalent circuit of the layout can be built. Fig. 3 shows the equivalent circuit of the Design-A layout. To make graphics tidy, complex and the large number of mutual inductance and resistors are not marked. AC self-inductances of the circuit are shown in Table. I.

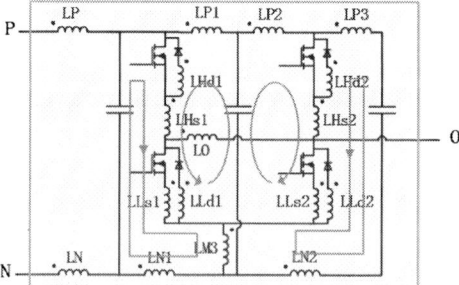

Fig. 3. Equivalent circuit of the Design-A layout with stray inductance.

With split damping capacitors inside, the commutation loop during switching is much smaller, this will make the loop inductance much smaller.

TABLE I.   SELF-INDUCTANCE FOR EQUIVALENT CIRCUIT OF THE DESIGN-A LAYOUT

Name	Value(nH)	Name	Value(nH)	Name	Value(nH)
LP	1.3969	LHs1	4.6638	LLd2	1.8964
LP1	0.1590	LHs2	4.6329	LN	0.2363
LP2	0.1689	LO	3.2017	LN1	1.2591
LP3	1.3201	LLs1	1.6027	LN2	1.8538
LHd1	1.3751	LLs2	1.6234	LN3	1.6447
LHd2	1.3791	LLd1	1.8704		

From Table I, we can find that wire bonding connections make up the dominant parasitic inductance of commutation loop. Long wire bonding connections form upper MOSFET to lower MOSFET contribute approximate 4.6nH inductance. The self-inductance of single commutation loop is 6.2nH which is a quite small value. It should be noted that this value is only the self-inductance of the specific inner loop including split damping capacitor. This value does not equal to the L_{Loop} mentioned in the (1) which represent the inductance of the loop include not only inner loop including split damping capacitor but also outside loop including bulk DC-link capacitor. The certain value of L_{Loop} can only be calculated from the experimental such as double-pulse-test.

IV.   SIMULATION APPROACH AND RESULTS

In power module, parasitic parameters are inevitable, among these parameters, stray inductance play a most important role on switching performance. To verify our design greatly decrease stray inductance of commutation loop which lead to smaller voltage spike, a comparative simulation is performed. The parasitic parameters model extracted by Ansys Q3D extractor can be saved as Multi-port network Spice model [12], [13]. Then, the device model provided by the semiconductor manufacture and extracted model are combined together in LTspice software with some other electrical line and electrical components. The simulation circuit is synchronous buck circuit. The series stray inductance of outside bus capacitor is 27.8nH from literature [14] which already optimizes the structure to decrease the inductance. In the simulation of Design A, three inside capacitors are all 4.7nF.

Fig.4 (a), (b) show the turn-off voltage waveforms of Design-A and Design-B. Fast turn-off transient in Design B causes 71V spike thanks to the optimized layout. Turning-off transient in Design A cause 55V spike due to both the optimized layout and split damping capacitors inside. Fig. 4 (c), (d) shows the well current consistency through upper two parallel SiC

MOSFETs of Design-A during turning on transient and current rising phase.

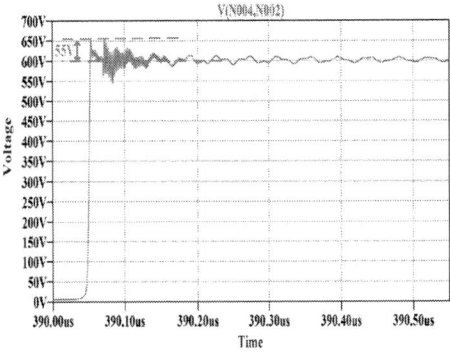

(a)

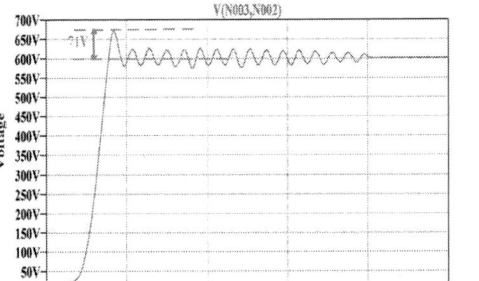

(b)

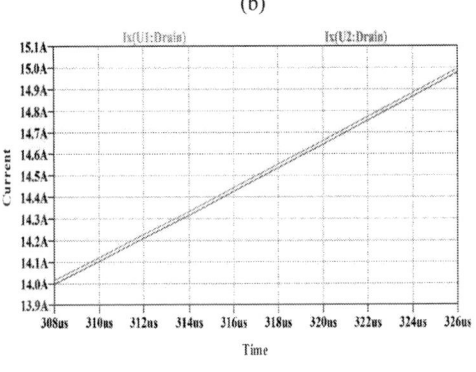

(c)

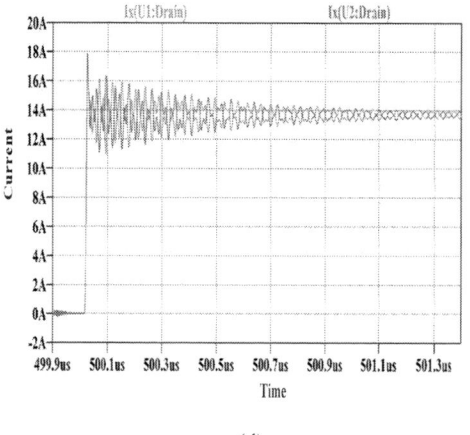

(d)

Fig. 4. Turn-off voltage waveforms of Design-A (a) and Design-B (b); Current through upper two parallel SiC MOSFETs of Design-A during turning on transient (c) and current rising phase (d).

The relationship between the capacitance of inside capacitors and voltage spike is shown in Fig.5. It indicates that when the capacitance is too small like less than 0.01nF, the voltage spike is larger than the Design B since the resonant phenomenon. The voltage spike decrease with increase of capacitance. The value of capacitance is a tradeoff between the size of capacitor and voltage spike.

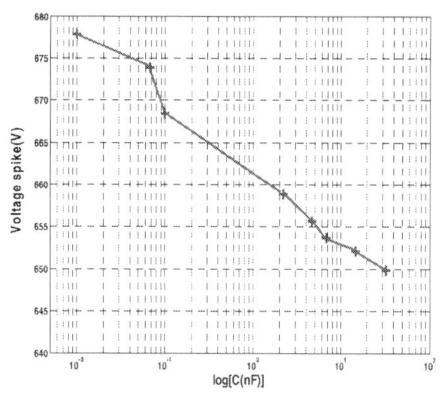

Fig. 5. The relationship between the capacitance of inside capacitors and voltage spike

V. EXPERIMENTAL RESULTS

Two comparative 1.2KV, 40A prototypes shown in Fig. 6 (a), (b) based on regular wire-bond structure were fabricated to valid the low inductance layout design. Since the larger capacitance value means the smaller voltage spike, we choose larger capacitor when the prototypes were fabricated. Three capacitors in Design A are KEMET 1KV, 56nF multilayer Ceramic Chip Capacitors. It is known that the gate resistor decide the driving speed of the gate driver, moreover, the lower gate loop

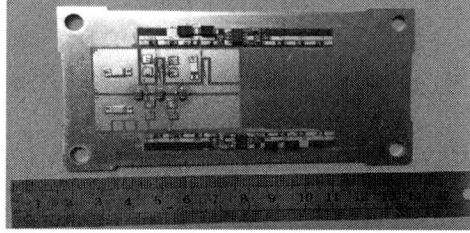

(a)

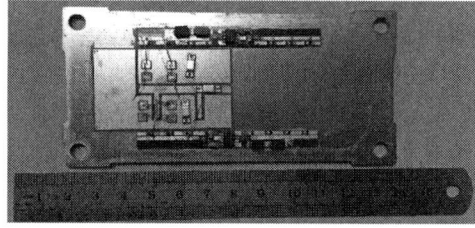

(b)

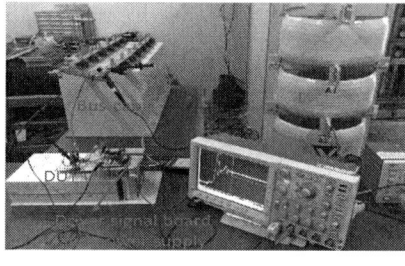

(c)

Fig. 6. 1.2KV, 40A SiC MOSFET experimental prototypes and experimental facility: (a) Design-A with split damping capacitors; and (b) Design-B without split damping capacitors; (c) experimental facility.

inductance is the smaller gate resistor can be. In this study, a gate driving board is designed specially for this layout, and the driving boards were placed near the DBC baseplate to reduce the stray inductance of driving loop. In order to evaluate the switching performance of the proposed design, two comparative prototypes were measured under double-pulse-test. Fig. 6 (c) show the double-pulse-test platform for the modules.

Each module is tested with only 50V DC input due to the absence of insulated silica gel. Each device has 2.5ohms external driver resistor and 4.6 ohms internal gate resistor. The

978-1-4673-9551-9/16 $31.00 © 2016 IEEE   2105

results shown in Fig.7 reveal 37V voltage spike of Design B and only 7.9V of Design A. It's worth noting that the high voltage spike is relevant to the stray inductance of outside DC-bus loop. In this test, the outside Bus-cap loop is not optimized and thus the voltage spike is bigger than the value in simulation. Despite all this, the proposed layout show the much lower internal inductance, thus the layout has higher switching frequency potential.

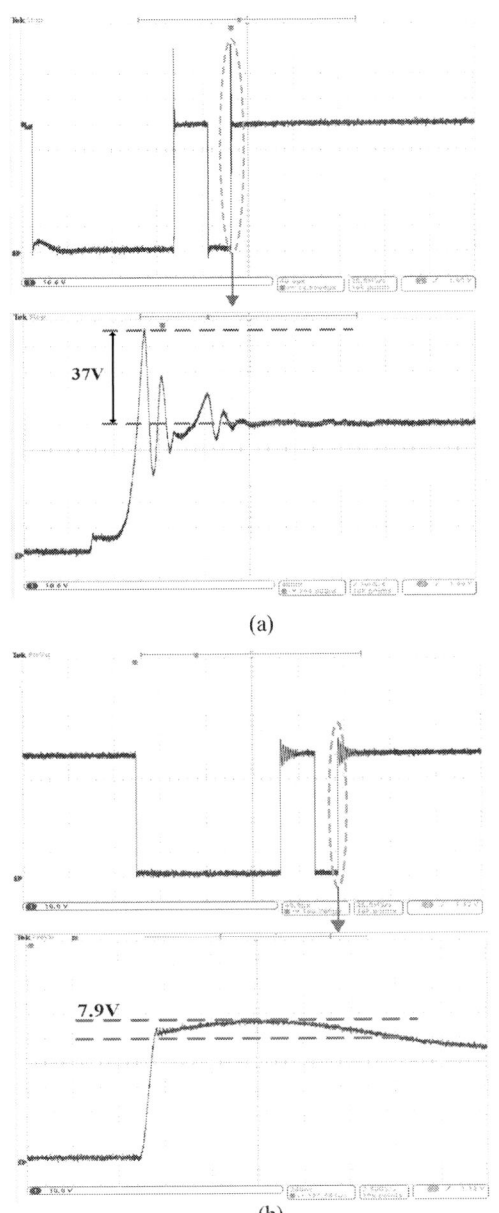

(a)

(b)

Fig. 7.   Double-pulse-Test experimental results: (a) Design-A; (b) Design-B

VI.   CONCLUSIONS AND FUTURE WORK

A novel packaging layout of Full-SiC MOSFET module embedding split damping capacitors has been proposed. Two comparative 1.2KV, 40A prototypes based on regular wire-bond structure were fabricated to valid the low inductance layout design. The total self-inductance of the single commutation loop is only 6.2nH and the experimental and circuit simulation results show the lower voltage spike at turn-off. The test voltage will be increased with the help of insulated silica gel in the future. Outside DC-bus loop will be optimized to decrease external inductance. The relationship between the voltage spike with internal capacitance, internal inductance and external DC-bus stay inductance will also be analyzed.

ACKNOWLEDGMENT

This work was supported by the National Key Basic Research Program of China (973 Program) under project 2015CB251001 and 2015CB251004.

REFERENCES

[1] A. Emadi, Y. J. Lee, and K. Rajashekara, "Power electronics and motordrives in electric, hybrid electric, and plug-in hybrid electric vehicles," IEEE Trans. Ind. Electron., vol. 55, no. 6, pp. 2237–2245, Jun. 2008.

[2] J. Millan, "A review of WBG power semiconductor devices," in Proc. 2012 Int. Semicond. Conf. (CAS), pp. 57–66, 2012.

[3] J. Millan, P. Godignon, X. Perpina, A. Perez-Tomas, and J. Rebollo, "A survey of wide bandgap

[4] A. Argawal, "Advances in SiC MOSFET Performance", in ECPE SiC & GaN Forum Potential of Wide Band gap Semiconductors in Power Electronic Applications, 2011.

[5] K. Xing, F. C. Lee, and D. Boroyevich, "Extraction of parasitics within wire-bond IGBT modules," in Proc. IEEE Appl. Power Electron. Conf. Expo., pp. 497–503, 1994.

[6] W. Kangping, Y. Xu, L. Hongchang, M. Huan, Z. Xiangjun, and C. Wenjie, "An Analytical Switching Process Model of Low-Voltage eGaN HEMTs for Loss Calculation," Power Electronics, IEEE Transactions on, vol. 31, pp. 635-647, 2016.

[7] Zhenxian Liang ; Puqi Ning ; Wang, F. "Development of Advanced All-SiC Power Modules," IEEE Transactions on Power Electronics , Vol. 29, no. 5, pp. 2289 – 2295, 2014.

[8] Zhenxian Liang ; Puqi Ning ; Wang, F. ; Marlino, L. "A Phase Leg Power Module packaged with Optimized Planar Interconnections and Integrated Double-Sided Cooling." IEEE Journal of Emerging and Selected Topic sin Power Electronics, Vol. 2, no. 3, pp. 443 – 450, 2014.

[9] E. Hoene, A. Ostmann, B. T. Lai, C. Marczok, A. Musing, and J. W. Kolar, "Ultra-Low-Inductance Power Module for Fast Switching Semiconductors, " in Proc. PCIM 2013, 2013.

[10] R. Wang, D. Boroyevich, P. Ning, Z. Wang, F. Wang, P. Mattavelli, K. D. T. Ngo, and K. Rajashekara, "A high-temperature SiC three-phase AC-DC converter design for >100 ◦C ambient temperature," IEEE Trans-actions on Power Electronics, vol. 28, no. 1, pp. 555–572, Jan. 2013.

[11] Z. Chen, Y. Yao, D. Boroyevich, K. D. T. Ngo, P. Mattavelli, and K. Rajashekara, "A 1200-V, 60-A SiC MOSFET multichip phase-leg module for high-temperature, high-frequency applications," IEEE Transcations on Power Electronics , Vol. 29, no. 5, pp. 2307–2320, May 2014.

[12] Ansys., Canonsburg, PA . [Online]. Available: http://www.ansys.com.

[13] J. Z. Chen, L. Yang, D. Boroyevich, and W. G. Odendaal, "Modeling and measurements of parasitic parameters for integrated power electronics modules", in Proc. IEEE-APEC 2004, vol. 1, pp. 522-525.

[14] Yamamoto, T. ; Hasegawa, K. ; Ishida, M. ; Takao, K. "Switching simulation of SiC high power module with low parasitic inductance," 2014 International Power Electronics Conference (IPEC-Hiroshima 2014- ECCE-ASIA), pp: 3707 – 3711, 2014.

Comprehensive Parametric Analyses of Thermally Aged Power MOSFETs for Failure Precursor Identification and Lifetime Estimation Based on Gate Threshold Voltage

Serkan Dusmez, *Student Member*, and Bilal Akin, *Senior Member, IEEE*
Electrical and Computer Science Department, Power Electronics and Drives Laboratory
University of Texas at Dallas, Richardson, TX, USA
serkan.dusmez@utdallas.edu, bilal.akin@utdallas.edu

Abstract— **Thermal/power cycles are widely acknowledged methods to accelerate the package related extrinsic failures. Many studies have focused on particular failure precursor at a time and continuously monitored it using custom-built circuits. Due to the difficulties in taking sensitive measurements, the reported findings are more on the quantities requiring less sensitive measurements such as on-state resistance. In this paper, a custom-designed test bed is used to age a number of power MOSFETs and an automated curve tracer is utilized to capture parametric variations in I-V curves, transfer capacitances and gate charges at certain time intervals throughout the aging. The results suggest that the failure precursors which exhibit continuously increasing trend are the on-state resistance, body diode voltage drop, parasitic capacitances and threshold voltage. Based on the results, an exponential empirical model for the gate threshold voltage that fits successfully with the experimental data is proposed. Furthermore, Kalman Filter is employed to filter out the measurement noise and model uncertainties, which is also used to estimate the remaining useful lifetime of the degraded switches.**

Keywords—failure diagnosis, power MOSFET, on-state resistance, gate threshold voltage, health monitoring, remaining useful lifetime.

I. INTRODUCTION

Reliability of power semiconductor devices is of great importance particularly for mission critical systems, and has been exhaustively investigated in recent literature [1]-[7]. Researchers have focused on the identification of failure precursors that can be evaluated to engage on protective circuitry, prevent costly shutdowns, and bring down maintenance costs. To accelerate the time duration of the reliability tests, custom test-beds that are capable of applying thermal/power cycling have been designed [4],[6]-[8]. Auxiliary circuits are employed to continuously monitor the failure precursors. Majority of the work has been devoted to the reliability of IGBT modules where multiple active chips are paralleled through multiple bond-wires [1],[3]-[5],[7]. The reported findings in the literature are; variations in saturation voltage [4],[6],[9], thermal impedance [10], turn off time [11], change in phase and amplitude of ringing during turn-off [12], input, output and reverse transfer capacitances [3], threshold voltage [6], [14]. On the other hand, a few studies have investigated the failure precursors observed in power MOSFETs [8], [15]-[17]. The failure precursors identified for power MOSFETs are the on state-resistance and gate threshold voltage.

Although some of the failure precursors of thermal aging have been identified in literature, the effects on the MOSFET's current/voltage characteristics have not been fully explored comprehensively in a single study. One important reason is the necessity of high resolution source and measurement units for these tests. For instance, the gate current or drain-source leakage currents are in the order of several pA which can easily be interfered by the noise present in the circuit, if the measurement interface board is not well designed.

In this work, 400V/11A power MOSFETs are thermally aged on a custom-designed test-bed presented in [8]. It is aimed to stress the gate oxide and solder joints by applying thermal cycle. At certain time intervals (50 or 200 thermal cycles), the switches are removed from the test-bed and placed on the Keysight B1506A automated curve tracer as shown in Fig. 1. The following parameters are analyzed in this work; 1) threshold voltage, 2) breakdown voltage, 3) leakage current, 4) body diode avalanche voltage, 5) gate charge, 6) parasitic capacitances, 7) on-state resistance, 8) body diode voltage drop. In earlier studies, on-state resistance variation has been continuously monitored and identified as a failure precursor and evaluated to estimate the remaining useful lifetime [17]. In this study, all parametric changes in the thermally aged power switches are investigated based on the data collected in the laboratory environment. Furthermore, as a result of the findings, the threshold voltage is identified as a feasible failure precursor that increases consistently throughout aging, which can be used to assess the state of health of the switches.

978-1-4673-9551-9/16 $31.00 © 2016 IEEE

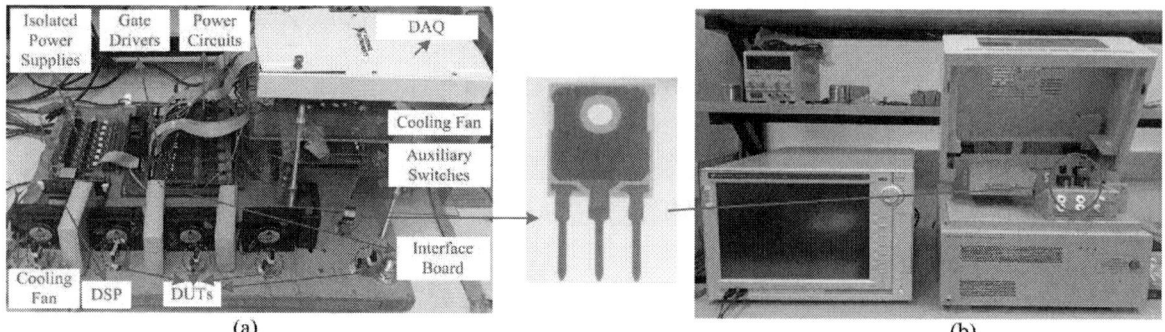

Fig. 1. Illustration of the test-bed; (a) custom designed aging setup, (b) curve tracer.

Based on the experimental data, an empirical model in the form of an exponential function is established and Kalman Filter (KF) is used to filter out the noise and deal with the model uncertainties.

II. PARAMETRIC ANALYSIS OF THERMALLY AGED POWER MOSFETs

The conducted thermal aging tests intend to cycle the power MOSFETs with a junction temperature swing between 40°C and 180°C, at a constant drain current of 5.2A, where rated continuous drain current of the switch at room temperature is 11A. At every 200 thermal cycles, the devices are removed from the setup shown in Fig. 1(a), and then plugged into the curve tracer for parametric analyses, as shown in Fig. 1(b). The tests have been performed on multiple samples, but only results of two samples that can represent the general behavior of the power MOSFETs have been shown here. There are two failure modes observed during the tests. Most commonly, the gate control was lost. Secondly, the drain-source terminal was shorted. Basically, there are three package related degradation mechanisms taking place in a semiconductor power device.

The first degradation mechanism is the aluminum reconstruction, which happens at relatively high junction temperatures. The thermal cycles at high temperatures result in tensile stresses on the upper metallization layer due to mismatch of the coefficients of the thermal expansions (CTE) of Al (22.2) and Si (3), where unit of CTE is $10^{-6}m/(mK)$. This induced stress can be much higher than the elastic limit, which results in extrusion of the aluminum grains at the grain boundaries. This causes the sheet resistance to increase, which increases the on-state resistance measured between drain to source terminals, and also degrades the gate-oxide layer as the increased sheet resistance of a cell increases the current densities at adjacent cells [18].

Another degradation/damage is typically observed at the wire-bonds. Being close to the active Si chip area, particularly the bond pad is exposed to the full thermal swing under power cycling. In addition, due to the skin effect becoming more dominant at higher switching frequencies, the current distribution or in other words power dissipation is not homogeneous. Most commonly, a crack is forming up at the tail of the bond, which then propagates towards the Al surface where the bond-wire gradually lifts off [18]. The root of this failure is again mismatch of the CTE between Al and Si. Bond-wires experience thermal swings larger than the ones experienced by the die.

Thirdly, the solder joint between Cu base plate and Direct Bounded Copper (DBC) substrate degrades with respect to thermal swings. Typically, the solder material is lead-tin alloy bonded to the base copper plate with large lateral dimension. When this interface experiences thermal cycles, the cracks are formed at the border of the joint on the copper plate side as maximum shear stress occurs at the border [18]-[19]. The degradation on this layer affects the thermal impedance as heat flow within the device is from Si to the base plate.

The DBC layer mostly helps reducing the shear stress on the die attach solder as CTE of DBC ranges from 4 to 7.1 while it is 3 for Si. However, DBC layer is not preferred in discrete packages due its high cost. Moreover, isolation between base plate and drain/source plate is not required in discrete switches. When die is attached to a Cu base plate, the CTE mismatch of die and base plate becomes considerably large with relatively larger lateral dimension in comparison to bond-wires. In addition, the thermal time constant of a discrete package is significantly less than that of the power module. Because of this reason, it is likely that die attach solder joint degrades in discrete packages at lower frequencies faster than the bond-wires. Die attach degradation increases both the electrical resistance and thermal impedance.

In accelerated tests, these three failure mechanisms occur simultaneously. Some studies have also tried to trigger only one type of failure. For instance, in the bond-wire failure tests, the time duration of the applied thermal swing is less than a couple of seconds with a maximum junction temperature not exceeding 110°C, where the switches are driven with PWM under rated load. The purpose is to only increase the junction temperature through heat generated as a result of power loss, but meantime not heating the case temperature too much so that solder joints or DCB substrate do not experience significant temperature swing. This type of test reflects a real condition where the operation frequency is higher than 1-10Hz

978-1-4673-9551-9/16 $31.00 © 2016 IEEE   2109

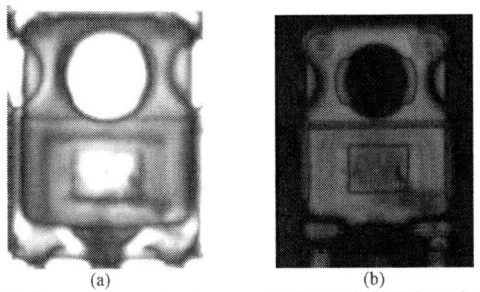

(a)   (b)

Fig. 2. Failure analysis results of a power MOSFET after 300 thermal cycles; (a) T-SAM, (b) C-SAM.

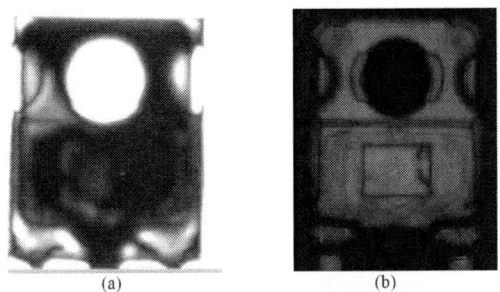

(a)   (b)

Fig. 3. Failure analysis results of a power MOSFET after 3300 thermal cycles; (a) T-SAM, (b) C-SAM.

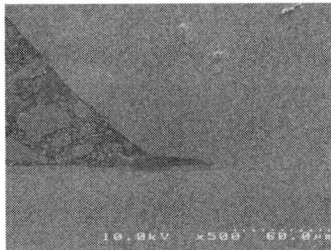

Fig. 4. Bond wire image after cross-sectioning.

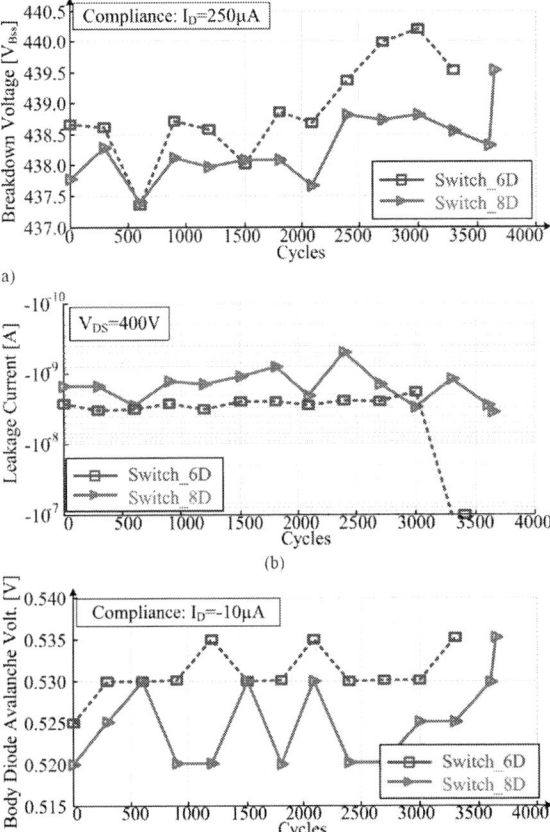

Fig. 5. Parameter variations according to the thermal cycles; (a) Breakdown voltage, (b) Leakage current, (c) Body diode forward avalanche voltage.

and it is more suitable for power modules where there is a significant thermal capacity difference between the die and DBC substrate. In discrete packages, it is more likely to observe die attach degradation under 60Hz as CTE mismatch of Cu and Si is large, the lateral dimension of die attach is larger than bond-pad, and thermal time constants of a discrete device is much smaller than power modules.

In our custom-designed test bed, the devices are actively heated without attaching a heat-sink and cooled down by forced air-cooling through fans. Thus, the junction temperature is increased up to very high degrees (180°C which triggers all three failure mechanisms. Yet, as pointed out earlier, the large thermal swings together with long relaxation and heating times in fact stress the die attach solder joints, and gate oxide more than the bond-wires due to the larger lateral dimensions. In our tests, one thermal cycle lasts for approximately one minute (20 sec heating, 40 sec cooling). Thus, we observed the consequences of these two aging mechanisms before there was any damage on the bond-wires.

The T-SAM and C-SAM analyses are conducted to observe possible delamination and voids on the surfaces. The dark spots in T-SAM and the red spots in C-SAM reveal possible delamination. The C-SAM and T-SAM results of a power MOSFET cycled for 300 thermal cycles shown in Fig. 2. It is seen that the images are clear and shows no sign of voids/delamination. The failure analyses of the same sample after 3300 thermal swings are shown in Fig. 3. The T-SAM result suggests possible delamination on the die attach, and C-SAM results indicate voids on several spots. The device has been cross-sectioned and bond-wire has been analyzed as shown in Fig. 4. As it can be seen, there is no significant damage on the bond-wire.

Some of the I-V curves under the aforementioned aging conditions are given in Fig. 5-7. The breakdown voltage has been measured as shown in Fig. 5(a) under the compliance of I_{DS}=250μA. It is seen that it increases by 2V at the end of the aging cycles; however, the increase is relatively small and does not exhibit a consistently increasing trend. As seen from Fig. 5(b), the leakage currents do not change significantly till the complete failure. The sample 6D has failed under short-

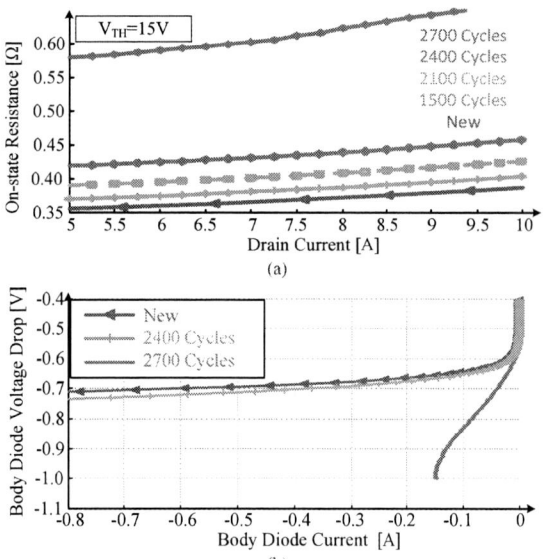

Fig. 6. Switch characteristics with respect to aging; (a) On-state resistance, (b) body diode voltage drop.

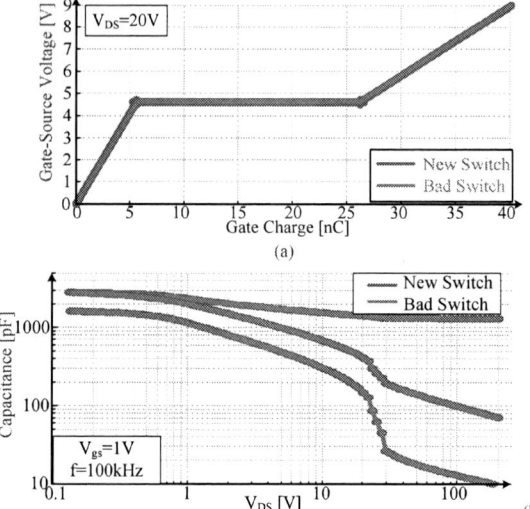

Fig. 7. Switch characteristics with respect to aging; (a) Gate charge, (b) Parasitic capacitances.

circuit fault. This failure can be observed from the increased leakage current around 3300 cycles. Fig. 5(c) shows the body diode forward avalanche voltages of the two samples. As it is seen, the p-n junction characteristics do not change which suggests that the failure is rather extrinsic

Fig. 6 presents the on-state resistance and body diode voltage drop at different aging intervals. As expected, the on-state resistance increases while the increase in the body diode voltage drop verifies the increased resistance between drain to source terminals. From electrical point of view, the body diode voltage drop of a switch is expressed as

$$V_{SD} = V_F + r_L I_{SD} \qquad (1)$$

where, V_{SD} is the voltage drop between source and drain terminals, r_L is the internal resistance between source to drain terminals, V_F is the forward voltage drop of the p-n junction, and I_{SD} is the current passing through the body diode. As it has been shown in Fig. 5(c), V_F does not change throughout the aging, which suggests that the increase in V_{SD} is due to the increased r_L. It is found out that the increment in the r_L is in fact equal to increment in the on-state resistance, $R_{ds,on}$.

In Fig 7, the gate charge and parasitic capacitances variations have been plotted for healthy and aged states. The curves suggest very small change in these characteristics. As shown in Fig. 8, the threshold voltage increases correspondingly. It has been reported in literature that the bond-wire liftoffs in a power module reduces the overlap surface of the gate polysilicon with doped N and P regions, which decreases the gate-source capacitance [3]. However, the device under test is in a discrete package with only one drain-to-source bond-wire. The solder degradation is not expected to have an effect on the voltage threshold, as thermal impedance

degradation at the joints does not have any physical interference with Si chip. Considering that compliance for threshold voltage test is 250µA, the junction would not be heating up. In fact, threshold voltage is used to correct the thermal impedance model of a degraded switch in [20], as it is not affected by the solder joint degradation.

One of the anticipated reasons for this change is the gate oxide degradation due to damaged source metallization and trapped electrons and holes in the gate-oxide. The devices had been decapsulated after the failure, which revealed source metallization damage. This partially degraded gate-oxide layer causes the changes in the gate oxide capacitance, which has the following relations with gate-source and gate-to-drain capacitances [3].

$$C_{gs} = \frac{C_{oxs}C_s}{C_{oxs} + C_s} + \frac{C_{oxc}C_c}{C_{oxc} + C_c} \qquad (2)$$

$$C_{gd} = \frac{C_{ox}C_{dep}}{C_{ox} + C_{dep}} \qquad (3)$$

where C_{oxs} and C_{oxc}, are the capacitances between gate oxide and $N+$ region, gate oxide and $P+$ region, respectively. C_s and C_c represent the capacitances of the depletion regions for $N+$ and $P+$ regions, and C_{dep} denotes the capacitance of the depletion region under the gate oxide. It is clear that decrease of gate oxide capacitances reduces both C_{gs} and C_{gd}. This decrease can also be observed in the gate threshold voltage variation as plotted in Fig. 8. The threshold voltage value has shifted by around 55mV at the end of the aging period. The threshold voltage is expressed as [21]

$$V_{Th} = V_{FB} + 2\phi_F + \frac{\sqrt{4\varepsilon_s q N_a \phi_F}}{C_{ox}} \qquad (4)$$

978-1-4673-9551-9/16 $31.00 © 2016 IEEE   2111

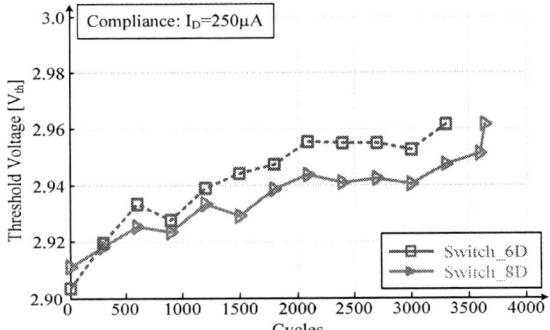

Fig. 8. Gate threshold voltage variation captured from the first set of test sampled at every 200 thermal cycles.

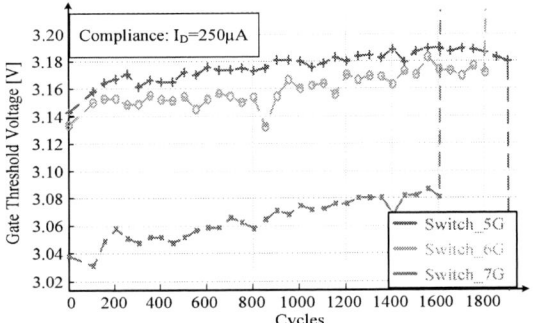

Fig. 9. Gate threshold voltage variation captured from the second set of test sampled at every 50 thermal cycles.

where, V_{FB} is the flatband voltage, $\square_F$ is the half contact potential, q is the charge of an electron, N_a is the doping density, ε_s is the dielectric constant of silicon. As Eq. (4) dictates, reduced gate oxide capacitance increases the threshold voltage value. Since the degradation of gate oxide is not severe in these samples, both the parasitic capacitances and threshold voltage variations are rather small. Yet, it is pertinent that threshold voltage steadily increases during aging, which makes it a potential failure precursor for remaining useful lifetime estimation.

To confirm this conclusion, another set of experiments were carried out and the results are presented in Fig. 9. The data is captured at every 50 thermal cycles rather than 200 thermal cycles to obtain more data points for evaluation. The resultant experimental data can be modeled empirically through exponential fit function. Based on the data, the threshold voltage can be expressed as in Eq. (5).

$$V_{Th}(t) = \alpha\left(1 - e^{-\beta t}\right) + V_{Th,init} \qquad (5)$$

For constant α and β, Eq. (5) can be written in state-space form as

$$\dot{V}_{Th}(t) = -\beta V_{Th}(t) + \beta\left(\alpha + V_{Th,init}\right) \qquad (6)$$

Discretizing the system yields

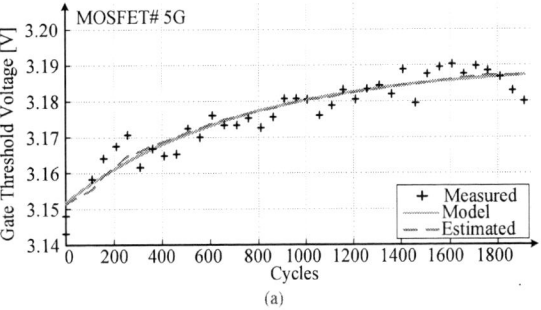

(a)

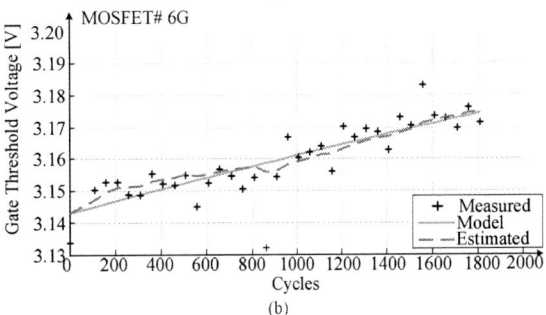

(b)

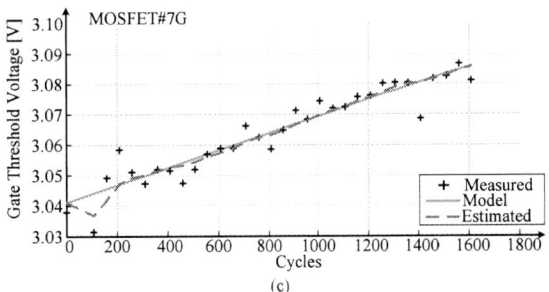

(c)

Fig. 10. State estimation with Kalman Filter for samples (a) MOSFET #5G, (b) MOSFET #6G, (c) MOSFET #7G.

$$V_{Th}(k+1) = V_{Th}(k)\cdot\left(1 - \Delta t\beta\right) + \beta\Delta t\left(V_{Th,init} + \alpha\right) \qquad (7)$$

It is possible to estimate the posterior state by applying KF to the empirical model given in Eq. (7). The implementation steps of the KF is similar to [17], in which on-state resistance variation had been identified as a failure precursor and the empirical model is filtered with KF. At every time step, the new state is estimated in a recursive manner using the previous posterior estimate that is found through the time and measurement updates. The estimated state obtained from the KF is compared with the actual measurement data in Fig. 10. Normal prediction horizon of EK is one step. However, EK can also be used for predicting n steps ahead, with current error covariance. As the real measurements are taken, the error covariance is updated and trajectory converges to the one obtained from the model. Fig. 11 presents the estimated trajectory of the state estimate after 200 and 400 thermal cycles for device 5G.

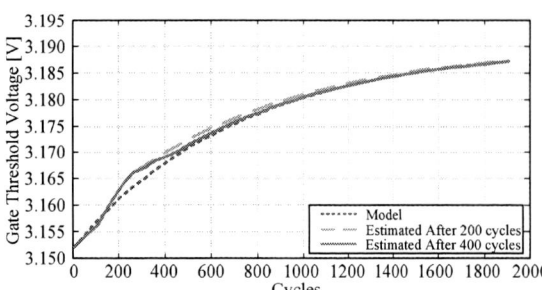

Fig. 11. RUL prediction for MOSFET #5G.

III. CONCLUSION

In this paper, comprehensive analyses of thermally aged power MOSFETs based on parametric current/voltage, parasitic capacitances, and gate charge variations with respect to aging cycles are presented. The power switches are exposed to thermal cycling, and the parametric variations are captured using an automated curve tracer at every determined number of cycles. It has been shown that on-state resistance, body diode voltage drop, parasitic capacitances and gate threshold voltage are the only viable failure precursors for die attach solder joint and gate oxide degradation mechanisms. The other parameters either do not consistently change or exhibit sudden change before failure, which can be used for instantaneous failure detection. Based on the findings, the variation in the gate threshold voltage is modeled with an exponential function, and Kalman Filter is used to filter out the measurement noise and model errors, and estimate the remaining useful lifetime. The experimental results on more than 20 samples suggest that threshold voltage variation is a feasible precursor to monitor.

IV. ACKNOWLEDGEMENT

This work was supported in part by the Office of Naval Research (ONR) under Award Number N00014-15-1-2325, and TXACE/SRC under TASK 1836.154. Authors would like to thank Priority Labs for providing the failure analysis results.

REFERENCES

[1] K. Li, G.Y. Tian, L. Cheng, A. Yin, W. Cao, and S. Crichton, "State Detection of Bond Wires in IGBT Modules Using Eddy Current Pulsed Thermography," *IEEE Trans on Power Electron.*, vol. 29, no. 9, pp. 5000-5009, Sept. 2014.

[2] S. Yang, D. Xiang, A. Bryant, P. Mawby, L. Ran, and P. Tavner, "Condition Monitoring for Device Reliability in Power Electronic Converters: A Review," *IEEE Trans on Power Electron.*, vol. 25, no. 11, pp. 2734-2752, Nov. 2010.

[3] W. Kexin, D. Mingxing, X. Linlin, and L. Jian, "Study of Bonding Wire Failure Effects on External Measurable Signals of IGBT Module," *IEEE Trans on Device and Materials Reliability,* vol. 14, no. 1, pp. 83-89, March 2014.

[4] V. Smet, F. Forest, J.-J. Huselstein, F. Richardeau, Z. Khatir, S. Lefebvre, and M. Berkani, "Ageing and Failure Modes of IGBT Modules in High-Temperature Power Cycling," *IEEE Trans on Ind. Electron.*, vol. 58, no. 10, pp. 4931-4941, Oct. 2011.

[5] H. Oh, B. Han, P. McCluskey, C. Han, and B.D. Youn, "Physics-of-Failure, Condition Monitoring, and Prognostics of Insulated Gate Bipolar Transistor Modules: A Review," *IEEE Trans on Power Electron.*, vol. 30, no. 5, pp. 2413-2426, May 2015.

[6] N. Patil, J. Celaya, D. Das, K. Goebel, and M. Pecht, "Precursor Parameter Identification for Insulated Gate Bipolar Transistor (IGBT) Prognostics," *IEEE Trans on Reliability,* vol. 58, no. 2, pp. 271-276, June 2009.

[7] B. Ji, V. Pickert, W. Cao, and B. Zahawi, "In Situ Diagnostics and Prognostics of Wire Bonding Faults in IGBT Modules for Electric Vehicle Drives," *IEEE Trans on Power Electron.*, vol. 28, no. 12, pp. 5568-5577, Dec. 2013.

[8] S. Dusmez and B. Akin, "An accelerated thermal aging platform to monitor fault precursor on-state resistance", in *Proc. IEEE International Electric Machines and Drives Conference (IEMDC)*, pp. 1-6, 10-13 May 2015.

[9] M. Held, P. Jacob, G. Nicoletti, P. Scacco, M.-H. Poech, "Fast power cycling test of IGBT modules in traction application," in *Proc. Power Electronics and Drive Systems*, pp. 425-430, 26-29 May 1997.

[10] W. Sleszynski, J. Nieznanski, A. Cichowski, J. Luszcz, A. Wojewodka, "Evaluation of selected diagnostic variables for the purpose of assessing the ageing effects in high-power IGBTs," in *Proc. IEEE International Symposium on Industrial Electronics (ISIE)*, pp. 821-825, 4-7 July 2010.

[11] Brown, D., M. Abbas, A. Ginart, I. Ali, P. Kalgren, and G. Vachtsevanos, "Turn-off Time as a Precursor for Gate Bipolar Transistor Latch-up Faults in Electric Motor Drives," in *Proc. Annual Conference of the Prognostics and Health Management Society*, pp. 1-8, 2010.

[12] Sonnenfeld, G., K. Goebel, and J.R. Celaya, "An agile accelerated aging, characterization and scenario simulation system for gate controlled power transistors," in *Proc. IEEE AUTOTESTCON*, pp. 205-215, 2008. Turn-Off Time as an Early Indicator of Insulated Gate Bipolar Transistor Latch-up.

[13] A. Ginart, M. J. Roemer, P. W. Kalgren, and K. Goebel, "Modeling aging effects of IGBTs in power drives by ringing characterization," in Proc. Annual Conference of the Prognostics and Health Management Society, pp. 1-7, 2008.

[14] Celaya, J.R., P. Wysocki, V.Vashchenko, S. Saha, and K. Goebel, "Accelerated aging system for prognostics of power semiconductor devices," in *Proc. IEEE AUTOTESTCON*, pp. 1-6, 2010.

[15] Celaya, J., A. Saxena, P. Wysocki, S. Saha, and K. Goebel, "Towards Prognostics of Power MOSFETs: Accelerated Aging and Precursors of Failure", in *Proc. Annual Conference of the Prognostics and Health Management Society*, pp. 1-10, 2010.

[16] Celaya, J.R., N. Patil, S. Saha, P. Wysocki, and K. Goebel, "Towards Accelerated Aging Methodologies and Health Management of Power MOSFETs (Technical Brief)", in *Proc. Annual Conference of the Prognostics and Health Management Society*, pp. 1-8, 2009.

[17] S. Dusmez, and B. Akin, "Remaining Useful Lifetime Estimation For Degraded Power MOSFETs Under Cyclic Thermal Stress", in *Proc. IEEE Energy Conversion Conference & Expo.*, pp. 3846-3851, 2015.

[18] M. Ciappa, "Selected Failure Mechanisms of Modern Power Modules," Microelectronics Reliability, vol. 42, no. 4–5, pp. 653-667, 2002.

[19] J.-M. Thebaud, E. Woirgard, C. Zardini, S. Azzopardi, O. Briat, J.-M. Vinassa, "Strategy for designing accelerated aging tests to evaluate IGBT power modules lifetime in real operation mode," *IEEE Trans on Components and Packaging Technologies*, vol. 26, no. 2, pp. 429-438, June 2003.

[20] H. Chen, B. Ji, V. Pickert, W. Cao, "Real-Time Temperature Estimation for Power MOSFETs Considering Thermal Aging Effects," *IEEE Trans on Device and Materials Reliability*, vol. 14, no. 1, pp. 220-228, March 2014.

[21] Bart Van Zeghbroeck, *Principles of Semiconductor Devices and Heterojunctions.* Prentice Hall , 2007.

Modeling and Design Guidelines of High Density Power Inductor for Battery Power Unit

Zhigang Dang, *Student Member, IEEE* and Jaber A. Abu Qahouq, *Senior Member, IEEE*

The University of Alabama
Department of Electrical and Computer Engineering
Tuscaloosa, Alabama 35487, USA

Abstract—This paper presents the modeling, simulation and design guidelines of a high density power inductor for battery power unit (BPU) in order to reduce weight and size especially in automotive applications. A ~4.5μH 30A high current density low profile permanent magnet power inductor (PMPI) design is developed and evaluated. By placing a small piece of NdFeB permanent magnet in the air gap of conventional power inductor (PI), PMPI achieves doubled saturation current while maintaining the same size and the same inductance value. Alternatively, PMPI achieves size reduction to half while maintaining the same inductance and the same saturation current. ANSYS®/Maxwell® 3-D physical model simulation results are used to illustrate the saturation current doubling (from 15A to 30A) and 55.5% size reduction (from 7.94cm³ to 3.53cm³) of the PMPI compared with PIs.

Keywords—battery management system; battery power unit; dc-dc; electric vehicles; ferrite; finite element analysis (FEA); magnetic field; power converter; power inductor; power magnetics; 3-D modeling and simulation.

I. INTRODUCTION

Switching power converters size and weight reduction while maintaining high efficiency and other steady-state and dynamic performance requirements are critical in many applications. These applications include, but are not limited to, automotive and mobile electronics [1-15]. In automotive applications, especially Electric and Electric-Hybrid vehicles (EVs and HEVs), weight and size affect driving range and efficiency [1, 5, 7].

Power inductor is one of the most important parts in a variety of switched mode power converters ranging from basic non-isolated DC-DC converter topologies such as buck converter, boost converter, buck-boost converter to isolated topologies such as forward converter, full bridge or half bridge converters, among others [1-15]. In a boost DC-DC power converter, for example, power inductor stores energy in the form of magnetic field during the time interval when the control switch is turned on and it discharges the stored energy to support the current flow to the load during the time interval when the synchronous switch (or diode) is on. The desired inductance value for a power converter is normally determined based on the switching frequency, maximum inductor current ripple, input voltage and output voltage, among others [16]. Power converter with higher switching frequency requires a smaller inductance value. Therefore, one of the most effective approaches to shrink the size of power inductor is to push the switching frequency higher. However, due to the limitations of the switching devices such as power MOSFETs, switching frequency is also limited by switching power losses and temperature rise [1]. Therefore, a reasonably sufficient inductance value for a given application and requirements is still required. Moreover, due to the nonlinearity of the magnetic material used for the power inductor core, a reasonably large size is required to prevent the core from saturation at heavy load conditions [2, 16]. Due to the requirements of inductance and saturation current values, power inductor is one of the largest and heaviest devices in a power converter, which impedes the higher level integration of switching power electronics.

In an EV or HEV electric power drive train, smaller power converters size and weight are critical to meet size and weigh requirements and targeted driving range increase. The battery pack voltage can be regulated by a high power large power converter with large power inductor or when the distributed battery energy storage system presented in [5] is used there will be the need for several power converters with several power inductors. This is because in the distributed battery energy storage system, rather than connecting battery cells in series to form a battery string and regulating the voltage of the battery string through a high power DC-DC power converter, battery cells are decoupled from one another by connecting each cell with a lower power boost power converter as illustrated in Fig. 1a [5]. Each battery power unit (BPU) consists of one battery cell and one boost power converter. Fig. 1b shows the schematic of boost power converter used in BPU, which include input and output capacitors C_{in} and C_o, control and synchronous power MOSFETs S_u and S_l and a power inductor. A small size and weight reduction in each boost power inductor will contribute the to the size and weight reduction in the whole battery pack.

978-1-4673-9551-9/16 $31.00 © 2016 IEEE

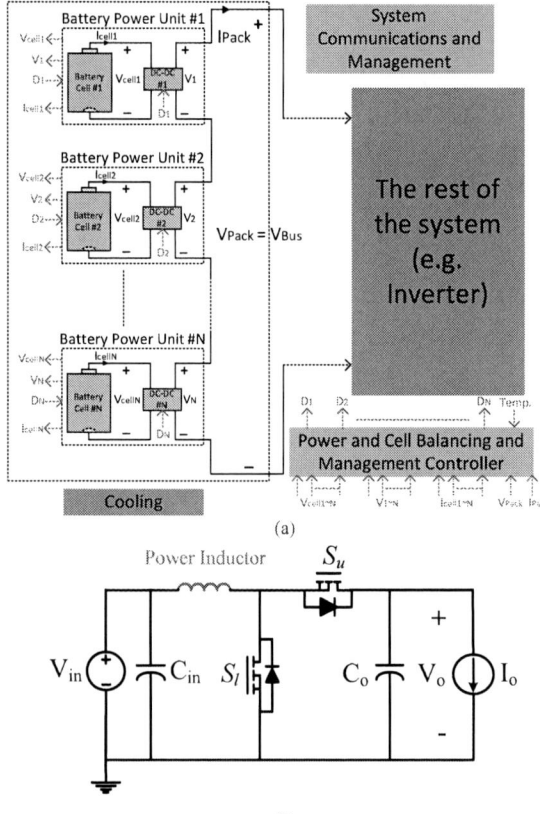

(a)

(b)

Fig. 1. (a) The simplified block diagram of the distributed battery energy storage system architecture [5] and (b) schematic of DC-DC boost power converter used in each BPU

In order to achieve lower profile, larger inductance density, higher saturation current and higher power density for power inductors, different levels of efforts have been taken over the years [17-29]. These efforts include 1) obtaining higher permeability, higher saturation magnetization, higher frequency and lower loss magnetic materials for inductor core [17-20], 2) improving the magnetic core structure such as the use of the multi-permeability core to make the better utilization of the magnetic core material [21-25], 3) improving the winding structure to reduce the winding losses and increase the inductor power density [26-27] and 4) new types of inductors such as LTCC based inductors to achieve higher level of system integration [22, 28-29]. Even after all these efforts have been taken, the intrinsic nonlinearity of magnetic material cannot be eliminated and the current carrying capabilities are still limited by the saturation of magnetic material.

To further increase the saturation current of the power inductor (PI), one or more pieces of permanent magnets (PM) can be placed in its airgap to partially cancel the winding fluxes [2, 30-32]. It is known that magnetic core of conventional power inductor normally operates in the first

quadrant on the BH curve. The addition of the PM could achieve better utilization of the ferrite magnetic core because the PM is able to push the magnetic core to operate in the third quadrant of the BH curve when DC current is relatively low. As a result, permanent magnet power inductor (PMPI) can achieve significant saturation current increase or effectively significant size and weight reduction [2, 30-32].

This paper presents design guidelines and the modeling and simulation results of a high density power inductor for battery power unit (BPU) based on the concept of PMPI. A ~4.5μH 30A high density PMPI, which reduces the size of the conventional power inductor (PI) to less than half, is designed and evaluated. The size reduction is achieved by placing a small piece of the fabricated NdFeB-N35EH magnet in the air gap of PI. Next section reviews the operation principle and presents design guidelines and procedure for the PMPI of a BPU. Section III presents the 3-D physical modeling and simulation results of the PMPI using ANSYS®/Maxwell®.Section IV gives the paper conclusion.

II. DESIGN GUIDELINES AND STEPS OF THE PMPI

The operation principle of the toroid core, EE core and EI core PMPIs are described in [2, 30-32]. A PMPI utilizes a small piece of PM in the airgap of a PI core to cancel the magnetic flux generated by the current carrying inductor's winding such that the saturation current can potentially be doubled. Alternatively, PMPI can also achieve significant size and weight reduction compared with conventional PI. Major design characteristics of interest of PMPI including inductance and saturation current calculation, and PM material selection and dimension calculations have been discussed in [2, 30, 32].

This section provides a design procedure of a PMPI as summarized in the flowchart of Fig. 2a. The first step is to determine the inductance value (L_{PMPI}) and saturation current (I_{sat_PMPI}) based on the specs of the power converter. Step 2 determines the specs of the corresponding PI, i.e. L_{PI} = L_{PMPI} and I_{sat_PI} = $0.5 \times I_{sat\_PMPI}$. Note that the saturation current of PI (I_{sat_PI}) is a half of I_{sat_PMPI}. Based on these specs, the conventional PI is designed in step 3, the detailed procedure for PI design is illustrated in Fig. 2b. Then in step 4, a check is performed to ensure that the PI meets the design specs. In step 5, a PM is designed and placed in the airgap of PI to obtain an initial prototype of PMPI, which is detailed in Fig. 2c. In step 6, a check is performed to ensure that the PMPI achieves the targeted I_{sat_PMPI}. Then, the PMPI needs to be tested in a designed power converter in step 7 in order to evaluate in step 8 that all of the performance parameters of interest including efficiency, power losses, temperature rises and/or radiated EMI. If acceptable PMPI performance parameters are obtained in step 9, the PMPI design can be considered complete. Otherwise the process might need to be repeated until satisfactory performance parameters are achieved.

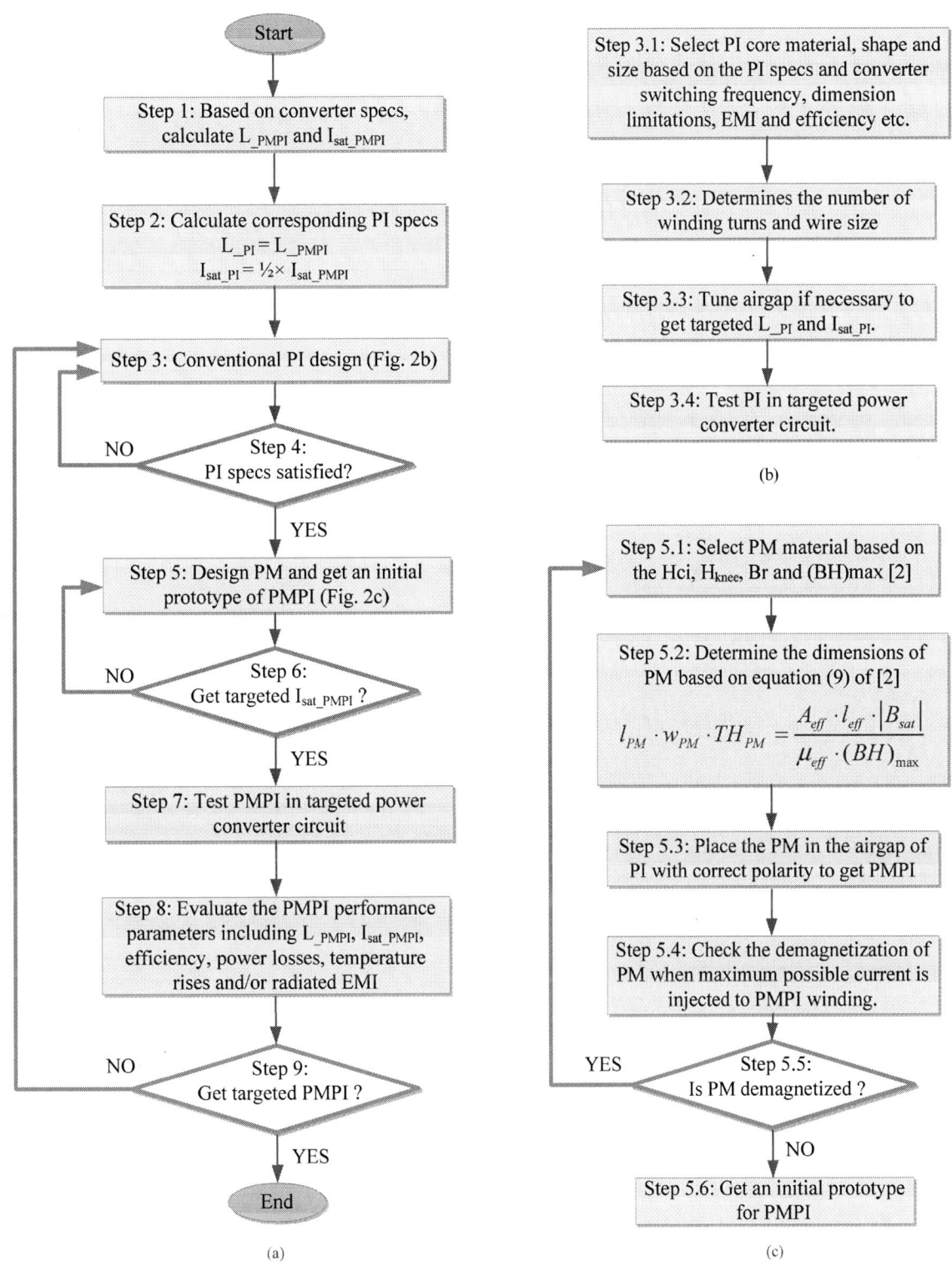

Fig. 2. Design procedure for the PMPI: (a) the main design flow, (b) conventional PI design flow and (c) PM design flow

Fig. 2b describes the design flow for the conventional PI. In step 3.1, PI core material, shape and size are determined based on the PI specs, power converter switching frequency, dimension limitations, EMI and efficiency. In step 3.2, the number of winding turns and wire size is determined based on the PI specs. Airgap of PI is tuned if necessary in step 3.3 to obtain the targeted L_{PI} and I_{sat_PI}. Then in step 3.4, the PI prototype is tested in a power converter circuit for the evaluation of performance parameters for step 4.

Fig. 2c describes the design flow for the PM. In step 5.1, the PM material type is selected based on the requirements for Hci, H_{knee}, Br and (BH)max which are detailed in [2]. In Step 5.2, the dimensions of PM is determined based on the method developed in [2] which is based on the equivalent energy stored in a gapped magnetic core and energy supplied by the PM. Note that the thickness of designed PM must be smaller than thickness of the airgap of magnetic core. In step 5.3, the PM needs to be carefully placed in the airgap with the correct direction of polarity. While this might be a relatively straightforward step, it is very critical to the performance of the PMPI to place the PM in the correct direction within the airgap. The right PM polarity can result in doubling of the saturation current, while the wrong PM polarity can reduce the saturation current to zero, which is illustrated in the simulation results in the next section. Step 5.4 is to check the demagnetization of the PM when maximum possible current is injected to PMPI winding. This is a very important step because designer needs to make sure that the PM is never demagnetized under the maximum input current, otherwise, all advantages achieved by PMPI will be lost. If the PM is not demagnetized, then an initial prototype for PMPI is obtained, otherwise, PM design flow needs to be repeated.

III. ANSYS®/MAXWELL® 3-D MODELING AND SIMULATION RESULTS OF THE PMPI

This section presents the design (using the design guidelines of Section II), 3-D physical models and simulation results for the PMPI design using ANSYS®/Maxwell® (an industrial standard FEA software package used for electromagnetics [2, 30-39]) and illustrates the saturation current doubling and size reduction of the PMPI. The design objective is to obtain ~4.5µH 30A power inductor, which operates in a DC-DC Boost Power Converter with 4V input and 8V output for a Battery Power Unit. The switching frequency of the BPU is 150kHz and EE core is utilized in this design. Core material used is 3F3 ferrite, which has the relative permeability of μ_r=4000, saturation flux density B_{sat} in the first quadrant of 0.35T at 25 °C and resistivity of 2Ωm at 25 °C [40]. The PM used in the PMPI is NdFeB-N35EH with the intrinsic coercivity H_{ci}= 2244.8kA/m, the coercivity H_c = 574.8kA/m, the residual flux density B_r=1.06T and the maximum energy product (BH)$_{max}$=188kJ/m³ [2].

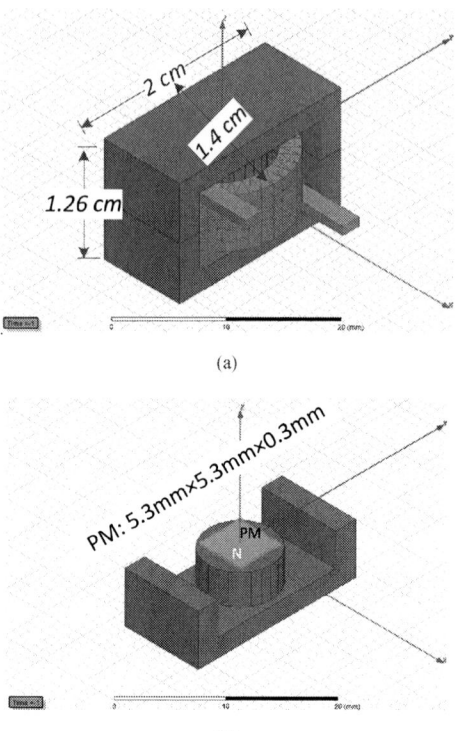

(a)

(b)

Fig. 3. 3-D model of (a) PI and (b) part of PMPI core developed in ANSYS®/Maxwell®

Fig. 3 shows the 3-D models developed by using ANSYS®/Maxwell®. After the 3-D model is built, material properties are assigned for ferrite cores, copper winding and NdFeB PM. One important step for PMPI simulation is to assign the winding excitation current and PM polarity. They both need to be in the right directions, as illustrated in Fig. 4b, in order to achieve flux cancelation. Meshes for each part need then to be assigned and the FEA simulation for inductor design is run.

Five developed power inductor physical models for comparative study are shown in Fig. 4. All five power inductors have the same number of winding turns and almost the same inductance values. PMPI and PI have the same size but PMPI has a PM in its airgap as illustrated in Fig. 4b. The Big-PI-1 is obtained by scaling up the dimensions of PI (including winding and core) 1.2 times in x, y and z directions, which makes its volume and weight 1.728 times of PI's. The airgap of Big-PI-1 is then tuned to obtain the same inductance value. The Big-PI-2 is obtained by scaling up the dimensions of PI 1.2 times in x and y directions and 1.5 times in z direction, which makes its volume and weight 2.16 times of PI's. The airgap of Big-PI-2 is also tuned to obtain the same inductance value. The Big-PI-3 is obtained by scaling up the dimensions of PI 1.5

978-1-4673-9551-9/16 $31.00 © 2016 IEEE   2117

times in x and y directions (while keeping z direction same), which makes its volume and weight 2.25 times of PI's. The airgap of Big-PI-3 is also tuned to obtain the same inductance value.

Comparison between the major characteristics of the five different power inductor designs are summarized in Table I. It can be observed that the PMPI achieves the highest inductance density and current density. Compared to the PI with the same size, weight and inductance, PMPI doubles the saturation current from 15 A to 30 A. Compared to the Big-PI-3 with the same inductance and saturation current, PMPI reduces the size and weight by 55.5%. The absolute value of weight reduction achieved by the PMPI is 14.51 g. The high inductance density and high current density PMPI can contribute to the size and weight reduction of BPU and the entire battery pack which consists of many BPUs.

Fig. 5 illustrates the B field of the different power inductors when the DC input current increases from 0 to 30A. For PI, the magnitude of B increases with the increase of the DC input current and the PI core is partially saturated (B_{sat}=0.35T) when the DC input current is 15A, i.e. I_{sat_PI}=15A. For PMPI, when the input current is 0, the ferrite core is at close point to be saturated by PM itself. When the input DC current increases from 0 to 30A, the net B value first decreases to zero at ~15A and the PMPI core is at close point to be saturated at 30A. When the DC input current is 15A, the flux density of winding and PM have the same magnitudes but are in opposite directions, which makes the net flux inside of the PMPI core equals to/near zero. These simulation results show that saturation current of the PMPI is twice as large as the saturation current of the PI, i.e. I_{sat_PMPI}=30A=2×I_{sat_PI}.

The designer of the PMPI has to ensure that the PM is never demagnetized under the maximum input current. It has been analyzed in [2] that as long as the reverse H field (resulted from the current carrying winding of the PMPI) is less than 1989kA/m, the NdFeB-N35EH magnet will not be irreversibly demagnetized. Fig. 6 shows the H field distribution of the PM when the DC input current is 30A. It can be observed that the maximum H value is 692.8 kA/m, which is much smaller than 1989 kA/m. This indicates that the PM used in the PMPI design is far from being irreversibly demagnetized when the input current is as high as 30A.

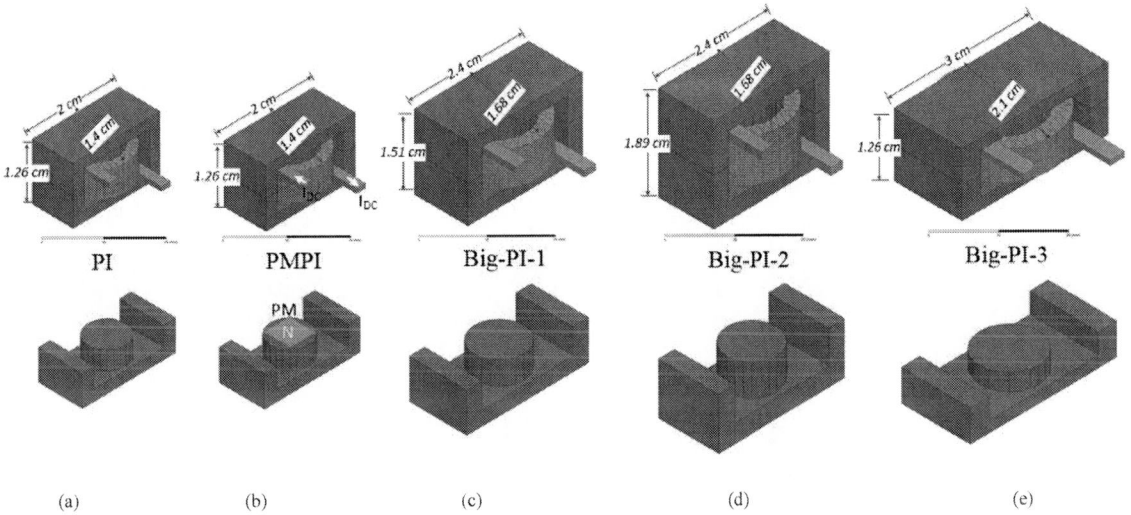

Fig. 4. Physical models of power inductor designs for BPU

Table I. Comparison between major characteristics of the different power inductor designs

	PI	PMPI	Big-PI-1	Big-PI-2	Big-PI-3
L (µH)	4.65	4.80	4.69	4.67	4.68
Isat (A)	15	30	25	25	30
Volume (cm³)	3.53	3.53	6.10	7.62	7.94
Weight (g)	11.61	11.61	20.06	25.08	26.12
Inductance density (µH/cm³)	1.32	1.36	0.77	0.61	0.59
Current density (A/ cm³)	4.25	8.50	4.10	3.28	3.78

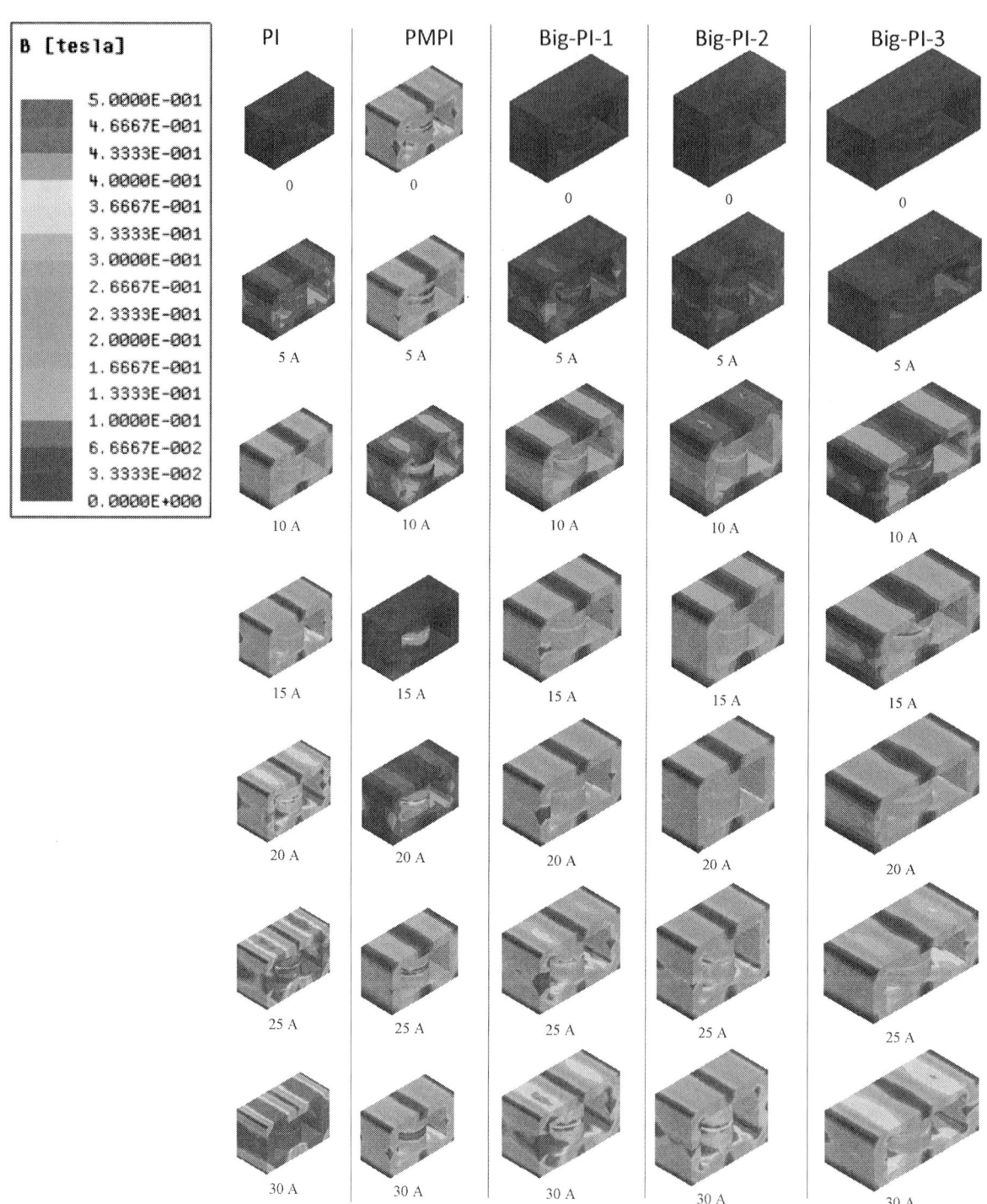

Fig. 5. B field of the different power inductors when the DC input current increases from 0 to 30A (fixed scaling)

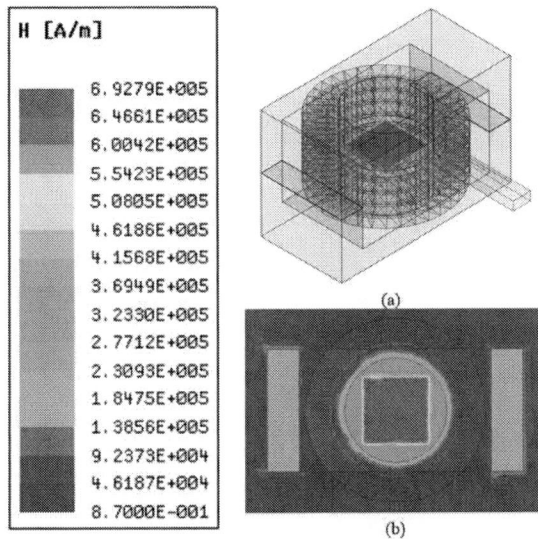

Fig. 6. Demagnetizing field of the PM in PMPI (a) 3-D view and (b) x-y plane cross sectional view (auto scaling)

IV. CONCLUSION

This paper presents the modeling and design guidelines in order to optimize the design of a high density power inductor with ~4.5µH, 30A for battery power unit (BPU). The design of PMPI achieves doubled saturation current compared to the same size and same inductance PI. PMPI also achieves 55.5% of size reduction compared to the bigger sized PI with the same saturation current and the same inductance. 3-D physical model simulation results using ANSYS®/Maxwell® visualize saturation current doubling and size reduction of the PMPI.

ACKNOWLEDGEMENT

This work is supported in part by The Center for Advanced Vehicle Technologies (CAVT) at The University of Alabama – Tuscaloosa (UA) and in part by the Alabama NASA EPSCoR Research Infrastructure Development grant no. NNX13AB09A. Any opinions, findings, and conclusions or recommendations expressed in this material are those of the author(s) and do not necessarily reflect the views of these entities.

REFERENCES

[1] J.D. van Wyk and F.C. Lee, "On a Future for Power Electronics," IEEE J. of Emerging and Selected Topics in Power Electron., vol.1, no.2, pp. 59-72, June 2013.

[2] Z. Dang and J. Abu Qahouq, "Evaluation of High-Current Toroid Power Inductor with NdFeB Magnet for DC-DC Power Converters," IEEE Transaction on Industrial Electronics, vol. 62, no. 11, pp. 6868-6876, Nov. 2015.

[3] J. Abu Qahouq and V. Arikatla, "Online Closed-Loop Auto-Tuning Digital Controller For Switching Power Converters," IEEE Transaction on Industrial Electronics, vol. 60, no. 5, pp. 1747-1758, May 2013.

[4] W. Huang and J. Abu Qahouq, "An Online Battery Impedance Measurement Method Using DC-DC Power Converter Control," IEEE Transactions on Industrial Electronics, vol. 61, no. 11, pp. 5987-5995, Nov. 2014.

[5] W. Huang and J. Abu Qahouq, "Energy Sharing Control Scheme for State-of-Charge Balancing of Distributed Battery Energy Storage System," IEEE Transaction on Industrial Electronics, vol. 62, no. 5, pp. 2764-2776, May 2015.

[6] Y. Jiang and J. Abu Qahouq, "Adaptive Step Size With Adaptive-Perturbation-Frequency Digital MPPT Controller for a Single-Sensor Photovoltaic Solar System," IEEE Transaction on Power Electronics, vol. 28, no. 7, pp. 3195-3205, July 2013.

[7] C. Deng, D. Xu, P. Chen, C. Hu, W. Zhang, Z. Wen and X. Wu, "Integration of Both EMI Filter and Boost Inductor for 1-kW PFC Converter," IEEE Transaction on Power Electronics, vol. 29, no. 11, pp. 5823-5834, Nov. 2014.

[8] H. Wu, J. Zhang and Y. Xing, "A Family of Multiport Buck–Boost Converters Based on DC-Link-Inductors (DLIs)," IEEE Transaction on Power Electronics, vol. 30, no. 2, pp. 735-746, Feb. 2015.

[9] H. Wu, T. Mu and Y. Xing, "Full-Range Soft-Switching-Isolated Buck-Boost Converters with Integrated Interleaved Boost Converter and Phase-Shifted Control," IEEE Transaction on Power Electronics, vol. 31, no. 2, pp. 987-999, Feb. 2016.

[10] P. Malcovati, M. Belloni, F. Gozzini, C. Bazzani and A. Baschirotto, "A 0.18-µm CMOS, 91%-Efficiency, 2-A Scalable Buck-Boost DC–DC Converter for LED Drivers," IEEE Transaction on Power Electronics, vol. 29, no. 10, pp. 5392-5398, Oct. 2014.

[11] H. Wu and Y. Xing, "Families of Forward Converters Suitable for Wide Input Voltage Range Applications," IEEE Transaction on Power Electronics, vol. 29, no. 11, pp. 6006-6017, Nov. 2014.

[12] X. Xie, J. Li, K. Peng, C. Zhao and Q. Lu, "Study on the Single-Stage Forward-Flyback PFC Converter With QR Control," IEEE Transaction on Power Electronics, vol. 31, no. 1, pp. 430-442, Jan. 2016.

[13] J. Lee, J.K. Kim, J.H. Kim, J. Baek and G. Moon, "A High-Efficiency PFM Half-Bridge Converter Utilizing a Half-Bridge LLC Converter Under Light Load Conditions," IEEE Transaction on Power Electronics, vol. 30, no. 9, pp. 4931-4942, Sept. 2015.

[14] A. Safaee, P.K. Jain, A. Bakhshai, "An Adaptive ZVS Full-Bridge DC–DC Converter With Reduced Conduction Losses and Frequency Variation Range," IEEE Transaction on Power Electronics, vol. 30, no. 8, pp. 4107-4118, Aug. 2015.

[15] M. Pahlevani, S. Eren, A. Bakhshai and P. Jain, "A Series–Parallel Current-Driven Full-Bridge DC/DC Converter," IEEE Transaction on Power Electronics, vol. 31, no. 2, pp. 1275-1293, Feb. 2016.

[16] R. W. Erickson and D. Maksimovic, "Fundamentals of Power Electronics", Springer, 2000.

[17] H. Jia, J. Lu, X. Wang, K. Padmanabhan and Z. J. Shen, "Integration of a Monolithic Buck Converter Power IC and Bondwire Inductors With Ferrite Epoxy Glob Cores," IEEE Transaction on Power Electronics, vol. 26, no. 6, pp. 1627-1630, June 2011.

[18] P. Herget, N. Wang, E.J. O'Sullivan, B.C. Webb, L.T. Romankiw, R. Fontana, X. Hu, G. Decad and W.J. Gallagher, "A Study of Current Density Limits Due to Saturation in Thin Film Magnetic Inductors for

On-Chip Power Conversion," IEEE Transaction on magnetics, vol. 48, no. 11, pp. 4119-4122, Nov. 2012.

[19] M. S. Rylko, K. J. Hartnett, J. G. Hayes, and M. G. Egan, "Magnetic material selection for high power high frequency inductors in dc-dc converters," The 2009 IEEE Applied Power Electronics Conference and Exposition, APEC'2009, pp. 2043-2049, Feb. 2009.

[20] W. Zhang, Y. Su, M. Mu, D. Gilham, Q. Li and F.C. Lee, "High-Density Integration of High-Frequency High-Current Point-of-Load (POL) Modules With Planar Inductors," IEEE Transaction on Power Electronics, vol. 30, no. 3, pp. 1421-1431, Mar. 2015.

[21] M.S. Perdigao, J. Trovao, J.M. Alonso and E.S. Saraiva, "Large-Signal Characterization of Power Inductors in EV Bidirectional DC–DC Converters Focused on Core Size Optimization," IEEE Transaction on Industrial Electronics, vol. 62, no. 5, pp. 3042-3051, May 2015.

[22] L. Wang, Z. Hu, Y. Liu, Y. Pei and X. Yang, "Multipermeability Inductors for Increasing the Inductance and Improving the Efficiency of High-Frequency DC/DC Converters," IEEE Transaction on Power Electronics, vol. 28, no. 9, pp. 4402-4413, Sept. 2013.

[23] Y. Su, Q. Li, F.C. Lee, D. Hou and S. She, "Planar inductor structure with variable flux distribution - A benefit or impediment?" The 2015 IEEE Applied Power Electronics Conference and Exposition, APEC'2015, pp. 1169-1176, Mar. 2015.

[24] X. FAng, R. Wu, Lu. Peng and J. Sin, "A Novel Integrated Power Inductor With Vertical Laminated Core for Improved L/R Ratios," IEEE Electron Device Letters, vol. 35, no. 12, Dec. 2014.

[25] T. Tera, H. Taki and T. Shimizu, "Loss Reduction of Laminated Core Inductor used in On-board Charger for EVs," 2014 International Power Electronics Conference (IPEC-Hiroshima 2014 – ECCE – Asia), pp. 876-882, 2014.

[26] A. Stadler, "The Optimization of High Frequency Inductors with Litz-Wire Windings," 8th International Conference on Compatibility and Power Electronics (CPE), pp. 209-213, 2013.

[27] V. Sung and W.G. Odendaal, "Litz Wire Pulsed Power Air Core Coupled Inductor," IEEE Transactions on Industry Applications, vol. 51, no. 4, pp. 3385-3393, July/Aug. 2015.

[28] L. Wang, Z. Hu, Y. Liu, Y. Pei and Z. Wang, "A Horizontal-Winding Multipermeability LTCC Inductor for a Low-profile Hybrid DC/DC Converter," IEEE Transaction on Power Electronics, vol. 28, no. 9, pp. 4365-4375, Sept. 2013.

[29] Y. Su, Q. Li and F.C. Lee, "Design and Evaluation of a High-Frequency LTCC Inductor Substrate for a Three-Dimensional Integrated DC/DC Converter," IEEE Transaction on Power Electronics, vol. 28, no. 9, pp. 4354-4364, Sept. 2013.

[30] Z. Dang and J. Abu Qahouq, "Permanent Magnet Power Inductor Circuit and Physical Modeling," International Review on Modelling and Simulations (I.RE.MO.S.), vol. 5, no. 5, pp. 2001-2006, Oct. 2012.

[31] Z. Dang and J. Abu Qahouq, "Permanent Magnet Toroid Power Inductor with Increased Saturation Current," The 2013 IEEE Applied Power Electronics Conference and Exposition, APEC'2013, pp. 2624-2628, Mar. 2013.

[32] Z. Dang and J. Abu Qahouq, "Permanent magnet power inductor with EE core for switching power converters," The 2015 IEEE Applied Power Electronics Conference and Exposition, APEC'2015, pp. 1073-1077, Mar. 2015.

[33] Z. Dang and J. Abu Qahouq, "On-Chip Three-Phase Coupled Power Inductor for Switching Power Converters," The 2015 IEEE Applied Power Electronics Conference and Exposition, APEC'2015, pp. 1045-1050, Mar. 2015.

[34] Z. Dang and J. Abu Qahouq, "On-chip coupled power inductor for switching power converters," The 2014 IEEE Applied Power Electronics Conference and Exposition, APEC'2014, pp 2854-2859, Mar. 2014.

[35] Z. Dang, Y. Cao and J. Abu Qahouq, "Reconfigurable Magnetic Resonance-Coupled Wireless Power Transfer System," IEEE Transactions on Power Electronics, vol. 30, no. 11, pp. 6057-6069, Nov. 2015.

[36] Z. Dang and J.A. Abu Qahouq, "Modelling and Simulation of Magnetic Resonance Coupled Wireless Power Transfer Systems," International Review of Modelling and Simulation (I.RE.MO.S), vol. 6, no. 5, pp. 1607-1617, Oct. 2013.

[37] Z. Dang and Jaber Abu Qahouq, "Range and Misalignment Tolerance Comparisons between Two-coil and Four-coil Wireless Power Transfer Systems," The 2015 IEEE Applied Power Electronics Conference and Exposition, APEC'2015, pp. 1234-1240, Mar. 2015.

[38] Z. Dang and Jaber Abu Qahouq, "Elimination Method for the Transmission Efficiency Valley of Death in Laterally Misaligned Wireless Power Transfer Systems," 2015 IEEE Applied Power Electronics Conference and Exposition, APEC'2015, pp. 1644-1649, Mar. 2015.

[39] Z. Dang and J. Abu Qahouq, "Modeling and Investigation of Magnetic Resonance Coupled Wireless Power Transfer System with Lateral Misalignment," The 2014 IEEE Applied Power Electronics Conference and Exposition, APEC'2014, pp 1317-1322, Mar. 2014.

[40] [Online]: http://www.ferroxcube.com/FerroxcubeCorporateReception/datasheet/3f3.pdf, accessed on Oct. 26, 2015.

Degradation of Low Voltage Metal Oxide Varistors in Power Supplies

Dawood Talebi Khanmiri [1] *, Roy Ball [2], Jerry Mosesian [2], Brad Lehman [1]

[1] Electrical & Computer Engineering Department
Northeastern University
Boston, MA, USA
Talebikhanmiri.d@husky.neu.edu

[2] Mersen USA Newburyport LLC
Newburyport, MA, USA

Abstract— The unprecedented proliferation of electronic devices in critical applications, such as smart grid and communication systems, has brought new challenges in system reliability. Electronic circuits and components are especially sensitive to transient overvoltages and lightning surges, and design must consider operation under these worst case situations. To address these transients, Metal Oxide Varistors (MOVs) are widely used in electronic circuits and power supplies. They absorb surges in the circuit input and maintain the voltage within acceptable limits. However, these surges and overvoltage transients can degrade MOVs over time. Reliability of an MOV depends on maintaining its Energy Absorption Capability (EAC) in an acceptable range. In this paper, for the first time, the effect of degradation is investigated on Energy Absorption Capability (EAC) of low voltage MOVs by experimental test in a UL-certified lab. The results show that EAC is decreasing with degradation due to surges. Furthermore, it is shown that Time-to-Failure (TtF) of the degraded MOVs is longer than that of new MOVs in some transient overvoltage (TOV) incidents. This helps explain the actual mechanism of MOV failures in real life applications and questions whether the widely accepted design guideline that a 10% change in varistor voltage at 1 mA is a degradation indicator. Finally, a nonlinear resistive – capacitor model is proposed for degraded MOVs.

Keywords— *Metal Oxide Varistor (MOV), Surge Protection, Power supply, Surge Protective Devices (SPD), Degradation, Energy Absorption Capability, Time to Failure.*

I. INTRODUCTION

A reliable power supply is a key element in today's communications, monitoring, and control systems. Particularly, surge protection is a critical part to safeguard in power supplies [1-5]. Often, electronic equipment must be installed on poles and towers, close to high voltage and medium voltage networks, and they are expected to work failure-free for a long time with minimum maintenance. On the other hand, considering various geographical areas, these electronics might be exposed to severe weather conditions. Lightning flash density varies from 0 to 16 flashes/sq. km/year in the United States, while in tropical areas this number can reach 70 flashes/sq. km/year. A lightning flash can discharge up to 200

kA, with a typical value of 20 kA, current into network, which causes a severe transient overvoltage [6]. Inductive coupling of lightning can cause severe overvoltages, too. A one kilometer distant lightning strike can induce about 200 volts in 1 meter of wire [6]. However, lightning is not the only source of surges. Utility switching and faults can also cause overvoltage transients in the network. Furthermore, internal sources, such as air conditioning units and induction motors are responsible for 70% of surges [7]. These surges are low amplitude, and degrade appliances over time.

Fig. 1 shows the spikes in otherwise normal output voltage of 120 V power supply [8]. These impulse overvoltages either go beyond the capability of the components and damage them permanently, or cause extra heating and degrade devices over time. Three decades ago researchers found that 75% of field failures of microprocessor-based electronics are caused by electrical overstresses and expressed the need for better and more reliable surge protection for electronic devices [6]. Since then, the miniaturization of electronic circuits, closer traces, and denser circuits has decreased insulation withstand and energy absorption capabilities of electronic components and made them more vulnerable to overvoltage transients and surges.

Metal Oxide Varistors (MOVs) are nonlinear components widely used for protection against surges and overvoltage transients in power supplies. New applications of MOVs have also recently emerged such as a snubber circuit in solid-state

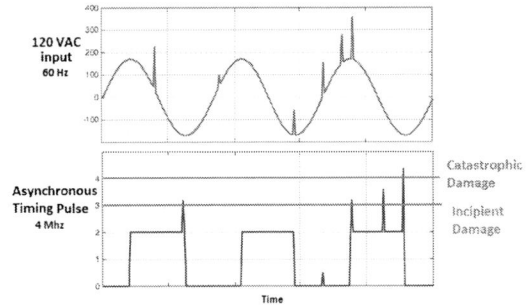

Fig. 1. Typical transients in power supply input and timing pulses

---

* The author gratefully acknowledges the support through grant by Mersen USA.

978-1-4673-9551-9/16 $31.00 © 2016 IEEE

breakers [9]. MOVs divert the surge current away from the sensitive load by providing a low resistance path to the ground and maintaining the voltage within acceptable limits. These discharge activities are known to change the microstructure of the MOVs and affect their voltage-current characteristics over time. Whether this "degradation" has effects on their Energy Absorption Capability (EAC) or not, has not been published in the literature so far.

Some studies have been conducted previously on EAC of new MOVs [10-13]. Also, degradation process was the subject of many papers and changes in electrical parameters [14-15], microstructure [16-17], and methods of identification [18-19] have been investigated. However, they do not comment on possible changes in EAC due to degradation in low voltage MOVs. Recently, Tuczek and Hinrichsen [10] have studied several types of high voltage station-class MOVs and found that EAC does not change after multiple AC or 2-ms long duration impulse stresses.

In this paper, we investigate the effect of degradation due to surges on EAC of low voltage MOVs by experimental tests on thirty two 150-volt MOV. Twenty one MOVs are degraded by 8/20 μs impulse currents. Then EAC test is performed in 60Hz AC current until their destruction. The results of the degraded MOVs are compared to EACs of new MOVs.

The main contribution of this paper can be summarized as below:

• Energy Absorption Capability of degraded low voltage MOVs are measured and compared to those of new ones. This is the first time that a paper reports on EAC of degraded low voltage MOVs.

• The results show EAC is decreasing with degradation due to surges. However, this decrease is not linear with varistor voltage or resistive leakage current.

• The paper shows that the degraded MOVs' average Time-to-Failure (TtF) is longer than that of new MOVs at 240 volts. This can challenge the credibility of using widely accepted 10% change in varistor voltage as an indication of degraded MOV.

• A nonlinear resistive-capacitor model for the degraded MOV is proposed.

The rest of this paper is organized as below. Section II provides a background on MOVs and their degradation. Section III describes the test specimens and procedure. Section IV presents the test results. In Section V we propose a simple nonlinear model for a degraded MOV. Finally, Section VI provides summary and conclusions.

II. DEGRADATION OF MOVS

Metal oxide varistors are sintered ceramic blocks, often composed of Zinc oxide (ZnO) grains in a matrix of small amounts of other metal oxides, showing a highly non-linear current-voltage characteristics. This non-linear behavior is the accumulated behavior of many micro-scale Schottky-barriers at grain boundaries that act like back-to-back Zener diodes in a randomly connected network. A typical current-voltage

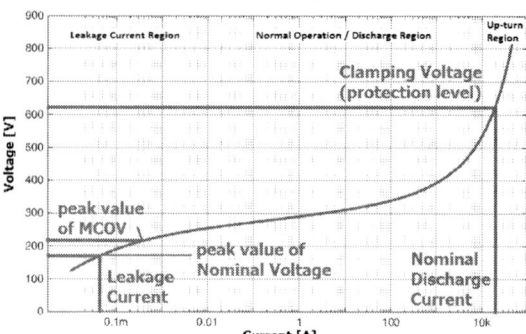

Fig. 2. Typical current-voltage characteristic of an MOV

characteristic of a varistor is shown in Fig. 2. The curve can be divided into 3 regions. In voltages below the Maximum Continuous Operating Voltage (MCOV), a small current flows through the MOV and the MOV is in high resistance mode. This region which sometimes is referred to as linear region shows the operating point of an MOV in normal system voltage and is called leakage current region. When the system voltage increases above MCOV, the MOV conducts a current larger than its leakage current and the voltage remains almost constant. This region is referred to as discharge region or normal operation region. In currents above the nominal discharge current, the voltage is enough to break all barriers and the only current limiting factor is grain's resistance. This region in which the voltage linearly increases with current is called up-turn region. Non-linearity factor, α, is defined as the slope of the V-I characteristic in normal operation region and often is calculated by measuring varistor voltage at 0.1 and 1 mA DC currents, as below:

$$\alpha = (\log 1 - \log 0.1) / (\log V_{1mA} - \log V_{0.1mA}) \qquad (1)$$

Metal oxide varistors are known to degrade over time when they experience surges and overvoltage transients. The leakage current of the MOV increases with degradation, thus the power dissipation increases within the varistor. Other parameters of the MOV like varistor voltage and capacitance change as well, more or less, with degradation [12-15].

Degradation in MOVs occurs because of the changes in their microstructure. A high magnitude current surge can change the electrical properties of Shottky-barriers at grain boundaries. High temperatures generated by the leakage current or applied externally can change the microstructure, too. These changes can be either asymmetrical (polarized) and change the barrier in one direction more than the opposite direction, or symmetrical and change the both sides equally [16]. Their collective effect reflects in changes in the current-voltage characteristics. It is believed that leakage current region of the V-I curve is more affected by degradation.

Furthermore, it should be noticed that these changes are stochastic processes and each MOV changes in a various random way. Thus, it is necessary to use multiple test samples in MOV degradation tests to reduce the possibility of

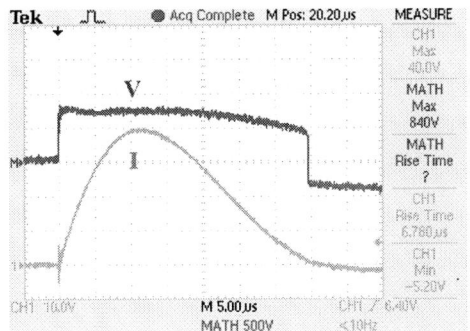

Fig. 3. Typical 8/20 μs surge current waveform and residual voltage on the 150-volt MOV

neglecting a particular behavior that might be seen only in a fraction of each production batch.

Degradation, or aging, of MOVs due to surges depends on the number of the surges, current magnitude discharged by the varistor, the duration of the current and its wave shape, the temperature of the varistor, the formulation of the varistor, and the polarity of the surge and varistor's previous polarization.

III. TEST PROCEDURE AND SPECIMENS

Thirty two low voltage 150-volt MOVs are used in this test in 3 different sets. Set 1 is new. Set 2 is degraded until their varistor voltage at 1 mA DC (V_{1mA}) decreases 10% of its initial value in the opposite direction of the applied surge. V_{1mA} is generally used by manufacturers in Quality Control (QC) tests. Usually, when V_{1mA} of an MOV changes more than 10% of its initial value, the MOV is considered as failed.

Different degradation methods are used in literature and manufacturer's QC tests. Surge degradation is designed to investigate the ability of the MOV to maintain an acceptable performance after tolerating a given number of impulse current surges. The standard 8/20 μs current surge waveform is the most widely accepted and used for test of Surge Protective Devices (SPDs) and other equipment for lightning protection. In this paper, for degradation of Set 2 and Set 3, 8/20 μs current surges are applied to MOVs in a UL-certified lab. After each surge, the MOV is investigated visually in order to detect any mechanical destruction. Also, their V_{1mA} are measured in order to identify the electrical break down. Fig. 3 shows a sample waveform of the applied impulse current and the residual voltage on a 150-volt MOV.

TABLE I. TEST SAMPLES

Set	Condition	Number of samples
1	New	11
2	10 % change in V	11
3	6 impulse of 40 kA	10

Eight out of eleven samples of Set 2 reached the 0.9 V_{1mA} only after one 40 kA impulse current surge of 8/20 μs. In fact, the manufacturer has declared it as maximum surge capability in the derating curve. Set 3 are subjected to 40 kA impulse surges until at least one failure is seen in the set. In the 6th surge, three samples are failed and the remaining 8 samples are saved for the EAC test. The failed MOVs tolerated the impulse currents, but in the post surge assessments after the 6th surge, they showed very low V_{1mA}, thus indicates they had developed short circuits through the MOVs. Table I summarizes the specimens' information.

EAC test is also conducted in the UL-certified high power lab. The MOVs are subjected to a 60Hz, 240 V AC voltage until their failure. The generator impedance is adjusted to deliver a short circuit current of 5 kA. This is a standard Short Circuit Current Rating (SCCR) according to UL-1449 standard [21]. Voltage and current waveforms are captured and are used to calculate the energy deposited in the MOV during the test. The MOV's heat dissipation to the environment during the test is neglected in these calculations. Since the current densities of each MOV depends on its voltage-current characteristics and differs from those of other MOVs, differences due to different current densities in different sets are not taken into account.

IV. TEST RESULTS

Energy Absorption Capability of 32 samples are measured and summarized in Table II. The average EAC and the minimum EAC is shown. Also Time to Failure of the samples are presented in the table. It is seen from the results that average EAC has decrease almost 40 percent from 962.3 Joules for new MOVs to 582.7 Joules for the MOVs after 6 surges. Also, the minimum EAC in the set has reached 220 Joules, which is almost one third of the minimum EAC for the new MOVs. Furthermore, it is seen that although the average EAC has not decreased significantly for the Set 2, there are some samples in this set in which the EAC decrease is noticeable. The minimum EAC for Set 2 is 538.1 Joules which is 17% less than minimum EAC for the new MOVs.

Decrease in EAC due to degradation is of importance in health monitoring and lifetime estimation of MOVs. However, in the real life application TtF might be more important in the design of a SPD. Although there are some debates on the main cause of MOV failures [19], transient overvoltages (TOVs) are one of the most commonly known causes. MOVs conduct high currents during a TOV and this may lead to its thermal runaway and failure.

TABLE II. TEST RESULTS

Set	Average EAC [J]	Minimum EAC [J]	Average TtF [sec]	Minimum TtF [sec]
1	962.3	649.7	0.597	0.178
	100%	100%	100%	100%
2	950.9	538.1	0.813	0.392
	98.8%	82.8%	136%	220%
3	582.7	220.6	1.473	0.500
	60.5%	33.9%	247%	280%

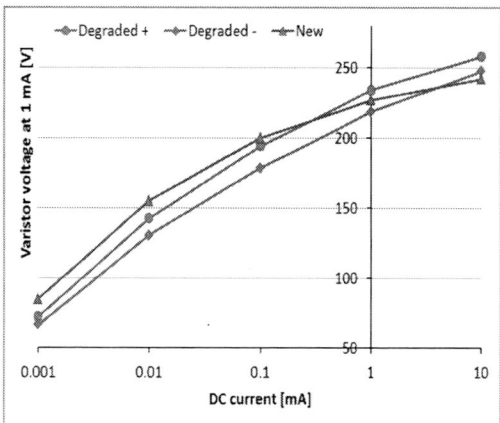

Fig. 4. V-I characteristics of a new and a degraded MOV

The test results show that during a TOV of 240 volts, a degraded MOV conducts less current than a new MOV, and it can tolerate the transient overvoltage for longer period of time, even though it absorbs less energy during this period. It is, perhaps, surprising that an MOV with higher leakage current conducts less current during a TOV. However, the voltage-current characteristics of a degraded MOV explain this phenomenon. Fig. 2 shows V-I characteristics for a new and a degraded MOV.

It is shown from Fig. 4 that when an MOV degrades, V-I characteristics move toward higher currents in leakage current region. However, in TOV region, it moves toward smaller currents. That means in higher voltages, which is typically TOV region, a new MOV's current is higher than a degraded MOV. This higher current generates more power and heats up the MOV faster. Thus, the new MOVs (at least these types that are tested) are more susceptible to TOVs than some degraded MOVs. It is worthy to mention that if the degradation process continues, it will eventually move the characteristics toward higher currents, even in TOV region. However, there is no clear evidence whether this much degradation occurs in real life applications.

Fig. 5 shows EAC of the Set 1, Set 2, and Set 3 versus V_{1mA} and resistive leakage current. As it is seen, V_{1mA} decreases with the first surge, as we expect. However, next surges can increase the V_{1mA} and decrease the resistive leakage current. The average Energy Absorption decreases from 962 joules to 950 joules, while the average V_{1mA} decreases from 248 volts (Set 1) to 218 volts (Set 2). Further surges increase the average V_{1mA} to 234 volts (Set 3), where the average EAC decreases to 582 joules. Similarly, the resistive leakage current and consequently power increases dramatically by the first surge, but then, following surges decrease both. Sometimes, this is referred to as "formation". It depends on the surge amplitude and polarity and somehow acts randomly and differently on different MOVs.

Comparing Set 1 and Set 3 in Fig. 5(b) shows that with degradation, resistive leakage current increases and EAC

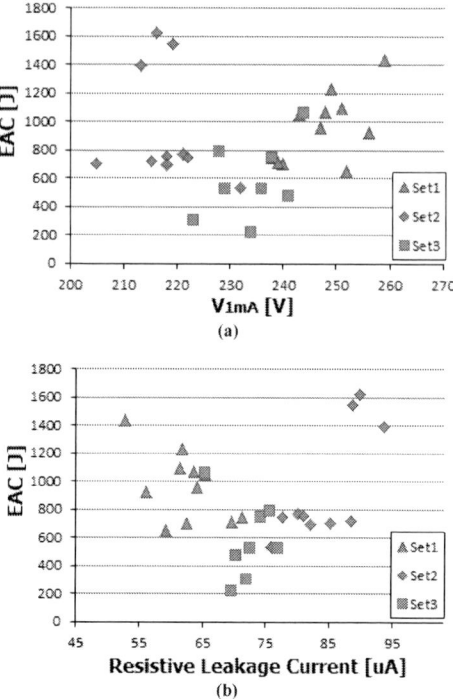

Fig. 5. Energy Absorption Capability vs. (a) Varistor Voltage (b) Resistive Leakage Current

decreases. However, Set 2 shows this change is not linear at least for degradation due to high surges, about nominal discharge current. It is possible that degradation due to smaller surges and over-voltages during long time show more linear characteristics. The same conclusion applies to Fig. 5(a), too. It is shown that although the average V_{1mA} of the Set 3 is higher than that of Set 2, their EAC is lower.

V. DEGRADED MOV MODEL

Experimental results in this paper is accordance with the commonly known fact that α, nonlinearity factor, decreases by degradation [16]. Typically, nonlinearity factor is in the range of 20 to 30 for new metal oxide varistors. Fig. 7 shows V-I characteristics of a new and a degraded MOV. Nonlinearity factor has decreased from 21.5 to 11.4 for this MOV.

Another feature of degradation, seen in these tests, that acknowledges the previously reported results [10], is that

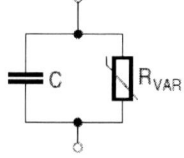

Fig. 6. Simple MOV model

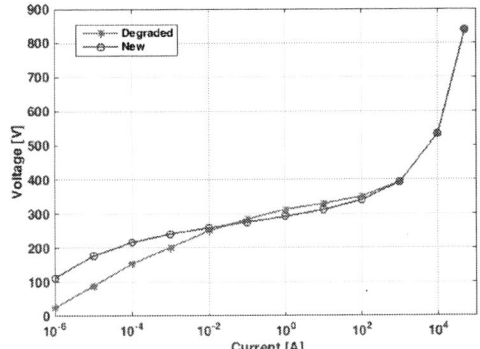

Fig. 7. Degradation effect on V-I characteristics of an MOV

clamping voltage is not affected by degradation. This means the discharge currents region of the V-I characteristics will remain unchanged. Finally, experimental tests in this paper show that changes in the capacitance of the MOV due to degradation is negligible.

A commonly used simple and effective model for the MOV [22] includes a nonlinear resistor in parallel with a capacitor, seen in Fig. 6. Nonlinear resistor is defined by:

$$\text{Log } V = b_1 + b_2 \log I + b_3 \exp(-\log I) + b_4 \exp(\log I) \quad (2)$$

Where V is the voltage across the MOV and I is the resistive component of the leakage current.

This research proposes that the same model can be used for the degraded MOVs, provided we keep the discharge currents region of V-I characteristics the same as before and replacing the leakage current and TOV regions by measured values. Then, the b1 to b4 factors are recalculated for the degraded MOV. Polarization can be either considered or neglected.

VI. CONCLUSION

In this paper the EAC of the new and degraded MOVs are measured by experimental tests in a UL-certified lab. Thirty two 150-volt MOVs are used in 3 different groups; two groups of differently degraded MOVs are compared to a group of new ones. It is found that EAC of low voltage MOVs decreases with degradation due to surges. However, it is seen that Time to Failure of degraded MOVs are longer than that of new MOVs at 240 V. This means used MOVs might tolerate TOVs better than new MOVs. Also, it is shown that change in varistor voltage and resistive leakage current might not show the level of degradation. Finally, a nonlinear model is proposed for degraded MOVs. Behavior of a degraded MOV can be of importance for reliability of the power supplies, as well as, simulation purposes.

REFERENCES

[1] Lin Du, et al., "A Novel Power Supply of Online Monitoring Systems for Power Transmission Lines," IEEE Transactions on Industrial Electronics, Vol. 57, No. 8, 2010.

[2] Emi Gohara, et al., "A Practical Approach of Lightning Protection Measures for Power Receiving Facilities in Telecom Building," IEEE 36th International Telecommunications Energy Conference (INTELEC), 2014.

[3] Eduard Shulzhenko, et al., "Applying of Surge Arresters in Power Electronic Network Components," 2014 International Conference on Lightning Protection (ICLP), China, 2014.

[4] Pu Xie, et al., "Research on Indirect lightning Protection for Military Power Supply," Second International Conference on Mechanic Automation and Control Engineering (MACE), 2011.

[5] Chai Yajing, et al., "Test on Lightning Characteristics of Electronic Equipment's Power Supply," International Symposium on Microwave, Antenna, Propagation and EMC Technologies for Wireless Communications, 2007.

[6] O.M. Clark, et al., "Lightning protection for microprocessor based electronic systems," IEEE Transaction on Industry Applications, Vol. 26, No. 5, 1990.

[7] Surge Protection Reference Guide, Emerson Network Power, 2011.

[8] Keith Brashear, "Lightning and Surge Protection Of Modern Electronic Systems," ILD Technologies, LLC, San Antonio, TX.

[9] J. Magnusson, et al., "Separation of the Energy Absorption and Overvoltage Protection in Solid-State Breakers by the Use of Parallel Varistors," IEEE TRANSACTIONS ON POWER ELECTRONICS, VOL. 29, NO. 6, JUNE 2014.

[10] M.N. Tuczek, V. Hinrichsen, "Recent Experimental Findings on the Single and Multi-Impulse Energy Handling Capability of Metal–Oxide Varistors for Use in High-Voltage Surge Arresters," Power Delivery, IEEE Transactions on, Vol. 20, No. 5, 2014.

[11] M. Bartkowiak, et al., "Failure Modes and Energy Absorption Capability of ZnO Varistors," IEEE Transactions on Power Delivery, Vol. 14, No. 1, January 1999

[12] D. Talebi Khanmiri, R. Ball, C. Mckenzie, B. Lehman, "Energy Absorption Capability of Low Voltage Metal Oxide Varistors in AC and Impulse Currents," *2016 IEEE Applied Power Electronics Conference and Exposition (APEC)*, in press.

[13] Jinliang He, et al, "Discussions on Nonuniformity of Energy Absorption Capabilities of ZnO Varistors," IEEE TRANSACTIONS ON POWER DELIVERY, VOL. 22, NO. 3, JULY 2007

[14] C. Salles, et al., "Ageing of Metal Oxide Varistors due to Surges," 2011 International Symposium on Lightning Protection (XI SIPDA), Fortaleza, Brazil, October 3-7, 2011

[15] M. A. Ponce, et al., "Influence of degradation on the electrical conduction process in ZnO and SnO2-based varistors," J. Appl. Phys. 108, 074505 (2010)

[16] Jinliang He, et al., "Non-uniform ageing behavior of individual grain boundaries in ZnO varistor ceramics," Journal of the European Ceramic Society 31 (2011) 1451–1456

[17] Mardira, et al., "The Effects of Electrical Degradation on the Microstructure of Metal Oxide Varistor," Transmission and Distribution Conference and Exposition, 2001 IEEE/PES, vol. 1, pp. 329-334, Atlanta, GA.

[18] X. Yan, et.al., "Study on the resistive leakage current characteristic of MOV surge arresters," Transmission and Distribution Conference and Exhibition 2002: Asia Pacific. IEEE/PES, Vol. 2, pp. 683 – 687, 2002.

[19] G. Lira, "MOSA Monitoring Technique Based on Analysis of Total Leakage Current," IEEE TRANSACTIONS ON POWER DELIVERY, VOL. 28, NO. 2, APRIL 2013.

[20] D. Kladar, F. Martzolff, D. Nastasi, "TOV Effects on Surge-Protective Devices," Power Quality Exhibition and Conference, Baltimore, October 25-27, 2005.

[21] "Standard for Surge Protective Devices," UL 1449 standard, 4th edition, 2014.

[22] M. Holzer, W. Zapsky, "Modeling varistors with PSpice: Simulation beats trial and error," Siemens-Matsushita Components.

978-1-4673-9551-9/16 $31.00 © 2016 IEEE

Characterization and Modeling of SiC MOSFET Body Diode

Kang Peng, Soheila Eskandari, Enrico Santi
Department of Electrical Engineering
University of South Carolina,
Columbia, SC, U.S.A
santi@engr.sc.edu

Abstract—In this paper, the static and switching characterizations of a SiC MOSFET's body diode are presented. The static characterization of SiC MOSFET's body diode is carried out using a curve tracer and a double pulse test bench is built to characterize the inductive switching behavior of SiC MOSFET's body diode. The reverse recovery of SiC MOSFET's body diode is shown at different forward conduction currents, junction temperatures and current commutation slopes. In order to evaluate the performance of SiC MOSFET's body diode in different applications, an accurate physics-based diode model is introduced to perform simulations of SiC MOSFET's body diode. The parameter extraction procedure for this body diode model is given. The validation of the body diode model shows good agreement between simulation and experimental results, which proves the accuracy of the model.

Keywords—SiC MOSFET body diode; characterization; modeling; parameter extraction; simulation

I. INTRODUCTION

SiC power MOSFET is a very good candidate for high-switching-frequency and low-loss power conversion applications. The lower on-resistance makes SiC power MOSFETs an ideal choice in high power applications, offering similar conduction loss as Si IGBTs while operating at a much higher switching frequency [1]-[3]. The switching loss of a SiC power MOSFET is much lower than that of a Si IGBT or Si GTO for the same voltage and current ratings, due to its lower device capacitance [4] [5]. In inductive hard switching, SiC MOSFET's body diode might be used if no external anti-parallel diode is connected [6] [7]. For example, in a synchronous buck converter the inductor current flows through the lower MOSFET's body diode during the dead time periods [8]. In order to utilize the body diode of SiC MOSFET, a complete characterization (static and dynamic) of SiC MOSFET's body diode is required. In addition, a circuit-oriented device model is needed to evaluate the performance of SiC MOSFET's body diode in power converter design.

The body diode in a SiC power MOSFET is a p-i-n diode, as shown in Fig.1. The low-doped drift region is sandwiched between drain and source, creating a vertical diode structure. This p-i-n diode can be utilized to conduct current through the SiC power MOSFET in third quadrant operation. It is desirable to utilize MOSFET body diodes to avoid additional

This work was supported by the Office of Naval Research under grant N00014-14-1-0165

cost of external anti-parallel diodes. However, a significant issue of body diode utilization is its reverse recovery. The reverse recovery is due to high concentrations of injected carriers stored in the drift region during conduction. During the switch turn-off transition, some carriers are swept away from the drift region, resulting in reverse recovery current. The reverse recovery current leads to additional switching loss in the complementary power switch [9].

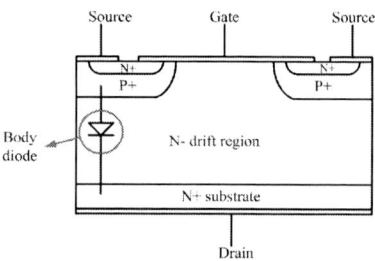

Fig. 1. Cross-sectional structure schematic of SiC power DMOSFET

In this paper, a complete performance characterization of SiC MOSFET's body diode is carried out. The study is conducted for a 1200V/36A SiC MOSFET from Cree Inc. The static characterization is done using a curve tracer. For dynamic switching characterization, a double pulse tester (DPT) printed circuit board (PCB) with an inductive load is built. The reverse recovery of SiC MOSFET's body diode is shown at different current commutation slopes, forward conduction currents, and junction temperatures. In addition, a Fourier-based-solution physics-based model for SiC MOSFET's body diode is proposed. The parameter extraction procedure for SiC MOSFET's body diode model is presented. Finally, the model is validated by comparing simulated results with experimental results under inductive switching condition.

II. CHARACTERIZATION OF SiC MOSFET BODY DIODE

A. Static characterization

In this section, static characterization of SiC MOSFET's body diode is described. Static characteristics (I-V) are measured with a Tektronix 371A curve tracer. The device under test (DUT) is a SiC MOSFET C2M0080120D from Cree Inc. rated at 1200V/36A.

The static characterization of SiC MOSFET's body diode is carried out with different gate-source voltages. Fig.2 shows the measured static characteristics (I-V) of SiC MOSFET's body diode at room temperature, when gate-source voltage V_{gs}=0,5,10,15,20V.

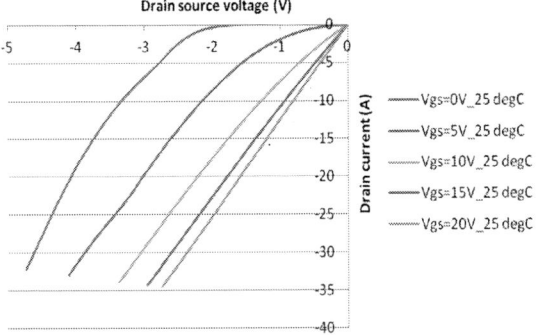

Fig. 2. Static characteristics of SiC MOSFET body diode with positive gate-source voltages

As shown in Fig.2, when the gate-source voltage increases, more current flows through the MOSFET channel. As a result, the voltage between source and drain is reduced. When gate-source voltage reaches a value V_{gs}=20V, MOSFET channel is fully turned on, and MOSFET conducts in the third quadrant in a manner similar to forward conduction in the first quadrant.

Fig.3 shows the measured static characteristics (I-V) of SiC MOSFET's body diode at room temperature, when gate-source voltage V_{gs}=0, -1, -2, -3, -4, -5V. As seen in Fig.3, when the gate-source voltage decreases, the voltage drop between source and drain increases.

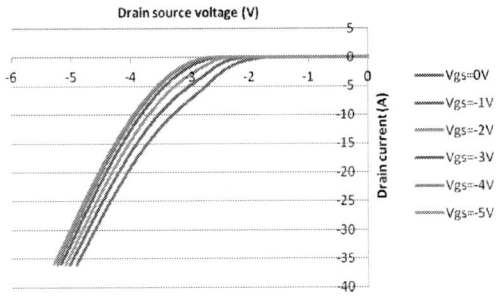

Fig. 3. Static characteristics of SiC MOSFET body diode with negative gate-source voltages

Fig.4 shows the measured static characteristics (I-V) of SiC MOSFET's body diode at junction temperatures 25°C and 125°C, when gate-source voltage V_{gs}=0, 5, 10, 15, 20V. Fig.5 shows the measured static characteristics (I-V) of SiC MOSFET's body diode at junction temperatures 25°C and 125°C, when gate-source voltage V_{gs}=0, -2, -5V.

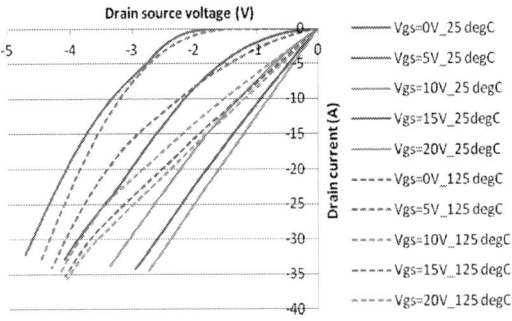

Fig. 4. Static characteristics of SiC MOSFET body diode with positive gate-source voltages at 25°C and 125°C

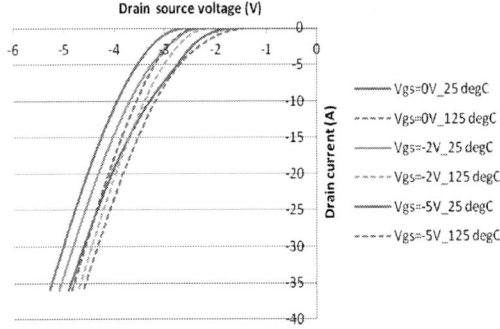

Fig. 5. Static characteristics of SiC MOSFET body diode with negative gate-source voltages at 25°C and 125°C

From Fig.4 and Fig.5, it can be seen that the MOSFET on-state resistance at V_{gs}=20V increases with junction temperature, due to lower carrier mobility at a higher junction temperature. In contrast, MOSFET's body diode built-in voltage potential decreases with junction temperature, due to higher intrinsic carrier concentration at a higher junction temperature. The body diode series resistance also decreases with junction temperature, because of higher minority carrier lifetime in drift layer at a higher junction temperature.

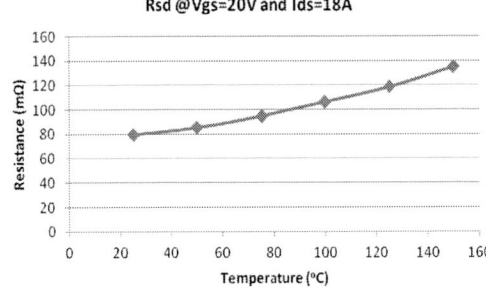

Fig. 6. Reverse conduction on-state resistance of SiC MOSFET as a function of junction temperature at V_{gs}=20V

Fig.6 shows the on-state resistance curve as a function of junction temperature, when V_{gs}=20V, and I_{ds}=18A. The on-state resistance is 79.6 mΩ at 25 °C junction temperature, while on-state resistance increases to 134.8 mΩ at 150 °C junction temperature.

Fig.7 shows body diode built-in potential curve as a function of junction temperature, when V_{gs}= -5V. The body diode series resistance as a function of junction temperature is shown in Fig.8, when V_{gs}= -5V.

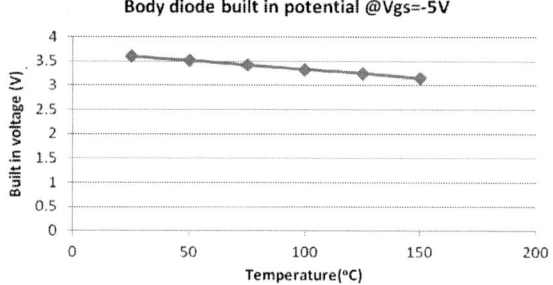

Fig. 7. Body diode built-in potential of SiC MOSFET body diode as a function of junction temperature at V_{gs}= -5V

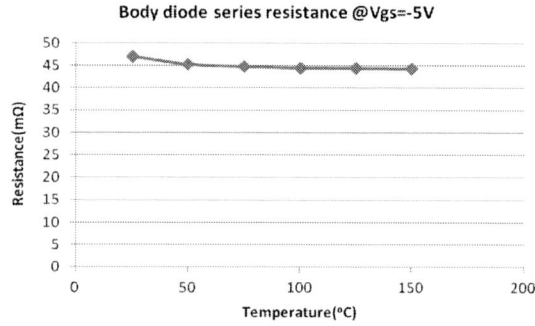

Fig. 8. Body diode series resistance of SiC MOSFET body diode as a function of junction temperature at V_{gs}= -5V

B. Switching characterization

In order to study switching behavior of SiC MOSFET body diode, a printed circuit board (PCB) test-bench was built to conduct the inductive switching experiments on SiC power devices. The parasitic inductances from the PCB layout were minimized, when the PCB was designed. Fig.9 shows the schematic of double pulse tester for SiC MOSFET body diode switching characterization. Fig.10 shows the experimental setup of inductive switching.

The test-bench includes a test socket for high-side SiC MOSFET, a test socket for low-side SiC MOSFET, gate drive circuit, input capacitor bank, a load inductor, probe-tip-adapters, and a Pearson coil for body diode current measurement. The MOSFET under test is a SiC MOSFET C2M0080120D from Cree Inc. rated at 1200V/36A. A gate driver IC IXDD609SI based on the totem-pole structure from

IXYS Corporation is used as the SiC MOSFET gate driver with 9A maximum source/sink drive current. A Pearson coil (model 2878) is used to measure the body diode current I_d of high-side SiC MOSFET. A 250 µH ferrite-core inductor is used as the load inductor for inductive switching experiments.

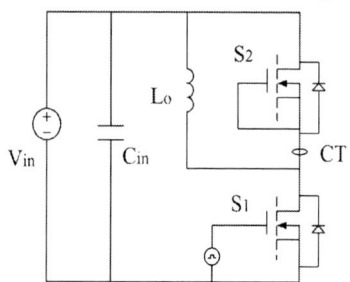

Fig. 9. Schematic of double pulse tester

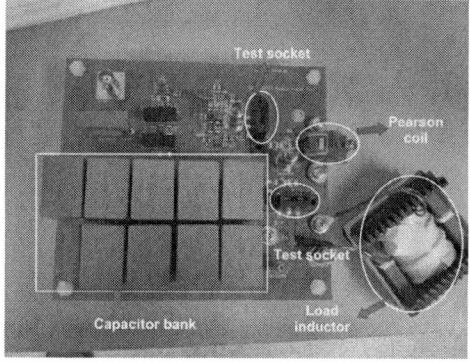

Fig. 10. Picture of double pulse tester

SiC MOSFET's body diode is based on a p-i-n diode structure, a lightly n- doped layer is inserted between n+ drain and p body. During the turn-off transition of MOSFET's body diode, the reverse recovery is observed, because the minority carriers in the drift layer must be removed or recombined before the body diode starts to block a reverse voltage. Reverse recovery is the foremost characteristic of MOSFET's body diode. In this section, the reverse current waveforms of SiC MOSFET's body diode with varied current commutation slopes, forward conduction currents and junction temperatures are measured to evaluate the switching performance of SiC MOSFET's body diode.

1) Varied current commutation slopes

Fig.11 shows the experimental body diode current waveforms with varied low-side MOSFET gate resistances at room temperature. The reverse recovery charge decreases with a large gate resistance, because more minority carriers recombine in a longer reverse recovery time.

Fig.12 shows the reverse peak currents and reverse recovery charges with varied low-side MOSFET gate resistances at room temperature. Both reverse peak current and reverse recovery charge decrease, when low-side MOSFET gate resistance increases.

978-1-4673-9551-9/16 $31.00 © 2016 IEEE   2129

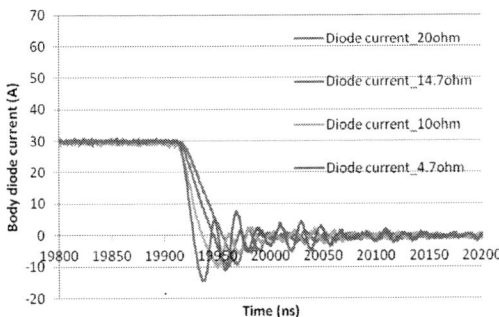

Fig. 11. Body diode current waveforms at different low-side MOSFET gate resistances (block voltage: 500V, forward conduction current: 30A, room temperature)

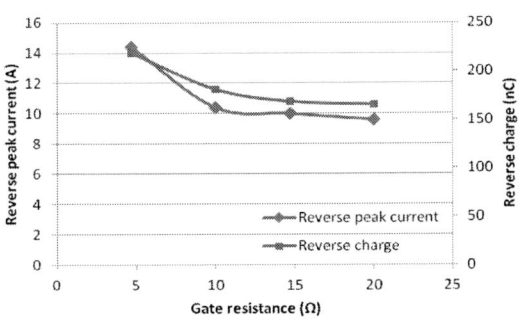

Fig. 12. Body diode reverse peak currents and reverse recovery charges at different low-side MOSFET gate resistances (block voltage: 500V, forward conduction current: 30A, room temperature)

Fig.13 shows the reverse recovery times and current commutating slopes (di/dt) with varied low-side MOSFET gate resistances at room temperature. As the low-side gate resistance increases, the current commutating slope (di/dt) decreases. By contrast, the reverse recovery time increases with the low-side MOSFET gate resistance.

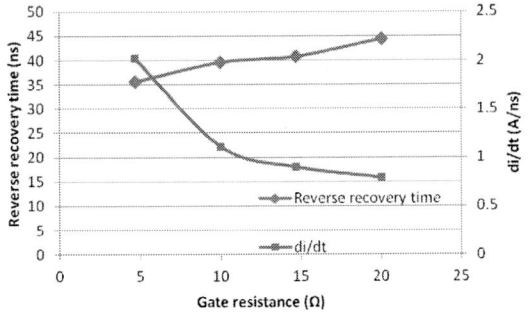

Fig. 13. Body diode reverse recovery times and di/dt at different low-side MOSFET gate resistances (block voltage: 500V, forward conduction current: 30A, room temperature)

2) Varied forward conduction currents

The body diode current waveforms in Fig.14 are measured with different forward conduction currents at 20 Ω gate resistance and room temperature. The reverse peak current and reverse charge increase with forward conduction current. A higher forward conduction current requires more free charges in the drift region, which results in a larger reverse recovery.

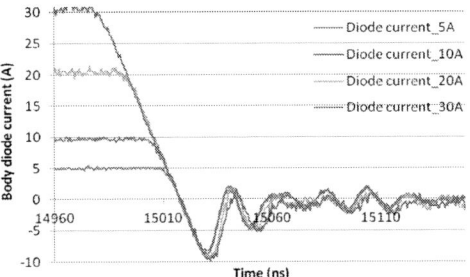

Fig. 14. Body diode current waveforms at different forward conduction currents (block voltage: 500V, low-side gate resistance: 20Ω, room temperature)

Fig.15 shows the reverse peak currents and reverse recovery charges with varied forward conduction currents at room temperature. Both reverse peak current and reverse recovery charge increase, when forward conduction current increases. However, the influence of forward conduction current on reverse peak current is weak. The forward conduction current increases from 5A to 30A, while the reverse peak current changes from 8A to 9.6A.

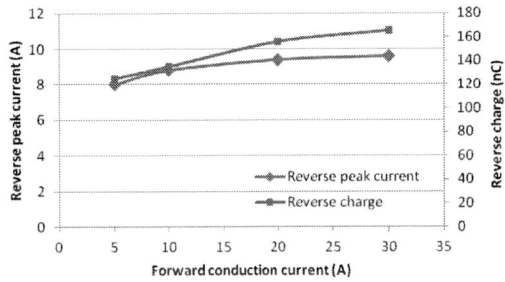

Fig. 15. Body diode reverse peak current and reverse recovery charge at different forward conduction currents (block voltage: 500V, low-side gate resistance: 20Ω, room temperature)

Fig.16 shows the reverse recovery time as a function of forward conduction current at room temperature. As forward conduction current increases, the reverse recovery time of body diode also increases.

978-1-4673-9551-9/16 $31.00 © 2016 IEEE

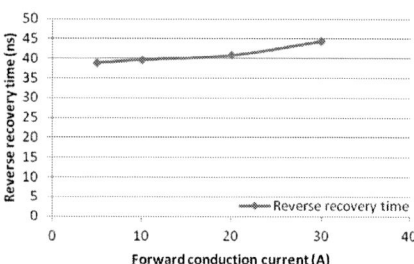

Fig. 16. Body diode reverse recovery times at different forward conduction currents (block voltage: 500V, low-side gate resistance: 20Ω, room temperature)

3) Varied junction temperatures

The DUT is heated by attaching it to a heat spreader, whose temperature is controlled by a thermal controller Eurotherm 94. At each temperature operating point, the device is heated for a long enough time to ensure that MOSFET's junction temperature is the same as the case temperature. Fig.17 shows the experimental waveforms at varied junction temperatures. Both the reverse peak current and reverse recovery charge increase with junction temperature.

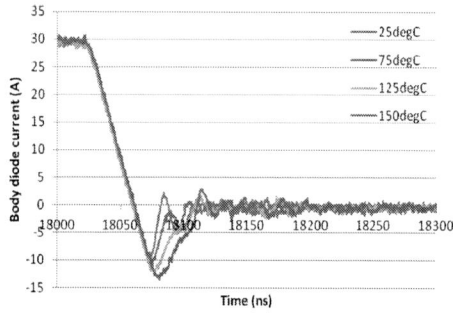

Fig. 17. Body diode current waveforms at different junction temperatures (block voltage: 500V, forward conduction current: 30A, gate resistance: 20Ω)

Fig.18 shows the reverse peak current and reverse recovery charge as a function of junction temperature. Fig.19 shows the reverse recovery time as a function of junction temperature.

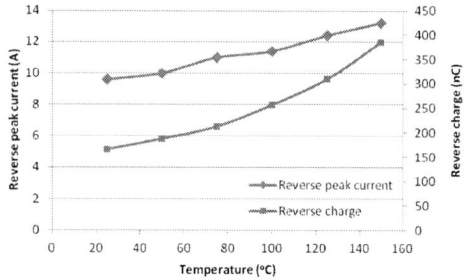

Fig. 18. Body diode reverse peak currents and reverse recovery charges at different junction temperatures (block voltage: 500V, forward conduction current: 30A, gate resistance: 20Ω)

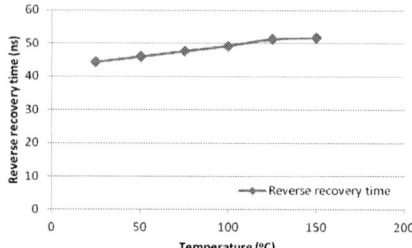

Fig. 19. Body diode reverse recovery times at different junction temperatures (block voltage: 500V, forward conduction current: 30A, gate resistance: 20Ω)

III. DEVELOPMENT OF BODY DIODE MODEL

The body diode model uses a Fourier series solution of the ambipolar diffusion equation (ADE) in the drift layer to find the carrier distribution in that region. This carrier distribution is used to find the conductive voltage drop in the drift region, accounting for conductivity modulation.

Fig.20 shows the general arrangement of the carrier distribution in n- drift region, including an un-depleted carrier storage layer and two depletion layers [10]. The carrier storage layer is sandwiched between two depletion layers. When the body diode is on, the two depletion layers shrink, and the carrier storage layer occupy the whole drift region. When the body diode is off, the two depletion layers expand from the two ends of the drift region, and some free carriers are swept from the carrier storage region. The depletion layers start to support a voltage and the body diode becomes reverse-biased.

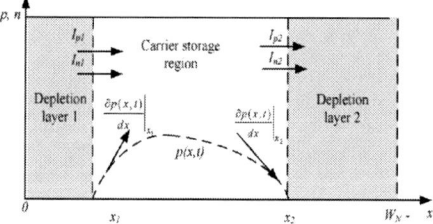

Fig. 20. Undepleted carrier storage layer and depletion layers in n-drift region

Under high level injection, the ambipolar diffusion equation (ADE) describes the carrier dynamics as follows:

$$D \frac{\partial^2 p(x,t)}{\partial x^2} = \frac{p(x,t)}{\tau} + \frac{\partial p(x,t)}{\partial t} \qquad (1)$$

where D is the ambipolar diffusion coefficient, τ is high-level carrier lifetime in the drift region, and $p(x,t)$ is the carrier concentration as a function of space x and time t. The ambipolar diffusion equation is a 2nd order partial differential equation, which describes the minority carrier distribution profile in the drift region of bipolar devices, as a function of time and space.

A Fourier-series based solution to ADE is proposed, which converts the 2nd order partial differential equation into an infinite set of 1st order ordinary differential equations with coefficients $p_0 \ldots p_k$. The Fourier-series based solution is given by:

$$p(x,t) = p_0(t) + \sum_{k=1}^{\infty} p_k(t) \cos\left[\frac{k\pi(x-x_1)}{(x_2-x_1)}\right] \qquad (2)$$

where x_1 and x_2 are the boundaries of the undepleted region.

The Fourier series coefficients p_k are given as follows:

$$p_0(t) = \frac{1}{x_2 - x_1} \int_{x_1}^{x_2} p(x,t) dx \qquad (3)$$

$$p_k(t) = \frac{2}{x_2 - x_1} \int_{x_1}^{x_2} p(x,t) \cos\left[\frac{k\pi(x-x_1)}{(x_2-x_1)}\right] dx \qquad (4)$$

By substituting equations (3) and (4) into equation (1), the Fourier-series coefficients p_k are determined in an infinite set of 1st order linear differential equations. The boundary conditions at the boundaries of the undepleted region (x_1 and x_2) are required, which give the gradients of the carrier densities. The boundary conditions are given by:

$$\frac{\partial p}{\partial x}\Big|_{x_1} = \frac{1}{2qA}\left(\frac{I_n}{D_n} - \frac{I_p}{D_p}\right)\Big|_{x_1} \qquad (5)$$

$$\frac{\partial p}{\partial x}\Big|_{x_2} = \frac{1}{2qA}\left(\frac{I_n}{D_n} - \frac{I_p}{D_p}\right)\Big|_{x_2} \qquad (6)$$

where D_n and D_p are electron and hole diffusion coefficients, I_n and I_p are electron and hole currents, and A is the device active chip area.

The infinite set of 1st order linear differential equations are given:

k=0:

$$\frac{D}{x_2-x_1}\Big[\frac{\partial p(x,t)}{\partial x}\Big|_{x_2} - \frac{\partial p(x,t)}{\partial x}\Big|_{x_1}\Big] = \frac{dp_0(t)}{dt} + \frac{p_0(t)}{\tau} + \frac{1}{x_2-x_1}\sum_{n=1}^{\infty}\Big[\frac{dx_1}{dt} - (-1)^n\frac{dx_2}{dt}\Big]p_n(t) \qquad (7)$$

k>0:

$$\frac{2D}{x_2-x_1}\Big[\frac{\partial p(x,t)}{\partial x}\Big|_{x_2}(-1)^k - \frac{\partial p(x,t)}{\partial x}\Big|_{x_1}\Big] = \frac{dp_k(t)}{dt} + \Big[\frac{1}{\tau} + \frac{Dk^2\pi^2}{(x_2-x_1)^2}\Big]p_k(t)$$
$$+ \frac{2}{x_2-x_1}\left(\sum_{\substack{n=1 \\ n \neq k}}^{\infty}\Big[\frac{dx_1}{dt} - (-1)^{n+k}\frac{dx_2}{dt}\Big]p_n(t)\frac{n^2}{n^2-k^2} + \frac{p_k}{4}\Big(\frac{dx_1}{dt} - \frac{dx_2}{dt}\Big)\right) \qquad (8)$$

The even harmonics and odd harmonics of the Fourier terms for the stored carrier charge can be represented using the electrical equivalent circuit shown in Fig.21.

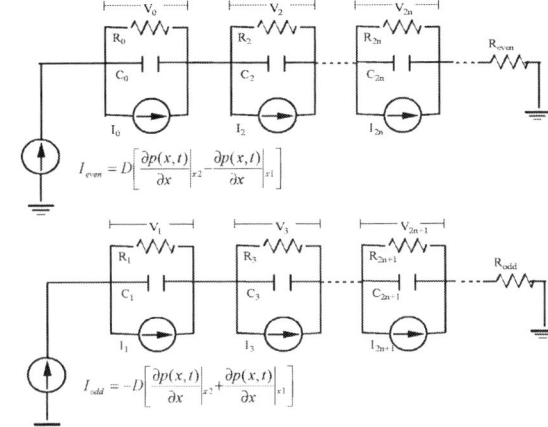

Fig. 21. Equivalent circuits used to calculate the carrier density representing the coefficients of the Fourier series solution to ADE

A. Voltage components

The voltage drop across the body diode V_{ak} is comprised of several components, including the voltages V_{j1} and V_{j2} across junctions J_1 and J_2, the voltages V_{d1} and V_{d2} across two depletion layers, and the voltage V_{n-} across the n- drift region.

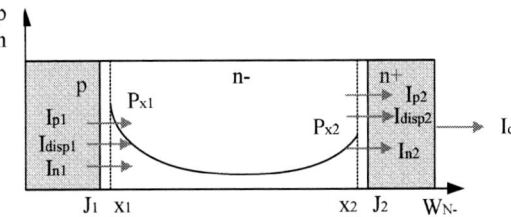

Fig. 22. Schematic structure and carrier densities of drift region of a SiC MOSFET body diode

The junction voltage of J_1 is given by:

$$V_{j1} = V_T \ln\left(\frac{P_{x1}N_{N-}}{n_i^2}\right) \qquad (9)$$

where n_i is the intrinsic carrier concentration in SiC, N_{N-} is the doping concentration in n- drift region and P_{x1} is the carrier density at the boundary x_1.

The junction voltage of J_2 is given by:

$$V_{j2} = V_T \ln\left(\frac{P_{x2}}{N_{N-}}\right) \qquad (10)$$

where P_{x2} is the carrier density at the boundary x_2.

In order to simplify the calculation of the drift region voltage V_{n-}, the discretized carrier distribution shown in Fig.23 is used, and the carrier storage region is divided into several segments of equal width. In the carrier profile, a straight line is used to connect two adjacent points. The tradeoff between accuracy and simulation speed is made by selecting the number of segments.

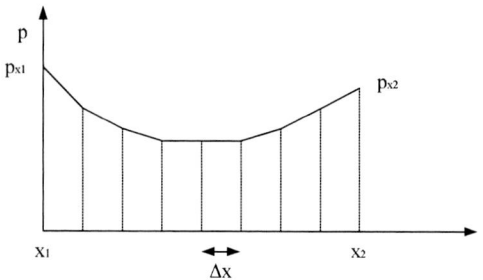

Fig. 23. Discretized carrier profile for simulation of $V_{n\text{-}}$.

The drift region voltage $V_{n\text{-}}$ in carrier storage region is calculated by:

$$V_{n\text{-}} = \frac{I_d}{qA(\mu_n + \mu_p)} \frac{x_2 - x_1}{M} \sum_{k=0}^{M} [\frac{1}{P_{T(k)} - P_{T(k-1)}} \ln(\frac{P_{T(k)}}{P_{T(k-1)}})] - V_T (\frac{\mu_n - \mu_p}{\mu_n + \mu_p}) \ln(\frac{P_{x2}}{P_{x1}})$$

(11)

where I_d is the body diode conduction current, A is the active chip area, and M is the number of segments in carrier storage region. μ_n and μ_p are the electron mobility and hole mobility, respectively. $P_{T(k)}$ is the carrier density at segment boundary points.

The depletion layer voltages V_{d1} and V_{d2} are derived using feedback from the carrier densities P_{x1} and P_{x2} at boundaries:

$$V_{d1} = \begin{cases} 0 & if \quad P_{x1} > 0 \\ -K_F P_{x1} & otherwise \end{cases}$$

(12)

$$V_{d2} = \begin{cases} 0 & if \quad P_{x2} > 0 \\ -K_F P_{x2} & otherwise \end{cases}$$

(13)

where K_F is the feedback constant.

B. Current components

The electron current I_{n1} in the p end region is given by:

$$I_{n1} = qAh_p P_{x1}^2$$

(14)

where h_p is the recombination parameter, and P_{x1} is the carrier density at the boundary x_1.

The displacement current I_{disp1} in the p end region is calculated by:

$$I_{disp1} = \varepsilon A \frac{1}{W_{d1}} \frac{dV_{d1}}{dt}$$

(15)

where W_{d1} is the depletion width at junction J_1.

The hole current I_{p2} in n+ end region is given by:

$$I_{p2} = qAh_n P_{x2}^2$$

(16)

where h_n is the recombination parameter, and P_{x2} is the carrier density at the boundary x_2.

The displacement current I_{disp2} in n+ end region is calculated by:

$$I_{disp2} = \varepsilon A \frac{1}{W_{d2}} \frac{dV_{d2}}{dt}$$

(17)

where W_{d2} is the depletion width at junction J_2.

IV. PARAMETER EXTRACTION METHOD

The body diode model parameters are listed in Table I. Only six parameters are needed for this body diode model, and they can be estimated from manufacturer's datasheets or from diode's turn-off waveforms.

TABLE I. BODY DIODE MODEL PARAMETER LIST

Symbol	Description
A (cm^2)	Active chip area
τ (μs)	High-level minority carrier lifetime
$W_{N\text{-}}$ (μm)	Drift region width
$N_{N\text{-}}$ (cm^{-3})	Doping concentration in drift region
h_p (cm^4/s)	Recombination parameter in P region
h_n (cm^4/s)	Recombination parameter in n+ region

The parameter extraction procedure includes an initial parameter estimation from the manufacturer's datasheets, and a parameter refinement based on the measured waveforms [11].

A. Initial parameter estimation

1) Active chip area A:
The active chip area can be obtained from datasheets or be roughly estimated from the maximum current density J (about 300A/cm^2) and current rating I_d. The active chip area A is calculated by:

$$A = \frac{I_d}{J}$$

(18)

2) High-level minority carrier lifetime τ:
The high-level minority carrier lifetime is estimated by:

$$\tau = \frac{Q_{rr}}{I_F}$$

(19)

where Q_{rr} is the reverse recovery charge in device datasheet, and I_F is the forward conduction current in the datasheet.

3) Low-doped drift region width $W_{N\text{-}}$:
The ionization coefficients for electrons and holes, which are electric field dependent, are given by:

$$\alpha_{n,p} = a \exp(-b / E)$$

(20)

where a and b are the constants, and E is the electric field.

Assuming avalanche breakdown in an abrupt junction, the equation for the breakdown voltage V_{BD} as a function of the constants a and b, and n- drift region width W_{N-}, can be given by:

$$V_{BD} = \frac{bW_{N-}}{\ln(aW_{N-})} \tag{21}$$

The drift region width W_{N-} can be derived from equation (21), using a=3.15×10^6 cm^{-1}, and b=1.04×10^7 V/cm. The breakdown voltage V_{BD} is the voltage rating in the manufacturer's datasheet plus some safety margin.

4) Doping concentration N_{N-} in n- drift region:

From the empirical effective impurity doping concentration in n- drift region, the doping concentration N_{N-} is assumed to be 6×10^{15} cm^{-3}.

5) Recombination parameters h_n and h_p:

The recombination parameters h_n and h_p control the carrier charge in the carrier stored region, and the amount of carrier charge in the drift region is reduced with higher recombination parameters. An initial estimate of 10^{-14} cm^4/s is made for both recombination parameter h_n and h_p.

B. Refinement of parameter values

With the values of Q_{rr} and I_F obtained from switching tests, the high-level minority carrier lifetime τ is refined. The high-level minority carrier lifetime τ is the critical parameter affecting the reverse recovery current waveforms. A better match can be achieved by refinement of high-level minority carrier lifetime τ. After this, the low-doped drift region width W_{N-} and doping concentration N_{N-} can be altered to improve the matching of diode voltage waveforms.

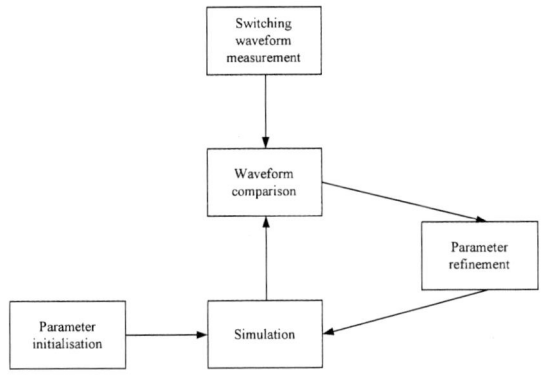

Fig. 24. The procedure of parameter extraction

C. Temperature dependence

The temperature dependent equation for high level minority carrier lifetime τ is given by:

$$\tau = \tau_0 \left(\frac{T}{300}\right)^{1.5} \tag{22}$$

where τ_0 is high level minority carrier lifetime at room temperature (300 K).

The electron mobility μ_n with temperature dependence is given by:

$$\mu_n = \mu_{n0}\left(\frac{300}{T}\right)^{2.7} \tag{23}$$

where μ_{n0} is electron mobility at room temperature (300 K).

The hole mobility μ_p with temperature dependence is given by:

$$\mu_p = \mu_{p0}\left(\frac{300}{T}\right)^{2.7} \tag{24}$$

where μ_{p0} is hole mobility at room temperature (300 K).

The intrinsic concentration n_i with temperature dependence is given by:

$$n_i = 1.70\times10^{16} T^{1.5} / \exp(-\frac{20800}{T}) \tag{25}$$

The junction temperature during operation is determined using thermal RC equivalent circuits. The junction temperature calculated from the thermal equivalent circuit is used to update temperature-dependent parameters in the model. The new values of temperature-dependent parameters are used to calculated body diode current and voltage.

V. MODEL VALIDATION

The extracted gate-to-source switching loop and drain-to-source switching loop parasitic inductance of the PCB layout are used in Pspice simulation together with SiC MOSFET model and SiC MOSFET body diode model to validate the device model for SiC MOSFET body diode. The SiC MOSFET model used in simulation is from the manufacturer Cree Inc., and the parasitic inductances of the PCB layout are extracted using FastHenry, which is a finite difference software tool for PCB parasitic element extraction [12]. Fig.25 shows the simulation circuit, including the extracted parasitic elements (red).

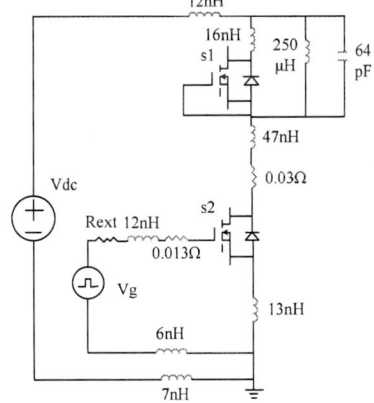

Fig. 25. Equivalent circuit used for inductive switching

Fig.26 shows the comparison of body diode turn-off voltage and current waveforms between experiment (dashed) and simulation (solid) at room temperature 25°C. The DC supply voltage is 500V, and forward conduction current of body diode is 30A. The results illustrates a very good matching between simulation and experiment. In the experimental diode voltage waveform, the diode voltage starts increasing and reaches a small value (about 40V) before the diode current reaches the reverse peak current. This voltage drop is caused by parasitic inductances from PCB layout and device packages. The diode begins to block reverse voltage when diode current reaches the reverse peak current.

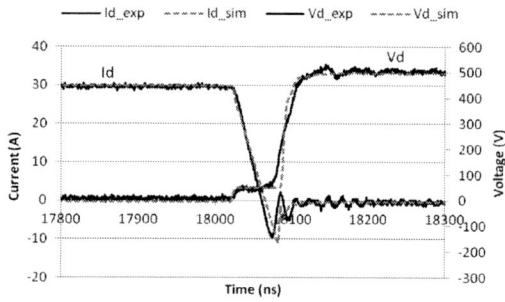

Fig. 26. Comparison of body diode turn-off voltage and current waveforms between experiment (solid) and simulation (dashed) at 25°C

Fig.27 illustrates the comparison of body diode turn-off voltage and current waveforms between experiment (solid) and simulation (dashed) at 150°C. The DC supply voltage is 500V, and forward conduction current of body diode is 30A. A good matching between experiment and simulation proves the accuracy of the model over a wide junction temperature range.

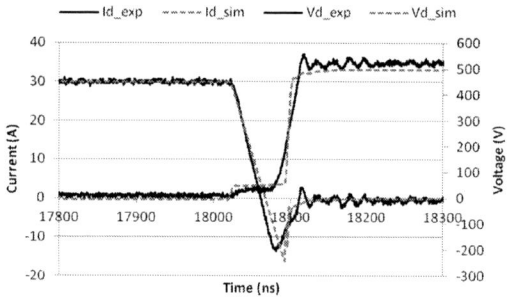

Fig. 27. Comparison of body diode turn-off voltage and current waveforms between experiment (solid) and simulation (dashed) at 150°C

VI. CONCLUSIONS

The static and dynamic characterizations of SiC MOSFET's body diode are provided. To our knowledge, this is the first complete characterization of SiC MOSFET's body diode in the literature. The static characterization of SiC MOSFET's body diode is carried out using a curve tracer. The

I-V curves of SiC MOSFET's body diode at varied junction temperatures are given. The dynamic characteristics of SiC MOSFET's body diode are tested based on a double pulse test bench. The switching behavior of SiC MOSFET's body diode at different current commutation slopes, forward conduction currents and junction temperatures is demonstrated. The device model of body diode is described in detail. The parameter extraction procedure for this model is introduced, which only requires data from the manufacturer's datasheets and one simple switching waveform measurement. Finally, the comparison between simulation and experiment proves the accuracy of the body diode model and the parameter extraction method over a wide junction temperature.

REFERENCES

[1] H.A.Mantooth, K.Peng, E.Santi, and J.L.Hudgins. "Modeling of wide bandgap power semiconductor devices-part I", IEEE Transactions on Electron Devices,vol.62, no.2,pp.423-433, Feb.2015.

[2] E.Santi, K.Peng, A.Mantooth and J.L.Hudgins. "Modeling of wide band-gap power semiconductor devices-part II", IEEE Transactions on Electron Devices,vol.62, no.2,pp.434-442, Feb.2015.

[3] R.Fu, A.E.Grekov, K. Peng and E.Santi, "Parameter extraction procedure for a physics-based power SiC Schottky diode model", IEEE Transactions on Industry Applications,vol.50, no.5,pp.3358-3568,Sep-Oct.2014.

[4] K.Peng, and E.Santi, "Performance projection and scalable loss model of SiC MOSFETs and SiC Schottky diodes", IEEE Electric Ship Technologies Symposium, ESTS 2015, pp.281-286, Jun.2015.

[5] K.Peng, S.Eskandari, and E.Santi, "Analytical loss model for power converters with SiC MOSFET and SiC Schottky diode pair," IEEE Energy Conversion Congress and Exposition, ECCE 2015, pp.6153-6160, Sep.2015.

[6] J. Jordan, V.Esteve, E.Sanchis-Kilders, E.J.Dede, E.Maset, J.B.Ejea, and A. Ferreres, "A comparative performance study of a 1200V Si and SiC MOSFET intrinsic diode on an induction heating inverter", IEEE Transactions on Power Electronics, vol.29, no.5, pp.2550-2562, May. 2014.

[7] A. Bolotnikov, J.Glaser, J.Nasadoski, P.Losee, S.Klopman, A.Permuy and L.Stevanovic, "Utilization of SiC MOSFET body diode in hard switching applications", Materials Science Forum, Vol. 778-780, pp. 947-950, Feb.2014.

[8] T.Funaki, M.Matsushita, M.Sasagawa, T.Kimoto, and T. Hikihara, "A study on SiC devices in synchronous rectification of DC-DC converter", Applied Power Electronics Conference and Exposition, APEC 2007, pp.339-344, Mar.2007.

[9] Z.Wang, J.Ouyang, J.Zhang, X.Wu and K.Sheng, "Analysis on reverse recovery characteristics of SiC MOSFET intrinsic diode", IEEE Energy Conversion Congress and Exposition, ECCE 2014, pp. 2832-2837, Sep.2014.

[10] T. Gachovska, J.L.Hudgins, A.Bryant, E.Santi, H.A.Mantooth and A. K.Agarwal, "Modeling, simulation, and validation of a Power SiC BJT," IEEE Transactions on Power Electronics,vol.27,no.10,pp.4338-4346, Apr.2012.

[11] X.Kang, A.Caiafa, E.Santi, J.L.Hudgins, and P.R.Palmer, "Parameter extraction for a power diode circuit simulator model including temperature dependent effects," Applied Power Electronics Conference and Exposition, APEC 2002, pp.452-458, Mar.2002.

[12] R.Fu, A.E.Grekov, K. Peng and E.Santi, "Parasitic modeling for accurate inductive switching simulation of converters using SiC devices," IEEE Energy Conversion Congress and Exposition, ECCE 2013,pp.1259-1265, Sep. 2013

A Simple Behavioral Electro-thermal Model of GaN FETs for SPICE Circuit Simulation

Liyao Wu
School of Electrical and Computer Engineering
Georgia Institute of Technology
Atlanta, Georgia 30332–0250
Email: lwu49@gatech.edu

Maryam Saeedifard
School of Electrical and Computer Engineering
Georgia Institute of Technology
Atlanta, Georgia 30332–0250
Email: maryam@ece.gatech.edu

Abstract—This paper develops a behavioral electro-thermal model of GaN FETs in SPICE environment for power-electronic circuit simulation. The model couples an available GaN FET electrical model with a thermal RC network for junction temperature estimation, while using modification circuits between the gate driver and device to consider thermal impacts on device performance. Both static and switching characteristics of the developed model are compared with those from the original electrical model. To demonstrate the performance and functionality of the developed electro-thermal model in circuit simulation studies, it is implemented in a boost converter as a benchmark system. Finally, a Double Pulse Tester (DPT) circuit is built and the performance of the developed model is compared with experimental results.

I. INTRODUCTION

The Wide Band-Gap (WBG) switching devices, particularly GaN devices, have become attractive switching devices for power electronic systems due to their unique electrical and thermal capabilities/characteristics compared to their Si counterparts, enabling considerable reduction of the size of passive components and thermal management effort and improvement of the efficiency and power density [1]–[3]. However, as the devices and power electronic systems become smaller, new challenges are imposed on thermal management as the heat dissipation becomes more difficult [4]. Therefore, at the design stage of GaN-based power electronic systems, it is necessary to determine the junction temperature profile of the switching devices to ensure that they operate within their safe operation area. The junction temperature also has significant impacts on the electrical performance of GaN FET devices, and therefore, for accurate temperature estimation, the device model needs to take the thermal impacts into account [5], [6]. Consequently, electro-thermal modeling of GaN devices is one of the important steps for system design.

Despite the importance of electro-thermal modeling of GaN devices, as mentioned in [7], an electro-thermal model for the GaN FET that considers the device self-heating is not yet readily available. Efforts have been made to develop electro-thermal models for GaN HEMTs, which considered nonlinear thermal effects, self-heating effects and bias dependence [5], [8], [9], for Radio Frequency (RF) applications. Reference [6] presents electro-thermal models for both GaN vertical MOSFETs and GaN lateral HEMTs based on device physics.

However, physical models generally involve many device parameters with mutual couplings, which make them difficult to implement for circuit simulation [10]. On the other hand, the results obtained from physics-based models may not be as accurate as those from behavioral models [11]. Thus, simple and accurate behavioral models are considered more favorable for circuit simulations. In addition, the existing electro-thermal models focus on the effect of junction temperature on the I_{ds}-V_{ds} relationship during the device conduction period. The thermal effects on the device switching characteristics, however, are only explored in terms of switching losses instead of parameters like $\frac{dv}{dt}$ and $\frac{di}{dt}$, which are important for loss evaluation, system reliability and EMI issues [5]–[9]. After all, to the best of the authors knowledge, no behavioral electro-thermal model for circuit simulation that considers both device static and switching characteristics has been developed for GaN FETs.

In this paper, a simple behavioral electro-thermal model for EPC2010C GaN FET is developed in the LTSPICE software environment. The model couples an available electrical model of the device with a thermal RC network to count the self-heating effects. The developed model models the impacts of junction temperature on the device static and switching characteristics by using a gate voltage modification circuit. The accuracy of the developed model is validated against the original fixed-temperature electrical model. To demonstrate the capability of the developed model in estimating the junction temperature profile of the device in the SPICE circuit simulation, it is investigated for a boost converter benchmark system. Finally, the simulation results with the developed model are compared with experimental results obtained from a Double Pulse Tester (DPT) circuit built with EPC2010C GaN FETs.

This paper is organized as follows: in Section II, the proposed electro-thermal model of GaN FET, including the operating principles of the gate voltage modification circuit during conduction periods and switching transients, is introduced. In Section III, performance of the developed electro-thermal model is validated against the original electrical model for both static and switching characteristics. Performance of the developed model in circuit simulation is also demonstrated in Section III. Experimental results from a DPT circuit built with EPC2010C GaN FETs are presented and used to validate the

978-1-4673-9551-9/16 $31.00 © 2016 IEEE

developed model in Section IV. Finally, Section V concludes the paper.

II. THE PROPOSED ELECTRO-THERMAL MODEL OF THE GaN FET

In general, an electro-thermal model consists of an electrical model of the device coupled with a thermal RC network [5], [12]. The thermal network, as shown in Fig. 1, takes the dissipated power in the device as the input and provides the voltage across the network, which represents the junction temperature, as the output. The junction temperature is fed back into the electrical model, impacting the device parameters and characteristics. Although the available electrical models have built-in functions to calculate device parameters at various junction temperatures, they are not capable of dynamically updating the parameters with a variable temperature profile during simulation. In this paper, a gate voltage modification circuit is developed, which, as shown in Fig. 2(a), modifies the gate voltage of the device based on junction temperature profile, thereby attaining an accurate electro-thermal model. The operation principles of the modification circuit applied to the gate terminal are explained in details in the following sections.

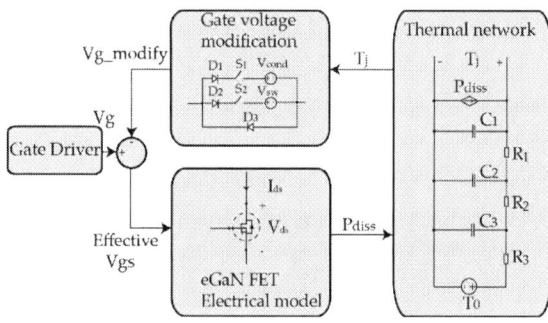

Fig. 1: Block diagram of the proposed electro-thermal model.

A. Modeling of Static Characteristic

The key parameter for the static characteristic of the device, which refers to the I_{ds}-V_{ds} relationship during conduction period, is the on-state resistance $R_{ds,on}$. The on-state resistance of GaN FET is a near-linear function of its junction temperature as shown in Fig. 3 [13]. On the other hand, $R_{ds,on}$ is also a function of gate voltage V_{gs} as in Fig. 3 [13]. Thus, by using an additional voltage source V_{cond} controlled by the junction temperature during conduction period as shown in Fig. 2(b), $R_{ds,on}$ can be changed based on the junction temperature profile. Due to the nonlinear relationship between $R_{ds,on}$ and V_{gs}, the value of V_{cond} is determined by a look-up table reflecting $R_{ds,on}$-V_{gs} relationship.

B. Modeling of Switching Characteristics

To model the temperature-dependent switching characteristics of the device, the impacts of junction temperature on the device turn-off and turn-on transients need to be identified. By

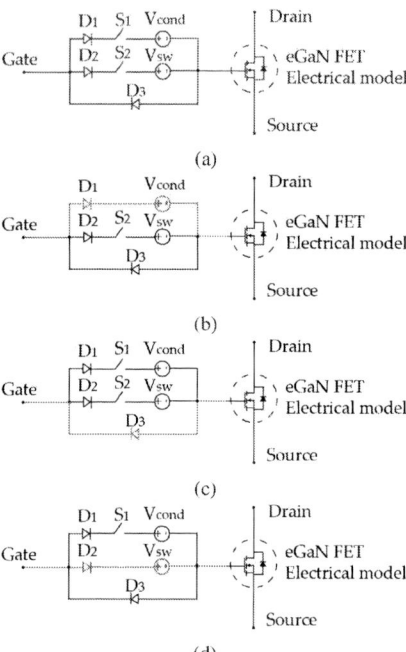

(a)

(b)

(c)

(d)

Fig. 2: (a) The proposed gate voltage modification circuit and (b), (c) and (d) its operation principles during device conduction period and turn-off and turn-on transients, respectively.

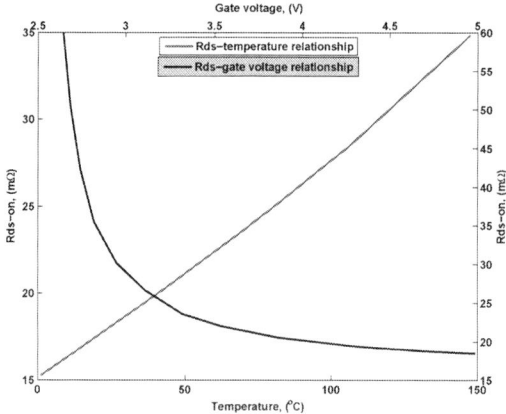

Fig. 3: Relationship between the on-state resistance $R_{ds,on}$, junction temperature and gate voltage.

using a conventional DPT circuit in LTSPICE, it is observed that the turn-off transient is almost independent of junction temperature, while $\frac{dV_{ds}}{dt}$ during the turn-on transition has a near-linear relationship with the junction temperature. Thus, during turn-off transition, the device is connected directly to the gate driver through the diode D_3, as shown in Fig. 2(c). During turn-on transition, $\frac{dV_{ds}}{dt}$ is determined by [14]:

$$\frac{dV_{ds}}{dt} = -\frac{V_{gs,on} - V_{plateau}}{R_g C_{gd}}, \tag{1}$$

where V_{ds} is the drain-source voltage, $V_{gs,on}$ is the turn-on voltage provided by the gate driver, $V_{plateau}$ is the Miller plateau voltage, R_g is the gate resistance and C_{gd} is the gate-drain capacitance of the device. Therefore, by using an additional voltage source V_{sw} in the gate loop during the turn-on transient as shown in Fig. 2(d), the effective $V_{gs,on}$ of the device can be modified. This, accordingly, changes $\frac{dV_{ds}}{dt}$. The value of V_{sw} is determined by a look-up table reflecting $\frac{dV_{ds}}{dt}$-$V_{gs,on}$ relationship.

III. MODEL VALIDATION

A. Static Characteristic

To validate the temperature-dependent static characteristic, the developed model is coupled with a time-varying temperature profile generated from a constant power source and a thermal RC network derived from the device datasheet [13], being under DC bias to extract the on-state resistance. As shown in Fig. 4(a), the effective gate-source voltage observed by the electrical device model is changed according to the varying junction temperature, and the modeled on-state resistance achieves good consistency with the resistance estimated based on datasheet throughout the time period as in Fig. 4(b).

B. Switching Characteristics

For switching transients, the developed model is implemented in a DPT circuit at various fixed temperatures, assuming that the time constant of the thermal system is sufficiently large so the temperature remains constant during a switching event [10]. The drain-source voltage V_{ds} and current I_{ds} waveforms during switching transients are shown in Fig. 5 for a junction temperatures of 85 ^{o}C. By comparing the switching characteristics of the device based on the original electrical model at 25 ^{o}C and 85 ^{o}C, the impacts of junction temperature on the switching characteristics can be observed in Fig. 5. As shown in Fig. 5(a), the estimated turn-off transient of the device based on the developed model at 85 ^{o}C is similar to the one from the original electrical model. The estimated turn-on transients of the device are also closely matched with the one from the original electrical model. Nevertheless, the overshoot current waveform of I_{ds} during the turn-on transition is not modeled quite accurately because under fixed-temperature simulation, the simulation temperature also impacts other devices such as diodes in the circuit while the proposed electro-thermal model only accounts for the GaN FET. However, the accuracy of the developed model is illustrated by $\frac{dI_{ds}}{dt}$ prior to the overshoot and the end of turn-on transient.

C. Circuit Simulation

To demonstrate performance of the developed model for circuit simulation studies, a 75-150 V boost converter is constructed in the LTSPICE software environment, using the developed electro-thermal model for the main switch, as shown

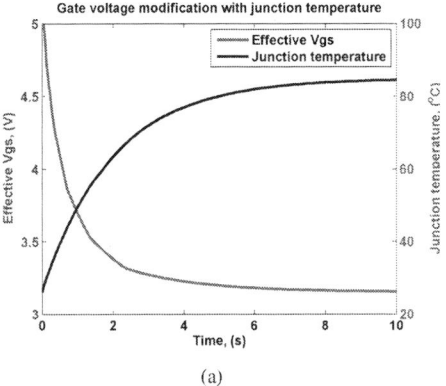

(a)

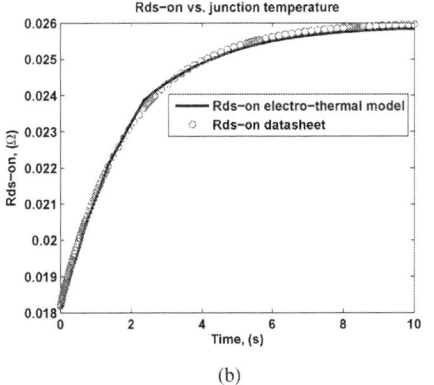

(b)

Fig. 4: (a) Gate voltage modification of the electro-thermal model during conduction mode with the junction temperature profile and (b) the modeled temperature-dependent on-state resistance and the resistance calculated based on datasheet.

in Fig. 6(a). The coupled thermal RC network is developed based on the thermal SPICE model provided by EPC [15]. The power loss profile of the main switch and the corresponding junction temperature variations during the first 1 ms of simulation time are shown in Fig. 6(b). As shown in Fig. 6(b), the developed model is capable of dynamically estimating the junction temperature transients during switching cycles.

In order to measure the simulation speed, the time to run the same boost converter simulation for 3000 cycles with and without the developed model is recorded. With the developed model implemented, the simulation time is about 1 minute, while the simulation time with only the original model is 26.71 seconds, both with a PC using Intel 3.40 GHz processor with 16 GB RAM.

IV. EXPERIMENTAL VALIDATION

In order to validate the developed electro-thermal model with experimental results, a DPT circuit is built based on an EPC2010C half-bridge evaluation board from EPC shown in Fig. 7. The board is placed on a heatplate to conduct experiments under different controlled temperatures.

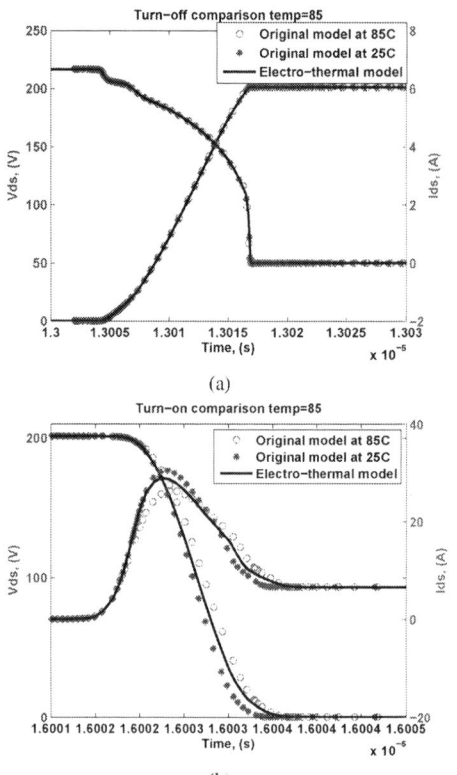

Fig. 5: (a) Turn-off and (b) turn-on switching transients.

Comparison of the experimental results and simulation results obtained with the developed electro-thermal model of V_{ds} of the lower switch at 50 ^oC during both turn-on and turn-off transients is presented in Figs. 8(a) and (b), and at 70 ^oC in Figs. 8(c) and (d), respectively. Based on Figs. 8(a) and (c), the simulated $\frac{dV_{ds}}{dt}$ during the transient is closely matched with the experimental result at different temperatures. On the other hand, the turn-off transient with the developed model also shows $\frac{dV_{ds}}{dt}$ that is close to the experimental result as in Figs. 8(b) and (d). However, the negative overshoot in V_{ds} present in the experimental result as observed in Figs. 8(a) and (c), as well as the positive voltage overshoot in Figs. 8(b) and (d), are not present in the simulation results. This is due to the circuit parasitics in the circuit [16], which are not included in simulation. Thus, the simulation results could be further improved if these circuit parasitics are accurately determined and included into the simulation.

V. CONCLUSION

In this paper, a new behavioral electro-thermal model for GaN FETs that considers the self-heating and thermal effects on the device characteristics is developed. The model, which is developed in the LTSPICE software environment, is validated under both conduction and switching conditions and is com-

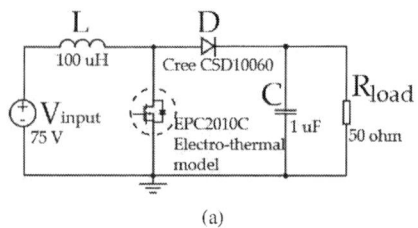

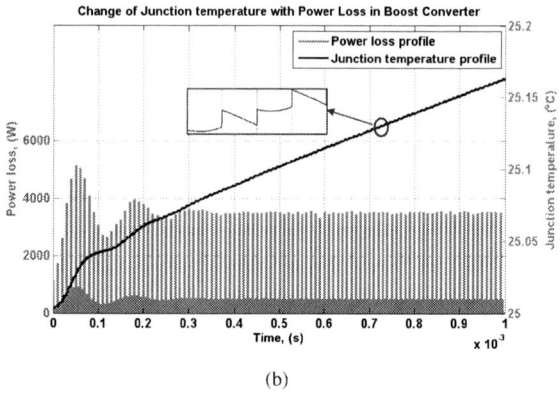

Fig. 6: (a) The boost converter benchmark system and (b) junction temperature and power loss profile of the main switch based on the proposed electro-thermal model.

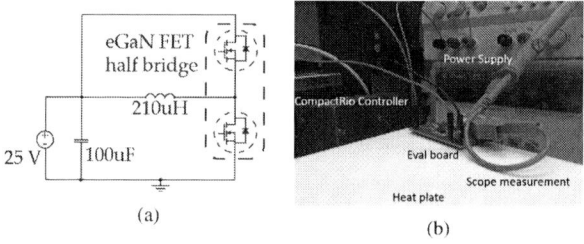

Fig. 7: (a) Block digram and (b) experimental setup of the DPT circuit to validate the proposed model.

pared against an available electrical model. To demonstrate the capability of the developed model in estimating the junction temperature transients, simulation studies are conducted on a boost converter benchmark system. The results from the proposed model are closely matched with the experimental results obtained from a DPT circuit. Further improvement could be expected by considering the parasitics in the circuit in the simulation.

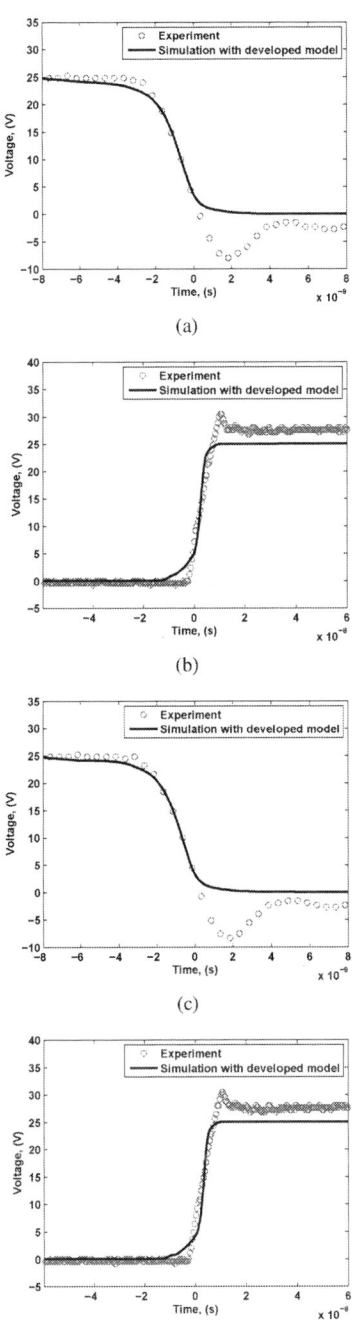

Fig. 8: Comparison of experimental and simulation results of V_{ds} of the lower switch at 50 ^{o}C during (a) and (b) lower switch turn-on and turn-off transients, respectively, and at 70 ^{o}C during (c) and (d) lower switch turn-on and turn-off transients.

REFERENCES

[1] A. Lidow, J. Strydom, M. D. Rooij, and D. Reusch, *GaN transistors for efficient power conversion*, 2nd ed. John Wiley & Sons, 2014.

[2] J. Millan, P. Godignon, X. Perpina, A. Prez-Toms, and J. Rebollo, "A survey of wide bandgap power semiconductor devices," *IEEE Trans. Power Electron. on*, vol. 29, no. 5, pp. 2155–2163, 2014.

[3] H. A. Mantooth, M. D. Glover, and P. Shepherd, "Wide bandgap technologies and their implications on miniaturizing power electronic systems," *IEEE Trans. Emerg. Sel. Topics Power Electron.*, vol. 2, no. 3, pp. 374–385, 2014.

[4] J. A. Ferreira, J. Popovic, J. D. van Wyk, and F. Pansier, "System integration of GaN technology," in *IEEE International Power Electronics Conference (IPEC)*, 2014, pp. 1935–1942.

[5] J. B. King and T. J. Brazil, "A comprehensive electrothermal GaN HEMT model including nonlinear thermal effects," in *IEEE MTT-S International Microwave Symposium Digest (MTT)*, 2012, pp. 1–3.

[6] Y. Zhang, M. Sun, Z. Liu, D. Piedra, H.-S. Lee, F. Gao, T. Fujishima, and T. Palacios, "Electrothermal simulation and thermal performance study of GaN vertical and lateral power transistors," *IEEE Trans. Electron Devices*, vol. 60, no. 7, pp. 2224–2230, 2013.

[7] E. Santi, K. Peng, H. A. Mantooth, and J. L. Hudgins, "Modeling of wide-bandgap power semiconductor devices – Part II," *IEEE Trans. Electron Devices*, vol. 62, no. 2, pp. 434–442, 2015.

[8] M. Thorsell, K. Andersson, H. Hjelmgren, and N. Rorsman, "Electrothermal access resistance model for GaN-based HEMTs," *IEEE Trans. Electron Devices*, vol. 58, no. 2, pp. 466–472, 2011.

[9] A. Xiong, C. Charbonniaud, E. Gatard, and S. Dellier, "A scalable and distributed electro-thermal model of AlGaN/GaN HEMT dedicated to multi-fingers transistors," in *IEEE Compound Semiconductor Integrated Circuit Symposium (CSICS)*, 2010, pp. 1–4.

[10] B. Du, J. L. Hudgins, E. Santi, A. T. Bryant, P. R. Palmer, and H. A. Mantooth, "Transient electrothermal simulation of power semiconductor devices," *IEEE Trans. Power Electron.*, vol. 25, no. 1, pp. 237–248, 2010.

[11] L. Starzak, M. Zubert, M. Janicki, T. Torzewicz, M. Napieralska, G. Jablonski, and A. Napieralski, "Behavioral approach to SiC MPS diode electrothermal model generation," *IEEE Trans. Electron Devices*, vol. 60, no. 2, pp. 630–638, 2013.

[12] S. Stoffels, H. Oprins, D. Marcon, K. Geens, X. Kang, M. V. Hove, and S. Decoutere, "Coupled electro-thermal model for simulation of GaN power switching HEMTs in circuit simulators," in *18th International Workshop on THERMal INvestigation of ICs and Systems*, 2012.

[13] EPC. (2015) "EPC2010C datasheet". [Online]. Available: http://epc-co.com/epc/Portals/0/epc/documents/datasheets/EPC2010C_datasheet.pdf

[14] Z. Chen. (2009) "Characterization and modeling of high-switching-speed behavior of SiC active devices". [Online]. Available: https://vtechworks.lib.vt.edu/bitstream/handle/10919/30778/Chen_Z_T_2009.pdf?sequence=1&isAllowed=y

[15] EPC. "EPC2010 SPICE thermal model". [Online]. Available: http://epc-co.com/epc/Portals/0/epc/documents/models/thermal/EPC%202010%20Thermal%20SPICE%20Model%20for%20web.pdf

[16] D. Reusch and J.Strydom, "Understanding the effect of PCB layout on circuit performance in a high-frequency gallium-nitride-based point of load converter," *IEEE Trans. Power Electron.*, vol. 29, no. 4, pp. 2008–2015, 2014.

Decomposition and Electro-Physical Model Creation of the CREE 1200V, 50A 3-Ph SiC Module

Adam J Morgan, Yang Xu, Douglas C Hopkins, Iqbal Husain, Wensong Yu

PowerAmerica
North Carolina State University
Raleigh, NC, USA
ajmorga4@ncsu.edu, yxu17@ncsu.edu, dchopkins@ncsu.edu, ihusain2@ncsu.edu, wyu2@ncsu.edu

Abstract—The CREE 1200V/50A, 25mΩ 6-Pack SiC MOSFET module (CCS050M12CM2) is decomposed into a full 3D CAD model, and materials identified, for use in electrical circuit and multi-physics simulations. A reverse engineering technique is first developed, outlined, and then demonstrated on the CREE module. The ANSYS Q3D Extractor is applied to the 3D CAD model where electrical, lumped parameter, parasitic circuit elements are determined. The model is also analyzed with a multi-physics simulator to provide in-situ thermal maps of the baseplate surface for application scenarios, e.g. with a thermal interface material and pin fin heat sink to capture the thermal spreading from junction to case. The complete model is made open source and freely distributed for use by the reader.

Keywords – *Parasitic Impedances; Lumped Parameter Model; Thermal Model; CREE Module*

I. INTRODUCTION

Fast adoption of wide-bandgap (WBG) semiconductor technology in today's power electronics is currently being heavily pursued by the US Department of Energy (DoE) in order to combat climate change [1]. Therefore, if design engineers are more easily able to simulate WBG-based power modules within their systems to witness multi-physical performance improvement, then proliferation of WBG technology into the market's various commercial and industrial sectors will occur at a faster rate [2]-[3]. The electrical-physical model is applied in the design of a 35kW EV motor drive, and provides accurate analysis of inverter circuit operation. The complete process is described from module modeling through parameter and operating condition values in line with a full inverter drive application.

II. TEAR-DOWN APPROACH OF THE CREE MODULE

The module was opened to expose the internal topology. A Hesse Mechatronics BJ939 Heavy Wire/Ribbon Bonder digital camera system, accurate to within 10µm, was used to measure die, interconnect, and substrate length and width dimensions; see Fig. 1.

Initial heights and thicknesses were measured through a 45° perspective holder, as shown in Fig. 2. Trigonometric relationships in conjunction with Snell's Law were used for determining the internal layer thicknesses of the substrate and

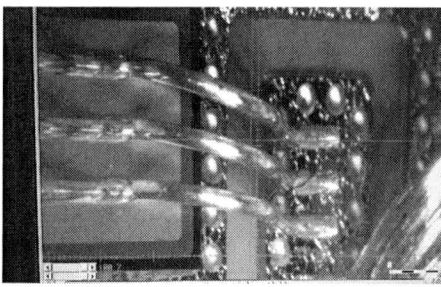

Figure 1 Internal view of CREE module

die. The silicone gel's index of refraction was ~1.33 at room temperature, due to its similar density to water [4]-[5].

The nondestructive measurement procedure using the BJ939 digital camera system is enumerated as follows:

1. Remove the module housing lid and place the opened module onto the known-angle perspective holder.

2. Place the module and perspective holder onto the center of the BJ939 work holder.

3. Center the BJ939 bond head and camera system over the opened power module, lowering the bond head until the internal power stage of the power module is in focus.

4. Use the BJ939 camera system target reticle cross-hairs to measure the horizontal distance, parallel to the work holder, between the top and bottom edges of each material layer within the power module.

5. Assuming the index of refraction of silicone gel is ~1.33, calculate individual layer thicknesses using the trigonometric relationships and Snell's Law, as shown in Fig. 2.

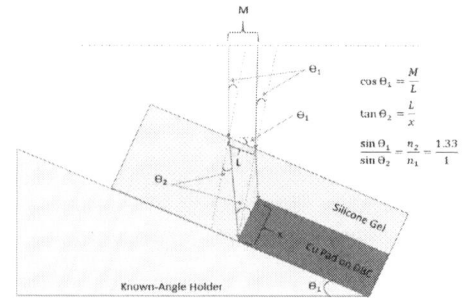

Figure 2 Power stage thickness calculations

Internal layer dimensions can be measured and calculated by repeating the procedure for all parts of the internal power stage. To validate the measurement procedure, the CREE module was disassembled (Fig. 3) and manually measured to within ±25.4μm (1mil). It should be noted that accurate measurement of the solder layers using a technique, such as SEM, was not used. However, solder layers were assumed to be 50μm for this study, as die attach thickness variations did not dominate the thermal model. Table 1 shows several examples of the accuracy of the nondestructive measurements. Power stage dimensions obtained using the nondestructive method are tabulated below in Fig. 4 and 5.

Figure 3 Destructive tear-down of CREE module

A complete 3D CAD model of the power stage, based on the measured dimensions, was created in SolidWorks; see Fig. 6, and includes the module housing, baseplate, die attachment, DBC substrate, wire bonds, SiC MOSFETs and diodes, and terminals.

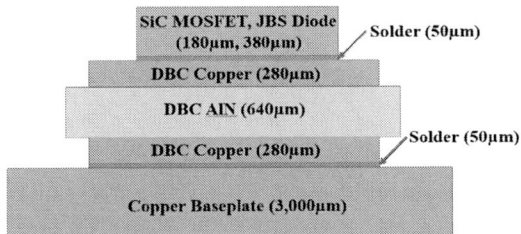

Figure 4 Power Stage Vertical Dimensions

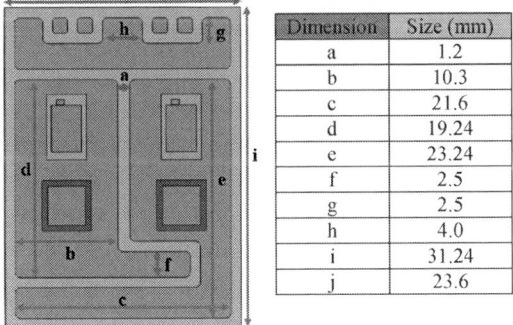

Dimension	Size (mm)
a	1.2
b	10.3
c	21.6
d	19.24
e	23.24
f	2.5
g	2.5
h	4.0
i	31.24
j	23.6

Figure 5 Power Stage Top View Dimensions

Table 1 Accuracy of nondestructive dimension measurements

DBC AlN	Nondestructive	Destructive	% Error
Thickness	24.2 mils	25 mils	3.2 %
Length	1210 mils	1236 mils	2.1 %
Width	885 mils	902 mils	1.9 %

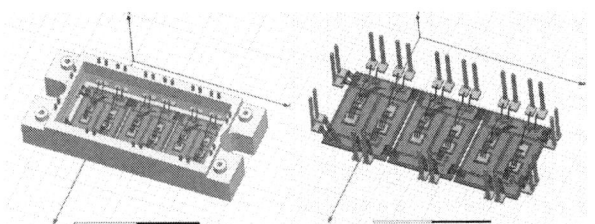

Figure 6 Full 3D CAD model of CREE module

III. PARASITIC MODEL EXTRACTION

The power stage 3D CAD model was imported into ANSYS Q3D Extractor to determine the lumped parameter SPICE-based parasitic circuit model. Each piece of the CREE module was assigned material and parameters; a circuit loop source and sink defined, and parasitic inductances and resistances extracted for each phase-leg using a discrete, single point frequency sweep: 0 Hz to 100 MHz. The CREE module can be connected to external circuitry in several configurations. Parasitic impedances for connection configurations A and B, portrayed in Fig. 7 and 8, are shown in Tables 2 and 3, respectively. Connection configuration A utilizes only one set of DC +/- terminals of the CREE module. Whereas configuration B utilizes both sets of DC +/- terminals of the CREE module.

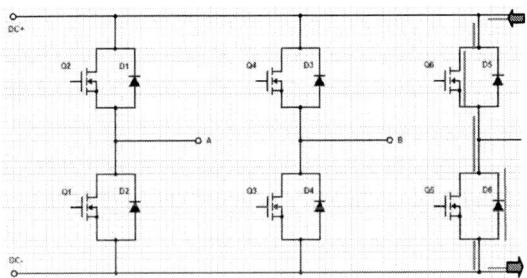

Figure 7 Connection configuration A and Q3D conduction path of CREE module terminals

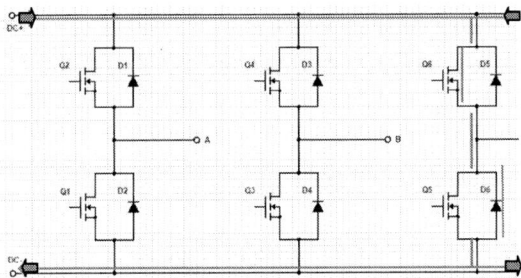

Figure 8 Connection configuration B and Q3D conduction path of CREE module terminals

Table 2 Configuration A extracted parasitic L and R values

Phase	Freq. (Hz)	DC	10k	100k	1M	10M	100M
A	L(nH)	115.7	102.2	83.66	78.91	74.16	73.43
B	L(nH)	81.73	74.8	61.9	56.52	54.78	54.22
C	L(nH)	47.39	46.49	40.89	37.41	36.23	35.86
A	R(mΩ)	7.33	8.13	12.49	47.26	73.16	219.1
B	R(mΩ)	6.11	6.61	9.83	20.76	55.55	165.6
C	R(mΩ)	4.54	4.71	6.56	13.44	35.48	105.3

Table 3 Configuration B extracted parasitic L and R values

Phase	Freq. (Hz)	DC	10k	100k	1M	10M	100M
A	L(nH)	43.37	42.20	36.21	32.76	31.61	31.24
B	L(nH)	51.52	48.72	40.46	36.54	35.25	34.85
C	L(nH)	41.56	40.44	34.80	31.56	30.48	30.13
A	R(mΩ)	4.83	4.98	6.90	14.12	37.28	110.6
B	R(mΩ)	5.04	5.31	7.59	15.64	41.34	122.7
C	R(mΩ)	4.76	4.90	6.79	14.13	37.72	112.5

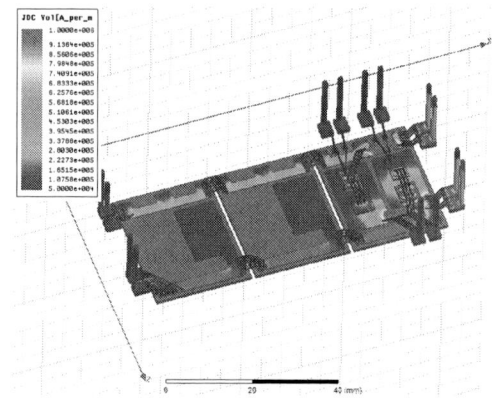

Figure 10 Configuration B current density

Comparing Tables 2 and 3, the parasitic values for each phase leg, over a discrete range of frequencies, are less when two sets of DC +/- terminals are connected to external circuitry. Lower parasitics are due to improved current symmetry within the CREE module. Therefore, the switching behavior of the SiC power semiconductor devices within the CREE module will also benefit from the more symmetrical connection configuration B.

IV. CREE MODULE TRANSIENT CHARACTERIZATION

A double pulse test circuit simulation was developed in LT-SPICE around the CREE module SiC MOSFET (C2M0025120D) and JBS diode (CPW51200Z050B) device models. Major extracted parasitic parameters from ANSYS Q3D Extractor were included in the double pulse test circuit in order to observe how the packaging of the CREE module influences the transient performance; see Fig. 11. For example, the series parasitic inductance extending from the MOSFET has been modeled as 15nH on each side of the source and drain, matching the stray inductance parameter value given in the CREE module datasheet.

The V_{DS} and I_{DS} switching waveforms for the turn-on and turn-off transients are displayed in Fig. 12. Parasitic effects are

The Q3D current density plots for configurations A and B are shown in Fig. 9 and 10 to further highlight the simulated conduction paths. Validity is given to the extracted parameters by comparing the values obtained through simulation with the values provided in the CREE module datasheet. For example, stray inductance within the datasheet is given as 30nH, and the parasitic inductance obtained through simulation, for configuration B, are within 5nH for all three phase cases. Choosing to connect the CREE module using either connection configuration A or B will alter the power loop inductance and resistance.

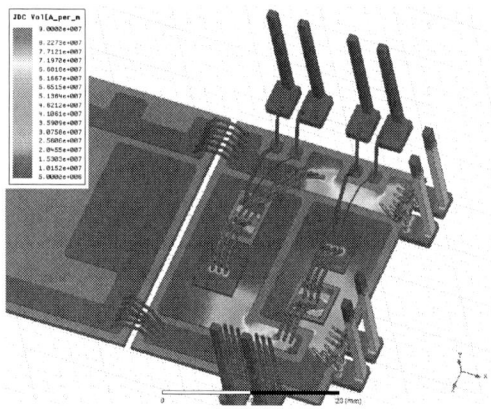

Figure 9 Configuration A current density

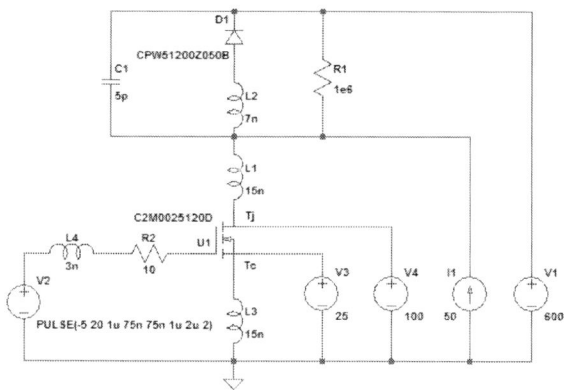

Figure 11 Double pulse test circuit of CREE module devices

978-1-4673-9551-9/16 $31.00 © 2016 IEEE   2143

observed contributing to the simulation switching output. A voltage plateau is seen during turn-on due to the parasitic L_{DS} adding an "L(di/dt)" voltage drop. I_{DS} contains undesirable high frequency ringing during both turn-on and turn-off contributed by the resonance between L_{DS} and C_j of the diode. Voltage and current overshoots occur by means of charging and discharging the parasitic inductances and capacitances, respectively, inherent to the CREE module packaging [6].

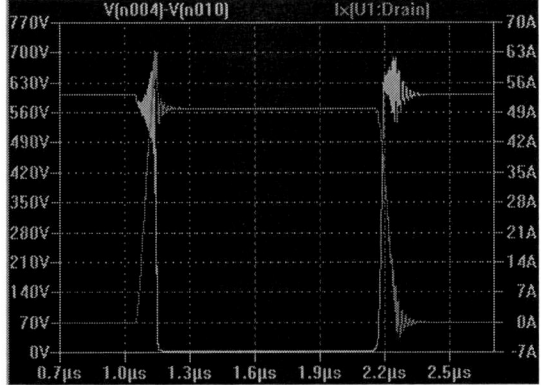

Figure 12 Switching transient of CREE module devices

V. THERMAL MODEL EXTRACTION

COMSOL was used to determine the thermal profile from junction to case of the CREE module. A vertical thermal profile was then extracted that displays thermal spreading throughout the CREE module from junction to case. Fig. 13 shows the thermal profile for operating points under rated conditions.

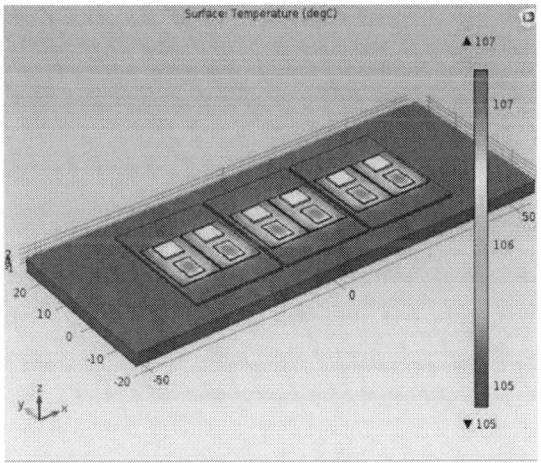

Figure 13 COMSOL CREE module power stage thermal profile

CREE module thermal performance characterization, using a thermal interface material (TIM) of 100μm thick thermal grease having 1 W/m-K thermal conductivity, and a 65°C water velocity of 0.1L/s at the inlet of a pin fin heat sink, was simulated on a per phase basis. The average MOSFET and diode dissipated power were 60W and 30W, respectively. Fig. 14 shows the surface temperature of a single phase leg of the inner power stage of the CREE module under the above conditions [7]-[9]. The ΔT_{jc} of approximately 35°C yields a junction temperature well below specified $T_{j,max}$ for the CREE MOSFET and diode, particularly under conditions typical of an EV motor drive [10]-[11].

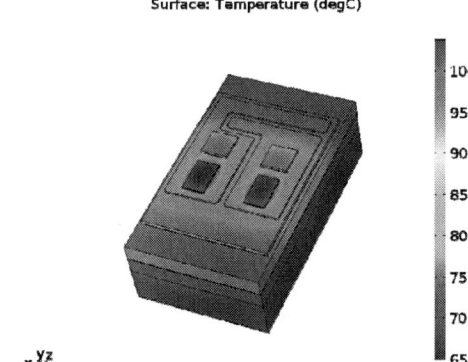

Figure 14 Phase leg surface temperature

Horizontal slices of the internal temperature within the baseplate and water velocity on a logarithmic scale, 5mm below the baseplate, inside the pin fin heat sink are shown in Fig. 15 and 16, respectively.

Slice: Temperature (degC)

Figure 15 Baseplate temperature distribution

The in-situ thermal map of the baseplate highlights how potential hotspots develop depending on the cooling approach used. The hottest baseplate temperatures occur where the water flow velocity is the lowest. Whereas the combined, precise placement of the heat sink pin fins directly beneath the MOSFETs and diodes of the power stage, along with the higher water flow velocity that contacts these pin fins, enables a much cooler baseplate surface temperature. The extracted thermal model portrays how the design of the heat sink for the CREE module is important for extracting the required heat from the power stage in order to allow MOSFET and diode device operation safely below the maximum allowable junction temperature.

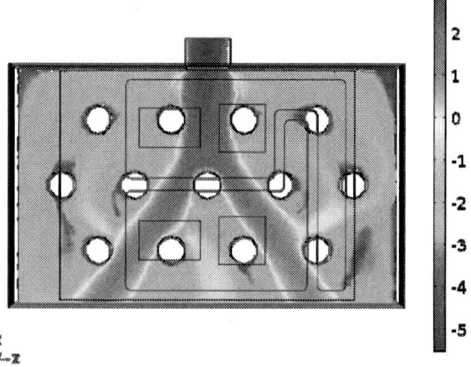

Slice: Velocity magnitude

Figure 16 Water velocity 5mm below module baseplate

The location of the inlet to the pin fins of the heat sink is also seen directly affecting not only the thermal behavior of the pin fins, but more importantly the MOSFETs and diodes within the power stage. Heat removal methods that allow for localized cooling of the CREE module baseplate are highly recommended, if operating conditions above rated values are desired. Therefore, this is one design example where the created physical model of the CREE module is capable of yielding a detailed thermal map for a given TIM and heat sink application. Other thermal scenarios can similarly be simulated using the model to give designers in-depth insight into how generated heat, under a given operating point, is spreading throughout the entire power stage.

VI. SUMMARY

A reverse engineering technique is first developed, outlined, and then demonstrated on the CREE module. ANSYS Q3D Extractor is applied to the 3D CAD model where electrical, lumped parameter, parasitic circuit elements are determined. The generated electrical, lumped parameter, parasitic circuit models enable simulators, such as LT-SPICE, to utilize a fully characterized device model; a model that is validated by the CREE module datasheet. Design engineers are provided the basis to a model that more closely represents actual module operation within a system. The model is also analyzed with a multi-physics simulator to provide in-situ thermal maps inside the CREE module structure to the

baseplate surface for a typical application scenario for an EV motor drive, e.g. with appropriate TIM and heat sink. Vertical thermal profiles are decomposed to capture the thermal spreading from junction to case. Design engineers are provided useful insight into what cooling method to use based upon the applied conditions to the back side of the CREE module. The complete model is freely distributed for use by the reader.

REFERENCES

[1] [1] U.S. Department of Energy, Office of Energy Efficiency & Renewable Energy. (2015, July 21). Power America [Online]. Available: http://energy.gov/eere/amo/power-america

[2] The White House Blog. (2014, January 15). Wide Bandgap Semiconductors: Essential to Our Technology Future [Online]. Available:https://www.whitehouse.gov/blog/2014/01/15/wide-bandgapsemiconductors-essential-our-technology-future

[3] McKinsey&Company, GSA Semiconductor Leaders Forum Taiwan. (2012, November 7). Unleashing Growth in Wide Bandgap: The upcoming disruptions in power electronics [Online]. Available: http://www.gsaglobal.org/events/2012/1107/docs/slft2012_wiseman.pdf

[4] NDT Resource Center. (2015, April 13). Refraction and Snell's Law [Online].Available:https://www.nde-ed.org/EducationResources/CommunityCollege/Ultrasonics/Physics/refractionsnells.htm

[5] The Engineering Toolbox. (2015, April 5). Refractive Index of Some Common Liquids, Solids, and Gases [Online]. Available: http://www.engineeringtoolbox.com/refractive-index-d_1264.html

[6] Z. Chen, "Characterization and Modeling of High-Switching-Speed Behavior of SiC Active Devices," M.S. thesis, Dept. Elec. Eng., VA Polytechnic Inst., Blacksburg, VA, 2009.

[7] Modeling and experimental study of thin bond line thermal interface material failure Shidong Li (IBM Corp., Hopewell Junction, NY, United States); Sinha, T.; Davis, T.J.; Sikka, K.; Bodenweber, P. Source: 2013 IEEE 63rd ECTC, p 803-6, 2013

[8] In-situ thickness method of measuring thermo-physical properties of polymer-like thermal interface materials [microelectronics cooling applications] Smith, R.A. (Micro-Electron. Heat Transfer Lab., Waterloo Univ., Ont., Canada); Culharn, R.J. Source: Twenty First Annual IEEE Semiconductor Thermal Measurement and Management Symposium (IEEE Cat. No.05CH37651), p 53-63, 2005

[9] Effect of the thickness of a thermal interface material (solder) on heat transfer between copper surfaces Xiangcheng Luo (Composite Mater. Res. Lab., State Univ. of New York, Buffalo, Buffalo, NY, United States); Chung, D.D.L. Source: International Journal of Microcircuits and Electronic Packaging, v 24, n 2, p 141-7, 2001

[10] Burress, T.; Campbell, S., "Benchmarking EV and HEV power electronics and electric machines," Transportation Electrification Conference and Expo (ITEC), 2013 IEEE , vol., no., pp.1,6, 16-19 June 2013, doi: 10.1109/ITEC.2013.6574498

[11] Burress, T.; Campbell, S., "Benchmarking EV and HEV power electronics and electric machines," Transportation Electrification Conference and Expo (ITEC), 2013 IEEE , vol., no., pp.1,6, 16-19 June 2013, doi: 10.1109/ITEC.2013.6574498

[12] M. Arun Noyal Doss, V.Ganapathy, R. Sridhar, S.S. Dash, D. Mahesh. "Analytical and Simulation Analysis of Stator Tooth on Cogging Torque of Brushless DC Motor Using Finite Element Analysis," in Proc. ICPERES 2014, pp. 1-8.

[13] M. Valan Rajkumar, P.S. Manoharan. "Modeling and Simulation of Three-Phase DCMLI Using SVPWM for Photovoltaic System," in Proc. ICPERES 2014, pp. 39-45.

[14] Schulz, M.; Allen, S.T.; Pohl, W., "The crucial influence of thermal interface material in power electronic design,"Semiconductor Thermal Measurement and Management Symposium (SEMI-THERM), 2013 29th Annual IEEE , vol., no., pp.251,254, 17-21 March 2013, doi: 10.1109/SEMITHERM.2013.6526839

[15] Gautam, D.; Wager, D.; Edington, M.; Musavi, F., "Performance comparison of thermal interface materials for power electronics

978-1-4673-9551-9/16 $31.00 © 2016 IEEE   2145

applications," Applied Power Electronics Conference and Exposition (APEC), 2014 Twenty-Ninth Annual IEEE , vol., no., pp.3507,3511, 16-20 March 2014, doi: 10.1109/APEC.2014.6803814

[16] Yafan Zhang; Belov, I.; Sarius, N.G.; Bakowski, M.; Nee, H.; Leisner, P., "Thermal evaluation of a liquid/air cooled integrated power inverter for hybrid vehicle applications," Thermal, Mechanical and Multi-Physics Simulation and Experiments in Microelectronics and Microsystems (EuroSimE), 2013 14th International Conference on , vol., no., pp.1,8, 14-17 April 2013, doi: 10.1109/EuroSimE.2013.6529944

[17] Mantooth, H.A.; Kang Peng; Santi, E.; Hudgins, J.L., "Modeling of Wide Bandgap Power Semiconductor Devices—Part I," Electron Devices, IEEE Transactions on , vol.62, no.2, pp.423,433, Feb. 2015, doi: 10.1109/TED.2014.2368274

[18] Santi, E.; Kang Peng; Mantooth, H.A.; Hudgins, J.L., "Modeling of Wide-Bandgap Power Semiconductor Devices—Part II," IEEE Trans.

on Electron Devices, , vol.62, no.2, pp.434,442, Feb. 2015, doi: 10.1109/TED.2014.2373373

A Three-Legged MATLAB/SIMULINK Transformer Model Using A Fictitious Delta Winding

Thomas A. Nondahl, *Fellow, IEEE*, Jingbo Liu, *Senior Member, IEEE,* Peter B. Schmidt, *Member, IEEE*

Rockwell Automation
1201 South Second Street, Milwaukee, WI 53204, USA
tanondahl@ra.rockwell.com, jliu2@ra.rockwell.com, pbschmidt@ra.rockwell.com

Abstract— **This paper proposes a simple MATLAB/ SIMULINK model of a three-phase, three-legged, saturated transformer. A fictitious delta winding is added to three single-phase saturable transformers. The proposed model is simple and fairly accurate to model the nonlinear characteristics of a saturable three-legged transformer including mutual coupling.**

Keywords— MATLAB; SIMULINK; Fictitious; Three-Legged Transformer; Model

I. INTRODUCTION

A transformer is a device utilized to change the voltage of an alternating current in one circuit to a different voltage in a second circuit, or to partially isolate two circuits from each other [1]. Three-legged transformers are the most commonly used transformers in motor drive systems. For a motor drive system, some applications have a transformer between a drive and motor. For example as shown in Fig. 1, a motor drive with output filter and transformer are used for some applications like Electric Submersible Pumps (ESP) for oil pumps. In such an ESP drive system, an output transformer is commonly utilized to increase the low voltage drive output voltage to the levels required to power a medium voltage motor through a long cable.

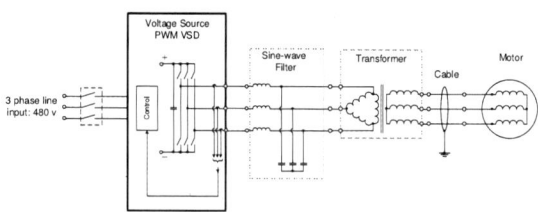

Figure 1.   An ESP drive with a three phase transformer

As is well known, a transformer will not pass DC voltages or currents to the motor. The magnetic saturation distorts the voltages applied to the motor. Meanwhile, magnetic saturation distorts the currents from the drive. Drive current and motor current can be very different. Therefore, motor control methods which use drive currents and/or drive voltages can fail. An accurate model of a transformer is crucial to the development of effective drive control algorithms.

The main challenges for three-phase transformer modeling have always been how to address three-phase core topologies and to determine nonlinear parameters [4]. There has been research effort focused on: coupled electric and magnetic equivalent circuits [1]-[3], nonlinear reluctances [6], hysteresis models [4]-[5], transient disturbances and DC bias [7]-[10], etc. Pedra et al. in [2] proposed a PSPICE model which utilizes a coupled electric and magnetic equivalent circuit. Although magnetic equivalent circuit models magnetic saturation and coupling between windings, flux from each leg are modeled to split equally between the other two legs [2]. Unfortunately, unequal splitting of flux is particularly hard to model and most existing transformer models in literature assume symmetric coupling. Furthermore, in most existing transformer models considering transformer nonlinearities is complicated. Those models might not be suitable for a high level motor drive system simulation.

A novel three-legged transformer SIMULINK model using a fictitious delta winding is presented in this paper. A fictitious delta winding is added to three single-phase saturable transformers. First, limitations of conventional 3-legged transformer models are summarized. Second, the proposed modeling strategy using a fictitious delta winding is analyzed. Third, single-phase parameters of transformers have been obtained from laboratory tests because accurate modeling of an E-Core three-phase transformer requires single-phase measurements of the transformer parameters for each leg. Last, experimental results with a 21-kVA 3-legged transformer are provided in the paper to validate the effectiveness of the proposed modeling method.

The proposed three-legged model is not only very simple but also fairly accurate to model the nonlinear characteristics of a saturable three-legged transformer including mutual coupling.

II. EXISTING MATLAB/SIMULINK TRANSFORMER MODELS

Fig. 2 illustrates the flux distribution in a 3-legged transformer. The inductance is calculated by: $L = \phi/I = N/R$. We define: Φ is flux, N is turns of winding, I is current, R is reluctance and L is inductance.

It can be seen that with only phase B excited, $\phi_a = \phi_c \approx -\frac{1}{2}$ ϕ_b. With only phase A excited, $\phi_b > \phi_c$. Fig.2 clearly shows

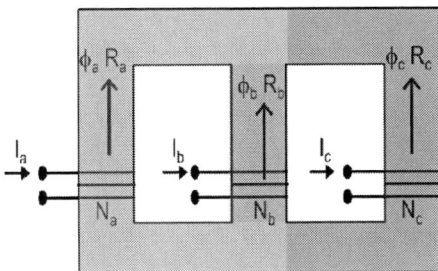

Figure 2. Flux distribution in 3-legged transformers

that the reluctance R_a equals R_c by symmetry, but R_b is smaller because the flux path is shorter.

This unequal splitting of flux in a 3-legged transformer is particularly difficult to model. Most transformer models in literature assume symmetric coupling. One major contribution of this paper is to provide a simple three-legged transformer model that adequately models the unequal splitting of flux.

A. SimPowerSystems™ Transformer Models

MATLAB/SIMLINK SimPowerSystems™ is one popular simulation tool that provides component libraries and analysis tools for modeling and simulating electrical power systems. The SimPowerSystems™ libraries offer models of electrical power components, including three-phase machines, flexible AC transmission systems, electric drives, and components for various applications, etc.

There are many models in SimPowerSystems which ignore magnetic saturation, such as models of grounding transformers, linear transformers, mutual inductance, three-phase transformer 12 terminals, three-phase transformer inductive matrix type (three windings), and three-phase transformer inductive matrix type (two windings).

On the other hand, SimPowerSystems have models which include magnetic saturation such as multi-winding transformer, saturable transformer (single phase model), three-phase transformer (three windings), three-phase transformer (two windings), and Zig-Zag phase shifting transformer. However, all models which include saturation are limited to one saturation curve per model. This is not adequate for 3-leg transformers.

B. Mutual Inductance Transformer Model

An inductance matrix can be calculated from voltage measurements. A known sinusoidal voltage is applied to one phase at a time and the open circuit voltages are measured on all other phases. Transformer voltage can be expressed in terms of mutual inductances, i.e.

$$[v] = [L]\frac{d}{dt}[i] \tag{1}$$

Where the inductance matrix is

$$[L] = \begin{bmatrix} L_{11} & L_{12} & \dots & L_{1N} \\ L_{21} & L_{22} & \dots & L_{2N} \\ \dots & \dots & \dots & \dots \\ L_{N1} & L_{N2} & \dots & L_{NN} \end{bmatrix} \tag{2}$$

The first step of the inductance matrix calculation is to measure open circuit voltages with single phase excitation. Each phase of the primary side and secondary is excited separately. Measure voltage and current of the phase being excited and also measure the induced voltages on all unexcited phases of the primary and secondary sides. For a 3-phase transformer with six windings, it requires a total of 30 voltage measurements to obtain the inductance matrix. Resistance matrix can be constructed from DC measurements of the resistance of each winding.

The second step of the inductance matrix calculation is to compute the coupling coefficient matrix [3]. Coupling measurements are made for each pair of windings. For example, the coupling coefficient (K_{ij}) can be calculated from test data:

$$k_{12} = \sqrt{\frac{v_{oc21}}{v_{d12}} \frac{v_{oc12}}{v_{d21}}} \tag{3}$$

Where v_{ocqr} is the voltage measured at winding q when winding r is driven. V_{drq} is the voltage measured at winding r when winding q is driven. Thus the mutual inductances are calculated by:

$$L_{12} = k_{12} \times \sqrt{L_{11}L_{22}} \tag{4}$$

Therefore the complete inductance matrix can be calculated from coupling coefficients and self-inductances. Conventional inductance measurement works well when the voltage and current are sinusoidal. However the transformer current is only sinusoidal for low levels of excitation. Once the transformer is saturated, it becomes difficult to measure the inductances because the current is hard to measure accurately. Another issue of the coupled inductor model is that it does not model magnetic saturation. The iron loss can be added by adding line-neutral resistors to the model. Nevertheless, the magnetic saturation has to be modeled by putting a saturating transformer model in parallel.

In summary, the coupled inductor model can model the unsaturated E-Core three-phase transformer. Magnetic saturation has to be modeled by putting the saturating transformer model in parallel. The coupled inductor model gives very good results at low voltages. However, the parameters of the coupled inductor model are extremely hard to measure. Small errors in parameters lead to an unstable model.

C. Inductance Matrix Transformer Model

Three-phase transformer inductance matrix type (two windings) notation is shown in Fig. 3. Three-phase winding 1 is composed of single-phase windings 1, 2, and 3. R1 is the resistance of single-phase winding 1, 2, or 3. We will assume three-phase winding 1 connects to a source of 3-phase electrical power. Similarly, three-Phase Winding 2 is composed of single-phase windings 4, 5, and 6. R2 is the resistance of single-phase winding 4, 5, or 6. Three-phase

978-1-4673-9551-9/16 $31.00 © 2016 IEEE

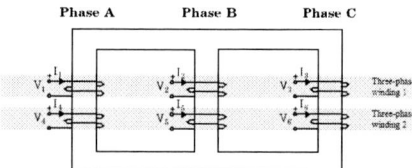

Figure 3 Three-phase transformer inductance matrix type (two windings) notations

winding 2 is assumed to connect to a load. Single-phase windings 1 and 4 are on the same leg of the transformer and designated as Phase A. Similarly, 2 and 5 are Phase B and 3 and 6 are Phase C.

Simulations based on the inductance matrix transformer model with a 21.5-kVA transformer (parameters shown in table 1) have been performed. Three-phase rated voltages are applied to three-phase windings of the transformer. All other windings remain open circuited. Measure RMS current and power in single-phase winding 1, 2, or 3. Simulation and corresponding test results are summarized in table 2.

Note that the zero sequence parameters required by the inductance matrix transformer model are not commonly used. It might take considerable effort to find references that described how to measure them.

TABLE I. SIMULATION PARAMETER SETS WITH THREE-PHASE TRANSFORMER INDUCTANCE MATRIX TYPE (TWO WINDINGS)INDUCTANCE MATRIX TYPE (TWO WINDINGS)

Parameter	21.5 kVA (Tested)
Connection	3-limb Y/Y
Nominal Power & Frequency (VA, Hz)	21.5e3 60
Nominal Line-Line Voltages (Vrms)	360 480
Winding Resistances (pu)	0.018 0.015
Positive-Sequence No-Load Excitation Current (% of Inom)	4.6
Positive-Sequence No-Load Losses (W)	133
Positive-Sequence Short-Circuit Reactance (pu)	0.036
Zero-Sequence No-Load Excitation Current (% of Inom)	322
Zero-Sequence No-Load Losses (W)	24500
Zero-Sequence Short-Circuit Reactance (pu)	0.29

TABLE II. THREE-PHSER TRANSFORMER INDUCTANCE MATRIX TYPE (TWO WINDINGS) RESULTS

Parameter	X1Y	X2Y	X3Y
Tested Current (Amps Pk)	0.15	0.096	0.15
Simulated Current (Amps Pk)	1.07	1.07	1.07
Tested Power (W)	7.01	5.30	9.43
Simulated Power (W)	10.06	10.06	10.06

Simulation results have shown that the three-phase transformer inductance matrix type (two windings) does not model the asymmetrical flux paths of a 3-leg (E-core) transformer. In other words, the coupling is not modeled correctly in the inductance matrix model.

III. NEW SATURABLE TRANSFORMER MODEL WITH A FICTITIOUS DELTA WINDING

Accurate modeling of an E-Core three-Phase transformer with three saturating transformer models requires two things: (i) single-phase measurements of the transformer parameters for each leg; and (ii) coupling between the transformer models used for each phase for unequal splitting of flux in a 3-legged transformer.

An example of a three-phase transformer modeled by three single-phase saturable transformers is shown in Fig. 4. Simulation results (omitted due to space limit) with the 3-phase transformer model in Fig. 4 have excellent match to the excitation current, but, almost no coupling between the three single-phase transformers. Therefore, three single-phase saturable transformers in Fig. 4 provide a good base model, however coupling needs to be considered.

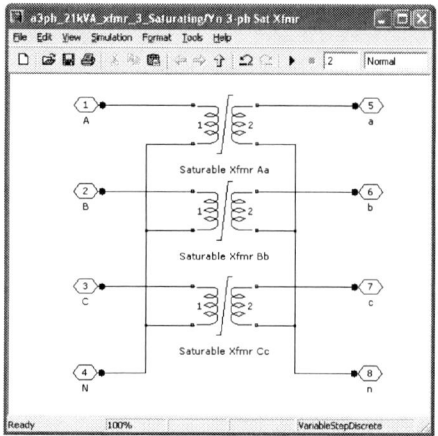

Figure 4. Three-phase Yn/Yn transformer modeled by three single-phase saturable transformers

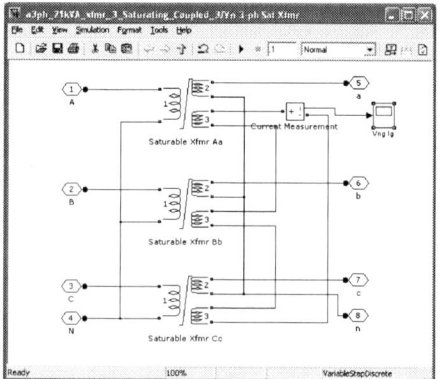

Figure 5. New three-phase transformer model: flux can be coupled through a fictitious delta winding

In this paper, we propose a new simple three-legged transformer model that improves the accuracy of the model to unequal splitting of flux through a fictitious delta winding. A fictitious delta winding is added to a base model which has three single-phase saturable transformers. As shown in Fig.5, a third winding (winding 3) is added to each transformer in Fig. 4 and all three windings are connected in series. Each winding 3 has the same rated voltage and very small impedance (for example 1e⁻⁶ Ω). Through a fictitious delta winding, current flows to always make the sum of the three voltages zero (voltage drop across impedance is less than 1 mV). When the sum of voltages is zero, the sum of fluxes in the three transformers is close to zero. Therefore the fictitious delta winding provides a simple solution to address the asymmetrical coupling between windings in the three-legged transformer model.

IV. EXPERIMENTAL AND SIMULATION RESULTS

An experimental setup as shown in Fig. 6 is utilized to verify the proposed transformer model using the fictitious delta winding. A 21 kVA , 360V/480V , Y/Y , 3-legged transformer is used and a 1-km long cable is short circuited. Parameters of the 21-kVA transformer are shown in Table 1. A drive which has closed loop current control determines the magnitude and frequency of currents flowing into the primary side of the transformer.

MATLAB/SIMULINK simulations have been performed to compare with the experimental results. In particular, a three-legged transformer is modeled using a saturable transformer model with a fictitious delta winding as shown in Fig. 7. The long cable is modeled as a resistor to simplify the system simulation, especially at low excitation frequencies. The data entry for each saturable transformer in SIMULINK is obtained

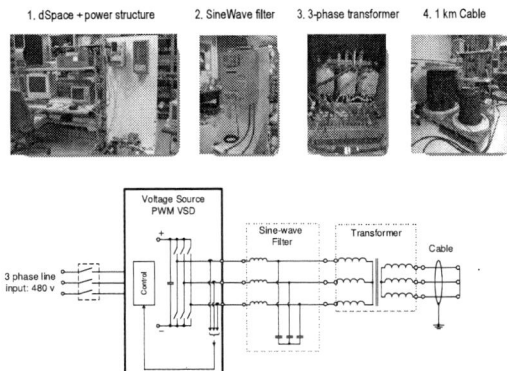

Figure 6.   Experimental setup: short circuit long cable, 21 kVA transformer Y/Y

via open circuit test, short circuit and DC resistance test, respectively.

Experimental results with transformer secondary short circuited and primary current reference at 1Hz and 1A are shown in Fig. 8 and Fig. 9, respectively.   Similarly, experimental results with transformer secondary short circuited and primary current reference at 1Hz & 10A, 8Hz & 25A, 12Hz & 20A, and 16Hz & 30A are shown in Fig. 10 through Fig. 17, respectively.

It is clear that the proposed model with a fictitious delta winding gives a decent match to experimental data.

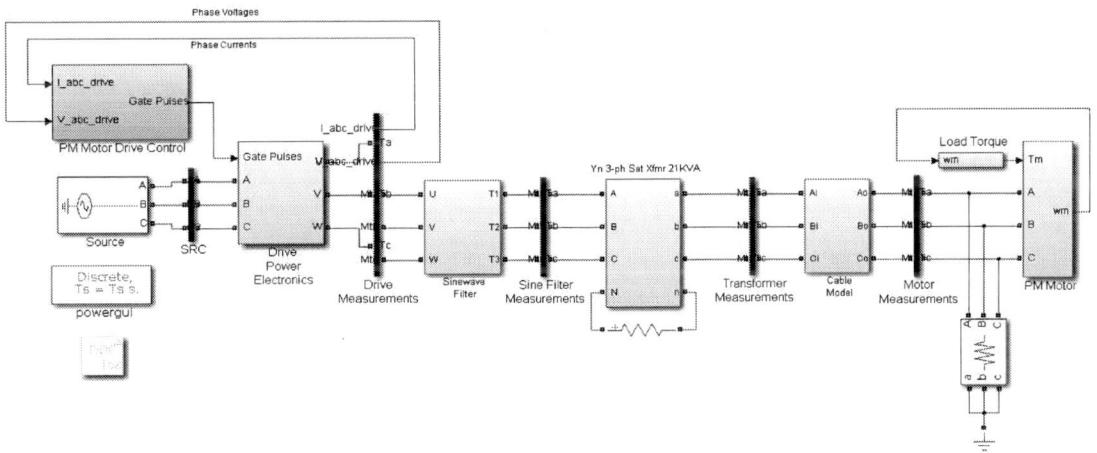

Figure 7.   Simulation model: short circuit long cable (motor disconnected), 21 kVA transformer Y/Y using a fictitious delta winding

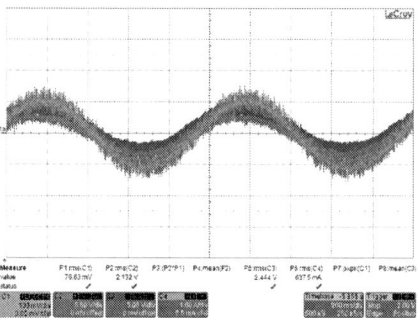

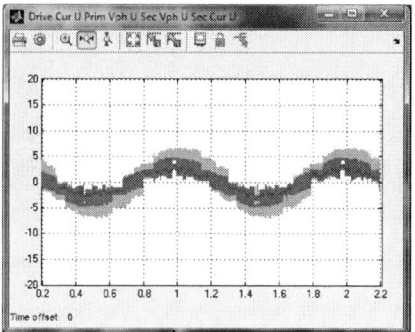

Figure 8. Experimental results with transformer secondary short circuited and primary current reference at 1Hz, 1A, Ch1: drive phase u current, 1A/div; Ch2: transformer primary voltage (L-N, phase U), 5V/div; Ch3: transformer secondary voltage (L-N, phase U), 5V/div; Ch4: transformer secondary current (phase u), 1A/div

Figure 9. Simulation results with transformer secondary short circuited and primary current reference at 1Hz, 1A, Ch1: drive phase u current, 1A/div; Ch2: transformer primary voltage (L-N, phase U), 5V/div; Ch3: transformer secondary voltage (L-N, phase U), 5V/div; Ch4: transformer secondary current (phase u), 1A/div

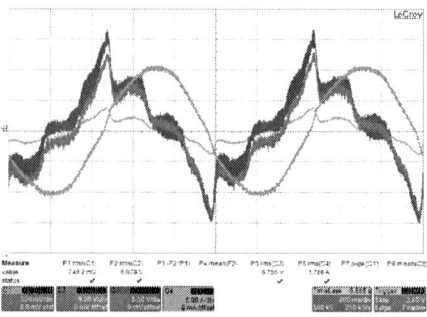

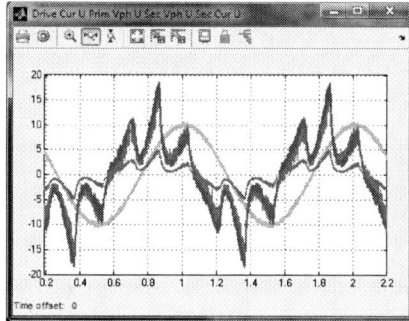

Figure 10. Experimental results with transformer secondary short circuited and primary current reference at 1Hz, 10A, Ch1: drive phase u current, 5A/div; Ch2: transformer primary voltage (L-N, phase U), 5V/div; Ch3: transformer secondary voltage (L-N, phase U), 5V/div; Ch4: transformer secondary current (phase u), 5A/div

Figure 11. Simulation results with transformer secondary short circuited and primary current reference at 1Hz, 10A, Ch1: drive phase u current, 5A/div; Ch2: transformer primary voltage (L-N, phase U), 5V/div; Ch3: transformer secondary voltage (L-N, phase U), 5V/div; Ch4: transformer secondary current (phase u), 5A/div

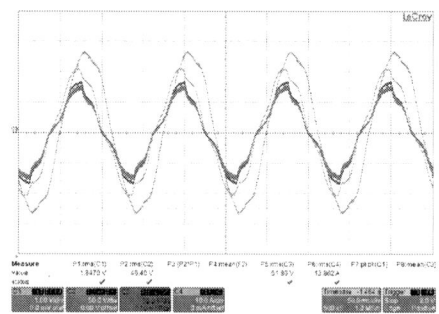

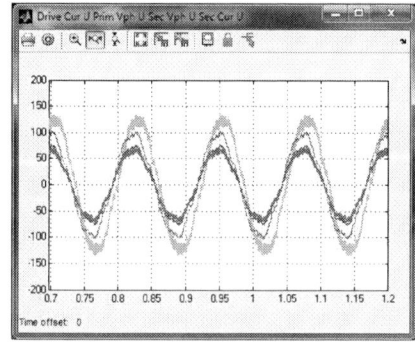

Figure 12. Experimental results with transformer secondary short circuited and primary current reference at 8Hz, 25A, Ch1: drive phase u current, 10A/div; Ch2: transformer primary voltage (L-N, phase U), 50V/div; Ch3: transformer secondary voltage (L-N, phase U), 50V/div; Ch4: transformer secondary current (phase u), 10A/div

Figure 13. Simulation results with transformer secondary short circuited and primary current reference at 8Hz, 25A, Ch1: drive phase u current, 10A/div; Ch2: transformer primary voltage (L-N, phase U), 50V/div; Ch3: transformer secondary voltage (L-N, phase U), 50V/div; Ch4: transformer secondary current (phase u), 10A/div

978-1-4673-9551-9/16 $31.00 © 2016 IEEE

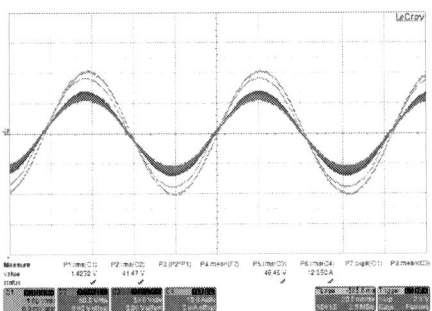

Figure 14. Experimental results with transformer secondary short circuited and primary current reference at 12Hz, 20A, Ch1: drive phase u current, 10A/div; Ch2: transformer primary voltage (L-N, phase U), 50V/div; Ch3: transformer secondary voltage (L-N, phase U), 50V/div; Ch4: transformer secondary current (phase u), 10A/div

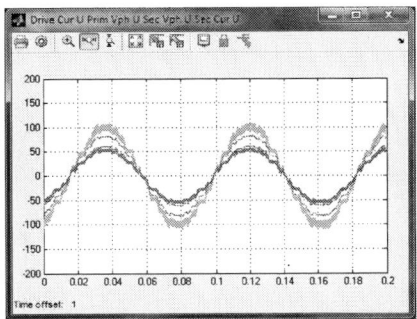

Figure 15. Simulation results with transformer secondary short circuited and primary current reference at 12Hz, 20A, Ch1: drive phase u current, 10A/div; Ch2: transformer primary voltage (L-N, phase U), 50V/div; Ch3: transformer secondary voltage (L-N, phase U), 50V/div; Ch4: transformer secondary current (phase u), 10A/div

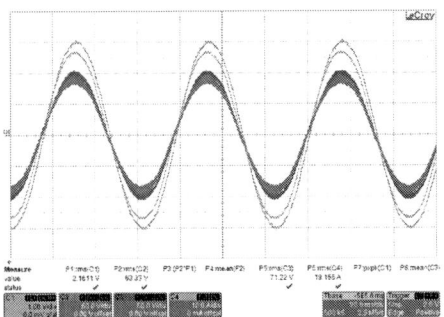

Figure 16. Experimental results with transformer secondary short circuited and primary current reference at 16Hz, 30A, Ch1: drive phase u current, 10A/div; Ch2: transformer primary voltage (L-N, phase U), 50V/div; Ch3: transformer secondary voltage (L-N, phase U), 50V/div; Ch4: transformer secondary current (phase u), 10A/div

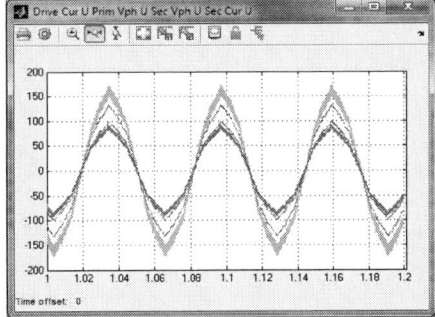

Figure 17. Simulation results with transformer secondary short circuited and primary current reference at 16Hz, 30A, Ch1: drive phase u current, 10A/div; Ch2: transformer primary voltage (L-N, phase U), 50V/div; Ch3: transformer secondary voltage (L-N, phase U), 50V/div; Ch4: transformer secondary current (phase u), 10A/div

Although a fictitious short-circuited delta winding greatly improves the accuracy of the simulated waveforms, it is discovered by simulations that the model with three uncoupled transformers fail to simulate the neutral to ground voltage properly. This explains the minor discrepancies between the simulation results and experimental results.

V. CONCLUSIONS

In general, accurate modeling of an E-Core three-phase transformer with three saturating transformer models requires two things: (i) single-phase measurements of the parameters for each leg; and (ii) coupling between the transformer models used for each phase.

This paper presents a novel three-legged transformer model using a fictitious delta winding. A fictitious delta winding is added to model unequal splitting of flux in a 3-legged transformer. The proposed model is not only very simple but also accurately models the nonlinear characteristics of a saturable three-legged transformer including mutual coupling.

Experimental results have verified that using a fictitious short-circuited delta winding greatly improves the accuracy of the simulated waveforms.

ACKNOWLEDGMENT

Thanks to Mr. Christian Höft, for his work during his internship on the 3-legged transformer modeling.

REFERENCES

[1] http://www.denverpels.org/Downloads/Denver_PELS_20070410_Hesterman_Magnetic_Coupling.pdf

[2] J. Pedra, L. Sainz, F. Córcoles, R. Lopez, and M. Salichs, "PSPICE Computer Model of a Nonlinear Three-Phase Three-Legged Transformer", IEEE Trans. on Power Delivery, Vol. 19, No. 1, pp. 200-207, Jan. 2004

[3] G. W. Ludwig, and S.-A. El Hamamsy "Coupled Inductance and Reluctance Models of Magnetic Components," IEEE Trans. on Power Electronics, Vol. 6, No. 2, pp. 240-250, April 1991

[4] P. S. Moses, M. A. S. Masoum, and H. A. Toliyat, "Dynamic Modeling of Three-Phase Asymmetric Power Transformers With Magnetic Hysteresis: No-Load and Inrush Conditions", IEEE Trans. on Engergy Conversion, Vol. 25, No. 4, pp. 1040-1046, Dec. 2010

[5] A. D. Theocharis, J. Milias-Argitis, and T. Zacharias, "Three-phase transformer model including magnetic hysteresis and eddy currents effects," IEEE Trans. Power Del., vol. 24, no. 3, pp. 1284–1294, Jul. 2009.

[6] X. Chen, "A three-phase multi-legged transformer model in ATP using the directly-formed inverse inductance matrix," IEEE Trans. Power Delivery, vol. 11, pp. 1554–1562, July 1996.

[7] C. M. Arturi, "Transient simulation and analysis of a three-phase five-limb step-up transformer," IEEE Trans. Power Delivery, vol. 6, pp. 196–207, Jan. 1991.

[8] A. Narang and R. H. Brierley, "Topology based magnetic model for steady-state and transient studies for three-phase core type transformers," IEEE Trans. Power Syst., vol. 9, pp. 1337–1349, Aug. 1994.

[9] E. F. Fuchs, Y. You, and D. J. Roesler, "Modeling and simulation, and their validation of three-phase transformers with three legs under DC bias," IEEE Trans. Power Delivery, vol. 14, pp. 443–449, Apr. 1999.

[10] X. Li, X. Wen, P. N. Markham and Y. Liu, "Analysis of Nonlinear Characteristics for a Three-Phase, Five-Limb Transformer Under DC Bias", IEEE Trans. Power Delivery, vol. 25, No. 4, pp. 2504-2510, Oct. 2010

A Lifetime Prediction Method for LEDs Considering Mission Profiles

Xiaohui Qu[1,2], Huai Wang[2], Xiaoqing Zhan[3], Frede Blaabjerg[2], and Henry Shu-Hung Chung[3]

[1] School of Electrical Engineering, Southeast University, China.
[2] Center of Reliable Power Electronics (CORPE), Department of Energy Technology, Aalborg University, Denmark.
[3] Center for Smart Energy Conversion and Utilization Research, City University of Hong Kong, Hong Kong.
Email: xhqu@seu.edu.cn

Abstract—**Light-Emitting Diodes (LEDs) has become a very promising alternative lighting source with the advantages of longer lifetime and higher efficiency than traditional ones. The lifetime prediction of LEDs is important to guide the LED system designers to fulfill the design specifications and to benchmark the cost-competitiveness of different lighting technologies. The existing lifetime data released by LED manufacturers or standard organizations are usually applicable only for specific temperature and current levels. Significant lifetime discrepancies may be observed in field operations due to the varying operational and environmental conditions during the entire service time (i.e., mission profiles). To overcome the challenge, this paper proposes an advanced lifetime prediction method, which takes into account the field operation mission profiles and the statistical properties of the life data available from accelerated degradation testing. It identifies also the key variables (e.g., heat sink parameters and lifetime-matching of LED drivers) that can be designed to achieve a specified lifetime and reliability level. Two case studies of an indoor residential lighting and an outdoor street lighting application are presented to demonstrate the prediction procedures and the impact of different mission profiles on the lifetime of LEDs.**

Index Terms—**LED lighting, lifetime prediction, reliability, mission profile.**

I. INTRODUCTION

Power Light-Emitting Diodes (LEDs) are increasingly applied for indoor and outdoor lighting applications due to their higher efficiency and longer lifetime compared to the traditional lighting sources. The lifetime of LED lamps involving LED drivers and source packages is routinely quoted as 25,000 to 50,000 hours in the market [1], [2]. However, the customer experiences may be different and some of the LED lamps can fail in a considerable time ahead of the claimed life [3]. The failure could be induced either by the LED drivers or by the LED packages. The discrepancies between the claimed lifetime and the field operation experiences are mainly due to the following reasons:

1) The definition of the specified lifetime of LED lamps is vague. A necessary lifetime definition should include at least four aspects: a) operation conditions; b) end-of-life criteria; c) required minimum reliability at the

end of the specified lifetime; d) confidence level of the specified lifetime.

2) The field environmental and operational conditions may vary within the specifications of the LED lamps, or even exceeding the specifications for severe users.

3) The lifetime mismatch between the LED drivers and the LED packages may occur. Sometimes, the lifetime of LED packages is misused as the claimed lifetime of LED lamps.

4) The thermal design of LED lamps is one of critical aspects that affect the lifetime of both drivers and sources.

Besides of the catastrophic failure due to the sudden breakdown of LED drivers, the two main concerns of LED failures are of source packages with the lumen depreciation and color shift due to degradation. The level of lumen depreciation is usually used as an end-of-life criteria. For color quality critical applications, the color shift level is also used as an additional criteria. Fig. 1 illustrates the definition of the lifetime L_p of LED individuals. For example, L_{70} is the time when the lumen is maintained at 70% of its initial value. With a more stringent requirement on lumen maintenance, the lifetime is shortened (e.g., L_{90} is less than L_{70} for a specific application). Currently, the L_{70} or L_{85} criteria are usually used for commercial and residential outdoor application and L_{90} is for residential indoor application. In some applications without the stringent lumen requirement, L_{50} is also taken into account and used as a design criteria.

It is well known that the L_p lifetime varies among LED samples with the same part number from the same manufacturer due to the variances in materials, process control, etc. Therefore, the percentile lifetime B_X for a population of LEDs is of more interest. Fig. 2 shows the definition of B_X lifetime based on the required minimum reliability at the end of life. For example, B_{10} lifetime means the time when 10% of the LEDs fail (i.e., with a reliability $R = 0.9$), and B_1 lifetime means the time when 1% of the LEDs fail. Accordingly, L_p/B_X lifetime refers to the time when $X\%$ of the LEDs have the lumen output below $p\%$ with respect to its initial values. The choices of p and X are application-dependent. Moreover, for the data not to be normally distributed, the

This work is supported by the National Science Foundation of China (Grant No. 51107009) and China Scholarship Council.

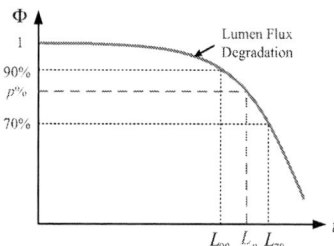

Fig. 1. L_p lifetime for LED individuals. L_p is defined as the time when $p\%$ of the initial output lumen of an LED is maintained.

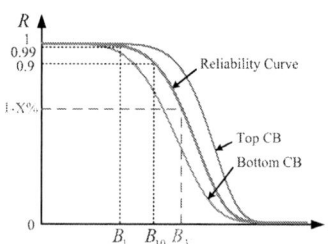

Fig. 2. B_X lifetime for a population of LEDs. B_X is defined as the time when $X\%$ of the LEDs have the lumen output below $p\%$ of their initial values.

median rank method is usually used in the reliability curve plotting to define the cumulative percentage of the population with a 50% confidence level. It is also possible to obtain the reliability range with a certain Confidence Bounds (CBs) as shown in Fig. 2. For example, the 2-sided 90% CBs have the top CB and the bottom CB curve to provide 5% and 95% confidence respectively.

The claimed long lifetime L_p is usually not measured all the time. The Illuminating Engineering Society (IES) [4] specifies the measurement procedures of lumen maintenance in the industry standard IES LM-80 [5] only for LED light sources, where a minimum 6,000 hours testing is carried out with enough sample populations and data. Most tests are provided by the LED manufacturers and last below 10,000 hours. Then longer lifetime is predicted by projecting the average data measured in IES LM-80 via an exponential curve-fitting extrapolation described in standard IES TM-21 [6]. Currently, the IES LM-80 test report is widely required for all LED products in the market to facilitate the assessment of LED performance and to determine the life expectancy of their products. However, the testing in this report is usually performed under several specific conditions, that are typical constant driving currents and at least three cases of ambient temperatures (55°C, 85°C and one selected by the manufacturers). In practise, the driving current depends on the user profiles and driving schemes. The ambient temperature may vary with time daily, monthly or even yearly and also depends on the place location in the world. Therefore, there are still gaps between the degradation testing data and the practical

applications as follows:

1) The specific reliability information (with a certain confidence level or confidence bounds) and the corresponding lifetime model are not readily available. A comprehensive analysis on those testing data is needed.

2) The mapping of the reliability under the specific accelerated testing conditions to the lifetime under field conditions (i.e., long-term mission profiles) is missing.

To overcome the above issues, this paper proposes an advanced lifetime prediction method concerning the long-term field operation mission profiles and the statistical properties of the life data available from accelerated degradation testing. With this method, some key variables for thermal design and lifetime-matching of LED drivers in the different field condition can easily be identified to achieve a specified lifetime and reliability level. Two case studies of an indoor residential lighting application and an outdoor street lighting application are presented to demonstrate the prediction procedures and the impact of different mission profiles on the lifetime of LEDs. The proposed method can also be extended to the prediction of the LED drivers and the entire LED lighting systems.

In this paper, Section II introduces a comprehensive lifetime model involving the statistical properties of life data and operational mission profiles. Based on this model, an advanced lifetime prediction method is then detailed in Section III. Two study cases are demonstrated and evaluated the performance of the proposed method in Section IV. Section V concludes the paper.

II. LIFETIME MODELS

Since LEDs are basically p-n junctions, the emitted lumen flux and intensity are proportional to the carriers' concentration [7]. The carrier's concentration depends on current density and junction temperature, which results in LED output lumen, color chromaticity and the forward voltage characteristics also varying with these two stresses. Hence, a generally accepted Black model in (1) can be used to describe the time to failure under different stresses [8], [9].

$$\text{Time to failure} = A_0 J^{-n} e^{\frac{E_a}{k_B T}}, \tag{1}$$

where A_0 is a constant, J is the current density, n is a scaling factor, E_a is the activation energy in unit of eV, k_B is the Boltzmann constant, and T is the absolute temperature in Kelvin.

The model shown in (1) describes the impact of current and temperature on the lifetime of LEDs. Therefore, L_p lifetime, defined as time-to-failure for LED individuals, can use this model. B_X lifetime based on a population of L_p lifetime data can also follow this equation to specify the reliability of an LED population. The parameters of A_0, n and E_a are usually obtained according to accelerated testing data. For an LED population with the same product series and number, the identical physical materials and degradation mechanism make the factors n and E_a constant under different stresses and end-of-life criteria. Hence, (1) can be rearranged as (2) and (3).

$$L_p(I_F, T_J) = A_p I_F^{-n} e^{\frac{E_a}{k_B T_J}}, \text{ and} \tag{2}$$

$$B_X(I_F, T_J) = A_X I_F^{-n} e^{\frac{E_a}{k_B T_J}}, \tag{3}$$

where I_F is the LED driving current proportional to current density, and T_J is the junction temperature of LEDs. Although A_p and A_X are dependent of the different L_p and B_X criteria, (2) and (3) have the same Acceleration Factors (AF).

$$AF(n, E_a) = \left(\frac{I_F}{I_{F0}}\right)^{-n} \cdot e^{\frac{E_a}{k_B}\left(\frac{1}{T_J} - \frac{1}{T_{J0}}\right)}, \tag{4}$$

and (I_{F0}, T_{J0}) is the initial stress level, whilst (I_F, T_J) is the accelerated stress level. To solve factors of n and E_a in (4), at least three different stress levels are required.

With this information, a study case based on an LM-80 test report [10] for Lumileds Luxeon Rebel LEDs [11] will show how to estimate the models in (2) and (3). Advanced Weibull distribution is the most widely used technique to process the lifetime data and then adopted here to analyze the reliability information. The report [10] provides multiple test conditions with stress levels of I_F from 0.35 A, 0.5 A, 0.7 A, to 1 A and air temperature T_a from 55°C, 85°C, 105°C, to 120°C. Each test condition has 25 samples with at least 9,000 hours testing. To solve n and E_a, three test conditions are randomly selected as shown in Fig. 3 and process the data using software tool *ReliaSoft*. With the exponential extrapolation curves, the L_{70} and L_{90} criteria can be applied to generate two groups of L_p lifetimes easily for each stress level using *Weibull++* degradation. These L_p lifetime data can be arranged in sequence and then ranked by the algebraic approximation of the Median rank in (5) [12].

$$\text{Median rank } r_j = \frac{j - 0.3}{N + 0.4}, \tag{5}$$

where j is the order number of the sequenced L_p data, $j \in [1, N]$ and N is the number of failure (i.e., the size of L_p data). The rank r_j is actually the probability to failure for the jth LED. With these ranks and corresponding L_p group at one stress level, the probability line for this stress level can be generated via *ALTA* (Accelerated Life Testing Analysis) degradation. Figs. 4 (a) and (b) illustrate the unreliability function $F(t)$ (i.e., probability to failure) at each operating stress level by median rank with 50% confidence level. Therefore, B_X satisfying $F(B_X) = X\%$ can easily be obtained in these lines. The probability lines follow the two parameter Weibull distribution and the cumulative failure $F(t)$ is described as

$$F(t) = 1 - R(t) = 1 - e^{-\left(\frac{t}{\eta}\right)^\beta}, \tag{6}$$

where t is time, β is the shape parameter, and η is the scale parameter of characteristic life $B_{63.2}$ (i.e., the life at which 63.2% of the tested samples fail) at each stress condition. For wear-out failure, $\beta > 1$. With the same failure mechanism, β is assumed constant under different stress levels within the physical limits [12]. In Figs. 4 (a) and (b), six well fitted curves

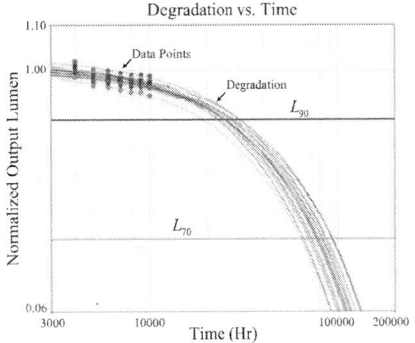

(a) I_F=0.35 A, T_a=120°C, and T_J=129°C

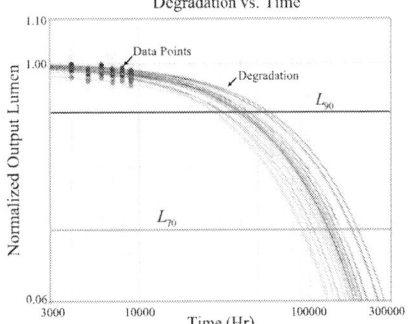

(b) I_F=0.7 A, T_a=55°C, and T_J=74°C

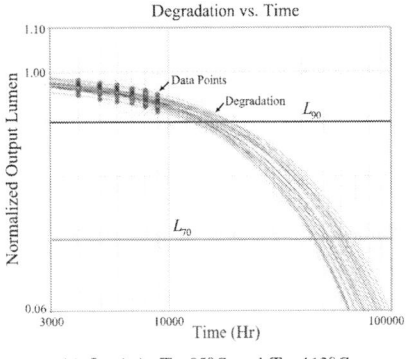

(c) I_F=1 A, T_a=85°C, and T_J=112°C

Fig. 3. Lumen degradation curves with L_{70} and L_{90} lifetime criteria under different LED operating conditions of (a), (b) and (c). X-axis is time in hour and Y-axis is the normalized output lumen to the initial lumen at t=0.

show good consistence on β, n and E_a, where small errors are caused by the distribution variation.

With known n and E_a in each figure, substitute any B_X value at one stress (I_F, T_J) into (3), A_X can be solved. Here, B_{10} and B_1 distributions based on L_{70} and L_{90} criteria are given in (7) and Fig. 5 as examples, which are valuable for further mapping the reliability information under field

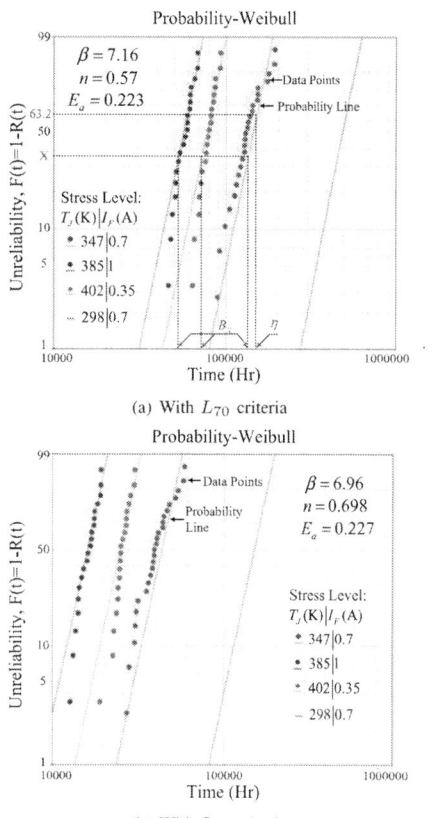

(a) With L_{70} criteria

(b) With L_{90} criteria

Fig. 4. Unreliability curves under different stress levels based on the L_p data given in Fig. 3. Y-axis is in %.

$$\begin{cases} \ln B_{10\_L70}(I_F, T_J) = 3.956 - 0.57 \ln I_F + \dfrac{2588}{T_J} \\[2mm] \ln B_{1\_L70}(I_F, T_J) = 3.628 - 0.57 \ln I_F + \dfrac{2588}{T_J} \\[2mm] \ln B_{10\_L90}(I_F, T_J) = 2.558 - 0.698 \ln I_F + \dfrac{2636}{T_J} \\[2mm] \ln B_{1\_L90}(I_F, T_J) = 2.221 - 0.698 \ln I_F + \dfrac{2636}{T_J} \end{cases} \quad (7)$$

III. PROPOSED MISSION-PROFILE BASED LIFETIME PREDICTION METHOD

From Fig. 5, I_F and T_J are the key factors to affect the lifetime and reliability in the LED lighting applications. In the field operation, LED driving current I_F depends on the user profiles (e.g., indoors or outdoors occasion for different lumen requirements, periodical operational hours per day, month or year, etc.) The control schemes including short-term dimming with a period of several milliseconds also affect I_F and then the lifetime and reliability. The junction temperature T_J, which is easily affected by the ambient temperature T_A, power loss in

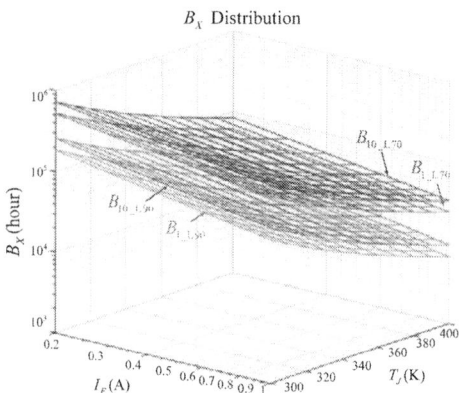

Fig. 5. B_X distributions versus forward current I_F and junction temperature T_J.

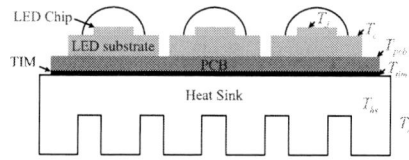

Fig. 6. The typical LED package structure.

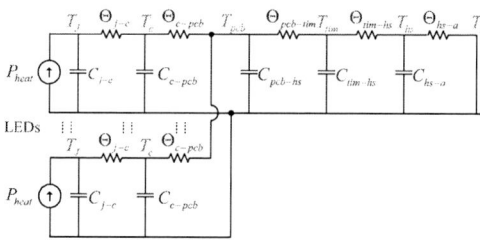

Fig. 7. The equivalent LED thermal circuit using Cauer model.

chip and thermal distribution of materials, cannot be measured directly due to the manufacture process. Although there are some methods to reflect T_J using easy-measured electrical variables, the online signal sensing, feedback, translation and calibration are costly and complicated in the long-term LED lifetime prediction [13]. Here, an accurate LED thermal model is more suitable to estimate T_J incorporating the complicated mission profiles.

A. LED Thermal Model

It is known that most energy of LEDs is converted to heat and the left energy is converted into visible radiant light energy. The heat in the LED die can be diffused only by heat sink or external cooling whereas the conventional light sources can emit most heat by radiation. Fig. 6 shows a typical LED package structure, where multiple LED dies are soldered on the same PCB substrate for electrical connection,

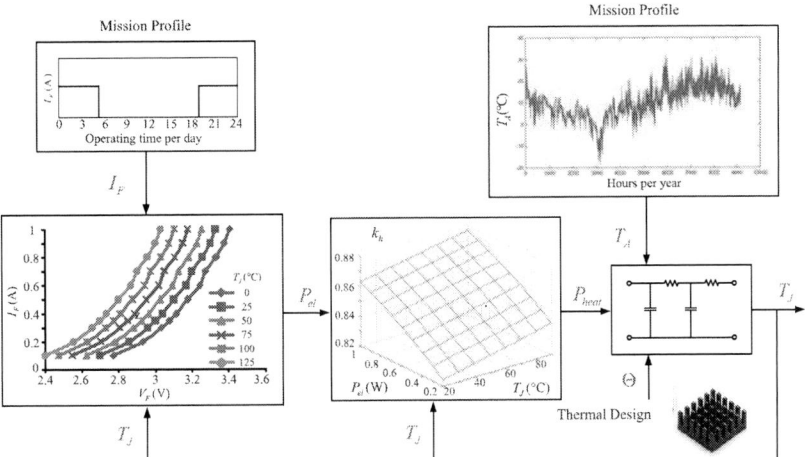

Fig. 8. The real-time thermal feedback model system of the LEDs. The mission profiles are for an outside street LED lamp located at Aalborg, Denmark. The VI relation and k_h are measured for Lumiled Rebel White LEDs as an example [11].

and the PCB is then usually attached to a heat sink for further heat conduction. To maximize the heat transfer between the heatsink and PCB, a good Thermal Interface Material (TIM) is needed to fill the air voids. Heat contributed to the PN junction, die case, cladding layer and heatsink can be described in Fig. 7 using Cauer thermal model, where P_{heat} is the dissipated heat power by each LED die, Θ is the thermal resistor of each layer in $\frac{°C}{W}$ and C is the thermal capacitor in $\frac{J}{°C}$. Both Θ and C are determined by the material and geometry of this layer.

Assuming there are m LEDs series connected in one package and having uniform heat dissipation performance, it follows (8) in the steady state.

$$
\begin{aligned}
T_J(T_A, P_{\text{heat}}, \Theta_{\text{j-a}}) &= T_A + P_{\text{heat}} \cdot \Theta_{\text{j-a}} \\
&= T_A + P_{\text{heat}} \cdot (\Theta_{\text{j-c}} + \Theta_{\text{c-pcb}} \\
&\quad + m\Theta_{\text{pcb-tim}} + m\Theta_{\text{tim-hs}} + m\Theta_{\text{hs-a}}),
\end{aligned} \tag{8}
$$

where $\Theta_{\text{j-c}}$ comes from the LED instinct characteristics, $\Theta_{\text{c-hs}}$ and $\Theta_{\text{hs-a}}$ are decided by designers. P_{heat} is caused by non-radiative electron-hole recombination and counts for a large proportion k_h of the input electrical power P_{el}. P_{el} is the product of driving current I_F and diode forward voltage V_F. Thus,

$$
P_{\text{heat}} = k_h P_{\text{el}} = k_h I_F V_F. \tag{9}
$$

As a semiconductor p-n junction, the VI-characteristic follows

$$
I_F = I_S \left(e^{\frac{eV_F}{Nk_B T_J}} - 1 \right). \tag{10}
$$

k_h is not constant for a single LED, varying slighting with T_J and P_{el} [14]. As P_{el} is also the function of I_F and T_J, P_{heat} can be calculated as

$$
P_{\text{heat}}(I_F, T_J) = k_h(I_F, T_J) \cdot I_F \cdot V_F(I_F, T_J). \tag{11}
$$

B. Acquisition of Operation Points

Comparing (11) and (8), a real-time feedback system is readily generated incorporating the thermal design and mission profiles of T_A and I_F to acquire the accurate operation point as shown in Fig. 8. The mission profiles of I_F and T_A describe the field operation condition of LEDs. Here, it is a typical example of an outside street lamp, whose ambient temperature profile is collected from yearly recorded data (1 hour per sampling data) located at the Aalborg University, Denmark. The street lamp is working at a constant driving current every day from 19:00 pm to the next day 5:00 am continuously. Dimming is not considered here. The relations of LED VI-curve, the dissipated heat coefficient k_h to I_F and T_J, and the thermal resistance of each layer can be obtained from the manufacture datasheets. For accuracy, it is better to measure them experimentally. $k_h = \frac{P_{\text{heat}}}{P_{\text{el}}} = \frac{1 - P_{\text{opt}}}{P_{\text{el}}}$, where P_{opt} is the radiated optical power and can be measured by Thermal Transient Tester (T3Ster). The thermal resistance $\Theta_{\text{j-c}}$ of LED chip, $\Theta_{\text{c-pcb}}$, $\Theta_{\text{pcb-tim}}$, $\Theta_{\text{tim-hs}}$ and $\Theta_{\text{hs-a}}$ of each layer can be calculated from FEM simulation or experimental measurement from T3Ster as shown in Section IV. So far, the real-time T_J in one period of the mission profiles can be acquired by the system shown in Fig. 8.

C. Mapping and Evaluation of Reliability

With a designated reliability $(1 - X\%)$, the B_X lifetime at a certain operating point of (I_F, T_J) can be calculated in (3). Mapping all the operating points in one mission profile period will produce a considerable amount of B_X data. How to evaluate the reliability of such products with different B_X in one period? Here, the consumed B_X lifetime in one period is defined in (12), based on the Palmgren-Miner linear

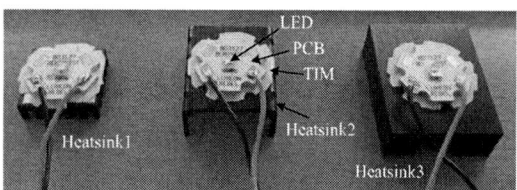

Fig. 9. The experimental LED setups for validating model.

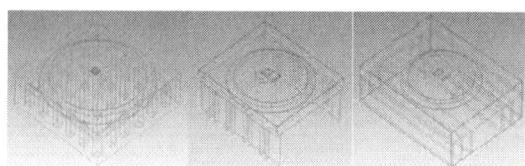

Fig. 10. The FEM simulation models with heatsinks 1, 2, 3 in Fig. 9.

cumulative damage model [15].

$$CL = \sum_{i=1}^{k} \frac{t_i}{B_{Xi}}, \qquad (12)$$

where k is the size of different stress levels, t_i is the duration of the accumulated at stress $(I_F, T_J)_i$ and B_{Xi} is the B_X lifetime at stress $(I_F, T_J)_i$ (e.g., for the case of Fig. 8, $k=24\times365=8760$ hours, $t_i=1$ hour, and B_{Xi} is calculated by (3)). A larger CL means the LED is more close to failure. When CL reaches 1, failure will occur. On the contrast, assuming the stress levels in the whole period are uniform, then the average B_X lifetime considering these periodic mission profiles can be defined as

$$B_{X\_avg}(\text{hour}) = \frac{\text{Period(hour)}}{CL}. \qquad (13)$$

Thus, the B_{X_avg} is the lifetime when $X\%$ of LEDs fail under the specified mission profiles.

D. Instruction to Thermal Design

With a given thermal design, B_{X_avg} can be calculated in (12) and (13) for LED quality assessment and lifetime-matchable driver design. With enough thermal data, a B_{X_avg} versus Θ_{hs-a} lookup table can be built. Consequently, for a designated reliability requirement, the maximum thermal resistor of the heat sink can be determined. A design example will be demonstrated in the Section. IV.

IV. DESIGN AND VALIDATION

To demonstrate the prediction procedures and the impact of different heat sinks and mission profiles on the lifetime of LEDs, two study cases for an indoor residential lighting application and an outdoor street lighting application at the two cities Aalborg in Denmark and Singapore with very different ambient temperatures are discussed here. The same Lumiled Rebel white LED [11] mounted on the three different heat sinks is used as shown in Fig. 9. The PCB substrate uses thermally conductive insulated metal substrate from Berquistand [16] and TIM uses very thin thermal pad from t-Global Technology [17]. The VI-curve and k_h are measured and given in Fig. 8.

A. Determination of Thermal Resistance

With the material characteristics and geometry of the LED package and heat sink, the model using Finite Element Method (FEM) simulation can be constructed in Fig. 10. The temperature distribution in the LED package with heatsink 1 can be

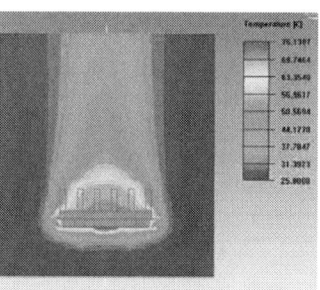

Fig. 11. Example temperature distribution in LED package with heatsink 1.

seen as example in Fig. 11. The other two models have the similar distribution. Because $\Theta = \frac{\Delta T}{P_{\text{heat}}}$, Θ of each layer can be calculated. Here, Θ_{j-hs} and Θ_{hs-a} for three cases are listed in Tab. I. With the same LED package, the calculated Θ_{j-hs1}, Θ_{j-hs2} and Θ_{j-hs3} are close and verify the simulation results. These thermal resistances can be used to predict junction temperature of LEDs.

B. Indoor Residential Lighting Case

The indoor residential lighting works in nearly constant room temperature, $22°C$ in the cold half year and $26°C$ in the warm half year from 19:00 pm to 24:00 pm every day. L_{90} critria runs here. For the lumen requirement, I_F is designed as 0.35 A and 18 LEDs work together. To simplify the simulation model, 18 LEDs are assumed to mount on a single heatsink. Then the equivalent Θ_{hs-a} should be $\Theta_{hs-a1,2,3}/18$ for heatsink 1,2,3. With three different Θ_{hs-a}, the average lifetime B_{10_avg} can be predicted following the flow-chart in Fig. 8 and Eq. (12),(13). With more Θ_{hs-a}, the B_{X_avg} versus Θ_{hs-a} can be drawn in Fig. 12. Obviously, the better heat radiation performance will make the LED to work longer If a minimum 50,000 hours lifetime with 90% reliability is required, the heat sink with a maximum Θ_{hs-a} of 2.96 °C/W is required and heatsinks 1,2,3 can satisfy the requirement.

C. Outdoor Street Lighting Case

Unlike the indoor application, the street lighting endures the variable outdoor temperatures at different locations. The mission profile in Fig. 8 shows the yearly ambient temperature in Aalborg, Denmark, whilst Singapore has a relatively high temperature all the year in Fig. 13. With the same LED street lamp working at $I_F=0.7$ A from 19:00 pm to the next day 5:00 am per day, Fig. 14 gives comparison of B_{10_avg} based

TABLE I
CALCULATED THERMAL RESISTANCES (°C/W) FROM THE FEM SIMULATION.

Θ_{j-hs1}	Θ_{hs-a1}	Θ_{j-hs2}	Θ_{hs-a2}	Θ_{j-hs3}	Θ_{hs-a3}
5.2591	45.8615	5.6043	31.583	5.3423	23.4623

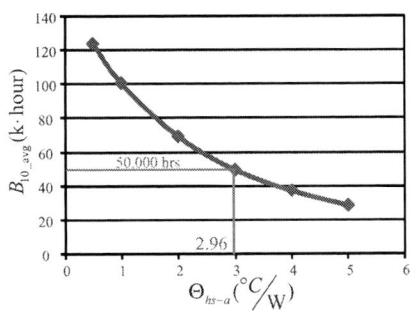

Fig. 12. Average lifetime B_{10_avg} versus Θ_{hs-a} of heat sink for indoor residential LED lighting with L_{90} model.

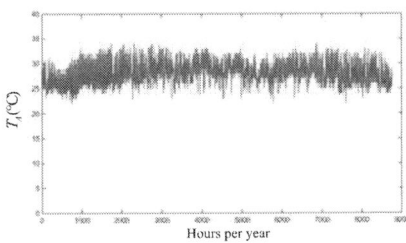

Fig. 13. The yearly ambient temperature distribution of Changi Airport in Singapore. The data is recorded per hour.

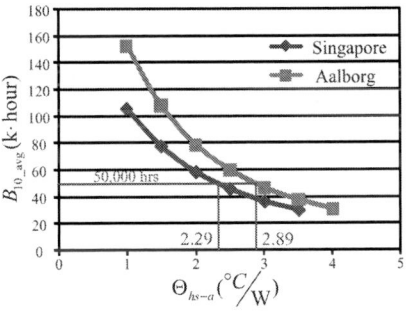

Fig. 14. Average lifetime B_{10_avg} versus Θ_{hs-a} of heat sink for Outdoor street lighting with L_{70} model.

on L_{70} model versus Θ_{hs-a} at these two cities. To have the same 50,000 hours lifetime, a better heat sink in Singapore with Θ_{hs-a} not exceeding 2.29 °C/W is required, which is reduced from the maximum Θ_{hs-a} of 2.89 °C/W in Aalborg. In this case, heatsink 1 with the equivalent Θ_{hs-a} of 2.548 can only be used in Aalborg.

V. CONCLUSION

A mission profile based lifetime prediction method is proposed to estimate the lifetime and reliability performance of LEDs in field operations. It is capable to take into account the impact of long-term electro-thermal loading stresses. Moreover, the statistic properties of life data from accelerated degradation testing are considered through Weibull analysis. The study case of an indoor residential lighting application reveals that the LED B_{10} lifetime could vary with different heat sink thermal resistances. Another specific case of an outdoor street lighting application shows the different predicted lifetime curves under mission profiles in Aalborg and Singapore. The outcomes of the study provide a guideline on the thermal design of LED lighting systems and a procedure to benchmark different design solutions. The experimental study on the junction temperature measurements under different thermal design solutions demonstrates the estimation method of key parameters in the thermal model.

REFERENCES

[1] Cree LED lamps,[Online] Available: http://creebulb.com/products.
[2] GE LED lamps.[Online] Available: http://catalog.gelighting.com/lamp/led-lamps/.
[3] Daily Mail Reporter, "The great LED lightbulb rip-off: One in four expensive 'long-life' bulbs doesn't last anything like as long as the makers claim," 2014. [Online] Available: http://www.dailymail.co.uk/news/article-2546363/.
[4] Illuminating Engineering Sciety, [Online] Available: http://www.ies.org/.
[5] Illuminating Engineering Sciety, "Approved Method: Measuring Lumen Maintenance of LED Light Sources," 2008. [Online] Available: https://www.ies.org/store/product/approved-method-measuring-lumen-maintenance-of-led-light-sources-1096.cfm.
[6] Illuminating Engineering Sciety, "Projecting Long Term Lumen Maintenance of LED Light Sources," 2011. [Online] Available: https://www.ies.org/store/product/projecting-long-term-lumen-maintenance-of-led-light-sources-1253.cfm.
[7] S. Koh, W. V. Driel, and G. Q. Zhang, "Degradation of light emitting diodes: a proposed methodology," *Journal of Semiconductiors*, vol. 32, no. 1, pp. 014004-1–4, Jan. 2011.
[8] International Sematech Inc., "semiconductor device reliability failure models," 2000. [Online] Available: http://www.sematech.org/docubase/document/3955axfr.pdf.
[9] C. J. Foo, "White Paper: Calculate the LED lifetime performance in optocouplers to predict reliability," 2014. [Online] Available: http://www.avagotech.com/docs/AV02-3401EN.
[10] Lumileds Inc., "IESNA LM-80 test report," 2012. [Online] Available: http://www.philipslumileds.com/uploads/362/DR05-1-pdf.
[11] Lumileds Inc., "Luxeon Rebel Gerneral purpose white," 2015. [Online] Available: http://www.lumileds.com/uploads/28/DS64-pdf.
[12] P. O'Conneor and A. Kleyner, "Practical reliability engineerign, 5th Ed.," *John Wiley & Sons*, 2012.
[13] X. Qu, S. C. Wong, and C. K. Tse, "Temperature measurement technique for stabilizing the light output of RGB LED lamps," *IEEE Transactions on Instrumentation and Measurement*, vol. 59, no. 3, pp. 661–670, Mar. 2010.
[14] X. Tao, H. Chen, S. Li, and S. Y. R. Hui, "A new noncontact method for the prediction of both internal themal resisitance and junction temperature of white light-emitting diodes," *IEEE Transactions on Power Electronics*, vol. 27, no. 4, pp. 2184–2192, Apr. 2012.
[15] M. Miner, "Cumulative damage in fatigue," *Journal of Applied Mechanics*, vol. 12, pp. A159–A164, 1945.
[16] Bergquist Inc., "Power LED IMS substrates," 2015. [Online] Available: http://www.bergquistcompany.com/pdfs/LEDConfig20Sht_Re204.pdf.
[17] T-Global Technology. "Lumi-Pad for LED configuration," 2015. [Online] Available: http://www.tglobaltechnology.com/LED-die-cuts/LP0005v01-Li98-0.25.PDF.

Enhanced Li-ion Battery Modeling Using Recursive Parameters Correction

Jae-Gu Kim, Jung-Hoon Ahn, and Byoung-Kuk Lee [†]

Department of Electrical and Computer Engineering
Sungkyunkwan University
Suwon, Korea
E-mail:bkleeskku@skku.edu

Abstract— **In this paper, an enhanced modeling method for accurate equivalent circuit model (ECM) of Li-ion battery is proposed. Modeling with curve fitting is analyzed and consequently, adaptive experimental conditions and appropriate fitting methods are proposed. An inaccurate phenomenon of the existing modeling method caused by inner capacitive impedance is analyzed. For matching model output and battery cell data during charge or discharge, model parameters are corrected recursively. Informative experimental single cell tests are carried out to verify improved accuracy of the proposed battery modeling.**

Keywords—Li-ion battery, battery modeling, 2nd RC Ladder model, offline parameterization

I. INTRODUCTION

Recently, Li-ion batteries are preferred because of their high energy densities, long lifetime, and etc. [1]. For the safe battery usage, it is necessary to predict a battery state such as state of charge (SOC) and state of health (SOH) through analyzing dynamic characteristics of the battery. The dynamic characteristics of the battery are analyzed through a battery model, hence the exact battery model is important [2]. In literature, electrochemical model, mathematical model, and equivalent circuit model (ECM) have been mostly researched. Among these models, ECM has advantages to represent chemical elements of the battery intuitively and has easy accessibility for users [3]. According to experimental analyses, the Li-ion battery is considered as an electrical model consisting of resistance and capacitance [4]. For estimating these parameters of the model, rest time voltage curves with pulse currents are fitted and the parameter values are calculated through the fitting results [5-7]. Therefore the accurate curve fitting and parameter calculation lead to exact battery model. However, there are no detail explains about mathematical method for curve fitting in literature and the curve fitting results are influenced by rest time length because scale of voltmeter crushes voltage data which leads to fitting error. Furthermore, an existing method for parameter estimation assumes steady state voltage of the battery during charge or discharge, but the voltage of battery is not in the steady state because of its large time constant and variations of RC

impedance, so that this assumption increases an error of the parameter estimation.

In this paper, the processes of adapting experimental conditions for the accurate curve fitting and enhancing parameter extraction methods are proposed in order to increase accuracy of the battery ECM. Experiment profiles and mathematical fitting methods leading to the accurate curve fitting are analyzed. Transient state voltage of RC networks taking account of variation of RC parameters during charge or discharge is calculated. And next, the model parameters are recalculated to correct the transient state voltage. Finally, a theoretical limit of the battery model is analyzed and complemented through comparing with a real battery cell at last. Experimental single cell test is performed to verify improvement of modeling accuracy.

II. ANALYZING PROBLEMS OF CONVENTIONAL METHOD FOR PARAMETER ESTIMATION

Li-ion batteries are commonly considered as 2nd RC Ladder model shown in Fig. 1 [4]. The 2nd RC Ladder model has one resistance component and two resistance-capacitance impedance networks which represent polarization and diffusion of the battery. The series resistance r_s represents step response, Vs, of the battery terminal voltage. In addition, the first RC network r_1-c_1 represent short-term response V_1 and the other RC network r_2-c_2 represent long-term response V_2 in the model. The battery is charged or discharged by pulse currents and voltage curves during rest time of the pulse currents are used to estimate the model parameters at fixed SOC. Equation (1) is the voltage curve of the 2nd RC ladder model during rest time. Fig. 2(a) shows the voltage curve consisting of V_s, V_1, V_2 and open circuit voltage (OCV) which are estimated through fitting the curve. Terminal voltage of the model V_t contains these four

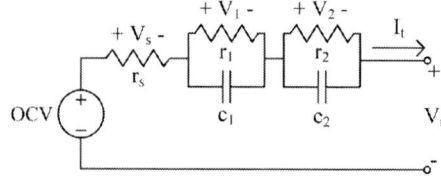

Fig. 1. 2nd RC ladder battery model.

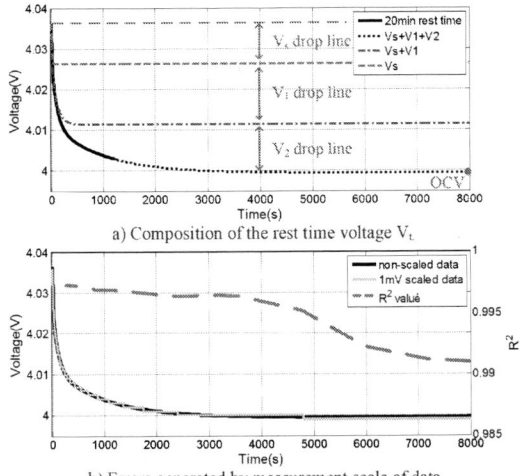

a) Composition of the rest time voltage V_t

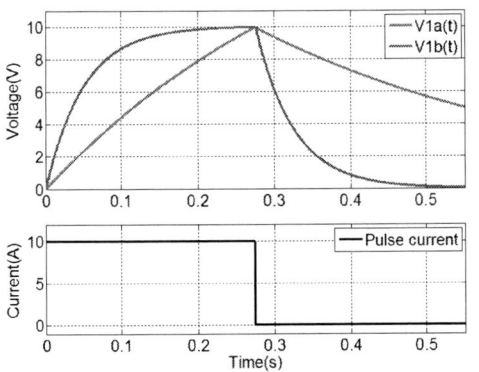

b) Errors generated by measurement scale of data

Fig. 2. Analysis of the rest time voltage curve.

Fig. 3. Curves of different RC ladder time constant $(r_{1a}=2\,\Omega,\ r_{1b}=1\,\Omega)$.

kinds of voltage and at last, only OCV is appeared in V_t. By fitting the rest time voltage curve with (1), values of V_s, V_1, V_2, τ_1, τ_2 and OCV are estimated and parameters r_s, r_1, r_2, c_1, c_2 are calculated by (2) through the curve fitting results on the assumption that capacitors in RC networks are saturated.

$$V_t(t) = OCV + V_s(1-u(t)) + V_1 e^{-\frac{t}{\tau_1}} + V_2 e^{-\frac{t}{\tau_2}} \quad (1)$$

$$r_1 = \frac{V_1}{I_t}, \ c_1 = \frac{\tau_1}{r_1} \ (I_t = input\ current) \quad (2)$$

A. Determination of profile for appropriate cell data

Cell voltage data for curve fitting is scaled by voltmeter in the process of measurement, hence the voltage data for the curve fitting contains errors compared with real value of battery voltage. This distortion of curve becomes deeper as length of the rest time becomes longer. R^2 values, which represents matching of data, are decreased through the increasing rest time shown in Fig. 2(b). The distortion of measured data makes it difficult to obtain exact fitting result. On the other hand, long rest time is necessary for the curve fitting to observe long-term response V_2. Short rest time means lack of data, hence it also leads to inaccurate curve fitting results. Therefore determination of length of the rest time is necessary with considering both the lack of data and the accumulated error.

B. Parameter extraction through exact curve fitting results

For the exact battery model, accurate curve fitting results are necessary. Among the elements of rest time voltage, V_s and OCV are especially estimated differently according to fitting methods. Because V_s is practically very short- term response in real battery, its value is changed by conversing method. OCV value is also changed by calculating point considering the profile rest time and the long-term effect of the battery. Moreover, the curve fitting has to be conducted in consideration of accuracy criterion such as error rate, RMSE and R^2 which are needed to compare fitting results with measured data. However, determination of these fitting methods and accuracy criteria for the exact curve fitting results is not researched sufficiently.

As shown in Fig. 3, although the value of r_{1a} is higher than r_{1b}, r_{1a} and r_{1b} are calculated equally if V_{1a} and V_{1b} are equally estimated. This phenomenon is because the parameter values are extracted directly from curve fitting due to conventional assumption that capacitance in RC networks is saturated during charge or discharge. However, large time constant and impedance variations of the RC ladders result in unsaturation state of the inner capacitance. Therefore, current flowing through resistors of the RC ladders is lower than input current I_t and real values of the resistance parameters of the RC ladder are calculated lowly.

III. PROPOSED ENHANCED METHOD FOR BATTERY ECM ACCURACY

A. Methods for getting the quality rest time voltage data

Rest time length of pulse wave has to consider the data scale of the measured voltage and the time constant of the battery because the rounded data occurs fitting errors. Fig. 4 indicates that long rest time results in fitting errors caused by low data scale and short rest time results in fitting errors caused by a lack of data. There is a problem that the time constant values of the battery are unknown before conducting curve fitting. Thereby reliable curve fitting on controlled conditions has to be conducted after the primarily measuring time constant. Additionally, fitting errors are occurred by low C-rate currents because low values of V_s, V_1, and V_2 are easy to be influenced by the data scale. High C-rate currents also occur fitting errors because of relatively short charge or discharge time compared with large time constant values, so the voltage values are also low. As a result, the current profile including 1.5C currents and 30min rest time is suitable for the curve fitting.

B. Methods for the exact fitting of curve

Methods of the curve fitting based on (1) are analyzed. For OCV estimation, the end point of the rest time curve is judged to OCV if rest time of a profile is long enough. This method is most precise to estimate OCV but the rest time is required to be long and the fitting error can be occurred because of long rest time. In this paper OCV is induced by (1) with relatively shorter rest time than the battery time constant. This method does not offer accurate OCV but facilitates reliable fitting accuracy. Secondly, a method of V_s estimation is analyzed. The voltage response of the real battery with small time constant has to be substituted for the step response V_s which can be estimated to response during fixed duration. This method is simple but not precise because responses of V_1 and V_2 occur simultaneously. In this paper, V_2 response which appeared on last part of the curve is measured first with estimating OCV. Then, V_1 is measured by analyzing middle part of the curve and lastly V_s is calculated inversely for the accurate curve fitting. Finally in order to get accurate curve fitting results,

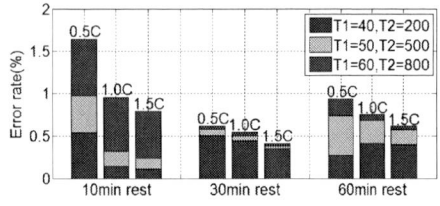

a) Error rates of a 1mV scale.

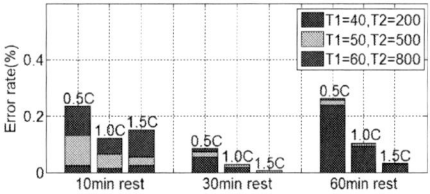

b) Error rates of a 0.1mV scale.

Fig. 4. Curves fitting error rates according to the experiment data conditions.

TABLE I.   PRECISION COMPARISON OF FITTING CRITERIA

	Error rate (%)	RMSE (%)	R^2(%)
V_2/V_1=1.0, T_2/T_1=10	99.961	99.961	99.961
V_2/V_1=0.7, T_2/T_1=12	99.979	99.979	99.979
V_2/V_1=1.3, T_2/T_1=8	99.987	99.987	99.987

TABLE II.   ESTIMATION RESULTS OF THE VIRTUAL MODEL

	r_s(%)	r_1(%)	r_2(%)	c_1(%)	c_2(%)
Before Correction	0.49	7.21	46.12	6.63	107.59
After Correction	0.49	0.75	3.68	0.72	3.28

criterion selection for curve fitting is analyzed. Error rate, RMSE, and R^2 value are set for the criteria of curve fitting and there is no difference of precision as written in Table 1. In this paper curve fitting is conducted by the R^2 criterion which is commonly used.

C. Methods for accurate paramters extraction from fitting results

For accurate calculation of model parameter, state estimation of the RC networks is needed. Transient state of the RC ladder voltage during charge or discharge is analyzed in order to correct the parameter calculation. The RC network is stayed in transient state by variation of resistance and capacitance during charge or discharge, as well as large time constant. Therefore changes of resistance and capacitance during charge of discharge have to be assumed as

$$r_1(t) = at + b, \; c_1(t) = pt + q. \tag{3}$$

Equation (3) represents the variations of the RC ladder through SOC changing. After the assumption as (3), the differential equation of the RC ladder voltage is expressed as

$$V_1(t) = r_1\left(I_T - \frac{d}{dt}\left(c_1(t) \cdot V_1(t)\right)\right). \tag{4}$$

By solving the differential equation, the transient state equation of the RC ladder voltage is expressed as

$$V_1(t) = ke^{A(t)} + B(t) \cdot C(t), \tag{5}$$

$$A(t) = \frac{1}{bp - aq}\ln\left(\frac{at+b}{pt+q}\right) - \ln\left(\frac{1}{bp-aq}\right), \tag{6}$$

$$B(t) = I_T\left(\frac{at+b}{bp-aq+1}\right)\left(\frac{p(at+b)}{bp-aq}\right)^{\frac{1}{bp-aq}-1}, \tag{7}$$

$$C(t) = {}_2F_1\left(\frac{1}{\alpha}, 1 + \frac{1}{\alpha}; 2 + \frac{1}{\alpha}; \frac{-a(pt+q)}{\alpha}\right) \cdot (\alpha = bp - aq) \tag{8}$$

Equation (8) is a hypergeometric function and k is coefficient for satisfying initial condition. First, the parameters r_s, r_1, r_2, c_1, and c_2 are calculated primarily by the existing method using (1) and (2) through SOC changing. The parameter values are sampled by SOC so values of a, b, p and q are calculated by (3) with the adjacent parameter values. For complementing the existing method, voltages of RC ladders V_1',$_2'$ are calculated by (5)-(8). Error of the voltages are calculated by considering curve fitting results $V_{1,2}$ as reference values. The parameters are recalculated by multiplying error between the fitting results $V_{1,2}$ and the calculated voltage values V_1',$_2'$. Fig. 5 shows the matching process of terminal voltage V_t and short-term response V_1 through repeating recalculations. According to recursive calculation loop, the voltage of RC ladder becomes matching with reference value. To verify the proposed method, parameter values of virtual 2nd

978-1-4673-9551-9/16 $31.00 © 2016 IEEE   2163

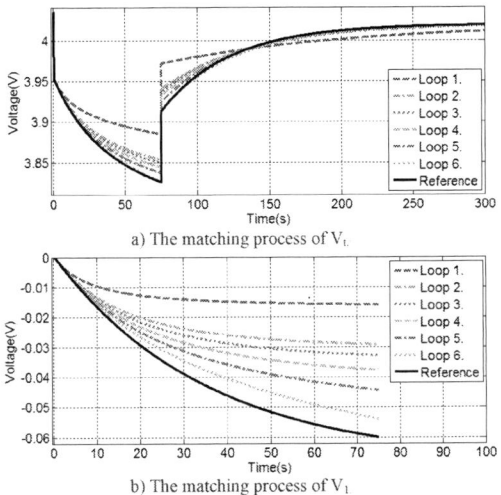

a) The matching process of V_t.

b) The matching process of V_1

Fig. 5. Matching processes of the reference voltage.

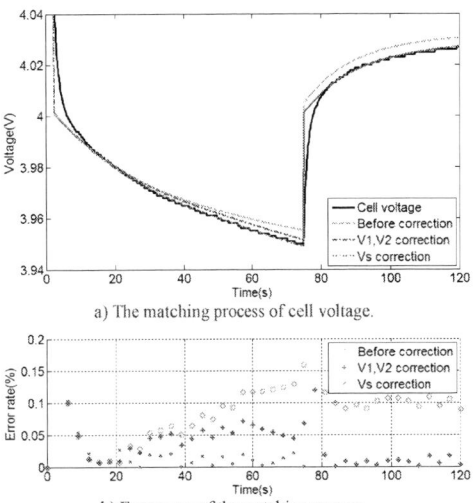

a) The matching process of cell voltage.

b) Error rates of the matching process.

Fig. 6. Step results of cell voltage curve correction.

RC ladder battery model is recalculated by both the existing method and the proposed method. Table 2 shows the result and accuracy of each parameter estimation is increased. As written in Table 2, r_s value is not changed through the recursive parameter correction and the reason is that the virtual 2nd RC ladder battery model has clear step response value V_s so the r_s value is exactly calculated even at the primary calculation.

D. Parameters Extraction From Battery Cell Data

Fig. 6(a) shows the result of a cell voltage test and the voltage curve after the correction is not matched with the cell voltage curve. This phenomenon is resulted from series resistance of the battery model. The series resistance r_s of the 2nd RC ladder ECM is practically parallel RC impedance with very small time constant in the real battery. The cell correction errors come from transient state of V_s, so r_s has to be recalculated through matching the end point of both curves. Fig. 6(a) shows the matching process of the real cell test through complementing the theoretical limit of the model and Fig. 6(b) show the decreased error rate. Fig. 8 shows the total algorithm of the proposed parameter extraction. Step 1 is the existing method which contains fitting the voltage data and

a) Battery cell   b) Experiment equipment for single cell test

Fig. 7. Matching processes of the reference voltage.

primarily calculating model parameters. Step 2 is the correction based on the 2nd RC model through recursive parameters calculation which leads to match transient curve with model output. Finally, Step 3 is the complementation of the limit of the model through converting RC impedance with very small time constant into series resistance.

IV. EXPERIMENTAL VERIFICATION OF PROPOSED BATTERY MODELING

A Kokam 75175280PS battery cell with a nominal voltage 3.7V and a nominal capacity 27Ah is used for parameters extraction and Single cell tests. The cell modeling and tests were done at 25°C and The battery cell and experiment equipment for these works are shown in Fig. 7. For parameters extraction, the cell is discharged 48 seconds with 1.5C current (ΔSOC=2%) and 30 minutes of rest time as mentioned earlier so the parameters are estimated per 2% SOC. The cell is discharged from SOC 100% to 0% with 3 cycles and Fig. 9 is the results of the battery parameters extraction. Parameter values of r_1 and r_2 are increased after correction, whereas c_1 and c_2 are decreased. Real values of current flowing through resistors are calculated by estimating transient state of the RC ladders and calculating voltage across the RC ladders. According to the correction, parameters are calculated taking account of unsaturated inner capacitance. Battery parameters are defined by mathematical function which is simpler than look-up table and more suitable for BMS [8]. The lines shown in Fig. 9 are fitting curves of model parameters using R^2 fitting criterion which is most accurate parameter fitting method [8]. The mathematical functions of parameter fitting curves are written in Table 3. As a result, error rates between the cell voltage curve for modeling and model output is decreased through parameters correction.

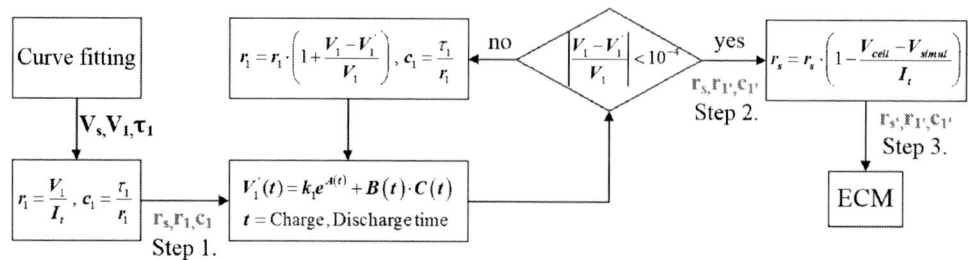

Fig. 8. Algorithm of the parameter calculation complementing.

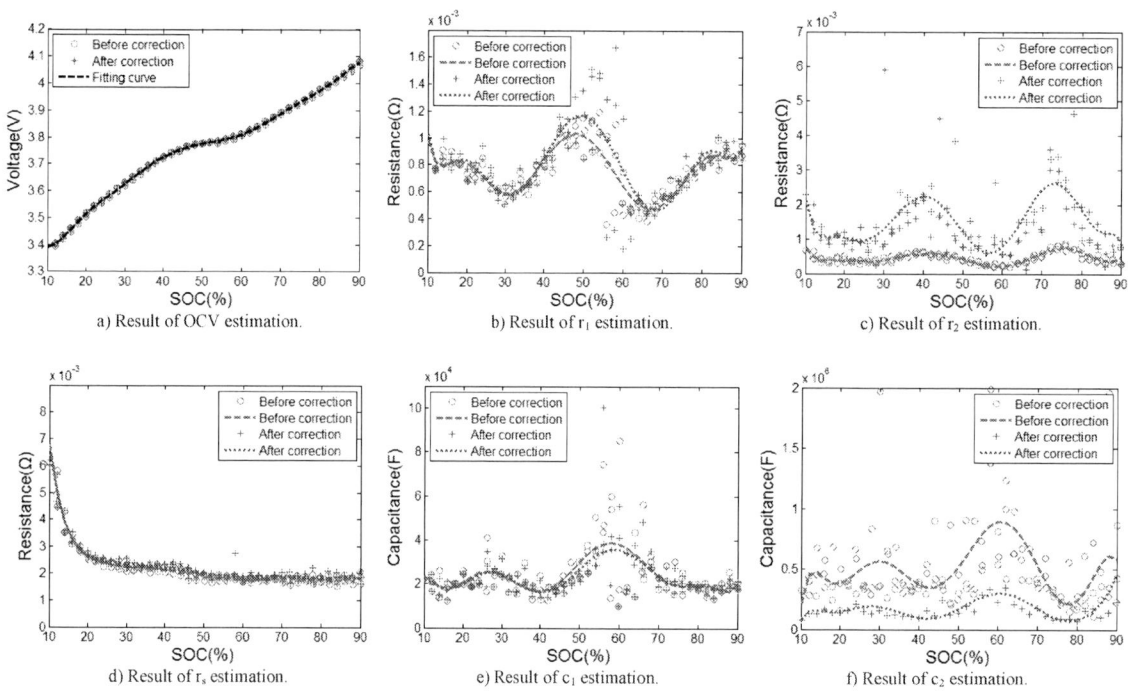

Fig. 9. Results of the battery parameters extraction.

TABLE III. RESULTS OF THE PARAMETERS EXTRACTION

| | Mathematical functions of parameters using R^2 fitting criterion | Error rate (%) | |
		Max	Mean
Before correction	$x = (SOC-50)/23.76$ $OCV = -0.00887x^9 + 0.00759x^8 + 0.0658x^7 - 0.0509x^6 - 0.175x^5 + 0.116x^4 + 0.216x^3 - 0.11x^2 + 0.0744x + 3.763$ $r_s = -1.62 \times 10^{-5}x^9 + 1.49 \times 10^{-5}x^8 + 6.21 \times 10^{-5}x^7 - 3.65 \times 10^{-5}x^6 - 1.02 \times 10^{-4}x^5 + 4.61 \times 10^{-5}x^4 + 8.60 \times 10^{-5}x^3 + 6.81 \times 10^{-5}x^2 - 2.86 \times 10^{-4}x + 0.00165$ $r_1 = -2.76 \times 10^{-5}x^9 + 1.46 \times 10^{-4}x^8 + 2.53 \times 10^{-4}x^7 - 9.99 \times 10^{-4}x^6 - 8.29 \times 10^{-4}x^5 + 2.27 \times 10^{-3}x^4 + 1.11 \times 10^{-3}x^3 - 1.79 \times 10^{-3}x^2 - 4.99 \times 10^{-4}x + 0.000999$ $r_2 = -9.78 \times 10^{-5}x^9 + 1.11 \times 10^{-4}x^8 + 7.82 \times 10^{-4}x^7 - 5.39 \times 10^{-4}x^6 - 2.25 \times 10^{-3}x^5 + 7.09 \times 10^{-3}x^4 + 2.57 \times 10^{-3}x^3 - 1.50 \times 10^{-4}x^2 - 8.32 \times 10^{-4}x + 0.000417$ $c_1 = 4.92 \times 10^3x^9 + 3.26 \times 10^3x^8 - 3.77 \times 10^4x^7 - 1.89 \times 10^4x^6 + 1.03 \times 10^5x^5 + 3.67 \times 10^4x^4 - 1.15 \times 10^5x^3 - 3.03 \times 10^4x^2 + 4.18 \times 10^4x + 31700$ $c_2 = 1.18 \times 10^5x^9 - 2.04 \times 10^5x^8 - 9.12 \times 10^5x^7 + 1.17 \times 10^6x^6 + 2.54 \times 10^6x^5 - 2.04 \times 10^6x^4 - 2.91 \times 10^6x^3 + 9.73 \times 10^5x^2 + 1.06 \times 10^6x + 514000$	0.588	0.137
After correction	$r_s = -2.94 \times 10^{-5}x^9 + 3.02 \times 10^{-5}x^8 + 1.42 \times 10^{-4}x^7 - 1.29 \times 10^{-4}x^6 - 2.37 \times 10^{-4}x^5 + 2.08 \times 10^{-4}x^4 + 1.35 \times 10^{-4}x^3 - 1.65 \times 10^{-5}x^2 - 2.82 \times 10^{-4}x + 0.00173$ $r_1 = -1.83 \times 10^{-6}x^9 + 2.03 \times 10^{-4}x^8 + 6.84 \times 10^{-5}x^7 - 1.37 \times 10^{-3}x^6 - 3.60 \times 10^{-4}x^5 + 3.09 \times 10^{-3}x^4 + 6.48 \times 10^{-4}x^3 - 2.46 \times 10^{-3}x^2 - 3.51 \times 10^{-4}x + 0.00116$ $r_2 = -6.65 \times 10^{-4}x^9 + 1.42 \times 10^{-4}x^8 + 4.83 \times 10^{-3}x^7 + 3.32 \times 10^{-4}x^6 - 1.24 \times 10^{-2}x^5 - 1.71 \times 10^{-3}x^4 + 1.27 \times 10^{-2}x^3 + 1.95 \times 10^{-3}x^2 - 3.61 \times 10^{-3}x + 0.00122$ $c_1 = 4.41 \times 10^3x^9 + 2.11 \times 10^3x^8 - 3.40 \times 10^4x^7 - 1.13 \times 10^4x^6 + 9.30 \times 10^4x^5 + 2.02 \times 10^4x^4 - 1.04 \times 10^5x^3 - 1.67 \times 10^4x^2 + 3.82 \times 10^4x + 28200$ $c_2 = 5.64 \times 10^4x^9 - 3.74 \times 10^4x^8 - 3.97 \times 10^5x^7 + 2.46 \times 10^5x^6 + 1.05 \times 10^6x^5 - 4.69 \times 10^5x^4 - 1.16 \times 10^6x^3 + 2.42 \times 10^5x^2 + 4.16 \times 10^5x + 168000$	0.238	0.0846

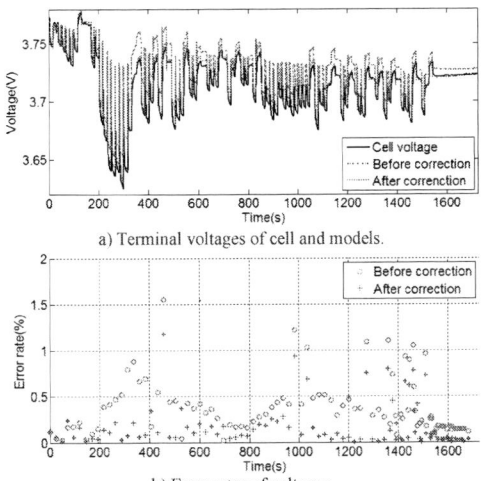

a) Terminal voltages of cell and models.

b) Error rates of voltages.

Fig. 10. Results of the single cell test by UDDS profile.

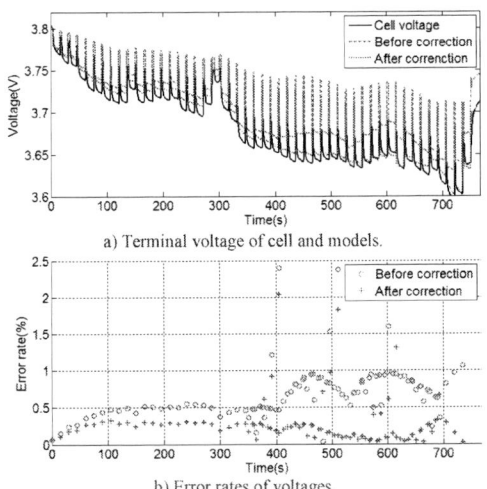

a) Terminal voltage of cell and models.

b) Error rates of voltages.

Fig. 11. Results of the single cell test by HWFET profile.

TABLE IV. ERROR RATES OF THE SINGLE CELL TESTS

	UDDS profile		HWFET profile	
	Before (%)	*After (%)*	*Before (%)*	*After (%)*
Max	2.918	2.421	3.236	2.487
Mean	0.365	0.140	0.698	0.351

Single cell tests are carried out through UDDS (Urban Dynamometer Driving Schedule) and HWFET (Highway Fuel Economy Test) profiles. Model output of both before and after correction is compared with cell outputs and the test results are shown in Figs. 10 and 11. Model outputs are matched better after parameter correction. Resistance values of the 2nd

ladder model are increased through parameter correction and they especially lead to the increase of model accuracy as shown in Fig 10 and 11. The maximum error rate of a single cell test decreases from 3.23% to 2.48% and the mean error rate decreases from 0.69% to 0.35% as written in Table 4.

V. CONCLUSIONS

An improved modeling method for battery 2nd RC ladder model is proposed. The appropriate current profile for accurate fitting results is dependent on the scale of measured data and the battery time constant so the rest time and C-rate of current profile have to be taken into account them. The suitable fitting method for measuring each value of voltage and time constant is determined. Model parameters are corrected through matching with the voltage curves of charge or discharge through adjusting impedance of the RC networks. At last, the limit of the 2nd RC ladder model is complemented through correcting the series resistance. The parameter extraction results are defined by mathematical function and model output becomes matching better through parameters correction. Maximum and mean error rate values of the single cell tests are all decreased.

ACKNOWLEDGMENT

This work was supported by the Industrial Strategic Technology Development Program (No.10053710) funded by the Ministry of Trade, Industry & Energy (MI, Korea).

REFERENCES

[1] L. Gao, S. Liu and R. A. Dougal, "Dynamic Lithium-Ion Battery Model for system simulation", IEEE Transactions on Components and Packaging Technologies, vol. 25, no. 3, pp. 495-505, Sep. 2002.

[2] K. W. E. Cheng, B. P. Divakar, H. Wu, K. Ding, and H. F. Ho, "Battery-Management System (BMS) and SOC Development for Electrical Vehicles", IEEE Transactions on Vehicular Technology, vol. 60, no. 1, pp. 76-88, Jan. 2011.

[3] E. Barsoukov, J. H. Kim, C. O. Yoon and H. S. Lee, "Universal battery parameterization to yield a non-linear equivalent circuit valid for battery simulation at arbitrary load", Journal of Power Sources, vol. 83, no. 1-2, pp. 61-70, Oct. 1999.

[4] J. M. Lee, J. H. Lee, O. Y. Nam, J. H. Kim, B. H. Cho, H. S. Yun, S. S. Choi, K. H. Kim, J. H. Kim and S. N. Jun, "Modeling and Real Time Estimation of Lumped Equivalent Circuit Model of a Lithium Ion Battery", Power Electronics and Motion Control Conference, Aug. 2006, pp.1536-1540

[5] J. Hafsaoui, J. Scordia, F. Sellier, and P. Aubret, "Development of an electrochemical battery model and its parameters identification tool", International Journal of Automotive Engineering, vol. 3, no. 1, pp. 27-33, Feb. 2012

[6] M. Chen and G. A. Rinc´on-Mora, "Accurate Electrical Battery Model Capable of Predicting Runtime and I-V Performance", IEEE Transactions on Energy Conversion, vol. 21, no. 2, pp. 504-511, Jun. 2006.

[7] H. G. Schweiger, O. Obeidi, O. Komesker, A. Raschke, M. Schiemann, C. Zehner, M. Gehnen, M. Keller and P. Birke, "Comparison of Several Methods for Determining the Internal Resistance of Lithium Ion Cells", Sensors, vol. 10, no. 6, pp. 5604-5625, Jun. 2010.

[8] A. S. Garcíaa, V. Alfonsina, S. Urréjolaa, Á. Sánchezb, "Optimal parametrization of electrodynamical battery model using model selection criteria", Journal of Power Sources, vol. 285, no. 1, pp. 119-130, Jul. 2015

Robust Sensorless Control of Grid Connected Converters with LCL Line Filters Using Frequency Adaptive Observers as AC Voltage Estimators

Vlatko Miskovic[†*], Vladimir Blasko[‡], Thomas Jahns[*], Robert Lorenz[*], Haojiong Zhang[†]

[†]Danfoss Drives, Loves Park, IL, 61111 USA, vlatko_m@danfoss.com, haojiongz@danfoss.com
[*]University of Wisconsin-Madison, Madison, WI, 53706 USA, jahns@engr.wisc.edu, lorenz@engr.wisc.edu
[‡]United Technologies Research Center, East Hartford, CT 06108, blaskov@utrc.utc.com

Abstract—This paper presents robust control of grid connected voltage source converters (VSC) with LCL line filter without grid voltage sensors, measuring only converter inductor currents and DC link voltage. As grid impedance and step-down transformer inductance values are usually not known, proposed control algorithm relies only on converter side inductor value for control parameter, making it robust against grid variations. Proposed method uses frequency adaptive observers as filter capacitor voltage estimators, which enables the use of well-known highly dynamic performing voltage oriented control (VOC). Two versions of the frequency adaptive observers are presented, one in synchronous frame and one in stationary frame. Theoretical results are verified in simulation environment and experimentally.

Index Terms—Active front end drive, LCL filter, robust, sensorless, grid synchronization, observers, frequency adaptive

I. INTRODUCTION

VSC with LCL line filters have become industry standard in active front end (AFE) drives in applications where bi-directional power flow, adjustable DC link voltage, adjustable power factor and low harmonics are required. Conventional VOC [1] uses measured grid voltages to synchronize with grid which is required for accurate power flow control. In order to reduce cost of VSC, it is desired to replace grid voltage sensors with estimators.

In [2-3] virtual flux estimator (VFE) is used to estimate grid voltage, which in order to avoid differentiation of the line currents, calculates integral of the grid voltage (i.e. virtual flux). Since pure integral has pole at origin, in practice, in order to avoid DC wind-up pure integrator has to be replaced with low pass filter, which introduces phase lag of voltage (and flux) estimate.

Another sensorless technique is introduced in [4] which uses modified proportional-integral (PI) control in the active *d* axis of the current controller to obtain the angle error signal, which drives phase locked loop (PLL) to estimate grid voltage angle. Estimator has zero lag properties at fundamental component of the grid voltage estimate, but is not independent of current control.

Augmented state-space observer is used in [5-6] to estimate multiple harmonics (both positive and negative sequences) and estimated harmonic voltage components are used for disturbance input decoupling (DID). In this work authors do not claim sensorless control, and observers are not adaptive to grid frequency.

In motor control applications observers are used extensively as motor back-emf estimators for sensorless control [7-11]. Those observers are implemented in stationary reference frame and have inherently lagging properties which depend on observer bandwidth [8].

Frequency adaptive observers in synchronous reference frame introduced by authors in [12] can estimate back-emf with zero phase lag at any frequency regardless of the observer bandwidth, but in that work they were not applied for sensorless control. Authors have shown in [13] that synchronous reference frame observers as back-emf estimators work well with model predictive control of VSC with L line filter, but grid frequency was assumed to be constant and known. Recent work in [14] developed sensorless control of VSC with LCL filter using adaptive observers based on small signal linearization of estimation-error, where converter side inductor and capacitor are assumed to be well known.

With accurate knowledge of LCL parameters and grid voltage sensors, high performance control algorithm can be achieved as in [15], however where voltage sensors and accurate LCL parameters are not available, there is a need for robust sensorless control of VSC.

Often it is the case that grid and step-down transformer inductances are not known, so this paper presents sensorless control using only one parameter, converter's side inductance, for current control and filter capacitor voltage estimation, thus making it robust against grid variations. The major constraint for proposed control algorithm to work is that unknown grid side inductance and possibly unknown filter capacitance need to be in the range such that effective resonant frequency of LCL is less than one third of switching frequency in order not to cause resonance problems. Frequency adaptive observers are used for filter capacitor voltage estimation, fundamental and harmonic components. Proposed method is shown to be robust against grid frequency variations, as well as converter side inductor parameter uncertainty. Control performance is verified in simulation environment and experimentally.

II. MATHEMATICAL MODEL OF THE SYSTEM

Fig .1 shows power circuit of VSC connected to grid with LCL filter with only converter inductor currents and DC link voltage measured.

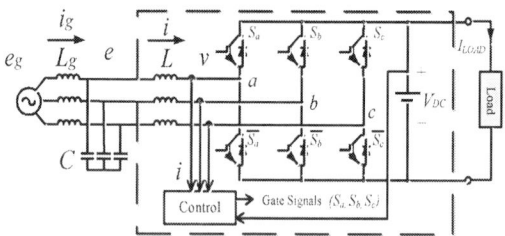

Fig. 1. Power circuit of grid-connected VSC with LCL line filter

Since only converter side inductance L is used for control, partial mathematical model of the system in Fig. 1 is represented by (1), where subscripts $\alpha\beta$ represent stationary reference frame.

$$L\frac{di_{\alpha\beta}}{dt} = -v_{\alpha\beta} + e_{\alpha\beta} - Ri_{\alpha\beta} \tag{1}$$

Stationary frame model in (1) rotating in ω reference frame is shown in (2), where subscripts dq represent synchronous (ω) reference frame.

$$L\frac{di_{dq}}{dt} = -v_{dq} + e_{dq} - R\cdot i_{dq} - j\omega L\cdot i_{dq} \tag{2}$$

Fig. 2 shows top level sensorless current control structure. Current reference i^*_{dq} is DC voltage regulator output, where d axis is aligned with active power. Filter capacitor voltage $e_{\alpha\beta}$, its frequency ω and position θ is estimated using frequency adaptive observers described in detail in Sections III and IV. Control is PI type controller and includes cross-coupling decoupling from current references and estimated capacitor voltage ($e_{\alpha\beta}$) feed-forward with delay compensations. Converter voltage input to the frequency adaptive observer $v_{\alpha\beta}$ is estimated using PWM duty cycles and DC link voltage.

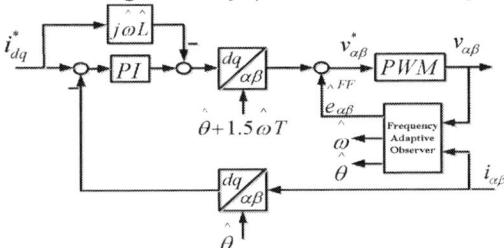

Fig. 2. Sensorless current controller structure

If filter capacitor C is known, reactive q axis current reference can be modified as in (3) such that desired power factor is achieved at filter capacitor voltage point.

$$i^*_q = i^*_{g\_q} - \hat{\omega}\hat{C}\hat{e}_D \tag{3}$$

In similar way if grid inductance L_g is known, i^*_q can be modified to achieve desired power factor on grid side.

III. FREQUENCY ADAPTIVE SYNCHRONOUS OBSERVERS (FASO)

A. Synchronous Observers (SO)

With synchronous reference frame plant (2), a continuous time synchronous observer (SO), introduced in [12], can be constructed as shown in Fig. 3. Synchronous frame frequency dependent cross-coupling term, $j\omega L i_{dq}$, in the observer uses measured states i_{dq} for faster convergence of the observer. Since in synchronous rotating reference frame, the synchronous states are DC quantities, PI controller can be used to achieve zero steady-state error. Transfer function of estimated filter capacitor voltage in synchronously rotating reference frame is shown in (4). For 50 or 60Hz frequency grid, impedance of inductor is mostly reactive (i.e. $\omega L \gg R$), so

it safe to assume that estimated inductor resistance $\hat{R} = 0$.

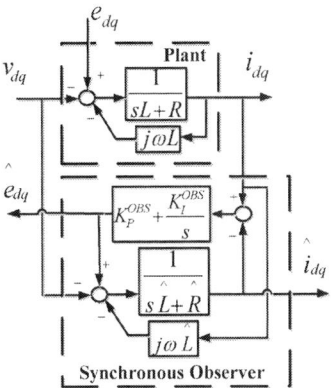

Fig. 3. Synchronous Observer (SO)

$$\frac{\hat{e}_{dq}(s)}{e_{dq}(s)} = \frac{sK^{OBS}_P + K^{OBS}_I}{s^2\hat{L} + s(R + K^{OBS}_P) + K^{OBS}_I}\left[\underbrace{\frac{(s+j\omega)\hat{L}+\hat{R}}{(s+j\omega)L+R}}_{\approx 1} + \frac{v_{DQ}}{e_{DQ}}\underbrace{\left(1 - \frac{(s+j\omega)\hat{L}+\hat{R}}{(s+j\omega)L+R}\right)}_{\approx 0}\right]$$

$$\approx \frac{sK^{OBS}_P + K^{OBS}_I}{s^2\hat{L} + s(R + K^{OBS}_P) + K^{OBS}_I} \tag{4}$$

With accurate parameters transfer function (4) at the synchronous frequency, i.e. $s = 0$, becomes:

$$\left.\frac{\hat{e}_{dq}(s)}{e_{dq}(s)}\right|_{s=0} = \left.\frac{sK^{OBS}_P + K^{OBS}_I}{s^2\hat{L} + sK^{OBS}_P + K^{OBS}_I}\right|_{s=0} = 1 \tag{5}$$

In the synchronous reference frame (which corresponds to frequency ω in stationary reference frame), the sinusoidal voltage can be estimated with zero phase and amplitude error, assuming accurate parameters. In stationary reference frame transfer function (4) becomes:

$$\frac{\hat{e}_{\alpha\beta}(s)}{e_{\alpha\beta}(s)} = \frac{(s-j\omega)K^{OBS}_P + K^{OBS}_I}{(s-j\omega)^2\hat{L} + (s-j\omega)K^{OBS}_P + K^{OBS}_I}\left[\underbrace{\frac{s\hat{L}+\hat{R}}{sL+R}}_{\approx 1} + \frac{v_{\alpha\beta}}{e_{\alpha\beta}}\underbrace{\left(1 - \frac{s\hat{L}+\hat{R}}{sL+R}\right)}_{\approx 0}\right] \tag{6}$$

Transfer function (6) at steady state (i.e. $s=j\omega$) becomes (7) which suggests that even if there is inductance value mismatch phase error of the estimate is negligible. It is important to note from (4) and (6) that estimates do not depend on current magnitude.

$$\left.\frac{\hat{e}_{\alpha\beta}(s)}{e_{\alpha\beta}(s)}\right|_{s=j\omega} \approx \frac{\hat{L}}{L} + \frac{v_{\alpha\beta}}{e_{\alpha\beta}}\left(1 - \frac{\hat{L}}{L}\right) \tag{7}$$

Industry standard method for estimating $e_{\alpha\beta}$ is so called virtual flux estimator (VFE) introduced in [2-3], which basically integrates (1) to obtain integral of $e_{\alpha\beta}$ (i.e. virtual flux $\lambda_{\alpha\beta}$), shown in (8). Transfer function of filter capacitor voltage estimate with VFE is shown in (9).

$$\hat{\lambda}_{\alpha\beta}(s) = \frac{1}{s}\left(v_{\alpha\beta} - \hat{R}i_{\alpha\beta}\right) - \hat{L}i_{\alpha\beta} \tag{8}$$

$$\frac{\hat{e}_{\alpha\beta}(s)}{e_{\alpha\beta}(s)} = \frac{j\omega\hat{\lambda}_{\alpha\beta}(s)}{e_{\alpha\beta}(s)} = \frac{j\omega}{e_{\alpha\beta}}\left[\frac{1}{s}\left(v_{\alpha\beta} - \hat{R}i_{\alpha\beta}\right) - \hat{L}i_{\alpha\beta}\right] = \frac{j\omega}{s} \tag{9}$$

Fig. 4 shows transfer functions of the filter capacitor voltage estimation with VFE (9) and SO (6) with detuned parameters $\{K_P^{OBS}, K_I^{OBS}\} = \{16L, 4L\}$ (detailed parameters of the plant are in Table I). In comparison with VFE, SO does not amplify low frequencies (below fundamental) and at steady state it can be detuned arbitrarily low to attenuate any higher order harmonics, and still have zero error estimate at fundamental component, as seen in Fig. 4.

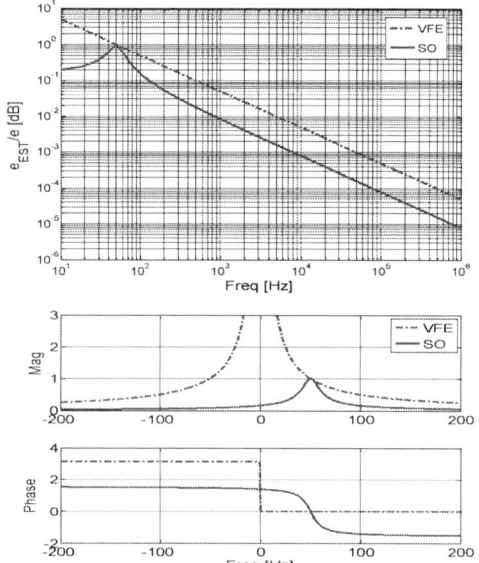

Fig. 4. Transfer function of VFE and SO, log scale (top) and linear scale (bottom)

B. Frequency Adaptive Synchronous Observers (FASO)

Since frequency of the capacitor voltage is usually not known a priori and can change over time, there is a need to track the frequency of the estimated capacitor voltage. Phase locked loop (PLL) shown in Fig. 5 [16] with estimated filter capacitor voltage as an input is used to track estimated capacitor voltage frequency ω and position θ.

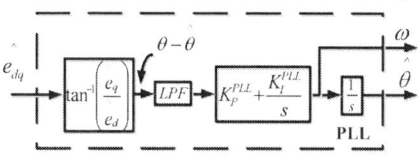

Fig. 5. Phase-locked loop (PLL)

Transfer function of the frequency estimate is shown in (10). Low pass filter (LPF), seen in Fig. 5, can be inserted in series with PI controller to further smoothen frequency estimates; however PLL performance was sufficient without optional LPF.

$$\frac{\hat{\omega}(s)}{\omega(s)} = \frac{s K_P^{PLL} + K_I^{PLL}}{s^2 + s K_P^{PLL} + K_I^{PLL}} \tag{10}$$

Finally SO together with PLL forms frequency adaptive synchronous observer (FASO) as capacitor voltage estimator which is shown in Fig. 6. FASO estimates filter capacitor voltage with zero error and adapts dynamically to its frequency and magnitude changes.

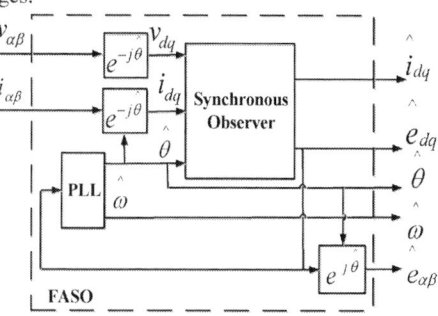

Fig. 6. Top level structure of FASO

All frequency adaptive observers in the paper are discretized using simple Euler forward approximation, including model of the plant in (1) and (2), PI controllers and integrator of PLL. FASO tuning and detailed parameters of the system are outlined in Table I.

To evaluate FASO in Fig. 6 performance, the VSC from Fig. 1 is simulated, using control structure in Fig. 2. Fig. 7 shows dynamic performance of FASO under ±20% step changes in the amplitude of grid voltage e_g. Estimated capacitor voltage converges fast in this transient event and tracks fundamental component of the voltage with zero lag at steady state.

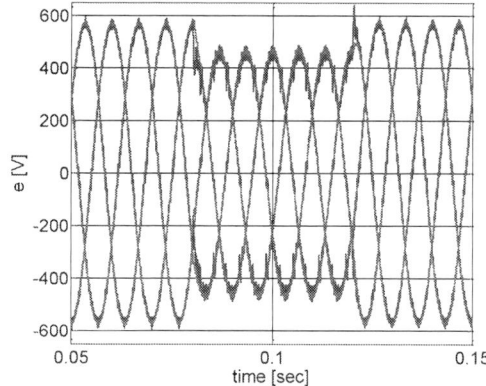

Fig. 7. Simulated (blue) and estimated using FASO (red) capacitor voltage e under ±20% step change in amplitude of e_g

Fig. 8 and 9 shows FASO performance when step change in grid voltage e_g frequency occurs. Top graph in Fig. 8 and 9 shows overlaid simulated (blue) and estimated using FASO (red) capacitor voltage e, while bottom graph shows estimated capacitor voltage frequency ω, also output of FASO. Fig. 8 shows FASO response with step change in frequency from 50Hz to 100Hz, while Fig. 9 shows the response with step change in frequency from 50Hz to 25Hz. In both cases FASO converges fast and is able to track new capacitor voltage at new frequency with zero amplitude and phase error at steady state.

978-1-4673-9551-9/16 $31.00 © 2016 IEEE

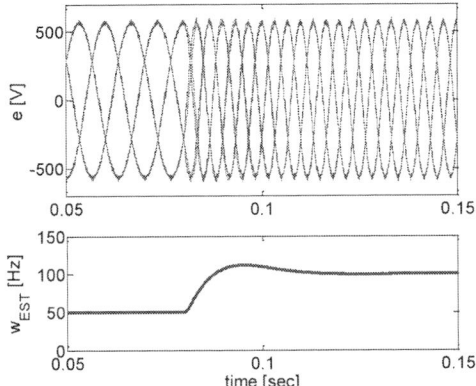

Fig. 8. Simulated (blue) and estimated using FASO (red) capacitor voltage e under +100% step change in frequency of e_g

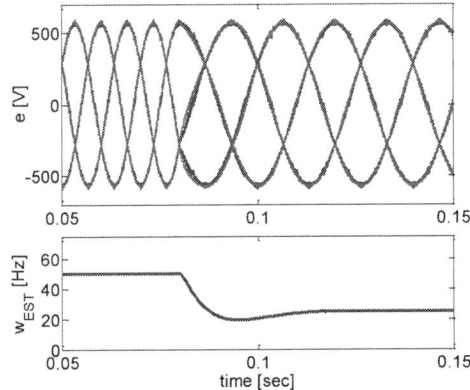

Fig. 9. Simulated (blue) and estimated using FASO (red) capacitor voltage e under -50% step change in frequency of e_g

IV. SYNCHRONOUS INTEGRATORS BASED FREQUENCY ADAPTIVE OBSERVERS (SI-FAO) IN STATIONARY REFERENCE FRAME

FASO performed well tracking filter capacitor voltage at fundamenatal frequency which can change in time as shown in Section III. However, if there is a need that multiple harmonics besides fundamental (i.e. negative sequence, 5th, 7th...) need to be estimated and decoupled as in [5-6], there is a need for frequency adaptive observer that estimates mutiple harmonics with zero error.

A. Frequency Adaptive Synchronous Integrators (FASI)

Fig. 10 shows integrator in arbitrary ω reference frame, i.e. synchronous integrator (SI). Input $x(t)$ to SI is transformed to ω reference frame, integrated and the result is transformed back to stationary frame providing SI output $y(t)$, as seen in Fig. 10.

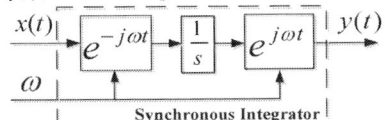

Fig. 10. Synchronous Integrator (SI)

Transfer function of SI [17] when frequency ω is constant is shown in (11):

$$\frac{Y(s)}{X(s)} = \frac{1}{s - j\omega} \qquad (11)$$

Clearly (11) has one pole at s=$j\omega$. When frequency ω of SI adapts to the frequency of estimated disturbance (i.e. capacitor voltage) we get the concept of frequency adaptive synchronous integrator (FASI).

B. Synchronous Integrators Based Frequency Adaptive Observers (SI-FAO)

Using the concept of SI, synchronous integrators based frequency adaptive observers (SI-FAO) in stationary ($\alpha\beta$) reference frame can be constructed as in Fig. 11. All inputs and the plant are in $\alpha\beta$ reference frame, and the observer controller is constructed using proportional gain K_P and multiple SI in parallel, each with gain K_h at the output, where h is a harmonic number (i.e. h=1 for fundamental, h =-5 for 5th, h =7 for 7th harmonic...). PLL is used to synchronize to fundamental frequency as in FASO.

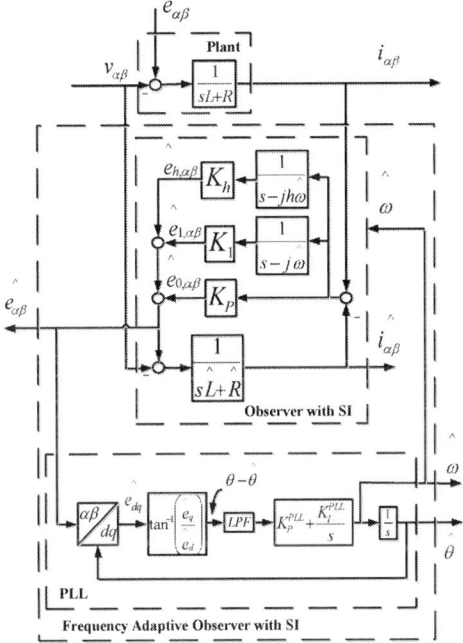

Fig. 11. Synchronous integrators frequency adaptive observer (SI-FAO)

Transfer function of the observer using SI with only fundamental h=1 is shown in (12):

$$\frac{e_{\alpha\beta}(s)}{\hat{e}_{\alpha\beta}(s)} = \frac{(s - j\omega)K_P + K_1}{s(s - j\omega)L + (s - j\omega)K_P + K_1} \qquad (12)$$

SI-FAO estimates $e_{\alpha\beta}$ with zero error at s=$j\omega$ which follows from (12). In similar way, without loss of generality, it can be easily shown that for any harmonic h in Fig. 11, SI-FAO estimates $e_{\alpha\beta}$ with zero error at $s=jh\omega$, because controller of the observer has a pole at that frequency.

Output from each of the SI in Fig. 11 is properly compensated for transport and PWM delays as shown in (13), and is used as feedforward for control as in Fig. 2.

$$\hat{e}_{\alpha\beta}^{FF} = \hat{e}_{0,\alpha\beta} + \hat{e}_{1,\alpha\beta} \cdot e^{j1.5\omega T} + \sum_h \hat{e}_{h,\alpha\beta} \cdot e^{j1.5h\omega T} \quad (13)$$

Fig. 12 and 13 show SI-FAO performance in the case when grid was polluted with higher order harmonics. SI-FAO is tuned in similar way to FASO ($K_h = K_1 = K_1$), and PLL is tuned to same. Fig. 12 shows the case where 5th harmonic was superimposed to the fundamental of e_g, and h was set to -5. SI-FAO tracks both fundamental and 5th harmonic of capacitor voltage with zero error. Similarly in Fig. 13, 7th harmonic was superimposed to the fundamental of e_g. When h was set to 7, SI-FASO is able to track both fundamental and 7th harmonic.

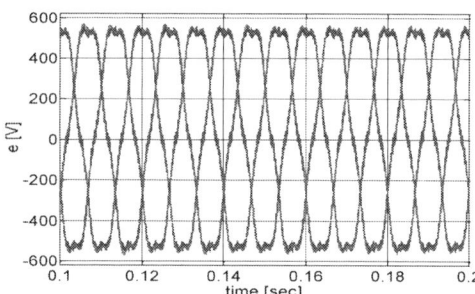

Fig. 12. Simulated (blue) and estimated with SI-FAO (red) capacitor voltage e with h=-5, when 5th harmonic is superimposed on e_g

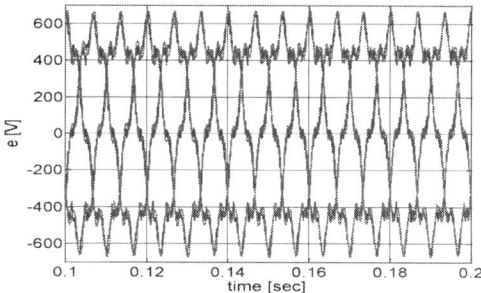

Fig. 13. Simulated (blue) and estimated using SI-FAO (red) back-emf e (filter capacitor voltage) with h=7, while 7th harmonic is superimposed on grid voltage e_g

V. SIMULATION RESULTS

All simulations in this paper are performed in MATLAB with SimPowerSystems Simscape™ toolbox. Same parameters are used in simulation as in Danfoss 800kW AFE drive used in experiment. Switching frequency F_{SW} is 4 kHz, while control frequency F_{CON} is 8 kHz. System parameters are shown in Table I. Maximum resonant frequency (at stiffest grid) is around 1378 Hz which is close to one third of switching frequency, so with proper delay compensations described above there were no resonance excitations. Simulation Figures 14-17

contain overlaid simulated (blue) and estimated with FASO (red) filter capacitor voltages (top), DC link voltage (middle) and AFE output current (bottom).

TABLE I
SYSTEM PARAMETERS

Symbol	Value	Symbol	Value
L	100μH	F_{SW}	4kHz
C	300μF	F_{CON}	8kHz
L_g	80μH	K_P^{OBS}	800 L
C_{DC}	32mF	K_I^{OBS}	40000 L
e_g (l-l)	690 V	K_P^{PLL}	200
V_{DC}^{*}	1030 V	K_I^{PLL}	20000

Proposed sensorless control algorithm is tested in simulation with following test cases:

A. 100% step load transients in both motoring and generating case under nominal inductance L

B. 100% motoring step load transients with ±20% error in inductance L

C. Dynamics performance under ±20% step change in grid frequency ω, under full motoring load.

A. Step Load Transients with Nominal Inductance

Fig. 14 and 15 show step load transients ON and OFF with nominal inductance L, in motoring and generating case respectively. In both cases the response is fast and well behaved, while estimated filter capacitor voltage has zero error and it has better filtering of switching harmonics.

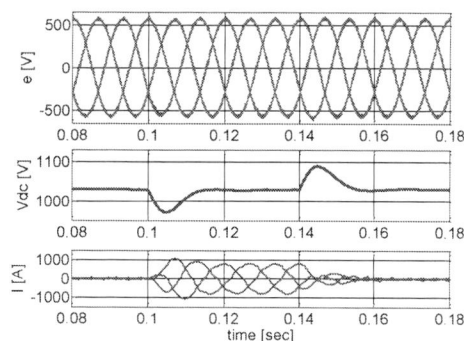

Fig. 14. 100% motoring step load transients with L=L_NOM

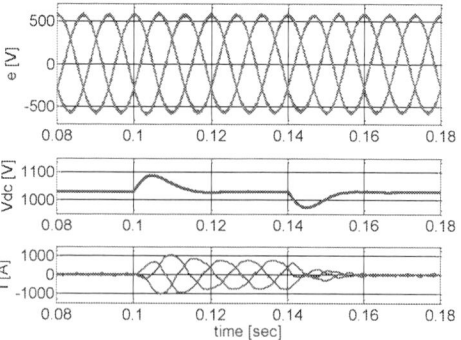

Fig. 15. 100% generating step load transients with L=L_NOM

B. Step Load Transients with 20% Error in Inductance

LCL parameters L_g and C are not used for control, except to possibly modify reactive current references to achieve desired power factor, which does not affect fast dynamics. To further show robustness of proposed algorithm, converter side inductor parameter variations are taken into account.

Fig. 16 shows motoring step load transients ON and OFF when converter inductor is 20% larger than nominal L, while nominal L is still being used for control. Fig. 17 shows similar transient response when converter inductor is 20% smaller than nominal L, while nominal L is used for control. In both cases the response is almost identical to the response with nominal parameters in Fig. 14, fast and well behaved.

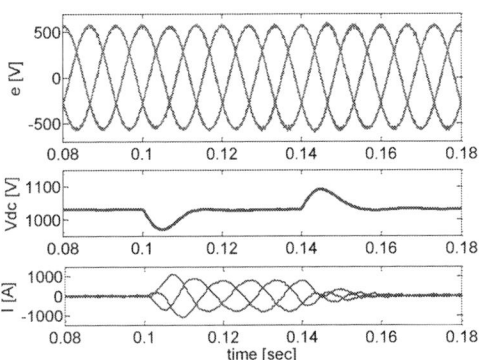

Fig. 16. 100% motoring step load transients with L=1.2L_{NOM}

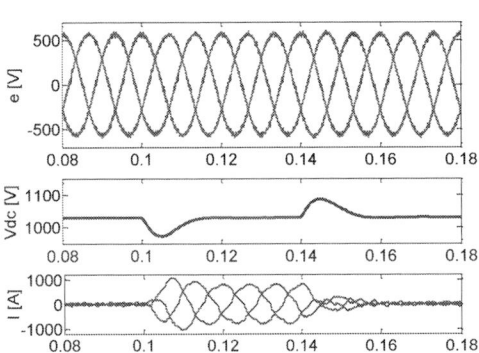

Fig. 17. 100% motoring step load transients with L=0.8L_{NOM}

C. Dynamic Performance with Step Changes in Grid Frequency under Full Load

In this section proposed algorithm performance is tested for cases of step changes of grid frequency. This is a hard test for sensorless control (as well as control with sensors), because if step (or fast) change in supply frequency occurs, power flow is no longer possible to control (as the angle of the grid voltage is drifting) and if control does not resynchronize quickly, AFE would likely trip due to overcurrent or overvoltage, especially under high load. Simulation Figures 18-19 contain overlaid

simulated (blue) and estimated with FASO (red) filter capacitor voltages (top), estimated frequency ω from PLL (middle) and AFE output current (bottom).

Fig. 18 shows dynamics performance of proposed control algorithm when grid frequency ω changes from 50Hz to 60Hz instantaneously under full motoring load, while Fig 19 shows the case when ω changes instantly from 50Hz to 40Hz also under full motoring load. In both cases PLL converged fast and filter capacitor voltages are estimated quickly with almost no phase lag, and overall operation of the proposed control was stable and well behaved.

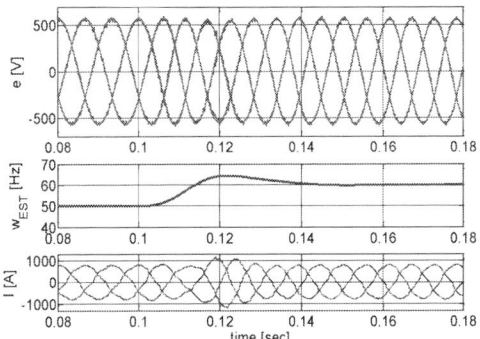

Fig. 18. +20% step change in grid frequency under nominal load

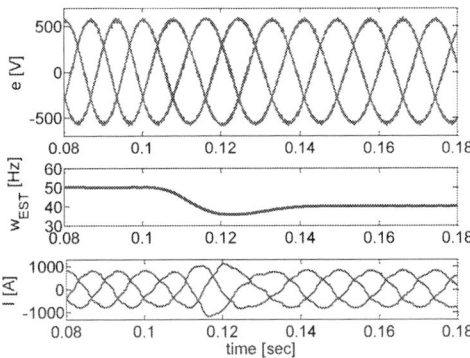

Fig. 19. -20% step change in grid frequency under nominal load

VI. EXPERIMENTAL RESULTS

Fig. 20 shows 800kW Danfoss AFE drive setup where proposed control algorithm is implemented. Parameters are listed in Table I. AFE is connected to the grid through 2.5MVA transformer, and grid voltage at low side is 690V. DC link voltage is regulated to 1030V, which results in about 5.5% boost on the rectified voltage, which is reasonable for high power AFE drive. Experimental results are performed on 60Hz grid and nominal parameter L is used for control.

For experimental results, only step load transients ON and OFF are performed since it was not possible to change supply frequency for high power AFE drive in the lab.

Fig. 20. Experimental setup

Step load transient ON is created by placing ~1.7Ω high power resistor in parallel with DC link. ~1.7 Ω resistor results in approximately 624kW step load transient. Step load transient OFF is created by running the induction motor coupled with dynamometer at load of approximately 624kW motoring, and suddenly removing the load by activating the safety function which opens all gates of motor drive inverter creating the step transient.

Fig. 21 shows ~624kW motoring step load transient ON, while Fig. 22 shows about the same size step load transient OFF.

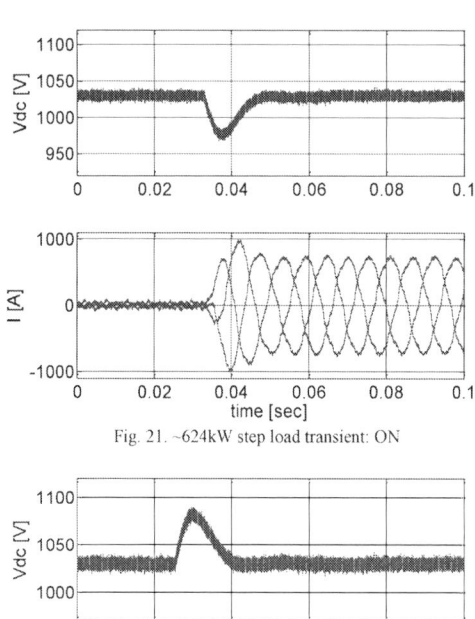

Fig. 21. ~624kW step load transient: ON

Fig. 22. ~624kW step load transient: OFF

In both experimental test cases dynamic response is well behaved. Similar response is seen in experimental and simulation results from similar test cases in Fig. 14. Differences between experimental and simulation results are: experimental results show bigger DC link ripple due to switching; simulation step load transient are performed on 50Hz grid while experiment could only be performed on 60Hz grid; and step loads were slightly bigger in simulation due to lab limitation.

VII. CONCLUSION

This paper presented novel control algorithm of grid connected VSC with LCL line filter without grid or filter capacitor voltage sensors, measuring only converter inductor currents and DC link voltage, and using only converter side inductor value for control. Two versions of frequency adaptive observers that estimate filter capacitor voltage, its frequency and angular position are presented.

Frequency adaptive synchronous observer (FASO) in synchronous reference frame was presented first, whose goal is to estimate only fundamental component of the filter capacitor voltage with zero error (phase and amplitude). Secondly, synchronous integrator based frequency adaptive observer (SI-FAO) is introduced which is able to estimate multiple harmonics of filter capacitor voltage with zero error.

The algorithm performed excellent in cases of: step load transient conditions; converter side inductor parameter uncertainties; step changes in grid frequency under full load. Presented method provides robust, low cost, with minimal number of sensors and minimal parameter knowledge, control solution for grid connected VSC.

REFERENCES

[1] V. Blasko, V. Kaura, "A New Mathematical Model and Control of a Three-Phase AC-DC Voltage Source Converter," *IEEE Transactions on Power Electronics*, vol. 12, no. 1, pp. 116-123, January, 1997.

[2] M. Malinowski, M.P Kazmierkowski, S. Hansen, F. Blaabjerg, G.D. Marques, "Virtual-Flux Based Direct Power Control of Three-Phase PWM Rectifiers," *IEEE Transactions on Industry Applications*, vol. 37, no. 4, pp. 1019-1027, July/August 2001.

[3] M. Malinowski, M. Jasinski, M.P Kazmierkowski, "Simple Direct Power Control of Three-Phase PWM Rectifier Using Space-Vector Modulation (DPC-SVM)," *IEEE Transactions on Industrial Electronics*, vol. 51, no. 2, pp. 447-454, April 2004.

[4] I. Agirman, V. Blasko, "A Novel Control Method of VSC Without AC Line Voltage Sensors," *IEEE Transactions on Industry Applications*, vol. 39, no. 2, pp. 519-524, March/April 2003.

[5] K. Lee, T.M. Jahns, T.A. Lipo, V. Blasko, R.D. Lorenz, "Observer-Based Control Metods for Combined Source-Voltage Harmonics and Unbalance Disturbances in PWM Voltage-Source Converters," *IEEE Transaction on Industry Application*, vol. 45, no. 6, pp. 2010-2021, Nov/Dec, 2009.

[6] K. Lee, T. M. Jahns, T. A. Lipo, and V. Blasko, "New control method including state observer of voltage unbalance for grid voltage-source converters," *IEEE Transactions on Industry Electronics*, vol. 57, no. 6, pp. 2054–2065, Jun. 2010.

[7] R. D. Lorenz, "Observers and state filters in drives and power electronics," *J. Elect. Eng.*, vol. 2, pp. 4–12, 2002.

[8] H. Kim, M. C. Harke, and R. D. Lorenz, "Sensorless control of interior permanent magnet machine drives with zero-phase lag position estimation," *IEEE Transactions on Industry Applications*, vol. 39, no. 6, pp. 1726–1733, Nov./Dec. 2003.

[9] M. C. Harke, L. A. de S. Ribeiro, and R. D. Lorenz, "Disturbance rejection limitations of back-EMF based sensorless PM drives," *in Proc. Eur. Conf. Power Electron. Appl.*, Aalborg, Denmark, 2007, pp. 1–10.

[10] B. Hafez, A. S. Abdel-Khalik, A. M. Massoud, S. Ahmed, and R. D. Lorenz, "Single-Sensor-Based Three-Phase Permanent-Magnet Synchronous Motor Drive System With Luenberger Observers for Motor Line Current Reconstruction," *IEEE Transactions Industry Applications*, vol. 50, no. 4, pp. 2602–2613, Jul./Aug. 2014.

[11] R. W. Hejny and R. D. Lorentz, "Evaluating the practical low-speed limits for back-EMF tracking-based sensorless speed control using drive stiffness as a key metric," *IEEE Transactions on Industrial Electronics*, vol. 47, no. 3, pp. 1337–1343, May/Jun. 2011.

[12] V.Miskovic, V.Blasko, T.M.Jahns, R.D. Lorenz, C.J. Romenesko, H. Zhang, "Synchronous Frame and Resonant Adaptive Observer as Disturbance Estimators and Their Applications in Power Electronics," *Proc. Energy Conversion Congress and Exposition (ECCE)*, Pittsburgh, PA, 2014, pp. 1248 – 1255.

[13] V.Miskovic, V.Blasko, T.M.Jahns, R.D. Lorenz, "Model Predictive Current Control of Voltage Source Converters with Back-Emf Estimation via Synchronous Reference Frame Observer," *Proc. International Conference on Industrial Technology (ICIT)*, Seville, Spain, 2015, pp. 2302 – 2307.

[14] J. Kukkola, M. Hinkkanen, "State Observer for Sensorless Control of a Grid-Connected Converter Equipped With an LCL Filter: Direct Discrete-Time Design," *Proc. Energy Conversion Congress and Exposition (ECCE)*, Montreal, QC, 2015, pp. 5511 – 5518.

[15] V. Miskovic, V. Blasko, T.M. Jahns, A.H. Smith and C. Romenesko, "Observer-Based Active Damping of LCL Resonance in Grid-Connected Voltage Source Converters," *IEEE Transactions on Industry Applications*, vol. 50, no. 6, pp. 3977–3985, Nov./Dec. 2014.

[16] V. Kaura, V. Blasko, "Operation of a phase locked loop system under distorted utility conditions," *IEEE Transactions on Industry Applications*, Vol.33, No.1 pp. 58-63, 1997.

[17] V. Blasko, L. Arnedo, P. Kshirsagar, S. Dwari, "Control and Elimination of Sinusoidal Harmonics on Power Electronics Equipment: System Approach," *Proc. Energy Conversion Congress and Exposition (ECCE)*, Phoenix, AZ, 2011, pp. 2827 – 2837.

Active stabilization of direct matrix converter input side filter through grid current control

Martin Leubner, Nico Remus, Marc Stübig and Wilfried Hofmann

Chair of Electrical Machines and Drives, Institute of Electrical Engineering, Technical University Dresden

Helmholtzstrasse 9, 01069 Dresden, Germany

Telephone: +49 (351) 463–39298; Fax: +49 (351) 463–33655

Email: martin.leubner@tu-dresden.de; Website: http://eeiema.et.tu-dresden.de

Abstract—**This paper proposes a combined control strategy for direct matrix converters (DMC) to stabilize the input side filter without any passive damping, simultaneously realizing the field oriented control (FOC) of an induction machine (IM) at the output. The speed control of the FOC with subsidiary torque control is left untouched. The rotor flux control loop is extended to a cascaded structure that additionally serves the needs for stabilization of the input filter, especially the active grid current. The reactive grid current is controlled by separate control loops that depend on the output side conditions created by the FOC. Linear controllers are used, which demand linearization of the instantaneous power based control loops. The functionality is proven by experimental results.**

I. Introduction

Since the DMC was shown first in 1976 [1] as "force commutated cyclo converter", it remains mainly a topic of academic research. Like back to back converters with DC link, it features bidirectional power flow, controllable input power factor as well as sinusoidal grid and output currents with arbitrary frequency and magnitude. The characteristic advantages are the high power density and the theoretical low failure probability due to the nonexistent DC link storage. For example, in [2] the realization of a SiC-based DMC is shown. The author states the potential of a high power density of $20\,\text{kW}$ per dm^3.

When the DMC is connected to the power grid, a filter is necessary between the switch matrix and the grid to serve the local codes for grid power restrictions. In [3] a LC filter is figured as the most economical solution and so it will be used for further investigations like shown in fig. 1. The LC filter is a second order system with a distinct resonance frequency. In order to ensure only allowed harmonic distortion in the resulting grid current [4], this frequency has to be damped. As shown in [5] a resistor in parallel to the inductor would increase the damping of the resonance frequency, but also would lead to a reduced damping of the switching frequency. A resistor in parallel with the filter capacitor would damp only the resonance frequency, but also leads to unacceptable losses in the filter. To realize the advantages of such parallel resistor without increasing overall losses, an extended control scheme can be used.

The recent works about damping oscillations of input side LC filter with a DMC are shown in [6] and [7]. There, a virtual parallel resistor to the filter capacitor is used to realize an active damping strategy which reduces harmonics of a damped or undamped input filter. This idea originates from [8] where it is used for a PWM driven current source inverter. In this paper the idea of the virtual resistor is discarded and instead a control of the grid current itself will be used. The similar idea of stabilization through grid power control is realized for a DMC with direct torque control already and shown in [9].

II. Basic assumptions

The considered DMC realizes a FOC of a squirrel cage IM like described in [10]. This is done with the symmetric space vector modulation (SSVM) mentioned in [11] together with the commutation strategy specified in [12]. Using the power-variant transformation in [13], the machine side values are oriented on the rotor flux angle φ_s (index: dq), the grid side will be oriented on the grid voltage angle φ_g (index: DQ) and the input of the switch matrix is oriented on the input voltage angle φ_i (index: D'Q'). The difference between D'Q' and DQ from (1) might be negligible, but is required to make some necessary simplifications using the linear control loops.

$$\varphi_{gi} = \varphi_g - \varphi_i \tag{1}$$

All equations in this paper are shown in the Laplace domain with s as Laplace variable. To design the controllers, the control loops are linearized using Taylor linearization.

III. Active grid current control

A. Input voltage control

The filter input voltage control uses the IM magnetizing current i_{sd} to achieve transient voltage changes of the input voltage $u_{iD'}$ at filter capacitor C_F. The correlation between the electrical values of the machine and input side can be deduced from the instantaneous power theory described in [14]–[16]. Neglecting converter losses the IM stator power p_s is equal the input power of the DMC p_i. This way energy can be transferred controlled between the IM stator flux and C_F.

$$p_s = \underbrace{\frac{3}{2}u_{sd}i_{sd}}_{p_{sd}} + \underbrace{\frac{3}{2}u_{sq}i_{sq}}_{p_{sq}} = \frac{3}{2}\left(u_{iD}i_{iD} + u_{iQ}i_{iQ}\right) = p_i \tag{2}$$

978-1-4673-9551-9/16 $31.00 © 2016 IEEE

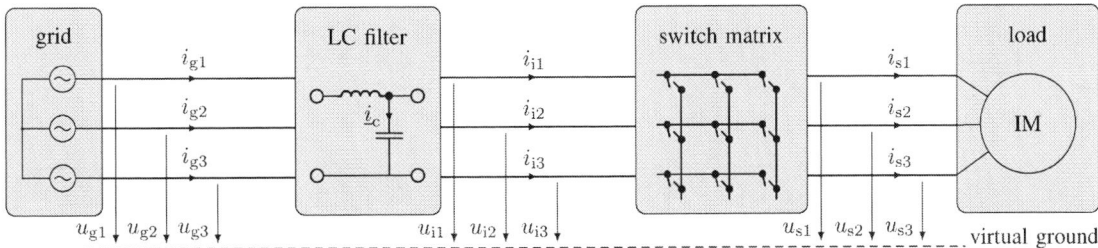

Fig. 1. General setup of the DMC installation

A control path can be achieved starting with the squirrel cage induction machine equations in rotor flux oriented coordinates (3) - (6) according to [17].

$$\underline{u}_s = R_s\underline{i}_s + j\omega_s\underline{\psi}'_s + s\underline{\psi}'_s \quad (3)$$

$$\underline{u}_r = R_r\underline{i}_r + j\omega_r\underline{\psi}_r + s\underline{\psi}_r = 0 \quad (4)$$

$$\underline{\psi}'_s = L_s\underline{i}_s + L_m\underline{i}_r \quad (5)$$

$$\underline{\psi}_r = L_m\underline{i}_s + L_r\underline{i}_r \quad (6)$$

Replacing the stator flux linkage $\underline{\psi}_s$ and the rotor current $\underline{i}_r$ in the stator voltage $\underline{u}_s$ results in (7) where the machine's stator resistance R_s, stator inductance L_s, inductance dispersion σ, rotor inductance L_r and mutual inductance L_m are supposed to be constant.

$$u_{sd} = \underbrace{i_{sd}(R_s + s\sigma L_s)}_{u_{sd,c}} \underbrace{-\omega_s\sigma L_s i_{sq} + s(1-\sigma)\frac{L_s}{L_m}\psi_{rd}}_{u_{sd,e}} \quad (7)$$

Similar to the traditional FOC the equation has to split up in the part which will be considered for the linear control loop $u_{sd,c}$ and the additional part $u_{sd,e}$. $u_{sd,e}$ contains the transformation based cross coupling as well as the derivation of the rotor flux linkage $\underline{\psi}_r$ which is assumed to change slowly. The machine side orientation is defined through (9). There, the rotor time constant T_r is introduced. The angular speed ω_s of the machine side vector space is estimated via the i_s-ω-model as shown in (10). This follows from (4) being set equal to zero. Besides the already mentioned parameters it takes the mechanical angular speed of the rotor ω_{mec}, the number of pole pairs z_p and the slip frequency ω_r of the IM into account.

$$\sigma = 1 - \frac{L_m^2}{L_s L_r} \quad (8)$$

$$\psi_{rd} = \frac{L_m i_{sd}}{1 + sT_r}; \quad \psi_{rq} = 0 \quad (9)$$

$$\omega_s = s\varphi_s = z_p\omega_{mec} + \underbrace{\frac{L_m i_{sq}}{T_r\psi_{rd}}}_{\omega_r} \quad (10)$$

Neglecting the cross coupling to the torque building current i_{sq} and the derivation of ψ_{rd}, the linear behavior in the Laplace domain between i_{sd} and $u_{sd,c}$ is shown in fig. 2a. There, i_{sd} is given by the subsidiary stator current control loop

of the original FOC with its substitute time constant T_{CC}. In turn the stator current control loop gets its reference from the PI input voltage controller. Considering (2), p_s leads to an appropriate input current of the switch matrix $i_{iD'}$. This can be done looking at the input voltage $\underline{u}_i$ in the $D'Q'$ vector space. Because of the chosen orientation the imaginary part of the input voltage $u_{iQ'}$ is equal to zero. The last part of the control path is given by the input voltage equation (13) where the active grid current $i_{gD'}$ is seen as a disturbance.

In absence of an optimization criterion in common literature, the PI controller for the linearized D' voltage control loop has to be optimized in the frequency domain. This can be done using a Bode diagram, which exemplary can be found in [18]. The methods of liniearization are listed there as well. To achieve a robust controller it is assumed that i_{sd} is equal to the machines rated current $\hat{I}_n$ and $u_{iD'}$ is equal to the nominal grid phase voltage $\hat{U}_{g,ph}$. Trying to achieve the ideal command response $|F(j\omega)| = 1$ by a coefficient comparison only leads to (12). The calculated dependency between the controller parameters can be used as an orientation for bode optimization. A common design of a PI controller is shown in (11). Corresponding parameters are defined for each control loop by alternating the indices.

$$PI(s) = \frac{K_{Rx}(1 + sT_{Nx})}{sT_{Nx}} \quad (11)$$

$$K_{RUID} = \frac{T_{NUID}R_S C_F \hat{U}_{g,ph}}{2\hat{I}_n(\sigma T_s T_{NUID} - T_{NUID}T_{CC} - \sigma T_s T_{CC})} \quad (12)$$

Although the already described elements would lead to a stable control loop an additional high-pass filter with the time constant T_{HP} has to be placed. It damps low frequency distortions in the input voltage error, which will result out of non ideal symmetric grid and input voltages, LC filter element parameters as well as non ideal measurements of voltages and currents.

B. Grid current control

Being able to regulate u_{iD}, a control of the active grid current i_{gD} must be achieved to stabilize the input side LC filter and reduce the harmonics caused by the resonance behavior. In fig. 3 the equivalent circuit of the filter in the grid voltage referenced DQ vector space is shown.

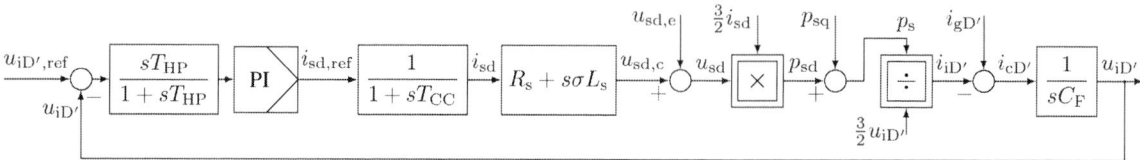

(a) D′ input voltage control loop

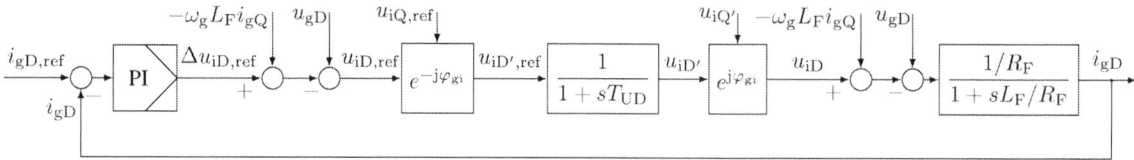

(b) Active grid current control loop

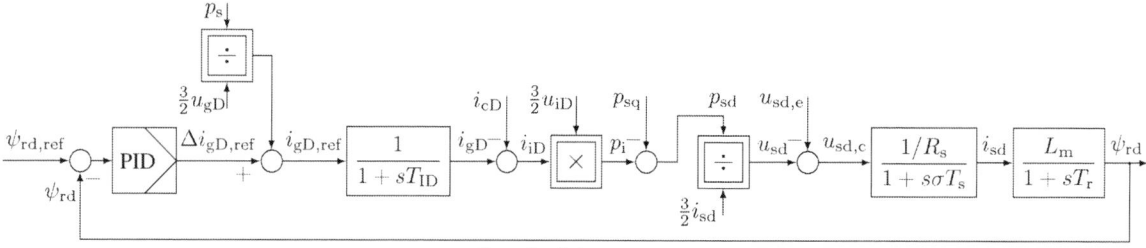

(c) Rotor flux linkage control loop

Fig. 2. Control loops for active current stabilization

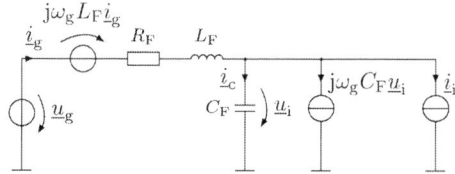

Fig. 3. LC filter in grid voltage referenced vector space

The filter coil is represented by its inductance L_{F} and its resistance R_{F}. The grid voltage referenced vector space is moving with the angular speed ω_{g}. Together with (13), (15) and (16) all this leads to the control loop in fig. 2b.

$$u_{\mathrm{iD}} = \frac{1}{sC_{\mathrm{F}}}\left(i_{\mathrm{gD}} - i_{\mathrm{iD}}\right) \tag{13}$$

$$u_{\mathrm{iQ}} = \frac{1}{sC_{\mathrm{F}}}\left(i_{\mathrm{gQ}} - i_{\mathrm{iQ}}\right) \tag{14}$$

$$\omega_{\mathrm{g}} = s\varphi_{\mathrm{g}} \tag{15}$$

$$i_{\mathrm{gD}} = \frac{1/R_{\mathrm{F}}}{1 + sL_{\mathrm{F}}/R_{\mathrm{F}}}\left(u_{\mathrm{gD}} - u_{\mathrm{iD}} + \omega_{\mathrm{g}}L_{\mathrm{F}}i_{\mathrm{gQ}}\right) \tag{16}$$

$$i_{\mathrm{gQ}} = \frac{1/R_{\mathrm{F}}}{1 + sL_{\mathrm{F}}/R_{\mathrm{F}}}\left(-u_{\mathrm{iD}} - \omega_{\mathrm{g}}L_{\mathrm{F}}i_{\mathrm{gD}}\right) \tag{17}$$

The grid current is controlled by a PI controller too, and the subsidiary voltage control loop in 2a is replaced by a PT$_1$

element with the substitute time constant T_{UD}. Knowing the transfer function of the linearized control loop, a simplified substitute time constant can be defined by the time where a step response reaches the 63 % value. The resulting control loop can be optimized using the in [19] mentioned amplitude optimum (AO) shown in (18) and (19).

$$K_{\mathrm{RIGD}} = \frac{L_{\mathrm{F}}}{2\,T_{\mathrm{UD}}} \tag{18}$$

$$T_{\mathrm{NIGD}} = \frac{L_{\mathrm{F}}}{R_{\mathrm{F}}} \tag{19}$$

C. Rotor flux linkage control

At least one additional control loop is necessary to control ψ_{rd} and generate an appropriate active grid current reference value $i_{\mathrm{gD,ref}}$. As shown in fig. 2c, $i_{\mathrm{gD,ref}}$ is primary defined by p_{s}. The additional control loop is needed because of unpredictable losses in semiconductors or the input side filter elements. It is based on (2) to (10) as well as the active grid current control substitute time constant T_{ID}. If more or less power is retrieved by the grid side as demanded by the IM operation point, the rotor flux changes. This results in the PID controlled additional active grid current reference $\Delta i_{\mathrm{gD,ref}}$. A common design of a PID controller is shown in (20). Here, u_{gD} and u_{iD} are approximated with $\hat{U}_{\mathrm{g,ph}}$. But to achieve a robust controller i_{sd}, different to the voltage control loop, is

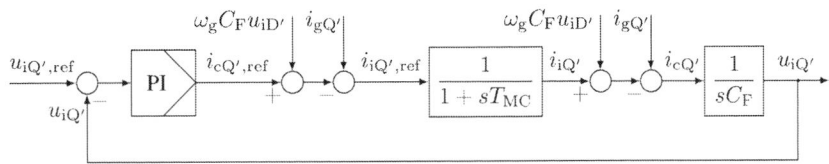

(a) Q' input voltage control loop

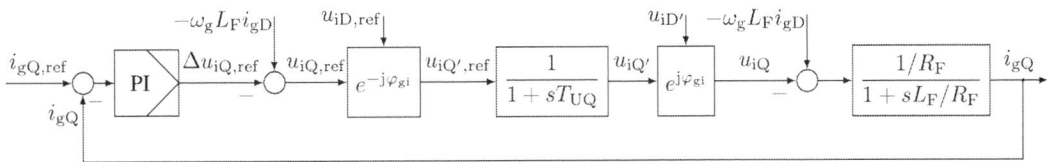

(b) Reactive grid current control loop

Fig. 4. Control loops for reactive current stabilization

considered at rated operation $i_{\mathrm{sd,n}}$. The controller parameters resulting from the AO are listed in (21) to (23).

$$\mathrm{PID}\,(s) = \frac{K_{\mathrm{R}x}\,(1 + sT_{\mathrm{N}x})\,(1 + sT_{\mathrm{V}x})}{sT_{\mathrm{N}x}} \tag{20}$$

$$K_{\mathrm{R}\psi} = \frac{T_{\mathrm{r}}\,i_{\mathrm{sd,n}}\,R_{\mathrm{s}}}{2\,\hat{U}_{\mathrm{g,ph}}\,L_{\mathrm{m}}\,T_{\mathrm{ICD}}} \tag{21}$$

$$T_{\mathrm{N}\psi} = T_{\mathrm{r}} \tag{22}$$

$$T_{\mathrm{V}\psi} = \sigma T_{\mathrm{s}} = \frac{\sigma L_{\mathrm{s}}}{R_{\mathrm{s}}} \tag{23}$$

D. Limitations

Because of the power based perspective it can not be allowed to emit a negative magnetizing current reference $i_{\mathrm{sd,ref}}$ by the input voltage controller in fig. 2a. Since a negative current would not lead to a negative input power the control loop would become instable. This also shows another limitation that only the amount of energy in the transient inductance σL_{s} can be used to stabilize the input side filter.

IV. REACTIVE GRID CURRENT CONTROL

A. Input voltage control

In contrast to the D' input voltage control of the active grid current, no adjustment of a IM side value is needed here. Nevertheless, the control of input voltage $u_{\mathrm{iQ'}}$ still depends on the machines operating point. To generate an appropriate input current $i_{\mathrm{iQ'}}$ a defined phase displacement φ_{ui} between input voltage and modulated input current has to be assigned to the SSVM. Based on [20] φ_{ui} can be calculated through (24) to (26) considering the IM stator current $\underline{i}_{\mathrm{s}}$ and the phase displacement φ_{s} between $\underline{u}_{\mathrm{s}}$ and $\underline{i}_{\mathrm{s}}$.

$$p_{\mathrm{i}} = \frac{3}{2}\,|\underline{u}_{\mathrm{i}}|\,|\underline{i}_{\mathrm{i}}|\,\cos\,(\varphi_{\mathrm{ui}}) = \frac{3}{2}\,|\underline{u}_{\mathrm{s}}|\,|\underline{i}_{\mathrm{s}}|\,\cos\,(\varphi_{\mathrm{s}}) = p_{\mathrm{s}} \tag{24}$$

$$\varphi_{\mathrm{ui}} = \arccos\left(\frac{|\underline{u}_{\mathrm{s}}|\,|\underline{i}_{\mathrm{s}}|}{|\underline{u}_{\mathrm{i}}|\,|\underline{i}_{\mathrm{i}}|}\cos\,(\varphi_{\mathrm{s}})\right) \tag{25}$$

$$\varphi_{\mathrm{ui}} = \arccos\left(\frac{|\underline{u}_{\mathrm{s}}|\,|\underline{i}_{\mathrm{s}}|}{u_{\mathrm{iD'}}\sqrt{i_{\mathrm{iD'}}^{2} + i_{\mathrm{iQ'}}^{2}}}\cos\,(\varphi_{\mathrm{s}})\right) \tag{26}$$

Like shown in (26) the modulation works in the input voltage vector space D'Q'. While $i_{\mathrm{iQ'}}$ will be given by its reference $i_{\mathrm{iQ',ref}}$, $i_{\mathrm{iD'}}$ is defined by p_{s} and u_{iD} represented by (28).

$$p_{\mathrm{i}} = \frac{3}{2}\,u_{\mathrm{iD'}}\,i_{\mathrm{iD'}} = \frac{3}{2}\,|\underline{u}_{\mathrm{s}}|\,|\underline{i}_{\mathrm{s}}|\,\cos\,(\varphi_{\mathrm{s}}) = p_{\mathrm{s}} \tag{27}$$

$$i_{\mathrm{iD'}} = \frac{|\underline{u}_{\mathrm{s}}|\,|\underline{i}_{\mathrm{s}}|}{u_{\mathrm{iD'}}}\cos\,(\varphi_{\mathrm{s}}) \tag{28}$$

The realization of $i_{\mathrm{iQ'}}$ is summarized in a PT$_1$ element with the converter time constant T_{MC} as shown in fig. 4a. T_{MC} contains the time delay of measuring and calculation as well as realizing the converter output voltage as a sum of pulses. The shown control loop will be stabilized by the in [21] mentioned PI controller with symmetrical optimum (SO). The parameters of the controller are shown in (30) and (31). According to [18] a is changeable factor to tune the phase reserve ϕ_{RSV} of the IT$_1$ control loop, which is shown in (29).

$$\phi_{\mathrm{RSV}} = \arctan\,(a) - \arctan\left(\frac{1}{a}\right)\,;\quad a > 1 \tag{29}$$

$$K_{\mathrm{RUIQ}} = \frac{C_{\mathrm{F}}}{a_{\mathrm{UIQ}}\,T_{\mathrm{MC}}} \tag{30}$$

$$T_{\mathrm{NUIQ}} = a_{\mathrm{UIQ}}^{2}\,T_{\mathrm{MC}} \tag{31}$$

Comparing the input voltage control loops for $u_{\mathrm{iD'}}$ and $u_{\mathrm{iQ'}}$, the missing cross coupling in fig. 2a stands out. But because of the chosen orientation $u_{\mathrm{iQ'}}$ is always equal to zero. Consequently, the cross coupling at the u_{iD} control loop is equal zero too.

B. Grid current control

The last control loop to be defined is the one for the reactive grid current i_{gQ} in fig. 4b. Due to the chosen orientation $u_{\mathrm{gQ}} = 0$ and is not shown in fig. 4b. Here, the SO is used to tune the controller instead of AO. This is allowed as

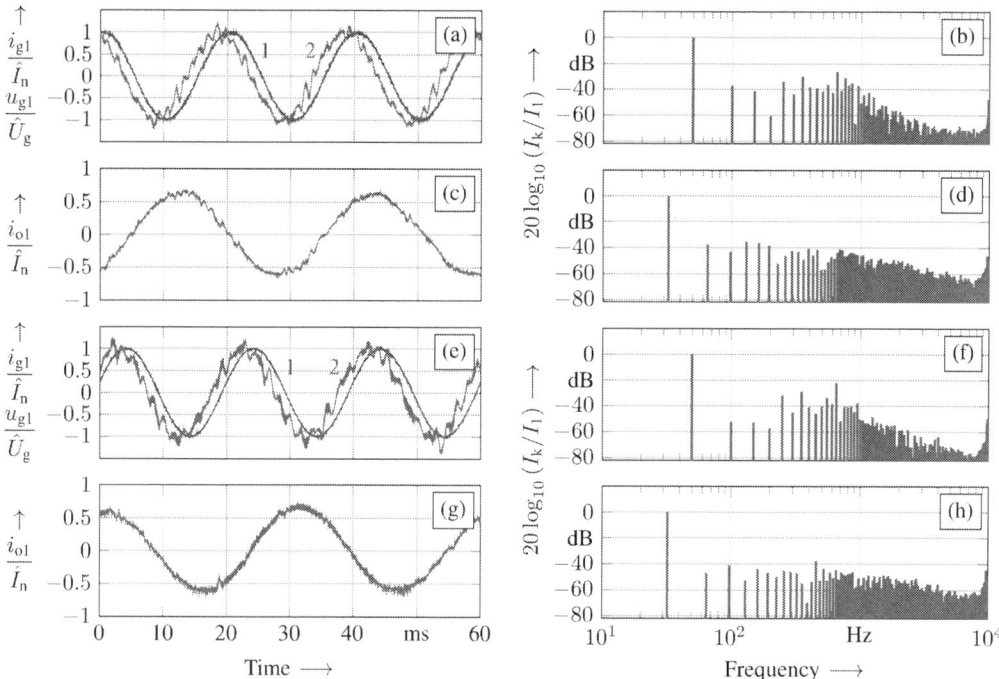

Fig. 5. Experimental results of the grid and output current with electric torque $m_\delta = \frac{1}{2}m_n$ and $n = \frac{2}{3}n_n$ for active stabilization ((a) to (d)) and classic FOC ((e) to (h)) with passive damping resistor $R_D = 50\,\Omega$ in parallel to L_F: (a,e)-1 grid voltage u_{g1}; (a,e)-2 grid current i_{g1}; (b,f) frequency spectrum of i_{g1}; (c,g) output current i_{o1}; (d,h) frequency spectrum of i_{o1}.

long as (32) is valid and done because of the SO's shorter disturbance settling time.

$$L_F/R_F \gg a_{IGQ}^2 T_{UQ} \tag{32}$$

$$K_{RIGQ} = \frac{L_F}{a_{IGQ} T_{UQ}} \tag{33}$$

$$T_{NIGQ} = a_{IGQ}^2 T_{UQ} \tag{34}$$

Limitations

According to [20] the output side modulation index h can be calculated through (35).

$$h = \frac{2\,|u_s|}{\sqrt{3}\,|u_i|\cos(\varphi_{ui})} \tag{35}$$

It shows that h and $\cos(\varphi_{ui})$ are indirectly proportional. This leads to a decision which of both should be prioritized. Here, the output side modulation index is first calculated assuming $\varphi_{ui} = 0$. Then the result is taken to calculate the limits of the input current $i_{iQ'}$. However, a certain amount of $i_{iQ'}$ is always needed for LC filter stabilization, which leads to reduced maximum output voltage of the DMC.

V. EXPERIMENTAL RESULTS

The introduced control strategy is implemented in a test bench. There, the structure shown in fig. 1 is coupled with

Fig. 6. DMC used in the test bench; (1) - control signals; (2) - gate driver; (3) - switch matrix connection; (4) - power semiconductor moduls; (5) - heat sink; (6) - optional passive damping resistor; (7) - filter inductor; (8) - filter capacitor

an additional IM with its own converter to control. This second IM generates the torque or speed to realize specific points of operation as well as transient changes in torque or speed. The parameters of the test bench equipment, necessary to reconstruct the measured values, are listed in tab. I, II and III. The DMC used is shown in fig. 6. A TMS320F28335

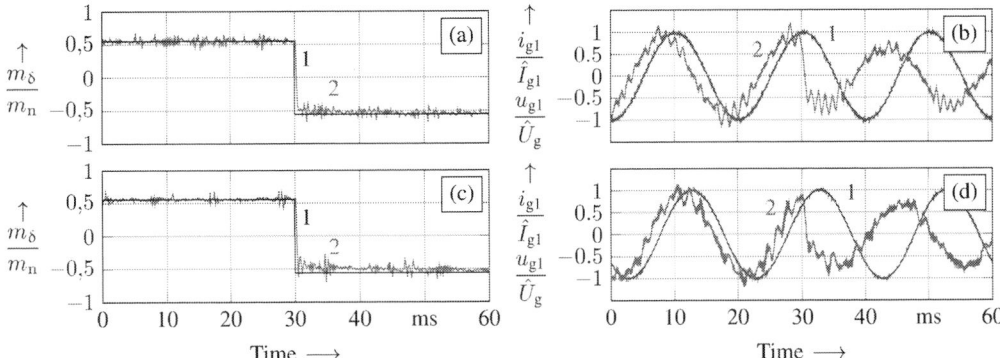

Fig. 7. Experimental results at electrical torque step $m_\delta = \frac{1}{2}m_n$ to $m_\delta = -\frac{1}{2}m_n$ at $n = \frac{2}{3}n_n$ for active stabilization ((a) and (b)) and classic FOC ((c) and (d)) with passive damping resistor $R_D = 50\,\Omega$ in parallel to L_F: (a,c)-1 electrical torque reference $m_{\delta\,ref}$; (a,c)-2 electrical torque m_δ; (b,d)-1 grid voltage u_{g1}; (b,d)-2 grid current i_{g1}

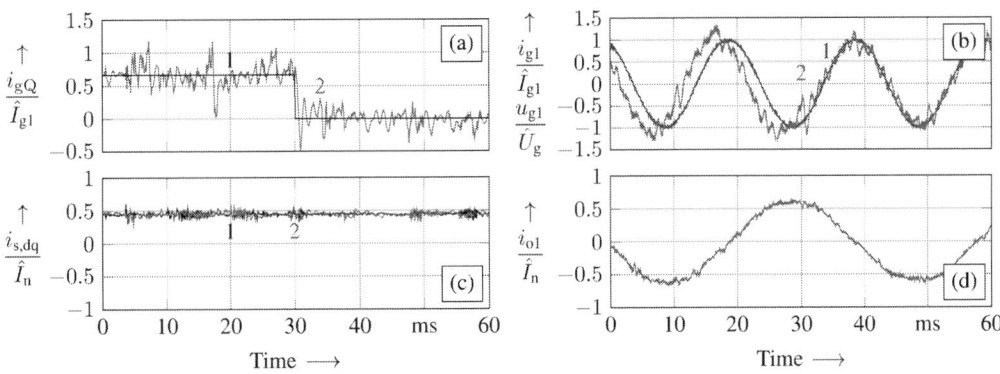

Fig. 8. Experimental results at reactive current step $i_{gQ,ref} = 2\,A$ to $i_{gQ,ref} = 0\,A$ at $n = \frac{1}{2}n_{rated}$: (a)-1 reactive grid current reference $i_{gQ,ref}$; (a)-2 reactive grid current i_{gQ}; (b)-1 grid voltage u_{g1}; (b)-2 grid current i_{g1}; (c)-1 torque building current i_{sq}; (c)-2 magnetizing current i_{sd}; (d) output current i_{o1}

TABLE I
PARAMETERS OF THE GRID, THE LC FILTER AND THE SWITCH MATRIX

$\hat{U}_{g,ph}$	$\sqrt{2}\,230\,V$	R_F	$0.05\,\Omega$	C_F	$19.5\,\mu F$
ω_g	$2\pi\,50\,Hz$	L_F	$1.9\,mH$	f_s	$10\,kHz$

TABLE II
PARAMETERS OF THE INDUCTION MACHINE

R_s	$0.694\,\Omega$	R_r	$0.612\,\Omega$	L_m	$146\,mH$
L_s	$151\,mH$	L_r	$151\,mH$	J_m	$39.7\,gm^2$
P_n	$3.6\,kW$	U_n	$330\,V$	I_n	$8.7\,A$
n_n	$1500\,min^{-1}$	m_n	$23\,Nm$	z_p	2
f_n	$51.7\,Hz$	$\cos\varphi$	0.86	$i_{sd,n}$	$5.5\,A$

TABLE III
PARAMETERS OF THE USED CONTROLLERS

K_{RUID}	$-20\,mS$	T_{HP}	$1.06\,ms$	$i_{sd,n}$	$5.5\,A$
T_{NUID}	$100\,\mu s$	T_{MC}	$100\,\mu s$	T_{CC}	$2\sqrt{2}\,T_{MC}$
a_{UIQ}	4	T_{UD}	$280\,\mu s$	T_{ID}	$600\,\mu s$
a_{IGQ}	3	T_{UQ}	$377\,\mu s$	T_r	$247\,ms$

TABLE IV
THD TILL THE 39TH OF THE CURRENTS IN FIG. 5

THD_{40}	i_{g1}	i_{o1}
without R_D	$7.8\,\%$	$4.2\,\%$
with R_D	$9.7\,\%$	$2.7\,\%$

micro controller from TI handles the control, from measuring till PWM, realizing the switching frequency f_s. Thereafter, a Spartan-3E FPGA from Xilinx realizes the mentioned commutation strategy between the phases with a dead time of $2\,\mu s$.

Fig. 5 shows measured values for the new control strategy without any passive damping and the classic FOC with a damping resistor R_D parallel to L_F. These measurements were taken at a stationary point of operation. There $i_{gQ\,ref}$ was fixed at 2 A to be comparable with the passive damped FOR without reactive current control. At first, fig. 5a and fig. 5c prove that the described control strategy leads to stabilization of the input side filter without using an additional damping resistor. The comparison of the THD_{40} values can be seen in tab. IV. The measuring shows that the saving of R_D results in a nearly equal THD_{40} of the grid current but leads to a slightly more distorted output current, which has to be expected by retracing the control loops. Comparing fig. 5b and 5f the steeper decline of the harmonics amplitude stands out because of the improved damping without R_D.

Fig. 7 and 8 show the behavior during a torque step and a reactive grid current step. In the first case the second IM provides the speed control. This way the fastest change of the i_{sq} proportional electrical torque m_δ at the DMC driven IM can be generated. It is shown that the stability remains while doing fast torque and reactive grid current changes.

VI. CONCLUSION

The presented stabilization strategy shows, that the input filter can operate without a passive damping resistor. The stability of the filter is then guaranteed by the introduced grid current control. Applying the new strategy will lead to comparable grid current distortion and a slightly higher distortion of the output current. This follows from the linear transmission of grid disturbances to the output side that can't be stabilized by the machines magnetizing current. Here, a first approach decouples those effects from the machine side with a simple first order high-pass filter. With respect to the input voltage control loop, a more advanced filter could reduce the distortions even more. Improvements in the strategy itself will come by reducing the converter time constant or the optimization of the input LC filter design with respect to the control strategy.

REFERENCES

[1] L. Gyugi and B. Pelly, *Static Power Frequency Changers: Theory, Performance and Applications*. New Yorker: Wiley, 1976.

[2] L. de Lillo, L. Empringham, M. Schulz, and P. Wheeler, "A high power density SiC-JFET-based matrix converter," in *Power Electronics and Applications (EPE 2011), Proceedings of the 2011-14th European Conference on*, 2011, pp. 1–8.

[3] P. Wheeler and D. Grant, "Optimised input filter design and low-loss switching techniques for a practical matrix converter," *IEE Proceedings-Electric Power Applications*, vol. 144, no. 1, pp. 53–60, 1997.

[4] DIN EN 61000-3-2 (IEC 61000-3-2:2005), Electromagnetic compatibility (EMC) - Part 3-2: Limits - Limits for harmonic current emissions (equipment input current $\leq$ 16 A per phase), 2006.

[5] H. She, H. Lin, X. Wang, and L. Yue, "Damped input filter design of matrix converter," in *Power Electronics and Drive Systems, 2009. PEDS 2009. International Conference on*, 2009, pp. 672–677.

[6] M. Rivera, C. Rojas, J. Rodríguez, P. Wheeler, B. Wu, and J. Espinoza, "Predictive current control with input filter resonance mitigation for a direct matrix converter," *IEEE Trans. Power Electron.*, vol. 26, no. 10, pp. 2794–2803, 2011.

[7] J. Lei, B. Zhou, X. Qin, J. Wei, and J. Bian, "Active damping control strategy of matrix converter via modifying input reference currents," *IEEE Trans. Power Electron.*, vol. 30, no. 9, pp. 5260–5271, 2015.

[8] J. Wiseman and B. Wu, "Active damping control of a high-power PWM current-source rectifier for line-current THD reduction," *IEEE Trans. Ind. Electron.*, vol. 52, no. 3, pp. 758–764, 2005.

[9] N. Remus and M. Leubner, "Direct control method for matrix converter with stabilisation of the input current," in *European Power Electronics and Applications - EPE. Geneva, Switzerland*, 2015.

[10] U. Riefenstahl, *Elektrische Antriebstechnik (Leitfaden der Elektrotechnik)*. Teubner, 2000.

[11] D. Casadei, G. Serra, A. Tani, and L. Zarri, "Matrix converter modulation strategies: a new general approach based on space-vector representation of the switch state," *IEEE Trans. Ind. Electron.*, vol. 49, no. 2, pp. 370–381, 2002.

[12] S. Krauss, N. Schwingal, and W. Hofmann, "Investigation of a 2/3-step voltage-based commutation method for matrix converters," in *Applied Power Electronics Conference and Exposition (APEC), 2012 Twenty-Seventh Annual IEEE*, 2012, pp. 397–404.

[13] IEC 62428:2008, Electric power engineering - Modal components in three-phase a.c. systems - Quantities and transformations, 2008.

[14] DIN 40110-1, Quantities used in alternating current theory; two-line circuits, 1994.

[15] DIN 40110-2, Quantities used in alternating current theory - Part 2: Multi-conductor circuits, 2002.

[16] H. Akagi, E. H. Watanabe, and M. Aredes, *Instantaneous Power Theory and Applications to Power Conditioning*. Wiley Interscience, 2007.

[17] K. Kovács, I. Rácz, and M. Kuzniarski, *Transiente Vorgänge in Wechselstrommaschinen. Bd. 2*. Akademiai Kiadó, Verl. d. Ungar. Akademie d. Wissenschaften, 1959.

[18] H. Lutz and W. Wendt, *Taschenbuch der Regelungstechnik*. Harri Deutsch, 2007.

[19] C. Kessler, *Über die Vorausberechnung optimal abgestimmter Regelkreise - Teil 3: Die optimale Einstellung des Reglers nach dem Betragsoptimum*. Regelungstechnik 3, pp. 16 - 22, 1955.

[20] L. Huber and D. Borojevic, "Space vector modulated three-phase to three-phase matrix converter with input power factor correction," *IEEE Trans. Ind. Appl.*, vol. 31, no. 6, pp. 1234–1246, 1995.

[21] C. Kessler, *Das symmetrische Optimum*. Regelungstechnik, part I pp. 395/400; part II pp. 432/436, 1958.

Impedance-Based Stability Analysis of Single-Phase Inverter Connected to Weak Grid with Voltage Feed-Forward Control

Jiangfeng Wang[1], Jianhui Yao[1], Haibing Hu[1], Yan Xing[1], Xiaobin He[2], Kai Sun[3]

1 Jiangsu Key Lab of New Energy and Power Conversion, Nanjing University of Aeronautics and Astronautics, Nanjing, China
2 Shanghai Institute Space Power Sources, Shanghai, China
3 State Key Lab of Power Systems, Department of Electrical Engineering, Tsinghua University, Beijing, China
wangjiangfeng@nuaa.edu.cn

Abstract—**Voltage feed-forward control (VFFC) is widely used due to its good low-frequency-harmonics suppression and easy implementation, however, it will worsen system stability under the weak grid condition. In this paper, the stability of single-phase grid-connected inverter with VFFC is studied based on impedance analysis method. With this method, the instability mechanism and the unstable area caused by VFFC are easily obtained. Based on this, system can be stabilized simply by adding phase-lead compensation, avoiding the real-time detection of grid impedance. Simulation and experimental results verify the analysis and the proposed control method.**

Keywords—voltage feed-forward control; system stability; weak grid; impedance analysis ; phase-lead compensation

I. INTRODUCTION

In a weak grid system, the stability of grid-connected inverters will be significantly influenced by grid impedance variations [1]-[3]. The impedance-based stability-assessment method has turned out to be an effective approach [4]-[8]. The well-known VFFC is widely used due to its sufficient grid disturbance rejection capability, such as sags, swells, flickers and harmonic pollution [9][10]. However, some literature reports that VFFC has negative impact on system stability under the weak grid condition [11]-[13]. The impact of VFFC is discussed in the three-phase L-type grid-tied inverter based on Generalized Nyquist stability Criterion, and a lower bandwidth of PLL is used to avoid the system instability [11]. However, the stability problem and this method have not been evaluated in a single-phase system. Ref [13] studies the stability problem introduced by VFFC in the single-phase system based on inverter control model, but the influence of delay time and phase-locked loop (PLL) are both ignored in the model. Possible solutions to deal with this problem are available in [6][13]. Ref [6] presents three methods to shape the inverter output impedance, in which a lead-lag compensator is added to the current regulator. This method can mitigate adverse impacts caused by VFFC and improve system robustness, however, the trade-off in the parameter design of current regulator and lead-lag compensator seems complicated. On the other hand, an adaptive current control is proposed in [13] to improve the control performance, but the real-time

detection of grid impedance is needed and unexpected noise may be introduced by the derivative feed-forward of measured grid impedance. In this paper, the stability of single-phase grid-connected inverter with VFFC is analyzed based on impedance stability criterion. The impedance model is established, with the delay time and PLL taken into account. Then the instability mechanism is studied based on impedance analysis. Finally, in order to improve the stability, a new phase-lead compensation method is proposed.

II. IMPEDANCE MODELING

Fig. 1 shows the main circuit of a single-phase LCL-type grid-connected inverter. The LCL filter consists of an inverter-side inductor L_1, a filter capacitor C_f, and a grid-side inductor L_2. U_{in} is the input dc voltage, u_{inv} is the inverter output voltage, i_L is the inverter-side current sampled to realize the active damping, i_g is the current injected to the grid. The weak grid is modeled as an ideal grid voltage source u_g in series with grid impedance Z_g. The PCC voltage u_{pcc} is sampled for the feed-forward and synchronizing purpose.

Fig. 2 presents the scheme of the dual-loop current control with grid voltage feed-forward. As seen in Fig. 2, I_{ref} is the current amplitude command, and the phase angle θ of the PCC voltage fundamental is determined by synchronous reference frame PLL (SRF-PLL) algorithm executed with the sampling (control) frequency of 40 kHz. The digital delay $G_d(s)$ consists of computation delay and PWM generation delay, which are equal to one sampling period and a half sampling period respectively. K_{pwm} is the transfer function of inverter bridge, H_{ic} is the feedback coefficient of capacitor current, $G_i(s)$ is the PI

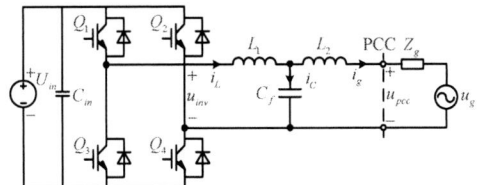

Fig. 1. Main circuit of single-phase LCL-type grid-connected inverter

---

This work is supported by the National Natural Science Foundation of China (51577088), the Industry-academic Joint Technological Innovations Fund Project of Jiangsu (BY2015003-008).

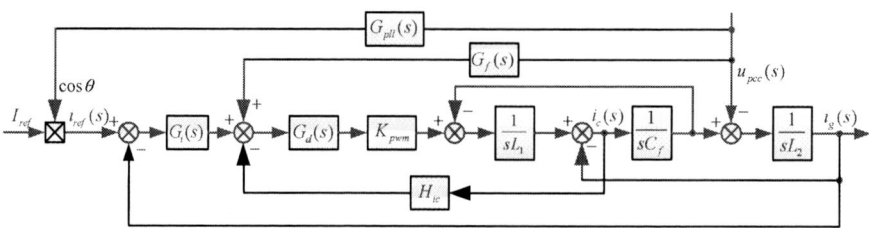

Fig. 2. Control block diagram of single-phase LCL-type grid-connected inverter

current regulator, $G_{pll}(s)$ is the transfer function of PLL, and the feed-forward function $G_f(s)$ equals to $1/K_{pwm}$.

According to Fig. 2, the current loop gain $T(s)$ can be derived as

$$T(s) = \frac{K_{pwm}G_i(s)G_d(s)}{L_1L_2C_fs^3 + H_{ic}K_{pwm}L_2C_fG_d(s)s^2 + (L_1 + L_2)s} \quad (1)$$

Furthermore, through equivalent transformation method, the injected grid current $i_g(s)$ can be written as

$$i_g(s) = i_s(s) - \frac{u_{pcc}(s)}{Z_{con}(s)} - \frac{u_{pcc}(s)}{Z_f(s)} \quad (2)$$

where $i_s(s)$ is the Norton equivalent current source. $Z_{con}(s)$ and $Z_f(s)$ are the impedance introduced by the current control loop and VFFC. They can be obtained as

$$i_s(s) = \frac{T(s)}{1+T(s)} I_{ref}G_{pll}(s)u_{pcc}(s) \quad (3)$$

$Z_{con}(s)$

$$= \frac{L_1L_2C_fs^3 + H_{ic}K_{pwm}L_2C_fG_d(s)s^2 + (L_1 + L_2)s + K_{pwm}G_i(s)G_d(s)}{L_1C_fs^2 + H_{ic}K_{pwm}C_fG_d(s)s + 1} \quad (4)$$

$Z_f(s)$

$$= -\frac{L_1L_2C_fs^3 + H_{ic}K_{pwm}L_2C_fG_d(s)s^2 + (L_1 + L_2)s + K_{pwm}G_i(s)G_d(s)}{K_{pwm}G_f(s)G_d(s)} \quad (5)$$

Considering the effect of PLL, the equivalent current source can be replaced by the negative parallel-impedance $Z_{pll}(s)$ [6][14], expressed as

$$Z_{pll}(s) = -\frac{1+T(s)}{I_{ref}G_{pll}(s)T(s)} \quad (6)$$

where the PLL transfer function $G_{pll}(s)$ is

$$G_{pll}(s) = \frac{1}{2} \frac{k_{p\_pll}(s - jw_0) + k_{i\_pll}}{(s - jw_0)^2 + U_m[k_{p\_pll}(s - jw_0) + k_{i\_pll}]} \quad (7)$$

In formula (7), U_m is the maximum value of grid voltage, k_{p_pll} and k_{i_pll} are parameters of the PLL regulator. Based on impedance modeling, the equivalent circuit of single-phase grid-connected inverter can be derived as Fig. 3. Then the inverter output impedance before and after adding VFFC can

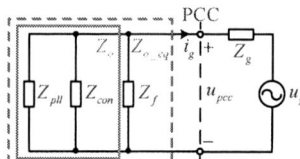

Fig. 3. Equivalent circuit of single-phase grid-connected inverter

be easily obtained, which are defined as $Z_o(s)$ and $Z_{o\_eq}(s)$ respectively, expressed as

$$Z_o(s) = \frac{Z_{pll}(s)Z_{con}(s)}{Z_{pll}(s) + Z_{con}(s)} \quad (8)$$

$$Z_{o\_eq}(s) = \frac{Z_o(s)Z_f(s)}{Z_o(s) + Z_f(s)} \quad (9)$$

On the right side of Fig. 3, considering the worst case, the grid impedance is regarded as a pure inductive component.

III. INFLUENCE OF VOLTAGE FEED-FORWARD CONTROL

Impedance-based stability analysis method is used in this paper due to its clear division of unstable area caused by VFFC, which will be analyzed in this section. Based on the theory proposed in [4], the system is stable only if the current-controlled inverter is stable with proper design of the current loop and PLL, and the impedance ratio $Z_g(s)/Z_o(s)$ should satisfy the Nyquist criterion. That means if $Z_g(s)$ and $Z_o(s)$ intersect at f_i, the phase margin (PM) must be a positive one. Here PM can be expressed as

$$PM = 90^0 + \arg[Z_o(j2\pi f_i)] \quad (10)$$

Based on (10), the critical frequency f_x is defined for the stability-boundary division, and the grid impedance corresponding to this frequency can be obtained by solving equations as follows

$$\begin{cases} |Z_o(f_x)| = |Z_g(f_x)| \\ \arg[Z_o(f_x)] = -90^0 \end{cases} \quad (11)$$

Before adding VFFC, different impedance of the equivalent circuit is depicted in Fig. 4. It can be observed that, when $Z_g(s)$ and $Z_o(s)$ intersect at frequencies above the critical frequency f_x, the phase gain is a positive one which means the system is always stable. On the other hand, with a low bandwidth of PLL, both the magnitude and phase of $Z_o(s)$ overlaps $Z_{con}(s)$ at

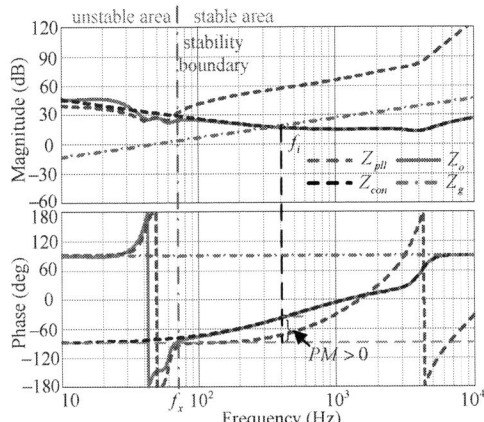

Fig. 4. Bode plot of impedance introduced by PLL (blue) and current loop (black), the inverter output impedance without VFFC (red) and the grid impedance (pink)

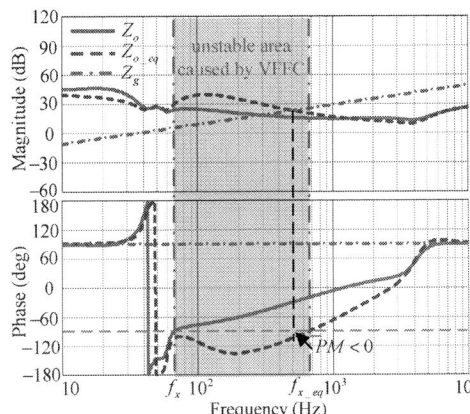

Fig. 5. Bode plot of inverter output impedance without VFFC (red) and with VFFC (blue)

frequencies above the critical frequency, meaning that $Z_{pll}(s)$ only affects the inverter output impedance within low-frequency range. Therefore, the low bandwidth PLL hardly influences the performance of system stability.

After adding VFFC, the shaped output impedance becomes $Z_{o\_eq}(s)$. Fig. 5 depicts the bode plot of $Z_o(s)$ and $Z_{o\_eq}(s)$ for comparison. The magnitude of $Z_{o\_eq}(s)$ is higher than $Z_o(s)$ in low-frequency range, indicating that VFFC improves the low-frequency-harmonics suppression ability. However, system instability occurs due to a large phase-lag introduced by VFFC, which results in a negative phase gain at intersection frequency. After adding VFFC, the critical frequency becomes

f_{x_eq}, and it is straightforward to show the unstable area caused by VFFC in Fig. 5 with the impedance-based analysis method, while it cannot be obtained easily based on inverter control model.

IV. PHASE-LEAD COMPENSATION METHOD

Due to the large phase-lag caused by VFFC, phase-lead compensation is used to improve system robustness. The phase-lead compensation can be added either in the forward path or feedback path of the outer current loop, as shown in Fig. 5(a) and (b). The compensation function $G_c(s)$ can be expressed as

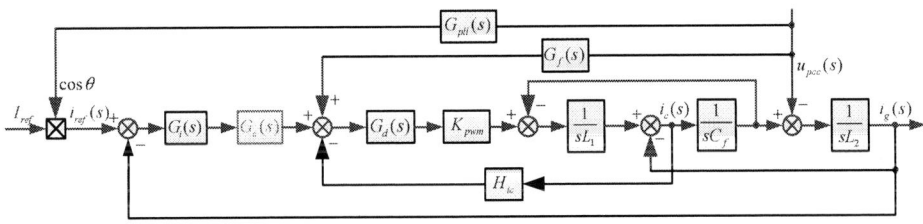

(a) Phase-lead compensation added in the forward path

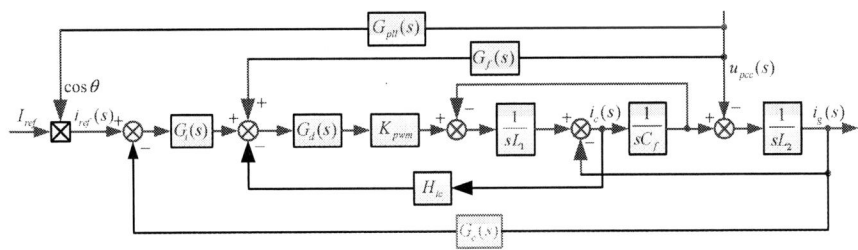

(b) Phase-lead compensation added in the feedback path

Fig. 6. Control block diagram of single-phase LCL-type grid-connected inverter with phase-lead compensation

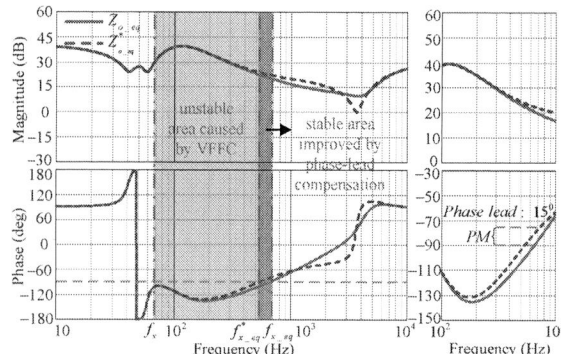

Fig. 7. Bode plot of inverter output impedance without phase-lead compensation (red) and with phase-lead compensation (blue)

$$G_c(s) = \frac{aTs+1}{Ts+1} \qquad (12)$$

The phase-lead compensation introduces a maximum phase boost φ_{max} at a certain frequency w_x. Different from the lead-lag compensator presented in [6], phase-lead compensation is added in this paper without any changes of current regulator parameters, considering that the parameters are properly designed under the ideal gird condition. Then the trade-off in the selection of w_x and φ_{max} is that, both the current controller bandwidth and phase gain are changed with the phase boost. To ensure good dynamic response and stability, a maximum phase boost of 20^0 is introduced at the cut-off frequency of the uncompensated current loop.

After adding phase-lead compensation, the shaped output impedance is defined as $Z^*_{o\ eq}(s)$. Fig. 7 gives the bode plot of $Z_{o\ eq}(s)$ and $Z^*_{o\ eq}(s)$. It is clearly observed that the unstable area caused by VFFC is narrowed with phase-lead compensation. The frequency range between 100Hz and 1kHz zoomed in the right side of Fig. 6 clearly shows a 15^0 phase margin is introduced to improve the robustness against grid impedance variations. By solving equations (11), the critical frequency $f_{x\ eq}$ equals to 665Hz and the grid impedance corresponding to this frequency is 2.4mH. After adding phase-lead compensation, the shaped critical frequency $f^*_{x\ eq}$ equals to 548Hz and the grid impedance corresponding to this frequency is 4.4mH. That means this method can nearly expand the maximum grid impedance range twice in comparison with the case without the phase-lead compensation.

V.   SIMULATION AND EXPERIMENTAL RESULTS

A 4-kW single-phase LCL-type grid-connected inverter is built and tested in both simulation and experiments to verify the analysis and the proposed control method. In order to mitigate the influence of PLL on the system stability, a low bandwidth of PLL is designed, whose cut-off frequency approximately equals to 20Hz. Meanwhile, the cut-off frequency of the current loop is set about 1.1kHz and the phase margin is 48^0 to ensure good dynamic response and stability. After adding phase-lead compensation，the cut-off

TABLE I.   PARAMETERS OF THE INVERTER

Parameters	Value	Parameters	Value
U_{in} / V	400	L_1 / uH	750
U_g / V	220	L_2 / uH	350
P_o / kW	4	C_f / uF	10
f_o / Hz	50	f_s / kHz	40
K_{pwm}	400	H_{ic}	0.027
K_{p_c}	0.018	K_{i_c}	30
K_{p_pll}	0.2	K_{i_pll}	12.8
a	2	T	1/16000

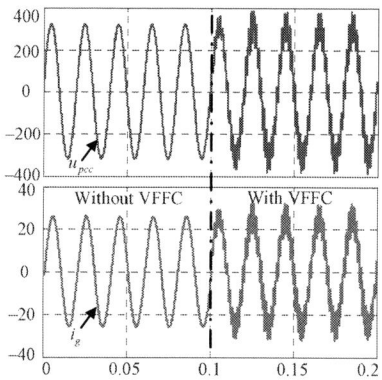

Fig. 8. Simulation waveforms before and after adding VFFC when L_g is 2.6mH

frequency and phase margin are changed to 1.5kHz and 60^0, respectively. The main parameters are listed in Tab. 1.

The inverter system before and after adding VFFC is compared by simulation firstly. Fig. 8 presents the simulation waveforms of u_{pcc} and i_g when L_g is 2.6mH. It can be seen that the inverter system is stable without VFFC due to enough PM as analyzed above, and it begins unstable with the seriously distorted current when VFFC is added.

The experiments are carried out when L_g are 0, 1.2 and 2.4mH respectively. The steady waveforms shown in Fig. 8 and Fig. 9 are compared before and after adding phase-lead compensation at half load. Fig. 9 presents experimental waveforms with VFFC before adding phase-lead compensation. The increasing grid impedance results in aggravated distortion of u_{pcc} and i_g, or even instability.

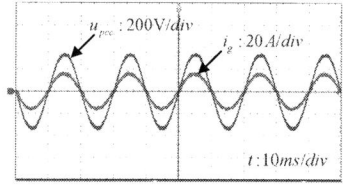

(a) Stiff grid with L_g=0

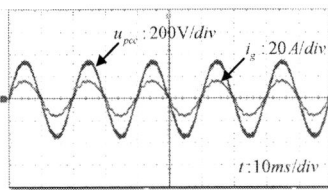

(b) Weak grid with L_g=1.2mH

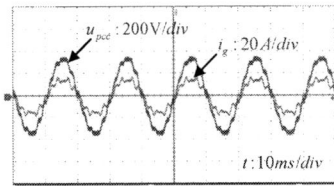

(c) Weak grid with L_g=2.4mH

Fig. 9. Experimental waveforms with VFFC before adding phase-lead compensation

The comparative experimental waveforms after adding phase-lead compensation are given in Fig. 10. Obviously, both u_{pcc} and i_g are improved by this method, particularly in large grid impedance cases, which agrees with the analysis pretty well.

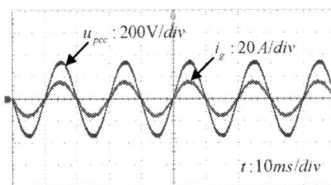

(a) Stiff grid with L_g=0

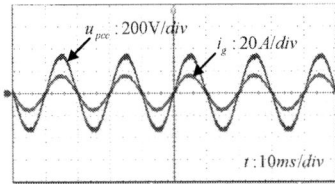

(b) Weak grid with L_g=1.2mH

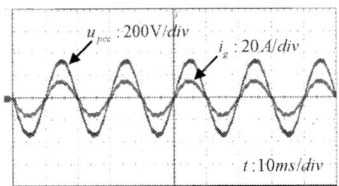

(c) Weak grid with L_g=2.4mH

Fig. 10. Experimental waveforms with VFFC after adding phase-lead compensation

VI. CLONCLUSIONS

This paper studies the stability of single-phase grid-connected inverter with VFFC using impedance-based analysis method. With this method, the unstable area caused by VFFC is clear and the maximum grid impedance range within which the inverter keeps stable can be easily obtained. Besides, due to the large phase-lag caused by VFFC, a new phase-lead compensation method is proposed to improve system robustness. The compensation can be added either in the forward path or feedback path of the outer current loop. Simulation and experimental results has verified the analysis and method.

REFERENCES

[1] M. Liserre, R. Teodorescu, and F. Blaabjerg, "Stability of photovoltaic and wind turbine grid-connected inverters for a large set of grid impedance values," *IEEE Trans. Power Electron.*, vol. 21, no. 1, pp. 263–272, Jan. 2006.

[2] M. Cespedes and J. Sun, "Renewable energy system instability involving grid-parallel inverters," in *Proc. IEEE Appl. Power Electron. Conf.*, 2009, pp. 1971–1977.

[3] J. L. Agorreta, M. Borrega, J. López, and L. Marroyo, "Modeling and control of N-paralleled grid-connected inverters with LCL filter coupled due to grid impedance in PV plants," *IEEE Trans. Power Electron.*, vol. 26, no. 3, pp. 770–785, Mar. 2011.

[4] J. Sun, "Impedance-Based Stability Criterion for Grid-Connected Inverters," *IEEE Trans. Power Electron.*, vol. 26, no. 11, pp. 3075-3078, Nov. 2011.

[5] B. Wen, D. Boroyevich, P. Mattavelli, R. Burgos, and Z. Shen, "Modeling the output impedance negative incremental resistance behavior of grid-tied inverters," in *Proc. IEEE Appl. Power Electron. Conf.*, 2014, pp. 1799-1806.

[6] M. Cespedes, J. Sun, "Impedance shaping of three-phase grid-parallel voltage-source converters," in *Proc. IEEE Appl. Power Electron. Conf.*, 2012, pp. 754-760.

[7] B. Wen, D. Dong, D. Boroyevich, R. Burgos, P. Mattavelli, Z. Shen, "Impedance-Based Analysis of Grid-Synchronization Stability for Three-Phase Paralleled Converters," *IEEE Trans. Power Electron.*, doi:10.1109/TPEL.2015.2419712.

[8] D. Yang, X. Ruan, H. Wu, "Impedance Shaping of the Grid-Connected Inverter with LCL Filter to Improve Its Adaptability to the Weak Grid Condition," *IEEE Trans. Power Electron.*, vol. 29, no. 11, pp. 5795-5805, Nov. 2014.

[9] S. Y. Park, C. L. Chen, J. S. Lai, and S. R. Moon, "Admittance compensation in current loop control for a grid-tie LCL fuel cell inverter, " *IEEE Trans. Power Electron.*, vol. 23, no. 4, pp. 1716–1723, Jul. 2008.

[10] X. Wang, X. Ruan, S. Liu, and C. K. Tse, "Full feed-forward of grid voltage for grid-connected inverter with LCL filter to suppress current distortion due to grid voltage harmonics," *IEEE Trans. Power Electron.*, vol. 25, no. 12, pp. 3119–3127, Dec. 2010.

[11] X. Zhang, F. Wang, W. Cao, Y. Ma, "Influence of voltage feed-forward control on small-signal stability of grid-tied inverters," in *Proc. IEEE Appl. Power Electron. Conf.*, 2015, pp. 1216-1221.

[12] J. Xu, S. Xie, T. Tang, "Evaluations of current control in weak grid case for grid-connected LCL-filtered inverter," *IET Power Electron.*, vol.6, no.2, pp. 227–234, Feb. 2013.

[13] J. Xu, S. Xie, T. Tang, "Improved control strategy with grid-voltage feedforward for LCL-filter-based inverter connected to weak grid," *IET Power Electron.*, vol.7, no.10, pp. 2660-2671, Oct. 2014.

[14] H. Wu, X. Ruan, D. Yang, "Research on the Stability Caused by Phase-locked Loop for LCL-type Grid-connected Inverter in Weak Grid Condition," *Proceedings of the CSEE*, vol.34, no.30, pp. 5259-5268, Oct. 2014.

New Configuration of Dynamic Voltage Restorer for Medium Voltage Application

Arash Khoshkbar Sadigh

Extron Electronics

Anaheim, USA

Vahid Dargahi and Keith Corzine

Microgrid and Power Electronics Laboratory

Holcombe Department of Electrical and Computer Engineering

Clemson University, Clemson, USA

Abstract— One of the undesirable power quality phenomenon in the distribution systems is voltage sag which risks the operation of sensitive loads. Dynamic voltage restorer (DVR) is well-known and reliable solution to mitigate this phenomenon and protect the sensitive loads. This paper proposes new configuration of DVR for medium-voltage applications. The proposed configuration contains diode rectifier to converter power grid voltage to medium-voltage dc link which is split between capacitors working at low-voltage level. Next stage is couple of full bridge (FB) dc-dc converters which are stacked together and each is fed with low voltage. Each FB dc-dc converter contains high-frequency (HF) transformer to make isolated dc link for each phase since DVR needs to inject isolated voltage in each phase individually. Last stage of proposed configuration is the cascaded multicell inverter (CMI) which works as series converter of DVR to inject required voltage. The output of each FB dc-dc converters are fed to one cell of CMI. CMI is controlled with phase shifted sinusoidal pulse width modulation to absorb the same power from each cell causing to split the voltage of main dc link equally between low-voltage capacitors. Simulation results of the proposed configuration are presented to show the performance and effectiveness of the circuit.

Keywords— *DVR; full-bridge dc-dc converter; multilevel cascaded inverter; power quality; SPWM; voltage sag.*

I. INTRODUCTION

In recent years, the number of sensitive loads integrated to power grid has been increased [1]–[3]. Consequently, the demand for high power quality and voltage stability becomes a significant issue. In the present power grids, voltage sags are recognized as a serious threat and a frequently occurring power-quality problem and have costly consequence such as sensitive loads tripping and production loss [4]–[7].

Voltage sags are results of transient phenomenon in power grid such as short circuits in the upstream power transmission line or parallel power distribution line connected to the point of common coupling (PCC), inrush currents involved with the starting of large machines, sudden changes of load, energizing of transformers or switching operations in the grid [8]–[10]. According to the IEEE STD 1159-2009, voltage sag (also called voltage dip in the IEC terminology) is defined as a decrease of 0.1 to 0.9 p.u. in the rms voltage at system frequency and with the duration of half cycle to one minute [11].

Due to the above mentioned effects of voltage sags on sensitive loads, compensating voltage sags and minimizing

their effects is necessary. Traditional methods of suppressing voltage variations include tap-changing transformers and uninterruptible power supplies (UPS) [12]. However, tap-changing transformer is bulky, costly and not fast enough to eliminate the voltage sag effects at load side. On the other hand, UPS is bulky and expensive device whose power rating should be same as load power rating [13]. Furthermore, there are custom power devices such as static synchronous compensator (STATCOM), distribution-STATCOM (D-STATCOM), unified power-quality conditioner (UPQC), and dynamic voltage restorer (DVR) as power electronics based solutions to minimize costly outcomes of voltage sags [12]. In comparison, DVR is more effective and direct solution for "restoring" the quality of voltage at its load-side terminals when the quality of voltage at its source-side terminals is disturbed [14]–[17].

DVRs compensate voltage sags by injecting the proper amount of voltages in series with the supply voltage, in order to maintain the load side voltage within the specification [18], [19]. Typically, a DVR consists of an energy storage device and an inverter which is coupled via a series transformer to the grid. The purpose of inverter is injecting the series voltage with a controlled magnitude and phase angle to restore the quality of load voltage and avoid load tripping [19], [20]. It is worth mentioning that DVR needs to produce the required voltage in each phase while it is isolated from other phases. There are several options to make a required isolation between phases. One option is to use line-frequency transformer at the output of series converter to couple it with power grid and use a common dc link [1]–[9], [11], [14], [15], [17], [19], [20]. Another way is to use isolated dc link individually in each phase. The later approach itself can be achieved in two ways. The first one is using line-frequency transformer at shunt converter side to make an isolated dc link for each phase [4], [16], [18]. The second one is using common dc link between phases and convert it to isolated dc link by utilization of isolated dc-dc converters [13], [21] in which isolation is done with HF transformer. Due to small size and low weight, HF transformer based isolation is more suitable for DVR application rather than line-frequency transformer based isolation. Another advantage of HF transformer is that it is easier to avoid any saturation during the startup transient since HF transformer fundamental cycle is much shorter than line cycle; thus, any delay, which can be a portion of HF transformer fundamental cycle, in startup to avoid saturation does not

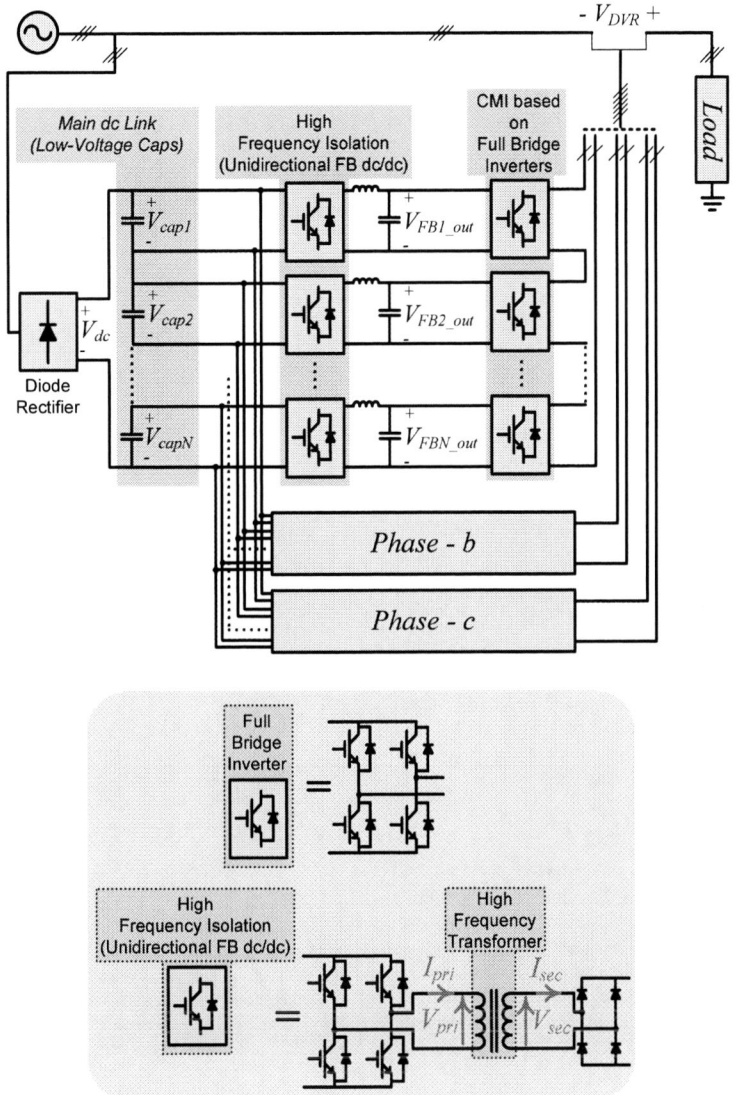

Fig. 1. Proposed configuration of DVR based on HF transformer for medium voltage application.

cause any noticeable delay in operation of DVR. On the other hand, saturation is a real issue in line-frequency transformer during the startup transient and in order to avoid it, it may be needed to have a delay up to quarter of fundamental cycle of line-frequency transformer. However, a few articles investigate and implement HF transformer based isolation in DVR and all of them just investigated low-voltage applications [13], [21].

This paper proposes a new configuration of DVR based on HF transformer isolation for medium voltage applications. In Section-II, the proposed configuration is illustrated and explained in detail. The control method of DVR is explained in Section-III. Furthermore, the simulation results of proposed DVR configuration are presented in Section IV.

II. PROPOSED DVR CONFIGURATION BASED ON HF TRANSFORMER FOR MEDIUM VOLTAGE APPLICATION

Fig. 1 illustrates the circuit schematic of proposed DVR whose first stage a diode rectifier to convert the grid voltage to dc link with medium-voltage rating. Just phase a is illustrated in detail for simplicity and other two phases are the same as phase a. To avoid the line-frequency transformer and related disadvantages mentioned in previous section, it is better to use HF transformer based dc-dc converter to make isolated dc links. However, due to high voltage rating of dc link made by diode rectifier [21], it is not practical to use one level FB dc-dc converter fed from main dc link. Therefore, it is proposed to use several FB dc-dc converters stacked together in order to split the voltage of main dc link between several low-voltage capacitors; and each capacitor feeds low-voltage FB dc-dc converter, as shown in Fig. 1. Each FB dc-dc converter works individually which means that each has its own output feedback to control its output voltage (V_{FBx_out}) at a desired value. Moreover, it is highly essential to distribute the required output power in each phase between all FB inverters of CMI evenly to split the main dc link voltage equally between low-voltage capacitors. Otherwise, if the output power of one FB inverter of CMI is higher than others, the related FB dc-dc converter needs to send out more power. Therefore, that specific FB dc-dc converter gets more input power from its own input low-voltage capacitor. Consequently, the voltage of that specific low-voltage capacitor in main dc link starts to decrease and others start to increase which will cause to damage the components. To overcome this phenomenon, it is proposed to use phase shifted sinusoidal pulse width modulation (PS-SPWM) method to control the CMI working as series converter. This approach equalizes the output power of each FB inverter in CMI which is the same as input power of each FB dc-dc converter. The PS-SPWM method to control 2-cell CMI is shown in Fig. 2.

III. CONTROL STRATEGY OF DVR

The control strategy of DVR has two main parts which are detection of disturbance, *i.e.* voltage sag, and reference determination of series injected voltage. Without the proper operation of each of these parts, DVR can't implement its suitable and required actions. In the following, each of these parts is described.

A. Voltage Sag Detection:

Here, the rms calculation which is an accurate and reliable method in spite of slow detection is implemented to detect the voltage sag. As the first step of voltage sag detection, the line-neutral grid voltages are measured in each phase individually and afterwards, the rms value of each phase voltage is calculated. If the calculated rms value in any of phases is less than predetermined threshold value, the DVR is activated and starts to inject the required amount of voltage.

B. Reference Determination of Series Injected Voltage:

A reliable method for compensating voltage sags is restoring the load voltage to the level and condition before the sag, called pre-sag compensation strategy. Therefore, the amplitude and the phase of the voltage before the sag have to be exactly restored. The phasor diagram of the pre-sag compensation strategy is shown in Fig. 3. In this figure, the dashed quantities (V'_{grid}, V'_{load}, V'_{dvr} and I'_{load}) indicate variables after the sag. The phasors prior to the sag are represented by V_{grid}, V_{load} and I_{load}. Moreover, angle of φ is phase angle difference between the load voltage and load current phasors and angle of δ is phase jump of grid voltage during the voltage sag. This strategy is able to compensate any kind of voltage sags including balanced or unbalanced voltage sags with or without any phase-variations in each phase of grid voltages. The

magnitude of injected voltage is:

$$V'_{DVR,k} = \sqrt{2} \cdot \sqrt{\left(V_{load}\right)^2 + \left(V'_{grid,k}\right)^2 - 2 \cdot V_{load} \cdot V'_{grid,k} \cdot \cos\left(\delta_k\right)} \quad (1)$$

and the phase angle of injected voltage phasor is:

$$\angle V'_{DVR,k} = \arctan\left(\frac{V_{load} \cdot \sin\left(\varphi\right) - V'_{grid,k} \cdot \sin\left(\varphi - \delta_k\right)}{V_{load} \cdot \cos\left(\varphi\right) - V'_{grid,k} \cdot \cos\left(\varphi - \delta_k\right)}\right) \quad (2)$$

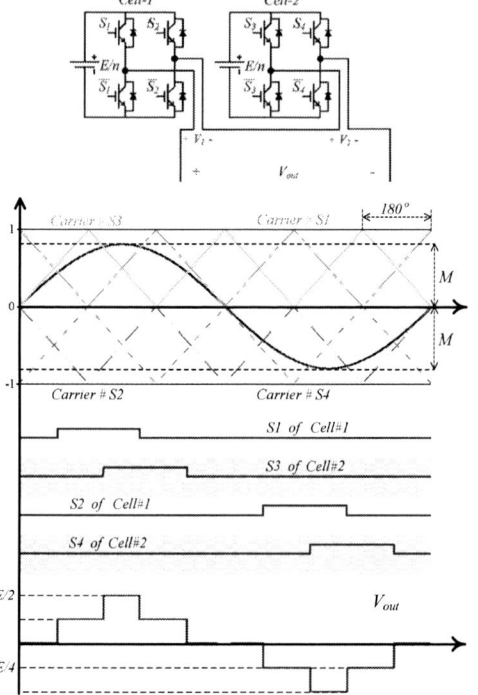

Fig. 2. PS-SPWM method to control general 2-cell 4-level CMI.

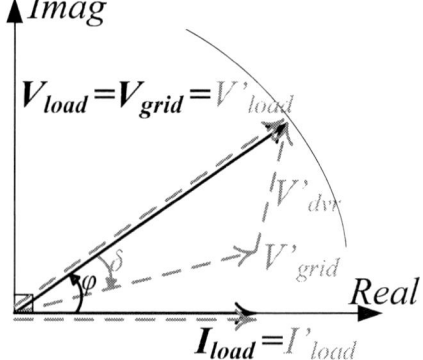

Fig. 3. Phasor diagram of pre-sag compensation strategy.

IV. SIMULATION RESULTS

Computer simulation is provided to verify the performance of the proposed DVR configuration. The main parameters used in the simulations are given in Table-I. Voltage sag is occurred at the power grid at $t = 0.06s$ and grid rms voltage of phases a drops to 70% of its nominal values, respectively. As shown in Fig. 4, the detection and determination methods are able to detect the voltage sag and determine the reference of injected voltage and as a result, DVR compensates the voltage sags within half cycle. As it is illustrated in Fig. 5, the main dc link voltage (Vdc_main) is split between 5 capacitors (Vcap1 to Vcap5) equally thanks to implemented PS-SPWM method to control CMI. As it can be pointed out, the main dc link capacitors can have low capacitance and therefore, more ripple (with ripple of ±25%) since each FB dc-dc converter regulates the output dc link (with ripple of ±6%) feeding each cell of CMI. This fact causes to use small capacitors at main dc link resulting in decrease the size and cost of dc link capacitors. It is worth mentioning that due to diode rectifier utilization, the grid current is distorted, as shown in Fig. 6. However, this phenomenon happens just for several cycles during the voltage sag compensation and according to standards (such as IEC 61000-3-2.) related the Electromagnetic Compatibility

(EMC) of equipment, it is considered as a transient. Furthermore, primary, secondary and output of FB#1 is illustrated in Fig. 7 showing HF operation f FB dc-dc converter. Moreover, primary and secondary current of HF transformer at FB dc-dc converter #1 is shown in Fig. 7. Other FB dc-dc converters have similar waveforms which is are provided here.

TABLE-I: MAIN PARAMETERS OF SIMULATED SYSTEM

System Parameters	Values
Nominal rms voltage (line to ground), V_{grid} (kV)	6.6
System frequency (Hz)	60
Capacitors used in main dc link (µF)	680
Ratio of HF transformer	0.6
Number of FB dc-dc converters stacked together	5
Switching frequency of FB dc-dc converter (kHz)	50
FB dc-dc converter output filter L (mH), C (mF)	0.2 ; 2.2
Switching frequency of CMI (kHz)	5
CMI output filter L (mH), C (µF)	0.1 ; 30
Resistance & inductance of load R (Ω), L (H)	12; 0.03

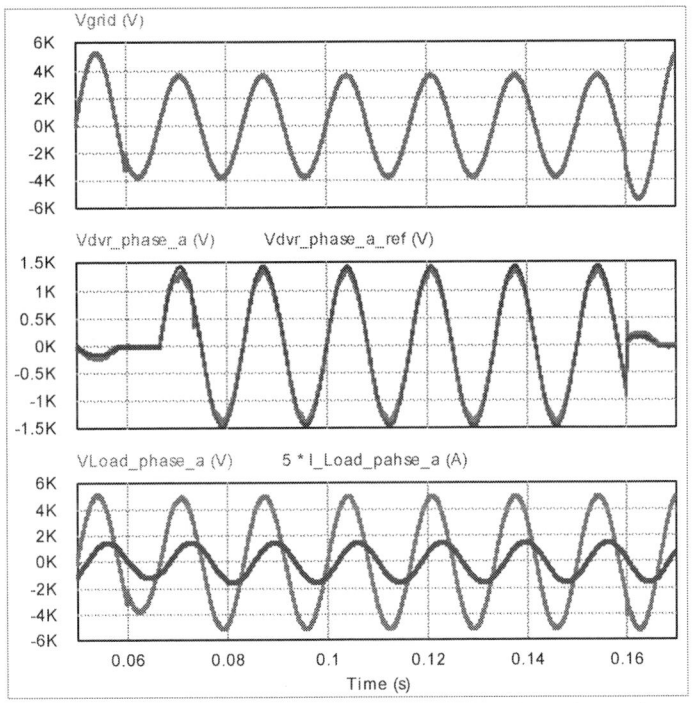

Fig. 4. Simulation results: (a) grid voltage; (b) DVR injected and reference voltage; (c) load voltage and current (increased 5 times).

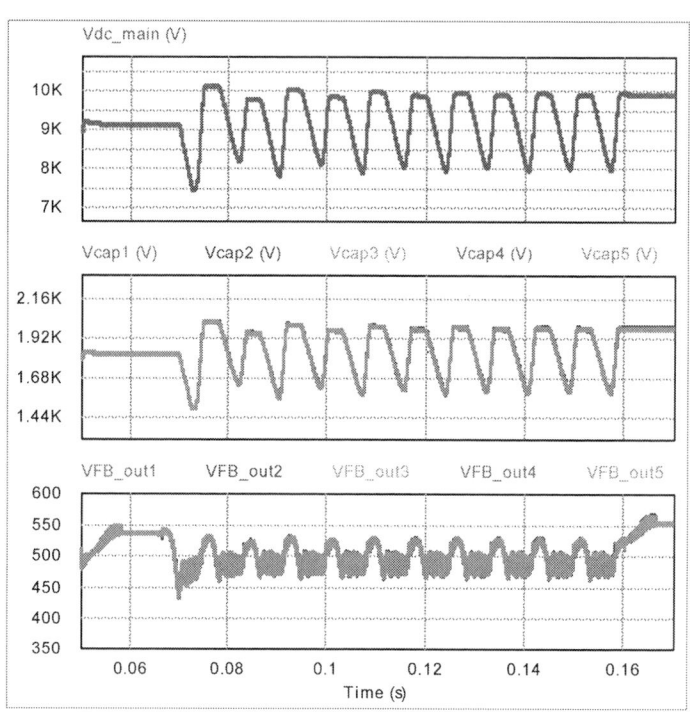

Fig. 5. Simulation results: (a) rectified main dc voltage; (b) capacitors voltages connected in series at dc link; (c) output voltage of dc-dc FB converters.

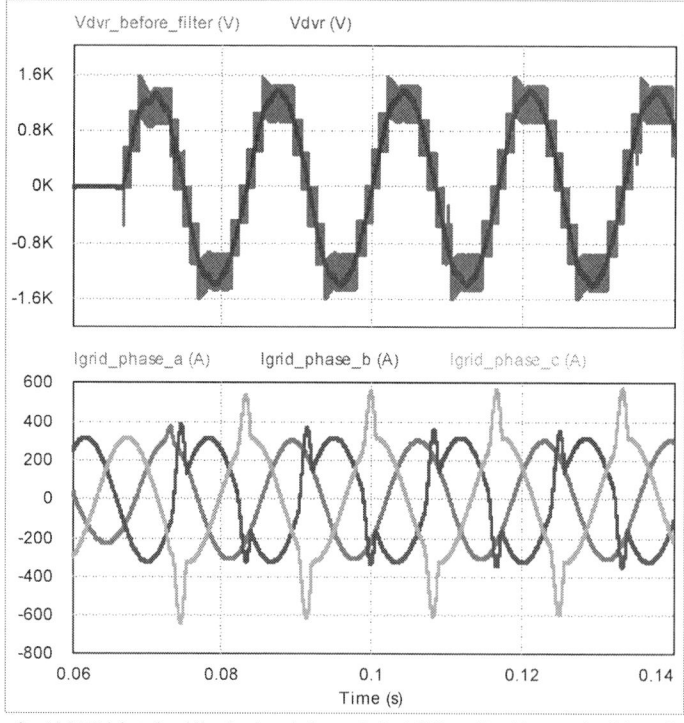

Fig. 6. Simulation results: (a) DVR injected multilevel voltage before and after LC filter; (b) grid current in all three phases.

978-1-4673-9551-9/16 $31.00 © 2016 IEEE

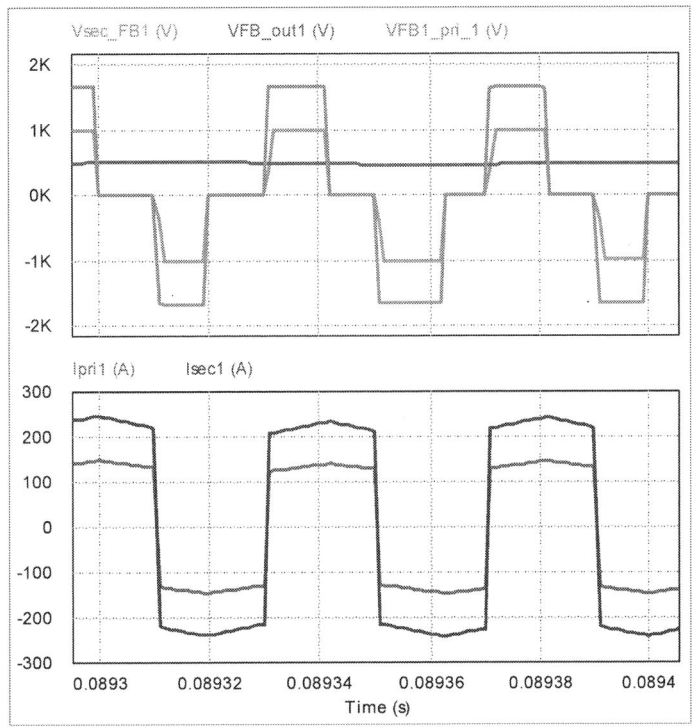

Fig. 7. Simulation results: (a) primary, secondary, and output voltage of dc-dc FB converter #1; (b) primary and secondary current of dc-dc FB converter #1.

V. CONCLUSION

In this paper, new configuration of DVR for medium-voltage application was proposed. The first stage of proposed configuration is diode rectifier to convert grid voltage to medium-voltage dc link. Moreover, it was proposed to utilize HF transformer based isolation to overcome large volume and size of line-frequency transformer as well as its saturation issue during startup transient. The HF transformer based isolation is done by utilization of FB dc-dc converters; and several of them were stacked together to split medium-voltage dc link into several low-voltage dc-links. This approach makes it possible to implement the FB dc-dc converters at low-voltage level in order to use low-voltage semiconductors and capacitors. To equalize the voltage between capacitors in main dc link, it was proposed to use PS-SPWM method to absorb the same amount of power from each FB dc-dc converter. The proposed DVR was simulated and obtained simulation results showed the well-performance of circuit in voltage sag detection and compensation.

REFERENCES

[1] F. Badrkhani Ajaei, S. Farhangi, and R. Iravani, "Fault Current Interruption by the Dynamic Voltage Restorer," *IEEE Trans. Power Deliv.*, vol. 28, no. 2, pp. 903–910, Apr. 2013.

[2] P. Roncero-Sanchez, E. Acha, J. E. Ortega-calderon, V. Feliu, A. Garcia-Cerrada, P. Roncero-sánchez, and S. Member, "A Versatile Control Scheme for a Dynamic Voltage Restorer for Power-Quality Improvement," *IEEE Trans. Power Deliv.*, vol. 24, no. 1, pp. 277–284, Jan. 2009.

[3] B. Wang and G. Venkataramanan, "Dynamic Voltage Restorer Utilizing a Matrix Converter and Flywheel Energy Storage," *IEEE Trans. Ind. Appl.*, vol. 45, no. 1, pp. 222–231, 2009.

[4] T. Jimichi, H. Fujita, and H. Akagi, "Design and Experimentation of a Dynamic Voltage Restorer Capable of Significantly Reducing an Energy-Storage Element," *IEEE Trans. Ind. Appl.*, vol. 44, no. 3, pp. 817–825, 2008.

[5] E. Babaei, M. F. Kangarlu, and M. Sabahi, "Mitigation of Voltage Disturbances Using Dynamic Voltage Restorer Based on Direct Converters," *IEEE Trans. Power Deliv.*, vol. 25, no. 4, pp. 2676–2683, Oct. 2010.

[6] F. M. Mahdianpoor, R. A. Hooshmand, and M. Ataei, "A New Approach to Multifunctional Dynamic Voltage Restorer Implementation for Emergency Control in Distribution Systems," *IEEE Trans. Power Deliv.*, vol. 26, no. 2, pp. 882–890, Apr. 2011.

[7] M. Moradlou and H. R. Karshenas, "Design Strategy for Optimum Rating Selection of Interline DVR," *IEEE Trans. Power Deliv.*, vol. 26, no. 1, pp. 242–249, Jan. 2011.

[8] P. Kanjiya, B. Singh, A. Chandra, and K. Al-Haddad, "'SRF Theory Revisited' to Control Self-Supported Dynamic Voltage Restorer (DVR) for Unbalanced and Nonlinear Loads," *IEEE Trans. Ind. Appl.*, vol. 49, no. 5, pp. 2330–2340, Sep. 2013.

[9] J. D. Barros and J. F. Silva, "Multilevel Optimal Predictive Dynamic Voltage Restorer," *IEEE Trans. Ind. Electron.*, vol. 57, no. 8, pp. 2747–2760, Aug. 2010.

[10] A. M. Massoud, S. Ahmed, P. N. Enjeti, and B. W. Williams, "Evaluation of a Multilevel Cascaded-Type Dynamic Voltage Restorer Employing Discontinuous Space Vector Modulation," *IEEE Trans. Ind. Electron.*, vol. 57, no. 7, pp. 2398–2410, Jul. 2010.

[11] C. N.-M. Ho and H. S.-H. Chung, "Implementation and Performance Evaluation of a Fast Dynamic Control Scheme for Capacitor-Supported Interline DVR," *IEEE Trans. Power Electron.*, vol. 25, no. 8, pp. 1975–1988, Aug. 2010.

[12] A. Prasai and D. Divan, "Zero Energy Sag Correctors - Optimizing Dynamic Voltage Restorers for Industrial

Applications," in *2007 IEEE Industry Applications Annual Meeting*, 2007, pp. 1585–1592.

[13] A. Y. Goharrizi, S. H. Hosseini, M. Sabahi, and G. B. Gharehpetian, "Three-Phase HFL-DVR With Independently Controlled Phases," *IEEE Trans. Power Electron.*, vol. 27, no. 4, pp. 1706–1718, Apr. 2012.

[14] N. H. Woodley, A. Sundaram, T. Holden, and T. C. Einarson, "Field experience with the new platform-mounted DVR," in *PowerCon 2000. 2000 International Conference on Power System Technology. Proceedings (Cat. No.00EX409)*, vol. 3, pp. 1323–1328.

[15] Y. W. Li, P. C. Loh, F. Blaabjerg, and D. M. Vilathgamuwa, "Investigation and Improvement of Transient Response of DVR at Medium Voltage Level," *IEEE Trans. Ind. Appl.*, vol. 43, no. 5, pp. 1309–1319, 2007.

[16] C. S. Lam, M. C. Wong, and Y. D. Han, "Voltage Swell and Overvoltage Compensation With Unidirectional Power Flow Controlled Dynamic Voltage Restorer," *IEEE Trans. Power Deliv.*, vol. 23, no. 4, pp. 2513–2521, Oct. 2008.

[17] Y. W. Li, F. Blaabjerg, D. M. Vilathgamuwa, and P. C. Loh, "Design and Comparison of High Performance Stationary-Frame Controllers for DVR Implementation," *IEEE Trans. Power Electron.*, vol. 22, no. 2, pp. 602–612, Mar. 2007.

[18] T. Jimichi, H. Fujita, and H. Akagi, "An Approach to Eliminating DC Magnetic Flux From the Series Transformer of a Dynamic Voltage Restorer," *IEEE Trans. Ind. Appl.*, vol. 44, no. 3, pp. 809–816, 2008.

[19] S. R. Naidu and D. A. Fernandes, "Dynamic voltage restorer based on a four-leg voltage source converter," *IET Gener. Transm. Distrib.*, vol. 3, no. 5, pp. 437–447, May 2009.

[20] Y. W. Li, D. M. Vilathgamuwa, P. C. Loh, and F. Blaabjerg, "A Dual-Functional Medium Voltage Level DVR to Limit Downstream Fault Currents," *IEEE Trans. Power Electron.*, vol. 22, no. 4, pp. 1330–1340, Jul. 2007.

[21] T. Jimichi, H. Fujita, and H. Akagi, "A Dynamic Voltage Restorer Equipped With a High-Frequency Isolated DC–DC Converter," *IEEE Trans. Ind. Appl.*, vol. 47, no. 1, pp. 169–175, Jan. 2011.

Studies on the Clustered Voltage Balancing Mechanism for Cascaded H-Bridge STATCOM

Daorong Lu[1], Haibing Hu[1], Yan Xing[1], Xiaobin He[2], Kai Sun[3], Jianhui Yao[1]

1 Jiangsu Key Lab. of New Energy and Power Conversion, Nanjing University of Aeronautics and Astronautics, Nanjing, China
2 Shanghai Institute Space Power Sources, Shanghai, China
3 State Key Lab of Power Systems, Department of Electrical Engineering, Tsinghua University, Beijing, China
Email: tcludaorong@nuaa.edu.cn

Abstract—To study the clustered voltage balancing mechanism for cascaded H-bridge STATCOM, the active power from the grid is decomposed of by using positive and negative sequences in *dq* frame. Based on the detailed analysis, portion of clustered active power generated by negative-sequence voltages and currents, referred as negative-sequence clustered active power (NCAP), can redistribute the active power among three clusters, which implies it can be utilized to balance the three clustered voltages. Then, the relationship between NCAP and control variables-duty cycle is built. The relationship reveals that three clustered voltages are capable of converging to stable voltages without any clustered balancing control, indicating the cascaded H-bridge STATCOM having the clustered voltage self-balancing feature. Finally a balancing control method is proposed to regulate NCAP. The effectiveness of the proposed control method is verified by the experiments.

Keywords—*cascaded H-bridge STATCOM; clustered active power; self-balancing feature; clustered balancing control*

I. INTRODUCTION

STATCOM has been a hotspot for power system to control the power factor and stabilize the power system. Cascaded H-bridge converter is often adopted for implementing STATCOM in high-voltage and high-power application due to its simple structure and modularity [1], [2]. However, the imbalance of each dc capacitor voltage has become a very critical issue for cascaded H-bridge converter [3].

A hierarchical control method is proposed to solve the imbalance problem [4]. Overall voltage control, clustered balancing control, and individual balancing control constitute three hierarchies of dc voltage control. For overall voltage control, *dq* decouple control is usually adopted [4], [7]. For individual balancing control, much research has been conducted [5]-[7]. Among them, the control method based on active voltage vector superposition has been analyzed [7] and it balances the individual voltages perfectly. For clustered balancing control, the input average active power of one cluster, which is named as clustered active power, should be equal to the power loss. Generally, the clustered balancing control methods can be categorized into following three approaches: (1) Active current control approach [4]. (2) Zero-sequence voltage injection approach [8], [9]. (3) Negative-sequence current injection approach [10], [11]. In active

current control approach, clustered active power is regulated by controlling the active current of each cluster. The control method does not need complicated calculation. But the effect of balancing three clustered voltages is not satisfying under unbalanced power system. In zero-sequence voltage injection approach, injecting zero-sequence voltage can regulate clustered active power and does not influence overall active power. However, a large voltage margin of dc capacitor is required. Compared with zero-sequence voltage injection approach, negative-sequence current injection approach does not have this issue. In [10] and [11], the model of clustered active power versus negative-sequence current is established. Then, based on the model, clustered active power can be regulated by controlling negative-sequence current, which is considered as indirect control. For the indirect control, complicated calculation is required to obtain the reference of negative-sequence current. Besides, the reference is susceptible to system parameters.

To clearly explain how the active power effects on the clustered voltages, we deliberately decompose the clustered active power into two parts based on positive and negative sequence decomposition. The first part, referred as positive-sequence clustered active power (PCAP), relies on the positive-sequence voltages, and the second part, referred as negative-sequence cluster active power (NCAP), relies on the negative-sequence cluster voltages. According to the detailed analysis (which will be given in following sections), the PCAP determines the overall active power absorbed from the grid, while the NCAP has this special function to redistribute the active power among three-phase clusters, which can be utilized to balance three clustered voltages. To regulate the NCAP to achieve the clustered voltage balanced, the relationship between control variables-duty cycle and the NCAP has to be derived. Similar to the clustered active power, duty cycle can also be divided into positive and negative-sequence components. The negative-sequence duty cycle can be utilized to regulate NCAP. Therefore, the relationship between NCAP and negative-sequence duty cycle is established based on the relationship between the negative-sequence voltage and duty cycle. Based on this analysis, the negative-sequence duty cycle control method is proposed naturally to regulate the NCAP and thus to balance three clustered voltages. The effectiveness of the proposed control method is verified by the experiments.

This work is supported by the National Natural Science Foundation of China (51577088), the Industry-academic Joint Technological Innovations Fund Project of Jiangsu (BY2015003-008).

II. Circuit Configuration of 10-kV Statcom

The circuit configuration of the star-connected STATCOM is shown in Fig. 1, which cascades n H-Bridge PWM converters in every cluster. u_{sk} are the phase voltages of the grid (k=a, b, c, similarly hereinafter), u_k and i_k are the phase voltages and currents, and u_{dckn} (n=1...12) are the capacitor voltage of three-phase H-Bridge modules.

III. Relationship Between Active Power and Clustered Voltages

A. Cluster Active Power Decomposition

Based on the circuit in Fig. 1, three clustered active power can be given in *abc* frame as follows:

$$P_a = \frac{1}{T}\int_0^T u_a \cdot i_a \, dt \quad P_b = \frac{1}{T}\int_0^T u_b \cdot i_b \, dt \quad P_c = \frac{1}{T}\int_0^T u_c \cdot i_c \, dt \quad (1)$$

The phase voltages and currents of STATCOM can be decomposed into positive and negative-sequence components as shown in (2) where subscripts p and n indicate the positive and negative-sequence components respectively, while ω stands for the fundamental frequency of the grid voltage. Furthermore, the phase currents i_k can be expressed in (3). Substituting (2) and (3) into (1), expression (4) in *dq* frame can be obtained, where U_{sd} represents the symmetric grid voltage in *dq* frame.

As seen in equation (4), each clustered active power consists of two parts. One part is the PCAP (P_p), which is same for three phases. The other is NCAP (P_{na}, P_{nb}, P_{nc}). The sum of three-phase NCAP is zero ($P_{na} + P_{nb} + P_{nc} = 0$), which reveals that NCAP can redistribute three clustered active power and does not influence overall active power. Therefore, the regulating of NCAP can be applied to balance three clustered voltages. Due to $P_{nc} = - P_{na} - P_{nb}$, only P_{na} and P_{nb} are analyzed to establish the relationship between NCAP and control variables-duty cycle. Similar to the clustered active power, duty cycle can be decomposed into positive and negative-sequence components (d_{dp}, d_{qp}, d_{dn}, d_{qn}). d_{dp} and d_{qp} are the output of *dq* decouple controller to realize the control of overall dc voltage and output reactive current. Thus, only d_{dn} and d_{qn} can be applied to regulate NCAP.

$$u_{sk} - u_k = jwL \cdot i_k \quad k = a, b, c \quad (3)$$

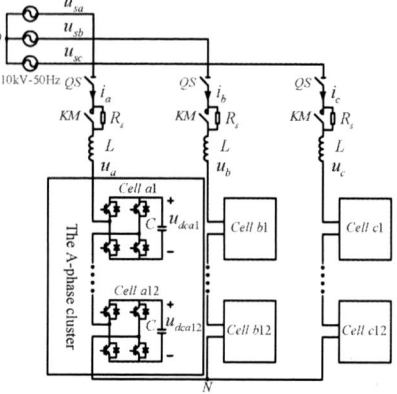

Fig. 1 Circuit configuration of 10-kV STATCOM

$$
\begin{bmatrix} P_a \\ P_b \\ P_c \end{bmatrix} = \begin{bmatrix} P_p \\ P_p \\ P_p \end{bmatrix} + \begin{bmatrix} P_{na} \\ P_{nb} \\ P_{nc} \end{bmatrix} = \underbrace{\begin{bmatrix} \dfrac{-U_{sd}}{2\omega L}\cdot U_{qp} \\[2mm] \dfrac{-U_{sd}}{2\omega L}\cdot U_{qp} \\[2mm] \dfrac{-U_{sd}}{2\omega L}\cdot U_{qp} \end{bmatrix}}_{PCAP} + \underbrace{\begin{bmatrix} \dfrac{U_{sd}}{2\omega L}\cdot U_{qn} \\[2mm] \dfrac{\sqrt{3}}{4}\dfrac{U_{sd}}{\omega L}\cdot U_{dn} - \dfrac{1}{4}\dfrac{U_{sd}}{\omega L}\cdot U_{qn} \\[2mm] -\dfrac{\sqrt{3}}{4}\dfrac{U_{sd}}{\omega L}\cdot U_{dn} - \dfrac{1}{4}\dfrac{U_{sd}}{\omega L}\cdot U_{qn} \end{bmatrix}}_{NCAP} \quad (4)
$$

B. Relationship Between NCAP and Negative-Sequence Duty Cycle

The second part of equation (4) shows NCAP is related to the negative-sequence voltages (U_{dn}, U_{qn}). To further establish the relationship between NCAP and negative-sequence duty cycle, it is desired to obtain the relationship between (U_{dn}, U_{qn}) and negative-sequence duty cycle. Before deriving this relationship, we assume that the positive-sequence duty cycle has been calculated to control the overall dc voltage and reactive current to their references through *dq* decouple controller. Fig. 2 illustrates the block diagram of the relationship between the negative-sequence voltage and the output duty cycle. In Fig. 2, d_a, d_b, d_c are the output duty cycles in *abc* frame which is composed of positive and negative-sequence components (d_{dp}, d_{qp}, d_{dn}, d_{qn}). The positive and negative-sequence *dq* to *abc* transformation come from equation (2). And U_{dca}, U_{dcb}, U_{dcc} are the three-phase clustered voltages, the average of which is regarded as their reference. To extract the positive and negative-sequence components of the output voltage (u_a, u_b, u_c), symmetrical component method is applied. Thereby, the relationship between (U_{dn}, U_{qn}) and the output duty cycle (d_{dp}, d_{qp}, d_{dn}, d_{qn}) can be expressed as (5). Furthermore, the output negative-sequence duty cycle can be achieved by superimposing the new controlled negative-

$$
\begin{bmatrix} x_a \\ x_b \\ x_c \end{bmatrix} = \begin{bmatrix} x_{ap} \\ x_{bp} \\ x_{cp} \end{bmatrix} + \begin{bmatrix} x_{an} \\ x_{bn} \\ x_{cn} \end{bmatrix} = \begin{bmatrix} \cos(\omega t) & -\sin(\omega t) \\ \cos\left(\omega t - \dfrac{2}{3}\pi\right) & -\sin\left(\omega t - \dfrac{2}{3}\pi\right) \\ \cos\left(\omega t + \dfrac{2}{3}\pi\right) & -\sin\left(\omega t + \dfrac{2}{3}\pi\right) \end{bmatrix} \cdot \begin{bmatrix} X_{dp} \\ X_{qp} \end{bmatrix} + \begin{bmatrix} \cos(\omega t) & \sin(\omega t) \\ \cos\left(\omega t + \dfrac{2}{3}\pi\right) & \sin\left(\omega t + \dfrac{2}{3}\pi\right) \\ \cos\left(\omega t - \dfrac{2}{3}\pi\right) & \sin\left(\omega t - \dfrac{2}{3}\pi\right) \end{bmatrix} \cdot \begin{bmatrix} X_{dn} \\ X_{qn} \end{bmatrix} \quad (2)
$$

$x = u, i, \quad X = U, I$

$$U_{dn} = U_{dcref} \cdot d_{dn} + \left(\frac{1}{2}d_{dp} - \frac{\sqrt{3}}{6}d_{qp} \right)\left(U_{dca} - U_{dcref} \right) - \frac{\sqrt{3}}{3}d_{qp}\left(U_{dcb} - U_{dcref} \right)$$

$$U_{qn} = U_{dcref} \cdot d_{qn} + \left(-\frac{1}{2}d_{qp} - \frac{\sqrt{3}}{6}d_{dp} \right)\left(U_{dca} - U_{dcref} \right) - \frac{\sqrt{3}}{3}d_{dp}\left(U_{dcb} - U_{dcref} \right)$$

(5)

sequence duty cycle (d_{dnc}, d_{qnc}) upon the original (d_{dno}, d_{qno}), as shown in Fig. 2. The original negative-sequence duty cycle roots in the negative-sequence currents through dq decouple controller. When going through the dq decouple controller, the negative-sequence currents in abc frame are converted to 100Hz ac signals in positive-sequence dq frame by positive-sequence abc to dq transformation. Since the integral effect of PI regulator on 100Hz signals can be negligible and the decouple gains is far less than the proportion of PI regulator, the output negative-sequence duty cycle can be given by

$$d_{dn} = d_{dnc} + K_p \cdot \frac{U_{qn}}{\omega L}$$

$$d_{qn} = d_{qnc} - K_p \cdot \frac{U_{dn}}{\omega L}$$

(6)

where K_p is the current proportion of PI regulator. After substituting (5) and (6) into (4) and performing some algebraic manipulation, the relationship between NCAP and the controlled negative-sequence duty cycle can be expressed by (7), where

$$G_{AA} = \frac{U_{sd}}{2\omega L(1+K_m^2)}\left[-\left(\frac{1}{2}K_m + \frac{\sqrt{3}}{6} \right)d_{dp} - \left(\frac{1}{2} - \frac{\sqrt{3}}{6}K_m \right)d_{qp} \right]$$

$$G_{AB} = \frac{U_{sd}}{2\omega L(1+K_m^2)}\left[-\frac{\sqrt{3}}{3}\left(d_{dp} - K_m d_{qp} \right) \right]$$

$$G_{BB} = \frac{U_{sd}}{2\omega L(1+K_m^2)}\left[\left(\frac{\sqrt{3}}{6} - \frac{1}{2}K_m \right)d_{dp} - \left(\frac{\sqrt{3}}{6}K_m + \frac{1}{2} \right)d_{qp} \right]$$

$$G_{BA} = \frac{U_{sd}}{2\omega L(1+K_m^2)}\left[\frac{\sqrt{3}}{3}\left(d_{dp} - K_m d_{qp} \right) \right], \quad K_m = \frac{K_p \cdot U_{dcref}}{\omega L}$$

(8)

From (7), it can be observed that NCAP contains two parts. The first part (P_{na_0}, P_{nb_0}) comes from the difference between the clustered voltages and their reference. The other part (ΔP_{na}, ΔP_{nb}) is determined by the controlled negative-sequence duty cycle d_{dnc} and d_{qnc}. Let's take a close look at the first part P_{na_0} (P_{nb_0}) whose block diagram is illustrated in Fig. 3. Considering that the absorbed active power of the STATCOM

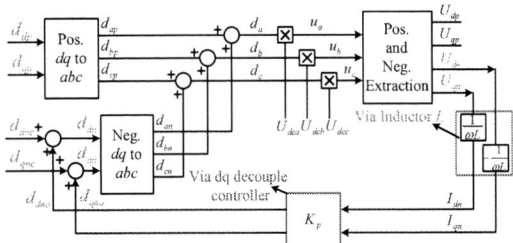

Fig.2 Relationship between negative-sequence voltage and duty cycle

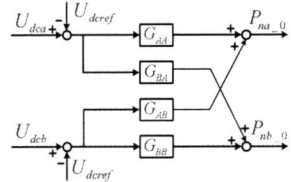

Fig. 3 The block diagram of P_{na_0}, P_{nb_0}

$$U_{dca} \uparrow \longrightarrow P_{na\_0} \downarrow \longrightarrow U_{dca} \downarrow$$

$$U_{dca} \uparrow \longrightarrow P_{nb\_0} \uparrow \longrightarrow U_{dcb} \uparrow \longrightarrow P_{na\_0} \downarrow U_{dca} \downarrow$$

Fig. 4 closed-loop regulating process of the clustered voltages

is very small in practical application, it is reasonable to assume $d_{dp} \approx 1$, $d_{qp} \approx 0$, and $d_{dp} >> |d_{qp}|$. For the sake of simplifying the following calculation, d_{dq} is equal to 1 and d_{qp} is equal to 0. Thereby, the gain parameters G_{AA}, G_{BB}, G_{AB} and G_{BA} meet the inequalities (9).

$$\Delta G_{AA} < 0, \ \Delta G_{BB} < 0, \ \Delta G_{AB} < 0, \ \Delta G_{BA} > 0 \qquad (9)$$

According to the block diagram and aforementioned assumption, we can qualitatively analyze the relationship between clustered voltages (U_{dca} and U_{dcb}) with the P_{na_0} (P_{nb_0}) as follows. Suppose phase A clustered voltage U_{dca} is larger than the reference U_{dcref}. In this scenario, its positive error multiplying G_{AA} will obtain negative P_{na_0}, and negative P_{na_0}

$$P_{na} = \underbrace{G_{AA} \cdot (U_{dcA} - U_{dcref}) + G_{AB} \cdot (U_{dcB} - U_{dcref})}_{P_{na\_0}} + \underbrace{\frac{U_{sd}U_{dcref}}{2\omega L(1+K_m^2)}(d_{qnc} - K_m \cdot d_{dnc})}_{\Delta P_{na}}$$

$$P_{nb} = \underbrace{G_{BA} \cdot (U_{dcA} - U_{dcref}) + G_{BB} \cdot (U_{dcB} - U_{dcref})}_{P_{nb\_0}} + \underbrace{\frac{U_{sd}U_{dcref}}{2\omega L(1+K_m^2)}\left[\frac{1}{2}(\sqrt{3}+K_m)d_{dnc} + \frac{1}{2}(\sqrt{3}K_m - 1)d_{qnc} \right]}_{\Delta P_{nb}}$$

(7)

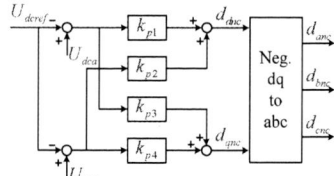

Fig. 5 Clustered balancing control method

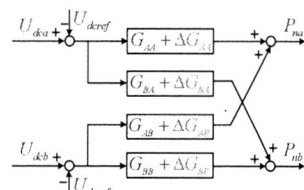

Fig. 6 The block diagram of the NPCA

will result in reducing the clustered voltage U_{dca}. Meanwhile, this positive error going through branch G_{BA} will lead to positive P_{nb_0}, resulting in the increase of the phase B clustered voltage U_{dcb}, and thus leading to reducing P_{na_0} via G_{AB} branch. With the help of two combining effects, phase A clustered voltage will come to a stable voltage, so is the same for phase B voltage. The diagram in Fig.4 illustrates closed-loop regulating process of the clustered voltages. The analysis clearly reveals that these clustered voltages can be regulated automatically to be stable, even without any additional clustered balancing control.

IV. PROPOSED CLUSTERED BALANCING CONTROL METHOD

Although three clustered voltages can maintain stable without any balancing control, they will deviate from the clustered voltage reference U_{dcref}. To keep three clustered voltages balanced, another control variable freedom (d_{dnc} and d_{qnc}) can be utilized to regulate ΔP_{na}, ΔP_{nb}. In line of this thinking, the clustered balancing control method is proposed as shown in Fig. 5, where d_{dnc}, d_{qnc} can be expressed as (10).

$$
\begin{aligned}
d_{dnc} &= k_{p1} \cdot (U_{dca} - U_{dcref}) + k_{p2} \cdot (U_{dcb} - U_{dcref}) \\
d_{qnc} &= k_{p3} \cdot (U_{dcb} - U_{dcref}) + k_{p4} \cdot (U_{dcb} - U_{dcref})
\end{aligned}
\tag{10}
$$

Substituting (10) into (7), ΔP_{na} and ΔP_{nb} can be expressed as (11), where ΔG_{AA}, ΔG_{AB}, ΔG_{BA}, ΔG_{BB} are shown in (12).

$$
\begin{aligned}
\Delta P_{na} &= \Delta G_{AA} \cdot (U_{dca} - U_{dcref}) + \Delta G_{AB} \cdot (U_{dcb} - U_{dcref}) \\
\Delta P_{nb} &= \Delta G_{BA} \cdot (U_{dca} - U_{dcref}) + \Delta G_{BB} \cdot (U_{dcb} - U_{dcref})
\end{aligned}
\tag{11}
$$

As illustrated in Fig.6, the gain parameters ΔG_{AA}, ΔG_{AB}, ΔG_{BA}, and ΔG_{BB} are added to regulate NCAP. Similar to the analysis of Fig. 4, every difference between clustered voltage and its reference has two branches to its NCAP. However, the gain parameters of two branches should be negative and adjustable so as to generate more NCAP to reduce the clustered

voltage difference when the clustered voltage deviates from its reference. Thereby, the new gain parameters ΔG_{AA}, ΔG_{BB}, ΔG_{AB}, and ΔG_{BA} should meet the inequalities (13). With appropriate design of the control parameters (k_{p1}, k_{p2}, k_{p3}, k_{p4}) by realizing the inequalities (13), the clustered voltage self-balancing mechanism can be achieved.

$$
\begin{aligned}
\Delta G_{AA} &= \frac{U_{sd} U_{dcref}}{2\omega L(1+K_m^2)}(k_{p3} - k_{p1} \cdot K_m) \\
\Delta G_{AB} &= \frac{U_{sd} U_{dcref}}{2\omega L(1+K_m^2)}(k_{p4} - k_{p2} \cdot K_m) \\
\Delta G_{BA} &= \frac{U_{sd} U_{dcref}}{2\omega L(1+K_m^2)}\left[\frac{1}{2}(\sqrt{3}+K_m) \cdot k_{p1} + \frac{1}{2}(\sqrt{3}K_m-1) \cdot k_{p3}\right] \\
\Delta G_{BB} &= \frac{U_{sd} U_{dcref}}{2\omega L(1+K_m^2)}\left[\frac{1}{2}(\sqrt{3}+K_m) \cdot k_{p2} + \frac{1}{2}(\sqrt{3}K_m-1) \cdot k_{p4}\right]
\end{aligned}
\tag{12}
$$

$$
\Delta G_{AA} < 0, \ \Delta G_{BB} < 0, \ \Delta G_{AB} < 0, \ \Delta G_{BA} > 0
\tag{13}
$$

V. EXPERIMENTAL RESULTS

As illustrated in Fig. 7, the hard-in-the-loop (HIL) and the controller were developed by our lab to verify the proposed clustered balancing control. TABLE I lists the main parameters of the circuit shown in Fig. 1 and TABLE II shows dq decouple controller parameters. Therefore, based on circuit parameters and controller parameters, the gain parameters of NCAP can be calculated, which are given by

$$
\begin{aligned}
G_{AA} &= -14.4, \\
G_{AB} &= -1.04, \\
G_{BB} &= -13.41, \\
G_{BA} &= 1.04, \\
\Delta G_{AA} &= k_{p3} - 15.48 k_{p1}, \\
\Delta G_{AB} &= k_{p4} - 15.48 k_{p2}, \\
\Delta G_{BA} &= 8.6 k_{p1} + 12.9 k_{p3}, \\
\Delta G_{BB} &= 8.6 k_{p2} + 12.9 k_{p4}
\end{aligned}
\tag{14}
$$

From equation (14), we can design the proportion parameters of the clustered balancing controller to meet inequalities (13), which are shown in TABLE III. Actually, the integral is added to generate enough NCAP and accordingly to eliminate the error between the clustered voltage and the reference.

Virtual oscilloscope in LABVIEW was used to display the operation waveforms. Besides, to emulate the different power

TABLE I. CIRCUIT PARAMETERS

Variables	Symbol	Value
Rated reactive power	Q	1 MVA
Line to line rms voltage	U_s	10 kV
Cascaded cell number	N	12
AC filter inductor	L	30 mH
Nominal dc voltage	U_{dc}	800 V
DC capacitor	C	1 mF

TABLE II. Dq Decouple Controller

Controller	Symbol	Value
Overall voltage controller	K_{vp}	0.01
	K_{vi}	0.3
Current controller	K_{cp}	0.0152
	K_{ci}	1

TABLE III. Clustered Balancing Controller

Controller	Symbol	Value
Clustered balancing controller	K_{p1}	0.0004
	K_{i1}	0.003
	K_{p2}	0
	K_{i2}	0
	K_{p3}	0.0004
	K_{i3}	0.003
	K_{p4}	-0.0004
	K_{i4}	-0.003

losses among H-bridge modules, power resistors with different values were deliberately connected to the dc link of each H-bridge modules. Fig. 8 shows the three clustered voltages under the symmetric power grid. During first stage, the proposed clustered balancing control was not activated. Three clustered voltages were imbalanced but stable, which proves the natural feature of cascaded H-bridge. During the second stage where the balancing control was activated, the average value of every clustered voltage began to converge to its reference and reached desired value 9600V in steady state.

However, when the power grid was asymmetric, the imbalanced clustered voltage would be much more severe. Fig. 9 gives the experimental results when the grid voltage of A-phase dropped 30%. Fig. 9 (a) shows the asymmetric line voltage of grid. From Fig. 9 (b), before the clustered voltage balancing control was activated, the three clustered voltage were imbalanced, especially for the A-phase. After the proposed balancing control was activated, the clustered voltage balancing was achieved. Both the experimental results under symmetric and asymmetric grids validate the analysis and the proposed balancing control method.

VI. CONLUSTIONS

This paper established the relationship between NCAP and negative-sequence duty cycle. Based on the relationship, three clustered voltages are analyzed to be stable naturally even without any clustered balancing control. Then, a novel clustered balancing control method based on NCAP is proposed to balance three clustered voltages. With the proposed control, clustered voltage balancing has been achieved under both symmetric and asymmetric grids.

REFERENCES

[1] Rodriguez, José, et al. "Multilevel inverters: a survey of topologies, controls, and applications." IEEE Transactions on Industrial Electronics49.4 (2002):724-738.

[2] Peng, Fang Zheng, et al. "A multilevel voltage-source inverter with separate dc sources for static var generation." Industry Applications IEEE Transactions on 32.5(1995):1130 - 1138.

Fig. 7 Experimental setup

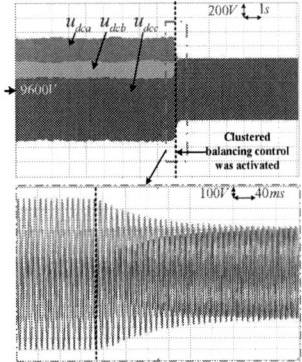

Fig. 8 Experimental waveforms under symmetric grid

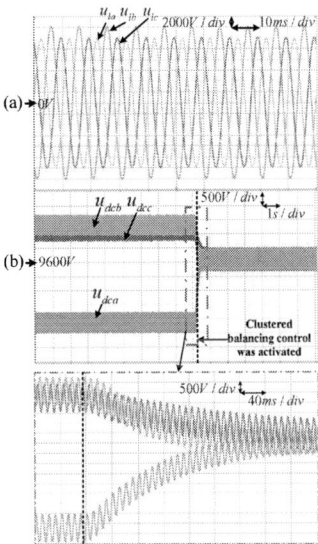

Fig. 9 Experimental waveforms under asymmetric grid

[3] S. Sirisukprasert, "The modeling and control of a cascaded-multilevel converter-based STATCOM." Ph.D. dissertation, Dept. Electr. Eng., Va. Polytechnic Inst. State Univ., Blacksburg, VA, Feb. 2004.

[4] Akagi, Hirofumi, Shigenori Inoue, and Tsurugi Yoshii. "Control and performance of a transformerless cascade PWM STATCOM with star configuration." Industry Applications, IEEE Transactions on 43.4 (2007): 1041-1049.

[5] Li, Yidan, and B. Wu. "A Novel DC Voltage Detection Technique in the CHB Inverter-Based STATCOM." Power Delivery IEEE Transactions on23.3 (2008):1613-1619.

978-1-4673-9551-9/16 $31.00 © 2016 IEEE

[6] Barrena, J.A., et al. "Individual Voltage Balancing Strategy for PWM Cascaded H-Bridge Converter-Based STATCOM." Industrial Electronics, IEEE Transactions on 1(2008):21-29.

[7] Liu, Zhao, et al. "A Novel DC Capacitor Voltage Balance Control Method for Cascade Multilevel STATCOM." Power Electronics IEEE Transactions on 27.1(2012):14 - 27.

[8] Song, Qiang, and Wenhua Liu. "Control of a cascade STATCOM with star configuration under unbalanced conditions." Power Electronics, IEEE Transactions on 24.1 (2009): 45-58.

[9] Ota, Yutaka, et al. "A Phase-Shifted-PWM D-STATCOM Using a Modular Multilevel Cascade Converter (SSBC)—Part I: Modeling, Analysis, and Design of Current Control." Industry Applications, IEEE Transactions on 51.1 (2015): 279-288.

[10] Hatano, N., and T. Ise. "A configuration and control method of cascade H-bridge STATCOM." Power and Energy Society General Meeting-Conversion and Delivery of Electrical Energy in the 21st Century, 2008 IEEE. IEEE, 2008.

[11] Lee, Chia-tse, et al. "Average Power Balancing Control of a STATCOM Based on the Cascaded H-Bridge PWM Converter With Star Configuration."Industry Applications, IEEE Transactions on 50.6 (2014): 3893-3901.

Design of a Fast Response Time Single-Phase PLL with DC Offset Rejection Capability

Abhijit Kulkarni and Vinod John
Department of Electrical Engineering,
Indian Institute of Science, Bangalore - 560012, India
Email: abhijitk@ee.iisc.ernet.in, vjohn@ee.iisc.ernet.in

Abstract—Second-order generalized integrator (SOGI) based phase-locked loops (PLLs) are commonly used for grid voltage synchronization in single-phase grid-connected power converters. SOGI-PLLs are attractive because of their simple structure that makes them suitable for implementation even in low-end digital controllers. In this paper, an SOGI based fixed-parameter PLL structure with full dc offset rejection capability is presented. This PLL uses two cascaded SOGI structures and it is termed as cascaded generalized integrator PLL (CGI-PLL). A systematic design procedure is proposed for the CGI-PLL minimizing the response time and unit vector harmonic distortion. This design achieves minimum settling time for any given worst-case frequency deviation in the grid voltage and ensures that the unit vector THD is less than 1%. The PLL designed using the proposed method has sufficient harmonic attenuation capability. The steady-state and transient response of this PLL have been validated experimentally and are found to agree with the theoretical analysis.

Index Terms—Phase-locked loops, second-order generalized integrator, inverters, dc offsets, current control, harmonic distortion.

I. INTRODUCTION

Phase-locked loops (PLLs) are used in the synchronization and closed-loop control of grid-connected power converters. PLLs estimate the frequency and phase of the grid voltage. Unit amplitude sine and cosine signals are generated from the estimated phase angle of the PLLs. These signals are commonly known as unit vectors and are used for closed-loop control reference generation.

There are different types of single-phase PLLs described in literature. In [1], three types are described namely pPLL, park-PLL and an enhanced PLL (EPLL). Second-order generalized integrator (SOGI) based PLL is described in [2]. The SOGI-PLL and parkPLL are based on the commonly used three-phase synchronous reference frame PLL (SRF-PLL) [3], [4]. There are many advanced single-phase and three-phase PLL structures reported [5]–[8]. They are computationally intensive and hence it can be difficult to implement them in low-end digital controllers.

SOGI-PLL has a simple implementation. It consists of a basic SOGI block to produce quadrature signals from the input voltage [2]. These quadrature signals are input to an embedded SRF-PLL [9]. The systematic design of basic SOGI-PLL is discussed in [9] in some detail. However, the basic SOGI-PLL is affected by the presence of dc offsets [10] in the input voltage. The dc offsets can occur due to the voltage sensor

offsets and dc offsets in A/D converters. The dc offsets can also occur due to mismatch in the semiconductor devices in a practical power converter [11]. If there is dc offset in the input voltage, the signal v_β in the basic SOGI [2], [9] will also contain dc offset [2]. Thus, the embedded SRF-PLL in the basic SOGI-PLL will have a dc offset in its input. This can result in dc offsets in the unit vectors [12]. This is highly undesirable as it can cause dc injection to the grid [12]. The grid interconnection standards such as IEEE 1547-2003 [13] impose a stringent limit of 0.5% dc injection to the grid.

To have adequate performance when the input to the PLL contains dc offsets, two approaches can be followed. One approach is to further reduce the bandwidth of the embedded SRF-PLL [12]. The second approach is to modify the SOGI structure to cancel the dc offsets. As the basic SOGI-PLL inherently contains a filtering SOGI block, the latter approach is preferred. This approach would also be advantageous in improving the overall settling time as very low bandwidths of the embedded SRF-PLL will not be necessary. For SOGI-PLL, a dc offset compensation method is proposed in [10] by modifying the structure of the SOGI-PLL. However, its design procedure relies on heuristic approach which is not easily systematised. The work reported in [14] considers multiple cascading of the SOGI blocks in a frequency-locked loop (FLL). The SOGI blocks are designed with different gains to achieve a desired input voltage harmonic attenuation. In such an approach, different combinations of the gains of SOGI are possible for the required harmonic attenuation but the possibility of settling time minimization is not considered.

A systematic design procedure is required for the PLLs with dc offset rejection based on SOGI structure, to have fast settling time and low unit vector distortion. These PLLs have low implementation complexity and can be preferred over the advanced PLLs which may be difficult to implement in low-end digital controllers.

In this paper, a single-phase PLL based on a cascaded-SOGI structure is discussed which can reject the dc offset present in the input. This PLL is referred to as cascaded generalized integrator PLL (CGI-PLL) in the paper. This is a fixed parameter PLL. In other words, adaptation of SOGI parameters in CGI-PLL is avoided to ensure stability and simple implementation.

It is known that input frequency deviations in fixed parameter SOGI-PLLs cause second harmonic ripple in the estimated

978-1-4673-9551-9/16 $31.00 © 2016 IEEE

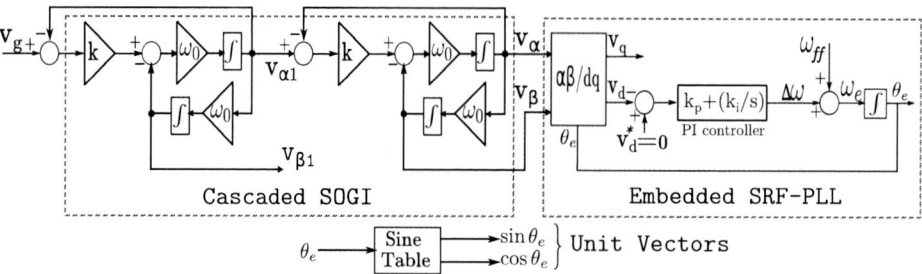

Fig. 1. Structure of cascaded generalized integrator PLL (CGI-PLL).

frequency [2]. This ripple results in harmonic distortion in the unit vectors [9]. The CGI-PLL is also affected by the frequency deviations in the input and hence its design must limit the unit vector THD under worst-case frequency deviations in the input voltage.

A systematic design method is proposed for the CGI-PLL which achieves minimum settling time for a given worst-case frequency deviation in the input voltage. The proposed design procedure can be summarized in the following two steps:

1) The parameter k in SOGI transfer functions is selected such that SOGI blocks have the fastest response to step change in the input.
2) A highest possible bandwidth (ω_{bw}) of the embedded SRF-PLL structure is chosen such that the unit vector THD is less than $u\%$ even for a frequency deviation of upto $\pm F\%$ in the grid voltage.

The grid frequency deviation range of $\pm F\%$ and the limit on unit vector THD $u\%$ can be specified by the designer. In this paper, a frequency deviation of upto $F = 8\%$ is considered. This means that a frequency range of $46Hz - 54Hz$ is considered for the design in a $50Hz$ system. The limit on unit vector THD is chosen to be $u = 1\%$. The proposed design ensures that even for the given worst case frequency deviation of $\pm 8\%$, the unit vector THD is within 1%. The actual frequency deviation is normally much smaller and hence the PLL will perform even better. It is also shown that the proposed design can give adequate unit vector performance even when input voltage contains considerable harmonic distortion.

The various performance measures such as dc offset rejection capability, frequency tracking, harmonic attenuation capability for the CGI-PLL under the proposed design are validated experimentally and agree with the analytical predictions.

This paper is organized as follows. Section II discusses the structure and operation of CGI-PLL in ideal and practical grid conditions. The proposed systematic design of the CGI-PLL is detailed in Section III. The comparison of CGI-PLL with state-of-the-art SOGI based single-phase PLLs is also included in Section III. The experimental validation of the performance of the CGI-PLL is discussed in Section IV. Conclusions are provided in Section V.

II. STRUCTURE AND OPERATION OF CGI-PLL

The CGI-PLL consists of two identical cascaded SOGI blocks followed by an embedded SRF-PLL. The structure of cascaded generalized integrator PLL (CGI-PLL) is shown in Fig. 1. Input voltage is fed to a standard SOGI block. Its output $v_{\alpha 1}$ is input to another SOGI block. The outputs of the second SOGI block are termed as v_α and v_β which are fed to the embedded SRF-PLL structure.

The transfer functions for v_α and v_β for this PLL topology are as follows,

$$G_{\alpha,c}(s) = \frac{v_\alpha}{v_g}(s) = \frac{(k\omega_0 s)^2}{(s^2 + k\omega_0 s + \omega_0^2)^2} \tag{1}$$

$$G_{\beta,c}(s) = \frac{v_\beta}{v_g}(s) = \frac{k^2\omega_0^3 s}{(s^2 + k\omega_0 s + \omega_0^2)^2} \tag{2}$$

The bode plots of the transfer functions in (1) and (2) are shown in Fig. 2. It can be seen from the bode plot and from the

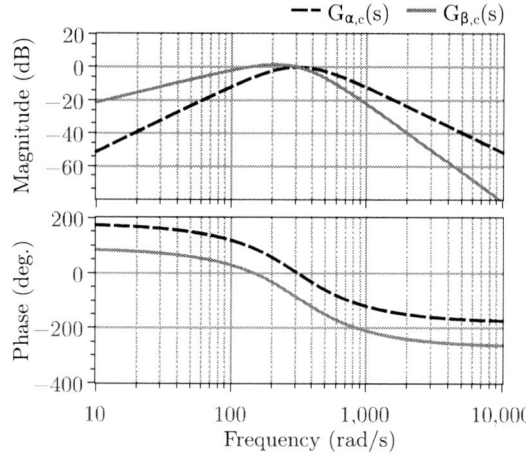

Fig. 2. Bode plot of the transfer functions $G_{\alpha,c}(s)$ and $G_{\beta,c}(s)$ of CGI-PLL.

transfer functions that the dc gain of both the transfer functions is zero. Hence, the input dc offset will not be reflected in the v_α and v_β which are input to the embedded SRF-PLL. As a result, the CGI-PLL will perform without any errors when the input contains dc offsets. For purely sinusoidal grid voltage with frequency ω_0, the outputs of the transfer functions can be observed to be balanced quadrature signals. Hence, the embedded SRF-PLL shown in Fig. 1, will have the estimated frequency $\omega_e = \omega_0$ and $v_d = 0$ in steady state.

The design parameters for this PLL are:

(i) The parameter k in SOGI transfer functions. Its value affects the transient response as well as the harmonic attenuation capability of the cascaded SOGI filtering block.
(ii) The PI controller parameters of the embedded SRF-PLL. Their values are directly related to the bandwidth of the embedded SRF-PLL [3], [4]. Similar to k, they affect the transient response and harmonic attenuation capability.

The various practical or non-ideal input conditions that affect the performance of the CGI-PLL are discussed in the following subsection.

A. Effect of Non-Ideal Grid Conditions

1) Transient amplitude and phase changes: The sensed grid voltage can have transient amplitude and phase changes. During these changes, the filtering SOGI blocks will give outputs that settle only after a known settling time. Hence, till the SOGI outputs settle, there will be transient error in the estimated frequency and phase. The parameter k mainly affects this settling time.

2) Frequency changes: Consider that the grid frequency changes from nominal $\omega_0 = 2\pi 50 \ rad/s$. From the bode plots in Fig. 2, it can be observed that the gains of the two transfer functions are different when the frequency deviates from ω_0. Similar observation is made for the basic SOGI-PLL in [9]. As discussed in [9], the different gains in the transfer functions result in different amplitudes of v_α and v_β. This means that these signals are not balanced quadrature signals. Hence, the embedded SRF-PLL will have a ripple at $2\omega_0$ in the estimated frequency and phase. This causes harmonic distortion in the unit vectors [4]. In order to limit the unit vector THD, the bandwidth of the embedded SRF-PLL is to be selected appropriately. It must be noted that only the adjustment of this bandwidth can attenuate the unit vector THD due to grid frequency changes. The method of parameter adaptation in the SOGI blocks [2], [15], [16] is not used to ensure system stability and to obtain simple and systematic design for the CGI-PLL. The parameter adaptation increases the implementation complexity also because in Fig. 1, ω_0 will not be a constant but will be a variable updated based on estimated PLL frequency.

3) Harmonic distortion: The cascaded SOGI transfer functions offer attenuation to the lower order harmonics as can be seen from the bode plot in Fig. 2. In addition to the SOGI blocks, the embedded SRF-PLL will also provide some attenuation depending on its design bandwidth. It is discussed in [9] that when the input contains harmonics, the basic SOGI-PLL will have harmonic distortion in the unit vectors. Similar observation can be made for the CGI-PLL. However, the attenuation to harmonics is expected to be higher in CGI-PLL

compared to the basic SOGI-PLL because of the cascading of the SOGI transfer functions in CGI-PLL.

The design of the PLL must consider all the above points while achieving the fast response and sufficient harmonic attenuation. The design must be systematic and must avoid trial-and-error approach. The proposed design achieves these requirements and is described in the following section.

III. Design of CGI-PLL

The design of CGI-PLL involves the selection of k and bandwidth ω_{bw} of the embedded SRF-PLL. The selection ensures fast transient response and unit vector THD less than 1% for upto 8% frequency deviation in the input.

A. Selection of the Parameter k

The parameter k for CGI-PLL is determined by considering the variation of the 2% settling time [17] of the cascaded SOGI blocks with the value of k. The design value of k is selected such that the step response settling time is the least. Fig. 3 shows the variation of settling times of the transfer functions $G_{\alpha,c}$ and $G_{\beta,c}$ with k. As it can be observed, there exists a value of $k = k_{opt}$ minimizing the overall settling time. This value is selected as the design value of k in CGI-PLL.

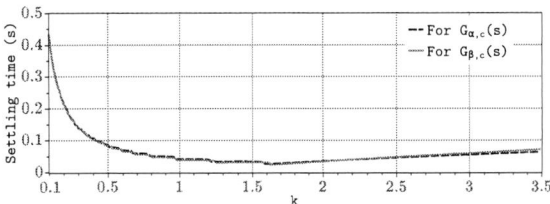

Fig. 3. Variation of 2% settling time versus k for CGI-PLL.

The design value of k from Fig. 3 is given by,

$$k = k_{opt} = 1.63 \tag{3}$$

For this value of k_{opt}, the settling times for v_α and v_β are observed to be $26.9ms$ and $23.5ms$ respectively. Thus the combined worst-case settling time ($t_{cgi,max}$) would be the maximum of the two settling times, that is,

$$t_{cgi,max} = 26.9ms \tag{4}$$

The step response of the transfer functions in (1) and (2) for $k_{opt} = 1.63$ is shown in Fig. 4. It can be observed that the overall settling time agrees with the value in (4).

B. Selection of the Bandwidth of Embedded SRF-PLL

As explained in Section III-A, the frequency deviation in the input results in harmonic distortion in the unit vector. To limit the unit vector THD to $u \leq 1\%$, for a maximum frequency deviation of $F = \pm 8\%$, the bandwidth of the embedded SRF-PLL must be chosen appropriately. This is done using

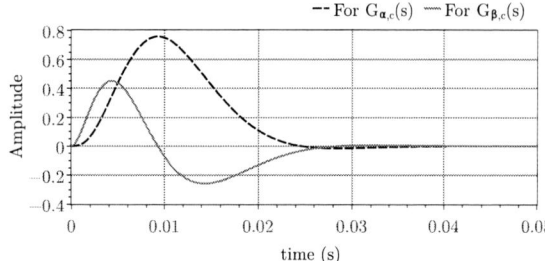

Fig. 4. Step response of the cascaded SOGI blocks in CGI-PLL for the design value of $k_{opt} = 1.63$.

the following steps. For a $50Hz$ system, the frequency range considered is,

$$f_{max} = 54Hz, \quad f_{min} = 46Hz \tag{5}$$

1) Compute v_α and v_β using the transfer functions given in (1) and (2) for a sinusoidal input voltage with frequency $f \in [f_{max}, f_{min}]$ and $f \neq 50Hz$. Note that both the amplitude and phase of v_α and v_β are to be determined.
2) Use the analytical expressions derived in [4] to compute the unit vector THD versus the bandwidth (ω_{bw}) of the embedded SRF-PLL for the pair of v_α and v_β determined in Step − 1 above.
3) Determine the highest bandwidth ($\omega_{bw,h}$) that results in unit vector THD to be just within $u = 1\%$.
4) Repeat steps 1 − 3 for different frequency deviations in the range $[f_{max}, f_{min}]$. For each frequency deviation, a corresponding highest bandwidth $\omega_{bw,h}$ is determined.
5) The design bandwidth is selected to be the lowest of the set of $\omega_{bw,h}$ determined in Step − 4. This ensures that for the whole range of the frequency deviations, the unit vector THD is limited to be at most 1%.

This procedure is illustrated graphically in Fig. 5. Unit vector THD versus bandwidth is plotted for CGI-PLL for frequency deviations from $46Hz$ to $54Hz$ in the steps of $2Hz$ and is shown in Fig. 5.

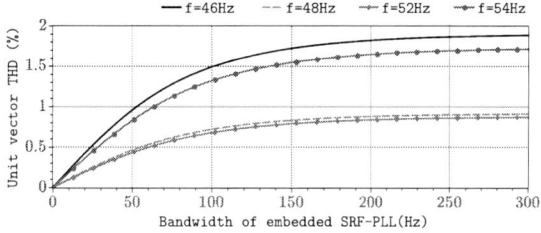

Fig. 5. Unit vector THD versus bandwidth for different frequency deviations in the input for CGI-PLL.

It can be observed that for frequencies of $48Hz$ and $52Hz$, the unit vector THD is well within the chosen limit of 1%. However, for an input frequency of $54Hz$, the highest bandwidth that can be chosen is about $65Hz$ or $408.4rad/s$.

Similarly, the highest bandwidth is $52Hz$ or $326.7rad/s$ when the input frequency is $46Hz$. To limit the unit vector THD to be within 1% for the entire range from $46 - 54Hz$, the bandwidth to selected is,

$$\omega_{bw} = 326.7rad/s \tag{6}$$

Once the bandwidth of the embedded SRF-PLL is determined, the PI controller parameters can be determined. The conventional small-signal model based design equations in [3] can be used. The design equations in [3] can be rewritten as

$$k_p = \frac{\omega_{bw}}{V_m} \tag{7}$$

$$k_i = k_p T_s \omega_{bw}^2 \tag{8}$$

k_p and k_i are the PI controller parameters used in Fig. 1. In (8), the parameter T_s is the sampling time used in the digital implementation of the PLL. The nominal sensed grid voltage peak is V_m in (7). For $T_s = 50\mu s$, $V_m = 5V$ and designed bandwidth $\omega_{bw} = 326.7rad/s$ from (6), the PI controller parameters are

$$k_p = 65.3 \quad \& \quad k_i = 348.5 \tag{9}$$

C. Harmonic Attenuation Capability of CGI-PLL

The design parameters for CGI-PLL are determined in Section III-A and III-B considering worst case frequency deviation in the input and the constraint on unit vector THD. However, it is important to determine the harmonic attenuation capability of CGI-PLL as the grid voltage normally contains harmonic distortion. The effect of harmonic distortion of the input voltage on the unit vector THD for CGI-PLL is quantified analytically. The procedure is explained as follows.

1) Let input voltage contain a known amount of dominant lower order harmonics with a known THD. Compute the harmonics in v_α and v_β using the transfer functions (1) and (2).
2) Analytical expressions derived in [4] are used to determine the unit vector THD for given design parameters of CGI-PLL.
3) Steps 1 − 2 are repeated for a range of harmonic distortion in the input.
4) The variation of unit vector THD is plotted versus input THD.

Fig. 6 shows the analytical plot of unit vector THD versus input THD for the design parameters of CGI-PLL given in (3) and (6).

The input harmonics considered are third, fifth, seventh and ninth. The harmonic content is considered to be with a relative magnitude as per the following equation.

$$V_x/V_y = y/x \tag{10}$$

That is, the third harmonic is considered to have highest amplitude which is $5/3$ times the fifth harmonic amplitude. Similarly the other odd harmonics upto ninth harmonic are considered. Using this relation, the individual harmonics for

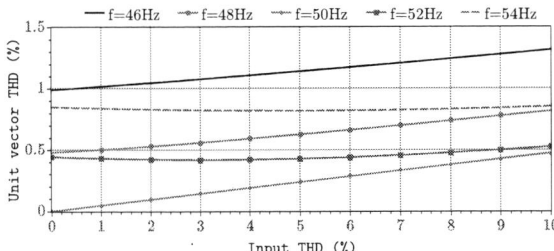

Fig. 6. Variation of unit vector THD with input THD for the proposed design of CGI-PLL under different grid frequency conditions.

any given net input THD can be determined considering the four dominant odd harmonics.

As it can be observed from Fig. 6, the unit vector THD is less than 1% even when the input voltage has a THD of upto 10% in addition to the grid frequency deviation in the range of $48Hz$ to $54Hz$. However, when the input frequency is $46Hz$, the unit vector THD is observed to be always slightly higher than 1%. For example, when input THD is 5% and frequency is $46Hz$, the unit vector THD is about 1.1%. This can be considered acceptable as it marginally exceeds the limit of 1% THD on the unit vector. It is possible to include the effect of input THD in the design of CGI-PLL. In that case, the input conditions to consider would be – the frequency deviations and input THD. The constraint on unit vector THD will be the same. As the input harmonics are attenuated by both the cascaded SOGI and the embedded SRF-PLL, the design cannot independently compute k in SOGI transfer functions as done in Section III-A. Since the unit vector THD is still close to 1% even when the input contains a THD of about 5% and a very high frequency deviation of 8%, the modified design method is not considered in this paper.

D. Worst-Case Additive Settling Time of CGI-PLL

With the proposed design, $k = 1.63$ and $\omega_{bw} = 2\pi52$ rad/s. The settling time due to the cascaded SOGI is given by $26.9ms$ from (4). For a bandwidth of ω_{bw}, the settling time for the embedded SRF-PLL is given by [18],

$$t_{srf} = 4/\omega_{bw} \tag{11}$$

Hence, the settling time due to the embedded SRF-PLL with a bandwidth of $\omega_{bw} = 326.7rad/s$ is $t_{srf} = 12.2ms$. The worst case additive settling time is equal to,

$$t_{sd} = t_{cgi} + t_{srf} = 39.1ms \tag{12}$$

Thus the CGI-PLL will settle within two fundamental cycles ($40ms$) for any input transients as per the value of t_{sd} determined in (12). Note that this is an additive worst case estimation and normally the settling time will be less than this value as both the transients in cascaded SOGI and embedded SRF-PLL occur simultaneously. This can be observed in the experimental results in Section IV.

E. Comparison of CGI-PLL with Popular SOGI based Single-Phase PLLs

The CGI-PLL presented in this paper is compared with popular SOGI based single-phase PLLs [2], [9], [10], [15]. The performance parameters used for comparison are – dc cancelling capability, design method used, number of design parameters and resource utilization in the implementation in a digital controller. Table I shows the summary of the comparison.

The basic SOGI-PLLs have lower resource utilization. However, they do not have the dc offset rejection capability. Hence, as it can be observed from Table I, the CGI-PLL with proposed design has better overall performance in all the aspects considered.

IV. EXPERIMENTAL VALIDATION

The experimental results in this section validate the performance of CGI-PLL for the following cases.

1) Validating the offset rejecting performance.
2) Validating the transient response.
3) Validating the unit vector THD when the input contains harmonic distortion.

The implementation of the CGI-PLL is done in Altera Cyclone EP1C12Q240C8 FPGA controller board using VHDL.

Fig. 7(a) shows the effect of presence of 10% dc offset in the input voltage of basic SOGI-PLL [2], [9]. The large dc offset is considered to clearly show the presence of the dc offset in the input voltage in Fig. 7(a). As it can be observed

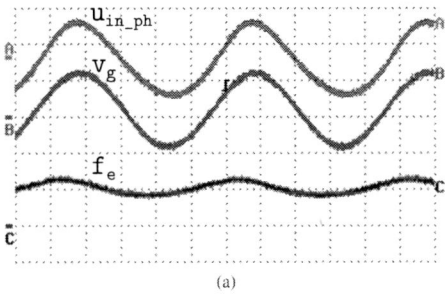

(a)

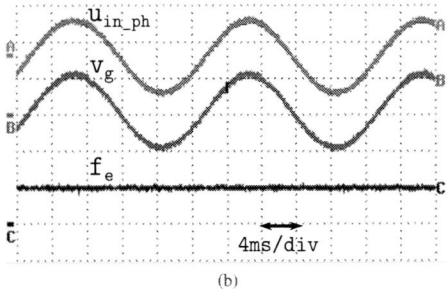

(b)

Fig. 7. Effect of 10% dc offset on (a) basic SOGI-PLL and (b) CGI-PLL. Ch. A = In-phase unit vector u_{in_ph} (1pu/div), Ch. B = Input voltage v_g (5V/div), Ch. C = Estimated frequency f_e(50Hz/div). Horizontal scale = 4ms/div.

TABLE I
COMPARISON OF CGI-PLL WITH POPULAR SOGI BASED SINGLE-PHASE PLLs.

PLL Type	DC cancelling capability	Design Parameters	Design Method	Resource Utilization* (Multiplications - M and additions - A)
Basic fixed SOGI-PLL [9]	No	2	Systematic	3M, 4A
Basic adaptive† SOGI-PLL [2], [15]	No	2	Heuristic	5M, 4A
Modified adaptive† SOGI-PLL [10]	Yes	3	Heuristic	7M, 8A
CGI-PLL (with proposed design)	Yes	2	Systematic	6M, 8A

*In addition to the SOGI blocks, the resource utilization in the embedded SRF-PLL is 7M, 6A for all the SOGI based PLLs. The resource utilization is computed considering forward Euler implementation. Trapezoidal or other discretization methods can also be used. However, the Euler method gives the least resources which may be important when the implementation is done on a low-end digital controller.

† The adaptive SOGI-PLLs update the gain ω_0 (as in Fig. 1) by the estimated frequency. Fixed SOGI-PLL and CGI-PLL use ω_0 as a constant equal to nominal grid frequency in rad/s.

from Fig. 7(a), the estimated frequency f_e contains a ripple error at fundamental frequency. This results in the presence of dc offsets and even harmonics in the unit vector [12]. The performance of CGI-PLL for the same 10% input dc offset conditions is shown in Fig. 7(b). The estimated frequency is a purely dc quantity for CGI-PLL indicating that the input dc has been rejected by the modified SOGI structure in CGI-PLL.

The transient response of CGI-PLL is observed by giving a step-phase-change. When the enable signal (En) in Fig. 8 goes high, the phase of the input voltage is given a step change of $60°$. The resulting the performance of CGI-PLL is shown in Fig. 8. It can be observed that the estimated frequency settles

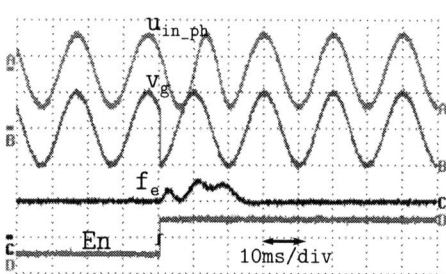

Fig. 8. Transient response of CGI-PLL to step-phase-change in the input. Ch. A = In-phase unit vector u_{in_ph} (1pu/div), Ch. B = Input voltage v_g (5V/div), Ch. C = Estimated frequency f_e(50Hz/div), Ch. D = Enable (En) signal. Horizontal scale = 10ms/div.

to steady value within $30ms$. Hence, the actual settling time of CGI-PLL is within the worst case value which was estimated to be $39.1ms$ in Section III-D.

The performance of CGI-PLL when the input contains harmonics is shown in Fig. 9. The input has a THD of 5% and frequency of $50Hz$. The unit vector THD is determined for the experimental waveform to be 0.3% which agrees with the analytical prediction in Fig. 6. The ripple in estimated frequency is practically negligible.

V. CONCLUSION

In this paper, a systematic design method is proposed for a low-complexity SOGI based single-phase PLL with

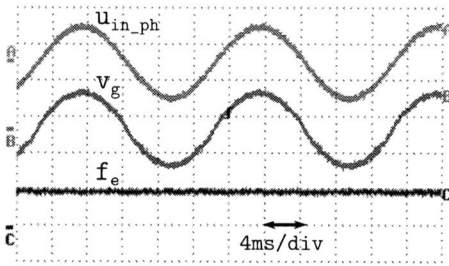

Fig. 9. Effect of harmonic distortion in the input voltage with THD=5%. Ch. A = In-phase unit vector u_{in_ph} (1pu/div), Ch. B = Input voltage v_g (5V/div), Ch. C = Estimated frequency f_e (50Hz/div). Horizontal scale = 4ms/div.

dc offset rejection capability. This PLL uses cascaded SOGI structures with fixed gain parameters and is termed as cascaded generalized integrator PLL (CGI-PLL). The proposed design selects the parameter in SOGI blocks such that the settling time for the cascaded SOGI is the least. Highest possible bandwidth of the embedded SRF-PLL is selected such that the unit vector THD due to worst-case input frequency deviations is limited to 1%. The proposed design achieves the fastest settling time for given constraints in the input voltage and unit vector THD. The steady-state, transient and harmonic attenuation performance of the CGI-PLL with the proposed design are validated experimentally. The overall performance of the CGI-PLL is shown to be better compared to popular SOGI based single-phase PLLs in terms of resource utilization in digital controller implementation and the use of systematic design method. Hence, the CGI-PLL can be preferred over advanced single-phase PLLs to achieve very good performance in addition to reduced resource utilization. This will be advantageous when low-end digital controllers are used for control implementation of grid-connected power converters in low-cost applications.

REFERENCES

[1] R. Santos Filho, P. Seixas, P. Cortizo, L. Torres, and A. Souza, "Comparison of three single-phase pll algorithms for ups applications," *IEEE Transactions on Industrial Electronics*, vol. 55, pp. 2923–2932, Aug 2008.

[2] M. Ciobotaru, R. Teodorescu, and F. Blaabjerg, "A new single-phase pll structure based on second order generalized integrator," in *37th IEEE Power Electronics Specialists Conference (PESC)*, pp. 1–6, June 2006.

[3] V. Kaura and V. Blasko, "Operation of a phase locked loop system under distorted utility conditions," *IEEE Transactions on Industry Applications*, vol. 33, pp. 58–63, Jan 1997.

[4] A. Kulkarni and V. John, "Analysis of bandwidth-unit-vector-distortion tradeoff in pll during abnormal grid conditions," *IEEE Transactions on Industrial Electronics*, vol. 60, pp. 5820–5829, Dec 2013.

[5] Q. Zhang, X.-D. Sun, Y.-R. Zhong, M. Matsui, and B.-Y. Ren, "Analysis and design of a digital phase-locked loop for single-phase grid-connected power conversion systems," *IEEE Transactions on Industrial Electronics*, vol. 58, pp. 3581–3592, Aug 2011.

[6] I. Carugati, P. Donato, S. Maestri, D. Carrica, and M. Benedetti, "Frequency adaptive pll for polluted single-phase grids," *IEEE Transactions on Power Electronics*, vol. 27, pp. 2396–2404, May 2012.

[7] Y. F. Wang and Y. W. Li, "Grid synchronization pll based on cascaded delayed signal cancellation," *IEEE Transactions on Power Electronics*, vol. 26, pp. 1987–1997, July 2011.

[8] F. Gonzalez-Espin, E. Figueres, and G. Garcera, "An adaptive synchronous-reference-frame phase-locked loop for power quality improvement in a polluted utility grid," *IEEE Transactions on Industrial Electronics*, vol. 59, pp. 2718–2731, June 2012.

[9] A. Kulkarni and V. John, "A novel design method for sogi-pll for minimum settling time and low unit vector distortion," in *39th Annual Conference of the IEEE Industrial Electronics Society-IECON 2013*, pp. 274–279, Nov 2013.

[10] M. Ciobotaru, R. Teodorescu, and V. Agelidis, "Offset rejection for pll based synchronization in grid-connected converters," in *Twenty-Third Annual IEEE Applied Power Electronics Conference and Exposition (APEC)*, pp. 1611–1617, Feb 2008.

[11] G. He, D. Xu, and M. Chen, "A novel control strategy of suppressing dc current injection to the grid for single-phase pv inverter," *IEEE Transactions on Power Electronics*, vol. 30, pp. 1266–1274, March 2015.

[12] A. Kulkarni and V. John, "Design of synchronous reference frame phase-locked loop with the presence of dc offsets in the input voltage," *IET Power Electronics*, 2015. Accepted for publication.

[13] "IEEE standard for interconnecting distributed resources with electric power systems," *IEEE Std 1547-2003*, 2003.

[14] J. Matas, M. Castilla, J. Miret, L. Garcia de Vicuna, and R. Guzman, "An adaptive prefiltering method to improve the speed/accuracy tradeoff of voltage sequence detection methods under adverse grid conditions," *IEEE Transactions on Industrial Electronics*, vol. 61, pp. 2139–2151, May 2014.

[15] Y. Yang and F. Blaabjerg, "Synchronization in single-phase grid-connected photovoltaic systems under grid faults," in *3rd IEEE International Symposium on Power Electronics for Distributed Generation Systems (PEDG)*, pp. 476–482, June 2012.

[16] P. Rodriguez, R. Teodorescu, I. Candela, A. Timbus, M. Liserre, and F. Blaabjerg, "New positive-sequence voltage detector for grid synchronization of power converters under faulty grid conditions," in *37th IEEE Power Electronics Specialists Conference (PESC)*, pp. 1–7, June 2006.

[17] K. Ogata, *Modern Control Engineering*. Prentice Hall, 5th ed., 2008.

[18] S. Golestan and J. Guerrero, "Conventional synchronous reference frame phase-locked loop is an adaptive complex filter," *IEEE Transactions on Industrial Electronics*, vol. 62, pp. 1679–1682, March 2015.

Four New Applications of Second-Order Generalized Integrator Quadrature Signal Generator

Zhen Xin[1], Rende Zhao[2], Xiongfei Wang[1], Poh Chiang Loh[1], Frede Blaabjerg[1]

Dept. of Energy Technology[1]
Aalborg University
Aalborg, Denmark
zxi@et.aau.dk, xwa@et.aau.dk, pcl@et.aau.dk,
fbl@et.aau.dk

Electrical Engineering Department[2]
China University of Petroleum (Eastern China)
Qingdao, China
zhaorende@126.com

Abstract— **The Second-Order Generalized Integrator (SOGI) was used as a building block for the SOGI-Quadrature-Signal Generator (SOGI-QSG) which has been widely used for grid synchronization, frequency estimation, and harmonic extraction over the past decade. This paper further investigates its integration and differentiation characteristics, with four new integrators and differentiators proposed. Theoretical analysis shows that the proposed SOGI-QSG based integration and differentiation methods can effectively overcome the drawbacks of the pure integrator and differentiator. The proposed four new methods are the frequency-adaptive integrator, the frequency-adaptive differentiator, the frequency-fixed integrator and the frequency-fixed differentiator respectively. The effectiveness of the proposed methods and the correctness of the theoretical analysis are finally evaluated by experimental results.**

Keywords—second-order-generalized integrator; pure integrator; digital differentiator; quadrature signal generator

I. INTRODUCTION

The Second-Order Generalized-Integrator Quadrature Signal Generator (SOGI-QSG) has previously been used in a single-phase PLL structure for grid synchronization [1] or in estimation of three-phase grid-voltage instantaneous-symmetrical components under unbalanced and distorted grid conditions [2], [3]. Recently, it has also been used for many other applications, such as delay estimation [4] and electromechanical oscillation estimation [5].

For the typical applications of SOGI-QSG, an early usage was for the frequency estimation [6]. Different from [1] where a PLL was used with the SOGI-QSG, reference [6] proposed a simpler Frequency-Locked Loop (FLL) structure which can achieve frequency estimation and thus make the SOGI-QSG frequency-adaptive. The gain of the FLL was normalized in [7] in order to guarantee a constant settling time independently of the input-signal characteristics. Besides, to realize an accurate grid synchronization under distorted grid conditions, the multiple SOGI structure was proposed in [8], which can not only estimate the sequence components at fundamental frequency but also other sequence components at the harmonic frequencies. Moreover, to eliminate the influence of the dc component in the input signal, dc-rejection methods were then proposed in [9] based on adding a Low-Pass Filter (LPF) and in [10] based on adding a new loop inside the SOGI-QSG, respectively.

For the recent development of SOGI, an interesting modified structure was proposed in [5] for the estimation of the frequency, magnitude, and damping of an oscillation. This algorithm is based on applying simple modifications to the existing SOGI-QSG structure, which can provide vital information about the stability condition of the power system. Another application is the time delay and frequency estimation of two sinusoidal signals received at two spatially separated sensors [4]. This function is realized by adding a simple new mechanism to the existing structure of the SOGI-QSG. Also it does not require any prior knowledge on the input signal frequency and therefore works at various sampling frequencies [4].

Apart from the above applications, there is also literature focusing on improving the performance of the SOGI-QSG. In [11], the cascaded SOGI-QSG structure was proposed, which actually changes the SOGI-QSG from a second-order system to a fourth-order system. The fourth-order structure can thus have a faster time response and higher harmonic rejection than the traditional SOGI-QSG structure. In [12] and [13], the third-order structure was proposed based on the SOGI-QSG, which shows better harmonic attenuation than the SOGI-QSG.

The SOGI-QSG has thus been used in many fields due to its flexible configurations. This paper further investigates the abilities of SOGI-QSG for integration and differentiation. The integrator and differentiator play critical roles in control systems, the principles of which are quite simple, yet their implementations are challenging. In this paper, four new approaches of SOGI-QSG are presented for the implementation of the integration and differentiation. The proposed methods are either frequency-adaptive or frequency-fixed integrators and differentiators, which provide multiple choices for the practical usages. The principles of the presented methods are explained with the bode diagrams and the effectiveness of two of the proposed methods are verified by experimental results.

This work was supported by European Research Council (ERC) under the European Union's Seventh Framework Program (FP/2007-2013)/ERC Grant Agreement [321149-Harmony], and China Scholarship Council (CSC).

978-1-4673-9551-9/16 $31.00 © 2016 IEEE

II. APPLICATIONS AND PROBLEMS OF INTEGRATOR AND DIFFERENTIATOR

Integrator and differentiator can be found in many applications. In this section, several typical applications of the integrator and differentiator are listed, along with the analysis of their typical problems.

A. Applications and Problems of Integrator

One typical application of the pure integrator is in the flux estimation of the induction motors [14], [15]. The voltage-model flux estimation method is used widely since the stator flux can simply be calculated by integrating the back Electro-Motive Force (EMF). Further, the rotor flux can easily be calculated from the stator flux, which is usually used for the field-oriented control. In the stationary reference frame, the stator flux estimation based on the voltage model can be expressed as

$$\psi_s = \int (u_s - R_s i_s) dt = \int e_s dt \qquad (1)$$

where ψ_s, u_s, i_s, e_s are the vectors of the stator flux, the stator voltage, the stator current and the back EMF respectively. R_s is the stator resistance. It is noted from (1) that a pure integrator is needed to realize the stator flux estimation.

The other typical application is the realization of the Rogowski-coil current sensor which has various advantages over the conventional magnetic current transformers, such as wide bandwidth and measuring range, without suffering from magnetic saturation, and light weight [16]. The principle of the Rogowski-coil current sensor is shown in Fig. 1 [17]. The voltage induced by the Rogowski coil is proportional to the rate change of the current enclosed by the coil-loop. It is thus necessary to integrate the coil voltage in order to produce a voltage proportional to the current. Hence, an integrator is required at the output of the Rogowski coil to obtain the measured current as shown in Fig. 1, which can be realized either analogly or digitally [18].

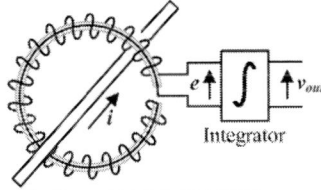

Fig. 1. Principle of the Rogowski coil current sensor.

Although the principles of the integration in these two applications are simple, the pure integrator suffers from the dc drift and the dc offset problems which makes it difficult to be implemented perfectly in practice. For a sinusoidal input signal with a dc component A_{dc} and an initial phase φ, $v_{ac}(t) = A\sin(\omega t+\varphi) +A_{dc}$, the output of the pure integrator will not only contain the integration of the sinusoidal component, but also contain the integration of this dc component and an additional dc offset, as shown in (2) and Fig. 2. The integration of the dc component will lead the integral output into saturation, while the dc offset is caused by the initial phase of the pure integrator.

$$v_{ac}(t) = -\frac{A}{\omega}\cos(\omega t + \varphi) + A_{dc}t + \frac{A}{\omega}\cos\varphi \qquad (2)$$

where A, ω, and φ is the magnitude, frequency, and initial phase of the input signal. A_{dc} is the dc component in the input sinusoidal signal.

In practice, the input dc component is often inevitable in the measured signal, which can finally drive the pure integrator into saturation. On the other hand, the dc offset in the output is caused by the initial phase of the sinusoidal signal to be integrated. The rapid change of the input sinusoidal signal can also introduce a dc offset in the integral output [15]. Usually, the pure integrators are replaced by Low-Pass Filters (LPF) to avoid these two problems. However, the LPF may produce magnitude error and phase error in the estimated flux, especially when the motor runs at a frequency lower than the cutoff frequency of the LPF. An Additional compensation algorithm should be added which in return makes the overall flux estimation algorithm more complicated. As for the Rogowski-coil usage, the low-frequency noise attenuation ability of the LPF is not strong enough which challenges the performance of the Rogowski-coil current sensor. Moreover, the LPF is usually implemented by the analog circuit, and thus the gain-frequency characteristic of the op-amp IC used for the integration will influence the bandwidth of the current sensor, which complicates the design of the integration circuit.

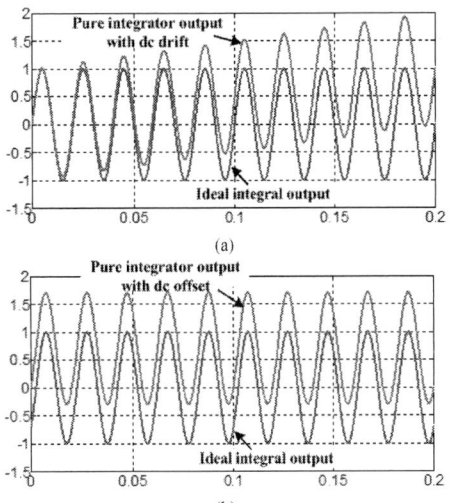

Fig. 2. Problems of pure integrator, (a) dc drift problem, (b) initial condition problem.

B. Applications and Problems of Differentiator

Differentiators are used widely in many applications as well. For example, one application of the differentiator can be found in the active damping of the LCL-filter resonance [19]–[22], where the capacitor-voltage is differentiated and fed back to achieve the damping function. Another recent application of the pure differentiator is reported in [23], where the digital differentiator is used in a newly proposed FLL structure for the grid synchronization. Applications of differentiator can also be found in the virtual inertia emulation in a doubly fed induction generator system [24], and synthesis of virtual impedance for a grid-tied converter [25], [26]. In these applications, the differentiator should be implemented digitally. However, digital implementation of a differentiator is not trivial due to the possible noise amplification and large phase error. The backward Euler method has usually been used to discretize the differentiator [27]. Nevertheless, it faces the problem of increasing large phase error at high frequency, which will eventually cause the differentiator loosing its effectiveness. The phase error caused by the backward Euler discretization can be seen from Fig. 3, where the frequency of the sinusoidal signal is 50 Hz, and the sampling frequency is 1 kHz.

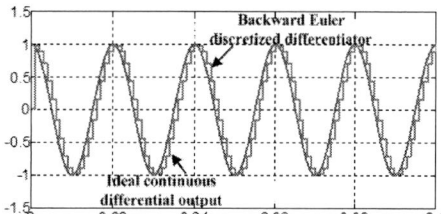

Fig. 3. Phase error caused by the backward Euler discretization of the differentiator.

C. Summary

It can be seen from the above analysis that the problems of the integrator and differentiator make them impossible to be used directly in practice. Therefore, additional methods should be researched to replace the integrator and differentiator. Considering the flexibility of the SOGI-QSG structure, the integration and differentiation characteristics of the SOGI-QSG will investigated in the following section with four new integrators and differentiators proposed.

III. BASIC PRINCIPLE OF SOGI-QSG

The SOGI-QSG is built by SOGI, whose transfer function can be written as (3), and the structure realization of SOGI is shown in the shaded part of Fig. 4. The structure of SOGI-QSG can then be realized by adding a unit feedback based on the structure of SOGI. In addition, the error signal and the in-quadrature signal can be used to realize a Frequency-Locked Loop (FLL), which was proposed in [6] as an effective mechanism to make the SOGI-QSG frequency adaptive. The overall structure is thus named as SOGI-FLL [6]. Usually, the two in-quadrature output signals of SOGI-QSG are used for the grid synchronization, which are defined by (4) and (5).

$$SOGI(s) = \frac{\omega's}{s^2 + \omega'^2} = \frac{\omega'/s}{1 + \omega'^2/s^2} \qquad (3)$$

$$D(s) = \frac{v'}{v} = \frac{k\omega's}{s^2 + k\omega's + \omega'^2} \qquad (4)$$

$$Q(s) = \frac{qv'}{v} = \frac{k\omega'^2}{s^2 + k\omega's + \omega'^2} \qquad (5)$$

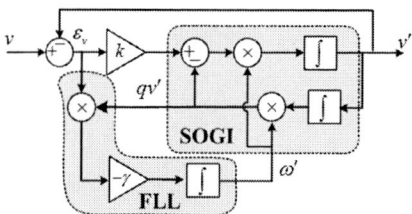

Fig. 4. Structure of the Second-Order Generalized Integrator Frequency Locked Loop (SOGI-FLL).

IV. FREQUENCY-ADAPTIVE INTEGRATOR AND DIFFERENTIATOR

A. Frequency-Adaptive Integrator

As mentioned previously, both the pure integrator and the LPF have problems when they are used for flux estimation. To solve their problems, a new flux integraion method is proposed based on SOGI-FLL, which can also be used in other applications. Fig. 5 illustrates the structure of the proposed flux estimator. Different from the traditional usage of the SOGI-FLL [19]-[24], the integral output is used here for the stator flux estimation, and in this case, the input signal will become the stator back EMF. It should be pointed out that the gain-normalization block of the FLL in [21]-[23] is essential here in order to ensure an unaltered time constant during the variation of the back EMF at different speeds. Considering ψ_s and e_s as the output and input signals respectively, the transfer function of the frequency-adaptive integrator is given by

$$\psi_s(j\omega) = \frac{\psi_s}{e_s} = \frac{k\omega'}{s^2 + k\omega's + \omega'^2} \qquad (6)$$

where ω' is the back-EMF frequency in rad/s detected by the FLL and k is the gain of SOGI-QSG which is set to $\sqrt{2}$ in order to achieve an optimal trade-off between the dynamic response and the overshoot [3].

It is worth to notice from (6) that it is equal to the pure integrator when the estimated back-EMF frequency ω' is equal to the actual back-EMF frequency ω, i.e. the integral characteristic of the SOGI-FLL is absolutely equal to the pure integrator when the frequency estimated by FLL is accurate. This is a very interesting feature for implementing integration. Bode plots of the transfer function (6) is shown in Fig. 6, which further describes the integral characteristic of SOGI-FLL with three different freqencies. It can be seen that SOGI-FLL has no magnitude/phase errors at the back-EMF frequency.

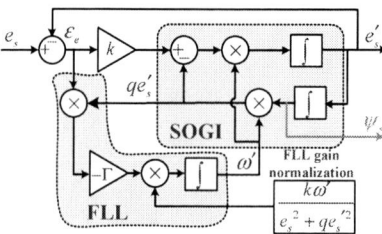

Fig. 5. Structure of the SOGI-based frequency-adaptive integrator.

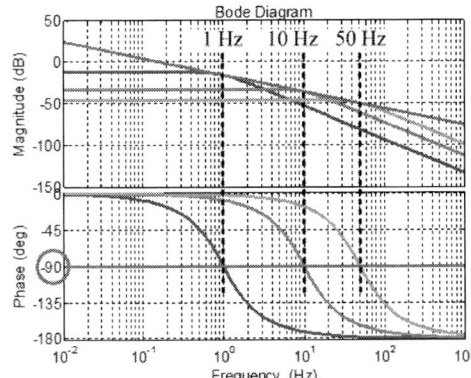

Fig. 6. Bode diagrams of the SOGI-based frequency-adaptive integrator.

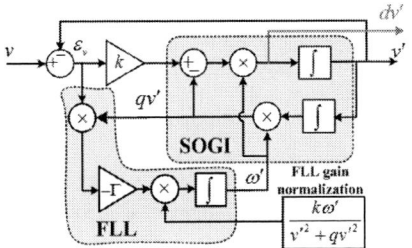

Fig. 7. Structure of the SOGI-based frequency-adaptive differentiator.

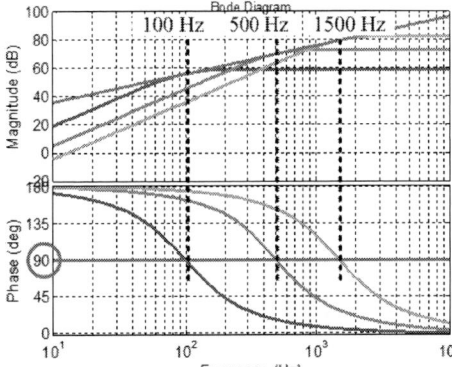

Fig. 8. Bode diagrams of the SOGI-based frequency-adaptive differentiator.

Besides, this integration algorithm can avoid the dc drift and the initial condition problems of a pure integrator as well. Considering an input back EMF signal $e_s(t) = A\sin(\omega t+\varphi) +A_{dc}$, the dc offset in the back EMF can be roughly considered as a step signal with a magnitude of A_{dc}, while the initial phase of the back EMF can be thought as an impulse signal with a magnitude of $A\cos(\varphi)/\omega$. Thus, the response of SOGI-FLL for the dc offset and initial condition can be described by the step response and impulse response of a second-order system respectively. The results are shown as

$$\psi_{dc}(t) = \frac{k}{\omega'}\cdot A_{dc} - \frac{2kA_{dc}}{\omega'\sqrt{4-k^2}}e^{-\sigma t}\cos(\omega_d t - \beta) \quad (7)$$

$$\psi_{ini}(t) = \frac{A\cos\varphi}{\omega'}\cdot \frac{2k}{\sqrt{4-k^2}}e^{-\sigma t}\sin(\omega_d t) \quad (8)$$

where $\sigma = \zeta\omega'$, $\beta = \tan^{-1}\left(k/\sqrt{4-k^2}\right)$ and $\omega_d = \omega'\sqrt{4-k^2}/2$.

It can be seen from (7) and (8) that both of them consist of an exponential-decay term which tends toward 0 with the time going on. Thus, the dc offset in the back EMF will be not integrated but reduced to k/ω' times of the value, and the effect of the initial condition can be eliminated. Furthermore, another remarkable aspect of the SOGI-based frequency-adaptive integrator is the high-order harmonics attenuation ability which can be found in Fig. 6. This is a very convenient feature to make the method more robust in front of back-EMF distortion due to the pulse width modulation control and measurement noise.

B. Frequency-Adaptive differentiator

Similar with the frequency-adaptive integrator, the SOGI-QSG can also be used as a frequency-adaptive differentiator. The principle of this differentiator is shown in Fig. 7. To illustrate, the transfer function for relating dv to v can be determined as (9) and the related bode diagrams are shown in Fig. 8. It can be seen that the SOGI-based frequency-adaptive differentiator has no magnitude/phase error at the frequency to be differentiated as long as the frequency is accurate. The proposed frequency-adaptive differentiator has a rapid gain roll-off below the differentiated frequency and a constant gain

above it. These features contribute to the selectiveness of the proposed differentiator, which can be helpful if only input of a certain frequency needs to be differentiated.

One application for this differentiator has been published in [28] where the selective differentiator is used to estimate the capacitor current from the capacitor voltage for the realization of active damping. However, for this application, the input frequency of SOGI-QSG should be tuned to the LCL-filter resonance frequency which is obtained from the calculation of LCL parameters rather than a FLL. For other applications, the FLL can also be used to estimate the input-signal frequency and makes the differentiator frequency adaptive. As for the digital implementation, it should be discretized by the Tustin method prewarped at the resonance frequency according to the analysis in [28].

$$G_{v_d'}(s) = \frac{v_d'}{v} = \frac{k\omega' s^2}{s^2 + k\omega' s + \omega'^2}. \quad (9)$$

V. FREQUENCY-FIXED INTEGRATOR AND DIFFERENTIATOR

The existing usages of SOGI-QSG are mainly based on its characteristic at the center frequency, such as the grid synchronization, and also the proposed frequency-adaptive integrator and differentiator. Actually, the characteristic of the SOGI-QSG at other frequency regions can also be used for the integration and differentiation as long as the input frequency and the damping gain k are tuned properly.

A. Frequency-Fixed Integrator

The structure of the SOGI-based frequency-fixed integrator is shown in Fig. 9. The gain at the output is to ensure the same gain with a pure integrator. To illustrate this, the transfer function for relating iv' to v can be given by

$$G_{iv'}(s) = \frac{iv'}{v} = \frac{s}{s^2 + k\omega's + \omega'^2} . \tag{10}$$

Substituting $s=j\omega$ to (10), its magnitude and phase expressions can be given by

$$\left| G_{iv'}(j\omega) \right| = \frac{\omega}{\sqrt{\left(\omega'^2 - \omega^2\right)^2 + \left(k\omega\omega'\right)^2}} \tag{11}$$

$$\angle G_{iv'}(j\omega) = \arctan\left(\frac{\omega'^2 - \omega^2}{k\omega\omega'}\right) \tag{12}$$

With $\omega' \ll \omega$ now considered, (11) and (12) will be approximately equal to $1/\omega$ and $-90°$. This characteristic can be seen from Fig. 10, where (10) has been plotted with an ideal integrator. The frequency ω' is set to 10π rad/s and a gain k is set to 1.414. For the Rogowski-coil current sensor application, traditional analog-LPF based integrator cannot ensure a satisfactory low-frequency noise attenuation ability, which influences the performance of the Rogowski-coil current sensor. Moreover, the elements used in the analog integrator circuits are nonideal, therefore, they may add some extra errors to the measured result. Realizing the integrator digitally is thus a good choice especially for the cases where a digital controller has already been added for the power quality analysis or other purpose. Unlike the LPF, the SOGI-based frequency-fixed integrator has a stronger attenuation than the LPF for the low-frequency noise, which can avoid the dc-drift and initial condition problems of the pure integrator, and improve the noise-attenuation performance of the current sensor.

On the other hand, similar to the LPF, the SOGI-based frequency-fixed integrator contains magnitude and phase errors as well. For example, in Fig. 10, the phase at 50 Hz, 150 Hz, and 250 Hz are -81.8°, -87.3° and -88.4° respectively, which are not equal to 90°. The phase error can be reduced by decreasing k but it cannot be eliminated, and a too large gain may deteriorate the low-frequency attenuation characteristic of the proposed differentiator. Besides, the phase error can also be reduced by decreasing the frequency ω', but the magnitude at the low-frequency region may increase at the same time. In this case, the compensation filter should be added, which can be implemented digitally, and thus its characteristic can be designed more flexibly than the analog compensation circuit according to the specific requirements.

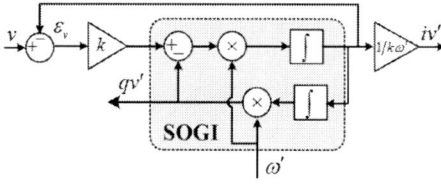

Fig. 9. Structure of the SOGI-based frequency-fixed integrator.

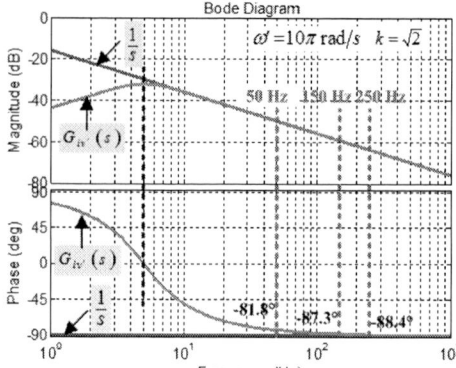

Fig. 10. Bode diagrams of the SOGI-based frequency-fixed integrator.

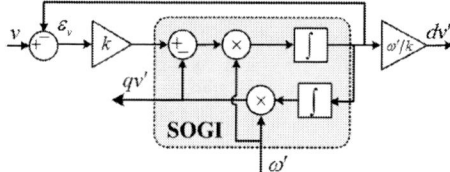

Fig. 11. Structure of the SOGI-based frequency-fixed differentiator.

B. Frequency-Fixed Differentiator

Similiar with the frequency-fixed integrator, the structure of the SOGI-based frequency-fixed differentiator is shown in Fig. 11, and the transfer function for relating dv' to v can be given by

$$G_{dv'}(s) = \frac{dv'}{v} = \frac{\omega'^2 s}{s^2 + k\omega's + \omega'^2} . \tag{13}$$

Substituting $s=j\omega$ to (13), its magnitude and phase expressions is shown below.

$$\left| G_{dv'}(j\omega) \right| = \frac{\omega\omega'^2}{\sqrt{\left(\omega'^2 - \omega^2\right)^2 + \left(k\omega\omega'\right)^2}} \tag{14}$$

$$\angle G_{dv'}(j\omega) = \arctan\left(\frac{\omega'^2 - \omega^2}{k\omega\omega'}\right) \tag{15}$$

With $\omega' \gg \omega$ now considered, (14) and (15) will be approximately equal to ω and 90°. In order to obtain the differentiation characteristic in a wide frequency range, the frequency ω' is set to the Nyquist frequency in a digital system. The resulted bode diagram is shown in Fig. 12, where (14) has been plotted with an ideal differentiator and the frequency ω' is set to the Nyquist frequency. It is noted that the SOGI-based frequency-fixed differentiator is actually the same with the nonideal GI based differentiator proposed in [29] and [28] as long as $k\omega'$ equals to the parameter ω_c of the nonideal-GI transfer function. Therefore, the parameter design of the SOGI-based frequency-fixed differentiator will also be the same with the non-ideal GI differentiator. Moreover, it is advised to be discretized with the first-order hold discretization technique according to the analysis in [28].

978-1-4673-9551-9/16 $31.00 © 2016 IEEE

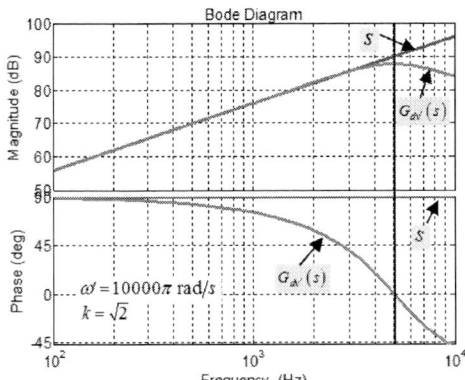

Fig. 12. Bode diagrams of the SOGI-based frequency-fixed differentiator.

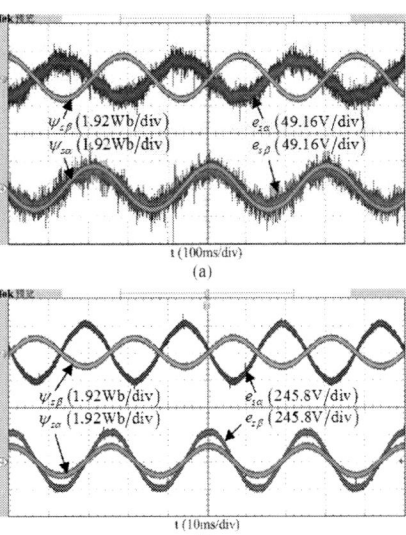

(a)

(b)

Fig. 13. Experimental results of the estimated stator fluxes in a motor drive which are compared with the back EMFs at 100 r/min and 1200 r/min respectively. (a) At 100 r/min. (b) At 1200 r/min.

VI. EXPERIMENTAL RESULTS

Two experimental setups are built to evaluate the performance of the proposed methods.

The first experiment is based on an 11 kW induction-motor rotor-flux-oriented vector control system. The rotor flux is calculated from the stator flux based on the voltage-model according to (16).

$$\psi_r = L_r \left(\psi_s - \sigma L_s i_s \right) \big/ L_m \qquad (16)$$

where ψ_s, ψ_r are the vector of the stator flux and the rotor flux. The stator flux ψ_s is calculated from (1) where an integrator is needed. L_s is the stator leakage inductance. $\sigma = \left(L_s L_r - L_m^2 \right) \big/ L_m^2$ is the leakage flux coefficient. L_m and L_r are the magnetizing and the rotor leakage inductances, respectively. The integration in the voltage model is implemented with the proposed SOGI-based frequency-adaptive integrator whose structure has been shown in Fig. 5.

Fig. 13 (a) and (b) show the stator fluxes estimated by the proposed SOGI-based frequency-adaptive integrator, which are compared with the back-EMFs at 100 r/min and 1200 r/min respectively. It can be seen that the estimated stator fluxes are stable sine waves despite the distortion of the back-EMFs. The fifth and seventh harmonics exist in the back EMFs due to the dead-time effect which is much more severe at low speed. Although the dead-time effect can be compensated, the compensation algorithm is not used in this control system, in order to validate the high-order harmonics attenuation performance of the SOGI-based frequency-adaptive integrator. The Fast Fourier Transform (FFT) of the α- axis back EMF at 100 r/min is shown in Fig. 14 (a), where it can be seen that the fifth and seventh harmonics are evident. The FFT of the estimated α- stator flux is shown in Fig. 14 (b), where it can be seen that the fifth and seventh harmonics are effectively eliminated with the proposed flux estimation method. The high-order harmonics attenuation performance of the proposed method can also be explained through the bode plot in Fig. 6.

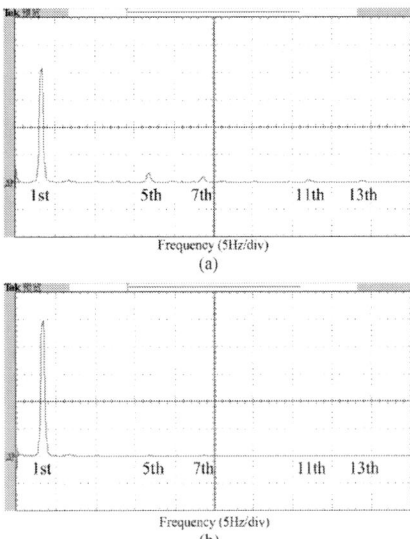

(a)

(b)

Fig. 14. Harmonic spectra of the back-EMF and the estimated stator flux at 100 r/min. (a) Harmonic spectra of the α- axis back EMF. (b) Harmonic spectra of the α- stator flux.

The second experiment is based on a 15-kW grid-connected LCL-filtered voltage source inverter system. The grid current is controlled to regulate the power injected to the grid. For synchronization and active damping of the LCL-filter resonance, only the filter capacitor voltage was measured. The capacitor voltage was differentiated by a backward Euler, HPF or SOGI-based frequency-fixed differentiator for comparison. Detailed discussion about the backward Euler and HPF differentiators can be found in [29] and [28].

978-1-4673-9551-9/16 $31.00 © 2016 IEEE

Fig. 16 - Fig. 18 show the experimental results of the capacitor-voltage-feedback active damping with the frequency-fixed differentiator applied. As mentioned, the SOGI-based frequency-fixed differentiator is the same with the nonideal GI based differentiator. Therefore, it is actually an adjustable differentiator and can be configured manually by regulating the value of k. This feature can be found in Fig. 15 where the proposed differentiator can become the same with the backward Euler differentiator and the HPF differentiator by choosing a proper k. To validate, three values of $k\omega'$ (ω' is set to the Nyquist frequency) are selected to test the effectiveness of the proposed method. First, the value of $k\omega'$ is set to 26000, in which case, the performance of the proposed differentiator should be the same with the HPF differentiator according to Fig. 15. The experimental results are shown in Fig. 16(a) with the result of HPF differentiator shown in Fig. 16(b). It can be seen that the resonance occurs during the step response of the grid current. Moreover, the magnitude of the resonance in these two figures are almost the same, which proves the correctness of the theoretical analysis. Further, the value of $k\omega'$ is tuned to 42000, in which case the performance of the proposed differentiator should be the same with the backward Euler differentiator according to Fig. 15. The corresponding results are shown in Fig. 17 where more severe resonances are activated due to a larger phase error of the differentiator when $k\omega'$ is tuned to 42000. Quite similar results in Fig. 17(a) and Fig. 17(b) again prove the adjustable characteristic of the proposed differentiator.

Finally, the value of $k\omega'$ is tuned to 5000. The SOGI-based frequency-fixed differentiator can thus have a more accurate phase characteristic than the backward Euler and the HPF differentiators. Fig. 18 shows the corresponding results in which the resonance is effectively damped during the step response of the grid current due to a more accurate differentiation characteristic of the proposed differentiator used for the capacitor-voltage-derivative-feedback active damping.

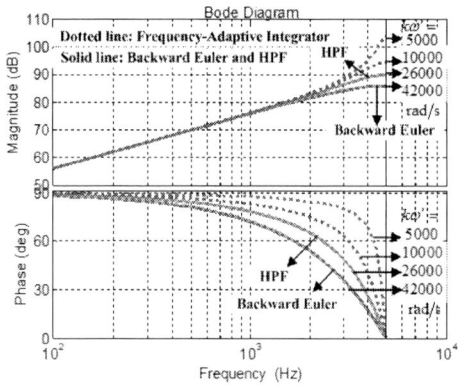

Fig. 15. Frequency responses of the first-order-hold discretized nonideal GI differentiator with four different parameters, and the backward Euler, HPF differentiator.

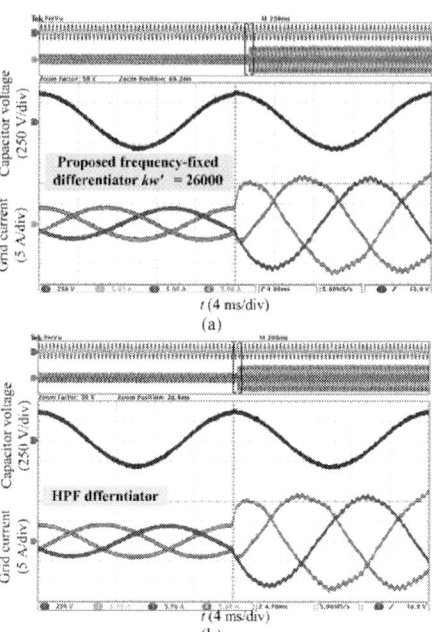

Fig. 16. Measured Capacitor voltage and grid current when the current command is stepped from 5 A to 10 A. (a) when the proposed frequency-fixed differentiator is used for active damping and $k\omega'$ is set to 26000. (b) when the HPF differentiator is used for active damping.

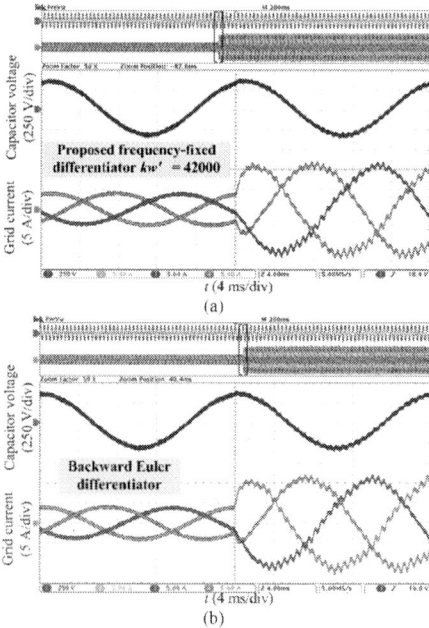

Fig. 17. Measured Capacitor voltage and grid current when the current command is stepped from 5 A to 10 A. (a) when the proposed frequency-fixed differentiator is used for active damping and $k\omega'$ is set to 42000. (b) when the HPF differentiator is used for active damping.

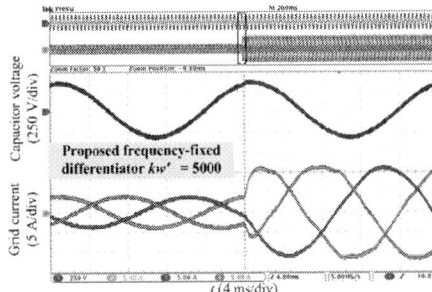

Fig. 18. Measured Capacitor voltage and grid current when the current command is stepped from 5 A to 10 A, and the proposed frequency-fixed differentiator is used for active damping and $k\omega'$ is set to 5000.

VII. CONCLUSIONS

In this paper, the integration and differentiation characteristics of the SOGI-QSG are investigated. Four new applications of SOGI-QSG are proposed to implement the pure integrator and digital differentiator. The proposed methods have quite simple structures and excellent performance. Besides, the proposed methods are also very easy to be implemented in practice. The proposed integrators and differentiators are either frequency-adaptive or frequency fixed algorithms, and thus they can be selected according to specific applications.

VIII. REFERENCES

[1] M. Ciobotaru, R. Teodorescu, and F. Blaabjerg, "A new single-phase PLL structure based on second order generalized integrator," in *Proc. IEEE PESC'06*, Jun. 2006, pp. 1–6.

[2] P. Rodriguez, R. Teodorescu, I. Candela, A. V. Timbus, M. Liserre, and F. Blaabjerg, "New positive-sequence voltage detector for grid synchronization of power converters under faulty grid conditions," in *Proc. IEEE PESC'06*, Jun. 2006, pp. 1–7.

[3] P. Rodriguez, A. Luna, R.S. Munoz-Aguilar, I. Etxeberria-Otadui, R. Teodorescu, F. Blaabjerg, "A stationary reference frame grid synchronization system for three-phase grid-connected power converters under adverse grid conditions," *IEEE Trans. Power Electron.*, vol. 27, no. 1, pp. 99–112, Jan. 2012.

[4] M. Ghadiri-Modarres, M. Mojiri, and M. Karimi-Ghartemani, "New adaptive algorithm for delay estimation of sinusoidal signals with unknown frequency," *IEEE Trans. Instrum. Meas.*, vol. 64, no. 9, pp. 2360–2366, Sep. 2015.

[5] M. Mansouri, M. Mojiri, M. Ghadiri-Modarres and M. Karimi-Ghartemani, "Estimation of electromechanical oscillations from phasor measurements using second-order generalized integrator," *IEEE Trans. Instrum. Meas.*, vol. 64, no. 4, pp. 943–950, Apr. 2015.

[6] P. Rodriguez, A. Luna, M. Ciobotaru, R. Teodorescu, and F. Blaabjerg, "Advanced grid synchronization system for power converters under unbalanced and distorted operating conditions," in *Proc. 32nd IEEE IECON, Paris, France*, Nov. 6–10, 2006, pp. 5173–5178.

[7] P. Rodriguez, A. Luna, I. Candela, R. Teodorescu, and F. Blaabjerg, "Grid synchronization of power converters using multiple second order generalized integrators," in *Proc. 34th Annu. Conf. IEEE Ind. Electron.*, Nov. 2008, pp. 755–760.

[8] P. Rodriguez, A. Luna, I. Candela, R. Mujal, R. Teodorescu, and F. Blaabjerg, "Multiresonant frequency-locked loop for grid synchronization of power converters under distorted grid conditions," *IEEE Trans. Ind. Electron.*, vol. 58, no. 1, pp. 127–138, Jan. 2011.

[9] M. Ciobotaru, R. Teodorescu, and V. Agelidis, "Offset rejection for PLL based synchronization in grid-connected converters," in *Proc. Appl. Power Electron. Conf. Expo.*, Austin, TX, Feb. 2008, pp. 1611–1617.

[10] M. Karimi-Ghartemani, S. Khajehoddin, P. Jain, A. Bakhshai, and M. Mojiri, "Addressing DC component in PLL and notch filter algorithms," *IEEE Trans. Power Electron.*, vol. 27, no. 1, pp. 78–86, Jan. 2012.

[11] J. Matas, M. Castilla, J. Miret, L. García de Vicuña, and R. Guzman, "An adaptive pre-filtering method to improve the speed/accuracy trade-off of voltage sequence detection methods under adverse grid conditions," *IEEE Trans. Ind. Electron.*, vol. 61, no. 5, pp. 2139–2151, May 2014.

[12] T. Ngo, Q. Nguyen, and S. Santoso, "Detecting positive-sequence component in active power filter under distorted grid voltage," in *IEEE PESGM 2015*, July 2015, pp. 1–5.

[13] W. Li, X. Ruan, C. Bao, D. Pan, and X. Wang, "Grid synchronization systems of three-phase grid-connected power converters: A complex vector- filter perspective," *IEEE Trans. Ind. Electron.*, vol. 61, no. 4, pp. 1855–1870, Apr. 2014.

[14] M. H. Shin, D. S. Hyun, S. B. Cho, and S. Y. Choe, "An improved stator flux estimation for speed sensorless stator flux orientation control of induction motors," *IEEE Trans. Power Electron.*, vol. 15, no. 2, pp. 312–318, Mar. 2000.

[15] J. Hu and B. Wu, "New integration algorithms for estimating motor flux over a wide speed range," *IEEE Trans. Power Electron*, vol. 1, no. 5, pp. 969–977, Sep. 1998.

[16] L. A. Kojovic, "Comparative performance characteristics of current transformers and Rogowski coils used for protective relaying purposes," in *Proc. IEEE PES Gen. Meeting*, Tampa, FL, 2007, pp. 1–6.

[17] L. A. Kojovic, "Application of Rogowski coils used for protective relaying purposes," *IEEE Power Syst. Conf. Exp.*, pp. 538–543, Oct.–Nov. 2006.

[18] M. Samimi, A. Mahari, M. Farahnakian, and H. Mohseni, "The Rogowski Coil Principles and Applications: A Review," *IEEE Sens. J.*, vol. 15, no. 2, pp. 651–658, Feb. 2015.

[19] J. Dannehl, F. W. Fuchs, S. Hansen, and P. B. Thogersen, "Investigation of active damping approaches for PI-based current control of grid-connected pulse width modulation converters with LCL filters," *IEEE Trans. Ind. Appl.*, vol. 46, no. 4, pp. 1509–1517, Jul. –Aug. 2010.

[20] L. Harnefors, A. G. Yepes, A. Vidal, and J. Doval-Gandoy, "Passivity-based controller design of grid-connected VSCs for prevention of electrical resonance instability," *IEEE Trans. Ind. Electron.*, vol. 62, no. 2, pp. 702–710, Feb. 2015.

[21] J. Xu, S. Xie, and T. Tang, "Active damping-based control for grid-connected LCL-filtered inverter with injected grid current feedback only," *IEEE Trans. Ind. Electron.*, vol. 61, no. 9, pp. 4746–4758, Sep. 2014.

[22] X. Wang, F. Blaabjerg, and P. C. Loh, "Grid-current-feedback active damping for LCL resonance in grid-connected voltage source converters," *IEEE Trans. Power Electron.*, vol. 31, no. 1, pp. 213–223, Jan. 2014.

[23] P. Kanjiya, V. Khadkikar, and M. Moursi, "A novel type-1 frequency-locked-loop for fast detection of frequency and phase with improved stability margins," *IEEE Trans. Ind. Electron.*, 10.1109/TPEL.2015.2435706.

[24] M. F. M. Arani and E. F. El-Saadany, "Implementing virtual inertia in DFIG-based wind power generation," *IEEE Trans. Power. Syst.*, vol. 28, no. 2, pp. 1373–1384, May. 2013.

[25] X. Wang, F. Blaabjerg, and P. C. Loh, "Virtual-impedance-based control for voltage-source and current-source converters," *IEEE Trans. Power Electron.*, vol. 31, no. 1, pp. 213–223, Jan. 2014.

[26] J. He and Y. W. Li, "Generalized closed-loop control schemes with embedded virtual impedances for voltage source converters with LC or LCL filters," *IEEE Trans. Power Electron.*, vol. 27, no. 4, pp. 1850–1861, Apr. 2012.

[27] X. Wang, F. Blaabjerg, and P. C. Loh, "Proportional derivative based stabilizing control of paralleled grid converters with cables in renewable power plants," in *Proc. IEEE of ECCE*, 2014, pp. 4917–4924

[28] Z. Xin, P. C. Loh, X. Wang, F. Blaabjerg and Y. Tang, "Highly accurate derivatives for LCL-filtered grid converter with capacitor voltage active damping," *IEEE Trans. Power Electron.*, 10.1109/TPEL.2015.2467313.

[29] Z. Xin, X. Wang, P. C. Loh, and F. Blaabjerg, "Realization of Digital Differentiator Using Generalized Integrator for Power Converters," *IEEE Trans. Power Electron.*, vol. 30, no. 12, pp. 6520–6523, Dec. 2015.

Three-Phase Multiple Harmonic Sequence Detection Based on Generalized Delayed Signal Superposition

Yong Lu, Guochun Xiao
School of Electrical Engineering and State Key Lab of
Electric Insulation and Power Equipment
Xi'an Jiaotong University
Xi'an, China
hearttoboat@stu.xjtu.edu.cn, xgc@mail.xjtu.edu.cn

Xiongfei Wang, Frede Blaabjerg
Department of Energy Technology
Aalborg University
Aalborg, Denmark
xwa@et.aau.dk, fbl@et.aau.dk

Abstract—**Grid synchronization has always been an important challenge for three-phase grid-connected converters under unbalanced and distorted grid conditions. Moreover, how to quickly and accurately extract multiple harmonic sequence information is essential for control systems. In this paper, a three-phase multiple harmonic sequence detection method is proposed for estimating both the fundamental and harmonic sequence components under adverse grid conditions. This detection method is denoted as MGDSS-PLL since it contains Multiple Generalized Delayed Signal Superposition operators and a Phase-Locked Loop. The proposed MGDSS-PLL can be flexibly tuned to extracting any harmonic components according to specific applications and it also exhibits great robustness to different grid disturbances. Simulations and experimental results are presented for verifying the performance of the MGDSS-PLL.**

Keywords—*Three-phase; Grid synchronization; Harmonic detection; Phase-locked loop; Delayed signal superposition*

I. INTRODUCTION

In recent years, Distributed Generation Systems (DGSs) based on renewable energy sources (RESs) have experienced rapid and continuous development due to the increasing demands for cleaner energy and also a dynamic grid system like the smart grid[1]-[4]. However, a high penetration level of RESs in turn challenges reliable, efficient and safe operation of the utility grid and therefore related grid codes become more and more rigorous [5]-[7]. Consequently, power-electronic converters, acting as energy conversion interface between the DGSs and electricity network, are required to provide stable performance and fault-ride-through capability in order to

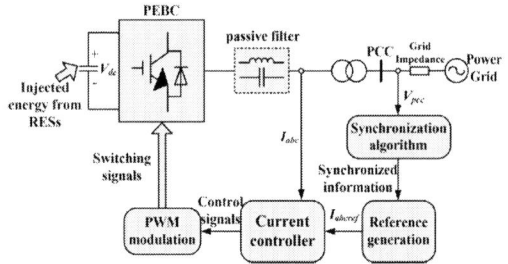

Fig. 1. General control system for power-electronic converters.

handle undesired grid perturbations (including voltage sags, system unbalance and severe distortions) [8]-[10]. In this scenario, one of the key issues in the design of reliable controllers for these converters is the grid synchronization at the Point of Common Coupling (PCC). Therefore, the synchronization algorithm should be accurate and robust under variety of grid conditions and plays a critical role in the control of converters [11]-[16], which is also demonstrated in Fig. 1.

In three-phase systems, Synchronous Reference Frame Phase-Locked Loop (SRF-PLL) is generally considered to be the most widely applied synchronization technique. The SRF-PLL exhibits satisfactory dynamic and accuracy in estimating the grid information under balanced and sinusoidal grid conditions, but its performance tend to be unacceptable when the grid is unbalanced and/or distorted [17]-[19]. To improve the behavior of SRF-PLL under abnormal grid conditions, various advanced approaches have been discussed in recent years [20]-[32]. A majority of these approaches are based on adding filtering stages, which can be further categorized into two types as extended-loop-filter-based techniques and prefilter-based techniques [24]. In practical implementation, the extended-loop-filter-based synchronization systems are often preferred when only the Fundamental Positive-sequence Component (FPC) of the grid is required, whereas the prefilter-based synchronization techniques are generally adopted to extract the fundamental and/or selected harmonic sequence components. Therefore, the prefilter-based techniques are normally considered to be more convenient for a flexible control strategy design under adverse grid conditions [25].

There are a variety types of prefilter-based synchronization techniques proposed, including Decoupled Synchronous Reference-Frame PLL (DSRF-PLL) [20], [26], Sinusoidal Signal Integrator PLL (SSI-PLL) [27], Signal Reforming PLL (SR-PLL) [28], Double Second Order Generalized Integrator PLL (DSOGI-PLL) [29], [30] and Delayed Signal Cancelation PLL (DSC-PLL) [31], [32], but these PLLs rarely gives satisfactory harmonic sequence detection performance considering their overall evaluation about the response time, steady-state error and robustness. In [33], [34], several types of prefilter-based synchronization algorithms are comparatively evaluated. In order to improve the harmonic detection abilities of the DSOGI-PLL, an improved solution based on Multiple

978-1-4673-9551-9/16 $31.00 © 2016 IEEE

Second Order Generalized Integrators and a Frequency-Locked Loop (MSOGI-FLL) is presented in [35]. The MSOGI-FLL consists of multiple DSOGIs tuned at different harmonics of the fundamental grid frequency and a Harmonic Decoupling Network (HDN) which decouples the interactions of different harmonic components on the input signal of these DSOGIs. Thus, the MSOGI-FLL can detect both fundamental and harmonic sequences even under highly polluted grid conditions. However, the number of DSOGI subsystems should be equal to the number of significant harmonics contained in the distorted grid and any unconsidered harmonics will affect the steady-state error of the HDN.

In this paper, a novel three-phase multiple harmonic sequence component detection method based on Multiple Generalized Delayed Signal Superposition operators and a Phase-Locked Loop (MGDSS-PLL) is proposed. The GDSS operators are derived according to the Delayed Signal Superposition (DSS) operators presented in [36] and it can extract any selected harmonic component out of the input signal with properly designed parameters. The proposed MGDSS-PLL structure consists of multiple GDSS operators and Positive-/Negative-Sequence Calculators (PNSCs), which allow extracting multiple desired harmonic components under adverse grid conditions. Unlike the aforementioned MSOGI technique, the number of GDSS operators only depends on the number of interested harmonics and these operators are insensitive to unconsidered harmonic components since they are rather selective. Moreover, the presented GDSS method is robust to small frequency variations and exhibits a fast transient response, i.e. the response time is no longer than one fundamental period. It should be mentioned that the GDSS is similar to the Cascaded Delayed Signal Cancellation (CDSC) introduced in [37] and they are both inspired by the concept of signal delay. However, differing from the CDSC, the proposed GDSS is derived through an algebraic recursive procedure, which provides an easier parameter design process and also a more intuitive harmonic elimination principle. Additionally, no magnitude or phase angle correction block is needed for the GDSS. Both simulation and experimental results are presented to verify the proposed GDSS-PLL.

II. GENERALIZED DELAYED SIGNAL SUPERPOSITION OPERATORS

Any voltage/current signal $u(t)$ can be regarded as a combination of the fundamental and harmonics. Then, $u(t)$ is described as:

$$u(t) = \sum_{h=1}^{H} U_h \cos(h\omega t + \varphi_h) \tag{1}$$

where h is the harmonic order (for fundamental component, $h=1$), H represents the maximum harmonic order considered, ω and φ_h are the fundamental frequency and initial phase of the h-order harmonics, respectively. Two different types of Generalized Delayed Signal ($GDS1$ and $GDS2$) operators for $u(t)$ are considered in this work:

$$GDS1[u(t)] = u(t - \frac{k}{h_s n}T)\cos\frac{2k\pi}{n} \tag{2}$$

$$GDS2[u(t)] = u(t - \frac{k}{h_s n}T)\sin\frac{2k\pi}{n} \tag{3}$$

where T is the fundamental period, h_s is the harmonic order to be detected and n, k are arbitrary integers. If $GDS1[u(t)]$ and $GDS2[u(t)]$ are added with k ranging from 0 to m ($m<h_s n$) and then multiplied by $2/(m+1)$, two types of GDSS operators can be constructed as:

$$GDSS1[u(t)] = \frac{2}{m+1}\sum_{k=0}^{m} u(t - \frac{k}{h_s n}T)\cos(\frac{2k\pi}{n}) \tag{4}$$

$$GDSS2[u(t)] = \frac{2}{m+1}\sum_{k=0}^{m} u(t - \frac{k}{h_s n}T)\sin(\frac{2k\pi}{n}) \tag{5}$$

When the h-order harmonic $u_h(t)=U_h\cos(h\omega t+\varphi_h)$ is considered, (4) can be further derived as:

$$GDSS1[u_h(t)] = \frac{2}{m+1}\sum_{k=0}^{m} U_h\cos(\alpha - \frac{2hk\pi}{h_s n})\cos(\frac{2k\pi}{n}) \tag{6}$$

where α represents the phase angle of $u_h(t)$ and $\alpha=h\omega t+\varphi_h$. Then (6) can be rearranged as:

$$GDSS1[u_h(t)] = \frac{U_h}{m+1}\sum_{k=0}^{m}[\cos(\alpha - 2kA) + \cos(\alpha - 2kB)] \tag{7}$$

$$A = \frac{(h-h_s)}{h_s n}\pi \tag{8}$$

$$B = \frac{(h+h_s)}{h_s n}\pi \tag{9}$$

According to certain sums of trigonometric functions provided in [38], (7) can be further simplified as:

$$GDSS1[u_h(t)] = \begin{cases} G(CD\csc A + EF\csc B) & h \neq h_i \\ U_h\cos\alpha + GEF\csc B & h = h_s(jn+1) \\ U_h\cos\alpha + GCD\csc A & h = h_s(jn-1) \end{cases} \tag{10}$$

where $C=\cos(\alpha-mA)$, $D=\sin[(m+1)A]$, $E=\cos(\alpha-mB)$, $F=\sin[(m+1)B]$, $h_i = h_s(jn\pm1)$ and j is a natural number. The same derivation process can be applied to analyze (5) and a similar expression can be obtained as:

$$GDSS2[u_h(t)] = \begin{cases} G(JD\csc A + KF\csc B) & h \neq h_i \\ U_h\sin\alpha + GKF\csc B & h = h_s(jn+1) \\ U_h\sin\alpha + GJD\csc A & h = h_s(jn-1) \end{cases} \tag{11}$$

where $J=\sin(\alpha-mA)$ and $K=\sin(\alpha-mB)$.

As it can be solved from (10), if D and F are set to zero with $m=h_s n-1$, $GDSS1$ operator will exhibit a zero gain for any harmonic at $h \neq h_i$ and has a unity gain and zero phase shift for harmonic components at $h = h_i$, including the selected harmonic order h_s. The same conclusion can be drawn for $GDSS2$ except that it will introduce an extra $-\pi/2$ phase shift to the harmonics at $h = h_i$. Therefore, $GDSS1$ and $GDSS2$ can be configured as a Quadrature Signal Generation (QSG) method and it can generate two pure quadrature signals at any targeted

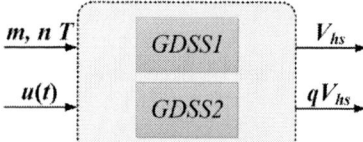

Fig. 2. Illustration of GDSS-based quadrature signal generation method.

TABLE I. RELATIONSHIPS AMONG h_s, m, n, h_i

h_s	m	n	h_i
1	14	15	1, 14, 16, 29, 31...
3	14	5	3, 12, 18, 27, 33...
5	14	3	5, 10, 20, 25, 35...
7	20	3	7, 14, 28...
9	26	3	9, 18, 36...
4	15	4	4, 12, 20, 28, 36...
8	23	3	8, 16, 32...

harmonic order h_s with properly designed m and n. Fig. 2 illustrates the QSG algorithm based on the GDSS operators (GDSS-QSG), where V_{hs} represents the h_s-order harmonic component targeted in the input signal V_{in} and $q = e^{-j\pi/2}$ denotes a $\pi/2$ phase delay operator. Typical parameter designs of m, n, h_s, and h_i can be calculated and they are as shown in Table I.

As it can be seen from (4) and (5), the maximum delay times for $GDSS1$ and $GDSS2$ are $mT/h_s n$, which enable a fast harmonic detection with response time as short as less than one fundamental period. Additionally, with further observation on (10) and (11), if even harmonic orders are not considered, m can be redesigned to $h_s n/2-1$ and the overall delay time will be significantly reduced to less than half a fundamental cycle.

In a practical implementation, GDSS-QSG will be realized with Digital Signal Processor (DSP) and thus discretization is needed for (4) and (5). If sampling points of $u(t)$ is set to N_S within a fundamental cycle, the discretized expression of the GDSS-QSG operator can be written as:

$$\begin{bmatrix} V_{hs}[i] \\ qV_{hs}[i] \end{bmatrix} = \begin{bmatrix} \dfrac{2}{m+1} \sum_{k=0}^{m} u[i - \dfrac{k}{h_s n} N_S] \cos(\dfrac{2k\pi}{n}) \\ \dfrac{2}{m+1} \sum_{k=0}^{m} u[i - \dfrac{k}{h_s n} N_S] \sin(\dfrac{2k\pi}{n}) \end{bmatrix} \quad (12)$$

According to (12), the $GDSS1$ and $GDSS2$ can be regarded as two Finite Impulse Responses (FIR) filters and its performance is similar to the comb filter discussed in the digital signal processing field [39]. In Fig. 3, frequency responses of the GDSS-QSG at $h_s = 4$ and 7 based on the parameters provided in Table I are demonstrated to verify the previous analysis. As also indicated by the presented frequency responses, interharmonics can be suppressed by the GDSS operators, although not completely eliminated.

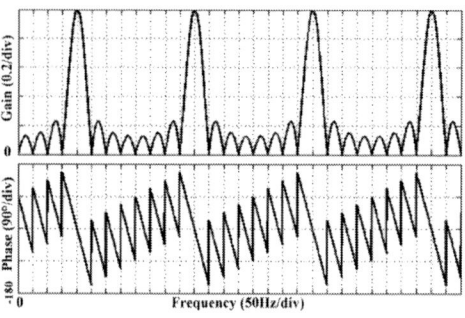

(a) Frequency response of $GDSS1$ at $h_s = 4$

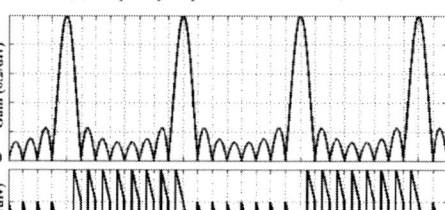

(b) Frequency response of $GDSS2$ at $h_s = 4$

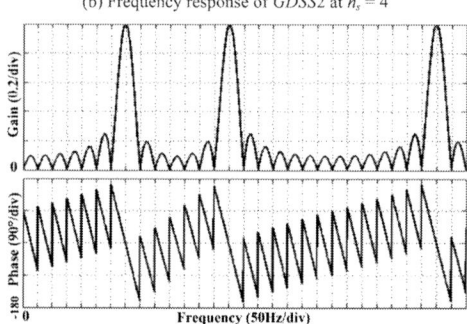

(c) Frequency response of $GDSS1$ at $h_s = 7$

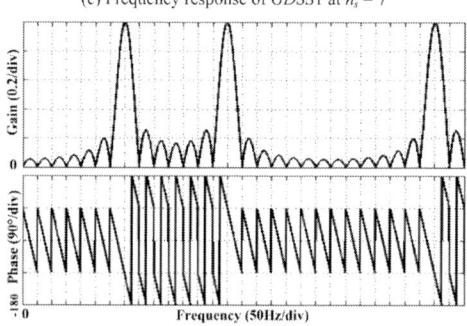

(b) Frequency response of $GDSS2$ at $h_s = 7$

Fig. 3. Frequency responses of GDSS-QSG at $h_s = 4$ and 7.

III. MGDSS-PLL AND SELECTED HARMONIC SEQUENCE DETECTION

In this section, the MGDSS-PLL structure is proposed for accurate estimation of the grid information under adverse grid conditions. The main configuration of the MGDSS-PLL is shown in Fig. 4, where multiple GDSS-QSGs tuned at different harmonic orders are adopted to generate interested sinusoidal quadrature signals and multiple PNSCs are applied to separate positive and negative components out of all the quadrature signals generated by theses GDSS-QSGs. Detailed descriptions about the MGDSS-PLL method are presented in the following part.

A. GDSS-QSG based Positive-/negative sequence calculator

The derivation process of the PNSC is introduced in [35] and the obtained expressions can be directly adopted in this paper. According to [35], the positive- and negative-sequence components in the $\alpha\beta$ reference frame can be calculated as:

$$\begin{bmatrix} V_\alpha^+ \\ V_\beta^+ \end{bmatrix} = \begin{bmatrix} T_{\alpha\beta+} \end{bmatrix} \begin{bmatrix} V_\alpha \\ V_\beta \end{bmatrix}; \quad \begin{bmatrix} T_{\alpha\beta+} \end{bmatrix} = \frac{1}{2} \begin{bmatrix} 1 & -q \\ q & 1 \end{bmatrix} \quad (13)$$

$$\begin{bmatrix} V_\alpha^- \\ V_\beta^- \end{bmatrix} = \begin{bmatrix} T_{\alpha\beta-} \end{bmatrix} \begin{bmatrix} V_\alpha \\ V_\beta \end{bmatrix}; \quad \begin{bmatrix} T_{\alpha\beta-} \end{bmatrix} = \frac{1}{2} \begin{bmatrix} 1 & q \\ -q & 1 \end{bmatrix} \quad (14)$$

where, V_α, V_β are signals transformed from the *abc* stationary frame by using the *Clark* transformation and V_α^+, V_β^+, V_α^-, V_β^- represent the positive- and negative-components under $\alpha\beta$ reference frame. Expression of the *Clark* transformation $T_{\alpha\beta}$ is written as:

$$\begin{bmatrix} T_{\alpha\beta} \end{bmatrix} = \sqrt{\frac{2}{3}} \begin{bmatrix} 1 & -\dfrac{1}{2} & -\dfrac{1}{2} \\ 0 & \dfrac{\sqrt{3}}{2} & -\dfrac{\sqrt{3}}{2} \end{bmatrix} \quad (15)$$

As it can be seen from (13) and (14), quadrature signals of V_α, V_β are required for the calculation of the desired V_α^+, V_β^+, V_α^-, V_β^-. Therefore, the aforementioned GDSS-QSG can be applied to provide these quadrature signals and the unwanted harmonic components will also be eliminated through this

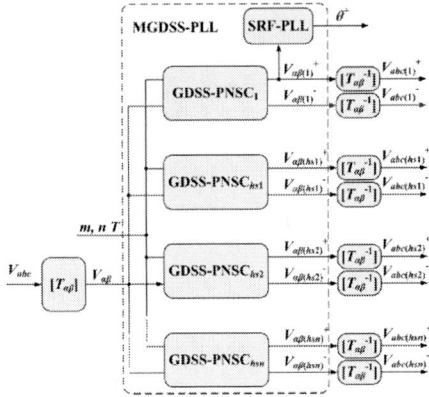

Fig. 4. Proposed configuration of the MGDSS-PLL.

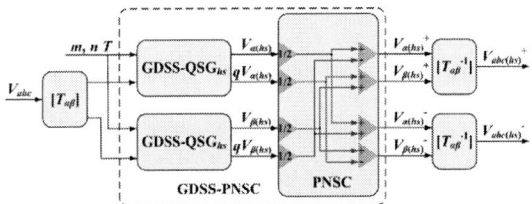

Fig. 5. Structure diagram of the GDSS-PNSC.

process. Fig. 5 illustrates the structure diagram of the discussed harmonic sequence components extraction algorithm which is marked as GDSS-PNSC.

In Fig. 5, V_{abc} represents the three-phase voltage/current signal to be estimated; $V_{\alpha(hs)}$, $V_{\beta(hs)}$ are the targeted h_s-order harmonic under the $\alpha\beta$ reference frame; $V_{\alpha(hs)}^+$, $V_{\beta(hs)}^+$, $V_{\alpha(hs)}^-$, $V_{\beta(hs)}^-$ denote the extracted positive and negative sequence components of $V_{\alpha(hs)}$ and $V_{\beta(hs)}$; $V_{abc(hs)}^+$ and $V_{abc(hs)}^-$ are respectively the computed positive and negative components of V_{abc}; GDSS-QSG$_{hs}$ stands for the GDSS-QSG operator tuned at the h_s-order harmonic. Moreover, $T_{\alpha\beta}^{-1}$ is the inverse *Clark* transformation and given as:

$$\begin{bmatrix} T_{\alpha\beta}^{-1} \end{bmatrix} = \sqrt{\frac{2}{3}} \begin{bmatrix} 1 & 0 \\ -\dfrac{1}{2} & \dfrac{\sqrt{3}}{2} \\ -\dfrac{1}{2} & -\dfrac{\sqrt{3}}{2} \end{bmatrix} \quad (16)$$

B. MGDSS-PLL

As shown in Fig. 4, multiple GDSS-PNSC subsystems aiming at fundamental and multiple harmonic frequencies are adopted to detect both the fundamental and targeted multiple harmonic sequences under highly unbalanced and distorted three-phase systems. Then, a standard SRF-PLL structure [17] is added to synchronize with the grid positive fundamental component. Based on the structure illustrated in Figs. 4 and 5, the signal processing procedure of the proposed MGDSS-PLL can be described as follows.

1) Three-phase signal V_{abc} is transferred to $V_{\alpha\beta}$ under $\alpha\beta$ reference frame using the *Clark* transformation and then fed to the multiple GDSS-PNSC subsystems.

2) Irrelevant harmonic orders contained in $V_{\alpha\beta}$ will be eliminated by the GDSS operators in each GDSS-PNSC subsystem and pure quadrature signals at targeted frequencies will be generated. Then these quadrature signals will be extracted into positive and negative sequence components under the $\alpha\beta$ reference frame by the PNSC blocks as shown in (13) and (14).

3) A standard SRF-PLL structure is implemented here to track the positive fundamental phase θ^+. Since the input signal $V_{\alpha\beta}^+$ is clean and balanced, the SRF-PLL control loop bandwidth can be designed to be very high for fast transient performance.

4) The obtained positive and negative harmonic sequence components at multiple desired orders are finally transferred to *abc* stationary reference frame.

As it can be seen from Fig. 4, the performance of the MGDSS-PLL is basically determined by the GDSS operators. Therefore, the MGDSS-PLL can be flexibly designed with a variable number of GDSS-PNSC subsystems for the specific applications and the control requirements. Furthermore, the parameters of the GDSS operators can also be optimized according to different grid conditions and control systems in order to achieve the most suitable synchronization performance in respect to the tracking accuracy and dynamic response. It is worth to mention that the MGDSS-PLL is immune to input frequency variations within a small range, since the gains and phases around the targeted frequencies are still close to their normal value as shown in Fig. 3.

IV. SIMULATION AND EXPERIMENTAL VERIFICATION

In order to verify the performance of the MGDSS-PLL under adverse grid conditions, simulations based on MATLAB/SIMULINK and experimental tests with dSPACE DS 1103 system are presented in this section. In the simulations and experiments, a highly unbalanced and distorted grid is applied and the grid parameters during normal and faulted conditions are specified in Table II. The MGDSS-PLL is configured to track the fundamental, 4th, 7th, 11th harmonic sequence components and thus there are only four GDSS-PNSC subsystems included. The grid fundamental frequency is 50 Hz during the simulation and experiments.

TABLE II. GRID PARAMETERS

Voltage component	Magnitude [V]	Phase [rad]
Fundamental positive-sequence	311	0
Fundamental negative-sequence	40	$\pi/3$
2nd harmonic positive-sequence	31	0
4th harmonic negative-sequence	31	$\pi/6$
5th harmonic positive-sequence	62	$\pi/6$
7th harmonic negative-sequence	62	$\pi/4$
8th harmonic positive-sequence	31	$\pi/3$
11th harmonic positive-sequence	62	$\pi/12$
13th harmonic negative-sequence	62	$\pi/9$

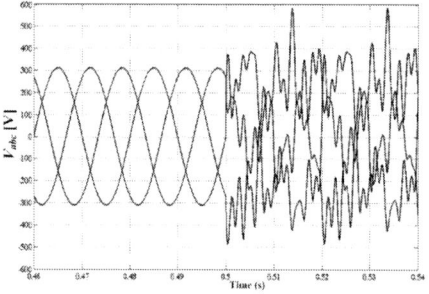

Fig. 6. Unbalanced and distorted grid.

A. Simulation results

The unbalanced and distorted grid mentioned is illustrated in Fig. 6, where the grid fault appears at *t*=0.5 s. Simulation results of the MGDSS-PLL under this extreme grid condition

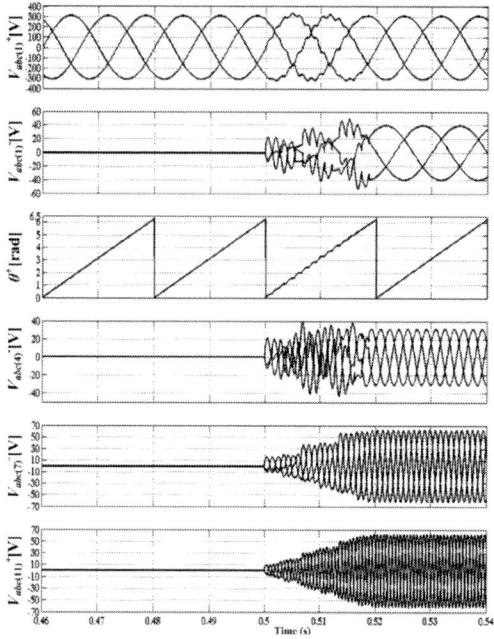

Fig. 7. Simulation of the MGDSS-PLL under abnormal grid condition.

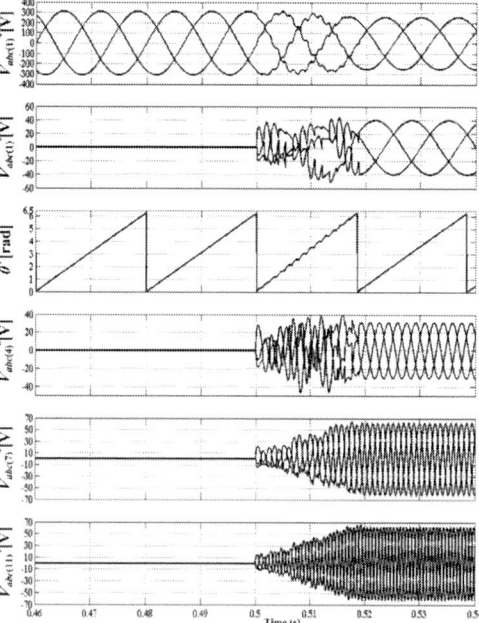

Fig. 8. Simulation of the MGDSS-PLL with voltage sag and phase jump.

978-1-4673-9551-9/16 $31.00 © 2016 IEEE

are given in Fig. 7. As indicated by the simulation waveforms, the proposed MGDSS-PLL can accurately extract all the interested harmonic sequence components within one fundamental period, which is consistent with the theoretical analysis. Then, the MGDSS-PLL is tested under more stringent scenarios with grid voltage sag, phase jump and frequency

variation. The obtained results are demonstrated in Figs. 8 and 9. In Fig. 8, the grid positive fundamental voltage drops from 311 V to 255 V with an extra $\pi/6$ phase jump at $t=0.5$ s. The other conditions remain the same as given in Table II. The same parameters are set for the grid voltage in Fig. 9 except that the positive fundamental frequency changes from 50 Hz to 50.5 Hz during the fault condition. As it can be seen from Figs. 8 and 9, the introduced MGDSS-PLL exhibits great robustness for abnormal grid conditions, which makes it an attractive synchronization technique.

To make it more clearly about the benefits of the derived MGDSS-PLL structure, the input voltage shown in Fig. 6 is also tested with the aforementioned MSOGI-FLL. The obtained simulation results are presented in Fig. 10, where $\Delta\omega^+$ = (ω-314.16) rad/s represents the error between detected positive fundamental frequency and its rated value. In the simulation, only the fundamental, 4th, 7th and 11th harmonics are considered to design the HDN of the MSOGI-FLL and other harmonic components contained are regarded as unknown components. According to Fig. 10, any unconsidered harmonics will affect the tracking performance of the MSOGI-FLL when the number of DSOGI subsystems is not equal to the number of significant harmonic components contained.

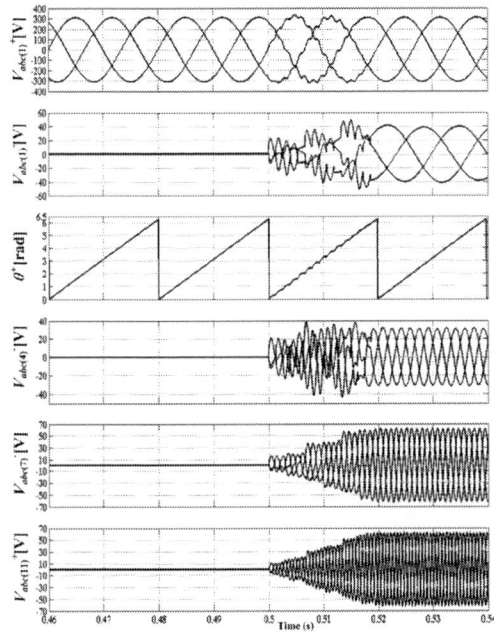

Fig. 9. Simulation of the MGDSS-PLL with frequency change.

B. Experimental results

The unbalanced and distorted input voltages used in the experiment is demonstrated in Fig. 11 and its important parameters are given in Table II. During the experiments, the input voltage is sampled by the analog-to-digital converters and fed to the detection algorithm. The waveforms recorded are all computed results of the detection algorithm and generated by the digital-to-analog converters of the control system. Sampling rate in the experiment is set to 15 kHz. Fig. 12 shows the experimental results of the proposed method under the input voltage given in Fig. 11. As it can be seen from the waveforms, all the positive- and negative sequence components contained in the input are accurately separated by the MGDSS-PLL and the settle time of the algorithm is about one fundamental period, which matches with the simulation and theoretical analysis. A more critical condition is also tested in the experiment and the results are demonstrated in Fig. 13, where the grid positive fundamental voltage drops from 311 V to 255 V with an extra $\pi/6$ phase jump and 0.5 Hz frequency change. As it can be confirmed by Fig. 13, the introduced MGDSS-PLL is immune to some typical grid perturbations like voltage sag, phase jump and small frequency variation, so it can maintain high tracking performance with great

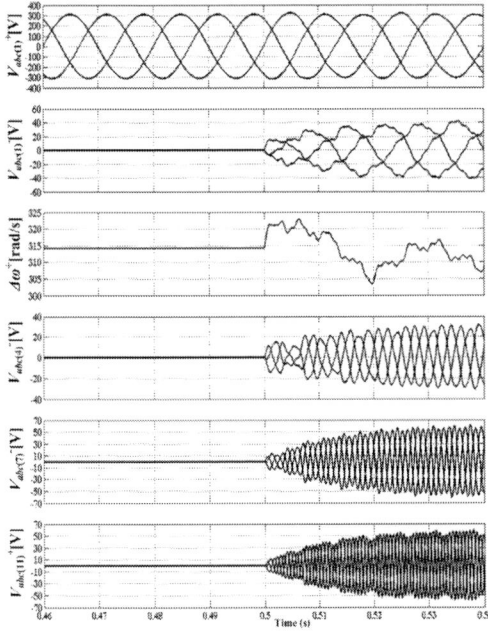

Fig. 10. Simulation of the MSOGI-FLL under abnormal grid conditions.

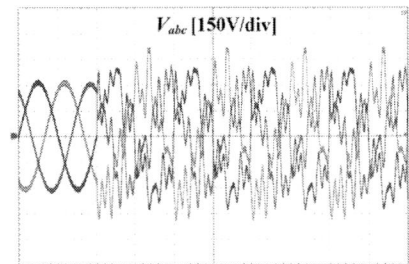

Fig. 11. Input voltage.

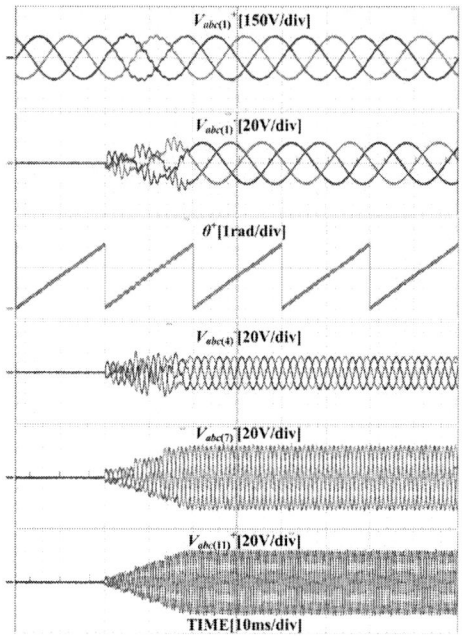

Fig. 12. Experimental results of MGDSS-PLL with abnormal grid.

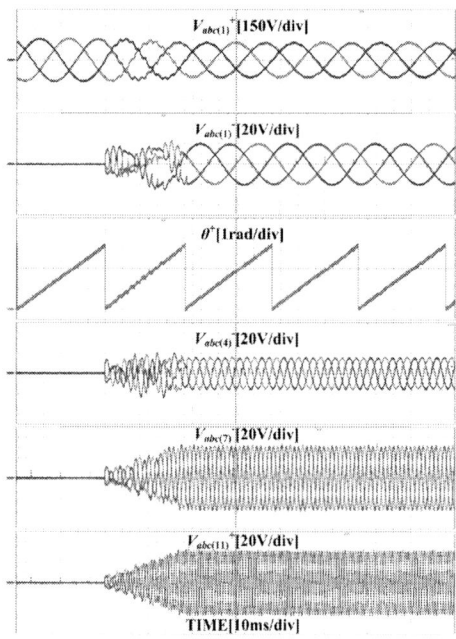

Fig. 13. Experimental results of MGDSS-PLL with grid perturbations.

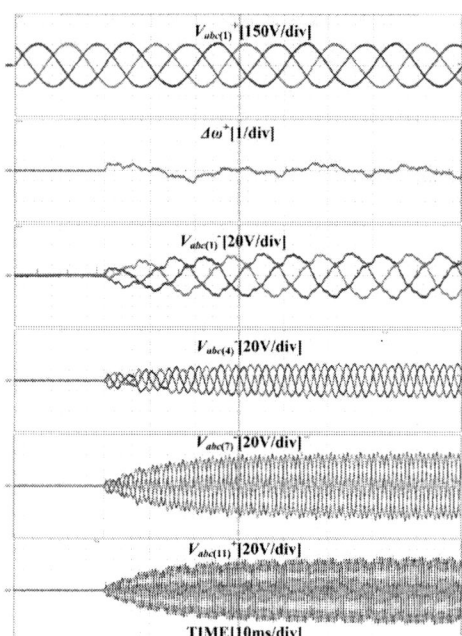

Fig. 14. Experimental results of MSOGI-FLL with abnormal grid.

performance of the MSOGI-FLL is rather sensitive to unknown harmonic components.

V. CONCLUSIONS

This paper has presented a novel three-phase multiple harmonic sequence components detection method named as MGDSS-PLL for extremely unbalanced and distorted grid conditions. The proposed MGDSS-PLL structure contains multiple GDSS-PNSCs, which enables accurate positive and negative components extraction at multiple interested harmonic orders within one fundamental period. A standard SRF-PLL structure is also adopted in the MGDSS-PLL for grid synchronization at the positive fundamental frequency. Moreover, the number of GDSS-PNSC subsystems is equal to the number of harmonics required to be detected and the MGDSS-PLL exhibits great robustness regarding to typical abnormal grid conditions including voltage sags, phase jumps and frequency variations. All these features make the proposed MGDSS-PLL to be a competitive synchronization algorithm. Simulation and experimental results are presented in the paper to verify the effectiveness of the proposed MGDSS-PLL.

ACKNOWLEDGMENT

This paper and its related research work are supported by national Natural Science Foundation of China (NSFC) (project no. 51277146).

REFERENCES

[1] F. Blaabjerg, Z. Chen, and S. B. Kjaer, "Power electronics as efficient interface in dispersed power generation systems," *IEEE Trans. Power Electron.*, vol. 19, no. 5, pp. 1184-1194, Sep. 2004.

robustness. Experimental results of the MSOGI-FLL under the grid voltage shown in Fig. 11 are given in Fig. 14. In the experiments, same HDN structure designed in the simulation is adopted and the obtained results indicate that the tracking

[2] J. Carrasco, L. Franquelo, J. Bialasiewicz, E. Galvan, R. Guisado, M. Prats, J. Leon, and N. Moreno-Alfonso, "Power-electronic systems for the grid integration of renewable energy sources: A survey," *IEEE Trans. Ind. Electron.*, vol. 53, no. 4, pp. 1002-1016, Jun. 2006.

[3] A. Timbus, M. Liserre, R. Teodorescu, P. Rodriguez, and F. Blaabjerg, "Evaluation of current controllers for distributed power generation systems," *IEEE Trans. Power Electron.*, vol. 24, no. 3, pp. 654-664, Mar. 2009.

[4] P. Acuna, L. Moran, M. Rivera, J. Dixon, and J. Rodriguez, "Improved active power filter performance for renewable power generation systems," *IEEE Trans. Power Electron.*, vol. 29, no. 2, pp. 687-694, Feb. 2014.

[5] *IEEE Standard for Interconnecting Distributed Resources with Electric Power Systems*, IEEE Std. 1547-2003, 2003.

[6] M. Tsili and S. Papathanassiou, "A review of grid code technical requirements for wind farms," *IET Renew. Power Gener.*, vol. 3, no. 3, pp. 308-332, Sep. 2009.

[7] B.-I. Craciun, T. Kerekes, D. Sera, and R. Teodorescu, "Overview of recent grid codes for PV power integration," in *Proc. OPTIM*, Brasov, Romania, 2012, pp. 959-965.

[8] K.-H. Kim, Y.-C. Jeung, D.-C. Lee, and H.-G. Kim, "LVRT scheme of PMSG wind power systems based on feedback linearization," *IEEE Trans. Power Electron.*, vol. 27, no. 5, pp. 2376-2384, May 2012.

[9] D. Santos-Martin, J. L. Rodriguez-Amenedo, and S. Arnaltes, "Direct power control applied to doubly fed induction generator under unbalanced grid voltage conditions," *IEEE Trans. Power Electron.*, vol. 23, no. 5, pp. 2328-2336, Sep. 2008.

[10] P. Rodriguez, A. Timbus, R. Teodorescu, M. Liserre, and F. Blaabjerg, "Flexible active power control of distributed power generation systems during grid faults," *IEEE Trans. Ind. Electron.*, vol. 54, no. 5, pp. 2583-2592, Oct. 2007.

[11] L. Asiminoaei, F. Blaabjerg, and S. Hansen, "Detection is key - Harmonic detection methods for active power filter applications," *IEEE Industry Applications Magazine*, vol. 13, pp. 22-33, Jul.-Aug. 2007.

[12] F. Blaabjerg, R. Teodorescu, M. Liserre, and A. Timbus, "Overview of control and grid synchronization for distributed power generation systems," *IEEE Trans. Ind. Electron.*, vol. 53, no. 5, pp. 1398-1409, Oct. 2006.

[13] X. Q. Guo, W. Y. Wu, X. F. Sun, and G. C. San, "Phase locked loop for electronically-interfaced converters in distributed utility network," *in Proc. 2008 ICEMS*, pp. 2346-2350, Oct. 2008.

[14] S. Alepuz, S. Busquets-Monge, J. Bordonau, J. A. Martinez-Velasco, C. A. Silva, J. Pontt, and J. Rodriguez, "Control strategies based on symmetrical components for grid-connected converters under voltage dips," *IEEE Trans. Ind. Electron.*, vol. 56, no. 6, pp. 2162-2173, Jun. 2009.

[15] K.-J. Lee, J.-P. Lee, D. Shin, D.-W. Yoo, and H.-J. Kim, "A novel grid synchronization PLL method based on adaptive low-pass notch filter for grid connected PCS," *IEEE Trans. Ind. Electron.*, vol. 61, no. 1, pp. 292-301, Jan. 2014.

[16] R. M. S. Filho, P. F. Seixas, P. C. Cortizo, L. A. B. Torres, and A. F. Souza, "Comparison of three single-phase PLL algorithms for UPS applications," *IEEE Trans. Ind. Electron.*, vol. 55, no. 8, pp. 2923–2932, Aug. 2008.

[17] S. Chung, "A phase tracking system for three phase utility interface inverters," *IEEE Trans. Power Electron.*, vol. 15, no. 3, pp. 431-438, May 2000.

[18] V. Kaura and V. Blasko, "Operation of a phase locked loop system under distorted utility conditions," *IEEE Trans. Ind. Appl.*, vol. 33, no. 1, pp. 58-63, Jan./Feb. 1997.

[19] M. A. Perez, J. R. Espinoza, L. A. Moran, M. A. Torres, and E. A. Araya, "A robust phase-locked loop algorithm to synchronize static-power converters with polluted AC systems," *IEEE Trans. Power Electron.*, vol. 55, no. 5, pp. 2185-2192, May 2008.

[20] P. Rodriguez, J. Pou, J. Bergas, J. Candela, R. Burgos, and D. Boroyevich, "Decoupled double synchronous reference frame PLL for power converters control," *IEEE Trans. Power Electron.*, vol. 22, no. 2, pp. 584-592, Mar. 2007.

[21] X. Guo, W. Wu, and Z. Chen, "Multiple-complex coefficient-filter-based phase-locked loop and synchronization technique for three-phase grid interfaced converters in distributed utility networks," *IEEE Trans. Ind. Electron.*, vol. 58, no. 4, pp. 1194-1204, Apr. 2011.

[22] A. V. Timbus, M. Liserre, R. Teodorescu, and F. Blaabjerg, "Synchronization methods for three phase distributed power generation systems. An overview and evaluation," in *Proc. IEEE PESC*, 2005, pp. 2474-2481.

[23] M. Karimi-Ghartemani, H. Karimi, and A. Bakhshai, "A filtering technique for three-phase power systems," *IEEE Trans. Instrum. Meas.*, vol. 58, no. 2, pp. 389-396, Feb. 2009.

[24] S. Golestan, M. Ramezani, J. M. Guerrero, and M. Monfared, "dq-frame cascaded delayed signal cancellation-based PLL: Analysis, design, and comparison with moving average filter-based PLL," *IEEE Trans. Power Electron.*, vol. 30, no. 3, pp. 1618-1632, Mar. 2015.

[25] W. Li, X. Ruan, C. Bao, D. Pan, and X.Wang, "Grid synchronization systems of three-phase grid-connected power converters: A complex vectorfilter perspective," *IEEE Trans. Ind. Electron.*, vol. 61, no. 4, pp. 1855-1870, Apr. 2014.

[26] P. Rodriguez, J. Pou, J. Bergas, I. Candela, R. Burgos, and D. Boroyevic, "Double synchronous reference frame PLL for power converters control," in *Proc. IEEE PESC*, 2005, pp. 1415-1421.

[27] R. Bojoi, L. Limongi, D. Roiu, and A. Tenconi, "Enhanced power quality control strategy for single-phase inverters in distributed generation systems," *IEEE Trans. Power Electron.*, vol. 26, no. 3, pp. 798-806, Mar. 2011.

[28] B. Liu, F. Zhuo, Y. Zhu, H. Yi, and F. Wang, "A Three-Phase PLL Algorithm Based on Signal Reforming Under Distorted Grid Conditions," *IEEE Trans. Power Electron.*, vol. 30, no. 9, pp. 5272-5283, sept. 2015.

[29] P. Rodriguez, A. Luna, M. Ciobotaru, R. Teodorescu, and F. Blaabjerg, "Advanced grid synchronization system for power converters under unbalanced and distorted operating conditions," in *Proc. 32nd Annu. Conf. IEEE Ind. Electron.*, Nov. 2006, pp. 5173-5178.

[30] P. Rodriguez, R. Teodorescu, I. Candela, A. V. Timbus, M. Liserre, and F. Blaabjerg, "New positive-sequence voltage detector for grid synchronization of power converters under faulty grid conditions," in *Proc. IEEE PESC*, 2006, pp. 1-7.

[31] Y. F. Wang and Y. W. Li, "Grid synchronization PLL based on cascaded delayed signal cancellation," *IEEE Trans. Power Electron.*, vol. 26, no. 7, pp. 1987-1997, Jul. 2011.

[32] F. A. S. Neves, M. C. Cavalcanti, H. E. P. de Souza, F. Bradaschia, E. J. Bueno, andM. Rizo, "A generalized delayed signal cancellation method for detecting fundamental-frequency positive-sequence three-phase signals," *IEEE Trans. Power Del.*, vol. 25, no. 3, pp. 1816-1825, Jul. 2010.

[33] L. R. Limongi, R. I. Bojoi, C. Pica, F. Profumo, and A. Tenconi, "Analysis and comparison of phase locked loop techniques for grid utility applications," in *Proc. IEEE Power Conversion Conf.*, 2007, pp. 674-681.

[34] N. Hoffmann, R. Lohde, M. Fischer, F. W. Fuchs, L. Asiminoaei, and P. B. Thogersen, "A review on fundamental grid-voltage detection methods under highly distorted conditions in distributed power generation networks," in *Proc. IEEE ECCE*, Phoenix, AZ, USA, 2011, pp. 3045-3052.

[35] P. Rodriguez, A. Luna, I. Candela, R. Mujal, R. Teodorescu, and F. Blaabjerg, "Multiresonant frequency-locked loop for grid synchronization of power converters under distorted grid conditions," *IEEE Trans. Ind. Electron.*, vol. 58, no. 1, pp. 127-138, Jan. 2011.

[36] Y. Lu, G. C. Xiao, L. F. Zang, X. L. Wu, F. W. Chen, "A novel synchronization method designed for single-phase distorted grid," in *Proc. IEEE APEC*, Charlotte, NC, USA, 2015, pp. 2866-2871.

[37] Y. Wang and Y. Li, "Three-phase cascaded delayed signal cancellation PLL for fast selective harmonic detection," *IEEE Trans. Ind. Electron.*, vol. 60, no. 4, pp. 1452-1463, Apr. 2013.

[38] I. S. Gradshteyn, and I. M. Ryzhik, "Table of integrals, series, and products," San Diego: Academy Press, 2007.

[39] P. Zahradnik and M. Vlcek, "Analytical design of optimal FIR combfilters," in *Proc. IEEE ICC*, May 2003, vol. 5, pp. 3590-3593.

978-1-4673-9551-9/16 $31.00 © 2016 IEEE

AUTHOR INDEX

A

Abdelmoaty, Ahmed	2437
Abdul Azeez, Najath	3140
Abe, Seiya	1640, 2422
Abedinpour, Siamak	3669
Abramov, Eli	111, 692
Abramson, Rose A.	1138
Abu Qahouq, Jaber A.	1868, 2114, 3611, 3684
Abu-Rub, Haitham	1214, 3663
Acero, J.	3020, 3026, 3566
Achanta, Prasanta K.	3273
Adhikari, Jeevan	9
Adragna, Claudio	564
Afridi, Khurram K.	1138, 1392, 1947, 2395
Afsharian, Jahangir	33, 899, 2312, 2320
Agamy, Mohammed	3403
Agelidis, Vassilios G.	236, 1702
Agostinelli, M.	339, 350
Agostini, Francesco	472
Agrawal, Neeraj	951
Aguilar, A.	2540
Ahmed, Emad M.	1505
Ahmed, Ibrahim	1882
Ahmed, Mahrous	1505
Ahmed, Shamim	1646
Ahmed, Shehab	936
Ahmed-Zaid, Said	2821
Ahn, Jung-Hoon	163, 1273, 2161
Ahsanuzzaman, S.M.	2497
Akagi, Hirofumi	1163
Akamatsu, Keiji	2607
Akin, Bilal	505, 1096, 1176, 2108, 2875
Alatise, Olayiwola	253, 2645
Al-Durra, Ahmed	1941
Alexandrov, Peter	2973
Al-Hallaj, Said	3128
Alharbi, Mahmood	3333
Ali, Kawsar	2491
Allard, Bruno	524
Allen, Scott	979
Allmeling, Jost	1108
Alonso, J. Marcos	1115
Alonso, R.	3020

Alonso, Rafael 3026
Alou, P. 2409
Al-Shyoukh, Mohammad 2437
Am, Sokchea 2401, 2700
Amaro, Mike 66
Ambacher, Oliver 2083
Amirabadi, Mahshid 3704
Amirahmadi, Ahmadreza 3333
Amon, Cristina 1350
Amouzandeh, Maryam S. 329
Andersen, Michael A.E. 1090, 1430, 1541, 1842, 2252, 2473
Andersen, Thomas 1430, 1842
Ando, Masato 2986
Aniruddhan, Sankaran 1878
Anthon, Alexander 1235, 2252
Antunes, Fernando L.M. 2592
Anwar, Saeed 424
Anwar, Usama 1947
Arafat, A.K.M. 1123
Arafat, Akm 2847
Arefifar, Ali 2561
Arias, Andrea 79
Arias, M. 1823
Arias, Manuel 822
Arnold, Cory 1597, 3273
Asa, Erdem 1323, 1756, 1767, 2587
Asensi, R. 1624
Ayers, Curtis 3529
Ayyanar, Raja 432, 3364
Azcondo, F.J. 2389
Azuma, Katsunori 283

B

Badawey, Mohammed 392
Baek, Jeihoon 3004
Bagawade, Snehal 544
Bahman, Amir Sajjad 261, 3012
Bahmani, M.A. 3043
Bai, Hua 529
Bai, Yongjiang 766, 3623
Baier, Thomas 2897
Bak, Claus Leth 3051
Bak, Yeongsu 2764, 3416
Baker, Michael W. 1597
Bakhshai, Alireza 460
Bakker, Cas 2457
Balasubramanian, Bharat 1868
Balda, Juan Carlos 143, 362, 651, 1387, 3712
Ball, Roy 2122, 3038

Bandyopadhyay, Santanu	3286
Banerjee, Arijit	2881
Baranwal, Rohit	2043
Barbosa, A.U.	3231
Bari, Syed	3259
Barlow, Matthew	1646
Barner, Alexander	106
Barrado, A.	2545, 3090
Barth, Christopher	1512
Barthelmebs, Clement	453
Batarseh, Issa	1381, 3333
Bawohl, Melanie	3069
Bayhan, Sertac	3663
Bazzi, Ali M.	2666
Bęczkowski, Szymon	704, 974, 3101
Bede, Lorand	1702, 2264
Beres, Remus	3051
Bergman, Joshua	79
Bermejo, M.	3090
Bermingham, Jack	2794
Berzoy, Alberto	928, 3200
Betz, Vaughn	1882
Bezdenezhnykh, Yevgeny	308
Bhalla, Anup	2973
Bhangu, Bicky	715
Bhardwaj, Manish	505
Bhattachaarjee, Parijat	1344
Bhattacharya, Subhashish	295, 601, 778, 886, 1497, 1632, 2076
Bhattacharya, Tanmoy	199
Bhowmik, Pankaj Kumar	2706
Biglarbegian, Mehrdad	2998
Biswas, Suvankar	1934
Bizjak, Luca	1663
Blaabjerg, Frede	221, 229, 261, 288, 370, 1253, 1872, 1941, 1995, 2011, 2154, 2207, 2215, 2264, 3012, 3051, 3416, 3431, 3500
Blalock, Benjamin J.	684, 893, 1569, 3255
Blasko, Vladimir	2167
Böcker, Joachim	1547
Bodano, Emanuele	1663
Bojarski, Mariusz	1756
Bonthu, Sai Sudheer Reddy	1131
Bonyadi, Roozbeh	253
Bonyadi, Yeganeh	253
Borges, Beatriz	3637
Born, Rachael	1148, 3243
Boroyevich, Dushan	177, 516, 524, 739, 1024, 1315, 1561
Botting, Chris	854
Boynuegri, Ali R.	207
Braga, A.P.S.	3231

Brandt, Tobias 3172
Brar, Berinder 79
Breaz, Elena 3476
Briggs, Roger 2927
Brohlin, Paul 838
Brothers, John A. 990, 1967
Brown, Alan 529
Burdío, J.M. 1040, 1762, 3020, 3026, 3566
Burgos, Rolando 177, 516, 524, 1024, 1561
Buticchi, Giampaolo 2449, 3493, 3629
Buttay, Cyril 524

C

Cai, Wen 1057, 1861, 2599
Campbell, S.L. 1307
Canacsinh, Hiren 3637
Canales, Francisco 472
Cao, Dong 3553
Cao, Jiankun 1371
Cao, Wenchao 2229
Cao, Yuan 1868, 3684
Carlos, Gregory A.A. 3641
Carretero, C. 663, 3020, 3026, 3566
Casady, Jeffrey 979
Castro, Ignacio 25, 822
Castro, Marcus R. 2592
Ceballos, Salvador 236
Cervera, Alon 111, 692, 2298
Cha, Hanju 511, 1708, 3598
Chae, Young-Ho 2801
Chakraborty, Shiladri 1954, 2652, 3389
Challingsworth, Mark 3069
Chang, C.-H. 951
Chatterjee, Urmimala 1183
Chattopadhyay, Ritwik 778
Chattopadhyay, Souvik 1954, 2652, 3389
Chawda, Pradeep 3266
Chee, Seung-Jun 1206, 2370
Chen, Alian 3453, 3465
Chen, Changdong 2981
Chen, Cheng-Po 3255
Chen, Chingchi 1554
Chen, Di 529
Chen, Fang 177
Chen, Guipeng 1450
Chen, Guodong 499
Chen, Guoliang 1227
Chen, Hao 1437
Chen, Hua 1947

Chen, Jie	3453
Chen, Min	138
Chen, Minjie	1138, 1443
Chen, Qianhong	2518
Chen, Runruo	1045
Chen, Weiqiang	2666
Chen, Wenjie	493, 766, 3115, 3623
Chen, Woei-Luen	3471
Chen, Xinwen	1358
Chen, Xuling	1788
Chen, Yang	899, 2304, 2312, 2320
Chen, Yang-Lin	558
Chen, Yaow-Ming	558
Chen, Ying	2071
Chen, Yuxiang	499, 1462
Chen, Zhe	3431
Cheng, Chun Sing	1795
Cheng, Kuang-Yao	118
Cheung, Chun	1616
Chi, Yongning	1462
Chinthavali, M.	1307
Cho, Bo-Hyung	487, 1416
Cho, Shin Young	3690
Choe, Songbaek	2051
Choi, Beomseok	1947
Choi, Byeung G.	1773
Choi, Hee-Su	3153
Choi, Seungdeog	631, 1123, 1131, 1748, 2847, 3004
Choi, Sewan	859
Choi, Sung-Jin	3153
Choi, Wooin	1416
Choi, Wooyoung	2679
Chou, Derek	1512
Chow, Jeff Po Wa	1795
Chowdhury, Md Asif Mahmood	207
Chub, Andrii	2533
Chun, Chang Yoon	3322
Chung, Henry Shu-Hung	1795, 1807, 2154
Chung, Steven	1350
Church, Ron	786
Ci, Song	3189
Ciobotaru, Mihai	1702
Cobos, J.A.	2409
Coelho, Ernane A.A.	3585
Colak, Kerim	1323, 1756, 1767, 2587
Colmenares, Juan	746, 1018
Comanescu, Mihai	2759, 2855
Connaughton, A.M.	355
Conway, Thomas	1670

Cook, M.	3537
Correa, Maurício B.R.	3641, 1032
Corzine, Keith	720, 1191, 1481, 2187, 2840
Cosetin, Marcelo	1115
Costa, Levy F.	2449
Costa, Louelson A.	1032
Costa, Paulo Junior Silva	2376
Costinett, Daniel	424, 872, 893, 1010, 1569, 2441, 3255, 3577
Craciun, Marian	854
Cui, Shenghui	2620
Cui, Yutian	893
Curuvija, Boris	3553
Cuzner, Robert	1577
Czarkowski, Dariusz	1323, 1756, 1767, 2587
Czwickla, Christoph	3069

D

Dahan, Nadav	802
Dai, Ke	1358
Dai, Zhiyong	3134
Dai, Ziwei	1358
Dally, William J.	86
Dang, Zhigang	2114, 3684
Daniel, Michael T.	1695
Dargahi, Vahid	720, 1191, 1481, 2187, 2840
Daryaei, Mohammad	2579
Das, Partha Pratim	2652
Das, Pritam	552, 2491
Dashmiz, Shadi	3297
Davari, Pooya	221, 229
Davletzhanova, Zarina	253
de Almeida Carlos, Gregory A.	2720
de Almeida, Bruno Ricardo	60
De Carne, Giovanni	3493
De Doncker, Rik W.	643
de Oliveira Pacheco, Juliano	3346
de Rooij, Michael	2292
de Souza Oliveira Jr., Demercil	60, 3231, 3346
De, Ankan	295, 1632
Debnath, Suman	1528
Deboy, Gerald	3570
Degner, Michael W.	241
Delhotal, J.	1926
Demerdash, Nabeel A.O.	1065, 2826
Deng, Cheng	143, 362, 651
Deng, Hao	816
Deng, Lu	3521
Deng, Yan	1450
Dias Jr., A.J.S.	3231

Diaz Reigosa, Paula	288
Diaz, Nelson	1227
Dimarino, Christina	516
Ding, Pengling	1371
Ding, Weisheng	440
Dinulovic, Dragan	3097
Ditze, Stefan	864
Divan, Deepak	1437
Dix, Jeffery	684
do Prado, Ricardo N.	1115
Dobmeier, Christian	1741
Domoto, Kazuhide	2422, 2465
Dong, Dong	3403
Dong, Zhou	73, 2518
Doolla, Suryanarayana	2245, 3376
Dorn-Gomba, Lea	453
dos Santos Jr., Euzeli C.	2720
Dos Santos, Gutemberg G.	1032
Dou, Manfeng	3134
Dou, Qingyun	2272
Dousoky, Gamal M.	2735
Driesen, Johan	1183
Driessen, Anton	2457
Drofenik, Uwe	472
Du, Weijing	1002
Du, Weijing	2334
Du, Xiong	2992
Duarte, J.L.	3158
Dujic, Drazen	156, 1108
Dumais, Alex	3219
Dusmez, Serkan	505, 1176, 2108
Dutta, Atanu	3012

E

Eberle, Wilson	1286
Ebrahimi, Mohammad	2579, 3207
Edpuganti, Amarendra	402, 943
Egan, Michael	2794
Ehrlich, Stefan	1741
Einspieler, Sascha	759
Ekhtiari, Marzieh	1430
Elrayyah, Ali	392, 2660, 2806
Elsayed, Ahmed T.	1267
El-Taweel, Nader A.	830
Emadi, Ali	453, 1300
Engelmann, Georges	643
Eni, Emanuel-Petre	974, 3101
Enjeti, Prasad	936, 1695, 2567, 3545
Enshaei, Hossein	2813

Enslin, Johan 2998
Erickson, Robert 1947
Ertl, H. 1
Escobar-Mejía, Andrés 362
Eskandari, Soheila 2127
Essakiappan, Somasundaram 2706
Eum, Hyunchul 2355
Evzelman, Michael 1603
Ezra, Ofer 308

F

Fabricio, Edgard L.L. 3641
Fan, Bo 334
Faraci, Eric 838
Fard, Miad 1403
Farhang, Peyman 733
Farley, Kathleen Blair 1737, 3526
Farnell, Chris 143
Faulkner, Bryan 54
Fayed, Ayman 2437
Fedison, J.B. 247
Fei, Chao 322
Feng, Junjie 1534, 2334
Ferdowsi, Mehdi 1962, 2687
Fernandes, B.G. 2245, 2673, 3376
Fernandes, Darlan A. 1032
Fernandez, A. 1823
Fernandez, C. 2545, 3090
Ferrieux, Jean-Paul 2700
Figge, Heiko 1547
Flankl, Michael 623
Foulkes, Thomas 1512
Francés, A. 1624
Francis, A. Matt 1646
Freitas, Antônio A.A. 2592
Freitas, Luiz C.G. 3585
Frey, David 2401, 2700
Friedrichs, Daniel 3577
Fröhleke, Norbert 1547
Fu, Lixing 1554, 1967
Fu, Shihang 1475
Furukawa, Keita 1336

G

Gaafar, Mahmoud A. 2735
Gafford, James 1577
Gajanayake, C.J. 2942, 3058
Gakhar, Vikram 1878
Galiano Zurbriggen, Ignacio 386

Gan, Yiliang	3560
Gandikota, Srikant	1051
Gao, Fei	3476
Gao, Feng	410, 536, 921, 2259, 2935
Gao, Mingzhi	138
Gao, Rui	3383
Gao, Sugu	2868
Gao, Xieping	3185
Gao, Yabiao	1737, 3526
Gao, Yikai	1861
Garces, Luis	3403
Garcia Rodriguez, Luciano Andres	651
Garcia, Jorge	3508
Garcia, O.	2409, 1624
Garcia, Pablo	3508
Garcia, Virginia	3069
Garcia-Rodriguez, Luciano A.	362, 3712
Gavagsaz-Ghoachani, Roghayeh	446, 3397
Ge, Baoming	1214
Ge, Hongjuan	1424
Ge, Ting	668
Ge, Xiongxuan	3080
Geng, Shengbao	3560
Georgious, Ramy	3508
Gerfer, Alexander	2553, 3097
Gerling, Dieter	215
Ghaffarzadeh, Hooman	3353
Ghandi, Reza	3255
Ghat, Mahendra B.	2342
Ghias, Amer M.Y.M.	236
Giezendanner, Florian	1018
Ginart, Antonio	1737, 3526
Glavanovics, Michael	759
Glover, S.	3537
Goetz, Stefan M.	2349
Gohil, Ghanshyamsinh	1702, 2264
Goktas, Taner	1096, 2875
Gong, Bing	33
Gonnet, Luc	2700
Gonzalez, S.	1926
Gonzalez-Llorente, Jesus	3712
Gorla, Naga Brahmendra Yadav	2491
Gotovac, Ante	1663
Gou, Ruifeng	2071
Gray, C. Thomas	86
Greer III, Thomas H.	86
Gritti, Giovanni	564
Grosse, Thorben	643
Gu, Bin	838

Gu, Dong-Jie 1243
Gu, Lei 2889
Guerrero, Josep M. 398, 1227, 1376, 3459, 3697
Gui, Han-Dong 1243
Guirguis, David 1350
Gulbudak, Ozan 3248
Gunasekaran, Deepak 1045, 2525
Gundel, Paul 3069
Gunter, Samantha J. 1138
Guo, Ben 1010
Guo, Feng 1682
Guo, Suxuan 2063
Guo, Xiaoqiang 398
Gupta, Amit K. 715, 2942, 3058
Gupta, Ankit 1344
Gupta, Mahima 2919
Gupta, Ranjan K. 1520
Gupta, Shalabh 2666
Gurusinghe, Nicoloy 2479

H

Ha, Jung-Ik 193, 487, 1398, 2801, 3717
Hadjidemetriou, Lenos 3500
Hafez, Bahaa 936
Halivni, Bar 111
Hameyer, Kay 643
Han, Di 2861, 2950
Han, Jung Kyu 3690
Han, Xiangyu 2957
Han, Yang 816
Hang, Lijun 3560
Hanrahan, Robert 3266
Hanson, Alex J. 98
Haque, Moinul Shahidul 3004
Hare, James 2666
Harfman-Todorovic, Maja 3403
Hariharan, K. 315
Hariya, Akinori 2430
Harris, Richard Kyle 3255
Harrison, M.J. 247
Hartmann, M. 1
Haryani, Nidhi 1024, 1561
Hasan, Iftekhar 638
Hasegawa, Kazunori 3032
Hata, Katsuhiro 1731
Hata, Yuki 468
Hatae, Shinji 468
Hattori, Yoshiyuki 3146
Haug, Martin 3097

Hayakawa, Seiichi	283
Hayes, John G.	2794
He, Dingyi	2599
He, Haibing	3185
He, Jiangbiao	1065, 1084, 2826
He, Jinwei	1249
He, Ruirui	766
He, Xiangning	499, 1450, 1462, 1475
He, Xiaobin	2182, 2194
Heckel, Thomas	864
Heldwein, Marcelo Lobo	2833
Henke, M.	700
Henkenius, Carsten	1547
Herbert, Joseph	1123, 3004
Hernando, Marta M.	822
Higaki, Yusuke	1713
Hilber, Patrik	746
Hinken, Reiner	303
Hitoshi, Ishii	676
Ho, Carl Ngai-Man	2905
Ho, Kwun Yuan Godwin	2328
Hofmann, Heath	1721, 1726
Hofmann, Wilfried	2175
Holmes, Grahame	2252
Hopkins, Douglas C.	295, 2141
Hori, Yoichi	1731
Hosseini, Rasoul	1577
Hou, Dongbin	657
Hou, Ruoyu	1300
Hsiehu, H.-C.	951
Hu, Haibing	2182, 2194
Hu, Ji	253
Hu, Sheng	3310
Hu, Xiaolei	1071, 3409, 3591
Hu, Zhiyuan	899, 2320
Huang, Alex Q.	132, 269, 983, 2063, 2365, 2727, 2927, 3383, 3648, 3677
Huang, J.-W.	1364
Huang, Kuohsien	2355
Huang, Qingjun	3521
Huang, Qingyun	2727, 3648
Huang, Xiucheng	1002, 1534, 1853, 2334
Huang, Yi	1616
Huang, Ying	1888, 1894
Huang, Yung-Ting	1900
Huang, Zhengrong	1847
Huber, Laszlo	38, 46
Huh, Sungjae	2370
Hui, S.Y. Ron	169, 913, 1169, 1888, 1894, 2328, 3302, 3481
Hui, Zhao	2019

Hull, Brett 979
Husain, Iqbal 2141, 2927, 3383
Hwu, K.I. 2415

I

Iannuzzo, Francesco 288
Ibrahim, Mahmoud 2401, 2700
Iijima, Ryuji 3722
Illa Font, Carlos Henrique 2376
Ilves, Kalle 276
Imura, Takehiro 1731
Inaba, Masamitsu 283
Inokuchi, Seiichiro 468
Inoue, Shuntaro 3146
Ishikuro, Hiroki 1802
Ishizuka, Yoichi 2422, 2430, 2465
Islam, Md. Zakirul 631
Islam, Rakib 3279
Isobe, Takanori 3722
Isurin, Alexander 880
Itagaki, Atsushi 2465
Itakura, Tetsuro 1907
Itoh, Jun-Ichi 1336, 1911
IV, Prasanna 9
Iyer, Vishnu Mahadeva 295
Izuka, Arata 468

J

Jacobina, Cursino B. 2720, 3641
Jahns, Thomas 2167
Jain, Parth 2553
Jain, Praveen 378, 460, 544
Jang, Yujin 3690
Jang, Yungtaek 595, 1292
Jayasinghe, Shantha Gamini 2813
Jedtberg, Holger 3629
Jensen, Scott 1947
Jerinic, Vladan 303
Ji, Junpeng 493, 3115
Ji, Lin 3446
Ji, Shiqi 1456
Jia, Xiaoyu 398
Jiang, Dan 1788
Jiang, Dong 3616
Jiang, Ling 872
Jiang, Qirong 1468
Jiang, W.Z. 2415
Jiang, Xinjian 907
Jiao, Ningfei 2776

Jin, Qian	2409
Jo, Hyunsik	511
Jo, Jongmin	1708
Joffe, Christopher	1741
John, Vinod	951, 2200, 3439
Johnson, J.	1926
Jones, David C.	3273
Jones, Edward A.	1010, 2441
Jones, Vinson	1387
Jourdan, Charlie	3333
Jovanović, Milan M.	38, 46, 1292
Jung, Jae-Jung	2620
Jung, Jee-Hoon	3213
Jung, Kyungsub	17

K

Kakitani, Hisao	2969
Kallfass, Ingmar	2969
Kang, Taeyong	3545
Kang, Yong	3521
Kapat, Santanu	315, 2504, 3224, 3237
Karbalaye Zadeh, Mehdi	446, 3397
Karimi-Ghartemani, Masoud	3165
Karki, Ujjwal	2525
Kashyap, Avinash	3255
Katzir, Liran	3655
Kawai, Yasufumi	2051
Kawajiri, Toru	1802
Kawase, Daisuke	283
Kazama, Taisuke	1585
Ke, Haotao	295
Ke, Xugang	94
Ke, Ziwei	241
Kelly, Anthony	1591, 1670
Kerekes, Tamas	974, 1702, 2264
Khajehoddin, S. Ali	2579, 3207
Khaligh, Alireza	54, 440
Kharezy, M.	3043
Khayat, Joseph	66
Khoshkbar Sadigh, Arash	720, 1191, 1481, 2187, 2840
Khurram, Adil	2782
Kieferndorf, Frederick	472
Kikuchi, Naoto	3146
Kim, Byeong-Heon	1220
Kim, Dong-Hee	1273
Kim, Hyeokjin	1947
Kim, Hyeon-Sik	1206
Kim, Jae-Gu	2161
Kim, Ji-Min	3690

Kim, Jin-Woong 1398
Kim, Jonghoon 1690, 3322
Kim, Minjae 859
Kim, Nari 1273
Kim, Youngjong 2355
Kim, Yun-Sung 163
Kirshenboim, Or 111, 802
Kirtley, James L. 2881
Knott, Arnold 1541, 1842
Ko, Youngjong 3629
Koga, Tomoya 2430
Kolar, Johann W. 615, 623, 1198
Kolluri, Sandeep 552, 2491
Koltsov, H. 339
Kondo, Ryota 1713
Kondo, Takeshi 2788
Konishi, Kyohei 1780
Konrad, Werner 3570
Kostov, Konstantin 1018
Kou, Lei 1489, 2278
Kouchaki, Alireza 2382
Krischan, K. 355, 759
Krishna Moorthy, Radha Sree 794
Krishnamurthy, Mahesh 880, 3128
Kshirsagar, Parag 2027, 3616
Kubendran, S. 663
Kubo, Hajime 2788
Kudva, Sudhir S. 86
Kularatna, Nihal 2479
Kulkarni, Abhijit 2200, 3439
Kulkarni, Onkar Vitthal 2245, 3376
Kulkarni, S. 663
Kulothungan, Gnana Sambandam 402, 943
Kumar, Ashish 1392, 1947, 2395
Kumar, Misha 38, 46, 1292
Kumar, Nikhil 1344
Kumar, V. Inder 3224, 3237
Kumar, V.V.S. Pradeep 2673
Kurokawa, Fujio 754
Kusaka, Keisuke 1336
Kusama, Fumito 2607
Kwak, Sangshin 1748
Kwon, Yong-Cheol 1206
Kyriakides, Elias 3500

L

Lago, Jackson 2833
Lai, Jih-Sheng 1148, 1974
Lai, Jih-Sheng Jason 3243

Lai, Wei-Han 1974
Lam, John 786, 830
Lamar, Diego G. 25, 822, 1823, 2545
Lamo, Paula 2389
Lamoureux, Carl 1882
Langham, Jeff 334
Langmaack, N. 700
Lashway, Christopher R. 1267
Lave, M. 1926
Lazaro, A. 2545, 3090
Lazaro, Orlando 66
Lazzarin, Telles Brunelli 2376
Le, Hoai Nam 1911
Lee, Albert T.L. 169
Lee, Byoung-Kuk 163, 1273, 2161
Lee, C.K. 913
Lee, Eun S. 1773
Lee, Fred C. 322, 343, 657, 1002, 1534, 1608, 1847, 1853, 2334, 3259
Lee, Hyun-jun 1690
Lee, Jaedo 3598
Lee, Jae-Hyun 3086
Lee, Jian-Hsing 1900
Lee, June-Seok 3416
Lee, Junwon 511
Lee, Kevin 2003
Lee, Kun Wang 2875
Lee, Kyo-Beum 2764, 3416
Lee, Kyu-Chan 487
Lee, Moonhyun 1416
Lee, Woongkul 2679, 2861
Lee, Yong-Duk 125
Lee, Yongjae 193
Lee, Younggi 2370
Leeb, Steven B. 2881
Lefranc, Pierre 2401, 2700
Lehman, Brad 417, 2122, 2286, 3038
Lei, Yang 2063
Lei, Yutian 1512
Lemmon, Andrew 1577
Leng, Mingzhi 1554
Leng, Siyu 1941
Lenz, Kevin 303
Leong, K.K. 355
Lequesne, Bruno 2927
Leubner, Martin 2175
Levy, Aron 138
Li, Chendan 3459
Li, Dan 595
Li, David K.W. 1350

Li, Guojie 3560
Li, He 990, 1554, 1967
Li, Helong 704, 3101
Li, Hongxu 1657
Li, Hui 1675, 2237
Li, Jie 2770
Li, Jun 2963
Li, Kai 2613
Li, Kaiyuan 3422
Li, Peide 3697
Li, Qiang 322, 343, 657, 1002, 1534, 1608, 1847, 1853, 2334, 3259
Li, River Tin-Ho 2905
Li, Rui 1675
Li, Sinan 169
Li, Tao 2573
Li, Tengfei 2992
Li, Virginia 343
Li, Wenyu 1450
Li, Wuhua 499, 1462
Li, Xing 728
Li, Xinlei 3623
Li, Xuan 2063
Li, Xueqing 2973
Li, Yalong 2637
Li, Yan 1462
Li, Yan-Cun 1853
Li, Yaohua 3080
Li, Yongdong 3317
Li, Yuan 417, 2525
Li, Yun Wei 1249
Li, Yungui 3560
Li, Yunwei 185
Li, Zhiqing 1329
Li, Zhongxi 2349
Li, Zhongyu 3697
Liang, Beihua 1249
Liang, Lin 2981
Liang, Tsorng-Juu 1900
Liang, Xinyu 2349, 2714
Liao, Yi-Hung 1831
Liao, Zitao 1512
Lidow, Alex 587
Lightbody, Gordon 2794
Liivik, Liisa 2533
Lim, Changjin 2370
Lim, Seungbum 98
Lima, Gustavo B. 3585
Lin, Hua 728, 1078
Lin, L.-C. 951, 1364

Lin, Ni	3189
Lin, P.-H.	1364
Liserre, Marco	2449, 3493, 3629
Lisi, G.	2540
Liu, Baojin	3328, 3370
Liu, Bing	843, 2754
Liu, Bo	966, 2441, 2637
Liu, Fuxin	1788
Liu, Gang	38, 46, 595, 1292
Liu, Haichun	1371
Liu, Haoyan	3180
Liu, Hongpeng	1253
Liu, Jingbo	2147
Liu, Jinjun	739, 2272, 3193, 3328, 3370
Liu, Liming	990
Liu, Pei-Hsin	343
Liu, Pengkun	132, 983
Liu, Sucheng	1489, 2278
Liu, Teng	2272
Liu, Tianshu	899
Liu, Tingting	1410
Liu, Weiguo	1726, 2748, 2776
Liu, Wenbo	899, 2095, 2320
Liu, Wen-Chuen	1512
Liu, Xianzhuo	3317
Liu, Xiaohu	3403
Liu, Xiaokang	2742
Liu, Yan-Fei	899, 1243, 1489, 2087, 2095, 2278, 2304, 2312, 2320
Liu, Yang	959
Liu, Yunting	1045
Liu, Yushan	1214
Liu, Yushi	1392
Liu, Yusi	143
Liu, Zeng	739, 2272, 3328, 3370
Liu, Zhengyang	1847, 1853
Liu, Zhichao	1155
Loh, Poh Chiang	229, 1253, 1872, 1995, 2011, 2207
Lomonova, E.A.	3158
Lope, I.	3020
López del Moral, D.	3090
López, Felipe	2389
Lopez, Ozzie	3065
Lorenz, Robert D.	215, 2055, 2167
Lotfi, Ashraf	1882
Lu, Daorong	2194
Lu, Fei	1721, 1726
Lu, Jie	1392, 1947
Lu, Juncheng	529
Lu, Minghui	1941

Lu, Sizhao 2613
Lu, Ting 1456
Lu, Yong 2215
Lu, Zhengang 2613
Lu, Zhengyu 2003
Lu, Zhigang 398
Lu, Zhouyu 1243
Lucia, Oscar 1040, 1762, 3566
Lukic, Srdjan M. 2349, 2714
Luna, Adriana 1227
Luo, Fang 709, 2981
Luo, Guangzhao 2748
Luo, Haoze 499
Luo, Min 1108
Luo, Tianyi 3065
Lynch, Brian 66

M

Ma, Cong 3279
Ma, Dongsheng 94
Ma, Hongbo 3243
Ma, Jun 499
Ma, Ke 261
Ma, Weizhong 816
Ma, Yingxian 2497
Ma, Yiwei 966, 1261, 2229, 3121
Madhusoodhanan, Sachin 886, 1497, 1632, 2076
Madsen, Mickey P. 1842
Magne, Pierre 453
Mahajan, Anirudh 601
Mahdavikhah, Behzad 329, 3297
Mahmoodzadeh, Zahra 3353
Mainali, Krishna 886, 1497, 1632, 2076
Maitra, Arindam 1974
Makhdoomi Kaviri, Sajjad 378
Makoschitz, M. 1
Maksimović, Dragan 580, 1392, 1947, 2292, 3273
Malcolm, Doug 2087, 2095
Mallik, Ayan 54
Manabe, Shinya 2465
Mandal, Arindam 3237
Mandi, Bipin Chandra 2504
Manjrekar, Madhav 2706
Mansour, Makram 3266
Mantooth, H. Alan 143, 1646, 3012, 3180
Mao, Shuai 2776
Marsili, S. 339
Martín, Kevin 25
Martineau, Donatien 524

Martínez, Gilberto 1115
März, Martin 864, 1741
Mátéfi-Tempfli, Stefan 733
Mathew, Dinto 3286
Matsumori, Hiroaki 676, 3051
Matsuura, Ken 2430
Mattavelli, Paolo 1315
Mauerer, M. 1198
Mawby, Philip 253, 2645
Mazhari, Iman 2998
Mazumder, Paromita 1344
Mazumder, Sudip K. 1344, 1989
Mazzola, Michael 1577
McAmmond, Matt 529
McCann, Roy A. 143
McCue, Benjamin M. 3255
McDonald, Brent 329, 334, 3297
McGrath, Brendan 2252
McHugh, Colin 1947
McIntrye, W. 2540
McKenzie, Craig 3038
McRae, T. 2540
Meder, Dirk 2083
Meere, Ronan 2629
Megyei, George 3255
Mehrizi-Sani, Ali 3353
Mehrotra, Vivek 79
Mekhilef, S. 1163
Méllo, João Paulo R. 2720
Meng, Jinhao 2748
Meng, Peipei 1102
Meng, Tao 2748, 2776
Meola, Marco 1591
Mertens, Axel 3172
Meyer, Jeffrey 1947
Mi, Chris 1721, 1726
Michihira, Masakazu 2607
Mikata, Atsushi 2969
Mikulla, Michael 2083
Miraoui, Abdellatif 3476
Mishima, Tomokazu 1780
Mishra, Richa 2342
Miskovic, Vlatko 2167
Mitra, Rakesh 3279
Miwa, Brett 1597, 3273
Miyazaki, Koutarou 1640
Miyazaki, Takayuki 1907
Modes, Christina 3069
Moeini, Amirhossein 2019

Mohamed, A.A.S. 928
Mohammad, Mostak 1748
Mohammadi, Danyal 2813, 2821
Mohammadi, Mehdi 848
Mohammadpour, Bahador 378
Mohammed, Osama 928, 1267, 3200
Mohan, Ned 1051, 1520, 1934, 1982, 2043
Mojab, Alireza 1989
Molinas, Marta 446, 3397
Mønster, Jakob D. 1842
Moon, Gun-Woo 3690
Moon, Intae 1512
Moon, Seung-Ryul 1974
Morgan, Adam 295, 2141
Moroto, Takahiro 1802
Morris, Casey 2861, 2950
Morsy, Ahmed 2567
Mosesian, Jerry 2122
Moss, Jim 668
Motto, Eric R. 468
Moury, Sanjida 786
Mu, Xianmin 1381
Muetze, A. 355, 759, 3570
Mukherjee, Subhajyoti 2687
Mukhopadhyay, Shayok 2782
Mukhopadhyay, Siddhartha 315
Munk-Nielsen, Stig 288, 704, 974, 1376, 3101
Murmann, Boris 1650
Musavi, Fariborz 772
Musumeci, Salvatore 3669
Muyeen, S.M. 1941

N

Na, Woonki 3322
Nadarajan, Sivakumar 715
Nademi, Hamed 3291
Nagai, Shuichi 2051
Nahid-Mobarakeh, Babak 446, 3397
Nakano, Toshiya 468
Nakao, Hiroshi 754
Nakaoka, Mutsuo 1780
Nakashima, Yoshiyasu 754
Nan, Chenhao 432
Narasimhan, Sneha 2043
Nasr, Miad 1350
Navarro, Angel 3508
Nawaz, Muhammad 276
Nee, Hans-Peter 746, 1018
Neely, J. 1926, 3537

Neft, Charles	79
Negoro, Noboru	2051
Ngo, Khai	668
Nguyen, Duy T.	1773
Ni, Tianheng	2754
Ni, Xijun	983
Niapour, S.A.Kh. Mozaffari	3704
Nikolaidis, Ilias	3069
Ning, Puqi	3080
Ninomiya, Tamotsu	2422, 2430
Nishizawa, Shin-Ichi	3032
Niu, He	2055
Niu, Ying	2770
Noh, Shinyoung	859
Nomura, Katsuya	3146
Nondahl, Thomas A.	2147
Noquil, Jonathan	3065
Norisada, Takaaki	2607
Norum, Lars Einar	3291
Nowak, Torsten	3069
Nymand, Morten	609, 2382

O

O'Donnell, Terence	2629
O'Donovan, Gerard	2794
Ogawa, Taichi	1907
Oh, Chang-Yeol	163
Oh, Jaeyoon	2370
Ojo, Olorunfemi	2035
Okubo, Hiizu	2465
Oliver, J.A.	2409
O'Loughlin, Cathal	2629
O'Mathuna, C.	663
Omura, Ichiro	1640, 3032
Oña, E.	2545
Onal, Yasemin	2693
Onar, O.C.	1307
Orabi, Mohamed	1505
Ordonez, Martin	386, 848, 854, 2561
Orikawa, Koji	1336, 1911
Orr, Ray	1350
Ortiz-Gonzalez, Jose	253
Ortiz-Rivera, Eduardo I.	3712
Otsuka, Masafumi	1350
Ouyang, Ziwei	2473
Ozimek, Patrick E.	1084
Ozpineci, Burak	3529

P

Padhee, Varsha	1982
Pagano, Rosario	3669
Pahlevani, Majid	378
Pala, Vipindas	979
Palaniappan, Vishal	1350
Palmour, John	979
Pam, Srikanth	3266
Pan, Bing	2613
Panda, S.K.	9, 552, 715, 2491
Park, Hwa-Pyeong	3213
Park, Joung-hu	1690
Park, Sung-Yeul	125
Parkhideh, Babak	2998
Parsa, Leila	2573
Parvez, M.	1163
Patel, Ankur	150
Pathan, Abrar Ahmed	2497
Patil, Devendra	2889
Patra, Amit	2504
Pavlick, Stephanie A.	1138
Pavlovic, Z.	663
Paz, Francisco	386
Pedersen, Jeppe A.	1541, 1842
Peixoto, Paulo P.	3346
Peng, Chang	132, 269, 983, 2927
Peng, Fang Z.	959, 1045, 2525
Peng, Hao	1450
Peng, Jichang	2748, 2776
Peng, Kang	2127
Peng, Li	1358
Perales, Mico	1967
Perdigão, Marina	1115
Peretz, Mor Mordechai	111, 308, 692, 802, 2298
Perez, Aday	215
Pérez-Tarragona, Mario	1762
Perreault, David J.	98, 1138
Perrin, Remi	524
Perry, Jeff	3266
Persons, Ryan	3069
Pervaiz, Saad	1947, 2395
Peterchev, Angel V.	2349
Pevere, Alessandro	1183
Phillips, Evan	3684
Piepenbreier, Bernhard	2897
Pierfederici, Serge	446, 3397
Pigazo, Alberto	2389
Pilawa-Podgurski, Robert C.N.	1512
Ping, Dinggang	38, 46

Piya, Prasanna ... 3165
Pong, M.H. Bryan ... 2328
Poshtkouhi, Shahab ... 1350, 1403
Pou, Josep ... 236
Praça, Paulo P. ... 60, 3231
Pramod, Prerit ... 3279
Prasai, Anish ... 1437
Preciat, Philippe ... 524
Prieto, R. ... 1624
Prodić, Aleksandar ... 329, 1597, 2497, 2540, 2553, 3297
Puukko, Joonas ... 990

Q

Qi, Feng ... 2912
Qian, Qiang ... 1919, 3446
Qiao, Wei ... 3514
Qin, Jiangchao ... 1528
Qin, Liang ... 2525
Qin, Shibin ... 1512
Qin, Xianhui ... 843
Qiu, Maohang ... 138
Qiu, Yajie ... 2320
Qu, Liyan ... 3279, 3514
Qu, Xiaohui ... 2154
Quan, Zhongyi ... 185
Quay, Rüdiger ... 2083
Quentin, Nicolas ... 524

R

Raciti, Angelo ... 3669
Rahnamaee, Arash ... 1989
Raizada, Shirish ... 1344
Rajashekara, Kaushik ... 2788
Ramachandran, Rakesh ... 609
Ramadass, Yogesh ... 838
Ramani, Ramanathan ... 66
Rambal-Vecino, Andres ... 3712
Ramezani, Medhi ... 2035
Ramos, Francisco ... 215
Ramu, Krishnan ... 2027
Ran, Li ... 253, 1443, 2645
Ranstad, Per ... 1018
Rao, Yuan ... 1585
Rashkin, L. ... 3537
Rathore, Akshay K. ... 402, 794, 943
Ravey, Alexandre ... 3476
Redondo, Luís M. ... 3637
Rehman, Habibur ... 2782
Reiner, Richard ... 2083

Reitz, Jessica	3069
Remus, Nico	2175
Ren, Hai-Peng	2770
Ren, Ren	2441
Ren, Xiaoyong	73, 2518, 3488
Ren, Xizhou	2912
Ren, Yu	2071, 2102
Rengifo, Johnny	3200
Renjit, Ajit A.	1682
Reusch, David	587
Riazmontazer, Hossein	1989
Rim, Chun T.	1773
Roasto, Indrek	2533
Robbins, William	1934
Roberts II, Charles	3255
Rodrigues, Danillo B.	3585
Rodriguez, A.	1823
Roehrs, Benjamin D.	3255
Rogers, Daniel J.	1650
Romero, David	1350
Rosahl, Thoralf	106
Roßkopf, Andreas	1741
Round, W. Howell	2479
Ruan, Xinbo	1788, 3488
Ruiz, Juan M.	1292

S

Sá Jr., Edilson M.	2592
Saasaa, Raed	1286
Sadik, Diane-Perle	746, 1018
Saeed, Sarah	3508
Saeedifard, Maryam	1528, 2136, 2957
Safaee, Alireza	460
Saha, Aparna	2806
Sahoo, Ashish Kumar	1982
Sahoo, Saroj Kumar	199
Saito, Katsuaki	283
Saito, Shoji	468
Saket, Mohammad Ali	854, 2561
Sakurai, Takayasu	1640
Salameh, Mohamad	3128
Salcines, Cristino	2969
Salem, Ahmed	1505
Sandoval, José Juan	3545
Sangwongwanich, Ariya	370
Sankman, Joseph	94
Santi, Enrico	2127, 3248
Santiago-González, Juan A.	98
Santos de Moura, Diogo Cesar	3553

Santos Guimarães, Jéssica	3346
Sanz, M.	2545, 3090
Sariri, Kouros	3255
Sarlioglu, Bulent	2679, 2861, 2950
Sarnago, Hector	1040, 1762, 3566
Satija, Yudhister	3266
Sato, Masaki	2465
Saublet, Louis-Marie	3397
Saur, Michael	215
Savaghebi, Mehdi	1227, 3697
Scandrett, Brad	417
Schmidt, Peter B.	2147
Schubert, Michael	643
Schweitzer, Ben	3128
Sebastián, Javier	25, 822, 1823
Seeman, Michael	838
Seltzer, Daniel	1947
Sen, Paresh C.	1489, 2278
Senanayake, Thilak	3722
Senol, Murat	643
Seo, Gab-Su	487
Sepahvand, Alihossein	580, 1947
Serrano, J.	3020
Setyawan, Leonardy	3409
Severson, Eric	2043
Shafiei, Navid	848, 854, 2561
Shagar, Viknash	2813
Shah, Neel	2998
Shahbazi, Caitlin	3069
Shamsi, Pourya	1962, 2687
Shang, Fei	880
Shao, Jianwen	3659
Sharkh, Suleiman M.	2223
Sharma, Ratnesh	1682
Shen, Ang	1962
Shen, Guangtong	2003
Shen, Zhiyu	516, 1561
Sheng, Su	417, 2286
Shenoy, Pradeep S.	66
Shi, Baoping	843
Shi, Jianjiang	1475
Shi, Xiaojie	2637
Shi, Yuxiang	1675
Shimizu, Toshihisa	676, 3051
Shiu, T.-H.	1364
Shmilovitz, Doron	3655
Shousha, Mahmoud	3097
Shoyama, Masahito	2735
Shrivastav, Ashish	601

Shu, Zhan 2223
Shukla, Anshuman 2342, 3286
Silva, J. Fernando 3637
Silva, Paulo R. 3585
Silva, Rafael V. 2592
Simanjorang, Rejeki 2942, 3058
Singh, Amandeep 3140
Singh, Shikhar 601
Singh, Surinder P. 1585
Sinha, Sreyam 1947
Siwakoti, Yam P. 1872
Sleik, Roland 759
Sohn, Hoon 3690
Soltani, Hamid 229
Somani, Apurva 1520
Somani, Utsav 3333
Son, Yeongrack 3717
Son, Young-Kwang 2370
Song, Xiaoqing 132, 269, 983
Soni, Harshit 1344
Sozer, Yilmaz 207, 392, 638, 2693, 2806
Srdic, Srdjan 2714
Srinivasan, Dipti 402, 943
Srivastava, Vineet 786
Stack, David 1670
Stamm, Thomas 3659
Steenis, Joel 3219
Steyn-Ross, D. Alistair 2479
Stillwell, Andrew 1512
Strydom, Johan 587, 2292
Stübig, Marc 2175
Styles, Julian 529
Su, Yipeng 118
Subotic, Ivan 623
Suh, Yongsug 17
Sul, Seung-Ki 1206, 1220, 2370, 2620
Sun, Bo 1376
Sun, Kai 1227, 1410, 2182, 2194
Sun, Lei 505
Sun, Lejia 810
Sun, Libing 1227
Sun, Pengju 2992
Sun, Wei 1462
Sung, Won-Yong 163
Sveum, Peter 3128

T

T T, Anandha Ruban 1878
Tabata, Osamu 2051

Tadano, Hiroshi 3722
Tadeparthy, Preetam 1878
Tai, Heng-Ming 2992
Takamiya, Makoto 1640
Takano, Koushi 676
Talebi Khanmiri, Dawood 2122, 3038
Tan, Kai 132, 983
Tan, Linlin 2102
Tan, Nadia M.L. 1163
Tan, Pingan 3185
Tan, Siew-Chong 169, 913, 1169, 1888, 1894, 3302, 3481
Tang, Yi 1071, 3591
Tang, Yichao 440
Tang, Yuan 1443, 2645
Tareilus, G. 700
Tayebi, S. Milad 1381
Teixeira, Carlos 2252
Tekgun, Burak 207
Teodorescu, Remus 974, 1702, 2264
Tewari, Saurabh 1520, 2043
Thiringer, T. 3043
Thone, Jef 3097
Tian, Shuilin 1608
Tian, Ye 499
Ting, Lo Pang-Yen 1900
Tkachov, Sergii 350, 1663
Tolbert, Leon M. 893, 966, 1261, 1307, 1569, 2637, 3121
Tomas-Manez, Kevin 1235
Tomioka, Satoshi 2430
Tong, C.F. 2942, 3058
Trabelsi, Mohamed 3663
Tran, Yan-Kim 156
Trento, Brad 3577
Trescases, Olivier 1350, 1403, 1882
Trintis, Ionut 1376
Tripathi, Awneesh 886, 1497, 1632, 2076
Tse, Zion Tsz Ho 1737, 3526
Tseng, King Jet 1071
Tseng, K.J. 2942, 3058, 3107, 3422, 3591
Tsukuda, Masanori 1640
Tung, Chung-Pui 1807
Tüysüz, Arda 615, 623, 1198

U

Uceda, J. 1624
Uddin, Md Wasi 638
Ueda, Tesuzo 2051
Ueno, Takeshi 1907
Ugur, Enes 1176

Urteaga, Miguel 79

V

Vasić, M. 2409
Vásquez, Juan C. 1227, 3459
Vázquez, A. 25, 2545
Vechalapu, Kasunaidu 295, 886, 1497, 1632, 2076
Vekslender, Timur 308
Venkataramanan, Giri 2919
Venkateswaran, Muthusubramanian 1878
Vermulst, B.J.D. 3158
Vermulst, Bas 2457
Vesti, S. 339
Vilathgamuwa, Mahinda 2813
Villarejo, J.A. 1823
Vinnikov, Dmitri 2533
Vitorino, Montiê A. 1032
Vrankovic, Zoran 1084
Vu, Trong Tue 1835, 2485
Vukadinović, Nenad 1597

W

Wada, Keiji 1640, 2986
Walsh, Ray 2794
Waltereit, Patrick 2083
Wang, Chengshan 1249
Wang, Fan 334
Wang, Feng 810, 2742
Wang, Fred 424, 893, 966, 1010, 1261, 1456, 1569, 2229, 2441, 2637, 3121
Wang, Gangyao 979
Wang, Guo-Xiang 1123
Wang, Haoyu 480, 1280, 1329
Wang, Hongliang 899, 1489, 2278, 2304, 2312, 2320
Wang, Huai 370, 2154
Wang, Jiangfeng 2182
Wang, Jin 990, 1554, 1967
Wang, Jun 516
Wang, Kui 3317
Wang, Kun 1450
Wang, Laili 899, 2087, 2095, 2320
Wang, Li 2365
Wang, Long 2754
Wang, Meilin 1078
Wang, Meng-Jie 3471
Wang, Mengqi 3648
Wang, Miao 709, 2912
Wang, Ming-Hao 3302
Wang, N. 663
Wang, Peng 1071, 3409, 3591

Wang, Qin 1657
Wang, Shike 3328, 3370
Wang, Shiliang 2889
Wang, Shuo 2019, 3603
Wang, Wei 1253
Wang, Xiaoping 572
Wang, Xingwei 1078
Wang, Xiongfei 704, 1253, 1941, 2011, 2207, 2215, 3051, 3101, 3431
Wang, Yanbo 3431
Wang, Yi 1815
Wang, Zhenxiong 3358
Wang, Zhiqiang 2637
Watanabe, Hiroki 1336
Watanabe, Yoshitoshi 3146
Wattes, J.L. 3231
Weber, Bastian 3172
Wei, Chun 3514
Wei, Jiadan 843, 2754
Wei, Lixiang 1065, 2826
Wei, Yingdong 1468
Weise, Nathan 1065
Weiss, Beatrix 2083
Wen, Changyun 3409
Wen, Lucheng 990
Wen, Xuhui 3080
Wens, Mike 3097
Wespel, Matthias 2083
Wicht, Bernhard 106
Wijnands, C.G.E. 3158
Wiktor, Wlodek 66
Williamson, Sheldon S. 3140
Wilson, D. 3537
Winterhalter, Craig 1084
Wittmann, Jürgen 106
Wu, Dalei 3189
Wu, Hongfei 1410, 1424
Wu, John 1967
Wu, Liyao 2136, 2957
Wu, Qunfang 1657
Wu, T.-F. 951, 1364
Wu, Teng 3328, 3370
Wu, Tong 3529

X

Xia, Yinglai 3364
Xiao, Guochun 2215
Xiao, Jianfang 3409
Xiao, Lan 1657
Xiao, Xi 1424, 3338

Xie, Shaojun	1371, 1919, 3446
Xie, Xiaogao	816
Xie, Yicong	2360
Xin, Zhen	1995, 2207, 3697
Xing, Xiangyang	3453, 3465
Xing, Yan	1410, 1424, 2182, 2194
Xiong, Liansong	2742
Xiong, Song	1888, 1894
Xu, Chen	1358
Xu, Dewei David	33
Xu, Dianguo	1253
Xu, Jialin	1657
Xu, Jing	990
Xu, Jinming	1919, 3446
Xu, Longya	709, 2912
Xu, Qianwen	3409
Xu, Tao	921
Xu, Wei	2868
Xu, Yang	2141
Xue, Fei	3677
Xue, Lingxiao	1315

Y

Yadav Gorla, Naga Brahmendra	552
Yadav, Akshat	2076
Yamada, Go	2607
Yamada, Masaki	1713
Yamaguchi, Koji	3075
Yamaguchi, Masahiro	2465
Yamamoto, Keiichi	283
Yamamoto, Yasuhiro	2788
Yan, Shuo	913
Yan, Xingda	2223
Yanagi, Hiroshige	2430
Yang, Ching-Chieh	558
Yang, Enxing	499
Yang, Heya	1462
Yang, Hongbin	1078
Yang, Jianwei	3134
Yang, Liu	1261, 3121
Yang, Pengzhi	1554
Yang, Qichen	2957
Yang, Shuitao	959
Yang, Shunfeng	1071, 3591
Yang, Tao	2629
Yang, Tianbo	913
Yang, Xu	493, 766, 2071, 2102, 3115, 3623
Yang, Yang	1243
Yang, Yong	1788

Yang, Yongheng 221, 370, 1253, 3500
Yang, Yuanyu 843
Yang, Yuchen 1534
Yang, Yun 1169, 3481
Yang, Zhihua 33, 899, 2312, 2320
Yao, Chengcheng 990, 1554
Yao, Jianhui 2182, 2194
Yao, Kai 572, 1815
Yao, Wenxi 2003
Yau, Y.T. 2415
Yazdanian, Mehrdad 3353
Ye, Qing 2237
Ye, Zichao 1512
Yeo, H.L. 3107
Yeong, Lee Meng 3409
Yi, Fan 1057, 1861, 2599
Yi, Hao 3358
Yin, Shan 2942, 3058
Yonezawa, Yu 754
Young, George 1835, 2485
Yu, Hualong 1456
Yu, Ruiyang 3677
Yu, Sheng-Yang 2511
Yu, Wensong 2063, 2141, 2365, 2714, 2727, 3648
Yu, Xinyu 1468
Yu, Yanqi 1286
Yuan, Huawei 907
Yuan, Liqiang 2613

Z

Zafarani, Mohsen 1096, 2875
Zagrodnik, Michael 1071, 3591
Zane, Regan 1603
Zare, Firuz 221, 229
Zefran, Milos 1989
Zeltser, Ilya 802
Zeng, Hulong 1045
Zeng, Xiangjun 2102
Zhak, Serhii M. 3273
Zhan, Xiaohai 1424
Zhan, Xiaoqing 2154
Zhang, Bin 1155
Zhang, Binfeng 1919
Zhang, Canhui 138
Zhang, Chengduo 2349
Zhang, Chenghui 3453, 3465
Zhang, Chi 2714
Zhang, Dan 766, 3623
Zhang, Fan 1947, 2071, 2102

Zhang, Hao 2973
Zhang, Haojiong 2167
Zhang, Hua 1721, 1726
Zhang, Hui 3338
Zhang, Jianqiu 595
Zhang, Julia 241
Zhang, Jun 2992
Zhang, Junfang 572
Zhang, Ke 3476
Zhang, Lanhua 1148, 1974, 3243
Zhang, Li 3488
Zhang, Li-Heng 2770
Zhang, Lin 1868
Zhang, Liqi 269, 2063, 2365
Zhang, Shuoting 966, 3121
Zhang, Weimin 424, 893
Zhang, Wenli 524, 1002
Zhang, Xiangming 1102
Zhang, Xuan 990, 1967, 2229
Zhang, Xuning 177, 1024, 1561
Zhang, Yongchang 2868
Zhang, Yuanzhe 580, 2292
Zhang, Yuzhi 3180
Zhang, Zhe 1090, 1235, 1430, 2252, 3514
Zhang, Zhen 3134
Zhang, Zheyu 684, 1010, 1569, 2441
Zhang, Zhigang 3358
Zhang, Zhiliang 73, 1243, 2518
Zhang, Zhiyu 1475
Zhang, Zicheng 3453, 3465
Zhao, Bo 2912
Zhao, Chongwen 3577
Zhao, Dongdong 3134
Zhao, Hengyang 1475
Zhao, Hui 3603
Zhao, Rende 1995, 2207, 3697
Zhao, Shuze 1882
Zhao, Tao 33
Zhao, Xiaonan 1148, 3243
Zhao, Xin 295
Zhao, Zhengming 1456, 2613
Zheng, Sheng 966
Zheng, Yue 417
Zheng, Zedong 3317
Zhi, Na 3338
Zhou, Bo 843, 2754
Zhou, Daming 3476
Zhou, Jinping 2360
Zhou, Keliang 1155

Zhou, Liwei	410, 2259, 2935
Zhou, Luowei	2992
Zhou, Min	2360
Zhou, Qi	536
Zhou, Sizhan	3193
Zhou, Yong	2567
Zhou, Yuan	73, 2518
Zhou, Zhe	2912
Zhu, Bohang	2788
Zhu, Donghai	3521
Zhu, Guorong	3310
Zhu, Ke	990
Zhu, Qianlai	2365
Zhu, Tianhua	810
Zhuo, Fang	810, 2742, 3358
Zojer, Bernhard	996
Zou, Juan	651
Zou, Ke	1554
Zou, Xudong	3521
Zou, Xuewen	73, 2518
Zou, Zhi-Xiang	3493
Zumel, P.	2545, 3090

IEEE
445 Hoes Lane
Piscataway, NJ 08854-4141

ISBN 978-1-4673-9551-9